EFM15 – 10th Anniversary International Conference on Experimental Fluid Mechanics 2015

EPJ Web of Conferences Volume 114 (2016)

Prague, Czech Republic
17 - 20 November 2015

Editors:

Petra Dancova
Martin Vesely

ISBN: 978-1-5108-2208-5

Printed from e-media with permission by:

Curran Associates, Inc.
57 Morehouse Lane
Red Hook, NY 12571

Some format issues inherent in the e-media version may also appear in this print version.

Printed by Curran Associates, Inc. (2016)

For additional information, please contact EDP Sciences – Web of Conferences at the address below.

EDP Sciences – Web of Conferences
17, Avenue du Hoggar
Parc d'Activité de Courtabœuf
BP 112
F-91944 Les Ulis Cedex A
France

Phone: +33 (0) 1 69 18 75 75
Fax: +33 (0) 1 69 28 84 91

contact-edps@webofconferences.org

Additional copies of this publication are available from:

Curran Associates, Inc.
57 Morehouse Lane
Red Hook, NY 12571 USA
Phone: 845-758-0400
Fax: 845-758-2633
Email: curran@proceedings.com
Web: www.proceedings.com

TABLE OF CONTENTS

EPJ Web of Conferences **114**, 00001 (2016)
DOI: 10.1051/epjconf/201611400001

Experimental Fluid Mechanics 2015

Prague, Czech Republic, 17th–20th November 2015

efm.kez.tul.cz

10th Anniversary International Conference „Experimental Fluid Mechanics 2015" was held at 17th–20th November 2015 in Kaserštejn Palace in Prague, Czech Republic.

The conference is focused on obtaining overviews of projects under investigation and experimental methods used in the field of fluid mechanic and thermodynamic. Its aim is to organize a network of research workers who are interested in experimental, and also theoretical, work in the field of fluid mechanic and thermodynamic. The conference should contribute the creation of closer contacts and sharing experience in application of various experimental methods, preparation and implementation of experiments, processing of results and numerical simulations of experiments.

The main topics of the conference are:

- Experimental methods used in fluid mechanics and thermodynamics.
- Preparation and implementation of experiments.
- Application of numerical simulation for preparation of experiments, comparison of the results obtained by numerical or analytical methods with experiments.
- Processing the results.
- Experiments in the field of compressible and uncompressible fluid flow, stability of flow, heat and mass transfer, cavitation etc.
- Experiments and measuring carried out on power and regulating equipment.
- Cryogenic fluid dynamics.

Members of Sciences Committee are:

- Assoc. Prof. Tomáš Vít – Head of the committee (TU Liberec, Czech Republic)
- Simone Ferrari (University of Cagliari, Italy)
- Jan Hrubý (Academy of Science of Czech Republic)
- Hendric de Lange (Eindhoven Univ. of Technology, Netherlands)
- Václav Kopecký (TU Liberec, Czech Republic)
- Robert Pastuszko (Kielce University of Technology, Poland)
- Kazimierz Peszynski (UTP University of Science and Technology, Poland)
- Ludvík Prášil (TU Liberec, Czech Republic)
- Jaromír Příhoda (Academy of Science of Czech Republic)
- Rudi Santbergen (Delft University of Technology, Netherlands)
- Ladislav Skrbek (Charles University, Czech Republic)
- Zdeněk Trávníček (Academy of Science of Czech Republic)
- Václav Uruba (Academy of Science of Czech Republic)
- Magda Vestfálová (TU Liberec, Czech Republic)
- An-Bang Wang (National Taiwan University, Taiwan)
- Yoshifumi Yokoi (National Defense Academy of Japan, Japan)

Steering Committee:

- Petra Dančová – Head of the committee (TU Liberec, Czech Republic)
- Jan Novosád (TU Liberec, Czech Republic)
- Martin Veselý (TU Liberec, Czech Republic)

Keynote lectures for 2015:

- Dr. Shan Zhong (Manchester University, United Kingdom)
  Flow control using synthetic jets
- Dr. Rudolf Dvořák (Czech Academy of Sciences, Czech Republic)
  Avian aerodynamics (Aerodynamics of bird flight)
- Assoc. Prof. Robert Pastuszko (Kielce University of Technology, Poland)
  Enhanced pool boiling heat transfer on mini- and micro-structured surfaces

List of Sponsors (in alphabetical order):

- DANTEC DYNAMICS GmbH (www.dantecdynamics.com)
- LAVISION (http://www.lavision.de/en/)
- LENAM (http://www.lenam.cz/)
- MECAS ESI s.r.o. (www.esi-group.com)
- MIT s.r.o. (www.mit-laser.cz)
- TSI GmbH (www.tsi.com)

Conference venue:

Every year, the organizers want to prepare the conference at nice and interesting place of the Czech Republic. In 2015, the conference venue was Prague.

Prague is the capital city of the Czech Republic. Founded during the Romanesque and flourishing by the Gothic and Renaissance eras, seat of two Holy Roman Emperors and thus also the capital of the Holy Roman Empire. It was an important city to the Habsburg Monarchy and its Austro-Hungarian Empire and after World War I became the capital of Czechoslovakia. The city played major roles in the Protestant Reformation. Main attractions include the Prague Castle, the Charles Bridge, Old Town Square and the Jewish Quarter. Since 1992, the historic center of Prague has been included in the UNESCO.

Prague

Dr. Shan Zhong (1st keynote lecture)

Poster section

Instructions for the crypto-game

edp sciences

EPJ Web of Conferences

Statement of Peer review

In submitting conference proceedings to EPJ Web of Conferences, I certify to the Publisher that I adhere to the **Policy on Publishing Integrity** of the journal in order to safeguard good scientific practice in publishing.

1. All articles have been subjected to peer review administered by the proceedings editors.
2. Reviews have been conducted by expert referees, who have been requested to provide unbiased and constructive comments aimed, whenever possible, at improving the work.
3. Proceedings editors have taken all reasonable steps to ensure the quality of the materials they publish and their decision to accept or reject a paper for publication has been based only on the merits of the work and the relevance to the journal.

Title, date and place of the conference

EXPERIMENTAL FLUID MECHANICS 2015
17 - 20/11/2015
PRAGUE, CZECH REPUBLIC

Proceedings editor(s):

PETRA DANČOVÁ
MARTIN VESELÝ

Date and editor's signature

17/2/2016

17, avenue du Hoggar - PA de Courtabœuf – BP 112 - 91944 Les Ulis Cedex A (France)
Tél. : 33 (0)1 69 18 75 75 - Fax : 33(0)1 69 07 45 17 – www.edpsciences.org

EPJ Web of Conferences **114**, 01001 (2016)
DOI: 10.1051/epjconf/201611401001

Aerodynamics of bird flight

Rudolf Dvořák[1,a]

[1]*Institute of Thermomechanics v.v.i.,Academy of Sciences of the Czech Republic, 18200 Prague 8,The Czech Republic*

Abstract. Unlike airplanes birds must have either flapping or oscillating wings (the hummingbird). Only such wings can produce both lift and thrust – two *sine qua non* attributes of flying.The bird wings have several possibilities how to obtain the same functions as airplane wings. All are realized by the system of flight feathers. Birds have also the capabilities of adjusting the shape of the wing according to what the immediate flight situation demands, as well as of responding almost immediately to conditions the flow environment dictates, such as wind gusts, object avoidance, target tracking, etc. In bird aerodynamics also the tail plays an important role. To fly, wings impart downward momentum to the surrounding air and obtain lift by reaction. How this is achieved under various flight situations (cruise flight, hovering, landing, etc.), and what the role is of the wing-generated vortices in producing lift and thrust is discussed.The issue of studying bird flight experimentally from *in vivo* or *in vitro* experiments is also briefly discussed.

1 Introduction

People have always been fascinated by the perfection of the bird flight, and in the past centuries many enthusiasts attempted to copy it, unfortunately, in most cases with no success. About the end of the fifteenth century Leonardo da Vinci systematically studied the flapping bird wings and observed and described very carefully the bird´s flight.

The first successfully man-operated airborne vehicle by Otto Lilienthal (about the end of the nineteenth century) had fixed wings. However, Lilienthal also studied in detail the bird wing structure and discovered the importance of cambered profiled wings. Horatio F.Phillips about the same time had investigated the aerodynamics of curved double-surfaced airfoils, and in 1884 had it even patented. All this knowledge was applied by Wright brothers, who in 1904 performed a first succcessful flight with a heavier-than-air-airplane of their own design. Their airplane had also fixed wings, and, fixed wings had all other man designed and operated airplanes in the whole century which followed. It was a magnificent century of hectic research and development of new aircrafts, and at its end people proudly and immodestly claimed that in flight abilities their airplanes had reached almost a bird-like perfection.

It is quite interesting that during the twentieth century people scarcely ever sought inspiration in the animal world. Rather, they used the most sophisticated instrumentation, available mainly towards the end of the century, to realize that what they have achieved, birds have possessed for millions of years.

However, towards the end of the last century a new interest has appeared in small autonomous or remotely controlled flying vehicles. The so called micro-air-vehicles (MAV) were expected to be used in aerial reconnaisance in open as well as confined spaces, in monitoring polluted areas which people cannot enter, and (last but not least!) in many military applications. Soon it has become clear that if these vehicles should also be able to carry a reasonable payload, they would have to have either rotating or moving wings. Since the MAVs operate at about the same Reynolds numbers as birds, bats and larger insects, it has recently opened a new interrest in aerodynamics and flight mechanics of all of them. In this paper, however, we will deal only with avian flight, i.e., with the flight of birds.

2 Bird wings

To fly, birds must have wings. Unlike airplanes they must have moving wings, i.e., they must have either flapping or, in one exceptional case (the hummingbird), even oscillating wings. Only such wings can produce both lift and thrust – two *sine qua non* attributes of flying.

First airplane designers, like Otto Lilienthal and Horatio Phillips, believed that it is enough to copy the bird wing shape and the wing profile. Unfortunately, the bird wing is not a simple curved compact surface like any of the vintage airplane wings.

The basic structure of a bird´s wing resembles the human hand, only proportions of the bones are different, as much as they are different in every bird species. The hand section of the wing provides the main dynamic

[a] Corresponding author: rudvorak@gmail.com

control for the bird. It represents about 80% of the wing length in small birds, who have to manoeuvre in constraint environment. In larger birds, using the wings mainly for soaring, gliding or slow flapping, the hand section is much smaller, and the wing is ruled by the arm bones extending from 40 to 60% of the wing span (Fig.1), [1], [2], [3].

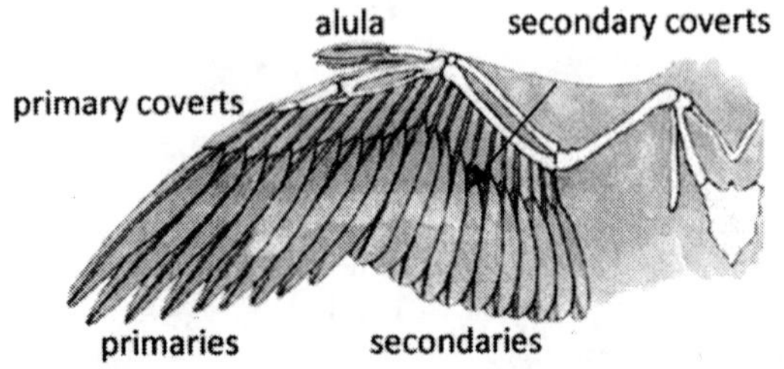

Figure 1. Flight feathers.

Primaries are connected to the bird's fingers. Primaries are the longest and narrowest of the outer feathers (the remiges), and they can be individually rotated. These feathers are the main source of thrust, mostly generated on the downstroke of flapping flight. On the upstroke the primaries are separated and rotated, reducing air resistance while still helping to provide some thrust. Remiges on the wingtips of large soaring birds like condors or vultures also allow for spreading the feathers, reducing thus the creation of wingtip vortices.

Secondaries are connected to the ulna. They remain close together in flight (they cannot be individually separated like the primaries) and help to provide lift by creating the airfoil shape of the bird's wing. Secondaries are usually shorter and broader than primaries.

The **alula feathers** are not flight feathers in the strict sense; however, they are very useful in slow flight. Attached to the bird's "thumb", they lie normally flush against the leading edge of the wing and detach only at higher angles of attack creating a gap between the alula and the rest of the wing (compare with slats on airplane wings). Birds can thus avoid stalling at low speeds or at landing.

The actual shape of the bird wing is made of two organized sets of feathers – in the first set are the **flight feathers** anchored in the digits (primaries) and the ulna (secondaries). There are also three sets of the so called **coverts** which act as a protective cover for all or part of the folded primaries and secondaries.

The vanes of each feather have hooklets that lock the feathers together, giving thus the wing – if locked – the necessary strength to withstand the lift force and to maintain its shape. Each feather has a bigger side and a lesser side. Moreover, the shaft bends slightly off its longitudinal axis with the lesser side to the front and the bigger side to the rear of the feather (Fig.2). The feather anatomy enables rotation of the feather in its follicle. During the up-stroke the bigger side is pressed down, opening thus the wing like a jalousie and allowing air to slip through the wing. This considerably reduces the upward resistance of the wing during upstroke (see again Fig.1). The integrity of the wing is reestablished during the downstroke, when the pressure acts from below and the feathers lock again to form a compact wing surface.

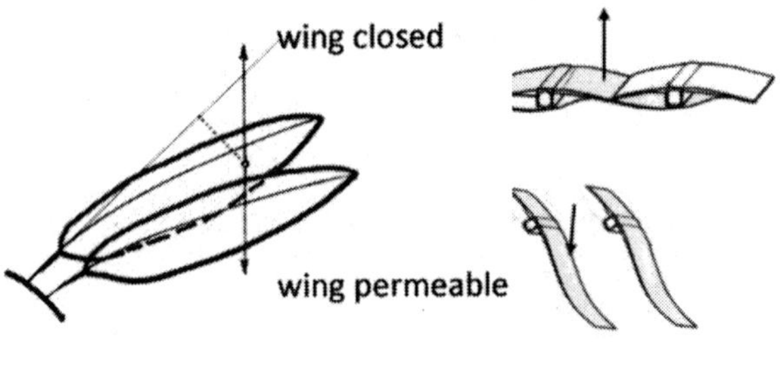

Figure 2. Forces on the flight feathers during downstroke (a) and upstroke (b).

The large flight feathers at the wing tip (primaries, outer remiges) contribute greatly to the production of lift. These feathers form the tip slots which can considerably reduce the induced drag by acting as winglets, i.e., by making wings effectively non-planar and eliminating thus the intensity of the tip vortex by spreading the vorticity vertically.

The flight feathers together with the covert feathers are responsible for morphing the wing shape. The many degrees of freedom of the wrist and elbow bones and all these feathers make the bird wing highly flexible in changing the wing chord and span, and in the spanwise twisting and bending of the wing, thus helping to maintain attached flow and reduce induced drag at the wing tip.

3 The bird flight

The command of flight is essential for the day-to-day struggle for survival, foraging, escaping from enemies, as well as for managing so demanding flight situations like landing and taking off. All this can be achieved only if the wings have the possibility of changing the lift in fractions of a second. These abilities are supported by a very short reaction time, which is almost by one order lower than in humans. It is important that birds have the capabilities of adjusting the shape of the wing (incl. camber and the wing area) according to what the immediate flight situation demands, as well as of responding almost immediately to conditions the flow environment dictates, such as wind gusts, object avoidance, target tracking, etc

There are two kinds of moving wings operating in two different modes –the **flapping mode** (birds, bats) and the **oscillating mode** (insects). Both are characterized by a **complex unsteady** and **threedimensional motion** of the wings, changing in every instant with the new position of the wings, [1],[2],[4],[5]. The wingbeat cycle in birds has two kinematically distinct phases – the **downstroke phase** (the wing goes usually down) and the **upstroke phase** (the wing returns to its initial position). The whole cycle highly depends on the purpose of the flight and has therefore various modifications even in the same species

The downstroke is often called the power stroke, as it provides the majority of lift and thrust, while the upstroke is a recovery stroke.

The motion of the wings is quite complex and depends on the aim of the flight or on the situation.

Close to the body, there is very little up and down movement of the wing, while movement of the outer part may be quite vigorous. The outer part of the wing has to twist and even change the profile so that each part of the wing can maintain the optimal angle of attack. During the downstroke, the angle of attack is increased and is decreased during the upstroke. As the wing twists, and as the outer part of the wing moves downward, the aerodynamic force is directed forward and the wing produces thrust. During the upstroke, the outer part of the wing must return with as minimum drag as possible, usually with zero angle of attack. The bird partially folds its wings, reducing thus the wingspan. As a rule, the primaries open, allowing passage of air through them, reducing further the drag. The inner part of the wing behaves in a different way. There is little up-and-down movement there so that this part of the wing continues to provide lift and functions more or less as a fixed wing. Thus lift does not fall down to zero between the downstroke phases of the wingbeat cycle.

Birds can very efficiently change the wing area. This is typical namely of seabirds who can almost instantly switch from the cruising flight with fully extended wings to a dive with wing close to the body. Even when descending for landing, the area of the wing is decreased by raising the inner part of the wing upwards (this at the same time increases the bird´s stability). By changing the wing area, birds can lower the aerodynamic load on the wing and thus decrease the risk of flow separation.

The primary feathers at the wing-tip of birds that soar over land spread out to form typical slotted tips (see, e.g., Fig.6). It has been shown by theoretical calculations as well as by wind tunnel measurements that these slotted tips can reduce the induced drag by acting as winglets on airplane wings, i.e.,by spreading vorticity both horizontally and vertically [1],[6].

The reduction of drag on wings with elliptical lift distribution as compared with planar wings without slotted tips amounts up to 25%.

Slotted wing tips are typical of birds that soar over land. Birds that soar over the sea (like gulls and albatrosses) usually have longer and narrower wings with pointed tips to reduce the induced drag. Slotted tips enable shorter wings than the narrow pointed wings without tip slots. Short wings are lighter than the narrow wings with pointed tips, producing the same amount of lift because they have lower root bending moment. They are also more manoeuvrable – they can roll and yaw at higher angular velocities for a given tip speed. All that is advantageous for birds flying in constraint environment like forests and soaring in unpredictably changing gusts over land. Sea birds are adapted to different living conditions. They operate at higher average speeds, and slotted tips may become a liability at these speeds, increasing profile drag more than reducing induced drag.

What are the means the bird wing has to increase instantly the lift or suddenly stop the flight at an perching point?

An airplane wing has a variety of flaps, slots and slats which effectively alter the shape of the wing, increase its curvature and allowing thus to divert more air, and thus create more lift [1],[5]. However, this alteration also increases drag, which helps a landing airplane to slow down, but necessitates more thrust during take-off. Slats perform the same function as flaps, changing temporarily the wing shape to increase lift on takeoff and landing. They deploy from the wing leading edge, like the flaps deploying from the trailing edge. Besides these, airplane wings have wing brakes to slow down the plane at landing. To minimize the induced drag at the wing tips modern airplanes have winglets to reduce the aircraft's drag by partial recovery of the tip vortex energy..

The bird wings have several possibilities how to obtain the same functions as the above described arrangements on airplane wings. All are realized by the system of flight feathers (see Fig.1), together with the ability to increase the camber, and also to exploit the aerodynamics function of the tail [7].

The tail is an integral part of the bird´s lifting system. Unlike the tail of an aircraft, it is very close to the trailing edge of the wing and helps to make the wing morphology more versatile. It has a variable area and a variable aspect ratio. It maintains the stability and balance, especially in slow flight, where the tail is widely spread to enhance the flow over the wing in a similar way as an extended slotted flap from the trailing edge of an aircraft wing.

The **underwing coverts** (clearly seen in Fig.3) open automatically along the leading edge at high angles of attack. They operate as an automatic high-lift device, analogous to a Krueger flap, increasing the wing´s effective camber [8].

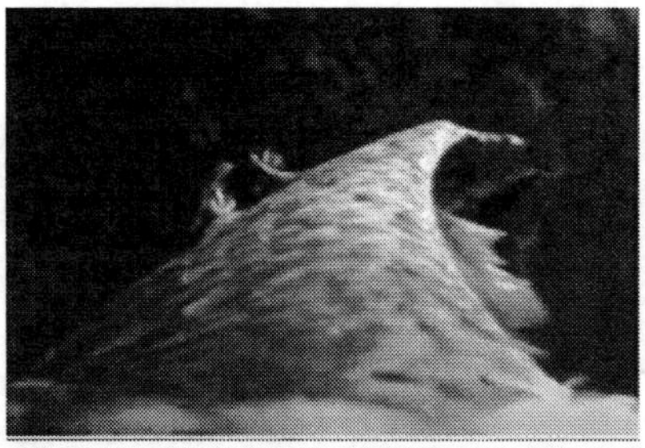

Figure 3. Deployed underwing coverts.

Take-off is one of the most energetically demanding manoeuvres of flight Small birds facilitate the take-off by jumping up, while larger birds take a run-up to receive the take-off velocity or simply try to receive it by facing into the wind. When perching on a branch or cliff all they need to start is to drop off into the air.

Landing is also a problem for many large birds with high airspeeds. In order to lose height and velocity rapidly prior to landing, large birds often use a series of sideslips.

The actual landing manoeuvre of all birds proceeds in a very similar way.

During the landing manoeuvre the tail is widely spread, extending thus the wing surface and its camber to generate maximum lift as possible. Close before perching, the tail moves up pushing the bird upwards (or even partly backwards) to the position where the wings stall and the bird slowly perches down. To enhance braking immediately before perching, the bird may use a few wingbeats with wings fully extended and set against the flow direction, the tail again lowered down to increase the lift (see, e.g., Fig. 4).

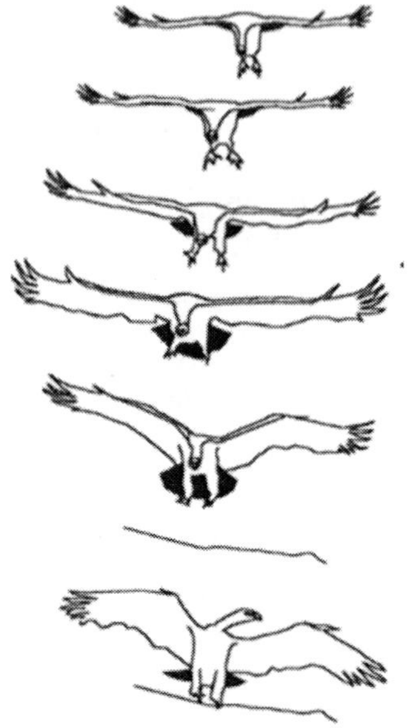

Figure 4. This picture is adapted from the ALR Thomas (1993) paper [7], and shows the last sequences of the landing manoeuvre of a white-backed vulture (*Gyps africanus).*Time between successive images is 1/5 of a second. Note especially the movements of the tail.

Figure 5. Picture of an Andean condor (*Vultur gryphus*) preparing for landing. Notice the raising up of the covert feathers on about one third of the wing span from the body and the widely spread tail.

Under normal flight conditions the tail acts independently of the wings as a kind of a control surface (both horizontal and vertical). Measurements in the wake have not detected the presence of separate wake of the bird´s body. This means that from the aerodynamics point of view the tail has to be treated as an integral part of the whole bird, at least together with the wings.

Flight at low speeds or flight with high angles of attack when high lift is required is inevitably accompanied by a risk of flow separation. Separation starts near the trailing edge but very quickly moves upstream to the low pressure peaks at the wing crest or near to the leading edge, leading ultimately to a complete stall. However, this reverse flow can be slowed down or even eliminated by self-activated flaps – spanwise fences on the wing upper surface [1] formed by covert feathers raising up automatically when necessary (see Fig.5).

The function of a leading edge slot in bird wings takes up the **alula** – feathers anchored in the first digit (see Fig.6).

Figure 6. Landing manoeuvre of the Northern gannet (*Morus bassanus)* documents well the use of the alula, the tail and wing morphing.

The wing-tip-slots do not eliminate completely the vortices at the wing tips (induced vortices), especially in lower aspect-ratio-wings, as, e.g., in passerines. They are generated during the downstroke phase and strengten by gradually engulfing the trailing edge vortex sheet. The strong downward-induced velocity thus effectively enhances the lift. Traditionally, tip vortices are considered as a nuisance increasing the drag, however, in low-aspect-ratio-wings they can favourably enhance the lift. The downward velocity behind the flyer is accompanied by an upward velocity field outside the wake, and in V-flight formation of migrating birds this can be used by the follower´s wing as an energy saving contribution to the generated lift (Fig.7) [1].

The increased downwash is also beneficial at near-to-ground flight (*ground effect*, skimming over the water surface), giving the bird considerable performance advantages, including a reduction in mechanical flight power of as much as 35%, compared with values for flight out of ground effect.

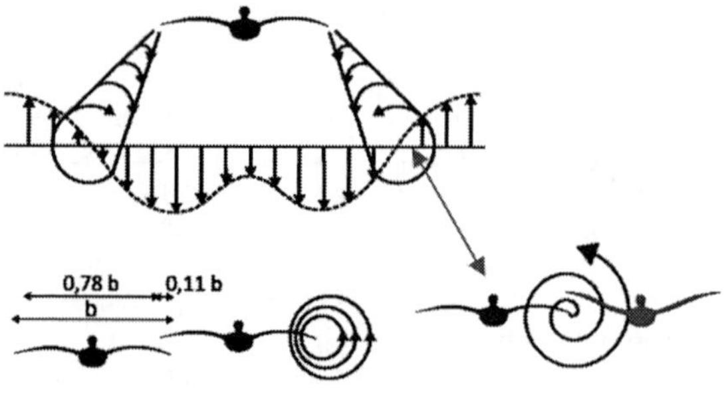

Figure 7. Flight in V-formation to save energy.

4 How do birds actually generate lift?

To fly, wings impart downward momentum to the surrounding air and obtain lift by reaction. This is Newton´s 3rd Law, which says that action and reaction are equal and opposite to each other. The lift force, which counterbalances the weight, is obtained from Newton´s 2nd Law.

To obtain the downward momentum, we have to know the downward component of the velocity deflected by the wing and the mass flow which is proportional to the air density, the wing area and the downward component of the velocity. This velocity component depends on the flight speed and on the angle of attack.The mass flow due to the deflected wing depends on the wing area, wing curvature, and in birds also on various lift-increasing-manipulations with the feathers. However, in principe, lift depends primarily on the angle of attack!

With the only exception of a hummingbird, birds generate lift and thrust by flapping the wings.This is a complex unsteady and threedimensional motion of the wings, changing at every instant with the new position of the wings. The aerodynamic analysis of bird flight was usually based on the **quasi-steady assumption,** according to which all instantaneous forces on a flapping wing in unsteady motion are assumed to be those corresponding to steady motion at the same instantaneous velocity and attitude. This assumption may be misleading (as is definitely the case when dealing with insect flight) and it is recommended to consider even the aerodynamics of the bird flight as unsteady.

Of course, wings not only generate lift and thrust, they cause drag as well. Optimally designed wings have to bring the air from upstream to downstream with as little loss of momentum, i.e. loss of drag, as possible and at the highest lift as possible.

Lift and drag are two components of the resulting aerodynamic force acting on the wing. The actual size of the components of the resulting aerodynamic force on vertical and horizontal axes depends primarily on the argument and magnitude of the wingbeat velocity (see, e.g., Fig.8)

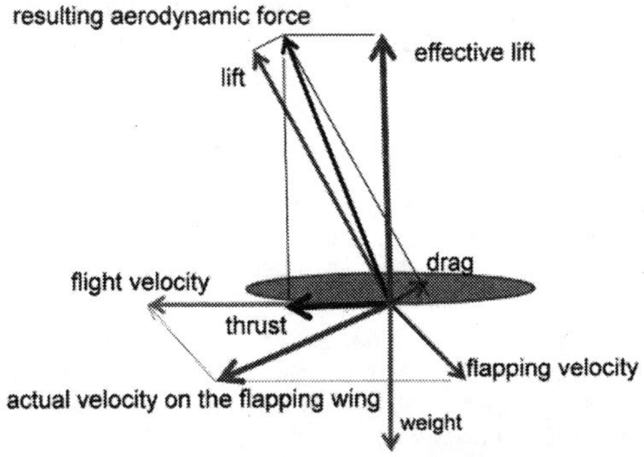

Figure 8. Aerodynamic forces on a flapping wing

The many facets of bird wing shapes and movements make a unifying theory of how lift is generated a difficult task. The only fact we can take for granted is that the mechanism of lift generation is ruled by vortices. The vortices drive in the surrounding air and impart downward momentum to this air [9]. The way how they do it depends on the wing shape, flapping velocity, as well as on the flight velocity. We will discuss here three possible models: high aspect ratio wings in cruising flight, high aspect ratio wings in more "energetic" flight and low aspect ratio wings.

The bird wings, however, are far from being twodimensional [10]. The tip vortices make the flow threedimensional and enter into nonlinear interaction with the spanwise shedding vortices. This interaction is, of course, significantly influenced by the wing shape, wing aspect ratio and the angle of attack. Especially on low aspect ratio wings the tip vortices can even completely destroy their original character of the vortex streets. The above described mechanism of producing thrust is thus valid only for cruising flight of birds with large aspect ratio wings with minimized tip vortices, like, e.g., condors, vultures or albatrosses.

During the downstroke phase of high aspect ratio wings typically two counterrotating vortices are generated – on the leading edge and on the trailing edge (Fig.9) [9], both being shed with the stream Fig.10). This vortex couple drives the air downward, generating thus lift in the opposite direction. On low aspect ratio wings the tip vortices on either wings dominate, forming again a couple to generate lift. Naturally, in this case interaction with the spanwise oriented vortices cannot be avoided (see, e.g., Fig. 12), as apparent from the complex wake structure [11].

Wake is a kind of footprint which any flying body leaves behind in the suroundin atmosphere. It represents not only the loss of momentum, comprising most of the losses generated by the flying body, but its structure also tells a lot about the behaviour of the body during the flight, especialy when the body contours are changing.

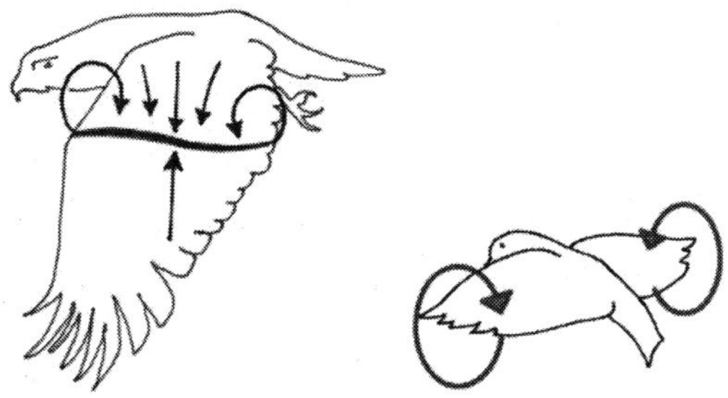

Figure 9. Vortex couples generating lift, a)high aspect ratio wings, b) low aspect ratio wings

Figure 10. Vortices shed off with the stream.

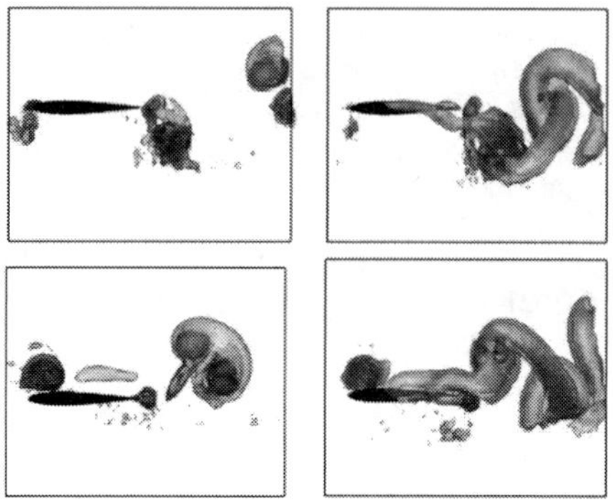

Figure 11. The left pictures show vortex shedding on a 2D profile at two different times, the right pictures refer to the ends of a3D profile at corresponding times and velocities and show interaction with the tip vortices.

The wakes behind fixed 2D wings with blunt trailing edges, or behind 2D blunt bodies, have a typical character of a Kármán vortex street (Fig.12a), [1].

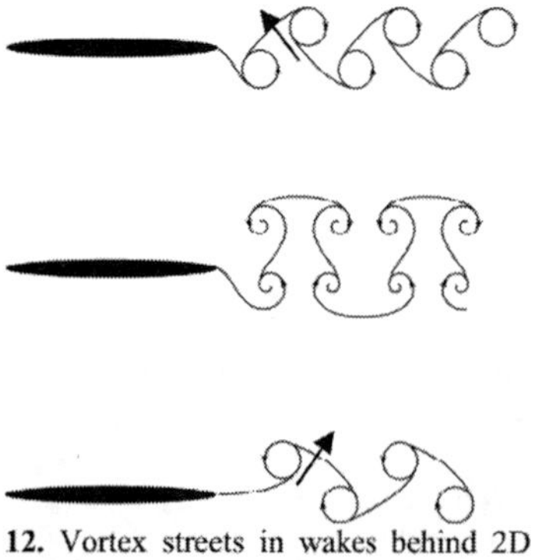

Figure 12. Vortex streets in wakes behind 2D fixed and oscillating airfoils.a) is a classical Kármán vortex street producing drag, b) mild lateral oscillations of the airfoil, the vortices are changing, c) higher frequency of oscillations generate a vortex street with opposite sence of vortex rotation to a), producing thus thrust.

However, provided the wing is heaving (moving in a direction perpendicular to the velocity of the oncoming flow), the vortex pattern in the wake may become different, being now dependent on the frequency of this movement. Behind a slowly heaving wing the wake is still similar to that behind the blunt trailing edge of a fixed wing.The sense of rotation of the vortices results from the momentum deficit behind the wing and represents net drag. By increasing the frequency the vortices in the wake gradually acquire opposite sense of rotation, the wake becomes "inverted" and the wing produces thrust (see, e.g., Fig.12c). This effect has been known since Knoller (1909) and Betz (1912) perceived that flapping a wing in a free stream flow resulted in an effective angle of attack with a normal force vector containing both lift and thrust components. This effect was experimentally demonstrated later by Katzmayr (1922) and revisited by Lai and Platzer in several papers published about the end of the twentieth century (see, e.g.,[12]. The effect of inverted Kármán vortex street is used by slowly soaring birds for occasional increase in lift. They slowly flap the wings up and down and generate thus the necessary lift.

Wakes behind flapping wings are much more complex than wakes behind fixed wings. Some time ago it was generally accepted that the downstroke is important in generating lift and thrust, and, therefore, it is sufficient to study only wakes in the downstroke phase. However, it has soon been realized that both the downstroke- and upstroke-generated wakes are equally important in building up the wake character. Moreover, it has also been proved that wake depends not only on the species in question, but in the same species it depends also on the particular gait (purpose of flight).

The wake develops while the wing flaps down and up and especially during downstroke the induced vortices play a considerable role. The tertials (the feathers nearest to the body) act in most cases as fixed wings and provide weight support throughout the whole cycle. Analysis of direct pressure measurements has suggested that the wingtip may be active even during the upstroke, providing some thrust. Fig.13 is just an example of a real wake to demonstrate its complexity [13].

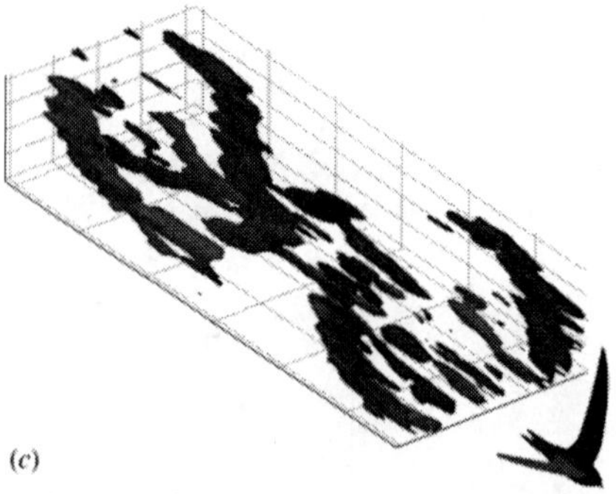

Figure 13. Iso-surfaces showing 1.5 wingbeats for the speed 7.7 m s^{-1}. Iso-value is 60 s^{-1}. The view is oriented so as if the bird was flying obliquely towards the observer from left to right. Red-coloured patches represent positive circulation (clockwise rotation as seen in direction of flight) and blue-coloured patches represent negative circulation (counterclockwise rotation as seen in direction of flight).Adapted from [13].

5 Experimental bird aerodynamics

This is a conference on experimental fluid dynamics so it is quite natural to ask whether it is possible to learn more about bird´s flight from laboratory or *in vivo* experiments.

Cruising flight has been filmed many times and these films, however interesting they may be, do not answer the most interesting (from the aerodynamics point of view) sequences of bird's flight like take-off and landing. This has changed in the last decades thanks to high speed cinematography and recently even to videorecording using miniaturized cameras placed directly on the bird's body (see, e.g., [8], or Fig.3 of this paper).

The wind tunnel tests were aimed at two groups of problems – direct measurement and evaluation of lift and drag from measuring in wakes, and *in vivo* measurements of the origin of these forces.

Wakes behind flying birds can be investigated experimentally in wind tunnels by application of several highly sophisticated experimental methods developed for experiments on airplane models.

It would be hardly possible to rely on any pressure measurements in the wake behind flapping wings. The only chance has been offered by the digital PIV (see, e.g.[14], or Fig.14 and Fig.15). However, a qualified interpretation of these results was first possible only after a CFD simulation of the same situation was available.

Experiments [14] were carried out using a closed-loop, low-turbulence wind tunnel designed for bird flight experiments. Four juvenile thrush nightingales *Luscinia luscinia* L. were caught, and after a period of acclimatization they were exposed to daily flight training for more than two months prior to experiments, under conditions that gradually resembled the experiment. Training started with low ambient light conditions, introduction and maintenance of fog particles, and occasional bursts of high intensity laser light.

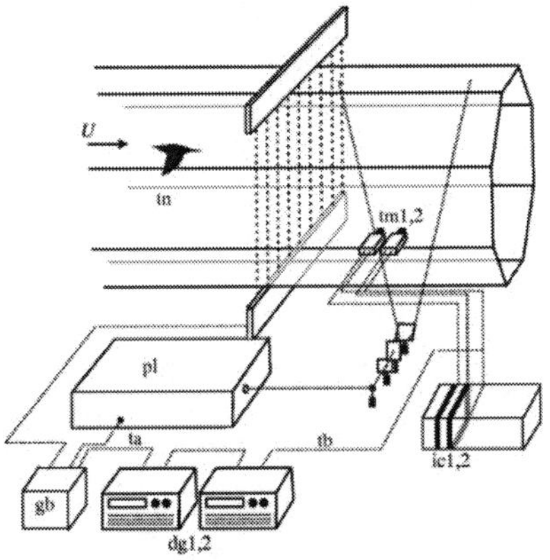

Figure 14. The bird (tn) is flying at constant speed, *U* is the independently controlled speed of the wind tunnel. Two delay generators (dg1,2), configured to run off a single crystal base timing clock, generate synchronised timing pulses to control the dual-head Nd:YAG (pl) laser and the asynchronous reset (tb) for the two CCD array cameras (tm1,2). Digital images are acquired at independent interface cards (ic1,2) and transferred directly to PC RAM. The laser timing pulses are gated (gb) with the summed output from an array of LED-photodiode pairs so that if any one or more beams are interrupted by the bird, laser output stops (Modified from[14]) .

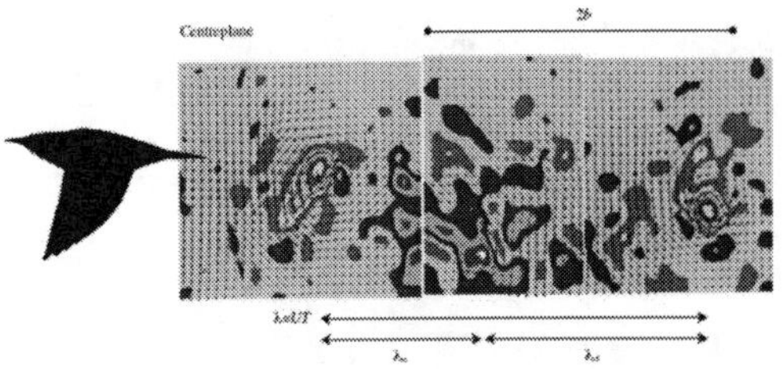

Figure 15. A reconstruction from three consecutive frames to show the vortex wake over slightly more than one wingbeat cycle. The wake is shown as if left in still air by the bird passing from right to left, from [14].

Recently, a special wind tunnel has been designed - the so called *Aerodynamic force platform* (AFP), [15], to measure directly the forces on bird wing during the bird flight. The AFP is a box instrumented with load cells enclosing the object (the bird) that generates the unsteady fluid force. According to Newton's 3rd law applied to a fluid, the unsteady net fluid force needs to be supported by an equal and opposite net force that acts on the control volume boundary.

Whereas the unsteady and convective terms are significant in the bulk fluid, they vanish at the surface of the AFP. The contour integral is thus simplified and when even the viscous shear forces acting on the wall can typically be ignored, to obtain the net aerodynamic force we can just integrate the pressure as measured on the wall.

The aerodynamic forces on the bird wing during take-off and landing were successfully measured *in vivo* by pressure transducers and accelerometers built in directly in the wing (see [16] and Figs. 16).

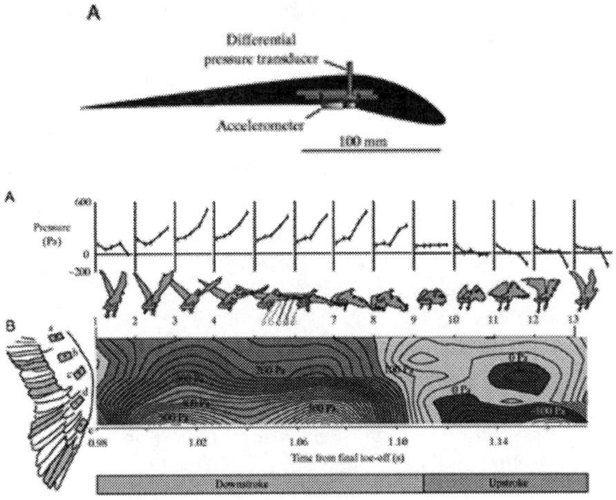

Figure 16. Pressure distribution related to inematice for a wing over a single flap during take-off flight, but five flaps after lift-off. The pressure distribution along the wing (A) show snapshots separated by 4/250 s. The contour plot (B) shows the pressure distribution through time as it would be mapped to an outstretched wing. Contours are calculated assuming a linear distribution of pressure with spanwise distance between the five measurement sites indicated on the wing image (labeled a–e). Positive pressures indicate a ventral-to-dorsal sense; the wing base maintains a weight-supporting sense throughout the wingstroke cycle.(adapted from[16]).

Also, further *in vivo* measurements were successfully performed with surgically implanted strain gauges of mechanical work and power output from the main muscles (see Fig.17, or[17]).

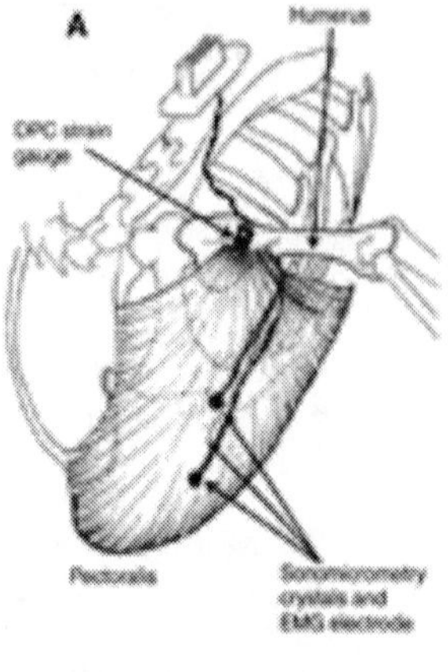

Figure 17. *In vivo* measurements of mechanical work and power output from the pectoralis.The primary downstroke muscle has surgically implanted strain gauges calibrated to measure force from bone strain on the deltopectoral crest of the humerus and sonimicrometry crystals to measure muscle length.

Closing remarks

Investigation into animal flight has its value not only for aircraft designers of the MAVs, but it can be inspiring for all those who do research in fluid dynamics.

The sole issue of generating lift and thrust by moving wings may even motivate us to reassess our knowledge on the same issue from the classical fixed wing aerodynamics.

References

1. Dvořák R.: *How they fly(*In Czech), Academia, Praha, 2015
2. Azuma A.: *The Biokinetics of Flying and Swimming*, Springer V., Tokyo, 1992
3. Šestakova G.S.: *Strojenije kryl'jev i mechanika poljota ptic*, Izd. Nauka, Moskva, 1971
4. Lighthill M.J.:Aerodynamic aspects of animal flight, In: *Swimming and flying in nature*, Plenum Press,pp.423-491
5. Norberg U.M.: *Vertebrate Flight*, Springer V., Berlin Heidelberg New York, 1990
6. Tucker V.A.: *J.Exp.Biology*, **180**, 1993, pp.285-310
7. Thomas A.L.R. : *Phil.Trans.Royal Soc., London, B*, Vol.340, pp.361-380
8. Carruthers A.C., Thomas A.L.R., Taylor G.K. *J. Exp. Biology,* **210**, 2007, pp.4136-4149.
9. Dvořák R.,*Applied and Computational Mechanics*, Vol.8, No 2, 2014, pp.147-156
10. Shyy W., Y. Lian, J. Tang, D. Viieru, and H. Liu: *Aerodynamics of Low Reynolds Number Flyers*, Cambridge Aerospace Series, No22, 2008
11. Calderon D. E., D. J. Cleaver, I. Gursul, Z. Wang *Physics of Fluids,* **26**, 071907 (2014)
12. Platzer M.F.,K.D.Jones,J.Young,J.C.S.Lai, AIAA Journal,**46**,2008,No 9,pp.2136-2149
13. Henningsson et al.,*J R Soc Interface*, 2011 June6, 8(59);pp.807-816
14. Spedding, G. R., Rosén, M. and Hedenström, A. *J. Exp. Biology,* **206**, 2003, pp.2313-2344
15. Lentink D., Haselsteiner A.F., Ingersoll R. *J R Soc Interface,* 2015, 12:20141283.
16. Usherwood J.R., T.L.Hedrick, A.A.Biewener, *J.Exp.Biology*, **206**, 2003, pp.4051-4056
17. Tobalske B.W., *J.Exp.Biology*, **210**, 2007, pp.3135-3146

EPJ Web of Conferences **114**, 01002 (2016)
DOI: 10.1051/epjconf/201611401002

Enhanced pool boiling heat transfer on mini- and micro- structured surfaces

Robert Pastuszko[1,a]

[1]*Kielce University of Technology, Department of Mechanics, Al. Tysiaclecia P.P.7, 25-314 Kielce, Poland*

Abstract. The surfaces used for investigating nucleate pool boiling for four working fluids had mini- and micro-fins of variable configurations, cross-sections and pitches, restrained by perforated foil or mesh cloth with various pore/opening diameters. Unique enhanced structures on these surfaces formed a system of interconnected horizontal and vertical tunnels. Four structured surfaces were proposed, each being a system of subsurface tunnels connected to 10 and 5 mm fins or 1 and 0.5 mm mini-fins. Measurement results for boiling water, ethanol, Fluorinert FC-72 and R-123 from more than 60 samples constituted the database used to verify the proposed theoretical models. These models were based on the results from the visualization studies, including internal visualization allowing observation of bubble nucleation, growth and displacement inside the tunnels, and on the analysis of existing boiling models for mini- and micro-structures.

1 Introduction

Traditional cooling methods based on natural or forced convection used in thermal management of electronic devices and systems are not capable of achieving the required performance in terms of effective heat removal. Harnessing boiling processes occurring on enhanced surfaces with an adequately formed system of subsurface tunnels is a very effective cooling option. The change of phase that accompanies a boiling process provides high heat fluxes at small temperature differences between the heating surface and the saturated fluid, thus increasing the heat transfer coefficient and reducing the size of heat exchanger.

The author of this paper and his collaborators at the Kielce University of Technology have been carrying out in-depth research devoted to determining pool boiling heat transfer coefficients for various working fluids from surface structures with mini- and micro-fins and/or subsurface tunnels and constructing theoretical models [1-7]. At the same time, the research team headed by Dr. M. Piasecka is studying flow boiling heat transfer for FC-72 in rectangular minichannels with enhanced heating surfaces [8,9,10].

This article summarizes the results of the analysis of pool boiling from four types of mini- and micro-structured surfaces.

2 Experimental research

2.1 Investigated surfaces

2.1.1 Tunnel structure (TS) surfaces

The specimens with the tunnel structures (TS) formed a square with sides of length 27 mm (w_f – figure 1).

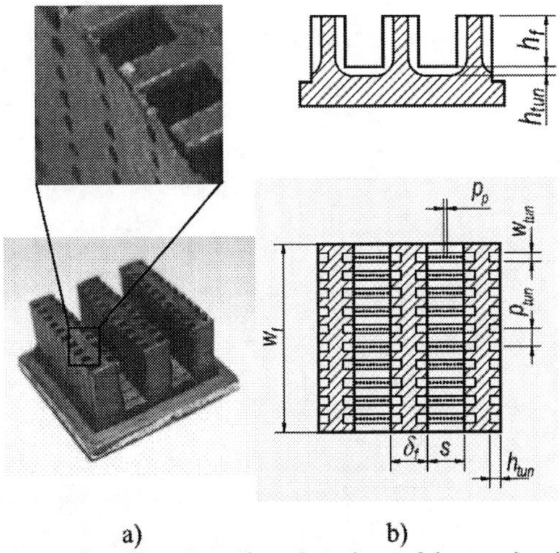

a) b)

Figure 1. TS surfaces investigated: a) views of the sample with 10 mm high fins, b) cross sections [1].

---

[a] Corresponding author: tmprp@tu.kielce.pl

Three main fins, 5 or 10 mm high (h_f) and 5 mm thick (δ_f), with inter-fin space of 5 mm (s), were modified to create tunnels on the vertical surfaces and in the horizontal inter-fin spaces. Perforated copper foil was sintered to the machined surfaces thereby forming a structure of combined U-shaped tunnels. The following variable parameters were used (figure 1):

- pore diameters: 0.3 – 0.4 – 0.5 mm (d_p),
- pore pitch: 0.6 – 0.8 – 1.0 mm (p_p=2d_p),
- tunnel pitch: 2.0 – 2.25 – 2.5 mm (p_{tun}).

2.1.2 Narrow tunnel structure (NTS) surfaces

The main fins, milled through, formed mini-fins with width corresponding to the thickness of the base fins. In addition, an array of interconnected narrow tunnels was obtained, closed with sintered perforated foil (figure 2). Constant parameters of the main fin geometry were identical to those on TS surfaces. Constant tunnel pitch of 2 mm (p_{tun}) was adopted. Variable parameters of the specimens were as follows:

- pore diameters (d_p): 0.3/0.32 – 0.4 – 0.5 mm,
- narrow tunnel width (w_{tun}): 0.6 – 1.0 – 1.5 mm.

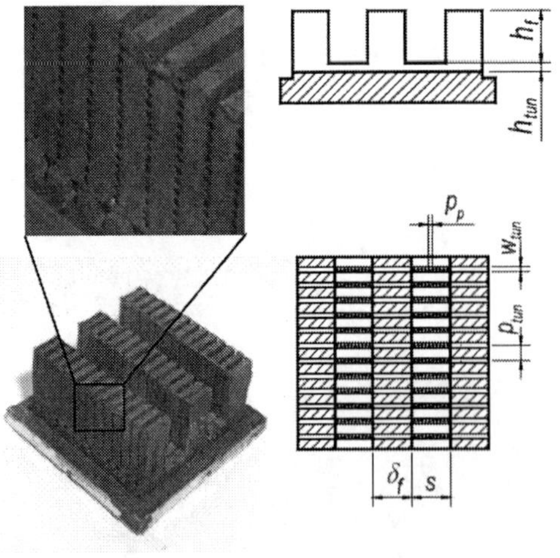

a) b)

Figure 2. NTS surfaces: a) views of the specimen with 5 mm high fins, b) cross sections [1,5].

2.1.3 Mini-fins with porous layer (MF+M/MF+F)

These are structural surfaces formed by sintering the copper wire mesh or perforated foil to the mini-fin tips. The copper specimens, square in shape with a side of 26.5 mm, have 112 mini-fins and form a system of tunnels (figure 3).

The following constant parameters of samples were used:

- mini-fin thickness: 2 mm (δ),
- inter-fin space width: 1.5 mm (s),
- tunnel pitch: 2 mm (p_{tun}).

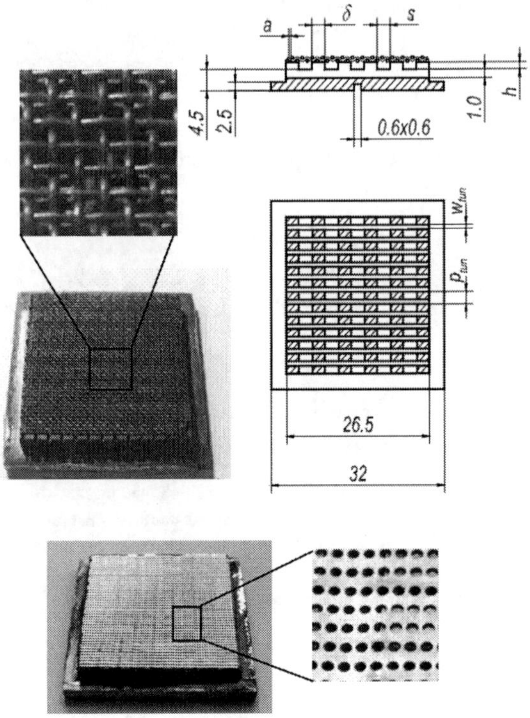

a) b)

Figure 3. MF+M (top) and MF+F (bottom) surfaces: a) views of the specimen, b) cross sections [3,6].

The following parameters were varied (figure 3):

- mesh aperture (a): 0.3 – 0.4 – 0.5 mm
- perforated foil hole diameters (d_p): 0.05 – 0.3 mm
- narrow tunnel width (w_{tun}): 0.6 – 1.0 – 1.5 mm,
- mini-fin height (h): 0.5 – 1.0 mm.

2.2 Experimental set-up

The experimental set-up designed for determining boiling curves and heat transfer coefficients was composed of the four modules and systems (figure 4).

2.3 Working fluids

The following boiling fluids were used:

- distilled water,
- ethyl alcohol,
- refrigerant R-123 – *dichlorotrifluoroethane*,
- fluorinert FC-72 – *perfluorohexane* or *tetradecafluorohexane* (C_6F_{14}).

Table 1 summarizes selected thermodynamic parameters of the liquid and vapour phases at saturation temperature.

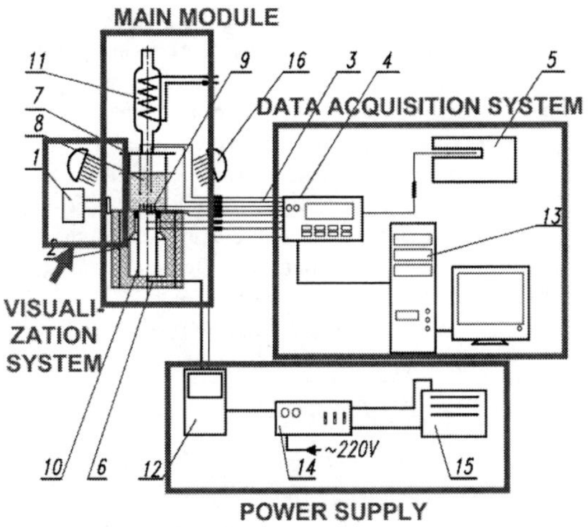

Figure 4. Measuring apparatus: 1 – digital camera, 2 – teflon casing, 3 – compensating leads, 4 – data logger, 5 – dry-well calibrator, 6 – insulation, 7 – glass vessel, 8 – boiling liquid, 9 – investigated sample, 10 – copper bar with cartridge heater, 11 – condenser, 12 – wattmeter, 13 – PC, 14 – power supply and fuses, 15 – autotransformer, 16 – lights.

Table 1. Thermodynamic parameters of the fluids

Saturation parameters (p=101.3 kPa)	Water	Ethanol	FC-72	R-123
T_{sat}, ^{0}C	100	78.3	56.4	27.5
ρ_l, kg/m^3	959	757	1602	1457
ρ_v, kg/m^3	0.597	1.43	13.24	6.39
μ_l, Pa s	$2.82\cdot10^{-4}$	$4.35\cdot10^{-4}$	$4.25\cdot10^{-4}$	$4.06\cdot10^{-4}$
c_l, J/kgK	4220	723	1103	1023
h_{lv}, kJ/kg	2251	963	94.9	170
λ_l, W/mK	0.68	0.169	0.055	0.077
σ_l, N/m	0.0589	0.0177	0.0081	0.015

3 Examples of results

3.1 TS surface – boiling of water

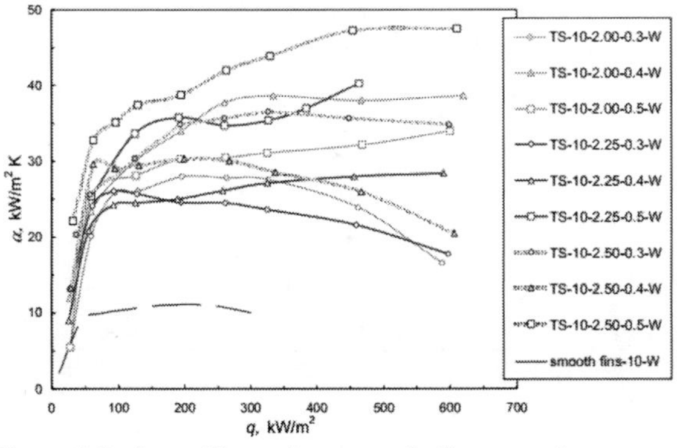

Figure 5. Surfaces with tunnel structure – boiling curves for water, TS surface code: h_f-p_{tun}-d_p; for smooth fins: h_f; W – water [5].

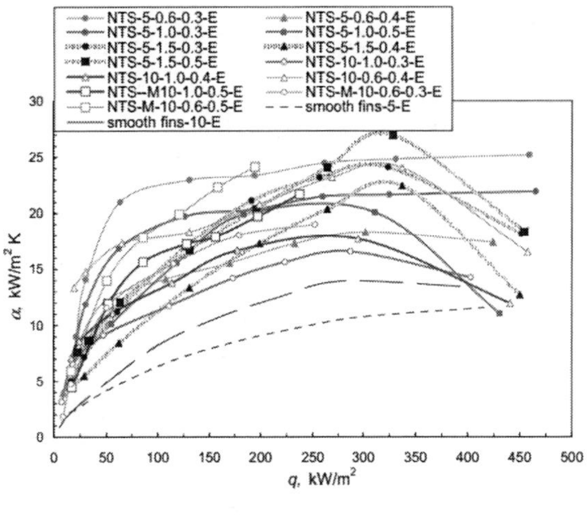

Figure 6. Surfaces with tunnel structure – boiling curves for ethanol, NTS surface code: h_f-p_{tun}-d_p, E – boiling of ethanol

Figure 5 shows the relationships between heat transfer coefficient and heat flux obtained for 10 mm high main fins. The highest heat transfer coefficients, about 47.5 kW/m^{2}K, were noted for the 10 mm fins with tunnels spaced at the highest pitch (2.5 mm) and pores with the largest diameter (0.5 mm).

3.2 NTS surfaces – boiling of ethanol

As for ethanol, NTS surfaces are especially advantageous when the main fins are 5 mm high (figure 6). An increase in heat transfer coefficients for smooth fins (α/α_{sf}) and the NTS-5-0.6-0.3 surface is 3.3 – 2.3 in the range of q = 100 – 300 kW/m^2.

The analysis of ethanol boiling curves for NTS surfaces demonstrates the advantages of coatings with pores 0.3 mm in diameter coupled with the narrowest tunnels (0.6 mm). Two-sided supply in the vertical tunnels compensates evaporation of the ethanol, even when the pores are 0.3 mm in diameter. When 1.5 mm wide tunnels are used, perforated foil with 0.5 mm holes provides the tunnel with suitable supply of fluid, precluding it from evaporating.

3.3 MF+M/MF+F surfaces

Figure 7 compares two types of micro-fin-structured surfaces. The surfaces with micro-fins covered with perforated foil with openings 0.3 mm in diameter can be used only when heat fluxes are low (below 20 kW/m^2). The values of the heat flux coefficient are similar in the range 10 – 100 kW/m^2 for micro-fins covered with foil with openings 0.3 mm in diameter and micro-fins with a mesh when the mesh aperture is 0.5 mm.

Above 40 kW/m^2, plain micro-fins produced the highest heat transfer coefficients, reaching 8 kW/m^{2}K at about 100 kW/m^2. In the range 100 – 150 kW/m^2, the coefficients decreased dramatically. This can be explained by the dry-out of the inter-micro-fin spaces, a condition that precludes boiling crisis.

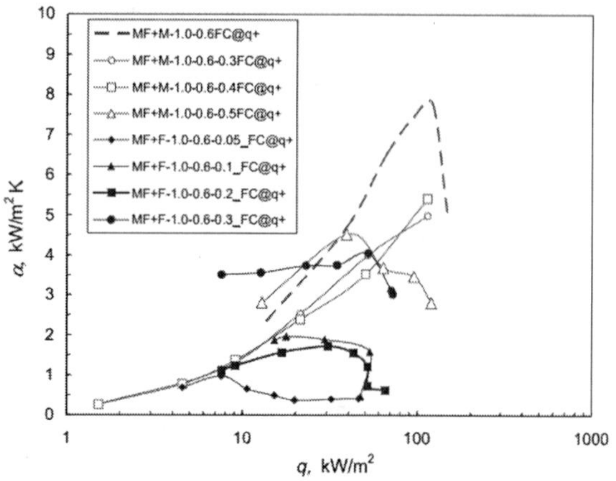

Figure 7. Surfaces with mini-fins with porous layer – boiling curves for FC-72, surface code: h_f-p_{tun}-d_p [3, 6].

4 Examples of boiling visualization

Identification of the phenomena that occur inside the confined space is the necessary condition to meet in order to describe theoretically the boiling process in tunnels, from the initiation to developed nucleate boiling to boiling crisis. To able to analyse nucleate boiling in the complex system of subsurface tunnels, the following data need to be found: the sites of vapour bubble nucleation, growth and departure, the sites of the influx of the fluid into the tunnels, and the sites of determining diameters of the departing bubbles together with their departure frequency.

The visual recording of boiling processes on the surfaces under investigation has been described by the author in several publications [2,4,6]. Examples below illustrate external visualization for the NTS surfaces with 10 mm high main fins.

Figure 8. Bubble growth at the vertical tunnel outlet, boiling water: NTS, h_f=10 mm, w_{tun}=1.0 mm, d_p=0.3 mm, ΔT=1.8 K, q=23 kW/m^2, 493 fps (every second frame is shown).

Figure 9. Bubble growth at the vertical tunnel outlet, boiling water, NTS, h_f=10 mm, w_{tun}=1.0 mm, d_p=0.3 mm, ΔT=2.1 K, q=37 kW/m^2, 493 fps (every second frame is shown).

At minor superheats (figure 8), there is a considerable difference in the number of active vertical tunnel outlets in each main fin. The appearing bubbles had spherical shapes; average bubble departure frequency at 1.8 K superheat can be estimated to have reached 16 Hz. The pores in the horizontal tunnels remained inactive.

With increasing superheat (figure 9), a larger number of vertical tunnel outlets become active and some of the departing bubbles are irregular in shape. The frequency of bubble departure increases to about 30 Hz. Marked differences in their diameters are observed.

5 Pool boiling models

Based on the own visualization studies and the existing boiling analytical models, the author proposed the models for boiling from four named above structured surfaces. These models have been thoroughly discussed in [2,4,6].

5.1 Model for TS surface [2]

In the semi-analytical model developed by the author, briefly described below, one of the mechanisms accounted for was the fluid evaporation in the corners of the tunnels, associated with the bubble nucleation cycle. This mechanism was considered together with non-isothermity of the main fin, i.e., assumed one-dimensional temperature distribution along the vertical tunnel walls. Another mechanism taken into account was that of the nucleation, movement and departure of vapour bubbles, different in vertical and horizontal tunnels. This model enables determining diameters of the departing bubbles, their frequency, density of nucleation sites and the overall heat flux for the system of interconnected subsurface tunnels confined with the perforated foil.

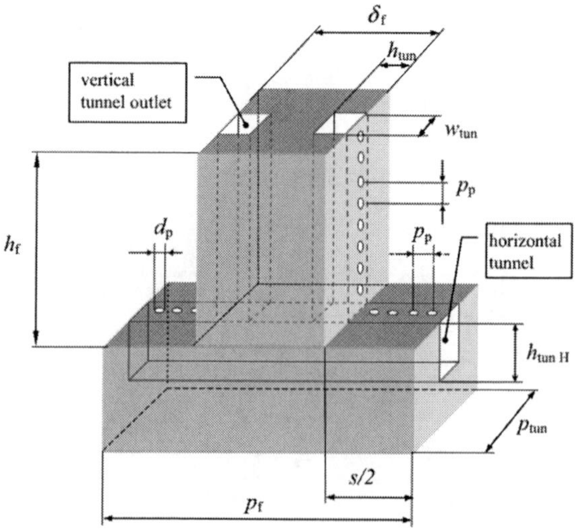

Figure 10. Interconnected vertical and horizontal tunnels of the TS segment in relation to the pitch of main fins and tunnels [2].

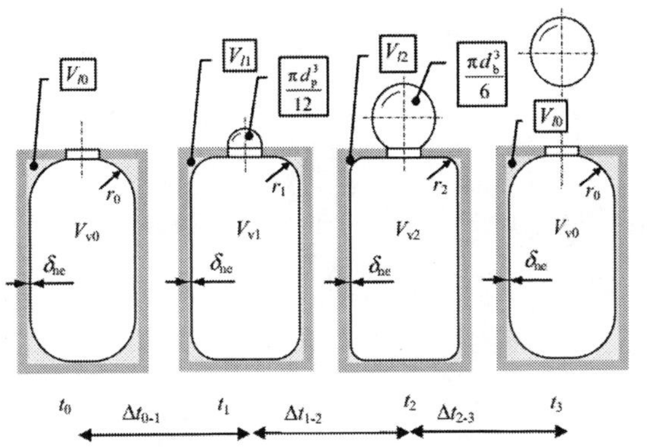

Figure 11. Fluid evaporation in the horizontal tunnel; bubble growth ($\Delta t_{0\text{-}1}$ – waiting period, $\Delta t_{1\text{-}2}$ – growth period, $\Delta t_{2\text{-}3}$ – period of filling with fluid).

5.1.1 Simplifying assumptions

The boiling mechanism in subsurface tunnels (TS – figures 10, 11) is defined by the following major simplifying assumptions:

- the holes in the foil act as elements supplying the surface tunnel structure;
- evaporation occurs from the menisci in the four corners of the tunnels;
- suction and evaporation occur independently for vertical and horizontal tunnels;
- the sites where the bubbles are released, that is, where vapour generated from the menisci in the tunnel corners outflows include: tunnels outlets in vertical tunnels and holes in the foil (pores) in the horizontal tunnels;
- bubble formation cycle in vertical and horizontal tunnels comprises a waiting period ($\Delta t_{0\text{-}1}$), a growth period ($\Delta t_{1\text{-}2}$) and a period of filling with fluid ($\Delta t_{2\text{-}3}$) – figure 11.

5.1.2 Heat flux calculation algorithm for the TS surface

These input data were used:

- Parameters: geometric, of the boiling fluid and the material of the structure: h_{tun}, w_{tun}, p_{tun}, h_f, δ_f, w_f, d_p, p_p, $\rho_{v\,sat}$, ρ_l, σ, λ_v, λ_l, i_{lv}, T_{sat}, λ_{Cu}.
- Literature-based constants: Hamaker's constant, growth period constant $C_{1\text{-}2}$.
- Experimental constants: exponent n in dependence $\alpha \sim \Delta T^n$

The following algorithm for heat flux determination is proposed:

1. Calculation of vapour density in the horizontal tunnel at the end of the waiting and growth periods.
2. Determination of the volume of fluid in the waiting and growth periods.
3. Calculation of the initial volume of fluid.
4. Calculation of the volume of fluid at the end of growth period according to the relationship.
5. Calculation of the evaporation heat transfer rate.
6. Calculation of the evaporation heat flux in the horizontal tunnel.
7. Determining the waiting and growth periods.
8. Calculation of bubble departure frequency in the horizontal tunnel.
9. Determination of the nucleation site densities in the horizontal tunnel.
10. Determination of the external and overall heat flux for the horizontal tunnel (q_H).
11. Calculation of the volume of fluid for the vertical tunnel according to items 1 – 4, after the introduction of hydraulic diameter.
12. Calculation of evaporation heat transfer rate from the four menisci of the vertical tunnel, using the height of the replacement cross-section.
13. Calculation of fin parameter m_{fm}.
14. Determination of the heat flux in the vertical tunnel ($q_{tun\,V}$).
15. Determination of nucleation site densities for the vertical tunnel (n_{sV}) and the frequency of the bubble departure (f_V).
16. Calculation of the external heat flux for the vertical tunnel (q_{exV}).
17. Calculation of the overall heat flux for the vertical tunnel (q_V).
18. Determination of the overall heat flux for the system of interconnected horizontal and vertical tunnel (q).

Figure 12 compares calculated heat flux and experimental data obtained during the boiling of water.

In the case of flat, homogeneous surfaces with horizontal tunnels, the model is simplified to the calculation of heat flux only according to the procedure for the surfaces with horizontal tunnels – exponent n necessary in the case of the TS complex structures, do not have to be calculated. The calculations were verified against the literature-based measurement data for tunnel structures for the boiling of R-123 and R-11 (figure 13).

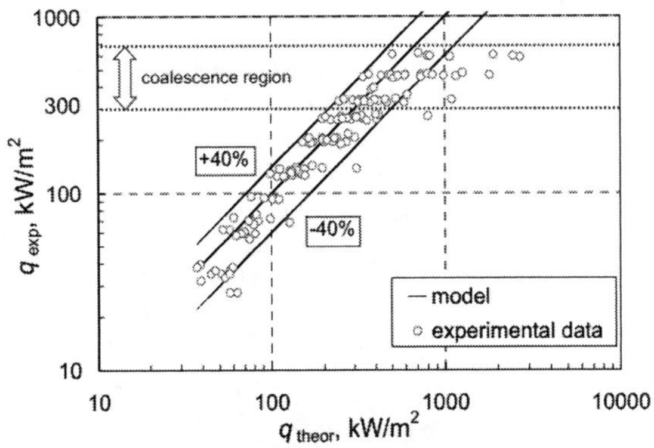

Figure 12. A comparison of experimental and theoretical heat fluxes for the boiling of water [2].

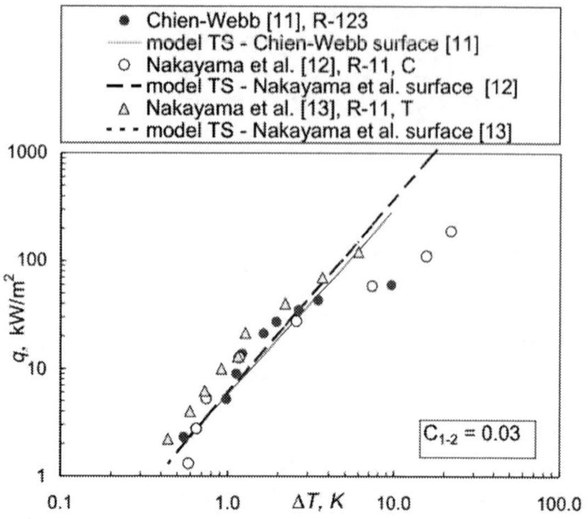

Figure 13. Verification of the model for experimental data from publications [11,12,13].

5.2 NTS surface model

This model, which is a modification of the one presented in [3], treats the area of the vertical tunnel and the area of adjacent horizontal tunnels jointly. One assumption remained – the holes in the foil are made only to supply the tunnels with fluid. Compared with the TS surface model, here the periods of waiting and growth are calculated in a different manner. To provide for the geometry of the structure, the author modified the Mikic and Rohsenow [14] relationship and introduced his own definition of the area of influence.

Additional simplifying assumptions include:

- pores in the horizontal tunnels remain inactive – the bubbles form only at the outlets of vertical tunnel,
- the overall heat flux, calculated jointly for the horizontal and vertical tunnel, comprises the evaporation heat flux inside the interconnected tunnels and the external heat flux associated with bubble departure.

Figure 14 shows the segment of the pitch of the main fins (p_f) and tunnels (p_{tun}) in relation to which the simplified model was formulated.

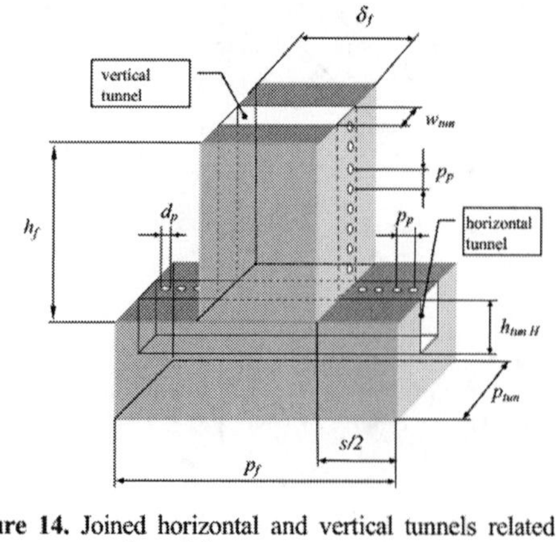

Figure 14. Joined horizontal and vertical tunnels related to the pitch segment volume [4].

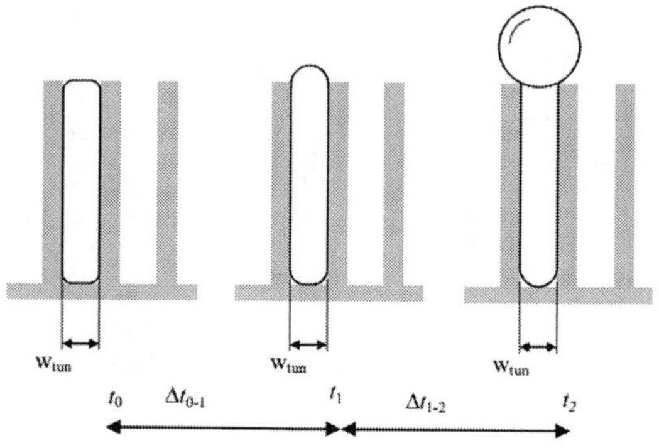

Figure 15. Liquid evaporation stages in the narrow vertical tunnel; bubble growth (Δt_{0-1} – waiting period, Δt_{1-2} – growth period).

The departing bubble diameter for NTS was calculated on the basis of buoyancy force and surface tension balance related to the hydraulic diameter of the vertical tunnel

$$d_{bV} = \left(\frac{6\sigma d_{hV}}{g(\rho_l - \rho_v)} \right)^{1/3} \tag{1}$$

The calculation of the waiting period was based on the modified Mikic and Rohsenow [14] relationship, assuming that this period starts at the moment when the area between the phases at the outlet of the vertical tunnel forms a horizontal plane, and ends when the bubble is formed having a hemispherical shape and the radius equal to the half of the tunnel width, i.e. $R_1 = w_{tun}/2$ (figure 15)

$$\Delta t_{0-1} = 0.19 \frac{1}{a_l} \frac{w_{tun}}{Ja} \tag{2}$$

where the Jakob number was defined by: $Ja = c_l \Delta T \rho_l / h_{lv} \rho_v$.

The modified Chien and Webb [15] relationship was used for the calculation of bubble growth period (Δt_{1-2}), taken that at the beginning of the growth period the bubble diameter $d_1 = w_{tun}$

$$\Delta t_{1-2} = \frac{1}{C_{1-2}} \left[\frac{7}{\pi} \frac{\rho_l T_{sat}}{h_{lv} \rho_v \Delta T} \frac{(d_{bV} + w_{tun})}{(d_{bV} - w_{tun})} \right]^{1/2} \left(\frac{d_{bV} - w_{tun}}{2} \right) \tag{3}$$

Apart for the fluid supply period, the frequency of bubble departure at the outlet of the vertical tunnel can be calculated as an inverse of the sum of waiting and growth periods:

$$f_V = \frac{1}{\Delta t_{0-1} + \Delta t_{1-2}} \tag{4}$$

The evaporation heat flux in the tunnel is expressed

$$q_{tun} = \rho_v h_{lv} n_{sV} \frac{\pi d_{bV}^3}{6} f_V \tag{5}$$

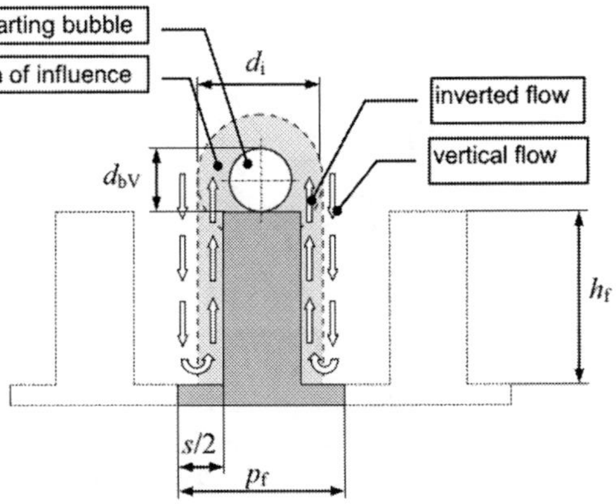

Figure 16. Area of influence for an NTS [4].

With the activity of all vertical tunnel outlets taken to be continuous, the density of nucleation sites remains constant and can be expressed as

$$n_{sV} = \frac{1}{p_{tun}(\delta_f + s)} \tag{6}$$

In relation to the main fin surface, the spatial domain of influence was defined as a surface covering the front of the fin and its lateral planes (figure 16). The dependence for the modified area of influence of the departing bubble has the following form

$$A_i = 2d_{bV}(2h_f + d_{bV} + \pi d_{bV}) \tag{7}$$

Convective (external) heat flux for defined area of influence is the following

$$q_{ex} = 2n_{sV} A_i \Delta T_{tip} \sqrt{\frac{\lambda_m c_l \rho_l f_V}{\pi}} \tag{8}$$

where ΔT_{tip} is the superheat at the tip of the mini-fin, corresponding to the superheat at the vertical tunnel outlet.

The overall heat flux for the surface with narrow tunnels is expressed by

$$q = q_{tun} + q_{ex} \tag{9}$$

Figure 17 presents a comparison of the experimental and theoretical data.

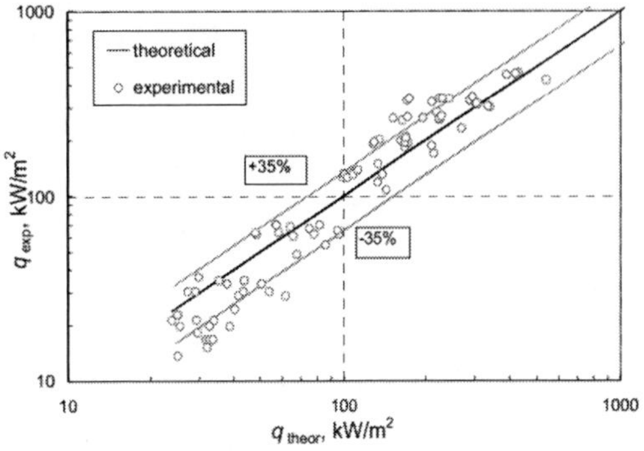

Figure 17. A comparison of experimental and theoretical heat fluxes for the boiling of ethanol [4].

5.3 Simplified model for surfaces MF+M/MF+F [6]

The calculation procedure is similar to that discussed in the previous simplified model (NTS). The diameter of the departing bubble was calculated according to (1). The way for the calculation of the waiting period was changed, assuming after Van Stralen et al. [16] that in pure liquids, the waiting period is related to the growth period by

$$\Delta t_{0-1} = 3\,\Delta t_{1-2} \tag{10}$$

To determine the growth period (Δt_{1-2}), d_p was substituted for w_{tun} in relationship (3). The value of C_{1-2} in this equation was selected through numerical simulation in terms of minimizing errors in the calculation of the frequency of bubble departure and heat flux.

Figure 18 presents the comparison of experimental data for Fluorinert with the values of q calculated from the presented model.

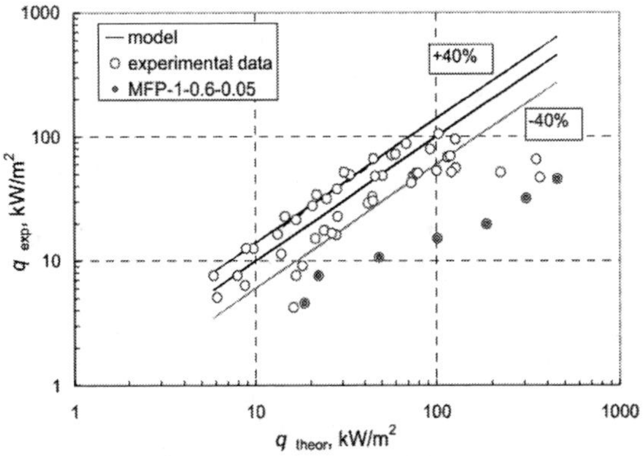

Figure 18. A comparison of experimental and theoretical heat fluxes for MFP surfaces at the boiling of FC-72 at $C_{1-2} = 0.0085$ [6].

Conclusions

- Compared with smooth surfaces, the mini and micro structured surfaces allow a substantial heat flux enhancement and up to seven times higher heat transfer coefficients.
- The presented models provide a prediction of heat flux within ±35% – ±40 % error margin for the boiling of water, ethanol and FC-72. Calculations require a selection of two (model for TS) or one experimental constant (NTS and MF+M/MF+F models) in the relationship for the growth period ($C_{1\text{-}2}$).
- These surfaces can be applied to the cooling of elements and systems which generate huge heat fluxes. They can be also used as evaporators of heat pipes or thermosyphons delivering heat to a Stirling engine heater.

References

1. R. Pastuszko, Exp. Thermal Fluid Science **32,** 1564-1577 (2008)
2. R. Pastuszko, M.E. Poniewski, Int. J. Thermal Sciences **47**, 1169-1183 (2008)
3. R. Pastuszko, Int. J. Thermal Sciences **49**, 2289-2298 (2010)
4. R. Pastuszko, Exp. Thermal Fluid Science **38**, 149-164 (2012)
5. R. Pastuszko, EPJ Web of Conferences **45,** 01020 (2013)
6. R. Pastuszko, T.M. Wójcik, Exp. Thermal Fluid Science **63**, 34-44 (2015)
7. R. Pastuszko, M. Piasecka, Journal of Physics: Conference Series **395,** 012137 (2012)
8. M. Piasecka, Annals of Nuclear Energy, **73**, 282-293 (2014).
9. M. Piasecka, Advanced Material Research, **874**, 95-100 (2014)
10. M. Piasecka, Heat Transfer Engineering **35**, 903-912 (2014)
11. L.-H. Chien, R.L. Webb, ASME-HDT **326,** 129-136 (1996)
12. W. Nakayama, T. Daikoku, T. Nakajima, 1982, J. Heat Transfer **104**, 286-291 (1982)
13. W. Nakayama, T. Daikoku, H. Kuwahara, T. Nakajima, Journal of Heat Transfer **102**, 445-450 (1980)
14. B.B.Mikic, W.M.Rohsenow, ASME Journal of Heat Transfer **91,** 245-250 (1969)
15. L.-H. Chien, R.L. Webb, Int. Journal of Heat and Mass Transfer **41**, 2183-2195 (1998)
16. S.J.D.Van Stralen, M.S.Sohal, R.Cole, W.M. Sluyter, Int. J. Heat Mass Transfer **18**, 453-467 (1975)

Nomenclature

a	thermal diffusivity, m^2/s
	mesh aperture, m
A	surface area, m^2
$C_{1\text{-}2}$	growth period constant
c	specific heat at constant pressure, J/kgK
d	diameter, m
f	frequency, Hz
h	height, m
h_{lv}	latent heat of vaporization, J/kg
Ja	Jakob number
MF+F	surface code (mini-fins + perforated foil)
MF+M	surface code (mini-fins + mesh)
NTS	surface code (narrow tunnel structure)
n	exponent in dependence $\alpha \sim \Delta T^n$
n_s	nucleation site density, m^{-2}
p	pitch, m
q	heat flux, W/m^2
R	radius, m
s	gap between fins, m
T	temperature, K
TS	surface code (tunnel structure)
t	time, s
w	width, m

Greek symbols

α	heat transfer coefficient referred to fin base surface, kW/m^2K
ΔT	wall superheat related to fin base surface, K
Δt	period, s
δ	thickness, m
λ	thermal conductivity, W/m K
ρ	density, kg/m^3
σ	surface tension, N/m

Subscripts

0	beginning of the growth period
1	beginning of the waiting period
0-1	waiting
1-2	growth
2-3	intake
b	departing bubble
ex	external
exp	experimental
f	fin
H	horizontal
h	hydraulic
i	influence
l	liquid
m	mean
p	pore
s	smooth
sat	saturation
theor	theoretical
tip	main fin tip
tun	tunnel
V	vertical
v	vapor

EPJ Web of Conferences **114**, 02001 (2016)
DOI: 10.1051/epjconf/201611402001

Effect of plasma actuator and splitter plate on drag coefficient of a circular cylinder

Hürrem Akbıyık[1,a], Yahya Erkan Akansu[2], Hakan Yavuz[1], and Ahmet Ertuğrul Bay[3]

[1]*Cukurova University, Mechanical Engineering Department, 01380 Adana, Turkey*
[2]*Nigde University, Mechanical Engineering Department, 51200 Niğde, Turkey*
[3]*Gaziantep University, Electrical and Electronics Engineering Department, 27310 Gaziantep, Turkey*

Abstract. In this paper, an experimental study on flow control around a circular cylinder with splitter plate and plasma actuator is investigated. The study is performed in wind tunnel for Reynolds numbers at 4000 and 8000. The wake region of circular cylinder with a splitter plate is analyzed at different angles between 0 and 180 degrees. In this the study, not only plasma actuators are activated but also splitter plate is placed behind the cylinder. A couple electrodes are mounted on circular cylinder at ±90 degrees. Also, flow visualization is achieved by using smoke wire method. Drag coefficient of the circular cylinder with splitter plate and the plasma actuator are obtained for different angles and compared with the plain circular cylinder. While attack angle is 0 degree, drag coefficient is decreased about 20% by using the splitter plate behind the circular cylinder. However, when the plasma actuators are activated, the improvement of the drag reduction is measured to be 50%.

1 Introduction

A circular cylinder has been studied by many researchers due to the fact that it underlies in flow around bluff body. There are many researches reporting on reduction of drag and wake region, vortex strength etc. and increase lift coefficient by using flow control methods.

As a passive flow control method, splitter plate can be used to reduce drag coefficient and vortex strength when it is placed behind the circular cylinder. Apelt et al. reported that they examined splitter plate behind circular cylinder in range of $L/D \leq 2$ at Reynolds numbers between 10^4 and $5x10^4$. They depicted that using a splitter plate reduces drag and produces narrower wake than that of bare cylinder. This means approximately 31% reduction in drag when L/D=1 [1]. Cimbala and Garg investigated plain circular cylinder, fixed/free rotatable splitter plate behind circular cylinder models in range of L/D=5 at $5x10^3<Re<2x10^4$. They reported that wake of free rotatable cylinder/splitter plate model is nearly same with plain cylinder. However, when the splitter plate was fixed at the 0 degree, the splitter plate changed wake. They also compared cylinder/rigid splitter plate with cylinder/free splitter plate in near wake about vortex formation length and Karman vortex street shedding frequency [2]. At $2x10^4<Re<8x10^4$ Cimbala and Leon exhibited that the change of drag coefficient of plain cylinder, cylinder/rigid splitter plate and cylinder/free splitter plate depending on L/D ratio. The change in drag was observed nearly the same in both splitter plate cases. However, when the splitter plate length was increased then drag was reduced [3]. Akansu et al. experimented the effect of the circular cylinder with splitter plate from 0 to 180 degree to observe pressure distribution, vortex shedding, drag and lift coefficient at $8x10^3<Re<6x10^4$. Drag coefficients were found at maximum and minimum values at 75 and 15 degree, respectively. While lift coefficient was observed to have two maximum values at the 15 and 165 degrees and minimum values were 0, 45, 158 and 180 degree [4].

As an active flow control method, plasma actuators are used to control flow around a circular cylinder on which they are mounted. A typical plasma actuator consists of two electrodes and a dielectric material between of them. One electrode is grounded and another one is connected to the high voltage. In literature, researchers widely used this application to reduce drag coefficient, noise level, wake region and vortex shedding. Thomas et al. mounted plasma actuators on a circular cylinder at ±90° and ±135° degrees for noise reduction and eliminating Karman shedding. They performed experiments at $Re=3.3x10^4$ [5]. Sung et al. altered flow separation point and flow field in the wake by using plasma actuators on a circular cylinder. At $Re=10^4$-$4x10^4$, six electrode pairs which were activated were placed on ±90°, ±120° and ±150° degrees of circular cylinder. They used forward and reversed electrode configuration to analyze flow around circular cylinder. When the forward electrode configuration was used, wake is reduced, whereas when reverse electrode

[a] Corresponding author: hakbiyik@cu.edu.tr

configuration was used, the wake significantly increased [6]. Tabatabaeian et al. demonstrated that the effects of electrodes were different configuration on flow structure around circular cylinder. Electrodes placed on 0 and 180 degree as first configuration. Then, they changed the position of electrodes on the rotating cylinder to different angles. They observed that using electrode changed pressure coefficient distribution, drag and lift coefficient of the flow around the cylinder [7].

In this study, in order to monitor the effects of flow around the circular cylinder, both flow control methods active (plasma actuator) and passive (splitter plate), is used. Force measurements are used to show the changes in drag and lift effects on a circular cylinder at Re=4x10^3 and 8x10^3. Wake region is also discussed for the experimental model's angle changed in every 15 degree. Flow visualization is made by using smoke-wire method.

2 Experimental setup

Experiments are conducted on an open loop suction type wind tunnel with a test section of 570x570x1000 mm^3. As seen in Figure 1, test model consist of circular cylinder, splitter plate, and plasma actuator. Circular cylinder which has 40 mm diameter and it is made of plexiglass. The length of the model is 400 mm. The splitter plate is also made from plexiglass. Length of splitter plate is 1D and thickness of splitter plate is about 4 mm. In order to produce the plasma one electrode is grounded and the other one is connected to a high voltage. Both electrodes have a width of 5mm. A kapton layer is placed between the two electrodes as a dielectric material. Symmetric electrode pairs are placed on the circular cylinder at ±90° degree angles. Homemade high voltage power supply was used for driving high voltage source. Plasma excitation frequency was set at 3500 Hz. Applied voltage was set at 6.5kV. Blockage ratio of model was 5% in the wind tunnel.

To measure drag and lift forces, ATI model six axis load cell is used. Measurements are taken 400 Hz sampling frequency and 50 values were averaged during 8 second. Therefore, 160 values were taken for each measurement every second and it has been repeated two times. Due to drag and lift forces of circular cylinder, it is examined in 2D: Drag and lift forces of endplates, holder rod, connector between model and holder rod are exiled from total drag and lift forces. In this way, the net drag forces are calculated.

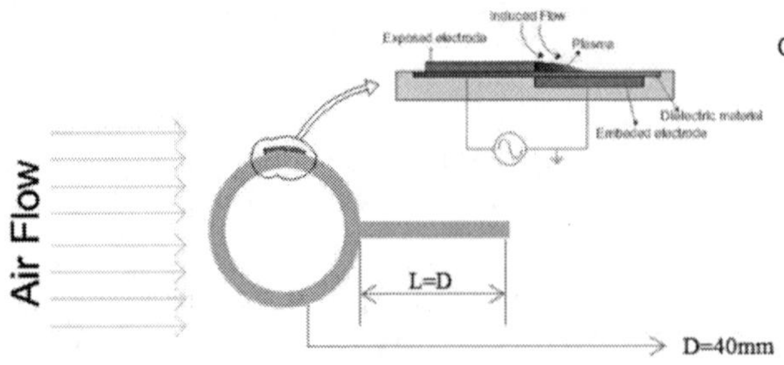

Figure 1. Schematic of the model.

Smoke wire method is used for monitoring the flow around the circular cylinder. Figure 2 show that the experimental setup model, connections of opponents of system and the wind tunnel. Home made two axis traverse system was used for positioning the rotary unit. ISEL rotary unit is used to set the angle of model.

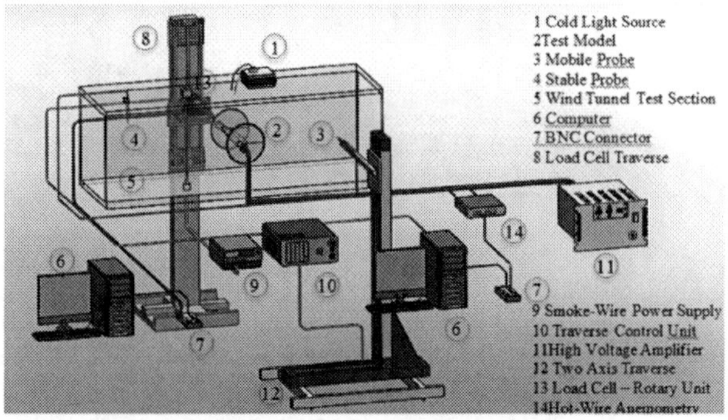

Figure 2. Schematic diagram of the experimental setup.

Multichannel CTA Dantec 54N81 model hot-wire anemometry is used to obtain free stream velocity. A stable hot-wire probe (55P16) is positioned 6D distance from front side of test model.

3 Results

3.1. Force measurement

In Figure 3, the drag coefficient versus attack angle is presented. Induced flow by means of the plasma is in the same direction via stream wise flow direction when the attack angle is between 0 and 90 degree. Thus, flow separation around circular cylinder is delayed by using plasma actuator. Also, the narrower wake region was observed and the reduction in drag was investigated. Induced flow by means of the plasma is reversed according to stream wise depending on the position of the placed electrodes when the attack angle is changed between 90 and 180. Therefore, wake region is widened and the drag coefficient of the circular cylinder is increased.

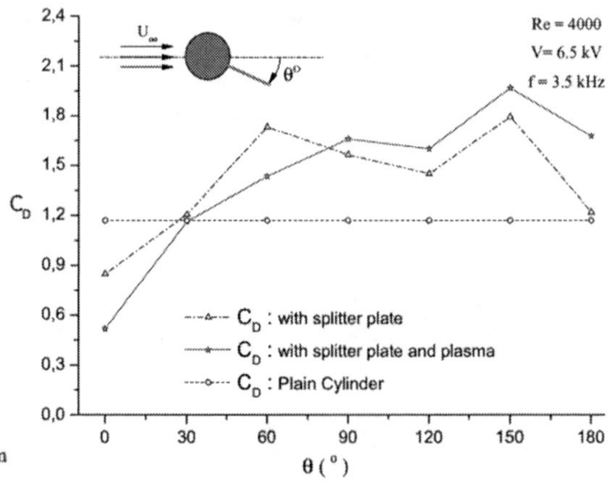

Figure 3. At Re=4000, different attack angle versus drag coefficient of circular cylinder with both plasma actuator and splitter plate.

In Figure 4, there is a very little change in drag coefficient. Because of increasing the effect of the plasma at the high Reynolds numbers, the voltage is needed to be increased [8].

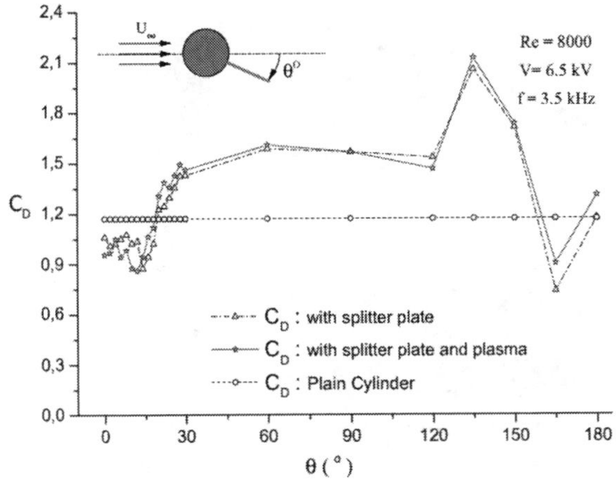

Figure 4. At Re=8000, different attack angle versus drag coefficient of circular cylinder with both plasma actuator and splitter plate.

3.2. Flow visualization

As it can be seen Figure 5, vortexes behind the circular cylinder are occurred regularly when the angle of attack is zero. If a splitter plate is placed behind the circular cylinder, vortex shedding behind the cylinder is occurred at a more far distance and in a smaller size. If both plasma actuators are placed on circular cylinder and splitter plate is used behind the circular cylinder, then both creations of vortexes are eliminated and shrinking of the wake region is observed.

If attack of angle is set to 180 degree, circular cylinder behind the splitter plate has narrower wake region than plain cylinder but vortexes are occurred more near at circular cylinder with splitter plate. When both plasma actuators are placed on circular cylinder and splitter plate is behind the circular cylinder, wake region is occurred that it is expanding. The reason of this is if circular cylinder's angle is changed, induced flow occurred by plasma actuators effect opposite direction to the flow. But, using plasma actuator, it is observed that it prevents occurrence of the vortex shedding.

As can be seen in Figure 6, for the angle of attack angle 30, 60, 90, 120, and 150, flow around the circular cylinder is experimented. When the angle of attack is 30 degrees, flow separation from the lower side of the cylinder delayed. Therefore, vortex shedding axis was deviated.

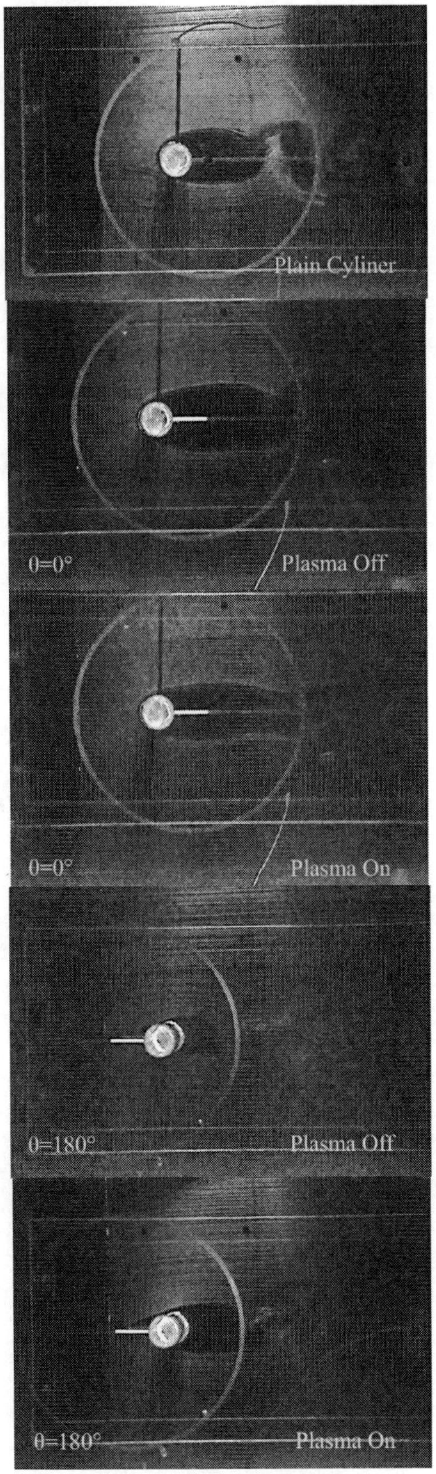

Figure 5. Visualized wake flows with applied flow control methods for V=6.5 kV_{pp} and f=3.5 kHz at Re=4000.

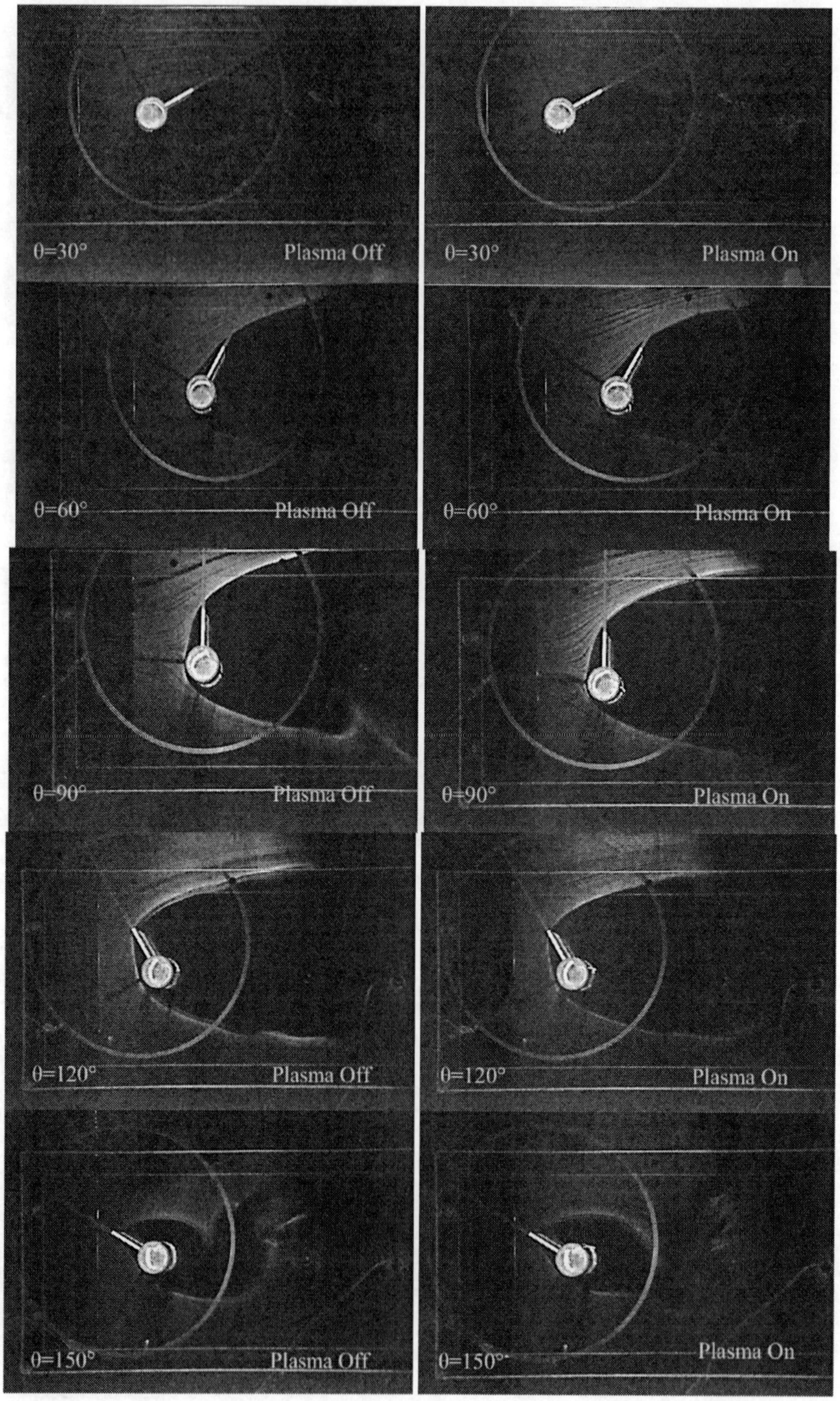

Figure 5. Visualized wake flows with applied flow control methods for $V=6.5\ kV_{pp}$ and f=3.5 kHz at Re=4000.

4 Conclusions

The flow around circular cylinder has been investigated experimentally by using both active and passive flow control method. At different angles of θ, between 0 and 180, the effects of plasma actuator and splitter plate were examined by considering the change of drag coefficient and flow visualization around the circular cylinder. As it can be seen from force measurement, there is a significant change depending on different angle of θ in drag at Re=4000. For θ=0, reduction in drag is %27 when the splitter plate was used behind the circular cylinder. But, reduction in drag is %56 when both splitter plate and plasma actuator are used together. Reduction in drag is more significant when both splitter plate and plasma actuator were used together than circular cylinder via splitter plate. For θ=0, 30, 60, reduction in drag are respectively %40, %36, and %17 when the splitter plate and plasma used together. But the same situation drag was increased %6, %10, %9, and %37 when the angle of θ was changed as 90, 120, 150, and 180, respectively. Effects of the using plasma actuator and splitter plate together give raise to narrower wake while circular cylinder with plasma actuator and splitter plate at 0 degree angle. But, circular cylinder wake region becomes larger while circular cylinder with splitter plate and plasma actuator at 180 degree angle. By using flow visualization, particularly wake region and vortex shedding are observed.

Acknowledgements

The authors would like to acknowledge the financial support of this work by the Scientific Research Projects (BAP) of the Çukurova University under the ID number of 5026.

Nomenclature

Re	Reynolds number
θ	attack angle
V	applied plasma voltage
f	applied plasma frequency
C_D	drag coefficient
U_∞	freestream velocity
D	diameter of the cylinder
L	length of the splitter plate

References

1. C.J. Apelt, G.S. West, A.A. Szewczyk, 1973, J. Fluid Mech., **61**, 187-198 (1973)
2. J.M. Cimbala, S. Garg , AIAA Journal, **29,** 6 (1990)
3. J.M. Cimbala, J. Leon, AIAA Journal, **34,** 11 (1996)
4. Y.E. Akansu, M. Sarioglu, T. Yavuz, AIAA Journal, **42**, 6 (2004)
5. F.O. Thomas, A. Kozlov, T.C. Corke, AIAA Journal, **46**, 8 (2008)
6. Y. Sung, W. Kim, M.G. Mungal, M.A. Cappelli, Exp. Fluids, **41**, 479-486 (2006)
7. S. Tabatabaeian, M. Mirzaei, A. Sadighzadeh, V. Damideh, A. Shadaram, Chinese J. of Aeronaut., **25**, 311-324 (2012)
8. H. Akbıyık, *Master Thesis*, Nigde University, Niğde/TURKEY (2014)

EPJ Web of Conferences **114**, 02002 (2016)
DOI: 10.1051/epjconf/201611402002

Experimental research of heterogeneous nuclei in superheated steam

Ondřej Bartoš[1,2,a], Michal Kolovratník[1], Bohuslav Šmíd[2] and Jan Hrubý[2]

[1]Czech Technical University in Prague, Department of Energy Engineering, Technická 4, Prague 6, 166 07, Czech Republic
[2]Institute of Thermomechanics of the CAS, v.v.i., Dolejškova 5, 182 00 Prague 8, Czech Republic

Abstract. A mobile steam expansion chamber has been developed to investigate experimentally homogeneous and heterogeneous nucleation processes in steam, both in the laboratory and at power plants using the steam withdrawn from the steam turbine. The purpose of the device is to provide new insight into the physics of non-equilibrium wet steam formation, which is one of the factors limiting the efficiency and reliability of steam turbines. The expanded steam or a mixture of steam with a non-condensable gas rapidly expands in the expansion chamber. Due to adiabatic cooling, the temperature drops below the dew point of the steam at a given pressure. When reaching a sufficiently high supersaturation, droplets are nucleated. By tuning the supersaturation in the so-called nucleation pulse, particles of various size ranges can be activated. This fact is used in the present study to measure the aerosol particles present in the air. Homogeneous nucleation was negligible in this case. The experiment demonstrates the functionality of the device, data acquisition system and data evaluation methods.

1 Introduction

The increasing global demand for electric power on the one hand, and strict requirements for reducing the emissions of pollutants and other environmental impacts of power production on the other, presents a new challenge of enhancing the efficiency of energy conversion systems. In most western countries – including the Czech Republic – most power is generated in nuclear and fossil-fuelled power plants. Because of the geographical conditions, other energy resources are limited and thermal power plants will continue their leading position in the coming decades as well. The requirement of efficiency and reliability enhancements also affects the steam turbine as an important part of the Rankine cycle of thermal power plants. In the low-pressure part (LP) of the steam turbine, the steam undergoes the process of nucleation and condensation into primary droplets [1]. Some of the primary droplets hit the blades and other surfaces and form a liquid film. Secondary droplets are formed by the disruption of the liquid film from blades and casing [2]. The non-equilibrium formation of the liquid phase significantly affects the budget of energy conversion efficiency. Despite the hundred-year development of steam turbines and investigations of the underlying physical processes, the microphysics of liquid phase formation is still only partially understood. Consequently, the mathematical models of liquid phase formation currently available to turbine designers and operators provide incomplete information on the effect of design and operation modifications on the efficiency and reliability of steam turbines. The rapid progress of computing hardware and computational fluid dynamics (CFD) allows for more and more accurate turbine-design computations. It has become a necessity to also bring the models of non-equilibrium condensation processes to a new level. The study of the effects of impurities in the super-heated steam is an important contribution for a better understanding of wet steam formation. The impurities in the super-heated steam are particularly important at the beginning of the nucleation and condensation processes. It can be expected that the impurities act as heterogeneous nucleation seeds.

2 Heterogeneous nucleation

Heterogeneous nucleation can be an important mechanism of early droplet formation (the so-called early condensate), because it requires a lower supersaturation to start [3, 4]. The actual value of supersaturation at which a heterogeneous particle is activated depends on its size and wetting properties for insoluble particles, size and solubility for soluble particles. Also the soluble-insoluble transition has been considered. It is assumed that the origin of most particles present in steam is in the droplets carried over by the steam from the boiler. In the superheater, the droplets dry out and low-volatile compounds (typically simple salts and oxides) dissolved in the droplets form fine particles. Consequently, such particles would be completely or partly soluble. Until

[a] Corresponding author: ondrej.bartos@fs.cvut.cz

recently, no data existed on heterogeneous particles in steam. Pioneering work [5] by the present team showed 3×108 to 6×109 particles per kilogram of steam in various power plants. The size distribution was characterized by the Sauter diameter of 30 nm. However, chemical analysis of condensed steam samples showed that only a small fraction of the low-volatile matter was captured by the present instrumentation, leading to the hypothesis that most of the matter is dispersed in a large number of particles smaller than 3 nm (the present detection limit).

3 Expansion chamber

The main motivation for the development of the expansion chamber was the lack of a suitable identification tool for identification of the particles size down to 10 μm. The classical aerosol devices like a condensing particle counter (CPC) has a detection limit of 3nm but the efficiency of the detection for small particles are low. The application of a scanning mobility particle sizer (SMPS) is difficult due to the insufficient number density of the particles and a high deposition rate in the sampling tubes [5]. The developed expansion chamber provides a new approach to particle identification. The measurement is performed as a series of expansions with different back pressures. The expansion ratio (ratio of the final to initial pressures) R and saturation S change together with the expansion and this means that different groups of heterogeneous particles are involved in the nucleation and condensation processes. In case heterogeneous particles are present in the sample, it can be expected that particles with a radius bigger than the critical droplet radius r* will grow and ultimately become droplets of a diameter on the order of 100 – 1000 nm, which can be detected optically. The critical radius can be estimated as

$$r^* = \frac{2\sigma}{\rho_l r T_g lnS} \quad (1)$$

Here, σ is surface tension, ρl is droplet density, r is the specific gas constant and Tg is the temperature of the gas, and S is supersaturation. Because the present experiments are performed at low pressures, the supersaturation can be computed as a ratio of the actual vapour pressure pv and the saturation vapour pressure ps corresponding to the actual temperature,

$$S = \frac{p_v}{p_s} \quad (2)$$

The purpose of the expansion chamber is to carry out the expansion of the sampled steam from the steam turbine in order to study homogeneous and heterogeneous nucleation processes. The entire device is mobile so that measurements can be performed on site at a power station. Usually the steam from the feed water heater or extraction pipes is used.

During its transport in the connecting tubing and in the expansion chamber, the diluted sample is kept at a constant temperature sufficiently high to avoid any condensation. A schema of the chamber is provided in Fig. 1.

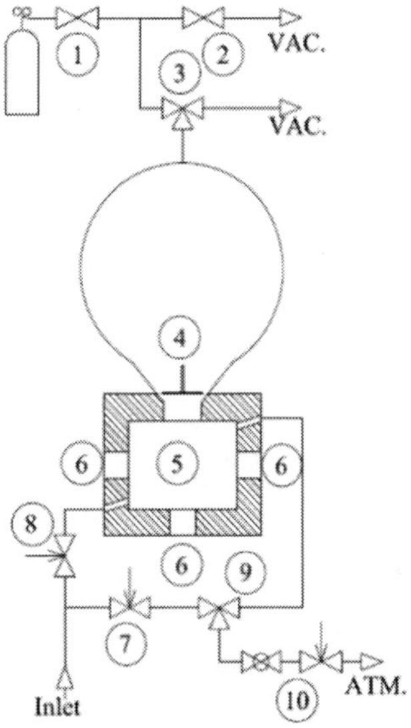

Figure 1. 1. High pressure valve, 2. Low pressure valve, 3. Evacuation valve, 4. Expansion valve, 5. Test section, 6. Optical window, 7. Bypass, 8. Chamber valve, 9. Bypass switching valve, 10. Outflow valves. Description is given in the text.

Solenoid valves 1 and 2 are switched by a PID controller to maintain the desired back pressure. Before the expansion, test section 5 of the chamber is flushed with the sample for a sufficient time to achieve adsorption equilibrium on the walls of the chamber and tubing. The specially designed fast-opening expansion valve 4 is actuated electro-pneumatically. Within a few milliseconds (see Fig. 2 showing the pressure trace), the vapour-gas mixture expands from the initial chamber pressure to the adjusted back pressure. During the expansion, the sample flow passes through the bypass with valve 7 to avoid pressure rise in the sampling tubing, valve 8 is closed. This valve presents a similar pressure drop to that of the test section and expansion valve. The 3-way valve 9 switches the sample flow from bypass or the test section. Throttling valve 10 is used to adjust the desired flow rate through the device. The laser light scattering and extinction measurements are performed through windows 6.

4 Methods

Special data acquisition and evaluation codes have been developed. The entire data processing procedure is not straightforward and requires several particular steps. The measurement sequence is triggered by the pressure drop in the chamber. The rapid pressure decrease is measured by a piezoelectric transducer with a Kistler charge amplifier. The temperature of the entire expansion chamber and the expansion valve is maintained at a constant value by a circulating bath thermostat. Together with the pressure trace, optical signals are also acquired: the power of the laser beam attenuated by the cloud of nucleated droplets is measured by a photodiode, the power of the incident beam is measured by another photodiode, the intensity of the scattered light measured with a photomultiplier.

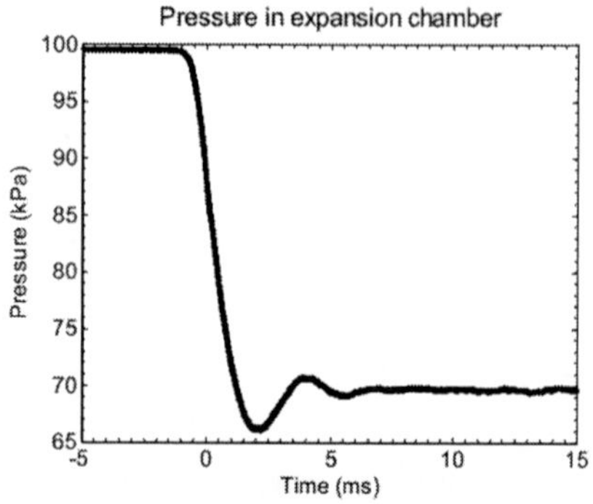

Figure 2. Pressure profile during the expansion.

Light extinction for determining the particle number density and the light scattering at 90 degrees for determining the droplets growth is acquired as a function of time. The determination of the number density of the droplets is based on the Lambert Beer law. The extinction ratio I/Io=f(λ) is measured as a function of time. Applying the Mie theory of light scattering on water droplets one can find the size distribution of the droplets φ(D) and number density. In general a system of ill-conditioned equations is obtained [6]:

$$\frac{1}{\ell}\ln\left(\frac{I_o}{I}\right)_i = \frac{\pi}{4}N_v\int_0^{D_\infty} Q(\pi D/\lambda_i)\cdot\varphi(D)\cdot D^2\mathrm{d}D\,. \quad (3)$$

Here, ℓ is the length of the measurement section, Nv is the volumetric number density of droplets, Q(πD/λi) is the extinction coefficient according to the Mie theory, D droplet diameter, φ(D) size distribution function of the droplets and i = 1,2,...*k* is the chosen number of the particular wavelengths $\lambda_{1,2,3...k}$ and this presents the number of the equation in the system.

Almost all droplets are nucleated within the so-called nucleation pulse, i.e. the pressure dip shown in Fig. 2. Consequently, the cloud of droplets can be considered as monodisperse and a monochromatic (675 nm) laser is used as the light source. Hence, equation (3) reduces simple form (4),

$$\frac{I}{I_0} = exp(-\pi N_v D\lambda QL) \quad (4)$$

The light scattering intensity as the function of the droplet radius is shown in Fig. 3.

The comparison of the peeks position or in general the theoretical and measured light intensity profile one can reach a dependence on droplets growth over time. Nevertheless, the desired result is the size distribution function of the heterogeneous seeds in the steam/gas at the moment the profile of the water droplet is determined. The application of this knowledge and the theory of droplet growth can help to achieve the final distribution function of the impurities in the gas.

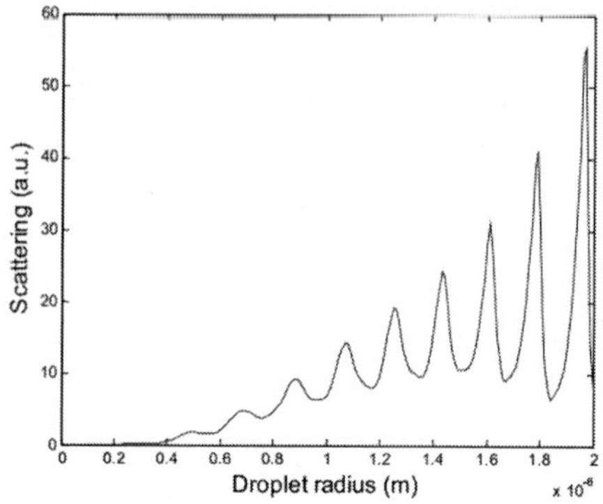

Figure 3. Dependence of the light intensity for light scattering at 90° for wavelength 675nm and variable droplet radius.

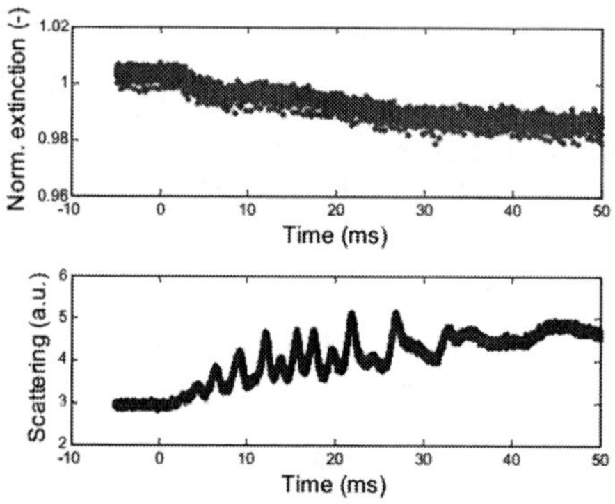

Figure 4. Profile of the light extinction and light scattering at 90°.

The growth of the droplet which start condensing on the foreign particle can be written with respect to some simplification [7,8]

$$r = \left(r_*^2 + \frac{2(S-1)}{C_1+C_2}(t-t_0)\right)^{0.5} \quad (5)$$

where S is the saturation of the steam, r* is the critical diameter or minimum size of the active foreign particle involved in the condensation. C1 and C2 are constants which respect diffusion of the water vapour in the gas, heat conductivity and physical properties of the gas and vapour.

Satisfactory results require a strong correlation between the measured diameters and the theoretical function of the droplet growth on the foreign particle over time. The whole data processing technique is sensitive to many variables and requires careful measurement.

5 Results

In this paper, an example of the results achieved by the measurement with the ambient air is presented. This kind of test proved useful because experiments at power plants are dependent on the operator's permission and can be performed on rare occasions. The unveiled test with the air was performed in the CTU laboratory. The measurement was used to evaluate the entire data acquisition and data processing method. The knowledge gained was used to enhance the measurement procedure. For example the fine adjustment of the control valves for maintaining the back pressure in the expanding vessel.

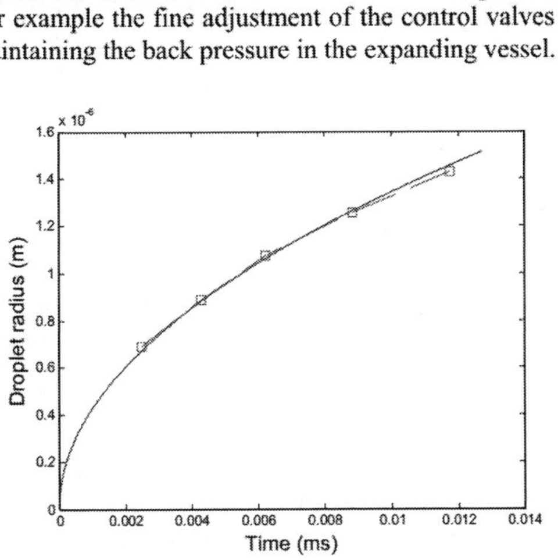

Figure 5. Example of the theoretical profile of the droplet growth compared with the measured values.

Fig.5 shows an example of the theoretical growth of the droplet over time together with the measured droplet size determined from the analysis of the light scattering. The expansion rate for the presented measurements was in the range 250÷350s-1 and the temperature of the gas was 20°C. The relative air humidity was 53%. Homogeneous nucleation was taken into account [9, 10] but for all measurements was considered as negligible because the saturation ratio (eq.5) never exceeded S=3 for all the expansions.

Measurement evaluations defined the requirements for precise control of the input parameters of the gas mixture. Precise modelling of the droplet growth and a model of the entire expansion process is necessary for determining the minimum starting diameter of the heterogeneous seeds in the gas mixture. In Fig. 6 the right part of the size distribution function is presented for the ambient air. The data correlates with that found in the literature [11].

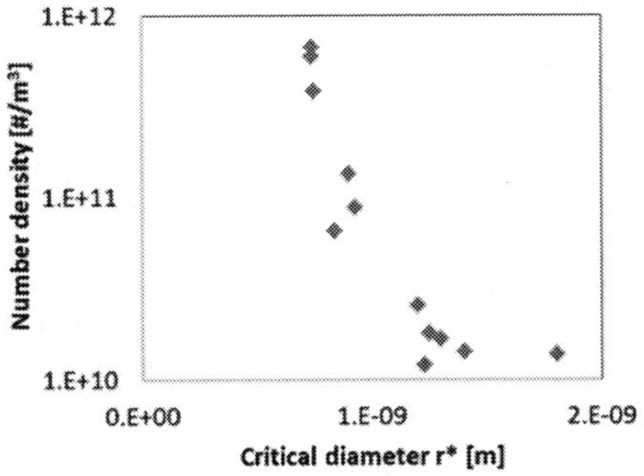

Figure 6. The determined heterogeneous particles size distribution function.

6 Conclusions

The evaluation measurement proves that the expansion chamber is a sufficient tool for the identification of foreign impurities in steam. Nevertheless, the entire process of the evaluation of the heterogeneous seeds size distribution function is very sensitive and requires a deep understanding of the nucleation process and precise measurement. Further enhancement of the determination method will focus on detailed knowledge of the input gas mixture properties and modelling of the expansion process. Other efforts will be focused on earlier detection of the growing particles through the use of light with a shorter wave length and by adjusting the entire optical system.

Acknowledgements

We gratefully acknowledge the support by Grant No. 101/11/1593 of the Grant Agency of the Czech Republic and by the Grant Agency of the Czech Technical University in Prague, grant No. SGS 14/183.

References

1. V. Petr, M. Kolovratník, Wet steam energy loss and related Baumann rule in low pressure steam turbine. Proceedings of the Institution of Mechanical Engineers Part A - Journal of Power and Energy. 2014, vol. 228, no. 2, p. 206-215. ISSN 0957-6509.
2. O. Bartoš, X. Cai, M. Kolovratník, EPJ Web of Conferences , 2014, vol. 67, ISSN 2100-014X.
3. W.T. Lindsay, Behavior of impurities in steam turbines, Power engineering, Vol. 5, pp. 68-72, 1975.
4. W.G. Steltz, et al, The verification of concentrated impurities in low-pressure steam turbines, TASME, J. of engineering for power, Vol. 105, pp. 192-198. 1983.
5. M. Kolovratník, J. Hrubý,V. Ždímal, O. Bartoš, I. Jiříček et al.: Nanoparticles found in superheated steam – A quantitative analysis of possible heterogeneous condensation nuclei. Proceedings of the Institution of Mechanical Engineers Part A - Journal of Power and Energy. 2014, vol. 228, no. 2, p. 186-193. ISSN 0957-6509.
6. C.F. Bohren, D.R. Huffman, Absorption and Scattering of Light by Small Particles. John Wiley & Sons, Inc., USA, 1998
7. H. Köhler, The nucleus in and the growth of hygroscopic droplets. Trans. Faraday Soc. 32 (1936) 1152-1161
8. W. E. Howell, The growth of cloud drops in uniformly cooled air. J. Meteor., 6, 134–149,1949.
9. V. Holten, Water nucleation. PhD Thesis, Eindoven: Eindhoven University of Technology 2009
10. J.Wolk, at al. Empirical function for homogeneous water nucleation rates J.Chem.Phys.117,4954(2002)
11. W.C. Hinds, Aerosol technology (Second Ed.). Wiley, New York, 1999.

EPJ Web of Conferences **114**, 02003 (2016)
DOI: 10.1051/epjconf/201611402003

Effect of the insulation by the mud on the convection in building in the Sahara of Algeria - case of Bechar City

Elhadj Benachour[1,a], Belkacem Draoui[1], Baachir Imine[2], Mohammed Hasnat[1], Lakhdar Rahmani[1]

[1]*Laboratory of Energy in Arid Regions, University of Tahri Mohamed - Bechar, P.O. P 417. 08000 , Road Kenedza , Bechar, Algeria*
[2]*Department of Mechanical Engineering, University of Sciences and Technology of Oran-Mohamed Boudiaf, P.O.1505 El-Mnaouar, Oran, Algeria*

Abstract. The thermal behavior of the buildings is a current problem which arouses the interest of many researchers. Indeed, the control of the loads of air conditioning or heating requires a thorough knowledge. Since the thermal quality of the buildings in the Maghreb is there generally very insufficient in particular in the Sahara of Algeria, the insulation proved that it is a very important parameter to minimize the thermal diffusion process inside on all when the insulator is a local material as the Mud which is in our region with a significant amount and almost free quantity. This work presents a study of numerical simulation aiming at the role and the influence of the insulation by the mud Reduced like a local material to the town of Bechar located at the south west of Algeria, or one is interested in a comparative study for pursued these goals. In this context, an analogy was used for the functions which are discretized by the finite difference method and integrated in the Fluent code which is based on the finite volume method. The validation of this procedure was confirmed while comparing some results. The results are presented in the form of distributions of the isotherms, the streamlines, local and average Nusselt of which the goal to study the influence on comfort.

1 Introduction

The thermal behavior of buildings is a current problem which arouses the interest of numerous and many researchers. Indeed, energy consumptions and greenhouse gas emissions are very high to the world. It is thus necessary to reduce the energy needs while improving the energy efficiency, in particular in the sector of the building. This one offers possibilities of considerable reduction of the energy consumption, in particular by improving the thermal insulation [1]. To reduce them, various strategies are organized, among which several aim at the improvement of the energy efficiency of buildings [2-5]. We indeed notice that numerous housing is not isolated yet. About 65 % of the walls of housing arrange no insulation. A significant portion and An important part of these not isolated external walls are massive solid brick walls or the thermal perceived in buildings was aimed by some research [6-9].Other works suggested studying the effect of the insulation on the energy consumptions and the energy performances in buildings. The insulation of the external walls is complex and several processes exist: insulation by filling of the slide, insulation by outside and insulation by the interior. In restoration, it very often arrives that the choice is restricted for technical reasons, economic and/or patrimonial reasons. Although it can generate many risks on the level of the hygrothermic and mechanical behavior of the walls, Then, since the thermal isolation is one of the investments of most profitable energy saving, in particular in the new building, Several numerical studies and experimental were realized[10-16]. The insulation by the interior then becomes often the only possible solution in particular with the buildings in the Sahara of Algéria. The hygrothermic parameters of the existing wall and reported materials must be well-known to design powerful and durable walls in time.

Our study consists in studying the thermal transfer through walls. Two types of walls are studied:
- A not isolated wall,
- A wall isolated inside by the mud.

This work presents a study of numerical simulation aiming at the role and at the influence of the insulation by the mud as a local material at the city of Bechar situated in the south west of Algeria, or we are interested in a comparative study for aimed these goals. In this context, an analogy was used for the used functions which are discretized by the finite difference method and integrated in the Fluent code which is based on the finite volume method. The validation of this procedure was confirmed by comparing some obtained results. The results are presented in the form of distributions of isotherms,

Stream lines and the local Nusselt number and average Nusselt number of whom (which) the purpose to study the influence on the comfort.

[a] Corresponding author: benachour_elhadj@yahoo.fr

2 Geometric configuration

By giving itself two configurations, the one being described the wall without insulation, and the other one with insulation by the mud. The studied configuration is represented on the figure 1. We chose the most widespread construction in the area. The studied room is a structure of a multi-house zone, but in is interested in only one, of dimension 16 m^3 (4 x 4 x 4).To simplify the calculation we are interested in a phenomenon of two dimensions. The outer walls are simple concrete walls of 20 cm, with the mortar coating outside surface of negligible thickness compared to the thickness of the outside wall. The roof is insulated [supposed or presumedly adiabatic]. And on the bottom floor earth (ground) is a concrete slab with a wooden frame [supposed adiabatic]. Temperatures imposed on interior and exterior walls are captured experimentally in one day of a summer of August the Bechar city to 15h00.Then to model this phenomenon. This is a square cavity of length L and a height h. The left and right vertical walls of the cavity are heated to a constant temperature equal to Tc 48 ° C (Tc = 1 for the dimensionless case) Tf equal to 25 ° C (Tf = 0.5 for the dimensionless case), and the other walls are maintained adiabatic . One will exploit the existence and the absence of the interior insulation to facilitate the comparative study. The thickness of concrete wall e_1 =L/20, and the thickness of the insulation interior by the Mud is e_2 =L/20. The factor of the form is A = 1.

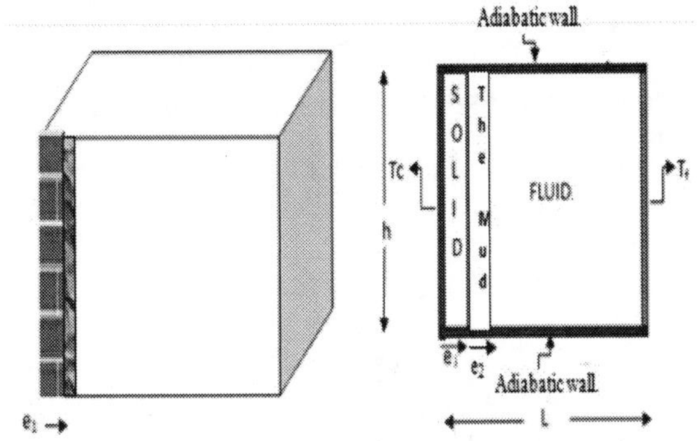

Figure 1. Schematic of the studied configuration

To simplify the problem, assume that:
- The fluid is Newtonian and incompressible.
- The heat dissipation by viscous friction is neglected.
- The Boussinesq approximation is considered

3 Objects

We can express our aim for this study in the following points:

- Study the Modelling of the coupling of convection - conduction.
- Study of the effect of the distribution of the heat inside buildings on the convection.
- Comparative study between the construction of buildings with and without insulation.
- Study the effect of the Rayleigh number on the convection in building.
- Test the reliability of the insulation with the Mud as local material.

4 Mathematical model

The fluid is assumed incompressible and obeys the Boussinesq approximation. In these cases, continuity in two dimensions and the equations governing the flow and energy is given by:

Continuity.

$$\frac{\partial U}{\partial X}+\frac{\partial V}{\partial Y}=0 \tag{1}$$

X-momentum.

$$\frac{\partial U}{\partial t}+U\frac{\partial U}{\partial X}+V\frac{\partial U}{\partial Y}=-\frac{1}{\rho}\frac{\partial P}{\partial X}+\upsilon\nabla^2 U \tag{2}$$

Y-momentum.

$$\frac{\partial V}{\partial t}+U\frac{\partial V}{\partial X}+V\frac{\partial V}{\partial Y}=-\frac{1}{\rho}\frac{\partial P}{\partial X}+\upsilon\nabla^2 V+g\beta\frac{\partial T}{\partial X} \tag{3}$$

Energy.

$$\frac{\partial T}{\partial t}+U\frac{\partial T}{\partial X}+V\frac{\partial T}{\partial Y}=\frac{\lambda}{\rho c_p}\nabla^2 T+\frac{1}{\rho c_p}\varphi \tag{4}$$

The derived equation of motion (2) over Y and the equation of motion (3) by contributing to x, then, after subtracting the two equations obtained, we obtain the equations dimensionless variables in writing Helmotz in terms of vorticity and stream function formulation are as follows:

$$\frac{\partial \omega}{\partial t}+U\frac{\partial \omega}{\partial X}+V\frac{\partial \omega}{\partial Y}=\Pr\nabla^2\omega+Ra\Pr\frac{\partial T}{\partial X} \tag{5}$$

$$\frac{\partial T}{\partial t}+U\frac{\partial T}{\partial X}+V\frac{\partial T}{\partial Y}=\frac{\partial^2 T}{\partial X^2}+\frac{\partial^2 T}{\partial Y^2}+\frac{\varphi}{\rho Cp}\frac{L^2}{a\Delta T} \tag{6}$$

$$\frac{\partial^2 \psi}{\partial X^2}+\frac{\partial^2 \psi}{\partial Y^2}=-\omega \tag{7}$$

The stream function and vorticity are related to the velocity components by the following expressions:

$$U=\frac{\partial \psi}{\partial Y},\ V=-\frac{\partial \psi}{\partial X}\ \ and\ \ \omega=\frac{\partial V}{\partial X}-\frac{\partial U}{\partial Y} \tag{8}$$

The dimensionless parameters in the equations above are defined as follows:

$$\begin{cases} X = \dfrac{x}{H}, \quad Y = \dfrac{y}{H}, \quad U = \dfrac{u}{u_i} \\ V = \dfrac{v}{u_i}, P = \dfrac{P}{\rho u_i^2}, T = \dfrac{T - T_i}{T_c - T_i}, t^* = \dfrac{t}{L^2 / a} \end{cases} \tag{9}$$

For solid, we are interested only in the following heat equation:

$$\frac{\partial T}{\partial t} = \frac{1}{a}\left(\frac{\partial^2 T}{\partial x^2} + \frac{\partial^2 T}{\partial y^2}\right) \tag{10}$$

5 Procedure of Simulation

The numerical calculation was conducted using the computer code "fluent" 6.2.13. The numerical procedure used in this work is that of finite volume.

It involves the integration of differential equations of mathematical model on finite control volumes for the corresponding algebraic equations.

The SIMPLE algorithm [17] was chosen for the coupling speed pressure in the Navier-Stokes equations on a staggered grid. The convective terms in all equations are evaluated using the schema upwind first order.

The discretization of the time term is made in a totally implicit scheme. The convergence of the solution is considered reached when the maximum relative change of all variables (u, v, w, p, t) between two successive time is not less than 10^4.

With an aim of following well any variation of the fields thermal and hydrodynamic, we used a uniform grid of 14241 nodes and 14480 Elements in non-stationary mode.

6 Validation of the Code of Calculation

6.1 Validation of the model

The computer code was validated on a natural convection problem of stale air in a square cavity with vertical walls differentially heated and adiabatic horizontal walls. Our results were compared with those obtained by De Vahl Davis (1983) [18] .The latter dealt with the same problem by adopting the finite difference method with the vorticity-stream function formulation (see Table 1).

Table1. Comparisons of the results of validation

	Ra = 10^3	**Ra = 10^4**	**Ra = 10^5**
Nu (average number) (hot wall)	**present study:** 1.2	**present study :** 2.257	**Present study :** 4.64
Nu (average number) (hot wall)	**De Vahl Davis [18]:** 1.12	**De Vahl Davis [18]:** 2.243	**De Vahl Davis [18] :** 4.52

7 Results and Discussion

The structure of the flow, the temperature field and heat transfer through the hot wall are discussed in this section.

In this study, to target the most important goal, we will show and studied the effect of the insulation by the Mud on dynamic and thermal behavior of the fluid in the cavity.

7.1 Isotherms

The isotherms are shown in Figures [2-9]. The heat distribution in the cavity is in accordance with the fluid circulation revealed by isotherms and iso currents. Indeed we find a heating fluid from the interface, if it causes the change of the heat distribution in the cavity (see figures) for different numbers of Rayleigh For a fixed value of the number of Prandtl equal 0.71.

Gradually, as the Rayleigh number has increased, the isotherms become increasingly wavy and heat transfer increases, so the flow intensifies and natural convection is expanding and predominates (natural convection is predominant).

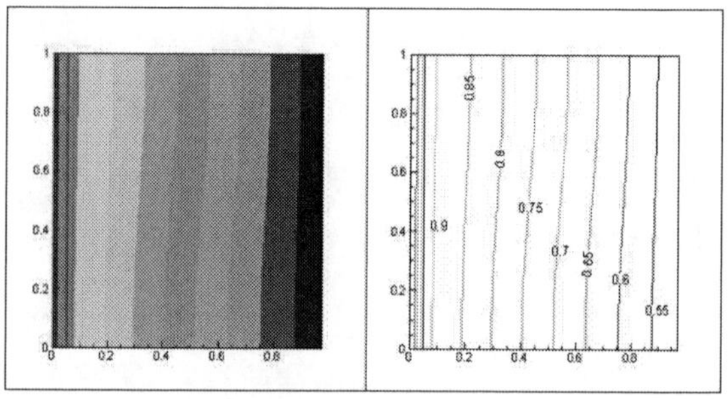

Figure 2. The Isotherms For uninsulated wall, Rayleigh Number Equal 10^3, Pr=0.71

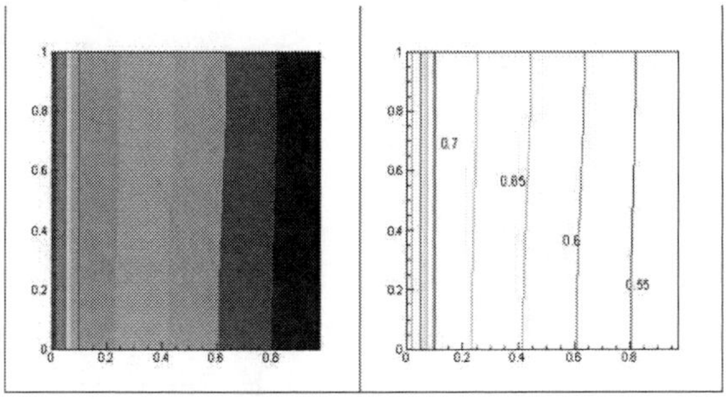

Figure 3. The Isotherms For Insulated Wall, Rayleigh Number Equal 10^3, Pr=0.71

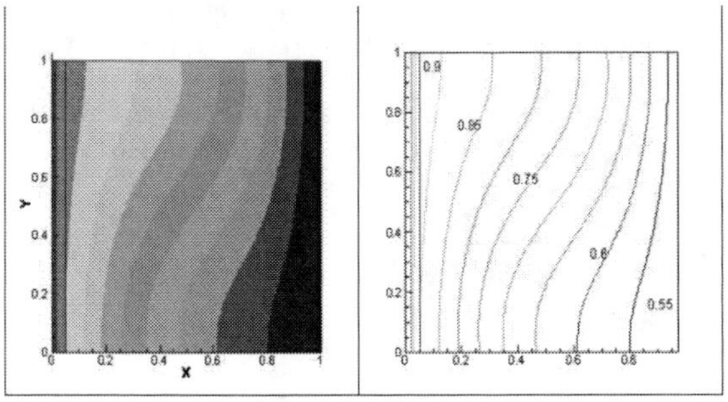

Figure 4. The Isotherms For uninsulated wall, Rayleigh Number Equal 10^4, Pr=0.71

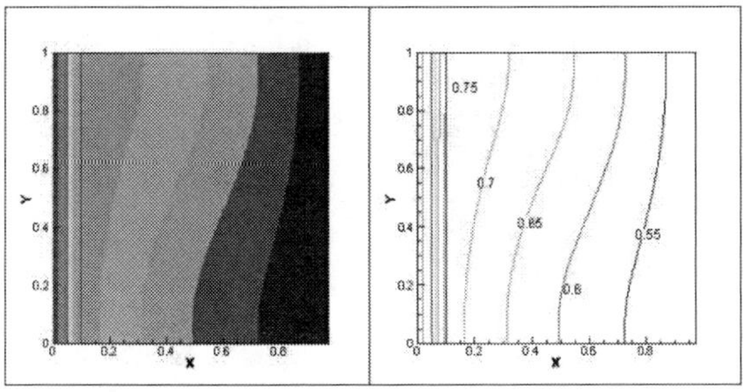

Figure 5. The Isotherms For Insulated Wall, Rayleigh Number Equal 10^4, Pr=0.71

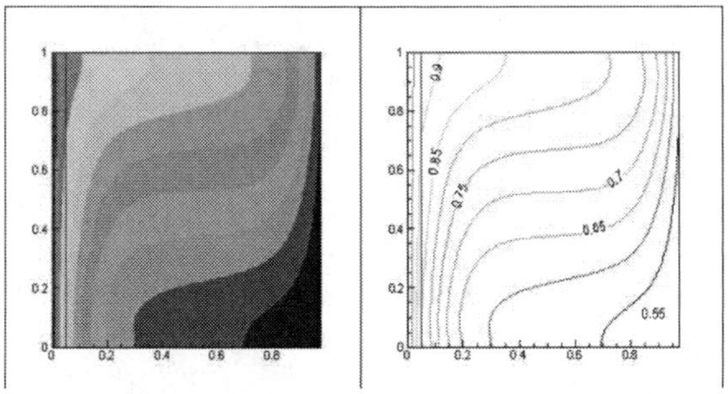

Figure 6. The Isotherms For uninsulated wall, Rayleigh Number Equal 10^5, Pr=0.71

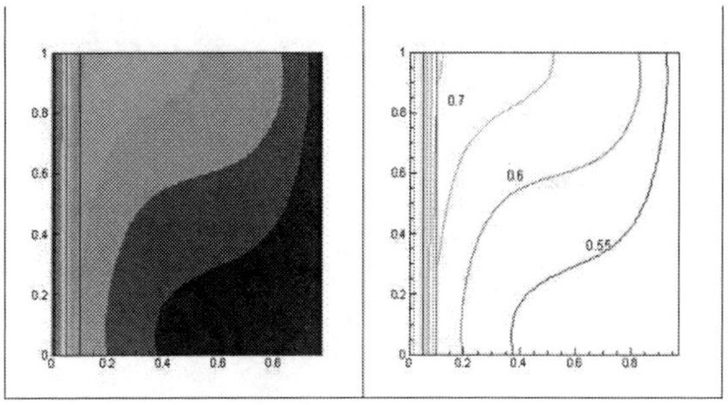

Figure 7. The Isotherms For Insulated Wall, Rayleigh Number Equal 10^5, Pr=0.71

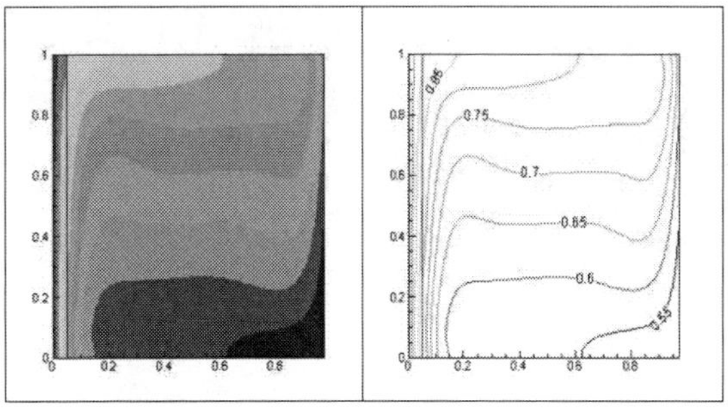

Figure 8. The Isotherms For uninsulated wall, Rayleigh Number Equal 10^6, Pr=0.71

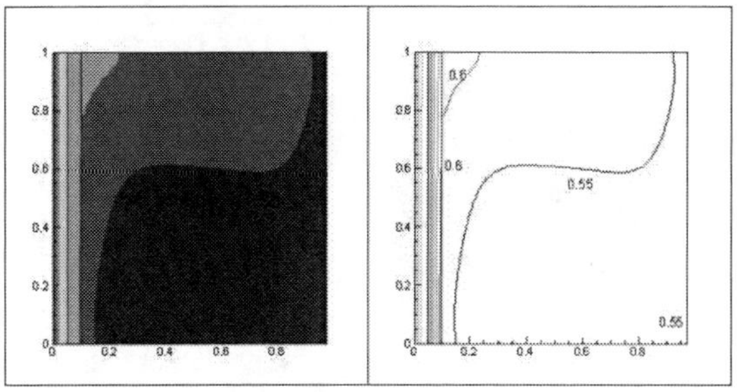

Figure 9. The Isotherms For Insulated Wall, Rayleigh Number Equal 10^6, Pr=0.71.

7.2 Streamlines

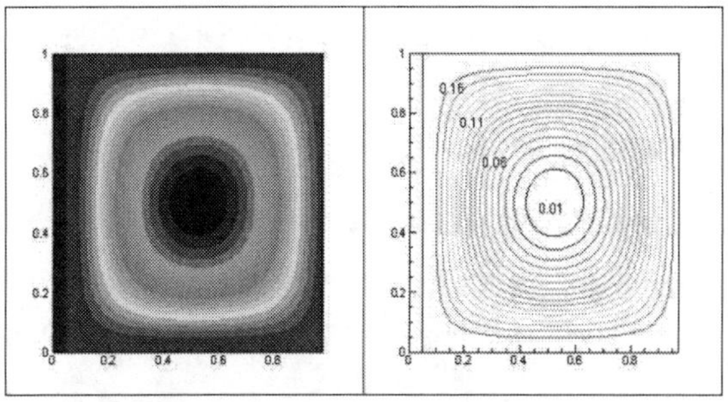

Figure 10. Streamlines obtained For uninsulated wall, Rayleigh Number Equal 10^3, Pr=0.71

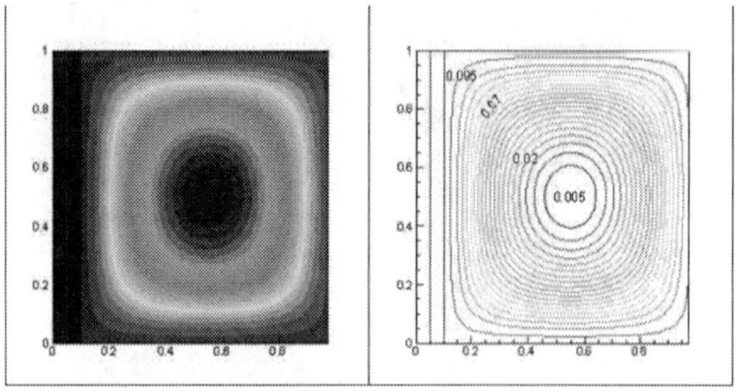

Figure 11. Streamlines obtained For Insulated wall, Rayleigh Number Equal 10^3, Pr=0.71

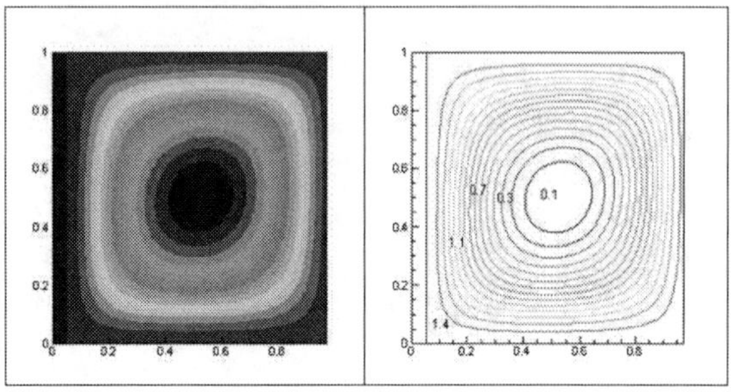

Figure 12. Streamlines obtained For uninsulated wall, Rayleigh Number Equal 10^4, Pr=0.71

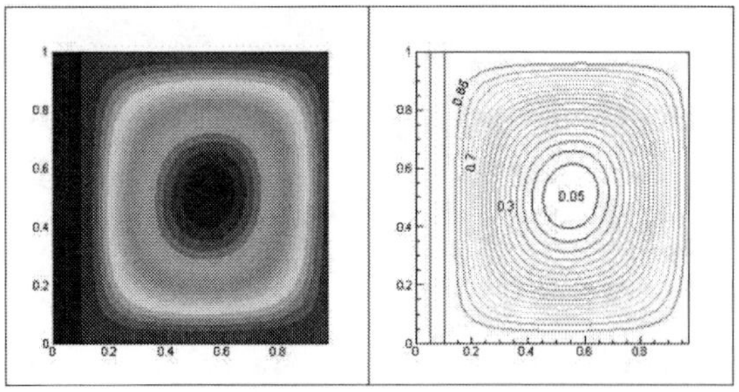

Figure 13. Streamlines obtained for Insulated wall, Rayleigh Number Equal 10^4, Pr=0.71.

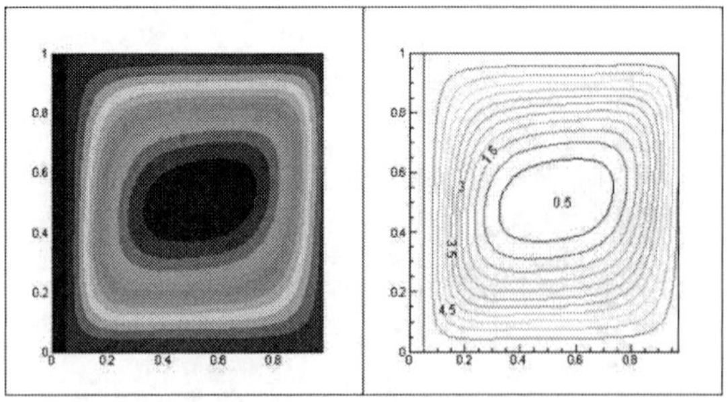

Figure 14. Streamlines obtained for uninsulated wall, Rayleigh Number Equal 10^5, Pr=0.71

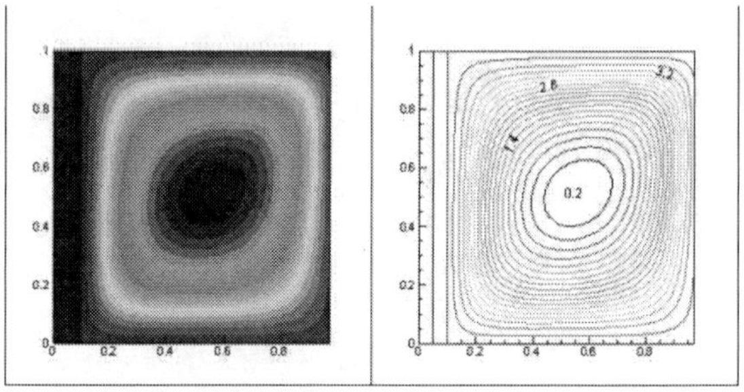

Figure 15. Streamlines obtained For Insulated Wall, Rayleigh Number Equal 10^5, Pr=0.71.

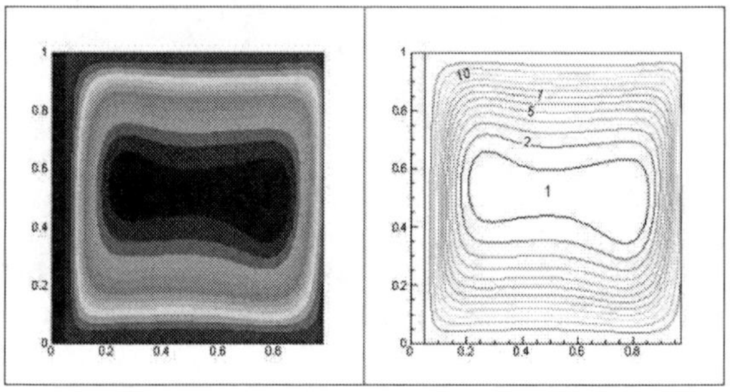

Figure 16. Streamlines obtained for uninsulated wall, Rayleigh Number Equal 10^6, Pr=0.71

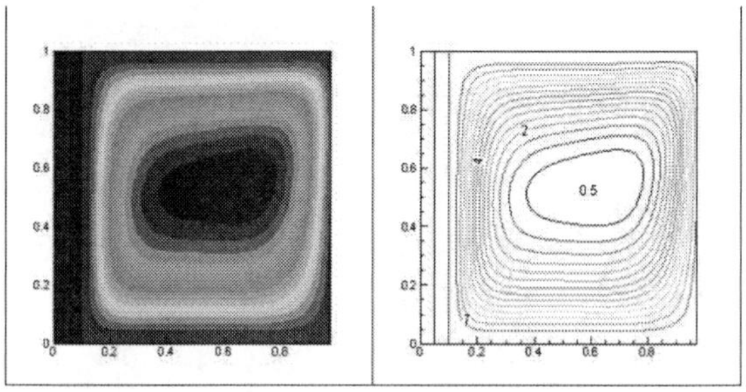

Figure 17. Streamlines obtained for Insulated Wall, Rayleigh Number Equal 10^6, Pr=0.71

Figures [10-17] shows the current lines corresponding to Rayleigh numbers: 10^3, 10^4,10^5 and 10^6 with a value of Prandt = 0.71 For uninsulated and Insulated Walls . When Ra = 10^3, the flow is almost symmetric, we see that there are cells that are closed on itself, the fluid motion was a motion block, which justifies the existence of recirculation zones closer the center of the cavities, the flow lines are substantially parallel to the vertical walls. So also for the representation of flow lines corresponding to the Rayleigh number equal to 10^4,10^5, we find that the current lines are tight at the vertical walls and in the middle, this is due to strong temperature gradient at the vertical walls and thus the flow is parietal which characterizes the large convective heat transfer to the vertical walls. We also observed the formation of two cells in the recirculation cavity for 10^6 same. It is also noted in the latter case, the expansion of two areas of recirculation of a cell in the clockwise direction and the other in the opposite direction, with the increase of Rayleigh.

7.3 Temperature profiles

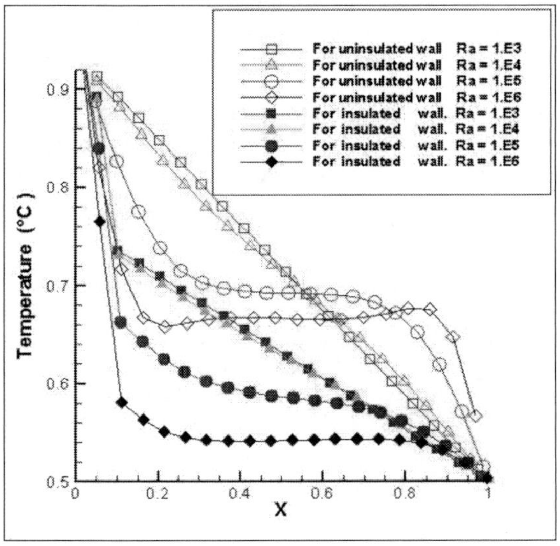

Figure 18. Dimensionless horizontal temperature profiles for different Rayleigh numbers for uninsulated and insulated walls, Pr = 0.71, Y = 0.5

Observation of the temperature profiles for different values of y = 0.5 (Figure 18) to Rayleigh leads us to confirm that the increase of temperature takes the form of the parabola is to say, this increase is condensed in the middle irrespective of the number Rayleigh, but it is sensitive, we can say there is a direct proportionality.

7.4 Nusselt numbers

We report in Figure 20 the influence of the Rayleigh number on average Nusselt number. We found a significant influence of the Rayleigh number. The increase in average Nusselt number is accompanied with the increase of Rayleigh, but after the Rayleigh number equal to 10^5, the heat transfer is almost the same, and then represents the better heat transfer in the cavity.

The local Nusselt number is presented in Figure 19, it is clear that it becomes important for every increase of Rayleigh because of the change in the exchange ratio between the wall and the air, a difference of around 16% on average, between insulated and uninsulated walls to Ra = 10^5, which can be doubled to Ra $=10^6$.

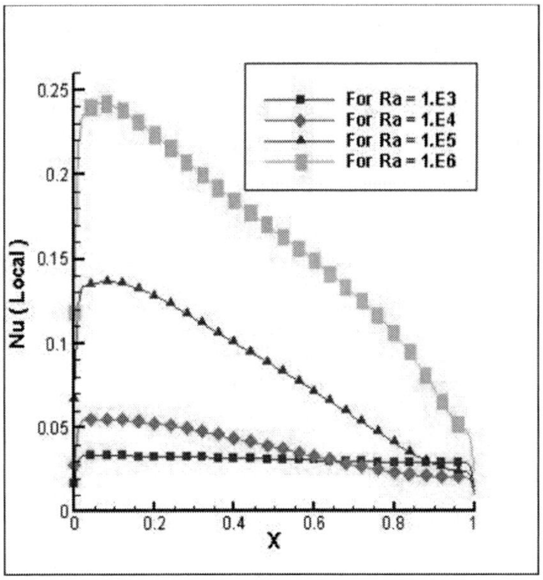

Figure 19. Profiles of Local Nusselt number, for different values of Ra=10^3,10^4,10^5 and 10^6, for insulated walls, Pr = 0.71.

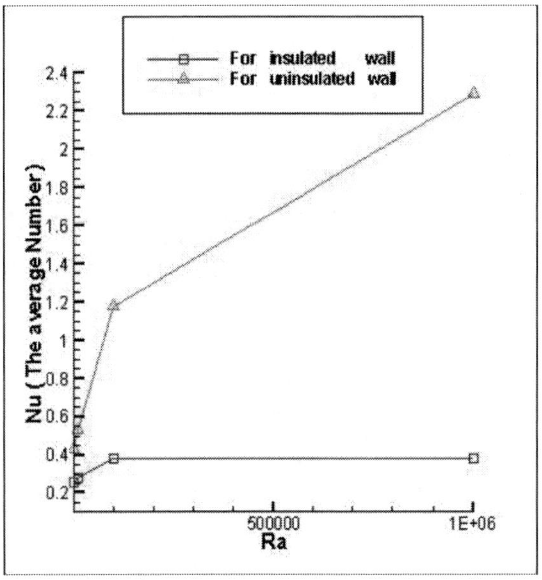

Figure 20. Profiles of average Nusselt number for different values of Ra=10^3,10^4,10^5 and 10^6, for uninsulated and insulated walls, Pr = 0.71.

8 Conclusion

The idea of this work is based on actual events; this is just the underfloor heating in buildings arides. The objective of our work was to study the behavior of the air in enclosed with two-dimensional numerical study of natural convection or (where) laminar flow is considered on is interested by conduction or convection coupling. Solving equations governing natural convection written Helmontz's variables ψ and ω which are discretized by the finite difference method. We validated the work in the literature for the hydrodynamic and thermal studies in the cavities where he showed good consistency. For our simulation, we presented the streamlines, isotherms and the influence of key parameters such as positioning after the interface and the number of Rayleigh convection. We have shown by this simulation, the following:

1. Convection affects the structure of the isotherms.

2. The temperature of the fluid increases dramatically with increase in Rayleigh number.

3. The Interior thermal isolation reduces the losses of heat and minimizes the energy needs (request of heating) and thus reduced in fact, at the same time: the invoice, consumption and pollution.

4. The Mud is insulating effective which to guarantee the comfort and the air quality inside the building

5. The interior thermal insulation helps to preserve a constant temperature in the integrality of your housing winter like summer.

6. The Inside thermal insulation has a promising future provided that all actors, around the sector of construction are justified and motivated and conscious of the ecological and financial issues.

References

1. R. Guechchati, M.A. Moussaoui, Ahm. Mezrhab and Abd. Mezrhab, Simulation of the effect of the thermal insulation of buildings psychoeducational center Cas du centre psychopédagogique Case of SAFAA in Oujda, Renewable Energy Review, Vol. 13 N°2, 223 – 232, (2010).
2. R. Belarbi, M. Spérandio et F. Allard, Evaluation of the Building industries Provided with Systems of Passive Cooling. Application to the Case of Adiabatique' Evaporation, General Review of Thermics,Vol. 36, N°7, pp. 547 - 561, (1997).
3. C. Inard, P. Depecker et J. Roux, A Model Simplified for the Prediction of the Field of Temperature in the Building industries, General Review of Thermics, Vol. 36, N°2, pp. 113 - 123,(1997).
4. G.A. Florides, S.A. Tassou, S.A. Kalogirou, L.C. Wrobel, 'Measures Used to Lower Building Energy Consumption and Their Cost Effectiveness', Applied Energy, Vol. 73, N°3, pp. 299 - 328,(2002).
5. F.F. Al-Ajmi and V.I. Hanby, 'Simulation of Energy Consumption for Kuwaiti Domestic Buildings', Energy and Buildings, Vol. 40, N°6, pp. 1101 – 1109, (2008).
6. M.S. Mohsen and B.A. Akash, 'Some Prospect of Energy Saving in Building', Energy Conversion Management, Vol. 42, N°11, pp. 1307 - 1315, (2001).
7. M.A. Aktacir, O. Büyükalaca and T. Yılmaz, 'A Case Study for Influence of Building Thermal Insulation on Cooling Load and Air-Conditioning System in the Hot and Humid Regions', Applied Energy, Vol. 87, N°2, pp. 599 - 607, (2010).
8. L. Peeters, R. de Dear, J. Hensen and W. D'haeseleer, 'Thermal Comfort in Residential Buildings: Comfort Values and Scales for Building Energy Simulation', Applied Energy, Vol. 86, N°5, pp. 772 - 780, (2009).
9. E. Benachour and al, Effect of positioning the heating element on natural convection in a square cavity (habitat type), international review of mechanical engineering, vol.4N.5, July (2010).
10. A. Mecke, I. Lee, J.R. Baker jr., M.M. Banaszak Holl, B.G. Orr, Eur. Phys. J. E **14**, 7 (2004).
11. U. Eicker, V. Fux, U. Bauer, L. Mei and D. Infield, 'Façades and Summer Performance of Buildings', Energy and Buildings, Vol. 40, N°4, pp. 600 - 611, (2008).
12. D. Medjelakh et S. Abdou, 'Impact of Thermal Inertia on Hygrothermic Comfort and the Energy consumption of the bâtiment', Review of Renewable energies, Vol. 11, N°3, pp. 329 - 341, (2008).
13. L. Peeters, R. de Dear, J. Hensen and W. D'haeseleer, 'Thermal Comfort in Residential Buildings: Comfort Values and Scales for Building Energy Simulation', Applied Energy, Vol. 86, N°5, pp. 772 - 780, (2009).
14. S.A. Klein, W.A. Beckman and J.A. Duffie, 'TRNSYS - A Transient Simulation Program', ASHRAE Transactions, Vol. 82, N°1, pp. 623 - 633, (1976).
15. S.A. Klein, W.A. Beckman, J.W. Mitchel, J.A. Duddie, N.A. Duffie and T.L. Freeman, 'TRNSYS Manual', University of Wisconsin, (2006).
16. N. Fezzioui, B. Draoui, M. Benyamine et S. Larbi, 'Influence of the Dynamic Characteristics of the Envelope of Building on Thermal Comfort in the South Algéria', Review of Renewable energies, Vol. 11, N°1, pp. 25 - 34, (2008).
17. S. V. Patankar, Numerical Heat Transfer and Fluid Flow, McGraw-Hill New York, (1980).
18. De Vahl Davis G, Natural convection of air in a square cavity. A bench mark numerical solution. (1983).

EPJ Web of Conferences **114**, 02004 (2016)
DOI: 10.1051/epjconf/201611402004

Visualization and evaluation of flow during water filtration: Parameterization and sensitivity analysis

Petr Bílek[1,a]

[1]*Technical University of Liberec, Studentská 1402/2, 461 17, Liberec 1, Czech Republic*

Abstract. This paper deals with visualization and evaluation of flow during filtration of water seeded by artificial microscopic particles. Planar laser induced fluorescence (PLIF) is a wide spread method for visualization and non-invasive characterization of flow. However the method uses fluorescent dyes or fluorescent particles in special cases. In this article the flow is seeded by non-fluorescent monodisperse polystyrene particles with the diameter smaller than one micrometer. The monodisperse sub-micron particles are very suitable for testing of textile filtration materials. Nevertheless non-fluorescent particles are not useful for PLIF method. A water filtration setup with an optical access to the place, were a tested filter is mounted, was built and used for the experiments. Concentration of particles in front of and behind the tested filter in a laser light sheet measured is and the local filtration efficiency expressed is. The article describes further progress in the measurement. It was carried out sensitivity analysis, parameterization and performance of the method during several simulations and experiments.

1 Introduction

Visualization and evaluation of flow is a suitable tool for investigation of flow, verification of numerical data and measurement of flow parameters. Various experimental methods were developed to measure important parameters of flow in different conditions. The most wide spread visualization and measuring methods are PLIF (planar laser induced fluorescence), PIV (particle image velocimetry), IPI (interferometric particle imagining) and shadow sizing [1, 2]. Even though the methods exist in many variants they have not to be suitable in some specials cases of use.

In this article the visualization of flow enables to see the filtration process in a laser sheet by a digital camera. This optical method helps to see, what is happening inside a filtration chamber and enables to measure the local filtration efficiency. The water filtration setup with an optical access is used for the experiments. Water is seeded by artificial submicron particles and the suspension is filtered by a tested filtration textile. In this article monodisperse non-fluorescent polystyrene particles with its diameter below one micron are used. The captured digital images of the filtration process are analysed and the local filtration efficiency is expressed. The measuring method is a relative method and it measures on the basis of a calibration. The calibration is carried out by known mass concentration of polystyrene particles. The most important issue is to keep the same conditions during the calibration and then during the measurement. More about the first experiments using PLIV and PIV method is in [3].

In this paper the new progress of the measuring method is discussed. The impact of a change of particle size, angle between the camera and the laser sheet, size of an evaluative area and flow velocity to the measurement were monitored. Also parameters of the method in various setting of the camera and the filtration setup were expressed.

2 Experimental setup

The filtration process is investigated in the filtration channel, where an optical access is enabled. The filtration channel is connected to the water filtration circuit, where two pressure sensors, a flow meter and a pump are mounted, Figure 1. In this article the water filtration setup is used only for the determination of the impact of flow velocity on measured results.

Due to the simplification of the other experiments a glass tank was used instead of the filtration chamber including the whole filtration setup. The glass tank was made from the same material as the filtration chamber, where a tested filter mounted is, so the optical parameters are the same. The picture of the setup with the glass container is shown in Figure 2. The glass container is usually used for calibration of the filtration channel. In this paper the glass tank is used for the sensitivity analysis.

[a] Corresponding author: petr.bilek@tul.cz

The flow is illuminated by the laser sheet generated by a 10° wobble angle Powell lens and a laser module. The laser module generates a 1 mm narrow green laser beam with the power of 92.8 mW (measured by LabMaster Ultima v. 2.35 with an attenuator 1:1000).

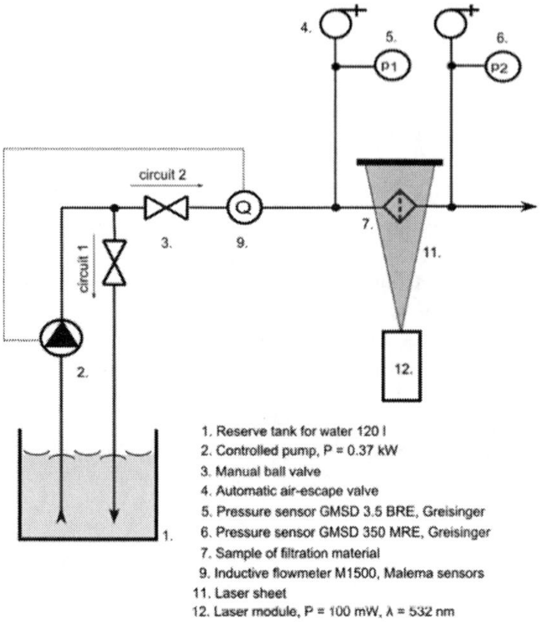

Figure 1. Scheme of the water filtration setup with the filtration channel and the laser sheet. The camera is installed at the place of a viewer.

Figure 2. Picture of the setup with the glass tank instead of the filtration chamber including the filtration setup.

The laser sheet is show in Figure 3. Unfortunately the laser sheet is not uniform and it includes brighter and darker strips about different light intensities. This is caused by defects of the cylindrical lens as well as the laser beam. The curve of a digital grey value across the laser sheet is shown in the graph in Figure 4. We can see the very uneven irradiation of the flow in the laser sheet. Nevertheless the defects of the laser sheet can be corrected by the calibration during evaluation of the filtration process. The camera is Pike F-210B/C with 1 inch size CCD chip with resolution 1920 × 1080 pixels. The camera offers plenty of settings and it is equipped by Samyang 35 mm F1.4 lens. The seeded particles are monodisperse made of polystyrene. The particles have excellent optical and mechanical features. The used sizes are 0.28, 0.42, 0.69, 0.96 and 1.7 micrometers.

Figure 3. The laser sheet in the glass tank full of water seeded by 0.69 μm polystyrene particles.

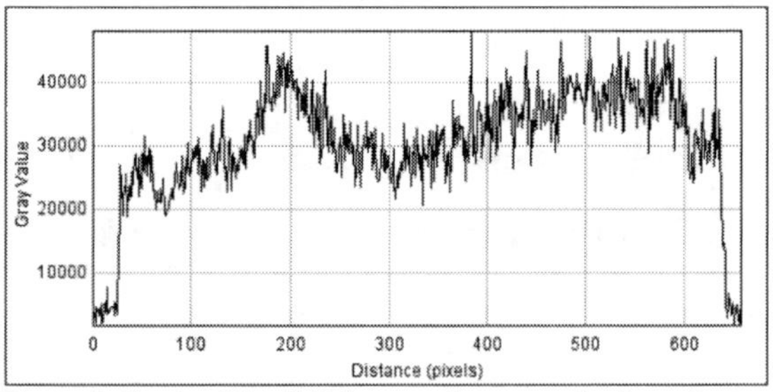

Figure 4. Graph of the digital grey value across the laser sheet, created by ImageJ software [4].

3 Experimental

The optical method for an evaluation of the local filtration efficiency is a relative method based on a calibration. The measuring conditions during the calibration have to be the same as during the measurement. In this article the impact of change of some chosen parameters between the calibration and the measurement was investigated. The parameters are scattering angle (the camera elevation against the laser sheet), particle size, flow rate and size of an evaluative area (the investigative area in an image, where the local concentration calculated is).

A part of the sensitivity analysis was carried out on the basis of simulations of the light scattering by small particles. A light intensity of the scattered light by small particles

$$\mathrm{E}(\theta) = E_0 \cdot \frac{C \cdot V \cdot |S(\theta)|^2}{k^2 \cdot d_1^2} \left[\frac{\mathrm{W}}{\mathrm{m}^2}\right] \quad (1)$$

is directly proportional to the scattering function $|S(\theta)|^2$ and to the concentration of particles C, where: E_0 is an incident light intensity, V is a small volume (characterized by the size of an evaluative area and the thickness of a laser sheet), θ is scattering angle, k is wave number and d_1 is the distance between an observer and particles [5]. In Figure 5, we can see the scattering function versus scattering angle for different sizes of particles. The scattering diagrams were created according to the Mie theory in software MiePlot [6]. The scattering diagrams were created for the used laser light and polystyrene particles suspended in distilled water.

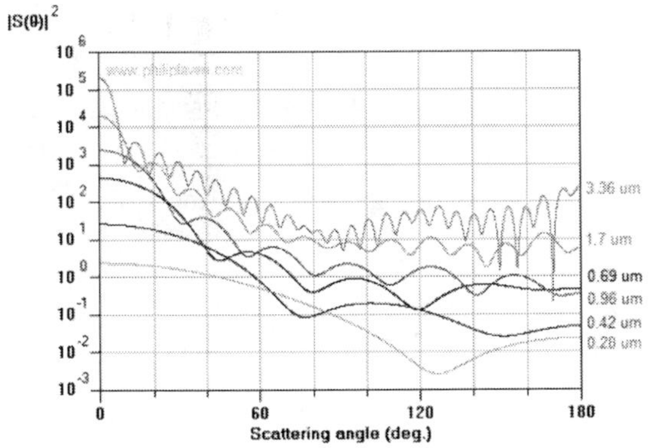

Figure 5. Scattering diagrams for polystyrene particles, generated by MiePlot [6].

According to (1) the change of a measured concentration value copies the change of a scattering function. Method sensitivity to the variation in a scattering angle is expressed by

$$S_\theta = \frac{\frac{|F(90.5°)-F(89.5°)|}{F(90°)} \cdot 100}{\Delta\theta} \left[\frac{\%}{°}\right], \tag{2}$$

where F is the scattering function and it is equal to $|S(\theta)|^2$ and $\Delta\theta = 90.5 - 89.5 = 1°$ is a small change of the scattering angle around the right angle. The scattering angle is the angle between a laser sheet and a camera and its value is usually right angle = 90°. The scattering angle is in fact the angle represeting a camera elevation against the laser sheet. The functions F were calculated by the MiePlot software.

Method sensitivity to the variation in a particle size is expressed by

$$S_\emptyset = \frac{\frac{|F(\emptyset+3\%)-F(\emptyset-3\%)|}{F(\emptyset)}}{\frac{\Delta\emptyset}{\emptyset}} [-], \tag{3}$$

where $\emptyset$ is size of a particle and $\Delta\emptyset$ is the small change 6 % from the nominal value $\emptyset$. The functions F were also calculated by the MiePlot software.

Method sensitivity to the variation in a flow rate is determined on the basis of an expeirment. The experiement was carried out on the water filtration setup. Water in rezervoir was seeded artificial particles and the flow through the filtration chamber was step by step increased from zero to 20 liter per minute. For every flowrate ten pictures of the flow were taken. The shift of particles in an image

$$d_{te} = \frac{Q_v \cdot t_e}{A_{ap} \cdot Z} < d_p, \tag{4}$$

should be smaller than the size of an active pixel d_p, where Q_v is volumetric flow rate, t_e is shutter time, A_{ap} is cross section of the filtration channel and Z is zoom of an objective. Relative error is calculated according to

$$\delta = \frac{H_m - H_s}{H_s} \cdot 100\ [\%], \tag{5}$$

where H_m is a measured value obtained from the measurement and H_s is the real value (calculated from the first 5 measurements during very low flowrate). The values can be expressed as digital grey values.

Method sensitivity to the variation in the size of an evaluative area is determined by another experiment, which was carried out with a glass tank instead of the filtration chamber and whole filtration setup. The concentration of artificial seeding particles was prepared to 100 µg per litre. The laser sheet was aimed to the glass tank and it was captured by the digital camera. Ten pictures were made. From these ten pictures a variation coefficient was calculated for the different sizes of an evaluative area and the different sizes of particles. The variation coefficient is expressed according to

$$C_v = \frac{\sigma_{Cm}}{\overline{C_m}}\ [\%], \tag{6}$$

where σ_{Cm} is a standard deviation and $\overline{C_m}$ is an average value of a mass cocnetration.

The last experiment was carried out with the glass tank as well. The concentration of artificial seeding particles was increased step by step in the glass tank and for every step ten pictures were taken. The concentration was increased from zero to 20 mg per liter. The calibration constant and the method sensitivity were calculated for various sizes of polystyrene particles. The sensitivity of the measuring method

$$S_{Cm} = \frac{\Delta Y_p}{\Delta C_m} = \frac{1}{K_T} \left[\frac{\mathrm{dm}^3}{\mu\mathrm{g}}\right] \tag{7}$$

is calculated from the change of the digital gray value ΔY_p and the change of the mass concentration ΔC_m. The smallest digital grey value is one, so the sensitivity is equal to the reciprocal calibration constant K_T.

4 Results

4.1 Sensitivity analysis

4.1.1 Angle between the camera and the laser sheet

The sensitivity of the measuring method to the angle between the camera and the laser cut is shown in Table 1. The sensitivity value was calculated according to (2) for monodisperse particles and for polydisperse as well. For very small particle sizes the impact of the scattering angle to the measurements is very low, but for particles bigger than micron is very significant. This is caused by a vibrating curve in the scattering diagram in Figure 5.

Nevertheless the sensitivity for the large particles is much smaller, if they are polydisperse. In fact no particles are absolute monodisperse and the majority of offered monodisperse particles are in fact polydisperse particles with a standard deviation. This implies that for the large particles the sensitivity is not such high.

Table 1. Method sensitivity to the scattering angle.

Particle size *Ø* **[μm]**	0.1	0.28	0.42	0.69	0.96	1.7	10
S_θ **[% / °]** Monodisperse particles	0.5	7.5	4	4.9	4.1	6.3	120
S_θ **[% / °]** Polydisperze, Std. Dev. 2 %	0.54	7.48	3.88	4.51	3.33	2.95	6.71

4.1.2 Size of seeding particles

The sensitivity of the optical method to particle size was calculated according to (3) and it is shown in Table 2. The particle size can be changed by dissolving or swelling of particles in water. The change of its size can cause significant error in measurement. For monodisperse particles the sensitivity is again higher for bigger particles. It is caused by its vibrating scattering diagram. In fact the sensitivity is much smaller due to standard deviation in size of the standard supplied particles.

Table 2. Method sensitivity to the size of seeding particles.

Particle size *Ø* **[μm]**	0.1	0.28	0.42	0.69	0.96	1.7	10
S_θ **[-]** Monodisperse particles	5.6	0.9	3.7	8.9	5.5	10.1	91.1
S_θ **[-]** Polydisperze, Std. Dev. 2 %	5.71	0.79	3.67	8.73	5.47	9.29	9.80

4.1.3 Flow rate in the filtration chamber

The optical method works during the wide range of flow rates. In Figure 6 we can see the influence of a relative error of the concentration measurement to the volumetric flow rate. The relative error was calculated according to (5).

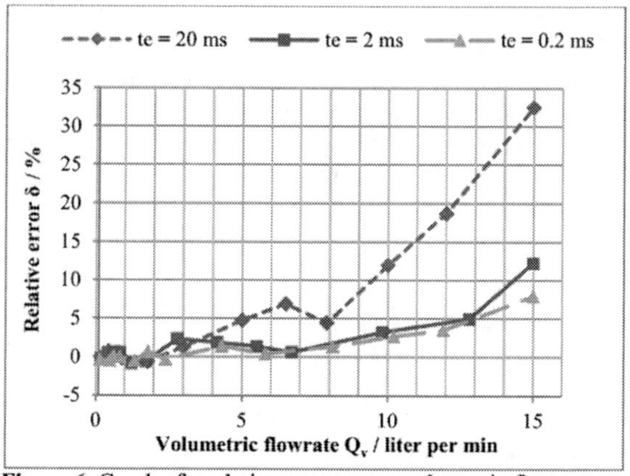

Figure 6. Graph of a relative error versus volumetric flow rate.

In the graph there are three curves for three exposition times. The relative error is the highest for the longest exposition time. This is caused by a blurring of an image according to (4). If the volumetric flow rate is high, the velocity of the particles in the filtration channel is high as well and the higher concentration of particles is measured in the blurred picture. The orange curve in the graph increases, though it is obtained during a very low shutter time. This is caused by air bubbles. The air bubbles are new sources of a scattered light and the method measures it as a higher mass concentration. In fact the mass concentration is still the same.

4.1.4 Size of the evaluative area

The graph in Figure 6 shows us the influence of the size of an evaluative area in an image. The variation coefficient was calculated according to (6). We can see high values for bigger particles. This is caused by distinguishability of the seeding particles in an image. In general the variation coefficient is the higher, the evaluative area is smaller.

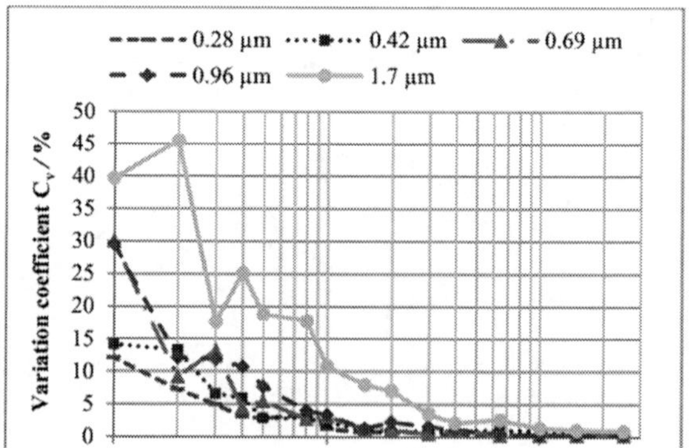

Figure 7. Graph of the variation coefficient versus size of an evaluative area.

4.2 Parameterization

The Table 3 shows the calibration constants and sensitivities of the optical method according to the used polystyrene particles. The sensitivity was calculated according to (7). We can see that the method is the most sensitive for the smallest particles. This is caused by their high number during the same mass concentration.

Table 3. Calibration constants for different particle sizes.

Particle size	K_T [μg/ dm^3]	S_{Cm} [dm^3/μg]	Range [μg/dm^3]
0.28 μm	0.03897	26	0.04 - 2620
0.42 μm	0.06104	17	0.06 - 3932
0.69 μm	0.07484	13	0.08 - 5242
0.96 μm	0.08098	12	0.08 - 5242
1.7 μm	0.10841	10	0.1 - 6553

The range of the measured mass concentration varies according to the sensitivity. Therefore the dynamic rage of the camera is still the same. The range of the optical method can be changed by setting of the camera, Table 4. If the binning is activated, the sensitivity is higher, but resolution is lower. If the evaluative area is bigger, the resolution of the method decreases. It is all about compromises and the optical method can be more powerful only in case of a better camera or a higher power of the laser sheet.

Table 4. Parameters of the optical method.

Size of the object [mm]	Size of the evaluative area in an image $O_x \times O_y$	Binning	Image resolution	Size of the evaluative area in an object [μm]
9.78 × 5.5	1 × 1	1 × 1	1920 × 1080	50.9 × 50.9
		2 × 2	960 × 540	101.9 × 101.9
	5 × 5	1 × 1	384 × 216	254.5 × 254.5
		2 × 2	192 × 108	509 × 509
	20 × 20	1 × 1	96 × 54	1019 × 1019
		2 × 2	48 × 27	2037 × 2037

5 Conclusion

The optical method for an evaluation of a filtration process discussed in the article is a strong tool for achievement new information about a filter. The method enables to visualize flow in the vicinity of the filter and characterize the filtration process. It is possible to measure the local filtration efficiency, which is connected to the geometrical structure of a filter. In this article a sensitivity analysis and a parameterization were carried out. Some of the measurements were simplified by a glass tank with the same optical performance as the filtration chamber. The glass tank was used instead of the filtration channel and water inside was seeded by polystyrene monodisperse particles.

The performance of the method is influenced by type of used particles, power of the laser module and settings of the digital camera. In general the method is capable to detect particles in range of 0.1 to 100 μm. The particles bigger than hundred of microns can be detected as well, but there is no requirement to measure these large particles. Smaller particles are detectable only in case of change of the camera setting to the most sensitivity mode at the expense of lower resolution or higher relative error. The small particles have to be in higher mass concentration as well. The maximum acceptable flow rate can be around 5 litres per minute for the used filtration chamber (5 cm). For higher values the relative error increases significantly because of blurring images and air bubbles occurring in the filtration chamber. The minimum size of an evaluative area in an image can be 10 × 10 pixels. The smaller values cause significant increase of a relative error. The method is very versatile and can measure wide range of particle sizes and their concentrations. However the versatility is limited by dynamic range of the digital camera. It is about compromises, the method can be very sensitive, but only in case of very low resolution and vice versa.

From the sensitivity analysis it ensures that the sensitivity of the optical method to the camera elevation (scattering angle) and size of particles is quite high. The calculations are propping upon Mie theory through the software MiePlot. The result is that the camera elevation against the laser sheet during the calibration has to be the same as during the measurement. To prevent size modification of seeding particles caused by swelling or dissolving, the particles have to be in water for the same time as during the calibration as during the measurement.

Acknowledgments

This work was supported by the Ministry of Education of the Czech Republic within the SGS project no. 21066/115 on the Technical University of Liberec.

References

1. C. Tropea, A. L. Yarin, J. F. Foss, *Springer handbook of experimental fluid mechanics*, Springer science and business media, p. 1557, ISBN: 978-3-540-25141-5, (2007).
2. J. E. Martin, M. H. García, *Combined PIV/PLIF measurements of a steady density current front*, Experimental fluids **45**, p. 265-276, (2009).
3. D. Jašíková, M. Kotek, P. Šidlof, J. Hrůza, M. Komárek, V. Kopecký : *Nanofilter evaluation using visualization methods*, 1st Nanocon International Conference, Tanger Ltd., pp. 186. (2009).
4. ImageJ. *Image analysis software* [online: 22.5.2015]. URL: <http://rsbweb.nih.gov/ij/index.html>.
5. H. G. van de HULST, *Light Scattering by Small Particles*. Willey, New York, pp. 470, ISBN: 0-486-64228-3, (1957).
6. P. Laven, *MiePlot software*, [online: 22.5.2015], <http://www.philiplaven.com/mieplot.htm>.

EPJ Web of Conferences **114**, 02005 (2016)
DOI: 10.1051/epjconf/201611402005

Velocity distribution measurements in a fishway like open channel by Laser Doppler Anemometry (LDA)

S.M.Sayeed-Bin-Asad[1,a], T. S. Lundström[1], A.G. Andersson[1] and J. G. I. Hellström[1]

[1]*Division of Fluid and Experimental Mechanics, Luleå University of Technology, SE-971 87 Luleå, Sweden*

Abstract. Experiments in an open channel flume with placing a vertical half cylinder barrier have been performed in order to investigate how the upstream velocity profiles are affected by a barrier. An experimental technique using Laser Doppler Velocimetry (LDV) was adopted to measure these velocity distributions in the channel for four different discharge rates. Velocity profiles were measured very close to wall and at 25, 50 and 100 mm upstream of the cylinder wall. For comparing these profiles with well-known logarithmic velocity profiles, velocity profiles were also measured in smooth open channel flow for all same four discharge rates. The results indicate that regaining the logarithmic velocity profiles upstream of the half cylindrical barrier occurs at 100 mm upstream of the cylinder wall.

1 Introduction

The turbulent flow around vertical cylinders in open channel has been extensively studied for many years; flow around semi-circular cylinders however has not been the subject of as many studies. One application for this type of structure is in fish migration where fishes swimming in the bow wake in front of the half cylinder used less energy to maintain their position in the flow [1]. It is then necessary to know how far the bow wake exists in front of the half cylinder and this can be identified measuring velocity profiles in different upstream region of the half cylinder. The area of bow wake ends where the velocity profile follows logarithmic law.

However, velocity distribution at or around any obstruction is one of important issues to obtain information of vortices. Different types of obstructions or barriers are installed in fish-ways to make proper passages to migrate fishes according to the situation. There are also some other application of open channel flow, such as, natural drainage systems in various creeks and rivers, rainwater in the channel of houses, flow in natural and human made canals, ditches of drainage, sewers, and gutters along roads, flow of small rivulets, and sheets of water across fields or parking lots and flow in the chutes of numerous water rides.

Many studies on open-channel phenomena have been performed since the 1970's [2-8]. Haro, A., et al. (2004) [9] studied the performance of swimming for upstream migrating fishes in open-channel flow for predicting passage through velocity barriers. They defined performance of upstream migrating fishes swimming through velocity barriers in a novel way and at a realistic scale in their study. Steffler, P. M. et al [10] conducted Laser Doppler Anemometry (LDA) measurements to measure mean velocity and turbulence for uniform subcritical flow in an open channel. The results indicated that mean velocity profiles follow logarithmic law. Ardiçhoğlu and Kirkgöz (1997) [11] conducted an experimental study on the progression of the flow starting from developing toward fully developed flow using LDA. They found that the extension of boundary layer occurs at fully developed turbulent flow axis for a certain aspect ratio. Madad, Reza, et al (2015) [12] conducted both experimental and numerical investigation for fully developed laminar and turbulent channel flow using an air–water interface. They noted that the higher energy is transferred to the water than the energy is transferred to a solid wall in moving condition. Balachandar and Patel (2002) [13] found that for a smooth surface, the logarithmic law (figure 1 is a typical velocity profile) is followed by the measured mean velocity profiles and for a rough surface, a suitable downward shifting occurs. Tachie et al. (2003) [14] measured velocities on two types of rough surfaces (different geometry) and a smooth surface in a channel and they found that the roughness effects on the velocity field were similar as found in turbulent boundary layer for a gradient of zero-pressure, although free surface influences the boundary layer in an open channel flow. They additionally found that wake parameters increased due to roughness of the surface as compared with a smooth surface. Afzal et al. (2009) [15] studied open-channel flows to find the effect of Reynolds number on the velocity distribution and they found that there is some extension of overlap with the log region which is affected by the variation of Reynolds number. However, Ghoma, Hussin [16] studied two-phase flows both experimentally and numerically in an

[a] Corresponding author: sayeed.asad@ltu.se

open channel flow and they observed the reasonable stream-wise velocity profiles distribution in the channel. They also found the reasonable agreement in the results between the numerical and the experimental studies.

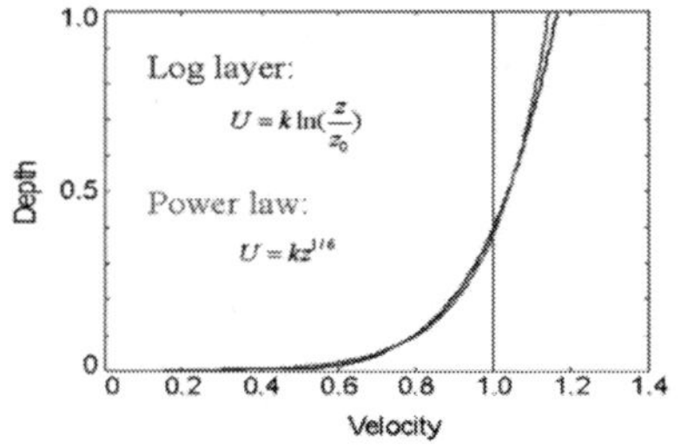

Figure 1. Typical logarithmic law and Power Law velocity profile, where U=Velocity ; k=Von Karman constant and z=Depth of water [17].

Laser Doppler velocimetry (LDV) is a well-proven non-invasive technique which accurately measures velocity of fluid flow at a point. An intersection of two laser beams is employed for this type of velocity measurement at a point. When a particle passes through the desired measuring point which is defined as probe volume, then the particle scatters light from the beams into a detector. The frequency of the resulted Doppler burst signal is directly proportional to the velocity of particle. Basic principle of a back scattering one-component LDA System *(Dantec Dynamics)* is shown in the figure 2. Velocity distribution in a channel flow can be accurately measured using the LDV which is, also known as LDA which measures the velocity of fluid based on the random sampling of individual velocity events which occur when particles pass through the measuring volume. Yeh and Cummins [18] first introduced LDA back in 1964, and since then it has been used broadly in experimental investigations of various fluid flow. LDA has been developed in the last 50 years as an enormously useful research instrument in the fluid dynamics area. Employing this LDA, accurate measurements all over the whole flow area can be obtained [19].

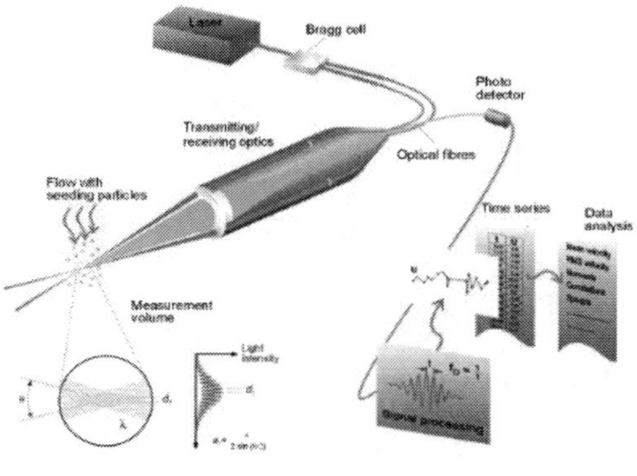

Figure 2. LDA principle [20].

A limited number of researchers have used the LDA technique for studying flow velocities in open channels. In the current study, the vertical velocity profiles in an open water channel with rectangular cross-section were experimentally investigated using the LDV technique for four different flow rates. The main focus is to investigate how steam-wise mean and RMS velocity profiles are varying in front of a half cylinder.

2 Experimental facility and method

The laboratory experimental investigations were carried out in a 7.5 meter long rectangular water flume with a cross-section of 295 mm x 310 mm shown in the Figure 3. A half cylindrical barrier (D-shaped) as *Liao, Beal [21]* used their study was used in the experiment. The length and diameter of this D-shaped half-cylinder are 390 mm and 100 mm respectively. Velocity distributions were measured vertically at four different upstream regions of this D-shaped cylinder for a constant water depth (D_w) of 180 mm (Height of the channel, $H= D_{w+} D_{e,}$) for four various flow discharge rates. An adjustable vertical gate was placed at the downstream end of the flume and a rail-mounted point gauges was installed on the top of the flume to control and measure the water depth in the channel. The sidewalls of the water flume were made of transparent 1.7 mm window glass to make possible velocity measurements using an LDA. A pump was used to re-circulate the water in the channel. A Danfoss MassFlo Coriolis flow meter (error <±0.5%) [22] was employed to control the discharge rate.

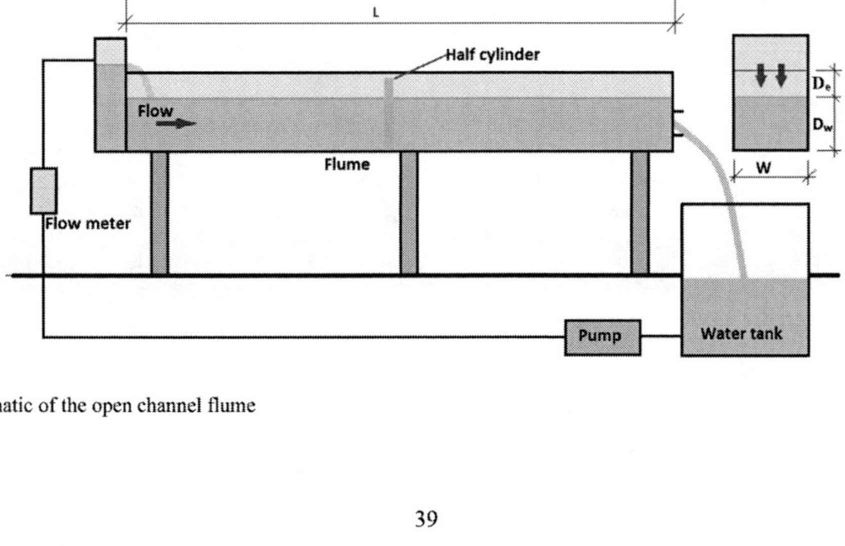

Figure 3. Schematic of the open channel flume

A one-component Power Sight Solid State Laser-based LDV system (TSI Inc.) was used to measure the velocity profiles. The LDV measuring system consists of a 500 mW laser, transmitting optics, a fiber probe equipped with a 350 mm focal length lens, a photo-detector, a signal processor and a computer. Installing half cylinder at middle of the channel, velocities at four different regions upstream of the cylinder such as close to cylinder wall, 25 mm, 50 mm and 100 mm ahead of the cylinder were measured for flow rate (Q)of 7650 liter/hour using 19 vertical points from channel bed to water surface. In the same way, velocities for 15000, 20000 and 25000 liter/min discharge rates were measured. However, similar steps of measurement were adopted for measuring velocities at only one region without installing cylinder (empty channel with water flow only) for each discharge rate. Table 1 shows the detail condition of these experiments.

Table 1. Experimental conditions

Discharge, (l/h)	Velocity distribution in the water flume
7650	Installing D-shaped cylinder (a)
	No D-shaped cylinder installed (a_0)
15000	Installing D-shaped cylinder (b)
	No D-shaped cylinder installed (b_0)
20000	Installing D-shaped cylinder (c)
	No D-shaped cylinder installed (c_0)
25000	Installing D-shaped cylinder (d)
	No D-shaped cylinder installed (d_0)

3 Results and discussion

3.1 Velocity Profiles

The comparison of stream-wise component of the mean velocity profiles in an open channel for different discharges is shown in figure 4. It is obvious that the velocity profiles in figure 4 are following logarithmic law for channel flow as comparing with the typical velocity profiles (figure 1).

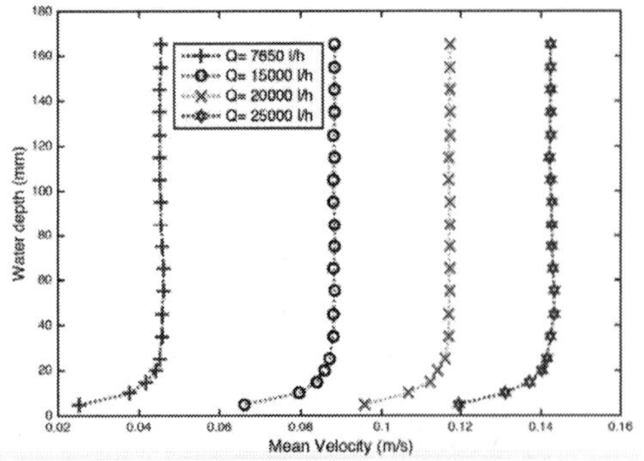

Figure 4. Velocity profiles for an open channel flow

Similarly, figure 5 depicts Root Mean Square (RMS) velocity or turbulent fluctuating velocity profiles for four flow discharge rates. All these profiles are comparable with profiles measured upstream of an installed half cylinder in the open channel.

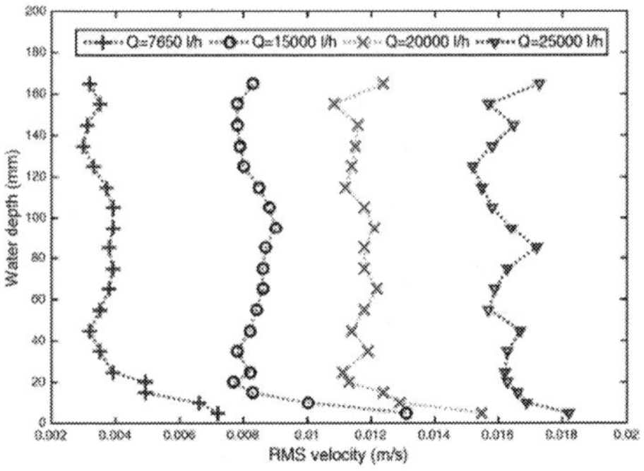

Figure 5. RMS velocity for an open channel flow

Figure *6a* depicts that mean stream-wise velocity profiles for flow both installing a half cylinder (four regions; close to wall, 25, 50 and 100 mm ahead of the cylinder) and with without installing a half cylinder (only open channel flow) for discharge rate (Q)of 7650 l/h. Similarly, figure *6b*, *6c* and *6d* depict the mean velocity profiles for discharge rates 15000, 20000 and 25000 l/h respectively. In the same way, figure 7 *(a-d)* show the Root Mean Square (RMS) velocity profiles for above mentioned all discharge rates. It is seen from figures 6 (a-d) that velocity profile is greatly affected at close to the cylinder wall.

For the case of close to wall (1 mm upstream of the wall), velocities close to the channel bed are low for all four discharge rates when the half cylinder is installed, while these velocities gradually increased as depth of water increased. This affect is gradually decreased as the distance from the cylinder wall is increased. Velocity profiles measured for all discharge rates are close to logarithmic profiles at 100 mm in front of the cylinder.

It is also found that velocities start increasing gradually for the case of close to wall (1 mm upstream of the wall) as height of water from channel bed increased and the maximum velocities are found close to water surface using a half cylinder installed in the channel but when no cylinder is installed (empty channel), velocities gradually increase up to water height of around 40 mm then these velocities keep almost constant up to water surface for all measured discharge rates.

Turbulent fluctuating or the Root Mean Square (RMS) velocity distribution for all four discharge rates are shown in figure 7 where it indicates that velocity fluctuation starts slowing down from channel bed to around 20 mm height of water for the case of close to wall but after that, the fluctuation starts rising rapidly. However, for other cases like 25, 50 and 100 mm away from wall as well as empty channel flow, maximum fluctuation occurred at channel bed which gradually

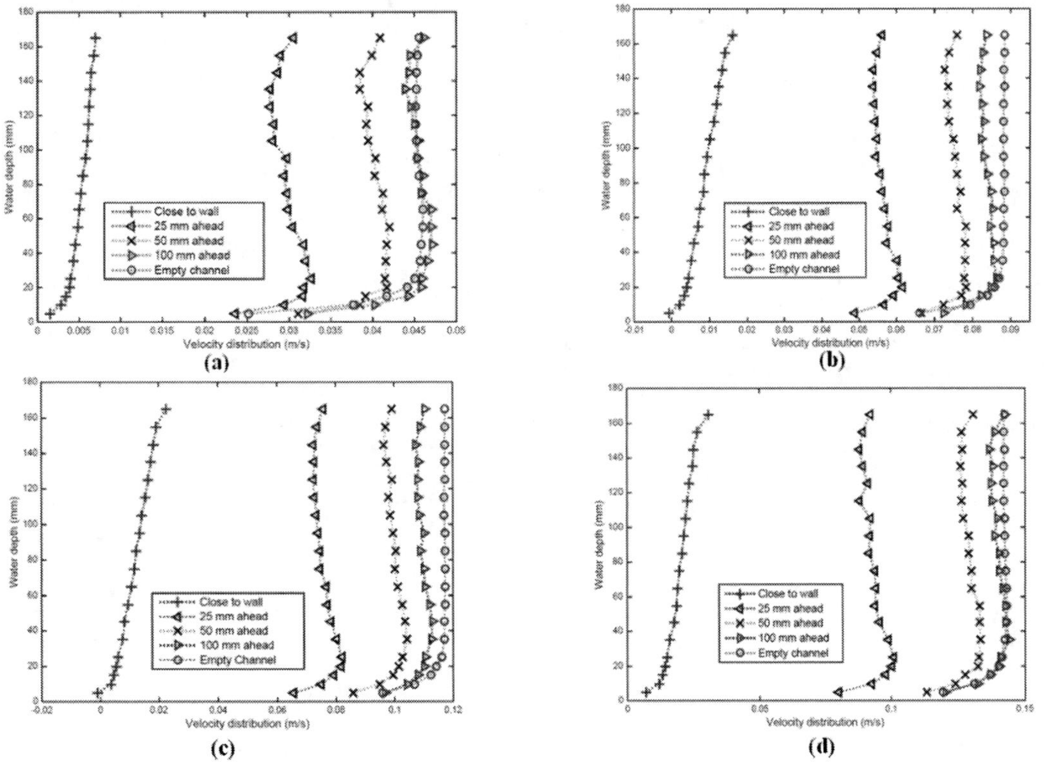

Figure 6. Velocity profiles for (a) Q=7650 l/h; (b) Q=15000 l/h; (c) Q=20000 l/h & (d) Q=25000 l/h

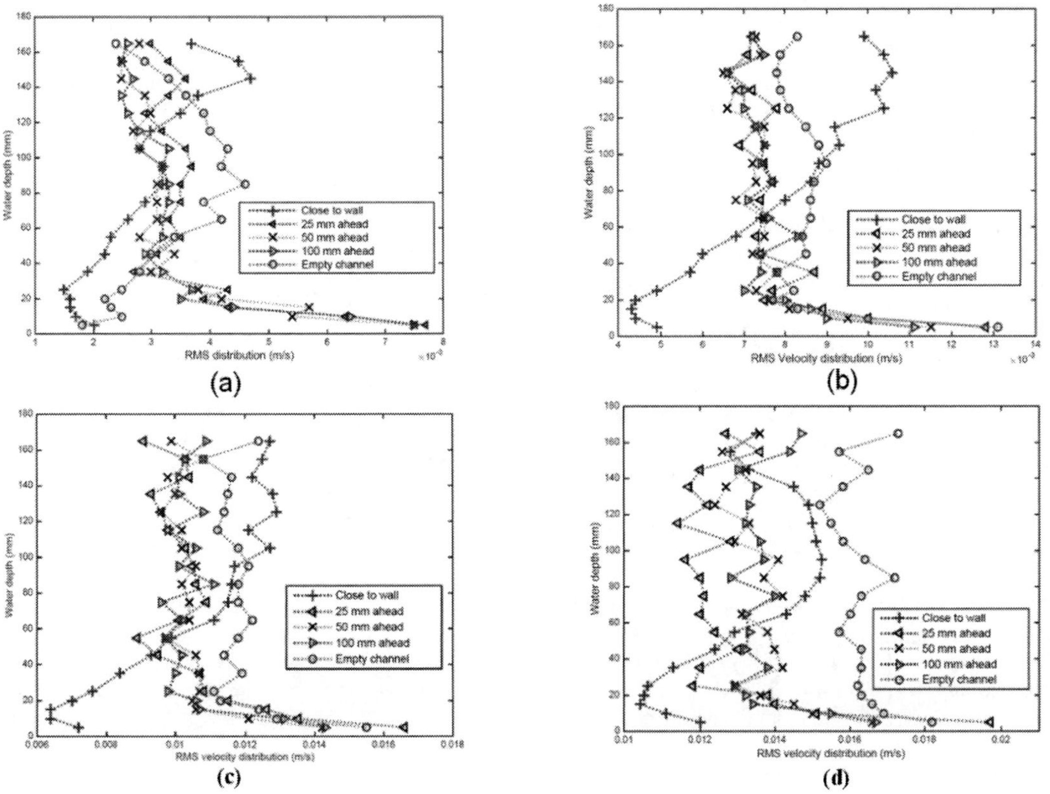

Figure 7. RMS Velocity profiles for (a) Q=7650 l/h; (b) Q=15000 l/h; (c) Q=20000 l/h & (c) Q=25000 l/

decreased up to 20 mm depth of water then the fluctuation started rising and dropping in zigzag way up to the water surface. The fluctuating velocities for almost all cases reach at roughly same value at approximately middle of water height but as discharge rate increases the deference between fluctuating velocities becomes larger, as it can be seen in the figure 7d for maximum discharge rate (Q=25000 l/h), RMS velocities do not reach at same point.

4 Conclusion

Velocity profiles were measured using LDV technique for four various discharge rates both installing a half-cylinder and without installing any half-cylinder in an open channel. Velocities at four different regions ahead of the cylinder were measured to examine the effect of the barrier on regaining logarithmic velocity profiles. The streamwise velocity profiles are regaining logarithmic velocity profiles at a distance of 100 mm (~1 diameter) upstream of the half cylinder for all four discharge rates which indicate that the bow wake exists up to 100 mm upstream of the half cylindrical obstruction. However, Maximum turbulent fluctuation has also been observed near the channel bed for most discharge rates.

Acknowledgement

This work has been funded by the collaboration initiative *StandUp for Energy*. The research program is a part of the Swedish governments commitment to high quality research in areas of strategic importance.

References

1. Liao, J.C., *A review of fish swimming mechanics and behaviour in altered flows.* Philosophical Transactions of the Royal Society B: Biological Sciences, 2007. 362(1487): p. 1973-1993.
2. Nezu, I. and W. Rodi, *Open-channel flow measurements with a laser Doppler anemometer.* Journal of Hydraulic Engineering, 1986. 112(5): p. 335-355.
3. Nezu, I., *Turbulence in open-channel flows.* 1993.
4. Kang, S., X. Yang, and F. Sotiropoulos, *On the onset of wake meandering for an axial flow turbine in a turbulent open channel flow.* Journal of Fluid Mechanics, 2014. 744: p. 376-403.
5. Fischer, H.B., *Longitudinal dispersion and turbulent mixing in open-channel flow.* Annual Review of Fluid Mechanics, 1973. 5(1): p. 59-78.
6. Yen, B.C., *Open channel flow resistance.* Journal of Hydraulic Engineering, 2002. 128(1): p. 20-39.
7. Muste, M. and V. Patel, *Velocity profiles for particles and liquid in open-channel flow with suspended sediment.* Journal of hydraulic engineering, 1997. 123(9): p. 742-751.
8. Kirkgöz, M.S., *Turbulent velocity profiles for smooth and rough open channel flow.* Journal of Hydraulic Engineering, 1989. 115(11): p. 1543-1561.
9. Haro, A., et al., *Swimming performance of upstream migrant fishes in open-channel flow: a new approach to predicting passage through velocity barriers.* Canadian Journal of Fisheries and Aquatic Sciences, 2004. 61(9): p. 1590-1601.
10. Steffler, P.M., N. Rajaratnam, and A.W. Peterson, *LDA measurements in open channel.* Journal of Hydraulic Engineering, 1985. 111(1): p. 119-130.
11. Kirkgöz, M.S. and M. Ardiclioglu, *Velocity profiles of developing and developed open channel flow.* Journal of Hydraulic Engineering, 1997. 123(12): p. 1099-1105.
12. Madad, R., et al., *An investigation of channel flow with a smooth air–water interface.* Experiments in Fluids, 2015. 56(6): p. 1-11.
13. Balachandar, R. and V. Patel, *Rough wall boundary layer on plates in open channels.* Journal of Hydraulic Engineering, 2002. 128(10): p. 947-951.
14. Tachie, M., D. Bergstrom, and R. Balachandar, *Roughness effects in low-Re θ open-channel turbulent boundary layers.* Experiments in Fluids, 2003. 35(4): p. 338-346.
15. Afzal, B., M.A. Faruque, and R. Balachandar, *Effect of Reynolds number, near-wall perturbation and turbulence on smooth open-channel flows.* Journal of Hydraulic Research, 2009. 47(1): p. 66-81.
16. Ghoma, M., K. Hussin, and S. Tait. *The Experimental and Numerical Approch of Two-phase Flows by a Wall Jets on Rough Beds in Open Channel Flow.* in *Proceedings of the World Congress on Engineering.* 2014.
17. Channelflow. *Logarithmic law and Power Law velocity profile.* 2015 [cited 2015 19-08-2015]; http://www.nortek.com.cn/lib/Tutorial%20-%20web%20content/channel-flow/channel-flow-fig1/image_view_fullscreen].
18. Yeh, Y. and H. Cummins, *Localized fluid flow measurements with an He–Ne laser spectrometer.* Applied Physics Letters, 1964. 4(10): p. 176-178.
19. Nezu, I., A. Tominaga, and H. Nakagawa, *Field measurements of secondary currents in straight rivers.* Journal of Hydraulic Engineering, 1993. 119(5): p. 598-614.
20. Dantecdynamics. *Laser Doppler Anemometer.* 2015 [cited 2015 18-08-2015].
21. Liao, J.C., et al., *Fish exploiting vortices decrease muscle activity.* Science, 2003. 302(5650): p. 1566-1569.
22. Green, T., *Particle image velocimetry in practice.* 2009: Luleå tekniska universitet.

EPJ Web of Conferences **114**, 02006 (2016)
DOI: 10.1051/epjconf/201611402006

Jet flow issuing from an axisymmetric pipe-cavity-orifice nozzle

Zuzana Broučková[1,a], Pavlína Pušková[2], Zdeněk Trávníček[1] and Pavel Šafařík[2]

[1]*Institute of Thermomechanics, the Czech Academy of Sciences, 182 00 Prague, Czech Republic*
[2]*Czech Technical University, Faculty of Mechanical Engineering, 166 07 Prague, Czech Republic*

Abstract. An axisymmetric air jet flow is experimentally investigated under passive flow control. The jet issues from a pipe of the inner diameter and length of 10 mm and 150 mm which is equipped with an axisymmetric cavity at the pipe end. The cavity operates as a resonator creating self-sustained acoustic excitations of the jet flow. A mechanism of excitations is rather complex – in comparison with a common Helmholtz resonator. The experiments were performed using flow visualization, microphone measurements and time-mean velocity measurements by the Pitot probe. The power spectral density (PSD) and the sound pressure level (SPL) were evaluated from microphone measurements. The jet Reynolds number ranged Re = 1600–18 000. Distinguishable peaks in PSD indicated a function of the resonator. Because the most effective acoustic response was found at higher Re, a majority of experiments focused on higher Re regime. The results demonstrate effects of the passive control on the jet behavior. Fluid mixing and velocity decay along the axis is intensified. It causes shortening of the jet transition region. On the other hand, an inverse proportionality of the velocity decay ($u \sim 1/x$) in the fully developed region is not changed. The momentum and kinetic energy fluxes decrease more intensively in the controlled jets in comparison with common jets.

1 Introduction

A jet flow is a type of fluid motion and one of basic types of shear flows. The jet flow is very important for theoretical studies in fluid mechanics – see Schlichting [1], Abramovich [2]. On the other hand, the jet flow is very important also from practical engineering point of view because it is frequently used in industrial applications such as fluid mixing, control of flow and thermal fields, thrust generation, convective heat transfer, and combustion and chemical processes.

This study is focused on turbulent submerged continuous jets. Despite the continuous (steady) jet's capability is unquestionable, unsteadiness effects can enhance applicability even more. It is worth noting here that unsteadiness is not a foreign feature to the continuous (steady) jets, quite the contrary, a tendency to form natural organized structures and oscillations are intrinsically incorporated in all jet flows – see, e.g., Crow and Champagne [3].

There are many ways how to generate/promote a pulsatile character of a jet flow. Generally, passive and active flow control methods can be distinguished – see Gad-El-Hak [4]. Relevant to this study are the passive methods which are based on geometry modifications and which can generate large vortex structures and/or significant acoustic excitations of the main flow. The most significant advantage of the passive control methods is a relative simplicity because they require neither actuators/transducers nor an external supply of energy.

A sensitivity of the jets to excitations is determined by instability and development processes either in the jet shear layers or in the entire jet columns, which are linked with the shear-layer and jet-column modes of the jet – e.g. Crow and Champagne [3] and Thomas [5]. The highest sensitivity of round jets to excitation, i.e. the most amplified mode under small artificial external excitations, is known to be at the most preferred mode linked with the Strouhal number of 0.3–0.35 – see Crow and Champagne [3] and Vlasov and Ginevski [6]. However, the Strouhal number of the jet-column mode of round jets themselves (natural passage frequency of vortices without any artificial excitation) can be higher, e.g. the value of about 0.5 concluded Crow and Champagne [3] and 0.61 was found by Liu and Sullivan [7]. Moreover, much wider range from 0.25 to 0.85 is considered by Thomas [5], who explained the reason of this wide range as an influence of the thickness of the boundary layer at the nozzle exit.

To create large acoustic excitations of jets by means of the passive control methods, Hill and Greene [8] designed the so called "whistler nozzle" consisting of an overhanging sleeve, which was co-axially mounted at the pipe exit. They concluded that the self-sustained acoustic oscillations can effectively enhance fluid mixing, which can be useful for various application such as design of ejectors, improvements in the flow beneath V/STOL

[a] Corresponding author: brouckova@it.cas.cz

aircrafts in ground effect including reduction of ground heating/erosion effects, and on/off suppression of exhaust plume detectability by reduction of infrared footprints of combat aircrafts. Note that the experiments [8] were made with the pipe nozzle of a diameter 25.4 mm at the jet exit velocity of 54 m/s, i.e. the Reynolds number was approximately 9×10^4. Hasan and Hussain [9] experimentally investigated the "whistler nozzle" phenomenon as an implication of a self-excited organ-pipe resonance in either the full-wave mode or the half-wave mode. In a follow-up paper, Hussain and Hasan [10] proposed more complex explanation showing that the whistler excitation results from the coupling of two independent resonance mechanisms, namely the organ-pipe resonance and the shear-layer tone resulting from an impingement of a pipe-exit shear layer onto the collar lip. Experiments in both studies [9, 10] were made with the pipe nozzles of diameters 25.4 mm and 7.62 mm at the jet exit velocity of 36 and 60 m/s, corresponding to the Reynolds number range from 6.2×10^4 to 3.1×10^5.

The so called orifice-cavity-orifice nozzle was proposed for generation of the self-sustained acoustic oscillations by Abell and Luxton [11]. As was described by Lee [12], fluid passes successively a pipe, orifice, resonance cavity and finally issues through the second, outlet orifice. The aim was to enhance mixing in combustion applications. As was explained by Nathan [13], the orifice-cavity-orifice nozzle can enhance stability and power of flames, but it is not easy to apply it in industry because flow rate variations can cause an abrupt drop-out effect from (rather narrow) resonance regime resulting in flame blowing off. To overcome this fact and keep advantages resulting from self-sustained oscillations (such as mixing intensification, flame stability increase, and NO*x* emissions decrease), another method based on fluidic rather than acoustic mechanism was developed by Nathan [13], Nathan and Luxton [14] and Nathan et al. [15]. Namely, they used self-sustained precession of entire jet flow column inside a cavity.

Page et al. [16] investigated a utilization of the "whistler nozzle" phenomenon for an enhancement of a heat transfer from impinging jets. The velocity and Reynolds number were 90–153 m/s and 6×10^4–1.5×10^5, respectively.

Other methods capable to generate jet pulsations use various no-moving-part fluidic elements to excite continuous fluid supply. For example, flapping jets based on fluidics can be designed either with external feedback loop (Viets [17], Raman et al. [18], Camci and Herr [19], Seifert and Pastuer [20], Tesař et al. [21]) or with an internal feedback action (Mi et al., [22]).

A kettle whistle consisting of two axially aligned orifice plates (i.e. the orifice-cavity-orifice nozzle as named by Abell and Luxton [11] and cited by Lee [12]) was investigated by Henrywood and Agarwal [23]. Their microphone experiments focused on aeroacoustics. Depending on the Reynolds number $Re = UD/\nu$ (where U is the average velocity through the exit orifice, D is the exit orifice diameter, and ν is the kinematic viscosity), two regimes were identified. The first regime is the constant frequency regime at low Reynolds numbers ($Re < 2000$). Behavior can be modeled by the Helmholtz resonator mechanism. Instead of fundamental zero-flow case, the resonator effective neck length is a function of the flow Mach number. The second regime is the constant Strouhal number regime at higher Reynolds numbers ($Re > 2000$). The driven mechanism is a vortex shedding at the exit orifice of the whistle.

As written above, Henrywood and Agarwal [23] focused on aeroacoustics behavior of the orifice-cavity-orifice nozzle. However, flow field behavior of that case was not investigated neither in study [23] nor in available literature. A motivation of this study is an evident lack of information related to the flow field behavior downstream of this type of whistler nozzles. The aims of this study are to design a whistler nozzle and to demonstrate its behavior. Namely, abilities to generate self-sustained acoustic oscillations and modify the time-mean jet-flow characteristics are investigated experimentally.

2 Experimental setup and methods

Figure 1(a) shows a schematic view of the present experimental setup. The airflow from the building compressor pass through a flexible plastic tube, ball valve (1), pressure regulator (2), throttle valve (3), passing the manometer (4) to the rotameter (5). In the case of flow visualization experiments, the airflow continues to the water fog generator (6); for velocity measurements this generator is not used. Thereafter the airflow continues through a T-shaped connecting tube (7) via the settling chamber (8) with a diameter/length of 25/117 mm. The settling chamber is equipped with a honeycomb to straighten flow and suppress a swirling component of the velocity. A size of cells is approximately 4.5 mm and their length is 27 mm. Thereafter the airflow continues to the investigated pipe-cavity-orifice nozzle (9), which is fixed on a traverse system (10) positioned manually along the horizontal axis; an accuracy of the adjustment is 0.1 mm.

Figure 1(b) shows the investigated nozzle in a detail including the used cylindrical coordinate system x, r. The nozzle is situated vertically and it consists of an duralumin pipe of D = 10.0 mm internal diameter. The pipe is 15 diameters long and a perpendicular cut at the end is machined making the sharp-edged nozzle lip. The end part is equipped with a resonant cylindrical cavity with an internal diameter of 45.6 mm. The cavity top is made out of an duralumin plate of thickens 0.4 mm with a sharp-edged hole of diameter D = 10.0 mm. The depth of the cavity h is can be adjusted in a range from h = 0 (i.e. the case without the cavity) to h = 40 mm. A more detailed description of the geometry was described by Pušková [24].

The volume flux through the apparatus thus the average velocity issuing from the nozzle are evaluated from measurements by rotameter (5) – see figure 1(a), with consideration of the pressure at the manometer (4). The rotameter is a multiple tube type with 150 mm tubes and built-in needle valves (Aalborg P41A6-BD1). An accuracy is ±2 % of full scale. The manometer is an electronic type with a resolution of 0.1 kPa (Greisinger GMH 3156 with GMSD 2 BR transducer). A thermistor probe is used to measure the air flow temperature.

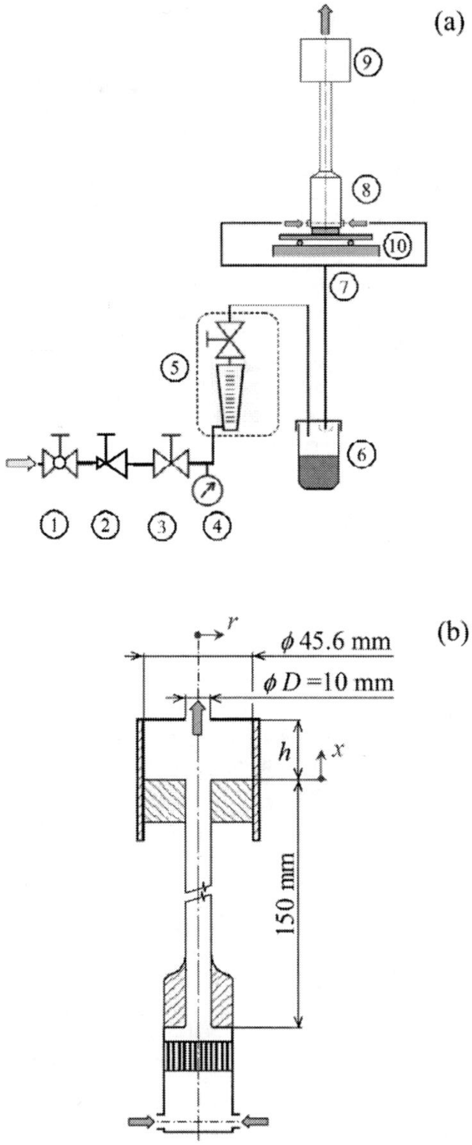

Figure 1. Experimental setup, 1: ball valve, 2: pressure regulator, 3: throttle valve, 4: manometer, 5: rotameter, 6: water fog generator, 7: T-shaped connecting tube, 8: settling chamber, 9: investigated pipe-cavity-orifice nozzle, 10: horizontal traverse system; (a) general scheme and (b) detail of the investigated nozzle.

Flow visualization used cold water fog produced by an ultrasonic piezoelectric nebulizer Mini Nebler – see figure 1(a), position 6. Contrasting white flow field pattern (streaklines, diffused by the flow turbulence) on black background were illuminated by a (left-hand) side photoflash, located at a distance of 0.8 m from the nozzle axis. The photoflash was equipped by an inhouse-built lens concentrating light into about 30 mm thick layer along the jet axis. Pictures were taken by a digital camera which was located in front of the nozzle, at a distance of 0.8 m from the nozzle axis. To obtain on-line control of the camera, it was connected to a PC via a USB cable. An exposure time of 1/125 s was used.

Frequencies of the sound effects were evaluated using an electret microphone of diameter 10 mm. The microphone was fixed perpendicularly to the nozzle axis in the place about x = 0, r = 0.6 m and its output was connected to the sound card of a computer. The microphone signal (without calibration) was on-line analyzed by means of Spectrum Lab V2.77 software to acquire the relative amplitude of the power spectral density (PSD).

The sound pressure level (SPL) was measured by means of the Sound Level Meter QUEST 2100 with a microphone of diameter 13.5 mm and accuracy of +/-0.7 dB. The device operates at the Frequency Weighting Networks A, which emulates the low level response of the human ear. The microphone was located close to the above mentioned electret microphone.

The time-mean flow velocity was evaluated by a Pitot tube (blunt-nosed impact tube with an outer diameter of 0.8 mm), connected to an electronic manometer (Greisinger GMH 3156) with the transducers GMSD 25 MR (resolution of 1 Pa) and GMSD 2,5 MR (resolution of 0.1 Pa). To evaluate time-mean velocity profiles, the Pitot tube probe was fixed on the traverse system positioned manually along the vertical axis. The accuracy of the adjustment was 0.5 mm.

Considering that the static pressure is equal to the barometric pressure, the measured time-mean total overpressure p can be converted into the time mean velocity as $U = c(2p/\rho)^{0.5}$, where c is the calibration factor (approximately c = 1.00 for the present probe) and ρ is the air density. An uncertainty analysis was performed according to the method of Kline and McClintock [25] for a single sample experiment. A typical centerline velocity magnitude in this study for x = 0-20D was above 10 m/s. For this magnitude, the uncertainty was evaluated within 7 %, based on the 95 % confidence level (i.e., ±2 standard deviation).

3 Results and discussion

3.1 Helmholtz resonator

To demonstrate possibility of the present measurements and validity of data processing, auxiliary experiments focusing on the cavity resonant frequency were performed. For this purpose, the cavity of the nozzle was adapted and the orifice from the pipe to the cavity was closed by an adhesive tape. To produce an acoustic response, an auxiliary excitation by air jets impinging obliquely onto the orifice lip (D = 10 mm, see figure 1(b)) was used. Note that the frequency response did not depend neither on a velocity of the excitation jet nor on details of the excitation arrangement.

Figure 2 presents examples of acoustic responses as the relative amplitude of the power spectral density (PSD). Figure 2(a) is a benchmark presenting a background noise in the laboratory before the test – no PSD peaks are identifiable.

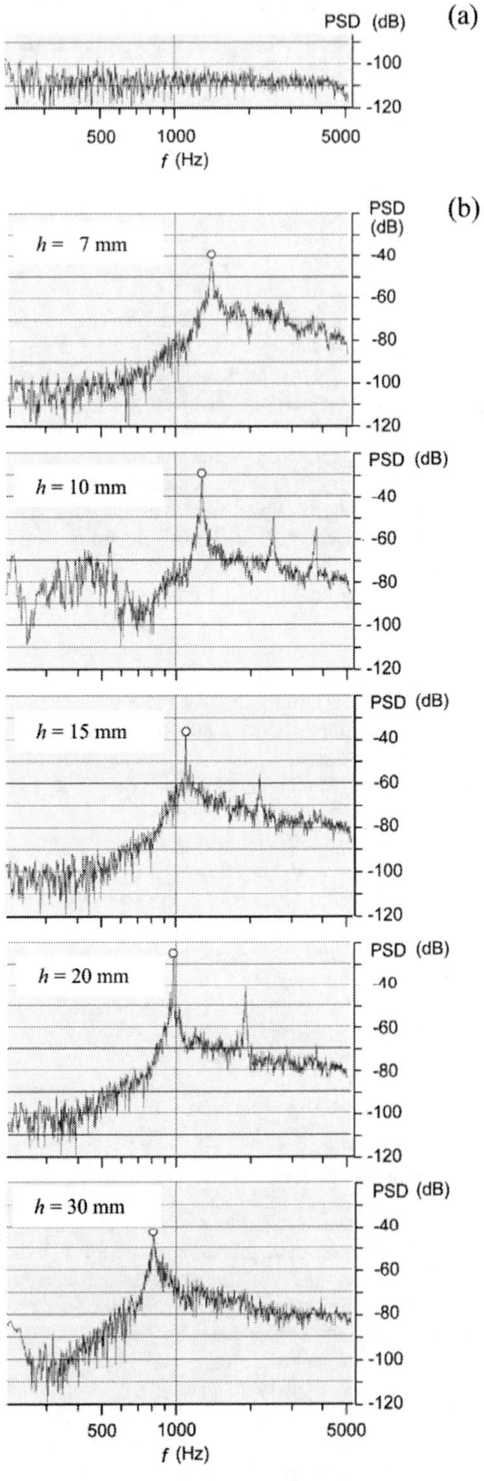

Figure 2. Power spectral density, (a) background noise, (b) acoustic response of the cavity for $h = 7$–30 mm.

Sequentially, excitation air jets was used and figure 2(b) shows an acoustic response for various cavity volumes expressed as cavity depths of $h = 7$–30 mm. A distinguishable PSD peaks clearly indicate the resonant

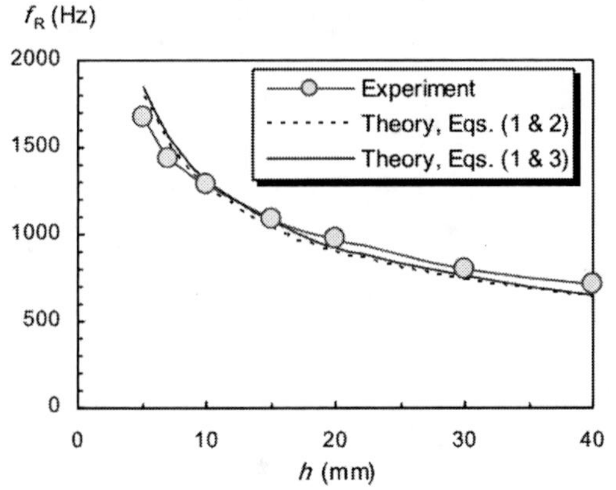

Figure 3. Resonant frequency – comparison of the present experimental data with theory.

frequencies f_R. Moreover, higher harmonics are manifested by other peaks with lower PSD level, see e.g. graph for $h = 10$ mm. As was expected (Kinsler [26]), an increase of the cavity volume, thus a depth increase in figure 2(b), causes a decrease of the resonant frequency. The present experimental data are plotted as a relationship between the resonant frequency f_R an the cavity depth h in figure 3. The result is well comparable with the natural frequency of the Helmholtz resonator (Kinsler [26]), which can be derived, considering the lumped acoustic system, isentropic expansion, ideal gas law, and compression during pressure oscillations, as

$$f_R = \frac{c}{2\pi}\sqrt{\frac{A}{L_{eff}V}}, \tag{1}$$

where c is the speed of sound, A is the cross-sectional area of the orifice ($A = \pi D^2/4$), V is the cavity volume ($V = Ah$), and L_{eff} is the effective length capturing mass of an oscillated fluid column (i.e., the slug flow model is assumed). Considering end correction for a flanged orifice (Kinsler [26]), the L_{eff} can be approximated as

$$L_{eff} = L + 8/(3\pi)\,D\,, \tag{2}$$

where L is the orifice length (thickness of the cavity top wall). Besides, Kinsler [26] suggests a bit shorter L_{eff} for a circular hole in a thin wall of a resonator:

$$L_{eff} = L + 0.8\,D\,. \tag{3}$$

Figure 3 shows that the present data reasonably agree with Eq. (1) with the L_{eff} from Eq. (2). The maximum deviation between the present data and values predicted by Eqs. (1 and 2) is less than 10 % in the whole investigated range. Moreover, using Eq. (3) decreases this deviation for a majority of experimental data points: the maximum deviations between the present data and values predicted by Eqs. (1 and 3) are within 7 % for $h = 10$-30 mm.

These measurements indicate that the tested cavity resonates as the Helmholtz resonator. The results demonstrate possibility of the experiments and validate the present approach.

3.2 Aeroacoustics of the pipe-cavity-orifice nozzle

Acoustic effects of the pipe-cavity-orifice nozzle on the jet flows was evaluated by means of the power spectral density (PSD). Investigated ranges were $Re = 1600$–18 000 and h = 7–40 mm. Figures 4–7 present typical results plotted as the relative amplitude of the PSD for jets without passive control (i.e. $h = 0$) and jets under the cavity control effect at h = 7–20 mm.

Figure 4 shows that the cavity control effect for $Re = 1600$ is relatively weak. Despite some redistribution in PSD can be identified, at h = 8 and 10 mm, no significant increase of sound pressure level (SPL) was found. In fact, for all cases at Re = 1600 (figure 4), the SPL was evaluated below a resolution of the instruments, i.e. below 40 dB.

Table 1 summarizes SPL measurements for all investigated jets at Re = 1600–18 000. The maximum obtained SPL for various Re by means of the cavity control effect is named SPL_{max}. Table 1 shows that an increase of the Re causes stronger sound generation. The highest SPL in this study was SPL_{max} = 79.3 dB; it was achieved at Re = 18 000 and h = 7 mm.

Table 1. SPL of jets without passive control (h = 0) and jets under the cavity control effect at h = 7–15 mm.

Re	No control (h = 0) SPL (dB)	With control SPL_{max} (dB)	h (mm)
1600	< 40	< 40	8
4400	41.2	50.8	15
8300	49.7	72.8	10
18 000	55.4	**79.3**	7

As is indicated by figures 5–7, distinct peaks in PSD appeared as results of the passive control at Re = 4400–18 000. This effect indicates significant changes in jet structures. The level of the PSD peaks is markedly higher than PSD of the uncontrolled reference jets, namely by 20 dB at Re = 4400 and even by 50 dB at Re = 8300 and 18 000. These peaks indicate a resonant behavior of the whole system. It can be deduced that the present nozzle design can be useful for a passive control of jet flows.

Moreover, a distinguishable cascade of peaks with lower PSD level can be identified in figures 5(b), 6(b), and 7(b), which resembles higher harmonics of the Helmholtz resonator. However, the aeroacoustic behavior of the present system with fluid flow is more complicate than the Helmholtz resonator. As was discussed by Henrywood and Agarwal [23], the excitation mechanism of a such system consists of a many particular sources such as (1) resonance behavior of the supplied piping system, (2) vortex shedding at the sudden expansion of the cross-sectional area (pipe-cavity interface), (3) cavity operating as the Helmholtz resonator, (4) sound propagation via edge tones at the jet impingement inside the cavity (cavity-orifice interface), (5) vortex shedding in the jet shear layer, and (6) turbulent noise of the jets inside as well as outside the cavity.

3.3 Flow visualization

Because the aeroacoustic effects described above are the most intensive at Re = 18 000, flow visualization experiments and velocity measurements focus on this Re case, when the control effects can be expected well distinguishable.

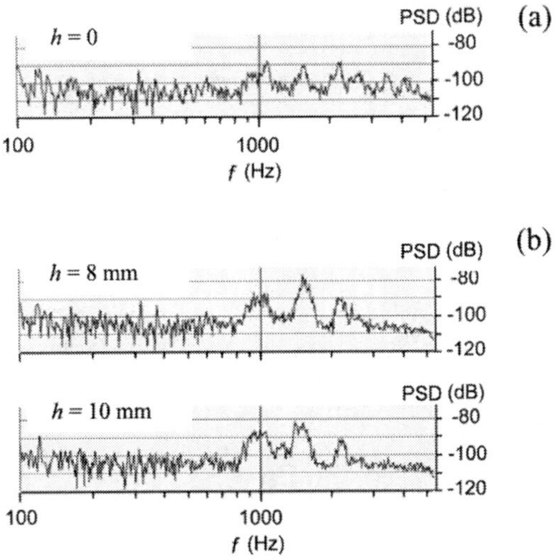

Figure 4. Power spectral density for Re = 1600, (a) jet without a passive control, (b) jet under the cavity control effect at h = 8 mm and 10 mm.

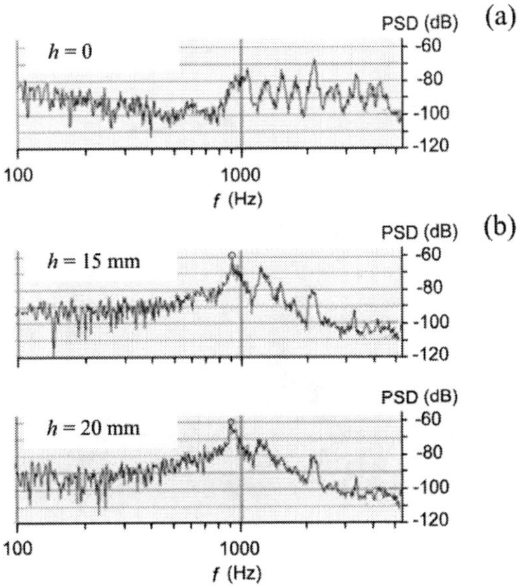

Figure 5. Power spectral density for Re = 4400, (a) jet without a passive control, (b) jet under the cavity control effect at h = 15 mm and 20 mm.

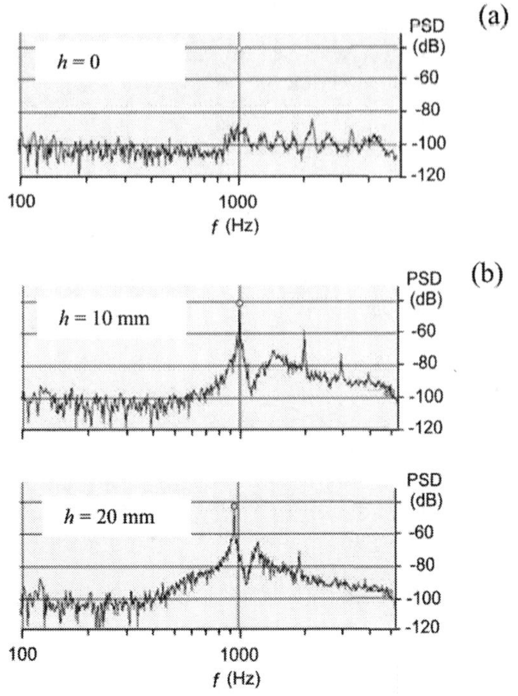

Figure 6. Power spectral density for $Re = 8300$, (a) jet without a passive control, (b) jet under the cavity control effect at $h = 10$ mm and 20 mm.

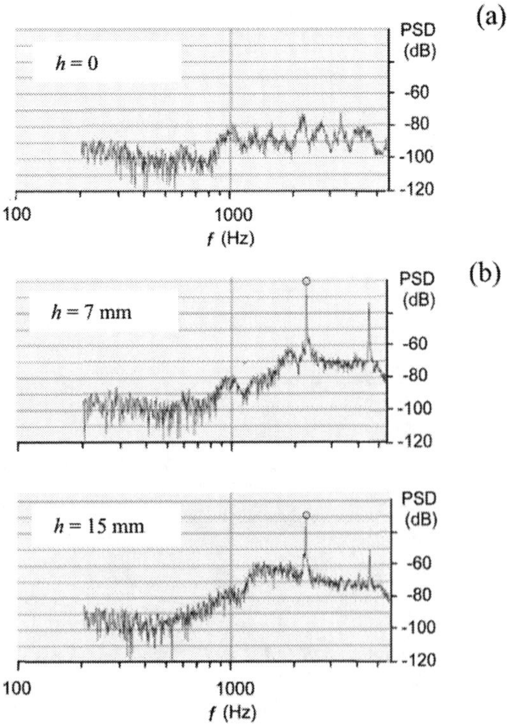

Figure 7. Power spectral density for Re = 18 000, (a) jet without a passive control, (b) jet under the cavity control effect at $h = 7$ mm and 15 mm.

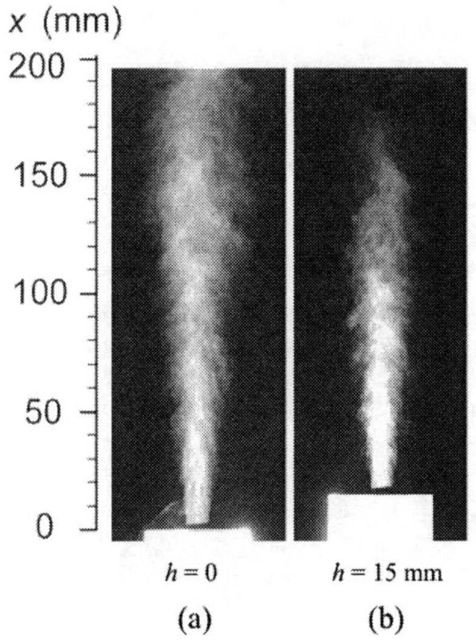

Figure 8. Flow visualization for $Re = 18\,000$; (a) without passive control ($h = 0$), (b) under the cavity control effect at $h = 15$ mm.

Figures 8(a, b) show typical flow visualization for $Re = 18\,000$, for the jet without a passive control (ie. at $h = 0$, see figure 8(a)) and for the jet under the cavity control effect at $h = 15$ mm, see figure 8(b). The control effect can be identified as a slight spreading and shortening of the visible jet flow area. However, flow visualization cannot capture unambiguously this effect. It will be quantified by the time-mean velocity measurement in the following text.

3.4 Velocity measurement of the flow field

Figure 9(a) shows the streamwise variation of the time-mean velocity along the jet centerline for $Re = 18\,000$ at various cavity depth from $h = 0$ (i.e. without a passive control) to $h = 20$ mm. A nondimensionalization of the velocity is made by the centerline velocity in the orifice u_0.

A behavior of a jet velocity decay is well comparable for all investigated jets. Firstly, the centerline jet velocity is approximately constant up to $x/D = 4$, i.e. the measurement is made in the jet core of in the initial region. Further downstream, the velocity decreases (the transition region) towards the final inverse proportionality of the velocity decay (fully developed region of the turbulent jet): $u/u_0 \sim (x/D)^p$, where the exponent p is approximately –1. This development is well-known from literature for fully developed axisymmetric turbulent jets – see Schlichting and Gersten [1] and Blevins [27]. A good agreement with literature is demonstrated by the relationship $u/u_0 = 6/(x/D)$ by Blevins [27], which is plotted as the dashed line in figure 9(a) too.

The control effect is well manifested by a shortening of the transition region and a sooner formation of the final inverse proportionality of the velocity decay indicating the fully developed region – the control effect is depicted by the arrow in figure 9(a). Note that the

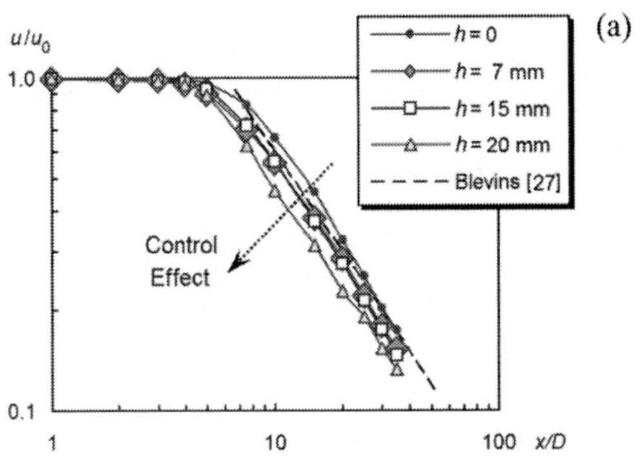

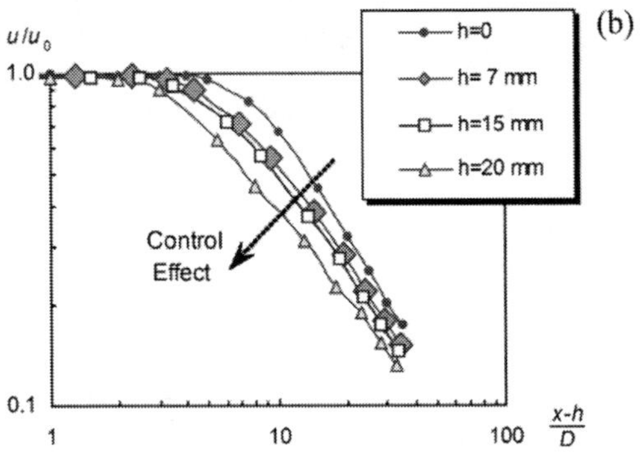

Figure 9. The centerline velocity decay at Re = 18 000 as a function of axial distance.

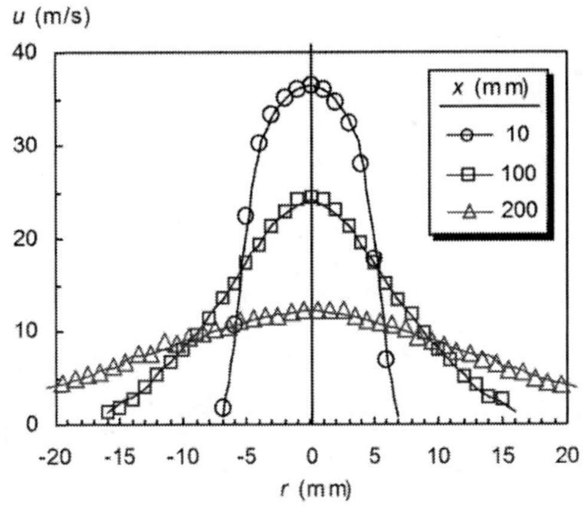

Figure 10. Velocity profiles at h = 0: points present experimental data, lines are symmetrical profiles obtained by the data symmetrization process.

exponent $p = -1$ of the exponential velocity decay was not changed by the control effect – see figure 9 too.

It is worthy of notice here that the streamwise coordinate x/D in figure 9(a) was purposely chosen in agreement with the paper by Hill and Greene [8]. On the other hand, another logical streamwise coordinate can originate in the outlet orifice cross section, thus the data presentation from figure 9(a) can be transformed into figure 9(b) with $(x\text{-}h)/D$ horizontal coordinate. Obviously, all the above described results demonstrating the control effect in figure 9(a) are evident in figure 9(b) too – in fact, the control effect is manifested in figure 9(b) even more clearly.

The present paper uses an axially symmetric setup thus we can expect an axially symmetric jet flows. Nevertheless, the expected symmetry of the jets has to be experimentally verified to exclude influences of manufacturing inaccuracies or random disturbances or suchlike. Figure 10 shows three typical velocity profiles at approximately Re = 18 000, which were measure for h = 0 (i.e. without a passive control) at x = 10, 100, and 200 mm. The profiles demonstrate the jet flow development, i.e. and decreasing of the centerline velocity and widening of the jet flow. Figure 10 demonstrates an acceptable symmetry of the velocity profiles. Obtained velocity profiles consisting of two halves for $r \leq 0$ and $r \geq 0$ were transformed into one symmetrical profile $u(r)$ by averaging the $u(+r)$ data and $u(-r)$ data for each data point of the traversing line. Figure 10 shows experimental data as well as the

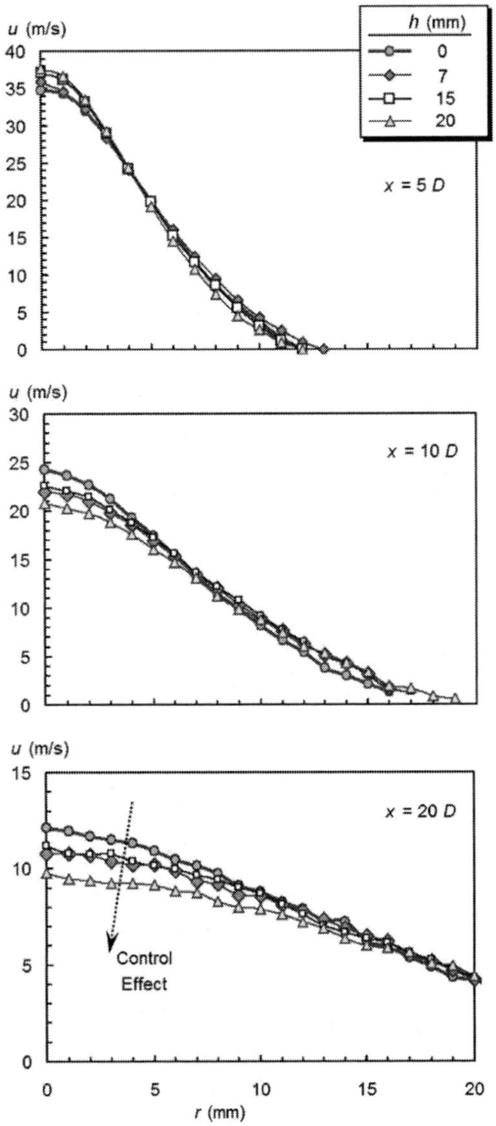

Figure 11. Velocity profiles at Re = 18 000.

symmetrical profiles resulting from the described data symmetrization process.

Figure 11 demonstrates a flow development for jets without/with the flow control: the centerline velocity decreases with increasing x and the jets are widened at the same time. The present control effect is well manifested by decreasing velocities in the central part, as is depicted by the arrow in figure 11 at $x = 20\,D$.

To compare jet flow development with available literature, the jet half-width was evaluated from experimental velocity profiles. The jet half-width (or half-radius for the present axisymmetric case) is the transverse distance from the axis to the point in which the axial velocity u falls to the one-half of the centerline value u_{max}. Figure 12 presents a development of the jet half-width for Re = 18 000 and a comparison with a formula by Schlichting and Gersten [1], $r_{0.5}/D = 0.09(x/D+1)$, which can be illustrated as the apical angle of 10.3° with the apex in the virtual origin of the jet at $x = -D$. Another relevant formula is $r_{0.5}/D = 0.086\ x/D$ with the apical angle of 9.8°, as proposed Blevins [27]. The both reference formulae are plotted in figure 12 too and an acceptable agreement of the present data with these references can be concluded.

Figure 13 shows the velocity profiles after transformation into a dimensionless form of $u/u_{max}(r/r_{0.5})$. Figure 13 demonstrates a similarity of the investigated profiles resulting from fully developed states of the jets, because they collapse quite well into one nondimensional profile. For comparison purposes, figure 13 shows also the reference profile by Blevins [27]: $u/u_{max} = e^{-94(0.086 r/r_{0.5})^2}$. A quite good agreement of the present data with this formula [27] can be concluded (except a very few points at the jet peripheral parts for x = 20 D, where a measurement of a small velocity u < 2 m/s is problematic).

3.5 Integral results

The velocity profiles can be integrated to obtain following integral quantities, namely the *volume flow rate Q*, the *flux of the axial momentum M*, and *the kinetic energy flux E*. The evaluations are defined as $Q = 2\pi\int_0^{+\infty} u\, r\, dr$, $M = 2\pi\rho\int_0^{+\infty} u^2 r\, dr$, and $E = \pi\rho\int_0^{+\infty} u^3 r\, dr$.

These integrations were applied with finite integration upper limits on the present experimental data for various cross sections from x = 0 to x = 20 D. The obtained results are presented in figure 14 in dimensionless forms of Q/Q_0, M/M_0, and E/E_0, where Q_0, M_0, and E_0 are the initial quantities evaluated at the nozzle.

Figure 14(a) shows a development of the volume flow rate along the jet axis, which increases because of fluid entrainments from surroundings. Figure 14(b) shows that the momentum flux along the jet axis slightly decreases, and this decrease is promoted by the control effect – see the arrow in figure 14(b) and 29% decrease at x/D = 20. Finally, the kinetic energy flux along the jet axis significantly decreases (see figure 14(c)) and the control

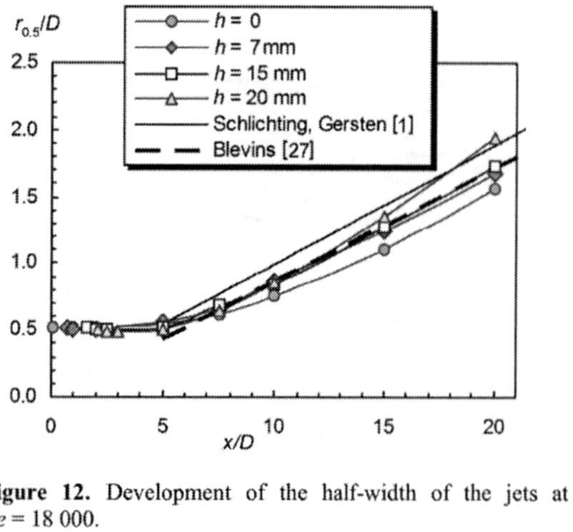

Figure 12. Development of the half-width of the jets at Re = 18 000.

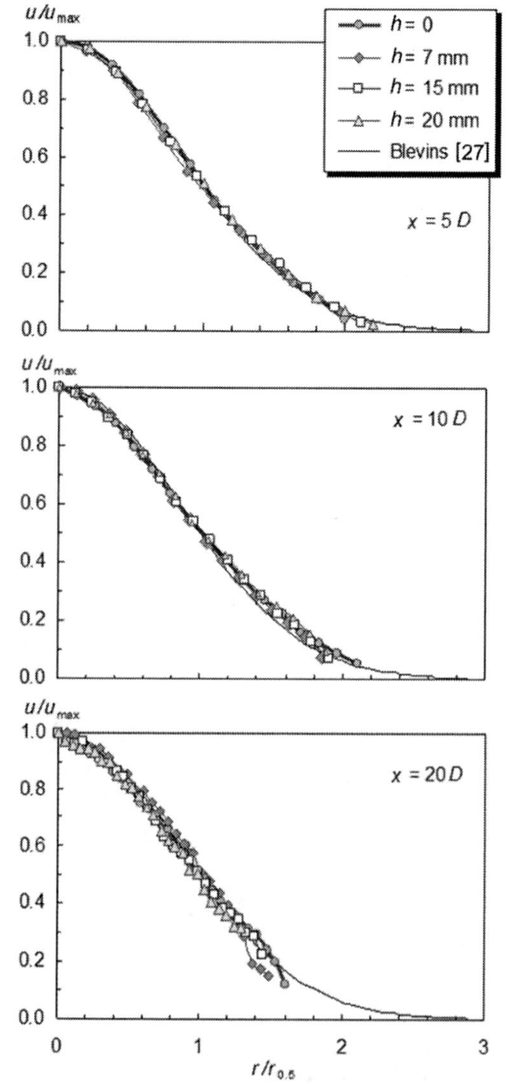

Figure 13. Similarity of dimensionless velocity profiles at Re = 18 000.

effect intensifies this process – see the arrow in figure 14(c) and 46% decrease at $x/D = 20$.

To assess these results, a comparison with available literature was performed. The volume flow rate increase resulting from the fluid entrainment can be expressed by the following formula (Blevins [27]):

$$Q/Q_0 = 0.32\, x/D\,. \qquad (4)$$

The momentum flux along the jet axis is very nearly constant, as was approximated by Schlichting and Gersten [1], and Blevins [27]. More realistic results were derived theoretically and confirmed experimentally by Kotsovinos [28] who described a small decrease of the momentum flux by the following formula:

$$M/M_0 = 0.983–0.0693\ \ln[x/(6D)]\,. \qquad (5)$$

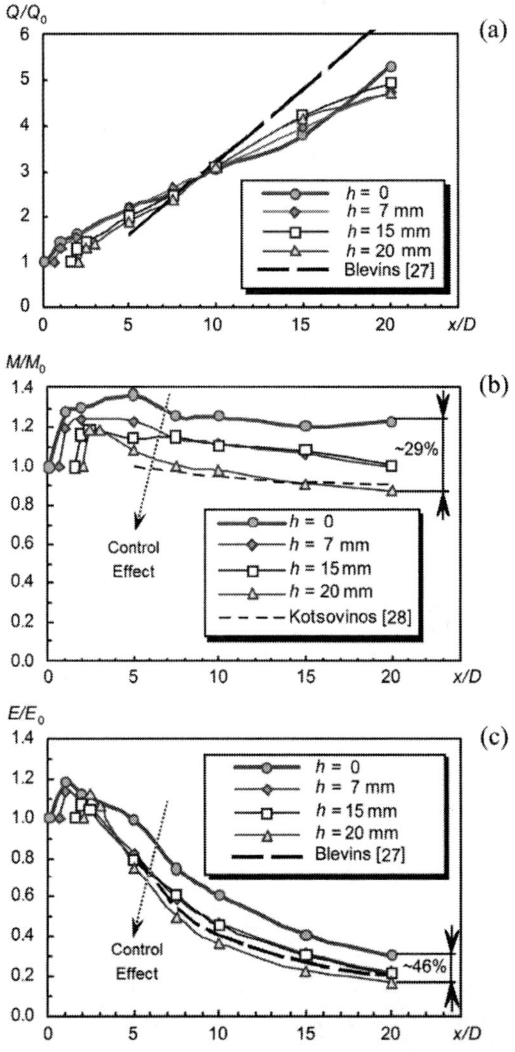

Figure 14. Integral characteristics of jets at $Re = 18\,000$; (a) volume flow rate, (b) momentum flux, and (c) kinetic energy flux.

Finally, the kinetic energy flux along the jet axis decreases as is described by the formula by Blevins [27]:

$$E/E_0 = 4.1\, D/x\,. \qquad (6)$$

Figure 14 shows a reasonable agreement of the present results with references [1, 27, 28]. Systematic shifts of the curves in figures 14(b and c), which is depicted by the arrows, indicate the revealed control effect. Namely, the fluid mixing is enhanced and velocity decay along the axis is promoted, hence the momentum and kinetic energy fluxes decrease more intensively in the controlled jets in comparison with common jets. The decreases by 29 % and 46 % are depicted in figures 14 too.

4 Conclusions

A pipe-cavity-orifice nozzle was designed. The nozzle consists of a pipe of the inner diameter of 10 mm and the length of 15 diameters, which is equipped with an axisymmetric cavity with an adjustable depth at the pipe end.

Firstly, the cavity resonance behavior was tested without jet flows. Frequencies of the sound effects were evaluated by means of a microphone measurement. A signal on-line analysis was made to obtain the power spectral density (PSD). The evaluated resonant frequencies well agree with a theoretical prediction of the natural frequency of the Helmholtz resonator. The maximum deviation between the present data and the theory is less than 7–10 %.

The nozzle was used for generation of axisymmetric air jet flows with a passive flow control by means of self-sustained acoustic excitations. The jets were experimentally investigated using flow visualization, microphone measurements and time-mean velocity measurement by the Pitot probe. The PSD and the sound pressure level (SPL) were evaluated from microphone measurements. The jet Reynolds number ranged Re = 1600–18 000. The most effective aeroacoustics response was found at Re = 18 000, therefore a majority of experiments was made at this Re regime.

The results demonstrate effects of the passive control on the jet behavior. In comparison with common jets, fluid mixing is enhanced and velocity decay along the axis is promoted by a shortening of the jet transition region. On the other hand, an inverse proportionality of the velocity decay ($u \sim 1/x$) in the fully developed region is not changed.

The control effect was also indicated from the integral quantities of the jet. The flux of the axial momentum and the kinetic energy flux decrease more intensively in comparison with common jets.

Acknowledgments

The authors gratefully acknowledge the support from the Grant Agency CR – the Czech Science Foundation (Project No. 14-08888S) and the institutional support RVO:61388998.

References

1. H. Schlichting, K. Gersten K. *Boundary-Layer Theory* (Springer-Verlag, Berlin, 2000)
2. G.N. Abramovich, *The Theory of Turbulent Jets* (MIT Press, Massachusetts, 1963) (original in Russian: *Teoriia turbulentnykh strui*, Moscow, GIFML, 1960).
3. S.C. Crow, F.H. Champagne, J. Fluid Mech. **48** (1971)
4. Mohamed Gad-el-Hak, *Flow Control: Passive, Active, and Reactive Flow Management* (Cambridge University Press, Cambridge, New York, 2007)
5. F.O. Thomas, Appl. Mech. Rev. **44** (1991)
6. Vlasov E.V., Ginevskii A.S., Sov. Phys. Acoust. **26**, 1 (1980)
7. T. Liu, J. P. Sullivan, Int. J. Heat Mass Transfer **39**, 17 (1996)
8. W. G. Hill Jr., P. R. Greene, Trans. ASME J. Fluids Eng. **99** (1977)
9. M. A. Z. Hasan and A. K. M. F. Hussain, J. Fluid Mech. **115** (1982)
10. A. K. M. F. Hussain and M. A. Z. Hasan, J. Fluid Mech. **134** (1983)
11. C.J. Abell, R. E. Luxton, *Mixing enhancement in combusting flow* (Technical Report 1979–1981, The University of Adelaide, Department of Mechanical Engineering, Adelaide, Australia) (cited by Lee [12])
12. S.-K. Lee, *Study of a naturally oscillating triangular-jet flow* (PhD. Thesis., University of Adelaide, School of Mech. Eng., 2009)
13. G.J. Nathan, *The enhanced mixing burner* (PhD. Thesis., University of Adelaide, Dept. Mech. Eng., 1988)
14. R.E. Luxton, G.J. Nathan, *Controlling the motion of a fluid jet* (Patent US 5060867 A, 1991)
15. G.J. Nathan, S.J. Hill, R.E. Luxton, J. Fluid Mech. **370** (1998)
16. R.H. Page, P.S. Chinnock, J. Seyed-Yagoobi, AIAA J. of Thermophysics and Heat Transfer **10**, 2 (1996)
17. H. Viets, AIAA J. **13**, 10 (1975)
18. G.Raman, E.J.Rice, D.M.Cornelius, Trans. ASME, Jour. Fluids Eng. **116** (1994)
19. C. Camci, F. Herr, Trans. ASME Jour. Heat Transfer **124** (2002)
20. A. Seifert, S. Pastuer, *Method and mechanism for producing suction and periodic excitation flow* (US Patent No. 7,055,541 B2, 2006)
21. V. Tesař, Z. Trávníček, J. Kordík, Z. Randa, Sens. Actuator A-Phys. **138**, 1 (2007)
22. J. Mi, G.J. Nathan, R.E. Luxton, Flow Turbul. Combust. **67** (2001)
23. R. H. Henrywood, A. Agarwal, Phys. Fluids **25** (2013)
24. P. Pušková, *Passive control of axisymmetric air flow* (*Pasivní řízení osově symetrického proudu vzduchu*) (Master thesis, Czech Technical University, Faculty of Mechanical Engineering, Prague, Czech Republic, 2015) (in Czech)
25. S. J. Kline, F.A. McClintock, Mechanical Engineering **75** (1953)
26. L.E. Kinsler, A.R. Frey, A.B. Coppens, J.V. Sanders, *Fundamentals of Acoustic* (Wiley, New York, 2000)
27. R. D. Blevins, *Applied Fluid Dynamics Handbook* (Florida: Krieger Publishing Company, 2003)
28. N.E. Kotsovinos, J. Fluid Mech. **87**, 1 (1978)

EPJ Web of Conferences **114**, 02007 (2016)
DOI: 10.1051/epjconf/201611402007

PIV and LIF study of slot continuous jet at low Reynolds number

Zuzana Broučková[1,a], Shu-Shen Hsu[2], An-Bang Wang[2] and Zdeněk Trávníček[1]

[1]*Institute of Thermomechanics, the Czech Academy of Sciences, 182 00 Prague, Czech Republic*
[2]*Institute of Applied Mechanics, National Taiwan University, Taipei, Taiwan R.O.C.*

Abstract. This study deals with a continuous jet issuing from a small narrow slot with a width of 0.36 mm. The experimental arrangement is based on the piezoelectric synthetic jet actuator studied previously for easy comparisons. The working fluid is water at room temperature. The experiments were performed using methods of particle image velocimetry (PIV) and flow visualization (laser induced fluorescence, LIF). The time-mean volume flux through the exit nozzle was quantified using precise scales. The mean velocity and the Reynolds number were evaluated as U_m = 0.12 m/s and Re = 90, respectively. The results of LIF and PIV techniques revealed the three-dimensional character of the flow field, namely the saddle-shape velocity profiles. This behavior is typical for steady jets from a rectangular nozzle. The obtained results were compared with previous measurements of the synthetic jet issuing from the same cavity and the slot nozzle.

1 Introduction

The turbulent continuous jets (CJ) issuing from planar and rectangular slots have been widely studied. The studies were focused on general characteristics of CJs (e.g. Gutmark and Wygnanski [1], Everitt and Robins [2], Tsuchiya et al. [3]), spatial developement of the velocity profile and the pressure distribution of the jet (Abramovich [4]), coherent structures in plane jets (e.g. Thomas and Goldschmidt [5]), axis-switching of the jet (Zaman [6]), saddle-back velocity profiles (Vouros et al. [7]), behavior of the jets at low Reynolds numbers (Sato and Sakao [8]) or the influence of an aspect ratio of the nozzle (Deo et al. [9]).

The synthetic jet (SJ) is a fluid flow generated from fluid oscillations during a periodical fluid exchange between an actuator cavity and surroundings. A meaningful advantage of SJs is a relative simplicity of an arrangement: neither blower nor fluid supply piping is required. SJs are investigated by means of experimental, theoretical and numerical methods – e.g. Smith and Glezer [10], Cater and Soria [11], Mallinson and Reizes [12].

This experimental study is focused on a continuous jet issuing at a low Reynolds number from a rectangular slot. The experimental setup is based on a previously studied piezoelectric SJ actuator (see [13], [14]) which was adapted to be able to work also in a continuous (steady) regime as is described thereinafter. The aim of this study is to verify the possibility of the device to work in continuous regime and also to obtain results which can be used as a benchmark for further experiments with SJs. For this reason the operating parameters of the CJ were chosen to be similar to those of SJ.

2 Parameters

The continuous rectangular jet is described by the dimensions of the exit slot (the width B and the length L_s) and by the Reynolds number:

$$Re = \frac{2BU_m}{\nu}, \tag{1}$$

where U_m is the mean exit velocity and ν is the kinematic viscosity.

The Reynolds number of SJs was defined [14] for the same rectangular slot ($B \times L_s$) as:

$$Re_{SJ} = \frac{2BU_0}{\nu}, \tag{2}$$

where U_0 is the time-mean exit velocity:

$$U_0 = \frac{1}{T}\int_0^{T_E} u_0(t)dt, \tag{3}$$

where $T = 1/f$ is the time period, f is frequency, T_E is the extrusion time and $u_0(t)$ is the centerline exit velocity (see e.g. Smith, Glezer [10]).

[a] Corresponding author: brouckova@it.cas.cz

3 Experimental investigation

3.1 Experimental setup

Figure 1 shows a scheme of the experimental device. The experimental arrangement is based on the piezoelectric synthetic jet actuator. The actuator body is made out of PMMA (polymethyl methacrylate). It is equipped with a piezoelectric diaphragm (1), which is the piezoceramic transducer KPS-100 with the diaphragm diameter of D_D = 39.5 mm. The cavity (2) diameter and depth are 44 mm and 6 mm, respectively. The jet issues from a slot (3) on the top of the actuator. The slot width and length are B = 0.36 mm and L_s = 40 mm, respectively, i.e. the aspect ratio of the slot is AR = L_s/B = 111. The length of the nozzle exit channel is L = 1.0 mm. The actuator coordinate system x, y, z is shown in figure 1 as well. The working fluid is water.

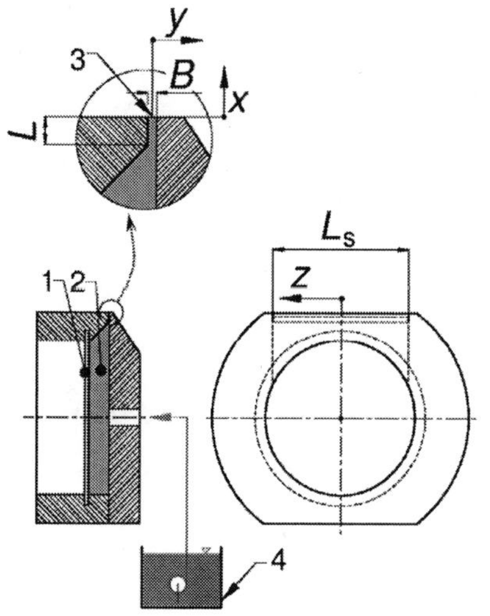

Figure 1. Investigated SJ actuator, 1: diaphragms, 2: cavity, 3: slot orifice, 4: pump.

The actuator can operate in two regimes: it can produce either the CJ or the SJ. When the device is operated in CJ regime, the fluid is fed into the actuator using centrifugal pump (4). The pump is connected with the actuator via the flexible pipe and the fluid enters the cavity opposite the center of the piezoelectric diaphragm – as is schematically shown in figure 1. The pump is placed either in the same tank as the actuator or in the separate tank. The former arrangement is used for PIV experiments, the latter for LIF visualization as well as for evaluation of the time-mean volume flux by means of precise scales (as will be described further). In the case of CJ regime, the piezoelectric diaphragm is switched off.

When the device operates as a SJ actuator, the inlet from the pump is closed and the piezoelectric diaphragm is driven by the sinusoidal current with a drive frequency f = 46 Hz and RMS voltage 18.0 V. A choice of the low frequency actuation at the near resonance frequency was refined recently by means of theoretical and experimental approaches – the details can be found in [13], [14]. Moreover, the time-mean exit velocity and the Reynolds number were evaluated in previous experiments [13], [14] as U_0 = 0.17 m/s and Re = 127, respectively.

3.2 Experimental methods

Two experimental methods were used: laser induced fluorescence (LIF) visualization and particle image velocimetry (PIV). The former experiments were made in the Institute of Thermomechanics (Prague, Czech Republic) and the latter one in the Institute of Applied Mechanics (Taipei, Taiwan R.O.C.).

3.2.1 LIF technique

The experiments were made in a glass water tank with the dimensions: height × length × depth = 150 × 245 × 140 mm.

The LIF technique is based on the fluorescence effect. When a fluorescence dye is under a light with suitable wavelength, the dye absorbs the energy and consequently reemits part of it at higher wavelengths (see e.g. Seuntiens et al. [15] for more details). The Rhodamine B was chosen as a fluorescence dye (absorption wavelength from 460 to 590 nm, emission wavelength from 560 to 670 nm) – see [16]. Rhodamine B is usually used for temperature measurements because it has a temperature-dependent emission intensity. However, in the present study this effect was not used, the present experiments were approximately isothermal.

The experimental device was placed into the glass tank with pure, demineralised water, prepared in four-stage reverse osmosis water purification system. The pump was placed in a separate smaller tank. The separate tank was filled with dye solution (highly concentrated solution of Rhodamine B). Subsequently, the pump was turned on. The dye-stained flow was lit by laser sheet (Nd:Yag pulsed laser, Litron, NANO S 65-15, with a wavelength of 532 nm and an output of 65 mJ, equipped with cylindrical optics). The laser sheet was perpendicular to the slot, i.e. in the plane of x–y, see figure 1. The thickness of the laser sheet was about 1 mm in the test section.

The pictures of the visible flow field pattern (streaklines) were taken using digital camera (HiSense Neo, 2560 × 2160 pixels, 16 bit). The camera was equipped with lens (Tokina 100 mm F2.8 Macro D). To remove the laser light from the pictures, the colour filter (cut-off wavelength 570 nm) was used. The camera was used to trigger the laser pulse. The measurement system was controlled using DynamicStudio v3.40 Software (Dantec Dynamics).

To evaluate the time-mean volume flux of the CJ, an auxiliary experiment was made using the precise scales (Mettler Toledo PR 8002 Delta Range). The actuator was placed into the separate tank and the amount of water discharged during the specified time was weighed.

The temperature of water was measured by a NTC thermistor sensor (Ahlborn AMR, Therm 2280-3). All experiments were made after thorough temperature equalization at room temperature. In this study, the temperature was 22°C.

3.2.2 PIV technique

The experiments were made in a cubic glass water tank with the side length of 390 mm. During PIV measurement the pump was placed in the same tank as the actuator. Polyamide seeding particles (Dantec, 5μm in diameter) were used. The light source was a Nd:YAG dual laser head system (NewWave, 120 mJ at 532 nm) with a cylindrical lens. The image pairs were acquired using a high resolution CCD camera (Kodak, MegaPlus ES1.0) with a spatial resolution of 1008 × 1016 pixels. The camera was equipped with the lens (Micro-Nikkor 105 mm). The measurement system was controlled using Dantec Dynamics PIV system (FlowMap1500). The velocity vectors were consequently calculated using DynamicStudio v3.40 Software. The PIV adaptive technique was used. The minimum interrogation area was 32 × 32 pixels, the maximum interrogation area was 64 × 64 pixels. The resulting velocity maps were averaged from a sequence of 100 images.

Previously, the velocity field of the SJ was captured at selected phases of the velocity cycle and the additional apparatus was used for this reason – the details can be found in [14]. The resulting pictures were also averaged from a sequence of 100 images in the same phase, [14].

4 Results and discussion

Firstly, the time-mean volume flux Q and the velocity U_m of the CJ was evaluated by means of the precise scales. The results are: $Q = 0.00167$ l/s and $U_m = 0.12$ m/s. The corresponding Reynolds number is $Re = 90$. These values are in a good agreement with the parameters of the previously studied SJ [13, 14].

Figure 2 shows a typical visualization of the jet (a) in the middle of the actuator slot (x-y plane, $z = 0$) and (b) along the z-axis (z-x plane, $y = 0$).

The visualization in the middle of the actuator slots (figure 2a) shows the evolution of the jet and its gradual spreading. The influence of the geometry of the exit part of the cavity and the nozzle is visible - since the exit part of the cavity is not symmetrical against the nozzle and the nozzle exit channel is relatively short, $L = 1.0$ mm (see figure 1), the issuing jet is slightly deflected from the vertical direction to the right hand side. From the figure 2b it is obvious that the distribution of the jet along the z-axis is not fully uniform. The flow in the surroundings of the midpoint of the slot ($z = y = 0$) is weaker then the flow at the off-axis parts (visible as darker area in the middle of the flow). This behavior is typical for 2D turbulent jets, the initially uniform jet creates two off axis peaks in proximity of the edges of the long size of the slot (here near the edges of the nozzle at $z = +L_s/2$ and $z = -L_s/2$), see e.g. Vouros et al. [7]. In the present experiment the behavior is probably also supported by the confined 3D geometry of the cavity (namely, the cylindrical cavity with an axial inlet and a radial outlet) and by the short nozzle exit channel.

Streamwise velocity maps at the distances $x/(2B) = 5$, 10, 15 and 20 are presented in figure 3. The maps were evaluated from the velocity profiles, which were measured at the planes of $z = (-15, -10, -5, 0, 5, 10, 15)$ mm. At the coordinates $z = -25$ mm and $z = 25$ mm the zero values were artificially added in the pictures. The measurement confirms the expected character of the jet – two off axis velocity maxima are visible at the distances $z = -15$ mm and $z = 15$ mm while the gradual decrease of the velocity in the center part of the jet is apparent. The resulting velocity profiles are known as saddle-back profiles of CJs (see e.g. Vouros et al. [7]). Note that this effect was identified for SJs too – see Smith and Glezer [10].

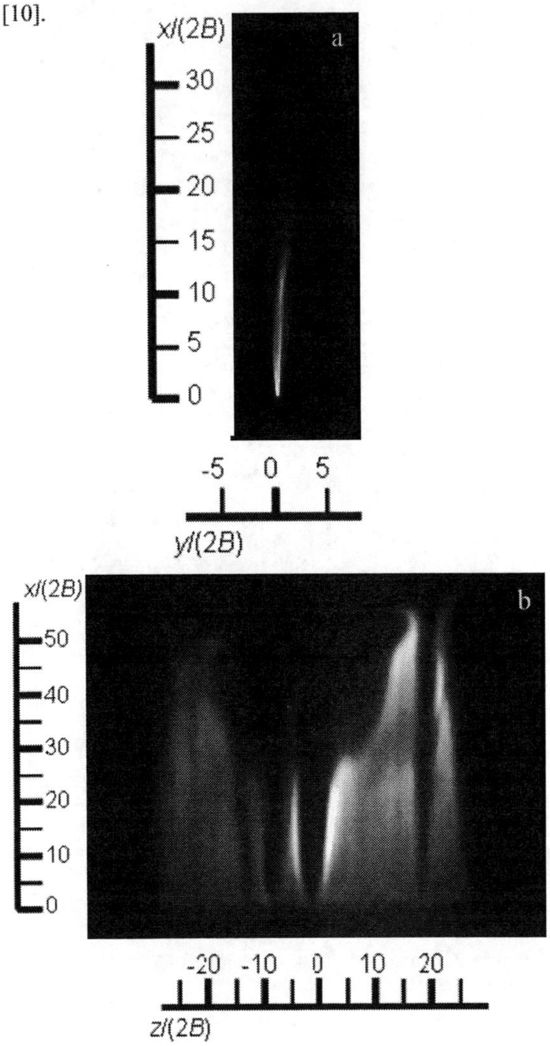

Figure 2. LIF visualization of the CJ
(a) In the middle of the actuator slot ($z = 0$),
(b) Along the z-axis.

It is worth noting here that the measurement using PIV in a close vicinity of a body surface is influenced by the unavoidable reflection of the laser beam from the surface (the near wall effect). Moreover, the diameter of the chosen particles is relatively large for the present geometry – namely, the particle diameter is only two orders lower than the lateral dimension of the slot (B), i.e. the initial lateral dimension of the jet. This fact can influence the ability of the particles to follow the flow properly. Moreover, the slot width B is approximately three times narrower that the laser sheet thickness. For these reasons the PIV measurement undervalues the velocity magnitude in the near field – approximately for the distance $x/(2B) < 6$. It is obvious from figure 3, the maximum velocity at the distance $x/(2B) = 5$ is unrealistically low – even lower than maximum velocities for $x/(2B) = 10$.

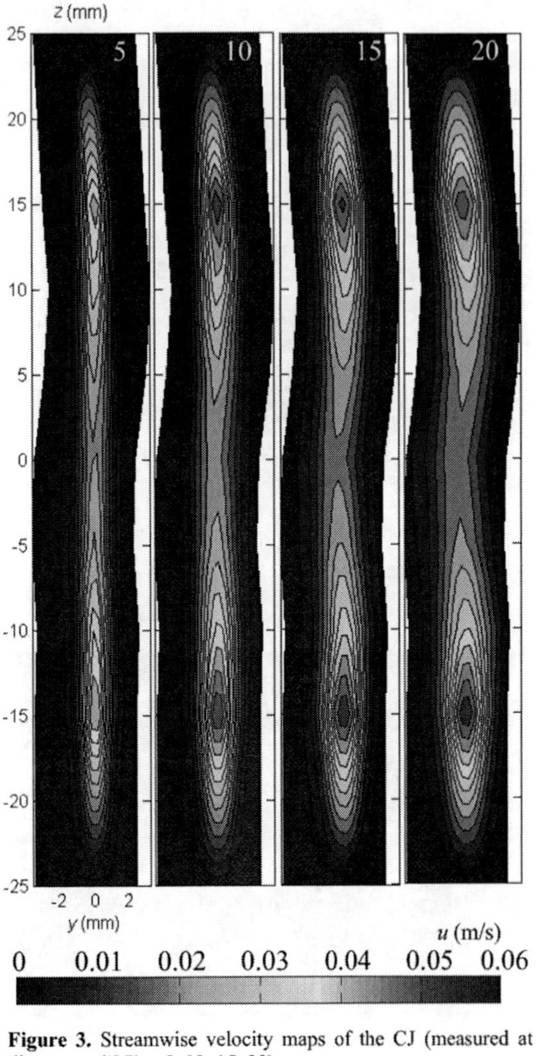

Figure 3. Streamwise velocity maps of the CJ (measured at distances $x/(2B) = 5, 10, 15, 20$).

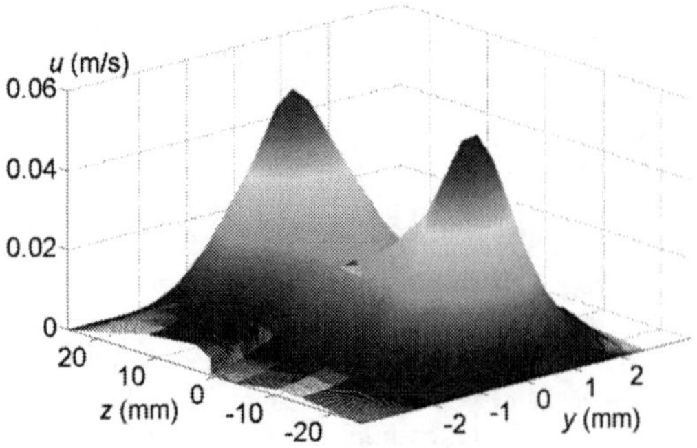

Figure 4. Streamwise velocity map at $x/(2B) = 15$.

For an illustrative purposes the streamwise velocity map at $x/(2B) = 15$ is also shown as three-dimensional contour plot in figure 4. The character of the jet is clearly visible – two off axis peaks near the ends of the nozzle slot ($z = +L_s/2$ and $z = -L_s/2$) and the lower velocity in the center of the jet create a typical saddle-shaped profile.

Figure 5 shows the velocity profiles (namely, the streamwise component of the velocity) in the dimensionless form. The horizontal coordinate is normalized by the half-width of the jet $y_{0.5}$ (it will be discussed in the following paragraph). The theoretical profile according to Blevins [17] is also shown, $u/u_{max} = \exp[-57\ (0.11\ y/y_{0.5})^2]$. The measured profiles collapse on one curve which agrees excellently with the theoretical one. The deviations of little import around the distance $y/y_{0.5} = -2$ are caused by the above mentioned deflection of the jet to the right hand side.

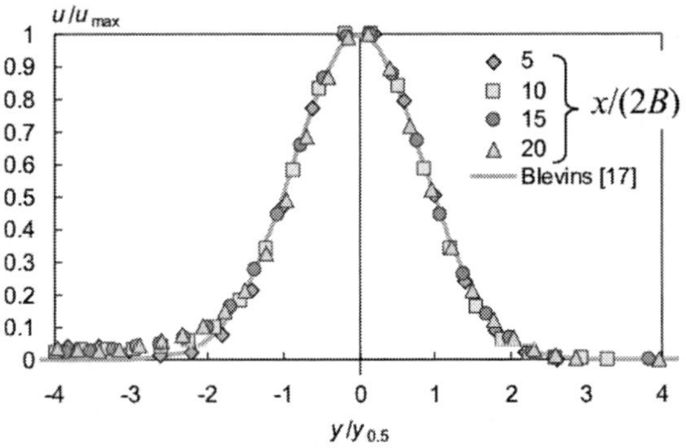

Figure 5. Streamwise velocity profiles.

Figure 6 presents the downstream evolution of the half-width of the jet. The curve for turbulent 2D jet according to Blevins [17] is plotted for comparison purposes, $y_{0.5}/(2B) = 0.11x/(2B)$. As expected the half-width increases gradually with the axial coordinate x as the surrounding fluid is entrained into the jet. It can be seen that the half-width depends on the spanwise location – the half-width decrease gradually with the coordinate z towards borders of the jet. The results for $x/(2B) \geq 10$ reasonably agree with the curve by Blevins. However the inclination of the present results is smaller. The main reason can be a relatively low Reynolds number of the examined jet with a transition from laminar to turbulent flow. Moreover, a relatively small aspect ratio can cause another difference from behavior of plane jets [17]. Note that the relationship by Blevins [17] is considered for fully turbulent flow ($Re > 3000$) and infinite length of the slot. The experimental data by Vouros et al. [7] ($Re = 46000$, AR = 10) and Deo et al. [9] ($Re = 3000$, AR = 61) are also shown in figure 6. The present data reasonably agree (for $x/(2B) \geq 10$) with these results [7, 9], although the operation parameters of [7] and [9] were quite different.

Note that the PIV results for $x/(2B) = 5$ are influenced by the near wall effect, as was explained in the text above. The region contaminated by the near wall effect is marked as a grey area in figure 6 (and similarly in the following figures).

Volumetric flow rate at the nozzle exit is defined as

$$Q_0 = \int_{-L_s/2}^{L_s/2}\int_{-B/2}^{B/2} u(y,z)\,dy\,dz = \int_{-L_s/2}^{L_s/2} Q_z(z)\,dz. \quad (4)$$

Figure 7 shows the volumetric flow rate relative to the z coordinate, Q_z. It is obvious that the entrainment of the flow is highly dependent on the spanwise location. Mixing with the surroundings is very intensive near the borders of the jet (z = -15 mm and z = 15 mm) because of fluid entrainment in both lateral and spanwise direction in the shear layers. Over against the entrainment around the centerline (i.e. around z = 0) is minimal.

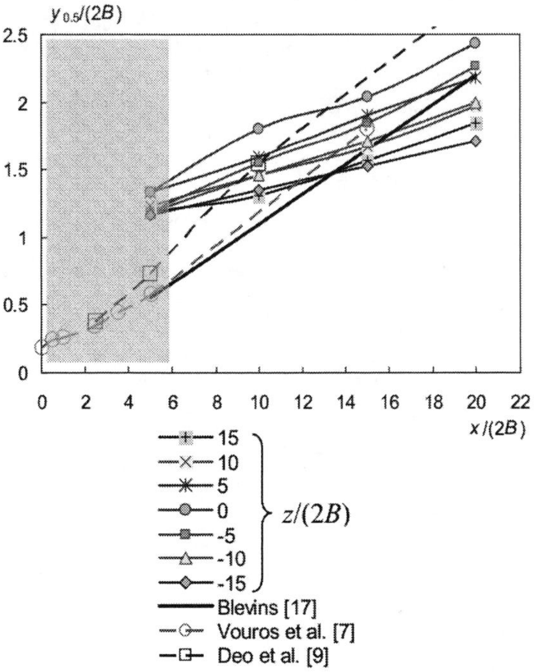

Figure 6. Half-width of the CJ; grey area depicts the near wall effect in PIV experiments.

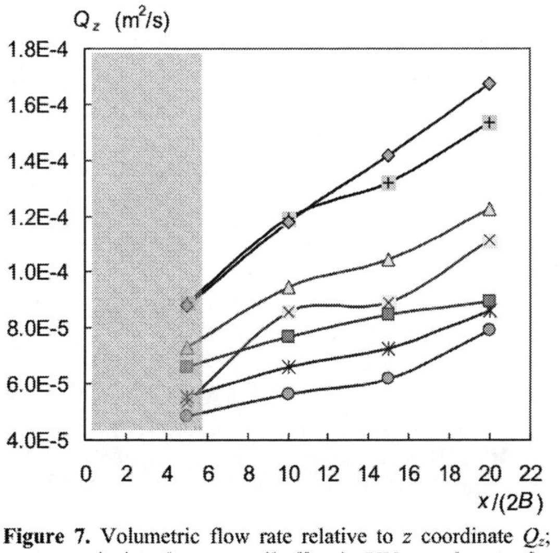

Figure 7. Volumetric flow rate relative to z coordinate Q_z; grey area depicts the near wall effect in PIV experiments; for legend see figure 6.

Figure 8 shows the comparison between the CJ and SJ velocity vector field (z = -15 mm). The SJ is shown in the blowing part of the cycle; more details were described previously – [13, 14]. The pulsatile character of the SJ is visible, mainly in the near field ($x/(2B)$ < 10). The SJ has a higher velocity at the axis and the jet is spread more than the CJ. The distinct deflections of the jets to the right hand side can be observed. This effect is stronger for SJ than for CJ. The reasons are a non-symmetrical geometry of the nozzle, as was written in the text above. Note the above discussed near wall effect is manifested in figure 8 as an undervaluation of the velocity magnitude in the near field for $x/(2B)$ < 5.

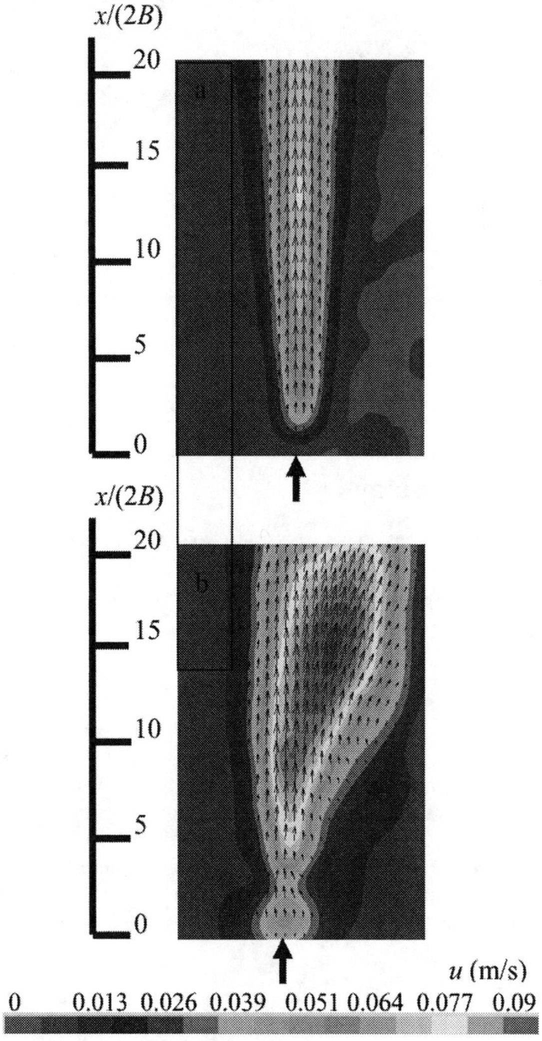

Figure 8. Velocity vector maps
(a) CJ
(b) SJ
The arrow depicts the position of the nozzle axis, y = 0.

Finally, figure 9 presents an evolution of the velocity magnitude along the centerline (y = z = 0). The velocity magnitude of the CJ and the time-mean velocity magnitude of the SJ are compared. The line for turbulent 2D jets according to Blevins [17] ($u \sim x^{-0.5}$) is plotted too.

The velocity of the CJ reaches its maximum approximately at $x/(2B)$ = 10. The reason of this unexpected result can be the above discussed near wall effect – see a grey area in figure 9.

More downstream, the velocity of CJ remains constant till $x/(2B)$ = 18 and than starts to gradually decrease. On the other hand, the SJ velocity reaches the maximum approximately at the same distance ($x/(2B)$ = 10) and than immediately drops. The SJ velocity decays at first with $x^{-0.5}$, further downstream the decay is faster ($x^{-0.78}$) – see figure 9.

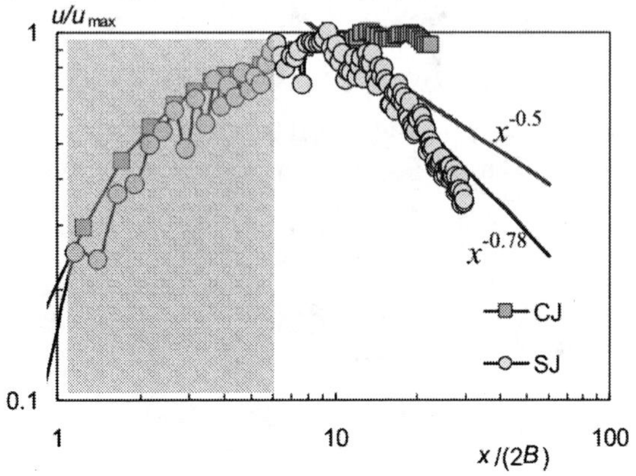

Figure 9. Centerline velocity decay; grey area depicts the near wall effect in PIV experiments.

5 Conclusions

The water continuous jet issuing from a small narrow slot (0.36 mm width) was experimentally studied. The experimental arrangement was based on the previously investigated piezoelectric synthetic jet actuator. The experiments using LIF and PIV techniques confirmed expected character of the jet: the three-dimensional character of the flow field was observed and the saddle-back velocity profiles were found. The comparison with previously studied synthetic jet was performed. The jet flows with the saddle-back velocity profiles are known for the continuous jets as well as for the synthetic jets. This effect is promoted by the present geometry, mainly by the 3D geometry of the cavity (the cylindrical cavity with an axial inlet and a radial outlet) and by a relatively short nozzle exit channel.

Acknowledgments

We gratefully acknowledge the support of the Grant Agency of the Czech Republic – Czech Science Foundation (project no. P101/14-08888S) and by the Czech-Taiwanese Joint Research Project under the Agreement on scientific cooperation between the Academy of Sciences of the Czech Republic and the National Science Council – Ministry of Science and Technology in Taiwan (PPP project no. MOST/14/001).

References

1. E. Gutmark, I. Wygnanski, J. Fluid Mech. **73** (1976)
2. K.W. Everitt, A.G. Robins, J. Fluid Mech. **88** (1978)
3. Y. Tsuchiya, C. Horikoshi, T. Sato, Exp. Fluids **4** (1986)
4. G.N. Abramovich, Int. J. Heat Mass Transfer **25**, 12 (1982)
5. F.O. Thomas, V.W. Goldschmidt, J. Fluid Mech. **163** (1986)
6. K.B.M.Q. Zaman, J. Fluid Mechanics **316** (1996)
7. A. P. Vouros, T. Panidis, A. Pollard, R. Schwab, Int. J. Heat Fluid Flow **51** (2015)
8. H. Sato, F. Sakao, J. Fluid Mech. **20** (1964)
9. R. C. Deo, J. Mi, G. J. Nathan, Phys. Fluids **20** (2008)
10. B.L. Smith, A. Glezer, Phys. Fluids **10** (1998)
11. J.E. Cater, J. Soria, J. Fluid Mech. **472** (2002)
12. S.G. Mallinson, J.A. Reizes, G. Hong, Aeronaut. J. **105**, 1043 (2001)
13. Z. Broučková, T. Vít, Z. Trávníček, EPJ Web of Conferences **92,** 02007 (2015)
14. Z. Broučková, S-S. Hsu, A-B. Wang, Z. Trávníček, Advanced Materiál Research **1104** (2015)
15. H.J. Seuntiëns, R. N. Kieft, C.C.M. Rindt, A.A. van Steenhoven, , Exp. Fluids **31**, 5 (2001)
16. Planar-LIF software, Installation & user's guide, Dantec Dynamics (2002)
17. R. D. Blevins, *Applied Fluid Dynamics Handbook.* (Florida: Krieger Publishing Company, 2003)

EPJ Web of Conferences **114**, 02008 (2016)
DOI: 10.1051/epjconf/201611402008

Simulation of dynamics of hydraulic system with proportional control valve

Adam Bureček[1], Lumír Hružík[1] and Martin Vašina[1,a]

[1]VŠB-Technical University of Ostrava, Faculty of Mechanical Engineering, Department of Hydromechanics and Hydraulic Equipment, 708 33 Ostrava, Czech Republic

Abstract. Dynamics of a hydraulic system is influenced by several parameters, in this case mainly by proportional control valve, oil bulk modulus, oil viscosity, mass load etc. This paper will be focused on experimental measurement and mathematical simulation of dynamics of a hydraulic system with proportional control valve, linear hydraulic cylinder and mass load. The measurement is performed on experimental equipment that enables realization of dynamic processes of the hydraulic system. Linear hydraulic cylinder with mass load is equipped with position sensor of piston. The movement control of piston rod is ensured by the proportional control valve. The equipment enables to test an influence of parameter settings of regulator of the proportional control valve on position and pressure system responses. The piston position is recorded by magnetostrictive sensor that is located in drilled piston rod side of the linear hydraulic cylinder. Pressures are measured by piezoresistive sensors on the piston side and the piston rod side of the hydraulic cylinder. The measurement is performed during movement of the piston rod with mass load to the required position. There is realized and verified a mathematical model using Matlab SimHydraulics software for this hydraulic system.

1 Introduction

Hydraulic systems are applied in various industrial sectors. They are applied mainly at manufacturing machines, transport and handling devices, logging machines, further in energetics, food-processing industry, aeronautical technics etc. The application of these systems increases due to increasing design qualities of single hydraulic elements in recent years. Advantageous drive characteristics, single continuous and adjustable control of parameters, positive ratio power/weight of an actuator and also a possibility of development into machine construction belong to general properties of hydraulic systems. It is not possible to replace these systems by other known types of mechanisms in the area of high-power amplifications. There are gradually increased requirements for technical performance, accuracy, complexity and control speed of hydraulic systems control. Combining hydraulics, electronics and computer technique are created much more accurate mechanisms that fulfill increasing requirements and thus affect the quality of production technology and increase the labour productivity.

It is possible to simulate dynamic behaviour of hydraulic elements and complete hydraulic systems due to increasingly more powerful computer technique and more sophisticated simulation programs. It is possible to simulate different operating conditions of a hydraulic system for different initial conditions and different element adjustment by means of a mathematical model, which represents a real hydraulic system. It is a big advantage of the mathematical simulation of hydraulic systems. Thus, we can identify design flaw or deficiency before the actual realization of a given system, but also to solve problems on the already operated equipment. It is necessary to experimentally verify the mathematical simulation because a simplification is an accompanying phenomenon of modelling. The fact that every mathematical model is always burdened with an error, does not diminish the importance of modelling and its results.

2 Control of hydraulic system

There is possible to control e.g. movement direction, speed, force and position of linear hydraulic cylinders in hydraulic systems. This paper is focused on position control of linear hydraulic cylinder by means of proportional control valve and control card. This card is used to comparison of the input desired quantity (i.e. the position for cylinder extension) with the specific position in which the cylinder is located at a given moment. The difference between the desired and the actual regulated quantities enters in the control card. The steering signal of the proportional control valve is created on basis of this difference. In this case it is control of spool position of the

[a] Corresponding author: martin.vasina@vsb.cz

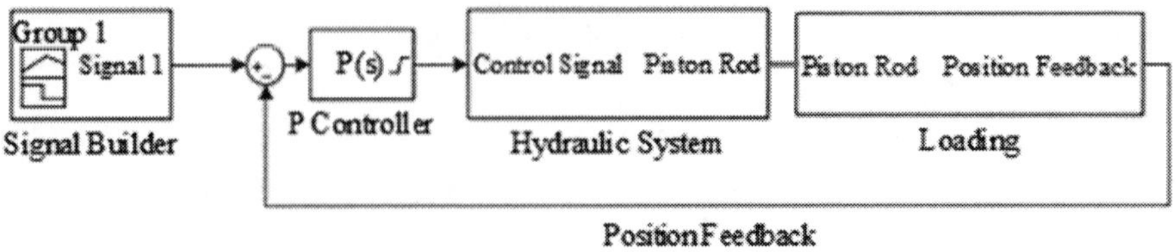

Figure 1. Schematic diagram of regulated system

proportional control valve. The control card PID00A-400 (Parker) was used in this case. The abbreviation PID indicates individual control components (i.e. proportional, integrative and derivative). Only the proportional control component was used to the regulation. This component is defined by the parameter *P*, which creates an output signal for the proportional control valve PV. The signal is proportional to the difference between the desired quantity and the controlled quantity. Parameter P represents the value of proportional gain of the controller.

Figure 1 shows the schematic diagram of the regulated system. A jump signal of the desired quantity (i.e. extension of piston road to the desired position) is adjusted by means of the Signal Builder block. The P Controller block represents the control card. The Hydraulic System block demonstrates the hydraulic system and the Loading block represents external load on the piston rod of the cylinder HC and the position measuring of the piston too.

3 Description of experimental equipment

The experimental hydraulic system was built on test stand from Parker company (see figure 2). The assembled hydraulic system allows control of the position of the hydraulic cylinder HC with mass load *m* in connection with position feedback. Movement control is realized by the proportional control valve PV and the PID00A-400 control card. The position feedback is ensured using the position sensor PoS.

The schematic diagram of the hydraulic system is shown in figure 3. The hydraulic pump HP is used as a source of pressure energy. Mineral oil of the VG46 type with the temperature t_t = 50 °C was used as the working liquid. The linear hydraulic cylinder HC is controlled by means of the proportional control valve PV. Control of the proportional control valve PV is realized by means of the PID00A-400 control card. The desired position of the hydraulic cylinder HC is generated using Matlab Simulink software by means of the AD612 card (Humusoft). The desired position comes up to the control card, where it is compared to the actual position of piston that is measured by the position sensor PoS. The mass load *m* is hanged on the piston rod of the hydraulic cylinder HC. That is connected to the proportional control valve PV by means of the hoses H-A and H-B. The proportional control valve PV is connected to the hydraulic pump HP by means of the hose H-P and to the tank T by the hose H-T. The hydraulic aggregate also consists of the check valve CV, the relief valve RV, the filter F and the cooler C. The measurement was performed as follows: The constant value of the pressure p = 8 MPa is adjusted at input of the proportional control valve PV. Furthermore, the control card receives a signal in the form of the desired position x = 0.15 m generated using Matlab Simulink software. The proportional control valve PV connects the channel P with the channel A and the channel T with the channel B at the same time. Hydraulic oil flows through the hydraulic hose H-A into the piston chamber of the hydraulic cylinder HC and the mass load *m* moves down. The extension speed of the piston rod of the hydraulic cylinder HC depends on the oil flow Q and on the size of the adjusted proportional component P of the control card. When the piston rod is approaching the desired position, there is a throttling and extension deceleration until its stop in the desired position. There are measured the pressure p_A on the piston side and the pressure p_B on the piston rod side by means of the pressure sensors PS A and PS B during this process. There were used the PR15 Hydrotechnik sensors (ranging from 0 to 40 MPa) with the measurement accuracy of 0.5% in this case. The piston rod position was measured by the Balluff BTL7 sensor (PoS) that is integrated in the body of the hydraulic cylinder HC. The signal from the sensors is recorded by means of the M5050 Hydrotechnik measuring equipment. Measured data are processed using Hydrowin software and subsequently compared with a mathematical model using MS Excel software [1-4].

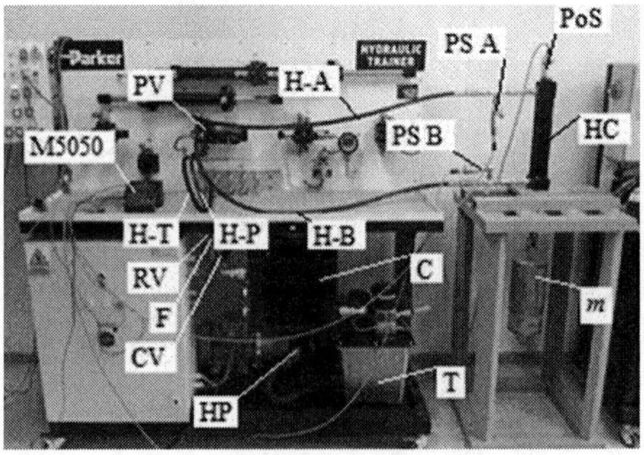

Figure 2. Experimental hydraulic system

Parameters of equipment:

- Hydraulic oil – density ρ = 870 kg m^{-3}, kinematic viscosity $\nu = 3.81 \cdot 10^{-5}$ m^2 s^{-1}.
- Hose H-A, H-B – length l = 2 m, outside diameter D = 0.018 m, inside diameter d = 0.01 m.
- Hose H-P, H-T – length l = 1.2 m, outside diameter D = 0.018 m, inside diameter d = 0.01 m.

[a] Corresponding author: martin.vasina@vsb.cz

- Hydraulic cylinder HC – piston stroke l = 0.3 m, piston diameter D = 0.04 m, piston rod diameter d = 0.028 m, mass load m = 50 kg.

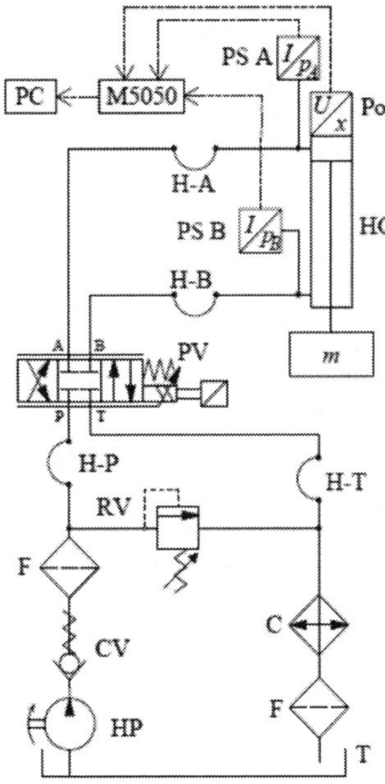

Figure 3. Schematic diagram of hydraulic system

The measured time dependencies of the piston position x for different values of the control component P are shown in figure 4. Furthermore, there is noticed the desired piston position in the form of the step change at the time t = 1 s to the position x = 150 mm. It is evident that the hydraulic cylinder HC reaches the desired position more quickly with the increasing value of the parameter P. The permanent control deviation is the smallest at the highest referred amplification. The proportional component P on the control card is in the range from 0 to 1000. In terms of clarity, only the selected values of the parameter P are shown in figure 4. The time dependencies of the position x in the range of the parameter P from 500 to 1000 were similar as in the case of the parameter value P = 500. For this reason these dependencies aren´t shown in figure 4.

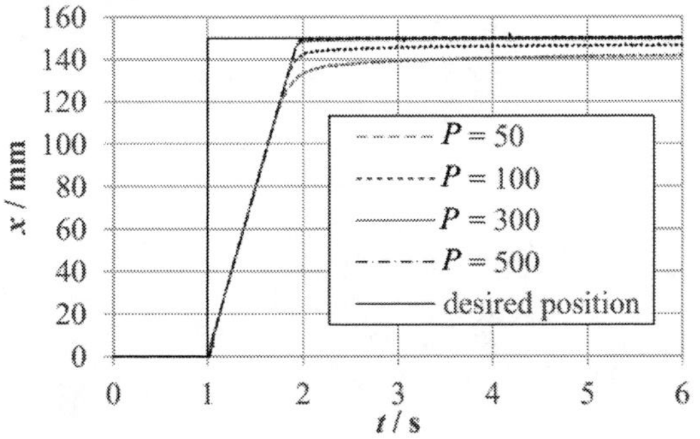

Figure 4. Measured time dependencies of the piston position x for different values of the parameter P

4 Description of measurement of Δp–Q characteristic of proportional control valve

It is necessary to know a flow characteristic depending on the pressure gradient through the proportional control valve PV for the accurate setting of the mathematical model of the proportional control valve PV. Therefore the hydraulic circuit (see figure 5) was assembled on the test stand in order to measure this characteristic. The schematic diagram of the hydraulic system for measurement of the Δp-Q characteristic of the proportional control valve PV is shown in figure 6. The measurement was performed as follows: The proportional control valve PV is gradually opened via the voltage signal from Simulink in the range of 0 to 10 V in steps of 1 V. The throttle valve TV is gradually closed until complete break of the flow Q for each opening of the valve PV. The pressure difference $\Delta p = p_A - p_B$ is measured for individual flows Q. The flow is measured by the flowmeter FM. The pressures are measured by means of the pressure sensors PS A and PS B. The resulting Δp-Q characteristic of the proportional control valve PV is shown in figure 7 [4].

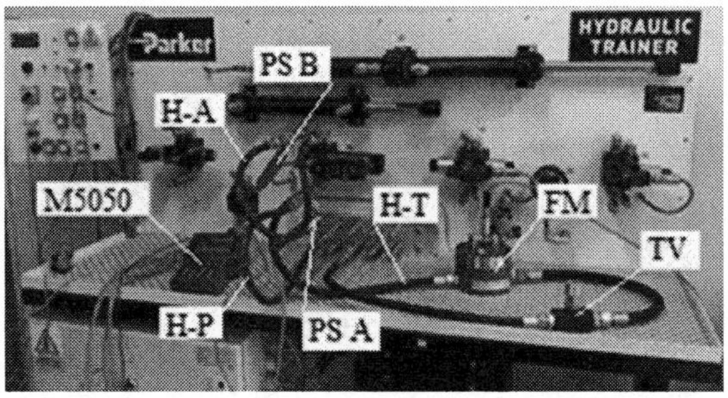

Figure 5. Experimental hydraulic system for measurement of the Δp-Q characteristic of the proportional control valve PV

It is evident (see figure 7) that hydraulic oil flows through the proportional control valve only from the voltage of 2 V. It can be caused by the positive overlap of the spool of the proportional control valve PV. It is also visible that the flow Q at a given pressure gradient Δp is increasing with increases of the voltage, which corresponds to the valve opening (i.e. to its flow area).

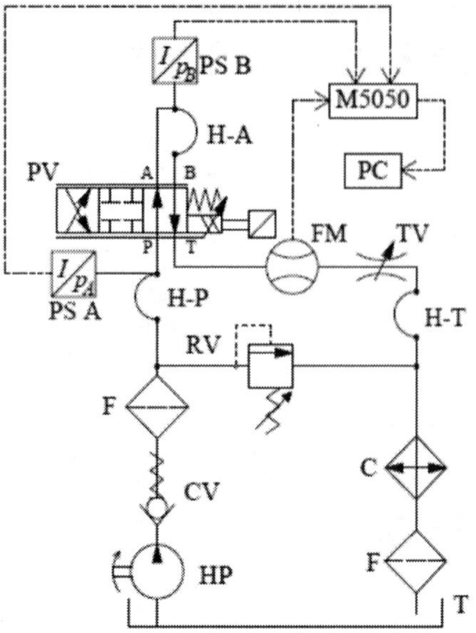

Figure 6. Schematic diagram of hydraulic system for measurement of the Δp-Q characteristic of the proportional control valve PV

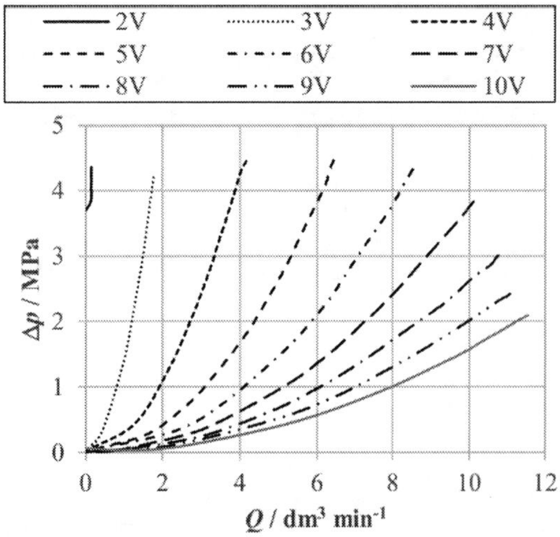

Figure 7. Measured Δp-Q characteristic of the proportional control valve PV

5 Description of mathematical model of experimental equipment

The mathematical model was created using Matlab SimHydraulics software (see figures 1, 8 and 9). This program allows solving the mathematical model based on the physical principle. The subsystems Hydraulic System (see figure 8) and Loading (see figure 9) were created in the model (see figure 1) for better clarity. The subsystem Hydraulic System represents the hydraulic system. The hydraulic pump HP is a source of oil pressure. The system pressure is limited by the relief valve RV to the value $p = 8$ MPa. Oil flows through the check valve CV, the hose H-P, the proportional control valve PV and the hose H-A to the piston side of the hydraulic cylinder HC. For this reason the mass load m moves down. Hydraulic oil flows from the piston rod space through the hose H-B, the proportional control valve PV and the hose H-T to the tank T. Furthermore, the model consists of the block Control Signal for control of the proportional control valve, the block Pressure Measuring for pressure measurement and the block Solver Configuration. The time step (i.e. $\Delta t = 0.005$ s) for scanning of the investigated quantities was adjusted similarly as in the case of the experimental measurement. The oil parameters are adjusted by the block Oil. The undissolved air content (in this case 1%) was defined in this model [5]. The block Fixed Orifice was used to adjust the resistance that was caused by elements on the real equipment, i.e. by the cooler C and the filter F. The subsystem Loading (see figure 9) represents the force and the inertia of the mass load m. Furthermore, the position measuring of the piston by means of the block PoS is located in this subsystem. The position feedback (Position Feedback) is generated from this block and is compared with the desired position before the control card P Controller (see figure 1) [2, 3, 6, 7].

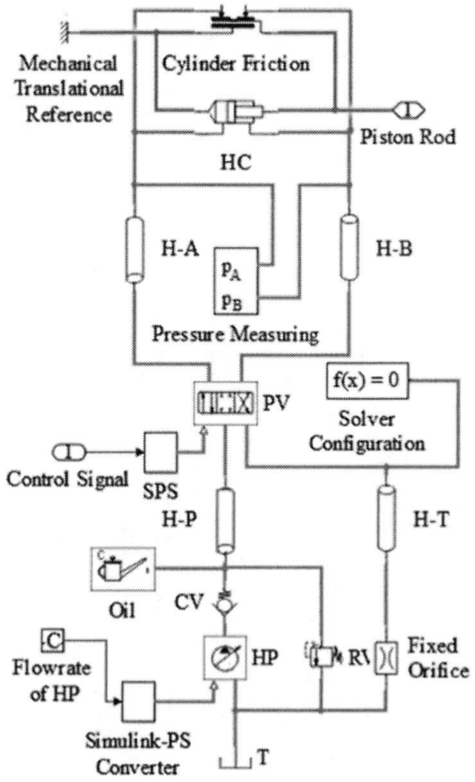

Figure 8. Schematic diagram of the subsystem Hydraulic System

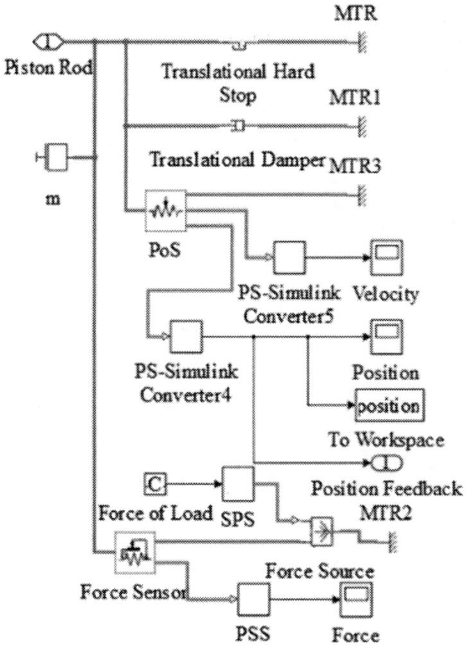

Figure 9. Schematic diagram of the subsystem Loading

6 Comparison of results of experimental measurement and mathematical model

The mathematical simulation was firstly performed for the optimum adjustment of the regulator P = 500. This adjustment corresponds to the quickest achieving of the desired piston position x with the minimum regulatory deviation without an overshoot of the desired position. The comparison of the simulated and the measured time dependencies of the pressures p_A and p_B and the position x is shown in figure 10. It is evident that the simulated and the measured time dependencies of the position x have a good agreement. Similarly the simulated and the measured time dependencies of the pressures p_A and p_B are consistent too. Their dependence is slightly different only when stopping the hydraulic cylinder HC in the desired position x. It can be caused by an unspecified hydraulic capacity in the hydraulic lines.

The simulations were also performed for further parameters, i.e. for P = 50, P = 100 and P = 300 (see figure 11). It is evident that the simulated dependencies of the position x for the individual parameters P correspond to the measured position dependencies. Similarly, the simulated dependencies of the pressures p_A and p_B for the individual parameters P correspond to the experiment and are not shown for clarity in this figure. The detail of the simulated and the measured dependencies of the piston position x during its arrival to the desired position x = 150 mm is shown in figure 12.

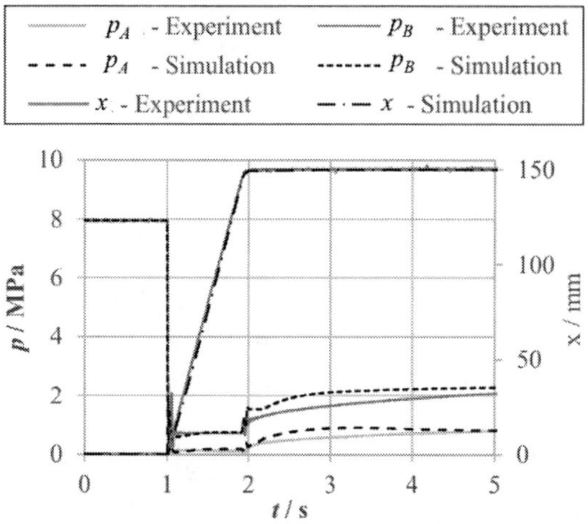

Figure 10. Comparison of measured and simulated time dependencies of pressures and position (proportional component of control P = 500)

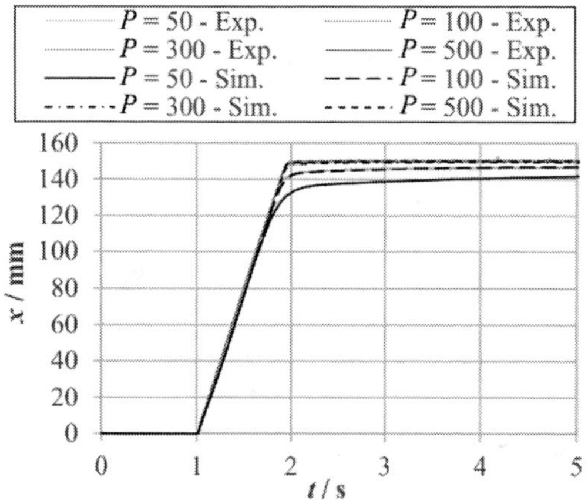

Figure 11. Comparison of measured and simulated time dependencies of the piston position x for different values of proportional component of control P

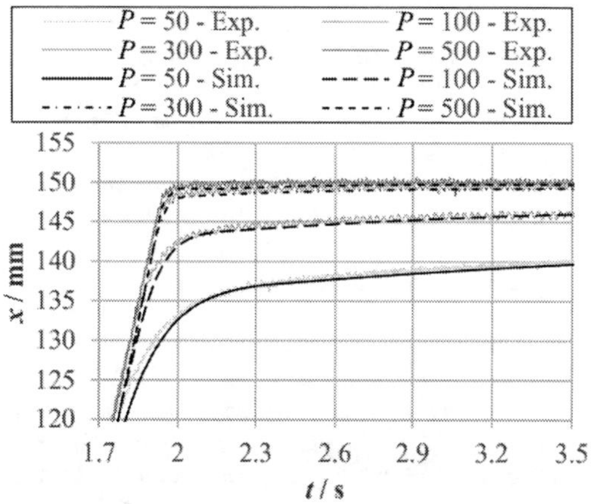

Figure 12. Detail of comparison of measured and simulated time dependencies of the piston position x during its arrival to the desired position

It was verified in the mathematical model that an overshoot of the desired position occurs when the parameter $P = 1100$. It was not possible to experimentally verify this fact because the applied control card has only the range $P = 0 \div 1000$.

7 Conclusions

This paper describes the experimental equipment with proportional control valve, linear hydraulic cylinder and mass load. This equipment allows control of the position with the feedback from the piston position. The Δp-Q characteristic of the proportional control valve for definition of its mathematical model was also measured. The mathematical model of the experimental equipment was created using Matlab SimHydraulics software. This model was verified by comparison of simulated and experimentally measured time dependencies of the piston position. The model was subsequently verified by comparison of simulated and experimentally measured time dependencies of pressures on both sides of the hydraulic cylinder. It was simulated the influence of the parameter of the proportional component P of the control card on time dependence of the piston position. The simulated time dependencies of the piston position for different parameters P correspond to the measured time dependencies of the piston position. It was found in the mathematical model that an overshoot of the desired position occurs when the parameter $P = 1100$. It is possible to debug the controller setting on PC due to the assembled mathematical model. This setting can be subsequently applied in real operation.

References

1. H. Dörr, R. Ewald, J. Hutter, D. Kretz, F. Liedhegener, A. Schmitt, *The Hydraulic Trainer Volume* 2 – Proportional and Servo Valve Technology (Mannesmann Rexroth GmbH, Lohr am Main, 1986)
2. A. Bureček, L. Hružík, *Simulation of Hydraulic Circuit with Linear Hydraulic Motor with Mass Load and Dynamic Characteristics*, J. of Appl. S. in the Term. and Fl. Mech. **5**, 6 (2011)
3. A. Bureček, L. Hružík, M. Vašina, EPJ Web of Conferences, EFM12, *Simulation of Dynamics of System with Hydraulic Lines and Linear Hydraulic Motor with Mass Load*, **45**, 01017, (2013)
4. Parker Hannifin Corporation, Communication Software ProPxD Software for Digital Valve Electronics
5. J. Jablonská, EPJ Web of Conferences, EFM13, *Compressibility of the fluid*, **67**, 02048, (2014)
6. P. Beater, *Entwurf Hydraulischer Maschinen: Modellbildung, Stabilitätsanalyse und Simulation Hydrostatischer Antriebe und Steuerungen* (Springer, Berlin, 1999)
7. The MathWorks, Matlab Simulink User´s Guide, SimHydraulics User´s Guide (USA, 2007)

Acknowledgements

The work presented in this paper was supported by a grant SGS "Investigation of dynamics of fluid systems" SP2015/95.

EPJ Web of Conferences **114**, 02009 (2016)
DOI: 10.1051/epjconf/201611402009

Comparison of methods for flow border detection in images of smoke visualization

Petr Caletka[1,a], Ondrej Pech[1], Jan Jedelsky[1], Frantisek Lizal[1] and Miroslav Jicha[1]

[1]*Brno University of Technology, Faculty of Mechanical Engineering, Technicka 2896/2,616 69 Brno, Czech Republic*

Abstract. A separation of the flow region from the surroundings is an essential step in the analysis of smoke visualization images. The separation can be performed using several detection methods from the image segmentation group. This paper deals with the border detection of the air flow downstream of a benchmark automotive vent using different threshold-based detection methods. An assessment of the methods on the basis of the resulting image quality is also addressed. The quality level depends on the quantity and brightness of disturbances in the background area. The disturbance is usually an isolated region of smoke, which naturally cannot be a part of the flow. Three representative images of different quality levels were selected for the detection, and three methods were used for the evaluation. Each of the methods was used to determine the threshold differently (by the level, by the ratio, and by the change of brightness). It is demonstrated that the change-based method with an appropriately selected parameter is the most convenient for images with the worst quality level while level- and ratio-based methods are only applicable for images of good quality.

1 Introduction

The smoke visualization is an experimental method used in fluid mechanics, especially for air flow studies. The main aim of the visualization is to get basic information about the flow field, e.g. its direction or shape, see Lezovic et al. [1]. One of the visualization outputs is the plane cut of the flow (of the image). There is a need to separate the flow area from the background by means of some segmentation technique. The flow border points are detected by the application of specific method of image segmentation, see Richter and Stastny [2].

Davies [3] divided techniques of image segmentation into three categories according to their principle (edge-based, region-based and thresholding). Spanel and Beran [4] described principles of the edge-based techniques and also global and local thresholding. McQuillan et al. [5] used the smoke visualization and the egde based technique for the indoor flow. Krska [6] visualized the flow downstream of the automotive vent by the smoke and then used the global thresholding for the border detection. Pech et al. [7] also visualized the flow downstream of the automotive benchmark vent using the smoke visualization, however the local thresholding was used for the border detection. Richter and Stastny [2] compared differences between the level-based (global thresholding) detection and the change-based (local thresholding) detection.

This paper deals with the detection of air flow shape from a benchmark automotive vent. This detection was done by thresholding in two planes perpendicular to each other. Three detection methods were used for the detection; each of them determines the threshold differently (by level, by ratio, and by change of brightness), see Richter and Stastny [2].

2 Experimental setup

According to Janotkova [8], the air flow behind the automotive benchmark vent at the volume rate 33.5 $m^3 h^{-1}$ has a character of the free isothermal turbulent jet (see Figure 1) with the Reynolds number of 10^4 and the Archimedes number of 1.2×10^{-3}.

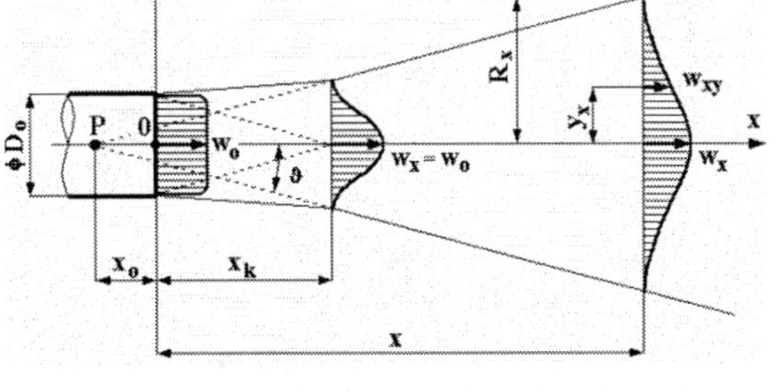

Figure 1. Typical shape of the free isothermal turbulent jet [8]

The jet was visualized by the smoke and illuminated with the use of a laser sheet. The quantity of the smoke fluid was represented by its consumption, which was set to 6 ml min^{-1}. The laser sheet output was set to 0.8 W.

[a] Corresponding author: petr.caletka@vutbr.cz

Horizontal and vertical planar cuts of the jet (see Figure 2) were taken with camera Canon EOS 300D with the lens Canon EF 17-40 mm f/4L USM. The EV value of the camera was set to 5.65. Setup of all of these components was documented by Pech et al. [7] during a complex optimization procedure.

3 Evaluated images

Each detection method was applied to a set of three images of different quality. The quality measure depends on the quantity and brightness of disturbed regions in the background area, see Figure 2 (white circles). The sensitivity to the disturbed regions is crucial for selection of the most appropriate method.

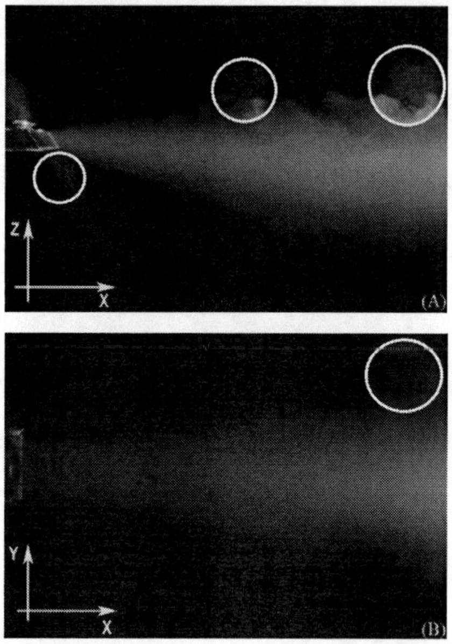

Figure 2. Disturbances in the images labelled by white circles: (A) side view; (B) top view

Caletka [9] assigned these images into three groups according to their quality:

a) with significant disturbed regions,
b) without significant disturbed regions,
c) without disturbed regions with computer retouching applied to background.

4 Detection methods

Richter and Stastny [2] documented that threshold-based methods for the jet border detection uses a number of cross-sections of the jet, which are perpendicular to the jet axis, see Figure 3. Each cross-section can be represented by the length (l, x-axis) – brightness (μ, y-axis) plot, see Figure 4.

Figure 3. Example of cross-section A-A

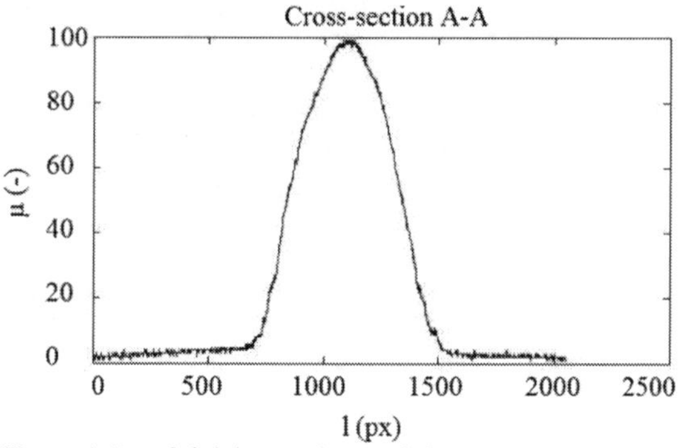

Figure 4. Length-brightness characteristic

A threshold $T(l, \mu)$, the border point of the jet, was determined in each cross-section by three different ways. First, the level-based detection from the global thresholding set of segmentation techniques was used. Then, two methods from the local thresholding set were used, namely the ratio- and change-based detection. The functionality and the character of each of the methods are controlled by adjustable parameters.

4.1 Level-based detection

Richter and Stastny [2] described a very simple principle of this detection method – global thresholding. There are border points T indicated in each cross-section depending on the selected level of adjustable parameter p_L, see Figure 5. In this case, p_L represents the jet border point, namely its brightness value μ. The p_L value was chosen as a specific percentile of the complete jet image brightness maximum $\mu_{i,max}$.

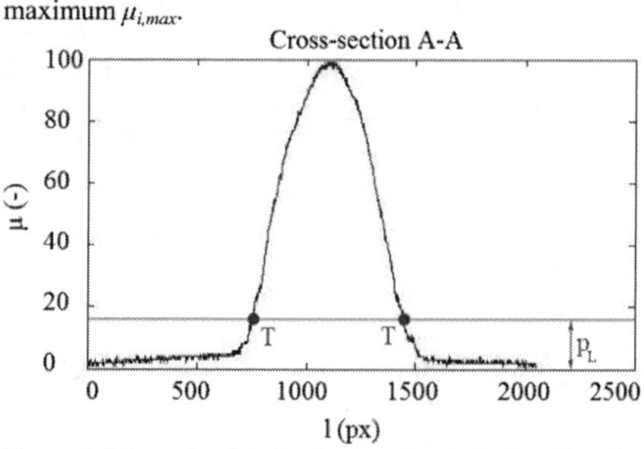

Figure 5. Schematic of the border detection in the level-based method

4.2 Ratio-based detection

This detection method based on local thresholding was described by Caletka [9]. There are border points T detected as the specific percentile p_R of the each cross-section brightness maximum $\mu_{c,max}$, see Figure 6. One of advantages of the method is the elimination of the uneven lighting of the visualized jet.

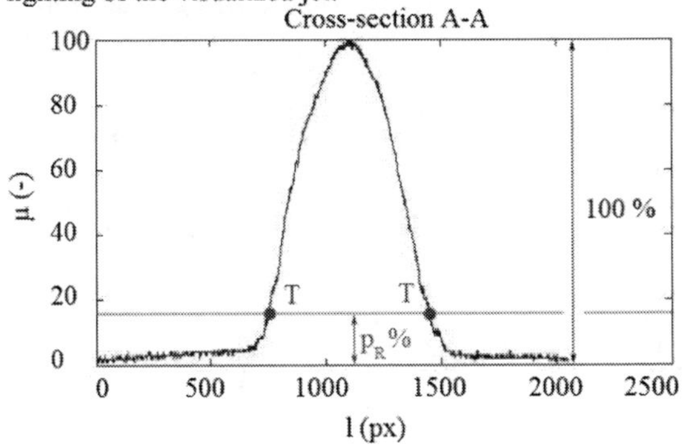

Figure 6. Schematic of the border detection in the ratio-based method

4.3 Change-based detection

This detection method belongs to the group of local thresholding methods and its principle (see Figure 7) is more complex than in case of the previous methods.

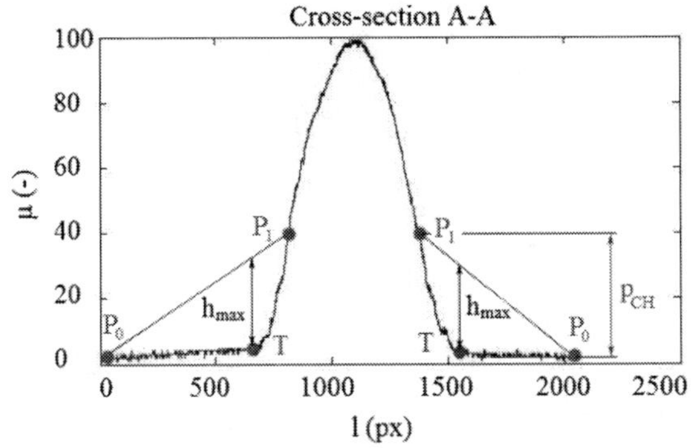

Figure 7. Schematic of the border detection in the change-based method

The extreme points p_0 of each cross-section are automatically assigned to the background, as documented by Richter and Stastny [2]. The points P_1 have the same brightness value μ as the chosen adjustable change parameter p_{CH}. There is also a presumption that these points are assigned to the flow area. The p_{CH} value is always selected as a specific percentile of brightness minimum in the axis of flow $\mu_{a,min}$. As can be seen in Figure 7, the height h between the segment line $|p_0p_1|$ and the current μ is traced. When the perpendicular line h_{max} is found, the point with current μ is indicated as the border point T.

5 Results and discussion

The approximate shape of the jet can be predicted on the basis of theory of free isothermal turbulent jet, see Janotkova [8] and Section 2. However, Seda [10] also studied and measured the jet properties, especially its velocity field by the CTA (constant temperature anemometry). The jet shape detected from his CTA measurement (see Figure 8) corresponds with the assumption presented in Janotkova [8].

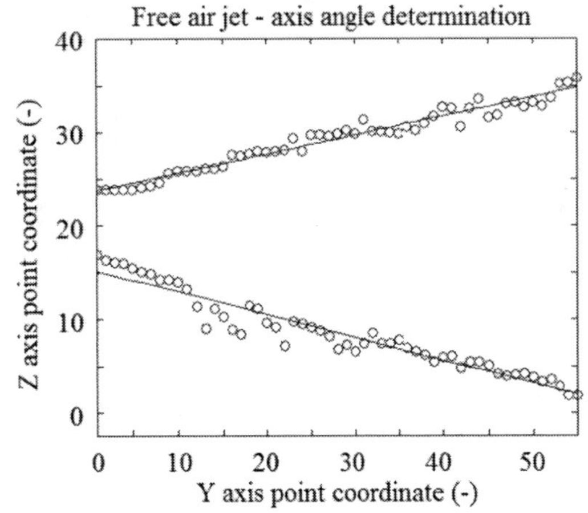

Figure 8. The jet shape indicated by CTA measurement [10]

A suitable detection method for jet border detection should be designed in order to not interpret the disturbed regions as a jet area. All the detected jet shapes in bottom-view images (X-Y plane) are presented in Figures 9-11. The specific values of adjustable parameters used in this work were selected as described in Caletka [9].

5.1 Image with significant disturbed regions

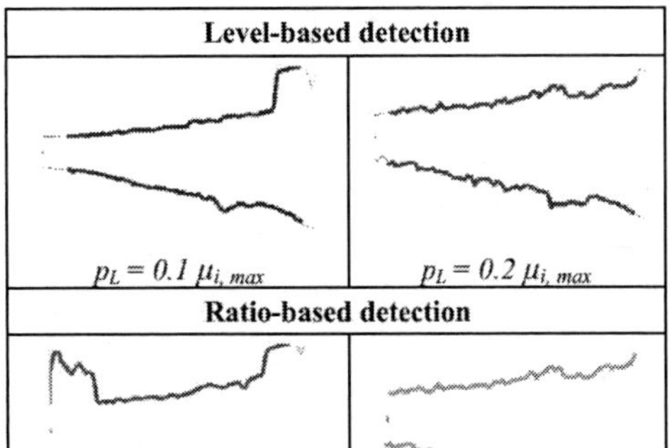

Figure 9. Jet shapes detected from the image with significant disturbed regions by all the methods

5.2 Image without significant disturbed regions

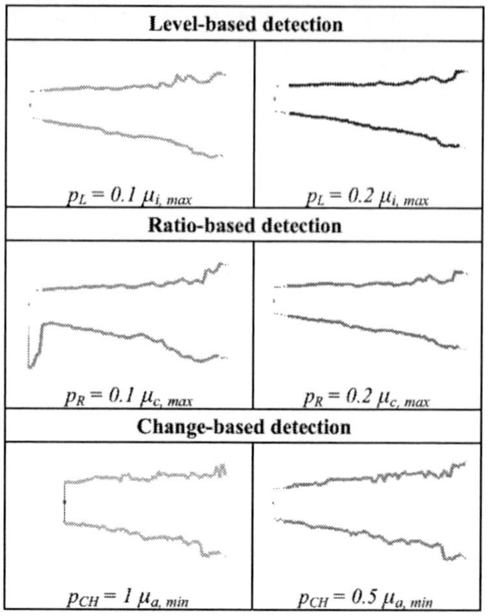

Figure 10. Jet shapes detected from the image without significant disturbed regions by all the methods

5.3 Image without the disturbed regions

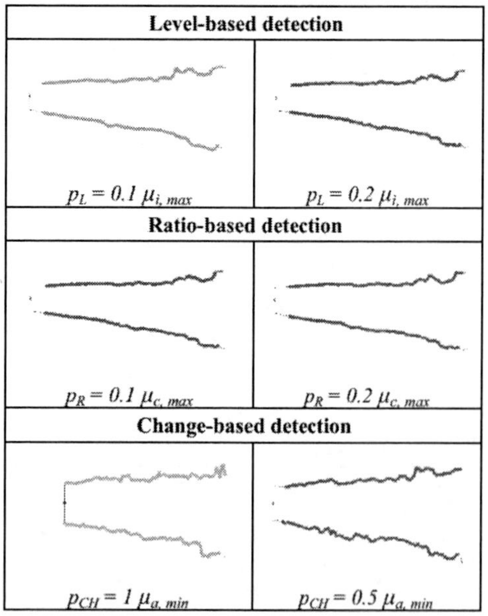

Figure 11. Jet shapes detected from the image without the disturbed regions by all the methods

6 Conclusions

It was found that the level-based method detection with a lower value of p_L is suitable especially in images without significant disturbed regions. When the quantity of disturbed regions increases it is possible to eliminate this issue by a higher value of p_L. However, this procedure is not optimal for correction of the border detection owing to the global character of this threshold method.

In the case of the ratio-based detection, it was found that the lower value of p_R is only suitable for images without the disturbed regions. The detection with the lower value of p_R is very sensitive to the disturbed regions, so when their quantity increases, the method does not work correctly. However, this issue can be eliminated by higher value of p_R. This procedure is the correct way to get relevant results due to the local character of this threshold method.

Finally, in case of the change-based detection, it was found that the lower value of p_{CH} is suitable for images with diverse quality levels. With a higher value of p_{CH}, the sensitivity to the presence of disturbed regions decreases. On the other hand there is a higher risk of wrong detection of jet shape immediately at the outlet of the benchmark automotive vent. The change-based method with an appropriately chosen parameter p_{CH} is the most universal. Flow borders detected using this method were not influenced by disturbed regions and this method was therefore found suitable for further application. Methods that find the border points according to the level or the ratio are suitable only for images with better quality, and in such cases usually with the p_L or p_R higher than 20.

Acknowledgements

The presented research was supported by the project LO1202 NETME CENTRE PLUS with the financial support from the Ministry of Education, Youth and Sports of the Czech Republic under the "National Sustainability Programme I" and by the project reg. number FSI-S-14-2355 of the Brno University of Technology. The authors also gratefully acknowledge the support from the "Josef Bozek Competence Centre for Automotive Industry", number TE01020020 of the Technology agency of the Czech Republic.

References

1. T. Lezovic, F. Lizal, J. Jedelsky, M. Jicha, HVAC&R Research, **19**, 9 (2013)
2. J. Richter, J. Stastny, *Advances in Computer Science*, 6, (2012)
3. E. R. Davies, *Computer & Machine Vision: Theory, Algorithms, Practicalities*, **4**, 912 (2012)
4. M. Spanel, V. Beran, *Obrazové segmentační techniky – přehled existujících metod, Brno University of Technology*; (2006)
5. B. McQuillan, J. Herzberg, L. D. Montoya, B&E **73**, 9 (2014)
6. L. Krska, *Master Diploma Thesis, Brno University of Technology*, 82, (2011)
7. O. Pech, J. Jedelsky, P. Caletka, M. Jicha, *EPJ Web of Conferences*, **67**, 4 (2014)
8. E. Janotkova, *Technika prostředí 1. část*, 58 (2011)
9. P. Caletka, *Bachelor Thesis, Brno University of Technology*, 66, (2014)
10. L. Seda, *Master Diploma Thesis, Brno University of Technology*, 72, (2015)

EPJ Web of Conferences **114**, 02010 (2016)
DOI: 10.1051/epjconf/201611402010

The effect of porosity on performance of phosphoric acid doped polybenzimidazole polymer electrolyte membrane fuel cell

Muhammet Celik[1], Gamze Genc[2,a], Gulsah Elden[2] and Huseyin Yapici[2]

[1]*Aksaray University, Dept. of Mechanical Engineering, 68100, Aksaray, Turkey*
[2]*Erciyes University, Dept. of Energy Systems Engineering, 38039, Kayseri, Turkey*

Abstract. A polybenzimidazole (PBI) based polymer electrolyte fuel cells, which called high temperature polymer electrolyte fuel cells (HT-PEMS), operate at higher temperatures (120-200°C) than conventional PEM fuel cells. Although it is known that HT-PEMS have some of the significant advantages as non-humidification requirements for membrane and the lack of liquid water at high temperature in the fuel cell, the generated water as a result of oxygen reduction reaction causes in the degradation of these systems. The generated water absorbed into membrane side interacts with the hydrophilic PBI matrix and it can cause swelling of membrane, so water transport mechanism in a membrane electrode assembly (MEA) needs to be well understood and water balance must be calculated in MEA. Therefore, the water diffusion transport across the electrolyte should be determined. In this study, various porosity values of gas diffusion layers are considered in order to investigate the effects of porosity on the water management for two phase flow in fuel cell. Two-dimensional fuel cell with interdigitated flow-field is modelled using COMSOL Multiphysics 4.2a software. The operating temperature and doping level is selected as 160°C and 6.75mol H3PO4/PBI, respectively.

1 Introduction

The proton exchange membrane (PEM) fuel cells are the most attractive alternative energy devices that can produce zero-emission power due to the fact that they do not emit harmful products after the electrochemical reactions. Nafion based PEM fuel cells operate under 100°C. But the operation conditions of these fuel cells have some disadvantages such as necessity humidifying membrane, water management problems etc. But Polybenzimidazole (PBI) based PEM fuel cells can solve many of these disadvantages by increasing the operation temperatures above 120°C. By running the PBI PEM fuel cells at 120°C-180°C, the water management becomes easier because the produced water will be in the gas phase.

PBI is an amorphous polymer which has strong thermochemical stability, and proton conductivity can be enhanced by phosphoric acid doping procedure. Because of this advantages, HT-PEMs have been attracted attentions and many studies such as its performance [1-5], CO contamination [6-8], numerical modelling [9-11] have been done. But in the literature the two phase flow in HT-PEMs are very scarce. Especially, the two phase flow at stop conditions should be investigated. Because when the operation of fuel cell finish, the water vapor will be condensed. So, the condensed water will accumulate in the porous structure. The phosphoric acid removal occurs in the presence of liquid water in the porous structure resulting low ionic conductivity of the PBI membrane. With this consideration the condensed water should be lessened inside the fuel cell. In this study the two phase flow at the stop conditions in a PBI based PEM fuel cell is investigated in order to observe how distribution of water change with the porosity of the cathode gas diffusion layer.

2 Mathematical Model

Because the water is produced at cathode, the cathode side of the fuel cell is selected as computational domain. The multiphase mixture (M^2) approach is used for modelling. The assumptions in numerical model are as follows;

- Gas phase is perfectly a mixture and ideal gas
- Hydrogen and oxygen is soluble in water
- Time-dependent regime

In addition, the fuel cell is in stop conditions when the fuel cell is not running and the electrochemical reactions are stopped. Hence, there is no hydrogen or oxygen entry to the fuel cell.

2.1 Governing equations

Mass conservation equation can be given as;

[a] Corresponding author: gamzegenc@erciyes.edu.tr

$$\varepsilon\frac{\partial\rho}{\partial t} + \nabla(\varepsilon\rho u) = 0 \quad (1)$$

where ε is porosity of the gas diffusion layer, u is velocity and ρ is the mixture density. In Eq(1) the ρu is defined the mixture velocity which can be expressed as;

$$\rho u = \rho_l u_l + \rho_g u_g \quad (2)$$

where ρ_l is the density of the liquid phase, ρ_g is the density of vapor phase, u_l is the liquid phase velocity and u_g is the vapor phase velocity.

The momentum conservation equation for two phase flow can be written as [12];

$$\frac{\partial(\varepsilon\rho u)}{\partial t} + \nabla(\varepsilon\rho uu) = \nabla(\varepsilon\mu\nabla u) - \varepsilon\nabla p + \varepsilon\rho_k g - \varepsilon^2\frac{\mu}{K}u \quad (3)$$

where the first term in the left hand side defines the variation of the momentum according to time. Second term defines the momentum according to convection. At the right hand side the variation of the momentum by viscous forces, pressure gradients, and gravity can be seen. The last term in the right hand side defines the flow in the porous structure. k subscript describes the phases (liquid and gas phase).

The viscosity of the two phase mixture can be expressed as [12];

$$\mu = \frac{\rho_l s + \rho_g(1-s)}{(k_{rl}/\nu_l) + (k_{rg}/\nu_g)} \quad (4)$$

where k_{rl} and k_{rg} describes relative permeability for liquid phase and gas phase, respectively. s defines the saturation of water. Relative permeabilities according the saturation can be defined as;

$$k_{rl} = s^3 \quad (5)$$

$$k_{rg} = (1-s)^3 \quad (6)$$

The saturation of water can be written as [13];

$$s = \frac{C^{H_2O} - C_g^{H_2O}}{C_l - C_g^{H_2O}} \quad (7)$$

where C_{H2O} is the mixture concentration, C_g^{H2O} is gas phase concentration and C_l is liquid concentration.

The species equation can be expressed as [14];

$$\varepsilon\frac{\partial C^{H_2O}}{\partial t} + \nabla\cdot(\varepsilon\gamma_c\vec{u}C^{H_2O}) = \nabla\cdot(D_g^{H_2O}\nabla C_g^{H_2O}) - \nabla\cdot\left[\left(\frac{mf_l^{H_2O}}{M^{H_2O}} - \frac{C_g^{H_2O}}{\rho_g}\right)\vec{j}_l\right] \quad (8)$$

where γ_c defines advection correction factor, $\vec{j}_l$ is diffusion flux and $mf_l^{H_2O}$ describes the mass fraction of liquid water. While the first term at the right hand side defines the diffusion of gas phase, the second term defines the motion according to capillary effect. The advection correction factor for water can be expressed as [12];

$$\gamma_{H_2O} = \frac{\rho(\lambda_l C_l^{H_2O} + \lambda_g C_g^{H_2O})}{\rho_l s C_l^{H_2O} + \rho_g(1-s)C_g^{H_2O}} \quad (9)$$

where λ_l and λ_g are the relative mobility for the liquid phase and gas phase, respectively. Relative mobilities can be defined as [12];

$$\lambda_l(s) = \frac{k_{rl}/\nu_l}{k_{rl}/\nu_l + k_{rg}/\nu_g} \quad (10)$$

$$\lambda_g(s) = 1 - \lambda_l(s) \quad (11)$$

Diffusion flux, $\vec{j}_l$, can be expressed as [13];

$$j_l = \frac{\lambda_l\lambda_g K}{\nu}\nabla P_c \quad (12)$$

Capillary pressure, P_c, can be expressed according to hydrophilic and hydrophobic surfaces. In this study the gas diffusion layer has hydrophobic structure. Hence, the capillary pressure for hydrophobic structure can be written as [13];

$$P_c = \sigma\cos\theta_c\left(\frac{\varepsilon}{K}\right)^{\frac{1}{2}}\left[1.417s - 2.120s^2 + 1.263s^3\right] \quad (13)$$

where σ describes the surface tension, θ_c describes the contact angle.

2.2 Numerical Procedure

In the numerical solutions, *COMSOL Multiphysics 4.2a* software which is based a finite element method solver is used. The computational domain and grid structure can be seen in Figure 1.

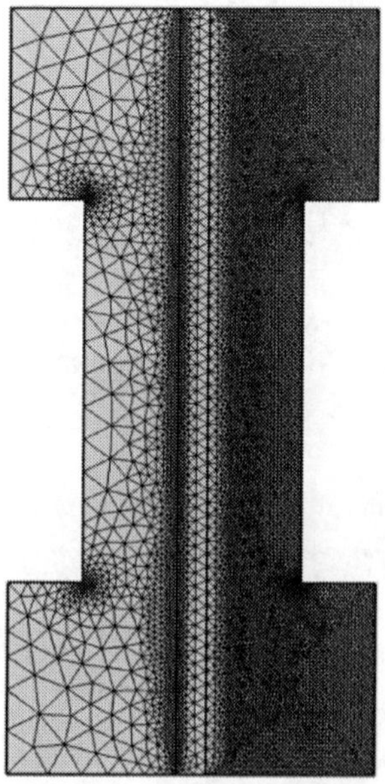

Figure 1. Computational domain and grid structure

As can be seen in Table 1, the parametric variations are observed according to numerical solutions for different element numbers. It is assumed that the parameter values are the same for 18522 and higher element numbers. So, the element number for the numerical model is chosen as 18522. Grid structure consists of 18522 triangular elements and the solver is selected as PARDISO. The parameters used in the modeling are listed in Table 2.

Table 1. Parametric variations according to element number

Element Number	**12772**	**18522**	**41097**
Saturation	2.553E-4	2.555 E-4	2.557 E-4
Temperature[K]	338.70	338.70	338.70

Table 2. Model Parameters.

Parameter	**Value**
Surface Tension	0.0625 N/m [14]
GDL porosity	0.4-0.7
GDL contact angle	110° [14]
Operation pressure	101325 Pa
Initial temperature	180°C
Acid doping level	6.75 mol H3PO4/PBI

3 Results and Discussions

The saturation according to time for GDL porosity of 0.4 is given in Figure 2. As can be seen in this Figure, the saturation increased with time and it became constant after 80 s. It means all of the gas phase condensed to the liquid phase in 100 s.

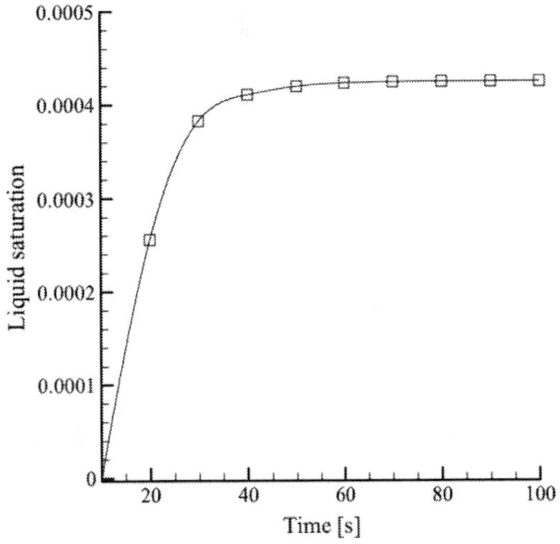

Figure 2. Saturation according to time (ε=0.5)

The saturation distribution according to time for GDL porosity of 0.4 is given in Figure 3. The saturation is higher at the centre of the gas diffusion layer due to the fact that the heat transfer from fuel cell to environment occurs at the interface of the gas diffusion layer and gas channel. As is seen in this figure, at the 100s, the saturation distribution became uniform in cathode gas diffusion layer and it means all of the gas phase condensed to liquid phase.

The diffusion of gas phase is relative to porosity of the medium. The gas molecules can effectively diffuse in the porous structure at higher porosity levels. In the Figure 4, the change of saturation according to porosity of the GDL at t=50s is given. While the porosity increases, the saturation level decreases. Because the gas molecules are transported effectively into the porous structure and they leave the cell before condensation, so it decreases the saturation level.

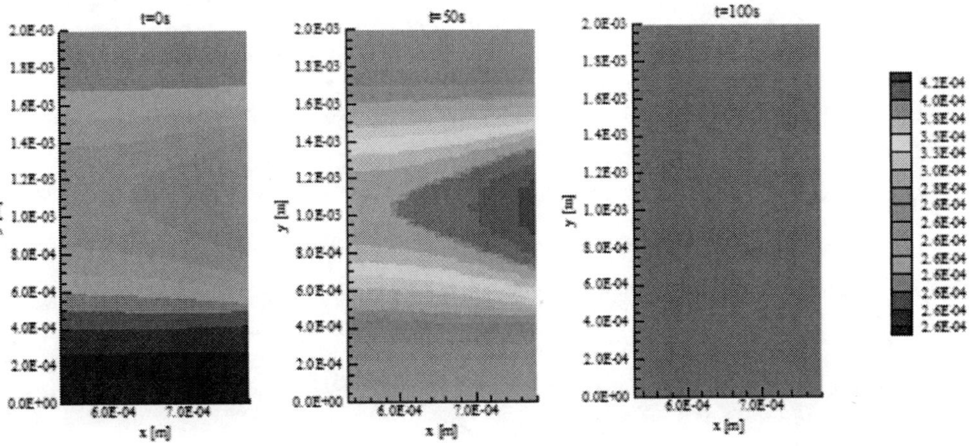

Figure 3. Saturation variation according to time (ε=0.5)

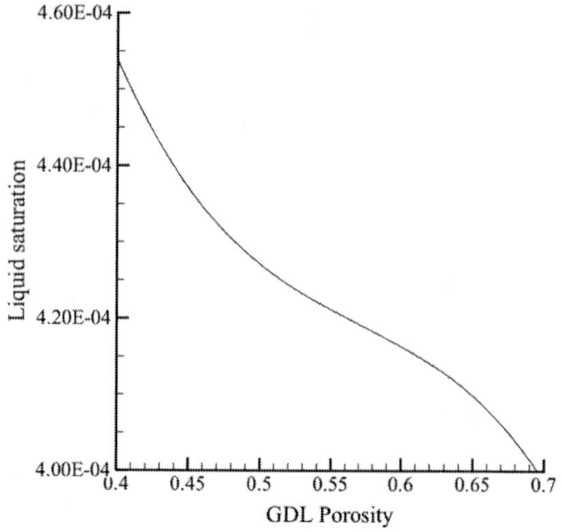

Figure 4. Saturation according to GDL porosity (t=50s)

In order to investigate the effect of GDL porosity in detail, a constant saturation level is selected as $4.2 \cdot 10^{-4}$. Time needed to reach saturation level of $4.2 \cdot 10^{-4}$ according to GDL porosity is given in Figure 5. As can be seen from this figure, time needed is increasing with increasing GDL porosity. When GDL porosity is 0.5, 36.2s needed to reach the saturation level of $4.2\ 10^{-4}$. It means while increasing the GDL porosity, the gas phase presence in the medium decreases. Therefore, the lower condensations occur at high GDL porosity.

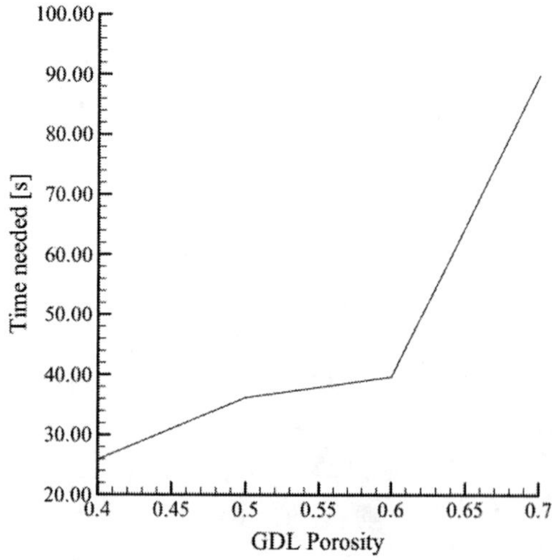

Figure 5. Time needed to reach desired saturation level according to GDL porosity

4 Conclusions

In this study, a transient, two dimensional, two phase numerical model for PBI based PEM fuel cell was made. Firstly, for the base case (ε=0.5), the saturation variation according to time was discussed. After that, for the various GDL porosity values (ε=0.4-0.7), the saturation level was investigated. Finally, the time needed to reach desired saturation level according to porosity was investigated. According to numerical solutions, it can be said as follows;

•The diffusion of gas molecules are highly related to porosity of gas diffusion layer.

•While porosity of GDL increases the saturation level decreases owing to the fact that the molecules leave the cell before condensation.

•Leaving the gas molecules from the cell facilitate the water management.

Acknowledgements

The authors would like to thank the Scientific Research Projects Unit of Erciyes University for funding supported by project coded FDK-2014-5442 and Scientific and Technological Research Council of Turkey (TÜBİTAK) for funding supported by project coded 111M071.

References

1. D.Cheddie, N. Munroe, Energ Convers Manage **47,** 11 (2006)
2. D.F. Cheddie, N.D.H. Munroe, J Power Sources **160,** 1 (2006)
3. J. Hu, H. Zhang, J. Hu, Y. ZhaiB. Yi, J Power Sources **160,** 2 (2006)
4. A. Su, Y. Ferng, J. HouT. Yu, Int J Hydrogen Energ **37,** 9 (2012)
5. R. Taccani, N. Zuliani, , Int J Hydrogen Energ **36,** 16 (2011)
6. S.J. Andreasen, J.R. Vang, S.K. Kær, Int J Hydrogen Energ **36,** 16 (2011)
7. A.R. Korsgaard, M.P. Nielsen, M. BangS.r.K. Kær. *Modeling of CO influence in PBI electrolyte PEM fuel cells*. in *ASME 2006 4th International Conference on Fuel Cell Science, Engineering and Technology*. 2006. American Society of Mechanical Engineers.
8. F. Seland, T. Berning, B. Børresen, R. Tunold, J Power Sources **160,** 1 (2006)
9. K. Jiao, X. Li, Fuel Cells **10,** 3 (2010)
10. C. Siegel, G. Bandlamudi, A. Heinzel, J Power Sources **196,** 5 (2011)
11. T. Sousa, M. Mamlouk, K. Scott, Chem Eng Sci **65,** 8 (2010)
12. Z.H. Wang, C.Y. Wang, K.S. Chen, J Power Sources **94,** 1 (2001)
13. U. Pasaogullari, C.Y. Wang, K.S. Chen, J Electrochem Soc **152,** 8 (2005)
14. U. Pasaogullari, C.Y. Wang, J Electrochem Soc **152,** 2 (2005)

EPJ Web of Conferences **114**, 02011 (2016)
DOI: 10.1051/epjconf/201611402011

Mathematical modeling of planar and spherical vapor–liquid phase interfaces for multicomponent fluids

David Celný[1,2,a], Václav Vinš[1], Barbora Planková[1,2], and Jan Hrubý[1]

[1] Institute of Thermomechanics, of the CAS, v. v. i., Dolejškova 1402/5, Prague, Czech Republic
[2] Faculty of Nuclear Sciences and Physical Engineering, Czech Technical University in Prague, Břehová 78/7, Prague, Czech Republic

Abstract. Development of methods for accurate modeling of phase interfaces is important for understanding various natural processes and for applications in technology such as power production and carbon dioxide separation and storage. In particular, prediction of the course of the non-equilibrium phase transition processes requires knowledge of the properties of the strongly curved phase interfaces of microscopic droplets. In our work, we focus on the spherical vapor–liquid phase interfaces for binary mixtures. We developed a robust computational method to determine the density and concentration profiles. The fundamentals of our approach lie in the Cahn-Hilliard gradient theory, allowing to transcribe the functional formulation into a system of ordinary Euler-Langrange equations. This system is then split and modified into a shape suitable for iterative computation. For this task, we combine the Newton-Raphson and the shooting methods providing a good convergence speed. For the thermodynamic properties, the PC–SAFT equation of state is used. We determine the density and concentration profiles for spherical phase interfaces at various saturation factors for the binary mixture of CO_2 and C_9H_{20}. The computed concentration profiles allow to the determine the work of formation and other characteristics of the microscopic droplets.

1 Introduction

The phase interface is the layer separating bulk thermodynamic phases, the liquid and the gas phases in the present case. The thickness of the phase interface is typically on the order of nanometres. Yet, understanding the physics of vapor-liquid phase interfaces is important for numerous natural processes and applications in technology. In particular, the present study aims at modeling the phase transition phenomena in systems relevant for carbon capture and storage technologies. The subject of interest of this study are microscopic droplets, whose diameter is comparable with the thickness of the phase interface. Such droplets occur in the process of nucleation of a liquid phase from the metastable supersaturated gas phase.

The phase equilibrium state of the system has to be investigated first. For this purpose, an equation of state (EoS) is needed. The usual approach is to choose one of the simple low parameter equation (Van der Waals, Redlich-Kwong, Peng-Robinson or their modifications). Here we employed an advanced PC-SAFT EoS. This equation belongs to a family of EoSs derived from the statistical fluid association theory (SAFT) [1]. The perturbed chain modification PC-SAFT was developed by Gross and Sadowski [2]. Various approaches are available to characterize the phase interface. The Gibbsian approach, treating the phase interface as infinitesimally thin, allows to properly describe the integral properties of the phase interface such as the surface tension, molecular adsorption and related thermodynamic properties. However, it does not provide a tool to predict the dependence of the surface tension on composition of the adjacent phases and on the curvature of the phase interface. The so-called capillarity approximation, which is one of the main assumptions of the classical nucleation theory (CNT) [3], assigns properties of the macroscopic planar phase interface to the strongly curved interface of the microscopic droplets. Modeling the effect of the surface curvature is possible with the gradient theory(GT) [4,5] or density functional theory (DFT) [6]. These theories enable to obtain the density profiles (dependence of the local density on the coordinate perpendicular to the surface) and other characteristics of the phase interface.

This study is a continuation of the research of phase interfaces using GT previously done by Hrubý *et al.* [7,8] and Planková [9]. In this work, we propose a new approach to the mathematical solution of the problem obtained from GT. This solution is iterative and offers a way of splitting the problem into two consecutive steps that can be solved separately.

2 The modeled system

We consider an n-component mixture (two components in this study) in the state of metastable supersaturated gas

[a] e-mail: : celnydav@fjfi.cvut.cz

phase. We consider a microscopic droplet in the thermal and chemical equilibrium with the gas phase, meaning that the temperature and chemical potentials are homogeneous throughout the system. The modeled droplet corresponds to the so-called critical cluster of the molecular theory. We note that the opposite situation with a bubble in the equilibrium with a superheated or stretched liquid would be very much alike.

When the system is composed of two components, the phase equilibrium state with a planar phase interface can be easily computed for the given temperature T and pressure p based on the conditions of thermodynamic equilibrium of the liquid and gaseous phases: equality of temperatures, chemical potentials for both components, and equality of pressures in both phases. As a result, we obtain the densities and compositions of both phases.

To proceed from the system with a planar phase interface to the system with a droplet, we increase the concentration of the less volatile component (here n-nonane) in the mother phase (gas) by multiplying it by the saturation ratio $\alpha > 1$, while keeping the concentration of the more volatile component (carbon dioxide) and temperature constant. The saturation ratio directly influences the size of the droplet: a bigger saturation ratio results in a smaller droplet. We note that choosing a too high saturation ratio might result into an unstable state (a state beyond the vapor-liquid spinodal). For the supersaturated gas phase obtained in this way, we determine the liquid phase in thermal and chemical equilibrium. From the condition of equality of chemical potentials of each components in both phases we determine the concentrations of the components in the liquid phase and the corresponding pressure of the liquid phase, which is bigger than for the mother gas phase. The magnitude of difference is also directly influenced by the saturation factor in a manner that higher α means higher pressure difference. The thermodynamic state of the liquid obtained in this way is the state of idealized liquid droplet in the Gibbsian picture. The difference of pressure is related to the Laplace pressure due to the surface tension of the curved interface. The "equilibration" of the droplet with the gas phase is iterative process. The iteration is continued even a few steps after the stop criterion is reached and the best result is chosen among the computed after-steps. The purpose of this measure is avoiding wrong results due to premature termination of the iteration.

The thermodynamic state obtained in this way is hypothetical because even the center of the droplet is influenced by the phase interface. However, the density and concentrations obtained in this way are used to provide the first guess of local conditions in the center of the droplet in the gradient theory modeling.

3 Gradient theory

The gradient theory, also called Cahn-Hilliard theory [4], was developed by Cahn and Hilliard in 1957, although its foundations were already given by van der Waals. The basis of the GT is an expansion of the Helmholtz energy into a Taylor series with respect to the gradient of density or gradients of concentrations. This expansion is then used to obtain the grand potential of the system, which is treated as a functional of the density or concentration profiles. These profiles are then determined by minimizing the functional, or more precisely, finding its stationary point by solving the corresponding Euler-Lagrange (E-L) differential equations.

In this study, we define the functional as the work of formation given by Eq. (1):

$$W = \Omega_{\text{inhom}} - \Omega_{\text{hom}} . \tag{1}$$

Here, Ω is the grand potential. The subscripts in this equation indicate homogeneous system (only the gaseous mother phase) and inhomogeneous system (system including a droplet with some gaseous surroundings at the same temperature and chemical potentials).

The grand potential Ω can be expressed in terms of the Helmholtz free energy of the inhomogeneous system. The form of the functional can be transformed into the spherical coordinates. The changed coordinate system allows using the symmetry of spherical droplet and simplifies the expression into the following form:

$$W = \int_0^R \left(\Delta\omega \left(\boldsymbol{\rho}(r)\right) + \frac{1}{2} \sum_{i,k=1}^{n} c_{i,k} \left(\nabla\rho_i(r)\right)\left(\nabla\rho_k(r)\right) \right) 4\pi r^2 dr . \tag{2}$$

We can see from Eq. (2) that integral is one dimensional and all concentrations ρ_i depend on the radial coordinate only. $c_{i,k}$ in Eq. (2) denotes the so-called influence parameter. And the term $\Delta\omega$ is a composite of the Helmholtz energy density and term $\sum_i \rho_i \mu_i$. Also this term depends on density profile function $\boldsymbol{\rho}(\boldsymbol{r})$. We denote density profile function in a vector shape because it is composed of concentration profiles for both components.

For given density profile the functional (2) returns the work needed to form an inhomogeneity represented by this density profile. Our task is now to obtain the most probable density profile for the given system conditions. Such demand according to the paper [4] requests a minimization of the formation work. The solution yields saddle point and minimal points of functional W. We are interested only in the saddle point because it corresponds to the required density profile for the droplet. The minimum points on the other hand represent a homogeneous system (gaseous or liquid) without any phase interface.

The saddle point is found using E-L system of two second order differential equations Eq. (3):

$$\frac{\partial W}{\partial \rho_k} - \frac{d}{dr}\frac{\partial W}{\partial \rho_k'} = 0, k \in 1 \dots n . \tag{3}$$

Where ρ' denotes differential of $d\rho/dr$. Expressing the work of formation in Eq. (3) in terms of the GT, we obtain Eq. (4):

$$\sum_{i=1}^{n} c_{i,k} \left(\frac{2}{r}\rho_i' + \rho_i'' \right) = \Delta\mu_k, k \in 1 \dots n . \tag{4}$$

Here, the right hand side of Eq. (4) represents differences between chemical potentials of a hypothetical homogeneous phase corresponding to local concentrations and the chemical potentials of the mother phase. In case that the interaction of the unlike molecules is not qualitatively different from the interactions of the like molecules, it is possible to approximate the influence parameter of mutual interaction of components $c_{i,k}$ as a geometric mean of the influence parameters for the pure components $\sqrt{c_{i,i} \cdot c_{k,k}}$:

$$\sum_{i=1}^{n} \sqrt{c_{i,i}} \left(\frac{2}{r} \rho_i' + \rho_i'' \right) = \frac{\Delta\mu_k}{\sqrt{c_{k,k}}}, k \in 1 \ldots n . \qquad (5)$$

Eq. (5) is the basis for the present computations. Various methods can be developed to solve this equation. The present method provides benefits of substantial speed-up, modifiability of computation and result precision.

4 Solution of the Euler-Lagrange equations

Our method uses elementary results of the gradient theory and modifies it into a two step iterative method. We notice the similarity in right hand side of equation within Eq. system (4) and subtract the first equation from the others. This operation gives us system composed of one differential equation (6) and $n - 1$ non-linear algebraic equations (7). For our case is $j \in 2 \ldots n$.

$$\sum_{i=1}^{n} \sqrt{c_{i,i}} \left(\frac{2}{r} \rho_i' + \rho_i'' \right) = \frac{\Delta\mu_1}{\sqrt{c_{1,1}}}, \qquad (6)$$

$$\frac{\Delta\mu_j}{\sqrt{c_{j,j}}} = \frac{\Delta\mu_1}{\sqrt{c_{1,1}}} . \qquad (7)$$

This system contains equations coupled through their right-hand sides. Our goal is to split the solution into solving algebraic part (7) and the differential part (6) separately. To ensure that these equations are coupled, the algebraic part (7) has to be solved first. Then, its results are used in the computation of equation (6). We perform a special substitution:

$$\tilde{\rho} = \frac{\sum_{i=1}^{n} \sqrt{c_{i,i}} \rho_i}{\sum_{i=1}^{n} \sqrt{c_{i,i}}} . \qquad (8)$$

The substitution transforms the components concentrations ρ_i into modified molar density $\tilde{\rho}$. Substitution also ensures simple shape of E-L equations and enable discretization. We can also see that shape of substitution (8) is invariable to differentiation d/dr. This feature of $\tilde{\rho}$ is used in differential equation. Secondly, a new artificial variable X is inserted. To clarify the dependence in system of Eq. (7, 6) we choose $X = -\Delta\mu_1 / \sqrt{c_{1,1}}$. For $j \in 2 \ldots n$ we then obtain a system:

$$\left(\sum_{i=1}^{n} \sqrt{c_{i,i}} \right) \left(\frac{2}{r} \tilde{\rho}' + \tilde{\rho}'' \right) = -X, \qquad (9)$$

$$\frac{\Delta\mu_j}{\sqrt{c_{j,j}}} = -X . \qquad (10)$$

This system (9,10) is basis of following computation. The solution of equation (9) is possible only with the knowledge of artificial variable X in the molar density interval. For that purpose X has to be computed. The dependence of X variable enables the discretization of $\tilde{\rho}$ across the molar density interval. At these discretized points we compute values of X.

4.1 Solution of the algebraic equations

As the first step we take only algebraic part (10). The artificial variable X allows to rewrite the algebraic part into a system of similar algebraic equations. As a result of adding new variable, one new equation also needs to be appended. The last equation is provided by definition (8). This completes the analytic system of equations (11). , where for $j \in 2 \ldots n$ we have:

$$\frac{\Delta\mu_1(\rho)}{\sqrt{c_{1,1}}} + X = 0$$

$$\vdots$$

$$\frac{\Delta\mu_n(\rho)}{\sqrt{c_{n,n}}} + X = 0$$

$$\frac{\sum_{i=1}^{n} \sqrt{c_{i,i}} \rho_i}{\sum_{i=1}^{n} \sqrt{c_{i,i}}} - \tilde{\rho} = 0. \qquad (11)$$

The system (11) needs to be solved for values of X and ρ at each point of discretization. This provides the function dependence sampled across the molar density interval. Results of each iteration $X(\tilde{\rho}_i)$, $\rho(\tilde{\rho}_i)$ are saved into data matrix M_1. The secondary computational data ($dX(\tilde{\rho}_i)/d\tilde{\rho}$, $d\rho(\tilde{\rho}_i)/d\tilde{\rho}$) are also saved into data matrix M_2. The data matrices are useful in post-processing phase for computation of approximate of requested functions X and ρ.

To solve this system the Newton-Raphson iterative method is used. For sufficiently fine division (500 points or more for best performance) the iteration is commenced. The Newton-Raphson method requires to compute the Jacobian of system (11) and sufficient first estimate. First guess for boundary points is provided by boundary conditions of system that we obtained from supersaturated equilibration. And points inside discretization use previous result as their first estimate. Owing to the fine discretization, the solution for each point is found in two to five Newton-Raphson iterations.

When the computation is completed, the obtained matrices M_1 and M_2 are used to compute piecewise cubic approximation of functions $X(\tilde{\rho})$, $\rho_1(\tilde{\rho})$, and $\rho_2(\tilde{\rho})$. Examples of these functions for saturation ratio $\alpha = 1.5$ are shown in figures 1 and 2. For each point of discretization except the first we compute coefficients for cubic polynomial with function values from M_1 and with derivatives from the M_2. The three piecewise cubic curves then provide a link between the two parts of the problem and are used for the solution of differential equation.

4.2 Solution of the differential equation

The differential part is the first equation (9). It is a second order differential equation which is now decoupled from the algebraic solution of the rest of the system. As input we use the piecewise cubic function $X(\tilde{\rho})$ computed from the solution of the algebraic system. The differential equation is accompanied with the boundary condition from the saturated system. From the mother phase end, we know the modified density of the gas phase $\tilde{\rho}_G$ and its derivative equal to zero. In the center of the droplet we know only that

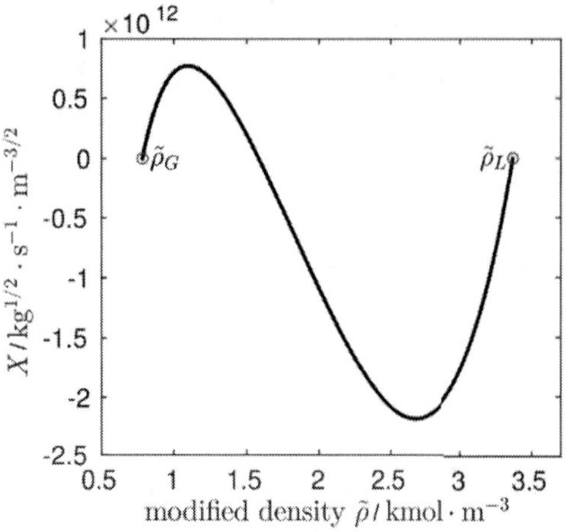

Fig. 1. Function $X(\tilde{\rho})$ computed by solving system of equations (11) for saturation ratio 1.5.

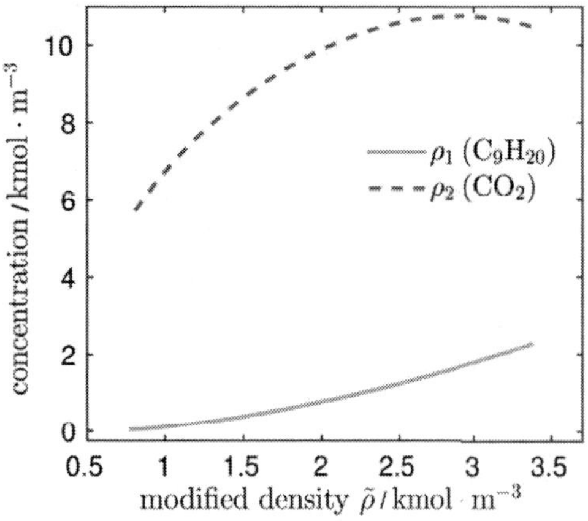

Fig. 2. Dependencies of the concentrations ρ_1 and ρ_2 on the modified density $\tilde{\rho}$ computed by solving system of equations (11) for saturation ratio 1.5

the derivative of the modified density is also zero as a consequence of the spherical symmetry.

The solution method is based on the decomposition of the second order equation into system of two first order equations. The boundary problem is then solved by a shooting method, converting the boundary problem into the initial one. Our shooting parameter is the initial modified liquid density $\tilde{\rho}_0$. The solution of the initial problem is computed with solving method ODE45, which is Runge-Kutta (4,5) method within MATLAB$^{(R)}$ libraries. As an option we requested to stop the computation either when the computed modified density reaches a negative value or its derivative equals zero.

The next value of shooting parameter is computed using the bisection method. At first the profile is computed for initial shooting parameter. The minimal value $\tilde{\rho}_{\min}$ of the profile is found and applied for the decision of the next shooting parameter $\tilde{\rho}_0$.

We use the robust decision criteria that takes only minimal value of the profile $\tilde{\rho}_{\min}$ into account. If the $\tilde{\rho}_{\min}$ value is below the $\tilde{\rho}_G$ level then lower shooting parameter $\tilde{\rho}_0$ is needed. Similarly, when $\tilde{\rho}_{\min}$ is above the gas density level $\tilde{\rho}_G$, the higher shooting parameter $\tilde{\rho}_0$ is chosen.

The iteration continues until specified precision is reached, or size of the interval is below numeric precision of double number format. When the iteration terminates successfully, the profile found this way is denoted as result profile and shooting parameter as $\tilde{\rho}_{0,\mathrm{res}}$.

5 Results

We implemented the splitting method and computed the density profiles for the system of carbon dioxide CO_2 and n-nonane C_9H_{20}. The phase equilibrium system conditions were set to $T = 350$ K and $p = 10$ MPa. The saturation factor alpha for this system was determined to range from $\alpha \in \langle 1.15, 4\rangle$.

The discretization constant for the algebraic solution was set to 1000 points inside interval $\langle \tilde{\rho}_G, \tilde{\rho}_L \rangle$. This option provided enough division points during the iteration and accurately use the interpolation. Performed analysis shows that the discretization interval can be chosen even ten times smaller, still providing accurate results. This fact is illustrated in table 1 and is caused by the efficiency of the piecewise cubic interpolation.

As problematic appear areas outside of the modified density interval. Here, extrapolation is used which is sensitive to errors in interpolation coefficients. However, extrapolation is only used for initial iteration or for unphysical profiles which might occur during the iteration. The result profile does not suffer from the extrapolated data.

Table (1) illustrates also the iteration dependency. Rough discretization have the average ratio of iteration per point around 3:1 but as the refinement increases the iteration ratio changes to 2:1. This nicely correspond to section section Algebraic solution where this behaviour was explained. The last column of table represent precision of result profile initial condition computation. The initial condition is our shooting parameter and we expect it to improve as the refinement of discretization improves. The space in values is inserted for clarity reasons at position of first different digit.

Table 1. Comparison of algebraic solution for different number of discretization points. N_{discr} is the number of discretization point, T is required time of computation, N_{it} show how many iterations were performed and $\tilde{\rho}_{0,res}$ is the found result profile initial condition

N_{discr}	T (s)	N_{it}	$\tilde{\rho}_{0,\mathrm{res}}$ (kmol · m^{-3})
2000	63.64	4042	3.31482151387 9655
1000	33.86	2165	3.31482151387 7961
500	19.65	1250	3.31482151 5545418
250	11.69	747	3.31482151 4526288
125	5.91	372	3.3148215 22776450
62	3.01	186	3.314821 671036016

The problem of precision is illustrated by following figure 3 that represents the behaviour of profiles for different initial conditions (shooting parameter) $\tilde{\rho}_0$. The figure depicts five initial conditions. But only two (red circle line and black full line) from figure 3 correspond to physical reality. The cyan square line depicts the convergence of solution to a thermodynamically unstable state that is also a valid solution. The other lines (green cross line and magenta triangle line) represent the non-physical solutions, where extrapolation of data from algebraic solution are used. These profiles quickly escapes the interesting region and are immediately truncated during computation.

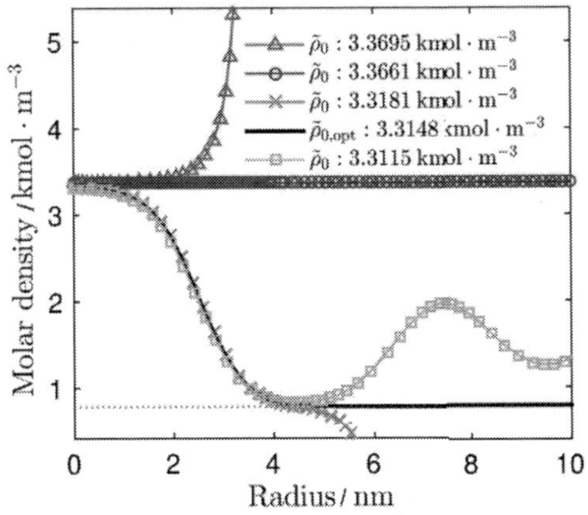

Fig. 3. Modified density profiles for different values of shooting parameter $\tilde{\rho}_0$ with saturation factor $\alpha = 1.5$. Black profile denoted as $\tilde{\rho}_{0,opt}$ is result profile.

The search for optimal starting condition exhibits interesting behavior when approaching the optimal value. This behavior is captured in figure 4. Compared to previous figure, we plot only the dependency of found minimum of the density profile on initial modified density. The black full line from figure 4 is composed of two segments. A linear segment and a "logarithmic" segment. The linear segment is of no importance because it only illustrate the incrementation of the initial condition $\tilde{\rho}_0$. The interesting situation happens for the logarithmic segment. This segment represents the search for optimal value of initial condition that is further used as result.

We know that the optimal profile is the one with minimum value equal to the modified gas density $\tilde{\rho}_G$ and, because of that the optimal profile is found at the intersection with the red dotted line on figure 4 representing $\tilde{\rho}_G$. Before this value we find profiles similar to cyan square line from previous figure 3 that snap to thermodynamically unstable solution. And after magenta circle we see rapid descend into negative region of figure that corresponds to the green cross line from previous figure 3. This is the reason for decision criteria to evaluate the minimal values of profiles. Because of this feature we are able to distinguish the small details, where the profile changes from cyan square line into green cross line. This is the moment we are looking for and even small variation in the initial condition influence the profile dramatically.

The common problem of such behaviour is the precision of double number format. Available range of representable numbers is not sufficient for capturing optimal profile. The

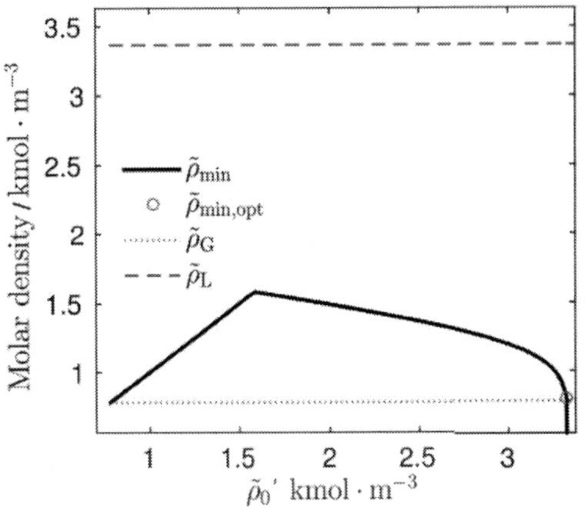

Fig. 4. Minimal value of modified density profile $\tilde{\rho}_{min}$ dependence on values of shooting parameter $\tilde{\rho}_0$ with saturation factor $\alpha = 1.5$. Magenta circle denoted as $\tilde{\rho}_{min,opt}$ is found minimum for result profile.

profile is suboptimal and display only small region where it satisfy the right boundary condition $\tilde{\rho}_{opt} = \tilde{\rho}_G$. The use of other number format would not solve the problem only extend the range of computable profiles. We have not performed the reimplementation into more precise number format because double is the greatest number type natively supported by MATLAB environment. The variable precision that is supported by minority of MATLAB function is also unacceptable due to is poor performance which would cause overall slowdown of computation.

For the reasons given above we chose the initial conditions in the safe region and perform computation there. We computed specified profile for three saturation factors of $\alpha \in \{4, 1.5, 1.15\}$ respectively to illustrate the growth of the droplet with decreasing saturation factor. With the modified density $\tilde{\rho}$ used in our computation and comparison of results we also computed components densities $\tilde{\rho}_1$ and $\tilde{\rho}_2$ that correspond to CO_2 and C_9H_{20} respectively. The boundary conditions were also included.

In our first system depicted in figure 5 we notice that only small droplet is formed. Due to the high saturation factor, the gas density is noticeably higher compared to figures 6 and 7. This figure also illustrates the droplet with no bulk liquid content. The droplet is so small in diameter that only phase interface is present. This fact is denoted by the density profile (black full line) that does not reach the blue dashed line $\tilde{\rho}_L$.

The second system in figure 6 captures a small droplet with nearly no liquid content. This figure corresponds to the saturation factor of $\alpha = 1.5$ and show interesting behavior of carbon dioxide. Carbon dioxide adsorbs at phase interface. It means that its molecules accumulate at the phase interface and density profile than have small peak within the phase interface. This behavior is noticeable also in figure 7.

The last figure shows an example of a large droplet. A significant portion of droplet is bulk liquid. If saturation factor would be lowered further, the droplet would grow until the planar phase interface is formed. This situation would happen at saturation factor $\alpha = 1$ but computationally it happens sooner due the finite machine

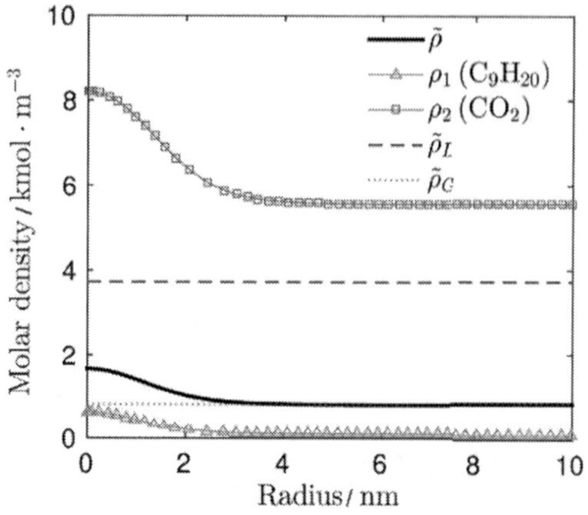

Fig. 5. Modified density profile and partial densities profiles (concentration) for the saturation factor $\alpha = 4$.

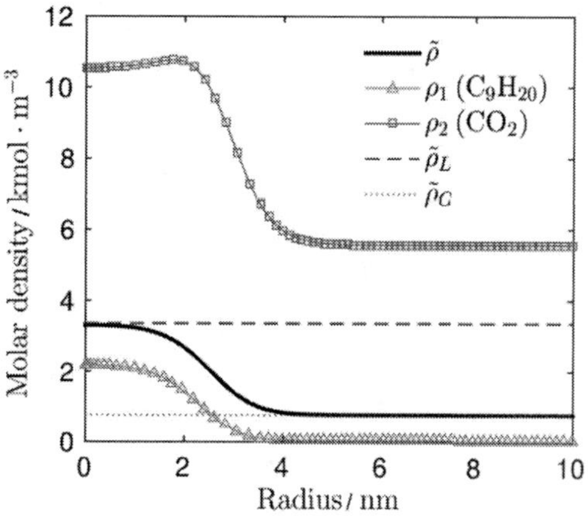

Fig. 6. Modified density profile and partial densities profiles (concentration) for the saturation factor $\alpha = 1.5$.

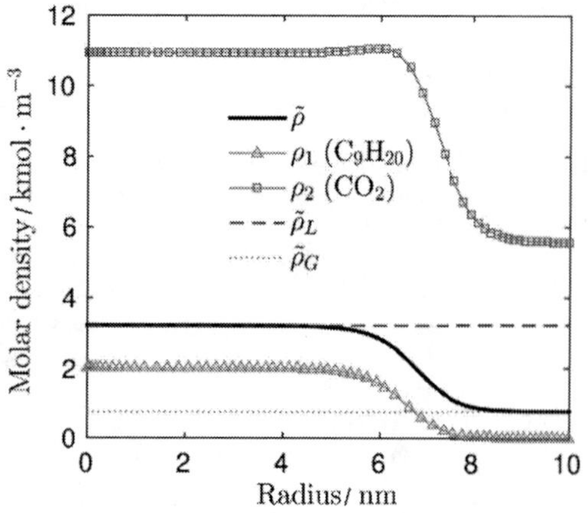

Fig. 7. Modified density profile and partial densities profiles (concentration) for the saturation factor $\alpha = 1.15$.

precision of computation. This error is most significant for bigger droplets with high ratio of pure liquid compared to phase interface. This problem is currently being solved with use of an analytical solution near zero radius.

6 Conclusion

Of the various approaches to the phase interface modeling, the gradient theory provides a good compromise between the physical detail and mathematical tractability. For this theory we developed a new solution method for multi-component mixtures and tested it for a binary system. In this work we presented the main steps of the method and offered information on variable parameters. We have also highlighted the problematic areas of computation which primarily concern very large droplets with almost "top hat" density profile.

Acknowledgement

We gratefully acknowledge the support by Grant No. GAP101/11/1593 of the Czech Science Agency, Grant No. 7F14466 of the Czech-Norwegian Research Programme CZ09 and institutional support RVO:61388998.

References

1. W. Chapman, K. Gubbins, G. Jackson, M. Radosz, Fluid Phase Equilib. **52**, 31 (1989)
2. J. Gross, G. Sadowski, Ind. Eng. Chem. Res. **40**, 1244 (2001)
3. Y. Drossinos, P.G. Kevrekidis, Phys. Rev. E: Stat., Nonlinear, Soft Matter Phys. **67** (2003)
4. J.W. Cahn, J.E. Hilliard, J. Chem. Phys. **28**, 258 (1958)
5. J.W. Cahn, J.E. Hilliard, J. Chem. Phys. **31**, 688 (1959)
6. J. Gross, J. Chem. Phys. **131** (2009)
7. J. Hrubý, D.G. Labetski, M.E.H. van Dongen, J. Chem. Phys. **127**, 164720 (2007)
8. J. Hrubý, B. Planková, V. Vinš, *Corrections to the classical work of formation of critical clusters*, in *Nucl. and Atmos. Aerosols: 19th Int. Conf.* (AIP Publishing LLC, 2013)
9. B. Planková, Master's thesis, CTU FNSPE (2011)

EPJ Web of Conferences **114**, 02012 (2016)
DOI: 10.1051/epjconf/201611402012

Finite volume - space-time discontinuous Galerkin method for the solution of compressible turbulent flow

Jan Česenek[1,a]

[1]Aerospace Research and Test Establishment, Beranových 130, 199 05 Praha - Letňany, Czech Republic

Abstract. In this article we deal with numerical simulation of the non-stationary compressible turbulent flow. Compressible turbulent flow is described by the Reynolds-Averaged Navier-Stokes (RANS) equations. This RANS system is equipped with two-equation k-omega turbulence model. These two systems of equations are solved separately. Discretization of the RANS system is carried out by the space-time discontinuous Galerkin method which is based on piecewise polynomial discontinuous approximation of the sought solution in space and in time. Discretization of the two-equation k-omega turbulence model is carried out by the implicit finite volume method, which is based on piecewise constant approximation of the sought solution. We present some numerical experiments to demonstrate the applicability of the method using own-developed code.

1 Introduction

During the last decade the space-time discontinuous Galerkin finite element method (ST-DG), which is based on piecewise polynomial discontinuous approximations of the sought solution, became very popular in the field of numerical simulation of the fluid flow. This method of higher order was successfully used for simulation of the Navier-Stokes equations ([2],[4],[5]). Unfortunately this method became unstable, when it was used for the whole system of the RANS equations with $k-\omega$ equations. Thus we solve this two systems separately. We use the ST-DG for the discretization of the RANS system of equations and the finite volume method for the equations of the $k-\omega$ turbulence model. As numerical experiments show this is good approach for the solution of this problem.

2 Formulation of the $k-\omega$ turbulence model

We consider compressible turbulent flow in a bounded domain $\Omega \subset R^d$, $d = 2, 3$. We assume that the boundary of Ω consists of three disjoint parts: $\partial\Omega = \Gamma_I \cup \Gamma_O \cup \Gamma_W$, where Γ_I is the inlet, Γ_O is the outlet and Γ_W is the impermeable wall.

The system of the RANS equations can be written in the following form

$$\frac{\partial \mathbf{w}}{\partial t} + \sum_{s=1}^{d} \frac{\partial \mathbf{f}_s(\mathbf{w})}{\partial x_s} = \sum_{s=1}^{d} \frac{\partial \mathbf{R}_s(\mathbf{w}, \nabla \mathbf{w})}{\partial x_s}, \tag{1}$$

where for $s = 1, ..., d$ we have

$$\mathbf{w} = (w_1, ..., w_{d+2})^T = (\rho, \rho v_1, ..., \rho v_d, E)^T \in R^{d+2},$$

$$\begin{aligned}
\mathbf{f}_s(\mathbf{w}) &= (f_{s,1}, ..., f_{s,d+2})^T \\
&= (\rho v_s, \rho v_1 v_s + \delta_{1s}(p + \tfrac{2}{3}\rho k), ..., \\
&\quad \rho v_d v_s + \delta_{ds}(p + \tfrac{2}{3}\rho k), (E + p + \tfrac{2}{3}\rho k) v_s)^T, \\
\mathbf{R}_s(\mathbf{w}, \nabla \mathbf{w}) &= (R_{s,1}, ..., R_{s,d+2})^T \\
&= \left(0, \tau_{s1}^V, ..., \tau_{sd}^V, \sum_{r=1}^{d} \tau_{sr}^V v_r + \left(\frac{c_p \mu_L}{Pr} + \frac{c_p \mu_T}{Pr_T}\right) \frac{\partial \theta}{\partial x_s}\right)^T,
\end{aligned}$$

where

$$\begin{aligned}
\tau_{sr}^V &= -\frac{2}{3}(\mu_L + \mu_T) \operatorname{div} \boldsymbol{v}\, \delta_{sr} + 2(\mu_L + \mu_T)\, d_{sr}(\boldsymbol{v}), \\
d_{sr}(\boldsymbol{v}) &= \frac{1}{2}\left(\frac{\partial v_s}{\partial x_r} + \frac{\partial v_r}{\partial x_s}\right).
\end{aligned}$$

We use the following notation: $\boldsymbol{v} = (v_1, ..., v_d)$ - velocity, ρ - density, p - pressure, θ - absolute temperature, E - total energy, γ - Poisson adiabatic constant, κ - heat conduction coefficient, c_v - specific heat at constant volume, c_p - specific heat at constant pressure, where $c_p = \gamma c_v$, Pr is the laminar Prandtl number, which can be express in the form $Pr = \frac{c_p \mu_L}{\kappa}$, Pr_T is the turbulent Prandtl constant number, μ_T is the eddy-viscosity coefficient, μ_L is the dynamic viscosity coefficient dependent on temperature via Sutherland's formula. The above system is completed by the thermodynamical relations

$$p = (\gamma - 1)\left(E - \frac{1}{2}\rho|\boldsymbol{v}|^2 - \rho k\right), \quad \theta = \frac{1}{c_v}\left(\frac{E}{\rho} - \frac{1}{2}|\boldsymbol{v}|^2 - k\right)$$

and is equipped with the initial condition

$$\mathbf{w}(x, 0) = \mathbf{w}^0(x), \quad x \in \Omega$$

and the following boundary conditions

a) $\rho|_{\Gamma_I} = \rho_D$,

[a] e-mail: jan.cessa@seznam.cz, cesenek@vzlu.cz

b) $\boldsymbol{v}|_{\Gamma_I} = \boldsymbol{v}_D = (v_{D1}, ..., v_{Dd})^T$, (2)

c) $\sum_{s,r=1}^{d} \tau_{sr}^V n_s v_r + \left(\frac{c_p \mu_L}{Pr} + \frac{c_p \mu_T}{Pr_T}\right)\frac{\partial \theta}{\partial n} = 0 \quad \text{on } \Gamma_I,$

a) $\boldsymbol{v}|_{\Gamma_W} = \mathbf{0}$,

b) $\left.\frac{\partial \theta}{\partial n}\right|_{\Gamma_W} = \mathbf{0}$,

a) $\sum_{s=1}^{2} \tau_{sr}^V n_s = 0,\ r = 1, 2, \quad \text{on } \Gamma_O,$ b) $\left.\frac{\partial \theta}{\partial n}\right|_{\Gamma_O} = \mathbf{0}$,

with given data $\mathbf{w}^0, \rho_D, \boldsymbol{v}_D$. It is possible to show that $\mathbf{f}_s(\alpha\mathbf{w}) = \alpha\,\mathbf{f}_s(\mathbf{w})$ for $\alpha > 0$. This property implies that

$$\mathbf{f}_s(\mathbf{w}) = \mathbf{A}_s(\mathbf{w})\mathbf{w}, \quad s = 1, ..., d, \tag{3}$$

where

$$\mathbf{A}_s(\mathbf{w}) = \frac{D\mathbf{f}_s(\mathbf{w})}{D\mathbf{w}}, \quad s = 1, ..., d,$$

are the Jacobi matrices of the mappings $\mathbf{f}_s$. The viscous terms $\mathbf{R}_s(\mathbf{w}, \nabla\mathbf{w})$ can be expressed in the form

$$\mathbf{R}_s(\mathbf{w}, \nabla\mathbf{w}) = \sum_{k=1}^{d} \mathbf{K}_{s,k}(\mathbf{w})\frac{\partial \mathbf{w}}{\partial x_k}, \quad s = 1, 2, \tag{4}$$

where $\mathbf{K}_{s,k}(\mathbf{w}) \in R^{(d+2)\times(d+2)}$ are matrices depending on $\mathbf{w}$.

The aforesaid system of the RANS equations has to be completed by a turbulence model for μ_T:

$$\frac{\partial \tilde{\mathbf{w}}}{\partial t} + \sum_{s=1}^{d} \frac{\partial \tilde{\mathbf{f}}_s(\tilde{\mathbf{w}})}{\partial x_s} = \sum_{s=1}^{d} \frac{\partial \tilde{\mathbf{R}}_s(\tilde{\mathbf{w}}, \nabla\tilde{\mathbf{w}})}{\partial x_s} + \tilde{\mathbf{s}}(\tilde{\mathbf{w}}), \tag{5}$$

where for $s = 1, ..., d$ we have

$$\tilde{\mathbf{w}} = (\tilde{w}_1, \tilde{w}_2)^T = (\rho\omega, \rho k)^T \in R^2,$$
$$\tilde{\mathbf{f}}_s(\tilde{\mathbf{w}}) = (\tilde{f}_{s,1}, \tilde{f}_{s,2})^T = (\rho\omega v_s, \rho k v_s)^T,$$
$$\tilde{\mathbf{R}}_s(\tilde{\mathbf{w}}, \nabla\tilde{\mathbf{w}}) = (R_{s,1}, R_{s,2})^T = \left((\mu_L + \sigma_\omega \mu_T)\frac{\partial \omega}{\partial x_s}, (\mu_L + \sigma_k \mu_T)\frac{\partial k}{\partial x_s}\right)^T,$$
$$\tilde{\mathbf{s}}(\tilde{\mathbf{w}}) = (P_\omega - \beta\rho\omega^2 + C_D, P_k - \beta^*\rho\omega k)^T.$$

Here ω is the turbulence dissipation, k is the turbulence kinetic energy and $\mu_T = \frac{\rho k}{\omega}$ is the eddy viscosity. We can write the production terms as

$$P_k = \sum_{s=1}^{d}\sum_{r=1}^{d} \tau_{sr}^T \frac{\partial v_s}{\partial x_r},$$
$$P_\omega = \alpha_\omega \frac{\omega P_k}{k},$$

where

$$\tau_{sr}^T = -\frac{2}{3}\mu_T \operatorname{div}\boldsymbol{v}\,\delta_{sr} - \frac{2}{3}\rho k\,\delta_{sr} + 2\mu_T\, d_{sr}(\boldsymbol{v}),$$
$$d_{sr}(\boldsymbol{v}) = \frac{1}{2}\left(\frac{\partial v_s}{\partial x_r} + \frac{\partial v_r}{\partial x_s}\right).$$

The cross-diffusion term C_D is defined as

$$C_D = \sigma_D \frac{\rho}{\omega} \max\left\{\sum_{s=1}^{d} \frac{\partial k}{\partial x_s}\frac{\partial \omega}{\partial x_s}, 0\right\}.$$

The coefficients $\beta, \beta^*, \sigma_k, \sigma_\omega, \sigma_D, \alpha_\omega$ are chosen by [9]:

$$\beta^* = 0.09, \quad \beta = \frac{5}{6}\beta^*, \quad \sigma_\omega = 0.5, \quad \sigma_k = \frac{2}{3},$$
$$\kappa = 0.41, \quad \sigma_D = 0.5, \quad \alpha_\omega = \frac{\beta}{\beta^*} - \sigma_\omega \frac{\kappa^2}{\sqrt{\beta^*}}.$$

This system is also equipped with the initial condition

$$\tilde{\mathbf{w}}(x, 0) = \tilde{\mathbf{w}}^0(x), \quad x \in \Omega$$

and the following boundary conditions

a) $\omega|_{\Gamma_I} = \omega_D$, b) $k|_{\Gamma_I} = k_D$, (6)

a) $\omega|_{\Gamma_W} = \omega_{wall}$, b) $k|_{\Gamma_W} = 0$, (7)

a) $\left.\frac{\partial \omega}{\partial n}\right|_{\Gamma_O} = \mathbf{0}$, b) $\left.\frac{\partial k}{\partial n}\right|_{\Gamma_O} = \mathbf{0}$, (8)

with given data $\tilde{\mathbf{w}}^0, \omega_D, k_D, \omega_{wall}$.

3 Space-time discretization of the RANS equations

3.1 Space discretization

By Ω_h we denote a polygonal (d=2) or polyhedral (d=3) approximation of the domain Ω. Let $\mathcal{T}_h$ be a partition of the domain Ω_h into finite number of closed elements with mutually disjoint interiors such that $\overline{\Omega_h} = \bigcup_{K\in\mathcal{T}_h} K$. In 2D problems, we usually choose $K \in \mathcal{T}_h$ as triangles or quadrilaterals. In 3D, $K \in \mathcal{T}_h$ can be, e.g., tetrahedrons, prisms or hexahedrons. By $\mathcal{F}_h$ we denote the system of all faces of all elements $K \in \mathcal{T}_h$. Further, we introduce the set of boundary faces $\mathcal{F}_h^B = \{\Gamma \in \mathcal{F}_h; \Gamma \subset \partial\Omega_h(t)\}$ and the set of inner faces $\mathcal{F}_h^I = \mathcal{F}_h \backslash \mathcal{F}_h^B$. Each $\Gamma \in \mathcal{F}_h$ is associated with a unit normal vector $\boldsymbol{n}_\Gamma$. For $\Gamma \in \mathcal{F}_h^B$ the normal $\boldsymbol{n}_\Gamma$ has the same orientation as the outer normal to $\partial\Omega_h$. For each $\Gamma \in \mathcal{F}_h^I$ there exist two neighbouring elements $K_\Gamma^L, K_\Gamma^R \in \mathcal{T}_h$ such that $\Gamma \subset \partial K_\Gamma^L \cap \partial K_\Gamma^R$. We use the convention that K_Γ^R lies in the direction of $\boldsymbol{n}_\Gamma$ and K_Γ^L lies in the opposite direction to $\boldsymbol{n}_\Gamma$. If $\Gamma \in \mathcal{F}_h^B$, then the element adjacent to Γ will be denoted by K_Γ^L.

Because of the space discretization we define the space of piecewise polynomial functions

$$\mathbf{S}_h^p = (S_h^p)^{d+2}, \quad S_h^p = \{v; v|_K \in P^p(K), \forall K \in \mathcal{T}_h\},$$

where $p > 0$ is an integer and $P^p(K)$ denotes the space of all polynomials on K of degree $\leq p$. A function $\Phi \in \mathbf{S}_h^p$ is, in general, discontinuous on interfaces $\Gamma \in \mathcal{F}_h^I$. By Φ_Γ^L and Φ_Γ^R we denote the values of Φ on Γ considered from the interior and the exterior of K_Γ^L, respectively, and set

$$\langle\Phi\rangle_\Gamma = \frac{1}{2}\left(\Phi_\Gamma^L + \Phi_\Gamma^R\right),$$
$$[\Phi]_\Gamma = \Phi_\Gamma^L - \Phi_\Gamma^R,$$

which denotes average and jump of Φ on Γ.

The discrete problem is derived in the following way: For arbitrary $t \in [0, T]$ we can multiply the system (1) by a test function $\Phi_h \in \mathbf{S}_h^p$, integrate over $K \in \mathcal{T}_h$, apply Green's theorem, sum over all elements $K \in \mathcal{T}_h$, use the concept of the numerical flux and introduce suitable terms mutually vanishing for a regular exact solution. Moreover, we carry out a suitable partial linearization of nonlinear terms.

In order to evaluate the integrals over $\Gamma \in \mathcal{F}_h$ in inviscid term we use the approximation

$$\mathbf{H}(\mathbf{w}_\Gamma^L, \mathbf{w}_\Gamma^R, \boldsymbol{n}_\Gamma) \approx \sum_{s=1}^{d} \mathbf{f}_s(\mathbf{w})(\boldsymbol{n}_\Gamma)_s,$$

where $\mathbf{H}$ is a numerical flux. For the construction of the numerical flux we use the properties (3) of $\mathbf{f}_s$. Let us define the matrix

$$\mathbf{P}(\mathbf{w}, \boldsymbol{n}) := \sum_{s=1}^{d} \mathbf{A}_s(\mathbf{w}) n_s,$$

where $\boldsymbol{n} = (n_1, ..., n_d)$, $n_1^2 + ... + n_d^2 = 1$. Then we have

$$\mathbf{P}(\mathbf{w}, \boldsymbol{n})\mathbf{w} := \sum_{s=1}^{d} \mathbf{f}_s(\mathbf{w}) n_s.$$

It is possible to show that the matrix $\mathbf{P}$ is diagonalizable. It means that there exists a nonsingular matrix $\mathbf{T} = \mathbf{T}(\mathbf{w}, \boldsymbol{n})$ and a diagonal matrix $\Lambda = \Lambda(\mathbf{w}, \boldsymbol{n})$ such that

$$\mathbf{P} = \mathbf{T}\Lambda\mathbf{T}^{-1}, \quad \Lambda = \mathrm{diag}(\lambda_1, ..., \lambda_{d+2}),$$

where $\lambda_i = \lambda_i(\mathbf{w}, \boldsymbol{n})$ are eigenvalues of the matrix $\mathbf{P}$. Then we can define the "positive" and "negative" parts of the matrix $\mathbf{P}$ by

$$\mathbf{P}^{\pm} = \mathbf{T}\Lambda^{\pm}\mathbf{T}^{-1}, \quad \Lambda^{\pm} = \mathrm{diag}(\lambda_1^{\pm}, ..., \lambda_{d+2}^{\pm}),$$

where $\lambda^+ = \max(\lambda, 0)$, $\lambda^- = \min(\lambda, 0)$. Using this concept, we introduce the so-called Vijayasundaram numerical flux

$$\mathbf{H}(\mathbf{w}^L, \mathbf{w}^R, \boldsymbol{n}) = \mathbf{P}^+\left(\frac{\mathbf{w}^L + \mathbf{w}^R}{2}, \boldsymbol{n}\right)\mathbf{w}^L + \mathbf{P}^-\left(\frac{\mathbf{w}^L + \mathbf{w}^R}{2}, \boldsymbol{n}\right)\mathbf{w}^R.$$

This numerical flux has suitable form for a linearization. Now we can define inviscid form in the following way:

$$\begin{aligned}
&b_h(\bar{\mathbf{w}}_h, \mathbf{w}_h, \Phi_h) := \\
&- \sum_{K \in \mathcal{T}_h} \int_K \sum_{s=1}^{d} \mathbf{A}_s(\bar{\mathbf{w}}_h)\mathbf{w}_h \cdot \frac{\partial \Phi_h}{\partial x_s}\, \mathrm{d}\boldsymbol{x} \\
&+ \sum_{\Gamma \in \mathcal{F}_h^I} \int_\Gamma \left(\mathbf{P}^+(\langle \bar{\mathbf{w}}_h \rangle, \boldsymbol{n}_\Gamma)\,\mathbf{w}_h^L + \mathbf{P}^-(\langle \bar{\mathbf{w}}_h \rangle, \boldsymbol{n}_\Gamma)\,\mathbf{w}_h^R\right) \cdot [\Phi_h]\, \mathrm{d}S \\
&+ \sum_{\Gamma \in \mathcal{F}_h^B} \int_\Gamma \left(\mathbf{P}^+(\langle \bar{\mathbf{w}}_h \rangle, \boldsymbol{n}_\Gamma)\,\mathbf{w}_h^L + \mathbf{P}^-(\langle \bar{\mathbf{w}}_h \rangle, \boldsymbol{n}_\Gamma)\,\bar{\mathbf{w}}_h^R\right) \cdot \Phi_h\, \mathrm{d}S
\end{aligned}$$

The boundary state $\bar{\mathbf{w}}_h^R$ is evaluated with the aid of the local linearized Riemann problem described in [8].

For the discretization of the viscous terms we use the property (4) and get the viscous form

$$\begin{aligned}
&a_h(\bar{\mathbf{w}}_h, \mathbf{w}_h, \Phi_h) := \\
&+ \sum_{K \in \mathcal{T}_h} \int_K \sum_{s=1}^{d} \sum_{k=1}^{d} \mathbf{K}_{s,k}(\bar{\mathbf{w}}_h) \frac{\partial \mathbf{w}_h}{\partial x_k} \cdot \frac{\partial \Phi_h}{\partial x_s}\, \mathrm{d}\boldsymbol{x} \\
&- \sum_{\Gamma \in \mathcal{F}_h^I} \int_\Gamma \sum_{s=1}^{d} \left\langle \sum_{k=1}^{d} \mathbf{K}_{s,k}(\bar{\mathbf{w}}_h) \frac{\partial \mathbf{w}_h}{\partial x_k} \right\rangle (\boldsymbol{n}_\Gamma)_s \cdot [\Phi_h]\, \mathrm{d}S \\
&- \sum_{\Gamma \in \mathcal{F}_h^B} \int_\Gamma \sum_{s=1}^{d} \sum_{k=1}^{d} \mathbf{K}_{s,k}(\bar{\mathbf{w}}_h) \frac{\partial \mathbf{w}_h}{\partial x_k} (\boldsymbol{n}_\Gamma)_s \cdot \Phi_h\, \mathrm{d}S \\
&- \Theta \sum_{\Gamma \in \mathcal{F}_h^I} \int_\Gamma \sum_{s=1}^{d} \left\langle \sum_{k=1}^{d} \mathbf{K}_{k,s}^T(\bar{\mathbf{w}}_h) \frac{\partial \Phi_h}{\partial x_k} \right\rangle (\boldsymbol{n}_\Gamma)_s \cdot [\mathbf{w}_h]\, \mathrm{d}S \\
&- \Theta \sum_{\Gamma \in \mathcal{F}_h^B} \int_\Gamma \sum_{s=1}^{d} \sum_{k=1}^{d} \mathbf{K}_{k,s}^T(\bar{\mathbf{w}}_h) \frac{\partial \Phi_h}{\partial x_k} (\boldsymbol{n}_\Gamma)_s \cdot \mathbf{w}_h\, \mathrm{d}S.
\end{aligned}$$

We set $\Theta = 1$ or $\Theta = 0$ or $\Theta = -1$ and get the so-called symmetric version (SIPG) or incomplete version (IIPG) or nonsymetric version (NIPG), respectively, of the discretization of viscous terms.

Further, we define the interior and boundary penalty form J_h^σ and the right-hand side form l_h in the following way:

$$\begin{aligned}
J_h^\sigma(\mathbf{w}_h, \Phi_h) &:= \sum_{\Gamma \in \mathcal{F}_h^I} \int_\Gamma \sigma[\mathbf{w}_h] \cdot [\Phi_h]\, \mathrm{d}S \\
&+ \sum_{\Gamma \in \mathcal{F}_h^B} \int_\Gamma \sigma \mathbf{w}_h \cdot \Phi_h\, \mathrm{d}S, \\
l_h(\bar{\mathbf{w}}_h, \Phi_h) &:= \sum_{\Gamma \in \mathcal{F}_h^B} \int_\Gamma \sigma \mathbf{w}_B \cdot \Phi_h\, \mathrm{d}S \\
&- \Theta \sum_{\Gamma \in \mathcal{F}_h^B} \int_\Gamma \sum_{s=1}^{d} \sum_{k=1}^{d} \mathbf{K}_{k,s}^T(\bar{\mathbf{w}}_h) \frac{\partial \Phi_h}{\partial x_k} (\boldsymbol{n}_\Gamma)_s \cdot \mathbf{w}_B\, \mathrm{d}S,
\end{aligned}$$

where $\sigma|_\Gamma = \frac{C_W}{Re\, d(\Gamma)}$, where $d(\Gamma)$ is the diameter of $\Gamma \in \mathcal{F}_h$, Re is the Reynolds number and $C_W > 0$ is a suitable sufficiently large constant. The boundary state $\mathbf{w}_B$ is defined on the basis of the Dirichlet boundary conditions and extrapolation:

$$\begin{aligned}
\mathbf{w}_B &= (\rho_D, \rho_D v_{D1}, ..., \\
&\qquad \rho_D v_{Dd}, c_v \rho_D \theta_\Gamma^L + \frac{1}{2}\rho_D |\boldsymbol{v}_D|^2 + \rho_D k_D) \quad \text{on } \Gamma_I, \\
\mathbf{w}_B &= \mathbf{w}_\Gamma^L \quad \text{on } \Gamma_O, \\
\mathbf{w}_B &= (\rho_\Gamma^L, 0, ..., 0, c_v \rho_\Gamma^L \theta_\Gamma^L + \frac{1}{2}\rho_\Gamma^L |z_D|^2) \quad \text{on } \Gamma_W.
\end{aligned}$$

In the vicinity of discontinuities or steep gradients nonphysical oscillations can appear in the approximate solution. In order to overcome this difficulty we employ artificial viscosity form, see [7]. They are based on the discontinuity indicator

$$g(K) := \frac{1}{d(K)\,|K|^{3/4}} \int_{\partial K} [\bar{\rho}_h]^2 dS, \quad K \in \mathcal{T}_h,$$

where $[\bar{\rho}_h]$ is the jump of the function $\bar{\rho}_h$ (= the first component of the vector function $\bar{\mathbf{w}}_h$) on the boundary ∂K, $d(K)$ denotes the diameter of K and $|K|$ denotes the area of the element K. Then we define the discrete discontinuity indicator

$$\begin{aligned}
G(K) &:= 0 \quad \text{if } g(K) < 1, \\
G(K) &:= 1 \quad \text{if } g(K) \geq 1, \ K \in \mathcal{T}_h,
\end{aligned}$$

and the artificial viscosity forms

$$\tilde{\beta}_h(\bar{\mathbf{w}}_h, \mathbf{w}_h, \Phi_h) :=$$
$$\nu_1 \sum_{K\in\mathcal{T}_h} d(K)G(K) \int_K \nabla \mathbf{w}_h \cdot \nabla \Phi_h \,\mathrm{d}x,$$
$$\tilde{J}_h(\bar{\mathbf{w}}_h, \mathbf{w}_h, \Phi_h) :=$$
$$\nu_2 \sum_{\Gamma\in\mathcal{F}_h^I} \frac{1}{2}\left(G(K_\Gamma^L) + G(K_\Gamma^R)\right) \int_\Gamma [\mathbf{w}_h]\cdot[\Phi_h]\,\mathrm{d}S,$$

with constants ν_1 and ν_2.

All these forms are linear with respect to $\mathbf{w}_h$ and nonlinear with respect to $\bar{\mathbf{w}}_h$.

Finally, we set

$$(\phi, \psi) = \int_{\Omega_h} \phi\psi \,\mathrm{d}x. \tag{9}$$

3.2 Full space-time DGM discretization

Let $0 = t_0 < t_1 < ... < t_M = T$ be a partition of the interval $[0, T]$ and denote $I_m = (t_{m-1}, t_m)$, $\tau_m = t_m - t_{m-1}$ for $m = 1, ..., M$. We define the space $\mathbf{S}_{h,\tau}^{p,q} = (S_{h,\tau}^{p,q})^{d+2}$, where

$$S_{h,\tau}^{p,q} := \left\{\phi\,;\,\phi|_{I_m} = \sum_{i=0}^{q} \zeta_i\phi_i, \text{ where } \phi_i \in S_h^p,\ \zeta_i \in P^q(I_m)\right\}$$

with integers $p, q \geq 1$. $P^q(I_m)$ denotes the space of all polynomials in t on I_m of degree $\leq q$. Moreover for $\Phi \in \mathbf{S}_{h,\tau}^{p,q}$ we introduce the following notation:

$$\Phi_m^\pm = \Phi(t_m^\pm) = \lim_{t\to t_m\pm} \Phi(t),$$
$$\{\Phi\}_m = \Phi_m^+ - \Phi_m^-.$$

Approximate solution $\mathbf{w}_{h\tau}$ of the problem will be sought in the space $\mathbf{S}_{h,\tau}^{p,q}$. Since the functions of this space are in general discontinuous in time, we ensure the connection between I_{m-1} and I_m by the penalty term in time

$$\left(\{\mathbf{w}_{h\tau}\}_{m-1}, \Phi_{h\tau}(t_{m-1}^+)\right).$$

The initial state $\mathbf{w}_{h\tau}$ is included by the $L^2(\Omega_h)$-projection of $\mathbf{w}^0$ on $\mathbf{S}_h^p$:

$$\left(\mathbf{w}_{h\tau}(t_0^+), \Phi_{h\tau}(t_0^+)\right) = \left(\mathbf{w}^0, \Phi_{h\tau}(t_0^+)\right) \quad \forall \Phi_{h\tau} \in \mathbf{S}_{h,\tau}^{p,q}.$$

Now we introduce a suitable linearization. We can use two possibilities.

1) We put $\bar{\mathbf{w}}_{h\tau}(t) := \mathbf{w}_h(t_{m-1}^-)$ for $t \in I_m$.

2) We prolong the solution from the time interval I_{m-1} to the time interval I_m.

We say that a function $\mathbf{w}_{h\tau} \in \mathbf{S}_{h,\tau}^{p,q}$ is the approximate solution of the problem (1) obtained by the ST-DG method, if it satisfies the following conditions

$$\sum_{m=1}^{M} \int_{I_m} \left(\frac{\partial \mathbf{w}_{h\tau}}{\partial t}, \Phi_{h\tau}\right) \mathrm{d}t$$
$$+ \sum_{m=1}^{M} \int_{I_m} \left(a_h(\bar{\mathbf{w}}_{h\tau}, \mathbf{w}_{h\tau}, \Phi_{h\tau}) + b_h(\bar{\mathbf{w}}_{h\tau}, \mathbf{w}_{h\tau}, \Phi_{h\tau})\right) \mathrm{d}t$$
$$+ \sum_{m=1}^{M} \int_{I_m} \left(\tilde{\beta}_h(\bar{\mathbf{w}}_{h\tau}, \mathbf{w}_{h\tau}, \Phi_{h\tau}) + \tilde{J}_h(\bar{\mathbf{w}}_{h\tau}, \mathbf{w}_{h\tau}, \Phi_{h\tau})\right) \mathrm{d}t$$
$$+ \sum_{m=1}^{M} \int_{I_m} J_h^\sigma(\mathbf{w}_{h\tau}, \Phi_{h\tau})\,\mathrm{d}t$$
$$+ \sum_{m=2}^{M} \left(\{\mathbf{w}_{h\tau}\}_{m-1}, \Phi_{h\tau}(t_{m-1}^+)\right) + \left(\mathbf{w}_{h\tau}(t_0^+), \Phi_{h\tau}(t_0^+)\right)$$
$$= \sum_{m=1}^{M} \int_{I_m} l_h(\bar{\mathbf{w}}_{h\tau}, \Phi_{h\tau})\,\mathrm{d}t + \left(\mathbf{w}^0, \Phi_{h\tau}(t_0^+)\right) \quad \forall \Phi_{h\tau} \in \mathbf{S}_{h,\tau}^{p,q}.$$

4 Finite volume discretization of the two-equation $k - \omega$ turbulence model

Similarly as in Section 3.1 let $\mathcal{T}_h = \{K_i\}_{i\in I}$ be a partition of the domain Ω_h into finite number of closed elements with mutually disjoint interiors such that $\overline{\Omega}_h = \bigcup_{i\in I} K_i$, where $I \subset Z^+ = \{0, 1, 2, 3, ...\}$ is a suitable index set. If two elements have a common face, then we call them neighbours and put $\Gamma_{ij} = \Gamma_{ji} = \partial K_i \cap \partial K_j$. For each $i \in I$ we define the index set $s(i) = \left\{j \in I; K_j \text{ is a neighbour of } K_i\right\}$. The boundary $\partial\Omega_h$ is formed by a finite number of sides of elements K_i adjacent to $\partial\Omega_h$. We denote all these boundary sides by S_j, where $j \in I_b \subset Z^- = \{-1, -2, -3, ...\}$ and set $\gamma(i) = \{j \in I_b; S_j \text{ is a side of } K_i\}, \Gamma_{ij} = S_j$ for $K_i \in \mathcal{T}_h$ such that $S_j \subset \partial K_i, j \in I_b$. For an element K_i, not containing any boundary side S_j, we set $\gamma(i) = \emptyset$. Moreover we define $S(i) = s(i) \cup \gamma(i)$ and $\boldsymbol{n}_{ij} = ((n_{ij})_1, ..., (n_{ij})_d)$ as the unit outer normal to ∂K_i on the side Γ_{ij}. We again consider a partition $0 = t_0 < t_1 < ... < t_M = T$ of the interval $[0, T]$ and denote $I_m = (t_{m-1}, t_m)$, $\tau_m = t_m - t_{m-1}$ for $m = 1, ..., M$.

Now we can integrate the system (5) over set $K_i \times (t_{m-1}, t_m)$, apply Green's theorem and approximate the integrals with the one-point rule and we get the following implicit finite volume scheme,

$$|K_i|\,(\tilde{\mathbf{w}}_i^m - \tilde{\mathbf{w}}_i^{m-1}) - \tau_m\,|K_i|\,\tilde{\mathbf{s}}(\tilde{\mathbf{w}}_i^m)$$
$$+ \tau_m \sum_{j\in S(i)} \left|\Gamma_{ij}\right| \tilde{\mathbf{H}}(\tilde{\mathbf{w}}_i^m, \tilde{\mathbf{w}}_j^m, \boldsymbol{n}_{ij})$$
$$- \tau_m \sum_{j\in S(i)} \left|\Gamma_{ij}\right| \sum_{s=1}^{d} \tilde{\mathbf{R}}_s(\tilde{\mathbf{w}}_{\Gamma_{ij}}^m, \nabla\tilde{\mathbf{w}}_{\Gamma_{ij}}^m)(n_{ij})_s$$
$$= 0 \qquad \forall i \in I.$$

Here $\tilde{\mathbf{w}}_{\Gamma_{ij}}^m$ resp. $\nabla\tilde{\mathbf{w}}_{\Gamma_{ij}}^m$ denote $\tilde{\mathbf{w}}^m$ resp. $\nabla\tilde{\mathbf{w}}^m$ at the center of the edge Γ_{ij} at time instant t_m. By $\tilde{\mathbf{w}}_i^m$ we shall denote approximate solution of $\tilde{\mathbf{w}}$ on the element K_i at time instant t_m

$$\tilde{\mathbf{w}}_i^m \approx \frac{1}{|K_i|} \int_{K_i} \tilde{\mathbf{w}}(x, t_m)\,\mathrm{d}x. \tag{10}$$

For the numerical flux $\tilde{\mathbf{H}}$ we again use the concept of the Vijayasundaram numerical flux in the following way

$$\tilde{\mathbf{H}}(\tilde{\mathbf{w}}_i^m, \tilde{\mathbf{w}}_j^m, \boldsymbol{n}_{ij}) = (\langle\mathbf{v}\rangle \cdot \boldsymbol{n}_{ij})^+ \tilde{\mathbf{w}}_i^m + (\langle\mathbf{v}\rangle \cdot \boldsymbol{n}_{ij})^- \tilde{\mathbf{w}}_j^m \quad j \in s(i),$$
$$\tilde{\mathbf{H}}(\tilde{\mathbf{w}}_i^m, \tilde{\mathbf{w}}_j^m, \boldsymbol{n}_{ij}) = (\mathbf{v} \cdot \boldsymbol{n}_{ij})^+ \tilde{\mathbf{w}}_i^m + (\mathbf{v} \cdot \boldsymbol{n}_{ij})^- \tilde{\mathbf{w}}_B \quad j \in \gamma(i),$$

where $\tilde{\mathbf{w}}_B$ denotes boundary conditions. One possibility of linearization of $\tilde{\mathbf{R}}_s$ and $\tilde{\mathbf{s}}$ is via Taylor expansion which is used in our case. For more details see [6].

Figure 1. The pressure distribution for the case 3.

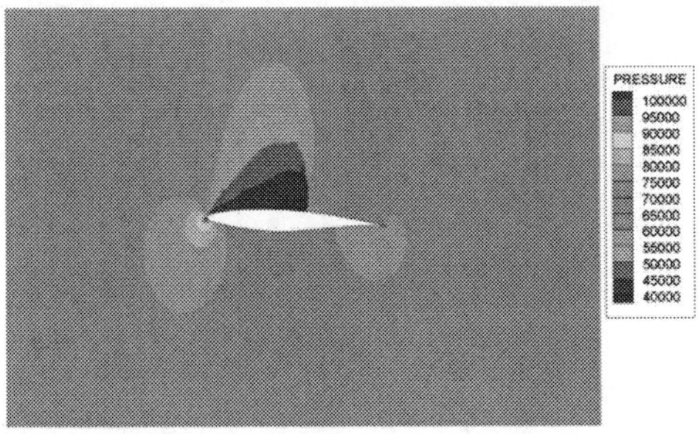

Figure 2. The pressure distribution for the case 9.

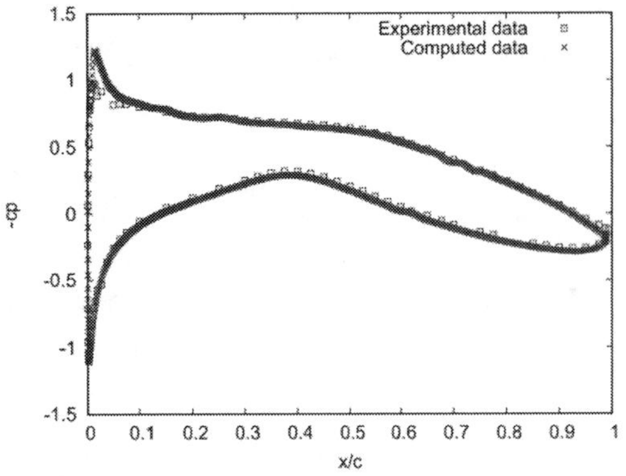

Figure 3. Comparison of the computed pressure coefficient c_p with the experimental data for the case 3.

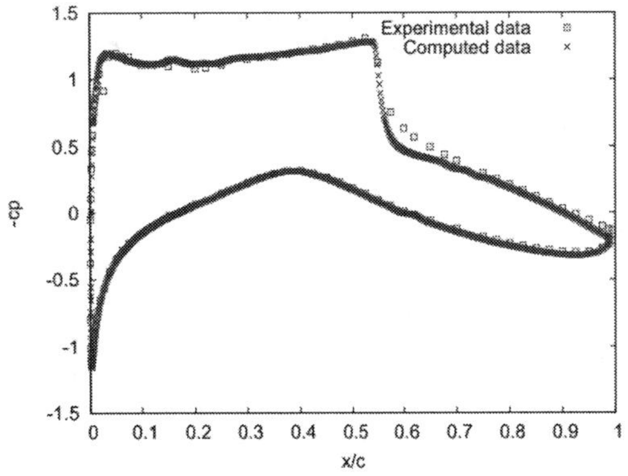

Figure 4. Comparison of the computed pressure coefficient c_p with the experimental data for the case 9.

5 Numerical experiments

In order to demonstrate the applicability of the developed method we shall present two numerical simulations of compressible turbulent flow in 2D. For this purpose we chose two regimes for the profile RAE2822 and compared it with experimental data, which can be found in [1]. First regime is characterized by the far-field Mach number $M_\infty = 0.6$ and the angle of the attack $\alpha = 2.57^o$ ('case 3' in [1]) and second case is characterized by the Mach number $M_\infty = 0.73$ and the angle of the attack $\alpha = 3.19^o$ ('case 9' in [1]). Figures 1 and 2 show the pressure distribution. Comparison of the computed pressure coefficient c_p with the experimental data is shown in Figures 3 and 4.

6 Conclusion

In this paper we dealt with the finite-volume - space-time discontinuous Galerkin method for the numerical solution of compressible turbulent flow. The applicability of the proposed method was demonstrated on the examples, which show, that the presented method is an efficient numerical scheme for the solution of the 2D compressible turbulent flow.

Acknowledgment

This result originated with the support of Ministry of Industry and Trade of the Czech Republic for the long-term strategic development of the research organisation. The authors acknowledge this support.

References

1. Cook P. H., McDonald M. A. and Firmin M. C. P. AGARD Advisory Report 138, (1979)
2. Česenek J. *Discontinuous Galerkin method for solving compressible viscous flow*. Doctoral Thesis, (2011)
3. Česenek J. et al., Appl. Math. Comput., doi:10.1016/1.amc.2011.08.077 (2011)
4. Česenek J., Feistauer M., Kosik A. ZAMM – Z. Angew. Math. Mech. **93**, 6 – 7, 387 – 402 (2013)
5. Česenek J. *Experimental fluid mechanics*,142-147 (2013)
6. Česenek J. *Combined finite volume - space-time discontinuous Galerkin method for the 2D compressible turbulent flow*. Technical report R-6154, VZLÚ, Beranových 130, Prague (2014)
7. Dolejší V., Feistauer M. *Discontinuous Galerkin Method*. Springer, Heidelberg (2015)
8. Feistauer M., Felcman J., and Straškraba I. *Mathematical and Computational Methods for Compressible Flow*. Clarendon Press, Oxford (2003)
9. Kok C.J. AIAA Journal, Vol. **38**. No. 7., (2000)
10. Sváček P., Feistauer M., Horáček J. J. Fluids Struc., **23**, 391-411 (2007)

EPJ Web of Conferences **114**, 02013 (2016)
DOI: 10.1051/epjconf/201611402013

Parametric model of ventilators simulated in OpenFOAM and Elmer

Václav Čibera[1,a], Richard Matas[1] and Jan Sedláček[1]

[1]*New Technologies – Research Centre, University of West Bohemia, Univerzitni 8, 306 14 Pilsen, Czech Republic*

Abstract. The main goal of presented work was to develop parametric model of a ventilator for CFD and structural analysis. The whole model was designed and scripted in freely available open source programmes in particular in OpenFOAM and Elmer. The main script, which runs or generates other scripts and further control the course of simulation, was written in bash scripting language in Linux environment. Further, the scripts needed for a mesh generation and running of a simulation were prepared using m4 word pre-processor. The use of m4 allowed comfortable set up of the higher amount of scripts. Consequently, the mesh was generated for fluid and solid part of the ventilator within OpenFOAM. Although OpenFOAM offers also a few tools for structural analysis, the mesh of solid parts was transferred into Elmer mesh format with the aim to perform structural analysis in this software. This submitted paper deals namely with part concerning fluid flow through parametrized geometry with different initial conditions. As an example, two simulations were conducted for the same geometric parameters and mesh but for different angular velocity of ventilator rotation.

1 Introduction

The work presented in this paper was predominantly focused on a design and simulation of ventilators. The main aim was to develop the tool as universal as possible to design shape and examine the working effectiveness of ventilators at the scale range starting at centimetres scale up to scale of metres. The part of the work included in the text bellow was mainly oriented on the part concerning the properties of fluid flow through the ventilator geometry. Further, the mesh preparation of the solid parts was considered and prepared for structural analysis.

Apart from the main aim of work, i.e. simulation of ventilators, the second main aim was to test and develop the simulation scripts within non-commercial open source software packages. Accordingly, all of the results presented here were achieved with freely available software. In particular, the mesh generation and simulations of fluid flow properties was conducted within OpenFOAM. The mesh of solid part was prepared for structural analysis in Elmer.

Because the ventilators are considered as rotating objects that rotate around their centres, the problem was simplified namely on the account of computing time. In this sense, only repeated segment of ventilator was considered and solved by scripts. Beside this, parallel computation is nowadays a must and matter of course, therefore almost every processes was parallelized. This was achieved namely due to wide range of possibilities in OpenFOAM and Elmer. Last but not least, for a post-processing was used program ParaView, which is also freely available.

[a] Corresponding author: vacibera@ntc.zcu.cz

The work and the scripts are still under development, but the simulations already work sufficiently for steady – states regardless of the values of input parameters.

The paper is organised as follows. After an introduction section, the methods section follows. In this section, the used software and other tools which were used are introduced. In this part of the text are also presented some effects of the various input parameters. This section contains scripts description as well. But according to the scope of this paper, this description is very simplified. Following section contains results of simulations. The last section is conclusion discussing the summary results.

2 Methods

The whole process of solution is a sequence of particulars commands arranged in the main bash script within Linux environment.

2.1 Preparation of meshes and script files

2.1.1 Generation of script files

First of all, it was necessary to spread out input parameters thorough the higher amount of predefined scripts. This was achieved by using m4 word pre-processor. Model inputs within m4 files are then: inlet inner radius, inlet outer radius, inlet height, rotor height, outlet length, blade length, blade width, width of solid part. The meaning of the parameters might be clear from the figures 1 and 2. Additional parameters might be set for the quality of mesh.

The coordinates of the mesh vertices were computed in Octave (also freely available software), when it was called within m4 word pre-processor. The meshes of the fluid part and solid part could be as well generated separately within particular tools of OpenFOAM and Elmer. Separately generated meshes might also avoid the problems with conversion between meshes format from OpenFOAM to Elmer. This might also lead to the more comfortable setup of initial and boundary conditions in Elmer. But since these two meshes were generated within one meshing tool – blockMesh, it satisfies that the nodal points at the common patches (faces) between solid and fluid parts have exactly the same coordinates. This leads to the simplification of transferring for example the data of the pressure on the walls obtained from fluid flow simulation to the structural simulation of the solid part.

Main bash script then contains sequence of commands something like: m4 files.m4, blockMesh –case fluid, createPatch, checkMesh, decomposePar, mpirun –np X simpleFoam – parallel, blockMesh – case solid, createPatch, foamMeshToElmer, ElmerSolver case.sif.

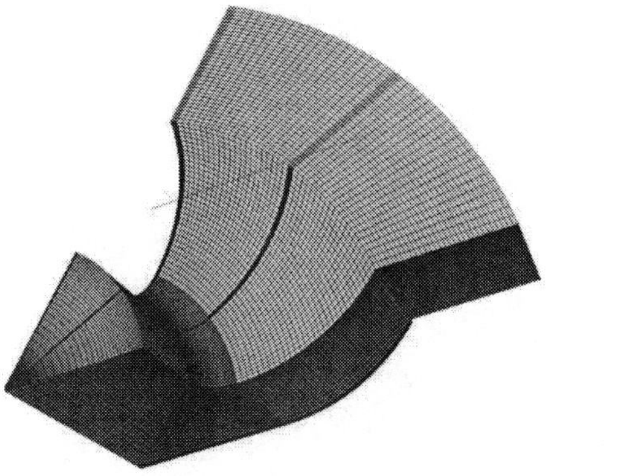

Figure 1. Mesh of the ventilator's fluid and solid parts loaded within ParaView. Depicted mesh represents a ventilator consisting of six main blades.

Figure 2. Mesh of the ventilator's solid part loaded within ParaView. The picture shows only rotor of solid part incorporated within fluid part.

2.2 Examples of the geometric parameters effects

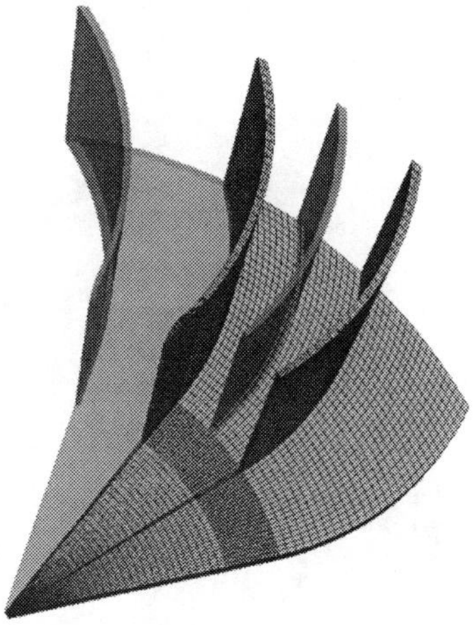

Figure 3. Effect of the number of blades in ventilator. The blue part depicts the ventilator with 9 main blades. The grey part depicts the ventilator with 6 main blades.

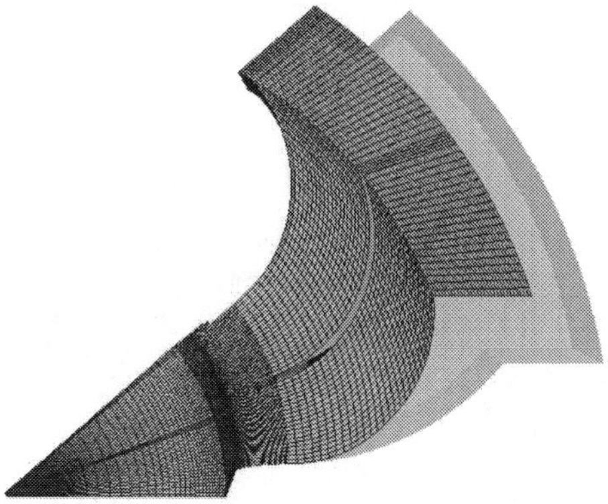

Figure 4. Effect of the blade´s radius change. The blades in blue meshed part have radius of 0.03 m and the radius of blades in grey area was set to 0.06 m.

2.3 Solution in OpenFOAM

Regarding to the use of the OpenFOAM, two "versions" can be found. This might be confusing. Namely it is OpenFOAM as accessible at [3] and secondly OpenFOAM Extend Project as available at [4]. Both of the versions have the most of the tools in common. But owing to the fact, that each version is developed by different groups, some required tools were missing either in first or the second version. Therefore at the end, both of the versions were used. The first noted version was used primarily to prepare meshes and for simulation of fluid parts of ventilator. The later noted version was predominantly used for preparation of the solid part of mesh. Specifically it was used for the conversion of the mesh from OpenFOAM format to Elmer mesh format.

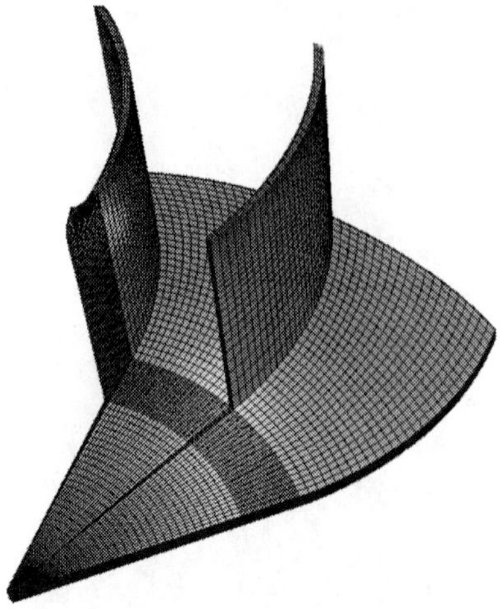

Figure 5. Mesh of the ventilator's solid part loaded within ElmerGUI after transferred from OpenFOAM mesh format to Elmer mesh format.

3 Results

As an example, two simulations were conducted. Both simulations were computed as the steady-state simulations. In both cases, the same geometric parameters of ventilator were used. Therefore the same mesh was generated. The best results reflecting reality were achieved with k-omega SST turbulence model. The fluid medium was considered as incompressible air. The boundary conditions for pressure were set as absolute atmospheric pressure at the inlet and outlet of ventilator. Consequently, flow velocity field and pressure field within ventilator geometry was computed.

3.1 Results for simulation for 1000 and 2000 rpm

The results as presented here were conducted for 1000 and 2000 rpm of ventilator. The tasks of steady-state solution converged within 3000 iteration to the acceptable values of residuals as well as for various rpm.

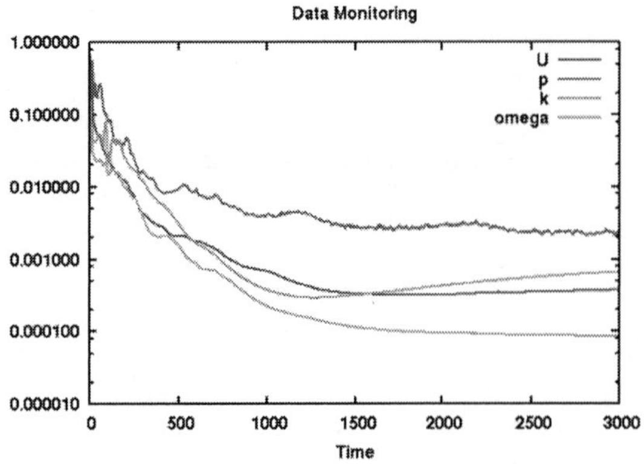

Figure 6. Convergence of simulation residuals. The picture is output from tool foamMonitor. The x-label Time means here in a fact a number of iteration (steady-state solution).

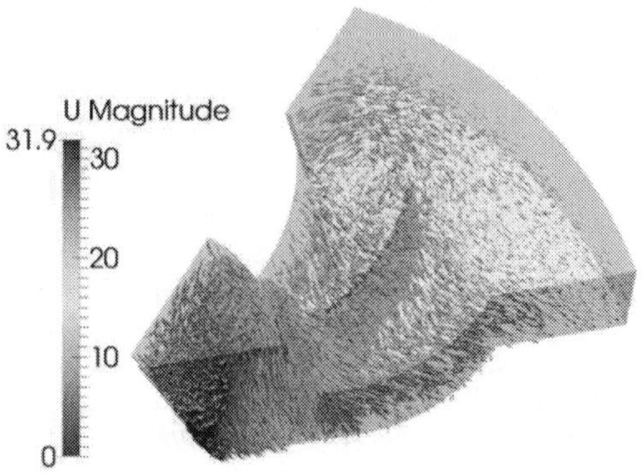

Figure 7. Velocity field obtained for the simulation of the 1000 rpm.

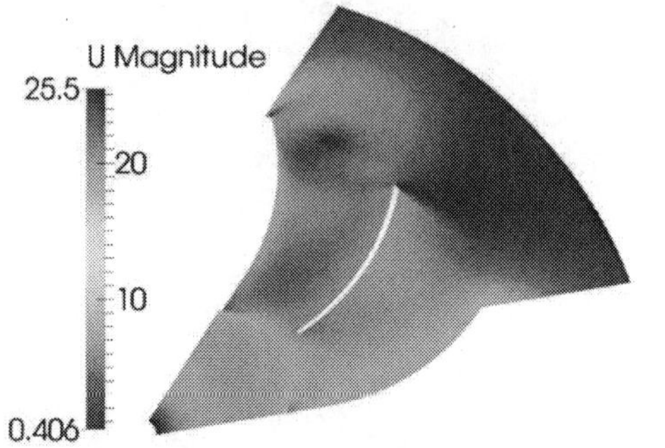

Figure 8. Velocity fields in the centre plane of the rotor achieved by 1000 rpm.

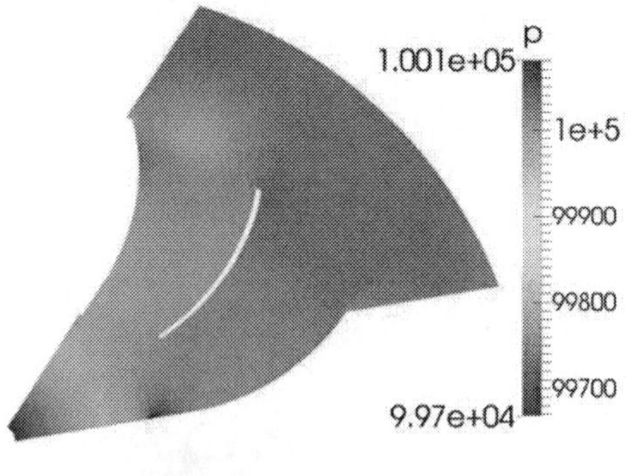

Figure 9. Absolute pressure in the centre plane of the rotor achieved by 1000 rpm.

The range of the velocity magnitudes as depicted in figure 7 does not represent the maximal value in the ventilator volume since the velocity vectors were generated randomly. Therefore the figure 7 is predominantly for illustration of the direction of the velocity fields. Hence this picture can be hard compared with the same kind of figure 10 for 2000 rpm in absolute values.

The values that can be compared on account of the change of the angular velocity of ventilator are depicted in the figures 8, 9 and 11, 12. In the figures 8 and 9 are shown velocity fields in the cross-section through the rotor. The cross-section is situated exactly in the middle of rotor height. As can be seen in these pictures, for the higher rpm of ventilator, the higher velocities were achieved which is theoretically in concordance with reality. But for the more accurate characteristic of ventilator, more states should be computed. Consequently, pressure fields were not changed significantly since the simulation supposes incompressible flow with atmospheric pressure at the outlet as well as inlet of ventilator.

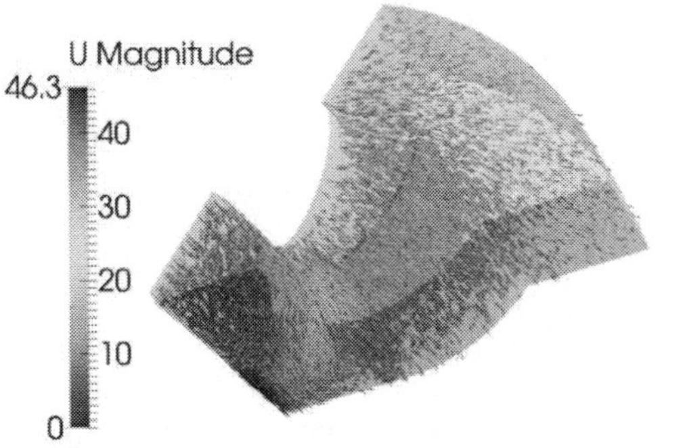

Figure 10. Velocity field obtained for the simulation of the 2000 rpm.

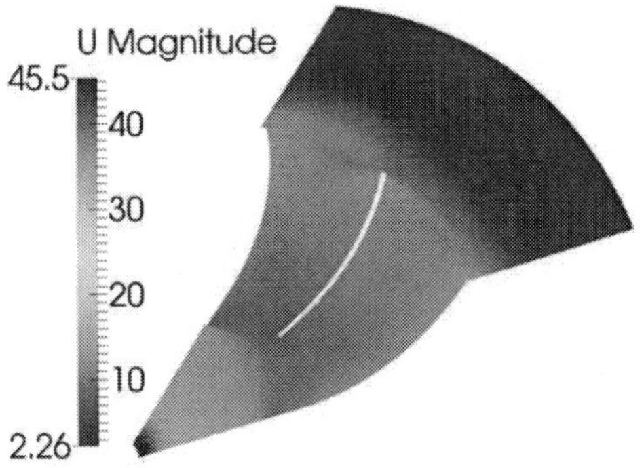

Figure 11. Velocity fields in the centre plane of the rotor achieved by 2000 rpm.

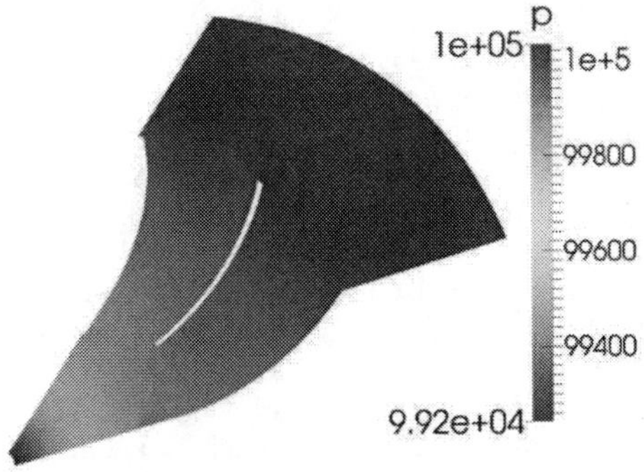

Figure 12. Absolute pressure in the centre plane of the rotor achieved by 2000 rpm.

4 Conclusions

Before all, it is worthy to notice that in comparison to commercial software, the pre-processing in open source programs as in OpenFOAM or Elmer might take longer time from various reason such as lack of documentation or lack of graphical user interfaces. This might be disadvantage namely in the cases, were it is demanded only a small number of simulations for small number of parameters variations.

The advantage of the use of the mentioned might be found in the cases as the solution introduced here. The reason for this is that since this software is open-source, new tool can be comfortably added, which facilitates scripting of the more extensive project. This way the solved project can be more comfortably prepared on account of the specific requests.

The results were compared with results achieved by using ANSYS FLUENT 15. The version that were used for OpenFOAM were 2.4, 2.3 respectively for OpenFOAM Extended Project. The same mesh, turbulent models etc. were used in both CFD solvers. The only parameters that were not set equally were the numerical solvers. Comparing the results of maximal velocities, the differences in absolute values were no higher than 3%.

Simulations usually converged to required accuracy within 3000 iteration. The results were achieved in the range of minutes for small meshes and in the range of tens of minutes.

According to the generation of the scripts by m4 word pre-processor, a little bit surprising was the slowness of generation. It was believed that it was caused by calling of Octave to compute the parameters of the mesh.

By using developed scripts, optimal working regime and characteristics properties of the ventilator with particular geometry can be found. This way, the various ventilators can be tune up for an ideal angular velocity, working regimes and manufacturing process.

The scripts are still being developed. The next step in the development is to include scripts, which would automatically compute modal analysis of solid parts. Last but not least, further the codes could be optimized for transient solution.

Acknowledgement

The work was supported by internal grant at University of West Bohemia: SGS-2014-070: "Zvyšování účinnosti, spolehlivosti a životnosti energetických strojů a zařízení 3".

References

1. J.H. Ferziger, M. Perić, Computationla Methods for Fluid Dynamics, third, rev. edition. Springer 2002.
2. C. Hirsch, Numerical computation of internal and external flows, volume 1, John Wiley and Sons, 1997
3. www.openfoam.com
4. http://www.extend-project.de/

EPJ Web of Conferences **114**, 02014 (2016)
DOI: 10.1051/epjconf/201611402014

Experimental and numerical research of synthetic jet array

Petra Dančová[1,2,a], Jan Novosád [1], Tomáš Vít [1] and Zdeněk Trávníček [2]

[1]Technical University of Liberec, Studentská 2, 46117 Liberec 1, Czech Republic
[2]Institute of Thermomechanics, CAS, v.v.i., Dolejškova 5, 18000 Praha, Czech Republic

Abstract. This paper describes the additional research of the synthetic jet array in a channel flow and continues the paper of authors Dančová, Trávníček and Vít, [1]. Numerical simulations support the experiments from [1] and the influence of the new slope of the synthetic jet orifices is studied here. This research will be used for preparation of the experiments with inclined orifices of the synthetic jet array.

1 Introduction

This work continues a larger group of experiments focused on the research of the synthetic jet (SJ) array interacting with a laminar channel flow [1 and 2]. This arrangement can be used in micro-scale application (e.g. a cooling of micro-electronics). In micro dimensions the flow is usually laminar with very small Reynold numbers. Therefore, the SJ actuator or the SJ array can be successfully used to disturb a laminar flow in these conditions.

Characteristics, parameterization and application of synthetic jet are described e.g. in [3 – 5]. Briefly described, SJ is generated by the periodical motion of an actuator oscillating membrane. In first half of the period of the membrane movement, the fluid is sucked into the sealed actuator cavity through the orifice. In the second half of the membrane movement, this fluid is extruded from the actuator cavity. The vortex rings are formed at the lip of the orifice during the fluid extrusion. These rings should move high enough from the orifice to prevent suction back into the cavity during the suction phase of the time period.

Based on the experimental results from [1], the adjustment for the numerical simulation in sw Ansys Fluent was prepared and tested. This adjustment will be used for new experiments preparation with inclined orifices of the synthetic jet array.

2 Experimental setup and method

2.1 Synthetic jet array experimental setup

The SJ array is made from four SJ actuators placed across the channel cross section. The actuator is made from Plexiglas and consists of a sealed cavity equipped with an emitting orifice and with a periodically moving piezoelectric membrane KINGSTATE KPS-100. Diameter of an actuator cavity is 36 mm; the cavity height is 5 mm.

For our research two kind of geometry for SJ actuators orifices were designed. Firstly, the orifices were placed upwards along the cavities axes and have diameter and height of 3 mm, 5 mm respectively (see figure 1). In the second geometry type, the orifices were inclined of 45 degrees from the SJ actuator axes (figure 2).

SJ array is placed on the Plexiglas with dimensions: length (length with start) x width x thickness = (860 (880) × 200 × 10) mm. The wall inlet was carefully rounded, and the SJ array was placed 520 mm from the beginning. The wall was put into the circulation water channel, which is described in [1 and 2].

2.2 Experimental method - PIV

The particle image velocimetry (PIV) method was used for the experiments. In this a non-contact laser optical method for research of flow, turbulence, microfluidics, and many others processes, where the measurement fluid has to be transparent enough. In the PIV, the velocity vectors are derived from sub-sections of the target area of the particle-seeded flow by measuring the movement of particles between two light pulses. For more description of this method see [6-9].

The settings of the PIV system was set the same as in [1]: a unit "Timer box" and DynamicStudio v.2.30 (DANTEC) software were used. All the control boards of the lasers, cameras, and timers of the input and output signals were placed directly into the main computer. DynamicStudio is used as a control and processing program for the unit Timer box. The computer takes the role of a measuring center. Data communication between the PC, laser, and cameras runs through the internal protocols of the DynamicStudio control software and are impossible to modify. Starting signals are taken out by means of a National Instruments PCI-6601 counting

[a] Corresponding author: petra.dancova@tul.cz

measuring card into the unit Timer box with eight BNC connector outputs and two inputs.

Signals for the start of every flash lamp and Qswitching of the laser are led through one cable, whose position it is necessary to define – to Timer box to connect the starting trigger signal into two inputs.

Figure 3 shows the principle of basic synchronization. From the SJ signal generator, the TTL signal goes into the PC (Timer box). The laser and camera start with the leading edge of the TTL. The starting point of measurement is changed during the SJ period and, therefore, the whole period is measured.

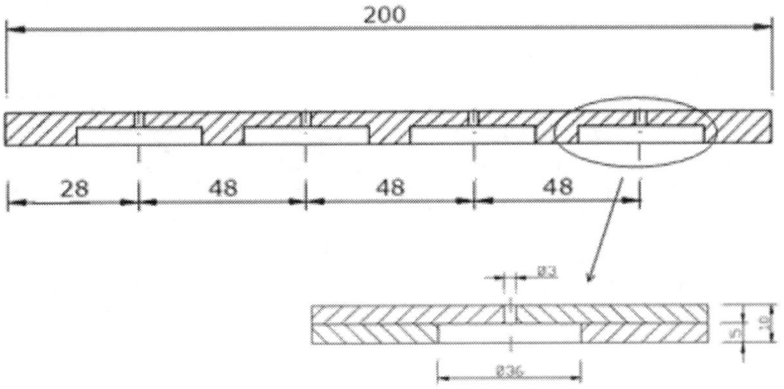

Figure 1. Schematic view of SJ array (straight orifices).

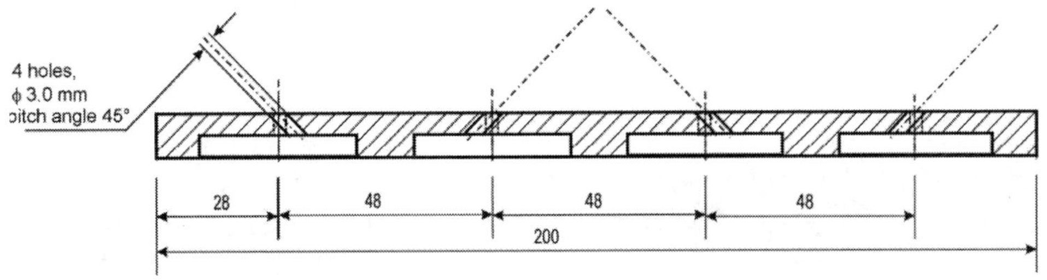

Figure 2. Schematic view of SJ array (inclined orifices).

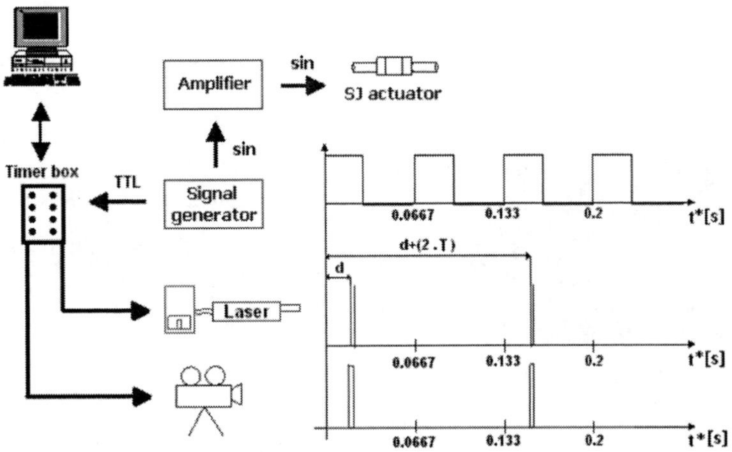

Figure 3. Schematic of basic synchronization for PIV measurements (T means the period).

The delay inside the electronic synchronization devices does not exceed 3 ms.

Polyamide balls with diameters of 20 mm (DANTEC) were used as the seeding particles. The particles were illuminated by a double pulse laser with a maximum 125 mJ per 10 ns pulse at a repetition rate of 2 × 15 Hz. The typical delay time between two pulses was 5 ms.

The laser beam was expanded into a light sheet of approx. 1 mm thick. The image pairs were acquired using the HiSense MKI camera (DANTEC) with a spatial resolution of 1280 × 1024 pixels.

The resulting vector maps were averaged over 100 PIV records. Velocity vectors were determined by adaptive correlation using interrogation windows of 32 × 32 pixels at a 25% overlap. Data processing was done using DynamicStudio 2.21 commercial software (DANTEC).

3 Numerical modelling

Numerical simulations were realized to promote the results of experiments. Preparation of model and mesh was carried out in Ansys Workbench environment. Ansys Design Modeler was used to create 3D model.

3.1 Model geometry and mesh

The model was based on the geometry of the SJ array. Each model presents only a half of the SJ array, because the symmetry boundary condition was used to simplify the simulations. Close region around the SJ array were modelled. Two models of channel were created as a tested area for both cases – straight and inclined SJ actuators orifices. Final models with basic dimensions are shown in figure 4.

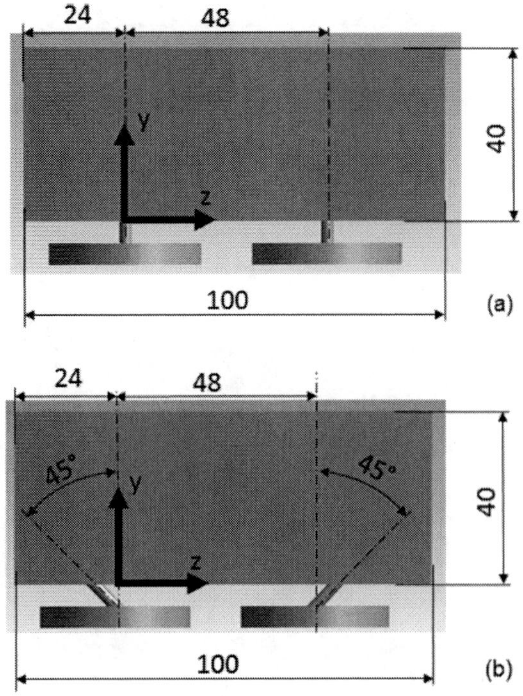

Figure 4. Model of SJ array – front view, (a) straight SJ orifices, (b) inclined SJ orifices.

Computational meshes for each model were generated in ANSYS Meshing. Automatic functions were used to create hybrid mesh with predominant hexahedral elements. Both meshes contain around 950 000 elements.

3.2 Boundary conditions

Boundary conditions are shown in figure 5. On unlabelled faces, the boundary condition „wall“ is applied. On the membrane faces, the BCs „velocity-inlet“ are applied. Oscillation of membranes is modeled with UDF function, which is defined as dependency of velocity on time (following equations and methodology in [2]). For the description of parameters see table 1.

Table 1. Boundary conditions.

Label	BC type	Parameters, description
Inlet	velocity-inlet	Inlet velocity 0.0067 m/s
Outlet	pressure-outlet	Channel outlet
Membranes	velocity-inlet	Velocity profile defined by UDF function
Symmetry	symmetry	Symmetry plane
Wall	wall	Outer walls of channel

3.3 Methodology of results assessment

To obtain velocity fields in close area to the SJ array several cutting planes were defined (for description see table 2) Global coordinate system is defined by triad X, Y, Z (see figure 4). To resolute the actuators, the actuator which is closer to symmetry plane is called „inner“ actuator, the another one is called „outer“ actuator.

Table 2. Cutting planes.

Label	Parameters, description
SJ-cross	Equal to YZ (x = 0) – through all actuators
In-PL	Equal to XY (z = 0) – through inner actuator
Out-PL	Parallel to XY (z = 48 mm) – through outer actuator

Velocity fields are obtained using Ansys FLUENT post-processing tools, especially contours of velocity magnitude and velocity vector fields.

4 Results

4.1 Straight SJ array orifices in a channel flow

Experiments of the straight SJ array with affection of the laminar channel flow are described in [1 and 2]. The delay between the two laser pulses was $t = 5$ms. The working frequency of SJ actuators was $f = 15$ Hz. Experimental results shown in figure 6 were performed in the plane x-y, for $z = (0$ and $48)$ mm according to the geometry used in figure 4. The results plotted in figure 6 are in the form of velocity magnitude contours (in discrete drawing style) in time $t/T = 0.75$ (maximum extrusion velocity). There is visible small bending of a SJ (approx. 10°) and SJ reached the opposite wall ($y = 40$ mm) at the location of approx. $x = 8$ mm. The time-mean velocity through the channel was evaluated approx. $U_C = 0.0067$ m/s. The synthetic jets' time-mean orifice velocity as $U_0 = 0.06$ m/s. These velocities are entered as the boundary conditions for the numerical simulations of this problem.

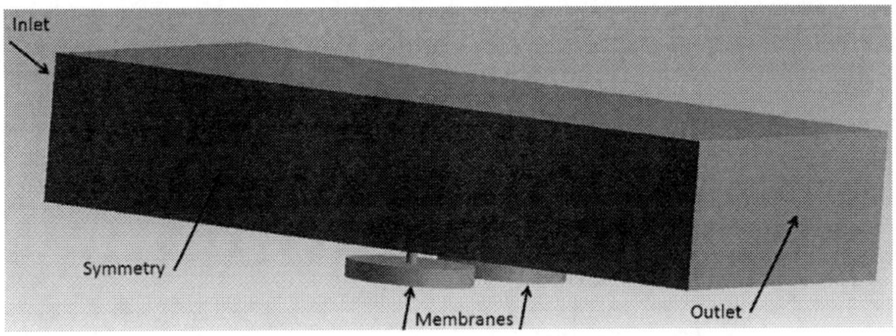

Figure 5. Boundary conditions used in numerical modelling.

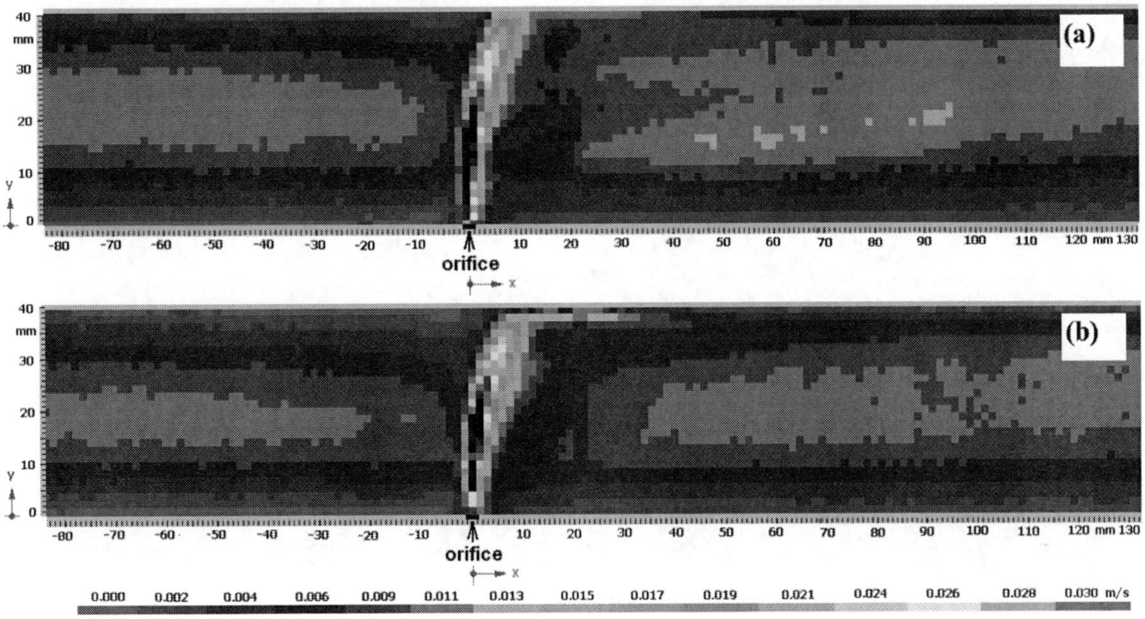

Figure 6. Straight SJ orifices – the experimental results (plotted in discrete drawing style), $t/T = 0.75$, (a) outer SJ actuator, (b) inner SJ actuator.

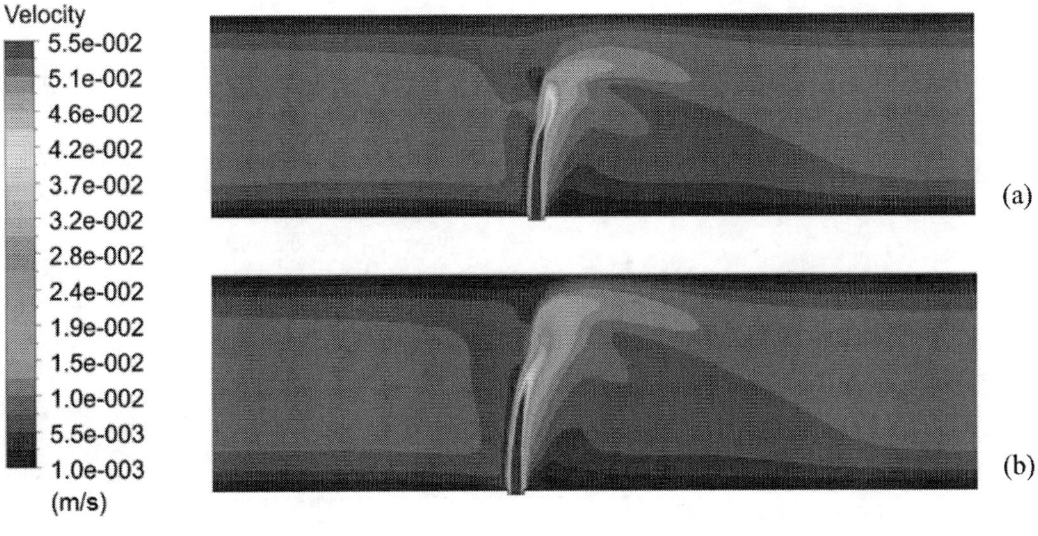

Figure 7. Straight SJ orifices – results of the numerical simulation, $t/T = 0.75$, (a) outer SJ actuator, (b) inner SJ actuator.

Figure 7 shows the numerical simulation of the same situation as is in the experiments: working frequency of SJ actuators was 15 Hz, the outer and inner SJ actuator was in the maximum extrusion time ($t/T = 0.75$). Comparison of the experiment and the numerical simulation in the plane x-y shows very good agreement (see figure 6 and 7).

The water channel arrangement used in [1 and 2] did not enable to measure the situation in the plane x-z. To make visible the behaviour of the SJ array in the channel flow in the plane x-z see figure 8. This figure shows the contours of velocity and velocity vector fields in time $t/T = 0.25$ and 0.75, i.e. maximum suction velocity and maximum extrusion velocity, respectively. Because the velocity in the SJ orifices is one order higher than the velocity in a channel, two scales are used in this figure. Figure 8 also shows detailed view into the SJ actuators. This view was not possible to investigate experimentally.

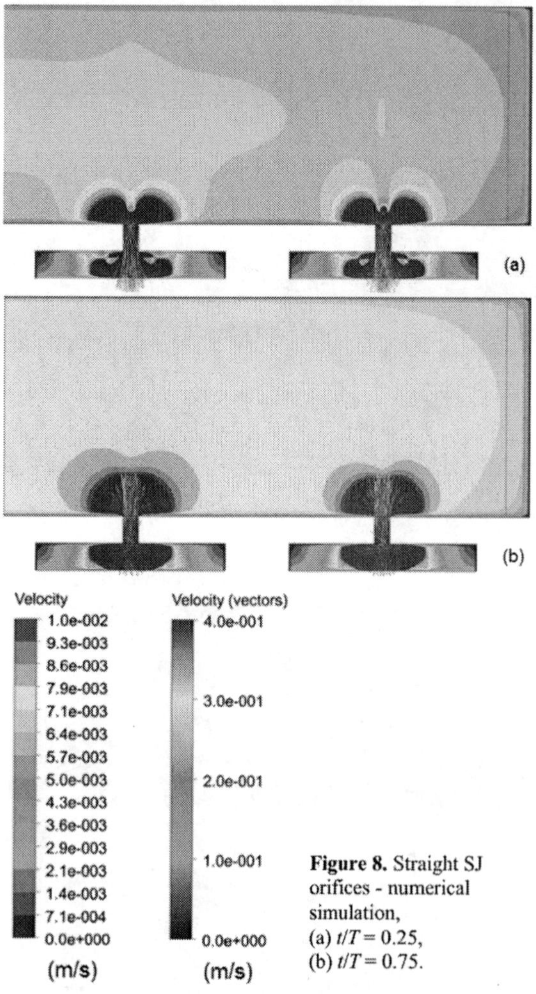

Figure 8. Straight SJ orifices - numerical simulation, (a) $t/T = 0.25$, (b) $t/T = 0.75$.

4.2 Inclined SJ array orifices in a channel flow

Because the results from the numerical simulation of the straight SJ orifices are in very good agreement with the experiment, the same adjustment for numerical simulation for inclined SJs was used.

Figure 9 shows the affection of inclined SJs with the channel flow in time of maximum suction and extrusion velocity ($t/T = 0.25$ and 0.75) in plane x-z. The SJ actuators working frequency was 15 Hz. Figure 9 also shows velocity distribution inside the SJ actuator.

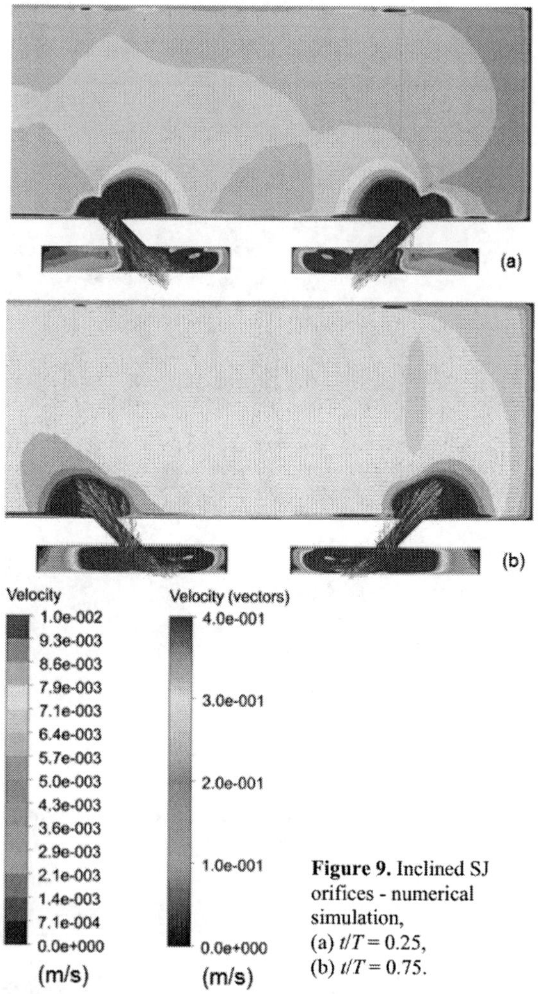

Figure 9. Inclined SJ orifices - numerical simulation, (a) $t/T = 0.25$, (b) $t/T = 0.75$.

For the experimental investigation - to set the laser sheet in a right position, the figure 10 will be used. This figure makes visible what is happened in a symmetry plane (a), how the inner and outer actuator influence a channel flow in the plane x-z (b), (c), and also shows behaviour of a channel flow between the inner and outer SJ actuator (d). The SJ orifices were plotted in time of maximum extrusion ($t/T = 0.75$).

5 Discussions

This paper brings results of a research of a channel flow influenced with a SJ array. According to the experiments from [1 and 2], the numerical simulations were carried out. Experimental data and results from [1 and 2] were used for adjustment of geometry, numerical model and boundary conditions. After a successful comparison, this model was used for next step of our research – for inclined SJ actuator orifices investigation and these results will be used to prepare the experiments.

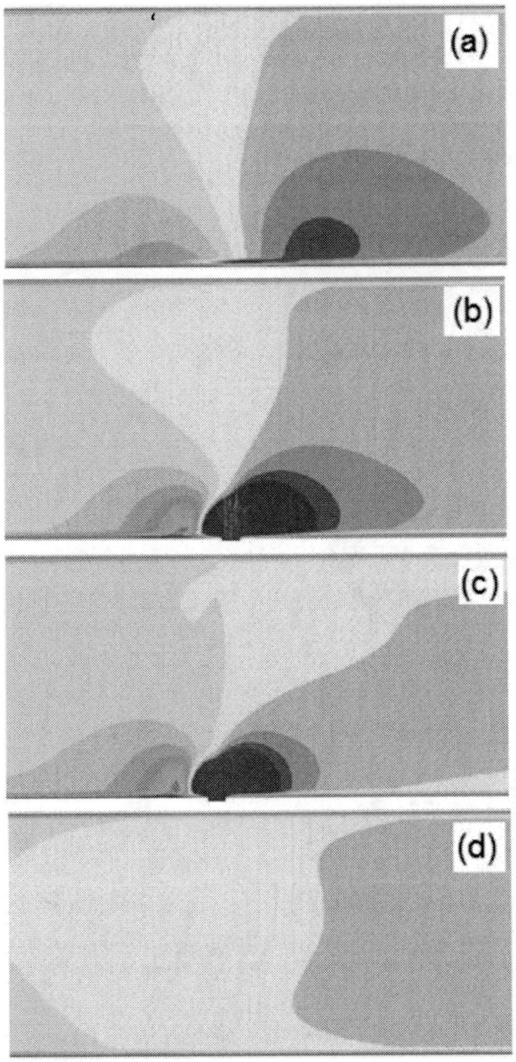

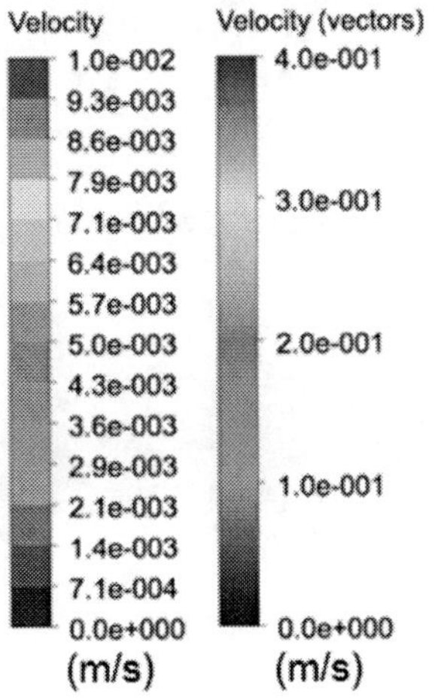

Figure 10. Inclined SJ orifices - numerical simulation, $t/T = 0.75$,
(a) symmetry plane,
(b) plane x-z of inner actuator,
(c) plane x-z of outer actuator,
(d) plane x-z between inner and outer actuator.

References

1. P. Dančová, Z. Trávníček, T. Vít, [Experimental investigation of a synthetic jet array in a laminar channel flow,] EPJ Web of Conferences, Vol. **45**, 01002 (2013)
2. P. Dančová, Ph.D. thesis (Technical university of Liberec, 2012)
3. B.L. Smith, A. Glezer, Phys. Fluids **10**, 2281 (1198)
4. J.E. Cater, J. Soria, J. Fluid Mech. **472**, 167 (2002)
5. A. Glezer, M. Amitay, Annu. Rev. Fluid Mech. **34**, 503 (2002)
6. http://www.lavision.de/en/techniques/piv.php
7. http://www.tsi.com/Fluid-Mechanic-Systems/
8. http://www.dantecdynamics.com/particle-image-velocimetry
9. http://www.openpiv.net/
10. S. Gavlas, P. Ďurčanský, R. Lenhard, J. Jandačka, [Mathematical simulation of heat exchanger working conditions,] EPJ Web of Conferences, Vol. **92**, 02019 (2015)
11. M. Kasanický, R. Lenhard, K. Kaduchová, M. Malcho, [An optical method for measuring the thickness of a falling condensate in gravity assisted heat pipe,] EPJ Web of Conferences, Vol. **92**, 02031 (2015)

Acknowledgements

We gratefully acknowledge the support of the Grant Agency of the Czech Republic - Czech Science Foundation (project no. 14-08888S).

EPJ Web of Conferences **114**, 02015 (2016)
DOI: 10.1051/epjconf/201611402015

Measurement of unsteady velocity flow field of pulsating flow

Petra Dančová[1,2,a], and Tomáš Vít[1]

[1]Technical University of Liberec, Studentská 2, 46117 Liberec 1, Czech Republic
[2]Institute of Thermomechanics, CAS, v.v.i., Dolejškova 5, 18000 Praha, Czech Republic

Abstract. This paper describes an investigation of the commercial synthetic jet actuator "SynJet ZFlow 90 Outdoor Cooler" (Nuventix). The velocity flow field was investigated using the thermoanemometry method in the constant temperature mode. The Eigen frequency of the used actuator was also calculated in this work.

1 Introduction

This paper deals with an investigation of unsteady velocity flow field which generates the actuator of synthetic jet. The commercial actuator "SynJet ZFlow 90 Outdoor Cooler" from Nuventix Company is used for our research, [1].

What the term "synthetic jet" (SJ) means and how the SJ actuators work is described in works e.g. [2 – 4]. A SJ is synthesized by the interactions within a train of vortex rings or counter-rotating vortex pairs in axis-symmetric or two-dimensional geometry, [2]. These vortex rings are generated by the periodical motion of an actuator oscillating membrane. This membrane enables fluid suction into the actuator cavity in the first part of the membrane motion. Then in the second part of the period, this sucked fluid is exhausted from the actuator cavity. One of the main advantages of a SJ is that the time-mean mass flux of the oscillatory flow in the orifice is zero; hence the other common expression is a zero-net-mass-flux (ZNMF) jet. A ZNMF eliminates the requirement of a blower and piping for the fluid inlet. Though the SJ actuator works with a ZNMF at the orifice, the momentum of the resultant SJ at a specific distance from the lip in y is non-zero [2-4].

2 Experimental setup and method

2.1 Experimental setup

For our experiments, the commercial synthetic jet actuator "SynJet ZFlow 90 Outdoor Cooler" (Nuventix) was used. For the schematic view and the actuator geometry see figure 1, for the technical specification see [1]. The SJ actuator orifice has the outer and inner diameter of 88 mm and 85 mm, respectively, and it is divided into four quarter circle parts (visible on figure 1 (c)). Figures 1 (c) and (d) also show the used coordinate system.

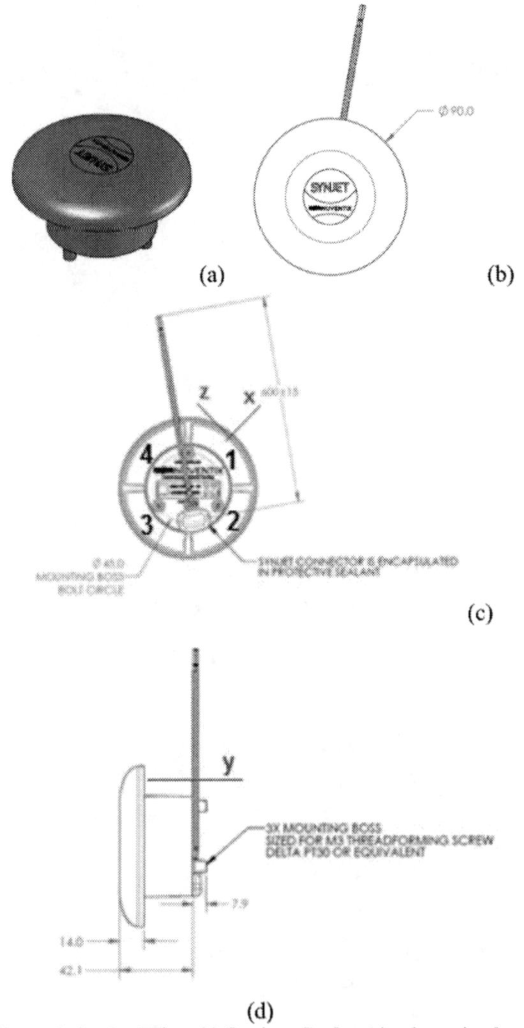

Figure 1. SynJet ZFlow 90 Outdoor Cooler, (a) schematic view, (b) view from the top, (c) platform, (d) side view.

[a] Corresponding author: petra.dancova@tul.cz

2.2 Experimental method - HWA

Hot wire anemometry (HWA) is one of the classical methods for measuring fluid velocities, including a fluctuating velocity component. It was used in this study to quantify SJ velocities in general and to evaluate the resonant frequency of SJ actuators in particular, [5].

The method is based on convective heat transfer from a heated body (hot wire or hot film probe) into a moving environment. The basic element of the measuring circuit is the probe, which is a wire fixed to the prongs of the carrier. The wire (typically tungsten) is heated by an electric current, [5]. The heat flux generated by the electric current in the wire element of length dx can be formulated according to Joule's law in the form:

$$d\dot{Q}_J = \frac{I^2 \chi_{\text{wire}}}{S_{\text{wire}}} dx \,, \qquad (1)$$

where I is the electric current, χ_{wire} is the specific resistance of wire material at the wire temperature T_{wire} and S_{wire} is the wire cross section area, [5].

The heat flux $d\dot{Q}_J$ generated on the wire element dx must be equal to the sum of the heat transferred into the surroundings, i.e. the heat transferred by convection $d\dot{Q}_{\text{conv}}$, the heat transferred by conduction into carriers $d\dot{Q}_{\text{cond}}$, the heat radiated into the surroundings $d\dot{Q}_{\text{rad}}$, and the accumulated heat $d\dot{Q}_{\text{ac}}$:

$$d\dot{Q}_J = d\dot{Q}_{\text{conv}} + d\dot{Q}_{\text{cond}} + d\dot{Q}_{\text{rad}} + d\dot{Q}_{\text{ac}} \,. \qquad (2)$$

The heat dissipated by radiation is in most cases negligible, and the heat dissipated from the wire into the carrier can be considered as independent of the parameters of the flowing medium, see [5]. The convective heat flux $d\dot{Q}_{\text{conv}}$ transferred into the flowing medium can be formulated as:

$$d\dot{Q}_{\text{conv}} = \pi d_{\text{wire}} \alpha (T_{\text{wire}} - T_\infty) dx \,, \qquad (3)$$

where d_{wire} is the wire diameter, α is the heat transfer coefficient, T_{wire} the wire surface temperature, and T_∞ is the fluid temperature.

If a constant temperature (CT) mode is assumed (this mode is used for velocity measurement), the thermal balance for an infinitely long wire, where the heat transfer into the holder is ignored, is given by:

$$I^2 R_{\text{wire}} = \pi d_{\text{wire}} L_{\text{wire}} \alpha (T_{\text{wire}} - T_\infty) \,, \qquad (4)$$

where R_{wire} is the wire resistance on a heated wire with the temperature T_{wire}, and L_{wire} is a length of the wire. After substitution, see [5], the equation (4) becomes:

$$\frac{E^2_{\text{wire}}}{R_{\text{wire}}} = (A + BU^n)(T_{\text{wire}} - T_\infty) \,, \qquad (5)$$

where A, B and n are the functions of wire diameter and the material properties of the wire, and of the surrounding fluid, whose magnitude are determined through calibration, [5]. E is the voltage measured on the wire.

Overheat ratio: An important parameter of probes working in a CT regime is the so-called "overheat ratio":

$$a_h = \frac{R_{\text{wire}}}{R_\infty} \,, \qquad (6)$$

and it is a guide for adjusting the wire temperature. For velocity measurement, it is recommended select an overheat ratio as high as possible (maximum values are limited by the material properties of the wire, especially by a tendency toward faster corrosion at higher temperatures). For experiments in fluids, the overheat ratio is limited by the boiling temperature of fluid.

Velocity calibration: Calibration is used to determine the constants in established calibration relations for velocity measurement $E = f(U)$. The most common relations for HWA calibration are described in [5]. For experiments published in this paper, the polynomial curve fit was used:

$$U = AE^0 + BE^1 + CE^2 + DE^3 + FE^4 \,, \qquad (7)$$

where A, B, C, D, and F are the calibration constants, and E is the voltage measured on the wire.

For experiments, the DANTEC system was used: two hot-film probes 55P11, anemometer 90C10, bus bar NI_CA1000, and A/D converter NI-PCI-MIO-16E-1. The measured data was analyzed in StreamWare 3.01 and Microsoft Excel.

3 Experimental results

The measurement was carried out with sampling frequency and number of samples 8 kHz and 16384, respectively. Overheat ration and the temperature on the probes wire were set as 0.80, and 230 K, respectively.

For the measurement it was necessary to use two hot-film probes 55P11; first was used as the reference (stable) probe and the second one (measuring probe) was used to receive data from the experiments. Because the reference probe was on the same position during whole measurement, the received signal on this probe was always the same. This is the reason why the signal from the reference probe was used for phase averaging of the signal received on the measuring probe. Principle of phase averaging can be found in [5 and 6].

Velocity calibration: The calibration constants were determined from forth order polynomial curve as:

$$U = 51.3514E^0 - 114.3149E^1 + 103.3519E^2 - 47.3964E^3 + 9.4725E^4$$

3.1 Eigen frequency of the SJ actuator

This kind of actuator can be not connected to the signal generator and the Eigen frequency cannot be determined experimentally (such as actuator frequency characteristic determination in [6]). This SJ actuator is directly connected to the power source where voltage and current values were set and thereby the SJ actuator working frequency is adjusted automatically.

The frequency was determined from data obtained when the actuator worked by voltage of 5 V and current of 0.2 A. The measured signal was interlaced with function $y = A + B\sin(2 \cdot \pi \cdot C + D)$, where C corresponds to the working frequency of 59 Hz, [7].

3.2 Velocity measurement

First experiment was focused to determine the effect of synthetic jet, because it was observed that an SJ far enough from the orifice has a character of a conventional steady jet. This is caused by the development and dissipation of vortexes. This experiment was carried out for voltage of 5 V and current of 0.2 A. Figure 2 demonstrates the dependence of the phase-averaged velocity U_P in a different instant of the period and the time-mean velocity U on the distance from the actuator orifice No. 1. There was a visible decrease in velocity as the distance increased along y direction. At a distance of $y = 30$ mm, the flow oscillation practically disappeared, and the flow can be considered as a steady jet flow. $t/T = 0.25$ represents phase maximum extrusion, $t/T = 0.75$ phase of maximum suction, respectively. Negative values of curves $t/T = 0.75$ and 0 in the orifice vicinity represents suction into the actuator cavity.

Dependence of the velocity distribution on time period in different distances from the SJ actuator orifice No. 1 shows figure 3. On this figure is also visible that the actuator works as the synthetic jet actuator until distance of approx. 30 mm above the orifice. According results plotted in figures 2 and 3, the next experiments were performed in distances (0-30) mm from the actuator orifice.

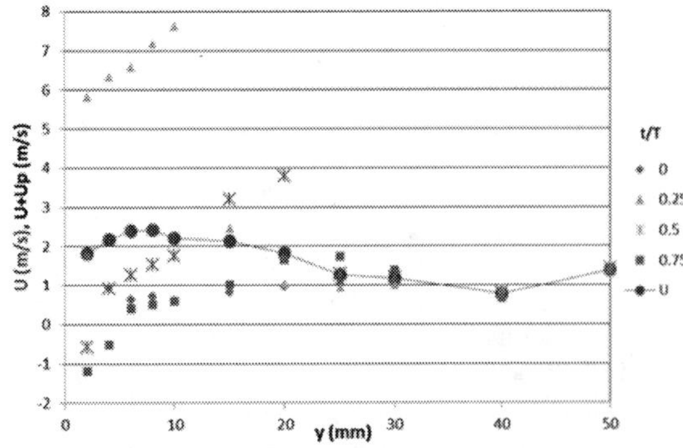

Figure 2. Dependence of the phase-averaged velocity magnitude at different instances in the period and of the time-mean velocity on the distance from the actuator orifice, orifice part No. 1, measured at 5 V, 0.2 A.

Due to the geometry of the SJ actuator, its symmetry was supposed. For the confirmation, the experiments presented in figure 4 were performed - the velocity distribution depending on the distance from the actuator orifices. The value of voltage and current was kept on 5 V and 0.2 A. The results demonstrated in figure 4 the assumption of the actuator symmetry does not confirmed. As it is visible, the shape of the velocity curves is very similar, but the values are different in range of approx. 20 % - 35 %, respectively. The results show that all parts of orifices do not work with the same power. This difference could be also caused with small changes of the electric power delivered into the actuator. But this power change was very small (in order of percent), so the actuator geometry can be supposed as non-symmetrical: small differences in the inner geometry, not identical geometry of the orifices, etc.

Figure 5 demonstrates the time mean velocity profiles in x axes measured in different distances y from the actuator orifice. With increasing distance along y, the velocity profile starts to be symmetrical in x direction. Near the orifice, $y = (0 - 5)$ mm, there is evident a role of influence of the orifice's shape and geometry.

Velocity profiles development during the time period in distance $y = 10$ mm demonstrates figure 6. There is visible that the vortexes of SJ are visible also in the distance ±6 mm from the axis of the orifice.

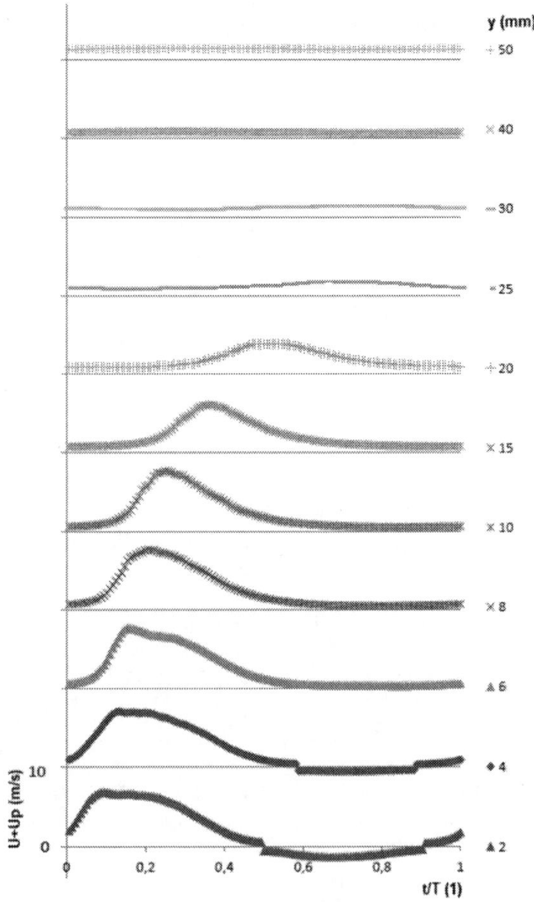

Figure 3. Phase-averaged velocity during the time period for different distances from the SJ actuator orifice, orifice part No. 1, measured at 5 V, 0.2 A.

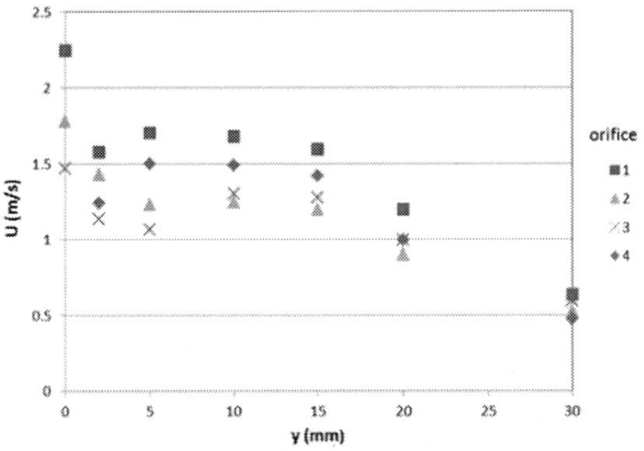

Figure 4. Time mean velocity distribution depending on the distance from the orifice, measured at 5 V, 0.2 A.

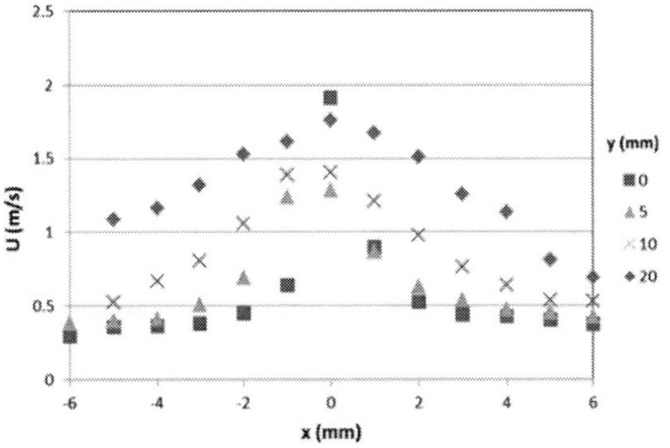

Figure 5. Time mean velocity profiles in different distances from the orifice, measured on orifice part No. 1, measured at 5 V, 0.2 A.

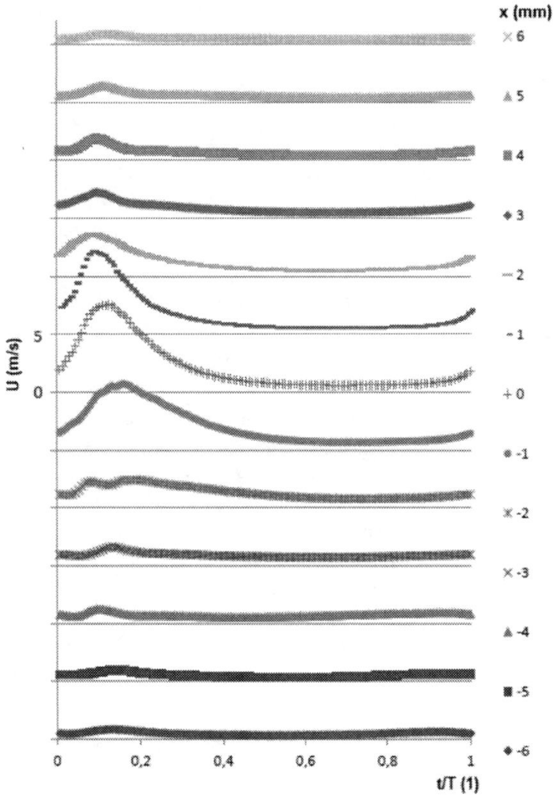

Figure 6. Phase-averaged velocity development during the time period for distance $y = 10$ mm from the SJ actuator orifice, orifice part No. 1, measured at 5 V, 0.2 A.

4 Discussions

The experiments performed on the commercial synthetic jet actuator from the company Nuventix; model SynJet ZFlow 90 Outdoor Cooler were described in this paper. Experiments were carried out using thermoanemometry method in the constant temperature mode. The dependence of the phase-averaged and time mean velocity on the distance from the orifice (y direction) were investigated. Time mean velocity profile distances from the orifice have been also described in this paper.

During the experiments small non-symmetry of the SJ actuator were detected.

All experiments were carried out with voltage of 5 V and current of 0.2 A; keeping of these values were monitored during the experiments.

In the next step of the experiments the changing of the electric power will be investigated, for these changes the working frequency will be determined. Also the cooling properties of this actuator will be studied.

References

1. http://www.aavid.com/sites/default/files/products/led/Aavid-ZFlow90-Outdoor-SynJet-April2015.pdf (cited 10/10/2015)
2. B.L. Smith, A. Glezer, Phys. Fluids **10**, 2281 (1198)
3. J.E. Cater, J. Soria, J. Fluid Mech. **472**, 167 (2002)
4. A. Glezer, M. Amitay, Annu. Rev. Fluid Mech. **34**, 503 (2002)
5. H.H. Bruun, Hot wire anemometry (Oxford Univ. Press, 1995)
6. P. Dančová, Ph.D. thesis (Technical university of Liberec, 2012)
7. M. Šimko, Bachelor thesis (Technical university of Liberec, 2015)

Acknowledgements

We gratefully acknowledge the support of the Grant Agency of the Czech Republic - Czech Science Foundation (project no. 14-08888S).

EPJ Web of Conferences **114**, 02016 (2016)
DOI: 10.1051/epjconf/201611402016

Numerical investigation of flow on NACA4412 aerofoil with different aspect ratios

Hacımurat Demir[1,2,a], Mustafa Özden[1], Mustafa Serdar Genç[1], Mücahit Çağdaş[1]

[1]*Wind Engineering and Aerodynamic Research Group, Department of Energy Systems Engineering, Erciyes University, Kayseri, Turkey*
[2]*Department of Mechanical Engineering, Aksaray University, Aksaray, Turkey*

Abstract. In this study, the flow over NACA4412 was investigated both numerically and experimentally at a different Reynolds numbers. The experiments were carried out in a low speed wind tunnel with various angles of attack and different Reynolds numbers (25000 and 50000). Airfoil was manufactured using 3D printer with a various aspect ratios (AR = 1 and AR = 3). Smoke-wire and oil flow visualization methods were used to visualize the surface flow patterns. NACA4412 aerofoil was designed by using SOLIDWORKS. The structural grid of numerical model was constructed by ANSYS ICEM CFD meshing software. Furthermore, ANSYS FLUENT™ software was used to perform numerical calculations. The numerical results were compared with experimental results. Bubble formation was shown in CFD streamlines and smoke-wire experiments at z / c = 0.4. Furthermore, bubble shrunk at z / c = 0.2 by reason of the effects of tip vortices in both numerical and experimental studies. Consequently, it was seen that there was a good agreement between numerical and experimental results.

1 Introduction

Low Reynolds (Re) number aerodynamics has gained more attention due to increasing applications of Micro Air Vehicle (MAV), Unmanned Air Vehicle (UAV) and wind turbine. At low Reynolds number flows, laminar separation bubble may cause negative effects, such as increasing on drag, decreasing on lift, reducing stability of the aircraft, vibration, and noise [1–7]. Wind turbine blades, unmanned air vehicles (UAV), micro air vehicles (MAV) and propellers are some examples of applications of low Reynolds number aerodynamics. In low Reynolds number flow regime, viscous forces are higher than inertia forces. The fluid flow is prone to separate, resulting in increases in drag and loss of efficiency under low Reynolds number regime.

Under such low Reynolds number regime, the maximum lift and stall angle are lower than high Reynolds number flow conditions. Due to the fact that the aerodynamic performance is lower, it is vital to control the flow and to achieve higher lift for this type of vehicles and devices. There are new methods being developed to ignore the negative effects of the separated flow for improving the aerodynamic performance. These methods called flow control methods and these methods could be categorized as active and passive [5-7]. Active flow control method can be made by adding energy to free stream or to boundary layer directly. On the other hand, passive flow control method can be executed by adding geometrical discontinuities. The active flow control methods can be suction and/or blowing [6] or high lift devices [7], the passive control methods can be vortex generators [8] or flexibility enhancer [9-10]. Lift can be increased, drag force can be reduced, stall can be delayed, vibrations and noise can be diminished and reattachment of the separated flow can be ensured by means of utilizing these flow control methods.

A series of studies have been executed to understand and determine the low Reynolds number aerodynamics around entirely translating low-aspect-ratio wings. [11-14]. R. Wahidi et al. [15] investigated to effects of laminar separation bubbles by measuring over NACA4412 airfoil model using volumetric three-component velocimetry (V3V) and particle image velocimetry (PIV) at Reynolds number 50000 and different angle of attack. They defined that onset of transition, location of separation and reattachment after they made comment relationships of different Reynolds normal and shear stress via the time-averaged V3V. Their results demonstrated that roll-up of spanwise caused vortices and these vortices inspired pairs of negative and positive wall-normal velocity. The pairs did an enormous act in the uncertainty of the reattachment of the separated shear layer. M. Agrawal and G. Saxena [16] aimed that making analysis for aerodynamic of wings. They selected NACA4412 airfoil as a model and achieved few results after completing wind tunnel tests sections. They observed from results that lift increased until a certain

[a] Corresponding author: hmdemir@aksaray.edu.tr

point if angle of attack increased. Their observation also included that drag started to become a dominant factor if angle of attack kept rising and stall occurred as a result of all these parts.

Karasu et al. [17] performed a numerical study on low Reynolds number flow over NACA2415 airfoil and compared the numerical results with the experimental results of their previous study [4, 18]. It was stated that the point of separation moved towards the leading edge due to increasing angle of attack. Moreover, the flow visualization results showed that as the angle of attack increased further, the bubble burst and the separated flow was not able to reattach to the airfoil surface, which indicated stall. Furthermore, in the numerical results, the transition models are shown to definitely predict the location of the separation bubble experimentally determined at lower angles of attack.

2 Numerical Modelling of Airfoil

For numerical model, manufacturing of NACA4412 airfoils with different aspect ratios were conducted by using SOLIDWORKS software. ANSYS Workbench module was utilized for numerical modelling of flow over the airfoil. The grid structures of test section and airfoil surface were constructed by means of ICEM CFD. The domain except the airfoil was selected as pressure far field and no slip condition was carried out to the airfoil surface. All calculations were performed on as density based, and steady-state solution.

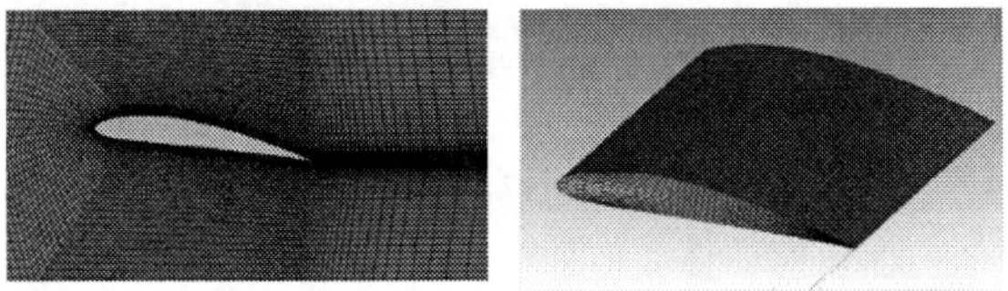

Figure 1. The detailed view of the mesh NACA4412 aerofoil.

ANSYS Fluent software was used for calculation and SST Transition model was used for the numerical solution method.

2.1 Experimental Study

2.1.1 Experimental apparatus and methods

The experiments were carried out in a low-speed, suction-type wind tunnel with a square working section of 500 mm x 500 mm located at the Department of Energy Systems Engineering, University of Erciyes via Wind Engineering and Aerodynamics Research Group (WEAR) as shown in Figure 2. Turbulence intensity of the tunnel is 0.35% and the operating range of tunnel speed was 1-45 m / s. The experiments were conducted at Reynolds numbers (Re) of 25000 and 50000.

NACA4412 airfoil was manufactured by means of 3D printer as seen in Figure 3.

Figure 2. Photograph of the experimental set-up.

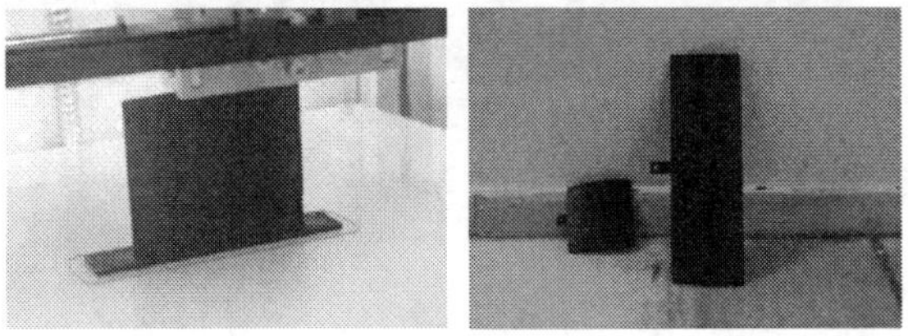

Figure 3. NACA4412 airfoils with different aspect ratios.

2.1.2 Smoke-wire and oil flow visualization

The smoke-wire technique allows for the introduction of fine smoke streamlines into the flow field through the electrical resistive heating of a very thin wire which has been coated with oil and which is located upstream from the leading edge of the related airfoil section. Smoke-wire tests have been conducted to visualize flows over NACA4412 airfoil surface at $z / c = 0.2$ and $z / c = 0.4$ and two different Reynolds numbers (Re = 25000 and Re = 50000) with various aspect ratios (AR = 1 and AR = 3) at 8° angle of attack. The expression of z / c is utilized for the location of smoke wire.

Oil flow visualization technique was employed for the flow visualization because this technique was simple like smoke-wire method to apply and effective to see flow conditions. In order to photograph the surface flow events using this method, the pigmented oils are painted onto NACA4412 airfoil surface (Figure 7). Once it has dried, the painted and patterned surface can be photographed. The mix should have the right consistency to efficaciously demonstrate boundary layer development. Additionally, the inertia forces of the moving oil should be lower than the viscous and surface tension forces [19] to not affect the conditions at the surface. Kerosene, light diesel oil and light transformer oil are most common oils, and titanium dioxide, china clay and fluorescent chrysene are most common pigments [19]. In addition to this, oleic acid can be added to the mixture to clearly see the pigment deposit on the oiled surface. The simplest mix to make was kerosene, titanium dioxide and a very small amount of oleic acid: the ratio of kerosene to oleic acid was roughly 20:1 [20] and this ratio of mixture was utilized in the present study.

3 Comparisons between Experimental and Numerical Results

3.1 Effects of tip vortex on separation bubble

When obtained experimental and numerical results are analyzed, bubble formation is observed in streamline and smoke-wire at z / c = 0.4. Furthermore, bubble shrunk at z / c = 0.2 due to the effects of tip vortices. The effects of tip vortices could be understood significantly with oil experiment.

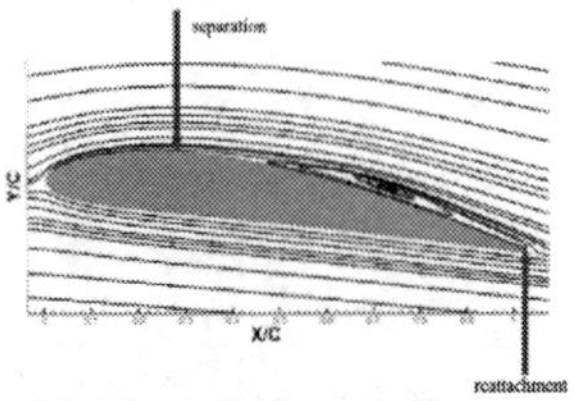

(a) Numerical stream lines on the NACA 4412

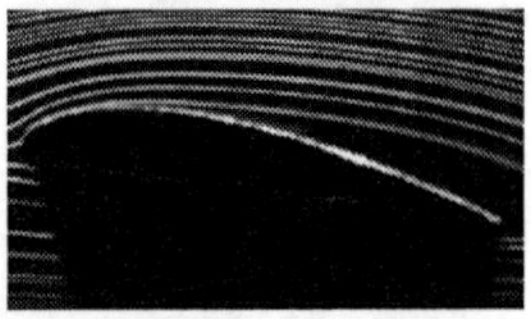

(b) Experimental smoke-wire visualization result

Figure 4. NACA4412 airfoil results for Re = 25000, AR = 3 and z / c = +0. 4 condition.

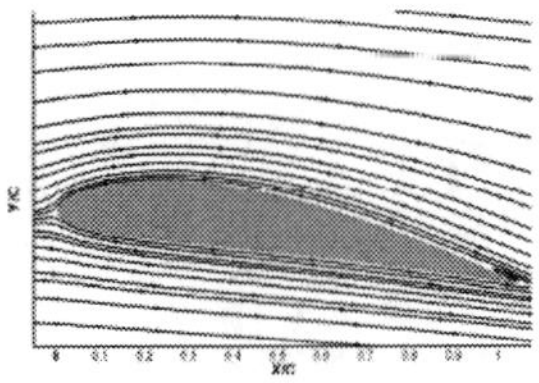

(a) Numerical stream lines on the NACA4412

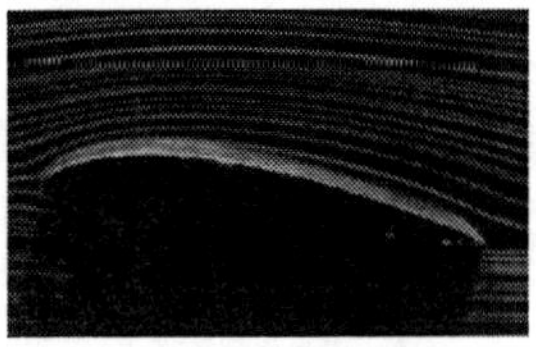

(b) Experimental smoke-wire visualization result

Figure 5. NACA4412 airfoil results for Re = 25000, AR = 3 and z / c = +0.2 condition.

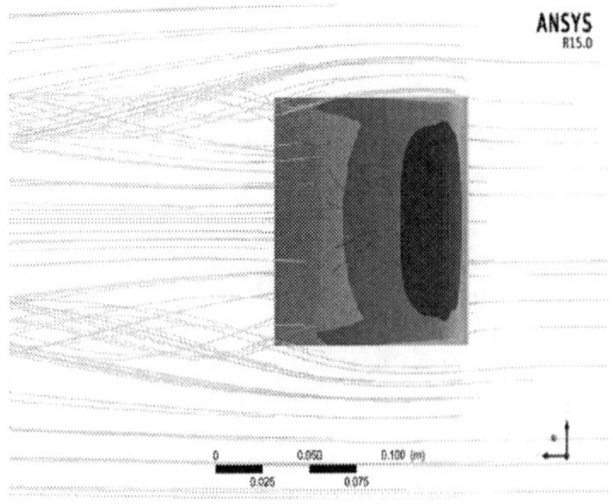

(a) Numerical stream lines on the NACA 4412.

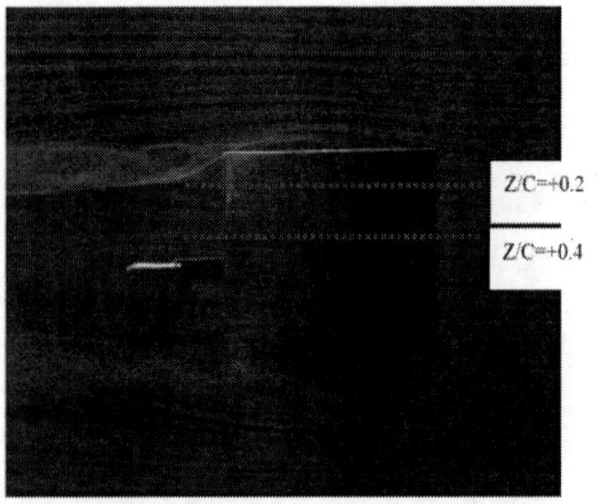

(b) Experimental smoke-wire visualization result.

Figure 6. Tip vortices at z / c = +0.4 and Re = 50000 for AR = 1 airfoil.

It is seen that tip vortices have an impact on the inner of the airfoil when comparing the smoke wire experimental and numerical results. It is observed that flow moved from lower side to the upper side of the airfoil due to the pressure side.

The effects of separation over the airfoil are seen after obtained oil flow visualization results. There is no change of oil mixture over the airfoil where flow separation is occurred but oil is swept at the location of the flow which is attached to the airfoil. When moving to the z direction on the airfoil, the size of separation increases in the middle location of the airfoil. On the other hand, oil mixture moves to inside of airfoil in the tip region. Tip vortices which are occurred on the airfoil tip affect the flow over the airfoil significantly. Furthermore, leading edge vortices combine with tip vortices and this situation causes to ensure 3D vortices. Flow obtains energy and bubble decreases due to these 3D vortices.

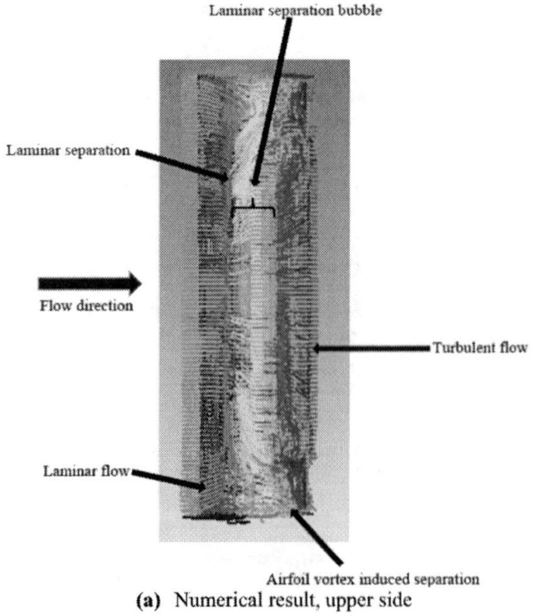

(a) Numerical result, upper side

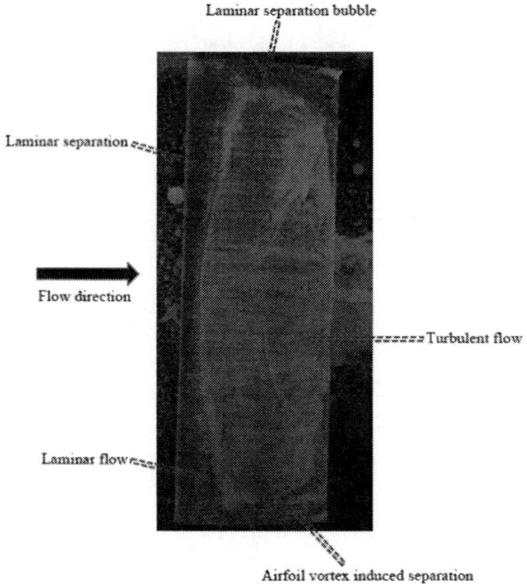

(b) Experimental oil flow visualization results, upper side

Figure 7. Numerical and experimental results for Re = 50000 and AR = 3.

4 Conclusion

In this study, the flow over NACA4412 airfoil which was manufactured by means of 3-D printer with AR = 1 and AR = 3 was investigated both numerically and experimentally at various Reynolds numbers. The experimental section was conducted in a low speed wind tunnel with various angles of attack and Re = 25000 and Re = 50000. In order to visualize the surface flow patterns of NACA4412 smoke-wire and oil flow visualization methods were utilized. Furthermore, the structural grid of numerical model was constructed by ANSYS ICEM CFD meshing software. Additionally, ANSYS FLUENT™ software was used to perform numerical calculations. The numerical results were compared with experimental results and it was seen that bubble formation occurred in streamline and smoke-wire at z / c = 0.4. In addition to this, by reason of the effects of tip vortices bubble shrunk at z / c = 0.2.

The effects of tip vortices could be understood significantly with oil experiment. There was no change of oil mixture over the airfoil where flow separation was occurred but oil was swept at the location of the flow which was attached to the airfoil. In the tip region, oil mixture moved to inside of airfoil. Tip vortices which were occurred on the airfoil tip affected the flow over the airfoil importantly. Additionally, leading edge vortices combined with tip vortices and this situation caused to ensure 3D vortices. Flow obtained energy and bubble decreased due to these 3D vortices. It was seen from the obtained results that there was a good agreement between numerical and experimental results.

Acknowledgments

The authors would like to acknowledge funding from the Scientific and Technological Research Council of Turkey (TÜBİTAK) under the project no: 213M329.

References

1. W. Zhang, R. Hain, C.J. Kahler, Scanning PIV investigation of the laminar separation bubble on a SD7003 airfoil. *Exp. Fluids.* **45(4)**, 725-743 (2008)
2. R. Ricci, S.A. Montelpare, Quantitative IR thermographic method to study the laminar separation bubble phenomenon. *Int J Therm Sci.* **44(8)**, 709-719 (2005)
3. M.S. Genc, Numerical Simulation of Flow over a Thin Aerofoil at High Re Number using a Transition Model. *Proc IMechE, Part C-J Mech Eng Sci,* **224(10)**, 2155-2164 (2010)
4. M.S. Genc, I. Karasu, H.H. Acıkel, An experimental study on aerodynamics of NACA2415 aerofoil at low Re numbers. *Exp Therm Fluid Sci.* **39**, 252-264 (2012)
5. M.S. Genc, I. Karasu, H.H. Acıkel, M.T. Akpolat, Low Reynolds number flows and transition, in: M. Serdar Genc (Ed.), Low Reynolds Number Aerodynamics and Transition, *Intech-Sciyo Publishing*, ISBN 979-953-307-627-9 (2012)
6. M.S. Genc, U. Kaynak, H. Yapıcı, Performance of transition model for predicting low Re aerofoil flows without/with single and simultaneous blowing and suction. *Eur J Mech B-Fluid.* **30(2)**, 218-235 (2011)
7. M.S. Genc, U. Kaynak, G.D. Lock, Flow over an Aerofoil without and with Leading Edge Slat at a Transitional Reynolds Number. *Proc IMechE, Part G: J Aerospace Eng.* **223(3)**, 217-231 (2009)
8. D. Lengani, D. Simoni, M. Ubaldi, P. Zunino, F. Bertini, Turbulent boundary layer separation control and loss evaluation of low profile vortex generators. *Exp Therm Fluid Sci.* **35(8)**, 1505-1513 (2011)
9. M.S. Genç, Unsteady aerodynamics and flow-induced vibrations of a low aspect ratio rectangular membrane wing with excess length. *Exp. Therm Fluid Sci.* **44**, 749-759 (2013)
10. P. Rojratsirikul, M.S. Genc, Z. Wang, I. Gursul, Flow-induced vibrations of low aspect ratio rectangular membrane wings. *J Fluid Struct.* **27**, 1296-1309 (2011)
11. G.E. Torres, T.J. Mueller, Low aspect ratio aerodynamics at low Reynolds numbers. *AIAA journal.* **42(5)**, 865-873 (2004)
12. A. Pelletier, T.J. Mueller, Low Reynolds number aerodynamics of low-aspect-ratio, thin/flat/cambered-plate wings. *J Aircraft.* **37(5)**, 825-832 (2000)
13. P. Freymuth, W. Bank, F. Finaish, Further visualization of combined wing tip and starting vortex systems. *AIAA journal.* **25(9)**, 1153-1159 (1987)
14. P.H. Cosyn, J. Vierendeels. Numerical investigation of low-aspect-ratio wings at low Reynolds numbers. *J Aircraft.* **43(3)**, 713-722 (2006)

15. R. Wahidi, W. Lai, J.P. Hubner, A. Lang. Volumetric three-component velocimetry and PIV measurements of Laminar Separation Bubbles on a NACA4412 Airfoil. *16th Int. Symp of Applications of Laser Techniques of Fluid Mechanics*. Lisbon, Portugal. 09-12 July, (2012)
16. M. Agrawal, G. Saxena, Analysis of wings using airfoil NACA4412 at different angle of attack. *IJMER*. **3**, 1467-1469 (2013)
17. I. Karasu, M.S. Genç, H.H. Açikel, Numerical study on low Reynolds number flows over an Aerofoil. *J. Appl. Mech. Eng*. **2**, 131 (2013)
18. I. Karasu, Experimental and numerical investigations of transition to turbulence and laminar separation bubble over aerofoil at low Reynolds number flows. Graduate School of Natural and Applied Sciences, Erciyes University, Kayseri, Turkey (2011)
19. Werzkirch, Flow Visualization, Academic Press Inc. Ltd., London, (1974)
20. D.F. Perrens, Flow visualisation in low speed wind tunnels. *Phys. Educ*. **5 (5)**, 262-265 (1970)

EPJ Web of Conferences **114**, 02017 (2016)
DOI: 10.1051/epjconf/201611402017

Identification of the noise using mathematical modelling

Josef Dobeš[1,a], Milada Kozubková[1] and Miroslav Mahdal[2]

[1]*Department of Hydromechanics and hydraulic equipment, Technical University of Ostrava, Ostrava, Czech Republic*
[2]*Department of Control Systems and Instrumentation, Technical University of Ostrava, Ostrava, Czech Republic*

Abstract. In engineering applications the noisiness of a component or the whole device is a common problem. Currently, a lot of effort is put to eliminate noise of the already produced devices, to prevent generation of acoustic waves during the design of new components, or to specify the operating problems based on noisiness change. The experimental method and the mathematical modelling method belong to these identification methods. With the power of today's computers the ability to identify the sources of the noise on the mathematical modelling level is a very appreciated tool for engineers. For example, the noise itself may be generated by the vibration of the solid object, combustion, shock, fluid flow around an object or cavitation at the fluid flow in an object. For the given task generating the noise using fluid flow on the selected geometry and propagation of the acoustic waves and their subsequent identification are solved and evaluated. In this paper the principle of measurement of variables describing the fluid flow field and acoustic field are described. For the solution of fluid flow a mathematical model implemented into the CFD code is used. The mathematical modelling evaluation of the flow field is compared to the experimental data.

1 The Possibilities of the Acoustic Task Solution

The numerical solutions of the generation and propagation of the acoustic waves help reduce the noise level of the individual components or the equipment altogether. Today the possibility of numerical modelling of the generation and propagation of acoustic waves is a very appreciated tool. These numerical simulations are compared with the data obtained by experimental measurements, which is, particularly for measurements of noise or sound intensity, highly expensive procedure as regards the experimental equipment and staff.

Noise can be generated by several different methods that, from physical perspective, can be divided into the following groups [1]:

- noise generated by the friction of solid parts - gearboxes,
- noise generated by the vibration of the components - engine covers,
- noise generated by the combustion - piston engines,
- noise generated by the shocks - explosions, cavitation,
- noise generated by the vortex shedding structures – aerodynamic noise.

For the numerical modelling of the fluid flow using CFD (Computational Fluid Dynamics) code and the subsequent modelling of propagation of acoustic waves the most interesting applications are in the field of cavitation behaviour of the fluid elements and in the field of the flown-around obstacles, known as aeroacoustic noise (CAA - Computational aeroacoustics).

2 The numerical solution of the aeroacoustic task

From the viewpoint of the numerical solution of the generation and the propagation of the acoustic waves through the fluid flow (cavitation, aeroacoustic noise) is the solution of the problem split into two particular parts, each of them is solved separately in the given order.

In the first part the problem of the description of the flow field of flowing fluid is solved through the Navier-Stokes equations, the continuity equation and additional equations for the appropriate turbulence model are numerically solved.

The second part of the task lies in dealing with the propagation of acoustic waves generated by the solved flow field in the given environment and their subsequent evaluation. On the simple task of the flown-around cylinder the method how to solve aeroacoustic task is shown. The cylinder is solved as a planar 2D task. The measuring section of the wind tunnel and inserted real flown-around object is the geometry.

The experimental work was focused on the measurement of the velocity profiles for the cylinder and the measurement of the sound intensity. Due to large complexity of the calculations and the experimental work

[a] Corresponding author: josef.dobes@vsb.cz

is the flown-around cylinder solved at one value of the velocity **u** and thus Reynolds number *Re*.

2.1 CFD tool OpenFOAM

OpenFOAM is a tool for solving CFD simulation for a wide range of engineering applications, which is based on the programming language C/C++. In the basic configuration it contains over 80 solvers and 170 applications enabling pre- and post-processing [2]. The numerical task is defined using the following directory structure. The main directory named "20_ms" has the following subdirectories:

• „system" – solver setup in the file "controlDict", tolerances and numerical schemes for solving of the partial differential equations,

• „constant" – contains information about the computational mesh, the physical properties of the flowing medium and the selected turbulent model,

• „0" - time directory contains information about boundary and initial conditions for solved tasks.

For this task the version of program OpenFOAM 2.3.x is applied. It is an air flow at a velocity $\mathbf{u} = 20\ m\ s^{-1}$ of unaffected field, we consider the flow of the liquid – air – incompressible flow and as the numerical solver pisoFoam is selected. The Reynolds number for the cylinder at diameter $d = 20\ mm$, at velocity $\mathbf{u} = 20\ m\ s^{-1}$ is $Re = 26\ 042$.

2.2 The creation of geometry and mesh

The geometry and the computational mesh was created in OpenFOAM software environment using the appropriate commands, checked in a graphical environment Paraview and subsequently modified or approved and then it went through the numerical calculation.

The geometry and mesh for this task is created directly by "blockMesh" command. The appropriate command creates geometry and mesh in a few steps:

• creates points placed in a 3D coordinate system,

• links these points into blocks,

• divides the blocks into individual cells according to parameters specified by the user,

• sets boundary conditions of the task.

OpenFOAM allows creating only a 3D object. For numerical solution of a 2D task you need to specify a special boundary condition - "empty".

2.3 The numerical solution of the fluid flow - k-epsilon turbulence Standard model

The partial differential equations describing fluid flow are defined for time averaging values, while in the equations Einstein summation rule is used. The continuity equation for incompressible fluid has the form [2, 3]:

$$\frac{\partial u_j}{\partial x_j} = 0 \tag{1}$$

where the member $\mathbf{u}_j$ is j-th component of velocity, the index j indicates the counting index.

For incompressible fluid the Navier - Stokes equation can be written as:

$$\frac{\mathrm{D}u_i}{\mathrm{D}t} = -\frac{1}{\rho}\frac{\partial p}{\partial x_i} + \nu\frac{\partial^2 u_i}{\partial x_j^2} + f_i \tag{2}$$

where the member $\frac{\mathrm{D}}{\mathrm{D}t}$ indicates substantial derivative, which is a kind of total derivatives following the movement. The member ν is the kinematic viscosity, $\mathbf{u}_i$ is i-th component of velocity vector, p denotes pressure, ρ density of fluid and, generally, the member f_i symbolizes i-th component of the external volume forces (gravitational, centrifugal force). The used turbulent k-epsilon Standard model is a two-equation turbulent model that adds next two transport equations to solving the equations mentioned above. One of them is the equation for turbulent kinetic energy k:

$$\frac{\partial}{\partial t}(\rho \cdot k) + \nabla \cdot (\rho k\vec{u}) = \nabla\left(\left(\mu + \frac{\mu_t}{\sigma_k}\right)\nabla k\right) + G_k + G_b - \rho \cdot \varepsilon - Y_M + S_k \tag{3}$$

the second one is the equation for the dissipation rate ε:

$$\frac{\partial}{\partial t}(\rho \cdot \varepsilon) + \nabla \cdot (\rho\varepsilon\vec{u}) = \nabla\left(\left(\mu + \frac{\mu_t}{\sigma_\varepsilon}\right)\nabla \varepsilon\right) + C_{1\varepsilon}\frac{\varepsilon}{k}(G_k + C_{3\varepsilon} \cdot G_b) - C_{2\varepsilon}\rho \cdot \frac{\varepsilon^2}{k} + S_\varepsilon \tag{4}$$

In the literature we can find a detailed description of this model [3]. Such modified differential equations are solved by finite volume method in program OpenFOAM.

2.4 Boundary conditions

The boundary conditions for the flown-around cylinder are specified in the files in the "0" directory. At the inlet into the geometry the velocity **u** and the turbulence quantities k and ε were assigned. The pressure $p = 0\ Pa$ was assigned at the outlet as relative. The gravity is set by the file „g" in the „constant" directory.

2.5 The numerical solution of acoustic waves

From the viewpoint of the possible solutions the aeroacoustic task the access using numerical methods can be grouped as follows [4]:

• a method of direct simulation of noise (CAA) – this method is the most demanding in terms of time step size and cell size of the mesh. Small fluctuations of pressure p and velocity **u** are solved directly which carries information of the noise generated by the flow. The noise source and the receiver must be placed in the solved area.

• a method when the solution of CFD simulation serves to define the noise sources for specialized acoustic programs - the acoustic waves propagation is solved.

• a method of modelling by the acoustic analogy – CFD simulation serves to calculate source members. In the selected parts of the calculation area, where the noise generation is anticipated, pressure p, velocity **u** and for compressible media the density ρ in each time step are recorded. From these records the propagation of acoustic waves in a "stationary" fluid is calculated.

2.5.1 The acoustic analogy

The acoustic analogy is used to describe the connection between the flow field and the acoustic field emerging due to flowing. It describes the relationship between the propagation of acoustic waves and its causes (sources). The acoustic analogy can be divided to acoustic analogy based on the fluid density ρ, acoustic analogy based on a logarithmic ratio of the pressures in the fluid, and acoustic analogy based on the enthalpy of the fluid H [5]. The Curle's analogy falls into analogy derived and subsequently expressed depending on the density of the fluid ρ. This is a special version of analogy Ffowcs Williams-Hawkings (abbreviated FW-H) that is derived from the basic Lighthill's analogy. The theory and the derivation of this analogy can be found in literatures [5]. If the surface S in the FW-H equation is assumed to be rigid and it does not move, the equivalent monopole surface source distribution q_{WS} and its substitute volume dipole and quadrupole source distributions f_{WVc} and T_{WVc} disperse as well as the corresponding density perturbation components ρ'_{vS}, ρ'_{fVc} and ρ'_{TVc} [5]. This leads to Curle's equation:

$$\frac{\partial^2 \rho'}{\partial t^2} - c^2 \nabla^2 \rho' = -\nabla \left[\vec{f}_{WS}\, \delta(w_1 - w_{10})\right] + \nabla\nabla : \left[\overline{\overline{T}}_L H(w_1 - w_{10})\right] \tag{5}$$

The detailed description of the individual variables can be found in the appropriate literature sources [1, 5].

2.5.2 The implementation of Curle's analogy

The Curle's analogy implementation was done according to literature [6] for the version of program OpenFOAM 2.3.x. The implementation itself is composed of several consecutive steps that guarantee the proper function of individual libraries for evaluation of acoustic quantities using Curle's acoustic analogy. It is necessary to implement a dynamic library to obtain the variables for the possibility of calculating the acoustic pressure, the library of analogy for the calculation of acoustic quantities and the library for the data evaluation.

2.5.3 The setup of Curle's acoustic analogy

The setup is done in the file named „CurleAnalogy" in the directory „system". The start and end time of sensing acoustic quantities, the source object – cylinder, the speed of sound in the medium **c**, the reference density ρ, pressure p and the reference sound pressure p_A are defined in this file. Furthermore, receivers' (the probes) positioning, the sampling frequency for evaluating a fast Fourier transform are defined. More information about this analogy can be obtained from the literature [6].

2.5.4 The results of aeroacoustic simulation using by program OpenFOAM

In the introduction of the Chapter 2 aeroacoustic task solution is generally described. In this section, the evaluation of numerical simulation task is indicated, more precisely the acoustic part. The aerodynamic part of the task is evaluated and also compared with the experimental data in Chapter 4.1.

The comparison of the numerical simulation with the experimental measurements for acoustic part has not been carried out yet. The experimental acoustic work is in the phase of research and development for the most accurate and appropriate measurement method, which will deal appropriately with the issue of the aeroacoustic task and verify the numerical simulations properly.

The result of the acoustic task's numerical simulation is plotting of sound pressure level depending on the frequency of individual receivers R1 until R5, which were located according to the drawing below (all dimensions in the drawings in this paper are in *mm*).

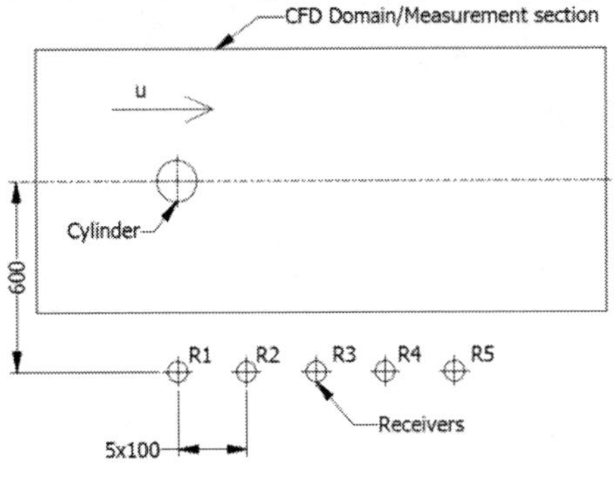

Figure 1. The drawing of receivers' location in acoustic simulation

The aeroacoustic simulation result is shown in the following chart. From the graphical evaluation it can be seen that the first significant peak of the sound pressure level (abbreviated SPL) is at the frequency 143 *Hz*. This value is lower than vortex shedding frequency of the flown-around cylinder. This value is recorded for all the receivers, when the values vary in the size of SPL where with increasing distance from the cylinder the SPL decreases. The highest sound pressure level at this frequency is 123 *dB* (for receiver R1), which is a very high value. The second significant peak is at the frequency 285 *Hz*. This value is approximately double compared to the first significant frequency. The highest value of SPL at this frequency is 80 *dB*. The results are not processed using any weighting filter (A-weighted or C-weighted).

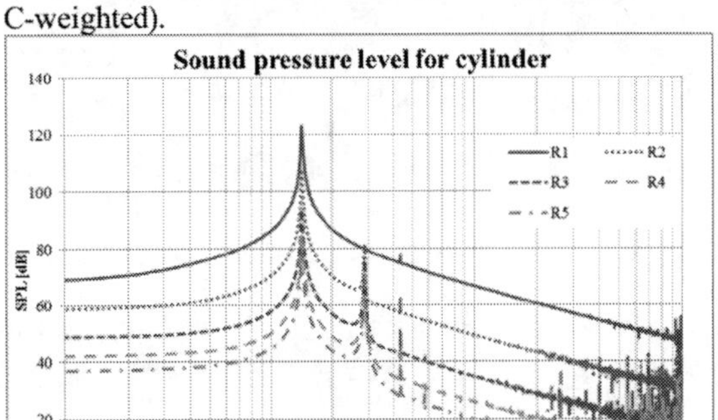

Figure 2. The evaluation SPL for acoustic receivers

3 The experimental part of acoustic task

The experimental part of task was performed using a wind tunnel on the Department Hydromechanics and hydraulic equipment of Technical university of Ostrava.

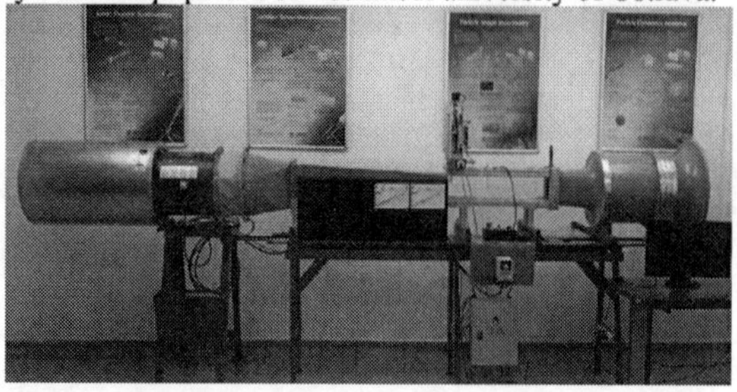

Figure 3. The experimental device for measuring task

In the first part the measurement was carried out to verify the numerically calculated flow field with the experimental data obtained with the hot-wire probe MiniCTA. In the second part, to obtain the data describing the acoustic field, the experimental work was performed by measuring the wind tunnel without obstacle by the intensity probe.

3.1 The measurement of the flow field

The measurement of velocity profiles in a defined distance behind the flown-around cylinder was done by the one-wire probe MiniCTA type 55P14. The probe was clamped into traversing device to ensure the most accurate position for measurement of velocity component in the flow direction.

The velocity profiles for the cylinder were measured in the distances: $2\times D$, $3\times D$, $4\times D$, $5\times D$, $6\times D$ a $7\times D$, when D is the cylinder diameter (20 *mm*). In these distances the velocity profile perpendicular to the main direction of the velocity vector was measured.

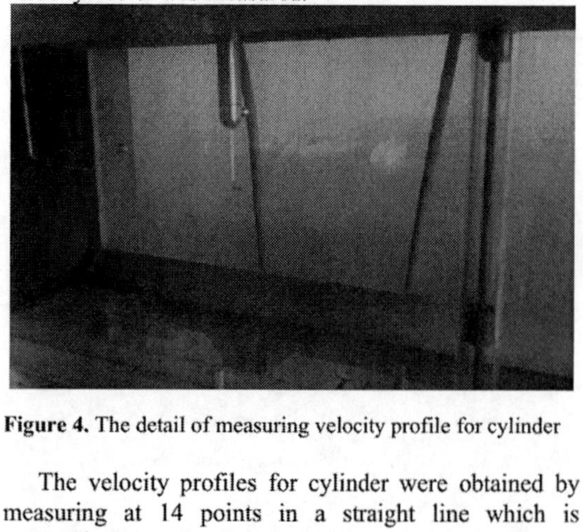

Figure 4. The detail of measuring velocity profile for cylinder

The velocity profiles for cylinder were obtained by measuring at 14 points in a straight line which is perpendicular to the direction of the velocity vector. The distance between two consecutive points was set at 6 mm. A finer measurement, the smaller the distance between two consecutive points, was rejected because of limited capacity of the disk storage. The drawing of the velocity profiles measurement is shown in the Figure 5.

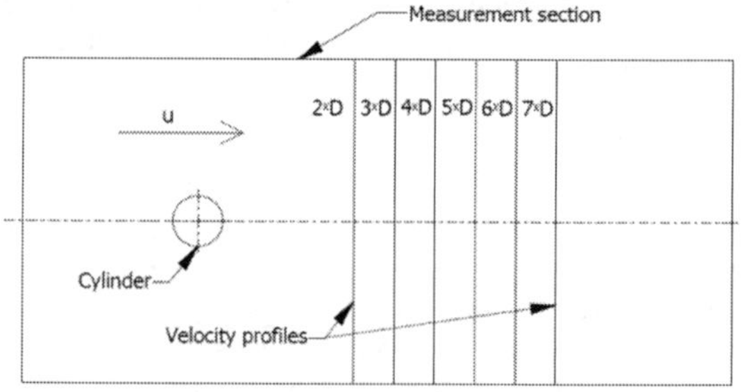

Figure 5. The drawing of velocity profiles measurement

Subsequently, all experimentally obtained velocity profiles were compared with the results of numerical simulation for checking of numerical CFD code setup.

Fast Fourier transform of the measured velocity data to determine the vortex shedding frequency f and the calculation of experimental Strouhal number St was also done. This value was compared to the results of numerical simulation and literature in the Table 1.

3.2 The measurement of the acoustic field

For the acoustic measurements so-called Intensity probes are suitable measuring instruments. These are based on the measurement of sound pressure in two nearby points, these probes are often termed as *p-p* probes or two-microphone probes. For the measurement the high quality condenser microphones are used and the most common setup of those is face to face. The distance between the microphones is defined using spacers.

Figure 6. The intensity probe for acoustic measurement

Subsequently, the measurement was evaluated using the relevant analyser and software. As intensity probe the probe type 3599 from the company Bruel & Kjaer was used. The drawing of the measurement is shown in the Figure 7. There are also shown the measuring points (A0 – A4) with the respective designation of these points.

4 The comparison of numerical and experimental data

In this part of the paper a verification of experimental data with data obtained by numerical simulation is done. In the first part the experimentally obtained velocity profiles were compared with numerically obtained velocity profiles and the numerically obtained drag coefficient C_d and Strouhal number St were compared with literature. In section 4.2 the results obtained by the intensity probe measuring are presented. For better clarity, this section is divided into comparison of the aerodynamic part - velocity profiles, drag coefficient C_d and Strouhal number St. For the acoustic part of the task measuring the tunnel without cylinder is done. To obtain data describing the acoustic behaviour of whole system, the acoustic measurement serves as a reference for the future measuring with obstacle

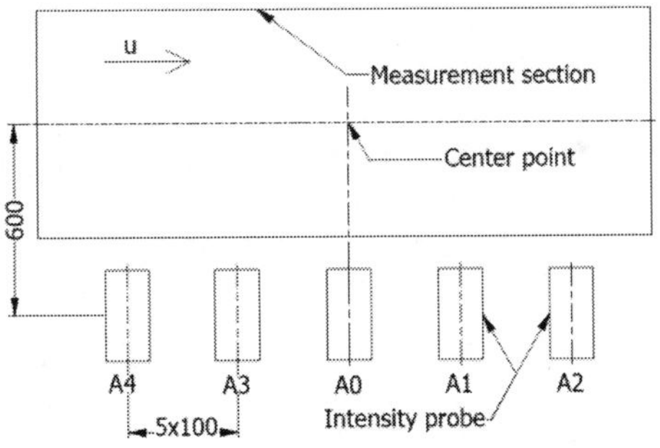

Figure 7. The acoustical measurement drawing for tunnel

4.1 The aerodynamic experimental part

To compare experimental and numerical data averaging velocity profiles for the flown-around cylinder were selected. As a complement, a comparison of the drag coefficient C_d and Strouhal number St with literature for cylinder is done [7]. In the Figure 8 experimentally obtained velocity profiles for the cylinder are shown.

In the graphic representation, the X-axis shows the measuring points, when point X8 is a measurement of the velocity profile past the cylinder axis. The Y-axis shows the arithmetic average of measured velocity. For the comparison the experimental data with the numerical results profiles 2×D and 3×D were selected. Parameters of the measurement were sampling frequency f = 480 Hz which corresponds to vortex shedding frequency of cylinder multiplied by 2.5 for compliance with the Nyquist- Kotelnikov theorem. The number of samples for all measurement was 4096.

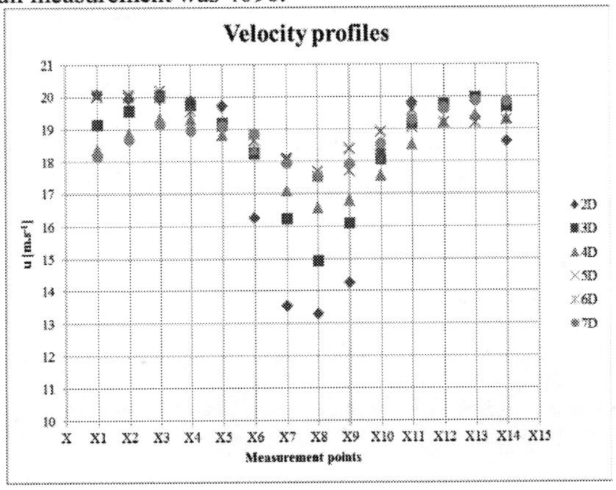

Figure 8. The experimental velocity profiles – cylinder

4.1.1 The comparison of velocity profiles for cylinder

The following chart shows comparing the velocity profiles between the numerical simulations indicated in the graph as "OpenFOAM 2D" and "OpenFOAM 3D", with experimental data, indicated in the graph as the "2D" and "3D" for the cylinder.

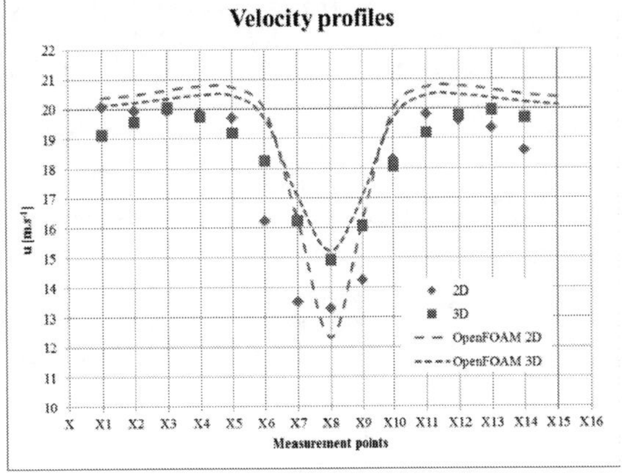

Figure 9. The velocity profiles comparison for cylinder

The comparison shows that the velocity profiles are identical in the field behind the flown-around obstacle. In the direction from the cylinder to the wall there are already differences which may be caused by short measuring record and by high sensitivity of the hot-wire probe on the turbulent fluctuations. These fluctuations arise already at the inlet to the measuring section of the tunnel and are not considered for the numerical simulation. To reduce this error in the future, for numerical simulation, the experimentally obtained velocity profile will be entered at the inlet to the measuring section. The technical solution by upgrading the suction side is also being prepared. The newer section should contain a suction sieve, together with the stabilizing honeycomb for calming flowing, and thus achieve better velocity profile of the measuring section.

4.1.2 The drag coefficient, Strouhal number

To calculate the drag coefficient C_d the library „libforces.so“ can be used. It uses the file „forceCoeffs“, in directory „system“. It allows calculating the drag forces (pressure, viscous) and the coefficients on the relevant, user defined, flown-around obstacle. Furthermore, in this file velocity **u**, density ρ, reference length L and cross sectional area S to perpendicular to the flow direction are defined. The drag coefficient C_d can be computed by the equation:

$$C_d = \frac{2 \cdot F_d}{\rho \cdot u^2 \cdot S} \qquad (6)$$

where F_d indicates total drag force which consists of a viscous and pressure drag forces, ρ indicates the density of fluid, **u** indicates velocity in unaffected field, S represents cross sectional area perpendicular to the velocity vector **u**.

The evaluation of Strouhal number St was carried out by calculating according to the following equation:

$$St = \frac{f \cdot l}{u} \tag{7}$$

where f indicates vortex shedding frequency, l represents the characteristic dimension (diameter of cylinder) and **u** represents the velocity. The frequency f was calculated using the inverse value of the difference between the two closest points of equal value of velocity **u** acquired from the graphical representation of t/**u** dependence. In the following table the comparison of numerically acquired data with data from the literature/experiment is done.

Table 1. The comparison of aerodynamic variables for cylinder

Parameter	**Cylinder Re = 26042**		
	Literature	**Numerical Results**	**Experiment**
Drag coefficient	1.009	1.066	-
Strouhal number	0.192	0.189	0.179

4.2 The acoustic experimental part of the task

As an example measurement evaluation for wind tunnel without inserted obstacles is shown. This measurement is used as a reference for measurement evaluation of acoustic variables for arbitrary obstacle.

The sampling frequency was set on the value 32.7 *kHz*, the length of the record was 10 *s*. The possibility of automatic averaging of measured data using by intensity probe software was used. Positive values were determined by the intensity sound in the direction from the measuring section of the wind tunnel, the negative intensity sound values were determined from the opposite side (room), where the measurement was carried out. This determination of positive/negative values was evaluated by a simple experiment - clap hands between the probe and the measuring section of the tunnel (for positive values), or behind the probe (for negative values) and then record this phenomenon and evaluate.

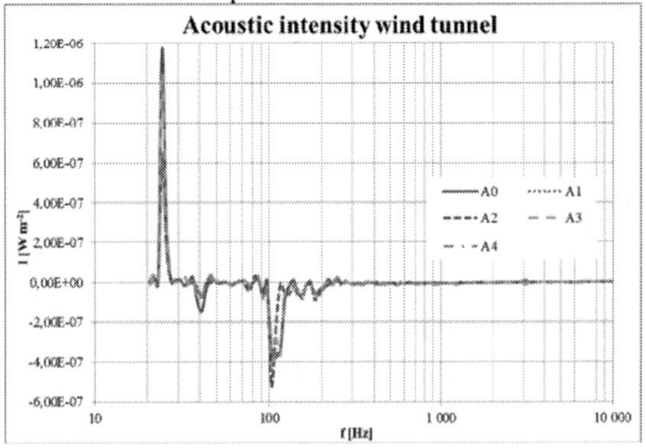

Figure 10. The evaluation of the intensity of the sound

At every point A0 - A4 three measurements were carried out. Subsequently, the average value of these three measurements was calculated and used to plot into the graph in the Figure 10.

The first significant (positive) peak is at the frequency of 25 *Hz*, which corresponds to the configuration of the frequency inverter. The value for the inverter was set to 24.8 *Hz*. The second significant (negative) peak is observed at the frequency of 109 *Hz* which can be attributed to the actual velocity of the 6-bladed fan. The silencer which is connected behind the fan causes the frequencies are reduced.

5 The final evaluation of task

In this paper a numerical solution of the noise generated by the cylinder was solved. As a CFD tool OpenFOAM was used. The turbulent model k-epsilon model Standard was selected. The measurement of velocity profiles using the hot-wire anemometer was done. Furthermore, measurement by the intensity probe in the tunnel without flown-around obstacle was carried out.

The comparison between the numerical methods and experiment was performed for velocity profiles. The comparison of velocity profiles is made in the Figure 9 for the cylinder. The comparison of the values (drag coefficient C_d, Strouhal number St) is provided in the Table 1. The comparison of numerically obtained drag coefficient and Strouhal number with the literature data indicates the values are in good compliance. In the second part of the task, after solving the aerodynamic task; the numerical solving of the acoustic wave propagation was done. To solve this problem the Curle's analogy through the implementation in the source code of the program OpenFOAM was used. The acoustic simulation results are shown in the Figure 2. The experimental part was carried out using the intensity probe. Significant peaks at two frequencies are presented.

The solving of the aeroacoustics issue will continue. In the numerical calculation area the solution of the acoustic pressure signal transition through a layer of plexiglass needs to be found. It is the border between the measuring section and the outside environment, which significantly influences the passing signal. In the experimental area the optimization of experiments will be made to reduce the noise, which is not the investigation subject and degrades the measurement for our needs.

Acknowledgement

This paper was funded by specific research on the Technical University of Ostrava named "Zkoumání dynamiky tekutinových systému", code SP2015/95.

References

1. Wagner, C. *Large-Eddy Simuation for Acoustics*. 470 (Cambridge University Press, 2012)
2. *OpenFOAM The Open Source CFD Toolbox User Guide, Version 2.3.1*. 221 (2014)
3. *Ansys Fluent Theory Guide, Release 15.0*. Ansys, Inc., 814 (2013)
4. Verhoeven, Olivier. *Trailing Edge Noise Simulations using IDDES in OpenFOAM*. 139 (Delft, 2011)

5. Uosukainen, S. *Foundations of acoustic analogies [Akustisten analogioiden perusteet].* 103 (VTT Technical Research Centre of Finland, 2011)
6. Kraposhin, M. V. Strizhak, S. V. *How to implement simple acoustic analogy in OpenFOAM.* 49 (8th OpenFOAM Workshop, Jeju, Korea 2013)
7. Roshko, A. *Experiments on the flow past a circular cylinder at very high Reynolds number.* 12 (Journal of Fluid Mechanics. 1961)
8. Ayachit, U. *The ParaView Guide*, 239 (2015)

EPJ Web of Conferences **114**, 02018 (2016)
DOI: 10.1051/epjconf/201611402018

Particle image velocimetry investigation of steady flow over a backward-facing step

Sharul Sham Dol[1,a]

[1]*Petroleum Engineering Department, Curtin University, Sarawak Campus CDT 250, Miri Sarawak, 98009 Malaysia*

Abstract. The backward-facing step (BFS) is a heuristic example, allowing for complex phenomena to arise in a simple geometry. Particle Image Velocimetry (PIV) investigations of mean-velocity distributions of backward-facing step flow with steady inlet condition were carried out and good agreement was obtained between current and previously published results for $50 \leq Re \leq 400$. This confirms that the current experimental capabilities can provide detailed and accurate velocity information. The flow behaviour downstream the step depends on the strength of separated shear layer, which the circulation depends on the bulk flow, recirculation zone length and vortex formation time. Since the vortex formation process is governed by the circulation flux convected along the wall layer from the step, for $Re \leq 400$, all of the circulation contained in the shear layer is drawn into the recirculation region. Thus, in a case where the shear layer characteristics are modified (e.g. in higher Reynolds number and unsteady flows), the balance of circulation is modified that would result in shedding.

1 Introduction

The backward-facing step (BFS) is a heuristic example, allowing for complex phenomena to arise in a simple geometry. The steady flow over a BFS has received much interest over the past decades (e.g. [1-6]), mainly due to the abundance of competing physical phenomena (e.g. internal flow separation, shear layer instability, boundary layer growth, flow recirculation, vorticity flow, vortex shedding and flow reattachment) of practical pertinence. The internal separated flow regions appear in many engineering and biomedical applications. Examples of these include flows in heat exchangers or through sudden expansions in pipe networks.

A BFS is also a suitable prototypical example for regions of flow recirculation-reattachment in physiological flows, such as in arterial bifurcation, stenoses and many biomedical devices including venous and urinary catheters, fluid transfer implants and artificial heart valves [7]. In these kinds of applications, the inlet flows are normally pulsatile and often characterized by zones of highly unstable and disturbed flows downstream of the step. Haidekker et al. [8] in their experimental studies used the BFS as an *in-vitro* model to examine the response of mammalian cells to spatial and temporal wall shear stress variations induced by disturbed flow and rapidly changing near-wall flow patterns. They showed that under pulsatile conditions, the BFS geometry causes local amplification of the shear stress in the recirculation region and that strong temporal gradients of the of flow in the BFS model also leads to a significant stimulation of endothelial cell growth. The flow pulsation induces hydrodynamic instabilities that result in flow unsteadiness downstream of the sudden expansion.

The flow over a BFS with steady inlet condition has been extensively studied both experimentally and numerically (e.g. [2, 5, 6, 9, 10, 11, 12, 13, 14, 15]). Armaly et al. [2] performed Laser-Doppler measurements of velocity distribution and reattachment length downstream of a BFS mounted in a 18:1 aspect ratio (step span/step height) channel for a wide range of Reynolds number, $70 < Re < 8000$. They reported that the length of the recirculating flow region in the immediate vicinity of the BFS (primary vortex) has strong, non-linear dependence on the Reynolds number. The primary recirculation length increases up to $Re \sim 1200$. An upper wall recirculation region (secondary vortex) appears for $Re > \sim 400$ due to the adverse pressure gradient created by the sudden expansion. The flow field downstream the step becomes unsteady due to the complex interaction between these primary and secondary vortices [6]. These vortices have different sizes and characteristics. While many questions remain, these basic flows are understood considerably better than the unsteady flows over a BFS.

Thus, the steady results will serve to verify the validity and reliability of the experimental set-up and technique used in this study. Moreover, a dynamic characterization of the strength (circulation) of the steady primary recirculation region, which has not been discussed in earlier work, provides a useful perspective

[a] Corresponding author: sharulsham@curtin.edu.my

on how to handle the different flow parameters in a case where circulation must shed (e.g. turbulent, pulsatile flow). For this purpose, steady inlet flows for $50 \leq Re \leq 400$ are experimentally investigated by using particle image velocimeter (PIV).

2 Experimental apparatus and technique

The experimental investigation was conducted in the BFS water channel as shown in Figure 1. This is a closed-loop flow chamber with a 250 mm-long working section. It incorporates backward-facing step (BFS) with an expansion ratio (H/h) of 2.0. The flow enters through a nozzle with an area contraction ratio of 10:1. This contraction was designed using the method of [16] to prevent over-speed separation. The outlet of the nozzle is connected to the inlet of the channel test section, which is 2 mm in height (h), 50 mm in width (W) and 100 mm in length up to the backward-facing step (l). These dimensions ensure a two-dimensional fully developed flow at the cross-section where the step is located. The downstream length (L) to the outlet of the channel is 150 mm. The outlet channel is long enough to allow the flow recovering into a fully developed channel profile. This BFS model has an aspect ratio of $W/h = 25$, which is sufficient to eliminate three-dimensional wall effects in the center of the span [17]. The channel walls were made of Plexi-glass for optical access with a laser measuring system. The laboratory coordinate system was taken with the origin centrally located at the sudden expansion plane. A Cartesian coordinate system is used with the x-axis aligned in the direction of the bulk flow, U_0, and the y-axis normal to the flow (transverse). A schematic of the BFS channel is shown in Figure 1 (b).

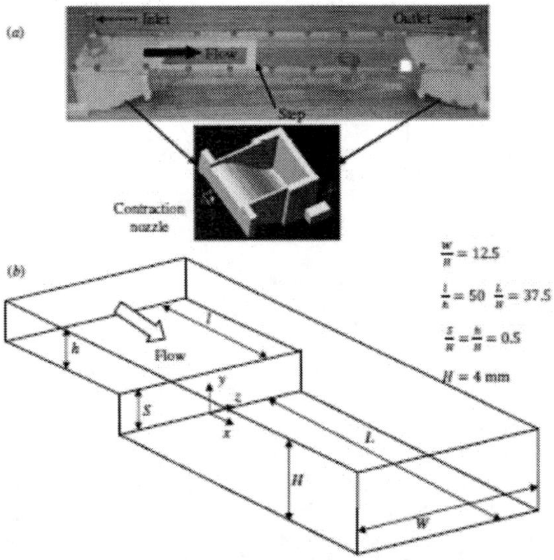

Figure 1. BFS channel; (a) Photograph of the channel; (b) Schematic of the BFS.

For steady flow experiments, a constant head gravity driven arrangement was selected, for which flow fluctuations could not be observed. Gravity driven flow for the steady flow experiments was achieved via hydrostatic pressure exerted by a difference in height between the supply reservoir and the BFS channel. Velocity measurements for steady flows were performed at $50 \leq Re \leq 400$ using high-frame-rate particle image velocimeter. The Reynolds number used in this study is defined based on $2h$ (hydraulic diameter of the inlet channel) and average bulk velocity upstream of the channel, U_0.

Flow field information was obtained by using a LaVision FlowMaster high-frame-rate PIV system. The ability of the system to simultaneously give very high spatial resolution velocimetry over a relatively large flow region and time resolved flow fields makes the technique essential to the study of evolving dynamics of coherent structures. Silicon-carbide particles with 8 μm nominal diameter were added to the water and used as tracers. The flow is illuminated by double-pulsed Nd:YLF laser (Photonics Industries DM20-527, λ = 527 nm) with a maximum energy of 20 mJ per pulse per laser head at a repetition rate of 1000 Hz.

In the present study the image recoding unit is a digital high-speed 10-bit CMOS camera of type HighSpeedStar 5 with maximum frame rate of 250 kHz. It has resolution of 1024 x 1024 pixels, captured at 3 kHz frame rates and the pixel size is 17 μm x 17 μm. The camera was fitted with an AF micro-Nikkor 60 mm f/2.8D lens and the object distance was adjusted to obtain a field-of-view of ~ 18.4 x 18.4 mm^2. The spatial resolution was 0.0179856 mm/pixel. Single-frame mode with recording rate of 125-1000 Hz was used and 500-1000 frames per acquisition were recorded in all the experiments. The vector analysis was done via DaVis 7.2 software in two steps; multi-pass with decreasingly smaller sizes interrogation windows of 64 x 64 and 16 x 16 pixel were used in the first and second step, respectively to achieve more precise results and higher spatial resolution. An overlap of 50% of the interrogation area was used together with a Gaussian window function to minimize the loss of pairs. A detailed description of the technique and the uncertainty estimation can be found in [18-19].

3 Results and discussions

Results of PIV investigations of steady laminar flows over an expansion ratio 2:1 backward-facing step (BFS) at $50 \leq Re \leq 400$ are summarized here.

3.1 Mean flow

Figure 2 (a-d) shows the mean vector flow fields and streamlines patterns for $50 \leq Re \leq 400$. The flow is laminar and steady. The results yield the expected primary zone of recirculating flow attached to the backward-facing step and also show no additional regions of flow separation downstream of the sudden expansion. This is consistent with the results reported in previous studies (e.g. [2, 4]. Armaly et al. [2] found that the secondary recirculation region (secondary vortex), due to an adverse pressure gradient created by the sudden expansion, on the opposite wall from the step develops for $Re >$ ~400.

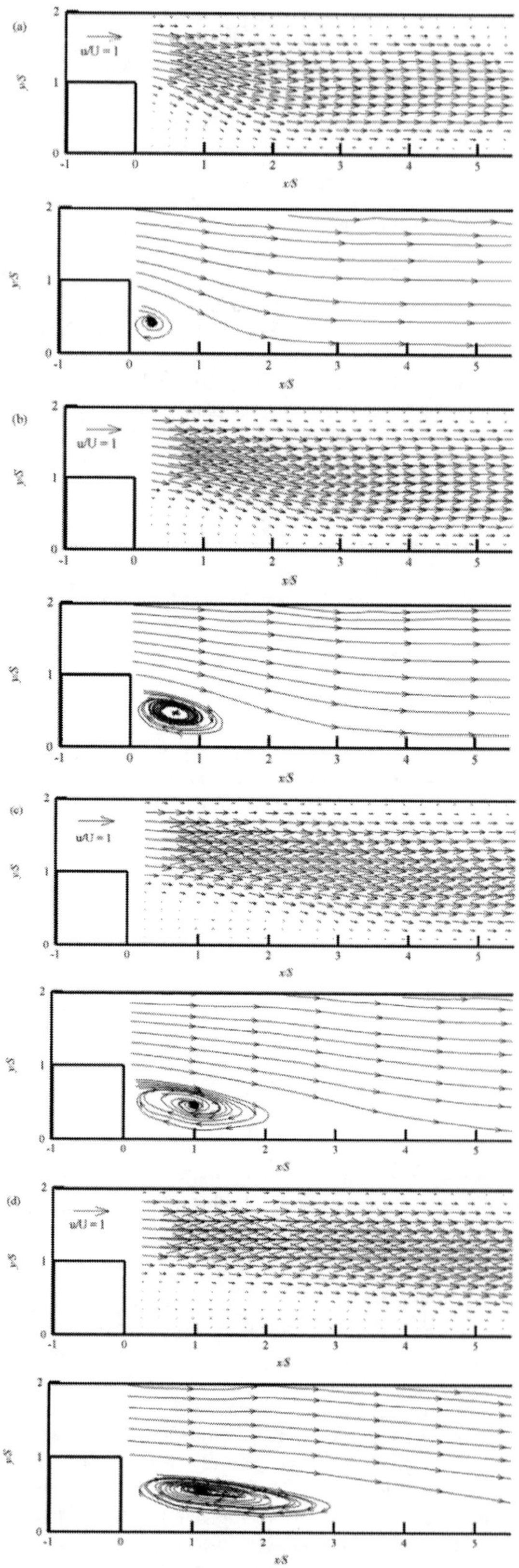

Figure 2. Mean flow fields and streamlines patterns; (*a*) Re = 50; (*b*) Re = 100; (*c*) Re = 200; (*d*) Re = 400.

Velocity profiles as shown in Figure 3 (*a-d*) demonstrate that the length of flow development downstream of the step increases with the Reynolds number. For $Re = 50$ and 100, the velocity profiles at $x/S = 4.5$ show that the velocity fields are close to that of a fully developed channel flow with a parabolic profile. However, deviations from the parabolic profile are present for higher Reynolds numbers ($Re = 200$ and 400) indicating the flow is still redeveloping at $x/S = 4.5$ due to a stronger separated shear layer, hence farther reattachment point.

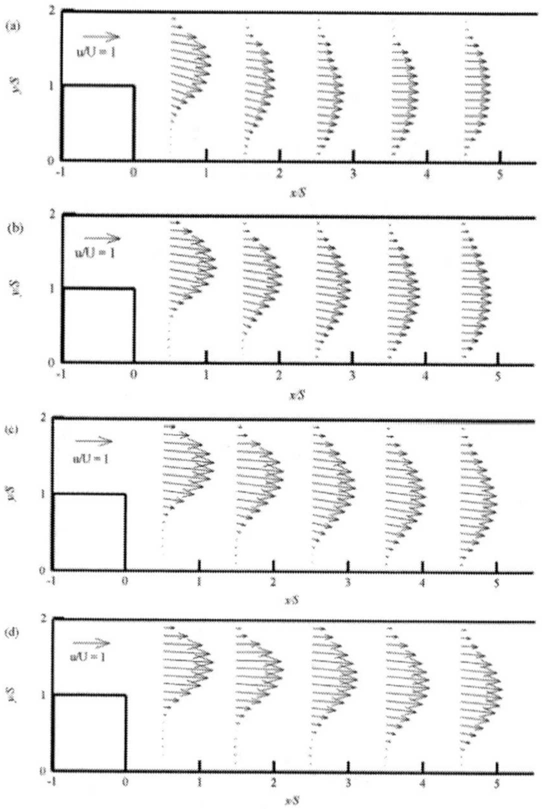

Figure 3. Velocity profiles at several locations downstream the sudden expansion; (*a*) $Re = 50$; (*b*) $Re = 100$; (*c*) $Re = 200$; (*d*) $Re = 400$.

3.2 Recirculation zone

Figure 4 shows the variation of recirculation zone lengths, X_R, with Reynolds numbers, Re. The results show that X_R increases almost linearly with Re for $50 \leq Re \leq 400$. The present results of X_R are consistent with the previous investigations [2, 14, 20]. Experimental results of [4] showed that the flow downstream of the step showed destabilization traits (e.g. transverse oscillations of the separated streamline) as soon as the secondary separation region appeared (Re ~400-500). The highly energetic shear layer becomes unstable which breaks, leading to the formation of isolated convective vortices [15] and for a steady flow over a BFS, this phenomena can only be seen for Re > ~1500 [6]. The nearly linear increase of X_R with Re suggests that the size of the recirculation region scales directly with the circulation flux along the separated shear layer as discussed in the next section.

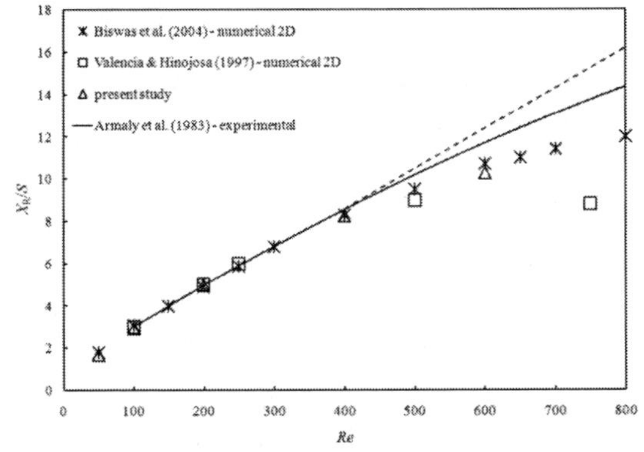

Figure 4. Variations of primary recirculation zone lengths with *Re*.

3.3 Vortex strength and formation time

Figure 5 shows the amount of circulation in the primary recirculation region (bound vortex) for $50 \leq Re \leq 400$. No vortex shedding occurs at this range of Reynolds numbers (i.e. there is no net convection of circulation downstream). The circulation flux must be balanced locally inside the steady recirculation region by the vorticity flux at the walls. Hence, it will be a good indication to see how the primary vortex characteristics (e.g. strength, formation time) changes with the bulk flow. The vortex strength (circulation), Γ_P was estimated according to vorticity integrated area of the primary vortex,

$$\Gamma_P = \oint \vec{\omega} \cdot d\vec{A} \approx \sum_{i,j} (\omega)_{i,j} \Delta A_{i,j} \tag{1}$$

where ω is the local vorticity, i and j are spatial indices, and A is the vortex area as defined by [21]. The circulation $\frac{\Gamma_P}{U_0 D}$ increases linearly, indicating that the strength of the primary recirculation region is linearly in proportion with U_0. The non-dimensional magnitude of the vorticity ($\omega = \frac{dv}{dx} - \frac{du}{dy}$) in the separated shear layer does not vary significantly with the Reynolds numbers (Figure 6), so that the circulation scales with the characteristic length X_R.

The rate of circulation feeding the primary vortex is assumed to be the rate at which circulation is advected along the wall boundary layer separating at the step:

$$\frac{D\Gamma}{Dt} \approx \int u \frac{\partial u}{\partial y} dy = \frac{{U_1}^2 - {U_2}^2}{2} \tag{2}$$

where U_1 and U_2 are the velocities at the edges of the shear layer, the strength of which is $\Delta U = U_1 - U_2$. Assuming that the shear layer behaves as a vortex sheet (no net vorticity flux from the shear layer), the initial circulation flux is preserved and should correspond to the circulation flux along the wall layer separating at the step (i.e. the wall no longer acts as a vorticity flux source). Hence, $U_1 = kU_0$ ($k = 1.5$ for a laminar flow) and and $U_2 = 0$ such that:

$$\frac{D\Gamma}{Dt} = \frac{k^2 {U_0}^2}{2} \tag{3}$$

which is shown in Figure 7. It shows the flux scales with U_0^2.

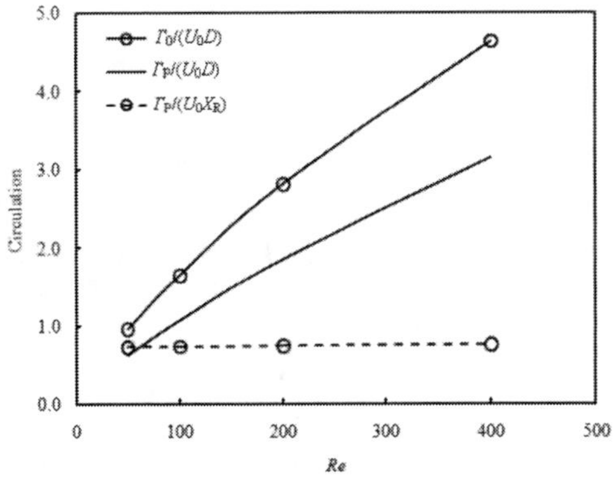

Figure 5. Variations of circulation with *Re*.

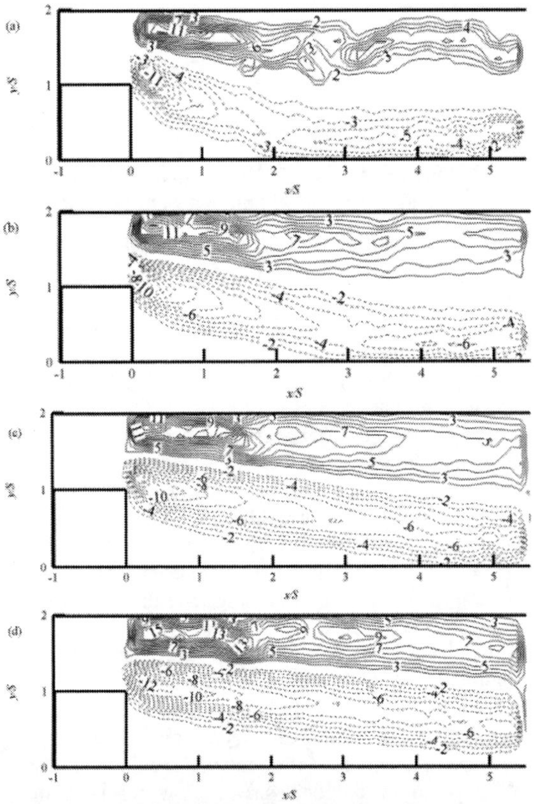

Figure 6. Mean vorticity contours ($\omega D/U_0$). Dashed (solid) lines represent constant negative (positive) vorticity values; (*a*) $Re = 50$; (*b*) $Re = 100$; (*c*) $Re = 200$; (*d*) $Re = 400$.

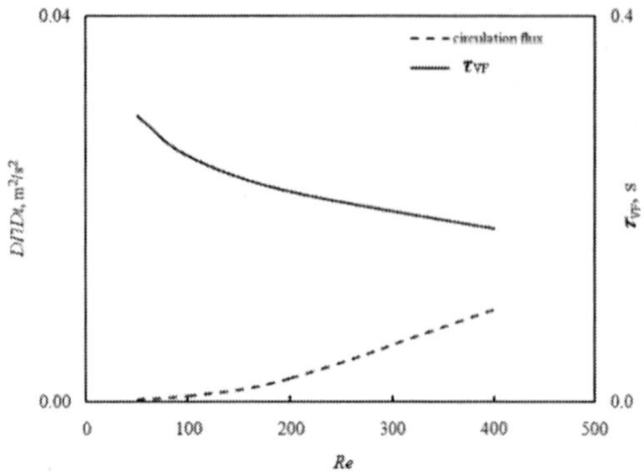

Figure 7. Variations of circulation flux and vortex formation time with *Re*.

If the shear layer is assumed to behave like a vortex sheet of intensity ΔU, then the total circulation contained in the shear layer should be:

$$\Gamma_0 = \int \Delta u ds \tag{4}$$

where s is the distance from the step along the vortex sheet. The distance s must scale with X_R such that $\Gamma_0 \propto \Delta u X_R = kU_0X_R$. Assuming that the circulation losses at the walls (vorticity flux along the walls due to the wall-parallel pressure gradient) also scale with U_0^2, one obtain that

$$\frac{\Gamma_P}{U_0X_R} \propto \frac{\Gamma_0}{U_0X_R} = \text{constant} \tag{5}$$

This assumption appears at least approximately valid for $Re \leq 400$, as shown in Figure 5, where $\frac{\Gamma_P}{U_0X_R} \approx 0.75$. This observation confirms that the formation process is governed by the circulation flux convected along the wall layer from the step. The implication of this observation is that for $Re \leq 400$, all of the circulation contained in the shear layer is drawn into the recirculation region. Noting that X_R in this range increases almost linearly with U_0, it also implies that Γ_P should scale with U_0^2.

As the Reynolds number increases beyond 400, it is noted that the rate of growth of X_R decreases significantly with *Re* (i.e. deviates from a linear progression), but the rate at which circulation is advected along the shear layer must continue to scale with U_0^2 (since the parabolic profile upstream of the step is independent of *Re* for laminar flow). Again, assuming that the wall vorticity flux scales with U_0^2, which seems reasonable since the vortex is vertically bound by the shear layer, an equilibrium with the circulation contained in the vortex cannot be sustained. In other words, some of the circulation in the shear layer must be advected downstream. This loss of circulation would thus result in perturbation of the downstream flow resulting in the observed instabilities [6, 11, 15]. For example, in case where the recirculation region may become unstable due to the breakdown of shear layer that leads to the formation of shed vortices, as described by [15], the balance of circulation maybe interrupted because the excess of circulation is released during the shedding period. Armaly et al. [2] found that the recirculation zone length, X_R decreased in the range of $1200 \leq Re \leq 6600$ (transitional regime). This infers an abruption to a steady vortex formation.

Another observation from the foregoing analysis is that there is a natural time-scale associated with the formation of the primary vortex. This vortex formation time, τ_{VF}, can be simply defined as

$$\tau_{VF} = \frac{\Gamma}{\frac{D\Gamma}{Dt}} \sim \frac{x_R}{U_0} \tag{6}$$

which is shown in Figure 7. This formation time may play a role in unsteady and pulsating flows. If the frequency of pulsation is very low (period of pulsation, T, much longer than τ_{VF}), it is expected that the influence of the pulsation will be small. Conversely, if $T \ll \tau_{VF}$, the pulsation will interfere with the vortex formation [18-19]. Examples of two cases of pulsatile flows with different period of pulsation are given below to see the effects of this vortex formation time.

The mean streamlines patterns for $St = fD/U_0 = 0.4$ where f is the pulsation frequency and for $Re = 100$ is shown in Figure 8. The inlet pulsation leads to different flow behaviour downstream of the sudden expansion compared to the steady flow field. The main different between the mean flow for the pulsatile and steady cases is the appearance of the upper wall recirculation region (secondary vortex). This upper wall separation only happens at $Re > \sim 400$ for the steady case. The upper wall vortex leads to a destabilization in the flow field through concave curvature instability of the boundary layer [6, 18, 19] and thus it is expected the circulation will be shed.

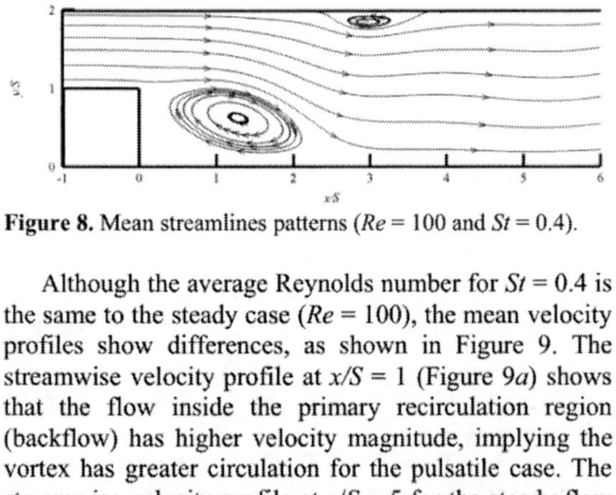

Figure 8. Mean streamlines patterns ($Re = 100$ and $St = 0.4$).

Although the average Reynolds number for $St = 0.4$ is the same to the steady case ($Re = 100$), the mean velocity profiles show differences, as shown in Figure 9. The streamwise velocity profile at $x/S = 1$ (Figure 9*a*) shows that the flow inside the primary recirculation region (backflow) has higher velocity magnitude, implying the vortex has greater circulation for the pulsatile case. The streamwise velocity profile at $x/S = 5$ for the steady flow shows that the velocity fields are close to that of a fully developed channel (parabolic profile) whilst the flow is still redeveloping for the pulsatile case (Figure 9*b*). The transverse velocity profiles also indicate the deviations from the steady flow behaviour (Figure 9*c*). These differences can be attributed to modification to the momentum transfer due to the flow pulsatility.

For $St = 0.035$, the formation of primary and secondary vortices is regular but no shedding is observed. Although the forming primary and secondary vortices are much stronger (e.g. compared to $St = 0.4$), these remain bound during the cycle and both the primary and secondary vortices lose their strength as soon as they enter the deceleration phase and eventually disappeared at the end of the pulsation cycle. Pulsatile flow with very small Strouhal number behaves more like a steady flow [22].

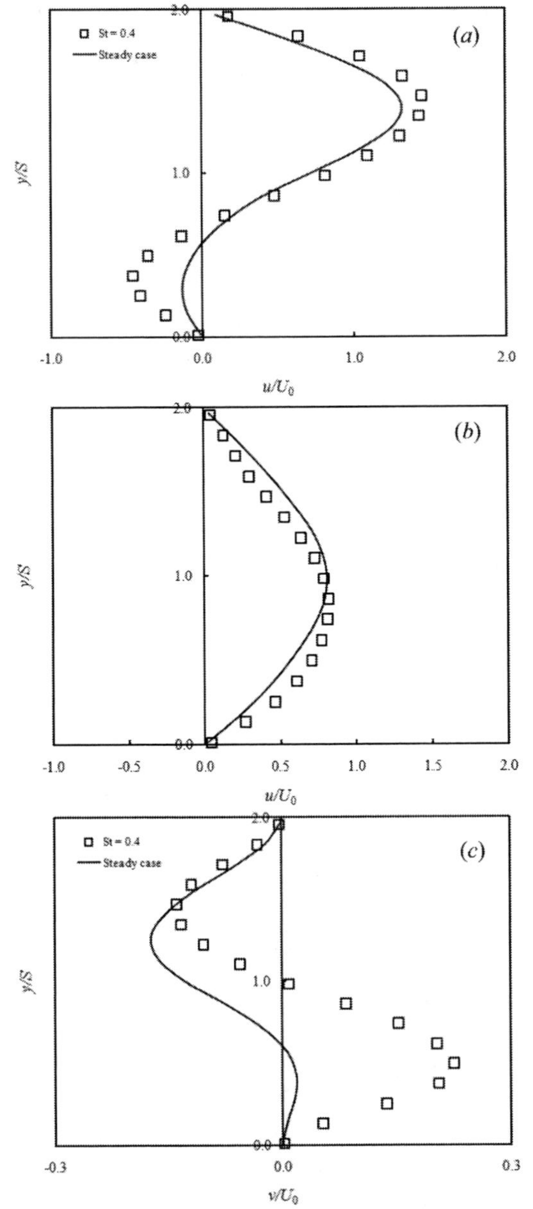

Figure 9. Mean velocity profiles for Re = 100; (*a*) Streamwise, x/S = 1; (*b*) Streamwise, x/S = 5; (*c*) Transverse, x/S = 1.

4 Conclusion

PIV investigations of mean-velocity distributions of backward-facing step flow with steady inlet condition were carried out by the author, and good agreement was obtained between current and previously published results for $50 \leq Re \leq 400$. This confirms that the current experimental capabilities can provide detailed and accurate velocity information. The flow behaviour downstream the step depends on the strength of separated shear layer, which the circulation depends on the bulk flow, U_0, recirculation zone length, X_R, and vortex formation time, τ_{VF}. Since the vortex formation process is governed by the circulation flux convected along the wall layer from the step, for $Re \leq 400$, all of the circulation contained in the shear layer is drawn into the recirculation region. Thus, in a case where the shear layer characteristics are modified (e.g. in higher Reynolds number or unsteady flow), the balance of circulation is modified that would result in shedding.

References

1. R. J. Goldstein, V. L. Eriksen, R. M. Olson, E. R. G. Eckert, Laminar separation reattachment, and transition of flow over a downward-facing step, *Trans. ASME D: J. Basic Engng.* **92**, 732 (1970).
2. B. F. Armaly, F. Durst, J. C. F. Pereira, B. Schönung, Experimental and theoretical investigation of backward-facing step flow, *J. Fluid Mech.* **127**, 496 (1983).
3. K. B. Chun, H. J. Sung, Visualization of a locally-forced separated flow over a backward-facing step, *Exp. Fluids* **25**, 133 (1998).
4. T. Lee, D. Mateescu, Experimental and numerical investigation of 2D backward-facing step flow, *J. Fluids Struct.* **12**, 703 (1998).
5. J. Kostas, J. Soria, M. S. Chong, Particle image velocimetry measurements of a backward-facing step flow, *Exp. Fluids* **33**(6), 838 (2002).
6. H. P. Rani, W. H. S. Tony, S. F. T. Eric, Eddy structures in a transitional backward-facing step flow, *J. Fluid Mech.* **588**, 43 (2007).
7. G. A. Truskey, K. M. Barber, T. C. Robey, L. A. Olivier, M. P. Combs, Characterization of a sudden expansion flow chamber to study the response of endothelium to flow recirculation, *ASME J. Biomech. Engg.* **117**, 203 (1995).
8. M. A. Haidekker, C. R. White, J. A. Frangos, Analysis of temporal shear stress gradients during the onset phase of flow over a backward-facing step, *J. Biomech. Engg.* **123**, 455 (2001).
9. J. Kim, and P. Moin, Application of a fractional step method to incompressible flows, *J. Comput. Phys.* **59**, 308 (1985).
10. I. Grant, E. Owens, Y. Yan, Particle image velocimetry measurements of the separated flow behind a rearward facing step, *Exp. Fluids* **12**, 238 (1992).
11. P. T. Williams, A. J. Baker, Numerical simulations of laminar flow over a 3D backward-facing step, *Int. J. Numer. Methods Fluids* **24**, 1159 (1997).
12. T. P. Chiang, T. W. H. Sheu, A numerical revisit of backward-facing step flow problem, *Phys. Fluids* **11**(4), 874 (1999).
13. D. Barkley, M. G. M Gomes, and Henderson, R. D. Three-dimensional instability in flow over a

backward-facing step, *J. Fluid Mech.* **473**, 190 (2002).
14. G. Biswas, M. Breuer, F. Durst, Backward-facing step flows for various expansion ratios at low and moderate Reynolds numbers, *J. Fluids Eng.* **126**, 362 (2004)
15. C. P. Schram, C., P. Rambaud, M. L. Riethmuller, Wavelet based eddy structure eduction from a backward facing step flow investigated using particle image velocimetry, *Exps. Fluids* **36**, 233 (2004).
16. P. A. Libby, H. R. Reis The design of two-dimensional contraction sections, *Quarterly of Applied Mathematics* **10**, 95 (1951)
17. V. de Brederode, P. Bradshaw, Three-dimensional flow in nominally two-dimensional separation bubbles, I. Flow behind a rearward-facing step, *Aero Report* 72-19, Imperial College of Science and Technology, London (1972).
18. S. S. Dol, M .M. Salek, R. J. Martinuzzi, Salek, R. J. Martinuzzi, Effects of Pulsation to the Mean Field and Vortex Development in a Backward-Facing Step Flow. *Journal of Fluids Engineering*, ***136***(1) (2014).
19. S. S. Dol, M .M. Salek, R. J. Martinuzzi, Energy Redistribution Between the Mean and Pulsating Flow Field in a Separated Flow Region. *Journal of Fluids Engineering*, ***136***(11), 111105 (2014).
20. A. Valencia, L. Hinojosa, Numerical solutions of pulsating flow and heat transfer characteristics in a channel with a backward-facing step, *Heat and Mass Transfer* **32**,143 (1997).
21. J. Jeong, F. Hussain, On the identification of a vortex, *J. Fluid Mech.* **285**, 69 (1994).
22. M .M. Salek, S. S. Dol, R. J. Martinuzzi, Analysis of Pulsatile Flow in a Separated Flow Region, *ASME Proceedings 5th Symposium on Fundamental Issues and Perspectives in Fluid Mechanics*, pp. 1429-1438 (2009).

EPJ Web of Conferences **114**, 02019 (2016)
DOI: 10.1051/epjconf/201611402019

Proper orthogonal decomposition analysis of vortex shedding behind a rotating circular cylinder

Sharul Sham Dol[1,a]

[1]*Petroleum Engineering Department, Curtin University, Sarawak Campus CDT 250, Miri Sarawak, 98009 Malaysia*

Abstract. Turbulence studies were made in the wake of a rotating circular cylinder in a uniform free stream with the objective of describing the patterns of the vortex shedding up to suppression of the periodic vortex street at high velocity ratios, λ. The results obtained in the present study establish that shedding of Kármán vortices in a rotating circular cylinder-generated wake is modified by rotation of the cylinder. Alternate vortex shedding is highly visible when $\lambda < 2.0$ although the strength of the separated shear layers differ due to the rotation of the cylinder. The spectral density in the wakes indicate significant changes at $\lambda = 2.0$. The results indicate that the rotation of the cylinder causes significant disruption in the structure of the flow. Alternate vortex shedding is weak, distorted and close to being suppressed at $\lambda = 2.0$. It is clear that flow asymmetries will weaken vortex shedding, and when the asymmetries are significant enough, total suppression of a periodic street occurs. Particular attention was paid to the decomposition of the flow using Proper Orthogonal Decomposition (POD). By analyzing this decomposition with the help of Particle Image Velocimetry (PIV) data, it was found that large scales contribute to the coherent motion. Vorticity structures in the modes become increasingly irregular with downstream distance, suggesting turbulent interactions are occurring at the more downstream locations, especially when the cylinder rotates.

1 Introduction

The study of vortices (i.e. vortex motions, vortex instability, vortex breakdown) will lead to an understanding of turbulent flow. Turbulence is rotational and characterized by high levels of fluctuating vorticity. For this reason, vortex dynamics plays an essential role in the description of turbulent flows. For example, the existence of energy transfer (energy cascade) from large eddies to small eddies is driven by vortex-stretching mechanism. This is an important concept to explain why turbulent flows are always dissipative.

The flow past a rotating cylinder has been studied extensively (e.g. [1-6]), yet much continues to be unknown about the flow structure and dynamics of this geometrically simple, turbulent flow. Vortical structures in the flow and their interactions are believed to play an important role in the generation of turbulent stresses, turbulent kinetic energy, wake entrainment and growth and mean flow development. However, the direct and correct identification and observation of vertical structures and their interactions when the cylinder rotates has received little attention to date. Dol [7] and Dol et al. [8-9] have conducted a detailed experimental study, using particle image velocimeter (PIV), of the structure and behaviour of vortex shedding and vortex street for a turbulent flow over a rotating cylinder.

The data obtained from PIV measurements is well suited to Proper Orthogonal Decomposition (POD) analysis. The aim is to capture most of the kinetic energy of the flow in the least number of modes possible. The snapshot POD technique is performed on instantaneous velocity fields. Proper orthogonal decomposition is performed on the velocity field data in an effort to identify and investigate time-dependent structures in the flow. The main objective of the present work is to study the breakdown of Kármán vortex street when the cylinder rotates. The interest is to examine the turbulent properties at the suppression of periodicity in vortex formation process.

2 Experimental apparatus and technique

The experiments were performed at the Boundary Layer Wind Tunnel Laboratory (BLWTL) of the University of Western Ontario (UWO) in a suction-type wind tunnel with a 1500 mm long working section and a 450 x 450 mm^2 cross section. Hot-wire anemometry (HWA) measurements across the test section showed a uniform free-stream and a turbulence intensity of less than 1%. A smooth steel circular cylinder of nominal diameter D = 24 mm was mounted horizontally in the midplane spanning the full width of the working section. A photograph of the set-up is shown in Figure 1. The

[a] Corresponding author: sharulsham@curtin.edu.my

cylinder aspect ratio was about 18.8 so that the statistical properties of the flow were effectively two-dimensional [2]. The blockage ratio was 5.3%, which is at the upper bound of the region of negligible blockage effects [10-11]; no correction was made for blockage effects. A definition sketch of the flow is given in [8]. Particle image velocimetry (PIV) experiments were performed at Reynolds number of 9000 and for speed ratios, λ, between 0 and 2.7. Operating parameters for the PIV set-up and measurements can be found in [7-9].

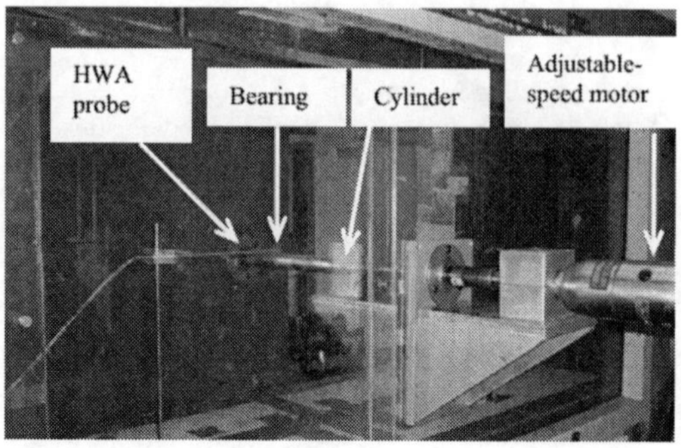

Figure 1. A photograph showing the mechanical set-up.

2.1 POD background theory

Let $f(x,t)$ be a random vector function on a finite spatial domain D and a finite time interval $0 \le t \le T$. Proper orthogonal decomposition involves solving the following Fredholm-type integral equation

$$\int_D R(x,x')\psi(x')dx' = \lambda\psi(x) \tag{1}$$

with $R(x,x')$ known as the kernel. $R(x,x')$ is the two-point averaged spatial correlation function. Solution of Eq. (1) gives $\psi^{(k)}(x)$ and λ_k which are the empirical eigenfunctions and eigenvalues respectively. The orthogonality condition implies that eigenfunctions of different order do not interact with each other in which case their scalar product,

$$\psi^{(k)},\psi^{(l)} \equiv \int_D \psi^{(k)}(x)\psi^{*(k)}(x)dx = \delta_{kl} \tag{2}$$

is zero (* represents the complex conjugate). The original data set can be reconstructed using

$$f(x,t) \approx \sum_{k=1}^{K} a^{(k)}(t)\psi^{(k)}(x) \tag{3}$$

which is known as the proper orthogonal decomposition of $f(x,t)$ .In practical situations, the data $f(x,t)$ are limited to a finite number of frames or snapshots, M. It can be shown that the matrix equation

$$C\phi^{(k)} = \lambda_k \phi^{(k)} \tag{4}$$

gives equivalent solutions to Eq. (1), [12]. Here C is the spatial correlation matrix

$$C_{m,n} = \frac{1}{M}\int_D f(x,t_m)f(x,t_n)dx, \quad m,n = 1,\ldots\ldots,M \tag{5}$$

with $\phi^{(k)}$ and λ_k being the eigenvectors and eigenvalues of C for the kth mode respectively. The empirical eigenfunctions $\psi^{(k)}(x)$ are constructed from a linear combination of the snapshots using

$$\psi^{(k)}(x) = \sum_{n=1}^{M} \phi_n^{(k)} f(x,t_n) \tag{6}$$

The problem has now been reduced to solving an M x M matrix eigenvalue problem instead of an N x N problem (where N is the number of spatial points in the domain and in most cases $N >> M$). This technique was first suggested by [12] and is commonly known as the method of snapshots. The direct method is computationally expensive for high-resolution data on a large spatial domain, while the snapshot method only becomes computationally expensive for large ensemble sizes (40 PIV snapshots were used in present analysis for each velocity ratio).

The energy of an instantaneous flow field variable, $f(x,t_n)$, represented by

$$E_{tot} = (f,f) = \int_D f(x,t_n)f(x,t_n)dx \tag{7}$$

The time average of the energy can thus be expressed as

$$\langle E_{tot}\rangle = \int_D \overline{f(x)}^2 dx + \left\langle \int_D f'(x,t_n)^2 dx \right\rangle = \langle E_{mean}\rangle + \langle E_{fluc}\rangle \tag{8}$$

which can be decomposed into a mean and fluctuating component. It can be shown [13] that the energy fraction associated with the kth mode, or eigenfunction $\psi^{(k)}(x)$ is

$$E_k = \frac{\lambda_k}{\sum_{k=1}^{M} \lambda_k} \quad (9)$$

Using the definition of the inner product given by Eq. (2), the energy of the flow based on the velocity, ***u*** yields the following results

$$\text{kinetic energy density} = \frac{1}{2}\rho(u,u) = \frac{1}{2}\rho\|u\|^2 \quad (10)$$

where ρ is the fluid density. Hence, applying POD to ***u*** will categorize structures according to their contribution of kinetic energy to the flow.

3 Results and discussions

The POD modes represent the most common events occurring in the turbulent fields. Modes 1, 2, 3 and 10 from the velocity decomposition are shown in Figure 2-5 for a stationary cylinder case. In agreement with [14], the modes contribute about 22%, 20%, 13%, 5% and 2% respectively to the total fluctuating kinetic energy in the flow (Figure 6). It is seen that the first 2 modes have almost similar levels of energy. This result is suggestive of a less diversity of flow structure in the velocity field and results in a distinct distribution of energy between the modes and hence, an energy decomposition. A trend noticeable in the progression towards higher modes is the increase in small-scale structure within them. This can be clearly seen in the comparison between vorticity plots (curl of velocity modes) between mode 1 and mode 10, shown in Figure 2 and 5. The small scales are found to contribute only a small fraction to the fluctuating energy of the flow. However, they still appear to represent physically meaningful flow structures.

Low order representations of the instantaneous *x-y* plane velocity fields $\omega(x,y,t_n)$, were reconstructed with the computed velocity modes, $\psi^{(k)}(x)$. Their corresponding vorticity distributions were compared with the distributions obtained from the PIV measurements. The reconstructions of the vorticity field using k = 1, 2, 4 and 10 modes are illustrated in Figure 7. The distribution obtained from the PIV for the same snapshot is also included. The figures illustrate the cumulative effect of using increasingly more modes in the data reconstruction. Good qualitative agreement of the vorticity contours is achieved with the measured distributions using approximately 4 modes in the reconstructions. As expected, the pattern of von Kármán vortices is well produced. Perrin et al. [14] found that taking $N > 2$ seems to lead an overestimation of the contribution of the coherent fluctuation. A comparison between the instantaneous PIV measurements and the POD modes shows how the location of flow structures in the POD modes coincide with the location of flow structures in the PIV measurements. Large-scale structures appear to be responsible for the majority of the vortex activity. The inclusion of high-order modes (small scale structures) contribute to the fine-scale structure in the reconstructed distributions. An exaggerated vorticity distribution becomes noticeable for the 10 modes reconstruction. It seems that once all vorticity levels are recovered, the use of increasingly more modes adversely affects the reconstructions. This suggests a poor convergence of the high order modes in the decomposition which arises from noise in the PIV measurements.

Modes 1, 2, 3 and 10 from the velocity decomposition are shown in Figure 8 for $\lambda = 2.0$ case. They contribute 27%, 6%, 5% and 3% respectively to the total fluctuating kinetic energy in the flow (Figure 9). It is seen that only the first mode has significant level of energy. This result is suggestive of a greater diversity of flow structure in the velocity field and results in a more even distribution of energy between the modes and hence a less efficient decomposition. The reconstructions of the vorticity field using k = 1, 2, 4 and 10 modes are illustrated in Figure 10. The distribution obtained from the PIV for the same snapshot is also included. Looking at the distribution of energy, where a constant decrease is observed after $N = 1$, and looking at the reconstruction of the instantaneous field, it seems that taking into account the first mode would represent a good picture of the coherent picture. Dol et al. [8] found that the vortex strength is only about 10% of the initial strength at this velocity ratio. The unequal strength of the separated shear layer disrupts the vortex formation process resulting in distortion of Kármán vortex street. In other words, coherent structures are suppressed. Probably this is the reason why the vorticity structures smear over the area in larger POD modes.

The results obtained in [7-9] establish that shedding of Kármán vortices in a rotating circular cylinder-generated wake is modified by rotation of the cylinder. Alternate vortex shedding is highly visible when $\lambda < 2.0$ although the strength of the separated shear layers differ due to the rotation of the cylinder. The statistics and turbulence quantities in the wakes show significant changes at $\lambda = 2.0$. The results indicate that the rotation of the cylinder causes significant disruption in the structure of the flow. It is clear that flow asymmetries will weaken vortex shedding, and when the asymmetries are significant enough, total suppression of a periodic street occurs.

The average centerline velocities represent the maximum velocity deficit of the wake, U_C. The velocity defect is smaller as λ increases. U_C/U_∞ is about unity when $\lambda \geq 2.0$. This is due to the reduced vortex activity in the formation region [9]. This also implies that the location at which the maximum velocity defects is shifted. The mean square value, the variance (or the rms value of the fluctuation, the standard deviation) becomes smaller at higher rotational rates (especially when $\lambda \geq 2.0$). This also means that the width of the probability density function becomes smaller (i.e. turbulent fluctuations are reduced into the vicinity of the mean value). Consequently, the turbulent intensities, T*i* are also getting smaller as λ increases due to suppression of coherent vortex structures [9].

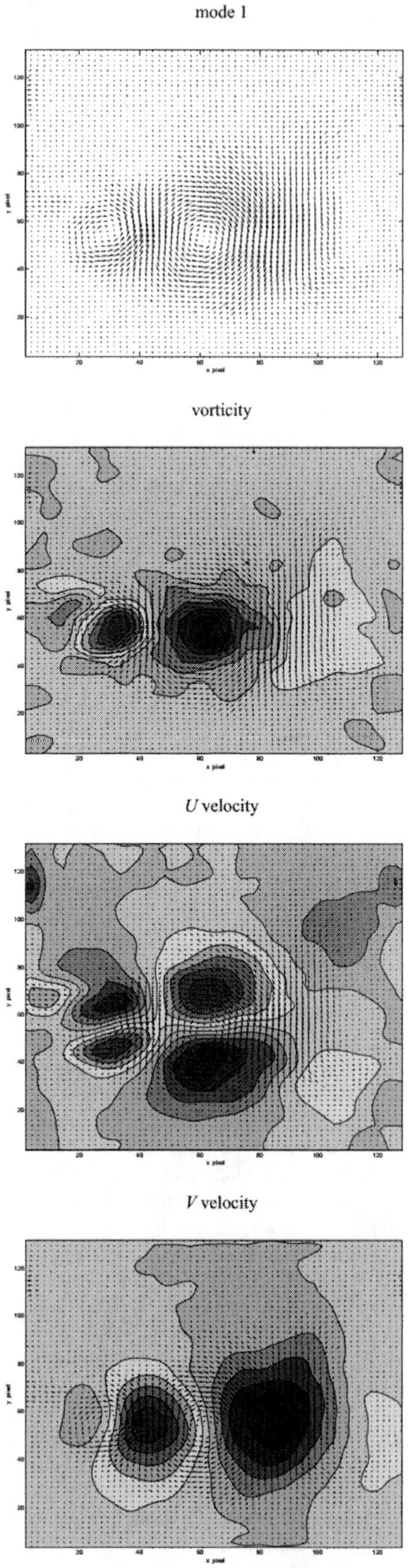

Figure 2. Mode 1 of velocity decomposition (center of the cylinder is at 0, 50).

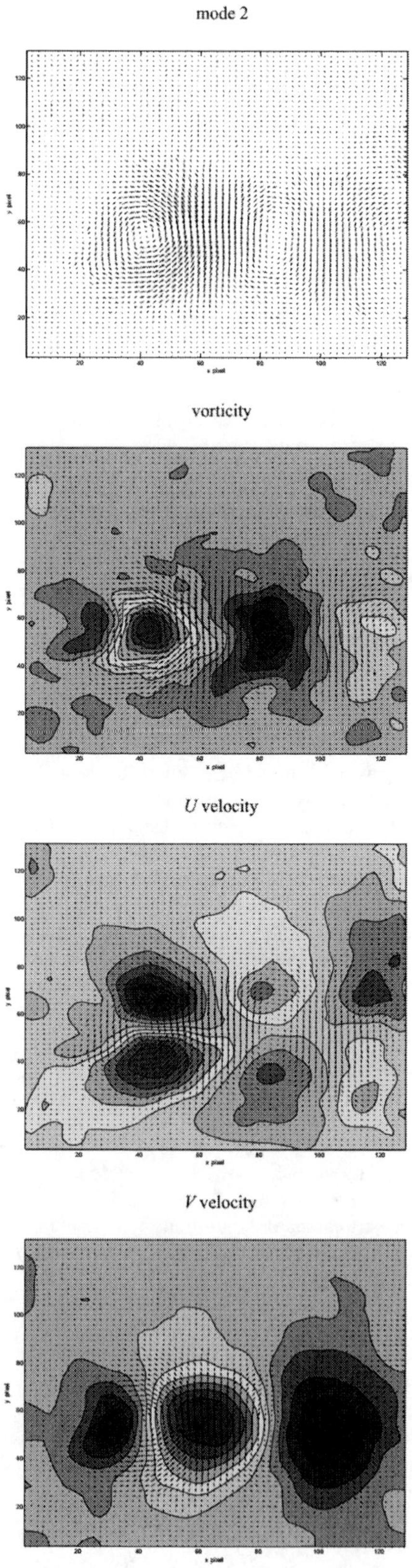

Figure 3 Mode 2 of velocity decomposition (center of the cylinder is at 0, 50).

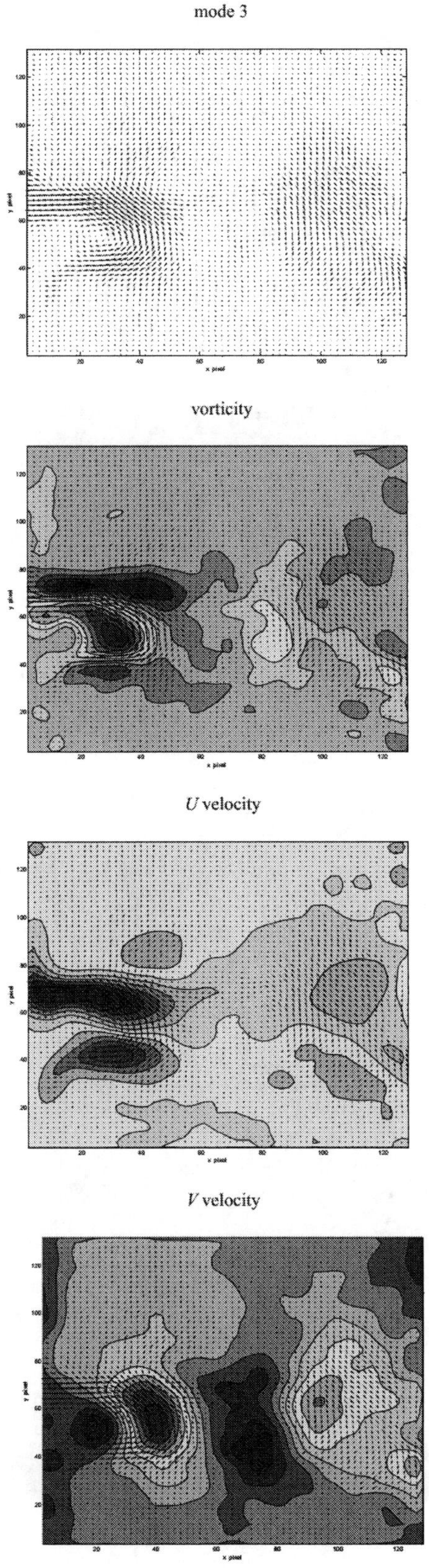

Figure 4 Mode 3 of velocity decomposition (center of the cylinder is at 0, 50).

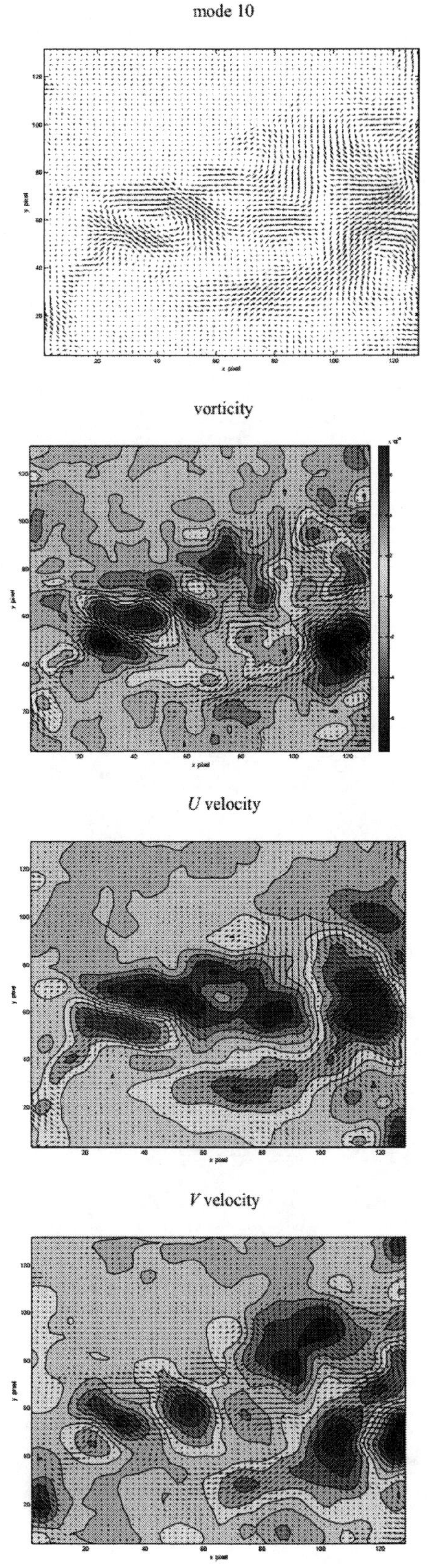

Figure 5 Mode 10 of velocity decomposition (center of the cylinder is at 0, 50).

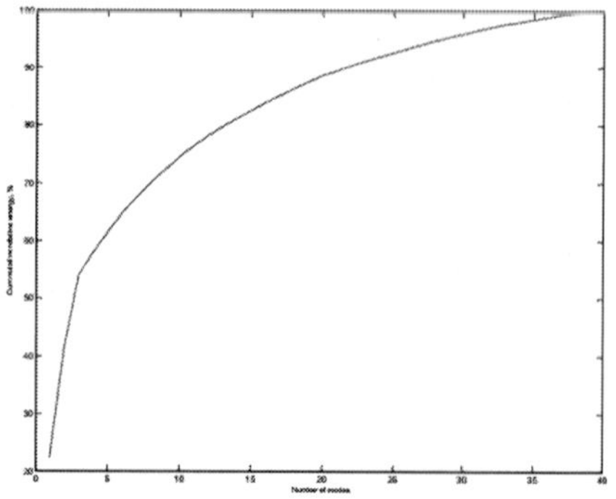

Figure 6 POD energy of the modes (in % of the total energy of the fluctuating motion on the domain.

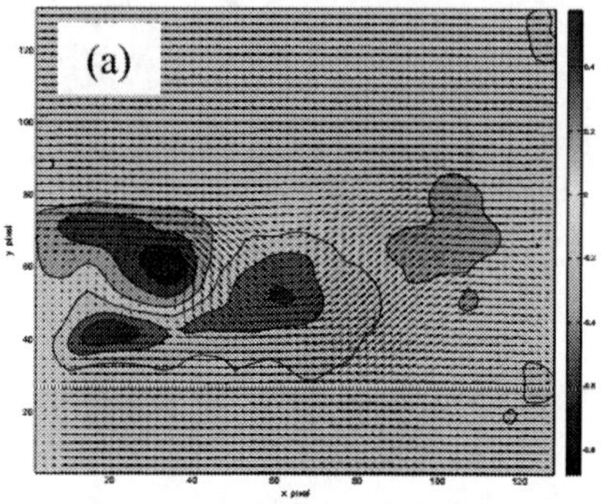

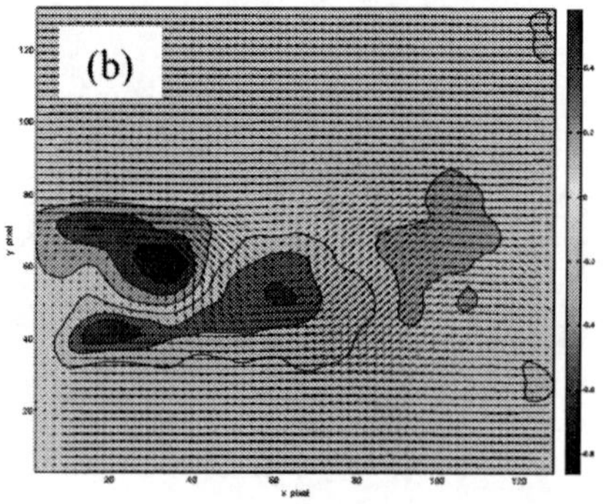

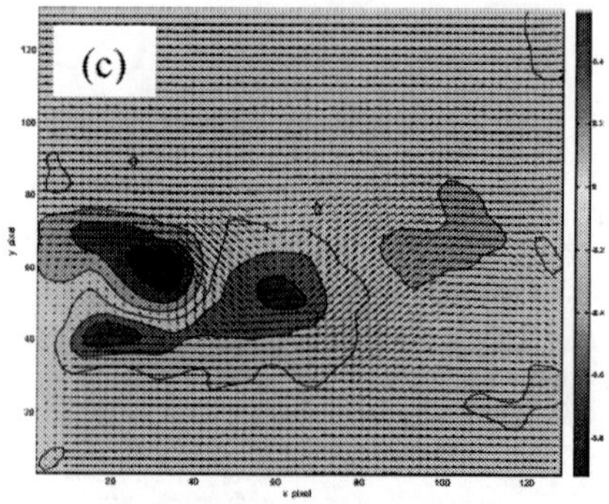

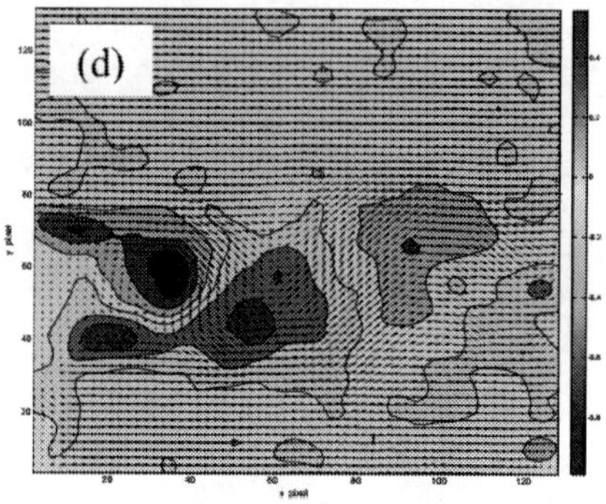

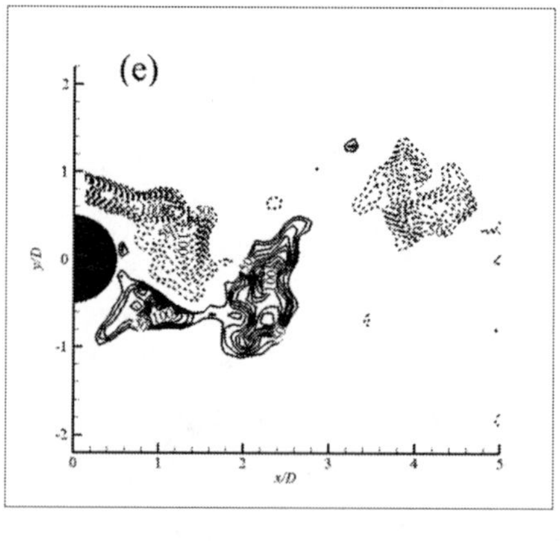

Figure 7 Vorticity distribution, obtained using (a) 1 (b) 2 (c) 4 (d) 10 modes in the reconstruction; and (e) from PIV data.

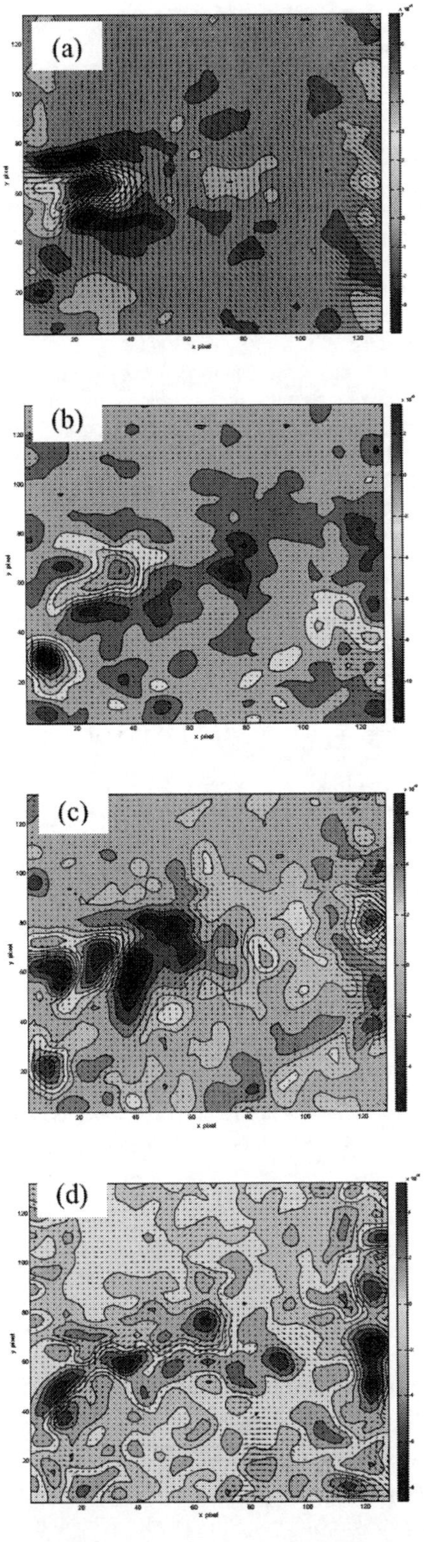

Figure 8. POD modes (a) 1 (b) 2 (c) 3 and (d) 10. Contours of vorticity are shown (center of the cylinder is at 0, 50).

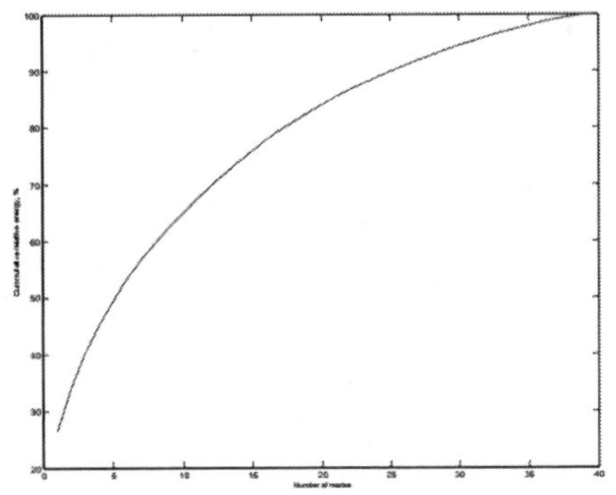

Figure 9. POD energy of the modes (in % of the total energy of the fluctuating motion on the domain.

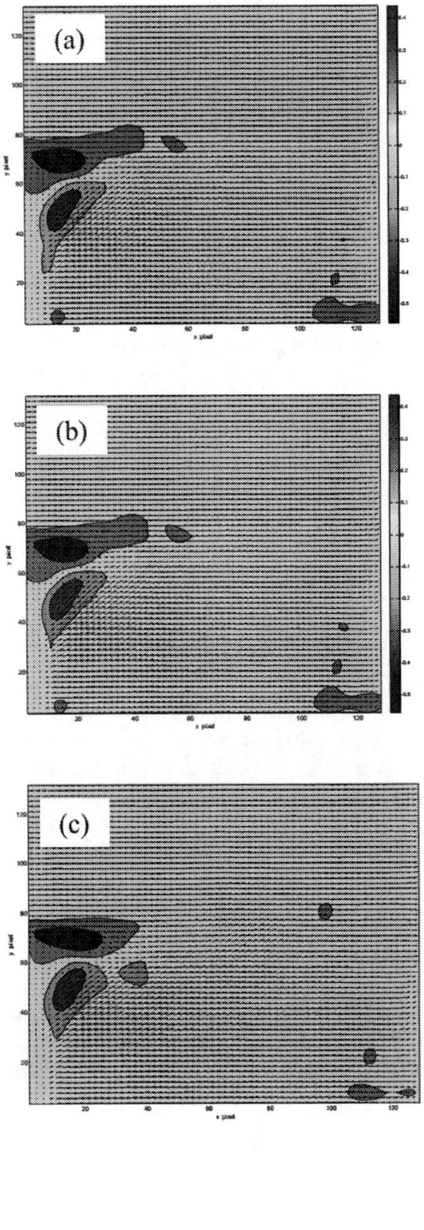

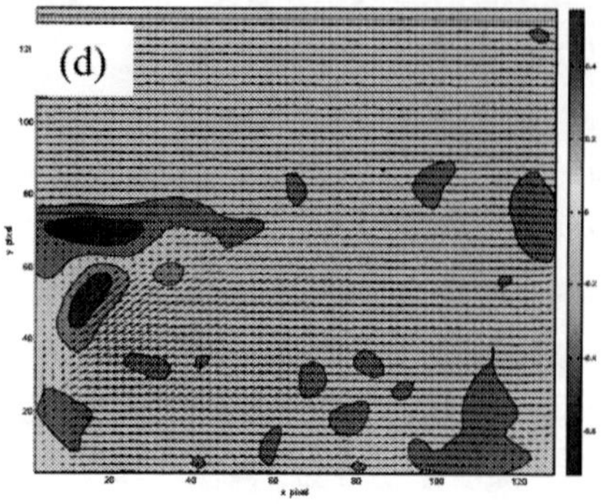

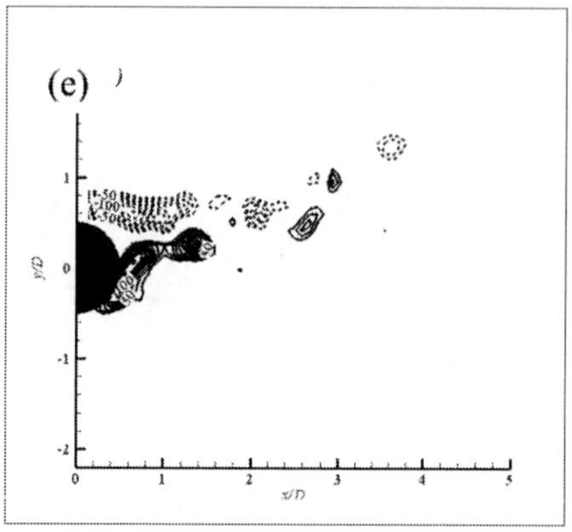

Figure 10 Vorticity distribution, obtained using (a) 1 (b) 2 (c) 4 (d) 10 modes in the reconstruction; and (e) from PIV data.

4 Conclusion

Turbulence studies were made in the wake of a rotating circular cylinder in a uniform free stream with the objective of describing the patterns of the vortex shedding up to suppression of the periodic vortex street at high velocity ratios. The results obtained in the present study establish that shedding of Kármán vortices in a rotating circular cylinder-generated wake is modified by rotation of the cylinder. Alternate vortex shedding is highly visible when $\lambda < 2.0$ although the strength of the separated shear layers differ due to the rotation of the cylinder. The results indicate that the rotation of the cylinder causes significant disruption in the structure of the flow. Alternate vortex shedding is weak, distorted and close to being suppressed at $\lambda = 2.0$. It is clear that flow asymmetries will weaken vortex shedding, and when the asymmetries are significant enough, total suppression of a periodic street occurs. Particular attention was paid to the decomposition of the flow using POD. By analyzing this decomposition with the help of PIV data, it was found that large scales contribute to the coherent motion. Vorticity structures in the modes become increasingly irregular with downstream distance, suggesting turbulent interactions are occurring at the more downstream locations, especially when the cylinder rotates.

References

1. F. Diaz, J. Gavalda, J. G. Kawall, J. F. Keffer, F. Giralt, Vortex shedding from a spinning cylinder, *Physics of Fluids*, **26**, pp. 3454 (1983).
2. J. Massons, X. Ruiz, F. Diaz, Image processing of the near wakes of stationary and rotating cylinders, *Journal of Fluid Mechanics*, **204**, pp. 167 (1989).
3. H. M. Badr, M. Coutanceau, S. C. R. Dennis, C. Menard, Unsteady flow past a rotating circular cylinder at Reynolds number 10^3 and 10^4, *Journal of Fluid Mechanics*, **220**, pp. 459 (1990).
4. C. C. Chang, R. L. Chern, Vortex shedding from an impulsively started rotating and translating circular cylinder, *Journal of Fluid Mechanics*, **233**, pp. 265 (1991).
5. P. T. Tokumaru, P. E. Dimotakis, Rotary oscillatory control of a cylinder wake, *Journal of Fluid Mechanics*, **224**, pp. 77 (1991).
6. Y. T. Chew, M. Cheng, S. C. Luo, A numerical study of flow past a rotating circular cylinder using a hybrid vortex scheme, *Journal of Fluid Mechanics*, **299**, pp. 35 (1995).
7. S. S. Dol, Weakened Vortex Shedding from a Rotating Cylinder. *International Journal of Mechanical, Industrial Science and Engineering*, **7**(10), pp. 330-337 (2013).
8. S. S. Dol, G. A. Kopp, R. J. Martinuzzi, The suppression of periodic vortex shedding from a rotating circular cylinder. *Journal of Wind Engineering and Industrial Aerodynamics*, ***96***(6), 1164-1184 (2008).
9. S. S. Dol, U. Azimov, R. J. Martinuzzi, Statistical Description in the Turbulent Near Wake of a Rotating Circular Cylinder. *World Academy of Science, Engineering and Technology*, 72, pp. 575-579 (2012).
10. G. S. West, C. J. Apelt, The effects of tunnel blockage and aspect ratio on the mean flow past a circular cylinder with Reynolds number between 10^4 and 10^5, *Journal of Fluid Mechanics*, **114**, pp. 361 (1982).
11. A. Laneville, Turbulence and blockage effects on two dimensional rectangular cylinders, *Journal of Wind Engineering and Industrial Aerodynamics*, **33**, pp. 11 (1990).
12. L. Sirovich, Turbulence and the dynamics of coherent structures. Part I: Coherent structures. Q Appl Mth XLV: 561-571 (1987).
13. P. Holmes, J. Lumley G. Berkooz, *Turbulence, coherent structures, dynamical systems and symmetry*. Cambridge University Press, Cambridge (1996).
14. R. Perrin, M. Braza, E. Cid, S. Cazin, A. Barthet, A. Sevrain, C. Mockett, F. Thiele, Obtaining phase averaged turbulence properties in the near wake of a circular cylinder at high Reynolds number using POD. *Exp Fluids* **43**:341-355 (2007).

EPJ Web of Conferences **114**, 02020 (2016)
DOI: 10.1051/epjconf/201611402020

Measurement of reed valve kinematics

Michael Fenkl[1,a], Václav Dvořák[1] and Tomáš Vít[1]

[1] *Technical University of Liberec, Department of Power Engineering Equipment, Studentská 1402/2, Liberec I, Czech Republic*

Abstract. The measurement of key kinematic parameters of a reed valve movement is necessary for the further development of the reed valve system. These parameters are dependent on the geometry and material properties of the valve. As they directly affect the quantity of air flowing around the valve, a simple and easy to implement measurement of various valve configuration based on the air flow has been devised and is described in this paper, along with its technical parameters and drawbacks when evaluating reed valves used in reciprocating air compressors. Results are presented for a specimen of a compressor under examination. All kinematic parameters, and timing of the opening and closing of the valve, obtained from the measurement are presented and discussed.

1 Introduction

Reed valve systems are thanks to their simplicity frequently used for regulation of suction and discharge strokes of reciprocating compressors. For a meaningful development of such systems, the kinematic parameters of the valves during a realistic operation, which depend on material properties and geometry of the system, need to be known. A suitable method for measuring the valve motion and subsequent data analysis is necessary for determination of such parameters. Given that the valvetrain is enclosed in the compressor head, any measurement requires certain modifications to the compressor. Measurement on a test stand, where the valve motion is simulated using compressed gas and the motion is recorded by a high-speed camera, is one of the simplest measurement methods [1]. This approach, however, does not reflect all effects present during real-world operation of the compressor. Another approach is a numerical solution of the problem using Computational Fluid Dynamics and Finite Elements Method (CFD/FEM) [2,3]. However, given the substantial complexity of the computations and quantity of the input data the interaction of a compressible fluid (gas) and a non-rigid body (valve) is greatly simplified, leading to less accurate results. Another approach is based on the use of an endoscope coupled to a high-speed camera to record a realistic reed valve movement [4], which requires, however, a rather complex data analysis. Yet another method is the examination of a combustion engine valve motion relative to the cam shape using a vibrometer [5].

2 Methods

2.1 Measurement setup

A Laser Doppler vibrometer Ometron VH-1000-D (Fig. 1), with a voltage output directly proportional to the velocity of the sensed object, was attached to the head of a tested compressor specimen using a custom-fabricated holder above a hole drilled above suction and discharge valves, fitted with a glass plate for non-contact sensing of the valve motion. The dynamic pressure in the cylinder and in the discharge duct downstream of the discharge valve was measured using a pair of optomechanical pressure sensors (Fig. 2, AutoPSI D82294-Q, Optrand, USA) (2). All three signals were synchronized with the crankshaft position sensed by an incremental encoder (Fig. 3, ARC 425/10KB) with a resolution of 512 points per revolution. All data were captured by a fast data acquisition board (Fig. 3, NI USB 6218). A Gray code to binary code converter was placed in between the incremental encoder and the data acquisition module.

The piston bottom dead center was chosen as the zero position of the incremental encoder. The output pressure was set by a needle valve capable of maintaing a constant pressure during a steady-state operation of the compressor. The control and data acquisition software working at a resolution of 512 data points per crankshaft revolution was written by the authors in LabView (National Instruments).

[a] Corresponding author: michael.fenkl@tul.cz

Figure 1. Vibrometer attachment.

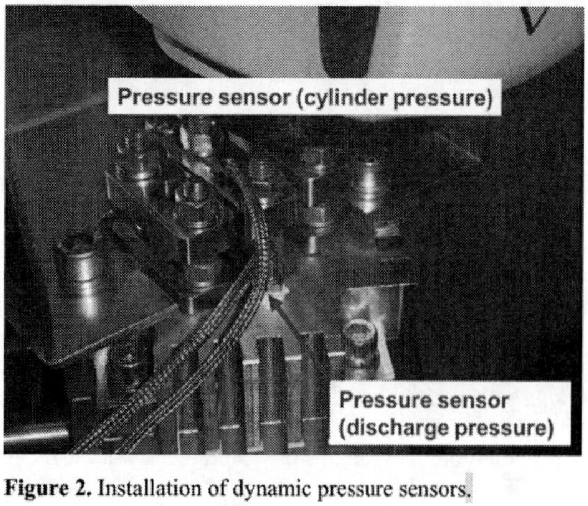

Figure 2. Installation of dynamic pressure sensors.

2.2 Drawbacks of the measurement method

Several minor issues were successfully resolved prior to the measurement. The own vibrations of the vibrometer, contributed to by its considerable mass, affected the measurement, which had to be compensated by subtracting vibrometer vibrations recorded during a series of measurements with the glass plate covered for each measured regime. This has greatly increased the accuracy of the measurement of the valve velocity, however, the measurement were still affected by the internal frequency of the vibrometer holder. The standard deviation of the valve velocity was up to 6.6% for the discharge valve and up to, in several cases, 20% for the suction valve affected by a considerable variability of the valve motion due to underpressure in the cylinder. During a comparable measurement of dynamic pressures in the cylinder and in the discharge duct, which was not affected by vibrations, the standard deviation did not exceed 0.2% for 10 thousands subsequent working cycles.

Figure 3. Data acquisition system.

3 Results and discussion

The measurement took place at the discharge pressure of 300, 500 and 800 kPa for four sets of valves A, B, C and D. Suction and discharge valves in sets A and B were identical except that valves B were thinner. In sets C and D, the valves were identical except that valves D were thinner. Valves C and D were longer than A and B. Ten thousands of working cycles were recorded, at 512 data points per cycle, and mean values were computed for each of the 512 data points.

In this work, only data from valves A for the working pressure of 5 bars are reported. Valve lift is reported in arbitrary units due to trade secret restrictions.

3.1 Discharge valve (DV).

Fig. 4 – 6 show a rapid increase in the discharge duct, followed by a small decrease caused by an opening of a one-way valve to the storage tank. This is in agreement with the in-cylinder pressure which also rapidly rises and later shows a small decrease, and with the reed valve movement, which partially closes after first peak of valve lift, as apparent from the velocity profile shown in Fig. 7. When the pressures are matched, the reed valve opens and remains open during most of the discharge stroke. Only during the last phase the valve opening increases slightly. This is due to the pressure differences across the valve, during which the acceleration of the flow in a narrow slot leads to a pressure decrease and to an aerodynamic paradox [6].

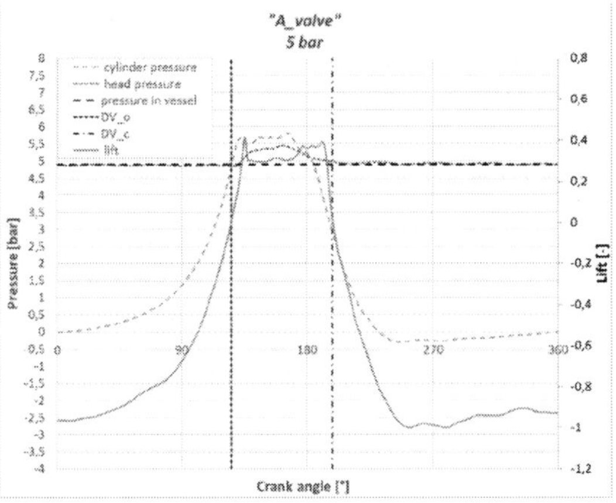

Figure 4. p-alfa diagram for 500 kPa – discharge valve lift.

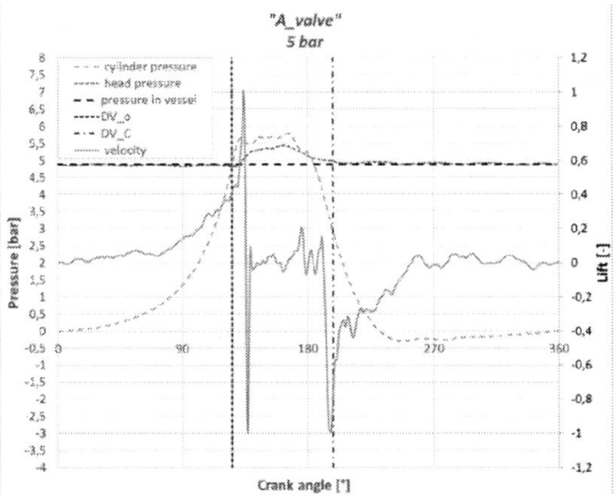

Figure 5. p-alfa diagram for 500 kPa – discharge valve velocity.

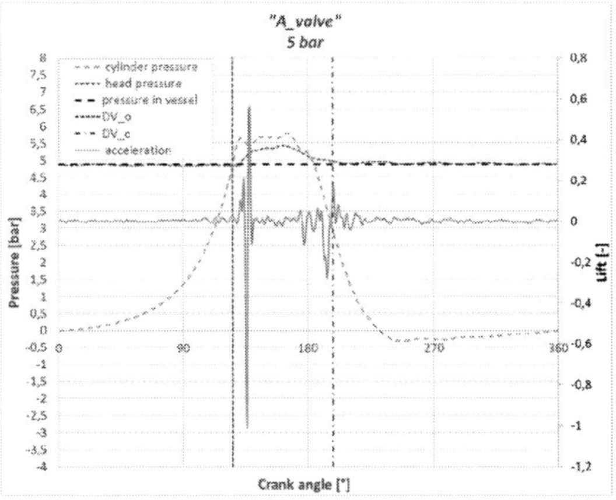

Figure 6. p-alfa diagram for 500 kPa – discharge valve acceleration.

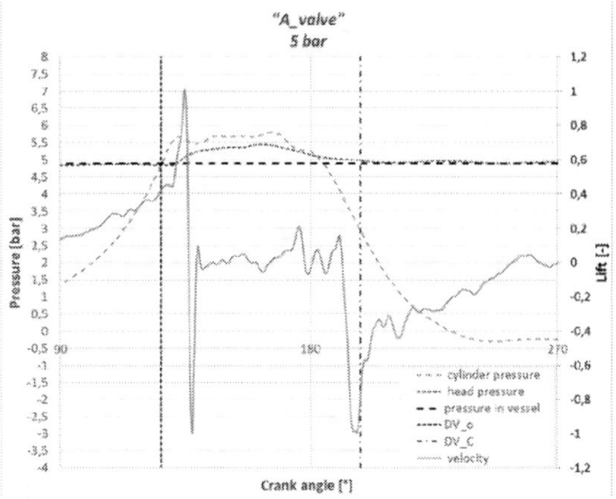

Figure 7. p-alfa diagram for 5 bar – detail of discharge valve velocity.

A numerical integration was performed on the velocity data to obtain the valve lift profile, using the boundary condition of zero lift $p_{cylindr} = p_{head}$ $<0°, CA\_p_{cylindr\_max}>$. This condition has ensured that the reference position is not skewed by the cylinder head overpressure. The acceleration of the valve was calculated as the numerical derivative of the velocity.

3.2 Suction valve (SV).

Fig. 8 – 10 show lift, velocity and acceleration of suction valve. A numerical integration was performed on the velocity data to obtain the valve lift profile. The integration constant could not be readily obtained using a boundary condition, as the suction valve is deformed by the pressure in the cylinder. Instead, zero lift was assigned to a time between crank angles of 0° and 90° during which:

1. there was a considerable decrease in valve acceleration (no major changes in acceleration),
2. there was at least a short period with a constant lift (no effect of in-cylinder pressure and valve flicker).

The zero lift position determined in this way was the compared with the lift at the beginning of the suction stroke and with acceleration following the suction valve opening, and corrected as necessary. The second point was not readily apparent, especially for valves „B“ and „D“, where flicker was leading into deformation caused by in-cylinder pressure. For these reasons, the actual beginning of the opening and end of the closing of the suction valve may differ by several crank angle degrees.

The results show that while using a vibrometer is seemingly simple and practical method with relatively easy analysis and interpretation of data, care should be taken to avoid creation of a measurement artefact originating out of the vibrations of the entire compressor unit, which in turn causes distortion and vibrations of the mounting of the vibrometer. The use of an optic fiber might be beneficial for resolving this problem.

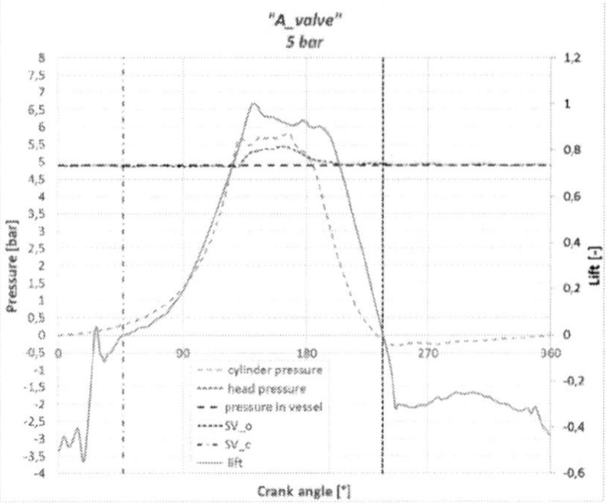

Figure 8. p-alfa diagram pro 5 bar - SV (lift).

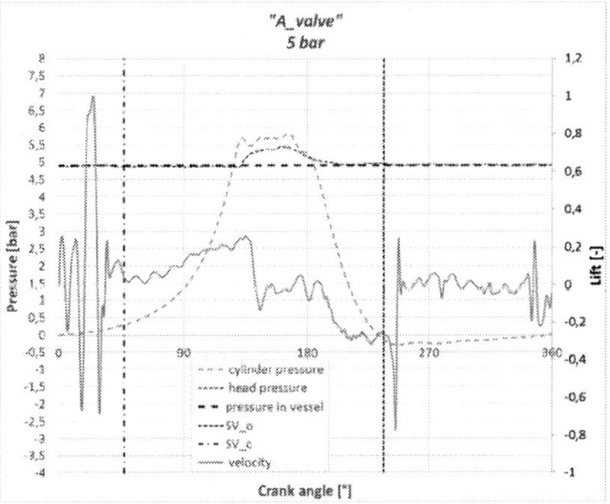

Figure 9. p-alfa diagram pro 5 bar - SV (velocity).

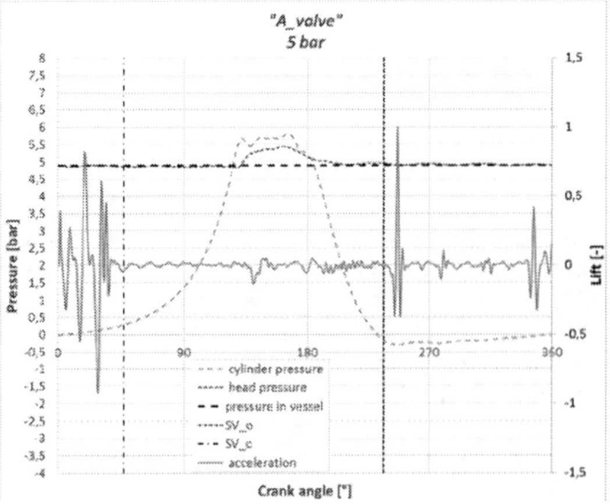

Figure 10. p-alfa diagram pro 5 bar - SV (acceleration).

References

1. A. Zimmermann, P. Hrnjak, "Visualization of the Opening Process of a Discharge Reed Valve in the Presence of Oil", International Compressor Engineering Conference, Paper 2369, (2014).
2. F. Lajus, C. Deschamps, M. Alves, "Numerical analysis of seat impact of reed type valves", 8th International Conference on Compressors and their Systems, London (2013)
3. Q. Tan, S. Pan, Z. Liu, X. Yu, Q. Feng, "Numerical study on transient effective flow and force areas of reed valve in a rotary compressor"; *Proceedings of the Institution of Mechanical Engineers Part A Journal of Power and Energy*, 228(8), (2014)
4. A. Ludu, A. Beto, G. Regner, "Endoscope Video of Compressor Valve Motion and Pressure Measurement Assist Simulations for Design Improvements", *International Compressor Engineering Conference.*, Paper 1420, (2000)
5. N. Paone, C. Santolini, E.P.Tomasini, "Application of a Laser Doppler Vibrometer to Evaluate Engine Poppet Valve Kinematics", *12° Int.Modal Analysis Conf.*,7, (1994)
6. W. Benenson, H. Stöcker, *Handbook of physics.* (2002)

EPJ Web of Conferences **114**, 02021 (2016)
DOI: 10.1051/epjconf/201611402021

N-butanol and isobutanol as alternatives to gasoline: Comparison of port fuel injector characteristics

Michael Fenkl [1,a], Martin Pechout[2], Michal Vojtisek[3]

[1]*Technical University of Liberec, Department of Power Engineering Equipment, Studentská 1402/2, Liberec I, Czech Republic*
[2]*Technical University of Liberec, Department of Vehicles and Engines, Studentská 1402/2, Liberec I, Czech Republic*
[3]*Czech Technical University in Prague, Center for Sustainable Mobility, Prilepska 1920, Roztoky u Prahy, Czech Republic*

Abstract. The paper reports on an experimental investigation of the relationship between the pulse width of a gasoline engine port fuel injector and the quantity of the fuel injected when butanol is used as a fuel. Two isomers of butanol, n-butanol and isobutanol, are considered as potential candidates for renewable, locally produced fuels capable of serving as a drop-in replacement fuel for gasoline, as an alternative to ethanol which poses material compatibility and other drawbacks. While the injected quantity of fuel is typically a linear function of the time the injector coil is energized, the flow through the port fuel injector is complex, non ideal, and not necessarily laminar, and considering that butanol has much higher viscosity than gasoline, an experimental investigation was conducted. A production injector, coupled to a production fueling system, and driven by a pulse width generator was operated at various pulse lengths and frequencies, covering the range of engine rpm and loads on a car engine. The results suggest that at least at room temperature, the fueling rate remains to be a linear function of the pulse width for both n-butanol and isobutanol, and the volumes of fuel injected are comparable for gasoline and both butanol isomers.

1 Introduction

Finite amount of fossil energy resources, energy security and concerns about climate changes are some of the main drivers of the quest to replace these energy source with renewable, locally produced alternatives. This process also includes utilization of various fuels for internal combustion engines used in transportation and mobile machinery as a substitute of mostly used oil-based in transportation sectors.

Currently, ethanol blended into gasoline at various concentrations is the most utilized alternative for spark ignition engines. Unfortunately, ethanol exhibits some negative attributes, such as significantly lower energy content, hygroscopic behaviour and chemical aggresivity [1-5]. Also, broader consequences of fuel production need to be considered, as the current ethanol feedstock crops may compete with food production.

One promising alternative to ethanol is butanol – a four carbon alcohol, present in four isomers (n-butanol, 2-butanol, isobutanol, terc-butanol), out of which n-butanol and isobutanol (2-methyl-propan-1-ol) have be used as a fuel in spark ignition engines [6-8].

Key properties of n-butanol and isobutanol are compared with ethanol and gasoline in Table 1.

Table 1. Selected properties of gasoline and alcohol alternatives

	Gasoline	N-butanol	Iso-butanol	Ethanol
Research octane number (RON)	95	94 [9-10] - 96 [11]	113	110
Density [kg·m^{-3}]	753 **)	810	806	790
Lower heating value [MJ·kg^{-1}]	42.9	33.3	33.3	26.8
Volumetric lower heating value [MJ·dm^{-3}]	32.3	27.0	26.8	21.2
Latent heat of vaporization [kJ·kg^{-1}]	380-500	716 [2]	579	904
Chemical formula	complex mixture	C_4H_9OH	C_4H_9OH	C_2H_5OH
Mass share of C [%]	86	65	65	52
Mass share of H [%]	14	13.5	13.5	13
Mass share of O [%]	0 **)	21.5	21.5	35
Viscosity [mPa·s]	0.4-0.8	2.57	3.33	1.08
Boiling point [°C]	199 (FBP)	118	108 [12]	78
Stoichimetric air/fuel ratio	14.7	11.2	11.2	9.0

[a] Corresponding author: michael.fenkl@tul.cz

Compared to ethanol, butanol has better miscibility with ordinary gasoline, is less corrosive, does not absorb water from ambient air, and is more compatible with the existing gasoline infrastructure. The energy density of butanol is significantly higher than ethanol. Potential drawbacks are that both butanol isomers exhibit significantly higher boiling temperature, viscosity and lover vapour pressure [13], which is important for running of cold engine.

On the other hand, production of both isomers – n-butanol and iso-butanol – from non-food crops such waste biomass, lignocelluloses etc. is more complicated to ethanol fermentation [2].

From the comparison of the properties of fuels it is apparent than the amount of injected fuel has to be increased when alcohol is added to maintain the desired – typically stoichiometric - air to fuel ratio, which is necessary for the proper operation of the engine including its exhaust gas aftertreatment devices. Production ECUs (engine control units) generally have the capability of some degree of short-term and long-term adjustment of the fuel injection pulse width, intended to serve as a compensation of variations in the composition and other properties of the fuel and in the condition and general working of the engine. Such capacity has not, however, been intended to cover, for example, a rather substantial increase in the fuel delivery rates when switching from gasoline to ethanol. The use of high concentrations of ethanol generally requires either an ECU designed for such fuel, or the installation of an auxiliary control unit, the function of which is often limited to the prolongation of the fuel injector pulse width.

The prolongation of the fuel injector pulse width is not necessarily simply a matter of multiplying the injector opening time with the ratio of the volumetric energy densities of traditional and alternative fuel. The steady-state flow through an open injector is also affected by the fuel viscosity. Also, rather complicated dynamic effect during injector opening and closing need to be considered when calculating the injector pulse width (the time for which the injector coil is energized) necessary to deliver the desired quantity of the fuel.

The effect on replacing gasoline with either n-butanol or isobutanol on the relationship between the fuel injector pulse width and the quantity of the fuel delivered is considered in this work.

In general, for a laminar flow, the effect of fuel viscosity on the injector flow rate, everything else being equal, is given by Hagen-Poiseuille equation [14]:

$$\dot{V} = \frac{\pi D^4 \Delta p}{128 \mu L} \tag{1}$$

where D is the diameter and L is the length of the pipe, p is pressure loss across the pipe, and μ is the dynamic viscosity of the fluid.

Unfortunately in injectors for engines with multi-point injection the flow passes through a thin orifice, which is too short to have purely laminar flow, and as a result, the Hagen–Poiseuille equation results in unphysically high flow rates. The flow is bounded by Bernoulli's principle, under less restrictive conditions [15]:

$$\dot{V}_{\max} = \pi R^2 \cdot \sqrt{\frac{2 \cdot \Delta p}{\rho}} \; . \tag{2}$$

The Reynolds number in a port injection spark ignition engine injector was reported to be on the order of 10-100 [16], depending on the fuel composition and temperature. While such a range of Re typically represents laminar flow through a pipe [17], not knowing the precise geometry of the injector, which is rather complex, the flow through the injector may lie in the transition region between laminar and turbulent flow [18-19]. Further uncertainties are expected during the injector opening and closing period.

Therefore, while no surprising non-linearities have been found with ethanol, and the volumes of the fuel injected during a given pulse width were reported not to differ more than 10% between ethanol and gasoline [20-21], the considerably higher viscosity of both n-butanol and isobutanol as compared to either gasoline or ethanol (see Table 1) complicates the development of a simple method describing the flow through the injector as a function of fuel properties.

Therefore, an experimental approach, also including potential effect of used fuel on pressure reducing valve and rail pressure, has been chosen to elucidate the effect of replacing gasoline with n-butanol or isobutanol on the relationship between the fuel injector pulse width and the energy content of the fuel delivered. As an added value, such determination could allow for determination of a correction factor for the ECU reported fuel consumption when using butanol as a fuel.

2 Experimental

2.1 Injection fuel system

The injector characteristics were examined on an in-house produced experimental apparatus consisting of a production fuel injector fed from a production fuel supply system. The injector was driven by a pulse generator and was injecting into a graduated cylinder. The scheme of this apparatus is shown on Figure 1.

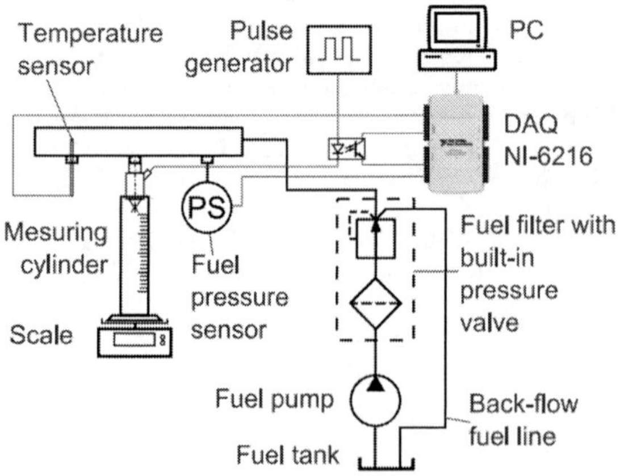

Figure 1. Scheme of injection and data acquisition system

Fuel is supplied by the fuel pump through filter holder with integrated output pressure regulator (4 bar) into a standard fuel rail, which has only one of the middle ports coupled to an injector. All parts were from a Volkswagen EA-111 series gasoline engine. The remaining ports were fitted with a thermocouple to measure temperature of the fuel in the rail, by a pressure sensor to record possible injection pressure changes due to different fuels properties, and by a plug.

Pressure sensor voltage output and thermocouple signal were sensed by data acquisition system (NI-6216). As the used pulse width generator does not allow setting of desired count of pulses, an A/D converter was connected using an optocoupler to ensure counting count of injection pulses. The mass accumulated over several thousands of pulses (each pulse corresponding to tens of mg of fuel) has been determined by a laboratory scale (Sartorius, resolution 0.1 g) and divided by the number of pulses to obtain the mean amount of injected fuel per one burst.

2.2. Injection conditions and data processing

Technical grade N-butanol and isobutanol, commonly sold for industrial use (such as solvents) were obtained from industrial distributors (Chemap, Dašice, CZ, and Chemlogistic, Pardubice, CZ). Non-oxygenated gasoline was obtained from the state-owned supplier (Cepro).

Combinations of injection pulse frequency corresponding to engine rpm 2000 to 5000 rpm with 1000 rpm step and fuel injection pulse width of 5, 10, 15 and 20 ms, and 3 ms at 1000 rpm to represent idling conditions, were selected based on real engine characteristics. All combinations were repeated three times to address the procedure repeatability.

Captured fuel injection voltage signal has been analysed and fuel injection pulses were counted using rising edge detection. The mean cycle based amount of injected fuel has been obtained as the ratio of total injected mass to the number of pulses.

3 Results and discussion

The mean mass of injected fuel as a function of frequency at a given fuel injection pulse width are displayed on Fig. 2 and the mean mass of injected fuel as a function of rpm are shown on Figure 3.

The data obtained show very good repeatability. It is clearly visible that the injected amount is nearly independent on the injection frequency. This is the case notably for both butanol isomers. For gasoline, where some dependency on the pulse frequency has been observed, the primarily lower fuel quantity at lower rpm (lower frequency) is possibly attributable to higher evaporation loses as a result of significantly higher vapour pressure of gasoline.

The amount of injected fuel has been close for both butanol isomers where amount of injected iso-butanol was up to 1% higher.

From the elementary Bernoulli equation based theoretical flow speed of

$$c = \sqrt{\frac{2 \cdot (p_{RAIL} - p_{IMA})}{\rho}} \tag{3}$$

based on pressure difference inside the fuel rail p_{RAIL}, pressure in intake manifold (or ambient pressure during this experiments) p_{IMA} and fuel density ρ a theoretical ratios of both injected volume and mass at a constant injection pressure, injection duration and duration of non-injecting par can be derived as a function of fuel density:

$$\frac{V_{Bu}}{V_{gas}} = \sqrt{\frac{\rho_{gas}}{\rho_{Bu}}};\ \frac{m_{Bu}}{m_{gas}} = \sqrt{\frac{\rho_{Bu}}{\rho_{gas}}}. \tag{4}$$

Using this hypothesis the ratio of injected butanol mass to gasoline mass should be, using the densities in Table 1, approximately 1.037 for n-butanol and 1.035 for isobutanol. The experimental results show mean values of 1.084 for isobutanol and 1.074 for n-butanol, with rather low differences among the three tests of about 1%, but more considerable differences among different pulse frequencies.

Using a purely volumetric metering hypothesis, under which the volume of the fuel injected is independent of the fuel properties, the ratios of fuel mass should be equal to the ratios of fuel densities, which are 1.096 for n-butanol and 1.088 for isobutanol. The density ratios are only about 1% lower than the observed ratios of fuel mass injected. It is therefore quite remarkable that metering of the all the fuels can be simplified as volumetric with error of total injected mass about 1%, which is within the experimental uncertainties.

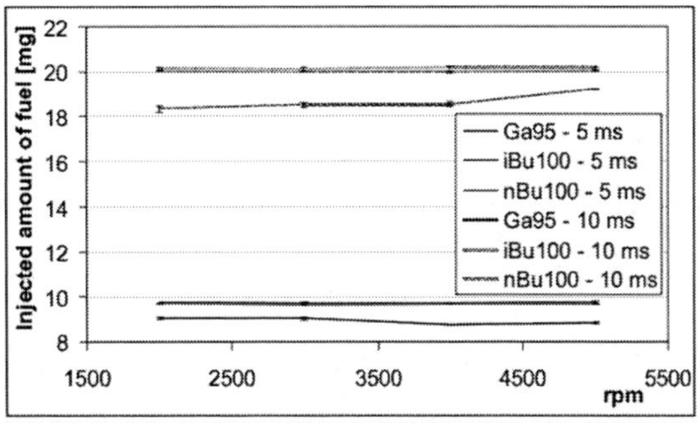

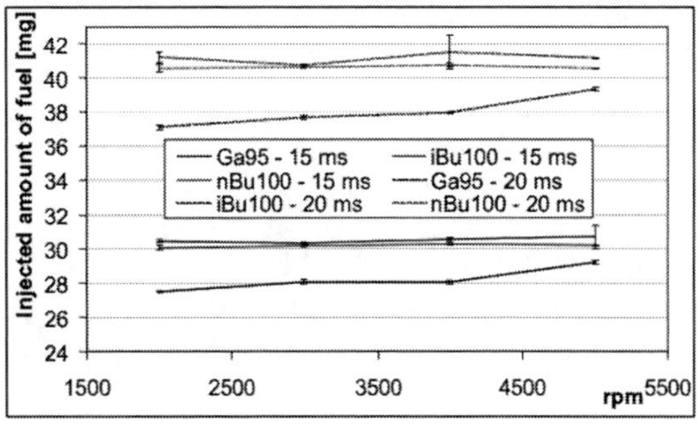

Figure 2. Effect of used fuel on mass amount of injected fuel for various fuel injection pulse width and rpm.

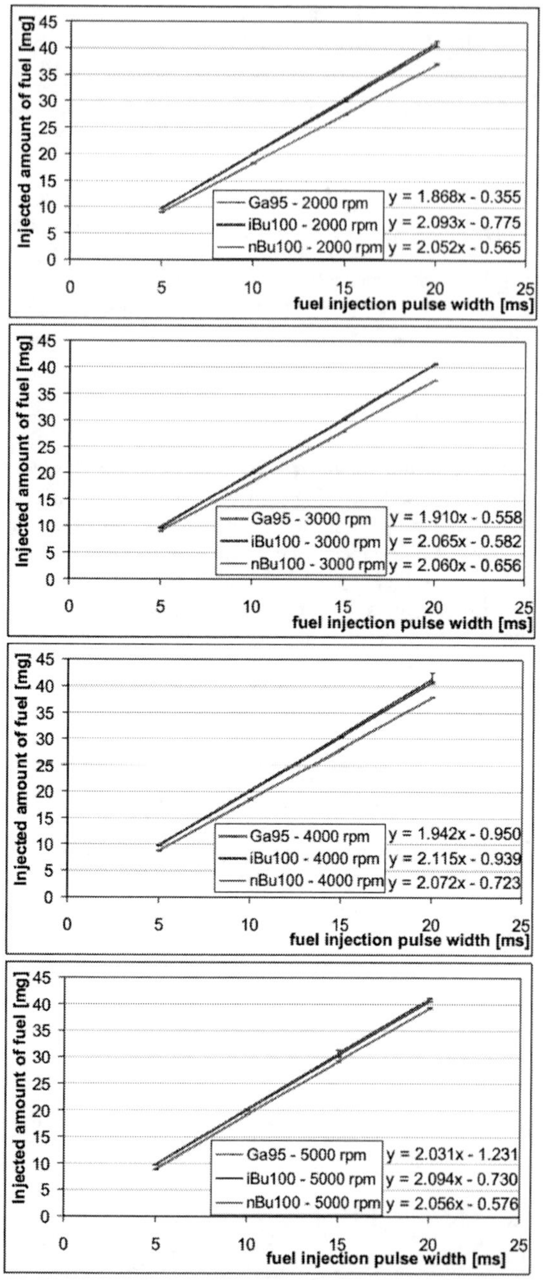

Figure 3. Effect of used fuel on mass amount of injected fuel for various fuel injection pulse width and rpm.

Somewhat higher increase of injected mass has been observed during injection conditions typical for engine idle (approx 3ms injection duration at frequency corresponding 1000 rpm, Figure 4). These increases were about 12.5% for iso-butanol and 11.4% for n-butanol.

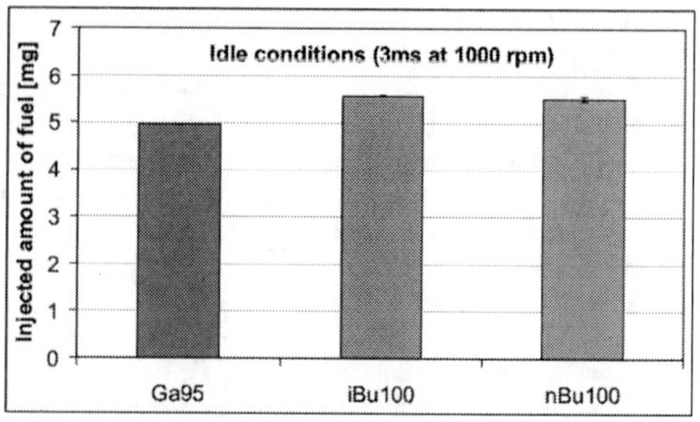

Figure 4. Effect of used fuel on mass amount of injected fuel at idling conditions.

To investigate this difference, the injection time "overhead", or the length of the non-effective period, was investigated. The overhead time was determined by extrapolation of the injected amount to the zero quantity of the fuel injected using linear regression coefficients (Fig. 3). This consideration resulted in values between 0.3 and 0.4 ms without remarkable dependency on rpm or the fuel used.

Based on stochiometric air to fuel ratio of gasoline and both butanol isomers (see Table 1), to maintain the same air-fuel ratio, the mass amount should be increased by about 31% when gasoline is replaced by butanol while 63% increase is necessary when neat ethanol is to be used. Considering that the injected mass increased by about 8.4% for n-butanol and 7.4% for isobutanol due to higher butanol density, an additional increase of only about 23% is needed, compared to approximately 52% for ethanol.

4 Conclusions

The relationship between the frequency and length of the fuel injector pulse width and the quantity of the fuel delivered has been tested for two candidate alternative fuels, n-butanol and isobutanol, and compared to gasoline. The testing was performed on a production port fuel injector coupled to production fuel system (fuel tank, fuel pump, fuel rail with pressure regulator), and driven by a pulse width generator.

The experimental data showed very good repeatability. Results showed minimum dependency of injected mass on rpm, especially for both butanol isomers, so the dependency of injected fuel mass on fuel injector pulse width is following a linear trend.

From extrapolation of the linear trend the non-effective part of injection period has been derived with no significant dependency on injection frequency and used fuel. As a result, this method poses a simple way how to determine the fuel consumption during engine non-steady state typical for real-world engine operation using on-board diagnostic interface.

Changes in the injected mass have shown that the simplified approach assuming volumetric metering generates inaccuracy about 1% when gasoline is replaced with neat butanol (either n-butanol or isobutanol).

Estimation of volumetric metering gives more accurate values for injected mass change compared to a simple model based on Bernoulli equation. The volumetric assumption gives a prediction error about 1% which is sufficiently small for practical usage.

References

1. R. Cascone, *Chem. Eng. Prog.* **104**, p. S4-S9, (2008)
2. P. Patakova, D. Maxa, M. Rychtera, L. Linhova et al., *M.A.D.S.*, ISBN 978-953-307-480-1, p. 243-266, InTech, Rijeka, Croatia, (2011)
3. B. Jones, G. Mead, P. Steevens, *"The Effects of E20 on Plastic Automotive Fuel System Components"* Tech. Rep. 20082/21/2008, *Minnesota Center for Automotive Research*, (2008).
4. B. Bailey, "Performance of Ethanol as a Transportation Fuel," In: *Handbook on Bioethanol: Production and Utilization.* Edited by Wyman C. E. Online at http://infohouse.p2ric.org/ref/36/35015.pdf (accessed October 10, 2015).
5. EPA (Environment Protection Agency U.S.), "1-Butanol as a Gasoline Blending Bio-component," BP, Mobile Sources, Technical Review Subcommittee, March 28, (2007), http://nsdi.epa.gov/air/caaac/mstrs/March2007/Wolf.pdf.
6. S. Szwaja, J. D. Naber, Fuel, **89**, Issue 7, p. 1573-1582, (2010), doi:10.1016/j.fuel.2009.08.043.
7. F. N. Alasfour, Int. J. of Energy Research, **21**, Issue 1, p. 21-30, (1997), doi:10.1002/(SICI)1099-114X(199701)21:1<21::AID-ER231>3.0.CO;2-K.
8. T. Wallner, A. Ickes, N. Shidore, "Impact of ethanol and butanol as oxygenates on SIDI engine efficiency and emissions using steady-state and transient test procedures," *DEER 2010*, Detroit, Michigan, September 27-30, (2010).
9. A. Cairns, A. Todd, P. Aleiferis, N. Fraser, et al., "A Study of Alcohol Blended Fuels in an Unthrottled Single Cylinder Spark Ignition Engine," *SAE Technical Paper* 2010-01-0618, 17, (2010), doi:10.4271/2010-01-0618
10. M. Thewes, M. Müther, A. Brassat, S. Pischinger et al., "Analysis of the Effect of Bio-Fuels on the Combustion in a Downsized DI SI Engine," *SAE Int. J. Fuels Lubr.* **5(1)**, p.274-288, (2011), doi:10.4271/2011-01-1991 .
11. S. Merola, C. Tornatore, G. Valentino, L. Marchitto, et al., "Optical Investigation of the Effect on the Combustion Process of Butanol-Gasoline Blend in a PFI SI Boosted Engine," *SAE Technical Paper* 2011-24-0057, 15, (2011), doi:10.4271/2011-24-0057.
12. Chemistry Webbook. National Institute of Standards and Technology. Online at http://webbook.nist.gov.
13. V. F. Andersen, J. E. Andersen, T. J. Wallington, S. A. Mueller, et al.,,"Vapor Pressures of Alcohol-Gasoline Blends," *Energy & Fuels*, **24**, p. 3647-3654, (2010), doi:10.1021/ef100254w
14. S. P. Sutera, R. Skalak, "The history of Poiseuille's law". *Annual Review of Fluid Mechanics* **25**: pp. 1–19 (1993).
15. C. O. Bennett, J. E. Myers, *Momentum, Heat, and Mass Transfer*, McGraw-Hill (1962).
16. P. G. Aleiferis, J. Serras-Pereira, Z. van Romunde, J. Caine, M. Wirth, Combustion and Flame **157,** pp. 735–756, (2010).
17. J. P. Holman, *Heat Transfer*, p. 207, (2002).
18. M. C. Potter, D. C. Wiggert, B. H. Ramadan, T. I. P. Shih, *Mechanics of fluids* (Fourth edition. ed.). Cengage Learning (2012).
19. D. Kolokotronis, Y. Hardalupas, A. Taylor, P. Aleiferis et al.: "Experimental Investigation of Cavitation in Gasoline Injectors", *SAE Technical Paper* 2010-01-1500, 23, (2010), doi:10.4271/2010-01-1500.
20. T. N. C. Anand, A. Madan Mohan, R. V. Ravikrishna, "Spray characterization of gasoline-ethanol blends from a multi-hole port.", *Fuel*, **102**, 10, (2012)
21. J. Gao, D. Jiang, Z. Huang, "Spray properties of alternative fuels: A comparative analysis of ethanol–gasoline blends and gasoline.", *Fuel,* **86,** 5, (2007)

Acknowledgements

The work of M.P. and material expenses were covered by the Student Grant Competition of the Technical University of Liberec, grant 21013 "Possibilities of utilizing higher alcohols for fast gasoline replacement." The work of M.V. was funded by the Czech Ministry of Education, Youth and Sports program NPU I (LO), project "Development of Vehicle Center of Sustainable Mobility" (LO1311).

The work of M.F. and material expenses were covered by the Student Grant Competition of the Technical University of Liberec, grant 21000.

EPJ Web of Conferences **114**, 02022 (2016)
DOI: 10.1051/epjconf/201611402022

A non-intrusive and continuous-in-space technique to investigate the wave transformation and breaking over a breakwater

Simone Ferrari[1,a], Maria Grazia Badas[1] and Giorgio Querzoli[1]

[1]*DICAAR - Dipartimento di Ingegneria Civile, Ambientale e Architettura, University of Cagliari,via Marengo 2, 09123 Cagliari, Italy*

Abstract. To design longshore breakwaters, the evaluation of the wave motion transformations over the structures and of the energy they are able to absorb, dissipate and reflect is necessary. To characterize features and transformations of monochromatic wave trains above a breakwater, both submerged and emerged, we have designed and developed a non-intrusive and continuous-in-space technique, based on Image Analysis, and carried out an experimental campaign, in a laboratory flume equipped with a wave-maker, in order to test it. The investigation area was lighted with a light sheet and images were recorded by a video-camera. The working fluid was seeded with non buoyant particles to make it bright and clearly distinct from dark background and breakwater. The technique, that is based on a robust algorithm to identify the free surface, has showed to properly work also in prohibitive situations for traditional resistive probes (e.g., very shallow waters and/or breaking waves) and to be able to measure the free surface all over the investigation field in a non-intrusive way. Two kind of analysis were mainly performed, a statistical and a spectral one. The peculiarities of the measurement technique allowed to describe the whole wave transformation and to supply useful information for design purposes.

1 Introduction

Longshore breakwaters are extensively used as littoral or harbour protection systems (see, for instance, [1]) but also, more recently, as devices for wave energy conversion ([2]). As a consequence, the estimation of the wave motion transformations over the breakwater and of the amount of energy this structure is able to absorb, dissipate and reflect, is necessary, as stated in [3]. For instance, the proper design of longshore breakwaters for littoral protection requires that wave motion in the sheltered area is within certain limits, so laboratory experiments can be carried out to determine reflection and dissipation of the energy of trains of monochromatic waves above a breakwater, as in [4]. Unfortunately, the traditional instruments employed for this kind of experiments, i.e. the classical resistive probes (e.g., see [5]), can successfully measure water level on the lee and weather side of the breakwater but are not able to properly work above and close to it, where the water level is very shallow and the waves are possibly breaking. As a consequence, using traditional probes the breakwater results as a "black box" and it is not possible a direct comprehension of the physical phenomena determining the wave evolutions.

To overcome these issues, we have designed and developed a non-intrusive and continuous-in-space technique, based on Image Analysis, and carried out an experimental campaign, in a laboratory flume equipped with a wave-maker, in order to test it.

2 Experiments and data elaboration

The experimental campaign was carried out in the 21-m long flume (see Figure 1) of the Hydraulic Laboratory of the DICAAR – University of Cagliari. The flume has glass walls and it is equipped on one side with a wave-maker and on the opposite side with an absorbing beach, designed in order to minimise the reflection of the incoming waves. The wave-maker is controlled by a pc and this system is able to generate both monochromatic regular wave trains or random waves from a given spectrum. More details on the flume and the wave-maker can be found in Ferrari and Querzoli, 2015 [6].

The model breakwater employed in this experimental campaign was a black painted trapezoidal obstacle made of Perspex: it is 26.0 cm height, with a lower side 124.0 cm long, an upper side 20.0 cm long and a cross section 30.0 cm wide (same width as the flume). As breakwaters for littoral protection or for energy conversion can be both submerged or emerged, the hydrostatic level in the experiments was of 29.0, 25.0 or 23.0 cm, so the obstacle was submerged in the first case, whist it was emerged in the other two. 27 experimental runs (9 with submerged breakwater and 18 with emerged breakwater) were recorded, resulting from the combination of 3 different

[a] Corresponding author: ferraris@unica.it

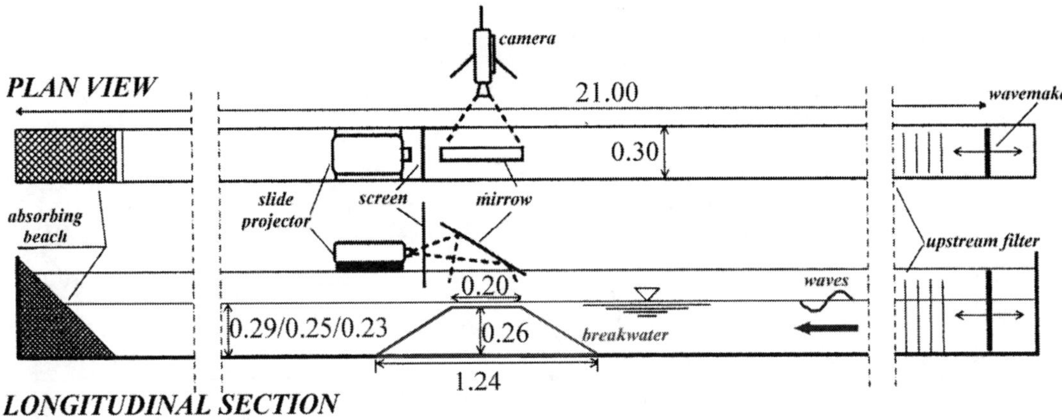

Figure 1. Experimental set-up (measures in m).

hydrostatic levels, wave periods T (0.8, 1.0 and 1.3 s) and amplitudes A. As breakwater and background were black, the working fluid was water seeded with non-buoyant particles and the area of interest was lighted with a light sheet, generated by two slide projectors, equipped with a black slide and a narrow vertical slit at its centre (see [7] for more details on this technique and on the characteristics of the light sheet). As a consequence, the water was clearly distinct from the background and the breakwater (see Figure 2) and the free surface can be identified as the interface between dark and lighted zones.

During the last years, various Image Analysis techniques have been developed by the staff of the DICAAR – University of Cagliari to measure concentrations ([7]) and velocities ([8], [9]) in a non-intrusive and continuous in space way: this gives the advantages of not modifying the flow inserting probes into it and of reducing the time spent in experiments, as the measures are obtained all over the investigation field. In order to develop and test a new Image Analysis technique able to estimate the wave motion transformations over a breakwater, experiments were recorded by a 3-CCD digital video camera, orthogonal to the light sheet, at the frequency of 25 images per second.

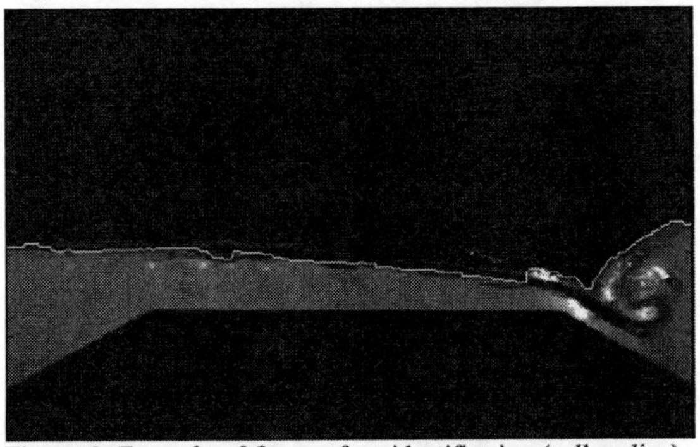

Figure 2. Example of free surface identification (yellow line): the weather side is on the right and the lee is side on the left.

The recorded images were digitalized at a 576x720 pixels resolution, then a median filter and afterwards a simple threshold segmentation were performed on them, with the target of identifying the free surface as the interface between water (white) from background (black). To avoid misleading (e.g., from small drops, small black areas in the lighted zone or in bright areas in the dark ones), the following procedure was employed:

1. each column in a given image was analyzed pixel by pixel, from the upper side of the image downwards, as far as a black/white interface was detected;
2. when a white pixel was detected, a certain number of successive pixels was checked;
3. if all of them were as well white, the first pixel was recognized as belonging to the free surface.

When the full image was elaborated and all the points of the free surface were found, another median filter was applied, to smooth the line representing free surface.

Following the described procedure, it is possible to identify the free surface at as many points as the horizontal resolution of the video-camera, also when the obstacle is emerged or with breaking waves. Consequently, this technique can identify the entire free surface, while traditional resistive probes are able to give information only in the points where they are used. Moreover, this method allows to define even a posteriori the number and the positions of the points to study. Eventually, this method works even in prohibitive situations for traditional resistive probes, as in the case of very shallow water above the breakwater or with breaking waves.

3 Results

As shown in Figure 1, the waves are generated by the wave-maker on the right side of the flume and they run from right to left, so on every figures the weather side is on the right and the lee is side on the left. Moreover, the origin of abscissa X is put in the middle of the obstacle longitudinal section; the origin of the wave height H is put at the hydrostatic level.

Two kind of analysis were mainly performed on the data, a statistical (on the wave mean height, the wave height variance, the wave mean height in phase, the wave height variance in phase and the probability density of the surface height.) and a spectral one (on the power spectral density, the fundamental frequency and the harmonic ones). Thanks to the above illustrated peculiarities of the measurement technique, the analyses were performed with continuity all over the top of the breakwater, so describing the whole transformation of the waves.

Before coming into the data analysis, a brief description of the phenomena that develop when the wave train comes over the obstacle will be done.

3.1 Phenomenological description

When the breakwater is submerged (hydrostatic level of 29.0 cm), upstream the obstacle a wave depression develops with the so-called plunging profile: it has the shape of a curl (see the right side of Figure 2) that grows up and then dives into the wave basis. This is the so-called wave breaking, that happens over the breakwater weather side, causing a strong turbulence with a consequent energy dissipation. In the experiments with wave period of 0.8 s, the profile is of surging type, which is characterized by a uniformly decreasing energy.

When the breakwater is emerged (hydrostatic level of 25.0 and 23.0 cm), the wave-breaking never happens and the water overflows the breakwater sliding on it, with no vortices on the weather side, while a hydraulic jump behaves on the lee side.

3.2 Statistical analysis

As before stated, the statistical analysis that will be shown hereafter are relative to the wave mean height (compared with hydrostatic level) and its variance, performed also phase by phase, and to the probability density of the free surface heights.

3.2.1 Overall mean value and variance of the wave height

In Figures 3 and 4, the wave mean heights H and the wave height variances σ^2 along the breakwater for the three experiments with submerged obstacle (hydrostatic level 29.0 cm) and period T of 0.8 s are plotted.

In Figure 3 the so-called wave set-up phenomenon is visible: downstream the breakwater, the water flow from weather side to lee one (that is closed) causes a raising of the free surface above the hydrostatic level. Upstream the breakwater, the phenomenon of the wave set-down is clear: the plunging profile behaviour determines a decreasing of the free surface under the hydrostatic level. Wave set-up and wave set-down are so accurately detected by the technique. Moreover, Figure 3 shows that both wave set-up and set-down tend to grow for increasing values of the wave amplitude A. From the comparison of Figure 3 with Figure 5 (same hydrostatic level but a different period T of 1.0 s), it is clear that wave set-up and set-down also tend to grow for increasing values of T.

In Figure 4 the wave height variance σ^2 is shown. σ^2 is proportional to A^2 and to the turbulence intensity, so the upstream peak, visible for all the amplitudes, points out the location X of the wave breaking: downstream this peak, the variance decreases almost uniformly, indicating that the profile is of plunging type. the wave height variance σ^2 tends to grow for increasing values of the wave amplitude A. From the comparison of Fig. 4 with Fig. 6 (same hydrostatic level but T = 1.0 s), it is clear that σ^2 tends to grow for increasing values of T.

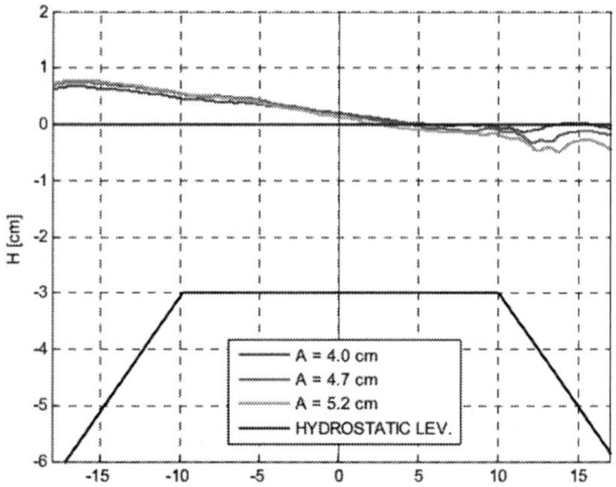

Figure 3. Wave mean height H along the breakwater for the runs with hydrostatic level of 29.0 cm, period T = 0.8 s and amplitude A of 4.0, 4.7 and 5.2 cm.

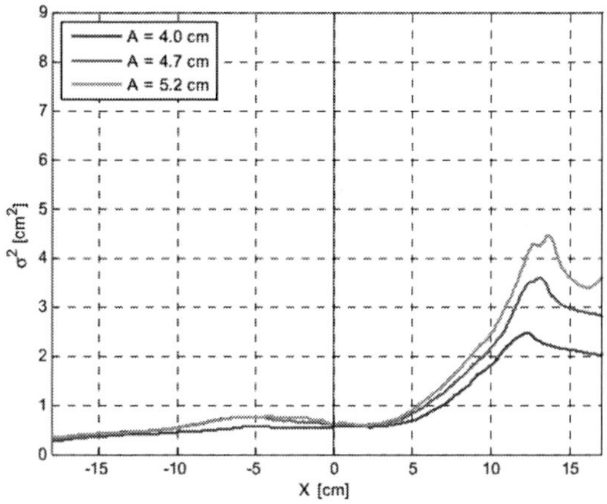

Figure 4. Wave height variance σ^2 along the breakwater for the same runs of Figure 3.

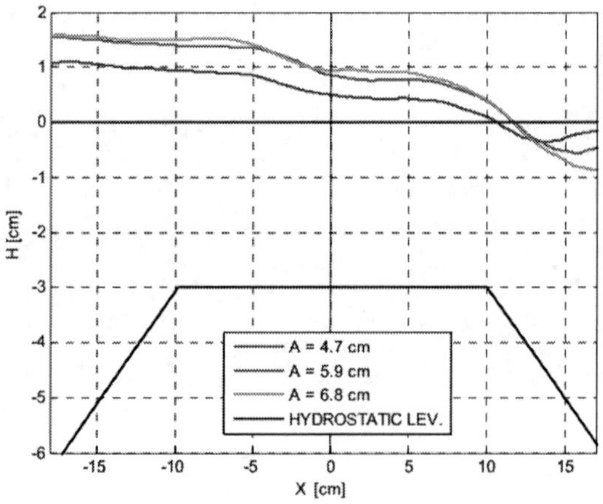

Figure 5. Wave mean height H along the breakwater for the runs with hydrostatic level 29.0 cm, period T = 1.0 s and amplitude A of 4.7, 5.9 and 6.8 cm.

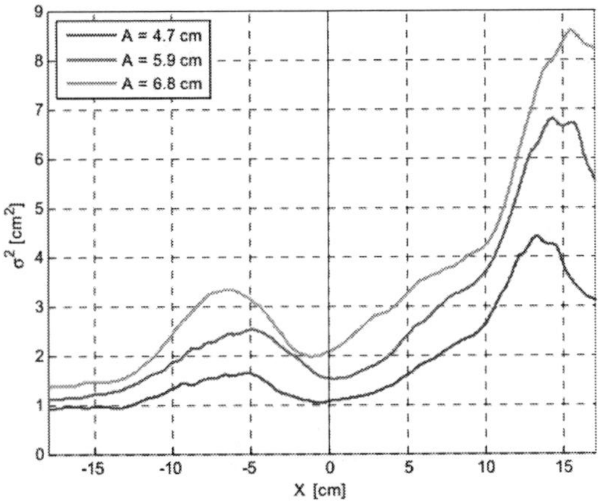

Figure 6. Wave height variance σ^2 along the breakwater for the same runs of Figure 5.

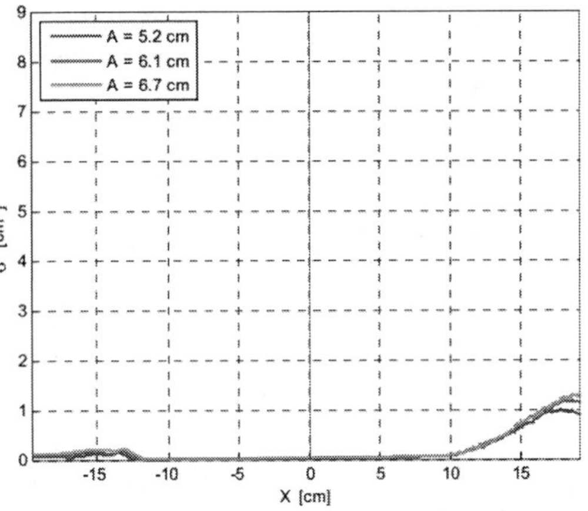

Figure 8. Wave height variance σ^2 along the breakwater for the same runs of Figure 7.

In Figure 6, there is instead a minimum value of σ^2 (in the middle of the breakwater) followed by another peak: the minimum value indicates the location X where the waves spill over themselves after the breaking, the following peak shows the consequent increase of the turbulence.

In Figures 7 and 8, the wave mean heights H and the wave height variances σ^2 along the breakwater for three of the runs with submerged breakwater (hydrostatic level 25.0 cm) and period T of 0.8 s are plotted. σ^2 in experiments with emerged breakwater has values that are very much lower than in ones with submerged breakwater, because there is not a breaking wave but only a raising followed by a regular decreasing of its height H. As a matter of fact, when the breakwater is emerged the water overtopping is just a slipping on the upper part of the breakwater, with a following hydraulic jump, visible in Figure 7. In this figure it is possible to note also that, near the downstream edge of the breakwater, the water height H is almost zero: these areas will be called "dry zones". A dry zone appears when the wave, during the overtopping of the breakwater, looses almost entirely its energy. The dry time DT represents how long, during a run, each part of the breakwater is a dry zone (i.e., how long each X experiences H $\cong$ 0 cm). In Figure 9 and 10, the dry time DT is plotted along the breakwater for the runs with hydrostatic level 25.0 and 23.0 cm.

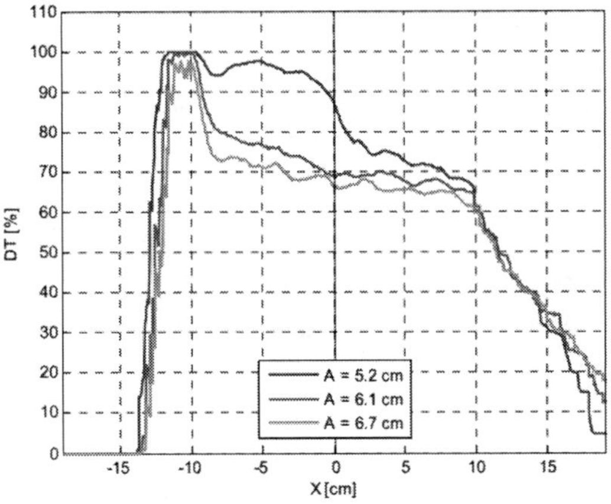

Figure 9. Dry time percentage DT along the breakwater for the runs with hydrostatic level of 25.0 cm, period T = 0.8 s and amplitude A of 5.2, 6.1 and 6.7 cm.

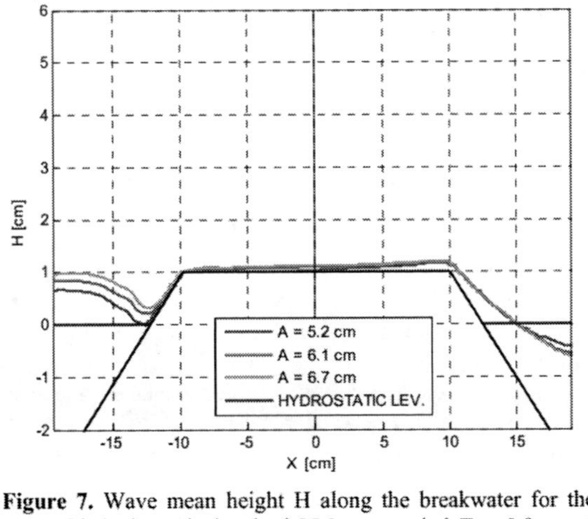

Figure 7. Wave mean height H along the breakwater for the runs with hydrostatic level of 25.0 cm, period T = 0.8 s and amplitude A of 5.2, 6.1 and 6.7 cm.

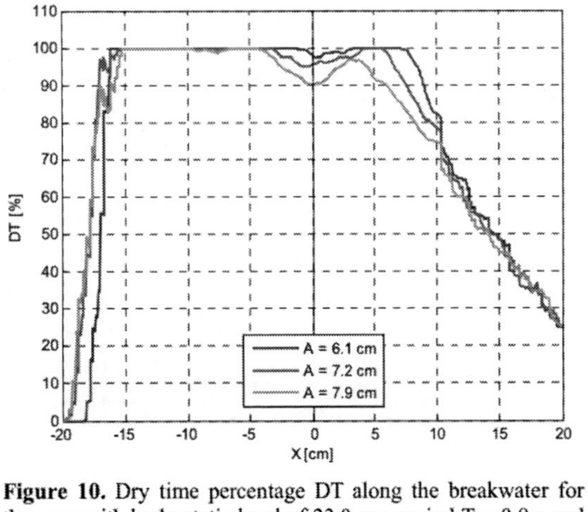

Figure 10. Dry time percentage DT along the breakwater for the runs with hydrostatic level of 23.0 cm, period T = 0.8 s and amplitude A of 6.1, 7.2 and 7.9 cm.

DT has a maximum on the downstream edge (Figure 9); moving downstream, DT tends to zero because H is always non zero in the lee zone of the breakwater. In the upstream breakwater edge, DT is non zero because the wave rundown makes the free surface go under hydrostatic level, leaving a zone of the breakwater uncovered. Comparing Figure 9 with Figure 10, it is possible to note that the dry zones and the dry time percentage DT increase with decreasing hydrostatic level, wave amplitude A and period T.

3.2.2 Mean value and variance of the wave height phase by phase

As already stated, the mean H and the variance σ^2 of the wave heights have been analyzed also phase by phase, in order, in particular, to highlight how σ^2 varies during the wave transit and transformation. In Figures from 11 to 14, four phases of the mean wave for the run with hydrostatic level 29.0 cm (submerged breakwater), T = 0.8 and A = 4.0 cm; above the wave mean height H in phase, compared with the mean one and the hydrostatic level; below, the variance σ^2 in phase.

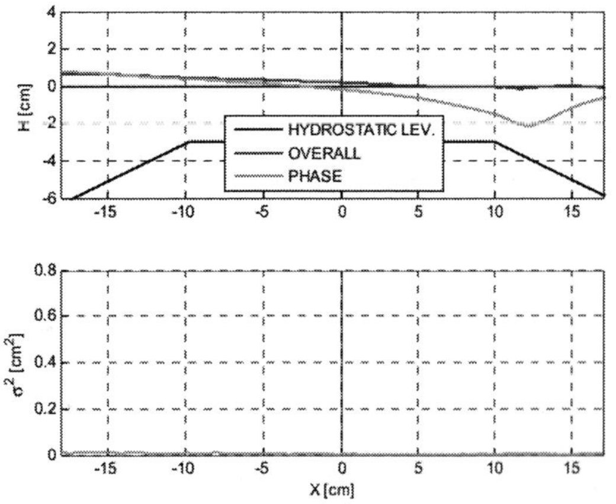

Figure 11. Wave mean height H (above) and variance σ^2 (below) along the breakwater at phase 1/5 for the run with hydrostatic level 29.0 cm, T = 0.8 s and A = 4.0 cm.

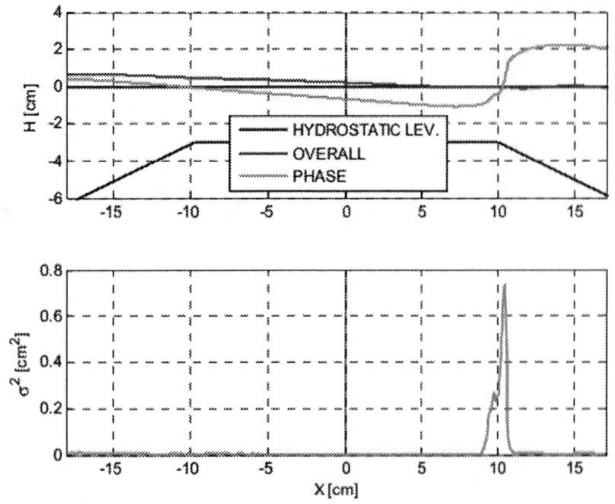

Figure 12. Wave mean height H (above) and variance σ^2 (below) along the breakwater at phase 2/5 for the same run of Figure 11.

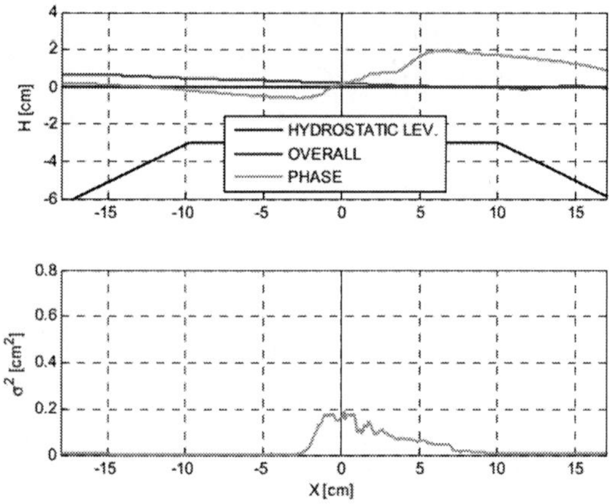

Figure 13. Wave mean height H (above) and variance σ^2 (below) along the breakwater at phase 3/5 for the same run of Figure 11.

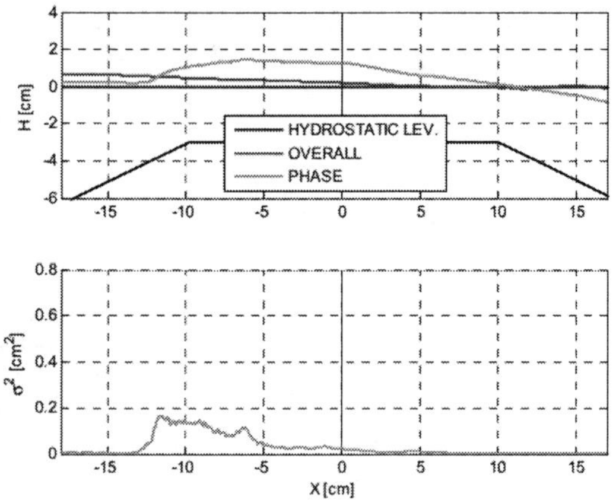

Figure 14. Wave mean height H (above) and variance σ^2 (below) along the breakwater at phase 4/5 for the same run of Figure 11.

During the wave set-down (Figure 11), the wave prepares for overtopping the breakwater, there is not yet any wave breaking or turbulence, so the variance σ^2 is almost zero along all the investigation area. When the wave breaks (Figure 12), the variance σ^2 reaches its maximum value, pointing out that the wave breaks, with a sudden increase of the turbulence level. This peak of variance σ^2 then moves along the obstacle, carried by the wave, with lower values.

The same analysis is shown in Figures from 15 to 18, for the run with hydrostatic level 25.0 cm (emerged breakwater), T = 0.8 and A = 6.1 cm. In the experiments with emerged breakwater, two peaks of variance σ^2 can be spotted: they have lower values than the one experienced with the submerged breakwater. The peak upstream the breakwater is caused by the wave that prepares for the overtopping of the obstacle; the peak downstream the breakwater point out the turbulence generated by the hydraulic jump that develops there. The peaks of variance σ^2 tend to increase with increasing values of amplitude A and period T, both in the case with submerged and emerged breakwater.

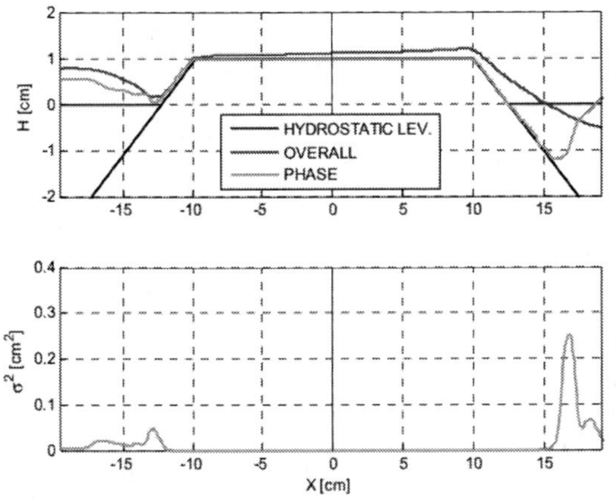

Figure 15. Wave mean height H (above) and variance σ^2 (below) along the breakwater at phase 1/5 for the run with hydrostatic level 25.0 cm, T = 0.8 s and A = 6.1 cm.

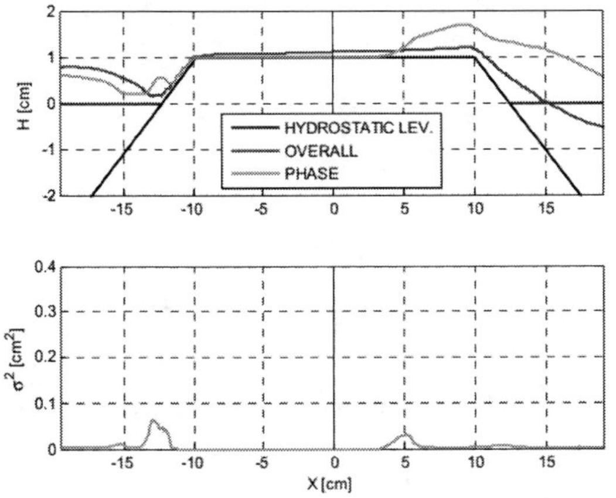

Figure 16. Wave mean height H (above) and variance σ^2 (below) at phase 2/5 for the same run of Figure 15.

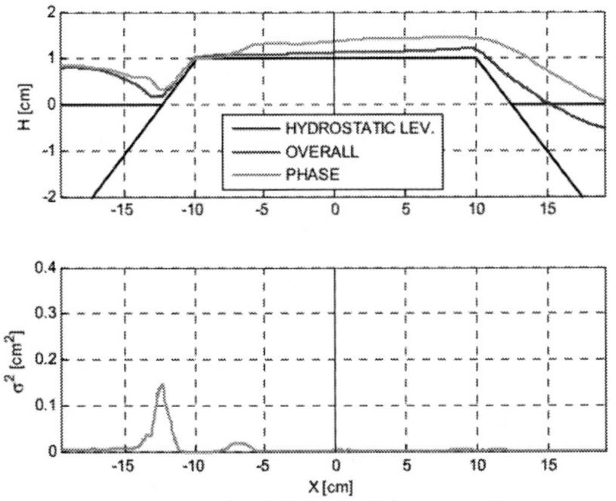

Figure 17. Wave mean height H (above) and variance σ^2 (below) at phase 3/5 for the same run of Figure 15.

3.2.3 Probability density of the free surface heights

As shown before, the Image Analysis technique employed in this work allows to perform measurements and, so, to analyse every single point in the investigation area.

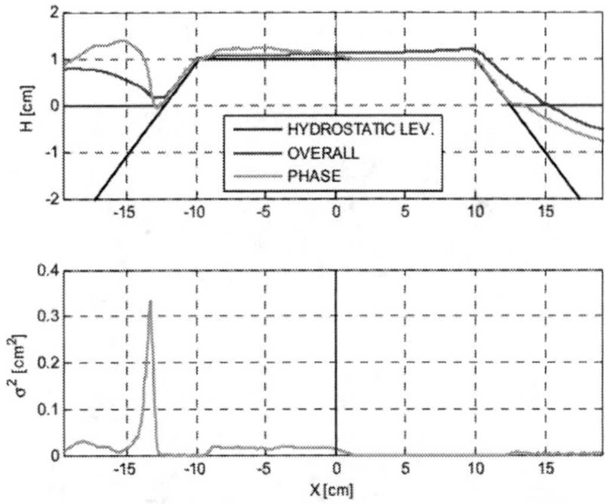

Figure 18. Wave mean height H (above) and variance σ^2 (below) along the breakwater at phase 4/5 for the same run of Figure 15.

We have taken advantage from this feature of the technique was to compute the Probability Density Function (PDF) of the wave heights H for the whole free surface, highlighting some wave characteristics that cannot be spotted from a simple statistical analysis of the wave mean height.

In Figure 19 the Probability Density Function (PDF) of the free surface height H in all the investigation area is shown for a hydrostatic level of 29.0 cm (breakwater submergence of 3.0 cm), with a wave amplitude A = 4.7 cm and a period T = 1.0 s; colours represent the probability density that the wave free surface was in a particular location (dark blue for a zero value, dark red for a high value). The point where the wave breaks corresponds to the upstream depression (X ≅ 0 cm). The net of blue lines visible on the right side of Figure 19 is due to the fact that the wave frequency F (1.0 Hz, being the inverse of the wave period T) is an exact divisor of the camera acquisition rate (25.0 Hz), so there is a high probability for the free surface to be in the same position after a fixed number of frames. The vertical lines of the net identify the positions occupied by the advancing wave front, while the horizontal lines identify the positions of the wave free surface during its withdrawal, up to the point where a new wave breaks. A multi-modal probability distribution is also apparent, describing the typical wave behaviour, both during the set-up (pointed out by red U in Figure 19) and set-down (white D) phases. Moreover, the occurrence of the blocking phenomenon is clear when looking at the upper boundary of the PDF: the wave behaves with time always in the same way, with blocked phases.

In Figure 20, the same analysis is performed on a run with an emerged breakwater: in this case, the Probability Density Function (PDF) of the free surface height H in all the investigation area is shown for a hydrostatic level of 25.0 cm, with a wave amplitude A = 5.6 cm and a period T = 1.0 s. In the PDF analysis with an emerged breakwater, the instants in which that point is in dry time are not taken into account in the statistics. When the obstacle is emerged, the above cited dry zones, the upstream raising and the downstream hydraulic jump are easily spotted.

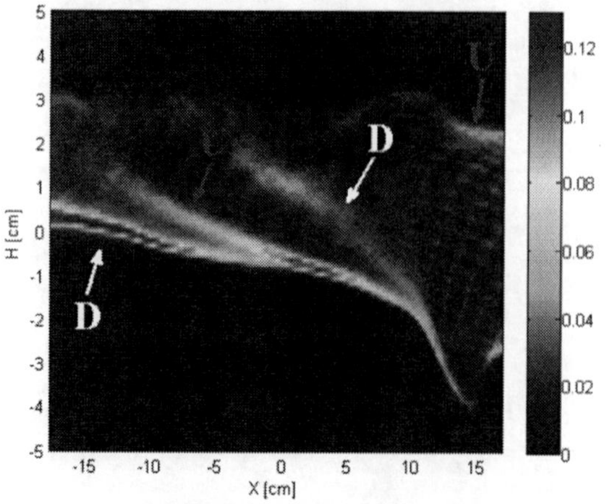

Figure 19. Probability Density Function (PDF) of the wave height H in the investigation area for the run with hydrostatic level 29.0 cm, T = 1.0 s and A = 4.7 cm.

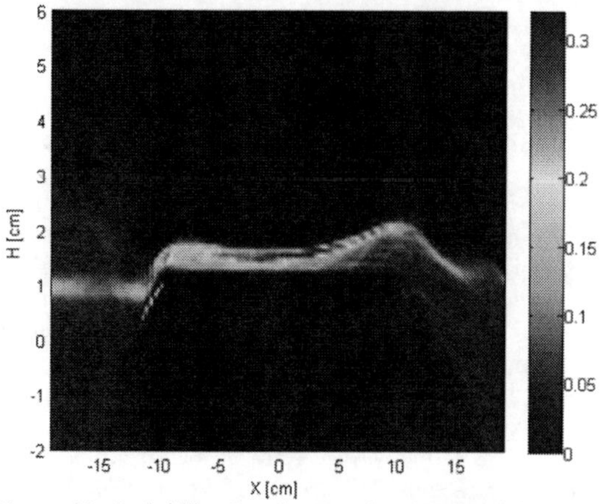

Figure 20. Probability Density Function (PDF) of the wave height H in the investigation area for the run with hydrostatic level 25.0 cm, T = 1.0 s and A = 5.6 cm.

3.3 Power spectral analysis

When a regular wave train, generated by the wave-maker, interact with the breakwater, it looses its linearity and, from this point downstream, it is no longer regular: in this case the waves can be studied with a power spectral analysis. The power spectral analysis performed continuously above the breakwater allows to assess the power density attenuation that each section of the breakwater causes. In Figure 21, 24 and 25 the log10 of the power spectral density PD (colours) is plotted in a chart with frequency F versus X.; log10 is employed to better highlight amounts of energy also very little. When the wave passes over the breakwater, there is an energy transfer from the fundamental frequency (the one carrying the highest fraction of energy) to the harmonic ones, multiples of the fundamental, so the wave energy decreases from the fundamental frequency to the harmonic ones, as it clearly visible in Figure 21. Indeed the highest harmonics are particularly noticeable only on the breakwater upstream side, showing the wave looses its linearity mainly over the breakwater, consistently with Ohyama and Nadaoka findings [10]. Moreover, this analysis confirms that the lower harmonic is the first to be generated, followed by all the other ones. In Figure 22, the fundamental frequency and the first harmonic are plotted along the breakwater: this plot allows to highlight that the greatest fraction of energy is dissipated on the upstream side of the breakwater which is, consequently, the most efficient from the energy dissipation point of view. Thanks to the technique, this density power dissipation can be quantified in of more than ten times of the initial value. Looking at Figure 23, where the fundamental frequencies of the three runs with hydrostatic level equals to 29.0 cm and period to 0.8 cm are plotted, it is possible to understand that the dissipation increases with increasing wave amplitude A.

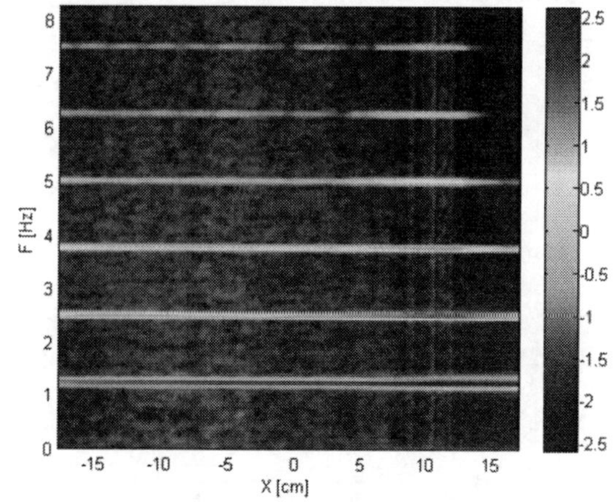

Figure 21. Log 10 of the Power Spectral Density (cm^2 s^2), for the various frequency F, along the breakwater for the run with hydrostatic level 29.0 cm, T = 0.8 s and A = 4.0 cm.

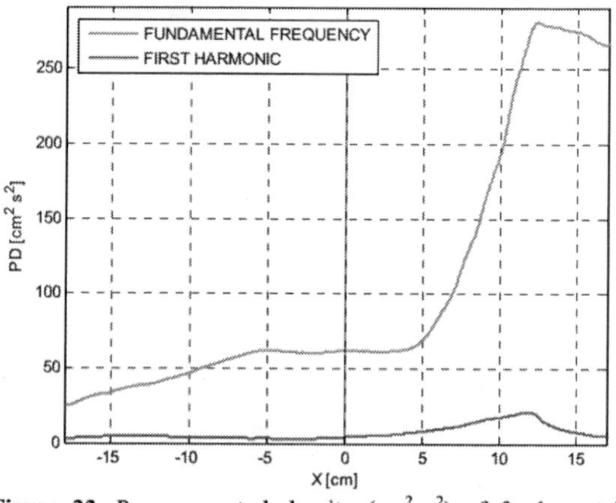

Figure 22. Power spectral density (cm^2 s^2) of fundamental frequency and first harmonic for the same run of figure 21.

In Figure 24, the previously mentioned blocking phenomenon is visible for the highest harmonics, as some of the harmonics do not have a constant energy along X but zones of higher energy are alternate to other of lower one. When, as in Figure 25, the breakwater is emerged, the power dissipation is almost complete already on the breakwater upstream side and a dry zone, with zero energy, arises (dark blue vertical stripe).

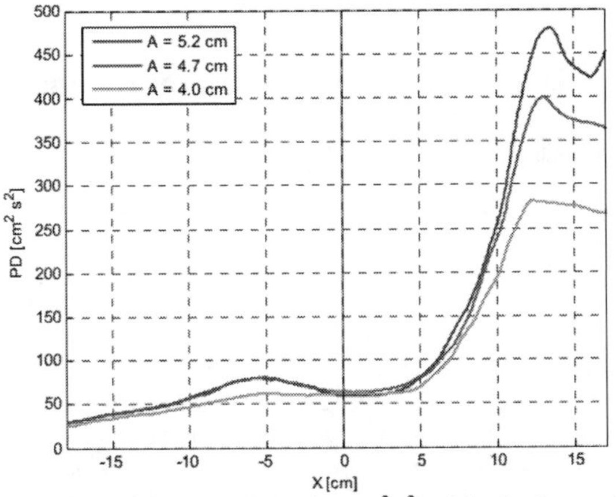

Figure 23. Power spectral density (cm^2 s^2) of the fundamental frequencies for the three runs with hydrostatic level 29.0 cm and period 0.8 s.

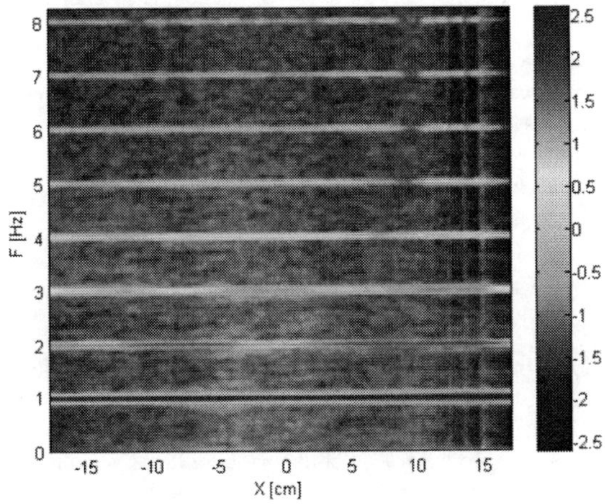

Figure 24. Log 10 of the Power Spectral Density (cm^2 s^2), for the various frequency F, along the breakwater for the run with hydrostatic level 29.0 cm, T = 1.0 s and A = 4.7.

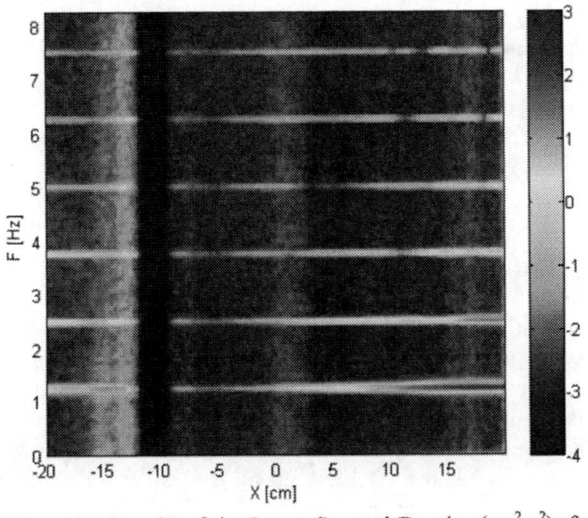

Figure 25. Log 10 of the Power Spectral Density (cm^2 s^2), for the various frequency F, along the breakwater for the run with hydrostatic level 25.0 cm, T = 0.8 s and A = 5.2 cm.

Downstream the dry zone, the wave slips down the breakwater recovering energy and forming a hydraulic jump: the energy is redistributed more uniformly among the various frequencies (pale blue area on the left of the dry zone) and the flow sharp separation in frequencies multiple of the fundamental is no more present.

4 Conclusions

A non-intrusive and continuous-in-space technique, based on Image Analysis, has been developed to overcome the limits of traditional resistive probes, usually employed to investigate wave transformation caused by a breakwater. This technique has proved to properly work also in situations that would be prohibitive for traditional probes (e.g. very shallow water and/or breaking waves) and has allowed performing quantitative analyses with continuity all over the investigation area. The wave transformations above a trapezoidal breakwater, both emerged and submerged, were studied in the laboratory by means of this method. The statistical analysis of wave height has allowed the observation of wave breaking and the measure of set-up and set-down: they increase with increasing wave amplitudes and periods. The measure of the variance of the wave height has allowed to identify if (and, in case, where) the wave breaks: also the variance increases with increasing wave amplitudes and periods. Dry zones (where the water level is almost zero) has been detected and quantified. The blocking phenomenon was spotted thanks to a power density analysis. The Power spectra have been also measured: they are able to quantitatively describe both the energy dissipation (and where it occurs above the breakwater) and the wave non-linear transformations from the weather to lee side of the breakwater, with the resulting energy transfer among the frequencies. The quantities measured thanks to this technique can be supply useful information for design purposes.

References

1. N.G. Rangel-Buitrago, G. Anfuso, A.T. Williams, A.T., Ocean Coast Manage **114**, 129-144 (2015).
2. M.E. McCormick, R.C. Murtha, J. Steinmetz, Mar Technol Soc J **47** (4), 187-192 (2013).
3. T. Sawaragi, *Proceedings of 23° International Conference on Coastal Engineering*, 351-368 (1992).
4. C.K. Ting Francis, K. Young-Ki, Coast Eng **24**, 23-49 (1994).
5. M. Di Risio, I. Lisi, G.M. Beltramiand, P. De Girolamo, Ocean Eng **37** (8-9), 777-789 (2010).
6. S. Ferrari, G. Querzoli, EPJ Web of Conferences **92**, 02018 (2015).
7. S. Ferrari, G. Querzoli, J Hydr Res **48**(5), 632-640 (2010).
8. L.A. Besalduch, M.G. Badas, S. Ferrari, G. Querzoli, EPJ Web of Conferences **67**, 02007 (2014).
9. L.A. Besalduch, M.G. Badas, S. Ferrari, G. Querzoli, EPJ Web of Conferences **45**, 01012 (2013).
10. T. Ohyama, K. Nadaoka, Coast Eng **24**, 1-22 (1994).

EPJ Web of Conferences **114**, 02023 (2016)
DOI: 10.1051/epjconf/201611402023

Airflow Measurement of the Car HVAC Unit Using Hot-wire Anemometry

Miloš Fojtlín[1,a], Michal Planka[1], Jan Fišer[1], Jan Pokorný[1] and Miroslav Jícha[1]

[1]*Department of Thermodynamics and Environmental Engineering, Energy Institute, Faculty of mechanical engineering, Brno University of Technology, Czech Republic.*

Abstract. Thermal environment in a vehicular cabin significantly influence drivers' fatigue and passengers' thermal comfort. This environment is traditionally managed by HVAC cabin system that distributes air and modifies its properties. In order to simulate cabin thermal behaviour, amount of the air led through car vents must be determined. The aim of this study was to develop methodology to measure airflow from the vents, and consequently calculate corresponding air distribution coefficients. Three climatic cases were selected to match European winter, summer, and spring / fall conditions. Experiments were conducted on a test vehicle in a climatic chamber. The car HVAC system was set to automatic control mode, and the measurements were executed after the system stabilisation—each case was independently measured three times. To be able to evaluate precision of the method, the airflow was determined at the system inlet (HVAC suction) and outlet (each vent), and the total airflow values were compared. The airflow was calculated by determining a mean value of the air velocity multiplied by an area of inlet / outlet cross-section. Hot-wire anemometry was involved to measure the air velocity. Regarding the summer case, total airflow entering the cabin was around 57 $l\,s^{-1}$ with 60 % of the air entering the cabin through dashboard vents; no air was supplied to the feet compartment. The remaining cases had the same total airflow of around 42 $l\,s^{-1}$, and the air distribution was focused mainly on feet and windows. The inlet and outlet airflow values show a good match with a maximum mass differential of 8.3 %.

1 Introduction

Europeans spend significant amount of time indoors in buildings and vehicles [1]. Consequently, proper indoor microclimate in relation to air quality, thermal comfort and other factors is important [2]. Regarding vehicular microclimate, inappropriate thermal environment has also a negative impact on drivers' fatigue and focus. Traditionally, microclimate in vehicles is managed via HVAC (Heating Ventilation and Air-conditioning) system. Besides others, the HVAC system also provides defogging or defrosting of windscreens, thus creating conditions for safer driving. These all HVAC features are achieved through smart distribution of the air with an appropriate air speed, temperature and humidity.

This work will analyse the air distribution within a passenger vehicle cabin with a focus on airflow through HVAC vents and calculation of corresponding air distribution coefficients. The HVAC will be set to automatic control mode and three interior temperatures will be studied. Both, system inlet and outlet airflow will be monitored to compare a match of these two values. The aim of this study is to develop a methodology to measure the airflow simply and relatively precisely. Next, three cases will be examined to match central European winter, summer and spring/fall ambient conditions. The experiments will be conducted in a laboratory of thermal comfort at the University of Technology Brno inside a climate chamber. Outcomes of this study will serve also as a reliable source of boundary conditions to numerical simulations of a car cabin.

2 Methods

To develop a simple and efficient methodology to measure airflow from the vents we needed to measure the airflow entering the HVAC system precisely. Thus, we were able to compare "the input" with the airflow leaving the vents in the car cabin. The vents were measured with a less precise and less time consuming method. Match of these two parameters was then studied.

2.1 HVAC system description

Nowadays, the HVAC systems in passenger vehicles are conceptually very similar. Firstly, fresh air is supplied from a front grille located below a windshield and the air is let through filters. Amount of the supplied fresh air is adjusted by an air recirculation door. Next, the airflow is driven by a centrifugal fan, the air passes heat exchangers and final air temperature is achieved in a mixing

[a] Corresponding author: milos.fojtlin@vut.cz

chamber. Now, the air is ready to be introduced to the cabin via air ducts and vents (Figure 1. and 2.). The whole process is controlled manually or by an HVAC control unit, when an automatic control mode is used. This study involved the automatic control mode that selects optimal settings based on ambient temperature, intensity of solar radiation and other numerous factors. Finally, the air leaves the cabin through openings with flaps mounted under a rear bumper.

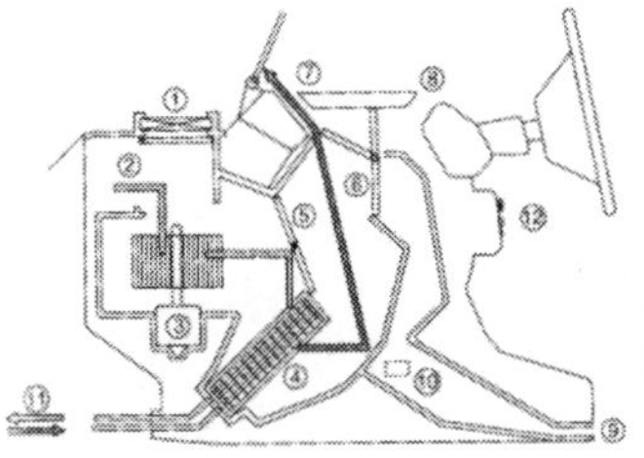

Figure 1. Schematic design of the vehicular HVAC system [3].

The test car possesses one inlet opening and 25 various vents in the interior. Vents arrangement of the test car shows Figure 2, complete description, surface area, and number of measurement points are shown in Table 1.

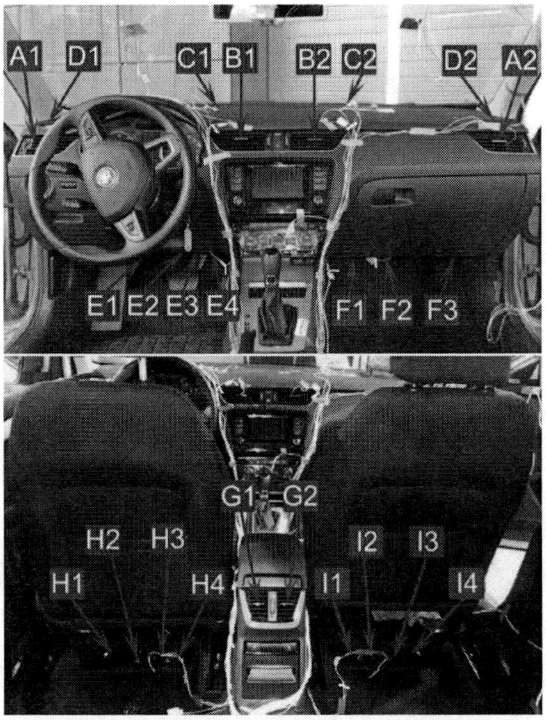

Figure 2. Arrangement of the vents in the tested vehicle.

Some of the vents are equipped with directional flaps and butterfly valves to adjust the flow direction and amount of the supplied air. During the experiments, all of the vents were fully opened and set to a neutral position. Next, it was found that an air pressure fluctuates inside the cabin when the cabin is closed. We ascribed this phenomenon to a cyclic opening and closing of the outlet flaps. Stability of the airflow entering the cabin and the measurement reproducibility was thus negatively influenced. Therefore, a trunk of the car was ajar. This helped to eliminate the fluctuations.

Table 1. Vents description, surface area, number of local air speed measurement points and proportional coefficients.

tag	description	S (cm^2)	no. of w_i	k_m
A1	L dashb.	50.4	4	0.58
A2	R dashb.	50.4	4	0.58
B1	L centr. dashb.	66.8	4	0.57
B2	R centr. dashb.	66.8	4	0.57
C1	L defrost	42.0	3	0.91
C2	R defrost	42.0	3	0.91
D1	L window	18.6	1	0.84
D2	R window	18.6	1	0.84
E1-4	L Feet fr.	11.3; 3.8; 4.3; 5.1	4 × 1	1
F1,2,3	R feet fr.	18.2; 17.8; 12.9	3 × 1	1
G1	L centr. bk.	26.9	2	0.65
G2	R centr. bk	26.9	2	0.65
H1-4	L feet bk	20.0; 9.8; 7.0; 12.6	4 × 1	1
I1-4	R feet bk	12.6; 7.0; 9.8; 20.0	4 × 1	1

2.2 Airflow measurement

Hotwire anemometry was used to measure local air speeds w_i at the system inlet (suction) and outlet (vents). Mean airspeed $\bar{w}$ was calculated using averaging method. The experiment was designed that w_i measurement points were evenly distributed across the examined area and each point represented the same area fraction. Thus, $\bar{w}$ is defined as the arithmetic mean of w_i. Number of measurement points per vent shows Table 1. For circular cross-section with radius R, w_i is determined at pre-calculated radial distances from the center point r_i using formula (1) – equidistant spacing using a transformed coordinate (r/R) in an interval from 0 to 1 [4],

$$r_i = R\sqrt{\frac{2i-1}{2n}} \tag{1}$$

where n is the number of measurement points; i is the sequence number of the measurement point from 1st to n^{th}. Next, to calculate airflow q formula (2) was applied

$$q = \bar{w}\,S\,k_m \tag{2}$$

where $\bar{w}$ is the corresponding mean air speed, m s^{-1}; S is the cross section of the vent or the inlet piping, m^2; k_m is a proportional coefficient to compensate solid cross-section area of the vents if there are the directional flaps present (Table 1). Mass flow of the air q_m was defined as

$$q_m = q\,\rho \tag{3}$$

where ρ is the air, kg m^{-3}. Since low air humidity (up to 20 %) was present during the test, modified equation of state was used

$$\rho = \frac{p}{r\,T} \tag{4}$$

where p is the air pressure, Pa (100 kPa was used); r is a specific gas constant of a dry air, J kg^{-1} K^{-1}; T is temperature, K.

2.2.1 Airflow at the HVAC system inlet

To determine the airflow at the inlet was a challenging task because of following conditions: 1. difficult access to the actual opening; 2. influence of the measurements due to blowing air from the engine fans; 3. unpredictable behavior of recirculation door that influences amount of fresh air entering the cabin.

Firstly, the car hood was disassembled and plastic covers of the suction were removed. Secondly, after several unsuccessful attempts to determine the inlet airflow by a handheld anemometric probe, we decided to mount an inlet tube with an adapter instead of the original grille (Figure 3. and 4.). The tube was 2 m long with diameter of 150 mm. Cross-section area of the tube and the grille was the same. It was also assumed that pressure losses in the tube are similar to the pressure losses induced by the grille, and the measurements are not significantly influenced by this substitution. Thirdly, the recirculation door was kept open to be able to compare the inlet and outlet airflow.

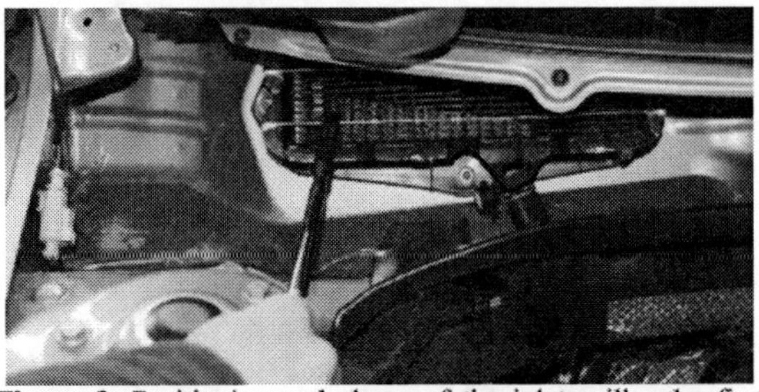

Figure 3. Positioning and shape of the inlet grille; the first attempts to measure the inlet airflow directly without the tube.

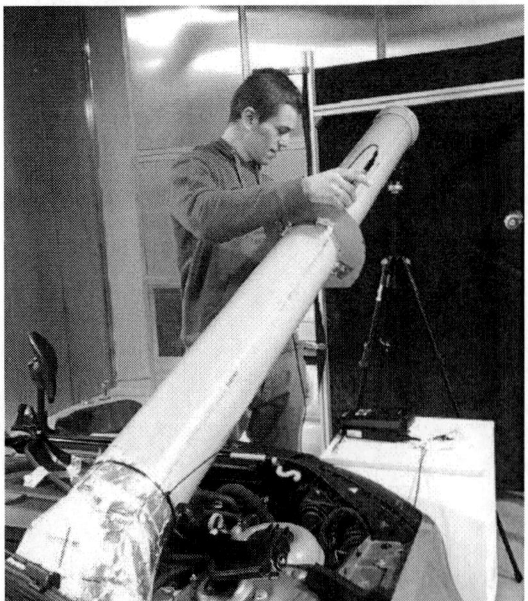

Figure 4. The suction tube with the adapter – a final assembly.

Omnidirectional anemometric and temperature probe Testo 350 M/XL-454 was used to examine airspeed in the tube. The local airspeed was determined in the middle of the tube length (that is 6.6 tube diameters) corresponding to Figure 5. It was also expected that the tube stabilizes the flow contributing to improve measurement precision. Each point was measured 10 times independently for each of the cases. Local airspeed of one point yielded from 2 s measurement average. At the same time, voltage of the HVAC fan was monitored. It was assumed that the fan voltage corresponds to the airflow and can serve as rough airflow estimation.

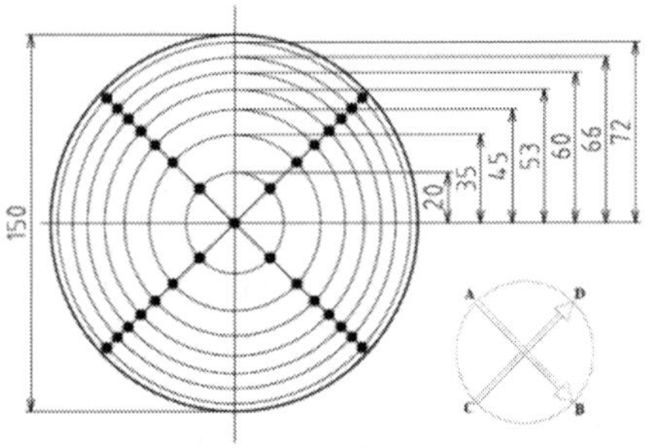

Figure 5. Pre-calculated points of w_i measurement according to the formula (1); quarters of measurement A-B, C-D.

2.2.2 Airflow at the HVAC system outlet

Airspeed was measured circa 2 cm in front of the vents at specified points by a handheld combined probe (airspeed, temperature and humidity) with a logger Testo 435. Number of the points varied due to diversity of vent shapes and final number of the measurement points per vent is indicated in Table 1. However, other methods to determine the airflow were considered, but as it could be seen in Figure 6., the vents pointing to the foot well are hardly accessible and have an irregular shape. This excludes most of the conventional airflow measurement methods.

Figure 6. Example – points of w_i measurement at the vent A1 (upper picture) and vents E1-4 located in the foot well.

The local airspeed was determined from 10 s average, and measurement in each point was repeated three times a case.

2.3 Experimental procedure

Firstly, the experimental vehicle was placed inside the climate chamber. Next, the climate chamber and the artificial solar system were initiated to reach desired ambient conditions, details presents Table 2. Typically, the preconditioning of the chamber with the car inside lasted one hour. Then, the vehicle was started up and the automatic HVAC system was set to desired interior temperature. The actual measurements started after around 45 minutes when the automatic control mode reached stable thermal conditions within the cabin.

Table 2. Boundary conditions of the experiment

case	t_{set} (°C)	$t_{chamber}$ (°C)	I_{solar} (W m^{-2})
summer	auto 23	30	800
spring / fall	auto 22	16	400
winter	auto 18	-2	0

Measurements were performed by two operators, one determining airflow through vents in the car, and the other examining the inlet airflow to the HVAC system outside. Performance of each case lasted around 40 min.

2.4 Statistical data evaluation

As long as no measurement is perfectly precise, there is a need to express uncertainty of the estimated value (measurement). This study dealt with the type A and B uncertainty evaluation of the airflow and the mass flow. Next, we adopted presumption that uncertainty of the air density calculation and the cross-section measurement had a small impact on the results. However, the main focus was on the airspeed uncertainty evaluation.

Type A uncertainty is represented by a statistically estimated sample standard deviation of the mean $u_A(x)$ (5) [5,6,7]. Formula (5) is valid under following conditions: 1. variable x follows a normal probability distribution; 2. the data origin from independent observations performed under the same conditions. Our previous experience proves normal probability distribution of the data.

$$u_A(x) = k_{uA} \left[\frac{1}{n(n-1)} \sum_{i=1}^{n} (x_i - \bar{x})^2\right]^{\frac{1}{2}} \quad (5)$$

where $u_A(x)$ is the type A uncertainty of the variable x; n is the number of the independent observations; x_i is the value of the ith variable x; $\bar{x}$ is the mean value of x; k_{uA} is the safety factor.

This study dealt with 10 observations in the case of the inlet measurements, thus safety factor $k_{uA} = 1$. Regarding the vents, $k_{uA} = 2.3$ was applied to compensate missing data because of the three observations a case [8].

The type B uncertainty $u_B(x)$ is an evaluation method by means other than the statistical data evaluation. We considered only the uncertainty due to limited precision of the anemometers (Table 3). Other sources of the B type uncertainty were omitted.

Next, expanded combined uncertainty was expressed as follows

$$U = k\,u = k\,\sqrt{u_A(x)^2 + u_B(x)^2} \quad (6)$$

where k is the coverage factor, $k = 2$ was applied to achieve 95 % level of confidence; u is the combined uncertainty (confidence level ca 68.3 %).

Table 3. Precision of the anemometers

anemometer	precision	note
Testo 350M/XL	± 0.02 m·s^{-1}	inlet
Testo 435	±0.02 m·s^{-1} + 4 % of est. value	outlet

Uncertainty of an indirect independent measurement is a function of each uncertainty from which the final uncertainty is calculated. If the final quantity x is a function of quantities a, b, c, …, the final uncertainty u_x is calculated as follows [8]

$$x = f(a, b, c \ldots) \quad (7)$$

$$u_x = \sqrt{\left(\frac{\partial f}{\partial a} u_a\right)^2 + \left(\frac{\partial f}{\partial b} u_b\right)^2 + \left(\frac{\partial f}{\partial c} u_c\right)^2 + \cdots} \quad (8)$$

3 Results and discussion

Results from the experiments are presented in form of air distribution coefficients d (%) that represent a fraction of the total airflow through the HVAC system. Each case is examined separately. Confidence level of the presented results is 95 %.

3.1 Summer case

An airspeed profile in the suction tube is depicted in Figure 7. Shape of the profile corresponds to an expected turbulent profile shape. It is obvious that the velocities are not evenly distributed across the tube. However, the peak (4.3 m s^{-1}) is caused due to the adapter shape which had to be customised to fit the inlet opening. Indication of diameters "*A-B, C-D*" corresponds to Figure 5.

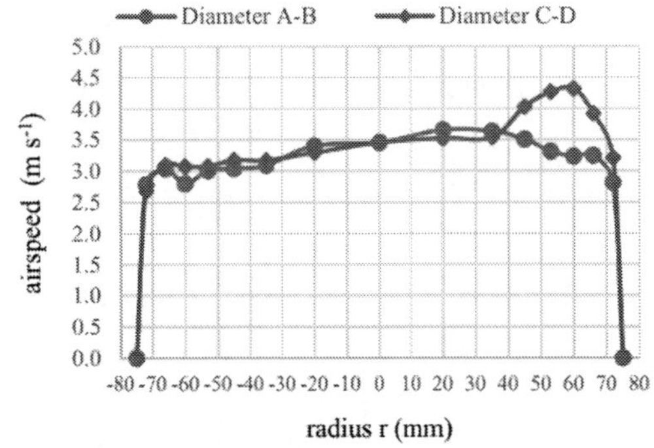

Figure 7. Summer case, the inlet velocity profile.

Voltage on the HVAC fan was stabilised at 4.6 V what meets approximately 57 l s^{-1} of airflow at the vents (sum of all airflows through the vents). Detailed information about the flows provides Table 4. Overall, good match of inlet and outlet airflow was achieved – the absolute difference of the mean values is 1.6 g s^{-1} (l s^{-1}).

Table 4. Summer case, the total airflow and mass flow.

flow	inlet	outlet	difference in-out	
			absolute	relative
(g s^{-1})	68.6 ± 0.3	70.2 ± 3.1	1.6	2.3 %
(l s^{-1})	59.0 ± 0.2	57.4 ± 2.6	1.6	2.7 %

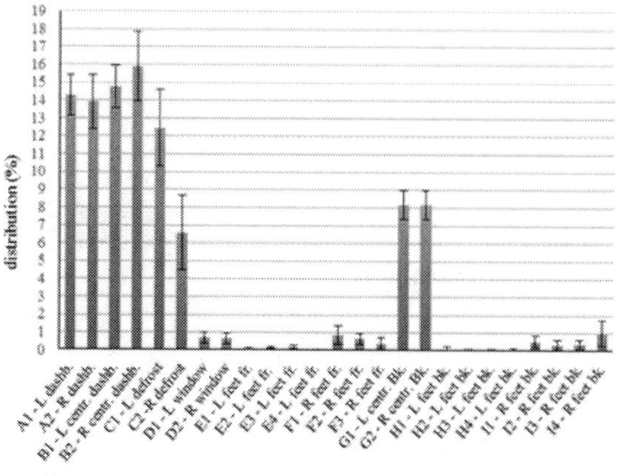

Figure 8. Summer case, distribution coefficients with 95% confidence error bars.

The most of the air, around 60 %, enters the cabin through dashboard vents (A1, A2, B1, B2). Next, about 20 % of the air is directed to the windshield and 16 % is conducted to the rear seats. Distribution is relatively symmetric, yet a defrost slot provided more air for the driver's section. However, relatively wide confident bars (± 2 %) indicate either a high airspeed fluctuation or distinct measurement points. Confidence intervals of other vents range around ± 1 %. As expected, the summer setup does not provide any air to the feet compartment. Residual airflow from the "feet vents" is up to 5 % in total, and it is ascribed to the distribution door leaks.

3.2 Spring / fall case

The airspeed profile in the suction tube is similar to the summer case one; however, the peak reaches only 3 m s^{-1}. In average, the voltage on the fan was 3.8 V what correlates with 42 l s^{-1} of airflow on the outlets (Table 5). That is 35 % less compared to summer case (57 l s^{-1}). Because of the lower heat load on the cabin, lower amount of the air is necessary to treat the thermal environment.

Table 5. Spring / fall case, the total airflow and mass flow

flow	inlet	outlet	difference in-out	
			absolute	relative
(g s^{-1})	49.6 ± 0.3	49.2 ± 1.8	0.4	0.8 %
(l s^{-1})	40.5 ± 0.2	42.1 ± 1.6	1.6	4.0 %

According to Figure 10, distribution of the air has changed in favour of the foot well (total ca 40 %), however the distribution here is not symmetric. The dashboard vents remained opened besides the central ones (B1, B2) with about 26 % of the total airflow. Around 20 % of air was distributed towards the windows.

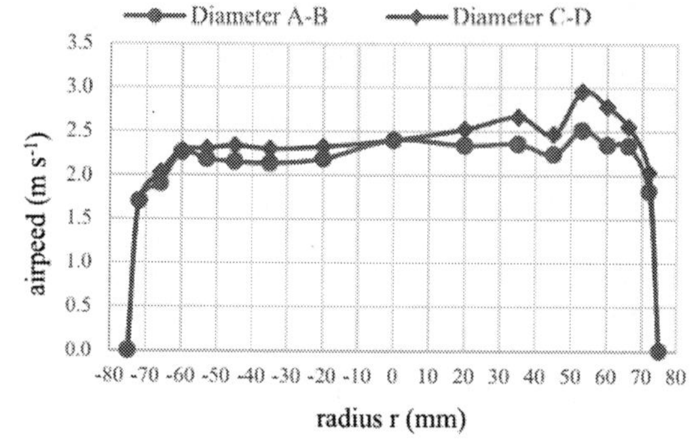

Figure 9. Spring / fall case, distribution coefficients with 95% confidence error bars.

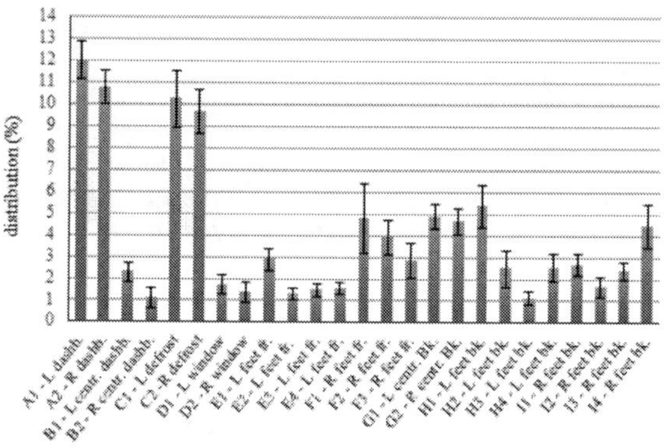

Figure 10. Spring / fall case, distribution coefficients with 95% confidence error bars.

3.3 Winter case

Theoretically, the total inlet airflow of spring / fall case and winter case should be the same because of the same voltage indicated on the HVAC fan during both experiments. Equal voltage should guarantee the same fan speed. This was assumed and the air mass flow was calculated with respect to a change of the air density with temperature (Table 6.). Although, estimated inlet airflow was used, there is 8.3% difference between the estimation and measured mass flow what is acceptable.

Table 6. Winter case, total flow comparison

flow	Inlet	outlet	difference in-out	
			absolute	relative
(g s^{-1})	52.7	48.3 ± 1.9	4.4	8.3 %
(l s^{-1})	40.5	40.6 ± 1.7	0.1	0.3 %

As shown in Figure 11, the air distribution is identical with the spring / fall one. However, this was expected, because majority of the air provides heating of the foot well primarily, and the rest of the air compensates radiative effect of cold windows and ensues their defogging or defrosting.

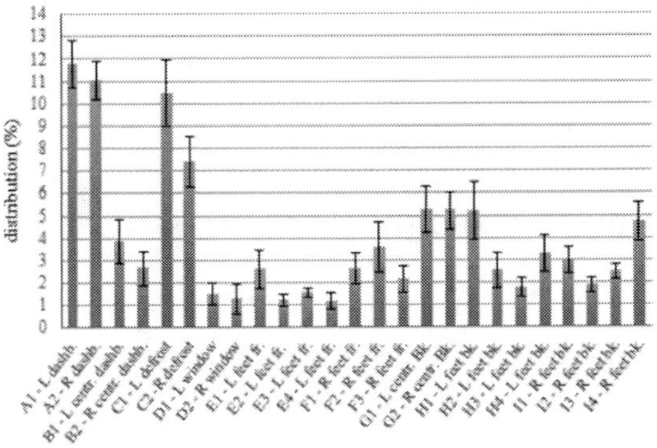

Figure 11. Winter case, distribution coefficients with 95% confidence error bars.

4 Conclusion

The aim of this study was to develop a methodology to measure airflow within a vehicle HVAC system simply and relatively precisely. Inlet and outlet HVAC airflows were studied exploiting hot wire anemometry. The test vehicle, however, remained static during the tests. The air mass flow was calculated under the prerequisite that the air is an ideal gas. Next, three climate cases were examined under steady thermal conditions. We arrived at following conclusions:

- To avoid airflow pulsations due to the cyclic opening of the air exhaust flaps, the cabin had to be connected to the ambient atmosphere (5th doors were ajar). This also means that under real driving conditions the airflow will be changed due to a rising ambient pressure with a driving speed. In the real driving conditions, this is partially solved by closing the recirculation door; however, the door was made fully opened during the tests.
- The mean values of the inlet and outlet flows match (the maximum difference is 1.6 $l\,s^{-1}$ or 4.4 $g\,s^{-1}$). Volumetric change of the air inside the HVAC unit was included in the calculations. Confidence intervals on the inlet are circa one order of magnitude smaller than on the outlet vents what is still acceptable. This could be improved, in the future, by special mounts to hold the probe precisely in its position.
- Summer case showed that the HVAC unit provides more air to compensate heat gains (ca 57 $l\,s^{-1}$) compared to winter and spring/fall case (ca 42 $l\,s^{-1}$).
- Air distribution in summer is focused on passengers' upper body (76 %) and windshield (20 %). Rest of the air leaks to the other vents insignificantly.
- Winter and spring / fall cases are foot well oriented (ca 40 %), the rest of the air is used to heat up the windows and improve passengers thermal comfort.
- Pair vents, besides the feet ones, provide equal amount of the air. This knowledge can be used to determine the airflow only from one vent of the pair in the future.

References

1. Office for Official Publications of the European Communities. *How Europeans spend their time.* Luxemburg: European Communities, pp. 132. ISBN 92-894-7235-9.A. (2004)
2. O. David, V. Ezratty. Energy Policy, Vol.**49**.pp. 116-121. ISSN 0301-4215,(2011)
3. S. Daly. *Automotive air-conditioning and climate control systems.* 1st ed. Oxford : Butterworth Heinemann. s. xvii, pp. 362. ISBN 07-506-6955-1. (2006)
4. Goodfellow, H. D. and E. Tähti. *Industrial Ventilation Design Guidebook. Part 12.* Rxperimental techniques. (2001)
5. R. Palenčár, et al., *Nejistoty v měření I: vyjadřování.* In: Automa. Brno: 7-8. pp. 50-54.(2001).
6. R. Palenčár, et al., *Nejistoty v měření II: nejisototy.* In: Automa. Brno: 10. pp. 52-56.(2001).
7. R. Palenčár, et al., *Nejistoty v měření III: vyjadřování.* In: Automa. Brno: 12. pp. 28-33.(2001).
8. P. Němeček. *Nejistoty měření.* Praha: Česká společnost pro jakost, o.s., (2008)

Acknowledgement

The research was supported by the project Reg. No. FSI-S-14-2355 of the Brno University of Technology; the project LO1202 Netme Centre Plus with the financial support from the Ministry of Education, Youth and Sports of the Czech Republic under the "National Sustainability Programme I" and the project of the Josef Božek Competence Centre for Automotive Industry TE01020020.

EPJ Web of Conferences **114**, 02024 (2016)
DOI: 10.1051/epjconf/201611402024

Operation in the turbulent jet field of a linear array of multiple rectangular jets using a two-dimensional jet (Variation of mean velocity field)

Shigetaka Fujita[1,a] and Takashi Harima[1]

[1]*National Institute of Technology, Tokuyama College, Dept. of Mechanical and Electrical Engineering, Gakuendai, Shunan, Yamaguchi, 745-8585, Japan*

Abstract. The mean flowfield of a linear array of multiple rectangular jets run through transversely with a two-dimensional jet, has been investigated, experimentally. The object of this experiment is to operate both the velocity scale and the length scale of the multiple rectangular jets using a two-dimensional jet. The reason of the adoption of this nozzle exit shape was caused by the reports of authors in which the cruciform nozzle promoted the inward secondary flows strongly on both the two jet axes. Aspect ratio of the rectangular nozzle used in this experiment was 12.5. Reynolds number based on the nozzle width d and the exit mean velocity U_e ($\cong$ 39 m / s) was kept constant 25000. Longitudinal mean velocity was measured using an X-array Hot-Wire Probe (l_h = 3.1 μm in diameter, d_h = 0.6 mm effective length : d_h / l_h = 194) operated by the linearized constant temperature anemometers (DANTEC), and the spanwise and the lateral mean velocities were measured using a yaw meter. The signals from the anemometers were passed through the low-pass filters and sampled using A.D. converter. The processing of the signals was made by a personal computer. Acquisition time of the signals was usually 60 seconds. From this experiment, it was revealed that the magnitude of the inward secondary flows on both the y and z axes in the upstream region of the present jet was promoted by a two-dimensional jet which run through transversely perpendicular to the multiple rectangular jets, therefore the potential core length on the x axis of the present jet extended 2.3 times longer than that of the multiple rectangular jets, and the half-velocity width on the rectangular jet axis of the present jet was suppressed 41% shorter compared with that of the multiple rectangular jets.

1 Introduction

The object of this experiment is to operate both the velocity scale and the length scale of the multiple rectangular jets using a two-dimensional jet.

Many examples of multiple rectangular jets in industrial situations were found in the combustion equipment [1], the high lift devices [2] and the high thrust ejectors [3], they were applied to the rapid mixing and the operation of the mixing rate. If the two-dimensional nozzle shape was divided into multiple rectangular nozzles, the improvement of the mixing rate is expected because the various jets arrangement will be thought. However, the systematic conjecture is in very difficult situation for the reason of that there are so many parameters to be investigated, for example, nozzle aspect ratio, nozzle arrangement pitch, Reynolds number and turbulent intensity. Furthermore, many difficulties are existing in both the method and the cost of the nozzle manufacturing.

So far, Krothapalli et al., [4] (Nozzle length: L = 50 mm, Nozzle width: d = 3.0 mm, Nozzle aspect ratio: AR (L / d) = 16.67, Nozzle arrangement pitch: S = 8 d, Channel type) and Marsters [5] (L = 38 mm, d = 3.96 mm, AR (L / d) = 9.6, S = 4.7 d, Channel type) reported about the multiple rectangular jets. The objective of both reports was to promote the thrust augmentation. In the report of the multiple rectangular jets by Krothapalli et al., [4], the streamwise variation of the characteristic scale in the CD (Characteristic Decay) region containing an initial region (20 times of nozzle width d) was almost the same with the result of the single rectangular jet by Krothapalli et al., [6] (L = 50 mm, d = 3.0 mm, AR (L / d) = 16.67, Channel type).

On the other hand, Yu [7] used the multiple rectangular jets of which those rectangular axes were arranged in a line (L = 2.0 mm, d = 1.0 mm, AR (L / d) = 2.0, S = 2.5 d) to make high quality gas wiping galvanized steel sheets with uniform thickness. In this case, an extent of the potential core length of the multiple rectangular jets is very important, but it is difficult to extent the potential core length effectively in the initial jet region. Therefore, a new method to improve the potential core length is expected. From these points of view, there are many subjects to be clarified, for example, the fundamental differences between the single

[a] Corresponding author: fujita@tokuyama.ac.jp

rectangular jet and the multiple rectangular jets, and the developing and decreasing properties for the case of handling the multiple rectangular jets as the object of operation.

From the mentioned above, the present study adapted a new method of running through transversely with a two-dimensional jet to change the property of the multiple rectangular jets. This adaption was obtained from the reports by authors [8, 9, 10] of which the cruciform shape nozzle jet produced the inward secondary flow toward the jet centre region on both nozzle axes.

The object of the present study is to delay the decreasing rate of the velocity scale near the jet centre region and to repress the developing rate of the half velocity width of the multiple rectangular jets field using a two-dimensional jet which runs through transversely perpendicular to the multiple rectangular jets. In this report, this present multiple rectangular jets run through transversely using a two-dimensional jet will be called the present 3-D jet.

In the case of comparing the results of the velocity and the length scales between the present 3-D jet and the multiple rectangular jets, the results of multiple rectangular jets by authors will be used, instead of the results of multiple rectangular jets by Krothapalli et al., [4], because the multiple rectangular jets by authors have the same cross section shape at the nozzle exit section, rectangular nozzle width d (= 10 mm), aspect ratio (AR = 12.5) and arrangement pitch (S = 12.5 d) with the present 3-D jet.

Nomenclature

The nomenclatures used in this study are as follows:

AR : Aspect ratio of the rectangular nozzle (L / d)

b_y : Half velocity width on the long axis (y axis) of the rectangular nozzle in the present 3-D jet and the multiple rectangular jets

D_2 : Length scale which is equivalent to the two-dimensional jet nozzle width made from the total nozzle exit area of each three-dimensional jet

d : Nozzle width (10 mm)

L : Length of long axis of the rectangular nozzle in the multiple rectangular jets (125 mm)

L_{ye} : Half-length of long axis of each rectangular nozzle (L / 2 = 62.5 mm)

l : Length of long axis of the two-dimensional nozzle in the present jet (990 mm)

S : Arrangement pitch of the rectangular nozzle (125 mm)

U : Longitudinal mean velocity in the x-direction

U_e : Longitudinal mean velocity at the nozzle centre ($x / d = 0$, $y / d = z / d = 0$) of the exit plane ($\cong$ 39 m / s)

U_{ox} : Longitudinal mean velocity along the jet centreline (x axis: $y / d = z / d = 0$) on the x axis

V : Lateral mean velocity in the y-direction

W : Spanwise mean velocity in the z-direction

x, y, z : Cartesian coordinate system with origin at the rectangular nozzle centre of the exit plane in the present 3-D Jet (See figure1)

Subscript

e : Value at the nozzle exit centre of the present 3-D jet ($x / d = 0$, $y / d = z / d = 0$)

m : Value at the middle location ($y / d = 0$, $z / d = 6.25$) which is the centre between the rectangular nozzles of the present 3-D jet

ox : Value along the jet centreline (x axis) of the present 3-D jet

rms : Value of root mean square

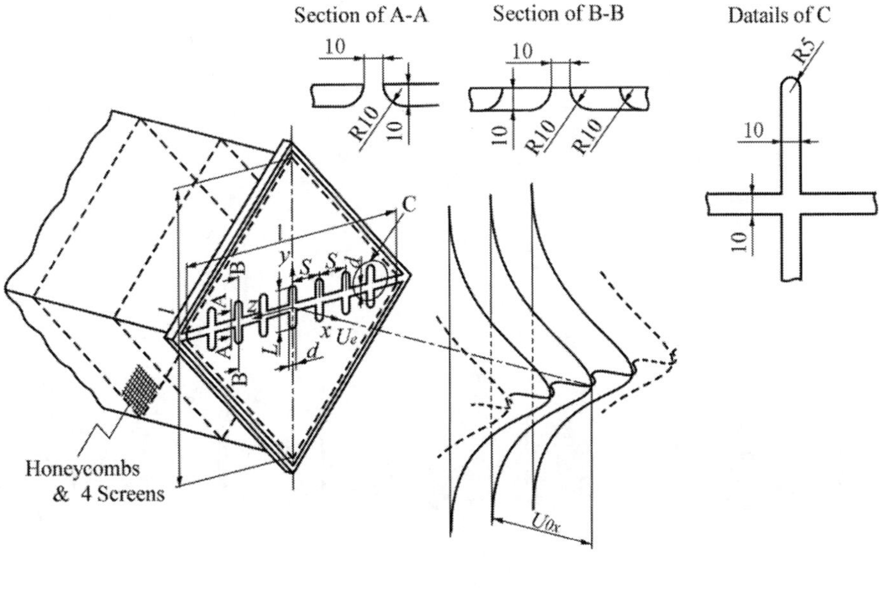

Fig. 1. Configuration of the flowfield and the coordinate system of the present 3-D jet.

2 Experimental equipment and procedure

The configuration of the flowfield and the coordinate system are presented in figure 1. The jet flow facility consists of a turbo fan, a settling chamber with a 990 mm diagonal line, honeycomb and 4 mesh wire screens. The present 3-D jet nozzle which is made from the acrylic acid resin of 10 mm thickness, is installed at the end of the settling chamber. The main axis length L and the width d of the rectangular nozzle are made to be 125 mm and 10 mm respectively, because the magnitude of the inward secondary flow on the rectangular long axis showed the maximum value for the case of cruciform nozzle aspect ratio (L / d) = 12.5 [10]. The seven multiple rectangular nozzles are arranged with same pitch S = 125 mm (S / d = 12.5). Here, the configurations of both the present 3-D jet nozzle and the multiple rectangular jet nozzles are shown in figures 2-(a) and (b) for comparison.

Longitudinal mean velocity were measured using an X-array Hot-Wire Probe (d_h = 3.1 μm in diameter, l_h = 0.6 mm effective length: l_h / d_h = 194) operated by the linearized constant temperature anemometers (DANTEC), and the spanwise and the lateral mean velocities were measured using a yaw meter. The signals from the anemometers were passed through the low-pass filters (10 kHz) and sampled using A.D. converter at 20 kHz. The processing of the signals was made by a personal computer. Acquisition time of the signals was usually 60 seconds (max acquisition time 100 seconds). The exit plane Reynolds number based on the nozzle width d and the exit mean velocity U_e, was kept constant 25000 throughout the present experiment. Furthermore, the value of the nozzle exit turbulent intensity u_{rms} / U_e was about 4×10^{-3}. In this experiment, the uncertainty associated with the longitudinal mean velocity U is estimated at ±3% of U_e, which includes calibration error of the linearized constant temperature anemometers, and that of the spanwise mean velocity V and the lateral mean velocity W is estimated at ±6% of U_e, which includes calibration error of the used yaw meter.

3 Experimental results and discussion

3.1 Streamwise variation of the longitudinal mean velocity

The streamwise variation of the longitudinal mean velocity profiles on both the y and z axes of the present 3-D jet (Black) are shown in figure 3, and the results of the multiple rectangular jets (Red) by authors are also shown for comparison in the region of $5 \leq x / d \leq 50$. The dotted line in figure 3 shows the z / d = 6.25 slice plane which is the middle location between the present 3-D multiple rectangular jets.

The profiles of the multiple rectangular jets on the y axis (Red) show the saddle-back shape which takes the minimum value at the jet centre and the maximum value near the jet end from the section of x / d = 5 to 30. On the other hand, the present 3-D jet (Black) shows a saddle-back shape of which the profile takes the maximum value at the jet centre and takes the local maximum value near the jet end, therefore, the saddle-back shape of the present 3-D jet is different from that of the multiple rectangular jets. Here, the reason of the value of $U / U_e > 1.0$ near the jet centre region within the region of $x / d \leq 20$, comes both from the active inward advection transport of the mean kinetic energy ($V \cdot \partial (U^2 / 2) / \partial y$ on the y axis, $W \cdot \partial (U^2 / 2) / \partial z$ on the z axis) [11] and the nozzle cross section shape which produces the inward secondary flow (as will be shown in figure 4).

The profiles on the z axis for the multiple rectangular jets (Red) show a monotonous decrease from the maximum value at the jet centre region to the zero value at z / d = 6.25 within the region of $x / d \leq 15$. In the region of $x / d \geq 20$, the profiles decrease monotonically from the jet centre region to the minimum value at z / d =

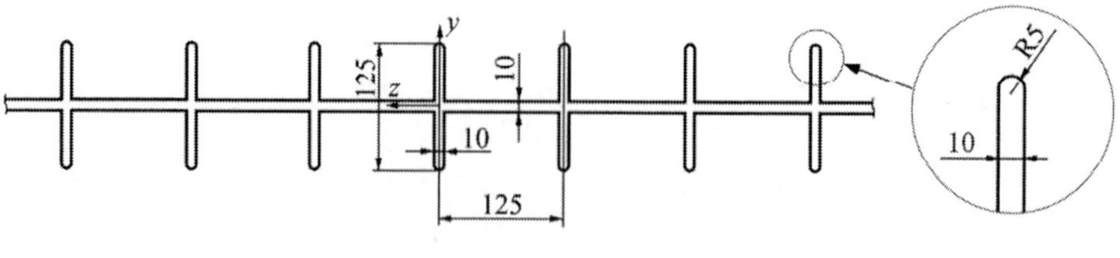

(a) Present 3-D jet

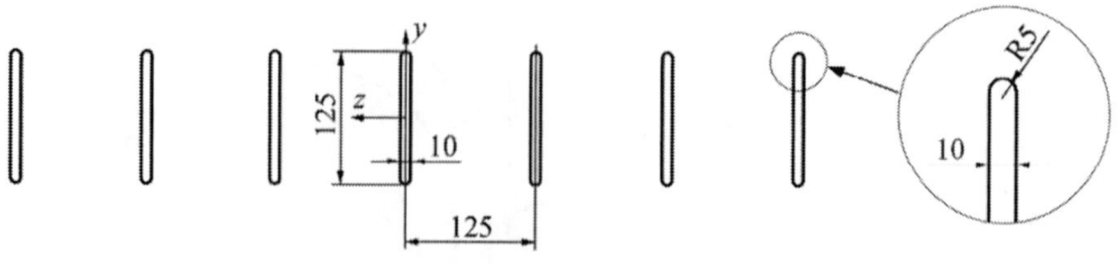

(b) Multiple rectangular jets

Fig. 2. Comparison of the nozzle exit shape between the present 3-D jet and the multiple rectangular jets.

6.25. On the other hand, the profiles on the z axis for the present 3-D jet (Black) at the sections of $x / d = 5$ and 10 take the maximum value at the jet centre region and decrease monotonically and then take the saddle-back shape. In the region of $x / d \geq 15$, the profiles show the monotonic decrease from the maximum value at the jet centre ($z / d = 0$) to the minimum value at $z / d = 6.25$. The decreasing rate of the value near the jet centre region of the present 3-D jet (Black) is smaller than that of the multiple rectangular jets (Red).

From this result, it is confirmed that the decreasing rate of the streamwise velocity scale near the jet centre region is repressed compared with that of the multiple rectangular jets.

3.2 Comparison of the secondary flow velocity

In this section, the confirmation of the secondary flow production and the magnitude of the inward secondary flow on both the y and z axes will be examined. Figure 4 shows the comparison of the streamwise variation of the secondary flow velocity profiles on both the y and z axes between the present 3-D jet (Black) and the single rectangular jet (Red) by authors [12]. Here, the secondary flow profiles on both the y and z axes for the multiple rectangular jets by authors were not measured. Therefore, the results which were acquired by the single rectangular jet [12] were used.

The strong inward secondary flow on the y axis is shown in the extent from the section of $x / d = 5$ to 15. In the spanwise region of $0.5 \leq y / d \leq 5.0$ on the y axis at the section of $x / d = 5$, it is seen that the inward secondary flow of the present 3-D jet is promoted compared with that of the single rectangular jet (Red). On the other hand, the inward secondary flow on the z axis is produced at the section of $x / d = 5$ and10, but it is not found in the profiles of the single rectangular jet.

From the mentioned above, it is confirmed that the magnitude of the inward secondary flow on the y axis in the present 3-D jet is remarkably promoted compared with that of the single rectangular jet.

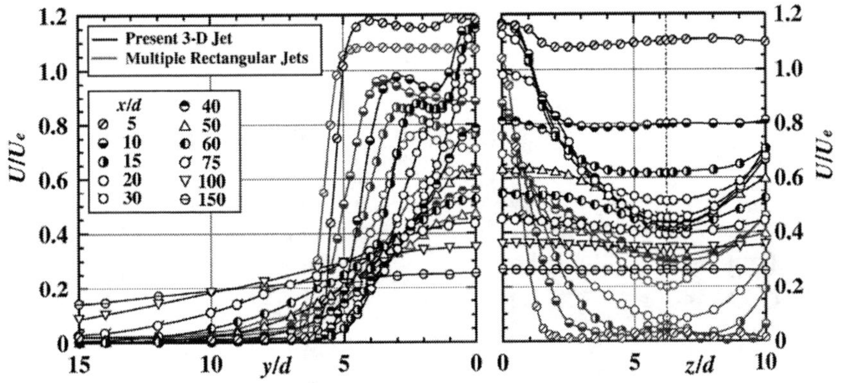

Fig. 3. Comparison of the streamwise variation of the longitudinal mean velocity profiles on both the y and z axes.

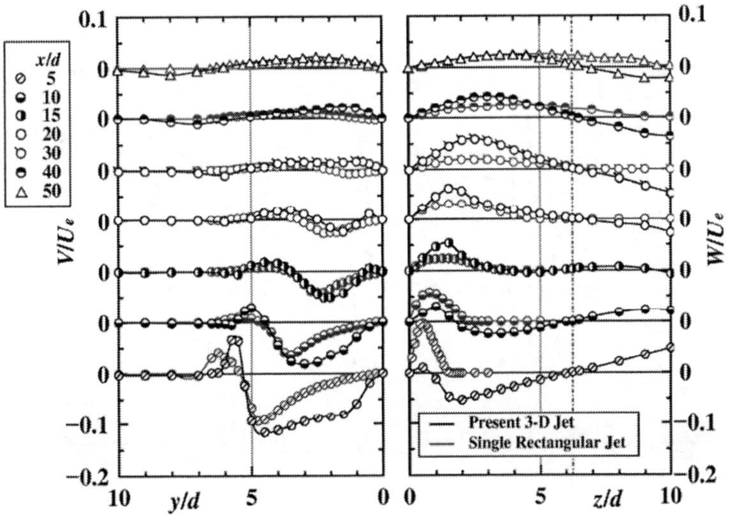

Fig. 4. Streamwise variation of the secondary flow velocity profiles on both the y and z axes.

3.3 Variation of the contour plots of the longitudinal mean velocity

In this section, the streamwise variation of the whole flowfield at each constant section of x / d will be investigated by comparison of velocity contour plots between the present 3-D jet and the single rectangular jet by authors [12]. Figures 5-(a), (b) and (c) show the streamwise variation of the velocity contour plots of the present 3-D jet, and figures 6-(a), (b) and (c) show the contour plots of the single rectangular jet for comparison.

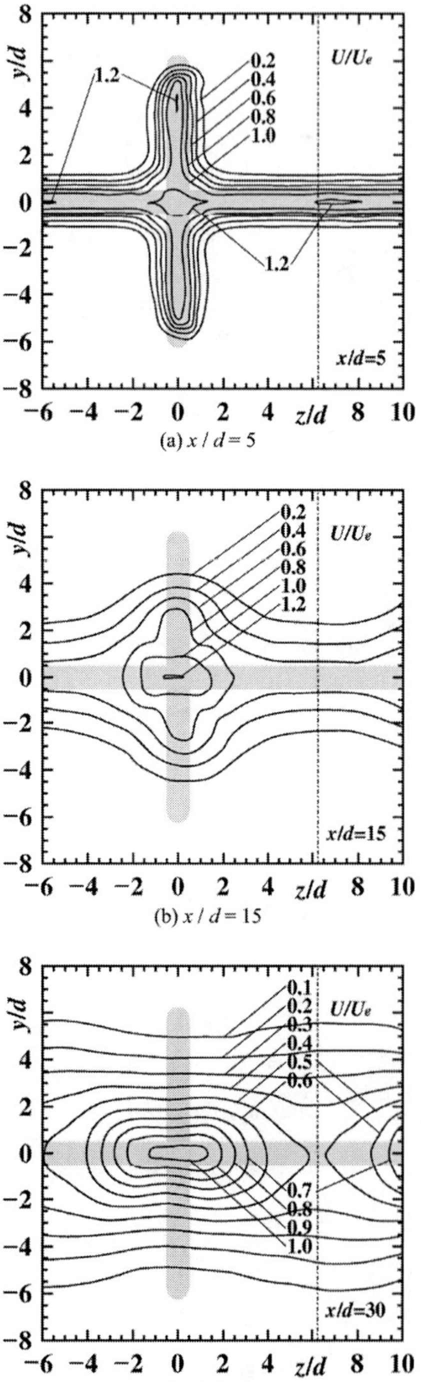

(a) $x / d = 5$

(b) $x / d = 15$

(c) $x / d = 30$

Fig. 5. Streamwise variation of the contour plots of longitudinal mean velocity of the present 3-D jet.

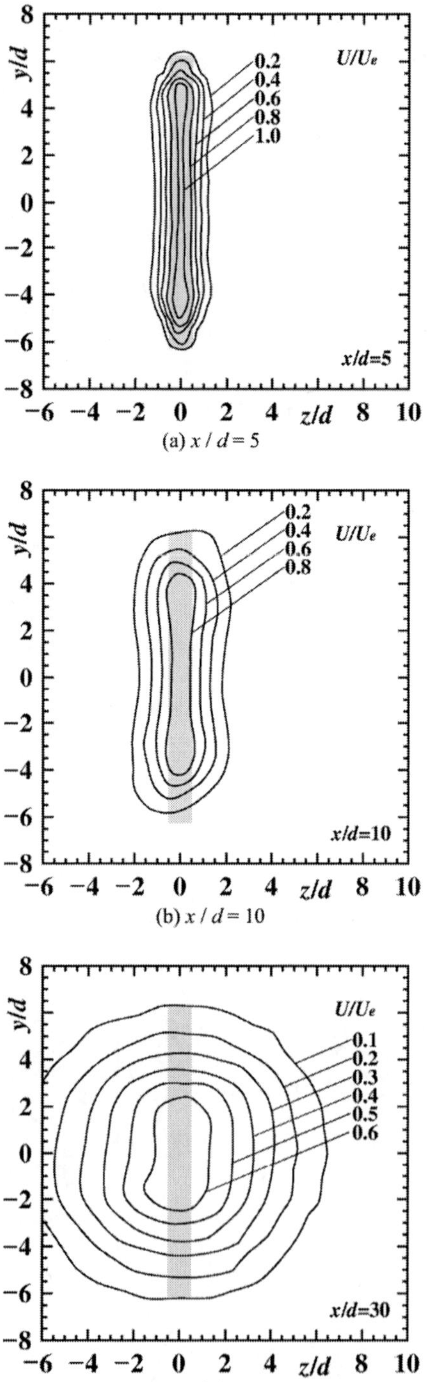

(a) $x / d = 5$

(b) $x / d = 10$

(c) $x / d = 30$

Fig. 6. Streamwise variation of the contour plots of longitudinal mean velocity of the single rectangular jet.

Furthermore, the nozzle shape at the nozzle exit plane is painted blue, and a dotted line in figure 5 is drawn on the $z / d = 6.25$ slice plane which is the middle location between the present 3-D multiple rectangular nozzles.

The contour plot of $x / d = 5$ in figure 5-(a) indicates the cruciform shape similar to the nozzle exit shape. The local maximum value ($U / U_e \cong 1.2$) on the y axis is located at $y / d = \pm 4.0$ due to the inward secondary flows, and the jet width of $U / U_e = 0.2$ on the y axis is smaller than the rectangular nozzle length (12.5 d). On the other

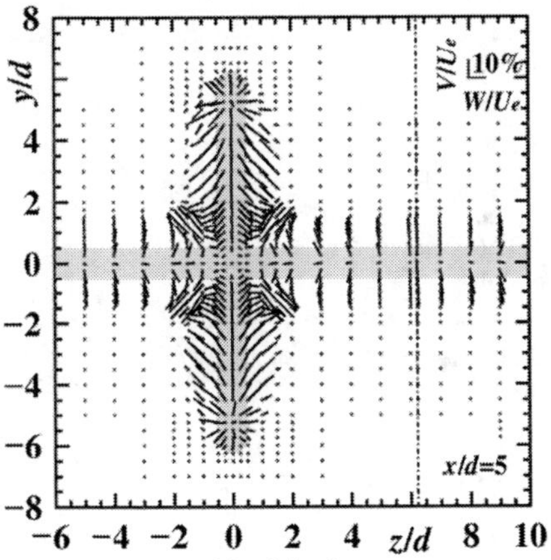

(a) $x / d = 5$

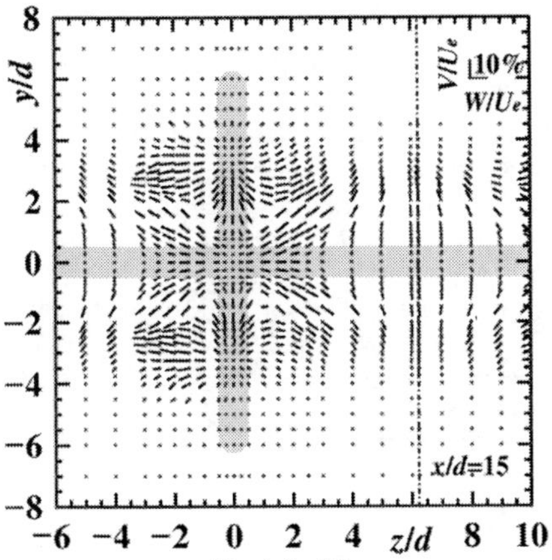

(b) $x / d = 15$

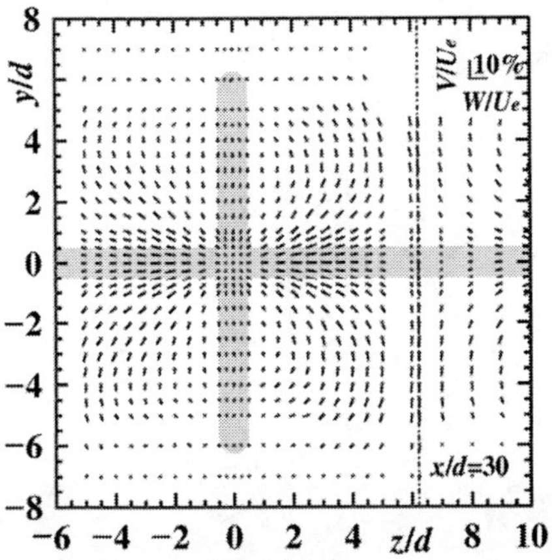

(c) $x / d = 30$

Fig. 7. Streamwise variation of the secondary flow vectors of the present 3-D jet at the three streamwise sections.

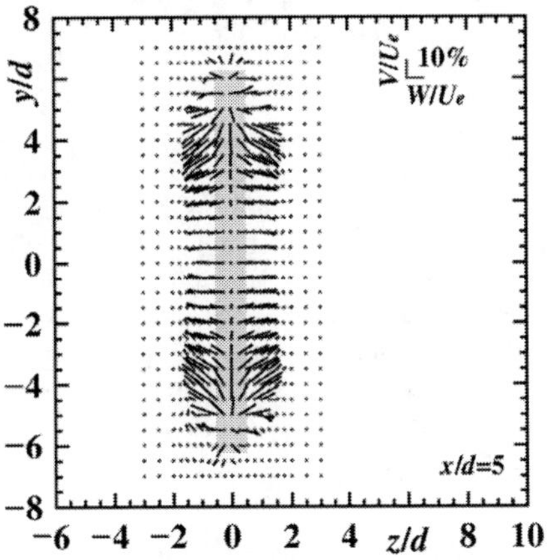

(a) $x / d = 5$

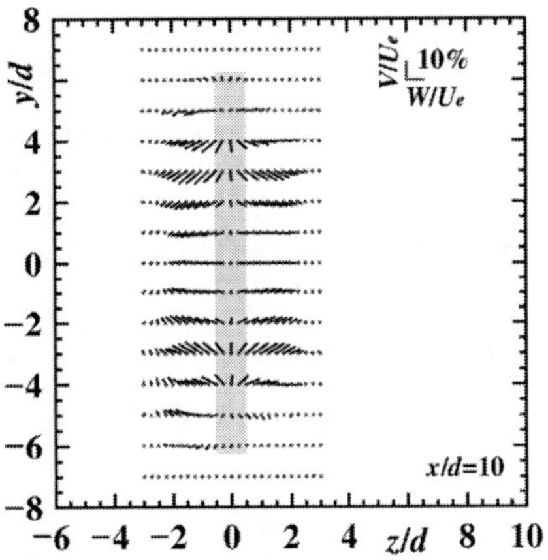

(b) $x / d = 10$

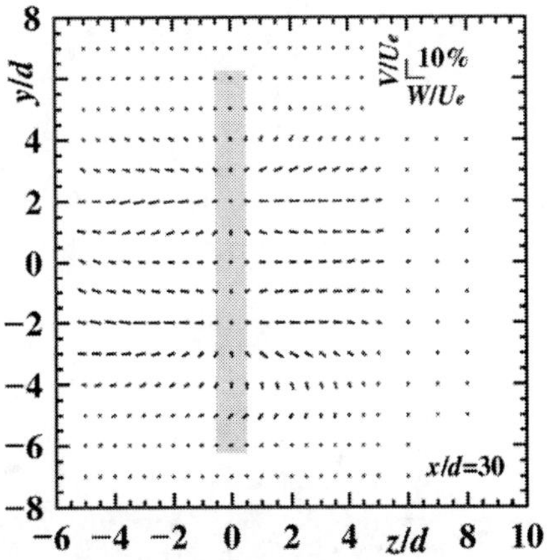

(c) $x / d = 30$

Fig. 8. Streamwise variation of the secondary flow vectors of the single rectangular jet at the three streamwise sections.

hand, the contour plot of the single rectangular jet at $x / d = 5$ also takes the local maximum value ($U / U_e > 1.0$) near $y / d = \pm 4.0$ on the y axis, and the jet width of $U / U_e = 0.2$ on the y axis is almost equal to the rectangular nozzle length.

The jet width ($U / U_e = 0.2$) on the y axis at $x / d = 15$ of the present 3-D jet is narrower than that at $x / d = 5$ because of the inward secondary flow, but the jet width in the y direction on the $z / d = 6.25$ slice plane is large compared with that at $x / d = 5$. The width in the z direction of the single rectangular jet at $x / d = 10$ expands compared with that at $x / d = 5$, but the width of $U / U_e = 0.2$ on the y axis is almost the same value.

Furthermore, the jet width ($U / U_e = 0.2$) of the present 3-D jet on the y axis at $x / d = 30$ is almost the same with that at $x / d = 15$ because there is no inward secondary flow on the y axis as shown in figure 4. However, the jet width ($U / U_e \leq 0.2$) on the $z / d = 6.25$ slice plane of the present 3-D jet at $x / d = 30$ is larger than that on the y axis.

3.4 Variation of the secondary flow vectors

Figures 7-(a), (b) and (c) show the secondary flow vectors of the present 3-D jet, and the results of the secondary flow vectors of the single rectangular jet by authors [13] are shown in figures 8-(a), (b) and (c) for comparison. In this section, the feature of the promoted secondary flow velocity of the present 3-D jet will be clarified.

At the section of $x / d = 5$ for the present 3-D jet, the inward secondary flows are confirmed in both the region of $y / d \leq 5.25$ on the y axis and of $z / d \leq 6.00$ on the z axis. As a result, the outward secondary flow is produced in the direction of $y = \pm z$ axes. The magnitude of the secondary flow velocity along the y axis is larger than that of the single rectangular jet, and it is seen that the inward secondary flow of the present 3-D jet is fairly promoted compared with that of the single rectangular jet as shown in figure 4.

The inward secondary flow velocity at the section of $x / d = 15$, is still found on the y axis near the jet centre region, but the extent of the inward secondary flow velocity becomes narrower than that at $x / d = 5$. The other hand, the inward secondary flow velocity in the region along the z axis is not found.

At the section of $x / d = 30$ of the present 3-D jet, there is no inward secondary flow velocity on both the y and z axes.

3.5 Variation of the velocity scale

The streamwise variation of the mean velocity scales along the x axis ($y / d = z / d = 0$) and the middle location ($y / d = 0$ and $z / d = 6.25$) for the present 3-D jet, is shown in figure 9. Here, the result of the multiple rectangular jets by authors is also indicated for comparison. The streamwise distance x of the upper section and the lower section of figure 9 was normalized by the nozzle width d and D_2, respectively. The reason of the adaption of the length scale D_2 is that this length scale is made from the total nozzle exit area of each three-

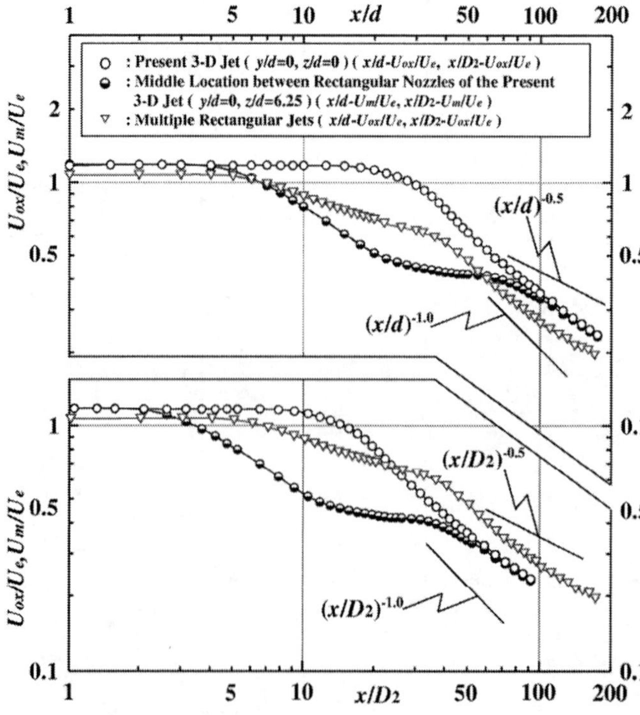

Fig. 9. Comparison of the streamwise variation of the mean velocity scales.

dimensional jet and is recognized as a suitable length scale for the three-dimensional jet which has each various shape [14]. The value of D_2 for the present 3-D jet is equal to 1.90 *d*, on the other hand, the value of the multiple rectangular jets ($S / d = 12.5$) is equal to 0.983 *d*.

In the upper section, the potential core length of the present 3-D jet (Black) exists until the section of $x / d = 28$. This result shows that the potential core length of the present 3-D jet is completely longer than that of the multiple rectangular jets. For the case using the length scale D_2, the velocity scale (U_{ox} / U_e) (Black) on the *x* axis for the present 3-D jet shows the potential core region until the section of $x / D_2 = 14$, this result is 2.3 times longer than that of the multiple rectangular jets ($x / D_2 = 6$). This result is recognized as that the inward secondary flow is produced on the both axes in the initial region (cross shaped nozzle), and the advection transport of the mean kinetic energy ($V \cdot \partial (U^2 / 2) / \partial y$ on the *y* axis, $W \cdot \partial (U^2 / 2) / \partial z$ on the *z* axis) is promoted on both the *y* and *z* axes by using the method of running through transversely with a two-dimensional jet perpendicular to the multiple rectangular jets.

From the result mentioned above, it is clarified that the decay rate of the velocity scale on the *x* axis is strongly repressed, and the potential core region is extended extremely, compared with the multiple rectangular jets.

3.6 Variation of the length scale

The streamwise variation of the half velocity width on the *y* axis (rectangular nozzle axis) of both the present 3-D jet and the multiple rectangular jets by authors, will be compared in figure 10. Here, the horizontal axis *x* and the vertical axis b_y are normalized by the nozzle width *d* in the upper section of figure 10. On the other hand in the lower section of figure 10, the horizontal axis *x* is normalized by D_2, and the vertical axis b_y is normalized by the half length of the rectangular nozzle length L_{ye}.

The half velocity width b_y / d (Black) on the *y* axis of the present 3-D jet decreases from the value ($b_y / d = 5.36$) at $x / d = 5$ to the minimum value ($b_y / d = 2.21$) at $x / d = 30$. On the other hand, the result of b_y / d (Red) of the multiple rectangular jets decreases from the value ($b_y / d = 6.15$) at $x / d = 2$ to the minimum value ($b_y / d = 3.74$) at $x / d \cong 30$. Comparing these results, the decreasing rate of b_y / d (Black) for the present 3-D jet is larger than that of the multiple rectangular jets (Red). From this result, it is confirmed that the development of the half velocity width on the *y* axis for the present 3-D jet, is suppressed remarkably compared with that of the multiple rectangular jets from the initial region to the CD region.

The minimum value of the half velocity width b_y which is normalized by the half-length L_{ye} (= 6.25 *d*), will

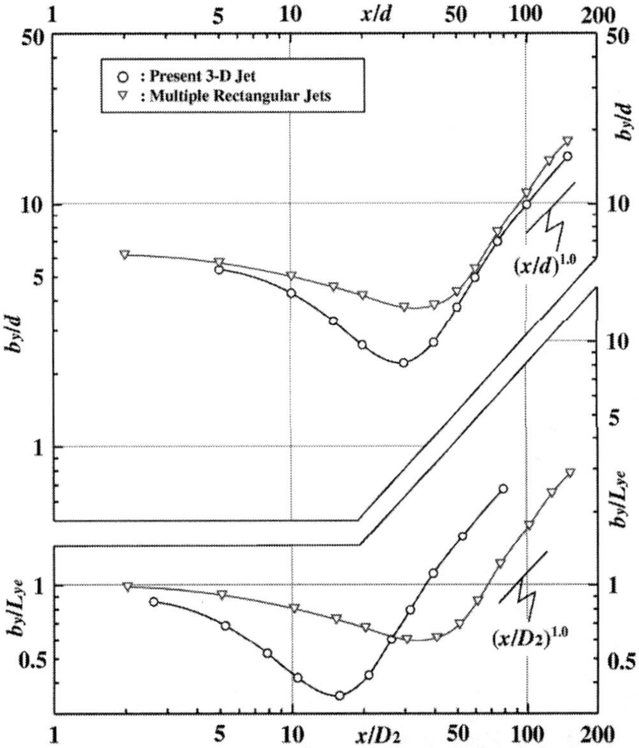

Fig. 10. Comparison of the streamwise variation of the length scales on the long axis of the rectangular jet.

be examined. The value of b_y / L_{ye} (Black) is 0.354 at x / D_2 = 15.7 for the present 3-D jet, and the value of b_y / L_{ye} (Red) is 0.60 at x / $D_2 \cong 30$ for the multiple rectangular jets. From this result, it is seen that the magnitude of the length scale on the y axis for the present 3-D jet is 41% smaller than that for the multiple rectangular jets.

From the mentioned above, it is clear that the developing rate of the half velocity width b_y as the length scale on the rectangular nozzle axis, from the initial region to the CD region for the present 3-D jet, is remarkably suppressed compared with that of the multiple rectangular jets, by using the method of running through transversely with a two-dimensional jet perpendicular to the multiple rectangular jets, even if the length scale was normalized by either nozzle width d or D_2.

4 Conclusion

The object of this experiment is to operate both the velocity scale and the length scale of the multiple rectangular jets. To resolve this problem, the method of running through transversely with a two-dimensional jet perpendicular to the multiple rectangular jets, was adapted. From the experiment, the results obtained are as follows:

(1) In this initial region of the present 3-D jet, the magnitude of the inward secondary flow on the rectangular nozzle axis (y axis) is remarkably promoted by a two-dimensional jet compared with that of the rectangular jet. Furthermore, the inward secondary flow in the initial region on the z axis of the present 3-D jet is produced.

(2) The decay rate of the velocity scale (U_{ox} / U_e) on the x axis near the nozzle initial region is remarkably suppressed, and the potential core length is extended 2.3 times longer than that of the multiple rectangular jets, even if the velocity scale was normalized by either nozzle width d or length scale D_2.

(3) The developing rate of the half velocity width b_y as the length scale on the rectangular nozzle axis from the initial region to the CD region for the present 3-D jet, is remarkably suppressed compared with that of the multiple rectangular jets.

References

1. Lilley, G. M., Aeronautical Journal, **88**-875, 213-223 (1984)
2. Bevilaqua, P. M., Journal of Aircraft, **11**-6, 348-354 (1974)
3. Lummus, J. R., Journal of Aircraft, **18**-4, 245-251 (1981)
4. Krothapalli, A., Baganoff, D. and Karamcheti, K., AIAA Journal, **18**-8, 945-950 (1980)
5. Marsters, G. F., AIAA Journal, **19**-2, 148-152 (1981)
6. Krothapalli, A., Baganoff, D. and Karamcheti, K., Journal of Fluid Mechanics, **107**, 201-220 (1981)
7. Yu, S., Japanese patent disclosure 2003-268523 (2003) (in Japanese)
8. Fujita, S. and Osaka, H., Transactions of the Japan Society of Mechanical Engineers, Series B, **53**-488, 1142-1149 (1987) (in Japanese)
9. Fujita, S., Harima, T. and Osaka, H., Transactions of the Japan Society of Mechanical Engineers, Series B, **65**-631, 905-911 (1999) (in Japanese)
10. Fujita, S., Harima, T. and Osaka, H., Transactions of the Japan Society of Mechanical Engineers, Series B, **72**-717, 1159-1165 (2006) (in Japanese)
11. Fujita, S., Harima, T. and Osaka, H., Transactions of the Japan Society of Mechanical Engineers, Series B, **70**-695, 1717-1724 (2004) (in Japanese)
12. Fujita, S., Harima, T. and Osaka, H., *Fluid Structure Interaction V*, WIT press, 61-70 (2009)
13. Fujita, S. and Harima, T., Transactions of the Japan Society of Mechanical Engineers, **80**-820 (2014) (in Japanese)
14. Luxton, R. E., Experiments in Fluids, **28**, 93-94 (2000)

EPJ Web of Conferences **114**, 02025 (2016)
DOI: 10.1051/epjconf/201611402025

The application of automatic differentiation techniques for the prediction of rotor-dynamic coefficients of labyrinth seals

Jiří Fürst[1,a]

[1] *Czech Technical University in Prague, Karlovo nám. 13, 121 35 Praha 2, Czech Republic*

Abstract. The article describes the development of bulk-flow code for the prediction of rotor-dynamic coefficients of labyrinth seals. The code is based on the so-called single control volume approach by Childs and Scharrer [1] and the the forces are evaluated using the automatic differentiation technique. The resulting code is very simple and provides reasonable predictions of stiffness and damping coefficients at short computational time.

1 Introduction

Labyrinth seals are commonly used in turbines and compressors to dissipate energy and reduce the amount of leakage flow. Unfortunately the rotor vibrations cause a non-uniformity in the circumferential pressure distribution which in turn produces forces acting on the rotor and can cause the rotor to become unstable. The prediction of rotor-dynamic coefficients and leakage flow for a straight-through teeth on stator labyrinth seal is the main subject of this work.

There are several methods for prediction of forces generated by the leakage flows based on different models ranging from single control volume bulk-flow code (see e.g. [1–3]), through more complicated two or three control volumes codes [4], to full 3D simulations using finite differences or finite volumes [5–7]. The later method based on the simulation of 3D turbulent flows gives the most complete description of the flow field and allows arbitrary geometric configurations of the seals but this comes at the price of an excessive amount of computational resources. On the other hand the single control volume bulk-flow method based on the solution of cavity-averaged flow quantities predicts the leakage mass flow rate as well as the rotor-dynamic coefficients in much simpler way at much lower price.

The original bulk-flow codes [1, 2] calculate the forces acting on disturbed rotor using perturbation technique, i.e. each relevant quantity is expressed as $\phi(t,\theta) = \phi^0 + \epsilon\phi^1(t,\theta)$ and a system of equation for zeroth order components ϕ^0 and first order components $\phi^1(t,\theta)$ is built and solved. In our code we simplify this method using automatic differentiation technique using complex extension of Clifford's dual numbers.

[a] e-mail: Jiri.Furst@fs.cvut.cz

2 Single control volume model

The swirling gas passing through the labyrinth seal enters the seal at high pressure through the clearance between the first tooth and the opposite wall and enters to first cavity. Its circumferential momentum is altered due to friction with the cavity walls and then it continues through second clearance to next cavity. Once the gas passes several cavities, it leaves the seal by the last clearance at much lower pressure. The flow in each labyrinth cavity is almost uniform in radial and axial directions with exceptions of boundary layers and "jets" downstream clearances. The non-uniformity of the flow in circumferential direction due to small rotor vibrations is however very important and causes the rotor-dynamic forces.

The single control volume code assumes that the gas pressure as well as the circumferential velocity in each cavity are independent of the radial and axial coordinates, the temperature is constant across the seal and the eccentricity of the rotor is small relative to radial clearance. Using this assumptions the following system of equations can be developed:

$$\frac{\partial(\rho_i A_i)}{\partial t} + \frac{1}{R_s}\frac{\partial(\rho_i V_i A_i)}{\partial\theta} + \dot{m}_{i+1/2} - \dot{m}_{i-1/2} = 0, \quad (1)$$

and

$$\frac{\partial(\rho_i V_i A_i)}{\partial t} + \frac{1}{R_s}\frac{\partial(\rho_i V_i^2 A_i)}{\partial\theta} + \dot{m}_{i+1/2}V_i - \dot{m}_{i-1/2}V_{i-1} = = -\frac{A_i}{R_s}\frac{\partial p_i}{\partial\theta} + \tau_{ri}ar_iL_i - \tau_{si}as_iL_i. \quad (2)$$

Here t is the time, θ is the angular coordinate denoting the position in the seal, ρ_i is the density in i-th cavity, $A_i = L_i(B+CR)$ is the transverse surface area, V_i is the circumferential velocity, $\dot{m}_{i+1/2}$ is the mass flow rate per circumferential length between i-th and $i+1$st cavity, p_i is the pressure, τ_{ri} and τ_{si} are the wall shear stresses at the rotor

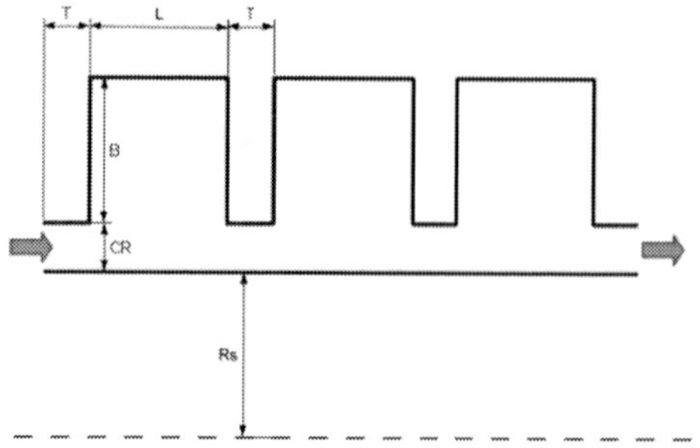

Figure 1. Geometry of straight through seal with stator teeth.

and stator part of the cavity, ar_i and as_i are the dimensionless rotor and stator lengths ($ar_i = 1$ and $as_i = (2B+L_i)/L_i$ for teeth on stator configuration, and $ar_i = (2B + L_i)/L_i$ and $as_i = 1$ for teeth on rotor). The geometry of the seal is given by the shaft radius R_s, teeth height B, teeth pitch L_i, clearance CR and by the number of teeth NT. The wall shear stresses are modeled as turbulent flow in a smooth pipe using Blasius correlation

$$\tau_{si} = \frac{\rho_i}{2} V_i^2 n_0 \left(\frac{|V_i| Dh_i}{\nu}\right)^{m_0} \operatorname{sgn}(V_i), \tag{3}$$

$$\tau_{ri} = \frac{\rho_i}{2}(R_s\omega - V_i)^2 n_0 \left(\frac{|R_s\omega - V_i| Dh_i}{\nu}\right)^{m_0} \operatorname{sgn}(R_s\omega - V_i), \tag{4}$$

where $Dh_i = 2L_i(B + CR)/(L_i + B + CR)$ is the hydraulic diameter. The constants m_0 and n_0 are given for turbulent flow between smooth surfaces as $m_0 = -0.25$ and $n_0 = 0.079$. The density ρ_i is related to the pressure p_i using the ideal gas law $p_i/\rho_i = rT$. The boundary conditions are as follows: the pressure p_0 and the swirl velocity V_0 are prescribed at the inlet ($i = 0$) and the outlet pressure p_{NT} is prescribed downstream the last tooth ($i = NT$).

The mass flow rate $\dot{m}_{i+1/2}$ is calculated using Neumann's model as

$$\dot{m}_{i+1/2} = C_0 \mu_{i+1/2} \frac{CR}{\sqrt{rT}} \sqrt{p_i^2 - p_{i+1}^2}. \tag{5}$$

The kinetic carry-over coefficient μ correcting the nonuniformity of axial velocity is taken as

$$\mu_{i+1/2} = \sqrt{\frac{NT}{(1 - J)NT + J}} \tag{6}$$

with $J = 1 - (1 + 16.6CR/L)^{-2}$ for $i > 0$ and $\mu_{1/2} = 1$ (see [2]). The orifice contraction coefficient C_0 which accounts for the vena contracta effect is given by Chaplygin's formula

$$C_0 = \frac{\pi}{\pi + 2 - 5S_{i+1/2} + 2S_{i+1/2}^2} \tag{7}$$

where $S_{i+1/2} = (p_i/p_{i+1})^{(1-1/\gamma)} - 1$.

2.1 Axisymmetric case

In the case of central position of the rotor and stationary uniform inlet and outlet parameters the equations (1,2) have a steady axisymmetric solution (i.e. $\partial/\partial\theta = 0$.). The solution is calculated with time marching using explicit Euler method:

$$\rho_i^{n+1} = \rho_i^n - \frac{\Delta t}{A_i}\left(\dot{m}_{i+1/2}^n - \dot{m}_{i-1/2}^n\right), \tag{8}$$

$$\begin{aligned}\rho_i^{n+1} V_i^{n+1} =& \rho_i^n V_i^n - \frac{\Delta t}{A_i}\left(\dot{m}_{i+1/2}^n V_i^n - \dot{m}_{i-1/2}^n V_{i-1}^n\right)\\ &+ \frac{\Delta t}{A_i}\left(\tau_{ri}^n ar_i L_i - \tau_{si}^n as_i L_i\right).\end{aligned} \tag{9}$$

In order to analyze the stability of the numerical method the equation (8) is rewritten using equation of state and the model for mass flow rate as

$$\begin{aligned}p_i^{n+1} =& p_i^n - \frac{\Delta t CR}{A_i\sqrt{rT}}\Big(C_0\mu_{i+1/2}\sqrt{(p_i^n)^2 - (p_{i+1}^n)^2} -\\ &-C_0\mu_{i-1/2}\sqrt{(p_{i-1}^n)^2 - (p_i^n)^2}\Big) = \mathcal{H}^c(p_{i-1}^n, p_i^n, p_{i+1}^n).\end{aligned} \tag{10}$$

The analysis will be further simplified by taking constant value of $C_0 = 0.716$ as proposed by Eser an Kazakia in [2]. The stability of the numerical method as well as the maximum principle for the pressure follows from the monotonicity of $\mathcal{H}^c$. Hence if

$$\frac{\partial\mathcal{H}^c(p_{i-1}, p_i, p_{i+1})}{\partial p_{i-1}} \geq 0, \tag{11}$$

$$\frac{\partial\mathcal{H}^c(p_{i-1}, p_i, p_{i+1})}{\partial p_{i+1}} \geq 0, \tag{12}$$

$$\frac{\partial\mathcal{H}^c(p_{i-1}, p_i, p_{i+1})}{\partial p_i} \geq 0, \tag{13}$$

then the method will be stable and produces monotone distribution of the pressure $p_i \leq p_{i-1}$. First two inequalities are satisfied always whereas the third one leads to stability condition for time step magnitude

$$\Delta t \leq \frac{A_i\sqrt{rT}}{p_i CR\, C_0} \frac{1}{\frac{\mu_{i+1/2}}{\sqrt{p_i^2 - p_{i+1}^2}} + \frac{\mu_{i-1/2}}{\sqrt{p_{i-1}^2 - p_i^2}}}. \tag{14}$$

2.2 Case with eccentric rotor

In order to evaluate the rotor-dynamic coefficients one assumes rotor with small eccentricity ϵ orbiting with precession speed Ω around its nominal position. The flow field is then no more axisymmetric and one has to include all terms in the eq. (1), (2). Codes developed by Childs and Scharrer [1], Eser and Kazakia [2], or by Malvano, Vatta, and Vigliani [3] use perturbation technique splitting all relevant quantities to its mean value corresponding to axisymmetrical case and to a perturbation due to rotor eccentricity $\phi(t, \theta) = \phi^0 + \epsilon\phi^1(t, \theta)$ and developing the system of equation for first order variables ϕ^1. In contrary to their approach we employ an automatic differentiation technique which greatly simplifies the code.

2.2.1 Automatic differentiation via dual numbers

The "classical" dual numbers extend the real numbers by adding a new element δ (similar to imaginary unit ι for complex numbers) with the property $\delta^2 = 0$. It means that each dual number x is represented by two real numbers: the main value x_v and the perturbation x_p, hence $x = \mathcal{D}(x_v, x_p) := x_v + \delta x_p$. Assume $x = x_v + \delta x_p$ and $y = y_v + \delta y_p$ are two dual numbers and α a real number. Then one can define the dual number arithmetics as follows

$$x + y := x_v + y_v + \delta(x_p + y_p) = \mathcal{D}(x_v + y_v, x_p + y_p), \quad (15)$$

$$\alpha x := \alpha x_v + \delta \alpha x_p = \mathcal{D}(\alpha x_v, \alpha x_p) \quad (16)$$

$$\begin{aligned} xy &:= x_v y_v + \delta(x_p y_v + x_v y_p) = \\ &= \mathcal{D}(x_v y_v, x_p y_v + x_v y_p), \end{aligned} \quad (17)$$

$$\begin{aligned} x/y &:= x_v/y_v + \delta(x_p y_v - x_v y_p)/y_v^2 = \\ &= \mathcal{D}(x_v/y_v, (x_p y_v - x_v y_p)/y_v^2). \end{aligned} \quad (18)$$

Next, one can extend all differentiable real functions using the following rule:

$$f(x) := f(x_v) + \delta f'(x_v) x_p, \quad (19)$$

where f and f' at the right hand side are the real function and its derivative.

Let's define two operators $\mathcal{V}(x) := x_v$ and $\mathcal{P}(x) := x_p$ giving the main value and perturbation part of a dual number. Then one can calculate function value and first derivative of any differentiable function as:

$$f(x_0) = \mathcal{V}\left(f(x_0 + \delta)\right), \quad (20)$$

$$f'(x_0) = \mathcal{P}\left(f(x_0 + \delta)\right). \quad (21)$$

Note that the formula holds even for compound functions

$$\begin{aligned} (f \circ g)'(x_0) &= \mathcal{P}\left(f\left(g(x_0 + \delta 1)\right)\right) = \\ &= \mathcal{P}\left(f\left(g(x_0) + \delta g'(x_0)\right)\right) = \\ &= \mathcal{P}\left(f(g(x_0)) + \delta f'(x_0) g'(x_0)\right) = f'(x_0) g'(x_0), \end{aligned} \quad (22)$$

and thus it can be used for calculation of derivatives of any (smooth) algorithms.

2.2.2 Dual numbers with complex rotating perturbations

However one can use the above mentioned technique for computing first derivatives, we propose an extension of dual numbers which is more appropriate for the calculation of rotor-dynamic coefficients. We assume real main value corresponding to centered position of the rotor and complex perturbation including rotation term $e^{\iota(\theta - \Omega t)}$:

$$\phi(t, \theta) = \mathcal{D}^*(\phi_v, \phi_p) = \phi_v + \delta \phi_p e^{\iota(\theta - \Omega t)}. \quad (23)$$

Here $\phi_v \in \mathbb{R}$ is the main value and $\phi_p \in \mathbb{C}$ is the complex perturbation quotient.

Note that this extension uses the same arithmetic operations as the "classical" dual numbers. Let $x = \mathcal{D}^*(x_v, x_p)$, $y = \mathcal{D}^*(y_v, y_p)$, and $\alpha \in \mathbb{R}$. Then

$$x + y = \mathcal{D}^*(x_v + y_v, x_p + y_p), \quad (24)$$

$$\alpha x = \mathcal{D}^*(\alpha x_v, \alpha x_p) \quad (25)$$

$$xy = \mathcal{D}^*(x_v y_v, x_p y_v + x_v y_p), \quad (26)$$

$$x/y = \mathcal{D}^*(x_v/y_v, (x_p y_v - x_v y_p)/y_v^2), \quad (27)$$

and any smooth real function f can be extended for duals as

$$f(x) := \mathcal{D}^*(f(x_v), f'(x_v) x_p). \quad (28)$$

Let's define the operator $\mathcal{V}(x) := x_v$ in the same way as for the "classical" dual numbers and the operator $\mathcal{P}(x) := x_p$ returns only the perturbation quotient.

2.2.3 Linearization using extended dual numbers

The above mentioned dual numbers with complex rotating perturbations can be easily used for automatic linearization of the code for the case of eccentric rotor making precession along a circular orbit with radius ϵ at the precession speed Ω. The (perturbed) clearance is then

$$CR(t, \theta) = CR_{s,v} - \epsilon \Re(e^{\iota(\theta - \Omega t)}) \quad (29)$$

and it can be represented by an extended dual number

$$\widetilde{CR} = \mathcal{D}^*(CR_{s,v}, -\epsilon) \quad (30)$$

hence the following identity holds

$$CR(t, \theta) = \mathcal{V}(\widetilde{CR}) + \Re(\mathcal{P}(\widetilde{CR}) e^{\iota(\theta - \Omega t)}). \quad (31)$$

Let's denote by tilde dual number representations of each relevant quantities, e.g. $\tilde{A}_i = L_i(B + \widetilde{CR})$. The extended dual number representation of unknown flow parameters $\tilde{\rho}_i$, $\tilde{p}_i$, $\tilde{V}_i$ can be obtained with time marching method similar to (8), (9) including corrections for $\partial/\partial t$ and $\partial/\partial\theta$ according to following rules:

$$\frac{\partial \mathcal{D}^*(\phi_v, \phi_p)}{\partial t} = \mathcal{D}^*\left(\frac{\partial \phi_v}{\partial t}, \frac{\partial \phi_p}{\partial t}\right) - \mathcal{D}^*(0, \iota \phi_p \Omega), \quad (32)$$

$$\frac{\partial \mathcal{D}^*(\phi_v, \phi_p)}{\partial \theta} = \mathcal{D}^*(0, \iota \phi_p). \quad (33)$$

The final time marching method is then

$$\begin{aligned} \tilde{\rho}_i^{n+1} &= \tilde{\rho}_i^n - \frac{\Delta t}{\tilde{A}_i}\left(\dot{\tilde{m}}_{i+1/2}^n - \dot{\tilde{m}}_{i-1/2}^n\right) - \\ &- \frac{\Delta t}{R_s \tilde{A}_i} \frac{\partial \tilde{\rho}_i^n \tilde{V}_i^n \tilde{A}_i}{\partial \theta} + \Delta t \frac{\mathcal{D}^*(0, \iota \mathcal{P}(\tilde{\rho}_i^n \tilde{A}_i)\Omega)}{\tilde{A}_i}, \end{aligned} \quad (34)$$

$$\begin{aligned} \tilde{\rho}_i^{n+1} \tilde{V}_i^{n+1} &= \tilde{\rho}_i^n \tilde{V}_i^n - \frac{\Delta t}{\tilde{A}_i}\left(\dot{\tilde{m}}_{i+1/2}^n \tilde{V}_i^n - \dot{\tilde{m}}_{i-1/2}^n \tilde{V}_{i-1}^n\right) + \\ &+ \frac{\Delta t}{\tilde{A}_i}\left(\tilde{\tau}_{ri}^n a r_i L_i - \tilde{\tau}_{si}^n a s_i L_i\right) - \frac{\Delta t}{R_s} \frac{\partial \tilde{p}_i^n}{\partial \theta} - \\ &- \frac{\Delta t}{R_s \tilde{A}_i} \frac{\partial \tilde{\rho}_i^n (\tilde{V}_i^n)^2 \tilde{A}_i}{\partial \theta} + \Delta t \frac{\mathcal{D}^*(0, \iota \mathcal{P}(\tilde{\rho}_i^n \tilde{V}_i \tilde{A}_i)\Omega)}{\tilde{A}_i}. \end{aligned} \quad (35)$$

The last two terms in eq. (34) and three terms in eq. (35) are the contributions from partial derivatives $\partial/\partial\theta$ and $\partial/\partial t$. The tilde denotes here extended dual numbers. Note that the mass flux rates, wall shear stresses, and isothermal gas law is also evaluated using extended dual arithmetics.

The steady solution is calculated using time-independent boundary conditions for $\tilde{p}_0 = \mathcal{D}^*(p_0, 0)$, $\tilde{V}_0 = \mathcal{D}^*(V_0, 0)$, and $\tilde{p}_{NT} = \mathcal{D}^*(p_{NT}, 0)$. The forces acting on the rotor are then evaluated in rotating coordinate frame neglecting wall shear stresses by integrating the pressure perturbation:

$$\mathbf{F} = -\int_{rotor} p\mathbf{n}\, dS =$$
$$= -\sum_{i=1}^{NT-1} L_i R_s \int_{-\pi}^{\pi} \Re(\mathcal{P}(\tilde{p}_i)e^{i\theta})(\cos\theta, \sin\theta)\, d\theta, \quad (36)$$

hence

$$F_r = -\Re\left(\pi R_s \sum_{i=1}^{NT-1} L_i \mathcal{P}(\tilde{p}_i)\right), \quad (37)$$

$$F_\theta = \Im\left(\pi R_s \sum_{i=1}^{NT-1} L_i \mathcal{P}(\tilde{p}_i)\right). \quad (38)$$

The above mentioned method has been implemented using operator overloading and polymorphic functions in Julia language [8].

2.3 Rotor-dynamic coefficients

The forces acting on the rotor are evaluated for several precession speeds and finally the rotor-dynamic coefficients (i.e. stiffness coefficients K, k, and damping coefficients D, d) are calculated using least-squares method using linear model (see [9])

$$F_r/\epsilon = -(K + c\Omega), \quad (39)$$
$$F_\theta/\epsilon = k - C\Omega. \quad (40)$$

Here F_r and F_t are the radial and tangential components of the force, ϵ is the rotor displacement, and Ω is the precession angular speed.

3 Validation

In order to validate our code a simulation of flow through a straight seal with stator teeth are performed. The geometrical parameters as well as the flow parameters are given in the table 1. The forces acting on the rotor are calculated for 10 precession speeds Ω in the range 0 rpm to 8000 rpm and the rotor-dynamic coefficients are calculated using least-square method.

The figures 2 and 3 show the cross-coupled stiffness and direct damping coefficients (hence the components of circumferential force F_θ) for several inlet swirl velocity ratios $V_0/(\Omega_r R_s)$. The results obtained with current code labeled by "Fuerst, one volume" are compared to experimental results of Childs and Scharrer [10], "classical" one control volume code [1], and with the two control volumes code [11]. One can see that the results for cross coupled stiffness and direct damping corresponds very well to experimental data as well as to results obtained with other codes.

On the other hand there is quite large discrepancy in direct stiffness K obtained with current code and experimental data (see figure 4). This discrepancy is caused probably by the fact that the radial force component shows non-linear dependency on the precession speed (see figure 5) and therefore both the direct stiffness K and cross-

Table 1. Parameters of test seal

R_s	shaft radius	72.5 mm
B	tooth height	3.175 mm
L	tooth pitch	3.175 mm
CR	clearance	0.4064 mm
NT	number of teeth	16
p_0	inlet pressure	0.822 MPa
p_{NT+1}	outlet pressure	0.1 MPa
T	gas temperature	298.2 K
r	specific gas constant (air)	287.06 J K^{-1} kg^{-1}
γ	specific heat ratio	1.4
Ω_r	rotor speed	8000 rpm

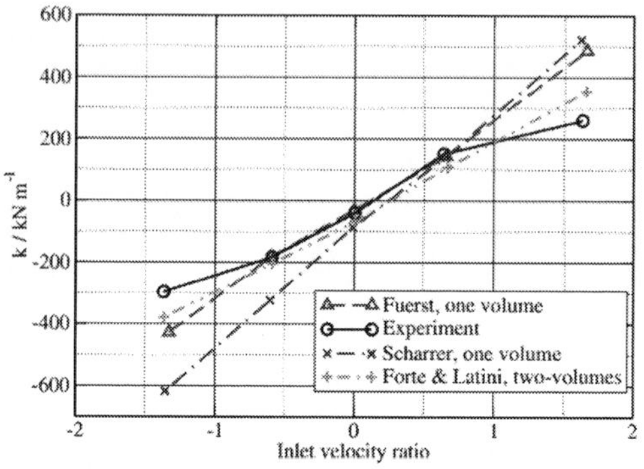

Figure 2. Cross coupled stiffness coefficient k, precession speeds 0 rpm to 8000 rpm.

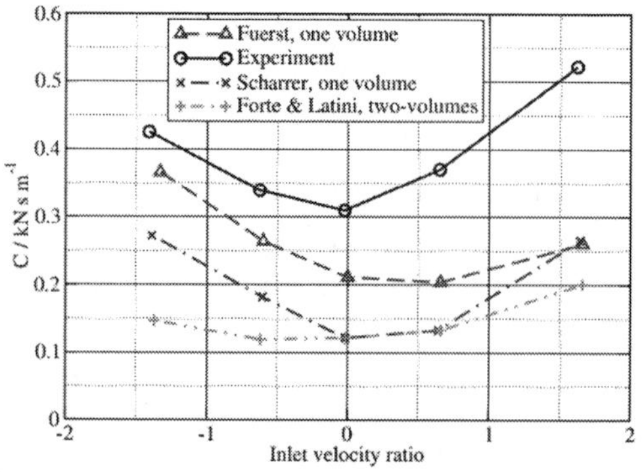

Figure 3. Direct damping coefficient C, precession speeds 0 rpm to 8000 rpm.

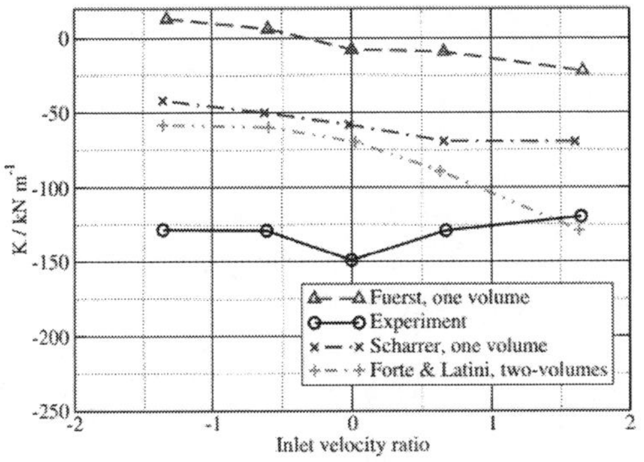

Figure 4. Direct stiffness coefficient K, precession speeds 0 rpm to 8000 rpm.

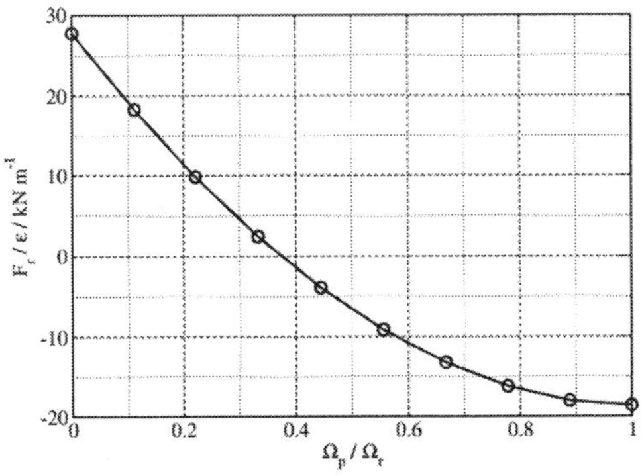

Figure 5. Radial force for precession speeds 0 rpm to 8000 rpm, ($V_0 = 1.65\Omega_r R_s$).

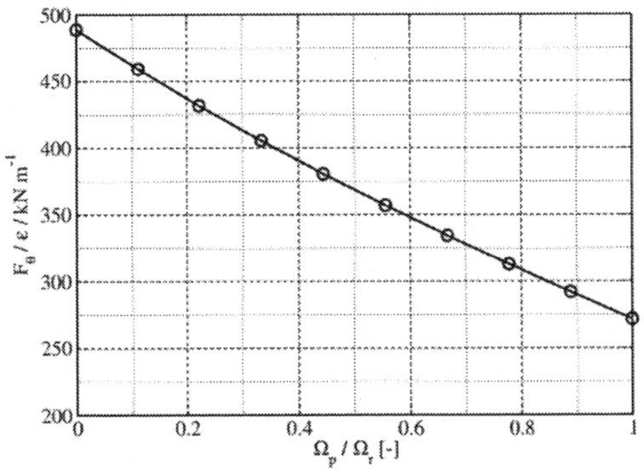

Figure 6. Circumferential force for precession speeds 0 rpm to 8000 rpm, ($V_0 = 1.65\Omega_r R_s$).

coupled damping c depends strongly on the precession assumed during computation. Note that this is not the case for the circumferential component of the force which is almost linear, see figure 6.

Unfortunately there is no or very little informations about precession speeds for the experimental data as well as for the results of Forte and Latini [11]. Nevertheless the description of the experimental apparatus states that

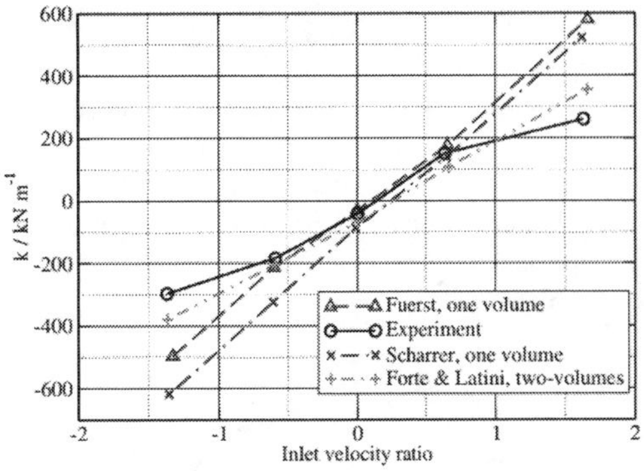

Figure 7. Cross coupled stiffness coefficient k, precession speeds −8000 rpm and 8000 rpm.

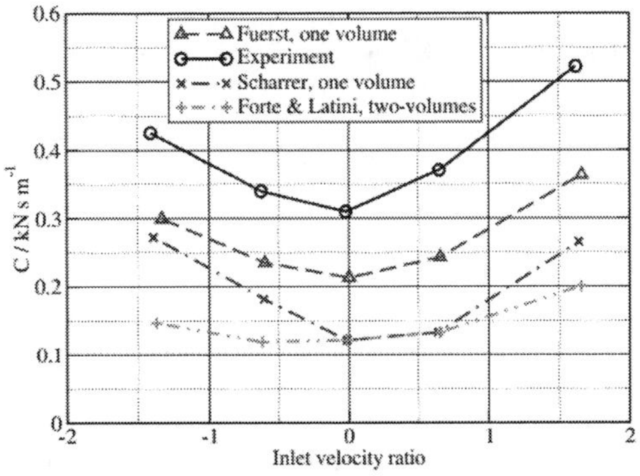

Figure 8. Direct damping coefficient C, precession speeds −8000 rpm and 8000 rpm.

the rotor makes linear side-to-side motion [10]. Similarly the "classical" perturbation based codes including those of [11] assume elliptical motion. In both cases the elliptical as well as the linear motion can be decomposed onto two circular orbits with opposite precession speeds.

Therefore the calculation is done here also for two opposite precession speeds Ω =−8000 rpm and 8000 rpm. The rotor-dynamic coefficients are then evaluated directly from two forces without using least-square method.

One can see that there is almost no difference in circumferential components k and C but the agreement with experimental data as well as with the computations of other authors is much better in this case.

4 Conclusion

The original automatic differentiation technique based on dual numbers with complex rotating perturbations allows rapid development of a code for dynamic analysis of flows in labyrinth seals without the need of tedious and error-prone linearization. The code is further simplified using modern programming techniques (operator overloading, polymorphic functions, ...).

The code is validated against experimental data [10] and numerical results of other authors using two variants of "classical" bulk-flow method for the case of a look-through seal with 16 stator teeth. The results show that the cross-coupled stiffness and direct damping coefficients are predicted in accordance to available data. Due to non-linear dependence of radial force on the precession speed the agreement of direct stiffness coefficient with available data depends highly on the precession speeds used during the computation.

Acknowledgements

The work was supported by the Grant Agency of the Czech Technical University in Prague, grant No. SGS 13/174/OHK2/3T/12.

References

[1] D.W. Childs, D.L. Rhode, Tech. rep., ASFOR (1988)

[2] D. Eser, J.Y. Kazakia, International Journal of Engineering Science **33**, 2309 (1995)

[3] R. Malvano, F. Vatta, a. Vigliani, Meccanica **36**, 731 (2001)

[4] R. Nordmann, P. Weiser, Rotordynamic Instability Problems in High-Performance Turbomachinery pp. 147–163 (1991)

[5] R. Nordmann, P. Weiser, Rotordynamic Instability Problems in High-Performance Turbomachinery pp. 161–176 (1989)

[6] J. Xi, D.L. Rhode, International Journal of Rotating Machinery **2006**, 1 (2006)

[7] A. Untaroiu, C.D. Untaroiu, H.G. Wood, P.E. Allaire, Journal of Engineering for Gas Turbines and Power **135**, 012501 (2013)

[8] J. Bezanson, A. Edelman, S. Karpinski, V.B. Shah, arXiv 1411.1607 (2014)

[9] J. Schettel, R. Nordmann, Eighth International Conference on Vibrations in Rotating Machinery (2004)

[10] D.W. Childs, J.K. Scharrer, Journal of Vibration Acoustics Stress and Reliability in Design **110**, 281 (1988)

[11] P. Forte, F. Latini, International Journal of Rotating Machinery **5**, 67 (1999)

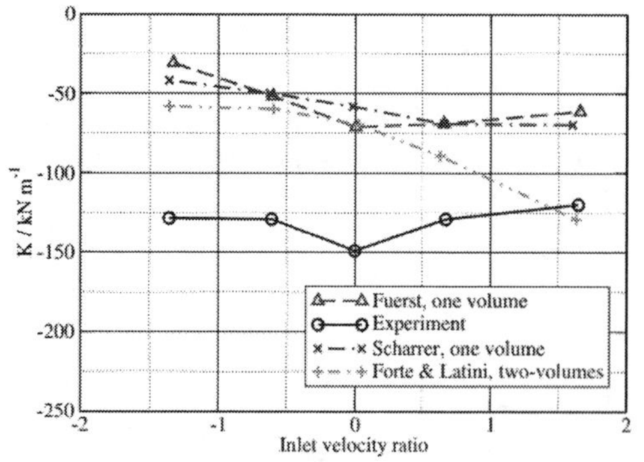

Figure 9. Direct stiffness coefficient *K*, precession speeds −8000 rpm and 8000 rpm.

EPJ Web of Conferences **114**, 02026 (2016)
DOI: 10.1051/epjconf/201611402026

Properties of Broezel static probe

Peter Gašparovič[1,a], Karol Semrád[1] and Miroslava Cúttová[1]

[1]*Technical University of Kosice, Faculty of Aeronautics, Rampova 7, 041 21, Kosice, Slovakia*

Abstract. The properties of flat static probe designed by Broezel and used in sailplanes are investigated for its planned use in low speed tunnel. Both the numerical CFD model and experiment in low speed wind tunnel confirm yaw insensitivity of the static pressure measured by the probe. The results indicate that the probe is sufficiently accurate for its planned use in wind tunnel measurements.

1 Introduction

The most common device for measurement of static pressure is a static tube. However, the static tube is not suitable when airflow direction is not known in advance. Examples of "standard" and Prandtl's pitot-static tubes are on the figure 1. The static tube is obtained from pitot-static tube by omitting total pressure port.

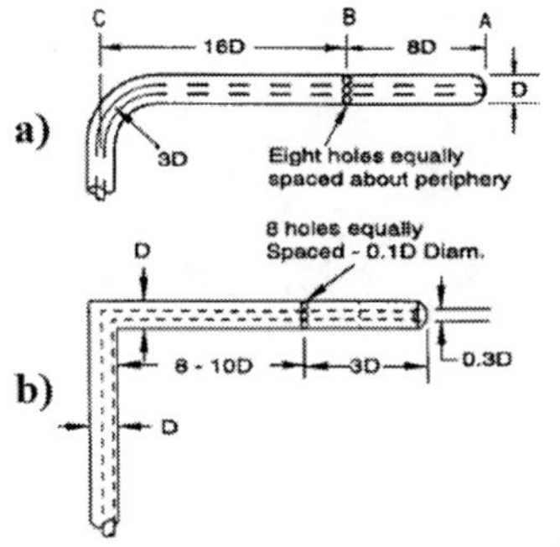

Figure 1. Pitot-static tubes a) N.P.L "Standard", b) Prandtl's design, [1]

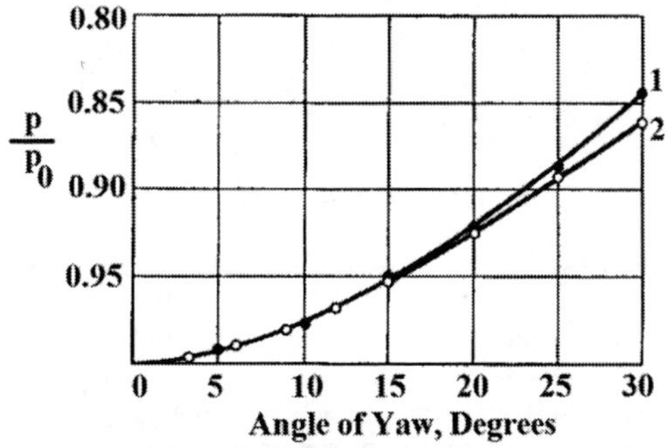

Figure 2. Influence of yaw on reading of static tube, [7]

Even when the dominant direction is known, strong directional sensitivity of the static tube makes accurate measurement very difficult an unreliable. On the figure 2 is shown influence of the angle of yaw on the static error measured by pitot-static tubes (1- N.P.L. Standard, 2- Round Nosed). Indeed, the directional dependency of the static pressure of a static tube has very narrow range of yaw angles, where the error is below 1% (typically less than plus/minus 5°). On the contrary, the total pressure of a decent pitot tube has usable range of plus/minus 10° and the error in most of this range is much lower and more gradual.

Some different types of the static probes, which are promised to be directionally insensitive, are on the following figures.

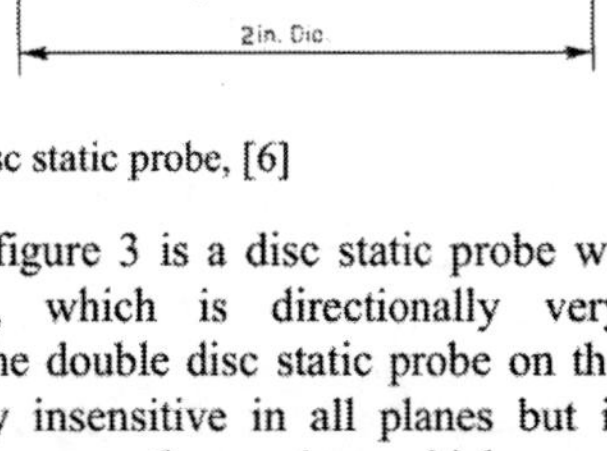

Figure 3. Disc static probe, [6]

On the figure 3 is a disc static probe with only one static port, which is directionally very sensitive. However, the double disc static probe on the figure 4 is directionally insensitive in all planes but its relatively bulky. There are other probes, which are directionally insensitive too, but are more practical for aeronautical applications. Very interesting probe was designed by A.M.O. Smith (figure 5). However, this probe has complicated geometry, which is not trivial to manufacture

[a] Corresponding author: peter.gasparovic@tuke.sk

(it must be smooth and accurate at the same time). The directional sensitivity of static probe by A.M.O. Smith, calculated by inviscid panel method, is on the figure 6.

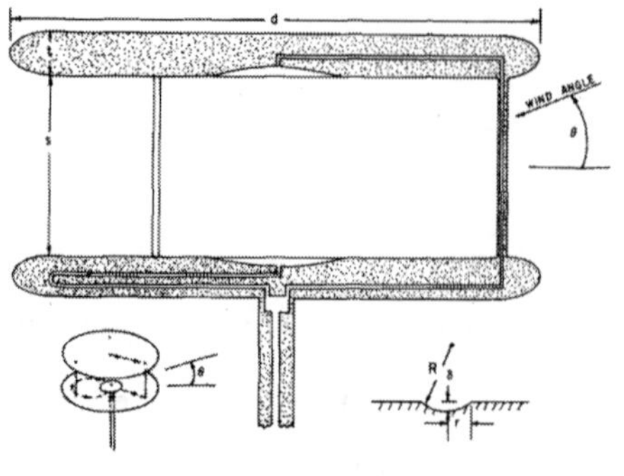

Figure 4. Double disc omnidirectional static probe, [5]

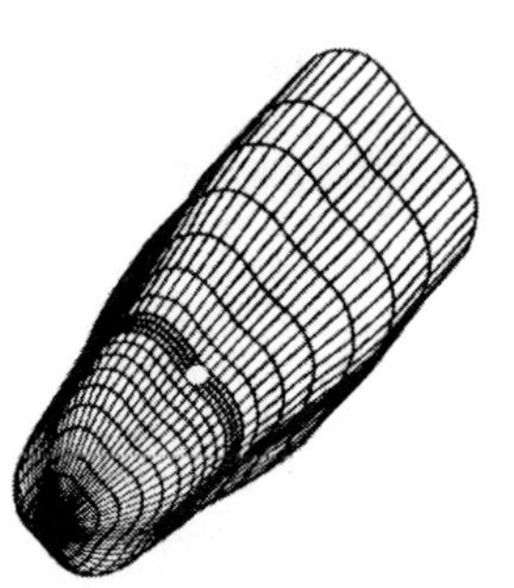

Figure 5. Static probe of A.M.O. Smith, [8]

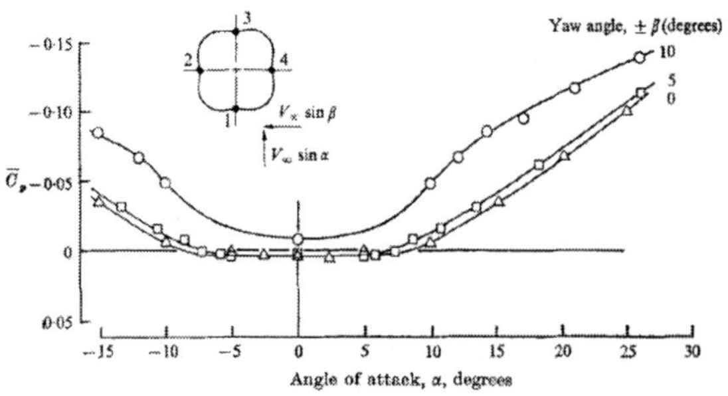

Figure 6. Characteristics of A.M.O. Smith static probe, [8]

The directional sensitivity of Smith probe is perfectly acceptable, but there are other solutions which are more easily manufacturable, for example wedge probe described in [4] and [3] or reversed delta wedge described in Patent [2]. These devices are expected for use at speeds approaching or exceeding sound of speed. However, for low speeds there is even simpler solution – Broezel flat static probe.

2 Broezel flat static probe and its CFD analysis

Sailplanes are flying without engines and to keep them at the safe altitude the pilot must use natural ascending motion of the air. To effectively use this motion, there is a need to differentiate the climbing motion of the air from the climbing motion of the sailplane at the expense of kinetic energy of the sailplane. One needs two inputs – total pressure at the speed of sailplane and static pressure. Whereas measuring total pressure is relatively simple task, to accurately measure the static pressure is almost unrealizable, because the sailplane is in the state of permanent change of its position relatively to surrounding airflow. That perpetual unrest creates dynamic changes of surface pressures over whole surface of the glider.

Sometime around year 1985, German inventor Rudolf Broezel, came with his own solution. He devised reliable static probe, which consisted of simple flat and narrow plate with static ports. Later he sold his design to the company ESA systems, which is still manufacturing it and offering under the name STATEK (STAtic probe + Total Energy Compensation), see figure 7.

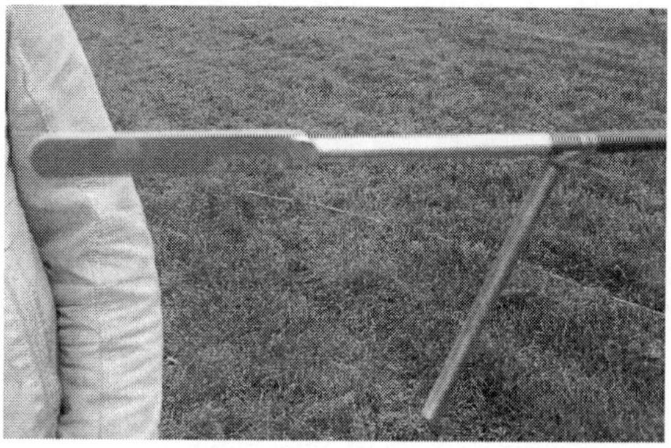

Figure 7. Static probe STATEK by R. Broezel

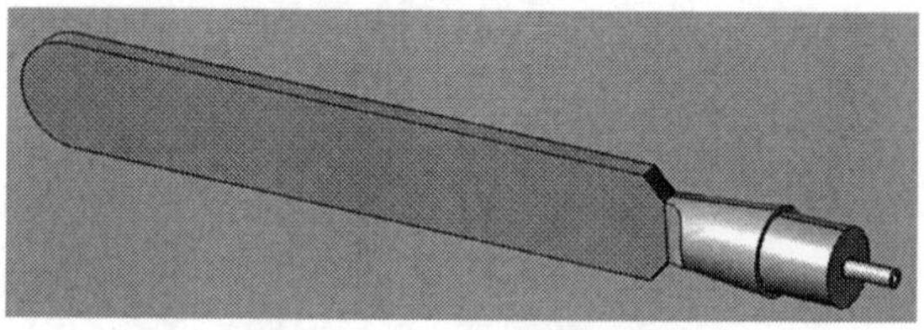

Figure 8. CAD geometry of Broezel static probe

In order to investigate its suitability for measurement of static pressure in our low speed wind tunnel, the geometry of the probe was reversely engineered from the photograph. Its CAD geometry is shown on the figure 8.

The computational mesh was created in program IcemCFD. The mesh is unstructured, contains 1.6 million cells and prism elements in boundary layer were scaled to achieve y+ < 1. The numerical solution of CFD analysis was performed in software Ansys CFX. The flow was modeled by the Reynolds Averaged Navier Stokes equations and SST turbulence model, with the ability to model the laminar-turbulent transition in boundary layer. The CFD analysis results are on the following pictures. The pressure coefficient at zero yaw of the Broezel static probe is on the figure 9. It can be seen, that the centre part of the plate has relatively mild pressure gradient.

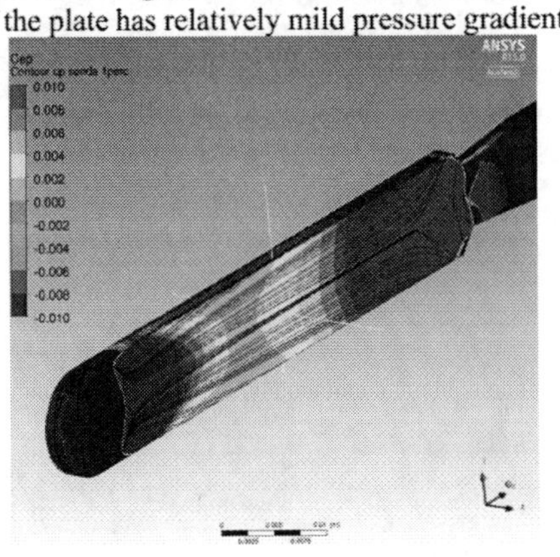

Figure 9. Pressure coefficient at zero yaw

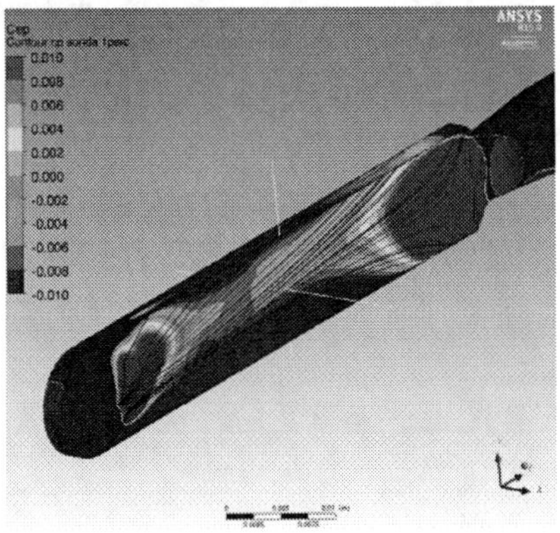

Figure 10. Pressure coefficient at angle of attack $\alpha = 10°$

An important feature of the probe is wider beam, which creates stagnated area with higher pressure in balance to lower pressure in front part. The figures 10 and 11 show very good insensitivity to lateral cross flow. Sensitivity to yaw angles in vertical plane (i.e. angle of attack) is higher, but better than sensitivity of static tubes.

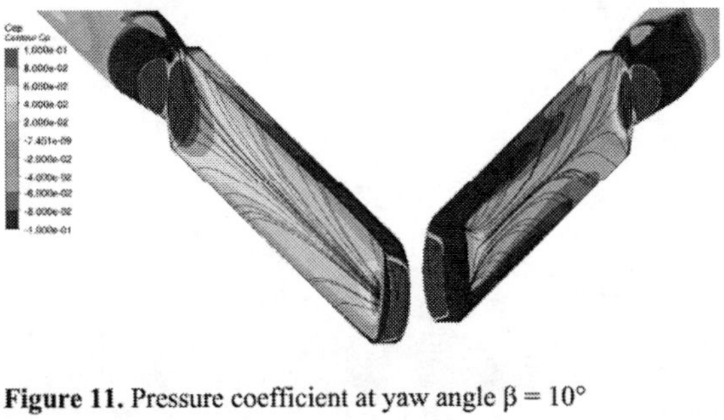

Figure 11. Pressure coefficient at yaw angle $\beta = 10°$

The static error comparison of the CFD analysis results and Prandtl's static probe is on the figure 12. It is apparent, that the Broezel flat static probe has a much wider range of yaw angles, where the static error is below 1% (more than 10°).

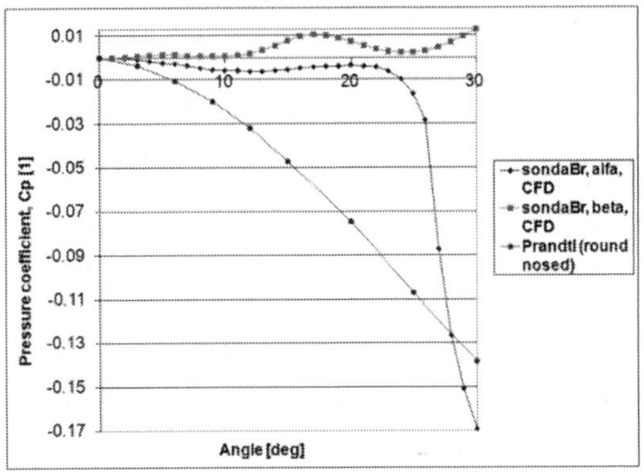

Figure 12. Comparison of the influence of yaw on static error of Broezel and Prandtl static probes

3 Experimental measurement of Broezel flat static probe in wind tunnel

For confirmation of insensitivity of the Broezel flat static probe the experimental measurement in wind tunnel was performed. First step was the manufacture of the probe using 3D printer. The material was PLA (polylactide) polymer and a surface of the probe was grinded in order to achieve its hydrodynamic smoothness.

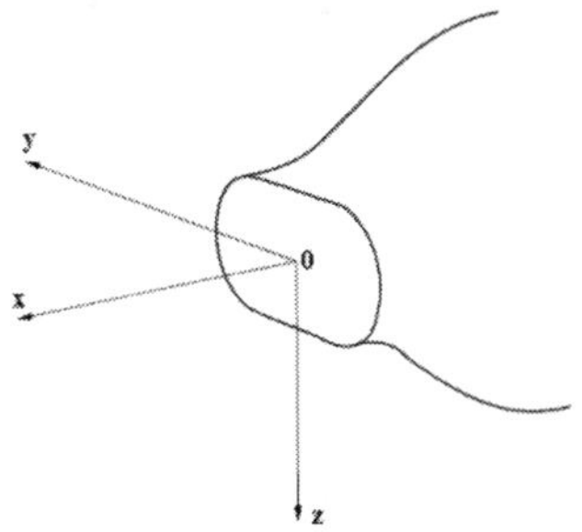

Figure 13. Coordinate system of the measurement section

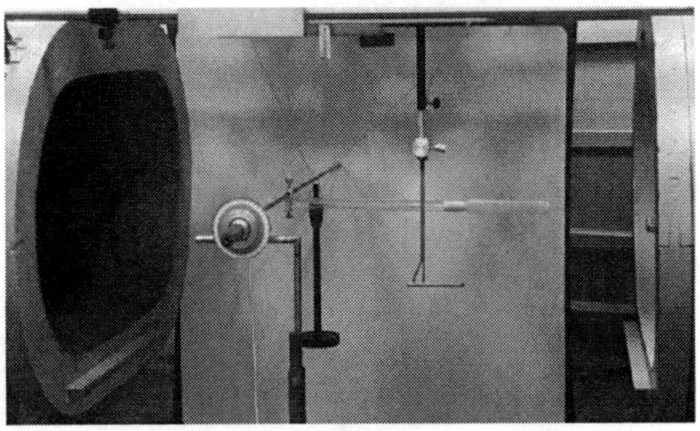

Figure 14. Position of the Broezel probe in a measurement section of the wind tunnel

The Broezel flat static probe was placed in a test section of the subsonic wind tunnel. The probe is positioned in the centerline of measurement section, with its static ports located on x-axis, 0.3 m from the origin by the figure 13. For resulting configuration of the measuring equipment in the measurement section see the figure 14.

The velocity of the airflow was approximately 25 m/s (90 km h^{-1}). The position of the probe relatively to the wind tunnel was in the same position during all measurements. The angle of attack and yaw angle of the probe was varied in the range ± 30°, always varying only one parameter at the same time and holding the other parameter at the zero value. The results of the wind tunnel experiment are shown on the following graphs.

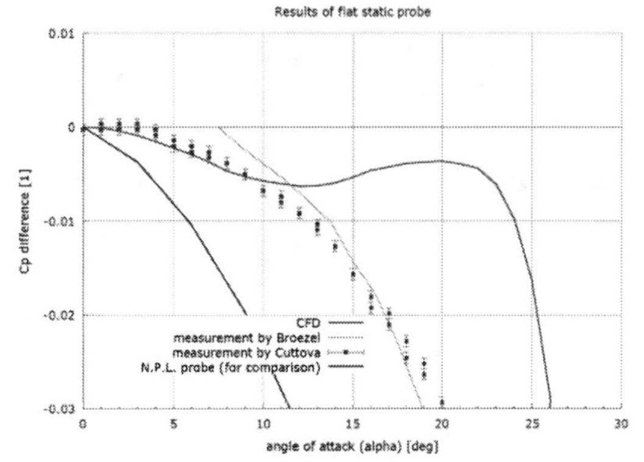

Figure 15. Angle of attack sensitivity of Broezel flat static probe

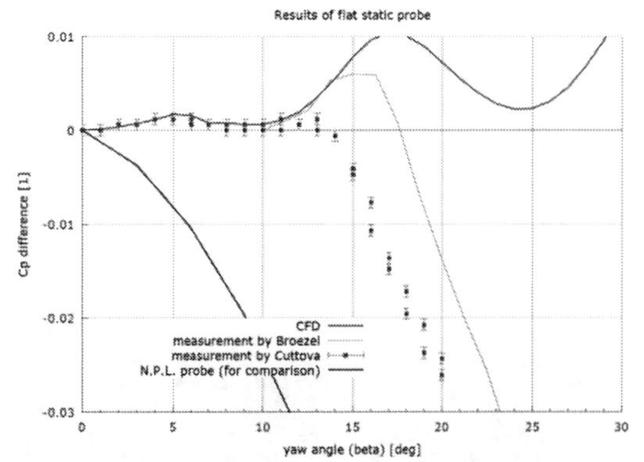

Figure 16. Yaw sensitivity of Broezel flat static probe

4 Conclusions

CFD results and experimental results in the range of angles up to 10° are similar and they have the same trend. At high angles, the measurement in wind tunnel has the same trend as the measurement by Broezel. The differences may be the result of relatively high turbulence in the wind tunnel. There is also another unknown – the value of the Reynolds number in Broezel's experiment.

Both the CFD results and experiment results confirm good properties of the probe. The Broezel flat static probe is suitable for accurate measurement of the static pressure with the yaw in the range of 3° with the static pressure error 0.1%. For comparison, Prandtl's probe must be set in the direction of the airflow 4 times more accurately.

References

1. J. B. Barlow, W. H. Rae, A. Pope, *Low-speed wind tunnel testing* (1999)
2. R. L. Beach, et al., US Patent 2876640 (1956)
3. J. P. Holman, *Experimental Methods for Engineers* (2011)
4. L. N. Krause, C. C. Gettelman, ISA Proc., **7**, 134–137 (1952)
5. R. W. Miksad, Appl. Meteor., **15**, 1215–1225 (1976)
6. E. Ower, R. C. Pankhurst, *The measurement of air flow* (1966)
7. R. C. Pankhurst, D. W. Holder, *Wind-tunnel technique* (1952)
8. A. M. O. Smith, A. B. Bauer, J. of Fluid Mech., **44**, 513–528 (1970)

Acknowledgement

This work has been supported by the grant KEGA 028TUKE-4/2013 (An integrated program of study "Unmanned Aircraft Systems").

EPJ Web of Conferences **114**, 02027 (2016)
DOI: 10.1051/epjconf/201611402027

Effect of tip vortices on flow over NACA4412 aerofoil with different aspect ratios

Mustafa Serdar Genç[1,a], Gökhan Özkan[1], Halil Hakan Açikel[1], Mehmet Sadık Kiriş[1], Rahime Yildiz[1]

[1]*Wind Engineering and Aerodynamic Research Group, Department of Energy Systems Engineering, Erciyes University, Kayseri, Turkey*

Abstract. Effect of tip vortices on flow and laminar separation bubble over NACA4412 aerofoil at low Reynolds numbers and different angles of attack was investigated in detail by performing force and flow visualization via smoke wire technique. Experiments have been done at Reynolds number of 50000 and the wing model of aspect-ratio was 1 and 3, respectively. From the experimental results, the flow visualization results showed that tip vortices effect on the laminar separation bubble and the bubble reduces over the wing with low aspect ratio as the angle of attack increased. Moreover, it was noticed that stall angles decreased as aspect-ratio increased at the same Reynolds number.

1 Introduction

The aerodynamic of low aspect-ratio wings have been extensively investigated by aerodynamic researches keen on low aspect ratio wings and low Reynolds numbers [1-19]. Viscous effects, separated and vortical flows are more prevalent and separation bubbles, transition, and reattachment processes are very attractive and also little understood at low Re number aerodynamics [1]. At low Reynolds number flow conditions, some negative effects are occurred such as decreasing on lift, increasing on drag, reducing stability of the aircraft, vibration, and noise due to the laminar separation bubble [4–11]. Genç et al. conducted an experimental [9] and numerical [11] studies over the NACA2415 aerofoil at various angles of attack and low Reynolds numbers. They noticed that due to the increasing Reynolds number and angle of attack the point of separation and reattachment moved towards to leading edge.

Recently, elliptical and rectangular wing model was thought to be hopefully for unmanned aerial vehicles (UAV), micro air vehicles (MAV) and other applications, since they were noticed that they could maximize the wing area for certain wing dimension. Thus, it can be said that is necessary to understand flow over the low aspect-ratio wings with these given wing models.

In these days, one study related to effects of aspect-ratio on the aerodynamic characteristics of wings at low Reynolds number has been presented by M. Mizoguchi and H. Itoh [12]. They firstly concluded that influences of aspect-ratio were not as important as they thought as aspect-ratio was 3 or more. But this situation was different at low aspect-ratios wings. Tip vortices had significantly important influences on aerodynamic characteristics of airfoil. Effect of tip vortices increased while aspect-ratio of wing decreased. Maximum lift coefficient and stall angle considerably increased as aspect-ratio was less or equal to 1. Okamoto and Azuma [13] investigated the effects of low Reynolds number and aspect ratios when Reynolds number was around 10000. Their results showed that stall angle increased significantly when aspect-ratio was less than or equal to 1. Torres and Mueller [14] studied low aspect-ratio wings and their characteristics at Reynolds number in the style of 100000. Their studies showed same results for low aspect-ratio wings. Since, aspect ratio reduced to 1 and accordingly maximum lift coefficient and stall angles increased considerably. Other numerical experimental results [15-19] have demonstrated that low aspect-ratio wing performance is affected by both linear and nonlinear sources of lift.

In this report, the effect of the aspect ratio on the aerodynamic characteristics is investigated experimentally. After all tests, results are compared and discussed with literature studies if these experimental results are in coherence.

2 Experimental Tests

First of all, NACA4412 airfoil model was produced via 3D printer and then it was rubbed to obtain a smoother surface of wing as shown in Figure 1.

[a] Corresponding author: musgenc@erciyes.edu.tr

Figure 1. NACA4412 aerofoil model produced by 3D printer.

2.1 Aerodynamic Force Coefficient for NACA4412 airfoil model

In this study, NACA4412 airfoil model was produced with the aspect ratio of 1 and 3 were chosen at Reynolds number of 50000 and at the angles of attack from 0° to 20°. Computer-controlled automatic angle changing force measurement system was used. Drag and lift forces were determined with strain-gauge which was connected to system. Then, these drag and lift forces were transformed into drag coefficient (C_D) and lift coefficient (C_L), respectively. Variations of drag and lift coefficients with angle of attack were given at Reynolds number of 50000 as shown in Figure 2.

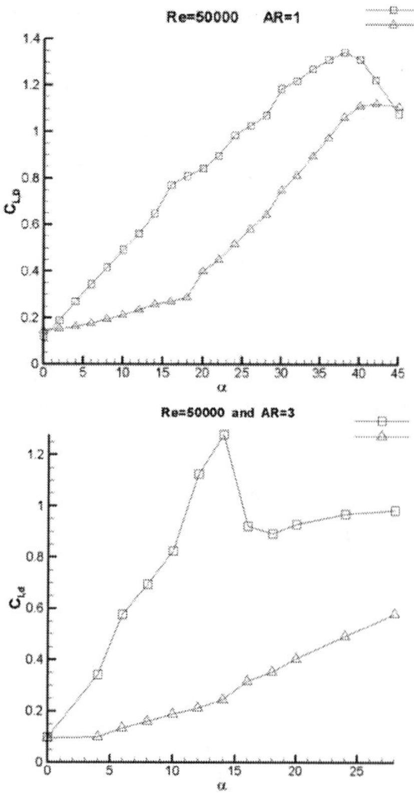

Figure 2. Aerodynamic force coefficient distributions for different aspect ratios at Re = 50000.

As seen from the figures, the stall angle is 38° and $C_{L,max}$ is 1.35 when the aspect-ratio is 1. Yet, stall angle is 14° and $C_{L,max}$ is 1.25 as the aspect-ratio has been increased as 3 at same conditions. The cause of this situation is that the effects of tip vortices over the wing are seen more at low aspect ratio. Thus, stall angle at the aspect-ratio of 1 is bigger than value of AR = 3. In addition to the results, it can be seen that stall at Reynolds number of 50000 for AR = 3 is sharper than stall for AR = 1. That is, it can be said that abrupt stall is seen at aspect ratio of 3 whereas mild stall is seen at AR = 1.

2.2 Flow Visualization Device

The smoke-wire test for flow visualization technique is employed in this study. In smoke-wire flow visualization, different smoke-wire tests are applied for visualization with AR = 1 and AR = 3. As shown in Figure 3, z / c was used as a term for location of smoke-wire. It can clearly be seen that this z / c term is located of the outer portion of wing if it is negative sign.

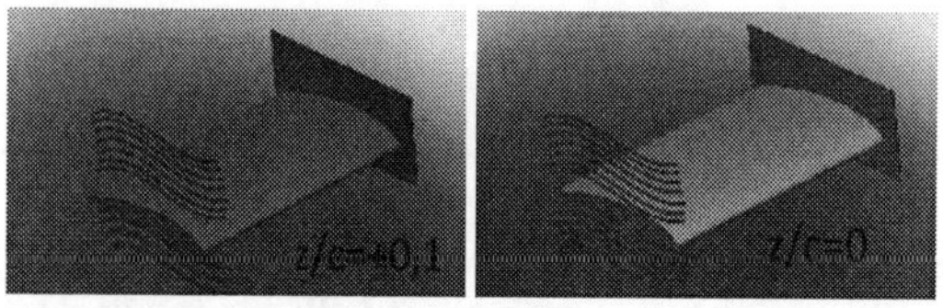

Figure 3. Different positions of smoke depending on the wire.

Flow visualization with smoke-wire test was presented for AR = 1 and AR = 3 at Re = 50000 as shown at Figure 4 and Figure 5. Moreover, flow visualization with smoke-wire test was also presented as top view of wing for AR = 1 and AR = 3 at Re = 50000 as shown in Figure 6 and Figure 7, respectively.

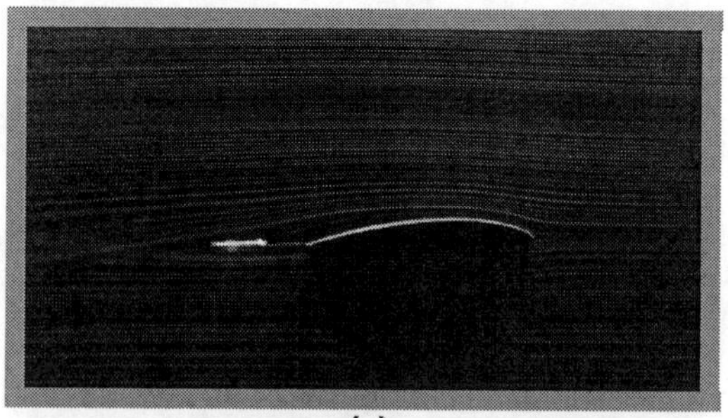

(a)

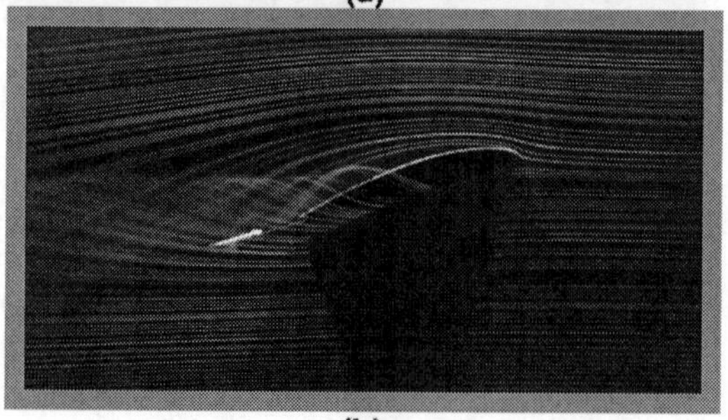

(b)

(c)

Figure 4. Flow visualization with smoke-wire test at Reynolds number of 50000 for AR = 1 and z /c = 0 a) 0°, b) 16°, c) 24°.

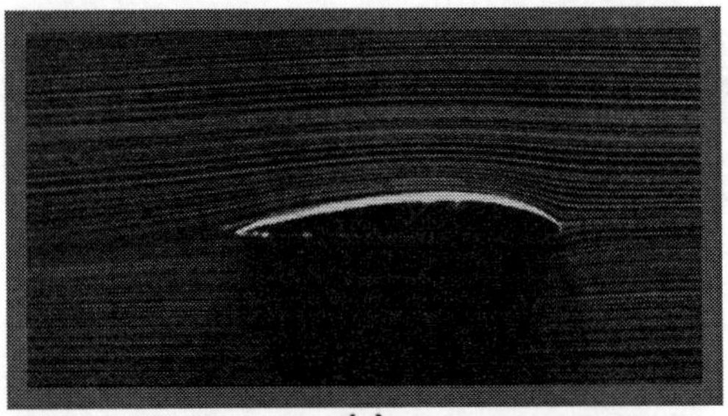

(a)

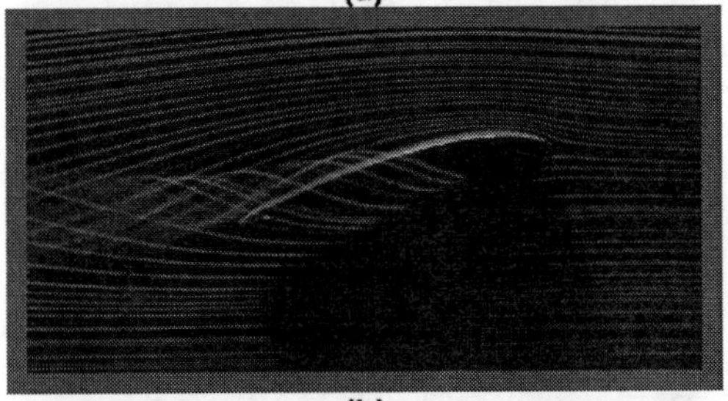

(b)

(c)

Figure 5. Flow visualization with smoke-wire test at Reynolds number of 50000 for AR = 3 and z / c = + 0.1 a) 0°, b) 16°, c) 24°.

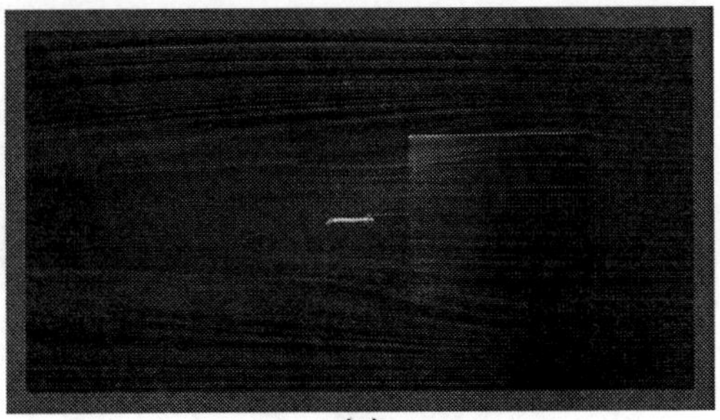

(a)

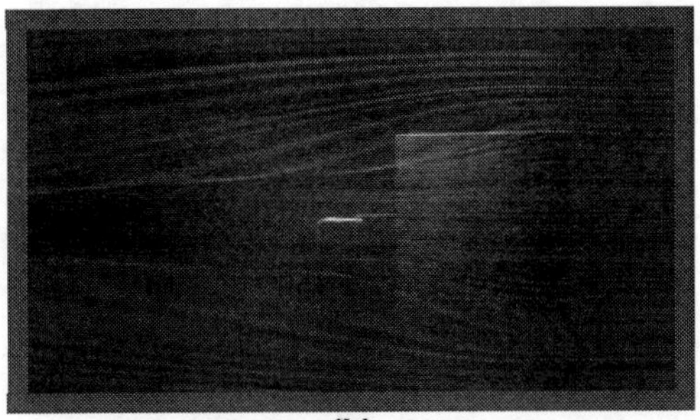

(b)

Figure 6. Flow visualization with smoke-wire test at Reynolds number of 50000 for AR = 1 and the top of view a) 4°, b) 16°.

(a)

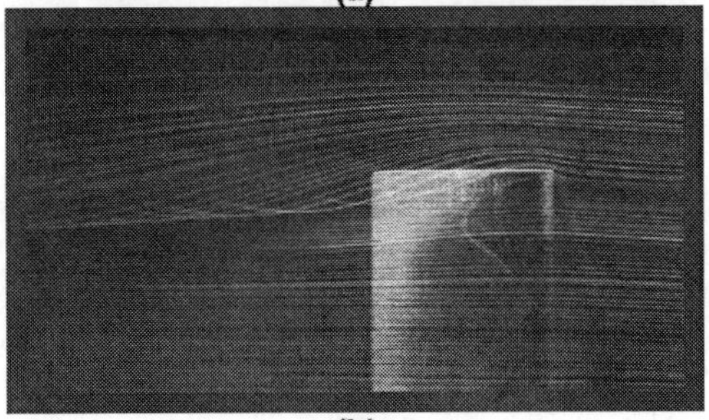

(b)

Figure 7. Flow visualization with smoke-wire test at Reynolds number of 50000 for AR = 3 and the top of view a) 4°, b) 16°.

In Figure (4a) and Figure (5a), flow over the wing model was in a normal way and was smooth. Tip-vortices became more dominant when angle of attack increased from 0° to 24°. In this type of flow regions for AR =1 and AR = 3, laminar separation bubbles were affected by tip-vortices. Tip-vortices over the wing model behaved like additional force which helped to attach for flow and separated flow at the middle of wing formed clinging flow at the tip of wing. Vortices coming from tips affected the majority of wing surface and instabilities increased. Also, rotational flow at wake region improved flows by doing vacuum effect with pressure drop.

For AR = 1, tip-vortices moved horizontally directed towards to middle of wing with increasing of angle of attack whereas they formed at the tip of wing at low angle of attack and continued to move towards to wake region of wing. Yet for AR = 3, movement of tip-vortices was limited as angle of attack increased.

3 Conclusion

The aerodynamic characteristics and corresponding flow structures happened by separated shear layer and tip-vortices over the wing upper surface at Reynolds number of 50000 for AR = 1 and AR = 3 have been experimentally investigated by using flow visualization technique like smoke-wire test and the force measurement tests. For low aspect-ratio wings, tip-vortices have a considerable on characteristics of NACA4412 wing model. As mentioned earlier, value of maximum lift coefficient ($C_{L,max}$) is 1.35 and stall angle is 38° for AR=1 at Reynolds number of 50000. On the other hand, $C_{L,max}$ is 1.25 and stall angle 14° when aspect-ratio has been increased from 1 to 3. It can be noticed that the effect of tip-vortices increases when the aspect-ratio decreases. The results in regards to the sample of flow visualization with separation laminar shear layer, flow reattached property and generation of tip-vortices on the upper region of wing can be demonstrated by the corresponding flow structure showed at figures in the present report. This type of flow structures moves from tip of wing to middle of the wing when angle of attack increases. Yet, they will vanish and they will be replaced by huge portion of wake region after a while.

Acknowledgments

The authors would like to acknowledge funding from the Scientific and Technological Research Council of Turkey (TÜBİTAK) under the project no: 213M329.

References

1. I. Gursul, Vortex flows on UAVs: issues and challenges. *Aeronautical Journal.* **108(1090)** 597–610 (2004)
2. M.S. Genç, Unsteady aerodynamics and flow-induced vibrations of a low aspect ratio rectangular membrane wing with excess length. *Exp. Therm Fluid Sci.* **44**, 749-759 (2013)
3. P. Rojratsirikul, M.S. Genc, Z. Wang, I. Gursul, Flow-induced vibrations of low aspect ratio rectangular membrane wings. *J Fluid Struct.* **27**, 1296-1309 (2011)
4. R. Ricci, S.A. Montelpare, Quantitative IR thermographic method to study the laminar separation bubble phenomenon. *Int J Therm Sci.* **44(8)**, 709-719 (2005)
5. W. Zhang, R. Hain, C.J. Kahler, Scanning PIV investigation of the laminar separation bubble on a SD7003 airfoil. *Exp. Fluids.* **45(4)**, 725-743 (2008)
6. M.S. Genc, U. Kaynak, G.D. Lock, Flow over an Aerofoil without and with Leading Edge Slat at a Transitional Reynolds Number. *Proc IMechE, Part G: J Aerospace Eng.* **223(3)**, 217-231 (2009)
7. M.S. Genc, Numerical Simulation of Flow over a Thin Aerofoil at High Re Number using a Transition Model. *Proc IMechE, Part C-J Mech Eng Sci*, **224(10)**, 2155-2164 (2010)
8. M.S. Genc, U. Kaynak, H. Yapıcı, Performance of transition model for predicting low Re aerofoil flows without/with single and simultaneous blowing and suction. *Eur J Mech B-Fluid.* **30(2)**, 218-235 (2011)
9. M.S. Genc, I. Karasu, H.H. Acıkel, An experimental study on aerodynamics of NACA2415 aerofoil at low Re numbers. *Exp Therm Fluid Sci.* **39**, 252-264 (2012)
10. M.S. Genc, I. Karasu, H.H. Acıkel, M.T. Akpolat, Low Reynolds number flows and transition, in: M. Serdar Genc (Ed.), Low Reynolds Number Aerodynamics and Transition, *Intech-Sciyo Publishing*, ISBN 979-953-307-627-9 (2012)
11. I. Karasu, M. S. Genç, H. H. Açikel, Numerical study on low Reynolds number flows over an Aerofoil. *J. Appl. Mech. Eng.* **2**, 131 (2013)
12. M. Mizoguchi, H. Itoh, Effect of aspect ratio on aerodynamic characteristics at low Reynolds numbers. *AIAA Journal.* **51(7)**, 1631-1639 (2013)
13. M. Okamoto, A. Azuma, Aerodynamic Characteristics at Low Reynolds Numbers for Wings of Various Planforms. *AIAA Journal.* **49(6)**, 1135-1150 (2011)
14. G. E. Torres, T. J. Mueller, Low Aspect-Ratio Wing Aerodynamics at Low Reynolds Numbers. *AIAA Journal.* **42(5)**, 865-873 (2004)
15. R. Bleischwitz, R. de Kat, B. Ganapathisubramani, Effects of aspect ratio on fluid-structure interactions in membrane wings. *52nd AIAA Aerospace Science and Technology Forum and Exposition, National Harbor, Maryland,* (2014)
16. P.H. Cosyn, J. Vierendeels, Numerical Investigation of low aspect-ratio wings at low Reynolds Numbers. *J. Aircraft.* **43(3)**, 713-722 (2006)
17. S. Kaplan, A. Altman, M. Ol. Wake vorticity measurements for low aspect-ratio wings at low Reynolds Numbers. *J. Aircraft.* **44 (1)**, 241-251 (2007)
18. E.V. Laitone, Wind tunnel tests of wings at Reynolds number below 70000. *Exp. Fluids.* **23(5)**, 405-409 (1997)
19. M. Shields, K. Mohseni, Experimental complications inherent to low Reynolds number wind tunnel testing. *AIAA Paper.* 2011-0873. Orlando, FL. January 2011.

EPJ Web of Conferences **114**, 02028 (2016)
DOI: 10.1051/epjconf/201611402028

Effect of tip vortices on membrane vibration of flexible wings with different aspect ratios

Mustafa Serdar Genç[1,a], Halil Hakan Açikel[1], Hacımurat Demir[1,2], Mustafa Özden[1], Mücahit Çağdaş[1], Iliasbek Isabekov[1]

[1]*Wind Engineering and Aerodynamic Research Group, Department of Energy Systems Engineering, Erciyes University, Kayseri, Turkey*
[2]*Department Of Mechanical Engineering, Aksaray University, Aksaray, Turkey*

Abstract. In this study, the effect of the aspect ratio on the aerodynamics characteristic of flexible membrane wings with different aspect ratios (AR = 1 and AR = 3) is experimentally investigated at Reynolds number of 25000. Time accurate measurements of membrane deformation using Digital Image Correlation system (DIC) is carried out while normal forces of the wing will be measured by helping a load-cell system and flow on the wing was visualized by means of smoke wire technic. The characteristics of high aspect ratio wings are shown to be affected by leading edge separation bubbles at low Reynolds number. It is concluded that the camber of membrane wing excites the separated shear layer and this situation increases the lift coefficient relatively more as compared to rigid wings. In membrane wings with low aspect ratio, unsteadiness included tip vortices and vortex shedding, and the combination of tip vortices and vortex shedding causes complex unsteady deformations of these membrane wings. The characteristic of high aspect ratio wings was shown to be affected by leading edge separation bubbles at low Reynolds numbers whereas the deformations of flexible wing with low aspect ratio affected by tip vortices and leading edge separation bubbles.

1 Introduction

Flexible membranes have been associated with a number of applications such as hang gliders, parachutes, paragliders, yacht sails, microlights, natural flyers such as bats and insects. Micro Air Vehicles (MAVs) and Unmanned Air Vehicles (UAVs), which are characterized by low flight speeds and small size, are of great interest to low Reynolds number applications. In low Reynolds number flow regime, viscous forces are higher than inertia forces. So the flow is separated from the wing and reattached over the wing, which is named as a laminar separation bubble. The flow separation is very important for MAVs or UAVs.

Gordnier and Attar [1] performed a numerical study on flexible membrane wings. They performed computations at Re = 24300 and angles of attack $\alpha = 10°$, $\alpha = 16°$ and $\alpha = 23°$ for wing with aspect ratio of 2. They noticed that there was a reasonable agreement between computational results with experimental PIV and surface deflection measurements. The flexible wing demonstrated a mean first mode structural response of increasing amplitude with increasing angle of attack. Because of favorable interactions between the flexible membrane wing and the unsteady flow over the wing, reduced separation and enhanced lift were obtained.

Tamai et al. [2] performed an experimental study in order to investigate the benefits of utilizing flexible membrane wings at low Reynolds numbers for MAV applications. Flow field measurements carried on different angle of attack for quantifying the transient behaviour of vortex and turbulent flow structures around the flexible membrane wings at by means of a high-resolution Particle Image Velocimetry (PIV) system. They noticed that flexible membrane wings ensured better aerodynamic performance comparing their rigid counterpart at low Reynolds numbers. In addition, it was seen that the aerodynamic performance of the membrane airfoils were effected the flexibility (or rigidity) of membrane skins remarkably. The PIV measurements clarified that flexible membrane wings altered their camber automatically in order to accommodate incoming flows for balancing the pressure differences applied on the upper and lower surfaces of the wings at different angle of attack.

Béguin and Breitsamter [3] investigated effects of membrane pre-stress on the aerodynamic characteristics of elasto-flexible morphing wing for unmanned air vehicle applications. The results of force and deformation measurements conducted at several dynamic pressures elucidated that varying the pre-stress effectively allowed to influence the passive deformation of the wing surface and regulated the wing camber within a certain range at a given flow condition. Moreover, it was seen that the lift and drag characteristics were affected such that an improvement of the maximum lift-to-drag ratio between

[a] Corresponding author: musgenc@erciyes.edu.tr

6% and 23% of the lowest values at a given flow condition could be achieved.

Genç [4] investigated unsteady aerodynamics and flow-induced vibrations of a low aspect ratio rectangular membrane wing with excess length. The experiments for the membrane and rigid wing models were conducted in a closed-loop open-jet wind tunnel with a circular working section of 760 mm in diameter at angles of attack ranging from 0° to 25° at Reynolds number of 48,700. Digital Image Correlation system was used for time-accurate measurements of membrane deformation. Furthermore, measurement of velocity field with Digital Particle Image Velocimetry system was conducted whereas normal forces of the wing were measured by means of a load-cell system. Obtained experimental results demonstrated that the camber of membrane wing with excess length excited the separated shear layer. Additionally, the membrane caused to generate a larger magnitude of the maximum camber in the both 2D and 3D rectangular membrane wings with excess length. On the other hand, dominant vibrational modes were obtained as a spanwise mode of two at the angles of attack of 6-10° and 19-22°. On the other hand, dominant spanwise modes were three at 11-18° and 23-25°.

Rojratsirikul et al. [5] performed an experimental study of a low aspect ratio rectangular wing in a low-speed, closed loop open-jet wind tunnel for a Reynolds number range of 2.4×10^4–4.8×10^4. Time accurate measurements of membrane deformation were added with the flow field measurements. Analysis of the fluctuating deformation detected chordwise and spanwise modes, which were due to the shedding of leading-edge vortices as well as tip vortices. It was seen that the second mode in the chordwise direction became dominant as the vortex shedding took place at higher angles of attack. The dominant frequencies of the membrane vibrations were similar to those of two-dimensional membrane airfoils. Furthermore, measured vortex shedding frequency from the low aspect ratio rigid wing suggested that membrane vibrations occurred at the natural frequencies close to the harmonics of the wake instabilities. Additionally, it was seen that vortex shedding frequency from rigid wings showed significantly small effect of aspect ratio even when it was as low as unity.

Hefeng et al. [6] investigated four types of segmented flexible membrane airfoil which were based on NACA0012 numerically. They applied fluid-structure interaction method in the numerical simulation for the aerodynamic characteristics and response between flow field and structure of the segmented flexible airfoils. They noticed that the segmented flexible airfoils carried out a higher maximum lift coefficient, and postponed the stall effectively. Furthermore, the deformation of the flexible thin membrane decreased the scale of the separation vortex at the higher angles of attack. Three-segment flexible airfoil had the best aerodynamic characteristics among the four kinds of segmented forms of wings. It created a profit of 39% on lift when comparing with the traditional rigid airfoil.

Song et al. [7] studied on aerodynamics of rectangular membrane wings with low aspect ratio for Reynolds numbers in the range of $0.7\text{-}2.0 \times 10^5$. The lift and drag coefficients were measured for wings with varying aspect ratio, compliancy and pre-strain values. Linear elastic theory was utilized in order to predict the natural frequencies of membrane. It was seen from the measurements that the membrane wings had a higher lift slope, maximum lift coefficients and delayed stall.

Hu et al. [8] investigated the aerodynamic performances of flexible membrane wings in flapping flight experimentally. They compared two flexible membrane wings with different skin flexibility (i.e., flexible nylon wing and a very flexible latex wing) with a conventional rigid wing for the development of flapping wing Micro Air Vehicles (MAVs). The measurement results demonstrated that flapping motion would bring significant aerodynamic benefits when the flapping flight was in unsteady state regime, and the flexibility had better aerodynamic performances for both soaring and flapping flights. It was found that the flexible membrane wings had better overall aerodynamic performance over the rigid wing for soaring flight, especially for high speed soaring flight or at relatively high angle of attack.

Ke et al. [9] studied on analysis and flexible structural modeling for oscillating wing using aeroelasticity. Furthermore, they detailed the design principle, design process and range of application of such oscillating wing analytical method. A flexible structural oscillating wing model was set up, and relevant time response analysis and frequency response analysis were carried out. It was found from the analytical results that adopting the new-type driving way for the oscillating wing would not have flutter problems and would be able to produce propulsive force. Additionally, it would consume much less power than the fixed wing for generating the same lift.

Bleischwitz et al. [10] investigated the aeromechanics of membrane wings in ground effect. They conducted the experiments in wind tunnel with a moderate Reynolds number of Re= 56,000 utilizing rectangular rigid and membrane wings in the vicinity of the ground. In order to allow correlating and identifying instantaneous membrane deformation, time-resolved forces and membrane dynamics were captured in parallel. In order to resolve membrane motions, Digital Image Correlation (DIC) was utilized. Furthermore, Proper Orthogonal Decomposition (POD) was applied to study the spatio-temporal structure of membrane vibrations. Decomposition (POD) modes of membrane fluctuations were sufficient to capture 90% of all deformation energy closes to stall. In addition to this, spanwise structural modes were found by means of POD in regions of strong lift increment.

Attar et al. [11] studied on experimental characterization of aerodynamic behaviours of membrane wings in low Reynolds number of Re = 13700, 22600 and 36300 and the angle of attack up to 27°. It was found that the pre-stall values were irresponsive to pre-strain level for the mean lift coefficient results. Furthermore, they noticed that the Reynolds number had a considerable effect on the pre-stall lift curve, with the lift-curve slope increasing with increasing Reynolds number for all the flexible cases and the rigid flat plate model. Reducing level of model pre-strain led to increase in the pre-stall drag coefficients generally. It was found that the flexible-model stall angles were similar to the rigid flat plate results with the exception of the highest Reynolds number. On the other hand, dynamic results for the flexible models appeared large RMS values of lift and drag at the largest Reynolds number, which mostly reduced with increasing angle of attack. The RMS values for the rigid flat plate were unremarkable comparing with the flexible models.

Wrist et al. [12] compared the aerodynamic properties for flexible membrane micro air vehicle (MAV) wings with flat and cambered frames. An Objet30 Pro 3D printer was used for construction of wing's frames. They tested the wings in a low speed wind tunnel at Reynolds number of 50000 with varying angle of attack from -4° to 24°. It was stated from the results that cambering the frames for the flexible MAV wings increased maximum lift and aerodynamic efficiency comparing with the flat frames. Furthermore, membrane wings with cambered frames resulted in more negative pitching moment than the flat frame, leading to the center of pressure located aft of the quarter-chord. An increase in frame camber caused to increase aero-induced membrane camber with regard to the adjusted chord line.

Viieru et al. [13] discussed recent progress in understanding the low Reynolds number unsteady fluid dynamics related to flapping wings, including pitching-up rotation, leading-edge vortices and wake-capturing mechanisms. They noticed that the structures of the leading-edge vortices changed with Reynolds number remarkably by investigating the flow structures at three Reynolds numbers, namely, 6000, 120 and 10. The leading edge vortex demonstrated an intense, conical structure, with a sustained spanwise flow at the vortex core, breaking down at nearly three-quarters of the span towards the tip at Reynolds number of 6000. The vortex break-down was lost, and it was found that the leading-edge vortices connected to the tip vortices at Reynolds number of 120. In addition to this, a vortex ring connecting the leading edge vortex, the tip vortex and the trailing vortex was observed at Re = 10.

Greenhalgh et al. [14] investigated aerodynamic properties of a two dimensional inextensible flexible airfoil. Measurement of the lift, drag and membrane tension forces acting on a single surface mylar membrane supported by elliptical rods at the leading and trailing edges were conducted in the experiments. The angle of attack at which separation occurs decreased with increasing excess length and the hysteresis range increased resulting in a narrow operating range for the highest excess lengths examined. They noticed that theoretical results were in a good agreement with experimental data for those angles of attack where the flow over the membrane was not separated. Furthermore, all of the membranes demonstrated maximum lift-to-drag ratios near zero angle of attack. They observed hysteresis of the membrane airfoil near zero angle of attack experimentally.

2 Experimental Apparatus and Measurement Methods

The experiments were carried out in a low-speed, suction-type wind tunnel with a square working section of 500 mm x 500 mm located at the Department of Energy Systems Engineering, University of Erciyes via Wind Engineering and Aerodynamics Research Group (WEAR) as shown in Figure 1. Turbulence intensity of the tunnel is 0.35% and the operating range of tunnel speed was 1-45 m / s. In the experiments, two aspect ratios, AR = 1 and AR = 3 were considered. For membrane aerofoils, a membrane which was made from a black latex rubber sheet with a thickness of t = 0.2 mm, Young's modulus of E = 2.2 MPa, and density of ρ_m = 1 g / cm^3 was used. The membrane was attached to the aerofoil-shaped leading and trailing edges for a membrane aerofoil. The wing used in the study was composed of a rigid stainless steel frame and attached membrane. This rigid frame was manufactured in order to have a cross-section of airfoil shape (as seen in Figure 2).

Figure 1. Photograph of the experimental set-up.

2.1 Experimental Studies for Membrane Wing

Details of experimental studies about the membrane wing are given below elaborately. First of all, the variations of aerodynamic force coefficients with respect to aspect ratio were investigated. After determination of the aerodynamic force coefficients, flow on the wing was visualized by means of smoke wire experiment.

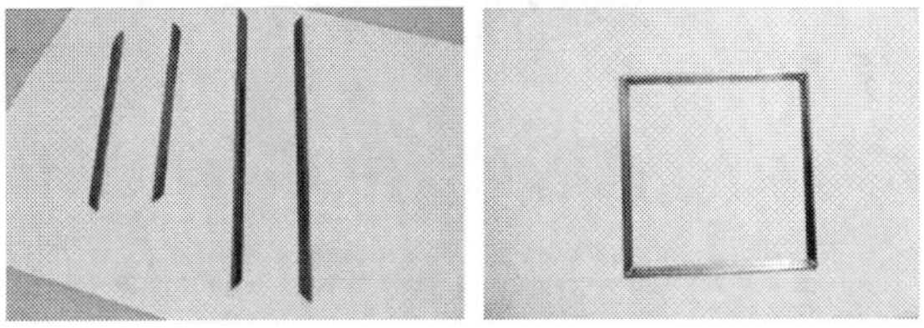

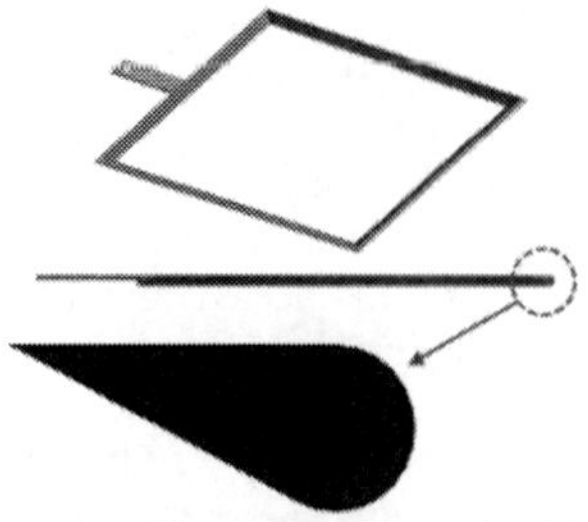

Figure 2. Schematic of the rectangular wing with cross-section of the rigid frame.

2.1.1 Aerodynamics performance measurement

Aerodynamic forces were measured at specific angles of attack from 0° to 45° by means of a computer controlled force measurement system having automatic change of angle. As seen from Figure 3, lift and drag forces are measured separately by means of strain gauges and these measured forces changed with coordinates according to the angle of attack of the wing. So, the measured values are converted to the lift (C_L) and drag (C_D) force coefficients in x and y axes by means of coordinate transformation.

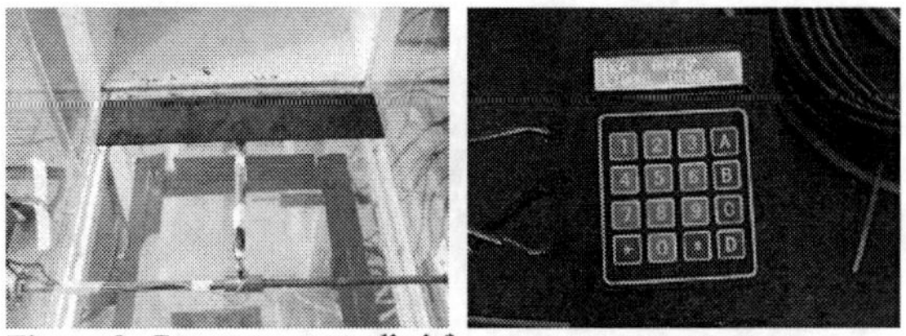

Figure 3. Computer controlled force measurement system.

2.1.2 Deformation measurement

Digital Image Correlation (DIC) system was used to measure the membrane deformations. The DIC system captures consecutive images and then calculates the displacement field over the specimen by following the deformation of spots (Figure 4) applied to the surface of the specimen using a cross-correlation method.

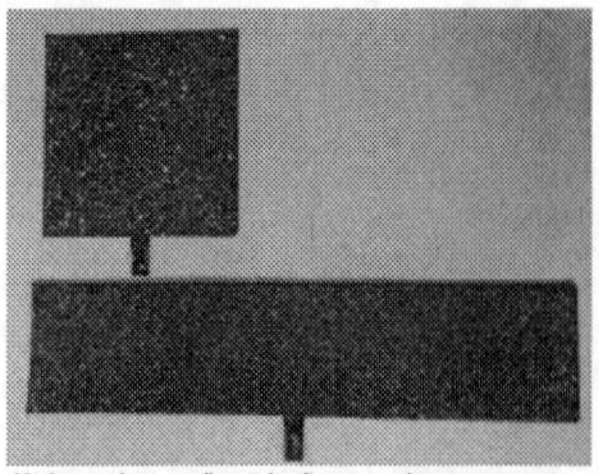

Figure 4. Flexible wings for deformation measurements.

2.1.3 Flow visualization via smoke-wire

Smoke-wire technique is preferred for the flow visualization due to the simplicity and reliability. A smoke-wire is utilized for heating the fluid and machine oil to mark the streamlines for flow visualization. Canon EOS-D1100 camera was used for taking related experimental photographs. The main problem in recording the smoke lines is reflection of images in the plexiglas wall test section of the tunnel. For this reason, lighting and camera arrangements are carried out carefully to ensure dark medium as seen in Figure 5a. The expression of z / c is utilized for the location of smoke wire as shown in Figure 5b. The behaviors of both laminar separation bubble and tip vortex are investigated in terms of various angles of attack, aspect ratios.

a)

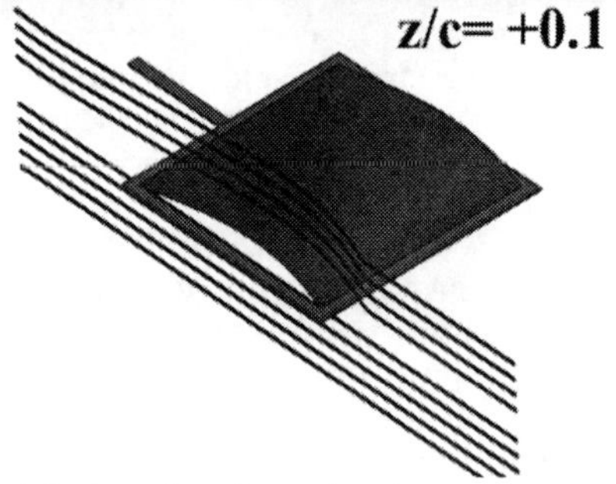

b)

Figure 5. a) Dark medium for smoke-wire measurement
b) Location of smoke wire on the membrane wing.

3 Experimental Results

3.1 Determination of aerodynamic coefficients of flexible membrane wing

The variation of C_L and C_D coefficients of membrane wing with respect to the angle of attack in various aspect ratios at Re = 25000 are given in Figure 6. According to the obtained results from these figures, stall angles of attack were 35° and 19° and the values of C_{Lmax} were 1.3 and 1.5 for AR = 1 and AR = 3, respectively. As the aspect ratio increased, stall angle and maximum lift decreased.

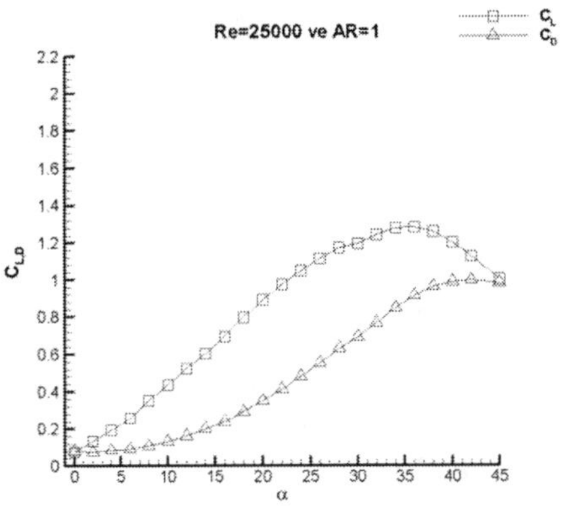

(a)

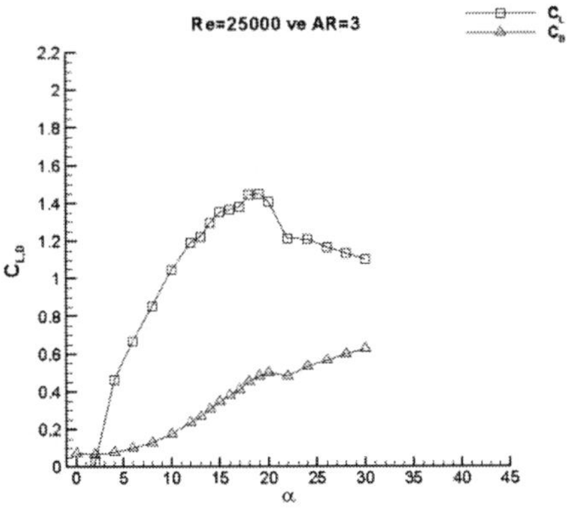

(b)

Figure 6. Aerodynamic force coefficients at various Re numbers.

3.2 Deformation results

Figure 7 and Figure 8 shows instantaneous membrane deformations for the wing with AR = 1 and AR = 3 at AoA = 12°. It is seen that the membrane deformations for the wing with AR = 1 change with time whereas for the wing with AR = 3 it has not instantaneous variations. Over the wing with AR = 1, tip vortices and leading edge separations are dominant, thus the flow over the wing are unsteady and the membrane deformations change instantaneously.

3.3 Smoke-wire visualization results

The variations of tip vortices with respect to the angle of attack are given from Figure 9 and Figure 10. Increase in angles of attack from 8° to 20° leads to improve tip vortices throughout the wake region at Re = 25000 and AR = 1 and AR = 3. These laminar separation bubbles moved towards to leading-edge and their size became smaller as angle of attack increased. Tip vortices delays stall for the wing at low aspect ratio, especially at low Re numbers due to affecting flow on the wing and separation bubble as seen in Figure 11 and 12. Separation bubble located near trailing-edge was occurred at low angle of attack.

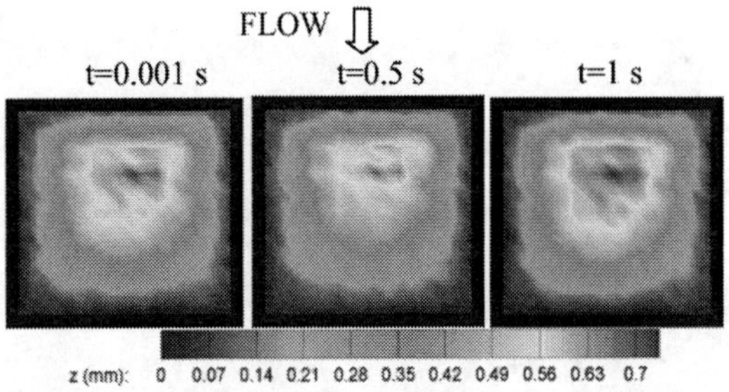

Figure 7. Instantaneous membrane deformations for the wing with AR = 1 at AoA = 12°.

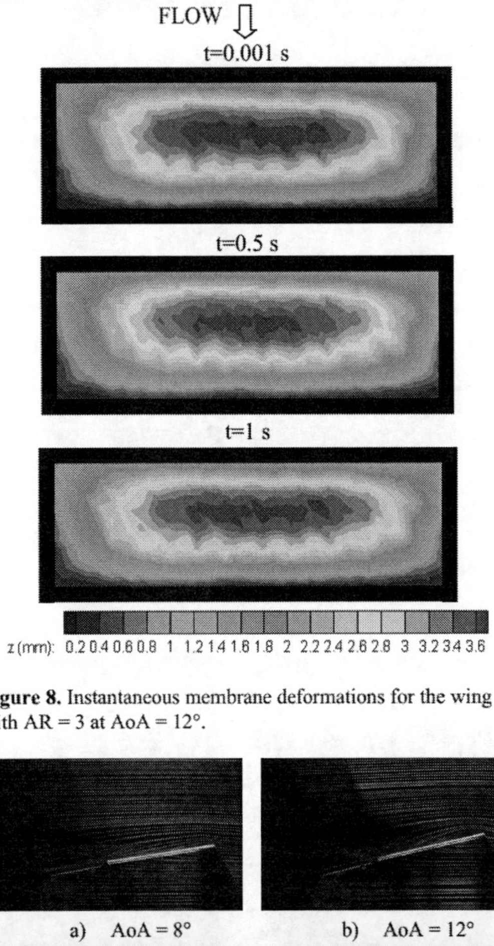

Figure 8. Instantaneous membrane deformations for the wing with AR = 3 at AoA = 12°.

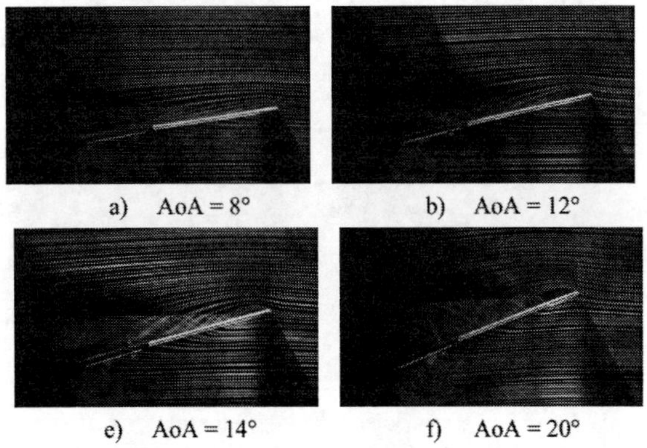

Figure 9. Smoke-wire visualization for tip vortices at z / c = +0.1 (Re = 25000 and AR = 1).

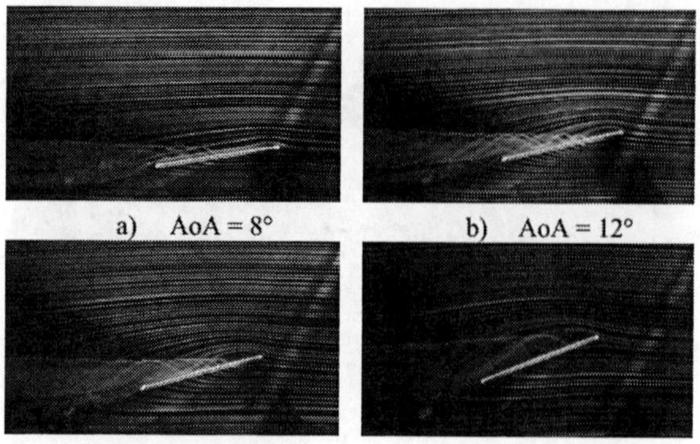

Figure 10. Smoke-wire visualization for tip vortices at z / c = +0.1 (Re = 25000 and AR = 3).

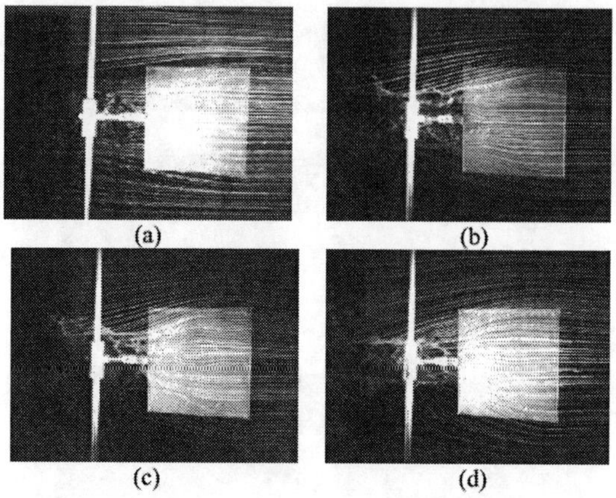

Figure 11. Top view smoke-wire visualization for tip vortices at z / c = +0.1 (Re = 25000 and AR = 1) for (a) α = 8°, (b) α = 16°, (c) α = 20° and (d) α = 24°.

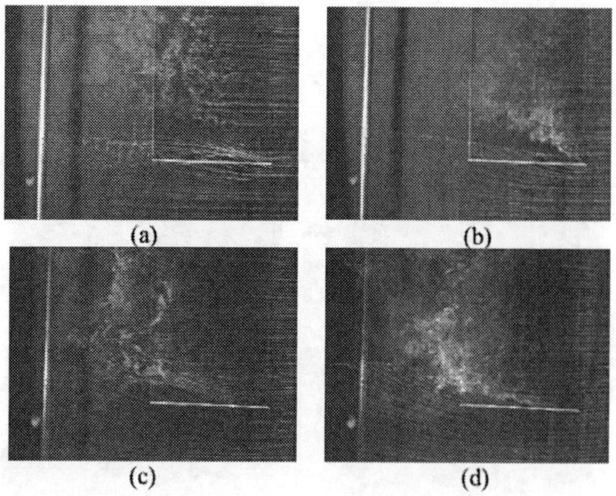

Figure 12. Top view smoke-wire visualization for tip vortices at z / c = +0.1 (Re = 25000 and AR = 3) for (a) α = 8°, (b) α = 12°, (c) α = 16° and (d) α = 20°.

4 Conclusion

The effect of tip vortices on membrane vibration of flexible wings with AR = 1 and AR = 3 was experimentally investigated at Re = 25000 in a low-speed, suction-type wind tunnel. In the force measurements, it was obtained that the stall angles of attack were 35° and 19° and the values of C_{Lmax} were 1.3 and 1.5 for AR = 1 and AR = 3, respectively. As the aspect ratio increased, stall angle and maximum lift decreased. In the DIC results, over the wing with AR = 1, tip vortices and leading edge separations are dominant, thus the flow over the wing are unsteady and the membrane deformations change instantaneously. In the flow visualization experiments showed that tip vortices delayed stall for the wing at low aspect ratio, especially at low Re numbers due to affecting flow on the wing and separation.

Acknowledgments

The authors would like to acknowledge funding from the Scientific and Technological Research Council of Turkey (TÜBİTAK) under the project no: 213M329, the Scientific Research Projects Unit of Erciyes University under the contract no: FDK-2015-6171, FBG-2014-5337 and FDA-2014-5273.

References

1. R.E. Gordnier, P.J. Attar, Impact of flexibility on the aerodynamics of an aspect ratio two membrane wing. *J. Fluids Struct.* **45**, 138-152 (2014)
2. M. Tamai, J. Murphy, H. Hu, An experimental study of flexible membrane airfoils at low Reynolds numbers. *AIAA Aerospace Sciences Meeting and Exhibit.* 7-10 (2008)
3. B. Béguin, C. Breitsamter, Effects of membrane pre-stress on the aerodynamic characteristics of an elasto-flexible morphing wing. *Aerosp. Sci. Technol.* **37**, 138-150 (2014)
4. M.S. Genç, Unsteady aerodynamics and flow-induced vibrations of a low aspect ratio rectangular membrane wing with excess length. *Exp. Therm Fluid Sci.* **44**, 749-759 (2013)
5. P. Rojratsirikul, M.S. Genc, Z. Wang, I. Gursul, Flow-induced vibrations of low aspect ratio rectangular membrane wings. *J Fluid Struct.* **27**, 1296-1309 (2011)
6. D. Hefeng, W. Chenxi, L. Shaobin, S.X. Zhen, Numerical Research on Segmented Flexible Airfoils Considering Fluid-structure Interaction. *Procedia Eng.* **99**, 57-66 (2015)
7. A. Song, X. Tian, E. Israeli, R. Galvao, K. Bishop, S. Swartz, K. Breuer, Aeromechanics of membrane wings with implications for animal flight. *AIAA journal.* **46(8)**, 2096-2106 (2008)
8. H. Hu, A.G. Kumar, G. Abate, R. Albertani, An experimental investigation on the aerodynamic performances of flexible membrane wings in flapping flight. *Aerosp. Sci. Technol.* **14(8)**, 575-586 (2010)
9. S. Ke, W. Zhigang, Y. Chao, Analysis and flexible structural modeling for oscillating wing utilizing aeroelasticity. *Chin. J. Aeronaut.* **21(5)**, 402-410 (2008)
10. R. Bleischwitz, R. Kat, B. Ganapathisubramani, Aeromechanics of Membrane Wings in Ground-Effect. *45th AIAA Fluid Dynamics Conf.* 2764, (2015)

11. P.J. Attar, B. J. Morris, W. A. Romberg, J. W. Johnston, R. N. Parthasarathy, Experimental Characterization of Aerodynamic Behavior of Membrane Wings in Low-Reynolds-Number Flow. *AIAA journal.* **50(7)**, 1525-1537 (2012)
12. A. Wrist, D. Pepley, Z. Zhang, J.P. Hubner, Aerodynamic Comparison of Flat and Cambered Frames for Flexible MAV Wings. *53rd AIAA Aerospace Sciences Meeting.* (2015)
13. D. Viieru, J. Tang, Y. Lian, H. Liu, W. Shyy, Flapping and flexible wing aerodynamics of low Reynolds number flight vehicles. *AIAA paper.* **503**, (2006)
14. S. Greenhalgh, H.C. Curtiss, B. Smith, Aerodynamic properties of a two-dimensional inextensible flexible airfoil. *AIAA journal.* **22(7)**, 865-870 (1984)

EPJ Web of Conferences **114**, 02029 (2016)
DOI: 10.1051/epjconf/201611402029

Flow characteristics over NACA4412 airfoil at low Reynolds number

Mustafa Serdar Genç[1,a], Kemal Koca[1], Halil Hakan Açıkel[1], Gökhan Özkan[1], Mehmet Sadık Kırış[1], Rahime Yıldız[1]

[1] *Wind Engineering and Aerodynamic Research Group, Department of Energy Systems Engineering, Erciyes University, Kayseri, Turkey*

Abstract. In this study, the flow phenomena over NACA4412 were experimentally observed at various angle of attack and Reynolds number of 25000, 50000 and 75000, respectively. NACA4412 airfoil was manufactured at 3D printer and each tips of the wing were closed by using plexiglas to obtain two-dimensional airfoil. The experiments were conducted at low speed wind tunnel. The force measurement and hot-wire experiments were conducted to obtain data so that the flow phenomenon at the both top and bottom of the airfoil such as the flow separation and vortex shedding were observed. Also, smoke-wire experiment was carried out to visualize the surface flow pattern. After obtaining graphics from both force measurement experiment and hot-wire experiment compared with smoke wire experiment, it was noticed that there is a good coherence among the experiments. It was concluded that as Re number increased, the stall angle increased. And the separation bubble moved towards leading edge over the airfoil as the angle of attack increased.

1 Introduction

In the past time and nowadays, aerodynamic researches concentrated on low Reynolds number aerodynamics, transition and laminar separation bubble for efficiency of wind power or flight regime [1-9]. As it was mentioned, these flight regime or efficiency of wind power can be improved before the analysis and design of aerofoil operating in the low Reynolds number. A better understanding of the associated flow phenomena must be obtained. The major concerns at specific conditions involve the wide region of separated flow can occur over the aerofoil for low Reynolds number incompressible flow. The various flow phenomena involved in the formation of a laminar separation bubble are especially of interest, because they can play an enormous role in the development of the boundary layer.

As previously discussed in the literatures, the flow separation and the laminar separation bubble affect the airfoils running with low Reynolds number [1-8]. The laminar separation bubbles are firstly separated from surface of airfoil and encounter with adverse pressure gradient [10]. It then passes to transition area including TS waves, spanwise vorticity, 3-dimensional breakdown, turbulent spots and subsequently reattaches onto surface [11, 12].

Genç et. al experimentally [7] and numerically [8] studied to the laminar separation bubble (LSB), transition and reattachment phenomenon over NACA2415 at low Reynolds number. After the experiments, they observed that the separation point moved towards the leading-edge of airfoil when angle of attack increased. Furthermore, it was concluded that short bubble burst at higher angle of attack caused long bubble to happen by using the oil-flow visualization experiment. Carmichael [13] explained low Reynolds number between 70000 and 200000 where size of laminar separation bubble, its formation widely affects aerodynamic performance of airfoil in his study. Wahidi et. al [14] investigated to effects of laminar separation bubbles over NACA4412 airfoil model using volumetric three-component velocimetry (V3V) and particle image velocimetry (PIV) at Reynolds number 50000 and different angle of attack. They described onset of transition, location of separation point and reattachment. Their results demonstrated that roll-up of spanwise caused vortices and these vortices inspired pairs of negative and positive wall-normal velocity. The pairs did an enormous act in the uncertainty of the reattachment of the separated shear layer. O'meara and Mueller [15] investigated behaviors and physique of laminar separation bubbles at low Reynolds number in their paper. They chose to NACA66$_3$-018 as airfoil model. Their range of Reynolds number was 50000-200000 and angle of attack was between 8°-12°. Their study showed that laminar separation bubble decreased as Reynolds number increased whereas the bubble increased in both thickness and length as angle of attack increased.

Agrawal and Saxena [16] aimed that making analysis for aerodynamic of wings. They chose to NACA4412 airfoil as a model and achieved few results after completing wind tunnel tests sections. They could observe from results that lift increased until a certain point if angle of attack increased. Their observation also

[a] Corresponding author: kemalkoca@erciyes.edu.tr

included that drag started to become a dominant factor if angle of attack kept rising and stall occurred as a result of all these parts.

In this study, a detailed experimental study was conducted to observe the flow phenomenon over NACA4412 airfoil. After the experiments completed, it was seen fine agreement between the obtaining the graphics and the flow visualization.

2 Experimental Studies

2.1. Experimental apparatus and methods

The experiments were performed in a low speed, suction type wind tunnel with a square working section of 500 mm x 500 mm at the laboratory located at the Department of Energy Systems Engineering, Erciyes University via Wind Engineering and Aerodynamics Research Group (WEAR) as shown in Figure 1. Turbulence intensity of the tunnel is 0.35% and the operating range of tunnel speed was 1-45 m / s [7, 17]. The experiments were conducted at Reynolds numbers of 25000, 50000 and 75000.

Figure 1. The experimental set-up and the wind tunnel.

A two-dimensional NACA4412 airfoil was produced via 3D printer as shown at Figure 2. After the two-dimensional NACA4412 airfoil was manufactured with 3D printer, it was rubbed with sandpaper to ensure smoothness of the surface and two plexiglas were used. The airfoil model has a chord (c) of 10 cm and a span of 10 cm.

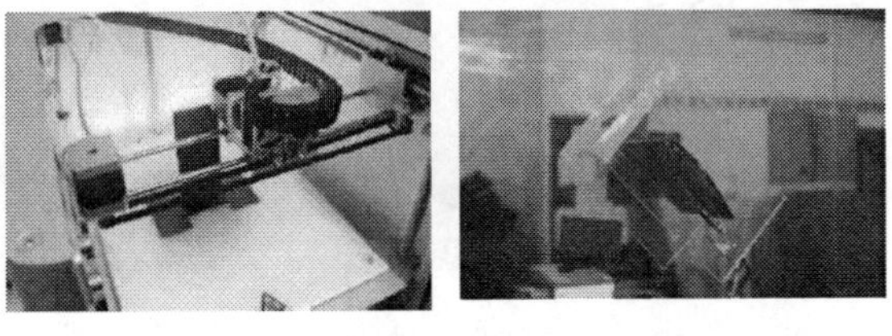

Figure 2. NACA4412 airfoil model produced via 3D printer.

2.2 Determination of the aerodynamic force coefficient for two-dimensional NACA4412 airfoil

To find out the aerodynamic force coefficients for two-dimensional NACA4412 airfoil, a computer-controlled automatic angle-changing force measurement system was used as demonstrated in Figure 3. Lift and drag forces were separately measured by strain-gauge. Then, these forces were converted to drag coefficient (C_D) and lift coefficient (C_L) where placed x-axis and y-axis at the coordinate system, respectively.

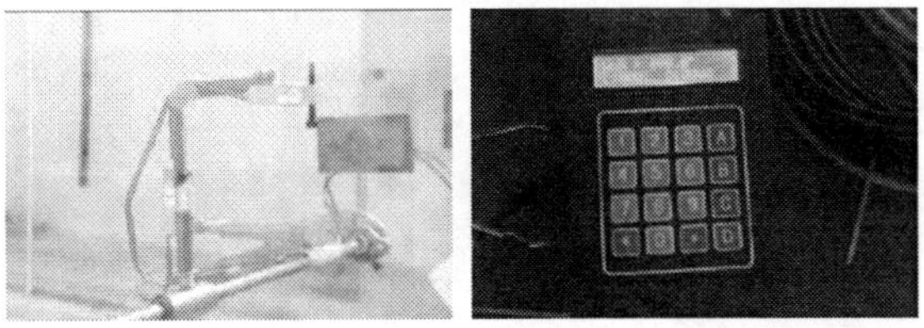

Figure 3. Automatic computer controlled force measurement system.

Figure 4 shows the experimental results of force measurements at Reynolds number of 25000. The stall angle is 12° and the maximum lift coefficient ($C_{L,max}$) is 0.9. The drag coefficient increased as angle of attack increased gradually. This situation for Reynolds number of 50000 is illustrated in Figure 5. The stall angle is 16° and the maximum lift coefficient is 1.25. It is seen that the swift climbing at the drag coefficient after the stall angle. The stall angle is 18° and the maximum lift coefficient is 1.35 for Reynolds number of 75000 as shown in Figure 6. The lift and drag coefficient increased dramatically until the stall was occurred. From these figures, it was concluded that as Re number increased, the stall angle increased.

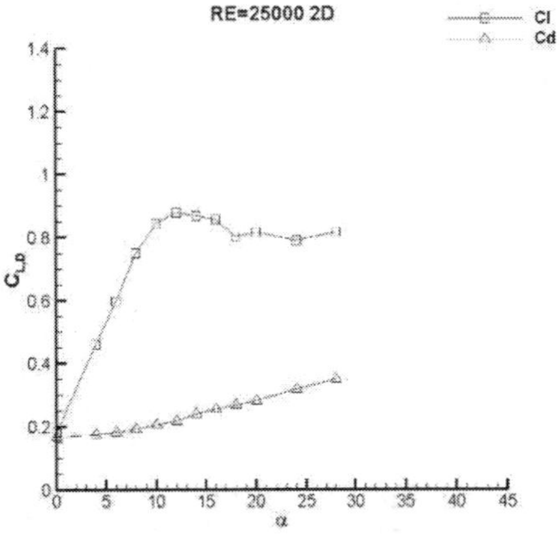

Figure 4. Aerodynamic force coefficients at Reynolds number of 25000 for NACA4412 airfoil.

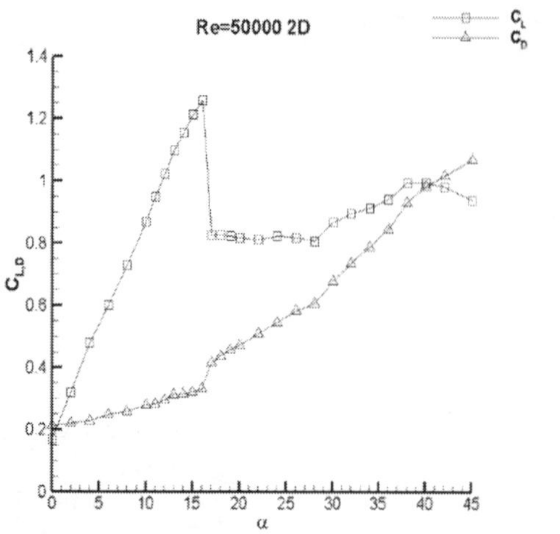

Figure 5. Aerodynamic force coefficients at Reynolds number of 50000 for NACA4412 airfoil.

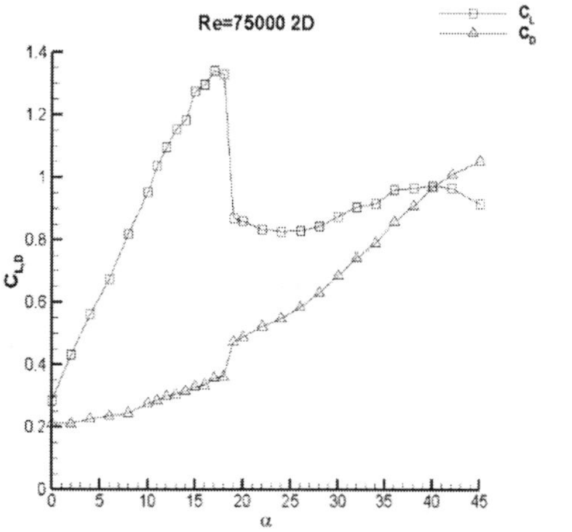

Figure 6. Aerodynamic force coefficients at Reynolds number of 75000 for NACA4412 airfoil.

2.3 Flow Visualization with smoke-wire experiment for NACA4412 airfoil model

Flow visualization with smoke-wire is an easier way in order to evaluate the structure of flow over the airfoil. The method consists of a slim wire positioned in the flow field of experiment chamber. The wire was located in front of leading-edge of airfoil. The vertical location of the wire was manually adjusted for proper positioning of the sheet of smoke. The small drops of the oil were gradually drained from the top of the wire by means of syringe. The electrical current was used through the wire in order to heat. The wire was heated as the small drops of the oil were drained. A compact camera was also used in order to visualize the sheets of smoke. It was almost positioned 50 cm forward of the experiment chamber and was focused to the airfoil. 30 frames were recorded at each Reynolds number. Two or three of fine frames were chosen and the corresponding flow structures were observed as demonstrated at Figure 7, Figure 8 and Figure 9, respectively.

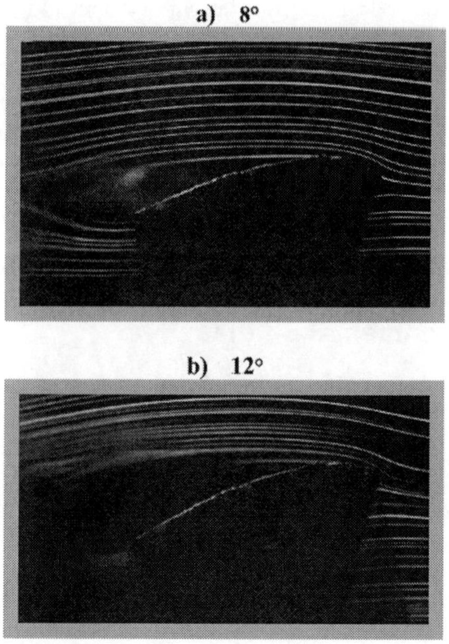

Figure 7. Flow visualization with smoke-wire experiments at Reynolds number of 25000.

The laminar separation bubble is seen clearly at Reynolds number of 25000 as shown in Figure 7(a). The separation bubble moved towards leading edge over the airfoil when the angle of attack increased from 8° to 12°. For Reynolds number of 50000, the leading edge flow separation happens at 16° even though the laminar separation bubble is observed at 12° as demonstrated in Figure 8(a) and 8(b), respectively. The flow separation can also observe at Reynolds number of 75000 as shown in Figure 9(b). Yet, the angle of the flow separation is 19° and the leading-edge separation is also seen at 19°.

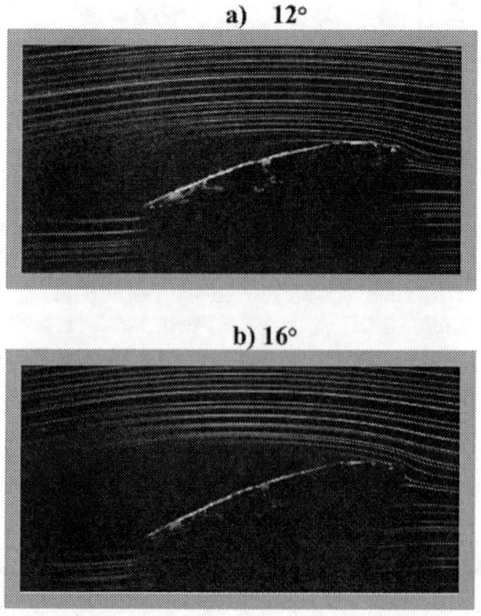

Figure 8. Flow visualization with smoke-wire experiments at Reynolds number of 50000.

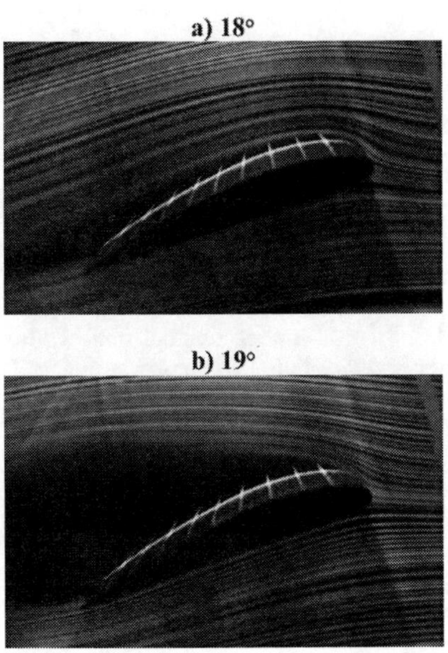

Figure 9. Flow visualization with smoke-wire experiments at Reynolds number of 75000.

2.4 Hot wire anemometry system

The hot wire system used in this experiment is a single sensor miniature wire probe. The miniature wire probe has a 5μm diameter, 1.25 mm long platinum-plated tungsten wire sensor. The system has the ability to measure one dimensional flows of low turbulence intensity. In velocity measurement experiments, the sample rate was 2 kHz and 20000 data was obtained for each experiment. DANTEC DYNAMICS mini-CTA software is used for the velocity measurement from 1C, which means 1 chord length of airfoil, forward of the trailing-edge of airfoil.

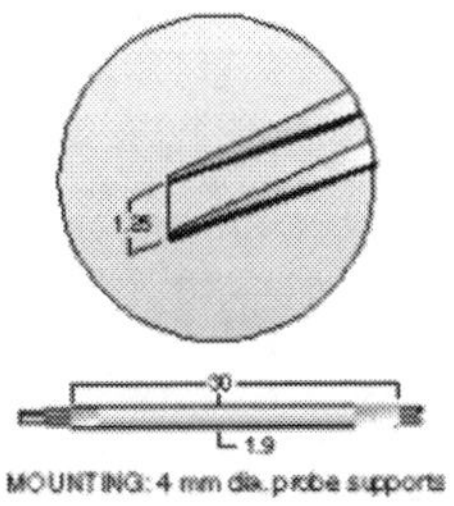

Figure 10. Hot-wire probe [18]

The hot wire probe needs to be calibrated before being used in the experiment. There is a relationship between the flow velocity and the measured voltages recorded the hot-wire anemometry system for hot-wire process. The traverse running both vertically and horizontally is conducted to move the hot-wire probe.

2.4.2 Hot wire anemometry data acquisition and analysis

Once the velocity calibration is completed, the hot wire probe is positioned 1C forward of trailing-edge of airfoil at the wake region of the airfoil. 2 cm above of the airfoil is chosen as starting point for measurement and the hot wire runs along the y-axis at the wake region of the airfoil. The hot wire continues to measure until 2 cm bottom of the airfoil. After the hot wire experiments are completed, the value of U/U_{∞} at the wake region is obtained at Reynolds number of 25000, 50000 and 75000 as shown in Figure 10, respectively. At Reynolds number of 25000, the velocity at the wake region decreased from 1.034 to 0.6 at 16°. The velocity similarly decreased from 1.074 to 0.37 at 20°. The similar decreasing is observed at Reynolds number of 50000 and 75000. After the flow separates from the surface of airfoil, it meets with the flow coming from the bottom of the airfoil at the wake region. For the separated flow over the airfoil is more dominant than the flow at the bottom of airfoil, velocity decreasing happens at the wake region. The flow separation moves to leading-edge of the airfoil when the angle of attack increases. The separated flow is more powerful than the separated flow at the former angle of attack. Thus, the velocity decreasing at the wake region increase dramatically when angle of attack increases as illustrated at figure 11.

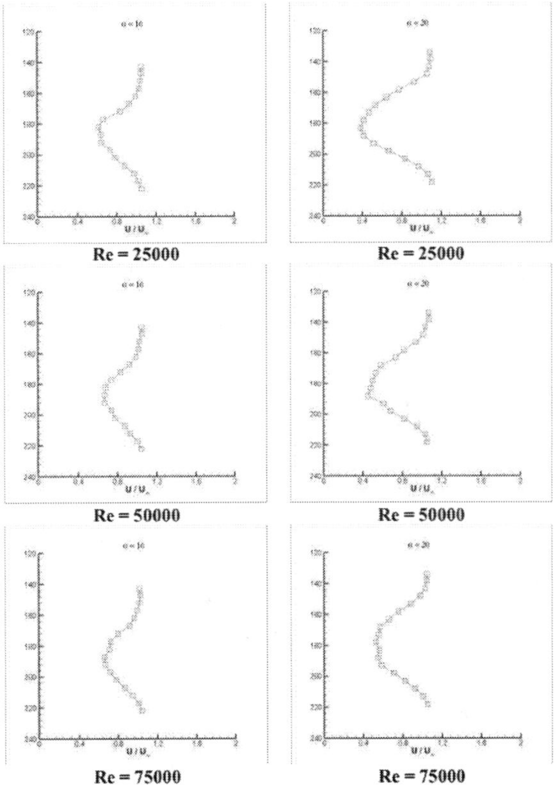

Figure 11. The value of U / U_{∞} at the wake region of NACA4412 airfoil

3 Conclusion

The experiments were conducted for the particular angle of attack in a low speed wind tunnel. The smoke-wire, the aerodynamic force coefficient and hot-wire experiments were carried out on an NACA4412 airfoil at Reynolds number of 25000, 50000 and 75000 where the leading-edge separation exists. The lift coefficient (C_L) increases gradually when the angle of attack increases at Reynolds number of 25000. Drag coefficient (C_D) also increases but there is no rapid climbing. The lift and drag coefficients also increase clearly at Reynolds number of 50000 and 75000 when the angle of attack increases. Unlike Reynolds number of 25000, there is a sudden climbing for the drag coefficient. The stall occurs at 16°, 19° at Reynolds number of 50000, 75000, respectively. An abrupt stall can be observed at Reynolds number of 50000 and 75000 whereas mild stall can be observed at Reynolds number of 25000. It is concluded that the stall angle increase when Reynolds number increase. Beside, the smoke-wire experiments at Reynolds number of 50000 and 75000 show a fine coherence with the aerodynamic force coefficient experiments. The velocity at the wake region of the airfoil decreased when angle of attack increased. It means that the flow separation headed to the leading-edge of the airfoil gradually when angle of attack increased.

Acknowledgments

The authors would like to acknowledge funding from the Scientific and Technological Research Council of Turkey (TÜBİTAK) under the project no: 213M329.

References

1. S.J. Schreck, N.N. Sørensen, M.C. Robinson, Aerodynamic structures and processes in rotationally augmented flow fields. *Wind Energy*, **10(2)**, 159-178 (2007)

2. M.S. Genc, U. Kaynak, H. Yapıcı, Performance of transition model for predicting low Re aerofoil flows without/with single and simultaneous blowing and suction. *Eur J Mech B-Fluid.* **30(2)**, 218-235 (2011)

3. M.S. Genc, I. Karasu, H.H. Acıkel, M.T. Akpolat, Low Reynolds number flows and transition, in: M. Serdar Genc (Ed.), Low Reynolds Number Aerodynamics and Transition, *Intech-Sciyo Publishing*, ISBN 979-953-307-627-9 (2012)

4. M.S. Genc, U. Kaynak, G.D. Lock, Flow over an Aerofoil without and with Leading Edge Slat at a Transitional Reynolds Number. *Proc IMechE, Part G: J Aerospace Eng.* **223(3)**, 217-231 (2009)

5. M.S. Genç, Unsteady aerodynamics and flow-induced vibrations of a low aspect ratio rectangular membrane wing with excess length. *Exp. Therm Fluid Sci.* **44**, 749-759 (2013)

6. P. Rojratsirikul, M.S. Genc, Z. Wang, I. Gursul, Flow-induced vibrations of low aspect ratio rectangular membrane wings. *J Fluid Struct.* **27**, 1296-1309 (2011)

7. M.S. Genc, I. Karasu, H.H. Acıkel, An experimental study on aerodynamics of NACA2415 aerofoil at low Re numbers. *Exp Therm Fluid Sci.* **39**, 252-264 (2012)

8. I. Karasu, M. S. Genç, H. H. Açikel, Numerical study on low Reynolds number flows over an Aerofoil. *J. Appl. Mech. Eng.* **2**, 131 (2013)

9. M.S. Genc, Numerical Simulation of Flow over a Thin Aerofoil at High Re Number using a Transition Model. *Proc IMechE, Part C-J Mech Eng Sci*, **224(10)**, 2155-2164 (2010)

10. White F.M. Fluid Mechanics, McGraw-Hill, Inc.4th edition (2004)

11. Schlicting H. Boundary Layer Theory. McGraw-Hill,Inc. /th edition (1979)

12. J. Ketz, A. Plotkin, Low-speed aerodynamics: from wing theory to panel methods. McGraw-Hill, Incorporated (1991)

13. B.H. Carmichael, Low Reynolds number airfoil survey. National Aeronautics and Space Administration, Langley Research Center (1981)

14. R. Wahidi, W. Lai, J.P. Hubner, A. Lang. Volumetric three-component velocimetry and PIV measurements of Laminar Separation Bubbles on a NACA4412 Airfoil. *16th Int. Symp of Applications of Laser Techniques of Fluid Mechanics*. Lisbon, Portugal. 09-12 July, (2012)

15. M. O'meara, T.J. Mueller, Laminar separation bubble characteristics on an airfoil at low Reynolds numbers. *AIAA journal*, **25(8)**, 1033-1041 (1987)

16. M. Agrawal, G. Saxena, Analysis of wings using airfoil NACA4412 at different angle of attack. *IJMER*. **3**, 1467-1469 (2013)

17. I. Karasu, Experimental and numerical investigations of transition to turbulence and laminar separation bubble over aerofoil at low Reynolds number flows. Graduate School of Natural and Applied Sciences, Erciyes University, Kayseri, Turkey (2011)

18. www.dantecdynamics.com

EPJ Web of Conferences **114**, 02030 (2016)
DOI: 10.1051/epjconf/201611402030

Unsteady flow over flexible wings at different low Reynolds numbers

Mustafa Serdar Genç[1], Mustafa Özden[1], Halil Hakan Açikel[1,a], Hacımurat Demir[1,2], Iliasbek Isabekov[1]

[1]*Wind Engineering and Aerodynamic Research Group, Department of Energy Systems Engineering, Erciyes University, Kayseri, Turkey*
[2]*Department of Mechanical Engineering, Aksaray University, Aksaray, Turkey*

Abstract. In this study, unsteady flow around flexible membrane wing which had aspect ratio of 1 (AR=1) was investigated experimentally at various Reynolds numbers (Re = 25000 and Re = 50000). Smoke-wire technique for flow visualization over the flexible membrane wing was utilized in the experiments. Digital Image Correlation system (DIC) was used for measuring deformation of AR = 1 flexible membrane wing. Instantaneous deformation measurements of membrane wing were combined with the flow field measurements. In low aspect ratio flexible membrane wings, unsteadiness includes tip vortices and vortex shedding, and the combination of tip vortices. In these types of wings, complex unsteady deformations occurred due to vortex shedding. The results showed that the increasing angle of attack results in increase of membrane deformation. Moreover, it was concluded that analysis of the instantaneous deformation revealed chordwise and spanwise, modes which were due to the shedding of leading-edge vortices as well as tip vortices. Consequently, vibrational mode decreased and maximum standard deviation location approached to the trailing edge by reason of increasing angle of attack.

1 Introduction

Recently, due to the advances in unmanned aerial vehicles (UAV), micro air vehicles (MAV) and wind turbines, aerodynamics researches concentrated on low Reynolds number aerodynamics, transition and laminar separation bubble (LSB) and its effects on aerodynamic performance [1-5]. In low Reynolds number aerodynamics, unsteadiness represents major problems. Furthermore, laminar separation bubble may cause negative effects, such as decreasing on lift, increasing on drag, reducing stability of the aircraft, noise and vibration in low Reynolds number aerodynamics. Membrane wings are used in many engineering applications such as microlight, yacht sails, parachutes, hang glider wings, paraglider and wings of small unmanned air vehicles go by the name of micro air vehicles (MAVs) [6]. Some studies show that flexible membrane wings may remarkably improve longitudinal static stability, delay stall and ensure a more favourable lift-to-drag ratio, and so enhance the overall aerodynamic performance when comparing with a rigid wing of similar geometry due to its aeroelastic effects and the adaptive inflation of the membrane skin.

Genç [7] investigated unsteady aerodynamics and flow-induced vibrations of a low aspect ratio rectangular membrane wing with excess length. The experiments for the membrane and rigid wing models are conducted in a closed-loop open-jet wind tunnel with a circular working section of 760 mm in diameter at angles of attack ranging from 0° to 25° at Reynolds number of 48,700. Digital Image Correlation system is used for time-accurate measurements of membrane deformation. Furthermore, measurement of velocity field with Digital Particle Image Velocimetry system was conducted whereas normal forces of the wing were measured by means of a load-cell system. Obtained experimental results demonstrated that the camber of membrane wing with excess length excited the separated shear layer. This situation leads to increase the lift coefficient relatively more as compared to membrane wings without excess length. The membrane caused to generate a larger magnitude of the maximum camber in the both 2D and 3D rectangular membrane wings with excess length. Dominant vibrational modes were obtained as a spanwise mode of two at the angles of attack of 6-10° and 19-22°. On the other hand, dominant spanwise modes were three at 11-18° and 23-25°.

Rojratsirikul et al. [8] investigated two dimensional flexible membrane airfoils experimentally with respect to the effects of membrane pre-strain and excess length on unsteady fluid structure interactions of membrane airfoils The largest camber and strain were observed for the airfoil with excess length. They noticed that the onset of the membrane airfoil vibrations was postponed to a higher angle of attack for the airfoils with excess length. Additionally, both the mode number and Strouhal number were in tendency to decrease with increasing angle of attack. Rojratsirikul et al. [9] performed an experimental study of a low aspect ratio rectangular wing for a Reynolds number range of $2.4x10^4$–$4.8x10^4$. Time

[a] Corresponding author: hhacikel@erciyes.edu.tr

accurate measurements of membrane deformation were added with the flow field measurements. Analysis of the fluctuating deformation detected chordwise and spanwise modes, which were due to the shedding of leading-edge vortices as well as tip vortices. It was seen that the second mode in the chordwise direction became dominant as the vortex shedding took place at higher angles of attack. The dominant frequencies of the membrane vibrations were similar to those of two-dimensional membrane airfoils. Additionally, it was seen that vortex shedding frequency from rigid wings showed significantly small effect of aspect ratio even when it was as low as unity.

Béguin and Breitsamter [10] studied on effects of membrane pre-stress on the aerodynamic characteristics of elasto-flexible morphing wing for applications of unmanned air vehicle. The results of force and deformation measurements conducted at several dynamic pressures explained that varying the pre-stress effectively permitted to influence the passive deformation of the wing surface and regulated the wing camber within a certain range at a given flow condition. In addition to this, it could be concluded that the characteristics of lift and drag were affected such that an improvement of the maximum lift-to-drag ratio between 6 % and 23 % of the lowest values at a given flow condition could be achieved.

Arbo´s-Torrent et al. [11] studied on leading- and trailing-edge effects on the aeromechanics of membrane aerofoils which was made of black latex rubber sheet. Two different cross-sectional geometries in two different sizes were used to in order to explore the effects of leading-edge (LE) and trailing-edge (TE) support designs on the performance of membrane aerofoils. They noticed that the mean camber as well as membrane vibrations changed with size and geometry of the LE and TE supports. Furthermore, the LE/TE supports with a rectangular cross-section continuously ensured higher lift forces and higher mean camber deformations comparing the support with circular cross-section. In addition to this, it was seen that membrane vibrations were higher for aerofoils with LE/TE supports with rectangular cross-section. LE/TE supports deviated under aerodynamic loading and thereby changed the performance of the aerofoil. Albertani et al. [12] investigated the aerodynamic coefficients and elastic deformations of low aspect ratio flexible wings. They conducted the experiments in a low-speed wind tunnel utilizing a visual image correlation system and a six-component strain gauge sting balance. It was seen that both lift and static longitudinal stability were increased and hence remarkable advantages in a flexible membrane wing performance were obtained.

Lian et al. [13] reviewed the aerodynamics of membrane and corresponding rigid wings under the micro air vehicle flight conditions. They noticed that the membrane wing both delayed stall as well as adapted to the unsteady flight environments, and that the proper orthogonal decomposition method was an economic tool in order to identify the flow structure around a wing and in order to facilitate flow and vehicle control. Aono et al. [14] investigated the effects of spanwise flexibility on flapping wing aerodynamics computationally. They concluded that the spanwise wing deformation and the increased effective angle of attack of the deformed wing increased thrust generation. On the other hand, it was noticed that when the flexibility was too high, the wing tip and root moved in inconsistent directions, which caused deteriorated aerodynamic performance. Kuo and Hsieh [15] studied on unsteady flow structure and vorticity convection over NACA0012 airfoil oscillating at high reduced frequency. They noticed that the leading edge separation vortices were combined one by one into a vortex near the trailing edge and then shed together downstream into the wake. The wavelength of unsteady vortex structure over the airfoil was found to be remarkably different from that at low reduced frequencies. Furthermore, they also noted that the effect of different pivoting axis was insignificant on the unsteady flow structure at such high reduced frequency oscillations. Ringuette et al. [16] investigated the role of the tip vortex in the force generation of low-aspect-ratio normal flat plates experimentally. They noticed that tip vortex resulted in a significant maximum in the plate force. Furthermore, decrease in aspect ratio caused to increasing of plate drag coefficient.

2 Experimental Set-up

The experiments were performed in a low-speed, suction-type wind tunnel (square working section of 500 mm x 500 mm) located at the Department of Energy Systems Engineering, University of Erciyes as shown in Figure 1. Turbulence intensity of the tunnel is 0.35% and the operating range of wind tunnel speed was 1-45 m / s. Flexible membrane wing and cross sectional area of the frames of the wing was shown in Figure 2. The wing had 125 mm chord length and 125 mm span length (AR = 1), was made from a black latex rubber sheet with a thickness of t = 0.2 mm, Young's modulus of E = 2.2 MPa, and density of ρ_m = 1 g / cm^3 was used. Experimental setup of DIC system for deformation measurement was given in Figure 3.

Figure 1. Photograph of the experimental set-up.

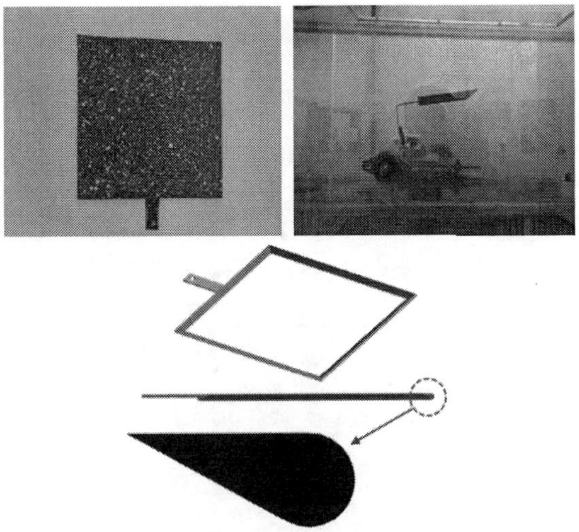

Figure 2. Schematic of the flexible membrane wing (AR = 1) with cross-section of the rigid frame in wind tunnel.

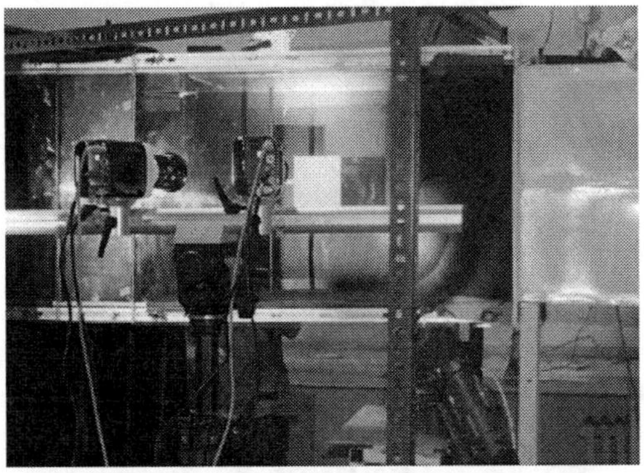

Figure 3. Photograph of the experimental set-up for DIC measurement.

Figure 4. Dark medium for smoke-wire measurement

2.1 Smoke-wire flow visualization and deformation measurement

2.1.1 Smoke-wire flow visualization

The smoke-wire visualization method was employed for flow visualization in order to mark the streamlines. The smoke experiments were conducted on by a heated wire and oil supply system which sent oil from top of the wire and burning oil moved with the free stream. Dark medium for smoke-wire measurement is seen in Figure 4.

2.1.2 Deformation measurement

Digital Image Correlation (DIC) system was used to measure the membrane displacements. Full-field measurements were obtained by means of using the DANTEC Dynamics 3D High-Speed Image Correlation System (Q-450: the number of pixels was 1280×800). The frame rate was 1 kHz. DANTEC Al-11-BMB_9x9 calibration target was utilized. During the deformation measurement, 1000 frame was captured each experiment.

3 Comparisons between Experimental and Numerical Results

3.1 Results of smoke-wire visualization

Smoke-wire visualization results for Re = 25000 and Re = 50000 are shown in Figure 5 and Figure 6. Firstly, smoke-wire experiment was carried out at constant Re = 25000 with angles of attack of $\alpha = 12^o$, 16°, 20° and 24°, respectively. As seen from the Figure 5, the formations of tip vortices were increased with increase in angle of attack from $\alpha = 12^o$ to $\alpha = 24^o$.

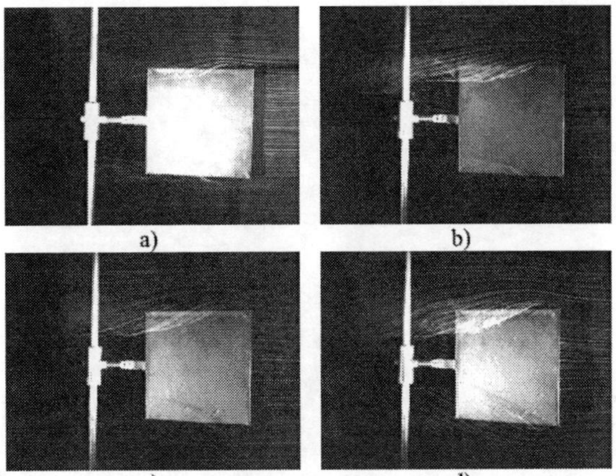

a) b) c) d)

Figure 5. Smoke-wire visualization results at Re = 25000 a) $\alpha = 12^o$, b) $\alpha = 16^o$, c) $\alpha = 20^o$ and d) $\alpha = 24^o$

Additionally, tip vortices grew up from leading edge to the trailing edge of membrane wing. Then, smoke-wire experiment was carried out at Re = 50000 for the same angles of attack. Results of experiments were given in Figure 6. It is seen from the Figure 6 that increasing angle of attack results in increase tip vortex. Due to this, tip vortices affected the unsteady aerodynamics of the membrane wing.

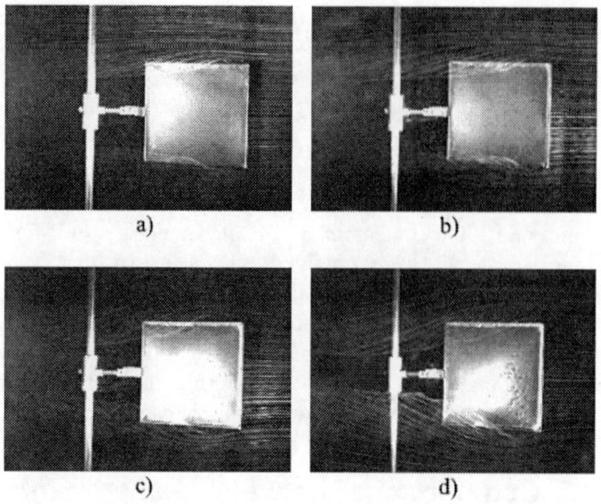

Figure 6. Smoke-wire visualization results at Re = 50000 a) α = 12°, b) α = 16°, c) α = 20° and d) α = 24°

3.1 DIC results

Due to the unsteady flow over the low aspect ratio (AR = 1) flexible membrane wing, deformations were observed over the related wing. Unsteady condition of membrane wing was investigated for Re = 25000 and Re = 50000 with different time intervals. Instantaneous displacement results for the flexible membrane wing at Re=25000 and Re = 50000 with various angles of attack were given from Figure 7 to Figure 12.

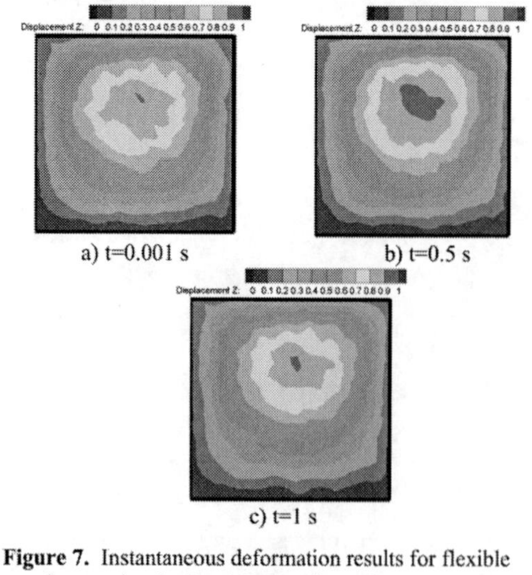

Figure 7. Instantaneous deformation results for flexible membrane wing for Re=25000 and α=16°

Deformation results for Re = 25000 with angles of attack α = 16°, α = 20° and α = 24° were demonstrated in Figure 7 to Figure 9. It was understood from obtained results that increase angle of attack caused to increase in maximum instantaneous deformation. At same Re number and angle of attack condition (i.e. Figure 7a), location of the maximum instantaneous deformation contour varied due to unsteady flow condition.

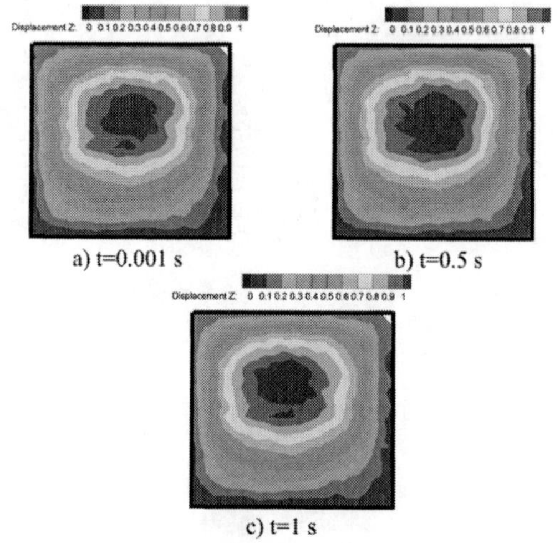

Figure 8. Instantaneous deformation results for flexible membrane wing for Re=25000 and α=20°

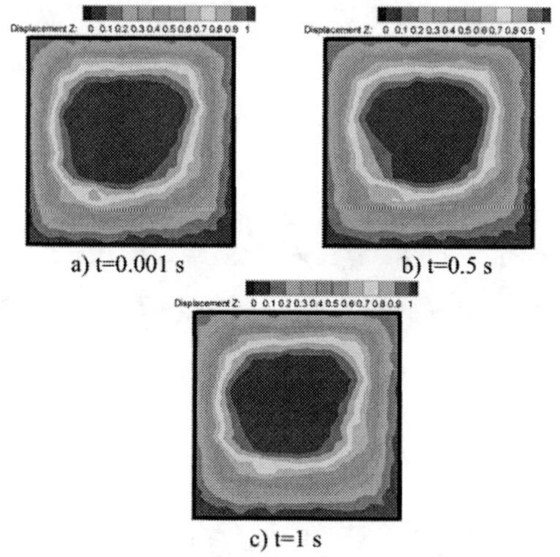

Figure 9. Instantaneous deformation results for flexible membrane wing for Re=25000 and α=24°

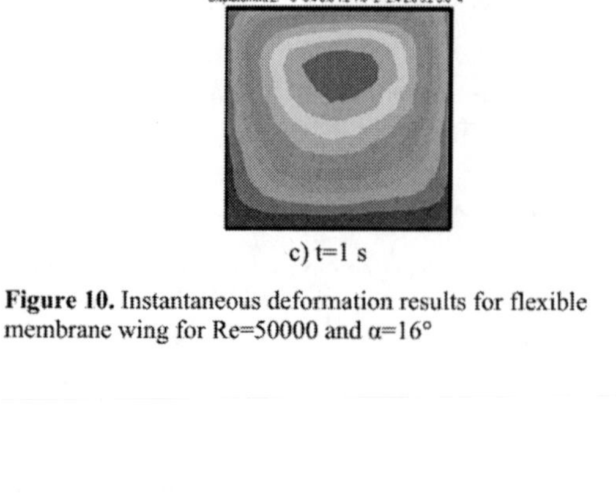

Figure 10. Instantaneous deformation results for flexible membrane wing for Re=50000 and α=16°

As seen from Figure 10 to Figure 12, deformation results for Re = 50000 with same angles of attack. The instantaneous displacements increase with increasing angles of attack as seen from Figure 10 to Figure 12. Shape of instantaneous deformation contour changed with formation of tip vortices. Deformations decreased towards to the trailing edge of membrane wing due to the effects of tip vortices.

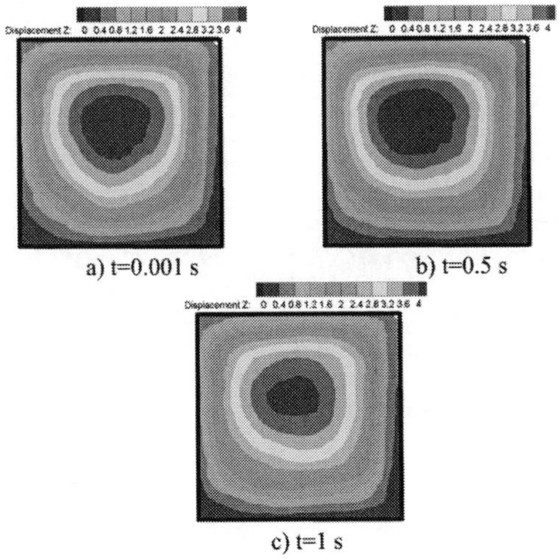

Figure 11. Instantaneous deformation results for flexible membrane wing for Re=50000 and α=20°

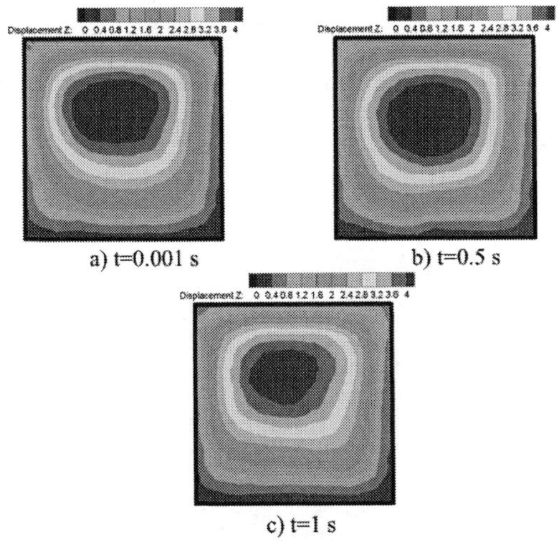

Figure 12. Instantaneous deformation results for flexible membrane wing for Re=50000 and α=24°

Figure 13 demonstrates the results of standard deviation of membrane wing in the direction normal to the chord-line. At angle of attack of 2°, mode of five was seen chordwise direction whereas second mode vibration was seen in spanwise especially at trailing edge of wing due to the effect of tip vortices. An increase in angle of attack caused to decrease in vibrational mode and the region of maximum standard deviation approached to the trailing edge.

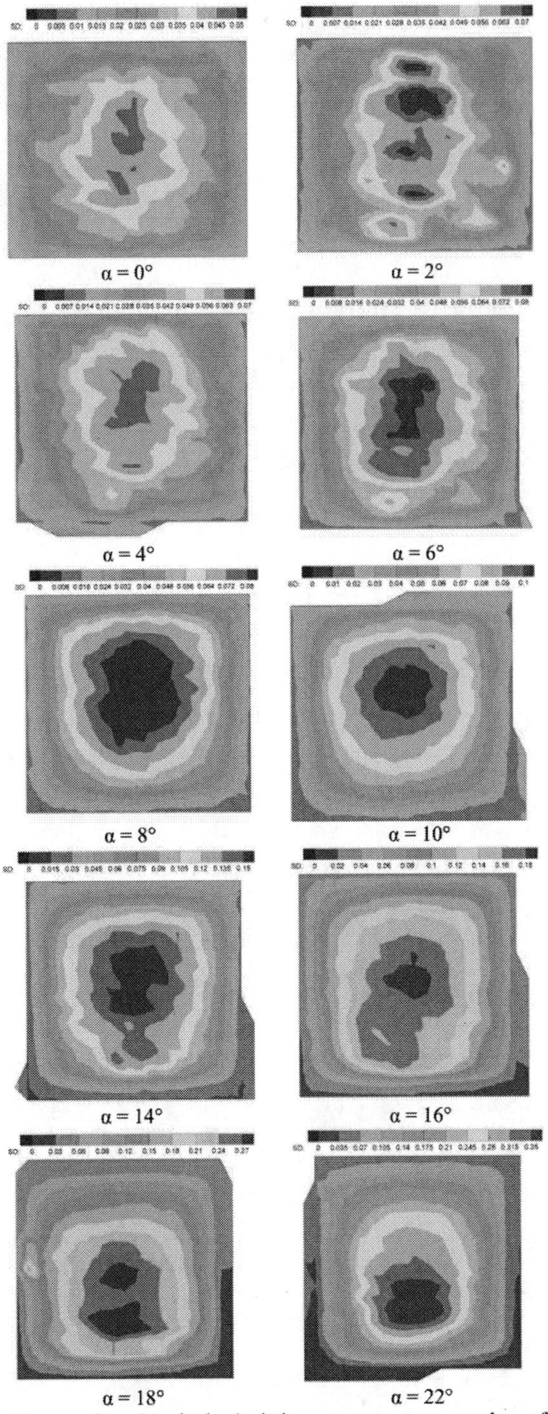

Figure 13. Standard deviation measurement results of membrane wing for AR=1 and Re=50000.

4 Conclusion

Unsteady flow around flexible membrane wings at low Reynolds numbers of 25000 and 50000 was investigated experimentally. Smoke-wire technique for flow visualization was utilized. Instantaneous deformation measurements of membrane wing which were performed by means of DIC system were combined with the flow field measurements. It was deduced that increasing angle of attack caused to increase in membrane deformation. Analysis of the instantaneous deformation revealed chordwise and spanwise, which were because of shedding of leading-edge vortices as well as tip vortices. Consequently, vibrational mode decreased and maximum standard deviation location approached to the trailing edge due to increasing angle of attack.

Acknowledgments

The authors would like to acknowledge funding from the Scientific and Technological Research Council of Turkey (TÜBİTAK) under the project no: 213M329, the Scientific Research Projects Unit of Erciyes University under the contract no: FDK-2015-6171, FBG-2014-5337 and FDA-2014-5273.

References

1. M.S. Genc, U. Kaynak, G.D. Lock, Flow over an Aerofoil without and with Leading Edge Slat at a Transitional Reynolds Number. *Proc IMechE, Part G: J Aerospace Eng.* **223(3)**, 217-231 (2009)
2. M.S. Genc, I. Karasu, H.H. Acıkel, M.T. Akpolat, Low Reynolds number flows and transition, in: M. Serdar Genc (Ed.), Low Reynolds Number Aerodynamics and Transition, *Intech-Sciyo Publishing*, ISBN 979-953-307-627-9 (2012)
3. M.S. Genc, U. Kaynak, H. Yapıcı, Performance of transition model for predicting low Re aerofoil flows without/with single and simultaneous blowing and suction. *Eur J Mech B-Fluid.* **30(2)**, 218-235 (2011)
4. M.S. Genc, I. Karasu, H.H. Acıkel, An experimental study on aerodynamics of NACA2415 aerofoil at low Re numbers. *Exp Therm Fluid Sci.* **39**, 252-264 (2012)
5. I. Karasu, M. S. Genç, H. H. Açikel, Numerical study on low Reynolds number flows over an Aerofoil. *J. Appl. Mech. Eng.* **2**, 131 (2013)
6. W. Shyy, M. Berg, D. Ljungqvist, Flapping and flexible wings for biological and micro air vehicles. *Prog Aerosp Sci.* **35**, 455-505 (1999)
7. M.S. Genç, Unsteady aerodynamics and flow-induced vibrations of a low aspect ratio rectangular membrane wing with excess length. *Exp. Therm Fluid Sci.* **44**, 749-759 (2013)
8. P. Rojratsirikul, Z. Wang, I. Gursul, Effect of pre-strain and excess length on unsteady fluid–structure interactions of membrane airfoils. *J Fluid Struct.* **26(3)**, 359-376 (2010)
9. P. Rojratsirikul, M.S. Genc, Z. Wang, I. Gursul, Flow-induced vibrations of low aspect ratio rectangular membrane wings. *J Fluid Struct.* **27**, 1296-1309 (2011)
10. B. Béguin, C. Breitsamter, Effects of membrane pre-stress on the aerodynamic characteristics of an elasto-flexible morphing wing. *Aerosp. Sci. Technol.* **37**, 138-150 (2014)
11. S. Arbós-Torrent, B. Ganapathisubramani, R. Palacios, Leading-and trailing-edge effects on the aeromechanics of membrane aerofoils. *J Fluid Struct.* **38**, 107-126 (2013)
12. R. Albertani, B. Stanford, J.P. Hubner, P.G. Ifju, Aerodynamic coefficients and deformation measurements on flexible micro air vehicle wings. *Exp. Mech.* **47(5)**, 625-635 (2007)
13. Y. Lian, W. Shyy, D. Viieru, B. Zhang, Membrane wing aerodynamics for micro air vehicles. *Prog. Aerosp. Sci.* **39(6)**, 425-465 (2003)
14. H. Aono, S.K. Chimakurthi, C.E. Cesnik, H. Liu, W. Shyy, Computational modeling of spanwise flexibility effects on flapping wing aerodynamics. *In 47th AIAA aerospace sciences meeting including the new horizons forum and aerospace exposition*, 18 (2009)
15. C.H. Kuo, J.K. Hsieh, Unsteady flow structure and vorticity convection over the airfoil oscillating at high reduced frequency. *Exp. Therm Fluid Sci.* **24(3)**, 117-129 (2001)
16. M.J. Ringuette, M. Milano, M. Gharib, Role of the tip vortex in the force generation of low-aspect-ratio normal flat plates. *J. Fluid Mech.* **581**, 453-468 (2007)

EPJ Web of Conferences **114**, 02031 (2016)
DOI: 10.1051/epjconf/201611402031

Experimental study of submillimeter droplets dynamics and breakup in continuous supersonic flow terminated by shock wave

Oleg Gobyzov[1,2], Yuriy Lozhkin[1,2], Mikhail Ryabov[1,2] and Dmitriy Markovich [1,2]

[1]*Kutateladze Institute of Thermophysics, Siberian Branch of Russian Academy of Sciences, 630090, Novosibirsk, Russia*
[2]*Novosibirsk State University, 630090, Novosibirsk, Russia*

Abstract. The present paper reports an application of optical methods, namely PIV, background-oriented-schlieren (BOS) and high-magnification imaging with background illumination to study of dynamics and breakup of 10-100 μm size droplets in continuous supersonic flow terminated by a normal shock wave. Flow diagnostics was performed by means of BOS and PIV. Shadow photography allowed to specify velocity ranges for different droplet sizes and to visualize droplets dynamics and breakup modes. Features of the experimental setup and certain details of implemented measurement system are considered. Results of velocity measurements and droplets behavior, including deformation and breakup, are presented and analysis of experimental conditions and dimensionless parameters affecting the droplets behavior is performed. Distinctive features of deformation and breakup processes of submillimeter scale droplets are revealed.

1 Introduction

Investigation of liquid drops deformation and breakup in high-speed flows has many scientific and technological applications: combustion of liquid fuels, erosion of steam turbine blades, cold spray processing and others. Through the last several decades a large number of experimental works, aimed on identification of droplets atomization regimes, systematization of amassed information and revealing of influencing parameters were performed. In this investigations it was shown that atomization of liquid droplets due to aerodynamic forces is a complex process that involves different physical mechanisms depending on the experimental conditions. Generally speaking it is considered that dynamics and breakup of liquid drops are governed by such dimensionless parameters as Weber number (*We*), Ohnesorge number (*Oh*) and Eötvös or Bond number (*Bo*), certain effect may also occur due to variation of Reynolds number (*Re*) [1]. Usually magnitudes of transitional Weber numbers considered to be quite stable for $Oh<0.1$, while for higher *Oh* increase in transitional *We* is observed [2]. Nevertheless, critical values of these parameters differ significantly from one paper to another.

Most part of the published works deal with the drops of size ≈ 1 mm (see e.g. [3, 4]). One of the main reasons for this is that study of smaller droplets is more challenging task, at least from the technical point of view. Besides that, one can usually choose a combination of dimensionless parameters to model the required conditions and to preserve separate parameters in the suitable range. At the same time there is still a place for presupposition that for small-scale droplets processes of deformation and breakup may have significant differences from those for millimetre-scale drops. First of all, such presuppositions are based on the fact that *We* and *Re* numbers cannot be preserved together with the decrease of the droplets size without variation of gas or liquid characteristics. More "intuitive" basis is that for the droplets of tens of micrometers size surface tension and capillary effects should play more important role. However, for the droplets less than 100 μm in diameter there is limited amount of experimental data on dynamics and breakup of droplets in gas flows and retrieving such data requires application of advanced experimental techniques.

In the present work several optical measurement methods, among which are particle image velocimetry (PIV), background-oriented schlieren (BOS) and high-magnification imaging with background illumination (shadow photography, SP) were applied to study dynamics and breakup of submillimeter droplets in continuous supersonic flow with normal shock wave. Results of flow and droplets velocity measurements and visualization of particle behaviour, including deformation and breakup, are presented.

2 Setup and measurement techniques

Continuous accelerating supersonic air flow was formed using a converging-diverging nozzle with rectangular cross section and throat size of 8×12 mm. Flat side walls of the nozzle were made of transparent acrylic glass to provide an optical access. Air was supplied from

[a] Corresponding author: oleg.a.g.post@gmail.com

the regenerative blower through the pressure regulator (fig. 1) at the rate of ≈ 35 l/s. Air flow conditions, namely pressure and temperature, were measured in the plenum chamber. The nozzle was operated in underexpanded regime to produce a normal shock wave (SW) in the diverging section of the nozzle.

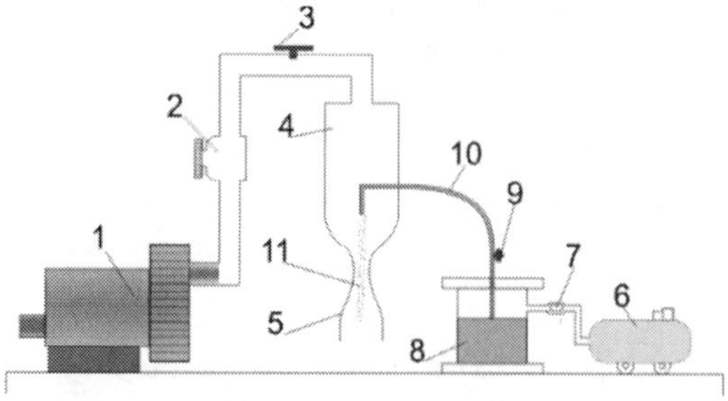

Figure 1. Scheme of the setup. 1 – air blower, 2 – pressure regulator, 3 – valve, 4 – plenum chamber, 5 – nozzle, 6 – compressor, 7 – electronic pressure regulator, 8 – water tank, 9 – valve, 10 – water supply tube, 11 - droplets.

Droplets were introduced into the flow by atomization of the coaxial liquid jet issued from the tube with diameter of 0.8 mm. Exit of the tube was fixed in the plenum chamber upstream of the converging part of the nozzle. Visualization of the jet atomization was performed to determine an atomization regime and size of the resulting droplets. So called fiber atomization regime [5] was observed with formation of large clusters at the distance of 3-5 mm from the tube exit. At the distance of 10-12 mm downstream of the tube exit an atomization was finished in general, although small amount of liquid ligaments still persisted in the flow. The size of resulting drops mainly varied in the range from 10 μm to 100 μm.

To fetch an image of a submillimeter droplet in a supersonic flow a high spatial resolution and short exposure of the recording system are required. Usually cameras, except for the ICCD, do not provide enough shutter speed and sensitivity to perform such visualization and thus high-energy and short duration (order of nanoseconds) light pulses are required for background lighting. An Nd:YAG pulse laser fits these requirements but due to coherent nature of emitted light it introduces small-scale speckles that deteriorate the image quality. In the current work a fluorescent background screen was used to provide an even background lighting without speckles. Fluorescent layer was excited by the 532 nm wavelength pulsed light beam generated by dual-head Nd:YAG laser. The size of the area excited by the laser light was adjusted using the diverging lens placed in an optical path of the laser beam.

To provide a required optical magnification an Infinity K2/SC long distance microscope was used as a camera lens. An optical magnification of ≈ 4.3:1, that is ≈ 1.6 μm/pix, was reached at the distance of ≈ 200 mm from the measurement area. Owing to 5 ns laser pulse duration and short Rhodamine B fluorescent paint lifetime (≈ 2-3 ns) droplet displacement during the exposure time didn't exceed 2 μm and practically «frozen» images of droplets were captured. Using dual-head laser and CCD-camera in double-frame exposure mode pairs of images with different interframe delay were captured in experiments. A certain disadvantage of this method of visualization was that the depth-of-field provided by the microscope was of the same order as the droplet size and therefore large number of fully or partially out-of-focus droplets were present in the images.

To measure gas and droplets velocity PIV technique was employed. The drawback of the technique for the case was that images of droplets were much brighter than the images of PIV tracers and therefore contributed most to the correlation function. No image filtering procedures were effective enough to eliminate droplets contribution to the correlation function, so droplets and gas velocity were measured in separate experiments. Water-glycerol tracer particles of the size ≈ 1 μm produced by the Laskin-nozzle type generator were introduced into the air supply line immediately after the pressure regulator. From the instant velocity field it was evaluated that tracer relaxation length at the shock wave front is about 2 mm, and thus velocity data immediately after the shock wave front was overestimated. For the measurements of droplets velocity no droplet size separation was made and only average liquid phase velocity was measured. Obviously, drops acceleration depends on its size and shape, so manual evaluation of velocity for droplets of different size was performed. At the distance of 3 mm upstream of the SW front, where velocity dispersion for droplets of different size should be close to maximum, average velocity of 40-μm droplets was ≈ 285 m/s, while for 15- μm droplets it amounted to ≈320 m/s. Gas and droplets velocity distribution is presented in fig. 3.

To correlate the droplets dynamics with a configuration of the flow a background oriented schlieren (BOS) technique was applied simultaneously with the shadow photography to determine the position of the shock wave. To provide simultaneous application of both techniques two cameras were used. First camera was installed normally to the measurement plane and was used for BOS imaging. A semi-transparent mirror was placed between the camera lens and the measurement area at the angle of 45° front (see fig 2). A second camera with microscope was installed alongside so that it observed the measurement area in the reflected light. A random black dot pattern was put on the fluorescent screen to use it as a background image for BOS. Because of the large difference in depth of field of the microscope and the lens of BOS camera fluorescent screen with black dot pattern was observed as practically even light-emitting surface by the microscope camera, while random dot pattern was clearly observed by the BOS camera.

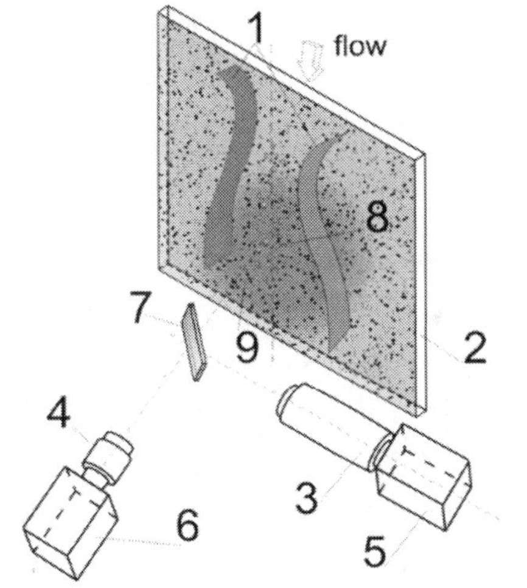

Figure 2. 1 – nozzle walls, 2 - fluorescent screen, 3 – LD microscope, 4 – camera lens, 5,6 – cameras, 7- half-transparent mirror, 8 – droplets, 9 – area of the screen excited by laser emission

In a case of shadow photography application for visualization of small-scale objects with short exposure time a proper distance between the fluorescent screen and measurement plane should be chosen: shorter distance yields brighter image but may lead to blurred edges in it, while longer distance yields sharper image but reduces an amount of light collected by the camera lens. Difference in image quality can be seen from fig. 3. In the experiments the distance between the fluorescent screen and measurement plane was ≈ 170 mm.

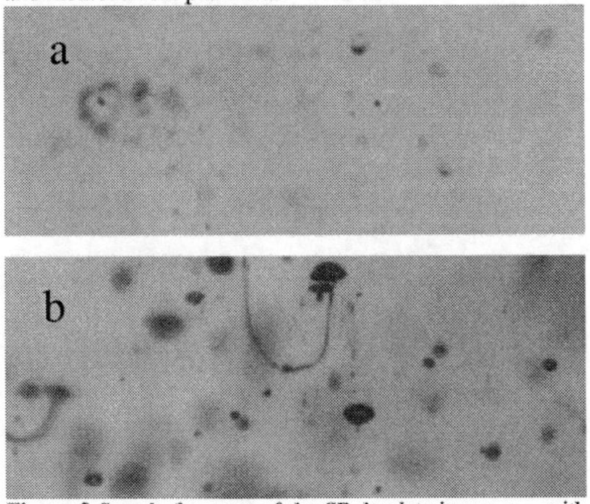

Figure 3 Sample fragment of the SP droplets image. a – with the screen at the distance of 25 mm from the measurement area. b – at the distance of 170 mm

3 Flow diagnostics

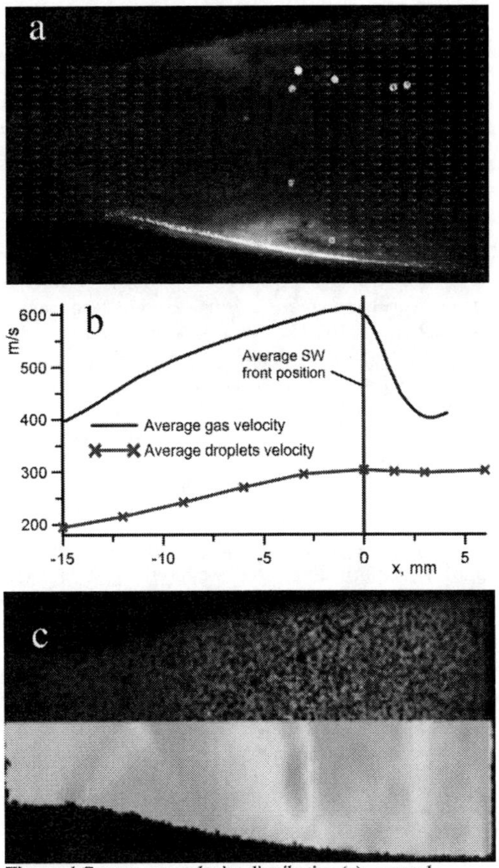

Figure 4. Gas average velocity distribution (a), gas and droplets velocity profile along axis of symmetry (b), background screen with random pattern and instantaneous result of BOS visualization (c)

PIV measurements have shown a smooth acceleration of the air flow with step deceleration (shock wave) inside the diverging part of the nozzle at the distance of approximately 20 mm from the nozzle throat (fig. 4a, 4b). Although the structure of the flow before the shock wave front remained the same during the experiment, oscillations of the shock wave front position and variations of the velocity were observed. BOS visualization (see fig. 4c) revealed that the shock wave front fluctuated within the range of ± 1.5 mm from its average position. Maximum velocity immediately before the step deceleration varied approximately from 570 to 620 m/s. Such instability is most likely due to non-stationary flow conditions downstream of the shockwave front: a secondary acceleration with a signs of flow separation and re-attachment near the wall was observed.

Droplets also exhibited smooth acceleration along the whole measurement area, but significantly lagged behind the gas flow. Eventually, droplets velocity before the SW front was close to the gas velocity after the step deceleration, and thus impact of the aerodynamic forces on droplets decreased abruptly immediately after the SW front.

3 Droplets deformation and breakup

Images of droplets were captured in the vicinity of the average shock wave front position and at the distances up to 15 mm upstream and downstream of it with a step of 3 mm. 300 pairs of images were obtained for each measurement area. Pairs of images were taken with interframe delay of 800 ns, 1500 ns and 2800 ns – 100 of image pairs for each delay value.

Due to the presence of out-of-focus particles and particles in a state of breakup it was virtually impossible to automate droplets images identification and sizing so image processing was performed manually. Due to this rather small number of droplets for each area of visualization (100-150 droplets) was accounted in statistics. Only drops retaining spherical shape or shape close to elliptical were taken into account. Eventually, droplets size and shape were evaluated from the images

Average droplets size, i.e. an average diameter of a sphere of equal volume

$$\langle d \rangle = \left\langle \sqrt[3]{a_x a_y^2} \right\rangle \quad (1)$$

where a_x is a streamwise ellipse dimension and a_y – transversal ellipse dimension, was ≈ 20-26 μm in a whole measurement area. Droplets exhibited streamwise contraction before the shock wave front and extension immediately after the shockwave front, yet, generally droplets extension and contraction rate for mentioned sizes did not exceed 30%, which is insufficient for transition to breakup.

From qualitative analysis of the captured images a few preliminary conclusions could be made: despite the fact that most intense impact of aerodynamic forces on droplets obviously should be immediately before the SW front, most part of the droplets in a state of disruption was observed at the distance of 15 – 9 mm upstream of the average SW front position. Closer to the SW front elliptical small droplets and "fibers" were mainly observed.

To interpret the obtained results range of dimensionless parameters affecting the droplets dynamics and breakup was estimated for several droplet sizes (see table. 1). Maximum droplets velocity lag $\Delta v_{max} \approx 300$ m/s was observed immediately before the shock wave front, so this value was substituted for estimation. A characteristic time of breakup induction,

$$t^* = \frac{d}{\Delta v}\left(\frac{\rho_d}{\rho_g}\right)^{\frac{1}{2}} \quad (2)$$

where Δv – velocity difference between droplet and gas, ρ_d – droplets density and ρ_g – density of gas, is also specified for different sizes of droplets in table 1.

Ohnesorge number for droplet sizes d>1 μm is below critical value of 0.1, above which Weber numbers demarking breakup mode boundaries tend to grow with increase of *Oh*. For droplets of 15 μm in diameter *We* was definitely below the critical value, while for droplets larger than 20 μm in diameter it was in the range, which is usually referred as a range of bag-breakup regime [3, 6]. However, it is considered that gradual acceleration leads to increase in critical *We* numbers demarking breakup modes [7], and thus We≈14 for 30-μm droplets should be regarded as «transitional», yet droplets with diameter d≈40÷50 μm, according to [8] should exhibit breakup.

Table 1. Typical values of characteristic parameters

droplet size [μm]	15	30	50
Weber number (We)	10,6	21.2	35.4
Ohnesorge number (Oh)	$2.8\cdot10^{-2}$	$2\cdot10^{-2}$	$15.6\cdot10^{-2}$
Laplace number(La)	$1.2\cdot10^{3}$	$2.5\cdot10^{3}$	$4.1\cdot10^{3}$
Reynolds Number (Re)	130	260	435
t*[s]	$2.1\cdot10^{-6}$	$4.3\cdot10^{-6}$	$7.1\cdot10^{-6}$

Contrary to these arguments, experiments have shown that droplets in a range of sizes from 10 μm to approximately 35 μm stably persisted in the flow upstream and downstream of the shock wave front. Such droplets exhibited shape oscillations, i.e. contraction in the continuous flow and elongation in the vicinity of the SW front, but did not display any clear tendency for breakup.

For larger droplets well-known breakup regimes – vibrational and bag breakup modes ([6, 7]) were detected in the area of flow acceleration (see fig. 5). Immediately after the shock wave front only final stages of such breakup regimes were observed. As the method of visualization allowed to fetch only pairs of subsequent "snapshots" of the breakup process but not the whole process in dynamics, interpretation of images with droplets in a state of disruption, especially evaluation of their initial size, was significantly hindered. Further to mentioned "classic" regimes in some rare cases distinct dual bag breakup, described by Cao et. al. [8] also occurred (see fig. 6).

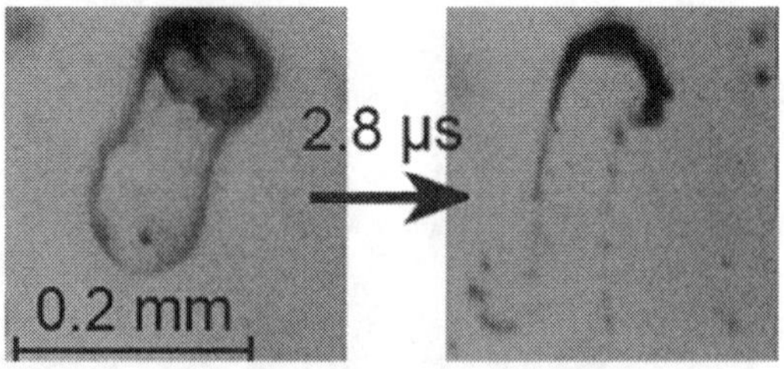

Figure 5 Example of developed bag breakup mode. Downward flow direction.

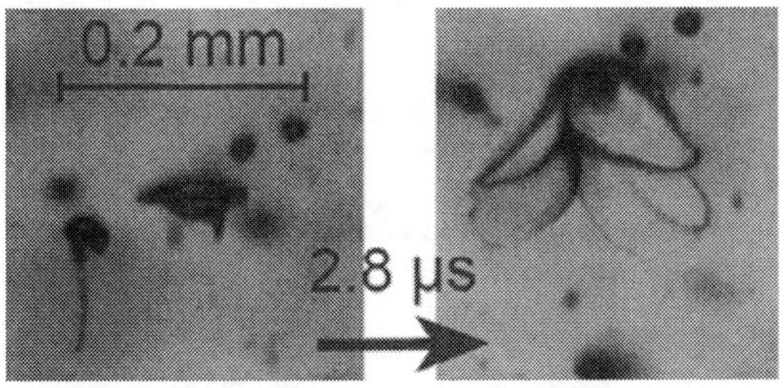

Figure 6 Example of dual-bag breakup regime. Downward flow direction.

Worth noting is the fact that final stages of the breakup observed in the experiments differ from those described in papers by Wierzba [7], Guildenbecher [6] and others. Although small number of large, fully developed "bags" that exhibited explosive type of disruption with formation of large number of small drops were detected (as in fig 5. and fig. 7), much more often after the rupture of the "bag" film pieces attached to the rim, formed elongated fibers that persist in a flow for a long time (see fig. 8).

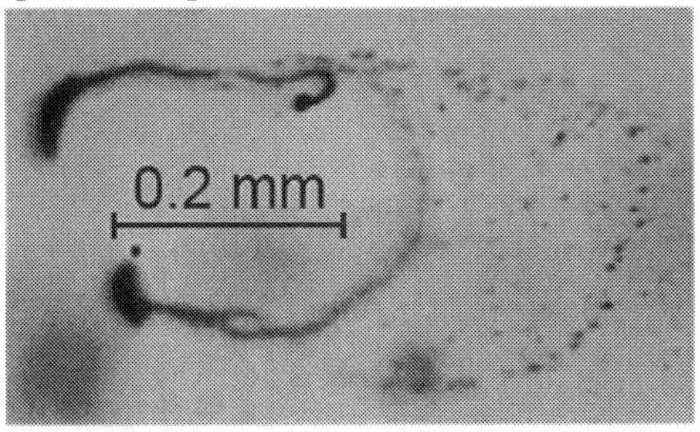

Figure 7 Example of bag breakup with formation of small droplets. Flow direction from left to right.

Another observation regarding the bag formation and breakup is that interruption of "bag" development was repeatedly observed in experiments. In the case of such interruption an undeveloped, usually asymmetric small bag breaks with formation of fiber-type structure An example of such process is shown in fig. 9. Generally speaking it is quite hard, if possible at all, to formally distinguish fully developed bag disruption from the interrupted one, but in the limiting cases difference is obvious and captured images give an impression that only very large bags disrupt in explosive manner. Presumably surface tension together with some local flow disturbances (turbulent pulsations) interrupts bag growth, yet there is no any strong evidence to support such explanation.

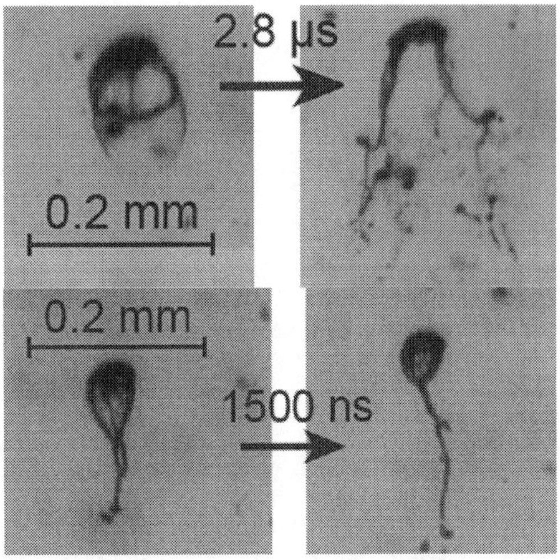

Figure 8. Example of bag rupture with formation of several fibers (up) and merging of fibers (down). Downward flow direction.

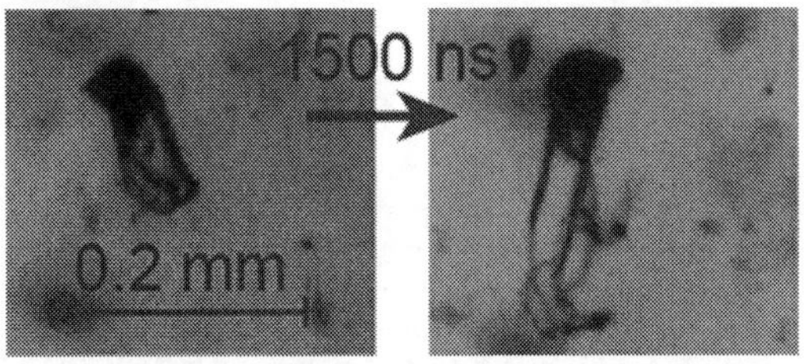

Figure 9. Example of vibrational breakup with formation of long connecting fiber. Downward flow direction.

Similar process occur for the vibrational breakup: droplet transforms into two smaller drops connected by a long fiber and then one of the smaller droplets detach from the formation, while another still persist in a flow with an attached fiber. In some cases such fibers can extend to the length close to 1 mm and, apparently, contain the considerable part of the droplet initial mass. Example of such breakup is shown in fig. 10.

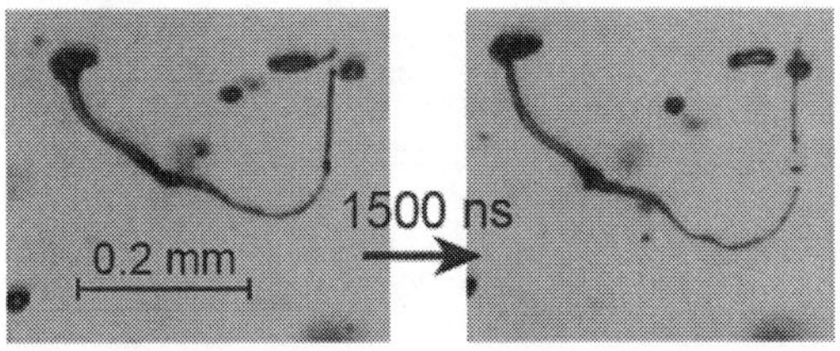

Figure 10. Example of vibrational breakup with formation of long connecting fiber. Downward flow direction.

Apart from modes mentioned above a less common breakup regime was also observed in experiments, example of such breakup mode is presented in figure 11. This mode features sheet stripping and bag formation at the same time. This regime, to the best of author's knowledge, doesn't resemble any of the "classical" ones. Most close visual similarity can be found in the recent numerical investigation by Kekesi et. al [9].

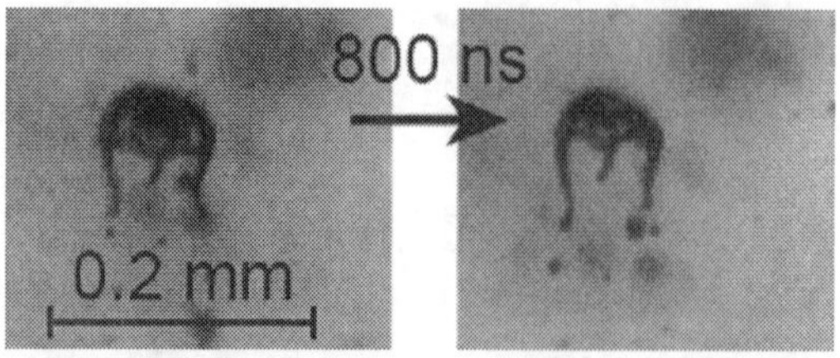

Figure 11 Example of transitional breakup mode. Downward flow direction.

As for the characteristic time of breakup, according to [1] it amounts to ≈ 5·t* for the bag-breakup mode, and ≈ 2·t* for vibrational breakup [1]. For such small droplets as were observed in the experiments t* is amounted to microseconds, and the displacement of the droplet during this time is of order of 1 mm. This means that all the breakup processes induced by aerodynamic forces should be finished upstream the shock wave front. Indeed, in the images taken at the distance of 6, 9 and 12 mm downstream of SW front droplets, on the average, retained spherical shape. Besides, breakup visualization in dual-frame regime confirms that this characteristic time estimation remains valid, at least in the order of magnitude, for submillimetre-scale droplets.

4 Conclusions

A behaviour of submillimetre-scale water droplets in a continuous transonic – supersonic flow terminated by normal shock wave was investigated using set of optical techniques, namely: PIV, PTV, BOS and shadow photography.

In general, it was found that small-scale droplets undergo breakup processes at slightly elevated *We* comparing to the "large" drops: no clear evidences of droplets breakup at the *We*<25 (diameter<35) were detected.

Larger droplets exhibited well-known vibrational and bag breakups and dual-bag disruption. Such breakup modes often resulted in formation of long "fibers" instead of large number of small drops. This effect is most likely due to the greater role of surface tension, which prevents immediate atomization of such "fibers".

Certain detected types of "transitional" breakup regimes could not be directly associated with any of regimes described in literature. Also different types of breakup realized for very similar droplet sizes. This conforms to a statement by Wierzba [7] that droplet behavior is extremely sensitive to the flow conditions and thus may vary significantly in one and the same flow configuration. Experimental results have also confirmed that characteristic breakup time estimation remains valid, at least in the order of magnitude, for the submillimetre-scale droplets.

Acknowledgements

The work was supported by Russian Foundation for Basic Research (grant number 14-08-31718).

References

1. R.I. Nigmatulin *Dynamics of Multiphase Media*, **1**. N.Y.: Hemisphere,. 507 p., (1991)
2. G.M. Faeth, L.P. Hsiang, P.K. Wu, Int J Multiphase Flow **21** (Suppl),:99 (1995)
3. V.M. Boiko, A.N. Papyrin, S.V. Poplavskii, Journal of Applied Mechanics and Technical Physics. **28**. I. 2, 263 (1987)
4. M. Pilch, C.A. Erdman, Int. J. Multiphase Flow **13,** 741 (1987)
5. J.C. Lasheras, E.J. Hopfinger, Ann. Rev. J. Fluid Mech. **32**, 275 (2000)
6. D.R. Guildenbecher; C.López-Rivera, P.E. Sojka, Exp. Fluids **46**, 371 (2009)
7. A. Wierzba, Exp. Fluids **9,** 59 (1990)
8. X.K. Cao, Z.G. Sun, W.F. Li, H.F. Liu, Z. H. Yu, Phys. Fluids **19 (5):** 057103, (2007)
9. T. Kekesi, G. Amberg, L.P. Wittberg, Journal of Multiphase Flow **66**, 1 (2014)

EPJ Web of Conferences **114**, 02032 (2016)
DOI: 10.1051/epjconf/201611402032

Turbulent transport measurements in a cold model of GT-burner at realistic flow rates

Oleg Gobyzov [1,2], Leonid Chikishev [1,2], Alexey Lobasov [1,2], Dmitriy Sharaborin [1,2], Vladimir Dulin [1,2], Artur Bilsky [1,2], Vakhtang Tsatiashvili [3], Valery Avgustinovich [3] and Dmitriy Markovich [1,2]

[1]*Kutateladze Institute of Thermophysics, Siberian Branch of Russian Academy of Sciences, 630090, Novosibirsk, Russia*
[2]*Novosibirsk State University, 630090, Novosibirsk, Russia*
[3]*JSC Aviadvigatel 614010, Perm, Russia*

Abstract. In the present work simultaneous velocity field and passive admixture concentration field measurements at realistic flow-rates conditions in a non-reacting flow in a model of combustion chamber with an industrial mixing device are reported. In the experiments for safety reasons the real fuel (natural gas) was replaced with neon gas to simulate stratification in a strongly swirling flow. Measurements were performed by means of planar laser-induced fluorescence (PLIF) and particle image velocimetry technique (PIV) at Reynolds number, based on the mean flow rate and nozzle diameter, ≈300 000. Details on experimental technique, features of the experimental setup, images and data preprocessing procedures and results of performed measurements are given in the paper. In addition to the raw velocity and admixture concentration data in-depth evaluation approaches aimed for estimation of turbulent kinetic energy (TKE) components, assessment of turbulent Schmidt number and analysis of the gradient closure hypothesis from experimental data are presented in the paper.

1 Introduction

Numerical codes for turbulent mixing calculations for GT burners require verification by comprehensive experimental data, which is often a challenging task to provide. Conventional single-point measurements usually have high temporal resolution but do not provide information about spatial features of the flow and turbulent transport. Contrary to that, planar optical methods, like PIV and PLIF can take advantage of capturing the instantaneous distributions of flow velocity and passive admixture concentration. Such data are usually more adequate for numerical code verification.

Fuel-air mixing plays one of key roles in flame stabilization and pollutant formation. Computational fluid dynamics code validation, using for complex swirling industrial flows demands on reliable experimental data. In particular, one of the key parameters for computational model verification is turbulent transport of passive admixture. Laser-induced fluorescence of acetone vapor (acetone PLIF) in combination with particle image velocimetry (PIV) can be used for quantitative measurements of local fuel/air ratio at constant temperature and pressure without chemical reaction [1]. There are several advantages of the acetone as a tracer in comparison with other polyatomic molecules: a high vapor pressure at room temperature, an absorption feature accessible with high-pulse-energy lasers, short emission lifetime, low toxicity, low cost for seeding at high flowrates, compatibility with air, insensitivity to effects of collisional quenching, and high fluorescence signal levels. At room temperature and atmospheric pressure acetone had a broadband absorption feature which extends from 225 to 320 nm with a peak near 275 nm. Fluorescence in the wavelength range between 350 and 550 nm is emitted from the first excited singlet state S1, following laser excitation from the ground electronic state S0. For example authors of paper [2] investigated turbulent mass-transport for transverse jet flows of laboratory scale.

Full cycle of PIV/PLIF experiment is not limited to the retrieving of raw experimental data, which can undergo certain distortions due to experimental conditions or limitations of the measurement technique. Such data requires proper evaluation to provide more complex and reliable basis for analysis and validation of CFD codes. Present paper is focused on flow structure measurements and mass-transfer investigation in premixing GT-burner. Details of the experimental techniques, description of the GT burner model apparatus, as well as methods of assessment and evaluation of the raw low-repetition experimental data are given.

[a] Corresponding author: oleg.a.g.post@gmail.com

2 Experimental setup and conditions

The measurements were carried out in cold combustion chamber model (see Fig. 1). The burner was provided by JSC Aviadvigatel (Perm, Russia). The burner with diameter d=80 mm has two channels for fuel supply, but in the present study fuel was supplied only through the central channel. Main air flow was supplied from compressed air line into the plenum chamber through four hoses (intake collector). Turbulence generating grid was installed. Inside the plenum chamber and the coaxial swirler of the burner generated high-swirl premixed flow. The model fuel (neon gas) was supplied to the central channel of the burner. Test section of combustion chamber was made of quartz cylinder (120 mm inner diameter). An optical section was 170 mm long.

The maximum air flow-rate was up to 0.5 kg/s, the temperature of the air was 17 °C. The Reynolds number based on the nozzle diameter, and the bulk velocity was about $3 \cdot 10^5$. The pressure inside the chamber was 2 atm. with pressure drop of 4% at the premixer. To model stratification in swirling flow and to keep experiment safe at the same time neon gas was used instead of methane. The gas was supplied from a 10 atm. Pressure tank. The neon flowrate was controlled using Bronkhorst flowmeters. Water-glycerol particles were used to provide PIV measurements (the main air flow was seeded). The seeder was connected to the main air pipe via bypass with 0.5 atm pressure difference between inlet and exit of the seeder. The temperature stabilized acetone bubbler was employed to seed the neon flow for PLIF measurements. Neon massflow rate was 5.4 g/s. Owing to temperature stabilization at constant flow-rate the acetone vapor concentration in neon flow also remained constant.

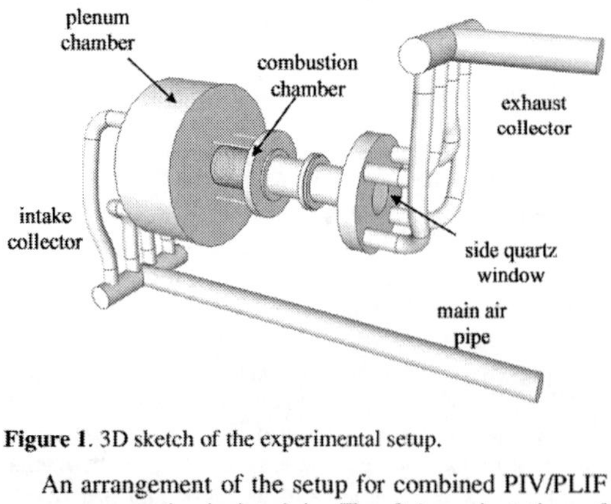

Figure 1. 3D sketch of the experimental setup.

An arrangement of the setup for combined PIV/PLIF measurements is depicted in Fig. 2.An orientation of coordinate axis is also shown in the fig. 2 and coordinates will be referred through the paper according to the figure. PIV system included double pulsed Nd:YAG laser (Quantel Ever Green 200) and 4M CCD camera ImperX IGV-B2020 supplied with the lens SIGMA AF 50 mm (F2,8 EX DG Macro) optical band-pass filter (Edmund Optics, 532±10 nm, 60% transmittance). To deliver the laser sheet inside the test section and to minimize scattering and reflections of the beam side quartz window (see fig. 1) was installed at the end of the cylindrical section. PLIF system included Quantel Brilliant B Nd:YAG laser and ICCD camera PI-MAX-4 (Princeton Instruments). Quantum efficiency of the s20 photocathode in spectral region 290-320 nm is about 25 %. LaVision UV-lens (f#2.8, 100 mm) with interference band-pass optical filter (280-600 nm) and multi-notch filter (1064, 532, 355 nm) was installed on the camera. Exposure time of ICCD camera in experiments was 200 ns. Fourth harmonic of Nd:YAG laser (266 nm) was used to excite fluorescence of the acetone vapor. Average pulse energy measured with Coherent LabMax Top was 70 mJ. To provide quantitative measurements collimated Sheet Optics (LaVision) was used. These two systems were synchronized using pulse generator (Berkeley Nucleonics BNC 575 TTL). Green laser pulse to pulse delay was 2 μs to prevent particle loss due to out-of-plane motion. ICCD camera exposition was triggered between PIV laser pulses.

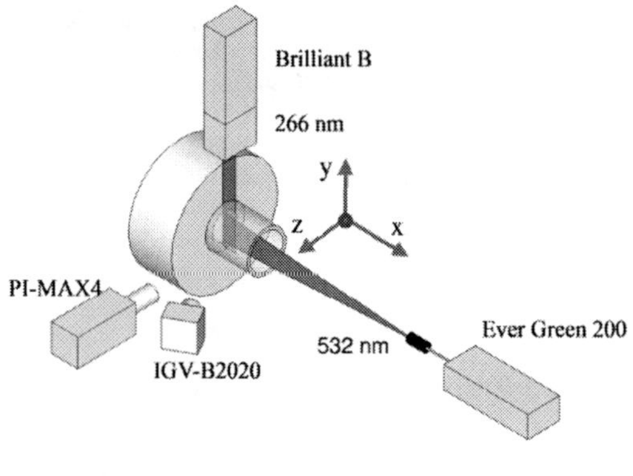

Figure 2. Scheme of the PIV/PLIF system arrangement.

A sheet of sensitive paper was placed into the measurement volume and two single shots of 266 and 532 nm lasers were done after each run to ensure that . PIV and PLIF measurement regions are well aligned. To measure magnification factor and the mapping functions of two cameras a calibration target was placed inside the cylinder.. In several minutes after seeding start the water-glycerol particles formed a film on a chamber wall making the cylinder non-transparent. Thus maximum statistics per one run was about 300 couples of images at 5 Hz frame-rate. The full statistics of 4000 fields of velocity and concentration respectively was collected in several consecutive experiments.

Besides the simultaneous PIV/PLIF experiments separate PIV experiments were performed in YZ plane at the distance of $0.1 \cdot d$ and $1 \cdot d$ from the mixing device exit to ensure that the flow is axisymmetric.

3 Images and data processing

3.1 PIV Image processing

Due to inclusions in side surface of the cylinder laser sheet used for PIV also had large number of stripy inhomogeneities. It resulted in distorted data retrieved from PIV evaluation: stripy structure emerged in average velocity fields, and especially in higher order momentums. An approach similar to light sheet correction in PLIF was adapted to correct an intensity of particle images. Since the inhomogeneities were produced by stationary defects in the cylinder material, intensity of each pixel was corrected by the time-averaged intensity:

$$I^*(X,Y) = \frac{I(X,Y)}{\langle I(X,Y)\rangle} I_0 \tag{1}$$

where I_0 is the mean of the time-averaged intensity:

$$I_0 = \frac{1}{n}\sum_x\sum_y \langle I(X,Y)\rangle \tag{2}$$

This procedure, although not fully compensating lighting flaws, substantially improved quality of particle images and reduced impact of "stripes" on velocity fields.

Since in simultaneous PIV/PLIF measurements PIV camera was inclined to the measurement plane a correction of the perspective distortions was applied to the PIV images. For calibration of the camera second order polynomial model [3] was used.

Spatial resolution, as reported in [4, 5] and many other works can have a substantial impact on the results of investigation. Apparently, PIV measurements in high-Reynolds number flow usually unable to fetch turbulent features at small scales, but the higher spatial resolution is, the larger part of the turbulent spectra would be resolved. One should also take into account that PIV resolution is bounded above due to spatial averaging over the light sheet thickness. In the present experiments light sheet thickness (≈1 mm) didn't exceed the size of interrogation window.

There are two ways to vary a spatial resolution in PIV experiment: variation of optical magnification factor and the size of interrogation window in PIV algorithm. In the experiments under consideration it was important to perform measurements across the whole test section, and therefore optical magnification was selected to satisfy this condition and was fixed during the series of measurements. As for the variation of interrogation window, excessive reduction of its size can result in emergence of intense data noise. If higher order statistical moments or spatial derivatives are to be retrieved from PIV velocity fields, the noise amplification issue should be considered.

To assess an effect of spatial resolution Δ, or rather of the interrogation window size, velocity field calculation was performed with final window size varied from 16×16 pix. to 128x128 pix. (see fig. 3). Even with qualitative analysis, it can be concluded that 16x16 pix. window interrogation introduces large amount of high-frequency noise, which contributes into the measured turbulence statistics, and if smoothed, distorts the profile in the area of low velocity pulsations. In its turn 128×128 pix. window interrogation results in substantial underestimation of spatial derivatives component, especially in the mixing layer. For 32×32 pix. and 64×64 pix. window almost no high-frequency noise component is observed, but larger interrogation window predictably yields a bit lower intensity of velocity pulsations. Same arguments can be applied to the second-order of velocity derivatives. In the final analysis PIV data were processed using final interrogation window size Δ of 32x32 pix. Peak validation and adaptive median filter (5x5) were employed to detect false vectors. In turbulent statistics false vectors were just excluded from evaluation, but for the calculation of spatial derivatives moving average interpolation of false vectors was applied to form a datasets on a regular grid. Since in all cases spatial overlap rate between the interrogation areas was 75%, the grid step h was four times smaller than Δ. The final spatial resolution was 2 vectors per 1 mm.

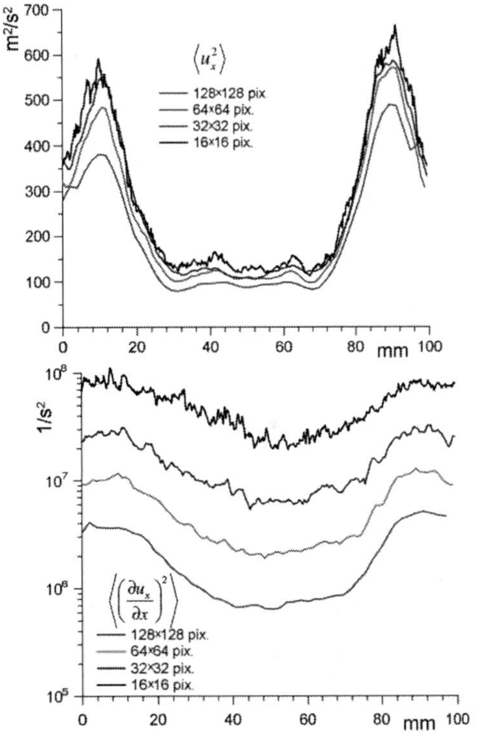

Figure 3 TKE component and second-order of velocity derivative profiles along Y axis at the distance of 22 mm from the mixing device exit.

Evaluation of more comprehensive descriptive parameters of turbulence and mixing processes, such as turbulent kinetic energy (TKE) dissipation rate and turbulent Schmidt number, both require calculation of velocity special derivatives. In its turn, numerical differentiation scheme for spatial derivatives calculation should be carefully chosen in order to keep proper balance between measurement noise amplification and truncation error.

For the case under consideration several numerical differential schemes were considered for calculation of spatial velocity derivatives (see. Fig 4). Among these schemes were: simple central differentiation (eq. 3), least-squares (eq. 4) scheme with different point spacing H, proposed in [3], and Richardson extrapolation scheme optimized for minimization of random error amplification (eq.5).

$$\frac{\partial u(x)}{\partial x}=\frac{u(x+H)-u(x-H)}{2H} \tag{3}$$

$$\frac{\partial u(x)}{\partial x}=\frac{2u(x+2H)+u(x+H)-u(x-H)-2u(x-2H)}{10H} \tag{4}$$

$$\frac{\partial u(x)}{\partial x}=\frac{16u(x+4H)+u(x+H)-u(x-H)-16u(x-4H)}{130H} \tag{5}$$

Profile of the second-order velocity derivative in a transversal cross-section of the measurement area at the distance of 1d from burner exit evaluated with different numerical derivation schemes is shown in fig.4. It can be seen, that simple central differentiation scheme and least-squares with $H=h$ introduce amount off noise comparable to the signal amplitude, while for least-square scheme with $H \geq 2h$ amount of high-frequency noise doesn't differ much, but truncation error increases significantly with increase of H. Note that least-square with $H=2h$ and Richardson scheme with $H=1h$ yield almost similar results because terms $u(x \pm 4h)$contribute most in both cases. According to this results, least-squares differential scheme with $H=2h$ was chosen for further evaluation.

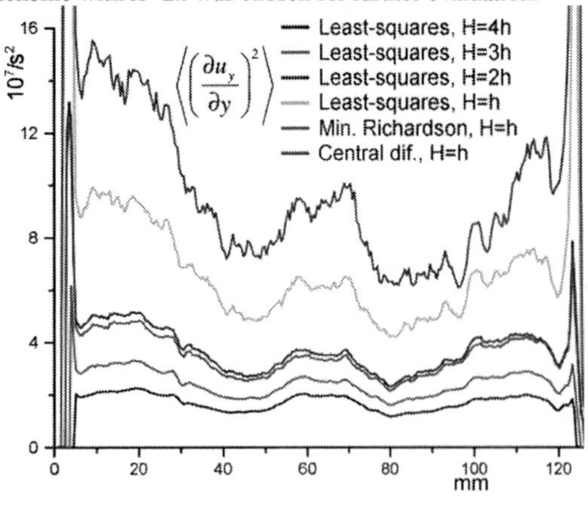

Figure 4 Second-order velocity derivative profile along Z axis at the distance of 1d from the mixing device exit evaluated with different numerical derivation schemes

3.2 PLIF Image processing

As it is known, in PLIF the linear dependency of the fluorescence signal and the acetone tracer concentration are provided only if the energy flux in the light sheet is below the saturation level and absorption can be neglected. Linear response of acetone vapor was previously reported by Bryant et. al. at maximum energy flux of 353 mJ/cm^3 when excited at 266 nm (10 ns pulse width) [6]. The maximum laser energy flux in the experiment under consideration was more than two times lower - 140 mJ/cm^3. The linearity of the signal was verified prior to measurements and no evidence of saturation was found.

The raw PLIF images were processed using background subtraction and laser sheet correction procedures. After background subtraction images were divided by a reference sensitivity image. The reference sensitivity image was recorded while the test section was flooded with an acetone tracer at homogeneous concentration (averaged over 200 images). This procedure accounts for the angular lens light collection efficiency, variations of sensitivity of the camera, the quantum efficiency of fluorescence and spatial variations of laser intensity. In order to correct for additional shot-to-shot fluctuations, each single image was corrected according to laser intensity deviation at the same time. Intensity variations were measured using energy meter. Fig. 5 shows an example of PLIF image evaluation (raw image and image after all correction procedures). The laser sheet direction is from top to bottom. The systematic error was significantly reduced using correction procedures.

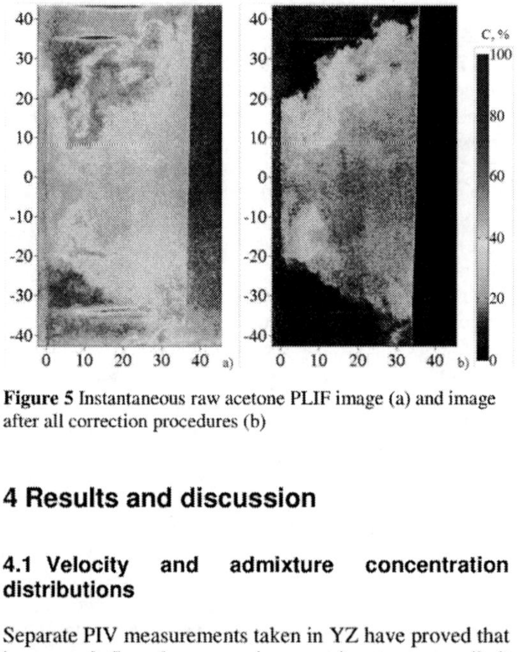

Figure 5 Instantaneous raw acetone PLIF image (a) and image after all correction procedures (b)

4 Results and discussion

4.1 Velocity and admixture concentration distributions

Separate PIV measurements taken in YZ have proved that in general flow has an axisymmetric structure, albeit some skewness, especially in radial component can be observed (see fig. 6).

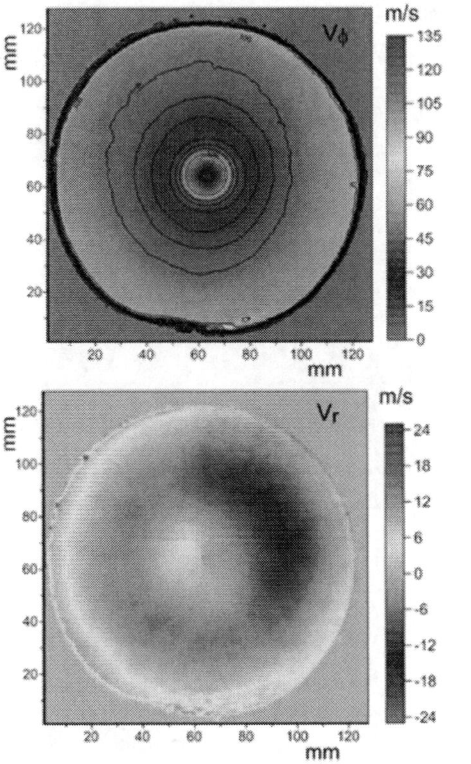

Figure 6. Average velocity components distribution in YZ plane at the distance of 1d from mixing device exit.

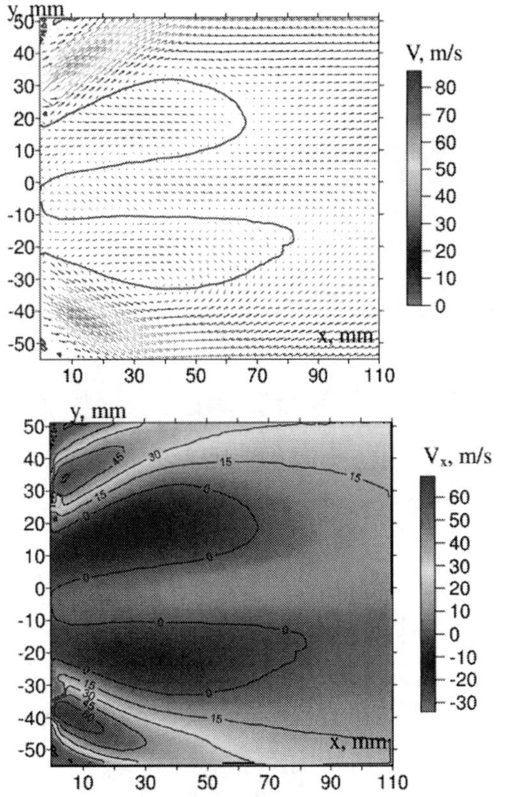

Figure 7. Averaged velocity field (upper) and axial velocity component field (lower).

Averaged velocity field in XY plane is presented in Fig. 7 (red line indicates the backflow region). As it is seen the flow is strongly swirled and after a sudden expansion backflow region appears. The maximum velocity at the burner exit is up to 80 m/s. Distinctive feature of the flow is a well pronounced recirculation region extending to the distance of one calibre. The central region of the flow exhibits jet flow due to model fuel supply at the center of the burner. The velocity maximum region reaches the wall at approximately 15 mm downstream of the burner. The corner recirculation region is not resolved due to poor signal/noise ratio at the chamber wall caused by laser reflections from the burner. More detailed quantitative description of the flow structure is available from the velocity profiles (fig. 8 and fig. 9). Axial and radial velocity components are presented for two distances downstream of the burner (0,1*d* and 0,5*d* respectively).

The prominent features of the mean axial velocity profiles are the inflow region with high velocities and pronounced recirculation zone with mean velocities of the order -10 m/s. The RMS values increases slightly with a distance from the axis of symmetry of the burner and have peaks in the inner and outer shear layers. Another feature of the flow is the central jet flow with axial velocity up to 10 m/s. Radial component of the velocity have its maxima only at the initial region of the flow. Downstream the burner mean radial velocity is approaches zero with RMS about 10 m/s.

As for the concentration distribution, it can be seen from figure 10 that downstream of the nozzle maximum mean concentration decreases 1.6 times. Another feature is that near the burner concentration profile has two well-pronounced maxima. This maxima indicates strong interaction of the central fuel jet with coaxial swirling flow. At the distance of $\approx 0.5d$ from the burner only one central maxima of concentration is present. The concentration pulsations are quite high near the burner exit (about 30 % of the maximum) and reaches its maximum in the inner shear layer of coaxial swirling flow.

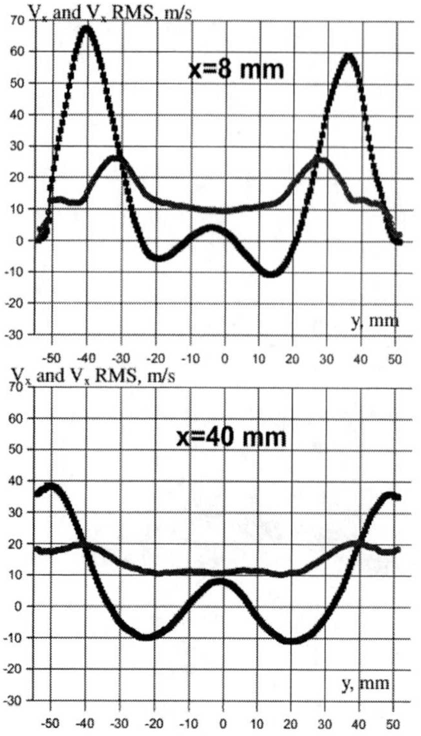

fig 8 Radial profiles of the mean (black) and RMS (blue) values of the axial *Vx*

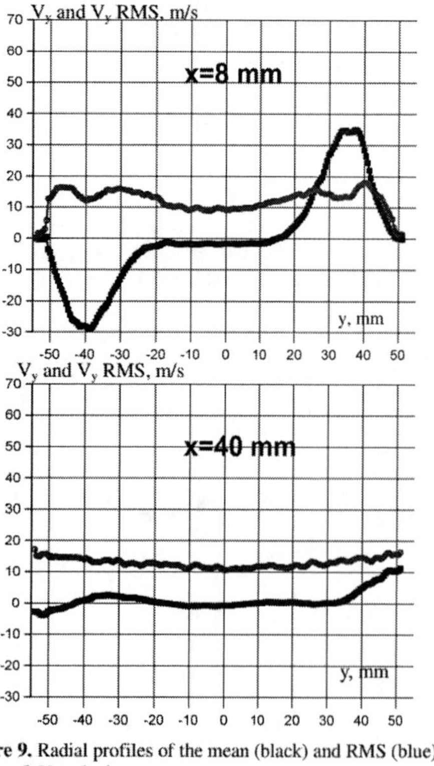

Figure 9. Radial profiles of the mean (black) and RMS (blue) values of *Vy* velocity component

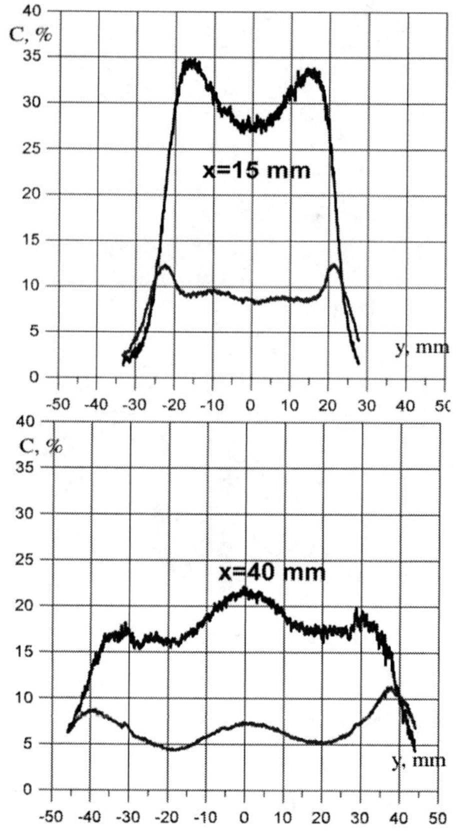

Figure 10. Concentration profiles (black line) and concentration RMS (blue line) at different distances from the burner exit

4.2 Turbulent kinetic energy dissipation rate

A turbulent kinetic energy (TKE) dissipation rate plays an important role in theories of turbulence and if recovered from experimental data can improve an overall understanding of the turbulent flow features.

TKE dissipation rate per unit volume depends on local kinematic viscosity of the fluid and strain rate tensor S_{ij}:

$$S_{ij} = \frac{1}{2}\left(\frac{\partial u_i}{\partial x_j} + \frac{\partial u_j}{\partial x_i}\right) \tag{6}$$

And TKE dissipation rate is:

$$\varepsilon = 2\nu\langle S_{ij} S_{ij}\rangle \tag{7}$$

Expressed in simple algebraic form, equation (7) will contain 12 second order-moments of velocity derivatives. Thus, in order to obtain TKE dissipation rate directly from the experiment, all 3 components of velocity vector should be measured, and measurement spatial resolution should be sufficient to resolve finest spatial scales in the flow without introducing truncation error in spatial derivatives calculation. Usually such experimental data is unavailable, especially for high-Reynolds number flows: velocity data may be incomplete, contain noise and lack spatial resolution. In some cases, proper physical assumptions can be applied to reduce number of velocity components needed to estimate TKE dissipation rate.

As PIV data in the experiments under consideration contained only two velocity components in XY plane, direct TKE dissipation rate calculation from eq.7 wasn't possible. At the same time, according to the statistical data yielded from measurements taken in YZ plane, an assumption of rotational symmetry about X axis was quite appropriate despite of some skewness. Under this assumption formulation proposed by Hussain et. al. in [7] for TKE dissipation rate calculation can be applied. In this formulation only four second order-moments of velocity derivatives are used:

$$\varepsilon = \nu\left[-\left\langle\left(\frac{\partial u_1}{\partial x_1}\right)^2\right\rangle + 2\left\langle\left(\frac{\partial u_1}{\partial x_2}\right)^2\right\rangle + 2\left\langle\left(\frac{\partial u_2}{\partial x_1}\right)^2\right\rangle + 8\left\langle\left(\frac{\partial u_2}{\partial x_2}\right)^2\right\rangle\right] \tag{8}$$

Eventually, if velocity derivatives calculated from experimental data using least-square scheme is substituted in eq. (8), ε yielded from this equation, here and below addressed as ε^m, will qualitatively represent distribution of TKE dissipation rate, but quantitatively will greatly underestimate the value. The latter statement can be indirectly confirmed by the following considerations: coarse estimation of Kolmogorov's scale η as $Re^{-3/4}$ yields $\eta \approx 50\mu m$, while PIV spatial resolution, taking into account interrogation window overlap, was ≈ 1.8 mm, besides, additional spatial filtering was introduced by numerical differentiation scheme. Consequently, for such measurement data large part of the spectrum falls into sub-grid scale, and, therefore, missed in TKE dissipation rate estimation.

To assess an «underestimation factor» a LES-based approach [8] together with appropriate model of turbulence can be applied. In this approach turbulence is assumed to conform to the selected model and limited spatial resolution of the measurement technique considered as a low-pass filter. Though PIV spatial filter is, in general, behaves like a cardinal sine function [9], for simplicity here it will be considered a sharp cut-off filter with $k_{max}=\pi/\Delta$, where spatial resolution is homogeneous in x and y directions $\Delta = \Delta_x = \Delta_y$. For the case, Δ was equal to $8h$, where h is a step between velocity vectors in a PIV velocity field. Factor 8 for the Δ was chosen because of 75% window overlap in PIV interrogation and decrease in effective spatial resolution for velocity derivatives due to selected numerical differentiation scheme.

The Smagorinsky model for turbulent viscosity ν_τ is:

$$\nu_\tau = C_S^2\Delta^2\sqrt{\left(2\overline{S_{ij}^* S_{ij}^*}\right)} \tag{9}$$

Where S_{ij} is strain rate tensor, Δ is the filter width, C_S is the Smagorinsky coefficient, and symbol * indicates instantaneous quantity. According to [10] and some other works ensemble averaging operator can be introduced on the norm of the strain-rate tensor without loss of the models performance:

$$\nu_\tau \approx C_S^2\Delta^2\left\langle 2\overline{S_{ij}^* S_{ij}^*}\right\rangle^{1/2} \approx C_S^2\Delta^2\left\langle 2\overline{S_{ij} S_{ij}}\right\rangle^{1/2} \tag{10}$$

Under the assumption of infinite Re TKE dissipation rate can be expressed from eq. 7 as

$$\varepsilon \approx 2\nu_\tau\left\langle\overline{S_{ij}^* S_{ij}^*}\right\rangle \approx 2\nu_\tau\left\langle\overline{S_{ij} S_{ij}}\right\rangle \tag{11}$$

For the inertial range spectrum, filtered by a sharp cut-off filter with width=Δ the value of C_S can be obtained as:

$$C_S = \frac{1}{\pi}\left(\frac{2}{3C}\right)^{3/4} \approx 0.173 \tag{12}$$

where $C \approx 1.5$ is Kolmogorov constant. In [11] other approximations, including an approximation for the case $L/\Delta >> 1$ (13) were proposed:

$$\varepsilon = 2\nu_\tau\left\langle\overline{S_{ij}^* S_{ij}^*}\right\rangle, \tag{13}$$

$$\nu_\tau = \sqrt{\left(C_{s,\infty}\Delta/\gamma\right)^4\left(2\overline{S_{ij}^* S_{ij}^*}\right) + \nu^2} - \nu \tag{14}$$

In this case, turbulent kinetic energy dissipation is given as a sum of sub-grid-scale and resolved dissipation, and γ depends on the filter shape and is equal to a unit for a sharp cutoff filter.

Recalling that resolved scales in this analysis corresponds to that measured with PIV, a relation (15) with classic Smagorinsky or modified model for turbulent viscosity can be applied to correct data for the amount of TKE dissipation rate at not resolved scales:

$$\begin{gathered}\varepsilon = \nu_\tau\varepsilon^m/\nu \text{, where} \\ \nu_\tau = C_S^2\Delta^2\left(\varepsilon^m/\nu\right)^{1/2} \\ \text{or} \\ \nu_\tau = \sqrt{\left(C_s\Delta/\gamma\right)^4\left(\varepsilon^m/\nu\right) + \nu^2} - \nu\end{gathered} \tag{15}$$

In accordance with the model described above a «correction factor» of the TKE dissipation rate over the measurement area was calculated and «corrected» TKE dissipation rate was assessed (fig. 11). Predictably, an area of the maximum TKE dissipation rate corresponds to the region of most intense large-scale eddy interaction with the small-scale turbulence in the mixing layer. A «correction factor» exceeds 600 over the whole measurement area and in the mixing region goes as high as 10^3. Eventually such results, in particular – extremely high «correction factor» mean that that only general features of the flow were fetched, and TKE dissipation rate assessment is valid to the order of magnitude at best. Generally it is not surprising as Re was high ($\approx 3\cdot 10^5$) and PIV spatial resolution was quite low.

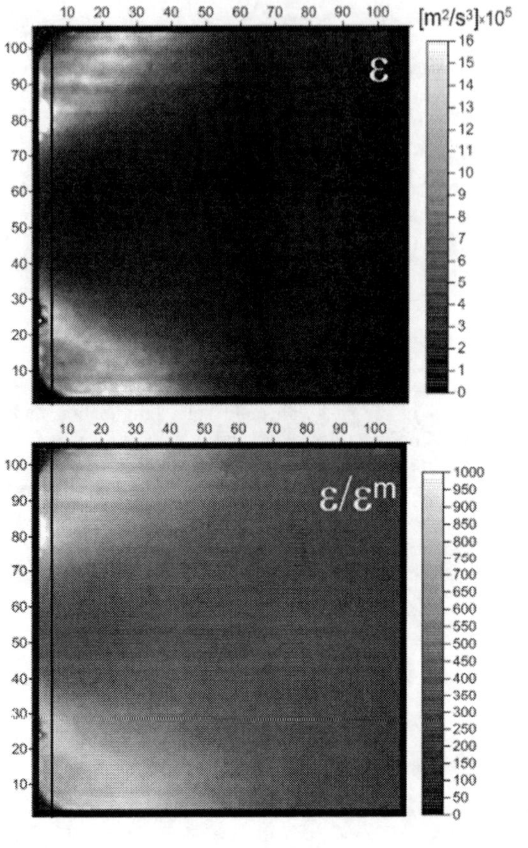

Figure 11. Distribution of TKE dissipation rate and «correction factor» over the measurement area in XY plane

4.3 Turbulent mixing characteristics

A conservation equation of species for turbulent flows can be written as:

$$\frac{\partial \bar{c}}{\partial t}+\bar{u}_j\frac{\partial \bar{c}}{\partial x_j}=\frac{D\partial^2\bar{c}}{\partial x_j^2}-\frac{\partial\langle u_j c\rangle}{\partial x_j} \tag{16}$$

where $\bar{c}$ is the time-averaged concentration, c - the concentration fluctuation, $\bar{u}_j$ - time averaged velocity component, u_j- velocity fluctuation component, D - molecular diffusion coefficient.

Right-hand part of the equation represents mass transport. In high-Reynolds number turbulent flows molecular diffusion term is usually negligible comparing to the turbulent diffusion term $\langle u_j c\rangle$ and can be omitted [12]. Than conservation equation takes following form:

$$\frac{\partial \bar{c}}{\partial t}+\bar{u}_j\frac{\partial \bar{c}}{\partial x_j}=-\frac{\partial\langle u_j c\rangle}{\partial x_j} \tag{17}$$

For this equation in CFD models a closure problem is usually solved by the assumption that turbulent mass transport is proportional to the mean concentration gradient - so called Fick's diffusion law assumption:

$$\langle u_j c\rangle=-Dt_j\frac{\partial \bar{c}}{\partial x_j} \tag{18}$$

Yet, to validate the model for certain type of the flow fidelity of this assumption should be performed. For the experiments under consideration validation of Fick's law assumption was performed using PIV and PLIF experimental data.

In fig. 12 a distribution of y-component of $\langle u_j c\rangle$ and y-component of the mean admixture concentration gradient are presented. An area of the reversed flow is bounded by the solid line in the figure. Using these two distributions a turbulent admixture diffusion coefficient Dt can be assessed.

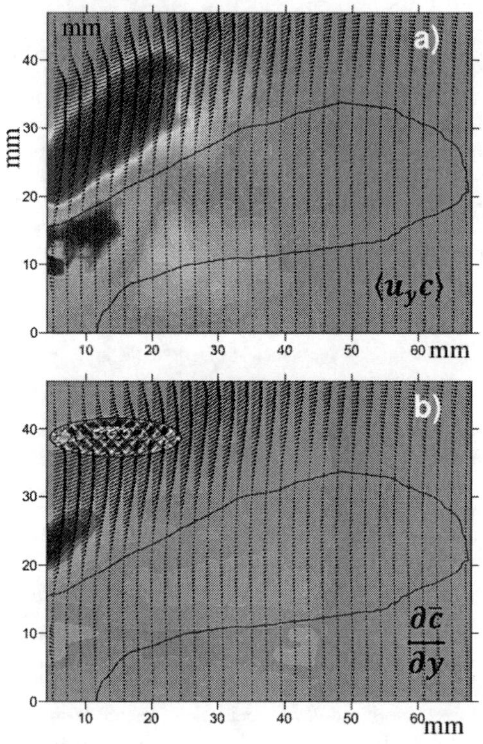

Figure 12. y-component of $\langle u_j c\rangle$ and y-component of the mean admixture concentration gradient

In fig.13 a profile of y-component of $\langle u_j c\rangle$ and y-component of the mean admixture concentration gradient across the measurement area is shown. It can be seen from the figure that directly measured turbulent transport is in good correspondence with the Fick's diffusion law. Turbulent diffusion coefficient Dt, that can be evaluated from such data by simple fitting of mean admixture concentration gradient profile to the $\langle u_j c\rangle$, was equal to 0.028 for the case.

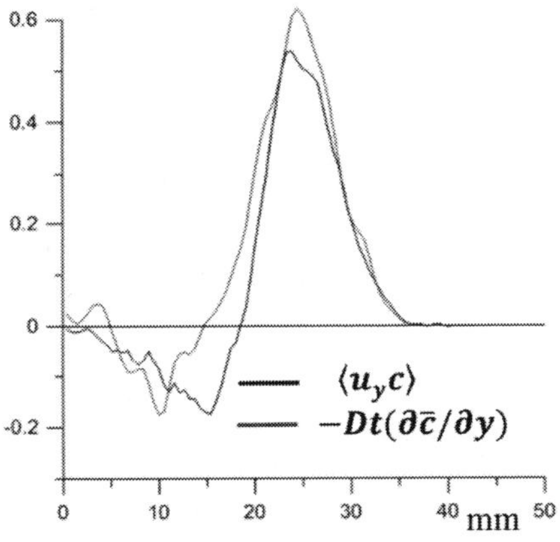

Figure 13. Profile of y-component of $\langle u_j c\rangle$ and y-component of the mean admixture concentration gradient across the measurement area.

Another representative characteristic that can be recovered from simultaneous PIV/PLIF measurements is turbulent Schmidt number, which represents the ratio between the rates of turbulent transport of momentum and the turbulent mass transport.

As Schmidt number $Sc_t = v_t/Dt$, a distribution of v_t is required to calculate Sc_t over the measurement area. Such distribution is available from PIV data through distribution of Reynolds stress and mean velocity gradient component, which are presented in fig. 14. As can be seen from the figure, maximum and minimum of both differ in sign and are reached within the shear layers of the swirling flow.

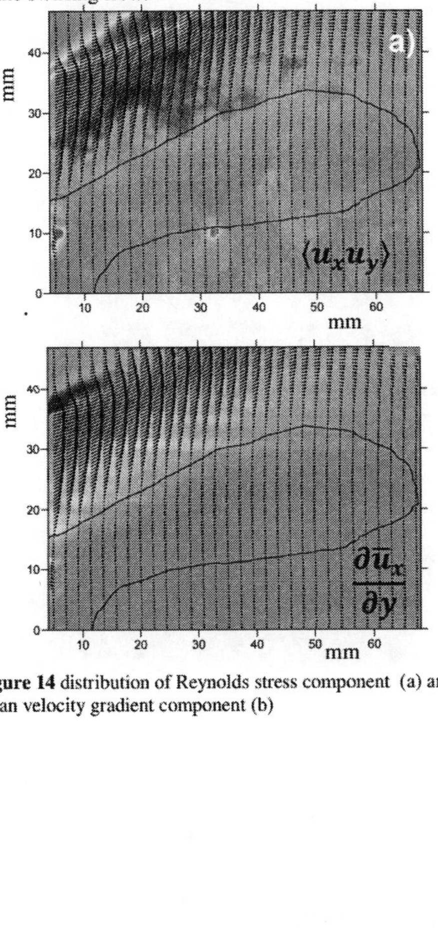

Figure 14 distribution of Reynolds stress component (a) and mean velocity gradient component (b)

Again, like in the previous case, turbulent viscosity v_t can be evaluated from experimental data by simple fitting of mean velocity gradient profile to the profile of Reynolds stresses (see fig. 15). Such evaluation yields $v_t \approx 0.01$, and corresponding Sc_t value of 0.35.

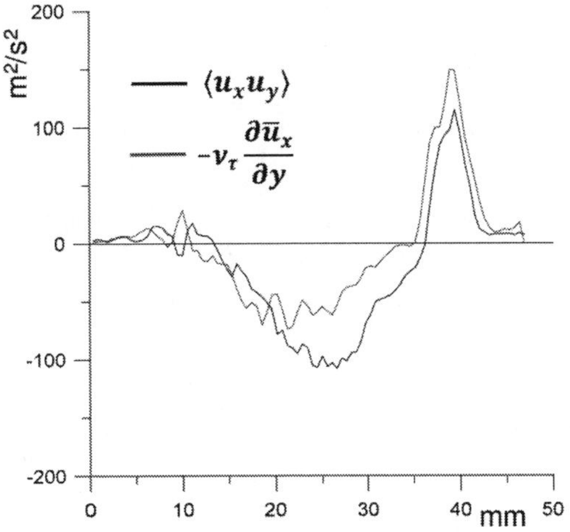

Figure 15. Profile of Reynolds stress component of $\langle u_x u_y\rangle$ and y-component of the mean streamwise velocity component gradient

5 Conclusion

In the present work combined PLIF/PIV experiments were carried out in a cold model of GT-burner at realistic flow-rates. The results include instantaneous simultaneously measured 2D concentration of passive admixture and velocity fields. The main goal of the investigation was a better understanding of the turbulent mixing process in an industrial burner and acquisition of experimental data for validation of numerical simulation. Despite of non-perfect quality of quartz glass cylinder and multiple laser sheet reflections data processing algorithms allowed to get reliable experimental data. The results indicate that there is a strong correlation between velocity and concentration fields.

In addition evaluation techniques for assessment of TKE dissipation rate and turbulent passive admixture transport for the case were considered. TKE dissipation rate was assessed from 2D-PIV data using an assumption of rotational symmetry, though assessment precision was rather low. Turbulent diffusion coefficient and turbulent viscosity, required for application of gradient closure hypothesis and evaluation of turbulent Schmidt number were retrieved from the PIV/PLIF simultaneous measurements. It was also shown, that in general, Reynolds stresses and turbulent passive admixture flux, both conform to the gradient closure hypothesis.

Acknowledgments

This work was funded by a grant from the Russian Science Foundation (project No. 14-29-00203 in NSU, supervised by Professor Kemal Hanjalic).

References

1. A. Lozano, S.H. Smith, M.G. Mungal, and R.K. Hanson, AIAA J., **32**, 218 (1993)
2. L.K Su., M.G. Mungal, *43rd Aerospace Science Meeting and Exhibit*, AIAA 2005-305, (2005)
3. M. Raffel, C. Willert, S. Wereley, J. Kompenhans, *Particle Image Velocimetry. A practical guide, Second Edition* (2007)
4. S. Alekseenko, A. Bilsky, V. Dulin, B. Ilyushin, D. Markovich, Non-intrusive determination of turbulent energy balance in free and confined jet flows. *4th International Symposium on Turbulence and Shear Flow Phenomena, Williamsburg, VA, USA* (2005)
5. J.M. Foucaut, B. Milliat, N. Perenne, M. Stanislas, Characterization of different PIV algorithms using the EUROPIV Synthetic Image Generator and real images from a turbulent boundary layer *EUROPIV 2 Workshop on Particle Image Velocimetry,*163 (2004)
6. R.A. Bryant, J.M. Donbar, J.F. Driscoll, Exp. Fluids **28**, **№. 5**, 471(2000).
7. W.K. George, H.J. Hussein, **233**, 1 (1991)
8. V.M. Dulin, D.M. Markovich, S.V. Alekseenko, Stereo PIV measurements of fine-scale turbulence statistics in a free jet flow *Proc. 6th Int. Symp. Turbulence, Heat Mass Transfer,* Begell House, Rome (2009).
9. J.M. Foucaut, J. Carlier, M. Stanislas, Meas Sci Technol **15**, 1046, (2004)
10. O. J. McMillan, J. H. Ferziger, AIAA J. **17**, 1340 (1979).
11. J. Meyers, P. Sagaut, J. Fluid Mech. **569**, 287, (2006)
12. P. J. W. Roberts, D. R. Webster, Turbulent diffusion, In: *Environmental Fluid Mechanics Theories and Application*, 7 (2002)

EPJ Web of Conferences **114**, 02033 (2016)
DOI: 10.1051/epjconf/201611402033

Processes of heat and mass transfer in straw bales using flue gasses as a drying medium

Wojciech Goryl[1,a], Mateusz Szubel[1] and Mariusz Filipowicz[1]

[1]*AGH University of Science and Technology, Faculty of Energy and Fuels, A. Mickiewicza 30 av., 30-059 Kraków, Poland*

Abstract. Moisture content is a main problem of using straw in form of bales for energy production. The paper presents possibility of straw drying in dedicated, innovative and patented in Poland straw dryers which using flue gasses as a drying medium. Paper presents an improved way of drying which proved to be very sufficient. Temperature and humidity of straw during the process of drying were measured. The measurements helped understand and perform numerical model of heat and mass transfer inside the straw bale. By using CFD codes it was possible to perform analysis of phenomenon occurring inside the dried straw bale. Based on the CFD model, proposals of the optimization and improvement process of drying have been discussed. Experimental and computational data have been compared in terms of convergence. A satisfying degree of agreement has been achieved. Applying improved drying method, homogenous field of moisture content and temperature in the straw bale is achieved in a very cost effective way.

1 Introduction

Currently, in the European Union, the share of energy from renewable sources in the complete balance of energy production increases from year to year. This has been initially dictated by the Directive 2001/77/EC, concerning the production of electrical energy from RES. Then, it was replaced by the Directive 2009/28/EC promoting not only electricity, but also the heat production from renewable sources. Poland after joining the European Union started to fulfil obligations of the Directive 2001/77/EC. Unfortunately, despite the new Directive entered into force in 2009 (2009/28/EC) Poland still tries to fulfil the obligations of the no longer applicable Directive, especially through the path of co-firing the biomass with coal. As a result, since 2004 we can observe a rapid increase of the biomass use to produce electricity, what is widely criticised by scientists and experts [1-3]. Until 2012 this was the twelvefold increase [4]. This involves a number of ecological, economic, transport and phytosanitary problems [2, 3]. The huge demand for biomass by professional coal power plants caused the drastic increase of its prices. In Poland, the greatest potential of agricultural biomass is in straw [4], therefore its price between the year 2005 and 2012 has almost doubled, reaching over 300 PLN/t [5].

Poland has large biomass resources in the form of straw. It is estimated that the annual production of straw is approx. 20 million tonnes [6] in which the technical potential of obtaining straw for energy purposes is approx. 7 million tonnes [7]. From year to year this potential will increase due to a decrease in the number of breeding animals [8], what will result in the lower demand for straw for animal and bedding purposes. Such large surpluses of straw will make it more common to meet the straw use for heating purposes by farmers, in public buildings, where the Municipality of Człuchów is a good example, where four schools use straw for heating buildings. Straw is also used in network heating in towns, like e.g. in Lubań in Lower Silesia.

Straw is characterised by low bulk density (approx. 10 times smaller than coal) and lower calorific value which on average equals 14-15 MJ/kg. In addition, it is quite a problematic fuel due to the heterogeneity of the material. The energy usefulness depends on the humidity, chemical composition and the straw type [9]. However, the basic problem of the energy use of straw in the form of bales as fuel is the moisture content. The main influence on it comes from the period of harvest and the conditions of its storage. Dry straw can reach humidity even under 10%, however the harvest during unfavourable weather conditions causes too severe moisture of the straw ranging between 60-70% of the relative humidity [10], what is the value often observed for fresh biomass [11]. In Polish climatic conditions and in many European countries the heating season is relatively long and lasts from October to May, while the harvest of straw takes place only in the summer period in July and August. Therefore, it is necessary to store straw, which will be used in the heating period. It should be stored in closed barns or haystacks covered with foils, tarps, or special non-water permeable materials. This allows to minimise

[a] Corresponding author: wgoryl@agh.edu.pl

the impact of weather conditions on the deterioration of the quality of stored fuel. Too much moisture of the straw has a negative impact on the burning value of the fuel and on the problems related to the burning of wet straw in boilers [10, 12]. It is best for the straw used in batch boilers to have approx. 10-20% humidity. This has a favourable effect on the combustion process due to the catalytic effect of water vapour to the combustion of the excessive amount of volatiles present in the biomass [10].

Harvest of straw in adverse weather conditions or the lack of possibility to store the straw in the place adapted for this determines the necessity to dry it in drying devices. The most often used drying devices in the industry are, e.g., rotary dryers, belt dryers, flash dryers or fluidised-bed dryers [13]. The main problem of this type of dryers is the need to obtain a loose form of the dried material. Straw used in batch combustion boilers is in the form of bales and it should also be dried in this form not to lose energy for its fragmentation before the drying process and the next balloting after the performed drying process. For this purpose special dryers should be used, which have the possibility to dry straw in the form of bales [10]. One of such devices is the innovative, prototype straw dryer of the MetalERG company, located by the school boiler room in the town of Wierzchowo-Dworzec in the municipality of Człuchów. It is used for drying straw bales using waste heat, which are the exhaust gases after the combustion process in the biomass boiler burned with straw or wood [14].

2 Description of the measurement stand

The main elements of the dryer used for studies of the possibility of drying straw include the drying chamber, transport table, discharge fan, fire prevention, complex of regulation and control of operation parameters, system of pre-extraction and separation of sparks and the exhaust introduction system.

Heat used during drying comes from the combustion process of biomass in the boiler. Such utilised waste heat increases the efficiency of the whole system and helps obtain a much more efficient fuel. The drying process involves the injecting of waste gas with the properly selected temperature using the specially designed nozzle with the diameters of 80cm into the bale of straw. Visualisation of the nozzle is in Figure 8.

For safety reasons exhaust gases coming from the boiler are pre-dusted in the sedimentation chamber (cyclone) due to the sling of large, sometimes incandescent stalks of unburned straw. Then, the gases are transported to the specially made mixer, where, if necessary, a process of mixing the exhaust gas with the ambient air takes place to obtain the desired temperature of the drying medium. After the mixing process, the fan introduces the pre-prepared mixture using the specially designed nozzle into the cylindrical bale of straw. The schematic diagram is presented in Figure 1.

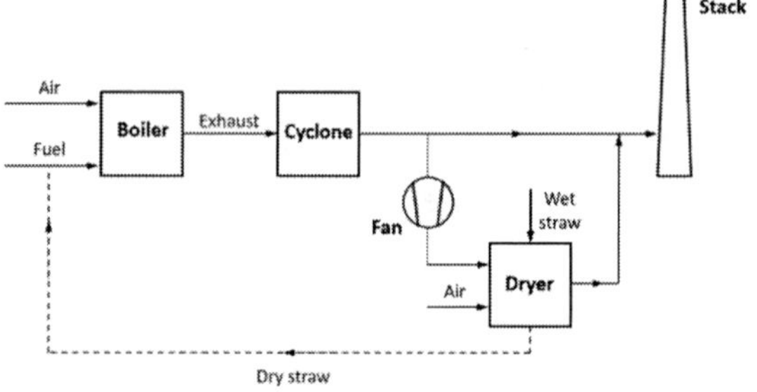

Figure 1. Schematic diagram of the installation used in experimental method.

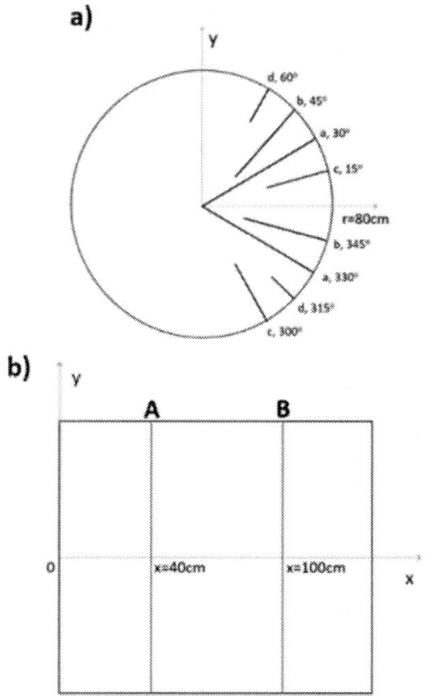

Figure 2. Visualisation of placing measurement probes: a) determining individual probes, b) coordinates of planes A and B.

The effectiveness of drying cylindrical straw bales (with dimensions: diameter 160 cm, length 140 cm) with exhaust gases was measured with a specially prepared recorder equipped with eight measurement probes. They were calibrated by producer on sample of straw with varied humidity. Capacitive relative humidity sensors and temperature sensors with the SHT11 device [15] were installed.

Measurement probes varied in length (80 cm, 60 cm, 40 cm and 20 cm denoted respectively by lowercase "a", "b", "c" and "d"), so it is possible to measure humidity and temperature in places with the preset distance from the axis of the straw bale. Figure 2 presents the visualisation of placement of the right probe in the bale during the measurement.

Measurements were carried out in two planes. They were distant from the beginning of the bale (x=0 – place where the drying mixture was pumped to the bale) by 40 cm (plane A, x=40) and by 100 cm (plane B, x=100). At first measurements were conducted in plane A, and then in plane B.

Experimental results are presented below. The curves show the change of temperature and humidity during the drying process. The prefix "A" and "B" at each label of the probes means the measurement plane according to Figure 2b, and the prefix "T" and "H" means the measurement of temperature and humidity, respectively.

This article presents the possibilities of drying straw in the modified drying process. After collecting a number of experiences in the straw drying process in the form of bales [16] some modifications of this process have been implemented, which involved the hollowing of a duct along the bale axis with the diameter of 63mm (the duct is marked as no. 3 in Figure 7) and applying the metal plate with the diameter of 1,000 mm to the opposite side of the bale (the plate marked as 4 in Figure 7) from the fitted drying medium. This modification was designed to allow the most radial propagation of the medium and to obtain the uniform temperature and humidity inside the bale in question.

3 Experimental stand

The curves in Figure 3-6 present the change of temperature and humidity for particular regions of the straw bale. "Combustion" – is the period, when the combustion of the insert took place in the form of straw in the biomass boiler. During this period the drying medium was supplied to the bale with the given temperature. After the combustion process the measurements were continued with the blowing air fan turned on. In this case the drying medium was the heated air after passing through the warmed but not operating boiler. As a result the medium had lower temperature comparing to the combustion period. Measurements were carried out in one straw bale with the potentially uniform initial humidity in the given measurement surface between 66-90% through the period of 280 min.

Changing temperature inside the dried straw bale in the A plane is presented in Figure 3. We can observe the rapid increase of temperature in the whole volume of the bale. Then, after approx. 30 min from beginning measurements the rapid temperature increase is reduced. In approx. 70 min of drying the highest temperature is achieved within the bale of about 45 °C. Only the "Aa30T" and "Aa330T" sensors located inside the bale recorded much higher temperatures of approx. 60 °C. After completion of the combustion process, temperature inside the bale started to slowly decrease. The bale layer represented by "Ab45T" sensors maintained a constant temperature after the completion of the combustion process. While the "Ab345T" sensor was characterised by the increase of the temperature from the moment of ending combustion until 170 min of the drying process. The temperature increase could be caused by the heat transfer from the layer from the bale axis to the more external layer described by the "Ab345T" sensor. Then, temperature started to slowly decrease until reaching the final value of 27 °C. The same final temperature has also been recorded in the location of installing the "Aa30T", "Aa330T" and "Ab45T" sensors. Table 1 presents the maximal temperatures for the given sensor in the plane obtained during experimental measurements and from CFD modelling described in the next chapter. One can observe a very good fit of the model to the experimental results in plane A. The best fit was achieved for the "Aa30T" sensor, from which 60 °C was achieved. The worst one characterised the "Ad60T" sensor, for which the difference of 6 °C was reached.

Table 1. The maximum temperatures reached for the location of the given sensor in individual planes in the experimental measurement and the CFD model.

	Plane A		Plane B	
Monitor Point	Experimental	CFD model	Experimental	CFD model
a30T	60 °C	60 °C	63 °C	63 °C
a330T	58 °C	60 °C	55 °C	63 °C
b45T	45 °C	42 °C	57 °C	60 °C
b345T	46 °C	42 °C	60 °C	62 °C
c15T	44 °C	40 °C	43 °C	41 °C
c300T	43 °C	41 °C	49 °C	44 °C
d60T	44 °C	38 °C	40 °C	41 °C
d315T	44 °C	39 °C	42 °C	42 °C

Figure 4 presents the change of humidity inside the bale of straw in plane A. The initial moisture value was approx. 66-90% and fluctuated within these limits to approx. 50 min of the drying process. At this point took place the rapid change of moisture in the place represented by "Aa30H" and "Aa330H" sensors. Until the completion of the combustion process in the boiler, the humidity in these places dropped until reaching 13%. The next layer represented by "Ab45H" and "Ab345H" sensors maintained at the constant level of 70-80% until 110 min of conducting data recording. Then the humidity started declining and reached the final value of 15%-20%, like for the "Aa30H" and "Aa330H" sensors. Humidity of more external layers during the whole drying process maintained at the permanent level. For "Ac15H" and "Ac300H" sensors it was 70-80%, and for "Ad60H" and "Ad315H" sensors 80-90%.

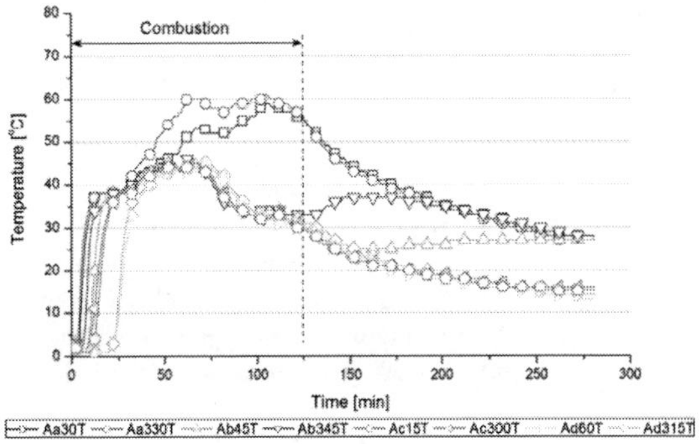

Figure 3. The change of temperature in the straw bale in the plane A.

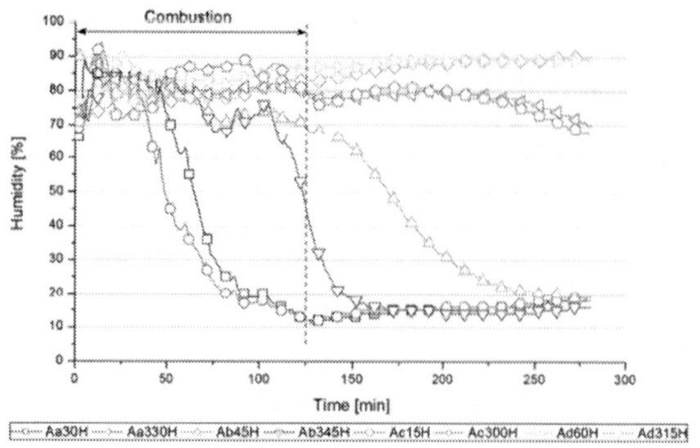

Figure 4. The change of humidity in the straw bale in the plane A.

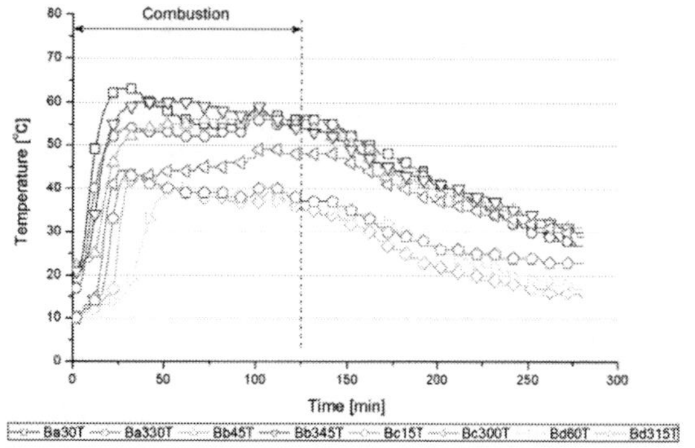

Figure 5. The change of temperature in the straw bale in the plane B.

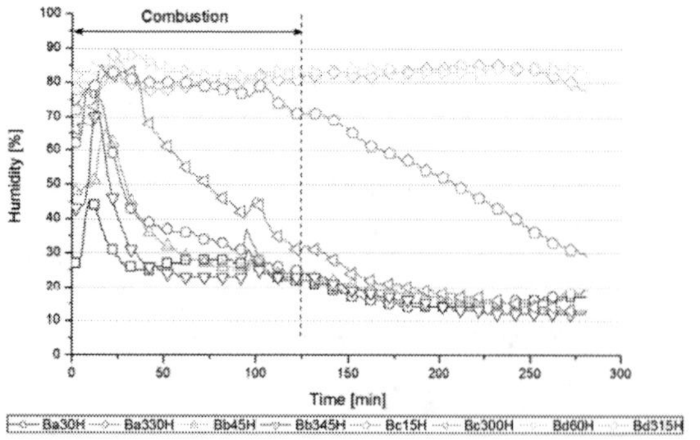

Figure 6. The change of humidity in the straw bale in the plane B.

After performing measurements in plane A, which lasted for 280 min, the observation of dynamics of straw bale drying has been started in plane B. Changing temperature inside the bale and straw humidity in plane B has been presented in Figure 5 and Figure 6. At first we observe the rapid increase of temperature in the bale axis layer, until reaching approx. 50-60 °C. Then, the next layers represented by proper sensors started heating later than the similar dynamics is observed in the case of the bale axis. Such heating of the individual layers in the straw bale is caused by the structure and physical properties of the straw. This is contributed by the dense packing of straw and the low coefficient of heat transfer [17, 18]. After achieving maximal temperatures by sensors representing the given layers in the period of 25-50 min from starting the measurement, the temperature maintained at the similar level until the end of the "Combustion" period. Then, temperatures started slowly dropping until reaching the value of 15-30°C. The highest temperatures and the fastest heating time was achieved by layers closest to the bale axis, and then the more external layers. Temperature distribution in case of plane B was slightly different than in case of plane A. This could be caused by the lower content of humidity or better radial flow of the drying factor caused by the application of the drying process modification, and as a result the higher pressures and better factor penetration occur closer to the place location than in the case of plane A.

Change of humidity during drying in plane B is presented in Figure 6. The initial bale humidity in plane B was between 27-82%. In the initial phase of the drying process the bale humidity rapidly increased, what was caused by the large increase of temperature in these places. After the period of approx. 20 min the humidity started declining. At the end of the "Combustion" process in the 125 min, the humidity for "Ba30H", "Ba330H", "Bb45H", "Bb345H" and "Bc300H" sensors was 22-30%. The "Bc15H" sensor recorded the value of 70%, and "Bd60H" and "Bd315H" sensors 80-85%. Moisture values at the end of the drying process for most sensors remained at the similar level as in the 125 min. Only the "Bc15H" sensor noted a significant decrease and at the end of the observation the humidity was 30%.

4 Numerical method

To perform series of the analyses of the straw dryer, a simple transient simulation of the heat and mass transfer inside the dryer has been performed. Vaporization from the straw bale pores initiates, when its temperature reaches the vaporization temperature of water. Due to this fact, it is reasonable to examine on the early stage of the numerical modelling the process of the heat transfer inside the bale.

Computational grid has been generated in steps - individually for each separated subregions. The subregions were created by division of the spatial geometry (subregions of the domains have been defined as so-called "parts") in ANSYS DesignModeler tool. This approach provides high fraction of the grid elements characterized by satisfying level of the quality. To evaluate mesh quality the "element metrics" coefficient has been used (equal 1 for the best elements and 0 for the worst). The fraction of grid elements of the quality larger than 0.75 is 57.5%. The total number of the elements in the computational grid is equal 525,000.

The model was divided into two domains: 1) fluid (air) and 2) solid (bale of straw). In the first case except the heat, mass transfer has been also considered due to necessity of simulation of the hot exhaust inflow (through the supplying pipes to first part of the dryer chamber).

The part of the domain representing the straw bale has been defined as a porous body, characterised by the porosity equal 0.5. Assumed parameters of the straw domain are summarised in the Table 2.

Table 2. Basic properties of the straw bale domain material

Parameter	Unit	Value
Density	$kg \cdot m^{-3}$	100
Specific heat	$J \cdot kg^{-1} \cdot K^{-1}$	1.51
Dynamic viscosity	$Pa \cdot s$	0.1
Thermal conductivity	$W \cdot m^{-1} \cdot K^{-1}$	0.03

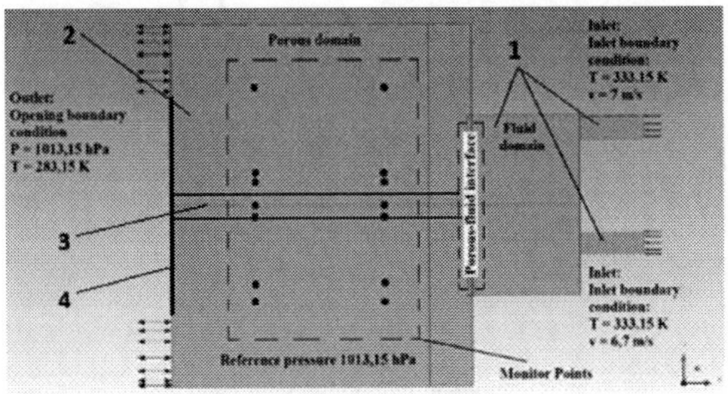

Figure 7. Chosen parameters and boundary conditions of the numerical simulation.

Gas heating the straw bale has been defined as an continuous fluid, characterized by the thermodynamic properties of air. Set of the boundary conditions has been defined. Values of the parameters, such as temperature and velocity of the inlet air (velocity of inlet air in each air duct was measured using anemometer), ambient pressure and temperature were determined based on the results of the experimental measurement. Detailed information about the preprocessing of the numerical model (boundary conditions and settings) are presented in Figure 7. Moreover, Figure 7 presents the air ducts and a special designed nozzle (1), straw bale (2), hole inside the bale (3) and metal plate (4).

Except the parameters, set of so-called monitor points have been defined (in Figure 7 – marked by dots in the crosses), which allowed to monitor temperature values calculated for the estimated points of location of the sensors used in the experimental tests of the dryer.

The k-epsilon model of turbulence has been implemented for the fluid domain. Such approach is reasonable in case of the fluid flow between hot exhaust gas inlets and the edge of the straw bale. Due to the high velocity of the fluid flow and small distance between the planes, the flow is turbulent. Only phenomena in central region of the straw bale were considered with special attention, so it was unnecessary to create advanced dense grid in regions near to the edges (lack of the dynamic changes of the pressure). Due to this fact, k-epsilon model of turbulence could be applied.

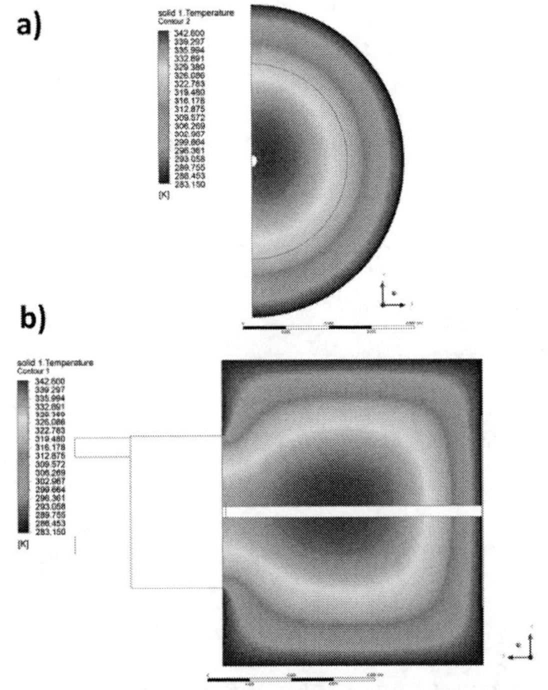

Figure 8. Temperature distribution in the straw bale in the agreed state: a) cross-section, b) longitudinal section.

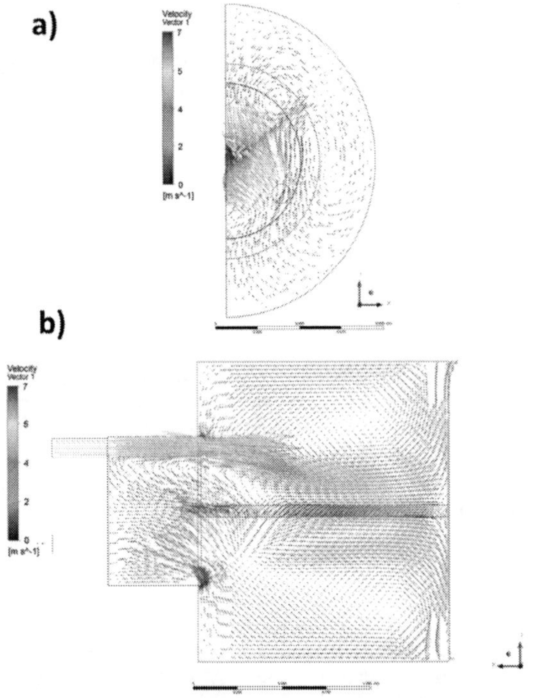

Figure 9. Distribution of the velocity vectors in the symmetry planes of the modelled domains: a) cross-section, b) longitudinal section.

Required residuals level for the solver has been set to 10^{-5}. Advection scheme has been defined as upwind (robust but less accurate than the high resolution mode). In first approach arbitrary chosen time scale were applied. Due to simplicity the time step was set as 1s. Using this time step dynamics of the temperature variation and other experimental parameters were well reflected in the simulation. Total time of the simulation was equal 1,500s (1,500 time steps). This time is sufficient to stabilize the temperature for each monitor points inside the straw bale. Longer simulation gives no essential improvements of the model output.

Figure 8 presents the cross-sections of the simulated solid domain (straw of bale). Colours on the planes correspond to appropriate value of the temperature in the last time step of the simulation, when state (temperature) of the straw bale was stable. The CFD simulation showed a very good temperature distribution around the straw bale perimeter. The highest temperatures were reached in the central part of the bale or in the plane presented in Figure 8a or Figure 8b, what also translates to the lowest humidity content. The further away from the bale axis, the lower the temperature. The same temperature distributions were achieved during experimental measurements.

Temperature distribution in the bale complies with the velocity vectors distribution, what is confirmed by the Figure 9. Relatively high temperature in the region of straw near to the nozzle is result of the turbulent movement of the fluid in the gas domain area.

The performance of modifications helped in the intensity of straw heating near the outlet of the dryer. In addition, the heat use has been improved, what translated into the more uniform heating of the straw bale and limitation of places where the heat was not supplied. After performing the CFD model, next modifications could be suggested, which should rely on optimising the hole inside the straw bale (its diameter, number and placement), a nozzle injecting the drying medium as well as the diameter of the metal plate.

5 Conclusions

The above article presented possibilities for drying straw in the form of bales. Experimental and numerical method was conducted for testing the effectiveness of drying and the transfer of heat and mass in the straw bale with the given boundary conditions. Experimental studies have shown that the internal layers of the bale are heated much faster. Additionally, the achieved temperatures are higher in the central part of the bale. After approx. 5 hours from starting the drying process, the satisfactory humidity of the straw was obtained in the bale axis of approx. 15-20%. Almost homogeneous humidity field was obtained after almost 10 hours of drying. After this time, only the most external layer maintained the humidity at the constant level of about 80%. The results obtained through the experimental and numerical way showed that the current drying process is satisfactory, because the bale heats evenly. The problem occurs for the external layer, where humidity remains at the constant, high level. After analysing the obtained results, the new, improved drying method was suggested, which will allow the most radial distribution of the drying medium. It will consist of optimising the hole inside the straw bale (its diameter, number and placement), a nozzle injecting the drying medium as well as the diameter of the metal plate. This will allow for more uniform heating of the bale in the whole volume, and thus its much shorter and more effective drying.

Acknowledgments

This work was financially supported by the Dean Grant of Faculty of Energy and Fuels at AGH University of Science and Technology and it was carried out by BioEcoMatic Project within KIC InnoEnergy Programme.

References

1. W. Goryl: Mater Thesis, AGH University of Science and Technology (Kraków, 2012)
2. W. Goryl, A. Guła, Zielona Planeta **112**, 16-19 (2014) [In Polish]
3. A. Guła, P. Wajss, W. Goryl, Przegląd Elektrotechniczny **5a**, 198-203 (2012)
4. M. Dakowski, S. Wiąckowski, *O energetyce dla użytkowników oraz sceptyków* (Fundacja Odysseum, Warszawa, 2005) [In Polish]
5. Central Statistical Office of Poland, *Procurement and prices of agri. prod. in 2014* (Warszawa, 2015)
6. Central Statistical Office of Poland, *Concise statistical yearbook of Poland 2015* (Warszawa 2015).
7. W. Goryl, *Proceedings of 11th International Conference ENEF*, Banska Bystrica (2014)
8. A. Ludwicka, A. Grzybek, Prob. of Agri. Eng., **2**, 101-111 (2010) [In Polish]
9. D. Kwaśniewski, Agric. Eng., **104**, 113-119 (2008)
10. A. Grzybek, *Straw – Implementation in Heating*, (Wydawnictwo ITP, Falenty, 2012)
11. V. Vladut, M. Chitoiu, A. Danciu, M. Militaru, C. Lehr, Bull. UASVM Agric., **67**, 292-300 (2010)
12. S. Kalinauskaite, A. Sakalauskas, E. Sarauskis, A. Jasinskas, Agron. Res., **11(2)**, 319-328 (2013)
13. H. Li, Q. Chen, X. Zhang, K. N. Finney, V. N. Sharifi, Appl. Therm. Eng., **35**, 71-80 (2012)
14. K. Sornek, M. Szubel, W. Goryl, E. Bożek, M. Filipowicz, Przem. Chem., **12(93)**, 2071-2076 (2012)
15. http://www.sensirion.com/fileadmin/user_upload/customers/sensirion/Dokumente/Humidity/Sensirion_Humidity_SHT1x_Datasheet_V5.pdf [access: 10 June 2015]
16. W. Goryl, *Proceedings of 2nd International Conference RENEWABLE ENERGY SOURCES engineering, technology, innovations - scientific conference*, Krynica (2015)
17. A. Thomson, P. Walker, Constr. Build. Mater., **68**, 135-141 (2014)
18. T. Ashour, H. Wieland, H. Georg, F.-J. Bockisch, W. Wu, Mater. Design, **31**, 4676-4685 (2010)

EPJ Web of Conferences **114**, 02034 (2016)
DOI: 10.1051/epjconf/201611402034

Clearance gap flow: extended pneumatic measurements and simulations by discontinuous Galerkin finite element method

Jindřich Hála[1,a], Martin Luxa[1], Ondřej Bublík[2], Helena Prausová[2], and Jan Vimmr[2]

[1] Institute of Thermomechanics AS CR, v.v.i., Academy of Sciences of the Czech Republic, Prague, Czech Republic
[2] NTIS - New Technologies for the Information Society, Faculty of Applied Sciences, University of West Bohemia, Pilsen, Czech Republic

Abstract. In the present paper, new results of measurements of the compressible viscous fluid flow in narrow channels with parallel walls under the conditions of aerodynamic choking are presented. Investigation was carried out using the improved test section with enhanced capability to accurately set the parallelism of the channel walls. The measurements were performed for the channels of the dimensions: length 100 mm, width 100 mm and for various heights in the range from 0.5 mm to 4 mm. The results in the form of distribution of the static pressure along the channel axis including the detailed study of the influence of the deviation from parallelism of the channel walls are compared with previous measurements and with numerical simulations performed using an in-house code based on Favre averaged system of Navier-Stokes equations completed with turbulence model of Spalart and Allmaras and a modification of production term according to Langtry and Sjolander. The spatial discretization of the governing equations is performed using the discontinuous Galerkin finite element method which ensures high order spatial accuracy of the numerical solution.

1 Introduction

The need to study compressible viscous fluid flow in narrow channels emerges from the occurrence of such flows in many engineering applications, for example in clearance gaps of the screw type compressors, during tip leakage flows in steam and gas turbine stages without shroud and in clearance gaps of the control valves of steam turbines. Such a widespread engineering application calls for need to explain and predict dissipative phenomena influencing thermodynamic losses and thus the efficiency of these machines.

The study follows earlier efforts of the authors, [1], and extends their most recent work, [2]. Similarly as in the case of previous experiments, the channel of the height varying from 0.5 to 4mm, length 100mm and width 100mm was chosen as the simplified model. However, the test section was improved to allow more accurate setting of the parallelism of the channel walls, because previous experiments showed that even small deviation have significant influence on the pressure distribution along the channel.

The aim of this study was to prove the increased capabilities of the new test section, namely to study the reproducibility of the experiments, and gather more reliable data for comparison with the results of numerical simulations, performed by an implicit solver based on the discontinuous Galerkin finite element method.

[a] e-mail: jindrich.hala@gmail.com

2 Experiment

2.1 Settings of the experiment

The experiments were carried out in the Aerodynamics Laboratory of the Institute of Thermomechanics AS CR, v.v.i. in Nový Knín. Scheme of the settings of the experiment can be seen in the Figure 1. The channel is formed by two parallel polished steel plates with the rounded leading edges (Figure 2) and by two glass windows. Unlike the test section used for the previous measurements, the current one was designed in order to have much higher stiffness to prevent the deformation of the walls. Moreover, eight adjusting screws were added (one in each corner of the lower and upper channel walls) to allow more precise setting of the parallelism of the gap.

During the measurement, the air continues from the laboratory through the channel and goes to the settling chamber in which the required back pressure p_b is maintained using the control nozzle downstream the settling chamber. Then the air continues through the pipes to the vacuum storage.

2.2 Methods of measurements

The upper wall of the channel was equipped with the static pressure taps, location of which can be seen in the Figure 3. For the pressure data acquisition, two 16 channel pressure scanners *PSI 9116* were used. The scanner integrates 16 silicon piezoresistive pressure sensors and microprocessor that performs a correction for sensor zero, span, linearity and thermal errors. The manufacturer guarantees system accuracy of up to ±0.05% of the full scale. The connection between the pressure scanners and pressure taps was

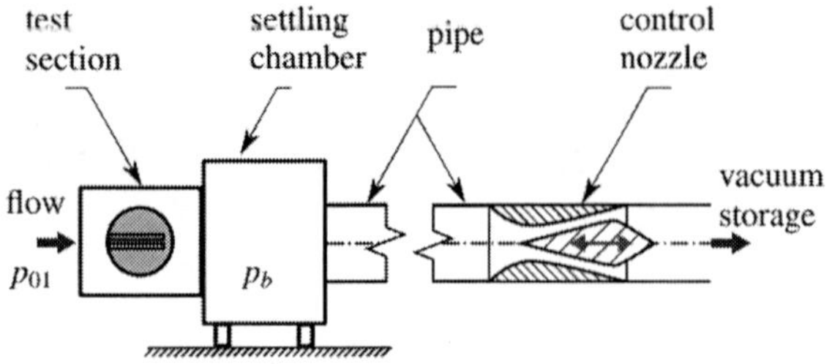

Fig. 1: Scheme of the set up of the experiment.

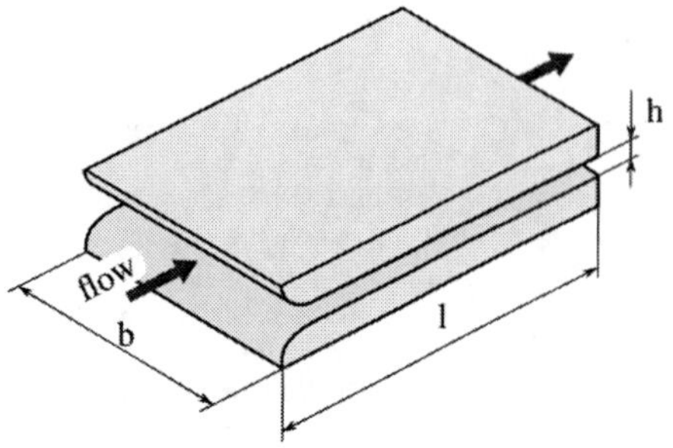

Fig. 2: Schematic view of the channel formed by two steel plates with marked main dimensions l, b and h (windows serving as a side walls are not depicted).

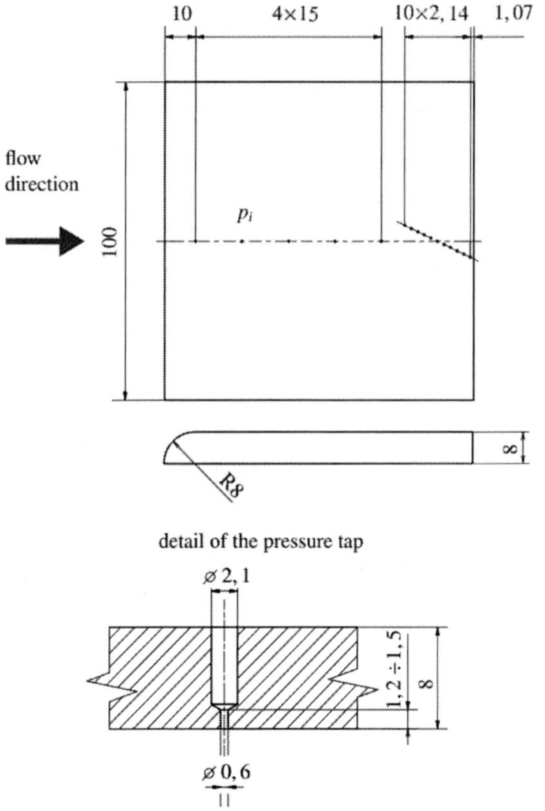

Fig. 3: Scheme of the upper wall of the minichannel of the lenght 100 mm with plotted positions of pressure taps and a detail view of the pressure tap.

realised using the plastic tubes of diameter 0.5 mm and length up to 0.5 m.

To obtain more information about the flow field in the channel, the optical methods were incorporated in the form of Mach-Zehnder interferometry, [2], [3]. This method allows to examine compressible fluid flow in the channel without intrusion of any probe, however, due to the small dimensions of the channel, quantitative evaluation of the interferograms is possible only in the case of gaps from height approximately $h = 2$ mm. Nevertheless, even for smaller gaps, flow field near the exit of the channel can be depicted in terms of isolines of index of refraction and used for qualitative assessment of the numerical simulation.

2.3 Results of the experiment

The experimental investigation was carried out for a channel of the height $h = 0.5, 1, 1.5, 2, 2.5, 3$ and 4 mm. Both optical and pneumatic measurements were performed for pressure ratios (ratio of the back pressure p_b to stagnation pressure p_{01}) $\pi = 0.189, 0.195, 0.308, 0.375$ and 0.457. Such pressure ratios were chosen in order to reach aerodynamic choking in the channel.

In the Figure 5 the comparison of the distribution of pressure ratio $\frac{p_i}{p_{01}}$ (ratio of the static pressure in each measuring point p_i to stagnation pressure p_{01}) along the channel axis for $\pi = 0.188$, obtained using pneumatic methods, can be seen for all investigated gap heights. For the gap heights up to 3 mm, we can observe steeper gradient of the pressure near the channel inlet than in the middle part of the channel. This might be caused by the abrupt change in the surface curvature on the interface of the rounded inlet and straight part of the channel, that is disturbing flow properties downstream such a point. In the middle part of the channel, the pressure distribution is linear and the slope of the line depends on the gap height, so that for narrow gap, the gradient of the pressure is higher due to more significant influence of the friction. With increasing gap height, the pressure gradient decreases towards the pressure distribution in the theoretical channel with negligible friction (zero pressure gradient). Following the linear part, there we can see significant pressure drop starting approximately at 80% of the channel length, where the flow accelerates towards sonic speed in the vicinity of the exit.

Pressure distributions for all other pressure ratios π are quite similar due to the phenomenon of aerodynamic choking. The only difference could be observed in the flow structure downstream the exit of the channel (Figure 4).

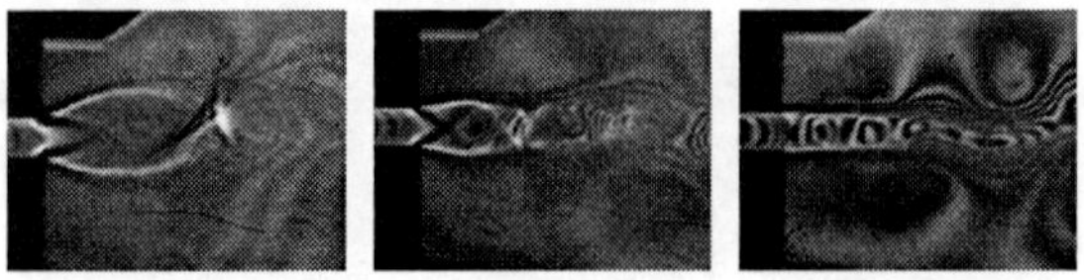

(a) (b) (c)

Fig. 4: Interferograms of the flow field downstream the exit of the channel of the height 4 mm for pressure ratios $\pi = 0.194$ (a), $\pi = 0.308$ (b) and $\pi = 0.458$ (c)

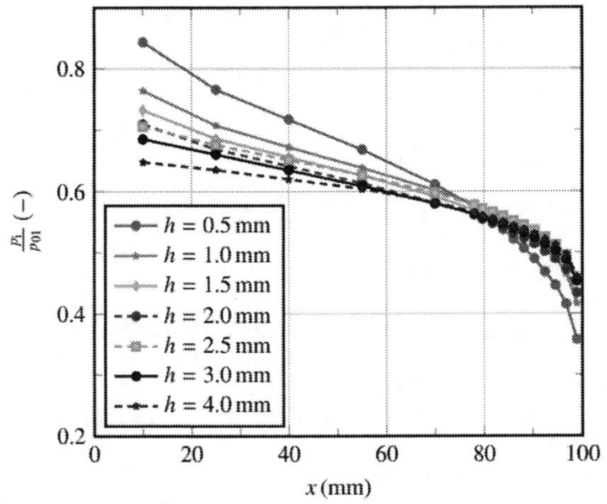

Fig. 5: Distribution of the pressure ratio $\frac{p_i}{p_{01}}$ along the minichannel axis for various channel heights obtained from pneumatic measurements, $\pi = 0.188$.

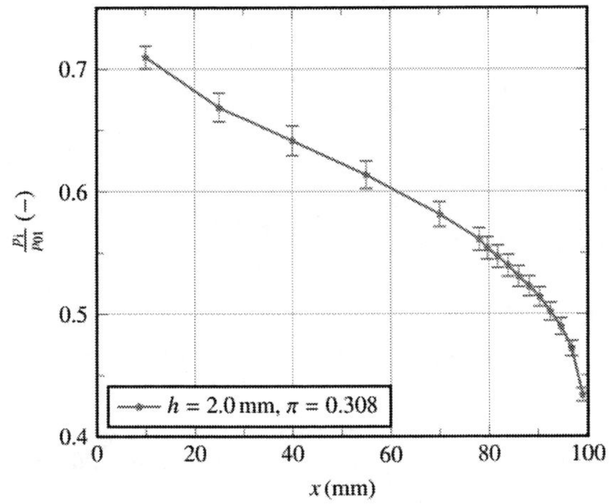

Fig. 6: Mean value of the ratio of the static pressure and stagnation pressure p_{01} along the channel axis for pressure ratio $\pi = 0.308$, obtained using 10 pneumatic measurements for channel height $h = 2$ mm, each with different adjustment. Error bars indicate the value of standard deviation.

The other part of the experiment was focused on the reproducibility of the results. For this purpose, the gap of the height 2 mm was used. Then 10 measurements were performed, each for slightly different adjustment of the adjustment screws. Deviation from the parallelism was measured in the locations of channel corners using feeler gauges and did not exceed 0.1 mm (5%). Both combinations of slightly converging and diverging channels were explored as well as channels with slight transverse slope. Results can be seen in the plot in the Figure 6, where the mean value of mentioned 10 measurements is plotted together with error bars indicating the value of standard deviation.

3 Numerical simulation

Numerical simulations of the air flow through the considered narrow channel were performed to further explore compressible fluid flow properties in this type of channels and demonstrate suitability of the in-house code based on the discontinuous Galerkin finite element method for that kind of computations. The width of the channel enables to simplify the geometry for numerical simulation into two dimensions.

3.1 Mathematical model

Mathematical model used to describe the problem is based on Favre-averaged system of Navier-Stokes equations written in dimensionless form as

$$\frac{\partial \bar{\varrho}}{\partial t} + \frac{\partial (\bar{\varrho}\tilde{u}_j)}{\partial x_j} = 0, \tag{1}$$
$$\frac{\partial}{\partial t}(\bar{\varrho}\tilde{u}_i) + \frac{\partial}{\partial x_j}(\bar{\varrho}\tilde{u}_i\tilde{u}_j + \bar{p}\delta_{ij}) = \frac{1}{Re}\frac{\partial}{\partial x_j}(\tilde{t}_{ij} + \tau_{ij}),$$
$$\frac{\partial}{\partial t}(\bar{\varrho}\tilde{e}) + \frac{\partial}{\partial x_j}(\bar{\varrho}\tilde{e}\tilde{u}_j + \bar{p}\tilde{u}_j) = \frac{1}{Re}\frac{\partial}{\partial x_j}\Big[(\tilde{t}_{ij} + \tau_{ij})\tilde{u}_i + $$
$$+ \frac{\kappa}{\kappa - 1}\left(\frac{\mu}{Pr} + \frac{\mu_t}{Pr_t}\right)\frac{\partial}{\partial x_j}\left(\frac{\bar{p}}{\bar{\varrho}}\right)\Big],$$

where $\bar{\varrho}$, $\tilde{u}_i$, $\bar{p}$, $\tilde{e}$ are dimensionless mass- or time-averaged values of density, velocity, pressure and energy. Mass-averaged stress tensor and Reynolds stress tensor are given by $\tilde{t}_{ij} = 2\mu\tilde{S}_{ij}$, $\tau_{ij} = 2\mu_t\tilde{S}_{ij}$, where $\tilde{S}_{ij} = \frac{1}{2}\left(\frac{\partial \tilde{u}_i}{\partial x_j} + \frac{\partial \tilde{u}_j}{\partial x_i}\right) - \frac{1}{3}\delta_{ij}\frac{\partial \tilde{u}_k}{\partial x_k}$.

A one-equation turbulence model of Spalart and Allmaras [5] is implemented to complete the Favre-averaged system of Navier-Stokes equations. It is given by transport equation for eddy viscosity $\tilde{\nu}$ written as

$$\frac{\partial (\bar{\varrho}\tilde{\nu})}{\partial t} + \frac{\partial (\bar{\varrho}\tilde{\nu}\tilde{u}_j)}{\partial x_j} = \frac{1}{Re}\frac{1}{\sigma}\left[\frac{\partial}{\partial x_j}\left((\mu + \bar{\varrho}\tilde{\nu})\frac{\partial \tilde{\nu}}{\partial x_j}\right) + \right. \tag{2}$$
$$\left. + \bar{\varrho}c_{b2}\frac{\partial \tilde{\nu}}{\partial x_j}\frac{\partial \tilde{\nu}}{\partial x_j}\right] + PTM\, c_{b1}\bar{\varrho}\tilde{S}\tilde{\nu} - \frac{1}{Re}c_{w1}f_w\frac{1}{\bar{\varrho}}\left(\frac{\bar{\varrho}\tilde{\nu}}{D}\right)^2,$$

which is completed by a set of auxiliary relations

$$\mu_t = \bar{\varrho}\tilde{\nu}f_{v1},$$
$$\tilde{S} = \bar{\Omega} + \frac{1}{Re}\frac{\tilde{\nu}}{\kappa^2 D^2}f_{v2},$$
$$f_w = g\left(\frac{1 + c_{w3}^6}{g^6 + c_{w3}^6}\right)^{\frac{1}{6}},$$
$$g = r + c_{w2}\left(r^6 - r\right),$$
$$r = \min\left(\frac{\tilde{\nu}}{Re\,\tilde{S}\kappa^2 D^2}, 10\right),$$
$$f_{v1} = \frac{\chi^3}{\chi^3 + c_{v1}^3}, \quad \chi = \frac{\tilde{\nu}}{\nu},$$
$$f_{v2} = 1 - \frac{\chi}{1 + \chi f_{v1}}$$

and constants $c_{b1} = 0.1355$, $c_{b2} = 0.622$, $\sigma = \frac{2}{3}$, $\kappa = 0.41$, $c_{w1} = \frac{c_{b1}}{\kappa^2} + \frac{1+c_{b2}}{\sigma}$, $c_{w2} = 0.3$, $c_{w3} = 2.0$, $c_{v1} = 7.1$, where Ω is the vorticity magnitude and D is the distance to the nearest wall.

The governing equations are then completed by a transitional model developed by Langtry and Sjolander (2002), [5]. Principle of this model is to restrict the production of turbulent kinetic energy in $k-\omega$ model by multiplying it by term PTM (production term modification) with values from 0 to 1. In the presented work the PTM term is applied to the eddy viscosity production term in transport equation (2). Value of PTM is computed from flow field variables and enables to switch gradually from laminar to turbulent flow regime. The model is based on the vorticity Reynolds number

$$Re_v = \frac{\varrho D^2}{\mu}\Omega.$$

The PTM term takes the form of

$$PTM = 1 - PTM_1\, F_3,$$

where function F_3 is defined as

$$F_3 = \exp\left[-\left(\frac{R_T}{6.5}\right)^4\right]$$

and viscosity ratio is simply $R_T = \frac{\mu_t}{\mu}$. PTM_1 term is given by two following polynomials

$$PTM_1 = \begin{cases} 1- \begin{bmatrix} (3.82E-04)Re_v - \\ (3.94E-07)Re_v^2 + \\ (1.43E-10)Re_v^3 \end{bmatrix}, & Re_v < 1000, \\ 1- [0.12 + (1.0E-05)Re_v], & Re_v \geq 1000. \end{cases}$$

System of averaged Navier-Stokes equations (1) and the transport equation of turbulence model (2) can be written together as

$$\frac{\partial \boldsymbol{w}}{\partial t} + \sum_{j=1}^{2} \frac{\partial}{\partial x_j} \boldsymbol{f}_j(\boldsymbol{w}) = $$
$$= \frac{1}{Re} \sum_{j=1}^{2} \frac{\partial}{\partial x_j} \boldsymbol{f}_j^v(\boldsymbol{w}, \nabla \boldsymbol{w}) + \boldsymbol{p}(\boldsymbol{w}, \nabla \boldsymbol{w}), \quad (3)$$

where $\boldsymbol{w}$ is a vector of conservative variables, $\boldsymbol{f}_j(\boldsymbol{w})$ are the inviscid numerical fluxes, $\boldsymbol{f}_j^v(\boldsymbol{w}, \nabla\boldsymbol{w})$ are the viscid numerical fluxes and $\boldsymbol{p}(\boldsymbol{w}, \nabla\boldsymbol{w})$ is a production term. The vectors are written as

$$\boldsymbol{w} = \begin{bmatrix} \bar{\varrho} \\ \bar{\varrho}\tilde{u}_1 \\ \bar{\varrho}\tilde{u}_2 \\ \bar{\varrho}\tilde{e} \\ \bar{\varrho}\tilde{\nu} \end{bmatrix}, \boldsymbol{f}_1(\boldsymbol{w}) = \begin{bmatrix} \bar{\varrho}\tilde{u}_1 \\ \bar{\varrho}\tilde{u}_1^2 + \bar{p} \\ \bar{\varrho}\tilde{u}_1\tilde{u}_2 \\ \tilde{u}_1(\bar{\varrho}\tilde{e} + \bar{p}) \\ \bar{\varrho}\tilde{\nu}\tilde{u}_1 \end{bmatrix}, \boldsymbol{f}_2(\boldsymbol{w}) = \begin{bmatrix} \bar{\varrho}\tilde{u}_2 \\ \bar{\varrho}\tilde{u}_1\tilde{u}_2 \\ \bar{\varrho}\tilde{u}_2^2 + \bar{p} \\ \tilde{u}_2(\bar{\varrho}\tilde{e} + \bar{p}) \\ \bar{\varrho}\tilde{\nu}\tilde{u}_2 \end{bmatrix},$$

$$\boldsymbol{f}_1^v(\boldsymbol{w}, \nabla\boldsymbol{w}) = \begin{bmatrix} 0 \\ \tilde{t}_{11} + \tau_{11} \\ \tilde{t}_{12} + \tau_{12} \\ \sum_{i=1}^{2} \tilde{u}_i(\tilde{t}_{i1} + \tau_{i1}) + \frac{\kappa}{\kappa-1}\left(\frac{\mu}{Pr} + \frac{\mu_t}{Pr_t}\right)\frac{\partial}{\partial x_1}\left(\frac{\bar{p}}{\bar{\varrho}}\right) \\ \frac{1}{\sigma}\left(\mu + \bar{\varrho}\tilde{\nu}\right)\frac{\partial \tilde{\nu}}{\partial x_1} \end{bmatrix},$$

$$\boldsymbol{f}_2^v(\boldsymbol{w}, \nabla\boldsymbol{w}) = \begin{bmatrix} 0 \\ \tilde{t}_{12} + \tau_{12} \\ \tilde{t}_{22} + \tau_{22} \\ \sum_{i=1}^{2} \tilde{u}_i(\tilde{t}_{i2} + \tau_{i2}) + \frac{\kappa}{\kappa-1}\left(\frac{\mu}{Pr} + \frac{\mu_t}{Pr_t}\right)\frac{\partial}{\partial x_2}\left(\frac{\bar{p}}{\bar{\varrho}}\right) \\ \frac{1}{\sigma}\left(\mu + \bar{\varrho}\tilde{\nu}\right)\frac{\partial \tilde{\nu}}{\partial x_2} \end{bmatrix},$$

$$\boldsymbol{p}(\boldsymbol{w}, \nabla\boldsymbol{w}) = \begin{bmatrix} 0 \\ 0 \\ 0 \\ 0 \\ \frac{1}{Re}\frac{1}{\sigma}\bar{\varrho}c_{b2}\left(\frac{\partial\tilde{\nu}}{\partial x_1}\frac{\partial\tilde{\nu}}{\partial x_1} + \frac{\partial\tilde{\nu}}{\partial x_2}\frac{\partial\tilde{\nu}}{\partial x_2}\right) + \\ PTM\ c_{b1}\bar{\varrho}\tilde{S}\tilde{\nu} - \frac{1}{Re}c_{w1}f_w\frac{1}{\bar{\varrho}}\left(\frac{\bar{\varrho}\tilde{\nu}}{D}\right)^2 \end{bmatrix}.$$

3.2 Discretization

The spatial discretization of equations (3) is performed using the discontinuous Galerkin finite element method (DGFEM), [6], [7], [8].

Let $\mathcal{T}_h = \{\Omega_k : \Omega_k \in \Omega, k = 1, \ldots, K\}$ be a triangulation of the computational domain $\Omega \in \mathcal{R}^2$ and $\Gamma = \Gamma^{int} \cup \Gamma^b$ be the set of inner Γ^{int} and boundary Γ^b edges. The solution of equations (3) is considered in the space of vector functions $\mathbf{S}_h = S_h \times S_h \times S_h \times S_h$, where

$$S_h = \{v(\mathbf{x}) : v(\mathbf{x}) \in \mathcal{P}^q(\Omega_k) \text{ if } \mathbf{x} \in \Omega_k,$$
$$v(\mathbf{x}) = 0 \text{ if } \mathbf{x} \notin \Omega_k, \forall\Omega_k \in \mathcal{T}_h\}$$

and $\mathcal{P}^q(\Omega_k)$ is the space of polynomials of q-th order at the most.

Multiplying the system of equations by a test function $\mathbf{v} \in \mathbf{S}_h$, integrating it over the element Ω_k and using the Green's theorem, the following integral identity is obtained

$$\int_{\Omega_k} \frac{\partial \boldsymbol{w}}{\partial t} \cdot \boldsymbol{v}\, \mathrm{d}\Omega - \int_{\Omega_k} \sum_{s=1}^{2} \boldsymbol{f}_s(\boldsymbol{w}) \cdot \frac{\partial \boldsymbol{v}}{\partial x_s}\, \mathrm{d}\Omega +$$
$$+ \oint_{\partial\Omega_k} \sum_{s=1}^{2} \boldsymbol{f}_s(\boldsymbol{w}) n_s \cdot \boldsymbol{v}\, \mathrm{d}l = \frac{1}{Re}\oint_{\partial\Omega_k} \sum_{s=1}^{2} \boldsymbol{f}_s^v(\boldsymbol{w}, \nabla\boldsymbol{w}) n_s \cdot \boldsymbol{v}\, \mathrm{d}l -$$
$$\frac{1}{Re}\int_{\Omega_k} \sum_{s=1}^{2} \boldsymbol{f}_s^v(\boldsymbol{w}, \nabla\boldsymbol{w}) \cdot \frac{\partial \boldsymbol{v}}{\partial x_s}\, \mathrm{d}\Omega + \int_{\Omega_k} \boldsymbol{p}(\boldsymbol{w}, \nabla\boldsymbol{w}) \cdot \boldsymbol{v}\, d\Omega.$$

The key part of the method is an approximation of curve integrals. They are evaluated in the same way as it is done in the case of finite volume method. The inviscid flux is evaluated as

$$\sum_{s=1}^{2} \boldsymbol{f}_s(\boldsymbol{w}) n_s \approx \mathcal{H}(\boldsymbol{w}^+, \boldsymbol{w}^-), \quad \text{on } \Gamma^{int} \in \partial\Omega,$$

$$\sum_{s=1}^{2} \boldsymbol{f}_s(\boldsymbol{w}) n_s \approx \mathcal{H}(\boldsymbol{w}^b, \boldsymbol{w}^-), \quad \text{on } \Gamma^b \in \partial\Omega,$$

where $\mathcal{H}$ is Lax-Friedrichs flux, [9], $\boldsymbol{n} = [n_1, n_2]$ is vector of outer normal, values $\boldsymbol{w}^-$ resp. $\boldsymbol{w}^+$ are the left resp. right limits of the vector $\boldsymbol{w}$ on the edge Γ and $\boldsymbol{w}^b$ is a boundary value of $\boldsymbol{w}$ computed from boundary conditions.

Similarly, the value of viscous flux and the value of $\boldsymbol{w}$ on the edge for gradient equation are evaluated using the appropriate numerical fluxes. In present paper, the approximation known as interior penalty (IP) method was chosen, [8].

Let $\boldsymbol{w}_k$ be a part of solution $\boldsymbol{w}$ corresponding to control element Ω_k and $\varphi_{k,i}$ be i-th basis function of the space $\mathcal{P}^q(\Omega_k)$. The m components w_k^m, $m = 1, 2, \ldots 4$, of vector $\boldsymbol{w}_k$ can be expressed as a linear combination

$$w_k^m = \sum_{i=1}^{q} w_{k,i}^m(t)\varphi_{k,i}(\boldsymbol{x}). \quad (4)$$

Inserting it into the integral identity and evaluating the volume and curve integrals using the Gauss integration rules of appropriate order, the following system of ordinary differential equations is obtained

$$\mathbf{M}\frac{\mathrm{d}\mathbf{W}}{\mathrm{d}t} = \mathbf{R}(\mathbf{W}), \tag{5}$$

where vector $\mathbf{W}$ consists of coefficients of linear combination (4) and $\mathbf{M}$ is a mass matrix. Linear Lagrange polynomials are chosen as the basis functions of the space $\mathcal{P}^q(\Omega_k)$.

Time integration of equation (5) is performed using the implicit backward Euler method in the form

$$\left[\frac{\boldsymbol{M}}{\Delta t} - \frac{\partial \boldsymbol{R}}{\partial \boldsymbol{W}}(\boldsymbol{W}^n)\right]\Delta \boldsymbol{W} = \boldsymbol{R}(\boldsymbol{W}^n), \qquad \boldsymbol{W}^{n+1} = \boldsymbol{W}^n + \Delta \boldsymbol{W}. \tag{6}$$

The system of linear equations (6) is solved iteratively using the GMRES method with block diagonal Jacobi preconditioner.

3.3 Numerical results and comparison with experimental data

Since the gap of the height 2 mm and pressure ratio π = 0.308 was chosen to perform repeated measurements to explore reproducibility of the results (Figure 6), air flow for the same conditions was simulated in the first place. Distribution of static to stagnation pressure ratio along the axis channel is displayed in Figure 7. Very good agreement of the numerical and experimental results can be seen as the numerical data fit into the range of the standard deviation. The transitional model as presented above was used.

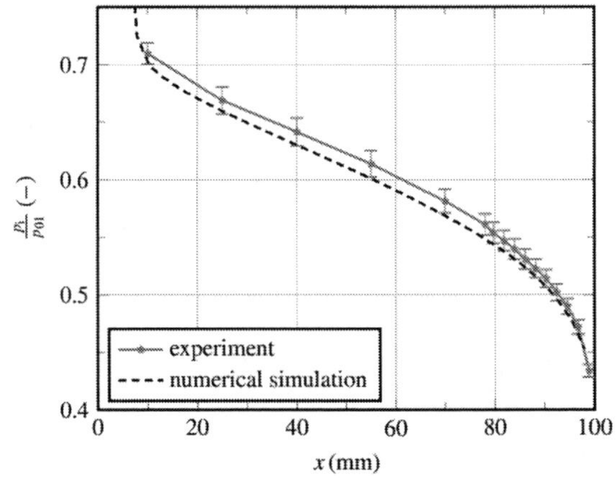

Fig. 7: Distribution of static to stagnation pressure ratio along the channel axis, h = 2 mm, π = 0.308: numerical and experimental results.

The aim of the numerical study is also to be able to compute properly the air flow behind the narrow channel. Qualitative comparison of computed density contour lines and the interferogram can be seen in Figures 8 and 9. The sequence of shock waves forming in the wake is captured well.

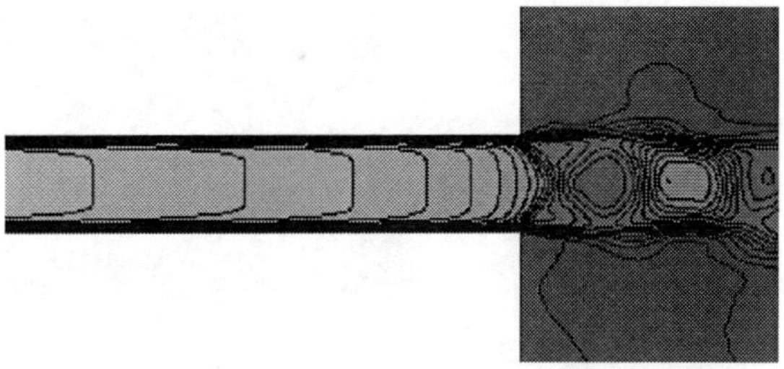

Fig. 8: Density isolines at the outflow, obtained from simulation with transitional model, h = 2 mm, π = 0.308.

Fig. 9: Interferogram at the outflow, h = 2 mm, π = 0.308.

Compressible flow in channels of the other heights was simulated as well. Distribution of static to stagnation pressure ratio along the axis channel of height 4 mm and pressure ratio π = 0.194 is displayed in Figure 10. Laminar, fully turbulent and transitional numerical results can be seen in comparison with experimental data.

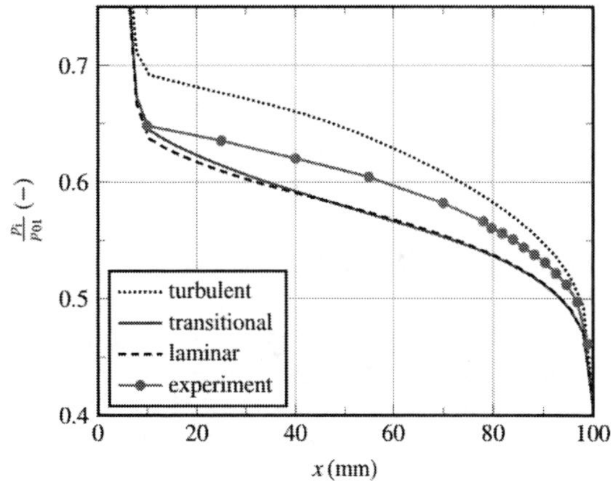

Fig. 10: Distribution of static to stagnation pressure ratio along the channel axis, h= 4 mm, π = 0.194: numerical and experimental results.

As in the case of 2 mm high channel, the air flow behind the channel of height 4 mm is also compared to experimental results, in Figures 11 and 12.

The good agreement in the pressure ratio at the beginning of the gap (position x = 0.01 m) in case of transitional modelling and the difference between experiment and sim-

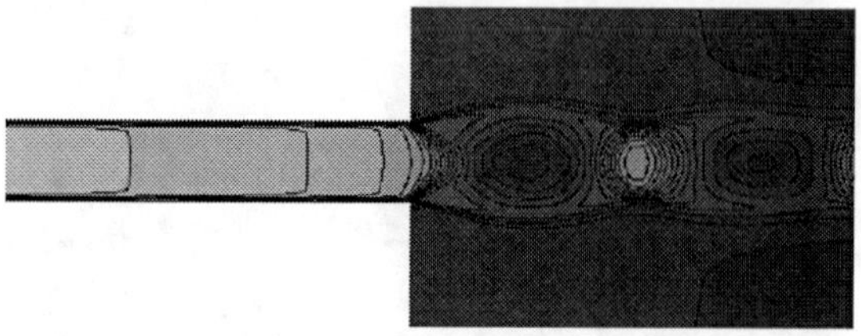

Fig. 11: Density isolines at the outflow, obtained from simulation with transitional model, h = 4 mm, π = 0.194.

ulation further downstream indicate the need to more appropriately capture the transition from laminar to turbulent flow in the channel.

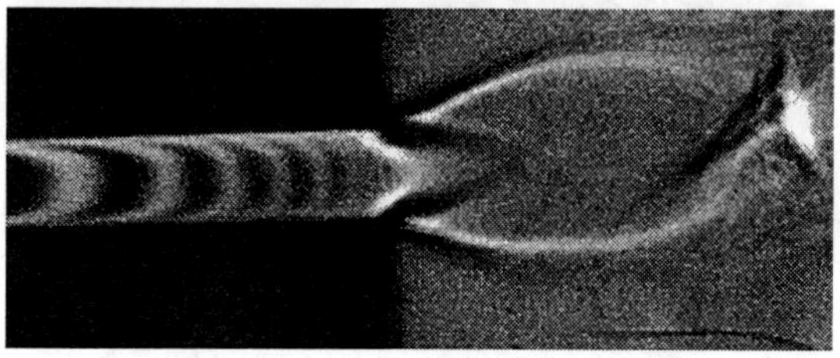

Fig. 12: Interferogram at the outflow, h = 4 mm, π = 0.194.

3.4 Summary

Viscous fluid flow in narrow gaps have been studied experimentally and the results were compared with numerical simulations. Improved test section proved to perform well during the experiments and allowed to gather new data for wider range of the gap heights. Moreover, the data obtained from reproduced measurements of the gap height $h = 2$ mm provided valuable data about the reproducibility of the measurements. Even though the results of the numerical simulations using transitional model of turbulence correspond well with experimental data, it must be said that the nature of the transition is still unexplored and will be subject to the following research as well as the study of wall friction.

Acknowledgement: Institutional support RVO: 61388998 is gratefully acknowledged. This publication was also supported by the project LO1506 of the Czech Ministry of Education, Youth and Sports.

References

1. Vimmr J., Klášterka H., Hajžman M., Luxa M., Dvořák R.: Journal of Thermal Science, 19(4), 289 -294 (2010).
2. Prausová H., Bublík O., Vimmr J., Luxa M., Hála J.: EPJ Web of Conferences 92, 02073 (2015).
3. M. Mokrý, *Aerodynamic measurements using Mach-Zehnder interferometer*. Scientific Report Z-239/67. Institute of Thermomechnics AS CR, v.v.i., Prague (1967).
4. Spalart P. R., Allmaras S. R.: A One-Equation Turbulence model for Aerodynamic Flows. AIAA Paper 92-439 (1992).
5. Langtry R. B., Sjolander S. A.: Prediction of Transition for Attached and Separated Shear Layers in Turbomachinery. AIAA Paper 2002-3641 (2002).
6. Reed W. H., Hill T. R.: Los Alamos Scientific Laboratory Report LA-UR-73-479 (1973).
7. Cockburn B., Shu C.-W.: Journal of Scientific Computing 16, 173 - 261 (2001).
8. Dolejší V., Holík M., Hozman J.: Journal of Computational Physics 230(11), 4176-4200 (2011).
9. Toro E. F.: *Riemann Solvers and Numerical Methods for Fluid Dynamics*, Springer, Heidelberg, pp. 325 (1999).

EPJ Web of Conferences **114**, 02035 (2016)
DOI: 10.1051/epjconf/201611402035

Signals features extraction in liquid-gas flow measurements using gamma densitometry. Part 1: time domain

Robert Hanus[1,a], Marcin Zych [2], Leszek Petryka[2], Marek Jaszczur[2] and Paweł Hanus[3]

[1]*Rzeszow University of Technology, 35-959 Rzeszow, Poland*
[2]*AGH University of Science and Technology, 30-059 Krakow, Poland*
[3]*Test & Measurement Solutions sp. z o.o., 53-025 Wrocław, Poland*

Abstract. The paper presents an application of the gamma-absorption method to study a gas-liquid two-phase flow in a horizontal pipeline. In the tests on laboratory installation two ^{241}Am radioactive sources and scintillation probes with NaI(Tl) crystals have been used. The experimental set-up allows recording of stochastic signals, which describe instantaneous content of the stream in the particular cross-section of the flow mixture. The analyses of these signals by statistical methods allow to determine the mean velocity of the gas phase. Meanwhile, the selected features of signals provided by the absorption set, can be applied to recognition of the structure of the flow. In this work such three structures of air-water flow as: plug, bubble, and transitional plug – bubble one were considered. The recorded raw signals were analyzed in time domain and several features were extracted. It was found that following features of signals as the mean, standard deviation, root mean square (RMS), variance and 4th moment are most useful to recognize the structure of the flow.

1 Introduction

Two-phase flow i.e. flow including gas and liquid phase commonly occurs in industry, especially in nuclear, chemical and petrochemical engineering. To fully describe such flow, it is required to determine variation of the velocity or flow rate for each fluid phase, as well as their mixing and distribution between the transported streams. Depending on the liquid phase velocity, viscosity and existence of turbulence, the same volume of gas may be transported in different ways, evolving from bubble up to annular flow. Selected structures in the case of liquid-gas flow in horizontal pipeline are presented in figure 1 [1].

Currently, such flows may be examined by using several methods, such as computer tomography (resistive, capacitive, optical and X-rays), Coriolis flowmeters, optical equipment, PIV, LDA and nuclear techniques [2-12]. These latter are also used by the authors of this article in studies of two-phase flows [13-21].

The nuclear methods are based on application of radiotracers or on a usage of a sealed gamma ray sources for the flow analysis. The last gamma-absorption method is particularly convenient for continuous non-invasive observation of two-phase flows, providing velocity of gas phase as well as a selected information about flow structure. That goal is achieved by statistical analysis of stochastic signals delivered by scintillation probes mounted outside the analysed stream as detectors of the radiation.

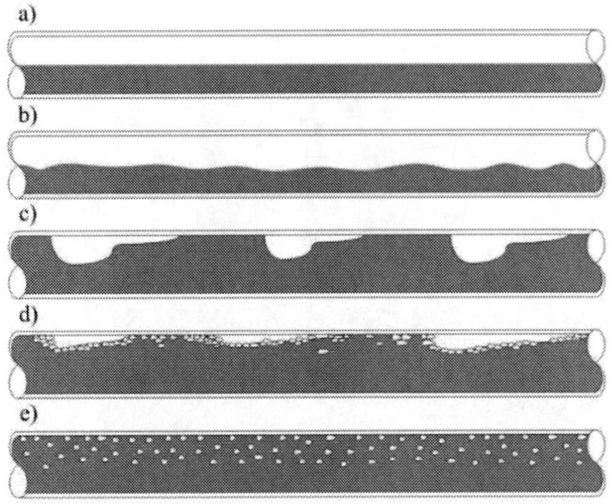

Figure 1. The selected liquid-gas flow patterns in horizontal pipeline: a) stratified smooth, b) stratified wavy, c) plug, d) transitional plug – bubble, e) dispersed bubble.

Flow regimes identification inspires many studies of two-phase flows. A quantitative description of flow regimes can be obtained by the analysis of the signals registered in radioisotope measurements. According to [22-31] the time-series analysis can be made in time domain, frequency domain (spectral analysis) or state-space domain (chaos analysis).

In this work the signals obtained by using gamma-absorption set were analyzed in time domain. Several

[a] Corresponding author: rohan@prz.edu.pl

features of signals, e.g., mean, standard deviation, kurtosis, skewness, root mean square (RMS), variance and 4th order moment were extracted. It was examined which features of signals are the most useful to recognize the flow structure.

2 Gamma-absorption in two-phase flow evaluation

The gamma absorption is based on exponential decreasing of a monoenergetic gamma beam in function of composition and geometry of absorbent [9]. The changes of the intensity of radiation are recorded by the scintillation detectors and converted into output electrical impulses [7].

The typical gamma-absorption measuring set is presented in figure 2.

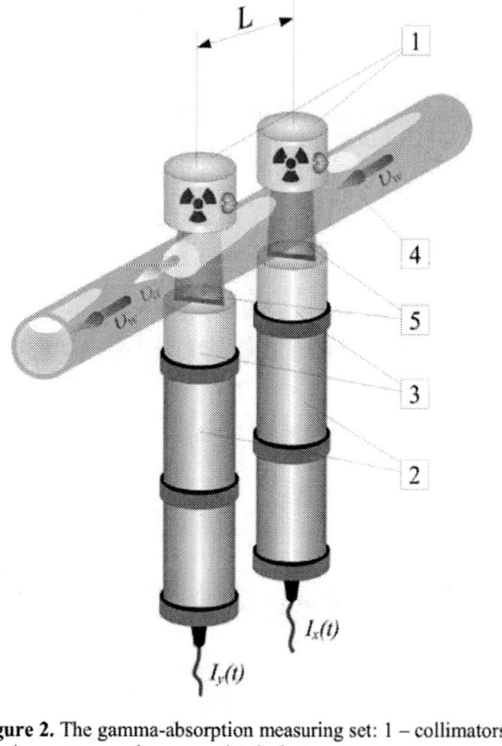

Figure 2. The gamma-absorption measuring set: 1 – collimators shaping gamma beam emitted by gamma ray sources, 2 – scintillation probes, 3 – detectors' collimators, 4 – pipeline, 5 – the main beam of gamma rays, υ_a – velocity of gas phase, υ_W – velocity of water.

Two sealed radioactive sources emit gamma radiation beams shaped by collimators (1). Photons pass through the pipeline with analyzed two-phase flow (air-water mixture) (4) and detector's collimator (3) before achieved scintillation probes (2). In the presented experiment two sets consisting of linear ^{241}Am sources with an activity of 100 mCi and probes with NaI(Tl) scintillation detectors are placed at a distance of $L = 97$ mm between them. At the outputs of two scintillation probes, count signals $I_x(t)$ and $I_y(t)$ are recorded for the first and second probe respectively.

3 Experimental set-up

The experimental set-up together with measuring equipment described above, was applied in the Industrial Radiometry Laboratory of the AGH University of Science and Technology in Krakow, Poland. Scheme of experimental set-up has been presented in figure 3.

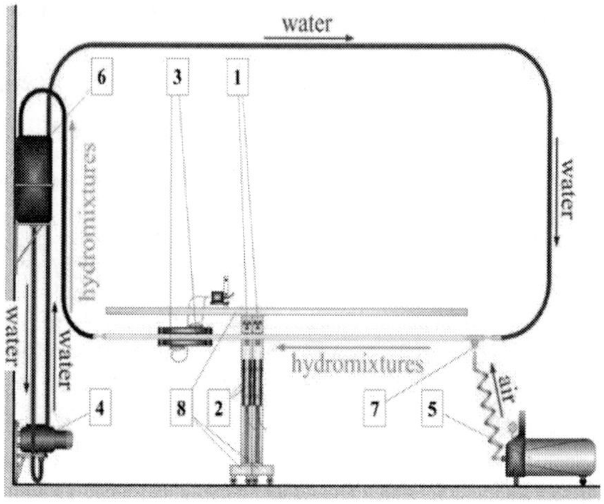

Figure 3. Scheme of experimental set-up: 1 - gamma-ray sealed source, 2 – scintillation probe, 3 – ultrasound flow meter, 4 – pump, 5 – compressor, 6 – air-removing container, 7 – air nozzle, 8 – shifting system of gamma-absorption set.

The test section consists of a transparent acrylic glass pipe with internal diameter of 30 mm and a length of 4.5 m. Water is pumped by a rotary pump (4) and air is provided by the compressor (5) by an injector nozzle (7). Due to this, the controlled mixture of water and air, fills the measuring pipe, and flows up to the air removing tank (6). The measuring system consisting of two γ-radioactive sources (1), two scintillation probes (2) and the system of data acquisition and analysis. The measuring set is mounted on a special trolley (8), which allows to move the set along the pipeline. The water flow rate being continuously measured by Uniflow 990 ultrasonic flowmeter (3). The rotary pump (4) controlled by an inverter enables selection of water velocity in the measuring pipe between 0.5 - 2.5 m / s. The transparent measuring section of the pipeline allows a photographic documentation of the observed two-phase flows [13].

The data acquisition equipment is comprised of PC with the dedicated counters card connected to a USB port. The pulses $I_x(t)$ and $I_y(t)$ obtained from the detectors counted within the sampling time $\Delta t = 1$ ms create stochastic signals $x(t)$ and $y(t)$. The analysis of these signals allows to determine the mean velocity of gas phase υ_a, its concentration and void fraction [18-21].

4 Examples of recorded signals

In this work there are three structures of air-water flow in horizontal pipeline, which are being considered: a plug flow, bubble flow, and transitional plug-bubble flow. For these types of flow the measurement signals were recorded in experiments denoted as: BUB6, BUB4, and BUB1 with a sampling rate of 1000 Hz. Obtained from the first probe data records length on 300,000 samples have been written to files:

- BUB601 - plug flow: $\upsilon_W = 0.90$ m / s, $\upsilon_a = 0.710$ m / s,
- BUB401 - transitional plug – bubble flow: $\upsilon_W = 1.36$ m / s, $\upsilon_a = 1.066$ m / s,
- BUB101 - bubble flow: $\upsilon_W = 1.92$ m / s, $\upsilon_a = 1.446$ m / s,

where υ_W – velocity of water measured by ultrasonic flowmeter Uniflow 990. Velocity of gas phase υ_a was determined using cross-correlation method [32-34]. Photos of particular types of structures performed in a laboratory and time records of the signals obtained in experiments BUB6, BUB4, and BUB1 are presented in figure 4.

5 Signals features extraction

The data records registered in the experiments BUB6, BUB4, and BUB1were divided into segments containing 10,000 samples. Developed in LabVIEW software allows to use moving average to smooth out short-term fluctuations, and overlapping (0 – 90%) of the segments [35]. Based on the analysis of these segments the following parameters ("features") were calculated for the three types of the flow regimes ("classes"): "plug flow", "transitional plug – bubble flow", and "bubble flow":

- mean value:

$$\mu = \frac{1}{N}\sum_{i=1}^{N} x_i \tag{1}$$

- variance:

$$\sigma^2 = m_2 = \frac{1}{N-1}\sum_{i=1}^{N}(x_i - \mu)^2 \tag{2}$$

- standard deviation (STD) σ,
- 4th order moment:

$$m_4 = \frac{1}{N}\sum_{i=1}^{N}(x_i - \mu)^4 \tag{3}$$

- root mean square (RMS):

$$RMS = \sqrt{\mu^2 + \sigma^2} \tag{4}$$

- skewness:

$$g_1 = \frac{m_3}{m_2\sqrt{m_2}} \tag{5}$$

where

$$m_3 = \frac{1}{N-1}\sum_{i=1}^{N}(x_i - \mu)^3 \tag{6}$$

- kurtosis:

$$g_2 = \frac{m_4}{m_2^2} - 3 \tag{7}$$

a)

b)

c)

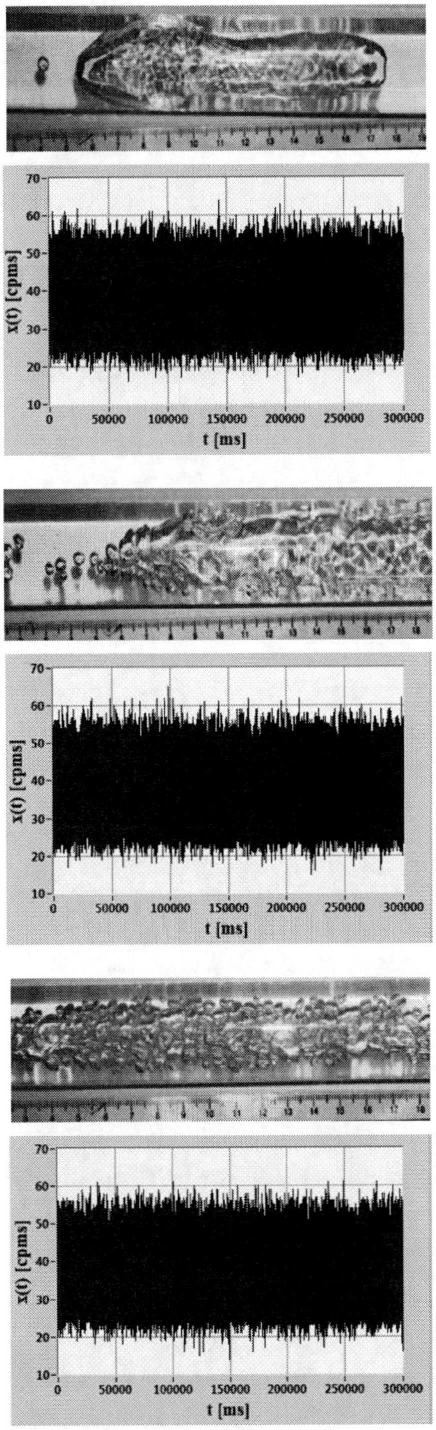

Figure 4. Pictures of the flow structures, and time records of signals $x(t)$ obtained in experiments (a) BUB6, (b) BUB 4 and (c) BUB1.

The data presented in figure 5 show values of the statistical parameters as a function of a segment number. In the analyses, the following settings, determined experimentally, were applied: width of moving average window: 20 samples, overlapping: 80% [35].

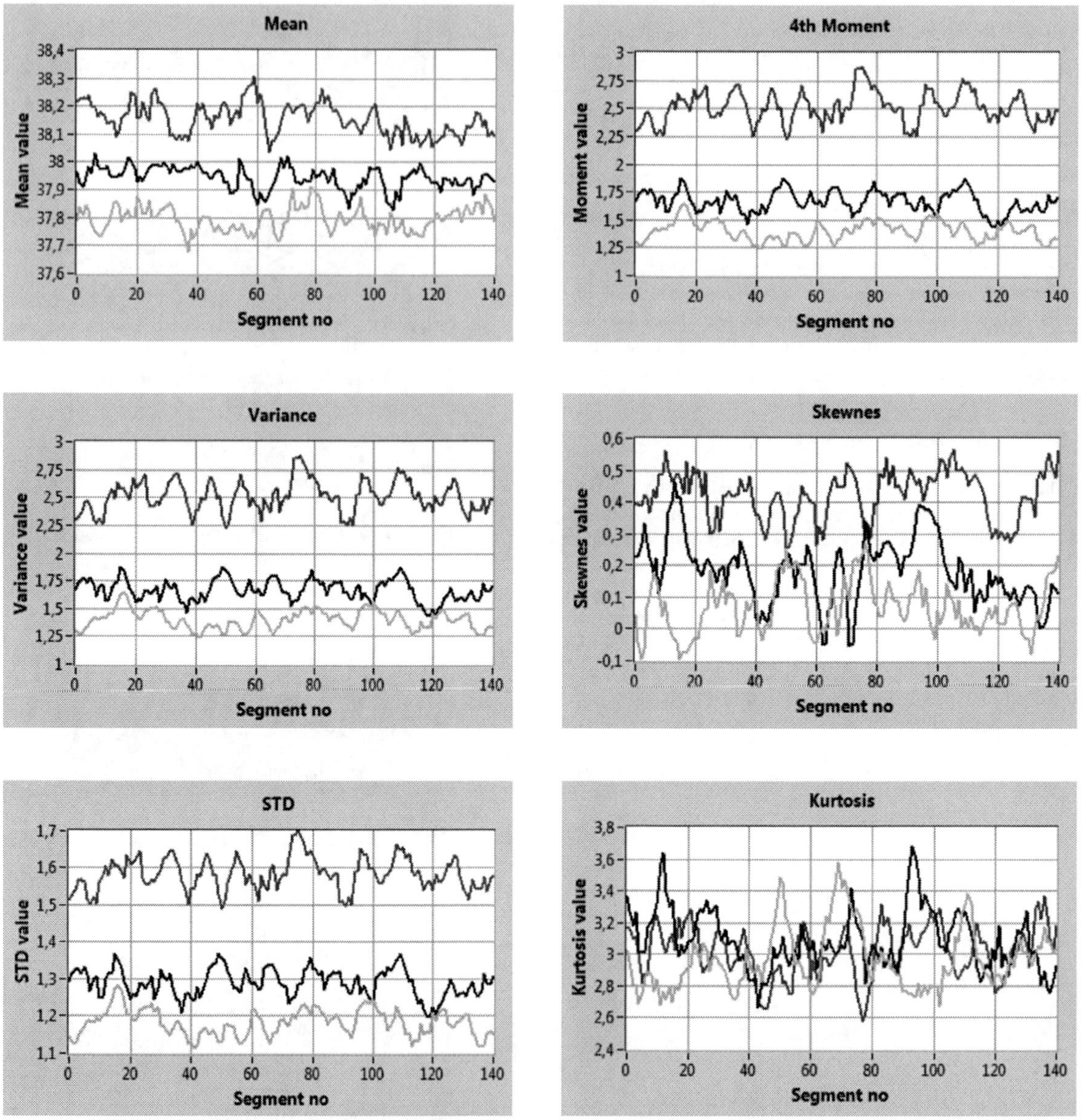

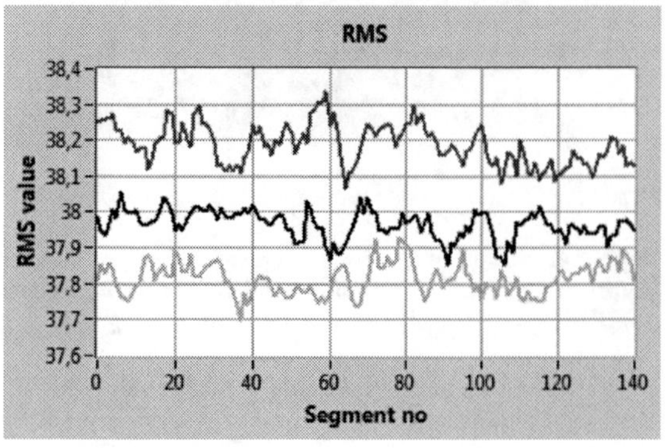

BUB601 - plug flow
BUB401 - transitional plug – bubble flow
BUB101 - bubble flow

Figure 5. Statistical parameters ("features") of BUB601, BUB401, BUB101 data records, obtained for segments length of 10,000 samples.

On the basis of the graphs presented in figure 5, it was found that the best separation for signals BUB601, BUB401 and BUB101 have mean, RMS, STD, variance and 4th order moment. The last three features are the best to separate the class "plug flow" (BUB601). It turned out that skewness and kurtosis do not allow the separation of flow structures.

6 Conclusions

In the paper an application of the gamma-absorption method to recognition of the gas-liquid flow regime in a horizontal pipeline has been presented. In the tests on laboratory installation the gamma-absorption set based on ^{241}Am radioactive sources and NaI(Tl) scintillation probes has been used.

Three structures ("classes") of air-water as plug, bubble, and transitional plug – bubble one have been considered. For these types of flow the signals from probes were recorded and analyzed. The selected parameters ("features") of signals in time domain were extracted to flow pattern recognition.

The best suitability for the recognition of flow structures are parameters for which the differences between classes are the greatest. On the basis of experiments it was found that the best indicator considered patterns have mean, RMS, standard deviation, variance and 4th order moment.

Further works will involve the application of the most useful features to flow recognition using machine learning methods.

References

1. M. Dziubiński, J. Prywer, *Mechanics of two-phase fluids* (WNT, Warsaw 2009) (in Polish)
2. G. Falcone, G.F. Hewitt, C. Alimonti, *Multiphase flow metering: principles and applications* (Elsevier, Amsterdam 2009)
3. M. Bottin, et al, Int. J. Multiphas. Flow **60,** 161-179 (2014)
4. T. Xue, et al, Flow Meas. Instrum. **27,** 29–36 (2012)
5. M.R. Rzasa, Nucl. Eng. Des. **239,** 699-707 (2009)
6. G.A. Johansen, P. Jackson, *Radioisotope gauges for industrial process measurements* (John Wiley, New York 2004)
7. Z. Yu, B. Qincheng, H. Richa, Appl. Therm. Eng. **60,** 398-410 (2013)
8. S.L. Soo (ed.), *Instrumentation for fluid-particle flow* (Noyes Publications, New Jersey, 1999)
9. W.A.S. Kumara, B.M. Halvorsen, M.C. Melaaen, Int. J. Multiphas. Flow **36,** 467-480 (2010)
10. V. Mosorov, Meas. Sci. Technol. **17,** 753-760 (2006)
11. S.H. Jung, et al, Appl. Rad. Isotop. **67,** 1254-1258 (2009)
12. P. Vlasak, et al, J. Hydrol. Hydrom. **62,** 241-247 (2014)
13. R. Hanus, Int. J. Multiphas. Flow **72,** 210–217 (2015)
14. M. Zych, et al, Flow Meas. Instrum. **35,** 11–15 (2014)
15. L. Petryka, et al, *EPJ WoC* **92,** 02065 (2015)
16. R. Hanus, M. Zych, L. Petryka, Flow Meas. Instrum. **40,** 58–63 (2014)
17. M. Zych, R. Hanus, L. Petryka, *MATEC WoC* **18,** 02004 (2014)
18. R. Hanus, M. Zych, L. Petryka, *JPCS* **530,** 012042 (2014)
19. M. Zych, et al., *EPJ WoC* **92,** 02122 (2015)
20. R. Hanus, et al., Math. Probl. Eng. **2014,** 475735 (2014)
21. R. Hanus, et al, *EPJ WoC* **92,** 02020 (2015)
22. S. Blaney, H. Yeung, Flow Meas. Instrum. **19,** 57–66 (2008)
23. O. Arjmandi-Tash, A. Zarghami, R. Zarghami, Particul. Sci. Technol. **32,** 498–505 (2014)
24. Y. Chen, et al, Chem. Eng. Sci. **129,** 156–167 (2015)
25. F. Johnsson, Int. J. Multiphas. Flow **26,** 663–715 (2000)
26. C.M. Salgado, Prog. Nucl. Energy **52,** 555–562 (2010)
27. Z. Sun, H. Zhang, Meas. Sci. Technol. **19,** 015401 (2008)
28. G.H. Roshani, Measurement **62,** 25–32 (2015)
29. C.M. Salgado, Appl. Radiat. Isot. **67,** 1812–1818 (2009)
30. E. Pawlowski, *20th International Conference on Methods and Models in Automation and Robotics (MMAR)*, 909-914 (2015)
31. M. Khorsandi, et al, Radiat. Meas. **59,** 183–187 (2013)
32. J.S. Bendat, A.G. Piersol, *Random data - analysis and measurement procedures* (John Wiley, New York, 2010)
33. M.S. Beck, A. Plaskowski, *Cross correlation flowmeters - their design and application* (Adam Hilger, Bristol, 1987)
34. W.Q. Yang, M.S. Beck, Flow Meas. Instrum. **8,** 77-84 (1998)
35. P. Hanus, MS thesis, AGH University of Science and Technology, Krakow 2014 (in Polish)

EPJ Web of Conferences **114**, 02036 (2016)
DOI: 10.1051/epjconf/201611402036

Signals features extraction in liquid-gas flow measurements using gamma densitometry. Part 2: frequency domain

Robert Hanus[1,a], Marcin Zych [2], Leszek Petryka[2], Marek Jaszczur[2] and Paweł Hanus[3]

[1]*Rzeszow University of Technology, 35-959 Rzeszow, Poland*
[2]*AGH University of Science and Technology, 30-059 Krakow, Poland*
[3]*Test & Measurement Solutions sp. z o.o., 53-025 Wrocław, Poland*

Abstract. Knowledge of the structure of a flow is really significant for the proper conduct a number of industrial processes. In this case a description of a two-phase flow regimes is possible by use of the time-series analysis e.g. in frequency domain. In this article the classical spectral analysis based on Fourier Transform (FT) and Short-Time Fourier Transform (STFT) were applied for analysis of signals obtained for water-air flow using gamma ray absorption. The presented method was illustrated by use data collected in experiments carried out on the laboratory hydraulic installation with a horizontal pipe of 4.5 m length and inner diameter of 30 mm equipped with two ^{241}Am radioactive sources and scintillation probes with NaI(Tl) crystals. Stochastic signals obtained from detectors for plug, bubble, and transitional plug – bubble flows were considered in this work. The recorded raw signals were analyzed and several features in the frequency domain were extracted using autospectral density function (ADF), cross-spectral density function (CSDF), and the STFT spectrogram. In result of a detail analysis it was found that the most promising to recognize of the flow structure are: maximum value of the CSDF magnitude, sum of the CSDF magnitudes in the selected frequency range, and the maximum value of the sum of selected amplitudes of STFT spectrogram.

1 Introduction

Knowledge of a two-phase flow structure is significant for the proper conduct a number of industrial processes, so flow regime identification inspires many studies. Recent works shows use for this purpose machine learning methods as e.g. artificial neural networks [1-12]. Generally computational intelligence methods utilize various features of signals in the time, frequency and state-space domain. Features of signals in the frequency domain may be determined using the Fourier Transform (FT), Short-Time Fourier Transform (STFT), wavelet transform and other methods [13-15].

In this work FT and STFT are applied to signals obtained in two-phase liquid-gas flow measurements using gamma-ray absorption, where data were collected in experiments carried out on a laboratory hydraulic installation with a horizontal pipe of 4.5 m length and inner diameter of 30 mm. Stochastic signals obtained for three structures of air-water flow as plug, bubble, and transitional plug – bubble one were considered. Laboratory stand and gamma absorption equipment are presented in the first part of this work and and are described in detail in [16-19].

For such features of recorded signals comparison in the frequency domain as maximum value of the autospectral density function (ADF), maximum value of the cross-spectral density function (CSDF) magnitude, sum of the ADF and sum of the CSDF in the selected frequency range, and the maximum value of the sum of selected amplitudes of STFT spectrogram were extracted.

2 Spectral analysis of signals using FT

The time records of stochastic signals obtained from gamma absorption set were presented in the first part of this work. These signals after pre-processing were employed for the analysis in the frequency domain. The one-sided autospectral density function $G_{xx}(f)$, and one-sided cross-spectral density function $G_{xy}(f)$ determined in practice for the frequency range $0 < f < \infty$ are defined by the following formulas [13, 20]:

$$G_{xx}(f) = 2\int_{-\infty}^{\infty} R_{xx}(\tau)e^{-j2\pi f\tau}d\tau \quad (1)$$

$$\underline{G}_{xy}(f) = 2\int_{-\infty}^{\infty} R_{xy}(\tau)e^{-j2\pi f\tau}d\tau \quad (2)$$

where: $R_{xx}(\tau)$, and $R_{xy}(\tau)$ are the autocorrelation and cross-correlation function respectively, f - frequency, τ – time delay.

[a] Corresponding author: rohan@prz.edu.pl

Table 1. ADF and |CSDF| of signals from the first scintillation probe obtained in BUB6, BUB 4 and BUB1 experiments.

Experiment	**ADF**	**|CSDF|**
BUB6	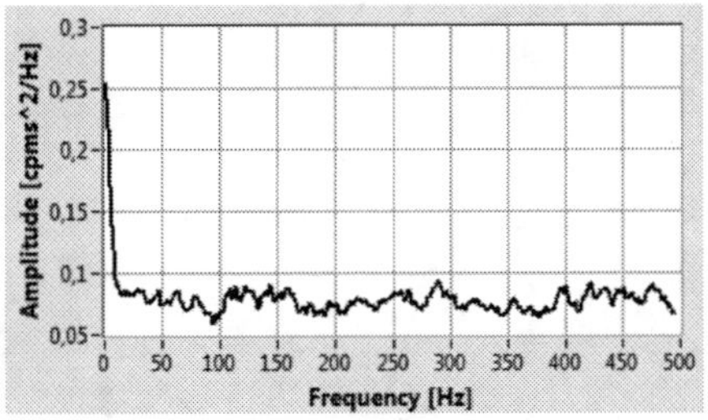	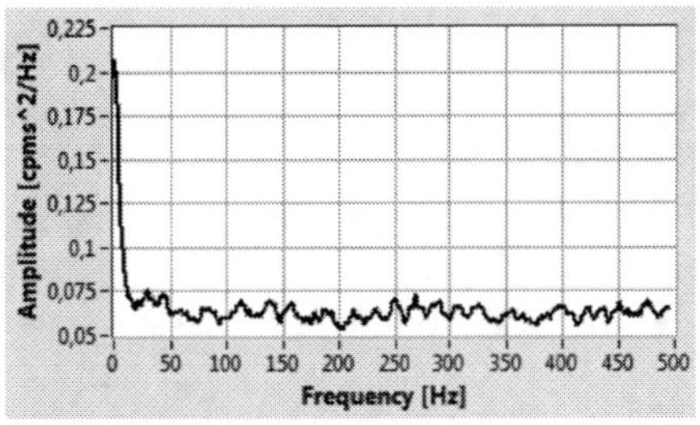
BUB4	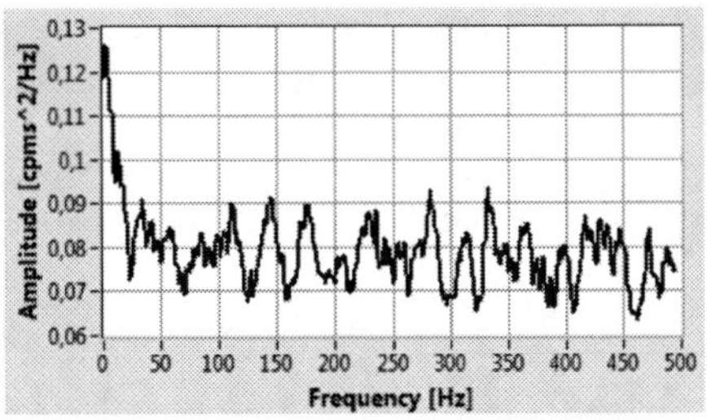	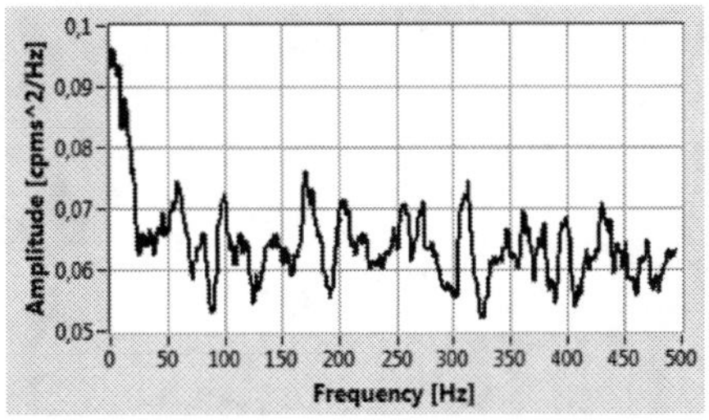
BUB1	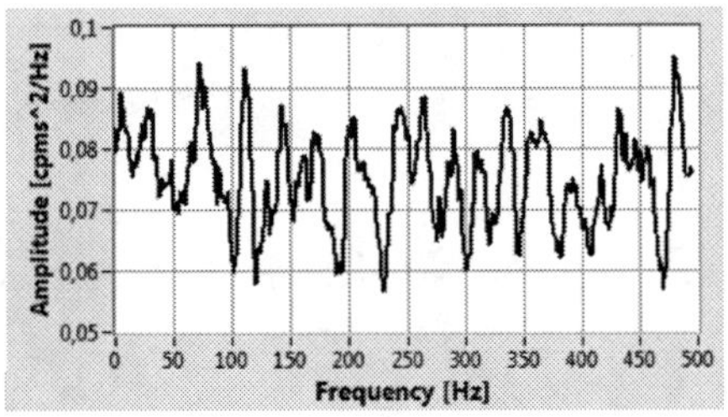	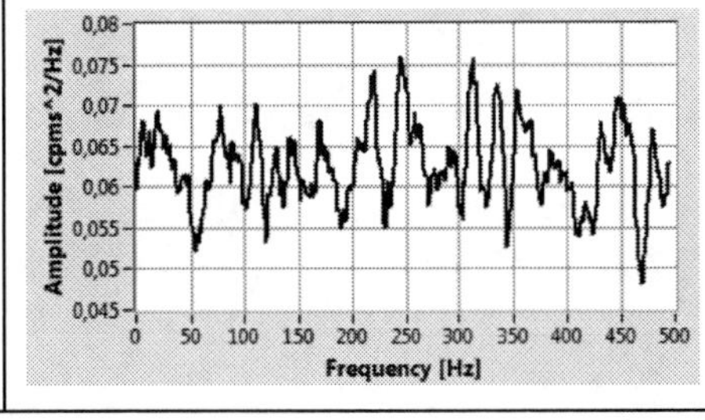

Since the cross-spectral density is a complex function, the module and phase of the CSDF are given as:

$$|CSDF| = \left\{ \mathrm{Re}\left[\underline{G}_{xy}(f) \right]^2 + \mathrm{Im}\left[\underline{G}_{xy}(f) \right]^2 \right\}^{0,5} \qquad (3)$$

$$\arg(CSDF) = arctg\left\{ \mathrm{Im}\left[\underline{G}_{xy}(f) \right] \Big/ \mathrm{Re}\left[\underline{G}_{xy}(f) \right] \right\} \qquad (4)$$

In consequence the phase distributions of CSDF (4) were used to measure the average velocity of the dispersed phase of the two-phase flow, like a cross-correlation [20-26].

Additionally the spectral analysis can be used to recognize the type of the gas bubbles transportation in the flow. The literature shows a use for this purpose ADF [27], however, according to [19], better results can be accomplished using the magnitude of cross-spectral density (3).

Table 1 shows the ADF and |CSDF| determined for signals in experiments BUB6 (plug flow), BUB 4 (transitional plug - bubble flow) and BUB1 (bubble flow). Windowing, overlapping of data segments and combined ensemble-averaging and frequency-averaging were used to ADF and CSDF estimation [21].

Based on the collected experiments, the following three parameters ("features") of spectral densities ADF and |CSDF| were identified:

- maximum of spectral density,
- area under the graph of spectral density for the selected frequency range (experimentally selected range from 0 Hz to f_g = 20 Hz).
- given above area for the selected frequency range related to the total area.

The graphs presented in figure 1 show the values of the descripting features as a function of the segment number. In the spectral analyses the following settings determined experimentally were used: segment width of 10,000 samples, Hamming window, 20 samples width of moving average window, 80% overlapping [21].

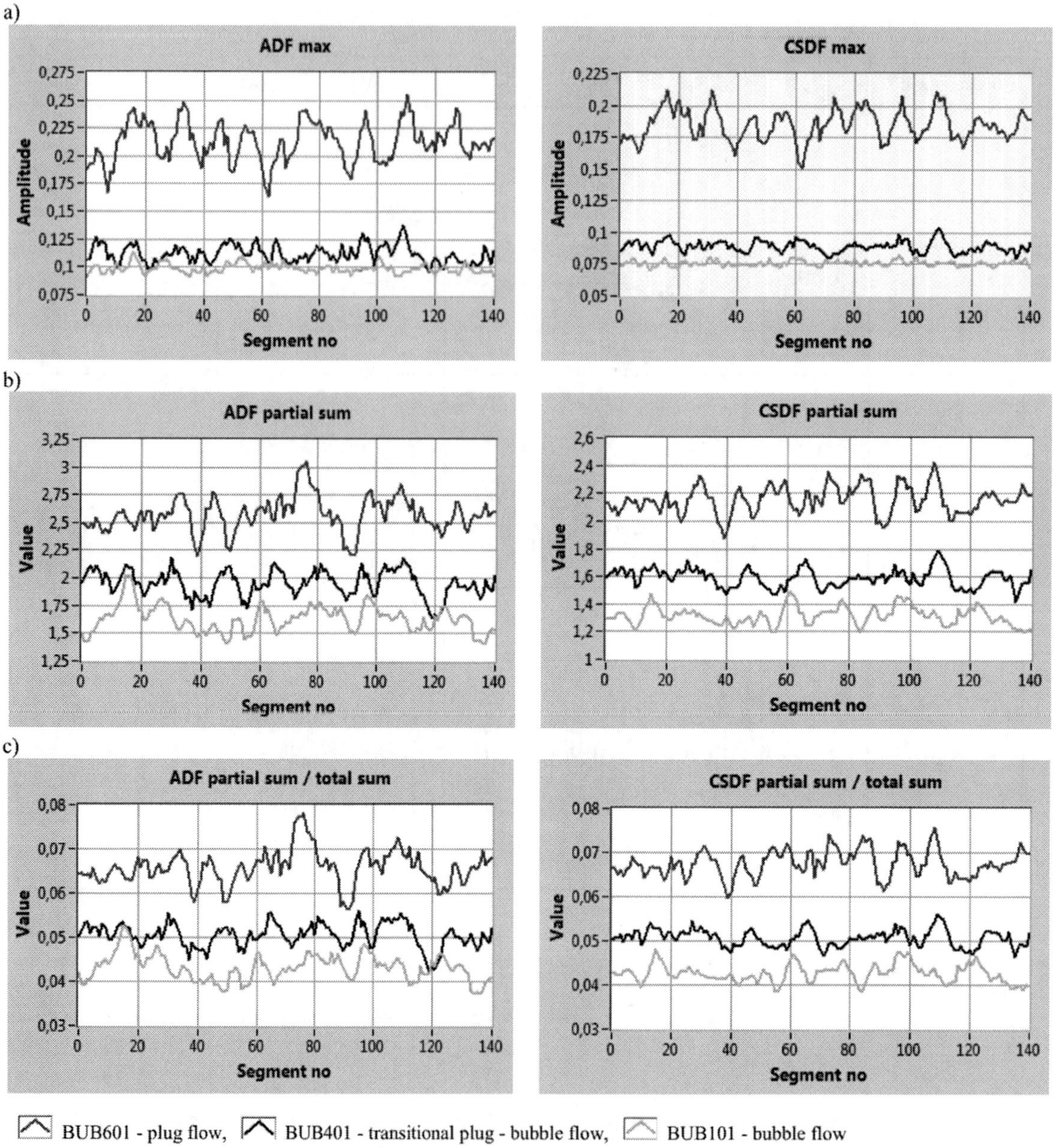

Figure 1. Features determined on the basis of ADF and |CSDF|: (a) maximum value of the ADF and |CSDF|, (b) sum of the ADF and |CSDF| in the selected frequency range, (c) relative sum of the ADF and |CSDF|.

3 The analysis of signals using STFT

STFT is time-frequency analysis method in which the signal $x(t)$ is divided into short, consecutive, possibly overlapping, segments. Each segment is windowed by a sliding time window and next is subjected to spectral analysis using FT. In this approach the window $w(t)$ defines each time segment to be analyzed (figure 2), thus resulting in a two-dimensional function STFT $X(t,f)$ defined by [28]:

$$X(t,f)=\int_{-\infty}^{\infty}x(\tau)w(\tau-t)e^{-jf\tau}d\tau \tag{5}$$

In the simplest case, the window $w(t)$ may be a square function, however, generally other types of windows, e.f. Hnning window are used.

The spectrogram $S_{xx}(t,f)$ of signal $x(t)$ is obtained by computing the squared magnitude of (5):

$$S_{xx}(t,f)=\left|X(t,f)\right|^2 \tag{6}$$

The spectrogram is a real-valued, nonnegative distribution which provides a signal representation in the time-frequency domain.

Graphs of spectrograms obtained using STFT for the beginning of signals BUB601, BUB401 and BUB101 are shown in figure 3.

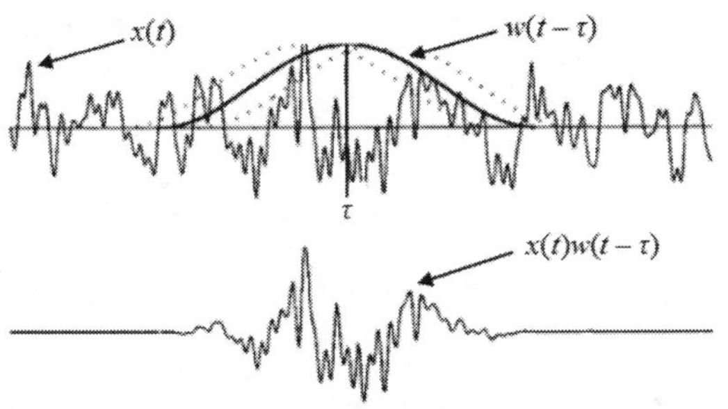

Figure 2. Application of sliding time window to $x(t)$ signal [28].

In the analyses the following settings determined experimentally were used: segment width of 10,000 samples, Hamming window, 80% overlapping.

It is easy to note that a spectrogram for three analyzed signals differ significantly. However, numerical parameters which allow separation of the classes (flow structures) were required. On the basis of many experiments carried out it was found that the following features may be proposed:

- maximum value of the sum of the spectrogram amplitudes above a fixed amplitude threshold Ag = 0.2. Graphs of such sums are shown on the right side of spectrograms in figure 3 ("Spectrogram sum"),
- area under the graph of "Spectrogram sum" for the first five components (frequency range from 0 to fg = 5 Hz). The value of fg = 5 Hz was determined experimentally.

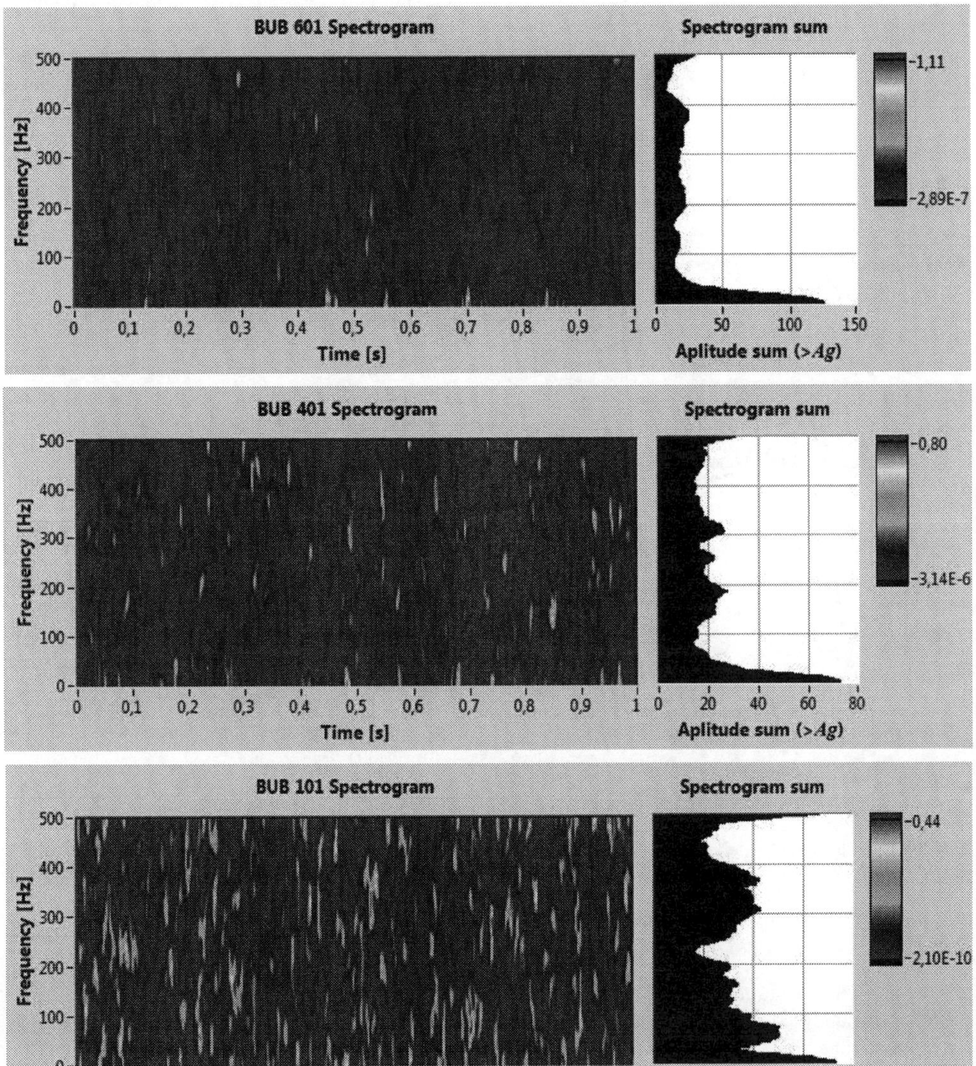

Figure 3. Spectrograms determined for signals BUB601, BUB401 and BUB101 using STFT.

- area under the graph of spectrogram sum in the frequency range from 0 to fg = 5 Hz, divided by the total area.

The graphs presented in figure 4 show the values of the mentioned above features as a function of the segment number.

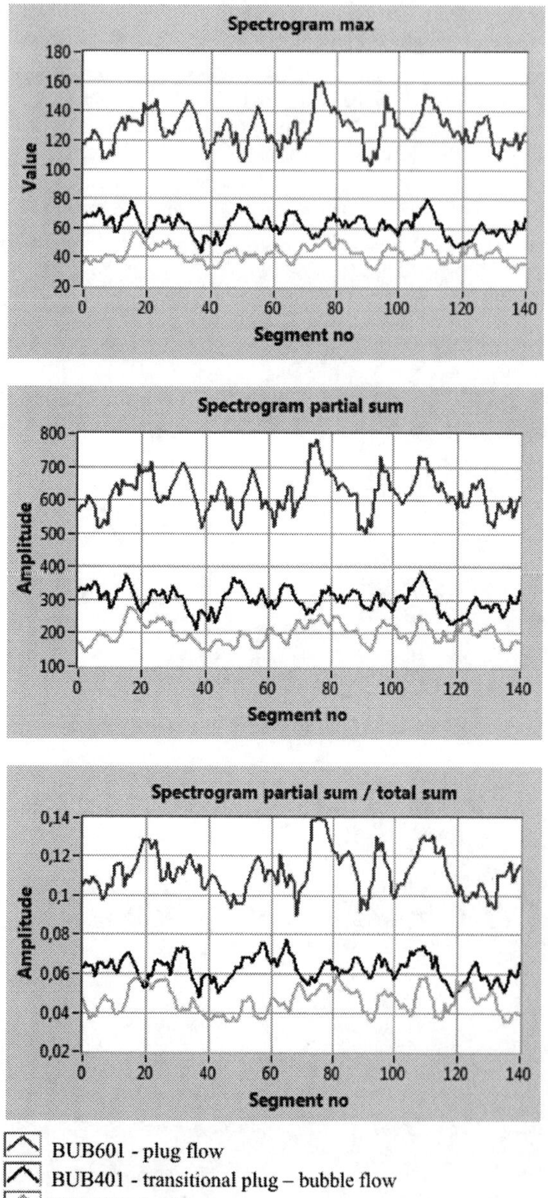

Figure 4. Features determined on the basis of STFT spectrograms.

The results presented in figure 1 suggests that the best separation for three analyzed structures of air-water flow have features based on the CSDF magnitude.

The STFT spectrogram may also be used for flow pattern recognition. As can be observed in figure 4 the best separation of patterns represented by signals BUB601, BUB401 and BUB101 has the maximum value of the sum of STFT spectrogram.

4 Conclusions

In this article stochastic signals recorded from scintillation detectors for three structures of water-air flow as plug, bubble, and transitional plug-bubble flows in horizontal pipeline were analyzed in the frequency domain. Based on the obtained results it can be stated that the selected parameters of ADF, CSDF and STFT spectrogram may be useful to the recognition of liquid-gas flow pattern.

According to the authors an improvement can be accomplished using the magnitude of the cross-spectral density, rather than the autospectral density function. On the basis of the carried out experiments, it was found that the best separation of the patterns give: maximum of the |CSDF| and relative area under the |CSDF| graph of recorded signals.

In the case of the STFT spectrogram the best separation for considered classes of flow has area under the graph of spectrogram sum in the selected frequency range of the transform.

It can be concluded, that the mentioned above features determined in the frequency domain would be applied to liquid-gas flow pattern recognition using machine learning methods.

References

1. S. Blaney, H. Yeung, Flow Meas. Instrum. **19,** 57–66 (2008)
2. O. Arjmandi-Tash, A. Zarghami, R. Zarghami, Particul. Sci. Technol. **32,** 498–505 (2014)
3. Y. Chen, et al, Chem. Eng. Sci. **129,**156–167 (2015)
4. F. Johnsson, Int. J. Multiphas. Flow **26,** 663–715 (2000)
5. C.M. Salgado, Prog. Nucl. Energy **52,** 555–562 (2010)
6. Z. Sun, H. Zhang, Meas. Sci. Technol. **19,** 015401 (2008)
7. G.H. Roshani, Measurement **62,** 25–32 (2015)
8. C.M. Salgado, Appl. Radiat. Isot. **67,** 1812–1818 (2009)
9. M. Khorsandi, et al, Radiat. Meas. **59,** 183–187 (2013)
10. H. Ding, et al, Flow Meas. Instrum. **18,** 37–46 (2007)
11. T. Elperin, M. Klochko, Exp. Fluids **32,** 674–682 (2002)
12. E.S. Rosa, et al, Int. J. Multiphas. Flow **36,** 738–754 (2010)
13. J.S. Bendat, A.G. Piersol, *Random data - analysis and measurement procedures* (John Wiley, New York, 2010)
14. J.H. McClellan, R.W. Schafer, M. A. Yoder, *Signal processing first* (Prentice Hall, New Jersey 2003)

15. A.V. Oppenheim, R.W. Schafer, J.R. Buck, *Discrete -Time Signal Processing* (Prentice Hall, New Jersey 1999)
16. R. Hanus, Int. J. Multiphas. Flow **72,** 210–217 (2015)
17. M. Zych, et al., *EPJ WoC* **92,** 02122 (2015)
18. R. Hanus, et al, *JPCS* **530,** 012042 (2014)
19. M. Zych, et al., *MATEC WoC* **18,** 02004 (2014)
20. M.S. Beck, A. Plaskowski, *Cross correlation flowmeters - their design and application* (Adam Hilger, Bristol, 1987)
21. P. Hanus, MS thesis, AGH University of Science and Technology, Krakow 2014 (in Polish)
22. M. Zych, et al, Flow Meas. Instrum. **35,** 11–15 (2014)
23. R. Hanus, L. Petryka, M. Zych, Flow Meas. Instrum. **40,** 58–63 (2014)
24. L. Petryka, et al, *EPJ WoC* **92,** 02065 (2015)
25. M. Zych, et al., *EPJ WoC* **92,** 02121 (2015)
26. R. Hanus, et al., Math. Probl. Eng. **2014,** 475735 (2014)
27. L. Liebenberg, J.P. Meyer, Exp Therm Fluid Sci. **31,** 127–140 (2006)
28. L. Sörnmo, P. Laguna, *Bioelectrical signal processing in cardiac and neurological applications* (Elsevier Academic Press, San Diego, 2005)

EPJ Web of Conferences **114**, 02037 (2016)
DOI: 10.1051/epjconf/201611402037

Condensation of the air-steam mixture in a vertical tube condenser

Jan Havlík[1,a], Tomáš Dlouhý[1]

[1]*Czech Technical University in Prague, Faculty of Mechanical Engineering, Department of Energy Engineering, Technická 4, 16607 Prague 6, Czech Republic*

Abstract. This paper deals with the condensation of water vapour in the presence of non-condensable air. Experimental and theoretical solutions of this problem are presented here. A heat exchanger for the condensation of industrial waste steam containing infiltrated air was designed. The condenser consists of a bundle of vertical tubes in which the steam condenses as it flows downwards with cooling water flowing outside the tubes in the opposite direction. Experiments with pure steam and with mixtures of steam with added air were carried out to find the dependence of the condensation heat transfer coefficient (HTC) on the air concentration in the steam mixture. The experimental results were compared with the theoretical formulas describing the cases. The theoretical determination of the HTC is based on the Nusselt model of steam condensation on a vertical wall, where the analogy of heat and mass transfer is used to take into account the behaviour of air in a steam mixture during the condensation process. The resulting dependencies obtained from the experiments and obtained from the theoretical model have similar results. The significant decrease in the condensation HTC, which begins at very low air concentrations in a steam mixture, was confirmed.

1 Introduction

Waste steam produced in many industrial processes has typically low parameters and it may contain infiltrated air limiting its following utilization. Therefore the steam is often exhausted to the ambient atmosphere. This paper deals with the influence of non-condensable gases (air) in a steam mixture on the utilization of its condensation heat. An evaluation of this effect is based on experiments with a condenser designed for the utilization of atmospheric pressure steam with a small content of infiltrated air. The dependence of the condensation heat transfer coefficient (HTC) on the air concentration in a steam mixture is given. The results are compared with the results of the theoretical calculation of this case. The influence of non-condensable gas on heat transfer is an important factor for the design of waste vapour heat condensers whose use may increase the energy efficiency of processes producing waste steam [1], [2].

2 Vertical tube condenser

Experiments are carried out on a vertical shell-and-tube heat exchanger in which the condensing water vapour flows downwards inside vertical tubes and the cooling water flows in a counter current in the outer shell (see Figure 1). This condenser design is used due to the potential presence of small solid particles in industrial waste steam (e.g. waste steam from fuel drying). These particles, which can stick to the tube wall, are spontaneously carried away from the tubes by the condensate blowing out. The condenser outlet on the vapour side is opened, so the steam condensation pressure is close to the atmospheric pressure. The tube bundle is formed by 49 tubes 865 mm in length, 28 mm in external diameter and 24 mm in internal diameter. The tubes are arranged in staggered arrays with a triangular tube pitch of 35 mm. The cross-section of the shell is rectangular in shape, 223 mm by 270 mm in size. Seven segmental baffles (223 x 230 mm) are used in the shell section. The tubes are made from stainless steel 1.4301 (AISI 304).

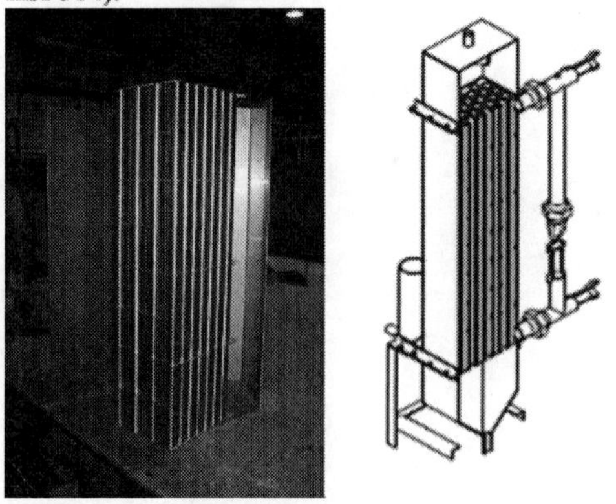

Figure 1. Vertical tube condenser

[a] Corresponding author: jan.havlik@fs.cvut.cz

3 Experimental set-up

The testing loop is shown in Figure 2. Steam is produced in a steam generator. Before the steam enters the vertical tube condenser, its parameters are reduced to the required value. Measured air flow is mixed to the steam through a controllable air fan. Recirculation of the hot water enables the cooling water temperature and the cooling water flow rate in the condenser to be regulated The measured parameters needed for the determination of the HTC from the heat balance are the inlet cooling water temperature T_{w1}, the temperature of water after recirculation T_{w2}, the outlet cooling water temperature T_{w3}, the cooling water flow rate Q_w, the volume of air added to the steam before the condenser V_a, the inlet steam pressure p_s, the inlet steam temperature T_s and the amount of steam condensate M_c.

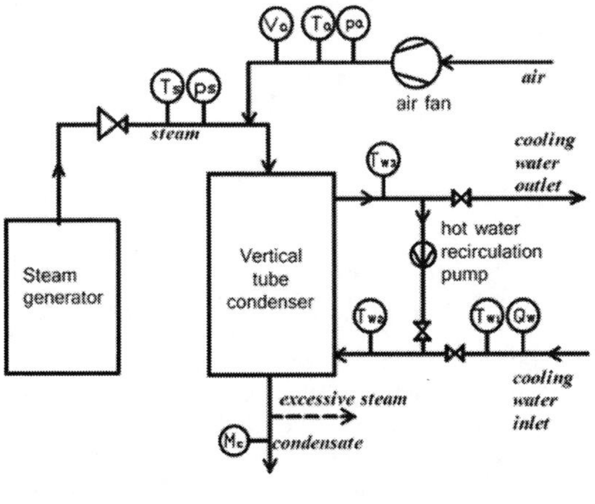

Figure 2. A schemata of the testing loop

4 Determination of the condensation HTC

Calculation of the HTC is based on the heat balance of the condenser [1]. The total heat power Q is given by the equation

$$Q = M_w \cdot c_w \cdot (T_{w,out} - T_{w,in}), \quad (1)$$

where M_w is the cooling water flow, c_w is the specific heat capacity of water, $T_{w,out}$ is the outlet cooling water temperature, and $T_{w,in}$ is the inlet cooling water temperature. The total heat power is also given by the equation

$$Q = U \cdot A \cdot \Delta T_{log}, \quad (2)$$

where U is the overall HTC, A is the heat transfer surface and ΔT_{log} is the logarithmic mean temperature difference. The tube-side HTC (vapour condensation) h_c, respectively the shell-side HTC (water) h_w, is possible to calculate from the term

$$U = \frac{1/D_o}{\frac{1}{d_i \cdot h_c} + \frac{1}{2k} ln\left(\frac{D_o}{d_i}\right) + \frac{1}{D_o \cdot h_w}}, \quad (3)$$

where d_i is the inside diameter of the tubes, k is the thermal conductivity of the tubes, and D_o is the outside diameter of the tubes.

4.1 Determination of the shell-side HTC

For the calculation of the condensation HTC h_c, it is necessary to fix the water HTC h_w for specific configuration of the heat exchanger. In the case of pure steam condensation, it is possible to calculate the condensation HTC h_c according to the Nusselt model of pure steam condensation on the vertical wall at distance x of the wall length [2], [3]

$$h_c(x) = \left[\frac{\rho_L g(\rho_L - \rho_p) h'_{fg} k_L^3}{4\mu_L \Delta T_{sat} L}\right]^{1/4}, \quad (4)$$

where ρ_L is the density of the condensate, ρ_p is the density of water vapor, h'_{fg} is the latent heat of condensation, k_L is the thermal conductivity of condensate, μ_L is the dynamic viscosity of condensate, ΔT_{sat} is the difference between the saturation temperature and the wall temperature, and L is the wall length.

4.2 Effect of non-condensable gas on the condensation HTC

When the value of $h_w = f(Re, \Pr) \approx f(M_w, T_w)$ [1], [3] is fixed, the value of h_c for various steam-air mixtures is provided by Equation (3). It is necessary to maintain a constant flow of cooling water M_w with minimal change of the water temperatures t_w, whose effect is negligible. Therefore the value of h_w is constant during measurement. Experimental determination of the shell-side HTC is described in [4].

5 Condensation theory of mixture containing non-condensable gas

If steam condenses in a mixture with inert gases, there is a difference in the concentrations of the steam in the bulk gas mixture C_∞ and at the condensation phase interface C_i. Steam concentrations decrease along the length of the tube equally with corresponding steam partial pressures p_∞ resp. p_i and temperatures T_∞ resp. T_i (see Figure 3). The molecules of steam diffuse through an inert gas towards the vapour-liquid interface and mass transfer occurs [1], [3]. On the contrary in the case of pure steam condensation, the temperatures T_∞ and T_i are identical and constant during the process. The heat balance of the total heat transfer process is shown in Figure 3.

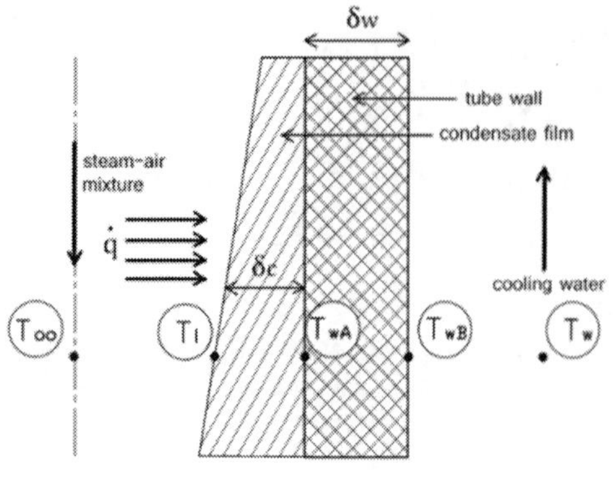

Figure 3. Heat transfer in a condenser tube

In the total heat transfer process, the value of heat flux $\dot{q}$ is constant.

- heat flux from condensing steam
$$\dot{q} = \dot{m}_c \cdot h'_{fg}, \quad (5)$$
where $\dot{m}_c\ [kg/sm^2]$ is the amount of condensing steam, and h'_{fg} is the latent heat of condensation.
- heat flux through condensate film
$$\dot{q} = h_{ci} \cdot (T_i - T_{wA}), \quad (6)$$
where h_{ci} is the condensation HTC, where T_i is the steam condensation temperature and T_{wA} is the wall temperature.
- heat flux through the tube wall
$$\dot{q} = \frac{k_w}{\delta_w} \cdot (T_{wA} - T_{wB}), \quad (7)$$
where k_w is the thermal conductivity of the tube wall, δ_w is the thickness of the tube, T_{wB} is the tube wall temperature.
- heat flux to cooling water
$$\dot{q} = h_w \cdot (T_{wB} - T_w), \quad (8)$$
where T_w is the cooling water temperature.

To determinate the condensation temperature at the phase interface T_i, it is necessary to take into account the mass transfer generated by the different partial pressures at the interface between the gas and the liquid phases p_i, and at bulk gas mixture p_∞ [1], [3]. In accordance with the analogy between mass transfer and heat transfer, a criterial equation for laminar flow is used. The Nusselt number *Nu* is replaced with the Sherwood number *Sh*, and the Prandtl number *Pr* with the Schmidt number *Sc*.

For laminar flow in a pipe ($Re \leq 2300$) in the entrance region for the boundary condition of a uniform heat flux, the local Sherwood number at distance x from the inlet is given by [1]

$$Sh_x = 1.302\left(ReSc\, {}^{D}/_{x}\right)^{1/3}. \quad (9)$$

The length of the entry region L_{er} is defined as

$$L_{er} = 0.05RePr. \quad (10)$$

For fully developed laminar flow in a pipe, the local Sherwood number Sh_x is constant and has the value of 4.36. The mass transfer coefficient β is given by equation [1]

$$\beta = \frac{Sh \cdot D_{12}}{D}, \quad (11)$$

where D_{12} is the diffusive coefficient of water vapour to air. The molar concentration at the phase interface y_{10} is calculated from Equation (12) derived from the Fick law for calculation of the molar flux of condensing steam $\dot{n}_1$ [1].

$$\dot{n}_1 = \beta C_T ln\left[\frac{1 - y_{1\delta}}{1 - y_{10}}\right], \quad (12)$$

where C_T is the total molar concentration, $y_{1\delta}$ is the molar concentration of steam at the boundary layer. The temperature at the phase interface T_i and the partial pressure of steam at the phase interface p_i are calculated from Equations (13) and (14), where p_{tot} is the total pressure in the bulk gas mixture.

$$p_i = y_{10} \cdot p_{tot}. \quad (13)$$

$$T_i = T_{sat}(p_i). \quad (14)$$

The HTC h_{ci} is given by Equation (4). The condensation HTC of the total process h_c is usually related to the temperature difference between the bulk gas mixture temperature T_∞ and the wall temperature T_{wA} [1] according to the term

$$h_c = h_{ci}\frac{(T_i - T_{wA})}{(T_\infty - T_{wA})}. \quad (15)$$

6 Results

Experiments were carried out for the laminar flow of the steam mixture with an inlet tube velocity of 1.5 m s^{-1} which corresponds with the Reynolds number of 1700. The range of tested mass concentration of air in the steam mixture was from 0 to 0.06. The cooling water flow rate and inlet temperature in the exterior shell-side was fixed for all cases. The values of the condensation HTC are evaluated according to the procedure described in Chapter 4. The shell-side HTC was fixed at the value of 1550 W m^{-2} K^{-1}. Experimental results are shown in Figure 4.

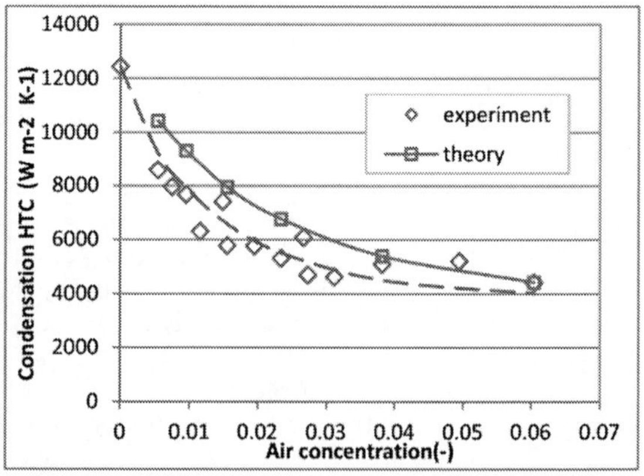

Figure 4. Effect of air content on the condensation process

The values obtained from the theoretical model are slightly higher. The deviation of the condensation HTC calculated using the theoretical formulas and evaluated by measurement ranges from 4% to 9% in the dependence on operating parameters. A significant decrease in the condensation HTC begins in both cases at very low air concentrations in the steam mixture. The values of condensation HTC (see Figure 4) are significantly higher in comparison with the value of the shell-side HTC. Therefore the effect of any potential inaccuracy of the HTC determination on the values of the overall HTC is relatively low for this design of a waste vapour condenser (see Figure 5).

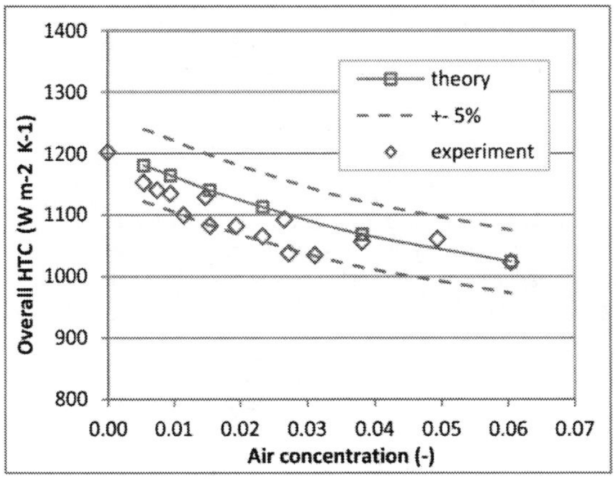

Figure 5. The deviation in the overall HTC

7 Conclusion

The influence of air content in waste steam on the condensation HTC value was investigated. A significant decrease in the condensation HTC value, in accordance with available literature, beginning at very low air concentrations in the steam mixture was experimentally verified. For the air concentration in the steam mixture of 1%, the condensation HTC decreases according to the experiments to 62% and according to the theoretical model even to 75% of the value for the condensation of pure steam.

The values obtained from the theoretical model are slightly higher, nevertheless the experimentally obtained dependence of condensation HTC on the air content in a steam mixture has a similar curve as the dependence calculated according to the theoretical model. Any potential inaccuracy in the HTC determination on the values of the overall HTC is relatively low for this design of waste vapour condenser. The sensitivity of the shell-side HTC value to the overall HTC is more significant than the condensation HTC. When calculating the overall HTC, it is of greatest importance to precisely determine the shell-side HTC.

Acknowledgement

This research has been supported by the Grant Agency of the Czech Technical University in Prague, grant No. SGS14/183/OHK2/3T/12.

References

1. G. F. Hewitt, G. L. Shires, T. R. Bott, *Process Heat Transfer* (Begell House, New York, 2000)
2. J. Šesták, F. Rieger, *Přenos hybnosti, tepla a hmoty* (ČVUT, Praha, 2005)
3. F. P. Incropera, D. P. Dewitt, *Introduction to Heat Transfer* (Wiley-Academy, New York, 1996)
4. J. Havlík, T. Dlouhý, *Proceedings of the 20th International Conference Engineering Mechanics 2014*, 216-219 (2014)

EPJ Web of Conferences **114**, 02038 (2016)
DOI: 10.1051/epjconf/201611402038

Check valve slam analysis in pumping station

D. Himr[a], V. Habán, and P. Dokoupil

Brno University of Technology, Faculty of Mechanical Engineering, Energy Institute, Victor Kaplan Department of Fluid Engineering, Technická 2, 616 69, Brno, Czech Republic

Abstract. Pumping station supplies water for technological process. The check valve in the station was replaced with a new one. The regular test of black out discovered the high pressure pulsations accompanied with noticeable pipeline movement of discharge pipe. It was caused by late check valve closing, probably, when the back flow reached the highest possible velocity. This statement was supported with analysis of results of pressure measurement near the check valve and with a numerical simulation of the flow in the pipeline system.

1 Introduction

Check valve is a necessary part of any pipeline delivering liquid from lower to upper place. It is usually placed immediately behind the pump. Its resistance should be low, when liquid flows in a positive direction and no reverse flow should be possible. It is important to choose an appropriate construction to get optimal performance of the check valve. There are many different constructions and each one has its advantages and disadvantages. For example, [1] compared properties of basic types (figure 1 to 4).

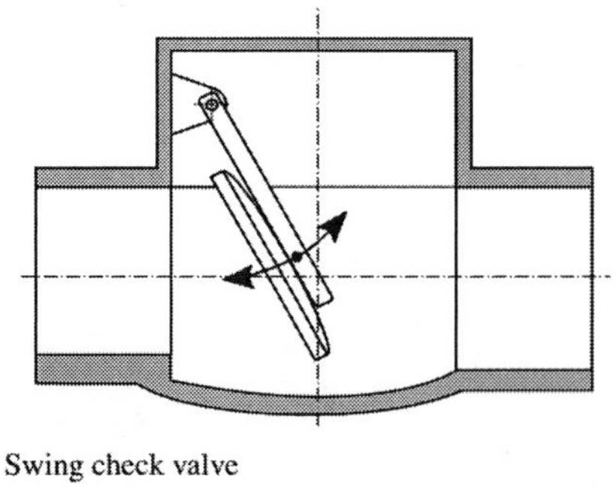

Fig. 1. Swing check valve

The ideal check valve is light with low inertia and low friction loss so it is easy to open it. The disc should have a short run or the closing should be accelerated (e. g. with springs).

There is a practice to slowdown the disc movement to avoid the water hammer, when it is not possible to close disc fast enough when flow reverses. Unfortunately, one has to accept a certain back flow in such case.

According to [2], following properties should be taken into account, while choosing the check valve:

- Closing speed with respect to duration and magnitude of the reverse flow,
- disc stability and its sensitivity to pressure pulsations,
- flow rate necessary to full opening and soft closing,

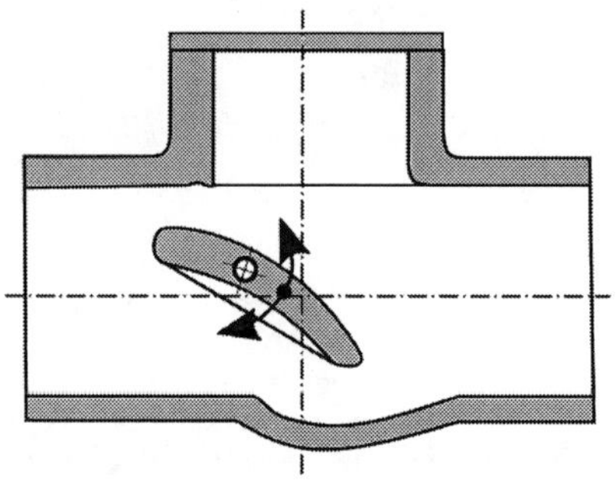

Fig. 2. Tilting check valve

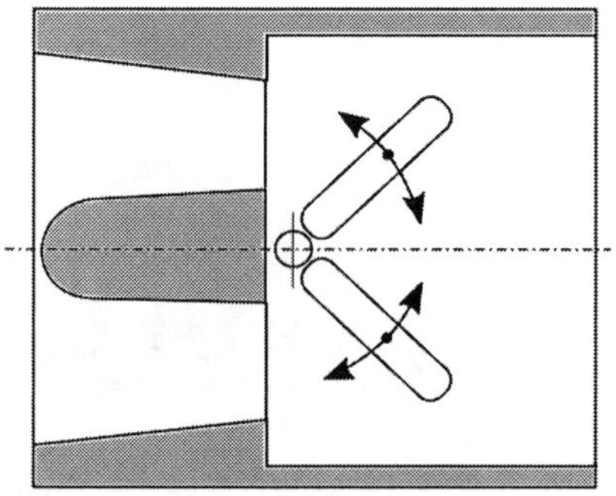

Fig. 3. Split disc check valve

- pressure loss for maximal flow rate,
- tightness and maintenance difficulty.

It means that check valve behaviour depends, besides its construction, on the pipeline system. Choice of a wrong dimension or a wrong type leads to pressure pulsations and check valve slam, which influences lifetime of the check valve and the whole pipeline system. Problems with check valve slam are very common in many industrial applications ([3] or [4]), also authors of this paper have already dealt with it [5]. The valve suffered from self-excited pressure pulsations, but in that case, the problems were not caused by check valve itself, but control valve in the different place of the system.

[a] e-mail: himr@fme.vutbr.cz

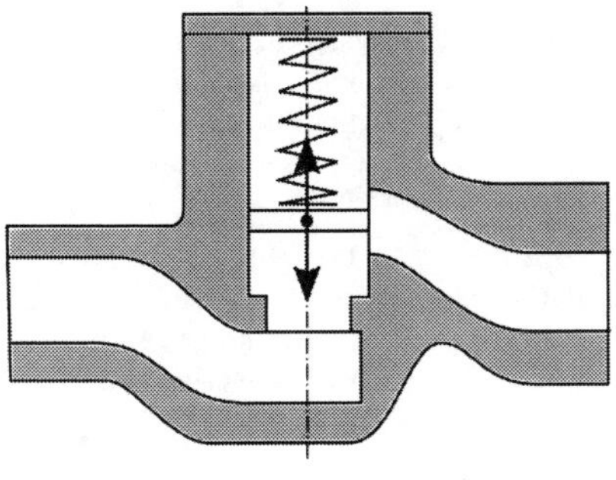

Fig. 4. Lift check valve

Pressure pulsations described in the present paper are consequence of wrong behaviour of the check valve.

2 Investigated system

Pumping system delivers water to a technological process, which depends on the continuous supplement of this liquid, see figure 5. One pump delivers water from suction

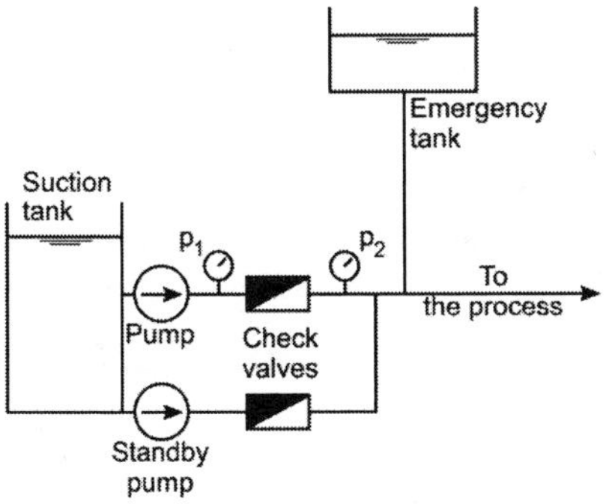

Fig. 5. Investigated system

tank to the discharge pipe. Since the technological process requires a continuous supply, there is another pump ready to start in case of failure of the first unit. Emergency tank connected to the discharge pipe is placed high enough to provide water using the gravity. It gives the time to start the standby pump. Pumps alternates in regular periods to secure the same wearing. Diameter of the discharge pipe is almost 1 m and when pump runs the steady flow velocity is $1.8\,\mathrm{m\,s^{-1}}$.

Back flow is prevented by check valves near to pumps, they have similar construction to figure 2. There are two counterweights on the shaft to make closing easier, see figure 6. These counterweights are placed symmetrically at both ends of the shaft.

There are also slide valves between pumps and check valves, but they are not drawn in the figure, because they are open and do not take part in any operating or transient event. These valves are closed just in case of pump revison, reconstruction or similar.

The tests of the system showed strong pressure pulsations accompanied with mechanical vibrations of the delivery pipe. The strength of pulsations led to destruction of operating pressure sensors. Computer interpreted loosing

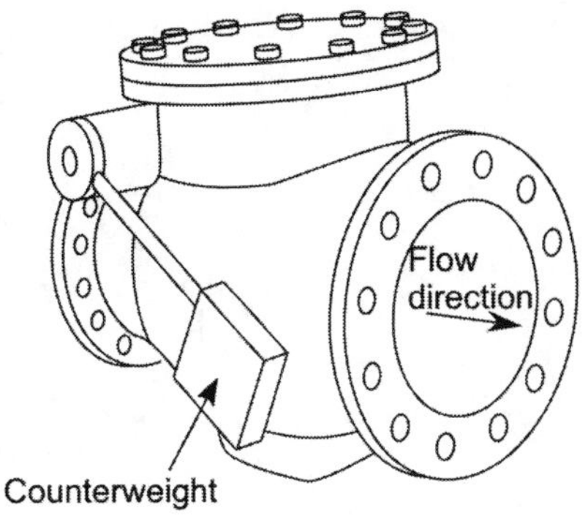

Fig. 6. Counterweight placement

the pressure signal as a pump failure and stopped all process. It was the first indication, that something was wrong.

The system was equipped only with pressure sensors with low sampling frequency. They are sufficient to find out whether the system works or not but they are unable to catch events faster than seconds. So the operator just replaced damaged sensors, but they did not have any suspicion that the problem was deeper. The new sensors failed again after several months.

The problems appeared after the replacement of check valves delivered by different manufacturer than the original ones. The manufacturer stated it had happened because of the wrong operation and it was not their problem, then stopped communication.

The operator took the check valve out of the system and found out it was damaged by the strong slams. The friction of the shaft made closing difficult. The damaged check valve was replaced with a new one from the store. Also this one had quite high friction of the shaft. The following test proved that slams were still on.

Operator tried to install dampers, but they were destroyed after several cycles as well. The operator had a theory that the check valve closed too early and fast when the flow reversed and asked for proving it, but in that case the pressure peak should be much lower than measured one. Description of the measurement follows in the next section.

3 Measurement evaluation

3.1 Sensors

Measurement conditions were quite difficult, because there is no flowmeter in the system and only two pressure sensors were allowed: downstream and upstream the check valve, see the figure 5. It was also possible to use accelerometer on the pipe wall at the check valve and the strain gauge to get actual disc position. Placement is shown in the figure 7. Range of pressure sensors was 0 to 5 Mpa with error ±0.5% from the range. Disc position was measured with error 0.2%. This error is for measurement of slow motion or steady position. When faster event is observed the error is greater.

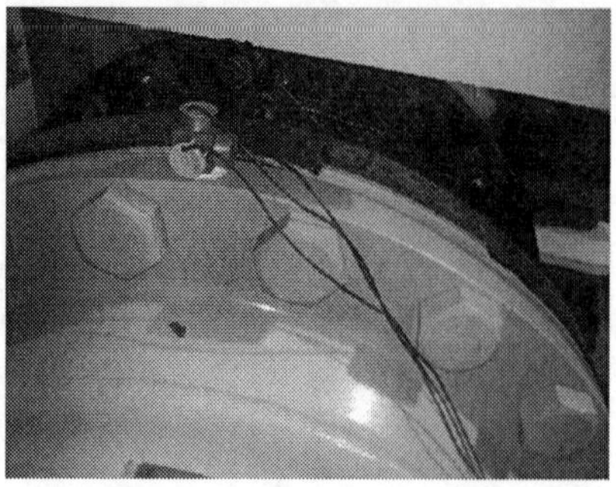

Fig. 7. Accelerometer placement

3.2 Test

Regular test starts when pump runs in steady conditions. The whole event is displayed in the figure 8. The pump was

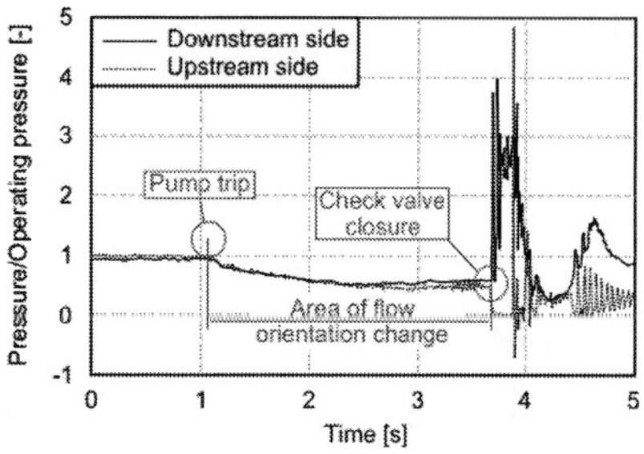

Fig. 8. Pressure measurement at the check valve

disconnected at time 1 s. Pressure went down in both observed places (upstream and downstream side of the check valve) the flow decelerated and reversed. The check valve closed after 2.8 s, which was connected with high pressure surge and strong slam.

The standby pump started running in 15 s after the the main pump disconnection so it could not influenced the measurement and it is not shown in the graph. Graph in the figure 9 shows what happened after the check valve closure. Point 1 indicates time when check valve started clos-

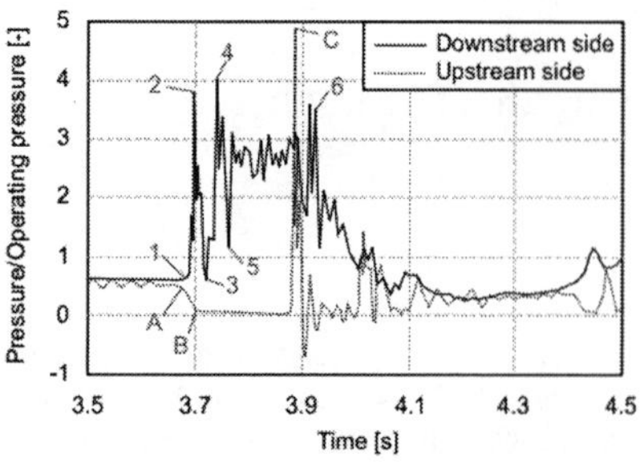

Fig. 9. Pressure measurement at the check valve – detail

ing. The end of closing corresponds to the maximal pressure downstream the check valve (point 2). Sudden drop to the point 3 is not connected with eigen frequency of the system, which is much lower, but when the disc hit the seat, it was bounced and partially opened and closed again (point 4). The bounce was repeated once more (point 5). Typical shock wave, which was induced when the check valve suddenly stopped the flow, travelled to the upstream tank and came back to the check valve. Point 6 marks the moment when the shock wave arrived and pressure dropped down.

The statement regarding the bounce of the disc of the check valve is supported with the graph in the figure 10. The position of the disc was measured with a position sen-

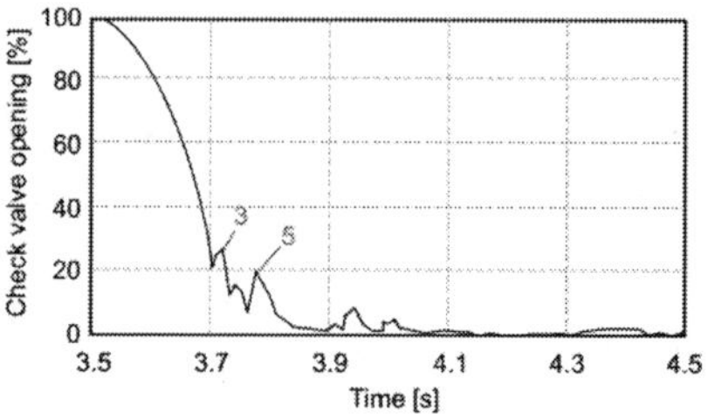

Fig. 10. Check valve opening

sor. This measurement is rather informative, because the sensor had quite great integrating constant, but it shows short periods when the disc was opening (point 3 and 5 in the figure 10, which correspond to the same points in the figure 9).

In fact, the disc started closing earlier than anything is visible on the pressure record, but its movement had the same velocity as the flow, so it did not make any pressure loss before the full closing.

Figure 9 also shows pressure upstream the check valve. Curve from A to B describes the disc closing, which is followed by cavitation formed at the check valve. Interruption of the back flow caused column separation. Collapsing cavitation is connected with high pressure peak (point C). Negative values do not have physical meaning. Pressure sensors might have been influenced by strong mechanical vibration of the pipe wall, which had acceleration 20 G at that moment, see figures 11 to 13 displaying records of the acceleration sensor at the check valve. Noticeable pipe-

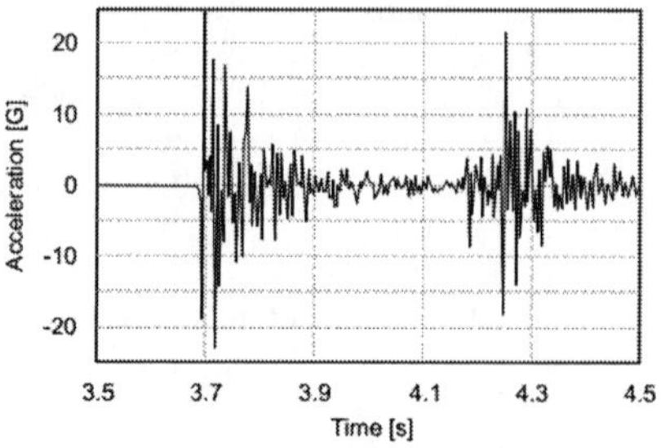

Fig. 11. Wall acceleration in axial direction

line movement was observed during the test and also first signs of damages in the building were observed. Specifically, ruptures in the wall, where the pipe line is going through, were found.

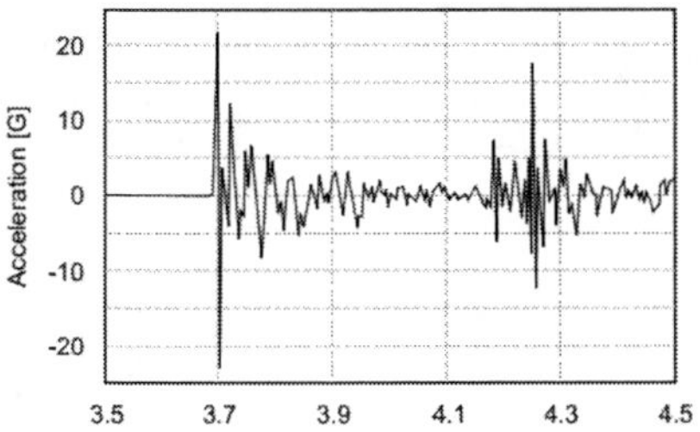

Fig. 12. Wall acceleration in radial (vertical) direction

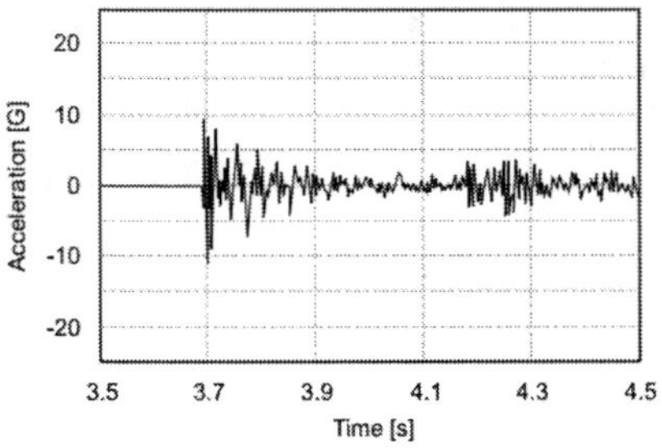

Fig. 13. Wall acceleration in tangential (horizontal) direction

3.3 Backflow

Pressure surge is always caused by sudden change in flow rate. There is no flow rate measurement in this case and it is not even possible to use Gibson method to compute time dependence of the flow rate, because the only pressure sensors were at the valve. When the valve closes, it splits the system into two independent parts and each part has only one pressure sensor, but it is possible to estimate back flow velocity using Zhukovsky equation (1).

$$\Delta v = \frac{\Delta p}{a \cdot \rho}. \tag{1}$$

Pressure surge Δp can be taken from the pressure record downstream the valve. It is the difference between points 2 and 1 (figure 9). Wave speed can be computed from period of pressure pulsations after the fifth second. The check valve was closed and the standby pump was not running yet. The wave speed can be evaluated from both pressure records (p_1 and p_2), see figure 5 and equation (2), which is valid for simple pipe with asymmetric boundary conditions (pressure boundary condition at the one side and flow rate boundary condition at the other side).

$$a = 4 \cdot f \cdot L. \tag{2}$$

The back flow velocity can be computed also from the record obtained with pressure sensor upstream the check valve. The process is more complicated than in case of Zhukovsky equation and it is described in [6]. Authors did extensive experiments with cavitating water hammer. Important inputs are initial pressure (point A in figure 9) and duration of cavitation (time between points B and C). Both approaches give similar value of back flow velocity $1.60\,\mathrm{m\,s^{-1}}$ and $1.52\,\mathrm{m\,s^{-1}}$ respectively, which are too great to say the check valve works in good way.

4 Computation

A numerical model of the system has been created to better understand the system behaviour after the pump trip. It is a standard numerical model of one-dimensional flow, where all elements are described with an appropriate mathematical equation.

Look at the figure 5. Suction tank, emergency tank and downstream end of the discharge pipe are conditions of constant pressure. In fact, the emergency tank starts emptying after the pump trip, but observed period is short and it is possible to neglect water level change in this tank. Pump is described by Q – p and Q – M curves, so, it is possible to simulate pump deceleration and loss of the head. The other pump does not work and its branch is not considered in the simulation, because it is short and does not have any influence on the event.

The check valve is considered as resistance so there is certain pressure loss (3) in the positive direction (direction of flow when pump is running) and backflow is not allowed.

$$\Delta p = R \cdot |Q| \cdot Q, \forall Q > 0. \tag{3}$$

Thus, when no solution with positive flow rate is found the flow rate is set to zero and appropriate pressure is computed.

One-dimensional flow in the pipeline follows momentum equation (4) and continuity equation (5). These equations are solved with Lax-Wendroff method.

$$\frac{\partial Q}{\partial t} + \frac{S}{\rho} \cdot \frac{\partial p}{\partial x} + \frac{\lambda}{2 \cdot D \cdot S} \cdot |Q| \cdot Q = g_p \cdot S, \tag{4}$$

$$\frac{\partial p}{\partial t} + \frac{K}{S} \cdot \frac{\partial Q}{\partial x} = 0. \tag{5}$$

As the model has to catch also cavitation at the valve, the dependence of wave speed on the pressure is used [7]. It comes from water – air mixture, where only little air is considered. The air is solved by the gas state equation (6) and it has great impact on wave speed.

$$\frac{p}{\rho} = r \cdot T. \tag{6}$$

When pressure goes down, the wave speed goes down as well. In an extreme case it can go near to zero.

The result of pump trip numerical simulation does not show any significant pressure fluctuation, see figure 14. Pressure at the check valve decreases when pump loses

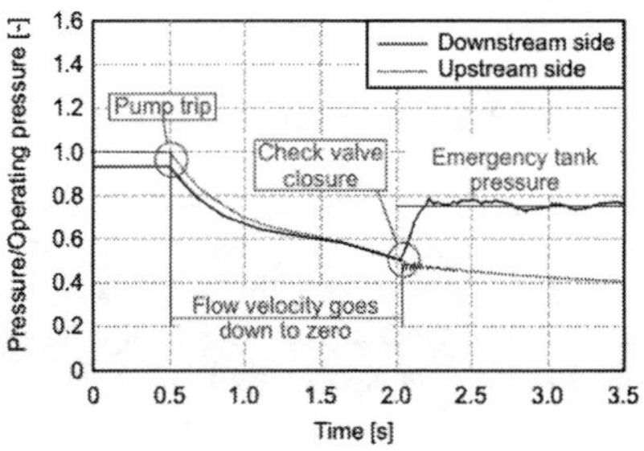

Fig. 14. Pump trip simulation without backflow through the check valve

energy. As soon as flow rate reaches zero value the check

valve closes and pressure at the downstream side increases to value given by emergency tank. No pressure pulsations appear and there is not any cavitation at the upstream side. Obviously, the result does not correspond to the measurement. It is a confirmation that check valve does not work correctly. The simulation predicts check valve closing approximately 1.5 s after the pump trip, but measured time is more than two times greater. Compare it with figure 8.

Figure 15 shows result with adjusted check valve condition. In this case the check valve allows back flow for

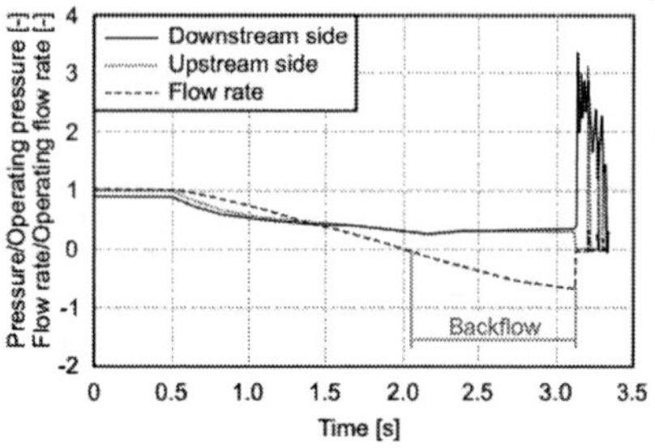

Fig. 15. Pump trip simulation with no backflow through the check valve

certain time and then suddenly closes (the dashed line in the graph). It causes pressure surge (solid and dotted line) fairly corresponding to the measured one. The back flow velocity when the check valve closes is 1.7 m s^{-1}. This value fairly correspond to the velocities obtained from pressure measurement analysis described in the previous chapter.

The model of the disc used in the simulation is the simplest one. Inertia, friction or damping are not considered and water does not influences disc behaviour thus the disc bouncing is not present in the result.

4.1 Discussion

Simulation showed that the system is designed well and pressure surge cannot appear with correctly working check valve (figure 14). Despite this fact, measured pressure surge indicates great backflow, which has almost the same velocity like the flow when pump is running. It is proved with three independent tools: Analysis of the pressure downstream the valve, upstream the valve and numerical simulation of the pump trip.

It leads to advice to replace the check valve with a different one made by another producer. The temporary solution is to increase the counterweight, which makes closing easier so the backflow will not evolve or it will evolve much lower. This adjustment increases pressure loss, but the pump has enough reserve to secure sufficient flow rate.

It should be stressed, that this solution is temporary. Increased counterweight causes greater strain in the shaft and could lead to its damage as well as the slams.

5 Conclusion

This paper describes investigation of real problem with the check valve slam. It appears in a pumping system when the pump loses power supply and it causes noticeable pipeline moving, which is definitely going to serious damage of the system. The problems appeared after the check valve replacement with a new one made by different producer.

The analysis of pressure records upstream and downstream the valve showed that the valve closes when too great back flow evolves. It is too late and it causes strong water hammer. The same result was obtained from results of numerical simulation of flow in pipeline system during the pump trip. The simulation also proved that the system is well designed and it can work without any problem, when good check valve is installed.

To increase the weight of the valve is a temporary solution, which makes its closing easier. The increased pressure loss of the valve due to this modification is acceptable, because the pump is able to cover it and secure the required flow rate.

Acknowledgement

This work has been supported by Technology Agency of the Czech Republic under the project Innovative research of check valves for extreme operating conditions in energetics TH01011352.

Nomenclature

a	(m s^{-1})	wave speed,
D	(m)	diameter,
f	(Hz)	frequency,
g_p	(m s^{-2})	projection of gravitational acceleration to pipe axis,
K	(Pa)	bulk modulus,
L	(m)	pipe length,
M	(N m)	torque,
p	(Pa)	pressure,
Q	(m^3 s^{-1})	flow rate,
R	(kg m^{-7})	resistance,
r	(J kg^{-1} K^{-1})	gas constant,
S	(m^2)	pipe cross-section,
T	(K)	temperature,
t	(s)	time,
x	(m)	longitudinal coordinate,
Δp	(Pa)	pressure difference,
Δv	(m s^{-1})	velocity change,
λ	(-)	coefficient of friction loss,
ρ	(kg m^{-3})	density.

References

1. K. L. McElhaney, Nuclear Engineering and Design **197**, 169 – 182 (2000)
2. D. Y. Goswami, *The CRC Handbook of Mechanical Engineering, Second Edition* (CRC Press, 2004), pp. 2688
3. C. Chiu, American Nuclear Society **54**, 289 – 291 (1987)
4. J. R. Travis, M. D. Torrey, *American Society of Mechanical Engineers winter annual meeting*, 1 – 9 (Miami 1985)

5. D. Himr, V. Habán, Applied Mechanics and Materials **630**, 375-382 (2014),
6. R. Autrique, E. Rodal, A. Sánchez, L. Carmona, IOP Conference Series: Earth Environmental Science *15* 022014 (2012)
7. D. Himr, Journal of Hydraulic Engineering **141** 04014080 (2015)

EPJ Web of Conferences **114**, 02039 (2016)
DOI: 10.1051/epjconf/201611402039

Diagnostics of gas turbines based on changes in thermodynamics parameters

Marián Hocko[1,a], Marek Klimko[2]

[1]*Technical University in Košice, Department of Aviation Engineering, Rampová 7, 041 21 Košice, Slovak Republic*
[2]*University of West Bohemia, Department of Power System Engineering, Univerzitní 8, 306 14 Pilsen, Czech Republic*

Abstract. This article is focused on solving the problems of determining the true state of gas turbine based on measured changes in thermodynamic parameters. Dependence between the real individual parts for gas turbines and changing the thermodynamic parameters were experimentally verified and confirmed on a small jet engine MPM-20 in the laboratory of the Department of Aviation Engineering at Technical University in Košice. The results of experiments confirm that the wear and tear of basic parts for gas turbines (turbo-compressor engines) to effect the change of thermodynamic parameters of the engine.

1 Introduction

Diagnostics of gas turbines (GT) based on changes in thermodynamics parameters is one of the most spread and efficient methods of their evaluation. This method is used to analyze mainly the capacity of GT to ensure their basic function – to supply required power.

Various types of malfunction and deterioration of the flow components are typical for different types of GT in the stage of their testing and operation.

Operational factors, among which are acceleration, deceleration, heat non-stationarity of construction components, friction parts wear, change in climatic conditions, sucking of foreign objects etc., influence GT during the operation. The influence of operational factors manifests in extension of radial clearance in blading machines, wear of interacting components, quality of sealing, erosive changes in a blade profile shape, appearance of clearance in the blade control system of compressor rectifier, clogging of an engine flow part and cooling system channels.

Wear of gas generator components is the basic cause of thermodynamic parameters impairment. Typical failures of the engine flow part are gas turbine failures – cracks on individual rectifying and rotor blades, erosive wear of stationary blades edges. Among other failures there are damage in compressor blades due to the sucking of foreign objects and erosive wear, fuel nozzles burnout, flue of combustion chambers, air leaks through flange joints of the compressor and combustion chamber at the points of fixing pipes, valves, pipe plugs and flanges of a gas turbine.

2 Requirements on checked parameters

An ideal case shall require measuring gas flow parameters (total pressure and total temperature) in all characteristic cross sections: at the inlet to the compressor P_{1C}, T_{1c}, combustion chamber p_{2c}, T_{2c} and gas chamber p_{3c}, T_{3c} for a perfect diagnosis of the GT flow part. The list of measured parameters is substantially lower in real life, although there is a trend to increase it. Total pressure p_{0c} and total air temperature T_{0c} at the engine inlet, and total pressure p_{2c}, total air temperature T_{2c} behind the compressor, and total pressure p_{4c} and total gas temperature T_{4c} behind the gas turbine, are the most frequently measured parameters of current engines. The essential measured parameters are engine rotor speed and those parameters, which characterize a change in controlled parameters of engine mechanization, such as a value of extracted air needed for a defroster and other systems. Most of listed parameters participate in an engine regulating process (engine regulating principle), which is a condition for their use. Total gas temperature behind a low-pressure gas turbine (LPT) p_{4c} represents a complex parameter, as for sensitivity, characterizing a condition of the engine flow component. Increase in sensitivity may be induced by various causes – a growth of air leaks from a high-pressure compressor (HPC) p_{2c}, impairment of efficiency (characteristics) of a compressor and a gas turbine. Air temperature around a combustion turbine enables to detect hot air leaks from HPC. A character of a temperature change behind HPC p_{2c} confirms the change of HPC efficiency, depending on the number of duty hours.

[a] Corresponding author: marian.hocko@tuke.sk

3 Algorithms of diagnosing the engine flow section

Currently, the processes running in GT (turbo-compressor engines), are sufficiently well controlled. Generally, they can be described by a system of non-linear equations, which are interrelated by status parameters (coefficients of total pressure preservation, efficiency of individual nodes, flow sections areas, values of gas and air leaks) and measured parameters (total pressures and total air and gas temperatures in various engine cross-sections, rotor speed, fuel consumption, power etc.).

An origin of failures (fluent or step ones) is usually subject to a change of efficiency (characteristics) of nodes, which cannot be measured, yet this change demonstrates immediately in a shift of an engine thermodynamic and gas dynamic parameters. An interaction between measured parameters and properties of nodes is described by thermodynamic and gas dynamic mathematical models on various complexity levels. The given models are utilized in all methods of malfunction detection.

Methods of detecting a flow section malfunction of GT (turbo-compressor engines) may be divided into two basic types:

- Detecting malfunction based on the analysis of measured parameters;
- Detecting malfunction on the basis of identifying a thermo-gas dynamic model, followed by malfunction detection based on the analysis of nodes properties parameters.

A selection of the model form, which will be the most suitable for realization in practice, shall depend on the ongoing life cycle of GT (turbo-compressor engines). Advantages of the first mentioned method prevail, depending on the operating time length and collected empirical information on the engine condition. These methods additionally provide a good illustration when implementing an expert analysis on the group of objects, diagnostics, and regression analysis of time series. When using the methods of the first type, parameter values are recalculated on standard atmospheric conditions and computational mode.

So called unrestrained parameters of nodes characteristics are utilized the most for the methods of the second type. Using a mathematical model (where a model identification is used), parameter values for various failures are calculated and a deviation from common parameter values is determined, for given external conditions and the engine operational mode. Obtained differences represent "separateness" of parameters. The minimum "separateness" of parameters is reached for such a state, which corresponds with the most probable malfunction defined in the model. The model is based on the model identification and a compilation of possible malfunctions with a small repeatability. A decision of a concrete turbo-compressor engines condition is carried out on the basis of the minimum "separateness" principle. Despite the positive properties of this method, there are flaws as well, which include a low illustration when performing an expert analysis on the group of diagnostics objects and, most of all, when analyzing time series.

Admissible values in characteristics changes of diagnosed nodes are defined on the basis of statistics data acquired during testing and operation of those engines, which showed malfunctions, and those, which worked off their technical life failure-free.

Random effects influence the computational results in the process of parameter analysis in operational conditions:

- Changes in air extractions or the performance, demanded by the engine or other devices,
- Inaccuracies of measurements and parameter registration,
- Inaccuracies in keeping the engine operation by a regulation system,
- Air flow non-uniformity at the engine inlet,
- Change in a generating set load,
- Technological scattering of geometrical dimensions, nodes characteristics during engine production,
- Wear of engine parts in the operational process,
- Humidity, air turbulence, wind direction and velocity.

Algorithms of mathematical models identification are traditionally used to diagnose the flow part failures, and “trend-analysis“ algorithms are used to detect a time change tendency of measured parameters in GT (turbo-compressor engines) automatized diagnostics systems. A mathematical model identification consists of the determination of differences between measured and computed parameter values. The use of “trend-analysis“ procedure enables, based on the deviation between measured and computed parameter values, to define tendencies of their change, depending on the operational time on the basis of indeterminate errors of measurements. It is difficult to guarantee the diagnostics of a flow part condition at changeable operation modes of the engine at a relatively low frequency of information registration by panel registration systems. The most suitable mode to diagnose the GT (turbo-compressor engines) part is the engine computational mode. General requirements for a selection of diagnostics modes are possible to define as follows:

- The engine operational mode must be high enough to differ changes in parameters, which were induced by the change of the flow part state from changes induced by random factors;
- Measuring parameters must be subject to the least amount of random non-measured effects due to the mode selection with the maximum volume of identical (similar) external conditions and parameters of the engine operational mode.

The engine has maximum parameter values at the maximum operational duty, which leads to the drop in measurement inaccuracies (faults). However, some problems appear, which are related to ensuring similar conditions of ongoing processes of massive rotor components overheating, valve closedowns etc.

Forming the optimum time and the selection size to determine the parameter values are chosen on the basis of the registered engine information oddness, as well as the length of ongoing processes in the engine.

A concept of a "reference point" is used fairly frequently in practical life. This concept enables to define a set up mode, as well as parameter mean values computed from selected values. It is necessary to recalculate the parameters onto the international standard atmospheric conditions (ISA) to eliminate the influence of meteorological conditions on the checked parameters values of GT (turbo-compressor engines). It is convenient to perform this recalculation on the standard atmospheric conditions, which correspond to the selected reference points, to decrease the inaccuracy influence during recalculating.

The ISA recalculation is performed using standard gas dynamic functions, which conform to the reference point of the given ISA type.

The values of acquired recalculated parameters on ISA, regarding the reference point of the given type, are related to close, though usually different, engine operational modes. Therefore, it is necessary to carry out their recalculation on the computational mode, to make the parameters value real. To be able to realize such a process, it is needed:

- To obtain the functional dependence course, which sufficiently describes a change in the parameter depending on the mode parameter,
- To define how the behaviour of this dependence changes for various engines,
- To determine the relation between a real dependence and dependence acquired via the engine mathematical model,
- To define the character of a functional dependence change for increased number of the engine operational hours (the shift of the characteristics, or its deviation),
- To define the character of the influence of extracted air flow rate, the position of compressor mechanization components etc.

The use of the method of parameter values recalculation towards standard atmospheric conditions, and consequently onto the computational mode, provides a possibility to create dependencies of parameter changes against the engine operational time. A graphic illustration of those dependencies is a useful tool to define a perfect visual formulation of parameter changes and long-term tendencies.

Possibilities of the flow part diagnostics by the time series method are open, with regard to the above mentioned. The easiest method is to determine basic parameter values at the beginning of the engine operation, and to consequently check up the deviation of recalculated parameters towards basic parameters.

4 Experimental verification of the influence of changes in the inlet device of the jet turbo-compressor engine on thermodynamic parameters

The experiment has focused on monitoring changes in thermodynamic parameters MPM-20 when the flow conditions in its inlet device have changed.

The theory of small changes, described for sub-critical flow conditions in an outlet nozzle for MPM-20, defines that a change in conditions in an inlet device will manifest itself as a change in all parameters of an operational process of a single-flow turbo-compressor engines. Four sets of experimental measurements have been prepared to verify this theory in practice; and their objective has been to stepwise throttle the air flow through the inlet device MPM-20. The measured values of parameters are compared with values acquired during a calibration measurement and theoretical calculation on the basis of the small changes theory. Four experimental measurements have been realized in stages as follows:

1. Air flow through the inlet device MPM-20 without an inlet sieve.
2. Air flow through the inlet device MPM-20 with a sieve of the aperture dimension of 1,5 x 1,5 mm.
3. Air flow through the inlet device MPM-20 with a doubled sieve (a sieve with the aperture dimension of 1,5 x 1,5 mm and a sieve with the aperture dimension of 2,5 x 2,5 mm).
4. Air flow through the inlet device MPM-20 with a fluorescent TV screen with holes of 0,4 mm in a diameter.

4.1 Air flow through the inlet device without an inlet sieve

An inlet sieve has been removed from the inlet device MPM-20 to follow the conditions of the given experiment.

Figure 1. View to the inlet sieve MPM-20.

10 measurements have been carried out during the experiment, whereas the same parameters as for the calibration measurement have been measured and analysed. The operational mode of MPM-20 has been set up onto such values, that correspond to the computational mode.

Differences for individual parameters are minimum and they range within -5,4% (static pressure behind the gas turbine p_{4c}) up to + 2,9% (total gas temperature behind the gas turbine T_{4c}), as implied from the data mentioned in Table 1, in which the measured values for parameters during experimental measurements without an inlet sieve and during the calibration measurement are compared. Most of observed parameters did not change much, even though the flow area at the inlet into MPM-20 enlarged by 35% by removing the inlet sieve.

The results obtained by direct measurements on MPM-20 without an inlet sieve confirm that a selection of a wire diameter as well as an aperture dimension have been optimized by the engine producer, so that the used sieve can provide a protection against the sucking of foreign objects at the minimum influence of MPM-20 parameters. A calculation of a force F_T and specific fuel consumption cm has been made, based on measured parameters.

4.2 Air flow through the inlet device without an inlet sieve

A fine inlet sieve has been implemented during the second experimental measurement. The total area of the inlet sieve was 0,078593 m^2. The wire area of the inlet sieve takes up 34,984% when using the wire sieve of 1,5 x 1,5 mm of the aperture dimension and 0,3 mm of the wire thickness.

The measured parameter values of MPM-20 listed in Tab. 1 are very near to the measured values of calibration measurements with the original sieve.

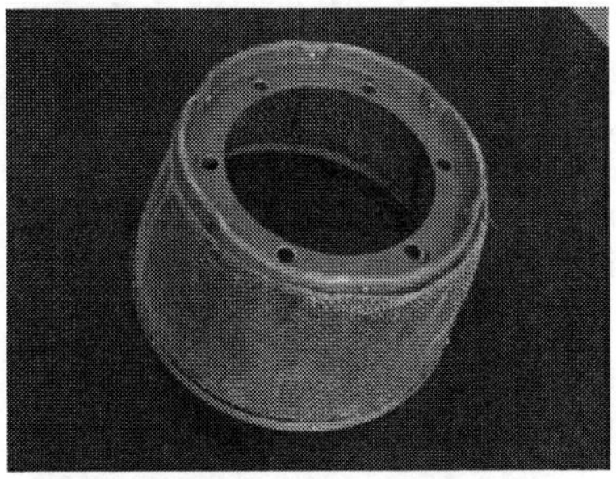

Figure 2. View to the fine inlet sieve of MPM-20.

4.3 Air flow through the inlet device with a doubled inlet sieve

During the third experimental measurement, a sieve with 1,5 x 1,5 mm of an aperture dimension and a wire thickness of 0,3 mm was mounted on the original inlet device of MPM-20, which decreased the flow area down to 58%. Based on this drop in the inlet device flow area, the measured parameters have changed, which led to the change in the dependent parameters. The biggest change has been observed in the stage of gas expansion on the gas turbine π_T (by 19,194%) and in the change of the total gas temperature behind the gas turbine T_{4c} (by 4,475%).

Figure 3. View to the inlet with the finest sieve from the inside.

Figure 4. View to the inlet device with the finest sieve from the outside.

4.4 Air flow through the inlet device with a doubled inlet sieve

A sieve with circle holes of 0,4 mm was implemented during the fourth experimental measurement instead of the original MPM-20 sieve, which resulted in a decrease of the flow area of the inlet MPM-20 device by 83%. A decrease in the flow area of the engine inlet device led to a significant degradation in measured parameters compared to calibration measured parameters (Δp_{2C} = -48 978,8 Pa, Δp_3 = 24198,5 Pa, Δp_4 = -24957,82 Pa, T_{2c} = -11,6 K, ΔT_{3c} = -92,4 K, ΔT_{4c} = -177,5, c_h = =-4,3465 kg.h^{-1}, $\Delta\pi_T$ = 0,6, $\Delta\pi_{Kc}$ = -0,509 a $\Delta\pi_{VD}$ = -0,257.

Table 1. Change in MPM-20 parameters when throttling the inlet system.

PARAMETER	$S_{vst.}$	p_H	T_H	Time of operation	$p_{2,K}'$	$p_{2c,K}'$
UNIT	%	Pa	K	second	Pa	Pa
Inlet system without a sieve	100	100525,3	288,75	51,0	178967,1	388941,8
Inlet system with a fine sieve	80	100365,6	290,95	49,7	179136,7	390028,1
Inlet system with a basic sieve	65	99885,82	290,65	51,7	180787,9	391185,7
Inlet system with a doubled sieve	58	100125,3	290,65	49,7	198319,4	397633,7
Inlet system with the finest sieve	17	99991,98	290,29	50,0	213852,3	428407,3
PARAMETER	p_{2c}	p_3	p_4	T_{2c}'	T_{2c}	T_{3c}
UNIT	Pa	Pa	Pa	K	K	K
Inlet system without a sieve	351178,4	351524,7	128716,48	458,35	459,55	1169,25
Inlet system with a fine sieve	351571,5	351462,6	128977,1	458,45	459,95	1170,05
Inlet system with a basic sieve	351971,5	351326,2	154655,78	458,55	460,25	1170,35
Inlet system with a doubled sieve	366122,2	340125,3	154192,9	460,65	461,95	1172,95
Inlet system with the finest sieve	400157,2	327326,2	153674,3	470,01	471,15	1261,65
PARAMETER	T_{4c}	Q_{pal}	G_{pal}	c_h	F_T	c_m
UNIT	K	$cm^3/cycle$	kg/cycle	$kg.h^{-1}$	N	$kg.h^{-1}N^1$
Inlet system without a sieve	1039,15	1407	1,09045	77,0065	732,5*	0,1051283
Inlet system with a fine sieve	1041,15	1405	1,0889	78,8700	731,6*	0,107811
Inlet system with a basic sieve	1043,45	1397	1,0819	75,2900	698,1*	0,107847
Inlet system with a doubled sieve	1085,65	1330	1,0312	74,7434	598,74*	0,125334
Inlet system with the finest sieve	1216,65	1293,3	1,0027	72,6600	541,11*	0,13428

Table 2. Change in dependent MPM-20 parameters when throttling the inlet system.

PARAMETER / $S_{vst.}$	π_T	π_{Kc}	π_{VD}	T_{2c}	T_{3c}	T_{4C}	c_h	F_T	c_m
	1	1	1	K	K	K	$kg.h^{-1}$	N	$kg.h^{-1}.N^{-1}$
100 %	2,730	3,49	1,280	459,55	1169,25	1039,15	77,01	732,5	0,105128
80 %	2,725	3,50	1,285	459,95	1170,05	1041,45	78,87	731,6	0,107811
65 %	2,717	3,52	1,548	460,25	1170,35	1043,45	75,29	698,1	0,107847
58 %	2,206	3,66	1,540	461,95	1172,95	1085,65	73,92	598,7	0,125334
17 %	2,130	4,00	1,537	471,15	1261,65	1216,65	72,66	541,1	0,134280

5 Conclusion

Following conclusions can be generalized from measured values of MPM-20 for individual experimental measurements:

1. The MPM-20 inlet device throttling leads to a change in engine parameters, whereas their character is the same like when changing the coefficient of total air pressure preservation in the inlet system σ_{vst}.

2. The following thermodynamic parameters react the most intensively to the inlet device throttling: static gas pressure behind the gas turbine p_4 (-19,39%), total gas temperature behind the gas turbine T_{4c} (-17,081%), total air pressure behind the compressor p_{2c} (- 13,95%), total gas pressure in front of the gas turbine T_{3c} (-7,903%), and total air temperature behind the compressor T_{2c} (-2,524%).

3. The MPM-20 inlet device throttling leads to intense approximation of the working point to the pumping border in the compressor characteristics. Despite higher resistance of the radial compressor MPM-20 against unstable work, 7th measurement with 83% shadowing of the inlet device resulted in its stalls and destroyed blades of the gas turbine, and the sealing mass burnout in the gas turbine body above rotor blades.

Figure 5. View at destroyed rotor blades of gas turbine MPM-20

Acknowledgement

The presented research is supported by the project CZ.1.05/2.1.00/03.0108 (Sustainable Energy, SUSEN) and student project SGS-2014-070 (Increasing the efficiency, reliability and durability of power system devices 3).

References

1. L. T. De Luca, *Propulsion physics* (EDP Sciences, Les Ulis, 2009)
2. G. G. Oľchovskij, *Energetičeskije gazoturbinnyje ustanovki* (Energoatomizdat, Moscow, 1985)
3. A. A. Inozemcev, V. L. Sandrackij, *Gazoturbinnyje dvigateli* (AOA Aviadvigateľ, Perm, 2006)
4. A. J. Čerkez, *Inžinernyje rasčoty gazoturbinnych dvigatelej metodom malych odklonenij* (Mašinostrojenije, Moscow, 1965)
5. M. Hocko, M. Komjáty, *Application of the method of small alterations to analyse funda-mental parameters of a small turbojet engine MPM-20* (Acta Avionica, 2014), ISSN 1335-9479
6. M. Hocko, *Hodnotenie stavu LTKM na základe zmeny termodynamických parametrov* (VLA M.R.Š., Košice, 2003)
7. T. Lazar, *Inovatívne výstupy z transformovaného experimentálneho pracoviska s malým prúdovým motorom* (Elfa, Košice, 2011) ISBN 978-80-8086-170-4

EPJ Web of Conferences **114**, 02040 (2016)
DOI: 10.1051/epjconf/201611402040

The use of wind tunnel facilities to estimate hydrodynamic data

Kristoffer Hoffmann[1,a], Johannes Tophøj Rasmussen[1], Svend Ole Hansen[1], Marit Reiso[2], Bjørn Isaksen[3], and Tale Egeberg Aasland[4]

[1] Svend Ole Hansen ApS, Copenhagen, Denmark
[2] Reinertsen AS, Trondheim, Norway
[3] Norwegian Public Roads Administration, Oslo, Norway
[4] Norwegian Public Roads Administration, Stavanger, Norway

Abstract. Experimental laboratory testing of vortex-induced structural oscillations in flowing water is an expensive and time-consuming procedure, and the testing of high Reynolds number flow regimes is complicated due to the requirement of either a large-scale or high-speed facility. In most cases, Reynolds number scaling effects are unavoidable, and these uncertainties have to be accounted for, usually by means of empirical rules-of-thumb. Instead of performing traditional hydrodynamic measurements, wind tunnel testing in an appropriately designed experimental setup may provide an alternative and much simpler and cheaper framework for estimating the structural behavior under water current and wave loading. Furthermore, the fluid velocities that can be obtained in a wind tunnel are substantially higher than in a water testing facility, thus decreasing the uncertainty from scaling effects.
In a series of measurements, wind tunnel testing has been used to investigate the static response characteristics of a circular and a rectangular section model. Motivated by the wish to estimate the vortex-induced in-line vibration characteristics of a neutrally buoyant submerged marine structure, additional measurements on extremely lightweight, helium-filled circular section models were conducted in a dynamic setup. During the experiment campaign, the mass of the model was varied in order to investigate how the mass ratio influences the vibration amplitude. The results show good agreement with both aerodynamic and hydrodynamic experimental results documented in the literature.

1 Introduction

Vortex-induced vibrations (VIV) of cylinders is a widely studied phenomenon due to its relevance in numerous engineering disciplines. Several reviews have been written on the matter of VIV of circular cylinders [1–3], as well as the more complicated cases of interference between two closely separated cylinders [4] and the structural instabilities of rectangular cylinders [5].

A series of wind tunnel test has been performed to investigate the response characteristics of different cylinder cross sections. The objective of the tests is to provide necessary key parameters for the initial design of a submerged floating tube bridge (SFTB) under consideration for the Bjørnafjord crossing south of Bergen, Norway. At the current stage in the design process, the use of wind tunnel testing, as opposed to hydrodynamic testing, is a great advantage.

Investigation of vortex-induced structural oscillations in hydrodynamic laboratories is time-consuming and expensive. In addition, the results will have an inherent uncertainty due to Reynolds number scaling effects. Similarity in Reynolds number is exceedingly hard to achieve in a water setup, as it requires either a large-scale setup or very high fluid velocities. The latter situation will also induce very large forces on the model structure. In a wind tunnel, these effects are less prominent, as the fluid velocities that can be obtained are substantially higher than in a water testing facility and the inertial forces are much smaller. Wind tunnel experiments allow Reynolds numbers in the range of the resonance wind velocities ($10^4 \leq Re \leq 10^6$), which also gives a good basis for predicting the behaviour in full scale. This, along with the relative ease of the setup and the reduction in cost, makes the use of wind tunnel testing to determine hydrodynamic data an attractive alternative to water facility testing.

Wind tunnel facilities are widely used in the investigation of vortex dynamics and the related structural issues, and numerous studies of VIV and flow instabilities of cylinders have been carried out in wind tunnels. What these studies have in common, is that they operate with comparatively large mass ratios. For instance, a recent study [6] of VIV on circular cylinders in the post-critical regime refers to a mass ratio of 35 as "quite low with respect to tests in air". A widely sited master's thesis work [7] employs a mass ratio of approximately 250 for wind tunnel experiments. However, for neutrally buoyant structures submerged in water the mass ratio is typically of order 1, and at such low mass ratios the scientific literature does not seem to cover wind tunnel experiments characterising the vortex-induced structural dynamics.

Mass and damping have a significant effect on VIV, and in order to obtain a realistic result for the present study, this must be taken into account. The effect of mass and damping on VIV for a circular cylinder has been investigated numerically, and it was found that variations of the mass and damping ratios have almost the same result on the system

[a] e-mail: kho@sohansen.dk

response. Increasing the mass ratio will reduce the maximum amplitude, and also the velocity range over which lock-in occurs [8]. This corresponds to earlier findings [9], where experiments with the same mass-damping parameter were carried out in both air and water. In the present investigation, lightweight helium-filled models have been used for the dynamic tests, in order to get model mass ratios that simulate neutral buoyancy in water.

2 Existing theories and methodologies

Vortex-induced vibrations are described mathematically by a number of models of the physics behind vortex formation, the associated oscillating loads and the structural dynamics, incorporating the associated coupling of these phenomena. The physical parameters and principles allow for the formulation of response models used to establish general design criteria primarily related to structural vibrations.

2.1 Determining parameters and basic principles

A number of basic non-dimensional parameters have been established to describe and characterise the main static and dynamic behaviour of structures subjected to in-line and cross-flow vortex-induced vibrations. These parameters depend on relevant structural and flow-specific properties.

The specific unique ratio of inertial and viscous forces is given by the Reynolds number, defined by the well-known expression

$$Re = \frac{UD}{\nu}, \tag{1}$$

where U is the mean velocity of the undisturbed flow, D is the cross-flow structural dimension and ν is the kinematic viscosity of the fluid.

The ratio between the oscillating mass of the structure, being in the model or full-scale configuration, and the mass of the equivalent displaced fluid can be expressed by the mass ratio

$$\lambda = \frac{m_{\text{osc,structure}}}{m_{\text{osc,fluid}}}. \tag{2}$$

Note that this is a purely structural parameter. In this expression $m_{\text{osc,structure}}$ denotes the complete mass of all oscillating parts. For both the numerator and the denominator, the oscillating mass of the surrounding fluid is in this paper approximated by the mass of the displaced fluid. For example, a structure having the same density as the surrounding fluid, will have a mass ratio of 1. A structure with twice the density of the surrounding fluid, will have a mass ratio of 3/2, since the density of the oscillating surrounding fluid reamins unchanged.

The Strouhal number is defined by the well-established relation

$$St = \frac{n_s D}{U}, \tag{3}$$

where n_s is the frequency of the lateral loads. A similar relationship is defined in oscillatory flow, where the relative orbital motion of fluid particles relative to the characteristic length of the immersed object plays a crucial role. This relationship is expressed by the Keulegan–Carpenter number [10].

For the present analysis, the structural susceptibility to vortex-induced vibrations is characterised by the general mass-damping parameter

$$Sc_G = \frac{2\delta_s m_e}{\rho L D}. \tag{4}$$

In the formula above, δ_s is the structural damping quantified by the logarithmic damping decrement in still fluid, m_e is the equivalent mass per unit length corresponding to the mode considered, ρ is the density of the surrounding fluid, and L is the in-line structural dimension. The presented expression is a generalisation of the Scruton number, taking into account structures having a non-circular cross-sectional geometry [11]. Note that in the case of a structure exposed to vortex shedding along its complete extension, the mass-damping parameter is simply a damping term multiplied by the ratio between the density of the oscillating structural mass and the density of the surrounding fluid.

The structural response is described in terms of a fundamental relation between the reduced fluid velocity and the reduced amplitude. The reduced fluid velocity is in relation to in-line vibrations defined by

$$U_r = \frac{U}{n_e D}, \tag{5}$$

where n_e is the in-line natural frequency of the structure in still fluid, and the reduced amplitude is defined by

$$A_r = \frac{A}{D}, \tag{6}$$

where A is the amplitude of the in-line oscillation.

It should be noted that the use of the reduced velocity as the parameter characterising the in-line vibrational response may not be optimal, at least for very low mass ratio structures. It seems reasonable that the forcing frequency caused by vortex shedding is dependent on the actual oscillation frequency and not the natural frequency of the structure in a still fluid. For cross-flow oscillations, the velocity normalised by the actual oscillation frequency has been shown to give a better description of the experimental data in some situations [12]. However, in the present paper, the reduced velocity is used as the descriptive parameter, since this is in accordance with the majority of the literature and design codes related to in-line oscillations.

The basic physical principles of vortex shedding induce an oscillating drag force which drive in-line structural oscillatory vibrations. Two different scenarios can occur, both corresponding to alternating low-pressure vortices on the downstream side of the object. Normal vortex shedding combined with a secondary, symmetric vortex shedding, which occurs as a result of in-line structural motion relative to the fluid, can create downstream pressure variations oscillating with a frequency of approximately three times the Strouhal frequency. This mechanism is responsible for the first fluid instability causing in-line vibrations. The second fluid instability causing in-line vibrations occur for slightly larger fluid velocities, where the forcing is entirely due to normal vortex shedding. Since the shedding is limited to low-pressure vortices on the downstream side of the object, the forcing frequency in the in-line direction is approximately twice the Strouhal frequency. It should also be mentioned, that since vortex-induced in-line vibrations in steady flow occur at relatively low fluid velocities, the oscillation amplitudes are in the order of one magnitude smaller than the corresponding cross-flow vibrations usually experienced at larger flow velocities [10].

2.2 In-line amplitude response models

Structural vibrations caused by vortex-induced oscillating loads can be described in a simplified fundamental mathematical framework, based on theoretical and experimental research in the oscillating fluid/structure interaction, combined with general accepted experience from the construction and operation of different types of structures susceptible to VIV. To a certain extent, experiments in fluid dynamic testing facilities has also helped to clarify and specify certain aspects of this description.

A fundamental description of the in-line reduced amplitude response for submerged cylindrical structures in flowing water can be given in terms of the reduced fluid velocity and a mass-damping parameter. Experiments, such as those performed by King [13], have verified the presence of two fluid instability regions and in broad terms defined the associated in-line response amplitude characteristics.

Det Norske Veritas (DNV) has developed application-based design criteria and recommendations for free-span submarine pipelines which prescribe in-line response curves based on the reduced velocity and a stability parameter similar to the mass-damping parameter, but taking into account to total modal damping [14]. The in-line response model covers both fluid instability regions, corresponding to symmetric and regular, asymmetric vortex shedding for a uniform mode shape. The DNV model is used as a reference guideline for the maximum in-line response of cylindrical low mass ratio structures submerged in steady flowing water; thus, together with the wind tunnel measurements it is used to compare the response characteristic measured for low mass ratio structures in steady air with those expected for similar low mass ratio structures submerged in steady flowing water.

In its general description, the maximum reduced in-line amplitude response for neutrally buoyant cylindrical structures submerged in steady flowing water is similar to the principal sketch in figure 1. In King's laboratory testing with an oscillation given by a free-end mode shape, the two instability regions are clearly visible. The DNV specifications are similar, but more general and the response covers are larger area, especially for $U_r > 3$. The DNV model is in figure 1 presented for a stability parameter of 0.2. At higher levels of total damping, the onset reduced velocity is increased slightly and the maximum response is decreased. The reduced velocity above which in-line vibrations are insignificant is also lowered. Note that the response models only describe the response caused by in-line VIV, and other phenomena, such as galloping or torsional structural instabilities, may change the actual response significantly.

The ultimate goal of the present analysis is to determine if the known dynamic behaviour of neutrally buoyant structures in steady flowing water, expressed by the presented in-line response models, is similar to the dynamics of neutrally buoyant model structures in air.

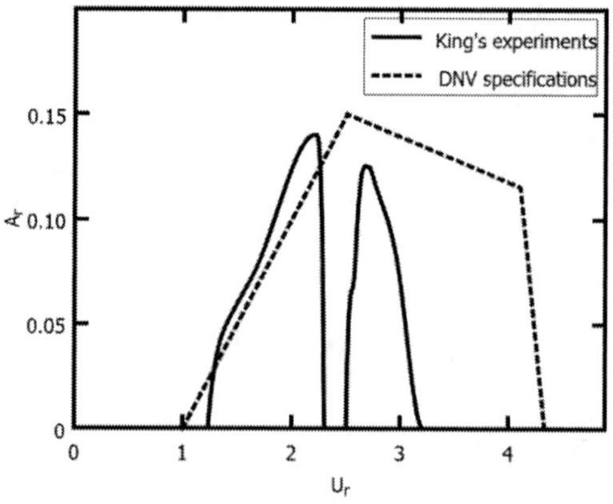

Figure 1: Principal sketch of the maximum reduced in-line amplitude response as a function of the reduced fluid velocity for low mass ratio cylindrical structures submerged in steady flowing water. The presented general response characteristics are based on published results from laboratory experiments in water for a free-end mode shape [13] ($Re = 6 \times 10^4$) and a response model from code specifications for a uniform mode shape, assuming a stability parameter of 0.2 [14].

2.3 Experiences and common practice when designing submerged marine structures

Vortex-induced vibrations are a key issue in many disciplines of marine engineering design. Currently defined response models, as those presented previously, are used to give overall indications of the onset and intensity of both in-line and cross-flow VIV. Certain structure-specific properties, such as the cross section geometry, inter-structural spacing and the distance to any flow boundaries may, however, influence the response significantly [15]. Hydrodynamical tests are as a natural consequence used to investigate the possibility of structural behaviour not covered by existing response models, such as for instance was the case with the initial studies performed at the Norwegian Marine Technology Research Institute (MARINTEK) in connection with the proposed Høgsfjord SFTB crossing and the Ormen Lange pipeline project.

The motivation for the present study is to investigate if one can use wind tunnel experiments, as an alternative to hydrodynamic experiments, to obtain information of the hydrodynamic behaviour of cross sections relevant for submerged floating tube bridges. The procedure will also provide valuable indications of how to utilise wind tunnel experiments to link the aerodynamic and hydrodynamic data base closer together and describe a fundamental methodology for the transfer of certain aerodynamic actions to equivalent hydrodynamic actions.

3 Experimental design and setup

The performed experimental testing includes measurements of static and dynamic flow-induced model responses performed in wind testing facilities at Svend Ole Hansen ApS, Copenhagen, Denmark and SOH Wind Engineering LLC,

Vermont, USA. In the following section, the experimental methodologies of the static and dynamic measurements are outlined and the geometric properties of the analysed cross section models are presented.

3.1 Testing procedure

All conducted model tests are twofold: Tests in the static rig provide data to estimate the drag coefficient C_D, and the lift and moment coefficients C_L, C_M, together with their angular derivatives, which enable a description of fluid-induced forces and the aeroelastic or hydroelastic characteristics of the structure. Spectral analyses of the drag and lift signals provide an additional insight into the oscillating flow mechanisms for the fixed structure. This is described by the Strouhal number for cross-flow vortex-induced oscillations. Tests in the dynamic rig provide data used to predict the risk of flow-induced vibrations, especially vortex-induced in-line vibrations of the structure. Different degrees of spring stiffness are tested, giving resonance wind velocities in the entire wind velocity range of the wind tunnel.

The analysis considers the following two cross section models:

- Circular cylindrical tube
- Rectangular 1:3 box

For the rectangular box model, the in-line dimension is three times larger than the cross-flow dimension.

3.2 Static wind tunnel rig

To measure the static response in a wind tunnel, a force transducer is placed on the upstream side of the model facilitating measurements of the drag force. Two force transducers are placed at the top to measure the wind-induced lift force and moment. A relatively large mass of the model ensures that the top wires achieve a sufficient pretension such that a positive force is always measured on the top force transducers. The setup is identical on the other side of the wind tunnel, with a similar force transducer placed on the upstream side of the model. A principal sketch of the measurement setup used for static testing is depicted in figure 2.

The output from the calibrated force transducers, knowledge of the model geometry and measurements of the flow velocity allow for the estimation of the static coefficients on the fixed structure. These are defined by the following expressions of the mean drag, lift and moment load per unit length acting on a model

$$F_D = q_m D C_D, \quad F_L = q_m L C_L, \quad F_M = q_m L^2 C_M, \quad (7)$$

where q_m is the mean wind velocity pressure.

Analysing the temporal drag and lift force readings on the nominally non-vibrating model, reveals that the spectral power of the two signals have excess power at certain frequencies. For the lift, this corresponds to the oscillating force caused by vortex shedding, expressed by the Strouhal number for a fixed structure. All results from the static measurements are presented in Sect. 4.1.

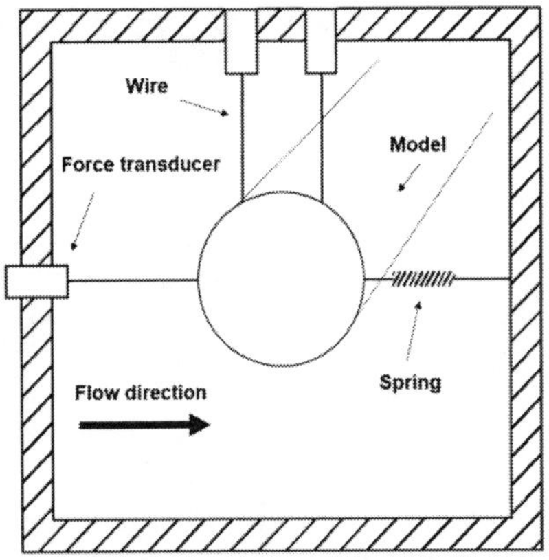

Figure 2: Principal sketch of the measurement setup used for static testing, looking from the side of the wind tunnel. The setup facilitates measurements of the wind-induced drag and lift force and moment. The setup is identical on the other side of the wind tunnel with a force transducer also placed on the upstream side of the model.

3.3 Dynamic wind tunnel rig

To measure the dynamic response, a lightweight model is mounted in the tunnel, constrained in the vertical direction, such that cross-flow deflections are restricted. Initial testing provided evidence that in-line VIV were initialised at wind velocities considerably below those where cross-flow VIV are of relevance, thus the restriction in the vertical direction is used only to provide a stable vertical position of the model in the low-turbulent wind tunnel flow. Any resonant cross-flow forcing can easily be identified by a corresponding loss of tension in the top wires, and all measurements performed under such conditions have been discarded. In the horizontal direction, the model is positioned between springs to achieve a desired oscillation frequency. One spring connects to a force transducer via a string that allows the deflection to be read, by a calibration done prior to beginning the experiment. Note that the US wind tunnel is also equipped with lasers to enable direct measurements of the model deflections. A principal sketch of the measurement setup used for dynamic testing is presented in figure 3. The dynamic measurements allows for a characterisation of the model-specific dynamic behaviour, especially linked to vortex-induced in-line vibrations.

The wind tunnel is set to run at a series of wind velocities whilst measuring the output of the force transducers. Between each measurement, an appropriate run-in time with constant wind velocity ensures that the oscillations have found a constant level. For each wind velocity, the standard deviations of the measured displacements are determined and by assuming a harmonic motion, the corresponding amplitudes are determined by multiplication by the square root of two.

In the dynamic rig, oscillations are measured as functions of different mass ratios, mode shapes and thereby different mass-damping parameters are simulated. Furthermore, decay tests are performed to measure the natural frequency and the damping at zero wind velocities. These data are relevant in the subsequent analysis of the dynamic behaviour of the full-scale structures. Note that the oscil-

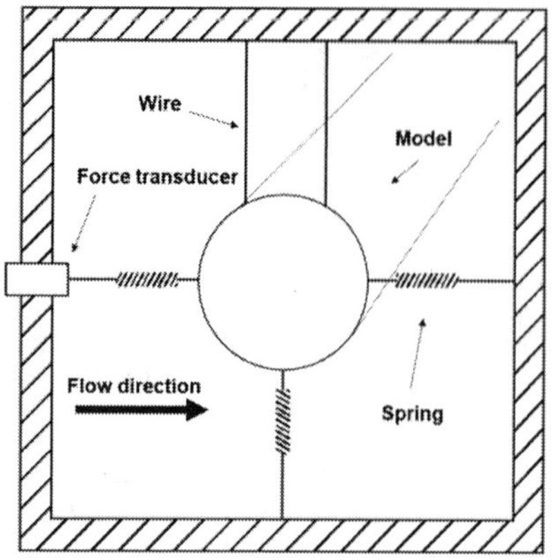

Figure 3: Principal sketch of the measurement setup used for dynamic testing, looking from the side of the wind tunnel. The setup facilitates measurements of an oscillating wind-induced in-line force. The US wind tunnel is also equipped with lasers to enable direct measurements of the model deflections.

lating mass is very different in still air relative to when the models are subjected to flow.

The results from the dynamic response measurements are presented in Sect. 4.2. Furthermore, pressure variations downstream to an in-line oscillating model have been captured using pitot tubes in the US wind tunnel for a single model configurations. This allows for the identification of the different fluid instability regions and therefore also the different physical forcing mechanisms responsible for in-line oscillations. These results are presented in Sect. 4.3.1.

3.4 Model design and construction

The measurements of the static load coefficients do not introduce mass restrictions on the model and readily available robust high-density materials can be in the model construction. In the dynamic wind tunnel rig, the section models are designed to be approximately neutrally buoyant. This may be established by using a low-mass hollow model construction, and replacing the air inside the model with a low-density gas such as helium. The model including suspension will hereby have a mass close to the mass of the displaced air.

The construction of the low-density models required significant engineering and experimentation to meet the structural requirements. The cylindrical models are based on helium-filled cylindrical containers, obtaining additional structural stiffness in part by the internal pressure, but also by using an appropriate chosen shell made by helically wrapped paraffin paper or aluminised polyester. Circular cylindrical models with diameters of 250 mm, 500 mm and 1000 mm have been used for the present analysis.

The rectangular model consists of a carbon fiber rod structure, covered with a light-weight paper-type fabric. Due to the sharp-edged geometry, the use of interior helium filled containers, which attain a cylindrical or spherical shape when inflated, were not found to be convenient in this case. A rectangular model with a cross-flow dimension of 200 mm has been used for the present analysis.

The width of the wind tunnel models are 1.70 m or 2.40 m, using close to the full width of 1.75 m and 2.50 m for the Copenhagen and Vermont wind tunnels, respectively. The height of the wind tunnels are 1.50 m and 3.00 m, respectively. Resonance wind velocities will occur at Reynolds numbers of approximately $10^4 - 10^6$. The possibility of covering both the subcritical and supercritical Reynolds number regime gives a good basis for predicting the behaviour in full scale.

3.5 Blockage effects

A flow occurring in the atmospheric boundary layer or open waters can only to some extent be recreated in an experimental testing facility, due to the finite extent of the generated stream. The spatial limitations may produce several boundary effects which could significantly influence the measurements, i.e. they do not resemble measurements performed in flow conditions without boundary effects. In that sense, it is usually important to consider boundary effects in order to provide a more accurate description of the aerodynamic or hydrodynamic behaviour. However, there is no single established method that takes account for a combination of flow contraction and wall constraints, which are often causing the main blockage effects. In this paper, the data is therefore presented without any correction for blockage, since applying a somewhat heuristic blockage correction expression would produce a data set with a significant level of uncertainty. Also, since most measurements are based on similar model scales, the influence from blockage effects are similar between the measurement series, and the results using different model configurations are therefore often directly comparable. For the 1000 mm diameter circular cylindrical model, the blockage is 33%.

Boundaries causing spatial limitations in the cross-flow direction are known to reduce the amplitude of vortex-induced cross-flow vibrations. This effect is not expected to cause a similar reduction for in-line vibrations, since similar spatial limitations do not exist in the main flow direction.

4 Main results

The static load coefficients and dynamic properties of several cylindrical models are documented in the following section. These originate from a number of different wind tunnel flow situations, where the in-line dynamics are evaluated exclusively for low mass ratio models. In this setting, the physical flow characteristics causing the two in-line vortex-induced fluid instability regions have been measured and identified using recordings of spatial and temporal downstream fluid pressure variations. Additionally, the risk of galloping and torsional structural instability have been assessed by evaluating the static load coefficients for different inclination angles.

4.1 Static load coefficients – nominally non-vibrating models

Drag, lift and moment coefficients for a circular cylindrical model without end plates have been measured in a wind

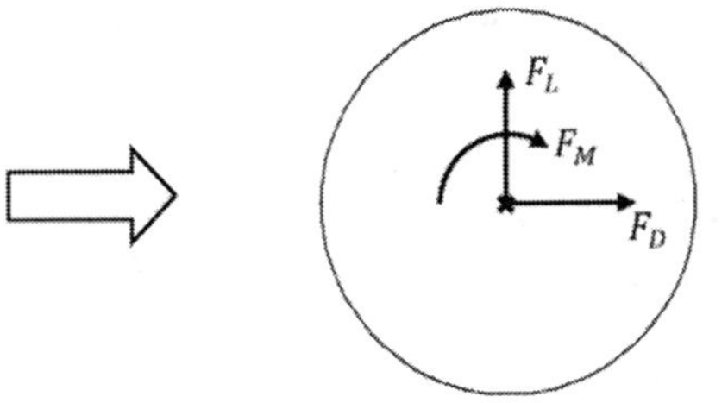

Figure 4: Sign convention of static forces and moments, here shown for the circular cross section. The flow is from left to right.

Table 1: Calculated static load coefficients of the two section models for a zero degree inclination of the approaching flow in the listed Reynolds number regime.

Cross section	C_D	C_L	C_M	St	Re
Circular	0.93	-0.07	-0.01	0.18	4×10^4
Rectangular	1.27	0.01	-0.02	0.05	1×10^5

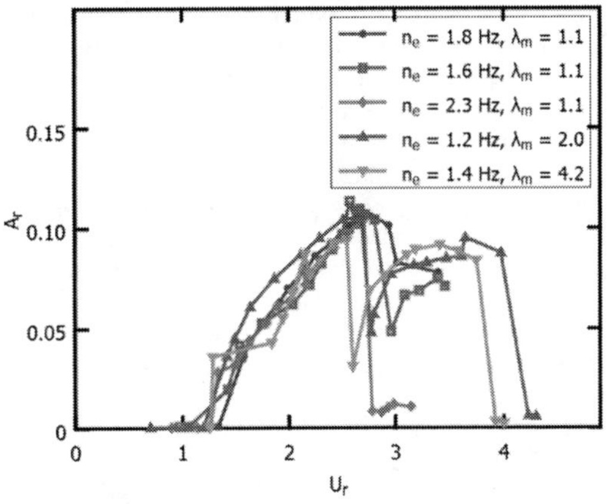

Figure 5: Measured response of five model cylinders configurations in steady air as function of the reduced wind velocity for different model configurations oscillating in a uniform mode shape. The flow conditions correspond to $1 \times 10^4 \leq Re \leq 3 \times 10^4$.

environment corresponding to $Re \approx 4 \times 10^4$ in the Copenhagen wind tunnel. In the wind tunnel in Vermont, similar measurements have been performed on the rectangular 1:3 cross section model in a flow corresponding to $Re \approx 1 \times 10^5$. The results are listed in table 1, where the static load coefficients are presented according to the definitions stated in Sect. 3.2. The diameter of the circular cylindrical model is 250 mm and the cross-flow and in-line dimensions are 500 mm and 1500 mm, respectively, for the rectangular model. The sign convention of static forces and moments is presented in figure 4.

The drag coefficient of a circular cylinder without free-end flow in the relevant Reynolds number regime is specified to be $C_{D,0} = 1.2$ [16]. In the wind tunnel setup, free-end flow exists, but is somewhat limited due to the close proximity of the side walls. The measured drag coefficient is approximately 20% smaller than this value, which is in good agreement with the presence of a limited free-end flow. Note that the drag coefficient of this section model depends on the Reynolds number regime.

The drag coefficient of a sharp-cornered rectangular cross section without free-end flow is stated in the Eurocode 1 specifications [17]. For a ratio of the cross-flow and in-line dimension of 1:3 a drag coefficient of $C_{D,0} = 1.36$ is specified. The measured drag coefficient is approximately 7% smaller than the specified value, which again is in good agreement with the presence of a limited free-end flow. Note that the drag coefficient for a sharp edged section does not depend on the Reynolds number regime.

The lift and moment coefficients are theoretically both zero due to the symmetry of the cross sections considered; thus, the measured responses of negligible magnitudes are entirely due to very small fluctuations in the wind flow, model imperfections and minor measurement uncertainties.

By analysing the spectral power density function, the lift force signals are found to have distinct peaks caused by vortex shedding. For each cross section model, this procedure allows for the estimation of the Strouhal number for the fixed structure, as listed in table 1.

4.2 Vortex-induced in-line vibrations in steady flow – vibrating models

Vortex-induced in-line vibrations in steady flow have been measured on several experiemental configurations, corresponding to differences in model mass ratio, natural frequency, mode shape and Reynolds number regime.

4.2.1 *Circular cylindrical model - Mass ratio and natural frequency*

The dynamic in-line response has been measured on five configurations of a 250 mm diameter cylinder model in flow conditions corresponding to $1 \times 10^4 \leq Re \leq 3 \times 10^4$, see figure 5. For these experiments, the model was set to oscillate in a uniform mode shape and the model was equipped with end plates to limit the disturbance caused by end effects.

A precise estimation of the mass-damping parameter is not possible, since the structural damping is not straightforward to determine. However, for these models the mass-damping parameter scales with the mass ratio. Assuming a structural damping of $\delta_s = 0.03$, the mass-damping parameter is $S c_G \approx 0.12$ for $\lambda_m = 1.1$.

The onset of the first fluid instability region is located at $U_r \approx 1.1$ for all five configurations, while the maximum amplitude, being approximately 11% of the cylinder diameter for the mass ratios $\lambda_m = 1.1$ and $\lambda_m = 2.0$, is only slightly reduced for the larger mass ratio of $\lambda_m = 4.2$. The shift between the first and second fluid instability seems to depend on the mass ratio as well. For larger mass ratios, corresponding to a larger stability parameter, the response drops at slightly smaller reduced wind velocities, which is in accordance with the DNV response model.

A change in the natural frequency of the model does not seem to influence the characteristics of the response in the first fluid instability region, justifying the use of the reduced fluid velocity as a descriptive parameter. For all configurations, the maximum response occurs in the range $2 < U_r < 3$, which fits well with the previously presented in-line response models. Also, by increasing the mass ratio, the maximum response seems to occur at slightly smaller reduced wind velocities.

The second fluid instability responses for $\lambda_m = 2.0$ and $\lambda_m = 4.2$ are similar, being slightly smaller in maximum amplitude for $\lambda_m = 4.2$. For the models with a mass ratio of $\lambda_m = 1.1$, the model begins to buckle and deform at reduced wind velocities above $U_r \approx 3$. When this happens, the tests are stopped because the mode shapes are no longer uniform. As a result, the second fluid instability region is not fully characterised by the models with the lowest mass ratio.

4.2.2 *Circular cylindrical model - Mode shape*

For a cylindrical 250 mm diameter model, the influence of the mode shape on the in-line vibrations has been investigated for two model configurations, see figure 6. In these measurements, the flow conditions correspond to $1 \times 10^4 \leq Re \leq 3 \times 10^4$.

The cylindrical model is oscillating with an amplitude of around 15% of the cylinder diameter for a free-end mode shape, somewhat larger than the amplitude found for models oscillating in a uniform mode shape, see figure 5. The difference in the amplitude between the mode shapes can likely be explained by a different scaling of positive and negative fluid dynamic damping effects along the structure. This phenomenon is actually well-known for cross-flow VIV [18].

Attaching additional mass to the free end of the cylinder increases the mass-damping parameter and decreases the maximum amplitude of the oscillations to around 10% of the cylinder diameter. Also, note that the second fluid instability region ($U_r \approx 3$) apparently does not give a distinct response in this mass-damping range for a free-end mode shape. The fact that the mass is added to the free end of the model implies that the mass-damping is increased relatively more than the mass ratio listed in the legend on figure 6. Assuming a structural damping of $\delta_s = 0.03$, the mass-damping parameters are $Sc_G \approx 0.17$ and $Sc_G \approx 0.30$ for the two presented model configurations.

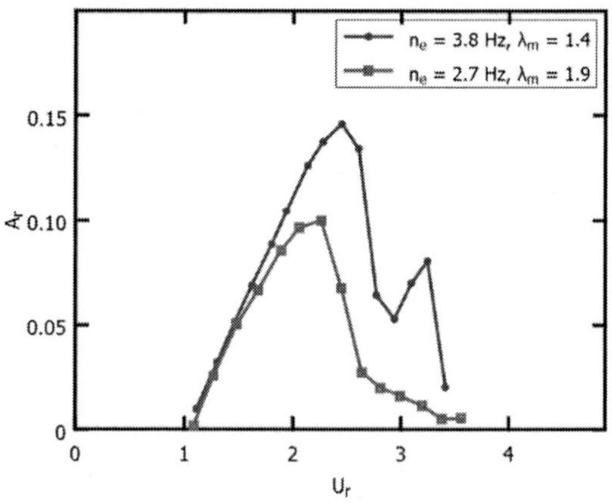

Figure 6: Measured in-line response of a cylindrical model as function of the reduced wind velocity for two different configurations oscillating in a free-end mode shape. The maximum response depends on the mode shape (compare to figure 5) and is reduced when a large mass is added. The flow conditions correspond to $2 \times 10^4 \leq Re \leq 4 \times 10^4$.

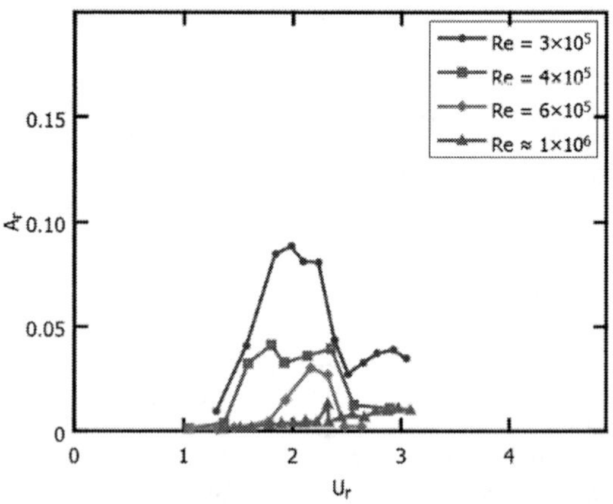

Figure 7: Measured in-line response of a 1000 mm diameter model cylinder in steady air as a function of the reduced wind velocity for different Reynold number flow regimes.

4.2.3 *Circular cylindrical model - Reynolds number*

Dynamic response characteristics in flow conditions corresponding to higher Reynolds numbers, can be obtained in a wind tunnel by increasing the model dimensions or by increasing the wind velocity. Both approaches imply considerable challenges for the model structure. A larger model dimension naturally implies increased sag and bending moments, and the present low mass ratio requirements would often result in a model which is less likely to withstand higher velocities without deformations or similar undesirable effects. Nevertheless, response measurements in even higher Reynolds number regimes have been obtained using a large inflatable scale model, having a diameter of 1000 mm. The responses of four such configurations in high Reynolds number regimes are presented in figure 7. For these four configurations, the mass ratio is held approximately constant and the model was equipped with end plates to limit the disturbance caused by end effects.

For $Re = 3 \times 10^5$, the model is seen to have an in-line response amplitude of up to 9% of the cylinder diameter and two visible fluid instability regions. A single fluid instability region with a much reduced response is observed for $Re = 4 \times 10^5$ and $Re = 6 \times 10^5$, and for the latter flow regime, non-uniform oscillations were observed for reduced wind velocities larger than $U_r \approx 2.2$. For the flow regime corresponding to $Re = 1 \times 10^6$, no clear indication of in-line fluid instabilities is observed and the response is just slowly increasing with wind velocity. The only exception is a single peak at $U_r \approx 2.3$ corresponding to an amplitude of around 1% of the cylinder diameter.

The maximum response, as well as the velocity interval in which oscillations occur, are seen to be significantly reduced for increasing Reynolds numbers. This general trend is most likely not entirely caused by the change in the flow regime characteristics with the increasing Reynolds number, but is also due to related side effects of the increased wind velocity, such as an increased deformation force of the wind tunnel model, and an associated increased energy dissipation in the complete oscillating system. However, it is reasonable to assume that no significant amplification of the overall response would happen as a result of the change in the flow regime for the Reynolds numbers considered. Note also that, as stated previously, wind tunnel blockage is not accounted for in the presented data; thus the reduced wind velocities, and thereby also the Reynold numbers, are slightly underestimated.

The physical mechanisms responsible for in-line vibrations is a certain combination of forcing due to vortex-shedding and a related change in the aerodynamic damping in the direction of the flow. Exactly the same physical mechanisms working in the cross-flow direction are responsible for cross-flow VIV. In the latter case, a change in aerodynamic damping is responsible for the transition range at which the smaller deflections governed by lift forces changes to larger deflections governed by motion-induced forces. The location of the transition is described by the aerodynamic damping factor for cross-flow amplitudes [11]. The presence of large cross-flow deflections governed by motion-induced forces therefore depends on the aerodynamic damping factor which in turn depends on Reynolds number. Since the physical mechanisms responsible for vibrations are similar in the in-line and cross-flow directions, it is expected that the maximum amplitudes of in-line oscillations show a similar dependence on the aerodynamic damping through the Reynolds number. For cross-flow oscillations, the aerodynamic damping factor attains its minimum at $Re \approx 6 \times 10^5$ according to Eurocode 1 specifications [17], which implies that in this range only small deflections governed by lift forces are to be expected. Apparently, a similar tendency is true for in-line oscillations.

4.2.4 *Rectangular model*

In contrast to the previously tested cylindrical models, the geometry of the rectangular cross section requires that the model is constructed using a lightweight skeleton structure, since the sharp-edged geometry cannot be obtained by pressurised balloons alone. This implies that a mass ratio very close to one is very difficult, if not impossible, to obtain with the model dimension of D = 200 mm related to the present wind tunnel testing. Therefore, the smallest mass ratios considered for the rectangular section model are larger than those related to the cylindrical models. A plot of the measured normalised response amplitudes as a function of the reduced wind velocity is presented in figure 8. The flow conditions correspond to $Re \approx 2 \times 10^4$. Again, for these models the mass-damping parameter scales with the mass ratio. Assuming a structural damping of $\delta_s = 0.03$, the mass-damping parameter is $S_{CG} \approx 0.20$ for $\lambda_m = 2.2$.

The presented results show that the rectangular cross section apparently is much less prone to in-line vortex-induced oscillations than the cylindrical cross section. The maximum amplitude is below 2% of the cross-flow dimension, and there is actually a slight increase in amplitude with increasing mass ratio for the four presented model configurations. Note that only a single fluid instability region is visible, located at approximately $3 \leq U_r \leq 4$.

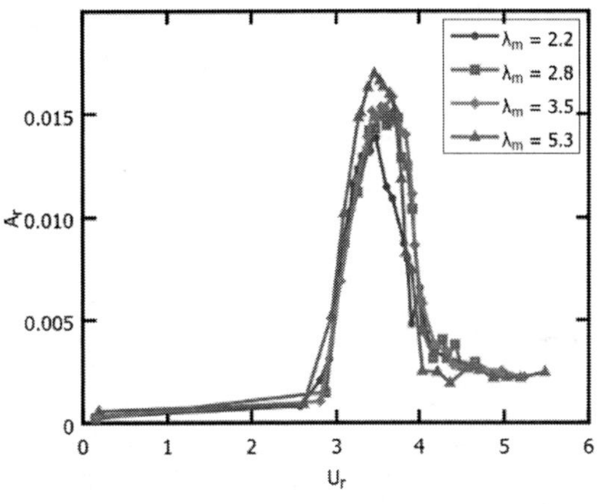

Figure 8: Reduced amplitudes of the rectangular section model for different mass ratios. The largest amplitude is below 2% of the cross-flow dimension. For all model configurations, a clear in-line model response is only seen for wind velocities corresponding to approximately $3 \leq U_r \leq 4$.

The aerodynamic damping for the rectangular cross section is larger than for circular cross sections, simply due to the larger drag coefficient and the fact that a higher reduced wind velocity is necessary to initialise the oscillations. This effect is also relatively more pronounced at small mass ratios at similar reduced wind velocities. It is deemed that this effect is responsible for the limited in-line response. The decrease in aerodynamic damping with increasing model mass could also account for the slight increase in maximum response for larger mass ratios.

The initialisation of the in-line vortex-induced forcing is related to the Strouhal number for the fixed structure. For the rectangular model and the cylindrical model, the ratio of the measured Strouhal numbers for the fixed structures are approximately 3.6, see table 1, implying an onset of the in-line forcing frequency relative to the structural eigenfrequency at three to four times higher wind velocities for the rectangular model. This is in agreement with the onset of the in-line vortex-induced forcing close to the structural eigenfrequency at $U_r \approx 3$, which is three times larger than the onset found for the cylindrical models, see Sect. 4.2.1.

As seen on figure 8, the in-line vibrational behaviour was investigated for reduced wind velocities up to approximately $U_r = 5$. For a similar 1:3 section model, torsional vortex-induced oscillations in air have been reported to have an onset near $U_r = 6.5$ [19]. This is approximately twice the reduced velocity at which the in-line oscillations were seen to attain the largest amplitudes, which suggests that the in-line oscillations measured on the rectangular section model are caused by a normal asymmetric vortex shedding pattern.

4.3 Additional investigations

Several experiments have been performed in addition to the investigations of model-specific static and dynamic load

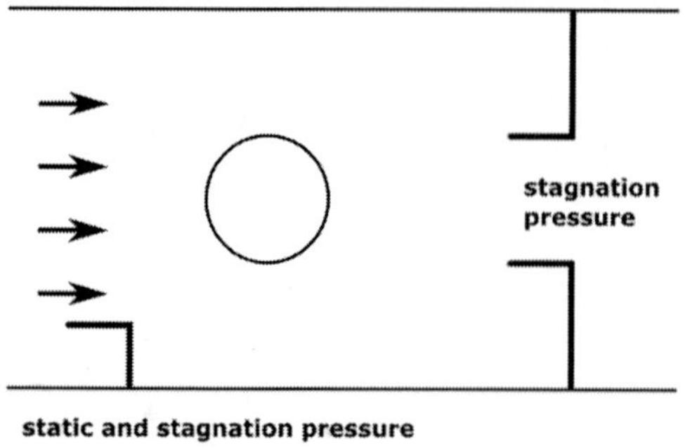

Figure 9: Sketch of the model setup capturing the downstream pressure variations due to vortex shedding. Two downstream pitot tubes are used to capture the spatial information needed to distinguish between asymmetric and symmetric vortex shedding.

characteristics. In this section some supplementary findings are listed, which confirms or enlightens the previously documented results linking in-line vortex-induced vibrations to basic flow characteristics, and the fluid-dynamical and structural parameters.

4.3.1 *Identification of fluid instability regions*

As mentioned in Sect. 2, the mechanism responsible for the first fluid instability causing in-line oscillations is regular, alternating vortex shedding combined with a secondary, symmetric vortex shedding pattern which occurs as a result of in-line structural motion relative to the fluid. In the second fluid instability region, the forcing is entirely due to regular vortex shedding from alternating sides of the cylinder. Therefore, the two types of in-line responses are caused by downstream pressure variations oscillating with three and two times the Strouhal frequency, respectively. Measurements of the spatial and temporal downstream pressure variations could therefore be used to identify the forcing mechanism; thus, identifying the corresponding fluid instability regions.

In the US wind tunnel, the downstream pressure variations have been captured using two pitot tubes located approximately 750 mm behind a 500 mm diameter cylindrical model oscillating in the in-line direction, having an eigenfrequency of $n_s \approx 2.7$ Hz. The pitot tubes are vertically aligned with the model top and bottom, see figure 9. The probes are 8 mm diameter NPL type pitot-static tubes with ellipsoidal head complying with ISO 3966. The dynamic pressure sensors have a response time of less than 10 ms, and the setup can easily capture the generated pressure fluctuations. Hot-wire anemometers which are often used to capture high frequency flow fluctuations are therefore not required in the present case.

When model oscillations couple with vortex shedding, the eigenfrequency of the model becomes visible in the power spectra of the measured time series of pressure variations at the two downstream pitot tubes. Figures 10 and 11 illustrate such spectral power densities at $U_r = 2.16$ and $U_r = 3.26$, respectively.

In the first fluid instability region, the vortex shedding pattern is complex as it is a mix of both alternating and

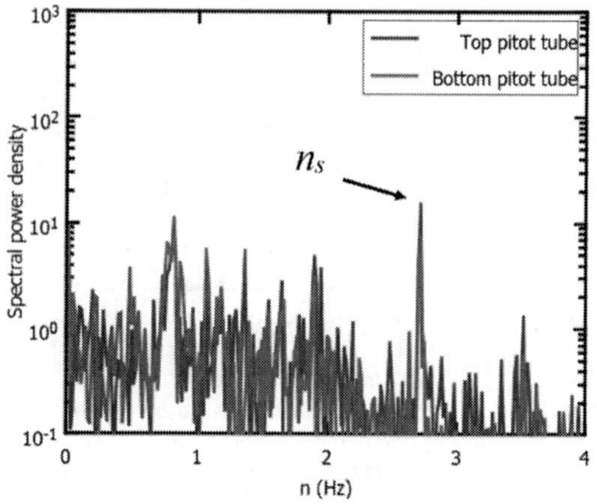

Figure 10: Spectral power density of the top and bottom pitot tube pressures at $U_r = 2.16$, in the first fluid instability region, indicating a clear peak at the models natural frequency n_s.

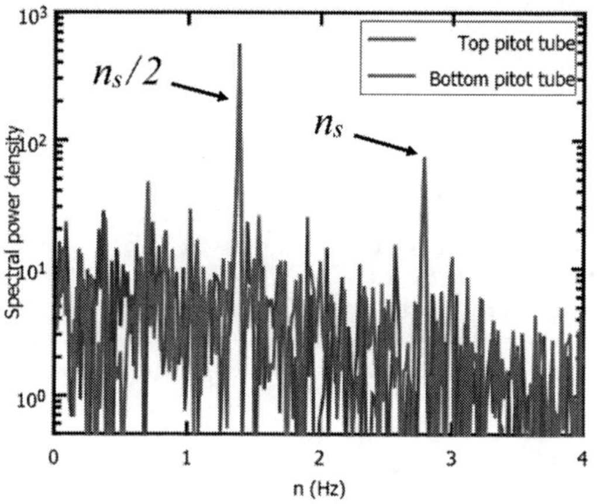

Figure 11: Spectral power density of the top and bottom pitot tube pressures at $U_r = 3.26$, in the second fluid instability region, indicating clear peaks at both the models natural frequency n_s and at $n_s/2$.

symmetric vortex shedding. In general terms, it can be divided in two sub-regions:

- For relatively low reduced wind velocities ($U_r \approx 1$), the cross-correlation of the two signals are dominated by correlation peaks of a period T, here defined as the inverse of the model eigenfrequency of approx. 2.7 Hz. This indicates that symmetric vortices, which are highly correlated in the wake, are shed with the motion of the model. Concurrently, correlation at zero time lag exists, but maximum correlation is not achieved here due to the occasional alternating vortex pair. By lowering the wind velocity, the vortex shedding becomes increasingly dominated by symmetric pairs, and the model eigenfrequency becomes visible while half the model eigenfrequency disappears from the spectra.
Depending of the model natural frequency, the regular asymmetric vortex shedding on a fixed cylinder corresponding to a Strouhal number of $St = 0.18$ may be-

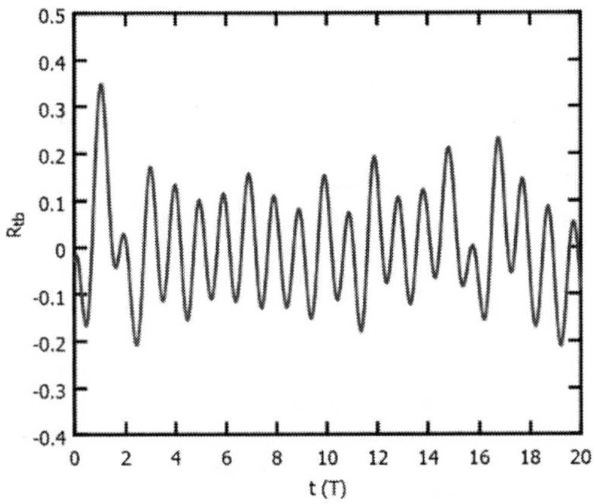

Figure 12: The cross-correlation between top and bottom pitot pressures at $U_r = 2.16$, indicating a mix of symmetric and regular, asymmetric vortex shedding.

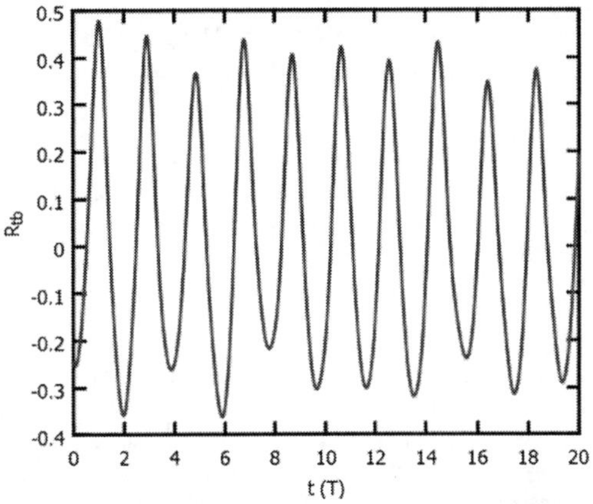

Figure 13: The cross-correlation between top and bottom pitot pressures at $U_r = 3.26$, indicating regular, asymmetric vortex shedding.

come apparent at reduced wind velocities below, and in, the first part of the first fluid instability region where very little or no response of the model is observed.

- At higher reduced wind velocities ($U_r \approx 2$), the cross-correlation is approximately sinusoidal, with some irregularities and a slight decay of the peak correlations for increased time lag, see figure 12. At $U_r = 2.16$, the first peak of correlation occurs at $t = T$ and reoccurs with period $2T$. This indicates that the vortex shedding is predominantly alternating. The correlation also show a smaller sinusoidal component, positive at $t = 0$ s with period T, indicating symmetric vortex shedding. Half the model eigenfrequency is not visible in the corresponding spectra, distinguishing this region from the second fluid instability region, see figures 10 and 11.

The second fluid instability region is due to vortex shedding from alternating sides as is familiar on fixed cylinders. This region is characterised by the appearance of peaks in the spectrum at the model eigenfrequency and at half the model eigenfrequency, see figure 11. This suggests that vortices are shed once per model oscillation, as the spectral footprint consists of the frequency by which each single vortex is shed, and also the frequency corresponding to the shedding of a full period of a pair of vortices. If the pitot pressures are subtracted from each other the resulting spectrum reveals no significant contribution at the frequency corresponding to the model eigenfrequency. If the signals are added a clear peak at the model eigenfrequency occurs in the corresponding spectrum. The cross-correlation on figure 13 of this shedding type is approximately sinusoidal with periodic peaks of constant value indicating a long time correlation. The period is $2T$, twice the period of the model oscillations, with the first peak correlation at T consolidating the alternating nature of the shedding.

Between the two fluid instability regions, the cross-correlation of the two pitot pressures remains sinusoidal with a period of $2T$, but the level of correlation decays quickly. The model eigenfrequency leaves very little footprint in the frequency spectra while half the eigenfrequency remains visible.

Measurements of the spatial and temporal downstream pressure variations on an in-line oscillating cylindrical model therefore allows for the identification of the first and second fluid instability region; thus, identifying the corresponding forcing mechanism causing in-line structural oscillations in stationary fluid flow. The investigation verifies that the physical phenomena responsible for vortex-induced vibrations in air are similar as those reported in water [10].

4.3.2 *Structural instability phenomena*

The angular dependence of the static load coefficients may be used not only to predict fluid-induced forces, but also to investigate if the cross section is susceptible to aeroelastic or hydroelastic structural instability phenomena such as galloping and flutter. For this, the Scruton number, of the more general non-dimensional mass-damping parameter, may be used in predictions of cross-flow galloping, assuming a driving force proportional to the in-line dimension. The rotational structural response can be modelled by a one degree of freedom system with a torsional forcing determined by the fluid-induced moment load. It is a standard approach to consider a linearisation argument to indicate that two types of structural instabilities can occur; one being torsional galloping and one being static divergence [20].

Due to the geometrical symmetry, galloping and torsional structural stability are not relevant for the circular cross section model. For the rectangular cross section, a negative slope of the aerodynamic lift was measured, implying that cross-flow galloping vibrations might exist for small inclinations of the flow due to unstable aerodynamic loading conditions. The aerodynamic moment has a negative slope for small inclinations. This implies that the structure is stable towards flow-induced excessive twist. However, a negative slope is known to increase the risk of torsional galloping, due to negative aerodynamic damping effects.

The slope of the moment coefficients for rectangular sections rotating about the geometric center have previously been reported for 1:2 and 1:4 cross sections in the literature, being -0.64 and -18, respectively, for α measured in radians, but using the cross-flow dimension as the characteristic dimension in the definition of the moment load per unit length [21]. For the analysed 1:3 cross section, the measurements suggest $\frac{dC_M}{d\alpha}|_{\alpha=0} \approx -1.1$, corresponding to a value of -9.9 using the cross-flow dimension as the characteristic dimension.

5 Discussion

For free-span circular cross-section pipelines submerged in water, the DNV in-line response model gives an empirical description in current dominated full-scale conditions. The presented results indicate that the maximum in-line response of a close to neutrally buoyant circular cylindrical model oscillating in a uniform mode shape in air is well-described by the DNV model, both in terms of maximum response and the reduced velocities at which in-line VIV is relevant. The dependence on the stability parameter expressed in the DNV model cannot be confirmed based on the conducted measurements, since a precise estimation of the total damping, or in particular the aerodynamic damping, is not straightforward. However, the overall tendency of a reduction in the maximum response occurring at a lower reduced flow velocity when increasing the stability parameter, or in the present case increasing the mass ratio, is indeed observed in the measured responses.

In the performed dynamic testing, models with a mass ratio in the range of, say, 1.0 – 2.0 of the surrounding fluid, apparently produce a comparable maximum response for a uniform mode shape. In other words, the related Scruton curve has a small slope near zero. This observation can be used as a guidance for future testing, since it implies that maximum responses might be correctly estimated using models which are only close to being neutrally buoyant.

Based on additional measurements in flow conditions corresponding to $10^5 \leq Re \leq 10^6$, it is reasonable to assume that no significant amplification of the overall response would happen as a result of the change in the flow regime for the Reynolds numbers considered. A compensation for the slightly lower Reynolds number in the wind tunnel environment compared to full scale conditions, is therefore likely not necessary to be included in the safety requirements. Nevertheless, subsequent wind tunnel measurements at even higher Reynolds numbers could potentially provide additional insight into the importance of this topic.

Measuring the spatial and temporal downstream pressure variations on an in-line oscillating cylindrical model allowed for the identification of the first- and second fluid instability region, thus identifying the forcing mechanism causing in-line structural oscillations in steady fluid flow. The investigation verified that the physical phenomena responsible for vortex-induced vibrations in air are similar as those reported in water [10].

The rectangular cross section is apparently much less sensitive to in-line VIV than the cylindrical cross sections, likely caused by a larger fluid-dynamic damping. The onset of in-line VIV happens at larger reduced velocities, probably due to the smaller Strouhal number for the structure.

6 Conclusion

The presented results provide a preliminary insight into a fundamental methodology for the transfer of certain aerodynamic actions to equivalent hydrodynamic actions, related to both static response characteristics and in-line vibrations caused by vortex shedding. A consistency between the results obtained in the wind testing facilities at Svend Ole Hansen ApS and SOH Wind Engineering LLC, and results published by others were observed; thus, verifying the understanding of the general dynamic properties of low mass ratio structures and the type of physical phenomena which are of relevance for the in-line structural oscillations in steady fluid flow. This consistency was found on both the qualitative and quantitative levels, giving a response of similar magnitude for experimental configurations corresponding to comparable basic flow characteristics and model parameters. In this sense, the wind tunnel testing has turned out to be a very economical and powerful tool for determining and verifying hydrodynamic data.

It is of fundamental interest that wind tunnel experiments have shown that in-line VIV caused by steady winds on model structures with a density comparable to air, and the response prescribed on neutrally buoyant structures in water are of very similar nature. As this is a very wide-ranging field, the presented work cover basic principles and demonstrates prevailing tendencies. The conversion of model-scale results to full-scale predictions should therefore be assessed accordingly and the interpretation of the presented model-scale data should focus on the fact that the basic physical mechanism generating the in-line vibrations in question are equivalent in model scale and full scale, and in air and water. Nonetheless, the presented model tests have provided clear indications of the static and dynamic properties of specific neutrally buoyant structures and should form a basis for future systematic investigations providing in-depth descriptions on the flow-induced structural behavior.

The authors would like to thank Ph.D. candidate Andreas Saur Brandtsegg from Dr.techn.Olav Olsen AS and the Norwegian Institute of Technology (NTNU), and Professor Odd Faltinsen from NTNU for valuable advice and expertise. A special thanks also goes to the employees at Svend Ole Hansen ApS, SOH Wind Engineering LLC, Reinertsen AS, Dr.techn.Olav Olsen AS and the Norwegian Public Roads Administration.

References

1. R. King, Ocean Engineering **4**, 141-171 (1977)
2. T. Sarpkaya, Journal of Applied Mechanics **46**, 241-258 (1979)
3. P. W. Bearman, Annual Review of Fluid Mechanics **16**, 195-222 (1984)
4. M. M. Zdravkovich, Journal of Fluids Engineering **99**, 618-633 (1977)
5. C. Mannini, A.M. Marra, G. Bartoli, Journal of Wind Engineering and Industrial Aerodynamics **132**, 109-124 (2014)
6. M. Belloni, S.Giappino, S. Morganti, S. Muggiasca, A. Zasso, Ocean Engineering **94**, 140-154 (2014)

7. C. C. Feng, *The measurement of vortex-induced effects in flow past stationary and oscillating circular and D-section cylinders* (Master's Thesis, University of British Columbia, Vancouver, B. C., Canada, 1968)
8. M.H. Bahmani, M.H. Akbari, Ocean Engineering **37**, 511-519 (2010)
9. S. E. Ramberg, O. M. Griffin, Proceedings of Hydrodynamics in Ocean Engineering, 1223-1245 (1981)
10. B. M. Sumer, J. Fredsøe, *Hydrodynamics Around Cylindrical Structures*, (World Scientific, 2006)
11. S. O. Hansen, Proceedings of the Institution of Civil Engineers - Structures and Buildings **166**, 560-571 (2013)
12. A. Khalak, C. H. K. Williamson, Journal of Fluids and Structures **13**, 813- 851 (1999)
13. R. King, *Vortex excited structural oscillations of a circular cylinder in flowing water* (PhD Thesis, Loughborough University, 1974)
14. Det Norske Veritas, *Recommended Practice - Free Spanning Pipelines DNV-RP-F105* (Det Norske Veritas, 2006)
15. H. Braaten, H. Lie, M. Søreide, S. Svardal, Proceedings of the 26th International Conference on Offshore Mechanics and Arctic Engineering, (2007)
16. C. Dyrbye, S. O. Hansen, *SBI-Anvisning 158 - Vindlast på bærende konstruktioner*, (Statens Byggeforskningsinstitut, 1989)
17. DS/EN 1991-1-4:2007: Eurocode 1, *Actions on structures - Part 1-4: General actions - Wind actions*, (Dansk Standard, 2007)
18. S. O. Hansen, Proceedings of the third Structural Engineers World Congress, (2007)
19. K. Washizu, A. Ohya, Y. Otsuki, K. Fujii, Journal of Sound and Vibration **72**, no. 4, 507-521, (1980).
20. C. Dyrbye, S. O. Hansen, *Wind loads on structures* (John Wiley & Sons Ltd., 1997)
21. M. P. Païdoussis, S. J. Price, E. de Langre, *Fluid-Structure Interactions: Cross-Flow-Induced Instabilities* (Cambridge University Press, 2010)

EPJ Web of Conferences **114**, 02041 (2016)
DOI: 10.1051/epjconf/201611402041

A combined experimental-numerical approach for two-phase flow boiling in a minichannel

Sylwia Hożejowska [1,a], Mirosław Grabowski [2]

[1]*Kielce University of Technology, Department of Informatics and Applied Mathematics, Al. 1000-lecia PP 7, 25-314 Kielce Poland*
[2]*Warsaw University of Technology, Institute of Mechanical Engineering, Plock Campus, ul. Łukasiewicza 17, 09-400 Plock, Poland*

Abstract. The paper addresses experimental and numerical modeling of the two-phase flows in an asymmetrically heated horizontal minichannel. Experimental measurements concerned flows of evaporating ethanol in a minichannel with rectangular cross section 1.8mm x 2 mm. In order to observe the flows, measuring system was designed and built. The system measured and recorded basic heat and flow parameters of flowing fluid, and the temperature of external surface of the heater by using infrared camera and recorded images of flow with high-speed camera. The second aim of the paper was to formulate appropriate flow boiling heat transfer model, which would minimises the use of experimentally determined constants. The procedure of calculating the temperature of the ethanol is coupled with concurrent process of determining the temperature distributions in the isolating foil and the heating surface. The two-dimensional temperature distributions in three subsequent domains were calculated with Trefftz method. Due to the Robin condition, heat transfer coefficient at the heating surface-ethanol interface was calculated based on the known temperature distributions of the foil and liquid. Additionally, the paper describes the relation between two sets of functions used in the calculation. Numerical calculations made by Trefftz method were performed with using experimental data.

1 Introduction

Progress in miniaturization of modern components and systems, driven by technological advances, has resulted in the development of methods that provide their functionality and proper performance. Reduced energy and material consumption are required for contemporary small-scaled heat exchangers, while maintaining or possibly improving their thermodynamic performance by transferring increased heat fluxes. One way to intensify heat transfer and meet the criteria above is connected with the phase transition of refrigerants, in particular in flow boiling in minichannels. This results in idea of the so-called compact heat exchangers (evaporators and condensers) used within cooling systems. Additionally, due to small internal volumes of the minichannels, their environmental impact shall be limited, because in the case of a leakage, the amount of the refrigerant released to the surroundings will be negligible. The main condition for proper operation of high-efficiency miniature heat exchangers is steady and stable flow. One of the key problems in design process is the selection of proper calculation procedures that will reflect the processes governing the heat transfer in the small-sized channels. The objective of this study is to build appropriate mathematical models describing heat transfer processes and to provide effective solutions for related problems.

Mathematical model building should be based on the experiment that both provides input data and verifies the results. This paper discusses the results of experimental study of flow boiling of ethanol in a horizontal minichannel with 1.8 mm × 2 mm cross-section. The design of the measurement system was based on the experience gained from earlier works reported by the authors and other researchers [1, 2]. The paper proposes a mathematical model describing heat transfer in the measurement module for single and two-phase flows of ethanol in the minichannel. To determine approximate temperature distributions for the insulating foil, the heater and the ethanol (two-phase flow), the Trefftz method was used. Known temperature distribution on the heating surface and the ethanol allowed determining the heat transfer coefficient in the heater-ethanol contact region. Measurement data from the experiment were used in the calculations.

2 Experiment

The investigations were conducted on the experimental setup, figure 1, which allowed:
1. pumping the working liquid through the minichannel at the set volumetric flow rate, inlet temperature and pressure.
2. heating the liquid in the minichannel to the boiling temperature.

[a] Corresponding author: ztpsf@tu.kielce.pl

3. measuring the basic thermodynamic and flow parameters: volumetric flow rate of the liquid flowing through the minichannel, temperature at the inlet and outlet, pressure at the inlet, pressure drop along the minichannel length, the power of the electrical current flowing through the heater; recording the temperature distribution on the surface of the insulating foil separating the heater from the ambient, 1.3 in figure 2; recording resulting two-phase flow structures with a video camera.

The main part of the setup, figure 1, comprises the measurement module with the minichannel 1. The module is fitted with pressure sensors (*p*) and temperature sensors (*T*) with index 1 for the inlet and 2 for the outlet.

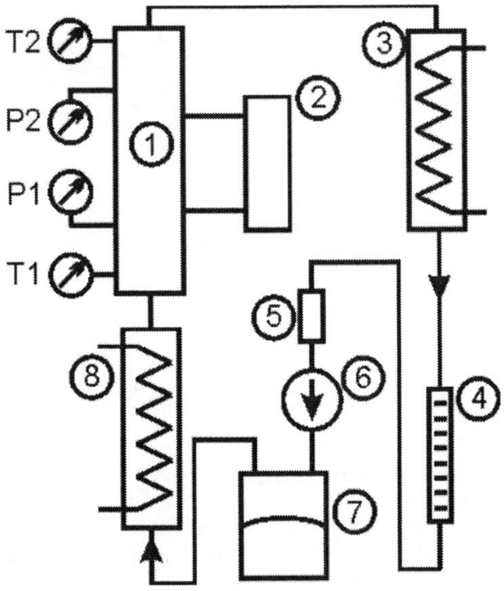

Figure 1. Flow loop: 1 – measurement module with the minichannel, 2 – DC power supply for the heater, 3 – condenser, 4 – rotameter, 5 – filter, 6 – pump, 7 – pressure control, 8 – preheater.

On one of the minichannel walls, a heater is mounted 1.2 figure 2, which is supplied from the DC power supply 2 figure 1. The boiling fluid leaving the minichannel is directed to the condenser 3. From the condenser, the fluid flows to the rotameter 4 and further to the filter 5 before it reaches the precision micropump 6. The micropump pumps the fluid to the pressure control, which is an expansion vessel. The fluid is on one side of it and on the other side, the supplied compressed air is used to force appropriate pressure value in the system. From the pressure control the fluid flows to the preheater, where it is preheated to the temperature close to the saturation temperature. The difference between the saturation temperature and that at the minichannel inlet is known as supercooling. Leaving the preheater, the fluid flows into the minichannel, figure 2. The measurement module is composed of the hub 1.1, in which the minichannel is milled 1.4. A lid with a transparent viewing window closes the minichannel on one side 1.5, and on the other side there is a heater 1.2 in the form of a flat strip made of Fecral 135 electrical resistance alloy with dimensions 2 mm× 0.1 mm, sealed with the insulating foil 1.3. Type K thermocouples with the cross-section of 0.5 mm, and pressure sensors 0 - 2.5 bar are mounted in the hub.

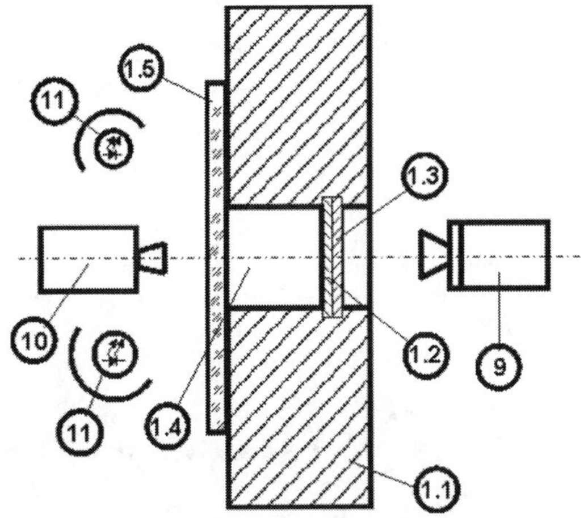

Figure 2. Cross-section of the measurement module with the minichannel and the location of the IR and video cameras and lighting (not to scale): 1.1 – hub, 1.2 – heater, 1.3– insulating foil, 1.4 – minichannel with boiling fluid, 1.5 – glass lid closing the channel, 9 – IR camera, 10 – video camera, 11 – LED lights.

The heater, 1.2 figure 2, is powered from the programmable high current DC power supply 2 figure 1, manufactured by TDK Lambda. During heating and phase change, the two-phase structures can be observed and recorded using the fast video camera, 10 figure 2, (Phantom 711 by Vision Research), with a resolution of 1280 × 800 pixels at 7530 fps. This camera to be efficient requires very intense lighting. Powerful incandescent lamps would heat the channel thus introducing an additional measurement error. To avoid this, LED light 64 W constructed by the authors were used, 11 figure 2. The lighting consists of two diode sections LED CLL 010 - 03 05 A1 - 27 3 M1 A2 with 7 diodes in each section arranged along the minichannel length and fitted with additional lenses to disperse light. A section of extra 9 LEDs can be placed on a flexible arm to precisely direct the luminous flux to the selected spot.

The minichannel can be divided into three regions associated with the ways of heat supply. The heat generated by the heater is supplied only to region I, figure 3.

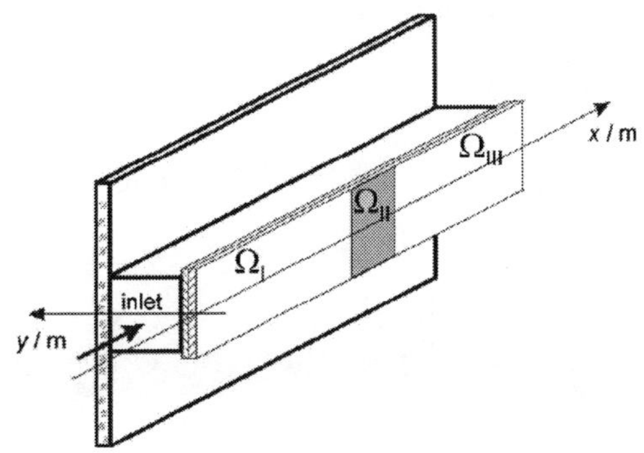

Figure 3. Minichannel division into the heated and the adiabatic parts.

For reasons related to the design, temperature field on the insulating foil cannot be recorded in region II. In region III, as in the case of region II, the heat is not supplied and all heat and flow parameters are recorded (as in region I). The design of the experimental stand is described more detailed in [3].

3 Experimental results

The experiments allowed recording the flow parameters, the heat flux supplied to the heater, the images of the two-phase structures occurring during boiling in the minichannel, figure 4, and the thermograms with temperature field on the insulating foil.

Figure 4. Two-phase flow structures in the flows of the following parameters: 1 – volumetric flow rate 1.41E-7 $m^3\,s^{-1}$, heat flux 143.3 kWm^{-2}, distance from the minichannel inlet 20 mm, 2 – volumetric flow rate 1.41E-7 $m^3\,s^{-1}$, heat flux 120.3 kW m^{-2}, distance from the minichannel inlet 150 mm, 3 –volumetric flow rate 5.88E-8 $m^3\,s^{-1}$, heat flux 120.3 kW m^{-2}, distance from the minichannel inlet 40 mm, 4 – volumetric flow rate 2.1E-8 $m^3\,s^{-1}$, heat flux 143.3 kW m^{-2}, distance from the minichannel inlet 40 mm.

The data above were used to prepare temperature profiles for each measurement instance. Figure 5 shows an example of temperature distribution along the channel length for the flow with the volumetric flow rate of 8.3E-8 $m^3\,s^{-1}$ and the heat flux of 153.3 kW m^{-2}.

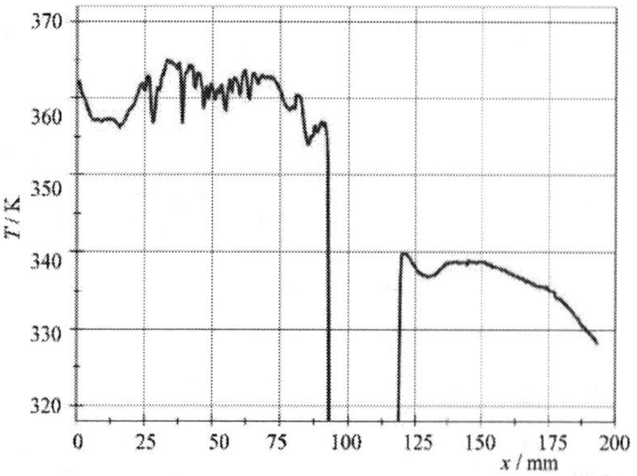

Figure 5. Temperature distribution recorded by the IR camera on the insulating foil for the flow with volumetric flow rate of 8.3E-8 $m^3\,s^{-1}$, and the heat flux of 153.3 kW m^{-2}.

The analysis of the photographs obtained using the fast video camera allowed determining the void fraction in the minichannel φ, defined as a ratio of the surface area of vapor bubbles to the surface area of the channel (projection area on the photo plane). The way to estimate void fraction was presented in [4]. An example of the void fraction along the channel length is shown in figure 6.

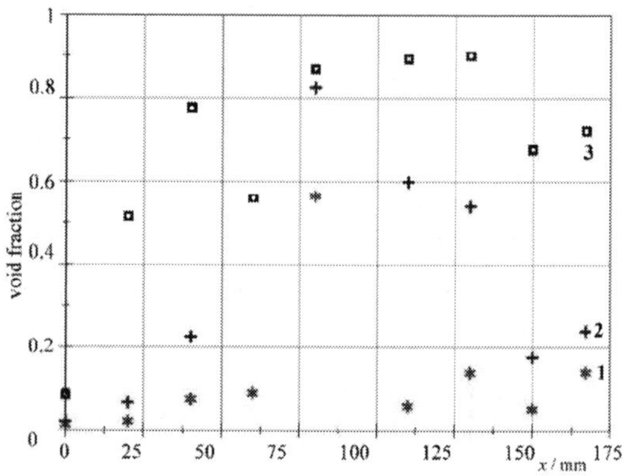

Figure 6. Void fractions relative to the distance from the minichannel inlet for selected thermal and flow parameters: 1 – volumetric flow rate 8.3E-8 $m^3\,s^{-1}$, heat flux 100.4 kW m^{-2}, 2 – volumetric flow rate 5.1 E-8 $m^3\,s^{-1}$, heat flux 99.2 kW m^{-2}, 3 –volumetric flow rate 2.1E-8 $m^3\,s^{-1}$, heat flux 102.6 kW m^{-2}.

The rapid void fraction changes seen in the diagram above are associated with the oscillations that occur at higher heat fluxes. The oscillations were characterized by cyclic violent evaporation (high φ) followed by an inflow of the fluid (low φ). Figure 7 illustrates the influence of the volumetric flow rate and the heat flux on the void fraction.

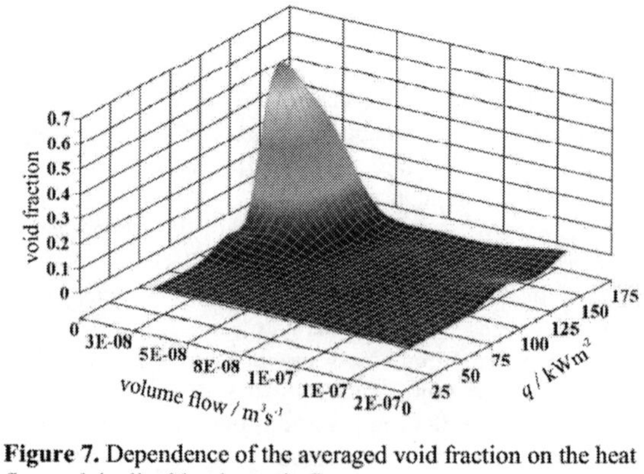

Figure 7. Dependence of the averaged void fraction on the heat flux and the liquid volumetric flow rate.

The measurement data were read and recorded by a control system based on specialized measurement modules manufactured by National Instruments. The basic part of the system is the main module NI cDAQ-9172, in which submodules are mounted for measuring process parameters and control parameters for the pump, light and cameras. To avoid large errors, high accuracy sensors are used in the measurement systems. Some of these sensors, for example, Czaki thermocouples cooperating with module NI 9211 were calibrated, which allowed obtaining the limiting temperature measurement error of up 0.55 K. The TDK Lambda adapter internal measurement system was used to measure the amperage and voltage of the current supplied to the heater. The limiting reading error in this was 0.1% measured value. To estimate the resultant measurement error for the heater output and the volumetric flow rate of the working liquid, the maximum uncertainty method was used. Table 1 shows the values of absolute errors. Error in measuring the void fraction is more difficult to estimate. The measurements are based on a computer graphic analysis of photographs obtained with a high-speed video camera. Graphical methods do not provide high accuracy, especially for cases with significant dispersion of vapour bubbles and their irregular pattern. These reasons result in difficulties to identify the interfacial boundary. The error of measurement series can be defined by the standard deviation. The median of standard deviation for measurements for the horizontal arrangement of the channel is 36.4% and 24.9% for the vertical.

Table 1. Absolute errors of the measured and calculated quantities.

Parameter	Error
Pressure (*p*)	0.01325 bar
Temperature (*T*)	0.55 K
Pressure drop (*Δp*)	1.725 Pa
Voltage (*U*)	0.006 V
Amperage (*I*)	0.006 A
Volumetric flow rate ($\dot{V}$)	1.04 E-9 $m^3 s^{-1}$
Heating power (*N*)	0.072 W

4 Mathematical model

Due to the measurement module design, figure 3, this mathematical model concerns heat transfer in only two regions: region I (Ω_I), where in the section with a length of H_I = 93 mm there is the heater heating the liquid flowing into the minichannel and region III (Ω_{III}) with adiabatic flow; this region concerns the 77 mm long end section of the measurement module.

The steady state in the test module was assumed for the mathematical model. The temperature changes in the insulating foil, heater and liquid along the minichannel width were assumed to be minor. Two dimensions were considered in this approach: dimension *x* parallel to the flow direction and perpendicular to it dimension *y* for the thickness of the insulating foil (δ_f), the heater (δ_h) and the depth of the minichannel (δ_m). Further considerations refer only to the central cross section of the measurement module. The parameters referring to the foil, heater and liquid were annotated by subscripts: *f* for the insulating foil, *h* for the heater and *L* for the liquid.

The temperature of the insulating foil T_f in $\Omega_{f,I} = \{(x,y) \in R^2 : 0 < x < H_I,\ \ 0 < y < \delta_f\}$ was assumed to fulfill Laplace's equation

$$LT_f = 0 \tag{1}$$

where $L = \frac{\partial^2}{\partial x^2} + \frac{\partial^2}{\partial y^2}$.

For equation (1) the following boundary conditions were adopted

$$T_f(x_k, 0) = T_k \text{ for } k = 1,2,\ldots, K_I \tag{2}$$

$$\frac{\partial T_f}{\partial y} = 0 \text{ for } y = 0,\ 0 \le x \le H_I \tag{3}$$

$$\frac{\partial T_f}{\partial x} = 0 \text{ for } 0 \le y \le \delta_f,\ x = 0 \tag{4}$$

where T_k denotes discrete foil external temperature measurement made with the IR camera at K_I measurement points $(x_k, 0)$.

The temperature of the heater, T_h, in $\Omega_{h,I} = \{(x,y) \in R^2 : 0 < x < H_I,\ \ \delta_f < y < \delta_f + \delta_h\}$ satisfies the Poisson equation

$$LT_h = -\frac{q_V}{\lambda_h} = -\frac{I\,\Delta U}{A_h\,\lambda_h} \tag{5}$$

where q_V denotes the volumetric flow rate supplied to the heating foil, λ_h is the heater thermal conductivity and A_h is the surface area of the heater.

The following boundary conditions were adopted for equation (5)

$$T_f(x, \delta_f) = T_h(x, \delta_f) \text{ for } 0 \le x \le H_I \tag{6}$$

$$\lambda_f \frac{\partial T_f}{\partial y} = \lambda_h \frac{\partial T_h}{\partial y} \text{ for } y = \delta_h,\ 0 \le x \le H_I \tag{7}$$

The distribution of ethanol temperature depends on the flow type. Different ways of temperature computation are used when the single-phase convection region predominates in the minichannel and ends with boiling incipience, and when two-phase flow occurs with developed bubbly or bubbly-slug boiling. The first of these flow types is still defined as single-phase flow, whereas the latter is just the two-phase flow. The assumptions for both flow types were as follows: (a) the flow of the liquid in the minichannel is laminar (Re < 2100) and stationary with constant mass flux,

(b) the velocity of the liquid in the minichannel has only one non-zero component $w(y)$ parallel to the heating surface, fulfilling the condition resulting from the momentum equation

$$-\frac{dp}{dx}+\mu_L\frac{\partial^2 w}{\partial y^2}=0 \tag{8}$$

where μ_L is the dynamic viscosity of the liquid.
(c) the temperature of the liquid at the inlet (T_{in}) and outlet (T_{out}) of the minichannel is known.

For the single-phase flow with boiling incipience, the liquid temperature was taken to change linearly in the minichannel: from the inlet temperature to the outlet temperature. Both of these temperatures are known from the measurements. For the two-phase flow, when limiting the considerations to the liquid phase, the temperature of the liquid in region $\Omega_{L,I}=\{(x,y)\in R^2: 0<x<H_I, \delta_f+\delta_h<y<\delta_f+\delta_h+\delta_m\}$ was assumed to fulfill the energy equation

$$\lambda_L LT_L = w(y)c_p\rho_L\frac{\partial T_L}{\partial x}-\mu_L\Phi-w(y)\frac{dp}{dx} \tag{9}$$

where $\frac{dp}{dx}\approx\frac{p_{out}-p_{in}}{H_{III}}=\frac{\Delta p}{H_{III}}$, H_{III} = 193 mm is the minichannel length (figure 3) and $\Phi=\frac{q_V}{\mu_L}=\left(\frac{dw}{dy}\right)^2$ is the Rayleigh dissipation function.

For equation (9) additional boundary conditions were adopted:
(d) the two-phase mixture resulting from the flow in the minichannel is characterized by void fraction φ. The heat flux generated in the heating foil is transferred to the liquid phase in the proportion dependent on the void fraction, i.e.

$$\lambda_L\frac{\partial T_L}{\partial y}=(1-\varphi(x))\lambda_h\frac{\partial T_h}{\partial y} \tag{10}$$

(e) the liquid temperature for x = H_I satisfies the condition

$$T_L(H_I)=T_{sat}(H_I) \tag{11}$$

where $T_{sat}(x)$ is the saturation temperature dependent on linearly variable pressure.
(f) the liquid temperature at the contact with the heater satisfies the condition

$$T_L=\begin{cases}T_h, & \text{for } T_h<T_{sat}\\ T_{sat}, & \text{for } T_h\geq T_{sat}\end{cases} \tag{12}$$

The known distributions of the heater and liquid temperatures allow determining the heat transfer coefficient $\alpha(x)$ from the Robin boundary condition

$$\alpha(x)=\frac{-\lambda_h\frac{\partial T_h}{\partial y}(x,\delta_h+\delta_f)}{T_h(x,\delta_h+\delta_f)-T_{L,0}(x)} \tag{13}$$

In (13), the reference temperature $T_{L,0}$ is computed from the depending on the flow type:
- for single-phase flows

$$T_{L,0}(x)=T_{in}+\frac{T_{out}-T_{in}}{H_{III}}x \tag{14}$$

- for two-phase flows

$$T_{L,0}(x)=\frac{1}{0.5\delta_m}\int_{\delta_f+\delta_r}^{\delta_f+\delta_r+0.5\delta_m}T_L(x,y)dy \tag{15}$$

In region III in the cross-section of the measurement module, figure 3, there is only the insulating foil and the minichannel with the ethanol. The foil temperature in $\Omega_{f,III}=\{(x,y)\in R^2: H_{II}<x<H_{III},\ 0<y<\delta_f\}$ was assumed to satisfy equation (1) with boundary conditions

$$T_f(x_k,0)=T_k \text{ for } k=1,2,\dots,K_{III} \tag{16}$$

$$\frac{\partial T_f}{\partial y}=0 \text{ for } y=0,\ H_{II}<x<H_{III} \tag{17}$$

where H_{II} = 116 mm and T_k are discrete temperature measurements made at K_{III} measurement points.

The ethanol temperature distribution is computed in $\Omega_{L,III}=\{(x,y)\in R^2: H_{II}<x<H_{III},\ \delta_f<y<\delta_f+\delta_h+\delta_m\}$ similarly as in the region $\Omega_{L,I}$. In the case of the two-phase flow, only boundary conditions are modified: conditions (d) and (f) do not occur, and condition (e) is subject to change, i.e. for x = H_{II}, the new condition (e') has the form

$$T_L(H_{II})=T_{sat}(H_{II}) \tag{18}$$

5 Trefftz method

Equations (1), (5), (9) with an adequate set of boundary conditions comprise a system of three subsequent (in region Ω_I) or two subsequent (in region Ω_{III}) inverse heat conduction problems. The approximate solutions to these equations were found using the Trefftz method.

Two sets of Trefftz functions were used in the calculations. To determine the approximate solution to equations (1) and (5), harmonic polynomials $u_{n,0}(x,y)$ were used, analogously as in [5, 6], defined by the formula

$$u_{n,0}(x,y)=\begin{cases}\frac{1}{\left(\frac{n}{2}\right)!}Re(x+iy)^{\frac{n}{2}} & \text{for } n=0,2,4,\dots\\ \frac{1}{\left(\frac{n+1}{2}\right)!}Im(x+iy)^{\frac{n+1}{2}} & \text{for } n=1,3,5,\dots\end{cases} \tag{19}$$

where i is the imaginary unit.

In each region, Ω_I and Ω_{III} the insulating foil temperature T_f is first determined, approximated with a linear combination of harmonic polynomials, [5, 6]

$$T_f(x,y)=\sum_{n=0}^{N_f} a_n u_n(x,y) \tag{20}$$

Coefficients of the linear combination (20) are determined in the way analogous to that in [5, 6], by minimizing the functional that describes the mean square error with which the approximate T_f satisfies boundary conditions (2) - (4) (in region Ω_I) or (16), (17) (in region Ω_{III}). Analogous procedure is used to determine the approximate temperature of the heater, T_h, in region Ω_I . In [7, 8] the Trefftz functions (19) were used for solving nonlinear direct and inverse heat transfer problems.

Trefftz functions $\theta_n(x,y)$ for the energy equation in the form

$$\lambda_L L T_L = w(y) c_p \rho_L \frac{\partial T_L}{\partial x} \tag{21}$$

are used to find the approximate temperature of ethanol.

As in [9], the solution to equation (9) is a sum of the linear combination of functions $\theta_n(x,y)$ and the particular solution to equation (9). The fashion of determining the coefficients of the linear combination has been described in detail in [9]. In [9], Trefftz functions $\theta_n(x,y)$ were determined using the expansion of the solution to equation (21) into the Taylor series. Also, several first functions $\theta_n(x,y)$ for the parabolic velocity profile $w(y)$ were presented. It is easy to prove that the n-th function $\theta_n(x,y)$ is the sum of the harmonic polynomial $u_{n,0}(x,y)$ and the polynomials depending on the form of the velocity profile $w(y)$, i.e.,

$$\theta_n(x,y)=u_{n,0}(x,y)+\sum_{i=1}^{m} u_{n,i}(x,y) \tag{22}$$

Functions $u_{n,i}(x,y)$ for $i=1,2,...,m=\left[\frac{n}{2}\right]$, where $[\cdot]$ denotes the greatest integer function, satisfy the relation

$$u_{n,i}(x,y)=\frac{1}{a}L^{-1}\left[w(y)\frac{\partial u_{n,i-1}}{\partial x}\right] \tag{23}$$

In (23), the inverse operator L^{-1} for monomials $\frac{x^n y^k}{n!k!}$, where $n, k = 0, 1, 2, ...$, is represented by, [10]

$$L^{-1}\left\{\frac{x^n y^k}{n!k!}\right\}=\sum_{j=0}^{m}(-1)^j \frac{x^{n-2j} y^{k+2j+2}}{(n-2j)!(k+2j+2)!} \tag{24}$$

Determined using the Trefftz method, temperature approximations T_f, T_h, T_L are the continuous functions that satisfy equations (1), (5) and (9) exactly, and fulfill the boundary conditions adopted for these equations approximately.

6 Results

The results of the calculations performed for the experimental data presented in figures 8 and 9 are shown below. For single-phase flows, the measurements from the IR camera of the foil surface for the region Ω_I are shown in figure 8a. First, the approximate 2D foil temperature distribution was determined using the Trefftz method. Then the same procedure was conducted for the heater. Figure 8 b and figure 8c show good agreement in satisfying conditions (6) and (7) by approximated temperatures T_f and T_h. The largest differences occur between the measurements and the temperature of the insulating foil, figure 12 a, said differences reaching maximum of 3.9 K (1.8 K on average).

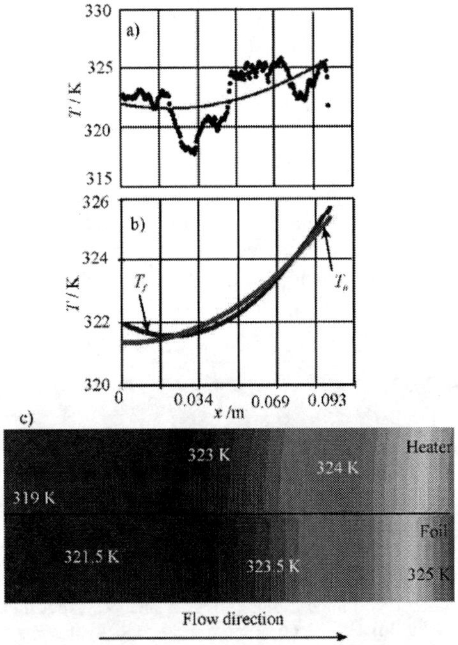

Figure 8. For the single-phase flow: a) foil temperature obtained from the IR camera (dotted line) and that calculated using the Trefftz method (solid line), b) foil (T_f) and heater (T_h) temperature at the foil-heater contact, c) 2D temperature distribution for the insulating foil and the heater obtained using the Trefftz method in region Ω_I ; experimental data: mean velocity 0.022 m s^{-1}, mass flux 134 kg m^{-2} s^{-1}, inlet pressure 6.83 kPa, heat flux: 23.71 kW m^{-2} .

For two-phase flows, the calculation in region Ω_I were performed at a small (~ 8%), medium (~ 60%), and large (~ 80%), volume of void fraction, figure 9. The void fractions experimentally determined at selected points were approximated with a cubic function, figure 9. The foil temperature distribution obtained using the IR camera is analogous to that shown in figure 8 a (dotted line).

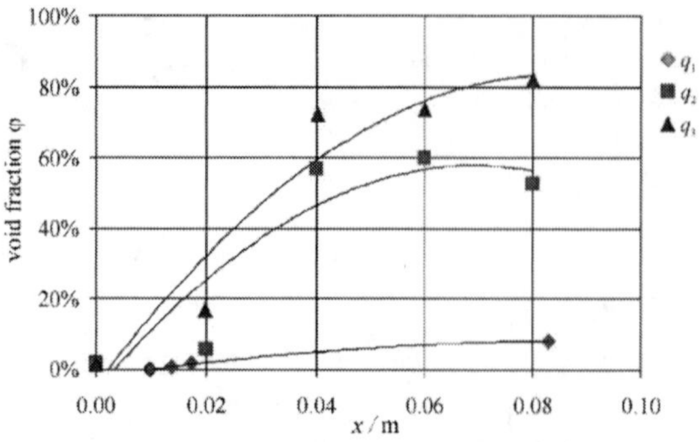

Figure 9. Experimentally determined void fractions approximated using a cubic function when: q_1= 73.66 kW m^{-2}, (the coefficient of determination R^2 = 0.9), q_2 = 122.96 kW m^{-2} (R^2 = 0.82), q_3 = 152.04 kW m^{-2} (R^2 = 1); experimental data: mean velocity 0.022 m s^{-1}, mass flux 134 kg m^{-2} s^{-1}, inlet pressure 6.83 kPa .

While determining the approximate temperature distributions in the measurement module using the Trefftz method, first the insulating foil temperature distribution was determined, then the temperature distribution of the heater was found and finally that of the ethanol. This procedure leads to the solution of three sequential inverse heat conduction problems in region Ω_I. The number of Trefftz functions was selected so that the calculated approximations were polynomials of the same degree. Figure 10 summarizes examples of 2D diagrams of approximated temperature distributions for the heater and ethanol determined in region Ω_I.

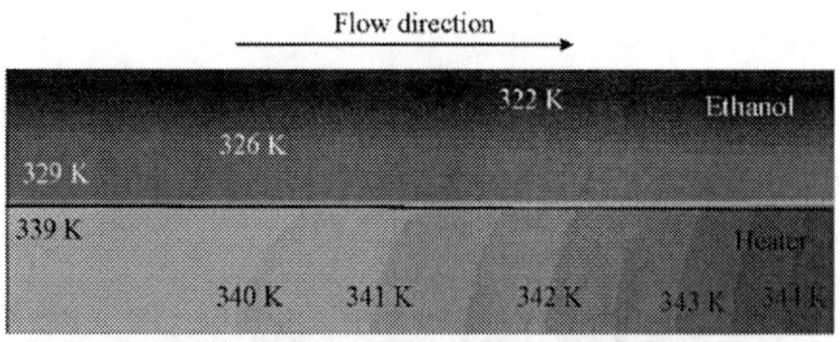

Figure 10. 2D temperature distributions of the ethanol and the heater determined using the Trefftz method; experimental data: heat flux q_1 = 73.66 kW m^{-2}, mean velocity 0.022 m s^{-1}, mass flux 134 kg m^{-2} s^{-1}inlet pressure 6.83 kPa.

The 2D temperature distribution for the ethanol in Ω_I and in Ω_{III} is shown in figure 11. Figure 11 shows, especially visible in Ω_I, the influence of the parabolic velocity profile on the liquid temperature distribution.

Known temperature distribution of the ethanol and the heater allows defining the heat transfer coefficient on the basis of boundary condition (13). Figure 12 shows heat transfer coefficients computed in Ω_I for the considered void fractions dependent on the heat fluxes supplied to the heater, figure 9.

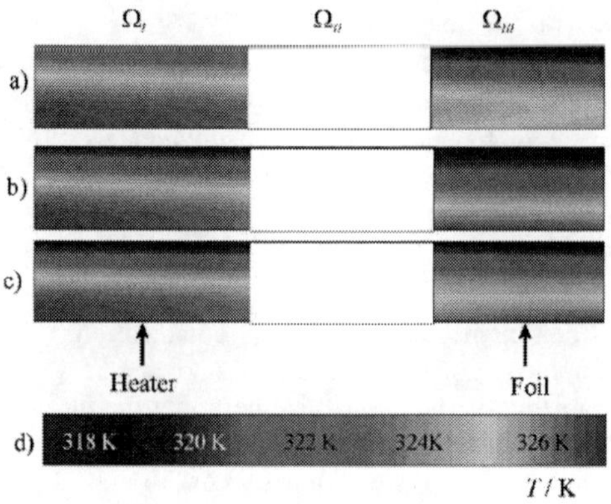

Figure 11. 2D ethanol temperature distributions for:
a) q_1 = 73.66 kW m^{-2}, b) q_2 = 122.96 kW m^{-2},
c) q_3 = 152.04 kW m^{-2},
d) scale; experimental data as in figure 10.

For small void fractions, during the flow of subheated liquid in the flow axis (few vapor bubbles appear only on the heater wall), the heat transfer coefficient decreases (figure 12 for q_1). Increasing the heat flux supplied to the heater leads to the development of boiling in the ethanol and the increase in the heat transfer coefficient value, (figure 12 for q_2 and q_3).

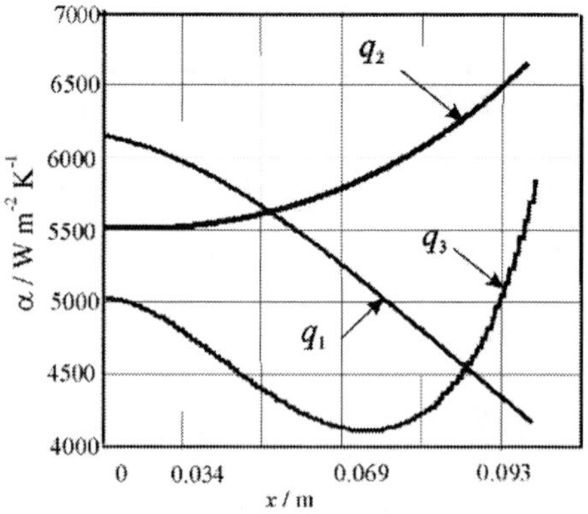

Figure 12. Heat transfer coefficients for q_1 = 73.66 kW m^{-2}, q_2 = 122.96 kW m^{-2}, q_3 = 152.04 kW m^{-2}; experimental data as in figure 10.

7 Conclusions

The paper describes the experiments that refer to flow boiling of ethanol in minichannels. The adopted heat and flow parameters (pressure and temperature at the minichannel inlet, mass flux, heat flux supplied to the heater) as well as the measurements of temperature, pressure and void fraction obtained from the experiments were the sources of data for further simulation calculations. The images from a fast video camera allowed determining void fractions in the minichannel. Thermograms of the heater surface and of the adiabatic part of the minichannel allowed obtaining temperature profiles along the minichannel length. The read and calculated parameters met the expected dependencies (especially the void fraction as a function of the volumetric flow rate and heat flux) and were used to solve inverse heat conduction problems.

The paper proposes a simple mathematical model to describe stationary heat transfer in the measurement module elements and in the minichannel. Single- and two-phase flows were considered in the model. The model was formulated so as to limit to minimum the occurrence of experimentally determined constants – here these are the external temperature measurements of the insulating surface and the void fraction. In the model, the calculation procedure of the temperature of the ethanol flowing through the minichannel is coupled with the process of determining temperature distributions in the elements of the measurement module containing the minichannel (including the heater). Known temperature distributions of the heater and ethanol allow defining local heat transfer coefficients at the heater/ethanol interface. The analysis of the results indicates that the proposed Trefftz method is a highly accurate and stable tool for solving inverse heat conduction problems leading to the searched function and its gradient. It is worth noting that in the instances discussed above a maximum of three subsequent inverse problems were solved.

References

1. R. Kaniowski, M.E. Poniewski, Arch. Therm. **34**, 1-19 (2013)
2. S. Witczak, *Flow boiling of natural refrigerants inside conventional size channel and minichannels* (in Polish), Opole University of Technology (2013)
3. M. Grabowski, M. Poniewski, S. Alabrudziński, K.Płaczkowski, App. Mech. Mater. (to be published)
4. K. Płaczkowski, M. Poniewski, M. Grabowski, S.Alabrudziński, App. Mech. Mater.(to be published)
5. M. Piasecka, S. Hożejowska, M.E. Poniewski, Int. J. Heat Fluid Flow **25**, 159-172 (2004)
6. S. Hozejowska, M. Piasecka, M.E. Poniewski, Int. J. Therm. Sci. **48**, 1049-1059 (2009)
7. K. Grysa, A. Maciąg, A. Pawińska A., Int. J. Heat Mass Transf. **55**, 7336 –7340 (2012)
8. K. Grysa, A. Maciąg, *APCOM&ISCM* (2013) Singapore
9. S. Hożejowska, R. Kaniowski, M.E. Poniewski, Int. J. Numer. Methods Heat & Fluid Flow **24**, 811-824 (2014)
10. M. Sokała, M.J. Ciałkowski, *TRECCOP'04*, 251-259 (2009)

EPJ Web of Conferences **114**, 02042 (2016)
DOI: 10.1051/epjconf/201611402042

Heat transfer coefficient for flow boiling in an annular mini gap

Sylwia Hożejowska[1,a], Tomasz Musiał[2] and Magdalena Piasecka[2]

[1]*Kielce University of Technology, Faculty of Management and Computer Modelling, Al. 1000-lecia PP 7, 25-314 Kielce, Poland*
[2]*Kielce University of Technology, Faculty of Mechatronics and Mechanical Engineering, Al. 1000-lecia PP 7, 25-314 Kielce, Poland*

Abstract. The aim of this paper was to present the concept of mathematical models of heat transfer in flow boiling in an annular mini gap between the metal pipe with enhanced exterior surface and the external glass pipe. The one- and two-dimensional mathematical models were proposed to describe stationary heat transfer in the gap. A set of experimental data governed both the form of energy equations in cylindrical coordinates and the boundary conditions. The models were formulated to minimize the number of experimentally determined constants. Known temperature distributions in the enhanced surface and in the fluid helped to determine, from the Robin condition, the local heat transfer coefficients at the enhanced surface – fluid contact. The Trefftz method was used to find two-dimensional temperature distributions for the thermal conductive filler layer, enhanced surface and flowing fluid. The method of temperature calculation depended on whether the area of single-phase convection ended with boiling incipience in the gap or the two-phase flow region prevailed, with either fully developed bubbly flow or bubbly-slug flow. In the two–phase flow, the fluid temperature was calculated by Trefftz method. Trefftz functions for the Laplace equation and for the energy equation were used in the calculations.

1 Introduction

In recent years, many experimental and theoretical works have been published to address the issues of heat transfer in flow boiling in mini- and microchannels. The use of minichannels in heat exchangers improves the effectiveness of heat transfer processes, which is the major advantage of this technology. An example of experimental studies conducted to investigate practical applications for heat exchangers with minichannels can be found in [1], where the use of mini-gaps in microchip cooling is discussed. In [2], the authors report heat transfer investigations of the initial flow boiling phases in mini-gaps for the ecological refrigerants which are superseding traditional thermodynamic agents due to strengthened requirements. In [3], the researchers present the studies and models for the boiling of refrigerant in the heat exchanger and determine the heat transfer coefficient for the channels with conventional and small diameters. In [4], boiling heat transfer for the counter-flow of R134a was studied. Experimental investigations [5] show that the heat transfer coefficient in compact heat exchangers with mini gaps is higher. Investigations often deals with frictional pressure drop in minichannels [6].

The use of microstructured heated surfaces in boiling research is known to intensify the heat transfer process. Enhanced surfaces are considered to be very efficient because of providing a large number of nucleation sites. The properties of a microstructured surface can be modified using chemical, thermal, mechanical or combined mechanical and thermal processes. The thermal processes include plasma spraying, electric arc spraying, sintering, laser surface texturing or electrical discharge texturing [7-13]. Very broad critical review of past achievements can be found in [14], where much room was devoted to discussing mathematical models and their application to numerical analysis. The most general model, i.e. the two fluid four-field model, which at least theoretically allows permanent modelling of flow structure changes is presented in [15]. Used mainly in numerical analysis to adiabatic/non-adiabatic, steady/ unsteady flows through the channels with different cross-sections and hydraulic diameters, the model gave satisfactory results [16, 17]. A major problem in solving this type of equations is to determine the appropriate set of closure equations. Solving such a complex system of equations, even when using CFD commercial software is not always possible, especially in the case of inverse problems.

In the present paper, one- and two-dimensional mathematical models are proposed to describe the stationary heat transfer in flow boiling of cooling liquid in the cylindrical annular gap. A set of experimental data governs the form of energy equations in the cylindrical coordinates and the boundary conditions. The two-dimensional model is formulated to minimize the number of experimentally determined constants. In addition to the fixed thermal-flow parameters, this model uses two quantities determined experimentally: the surface temperature of the thermal conductive filler layer and void fraction. The method for determining the fluid temperature distribution depends on the flow type. The

[a] Corresponding author: ztpsf@tu.kielce.pl

temperature of the fluid is calculated in one way when the domain of single-phase convection ended with the boiling incipience prevails in the gap and in a different way, when the domain with two-phase flow dominates along with developed bubbly or bubbly-slug flow. In the two-phase flow, the calculation of fluid temperature is related to the approximate determination of temperature distributions in the main elements of the measurement module. Paper [18] proposes Trefftz method to determine two-dimensional temperature distributions in the thermal conductive filler layer, in the layer with enhanced surface and in the fluid. The knowledge of temperature distributions in the heating surface and fluid helps to determine, using the Robin condition [19], the local heat transfer coefficients at the interface of the thermal conductive filler and the fluid.

2 Experimental setup

2.1 Experimental setup

The experimental setup (figure 1) consists of several systems: the test loop in which FC-72 Fluorinert circulates as the working fluid, the data and image acquisition system, the supply and control system, and the lighting system.

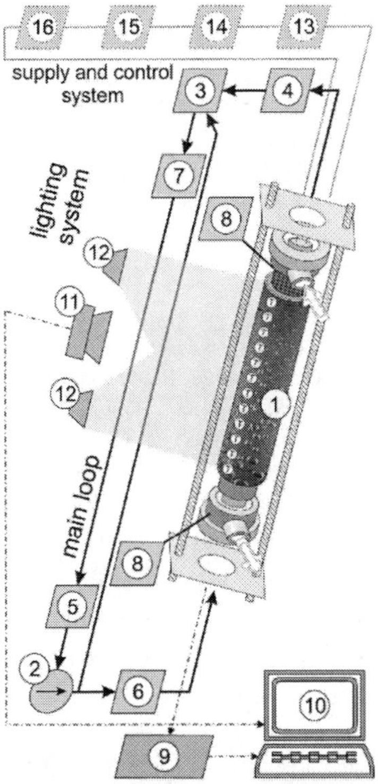

Figure 1. The schematic diagram of the main loops at the experimental setup, 1-measurement module with a mini annular gap; 2-rotary pump; 3-compensating tank/pressure regulator; 4-tube-type heat exchanger, 5-filter, 6-rotameters/flowmeter; 7-deaerator, 8-pressure transducer, 9-data acquisition station, 10- computer, 11-digital SLR camera, 12-halogen reflectors, 13-inverter welder, 14-voltmeter, 15-ammeter, 16-shunt.

The test loop is composed of a rotary pump, a heat exchanger, a compensating tank, rotameters, a filter, and a deaerator. The data and image acquisition system consists of two digital cameras, an infrared camera, a data acquisition station, a computer with special software, and the lighting. The experimental setup is similar to that described in [19-22].

2.2 The testing module with an annular mini gap

Figure 2 presents the testing module with an annular mini gap. The mini gap 1-3 mm wide will be created between the metal pipe (2) with enhanced exterior surface and the external tempered glass pipe (1) positioned along the same axis.

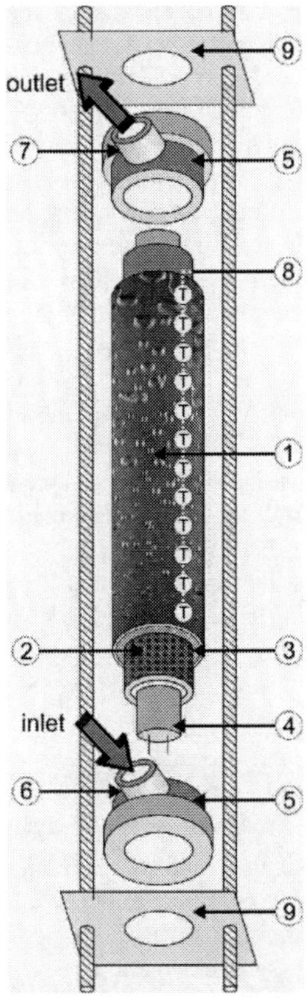

Figure 2. The measurement module layout with annular mini gap: 1-glass pipe for viewing flow patterns, 2-metal pipe with enhanced surface, 3-annular mini gap, 4-cartridge heater, 5-module header with sealing, 6-inlet port for the medium, 7-outlet port for the medium, 8-thermoelements between the heater (4) and the metal pipe (2) in the thermal conductive filler layer, 9-stirrup bolts.

Several types of microstructured heating surfaces are planned to be used: ‘microcavities’ (produced by laser surface texturing), ‘minicavities’ (produced by spark erosion) [22], and the surfaces with capillary metal fibrous structure or with sintered metallic powder. Inside the metal pipe (2), a cartridge heater (4) will be located axially and symmetrically, powered by autotransformer with adjusted current intensity. Longitudinal slots will be made in the inside surface of the pipe to fix the thermoelement sensors with a spacing of approximately 1 cm in the flow line. Thermoelement wires will run in the gap between the heater and the metal pipe in the thermal conductive filler layer. The module header will have a sealing (5) and the sensors of successive thermoelements and pressure transducers to measure the coolant temperature and positive pressure in the inlet/outlet to/from the annular mini-gap.

3 Mathematical models

The assumptions made in the mathematical model (see figure 3) included the steady state in the measurement module and the laminar flow of the fluid (Reynolds number < 2 100). Two dimensions were provided for in the proposed approaches: dimension z parallel to the flow direction and dimension r perpendicular to z, for the widths of the consecutive elements of the measurement module. The temperature distribution for the elements in the module is only dependent on variables r and z, (it is axi-symmetric). The temperature distribution of the coolant in the model is dependent on the flow type. Different techniques are used to calculate the temperature in the case when the single-phase convection region prevails in the annular gap, ended with boiling incipience, and in the case of two-phase flow with developed bubbly or bubbly-slug boiling. The first case mentioned above is further in the text referred to as single phase flow with boiling incipience, the latter case – simply two-phase flow. In both cases, the assumptions were as follows:

- the fluid flow in the minichannel is laminar and stationary, with a steady mass flux,
- fluid velocity has only one nonzero component $w(r)$ parallel to the heating surface, which satisfies the condition arising from the momentum equation:

$$\frac{1}{r}\frac{d}{dr}\left(r\frac{dw}{dr}\right) = -\mu_F \frac{\Delta p}{H} \tag{1}$$

where p is the pressure, μ_F is the dynamic viscosity of the fluid and H is the length of the minichannel, and

- the fluid temperature at the inlet ($T_{F,in}$) and outlet ($T_{F,out}$) of the minichannel is known.

Placing the thermoelement sensors on the external surface of the thermal conductive filler layer, figure 3, gives the surface temperature at M points: z_1, z_2, ..., z_M.

Two approaches are proposed below for describing heat transfer in particular elements of the measurement module with the annular mini-gap.

3.1 One-dimensional model

In the one-dimensional heat transfer model, for each cross-section (z_i = *constants*), the heater temperature T_h is assumed to satisfy the Poisson equation in the form

$$\frac{1}{r}\frac{d}{dr}\left(r\frac{dT_h}{dr}\right) = -\frac{q_V}{\lambda_h} \tag{2}$$

where q_V - is the volumetric heat flux generated in the heater, and λ_h - is the heat transfer conductivity of the heater.

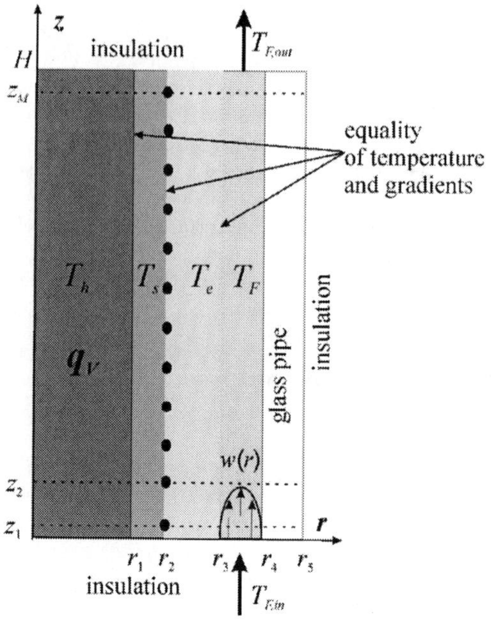

Figure 3. Diagram illustrating the assumptions adopted for the calculation methods (pictorial view, not to scale).

Equation (2) was solved assuming that:

- the temperature of the heater at each point of the cylinder is finite (including inside, where r=0).
- for $r = r_1$ at each point z_1, z_2, ..., z_M, the condition below is satisfied

$$\tfrac{1}{2} r_1 q_V = \lambda_h \frac{\partial T_h(r_1,z_i)}{\partial r} = \lambda_s \frac{\partial T_s(r_1,z_i)}{\partial r} \tag{3}$$

Taking for r=r_1 equal temperatures of the heater, T_h, and the thermal conductive filler T_s and considering that the width of the measurement module elements is small, we can write

$$\frac{\partial T_s(r_1,z_i)}{\partial r} \approx \frac{T_{s,i} - T_h(r_1,z_i)}{r_2 - r_1} \tag{4}$$

where $T_{s,i} = T_s(r_2,z_i)$ are the temperatures measured using the thermocouples at points (r_2,z_i), figure2.

Calculating temperature $T_h(r_1,z_i)$ from (3) and (4), the solution to equation (2) can be expressed as

$$T_h(r,z_i) = T_{s,i} - q_V\left(\frac{r^2 - r_1^2}{4\lambda_h} + \frac{(r_2 - r_1)r_1}{2\lambda_s}\right) \tag{5}$$

Assuming equal respective temperatures and heat fluxes at the boundaries $r = r_2$ and $r = r_3$, figure 3, at each measurement point z_1, z_2, ..., z_M, we have

$$T_e(r_3,z_i) = T_{s,i} + \frac{q_V r_1^2 (r_3 - r_2)}{2 r_2 \lambda_e} \tag{6}$$

$$\lambda_F \frac{\partial T_F(r_3,z_i)}{\partial r} = \frac{q_V r_1^2}{2r_3} \tag{7}$$

From (7) and from the Robin boundary condition, [23], we obtain the formula for the local heat transfer coefficients at points $z_1, z_2, ..., z_M$, at the interface between the enhanced surface and the fluid

$$\alpha(z_i) = \alpha_i = \frac{r_1^2}{2r_3} \frac{q_V}{T_{s,i} + \frac{q_V r_1^2 (r_3 - r_2)}{2r_2 \lambda_e} - T_F^{ref}} \tag{8}$$

In formula (8), the reference temperature T_F^{ref} is calculated depending on the flow type:
- for single phase flow with boiling incipience, the fluid temperature T_F^{ref} changes linearly in the minichannel, from the inlet temperature ($T_{F,in}$) to the outlet temperature($T_{F,out}$). Both temperatures are known from the measurements;
- for two-phase flow in this approach, we assume that

$$T_F^{ref} = \begin{cases} T_F & \text{if } T_{sat} > T_F \\ T_{sat} & \text{if } T_{sat} \le T_F \end{cases} \tag{9}$$

3.2 Two-dimensional model

In the two-dimensional approach we assume that the temperature of the thermal conductive filler layer T_s and of that of the enhanced surface T_e both satisfy respective equations

$$\text{a) } LT_s = 0 \qquad \text{b) } LT_e = 0 \tag{10}$$

where the Laplace operator in coordinates (r, z) is defined by $L = \frac{1}{r}\left(r\frac{\partial}{\partial r}\right) + \frac{\partial^2}{\partial z^2}$. The following boundary conditions, figure 3, were adopted for the temperature T_s

$$\lambda_s \frac{\partial T_s}{\partial r} = \frac{q_V r_1}{2} \quad \text{for } r = r_1 \tag{11}$$

$$T_s(r_2, z_i) = T_{s,i} \tag{12}$$

$$\frac{\partial T_s}{\partial z} = 0 \quad \text{for a) } z = 0 \text{ and b) } z = H \tag{13}$$

Analogous assumptions were adopted for the temperature of the thermal conductive filler layer, T_e, i.e.,

$$\lambda_e \frac{\partial T_e}{\partial r} = \frac{q_V r_1^2}{2r_2} \quad \text{for } r = r_2 \tag{14}$$

$$T_e(r_2, z_i) = T_{s,i} \text{ and b) } T_e(r_2, z) = T_s(r_2, z) \tag{15}$$

$$\frac{\partial T_e}{\partial z} = 0 \quad \text{for a) } z = 0 \text{ and b) } z = H \tag{16}$$

For two-phase flow, the fluid temperature T_F is assumed to satisfy the energy equation in the form

$$\lambda_F LT_F = w(r) c_{p,F} \rho_F \frac{\partial T_F}{\partial z} \tag{17}$$

where the fluid velocity in the mini-gap $w(r)$ is calculated from condition (1), and $c_{p,F}, \rho_F$ are the specific heat and liquid density, respectively. The following boundary conditions were adopted for equation (17)

$$\text{a) } T_F(r,0) = T_{F,in}, \quad \text{b) } T_F(r,H) = T_{F,out} \tag{18}$$

$$\lambda_F \frac{\partial T_F}{\partial r} = \lambda_e (1 - \varphi(z)) \frac{\partial T_e}{\partial r} \quad \text{for } r = r_3 \tag{19}$$

where $\varphi(z)$ is the experimentally determined void fraction. Also, on the boundary $r = r_3$, the fluid temperature T_F is assumed to satisfy condition (9).

Known temperature distributions of the enhanced surface, $T_e(r,z)$, and the fluid, $T_F(r,z)$, allows determining the heat transfer coefficient as in 3.1, from the Robin condition, which in this approach has the following form

$$\alpha(z) = \frac{-\lambda_F \frac{\partial T_e}{\partial r}(r_3, z)}{T_e(r_3, z) - T_F^{ave}(z)} \tag{20}$$

The reference temperature T_F^{ave} is calculated depending in the flow type – in the same way as in 3.1 for single phase flow with boiling incipience, and as an average liquid temperature in the gap – for two-phase flow, i.e.,

$$T_F^{ave} = \frac{1}{r_4 - r_3} \int_{r_3}^{r_4} T_F(r,z) dz \tag{21}$$

In the two-dimensional model, the form of energy equations and that of the boundary conditions depend on the set of the experimental data. The model was formulated in the way which minimizes the occurrence of experimentally determined constants – in this case, these are temperature measurements of the filler layer surface and void fraction. For single phase flow with boiling incipience, the solution of the differential equations leads to the solution of one direct heat transfer problem in the filler layer and one inverse problem in the enhanced layer. For two-phase flow, the procedure of calculating fluid temperature is coupled with the process of determining temperature distributions in two elements of the measurement module, which leads to a triple coupled heat transfer problem. This problem is composed of one direct problem (filler layer) and two successive inverse problems (filler layer and fluid).

3.2.1 *Trefftz method*

The differential equation written in section 3.2, together with adequate systems of boundary conditions, was solved using the Trefftz method. The choice of this method resulted from the fact that solving respective equations led to the solution of one or two inverse heat transfer problems. The Trefftz method, as confirmed by available research reports, [19,24], ensures stable results also for inverse problems. The essence of the Trefftz method lies in approximating an unknown solution of a differential equation using a linear combination of Trefftz functions, that is, the functions which satisfy the equation exactly. Two types of Trefftz functions were used in the numerical calculations: Trefftz functions for the Laplace equation (10) and Trefftz functions for energy equation (17). The Trefftz functions for equation (10) are defined by the following formula [25]

$$u_{n,0}(r,z)=\sum_{k=0}^{\left[\frac{n}{2}\right]}\frac{(-1)^k 2^{-2k}}{(k!)^2(n-2k)!}r^{2k}z^{n-2k} \tag{22}$$

where $[\cdot]$ denotes the greatest integer function.

Trefftz functions $h_n(r,z)$ for the energy equation (17) are the sum of functions $u_{n,0}(r,z)$ and polynomials dependent on the form of the velocity profile $w(r)$. This relationship can be expressed as

$$h_n(r,z)=u_{n,0}(r,z)+\sum_{i=1}^{n}u_{n,i}(r,z) \tag{23}$$

Assuming that $a=\frac{\lambda_F}{c_{p,F}\,\rho_F}$, functions $u_{n,i}(x,y)$, $i=1,2,\ldots,n$, satisfy the relationship

$$u_{n,i}(x,y)=\frac{1}{a}L^{-1}\left[w(r)\frac{\partial u_{n,i-1}}{\partial z}\right] \tag{24}$$

The inverse operator L^{-1} for monomial in the form $r^n z^k$ is defined for $n = 0, 1, 2\ldots$, as follows

$$L^{-1}\left(r^n z^k\right)=\begin{cases}\dfrac{r^{n+2}z^k}{(n+2)^2} & \text{for } k=0,1\\[2ex] \dfrac{r^{n+2}z^k}{(n+2)^2}-\dfrac{k(k-1)}{(n+2)^2}L^{-1}\left(r^{n+2}z^{k-2}\right) & \text{for } k\geq 2\end{cases} \tag{25}$$

Formula (25) applies when velocity $w(r)$ is a polynomial. If $w(r)$ is not a polynomial, the function is expanded as a Taylor series, which allows representing it in the form of a polynomial, to a random accuracy. The Trefftz functions given by formulas (22) and (23) will be used to determine approximate solutions to equations (10) and (17), respectively.

For example, the solution of equation (17) is approximated by the linear combination of the Trefftz functions $h_n(r,z)$

$$T_F(r,z)=\sum_{n=1}^{N}a_n h_n(r,z) \tag{26}$$

where N means the number of Trefftz functions (23). The unknown coefficients a_n of linear combination (26) are determined by minimizing the functional that describes the mean squared error with which function (26) satisfies conditions (9), (18) and (19). In this case, the minimized functional has the following form

$$\begin{aligned}J_F &= \int_{r_3}^{r_4}\left(\sum_{n=1}^{N}a_n h_n(r,0)-T_{F,in}\right)^2 dr+\\ &+\int_{r_3}^{r_4}\left(\sum_{n=0}^{N}a_n h_n(r,H)-T_{F,out}\right)^2 dr+\\ &+\int_{0}^{H}\left(\sum_{n=0}^{N}\lambda_F a_n\frac{\partial h_n(r_3,z)}{\partial r}-\lambda_e(1-\varphi(z))\frac{\partial T_e(r_3,z)}{\partial r}\right)^2 dz+\\ &+\int_{0}^{H}\left(\sum_{n=0}^{N}a_n h_n(r_3,z)-T_F^{ref}\right)^2 dz\end{aligned} \tag{27}$$

The minimum of the functional is determined by solving the linear equation system, where coefficients of linear combination a_n are the unknown quantities

$$\frac{\partial J_F}{\partial a_n}=0 \text{ for } n=1,2,\ldots,N \tag{28}$$

The temperatures of the filler layer, T_s, and the enhanced layer, T_e, are determined analogously using the Trefftz functions (22). The approximate temperature distribution in the thermal conductive filler layer, T_s, is calculated before calculating the temperature of the external enhanced surface, T_e, and that of the fluid, T_F. The Trefftz method- based approximate temperatures T_s, T_e and T_F (two-phase flow) are continuous functions, which satisfy equations (10a), (10b) and (17) exactly, with the adopted boundary conditions satisfied approximately.

4 Conclusions

The purpose of the paper was restricted only to presenting:

- the conceptions of two mathematical models which describe the heat transfer in flow boiling in annular gaps,
- the procedure for calculating the heat transfer coefficient at the interface between the enhanced surface and the fluid.

The two mathematical models of heat transfer processes occurring in annular mini-gaps proposed in this paper are simple, which is their major advantage. The analogous mathematical models of heat transfer in rectangular minichannels, developed and verified by the authors of the present paper, gave the results consistent with the physics of the problem. In the model presented here, two experimentally determined parameters were taken into account: the temperature distribution on the surface of the thermal conductive filler layer and void fraction. The node-less analytical numerical Trefftz method was used to solve the two-dimensional model. Owing to the application of that method, the two-dimensional direct and inverse heat transfer problems were solved in three successive adjacent regions with different physical parameters.

Further studies will focus on comparison of the results obtained from both proposed mathematical models and their verification by comparison to the existing theoretical models. The other task will be determining the thermal stress that occurs on the surfaces of a heat exchanger. Thermal strains have a significant influence on heat transfer conditions because in the case of the minichannel surface dimensions, this may lead to local narrowing in the mini-gaps, thus changing the physical conditions of heat exchanger operation. In extreme cases, thermal strains may cause a complete geometric deformation of the mini-gap.

Acknowledgments

The research has been financially supported by the National Scientific Center from funds granted by virtue of decision No. DEC-2013/09/B/ST8/02825.

References

1. L. A. Campbell S. G. Kandlikar, *Proc. 2nd Int. Conf. on Microchannels and Minichannels*, USA (2004)
2. T. Bohdal, Annual Set Env. Protect. (in Polish) **2**, 262–279 (2000)
3. D. Mikielewicz, J. Mikielewicz, J. Tesmar, Int. J. Heat Mass Tran. **50**, 3949–3956 (2007)
4. J. Kaew-On, K. Sakamatapan, S. Wongwises, Exp. Therm. Fluid Sci. **35**, 364–374 (2011)
5. T. Bohdal, Annual Set Env. Protect. **15**, 107–126 (2013)
6. K. Dutkowski, Heat Tran. Eng. **31**, 321–330 (2010)
7. B. Grabas, Arch. Metall. Mater. **60** (1), 33–39 (2015)
8. B. Grabas, Exp. Therm. Fluid Sci. **68**, 499–508 (2015)
9. B. Grabas, Adv. Mater. Research **874**, 71–75 (2014)
10. W. Depczyński, *Proc. 23rd Int. Conf. Metallurgy Materials METAL 2014*, Poland (2014)
11. W. Depczyński, S. Spadło, P. Młynarczyk, E. Ziach, *Proc. 24th Int. Conf. Metal. Mater. METAL 2015*, Poland (2015)
12. N. Radek, E. Wajs, M. Luchka, Powder Metallurgy Metal Ceramics **47** (3-4), 197–201 (2008)
13. N. Radek, B. Antoszewski, Kovove Materialy-Metallic Mater. **47** (1), 31–38 (2009)
14. Z. Guo, D.F. Fletcher, B.S. Haynes, J. Comp. Multiph. Flows **6**, 79–110 (2014)
15. R.T. Lahey Jr., *Proc. XIII US Nat. Con. Applied Mechanics*, USA (1998)
16. M.Z. Podowski, Int. J. Numer. Methods Heat & Fluid Flow **18**, 491–513 (2008)
17. M.Z. Podowski, R.M. Podowski, Sc. Techn. Nucl. Instal., 1-10; doi: 10.1155/2009/387020 (2009)
18. T. Trefftz, *Proc. 2nd Int. Con. App. Mech.*, 131-137 (1926)
19. S. Hożejowska, M. Piasecka, Heat Mass Tran. **50**, 1053–1063 (2014)
20. M. Piasecka, Heat Tran. Eng. **35**, 903–912 (2014)
21. M. Piasecka, Ann. Nucl. Energy **73**, 282–293 (2014)
22. M. Piasecka, Adv. Mater. Research **874**, 95–100 (2014)
23. S. Hożejowska, M. Piasecka, E. Poniewski, Int. J. Therm. Sci. **48**, 1049–1059 (2009)
24. S. Hożejowska, R. Kaniowski, M.E. Poniewski, Int. J. Numer. Methods & Heat Fluid Flow **24**, 811–824 (2014)
25. M.J. Ciałkowski, S. Hożejowska, L. Hożejowski, *Proc. XVII Conf. Therm.* (in Polish), 259–269 (1999)

EPJ Web of Conferences **114**, 02043 (2016)
DOI: 10.1051/epjconf/201611402043

Simplification of simulation processes at gravity heat pipes

Peter Hrabovský[1,a], Štefan Papučík[1], Richard. Lenhard[1]

[1]*Department of Energy Technology, Faculty of Mechanical Engineering, University of Zilina, Slovakia*

Abstract. Water heating by heat pipe is currently the object examined on the use in sphere of recovering heat from technological processes. The heat pipe is the device for water heating that provides us a very effective way of transferring heat from the heat source (combustion) to the place of consumption (water). For the draft proposal of such equipment is produced under the required conditions mathematical model of ANSYS that verifies the actual measurements the experiments. The paper deals with the possibility of simulations of heat pipes in the process of heat transport and apposite simplification of the simulation process by defining apposite the substitutes of the solid materials with its own thermal properties that ensure a similar heat transport as a heat transfer medium in the heat pipe.

1 Introduction

Heat pipes are now mainly used in refrigeration engineering, to remove heat from electrical components. They can be found mainly in computing, which is mounted on the processor (CPU, GPU, chipset ...) from which dissipates heat through active or passive cooler. The heat pipes are mainly used in the industry where it is necessary to comply with dust-proof, air tightness, water tightness. From such a closed space can removed heat by heat pipes [1].

Heat pipes are used in solar system for heating and heat transfer with heat exchangers. Heat pipes are devices that transfer heat efficiently. They can be used to transfer heat from the source at a higher energy potential for example built-in fireplaces in the home [2]. Of these, it is possible to capture the radiant heat component, which can be used for hot water or heating other rooms.

The aim of this experiment is to create a material that will serve as a substitute for the heat pipe. This material simplifies the mathematical models of CFD simulation, in which the problem is to define the overall heat transfer of heat pipe, respectively the actions that are going in the heat pipe. For example boiling, condensation, heat transfer.

2 Characteristic of solid material

The most commonly used electrical and thermal conductor is copper. We can find it in the electrical installation, heating cables, heating and cooling systems. In refrigeration engineering, is used in heat exchangers, which transfers heat to the ambient. Heating systems used copper pipes and heat exchangers that allow the exchange and transport of heat. Given its price and availability are replaced by other materials, for example stainless steel and plastic.

In this experiment, we used a full copper pipe, figure 1, with a diameter of 18 mm and length 400 mm.

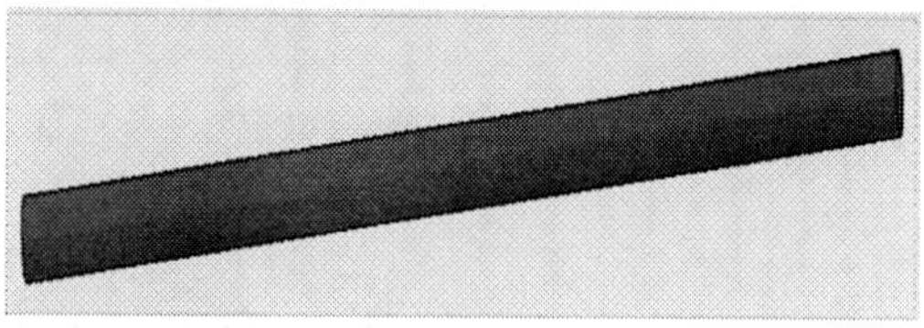

Figure 1. Model of full copper pipe

3 Characteristic of heat pipe

We used a heat pipe to transfer heat from the heat source to the point of consumption of heat with high efficiency and low temperature difference. It is a sealed device in which the net substance (water) exist in the liquid and the gas phase at a pressure below atmospheric pressure (vacuum). Cleanliness of the working fluid is an important parameter for the efficiency of the heat pipe, in terms of heat transfer [3, 4 and 5]. The actual heat pipe, figure 2, consists of evaporation, condensation, and the adiabatic section. In the evaporator the working fluid is evaporated at a temperature below normal conditions. In the adiabatic section is not exchanged heat transfer through the ambient area, due to the isolation of a few working fluid to the wall of the heat pipe condensation film of condensed working fluid. In the condenser the vapor of working fluid condenses on the walls of the heat pipe,

[a] Corresponding author: peter.hrabovsky@fstroj.uniza.sk

which transmit latent (condensing) heat through the walls of the heat pipe to the heat consumption point [6].

Filling of the heat pipe in this experiment is the distilled water. The production process of the heat pipe consists of the preparation of the material, the soldering of components, filling the working fluid and its evaporation to required volume. The final volume of the working fluid is about 20 % of the total volume of the heat pipe, which is about 16 ml of water. This volume has been established on the basis of previous experiments that compared the effectiveness of the heat pipe at various percentages volumes of the working fluid. The value 20 % of the working fluid in the heat pipe reached the best results. The diameter of the heat pipe is 18 mm and length 400 mm.

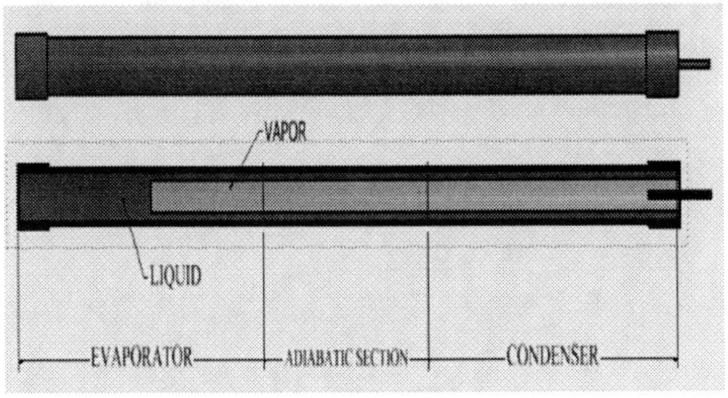

Figure 2. Model of heat pipe

4 Characteristic of test device

Device of experiment consists of two parts (cases), on which the fixed the full copper pipe or the heat pipe. In figure 3 is shown the experimental device.

A case situated in the upper part, (figure 4b, reference 3) is the part of a cooling device. It is connect to the cooling device, the cooling medium is water. The water circulates in a closed circuit between the cooling device and the case. In the cooling circuit included the safe cases, in which there are devices for measuring the temperature at entry and exit of water. In the circuit is included flow meter too that shows speed of flow of the refrigerant in a closed circuit. A case situated at the bottom part (Figure 4b ref. 2) is the heating of the device. It is connected to the heating device, the working medium is water. In a closed heater circuit included thermocouples for temperature measurement inlet and outlet water from the case, flow meter for flow measurement.. The circuit contents a valves that get disconnect the heater case when it is need for cooling with cold water, while avoiding the cooling of heating up the water in the heater circuit. Measured samples are mounted in the cases by caulked joints with seals, which are not being relaxed, possible leakage of cooling or heating of the working fluid.

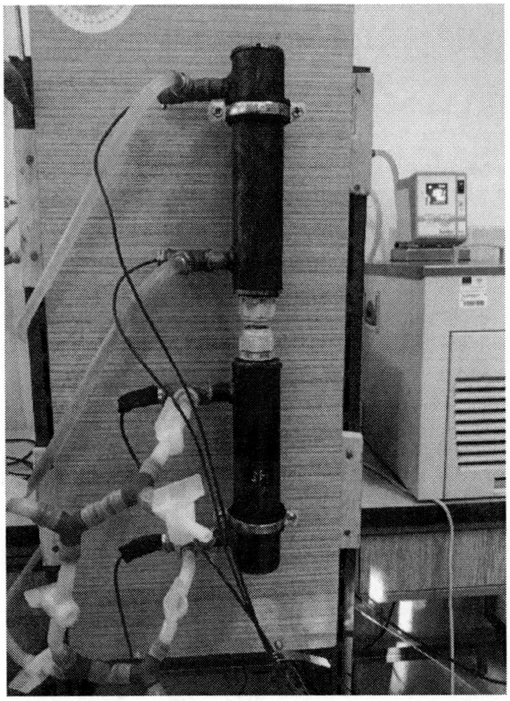

Figure 3 Real view on the experimental device

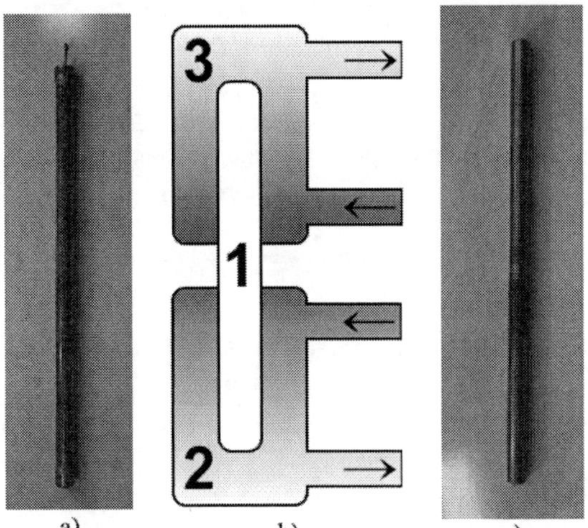

a) b) c)

Figure 4. a) heat pipe, b) scheme of experimental device, c) copper bar

5 Experiment

Experiment came through in normal conditions at 20 °C. As the first sample was measured by a full copper pipe. The cooling circuit was set to 10 °C at a flow rate of 1.72 dm^3 min^{-1}. The heating circuit was set to 80 °C at a flow rate of 4.79 $dm^3.min^{-1}$.

In Figure 65 we can see total progress of heat transfer at the full copper pipe. Equilibrium of heat transfer achieved in 500 s. The subsequent recording of data was found average performance 61.9 W.

After that came through the measurement of the heat pipe under the same conditions as the full copper pipe. The cooling circuit was set to 10 °C at a flow rate of 1.72 dm^3 min^{-1}. The heating circuit was set to 80 °C at a flow rate of 4.79 dm^3 min^{-1}.

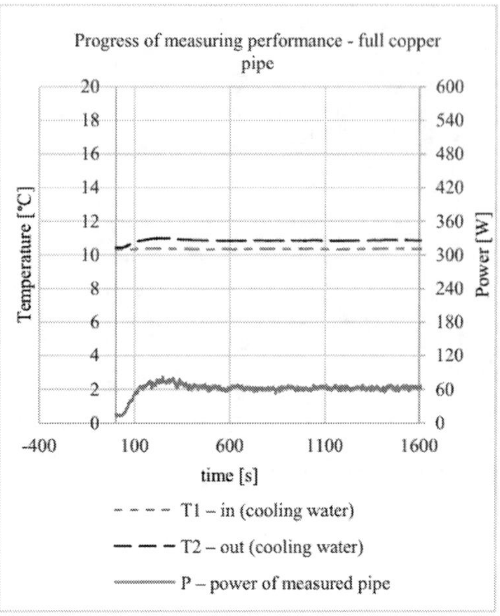

Figure 5. Graph result of full copper pipe

In Figure 6 we can see total progress of heat transfer at the heat pipe. Equilibrium of heat transfer achieved in 500 s. The subsequent recording of data was found average performance 516.1 W.

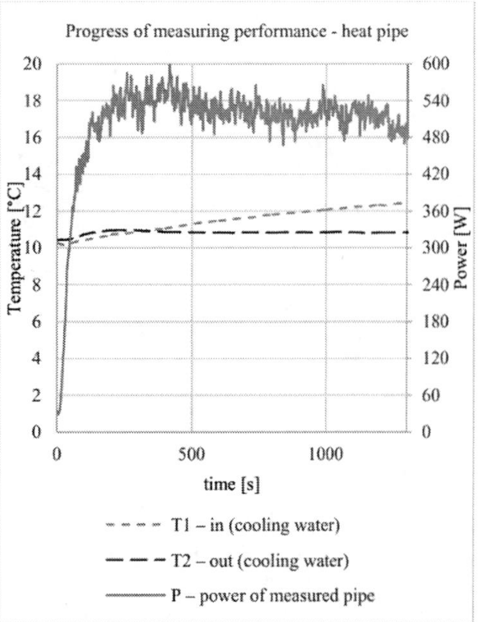

Figure 6. Graph result of heat pipe

6 CFD simulation

CFD simulation uses pc hardver to gain the results from thermodynamics systems. The CFD results we can use to comparing with real model of system. The experiment was carried out in a mathematical simulation of CFD ANSYS. The resulting mesh in Figure 7 is a simplified model of polyhedral and has parameters 239 302 cells, 1 552 487 faces, 1 306 224 nodes.

Individual volumes created appear cold water, hot water, copper rod, heat pipes filled with water-vapor – its thermal conductivity value was choose from set of numbers (500, 1000, 2500, 5000, 10000, 15000, 20000, 25000, 30000, 40000, 45000, 50000), the numbers of set are value of thermal conduction and unit is W m^{-1} K^{-1}.

Boundary conditions are the same as physical experiments (entered from physical experiment) – energy on, laminar flow – revealed by the Re f(v,d), time-independent model. In figure 7 left, the bottom part is an evaporator and the upper part is a condensation.

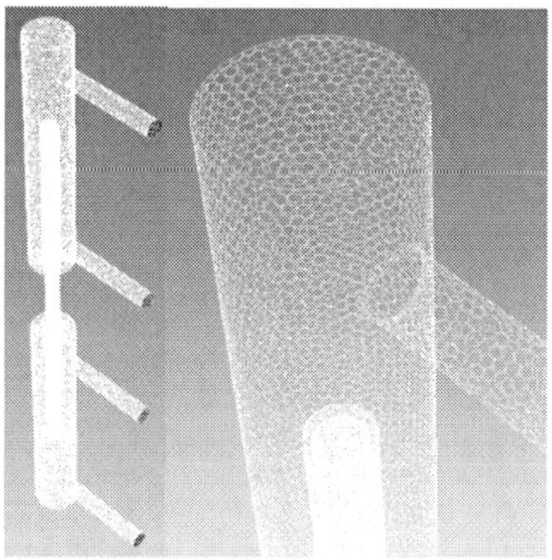

Figure 7. Mesh of CFD model

Input parameters of cooling water - temperature 10 °C, the mass flow rate 0.0287 kg s^{-1}. The outlet of cooling water was determined with parameter outflow. Input parameters of heater - temperature 80 °C, the mass flow rate 0.0781 kg s^{-1}. The outlet of the heating water was determined with parameter outflow.

Table 1 Result of simulation of heat pipe

Thermal coductivity	In cold water	Out cold water	Heat capacity	Power
W m^{-1} K^{-1}	K	K	J kg^{-1} K^{-1}	W
500	283.15	283.42	4197.62	32.92
2500	283.15	284.30	4197.15	138.52
5000	283.15	284.96	4196.80	217.71
10000	283.15	285.75	4196.39	313.40
15000	283.15	286.22	4196.14	369.55
20000	283.15	286.53	4195.98	406.78
25000	283.15	286.74	4195.87	433.04
30000	283.15	286.90	4195.79	451.71
40000	283.15	287.07	4195.70	472.79
45000	283.15	287.10	4195.68	476.53
50000	283.15	287.14	4195.67	480.26

The results of simulation of heat pipe are shown in the picture Figure 8. We can see, that the result of simulation are similarly as the result of physical model. When we use the water-vapor as the working fluid of heat pipe, the result of simulation are approaching to the result of physical model. We had changed one property of water-vapor, which is thermal conductivity.

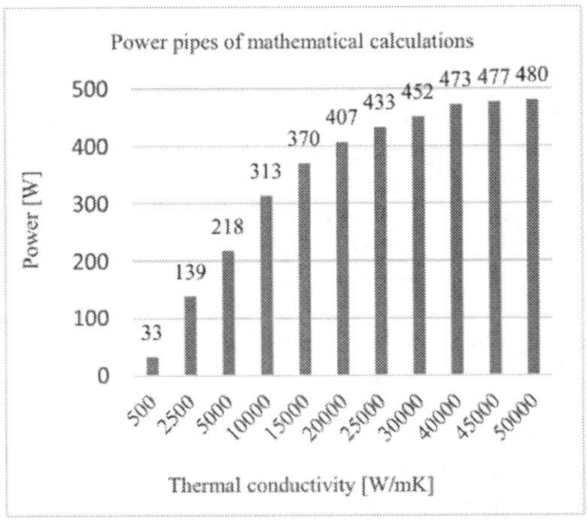

Figure 8. Depend the power of thermal conductivity

In Figure 9 we can see progress of water-vapor, which has changeable property of thermal conductivity. Efficiency of heat transfer is directly depending on the value of thermal conductivity.

In Table 1 are value of thermal conductivity, temperature of cooling water (in, out), heat capacity at temperature outer cooling water and power that the new material transferred in form of heat.

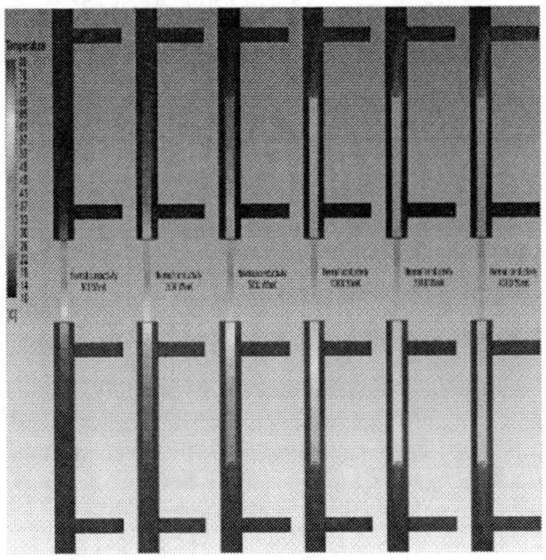

Figure 9. Heat transfer at various values of thermal conductivity

7 Conclusion

This model was computed as steady model. This model shown new material that is similarly as water-vapor. The changing is on one property now, which is thermal conductivity. Actually is in progress transient model that is depend of time. New model can show new results of progress from field of heat transfer.

Acknowledgment

This article was elaborated within the solution of project KEGA č.029ŽU-4/2015 „Recovering heat from technological processes“.

References

1. M. Kasanicky, S. Gavlas, M. Vantuch, M. Malcho, *Possibilities of using Carbon Dioxide as Fillers for Heat Pipe to Obtain Low-potential Geothermal Energy*. Experimental fluid mechanics, 2012, p. 332-336, ISBN 978-80-7372-417-7
2. M. Patsch, J. Labaj, *Influence of Glycerin Combustion in Experimental Burner on Production of Emissions*. Acta metallurgica slovaca conference. ISSN 1338-1660
3. P. Nemec, A. Caja, *Heat Transfer Intensification Applying Thermal Tubes*, Acta Metallurgica Slovaca, 2009, vol. 15, No. 1, pp. 215-220.
4. P. Nemec, A. Caja, *Heat Transport Visualization of Heat Pipe by Thermovision*, Experimental fluid mechanics, 2010, ISBN 978-80-7372-670-6, pp. 431-434.
5. P. Nemec, *Influence Heat Transfer Limitations of Heat Pipes on their Cooling Power,* Transcom, 2009, ISBN 978-80-554-0031-0, pp. 103-108.
6. P. Nemec, M. Malcho, M. Smitka, J. Matusov, *Perofrmance Parameters of Clodes Loop Thermosyphon.* Communications - Scientific Letters of the University of Zilina, vol. 14, No. 4a, 2012, pp. 53-57, ISSN 1335-4205.

EPJ Web of Conferences **114**, 02044 (2016)
DOI: 10.1051/epjconf/201611402044

CFD analysis of the two-phase bubbly flow characteristics in helically coiled rectangular and circular tube heat exchangers

Alamin Hussain[1], Andrew M. Fsadni[2,a]

[1,2]*School of Engineering, Kirkham Room 124, University of Central Lancashire, Preston, PR1 2HE, UK*

Abstract. Due to their ease of manufacture, high heat transfer efficiency and compact design, helically coiled heat exchangers are increasingly being adopted in a number of industries. The higher heat transfer efficiency over straight pipes is due to the secondary flow that develops as a result of the centrifugal force. In spite of the widespread use of helically coiled heat exchangers, and the presence of bubbly two-phase flow in a number of systems, very few studies have investigated the resultant flow characteristics. This paper will therefore present the results of CFD simulations for the two-phase bubbly flow in helically coiled heat exchangers as a function of the volumetric void fraction and the tube cross-section design. The CFD results are compared to the scarce flow visualisation experimental results available in the open literature.

1 Introduction

Helically coiled tube heat exchangers have seen extensive use in industrial applications such as in refrigeration, condensers and evaporators in the food industry, heat recovery systems, power generation, process plants and the nuclear industry where helically coiled heat exchangers are used for residual heat removal systems. This is due to the fact that helically coiled heat exchangers have a greater heat transfer rate as well as a more compact design compared to straight tube heat exchangers [1, 2].

Micro bubble formation is a phenomenon caused by water supersaturated with dissolved air, consequently, leading to bubble nucleation on the walls of the heat exchanger. Bubble nucleation occurs at low temperatures such as when the air can absorb the largest magnitude of gas [3, 4]. For example, in a domestic boiler system, bubble nucleation will occur when the boiler is switched on in the morning. Micro bubble formation affects many of the aforementioned industries. Computational investigations on simulating the effect of micro bubble formation in heat exchangers can lead to design improvements of deaerators which will, as a result, improve the efficiency of the heat exchangers [5]. Bubbles accumulating within the heat exchanger can result in cold spots, reducing the area in which heat transfer can occur thus reducing the efficiency of the heat exchanger and the overall heat transfer coefficient. Furthermore, a comprehensive knowledge of the behaviour of the bubbles such as the second phase distribution within the heat exchangers, is essential for an accurate representation of the heat exchanger's performance.

A bubbly flow in a heat exchanger is the result of the micro bubbles detaching from the walls of the heat exchanger and into the flow system. Two-phase bubbly flow is characterised by the presence of the bubbles with a maximum size much smaller than that of the containing vessel or duct. Furthermore the bubbles are dispersed in a continuous water phase [6].

Previous studies in this field, are mostly theoretical. However there have been some experimental investigations carried out. The lack of experimental data can be due to difficulties in obtaining reliable data on bubbly flow and bubble nucleation, as well as the inherent complexity associated with two-phase flows making the data difficult to analyse [5, 7].

In this paper we will investigate the phase distribution of bubbly flow in a helically coiled tube heat exchanger using computational fluid dynamics (CFD) at system conditions typical of domestic wet central heating systems. Due to the difficulties related to experimental methods and the cost and investment needed, the development and validation of a CFD simulation using existing experimental data, may prove beneficial for the development of helically coiled tube heat exchangers.

2 CFD Modelling

[a] Corresponding author: afsadni@uclan.ac.uk

The simulation was modelled using the ANSYS version 15 CFD software package. The mesh was developed using the ANSYS design modeller's meshing tools and the fluid flow was modelled using the FLUENT solver. The 3D model of the two helical pipes, one with a circular cross-section and one with a rectangular cross-section, were imported into ANSYS as a STEP file which was initially modelled in Pro/E wildfire. Three meshes of each 3D-model were produced with element counts of circa 2 e+6, 3.5 e+6 and 5 e+6 elements. The initial simulations of the rectangular pipes took between 1000 and 1300 iterations to converge, and the initial simulations of the circular pipes took between 600 and 1000 iterations to converge. The "initial simulations" refer to the first group of multi-phase flow simulations carried out on the pipes, the setup of which is specified below. The Reynolds number of the circular pipe was 2500 and the Reynolds number of the rectangular pipe was 3225, suggesting both flows through the pipes are transient flows. For both cases, the calculations were done under the assumption that the presence of air has a negligible effect on the overall fluid properties. Figures 1 & 2 illustrate the 3D models of the tubes used in the present study.

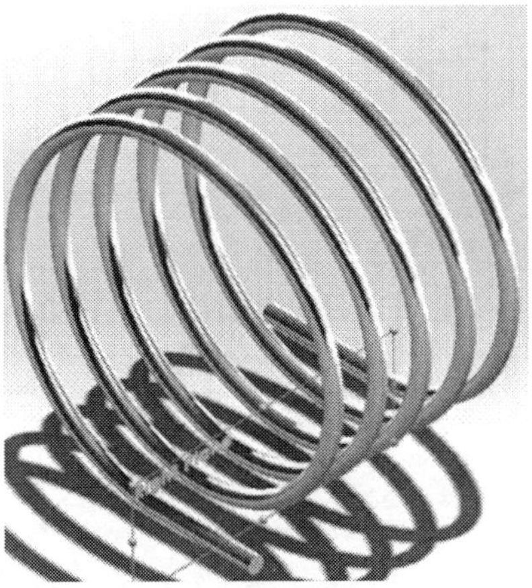

Figure 1. 3D model of helically coiled tube with circular cross-section

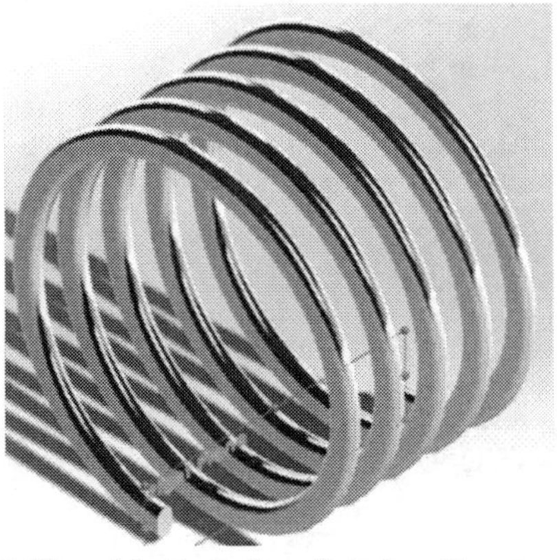

Figure 2. 3D model of helically coiled tube with rectangular cross-section

2.1 Model setup – FLUENT

The two-phase flow was simulated using one of the multiphase methods available in FLUENT, namely the mixture method. It was reported that for bubbly flows the mixture method was the most suitable, hence the use of this method in the study [8].

The solver type was a pressure based solver, which enabled the pressure based Navier Stokes solutions. The flow was assumed to be in a steady state, implying that the solution is independent to time, and a gravitational force was applied in the relevant direction in order to include the buoyancy of the bubbles in the simulation.

The turbulence model used was the realizable *k-ε* turbulence model with standard wall functions used as the near wall treatment. The model also assumes isotropic turbulence. The realizable *k-ε* turbulence model is a significant improvement to the standard *k-ε* turbulence model in almost all cases, thus suggesting that the realizable model will produce more accurate results [9]. The partial differential equation is closed using the realizable *k-ε* turbulence model where one equation is solved for the turbulence kinetic energy *k* as in equation (1) and another equation is solved for the turbulent dissipation *ε* as in equation (2).

$$\frac{\partial}{\partial t}(\rho\, k) + \frac{\partial}{\partial x_i}(p\, k\, u_j) = \frac{\partial}{\partial x_j}\left[\left(\mu + \frac{\mu_t}{\sigma_k}\right)\frac{\partial k}{\partial x_j}\right] + P_k + P_b - \rho\varepsilon - Y_M + S_k, \tag{1}$$

$$\frac{\partial}{\partial t}(\rho\, \varepsilon) + \frac{\partial}{\partial x_j}(p\, \varepsilon\, u_j) = \frac{\partial}{\partial x_j}\left[\left(\mu + \frac{\mu_t}{\sigma_\varepsilon}\right)\frac{\partial \varepsilon}{\partial x_j}\right] + \rho\, C_1\, S_\varepsilon - \rho\, C_2 \frac{\varepsilon^2}{k+\sqrt{v\varepsilon}} + C_{1\varepsilon}\frac{\varepsilon}{k}\, C_{3\varepsilon}\, P_b + S_\varepsilon, \tag{2}$$

where C_1 is calculated using equation (3), η is given by equation (4) and S is the deformation tensor found using equation (5).

$$C_1 = max\left[0.43, \frac{\eta}{\eta+5}\right], \tag{3}$$

$$\eta = S\,\frac{k}{\varepsilon}, \tag{4}$$

$$S = \sqrt{2\, S_{ij}\, S_{ij}}. \tag{5}$$

In these equations P_b is the generation of the turbulence kinetic energy calculated using equation (7), and P_k is the generation turbulence kinetic energy due to the mean velocity gradients calculated using equation (6).

$$P_k = \mu_t\, S^2, \tag{6}$$

$$P_b = \beta\, g_i\, \frac{\mu_t}{P\, r_t}\frac{\partial T}{\partial x_i}, \tag{7}$$

where Pr_t is the Prandtl number of a turbulent flow, which for the realizable model is 0.85, $\mathbf{g}_i$ is the vector of the gravitational component in the Ith direction, and β is the thermal expansion coefficient, which is defined by equation (8).

$$\beta = -\frac{1}{\rho}\left(\frac{\partial \rho}{\partial T}\right)_p, \quad (8)$$

The turbulent viscosity μ_t is found using equation (9).

$$\mu_t = \rho\, C_\mu \frac{k^2}{\varepsilon}, \quad (9)$$

where C_μ is a function of the mean strain and rotation rates, the angular velocity of the system rotation and the turbulence fields modelled using the following equations,

$$C_\mu = \frac{1}{A_0 + A_s \frac{k\, U^*}{\varepsilon}}, \quad (10)$$

$$U^* = \sqrt{S_{ij}\, S_{ij} + \tilde{\Omega}_{ij}\, \tilde{\Omega}_{ij}}, \quad (11)$$

$$\tilde{\Omega}_{ij} = \Omega_{ij} - 2_{\varepsilon_{ijk}\omega_k}, \quad (12)$$

$$\Omega_{ij} = \overline{\Omega_{ij}} - \varepsilon_{ijk}\, \omega_k, \quad (13)$$

where $\overline{\Omega_{ij}}$ is, in a rotating reference frame, the mean rate of rotation tensor viewed, with angular velocity ω_k. The model constants have the following values;

$C_{1\varepsilon} = 1.44$, $C_2 = 1.9$, $\sigma_k = 1.0$, $\sigma_\varepsilon = 1.2$,
$A_0 = 4.04$, $A_s = \sqrt{6}\cos\phi$

In this case A_s is found using the following equations;

$$\phi = \frac{1}{3}\cos^{-1}\left(\sqrt{6}W\right), \quad (14)$$

$$W = \frac{S_{ij}\, S_{jk}\, S_{ki}}{\tilde{S}^3}, \quad (15)$$

$$\tilde{S} = \sqrt{S_{ij}\, S_{ij}}, \quad (16)$$

$$S_{ij} = \frac{1}{2}\left(\frac{\partial u_j}{\partial x_i} + \frac{\partial u_i}{\partial x_j}\right). \quad (17)$$

[5, 10]

The primary phase was water with the secondary phase being air. The air bubble diameter was specified as 0.2 mm and the bubbles were assumed to be spherical in shape. The drag law selected was the Schiller-Neumann as this was the default method and is acceptable for general fluid to fluid multiphase simulations [11].

As both the air and water were specified to have the same initial velocity both were also assumed to have the same change in velocity as the fluids travelled through the pipe. If both fluids are assumed to have the same change in velocity a further assumption that can be made, is that there is no slip between the water and air. Therefore when specifying the slip model of the phase interaction the selection was changed from the standard option (manninen et al.) to none.

The velocity of both the water and the air at the inlet was 0.25 m s^{-1} and the volume void fraction was 1.05 e-4. The pressure of the system was 270 kPa, the turbulent intensity being 5% and hydraulic diameter 12.9 mm. The SIMPLE (Semi-Implicit Method for Pressure Linked Equations) scheme was used for the pressure-velocity coupling as this was the standard coupling method and for complicated flows such as turbulent flows the SIMPLE and SIMPLEC (SIMPLE-Consistent) schemes will converge at similar rates [12].

The first order discretization scheme used for the momentum, volume fraction, turbulent kinetic energy and turbulent dissipation rate, while the PRESTO and least squares cell based schemes were used for the pressure and the gradient respectively. The residual convergence was set at 1 e-5. The solution was initialised using the hybrid initialisation method.

Table 1. Simulation Parameters for both groups of Simulations

Parameters	Initial simulation	VF 0.05
V (both phases – no slip)(ms^{-1})	0.25	0.25
Volumetric void fraction of air	1.05 e-4	0.05
D_b (mm)	0.2	0.2
P (kPa)(abs)	270	270
Multiphase model	Mixture	Mixture
Convergence residual value	1 e-5	1 e-5
Simulation order scheme	First order	Second order

As can be seen from table 1 two sets of simulation parameters were modelled. The first being detailed above and the second seeing two changes. The first change was to increase the volumetric void fraction of the air, the second being changing the simulation method scheme from first order to, second order for the momentum, turbulent kinetic energy and turbulent dissipation rate, and the QUICK scheme was used for the volume fraction. The purpose for changing the scheme is that the second order and QUICK schemes should produce more accurate results from the simulations.

A further simulation was performed with a velocity increase from 0.25 m s^{-1} to 1 m s^{-1} changing the flow from transient to turbulent. As a result the number of iterations for the simulation to converge increased to a point where the simulation was taking too long to complete and thus was prematurely ended.

3 Results and Discussion

The results of the CFD simulations as detailed above are shown in figures 3 - 6. The figures all show the volume fraction of the air within the flow. The labels shown on figure 3 show, at what point of the pipe the cross-sections contours appear at. The inlet and outlet are at the point where the fluids flow into the pipe and out of the pipe respectively. Loop 1 - 4 refers to how many times the fluids have travelled a full loop within the pipe, returning to a point that is parallel to the inlet point. The results of the other simulations will be shown similarly to what is shown in figure 3.

Figure 3 illustrates the volume fraction distribution of the air within the circular helical pipe with mesh elements counts of 5 e+6 elements. The simulation parameters were that of the initial simulation shown in Table 1. The figures display cross sections of the pipe from the inlet to the outlet at every point the pipe loops. Figure 3 shows that the distribution of the volume fraction is rather evenly distributed across the cross section of the pipe, showing no tendency for the air to separate from the water or for the air bubbles to accumulate at any point within the pipe. Furthermore, the void fraction distribution suggests that the centrifugal force affects the air and water to a similar extent.

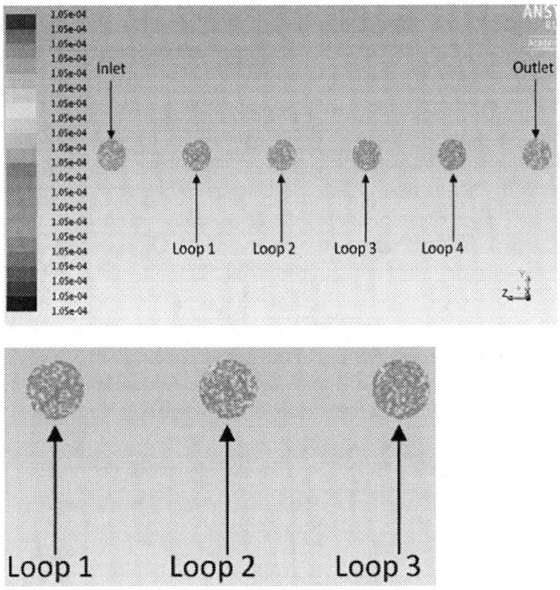

Figure 3. Volume void fraction distribution of air within the circular pipe under initial simulation conditions with enlarged sections for loops 1 – 3.

Figure 4 illustrates the same series of simulations as in figure 3 but the pipe model is different. Where figure 3 uses the circular cross-sectioned pipe, figures 4 uses the rectangular cross-sectioned pipe, and in both cases the element count is 5 e+6 elements.

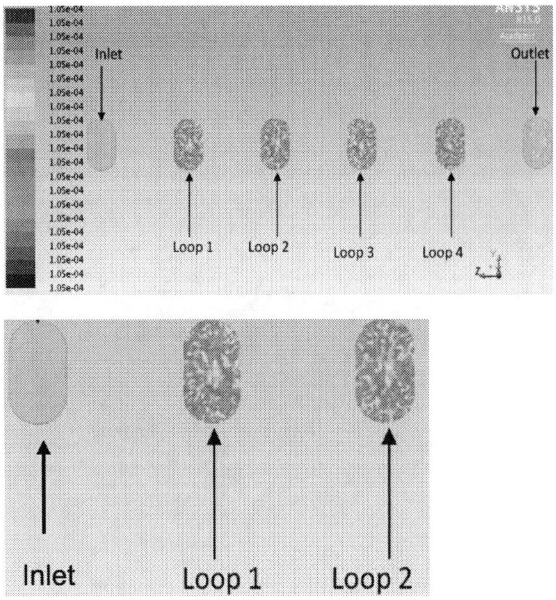

Figure 4. Volume void fraction distribution of air within the rectangular pipe under initial simulation conditions with enlarged sections for the inlet and loops 1 & 2.

The results shown in each series of simulations are very similar, in that the volume fraction distribution is spread throughout the pipe thus suggesting that the air and water phases are not separating.

A difference that is noticeable between the rectangular pipes results and the circular pipes results is that, the inlet and outlets on the rectangular pipes show a far more uniform distribution. This is likely because the plane at which the contours are applied to are much closer to the model's actual inlet and outlet points for the rectangular pipes, whereas the plane for the circular pipes lies a few millimetres into the pipe from the inlet and outlet. This is due to the fact that circular pipe has a short lead in before the pipe starts to coil and the rectangular pipe doesn't. The following figures illustrate the results from the simulations carried out with a volumetric void fraction of 0.05 and the solution methods being second order as well as having an element count of 5 e+6 elements for both figures.

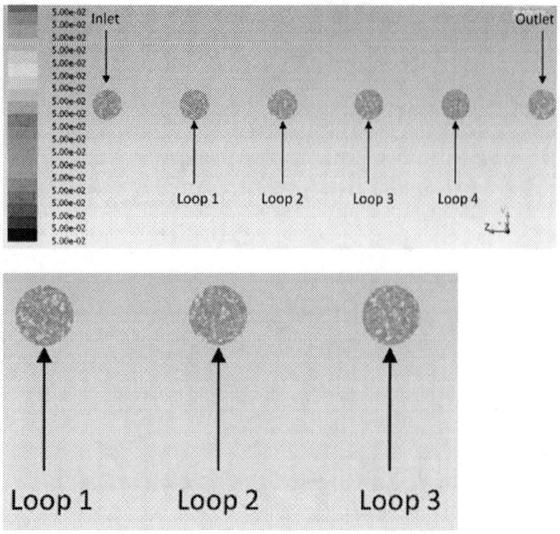

Figure 5. Volume void fraction distribution of air within the circular pipe with a volume fraction of 0.05 with enlarged sections for loops 1 – 3.

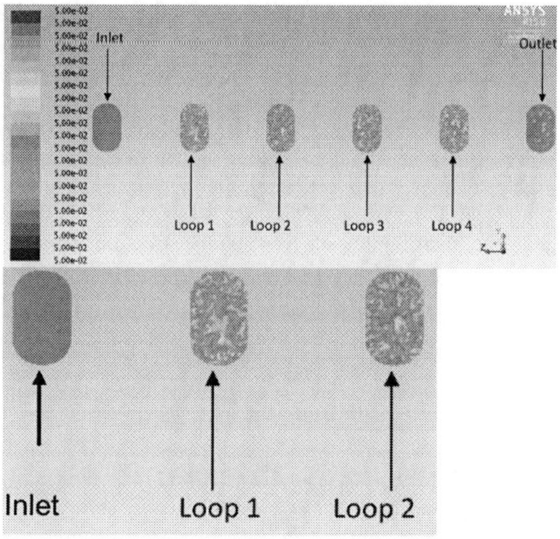

Figure 6. Volume void fraction distribution of air within the rectangular pipe with a volume fraction of 0.05 with enlarged sections for loops 1 – 3.

Figure 5 illustrates the results for the circular cross-section tube, where almost no change in the volume fraction distribution than what is seen in figure 3. This suggests that the increase in the volume fraction and the change of the solution method to a second order type have resulted in no significant change in the volume fraction distribution. Figure 6 illustrates the volume fraction distribution for the rectangular cross-sectioned pipe. This shows a very similar volume fraction distribution as in figure 4, further reinforcing the idea that changing the volume fraction magnitude has little effect on the volume fraction distribution. It was expected that the centrifugal force would cause the bubbles to shift towards the inner part of the pipe/centre of the coil. In figures 2 - 6 there would be a higher volume fraction to the top of the contoured cross-sections. However the volume fraction distribution seen in figures 2 - 6 may be due to the Reynolds number. As a recent experimental two-phase investigation reported, with increasing Reynolds number, the bubble distribution within the fluid becomes more homogenous [13]. As mentioned before the Reynolds number of the flows lied within the transitional state, the Reynolds numbers being calculated without taking the air bubbles into account. The presence of air would likely increase the Reynolds number as the dynamic viscosity of air is smaller than that of water [14].

Another factor that could affect whether the flow is turbulent or not, is torsion. It was found that torsion can destabilize the flow and reduce the Reynolds number at which turbulent flows occur, which could result in a flow that is transitional in a straight pipe, being turbulent in helically coiled pipe [1]. The scale for the contours shown on figures 2 - 6 do not change. This is because the scale is shown to three significant figures. This suggests that there is a less than 1% change in volume fraction, further reinforcing the idea that the volume fraction distribution is homogenous.

The reason figures 5 & 6 show similar volume fraction distributions to figures 3 & 4 is because changing the bulk fluids volumetric void fraction has a minimal effect on the actual volume fraction distribution in the tube [5]. A possible way of improving the CFD simulation and the results found could involve the application of a different two-phase modelling scheme, namely the Eulerian model. When using the mixture model, pseudo properties are used to solve a single set of conservation equations. But when using the Eulerian model each of the phases are solved using a complete set of conservation equations. Furthermore for a well-mixed gas-liquid region, it has been reported that using the mixture model will not yield correct results, due to the centrifugal and Coriolis forces. Therefore the Eulerian model should be used for future modelling [1].

Further improvements to the CFD simulations carried out would be to run simulations with varying Reynolds number from circa 100 to 10,000 to further validate the simulations using the idea that the phase distribution becomes more homogenous with increasing Reynolds number [13]. This would be carried out by changing either the velocity of the flow or the hydraulic diameter of the pipes.

4 Conclusions

This paper has presented a CFD simulation study on the volume fraction distribution of a two-phase bubbly flow for two helically coiled heat exchangers, one with a circular cross-section and the other having a rectangular cross-section under system conditions typical of a domestic wet central heating system. The present study has suggested that the volume fraction distribution found through the CFD simulation does resemble that found in experimental results assuming that there are factors that have affected the Reynolds number that were not taken into account when initially calculating the Reynolds number. This is further shown by the fact that the volume fraction distribution does not change when the air volume void fraction value is changed.

The importance of this study lies with the fact that having a thorough understanding of the volume fraction distribution for bubbly flows within heat exchangers should lead to enhanced deaeration system efficiencies and hence less energy would be needed for the heat exchangers to perform as desired. Furthermore having a validated CFD simulation for bubbly flow within heat exchangers will help develop heat exchangers as experimental methods are difficult, costly and time consuming. Finally, due to the widespread use of helically coiled heat exchangers having a validated CFD simulation model of such heat exchangers will greatly aid the numerous industries that utilise and depend on helically coiled heat exchangers. Further studies that could be performed to build upon what has been found in this paper include the application of a different multi-phase simulation model such as the Eulerian model to investigate how the Eulerian model and the mixture model differ as well as changing the velocity to alter the system Reynolds number.

Notation List

- C_μ Function of the mean strain and rotation rates, the angular velocity of the system rotation and the turbulence fields (-)
- D Pipe Diameter (m)
- D_b Bubble Diameter (m)
- G_b Generation of turbulent kinetic energy due to the buoyancy of the second phase (J / m^3 s)
- G_k Generation of the turbulent kinetic energy due to the mean velocity tensor (J / m^3 s)
- k Turbulent kinetic energy (m^2 / s^2)
- P System pressure (Pa)
- P_g Partial pressure of dissolved gas (Pa)
- S Deformation tensor (-)
- T Fluid temperature (°C)
- V Fluid Velocity (m s^{-1})
- VF Volumetric void fraction (-)
- g_i Vector of the gravitational component (-)
- Pr_t Prandtl number (-)
- P_b The generation of the turbulence kinetic energy (J / m^3 s)

Greek letters

- α Saturation ratio, (-)
- β Thermal expansion coefficient (K^{-1})
- Υ Surface tension, (N / m)
- ν Kinematic viscosity, (m^2 / s)
- ρ Density of fluid (kg / m^3)
- ε Turbulent dissipation (m^2 / s^3)
- σ_k Turbulent Prandtl numbers for k (-)
- μ_t The turbulent viscosity (kg / m s)
- σ_ε Turbulent Prandtl numbers for ε (-)
- $\overline{\Omega_{ij}}$ Rotating reference frame
- ω_k Angular velocity (rad s^{-1})

Acknowledgments

The authors would like to thank the University of Central Lancashire, for supporting this research work.

References

1. J.S. Jayakumar, Helically Coiled Heat Exchangers, Heat Exchangers - Basics Design Applications, Dr. Jovan Mitrovic (Ed.), ISBN: 978-953-51-0278-6, InTech, (2012), Available from: http://www.intechopen.com/books/heat-exchangers-basics-design-applications/helically-coiled-heatexchangers. Last accessed (12/08/2015)
2. R. Manna, J.S. Jayakumar, R.B. Grover, Thermal Hydraulic design of a condenser for a natural circulation system, *J. energy, heat and mass transfer*, **18**, 39-46 (1996)
3. Gerrard W, (1976). *Solubility of gases and liquids.* New York: Plenum Press.
4. C.L. Young, R. Battino, H.L. Clever, The solubility of gases and liquids – introductory information, *Nitrogen and Air, solubility data series* **10**, Oxford: Pergamon Press (1982).
5. A.M. Fsadni, Y.T. Ge, Experimental and CFD analysis of two phase distribution in horizontal pipes for enhanced system deaeration. In: International Conference on *Applied Energy*, 5-8th July 2012, Suzhou, China (2012)
6. R.H.S. Winterton, J.S. Munaweera, Bubble size in two-phase gas-liquid bubbly flow in ducts, *Chemical engineering and processing* **40**, p437-447 (2012).
7. N.J., Hepworth, J.W.R. Boyd, J.R.M. Hammond, J. Varley, Modelling the effect of liquid notion on bubble nucleation during beer dispense. *Chemical engineering sciences.* **58**, 4071-4084 (2006)
8. N. Basu, A. Troshko, G. Nurnberg, Modelling of two-phase flow and boiling with FLUENT (2006) Available from: http://www.scribd.com/doc/3754733/Basu . Last accessed 05/08/2015.
9. T.H. Shih, W.W. Liou, A. Shabbir, Z. Yang, J. Zhu, A new k-ε eddy viscosity model for high Reynolds number turbulent flows—model development and validation, *Computers and Fluids.* **24** (3), 227-238 (1995)
10. CFD-online, Realisable k-epsilon model. Available from: http://www.cfd-online.com/Wiki/Realisable_*k*-epsilon_model . Last accessed 25/08/2015 (2015)
11. ANSYS (2009). Setting Up the Eulerian Model, *ANSYS FLUENT 12.0 User's Guide.* United States, (2009)
12. Sharcnet, *Choosing the Pressure-Velocity Coupling Method.* Available from: https://www.sharcnet.ca/Software/Fluent6/html/ug/node1021.htm. Last accessed 25/08/2015 (2006)
13. H. Saffari, R. Moosavi, E. Gholami, N.M. Nouri, The effect of bubble on pressure drop reduction in

helical coil. *Experimental Thermal and Fluid Science.* **1** (51), p251-256 (2013)

14. Engineeringtoolbox, Dynamic, Absolute and Kinematic Viscosity. Available from: http://www.engineeringtoolbox.com/dynamic-absolute-kinematic-viscosity-d_412.html. Last accessed 04/09/2015 (2015)

EPJ Web of Conferences **114**, 02045 (2016)
DOI: 10.1051/epjconf/201611402045

Evaluation of Travelling Vortex Speed by Means of Dynamic Mode Decomposition and Residual Vorticity

Tomáš Hyhlík[1,a] and Hana Netřebská[1]

CTU in Prague, FME, Department of Fluid Dynamics and Thermodynamics, Technická 4, 166 07 Prague

Abstract. The article deals with the analysis of synthetic jet flow field by means of dynamic mode decomposition (DMD) method. The speed of travelling vortex ring which is connected with synthetic jet creation is evaluated using the wavelength identified from DMD modes. Vortices are identified in the DMD modes by using residual vorticity which allows to identify regions in the flow field where fluid particles perform rotational motion. The analysis is based on the data from the numerical simulation of synthetic jet into quiescent air by using ANSYS Fluent code. The regime of synthetic jet with Re = 329 and Stk = 19.7 is chosen. An increase in the vortex speed close to the orifice and then the decrease is observed with maximum reaching almost one and half of averaged blowing orifice centerline velocity.

1 Introduction

Current work is dealing with the analysis of unsteady flow field with travelling vortex ring and from the point of view of modal decomposition is continuation of e.g. the works [1,2]. From the point of view of analysis of synthetic jet flow it is indirect visualization method unlike the works [3,4], where is the subject of synthetic jet flow studied by using direct visualization.

The visualization of vortex structures in the DMD modes is problematic. There is a standard approach based on vector lines, but identification of vortex core is inaccurate. The aim is to refine identification of vortices in the DMD modes by using the concept of residual vorticity.

2 Parameters of Synthetic Jet

Velocity scale is typically defined as time averaged blowing orifice centerline velocity over an entire cycle

$$U_0 = \frac{1}{T}\int_0^{T_E} u_0(r=0,t)\,\mathrm{d}t, \qquad (1)$$

where T_E is blowing time and $T = 1/f$ is oscillation period. Reynolds number can be defined using time averaged blowing orifice centerline velocity U_0 as

$$\mathrm{Re} = \frac{2U_0 D}{\nu} \qquad (2)$$

where D is orifice diameter. Stokes number is defined as

$$\mathrm{Stk} = D\sqrt{\frac{2\pi f}{\nu}}, \qquad (3)$$

where f is frequency and ν is kinematic viscosity. The regime of synthetic jet with Re = 329 and Stk = 19.7 is chosen.

[a] e-mail: tomas.hyhlik@fs.cvut.cz

3 Numerical Simulation

Unsteady incompressible laminar flow simulation has been performed using commercial solver ANSYS Fluent. The flow is assumed to be axisymmetric and is simulated on rectilinear grid with 81550 cells using non iterative time advancement method with second order implicit scheme. Fractional step scheme is used for pressure velocity coupling. Convective terms are discretized using third order MUSCL scheme. Both the flow in the orifice and actuator cavity are included in the simulation to get more accurate results. The effect of oscillating diaphragm is replaced by velocity boundary condition which should guarantee achievement of required Reynolds number

$$u(t) = \frac{\pi}{2}\frac{\mathrm{Re}\eta}{\rho D}\frac{D^2}{D_L^2}\sin(2\pi f t), \qquad (4)$$

where η is dynamic viscosity, ρ is air density and D_L is inlet diameter i.e. diameter of oscillating diaphragm. Details of the numerical simulation can be found in reference [5].

4 Dynamic Mode Decomposition

Dynamic mode decomposition is a kind of modal decomposition method which was developed based on Koopman analysis of dynamical system [6]. DMD algorithm approximates Koopman modes from finite set of data. This section is mainly based on the reference [7]. Let the velocity field be represented by ensemble of snapshots sampling at interval Δt with a form of matrix $\mathbf{V}_1^N = \{\mathbf{v}_1, \mathbf{v}_2, \ldots, \mathbf{v}_N\}$, where the column vector $\mathbf{v}_j$ contains velocity field at jth time step. Let's consider linear mapping $\mathbf{A}$ that maps each snapshot to the next one in the following way

$$\mathbf{v}_{j+1} = \mathbf{A}\mathbf{v}_j. \qquad (5)$$

This mapping is assumed to be same over the full sampling interval $[0, (N-1)\Delta t]$. If the flow fields stem from a nonlinear process, this assumption corresponds to linear tangent approximation [7]. The assumption of a constant mapping between the snapshots $\mathbf{v}_j$ allows us to formulate the sequence of flow fields as a Krylov sequence

$$\mathbf{V}_1^N = \left\{\mathbf{v}_1, \mathbf{A}\mathbf{v}_1, \mathbf{A}^2\mathbf{v}_1, \ldots, \mathbf{A}^{N-1}\mathbf{v}_1\right\}. \tag{6}$$

The goal of DMD is the extraction of dynamic characteristics such as eigenvalues and eigenvectors which are described by linear operator $\mathbf{A}$ based on sequence $\mathbf{V}_1^N$. Supposing that it is possible to write last snapshot $\mathbf{v}_N$ as linear combination of previous $N-1$ snapshots, for sufficiently long sequence of the snapshots

$$\mathbf{v}_N = a_1\mathbf{v}_1 + a_2\mathbf{v}_2 + \ldots + a_{N-1}\mathbf{v}_{N-1} + \mathbf{r}, \tag{7}$$

where $\mathbf{a}^T = \{a_1, a_2, \ldots, a_{N-1}\}$ and $\mathbf{r}$ is residual vector. It is possible to write following relations

$$\mathbf{A}\{\mathbf{v}_1, \mathbf{v}_2, \mathbf{v}_3, \ldots, \mathbf{v}_{N-1}\} = \{\mathbf{v}_2, \mathbf{v}_3, \mathbf{v}_4, \ldots, \mathbf{v}_N\}, \tag{8}$$

$$\{\mathbf{v}_2, \mathbf{v}_3, \mathbf{v}_4, \ldots, \mathbf{v}_N\} = \{\mathbf{v}_2, \mathbf{v}_3, \mathbf{v}_4, \ldots, \mathbf{V}_1^{N-1}\mathbf{a}\} + \mathbf{r}\mathbf{e}_{N-1}^T, \tag{9}$$

or in matrix form

$$\mathbf{A}\mathbf{V}_1^{N-1} = \mathbf{V}_2^N = \mathbf{V}_1^{N-1}\mathbf{Z} + \mathbf{r}\mathbf{e}_{N-1}^T, \tag{10}$$

where $\mathbf{e}_{N-1}^T$ is unit vector of dimension $N-1$. Matrix $\mathbf{Z}$ is companion matrix

$$\mathbf{Z} = \begin{pmatrix} 0 & & & & a_1 \\ 1 & 0 & & & a_2 \\ & \ddots & \ddots & & \vdots \\ & & 1 & 0 & a_{N-2} \\ & & & 1 & a_{N-1} \end{pmatrix}. \tag{11}$$

The eigenvalues of $\mathbf{Z}$ are approximations of eigenvalues of linear operator $\mathbf{A}$. Decomposition based on companion matrix $\mathbf{Z}$ is mathematically correct, but more robust algorithm based on similar matrix $\tilde{\mathbf{Z}}$ is used in this work. Robustness is achieved by preprocessing step using singular value decomposition of data sequence [7] $\mathbf{V}_1^{N-1} = \mathbf{U}\mathbf{\Sigma}\mathbf{W}^H$. Substituting singular value decomposition into (10) and rearranging the resulting expression we have

$$\mathbf{U}^H\mathbf{A}\mathbf{U} = \mathbf{U}^H\mathbf{V}_2^N\mathbf{W}\mathbf{\Sigma}^{-1} = \tilde{\mathbf{Z}}. \tag{12}$$

Dynamic modes can be calculated using equation [7]

$$\mathbf{\Phi}_i = \mathbf{U}\mathbf{y}_i, \tag{13}$$

where $\mathbf{y}_i$ is ith eigenvector of $\tilde{\mathbf{Z}}$

$$\tilde{\mathbf{Z}}\mathbf{y}_i = \mu_i\mathbf{y}_i. \tag{14}$$

Eigenvalues μ_i approximate eigenvalues of full mapping matrix $\mathbf{A}$ and provide temporal dynamics of the flow field.

5 Vortex Identification

The work [8] is offering an impressive concept of triple decomposition of motion where the tensor of velocity gradient is splitted into three parts. The first part is representing the elongation of fluid portion, the second part is connected with rigid body rotation and the third part is effective pure shearing tensor. The reference [8] presents both the simple two dimensional decomposition and the general three dimensional approach. The aim of this section is to extend the two dimensional approach to be applicable to axisymmetric flow. The key idea of the mentioned work is to rotate the coordinate system to the so called basic reference frame where it is possible to clearly recognize the kind of fluid motion.

It is possible to express the continuum particle speed u_i when it changes its position from point x_j to point $x_j + \mathrm{d}x_j$ as

$$u_i(x_j + \mathrm{d}x_j) = u_i(x_j) + S_{ij}\mathrm{d}x_j + \Omega_{ij}\mathrm{d}x_j, \tag{15}$$

where S_{ij} is symmetrical part of velocity gradient tensor and Ω_{ij} is antisymmetric part of velocity gradient tensor. Symmetrical part is

$$\mathbf{S} = \begin{pmatrix} \frac{\partial v_r}{\partial r} & 0 & \frac{1}{2}\left(\frac{\partial v_r}{\partial z} + \frac{\partial v_z}{\partial r}\right) \\ 0 & \frac{v_r}{r} & 0 \\ \frac{1}{2}\left(\frac{\partial v_r}{\partial z} + \frac{\partial v_z}{\partial r}\right) & 0 & \frac{\partial v_z}{\partial z} \end{pmatrix} \tag{16}$$

in the case of axisymmetric flow with zero swirl velocity and antisymmetric part can be simplified as

$$\mathbf{\Omega} = \begin{pmatrix} 0 & 0 & \frac{1}{2}\left(\frac{\partial v_r}{\partial z} - \frac{\partial v_z}{\partial r}\right) \\ 0 & 0 & 0 \\ -\frac{1}{2}\left(\frac{\partial v_r}{\partial z} - \frac{\partial v_z}{\partial r}\right) & 0 & 0 \end{pmatrix}. \tag{17}$$

The tensor in Cartesian coordinates can be converted from cylindrical coordinates using

$$\begin{pmatrix} S_{xx} & S_{xy} & S_{xz} \\ S_{yx} & S_{yy} & S_{yz} \\ S_{zx} & S_{zy} & S_{zz} \end{pmatrix} = \mathbf{B}\begin{pmatrix} S_{rr} & S_{r\theta} & S_{rz} \\ S_{\theta r} & S_{\theta\theta} & S_{\theta z} \\ S_{zr} & S_{z\theta} & S_{zz} \end{pmatrix}\mathbf{C}, \tag{18}$$

where transformation matrices are

$$\mathbf{B} = \begin{pmatrix} \cos\theta & -\sin\theta & 0 \\ \sin\theta & \cos\theta & 0 \\ 0 & 0 & 1 \end{pmatrix}, \quad \mathbf{C} = \begin{pmatrix} \cos\theta & \sin\theta & 0 \\ -\sin\theta & \cos\theta & 0 \\ 0 & 0 & 1 \end{pmatrix}. \tag{19}$$

The angle θ is arbitrary in the case of axisymmetric flow. Let angle be $\theta = \pi/2$. We get

$$\mathbf{S} = \begin{pmatrix} S_{xx} & S_{xy} & S_{xz} \\ S_{yx} & S_{yy} & S_{yz} \\ S_{zx} & S_{zy} & S_{zz} \end{pmatrix} = \begin{pmatrix} S_{\theta\theta} & 0 & 0 \\ 0 & S_{rr} & S_{rz} \\ 0 & S_{rz} & S_{zz} \end{pmatrix} \tag{20}$$

We have to find principal values and principal axes of the matrix of kind

$$\begin{pmatrix} S_{xx} & S_{xy} & 0 \\ S_{yx} & S_{yy} & 0 \\ 0 & 0 & S_{zz} \end{pmatrix}, \tag{21}$$

The value of $S_{zz} \neq 0$ is in general nonzero in the case of axisymmetric flow. Eigenvalues of matrix (21) are

$$\lambda_1 = e_1 + s, \quad \lambda_2 = e_2 - s, \quad \lambda_3 = e_3, \tag{22}$$

where

$$s = \frac{1}{2}\sqrt{\left(S_{xx} - S_{yy}\right)^2 + 4S_{xy}^2} \tag{23}$$

and

$$e_{1,2} = e = \frac{1}{2}\left(S_{xx} + S_{yy}\right), \quad e_3 = S_{zz}. \tag{24}$$

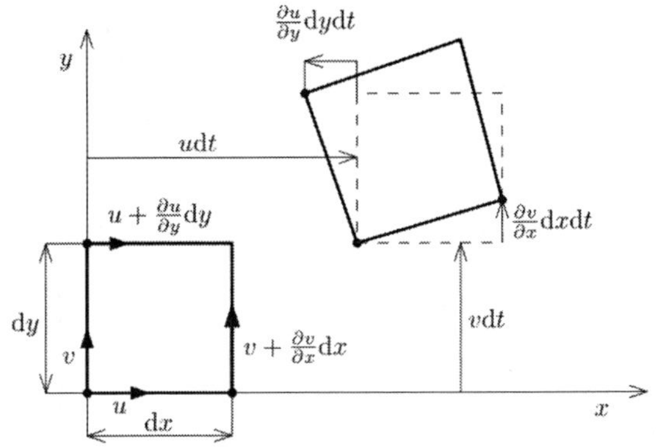

Fig. 1. Predominant rotation; the case where $\frac{\partial u}{\partial y} < 0$ and $\frac{\partial v}{\partial x} > 0$.

The velocity gradient tensor can be in principal axes expressed as

$$\nabla \mathbf{v} = \begin{pmatrix} e & 0 & 0 \\ 0 & e & 0 \\ 0 & 0 & e_3 \end{pmatrix} + \begin{pmatrix} s & 0 & 0 \\ 0 & -s & 0 \\ 0 & 0 & 0 \end{pmatrix} + \begin{pmatrix} 0 & -\omega & 0 \\ \omega & 0 & 0 \\ 0 & 0 & 0 \end{pmatrix}, \tag{25}$$

where

$$\omega = \frac{1}{2}\left(\frac{\partial v_y}{\partial x} - \frac{\partial v_x}{\partial y}\right). \tag{26}$$

An arbitrary rotation around z axis does not change the first and the third matrix in the relation (25). The first tensor in the relation (25) has no influence on shear deformation. It is possible to take into consideration only the second and the third part and rotate around z axis about angle $\alpha = \frac{3}{4}\pi$

$$\nabla \mathbf{v} \rightarrow \mathbf{D}\begin{pmatrix} s & -\omega & 0 \\ \omega & -s & 0 \\ 0 & 0 & 0 \end{pmatrix}\mathbf{F} = \begin{pmatrix} 0 & s-\omega & 0 \\ s+\omega & 0 & 0 \\ 0 & 0 & 0 \end{pmatrix}, \tag{27}$$

where transformation matrix $\mathbf{D}$ is

$$\mathbf{D} = \begin{pmatrix} \cos\alpha & \sin\alpha & 0 \\ -\sin\alpha & \cos\alpha & 0 \\ 0 & 0 & 1 \end{pmatrix} = \begin{pmatrix} -\frac{\sqrt{2}}{2} & \frac{\sqrt{2}}{2} & 0 \\ -\frac{\sqrt{2}}{2} & -\frac{\sqrt{2}}{2} & 0 \\ 0 & 0 & 1 \end{pmatrix} \tag{28}$$

and transformation matrix $\mathbf{F}$ is

$$\mathbf{F} = \begin{pmatrix} \cos\alpha & -\sin\alpha & 0 \\ \sin\alpha & \cos\alpha & 0 \\ 0 & 0 & 1 \end{pmatrix} = \begin{pmatrix} -\frac{\sqrt{2}}{2} & -\frac{\sqrt{2}}{2} & 0 \\ \frac{\sqrt{2}}{2} & -\frac{\sqrt{2}}{2} & 0 \\ 0 & 0 & 1 \end{pmatrix}. \tag{29}$$

Velocity gradient tensor is after transformation in so called basic reference frame [8]. It is possible to examine kinematics of the flow in the basic reference frame clearly.

All of us intuitively know that vortex is somehow connected with vorticity but on the other hand we know that there are cases like laminar shear flow where the value of vorticity is nonzero and there is not any vortex. The work [8] defines vortex as the case where the value of vorticity predominates the shear $|\omega| \geq |s|$. Let's assume only shear and vorticity parts of the velocity gradient tensor, then it has only two nonzero components in the basic reference frame $\frac{\partial u}{\partial y}$ and $\frac{\partial v}{\partial x}$ having opposite sign in the case of predominant vorticity, see figure 1. The residual shear parameter is in this case $s_{RES} = 0$ and shear parameter is $s_{SH} = s$.

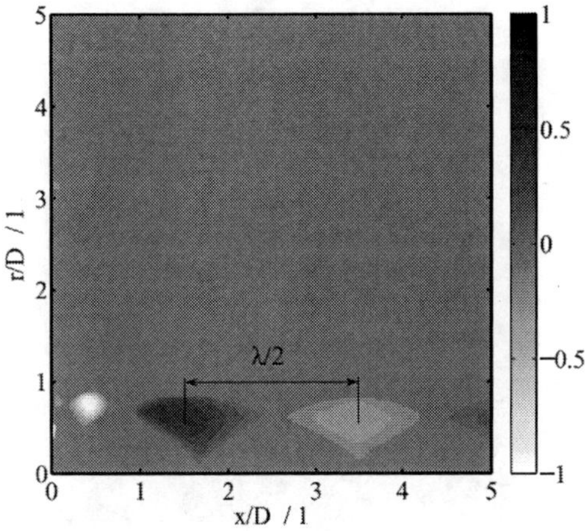

Fig. 2. Residual vorticity contours of imaginary part of the first DMD mode which is connected with frequency $f_1 = f$

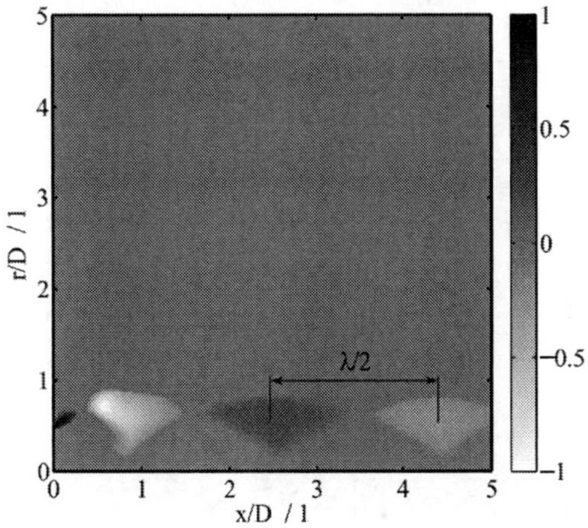

Fig. 3. Residual vorticity contours of real part of the first DMD mode which is connected with frequency $f_1 = f$

Let the residual vorticity be

$$\omega_{RES} = \mathrm{sgn}(\omega)\left[|\omega| - |s|\right]. \tag{30}$$

and shear vorticity be

$$\omega_{SH} = \mathrm{sgn}(\omega)|s|. \tag{31}$$

Examples of residual vorticity distributions in the DMD modes are depicted in the figures 2, 3, 4 and 5.

6 Results

There are basically two types of behaviour connected with spatial DMD modes. The first one is the case where the real part dominates over imaginary part. This kind of behaviour in the dynamic mode can indicate travelling wave like structure in the flow field [9]. The second kind of behaviour where the real and imaginary parts are similar, but there is space shift in the flow field. The space shift is typically one quarter of wave length which can be measured

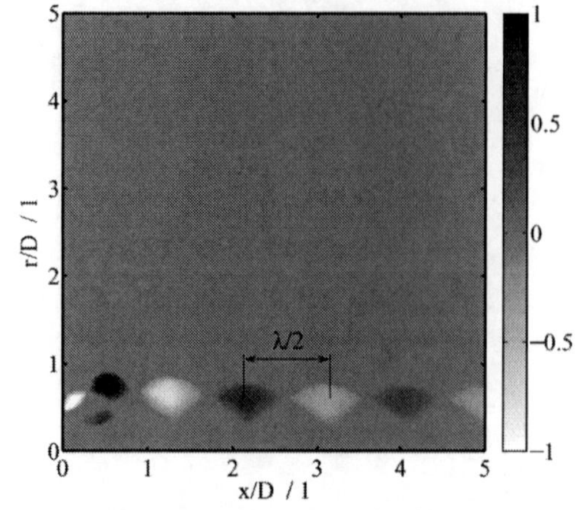

Fig. 4. Residual vorticity contours of imaginary part of the second DMD mode which is connected with frequency $f_2 = 2f$

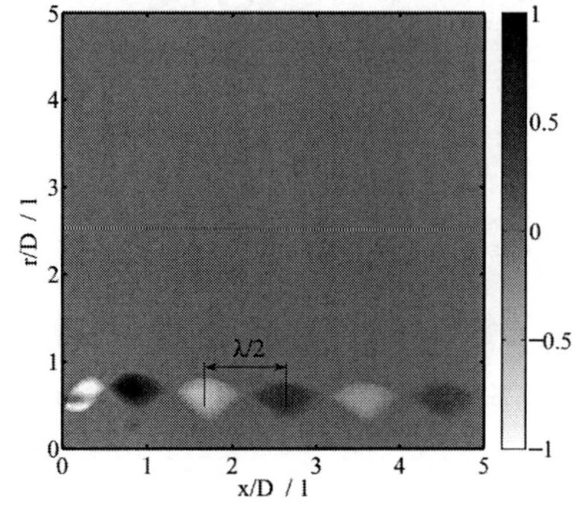

Fig. 5. Residual vorticity contours of real part of the second DMD mode which is connected with frequency $f_2 = 2f$

from the modes. This kind of behaviour is typical for the travelling vortex structures and is observed in the figures 2, 3, 4 and 5. There is a travelling vortex ring in the flow field depicted in the mentioned figures. The vortex ring is travelling from left to right in the x direction, where $x = 0$ corresponds to orifice exit. It is possible to observe opposite vortex rotation in the DMD modes in the figures, where vortices are visualized using residual vorticity. It is possible to directly measure wavelength λ in the case of every single DMD mode. The wavelength is changing with spatial position. The vortex speed is connected with wavelength as

$$u_c = \lambda f_m, \tag{32}$$

where f_m is frequency connected with given mode. There is an increase in the vortex speed close to the orifice and then is the vortex speed decreasing as shown in the figure 6. The maximum vortex speed is reaching almost one and half of averaged blowing orifice centerline velocity over an entire cycle U_0 in this case. The predicted behaviour is consistent with reference [10] but predicted maximal vortex speed is slightly higher.

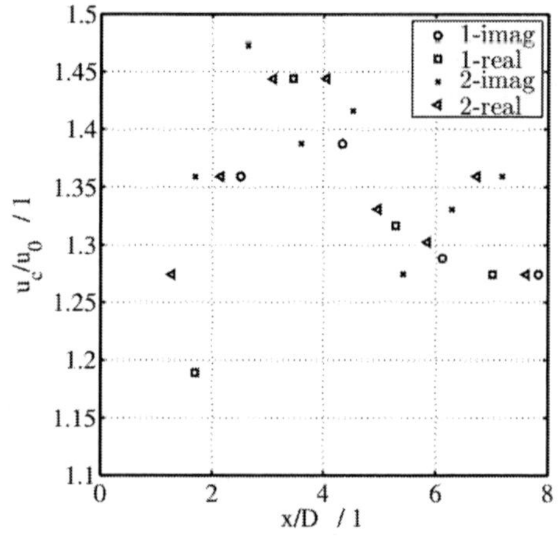

Fig. 6. Vortex speed normalized by averaged blowing orifice centerline velocity

7 Conclusions

The speed of travelling vortex ring is evaluated by using DMD modes, where the vortices are identified using residual vorticity. The vortex structure identification based on residual vorticity is modified to be applicable to axisymmetric flow. The contours of residual vorticity of individual modes are used to measure spatial change of wave length which together with the mode frequency allows to calculate speed of travelling vortex.

Acknowledgement

The support from Czech Science Foundation under Grant No. 14-08888S is gratefully acknowledged.

References

1. T. Hyhlík, P. Železný, J. Čížek, EPJ Web of Conferences, **45**, (2013), 01043
2. T. Hyhlík, H. Netřebská, EPJ Web of Conferences, **92**, (2015), 02026
3. Z. Broučková, Z. Trávníček, P. Šafařík, EPJ Web of Conferences, **25**, (2012), 01007
4. Z. Trávníček, Z. Broučková, J. Kordík, T. Vít, Journal of Visualization, (in press; online available March 2015)
5. T. Hyhlík, H. Netřebská, *33th Conference of Departments of Fluid Mechanics and Thermomechanics*, (Prague 2014)
6. C. W. Rowley, I. Mezić, S. Bagheri, P. Schlatter, D. S. Henningson, J. Fluid Mech., **641**, (2009), 115-127.
7. P. J. Schmid, J. Fluid Mech., **656**, (2010) 5-28.
8. V. Kolář, International Journal of Heat and Fluid Flow, **28**, (2007), 638-652.
9. O. Frederich, D. M. Luchtenburg, *Seventh International Symposium on Turbulence and Shear Flow Phenomena*, (Ottawa 2011)
10. B. L. Smith, A. Glezer, Physics of Fluids, **10(9)**, (1998), 2281-2297

EPJ Web of Conferences **114**, 02046 (2016)
DOI: 10.1051/epjconf/201611402046

Mathematical Model of Two Phase Flow in Natural Draft Wet-Cooling Tower Including Flue Gas Injection

Tomáš Hyhlík[1,a]

[1]CTU in Prague, FME, Department of Fluid Dynamics and Thermodynamics, Technická 4, 166 07 Prague

Abstract. The previously developed model of natural draft wet-cooling tower flow, heat and mass transfer is extended to be able to take into account the flow of supersaturated moist air. The two phase flow model is based on void fraction of gas phase which is included in the governing equations. Homogeneous equilibrium model, where the two phases are well mixed and have the same velocity, is used. The effect of flue gas injection is included into the developed mathematical model by using source terms in governing equations and by using momentum flux coefficient and kinetic energy flux coefficient. Heat and mass transfer in the fill zone is described by the system of ordinary differential equations, where the mass transfer is represented by measured fill Merkel number and heat transfer is calculated using prescribed Lewis factor.

1 Introduction

Various models of natural draft cooling tower flow, heat and mass transfer exist. Classical approach is described in reference [1], where simple algebraic draft equation is used to calculate moist air mass flow rate. An example of more complex approach based on commercial CFD code is the work [2]. As an example of other code can be mentioned e.g. work [3].

Current work can be understood as continuation of previous work of author, see e.g. [4], [5]. This work is leading to relatively complex CFD model of natural draft cooling tower flow, heat and mass transfer.

The main benefit of current model is possibility to simulate supersaturated moist air. In the case where ambient moist air is undersaturated or supersaturated is supersaturated moist air typically present in the cooling tower and the plume outside cooling tower exist. The present model provides data to estimate plume.

Another benefit of present model is the inclusion of flue gas injection. Flue gas injection has contradictory effect on natural draft wet-cooling tower mass flow rate and thereby on water cooling. From one point of view is flue gas injection reducing cross sectional area of cooling tower and thus reduces mass flow rate, but from the other point of view it creates ejector effect which increases mass flow rate. Last but not least, the inclusion of momentum and energy flux coefficients into the governing equation generalizes the model and allows to include multidimensional effect in quasi one-dimensional model.

2 Mathematical model of moist air flow

Lets assume homogeneous equilibrium model of two phase flow, where the two phases are well mixed and have the same velocity. Furthermore the thermal equilibrium of both phases is assumed. Thermodynamic equilibrium implies in the case of two phase flow, that mixture of water vapour and dry air is saturated. Water liquid has the temperature of the mixture and the pressure of mixture. The water is sub-cooled because mixture pressure is higher then water vapour partial pressure. This model is the kind of homogeneous equilibrium model [6] of two phase flow applied to case where the gas phase is the mixture of condensable and noncondensable components.

2.1 Mass fractions

A key step is the definition mass fractions. Overall density of two phase mixture can be expressed using void fraction of gas phase as [6]

$$\rho = \alpha\rho_g + (1-\alpha)\rho_l, \tag{1}$$

where α is volumetric void fraction at specific point x_i at time τ, i.e. the volume of gas phase V_g over the whole $V_g + V_l$

$$\alpha(x_i,\tau) = \frac{\mathrm{d}V_g(x_i,\tau)}{\mathrm{d}V_g(x_i,\tau) + \mathrm{d}V_l(x_i,\tau)}, \tag{2}$$

where ρ_l is the density of water liquid and ρ_g the density of gas phase which is a sum of dry air density ρ_a and the density of water vapour

$$\rho_g = \rho_a + \rho_v. \tag{3}$$

Specific humidity of supersaturated moist air is

$$x_s = \frac{\alpha\rho_v + (1-\alpha)\rho_l}{\alpha\rho_a} = \frac{\rho_v}{\rho_a} + \frac{1-\alpha}{\alpha}\frac{\rho_l}{\rho_a} \tag{4}$$

and the part corresponding to the gas phase is

$$x = \frac{\alpha\rho_v}{\alpha\rho_a} = \frac{\rho_v}{\rho_a}. \tag{5}$$

The case of supersaturated moist air is corresponding with $\alpha < 1$ and specific humidity of gas phase is equal to saturated specific humidity $x = x''$. The mass fraction of dry

[a] e-mail: tomas.hyhlik@fs.cvut.cz

air is in the case of supersaturated moist air equal to

$$w_a = \frac{\alpha\rho_a}{\rho} = \left[1 + \frac{\rho_v}{\rho_a} + \frac{1-\alpha}{\alpha}\frac{\rho_l}{\rho_a}\right]^{-1} = \frac{1}{1+x_s}, \quad (6)$$

water vapour mass fraction is defined as

$$w_v = \frac{\alpha\rho_v}{\rho} = \left[1 + \frac{\rho_a}{\rho_v} + \frac{1-\alpha}{\alpha}\frac{\rho_l}{\rho_v}\right]^{-1} = \frac{x}{1+x_s} \quad (7)$$

and water liquid mass fraction is

$$w_l = \frac{(1-\alpha)\rho_l}{\rho} = \left[1 + \frac{\alpha}{1-\alpha}\frac{\rho_a+\rho_v}{\rho_l}\right]^{-1}. \quad (8)$$

It is possible to express water liquid mass fraction also as

$$w_l = \frac{x_s - x}{1+x_s}. \quad (9)$$

Due to the utilization of void fraction in the definitions of individual mass fractions it is possible to show that

$$w_a + w_v + w_l = 1. \quad (10)$$

The ratio of gas mass fraction to liquid mass fraction can be expressed as

$$\frac{\alpha\rho_g}{(1-\alpha)\rho_l} = \frac{\alpha(\rho_a+\rho_v)}{(1-\alpha)\rho_l} = \frac{w_a+w_v}{1-w_a-w_v} \quad (11)$$

and

$$\frac{\alpha\rho_g}{(1-\alpha)\rho_l} = \frac{1+x}{x_s - x}, \quad (12)$$

where specific humidity is

$$x = \frac{w_v}{w_a}. \quad (13)$$

2.2 Governing equations

Mass balance equations of dry air, water vapour and water liquid are in this section written without diffusion term, which is mostly negligible. The overall mass of dry air in the two phase mixture can be expressed as

$$m_a = \int_V \left[\alpha\rho_a + (1-\alpha)\rho_l w_{a,l}\right] \mathrm{d}V = \int_V \varphi_a \mathrm{d}V, \quad (14)$$

where $w_{a,l}$ is mass fraction of dry air in the water liquid and φ_a can be expressed also as

$$\varphi_a = \alpha\rho_g w_{a,g} + (1-\alpha)\rho_l w_{a,l}, \quad (15)$$

where $w_{a,g}$ is mass fraction of dry air in the mixture of dry air and saturated water vapour

$$w_{a,g} = \frac{w_a}{w_a + w_v}. \quad (16)$$

Dry air mass balance can by written as

$$\frac{\partial\varphi_a}{\partial\tau} + \frac{\partial(\varphi_a v_i)}{\partial x_i} = 0, \quad (17)$$

where v_i is velocity vector. The solubility of dry air in the water liquid can be neglected (i.e. $w_{a,l} = 0$) and dry air mass balance can be simplified to

$$\frac{\partial(\alpha\rho_a)}{\partial\tau} + \frac{\partial(\alpha\rho_a v_i)}{\partial x_i} = 0, \quad (18)$$

or

$$\frac{\partial\left(\alpha\rho_g w_{a,g}\right)}{\partial\tau} + \frac{\partial\left(\alpha\rho_g w_{a,g} v_i\right)}{\partial x_i} = 0. \quad (19)$$

Water vapour mass balance can be expressed as

$$\frac{\partial(\alpha\rho_v)}{\partial\tau} + \frac{\partial(\alpha\rho_v v_i)}{\partial x_i} = \sigma_v(x_i), \quad (20)$$

where σ_v is density of water vapour mass source. Liquid phase mass balance is

$$\frac{\partial[(1-\alpha)\rho_l]}{\partial\tau} + \frac{\partial[(1-\alpha)\rho_l v_i]}{\partial x_i} = \sigma_l(x_i), \quad (21)$$

where σ_l is density of water liquid mass source and

$$\sigma_v(x_i) = -\sigma_l(x_i) \quad (22)$$

in the case of moist air flow without external sources of both water vapour and water liquid. Overall mass balance equation of supersaturated moist air can be obtained by adding of equations (18), (20) and (21)

$$\frac{\partial\rho}{\partial\tau} + \frac{\partial(\rho v_i)}{\partial x_i} = 0. \quad (23)$$

External sources of water vapour or water liquid are not included in the previous equation. Overall momentum balance can be written as

$$\frac{\partial(\rho v_i)}{\partial\tau} + \frac{\partial}{\partial x_k}\left[\rho v_i v_k - \sigma_{ik}\right] = -\rho g\delta_{i3} - \zeta\rho\frac{|v|^2}{2}\delta_{i3}, \quad (24)$$

where σ_{ik} is stress tensor, the first term on the right represents volume force connected with gravity in the direction $-z$ an the second term allows to add minor losses in the z direction. Overall energy equation is

$$\frac{\partial(\rho e)}{\partial\tau} + \frac{\partial}{\partial x_k}\left[\rho v_k e - \sigma_{kj}v_j + \dot{q}_k\right] = -g\rho v_3, \quad (25)$$

where total energy is the sum of internal energy and kinetic energy $e = u + \frac{v_k v_k}{2}$ and $\dot{q}_k$ is diffusion coefficient driven heat flux. The term on the right hand side is density of source term which is connected with gravity force.

We have the system of equations (18), (20), (21), (24) and (25) describing the flow of supersaturated moist air. The evaluation of source term σ_v appears to be one of the complication of the model, but it is possible to replace one of mass balance equations using overall mass balance equation (23) and evaluation of water vapour mass fraction should be based on thermodynamic relations for saturated moist air. Finally it is possible to formulate the system of governing equations where there are only overall continuity equation (23), overall momentum balance (24), dry air continuity equation (18) and overall equation of energy (25). The system of governing equations has to be closed using the definition of internal energy and using the equation of state.

2.3 Thermodynamic parameters

Internal energy u is defined as

$$u = \frac{1}{\rho}\left[\alpha\rho_g u_g + (1-\alpha)\rho_l u_l\right], \qquad (26)$$

where u_l is internal energy of water liquid and internal energy of the mixture of dry air and water vapour is

$$u_g = \frac{1}{\rho_g}(\rho_a u_a + \rho_v u_v) = u_a w_{a,g} + (1 - w_{a,g})u_v, \qquad (27)$$

where u_a is internal energy of dry air, u_v is internal energy of water vapour and $w_{a,g}(p,t)$ is mass fraction of dry air in the mixture of dry air and saturated water vapour. Finally can be internal energy u expressed as

$$u = c_{v_s}(p,t)t + w_v(p,t)\left(l_{TR} - \frac{p_{TR}}{\rho_{TR}}\right), \qquad (28)$$

where l_{TR} is heat of vaporisation at the triple point, p_{TR} is the triple point pressure and ρ_{TR} is the density at the triple point. Temperature in degrees of Celsius is t. Mixture specific heat capacity is

$$c_{v_s}(p,t) = w_a c_{v_a} + w_v(p,t)c_{v_v} + [1 - w_a - w_v(p,t)]c_{v_w}, \qquad (29)$$

where c_{v_a}, c_{v_v} and c_{v_w} are constant volume specific heat capacities of dry air, water vapour and water liquid respectively. Equation of state can be written as

$$p = r_s(p,t)\rho_g(p,t)T(p,t), \qquad (30)$$

where r_s is gas constant of the mixture of dry air and water vapour

$$r_s(p,t) = \frac{w_a}{w_a + w_v(p,t)}r_a + \frac{w_v(p,t)}{w_a + w_v(p,t)}r_v \qquad (31)$$

and temperature in Kelvins can be expressed as

$$T(p,t) = \frac{u - w_v(p,t)\left(l_{TR} - \frac{p_{TR}}{\rho_{TR}}\right)}{c_{v_s}(p,t)} + 273.15. \qquad (32)$$

The key part of this model of two phase flow is resolving the temperature and pressure based on the equations (28) and (30). The two equations for pressure and temperature are non-linear mainly because water vapour mass fraction is the function of saturation pressure at given temperature

$$w_v(p,t) = 0.622 w_a \frac{p_v(t)}{p - p_v(t)}. \qquad (33)$$

The saturation pressure at given temperature can be calculated by using IAPWS formulation [7].

2.4 Quasi one-dimensional model of natural draft wet-cooling tower including flue gas injection

The system of governing equations can be written for the case of quasi one-dimensional flow as

$$\frac{\partial(\mathbf{W}A)}{\partial\tau} + \frac{\partial(\mathbf{F}A)}{\partial z} = \mathbf{Q}_{nd} + \mathbf{Q}_{fg}, \qquad (34)$$

where $A(z)$ is cross sectional area of the cooling tower, z is spatial coordinate and τ is temporal coordinate. Lets assume the moist air flow in the direction of z axis. The first equation is overall continuity equation, the second is overall momentum equation, the third is mass balance of dry air and the last equation is balance of energy. Vector of conservative variables $\mathbf{W}$ and vector of fluxes $\mathbf{F}$ are

$$\mathbf{W} = \begin{bmatrix} \rho \\ \rho v \\ \rho w_a \\ \rho e \end{bmatrix}, \quad \mathbf{F} = \begin{bmatrix} \rho v \\ \rho v^2\beta + p \\ \rho w_a v \\ \left\{\rho(u + \gamma\frac{v^2}{2}) + p\right\} v \end{bmatrix}, \qquad (35)$$

where β is momentum flux coefficient and γ is kinetic energy flux coefficient taking into account non-uniformity of velocity profile and γ is kinetic energy flux coefficient

$$\beta = \frac{A\int_A v^2 \mathrm{d}A}{\left(\int_A v \mathrm{d}A\right)^2}, \quad \gamma = \frac{A^2\int_A v^3 \mathrm{d}A}{\left(\int_A v \mathrm{d}A\right)^3}. \qquad (36)$$

The utilization of momentum coefficient β and kinetic energy flux coefficient γ is important especially in the case of flue gas injection, where the non-uniformity of velocity profile is significant. The evaluation of β and γ coefficients is assumed to be based on CFD simulation of incompressible multidimensional flow with mass flow rate based on quasi one-dimensional simulation. Vector of sources responsible for natural draft and minor losses $\mathbf{Q}_{nd}$ is

$$\mathbf{Q}_{nd} = \begin{bmatrix} A(z)\sigma_m(z) \\ p\frac{\mathrm{d}A}{\mathrm{d}z} - A(z)\left(\rho g + \zeta(z)\rho\frac{v^2}{2}\right) \\ 0 \\ A(z)(\sigma_q(z) + \rho g v) \end{bmatrix}, \qquad (37)$$

where $\sigma_m(z)$ is the density of mass source which is connected with mass transfer in the fill zone and $\sigma_q(z)$ is the density of heat source connected with heat transfer between water film and moist air in the fill zone. Minor losses in the zones are represented by coefficient $\zeta(z)$, e.g. [8,9].

Flue gas injection corresponds to a reduction of effective cross-section by the cross-section of flue gas pipe A_{fg}. Flue gas mass flow rate is $\dot{m}_{fg}$, the temperature is t_{fg} and specific humidity of flue gas is x_{fg}. Let the pressure at flue gas injection is numerically solved, then it is possible to calculate density ρ_{fg} and velocity of flue gas v_{fg} there. Vector of sources responsible for flue gas effect $\mathbf{Q}_{fg}$ is

$$\mathbf{Q}_{fg} = \frac{\dot{m}_{fg}}{\Delta z_{fg}}\begin{bmatrix} 1 \\ v_{fg} \\ w_{a_{fg}} \\ e_{fg} \end{bmatrix}, \qquad (38)$$

where Δz_{fg} is height of control volume which represents flue gas injection, $w_{a_{fg}}$ is mass fraction of dry air in the flue gas and e_{fg} is total energy of flue gas.

3 Fill heat and mass transfer

The model of heat and mass transfer used in this section is modification of model [5]. Integration of the model equations in the fill zone is leading to the evaluation of sources in the governing equations of moist air flow.

3.1 Governing equations

It has been shown in the reference [5] that the change in water mass flow rate with height of the fill z is

$$\frac{d\dot{m}_w}{dz} = \alpha_m a_q A_{fr}(x''(t_w) - x_s), \quad (39)$$

where $\dot{m}_w$ is water mass flow rate, α_m is mass transfer coefficient, a_q is transfer area per unit volume, A_{fr} is cross sectional area of the fill and $x''(t_w)$ is specific moisture of saturated air at water temperature t_w. The change in water temperature can be expressed as

$$\frac{dt_w}{dz} = \frac{a_q A_{fr}}{\dot{m}_w c_{p_w}}\left[\alpha(t_w - t) + \alpha_m(x''(t_w) - x_s)l(t_w)\right], \quad (40)$$

where α is heat transfer coefficient and $l(t_w)$ is latent heat of vaporisation at water temperature. Further modifications of governing equations are based on the definition of Lewis factor [1]

$$Le_f = \frac{\alpha}{\alpha_m c_p}, \quad (41)$$

where unlike the reference [5] it is important to use dry air mass fraction w_a, water vapour mass fraction w_v and water liquid mass fraction w_l

$$c_p = w_a c_{p_a} + w_v c_{p_v} + w_l c_{p_w}. \quad (42)$$

Change in water temperature can be expressed using the definition of Lewis factor as

$$\frac{dt_w}{dz} = \frac{\alpha_m a_q A_{fr}}{\dot{m}_w c_{p_w}}\left[Le_f c_p(t_w - t) + (x''(t_w) - x_s)l(t_w)\right], \quad (43)$$

Merkel number [1] can be in general expressed as

$$Me = \int_0^A \frac{\alpha_m}{\dot{m}_w} dA = \int_0^{h_{fill}} \frac{\alpha_m a_q A_{fr}}{\dot{m}_w} dz, \quad (44)$$

where h_{fill} is height of the fill. Averaged value of non dimensional mass transfer coefficient is

$$\frac{\alpha_m a_q A_{fr}}{\dot{m}_w} = Me/h_{fill}. \quad (45)$$

Governing equations for water mass flow rate can be simplified using known value of Merkel number Me [8] as

$$\frac{d\dot{m}_w}{dz} = Me/h_{fill}\, \dot{m}_w(x''(t_w) - x_s), \quad (46)$$

$$\frac{dt_w}{dz} = \frac{Me/h_{fill}}{c_w}\left[Le_f c_p(t_w - t) + (x''(t_w) - x_s)l(t_w)\right]. \quad (47)$$

The main difference in comparison to previous model [5] is precise calculation of specific heat capacity, where parameters of supersaturated moist air are taken into account.

3.2 Calculation of source terms

Due to different definition of internal energy in this work unlike the work [5], the attention should be paid in expressions for source terms. Because the general mass balance between water and water vapour in infinitesimal control volume of height dz is

$$d\dot{m}_w = \dot{m}_a dx_s, \quad (48)$$

we can express mass source as

$$\sigma_m = \frac{\dot{m}_a}{A(z)}\frac{dx}{dz} = \frac{\alpha_m a_q A_{fr}}{A(z)}(x''(t_w) - x_s), \quad (49)$$

and after manipulation

$$\sigma_m = \frac{Me/h_{fill}}{A(z)}\dot{m}_w(x''(t_w) - x_s). \quad (50)$$

Heat source is

$$\sigma_q = \frac{\dot{m}_a}{A(z)}\frac{dh_{1+x}}{dz}. \quad (51)$$

Finally we get

$$\sigma_q = \frac{\alpha a_q A_{fr}}{A(z)}(t_w - t) + h_v(t_w)\sigma_m, \quad (52)$$

where $h_v(t_w)$ is enthalpy of water vapour at water temperature. After application of Lewis factor Le and Merkel number Me we have

$$\sigma_q = Le_f \frac{Me/h_{fill}}{A(z)}\dot{m}_w c_p(t_w - t) + h(t_w)\sigma_m. \quad (53)$$

4 Conclusions

Proposed model overcomes the drawbacks of previously developed models [4] and [5]. The applicability of the the model to the case of supersaturated moist air is significantly expanded. Finally, the model includes the possibility to simulate flue gas injection, where the approach based on momentum flux coefficient β, kinetic energy flux coefficient γ and additional source vector is used.

Acknowledgement

This work has been supported by *Technology Agency of the Czech Republic* under the project *Advanced Technologies for Heat and Electricity Production - TE01020036.*

References

1. D. G. Kröger, *Air-Cooled Heat Exchangers and Cooling Towers.* (Penn Well Corporation, Tulsa, 2004)
2. N. Williamson, M. Behnia, S. Armfield, Int. Journal of Heat and Mass Transfer, **51** (2008), 2227-2236
3. A. K. Majumdar, A. K. Singhal, D. B. Spalding, ASME J. Heat Transfer, **105(4)**, (1983), 728-735
4. T. Hyhlík, EPJ Web of Conferences, **67**, (2014), 02044
5. T. Hyhlík, EPJ Web of Conferences, **92**, (2015), 02027
6. S. M. Ghiaasiaan, *Two-Phase Flow, Boiling and Condensation in Convectional and Miniature Systems.* (Cambridge University Press, 2008)
7. *Revised Suplementary Release on Saturation Properties of Ordinary Water Substance.* (The International Association for the Properties of Water and Steam, 1992)
8. L. Dvořák, J. Nožička, EPJ Web of Conferences, **67**, (2014), 02024
9. J. Stodůlka, R. Vitkovičová, EPJ Web of Conferences, **92**, (2015), 02086

EPJ Web of Conferences **114**, 02047 (2016)
DOI: 10.1051/epjconf/201611402047

Short-term gas dispersion in idealised urban canopy in street parallel with flow direction

Hana Chaloupecká[1,2a], Zbyněk Jaňour[2] and Štěpán Nosek[2]

[1] *Charles University in Prague, Faculty of Mathematics and Physics, Czech Republic*

[2] *Institute of Thermomechanics Academy of Sciences of the Czech Republic, v.v.i. Dolejškova 1402/5, Prague 182 00, Czech Republic*

Abstract. Chemical attacks (e.g. Syria 2014-15 chlorine, 2013 sarine or Iraq 2006-7 chlorine) as well as chemical plant disasters (e.g. Spain 2015 nitric oxide, ferric chloride; Texas 2014 methyl mercaptan) threaten mankind. In these crisis situations, gas clouds are released. Dispersion of gas clouds is the issue of interest investigated in this paper. The paper describes wind tunnel experiments of dispersion from ground level point gas source. The source is situated in a model of an idealised urban canopy. The short duration releases of passive contaminant ethane are created by an electromagnetic valve. The gas cloud concentrations are measured in individual places at the height of the human breathing zone within a street parallel with flow direction by Fast-response Ionisation Detector. The simulations of the gas release for each measurement position are repeated many times under the same experimental set up to obtain representative datasets. These datasets are analysed to compute puff characteristics (arrival, leaving time and duration). The results indicate that the mean value of the dimensionless arrival time can be described as a growing linear function of the dimensionless coordinate in the street parallel with flow direction where the gas source is situated. The same might be stated about the dimensionless leaving time as well as the dimensionless duration, however these fits are worse. Utilising a linear function, we might also estimate some other statistical characteristics from datasets than the datasets means (medians, trimeans). The datasets of the dimensionless arrival time, the dimensionless leaving time and the dimensionless duration can be fitted by the generalized extreme value distribution (GEV) in all sampling positions except one.

1 Introduction

The presented paper deals with short-term gas releases. Such releases are dominant in emergency situations. By way of illustration, this type of emergency situation occurred in Catalonia, which is in north-eastern Spain, in February 2015. From Catalonia chemical plant, toxic orange gas cloud spread into six municipalities in central Catalonia. The gas cloud released when contents of delivery (including nitric oxide and ferric chloride) were mixed. The cloud was disappearing only very slowly because of very little wind. More details about this disaster can be found in [1]. The duration of such leakages is usually less than one hour. Therefore, they do not belong to mean field. If we conduct an experiment of a gas release once, we can see only one scenario from many possible ones that would happen with the same experimental set-up. That is the reason why we must repeat experiments many times to obtain statistically representative datasets. In this paper, we focus on dispersion from short-term gas releases in an urban area typical for European city inner parts.

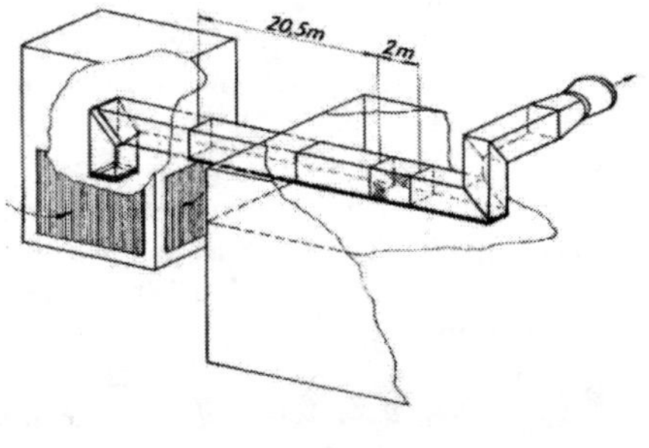

Figure 1. Scheme of an open low speed wind tunnel.

[a] Hana Chaloupecká: hana.chaloupecka@it.cas.cz

2 Experimental set-up

The experiments are conducted in the Laboratory of Environmental Aerodynamics of the Institute of Thermomechanics, Academy of Sciences of the Czech Republic. This laboratory has an open low speed wind tunnel specialised in boundary layer modelling. The boundary layer is developed in 20.5 m long section (see Figure 1). For the boundary layer development with neutral stratification similar to the one found in cities, spires and roughness elements are employed. Behind this section, the incoming flow comes to a test section, where measurements are conducted. To the bottom of the test section, models are fixed.

The model in 1:400 manufacture scale used in the presented measurements represents an idealised urban area found in city centres in Europe. It consists of houses with pitched roofs 63 mm high (50 mm houses bodies with 13 mm roofs). These houses are organised into closed courtyards of 150 x 300 mm dimensions. The courtyards are placed equidistant from each other (50 mm) in the model (see Figure 2).

Figure 2. Model of an idealised urban canopy.

A ground level point gas source is simulated by a pipe. It is placed into a middle of a street parallel with the incoming flow direction in the model. The pipe is fasten into an electromagnetic valve. By this electromagnetic valve (*3/2 electromagnetic valve of NORGREN*) operated by a timer relay (*Programmable Logic Controller Siemens LOGO!*), short duration gas leakages are created (see [2] for more details). The leakages duration in the presented experiments is 1 s in the model scale (a little less than 7 min in reality).

Concentrations are measured by a fast flame ionisation detector (*HFR400 fast flame ionisation detector of Cambustion*, FFID). The FFID recorded concentration time series are smoothed to 6 ms averages in model scale (2.4 s in reality). The FFID capillary, which sucks in the mixture of air and tracer gas, is placed into individual sampling places at the model. To prevent its oscillations in flow, the capillary is protected by two outer capillaries. Further, the FFID chamber is covered by a wooden casing to prevent its cooling by ambient flow. Moreover, the FFID flexible cable, which connects the control unit and the sample head, is insulated against cooling down from ambient flow. The FFID capillary end is placed into the high corresponding to the human breathing zone during the experiments. As the tracer gas, we use ethane. The simulations of the gas release for each measurement position are repeated many times (about 400 times) under the same experimental set up to obtain representative datasets. This enables determination of mean gas cloud arrival time with approximately 2%, mean cloud departure time with 4% and mean duration with 5% uncertainty at a sampling position. The uncertainties are counted from the repetitive measurements of 400 replicas at one measurement position.

3 Data analysis

In this paper, we present the results in dimensionless forms. To recalculate coordinates into the dimensionless form, we use the relation

$$x^* = \frac{x}{H}, y^* = \frac{y}{H}. \quad (1)$$

To recalculate the recorded concentration time series into the dimensionless form, we use the relation

$$C^* = \frac{C U_{ref} H^2}{Q}. \quad (2)$$

This relation is valid for the point source. The tests of this dimensionless concentration values independence on the Reynolds number and on the source intensity were conducted to set the appropriate experimental conditions. To recalculate time into the dimensionless form, we use the relation

$$t^* = \frac{t U_{ref}}{H}. \quad (3)$$

In these relations U_{ref} stands for the reference speed (correspondents to the free atmosphere), H the characteristic height (in this experiment houses height) and Q is the source intensity.

From the dimensionless concentration time series of each puff replica at each sampling place, puff characteristics are computed. The puff characteristics, which we discuss in this paper, are arrival time, leaving time and duration.

The data are analysed to obtain the puffs characteristics using the threshold criterion like in [3]. The threshold value is selected from the background concentration time series at individual sampling places. These time series are extracted from the recorded concentration time series of each puff replica (about 400 at each sampling place) as the last few seconds of them. From these sequences, 99% percentiles of concentrations are calculated. Finally, we calculate the maximum of the founded 99% percentiles at each sampling place.

3.1. Arrival (at*) & leaving (lt*) time

Definition of puff arrival (departure) time using the threshold method combats a problem with undesirable spikes in concentration time series. The spikes can be recognised as sudden high concentration values in concentration time series. This increase affects only a few measured values. The described situation is usually caused by a dust particle that is sucked into the thin FFID capillary.

If the definition of the puff arrival time requires only detection of a value of at least the threshold quantity in concentration time series (searched from the time when gas is released) spikes can cause false identification of gas cloud arrival. This situation happens when the spikes are presented within the time interval between the gas release and the true searched time of a cloud arrival at a sampling place in concentration time series. A very similar situation can occur during an identification of a gas cloud departure.

As the spikes affect only a few measured values, the problem could be solved as proposed Doran et al. in [4]. They require the detection of at least the threshold value in a specified time interval instead of only one value. But the requirement of at least the threshold quantity detection in a whole specified interval in these time series in the definitions of cloud arrival (departure) is problematic because of a huge intermittency in concentration signals. Choosing too long time interval causes delayed detection of a cloud arrival at a sampling place or the cloud do not have to be detected at all. On the contrary if we choose too short time interval the spikes can be falsely analysed as a cloud arrival at a sampling place. Therefore we overcome this problem requirement of detection of the value of at least the threshold quantity not in the whole but only in a specified percentage of cases within a specified interval in concentration time series in the definitions like in [3]. Concretely we require the detection of the value of at least the threshold quantity in 50% values from 100 ms time interval.

3.2 Duration (duration*)

Duration of a puff in a sampling place can be defined by the relation

$$duration = lt - at. \quad (4)$$

4 Results

The sampling places we present in this paper are chosen along the street parallel with flow direction (see Figure 3). The locations where the measurements took place are depicted by blue squares in Figure 3. The cross represents the gas source position. The arrow marks the flow direction.

In this paper, we show a mean value from datasets at each investigated position. Moreover, we display box plots to graphically depiction of the datasets at each sampling position. In the box plots, a median value (depicted by a square), 25% - 75% percentile (by a box) and a non-outlier range (by whiskers) are graphically represented. Furthermore, we display outliers (by circles) and extremes (by stars) in these figures. The outliers are defined as the values above (bellow) the 75% percentile (25% percentile) of the dataset plus (minus) 1.5 multiply of the difference between the 75% and 25% percentile. The extreme values are defined as the values above (bellow) the 75% percentile (25% percentile) of the dataset plus (minus) 3 multiply of the difference between the 75% and 25% percentile. The scheme of this graphical representation is displayed in Figure 4.

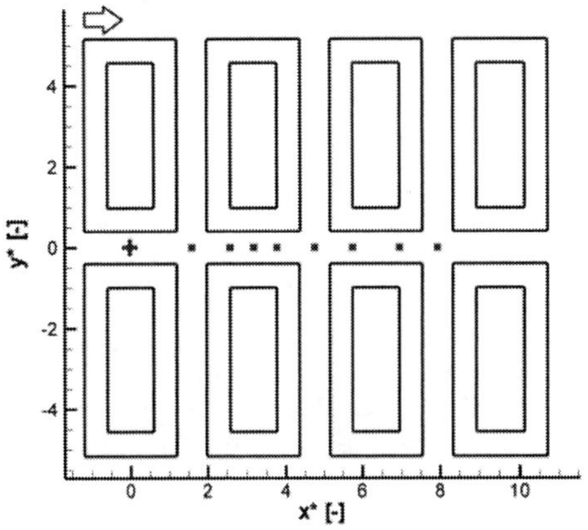

Figure 3. Scheme of investigated section.

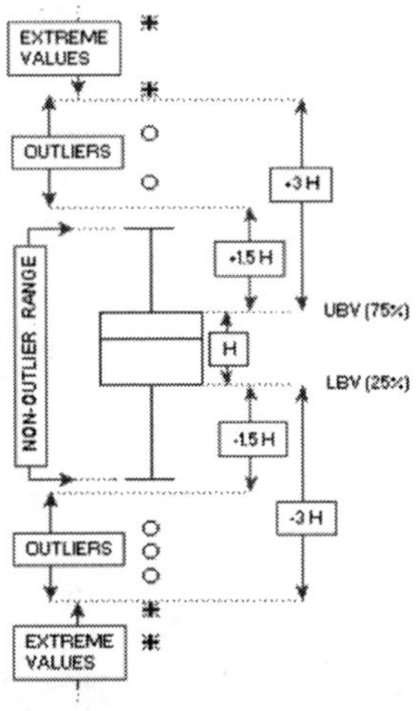

Figure 4. Scheme of graphical representation of datasets at each sampling place (adopted from [5]).

In Figure 5, dimensionless arrival time is displayed. The results indicate linear relationship between the dimensionless coordinate x and the mean value of the dimensionless arrival time. This relation can be described by the relation

$$at^*_{mean} = 2.65(\pm 0.11)x^* + 20.54(\pm 0.60). \quad (5)$$

It was found by the Least Squares approach. This linear model is proper according to an overall F-test, which tests a significance of regression, ($p = 0.000001$) as well as t-tests testing a statistical significance of individual coefficients (intercept $p = 0.000000$, independent value coefficient $p = 0.000001$). We also checked fulfilment of assumptions of the utilised method (independence of residuals and Gauss distribution of residuals with a constant mean value as well as variance). The found linear model has square of coefficient of determination R^2 0.99.

As well as the mean value of the dimensionless arrival time, also trimean of the dimensionless arrival time seems to be a linear function of the dimensionless coordinate x. Utilising the same analysis method as in the mean value of the dimensionless arrival time, we found the equation

$$at^*_{trimean} = 2.65(\pm 0.13)x^* + 20.13(\pm 0.65). \quad (6)$$

The results of the statistical tests are very similar to those found in the mean value of the dimensionless arrival time. We checked fulfilment of assumptions of the utilised method as well. The square of the coefficient of determination R^2 is also 0.99.

The datasets can be fitted by the generalized extreme value distribution (GEV) in all sampling positions except the location closest to the source. This hypothesis is not rejected according to the chi-square-goodness-of-fit test with $alpha = 0.05$.

In Figure 6, the puff characteristic dimensionless leaving (departure) time is demonstrated. The precise assessment of cloud departure time from a sampling location is very difficult. The reason is that the cloud is very torn shortly before it leaves the sampling location as reported in [3]. In comparison with the dimensionless arrival time, the range of datasets is wider. Moreover, the datasets contain more extreme values. The results of the dimensionless leaving time indicate that a linear relationship might be found between the dimensionless coordinate x and the mean value of the dimensionless leaving time. The proposed model equation is

$$lt^*_{mean} = 14.03(\pm 2.74)x^* + 164.55(\pm 13.71). \quad (7)$$

The equation (7) was found utilising the same approach as in the previous ones (F-test $p = 0.002170$, t-tests: intercept $p = 0.000020$, independent value coefficient $p = 0.002170$). The square of the coefficient of determination R^2 is 0.81.

The median of the dimensionless leaving time might be also estimated as a linear function of the dimensionless coordinate x. If we use the median value instead of the mean, the square of the coefficient of determination becomes better – 0.87. Other relevant values are: F-test $p = 0.000666$, t-tests: intercept $p = 0.000006$, independent value coefficient $p = 0.000006$. The found equation is

$$lt^*_{median} = 13.53(\pm 2.10)x^* + 156.43(\pm 10.52). \quad (8)$$

We could also use the trimean value of the dimensionless leaving time as in the equation (6).

$$lt^*_{trimean} = 13.69(\pm 2.59)x^* + 162.60(\pm 12.99). \quad (9)$$

In this case the relevant values are: $R^2 = 0.82$, F-test $p = 0.001875$, t-tests: intercept $p = 0.000016$, independent value coefficient $p = 0.001875$.

The datasets of the dimensionless leaving time can be fitted by GEV in all sampling positions except one ($x^* \approx 3.8$). This hypothesis is not rejected according to the chi-square-goodness-of-fit test with $alpha = 0.05$.

In Figure 7, the results of the dimensionless duration are depicted. In comparison with the equation (5) for the mean dimensionless arrival time, the line slope in the equation (7) for the dimensionless leaving time is higher. This denotes that the mean puff duration should increase with the increasing distance from the source along the the street parallel with flow direction. The described situation agrees with the results depicted in Figure 7. A linear function might be fitted to the data. However, this fit is worse than in all previous cases (equations (5) – (9)).The square of the coefficient of determination is only 0.73. The proposed model is

$$duration^*_{mean} = 11.38(\pm 2.85)x^* + 144.01(\pm 14.26). \quad (10)$$

Other relevant data are: F-test $p = 0.007167$, t-tests: intercept $p = 0.000055$, independent value coefficient $p = 0.007167$. This model can be also found utilising the equation (5) and (7) and the fact that $duration = lt - at$.

The median of the dimensionless duration might be also described as a linear function of the dimensionless coordinate x. This fact can be described by the relation

$$duration^*_{median} = 10.48(\pm 2.33)x^* + 138.72(\pm 11.68). \quad (11)$$

Other relevant data are: F-test $p = 0.004143$, t-tests: intercept $p = 0.000022$, independent value coefficient $p = 0.004143$. As well as in the dimensionless leaving time, this fit is better than in fitting the mean value because $R^2 = 0.77$.

A linear function might be also fitted between the trimean of the dimensionless duration and the dimensionless coordinate x. This situation can be described by the equation

$$duration^*_{trimean} = 10.99(\pm 2.70)x^* + 142.49(\pm 13.53). \quad (12)$$

Other relevant data are: F-test $p = 0.006613$, t-tests: intercept $p = 0.000043$, independent value coefficient $p = 0.006613$, $R^2 = 0.73$.

The datasets of the dimensionless duration can be fitted by GEV in all sampling positions except one ($x^* \approx 3.8$). This hypothesis is not rejected according to the chi-square-goodness-of-fit test with $alpha = 0.05$. This is exactly the same as in the dimensionless leaving time.

5 Conclusions

This paper deals with short-term gas releases (puffs). The puff characteristics definition of arrival and leaving time are crucial in the data analysis. We define them using the threshold criterion. These definitions must cope with the undesirable spikes in concentration time signals as well as the huge signal intermittency. We overcome this problem requirement of detection of the value of at least the threshold quantity not in the whole but only in a specified percentage of cases within a specified interval in concentration time series in the definitions of arrival (leaving) time.

The mean value of the dimensionless arrival time can be described as a growing linear function of the dimensionless coordinate x in the street with the gas source parallel with flow direction by the relation $at^*_{mean} = 2.65(\pm 0.11)x^* + 20.54(\pm 0.60), R^2 = 0.99$. A similar relation might be written for the mean value of the dimensionless leaving time, but the fit is worse ($R^2 = 0.81$). This fact can cohere with the difficult assessment of the precise departure time (discontinuity of the gas cloud tail). The proposed model is $lt^*_{mean} = 14.03(\pm 2.74)x^* + 164.55(\pm 13.71)$. In comparison with the equation for the mean dimensionless arrival time, the line slope is in the dimensionless leaving time higher. This denotes that the mean puff duration should increase with the increasing distance from the source along the the street parallel with flow direction. The described situation agrees with the found results. A linear function might be also fitted to the data. The proposed model is $duration^*_{mean} = 11.38(\pm 2.85)x^* + 144.01(\pm 14.26)$, but $R^2 = 0.73$. This model can be also found utilising the equation for the mean value of the dimensionless arrival time and the mean value of the dimensionless leaving time and the fact that $duration = lt - at$. Utilising a linear function, we might also estimate some other statistical characteristics from datasets than the means (medians, trimeans).

The datasets of the dimensionless arrival time, the dimensionless leaving time and the dimensionless duration can be fitted by GEV in all sampling positions except one. This hypothesis is not rejected according to the chi-square-goodness-of-fit test with $alpha = 0.05$.

Acknowledgement

Authors thank for support from the grant LD12007 of the Czech Ministry of Education and Sport and the grant 15-1894S of the Czech Science Foundation.

References

1. Toxic orange cloud spreads over Catalonia after chemical blast, http://www.theguardian.com/
2. H. Chaloupecká, Z. Jaňour, K. Jurčáková, L. Kukačka, Š. Nosek, EPJ Web of Conferences **67**, 020 45 (2014)
3. H. Chaloupecká, Z. Jaňour.9th International Conference on Urban Climate jointly with 12th Symposium on the Urban Environment (2015).
4. J.C. Doran, K.J. Allwine, J.E. Flaherty, K.L. Clawson, R.G. Carter. Atmospheric Environment **41**, 3440-3452 (2007).
5. STATISTICA Help, http://documentation.statsoft.com

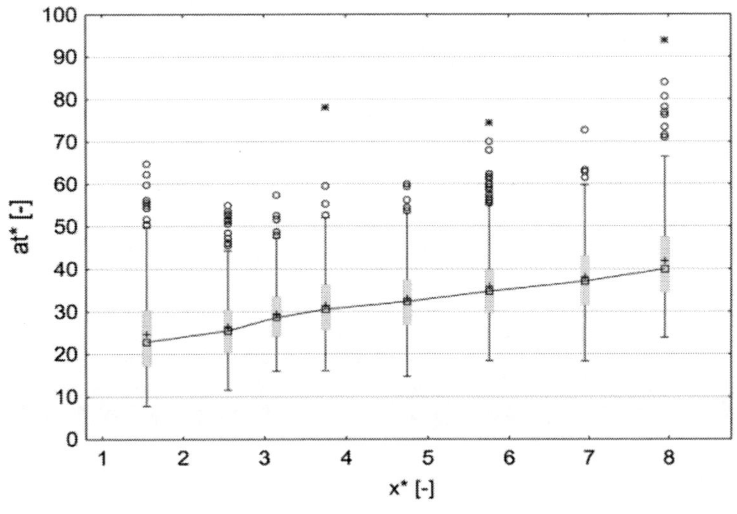

Figure 5. Dimensionless arrival time.

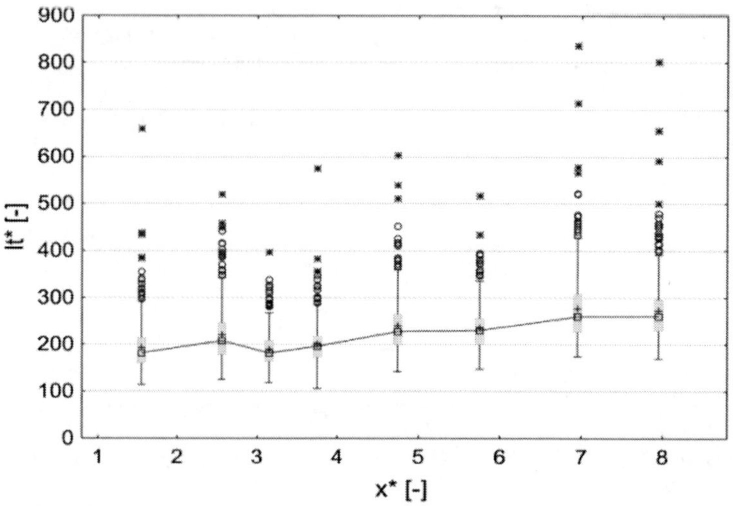

Figure 6. Dimensionless leaving time.

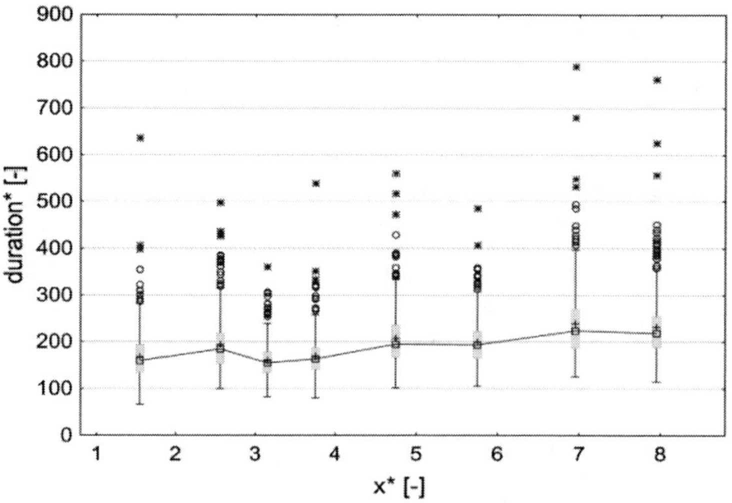

Figure 7. Dimensionless duration.

EPJ Web of Conferences **114**, 02048 (2016)
DOI: 10.1051/epjconf/201611402048

On the interaction between turbulence grids and boundary layers

Thomas Irps, Vasudevan Kanjirakkad

Thermo-Fluid Mechanics Research Centre, University of Sussex, Brighton, BN1 9QT, United Kingdom

Abstract. Turbulence grids are widely used in wind tunnels to produce representative turbulence levels when testing aerodynamic phenomena around models. Although the purpose of the grid is to introduce a desired turbulence level in the freestream flow, the wall boundary layers of the tunnel are subjected to modification due to the presence of such grids. This could have major implications to the flow around the models to be tested and hence there is a need to further understand this interaction. The study described in this paper examines wind tunnel wall boundary layer modification by turbulence grids of different mesh sizes and porosities to understand the effect of these parameters on such interaction. Experimental results are presented in the form of pressure loss coefficients, boundary layer velocity profiles and the statistics of turbulence modification.

1 Introduction

Grids of many shapes (round, square, structured, fractal, etc.), fabrication types (woven, welded/braced, perforated) and working principle (static, agitated/ pulsed, jet driven, active and passive) have been extensively used by aerodynamicists to introduce representative flow perturbations and turbulence into wind tunnel test sections. Since the turbulence produced by the grid decays with the flow downstream, the position of the grid installation becomes important. In many occasions this means that the grid needs to be positioned in relative proximity to the test section and the model. When the aerodynamic model to be tested is suspended away from the side walls of the test section, these grids are effective in providing a certain mainstream turbulence level with a uniform mean velocity profile in the transverse directions. However there are many test cases where the models to be tested need to be in close proximity to the wind tunnel walls (e.g., study of ground effects on moving objects, wind loading of static objects/buildings at ground level, turbomachinery blade cascade tests etc.). An interesting issue that is less discussed when one is encountered with the latter scenario is the interaction of the wind tunnel wall boundary layer upstream of the grid with the grid structure itself. Such interaction could produce unwanted localised flow variations both in an instantaneous and time averaged sense that could extend for an appreciable distance downstream and could affect the test conditions if the model to be tested is in close proximity.

Although only a handful of authors have discussed this in published literature, a common observation from those who did is the existence of a downstream velocity 'overshoot' and a thinning of the boundary layer over the wall surface that intersects the grid structure. To the authors' knowledge, the first observation of the boundary layer overshoot was described by Owen et al. [1] in an experiment which featured non-linearly distributed parallel rod grids. In that study which focused on the production of uniform shear flow, the authors describe the occurrence of local acceleration of the flow (relative to the mainstream flow) due to the pressure drop across the grid. This is explained by stating that the boundary layer suffers a lower total pressure loss due to the slower velocities encountered within it. Lau et al. [2] demonstrated the same effect and used the energy equation to show that a pressure drop coefficient larger than unity would create an overshoot with increasing size proportional to an increasing pressure drop. Mehta [3] suggested that the magnitude of the pressure drop coefficient (which thins the boundary layer) and the screen rigidity (i.e. possible curvature of the screen) plays a larger role in the generation of the overshoot due to the maximum streamline inclination towards the wind-tunnel surface.

When studying turbulence grid design it is important to understand the major parameters of relevance and how to use them to characterise the grid. Roach [4] provided a thorough summary of different geometric parameters that characterise a static turbulence grid (i.e. not featuring agitated bars, jet, aerofoil cascades and others). As shown by Mehta [3] and Roach [4] it is not the porosity alone that affects the performance of the flow but rather the imperfections and geometric parameters such as the grid type and curvature effects.

Kurian et al. [5] investigated flow through five different grids with the same porosity to characterize grid generated turbulence with a focus on energy spectra, correlation functions, turbulence length-scales, energy dissipation rates among others. The authors note that porosities should always be larger than 0.55 in order to avoid large-scale mean velocity profile fluctuations due to the merging of adjacent jets emanating from the holes.

Our intention in the on-going work, part of which is presented here in this paper, is to investigate the occurrence of the boundary layer overshoot, the extent and nature of its spatial existence and the cause of its origin. The paper tries to provide some insight into the boundary layer modification and the overshoot effect as brought about by the turbulence grid using the pressure drop data measured across the grid, velocity profiles resulting from the grid and the observed turbulence statistics upstream and downstream of the grid.

2 Methodology

Grid selection: To generate the overshoot five grids are employed in this study, three of which have the same porosity but varying mesh sizes and three having relatively similar mesh sizes but significantly varying porosities. These are manufactured from 2 mm thick aluminium sheets and are made following the design criteria described in Roach [4] for square-mesh arrays of square bars (SMS). All of the selected grids are expected to produce a pressure drop coefficient (K) larger than unity as per Roach's correlations. These employ empirically determined variables A and B in the form of

$$K = \frac{\Delta p}{q} = A\left(\frac{1}{\beta^2} - 1\right)^B, \tag{1}$$

where A is 0.98 and B is 1.09 and q is the upstream reference dynamic pressure

$$q = \frac{1}{2}\rho U_{ref}^{\;2} \quad . \tag{2}$$

The porosity is determined through

$$\beta = \left(1 - \frac{d}{M}\right)^2, \tag{3}$$

where, d is the strut width and M the mesh height (which is also the unit size). The resulting mesh characteristics for the five chosen grids (hereafter designated as G1, G2 G3, G4 and G5) are listed in Table 1.

Table 1. Grid parameters.

Grid	G1	G2	G3	G4	G5
Hole (mm)	4.5	8	9	10	18
Strut (mm)	1.5	4	3	2	6
Unit size (mm)	6	12	12	12	24
Porosity	0.563	0.444	0.563	0.694	0.563
K	2.27	4.51	2.27	1.05	2.27

Experimental setup and instrumentation: The test section is incorporated into the Sussex low-speed wind-tunnel which is a blower type tunnel driven by a centrifugal blower, handling a top range of 18.2 HP and 997 RPM with a resulting maximum flow rate of 6.32 m³/s. The flow through the tunnel is controlled via a mechanically regulated throttle at the inlet of the blower. Following the centrifugal blower is a diffuser section which consists of four screens. This is followed by a contraction section and the test section which is 2.2 m long and has a cross-sectional dimension of 0.15 m × 0.456 m.

Flow and boundary layer development is only measured along the lower horizontal wall of the duct and measurements are carried out only along the centreline of the duct thereby minimising the effects due to the sidewalls.

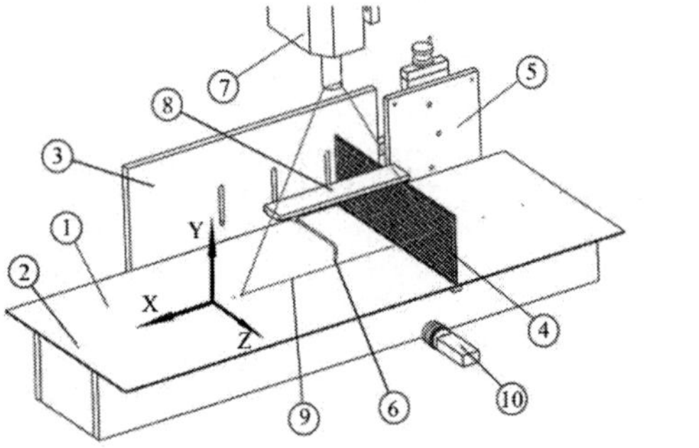

Figure 1. Test section instrumentation.

As shown in Figure 1 the setup consists of the fixed surface lower wall (1) where boundary layer flow is tracked. A series of static pressure tappings (2) are used along the centre line of the lower wall to measure pressure drop along the duct. A window (3) with multiple vertical slots is used to allow hot-wire (6) and pneumatic probe traverses in the wall normal direction (y-axis) at selected locations (see Figure 2) upstream and downstream of the grid (4). The linear actuator mechanism for traversing the probe is fixed on the outside of the wind-tunnel (5). The non-intrusive PIV setup consists of the non-reflective glass (8), the laser head (7), and the camera (10) for visualising the intended measuring area (9). The grids are held firmly in place by two u-shaped profiles that are cut into the upper and lower walls of the duct. It is important to note that care was taken to avoid close proximity between the lowest horizontal bar of the grid and the lower surface of the duct in order to prevent it from acting as an undesired boundary layer trip.

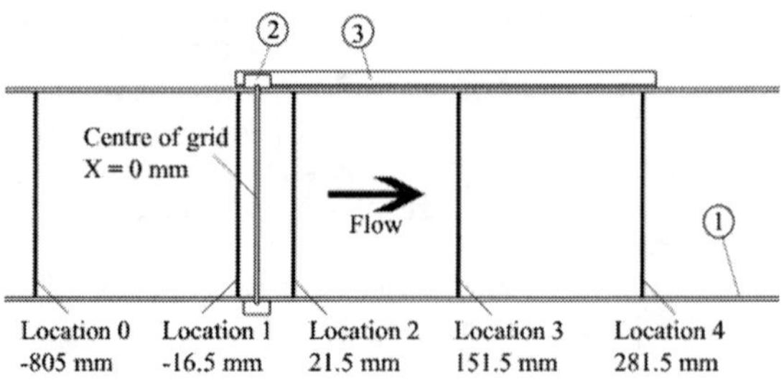

Figure 2. Measurement locations (not to scale).

Four measurement locations are chosen for gathering boundary layer data as shown in Figure 2. These locations allow to track both the flow development downstream of the grid as well as the boundary layer profile entering the grid. The number and the distance of the measurement locations with respect to that of the grid is also shown. In Figure 2, (1) denotes the test-section surface, (2) the junction securing grid in place and (3) the non-reflective glass for PIV measurements.

The various instrumentation types employed in this study include; hotwire (CTA), particle image velocimetry (PIV) and a range of steady pneumatic pressure measurements (for static and total pressure information).

The CTA setup consists of a single-wire SSP11 1-D all-purpose probe (5 µm diameter), held by a 90° support of the SSH21 type. A 4m long cable (A1863) connects the probe to the control system which is a Streamline 90N10 CTA frame from Dantec. The output from this is connected through NI USB-6251 A/D board to a laptop (which is configured via the Streamware 3.4 software) where data is logged using a LabView based software. All CTA signals were obtained at a frequency of 20 kHz.

The PIV laser system consists of a New Wave Research Class IV Solo PIV 120 Nd:Yag laser source. The beam diameter is 4.5 mm and has a pulse width of 3-5 ns. In order to generate sufficient amount of seeding particles the smoke machine "Le Maitre G-Force 1" is used which can generate up to 320 m^3/min of smoke with a particle diameter less than 5 µm. The camera used to capture the seeding is the Dantec Dynamics Flowsense Charged Coupled Device (CCD) camera with a resolution of 1600 x 1186 pixels and has a 10 bit analogue to digital converter. These two systems are linked to the Dantec Dynamics FlowMap processor which contains a correlator unit, input buffers and the synchronisation unit. The PIV measurements are currently ongoing and therefore only a limited amount of information would be presented in this paper.

The tunnel velocity is set using a pitot-static probe placed at mid height along the centre line of the duct 805 mm upstream of the turbulence grid, which is also used as reference (U_{ref}) velocity for all purposes. The above pitot-static tube is connected to a FC-034 500 Pa differential pressure transducer. A Druck DPI 740 barometer is used to measure the ambient pressure. A commercial thermocouple reader accurate to within ±1°C is used to track variation in room temperatures during the measurement. The wall static pressures along the duct lower surface and the total pressure data from a flattened pitot tube were captured using a 16 channel DSA 3217 transducer array. Tests were conducted for two different freestream velocities; 10.6 m/s and 17.6 m/s. However the results are only shown for the lower velocity case since initial analysis has produced very similar outcomes.

3 Results

In order to characterize the mean flow and the boundary layers several flow parameter definitions are used in this paper. To evaluate the static pressure variation inside the wind tunnel a static pressure coefficient is used as defined by:

$$C_{sp} = \frac{p_i}{q_{ref}}, \quad (4)$$

where, p_i is the local static pressure (gauge) along the test section and q_{ref} is the dynamic pressure of the flow at the reference location. The magnitude of C_{sp} is comparable to that of the grid pressure drop coefficient (K) as defined by Roach [4].

For the unperturbed velocity profiles, the boundary layer height (δ_{99}) is calculated as wall normal distance where the velocity reaches 99% of that in the freestream (U_∞). For cases with an overshoot the boundary layer height is calculated as the height at which the velocity reached 99% of the overshoot maxima.

The grid turbulence intensity has been empirically determined by Roach [4] to follow a decay law as represented by

$$Tu = C\left(\frac{x}{d}\right)^{-\frac{5}{7}}, \quad (5)$$

where, the value of the constant C for grids of the SMS type is 1.13

The measured turbulence intensity (Tu) is determined using the root mean square of the fluctuating components (u) and the mean freestream velocity (U) as per the expression,

$$Tu = \frac{(u)_{rms}}{U}. \quad (6)$$

For defining the turbulent length-scale (λ_x) we only consider the streamwise components of the velocity fluctuations and the derived energy spectrum $E(f)$:

$$\lambda_x = \left[\frac{E(f)\,U}{4u^2}\right]_{f\to 0}. \quad (7)$$

As explained by Roach [4] and by El-Gabry et al. [6] the frequency content above 10 kHz is negligible and only the energy content of the data asymptotically approaching a fixed value at frequencies close to zero is employed. This correlation assumes that the turbulent flow is truly homogenous and isotropic. The energy spectral density,$E(f)$, for a given frequency band as defined by:

$$E(f) = \frac{FFT(u)^2/L}{S}, \quad (8)$$

where, L is the sample length and S the sampling frequency.

To illustrate the boundary layer development through the duct and the interaction effect with the turbulence grid that is placed in the flow, Figure 3 compares the velocity profiles (hotwire) with and without a grid at four different locations along the duct on either side of the grid. Grid 3 (G3) is used as an example. Figure 3A compares the profiles at location-1 which is marginally upstream of the grid. Although this is an upstream location, it is clear that this is well within the potential field of the grid and hence there is a clear modification of the flow visible in the form of a thinning of the boundary layer when the grid is present as quantified in Table 3.

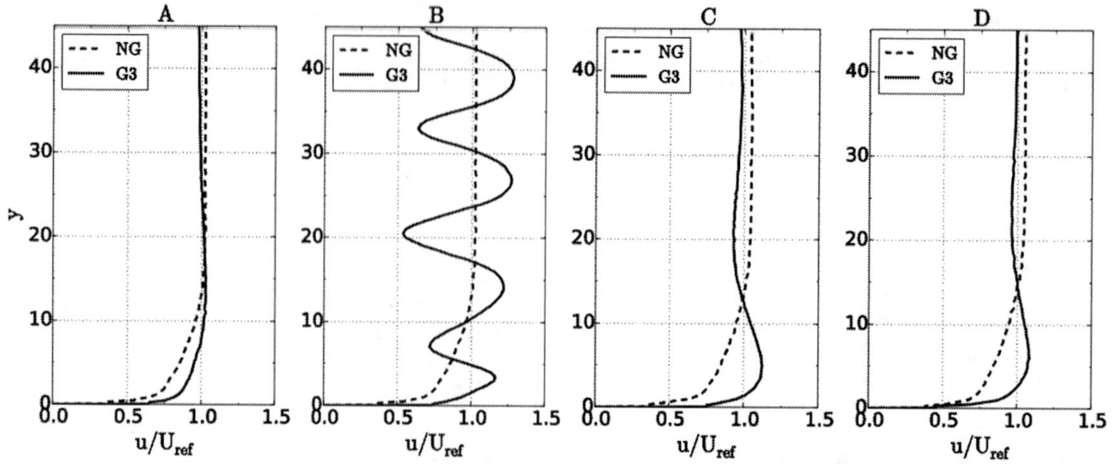

Figure 3. Boundary layer modification through grid interaction. A, B, C and D correspond to locations 1, 2, 3 and 4 respectively.

Figure 3B shows that at location 2 that is immediately downstream of the grid a more dramatic velocity distribution, where one cannot clearly discern the boundary layer due to the alternate wake and jet pattern, is emerging from the grid. Locations 3 and 4 are sufficiently downstream of the grid that the individual jets have now mixed and merged with one another but leaving a bump in the profile near the wall which is referred to as the velocity overshoot. Although the overshoot is present at both location-3 and location-4, it is clear from Figure 3 that the magnitude of the overshoot reduces as the flow marches downstream and the location of the velocity maxima moves further away from the wall. It is expected that the overshoot would disappear further downstream and a cleaner profile would result. However, it remains a fact that, turbulence intensities get damped and length-scales get larger further downstream. In many wind tunnel test setups this could mean that the model to be tested needs to be placed within the region of an active overshoot (for meeting a prescribed intensity and length-scale criteria) thus altering the 'ideal' test condition that was intended.

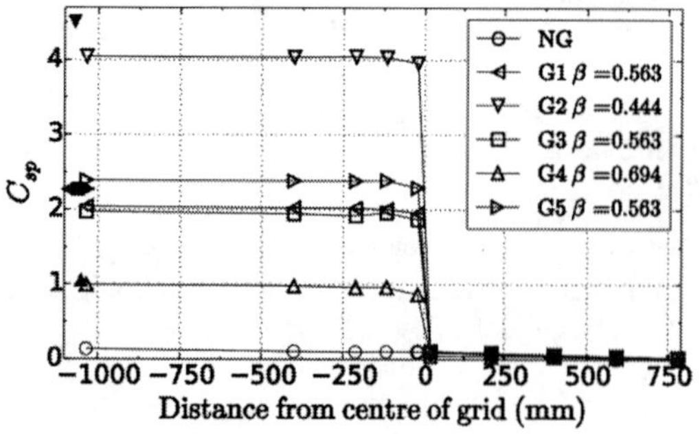

Figure 4. Wall static pressure coefficient variation in the test section.

One of the main consequences of placing a turbulence grid in the flow is the pressure drop that results due to the loss generated by the grid. The wall static pressure coefficient (C_{sp}) is presented in Figure 4, which shows the pressure drop effected by the placement of the grid in an otherwise clean duct test section (designated as NG, referring to a no-grid condition). Shown also alongside in closed symbols (of the same type) is the value of the pressure drop coefficient (K) for the SMS type grid from Roach's correlations. It is clear that the measurements largely agree with the results from the correlation. The slight disagreement as seen between grids of the same porosity (G1, G3 and G5) is more likely to be a result of manufacturing imperfections. It is possible that the coefficient of discharge of individual mesh openings could have been altered due to locally rounded edges of the strut. Otherwise these three grids exhibit similar pressure drops as expected.

Figure 5 presents the results from boundary layer traverses obtained with and without a grid at the four traverse locations for all five grids tested. As discussed previously, the flow immediately in front of all four grids (location-1) undergo a modification due to the potential interaction with the grid as seen in Figure 5A. The general tendency is to produce a thinner boundary layer, when a grid is present, as can be seen from the measured boundary layer thicknesses before (location-1) and after (location-2) the grid in Table 3. A small 'waviness' is also present on all of the profiles at location-1. This modification is thought to be related to the deceleration of the streamlines in the presence of the grid, but this needs to be understood further. Figure 5B shows the wall normal velocity distribution at location-2 which is immediately downstream of the grids and therefore strongly influenced by the alternate jets and wakes emerging from the openings and struts, respectively, of the grid. When compared with the no-grid case, it can be observed that the amplitude of the sinusoidal pattern is strongly linked to the mesh porosity such that the large amplitudes are associated with lower porosities. The strong local variations make the calculation of boundary layer integral parameters very difficult and hard to interpret at this location.

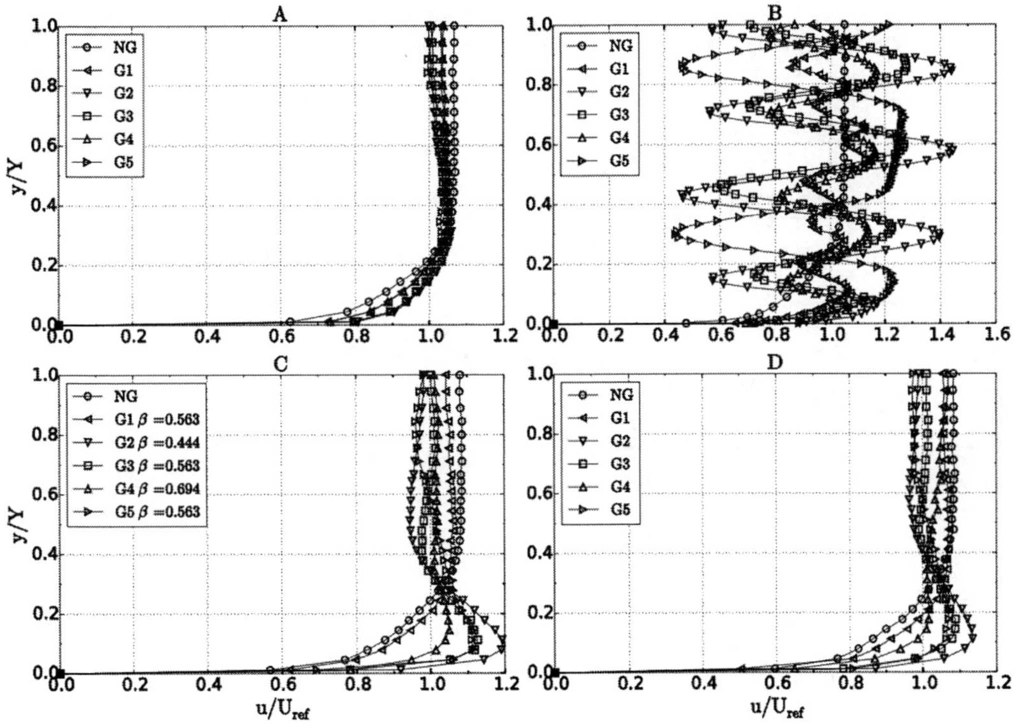

Figure 5. Boundary layer velocity profiles using hotwire measurements from the various traverse locations. Figures A, B, C and correspond to traverse locations 1, 2, 3 and 4 respectively. Here Y corresponds to y = 45 mm which is the last traverse coordinate normal to the surface.

Figure 5C and 5D show the velocity profiles further downstream of the grid at locations 3 and 4 respectively. At both locations the wake/jet related fluctuations (as seen at location-2) are absent due to the strong mixing and diffusion caused by the turbulent flow, which in turn is produced at the individual jet boundaries and from their subsequent merger with each other. The most notable feature of the flow in these locations is the lobbed structure closer to the wall that signifies the velocity overshoot as has been mentioned earlier in the paper. Similar to the sinusoidal variations in velocity, as found at location-2, the magnitude of the overshoot seems to be an inverse function of the grid porosity. The overshoot is larger for grids with lower porosities as seen by comparing those produced by grids G2, G3 and G4 in Figures 5C and 5D. When the mesh size is varied with a fixed porosity the overshoot is seen to increase first with the mesh size and then reduce afterwards as seen for the grids G1, G3 and G5 in Figures 5C and 5D. This suggests that, for a fixed porosity case, there is an 'optimum' mesh size for which the overshoot is maximised. Mehta's explanation of the overshoot as a simple function of the differential pressure loss effect across the incoming boundary layer seems to be inadequate to explain the actual physical mechanism that causes the formation of the overshoot since grids with relatively similar pressure drops are seen to produce dissimilar overshoots. The time-accurate and time-averaged data from the hotwire and PIV measurements during the present investigations will be further analysed to understand this phenomena better at a later stage when the PIV results are fully available. Note also from Figures 5C and 5D that an appreciable 'waviness' is present in the profiles in the wall normal direction at location-3 and location-4. This was also observed by Mehta [3] and was reiterated by Kurian [5] and has been explained to be associated with the random merging of adjacent jets amalgamating to form longitudinal vortices. In the present experiments this large scale non-uniformity is seen to get smoothened but not entirely even at location-4 that is almost two duct heights downstream of the grid.

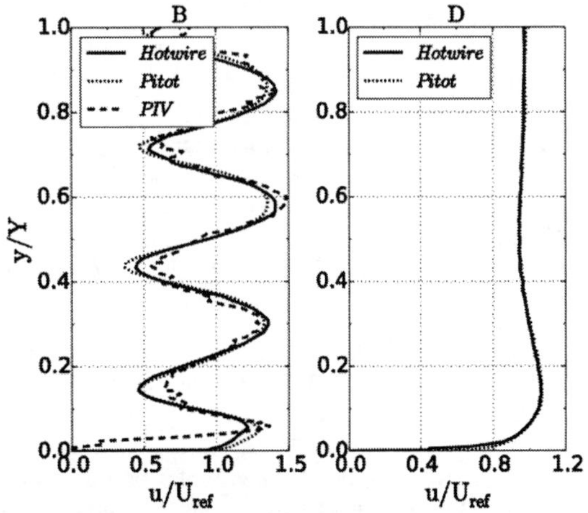

Figure 6. Comparison of velocity profiles from different measurement techniques. B) location-3 and D) location-4. Here Y corresponds to y = 45 mm which is the last traverse coordinate normal to the surface.

Different measurement techniques such as hotwire, PIV and flattened pitot are being used in this study and they provide results that are consistent with one another as shown in Figure 6. As mentioned elsewhere, the PIV measurements are currently incomplete.

Table 2. Comparison of freestream flow parameters; measured vs Roach's correlation.

	K		Tu		λ_x	
	Roach	Exp.	Roach	Exp.	Roach	Exp.
NG				1.96		11.65
G1	2.27	1.95	2.69	2.62	4.11	5.97
G2	4.51	3.99	5.41	4.87	6.71	4.65
G3	2.27	1.89	4.41	4.34	5.81	4.79
G4	1.05	0.92	3.29	3.54	4.75	5.09
G5	2.27	2.31	7.23	7.87	8.22	5.02

A comparison of the freestream derived pressure drop coefficient (between location-1 and location-4) and turbulence intensity and length-scale data for the measurements are shown in Table 2. The measured pressure drop coefficient (K) is well in agreement with Roach's correlation for SMS grids. The freestream turbulence values (Tu) also seem to agree quite well with the correlations presented by Roach. The present measurements provide a way of separating the effects of porosity from that of mesh size variation. For a fixed unit mesh size of 12mm, increasing the porosity (as seen from grids G2, G3 and G4) results in a reduction in turbulence intensity. The measured length-scales (λ_x) on the other hand are seen to increase with porosity. This tells us that, to a first order, the turbulence intensity scales with the strut width and the length-scales scale with the opening size (unlike what Roach's correlation shows). The decay and growth respectively of the freestream turbulence intensity and length-scale at locations 1, 3 and 4 are shown in Figures 7 and 8 respectively.

The freestream turbulence intensity (Figures 7) is seen to increase through the grid first (measured between locations 1 and 3) and then decay from location-3 to location-4. Although not shown here, the decay rate follows Roach's correlations adequately. The length-scales (Figure 8) are first seen to reduce as the flow goes past the grid and then to increase further downstream, as expected. The freestream values of turbulence intensity and length-scale are also tabulated in Table 3 for the three measurement locations 1, 3 and 4. Also shown is the boundary layer heights (δ_{99}) at these locations.

The turbulence intensity and length-scale variation at a height corresponding to 33% boundary layer edge velocity is plotted in Figures 7 and 8 respectively. For unperturbed velocity profiles the boundary layer edge velocity is the same as freestream velocity whereas for profiles with an overshoot the maximum velocity point is taken as the boundary layer edge. It is quite apparent that the strong coupling between porosity and the variation of the above parameters is non-existent. This suggest that there are different physical mechanism at work within the boundary layer such as an increased level of vorticity, in relation to the freestream flow, that could be causing a decoupling of local grid geometric parameters and the growth or decay of turbulence in this region. The values of turbulence intensity and length-scale at 33% boundary layer height is also tabulated in Table 4.

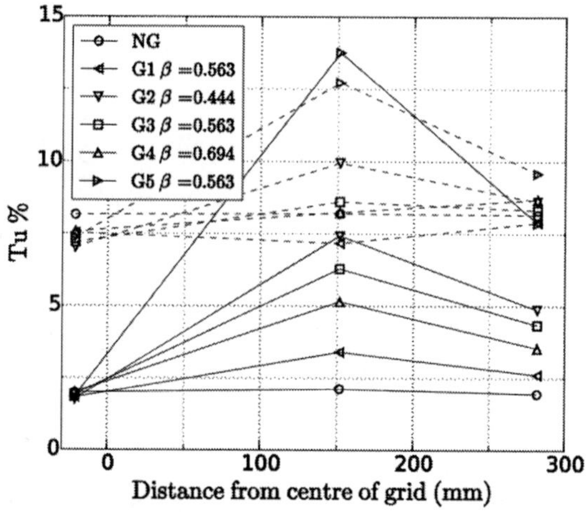

Figure 7. Turbulence decay with distance downstream of the grid: solid lines show freestream variation, dashed lines show variation inside boundary layer at 33% velocity location.

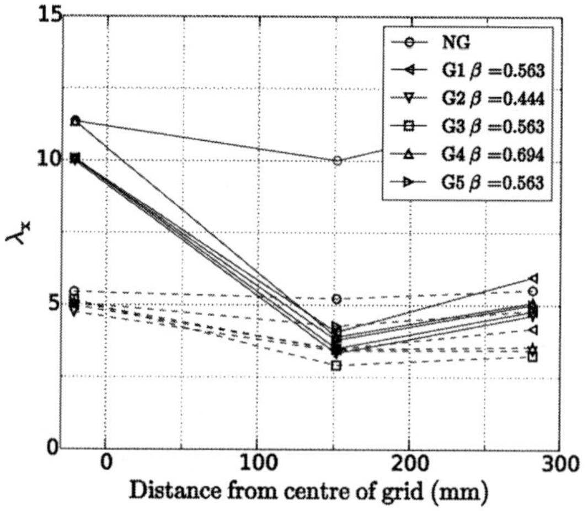

Figure 8. Turbulent length-scale growth with distance downstream of the grid: solid lines show freestream variation, dashed lines show variation inside boundary layer at 33% velocity location.

When the porosity is fixed and the unit mesh size is varied, the measured turbulence intensity values have been observed to increase with an increase in mesh size. This therefore scales with the strut size. However the measured length-scale values didn't seem to increase as per the correlations and as with the previous observation in the paper that this scales with the mesh opening size. This latter result needs to be revisited as the measurement of length-scales has proven difficult in some of the experiments due to the presence of a small bump in the spectra in the very low frequency region. No immediate source could be identified as a reason for its existence. The results however looked promising overall.

Table 3. Freestream turbulent flow statistics.

	NG	G1	G2	G3	G4	G5
			δ_{99}			
Loc-1	15.5	13.3	10.4	11.8	13.1	9.8
Loc-3	18.5	14.6	4.0	4.3	5.5	3.2
Loc-4	19	14.1	5.3	5.5	6.5	5.5
			Tu			
Loc-1	1.99	1.82	1.76	1.92	1.99	1.84
Loc-3	2.12	3.41	7.42	6.28	5.15	9.45
Loc-4	1.96	2.62	4.87	4.34	3.54	7.87
			λ_x			
Loc-1	11.35	10.1	9.99	10.1	11.4	10.1
Loc-3	10.01	4.07	3.33	3.5	3.19	3.79
Loc-4	11.65	5.97	4.65	4.79	5.09	5.02

Table 4. Turbulent flow statistics inside the boundary layer at 33% velocity location.

	NG	G1	G2	G3	G4	G5
			Tu			
Loc-1	8.16	7.53	7.01	7.19	7.58	7.35
Loc-3	8.2	7.16	9.95	8.61	8.24	12.8
Loc-4	8.16	7.89	8.62	8.32	8.69	9.59
			λ_x			
Loc-1	5.45	4.98	4.75	5.17	5.08	5.07
Loc-3	5.22	3.48	3.42	2.91	3.47	4.28
Loc-4	5.52	4.19	3.44	3.25	3.56	4.81

4 Conclusions

A set of experiments have been conducted in a low-speed wind tunnel with a modified cross section where an interchangeable turbulence grid could be placed. All measurements are conducted at 10.6 m/s. The flow past the grids is investigated for the nature of the interaction between the grid structure and the incoming boundary layer flow.

All the grids tested produced a pressure drop across it and the pressure drop magnitudes are in good agreement with the correlations for SMS grids by Roach.

For a fixed unit mesh size of 12 mm, varying the porosity has the effect that an increasing porosity produced a reduced level of freestream turbulence but at increased length-scales. This means that the turbulence intensity scales with the strut width and the length-scales scale with the opening size. Roach's correlation however suggests that both these parameters scale with the strut size alone. However, inside the boundary layer both the intensity and length-scale do not seem to have such a strong relationship with the porosity parameter.

A strong coupling has been observed between porosity and the strength of the velocity overshoot, for a fixed mesh size, in that lower the porosity higher is the overshoot found.

Small variations in pressure drop coefficient was observed between meshes of the same nominal porosity but this is very likely due to manufacturing imperfections resulting in local variation of the coefficient of discharge of the individual mesh openings. Turbulence intensity in the freestream has been observed to increase, at fixed porosity, with increased mesh size (therefore with increased strut width) whereas the measured freestream length-scale values did not show a coherent variation. The effect of increased mesh size for a fixed porosity on the overshoot effect is a notable result in that there seems to be an 'optimum' mesh size that generates the largest overshoot. This suggests that the simple way to look at the overshoot as a result of differential pressure drop across the boundary layer is insufficient in explaining the real physical mechanism that is causing the overshoot formation.

The turbulence decay and the length-scale growth in the freestream with distance downstream of the grid appear to follow the correlations of Roach although slightly elevated length scales were observed. The calculation of length-scales needs to be revisited to substantiate the accuracy of some of the findings.

Although the PIV measurements are still ongoing, the averaged results from the various techniques employed for boundary layer investigations (i.e. pneumatic traverse, hotwire traverses and PIV) have looked consistent and they complement each other well in providing physical insight.

References

1. P.R. Owen, H.K. Zienkiewicz, *J. Fluid Mechanics*, **2**, (1957)
2. Y.L. Lau, W.D. Baines, *J. Fluid Mechanics*, **33**, (1968)
3. R.D. Mehta, *AIAA Journal*, **23,** (1985)
4. P.E. Roach, *J. Heat and Fluid Flow*, **8** (1986)
5. T. Kurian, J.M. Fransson, *Fluid Dynamics Research*, **41** (2009)
6. L.A. El-Gabry, D.R. Thurman, P.E. Poinsatte, *NASA/TM – 2014-218403* (2014)

EPJ Web of Conferences **114**, 02049 (2016)
DOI: 10.1051/epjconf/201611402049

Flow characteristics of control valve for different strokes

Jana Jablonská[1,a] and Milada Kozubková[1]

[1] *VŠB-Technical University of Ostrava, Faculty of Mechanical Engineering, Department of Hydromechanics and Hydraulic Equipment, 17. listopadu 15/2172, 708 33 Ostrava – Poruba, Czech Republic*

Abstract. The article deals with the determination of flow characteristics and loss coefficients of control valve when the water flows in the interval of operating parameters, including the evaluation of vapour and air cavitation regime. The characteristics of the control valve are measured on the experimental equipment and subsequently loss coefficients are determined. Data from experimental measurements are used for creating of mathematical model with vapour and air cavitation and verification results. This validation will enable the application of methods of numerical modelling for valves of atypical dimensions e.g. for use in nuclear power industry. The correct knowledge of the valve characteristics and fundamental coefficients (e.g. flow coefficient, cavitation coefficient and loss coefficient) is necessarily required primarily for designers of pipe networks.

1 Cavitation in control valves

Hydrodynamic cavitation appears in hydraulic armatures during their current operation. This phenomenon is discovered for example in the narrowest gap between the regulating cone and the valve body. At this location the liquid is accelerated and the pressure is decreased. Vapour bubbles are formed in the liquid from cavitation nuclei (undissolved air and impurities). Growth of cavitation nuclei begins when the pressure falls below the value of saturation vapour pressure. It causes a sudden production of vapour bubbles, which are abducted by the liquid. Recover increase of pressure occurs behind the place of the smallest cross section. Vapour bubbles suddenly are diminished and disappeared by influence of the pressure. The gases contained in the bubbles partially diffuse into the surrounding liquid which is penetrated in their place. The remnants of gases in place of disappearing bubbles repeatedly decrease and increase their volume due to compressibility [1, 5]. Cavitation is influenced not only by vapour but also by dissolved air in the water, which is released at the narrowest section of the valve or by undissolved air sucked into the hydraulic circuit. In terms of the dynamics of bubbles it behaves like vapour. The maximum volume of the dissolved air in the water is 2% at normal conditions [3, 14, 15].

Cavitation in control valves is characterized by vibrations. Sound similar to the intermittent crackle in the field of the audible frequency can be observed in many cases [2, 4]. The occurrence of cavitation cloud in the flow field significantly influences the flow parameters of the valve [3, 5].

2 Mathematical models

Cavitation model assumes a minimum of two-phase mixture flowing - water and vapour. The release of dissolved air occurs at higher Reynolds numbers, and therefore model with three phases - water, vapour and air is considered. The dynamics of the gas comes from Rayleigh - Plesset equation for the dynamics of bubbles

$$\frac{p_{vap}(t)-p_{\infty}(t)}{\rho_l}=R_b\frac{d^2R_b}{dt^2}+\frac{3}{2}\left(\frac{dR_b}{dt}\right)^2+\frac{4\upsilon_l}{R}\frac{dR_b}{dt}+\frac{2S}{\rho_l R_b} \tag{1}$$

for certain value of local far pressure $p_{\infty}(t)$ the bubble radius $R_b(t)$ can be determined; p_{vap} is bubble surface pressure; σ is surface tension; ν_l is kinematic viscosity of liquid; ρ_l is density of liquid; see [9, 10, 11].

Transport equation of vapour volume fraction is taken into account.

$$\frac{\partial}{\partial t}(\rho\, f_{vap})+\nabla\left(\rho\, f_{vap}\,\vec{u}\right)=\nabla(\gamma\nabla f_{vap})+R_e-R_c \tag{2}$$

where f_{vap} is volume fraction of vapour; **u** is the velocity vector of the mixture; ρ is density of the mixture; R_e is efficient exchange factor for $p < p_{vap}$; R_c is efficient exchange factor for $p > p_{vap}$.

Individual cavitation models are different at the determination of R_e and R_c. For Schnerr Sauer cavitation model the follows equations are valid

$p \le p_{vap}$

$$R_e=\frac{\rho_l\rho_{vap}}{\rho}\alpha(1-\alpha)\frac{3}{R_b}\sqrt{\frac{2}{3}\frac{p_{vap}-p}{\rho_l}} \tag{3}$$

$p \ge p_{vap}$

$$R_c=\frac{\rho_l\rho_{vap}}{\rho}\alpha(1-\alpha)\frac{3}{R_b}\sqrt{\frac{2}{3}\frac{p-p_{vap}}{\rho_l}} \tag{4}$$

[a] Corresponding author: jana.jablonska@vsb.cz

Numerical simulation of flow in the valve is based on solution of Navier-Stokes equations by finite volume method and is performed by software Ansys Fluent 15. Model *k-ε* RNG has been used for modelling. This model is one of the modifications of the basic *k-ε* standard model. It is derived from the Navier-Stokes equations, using a mathematical technique called "renormalization group" (RNG) method. The model increases the calculating accuracy for the turbulent flow of lower Reynolds numbers and primarily for vortex flow. Therefore it is more reliable.

3 Description of the experiment

To test the control valve, a test circuit was built according to ČSN EN 60534-2-3 [6]. The test circuit is on figure 1.

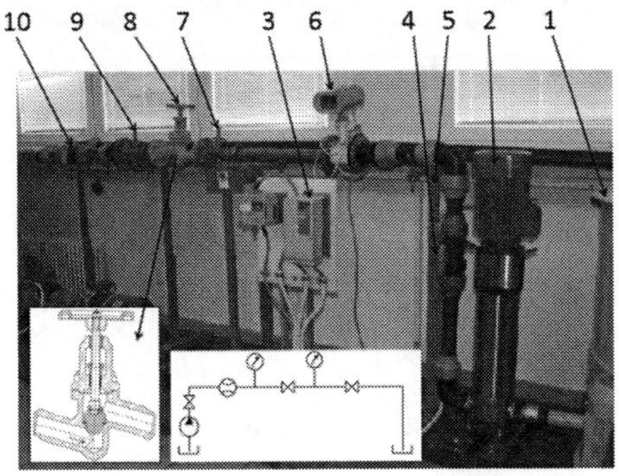

Figure 1. Circuit scheme – 1 - tank, 2 – pump, 3 – pump frequency converter, 4 – setting valve at inlet, 5 – frame, 6 – flowmeter, 7 – sampling point of pressure at inlet, 8 – test flow control valve DN40, 9 – sampling point of pressure at outlet, 10 – setting valve at outlet.

Measured element is the valve with a modified cone. The cone is adapted to be armature with control capabilities.

For measurement the water was used, i.e. an incompressible medium at a temperature of 20 °C. Accurate determination of the cone stroke was ensured by a system of lines around the perimeter of the valve [6, 7]. Accuracy of the cone stroke was ± 0.2 mm.

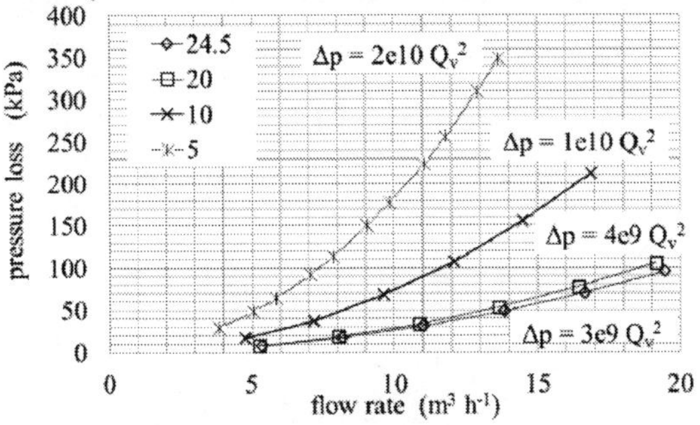

Figure 2. Control valve characteristic.

Basic characteristic of the valve, i.e. the dependence of pressure drop on the flow rate consists of individual characteristics that are designed for various valve strokes (5; 10; 20 and 24.5 mm). This characteristic is determined by measuring and evaluating for individual strokes of the control valve (see figure 2). The coefficients of these characteristics are determined by regression.

4 Evaluation of the experiment

Cavitation in the valve can arise in the area of the narrowest place of the valve. The increase of flow velocity and decrease of pressure (due to the Bernoulli equation) occurs due to narrowing of the flow cross section. The pressure can reach values of saturated vapour pressure.

Cavitation index is determined by design characteristics of the valve and expresses losses of the valve due to cavitation. It expresses comparison of the ability prevent the cavitation and susceptibility to cavitation.

$$\sigma = \frac{p_1 - p_v}{p_1 - p_2} \quad (5)$$

where p_1 is the absolute pressure at the inlet of the valve; p_v is the saturated vapour pressure for a given temperature; p_2 is the absolute pressure at the outlet of the valve.

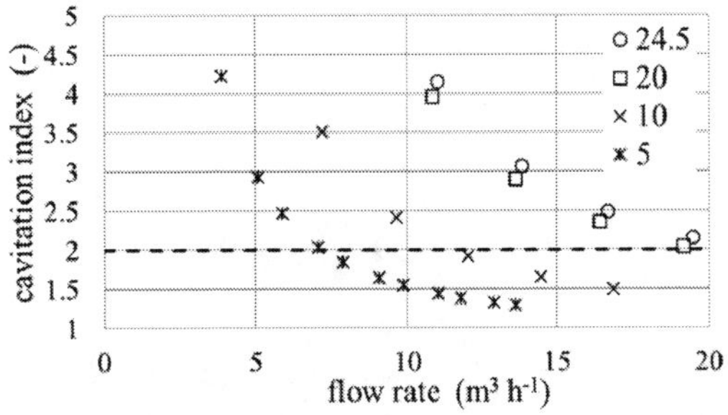

Figure 3. The relative flow characteristics - dependence of the cavitation index on the flow rate.

The critical value of cavitation index that determines the beginning of cavitation for the valve construction was determined based on a calculation from the experiment, see [8, 13, 15]. The value of cavitation index equals 2 and corresponds to the measured valve. If cavitation index was less than 2, then cavitation occurred.

Dependence of cavitation index on the flow rate is shown in figure 3. The graph shows that cavitation occurs already during flow rate 7.87 $m^3 h^{-1}$ for cone stroke 5 mm and for cone stroke 10 mm during flow rate 12.07 $m^3 h^{-1}$. Loss coefficient is typical for control valves used in European countries, which is derived from the equation

$$Q = \frac{1}{\sqrt{\zeta}} S \sqrt{\frac{2}{\rho}} \sqrt{\Delta p} \quad \Rightarrow \quad \zeta = \left[\frac{1}{Q} S \sqrt{\frac{2}{\rho}} \sqrt{\Delta p}\right]^2 \quad (6)$$

Decreasing of loss coefficient with increasing the valve stroke is shown in figure 4. Loss coefficient is determined from valve characteristic (see figure 2), where the equation $\Delta p = R \cdot Q^2$ is valid for turbulent flow.

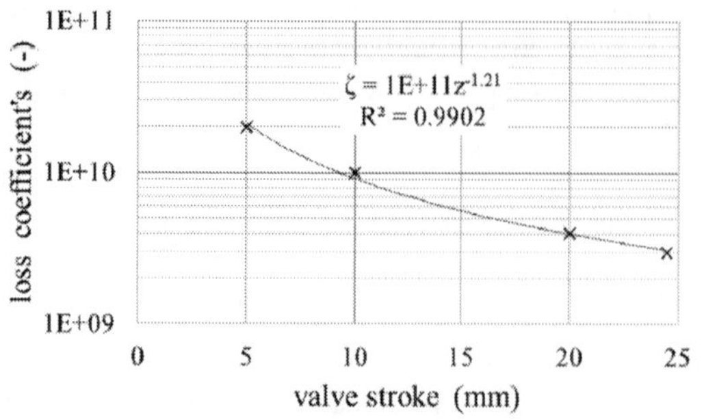

Figure 4. Loss coefficient dependence on the valve stroke.

5 Numerical modelling of flow through the valve

The task is modelled in a three dimensional space, using plane symmetry. The geometry of the created region, network and boundary conditions are seen in figure 5. Modelled region is identical with the measuring region (figure 1), inlet and outlet of the area are located in places of pressure sensors.

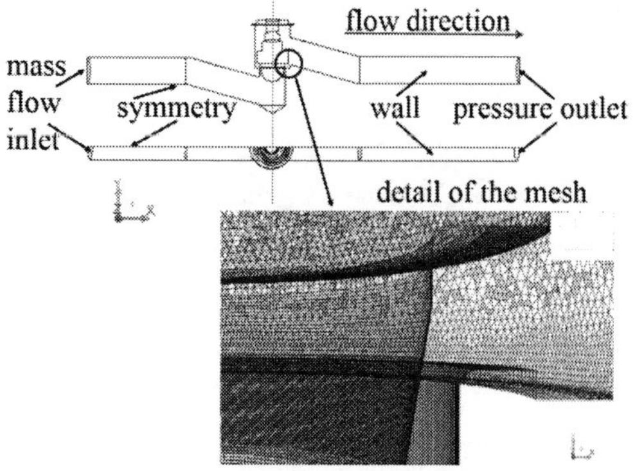

Figure 5. Geometry of the modelling region, grid and boundary conditions.

Tasks for cone strokes 5; 10; 20 and 24.5 mm were modeled. In the article only the task for cone stroke of 5 mm, where the cavitation at relatively low flow rates was occurred, is described in detail.

Two equation cavitation turbulent k-ε RNG model on the basis of the Reynolds number and recommendations of the literature [9, 12, 13] was selected. The third phase i.e. air was added at the inlet so that the pressure and flow rate agreed with the measured values. Volume fraction of air was specified by value of 2% (i.e. air mass flow rate is $3.958 \cdot 10^{-5}$ kg s^{-1}). The task was calculated as a time dependent due to the dynamics of cavitation area. Time step was 0.0001 s, which ensured the convergence of the task at each time step.

Table 1. The measured values for the cone stroke of 5 mm.

Q_{water}	Q_{water}	p_1	p_2
$m^3 s^{-1}$	kg s^{-1}	Pa	Pa
$3.278 \cdot 10^{-3}$	3.271	357537	101793

Three variants of the calculation for the cone stroke of 5 mm are compared in this article.

Variant A - water and vapour (two-phase model with cavitation)

Variant B - water and air (two-phase model without cavitation)

Variant C - water, air and vapour (three-phase model with cavitation)

Table 2. Boundary and cavitation conditions for the cone stroke of 5 mm.

	A	B	C
inlet - mass flow inlet* - mass flow rate kg s^{-1}			
water	3.271	3.271	3.271
vapour	0	0	0
air	-	$3.958 \cdot 10^{-5}$	$3.958 \cdot 10^{-5}$
outlet - pressure outlet Pa			
	101793	101793	101793
cavitation conditions	Schnerr – Sauer model	-	Schnerr – Sauer model
p_{vap}	2340 Pa	-	2340 Pa
n_b	10^{13}	-	10^{13}

* It is necessary to insert only half of the flow rate into inlet boundary condition because of the plane symmetry task.

Where p_{vap} is saturation vapour pressure; n_b is bubble number density.

Table 3. Physical properties.

	density kg m^{-3}	viscosity Pa s
water	998.2	0.001003
vapour	0.017	$1.34 \cdot 10^{-5}$
air	ideal-gas	$1.789 \cdot 10^{-5}$

Turbulent models and grid quality were tested in earlier tasks [11, 12].

6 Evaluation of flow modelling in control valve

Dependence of inlet pressure on the mass flow rate for numerical and experimental results for selected variants of cone stroke is evaluated in figure 6. It is evident a good agreement between calculated and measured values.

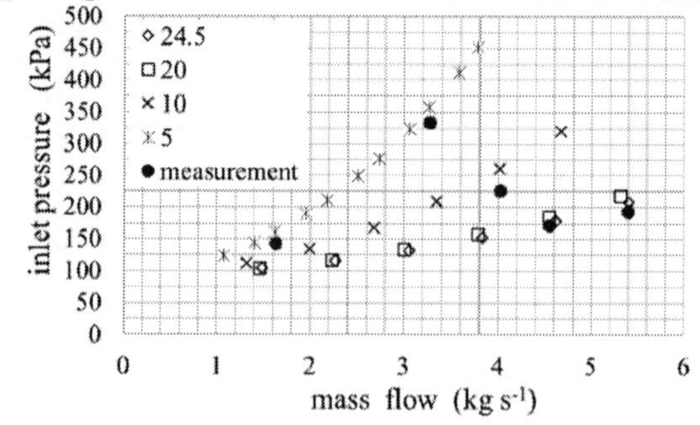

Figure 6. Dependence of pressure inlet (the modeled and measured) on the mass flow rate.

The dependence of inlet pressure and its deviations from the experiment for different variants of modelling is in figure 7. It is evident that deviation of the input pressure (as compared with measurement) is greatest for modelling of variant A (flow of water and vapour). The deviation was dramatically reduced by considering the flow of water and air (variant B). The air content in water is equal to 2%. If the cavitation model is added, i.e. the flow of water, vapour and air - variant C, the smallest deviation between calculated and measured values is observed.

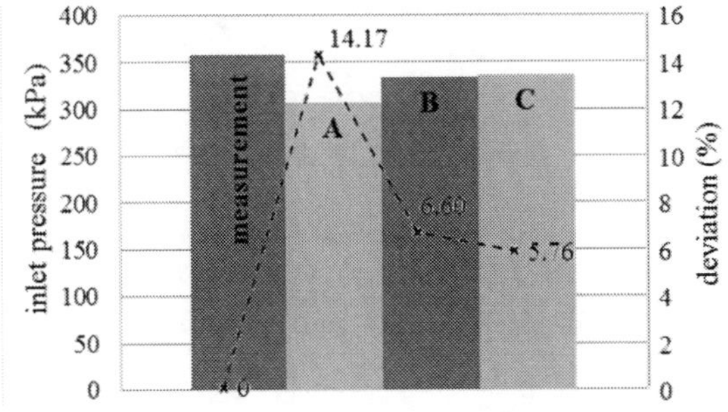

Figure 7. Dependence of inlet pressure and deviation for each variations of modelling for the cone stroke of 5 mm.

The pressure along the length of the valve for each modelled variants is shown in figure 8. The outlet is given by pressure boundary condition and is specified from experimental measurement, so the pressures on outlet are the same.

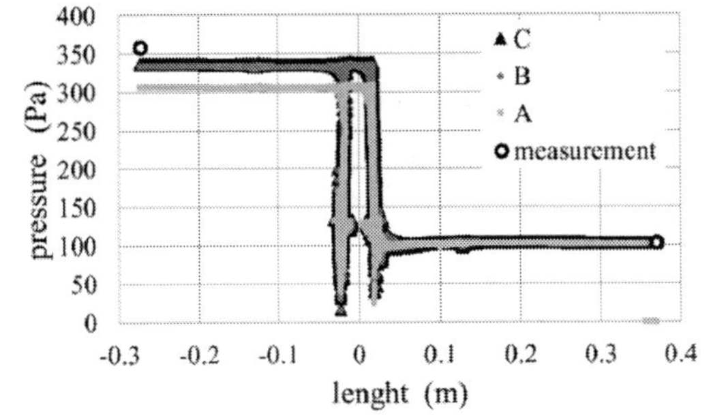

Figure 8. Pressure along the length of modelled variants for cone stroke of 5 mm.

It is obvious that the value of pressure at inlet depends on the choice of the multiphase model, see figure 8. Cavitation model of multiphase flow of water and vapour gives underestimated values of pressure drop in comparison with experiment. The air in variant A is not considered. By addition of air at the inlet the results are significantly more accurate what is confirmed by variants B and C.

From the literature [13, 14, 15] it is evident that air cavitation refines numerical calculation. The amount of air in the liquid is difficult to measure and it is only determined from the numerical experiments.

The volume fraction of vapour in variant A (water and vapour) is significantly higher than in variant C (water, vapour and air), see figure 9. Volume fraction of air in variant B is higher than in variant C. Evaluation of sum of vapour and air volume fraction is only possible for variant C. Comparison of the size of the cavitation area with the experiment is not possible because the valve is not transparent.

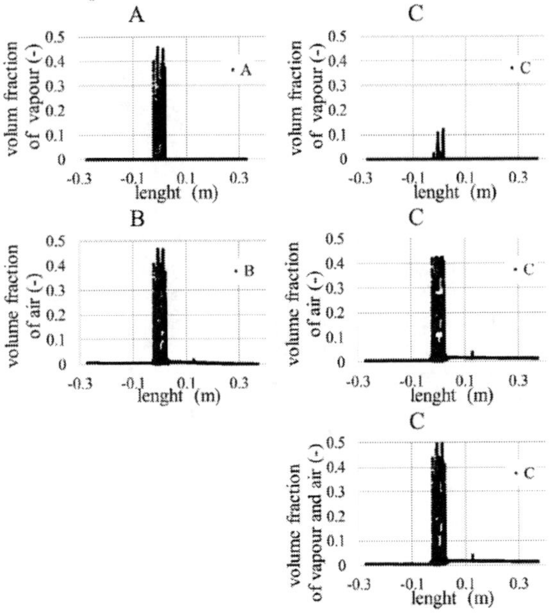

Figure 9. Volume fraction of vapour, air, vapour and air for individual variations vs. length of modelled area.

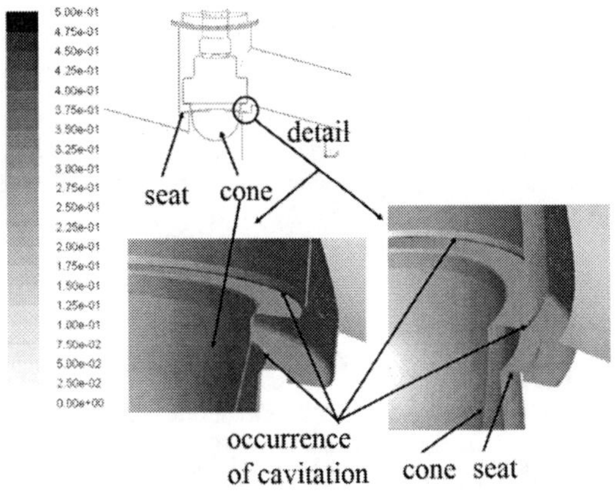

Figure 10. Cavitation area of the control valve.

Locations where cavitation occurs at the valve by the cone stroke of 5 mm are shown in figure 10. These locations of the valve are the most stressed and greatest wearing occurs on them.

7 Conclusion

The article deals with the formation of cavitation mathematical model for multiphase flow of water, air and vapour in the control valve for individual cone strokes. For verification and definition of the boundary conditions the characteristics and loss coefficients of were measured. Results of mathematical modelling were then compared with experimental results.

The evaluation of the cavitation number is an important result, which indicates that cavitation already arises at low flow rates for the cone stroke of 5 mm. For this reason the cone stroke of 5 mm was also selected for the evaluation of mathematical modelling.

The methodology of mathematical modelling of flow in control valve for each cone stroke was described. From evaluation of numerical solution and experimental measurements it was indicated, that to achieve consensus in the pressure boundary conditions, it was necessary to consider the addition of air, which moreover affects the size of the cavitation cloud.

Numerically it has been proven that vapour and air cavitation occurs at the edges of the valve. In practical point of view, presence of air cavitation reduces wearing in comparison with clean vapour cavitation.

The presence of air and vapour cavitation was also confirmed in other hydraulic elements [13, 14, 15].

References

1. C. E. Brennen, *An introduction to cavitation fundamentals*. (2011)
2. L. H. Yang, S. A. Liu, Y. L. Chen, X. H. Zhou, C. X. Wang, Y. M. Yao, J. L. Liu, *Applied Mechanics and Materials*, **548-549** (2014)
3. J. Noskievič, Kavitace. (1969)
4. S. C. LI, *Cavitation of Hydraulic Machinery*. (2000)
5. C. E. Brennen, *Fundamentals of multiphase flow*. (2005)
6. ČSN EN 60534-2-3 (1999)
7. ČSN EN 1267 (2012)
8. J. Parish, *Valve Magazine*. (2009)
9. Fluent 15 – User's guide. [Online]. Available to from: <URL: http://sp1.vsb.cz/DOC/Fluent_6.1/ html/ug/ /main_pre.htm>.
10. M. Kozubková, *Numerické modelování proudění – FLUENT I.* [Online]. Available to from: <URL: http://www.338.vsb.cz/seznam.htm>. (2005)
11. M. S. Plesset, R. B. Chapman, *J. of Fluid Mechanics*, **47** (02), (1971).
12. J. Dobeš, M. Kozubková, *EPJ Web of Conferences* **67**, p. 02019. EDP Sciences. (2014)
13. J. Jablonská, M. Kozubková, *Applied Mechanics and Materials*, **752-753**. (2015)
14. M. Kozubková, J. Rautová, *Proceedings of the 3rd IAHR International Meeting of the Workgroup on Cavitation and Dynamic Problems in Hydraulic Machinery and Systems*. (2009)
15. J. Jablonská, M. Kozubková, *J. of Computational Multiphase Flows*, **7** (2), (2015)

Acknowledgements

The work presented in this paper was supported by a grant SGS "Zkoumání dynamiky tekutinových systémů" SP2015/95.

EPJ Web of Conferences **114**, 02050 (2016)
DOI: 10.1051/epjconf/201611402050

Mathematical and experimental modelling of flow of air-saturated water through a convergent-divergent nozzle

Jana Jablonská[1,a] and Marian Bojko[1]

[1]*VŠB-Technical University of Ostrava, Faculty of Mechanical Engineering, Department of Hydromechanics and Hydraulic Equipment, 17. listopadu 15/2172, 708 33 Ostrava – Poruba, Czech Republic*

Abstract. In hydraulic elements an under-pressure is generated during fluid flow around sharp edges or changing the flow cross-section (e.g. for valves, switchgear, nozzles). In these locations air suction by leakages or release of air from the liquid during cavitation may occur. When flow modelling using classical mathematical model of cavitation at higher flow rates there is disagreement in the measured and calculated hydraulic variables before and behind hydraulic element. Therefore, it is necessary to use a mathematical model of cavitation applied to the three-phase flow (water, vapour, air). Nowadays it is necessary to look for mathematical approaches, which are suitable for quick engineering use in sufficiently precision numerical calculations. The article is devoted to theoretical investigation of multiphase mathematical model of cavitation and its verification using a laboratory experiment. At first case the k-ε RNG turbulent mathematical model with cavitation was chosen in accordance [9] and was applied on water flow with cavitation (water and vapour) in a convergent-divergent nozzle. In other cases a solution of water flow with cavitation and air saturation was investigated. Subsequently, the results of mathematical modelling and experimental investigation focused on monitoring of air content and its impact on the value of hydraulic parameters and the size of the cavitation area were verified.

1 Introduction

Various hydraulic elements (throttle valve, switchgear, measuring equipment, etc.), which are characterized by constriction of cross section, are much discussed from point of the fluid flow. It is a specific multiphase flow of fluid mixtures with cavitation (vapour, liquid). If the pressure is equal to the saturated pressure, the cavitation occurs. Moreover, at this place the air can be added by due to the release from the liquid and suction by leakages [13, 14]. The mixture is considered as the mixture of liquid, vapour and air.

Problem of mixture flow (gas - liquid) in convergent-divergent nozzle is described in the literature [5]. Authors determinate different values of the parameters such as pressure ratio, density ratio, Mach number and velocity ratio at different initial volume ratio. Measured values were taken from the experimental data while the calculated values came from theoretical relations. Error in the calculated and experimental values lay in the range of (10 - 25) % [5]. In reference [6] there are described experiments investigating the effects of air injection on the cloud cavitation of an oscillating hydrofoils. The article describes possibilities to reduce cavitation wear and noise levels during injection of air into the circuit.

The amount of air in the liquid has an influence on the size of the cavitation cloud and consequently on the distortion of the surfaces of various materials. Literature [7] generally described the influence of air and gases on cavitation (corrosion rate on the surface). The conclusions in the article show that increasing the oxygen level decreases the damage of the material surface. This decrease may be ascribed to the increased diffusion rate of air into cavitation bubbles during the growth cycle, leading to collapse cushioning [7].

Literature [8] studies effects of air content on cavitation and pressure fluctuations on the ship screw. From the results of the measurement, it is shown that air content has a great effect on the pressure fluctuations induced by the cavitating propeller. So air content has important effects on cavitation, especially on the bubble cavitation and the tip vortex cavitation and has little effect on the sheet cavitation range.

In this article we will deal with multiphase flow of water cavitation in the convergent - divergent nozzle (CD nozzle). From previous measurements authors know that at a given flow the pressure in the

[a] Corresponding author: jana.jablonska@vsb.cz

CD nozzle throat is decreasing to saturation pressure. To solve three-phase flow with cavitation the laboratory experiment for monitoring the size of the cavitation cloud and its changes at different air saturation was created. On the basis of this experiment the mathematical model was specified and the outcomes were verified with experimental data.

2 Cavitation and aeration

The fluid during cavitation is defined as multiphase mixture of liquid, vapour and air. Air is undissolved and takes the form of bubbles. The liquid containing of free air is not homogeneous medium (it is a suspension of gas in the liquid). Density of the liquid is significantly influenced by the content of gas, resulting in change in the dynamic properties of the fluid. The undissolved air exists in liquid always and like the vapour significantly affects the compressibility of the mixture. The undissolved air in the areas of low pressure combined with vapour creates a contiguous gas area. In the liquid there appears also dissolved air, which does not affect the compressibility of the fluid. Methods of detection of cavitation area and amount of gas are in figure 1.

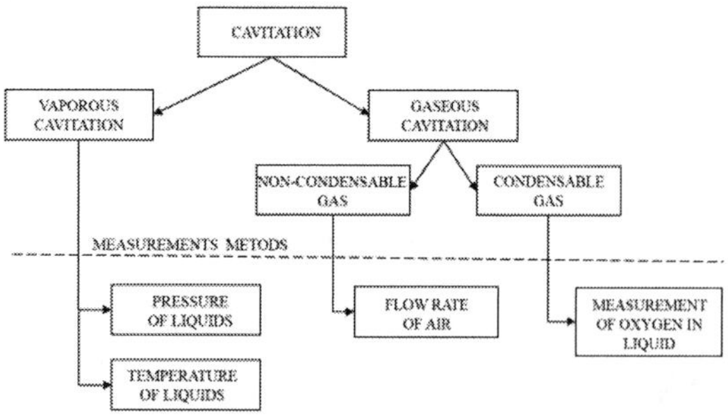

Figure 1. Specifications and methods of finding of cavitation area and gas quantities.

3 Description of measurement

The experimental equipment is located in the laboratories of the Technical University of Ostrava. The experimental equipment consists of a vertical centrifugal pump (figure 2, position 3), which is connected to water tank (figure 2, position 1). The centrifugal pump is driven by an electric motor (figure 2, position 4) with speed control by using the frequency converter (figure 2, position 5). Magnetic flowmeter is connected in the circuit below (figure 2, position 6). The circuit is supplemented by saturating air (figure 2, position 13 - 16), followed by transparent convergent divergent nozzle rectangular section - measured object (figure 2, position 8). Its basic dimensions are shown in figure 2. Sensors pressure (figure 2, position 9, 10) are disposed before and behind the CD nozzle. Time record of cavitation area (figure 2, position 17) is taken from each measurement using a camera.

Air is supplied to the circuit via the check valve and a mass flow regulator, which consists of a thermal mass flow sensor; control valve and accurate microprocessor PID (figure 2, position 13). Flow regulator adjusts quickly the required flow. The regulating device (figure 2, position 14) sets the desired volume flow rate of air, which then flows to the circuit.

The article deals with the influence of air amount on the size of the cavitation area at constant pump speed (the frequency converter is set at 26 Hz). The first investigation was done for variant A without added air into the circuit and further for the variants B - E with adding various amount of air. The size of the cavitation area was taken using the camera. Measuring of pressure before and after the CD nozzle, the volume flow rate of water (at 26 Hz) is evaluated in table 2.

Volume flow rates of air (see table 1) were recorded at each measurement. The measured values were averaged in time, and subsequently served as the boundary conditions for mathematical modelling. Operation range was limited by structural and material parameters of the equipment.

α_{inlet} is the volume fraction of air at the inlet of CD nozzle computed from measured pressure; Y is the mass fraction of air; pressures are in absolute values, the atmospheric pressure is equal 101104 Pa to the day of measurement.

Visualization of the cavitation cloud for each measured variant was used for further evaluation. The term „cloud cavitation" is used to refer to the process of coherent growth and collapse of clouds of cavitation bubbles [6].

Table 1. Indicated variants for different amount of air.

variant	Q_{air}	$Q_{m,\,air}$	α_{inlet}	Y
	$ml_N\ min^{-1}$	$kg\ s^{-1}$	-	-
A	0	0	0	0
B	100	$4.87\cdot10^{-6}$	$4.90\cdot10^{-4}$	$1.43\cdot10^{-6}$
C	300	$1.46\cdot10^{-5}$	$1.47\cdot10^{-3}$	$4.30\cdot10^{-6}$
D	400	$1.95\cdot10^{-5}$	$2.08\cdot10^{-3}$	$6.10\cdot10^{-6}$
E	500	$2.43\cdot10^{-5}$	$2.53\cdot10^{-3}$	$7.39\cdot10^{-6}$

Table 2. Measured value of water flow rate and pressure for different air flow rates.

variant	$Q_{m,\,water}$	p_{inlet}	p_{outlet}
	$kg\ s^{-1}$	Pa	Pa
A	3.393	188120	136539
B	3.391	191155	136535
C	3.380	192166	134516
D	3.360	194189	133505
E	3.325	195200	132493

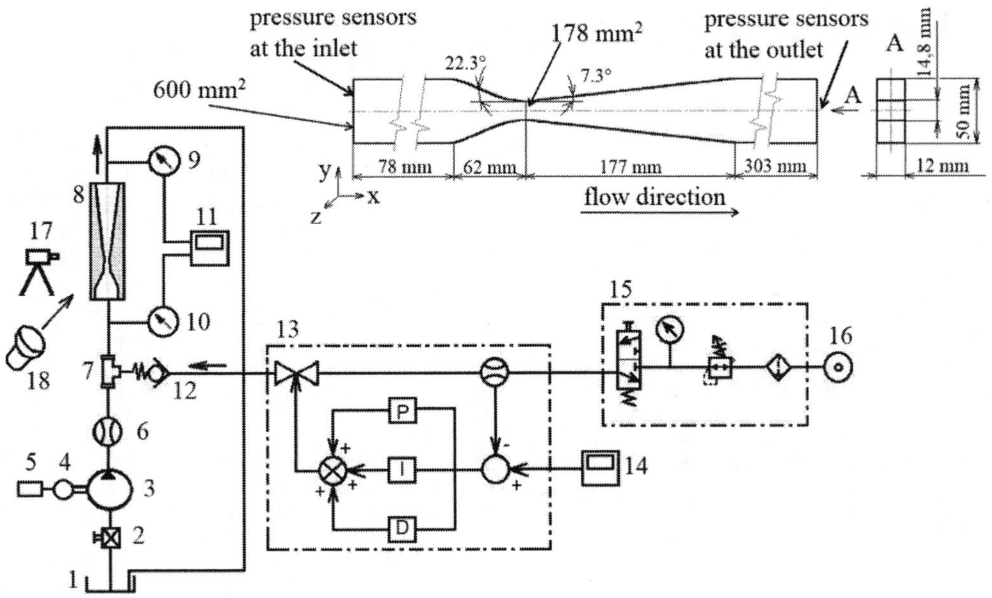

Figure 2. Circuit scheme – water and air segments (1 – tank, 2 – ball valve, 3 – hydro generator, 4 – motor, 5 – frequency converter, 6 – induction flowmeter, 7 – saturation with air, 8 – object measured, 9 –pressure sensor, 10 –pressure sensor, 11 – evaluation equipment, 12 – one-way valve, 13 – mass flow controller, 14 – control unit, 15 – pressure regulating valve with a filter unit and a condensate separator, 16 – compressed air source with a pressure gauge, 17 – video camera, 18 – light source).

4 Mathematical modelling

Flow in the CD nozzle is turbulent, so equation of continuity and Navier-Stokes equation which take into account turbulence were solved. To the solution turbulence models were used (k-ε standard, k-ε RNG, k-ε realizable, k-ω standard, k-ω SST, etc.). The system of differential equations was solved by finite volume method.

The mathematical model of turbulence cavitation takes into account the dynamics of gas bubbles, which is given generalized Rayleigh – Plesset equation for bubble dynamics (time-dependent pressure and size of bubbles) is

$$\frac{p_{vap}(t)-p_{\infty}(t)}{\rho_l}=R_b\frac{d^2R_b}{dt^2}+\frac{3}{2}\left(\frac{dR_b}{dt}\right)^2+\frac{4\upsilon_l}{R}\frac{dR_b}{dt}+\frac{2S}{\rho_l R_b} \quad (1)$$

for certain value of local far field pressure $p_{\infty}(t)$ the bubble radius $R_b(t)$ can be determined; p_{vap} is saturation pressure; σ is surface tension; ν_l is kinematic viscosity of liquid; ρ_l is density of liquid; see [1, 2, 3].

This differential equation has been too difficult to be applied into a multiphase flow model. Equation (1) was used in linearized form, for more details, please see [3, 4]

$$\frac{dR_b}{dt}=\sqrt{\frac{2}{3}\frac{p_{vap}(t)-p_{\infty}(t)}{\rho_l}} \quad (2)$$

Because the mathematical model should be taken into account evaporation of liquid and possibly air flow, the mathematical model is defined as multiphase flow model with cavitation by authors Schnerr and Sauer [2, 3, 4]. Cavitation model derives the exact expression for the mass transfer from liquid to vapour and condensation process equation for vapour volume fraction has the general form

$$\frac{\partial}{\partial t}(\rho_{vap}\alpha_{vap})+\frac{\partial}{\partial x_j}(\rho_{vap}\alpha_{vap}\overline{u}_j)=R \quad (3)$$

where α_{vap} is volume fraction of vapour; ρ_{vap} is density of vapour; R represents the vapour condensation or evaporation rate.

$$R=\frac{\rho_{vap}\rho_l}{\rho}\left(\frac{\partial\alpha_{vap}}{\partial t}+\frac{\partial(\overline{u}_j\alpha_{vap})}{\partial x_j}\right) \quad (4)$$

where density of mixture can be defined by

$$\rho=\alpha_{vap}\rho_{vap}+\alpha_g\rho_g+(1-\alpha_{vap}-\alpha_g)\rho_l$$

and the volume fraction of vapour is expressed in this model using the number of bubbles n_b in a unit of volume of liquid [1, 2].

$$\alpha_{vap}=\frac{n_b\frac{4}{3}\pi R_b^3}{1+n_b\frac{4}{3}\pi R_b^3} \quad (5)$$

after substituting

$$R=\frac{\rho_{vap}\rho_l}{\rho}\alpha_{vap}(1-\alpha_{vap})\frac{3}{R_b}\sqrt{\frac{2}{3}\frac{(p_{vap}-p)}{\rho_l}} \quad (6)$$

$$R_b=\left(\frac{\alpha_{vap}}{(1-\alpha_{vap})}\frac{3}{4\pi}\frac{1}{n_b}\right)^{\frac{1}{3}} \quad (7)$$

where R is mass transfer rate; R_b is bubble radius.

Also in this model, the only parameter which must be determined is the number of spherical bubbles per volume of liquid n_b. If you assume that no bubbles are created or destroyed, the bubble number density would be constant.

The main phase – water – contains small amount of non-condensable air, which can have significant influence on cavitation region due to compressibility of

[a] Corresponding author: jana.jablonska@vsb.cz

air by low pressure [1, 3, 4]. The calculation of the mixture density is modified as

$$\frac{1}{\rho}=\frac{f_{vap}}{\rho_{vap}}+\frac{f_g}{\rho_g}+\frac{1-f_{vap}-f_g}{\rho_l} \tag{8}$$

Non condensable gas density ρ_g is calculated by equation of state as

$$\rho_g=\frac{Mp}{RT} \tag{9}$$

Volume fractions of non-condensable gas and liquid are modified as

$$\alpha_g=f_g\frac{\rho}{\rho_g}, \qquad \alpha_l=1-\alpha_{vap}-\alpha_g \tag{10}$$

where α_{vap} is volume fraction of vapour; α_g is a volume fraction of gas; f_g is gas mass fraction.

5 Numerical modelling of flow in convergent divergent nozzle rectangular section

Task is solved as a three-dimensional, symmetrical area. The geometry of the modelled area identical with the measuring, grid and boundary conditions are shown in figure 3., input and output are in places located pressure sensors. The grid was created in the program Gambit, has 502500 hexahedrons cells. Due to assessing the size of cavitation area there is created in the model the evaluation plane with a constant distance of 0.01 m from the CD nozzle throat.

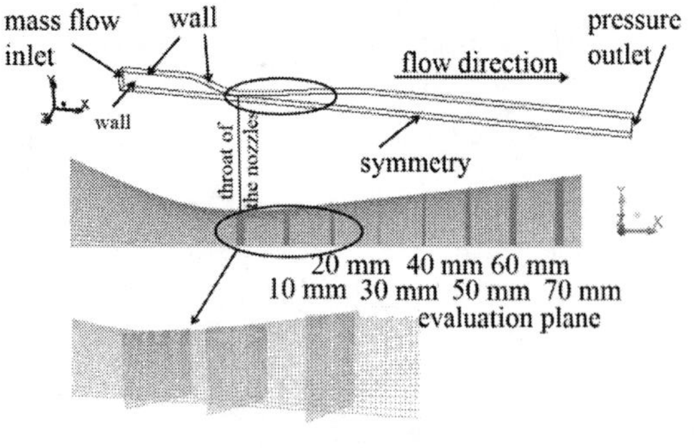

Figure 3. The geometry of the modelled area and boundary conditions.

K-ε RNG two-equation turbulent model and non-equilibrium wall functions were chosen based on the Reynolds number and recommendations references [10, 11, 12]. The task was calculated as the time dependent with a time step $1\cdot10^{-5}$ s, due to dynamics of cavitation cloud. Gravity acceleration in the negative direction of the x-axis was considered due to the location of the CD nozzle in the measurement.

The physical properties are defined for water and vapour as constant. Water density is 998.2 kg m^{-3} and viscosity of water is 0.001003 Pa s. Vapour density is 0.017 kg m^{-3} and viscosity of vapour is $1.34\cdot10^{-5}$ Pa s. Air density is defined as ideal-gas and viscosity of air is $1.789\cdot10^{-5}$ Pa s. Values of saturation vapour pressure and bubble number density 10^{+13} are necessary to define for the cavitation model. Saturation vapour pressure for a given constant temperature 25 °C (isothermal condition) is 2340 Pa. The measured data were averaged for the determination boundary conditions.

Table 3. Boundary and cavitation conditions.

inlet - mass flow inlet*	mass flow rate
water	see table 2, variant A - E
vapour	0 kg s^{-1}
air	see table 1, variant A - E
outlet - pressure outlet	see table 2, variant A - E
cavitation conditions	Schnerr – Sauer model
saturation vapour pressure	2340 Pa
bubble number density	10^{+13}

* Only half of the flow rate is necessary to insert into numerical simulation because of the symmetry task.

Boundary conditions are defined for different variants of the volume fraction of gas (air) (see table 1). The average mass flow of water, air at the inlet and the average outlet pressure is measured.

Testing of the turbulent model *k-ε* (standard, RNG, realizable) and *k-ω* (standard, SST) is presented in the publication [9]. Models were tested for the given geometry and the grid, but it showed some anomalies that are not compatible with the experiment. Model *k-ε* RNG is appropriate to use. This which follows the recommendations of the literature [10, 11, 12] and also from testing of the models. Evaluation of testing was performed by comparing of the modelling results and of the measurements.

Grid quality was evaluated by calculation, i.e. task was calculated with larger and smaller number of cells. Deviation of the average pressure value at the inlet was 1 %. The parameter y + was also evaluated for chosen model and grid, its value is 30 – 60, so the grid is suited.

6 Comparison of modelling and measurement

On figure 4 the cavitation cloud is visualized for each variant of the input boundary conditions. The figure shows that the area of vapour is approximately still the same. The cavitation cloud increases with increasing amount of air that is added to the circuit.

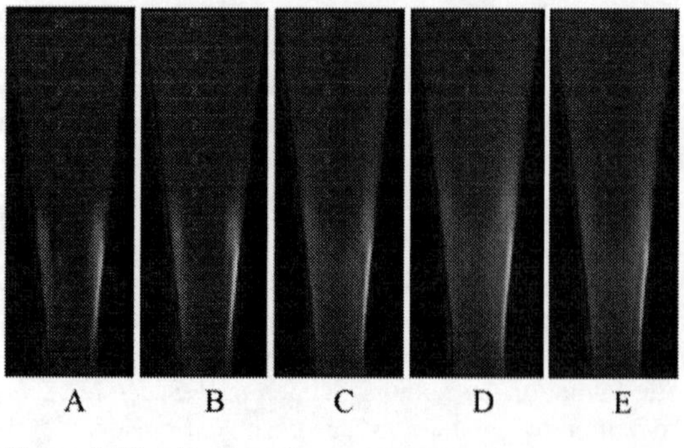

Figure 4. Visualization of measurement.

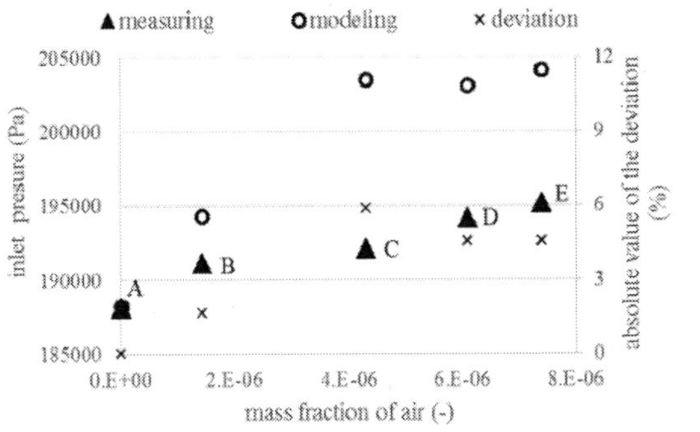

Figure 5. Inlet pressure dependence on the mass fraction of air.

Dependence of measured and mode led inlet pressure vs. mass fraction of air is evaluated on figure 5. Modelling and measurement error is 6 % (compared to the article mentioned in the introduction [5] where the error was (10 - 25) %. It is very good result.

From the graphical dependence of the volume fraction of vapour, air, vapour and air, and water on a mass fraction of air in the whole volume of the CD nozzle it is evident that the volume fraction of water decreases with increasing undissolved air. The volume fraction (or amount) of vapour can be evaluated as constant in whole the volume of the CD nozzle. It follows that the volume fraction of water (and consequently the size of the cavitation area) depends on the amount of undissolved air in the liquid.

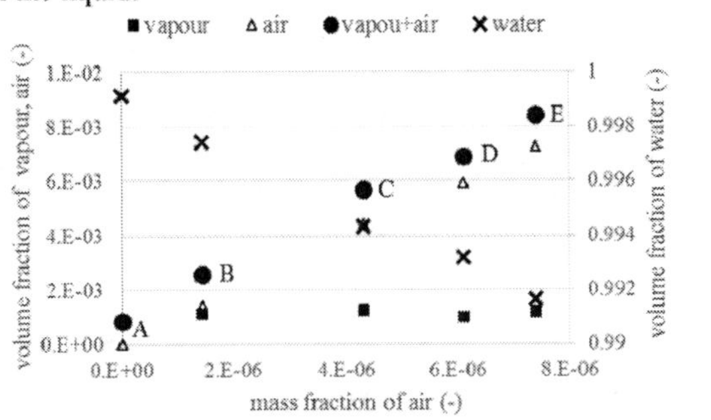

Figure 6. Dependence of volume fraction of vapour, air and water in the whole volume of the CD nozzle vs. mass fraction of air.

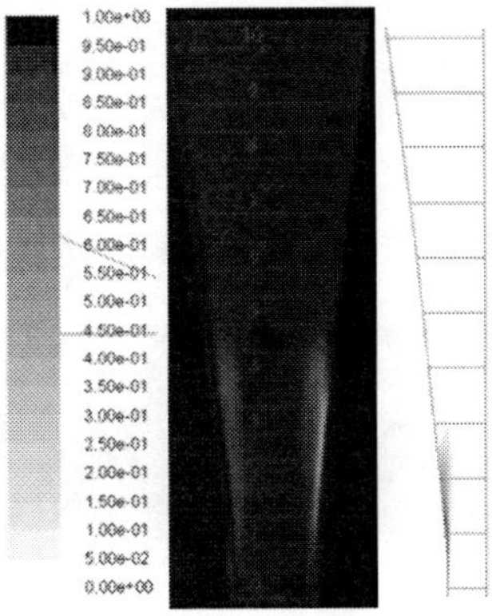

Figure 7. Variant A - comparison of the size of the cavitation cloud gained by experiment and modelling of water and vapour.

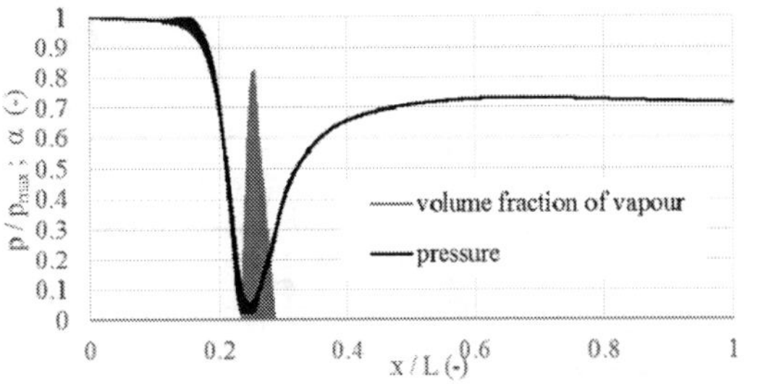

Figure 8. Variant A – distribution of dimensionless pressure ratio and volume fraction of vapour along the wall.

On figure 7 there is apparent good consensus of the size of the cavitation cloud when modelling variant A (water and vapour).

Change of pressure ratio along the length of the CD nozzle is apparent on the figure 8. Pressure minimum in the throat of the CD nozzle has value of 2340 Pa (saturated vapour pressure). At this place vapour is observed. The amount of vapour volume is declining with increasing pressure (along the length of the CD nozzle).

Vapour area is depicted at the left of figure 9 and vapour and air area is depicted at the right. When comparing figure 7 and figure 9 it is evident that the area of vapour was not changed. Increasing size of the cavitation cloud is in accordance with the growing amount of added air. Modelling shows that the vapour volume fraction changes only slightly depending on mass fraction of added air (see figure 6). The size of the area depends on the amount of undissolved air in the liquid.

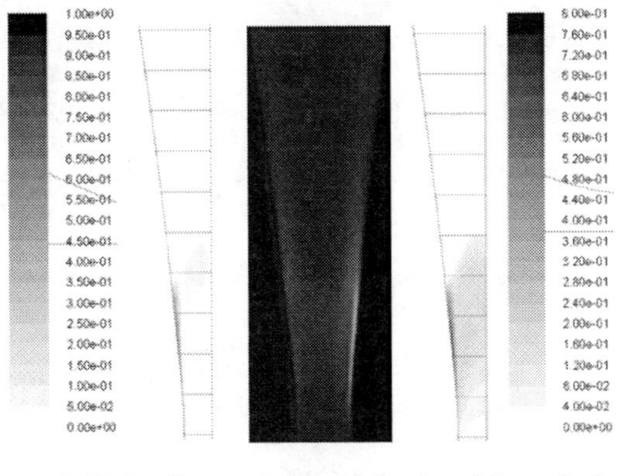

Figure 9. Variant E - comparison of the size of the cavitation cloud gained by experiment and modelling of water, vapour and air.

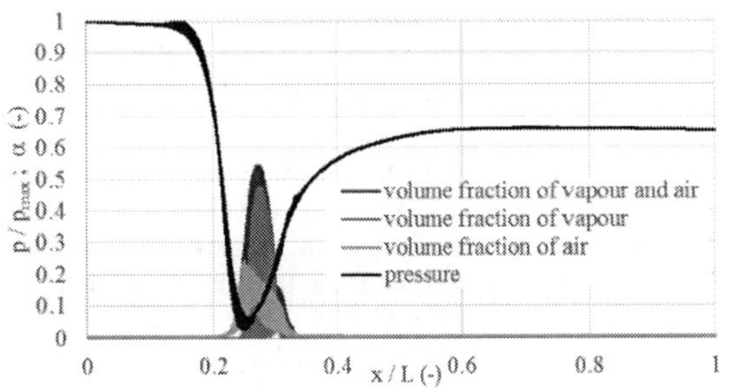

Figure 10. Variant E – distribution of pressure ratio and volume fraction of vapour along the wall.

Figure 10 shows that variation of pressure ratio along the length of the CD nozzle is not changed in comparison with the variant A but volume fraction of vapour decreased and the length of the area vapour and air is increased. The amount of vapour remains approximately constant in comparison with variant A, see figure 11, where the lower (dark) part of the column indicates the length of the cavitation vapour area. The whole length of the column indicates the size of the cavitation area of vapour and air.

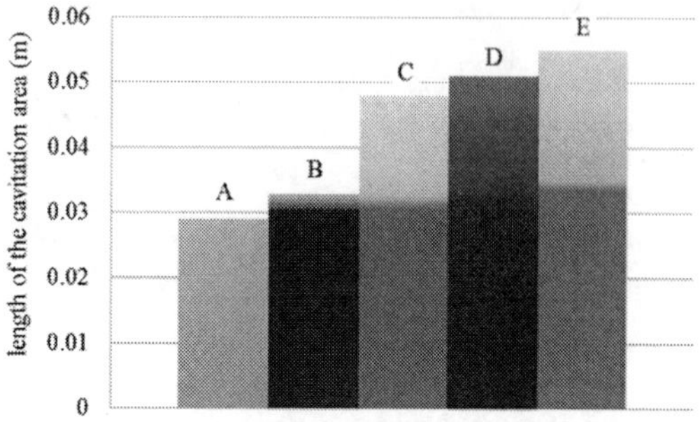

Figure 11. Length of the cavitation area on the wall (variant A - E).

7 Conclusion

The work is devoted to theoretical investigation of multiphase mathematical model of cavitation at higher flow rates where there is disagreement in the measured and calculated hydraulic variables before and behind hydraulic element. The article deals with the specification of the general mathematical model of flow with cavitation and its application to multiphase flows (water, vapour, air). The results were verified with experiments aimed on water flow with cavitation in the convergent-divergent nozzle and the air saturation.

- RNG k-ε mathematical model of turbulent flow with cavitation was defined for Reynolds number at the input about 100 000. Flowing medium was specified as a mixture of incompressible liquid (water) and vapour. Air was considered as a next compressible phase.
- The CD nozzle resistance was increased by adding air into the hydraulic circuit, therefore measured parameters as flow rate and pressure at the inlet and outlet at constant speed of the pump are changed. Hydraulic values obtained from experiment and numerical calculation differed by less than 6 %.
- Size of cavitation cloud depended on the amount of undissolved air, increasing amount of air increases the size of the cavitation cloud. This was confirmed by visualization and numeric calculation.
- Significant numerical result is that the amount of vapour is not dependent on the amount of air in the liquid.

We can conclude that the commonly used mathematical model of cavitation should be extended to phase of air when applied to liquid flow in hydraulic elements at higher operating parameters.

References

1. Fluent 15 – User's guide. [Online]. Available to from: <URL: http://sp1.vsb.cz/DOC/Fluent_6.1/ html/ug/ /main_pre.htm>.
2. M. Kozubková, *Numerické modelování proudění – FLUENT I.* [Online]. Available to from: <URL: http://www.338.vsb.cz/seznam.htm>. (2005)
3. M. S. Plesset, R. B. Chapman, *J. of Fluid Mechanics*, **47** (02), (1971).
4. A. K. Singhal, M. M. Athavale, H. Li, Y. Jiang, *J. of Fluids Engineering*, **124** (3), (2002).
5. A. Ullah, M. Nadeem, A. N. Ahmed, *Emerging Technologies, 2009. ICET 2009. International Conference on. IEEE*, (2009).
6. G. E. Reisman, M. E. Duttweiler, C. E. Brennen, *ASME Fluids Engineering Division Summer Meeting*, (1997)
7. J. G. Auret, O. F. R. A. Damm, G. J. Wright, F. P. A. Robinson, *Tribology international*, **26** (6), (1993).
8. J. M. Ye, Y. Xiong, L. I. Fang, S. Q. Chen, *J. of Hydrodynamics*, **Ser. B**, **22** (5), (2010).
9. J. Dobeš, M. Kozubková, *EPJ Web of Conferences* **67**, p. 02019. EDP Sciences. (2014)
10. S. Martynov, D. Mason, M. R. Heikal, S. S. Sazhin, M. Gorokhovski, *International Heat Transfer Conference 13.* (2006)
11. P. Roosen, O. Unruh, M. Behmann, *Report of the Institute for Technical Termodynamics.* (1996)
12. E. Winklhofer, E. Kull, E. Kelz, A. Morozov, *ILASS-Europe 2001, 17 International Conference on Liquid Atomization and Spray Systems.* (2001)
13. L. Hruzik, M. Vasina, A. Burecek, *EPJ Web of Conferences* **45**, 01041 (2013)
14. J. Jablonská, M. Kozubková, *J. of Computational Multiphase Flows*, **7** (2), (2015)

Acknowledgements

The work presented in this paper was supported by a grant SGS "Zkoumání dynamiky tekutinových systémů" SP2015/95.

EPJ Web of Conferences **114**, 02051 (2016)
DOI: 10.1051/epjconf/201611402051

Design of hydraulic recuperation unit

Pavel Jandourek[1,a], Vladimír Habán[1], Martin Hudec[1], Lenka Dobšáková[1] and David Štefan[1]

[1]*Brno University of Technology, Faculty of Mechanical Engineering, Energy Institute, Victor Kaplan Department of Fluid Engineering, Technicka 2896/2, 616 69 Brno, Czech Republic*

Abstract. This article deals with design and measurement of hydraulic recuperation unit. Recuperation unit consist of radial turbine and axial pump, which are coupled on the same shaft. Speed of shaft with impellers are 6000 1/min. For economic reasons, is design of recuperation unit performed using commercially manufactured propellers.

1 Introduction

This article deals with the study of recuperation unit, which would use excess pressure energy to propel the liquid. Recuperation device consists of a radial turbine and an axial pump, whose normal wheels rotates at the same speed and can be on a common shaft or separated by a magnetic coupling. Construction must be secured tightness of the turbine and pump circuit.

2 The suggestion of solution

According to set parameters are determined the dimensions of the hydraulic machines.

Table 1. Design parameters.

Design parameters of turbine	
Head H_T	10 m
Turbine flow Q_T	0.5 l/s
Design parameters of pump	
Specific energy Y	13.5 J/kg
Pump flow Q_P	1 l/s

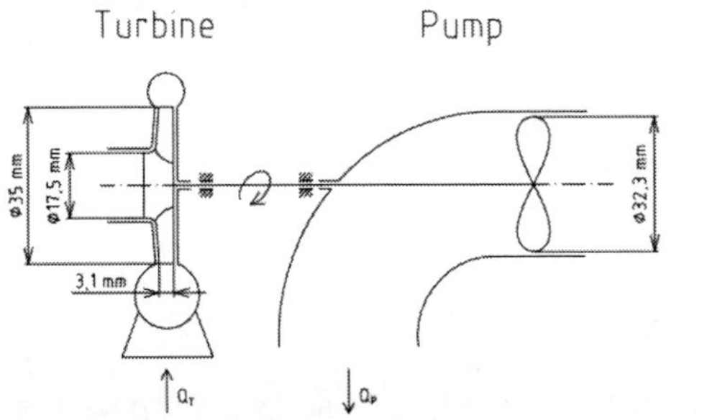

Figure 1. Diagram of recuperation unit.

2.1 Model of radial turbine

The turbine is designed to specified parameters. To better guide the water to the runner is used diffuser. The shaft is mounted in plain bearings, that are lubricated with water. Spiral case and runner are made using 3D printing. Spiral surface is painted, in order to avoid water seepage through the surfaces of turbine.

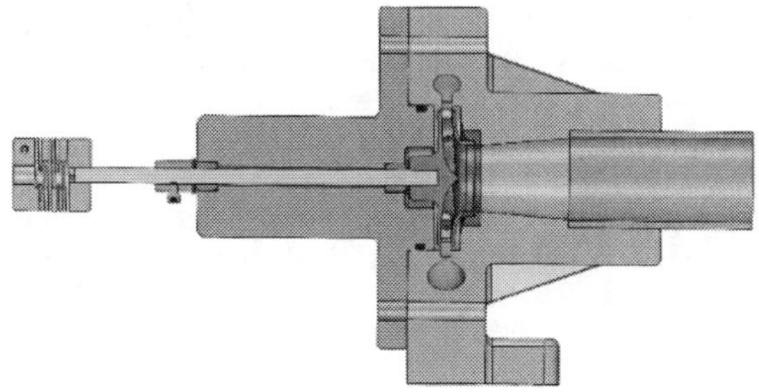

Figure 2. Cut of radial turbine.

2.2 Model of axial pump

The impeller is pressed onto the overhung shaft, which is led out of the elbow. They are used in plain bearings lubricated with water. Axial force on the impeller is captured by axial plain bearings below impeller and before the sleeve. Between a straight piece of pipe and elbow could be bet the diffuser, which directing swirling fluid behind the impeller.

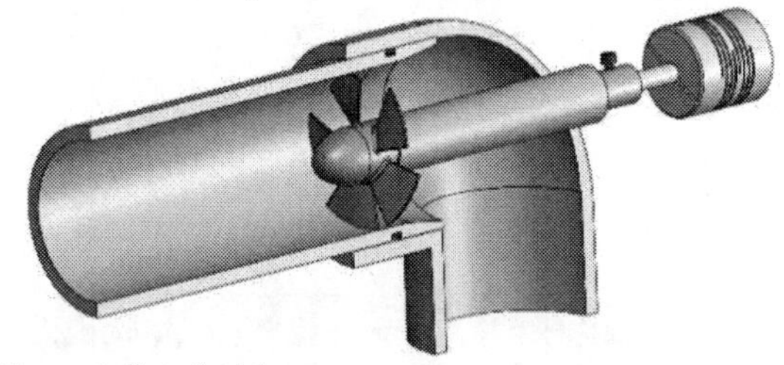

Figure 3. Cut of axial pump.

[a] Corresponding author: y101051@stud.fme.vutbr.cz

2.3 Recuperation unit

Recuperation unit is formed by the merger turbine and pump.

Figure 4. Cut of recuperation unit.

3 Development and measurement

3.1. Measurement of turbine

Measuring rig for turbine is composed of a suction tank, which serves as a liquid reservoir, the suction and discharge pipes and the feeding pump with variable speed drive. For feed pump is included a flowmeter and the measured turbine. On the discharge line is positioned throttle valve. Torquemeter for measuring output, torque and speed is attached bellows couplings between the pump and motor. The electric motor has a speed drive. There are two pressure sensors on the measurement rig. One is in suction pipe and second is in discharge piping. For donations of pressure and flow are observed length to calm the roiling of flowing liquid. Pressure sensors are powered by a stabilized source.

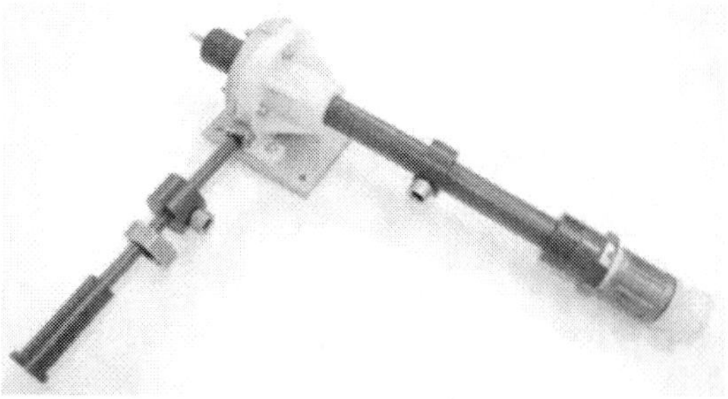

Figure 5. Turbine with stilling sections.

Table 2. Measured quantities at the turbine.

p_1	pressure in discharge tube	Pa
p_2	pressure in suction tube	Pa
Q	flow	m^3/s
M_k	torque	Nm
P_{out}	output	W
n	speed	1/s

Measurement procedure - after flooding rig and subsequent venting set the turbine speed. Feed pump settles to the required pressure drop. Let it steady and save the measured values. We continue adjusting other operating points with different speeds until to measure whole characteristic of the turbine. Shutting turbines perform gradually decreasing head and speed of the turbine. To determine the difference of values in sensors measure the pressure difference in the measurement rig in calm state and then drain the water from it. Saved mean values will be used to determine the functional dependence Q11 = f (n11), η = f (n11). The resulting characteristics create polynomial fitting the computed points.

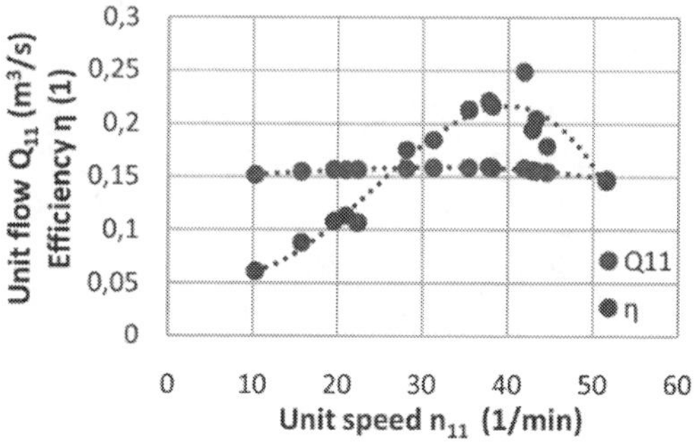

Figure 6. Dependence of unit flow and efficiency on unit speed.

Dependence of torque M_k on turbine flow is recalculated for different values speed of runner.

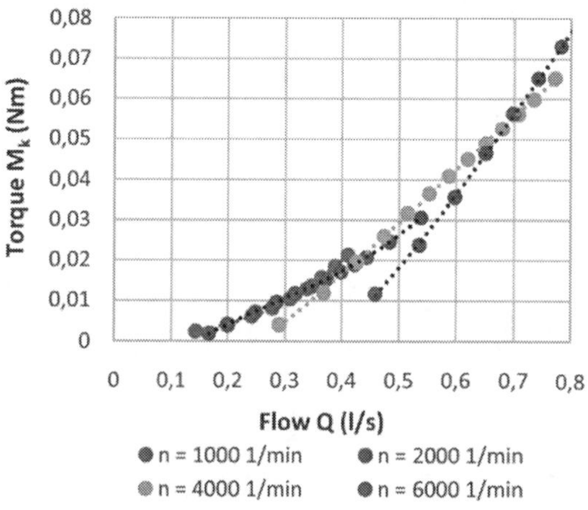

Figure 7. Torque of turbine in dependence on flow for values of speed between 1000 until 6000 1/rpm.

3.2. Measurement of pump

Measuring rig for pump is composed of a suction tank, which serves as a liquid reservoir, the suction and discharge pipes and the feeding pump with variable speed drive. For feed pump is included a flow meter and the measured axial pump. On the discharge line is positioned throttle valve. Torque meter for measuring output, torque and speed is attached bellows couplings between the pump and motor. The electric motor has a speed drive.

There are two pressure sensors on the measurement rig. One is in suction pipe and second is on discharge piping.

Impellers proposal was made by its own program operating on the principle of differential geometry. Due to the extremely poor quality of surface axial impellers made by using 3D printing, have been purchased for the development of pumps 3 propellers, designed for marine modellers.

Figure 8. 3,4 a 5 bladed ship impellers with diameter 40 mm.

Figure 9. A device for measurement of the ship propellers with a diameter 40 mm.

Table 3. Measured quantities at the pump.

p_1	pressure in suction tube	Pa
p_2	pressure in discharge tube	Pa
Q	flow	m^3/s
M_k	torque	Nm
P_{in}	power	W
n	speed	1/s

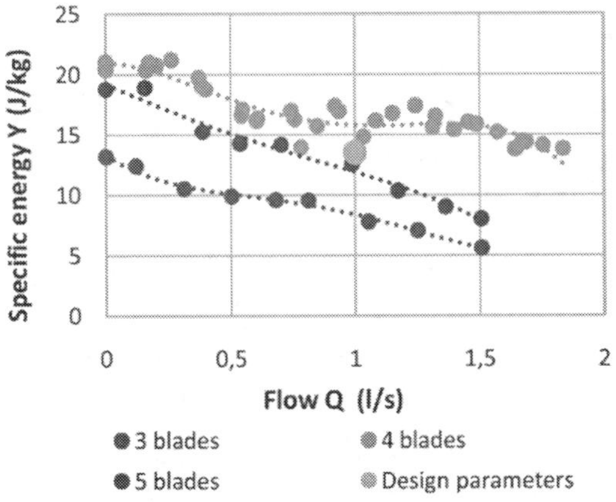

Figure 10. Dependence of specific energy on flow.

Best values while meeting the parameters of the design point reached propeller of diameter 40 mm with 4 blades.

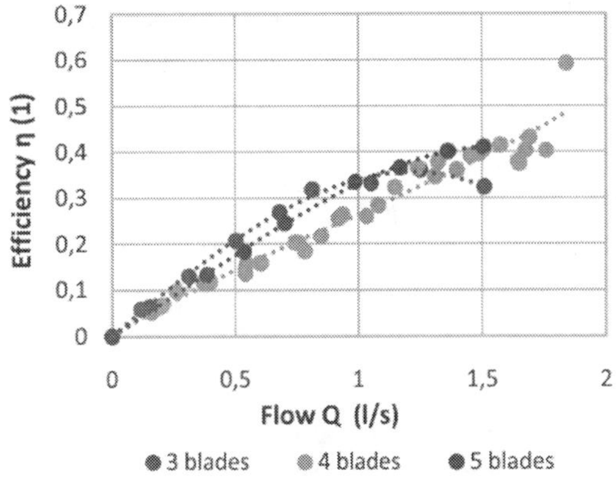

Figure 11. Dependence of efficiency on flow.

Propeller was machined from 40 to 33 mm, and was made a new hub. To be used propeller as a impeller of the axial pump, propeller must be additionally rotated. This will maintain the correct direction of flow through the pump.

Figure 12. Rotated ship propeller with drilled hub and machined on diameter 33 mm.

Figure 13. Adjusted propeller in the measuring circuit.

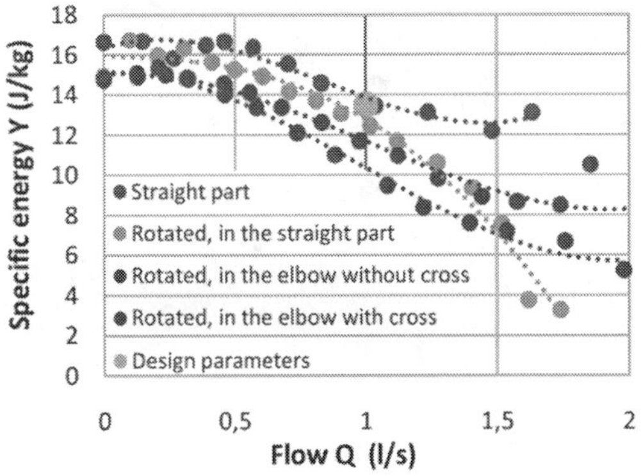

Figure 14. Dependence of specific energy on flow.

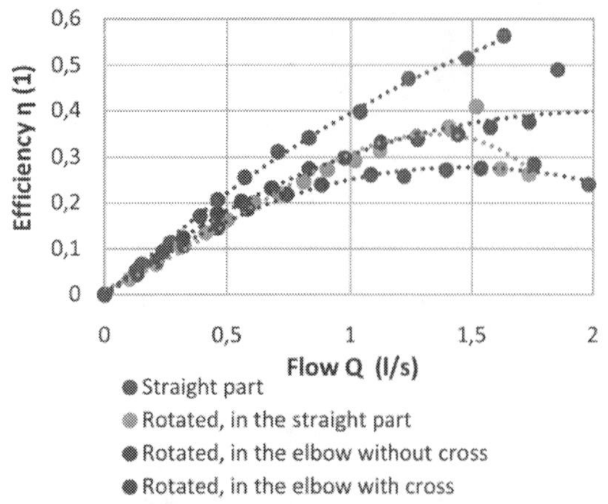

Figure 15. Dependence of efficiency on flow.

The best hydraulic profile of impeller for axial pump was created using the propeller machined from 40 to 33 mm. The specific energy characteristic is slightly smaller than the design point. This is caused by hydraulic losses in the piping system.

4 Recuperation unit

On the left side is circuit of turbine. The turbine is driven by the feeding pump. On the right side is the pump circuit, which is to pump water through the spillway. The turbine and the pump is fixed.

Figure 16. Test rig of recuperation unit.

The following graph shows the dependence of the flow on the pump flow turbine.

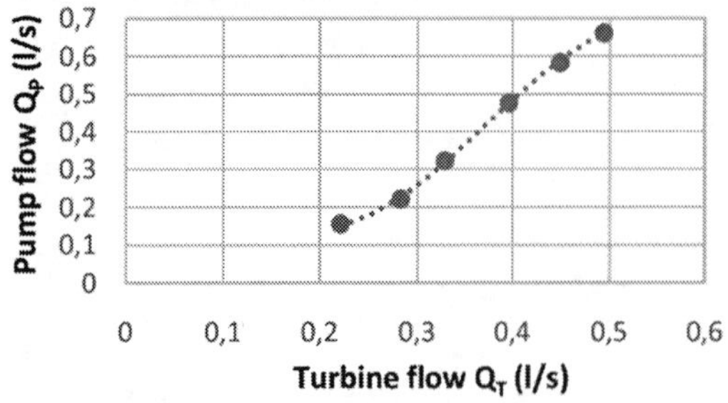

Figure 17. Dependence of pump flow on turbine flow.

Dependence of pressure losses on the turbine on the turbine flow shows resistive characteristics.

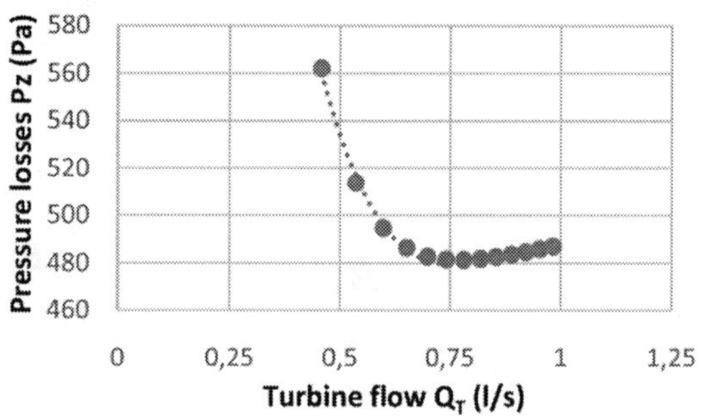

Figure 18. Resistive characteristics.

5 Conclusion

Hydraulic design of a turbine and a pump for the specified parameters recuperation unit was executed. These are the radial and axial turbine pump. These two machines work at the same speed. The turbine drives an axial pump and the pump pumps liquid.

According to the design has been using the method of 3D printing made turbine with these hydraulic parameters (Fig. 6. and 7.).

Axial pump was also made using 3D printing technology, but the results of multiple variants were not satisfactory. Especially the influence of roughness printed 3D models and the effect of the gap between the blades and stator, led to very low efficiency and failure to comply with the required hydraulic parameters. The final design consisted in the use of commercial hydraulic of the modelling propeller, which after tests achieved the required flow rate and efficiency

Fig. 7. shows the flow of water from the storage tank, depending on the water flow turbine calculated according to specified parameters. By entering the flow of the water pump 1.034 l/s. Hydraulics achieve with current flow 0.66 l/s. It could be assumed, that the efficiency (output) turbine produced by a other method than 3D printing, will be significantly higher and the pump will reach a higher flow rate. The models produced by 3D printing have high roughness value that the flow is causing high losses. Measured the efficiency of the turbine was up 24%. By reducing the roughness of the hydraulic space can be expected to increase the efficiency to 35-45%. Using hydraulics, i.e. hydraulic of pump and turbine can be designed a functional sample of recuperation unit. It can assume, that the parameters will be better than the corresponding Fig. 7., particularly with regard to the lower surface and smaller mechanical losses of prototype. In future prototype of recuperation unit would be appropriate to measure hydraulically again.

Acknowledgement

Presented research was supported by center of competence of Technology Agency of The Czech Republic TE02000232 “Rotary machines”.

EPJ Web of Conferences **114**, 02052 (2016)
DOI: 10.1051/epjconf/201611402052

A liquid interaction with ultrahydrophobic surfaces

Darina Jasikova[1,a], Michal Kotek[1], Simona Fialova[2] and Vaclav Kopecky[1]

[1] *Department of Physical Measurement, Technical University of Liberec, The Institute for Nanomaterials, Advanced Technology and Innovation, Studentska 1402/2 Liberec 1 461 17, Czech Republic*

[2] *Faculty of mechanical engineering, Brno University of Technology, Technicka 2896/0, Kralovo Pole, 61669, Brno, Czech Republic*

Abstract. The interaction of the liquid with ultra-hydrophobic surfaces was so far studied through estimation of static contact angles. It appears now that this interaction is more complex, and cannot be described only with static methods. Effect of ultra-hydrophobic surfaces and their advantages are also particularly in dynamic interaction with liquids. One of the parameters that determine the character of the dynamic interaction is presence of air film close to the surface. The thickness of air film can be measured with long distance microscopy and the interaction with the flow using micro PIV method. Here we present the results of measurements of the air film that is created close to ultra-hydrophobic surfaces and the dependence of its thickness on the Re number.

1 Introduction

The effect of the surface roughness on the skin friction is well known. There are not explored enough interactions between chemically modified surfaces and liquids. First, because the development of new surfaces is based on the influence of surface energy using cold plasma technique and applying special chemical structures. In any event, such surfaces macroscopically appear as ultra-hydrophobic (UH) and provide their use specifically in the interaction with liquids, i.e. in hydraulic machines, or a pipe system. It is expected that the change of the static contact angle (CA>150°) is reflected in the dynamic properties, e.g. skin friction coefficient and reduction of hydraulic losses.

Dynamic interaction of the liquid with UH surface is accompanied by formation of a layer of air film. The parameters and the quality of the air film are dependent on the quality and uniformity of UH surface and also its roughness, which results from the basic characteristics of skin friction.

Although, there were already published theoretical works on mechanism of air layer, there has not been presented satisfactory experimental evidence yet. [1]

The motivation of this project is the visualization of air film and its quantitative description. There has been compiled methodology and principles based on micro PIV and long distance microscopy methods as part of this project.

2 Experimental facilities

The UH surface is commonly coated on boards or entire blocks. Here there was manufactured the channel that should followed the requirement of the visualization methods.

Manufacturing of experimental facilities belongs to the one of research itself. It's a sector as important as the compilation of measurement technology and design of fluid section. There was designed a channel made from the modification of plastics (PCBs, PVC, PA) and traditionally manufactured by segments bonding.

Individual parts were designed so channel assembly was created (mini-channel) with dimensions of width w = 5 mm, height h = 3 mm, and functional length l = 150 mm. Each part of the channel was accurately machined and edges polished for a defined roughness Ra of 0.5 micron. Furthermore, parts were glued together as it is seen on figure 1.

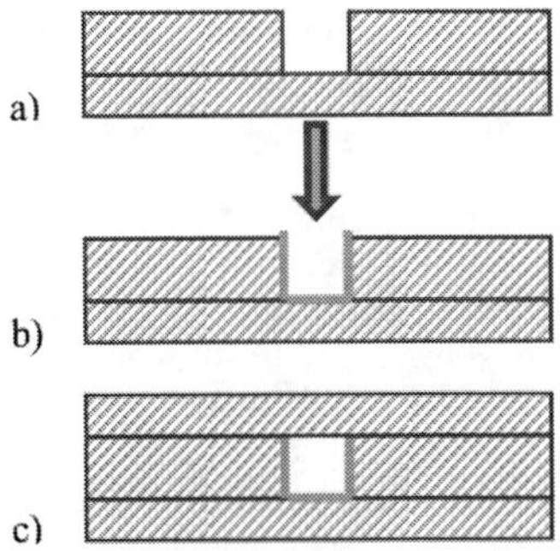

Figure 1. Process of manufacture of the mini-channel: A) opened channel prepared for the surface treatment procedure, B) surface treatment, C) closing the channel with the optically clear and untreated surface.

The inner surface of the mini-channel was treated with UH surface in cooperation with Masaryk University

[a] Corresponding author: darina.jasikova@tul.cz

of Brno. Before the completion of the experimental facility, the surface roughness and static contact angle was measured.

Figure 2. Construction of segments, their completion and set up with the joining reductions connected with the rest of hydraulic circuit.

The channel had the rectangular profile, so it was fitted with interfaces from circular to rectangular section for additional connection with hoses. The interfaces were designed in CAD software and produced by Rapid Prototyping - 3D printing technique. Connections were made of Vero White material, i.e. cross linked UV photopolymer used in 3D printers. The jumpers were optimized for Luer-lock connection system and a standard hose (6x4) mm.

3 Visualization techniques

The visualization of the air film came out from the micro PIV measurement technique. This technique allows the visualization, tracking and evaluation processes within a very small scale of nanometres to micrometres. The design of the measurement track and the samples were adapted to his method.

The micro PIV technique provided us the information about the air film thickness. This technique was suitable for studying small area of interest, such as 1,5mm. We used the long distance microscopy to cover the whole velocity profile across the channel – 5mm.

The basis of micro PIV technique is a laser beam, which illuminates the entire volume of the examined space. The investigated volume of liquid is enlarged with an inverted microscope, which is equipped with a magnifying lens. The magnifying lens also ensures the depth of field. The Technical university of Liberec is equipped with micro PIV system that is based on an inverted microscope Leica DMIL and fitted with a set of lenses: 5x, 10x, 20x and 50x. For illumination of the evaluated area NewWave Gemini pulsed laser is used Nd: YAG type, energy of 120mJ per pulse length of 10ns, wavelength 532nm.

The images are captured with Neo CMOS chip 5,5MPixel size of 6.5 microns camera. Laser and camera system is controlled from a computer and synchronized with the external TimerBox.

Unlike standard PIV method, where the depth of field depends on the distance to the measurement object from the camera lens, f-number, in the micro PIV method depth of field is dependent on the parameters of the microscope lens, index of refraction of the liquid, and the wavelength of used light. Other parameters are related to the system of lenses: NA - iris, M - total magnification microscope, and e - the distance between the pixels on the CCD camera.

The depth of field is calculated using the relation (1). The arrangement of micro PIV has a constant depth of field 4um at 10x magnification. This factor is important for choosing the size of seeding particles. Here in this investigating we used coated Rhodamine B 2 µm fluorescent particles.

$$\delta = \frac{n\lambda}{NA^2} + \frac{n.e}{NA.M} \qquad (1)$$

We used set of microscopic calibration targets for the calibration of the system to determine the magnitude coefficient. At 5x magnification dimensional coefficient was set at 0.1999 and at 10x magnification was set at 0.0998. The records of calibration grid also allow creating an image for further image dewarping to eliminate the image distortion. The figure 3 shows the calibration target with 50 µm grid.

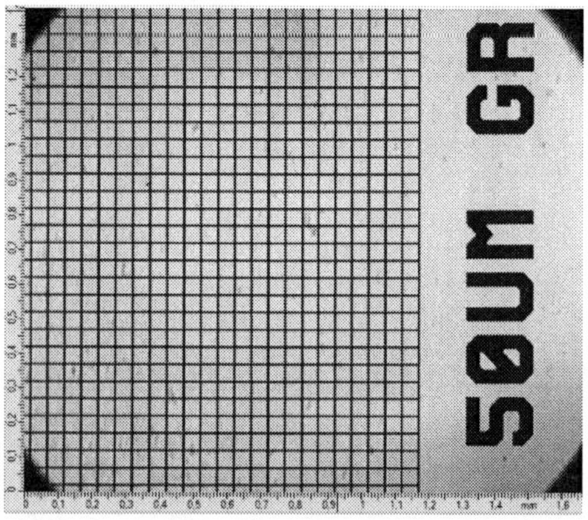

Figure 3. The example of the calibration target optimal for the magnification 10x.

The long distance microscopy method was used for the measurement of the velocity profile over the whole channel in one step. This optical setup was based on INFINIPROBE™ TS-160 universal macro/micro imaging system that enables 4x and 16x magnification. This probe was working in the set of compressor for the wide field of view, and the working distance was 100mm in the macro regime. The area of interest was illuminated with the laser light led with optical cable towards the sample.

The measurements run at 12Hz frequency with 60mW laser power at one flash. The light from the laser head was brought to the microscope lens system with optical cable. For each measurement at least 290 records were always acquired and were further processed by the correlation function, validation and filtering before being evaluated statistics and estimated velocity profiles.

4 Testing section

The assembling of hydraulic circuit went from lab equipment and requirements of the visualization technique on the organization of the experiment.

The investigation of the interaction between hydrophobic surfaces and liquids was run under Re number in laminar, transient and turbulent zone. Following this requirements we used Micropump gear pump. The advantage is almost zero slip, accurate dosing and zero pulsation. Faultless operation of the pump pulsation has been assessed by measurement. Pump speed was controlled by the applied voltage. The system was therefore enhanced with a linear precision voltage source KORADO KA3005P controllable from a PC and an adjustable voltage with an accuracy of hundredths. The flow was measured by the flowmeter FL 1440 to ensure the repeatability of measurement. The dependence of voltage, flow, and consequently Re numbers were necessary to provide for the conversion table.

Re number was determined according to the equation 2 and 3.

$$D = \frac{4A}{P} = \frac{4wh}{2w+2h} \quad (2)$$

$$Re = \frac{uD}{\nu}, \quad (3)$$

where u is the mean velocity on the outlet, D hydraulic parameter, and ν is kinematic viscosity of water ($1{,}004.10^{-6}$ at 20 °C). Characteristic hydraulic parameter of the mini channel is 3,75 mm. The calculation of the Re number was performed for the channel without the surface treatment, so the effect of the air film was neglected.

5 Results and discussion

5.1 The preparation of the samples

The inner walls of channel were treated with three different methods on polypropylene and polycarbonate to create UH deposited layer.

The sample A was spray coated with nanopolymeric UltraEverDry®, UED® in two layer was applied on the sample B and the sample C was continuously sprayed with UED® and cold Ar+O_2 plasma treated [4].

Surface roughness was measured before the channel chemical treatment of the inner surface and it was 0,5um. This surface roughness correspond to static contact angle 110°. We used the touch laboratory device Mitutoyo Surftest SV-2000 for the surface roughness estimation and it was measured and averaged over 100 mm. The average roughness of the treated surface sample A increased to 0,8 μm, on the sample B it increased to 1,5 μm, and on the sample C roughness increased to 0,9 μm.

The global increase of the surface roughness can be explained due the UED® nanoparticles coagulation during the process, as it is seen in figure 4. The surface roughness has the direct effect on the static contact angle: the surface on the sample A had the CA 150°, sample B 160°and the sample C 150°. Wet surface was also measured, but the changes in static contact angle were not significant. All measurements were also done on the reference sample. The reference sample was left untreated.

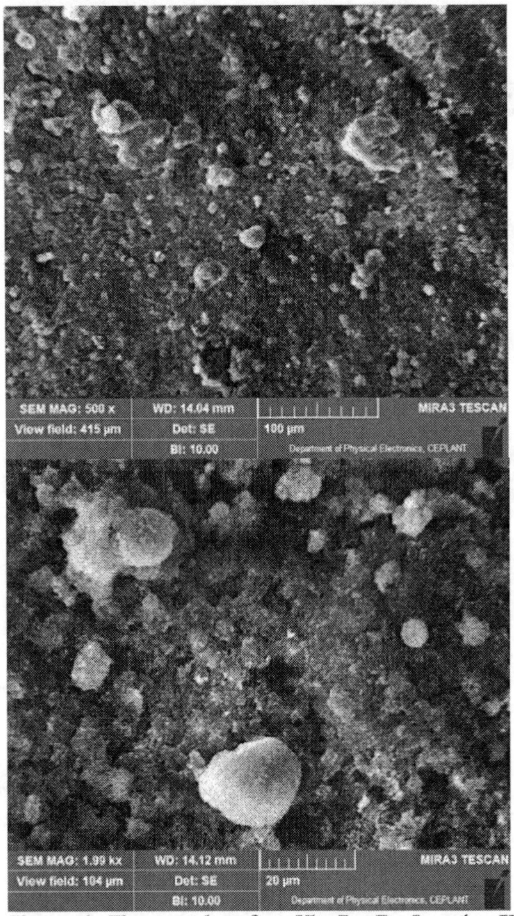

Figure 4. The treated surface UltraEverDry® under SEM microscope, magnification of 500x and 2000x.

5.2 Estimation of air film

The character of the air film comes from the principles of manufacturing of samples. There are three walls treated and the one that creates the cover is left clean. This clear wall allows optical access for the visualization and measurement.

Here we worked with distilled water. It is supposed that water contains certain amount of dissolved inert gases and micro-bubbles. The whole amount of present gas in water was 8 mg O_2/l at 20 °C. This value was measured with laboratory oximeter. There can be estimated total dissolved gas in the water with the knowledge of concentration of oxygen in air. The concentration of the dissolved air and amount of bubbles is strongly dependant on the partial pressure. It is said that degassed water contains about $0{,}911 \times 10^{9}$ nucleus of most probable size about 6 μm. [1]

Basically the growth of dissolved air on the surface can be divided into three categories: the first describe the situation, when the dissolved air is due the diffusion entrapped on the surface as the unique bubbles. These bubbles are said to be initialized in the turbulence layer in seeds. The seeds can be caused by local roughness, impurities or local change in pressure/density gradient. The growth of these single bubbles is time dependent and could grow up to 1,5 mm. Then these bubbles are changing character of the boundary layer and, of course, they active react on local pressure changes that lead to pulsations.

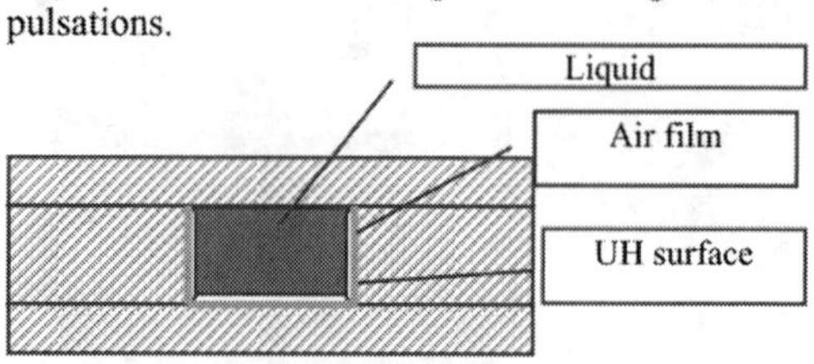

Figure 5. The geometry of the air film formation in the mini-channel.

In the second category the dissolved air is formed into more or less uniform air film. The formation of the air film is dependent on the flow pattern. This flow pattern is stratified into film–gas interface and could finally lead to multi-phase flow. The film structure is dependent on the geometry of the channel, roughness of the surface and it can be symmetrical or unsymmetrical.
It is said, that the air film thickness is dependant mostly on the channel geometry, that is represented by the hydraulic parameter D and α the amount of the dissolved air, so we can estimate the relation:

$$d=D(1-\sqrt{1-\alpha})/2 \tag{4}$$

, but according to previous formulation, that the air film is dependent on the flow pattern, this can be expressed using Re number.
Following this formulation the interaction of the liquid flow and the air film can be divided into two regions: Re<500, and Re>500. In the first are acting macroscopic eddies instead of the second one with microscopic eddies. Anyway, here the interaction moved from the liquid/surface to gas/liquid interface.

In the third category of formation of dissolved air, it overgrows the channel geometry and aerates the channel.

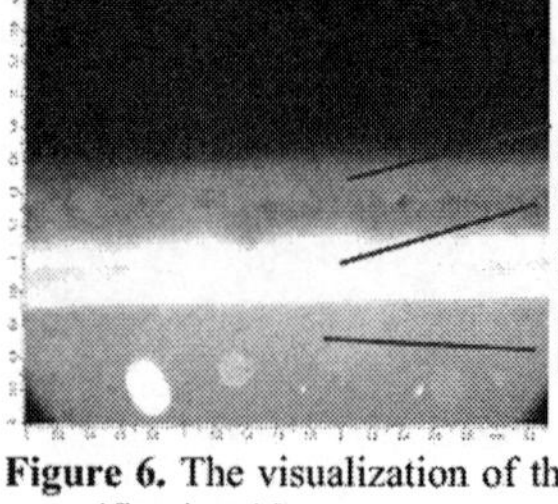

Figure 6. The visualization of the air film with the microscope magnification 10x.

Re 340

Re 1400

Re 3100

Figure 7. The visualization of the air film on the sample A – the situation in the whole channel. There are three examples selected: Re 340, Re 1400 and Re 3100.

This theoretical description of the existence of the air film on the UH surfaces had to be firstly proved its existence and demonstrated with the visualization technique. This would allow further description of the effect.
The existence of the air film formation was proved with microscopic view. The film is formed in the region Re<500 and can be seen as light stripe that reflects the light. This stripe narrows in dependence on the Re number, respectively of the flow velocity. The visualization of the air film is seen in figure 6.

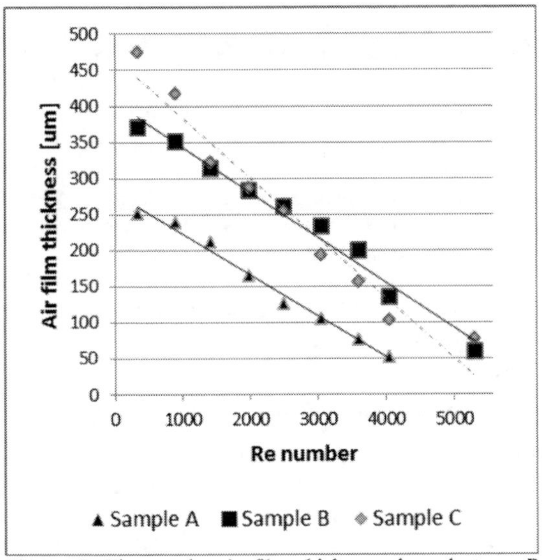

Figure 8. The graph: air film thickness dependant on Re number.

The microscopic view proved the local existence of the air film, but we were interested in the situation in the whole channel geometry. For this study we used the long distance microscopy setup. Figure 7 shows the complex situation of the air film formation depending on the Re number. There could be seen the decrease of the film thickness, when Ren number exceed the value for transient zone. The air film completely extinguished with high Re numbers.

Conclusiveness of this visualization was validated with the fluorescent method. The water was coloured with fluorescence dye (Rhodamine 6G), that is concentration sensitive and the air film was visualized in the green laser light (532 nm) for the excitation. The emitted light gave us information about local concentration and the relation between water and gas. The air film was seen as a very dark region. This means there was no concentration of the fluorescent dye in this layer.

5.3 The effect of air film on velocity profile

After this proof of the existence of symmetrical air layer with two different methods, that were repeatable, we studied the effect of the air film on the character of the velocity profile [5].
In the first step the velocity profile was measured. The air layer reduces the lumen cross section, resulting in maintaining a constant speed of the pump it leads to increase of the maximum velocity in the flow profile. The maximal differences could be seen in laminar zone. The thickness of air film for Re 340 reaches for sample A 250 µm that reduces the lumen cross section to 82%, for sample B 360 µm, that means 75%, and for sample C the air film thickness is 470 µm, so the reduces are to 68% of full cross section area. This corresponds with the maximal velocity that is seen in figure 9. There is added velocity profile for reference sample in figure 9, 10, and 11. This reference sample was untreated polycarbonate channel with (5x3) mm cross section. The velocity profiles were taken over the whole cross section and respect the air film on the boundaries. The profiles were analysed from the statistics that contained 290 dataset of vector maps taken with frequency 12Hz.

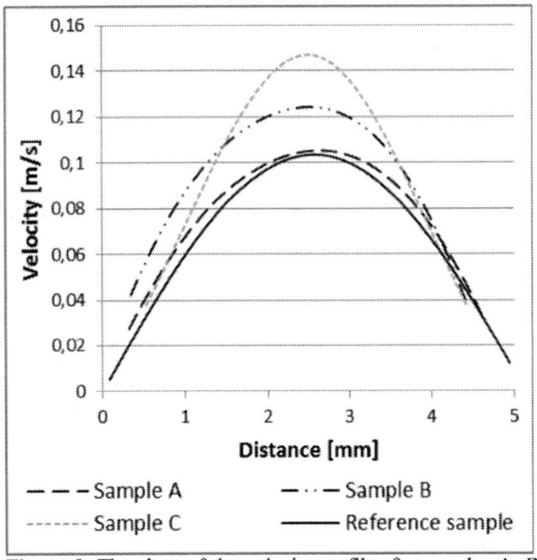

Figure 9. The chart of the velocity profiles for samples A, B, and C for Re 340.

In the figure 10 there is seen the situation of the velocity profile for the Re number 1400 that supoosed to be in the transient zone. There is significant decrease of velocity in the middle of the channel. This effect is caused by the increases of micro eddies in the middle of the cross section. The air film thickness for sample A was 200 µm, for sample B 314 µm and for sample C it was measured 322 µm. The effect of decrease of cross section area due the air film is not so significant and more reduce the transition on the boundaries close to the surface [2, 3].

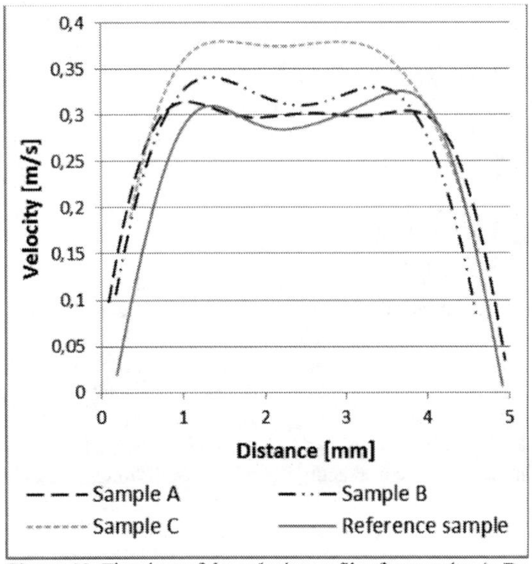

Figure 10. The chart of the velocity profiles for samples A, B, and C for Re 1400.

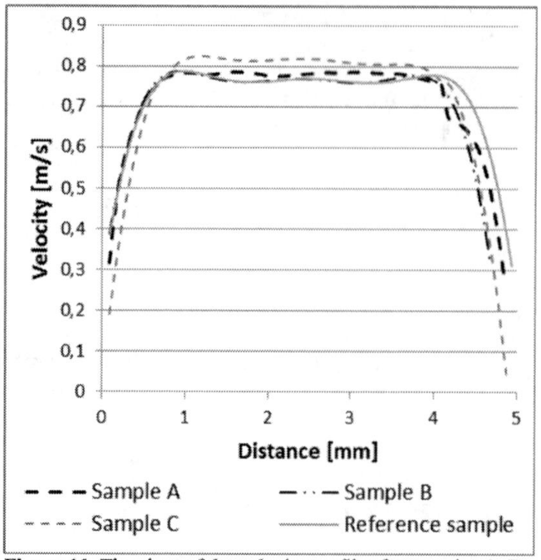

Figure 11. The chart of the velocity profiles for samples A, B, and C for Re 3100.

The figure 11 shows the chart of velocity profile for Re 3100 that is supposed to be in higher edge of transient zone. The air film disappears with the higher Re numbers. For Re 3100 the air film thickness reached round 50 μm for all samples. There is supposed to be no air film for the Re number in turbulence zone, for Re > 4000. For high Re number the surface structure is fully wet and the pores are filled with water. When this happens, there is no recovery of air film over the surface, even if the water flow stops. The measurement of the static contact angle proved the decrease of 20° and more, on the fully wet surface. This corresponds with nonexistence of air film on the surface that reach static contact angle round 110°. This value of static contact angle was typical for the untreated reference sample. The ultra-hydrophobicity can be restored after deep drying of the surface. Testing how many cycles' surface withstands (full wetting and drying) may be a novel method that will test the quality of the UH surface dynamic test. In this study all samples were strained three times and the result were comparable, this means there were no significant changes in the structure that would negatively influence its quality.

6 Conclusions

Dynamic interaction of the liquid with UH surface is accompanied by formation of a layer of air film. The proof of the quality of air film belongs to one of dynamic characteristics of ultra-hydrophobic surfaces. So far, the ultra-hydrophobic surfaces were described only by static method, measuring contact angle. However, sometimes the UH surfaces are not achieve the expected behaviour in the dynamic interaction with liquids, although their CA is higher than 150°. The parameters and the quality of the air film are dependent on the quality and uniformity of UH surface and also its roughness, which results from the basic characteristics of skin friction. The quality of the surface can be studied with microscope. Here we set the method to explore the quality of the air film depending on the surface roughness and the Re number. This relationship is mutual according to the results. As the dynamics of fluid flow is acting on the air film, the air film is influencing the character of liquid interaction as well as the maximal velocity of the flow. The qualities of UH surface are shown as the flow velocity is kept in transient area of Re number. In any event, the knowledge of this phenomenon and the way how it affects the flow expands the awareness of the dynamic interaction between the UH surfaces and liquids and has to be taken in account in the mathematical descriptions.

Acknowledgements

The results of this project LO1201 were obtained with co-funding from the Ministry of Education, Youth and Sports as part of targeted support from the „National Programme for Sustainability I " programme and the support of the Czech Science Foundation GA13-20031S - The Properties of hydrophobic surfaces in interaction with the fluid.

References

1. Nikolay I. Kolev, *Multiphase Flow Dynamics 3*, (Springer-Verlag Berlin Heidelberg, 2007)
2. Ch. J. Kahler, S.Scharnowski, Ch. Cierpka, Exp. Fluds, **52** (2012)
3. V. K. Natrajan, K.T. Christense, Microfluid Nanofluid, **9** (2010)
4. M. Klíma, etal., *Method of realisation of polyreactions, plasma-chemical polyreactions, their modification and modification of macromolecular substances by the plasma jet with a dielectric capillary enlaced by a hollow cathode*, EP 07466017, (2007)
5. F. Pochyly, et. al, International Journal of Fluid Machinery and Systems **3** (2010)

EPJ Web of Conferences **114**, 02053 (2016)
DOI: 10.1051/epjconf/201611402053

Development of a technique to evaluate the performance of cooling headers and its applications in POSCO plate mills

Jong-Hoon Kang[1,a] and Pil-Jong Lee[1]

[1] *Thermo-Fluid and Process Research Group, POSCO Technical Research Lab., Pohang 790-300, Korea*

Abstract. Hundreds of thousands of liquid jet nozzles are widely used in controling cooling equipment to accelerate the cooling of hot plates. The holes of these nozzles become clogged or the hole size becomes smaller as foreign substances like grease, scrap, and mud containing cooling water are flushed through. In this case, cooling water sprays abnormally through the partially clogged or restricted nozzle. This causes inhomogeneous cooling of hot plates. The objective of this study is to develop an evaluation system of cooling headers including nozzles. This paper presents the method to evaluate the conditions of spraying nozzles. The sensor developed in this research measures the liquid column of spraying nozzles using differential pressure between the inside sensor block and ambient air. In addition, the results of a field test in a real, large factory are introduced in this paper.

1 Introduction

Slabs manufactured through a continuous casting process and a process using a reheating furnace are produced as thick plates having a predetermined thickness through roughing mill (RM) and finishing mill (FM). As shown in Figure 1, a slab is heated to a target temperature - depending on the grade of steel - using a reheating furnace. The slab, having passed through the reheating furnace, is rolled to a predetermined thickness. Meanwhile, the thick plate is passed through an accelerated cooling unit so as to control grain refinement and transformation structure. It is necessary to remove factors causing the need for equipment management and non-uniformity in temperature in the rolled thick plate. The factors causing the non-uniformity in temperature are relevant to the controlling of the temperature of a material in the accelerated cooling unit. The accelerated cooling unit may be configured of several banks, and respective banks include dozens of cooling headers. A respective unit header includes thousands of nozzles forming columns of liquid and aligned in rows having uniform intervals.

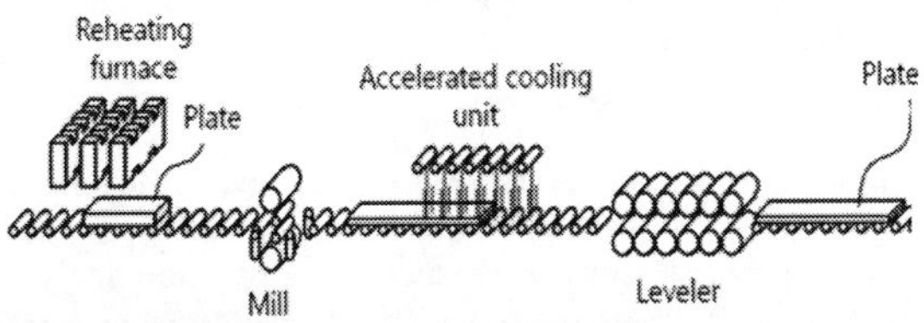

Figure 1. Schematic diagram of the thick plate manufacturing process.

[a] Corresponding author: engdr@posco.com

An impinging jet is a flow state that is widely prevalent in nature, such as the flow escaping from a water tap system, the falls of a waterfall from a high mountain, and so on. Because of high-efficiency heat transfer, impinging jets have been widely used to cool high temperature products in modem industrial production. For example, air impinging jets can be used to cool high temperature electronic elements in the electronics industry; water impinging jets can be used to accelerate the cooling of steel plates in the steel rolling process to achieve better mechanical properties and to improve productivity; likewise, impinging jet cooling has been widely used in the nuclear industry and the glass industry for a long time. However, just like the jet breakup of a tap water stream, jet breakup phenomenon is possible after the liquid impinging jet exited from the Accelerated Control Cooling (ACC) equipment had passed a certain.

The stability parameter is used to characterize jet breakup. Depending upon the point of view that is taken, can be characterized as the reciprocal of the amplification rate for disturbances or as the logarithm of the initial disturbance level. The stability of liquid jets has been studied for more than a century as a consequence of important practical applications such as cooling processes as well as fuel injection systems(citation needed). A number of experimental studies have been run to understand the mechanisms of the disintegration of liquid jets into droplets (citation needed). The measurement methods employed to study the stability of liquid jets allow us to classify them into two types that we will refer to as the local type and global type. Local type measurements are concerned with the temporal evolution of the interfacial displacement of the jet at a particular downstream position. These techniques commonly use a laser beam interacting with a small portion of the liquid jet.

Diffracted or transmitted light is analyzed to determine the diameter of the jet. For the global type measurement, the entire jet, or a significant part of it, is spatially analyzed. Global type techniques use imaging systems which record a shadow of the jet. Manual or automatic image analysis is used to determine the downstream variation of the jet diameter. Most experiments dealing with the measurement of the rate of growth of the disturbance use controlled jet destabilization.

The nozzles provided in the headers of the accelerated cooling unit are easily blocked in the inner portions of nozzles and outer portions thereof (jet holes) due to a difference in the ionization of constituent components thereof, heating and cooling. The blocking phenomenon in the nozzles of headers mostly occurs as holes are plugged with scales generated due to degradation. Therefore, in a case in which only portions of a considerable amount of nozzles in one unit header are blocked or nozzles are narrowed due to other foreign matter, corrosion, or the like, the cooling of the thick plate to its finish cooling temperature (FCT) may be difficult to attain or a cooling defects may be caused. Furthermore, when the nozzles are blocked or narrowed, internal pressure of the accelerated cooling unit increases and relevant components connected thereto, such as pipes, measuring instruments, pumps, and the like, are more apt to break down.

In this study, we developed the POCHEF (POSCO Cooling Header Evaluation Facility) for determining the condition of the nozzle using differential pressure between the inside sensor block and ambient air. In addition, the results of a field test in a real, large factory in POSCO are introduced in this paper.

2 Experimental methods

The spraying pattern varied with the condition of nozzles inside the cooling header. As shown in Figure 2, the spraying patterns can be classified into three types. If strange substances are adhered to the nozzle hole or the machining condition of nozzles are wrong, the liquid column of the impinging jet tilts forward-backward/ left-right. In addition, the cooling water does not spray if the nozzle is clogged. Figure 3 shows the condition of liquid column spraying at the cooling header in large factory.

It was observed prior to this study that the impinging pressure of the abnormal jet is lower than that of the normal jet. The collision force of the column of liquid indicates whether the column of liquid is normally dispensed through the nozzles, or whether the nozzles are blocked partially or in the entirety thereof so as not to allow the column of liquid to be dispensed in a normal amount. Alternatively, measurement values recorded by the pressure sensor may differ from one another depending on an extent of collision force which the column of liquid itself collides with the membrane member. In this study, the cooling header evaluation system was designed and made the pilot equipment using the principles of this phenomenon. Pilot equipment was installed in the cooling header used in the field.

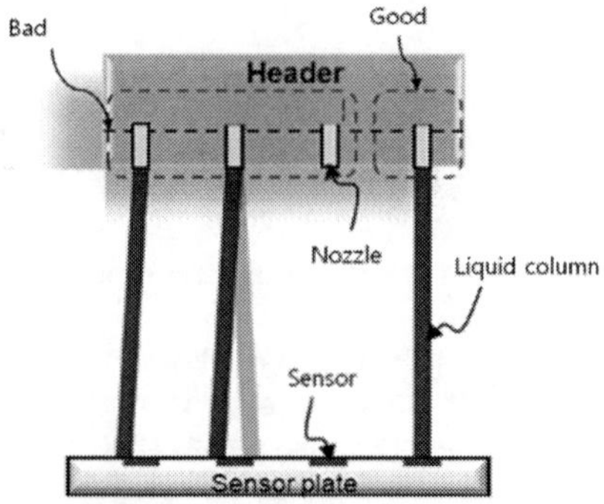

Figure 2. Schematic diagram of liquid column according to nozzle condition of multi-jet cooling header.

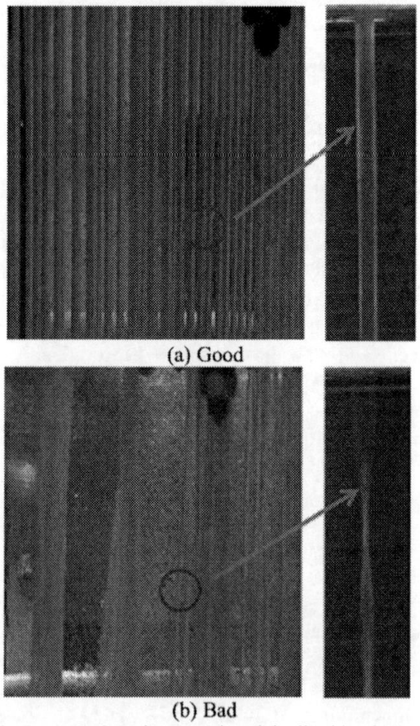

(a) Good

(b) Bad

Figure 3. Comparative photographs of the liquid column condition of the impinging jet.

A sensor device may include a housing including an air space; a membrane member provided with the housing and inducing a change in a volume of the air space at the time of a collision or contact of a pressure-applying material with the membrane member; and a sensor unit provided in the housing to communicate with the air space. The pressure sensor may measure a pressure value and then generate a voltage signal. The diameter of sensor membrane is 18mm. The sensor block has 12 sensors and 2 array. Figure 4 shows the schematic diagram of the sensor block.

An apparatus for evaluating performance of a cooling system, including a sensor device, may include: a sensor mounting unit provided on a movement path on which the cooling target material is moved and provided with a cooling header including nozzles, and the sensor device described above, provided in the sensor mounting unit to be aligned to correspond to the nozzles. The sensor structure allows for the generation of a quantitative signal as well as a qualitative signal corresponding to pressure applied at the time of contact of the liquid column, injection water dispensed or injected through nozzles, or the like, and allow for the establishment of data.

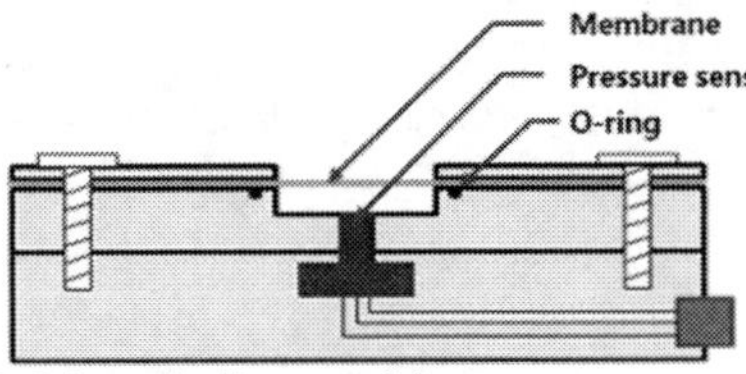

Figure 4. Schematic diagram of the sensor.

Figure 5 shows the schematic diagram of the measurement method using the cooling header evaluation system. The cooling jet system has 1,000 nozzles with nozzle hole 5mm in diameter. Total injection flow rate is $6m^3$/min per one cooling header; each nozzle 6l/min. The moving speed of the sensor plate is 1~3m/min. When the column of liquid dispensed from the upper header falls on and collides with the sensor, the water flow increases, and the pressure sensor may measure a pressure value (generate a voltage signal). This system measured automatically all of the nozzles in the width direction of the cooling header. The wireless transmission and reception system was used for transmitting measured data.

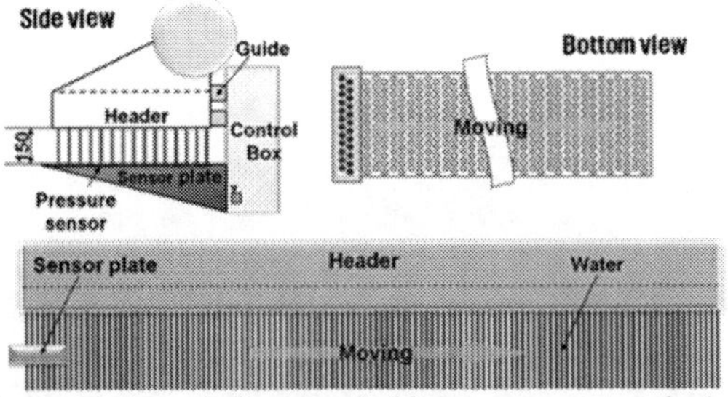

Figure 5. Schematic diagram of of the measurement method using the cooling header evaluation system.

3 Results and discussion

The membrane of the sensor device is formed using a material appropriate such as urethane, silicon, stainless steel, or the like, and may have a thickness appropriate in a range of from 0.5 to 1.5 mm. When the membrane has a reduced thickness of less than 0.5 mm, although reaction sensitivity of a sensor may be excellent, durability thereof may be decreased, and when the membrane has a thickness greater than 1.5 mm, although the durability thereof may be secured, a deflection amount of the membrane at the time of the generation of collision by the pressure-applying material may be relatively small to reduce a reaction speed or the sensitivity thereof. In addition, when the thickness of the membrane is less than 0.5 mm, the lifespan of the membrane or level of elasticity to return to its original position after the collision of the column of liquid may be decreased, such that the overall lifespan may be reduced. Meanwhile, in consideration for securing durability in terms of repetitive collisions of the column of liquid and the return of the membrane to the initial position after the collision of the column of liquid, the membrane may be formed using a stainless material.

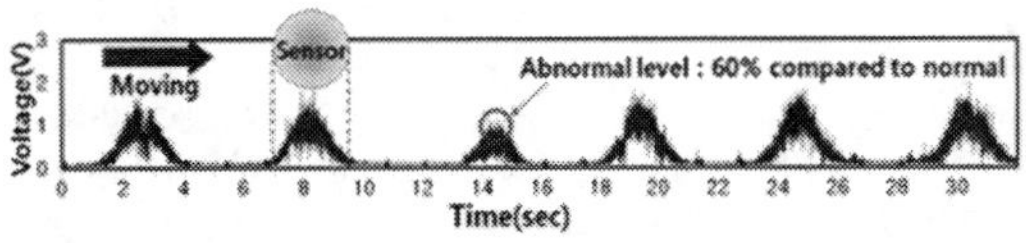

Figure 6. Analyzed data of the nozzle condition gathered by the POCHEF sensor.

Numerous data received through the POCHEF was processed in a short period of time. In addition, the optimal algorithm was necessary for removing noise caused by auxiliary actuating devices and water cooling installations in the factory building. In this case, when a reference pressure value indicating that pressure is not being applied to sensors is preset and a controlling unit may compare differential pressure with the measured pressure value, quantitatively measured data may also be established. Figure 6 shows the measured voltage signal. When the column of liquid (Figure 2 "Good") is normally dispensed from the upper header, the column of liquid continuously collides with the membrane of the sensor device. This may indicate that an electrical output signal generated by the pressure sensor may be represented the normal level of nozzles. However, when portions of the nozzles of the header are blocked, so as not to allow for the normal dispensing of liquid and columns of discontinuous liquid (Figure 2 "Bad") are dispensed, a voltage signal transferred by the pressure sensor may be short-circuited(ex. abnormal level in Figure 6).

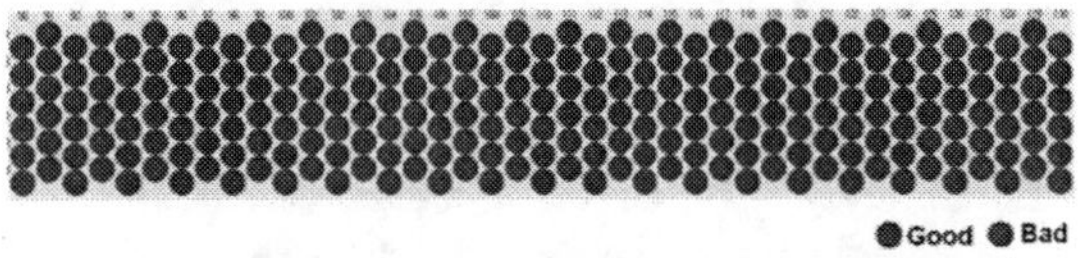

Figure 7. The measurement result using POCHEF in field.

The POCHEF may evaluate the column of normal liquid and the column of abnormal liquid dispensed through the nozzles of the cooling header. The extracted data of the field test result is illustrated in Figure 7. The green circle of the legend means the normal nozzle and the red one means the abnormal nozzle. From this results, the condition of cooling header can be evaluated. In addition, the cooling header may be frequently monitored using the POCHEF.

4 Conclusions

To improve maintainability and operability of cooling systems, we investigated a sensor device and an apparatus for evaluating the performance of a cooling system, and more particularly, a sensor device allowing for the generation of a quantitative signal as well as a qualitative signal corresponding to the pressure of liquid impacting a membrane member. The pressure of a column of water injected from a nozzle on a membrane member was measured allowing for the establishment of data and preventing the occurrence of errors in measurement due to changes in temperature, to precisely evaluate cooling system performance (in a pouring or injection state) at least quantitatively.

It is necessary to provide a sensor device allowing for the establishment of a sensor to generate a quantitative signal as well as a qualitative signal corresponding to pressure applied to a membrane component of the sensor at the time of collision (contact) of pressure-applying matter such as liquid, a column of liquid, or injection water dispensed or injected through a nozzle. A sensor device capable of preventing the occurrence of measurement errors by controlling a temperature of air transferring pressure applied to a sensor, to allow gas density to constantly change depending on coolant or ambient atmospheric temperature, and secure precision in measurement is required.

It is also necessary to provide a sensor device allowing for stabilized and quantitative evaluation by filling an operation space with incompressible fluid in which no change or a relatively small change in density according to temperature occurs, so as not to be affected by temperature.

Therefore, the POCHEF provides an apparatus for evaluating cooling system performance, capable of preventing the occurrence of cooling errors through quantitative, rapid and precise evaluation of cooling system performance, based on the sensor device according to the present disclosure, and allowing for easy maintenance and repair of equipment by determining the state of nozzles of a cooling system, based on automation in a poor environment confined to a relatively small workplace.

References

1. Lord Rayleigh, On the instability of jets, [in] Proc. 10th Math. Soc., London, (1878)
2. C. Weber, Zum zerfall eines flüssigkeitsstrahles, Z. Angew. Math. Mech., 11(1931)
3. D.B. Bogy, Drop formation in a circular liquid jet, Annu. Rev. Fluid Mech., 11(1979)
4. L. Crane, S. Birch, and P.D. McCormack, The effect of mechanical vibration on the breakup of a cylindrical water jet in air, Br. J. Appl. Phys., 15(1964)
5. R.J. Donnelly and W. Glaberson, Experiment on capillary instability of a liquid jet, [in] Proc. R. SOC., A290(1966)
6. E.F. Goedde and M.C. Yuen, Experiments on liquid jet instability, J. Fluid Mech, 40(1970)
7. A.L. Yarin, Free Liquid Jets and Films: Hydrodynamics and Rheology, Longman Scientific and Technical, Harlow, Essex, (1993)
8. P.-K. Wu, R.F. Miranda, and G.M. Faeth, Effects of initial flow conditions on primary breakup of nonturbulent and turbulent round liquid jet, At. Sprays, 5(1995)
9. Z. Dai, W.-H. Chou, and G.M. Faeth, Drop formation due to turbulent primary breakup at the free surface of plane liquid wall jets, Phys. Fluids, 10(1998)
10. T.A. Kowalewski, On the separation of droplets from a liquid jet, Fluid Dyn. Res., 17(1996)
11. A.W. Kerst, B. Judat, and E.-U. Schlunder, Flow regimes of free jets and falling films at high ambient pressure, Chem. Eng. Sci., 55(2000)
12. K.A. Sallam, Z. Dai, and G.M. Faeth, Liquid breakup at the surface of turbulent round liquid jets in still gases, Int. J. Multiphase Flow, 28(2002)

EPJ Web of Conferences **114**, 02054 (2016)
DOI: 10.1051/epjconf/201611402054

Influence of recirculation on Y-Q characteristic curve of hydrodynamic pump

Roman Klas[1,a], František Pochylý[1] and Pavel Rudolf[1]

[1]*Brno University of Technology, Faculty of Mechanical Engineering, Victor Kaplan Department of Fluids Engineering, Technická 2, 61669 Brno, Czech Republic*

Abstract. Contribution is focused on discussion of different design modifications of the volute, impeller and rotor-stator cavity in case of very low specific speed pump with recirculation channels. Amount of the liquid flowing through the recirculation channels has significant effect on delivery height, stability of the head curve and hydraulic efficiency. Analysis of these effects is based on the evaluation of the dissipated power in different internal parts of the pump and for different flow rates. It has already been proved in our previous research that volute has substantial impact on stability of the head curve. It is apparent that similar effect can also be attributed to distribution and shape of the recirculation channels. This fact is connected with the inflow into the channels and with magnitude of the flow rate through the recirculation channels. Influence of mentioned parameters on recirculation is discussed in present paper.

1 Introduction

If it is necessary to use the classic low specific speed hydrodynamic pumps for specific condition, we cannot avoid some issues during their design. These issues are mainly related to low efficiency and sometimes are connected with reaching of the required parameters. So called local eddy is formed within the blade channels, which is a special form of secondary flow [1]. Vortices fill the area of the blade channel, which causes the streamlines at the outlet of the impeller to bend. Angle, at which fluid leaves the outlet of the impeller (within the rotating frame of reference) to the spiral case, is smaller due to aforementioned phenomena. It is obvious, from Euler equations, that this fact will also bring the decrease of the pump head. There will also occur an increase in hydraulic losses. Both of those facts have a negative impact on the value of the hydraulic efficiency of the pump [2].

We can avoid the above mentioned drawbacks by using of specific designs of impellers. Careful analysis of the flow and the effort to suppress an initialization of the secondary vortex brought the idea of an impeller with thick trailing edges or alternatively an impeller with thick trailing edges with recirculation channels. This design will result in a different flux inside the impeller and primarily a different way of the spiral case filling.

The first designs of the impellers with thick trailing edges came out recently and their options are not yet sufficiently explored with respect to the design and optimization of the parameters. Especially the influence of recirculation channels on hydraulic efficiency and specific energy of pumps is not very well assessed. Therefore, the influence of modifications of the spiral case and the area between rotor and stator discs on flow rate through the recirculation channels will be examined. Flow rate through those channels has its significant influence especially on the hydraulic efficiency and total specific energy of the pump. Achievable specific energy will be observed also in terms of the Y-Q curve stability. As a tool for stability assessment of this curve, the dissipation function and its analysis with respect to power dissipated inside the individual parts of the interior of the pump was chosen. CFD methods were the main tool of the analysis. Results of CFD simulations were compared with experiments to prove its relevance [3, 4].

2 CFD simulation parameters

The basic tool for analysis of the interior of the pump and its parts was simulation software Ansys Fluent 14.5.7. Preprocessor GAMBIT 2.4.6 was used for modeling of the computational domain. The main parameters are listed in the Table 1.

3 Main parameters

Design parameters of tested pump are listed in the Table 2. Basic views on impellers with and without recirculation channels are shown in the Figure 1. There were used two types of spiral cases for all CFD

[a] Corresponding author: klas@fme.vutbr.cz

simulations, see the Figure 2. The appropriate notation is also contained within the figures.

Table 1. Basic design parameters and description of numerical model.

	B	T	L
B	Impeller with thick trailing edges		
T, L	Impeller with thick trailing edges and recirculation channel T,L - shape of the recirculation channels		
z - number of blades	5	7	
β_2 (°) – average value of outlet blade angle	32	63	
D_2 (m) – impeller diameter	0.320		
Q (l s^{-1}) – flow rate	6.94		
Y (J kg^{-1}) – specific energy	314		
n (1 s^{-1}) – rotational speed	24.167		
n_s (1 min^{-1}) – specific speed	33		
Number of computational cells	12 – 13 milions		
Turbulence model and near wall modeling	realizable k – ε non equilibrium wall function		
Boundary conditions	Inlet: velocity inlet Outlet: pressure outlet		
Calculation mode	unsteady, sliding mesh incompressible flow		

The flow areas in individual sections of the spiral case were preserved. The spiral case S1 was in fact designed for specific assembly conditions of the pump [5].

Both shown impellers differ from each other only in the built–in recirculation channels, but geometry of the main blade channels and also all main dimensions remain the same. Impeller without recirculation channels is marked as B. The recirculation channels could be T or L-shaped and are marked same as in the figures. All recirculation channels have the same diameter and in the direction perpendicular to the rotation axis they are drilled to the same depth. Number of the recirculation channels is the same for all designs to make the comparison easy. Correctness of the data, which were obtained from CFD simulation (e.g. dissipation) and could not be verified by appropriate experiments, was authenticated primarily by using integral characterictics like hydraulic efficiency and specific energy curves. There is correspondence between experimental data and data obtained from CFD in case of spiral case S1 [5].

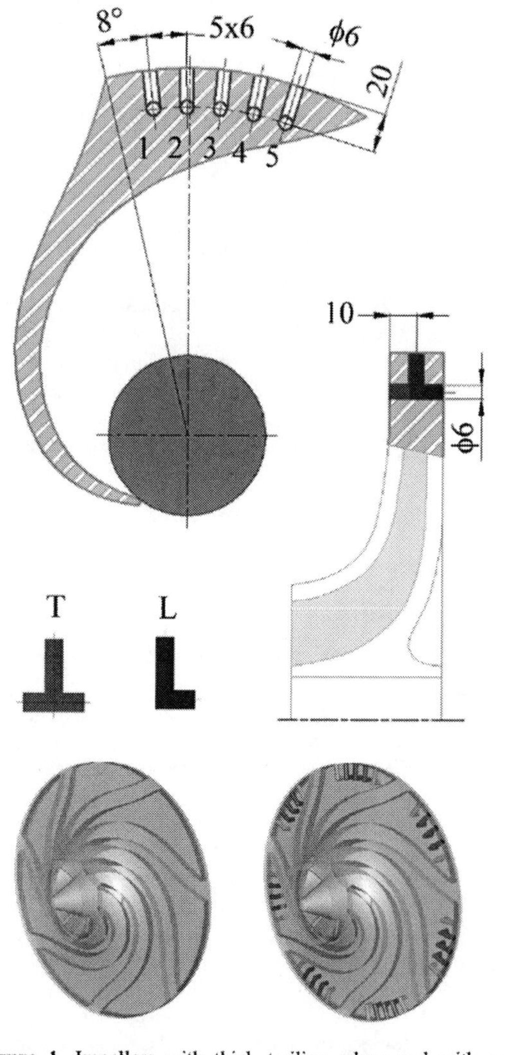

Figure 1. Impellers with thick trailing edges and with and without recirculation channels.

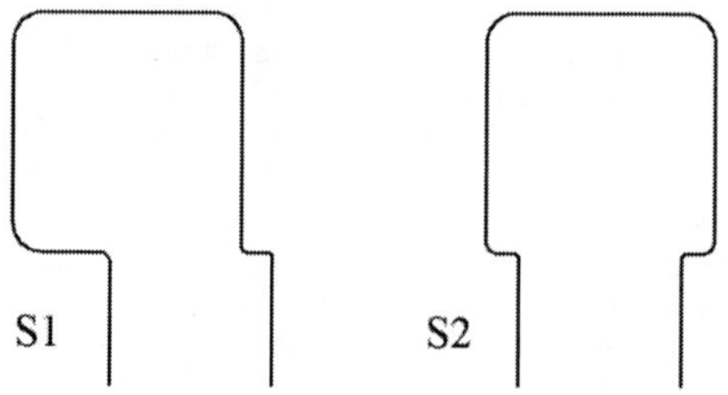

Figure 2. Cut through spiral cases.

In case of impeller with thick trailing edges with L-shaped recirculation channels there could possibly appear a moderate instability in the area of shut-off point. In case of the impeller without recirculation channels (variant B) we are nearly on the stability border.

Although the impeller with T-shaped recirculation channels reached the highest values of specific energy, it also reached the lowest values of hydraulic efficiency. These results are clearly illustrated in [5].

4 Dissipated power

With respect to the curves of specific energy or in other words the pump head, it is important to define criteria of stability appropriately, so that we can assess the impact of the recirculation on stability of those curves. As already mentioned, the main aim should be to assess the influence of changing the spiral case and the area between rotor and stator discs to curves of specific energy and their stability. A general criterion of stability (1) is defined by the following equation.

$$\frac{\partial Y}{\partial Q} < 0 \tag{1}$$

It turns out that it is beneficial to use the stability criterion based on the analysis of dissipated power $2D_H$. More information about this criterion is summarized in [6].

The most important advantage of so-defined criteria is the possibility to split the entire interior of the whole hydrodynamic pump into basic functional parts. Then a distribution of dissipated power in dependence on the flow through the pump can be determined within those parts.

This procedure enables to assess the contribution of individual components to the stability or instability of specific energy curves. Magnitude of the dissipated power is defined e.g. in [5].

Unfavorable fact is that the magnitude of the rate-of-strain tensor vij could not be determined precisely enough through CFD simulations, see [5]. Therefore, it is appropriate to use the difference of input and output powers instead to determine the magnitude of the dissipated power.

In addition to this, it is necessary to consider a significant fact. The required data, which are close to the pump shut-off point, cannot be obtained by CFD simulations.

The dependencies of specific energy and efficiency are plotted for different impellers with the spiral case S2. Figure 3 shows curves of specific energy and Figure 4 represents curves of hydraulic efficiency. However, comparison between CFD simulations and laboratory experiments for the spiral case S2 is not available.

Regarding to the maximum attainable specific energy or delivery height it is obvious, that a change of design parameters did not occur. There is one interesting fact connected with the curve of specific energy of the impeller with L-shaped recirculation channels. There can be seen some shift of this curve, which causes the curve of specific energy to appear more stable than it was shown in case of spiral case S1 [5]. Values of hydraulic efficiency remained preserved in the design point. The shift to an area of higher hydraulic efficiency occurred in the area of higher flow rates.

Distribution of the power dissipated within the pump interior is shown in the Figure 5 and 6. This interior will be separated into the following regions for further assessment: spiral case; blade channels; area between shroud and the static parts of the pump and finally an area between hub and the cover of the pump. The contribution of aforementioned regions to stability of specific energy curves will be assessed afterwards, at first the curves of dissipated power will be analyzed.

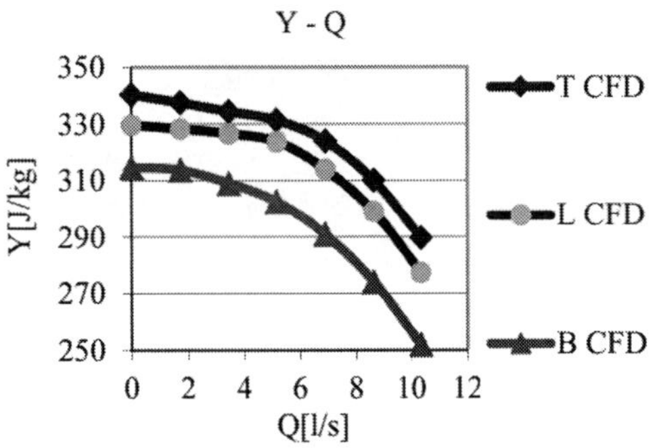

Figure 3. Curves of specific energy, spiral case S2.

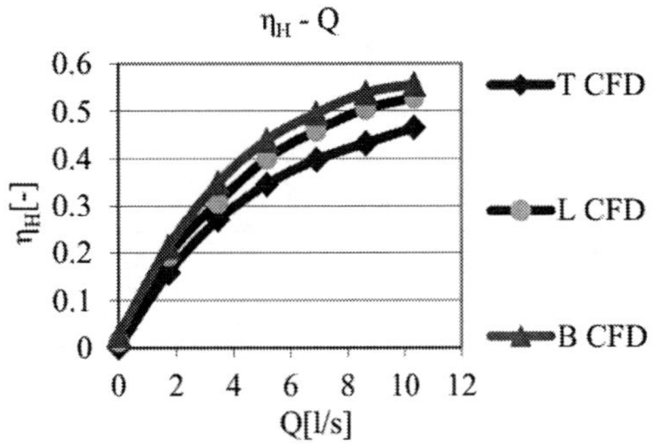

Figure 4. Curves of hydraulic efficiency, spiral case S2.

The reasons for modifications of the spiral case S1 to the spiral case S2 (see Figure 2) can be found in [5].

Regarding to the condition, there is a significant contribution to the instability of specific energy curves caused by the combination of spiral case S1 and impellers with thick trailing edges with or without recirculation channels (see Figure 6). Regardless of whether this curve is stable or not, recirculation channels are designed as L-shaped and T-shaped. In case of spiral case S2 and its combination with impeller with thick trailing edges without recirculation channels and impeller with thick trailing edges with L- shaped recirculation channels, there is slight contribution to the stability observable. This corresponds to conclusions in Figure 3. It appears that the correct design of the spiral case could have a decisive influence on the final quality of the whole design [8]. Significance of this is obvious from comparison of the power dissipated inside the spiral case S2. It is necessary to keep in mind, that the flow inside the spiral case is considerably affected by the flow in the impeller and conversely that individual effects are to some extent coupled [9]. Subsequently, dissipated powers in remaining regions of the pump will be plotted.

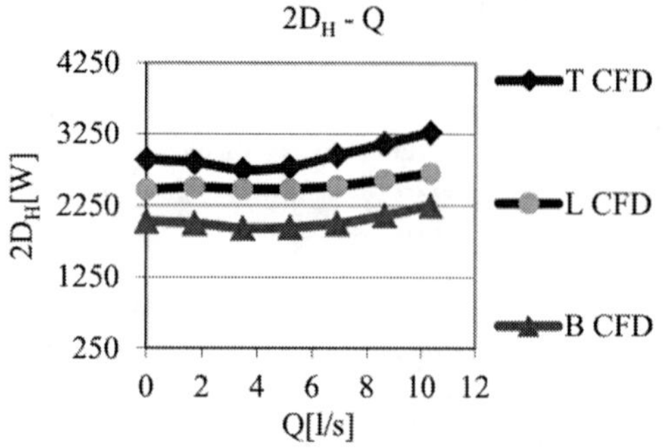

Figure5. Power dissipated within the pump with the spiral case S2.

The following section provides curves of power dissipated within spiral cases [8] S1 and S2 (see Figure 7).

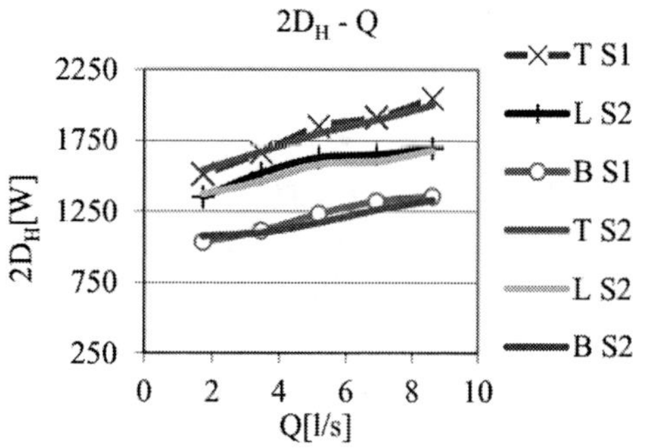

Figure 6. Power dissipated within spiral cases S1 and S2.

Contribution of blade channels to the stability of specific energy is very low for the impellers with thick trailing edges (see Figure 7). Differences in dissipated powers are negligible also for different spiral cases S1 and S2. It can be demonstrated that when discussing the classic impellers, the positive effect, which leads to stability of specific energy curves, is crucial. Negative impact of spiral case on stability is also obvious (see for example [7]).

It remains to assess the magnitude of dissipated power in the gap between impeller shroud and static parts of the pump (see Figure 8) and in the gap between impeller hub and the cover of the pump (see Figure 9).

The influence of the spiral case S2 on the impeller with thick trailing edges with L-shaped recirculation channels and the flow through this gap could obtained from the analysis of the dissipated power in the gap, see Figure 8. Changes were caused by an increase of recirculation inside of the spiral case. Similar effect can be observed also in T-shaped recirculation channels.

By contrast, the gap between the hub and static parts of the pump has a positive impact on stability for all types of impellers regardless of the shape of the spiral case.

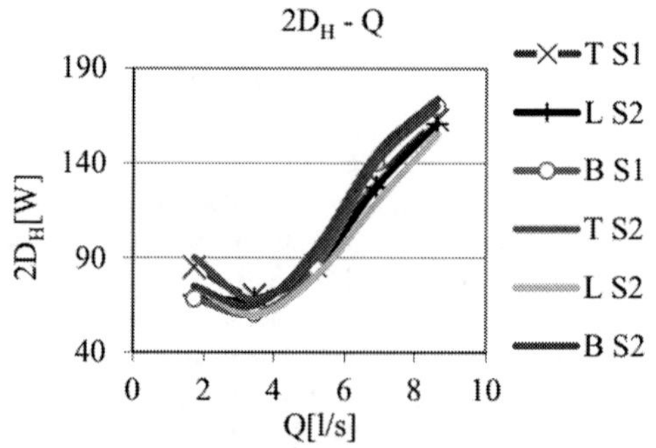

Figure 7. Powers dissipated inside blade channels of pump with spiral cases S1 and S2.

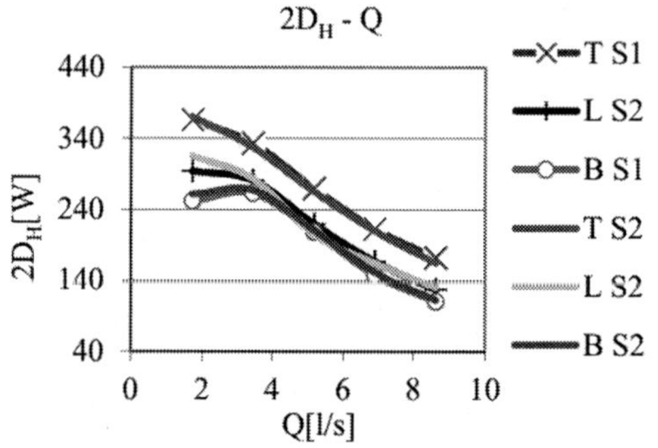

Figure 8. Power dissipated within the gap between impeller shroud and static parts of the pump with spiral cases S1 and S2.

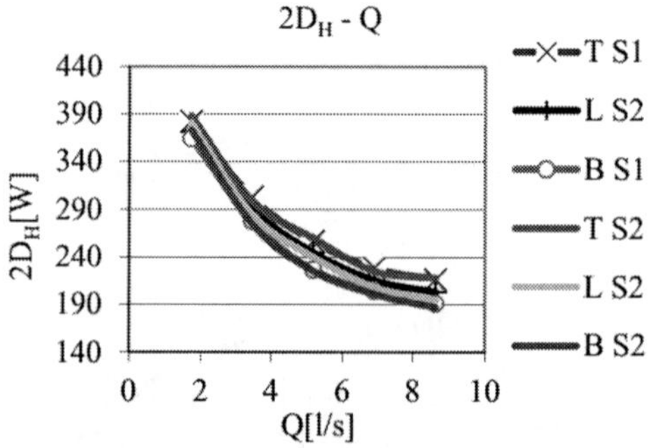

Figure 9. Power dissipated within the gap between hub and static parts of the pump with spiral cases S1 and S2.

5 Modifications of the spiral case on the side of hub

Another spiral case, S3, was designed to assess the effect of recirculation on curves of hydraulic efficiency and specific energy. This spiral case is slightly wider than the spiral case S2 (see Figure 10). Cross-sectional areas remain preserved in the majority of sections. Except these changes, the gap on the hub side was extended up to depth of 30 mm, see Figure 11. Original size of the gap was 2.5 mm and it corresponds to position no. 0. Subsequently, the gap was gradually enlarged always by 0.5 mm. The aim was to assist the recirculation of fluid in the impeller with T-shaped recirculation channels and then to verify possibilities of the influence on the size of the hydraulic efficiency. This impeller was also the only one which was tested in this part of work. Width of the gap could not be changed completely arbitrary due to its connection to the magnitude of the hydraulic losses [1].

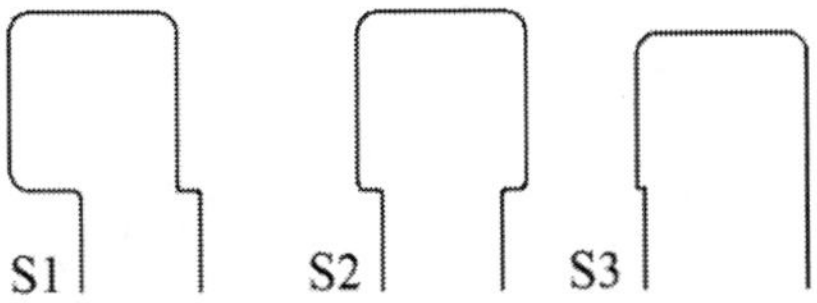

Figure 10. Sections of the spiral cases S1, S2 and S3.

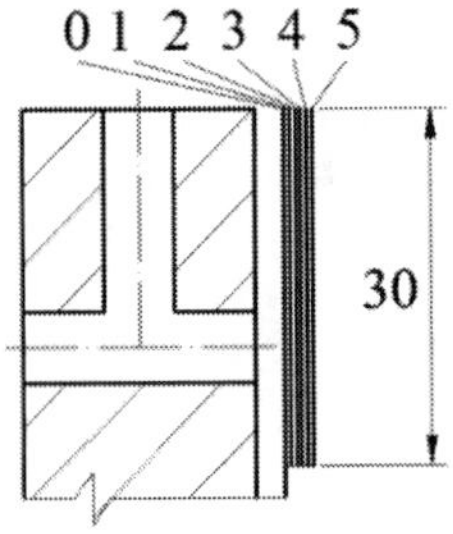

Figure 11. Gap between the hub and the cover of the pump.

Data were evaluated only on the basis of CFD simulations. First of all, there are the curves of specific energy for the impeller with T-shaped recirculation channels and the spiral case S3 presented. Labelling of the width of the gap corresponds to the Figure 11.

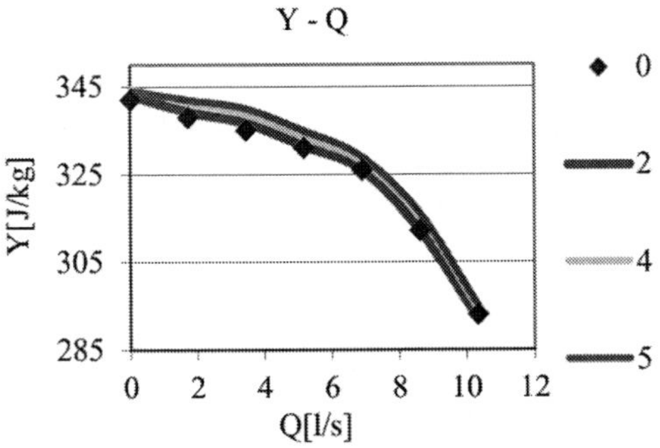

Figure 12. Specific energy inside the pump with the impeller with T-shaped recirculation channels and the spiral case S3.

According to curves shown in Figure 12 it appears that there is minimal impact of spiral case width on the specific energy. The only noticeable influence is connected to positions 4 and 5. There were observed some changes close to the shut-off point of the pump. These were connected with a decrease of hydraulic resistance on the inlet to the recirculation channels on the side of the hub. This caused the characteristics of pump with impeller with T-shaped recirculation channels to resemble more to the characteristics of the pump with impeller with L-shaped recirculation channels.

The similarity with described phenomena is also observable in curves of hydraulic efficiency, see Figure 14.

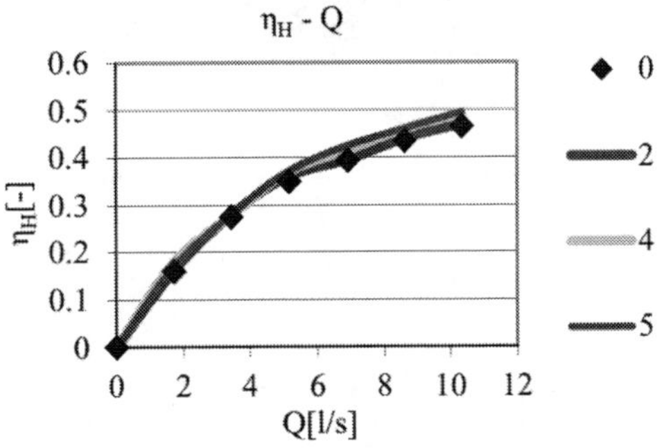

Figure 13. Hydraulic efficiency for the pump with impeller with T-shaped recirculation channels and with spiral case S3.

Increase of mechanical energy of the pump with gaps on the side of the hub 4 and 5 produces only slight increase in hydraulic efficiency compared to pumps with spiral cases S1 and S2 and original width of the spiral case.

6 Pressure pulsations within the pump

CFD simulations were used to obtain information about the size of pressure pulsations inside the pump. Pressure differences between suction and discharge were determined near the cut-off point and design point of the pump for spiral cases S1 [5] and S2.

Pumps with spiral case S2 also show similar results, see Figure 14 and Figure 15. The worst results are connected with impellers with T-shaped recirculation channels. While impellers with thick trailing edges without recirculation channels appears to be the best.

There is observable increase in the pressure pulsations nearby the design point of pumps with impellers with thick trailing edges without recirculation channels. On the contrary, impellers with thick trailing edges with recirculation channels shows significant decrease in static pressure pulsations.

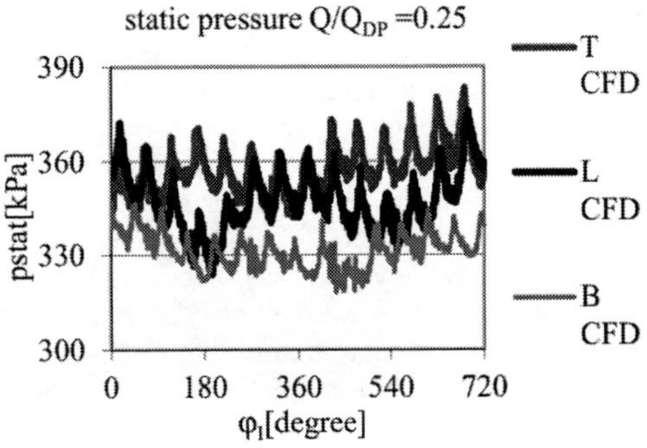

Figure 14. Pulsations of static pressure inside the pump with spiral case S2 in close proximity to shut-off point.

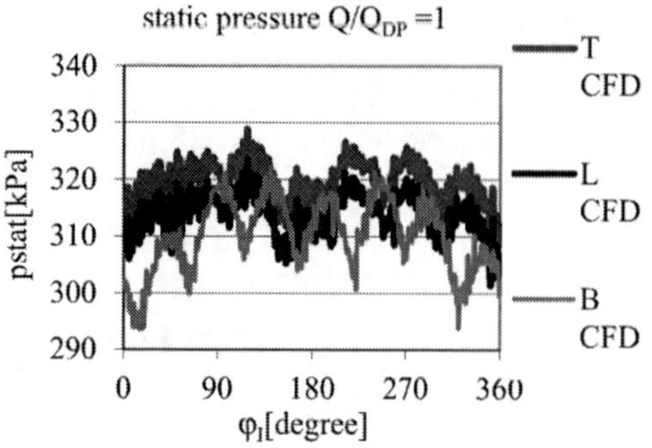

Figure 15. Pulsations of static pressure inside the pump with spiral case S2 in close proximity to design point.

Although, amplitudes of the static pressure are lower in these cases, conclusions about the properties of individual impellers near the cut-off point and around of the design point apply.

7 Conclusions

It can be summarized that regarding to stability of specific energy curves of the impellers with thick trailing edges, the best results were obtained for impellers with T-shaped recirculation channels. However, the total and hydraulic efficiency is smaller compared with other types of impellers. This fact was not significantly influenced even after some modifications were made to spiral cases S2 and S3, where the width of the gap between the impeller hub and the cover of the pump was increased. There was marginal increase of hydraulic efficiency for particular gap width and for enhanced inflow to recirculation channels, but the change was very small. Enlarging of the gap width on the side of the impeller hub improved the inflow into the recirculation channels, thus the impeller with T-shaped recirculation channels started behaving like the impeller with L-shaped recirculation channels in terms of specific energy curve stability.

Modification of spiral case S1 to the spiral case S2 brought us the improvement of the slight instability of specific energy curve for the impeller with L-shaped recirculation channels. The general significant contribution of the spiral case to possible instabilities of specific energy curves of hydrodynamic pumps was confirmed.

It was found that the gap between the impeller shroud and static parts of pump has a substantial impact on the stability. On the other hand, the impact of the gap between the impeller hub and the cover of the pump on the rate of recirculation (the shape of recirculation channels) corresponds to the course of specific energy (Euler turbomachinery equation). Recirculation enhances the stabilizing effect. The effect of stabilization is further reinforced by the optimization of the spiral case (change from S1 to S2), but the change is not significant from a quantitative point of view.

Effect of blade channels is only slight with respect to stability of the curves of specific energy.

Pulsations of static pressure in the interior of the pump close to the shut-off point are increasing while using the impellers with thick trailing edges with the highest rate of recirculation. This fact mainly concerns the T-shaped recirculation channels and it is related to the secondary flow rate within the spiral case.

References

1. J. F. Gülich, *Centrifugal Pumps* (Heidelberg Springer-Verlag, 2010)
2. A.Wilk, *International Journal of Mechanics* Issue **2**, Volume **4**, pp 33-41 (2010)
3. Y. Meng, Ch. Li, H. Yan, X. Liu, *Proc. Int. Conf. on Computer Eng. and Aplications (Manila)*, **2**, pp 179-183 (2009)
4. M.W. Heo, K. Y. Kim, S. B. Ma, I. S. Yoo, W. C. Choi, J. H. Kim, Y. S. Choi, *IOP Conf. Ser.: Earth and Enviromental Science*, **22**, (2014)
5. R. Klas, F. Pochylý, P. Rudolf, *EPJ Web of Conferences*, **92** (2015)
6. F. Pochylý, M. Haluza, S. Drábková, Engineering Mechanics 2009 – book of extended abstract, 208-215 (2009)
7. N. Tiwari, D. Kumar, *Afro – Asian Conf. on Sci., Engineering & Technology*, Paper ID ME00C504 (2015)
8. S. Yang, F. Kong, B. Chen, *Int. Journal of Rotating Machinery*, Article ID 137860 (2011)
9. E. C. Bacharoudis, A. E. Filios, M. D. Mentzos, D. P. Margaris, *The Open Mechanical Journal*, **2**, pp 75-83 (2008)

Acknowledgement

Project TA03020243 "Výzkum a vývoj čerpadel pro bezpečnostní systémy jaderných elektráren" is gratefully acknowledged for support of this research.

EPJ Web of Conferences **114**, 02055 (2016)
DOI: 10.1051/epjconf/201611402055

Experimental research of reaction blading on air turbine VT-400

Marek Klimko[1,a], Daniel Okresa[1]

[1]University of West Bohemia, Department of Power System Engineering, Univerzitní 8, 306 14 Pilsen, Czech Republic

Abstract. The article deals with testing a reaction blading on an experimental air turbine VT-400, which is situated at the Department of Power System Engineering at University of West Bohemia. Experiments were carried out in cooperation with an industrial partner Doosan Skoda Power. The outputs of these measurements are for example: results of the stage efficiency depending on the speed ratio u/c, the course of reaction, the input and output angles and profile losses along a blade.

1 Introduction

One of the important steps when designing a flow section of a steam turbine is to calculate efficiency, or more precisely, its preliminary estimation (efficiency prediction). This is a valuable and confidential know-how of every steam turbine producer. It is a long-term process carried out in research laboratories, including experience from real operation. Increasing speed and improving quality of CFD calculations enables to simulate the influence of a whole range of geometrical and operational factors, yet experimental testing still maintains its important role.

This paper represents a continuation of a long-term research carried out at the Department of Power System Engineering (KKE) in a close cooperation with the company DOOSAN ŠKODA POWER (DSPW). In the past, the research on an experimental air turbine was focused on secondary losses caused by flow in the stage of a steam turbine, as a result of 3Dflow in the channel endwall areas. The objective of the studies was to find a suitable solution, which would lead to the most significant decrease of those losses by a 3Dblade shaping method.

A simple construction layout of the test rig enables an effortless exchange of stator and rotor wheels according to measurement requirements. An input channel layout is possible to be changed as well. A large advantage of the experimental air turbine VT 400 is its easy and fast launch, and primarily its cheap operation, in comparison with a steam turbine. However, the use of this turbine is limited in some extent by the operational parameters range, especially by Mach and Reynolds numbers.

Two variants of reaction stages have been tested at the Department during the past two years. There are two types of stator (high aspect ratio blading: "Pretlak-1" and low aspect ratio blading: "Pretlak-2"), whereas the rotor part of the stage remains the same for both variants. There are basic geometric characteristics for both tested variants mentioned in the article, including a description of the measuring methods and evaluation of measured data, followed by results of graphic behaviour of selected parameters. Tested variants are compared from the point of view of the stage efficiency, reaction, Mach and Reynolds numbers, and also regarding the aspect of profile losses.

[a] Corresponding author: klimko@kke.zcu.cz

2 Basic description of the experimental air turbine VT-400

It is a single-stage air turbine located in the compressor suction. The turbine is a model of a high-pressure steam turbine component stage in the scale of 1:2. Apart from almost constant air inlet parameters, this layout also provides an easy access to the turbine and measuring points.

Figure 1. Experimental device.

A part of the test rig is a direct-current dynamometer, which serves for the determination of speed and torque. Air, volume of which is measured by a nozzle, leaves the blower pressure discharge and out of the laboratory.

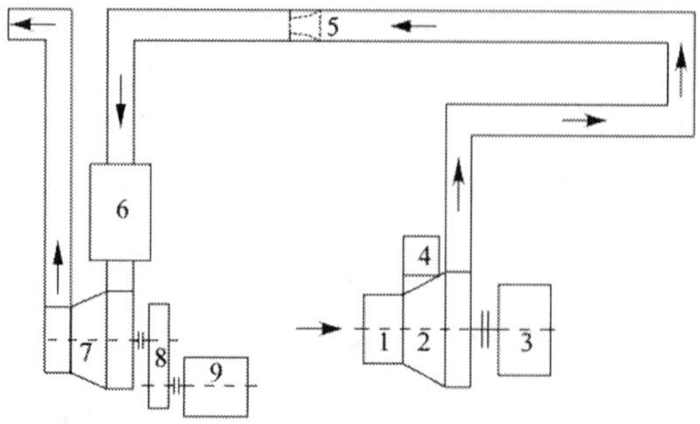

1 - Filter
2 - Turbine
3 - Dynamometer
4 - Traverser
5 - Nozzle
6 - Silencer
7 - Compressor
8 - Gearbox
9 - Electric motor

Figure 2. Scheme of the device.

The turbine is equipped by a traverser device. The traverser enables the probe to move radially (under the blade root section and above the blade tip), peripherally (across two stationary blade spacings), and it also allows its automatic turning in the flow direction. That enables a thorough measurement of a flow field behind stator and rotor blades.

There are several static pressure extractions on the turbine – in front of the stage, inside, and behind it, always on the root section and tip diameter. There are holes at the input, for inserting e.g. Prandtl probe and measuring the inlet flow. All pressures are scanned by a fast, 16-channel pressure transducer. Temperatures are scanned by a resistance thermometer in front of and behind the stage. The probe measurement and move is automatic; the utility program is created in LabView 7.

3 Tested variants of reaction blading

In 2013 DSPW proposed two versions of a stage designed for testing on the ZCU air turbine, in accordance with newly prepared methodologies and using new profiles for reaction blading. The aim is to verify design data concerning aerodynamics, and to obtain data for specification of a loss model of reaction stages. The variants differ in aspect ratio of a stationary blade – the main parameter influencing secondary losses development. The basic geometrical data are summarized in Tab. 1. The blading has been produced and tested on the air turbine VT-400.

4 Principle of the measurement course and basic procedure of data evaluation

Basically, there are 2 elementary types of measuring on the air turbine:

1) Turbine integral characteristics

The turbine is equipped with a set of static pressure extractions, which determines the pressure drop on the stage (pressures in front of and behind the stage) and also the stage reaction at the root section and the tip (according to the pressure extractions behind stator). The turbine is connected to a dynamometer, which provides directly measured values of speed and the moment of torsion. The third, and significant, value is a flow rate, which is detected by a precise nozzle in the pipeline between the turbine and the blower.

Figure 3. Locations of static pressure extractions on the nozzle.

2) Traversing of flow fields

By a pneumatic (5-hole) probe, which is fixed in the traversing device, velocity and pressure fields behind stator, and also behind rotor when relocated, can be examined in detail. These measurements do not directly provide the final data, but data determined for further proceeding – averaging, recalculating important points of the expanse line etc. The final output is, for instance, distribution of losses, angles, reaction etc. along the blade length.

Figure 4. Detail of a 5-hole probe.

Table 1. Basic geometrical characteristics of both variants.

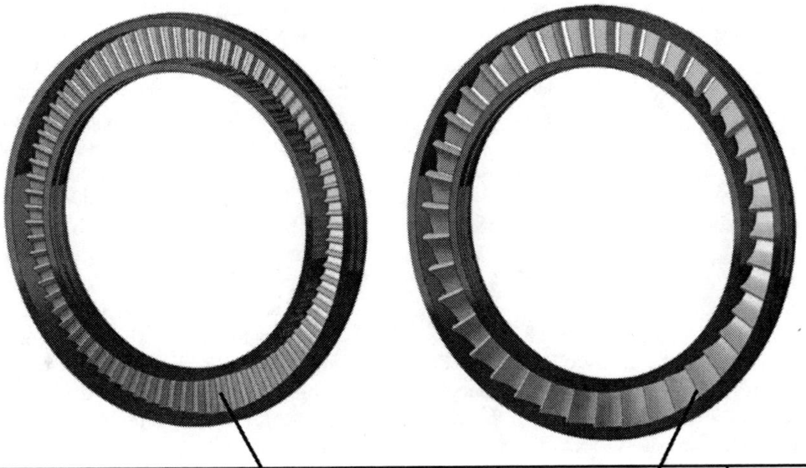

			"PRETLAK-1"	"PRETLAK-2"	ROTOR
			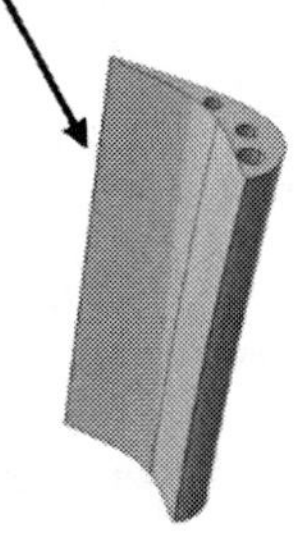	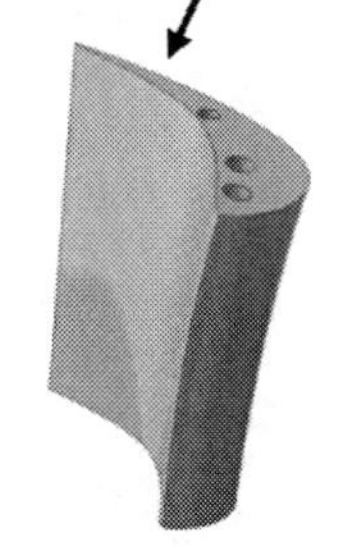	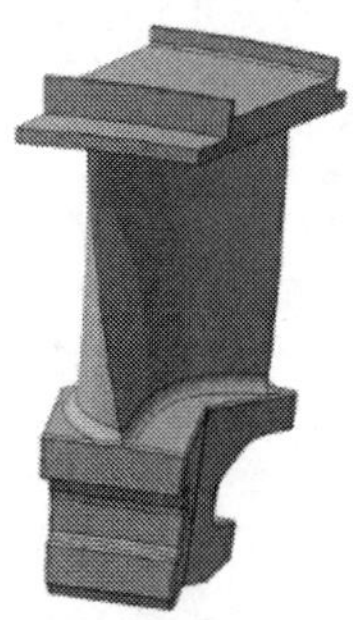
No. of blades	N	[1]	78	40	56
Blade length	L	[mm]	45,5		47
Hub diameter	D_P	[mm]	400		400
Relative length	L/D_p	[1]	0,114		0,118
Medium pitch	t_s	[mm]	17,94	34,99	25,08
Profile chord	c	[mm]	22,51	43,90	32,09
Blade width	B_{ax}	[mm]	15,85	30,47	21,43
Relative pitch	t_s/c	[1]	0,797	0,797	0,782
Aspect ratio	L/c	[1]	2,021	1,036	1,465

4.1 Turbine integral characteristics

The inlet pressure in front of the turbine is determined by atmospheric pressure, with the respect to the experimental device layout (an open suction tunnel), and only outlet pressure, and thus the pressure drop on the stage can be changed. The pressure drop is changed by the blower performance, or more precisely, by the performance of the powering engine. Changes are carried out in steps and manually, which makes concrete required mode settings complicated by maintaining a constant state.

Changing the pressure ratio at fixed turbine speed enables to change the velocity rate u/c and obtain the stage basic integral characteristics – dependence of efficiency on the u/c ratio. This characteristic can be measured for various turbine speed.

Basic integral data can be evaluated from each measuring point:

- barometric pressure and temperature (measured directly) + density recalculation,
- moment of torsion and speed (measured directly) + bearing and friction losses recalculation (dissipation),
- static pressure in front of and behind the stage (measured directly) + density recalculation,

[a] Corresponding author: klimko@kke.zcu.cz

- flow rate (recalculation from pressures on the nozzle and behind the stage temperature),
- reaction on the root section and the tip (recalculation from pressures behind stator, varied definitions),
- inlet and outlet velocity (recalculation from annular areas in front of and behind the stage, flow rate and other data),
- stage efficiency (recalculation according to varied definition (total-static, total-total),
- velocity ratio u/c (recalculation according to varied definition (total-static, total-total),
- another stage pure characteristic numbers (Mach and Reynolds numbers).

As mentioned above, we are able to change the velocity ratio u/c by changing the pressure ratio at fixed turbine speed, which results in one of the most essential stage characteristics – efficiency dependence on the u/c ratio. The turbine speed values have been set to 2000, 2300, 2700, 3000 and 3400 RPM during measuring both variants, which changed the pressure drop, Mach number, and partially Reynolds number as well. We have agreed with DSPW to focus on speed of 2300 rpm. The objective of the measurement was to obtain the turbine optimum operating point, i.e. such a point where the turbine reaches the highest efficiency value with the corresponding optimum velocity ratio u/c (Fig. 5).

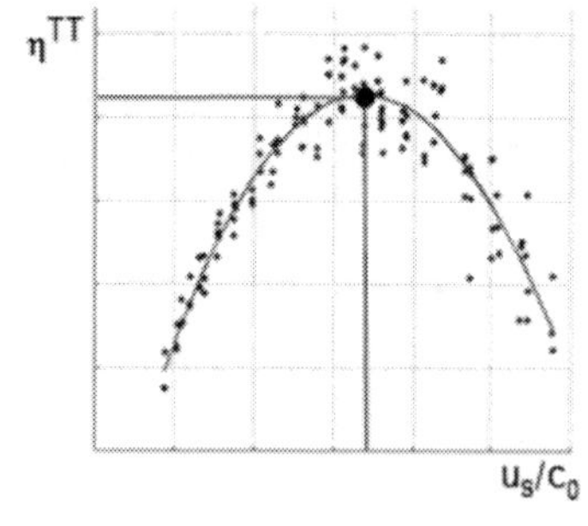

Figure 5. Determination of the turbine optimum operating mode

This optimum operating point was our input when executing stationary measurements of pressure fields behind stator and rotor.

4.2 Traversing flow fields

This measurement provides very important information about flow in the stage. It leads to a stationary pressure, or rather a velocity image of the flow field, and after suitable peripheral data averaging also a behaviour of quantities along the blade length, which is highly valued by designers. It is not only about loss values, but also velocities, or more precisely, angles for calculating of flow incidences and deviations; furthermore, and for instance, the reaction level or volumetric flow etc. Such data provides, when compared with the theoretical design, substantial information on behaviour and functioning of shape modifications of blades, or the influence of off-design modes.

Flow fields measurement takes place at a selected operating mode; in our case it was the turbine optimum operating duty, which was set on the basis of the pressure drop and the u/c ratio. This operating mode is maintained, as good as it can be, on a constant value in the tunnel. Measuring is performed by a 5-hole pneumatic probe of the size of 4 mm, which represents approximately 9% of the blade height, and thus the acquired data may be partially distorted. Due to the probe size, including its holder, an axial gap between stator and rotor was enlarged. The probe is fixed in the traversing device. It can be moved manually in two positions – for measuring behind stator or rotor. The traverser enables a peripheral probe shift (approximately 22°), radial shift (under the root section and above the tip), and also its flow-wise turn (automatically according to the comparison of pressures from the left and right probe extraction).

4.2.1 Traversing behind stator

Measuring is relatively accurate and its evaluation fairly simple. Isentropic expansion (velocity) can be determined from integral static pressure and static pressure from the probe, and, for example, losses in the stationary blades cascade can be calculated from the total pressures. Furthermore, an outlet angle from stator or stage reaction can be analyzed as well. An inlet flow angle into rotor or corresponding components of relative velocity can be analyzed by further recalculation (by including the peripheral velocity).

This can be implemented on the whole measured flow field (constant total pressure is considered in front of the stage), or more often the measured data are averaged after individual radials and one representative value is defined for a particular radial.

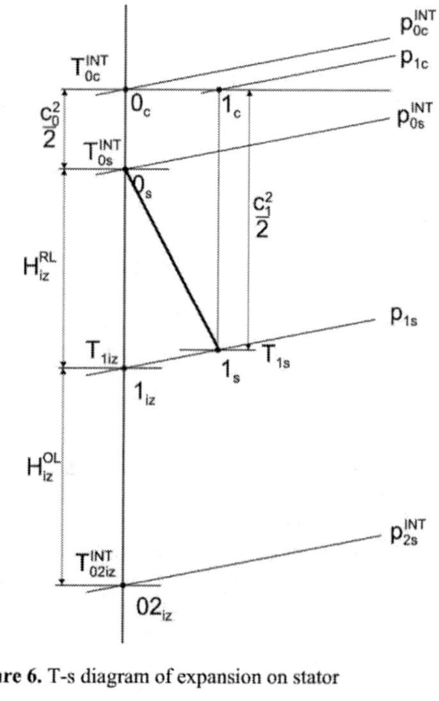

Figure 6. T-s diagram of expansion on stator

Stage reactions (static-static, total-static) are defined generally:

$$\rho^{SS} = \frac{H_{iz}^{OL}}{H_{iz}^{ST}} \tag{1}$$

$$\rho^{TS} = \frac{H_{iz}^{OL}}{H_{iz}^{ST} + \frac{c_0^2}{2}} \tag{2}$$

We are able to calculate isentropic drops H_{iz}^{ST}, H_{iz}^{OL} and H_{iz}^{RL} based on the difference of temperatures and isobaric specific heat capacity.

$$H_{iz}^{ST} = c_p \cdot \left(T_{0s}^{INT} - T_{02,iz}^{INT}\right) \tag{3}$$

$$H_{iz}^{OL} = c_p \cdot \left(T_{1,iz} - T_{02,iz}^{INT}\right) = H_{iz}^{ST} - H_{iz}^{RL} \tag{4}$$

$$H_{iz}^{RL} = c_p \cdot \left(T_{0s}^{INT} - T_{1iz}\right) \tag{5}$$

Temperature T_{0s}^{INT} is measured directly, temperatures $T_{02,iz}^{INT}$ and T_{1iz} are calculated as follows:

$$T_{02iz}^{INT} = T_{0s}^{INT} \cdot \left(\frac{p_{2s}^{INT}}{p_{0s}^{INT}}\right)^{\frac{\kappa-1}{\kappa}} \tag{6}$$

$$T_{1iz} = T_{0s}^{INT} \cdot \left(\frac{p_{1s}}{p_{0s}^{INT}}\right)^{\frac{\kappa-1}{\kappa}} \tag{7}$$

Pressures p_{0s}^{INT} and p_{2s}^{INT} are measured by extractions of static pressures in front of stator and behind rotor, and pressure p_{1s} is acquired from data from the 5-hole probe after recalculation according to calibration. We obtain from the velocity triangle:

$$T_{1s} = T_{0C}^{INT} \cdot \left(\frac{p_{1s}}{P_{1c}}\right)^{\frac{\kappa-1}{\kappa}} \tag{8}$$

$$T_{0c}^{INT} = T_{0s}^{INT} + \frac{c_0^2}{2 \cdot c_p} \tag{9}$$

$$c_2 = \sqrt{2 \cdot c_p \cdot \left(T_{0c}^{INT} - T_{1s}\right)} \tag{10}$$

$$c_{1a} = c_1 \cdot cos\vartheta \cdot cos(90 - \varphi) \tag{11}$$

$$c_{1u} = c_1 \cdot cos\vartheta \cdot sin(90 - \varphi) \tag{12}$$

$$c_{1r} = c_1 \cdot sin\vartheta \tag{13}$$

$$w_{1u} = c_{1u} - u_1 \tag{14}$$

$$w_1 = \sqrt{c_{1a}^2 + c_{1u}^2} \tag{15}$$

$$\beta_{1t} = acos \cdot \frac{w_{1u}}{w_1} \tag{16}$$

Stationary blade efficiency is then:

$$\eta^{RL} = \frac{T_{0c}^{INT} - T_{1s}}{T_{0c}^{INT} - T_{1iz}} = \frac{\frac{c_1^2}{2}}{H_{iz}^{RL} + \frac{c_0^2}{2}} \tag{17}$$

4.2.2 Traversing behind rotor

From the evaluation point-of-view, this measurement is far more demanding, as it is necessary to first analyze the probe measuring behind stator to be able to define the inlet state in front of rotor. Additionally, it is necessary to make sure that measuring takes place at the utmost identical modes of the turbine operation.

Again, the analysed data are blade cascade losses and outlet angle (both in relative coordinate and absolute systems). Data is then averaged and plotted as radial dependences on the relative blade length.

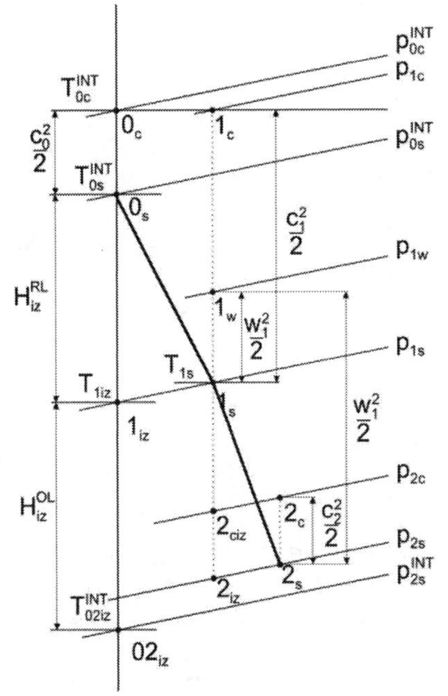

Figure 7. T-s diagram of stage expansion

As mentioned above, we need to know parameters of output state from stator to be able to determine the input state into rotor. Parameters $\rho^{SS}, \frac{c_1}{c_{02,iz}}, \frac{w_1}{c_{02,iz}}$ are used for integration and bonding of data behind stator and rotor. We obtain velocity $c_{02_{iz}}$ from the following relation:

$$c_{02iz} = \sqrt{2 \cdot H_{iz}^{ST}} \tag{18}$$

We shall also need parameters T_{02iz}^{INT}, H_{iz}^{ST}, T_{0c}^{INT}, which were described in the chapter 4.2.1.

Isentropic drop on stationary blades can be written in the following form:

$$H_{iz}^{RL} = (1 - \rho^{SS}) \cdot H_{iz}^{ST} \tag{19}$$

We will realize the calculation of parameters at the input into rotor on the basis of the following relations:

$$p_{1s} = p_{0s}^{INT} \cdot \left(\frac{T_{1iz}}{T_{0s}^{INT}}\right)^{\frac{\kappa}{\kappa-1}} \tag{20}$$

$$c_1 = \left(\frac{c_1}{c_{02iz}}\right) \cdot c_{02iz} \tag{21}$$

$$w_1 = \left(\frac{w_1}{c_{02iz}}\right) \cdot c_{02iz} \tag{22}$$

$$T_{1s} = T_{0c}^{INT} - \frac{c_1^2}{2 \cdot c_p} \tag{23}$$

We can consequently proceed to the calculation of expansion on rotor.

$$T_{1w} = T_{1s} + \frac{w_1^2}{2 \cdot c_p} \tag{24}$$

$$T_{2iz} = T_{1s} \cdot \left(\frac{p_{2s}}{p_{1s}}\right)^{\frac{\kappa-1}{\kappa}} \tag{25}$$

We can estimate the velocity $c_2$on the basis of the difference in isentropic temperatures:

$$\tilde{c}_2 = \sqrt{2 \cdot c_p \cdot (T_{2ciz} - T_{2iz})} \tag{26}$$

Where:

$$T_{2Ciz} = T_{1s} \cdot \left(\frac{p_{2c}}{p_{1s}}\right)^{\frac{\kappa-1}{\kappa}} \tag{27}$$

Individual parts of the estimated absolute velocity are determined by the following relations:

$$\tilde{c}_2 = \sqrt{2 \cdot c_p \cdot (T_{2ciz} - T_{2iz})} \tag{28}$$

$$\tilde{c}_{2a} = \tilde{c}_2 \cdot cos\vartheta \cdot cos(90 - \varphi) \tag{29}$$

$$\tilde{c}_{2u} = \tilde{c}_2 \cdot cos\vartheta \cdot sin(90 - \varphi) \tag{30}$$

Then the relative velocity from the estimated absolute velocity at the outlet from rotor is:

$$\widetilde{w}_{2u} = \tilde{c}_{2u} - u_2 \tag{31}$$

$$\widetilde{w}_2 = \sqrt{\tilde{c}_{2a}^2 + \widetilde{w}_{2u}^2} \tag{32}$$

Now we are able to define static and total temperatures at the outlet from rotor and calculate the real value of the outlet absolute velocity from rotor, including its components.

$$T_{2s} = T_{1w} - \frac{\widetilde{w}_2^2}{2 \cdot c_p} \tag{33}$$

$$T_{2c} = T_{2s} \cdot \left(\frac{p_{2c}}{p_{2s}}\right)^{\frac{\kappa-1}{\kappa}} \tag{34}$$

$$c_2 = \sqrt{2 \cdot c_p \cdot (T_{2c} - T_{2s})} \tag{35}$$

$$c_{2a} = c_2 \cdot cos\vartheta \cdot cos(90 - \varphi) \tag{36}$$

$$c_{2u} = c_2 \cdot cos\vartheta \cdot sin(90 - \varphi) \tag{37}$$

$$c_{2r} = c_2 \cdot sin\vartheta \tag{38}$$

Then the outlet relative velocity and the outlet flow angle from rotor are as follows:

$$w_2 = \sqrt{c_{2a}^2 + w_{2u}^2} \tag{39}$$

$$\beta_{2f} = acos\frac{w_{2u}}{w_2} \tag{40}$$

The final calculation is a determination of rotor efficiency, which we obtain from the following relation:

$$\eta^{OL} = \frac{T_{1w} - T_{2s}}{T_{1w} - T_{2iz}} \tag{41}$$

5 Results of measured data

It is necessary to mention at the beginning of this chapter that all the results from measuring on the experimental air turbine VT-400 belong to the property of DSPW, and therefore some courses of resultant parameters will be indicated in relative values.

The picture (Fig. 13) shows an efficiency comparison of both reaction blading tested variants of depending on the velocity ratio u/c. The efficiency value for variant "Pretlak-1"is slightly higher. The turbine optimum operating duty can be easily defined from both courses.

It is apparent from the Mach number behaviour on stationary blades (Fig. 8), that there is a deeply subsonic flow for both variants. Reynolds number ranges between 160 000 to 180 000 (Fig. 9) in the middle part of stationary blades.

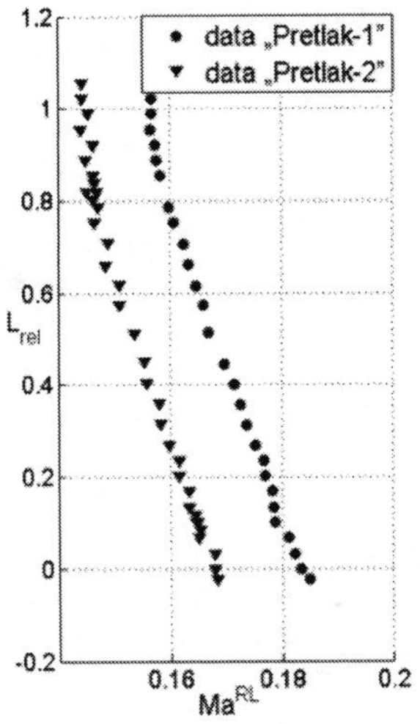

Figure 8. Mach number behaviour along a stator blade.

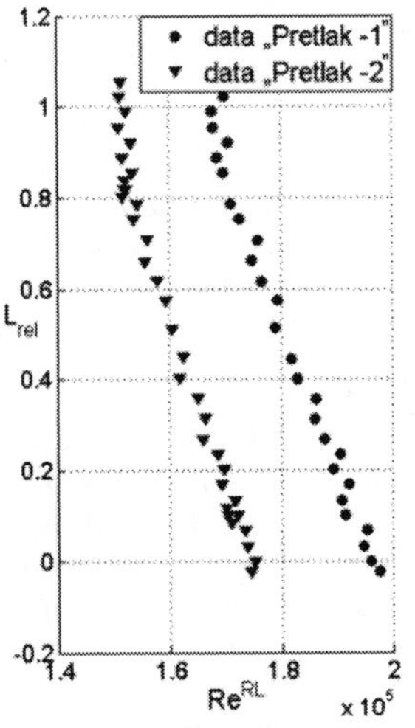

Figure 9. Reynolds number behaviour along a stator blade.

The following dependencies represent the reaction course (static-static) (Fig.10) and profile losses course separately on stationary and moving blades, depending on the blade relative length. (Fig. 11, Fig. 12). Distribution of profile losses along the stationary blades has an expected character. The losses value is low in the central part of the blade profile, but a significant losses increase occurs in the top of the blade, which we attribute to the influence of a space between the over-shroud (tip) seal and the stator. This space is relatively high due to the axial gap between the stator and rotor and there may be a swirl flow, which partially extends also to the output flow from the stator blades. Profile losses along the moving blades are obviously higher in comparison with the stationary blades.

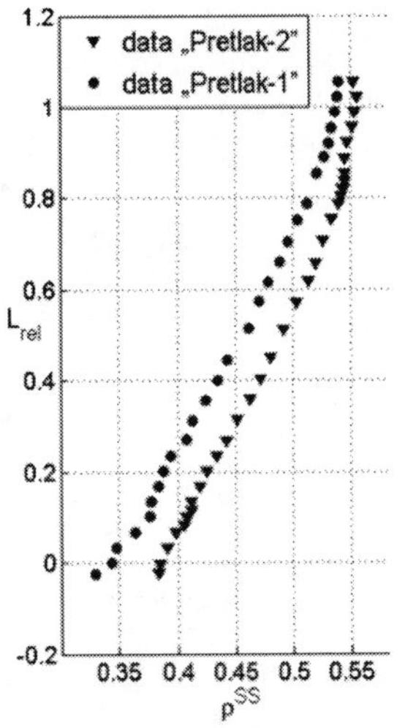

Figure 10. Reaction courses along the stator blade.

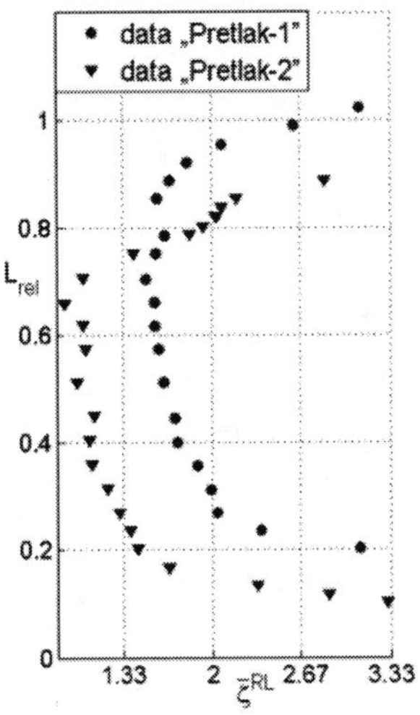

Figure 11. Profile losses courses along a stator blade.

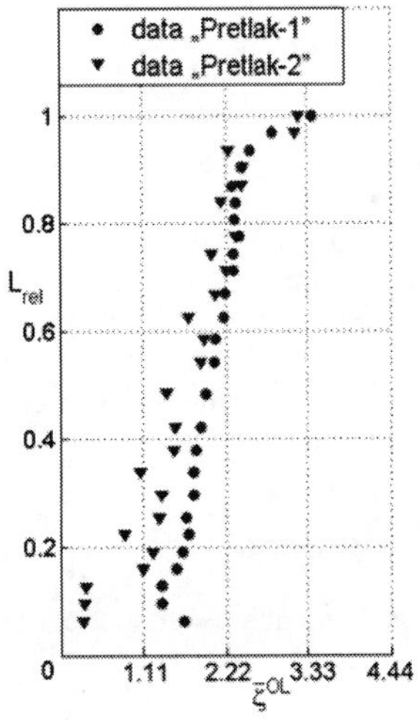

Figure 12. Profile losses courses along a rotor blade.

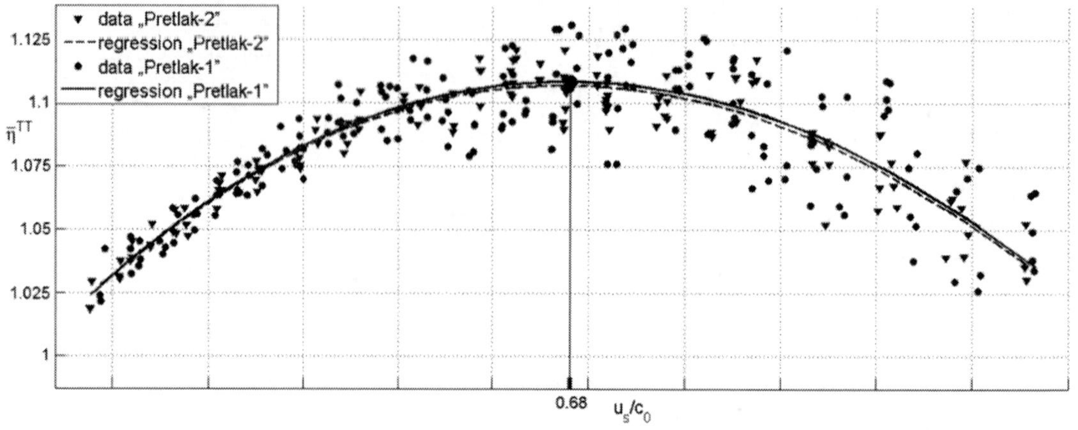

Figure 13. Efficiency comparison of tested variants.

Volume of profile losses is logically lower for stationary blades then for moving blades. As for the variants, the variant "Pretlak-2"demonstrates better results regarding losses in both cases, mainly due to better aspect ratio blade cascades. The middle part of the stationary bladefor the variant "Pretlak-1"displays far too high values of profile losses, which leads to the need of further testing.

6 Conclusion

At present, the experimental device is undergoing a fairly large reconstruction due to some outdated components. The compressor used up to now is sufficient as for its performance, yet its manual step regulation is inconvenient, likewise a possibility to maintain its constant operating duty.

Currently, pressure measurements are carried out by a 16-channel pressure transducer. To acquire more detailed information on stage integral parameters, the number of pressure measurements shall be doubled, and measurements will be extended by total states in front of and behind the stage using inserted probes.

As far as traversing behind stator and rotor is concerned, those measurements should carry on simultaneously after the reconstruction of the device, which shall simplify the evaluation process, and also reduce the measuring time.

Implementation of measuring system for non-stationary measurements by means of the measuring apparatus CTA (Constant Temperature Anemometry), which uses heated wire cooling for non-stationary measuring, is considered for the future, in addition to stationary measurements of tested stages.

Acknowledgement

The presented research is supported by the project CZ.1.05/2.1.00/03.0108 (Sustainable Energy, SUSEN) and student project SGS-2014-070 (Increasing the efficiency, reliability and durability of power system devices 3).

References

1. J. Vomela, *Experimental and numerical development of turbine blades of high efficiency* (5th European conference on turbomachinery, Praha, 2003)
2. J. Vomela, P. Milčák, *The development of high-efficiency turbine blades* (Cieplne maszyny przeplywowe. Turbomachinery, Lodz, 2005)
3. P. Milčák, M. Hoznedl, D. Krivánka, *Testování turbínového stupně s 3D lopatkováním – úvodní experiment* (Mechanical Engineering Journal, 2009), ISSN 1335-2938
4. B. R. Haller, *Full 3D turbine blade design in secondary and tip clearance flows in axial turbines* (von Karman Institute for Fluid Dynamics, 1997)
5. C. H. Sieverding, *Recent Progress in the understanding of basic aspects of secondary flows in turbine blade passages* (Journal of Engineering for Gas Turbines and Power, 1985).
6. P. Milčák, P. Žitek, M. Hoznedl, *Comparison of Resultsfrom Experimental Testing of Three Variants of Turbine Stage with Modern 3D Blades* (Experimental Fluid Mechanics, Liberec, 2012), DOI: 10.1051/epjconf/20134501063
7. P. Milčák, M. Hoznedl, P. Žitek, *Measurement on stages with 3D bladings and different relative width of stator blades* (European Physics Journal, 2011), DOI: 10.1051/epjconf/20122501054

EPJ Web of Conferences **114**, 02056 (2016)
DOI: 10.1051/epjconf/201611402056

Experimental investigation of the steam wetness in a 1000 MW steam turbine

Michal Kolovratník[1,a], Ondřej Bartoš[1]

[1]*Czech Technical University in Prague, Department of Energy Engineering, Technická 4, Prague 6, 166 07, Czechia*

Abstract. The aim of this paper is to introduce the experimental data of the wetness distribution obtained in the year 2015 in front of and behind the last stage of the 1000MW steam turbine in the power plant Temelín. Two different optical probes developed at Czech Technical University were used. For the first time in the Czech Republic pneumatic and optical measurement of the wet steam flow field in front of the last stage of a nuclear power-station steam turbine was provided. This unique measurement opportunity provided lots of new information for the manufacturer and operator of the steam turbine and valuable experimental data for the phase transition modelling in the wet steam flow. The experimental measurement was done in cooperation with Doosan Škoda Power s.r.o.

1 Introduction

Measurements of the wetness distribution along the last blades were obtained [1] by Czech Technical University in Prague (CTU) in 2014 as part of a guarantee measurement at block 1. of the power plant Temelín (ETE), newly equipped with an enhanced low pressure part (LP) of the steam turbine.

In 2015, with the support of all concerned partners we extended the measurement in front of the last stage. Doosan Škoda Power, s.r.o. (DSP) prepared the new measurement ports in front of the last stage of the LP for the insertion of the pneumatic and optical probes. At the turn of August/September in 2015 the measurement process began on the 3rd LP part of the ETE steam turbine while boosting the performance during the nominal operating conditions. DSP performed the initial measurement of the flow field with the pneumatic probe and CTU simultaneously carried out parallel optical measurements of the wet steam liquid phase structure.

For these measurements CTU used the original and modified concepts of the optical probes – the light extinction and photogrammetric probe. The aim of the measurement development, which is currently still in progress, is to obtain experimental data from a real operation to help to improve the operation of the existing turbines (especially to increase their efficiency and reliability) as well as the newly proposed turbines. Knowledge about the formation of liquid phase in the steam turbine flow field, which is gained from the measurements, is very important for modelling the phase transition. These measurements offer a unique opportunity to verify nucleation and condensation models.

2 The measurement methods

2.1 Light extinction

The extinction measurements data processing technique uses the known relationship between the number density and size distribution function of the examined particles and attenuation of the light intensity Io. The light beam passes through the layer of particles which reduces the light intensity to I.

$$\frac{1}{\ell}\ln\left(\frac{I_o}{I}\right)=\frac{\pi}{4}N_v\int_0^{D_\infty}Q(\pi D/\lambda)\cdot\varphi(D)\cdot D^2\,dD \qquad (1)$$

Where ℓ is the thickness of the measurement section of the probe, N_v is the volumetric number density of droplets, $Q(\pi D/\lambda)$ is the extinction coefficient according to the Mie theory, D the droplet diameter, and $\varphi(D)$ the size distribution function of the droplets.
The wet steam is a mixture of saturated steam and water droplets. One expects that the droplets are spherical in shape and form a system of polydisperse particles. For the reconstruction of this polydisperse system or to reach the size distribution function of $\varphi(D)$ it is suitable to get a ratio $I/I_o = f(\lambda_i)$ where $i = 1,2,...k$ is the chosen number of the particular wavelengths $\lambda_{1,2,3...k}$ and this presents the number of the equation in the system.

The measurement procedure with the extinction probe usually has two main stages. In the first stage the reference signal is acquired in order to know the

[a] Corresponding author: michal.kolovratnik@fs.cvut.cz

dependence of the light intensity I_o and wavelength λ without any water droplets in the measurement section. Then follows the second stage of measurement in the wet steam and again the dependence of I and λ is acquired. Due to the light scattering on the water droplets in the wet steam, the light intensity I is lower than I_o.

$$\frac{1}{\ell}\ln\left(\frac{I_o}{I}\right)_i = \frac{\pi}{4} N_v \int_0^{D_m} Q(\pi D/\lambda_i)\cdot\varphi(D)\cdot D^2 \mathrm{d}D \qquad (2)$$

By determining the extinction ratio $I/I_o=f(\lambda)$ and applying the Mie theory of light scattering on water droplets one can find the size distribution of the droplets φ(D) and the wetness of the wet steam y. In general the solution leads to a system of Fredholm integral equations of the first kind [2, 3].

2.2 Photogrammetric measurement

The measurement of the coarse droplets is not simple due to the wide range of diameters. Coarse droplets are formed by the disruption of the water films on the blades and casing of the steam turbine. The photogrammetry method was used to determine the object properties from images captured by a camera [4]. The head of the probe (Figure 1) is equipped with a camera with a telecentric lens on one side and an LED light source on the other. The expected size of the coarse droplets in steam turbines is in a wide range of 1 to 500 μm. The magnification of the lens is limited by the minimum exposure time of the camera when the global shutter is used. Higher magnification can bring better resolution for smaller droplets, but the displacement of the droplet may then be greater than the size of the observed volume when the picture is taken. The velocity of droplets can reach 300 m/s.

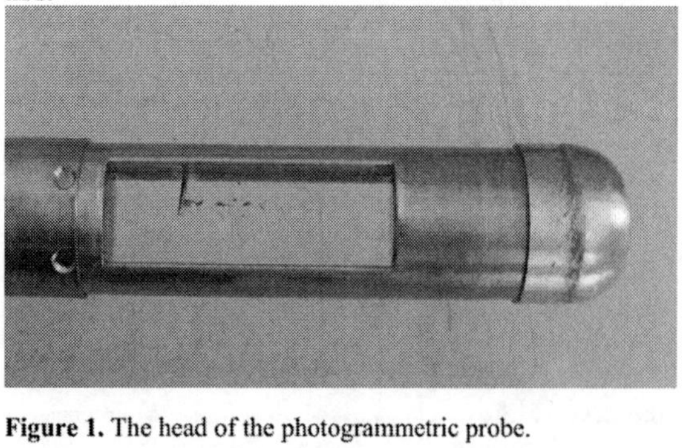

Figure 1. The head of the photogrammetric probe.

The path length of the droplet and the exposure time make it possible to compute the velocity and determine the expected direction of the droplet's trajectory. The telecentric lens provides the advantage that the image size is independent of the distance from the lens. Calibration is required for the correct pixel/dimension ratio [5].

3 Measurement probes

3.1 Light extinction probe

The probe holder with a diameter of 35 mm and a length of about 4.5 m was used for extinction measurements in the last stage LP part of the 1000 MW steam turbine with the length of the last rotating blade <1.2 m. The holder can be used to adjust the probe head at selected positions of measurement, as shown in Figure 2.

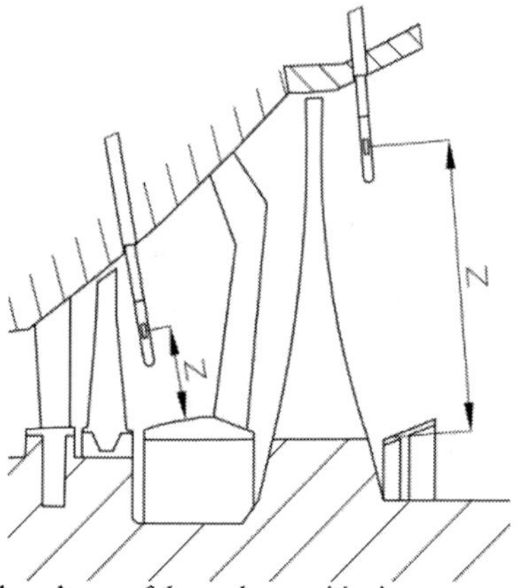

Figure 2. The schema of the probes positioning.

Two different versions of the probe head of the extinction probe were used, M-1 and M-2.

For the measurement behind the stage the tested probe head M-1 was used. Its basic dimensions are shown in Figure 3. In a cylindrical body with a diameter of 21 mm is fabricated a rectangular window measuring 50 x 13 mm through which the wet steam passes through. A white light beam is collimated perpendicular to the steam flow. As a result of light scattering on the droplets the light intensity is decreasing (see the description of the principle of measurement).

The light is drawn from an external deuterium-halogen light source to the probe head through a fibre optics cable to the measurement slot which is restricted by the optical windows. The fibre optics cable was used for the incoming light signal as well as for collecting the light which passes through the measurement slot. The fibre optics cable terminates in the collimators. Spectral analysis of the collected light is performed over the entire wavelength range $\lambda \sim 200 \div 1000$ nm.

In the results of the model calculations of the liquid phase nucleation and condensation within the passage of the LP part of the ETE's steam turbine processed at CTU [6, 7], it was expected that the measurement in front of the last stage would not be successful with the identical probe head. Nucleation starts several blade stages before the last stage and due to subsequent condensation the structure of the liquid phase is fully developed. Together with higher pressure levels, before the last stage there is a significantly higher optical density of wet steam. For that reason a probe with a shorter measuring slot has to be used in front of the last stage. A probe head with a length of 20 mm was developed. (Note: In steam turbines in fossil-fired power plants the condensation mainly occurs in the last stage, therefore the liquid phase structure is not as developed and the optical density before the stage is lower. Accordingly, the same probe head is generally used for measurement in both positions).

The concept these probes is the same which was already proven in 2014. The influence of shaking and fibre optic bending is reduced by the placement of a light source and a spectrophotometer on the outer end of the probe holder. The holder is also used as the collector for purging air of the optical window, together with the fibre optics cable and the rod for closing the head of the probe while measuring the reference intensity I_o. The probe head has a cylindrical holder reduced to 25 mm in diameter equal to the diameter of the collimator housing.

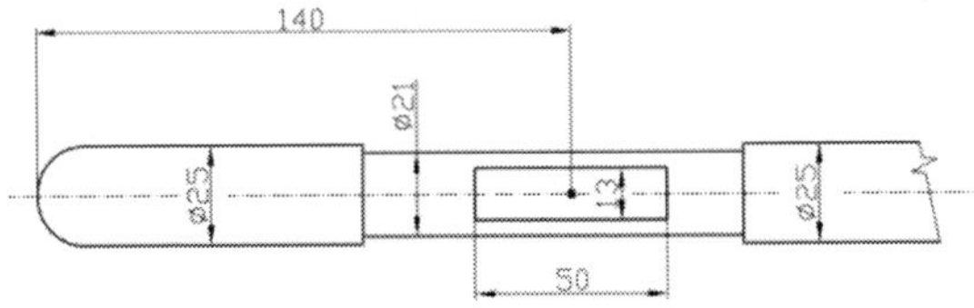

Figure 3. The main dimensions of the probe head M-1.

Due to the its length determined by the lens cover and the distance between the measuring gap and the tip, the M-1 probe cannot be used for measurements in the vicinity of the hub diameter. A new "single rib" probe head (M-2) was prepared and tested with modified optics and a reduced length. Its main dimensions are shown in Figure 4.

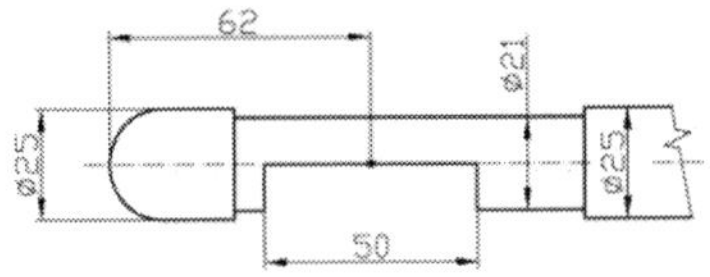

Figure 4. The main dimensions of the probe head M-2.

3.2 Photogrammetric probe

The measurement in the turbine is done together with the measurement with the extinction probe. The photogrammetric probe is enclosed in a stainless steel tube measuring 50 x 3 mm [8]. This probe is inserted to the LP part directly through the sealing and the gate valve. The camera and the lens are in the head of the probe on one side of the test section. The LED light source is fixed on the other side. The glass windows are purged by air. Previous experience has shown that the purging window is necessary to blow away the film of water deposited on the surface of the glass window. It is thought that the influence of the purging air curtain is negligible. When the probe is placed at the desired position in the main flow of the steam, the directions of the air and the steam flow are similar.

Due to the high optical density a new light source was developed and tested. The flash of the LED diode is synchronised with the camera. A 25W LED was used with nominal performance but for short exposures it was increased to approx. 100W. Thegoverning unit of the light source and the scope for determining the exposure time is presented in Figure 5.

Figure 5. The schema of the photogrammetric probe.

4 Data processing

4.1 Steam wetness by the extinction measurement

The size distribution functions of the droplets $\varphi(D)$ for the radial probe positions were determined by solving a system of equations (1) with the measured values of the light extinction $I/I_o = f(\lambda)$. The numerical solution of the equation system (determination of the function $\varphi(D)$ was done through the use of the regularization technique RNL developed at the Department of Energy Engineering at the Faculty of Mechanical Engineering CTU in Prague [3].

From the droplets size distribution function the flowing steam wetness was determined from the amount of fine droplets.

$$y = \frac{I}{1+I} \tag{3}$$

where $$I = \frac{\pi}{6}\frac{\rho_k}{\rho_p} N_v \int_{D_{min}}^{D_{max}} \varphi(D) D^3 \mathrm{d}D\,, \tag{4}$$

ρ_k and ρ_p is the density of liquid and vapour respectively.

The information about the liquid phase properties yielded by the processing of the extinction measurement is shown in Figure 7. The performance of the steam turbine during the measurement was 1090 MW.

4.2 Coarse droplets

Unlike light scattering, the photogrammetry method requires calibration. The calibration was carried out with the standard resolution target. This target was used for testing the spatial resolution of the new data processing method. The ratio for the probe is 180px/1mm.

5 Measurement on the ETE

The measurement with the extinction probe was carried out in front of and behind the last stage of the 3. LP part of the 1000 MW steam turbine on block 1 in ETE. The measurement ports are sealed by packing and closed through the use of a gate valve and it is possible to insert both probes inside the turbine when the turbine is under operation. The extinction probe required reduction housing and the photogrammetric probe can be inserted directly through the port.

Figure 6. The extinction probe inserted in to the turbine; the measurement ports on the left side for the position in front of and behind the last stage.

The positioning of the probe in the radial direction can be done in several positions in order to collect the wetness distribution along the length of the blade.

The position of the probe head is determined by the coordinate z (Figure 2) as the position of the measurement section of the probe centred with the hub diameter of the blade.

The radial distance between the hub and the tip of the blade is approx. 600 mm and 1200 mm for the position in front of and behind the last stage respectively. Due to the radial positioning of the probe it is possible to perform a series of measurements to determine the wetness along the last blade. The measurement usually proceeds from the tip to the hub of the blade. The direction of the measurement slot was set according to data about the flow field provided by DSP.

6 An example of the results

The new five-stage LP part of the ETE steam turbine manufactured by DSP is equipped with the last stage rotor blades with a length of 1220 mm. The main goal for the measurements with optical probes in 2015 was to gain more information about the wetness distribution in front of the last stage and thereby combine it with the data acquired in 2014.

For the measurements two types of optical probes were used, an extinction probe and a photogrammetric probe. The placement of the probes in the measurement position is shown in Figures 2 and 6.

Currently, the first series of measurements in front of the last stage has been done. For research of the liquid phase development within the expansion of the wet steam in the steam turbine the measurement behind the last stage will be carried out soon, very similar conditions in front and behind the last stage are expected.

In this paper the results of two measurements on the right side of the turbine are presented, which make it possible to compare the wetness distribution in both positions.

Duting the operating mode of the turbine a performance of 1060 MW makes it possible to determine the pressure range behind the last stage as 8 ÷ 10 kPa. The absolute value of moisture cannot be published. Therefore, the comparison of the wetness along the last blade is presented in Figure 7 in normalised values $y/y_{max}=f(z)$.

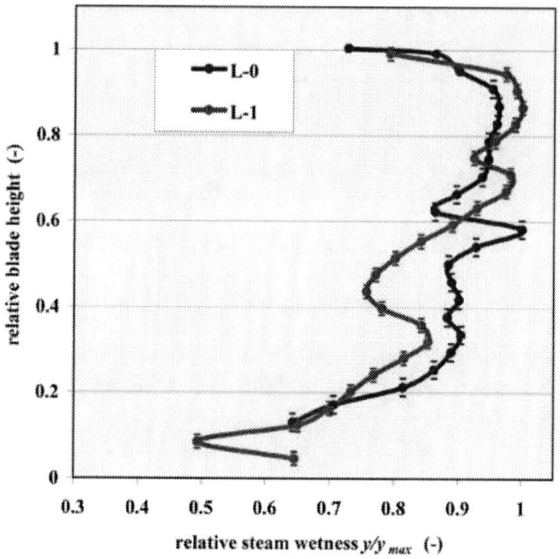

Figure 7. The wetness distribution along the last blade for the position in front of and behind of the last stage in the turbine.

The particular values of the wetness for the measurement positions are presented in Figure 7 at the position of the measurement section centre. The thin line presents the size of the measurement section (20/50 mm) of the probe and borders the region of signal acquisition by the extinction probe.

The photogrammetry method is very simple, the measurement in the steam turbine is still pushing the edge of the present abilities of the equipment being used. The main goal for the present measurement was to verify the new light source and attempt to repeat the successful measurement from 2014 when a few images of coarse droplets were captured.

Figure 8 presents an image taken at the position z = 1200 mm. In the image it is possible to see a coarse droplet, this is a static droplet probably produced by the probe head itself. This image proves the ability to measure and it is an example of an acquired picture.

The photogrammetric probe was placed in the turbine for a short time due to technical reasons and during that time no proven coarse droplets were captured. From ast year's results it is known that the number density of coarse droplets is not sufficient to obtain the correct statistical analysis of the size distribution function in the case of this measurement campaign. In our case approximately 1000 pictures were taken. The better results are due to the use of the new light source which is sufficient for the optical properties in the wet steam.

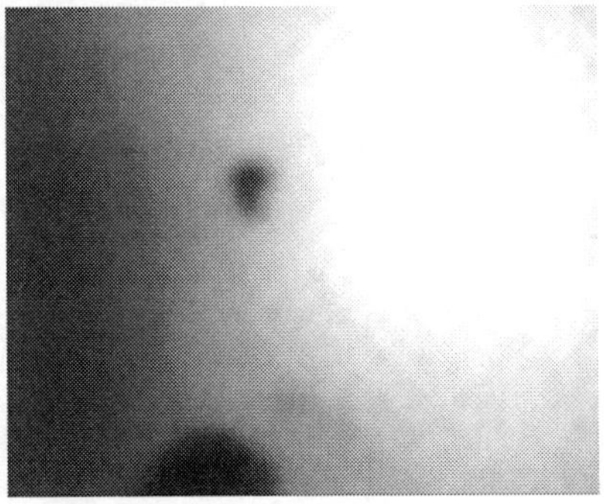

Figure 8. The coarse droplet.

7 Conclusions

The paper presents a brief description of the experimental research of wetness in the 1000 MW steam turbine in the 1. unit of the power station ETE.

A measurement system developed at CTU in Prague was used, which consists of extinction probes and a photogrammetric probe, auxiliary units for measurement support, and basic instruments for data acquisition and data processing. The parameters of the steam necessary for data processing such as the admission temperature and pressure, and the emission pressure are reached through the control and ETE measurement system and from data provided by the DSP measuring team. Experimental data from the extinction measurement was processed by the numerical method RLN.

The results are presented as an example of normalised wetness distribution along the blade on the left side of the steam turbine in the position L-1 in front and L-0 behind the last stage.

The photogrammetric method of coarse droplets detection was tested successfully. However, due to the low number density of coarse droplets it is still necessary to develop a new method for determining the coarse droplets size distribution function.

Acknowledgement

We gratefully acknowledge the support by Grant No. 101/11/1593 of the Grant Agency of the Czech Republic

The authors gratefully acknowledge the support of all colleagues from DSP and the ČEZ company who helped in the completion of this project.

We gratefully acknowledge the technical and measurement assistance of our colleague Petr Stříž.

References

1. M. Kolovratník, O. Bartoš, , *EPJ Web of Conferences* **92**, 02035, (2015)
2. Bohren, C.F., Huffman, D.R.: Absorption and Scattering of Light by Small Particles. John Wiley & Sons, Inc., USA, 1998
3. Kolovratník, M.: Extinkční sondy pro měření v parních turbínách. Habilitační práce, Praha, 2004
4. X. Cai,L. Wang, Y. Pan, X. Ouyan, J. Shen, J. Thermal Science. **10**: 123-126, (2001)
5. O. Bartoš, X. Cai, M. Kolovratník, The Cambridge Baumann Centenary Conference on Wet Steam Flows (Cripps Court Conference Centre, Magdalene College, Cambridge, UK; 10-11 September 2012)
6. V. Petr, M. Kolovratník, Diagnostic of wet steam in LP steam turbines, 5th European conf. on turbomachinery, fluid dynamics and thermodynamics, (687-698), Prague, (2003).
7. V. Petr, M. Kolovratník, Journal of power and energy, Vol. 228, Is.2, pp.206-2015,(2014)
8. O. Bartoš, X. Cai, M. Kolovratník, *EPJ Web of Conferences* **67**, 02005, (2014)

EPJ Web of Conferences **114**, 02057 (2016)
DOI: 10.1051/epjconf/201611402057

Reliability of fluid systems

Jaroslav Kopáček[1], Kamil Fojtášek[1,a] and Lukáš Dvořák[1]

[1]*VŠB-Technical University of Ostrava, Faculty of Mechanical Engineering, Department of Hydrodynamics and Hydraulic Equipment, 17. listopadu 15/2172, Ostrava 70833, Czech Republic*

Abstract. This paper focuses on the importance of detection reliability, especially in complex fluid systems for demanding production technology. The initial criterion for assessing the reliability is the failure of object (element), which is seen as a random variable and their data (values) can be processed using by the mathematical methods of theory probability and statistics. They are defined the basic indicators of reliability and their applications in calculations of serial, parallel and backed-up systems. For illustration, there are calculation examples of indicators of reliability for various elements of the system and for the selected pneumatic circuit.

1 Introduction

Each machine or technological device is formed from mechanical, hydraulic, pneumatic and electric elements or their combinations, usually coupled with sophisticated proceedings on the basis of electronics and computers. In this case we are talking about the mechanisms, consisting of systems.

Fluid Systems, which is now an established name for hydraulic and pneumatic systems, have in many practical cases the dominant or exclusive position. With the help of fluid systems is realized many manufacturing, assembly, transport, handling and packaging processes. These processes are not only in mechanical engineering, but largely in the food industry, electrical, pharmaceutical and many others. It should be noted that many applications are characterized by high work productivity and also high acquisition and operating costs. It must also consider that, the fluid systems in these applications have a dozens of working elements (linear and rotary actuators) and hundreds control elements with various functions (for pressure control, flow direction, speed, position, etc.), which relate with the function of hundreds of sensors and switches. The failure of a single member of this system leads to a halt operations of a whole manufacturing, assembly or packaging lines. This leads not only to loss of production, but in some cases to the irreversible material losses. And here starts the issue of reliable function of the individual components and the entire system. This issue has not been solved for fluid systems in general or a specific level, in particular in interaction with the electric control circuits.

The issue of reliability of elements and systems, according to available literature elaborated and continuously monitored in the strategic application in military industry (aerospace, missile), in radio engineering and electronics in general, and probably also in the area of large power plants (nuclear power stations). However, as discussed below, reliability is understood as a complex feature which has precisely defined terminology in our, European and world standards and extensive theory of reliability is based on the mathematical theory of probability and statistics.

The theory of reliability deals with three basic tasks:

- forecasting (prediction) of reliability,
- detection of reliability,
- control (diagnosing) of reliability.

For further explanation it is necessary to define the reliability according to universally accepted standards: „Reliability is a general property of an object consisting of the ability to perform the required functions, with maintaining the values specified operating parameters within limits and time set by a technical conditions.“

Emphasized the complex property of reliability depends on the inclusion of terms such as durability, repairability, failure rate and many others. This means that reliability cannot be precisely quantified. When providing guarantees of reliability, which is the content of economic contracts, it should be emphasized, that featured an indicator of reliability is for example predicted, anticipated, empirically validated or guaranteed.

The basic criterion for assessing the reliability is the failure of object. This failure is a random phenomenon and means ending the ability of an object (element or the whole system) to perform the desired function from any cause. Failures classification is quite complicated. In terms of reliability and in its time course may be a sudden failure, occasional or sequential. The time course of failure is most commonly demonstrated as bathtub curve

[a] Corresponding author: kamil.fojtasek@vsb.cz

Fig. 1, with three specified courses. The causes of failure are very complicated, and their share of system failure indicates Tab. 1 [1].

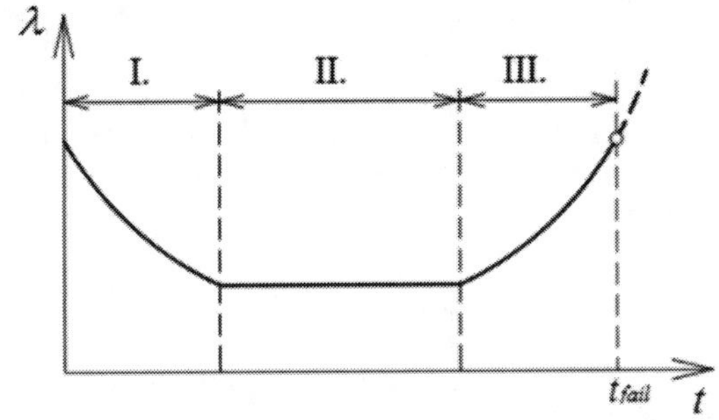

Figure 1. Idealized course of failure per time (bathtub curve)

Table 1. The reasons and causes of failures

The nature and causes of failure	The failure rate [%]
a) causes of damage	
construction	20
production	50
working	30
b) character of parameter changes	
sudden	40
progressive	60
predictable	20
unpredictable	80
c) signs of damage	
leak	45
bad settings	15
lack of functionality	15
violation of dynamic stability	10
destruction of the power components	15
d) elements of the hydraulic system	
pipes and hoses	35
solenoid coil of valves	20
automation component	10
filters	25
power elements	10

2 Theory

Quantitative evaluation of the reliability of the system is based on the evaluation of the reliability of individual elements. Evaluation is done through a series of characteristics and reliability indicators. These indicators are based on the application of the theory of probability and mathematical statistics.

Our interpretation can significantly simplify and point out the extensive literature in this area. We present four basic relations for evaluating the reliability:

Probability of failure

$$F(t) = P(\tau < t), \tag{1}$$

or

$$F(t) = \int_0^t f(t)dt.$$

Density of failures

$$f(t) = \frac{d}{dt}F(t), \tag{2}$$

the probability of failure-free operation

$$R(t) = 1 - F(t), \tag{3}$$

and intensity of failures

$$\lambda(t) = \frac{f(t)}{R(t)} = \frac{f(t)}{1 - F(t)}. \tag{4}$$

Intensity of failures is one of the most important indicators of reliability used in practice (the only published value). It should be noted that the λ value depends on many factors, especially on production technology, on operating conditions, on the size and complexity of the object and on time, as shown in Fig. 1 [2].

In the last column of Tab. 2 presents the basic theoretical equations, which are supplemented formulas to calculate statistical indicators. In these equations, n is the number of monitored objects, r is the number of damaged objects (objects in failure) and Δt is monitored time interval of total time t.

The literature indicates a number of values of reliability indicators $\lambda(t)$, apparently obtained empirically, as shown in Tab. 3. Computational application of theoretical equations from Tab. 2 for practical use should be further narrowed to law exponential distribution objects for value $\lambda = const.$, what can be in terms of technical experience to admit.

With these assumptions we obtain practically usable equations [3]:

The probability of failure-free operation

$$R(t) = e^{-\lambda \cdot t}, \tag{5}$$

density of probability of failure

$$f(t) = \lambda \cdot e^{-\lambda \cdot t}. \tag{6}$$

The median duration of failure-free operation

$$T_s = \frac{1}{\lambda}, \tag{7}$$

and also

$$R(t) = e^{-\frac{t}{T_s}}. \tag{8}$$

Table 2. Probabilistic definitions and statistical significance of reliability indicators

Indicator of reliability	Mark	Determination of probabilistic	Statistical estimation (empirical determination)
Probability of failure-free operation	$R(t)$	$R(t) = (\tau > t)$	$R(t) = \frac{n(t)}{n}$
Probability of failure	$F(t)$	$F(t) = P(\tau \leq t)$	$F(t) = \frac{r(t)}{n} = \frac{n - n(t)}{n}$
Density of probability of failure	$f(t)$	$f(t) = \frac{dF(t)}{dt}$	$f(t) = \frac{r(t)}{n \cdot \Delta t} = \frac{n - n(t)}{n \cdot \Delta t}$
Intensity of failure	$\lambda(t)$	$\lambda(t) = \frac{f(t)}{R(t)}$	$\lambda(t) = \frac{r(t)}{n(t) \cdot \Delta t}$

Equation (8) shows that when the median duration of failure-free operation T_s is equal to the duration of operation t, then $R(t) = e^{-1} = 0.37$. Which means that the probability of failure-free operation is reduced to 37%.

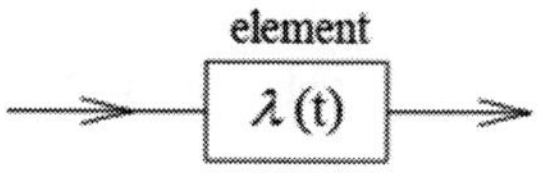

a) block representation

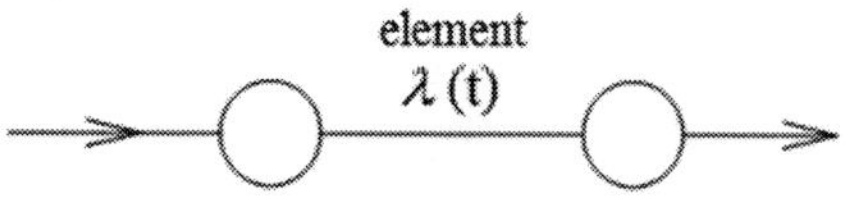

b) representation by oriented graph

Figure 2. Layout of boundary conditions for sliding mesh

For the calculation of the reliability are introduced models in the form of block diagrams or oriented graphs Fig. 2. The individual elements of these models can be connected in series, parallel or series-parallel. For serial connection Fig. 3 is probability of failure-free operation [4]

$$R(t) = \prod_{i=1}^{n} R_i(t) = \prod_{i=1}^{n} e^{-\lambda_i \cdot t_i}, \tag{9}$$

intensity of failure

$$\lambda(t) = \sum_{i=1}^{n} \lambda_i(t), \tag{10}$$

and probability of failure

$$F(t) = 1 - R(t) = 1 - \prod_{i=1}^{n} [1 - F_i(t)]. \tag{11}$$

Median duration of failure-free operation (duration to the failure)

$$T_s = \int_0^{\infty} R(t)dt = \frac{1}{\lambda(t)}, \tag{12}$$

and then

$$R(t) = e^{-\frac{t}{T_s}}. \tag{13}$$

For parallel connection we have more complicated equations:

probability of failure-free operation

$$R(t) = 1 - \prod_{i=1}^{m} \left(1 - e^{-\lambda_i(t)}\right), \tag{14}$$

probability of failure

$$F(t) = \prod_{i=1}^{m} \left(1 - e^{-\lambda_i(t)}\right), \tag{15}$$

intensity of failure

$$\lambda(t) = \frac{F(t)}{1 - F(t)} \cdot \sum_{i=1}^{m} \frac{f_i(t)}{F_i(t)}, \tag{16}$$

and median duration of failure-free operation

$$T_s = \int_0^{\infty} R(t)dt. \tag{17}$$

In parallel reliability models are certain elements simplified backup (redundancy), most often in their control part. Assuming that a backup is performed the same element (with intensity $\lambda_1 = \lambda_2 = \lambda_3 \ldots \lambda_0$), and if this element is also under load, then the probability of failure-free operation of the system [5]

$$R(t) = e^{-\lambda_0 \cdot t} \cdot \left(2 - e^{-\lambda_0 \cdot t}\right), \tag{18}$$

probability of failure

$$F(t) = 1 - R(t), \tag{19}$$

and median duration of failure-free operation

$$T_s = \frac{3}{2 \cdot \lambda_0}. \tag{20}$$

[a] Corresponding author: kamil.fojtasek@vsb.cz

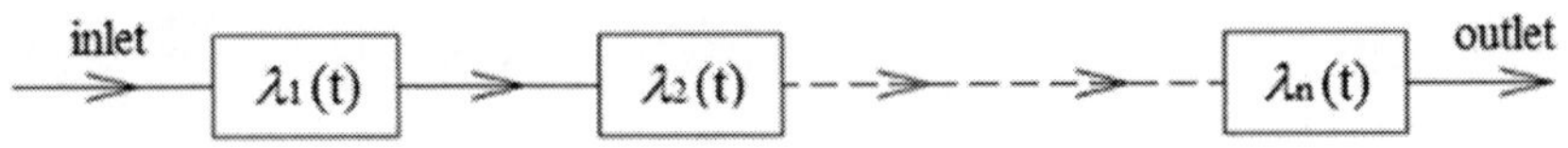

a) block representation

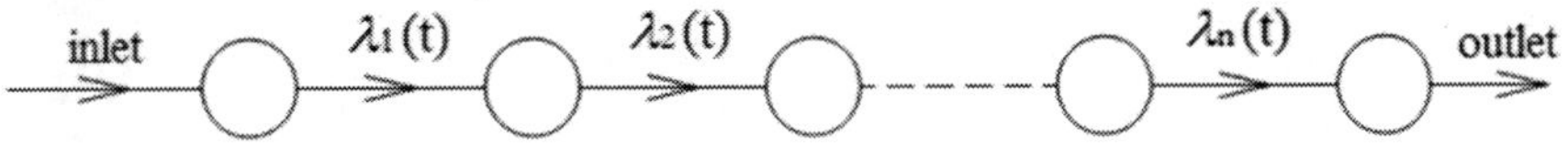

b) representation by oriented graph

Figure 3. Serial reliability diagram

Table 3. Description of the variant

Element	Intensity of failure $\lambda(t)10^{-6}\,h^{-1}$		
	maximum	median.	minimum
Gear pump	-	13	-
Axial piston pump	13	9	6
Rotary hydraulic actuator	7.15	4.3	1.45
Linear hydraulic actuator	0.12	0.08	0.005
Relief valve	14.1	5.7	0.224
Overflow valve	0.24	0.5	0.26
Electrohydraulic servo valve	56	30	16.8
Check valve	14.1	5.7	3.27
Spool valve	0.112	0.054	0.041
Accumulator	7.5	6.8	0.35
Rubber high-pressure hoses	3.27	2	0.05
Filtres	1.69	0.79	0.01
Hydraulic tank	2.52	1.5	0.48
Static seal	7.6	4.9	2.2
Dynamic seal	1.12	0.7	0.25
Pneumatic regulators	-	2.4	-
Pneumatic cylinder	0.013	0.004	0.002

If backup element isn not under load

$$R(t) = e^{-\lambda_0 \cdot t} \cdot (1 + \lambda_0 \cdot t), \qquad (21)$$

and

$$T_s = \frac{2}{\lambda_0}. \qquad (22)$$

For series-parallel connection are calculation equations more complicated and they are based on the progressive solution of serial and parallel parts. The resulting equations for the overall probability are multiplication of partial solutions.

For the calculation of reliability of complex systems are used mathematical methods and procedures. In literature are mentions about various calculation software. In literature are mentions about various calculation softwares.

3 Examples of calculation of reliability indicators

These calculations are performed in the newly designed systems, but also in existing systems (under the operating loads). Usually there is not enough partial information on individual elements, which may be best obtained experimentally (which relates to hydraulic and pneumatic systems). For this reason, calculation focuses only on some indicators of reliability. The most commonly is known the value of intensity of failure $\lambda(t)$. According to this value is verified probability of failure-free operation $R(t)$ and median duration of failure-free operation T_s.

Quite often in the individual elements of the system (electrical, hydraulic or pneumatic) it is given „lifetime“ value. This value is often confused with the parameter of reliability, expressed in hours of operation (pumps), in number of working cycles (directional valves), or like distance (fictitious stroke) in kilometers for linear motors.

Then time t is replaced to mark h and probability of failure-free operation will be

$$R(t) = e^{-\frac{h}{T_s}}, \qquad (23)$$

for example when fictious value h is equal median duration of failure-free operation T_s, then $R(t) = e^{-1} = 0.37$. In this case the element will be working without failure only for 37% of duration T_s, where $T_s$was calculated from equation (7).

The equation (23) could be modified by the help of known value $\lambda(t)$

$$R(t) = e^{-\frac{\lambda \cdot h}{0.37}} = e^{-2.7 \cdot \lambda \cdot h}. \quad (24)$$

If we choose intensity of failure $\lambda(t)$ for hydraulic cylinder from Tab. 3 $\lambda(t)_{max} = 0.12 \cdot 10^{-6}\, h^{-1}$ and if we assume a working operation 6000 hours per year, then probability of failure-free operation will be

$$R(t) = e^{-2.7 \cdot \lambda \cdot h} = e^{-2.7 \cdot 0.12 \cdot 6 \cdot 10^{-3}} = 0.998,$$

which is close to 100% reliability and this value should not dramatically reduced in next two or three years of working.

For electrohydraulic servo valve with maximal value of intensity of failure $\lambda(t) = 56 \cdot 10^{-6}\, h^{-1}$ Tab. 3, and with working operation $t = h = 6000\, h$

$$R(t) = e^{-2.7 \cdot \lambda \cdot h} = e^{-2.7 \cdot 56 \cdot 6 \cdot 10^{-3}} = 0.403,$$

and for redundant servo valve, whose backup control element (electrical, hydraulic) is also under load $\lambda_1(t) = \lambda_2(t) = \lambda_0$. Probability of failure-free operation of element with backup

$$R(t) = e^{-2.7 \cdot \lambda_0 \cdot h} \cdot \left(2 - e^{-2.7 \cdot \lambda_0 \cdot h}\right)$$

$$R(t) = e^{-2.7 \cdot 56 \cdot 6 \cdot 10^{-3}} \cdot \left(2 - e^{-2.7 \cdot 56 \cdot 6 \cdot 10^{-3}}\right)$$

$$R(t) = 0.643,$$

which is about 50% more value probability of failure-free operation than servo valve without backup.

And now one practical example. In the company ZTS Dubnica which produced axial piston convertors was processed sophisticated collection of information about their operation in various working machines. This is of the examples of evaluation that was performed in accordance with the then current standards.

Total production listed convertors was 55 773 pieces. Of that number 220 pieces was defective (unrepaired and eliminated). For the calculation will apply values and calculation formulas from Tab. 2:

Time interval of monitoring $\Delta t = 1\, year = 200\, days\ x\ 8\, hours\ of\ operation = 1600\, hours.$, the number of monitored convertors $n = 55\,773$, the number of defective and eliminated convertors in time interval $r(t) = 220\, pcs$, the number of convertors without failure in this time interval $n(t) = n - r(t) = 55\,553\, pcs$. Calculated indicators of reliability:

Probability of failure-free operation

$$R(t) = \frac{n(t)}{n} = \frac{55\,553}{55\,773} = 0.996,$$

probability of failure in time interval Δt

$$F(t) = \frac{r(t)}{n} = \frac{220}{55\,773} = 0.00394,$$

density of probability of failure in time interval Δt

$$f(t) = \frac{r(t)}{n \cdot \Delta t} = \frac{220}{55\,773 \cdot 1600} =$$

$$f(t) = 2.465 \cdot 10^{-6}\, h^{-1},$$

intensity of failures in time interval Δt

$$\lambda(t) = \frac{f(t)}{R(t)} = \frac{r(t)}{n(t) \cdot \Delta t} = \frac{220}{55\,553 \cdot 1600}$$

$$\lambda(t) = 2.474 \cdot 10^{-6}\, h^{-1},$$

which is significantly lower value than that mentioned in Tab. 3, and indicates a high level of production of the convertors.

We present an application example of calculating the reliability of the pneumatic system, shown in Fig. 4.

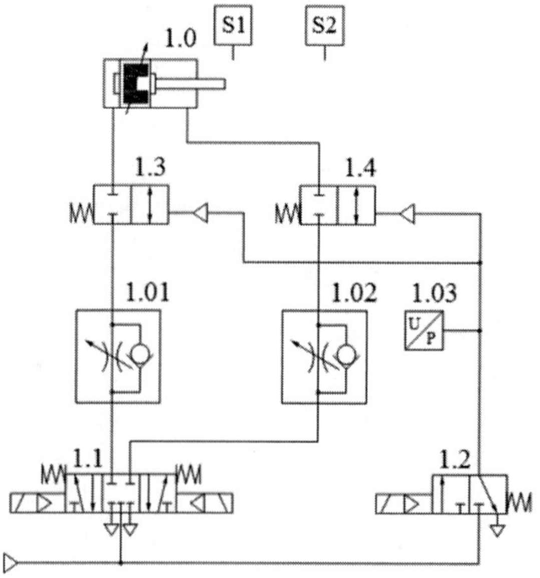

Figure 4. Pneumatic circuit

Table 4. Description of the variant

	Element	Value $\lambda(t) \cdot 10^{-6}\, h^{-1}$
$\lambda_1(t)$	rubber hose	3.27
$\lambda_2(t)$	spool valve	0.112
$\lambda_3(t)$	rubber hose	3.27
$\lambda_4(t)$	relief valve	14.1
$\lambda_5(t)$	rubber hose	3.27
$\lambda_6(t)$	spool valve	0.112
$\lambda_7(t)$	rubber hose	3.27
$\lambda_8(t)$	pneumatic cylinder	0.013
$\sum_1^8 \lambda(t) = \lambda_C$		$27.417 \cdot 10^{-6}\, h^{-1}$

The calculation was performed for input and output channel, both in series connection. From Tab. 3 we choose intensity of failure $\lambda_1(t) \div \lambda_8(t)$ for each individual element of input channel and determine the sum.

We choose the time to calculate the failure-free operation $t = h = 6000\, h$.
Probability of failure-free operation $R(t)$

$$R(t) = e^{-2.7 \cdot \lambda_C \cdot h} = e^{-2.7 \cdot 27.417 \cdot 6 \cdot 10^{-3}} = 0.641,$$

median duration of failure-free operation T_s

$$T_s = \int_0^\infty R(t)dt = \frac{1}{\lambda_C} = \frac{1}{27.417 \cdot 10^{-6}}$$

$$T_s = 36\,474\, h.$$

Similarly, for the output channel with a slightly different structure of circuit, Fig. 4 the sum is $\sum \lambda(t) = 24.147 \cdot 10^{-6}\, h^{-1}$, probability of failure-free operation $R(t) = 0.676$ and median duration of failure-free operation $T_s = 41\,413\, h$.

Prerequisite for the proposed pneumatic circuit that as a whole system will work more than six years without failure, with probability over 60%.

4 Conclusion

In the article it was pointed out to the importance of detection of reliability, especially in complex fluid systems for demanding manufacturing and other technologies. The default criterion is the failure of a component (object) of system. This failure is understood as a random variable, their values can be processed using mathematical methods of probability theory and statistics.

It was defined the basic indicators of reliability and their application in calculations of serial, parallel or backup systems. For illustration were shown examples of calculation of the reliability coefficients for the individual elements of the system and for chosen pneumatic circuit.

The aim of article was pointed out to the still-neglected issue of the reliability of hydraulics and pneumatics mechanisms.

Acknowledgements

This work was supported by the project SP2015/95 of VŠB Technical University of Ostrava.

References

1- L. Maixner, Z. Kolníková, *Spolehlivost automatických výrobních systémů*, (1984)
2- T.A. Syricyn, *Nadežnosť gidro-i-pnevmoprivodov*, (1981)
3- J. Kopáček, *Technická diagnostika hydraulických mechanismů*, (1996)
4- S.R. Calabro, *Základy spolehlivosti*, (1965)
5- B. Dodson, D. Nolan, *Reliability Engineering Handbook* (1999)

EPJ Web of Conferences **114**, 02058 (2016)
DOI: 10.1051/epjconf/201611402058

Experimental research of multiphase flow with cavitation in the nozzle

Milada Kozubkova[1], Marian Bojko[1], Jana Jablonska[1], Dorota Homa[2,a] and Jiří Tůma[3]

[1]*Department of Hydromechanics and Hydraulic Equipment, Faculty of Mechanical Engineering, VSB – Technical University of Ostrava, 17. listopadu 15, 708 33 Ostrava, Czech Republic*
[2]*Institute of Power Engineering and Turbomachinery, Faculty of Energy and Environmental Engineering, Silesian University of Technology, Konarskiego Street 18, 44 100 Gliwice, Poland*
[3]*Department of Controls Systems and Instrumentations, Faculty of Mechanical Engineering, VSB – Technical University of Ostrava, 17. listopadu 15, 708 33 Ostrava, Czech Republic*

Abstract. The paper deals with the problems of cavitation in water flow in the nozzle. The area of research is divided into two directions (experimental and numerical research). During the experimental research the equipment with the nozzle is under the measurement and basic physical quantities such as pressure and volume flow rate are recorded. In the following phase measuring of noise which is generated during flow through the nozzle in the area of cavitation is measured at various operating conditions of the pump. In the second part the appropriate multiphase mathematical model including the consideration of cavitation is defined. Boundary conditions for numerical simulation are defined on the basis of experimental measurements. Undissolved air in the flow is taken into account to obtain pressure distribution in accordance to measured one. Results of the numerical simulation are presented by means of basic current quantities such as pressure, velocity and volume fractions of each phase. The conclusions obtained from experimental research of cavitation were applied to modify the multiphase mathematical model.

1 Introduction

Cavitation is fairly common phenomenon in issues connected with hydrodynamics and pumps. Hydrodynamic cavitation, the one most common in hydraulic equipment, can be observed in areas of lowered pressure and higher velocity appearance. In pumps it is near the leading edge of rotor blades. For hydraulic equipment such as valves and rotor of pumps working under cavitation conditions can lead to their serious damage and loss of further working possibility [1]. Therefore it is crucial to be able to predict and model cavitation occurrence and development.

Cavitation is a process that includes both vaporization (creation of vapour bubbles) and condensation (sudden bubble collapse). At first, when pressure is lowered cavitation nuclei of diameter about 10^{-6} m occur in the flow. If the pressure is still lowered (which happens in flow through the nozzle), the nuclei start to grow due to the evaporation on their surface. This is called vapour cavitation [2, 3] If the lowering of pressure is taking place at slow rate the growth of the bubbles can be due to the diffusion on undissolved gases in the liquid. This is called gaseous cavitation [2, 3]. When bubbles flow to the area of higher pressure and lower velocity, they collapse. The vacuum formed around collapsing bubbles causes the liquid to flow in with high velocity. The liquid compresses the left gases to the very high pressure. The gases at high pressure expand again and the vaporization at the surface of the bubble takes place. The whole process of vaporization and condensation happens alternately at high frequency. The natural frequency of bubbles is mentioned to be about 2-5 kHz [4].

As the cavitation phenomenon is highly unsteady and includes complicated interaction between phases, there are different mathematical models describing it [5-7]. They consider mainly on vapour phase creation in the flow. In presented numerical simulation the undissolved air was taken into account to obtain the results in accordance with experiment. Aim of this paper is to confront results of the mathematical simulation with an experiment.

2 Description of experiment

2.1 Experimental methods

To widen the knowledge of the cavitation process and provide data to compare with mathematical simulation the following experiment was performed. Cavitating flow in circular converging-diverging nozzle installed in a hydraulic circuit was observed. The methods included visualisation of the flow with means of camera recording, measuring of the flow rate, pressure and noise [8].

[a] Corresponding author: dorota.homa@polsl.pl

2.2 Experimental equipment

Experimental equipment consisted of hydraulic circuit with circular converging-diverging nozzle, frequency converter, flowmeter, pressure and noise sensors and dissolved oxygen transmitter. The scheme of the circuit is shown in figure 1.

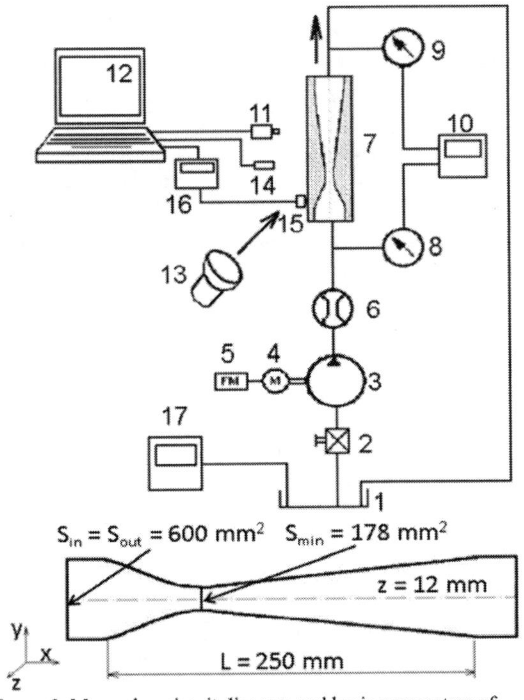

Figure 1. Measuring circuit diagram and basic parameters of the investigated object 1 – tank, 2 – ball valve, 3 – GRUNDFOS five-stage water pump, 4 – electric motor, 5 – YASKAWA VS mini J7 frequency inverter, 6 – ELIS PLZEŇ electromagnetic flow meter, 7 – measured object - nozzle, 8, 9 – Hydrotechnik pressure sensors, 10 – Hydrotechnik analyzer, 11 – Baumer HXC20c high-speed camera with CMOSIS CMV2000 sensor, 12 – PC with analysis software, 13 – light source, 14 – ½" TYPE 489 microphones, 15 – vibration sensor, 16 – Microlog analyzer, 17 - ENDRESS + HAUSER LIQUISYS-M dissolved oxygen transmitter

The flow in the circuit was forced by the pump connected to the tank, working liquid was water. Water passed through the flowmeter and reached the nozzle. With increasing pump motor's frequency more air was released in the nozzle. The change of oxygen content in the water was measured by the dissolved oxygen transmitter located in the tank. In the beginning and end of the nozzle the pressure sensors were installed. The nozzle is made of transparent plastic. The angle of extension in the diffuser part of the nozzle is equal to 7.3°, which is less than the critical value of 12° at which separated flow occurs [2]. The flow was regulated by changing the frequency of pump's motor by means of frequency converter. The measurements were performed at twelve different pump's motor frequencies, from 10 Hz to 32 Hz with step 2 Hz. For each frequency noise measurement was performed by means of sound intensity probe, shown in figure 2.

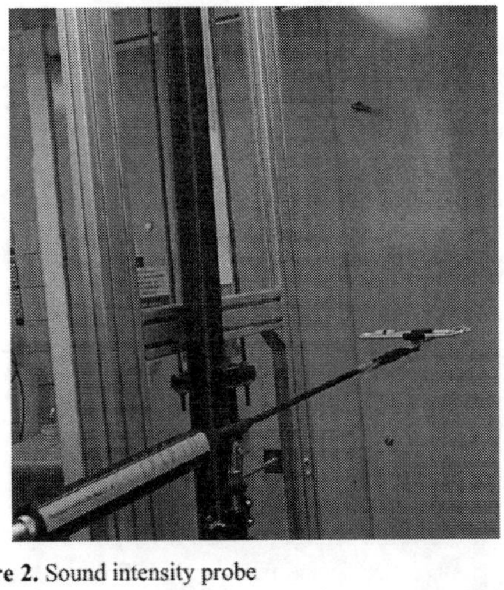

Figure 2. Sound intensity probe

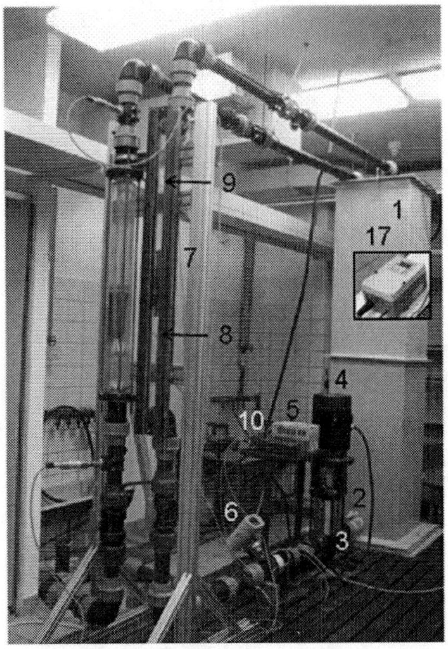

Figure 3. Measuring circuit photo

3 Evaluation of experiment

3.1 Visualisation of cavitation clouds

The cavitation could be observed from frequency equal to 18 Hz. As cavitation is not uniform in time, for each frequency the sequence of pictures were taken. After that, the mean value of cavitation cloud area at each frequency was prepared using MATLAB software. The results of measurements including *f* frequency of pump's motor; Q_v mean volumetric flow of water; p_{in} mean static pressure at inlet to the nozzle, p_{out} mean static pressure at the outlet to the nozzle; *L* length of cavitation cloud are shown in table 1. In figure 4 the clouds are plotted.

Table 1. Results of measurements

f	Q_v	p_{in}	p_{out}	L
Hz	l/s	Pa	Pa	cm
18	2.61	138982	96476	1.6
20	2.79	155301	96725	2.2
22	2.95	173085	97048	4.1
24	3.15	192279	97329	5.3
26	3.31	213556	97775	7.0
28	3.49	236471	98202	9.4
30	3.68	260885	98493	11.3
32	3.88	286609	97937	11.7

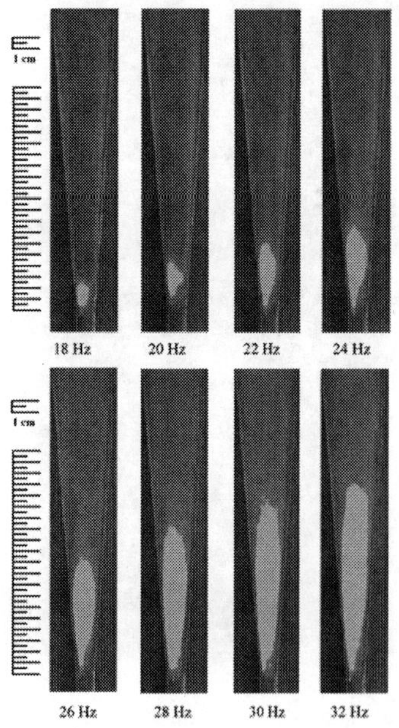

Figure 4. Cavitation clouds development

The clouds are slightly not axisymmetric as the nozzle was lit from the one side.

3.2 Results of noise measurements

For each frequency of the pump's motor noise generated by the flow in the nozzle was measured by sound intensity probe. As the sound probe also received signals from pump's motor, it was necessary to place it between the nozzle and the motor and reject the signal from motor's site. After collecting the data, the FFT analysis was performed to identify main noise frequencies. The results for motor's frequency from 18 Hz to 24 Hz are shown in figure 5.

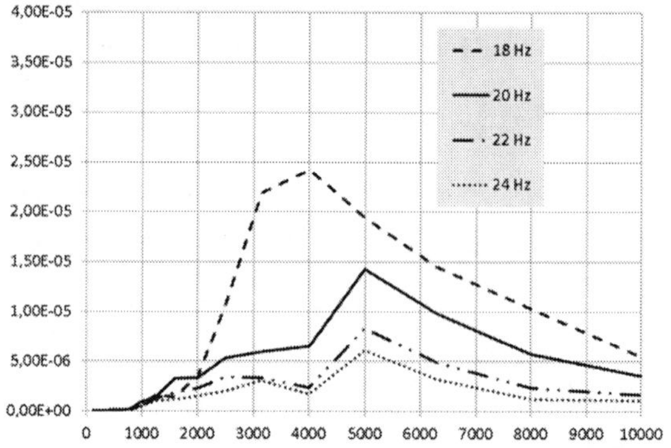

Figure 5. FFT analysis of noise measurements, frequency of motor from 18-24 Hz

For motor's frequencies of 18 Hz only one peak frequency can be noticed, equal to 4000 Hz. As the frequency and flow rate is increasing, the frequency is equal to about 5000 Hz and from 24 Hz (3.15 l/s) the two noise frequencies can be observed, 3000 Hz and 5000 Hz. For 3000 Hz magnitude is lower. However, from motor's frequency 28 Hz the higher magnitude appears at this frequency. Results of FFT analysis for further flow rates and motor's frequencies is shown in figure 6.

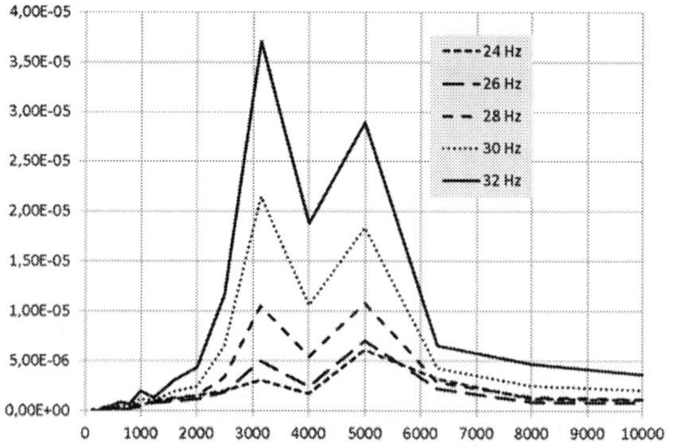

Figure 6. FFT analysis of noise measurements, frequency of motor from 24-32 Hz

With increase of motor's frequency (which indicates increase of flow rate) the magnitudes are getting higher. The highest ones are observed at motor's frequency 32 Hz.

The possible explanation of dividing the noise frequency into two from flow rate 3.15 l/s (24 Hz motor's frequency) is that cavitation at first is mainly of vapour type. It dominates at flow rate equal to 2.61 l/s (motor's frequency 18 Hz), then the dissolved air starts to play more significant role. It is possible that two noise frequencies stands for two gases – one is undissolved air and the other is water vapour, but further studies are needed to confirm that hypothesis.

4 Computer simulation of cavitation

On the basis of experimental investigation computer simulation was performed by means of ANSYS FLUENT software. The 2D case was calculated as the nozzle is axisymmetric. To create the proper mesh the flow without cavitation was investigated. The heights of the cells near the wall of the nozzle were set to the values that y* parameter (wall unit) varied between 8 and 50. The values about 8 occur only near inlet and outlet of the nozzle. It is sufficient to model boundary layer properly as far as k-ε turbulence model is concerned [6]. Finally, the 2D mesh has 21000 cells. The calculations were transient, with pressure-based solver. The gravity was taken into account. The inlet boundary condition was mass flow rate, the outlet – absolute static pressure. These parameters were known from experimental investigation.

Figure 7. Computing domain overview

4.1 Cavitation model

To model cavitation Schnerr & Sauer cavitation model was chosen. This model bases on Rayleigh-Plesset equation, which states as follows [4]:

$$r_B \frac{D^2 r_B}{Dt^2} + \frac{3}{2}\left(\frac{Dr_B}{Dt}\right) = \left(\frac{p_B - p}{\rho_l}\right) - \frac{2S}{\rho_l r_B} \quad (1)$$

Where : r_B radius of bubble, m; t – time, s ; p_B pressure in a bubble, Pa; p pressure in the flow, Pa; ρ_l density of liquid phase, kg/m^3; S surface tension, J/m^2.

In Schnerr & Sauer model the simplifier formulas is used, when neglecting surface tension and second order term the following equation is obtained [4]:

$$\frac{Dr_B}{Dt} = \sqrt{\frac{2}{3}\frac{p_B - p}{\rho_l}} \quad (2)$$

Cavitation models in general can be gathered into two main groups [5]:

- one-fluid models (mixture models)
- two-fluid models

Schnerr & Sauer cavitation model is one-fluid model. In these models the flow is threatened as the mixture of two phases and the conservation equations (mass and momentum) for the mixture are solved. In these models no slip between the phases is assumed. To calculate fraction of gaseous phase mass conservation equation of vapour is solved [5, 6].

$$\frac{\partial \alpha \rho_v}{\partial t} + \nabla(\alpha \rho_v u) = R_e - R_c \quad (3)$$

Where: α vapour volume fraction, ρ_v – vapour density, kg/m^3; u mixture velocity, m/s; R_e, R_c source terms for evaporation and condensation respectively, kg/m^{3}s.

The main concern is mixture models is to determine the source terms R_e and R_c. In Schnerr & Sauer model the authors use the following formula to connect vapour volume fraction α with bubble radius r_B and number of bubbles per volume of liquid n_B [9]:

$$r_B = \left(\frac{\alpha}{1-\alpha}\frac{3}{4\pi}\frac{1}{n_B}\right)^{\frac{1}{3}} \quad (4)$$

Then the final formulas for R_e and R_c:

$$R_e = \frac{\rho_l \rho_v}{\rho}\alpha(1-\alpha)\frac{3}{r_B}\sqrt{\frac{2}{3}\frac{p_s - p}{\rho_l}} \quad (5)$$

$$R_c = \frac{\rho_l \rho_v}{\rho}\alpha(1-\alpha)\frac{3}{r_B}\sqrt{\frac{2}{3}\frac{p - p_s}{\rho_l}} \quad (6)$$

The model parameters were set as follows [9]:

$n_B = 10^{13}$ m^{-3}, $r_B = 10^{-6}$ m.

Vapour consideration in the flow was not sufficient to obtain the pressure distribution in accordance with experiment (the pressure at the inlet was too low). Therefore it was necessary to add dissolved air at the inlet of the domain. This air due to the lower pressure in the nozzle is released.

4.2 Turbulence model

To choose the most suitable turbulence model to the further calculation the series of tests were performed. At one frequency of pump's motor equal to 32 Hz (maximum flow) four different turbulence models were tested:

- k-ε RNG
- k-ε realizable
- k-ω
- SAS

The area of vapour appearance for each turbulence model is shown in figure 8. In table 2 the outreach of vapour and gas (vapour + undissolved air) cloud is described.

Table 2. Results of turbulence models test

Turbulence model	Outreach of vapour cloud	Outreach of gas cloud
	cm	cm
k-ε realizable	12	9
k-ε RNG	7	11
SAS	7.5	9
k-ω	6	7

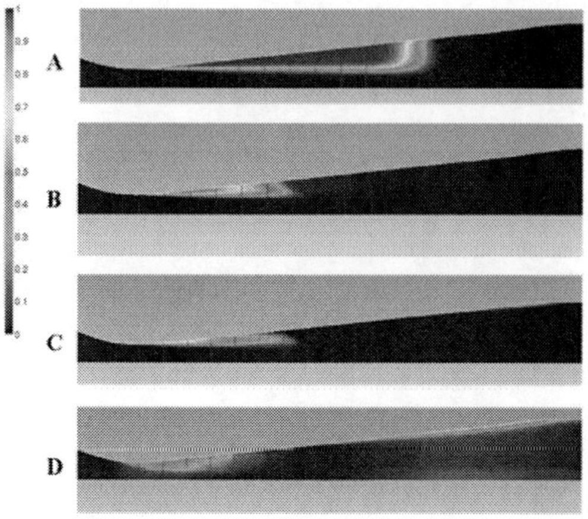

Figure 8. Vapor appearance outreach for different turbulence models: A) k-ε realizable B) k-ε RNG C) SAS D) k-ω

The analysis of the area occupied by the vapour for different turbulence models indicates that the greatest area of vapour appearance was observed while realizable k-ε model was used. For k-ω model the length of the vapour cloud is too small (compared to the experiment). Moreover, for SAS and k-ω model vapor appears far away from the critical cross section, which was not observed during the experiment.

Next the dissolved air was taken into consideration. The mass flow of the air was set to the value which allows obtaining pressure at inlet close to the measured one. The area of air and both gases appearance were compared after the calculations (figure 9, 10).

Comparing area of gas appearance for different turbulence models (figure 9) it can be seen that for SAS and k-ω models the area is too small in regards to experiment. For k-ε models, realizable and RNG the area of gas is larger, reaching up to 9 cm (realizable) and 11 cm (RNG). The shape of the gas cloud is different – for realizable model the outreach of gas is similar near wall and along the axis of symmetry. In case of RNG mode, the gas cloud near wall is significantly shorter that along the axis. However, the analysis of areas occupied by air for different turbulence models (figure 10) shows that for realizable k-ε model area of high air volume fraction occurs near wall at the length of 8-9 cm. To investigate the reason for that discrepancy between the two k-ε models the further analysis was performed.

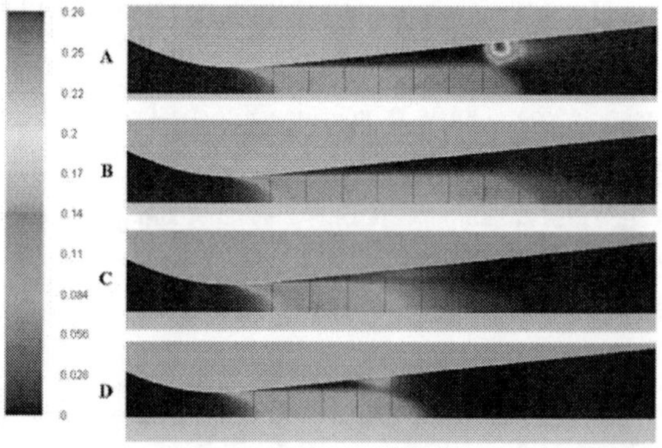

Figure 9. Undissolved air appearance outreach for different turbulence models: A) k-ε realizable B) k-ε RNG C) SAS D) k-ω

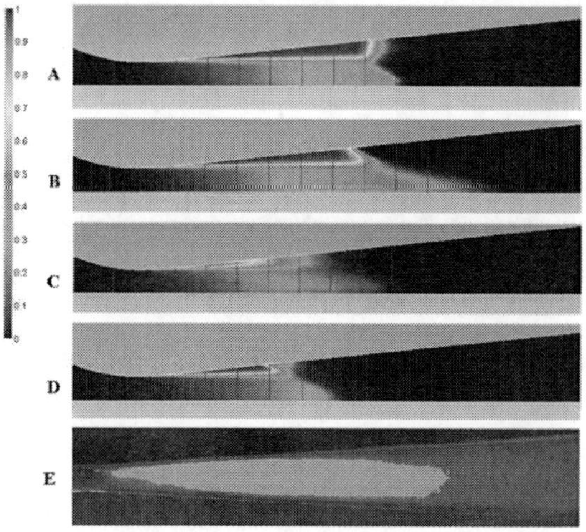

Figure 10. Gas (vapour+air) appearance outreach for different turbulence models: A) k-ε realizable B) k-ε RNG C) SAS D) k-ω E) gas area mean measured cavitation cloud

The main difference between the results obtained by RNG and realizable k-ε model is distribution of air along the walls of the nozzle. In figure 11 air volume fraction on the walls for both models is shown. In case of realizable k-ε model the sudden growth of air volume fraction is observed at x coordinate equal to 0.09 m, which corresponds to the bubble of air noticed in figure 10A. The next discrepancy between the both models is wall shear stress. The distribution of this value on the walls is shown in figure 12. The highest value of this variable occurs near 0 x coordinate. For realizable k-ε model at x coordinate equal to 0.09 the sudden peak of wall shear stress is observed. There is no physically based explanation for such a sudden change of this parameter. Finally, the k-ε RNG model was chosen for further calculation.

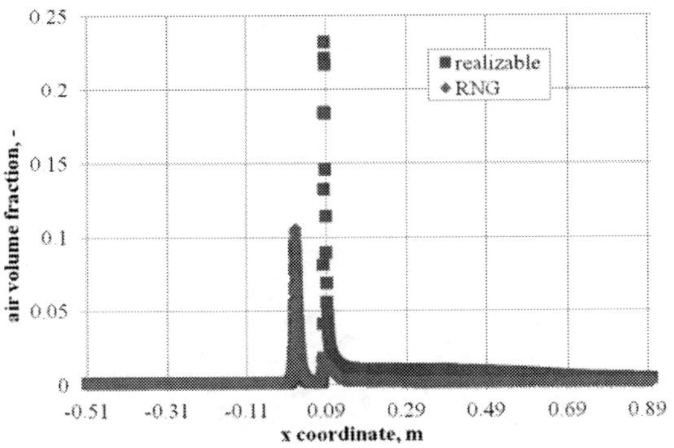

Figure 11. Distribution of air volume fraction on walls of the nozzle for RNG and realizable k-ε model

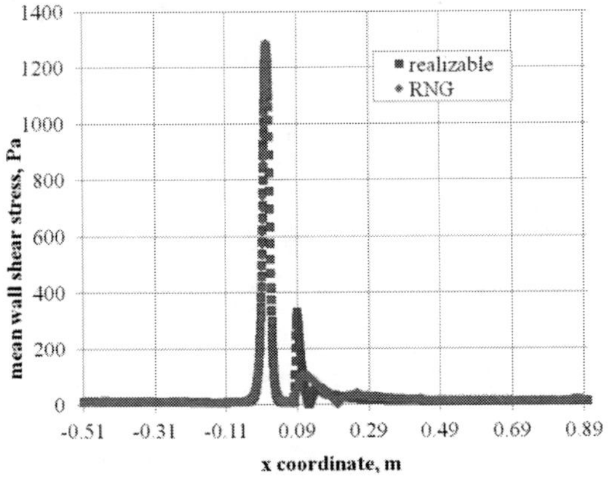

Figure 12. Distribution of mean wall shear stress for RNG and realizable k-ε model

5 Results of simulation

After choosing the turbulence model the calculations with smaller time step ($\Delta t = 10^{-5}$ s) with data saving from monitor points were performed. The monitor values points were volume integrals of each phase: water, vapour and air. The results for two flow rates are presented: 2.61 l/s and 3.88 l/s (18 Hz and 32 Hz pump's motor respectively). With no air in the flow the pressure at the inlet obtained in calculations (p_{in1}) was lower than the one from the measurements (p_{in}). The mass flow rate of air was added (Q_a) and set to the value which leaded to higher pressure at the inlet (p_{in2}), which is close to the measured values.

Table 3. Details of case investigated.

f	Q_v	Q_a	p_{in}	p_{in1}	p_{in2}
Hz	l/s	kg/s	Pa	Pa	Pa
18	2.61	1.87×10^{-6}	138982	127000	138500
32	3.88	5.15×10^{-6}	286609	272000	287500

The undissolved mass flow rate is increasing with the motor's frequency, which was also observed during the experiment. For flow rate equal to 3.88 l/s the flow of the air is maintained at level of 5.15×10^{-6} kg/s. In analysis of the noise frequency (section 3.2. of this paper) two main noise frequencies were observed (5000 Hz and 3000 Hz) at this flow rate. The changes of volume of air and of vapour in the domain in flow time are shown in figure 13 and 14, respectively. It can be observed that for vapour the one frequency dominates the others, in case of air volumes a few frequencies can be seen. The results of FFT analysis are shown in figure 15 (air) and 16 (vapour). The frequencies obtained by FFT analysis were 12.2 Hz, 24.4 Hz and 36.6 Hz.

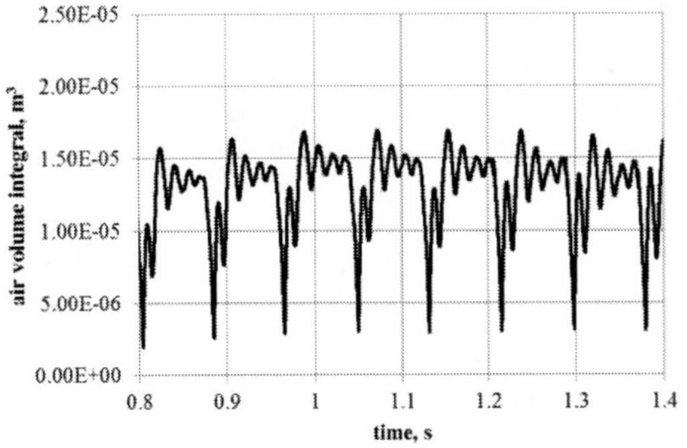

Figure 13. Volume integral of air phase in time, 32 Hz motor's frequency

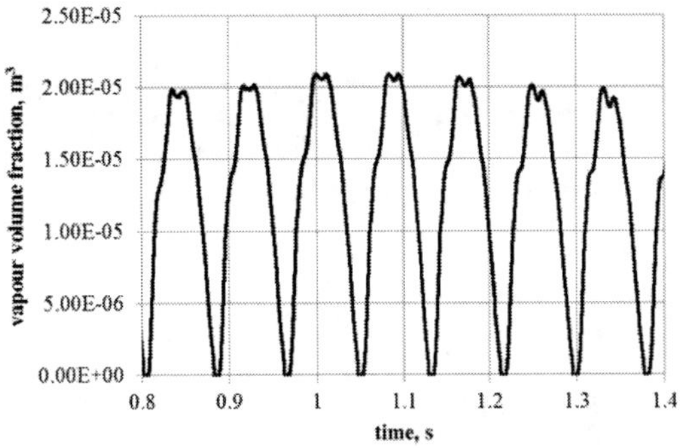

Figure 14. Volume integral of vapour phase in time, 32 Hz motor's frequency

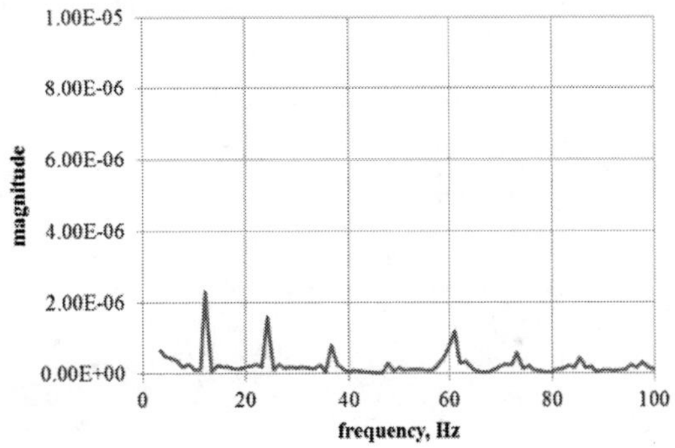

Figure 15. FFT analysis of volume integral of air phase, 32 Hz motor's frequency

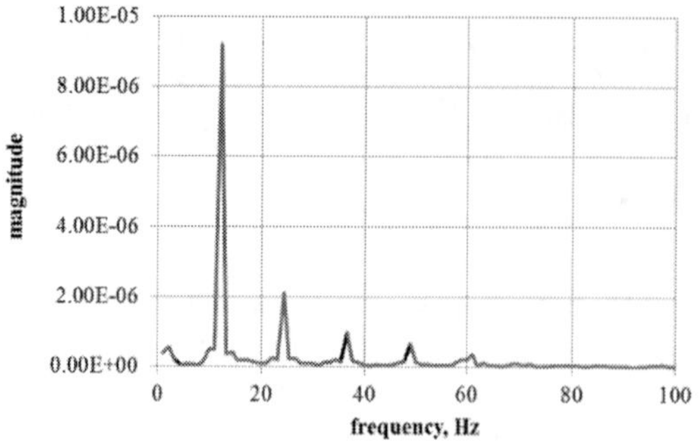

Figure 16. FFT analysis of volume integral of vapour phase, 32 Hz motor's frequency

Similar investigation was performed for flow rate 2.61 l/s (18 Hz pump's motor frequency). The changes of air and vapour volume in the domain in flow time are shown in figure 17 and 18, respectively. For vapour volume one frequency significantly dominates. For air volume the similar situation as for flow rate 3.88 l/s takes place – there are a few frequency of changes which covers one another. The results of FFT analysis are shown in figure 19 (air) and 20 (vapour). The frequencies obtained by FFT analysis were 37.5 Hz, 43.7 Hz.

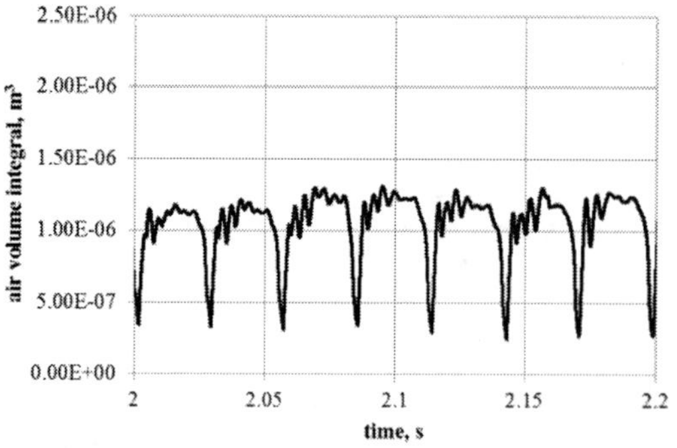

Figure 17. Volume integral of air phase in time, 18 Hz motor's frequency

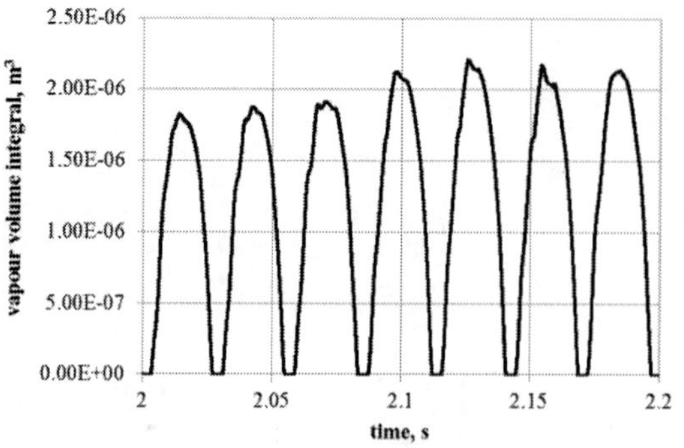

Figure 18. Volume integral of vapour phase in time, 18 Hz motor's frequency

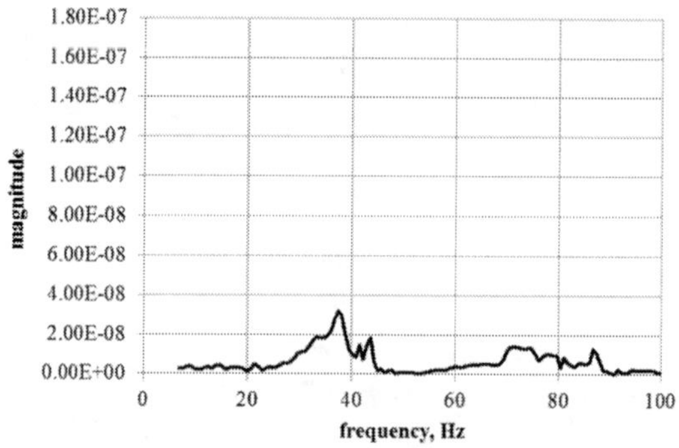

Figure 19. FFT analysis of volume integral of air phase, 18 Hz

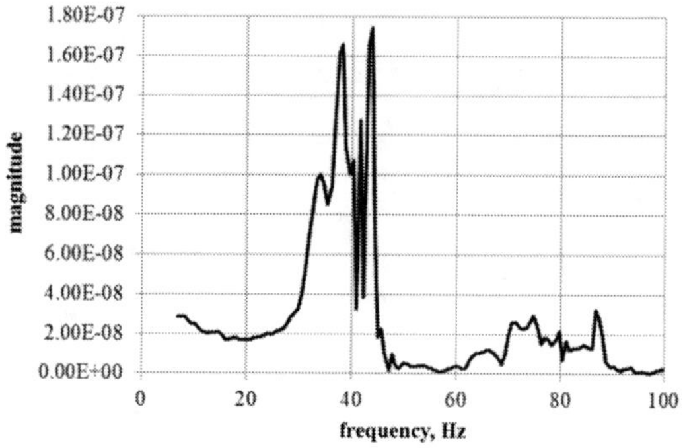

Figure 20. FFT analysis of volume integral of vapour phase, 18 Hz motor's frequency

After simulation performance it can be stated that in case of vapour phase there is one main growing and shrinking frequency. Meanwhile the air phase is changing with number of frequencies with no such clear domination. It also was observed that for vapour type of cavitation (which occurs at flow rate 2.61 l/s – 18 Hz motor's frequency) higher frequencies of phase changing appeared. When more air is observed in cavitation cloud, the frequency of phase growing and shrinking is decreasing. That indicates that air is responsible for lower frequencies in noise analysis described in chapter 3.2.

6 Results of 3D simulation

To study the influence of 2D flow assumption the 3D simulation was performed. The calculation domain includes ¼ of the nozzle. The calculated case was flow rate 3.88 l/s (32 Hz pump's motor frequency). The 3D mesh contained 292 000 cells. The same turbulence model was used as for 2D calculation (k-ε RNG). The results are shown in figure 21 and 22.

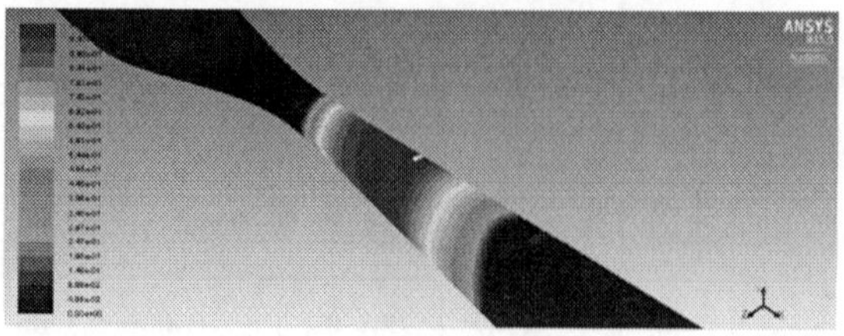

Figure 21. Vapour phase, 3D calculation

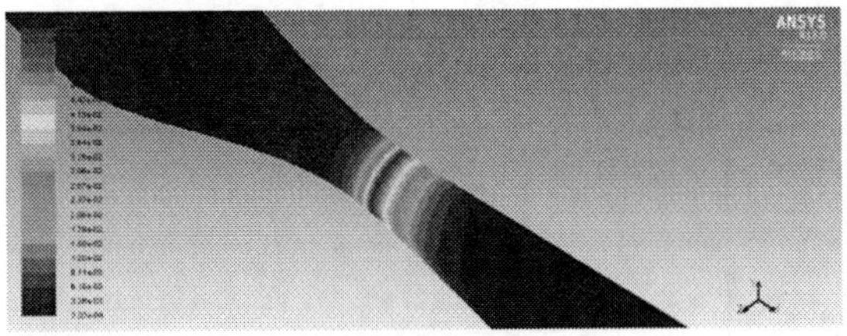

Figure 22. Air phase, 3D calculation

On basis of 3D calculation results it can be stated that the distribution of air and vapour volume fraction is slightly not uniform in circumferential direction on the wall. Therefore 3D calculation can provide some additional information about the shape and outreach of cavitation structures. However, this kind of calculations takes a lot more computational time than the 2D case. Next stage of investigation will include using LES turbulence model.

7 Conclusions

In the paper the experimental and numerical investigation of the cavitating flow in the water nozzle was described. The following conclusions can be provided:

First stage of cavitation was observed at pump's motor frequency equal to 18 Hz, which stands for water flow 2.61 l/s. In this case the length of cavitation structure is about 1.6 cm. With increase of flow rate cavitation structure expands. At flow rate 3.88 l/s (32 Hz motor's frequency) the structure enlarges to 11.7 cm. The evaluation of noise measurements results leads to conclusion that with increasing of flow rate the noise frequency split into two: instead of one frequency of 4000 Hz two frequencies equal to 3000 Hz and 5000 Hz are observed. This may be connected with the existence of two cavitation mechanism types: vapour and gaseous one.

Numerical investigation was performed in use of Schnerr & Sauer cavitation model. Several turbulence models were tested. Model k-ε RNG captured the best distribution of vapour and air volume fraction in the nozzle. FFT analysis of air and vapour volume changes in time in the whole domain was also performed. It showed that for first stage of cavitation higher frequency occurs than for developed one. Growing and shrinking of vapour structures appear with one main frequency, while air structures change their size with a few frequencies with similar magnitude. Analogous situation takes place for last stage of cavitation, at flow rate equal to 3.88 l/s.

Although the results of FFT analysis of phases volumes changes in time cannot be compared quantitatively with noise measurements results, they are in good agreement qualitatively: the developed gas cavitation can be characterized by lower frequency than the vapour one. The frequencies observed during noise measurement are connected with frequency of single bubble, the frequencies observed during numerical simulation are related to the frequency of cavitation clouds.

3D calculation with ¼ of the nozzle also was carried out. It showed that there is slightly non-uniform distribution of phases volume fractions in a circumferential direction. The further 3D calculation will be performed in the future to investigate in detail this issue.

Acknowledgements

This paper has been elaborated in the framework of the Project Intention Moravian-Silesian Region, Support for Science and research in the Moravian – Silesian Region 2014 subsidy Title1.

References

1. C. E. Brennen, *Hydrodynamics of Pumps*, Oxford University Press, (1994),
2. J. Noskievič, *Kavitace*, Praha : Ceska akademie ved, p. 509-21-875, (1969)
3. J. Jablonska, M. Bojko, Multiphase Flow and Cavitation – Comparison of Flow in Rectangular and Circular Nozzle, EPJ Conference, (2014)
4. C. E. Brennen, *Cavitation and Bubble Dynamics*, Oxford University Press, (1995)
5. B. Puffary, Numerical modelling of cavitation, design and analysis of high speed pumps. Educational Notes RTO-EN-AVT-143, paper 3, p. 3-1 – 3.54, (2006)
6. ANSYS FLUENT 2015 User's guide
7. J. P. Franc, J. M. Michel, *Fundamentals of Cavitation*, Fluid Mechanics and Its Application, Volume 76, Kluwe Academic Publischers, (2004)
8. J. Jablonska, Experimentalni zarizeni k vzniku a vyvoji kavitace w 2D Lavalove dyze, 135/18-12-2013_F. submitted 18. 12. (2013)
9. G. H. Schnerr, J. Sauer, Physical and Numerical Modelling of Unsteady Cavitation Dynamics, Fourth International Conference on Multiphase Flow, New Orleans, Louisiana, (2001)

EPJ Web of Conferences **114**, 02059 (2016)
DOI: 10.1051/epjconf/201611402059

Development of an Analytical Method for Predicting Flow in a Supersonic Air Ejector

Jan Kracik[1a], Vaclav Dvorak[1]

[1]*Department of Power Engineering Equipment, Faculty of Mechanical Engineering, Technical University of Liberec, Studentska 2, 46117, Liberec, the Czech Republic*

Abstract. The article deals with development of an analytical method for predicting flow in an ejector with twelve supersonic nozzles, which are located at the periphery of the mixing chamber of the ejector. Supersonic primary air stream makes the investigation more complex. The secondary air (atmospheric) is sucked in direction of the ejector axis. The shape of the mixing chamber is convergent – divergent and a throat is formed behind the primary nozzles. Each of the primary nozzles can be treated independently so there can be various number of nozzles under operation in the ejector. According to previous investigations, constant pressure mixing is assumed to occur inside a part of the mixing chamber. The method under investigation is considered for isentropic flow in the first approximation and after that the stagnation pressure corrections at the inlets are considered. Furthermore, the decrease in stagnation pressure in the mixing chamber is considered to take losses in the mixing chamber and diffuser into account. The numerical data of the stagnation pressure has been obtained from Ansys Fluent software. In addition, a comparison with previous experimental results is introduced.

1 Introduction

Nowadays, there are still higher and higher requirements for efficient operation of engineering devices and optimal operating regimes are required. The efficient working regimes are connected with losses and with properly working devices. Ejectors with their low efficiency, generally less than 30 per cent, require more attention and should be investigated in more details. The low efficiency is the most significant disadvantage of these devices. On the other hand, there are many advantages of ejectors. Among advantages especially belong absences of moving parts, ointment or seals. Other advantages are simply designs and relatively low manufacturing costs. They are usually used in places where sufficient amount of the working fluid is available or in industrial processes where some low-grade thermal energy is available, which is especially highly desirable in these days when new environment-friendly technologies are under development.

By the time that Keenan et.al. [1] performed the first comprehensive study of mixing, two cases of mixing were distinguished: the constant pressure mixing and the constant area mixing. From this follows usage of ejectors. Another distinction of ejectors is in terms of primary and secondary streams arrangement. The primary stream can either be supplied in the ejector axis and secondary stream is sucked at the periphery of the mixing chamber or primary stream enters the periphery of the mixing chamber and the secondary stream is sucked in the ejector axis. The latter design, which is usually used for wind tunnel propulsions, can be based on two arrangements. The first arrangement uses a slot primary nozzle. The second one uses several primary nozzles and is investigated in this paper, see figure 1. Many theoretical and experimental studies have been carried out to obtain the performance of ejectors so far. For more information, see [2-4].

The design of the air ejector under investigation was also investigated in more detail in works [5, 6 and 7]. The ejector is used for propulsion of a small experimental variable wind tunnel, see [8]. As can be seen in figure 1, the declination angle of the primary supersonic nozzles towards the ejector axis is 8.2°. The declination should prevent the occurrence of reversal flow in the ejector axis, as has been stated in work [9]. The method used in this paper for predicting ejector performance is based on assumptions which were first proposed by Christianovic and later simplified by Kiselev [10].

Four cases with two, four, eight and twelve activated primary nozzle have been selected for the analysis.

[a] jan.kracik@tul.cz

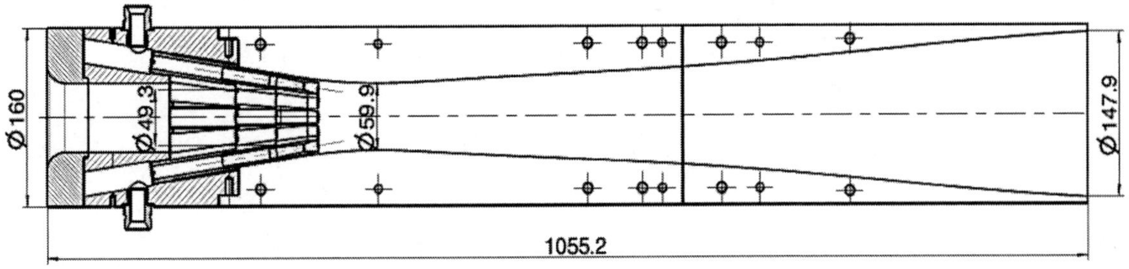

Figure 1. Longitudinal cross-section of the supersonic ejector (All dimensions given in millimeters) [8].

2 Theoretical analysis

The aim of this paper is to obtain output values of secondary (or sucking) mass flow rate and entrainment ratio, which is defined as ratio of mass flow rates, i.e.

$$\Gamma = \frac{\dot{m}_2}{\dot{m}_1}, \qquad (1)$$

where subscript *1* denotes primary and *2* secondary flow.

The entrainment ratio is considered as the most important indication of ejector performance in many supersonic ejector applications. Three operational regimes of supersonic ejectors are distinguished, see figure 2.

Due to availability of setting any number of the primary nozzles the situation inside the ejector is rather complex. Figure 3 shows the scheme of the investigated ejector. The section x-x, which is the ejector throat, is so called effective area where the flow is assumed to be choked.

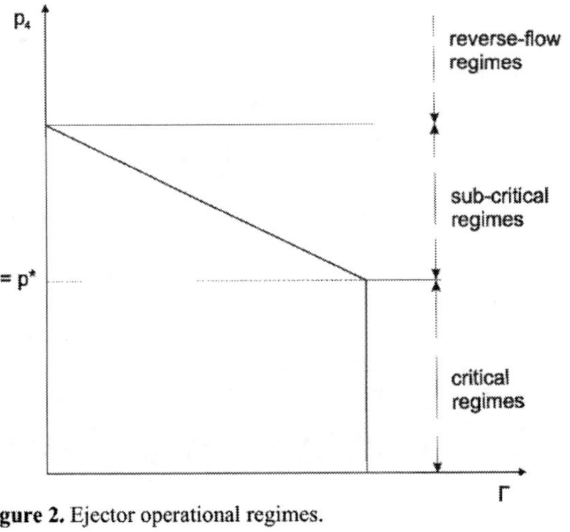

Figure 2. Ejector operational regimes.

The method used for the analysis is similar to the other methods that have been already investigated in previous works. These assumptions have been made:

1. The flow is considered to be steady and one-dimensional.
2. The primary and the secondary stream of air can be treated as ideal with constant κ and c_p.
3. The stagnation conditions are constant at the primary and secondary air inlets, as well as at the exit of the mixing chamber and diffuser.
4. The mixing chamber and the diffuser walls of the ejector are adiabatic.
5. Isentropic relations are used for calculations.
6. Shock waves are not considered.
7. Stagnation pressure coefficients are used for correction of the stagnation pressures at the boundaries (inlets) at section *12-12*.
8. Stagnation pressure correction has not been considered for the diffuser.
9. The mixed flow is considered as critical or sub-critical ($\lambda_3 \leq 1$).

The used method is based on the following relations, which are the law of conservation of mass

$$c_1\rho_1A_1 + c_2\rho_2A_2 = c_3\rho_3A_3, \qquad (2)$$

where A_1 and A_2 are cross section areas at section *12-12* and A_3 is the exit cross section area of the mixing chamber, *c* denotes the stream velocities and *ρ* the stream densities. The other relations are the balance of momentum.

$$\dot{m}_1 c_1 + p_1 A_1 + \dot{m}_2 c_2 + p_2 A_2 = (\dot{m}_1 + \dot{m}_2) c_3 + p_3 (A_1 + A_2), \tag{3}$$

where p denotes stream static pressure and the balance of energy

$$\dot{m}_1 \left(c_{p1} T_1 + \frac{c_1^2}{2} \right) + \dot{m}_2 \left(c_{p2} T_2 + \frac{c_2^2}{2} \right) = (\dot{m}_1 + \dot{m}_2) \left(c_{p3} T_3 + \frac{c_3^2}{2} \right), \tag{4}$$

which has been derived for adiabatic flow in the mixing chamber and where T is static temperature. The relations mentioned above have been derived with the assumptions that isobaric specific heat capacities are constant, i.e.

$$c_{p1} = c_{p2} = c_{p3}, \tag{5}$$

And also ratio of specific heats is constant, i.e.

$$\kappa_1 = \kappa_2 = \kappa_3 . \tag{6}$$

Stagnation pressure of the mixed flow after finished mixing is calculated by relation

$$p_{03} = p_{01} \frac{\sqrt{(1+\Gamma)(1+\Gamma\Theta_{21})}}{1 + \frac{p_{01}}{p_{02}} \Gamma \sqrt{\Theta_{21}} \frac{q(\lambda_1)}{q(\lambda_2)}} \frac{q(\lambda_1)}{q(\lambda_3)}, \tag{7}$$

where Θ_{21} is the ratio of the stagnation temperatures of both streams defined as

$$\Theta_{21} = \frac{T_{02}}{T_{01}} \tag{8}$$

and $q(\lambda)$ is the aerodynamic function of mass flow rate given by

$$q(\lambda) = \frac{\rho c}{(\rho c)*} = \left(1 - \frac{\kappa - 1}{\kappa + 1} \lambda^2 \right)^{\frac{1}{\kappa - 1}} \left(\frac{\kappa + 1}{2} \right)^{\frac{1}{\kappa - 1}} \lambda, \tag{9}$$

where λ is dimensionless speed obtained from ratio

$$\lambda = \frac{c}{c*}, \tag{10}$$

where c is the velocity at given cross section and $c*$ is the critical velocity at the same cross section. Stagnation temperature after mixing is derived directly from the balance of energy

$$T_{03} = T_{01} \frac{1 + \Gamma\Theta_{21}}{1 + \Gamma}. \tag{11}$$

Dimensionless speed of the mixed flow is obtained from relation

$$z(\lambda_3) = \frac{z(\lambda_1) + \Gamma \sqrt{\Theta_{21}} z(\lambda_2)}{\sqrt{(1+\Gamma)(1+\Gamma\Theta_{21})}}, \tag{12}$$

where $z(\lambda)$ is aerodynamic function given by

$$z(\lambda) = \lambda + \frac{1}{\lambda}. \tag{13}$$

The following relations have been used also during calculations. Since there is no exact distinction between the mixing chamber and the diffuser due to the fluent shape of the inner channel, following relation has been used to determine the mixing chamber cross section A_3 and consequently the position of the end of the mixing chamber

$$A_{1et} + A_2 = A_3, \tag{14}$$

where A_{1et} is the total exit cross section area of all activated nozzles, A_2 is the cross section area remaining for the secondary flow at section *12-12* and A_3 is the exit cross section area of the mixing chamber or the inlet cross section area of the diffuser, respectively.

The most significant design parameter of the ejector is the ratio of cross section areas of both streams defined as

$$\mu(A_{1et}) = \frac{A_{1et}}{A_2}. \tag{15}$$

This approach allows calculation of ejector efficiency, which is defined as

$$\eta_E = \frac{\dot{m}_2}{\dot{m}_1} \frac{\left(\frac{p_4}{p_{02}} \right)^{\frac{\kappa-1}{\kappa}} - 1}{1 - \left(\frac{p_4}{p_{01}} \right)^{\frac{\kappa-1}{\kappa}}} \frac{T_{02}}{T_{01}}, \tag{16}$$

where p_4 is back-pressure, i.e. the pressure at the diffuser exit.

The relations mentioned above assume one-dimensional isentropic flow of ideal compressible fluid. These relations allow to calculate all important variables of the mixed flow for a given regime of the ejector.

The stagnation pressure correction has been treated by coefficients obtained from numerical data

$$p_0' = p_{12} \left\{ 1 - \eta_{nozzle} \left[1 - \left(\frac{p_{12}}{p_0} \right)^{\frac{\kappa-1}{\kappa}} \right] \right\}^{\frac{\kappa}{1-\kappa}}, \tag{17}$$

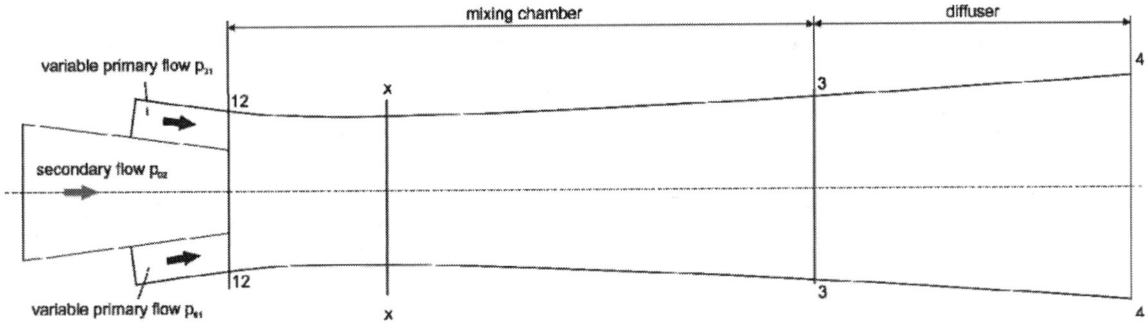

Figure 3. Analytical scheme of the ejector performance.

where η_{nozzle} is efficiency coefficient of the primary nozzle or the secondary inlet part, respectively, that has nozzle-like behavior as well. The efficiency has been calculated from resulta of numerical calculations using following equation

$$\eta_{nozzle} = \frac{1-\left(\frac{p}{p_0'}\right)^{\frac{\kappa-1}{\kappa}}}{1-\left(\frac{p}{p_0}\right)^{\frac{\kappa-1}{\kappa}}} = \varphi^2 , \qquad (18)$$

where ϕ is a velocity coefficient, p is the static pressure obtained from numerical results as an area-weighted-average value, p_0 is the stagnation pressure in the beginning of expansion, i.e. in the nozzle inlet, obtained from numerical results as a mass-weighted-average value and similarly p_0' is stagnation pressure after expansion, i.e. at the nozzle exit, again as mass-weighted-average, see figure 4.

The stagnation conditions of the primary air have been set $p_{01} = 397$ kPa (absolute) and $T_{01} = 295$ K. The Stagnation pressure of the secondary air has been $p_{02} = 97$ kPa (absolute) and the stagnation temperature has been considered the same as for the primary air, i.e. $T_{02} = T_{01}$. Static pressure p_{12}, which varied from value 96.9 kPa down to 30.0 kPa, has been considered as the only one independent variable.

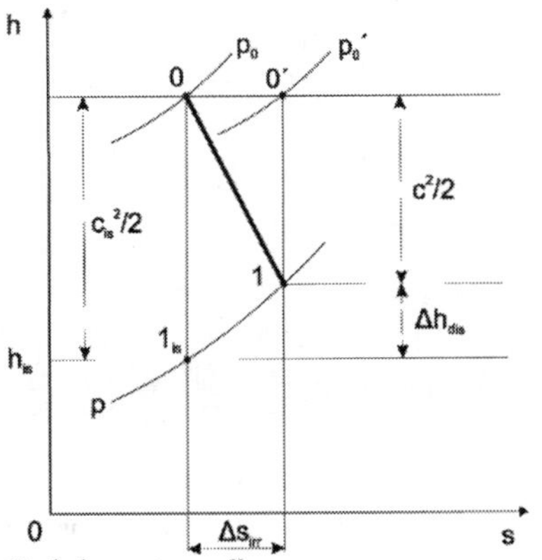

Figure 4. Enthalpy-entropy diagram.

3 Numerical analysis

A One fourth model of the ejector has been used to obtain numerical data of the stagnation and static pressures within the mixing chamber. The k-ω SST turbulence model was used to take for turbulence effects into account. This model is suitable for calculation of flows in supersonic ejectors as it was stated in previous works. For more information, see Ansys Fluent documentation [11].

4 Results

In figures 5-8 courses of efficiency and relative back-pressure for different number of activated primary nozzles can be seen. In figure 5, the results for ejector efficiency obtained from the analytical method are shown. It can be seen that the whole curve, i.e. the position of maximal efficiency and position of choking, moves to lower ejection ratios with increasing number of activated nozzles. These observations are in agreement with the previous experimental data as shown in the next text. The courses of relative back-pressure obtained from the analytical method can be seen in figure 6. The highest value of entrainment ratio obtained from analytical method was approximately of 7.7 for two activated nozzles, while the lowest ejection ratios but the highest back pressure ratios are obtained for twelve nozzles.

In order to verify proposed analytical model, experimental data obtained by Kracik (2014) on a supersonic air ejector with a circular cross section were used, see figure 1.

Figure 7 and 8 show comparison of curves of efficiency and relative back-pressure obtained by presented analytical model and by experiments [8].

As mentioned earlier the maximum of efficiency moves to the left with increasing number of activated nozzles in both the presented model and the experiments. There are obvious relatively large disagreement between both the efficiency and the relative back-pressure curves. The analytical method underestimates reached ejection rations, but overestimates maximums of efficiency. One of the possible ways to reduce these discrepancies is to include friction and others losses that take place within the ejector mixing chamber and ejector diffuser. Further, it is necessary to analyse experimental and numerical results more thoroughly to deeply understand flow processes inside the mixing chamber and include and describe all phenomena to predict the chocking conditions more precisely.

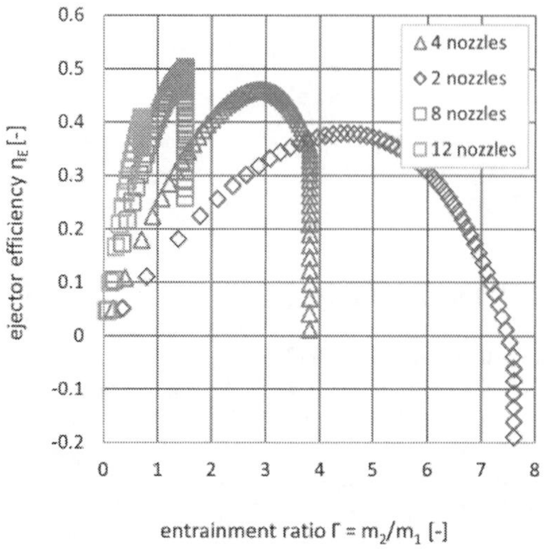

Figure 5. Comparison of ejector efficiency for various number of activated primary nozzles.

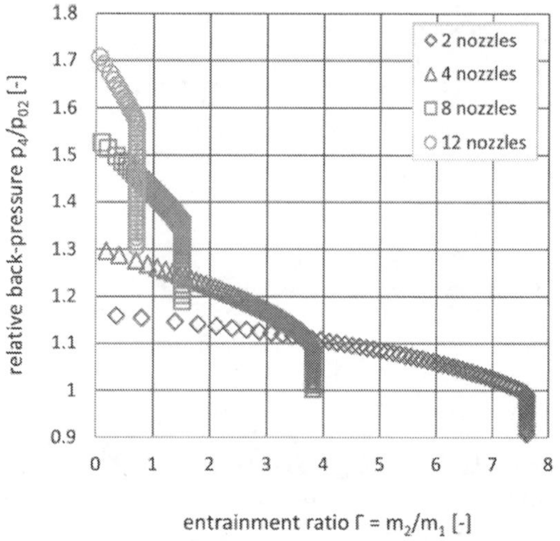

Figure 6. Comparison of relative back-pressure for various number of activated primary nozzles.

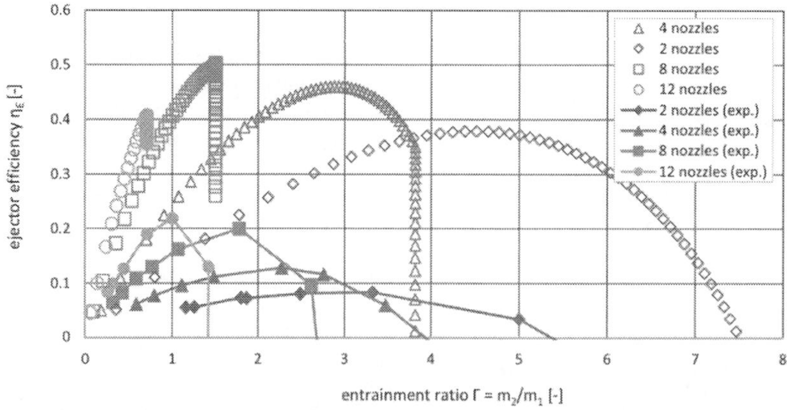

Figure 7. Ejector efficiency comparison of presented model and experimental data from [8] for various number of activated primary nozzles.

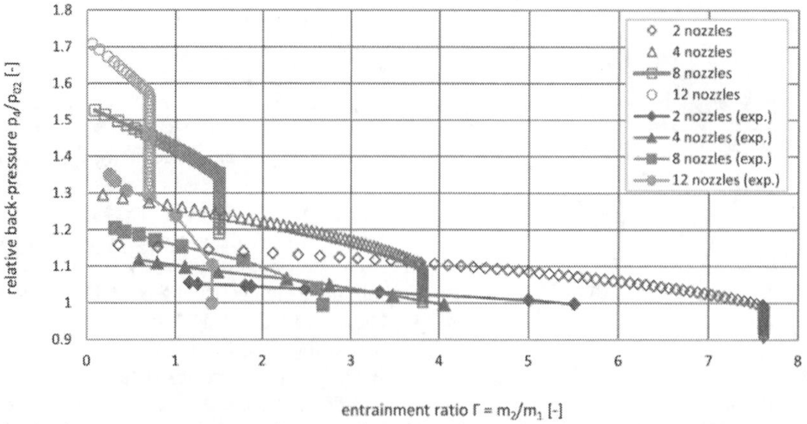

Figure 8. Relative back-pressure comparison of presented model and experimental data from [8] for various number of activated primary nozzles.

5 Summary and conclusions

The first attempt to predict ejector performance with twelve supersonic primary nozzles has been presented. Isentropic flow has been considered as the first approximation. Moreover, the correction coefficients have been taken into account for the primary and secondary inlet. It has been shown that the maximum of efficiency moves to lower ejection ratios with increasing number of activated nozzles, i.e. the primary mass flow rate is increased both in the presented model and experiments performed earlier. There are obviously significant discrepancies in both the efficiency and relative back-pressure curves between predicted and measured values.

It is obvious that friction and others losses that take place within the mixing chamber and diffuser must be considered in the method to improve agreement with experiments. Furthermore, a thorough analyse of experimental and numerical results is necessary to understand flow processes inside the mixing chamber and include and describe all phenomena and conditions to predict the chocking more precisely.

Acknowledgment

We gratefully acknowledge the support of the Grant Agency of the Czech Republic - Czech Science Foundation (project no. 14-08888S). We also gratefully acknowledge the support of the Student Grant Agency of the Technical University of Liberec (project no. 21000).

References

1. J.H. Keenan, E.P. Neumann and F. Lustwerk, *An Investigation of Ejector Design by Analysis and Experiment.* Journal of Applied Mechanics, Trans. ASME **17** (1950), pp 299 – 309.
2. J.H. Keenan, E.P. Neumann, *A simple air Ejector,* Journal of Applied Mechanics, Trans. ASME **64**, pp A75 – A81 (1942)
3. S. Aphornratana, I.W. Eames, *A small capacity steam-ejector refrigerator: experimental investigation of a system using ejector with movable primary nozzle,* International Journal Refrigeration, Trans. ASME **20**, pp 352 – 358 (1997)
4. B.J. Huang, J.M. Chang, C.P. Wang, V.A. Petrenko, *A 1-D analysis of ejector performance,* International Journal Refrigeration, Trans. ASME **22**, pp 354 – 364 (1999)
5. J. Kracik, V. Dvorak, *Experimental and numerical investigation of an air to air supersonic ejector for propulsion of a small supersonic wind tunnel,* EPJ Web of Conferences **92** (2015) 02038
6. J. Kracik, V. Dvorak, J. Kolar, *Development of air to air ejector for supersonic wind tunnel,* Web of Conferences **67** (2014) 02059
7. J. Kracik, V. Dvorak, J. Kolar, *Development of air to air ejector for supersonic wind tunnel,* Web of Conferences **67** (2014) 02059
8. J. Kracik, *Design of Supersonic Wind Tunnel* Diploma thesis, In Czech, (2014)
9. V. Dvorak, *Shape Optimization of Supersonic Ejector for Supersonic Wind Tunnel.* Applied and Computational Mechanics, Plzen, (2010)
10. G.N. Abramovich, *Applied Gas Dynamics,* Volume **2** Set edition (1973)
11. Ansys Fluent 15.0 *User's Guide*, Ansys Inc.
12. J. Kolar, V. Dvorak, *Interaction of Shock Waves in Supersonic Ejector,* Proceedings of the 27th Meeting of Thermodynamics and Fluid Mechanics Departments, Plzen (2008)
13. J. Simak, *Computation of the Flow and Interaction of Shock Waves in a 2D Supersonic Ejector,* Colloquium Fluid Dynamics 2009, Prague (2009)
14. Y. Bartosiewicz, Z. Aidoun, P. Desevaux, Y. Mercadier, *Numerical and Experimental Investigations on Supersonic Ejectors,* International Journal of Heat and Fluid Flow **26** (2005)
15. J. Kolar, *Design and Optimization of Ejector for Supersonic Wind Tunnel*, Diploma thesis, Liberec, In Czech (2007)
16. B.J. Huang, J.M. Chang, C.P. Wang, V.A. Petrenko, *A 1-D analysis of ejector performance,* International Journal Refrigeration, Trans. ASME **22**, pp 354 – 364 (1999)
17. B.M. Kiselev, *Rasčet odnorazmernych gazovych tečenij,* Prikladnaja matem. i mech, No **1** (1949)
18. W. Cheng, M. Liu, D. Chong, J. Yan, A.B. Little, Y. Bartosiewicz, *A 1D model to predict ejector performance at critical and sub-critical operational regimes,* International Journal of Refrigeration **36**, pp 1750 – 1761. (2013)

EPJ Web of Conferences **114**, 02060 (2016)
DOI: 10.1051/epjconf/201611402060

Development boiling to sprinkled tube bundle

Petr Kracík[1,a] and Jiří Pospíšil[1]

[1]*Institute of Power Engineering, Brno University of Technology, Faculty of Mechanical Engineering, Technická 2896/2, 616 69 Brno, The Czech Republic*

Abstract. This paper presents results of a studied heat transfer coefficient at the surface of a sprinkled tube bundle where boiling occurs. Research in the area of sprinkled exchangers can be divided into two major parts. The first part is research on heat transfer and determination of the heat transfer coefficient at sprinkled tube bundles for various liquids, whether boiling or not. The second part is testing of sprinkle modes for various tube diameters, tube pitches and tube materials and determination of individual modes' interface. All results published so far for water as the falling film liquid apply to one to three tubes for which the mentioned relations studied are determined in rigid laboratory conditions defined strictly in advance. The sprinkled tubes were not viewed from the operational perspective where there are more tubes and various modes may occur in different parts with various heat transfer values. The article focuses on these processes. The tube is located in a low-pressure chamber where vacuum is generated using an exhauster via ejector. The tube consists of smooth copper tubes of 12 mm diameter placed horizontally one above another.

1 Mathematical model of heat transfer of boiling liquid

Lorenz and Yung created a mathematical model published at **[1]** in 1979 for prediction of a heat transfer coefficient at the surface of a single sprinkled horizontal tube using a simplified geometry. The results of for instance **[2, 3, 4]** were compared with this model. The mathematical calculation considers a single vertical half of a horizontal tube that has been unrolled.

The mathematical model further takes into account that the temperature of the liquid falling on a heated tube equals the temperature of a saturated liquid. On these grounds two basic regions are formed at the tube during consequent heating. The first one is a transition convective region where boiling develops and the second one represents a fully developed boiling where evaporation occurs. The total heat transfer coefficient at the surface of a sprinkled horizontal tube equals a sum of partial heat transfer coefficients [at nucleate boiling effect (α_{ob}), at thermal developing region (α_{od}) and at fully developed region (α_{oc})] which are defined below and corrected by length for which their mean value has been set against the total studied length (of the unrolled area)

$$\bar{\alpha}_o = \alpha_{ob} + \alpha_{od} \cdot \frac{L_d}{L} + \alpha_{oc} \cdot \left(1 - \frac{L_d}{L}\right) \quad (1)$$

The heat transfer coefficient at nucleate boiling effect (also referred to as boiling nucleus) is based on a sufficient heat transfer at which the liquid starts to boil. According to **[1]** higher heat transfers with boiling temperature are achieved in thin liquid films than in steam environment and a conservative approach is the application of a relation defined by Rohsenow **[5]** who uses in his equation besides other things latent heat [enthalpy of vaporization l_v ($J \cdot kg^{-1}$)]

$$\alpha_{ob} = \frac{\mu \cdot l_v}{C_{sf}^3 \cdot \sqrt{\frac{g_0 \cdot \sigma}{g \cdot \rho}}} \cdot \left(\frac{c_p}{l_v \cdot Pr}\right)^3 \cdot \Delta T^2 \quad (2)$$

and parameter C_{sf} is a function of the surface tension of the liquid and of the surface which the liquid flows down from. Parameter C_{sf} for water and various surfaces is for example in **[6, tab. 9.1]**.

Cooper (1984) **[7]** created relatively simple relation for experimental data of heat transfer at nucleate boiling effect that is only a function of reduced pressure, molar mass and roughness of the surface of a sprinkled tube bundle. Cooper formed several identical functions. One of them applies to common logarithm

$$\alpha_{ob} = \left\{90 \cdot \frac{\Delta T_{ws}^{0.67}}{\sqrt{M}} \cdot \frac{p_r^m}{[-\log_{10}(p_r)]^{0.55}}\right\}^{3.0303} \quad (3)$$

where p_r (–) is a reduced pressure, i.e. pressure in the tube bundle's environment divided by critical pressure of the liquid that flows around the tube bundle. ΔT_{ws} (K) stands for the overheating temperature, i.e. the temperature of the tube wall is deducted by the temperature of saturation of the liquid flowing around it at a given environment pressure. The capital "M ($kg \cdot kmol^{-1}$)" represents molar mass and the exponent m (–) is dependent on the roughness of the tube surface

[a] Corresponding author: kracik@fme.vutbr.cz

$$m = 0.12 - 0.2 \cdot \log_{10}\left(\frac{Ra}{0.4}\right) \quad (4)$$

Ra (μm) stands for an average (mean) arithmetical deviation of the studied profile of the given surface calculated from the roughness profile. More in [8]

The thermal developing region where the heat is used for liquid overheating and where boiling does not occur yet, but evaporation does, is limited by the length L_d (m) [1]

$$L_d = \frac{c_p \cdot \sqrt[3]{\Gamma^4}}{4 \cdot \pi \cdot \rho \cdot \lambda} \cdot \sqrt{\frac{3 \cdot \mu}{g \cdot \rho^2}} \quad (5)$$

this equation is derived for laminar to slightly turbulent flow and the mean value of a heat transfer coefficient at the tubes' surface in this region is [1]

$$\alpha_{od} = \frac{3}{8} \cdot c_p \cdot \frac{\Gamma}{L_d} \quad (6)$$

which is based on energy balance.

For a fully developed region at a vertical smooth tube the following two equations are recommended by [1] for a calculation of a heat transfer coefficient at laminar (and also pseudo laminar) and at turbulent flow according to Chun and Seban [9]

$$\alpha_{oc,lam} = 0.821 \cdot \sqrt[3]{\frac{\lambda^3 \cdot g}{\nu^2}} \cdot \sqrt[0.22]{\frac{\mu}{4 \cdot \Gamma}} \quad (7)$$

$$\alpha_{oc,tur} = 0.003\,8 \cdot \sqrt[3]{\frac{\lambda^3 \cdot g}{\nu^2}} \cdot \sqrt[0.4]{\frac{4 \cdot \Gamma}{\mu}} \cdot \sqrt[0.65]{\frac{\nu}{a}} \quad (8)$$

The laminar mode considers the influence of ripple and waves as well that have an effect on the increase of heat transfer and the decrease of film's thickness. Both equations are a function of the Reynolds and Prandtl number and they should be applied for a condition at a constant heat transfer or at constant thermal conditions at the tube wall. Correlation between the equations (7) and (8) creates a transition boundary defined as a functional dependence of the Reynolds number on the Prandtl number

$$Re_{f,tr} > \frac{5\,800}{Pr^{1.06}} \quad (9)$$

This equation does not define a real boundary between pseudo laminar and turbulent mode, but the transition between the equations (7) and (8) only.

2 Mathematical model for two-phase flow

Chien and Cheng (2006) [3] state that the model of Chun and Seban (1971) [9] does not comply well with practically measured results for various coolants, i.e. shows a significant error. For these reasons they suggest applying this model, which draws on the fact that the calculation of heat transfer coefficient at a tube surface should also consider occurrence of bubbles, to a horizontal tube. The model is based on the superposition principle, i.e. the final heat transfer coefficient consists of multiple elements which add up to create its resultant. This model has been developed by Chen (1966) [10] and it is given by the basic equation

$$\alpha_o = S \cdot \alpha_{ob} + \alpha_{cv} = S \cdot \alpha_{ob} + E \cdot h_l \quad (10)$$

which is identical also for flow inside a channel, where S (–) represents the suppression factor of heat transfer coefficient at nucleate boiling effect (α_{ob}). The factor is determined by the equation (3). A modified form of the mentioned equation which is defined by Cooper [7] can be used which uses natural logarithm and the overheating temperature is replaced by the heat transfer density.

$$\alpha_{ob} = 93.5 \cdot \frac{\dot{q}^{0.67}}{\sqrt{M}} \cdot \frac{p_r^{0.12 - 0.434\,3 \cdot \ln\left(\frac{Ra}{0.4}\right)}}{[-0.434\,3 \cdot \ln(p_r)]^{0.55}} \quad (11)$$

For the equation (11) Chien and Cheng [3] state the deviation ±10% at the largest height of protrusions Ra=0.1 μm.

The quantity α_{cv} in the equation (10) is a convective heat transfer coefficient that equals the product of a two-phase convection enhancement factor E (–) and the heat transfer coefficient for a single-phase liquid flow. Chien and Cheng (2006) [3] suggest to use the following functions to calculate the suppression factor and the two-phase convection enhancement factor

$$S = \frac{1}{1 + 2.56 \cdot 10^{-6} \cdot \left(Re_f \cdot E^{1.25}\right)^{1.17}} \quad (12)$$

$$\left.\begin{aligned} E &= 1\,[-] && for\ \frac{1}{X_{tt}} < 1\,[-] \\ E &= 2.35 \cdot \left(0.213 + \frac{1}{X_{tt}}\right)^{0.736} && for\ \frac{1}{X_{tt}} \geq 1\,[-] \end{aligned}\right\} \quad (13)$$

where X_tt [–] is the Martinelli parameter

$$X_{tt} = \left(\frac{1-x}{x}\right)^{0.875} \cdot \left(\frac{\mu_f}{\mu_g}\right)^{0.125} \cdot \left(\frac{\rho_g}{\rho_f}\right)^{0.5} \quad (14)$$

3 Measurement apparatus

For the purposes of examination of the heat transfer at sprinkled tube bundles a test apparatus has been constructed; see the diagram at Fig. 1. A tube bundle at which a heat transfer from a heated water flowing inside the tubes into a falling film liquid is studied is placed in a vessel where low pressure is created by an exhauster through an ejector.

The test apparatus chamber is a cylindrical vessel of the length 1.2 m with three apertures in which the tube bundle of the examined length 940.0 mm is inserted. The tube bundle is installed in two fitting metal sheets which define the sprinkled area. The bundle consists of eight copper tubes of the diameter 12.0 mm situated horizontally one above another, with a distribution tube above them with apertures of the diameter from 1.0 mm to 9.2 mm. The bundle can be operated using only the first four or six tubes too.

Two closed loops are connected to the chamber. A heating one and a sprinkling one. The heating liquid flowing inside the tubes is intended for overpressure up to 1.0 MPa. The second loop contains flowing falling film liquid. There is a pump, a regulation valve, a flow meter and plate heat exchangers attached to both loops.

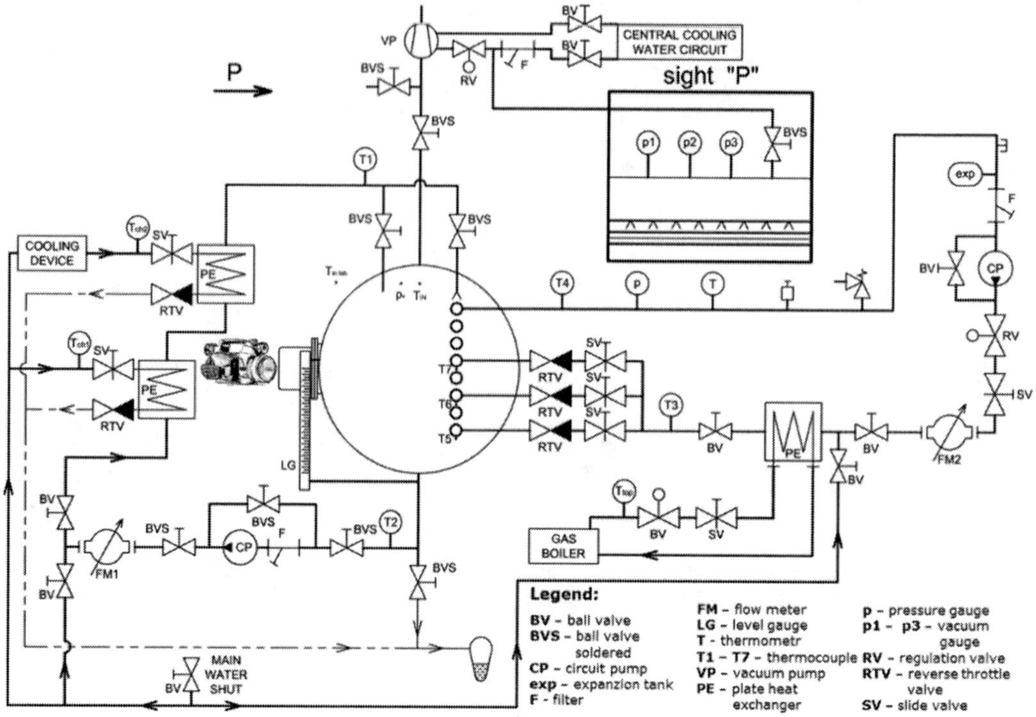

Figure 1. Measurement Apparatus Diagram

The plate heat exchanger at the heating loop is connected to a gas boiler which supplies heat to the heating liquid. The sprinkling loop uses two plate heat exchangers. In the first one the falling film liquid is cooled by a cold drinkable water from water mains and the falling film liquid is cooled in the second exchanger by a drinkable water cooled in a cooler which regulates the temperature up to 1.0 °C. In order to enable visual control the heating loop also includes a manometer and a thermometer. The thermal status in individual loops is measured by wrapped unearthed T-type thermocouples on the agents' input and output from the vessel. All thermocouples have been calibrated in the CL1000 Series calibration furnace which maintains a given temperature with the accuracy ±0.15 °C. None of the thermocouples has exceeded the error ±0.5 °C within the studied range from 28 °C to 75 °C. That is why the total error at temperature measurement is set uniformly for all thermocouples along the whole studied range ±0.65 °C.

There are three vacuum gauges measuring the low pressure. The first vacuum gauge is designed for visual control and it is a mercury meter, the second one is a digital vacuum gauge Baumer TED6 and enables measuring within the whole desired low-pressure range, but it is less accurate with lower pressure values. To allow a precise measuring of the low spectrum a third digital vacuum gauge of the range 2.0 kPa – 0 Pa is used. The accuracy of a vacuum gauge the results of which have been used for the assessment is 0.5% from the measured range, i.e. ±0.5 kPa.

Electromagnetic flow meters Flomag 3000 attached to both loops measure the flow rate. The flow meters' range is 0.007 8 ÷ 0.942 4 $l \cdot s^{-1}$, where the accuracy is 0.5% from the measured range, i.e. ± 0.004 67 $l \cdot s^{-1}$. All examined quantities are either directly (thermocouples) or via transducers scanned by measuring cards DAQ 56.

4 Methodology of data assessment

The assessment of the measured data is based on the thermal balance between the operation liquid circulating inside the tubes and a sprinkling loop according to the law of conservation of energy. Heat transfer is realized by convection, conduction and radiation. In lower temperatures the heat transferred by radiation is negligible, therefore it is excluded from further calculations. The calculation of the studied heat transfer coefficient is based on the Newton's heat transfer law and Fourier's heat conduction law that have been used to form the following relation

$$\alpha_o = \frac{1}{2 \cdot \pi \cdot r_o \cdot \left[\frac{1}{k_S} - \frac{1}{2 \cdot \pi \cdot \alpha_i \cdot r_i} - \frac{\ln\left(\frac{r_o}{r_i}\right)}{2 \cdot \pi \cdot \lambda_S} \right]} \qquad (15)$$

where α_o ($W \cdot m^{-2} \cdot K^{-1}$) is the heat transfer coefficient at the sprinkled tubes' surface, α_i ($W \cdot m^{-2} \cdot K^{-1}$) is the heat transfer coefficient at the inner side of a tube set for a fully developed turbulent flow according to [11, page 59 and 138], r_o; r_i (m) are outer and inner tube radii, λ_S ($W \cdot m^{-1} \cdot K^{-1}$) is thermal conductivity and k_S ($W \cdot m^{-1} \cdot K^{-1}$) is heat admittance based on the above mentioned laws governing heat transfer which is calculated from heat balance of the heating side of the loop, that is why the following must be valid:

$$\dot{Q}_{34} = k_S \cdot L \cdot \Delta T_{ln} = \dot{M}_{34} \cdot c_p\left(\frac{t_3+t_4}{2}\right) \cdot (t_3 - t_4) \qquad (16)$$

where M_{34} (kg·s^{-1}) is the mass flow of heating water, c_p (J·kg^{-1}·K^{-1}) is the specific heat capacity of water at constant pressure related to the mean temperature inside the loop, L (m) is the total length of the bundle and ΔT_{ln} (K) is a logarithmic temperature gradient where a counter-current exchanger was considered

The above mentioned calculation is valid only for heat transfer from water flowing inside the tubes into a falling film liquid which does not boil at the surface or where insignificant boiling may occasionally occur at the last tube. In case boiling occurs at a major part of the tube bundle, i.e. the tube bundle is sprinkled by a saturated water or higher at a corresponding pressure in a chamber p_V (Pa), two-phase flow should be considered. The amount of developed steam can be calculated according to the law of conservation of mass. I.e. the sum of the mass flow rate of steam M_v (kg·s^{-1}) and the mass flow rate of the water flowing down after the exchanger M_f (kg·s^{-1}) must equal the mass flow rate flowing from a distribution tube M_l (kg·s^{-1}). However, there are two unknown variables in this equation. That is why it is necessary to apply the law of conservation of energy as well. I.e. the sum of the energy necessary to develop steam Q_v (W) and the energy transferred to a falling film water Q_f (W) must equal the energy supplied by the heating liquid Q_{34} (W). Then the following is valid for the calculation of the amount of developed steam:

$$\dot{M}_v = \frac{\dot{Q}_{34} - \dot{M}_1 \cdot [i_2(t_2; x=0) - i_1(t_1; x=0)]}{i_{IN}(t_{IN}; p_v) - i_2(t_2; x=0)} \qquad (17)$$

under the condition that the temperature of the falling film water t_1 (°C) in a distribution tube is max. at the saturation boundary or slightly below it. In the mentioned equation, i_{IN} (J·kg^{-1}) stands for the enthalpy of a developed steam cushion at the temperature t_{in} (°C), and i_2 (J·kg^{-1}) stands for the enthalpy of a heated falling film water at the temperature t_2 (°C), after the last tube the test tube bundle.

The studied heat transfer coefficient can be determined on the basis of the Newton's law of heat transfer where the wall temperature is set as a mean temperature of the heating water and the temperature of the liquid flowing around it as a mean value of the temperature of a falling film water flowing out of the distribution tube and the temperature of a steam cushion.

$$\left.\begin{aligned} \dot{Q}_v &= \alpha_{o,v} \cdot S \cdot (T_w - T_\infty) \\ \Longrightarrow \alpha_{o,v} &= \frac{\dot{M}_v \cdot [i_{IN}(t_{IN}; p_v) - i_1(t_1; x=0)]}{S \cdot \left(\frac{t_3+t_4}{2} - \frac{t_{in}+t_1}{2}\right)} \end{aligned}\right\} \qquad (18)$$

Where S (m^2) stands for a heat-transfer region, i.e. the surface of the tube bundle.

5 Experiments results

Experiments published in this paper have focused on the influence of a heat-exchanging surface size on boiling development at a sprinkled tube bundle at two thermal levels. A tube bundle has been assembled consisting of eight tubes positioned horizontally one above another where only the first four, six or all eight tubes were heated. The input heating liquid temperature (t_3) was maintained at the level of approx. 40°C or 50°C with one degree centigrade deviation and at the flow rate (V_1) of 7.2 litres per minute with a deviation up to 0.1 litre per minute.

The falling film liquid flow rate was adjusted and maximum low pressure was created in the vessel by an exhauster for each experiment. The difference in pressures at individual thermal levels was given by the fact that with more vapour generation of a higher thermal level the exhauster's absorption capacity decreased. The heat source itself was regulated by the temperature of the falling film water at the distribution tube's output. Summary of basic data gained by measurement is provided in Table 1. Apart from the quantities mentioned above the table also contains falling film liquid mass flow rate related to the sprinkled zone length (Γ_1), Reynolds number of the liquid phase (Re_f), Prandtl number of the liquid phase (Pr_f), pressure in a chamber (p_v) during the whole experiment, specific heat transfer derived from the heating liquid for boiling (q“) that was determined on the basis of the law of conservation of energy and mass, and a total heat transfer derived from the heating liquid (q).

Due to the fact that the waveform of a falling film liquid heating on the bundle at tested flow rates is approximately linear, it was possible to apply linear approximation to find the borderline of saturated liquid at the bundle. Figure 2 shows four sets of points in relation to the Prandtl number of the liquid phase and on the falling film liquid temperature at the distribution tube's output, deducted by the temperature of saturation corresponding to the current pressure in a chamber. Points "0%" mark a state when the falling film water temperature flowing out of the exchanger reaches a saturation borderline. Points "50%" mean that the saturation state occurs between the 4th and 5th tube in case of an 8-tube bundle etc. The results for the Prandtl number up to approx. 4.7 come from experiments where the heating water at the bundle's input was approx. 50°C. Higher values of the Prandtl number correspond to the heating water of approx. 40°C.

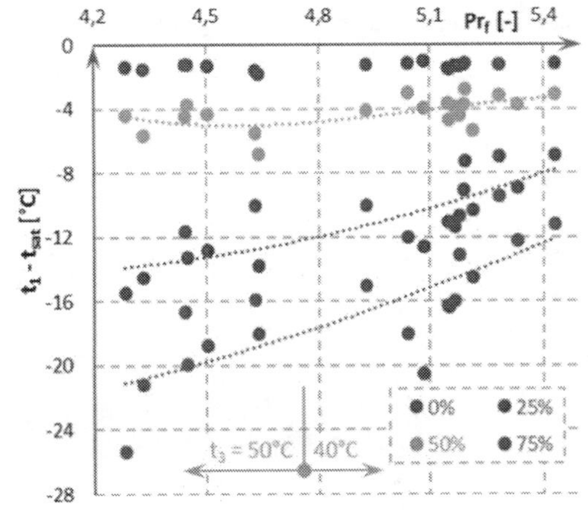

Figure 2. Boiling Development at Sprinkled Tube Bundle

Table 1. Basic information about the experiments

	t_3 (°C)	V_1 (l/min)	Γ_1 (kg/(s.m))	Re_f (-)	Pr_f (-)	p_v (kPa)	q" (kW/m²)	q (kW/m²)
4 tubes	40.4 ± 0.3	4.04 ± 0.03	0.0713 ± 0.0005	361.8	5.3	4.4 ± 0.1	3.8 ± 0.9	12.1 ± 2.5
	40.4 ± 0.3	6.03 ± 0.01	0.1063 ± 0.0003	544.5	5.3	4.5 ± 0.2	3.4 ± 1.0	12.4 ± 3.5
	40.3 ± 0.4	4.03 ± 0.00	0.0710 ± 0.0001	369.0	5.2	4.7 ± 0.3	4.4 ± 0.9	10.4 ± 2.0
	40.4 ± 0.2	6.03 ± 0.01	0.1063 ± 0.0002	531.0	5.4	4.2 ± 0.2	4.8 ± 1.2	14.5 ± 3.8
	50.3 ± 0.4	4.03 ± 0.01	0.0708 ± 0.0003	405.5	4.6	6.2 ± 0.4	8.0 ± 1.6	18.3 ± 3.9
	50.1 ± 0.3	6.03 ± 0.02	0.1062 ± 0.0005	610.5	4.6	6.2 ± 0.5	7.3 ± 1.4	19.8 ± 4.5
6 tubes	40.6 ± 0.2	4.03 ± 0.02	0.0710 ± 0.0003	367.3	5.2	4.7 ± 0.1	3.4 ± 0.6	7.6 ± 1.1
	40.5 ± 0.2	6.02 ± 0.02	0.1062 ± 0.0003	552.1	5.2	4.7 ± 0.2	3.5 ± 1.1	9.4 ± 2.7
	40.3 ± 0.3	4.06 ± 0.01	0.0715 ± 0.0001	374.4	5.1	4.8 ± 0.2	2.7 ± 0.5	6.8 ± 1.9
	40.5 ± 0.3	5.02 ± 0.01	0.0885 ± 0.0003	462.0	5.2	4.8 ± 0.2	3.2 ± 0.8	8.3 ± 2.5
	50.2 ± 0.4	4.03 ± 0.03	0.0710 ± 0.0006	416.7	4.5	6.7 ± 0.5	6.8 ± 1.6	12.6 ± 2.8
	50.1 ± 0.5	5.01 ± 0.02	0.0881 ± 0.0003	524.2	4.5	6.9 ± 0.3	6.5 ± 1.4	13.5 ± 2.7
	50.4 ± 0.2	6.05 ± 0.01	0.1064 ± 0.0002	633.9	4.4	7.0 ± 0.6	7.3 ± 1.6	15.2 ± 3.4
8 tubes	40.5 ± 0.2	4.02 ± 0.01	0.0708 ± 0.0001	370.2	5.2	4.8 ± 0.2	2.5 ± 0.7	6.6 ± 1.7
	39.9 ± 0.7	4.02 ± 0.04	0.0708 ± 0.0007	375.2	5.1	5.0 ± 0.2	1.9 ± 1.0	5.5 ± 2.1
	40.2 ± 0.5	5.04 ± 0.01	0.0887 ± 0.0001	473.8	5.0	5.1 ± 0.2	1.8 ± 0.8	5.3 ± 1.9
	40.4 ± 0.2	6.03 ± 0.02	0.1063 ± 0.0003	554.4	5.2	4.7 ± 0.2	2.1 ± 0.8	6.8 ± 2.4
	40.4 ± 0.5	8.08 ± 0.00	0.1423 ± 0.0001	776.5	4.9	5.4 ± 0.2	1.4 ± 0.6	5.4 ± 1.8
	50.2 ± 0.3	4.04 ± 0.04	0.0710 ± 0.0007	436.4	4.3	7.6 ± 0.3	4.4 ± 0.8	9.7 ± 1.6
	50.3 ± 0.3	6.02 ± 0.02	0.1058 ± 0.0003	645.9	4.3	7.4 ± 0.5	4.2 ± 0.8	10.7 ± 2.3

Figure 2 clearly shows that at the temperature difference of -3 and more, the boiling occurs at more than 60% of the tube bundle's height (considered from the last tube upwards). Simultaneously, the heat transfer coefficient was being stabilized at this value. That is why an average value and a standard deviation have been calculated starting at this borderline up for the purpose of further comparison. Figure 3 shows heat transfer coefficients of the vapour phase that have been determined according to the equation (18) in relation to the Reynolds number. Red tones represent an 8-tube bundle, green tones represent a 6-tube bundle and blue tones represent a 4-tube bundle. A lighter tone displays t_3 = 50°C and a darker tone displays t_3 = 40°C. This legend is applied to the following diagrams too, with the exception of the last one.

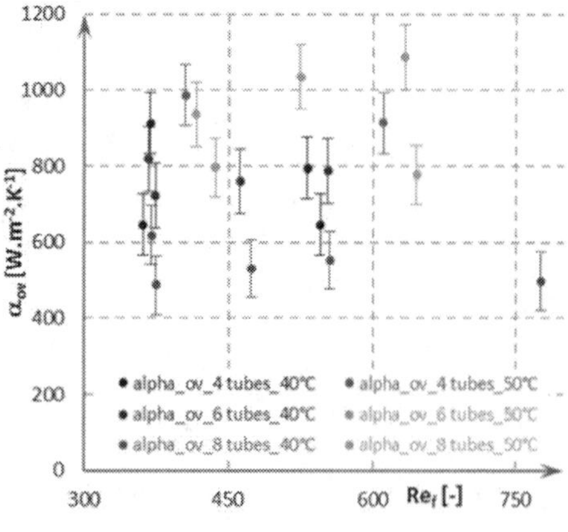

Figure 3. Heat Transfer Coefficient of Vapour Phase

A deviation is highlighted at individual points that is uniform at the given number of tubes. It was determined according to the average value of individual deviations increased by a standard deviation of this average value. Deviations ranged between 7 and 25%.

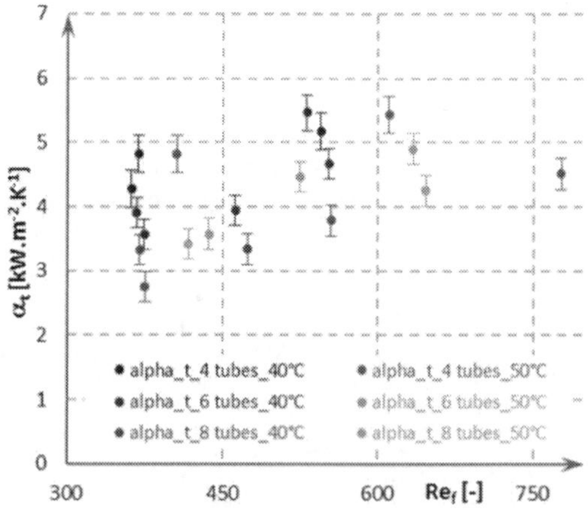

Figure 4. Total Heat Transfer Coefficient

The liquid phase coefficient given by the equation (15) as well as the vapour phase coefficient is derived from total energies at the input and output of the bundle. For this reason the coefficients are valid for the whole bundle and they represent its mean value. On the basis of the superposition principle it is therefore possible to total up both coefficients, thus creating a total mean heat transfer coefficient at the tube's surface which is displayed in Figure 4 also in relation to the Reynolds number of the liquid phase. Figure 5 shows the values of the generated vapour amount in relation to the Reynolds number.

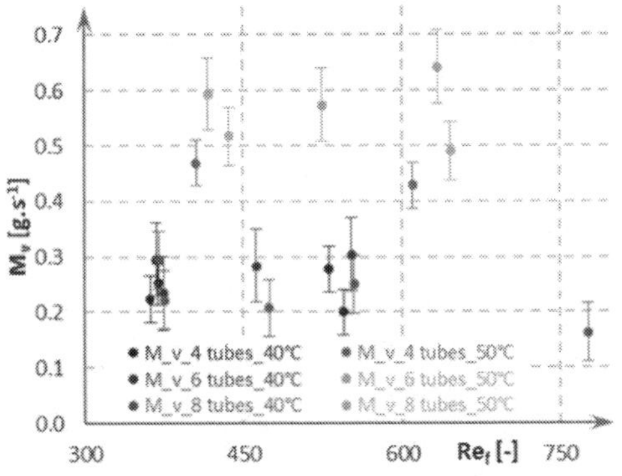

Figure 5. Generated Vapour Amount

The last Figure 6 displays all values of the total mean coefficient in single colour (green) and compares them with two mathematical models given in the theoretical part. Equations of these models were completed by the state values gained by our measurement. Blue values have been calculated according to the Lorenz and Yung model, i.e. by the basic equation (1), and red values have been calculated according to the Chien and Cheng model (2006), i.e. by the basic equation (10).

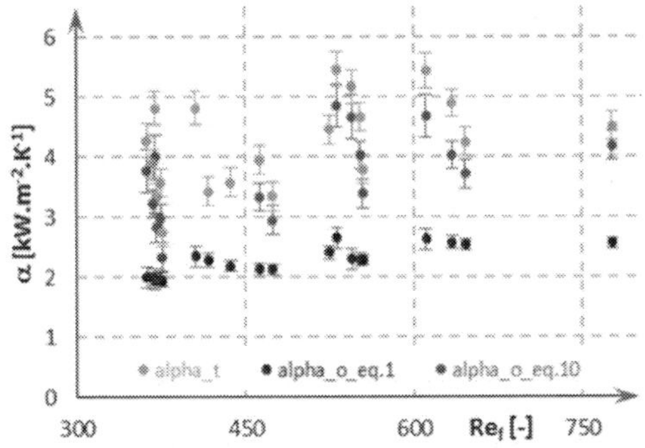

Figure 6. Results Comparison with Other Authors

Conclusion

The objective of this paper was to describe the influence of heat-exchanging surface size on boiling development at a sprinkled tube bundle at two thermal levels of a heating liquid that flows inside the tubes. Boiling development, i.e. its consequent testing on a bundle, was realized using the temperature of a falling film water that has been sprinkled on a tube bundle. It has been proved that the lowest values of the heat transfer coefficient of the vapour phase occurred at the 8-tube bundle and fluctuated around 550 W/(m^2.K). On the contrary, the highest values (almost double) have been reached at the 6-tube bundle. The vapour phase coefficient as well as the generated vapour amount showed to be functionally independent on the value of the total falling film liquid flow rate.

While the convenient number of tubes cannot be clearly determined at the thermal level of 40°C because the vapour amount fluctuated around the mean value of 0.25 g/s, the higher thermal level provides more clear-cut results. The highest values have been reached with six tubes where the vapour amount fluctuated around approx. 0.6 g/s. 8 tubes follow where the amount was approx. 0.5 g/s and the mean value at four tubes oscillated around 0.45 g/s. Higher amounts of generated vapour at six tubes can be justified by a better distribution of the temperature field at the bundle which is visible at the boiling development. Whereas the gradient increases between six and eight tubes, it falls sharply due to further reduction of the heat-exchanging surface (best visible at 50 and 75%). The total heat transfer coefficient at the tube surface consists mostly of the liquid phase coefficient. That is why it is also functionally dependent on the value of the falling film liquid flow rate. The highest values have been reached at a 4-tube bundle. Values of a 6-tube bundle were approx. 20% lower and the difference at an 8-tube bundle was up to approx. 40%.

Coefficients suggested by us have been compared with two other models. State values gained by our measurement have been put into these models. The first model by Chien and Cheng did not correspond very well, i.e. the calculated values were lower in the range of approx. 30 to 56%. Contrary to this, the second model showed a relatively good concordance. Values were lower only in the range of approx. 7 to 17%.

Acknowledgment

This work is an output of research and scientific activities of NETME CENTRE PLUS (LO1202) by financial means from the Ministry of Education, Youth and Sports under the „National Sustainability Programme I".

References

1. J. J. Lorenz and D. Yung. Journal of Heat Transfer, vol. 101, issue 1, pp. 178-180, (1979).
2. W. H. Parken, L. S. Fletcher, V. Sernas and J. C. Han. Journal of Heat Transfer, vol. 112, issue 3, pp. 744-750, (1990).
3. L.-H. Chien and Ch.-H. Cheng, HVAC, vol. 12, issue 1, pp. 69-87, (2006).
4. J. J. Lorenz and D. Yung, Proceedings, Fifth Ocean Thermal Energy Conversion Conference, Miami Beach, Florida, pp. 46-69, (1978).
5. W. M. Rohsenow, Journal of Heat Transfer, vol. 94, issue 2, pp. 255-256, (1972.
6. J. R. Thome, Wolverine Tube, Inc. <http://www.wlv.com/products/databook/db3/DataBookIII.pdf>, (2004-2010)
7. M. G. Cooper, 1st UK National Conference on Heat Transfer, vol. 86, pp. 785-793, (1984).
8. ČSN EN ISO 4287, Praha: ČSNI, 04/1999.
9. K. R. Chun and R. A. Seban, Journal of Heat Transfer, vol. 93, issue 4, pp. 391-396, (1971).

10. J. C. Chen, Industrial & Engineering Chemistry Process Design and Development, vol. 5, issue 3, pp. 322-329, (1966).
11. M. Jícha, M. Issue 1. 160 p. ISBN 80-214-2029-4, (2001).

EPJ Web of Conferences **114**, 02061 (2016)
DOI: 10.1051/epjconf/201611402061

Optimization of the heat output of wall convectors with using an unconventional slumped glass cover

Petr Kulhavy[1,a], Jitka Kleckova[2], Michal Petru[1] and Miroslav Havlicek[3]

[1] Faculty of Mechanical Engineering, Technical University of Liberec, Studentská 2, 461 17, Liberec 1, Czech Republic

[2] Department of Mechanics, Faculty of Applied Sciences, University of West Bohemia, Pilsen. Univerzitní 22, Pilsen, Czech Republic.

[3] Licon heat s.r.o., Construction department, Průmyslová zóna Sever, 460 11, Liberec 11, Czech Republic

Abstract. This paper deals with study of a new shape of the glass cover of heat convectors. Design of the heat convectors is characterized by the low water volume, highly variable geometry, adjustable heat output and the pressure losses that arise when the fluid flow through the exchanger. Based on a new concept of a slumped glass cover of the exchanger have been created some numerical models and also carried out experiments in order to find and optimize an appropriate shape. The glass materials are very specific mainly due to their specific shaping abilities and technological capabilities of manufacturing. The aim is to determine an appropriate shape and a curvature of the glass body and also the position and size of the exchanger. It has been found a significant dependence of the heat exchanger position on to the total heat output of an entire device. Simultaneously has been proved also the dependence of a shape of the cover plate, to the total heat output, that is more considerably for natural than for the forced convection.

1 Introduction

For a heating of houses, flats and industrial buildings are being used a lot of various devices. From the still available and most commonly used conventional cast-iron radiators to devices, supplemented by regulation and fans, so-called convectors. Classic radiators are used for the transmission and transformation of heat into the surroundings through radiation. Modern heating bodies, called convectors use principle of free or forced convection, when the air movement is accelerated by the fan [1]. The energy that we can obtain is usually from the electricity, fossil fuels or woods. The final conversion of the supplied energy to thermal has been carried out using our own equipment or from heating plants, such as long-haul networks or central distribution systems. In most cases is the thus obtained energy further transmitted to a fluid medium in the mostly number of cases to water, sometimes also oils etc.. This way is the thermal energy distributed through buildings and by using exchangers, is the heat transferred into the rooms. Today there is possible to choose from an enormous offer of heating elements and their variants. Other possibilities including e.g. the floor heating systems and the current top, we can say, are the individual central distribution systems with heat pump, often complimented by recuperation or air conditioning units. These devices instead of only a temperature control at the same time, providing adequate exchange, regulation of humidity and air quality. Practically in all industries, with regard to an efficient use of energies, we encounter reduction of the medium (water) volume, a total weight of the device and the input power. As mentioned earlier from the original cast iron radiators is the move over time towards their smaller, lighter and more powerful variants. Generally, heating convectors designed for the heating in the living spaces, have been gradually replacing the classic radiators [1,2]. The most important parameters indicating the competitiveness of products is definitely their thermal power and in the case of convectors with forced convection also an energy consumption and acoustic characteristics of the fans, namely the sound pressure level. In these parameters is already achieved almost a physical maximum and more important than a little bit exceed the others, is primarily to keep the pace with the top. Currently, a real fight is in this sector on the field of design. There are e.g. various micro lamellar systems [3], iron castings sculptures or other highly complicated shapes. Furthermore, for this sector completely new materials have been used like e.g. glass, mirror and polished stone or wooden board [4]. In our case we concentrated to a glass plate manufactured of technology

[a] Corresponding author: petr.kulhavy@tul.cz

so-called "slumping" with simultaneous usage of a fusing. The aim of this work is in cooperation with the manufacturer of heat convectors Licon Llc assess the dependency of a shape of the front cover plate and position of the heat exchanger on the overall thermal performance.

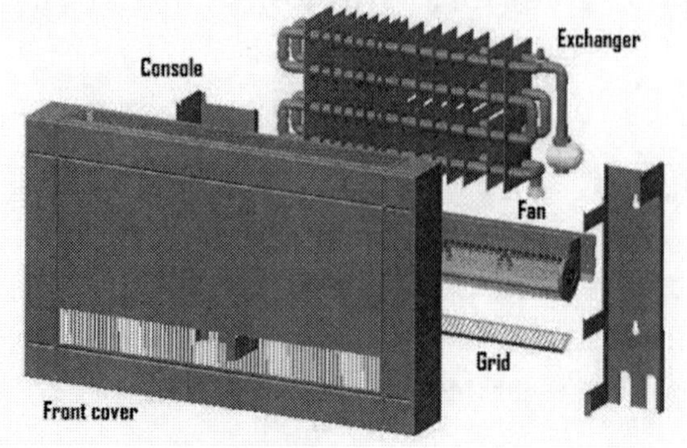

Figure 1. The original version of the wall-mounted convector.

2 The solved problem

About the advantages of devices based on convection instead of only radiation for heating in living rooms speak a lot of researchers. We can convince about it eg. at the work of Hamza [5], which speaks about energy savings in some model and simulated situations. This work is comparing two methods of heating (radiation, convection) in an experimental room at different ambient temperatures. A lot of international authors e.g. [6,7,8] deal with the question of an appropriate type of convection and they have been trying to find out a compromise between thermal comfort and the size of the supplied energy. The results showed that in energy terms is more advantageous to use the natural convection instead of the forced convection. This should be logically caused by the fact that the movement of warm air during heating arises naturally and there isn't need to supply any additional electric energy. Such a solution appears to be preferable, especially in terms of energy saving. The usage of the fan, the control and power required for the power supply, directly associate also with the end product and overhead cost [2,5]. From the other side, with the forced convection is generally reached significantly higher levels of heat output and faster achieved feeling of thermal comfort in the entire heated space and that is why this method is so widely preferred. The question and the subject of many numerical models and experiments, remains the subjective feeling and perception of thermal comfort. In our case the wall-mounted radiator, that standardly uses the forced convection, is tested and innovated. The volume of water in the individual types of used heat exchangers is in ranges from 0.18 to 1.2 liters per meter of length and the volumetric flow up to 5 l / min. The thermal output, of the tested wall mounted convectors, according to EN 442 at an ambient temperature 20 °C and at the operation mode 70/55 ° C, reaches the values in range from approximately 190 to 7800 W depending on length of the convector, flow rate and a revolutions of used fan.

2.1 The tested prototype

As is already mentioned above, the leading manufacturers of heating systems has already reached the real maximum of thermal performance and now is the competition especially in a field of design. In our case, the original steel body of convector (Fig.1) was replaced by a transparent front plate made from slumped glass (Fig. 2). This slumped cover is dominant not only for aesthetic reasons, but also due to a very specific shapes that can be achieved and also can be optimized with regard to the internal flow with the aim to increase heat output.

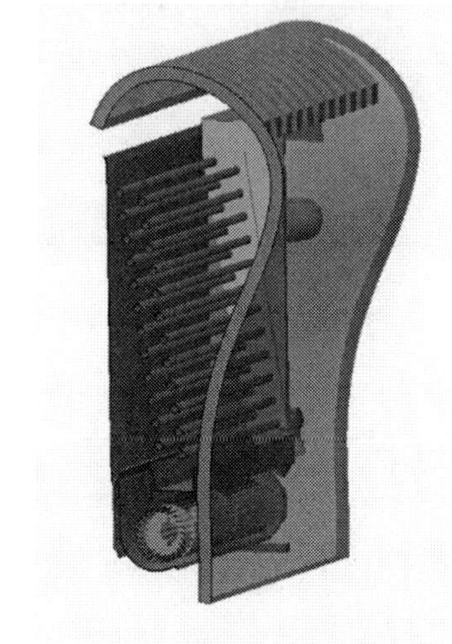

Figure 2. The first idea of a shape the slumped glass cover.

By research of an ideal shape of the fins fitted on tubes of a heat exchanger, their position and air flowing around them have dealt very thoroughly eg. Lemfeld and Frana [2,9,10], as shown in Figure 3.

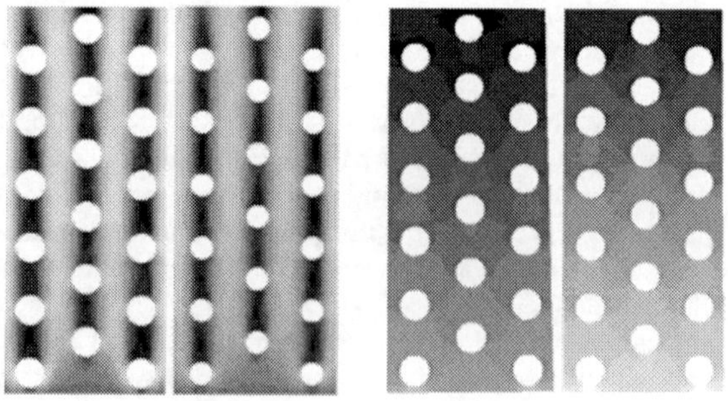

Figure 3. Example of distributions temperature fields (left) and static pressure (right) for different types of fins [9].

3 Experiments

Experiments were performed in thermodynamic laboratories of the manufacturer. The aim of the carried tests was to assess the effect of shape of the innovated cover to the thermal output and optimizing its geometry.

The thermal power was measured using a thermocouple system. Measurements were also conducted for the exchanger without any sheathing. Instead of the glass was at first used transversely formable plastic material on the inner side covered with thermally insulating and reflective foils. With that is possible to achieve changeable geometric shapes. To achieve the exact shape and repeatability of the experiment, it was used steel templates, created in accordance to the CAD drawings. The desired final shape (Fig. 4) was with using a CNC machined mold made of a slumped glass.

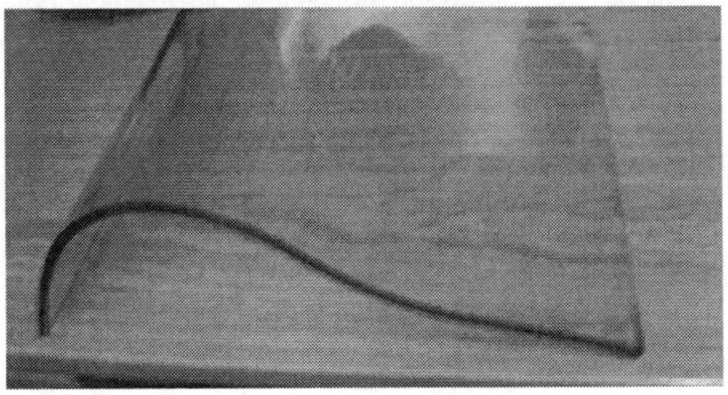

Figure 4. The first created prototype of a glass cover.

Wall convector completed by some changeable covering panels was connected to the water circuit in manufacturer's hydrodynamic laboratory. For testing liquid at a given temperature it is necessary to ensure a sufficiently powerful heating. It was used a thermally insulated tank with electric heating dimensioned over the maximum estimated thermal performance of the connected devices. Liquid pumped with an electronically controlled pump, fed from the tank through the control ultrasonic flowmeter into the exchanger of the tested wall convector and then is the liquid again returned to the tank [1]. The measurement was performed for 2-point temperature of water 65/ 55°C and the ambient temperature 20°C. The electronically controlled variable flow rate in a range 0.1-5 l×min-1 was joined to the connecting fitting of the heat exchanger (Fig. 5). This constitutes a speed from approximately 0.01 to 0.3 m / s when a size of the Re number reaches value 100-4800.

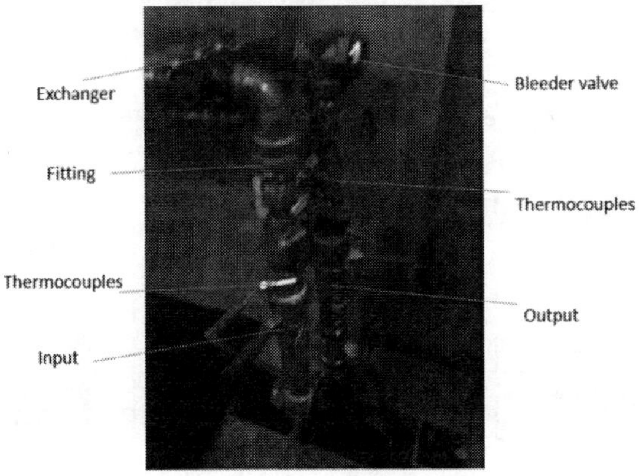

Figure 5. Joining parts of the exchanger with thermocouples.

Measurements were performed by using the interface National Instruments - NI cDAQ-9174 with additional modules. To the module are connected NiCr-Ni thermocouples. Their installation is 3 to input, 3 to output and the final value is their average. Several other couples measure the temperature in the heater, the pump and of course the ambient temperature. Based on these data, is the density of water calculated for the precise determination of the mass flow.

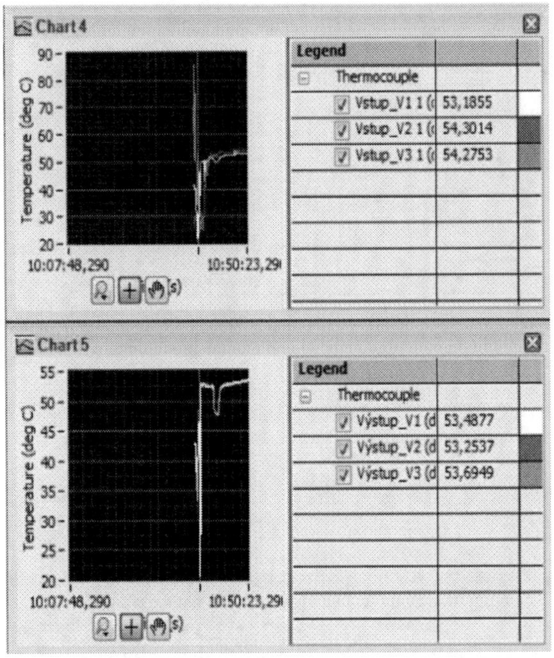

Figure 6. Interface of used NI® Lab View measurement software.

4 Numerical simulation

Experimental results were compared with numerical results obtained by using commercial CFD software ANSYS Fluent, which allows the modelling of flow of isotropic and anisotropic structures and other physical phenomenon. This software contains a wide variety of tools for simulation of the various problems, which are described by partial differential equations. CFD software ANSYS Fluent uses for the calculation the implicit algorithm, where individual states of the analyzed compression are updated gradually in time t to time $t+dt$ according to Eq. (1).

$$\delta u_{i+1} = u_{i+1}^{t+\Delta t} - u_i^{t+\Delta t} \tag{1}$$

where $u_i^{t+\Delta t}$ is vector of nodal displacements for i^th iteration in the time $t+\Delta t$.

4.1 Numerical model

Two cases of heat convectors were compared by using numerical simulations. The type with new glass cover and convector with standard plated cover. To simplify a problem, computational domain was considered as a two-dimensional section of the original model, as shown in Fig. 7. The computational mesh was thickened in the vicinity of walls of heat exchanger tubes. Numerical simulations were accomplished for the temperature of tubes equal to 65°C. Boundary conditions at the inlet area correspond to conditions of experimental measurement which were 20°C and atmospheric pressure.

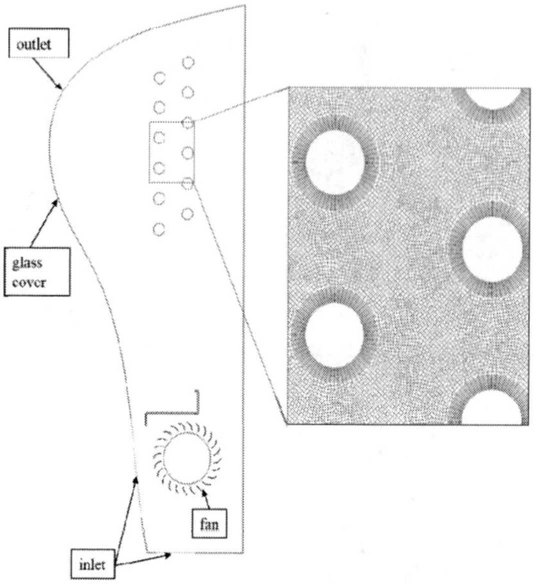

Figure 7. Two-dimensional computational domain (left), detail of computational mesh in vicinity of tubes (right).

Two-dimensional air flow was modelled by standard k-ε turbulence model which was first introduced by Launder and Spalding [14]. Standard k-ε turbulence model consists of two transport equations for turbulence kinetic energy k and its dissipation rate ε.

$$\frac{\partial(\rho k)}{\partial t}+\frac{\partial(\rho k u_i)}{\partial x_j}=\tau_{ij}\frac{\partial(\rho u_i)}{\partial x_j}+\frac{\partial}{\partial x_j}\left[\left(\mu+\frac{\mu_t}{\sigma_k}\right)\frac{\partial k}{\partial x_j}\right]-\rho\varepsilon \;, \quad (2)$$

$$\frac{\partial(\rho\varepsilon)}{\partial t}+\frac{\partial(\rho\varepsilon u_i)}{\partial x_j}= \quad (3)$$

$$=\frac{\varepsilon}{k}C_{\varepsilon 1}\tau_{ij}\frac{\partial(\rho u_i)}{\partial x_j}+\frac{\partial}{\partial x_j}\left[\left(\mu+\frac{\mu_t}{\sigma_\varepsilon}\right)\frac{\partial\varepsilon}{\partial x_j}\right]-\frac{\varepsilon^2}{k}C_{\varepsilon 2}\rho \;,$$

where ρ is fluid density, $\mathbf{u}_i$ are the components of velocity vector, $\mathbf{x}_j$ are components of position vector, t is time, $\boldsymbol{\tau}_{ij}$ is Reynolds stress tensor, μ is dynamic viscosity and μ_t is turbulence eddy viscosity defined as

$$\mu_t=C_\mu\rho\frac{k^2}{\varepsilon}\;, \quad (4)$$

where $C_{\varepsilon 1}$=1,44; $C_{\varepsilon 2}$=1,92; C_μ=0,09; σ_k=1; σ_ε =1,3 are empirically derived constants which could be set by testing of simple fluid flow problems or by numerical optimization.

For the spatial discretization of computational domain the finite volume method is implemented in computational software Fluent. The second order upwind scheme was used to discretize convective terms of the governing equations. In ANSYS Fluent, the fluid flow in rotating domain could be solved using the multiple reference frame (MRF) function. This method is useful for the problems where some parts of geometry are moving (fan) and other parts are stationary (cover and walls of conductor). The main advantage of using MRF is, that there is no need to use any deforming mesh algorithm. The computational domain was divided by a circular interface into the two parts - the annular domain which contains fan blades and domain of surrounding fluid, see Fig. 8.

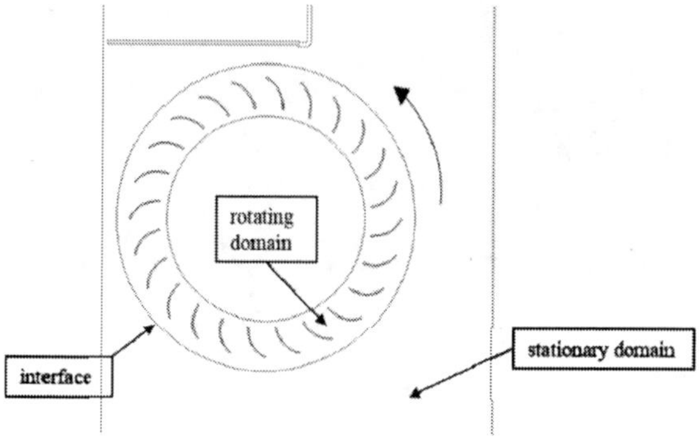

Figure 8. Division of domain into the moving and stationary part.

The blades are rotating with an angular velocity which is prescribed to whole rotating domain as a cell-zone condition in Fluent. Relative velocity must be applied into a numerical model to solve the fluid flow in rotating domain. This velocity is computed as Eq. (5).

$$\vec{v}_r=\vec{v}-(\vec{\omega}\times\vec{r})\;, \quad (5)$$

where $\vec{v}$ is the vector of absolute velocity, $\vec{\omega}$ is the angular velocity of rotating frame, $\vec{r}$ is the position vector in rotating frame.

4.2 Results of numerical simulation

Numerical simulations were provided for two fan velocities: 1000 and 1600 rpm. The comparison of convergence of solved cases is shown in the next Table 1.

Table 1. The comparison of solved cases

case	number of mesh elements	rpm	number of iterations	continuity residual value
Standard shape	210000	1000	10^4	$2,2 . 10^{-8}$
		1600	10^4	$5,8 . 10^{-7}$
New shape	250000	1000	10^4	$2,9 . 10^{-11}$
		1600	10^4	$1,5 . 10^{-9}$

Velocity vectors in computational domain in case of conventional steel cover are shown in the Fig. 9.

Temperature distribution in computational domain for the first case of fan velocity is shown in the Fig. 10.

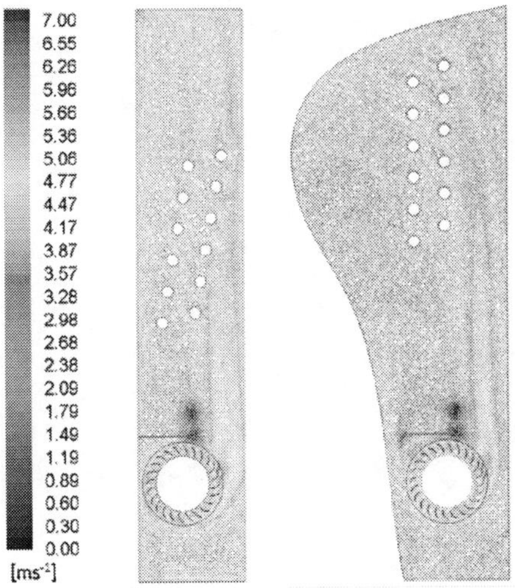

Figure 9. Velocity vectors for the 1000 rpm fan velocity with the convetional steel cover in comparison to the new one.

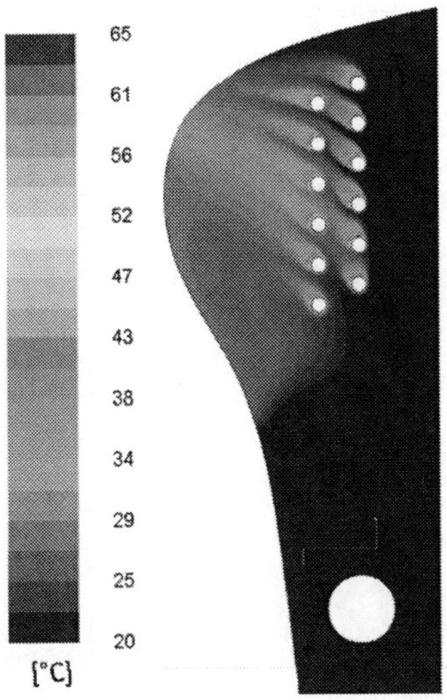

Figue 10. Temperature distribution for the 1000 rpm fan velocity with inovated shape of cover.

The heat output was computed for each of these cases by formula

$$P = Qc_P\Delta T\,, \tag{6}$$

where Q is the mass flow of air at the domain outlet, c_p is specific heat equal to value 1006,43 $Jkg^{-1}K^{-1}$ and ΔT is temperature difference between outlet and inlet area. The comparison of solved cases is shown in the Tab. 2 below.

Table 2. Heat output comparison

case	RPM	Q [kg s⁻¹]	ΔT [°C]	P [W]
Standard shape	1000	0,186	6,8	1273
	1600	0,260	5,9	1544
New shape	1000	0,198	7,2	1435
	1600	0,271	6,1	1664

5 Results and discussion

The experimental and numerical determination of heat outputs was performed in order to find a dependence on the shape of used covering front plate. The used exchanger optimized for forced convection achieves without using any fans only a very small thermal power. These quite small values could be measured only a very inaccurate with significant statistical variances, especially with a lowest Re numbers of the flowing medium in the exchanger. After turning on the fan even at very low speeds, the heat output starts to increase almost exponentially. Using of exchanger just without covering seems to be less appropriate. This methods are in practice used quite rarely, especially in some special customer applications like e.g. between the multi windows and facade systems.

As has been possible to observe, the ordinary rectangular metal sheeting, its geometry and dimensions are designed to reach the almost maximum possible power. In only one configuration of the curved cover with an upper convexity, we have succeeded in approximately 8% increasing of the heat power, compared to the conventional one. It can be concluded that the sheeting, its shape and parameters play a significant role in the construction of heating systems and to the heat output they have a very significant impact. This all is necessary to consider in a design process of new devices. It was produced, constructed and subsequently experimentally test a variety of shapes. These include position of the heat exchanger, its angle relative to the rear side and of course the shape of the used cover itself. The Fig. 11 and 12 show several variants where it was possible to observe the most fundamental differences. Figure 11 shows results of experimental measurement and Figure 12 shows results of numerical simulations.

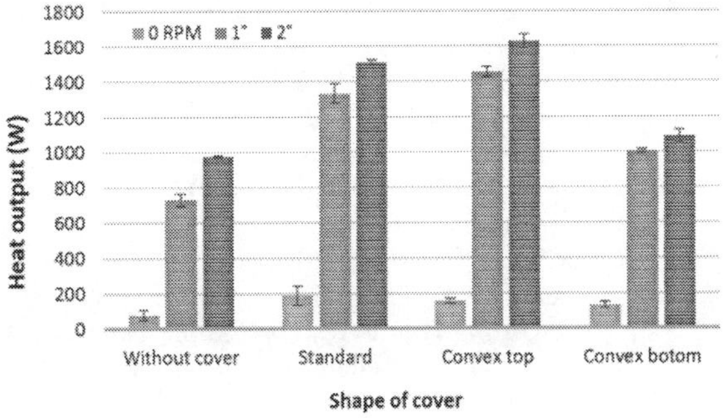

Figure 11. The selected measurement of heat outputs.

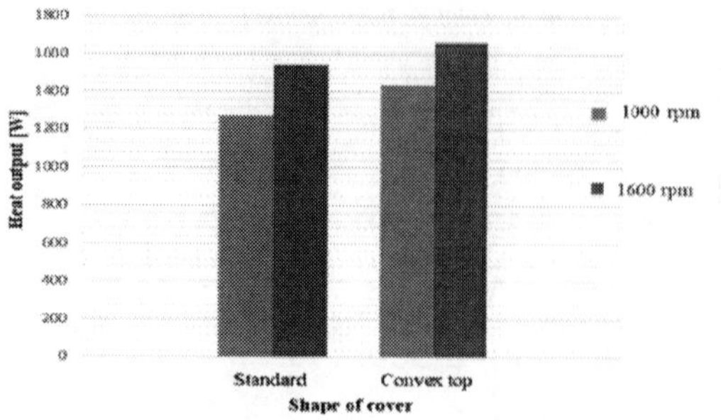

Figure 12. The simulation results of heat output dependence on angular velocity of the fan.

6 Conlusion

The differences between experimental and numerical results in values of heat output could be caused by simplification of the complex three-dimensional problem to a fluid flow in a two-dimensional domain. On the other hand, the simplified model was adequate to show the differences between standard and new glass cover. While considering the conventional shapes is in this solution achieved practically the maximal possible power. For consideration and as a subject for further work can be also clearly established the dependence of the heating medium and air flow velocity to the total heat output.

Acknowledgements

We give our thanks to the construction department and management of the company Licon heat s.r.o. for the collaboration and provided information.

This work was supported by the Ministry of Education of the Czech Republic within the SGS project no. 21007 on the Technical University of Liberec.

References

1. P. Kulhavy, M. Petru, P. Srb, G. Rachitsky: Eur. Phys. J. **92**, 02064 (2014)
2. K. Frana, M. Muller, F. Lemfeld: WAS, Engineering and Technology **6**, 5154 (2012)
3. Z. Chen, H. Guanren, M. Yuanle: Soc. for Industrial and Applied Math., xxix, **531** (2006)
4. LICON HEAT s.r.o., Směrnice technologie a konstrukce, Liberec (in Czech), (2014)
5. A. Hamza, G. M. Morsy, Energy efficiency and indoor thermal perception, **3** (2010)
6. T. Aicher, H. Martin, Intern. Journal of Heat and Mass Transfer, **40**, 3617-3626 (1997)
7. T. Kuppan, *Heat Exchangers Design Handbook*, Taylor & Francis, New York (2000)
8. P. Dancova, Z. Travnicek, T. Vit, EPJ Web of Conferences **45**, 01002 (2013)
9. F. Lemfeld, K. Frana, *The influence of geometry changes to the flow in a floor heating convector* In 30th conf. of Dept of Fluid Mechanic, Liberec (2011)
10. F. Lemfeld, K. Frana, Journal of Applied Science in the Thermodynamics, **5** (2011)
11. EN 442-1 ed. 2 *Radiators and convectors:* Technical specifications and requirements
12. M. Petru, O. Novak, D. Herak, S. Simanjuntak: Biosystems Eng., **111** (2012)
13. M. Petrů, O. Novák, L. Ševčík, P. Lepšík: Applied Mechanics and Materials, **157** (2014)
14. B. E. Launder, D. B. Spalding: *Mathematical Models of Turbulence*, Academic Press (1972)

EPJ Web of Conferences **114**, 02062 (2016)
DOI: 10.1051/epjconf/201611402062

Distribution of Acoustic Power Spectra for an Isolated Helicopter Fuselage

A.N. Kusyumov[1,a], S.A. Mikhailov[2], L.I. Garipova[3], A.S. Batrakov[4] and G. Barakos[5]

[1-4]Kazan National Research Technical University n.a. A. N. Tupolev (KNRTU-KAI), Kazan, Russia
[5]University of Liverpool, Liverpool, UK

Abstract. The broadband aerodynamic noise can be studied, assuming isotropic flow, turbulence and decay. Proudman's approach allows practical calculations of noise based on CFD solutions of RANS or URANS equations at the stage of post processing and analysis of the solution. Another aspect is the broadband acoustic spectrum and the distribution of acoustic power over a range of frequencies. The acoustic energy spectrum distribution in isotropic turbulence is non monotonic and has a maximum at a certain value of Strouhal number. In the present work the value of acoustic power peak frequency is determined using a prescribed form of acoustic energy spectrum distribution presented in papers by S. Sarkar and M. Y. Hussaini and by G. M. Lilley. CFD modelling of the flow around isolated helicopter fuselage model was considered using the HMB CFD code and the RANS equations.

1 Introduction

For a conventional helicopter, there are two fundamental elements that contribute to the generation of near-field and far-field noise, the main rotor and the tail rotor [1]. Engine and fuselage noise are typically of secondary significance. A helicopter main rotor generates primarily tonal and broadband noise. Additional sources, including Blade Vortex Interaction (BVI) noise and High Speed Impulsive (HSI) noise, that are dominant at specific flight regimes. Basic loading noise during hover is generally dominant in a conical region directed 30 to 40 degrees downward from the rotor plane, while broad-band noise radiates mostly out of the plane of the rotor [1]. So for some operating conditions and for specific directions of sound propagation, the broadband noise generated by all parts of helicopter (including fuselage) can be a significant contributor to the overall helicopter noise.

According to Proudman [2] the broadband noise can be considered, assuming isotropic flow turbulence and decay. The total acoustic power radiated from embedded finite regions of turbulence contained within an infinite volume of compressible fluid, is a function of the local time-averaged kinetic energy of the turbulence per unit volume, k, and the time averaged rate of dissipation of the kinetic energy per unit volume, ε.

Another aspect, is the broadband acoustic spectrum and the distribution of acoustic power over a range of frequencies. The acoustic energy spectrum distribution in isotropic turbulence is non monotonic. The power spectrum function has a maximum at some value of Strouhal number. In the present work the value of acoustic power peak frequency is determined using a prescribed form of acoustic energy spectrum distribution presented in papers by S. Sarkar and M. Y. Hussaini [3] and by G. M. Lilley [4].

CFD modelling of the flow around isolated helicopter fuselage model is considered in this paper using the HMB CFD code and the RANS equations.

2 Broadband isotropic turbulence noise (Proudman's formula)

A simple approach to estimate acoustic emission of flying vehicle for turbulent flows, assumes that the emitted noise does not have any distinct tones, and that the sound energy is continuously distributed over a broad range of frequencies. In this case the broadband noise power, can be estimated from RANS equations using the mean flow

[1a]Prof., department of aero-hydrodynamics, postbox7@mail.ru
[2]Prof., department of aero-hydrodynamics, sergey.mikhaylov@kai.ru
[3]PhD student, lyaysan_garipova@mail.ru
[4]PhD student, batrakov_a.c@mail.ru
[5]Prof., department of Engineering, George.Barakos@glasgow.ac.uk

field, turbulent kinetic energy k and the dissipation rate ε. Unlike the direct method of simulation and the integral methods, Proudman's [2] approach does not require unsteady CFD solutions and is based on Lighthill's acoustic analogy [5].

The intensity $I(\mathbf{x},t)$ of the radiated sound in the far-field is proportional to the square of the fluctuating pressure p and is defined by

$$I(\mathbf{x},t) = \frac{<(p-p_\infty)^2>}{\rho_\infty c_\infty},$$

were ρ_∞, p_∞ are the air density and pressure, c_∞ is the ambient sound speed. The intensity of radiated sound at any point **x**, and at time t can be determined as follows [4,6]:

$$I(\mathbf{x},t) = \frac{1}{16\pi^2\rho_\infty c_\infty^5 x^2} \iiint d\mathbf{y} \iiint d\mathbf{r} \frac{\partial^4}{\partial\tau^4} P_{xx,xx}(\mathbf{y},\mathbf{r},\tau).$$

Here ρ_∞ is the air density, c_∞ is the ambient sound speed, **y** is a position, **r** is the distance vector between two internal points **y** and **y′** inside the integration volume, and τ is the retarded time, $x=|\mathbf{x}|$. The space-time correlation function $P_{xx,xx}$ can be written in the form

$$P_{xx,xx} = \alpha_P u^4 \Phi(r/l)\Psi(\Omega\tau) \quad (1)$$

with the non-dimensional space correlation function, $\Phi(r/l)$, and the non-dimensional retarded time correlation function, $\Psi(\Omega\tau)$. Here u, l, and Ω are, respectively, reference values of the turbulent velocity, the turbulent length scale, and turbulent frequency; $\alpha_P = const$ is a scale coefficient ($\alpha_P = 1$ in [4]). The forms for the space and temporal correlation functions accepted in [4] are

$$\Phi(r/l) = exp\left(-\frac{\pi}{2}\frac{r^2}{L^2}\right), \; \Psi(\Omega\tau) = exp\left(-\frac{\pi}{2}\Omega^2\tau^2\right), \quad (2)$$

where L is the integral lenght scale.

Using the expression (2) for the space correlation function, Proudman obtained an estimate of the acoustic power per unit of volume in the time domain:

$$P_{at}=\alpha_\varepsilon \rho_\infty \varepsilon M_t^5, \quad (3)$$

where

$$M_t=(2k)^{1/2}/c_\infty$$

is the turbulent Mach number, k is the is turbulent kinetic energy per unit mass, and c_∞ is the speed of sound. The turbulent dissipation rate ε can be determined [7] by

$$\varepsilon = \frac{k^{3/2}}{L}. \quad (4)$$

Sarkar and Hussaini [3] recommended the re-scaled constant, α_ε=0.1, based on DNS data. It should be noted that there are different expressions for the reference turbulent velocity u (see, for example, paper [2], were $u = (2k/3)^{1/2}$). In this paper $u = (2k)^{1/2}$ following reference [8].

3 Total radiated acoustic power

An alternative approach to determine the acoustic power per unit of volume is based on the spectral transformation of the intensity $I(\mathbf{x},t)$ of the radiated sound. The spectral density $I(\mathbf{x},\omega)$ corresponding to the intensity $I(\mathbf{x},t)$ of the radiated sound is ([6])

$$I(\mathbf{x},\tilde\omega) = \frac{1}{2\pi}\int I(\mathbf{x},t)exp(i\tilde\omega t)dt =$$
$$= \frac{\pi\omega^4}{2x^2\rho_\infty c_\infty^5}\int_V P_{xx,xx}(\mathbf{y},\mathbf{k},\tilde\omega)d^3\mathbf{y}.$$

The frequency of the sound $\tilde\omega$ is the same as in the turbulence and **k** is the wavenumber vector of the sound. Taking into account (2) the space-time correlation function (1) may be written:

$$P_{xx,xx} = \alpha_P u^4 exp\left(-\frac{\pi}{2}\left(\frac{r^2}{L^2}+\Omega^2\tau^2\right)\right)$$

with four-dimensional wave-number/frequency spectrum function [4]:

$$P(k,\tilde\omega) = \frac{1}{4\pi^4}\frac{u^7 S_T^3}{\Omega^4} exp\left(-\frac{\tilde\omega^2}{2\pi\Omega^2}C\right). \quad (5)$$

and the acoustic spectral density

$$P_s(\tilde\omega) = 2\pi^2 \frac{\rho_\infty}{c_\infty^5}\omega^4 P(k,\tilde\omega). \quad (6)$$

Here

$$S_T = \Omega L/u \quad (7)$$

is the turbulent Strouhal number and

$$C = 1 + S_T^2 u^2/c_\infty^2 = 1 + S_T^2 M_t^2.$$

According to [4] the total radiated acoustic power per unit volume of turbulence is in the frequency domain:

$$P_{a\omega} = \alpha_P \int_0^\infty P_s(\tilde\omega)\, d\tilde\omega.$$

Substituting (5), (6) gives the total radiated acoustic power per unit volume of turbulence

$$P_{a\omega} = \alpha_P \rho_\infty \frac{3\pi}{\sqrt{2}C^{5/2}}\frac{u^3}{L} S_T^4 M_t^5. \quad (8)$$

4 Proudman's constant

The expression (3) for the total acoustic power radiated per unit volume of flow in the time domain can be written as

$$P_{at} = \alpha\rho_\infty \frac{u^3}{L} M_t^5, \quad (9)$$

where α is Proudman's constant. Taking into account (4) the re-scaled constant α_ε, and Proudman's constant are connected by the expression:

$$\alpha_\varepsilon = (2)^{3/2}\alpha. \quad (10)$$

The proposal that the total acoustic power radiated per unit volume in the time domain (3) is equal to the acoustic power in the frequency domain (9) can be written as

$$P_{at} = P_{a\omega}. \quad (11)$$

After substitution (3) and (8) to (11) we have

$$\alpha = \alpha_P \frac{3\pi}{\sqrt{2}C^{5/2}} S_T^4. \quad (12)$$

In this case, taking into account (10), we can rewrite Proudman's formula (3) in the form

$$P_{at}=\alpha_P \frac{6\pi}{C^{5/2}} S_T^4 \rho\varepsilon M_t^5$$

According to the approximate estimations [4] the turbulent Strouhal number can be accepted as

$$S_T \sim 1. \quad (13)$$

For low speed subsonic flow (M_t<<1 or $C\approx1$) we have the approximate value

$$\alpha_P \approx \alpha\frac{\sqrt{2}}{3\pi} = \frac{\alpha_\varepsilon}{6\pi} = 0.0053.$$

5 Turbulent reference frequency

The condition (13) and expression (4) can be used to determine the reference frequency Ω

$$\Omega \approx u/L = \sqrt{2}\varepsilon/k.$$

As $\varepsilon = \beta^* k\omega$ (ω is the specific dissipation parameter, $\beta^* = 0.09$) [8] we finally have

$$\Omega \approx \sqrt{2}\beta^*\omega. \quad (14)$$

According to [4] the reference frequency Ω can be used to determine a frequency peak in the acoustic power spectrum:

$$\omega_m = \sqrt{8}\Omega \quad (15)$$

and with formula (14) we can rewrite (15) as

$$\omega_m \approx 4\beta^*\omega. \quad (16)$$

Expression (3) can be rewritten in the form

$$P_{at} = \frac{1}{4}\alpha_\varepsilon \rho_\infty c_\infty^2 \beta^* \omega M_t^7 \quad (17)$$

or

$$P_{at} = \frac{1}{16}\alpha_\varepsilon \rho_\infty c_\infty^2 \omega_m M_t^7. \quad (18)$$

Formulas (16) – (18) allow estimation of the frequency peak and the acoustic power per unit volume spatial the steady computation of the flow using the turbulent Mach number M_t and the specific dissipation ω for the k-ω turbulence model [9].

Another estimation of the acoustic power and the acoustic intensity was given by Lighthill [5]:

$$P_a \sim \frac{\rho_\infty u^8}{c_\infty^5} L_{ref}^2, \quad (19)$$

$$I \sim \frac{\rho_\infty u^8}{c_\infty^5}. \quad (20)$$

Here L_{ref} is a reference length. From the comparison of expressions (19) and (20) we have that

$$I \sim \frac{P_a}{L_{ref}^2}.$$

Eventually we can write for the local sound pressure at the vicinity of an emitting body surface:

$$p_s = (I\rho_\infty c_\infty)^{1/2} = \left(\frac{P_a \rho_\infty c_\infty}{L_{ref}^2}\right)^{1/2}. \quad (21)$$

6 Numerical simulation of flow around an isolated helicopter fuselage

For simulations, an early version of the ANSAT helicopter produced by the JSC Kazan Helicopters is used. CFD computations and experiments were conducted for fuselage model with (without) skids and cross-beams. Multiblock grids for the CFD computations were constructed using the ANSYS ICEM-hexa software. The computational domain was resolved using hexahedral grids and the 3D steady incompressible Reynolds-Averaged Navier-Stokes (RANS) equations. Fully turbulent calculations were performed using the k-ω model [9]. The computational hexa-grid for this model (without skids and cross-beams) contained 964 blocks and about 13.500.000 cells. The computations were performed using the HMB solver of Liverpool University [10]. The results of the CFD modelling were compared to wind tunnel experimental data. Fuselage was tested at the low speed T-1K wind tunnel of KNRTU-KAI (2.25 m diameter).

The wind tunnel model fuselage was constructed using the same CAD model (figure 1), used for CFD modelling. The mesh topology and the surface grid near the area of the exhausts are presented in figure 2. The aerodynamic performances of this model and CFD code validation were considered in references [11-14], and were studied using the T-1K wind tunnel, that is equipped with a six-component Prandtl-type balances. The conditions of the wind tunnel experiment and CFD modelling corresponded to the free stream Mach number and the Reynolds numbers were of 0.1 and $4.4\cdot10^6$, respectively.

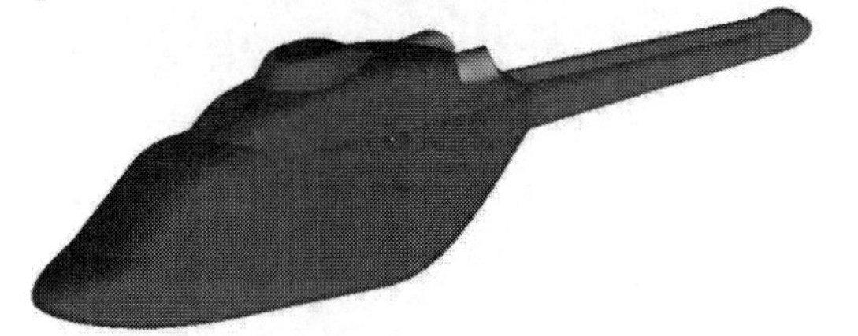

Fig. 1. Isolated helicopter fuselage model.

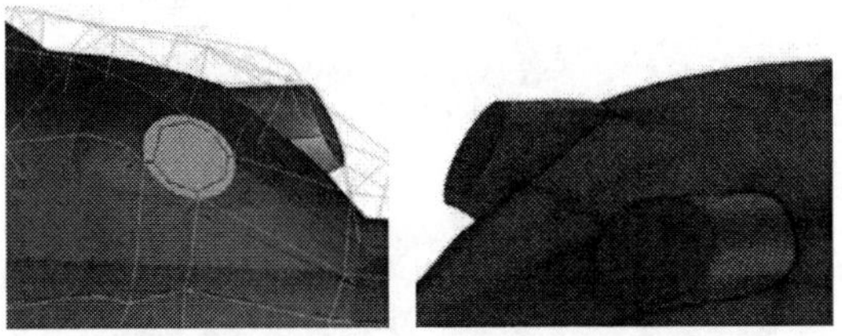

Fig. 2. Topology of blocks, and surface grid at the area of the engine exhausts.

Results of numerical simulation for the value of pitch angle of 0 degrees, including visualization of the Sound Pressure Level, the vector field and the frequency peak in the acoustic power spectrum are resented at sections of flow, according to the slices map in figure 3.

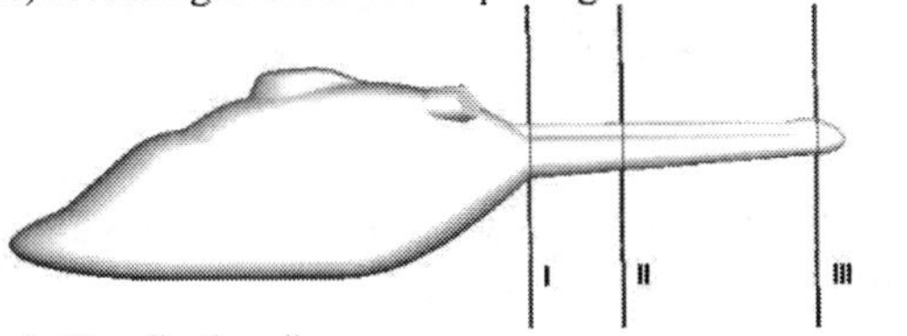

Fig. 3. Visualisation slices map.

The velocity field distribution scale is estimated according to the expression

$$\bar{V} = \frac{V}{V_\infty},$$

were V is a local velocity value, V_∞ is the free stream velocity. The Sound Pressure Level (SPL, dB) is evaluated according the formula

$$\text{SPL}=20log_{10}(p_s/p_{ref}).$$

Here p_s is the sound pressure, according to equations (21), and $p_{ref}=2\times10^{-5}$ Pa.

2D distribution of velocity field distributions and flow streamlines (figure 4) are given in the sections of the slices map. The vortex structures visualized at the cross flow sections are generated by the exhausts, the engine faring and the rear part of fuselage.

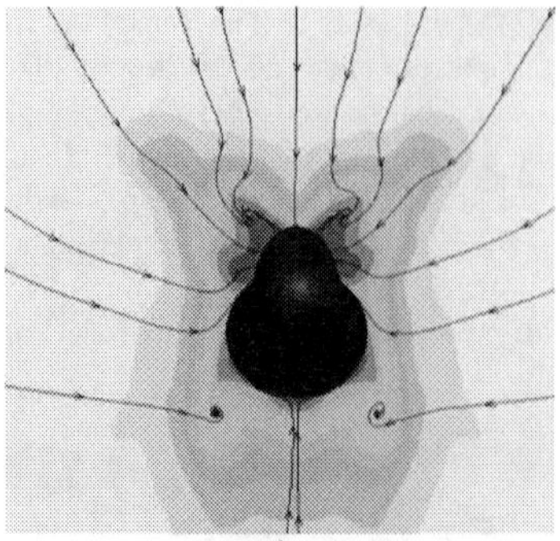

a)

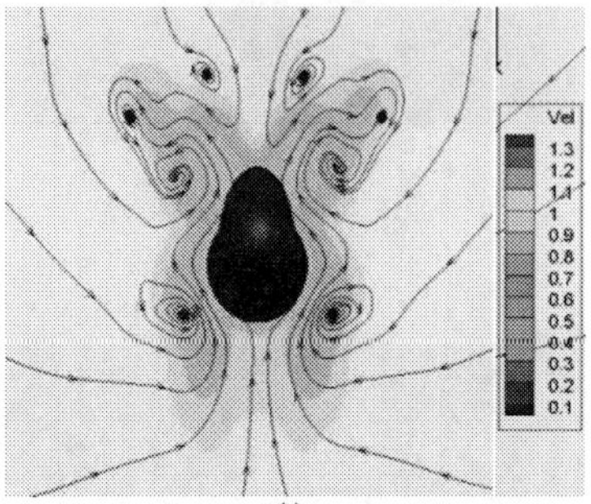

b)

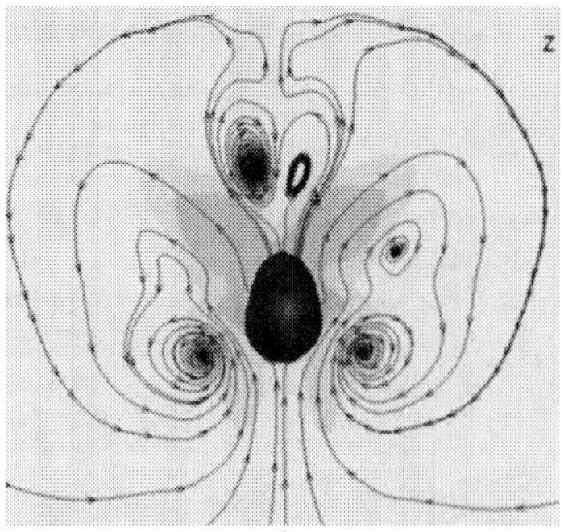

c)

Fig. 4. Velocity field distributions and flow streamlines for the sections a) I, b) II, c) III.

Localisation of the vortex structures in general corresponds to perturbations of the velocity field in figure 4. It can be expected that perturbations of the velocity field area correspond to the distribution of high SPL intensity, presented in figure 5.

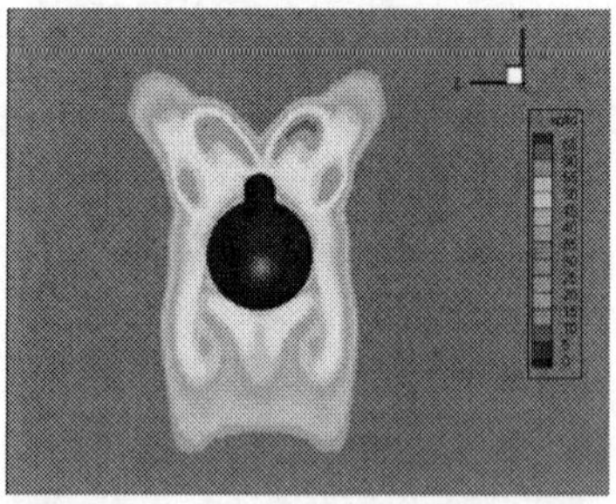

a)

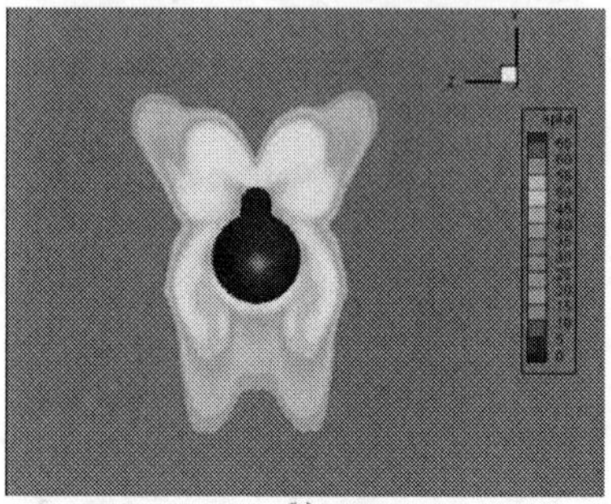

b)

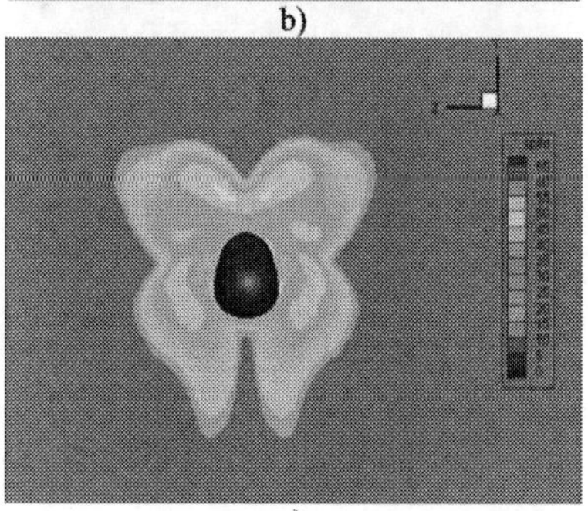

c)

Fig. 5. SPL distributions for the sections a) I, b) II, c) III.

Figure 6 presents the distribution of frequency peak in the acoustic power spectrum distribution (see equation (16)) for the same sections. Lowering of the vector field disturbances (figure 4) leads to lowering of the SPL intensity (figure 5) and to decreasing of the frequencies spectrum area (figure 6).

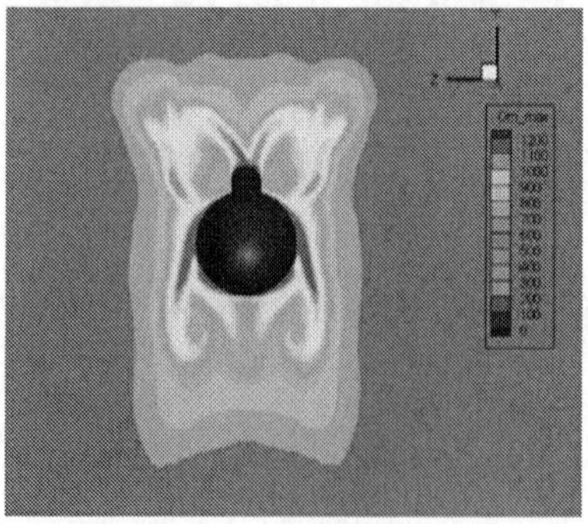

a)

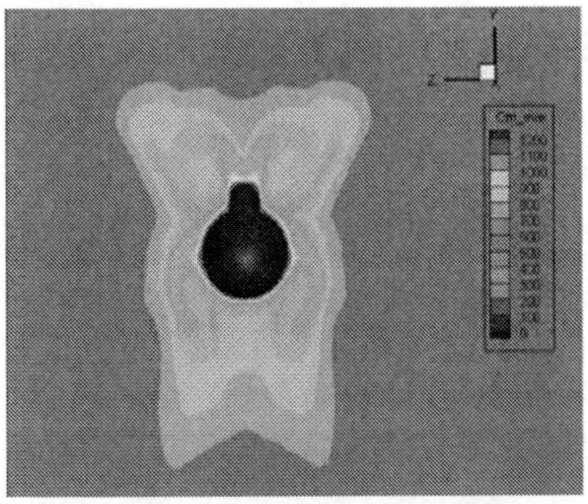

b)

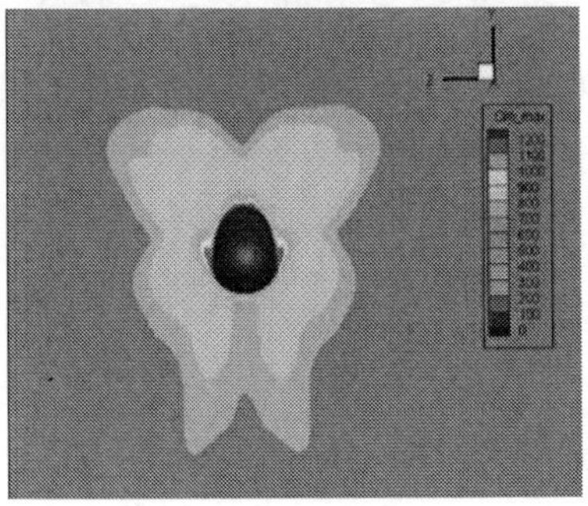

c)

Fig. 6. The frequency peak in the acoustic power spectrum distribution for the sections a) I, b) II, c) III.

7 Conclusions and future work

The acoustic properties (broadband noise) of the flow, and the frequency peak in the acoustic power spectrum distribution were estimated. Using Lighthill's acoustic analogy and the RANS equations, the broadband noise Sound Pressure Level was estimated based on the Proudman's formula. Proudman's approach was combined with Lilley's estimations to determine the frequency of the acoustic power spectrum distribution peak distribution. The structure and the acoustic properties of the flow around an isolated fuselage helicopter were examined in the in terms of mean flow field, turbulent kinetic energy and the dissipation rate.

In the future, the problem of the helicopter fuselage acoustic properties will be considered using CFD modelling based on the URANS equations and advances turbulence models.

This work is supported by the "Project part of state task in the field of the scientific activity" Grant (No 9.1694.2014/K).

References

1. Ch. Edwards Band Cox, NASA/CR-2002-211650, May (2002).

2. I. Proudman, *Proc. Roy. Soc.*, **214**, 119-132, August (1952).

3. S. Sarkar, M. Hussaini, ICASE Rep, **93-74** (1993).

4. G. Lilley, NASA Contract Report, **93-75**, NASA Langley Research Center, Hampton, VA (1993).

5. M. J. Lighthill, *Proc. Roy. Soc. London Ser*, **211**, 564–587. MR 13,879i. Zbl 049. 259 (1952).

6. G. Lilley, Center for Turbulence Research, Annual Research Briefs, 241 – 255 (1995).

7. H. K. Versteeg and W. Malalasekera, Pearson Education Limited, 503, (2007).

8. ANSYS FLUENT 12.0/12.1 Documentation.

9. D. C. Wilcox, *AIAA J.*, **26**, 11, 1299–1310 (1988).

10. G. Barakos, R. Steijl, K. Badcock and A. Brocklehurst, 31st European Rotorcraft Forum, 13-15 September 2005, Florence, Italy (2005).

11. A. Kusyumov, S. Mikhailov, E. Romanova, A. Garipov, E. Nikolaev, G. Barakos, EFM 2012 conference, Hradec Kralove, Czech. Republic (2012).

12. A. Batrakov, A. Kusyumov, S. Mikhailov, V. Pakhov, A. Sungatullin, V. Zherekhov, G. Barakos, 39 European Rotorcraft Forum, Moscow, Russia (2013).

13. A. Kusyumov, S. Mikhailov, A. Garipov, E. Nikolaev, G. Barakos, *Transactions on Control and Mechanical Systems*, **1**, 7, 318 – 324 (2012).

14. A. Batrakov, A. Kusyumov, S. Mikhailov, V. Pakhov, A. Sungatullin, V. Zherekhov, G. Barakos, ASME 2013 International Mechanical Engineering Congress, San Diego, California, USA (2013).

EPJ Web of Conferences **114**, 02063 (2016)
DOI: 10.1051/epjconf/201611402063

Optimization of ejector design and operation

Konstantin Kuzmenko[1,a], Nina Yurchenko[1], Pavlo Vynogradskyy[1], Yuriy Paramonov[1]

[1] *Laboratory for Advanced Aerodynamics, Institute of Hydromechanics, Nat. Academy of Sci.,8/4 Zheliabov St., Kiev 03680, Ukraine*

Abstract. The investigation aims at optimization of gas ejector operation. The goal consists in the improvement of the inflator design so that to enable 50 liters of gas inflation within ~30 milliseconds. For that, an experimental facility was developed and fabricated together with the measurement system to study pressure patterns in the inflator path.

1 Background

Supersonic gas ejectors are known to be applied in various fields of industry. Their operation is based on energy transfer from pressurized gas to a high-speed flow at lower pressure [1, 2]. The process entrains not only the gas from the high-pressure source but also the surrounding gas at normal pressure, thereby increasing a total amount (mass/volume) of gas which is to be pumped into a certain container. An aspiration coefficient is the main criterion of the system efficiency; it is defined by a ratio of the total amount of pumped gas to the amount of pressurized gas.

2 Experimental facility

An experiment facility was designed and fabricated for testing developed ejector system in a process of their characteristics improvement (Fig. 1). It consists of the following basic parts: high-pressure tank (250 bar), receiver for averaged and high pressure (150 bar), pressure gauges, two high-speed valves, data acquisition and processing system, small-size ejector for flowrate testing, high- and low-pressure compressors, etc. To model the ejection process needed air consumption with cold air two high-speed valves were used. The air is pumped into the 40 L high-pressure tank by high-pressure PE-100W compressor up to 100 bar. From this tank, air goes to the 40 L receiver through a reduction valve. The reduction valve is adjusted to the range of pressure values required in experiments. Pressure in the receiver is measured using two pressure gauges: the first one up to 40 bars, the second one up to 100 bars. Both gauges indications do not differ more than 0.5 bar during the experiment. The drainage crane is provided for air exhaust in case of arising overpressure in the receiver. From the receiver, air can flow in one of two pipes leading to a particular high-speed valve. The first KB-20 high-speed valve has an opening time of 2 ms in a pressure range of 1-70 bars. The second ARMA high-speed valve was found to have an opening time of 4-6 ms in the pressure range of 1-36 bars. Specified high-pressure sensors for 1-70 bars are mounted after each valve to measure pressure losses.

2.1 High-speed valves

To satisfy the task requirements of 50 L inflation within 30 ms, high-speed valve had to be used as a heart of the inflation system. Two high-speed valves were used for ejection process investigation with compressed air. First one KB-20 **(ISTA Inc., St. Petersburg, Russia, http://www.ista-pneumatics.ru/ru/kv.shtml)**.

Figure 1. Experimental facility

[a] Corresponding author: k_kuzmenko@ex.ua

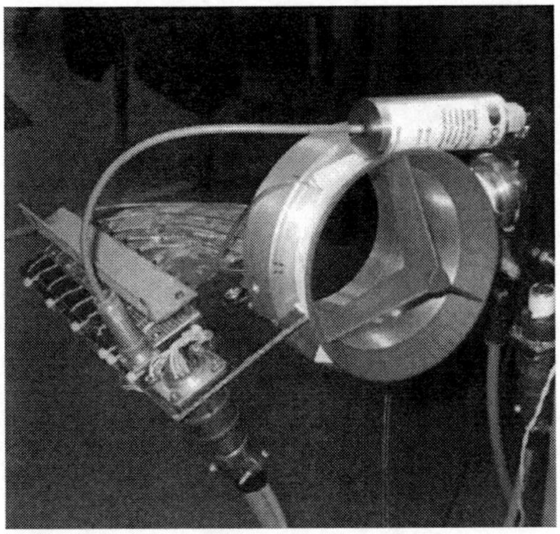

Figure 2. Inlet of the ejector.

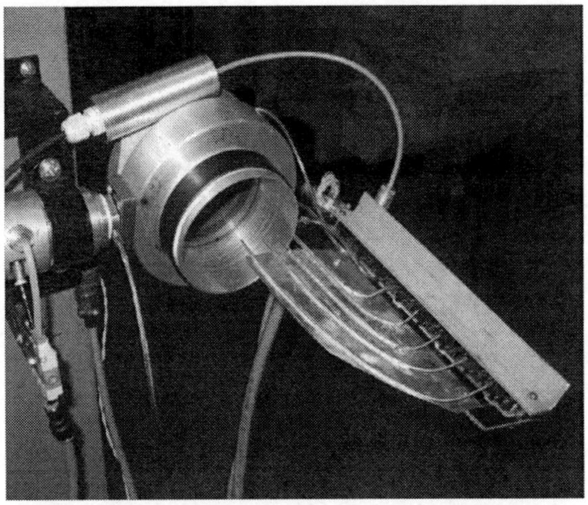

Figure 3. Pressure measuring comb.

The second valve is designed at the specialized ARMA enterprise (**Central design bureau of valves, Kiev, Ukraine**). Unlike the first valve, The ARMA valve makes its further development possible with parameters adjusted to the future tasks. At present, the key parameters are as follows.

Valve opening time is 4-6 ms;

system response time (between an initiating electric signal and the valve opening) is 0.01÷0.012s;

throat diameter is 12 mm;

operating pressure range is 0.1÷3.5 MPa.

This valve has an electrical control unit which provides both manual and remote control of its operation within 20 to 50 ms.

A basic difference of the two valves design is their throat diameter of 20 mm (KB-20) and 12 mm (ARMA valve). It has an essential impact on a flow rate in the inflator system that is important to seek for ways of the inflation system improvement. That is why both kinds of the high-speed valves were integrated into the facility and thoroughly studied. Valve opening time can be varied in a wide range of 20-50 ms. Present experiments were carried out at 30 ms. It was chosen from a viewpoint of the formulated problem, of technical characteristics of available high-speed valves as well as of prospective investigations to follow.

2.2 Measurement system

The available PCI-1747U 64-channel analog input board with up to 250 kS/s sampling rate provides sufficient measuring capabilities along with Advantech AIMB-782QG2-00A1E motherboard and Intel Core i7-4790K processor. Such configuration of the data acquisition and processing system enables measurements of up to 25 parameters with time resolution of 0.1 ms. It is considered quite sufficient providing up to 1000 measurements of each parameter during the ejection process cycle.

For the flowrate measurements, 6 sensors are mounted at the ejector outlet. The designed pressure sensor comb combines 6 sensor tubes. Three of them are located near a wall; one is situated on an ejector outlet axis. OPTrand sensor (pressure range 0÷70 bar) is placed between the valve outlet and the ejector inlet. Three sensors are flush-mounted in the ejector internal tract for static pressure measurement. For velocity field measurements, five more pressure probes (Fig. 3, 5) are mounted at the side to avoid their influence on exhaust flow from the ejector outlet. Full pressure probes are mounted on a holder under various steps. Analog and digital signals from sensors go to the centralized Data Acquisition and Processing Complex.

Fig. 4 shows typical data obtained from a high-pressure

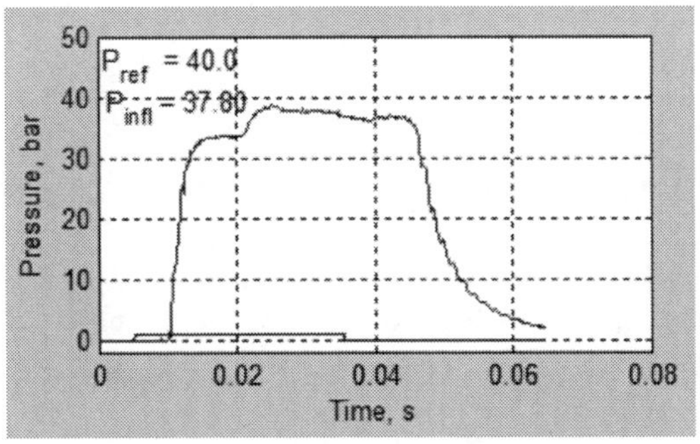

Figure 4. Pressure after valve (valve opening moment)

sensor. It demonstrates the high-speed valve operation and time when the pressure reaches its maximum.

Fig. 2 shows the ejector experimental model. Its inlet diameter is 13 mm, outlet diameter is 84 mm. Slit width can be changed within 0.095 – 0.25 mm. According to technical requirements, the overall ejector size (diameter) must be less than 150 mm, and the length should not exceed 120 mm.

Figure 5. Pressure comb. Position #1-3 is near wall pressure probes, probe #6 is on the axis of ejector outlet

3 Results

Tests were performed at different pressure ranges and cycle durations. A basic phenomenon which was studied under testing was jet separation. The separation restricts the arbitrary pressure growth aimed to increase higher flow rates of ejected air. Thus the work is focused on jet separation delay with increased pressure, because if it is so there is no ambient air sucked in ejector.

3.1 First experimental series

Experimental investigation started with a slit width of 0.25 mm decreasing to 0.12 mm in this test series. At this configuration, separation takes place at low pressure values (14.5 and 17.8 bar respectively). After that, the slit width was decreased to 0.08 mm. The latter resulted in jet separation started at 25 bar and ended up at 29 bar. It was called as 'Full separation' characterizing the situation of complete jet separation from the wall. The 'Full separation' velocity is registered with a sensor located at the ejector outlet axis. To ensure accuracy and repeatability of measurement results, the pressure sensor #3 was connected to the third near-wall tube. Fig. 6 shows the experimental ejector arrangement.

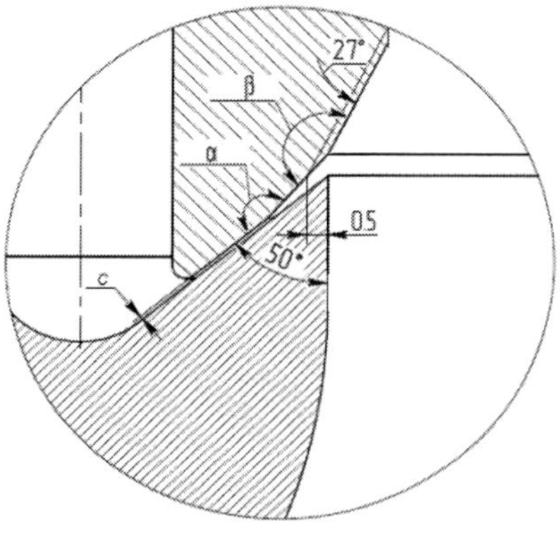

Figure 6. Initial slit dimensions

3.2 Jet separation determination

Jet separation was detected by pressure drop at near wall sensors and simultaneously pressure increasing at sensors located at the ejectors outlet axis. For example separation occurrence at small pressure increasing (from 40 to 40.5 bars) is shown at Fig. 8. and 9. Simultaneously with this process sensors located near at ejector axis detect air jet. This is shown at Fig. 10 and Fig. 11. Static pressure measurement sensor located at ejector inlet also detect jet separation this results are shown in Fig. 12-14.

3.3 Second experimental series

The second cycle of measurements was carried out with the changed slit configuration. Two slit angles were changed to values of α = 20°, β = 23°. The sharp edge of the slit projecting part has been reduced to 0.5 mm, so sharp edge get lower than bending point of β angle. Experiment test was performed for three slit width 0.1mm, 0.14 and 0.25mm. At slit width 0.1mm and new slit shape contour the jet separation appears at pressure values more than 30 bars (at pressure equal to 30.36 bar

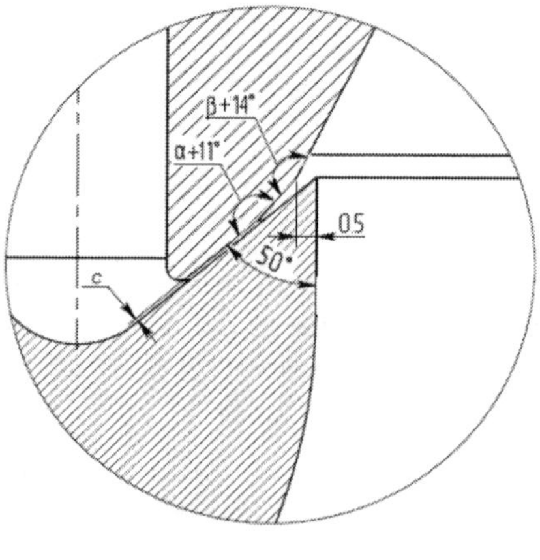

Figure 7. Second version of ejector slit

near-wall jet is already attached to wall) also vacuum pressure at the ejector inlet (sensor #1) reaches 0.08 bar. But small pressure increasing to 30.57 bar leads to separation phenomena and full separation happens at 31-32 bar. Slit dimension after first changing is shown in Fig. 7.

3.4 Third experimental series

Second ejector version was made based on previous results. Main changes were: increased angles: α to 20°, β to 5°. Experiment series were made at slit width 0.1 mm and 0.21mm. Under this geometry configuration (at slit width 0.1 mm) the separation occurs in range from 40 to

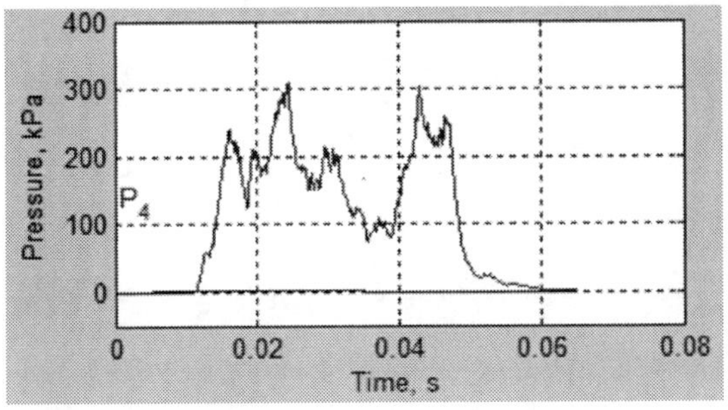

Figure 8. Near-wall sensor data at pressure 40 bar

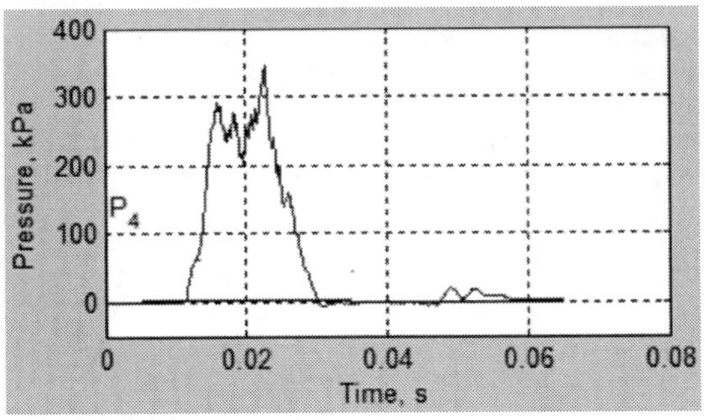

Figure 9. Near-wall sensor data at pressure 40.5 bar

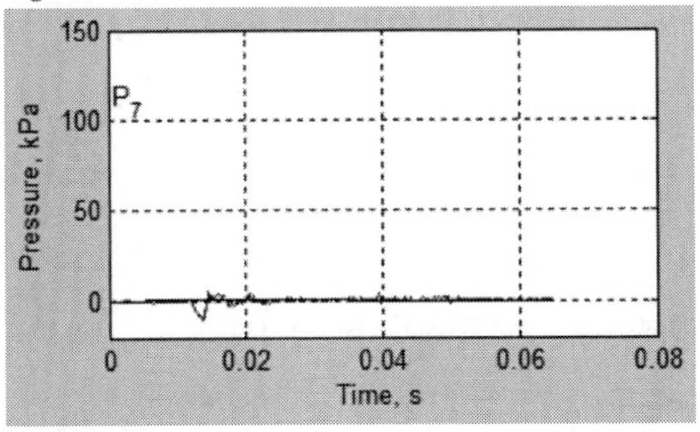

Figure 10. Axis sensor data at pressure 40 bar

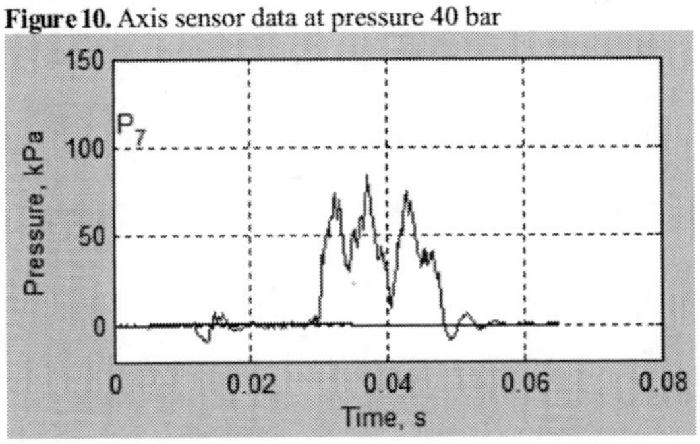

Figure 11. Axis sensor data at pressure 40.5 bar

40.5 bars. Experiment was made also with slit width 0.21mm and showed high vacuum pressure value at the ejector inlet. Sucked jet velocity in ejector reached 160 m/s at the vacuum pressure about 0.17 bar.

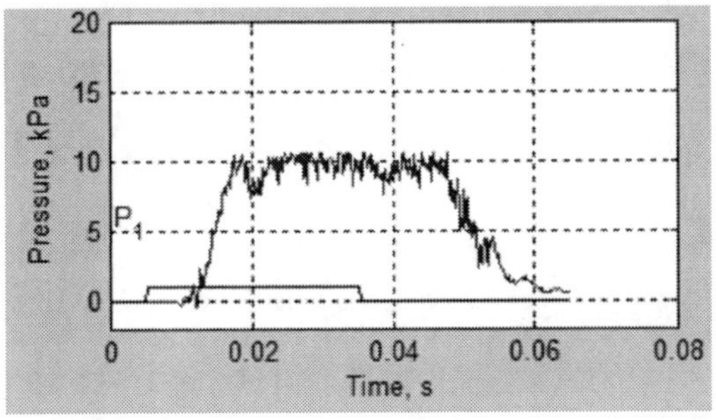

Figure 12. Pressure at the ejector inlet, P = 30 bar

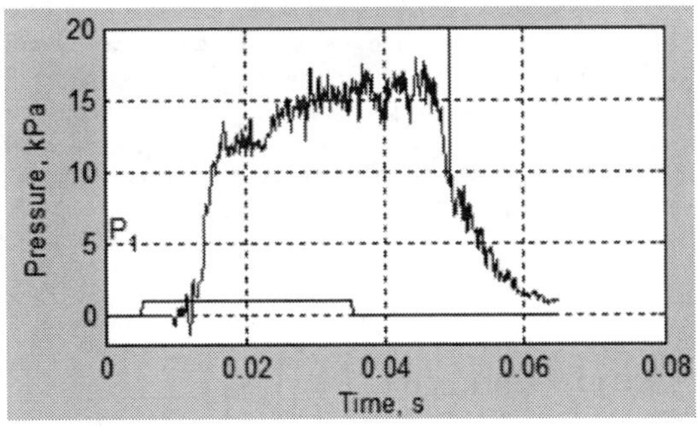

Figure 13. Pressure at the ejector inlet, P = 39 bar

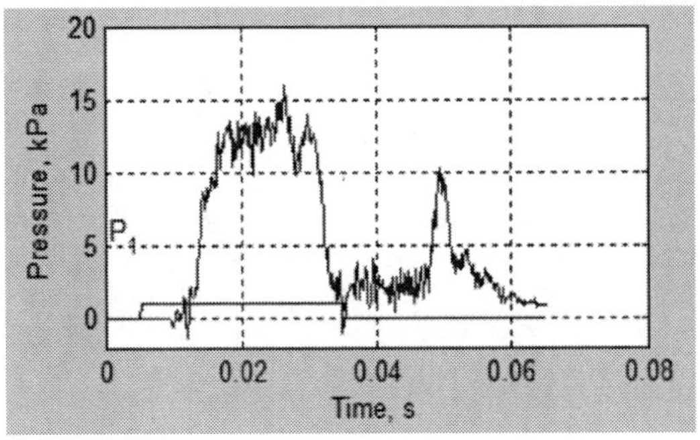

Figure 14. Pressure at the ejector inlet, P = 40.5 bar

Conclusions. To optimize gas ejector operation, the experimental facility was designed together with the remote data acquisition and processing system. The ejector test model was designed under guidance of numerical results modeling the ejection process. Three experimental test series resulted in the successive ejector geometry optimization. As a result, jet separation occurred at ~3 times higher pressure (40.5 bar) compared to the initial case (17 bar). The work is in progress.

References

1. E.Y.Sokolov, H.M.Zinger, *Jet devices* (Government energetic publishing, Moscow, 1960) In Russian
2. Y.K.Arkadov, *New gas ejector* (Phys.-Math lit., Moscow, 2001) In Russian.

EPJ Web of Conferences **114**, 02064 (2016)
DOI: 10.1051/epjconf/201611402064

Numerical simulation of the passive gas mixture flow

Martin Kyncl[1,a] and Jaroslav Pelant[1,b]

[1] Aerospace Research and Test Establishment (VZLÚ),
Beranových 130, 199 05 Praha - Letňany, Czech Republic

Abstract. The aim of this paper is the numerical solution of the equations describing the non-stationary compressible turbulent multicomponent flow in gravitational field. The mixture of perfect inert gases is assumed. We work with the RANS equations equipped with the k-omega and the EARSM turbulence models. For the simulation of the wall roughness we use the modification of the specific turbulent dissipation. The finite volume method is used, with thermodynamic constants being functions in time and space. In order to compute the fluxes through the boundary faces we use the modification of the Riemann solver, which is the original result. We present the computational results, computed with the own-developed code (C, FORTRAN, multiprocessor, unstructured meshes in general).

1 Introduction

The aim of this work is to simulate the complicated behaviour of the perfect gas mixture. In this contribution we consider the Reynolds- Averaged Navier-Stokes equations with the k-ω model of turbulence. This system is equipped with the equation of state in more general form, and with the mass conservation of the additional gas specie. We choose the well-known finite volume method to discretize the analytical problem, represented by the system of the equations in generalized (integral) form. In order to apply this method we split the area of the interest into the elements, and we construct a piecewise constant solution in time. The crucial problem of this method lies in the evaluation of the so-called fluxes through the edges/faces of the particular elements. We use the analysis of the exact solution of the Riemann problem for the discretization of the fluxes through the boundary edges/faces. Own algorithms for the solution of the boundary problem were coded, and used in the numerical examples.

2 Formulation of the Equations

We consider the conservation laws for viscous compressible turbulent flow of ideal gas with the zero heat sources in a domain $\Omega \in \boldsymbol{R}^N$, and time interval $(0, T)$, with $T > 0$. The system of the Reynolds-Averaged Navier-Stokes equations in 3D has the form

$$\frac{\partial \boldsymbol{w}}{\partial t}+\sum_{s=1}^{3}\frac{\partial \boldsymbol{f}_s(\boldsymbol{w})}{\partial x_s}=\sum_{s=1}^{3}\frac{\partial \boldsymbol{R}_s(\boldsymbol{w},\nabla \boldsymbol{w})}{\partial x_s}+\boldsymbol{S} \quad \text{in } Q_T=\Omega\times(0,T). \tag{1}$$

Here x_1, x_2, x_3 are the space coordinates, t the time, $\boldsymbol{w} = \boldsymbol{w}(x,t) = (\varrho, \varrho v_1, \varrho v_2, \varrho v_3, E)^T$ is the state vector, $\boldsymbol{f}_s = (\varrho v_s, \varrho v_s v_1 + \delta_{s1} p, \varrho v_s v_2 + \delta_{s2} p, \varrho v_s v_3 + \delta_{s3} p, (E + p) v_s)^T$ are the inviscid fluxes, $\boldsymbol{R}_s = (0, \tau_{s1}, \tau_{s2}, \tau_{s3}, \sum_{r=1}^{3} \tau_{sr} v_r + C_k \partial\theta/\partial x_s)^T$ are the viscous fluxes, $\boldsymbol{S}$ are additional sources. $\boldsymbol{v} = (v_1, v_2, v_3)^T$ denotes the velocity vector, ϱ is the density, p the pressure, θ the absolute temperature, $E = \varrho e + \frac{1}{2}\varrho \boldsymbol{v}^2 + \varrho k$ the total energy. Further $\tau_{ij} = \begin{cases} (\mu+\mu_T)S_{ij}, & i \neq j \\ (\mu+\mu_T)S_{ij} - \frac{2}{3}\varrho k, & i = j \end{cases}$, with $S_{11} = \frac{2}{3}\left(2\frac{\partial v_1}{\partial x_1} - \frac{\partial v_2}{\partial x_2} - \frac{\partial v_3}{\partial x_3}\right), S_{12} = \frac{\partial v_1}{\partial x_2} + \frac{\partial v_2}{\partial x_1}, S_{13} = \frac{\partial v_1}{\partial x_3} + \frac{\partial v_3}{\partial x_1}, S_{21} = S_{12}, S_{22} = \frac{2}{3}\left(-\frac{\partial v_1}{\partial x_1} + 2\frac{\partial v_2}{\partial x_2} - \frac{\partial v_3}{\partial x_3}\right), S_{23} = \frac{\partial v_2}{\partial x_3} + \frac{\partial v_3}{\partial x_2}$, $S_{31} = S_{13}, S_{32} = S_{23}, S_{33} = \frac{2}{3}\left(-\frac{\partial v_1}{\partial x_1} - \frac{\partial v_2}{\partial x_2} + 2\frac{\partial v_3}{\partial x_3}\right)$, where μ is the dynamic viscosity coefficient dependent on temperature, μ_T is the eddy-viscosity coefficient. For the specific internal energy $e = c_v\theta$ we assume the caloric equation of state $e = p/\varrho(\gamma - 1)$, c_v is the specific heat at constant volume, $\gamma > 1$ is called the *Poisson adiabatic constant*. The constant C_k denotes the heat conduction coefficient $C_k = (\frac{\mu}{P_r} + \frac{\mu_T}{P_{r_T}})c_v\gamma$, and P_r is laminar and P_{r_T} is turbulent Prandtl constant number. In our application of flow in the gravitational field we set the source terms to $\boldsymbol{S} = (0, \varrho g_1, \varrho g_2, \varrho g_3, \varrho \boldsymbol{g} \cdot \boldsymbol{v})$, where $\boldsymbol{g} = (g_1, g_2, g_3)$ is the gravity vector. For the gas mixture with two species we use the Dalton's law for the total mixture pressure

$$p = p_1 + p_2,$$

where p_1 and p_2 are the partial pressures of the first and second component gas. Let ϱ_1 and ϱ_2 be the mass density of these components. Then the total mass density of mixture is

$$\varrho = \varrho_1 + \varrho_2.$$

Temperature θ is same for all gases in the mixture, and the equation of state holds

$$p_i = \varrho_i R_i \theta, \quad R_i = \frac{R_g}{m_i},$$

where $R_g = 8.3144621$ is universal gas constant, and m_i denotes the mollar mass of the ith specie. We can introduce

[a] e-mail: kyncl@vzlu.cz
[b] e-mail: pelant@vzlu.cz

the species mass fractions Y_1, Y_2, with $Y_1 = \frac{\varrho_1}{\varrho}, Y_2 = \frac{\varrho_2}{\varrho}$, it is $Y_1+Y_2 = 1$. The thermodynamic constants of the mixture satisfy (using the decomposition of the internal specific energy and enthalpy)

$$\varrho c_p = \sum \varrho_i c_{pi}, \quad \varrho c_V = \sum \varrho_i c_{Vi},$$

then the adiabatic constant γ, needed in the solution of (1), can be written as

$$\gamma = \frac{c_p}{c_V} = \frac{\sum Y_i c_{pi}}{\sum Y_i c_{Vi}}.$$

The system (1) is then extended with the conservation law of the mass for one gas component (specie)

$$\frac{\partial \varrho Y_1}{\partial t} + \frac{\partial \varrho Y_1 v_1}{\partial x_1} + \frac{\partial \varrho Y_1 v_2}{\partial x_2} + \frac{\partial \varrho Y_1 v_3}{\partial x_3} =$$
$$\frac{\partial}{\partial x_1}\left(\sigma_C \mu_T \frac{\partial Y_1}{\partial x_1}\right) + \frac{\partial}{\partial x_2}\left(\sigma_C \mu_T \frac{\partial Y_1}{\partial x_2}\right) + \frac{\partial}{\partial x_3}\left(\sigma_C \mu_T \frac{\partial Y_1}{\partial x_3}\right) \quad (2)$$

Here σ_C is diffusion coefficient. The mass conservation for the second specie is automatically satisfied via the system (1).

Here we assume the system (1),(2) equipped with the two-equation turbulent model $k-\omega$ (Kok), described in [1]. The effective turbulent viscosity is $\mu_T = \varrho k/\omega$.

$$\frac{\partial \varrho k}{\partial t} + \frac{\partial \varrho k v_1}{\partial x_1} + \frac{\partial \varrho k v_2}{\partial x_2} + \frac{\partial \varrho k v_3}{\partial x_3} = P_k - \beta^* \varrho \omega k +$$
$$+ \frac{\partial}{\partial x_1}\left((\mu + \sigma_k \mu_T)\frac{\partial k}{\partial x_1}\right) +$$
$$+ \frac{\partial}{\partial x_2}\left((\mu + \sigma_k \mu_T)\frac{\partial k}{\partial x_2}\right) + \frac{\partial}{\partial x_3}\left((\mu + \sigma_k \mu_T)\frac{\partial k}{\partial x_3}\right), \quad (3)$$

$$\frac{\partial \varrho \omega}{\partial t} + \frac{\partial \varrho \omega v_1}{\partial x_1} + \frac{\partial \varrho \omega v_2}{\partial x_2} + \frac{\partial \varrho \omega v_3}{\partial x_3} = P_\omega - \beta \varrho \omega \omega +$$
$$C_D + \frac{\partial}{\partial x_1}\left((\mu + \sigma_\omega \mu_T)\frac{\partial \omega}{\partial x_1}\right) +$$
$$+ \frac{\partial}{\partial x_2}\left((\mu + \sigma_\omega \mu_T)\frac{\partial \omega}{\partial x_2}\right) + \frac{\partial}{\partial x_3}\left((\mu + \sigma_\omega \mu_T)\frac{\partial \omega}{\partial x_3}\right), \quad (4)$$

where k the specific turbulent kinetic energy and ω the turbulent dissipation are functions of time t and space coordinates x_1, x_2, x_3. The production terms P_k and P_ω are given by formulas

$$P_k = \tau_{11}\frac{\partial v_1}{\partial x_1} + \tau_{12}\frac{\partial v_1}{\partial x_2} + \tau_{21}\frac{\partial v_2}{\partial x_1} + \tau_{22}\frac{\partial v_2}{\partial x_2} + \tau_{13}\frac{\partial v_1}{\partial x_3} +$$
$$+\tau_{31}\frac{\partial v_3}{\partial x_1} + \tau_{23}\frac{\partial v_2}{\partial x_3} + \tau_{32}\frac{\partial v_3}{\partial x_2} + \tau_{33}\frac{\partial v_3}{\partial x_3},$$
$$P_\omega = \frac{\alpha_\omega \omega P_k}{k},$$

where functions τ are defined in (1) with $\mu = 0$, $\alpha_\omega = \frac{\beta}{\beta^*} - \frac{\sigma_\omega \kappa^2}{\sqrt{\beta^*}}$ and $\sigma_k = \frac{2}{3}$, $\beta^* = 0.09$, $\beta = \frac{5}{6}\beta^*$, $\sigma_\omega = 0.5$, $\kappa = 0.41$. The cross-diffusion term C_D is defined as

$$C_D = \sigma_d \frac{\varrho}{\omega} \max\left\{\frac{\partial k}{\partial x_1}\frac{\partial \omega}{\partial x_1} + \frac{\partial k}{\partial x_2}\frac{\partial \omega}{\partial x_2} + \frac{\partial k}{\partial x_3}\frac{\partial \omega}{\partial x_3}, 0\right\},$$

where $\sigma_d = 0.5$ is constant.

3 Numerical Method

For the discretization of the system we proceed as described in [6]. We use either explicit or implicit finite volume method to solve the systems sequentionally. Here we present the discretization of the system (1). By Ω_h let us the denote the polyhedral approximation of Ω. The system of the closed polyhedrons with mutually disjoint interiors $\mathcal{D}_h = \{D_i\}_{i\in J}$, where $J \subset Z^+ = \{0, 1, \ldots\}$ is an index set and $h > 0$, will be called a *finite volume mesh*. This system $\mathcal{D}_h$ approximates the domain Ω, we write $\overline{\Omega}_h = \bigcup_{i\in J} D_i$. The elements $D_i \in \mathcal{D}_h$ are called the *finite volumes*. For two neighboring elements D_i, D_j we set $\Gamma_{ij} = \partial D_i \cap \partial D_j = \Gamma_{ji}$. Similarly, using the negative index j we may denote the boundary faces. Here we will work with the so-called *regular* meshes, i.e. the intersection of two arbitrary (different) elements is either empty or it consists of a common vertex or a common edge or a common face (in 3D). The boundary ∂D_i of each element D_i is

$$\partial D_i = \bigcup_{\Gamma_{ij}\in\Gamma_{D_i}} \Gamma_{ij}. \quad (5)$$

Here the set $\Gamma_{D_i} = \{\Gamma_{ij}; \Gamma_{ij} \subset \partial D_i\}$ forms the boundary ∂D_i. By $\boldsymbol{n}_{ij}$ let us denote the unit outer normal to ∂D_i on Γ_{ij}. Let us construct a partition $0 = t_0 < t_1 < \ldots$ of the time interval $[0, T]$ and denote the time steps $\tau_k = t_{k+1} - t_k$. We integrate the system (1) over the set $D_i \times (t_k, t_{k+1})$. With the integral form of the equations we can study a flow with discontinuities, such as shock waves, too.

$$\int_{D_i}\int_{t_k}^{t_{k+1}} \frac{\partial \boldsymbol{w}}{\partial t} dx\, dt + \int_{t_k}^{t_{k+1}}\int_{D_i}\sum_{s=1}^{3}\frac{\partial \boldsymbol{f}_s(\boldsymbol{w})}{\partial x_s} dx\, dt =$$
$$= \int_{t_k}^{t_{k+1}}\int_{D_i}\sum_{s=1}^{3}\frac{\partial \boldsymbol{R}_s(\boldsymbol{w}, \nabla\boldsymbol{w})}{\partial x_s} dx\, dt \quad (6)$$

Using the Green's theorem on D_i it is

$$\int_{D_i}\sum_{s=1}^{3}\frac{\partial \boldsymbol{f}_s(\boldsymbol{w})}{\partial x_s} dx = \int_{\partial D_i}\sum_{s=1}^{3}\boldsymbol{f}_s(\boldsymbol{w}) n_s\, dS, \quad (7)$$
$$\int_{D_i}\sum_{s=1}^{3}\frac{\partial \boldsymbol{R}_s(\boldsymbol{w}, \nabla\boldsymbol{w})}{\partial x_s} dx = \int_{\partial D_i}\sum_{s=1}^{3}\boldsymbol{R}_s(\boldsymbol{w}, \nabla\boldsymbol{w}) n_s\, dS.$$

Here $\boldsymbol{n} = (n_1, n_2, n_3)$ is the unit outer normal to ∂D_i. Further we use (5), and we rewrite (6)

$$\int_{D_i} (\boldsymbol{w}(x, t_{k+1}) - \boldsymbol{w}(x, t_k))\, dx + \quad (8)$$
$$\int_{t_k}^{t_{k+1}} \sum_{\Gamma_{ij}\in\Gamma_{D_i}} \int_{\Gamma_{ij}} \sum_{s=1}^{3} (\boldsymbol{f}_s(\boldsymbol{w}) - \boldsymbol{R}_s(\boldsymbol{w}, \nabla\boldsymbol{w}))\,(n_{ij})_s\, dS\, dt = 0$$

We define a finite volume approximate solution of the system studied (1) as a piecewise constant vector-valued functions $\mathbf{w}_h^k$, $k = 0, 1, \ldots$, where $\boldsymbol{w}_h^k$ is constant on each element D_i, and t_k is the time instant. By $\boldsymbol{w}_i^k$ we denote the value of the approximate solution on D_i at time t_k. We approximate the integral over the element D_i

$$\int_{D_i} \boldsymbol{w}(x, t_k)\, dx \approx |D_i| \boldsymbol{w}_i^k. \quad (9)$$

Further we proceed with the approximation of the fluxes. Usually the flux $\sum_{s=1}^{3} \boldsymbol{f}_s(\boldsymbol{w})(n_{ij})_s|_{\Gamma_{ij}}$ is being approximated by a *numerical flux* at suitable time instant t_l

$$\sum_{s=1}^{3} \boldsymbol{f}_s(\boldsymbol{w})(n_{ij})_s|_{\Gamma_{ij}} \approx \boldsymbol{H}(\boldsymbol{w}_i^l, \boldsymbol{w}_j^l, \boldsymbol{n}_{ij}),$$

with $\boldsymbol{w}_i^l$, $\boldsymbol{w}_j^l$ denoting the approximate solution on the elements adjacent to the edge Γ_{ij} at the time instant t_l. In the case of a boundary face the vector $\boldsymbol{w}_j^l$ has to be specified. Here we show the numerical flux based on the solution of the Riemann problem for the split Euler equations. By $\boldsymbol{w}_{\Gamma_{ij}}^l$ let us denote the state vector $\boldsymbol{w}$ at the center of the edge Γ_{ij} at the time instant t_l, and let us suppose $\boldsymbol{w}_{\Gamma_{ij}}^l$ is known. Evaluation of these values will be a question of the further analysis, here we use them to approximate the integrals with the one-point rule

$$\int_{\Gamma_{ij}} \sum_{s=1}^{3} \boldsymbol{f}_s(\boldsymbol{w}(x,t_l))(n_{ij})_s \, dS \approx |\Gamma_{ij}| \sum_{s=1}^{3} \boldsymbol{f}_s(\boldsymbol{w}_{\Gamma_{ij}}^l)(n_{ij})_s. \tag{10}$$

$$\int_{\Gamma_{ij}} \sum_{s=1}^{3} \boldsymbol{R}_s(\boldsymbol{w}(x,t_l), \nabla \boldsymbol{w}(x,t_l))(n_{ij})_s \, dS \approx |\Gamma_{ij}| \sum_{s=1}^{3} \boldsymbol{R}_s(\boldsymbol{w}_{\Gamma_{ij}}^l, \nabla \boldsymbol{w}_{\Gamma_{ij}}^l)(n_{ij})_s.$$

Here $\nabla \boldsymbol{w}_{\Gamma_{ij}}^l$ denotes the $\nabla \boldsymbol{w}$ at the center of the edge Γ_{ij} at time instant t_l. Now it is possible to approximate the system (8) by the following explicit finite volume scheme

$$\frac{|D_i|}{\tau_k}(\boldsymbol{w}_i^{k+1} - \boldsymbol{w}_i^k) + \tag{11}$$

$$\sum_{\Gamma_{ij} \in \Gamma_{D_i}} |\Gamma_{ij}| \sum_{s=1}^{3} \left(\boldsymbol{f}_s(\boldsymbol{w}_{\Gamma_{ij}}^k) n_s - \boldsymbol{R}_s(\boldsymbol{w}_{\Gamma_{ij}}^k, \nabla \boldsymbol{w}_{\Gamma_{ij}}^k) n_s \right) = 0$$

With this finite volume formula one computes the values of the approximate solution at the time instant t_{k+1}, using the values from the time instant t_k, and by evaluating the values $\boldsymbol{w}_{\Gamma_{ij}}^k$ at the faces Γ_{ij}. In order to achieve the stability of the used method, the time step τ_k must be restricted by the so-called CFL condition, see [4]. The crucial problem of this discretization lies with the evaluation of the edge values $\boldsymbol{w}_{\Gamma_{ij}}^k$. Or one deals with the problem of finding the face fluxes $\boldsymbol{H}(\boldsymbol{w}_i^k, \boldsymbol{w}_j^k, \boldsymbol{n}_{ij})$. It is also possible to use the **implicit scheme**

$$\frac{|D_i|}{\tau_k}(\boldsymbol{w}_i^{k+1} - \boldsymbol{w}_i^k) + \tag{12}$$

$$\sum_{\Gamma_{ij} \in \Gamma_{D_i}} |\Gamma_{ij}| \left(\boldsymbol{H}(\boldsymbol{w}_i^{k+1}, \boldsymbol{w}_j^{k+1}, \boldsymbol{n}_{ij}) - \sum_{s=1}^{3} \boldsymbol{R}_s(\boldsymbol{w}_{\Gamma_{ij}}^{k+1}, \nabla \boldsymbol{w}_{\Gamma_{ij}}^{k+1}) n_s \right) = 0$$

The crucial problem of this discretization lies with the evaluation of the face fluxes $\boldsymbol{H}(\boldsymbol{w}_i^{k+1}, \boldsymbol{w}_j^{k+1}, \boldsymbol{n}_{ij})$. One possibility is to use the linearization via the Taylor expansion of the vector function $\boldsymbol{H}(\boldsymbol{w}\boldsymbol{I}, \boldsymbol{w}\boldsymbol{J}, \boldsymbol{n})$, this was shown in [6]. One possibility of the face flux evaluation is to approximate the face values $\boldsymbol{w}_{\Gamma_{ij}}^k$ and then compute the numerical flux

$$\boldsymbol{H}|_{ij}^k = \sum_{s=1}^{3} \boldsymbol{f}_s(\boldsymbol{w}_{\Gamma_{ij}}^k)(n_{ij})_s. \tag{13}$$

To approximate the face values $\boldsymbol{w}_{\Gamma_{ij}}^k$ at time instant t_k we solve the simplified system (14) in the vicinity of the face Γ_{ij} in time with the initial condition formed by the state vectors $\boldsymbol{w}_i^k$ and $\boldsymbol{w}_j^k$. Using the rotational invariance of the Euler equations, the system is expressed in a new Cartesian coordinate system $\tilde{x}_1, \tilde{x}_2, \tilde{x}_3$ with the origin at the center of the gravity of Γ_{ij} and with the new axis $\tilde{x}_1$ in the direction of $\boldsymbol{n} = (n_1, n_2, n_3)$, given by the face normal $\boldsymbol{n} = \boldsymbol{n}_{ij}$.

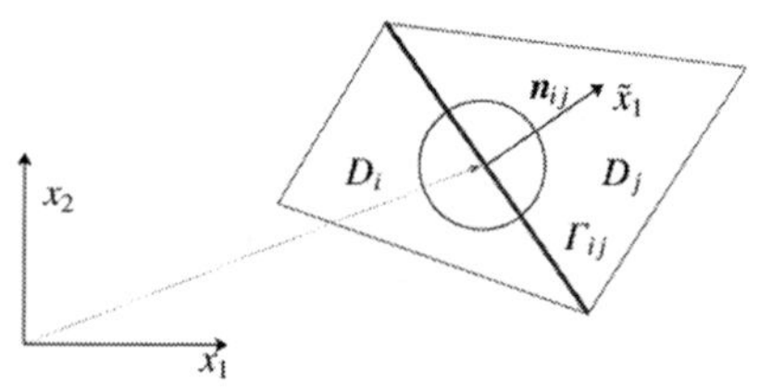

Fig. 1. Coordinate transformation for the inner edges in 2D.

Then the derivatives with respect to $\tilde{x}_2, \tilde{x}_3$ are neglected and we get the so-called split 3D Euler equations, see [4, page 138]:

$$\frac{\partial \boldsymbol{q}}{\partial t} + \frac{\partial \boldsymbol{f}_1(\boldsymbol{q})}{\partial \tilde{x}_1} = 0. \tag{14}$$

The values $\boldsymbol{w}_i^k$ and $\boldsymbol{w}_j^k$ adjacent to the face Γ_{ij} are known, forming the initial conditions

$$\boldsymbol{q}(\tilde{x}_1, 0) = \boldsymbol{q}_L = \mathbb{Q}\, \boldsymbol{w}_i^k, \qquad \tilde{x}_1 < 0, \tag{15}$$

$$\boldsymbol{q}(\tilde{x}_1, 0) = \boldsymbol{q}_R = \mathbb{Q}\, \boldsymbol{w}_j^k, \qquad \tilde{x}_1 > 0. \tag{16}$$

The transformation matrix $\mathbb{Q}$ is defined in [6]. The problem (14), (15), (16) has a unique "solution" in $(-\infty, \infty) \times (0, \infty)$, the analysis can be found in [4, page 138]. Let $\boldsymbol{q}_{RS}(\boldsymbol{q}_L, \boldsymbol{q}_R, \tilde{x}_1, t)$ denote the solution of this problem at the point $(\tilde{x}_1, t)$. We are interested in the solution of this local problem at the time axis, which is the sought solution in the local coordinates $\boldsymbol{q}_{\Gamma_{ij}} = \boldsymbol{q}_{RS}(\boldsymbol{q}_L, \boldsymbol{q}_R, 0, t)$. The backward transformation of the state vector $\boldsymbol{q}_{\Gamma_{ij}}$ into the global coordinates is

$$\boldsymbol{w}_{\Gamma_{ij}}^k = \mathbb{Q}^{-1} \boldsymbol{q}_{\Gamma_{ij}} = \mathbb{Q}^{-1}\, \boldsymbol{q}_{RS}(\mathbb{Q}\, \boldsymbol{w}_i^k, \mathbb{Q}\, \boldsymbol{w}_j^k, 0, t). \tag{17}$$

The constructed numerical flux can be written as

$$\boldsymbol{H}(\boldsymbol{u}, \boldsymbol{v}, \boldsymbol{n}) := \sum_{s=1}^{3} \boldsymbol{f}_s\left(\mathbb{Q}^{-1}\, \boldsymbol{q}_{RS}(\mathbb{Q}\, \boldsymbol{u}, \mathbb{Q}\, \boldsymbol{v}, 0, t)\right) n_s. \tag{18}$$

Let Γ_{ij} be the face of the element D_i laying at the boundary of the computational area. To approximate the face values $\boldsymbol{w}_{\Gamma_{ij}}^k$ at time instant t_k we solve the incomplete simplified system (19) in the vicinity of the face Γ_{ij} in time with the initial condition (20) determined by the state vector $\boldsymbol{w}_i^k$. By adding properly chosen equations into the system (19),(20) it is possible to reconstruct the boundary state $\boldsymbol{q}_B$ such that the system (19),(20) has a unique solution at the boundary, see [3]. We will refer to these added equations as to **complementary conditions**. Several choices of the complementary conditions were discussed in [3], [5].

4 The Riemann Problem for the Split Euler Equations

For many numerical methods dealing with the two or three dimensional equations, describing the compressible flow, it is useful to solve the Riemann problem for the *3D split* Euler equations. We search the solution of the system of partial differential equations in time t and space (x, y, z)

$$\begin{aligned} \frac{\partial \varrho}{\partial t} + \frac{\partial \varrho u}{\partial x} &= 0 \\ \frac{\partial \varrho u}{\partial t} + \frac{\partial (p+\varrho u^2)}{\partial x} &= 0 \\ \frac{\partial \varrho v}{\partial t} + \frac{\partial \varrho u v}{\partial x} &= 0 \\ \frac{\partial \varrho w}{\partial t} + \frac{\partial \varrho u w}{\partial x} &= 0 \\ \frac{\partial E}{\partial t} + \frac{\partial u(E+p)}{\partial x} &= 0, \end{aligned} \tag{19}$$

equipped with the initial conditions

$$\varrho(x,t) = \varrho_L, \boldsymbol{v}(x,t) = \boldsymbol{v}_L, p(x,t) = p_L, \quad x < 0, \tag{20}$$

$$\varrho(x,t) = \varrho_R, \boldsymbol{v}(x,t) = \boldsymbol{v}_R, p(x,t) = p_R, \quad x > 0. \tag{21}$$

Vector $\boldsymbol{v} = (u, v, w)$ denotes the velocity, ϱ density, p pressure, $E = \varrho e + \varrho|\boldsymbol{v}|^2$ is the total energy, with e denoting the specific internal energy. We assume the equation of state for the calorically ideal gas

$$e = \frac{p}{\varrho(\gamma - 1)}.$$

'Split' means here that we still have 5 equations in 3D, but the dependence on y, z (space coordinates) is neglected, and we deal with the system for one space variable x. The system (19) is considered in the set $Q_\infty = (-\infty, \infty) \times (0, +\infty)$.

The solution of this problem is fundamentally the same as the solution of the Riemann problem for the 1D Euler equations, see [4, page 138]. In fact, the solution for the pressure, the first component of the velocity, and the density is exactly the same as in one-dimensional case. It is a characteristic feature of the hyperbolic equations, that there is a possible raise of discontinuities in solutions, even in the case when the initial conditions are smooth, see [8, page 390]. Here the concept of the classical solution is too restrictive, and therefore we seek a weak solution of this problem. To distinguish physically admissible solutions from nonphysical ones, entropy condition must be introduced, see [8, page 396]. By the solution of the problem (19),(20),(21) we mean the *weak entropy solution* of this problem in Q_∞. The analysis to the solution of this problem can be found in many books, i.e. [4], [8], [9]. The general theorem on the solvability of the Riemann problem can be found in [4, page 88].

The solution is piecewise **smooth** and its general form can be seen in Fig. 2, where the system of half lines is drawn.

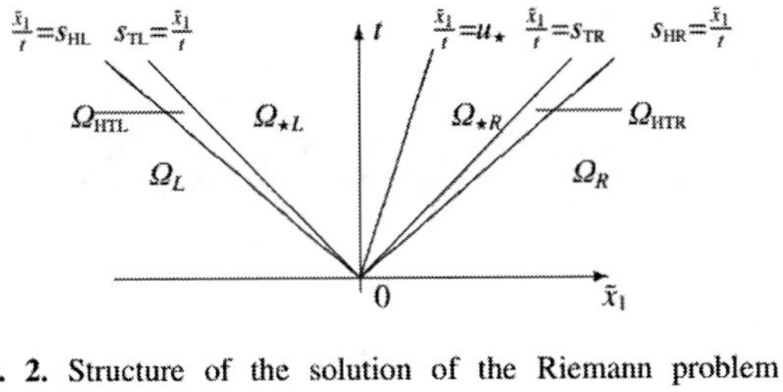

Fig. 2. Structure of the solution of the Riemann problem (19),(20),(21)

These half lines define regions, where solution is either constant or given by a smooth function. Let us define the open sets called **wedges** (see Fig. 2):

$$\begin{aligned} \Omega_L &= \text{wedge}(L_{-\infty}, s_{HL}), & \Omega_{\star R} &= \text{wedge}(u_\star, s_{TR}), \\ \Omega_{HTL} &= \text{wedge}(s_{HL}, s_{TL}), & \Omega_{HTR} &= \text{wedge}(s_{TR}, s_{HR}), \\ \Omega_{\star L} &= \text{wedge}(s_{TL}, u_*), & \Omega_R &= \text{wedge}(s_{HR}, L_{+\infty}). \end{aligned}$$

Using the theory in [4], [8], [9], we can write the solution for the primitive variables as

$$\begin{aligned} (\varrho, u, v, w, p)|_{\Omega_L} &= (\varrho_L, u_L, v_L, w_L, p_L), \\ (\varrho, u, v, w, p)|_{\Omega_{\star L}} &= (\varrho_{\star L}, u_\star, v_L, w_L, p_\star), \\ (\varrho, u, v, w, p)|_{\Omega_{\star R}} &= (\varrho_{\star R}, u_\star, v_R, w_R, p_\star), \\ (\varrho, u, v, w, p)|_{\Omega_R} &= (\varrho_R, u_R, v_R, w_R, p_R). \end{aligned}$$

The folowing relations for these variables hold:

$$u_\star = u_L + \begin{cases} -(p_\star - p_L)\left(\dfrac{\frac{2}{(\gamma+1)\varrho_L}}{p_\star + \frac{\gamma-1}{\gamma+1}p_L}\right)^{\frac{1}{2}}, & p_\star > p_L \\ \frac{2}{\gamma-1}a_L\left[1 - \left(\frac{p_\star}{p_L}\right)^{(\gamma-1)/2\gamma}\right], & p_\star \le p_L \end{cases} \tag{22}$$

$$\varrho_{\star L} = \begin{cases} \varrho_L \dfrac{\frac{\gamma-1}{\gamma+1}\frac{p_L}{p_\star} + 1}{\frac{p_L}{p_\star} + \frac{\gamma-1}{\gamma+1}}, & p_\star > p_L \\ \varrho_L\left(\frac{p_\star}{p_L}\right)^{\frac{1}{\gamma}}, & p_\star \le p_L \end{cases} \tag{23}$$

$$s^1_{TL} = \begin{cases} u_L - a_L\sqrt{\frac{\gamma+1}{2\gamma}\frac{p_\star}{p_L} + \frac{\gamma-1}{2\gamma}}, & p_\star > p_L \\ u_\star - a_L\left(\frac{p_\star}{p_L}\right)^{(\gamma-1)/2\gamma}, & p_\star \le p_L \end{cases} \tag{24}$$

$$u_\star = u_R + \begin{cases} (p_\star - p_R)\left(\dfrac{\frac{2}{(\gamma+1)\varrho_R}}{p_\star + \frac{\gamma-1}{\gamma+1}p_R}\right)^{\frac{1}{2}}, & p_\star > p_R, \\ -\frac{2}{\gamma-1}a_R\left[1 - \left(\frac{p_\star}{p_R}\right)^{(\gamma-1)/2\gamma}\right], & p_\star \le p_R \end{cases} \tag{25}$$

$$\varrho_{\star R} = \begin{cases} \varrho_R \dfrac{\frac{p_\star}{p_R} + \frac{\gamma-1}{\gamma+1}}{\frac{\gamma-1}{\gamma+1}\frac{p_\star}{p_R} + 1}, & p_\star > p_R \\ \varrho_R\left(\frac{p_\star}{p_R}\right)^{\frac{1}{\gamma}}, & p_\star \le p_R \end{cases} \tag{26}$$

$$s^3_{TR} = \begin{cases} u_R + a_R\sqrt{\frac{\gamma+1}{2\gamma}\frac{p_\star}{p_R} + \frac{\gamma-1}{2\gamma}}, & p_\star > p_R \\ u_\star + a_R\left(\frac{p_\star}{p_R}\right)^{(\gamma-1)/2\gamma}, & p_\star \le p_R \end{cases} \tag{27}$$

Here $a_L = \sqrt{\gamma p_L/\varrho_L}$, $a_R = \sqrt{\gamma p_R/\varrho_R}$, and γ denotes the adiabatic constant. Further s^1_{TL} denotes "unknown left wave speed", s^3_{TR} "unknown right wave speed". Note, that the system (22) - (27) is the system of 6 equations for 6 unknowns $p_\star$,$u_\star$,$\varrho_{\star L}$ $\varrho_{\star R}$,s^1_{TL},s^3_{TR}.

Solution for the Pressure $p_\star$

The combination of the equations (22), (25) gives the implicit equation for the unknown pressure $p_\star$

$$u_L + F_{1L}(p_\star) = u_R + F_{3R}(p_\star),$$

with

$$F_{1L}(p) = \begin{cases} -(p - p_L)\left(\dfrac{\frac{2}{(\gamma+1)\varrho_L}}{p + \frac{\gamma-1}{\gamma+1}p_L}\right)^{\frac{1}{2}}, & p > p_L \\ \frac{2}{\gamma-1}a_L\left[1 - \left(\frac{p}{p_L}\right)^{(\gamma-1)/2\gamma}\right], & p \le p_L. \end{cases},$$

$$F_{3R}(p) = \begin{cases} (p - p_R)\left(\frac{\frac{2}{(\gamma+1)\varrho_R}}{p+\frac{\gamma-1}{\gamma+1}p_R}\right)^{\frac{1}{2}}, & p > p_R, \\ -\frac{2}{\gamma-1}a_R\left[1-\left(\frac{p}{p_R}\right)^{(\gamma-1)/2\gamma}\right], & p \le p_R. \end{cases}$$

This is a nonlinear algebraic equation, and one cannot express the analytical solution of this problem in a closed form. The solution $p_\star$ can be found as the root of the function $F(p)$ defined as

$$F(p) = F_{3R}(p) - F_{1L}(p) + u_R - u_L. \tag{28}$$

The analysis of this function can be found in [4]. Here we state, that $F(p)$ is monotone and concave. Also the analytic expression for its derivative is available. For a positive solution for the pressure $F(0) < 0$ is required. This gives the *pressure positivity condition*

$$u_R - u_L < \frac{2}{\gamma - 1}(a_L + a_R). \tag{29}$$

The Newton method can be used to find the root of $F(p) = 0$. With the pressure $p_\star$ known, we use the equations (22)-(27) to obtain the whole solution.

Remarks

- Once the pressure $p_\star$ is known, the solution on the left-hand side of the contact discontinuity depends only on the left-hand side initial condition (20). And similarily, with $p_\star$ known, only the right-hand side initial condition (21) is used to compute the solution on the right-hand side of the contact discontinuity.
- The solution in $\Omega_L \cup \Omega_{HTL} \cup \Omega_{\star L}$ (across 1 wave) (solvability in general case)
  The solution components in $\Omega_L \cup \Omega_{HTL} \cup \Omega_{\star L}$ region must satisfy the system of equations (22)-(24). It is a system of three equations for four unknowns ($\varrho_{\star L}$, $p_\star$, $u_\star$, s^1_{TL}). We have to add another equation in order to get the uniquely solvable system in $\Omega_L \cup \Omega_{HTL} \cup \Omega_{\star L}$. This is the key problem for the outlet boundary condition.
- The equation (22) can be written as the *equation for pressure*, see [3]

$$p_\star = E_1(u_\star), \tag{30}$$

$$E_1(u) = \begin{cases} \frac{2p_L+\frac{\gamma+1}{2}\varrho_L(u_L-u)^2+(u_L-u)\sqrt{4\varrho_L\gamma p_L+\varrho_L^2(\frac{\gamma+1}{2})^2(u_L-u)^2}}{2}, & u < u_L, \\ p_L\left(\frac{-u+u_L+\frac{2}{\gamma-1}a_L}{\frac{2}{\gamma-1}a_L}\right)^{\frac{2\gamma}{\gamma-1}}, & u_L \le u < u_L + \frac{2}{\gamma-1}a_L. \end{cases}$$

In our work we use the analysis of this problem also for the solution of the initial-boundary value at the boundary faces.

5 Boundary Condition by Preference of Pressure

Using this boundary condition, we try to prescribe given static pressure at the face. This boundary condition corresponds to the real-world problem, when we deal with the experimentally obtained pressure distribution at the boundary. The conservation laws must be satisfied in close vicinity to the boundary face. We use the analysis of the incomplete Riemann problem to construct the values for the density and the pressure at each boundary face. The following notation is used (some values are used only for an INLET case)

ϱ_B	static density at the boundary (unknown)
p_B	static pressure at the boundary (unknown)
$\boldsymbol{v}_B$	velocity vector at the boundary (in global coordinates) (unknown)
u_B	normal component of velocity at the boundary (local coordinates) (unknown)
$\boldsymbol{n} = (n_1, n_2, n_3)$	unit outer normal to face
ϱ_L	static density near the wall, time 0
p_L	static pressure near the wall, time 0
$\boldsymbol{v}_L$	velocity vector, state near the wall at time 0
u_L	normal component of velocity, state near the wall at time 0 $u_L = \boldsymbol{v}_L \cdot \boldsymbol{n}$
p_{GIVEN}	given value for the pressure (=desired pressure at the face)
ϱ_{GIVEN}	given value for the density (=desired INLET density at the face)
$\boldsymbol{d} = (d_1, d_2, d_3)$	given INLET velocity direction

At the vicinity of the boundary we solve the modified Riemann problem for the split Euler equations (19), introduced in Section 4, with the left-hand side initial condition (20) and the **complementary conditions** (prescribed pressure).

$$p_\star = p_{\text{GIVEN}}, \tag{31}$$

$$p_R = p_\star, \tag{32}$$

Here $p_\star$ denotes the pressure in $\Omega_{\star L} \cup \Omega_{\star R}$ part of the solution of the Riemann problem for the split Euler equations, shown in Section 4. This way we have prescribed the desired pressure value whenever it is possible (see the general form of the solution in Section 4). Now it is possible to use (22),(23),(24) Further we must discuss all the possibilities of the left (shock, rarefaction) and the middle (discontinuity) wave in the solution, which is shown in figure 3. In the case of **outlet** ($u_\star \ge 0$), the transformation of the velocity into the global coordinates is

$$\boldsymbol{v}_B = \boldsymbol{n}(u_B - u_L) + \boldsymbol{v}_L.$$

Here u_B denotes the computed velocity in the normal direction. In the case of an **inlet** ($u_\star < 0$) we must **prescribe another conditions** in order to obtain a uniquely solvable system. Here we may choose (for example) arbitrary density ϱ_{GIVEN} and the direction of the velocity $\boldsymbol{d}$.

$$\varrho_{\star R} = \varrho_{\text{GIVEN}}, \quad \boldsymbol{v}_B = \boldsymbol{d}|\boldsymbol{v}|, \text{ with } \boldsymbol{v}_B \cdot \boldsymbol{n} = u_\star \tag{33}$$

Then it is

$$p_B = p_{\text{GIVEN}}, \; \varrho_B = \varrho_{\text{GIVEN}}, \; \boldsymbol{v}_B = \frac{du_\star}{\boldsymbol{d} \cdot \boldsymbol{n}}.$$

The analysis of this problem was shown also in [3].

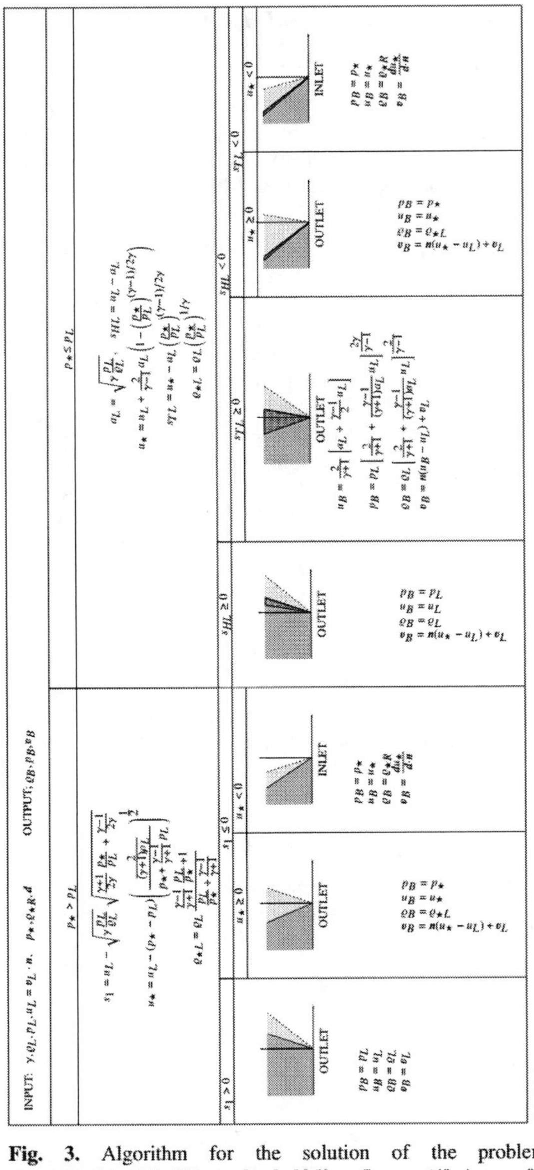

Fig. 3. Algorithm for the solution of the problem (19),(20),(31),(32),(33) at the half line $S_B = \{(0,t);\ t > 0\}$. The value of the pressure $p_\star$ is prescribed. Possible situations are illustrated by the pictures showing $\Omega_L \cup \Omega_{HTL} \cup \Omega_{\star L}$ region. The sought boundary state located at the time axis, marked red.

6 Boundary Condition by Preference of Velocity

Here we show the boundary condition prefering the prescribed given velocity at the face. The conservation laws must be satisfied in close vicnity to the boundary face. Therefore we use the analysis of the incomplete Riemann problem to construct the values for the density and the velocity at each boundary face. The following notation is used (some values are used only for an INLET case)

ϱ_B	static density at the boundary (unknown)
p_B	static pressure at the boundary (unknown)
$\boldsymbol{v}_B$	velocity vector at the boundary (in global coordinates) (unknown)
u_B	normal component of velocity at the boundary (local coordinates) (unknown)
$\boldsymbol{n} = (n_1, n_2, n_3)$	unit outer normal to face
ϱ_L	static density near the wall, time 0
p_L	static pressure near the wall, time 0
$\boldsymbol{v}_L$	velocity vector, state near the wall at time 0
u_L	normal component of velocity, state near the wall at time 0 $u_L = \boldsymbol{v}_L \cdot \boldsymbol{n}$
$\boldsymbol{v}_{\text{GIVEN}}$	given velocity vector (=desired value at the face)
u_{GIVEN}	given value for the normal vcomponent of velocity (=desired normal velocity at the face) $u_{\text{GIVEN}} = \boldsymbol{v}_{\text{GIVEN}} \cdot \boldsymbol{n}$
ϱ_{GIVEN}	given value for the density (=desired INLET density at the face)

At the vicinity of the boundary we solve the modified Riemann problem for the split Euler equations (19), introduced in Section 4, with the left-hand side initial condition (20) and the **complementary conditions** (prescribed normal component of velocity).

$$u_\star = u_{\text{GIVEN}}, \tag{34}$$

$$u_R = u_\star, \tag{35}$$

Here $u_\star$ denotes the normal component of velocity in $\Omega_{\star L} \cup \Omega_{\star R}$ part of the solution of the Riemann problem for the split Euler equations, shown in Section 4. This way we have prescribed the desired normal component of velocity whenever it is possible (see the general form of the solution in Section 4). Now it is possible to use (30),(23),(24). Further we must discuss all the possibilities of the left (shock, rarefaction) and the middle (discontinuity) wave in the solution, which is shown in figure 4. The tangential velocities are conserved (same as "left" values) in the case of **outlet** ($u_\star \geq 0$), and the velocity vector in the global coordinates is

$$\boldsymbol{v}_B = \boldsymbol{n}(u_B - u_L) + \boldsymbol{v}_L.$$

Here u_B denotes the computed velocity in the normal direction. In the case of an **inlet** ($u_\star < 0$) we must **prescribe another conditions** in order to obtain a uniquely solvable system. Here we may choose (for example) arbitrary density ϱ_{GIVEN} and the whole vector of velocity

$$\varrho_{\star R} = \varrho_{\text{GIVEN}}, \quad \boldsymbol{v}_R = \boldsymbol{v}_{\text{GIVEN}}. \tag{36}$$

For example, it is possible to choose the density $\varrho_{\star R}$ using the given entropy index $\frac{p_o}{\varrho_o{}^\gamma}$ (p_o, ϱ_o are some reference values for the pressure and the density)

$$\varrho_{\star R} = p_\star^{1/\gamma} / \left(\frac{p_o}{\varrho_o^\gamma}\right)^{1/\gamma} = \varrho_o \left(\frac{p_\star}{p_o}\right)^{1/\gamma}$$

Another possibility is to conserve (choose) the total temperature θ_0 for the inlet case. Then, using thermodynamic relations (energy equation for a perfect gas), it is

$$\varrho_{\star R} = \frac{p_\star}{R\theta_0\left(1 - \frac{\gamma-1}{2}\frac{|\boldsymbol{v}_{\mathrm{GIVEN}}|^2}{\gamma R\theta_0}\right)}. \tag{37}$$

Let $\theta_0 = \theta_{0R} = p_R/(R\varrho_R\left(1 + \frac{\gamma-1}{2}\frac{|\boldsymbol{v}_R|^2}{\gamma p_R/\varrho_R}\right))$ and $\boldsymbol{v}_{\mathrm{GIVEN}} = \boldsymbol{v}_R$. Then, using (37), it is $\varrho_{\star R} = \varrho_R \frac{p_\star}{p_R}$.

The algorithm for the construction of the primitive variables $\varrho_B, \boldsymbol{v}_B, p_B$ at the boundary face is shown in Fig. 4. The complete analysis of this problem is shown in [3].

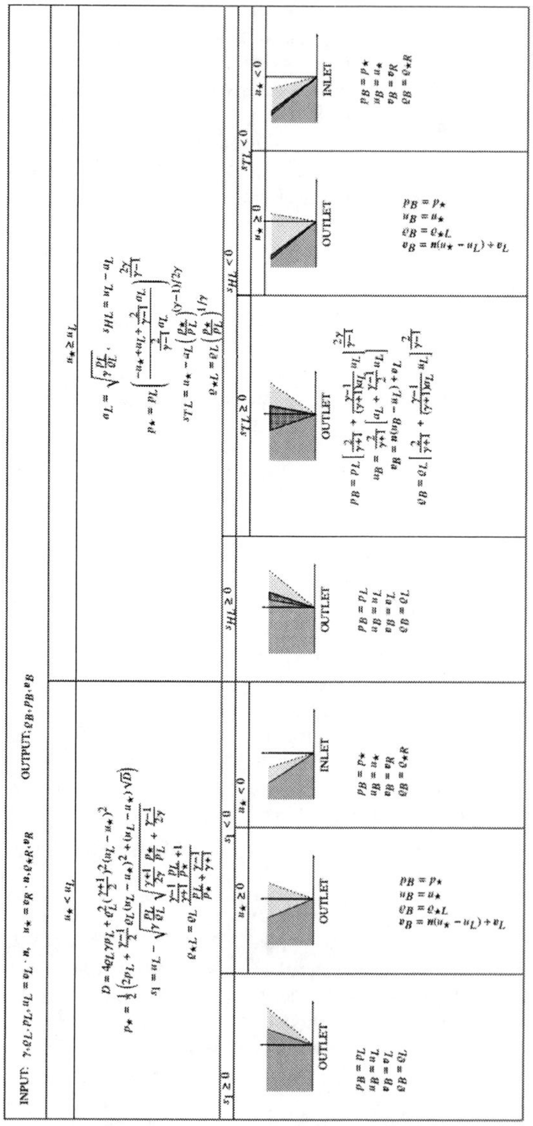

Fig. 4. Algorithm for the solution of the problem (19),(20),(34),(35),(36) at the half line $S_B = \{(0,t);\ t > 0\}$. Possible situations are illustrated by the pictures showing the region $\Omega_L \cup \Omega_{HTL} \cup \Omega_{\star L}$ with the sought boundary state located at the time axis.

7 Boundary Condition for the Imermeable Wall

Slip wall (Eulerian, inviscid, symmetric boundary)
Using this boundary condition, we set the normal component velocity at the wall to zero. The conservation laws must be satisfied in close vicnity to the boundary face. We use the analysis of the incomplete Riemann problem to construct the values for the density and the pressure at each boundary face. The following notation is used

ϱ_w	static density at the wall
p_w	static pressure at the wall
$\boldsymbol{v}_w$	velocity vector at the wall (in global coordinates)
$\boldsymbol{n} = (n_1, n_2, n_3)$	unit outer normal to face
ϱ_L	static density near the wall, time 0
p_L	static pressure near the wall, time 0
$\boldsymbol{v}_L$	velocity vector, state near the wall at time 0
u_L	normal component of velocity, state near the wall at time 0 $u_L = \boldsymbol{v}_L \cdot \boldsymbol{n}$

At the vicinity of the boundary we solve the following problem (modified Riemann problem for the split Euler equations), with the left-hand side initial condition and the complementary condition (zero normal velocity at the boundary)

$$u_\star = 0. \tag{38}$$

Using the analysis shown above (Section 4), the solution of the system (19),(20),(38) lead us to the equations (30), (23). There are two possibilities.

- $u_L > 0$ **(1-shock wave)**

$$p_w = p_\star = \frac{2p_L + \frac{\gamma+1}{2}\varrho_L u_L^2 + u_L\sqrt{4\varrho_L\gamma p_L + \varrho_L^2(\frac{\gamma+1}{2})^2 u_L^2}}{2},$$

$$\varrho_w = \varrho_L \frac{\frac{\gamma-1}{\gamma+1}\frac{p_L}{p_\star} + 1}{\frac{p_L}{p_\star} + \frac{\gamma-1}{\gamma+1}}.$$

- $u_L \le 0$ **(1-rarefaction wave)**

$$p_w = p_L\left(1 + \frac{\gamma-1}{2}\frac{u_L}{a_L}\right)^{\frac{2\gamma}{\gamma-1}}, \text{ with } a_L = \sqrt{\gamma p_L/\varrho_L},$$

$$\varrho_w = \varrho_L\left(1 + \frac{\gamma-1}{2}\frac{u_L}{a_L}\right)^{\frac{2}{\gamma-1}}.$$

The velocity in local coordinates is $(0, v_L, w_L)$. The backward transformation of the computed values back into global coordinates (transformation of the velocity components) gives the sought velocity vector at the boundary

$$\boldsymbol{v}_w = \boldsymbol{v}_L - u_L\boldsymbol{n} = (v_{L1} - u_L n_1, v_{L2} - u_L n_2, v_{L3} - u_L n_3).$$

Remark. It is possible to solve this boundary problem by prescribing the special right-hand side initial condition to the original Riemann problem, described in Section 4., i.e. $\varrho_R = \varrho_L,\ u_R = -u_L,\ p_R = p_L$.

Non-slip wall
For the non-slip wall (viscous flow) we set the velocity vector $\boldsymbol{v}_w = 0$ at the boundary. The turbulent kinetic energy at such non-slip wall is $k = 0$. For the evaulation of the specific dissipation at the boundary ω we use the statements from [1]. Let us use the following notation and relations

θ_w	static temperature at the wall
μ_w	dynamic viscosity at the wall, using Sutherland's law $\mu_w = \mu_Z \left(\frac{\theta_w}{\theta_Z}\right)^{3/2} \frac{\theta_Z+\theta_S}{\theta_w+\theta_S}$, $\theta_S = 110$, $\theta_Z = 273.15$, $\mu_Z = 1.716 \cdot 10^{-5}$
k_w	turbulent kinetic energy at the wall, $k_w = 0$
ω_w	turbulent specific dissipation at the wall
$\boldsymbol{n} = (n_1, n_2, n_3)$	unit outer normal to face
$\boldsymbol{F} = (F_1, F_2, F_3)$	force (vector) acting on surface

$$F_1 = pn_1 - (\tau_{11}n_1 + \tau_{12}n_2 + \tau_{13}n_3)$$
$$F_2 = pn_2 - (\tau_{12}n_1 + \tau_{22}n_2 + \tau_{23}n_3)$$
$$F_3 = pn_3 - (\tau_{13}n_1 + \tau_{23}n_2 + \tau_{33}n_3)$$

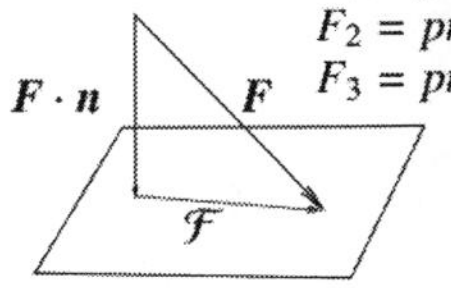

$$\tau_{11} = \tfrac{2}{3}\mu_w\left(2\frac{\partial v_1}{\partial x_1} - \frac{\partial v_2}{\partial x_2} - \frac{\partial v_3}{\partial x_3}\right) \qquad \tau_{12} = \mu_w\left(\frac{\partial v_1}{\partial x_2} + \frac{\partial v_2}{\partial x_1}\right)$$
$$\tau_{22} = \tfrac{2}{3}\mu_w\left(-\frac{\partial v_1}{\partial x_1} + 2\frac{\partial v_2}{\partial x_2} - \frac{\partial v_3}{\partial x_3}\right) \qquad \tau_{13} = \mu_w\left(\frac{\partial v_1}{\partial x_3} + \frac{\partial v_3}{\partial x_1}\right)$$
$$\tau_{33} = \tfrac{2}{3}\mu_w\left(-\frac{\partial v_1}{\partial x_1} - \frac{\partial v_2}{\partial x_2} + 2\frac{\partial v_3}{\partial x_3}\right) \qquad \tau_{23} = \mu_w\left(\frac{\partial v_2}{\partial x_3} + \frac{\partial v_3}{\partial x_2}\right)$$

$\mathcal{F}$	surface shear stress, friction tension (force magnitude in surface direction) $\mathcal{F}^2 = F_1^2 + F_2^2 + F_3^2 - (\boldsymbol{F} \cdot \boldsymbol{n})^2$
u_τ	friction velocity (velocity scale) $u_\tau^2 = \frac{\mathcal{F}}{\varrho_w}$

In the real application it is necessary to prescribe the static temperature of the surface θ_w.

The law of wall, [1, pages 13-18]
The law of the wall is empirically-determined relation between the streamwise velocity U and distance from surface y, see [1, page 15]

$$U^+ = \frac{1}{\kappa} ln\,(y^+) + C, \text{ with } U^+ = \frac{U}{u_\tau},\ y^+ = \frac{u_\tau \varrho}{\mu} y. \quad (39)$$

Measurements indicate $C \approx 5.0$ for smooth surfaces, $\kappa \approx 0.41$ (Karman's constant) for smooth and rough surfaces, y denotes the distance from surface. For the rough surface, with elements of average height k_S, experiments show that the C is a function of k_S,

$$C \rightarrow 8.5 - \frac{1}{\kappa} ln\, k_S^+, \text{ where } k_S^+ = \frac{u_\tau \varrho_w}{\mu_w} k_S.$$

Here k_S^+ is dimensionless roughness height. The the law of the wall (39) then yields

$$\frac{U}{u_\tau} = \frac{1}{\kappa} ln\left(\frac{y}{k_S}\right) + 8.5 \quad \text{(completely rough wall)}$$

There is no viscosity in this relation, corresponding with the idea that the "shear stress" is due to pressure drag on the roughness elements. The last relation can be rewritten as

$$\frac{U}{u_\tau} = \frac{1}{\kappa} ln\left(\frac{y}{z_S}\right), \text{ with } z_S = k_S e^{-8.5\kappa}.$$

Condition for ω, smooth wall
In the case of smooth wall we use the relation

$$\omega_w = 120 \frac{\mu_w}{\varrho_w y_S^2},$$

where y_S is the distance of the face from the center of the adjacent element.

Condition for ω, completely rough wall
We simulate the surface roughness by modification of the turbulent dissipation ω at the wall, as described in [1][page 175-177,308].

$$\omega_w = \frac{u_\tau^2 \varrho_w}{\mu_w} S_R$$

The coefficient S_R is defined with the use of the roughness height $k_S = z_S\, e^{8.5\kappa}$

$$S_R = \begin{cases} (50/k_S^+)^2, & k_S^+ \leq 25 \\ 100/k_S^+, & k_S^+ > 25 \\ \left([1][\text{page } 308] \quad 500/(k_S^+)^{\frac{3}{2}}\right., & \left. k_S^+ > 25\right) \end{cases}.$$

Here k_S represents the roughness height. S_R is a continuous function of k_S^+.

7.1 The Flat Plate Example

Here we show example of the compressible viscous flow along the flat plate. The velocity regime $u = 15\ m\ s^{-1}$. The total pressure $p_o = 101325\ Pa$, total temperature $\theta_o = 273.15\ K$, and properties $R = 287.04\ J\ K^{-1}\ mol^{-1}, \gamma = 1.4$, gas flows from the left to the right side. Initial condition for the computation is given by isentropic approximation (and the use of energy equation for a perfect gas)

$$p = p_o\left(1 - \frac{\gamma - 1}{2} u^2/(\gamma R \theta_o)\right)^{\frac{\gamma}{\gamma-1}} \approx 101179.69\ Pa,$$

$$\varrho = \varrho_o\left(1 - \frac{\gamma - 1}{2} u^2/(\gamma R \theta_o)\right)^{\frac{1}{\gamma-1}} \approx 1.291\ kg\ m^{-3},$$

The initial conditions for the free stream turbulent kinetic energy k and the specific turbulent dissipaton ω are given by the turbulent intensity $I = 0.1$ and $\mu_T/\mu = 0.01$:

$$k = \frac{3}{2} v^2 I^2, \quad \omega = \frac{\varrho k}{\mu\left(\frac{\mu_T}{\mu}\right)}.$$

($k = 3.375\ m^2\ s^{-2}, \omega = 2.53995e + 07\ s^{-1}, \mu = 1.716e - 05\ kg\ m^{-1}\ s^{-1}$). The computation was made for the simulation of the rough wall (bottom channel wall) with the roughness parameter $z_S = 0.0002$, see figures 5 and 6, the formula for ω at wall was chosen accordingly to [1][page 177].

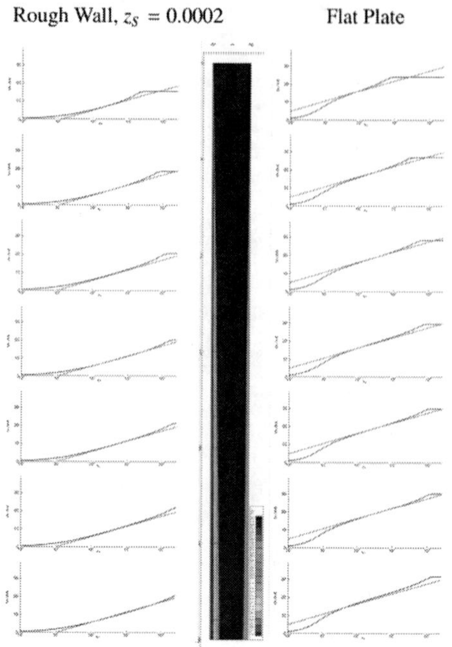

Fig. 5. Computational simulation, the flow along the flat plate and along the rough wall ($z_S = 0.0002$) simulation, regime 15 $m\ s^{-1}$, isolines of k turbulent kinetic energy. The $Y^+ - U^+$ graph (log scale) at chosen line cuts (marked by the lines in the central picture). The dotted line shows U+ given by (39), full line shows the computed results.

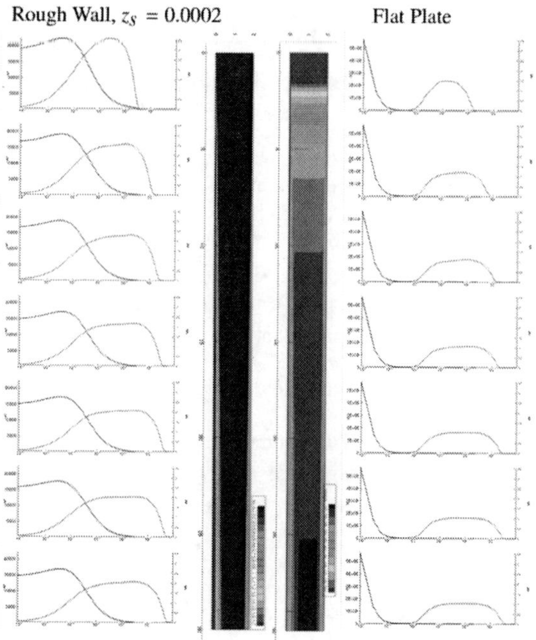

Fig. 6. Computational simulation, the flow along the flat plate and along the rough wall ($z_S = 0.0002$) simulation, regime 15 $m\ s^{-1}$, isolines of the turbulent kinetic energy k and dissipation ω. The wall distance - k (green) and ω (red) graphs at the chosen line cuts (marked by the lines in the central picture).

7.2 Numerical Tests on Shortened Domains

In order to achieve faster computations it is desired to use smaller computation areas. In such case, the boundary conditions must be chosen appropriately. Here we present numerical experiments with the boundary conditions on smaller domains. At first we assume the compressible viscous flow in the long channel (x-coordinate from -13 to 4), with only small area of interest (x-coordinate from -1 to 2). The velocity regime $u = 15\ m\ s^{-1}$. The total pressure $p_o = 101325\ Pa$, total temperature $\theta_o = 273.15\ K$, and the gas constants set to $R = 287.04\ J\ K^{-1}\ mol^{-1}$, $\gamma = 1.4$, gas flows from the left to the right side. Initial condition for the computation is given by isentropic approximation (and the use of energy equation for a perfect gas). The initial conditions for the free stream turbulent kinetic energy k and the specific turbulent dissipaton ω are given by the turbulent intensity $I = 0.1$ and $\mu_T/\mu = 0.01$, see also section 7.1. The results are depicted in figure 7.

The computed results were used in another series of computations on shortened domain (with y-coordinate -1 to 2), in order to test various boundary conditions. The aim was to test various boundary and initial conditions in order to match previously computed data on the original (longer) channel. Here we present the results using the boundary conditions given by the preference of pressure at the outlet, and with the preference of total quantities and the direction of velocity at the inlet. The input values for these boundary conditions were given from the previous computation on larger domain (different values at each boundary face). Results are shown in figures 8 - 10.

Further we show the simulation of emission of another gas specie into the area. We choose the additional gas with

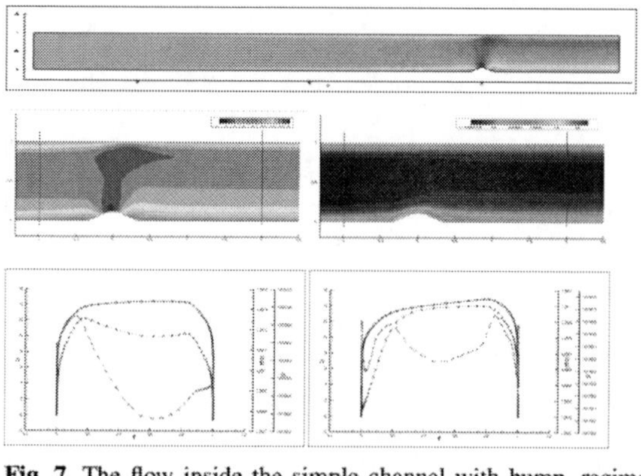

Fig. 7. The flow inside the simple channel with bump, regime 15 $m\ s^{-1}$, isolines of u velocity component and k turbulent kinetic energy. The $y - \varrho, u, p$ graphs at chosen line cuts (shown red).

the properties $R = 461\ J\ K^{-1}\ mol^{-1}, c_V = 1450\ J\ K^{-1}\ mol^{-1}$, which enters the area. This emission is realized by introducing the fixed source located at the given cell (with cell center coordinates [0.092,0.0997,0.5]). The mass fraction of additional gas component at this cell was fixed to $Y_1 = 1$, and density was set accordingly to equation of state, using the computed pressure and temperature. Figure 11. shows the propagation of mixture in time.

Conclusion

This paper shows the formulation of the equations describing the mixture of two inert perfect gases in 3D, it is further focused on the boundary conditions and simulation of

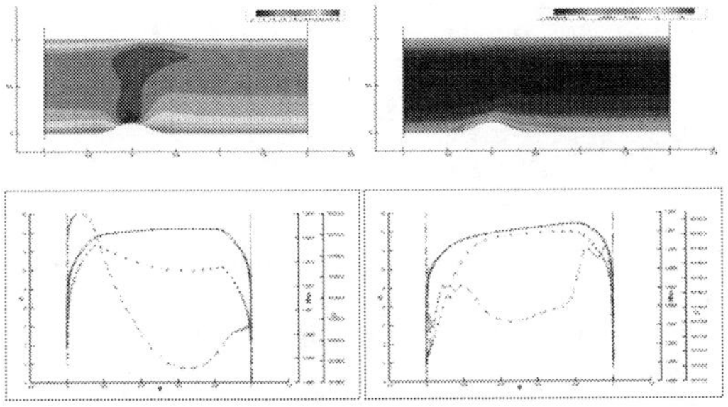

Fig. 8. Computational results on shortened domain. Inlet: fixed "ghost cell" values, outlet: fixed "ghost cell" values.

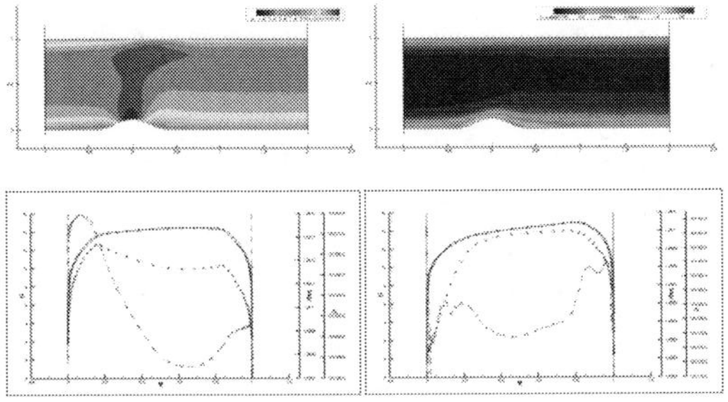

Fig. 9. Computational results on shortened domain. Inlet: fixed "ghost cell" values, outlet: preference of velocity

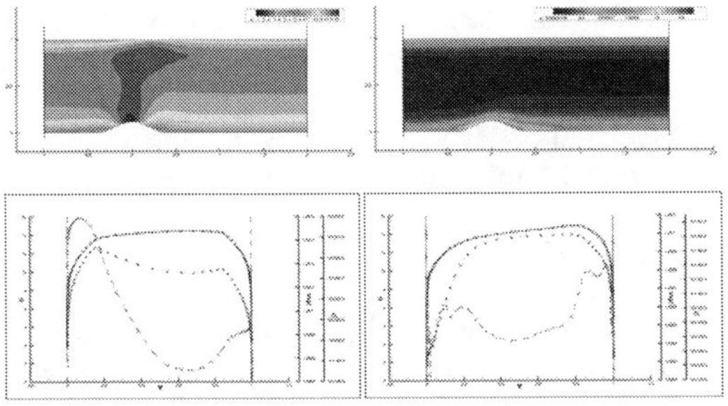

Fig. 10. Computational results on shortened domain. Inlet: preference of total quantities, outlet: preference of pressure.

the wall roughness. The numerical method (finite volume method) is applied for the solution of these equations. Own software was programmed. The modification of the Riemann problem is used at the boundaries.

Acknowledgment
This result originated with the support of Ministry of the Interior of the Czech Republic, project SCENT. The authors acknowledge this support.

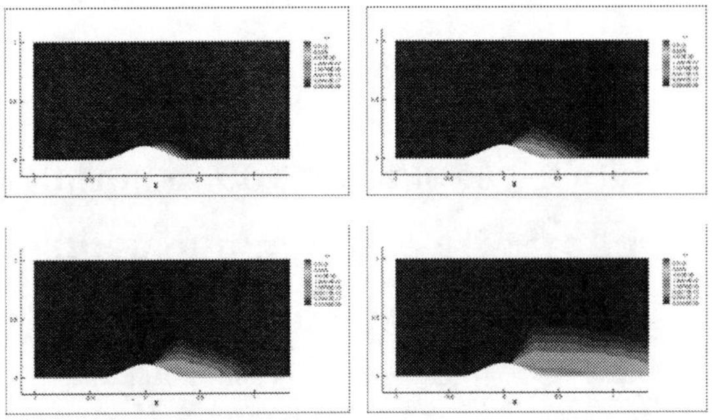

Fig. 11. Computational results on shortened domain. Isolines of 1st specie mass fraction Y_1 in time.

References

1. C. J. Kok. AIAA Journal, Vol. **38.**, No. 7., (2000).
2. M. Kyncl and J. Pelant. *Applications of the Navier-Stokes equations for 3d viscous laminar flow for symmetric inlet and outlet parts of turbine engines with the use of various boundary conditions.* Technical report R3998, VZLÚ, Beranových 130, Prague, (2006).
3. M. Kyncl. *Numerical solution of the three-dimensional compressible flow.* Master's thesis, Prague, (2011). Doctoral Thesis.
4. M. Feistauer, J. Felcman, and I. Straškraba. *Mathematical and Computational Methods for Compressible Flow.* Oxford University Press, Oxford, (2003).
5. M. Kyncl and J. Pelant. *Implicit method for the 3d Euler equations.* Technical report R5375, VZLÚ, Beranových 130, Prague, (2012).
6. M. Kyncl and J. Pelant. *Implicit method for the 3d RANS equations with the k-w (Kok) Turbulent Model.* Technical report R5453, VZLÚ, Beranových 130, Prague, (2012).
7. H. W. Liepmann and A. Roshko. *Elements of Gasdynamics.* John Wiley & Sons, Inc., New York, (1957).
8. M. Feistauer. *Mathematical Methods in Fluid Dynamics.* Longman Scientific & Technical, Harlow, (1993).
9. E. F. Toro. *Riemann Solvers and Numerical Methods for Fluid Dynamics.* Springer, Berlin, (1997).

EPJ Web of Conferences **114**, 02065 (2016)
DOI: 10.1051/epjconf/201611402065

Numerical models and experiment of air flow in a simulation box for optical wireless communications

Jan Latal[1], Lukas Hajek[1], Marian Bojko[2], Jan Vitasek[1], Petr Koudelka[1], Stanislav Kepak[1], Ales Vanderka[1], Vladimir Vasinek[1]

[1]*Department of Telecommunications, Faculty of Electrical Engineering and Computer Science, VSB-Technical University of Ostrava, 17.listopadu 15, Ostrava-Poruba, 708 33, Czech Republic*

[2]*Department of Hydrodynamics and Hydraulic Equipment, Faculty of Mechanical Engineering, VSB-Technical University of Ostrava, 17.listopadu 15, Ostrava-Poruba, Czech Republic*

Abstract. In this article, the authors focused on real measurements of mechanical turbulence generated by ventilators in the simulation box for Optical Wireless Communications. The mechanical turbulences disturb the optical beam that propagates along the central axis of the simulation box. The aim of authors is to show the effect of mechanical turbulence on optical beams at different heights in the simulation box. In the Ansys Fluent, we created numerical models which were then compared with real measurements. Authors compared the real and numerical models according to statistical methods.

1 Introduction

Today communication has become a necessity for people. It is necessary to be mobile and accessible on mobile devices that are connected to the Internet. This has resulted in increasing demands from internet service providers (ISPs) that must be able to cover large areas with sufficient signal levels. To satisfy this demand, ISPs are using all available communication systems including optical fiber-free connections. They are characterized by the fact that they do not need a licensed band for communication and are also easily installed on rooftops. Among the disadvantages that reduce the deployment of these systems is the need for line of sight between the transmitter and receiver and the influence of atmospheric phenomena.

For fiber-free optic communications the data and signal transmission uses an optical beam that is subject to atmospheric conditions, such as wind, rain, snow, turbulence, absorption, mist, fog, insects or birds, that can cause a decrease or loss of signal power level, or can change the spatial coherence of the optical beam. Before installing fiber-free links in a location, the implementation of a model behavior of the atmosphere and its influence on the fiber-free link is performed to reduce the overall degradation.

The aim of this article is to carry out a thorough analysis and real measurements of the air flow in a closed plexiglass box with dimensions of 250x50x50 cm and to create a numerical model in the ANSYS FLUENT 15 software, which is used for the simulation of fluid dynamics using the finite volume method (FVM) and for numerical modeling of the turbulent flow. Measured values will be compared to simulated values.

2 Atmospheric turbulence

The airflow in the atmosphere is divided into two kinds of flow: laminar flow and turbulent flow. During laminar flow the parallel motion of particles occurs. The particle motion is considered to be in layers that are mutually unmixed while friction among these layers is considered negligible.

However, for turbulent flow particle tracks are mixed. Particles do not obey laminar only motion, but also perform their own complex movement with irregular direction and speed in such a way that the flow can be considered to be stochastic and non-stationary. As the particles travel through different layers, these layers mix leading to the formation of vortices.

Atmospheric turbulence is divided into three types, namely thermal turbulence, turbulence and dynamic mechanical turbulence. In dynamic mechanical turbulence, a large wind shear occurs due to friction between the layers of air, each having different velocity and direction. In atmospheric optical links dynamic turbulence occurs only at altitudes above 4 km and could therefore affect satellite communication with the ground station for instance.

The airflow results in random fluctuations in the signal transmission quality, changing the power density incident on the detector. Random beam deflections lead to an increase in bit error rate or even to a complete loss of signal [1-6].

[a] Corresponding author: jan.latal@vsb.cz

2.1 The effects of turbulence on the optical beam

Turbulence in the atmosphere are very unstable pockets of different sizes and densities unstable in time, continuously forming, changing and disappearing. They cause amplitude and phase fluctuations of coherent waves with time. Turbulence can be divided according to a relationship between the size of the light beam and the level of turbulence. All types of turbulence can impact transmission through the fluctuations in received power level at a frequency in the tens of Hz.

2.1.1 The diameter of the beam is smaller than the size of the turbulence

If the optical beam diameter is smaller than the diameter of the turbulent vortex a deflection of the entire optical beam from its path takes place. The optical beam exits at a different angle from the turbulence as a refracted beam caused by the refraction index gradient of the turbulence. This may lead to a decrease in received power level to the point where only minimal optical power is detected.

2.1.2 The diameter of the beam is the same size as turbulence

If the optical beam has the same diameter as the turbulent vortex, turbulence acts as a lens which does not change the direction of the optical beam but increases its divergence. Additionally, this turbulence causes a constant change of attenuation and phase delay.

2.1.3 The diameter of the beam is larger than the turbulence

If the optical beam has a larger diameter than the turbulent vortex diameter, the beam is split into separate beams that exit the turbulent flow at different angles in addition, to the occurrence of intensity scattering and interference. The beams propagate along different paths resulting in a variation in the intensity of the optical beam called scintillation.

The relative variations in the optical intensity σ_I^2 are defined as:

$$\sigma_I^2 = \frac{\langle I^2\rangle - \langle I\rangle^2}{\langle I\rangle^2} = \frac{\langle I^2\rangle}{\langle I\rangle^2} - 1 \qquad (1)$$

where I is the optical intensity of the optical beam. If $\sigma_I^2 << 1$, we can use Rytovov approximation which works even with the structural parameter of the refractive index C_n^2:

$$\sigma_I^2 = K \cdot C_n^2 \cdot k^{7/6} \cdot L^{11/6}, \qquad (2)$$

where K represents a constant for a plane wave equal to 1.23 and for a spherical wave equal to 0.5, k is the wave number and L is the distance between the transmitter and receiver.

If the relative variations change of the refractive index of is less than 1 $\sigma_I^2 < 1$ it is a weak turbulence. If this condition is not met, it is a strong turbulence.

Scintillation can be eliminated by using multiple transmitters or using large area receivers. The amplitude and frequency of the scintillation depends on the cell size compared to the size of the light beam. The amplitude and frequency of the scintillation increases with the increasing frequency of light.

For a plane waves in a low turbulence environment and for a specific receiver the scintillation variance can be expressed by the following formula:

$$\alpha_{\sin} = 2\sqrt{23{,}17 \cdot \left(\frac{2\pi}{\lambda}10^9\right)^{\frac{7}{6}} \cdot C_n^2 \cdot L^{\frac{11}{6}}} \quad [dB], \qquad (3)$$

where the individual variables can be described as follows λ transmission wavelength [nm], L transmission channel distance [m], C_n^2 structural parameter of the refractive index [$m^{-2/3}$].

According to Andrews, the optical intensity variation equation is expanded to include the diameter of the receiving lens D_{RXA}. This idea is based on the assumption that turbulence is not absorbing optical power as in the case of absorption, but merely diverting the direction of the beam. These diverted beams are then focussed on the lens surface of the detector.

$$\sigma_I^2(D_{RXA}) = \exp\left(\frac{0{,}49 \cdot \beta^2}{\left(1 + 0{,}18 \cdot d^2 + 0{,}56 \cdot \beta^{12/5}\right)^{7/5}} + \right.$$

$$\left. + \frac{0{,}51 \cdot \beta\left(1 + 0{,}69 \cdot \beta^{12/5}\right)^{-5/6}}{\left(1 + 0{,}90 \cdot d^2 + 0{,}62 \cdot d^2 \cdot \beta^{12/5}\right)^{7/5}} - 1\right). \qquad (4)$$

Coefficients β and d are determined as:

$$\beta = 0{,}5 \cdot C_n^2 \cdot k^{7/5} \cdot L^{11/6}, d = \sqrt{\frac{2\pi}{4\lambda L}} D_{RXA} . (5)$$

Attenuation expressed using the Andrews approximation and can be written as:

$$\alpha_{turbulence} = 10\log\left|1 - \sqrt{\sigma_I^2(D_{RXA})}\right| \qquad (6)$$

The Andrews expression of attenuation due to the turbulent atmosphere envisages the lowest possible value of attenuation. Another method which is focused on the determination of turbulent attenuation is called the method of available power. It is used for the sources with a Gaussian distribution of the optical intensity in the plane perpendicular to the direction of propagation [1-6].

3 Characteristics of the mathematical model for fluid flow

The flow of fluids (gases and liquids) is divided into laminar and turbulent flow. To determine the nature of flow is used dimensionless parameter called the Reynolds number:

$$Re = \frac{v \cdot d_h}{\nu}, \quad (7)$$

where (d_h) [m] describes the characteristic diameter equal to the diameter of the inlet hole, (v) [m·s^{-1}] velocity of air and (ν) [m^2·s^{-1}] kinematic viscosity of flowing media. Type of flow can be determined from the size of the Reynolds number. In the determination of the boundaries between laminar and turbulent flow is discovered many values which may vary by pressure, altitude, or temperature. The approximate values of $2320 < Re < 4000$, where $Re < 2320$ is the laminar flow and at $Re > 4000$ is the turbulent flow are taken into consideration. The area between them is known as the transition region [10]. Laminar flow is considered to be a fluid flow of such a viscosity that the streamlines does not mix and are therefore parallel to the propagation direction of the medium. This type of flow can be observed for example in liquids at low speeds and high viscosity. On the contrary, turbulent flow shows chaotic changes (fluctuations) in the case of temperature, pressure, speed and direction of flow [5, 8]. The modelling of turbulent flow can be made using many methods made possible by the computing power of today's computers. The basic techniques include simulations of large eddies (LES) or direct numerical simulation (DNS). However, the DNS method becomes technically unfeasible because of the growing complexity of models, size of computational net and Reynolds number. One of the most used methods today is the Reynolds averaging of the Navier-Stokes equations (RANS). This method is based on the time-averaged values of turbulent flow and time-averaging procedure at balance equations [5, 9].

$$\zeta = \overline{\zeta} + \zeta' \quad (8)$$

$$\overline{\zeta} = \frac{1}{T}\int_0^T \zeta dt, \ \overline{\zeta'} = 0 \quad (9)$$

Individual immediate values of the turbulent flow (ζ) are divided into the sum of the median value ($\overline{\zeta}$) and fluctuating value (ζ').

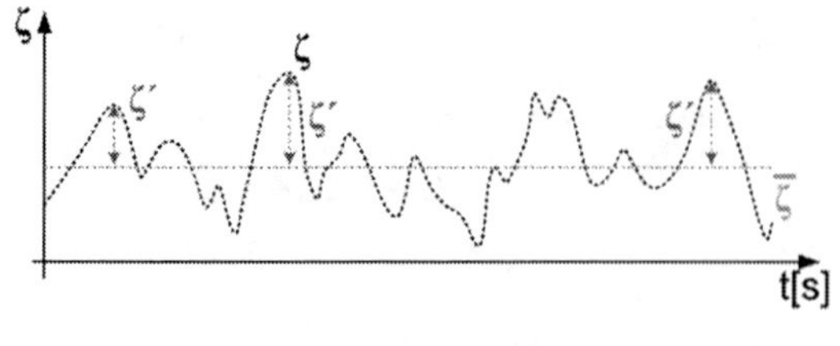

Figure 1. Mean, fluctuation, immediate value [7].

The median value is the arithmetic average in a specific period *T* and the average value of fluctuations is equal to zero (see equation (9)). Based on equation (8) and (9) the equation of continuity (10) and N-S equation (11) of incompressible, time-independent flow can be defined into the form:

$$\frac{\partial \overline{u}_j}{\partial x_j} = 0, \quad (10)$$

$$\frac{\partial(\overline{u}_i \overline{u}_j)}{\partial x_j} + \frac{\partial(\overline{u'_i u'_j})}{\partial x_j} = -\frac{1}{\rho}\frac{\partial(\overline{p})}{\partial x_i} + \nu \frac{\partial^2(\overline{u}_i)}{\partial x_j^2} + \overline{f_i} \quad (11)$$

After adjustment of the N-S equations, the tensor of Reynolds (turbulent) stress arises in the link (11), the tensor must be numerically modelled. In order to limit the computing demand on the calculation of the turbulent flow, the Boussinesq hypothesis was introduced. The hypothesis uses the Newtonian relation to replace the tensor of nine shear stresses with one quantity of the same unit as the dynamic viscosity. The result of the Boussinesq hypothesis is the modified N-S equation for noncompressible, time-independent flow:

$$\frac{\partial(\overline{u}_i \overline{u}_j)}{\partial x_j} = -\frac{1}{\rho}\frac{\partial(\overline{p})}{\partial x_i} + \nu_{eff} \frac{\partial^2(\overline{u}_i)}{\partial x_j^2} + \overline{f_i}. \quad (12)$$

The Boussinesq hypothesis is the basis of numerical models that have lower computational requirements in calculating the turbulent viscosity. The division into zero, one and two-linear equation models is represented. The Zero-linear equation model refers to the fact that apart from the equation of conservation of energy, momentum and mass no transport equations are needed. Next is the single linear equation model, which uses a transport equation for modeling the turbulent kinetic energy that is a function of the turbulent viscosity. However it is necessary to specify the turbulent rule that is dependent on the qualities of the flow. The zero-linear and one-linear equation models are called incomplete models because it is necessary to know the character of the flow during the calculation of these models. When modeling the turbulent flow, the two-linear equation models k-ω a k-ε [2] are defined the most often. The models which are used in this work are the two-linear equation models: k-ε Standard, k-ε RNG, k-ε Realizable, k-ω Standard and k-ω SST.

4 Simulation of mechanical turbulence

CFD simulations of air flow in the box were made using the ANSYS Fluent 15 software. To create a numerical model, the geometric model must be made first. The computational grid with the appropriately set boundary conditions is then applied to the geometric model. The turbulent models which are used in this work are the two-linear equation models k-ε (Standard, RNG, Realizable) and k-ω (Standard and SST). For solution of turbulent flow are many models. Models are suitable for different characters of flow and different conditions of flow. In this article we focus on experimental measurements and mathematical modelling. Therefore, the use of powerful CFD model of turbulence will be the content follow research turbulent flow. Some short information about characteristics of others turbulent models. DES models have been specifically designed to address high Reynolds number wall bounded flows, where the cost of a near-wall resolving Large Eddy Simulation would be prohibitive. The difference with the LES model is that it relies only on the required RANS resolution in the boundary layers. The application of DES, however, may still require significant CPU resources and therefore, as a general guideline, it is recommended that the conventional turbulence models employing the Reynolds-averaged approach be used for most practical calculations. Resolving only the large eddies allows one to use much coarser mesh and larger time-step sizes in LES than in DNS. However, LES still requires substantially finer meshes than those typically used for RANS calculations. In addition, LES has to be run for a sufficiently long flow-time to obtain stable statistics of the flow being modeled. As a result, the computational cost involved with LES is normally orders of magnitudes higher than that for steady RANS calculations in terms of memory (RAM) and CPU time. Therefore, high-performance computing (for example, parallel computing) is a necessity for LES, especially for industrial applications. Next turbulent models are Scale-Adaptive Simulation (SAS), Embedded Large Eddy Simulation (ELES), Reynolds Stress Model (RSM), more information [11], [12]. For application these models will be necessary to ensure performance computing serve.

4.1 CFD model for simulation in plexiglass box

As stated previously, the actual measurement was carried out in plexiglass boxes with dimensions of 2.5 × 0.5 × 0.5 m. In ANSYS Fluent 15 the geometric model of the box was created. Then the boundary conditions derived from real measurements were applied to the model. Using the Extrude method a 0.5 m height box was made. Four inlet and outlet holes with the shape of a circle with 11 cm diameter were created in the model. Because fan comprises of a rotor and a stator it was necessary to create another circle with 5.4 cm diameter inlets these circles, which then represents the stator. These prepared inlets to the plexiglass box were modeled like in the real box. The geometric model of the box can be seen in Figure 2.

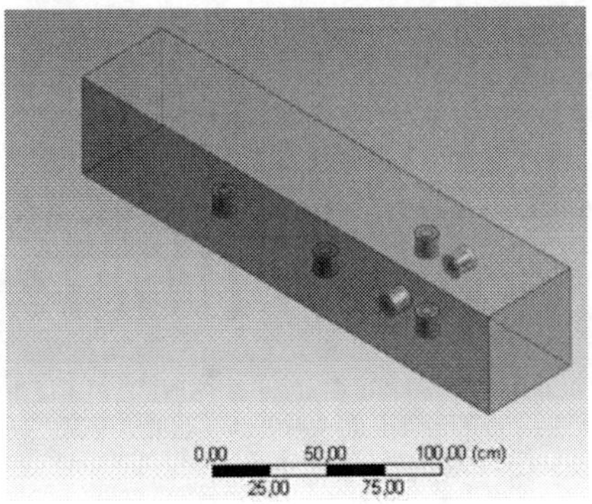

Figure 2. Geometric model of the box.

After the creation of the geometric model and computational network the boundary conditions were set. These settings were based on the real measurement data. The velocity for individual fans and outlets were measured, as shown in Figure 3.

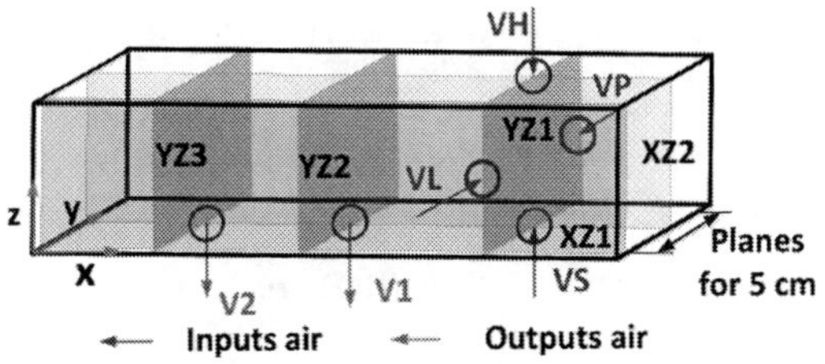

Figure 3. Nine measured planes for 5cm in the plexiglass model box.

Whirling flow behavior of the air media was generated using the inlet fan rotor blades mounted on the bottom side at the center hole. A SUNON A2123-HBT fan was used. The average measured velocity at the individual inlets and outlets are listed in Table 1. The most essential part of the boundary conditions setting has been given to the inlet. It was important to stick with the same parameters with the actual model as closely as possible. After the setting of these values in the simulation, it was found that the average velocity at the inlets was around 11 m·s^{-1} and 10 m·s^{-1} at the outlets. This could be mainly due to the fact that the simulation is actually an expression of the ideal state, which does not calculate fan blades that must be modeled and inserted into a simulation separately. By gradually reducing the velocity it was discovered that it is best to count with the inlet velocity at about 1.7 m·s^{-1} lower. After using this reduction in speed method, it was found that the average velocity reached values that were read during real measurements at the inlet and outlets.

Table 1. Measured velocity at the box inlets and outlets.

Box inlets and outlets	Velocity [m·s^{-1}]
Left inlet	7,00
Bottom inlet	6,75
Right inlet	6,65
Top inlet	6,25
Outlet	8,00
Outlet	8,00

Subsequently, the mass flow rate was calculated by the following formula:

$$Q_m = S \cdot v \cdot \rho = \pi \cdot r^2 \cdot v \cdot \rho, \qquad (13)$$

where S represents the content of the inlet area, v is velocity, and ρ is the air density. Since we know that the fans are formed by two parts the rotor and stator, it is necessary to obtain an area value which defines the area of flow. Therefore, the surface of the rotor and stator are subtracted to obtain the total inlet area. The parameter of turbulent intensity and the hydraulic diameter at the inlet can be set for the boundary conditions within the simulation. The hydraulic diameter is defined as a difference between the internal and external diameter of the box inlet and outlet holes. The turbulent intensity is derived from Reynolds numbers [7-9] and can be calculated using the following equation:

$$I = 0.16 \cdot (\mathrm{Re})^{-\frac{1}{8}}, [\%]. \quad (14)$$

To make the turbulent intensity calculation for the right intake fan it was first necessary to calculate the inlet area contents:

$$S = S_1 - S_2 \quad (15)$$

$$S_1 = \pi \cdot r^2 = 3.14 \cdot 5.5\ 2 = 94.98 \mathrm{cm}^2 \quad (16)$$

$$S_2 = \pi \cdot r^2 = 3.14 \cdot 2.7\ 2 = 22.8906 \mathrm{cm}^2 \quad (17)$$

$$S_{cel} = 94.985 - 22.8906 = 72.0944 \mathrm{cm}^2 = 0.0072 \mathrm{m}^2 \quad (18)$$

Air density was 1,184 kg·m^{-3} at a temperature of 25 °C. Based on these values, we can calculate the mass flow rate defined as:

$$Q_m = S \cdot v \cdot \rho = 0.0072 \cdot 5.3 \cdot 1.184 = = 0.04518144\, kg \cdot s^{-1} \quad (19)$$

The hydraulic diameter is given by:

$$d_h = Average\ S_1 - Average\ S_2 = 11 - 5.4 = = 5.6 \mathrm{cm} = 0.056 \mathrm{m} \quad (20)$$

Then using a formula to calculate the Reynolds number, we compute the value of R_e = 1658. This value is substituted into the formula for the turbulent intensity calculation, leading to a value of I = 4.7% for the left intake fan. The following table 2 lists the overview of the calculations for individual inlets.

Table 2. Calculated values for mass flow rate, turbulent intensity and hydraulic diameter.

Fan type	Quantity	Value
VP (right)	Q_m [kg·s^{-1}]	0.04219776
	I [%]	4.7
	d_h [m]	0.056
VS (bottom)	Q_m [kg·s^{-1}]	0.04305024
	I [%]	4.7
	d_h [m]	0.056
VH (top)	Q_m [kg·s^{-1}]	0.03878784
	I [%]	4.8
	d_h [m]	0.056

The pressure–outlet boundary condition at the inlets and mass–flow–rate–inlet boundary condition at the outlets was used in the simulation.

5 The real measurements of mechanical turbulence

When measuring mechanical turbulence, four fans were mounted on the right side of each wall and two bottom outlets were exposed. Measurements were performed using an anemometer (type Almemo 2290-2/3) capable of real-time air velocity measurement. The box was imaginarily divided into 9 vertical planes spaced apart by 5 cm, see Figure 3. In each plane 459 notional points were spaced 5 cm apart resulting in 4131 measured points. Because at every point the air flows from different directions, the measurement has to be made in three axes x, y and z. Overall we get 12393 measurements leading to the air velocity values. In this way, the whole box was measured. To obtain the actual value of the velocity at that point, we have to use following formula:

$$\|v\| = \sqrt{v_x^2 + v_y^2 + v_z^2} \quad (21)$$

where v is the velocity magnitude at a given point and v_x, v_y, v_z represents the measured values in different mutually perpendicular axes. These calculated velocity values were then processed and plotted in graphs using Matlab R2014a. In Figure 4 the air velocity in XZ1 plane can be seen. This plane is 5 cm distant from the beginning of the box. It can be seen that the largest velocity occurs in the area of air inlet VL, see Figure 3, with successively decreasing air velocity occuring in the direction of V1 and V2 outlets.

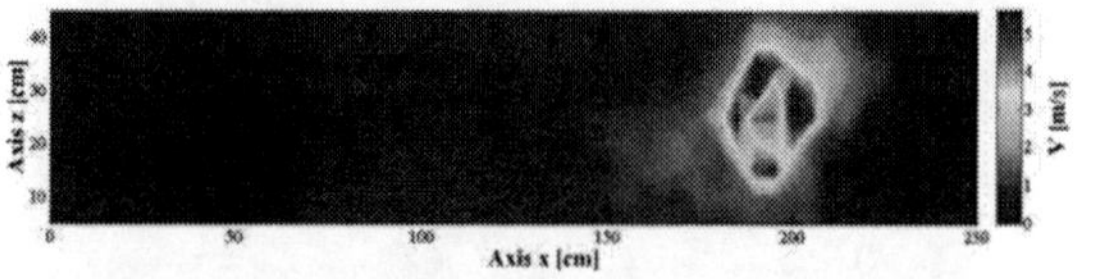

Figure 4. The air velocity in XZ1 plane

In the XZ2 plane which is depicted in Figure 5, the largest air velocity is apparent in the area of intake fans VS and VH. The measured velocity averaged at 7 m·s^{-1} and gradually declined. A smaller portion of the energy is located in the area of the outlets V1 and V2. This plane goes exactly through the center of the box which is also 25 cm away from the outer wall.

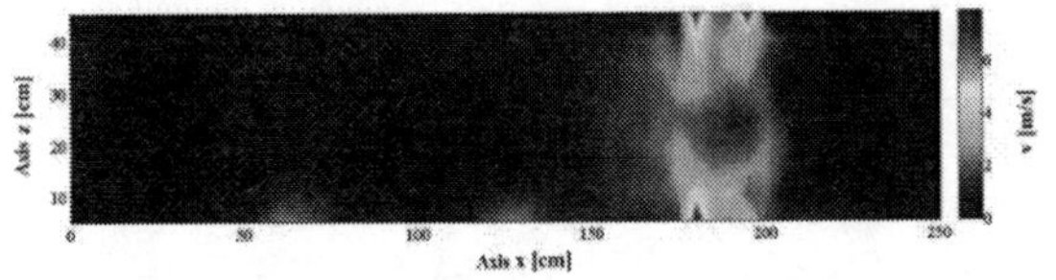

Figure 5. The air velocity in XZ2 plane.

Figure 6 shows three planes, the first two planes YZ3 and YZ2 pass through the middle of the V1and V2 outlets and YZ1 plane intersects the center of the fan inlets VL, VS, VP and VH. Visible difference in air velocity are between both outlets, where the V2 outlet shows larger decrease (up to 2.4 m·s^{-1}) than the V1 outlet where maximum air velocity was 2.8 m·s^{-1}. In the last plane YZ1, the air pumping from the fans is apparent and also the pumped air profile formed by the two turbulent lobes. It is caused due to the fact that the fan is formed by a rotor and stator whereby air can only pass through the rotor blades.

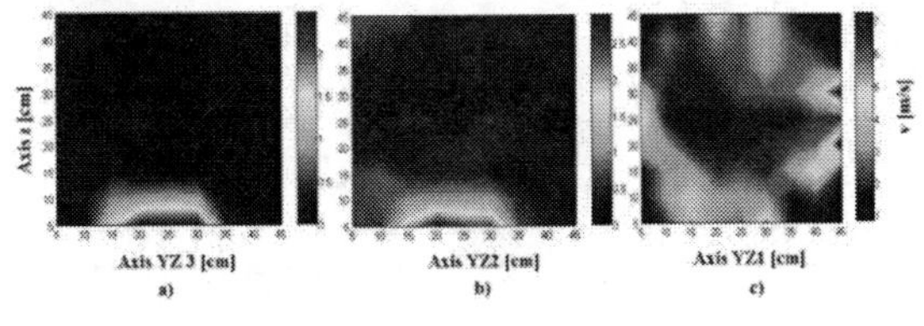

Figure 6. The air velocity: a) in YZ3 plane, b) in YZ2 plane, c) in YZ1 plane.

Individual velocity profiles at the fan inlets and outlets can be seen in Fig. 7 and 8. The maximum fan air velocity was determined to be around 10 m·s^{-1}, the average being however around 7 m·s^{-1}. This average value was then set in simulation software. The maximum air velocity at the outlets was around 8 m·s^{-1}.

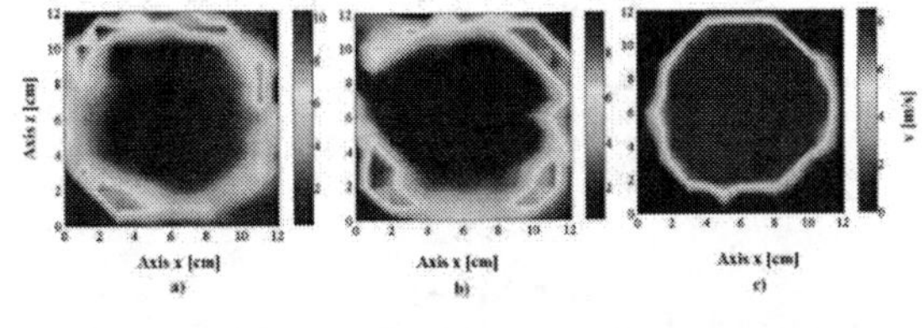

Figure 7. The air velocity a) VL fan, b) VS fan, c) V1 outlet.

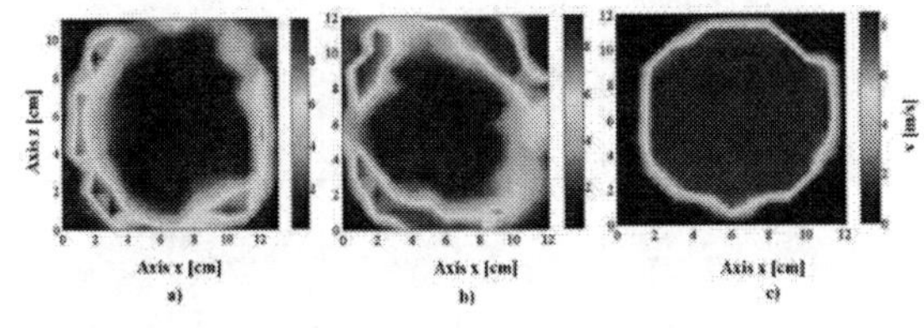

Figure 8. The air velocity a) VP fan, b) VH fan, c) V2 outlet.

6 Statistical comparisons of the numerical models

To compare different numerical models, three different methods were chosen. The first method used an averaging of the air velocity in individual planes. Another method used in the numerical model comparison used MSE, which is the root mean square error, and the last method used the NSE (Nash-Sutcliffe Efficiency). In Table 3 we can see the comparison of numerical models with real measured data by averaging the velocity in defined planes (see Figure 3). The highlighted models are the ones that most closely match the measured data. From Table 3 it is apparent that the k − ε RNG numerical model is the most similar to the measured values.

Table 3. Velocity averages comparison in defined planes. Velocity are in m·s^{-1}.

Inlets / Outlets	**Measured values**	**k − ε Standard**	**k − ε RNG**	**k − ε Realizable**	**k − ω Standard**	**k − ω SST**
VL	7	7.28	7.28	7.28	7.28	7.28
VS	6.75	6.94	6.94	6.94	6.94	6.94
VP	6.65	6.8	6.8	6.8	6.8	6.8
VH	6.25	6.25	6.25	6.25	6.25	6.25
V1	8	7.74	7.99	7.94	7.75	7.97
V2	8	7.8	8.5	8	7.95	8
			Plane			
XZ1	0.60	0.462	0.58	0.433	0.613	0.058

As part of the evaluation of the numerical model that most closely resembles the measurements, k − ε RNG model was chosen (see Figure 9). The simulation results in individual cross sections or planes are shown in the following figures. The numerical model k − ε RNG is an extension of k − ε Standard model, when it is apparent that k − ε RNG in greater extent calculates the turbulent flow at inlets and outlets, see Figure 10 and at the individual planes (see Figure 11) of the simulated box.

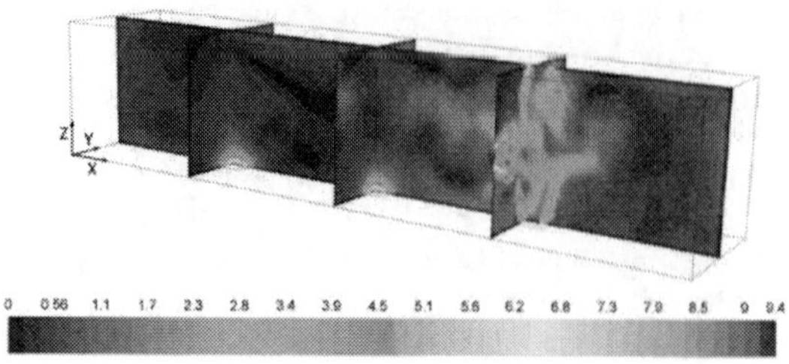

Figure 9. The velocity for different planes calculated using k – ε RNG model.

Figure 10. Section view of the velocity in XZ1 plane in the simulated box and for k – ε RNG model.

Figure 10 depicts the air velocity around the simulated box intake fan in XZ1 plane and calculated using k – ε RNG model. In XZ2 plane, the same simulation can be seen from a different perspective (Figure 11). The individual planes YZ1, YZ2 and YZ3 represent the simulated box cross sections according to Figure 3.

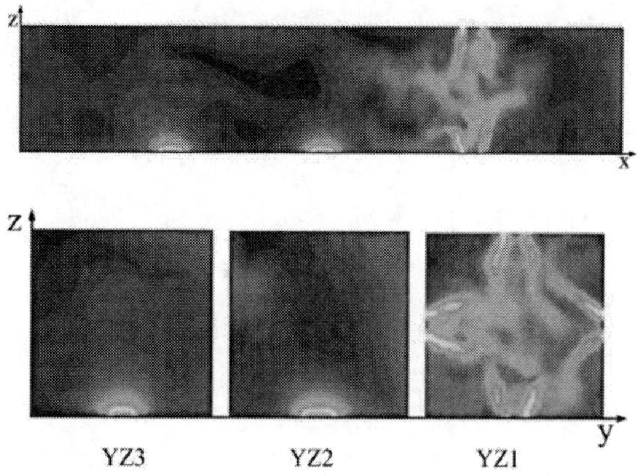

Figure 11. Section view of the velocity in XZ2, YZ1, YZ2 and YZ3 planes in the simulated box and for k – ε RNG model.

6.1 XZ1 plane evaluation

To determine the influence of turbulence on the optical beam, the XZ1 sectional evaluation was selected. The MSE and NSE methods were used to calculate the individual coefficients on selected layers, see Tables 4 and 5.

Table 4. Calculated coefficients for MSE method.

	Models				
cm	k – ε Standard	k – ε RNG	k – ε Realizable	k – ω Standard	k – ω SST
5	0.140	0.138	0.063	0.053	0.143
10	0.222	0.120	0.134	0.072	0.143
15	1.085	0.737	0.831	0.550	0.831
20	0.679	0.695	0.634	0.549	0.534
25	1.402	1.190	1.334	1.256	1.256
30	1.025	1.052	0.975	0.888	1.067
35	1.265	1.412	1.350	1.277	1.355
40	0.317	0.387	0.320	0.355	0.290
45	0.134	0.188	0.077	0.069	0.192

Table 5. Calculated coefficients for NSE method.

	Models				
cm	k – ε Standard	k – ε RNG	k – ε Realizable	k – ω Standard	k – ω SST
5	-0.125	-0.111	0.497	0.574	-0.149
10	0.291	0.359	0.289	0.618	0.239
15	-0.136	0.228	0.130	0.423	0.129
20	0.295	0.278	0.342	0.430	0.445

25	0.152	0.280	0.194	0.241	0.241
30	0.408	0.393	0.437	0.488	0.385
35	0.101	0	0.040	0.092	0.036
40	0.118	-0.080	0.108	0.010	0.192
45	0.051	-0.333	0.456	0.508	0.460

The highlighted areas show the greatest agreement with the measured values. The XZ1 plane closest to the box wall (5 cm) was simulated best using the k − ω Standard model. The simulated vs real measured velocity values of individual models at 25 cm level can be seen in Figure 12. The blue line represents the measured values while the dashed line the most realistic model. The measured values of velocity were up to 5.5 $m \cdot s^{-1}$ and the best approximation to that value was using the k − ε RNG model (solid red line in Figure 12).

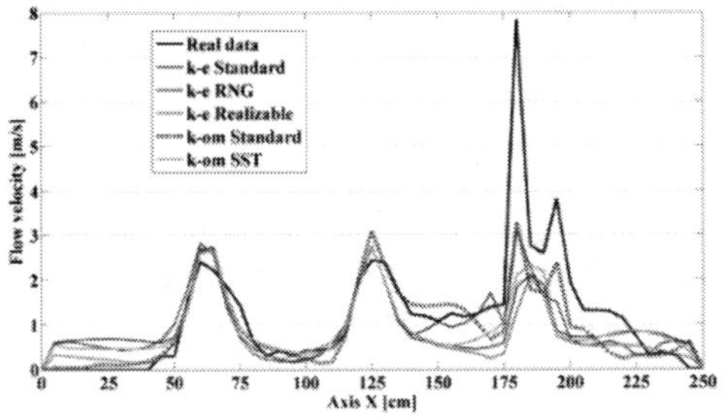

Figure 12. Model comparison at 25cm level.

7 Conclusion

In this article the team of authors have focused on the modelling of air flow in a simulation box used for modelling atmospheric conditions. From the measured data the realistic model was determined using different types of numerical models in the ANSYS Fluent 15 software. These models were compared and evaluated. Based on the NSE and MSE statistical methods, the numerical model that most closely corresponded to the actual measurements was chosen. The selected model can therefore be used to the model the air flow in the box.

Acknowledgement

The research described in this article was undertaken thanks to the active support of the Ministry of Education, Youth and Sports of the Czech Republic through grant project no. CZ.1.07/2.3.00/20.0217 within the frame of the operation programme Education for competitiveness financed by the European Structural Funds and from the state budget of the Czech Republic. This article was also supported by project Technology Agency of the Czech Republic TA03020439 and TA04021263. The research has been partially supported by the projects no. SP2015/130, SP2015/182 and VI20152020008.

References

1. T. David, J. Latal, F. Hanacek, P. Koudelka, J. Vitasek, J., P. Siska, J. Skapa, V. Vasinek, Advances in Electrical and Electronic Engineering **9**, 4 (2011)
2. O. Bouchet, J.C. Ricklin. Wireless optical telecommunications: principles and advances, John Wiley, Hoboken (2012)
3. R. Ramirez-Iniguez, S.M. Idrus, Z. Sun. Optical wireless communications: IR for wireless connectivity. CRC Press, Boca Raton (2008)
4. A. Vanderka, L. Hajek, J. Latal, J. Vitasek, P. Koudelka, Advances Electrical and Electronic Engineering, **12**, 6 (2014)
5. L. Hajek, J. Latal, M. Bojko, R. Poboril, P. Koudelka, J. Vitasek, P. Siska, V. Vasinek. EPJ Web of Conferences **67**, 02032 (2014)
6. A.K. Majumdar, J.C. Ricklin. Free Space Laser Communications Principles and Advances, Springer ISBN 978-0-387-28652-5 (2008)
7. M. Kozubkova, Modeling of fluid flow: FLUENT, CFX. Ostrava: Technical University of Ostrava, (2008)
8. T. Blejchar, Turbulence – Flow modeling - CFX. Ostrava: Technical University of Ostrava, ISBN 978-80-248-2606-6 (2010)
9. M. Kozubkova, S. Drabkova. Numerical modeling of flow: FLUENT I. Ostrava: Technical University of Ostrava, ISBN 80-248-0525-1 (2003)
10. M. Kozubkova, T. Blejchar, M. Bojko, Modelling of heat, mass transfer and momentum. Ostrava: Technical University of Ostrava, ISBN 978-80-248-2491-8 (2013)
11. FLUENT: Fluent 15.0 - ANSYS FLUENT, ANSYS, Inc., (2013)
12. D. C. Wilcox. Turbulence Modeling for CFD. D C W Industries. 522p. (2006)

EPJ Web of Conferences **114**, 02066 (2016)
DOI: 10.1051/epjconf/201611402066

Investigation of heat transfer in the petal detector structure

David Lávička[1,a] and Jindřich Kňourek[1]

[1]*University of West Bohemia, New Technologies – Research Centre, Univerzitní 8, 306 14 Plzeň, Czech Republic*

Abstract. This paper describes the CFD numerical simulation of heat transfer in the petal complex structure. Petal consists of core structure and detectors designed to improve the detection capabilities of the ATLAS detector and take advantage of operating at the High Luminosity LHC (HL-LHC). New generation silicon detectors in the petal require more efficient cooling of the front-end electronics and the silicon sensors themselves. To minimize reverse annealing of the silicon sensors the operating temperatures need to be reduced. In this case the evaporative CO2 cooling medium is used as it is ideal for this purpose. This work describes the flow simulation setup and results for the current prototype and heat transfer behavior for different operating states.

1 Introduction

The ATLAS detector is one of the seven particle detector experiments constructed at the Large Hadron Collider (LHC) in a particle accelerator at CERN in Switzerland. Currently, the development is focused in the Phase-II Upgrades of the ATLAS Detector and this development is coordinated by the DESY ATLAS group.

The DESY group is involved in a wide area of activities ranging from simulation studies, sensor assembly and testing, the overall system design (both mechanical and electrical) to the readout electronics. In our case, we will deal with petal component. The petal is the detector designed to improve the detection capabilities of the ATLAS detector and take advantage of operating at the High Luminosity LHC (HL-LHC). The increased luminosity and the accumulated radiation damage will render the current Inner Tracker no longer suitable for long term operations. It is one of the many reasons for replacing a current detector by new generation of silicon detectors. The new generation silicon detectors in the petal require more efficient cooling of the front-end electronics and the silicon sensors themselves. Other important requirements of the new generation cooling systems for the petal are the reduced mass and maintenance free operation.

In this case the evaporative CO2 cooling system is used as it is ideal for this purpose. It is known that two phase cooling is more efficient than single phase cooling. For high cooling and good evaporative capability the smaller diameters of tubes are used. The diameter of tubes is about 2 mm.

Our workplace New Technologies - Research Centre participates in the petal development in collaboration with DESY group. Our work is to focus to heat transfer numerical simulations. These numerical simulations describe the heat transfer and flow behavior in the the current prototype and also the behavior for different operating states. In this article we will focus on verifying the accuracy of the calculation model, the correct identification of material and first results describing the cooling capacity of petal component.

2 The Petal

The Figure 1 shows the assembly of all petal pieces in the detector. Turbo-fan end-cap support structure consists of seven disks with 32 petals each and a total of 50 000 readout chips.

Figure 1. Turbo-fan end-cap support structure. (source: http://wiki.nikhef.nl/atlas/Upgrade)

[a] Corresponding author: dlavicka@ntc.zcu.cz

The end-cap petals follow very closely the design of the barrel staves. However, the design of the petal is more complex in comparison to the barrel. The petal itself hosts 9 sensor modules per side and it is staffed from both sides (18 modules per petal in total). These modules are not identical within each petal: there are in total 6 different types of sensors, and 8 different types of hybrids. The petal can be divided in 6 different regions, or rings, depending on the type of sensor each ring hosts. This division is shown in Figure 2.

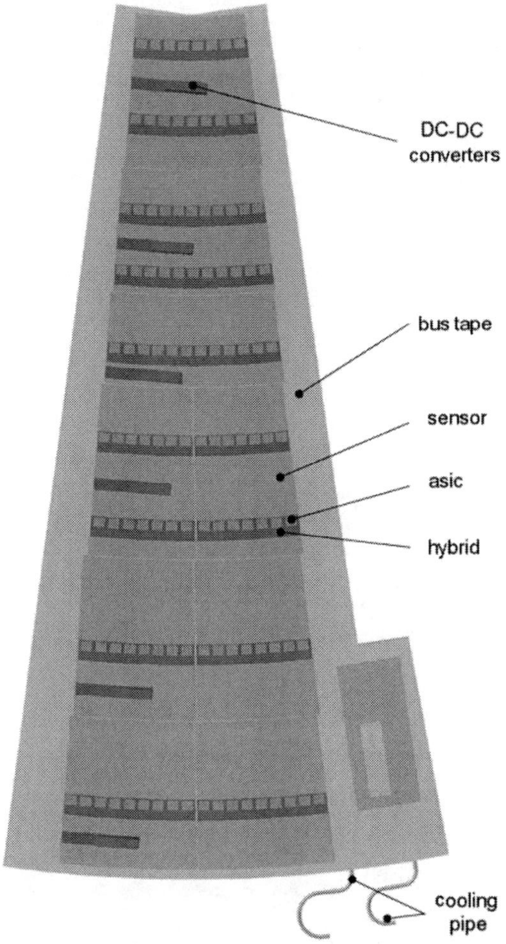

Figure 2. Description of the Petal

The Petal consist of the inner and outside parts. The base core box of petal is made from carbon fibre. Inner parts contain tubes for coolant medium. In our case the coolant medium is liquid CO_2, which is ideal for this purpose. The currently preferred material of tubes is titanium. In this work thin-walled tubes are used with the wall thickness of about 0.1 mm. The tube diameter is 2 mm. In future work, the investigation of other materials (for example stainless steel, carbon fibre, glass and more) is planned. The cut through the petal core structure is presented at Figure 3.

The space around the tubes is filled with high thermal conductive foam and the honeycomb. The honeycomb is currently replaced by air as its function is to isolate the rest of the inner part of petal. Inner part has a thickness of about 5 mm.

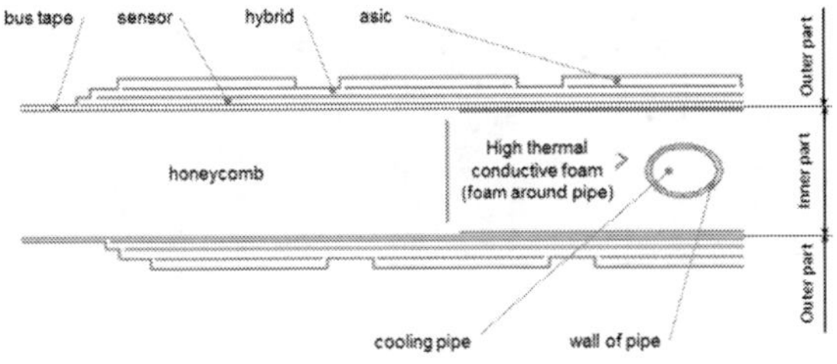

Figure 3. Inner and outer part of petal

The petal outer parts are made of silicon detectors of different sizes and in different layers. The thickness of silicon detectors is around 200 microns, which is a very small thickness. Some layers are described at Figure 2 and Figure 3. Each silicon layer is joined by glue to petal core or another layer. The presence of the glue appears to be a major obstacle for the precise definition of material properties to be used in the numerical model.

This dividing of model to the inner and outer parts is advantageous, because it allows to quickly respond to changing geometry of petal core inner part, for instance for changing the cooling tube shape.

3 Computational mesh

The petal structure geometry is exported from IGES files. The model needs to be to simplified and modified for easier and correct meshing. For instance, the bus cables connecting the silicon detectors to asic modules has to be neglected. Also, it was necessary to join and merge inner and outer parts as describing in Figure 3. The Figure 4 shows the diversity of the mesh density on the inner and outer parts.

Figure 4. The surface mesh on the inner and outer parts.

The CFD preprocessing software was used to modify the model geometry and to prepare the computational mesh. Volumetric mesh was obtained by application of the "cooper" mesh scheme (similar to "sweep") in that software. The vertical number of elements is not distributed uniformly. The volumetric mesh contains approximately 31.35 million of 3D mainly hexagonal cells. The inner part has approximately 13.03 million cells and outer parts has approximately 18.32 million cells.

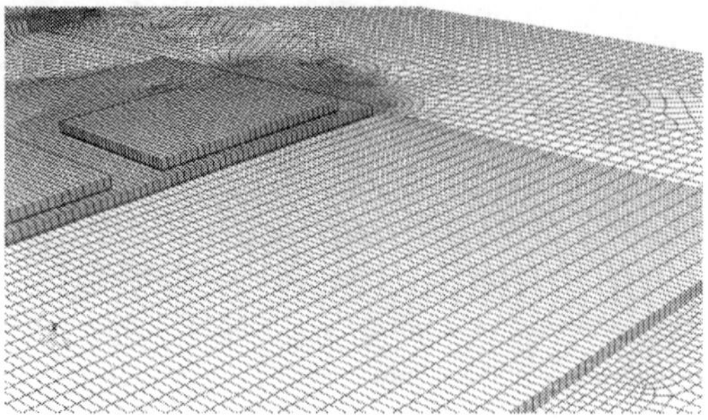

Figure 5. The surface mesh on silicon detectors.

The size of mesh is very big, because the simulations are focused mainly on the heat transfer to the surroundings. In this case, heat transfer simulations are very sensitive to the cell size and other characteristics such as skewness, expansion factor and more.

The volumetric mesh of tubes contains approximately 2.26 million of 3D cells and it is set as the fluid for flow of the cooling medium. The rest of cells is set as solid with different material properties to handle the simulation of heat transfer. The mesh detail is shown in Figure 6.

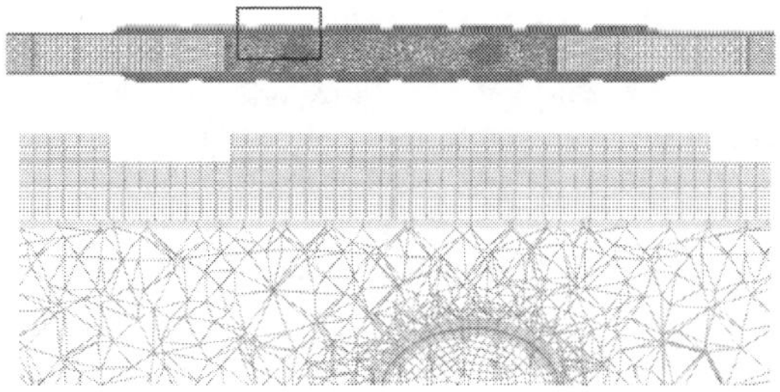

Figure 6. Detailed to boundary layer mesh of the tube.

4 Definition material and cooling fluid

The material definitions and material properties are the key issue for obtaining the correct results from numerical simulations. In our case, the material properties were supplied by DESY Atlas group from appropriate measurements, for instance for silicon detectors or for interlayer, where the material called polyimide is used. The honeycomb core structure inner part is replaced by air, because the material acts as insulation there. In numerical simulations, the adhesive layer between silicone detectors is not considered. This simplification may have a noticeable effect on the results.

Another important issue is the definition of cooling fluid. In fact, the used coolant is in two-phase mixture. In our case, the numerical simulation treats this mixture as a liquid and the evaporation effect of the liquid is not considered. But the material properties of liquid are set according to values for the material representing CO2 mixture. This mixture consists of liquid and vapor phases for which values were estimated.

5 Numerical simulation

The boundary conditions for numerical simulation are described in Figure 7. The blue tube (on right in Figure 7) has the boundary condition defined to the mass-flow-inlet. The mass flow rate has been set variably to values 1 g/s and 5 g/s. The red tube on left is set to pressure-outlet. The initial state before the start of calculation has been set for solid and fluid zones to 248 K (-25°C).

The numerical simulation and postprocessing was performed in the commercial software package ANSYS FLUENT v.15.0. The flow model included the turbulence model "Standard k-epsilon" in the second order of accuracy. The numerical simulation was performed in the stationary mode in the second order of accuracy in 6000 iterations.

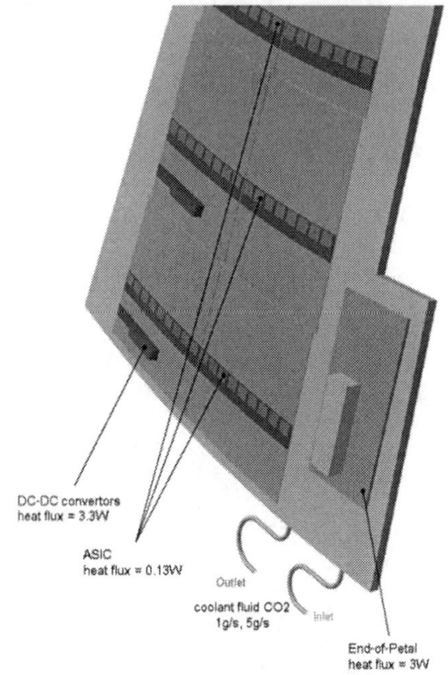

Figure 7. Boundary conditions and heat sources.

The heat sources (see Figure 7) has been set at appropriate components: in the DC-DC convertors to 3.3 W, in the ASIC to 0.13 W (for each silicon detector) and in the End-of-Petal component to 3 W. The total heat source through all components and for both sides is approximately 49 W. Heat transfer coefficient to surrounding medium is set to 15 W/m^{2}K and the ambient temperature is set to 248 K (-25°C).

6 Results

The simulation results are presented for 2 operating states for mass flow rates at 1 g/s and 5 g/s with the same initial conditions. Figure 8 shows the surface temperature distribution for coolant flow 5 g/s. Some temperature values are pinned for both flow rates. From the results it can be seen the largest heat source are the components of the DC-DC convertors, and they significantly affect the temperature of the whole petal structure. The bottom DC-DC convertor has higher temperature than the other DC-DC convertors. The temperature augmentation is there because the cooling pipe is not present in the core structure under this DC-DC convertor. This situation is same for both petal sides.

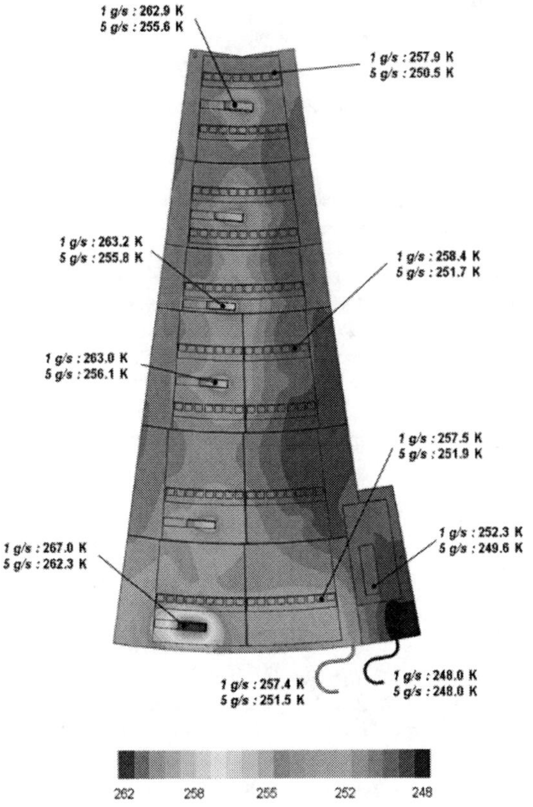

Figure 8. Surface-temperature distribution for 5 g/s.

Figure 9 shows the temperature situation in the cooling tube also for both operating conditions. The influence of the size of the flow rate and velocity in the pipe there is significant to temperature augmentation in the upper part of the device for a flow rate of 5 g/s. For the flow rate of 1 g/s the corresponding coolant velocity in the tube is approximately 0.26 m/s and Reynolds number is 1022. For the flow rate 5 g/s, it is almost 1.32 m/s and Reynolds number is 5190.

These calculations are characterized by Prandtl number 9.9. This number is only dependent on the material properties of the fluid.

The increased coolant velocity in the tube significantly affects the heat transfer mainly in the upper part of the petal. The surface temperature difference between the two variants has the value of 6 °C.

7 Conclusion

The aim of these simulations is to describe the heat transfer and flow behavior in the petal structure. The numerical model for CFD simulations and the definition of basic boundary conditions was set up. The results showed in the previous chapter are consistent with the expected results.

In future work, we will focus on improvement of material properties for solids and also to focus on more detailed description of the coolant mixture together with the evaporating process. In the next steps, we will also deal with the optimization of the cooling duct and the location of silicon detectors in the petal.

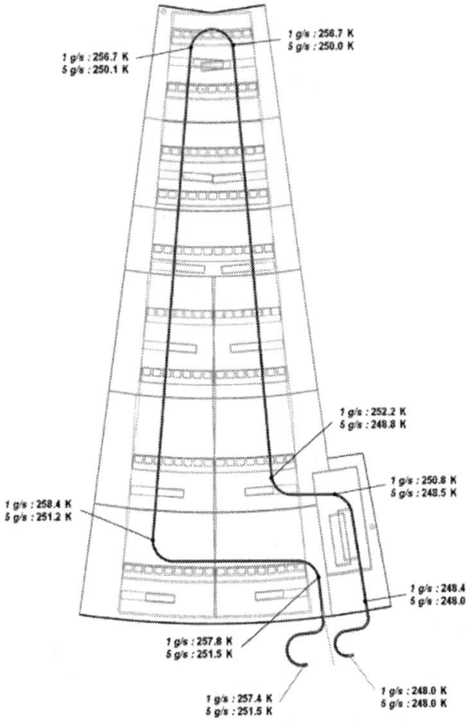

Figure 9. Temperature situation in cooling duct

Acknowledgment

The result was developed within the CENTEM project, reg. no. CZ.1.05/2.1.00/03.0088, cofunded by the ERDF as part of the Ministry of Education, Youth and Sports OP RDI programme and, in the follow-up sustainability stage, supported through CENTEM PLUS (LO1402).
We would also like to thank the DESY Atlas group for cooperation in R&D on the Petal.

References

1. N.P. Hessey, H. Schuilenberg, E. Hennes: ATLAS Endcap Stave Upgrade Plans at Nikhef, technical report, 2009.
2. B. Verlaat, M. V. Beuzekom, A. V. Lysebetten: CO2 cooling for HEP experiments, TWEPP-08, 2008.
3. N.P. Hessey, H. Schuilenberg, E. Hennes: ATLAS Upgrade R&D Plans at Nikhef, technical report, 2008.
4. http://atlas.desy.de/activities/
5. S. Diez: Silicon strip staves and petals for the ATLAS Upgrade tracker of the HL-LHC, 2012.

EPJ Web of Conferences **114**, 02067 (2016)
DOI: 10.1051/epjconf/201611402067

Experimental verification of boundary conditions for numerical simulation of airflow in a benchmark ventilation channel

Frantisek Lizal[1,a], Pavel Niedoba[1], Libor Seda[1], Jan Jedelsky[1], Miroslav Jicha[1]

[1]*Brno University of Technology, Faculty of Mechanical Engineering, Energy Institute, Technicka 2896/2, Brno 616 69, Czech Republic*

Abstract. Correct definition of boundary conditions is crucial for the appropriate simulation of a flow. It is a common practice that simulation of sufficiently long upstream entrance section is performed instead of experimental investigation of the actual conditions at the boundary of the examined area, in the case that the measurement is either impossible or extremely demanding. We focused on the case of a benchmark channel with ventilation outlet, which models a regular automotive ventilation system. At first, measurements of air velocity and turbulence intensity were performed at the boundary of the examined area, i.e. in the rectangular channel 272.5 mm upstream the ventilation outlet. Then, the experimentally acquired results were compared with results obtained by numerical simulation of further upstream entrance section defined according to generally approved theoretical suggestions. The comparison showed that despite the simple geometry and general agreement of average axial velocity, certain difference was found in the shape of the velocity profile. The difference was attributed to the simplifications of the numerical model and the isotropic turbulence assumption of the used turbulence model. The appropriate recommendations were stated for the future work.

1 Introduction

Computational fluid dynamics (CFD) is an efficient tool, which is used in many applications for calculation of airflow characteristics. In a frame of broader research task, we investigate the air flow field downstream of a benchmark automotive vent. The research is performed both numerically and experimentally. At first, the experimental test rig was designed, and the geometry of the channel was modelled in a simplified fashion and is used for the numerical simulation. The main goal of the broader project is to determine the influence of the upstream channel, adjustable grilles and design of the closing valve (these parts are hereafter referred to as the main section) on the flow.

The precise setting of boundary conditions at the inlet of the critical section, namely velocity profile and turbulence intensity, is necessary for a correct simulation of the flow. There are two ways to determine the boundary conditions. The first option is the experimental measurement performed directly upstream of the main section, the alternative way is the numerical simulation of a sufficiently long entrance section.

In this paper we compare the results acquired experimentally using constant temperature anemometry and numerically at the inlet boundary of the main section to determine the possible error in the velocity and turbulence intensity settings.

[a] Corresponding author: lizal@fme.vutbr.cz

2 Methods

The whole test rig for measurement of the flow downstream of the benchmark automotive vent (Figure 1) consists of a regular automotive ventilator connected to a 12 V laboratory electric supply. The flow rate is calculated from a measured pressure loss of an orifice plate according to EN ISO 5167. 3D fittings and stilling pipes are located upstream and downstream of the orifice.

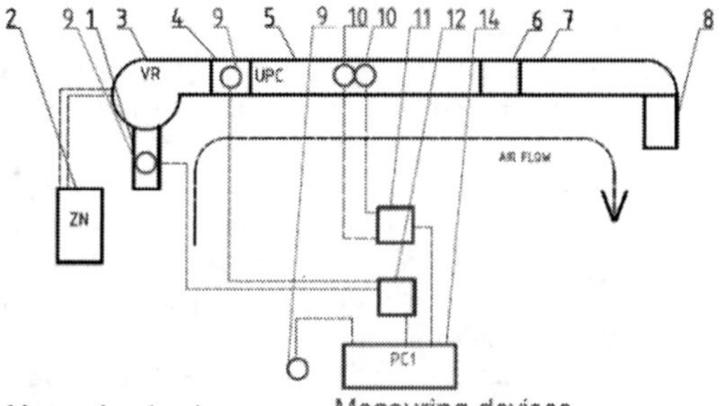

Measuring track
1 Suction adapter
2 Voltage supply
3 Ventilator unit
4 3D fitting
5 Stilling pipe
6 3D fitting
7 Exchangeable conduit
8 Benchmark air vent

Measuring devices
9 Resistance sensor Pt100
10 Pressure measurement
11 Pressure transducer
12 Temperature transducer
14 Computer (PC1)

Figure 1. Schematic of the benchmark automotive vent measuring rig [1].

The conduit directly upstream of the benchmark vent is interchangeable and either straight or bent channel can be inserted. Temperature is measured upstream and downstream of the ventilator and in the surrounding air. Absolute pressure is measured upstream of the orifice and atmospheric pressure is measured in the surroundings.

The velocity and turbulence measurement at the boundary of the main section (upstream of the interchangeable conduit) was performed in three vertical and five horizontal lines 272.5 mm upstream of the upper edge of the benchmark vent (see Figure 2). Both the measurement and simulation were performed for flowrate of 87.7 m^3/h. Dimensions of the rectangular channel in the measuring cross-section was 42 × 95 mm. Reynolds number was approximately 18,000, therefore the turbulent velocity profile had been expected.

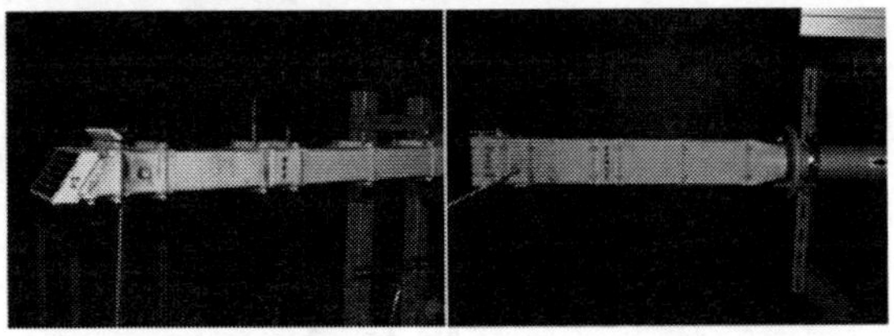

Figure 2. Photographs of the measuring rig (side view on the left and bottom view on the right) with drilled holes for insertion of the velocity probe [1].

2.1 Numerical simulation

The three-dimensional incompressible steady-state numerical simulation was carried out with the open-source code OpenFOAM. The governing Reynolds-averaged Navier-Stokes (RANS) equations were closed with the *k-ω* SST turbulence model [2]. The numerical code is based on the finite volume method using the SIMPLE scheme for the incompressible steady-state flow.

Figure 3 shows the simplified numerical model of the benchmark vent. In the process of the simplification, some geometry details had to be neglected to reduce the computational costs. The mesh was generated using the open-source mesh generator SALOME and consists of 1,158,134 tetrahedral cells. The case requires initial and boundary conditions settings for all the involved fields, i.e. velocity U or volumetric flowrate V, pressure p, turbulent kinetic energy k and specific rate of dissipation ω. The inlet turbulence properties were set to $k = 2.92$ m^2s^{-2} and $\omega = 637\ s^{-1}$.

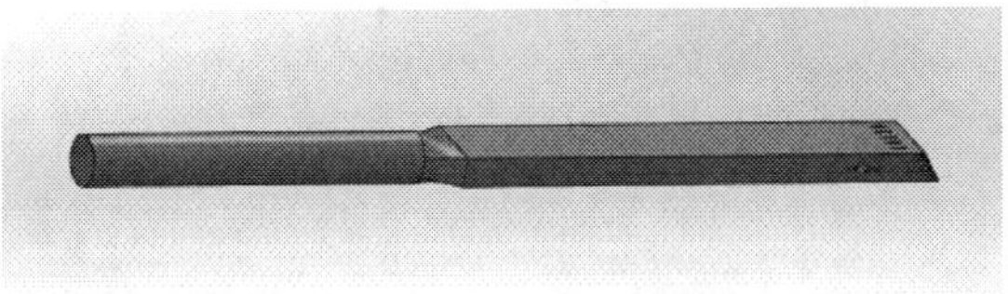

Figure 3. Simplified numerical model of the benchmark vent.

The calculation of the turbulence intensity (Tu) from numerical data was based on the turbulent kinetic energy, i.e.

$$Tu = \frac{\sqrt{\frac{2}{3}k}}{\overline{U}} \cdot 100\ (\%), \tag{1}$$

where $\sqrt{\frac{2}{3}k}$ represents the root mean square value of the turbulent velocity field u'. Note that the turbulent kinetic energy assumes isotropic fluctuations, which does not comply with our case of channel flow, where the axial turbulent velocity component dominates. Hence, the difference between numerical and experimental values of turbulent intensity is expected, i.e. the numerical turbulent intensity underestimates the experimental turbulent intensity in the axial direction.

2.2 Experiments

The measurement of velocity was performed by Constant Temperature Anemometry (CTA) system Dantec Streamline with single wire probe 55P04 (Dantec Dynamics A/S, Skovlunde, Denmark). The velocity calibration was performed in the range of 0.5 to 20 m/s in the Dantec StreamLine® Pro Automatic calibrator. Measuring frequency was 1 kHz and 1024 samples were taken in each measuring point. Traversing step of the velocity probe was 2 mm and shifting of the probe was provided by the automatic ISEL 3D traverse system (ISEL, Eiterfeld, Germany).

The turbulence intensity was calculated as follows:

$$Tu = \frac{u'_{rms}}{\overline{U}} \cdot 100\ (\%), \tag{2}$$

where u'_{rms} was determined as the root mean square value from the turbulent velocity samples u' recorded in one measuring point.

The uncertainty of the velocity measurement was calculated according to [3]. The uncertainty for the average velocity of 10 m/s was 3.1 %. The major source of uncertainty is the calibration process, other sources, such as linearization, A/D conversion, probe positioning, temperature variation of a wire and of ambient air, variation of ambient pressure and humidity have lower effect, but were included in the calculation of the velocity uncertainty.

The uncertainty of the flow rate measurement was calculated according to EN ISO 5167-2. The flow rate coefficient, piping dimensions, expansion coefficient, orifice dimension, pressure difference measurement, and density measurement were taken into account. The type B uncertainty of flow rate measurement reached 3.4 %. Repeatability of the flow rate setting, which depends mainly on the ventilator power fluctuations, was experimentally verified. Ventilator power is set by the voltage on the electric power supply. Setting of the flowrate was repeated 15 times and recorded. The standard deviation calculated from this data set, which corresponds to type A uncertainty, reached 0.230 m^3/h. The combined uncertainty of the flow rate was then 2.761 m^3/h.

3 Results and discussion

The numerical simulation provides detailed space-resolved information about the flow in the channel, while the CTA is only capable of a pointwise measurement. Therefore the axial velocity is displayed in the whole measuring cross-section (Figure 4) based on the simulation data, but only three horizontal and five vertical measuring lines can serve for the direct comparison of CFD and experiments (Figures 5 to 8).

Figure 4 shows the axial velocity component in the measured cross-section with lines on which the experimental measurement was carried out. It is obvious that there are no disturbances in the flow and almost ideally developed flow appears there.

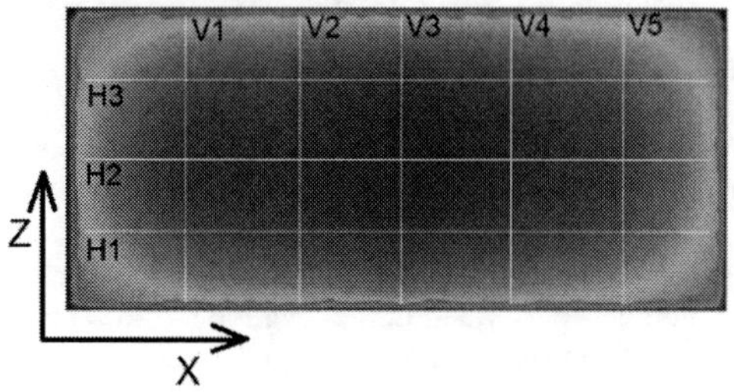

Figure 4. The mean axial velocity component calculated by CFD in the measured cross-section. The white lines correspond with the measuring lines and the labels correspond with labelling in Figures 5 to 8.

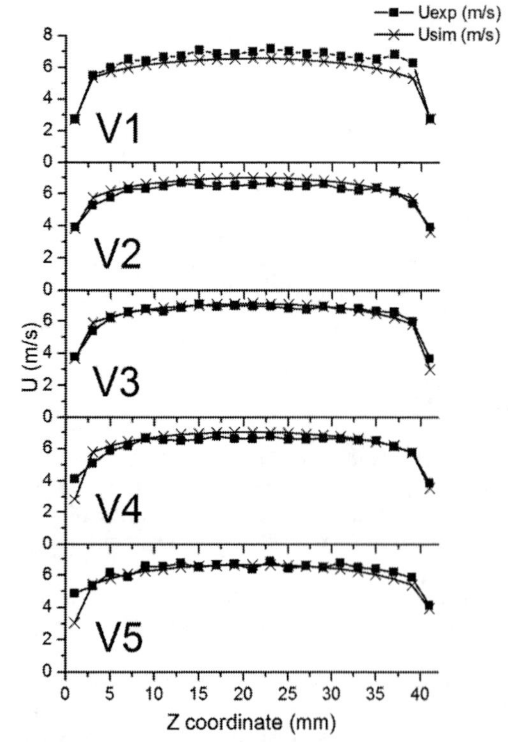

Figure 5. Axial velocity profiles in the vertical measuring lines. The top graph represents the left measuring line in the streamwise view on the measured cross-section.

Comparison of the velocity on the middle vertical measuring line V3 (Figure 5) shows very good agreement of CFD and experiments. Some slight differences can be seen in the V4 and V5 lines close to the probe insertion hole (position Z = 0 mm). These differences could arise from the vicinity of the hole, which can cause flow disturbances. These disturbances were not present in the CFD, as no holes were modelled.

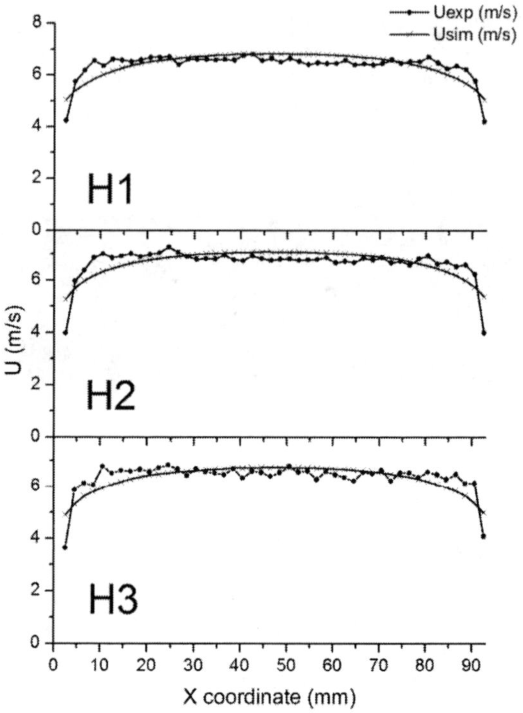

Figure 6. Axial velocity profiles in the horizontal measuring lines. The top graph represents the lowest measuring line.

It should be noted, that the bent probe was used and therefore the heated wire was approximately 5 mm upstream of the insertion holes plane and therefore only limited effect of the holes was expected. Only minor deviations can be observed in measuring lines V1 and V2.

Horizontal scans (Figure 6) differ significantly. Experimentally acquired profiles appear more turbulent than the numerically simulated. As the flowrate is identical in both the cases, the logical explanation is, that the numerical simulation does not capture the realistic development of the turbulence.

This assumption is partially supported by the comparison of the turbulence intensity presented in Figures 7 and 8. However, as was already mentioned, underestimation of the turbulence intensity in CFD was expected due to the invalidity of the isotropic turbulence assumption of the turbulence model and the geometry simplifications. Apart from that, the turbulence intensity profiles differ in the vertical lines in the vicinity of the insertion holes, which supports the supposition, that the holes could have influenced the measurement close to Z = 0 mm coordinates. The cause of asymmetry of the turbulence intensity profiles H1 and H3 is not known. The additional measurement will be performed to verify that the first measurement was correct. However, the result of the repeated measurement is not available at the time of finalizing this contribution.

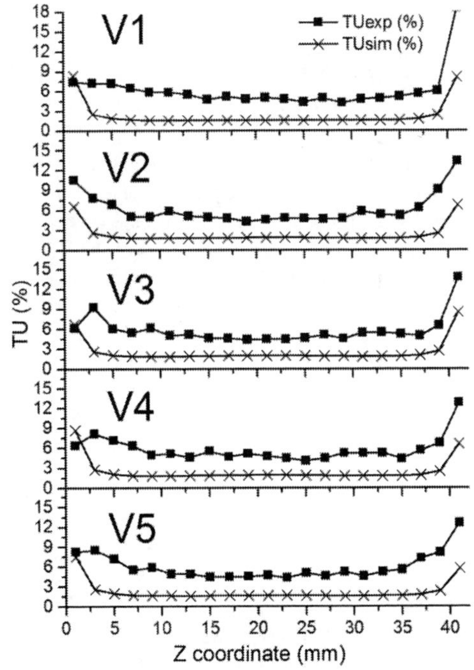

Figure 7. Turbulence intensity in the vertical measuring lines. The top graph represents the left measuring line in the streamwise view on the measured cross-section.

4 Conclusions

The comparison of experimentally and numerically acquired axial velocity and turbulence intensity at the boundary of a main section of our measuring rig showed, that despite the simple geometry and general agreement of the average cross-sectional velocity, certain difference was found in the shape of the velocity profile. The difference can be attributed to inaccurately simulated turbulence. Two reasons can be found, that may have influenced that. The first is the isotropic turbulence assumption of the k-ω SST model. However, more probable explanation is the second one, i.e. that the lower turbulence intensity in the simulation was caused by the absence of geometrical details, such as channel pieces connections, wall roughness and imperfections of the wall surface. These disturbances are transferred in the real flow and affect the development of the turbulence.

Our study indicates that researchers should resist the commonly encountered inclination to restrict the experimental measurement of boundary conditions and the verification of CFD results. It is important to combine both the approaches to achieve valuable results.

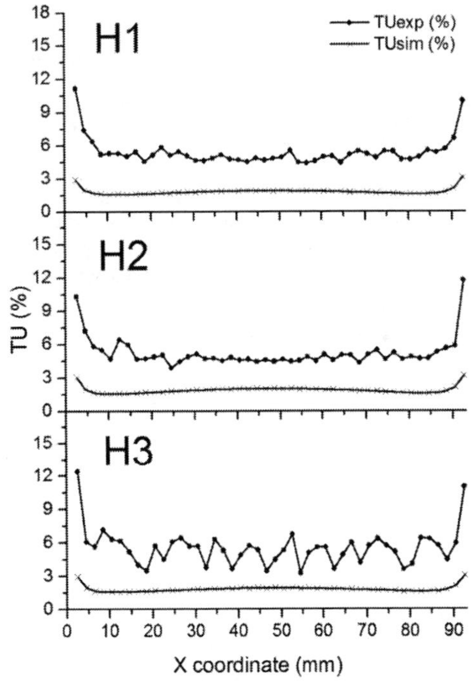

Figure 8. Turbulence intensity in the horizontal measuring lines. The top graph represents the lowest measuring line.

Acknowledgement

The presented research was supported by the project LO1202 NETME CENTRE PLUS with the financial support from the Ministry of Education, Youth and Sports of the Czech Republic under the "National Sustainability Programme I" and from the project reg. number FSI-S-14-2355 of the Brno University of Technology.

References

1 L. Šeda, Master Diploma Thesis, Brno University of Technology, 2015.
2 F. R. Menter, AIAA journal 32 (8), 1598 (1994).
3 F. Jorgensen, How to measure turbulence with hot-wire anemometers - a practical guide. (Dantec Dynamics, Skovlunde, Denmark, 2002).

EPJ Web of Conferences **114**, 02068 (2016)
DOI: 10.1051/epjconf/201611402068

Determination of the heat transfer coefficient from IRT measurement data using the Trefftz method

Beata Maciejewska[1,a], Kinga Strąk[2] and Magdalena Piasecka[2]

[1]*Kielce University of Technology, Faculty of Management and Computer Modelling, 25-314 Kielce*
[2]*Kielce University of Technology, Faculty of Mechatronics and Mechanical Engineering, 25-314 Kielce*

Abstract. The paper presents the method of heat transfer coefficient determination for boiling research during FC-72 flow in the minichannels, each 1.7 mm deep, 24 mm wide and 360 mm long. The heating element was the thin foil, enhanced on the side which comes into contact with fluid in the minichannels. Local values of the heat transfer coefficient were calculated from the Robin boundary condition. The foil temperature distribution and the derivative of the foil temperature were obtained by solving the two-dimensional inverse heat conduction problem, due to measurements obtained by IRT. Calculations was carried out by the method based on the approximation of the solution of the problem using a linear combination of Trefftz functions. The basic property of this functions is they satisfy the governing equation. Unknown coefficients of linear combination of Trefftz functions are calculated from the minimization of the functional that expresses the mean square error of the approximate solution on the boundary. The results presented as IR thermographs, two-phase flow structure images and the heat transfer coefficient as a function of the distance from the channel inlet, were analyzed.

1 Introduction

Boiling heat transfer during fluid flow in minichannels is process, which it exploited the latent heat of vapourization. Due to using flow boiling process, it is possible to remove more amounts of thermal energy than in single-phase flow. Flow boiling heat transfer is often used as cooling devices or compact heat exchangers. Channels of small dimensions ensure a greater contact surface in compact heat exchangers. It also enabled work under high pressure [1]. Process of flow boiling in small channel was susceptible for various kinds of instability than in traditional size channels. In [2] the two-phase flow instability of nanofluids in a minichannel have been presented.

In literature different classifications of small channels can be found. One of the most popular it was the classification proposed by Kandlikar [3], based on hydraulic diameter of channels. According to this classification microchannels have hydraulic diameter d_h: 10 μm ≤ d_h ≤ 200 μm, minichannels: 200 μm ≤ d_h ≤ 3 mm and conventional channels d_h > 3 mm. Another classification was presented by Shah, who divides channels in compact heat exchangers according to their hydraulic diameter into two groups: conventional channels (d_h > 6 mm) and minichannels (d_h < 6 mm) [4].

This paper presents calculations of local heat transfer coefficients between the heated foil and the cooling liquid FC-72 flowing in a vertical minichannel, by means of the Trefftz method. This method was proposed in 1926 by E. Trefftz [5]. The Trefftz method involves approximating the solution of a problem using a linear combination of Trefftz functions. This functions strictly satisfy the governing differential equation. Additional information on the Trefftz method can be found in [6-18].

2 Experiment

2.1 Experimental stand

Experimental stand consists of several systems: the flow loop with FC-72 as a working fluid, the supply and control system, the data and image acquisition system and the lighting system. Figure 1 presents the flow loop of the experimental stand. The main elements of the this loop are: a rotary pump, a heat exchanger, a compensating tank, rotameters, a filter, and a deaerator. The data and image acquisition system consists with: a digital camera, an infrared camera, a data acquisition station, a computer with special software, and lighting systems. Experimental stand are discussed in detail in [19-21].

[a] Corresponding author: beatam@tu.kielce.pl

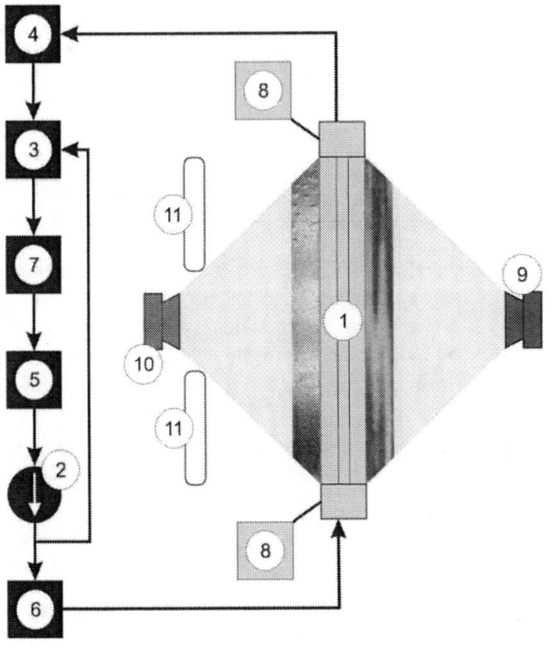

Figure 1. Schematic diagram of the flow loop with main elements of image acquisition and lighting systems of the experimental stand, 1 - testing module with a vertical minichannel; 2 - rotary pump; 3 - pressure regulator; 4 - tube-type heat exchanger, 5 - filter, 6 - rotameters; 7 - deaerator, 8 - pressure converter, 9 - digital SLR camera, 10 - infrared camera, 11 - halogen lamp.

2.2 The testing module

The testing module consists of two parallel rectangular, vertically-oriented minichannels, each 1.7 mm deep, 24 mm wide and 360 mm long, figure 2. The heating element for FC-72 flowing in each minichannel (1) is the thin foil (2). It is possible to measure the two-dimensional temperature distribution of the heating foil in each minichannel by two contactless temperature methods, separately. In one minichannel the foil temperature is obtained by applying infrared thermography (IRT minichannel), in the other minichannel - by using liquid crystal thermography (LCT minichannel). In this study only data for IRT minichannel are taken account. In the IRT minichannel, the outer surface of the foil is coated with black paint (3) to gain an emissivity of 0.83 [22]. Foil temperature is measured by IR camera in the central, axially symmetric, part of the channel (approx. 10 mm x 350 mm). The IRT channel is reinforced on both sides with glass panels to prevent the heating foil from deforming. The other side of the minichannel is observed through the glass pane (4). In this channel occur a two-phase flow during increasing the heat flux supplied to the foil. K-type thermocouples (7) and pressure converters are installed at the inlet and outlet of the minichannel.

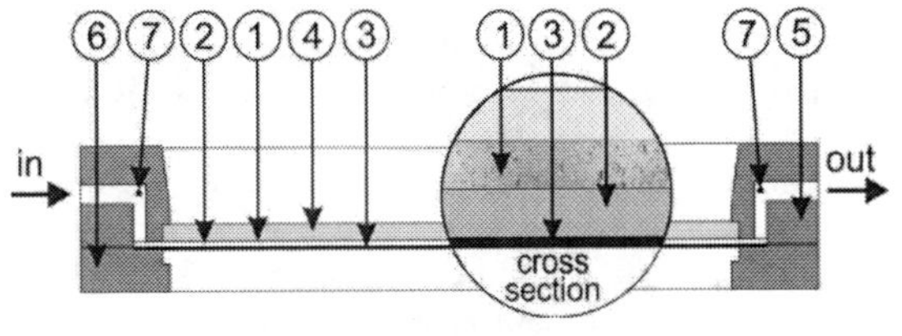

Figure 2. Schematic diagram of the testing module: 1 - IRT minichannel, 2 - heating foil, 3 - black paint layer, 4 - glass pane, 5 - channel body, 6 - front cover, 7 - thermocouple.

2.3 Experimental methodology

During experimental series, there is a laminar flow of FC-72 in the minichannel. When the desired pressure and flow rate are reached, there is a gradual increase in the electric power supplied to the heating foil followed by an increase in the heat flux transferred to the fluid in the channel. Data for four settings of the increasing heat flux supplied to the foil are analysed.

2.4 Experimental uncertainties

The images taken with a FLIR SC640 infrared camera system operating in the spectral range from 7.5 μm to 13 μm were recorded in a digital system with an image frequency of 30 Hz. The accuracy of the camera is assumed as ± 2% of reading [23].The mean error in the measurement of the heating foil temperature using infrared thermography was estimated as 0.50 K.

3 Local heat transfer coefficient determination

The local heat transfer coefficients were calculated from the following formula:

$$\alpha(x)=\frac{-\lambda\dfrac{\partial T(x,\delta)}{\partial y}}{T(x,\delta)-T_f(x)} \tag{1}$$

where λ – coefficient of thermal conductivity of the foil, δ – foil thickness, T – foil temperature, $T_f(x)$ – fluid temperature determined on the basis of the linear distribution of fluid temperature along the length of the minichannel from the inlet to the outlet. The foil temperature T was determined by solving the inverse problem [24-28] of the heat transfer in the heating foil.

The foil temperature T satisfies Poisson's equation:

$$\nabla^2 T=-\frac{q_V}{\lambda} \text{ for } (x,y)\in\Omega \tag{2}$$

where $\Omega=\{(x,y)\in R^2: 0<x<L,\ \ 0<y<\delta\}$, L –length of the minichannel and λ, δ defined as for Eq.(1).

The volumetric heat flux q_V was calculated according to the formula:

$$q_V=\frac{I\cdot\Delta U}{A_F\cdot\delta} \tag{3}$$

where: I – current, ΔU – voltage drop, A_F – surface area of the heating foil determined for IRT minichannel, δ – defined as for Eq. (1).

The boundary conditions as shown in figure 3 where: P - number of measurements obtained using infrared thermography (IRT) on the outer surface of the heating foil, T_p^{IRT} – foil temperature measured by infrared thermography at the boundary $y = 0$.

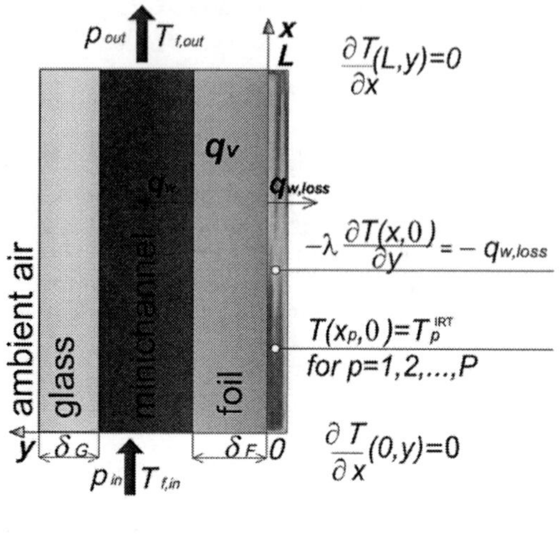

Figure 3. Boundary conditions for a two-dimensional approximation.

Heat loss to the surroundings was calculated as follows:

$$q_{wloss} = \alpha_s \left(\max_{p=1,2,\dots,P} T_p^{IRT} - T_s \right) \tag{4}$$

where: α_s – local values of the heat transfer coefficient at the interface between the heating foil and the surroundings [29], T_s – ambient temperature, T_p^{IRT} and P defined as for Eq.(2).

In domain Ω, the temperature T was approximated by means of the linear combination of Trefftz functions:

$$T(x,y) = u(x,y) + \sum_{i=1}^{N} a_i v_i(x,y) \tag{5}$$

where: $u(x,y)$ - particular solution of Eq. (2), $v_i(x,y)$ - the Trefftz functions.

The unknown coefficients a_i were calculated by minimizing the functional:

$$\begin{aligned} J = & \int_0^{\delta} \left((\frac{\partial T}{\partial x}(0,y))^2 + (\frac{\partial T}{\partial x}(L,y))^2 \right) dy + \\ & + \int_0^{\delta} (\frac{\partial T}{\partial y}(x,0) - \frac{q_{w,loss}}{\lambda})^2 dx + \\ & + \sum_{p=1}^{P} (T(x_p,0) - T_p^{IRT})^2 \end{aligned} \tag{6}$$

4 Results

Figure 4a shows the infrared thermographs of the complete experimental series. Four heat flux settings were selected for further analysis. Figure 4b,c shows the data for these settings in the form of IR thermograms of the heating foil surface and two-phase flow images, respectively.

Heat transfer processes were analysed using the experimental data - the temperature of the heating foil, two-phase flow structure images, the temperature of the working fluid, the pressure in the channel - and the plots of the heat transfer coefficients (figure 5) as a function of the distance from the minichannel inlet. The heat transfer coefficients calculated using the two-dimensional model were discussed. The calculation results were presented for four settings of the increasing heat flux supplied to the heating foil.

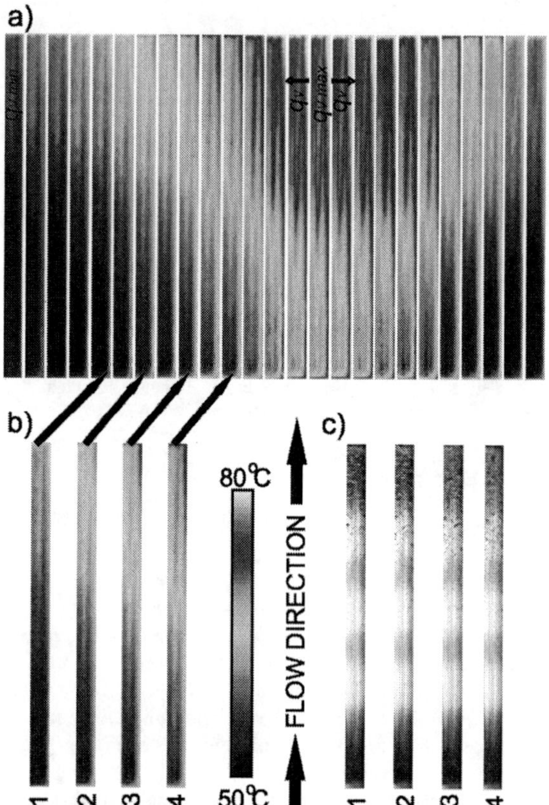

Figure 4. a,b) Thermographs of the heating foil surface measured by infrared thermography, data for: the complete experimental series (a), selected settings of the heat flux (b); c) two-phase flow structure images - data for selected settings of the heat flux; experimental parameters: average mass flux of 164 kg m^{-2}s^{-1}, average inlet pressure of 165 kPa, average inlet liquid subcooling of 45.8 K, volumetric heat fluxes: #1) q_V= 9.60·10^4 kW m^{-3}; #2) q_V= 1.06·10^5 kW m^{-3}; #3) q_V= 1.17·10^5 kW m^{-3}; #4) q_V= 1.28·10^5 kW m^{-3}.

In 2D approach, for approximation of the foil temperature, 12 Trefftz functions defined by the following formulas were used [12]:

$$v_n(x,y) = \sum_{k=0}^{[\frac{n}{2}]} \frac{(-1)^k x^{n-2k} y^{2k}}{(2k)!(n-2k)!}, \quad n = 0,1,\dots,7 \tag{7}$$

$$v_m(x,y)=\sum_{k=0}^{[\frac{m-1}{2}]}\frac{(-1)^k x^{m-2k-1} y^{2k+1}}{(2k+1)!(m-2k-1)!}, \quad m=1,2,\ldots,8 \qquad (8)$$

as a particular solution of Eq. (2)

$$u(x,y)=-\frac{q_V(x^2+y^2)}{4\lambda_F} \qquad (9)$$

was accepted.

The IRT temperature measurements allowed determining the foil temperature over the entire minichannel length, which was not the case in the authors' earlier works [20,21,30,31], where temperatures were measured using liquid crystal thermography (LCT) limited by the active band of the liquid crystal mixture.

The values of local heat transfer coefficients (see figure 5) indicate their rapid increase in the boiling region. Increasing values of the vapour quality were usually obtained by increasing the applied heat flux. Local vapour quality increases with increasing distance from the minichannel inlet [20]. The highest heat transfer coefficient for all heat flux settings was achieved at the minichannel outlet for which vapour quality was the highest. This observation is consistent with some of the findings by other researchers [32,33] but not with the previous research results reported by the authors of this paper. The onset of boiling was accompanied by a sharp increase in the heat transfer coefficient. In fully developed nucleate boiling, the heat transfer coefficient was the highest but then it decreased sharply with an increase in the vapour quality. Very high values of the local heat transfer coefficient at the minichannel outlet were typically determined for the saturated boiling region. In the subcooled boiling region, significantly lower heat transfer coefficients were obtained [20,21,30]. In the present study, for the experimental data, a sharp increase in heat transfer coefficient was observed at the distance of 0.25 m to 0.35 m from the minichannel inlet, but the experimental data do not indicate that the saturated boiling region developed (except the narrow region near the minichannel outlet). However, the analysed region overlapped with the area where an increase in the number and size of the generated bubbles was detected (see figure 4c). Furthermore, no 'nucleation hysteresis' was observed while increasing the heat flux to the heating foil. It is known that under certain conditions, a considerable rise in the wall temperature can occur above the saturation point before the boiling begins. This "temperature overshoot" also referred to as "superheated excursion" or "nucleation hysteresis" is characteristic for highly wetting dielectric fluids (e. g. refrigerants) [33-38]. The nucleation hysteresis is mainly due to the local distribution of nucleation sites and the wetting characteristics of the used refrigerant [33]. The abrupt decrease in heating surface temperature results from the vapor bubbles spontaneous formation in the wall adjacent layer. They function as internal heat sinks, absorbing significant amount of energy transferred to the liquid [35-38].

The untypical heat transfer coefficient plot and the fact that 'nucleation hysteresis' did not occur may have been due to large pressure fluctuations recorded for the increasing heat flux. During the experimental series, extremely large pressure fluctuations were measured at the outlet of the channel. Increasing local pressure may have caused disturbance and turbulent flow. This probably increased the local bulk fluid temperature, resulting in high local heat transfer coefficients at the outlet of the minichannel.

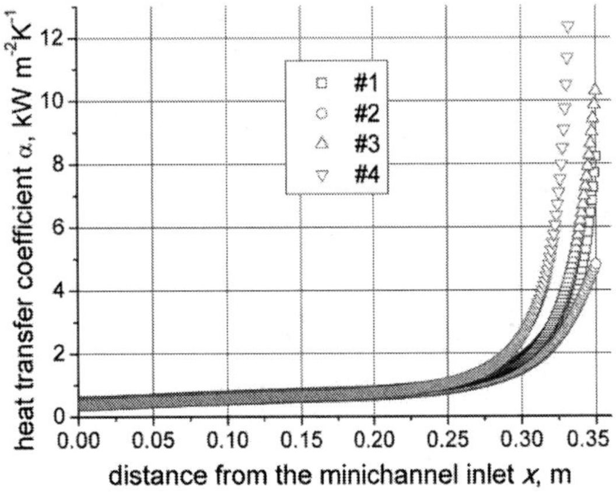

Figure 5. Heat transfer coefficient as a function of the distance from the minichannel inlet, obtained by using IRT.

5 Conclusions

The paper presents the method of heat transfer coefficient determination for boiling research during FC-72 flow in a minichannel. Local values of the heat transfer coefficient were calculated from the Robin boundary condition. The foil temperature distribution and the derivative of the foil temperature were obtained by solving the two-dimensional inverse heat conduction problem, due to measurements obtained by infrared thermography. Calculations was carried out by the method based on the approximation of the solution of the problem using a linear combination of Trefftz functions.

The results were presented as IR thermographs and two-phase flow structure images. The heat transfer coefficient were given as a function of the distance from the channel inlet. The phenomenon of "nucleation hysteresis" was not observed. Local heat transfer coefficients increased rapidly in the developed boiling region. The greatest value of the coefficients was achieved at the minichannel outlet for all heat flux settings under analysis. Large pressure fluctuations recorded while increasing the heat flux may have been a factor in the non-occurrence of "nucleation hysteresis" and may have contributed to the untypical plot of the heat transfer coefficient.

Acknowledgements

The research reported herein was supported by a grant from the National Scientific Center (No. DEC-2013/09/B/ST8/02825).

References

1. S. Saisorn, J. Kaew-On, S. Wongwises, Exp. Therm. Fluid Sci. **46**, 232-244 (2013)
2. L. Yu, A. Sur, D. Liu, J. Heat Tran. **137** (5), (2015)
3. S. G. Kandlikar, Exp. Therm. Fluid Sci. **26**, 389-407 (2002)
4. R. K. Shah, *Classification of Heat Exchangers, Heat Exchangers–Thermal/Hydraulic Fundamentals and Design* (Hemisphere Publishing Washington DC, 9-46, 1981)
5. E. Trefftz, *Proc. Int. Kongress für Technische Mechanik*, Zürich, 131-137 (1926)
6. E. Kita, N. Kamiya, Adv. Eng. Softw. **24**, 3-12 (1995)
7. V. Kompis, F. Konkol, M. Vasko, Comp. Assisted Mech. Eng. Sci. **8**, 385-395 (2001)
8. I. Herrera, Numer. Meth. Partial Diff. Eq. **16**, 561-580 (2000)
9. S. Hożejowska, R. Kaniowski, M. E. Poniewski, Int. J. Numer. Method Heat Fluid Flow **24**, 811-824 (2014)
10. K. Grysa, A. Maciąg, A. Pawińska, Int. J. Heat Mass Tran. **55**, 7336-7340 (2012)
11. B. Movahedian, B. Boroomand, S. Soghrati, Eng. Anal. Bound. Elem. **37**, 868-883 (2013)
12. M. J. Ciałkowski, A. Frąckowiak, J. Therm. Sci. **11**, 148-162 (2002)
13. M. J. Ciałkowski, K. Grysa, J. Inv. Ill-Posed Problems **18**, 595-616 (2010)
14. M. J. Ciałkowski, A. Frąckowiak, K. Grysa, J. Inv. Ill-Posed Problems **15**, 347-364 (2007)
15. M. J. Ciałkowski, A. Frąckowiak, K. Grysa, Int. J. Heat Mass Tran. **50**, 2170-2181 (2007)
16. A. P. Zieliński, Afdv. Eng. Softw. **24**, 147-155 (1995)
17. A. Frąckowiak, J. V. Wolfersdorf, M. Ciałkowski, Int. J. Heat Mass Tran. **54**, 1236-1243 (2011)
18. M. S. Karaś, A. P. Zieliński, Commun. Num. Methods Eng. **24**, 605-625 (2008)
19. M. Piasecka, Heat Tran. Eng. 35 (10), 903-912 (2014)
20. M. Piasecka, Int. J. Refrig. **56**, 198-212 (2015)
21. M. Piasecka, Ann. Nucl. Energy **73**, 282-293 (2014)
22. T. Orzechowski, *Heat transfer on ribs with microstructured surface*, Monographs, studies, hearings 39 (in Polish) (the Publishing House of the Kielce Univ. Techn., Kielce, Poland, 2003)
23. *User's manual ThermaCam B640, P640, SC640*, Publ. No 155850 Rev.a 201-ENGLISH (EN) (2007)
24. J. V. Beck, B. Blackwell, C. R. St Clair, *Inverse heat conduction. Ill-posed Problems* (Wiley – Interscience Publ., New York, 1985)
25. O. M. Alifanow, *Inverse heat transfer problems* (Springer-Verlag, Berlin, 1994)
26. K. Kurpisz, A. J. Nowak, *Inverse Thermal Problems, International Series on Computational Mechanics Publications* (Southampton, UK and Boston, USA, 1995)
27. M. N. Ozisik, H. R. B Orlande, *Inverse heat transfer: fundamentals and applications* (Taylor & Francis, New York, 2000)
28. A. N. Tikhonov, V. Y. Arsenin, *Solution of ill-posed problems* (Wiley, New York, 1977)
29. Y. A. Cengel, and R.H. Turner, *Fundamentals of thermal-fluid sciences* (McGraw-Hill Higher Education, International Edition 2001)
30. M. Piasecka, B. Maciejewska, Exp. Therm. Fluid Sci. **44**, 23-33 (2013)
31. M. Piasecka, Int. J. Heat Mass Tran. **81**, 114-121 (2015)
32. Y.M. Lie, F.Q. Su, R.L. Lai, T.F. Lin, Int. J. Heat Mass Tran. **49**, 207-218 (2006)
33. L. Saraceno, G.P. Celata, M. Furrer, A. Mariani, G. Zummo, Int. J. Therm. Sc. **53**, 35-41 (2012)
34. G. P. Celata, M. Cumo, T. Setaro, Exp. Heat Tran. **5**, 253-275 (1992)
35. M. Piasecka, M. E. Poniewski, Heat Tran. Eng. **25** (3), 44-51 (2004)
36. S. Hożejowska, M. Piasecka, Heat Mass Tran. **50**, 1053-1063 (2014)
37. Z. Bilicki, Int. J. Heat Mass Tran. **26**, 559-565 (1983)
38. T. Bohdal, Exp. Heat Tran. **14**, 199-215 (2001)

EPJ Web of Conferences **114**, 02069 (2016)
DOI: 10.1051/epjconf/201611402069

Numerical study of flow around NACA0015 in ground effect

Khadidja Malti[1], Hanaa Hebow[2] and Bachir Imine[1,a]

[1]*LASP Laboratory, Faculty of Mechanical Engineering, USTO, Algeria.*
[2]*University of Aleppo, Faculty of mechanical Engineering, Syria.*

Abstract. The aim of this work is to present a numerical simulation of flow around a wing profile NACA0015 under the ground effect. CFD software has been used to determine the aerodynamic performance for different angles of incidence. The flow is considered two-dimensional and the adopted meshing considered the effects of the boundary layer. The Spalart-Almaras turbulence model was adopted for the investigation of complex flow around the profile. The results obtained by CFD were compared to those obtained by the literature.

1 Introduction

When a wing approaches the ground two phenomena are actually involved in the increase of the lift force and reduction of the drag. The effects of proximity to the ground on the aerodynamic performances of the wing were investigated since 1920. Previous studies have shown that the decrease in altitude generates a rapid increase in lift and drag reduction. So favorable aerodynamic characteristics are obtained when a wing is close to the ground. For an open-wheel racing car, the flow around a number of components including the front wings, distributor, and wheels are subjected to the direct influence of the ground effect. The improved aerodynamic response can have a significant effect on the performance of the down force. For example, the down force can reach three times the weight of the car, one-third from the front wing, a third fairing and the rear diffuser and the rest of the rear fender.

During the first decades of the twentieth century [1], aerodynamic research has been carried in the automotive field and in particular to reduce drag to achieve the performance of advanced vehicle speed. During the second half of this century, development continued in the same direction, with reduction or reversal of the lift motor to increase their roll stability.

The flow characteristics over a symmetrical airfoil—NACA 0015—are studied experimentally in a low speed wind tunnel by Ahmed an al.[2]. The pressure distribution on the airfoil surface was obtained, lift and drag forces were measured and mean velocity profiles were obtained over the surface. As a result, they obtained a higher values of lift coefficient when the airfoil is close to the ground. Also, the wing in ground effect has been studies by Moure and al.[3]., they detailed experimental results from wind tunnel studies of a DHMTU and NACA 0012 section operating in ground effect. It was found that the drag of the DHMTU 12-35.3-10.2-80.12 increased with decreasing altitude, contrary to expectations and superior lift performance to the NACA 0012 baseline section was generated. In a numerical tier, Abramowski [4] proposed numerical simulations of the ground effect which affects an air flow when a wing approaches the ground or sea surface. The two-dimensional viscous flow problem is solved and the results of CFD are presented. On the basis of numerical calculations, an empirical formula has been proposed for quantitative assessment of the chord dominated ground effect.

In 2006, Hebow conducted an experimental investigation on the ground effect on a turbulent flow around a profile NACA0015 with a negative incidence. It has been proven that the aerodynamic performances of the airfoil degrade when clearance decreases.

In this present study, a NACA0015 profile type was chosen in order to simulate the operating conditions of the work of Hebow [5] to validate the numerical model used. As it shown in figure 1, the incidence α is defined as an angle between the direction of the undisturbed flow profile and the axis of symmetry, H is the height between the floor and the underside: determined from the extreme point of the edge leak. The variation of the angle of attack α and the parameter H to estimate the coefficient of drag and lift in several cases studied. Also in this study, a negative angle of attack are considered in order to investigate the roll stability.

[a] Corresponding author: imine_b@yahoo.fr

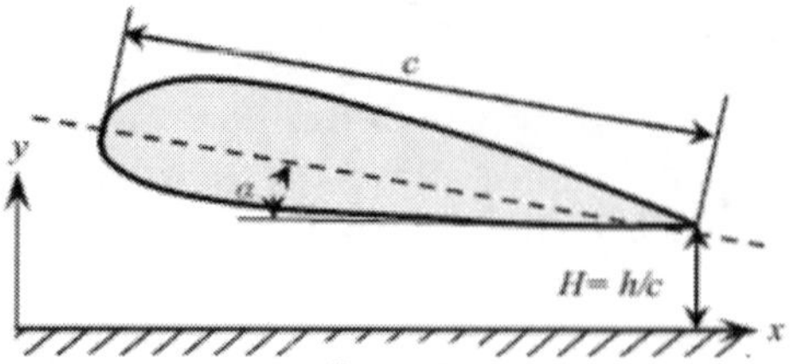

Figure 1. Geometry of physical problem

2 Transport equations and turbulence model

The flow around the profile is considered turbulent symmetrical. The general forms of the transport equations can be written in Cartesian coordinates as:

Continuity equation

$$\frac{\partial}{\partial x_j}\left(\bar{\rho}\tilde{U}_j\right)=0 \qquad (1)$$

Momentum conservation equation

$$\frac{\partial}{\partial x_j}\left(\bar{\rho}\tilde{U}_i\tilde{U}_j\right)=\bar{\rho}g_i-\frac{\partial\bar{p}}{\partial x_i}-\frac{\partial}{\partial x_j}(\rho u_i'' u_j'')$$
$$+\frac{\partial}{\partial x_j}\left[\bar{\mu}\left(\frac{\partial\tilde{U}_i}{\partial x_j}+\frac{\partial\tilde{U}_j}{\partial x_i}\right)-\frac{2}{3}\bar{\mu}\frac{\partial\tilde{U}_k}{\partial x_k}\delta_{ij}\right] \qquad (2)$$

Turbulence model

The model of turbulence used in this present work is a model with a transport equation for the v quantity suggested by Spalart and Allmaras (3):

$$\frac{\partial(\bar{\rho}\tilde{v})}{\partial_t}+\vec{\nabla}.\left(\bar{\rho}\vec{\tilde{V}}\tilde{v}\right)=\vec{\nabla}.\left[\frac{\bar{\mu}+\bar{\rho}\tilde{v}}{\partial_{SA}}\vec{\nabla}\tilde{v}\right]+P_v-D_v \qquad (3)$$

Or the terms of production and destruction are defined as follows:

$$P_v=C_{b1}C_\mu\tilde{S}\bar{\rho}\tilde{v}+C_{b2}\frac{\bar{\rho}(\vec{\nabla}\tilde{v})^2}{\partial_{SA}} \qquad (4)$$

$$D_v=C'_{\omega1}f_\omega\frac{\bar{\rho}(\vec{\nabla}\tilde{v})^2}{\rho d^2} \qquad (5)$$

With

$$f_x=g\left(\frac{1+C_{\omega3}^6}{g^6+C_{\omega3}^6}\right)^{1/6} \text{ and } g=r+C_{\omega2}\left(r^2-r\right) \qquad (6)$$

And

$$M_t=\bar{\rho}\tilde{v}\frac{\chi^3}{\chi^3+C_{v1}} \qquad \chi=\frac{\bar{\rho}\tilde{v}}{\mu} \qquad (7)$$

And the coefficients of closing are given by the following values:

$$\sigma_{SA}=\frac{2}{3};\quad C_{b1}=0.1355;\quad C_{b2}=0.622;$$
$$C_{v1}=7.1;\ C_{\omega1}=0.3;\ C_{\omega3}=2. \qquad (8)$$

(For more information see reference 5)

3 Meshing

For the airfoil case, NACA 0015 airfoil with 0.1 m of chord length c and as it shown in figures 2 and 3, the 2D structured mesh are generated The geometry of our model being variable according to the parameters h / c and the angle of attack α. Our main concern was to develop a grid that ensures resolutions in agreement with the available literature [5]. The geometry considered is composed of four distinct surfaces and a profile NACA 0015. The separation into four areas is to enable to refine the mesh without imposing a total number of items too high.

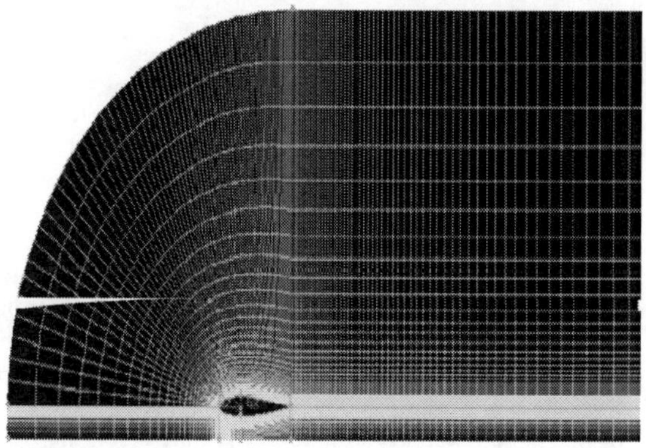

Figure 2: Meshing of computational domain

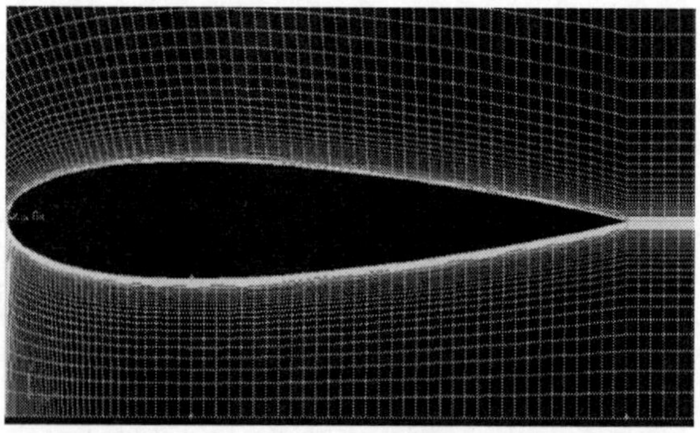

Figure 3: Meshing around the profile

Table I: Quality of meshing

h/c	Angle of attack α	Elements
0.25	0°	14475
	-5°	15349
	-10°	16683
0.35	0°	14475
	-5°	15073
	-10°	16407
0.45	0°	14475
	-5°	14475
	-10°	16177

4 Results

Several compilations were carried out for different negative angles of attack and different ground clearance to determine the lift and drag. The negative angles of attack were used in order to simulate the downforce. Each case was carried out at the same speed: 20 meters per second. The angle of attack extended from the following values: α =0, -5 and -10 degrees.

Figure 4 compares the computational evolution of the lift according to the angles of attack α with the experimentally data [5] for different cases of ground clearance h/c: A,B and C. An almost agreement of the two curves is noted in the linear evolution and the lift coefficient becomes more negative with decreasing angle of attack. However, examining the angle of attack values from α = - 5 to - 10, the differences between the numerical study and deviations from the measured curve become visible for all the ground clearance cases. It is clear that the numerical simulation is far from predicting the negative incidence, because the models currently used are not accurate for flows with separate streams. As it shown in the A,B and C cases, the downforce increases when the ground clearance h/c decreases.

a-

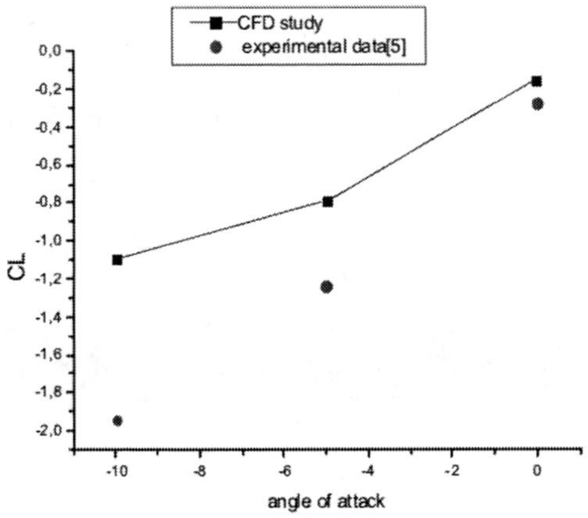

b-

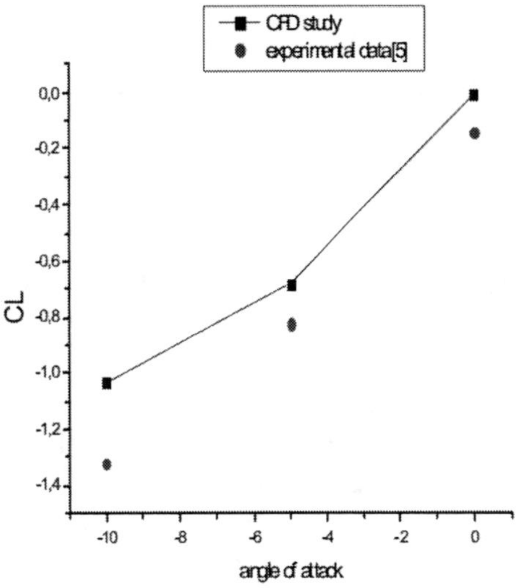

c-

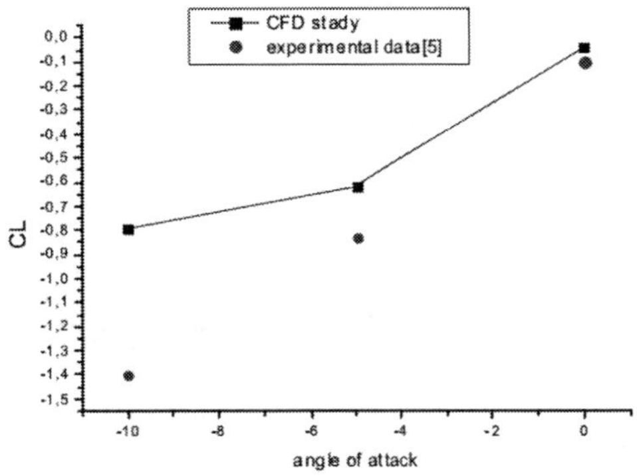

Figure 4: Variation of the lift coefficient with different ground clearance (h/c) **a** : h/c=0.25, **b** :h/c= 0.35 and **c**: h/c=0.45

The evolution of the drag coefficient according to the negative angle of attack with increasing ground clearance h/c=0.25, 0.35 and 0.45 was observed as illustrated in Figure 5. All curves are quasi linear type tendency similar to those found on the literature [5]. In particular, one notes the increase in the drag with the increase in the angle of attack. However, a slight conformity between the present study and experimental one was only partial. It was shown in the two graphs a reduction of the drag coefficient versus the angle of attack α, but depending on the ground clearance 'h / c', this analogy exists only at α = -10 °.

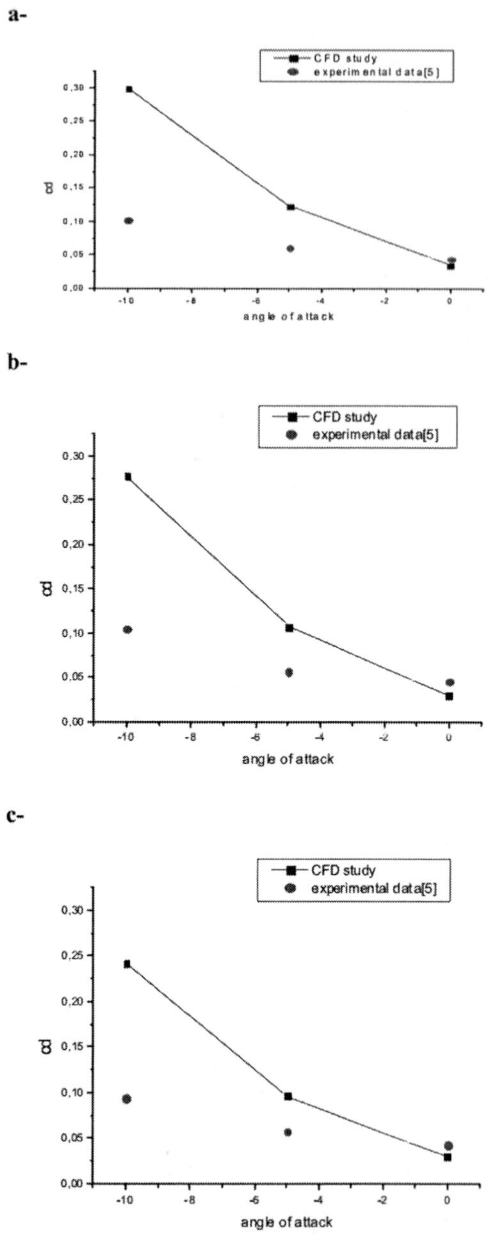

Figure 5 The evolution of the drag coefficient with different ground clearance. **a**: :h/c=0.25, **b**: h/c=0.35 and **c** : h/c = 0.45.

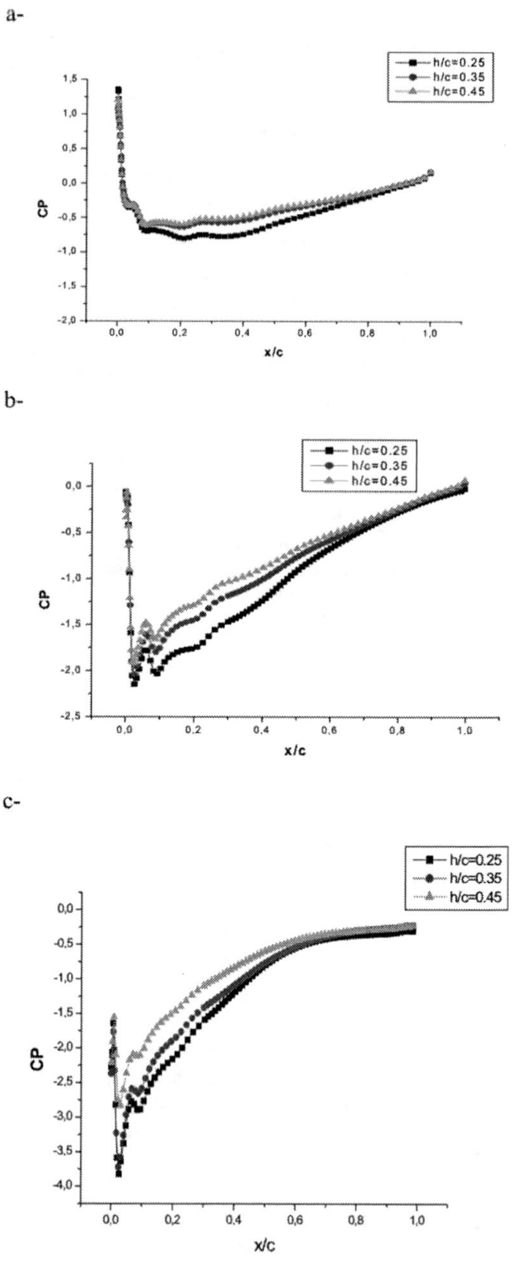

Figure 6: Pressure distribution on the surface of the airfoil for an angle of attack of (a) $\alpha = 0°$, (b) $\alpha = -5°$, (c) $\alpha = -10°$.

Figures 6a, 6b and 6c shown the variation in the pressure on the airfoil surface for different values of attack angle: $\alpha = 0$ °, - 5 ° and -10 ° and with different ground clearances h/c= 0.45, 0.35 and 0.25, For all cases, the pressure is high on the upper surface at points close to the leading edge. It was noted that for $\alpha = -5°$ and -10°, the pressure coefficient is negative. This is because the area between the airfoil and the ground forms a divergent–convergent passage, and its effect is more prominent for lower ground clearances.

Conclusion

To study the performance of a NACA 0015 wing in ground effect, a detailed numerical investigation was carried out using CFD code with the Spalart-Allmaras turbulence model. It was found from the present numerical study that both the ground clearance and the angle of attack have a good effect on the aerodynamic performances of the used configuration. The comparison between the experimental and numerical fields, gives a reasonably good correlation.

References

1. Florent Mortel (2002) course 'l'aerodynamique dans l'automobile' Belford technology university, 2002.
2. M.R. Ahmed , S.D. Sharma , An investigation on the aerodynamics of a symmetrical airfoil in ground effect. International Journal of Mechanical, Aerospace, Industrial, Mechatronic and Manufacturing Engineering Vol:8, No:7, 2014.
3. Lange R, Moore J, "Large Wing-in-Ground Effect Transport Aircraft", Journal of Aircraft, Vol.17, No.4, April 1980.
4. Tomasz Abramowski; Numerical investigation of airfoil in ground proximity. Journal of theoretical and applied mechanics 45, 2, pp. 425-436, Warsaw 2007.
5. Hebow. H, (2006). 'Interferences between boundary layers on an Airfoil and a flate plate'. Thesis master degree in Mechanical Engineering. Aleppo University. 123p.

EPJ Web of Conferences **114**, 02070 (2016)
DOI: 10.1051/epjconf/201611402070

The influence of spill-line geometry on a spray generated by a pressure-swirl atomizer

Milan Malý[1,a], Lada Janáčková[1], Jan Jedelský[1] and Miroslav Jícha[1]

[1]*Brno University of Technology, Faculty of Mechanical Engineering, Energy Institute, Technická 2896/2 Brno 616 69, Czech Republic*

Abstract. An experimental investigation of characteristics of spray generated by a pressure-swirl atomizer (spill-return type) was performed using shadowgraphy and Phase-Doppler Anemometry (PDA). Several different geometries of the spill-return orifice were tested in terms of a spray stability and quality on a cold test bench. PDA measurement yields a drop-size distribution and velocity data while the shadowgraphy unveils a break-up process in detail. Performed measurements reveal significant differences in spray characteristics as well as differences in spray stability. The results suggest that the air core, formed inside the swirl chamber, passes through the spill orifice, which causes instability of the inner flow. These instabilities lead to a chaotic state of sheet breakup resulting in shortening of breakup distance. Obtained findings are used to propose possible changes in the atomizer design for improvement of its performance.

1 Introduction

A spill-return atomizer is basically a pressure-swirl atomizer that contains a bypass in the swirl chamber known as a spill-line (SL) orifice. The liquid is injected via tangential ports to the swirl chamber, where it is divided into two streams, one of them is discharged outside and atomized, the second one is "spilled" back to the reservoir throught the SL orifice. The amount of spilled fuel is regulated by a control valve in a spill-line. Fuel is always supplied to the swirl chamber at high pressure providing good atomization over a wide span of injection flow rate. This feature allows using the spill-return atomizers in applications requiring the wide regulation range such as industrial burners and gas turbine combustors [1].

A number of studies focused on improvement of spray characteristics of pressure-swirl atomizers, but only a few of them were investigating spill-return atomizer and even less examined the influence of the spill-return orifice on the spray quality and stability. Nasr et al. in [2] tested several different diameters of spill orifice in terms of Sauter mean diameter (SMD) and flow rate. However, their atomizers were run with open spill line without any regulation, and the spill-feed ratio was controlled by a cross section of SL orifice. Our research is concentrated on test with regulated spill line along with closed spill line to simulate the full power of the atomizer.

Several small sized spill-return atomizers were chosen for this study varying only in the SL orifice geometry. Two different diameters of SL orifice were investigated with or without an obstacle blocking direct axial flow through the SL orifice. Also, several axial positions of the obstacle in the SL orifice were taken into account.

2 Experimental setup

Experiments were performed at a specially designed facility for cold spray generation under controlled conditions in the Spray laboratory at the Brno University of Technology.

2.1 Cold test bench

A schematic layout of the cold test bench is shown in Figure 1. It consists of a gear feed pump (3) that supplies fuel from a main tank (1) through filters (2), a flow meter (4), a pressure sensor (5), a temperature meter (7) and a control valve (8) into the atomizer (9). The spray falls into a collector and it is then returned to the main supply tank. The flow rate is controlled by a bypass needle valve (9). Spill-return line consist of a pressure sensor (6), a needle regulating valve (11) and a gear flow meter (12)

Flow rate in the fuel line is metered by Siemens Mass 2100 Di3 Coriolis mass flow meter fitted with a Mass 6000 transmitter. Flow rate uncertainty is 0.2 % of its actual value. The uncertainty of pressure sensors (BD Senzor DMP 331i) is 0.35 % of the actual value. The error of temperature sensor Omega PR-13 is 0.2 °C.

Operation regimes of the atomizer were controlled by setting the inlet pressure to 0.5, 1 and 1.5 MPa with the closed spill line by the valve near the atomizer. Spill mode of the atomizer was performed at an inlet pressure of 1 MPa with the spill-feed ratio of 0.4 and 0.8. The

[a] Corresponding author: milan.maly@vutbr.cz

spill-fill ratio is defined as the ratio of the amount of

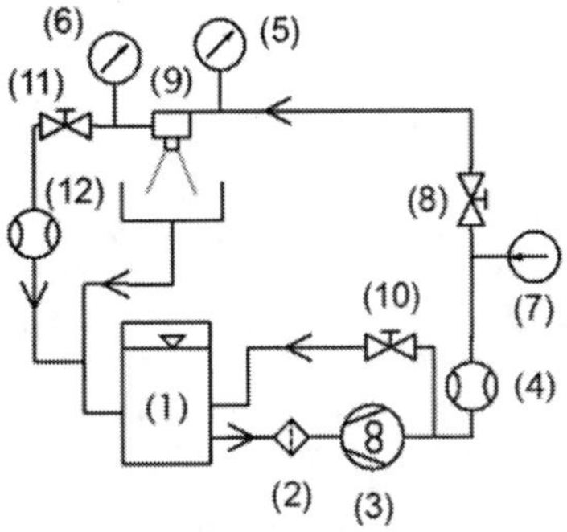

Figure 1. Schematic layout of the experimental facility

spilled fuel to the amount of pumped fuel to the atomizer

Kerosene Jet A-1 was used as the testing liquid. Physical properties of Jet A-1 at room temperature are: surface tension $\sigma = 0.029$ kg/s^2, dynamic viscosity $\mu_l = 0.0016$ kg/(m·s) and density $\rho_l = 795$ kg/m^3. All tests were done with one batch at 25 °C.

2.2 Phase Doppler anemometry

Droplet sizes and velocities were probed by two component Phase Doppler anemometer (PDA) Dantec Dynamics A/S, see a schematic layout in Figure 2. PDA measurement was made at an axial distance Z = 25 mm from the exit orifice. One radial profile with 31 positions was measured. In each position, 35,000 samples or 15 second long measurement were taken. Detailed setup information is summarized into Table 1.

Spray images were taken by Canon EOS D70 with Canon EF 100 mm f/2.8 UMS Macro lens fitted to 68 mm long extension tube. The spray was illuminated by Nd:YAG NewWave Research Gemini laser with a pulse duration of 5 ns and pulse energy of 5 mJ. The imaging system was configured to capture a shadowgraphy images.

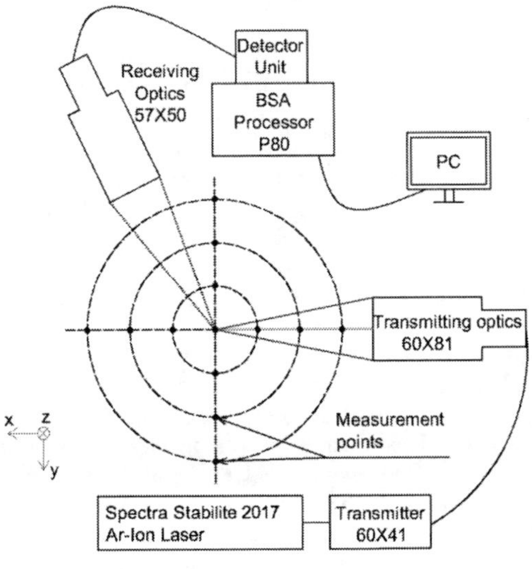

Figure 2. A setup of the PDA measurement

Table 1. PDA setup

Parameter	Value
Laser power output	1 W
Wavelength	514.5 nm
Front focal length of transmitting optics	310 mm
Front focal length of receiving optics	800 mm
Scattering angle	70°
Mask	B
Spatial filter	0.050 mm
Velocity	Axial
Velocity center	8 m/s
Velocity span	32 m/s
Sensitivity	800 V
SNR	0 dB
Signal gain	20 dB
Level validation ratio	8

2.3 Tested geometries

Several small pressure-swirl atomizers with internal dimensions corresponding to
Figure 3 were investigated. Tested variations are listed in Table 2, where the B dimension represents a distance of the blocking obstacle from the top of the swirl chamber and d_s a diameter of the SL orifice.

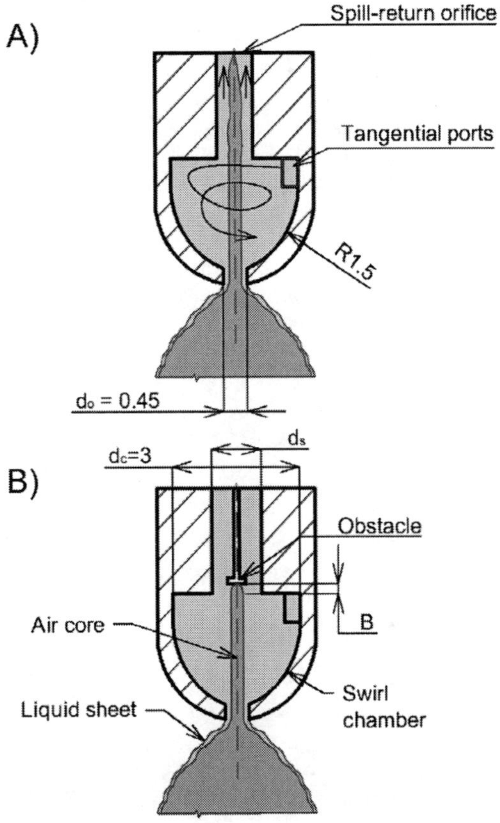

Figure 3. The geometry of the tested atomizer. A) Standard design (the SL orifice in the centre of the swirl chamber), B) Modified spill-line geometry with the obstacle

Table 2. Tested atomizers configurations

Configuration	B [mm]	d_s [mm]
S1	NA*	0.8
S1B0	1	0.8
S1B2	3	0.8
S1B4	5	0.8
S2	NA	1.09
S2B0	1	1.09
S2B2	3	1.09
S2B4	5	1.09

*NA = Not applied

3 Results and discussion

The first part of this chapter focuses to the flow characteristics such as mass flow rate, discharge coefficient and overpressure in the spill line, the second one deals with the liquid breakup and the third one is concerning PDA measurements.

3.1 Discharge characteristics

Measuring of liquid mass flow through the atomizers exit orifice and pressure on the spill line (with closed spill line) shows unexpected results. As it is clear from Figure 4, liquid mass flow rate depends on the geometry of SL orifice even if the other dimensions remain unchanged. The lowest flow rate is almost the same for both SL orifice diameters (0.8 and 1.09 mm) with the obstacle closest to the swirl chamber. Moderate flow rate is achieved by 0.8 mm SL orifice, where the flowrate is the same for all positions of the obstacle (expect the closest one) as well as for atomizer without the obstacle. Only at 1.5 MPa, the S1B4 show much higher flow rate than it would be expected and the reason of it is still unexplained. The 1.09 SL orifice (except the S2B0) have the highest flow rates. A difference about 40 % in the mass flow rate between atomizers with and without obstacle shows a strong influence of SL orifice to the internal flow.

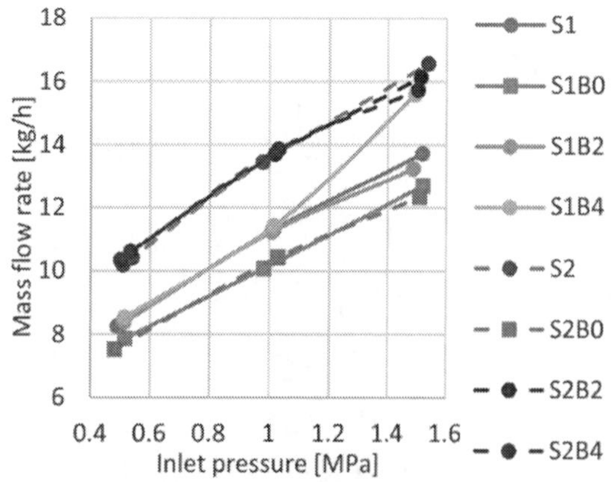

Figure 4. Mass flow rate with closed spill-line

Coefficient of discharge (C_D) is a ratio between the actual and the theoretical mass flow rate through the exit orifice, defined as:

$$C_D = \frac{\dot{m}_l}{A_0\sqrt{2\rho_l \Delta p}} \tag{1}$$

The C_D is for pressure swirl atomizers usually low due to the presence of air core blocking off the central portion of the exit orifice and it is roughly independent on the inlet pressure. The C_D values as a function of inlet pressure are shown in Figure 5 for all tested atomizers. Figure 4 indicates main trends of C_D distribution over tested geometries. Atomizers with the obstacle closest to the swirl chamber have the lowest C_D varying from 0.47 to 0.44. The highest C_D was reached with 1.09 mm SL orifice diameter, changing from 0.64 to 0.57.

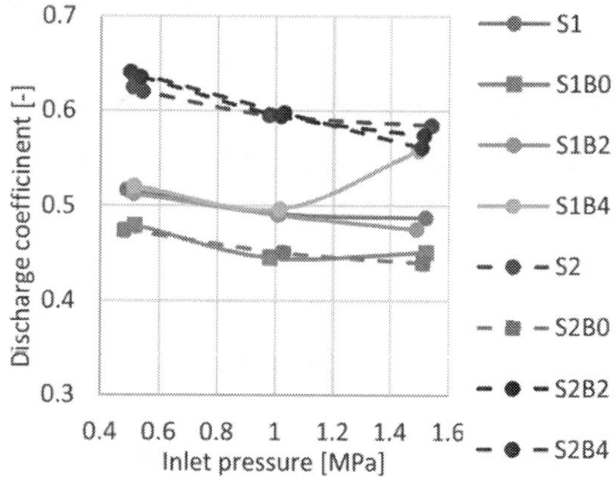

Figure 5. Discharge coefficient

A prediction of the discharge coefficient based on atomizer dimensions published by Rizk and Lefebvre in [3]:

$$C_D = 0.35\left(\frac{A_p}{d_s d_o}\right)^{0,5}\left(\frac{d_s}{d_o}\right)^{0,25} \tag{2}$$

gives $C_D = 0.39$ and it is independent on inlet pressure and SL orifice dimensions.

Lee et al. [4] report C_D as a function of Reynolds number varying from 0.45 for high Re to 0.68 for low Re. They found a dependecy of air core stability on Re. Stable air core means a drop in the fuel flow rate as the center of the exit orifice is occupied by the air core.

Assuming that the air core created by internal swirl motion of tangentially injected liquid passes through the SL orifice, spill-return atomizer can be simplified into simple pressure-swirl type with extended swirl chamber. Influence of dimensions of the swirl chamber on internal flow was investigated by Kim et al. in [5]. They described a dependency of stability of the air core on the length of the swirl chamber, where longer swirl chamber lowers the air core stability. Taking these findings into account, we can assume that the geometries with the obstacle closest to the swirl chamber have stabilized the air core, while in the different configuration the air core is longer than its critical length and behave unstably, hence the C_D and the mass flow rate increases.

The influence of spill-line pressure on the inlet pressure is revealed in Figure 6. As expected, SL pressure depends on the SL orifice diameter. The orifice with larger cross section has lower pressure losses and vice versa. But the obstacle in the spill line strongly increases the SL pressure (pressure drop is smaller). This behaviour is observed for both the SL orifice diameters it is attributed to the low C_D value.

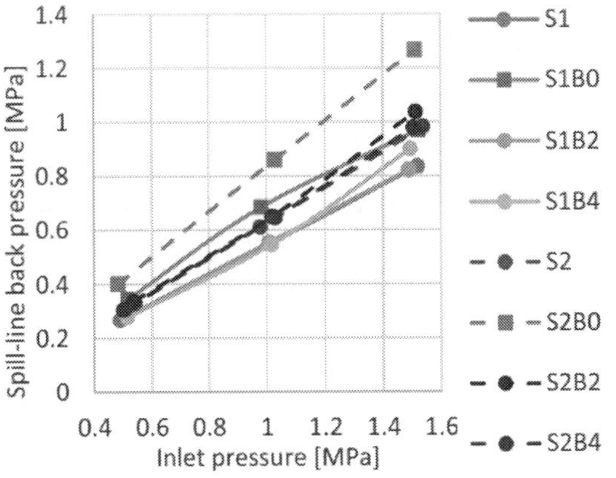

Figure 6. Spill-line back pressure

3.2 Liquid breakup

Droplet formation is a chaotic process, where discharged liquid sheet disintegrates due to aerodynamic forces between the liquid and stagnant surrounding air into filaments and ligaments and finally into drops in the form of a hollow cone spray. Longer breakup length would allow the liquid sheet to be thinned before the breakup and would result in smaller droplets. Entire breakup process is visualized in Figure 7. Images of the breakup process reveal how compact the liquid sheet is and if there are any long time instabilities.

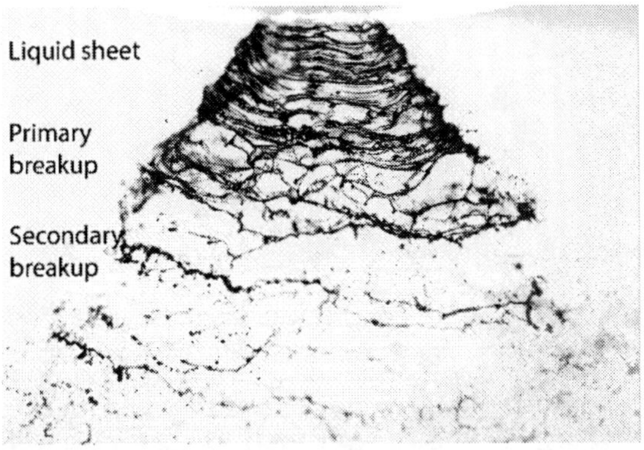

Figure 7. Near-nozzle spray structure

For detailed breakup, images see Figure 9. All tested configurations have in common that the breakup length strongly shortens with increasing inlet pressure, the atomizer S2B0 has the longest breakup length, which is most evident at $\Delta p = 0.5$ MPa and 1.0 MPa, while the S1B0 does not appear to have similar nature and behave the same way as the configurations without the obstacle. This is in contrary to the findings from the previous chapter, where both the atomizer have exactly the same C_D. Thus the same mode of the liquid break up was expected. Other atomizers (even those not shown in Figure 8) feature a chaotic breakup process where the liquid sheet is almost undetectable and very short. Short break up length might be linked to the instabilities of the air core that was predicted in the previous chapter.

3.3 Spray characteristics

Basic spray characteristics such as axial velocity profile, Sauter mean diameter profile or liquid mass distribution within the spray are evaluated at inlet overpressure 1 MPa with closed spill line. .The configurations with the obstacle placed further from the exit orifice were omitted from the PDA measurements.

The radial distribution of the mean droplet axial velocity corresponds well to other observations of hollow cone sprays for the case of atomizers with the obstacle (see Figure 8), where axial velocity reaches a local maximum at approx. 16 mm from the spray centre. Geometries without the obstacle have this maximum much less distinct. Velocities in the spray centre are divided by the diameter of SL orifice, where 1.09 mm orifice (both, with and without the obstacle) have by 25% higher axial velocity than 0.8 mm orifice.

The SMD is lower for atomizers with the obstacle in the spray centre (see Figure 10). The difference between S1 and S1B0 is modest in all radial distances, however S2 have 2.5 times higher SMD in the centre compared to the same geometry with the obstacle (S2B0) which has the lowest SMD in almost every position.

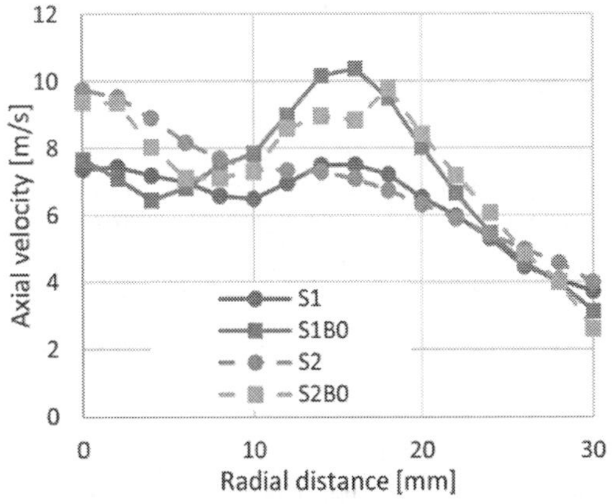

Figure 8. Axial velocity for $\Delta p = 1$ MPa with closed spill line

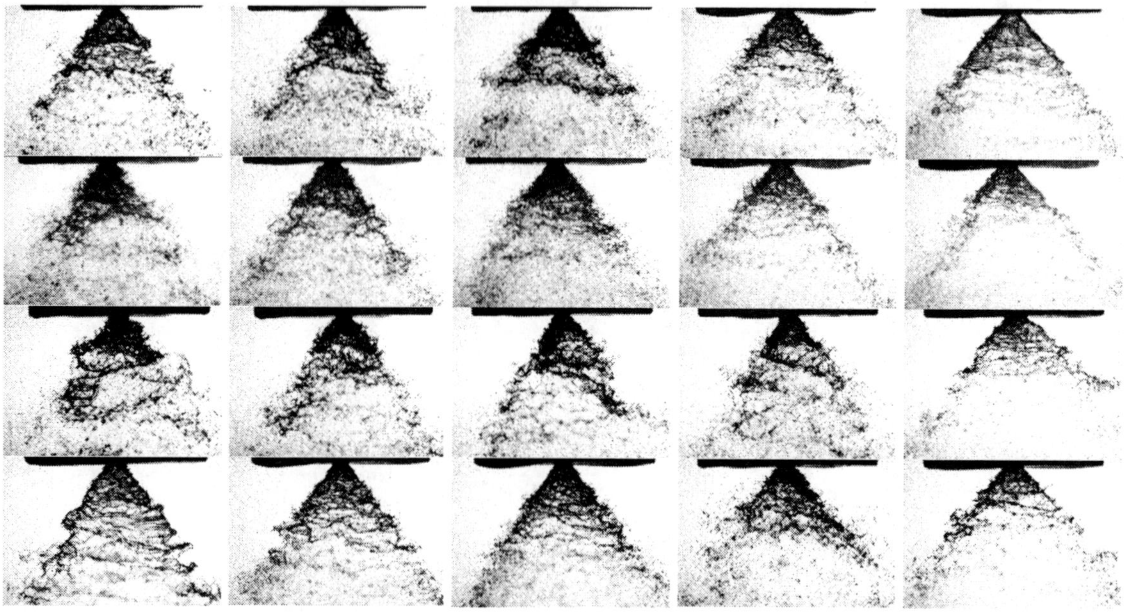

Figure 9. Liquid breakup visualization. Columns: from left Δp = 0.5, 1, 1.5 MPa with SFR 0; 1 MPa SFR 0.4 and 1 MPa SFR 0.8. Rows from top: S1, S1B0, S2, and S2B0. The size of each photograph is 7x10 mm.

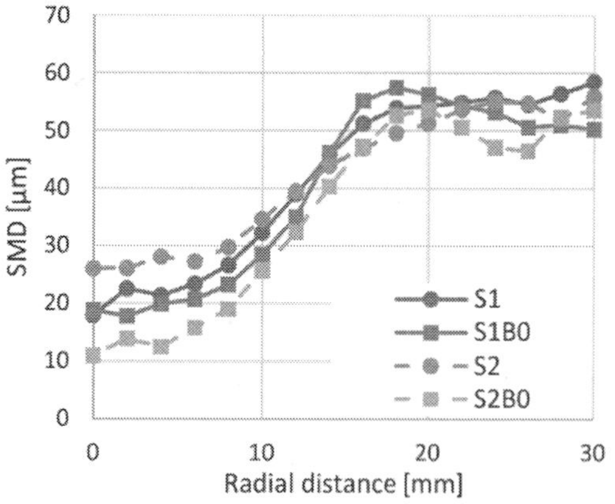

Figure 10. SMD for Δp = 1 MPa with closed spill line

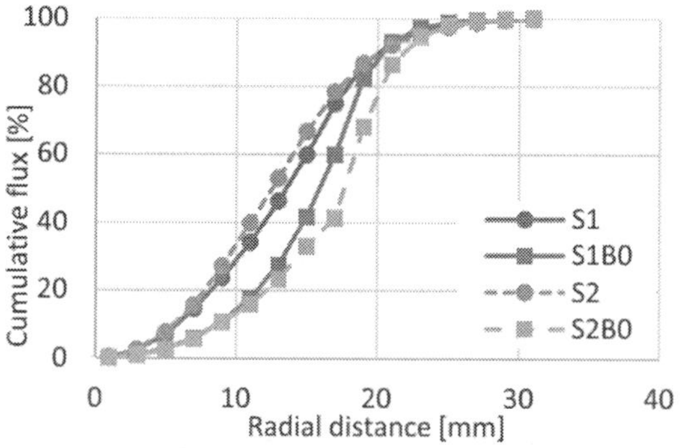

Figure 11. Liquid mass distribution Δp = 1 MPa

Cumulative liquid mass distribution in Figure 11 reveals significant divergence between the configurations with and without the obstacle. S1B0 and S2B0 have more liquid concentrated at positions further from spray centreline – where an annular liquid sheet is expected. Atomizers without the obstacle have concentrated about 37% of entire liquid mass flow at radial distance of 11 mm while the atomizers with the obstacle only 16% thus the mass distribution of S1 and S2 does not fully correspond to the idea of hollow cone spray.

This observation together with findings from Figure 8 indicate a problem of the liquid sheet stability that was confirmed by the breakup visualization in Figure 9.

For evaluation of inlet pressure influence on SMD a global form of the SMD was calculated:

$$ISMD = \frac{\sum_{i=1}^{N_i} n_i f_i D_i^3}{\sum_{i=1}^{N_i} n_i f_i D_i^2} \qquad (3)$$

As shown in Figure 12, with increasing pressure the ISMD decreases for all tested atomizers. The highest ISMD in the whole tested range is attributed to the S1B0 while the S2B0 has about 10% lower ISMD. Both atomizers have the same C_D, but the disparity in their ISMD confirms a different breakup mode showed in Figure 9. The lowest ISMD at 0.5 MPa was reached by the S2B0 as it was predicted by the visualization (the longest breakup) however the S2 has the lowest ISMD at the rest of the tested regimes.

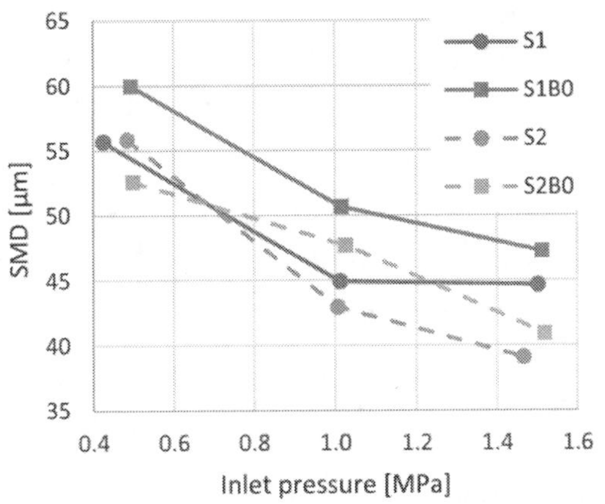

Figure 12. Global SMD as function of inlet pressure

4 Conclusion

The study of the influence of various SL orifice on spray was focused on discharge characteristics, near nozzle liquid break up and spray quality in terms of liquid distribution and SMD. Discharge coefficient of the flow through the exit orifice was found dependent on the diameter of SL orifice and also on the obstacle presence. A question of the air core instabilities caused by air core passing through the SL orifice were discussed with other related publications.

Spray visualization shows very short and chaotic breakup except the atomizer with the obstacle close to the swirl chamber and large diameter of SL orifice.

The spray structure was described based on spatially resolved velocity, drop size and mass flux. Different radial distributions of axial velocity indicate a problem with the hollow cone liquid sheet formation in the case of atomizers without the obstacle which is also confirmed by the analysis of liquid mass distribution across the spray diameter.

This study is a brief introduction to the problem of the influence of SL orifice geometry on the spray characteristics and reveals the importance of proper SL orifice design. Further investigation with a focus on the internal flow shows its necessity due to a lack of published papers dealing with phenomena.

Acknowledgement

This work has been supported by the project No. GA15-09040S funded by the Czech Science Foundation, the project LO1202 NETME CENTRE PLUS with the financial support from the Ministry of Education, Youth and Sports of the Czech Republic under the "National Sustainability Programme I" and project Reg. No. FSI-S-14-2355 funded by the Brno University of Technology.

References

1. A. H. Lefebvre, *Atomization and sprays* (1989)
2. G. Nasr, A. Yule, and S. Lloyd, *ILASS* (2007)
3. N. K. Rizk, and A. H. Lefebvre, Journal of Propulsion and Power, **1**, 193-199 (1985)
4. E. J. Lee, S. Y. Oh, H. Y. Kim., Experimental thermal and fluid science, **34**, 1475-1483 (2010)
5. S. Kim, T. Khil, D. Kim, Measurement Science and Technology, **20** (2009)

EPJ Web of Conferences **114**, 02071 (2016)
DOI: 10.1051/epjconf/201611402071

Preliminary study to the temperatures of the thermocouple probes affected by the environment of heated walls

Richard Matas[1,a], Jan Sedláček[1] and Václav Čibera[1]

[1]*New Technologies – Research Centre, University of West Bohemia, Univerzitni 8, 306 14 Pilsen, Czech Republic*

Abstract. Measuring of the total temperatures of the flowing medium is one of the pivotal points in the development when a new compressor stages are tested. Measurements conducted on the test compressor indicated that in the specific cases the measured temperature can be affected by the temperature of the walls which hold thermocouple probe. By implication of this observed fact, the preliminary study about this phenomenon was investigated. It was considered the influence of the heat conduction in the body of the probe and in the thermocouples. The main aim of the study was to examine the properties and the behaviour of the thermocouple probes in various working regimes. The preliminary study was focused on the development of the tools that can be used for complex solution of the problem by using of measured data. The results of this study might help to correct measured data in the case of skewed measurements affected by surroundings of the thermocouple probes.

1 Introduction

The measurements of the total temperatures are one of the crucial steps within verifying of the parameters of the new compressor stages [1]. The accurate values of the measured temperatures are necessary for the evaluation and determining the efficiency and other properties. Further, the correct measured temperatures are important for design documents as well as for comparison with results of CFD simulations [2].

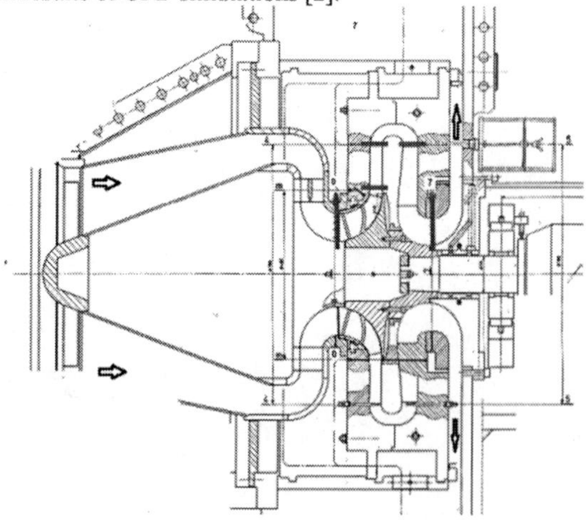

Figure 1. The arrangement of the examined planes in the tested compressor stage.

The probes in different variations have been used to measure total temperatures and total pressure for the testing of the radial compressors. The first type of the probe is a multiple sensors probe containing more than one sensor arranged on the same shank. This kind of probe is used at the inlet and outlet planes of compressor. The second type of probe is the probe containing only a single pressure sensor or a single temperature sensor. This kind of probe is used in inner planes of compressor. The figure 1 shows the arrangement of probes in examined compressor.

The probes are designed and delivered by specialized company to conform demanded high quality. Individual probes are produced to satisfy specific flow cross-section of a compressor stage.

To accomplish complete measurements of temperatures, the temperatures of the inlet and outlet air at orifices have been measured. The temperature of the oil was measured as well by the use of resistance thermometers.

2 Measuring of the temperatures in a single compressor stage

The figure 2 shows the real design and details of the used thermocouples of considered probes, which are installed into measuring planes inside compressor. The signals received from these probes are processed and transmitted into control computer. After the correction regarding to the flow rate of air, the signals are used to evaluate the parameters of the compressor stage.

During the measurements it has been assumed that the temperature of the air at the inlet and the outlet is approximately constant across the cross-section of channel. The temperatures across the inner measuring planes might be unequal. And then on the end of the

[a] Corresponding author: mata@ntc.zcu.cz

channel the temperatures across the channel cross-section should be partially equalized. But the measured results for the various conditions seem to indicate that for the various temperatures of the inlet atmospheric air and for the differently warmed parts of the compressor the measured temperature at the inlet cross-section is inhomogeneous. During the observation of this phenomenon, it was confirmed that the temperature measured by the multiple probe was always higher for the sensors closer to the wall of compressor. Contrary, the temperature decreased in the direction to the end of the probe. This effect was amplified when the colder air was sucked in in the course of the measurements during the winter months. In this case the average difference of the temperature at the sensors arranged along the same probe was sometimes more than 1°C at the inlet plane. This value can already negatively affect the measured efficiency of the investigated compressor stage. On the other hand, during the summer months, when the warmer air was sucked in, this phenomenon was almost negligible.

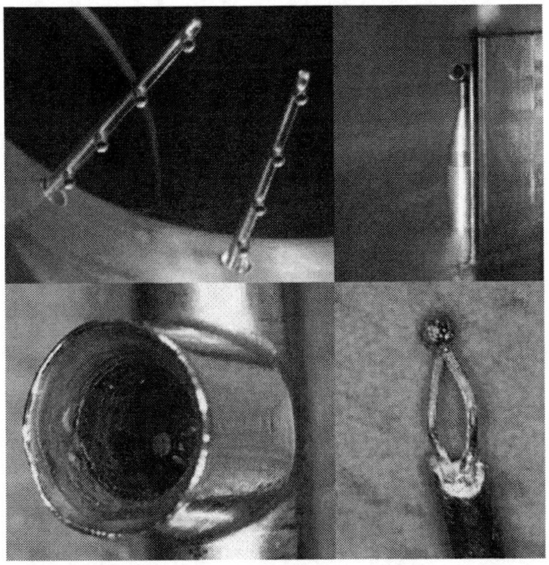

Figure 2. Examples of the probes containing a multiple number of thermocouples and example of the probes with a single thermocouple.

3 Model of the probe containing three thermocouples

3.1. Measured values

As the reference state served the case with the measurements by using multiple sensors probe containing three thermocouples - the similar with the four thermocouples can be seen in figure 2 up and left. From the recorded measured points, two cases were chosen. These two cases were characterized by the different values for the temperature as well as for the flow rate and so for average inlet velocities 23 and 105 m/s. The point values of the temperature were the inlet air at the inlet aperture T_a, the value of the temperature of the wall T_w. Further, the velocity of the air flow was used as boundary conditions. The average values of the temperatures achieved by the probes on the particular radial positions are used for comparison with the results of simulations.

3.2 Model of the thermal network

The first used method for the simulation was the method of thermal network. In this approach, the thermal network is solved as analogy of an electrical circuit. The examined subject is substituted by the elements of electrical circuit. The part of the system is then replaced by elements, usually by π elements. The parameters of these elements conform to the physical properties of investigated objects and their relationships to the surroundings environment. This method is often used for design of electrical machines, electronical devices as well as in various other research areas. The advantage of this method is that it provides quick analysis of objects warming in steady-states as well as in transient states.

By use of this method, the multiple sensors probe was divided into 13 sections along its longest directions. Single thermocouples and deflectors we considered. The parameters as thermal resistances of individual sections were computed as substitute values taking into account material properties of appropriate parts and their proportional representation. The coefficient of the heat transfer among the fluid and solid was considered as value depending on thermal gradient among probes and air. Further this coefficient was considered reliant on the velocity of air flow. The computation of the heat transfer coefficient was conducted according to the method introduced in [3].

The computation of thermal network was realized within MATLAB and Simulink environment. Namely, the tool PLECS designed for the analysis of the electrical circuits was used. The solution was computed for different thermal proportions as the temperatures of the walls, thermal gradient according to temperature of flowing air and further for the various velocities of flow.

The results of the simulations are the distribution of the temperature along the probe body, heat flux through the probe, temperatures at the end of the probe and the deviations of the expected values. The warming along the probe for case 1 is showed in the figure 3.

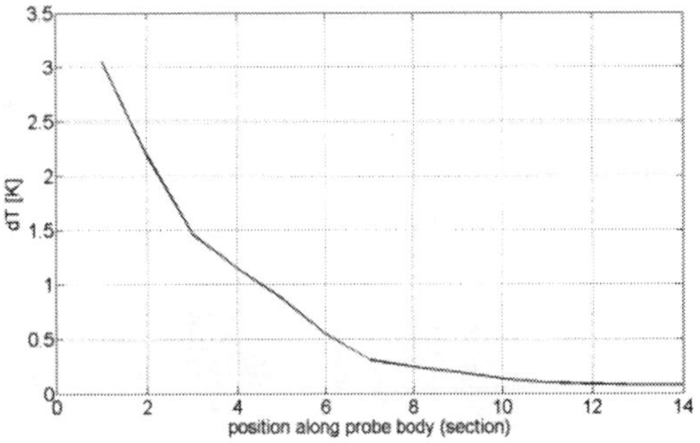

Figure 3. The warming of the probe along its body. The result was achieved by the method of thermal network.

3.3 CFD model of the probes with multiple numbers of thermocouples

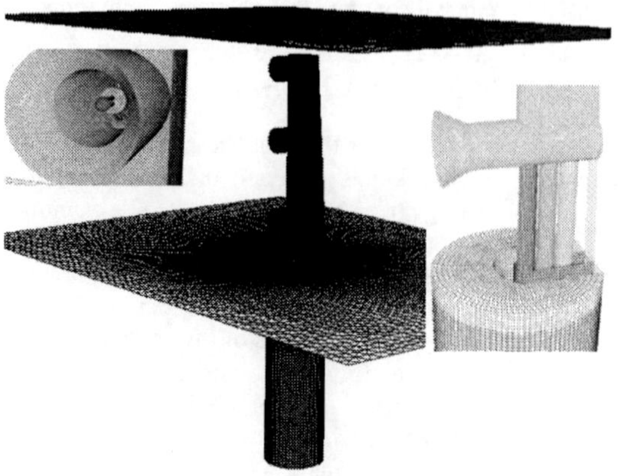

Figure 4. Surface mesh of the probe model containing three thermocouples and the details of the mesh.

The model for the CFD simulations was assembled for the more detailed investigation of the heat and flow proportions in the probes and in its surroundings. This model contains approximately 4.7 million of the mesh cells. The mesh can be seen in the figure 4. The model was prepared for the parameters of the real thermocouple probe. The model includes all the geometric properties including the geometry of the inner part of probe and material properties.

The numerical simulation was conducted within the software ANSYS FLUENT 15. The flow of the air was considered as turbulent, rng k-ε model was used for first simulation. The flowing medium was set as incompressible. The solved case was computed as a steady state. Further, radiation was for the first preliminary results neglected.

The result of the temperature distribution for case 1 is in figure 5. As can be seen from this picture, the warming due to higher temperature of the walls is considerable.

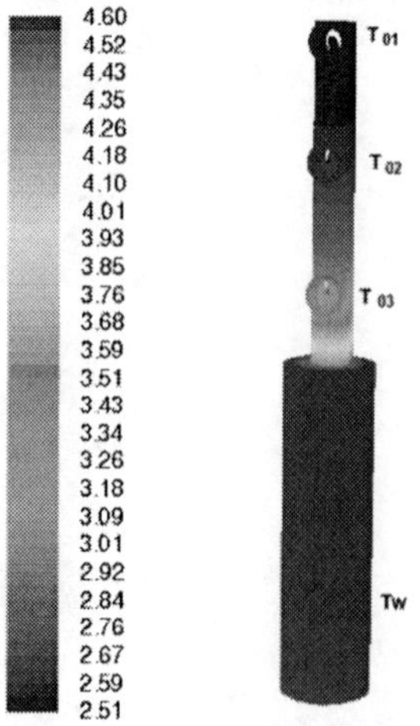

Figure 5. Distribution of the temperature along the probe body.

The figure 6 depicts the cross-section through the probe with surrounding velocity fields. The picture illustratively shows the inhomogeneous velocity fields in the surroundings of the probe body and especially near the thermocouple sensors. The considerable decrease of the flow rate through the deflectors of thermocouple in the direction to the wall of channel can be observed in results of simulation. This might have other negative effects on the measurement of temperature.

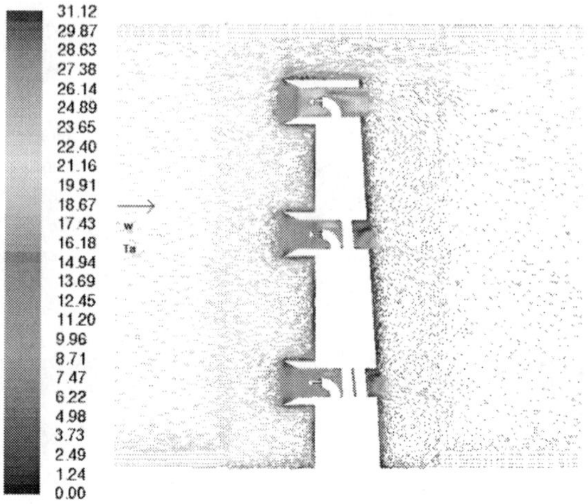

Figure 6. Distribution of the velocity fields near the cross-section of probe – case 1.

The preliminary results are summarized in table 1. The table contains the measured data as well as results of simulation for the chosen points T_a, T_w and the differences of the temperatures related to the end of probe, where the thermocouple is surrounded with the highest flow rates.

Table 1. Summary of the results achieved by the described methods for both solved cases.

	Measured data		Thermal network results		CFD results	
w [m/s]	23	105	23	105	23	105
T_a [°C]	3.05	5.6	3.05	5.6	3.05	5.6
T_w [°C]	4.6	22.6	4.6	22.6	4.6	22.6
T_{01} [°C]	3.20	6.15	3.09	5.61	3.05	5.60
ΔT_{12} [°C]	0.02	0.33	0.08	0.08	0.03	0.07
T_{02} [°C]	3.22	6.48	3.17	5.69	3.08	5.65
ΔT_{13} [°C]	0.22	1.98	0.54	1.07	0.20	0.68
T_{03} [°C]	3.42	8.13	3.63	6.67	3.25	6.28

Besides already mentioned, the next conducted study was the examination of the effect of the used turbulence model. The goal was the determination of the better approach to the comparison of experimental values, because the differences for the case 2 are relative high. The results for case 2 are summarized in the table 2.

Table 2. Effect of the turbulence model on the results.

	rng k-ε	SST k -ω	RSM
w [m/s]	105		
T_a [°C]	5.6		
T_w [°C]	22.6		
T_{01} [°C]	5.60	5.60	5.61
ΔT_{12} [°C]	0.05	0.07	0.07
T_{02} [°C]	5.65	5.67	5.68
ΔT_{13} [°C]	0.68	0.86	0.9
T_{03} [°C]	6.28	6.46	6.51

4 Model of the single thermocouple probe for sensitive analysis

Since the results achieved at the multiple sensors probe were interesting, it was decided to proceed to the more complex solution including the modelling and simulation of simple probes containing single thermocouples. Other models of probes as depicted in figure 2 were prepared for simulations. The models were prepared for the cases with and without deflectors.

One of the easier cases of the probe with single thermocouple without deflector was simulated within CFD software OpenFOAM 2.4. An example of the simulation results is depicted in figure 7. The figure shows flow around the thermocouple with heat transfer between solid and fluid, where the fluid warms probe.

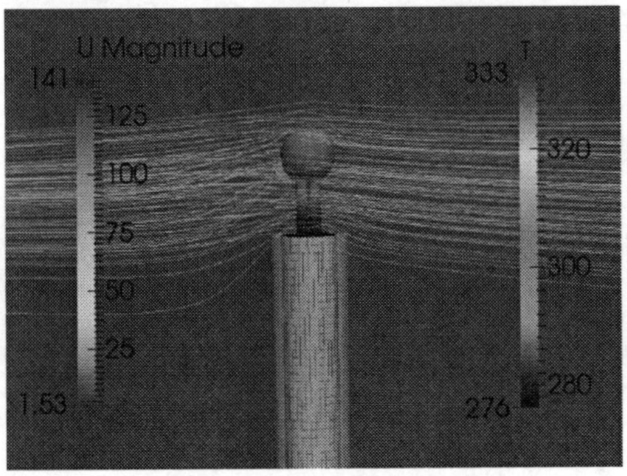

Figure 7. Model of the probe consisting of a single thermocouple. The presented model was prepared in OpenFOAM.

The simulation in OpenFOAM includes radiation. As a radiation model was used view factor model.

5 Conclusions

During the measurements of the compressor´s stages, the unrealistic and inhomogeneous temperatures were measured in the specific cases. Skewed results were obtained namely at inlet plane of the compressor.

To accomplish the measurements, two different kinds of simulations were conducted. The achieved results of the simulations are in concordance with observed data. That is that the temperatures of the walls affect measurements of total temperatures in the special cases.

The results of the used models unambiguously indicate that there is an impact on the measured total temperatures caused by temperatures of the nearest walls. But it must be admitted that the values achieved by simulation does not correspond with the measured data exactly. On account of this fact, the models still need to be further developed and tuned up to describe the reality more precisely.

Therefore, the simulation models were prepared with the aim for more detailed sensitive analysis of phenomenon. The models include radiation with the aim to examine the impact of the probe surroundings on the particular thermocouple probes.

In the nearest future, the experiments of controlled measurements in the presence of various surrounding conditions are going to be performed. The simulations models are further being developed more sophistically to identify all effects that could have skewed measurements of temperatures of fluid medium through the compressor stages.

Acknowledgement

These results were achieved in the framework of the FR-TI1/126 project of the programme "TIP", Ministry of Industry and Trade of the Czech Republic programme and in specific research by the internal grant at University of West Bohemia: SGS-2014-070.

References

1. P. C. Hanlon, *Compressor handbook* McGraw-Hill, USA (2001)
2. T. Syka, R. Matas, J. Kňourek *Hub and shroud fillets influence on the radial compressor stage efficiency*, EPJ Web of Conferences, **92** (2015)
3. *VDI Heat Atlas 2nd edition,* Springer-Verlag Berlin Heidelberg (2010)
4. www.ansys.com
5. www.openfoam.com

EPJ Web of Conferences **114**, 02072 (2016)
DOI: 10.1051/epjconf/201611402072

Measurement of Liquid Film Thickness via Light Absorption and Laser Tomography

M.A. Mendez[1,2,a], L. Németh[1], and J-M. Buchlin[1]

[1] *Von Karman Institute for Fluid Dynamics, Environmental and Applied Fluid Dynamics Department*
[2] *Université Libre de Bruxelles, Département d' Aéro-Thermo-Mécanique, Faculté des Sciences Appliquées*

Abstract.
This paper presents film thickness measurement over a vertical falling liquid film, carried out with Light Absorption (LAbs) and a 2D laser tomography techniques known as Level Detection and Recording (LeDaR). The presented test cases are in the low Reynolds regime ($Re = 1-8$) and include undisturbed conditions, i.e. constant flow rate, and inlet-forced conditions, i.e. flow rate pulsing at frequencies in the range $f_p = 12-18Hz$. The liquid used is Dipropilene Glycol (DPG), with Kapitza number $Ka = 4.6$. The results include average film thickness in the steady case, wave celerity and wave profile in the forced conditions. All the measurements are scaled using the Skhadov scaling, to facilitate the comparison with analytical low-dimensional models. Both LAbs and LeDaR techniques are presented in detail, including calibration, uncertainties and data processing aspects.

1 Introduction

Falling liquid films are encountered in many applications, including the coating processes of paint, paper or galvanizing industries, and the heat and mass transfer equipment of process industries. These flows are inherently unstable and prone to develop interface instabilities which are crucial in any application: they can limit the quality of coated products or enhance heat and mass transfer.

Besides the industrial concerns, falling liquid films motivated a large amount of fundamental research since the seminal work of Kapitza [1], as liquid films are in fact laminar flows amenable to analytical treatment. Several simplified models have been developed over the years [2–4]. These models aim to reduce the complexity of the governing Navier-Stokes (N-S) equations by identifying a stream-wise length scale ($[x]$) much larger than the thickness scale ($[h]$), to define a reordering parameter $\varepsilon = [h]/[x] \ll 1$ which weights the forces governing the film flow: gravity, viscosity, inertia and surface tension. Neglecting terms above a given order p (i.e. $\propto \varepsilon^p$) leads to model analytically tractable, yet capable of describing the film dynamics. This separation of scales ($[x] \gg [h]$), achieved via the Shkadov scaling presented in Sec.5, is the fundamental framework of any analytical model.

Little attention, however, has been paid to a clear definition of the conditions under the separation of scales is possible. As discussed in Sec.5, this depends on the liquid properties. In particular, on the ratio between the capillary length scale $l_\sigma = \sqrt{\sigma/\rho g}$ and the viscosity length scale $l_\nu = \sqrt[3]{\nu^2/g}$, regrouped in a dimensionless number known as Kapitza number ($Ka = l_\sigma^2/l_\nu^2$). For low Ka liquids (e.g. organic oils or paints, with $Ka = O(1)$) separation of scales is not possible, since $\varepsilon \approx 1$. An experimental validation of these analytical models, however, requires 2D or 3D film thickness measurements with high spatial and temporal resolution [5–8], and few authors have reported detailed thickness measurements at low Ka.

Krantz & Goren ([9], $Ka = 3.4$) focused on the experimental prediction of natural instability inception, by measuring whether introduced disturbances were amplified or damped. Moran *et al* ([10], $Ka = 18.4$) proposed an extensive characterization of a Nusselt flow down to an inclined plane, in terms of averaged thickness and averaged velocity profiles. Nguyen & Balakotaiah ([11], $Ka = 6-22$) analyzed large amplitude waves on a low Kapitza liquid, but their single-point measurements do not allow for a complete insight into the travelling wave profile. Several cases at $Ka = 14$ and $Ka = 85$ have been presented by Charogiannis *et al* [12], who combined the thickness measurement with velocity measurement within the liquid.

This paper focuses on the experimental characterization of 2D waves traveling over a vertical falling liquid film of Dipropilene Glycol (DPG, $Ka = 4.6$), having $\rho = 1023 \pm 0.1\%[kg/m^3]$, $\mu = 87.5 \pm 1\%[cSt]$ $\sigma = 32 \pm 2\%[mN/m]$, in the range of Reynolds number $Re = q/\nu = 1-8$, being q the flow rate per unit width. The conditions analyzed include undisturbed regime and inlet forced regime, with superimposed pulsing flow rate at frequencies $f_p = 12-18Hz$. Time resolved, 2D and 3D film thickness measurements were performed using two measurement techniques. The first technique is the light absorption (LAbs), implemented in the 3D version used in

[a] e-mail: mendez@vki.ac.be

[17–19]; the second technique is an adaptation of a visualization method based on planar laser-induced fluorescence (PLIF) and image processing, referred to as level detection and recording (LeDaR, [23–25]). In the LAbs technique the film thickness is retrieved by measuring the intensity of the light transmitted through the liquid film. Developed as a point-wise technique, using photodiods or photomultipliers as light receivers [14, 15], the LAbs has later become a full field technique using digital video cameras and basic image processing operations [17–19]. While this method can provide high spatial and temporal resolution, its main limitation is in the assumption that the light transmittance is solely linked to light absorption, disregarding spurious effects such as light refraction, reflection or scattering.
In the LeDaR technique the film thickness is retrieved via digital processing of flow visualizations images. In these images the contrast liquid-background is enhanced using planar laser-induced fluorescence (PLIF) of a dye diluted in the liquid and an optical band-pass filter on the camera's objective. The key difference between this approach and other LIF-based methods such as in [20–22] is that the intensity of the fluorescence emission is not related to the film thickness, but only serves visualization purposes. As a result, the measurement requires no calibration. The image processing technique used to detect the interface varies according to the complexity of the analysed configuration. Examples of these approaches are proposed in [27–31] and in [12], where it is successfully combined with PIV velocity measurements within the liquid.

The next section presents the VKI facility and its operation in steady and unsteady condition. Sec.3 and Sec.4 reviews, respectively, the LAbs and the LeDaR techniques. Sec.5 briefly presents the measurement scaling according to the most popular scaling approach used in the literature of liquid film modeling: the Shkadov scaling. Sec.6 presents the experimental results, Sec.7 the conclusions.

2 Liquid film facility

The VKI liquid film facility is sketched in figure 1. It consists of a closed loop where the liquid falls along a vertical test section (1), is collected in a tank (2) and re-circulated into a pneumatic chamber (5) through an inlet line (4), by a volumetric pump (3, Pollard PR40). The pump is a rotative vane pump, which introduces a negligible temperature rise, avoiding the need of a heat exchanger. The liquid temperature was monitored in the test section and in the tank, and remained constant at $21.7^oC \pm 1\%$ during the tests presented in Sec.6.

In the pneumatic chamber, the liquid motion is minimized splitting both the input and the by-pass flow rates into four parts, positioning the inlet line $15cm$ below the liquid level, and hosting a large liquid volume ($18l$) during the experiment. This allows to maintain the liquid surface flat, and to control the hydrostatic head of the liquid flow by monitoring the rotational speed of the pump and the opening of the by-pass line. The volume of the liquid gives to the system a high inertia, to simplify the regulation of the facility and allow the user to set a constant level during the tests. Moreover, the chamber is pressurized from an inlet line (7), equipped with a pressure regulator, and discharges in atmosphere through an outlet line (8). A deflecting frame (9) avoids the impingement of the incoming air on the liquid interface within the chamber. The opening of the lines 7-8 is controlled by two butterfly valves (detail A), mounted with 90° phase delay on the shaft of an electric motor (Maxon EC-4pole 45).

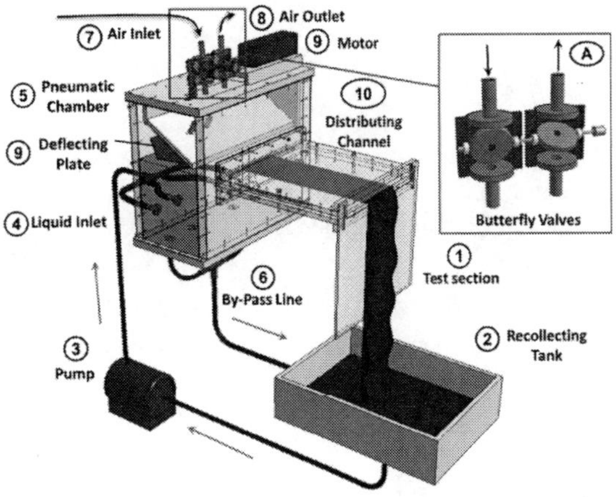

Figure 1: Sketch of the VKI liquid Film Facility All components are manufactured in ®PLEXIGLAS, with $1^{1/4}inch$ copper pipes for the connections 4-6.

In the steady operation mode, the motor is off, both valves are open, and a constant chamber pressure is settled by the upstream pressure regulator. In the unsteady operation mode, the motor opens and closes lines 7 and 8 periodically, producing two charge and discharge phases at every round, thus a pressure pulsation in the chamber.

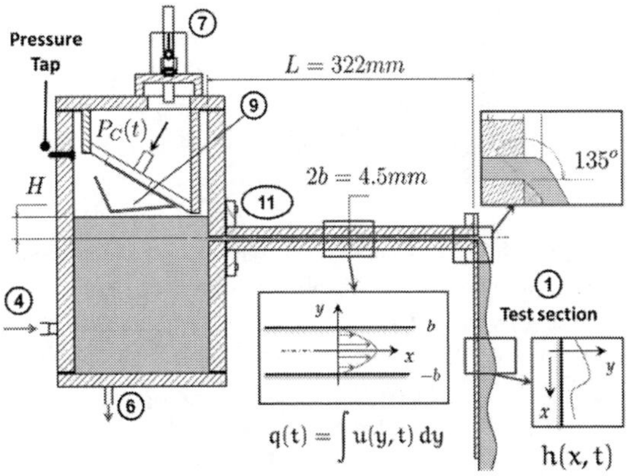

Figure 2: Facility Cross Section. The pressure pulsation $P_C(t)$ produces a flow rate pulsation $q(t)$, resulting in a wavy liquid film $h(x, t)$. A smooth connection between channel and test section prevents liquid detachment.

As the film thickness adhering the walls is considerably thinner than in the investigated areas, this interruption has no appreciable effects on the film thickness. This can

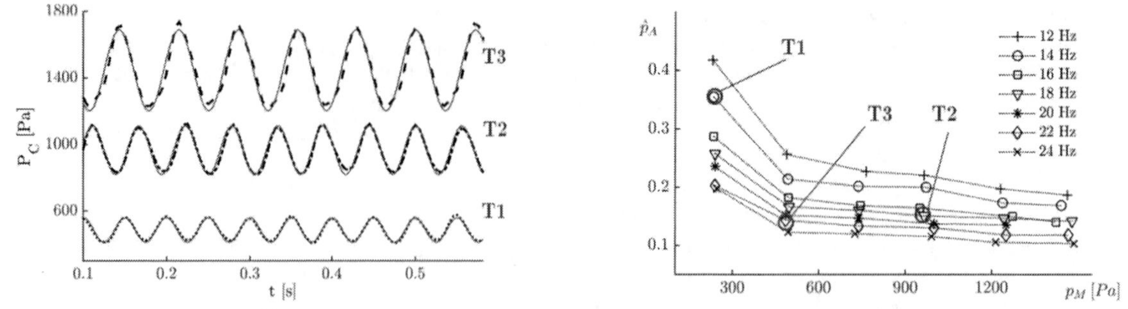

(a) (b)

Figure 4: Pressure measurements in the pneumatic chamber. Figure 4a: examples of pressure signals $P_C(t)$ acquired from the pressure transducer at sampling $f_s = 800Hz$. Figure 4b: relative pressure amplitude $\hat{p}_A(p_M, f)$ curves for seven frequencies, six mean pressures.

be observed in the shape of the 3D liquid edges forming in the two sides of film, shown in figure3: no appreciable difference appears between the open (right) and closed (left) sides.

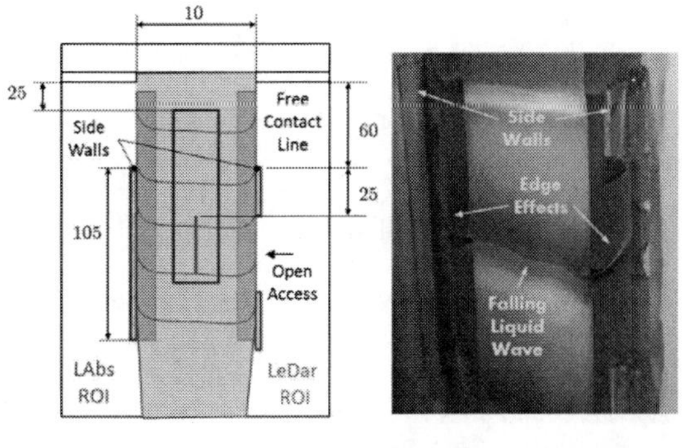

Figure 3: Test section front view, with measurements ROIs, and picture of a traveling wave during the test. The test section is illuminated by diffused red light, as described in section 3. The 3D lateral edges are unaffected by the open access in one of the side walls.

figure4a shows three examples of pressure signals $P_C(t)$, measured in the pneumatic chamber. On each of these signals it is fitted, in red, a harmonic pulsating pressure having mean value p_M, frequency f_p and amplitude p_A. The user controls the mean pressure p_M, through the pressure regulator in the inlet line, and the pulsation frequency f_p, through the motor speed $\omega_m = 4\pi f_p$. The pressure amplitude p_A, on the other hand, depends on the facility line-cavity behavior. As shown in figure 4b, the relative pressure amplitude $\hat{p}_A = p_A/p_M$ decreases asymptotically when increasing p_M and f_p.

Since the liquid flow in the channel is incompressible, 2D, and fully developed, the governing equation is linear thus the liquid flow rate per unit width q is the superimposition of a mean flow q_M, sustained by the mean pressure p_M, and an oscillatory component $q_A \sin(2\pi f_p - \phi)$, sustained by the oscillating pressure $p_A \sin(2\pi f_p)$. Referring to the sketch in figure 2, the liquid flow reaching the test section is governed by the x-wise momentum equation:

$$\frac{\partial u}{\partial t} = -\frac{1}{\rho}\frac{\partial p}{\partial x} + \nu\frac{\partial^2 u}{\partial y^2}\,. \tag{1}$$

As the total pressure is $p(t) = P_C(t) + \rho g H$ and the pressure drop is linear, the pressure gradient term reads:

$$-\frac{\partial p(t)}{\partial x} = -\left(\frac{P_M}{L} + \frac{p_A}{L}\cos(2\pi f_p t)\right), \tag{2}$$

where $P_M = p_M + \rho g H$ is the mean pressure head.
The momentum equation in eq.1 is conveniently rescaled via the following reference quantities: $\hat{y} = y/b$, $\hat{x} = x/L$, $\hat{u} = u/(P_M b^2/\mu L)$, $\hat{P} = P/P_M$, $\hat{t} = t/(2\pi f_p)$, obtaining:

$$\mathcal{W}^2\frac{\partial \hat{u}}{\partial \hat{t}} = -\frac{\partial \hat{p}}{\partial \hat{x}} + \frac{\partial^2 \hat{u}}{\partial \hat{y}^2} = -\left(1 + \hat{P}_A\cos(\hat{t})\right) + \frac{\partial^2 \hat{u}}{\partial \hat{y}^2}\,, \tag{3}$$

where eq.2 has been introduced. The only dimensionless number controlling the channel flow response is the Womersley number $\mathcal{W} = b\sqrt{\omega/\nu}$ ([32–34]).
The steady solution of eq.3, related to the mean flow, is the steady Poiseuille flow:

$$\hat{u}_M(\hat{y}) = \frac{1}{2}\left(1 - \hat{y}^2\right) \Longrightarrow \hat{q}_M = \int_{-1}^{1}\hat{u}_M(\hat{y})d\hat{y} = \frac{2}{3} \tag{4}$$

from which, reversing the scaling, the mean Reynolds number Re_M reads:

$$Re_M = \frac{q_M}{\nu} = \frac{2b^3}{3\rho\nu^2}\left(\frac{\rho g H + p_M}{L}\right) \tag{5}$$

The harmonic component of eq.3 can be obtained analytically in terms of a periodic, variable separated, complex solution $\tilde{u}(\hat{y}, \hat{t}) = \tilde{f}(\hat{y})e^{i\hat{t}}$ which trasforms eq.3 into a simple ordinary differential equation. The results, as shown in [35, 36], is the real part $\hat{u}_A = \mathbb{R}(\tilde{u})$:

$$\hat{u}_A(\hat{y}, \hat{t}) = \mathbb{R}\left\{\left(1 - \frac{\cosh(\mathcal{W}\sqrt{i}\,\hat{y})}{\cosh(\mathcal{W}\sqrt{i})}\right)\frac{i\hat{P}_A}{\mathcal{W}^2}e^{i\hat{t}}\right\}\,. \tag{6}$$

The corresponding flow rate amplitude can therefore be linked to a modulation term $\mathcal{T}$ which only depends on the Womersley number and the dimensionless pressure oscillation $\hat{P}_A$:

$$\hat{q}_A = \frac{1}{\cos(\hat{t}-\hat{\phi})}\int_{-1}^{1}\hat{u}_A(y,t)d\hat{y} = \mathcal{T}(\mathcal{W})\hat{P}_A \quad (7)$$

from which, reversing the scaling, the amplitude of the Reynolds pulsation Re_A reads:

$$Re_A = \frac{q_A}{\nu} = \frac{3}{2}\mathcal{T}(\mathcal{W})\hat{P}_A Re_M \quad (8)$$

figure 5 shows the modulation function $\mathcal{T}$ obtained analytically from eq.6 (Th), together with the results of CFD laminar simulations for three relative pressure amplitudes ($\hat{P}_A = 0.2-0.3-0.6$) and frequency in the range $10-30Hz$ [36]. The strength of the flow rate pulsation is hereinafter measured in terms of $\zeta = Re_A/Re_M$.

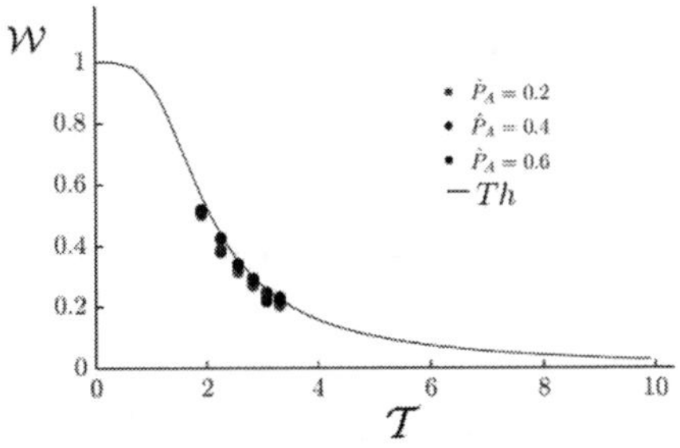

Figure 5: Modulation function $\mathcal{T}$ as a function of the Womersley number $\mathcal{W}$ produced by the pressure pulsation in the VKI Facility in figure 1. Comparison of CFD results and analytical derivation [36].

3 Light Absorption (LAbs)

The light absorption relates the instantaneous thickness distribution $h(x,y,t)$ with the amount of light absorbed from a back-lighting source, using the well known Beer-Lambert's law:

$$h(x,z,t) = \frac{1}{\gamma_g}\, ln\left(\frac{I_0(x,z)}{I(x,z,t)}\right) = \frac{A(x,z,t)}{\gamma_g} \quad (9)$$

Where γ_g is the light extinction coefficient of the liquid, determined via calibration, $I_0(x,z)$ is the back-light intensity, referred to as *Reference Image*, $I(x,z,t)$ is the transmitted light intensity trough the liquid film, referred to as *Trasmittance Image*. The dimensionless quantity $A = ln(I_0/I)$ is referred to as *Absorbance*. The absorption coefficient γ_g is the average over the total spectrum of light captured by the receiver, and it is a function of the temperature. Light absorption is typically enhanced diluting in the liquid small concentration of colourant dye whose absorbance spectrum matches the emission spectrum of the light source. The colourant concentration c is thus a key variable affecting the technique's sensitivity and its influence is hereinafter analyzed. As for the temperature influence, its effects are canceled by performing calibration and measurements in the same conditions.

3.1 LAbs Test Section

The light absorption (LAbs) test section is sketched in figure 6. The light source (1) consists of two arrays with 4x4 LEDs of $\approx 0.7\,W$ each, with 90% emittance in the range $636 \pm 26\mu m$. They are placed $10\,cm$ behind a $3mm$ thick screen Opal ®PLEXIGLAS (2), which diffuses the light over the test section (3), to which it is attached. The receiver (4) is a rolling shutter $16bits$ CMOS camera (Hamamatsu ORCA-Flash4.0), synchronized with the LEDs, to acquire at $200Hz$ with a resolution of $500x2048pixels$.

The synchronization is achieved operating the camera in free running mode, with the camera controlling the LEDs so that the light flash occurs when the rolling shutter exposes the whole sensor. As a result, the *Reference* image $I_0(x,z)$ is determined with an uncertainty of 0.5% and no light flickering problems arise. A red absorbing colourant, the blue methlyene, is diluted in the DPG and an optical band-pass filtered ($640 \pm 40\mu m$) is installed on the $50mm$ camera's objective, to limit the camera sensitivity to the portion of spectra in which absorption takes place.

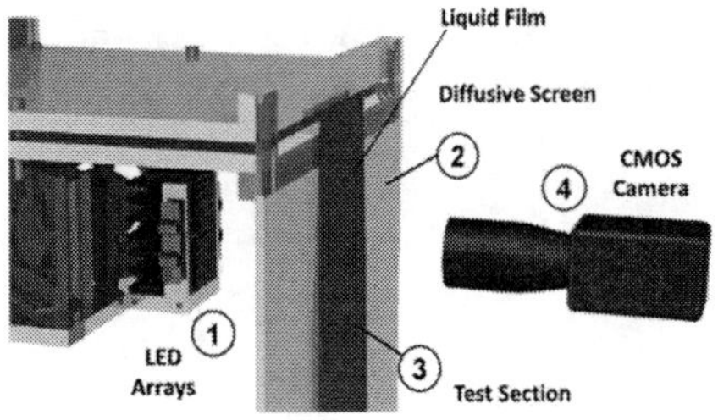

Figure 6: Light Absorption (LAbs) Test Section: LEDs (1) and screen (2) to back-light the test section (3) with diffuse, monochromatic light. A CMOS camera (4), to measure the attenuation of light produced by the liquid film.

3.2 LAbs Calibration and Data Processing

The LAbs calibration consists in evaluating the scaling factor $M[pixel/mm]$ and measuring the global extinction coefficient γ_g. The scaling factor is obtained by binarizing an image containing pattern of circles and extracting the diameter via standard morphology operations. In the LAbs measurement, it is typically $M = 19.6pixel/mm \pm 1\%$. The extinction coefficient is measured using a calibrating vat and positioning the set up horizontally. Figure 7 sketches the calibration step and shows a typical absorbance profile. The absorbance profile is computed from the gray scale reference video $I_0(x,y,t)$, taken with an empty vat, and the gray scale transmittance video $I(x,y,t)$, taken with the vat filled with liquid. The corresponding absorbance image $A = ln(\overline{I}_0/\overline{I})$ is obtained from the time averages of the videos $\bar{I}_0$ and $\bar{I}$. The absorption image is spatially filtered with a low-pass Gaussian and the average profile is fitted via linear regression. The camera alignment error can be corrected by rotating the image so as to have a zero gradient component $\partial A/\partial y = 0$.

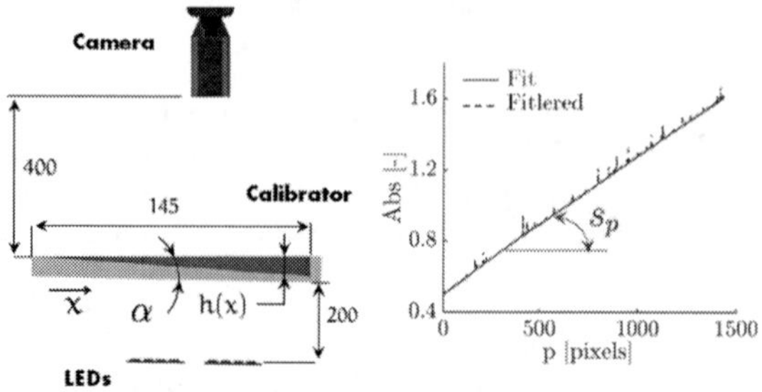

Figure 7: Calibrator of the LAbs measurement: calibrating vat and typical absorbance curve.

The absorption coefficient is retrieved from the vat slope $s_x = \tan(\alpha)$, the slope of the absorption profile in the image s_p, and the image scaling factor $M = x/p$, using eq. 9:

$$\gamma_g = \frac{A(x)}{h(x)} = \frac{s_p \cdot x/M}{x \cdot s_x} = \frac{s_p}{M \cdot s_x} \tag{10}$$

Figure 8 shows the linear dependency of the global absorption coefficient for the low colourant concentration used.

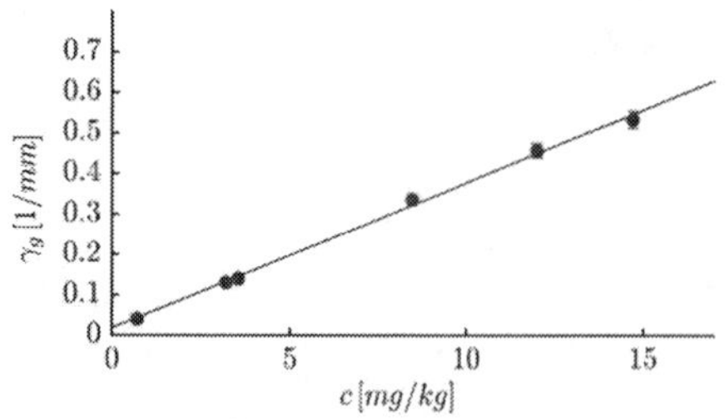

Figure 8: Extinction coefficient γ_g as a function of blue methylene diluted in DPG in mg/kg. The concentration range selected is sufficiently low to let the linearity in eq.9 hold.

3.3 LAbs Uncertainties and Sensitivity Analysis

The sensitivity–and the uncertainty– analysis of the LAbs measurement $\mathcal{U}_h(\gamma_p, A)$, as a function of the extinction coefficient and the absorbance, can be obtained from eq.9:

$$\mathcal{U}_h = \sqrt{\frac{\mathcal{U}_A^2}{\gamma_p^2} + \frac{A^2\,\delta\mathcal{U}_{\gamma_g}^2}{\gamma_g^4}}\,, \tag{11}$$

and depends on the absorbance and the extinction coefficient uncertainty $\mathcal{U}_A$ and $\mathcal{U}_{\gamma_g}$. In particular, the absorbance uncertainty δA depends on the measured gray scale transmittance I and reference I_0 images. From eq.9 it is:

$$\mathcal{U}_A = \sqrt{\left(\frac{\mathcal{U}_I}{I}\right)^2 + \left(\frac{\mathcal{U}_{I_0}}{I_0}\right)^2} \tag{12}$$

$$\approx \sqrt{\frac{\mathcal{U}_I^2}{I_0^2}\left(e^{2A} + 1\right)} \tag{13}$$

using eq. 9 and assuming that $\mathcal{U}_I \approx \mathcal{U}_{I_0}$. This equation highlights an exponential growth of the uncertainty as the absorbance –thus the liquid thickness or the concentration– increases. The relative uncertainty of the light intensity $\mathcal{U}_I/I_0$ modulates this growth and represents the most important parameter. The synchronization camera-LED in this works gives $\mathcal{U}_I/I_0 = 0.15\%$. The uncertainty of the extension coefficient $\mathcal{U}_{\gamma_p}$, on the other hand, does not depend on the operating condition but only on the calibration. From eq.10 it is:

$$\mathcal{U}_{\gamma_p} = \sqrt{\left(\frac{\mathcal{U}_{s_p}}{M s_x}\right)^2 + \left(\frac{s_p\,\mathcal{U}_M}{M^2 \cdot s_x}\right)^2 + \left(\frac{s_p\,\mathcal{U}_{s_x}}{M \cdot s_x^2}\right)^2} \tag{14}$$

and leads to a relative uncertainty of $\mathcal{U}_{\gamma_g}/\gamma_g = 2\%$. Introducing eq.14 and eq.12 in eq.11 gives the thickness uncertainty as a function of the thickness and the extinction coefficient. The results are collected in figure 9.

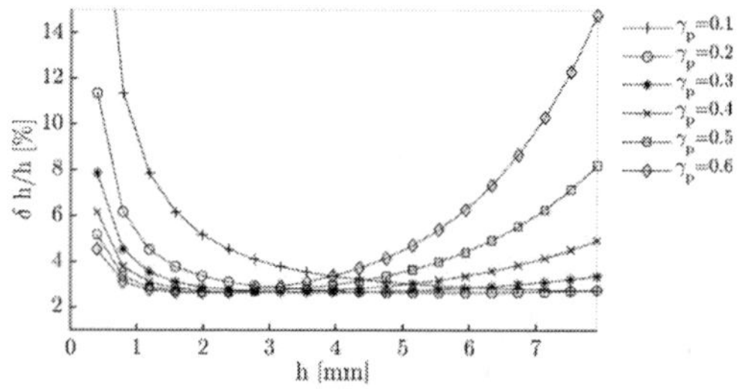

Figure 9: Relative uncertainty $\delta h/h$ as a function of the liquid thickness h for the different extinction coefficients γ_g, thus colourant concentrations.

It is evident that the choice of the extinction coefficient, thus the colourant concentration, depends on the film thickness range of interest: high concentrations are suited for thinner films while the reverse is true for low concentrations. As this work considers film thickness ranges $h = 1 - 6mm$, a colourant concentration of $c = 15mg/kg \pm 1\%$ was chosen, leading to $\gamma_g = 0.55mm^{-1} \pm 2\%$.

An example of LAbs measurement and processing is detailed in figure 10. The image processing only consists in converting the reference image I_0 and the transmittance image I into double precision array and use equation 9. The resulting thickness maps are low pass filtered using the fast smoothing spline method from [37], in order to remove the high frequency noise due to entrained micro bubbles without smoothing the liquid film thickness gradients.

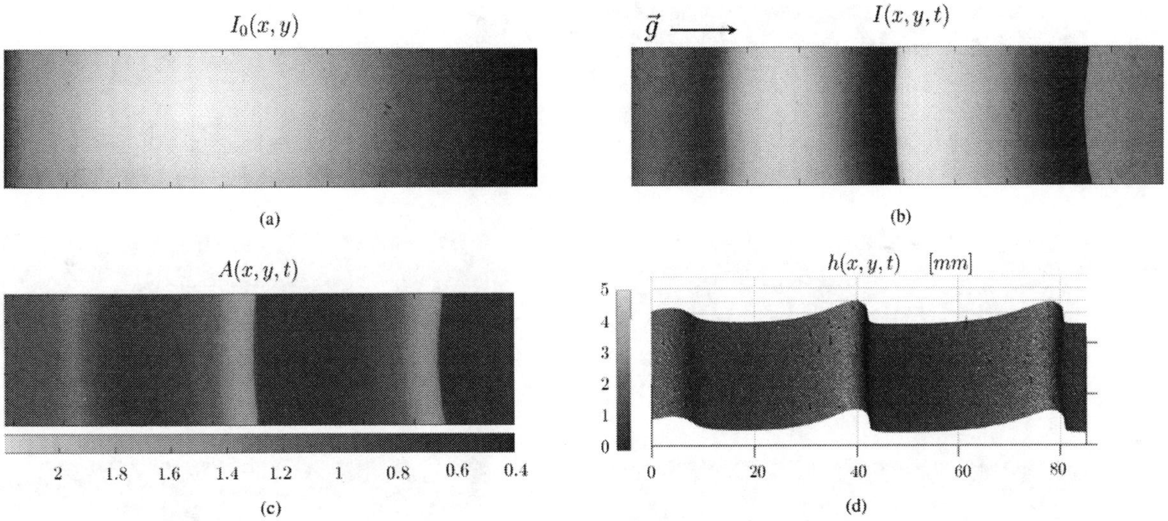

Figure 10: Image processing for LAbs measurement: the reference image $I_0(x,y)$ (in a), taken with the dry test section, is divided by the transmittance image $I(x,y,t)$ (in b) to compute the absorbance image $A(x,y,t) = ln(I_0/I)$. The corresponding instantaneous thickness contourmap, computed from Beer-Lambert's law (in d).

4 Level Detection and Recording (LeDaR)

The Level Detection and Recording (LeDaR) is an image processing technique developed at the von Karman Institute to extract air-liquid interfaces from laser tomography images ([23–25]). Other versions of the method have been independently developed in the literature ([27–31]). This method uses the planar laser induced fluorescence as a contrast enhancement tool and detects the liquid interface via image processing techniques, either via gradient-based methods (typically convolution with Sobel-like kernels) or via segmentation-based methods (typically Otsu's binarization preceded by histogram equalization). However, as the liquid waves observed in this work have larger amplitude and curvature than those measured in literature with this method, none of the two approaches proved successful. The main challenge is that the high interface curvature acts as a lens focusing light within a narrow region below the wave crest, hindering the uniformity of the illumination in the liquid domain. To overcome the problem, the routine presented combines directional image re-contrasting with standard gray scale gradient analysis.

4.1 LeDaR Test Section

The LeDaR test section is sketched in figure 11. An optical bench (1) produces and direct a laser sheet towards the test section (2). A CMOS camera (3, Hamamatsu ORCA-Flash 4.0), equipped with $350mm$ zoom lens and an optical high pass filter ($> 570\mu m$) allows to visualize the laser induced fluorescence. The laser source is a continuous argon laser (Stability 2017, Spectra Physics); the diluted die is Rhodamine B with a concentration of $20mg/kg$.

As for the light absorption measurement, the liquid flow lateral edges are defined by the lateral contact lines: no wall nor boundaries are placed between the focal plane and the camera's objective which visualizes the illuminated plane through the liquid film. Contrary to the set ups in [23–25], no index-matching methods are therefore used. An example snapshot is shown in figure 12a. The video consists of a sequence of $n_T = 1000$, $16bits$ gray scale images with $400 \times 2048 pixels$ resolution, sampled at $f_s = 200Hz$.

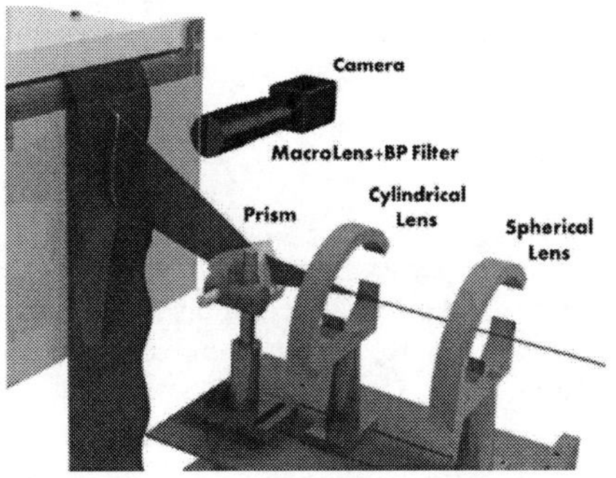

Figure 11: LeDaR setup: plane laser sheet, fluorescent dye diluted in the liquid, video camera. As for the LABS, the test section wall is wet before the test, to let the lateral flow boundaries defined by the liquid contact-line.

4.2 LeDaR Calibration and Data Processing

The calibration, performed in the dry test section, consists in evaluating the wall position and the scaling factor, using a pattern of circles. The scaling factor in the images was $M = 61 pixel/mm \pm 1\%$; the wall line was set as zero thickness with a typical alignment error of 1°. No

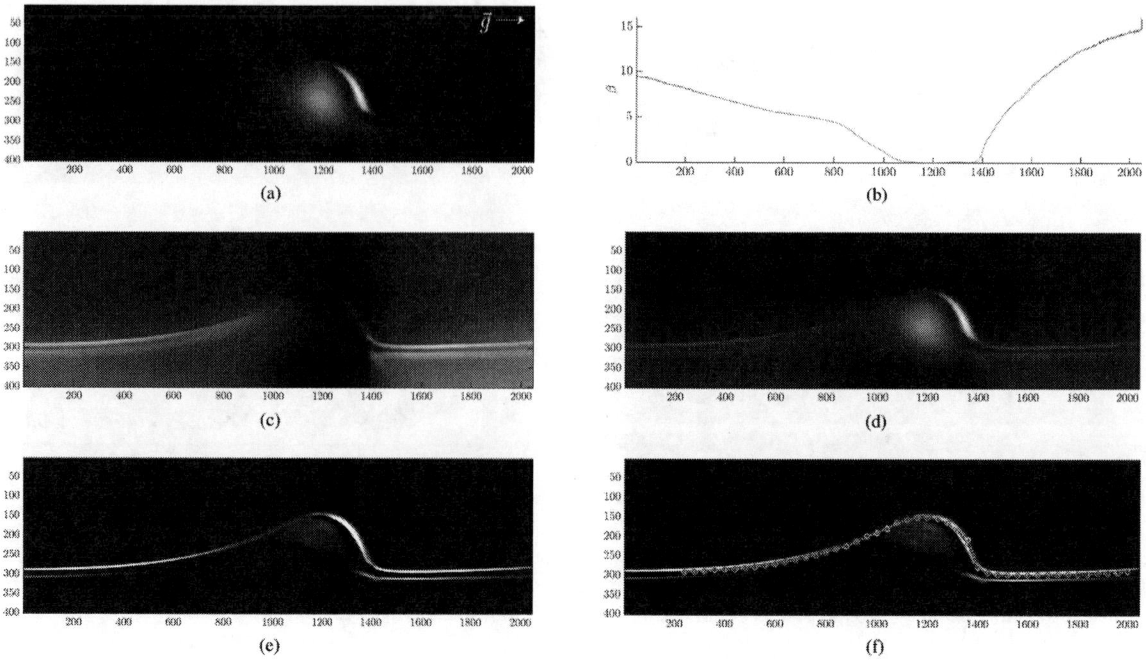

Figure 12: Image processing steps for LeDaR: From the raw image (in a), the algorithm computes the enhancement factor β (plotted in b), and use it to directionally recontrast the image. The results (in c), is added to the original image, obtaining the figure in d. Finally, a sobel-like kernel enhances the liquid interface (in e) and the column-wise peak finder method locates the interface in the image (in f).

appreciable perspective error was observed. The image processing technique proposed in this work is divided in three steps: gray scale re-contrasting, edge enhancement and edge location. The re-contrasting is used to uniform the light intensity within the liquid film, which is reduced by large waves such as in figure 12a. This step is based on a directional gray scale averaging. The idea is sketched in figure 13. Let $A(x_i, y_i) \in \mathbb{R}^2$ be the gray scale image with $x_i \in [0, n_x]$ columns and $y_i \in [0, n_y]$ rows having a mean gray scale level μ_A. Let $D = f(x_i) \in \mathbb{R}^{1 \times n_p}$ be a user defined line, passing through n_p image pixels, each of which identifying an orthogonal line $P_n(x_i)$ and a gray scale profile $A(x_i, P_n(x_i))$ having average μ_n. In the directional re-contrasting step, each of the n_p gray scale profile $A(x_i, P_n(x_i))$ is multiplied by an enhancement factor of the form $\beta_n = (\mu_n/\mu_A)^k$.

In this work the stretching direction D is chosen to be horizontal, thus the gray scale profile are taken column-wise. For the image snapshot in figure 12a, the enhancing factor with $k = 2$ is plotted as a function of the image column in figure 12b. The stretching result, shown in figure 12c, improves the light uniformity in the darkest regions while causes a darkening of the brightest ones. This image is added to the original one, obtaining the image B, in figure 12d. The second step of the processing is the edge enhancement. This is done by convolving the image B with an extended Sobel-like [9x3] kernel K, which is a reduced version of the maximum forward [21x3] kernel presented in [25]:

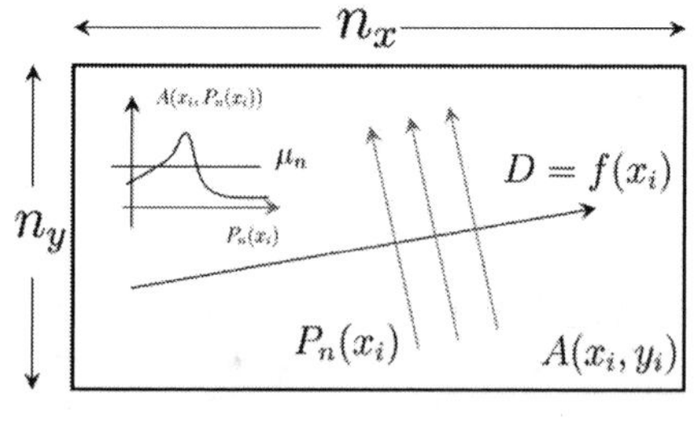

Figure 13: Working principle of the directional re-contrasting used as first step of the image processing.

$$C(x_n, y_n) = B(x_n, y_n) * \begin{pmatrix} -E \\ C \\ E \end{pmatrix} \quad (15)$$

where:

$$C = \begin{pmatrix} 0 & -1 & 0 \\ 0 & 0 & 0 \\ 0 & 1 & 0 \end{pmatrix} \quad E = \begin{pmatrix} 1 & 2 & 1 \\ 1 & 2 & 1 \\ 1 & 2 & 1 \end{pmatrix} \quad (16)$$

This operation enhances the horizontal edges of the image and results in a bright interface region with thickness varying between 4-16 pixels. The results, shown in figure 12e, is used to locate the interface position $h_p(x_i)$ in each column from the position of the peak in the profile $C(x_i, y_n)$. In particular, the peaks are detected using standard prominence method on a smoothed version of the profiles, ensuring that the interface positioned is in the middle of the white interface band as shown in figure 12f.

4.3 LeDaR Uncertainty

The measurement uncertainty is assumed to be due to the interface location h_p in the pixel space and the scaling factor M used to convert the results in the physical units. The small and purely random uncertainty in the detection of the scaling factor M and the absence of perspective deformation during the calibration allows to neglect camera alignment uncertainties.

5 Data Scaling and Parametrization

The dynamics of a vertical falling liquid film is controlled by two dimensionless number. The natural scaling of the N-S equations leads to the Reynolds number $Re = [u][h]/\nu$ and the Weber number $We = \sigma/\rho[u]^2[h]$ were $[u]$ and $[h]$ are the reference velocity and thickness respectively. The selection of these quantities is typically based on the definition of the *Nusselt* flow, i.e. the liquid film flow obtained in steady condition assuming that the interface remains flat and constant. These two assumptions suppress the role of inertia and surface tension and the flow results from the balance of viscosity and gravity. The film thickness h_N and the corresponding mean velocity u_N are consequently linked by the liquid properties and the flow rate per unit width $q_N = h_N u_N$, thus the Reynolds number Re:

$$h_N = \left(\frac{3\nu q}{g}\right)^{1/3} = \left(\frac{3\nu^2}{g} Re\right)^{1/3} = l_\nu \left(3Re\right)^{1/3} \quad (17)$$

$$u_N = \frac{g h_N^2}{3\nu} = \frac{1}{3}(\nu g)^{1/3}\left(3Re\right)^{2/3} = \frac{l_\nu}{3t_\nu}\left(3Re\right)^{2/3} \quad (18)$$

where $l_\nu = (\nu^2/g)^{1/3}$ and $t_\nu = (\nu/g^2)^{1/3}$ are the viscous-gravity length and time scales. The Nusselt thickness h_N is chosen as reference length $[h] = h_N$ in the cross-stream direction whereas it is worth observing that the definition of the velocity scale $[u]$ is not unique in the literature of liquid film. In the definition of the Reynolds number it is $[u] = u_N$, whereas it is usually considered as $[u] = 3u_N$ when scaling the N-S equation (see [3] and [4]). This approach, responsible for the factor $[u]/u_N = 3$ appearing in various term of the scaled N-S, and in eq.s 17 and 18, is motivated by the willing of maintaining the standard definition of $Re = q_N/\nu$ all the while scaling the liquid film velocity with respect to the fastest waves arising from the natural interface instabilities. These are the kinematic waves arising when neglecting the non linear inertial and surface tension effects, traveling at $c_w = 3u_N$ ([2]).

Regarding the stream-wise reference length scale $[x]$, as waves in liquid films are typically long wavelength $\lambda >> h_N$, analytical models [7, 8] compress the stream-wise coordinate by a factor $k = 1/\varepsilon$, so that $\varepsilon = [h]/[x] <<$ 1. After Shkadov [13], this factor is typically identified by selecting the stream-wise reference length $[x]$ from the balance of capillary pressure gradient ($\approx \sigma\partial_{xxx}h$) and gravity volume force (ρg):

$$\sigma\,\partial_{xxx}h \approx \sigma\,\frac{h_N}{[x]^3} \approx \rho g \rightarrow [x] = We^{1/3}h_N\,, \quad (19)$$

from which $\varepsilon = We^{-1/3}$. This reordering parameter is conveniently rewritten by replacing the Weber number We with a dimensionless number which solely depends on the liquid properties. This is the Kapitza number Ka:

$$Ka = \frac{9\,We}{(3Re)^{5/3}} = \frac{\sigma}{g^{1/3}\nu^{4/3}\rho} = \left(\frac{l_\sigma}{l_\nu}\right)^2, \quad (20)$$

being $l_\sigma = \sqrt{\sigma/(\rho g)}$ the capillary length. This number compares the importance of surface tension to viscous dispersion. Introducing eq.20 in eq.19 yields:

$$\varepsilon = \frac{(3Re)^{2/9}}{Ka^{1/3}} \quad (21)$$

For a laminar liquid film with $Re = O(10)$, the assumption of $\varepsilon \ll 1$ implies that $Ka \gg 1$. This is true for liquid such as water or alcohols ($Ka = O(10^3)$), but not for liquids such as the Dipropilene Glicole used in this work, for which $Ka = 4.61$. As result, the stream-wise length scale is $[x] \approx [h]$ and the analytical models described in [3, 4] lose their fundamental framework. The Shkadov scaling remains nevertheless the best attempt to separate the "fast" cross-stream scale from the "slow" stream-wise scale, and it is therefore used to scale the thickness measurement in this work. In particular, the flow regime is defined in terms of a *Reduced Reynolds number* δ and the *Viscous Dispersion Parameter* η:

$$\delta = 3\varepsilon Re = \frac{(3Re)^{11/9}}{Ka^{1/3}}\,,\ \eta = \varepsilon^2 = \frac{(3Re)^{4/9}}{Ka^{2/3}}\,, \quad (22)$$

while the scaling of the film thickness variables reads:

$$\hat{h} = \frac{h}{h_N}\ \ \hat{x} = \frac{x\varepsilon}{h_N}\ \ \hat{c}_w = \frac{c}{3u_N} \quad (23)$$

6 Results

For a set of steady tests at different mean pressures, figure 14a compares the dimensionless mean film thickness measured via LAbs (red, dotted line) and LeDaR (blue, dashed line) with the theoretical Nusselt film thickness ($\hat{h} = 1$) in eq.17, where eq.5 is used to compute the Reynolds number Re from the measured mean pressure P_M. The error bars are computed as the least square of the systematic uncertainties from eq.11 and the standard deviation resulting from the spatial and temporal averaging. The amplitude of the long wavelength waves naturally developing over the mean flow remains within the systematic measurement uncertainties and do not produce increasing error bars. The good agreement between measurement and theoretical prediction validates the derivations in eq.17 and

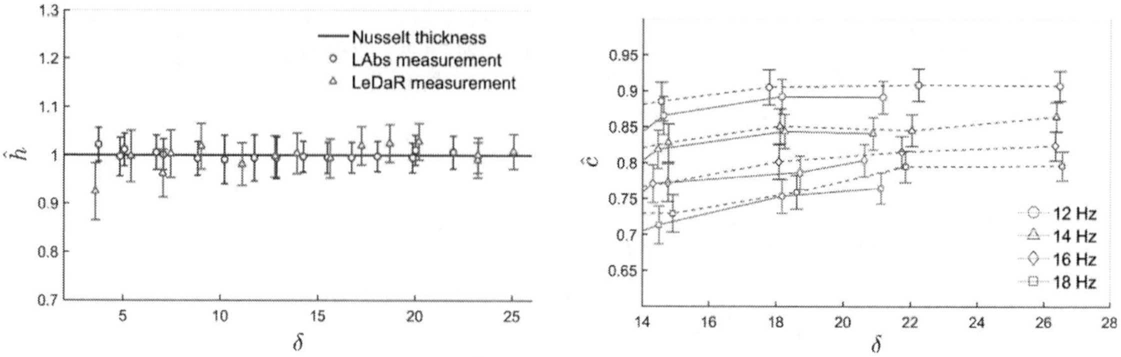

(a) (b)

Figure 14: Deviation from Nusselt film thickness ($\hat{h}$ = 1,figure 14a) as a function δ. Figure 14b: wave celerity $\hat{c}$ as a function of δ and perturbation frequencies f_p, for LeDaR (blue, dashed line) and LAbs (red, continuous line) measurements.

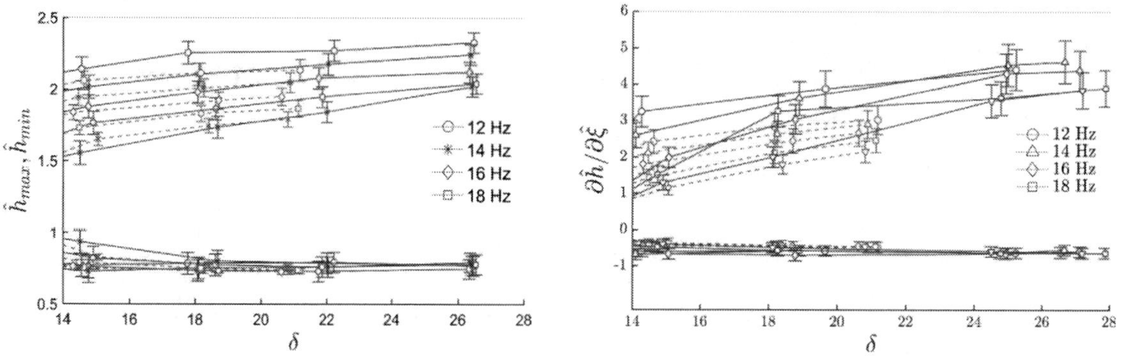

(a) (b)

Figure 15: Minimum and maximum thickness ($\hat{h}_{min}/\hat{h}_{max}$,figure 15a) and maximum slopes ($\partial\hat{h}/\partial\hat{\xi}$, figure 15b) in the wave front and tail as a function of δ for different perturbation frequencies f_p, from LeDaR (blue, dashed line) and LAbs (red, continuous line) measurements.

eq.5, which are hereinafter used to non-dimensionalize the test cases with unsteady flow rate. The comparison between LAbs and LeDaR for the unsteady cases is carried out in the same spatial location, thus in the final portion of the LAbs measurement (see ROIs in figure 3). Figure 14b presents the wave celerity, obtained by cross-correlating the thickness profiles in two subsequence images. In the range of reduced Reynolds δ considered, both techniques show a weak dependency of the dimensionless wave celerity as a function of δ, and a more evident dependency on the perturbation frequency f_p. This is due to amplitude of the perturbation produced in the facility, as higher frequency, thus high $\mathcal{W}$, leads to stronger inertial damping, thus smaller perturbation strength $\zeta = Re_A/Re_M$, as shown in figure 5. Figure 15a shows the evolution of wave maxima and minima as function of δ for several perturbation frequencies f_p. While the wave peaks grow monotonically, the wave minima exhibit an asymptotic behavior towards $\hat{h}_{min} \approx 0.75$, regardless of perturbation amplitude and frequency. A fair agreement in both techniques is observed.

Figure 15b shows the evolution of the maximum thickness slope $\partial_{\xi}\hat{h}$ in the front and in the tail of the waves. These gradients are computed from the wave profiles extracted in the moving reference frame ($\hat{\xi} = \hat{x} - \hat{c}\hat{t}$), shifted to have the wave peak at $\hat{\xi} = 0$. For plotting purposes, we assign a positive sign to the tail's slope, a positive sign to the front's slope. In the latter, a systematic discrepancy between the two techniques is evident, with the LeDaR detecting higher values than the LAbs. This discrepancy is due to a different prediction of the wave profile in the wave front. Figure 16 and figure17 show several dimensionless profiles for LAbs and LeDaR, at $12Hz$ and $18Hz$ respectively, for different mean Reynolds number. Besides the difference in the peaks, due to the slight difference in the mean Reynolds, the profiles detected in the LeDaR have a curvature in the front which is not observed in those detected in the LAbs. Such difference increases with the wave peak, thus the interface curvature, and it is due to light refraction in the wave front, which biases the LeDaR measurements. Finally, it is worth noticing, in the profiles obtained with both techniques, the absence of capillary ripples in the front wave. These ripples, typical of high Ka liquids, are damped in low Ka liquids by the high

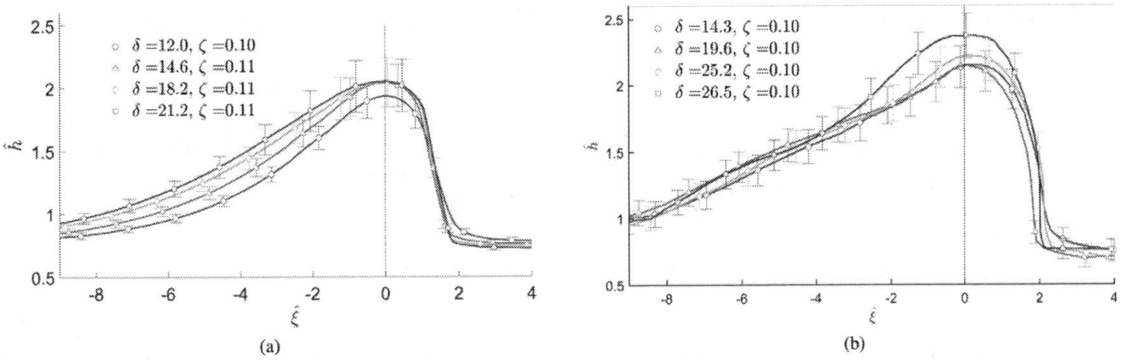

Figure 16: Wave profiles obtained via LAbs (figure 17a) and LeDaR (figure 17b) at different Reynolds number δ and perturbation strength ζ, for a perturbation frequency $f_p = 12Hz$.

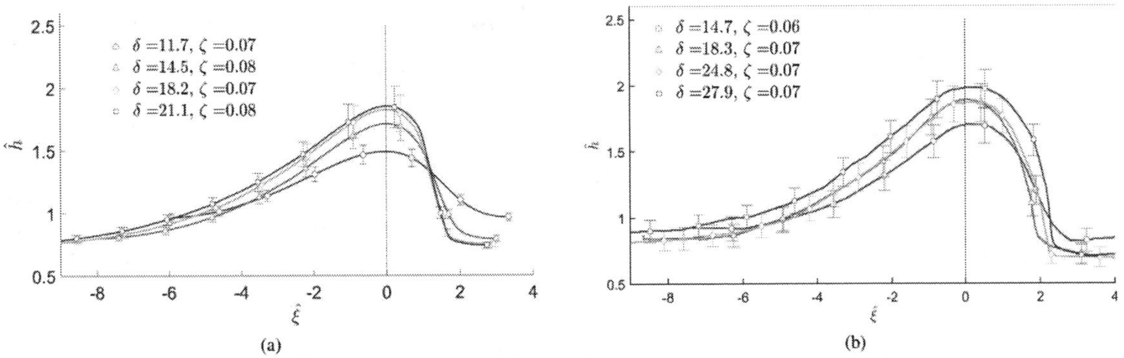

Figure 17: Wave profiles obtained via LAbs (figure17a) and LeDaR (figure 17b) at different Reynolds number δ and perturbation strength ζ, for a perturbation frequency $f_p = 18Hz$.

viscous dispersion, which is of the same order of magnitude of surface tension ($\varepsilon \approx \eta \approx 1$).

7 Conclusions

This paper has presented film thickness measurements, using Light Absorption and Laser Tomography, on a vertical falling liquid film with low Kapitza number. Unperturbed (steady) and perturbed (pulsing) flow rates have been considered. For both techniques, calibration, uncertainty, and processing aspects have been discussed in detail. In the steady tests, the results show excellent agreements between the techniques in terms of mean film thickness, recovering the Nusselt film theory. In the unsteady tests, the techniques agrees in the measurement of wave celerity and thickness maxima and minima, with a slight discrepancy in the wave profiles, particularly in wave front.

Acknowledgments

M. A. Mendez is supported by an F.R.S.-FNRS FRIA grant and the research program is funded by Arcelor-Mittal. The authors gratefully acknowledge the contributions of Dr. Flora Tomasoni in the design and the preparation of the experiments and the fruitful discussions with Prof. B. Scheid, from ULB University, concerning the modeling and the scaling of liquid film flows.

Nomenclature

Notation

$[x]$	Characteristic scale for x
$\hat{x}$	Nondimensionalized x
x_A	Oscillating amplitude of x
x_M	Mean component of x
$\mathcal{U}_x$	Uncertainty of x

Dimensionless Quantitites

Re	Reynolds number
Ka	Kapitza number
$\mathcal{W}$	Womersley number
$\mathcal{T}$	modulation factor
ζ	perturbation strength
δ	reduced Reynolds number
ϵ	ordering parameter
η	viscous dispersion parameter

Dimensional Quantities

A	Absorbance coefficient	-
b	channel half opening	m
β_n	contrast enhancing factor	-
c	colourant concentration	mg/kg
c_w	wave celerity	m/s
f_p	perturbation frequency	Hz
f_s	sampling frequency	Hz
γ_g	global absorption coefficient	$1/mm$
H	hydrostatic head	m
h	film thickness	m
I_0	reference image	$graycounts$
I	transmittance image	$graycounts$
L	channel length	m
l_σ	capillary length scale	s
l_v	viscous length scale	s
M	image scaling factor	$pixel/mm$
μ	dynamic viscosity	$Pa\,s$
μ_m	gray scale average along P	$Pa\,s$
M	image scaling factor	$pixel/mm$
n_t	number of images	
$n_{x/y}$	image size	$pixels$
ν	kinematic viscosity	m^2/s
ω	angular frequency	rad/s
P	total pressure	Pa
p	gauge pressure	Pa
P_n	image profile direction	$pixel$
ρ	density	kg/m^3
q	flow rate per unit width	m^2/s
σ	surface tension	mN/m
t_v	viscous time scale	s
u	stream-wise velocity component	m/s
ξ	traveling reference frame	m

References

[1] Kapitza, P.L., Kapitza, S.P, Collected Papers of P.L. Kapitza, **II**, (1938-1964), Pergamon Press, pp. 662–709.

[2] Alekseenko, S.V., Nakoryakov, V.E., Pokusaev, B.G., *Wave Flow of Liquid Films* (Begell House,1994).

[3] S. Kalliadasis, S., Ruyer-Quil, C., Scheid, B., Velarde, M.G., *Falling Liquid Films*, Vol. 176 (Springer, 2012).

[4] Ruyer-Quil, C., Kofman, N., Chasseur, D., Mergui,S., Eur. Phys. J. E **37** (30), (2014).

[5] Alekseenko, S.V., Nakorykov, V.E., Pokusaev, B.G., Int.J.Multiphase Flow, **11** (5), (1985) pp. 607-627.

[6] Chang, H.-C., Annu. Rev. Fluid Mech. **26**, (1994) pp. 103-136.

[7] Scheid, B., Ruyer-Quil,C., Manneville, P. , J. Fluid. Mech **531**, (2005) pp. 181-190.

[8] Scheid, B., Ruyer-Quil,C., Manneville, P. , J. Fluid. Mech **562**, (2006) pp. 183-222.

[9] W.B. Krantz, S.L. Goren, Ind. Eng. Chem. Fundam. **10**(91) (1971).

[10] Moran, K., Inumaru, J., Kawaji, M., Int. J. Multiphase Flow **28**(5) (2002).

[11] L. T. Nguyen, V. Balakotaiah, Phys.Fluids, **12**(2236) (2000)

[12] Charogiannis, A., An, J.S., Markides, C.N., Exp. Therm. Fluid Sci., 68, (2015).

[13] Shkadov, V. Y., Izv. Akad. Nauk SSSR, Mekh. Zhidk Gaza **1**(63), (1977) pp. 29-34.

[14] Lilleleht, L.U., Hanratty T.J., J. Fluid Mech **11**, (1961), pp. 65-81.

[15] Stainthorp,F.P., Allen,J.M. Trans.Inst.Chem.Engrs **43**(3), (1965), pp. 85-91.

[16] Portalski,S., Chem.Eng.Sci. **18**, (1963), pp. 787-804.

[17] Özgen, S., Carbonaro, M., Sarma, G.S.R, Phys. Fluids **14**, (2002), pp. 3391-3420.

[18] Gosset, A., in *Thermohydraulic Instabilities*, VKI Lecture Series 2006-07, pp 1 - 25.

[19] Njifenju, A.K., Bico, J., Andrés,E., Jenffer,P., Fermigier, M. Exp. Fluids **54** (1506), (2013), pp. 3391-3420.

[20] Driscolli,D.I., Schmitt,R.M., Stevenson, W.H. J.Fluids Eng. **114**, (1992), pp. 107-112.

[21] Liu,J., Schneider J.B., Gollub J.P., Phys.Fluids **7**(1), (1994), pp. 55-67.

[22] Liu,J., Gollub J.P., Phys.Fluids **6**(5), (1994), pp. 1702-1712.

[23] D. Bouchez, D., Zimmer, L., Reithmuller, M.L. Proceedings of the 9th Symp. of Flow Visualization, (2000), Edinburgh, SCT.

[24] Planquart, Ph., Proceedings of the 4th European continuous casting conference, (2002), Birmingham, UK.

[25] Toth, B., Anthoine, J., Reithmuller, M.L. Congres Francophone de Techniques Laser, (2006), Touluse, FR.

[26] Toth, B., Ph. D. thesis, Free University of Bruxelles, Brussels, Belgium, (2008).

[27] Hewitt, G.F., Jayanti, S., Hope, C.B., Gingell, D., Int. J. of Multiphase Flow **16**(6), (1990), pp. 951-957.

[28] Ursenbacher, T., Wojtan, L., and Thome., J. R., Int. J. of Multiphase Flow **30**(2), (2004), pp. 107-124.

[29] Wojtan, L., Ursenbacher, T., and Thome., J. R., Int. J. of Multiphase Flow **30**(2), (2004)(2), pp. 125-137.

[30] Rodriguez, D. J., Shedd, T.A., ASME Heat Transfer/Fluids Eng. Summer Conference, (2006) Charlotte, NC, USA.

[31] Farias, P.S.C., Martins, F.J.W.A., Sampaio, L.E.B., Serfaty, R., Azevedo, L.F.A., 15th Int. Symp. on Applications of Laser Techniques to FLuid Mechanics, (2010) Lisbon, PT

[32] Womersley, J.R., J. Physiol., 127 (1955).

[33] Rohlf, K., Tent, G., Journal of biomechanics, 34(1) (2000)

[34] Loudon, C., Tordesillas, A., J. Theor. Biol., 191(1) (1998)

[35] Karagoz, I., Uludag Universitesi MMFD, 7 (2002)

[36] Mendez, M.A., von Karman Institute, TN 215 (2015).

[37] Garcia, D., Comput. Stat. Data Anal. 54 (2010).

EPJ Web of Conferences **114**, 02073 (2016)
DOI: 10.1051/epjconf/201611402073

Ionic wind measurements on multi-tip plasma actuators

F. Messanelli[1,a] and M. Belan[1]

[1]Politecnico di Milano, Dipartimento di Scienze e Tecnologie Aerospaziali, via La Masa 34, 20156, Milano, Italia

Abstract. This work presents an experimental investigation about the effects of triangular tips on the active electrodes of plasma actuators. The tests are performed on two sets of actuators, corona and DBD, parameterized by means of the tip sharpness and the tips number per unit length. A total number of 30 actuators is considered. The devices are evaluated on the basis of the far field ionic wind velocity, that has been chosen as a representative test common to both kinds of actuator. The dataset includes velocity profiles and maps, that can be integrated to give mass flows and electromechanical efficiencies. Some results are also presented in the parameter space defined by tip sharpness and tips number per unit length: this gives the chance of defining optimal electrode shapes within each set. In general, the longitudinal velocity of the gas increases downstream of the tips in all the actuators tested, but the velocity field is modified to different extents in the two kinds of actuators, and is more complicated for the DBDs than for the coronas. The tips also increase the efficiency of all the actuators, particularly for the corona set, where even the stability is remarkably improved.

1 Introduction

The present work deals with some properties of the plasma actuators, widely studied because of their capacity to produce a ionic wind. These devices may be operated with direct (corona actuators) or alternate (DBD actuators) current. The induced gas velocity can be used for several flow control applications, as described in the review by Moreau [1], and specifically for the DBD by Corke et al. [2], Wang et al. [3] and Benard et al. [4].

In particular, this work describes an experiment about plasma actuators, both corona and DBD types, in order to characterize the flow field induced by a multi-tip geometry of the upstream electrodes. The tips are triangular, so that the geometry of the electrodes under study can be considered as a generalized form of the serrated edge shape. This shape has already been studied in some previous works for DBDs [5, 6], but a proper parametric study of the relevant geometry has never been performed, while it is completely innovative for corona devices, to the best of our knowledge.

Two sets of actuators, corona and DBD, including 30 different devices, are considered in this experiment. A parametric study is carried out identifying the different actuators by their tip sharpness r (defined as the ratio between tip length and width) and tips number per unit length n. Straight edge reference actuators (wire-to-plate for corona, plate-to-plate for DBD) are also tested in order to have a term of comparison. The actuators are evaluated on the basis of the far field ionic wind velocity, measured by means of a pitot probe. The devices under test are located on a flat plate in still air and the induced longitudinal velocity is measured across two planes normal to the ionic wind, at distances far enough to avoid the strong tridimensionality of the near field. Current and voltage waveforms for every actuator tested are also acquired, in order to evaluate the power consumption and the electro-mechanical efficiency.

The available data make possible several comparisons among different actuators, within each set and also between actuators of different sets (corona vs DBD). The results exposed below indicate that in general the multi-tip geometry gives better performance with respect to the straight geometry, both for corona and DBD, considering the intensity of the induced ionic wind and the mass flow. Power consumption considerations indicate that corona actuators are more efficient than DBDs. Furthermore, the multi-tip geometry increases corona stability, reducing glow-to-arc transition and favouring the ignition of the discharge.

2 Experiment setup

The corona actuators under study are represented in figure 1 and listed by their geometrical parameters in table 1: the set includes a wire-to-plate actuator C1 (figure 1a) as a reference, several actuators with serrated anodes C2...C12 (figure 1b) and an actuator with isolated tips C13 (figure 1c). Gap g and cathode size c are common to all the actuators, as reported in figure 1. The specific geometrical parameters are the tip height h and the tip width w, listed in table 1 together with the tip sharpness $r = h/w$ and

[a] e-mail: federico.messanelli@polimi.it

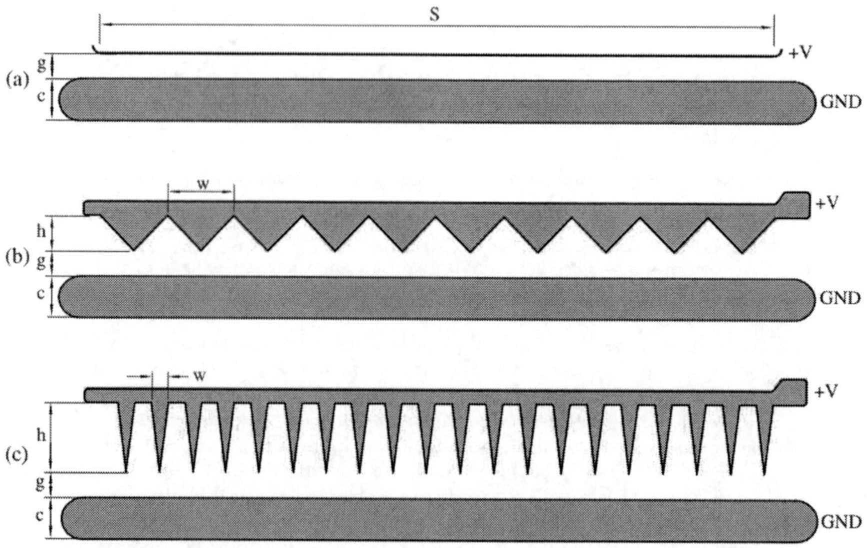

Fig. 1: Geometry of the corona actuators under study: (a) reference wire-plate; (b) multi-tip, serrated; (c) multi-tip, isolated tips. All the actuators have gap $g = 15$ mm and cathode size $c = 24$ mm. The span is $S = 400$ mm for all the actuators except the reference (a), which is 35mm wide to ensure its operational stability.

the tips number per unit length $n = N/S$ = (number of tips)/span. The span is $S = 400$ mm for all the actuators except the reference wire-plate (figure 1a), that can not be operated in still air over such a span because of its poor stability. This actuator has $S = 35$ mm to avoid excessive transitions to spark, so that in figure 1a is sketched out of scale. The 2 actuators labeled Cx and Cxx in table 1 have been prepared for the tests but later excluded because of their poor stability, similar to the wire-plate case C1. All the corona actuators are operated in the streamer regime, thanks to a suitable high voltage driver. The operating voltage is $V_0 = 17.2 \pm 0.2$ kV, except for the wire–plate device, that requires $V_0 = 16 \pm 0.1$ kV to ignite correctly with the same gap width g as the other actuators. The HV driver–actuator connection is made through a low-pass capacitive filter ($R = 4.4$ MΩ, $C = 0.9$ nF) and a ballast resistor $R_b = 4.4$ MΩ. Voltage measurements are obtained by means of a 1:1000 voltage divider (load 47 MΩ), current measurements by a shunt resistor ($R_s = 100\Omega$) on the ground electrode of the actuator under test.

The DBD actuators are represented in figure 2 and listed by their geometrical parameters in table 2: the set includes a plate-to-plate actuator D1 (figure 2a) as a reference, several actuators with serrated anodes D2...D14 (figure 2b) and an actuator with isolated tips D15 (figure 2c). Span S and ground plate size c are common to all the actuators, as reported in figure 2. The ground plate is always aligned with the roots of the tips. As for the corona actuators, table 2 lists the specific geometrical parameters, namely tip height and width h and w, tip sharpness $r = h/w$ and tips number per unit length $n = N/S$. The DBD actuators are operated by an oscillator–amplifier–output transformer chain, purposely built. All measurements have

Table 1: Parameters of the corona actuators under test.

Name	h (mm)	w (mm)	$r = h/w$	$n = N/S$ (m^{-1})
C1	0	–	0	–
Cx	3	10	0.3	100
Cxx	5	10	0.5	100
C2	8	4	2	250
C3	8	5	1.6	200
C4	10	10	1	100
C5	15	10	1.5	100
C6	20	4	5	250
C7	20	5	4	200
C8	20	8	2.5	125
C9	20	10	2	100
C10	20	20	1	50
C11	30	10	3	100
C12	40	10	4	100
C13	40	10	4	50

been carried out with a sinusoidal waveform having a peak-to-peak voltage $V_{pp} = 20 \pm 0.5$ kV with frequency $f = 815 \pm 5$ Hz. The measurement section (voltage divider and current shunt) is the same as for corona actuators.

All the voltage and current waveforms have been acquired at sampling frequencies of 100 Msamples/s. The accuracy of the electrical measurements, reported for the power consumption of the devices, is in the order of 0.4 W (or 1 W/m) for the coronas and 1.2 W (or 3 W/m) for the DBDs.

Sizes and operating conditions for the two sets have been selected after preliminary tests in order to give ionic wind velocities in the same order of magnitude, thus comparisons within each set are possible as well as comparisons between actuators in different

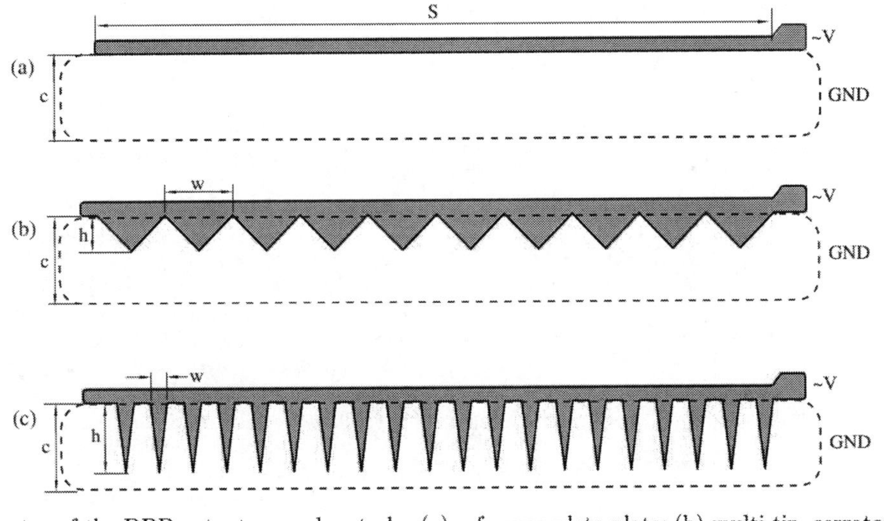

Fig. 2: Geometry of the DBD actuators under study: (a) reference plate-plate; (b) multi-tip, serrated; (c) multi-tip, isolated tips. All the actuators have span $S = 400$ mm and ground plate size $c = 50$ mm.

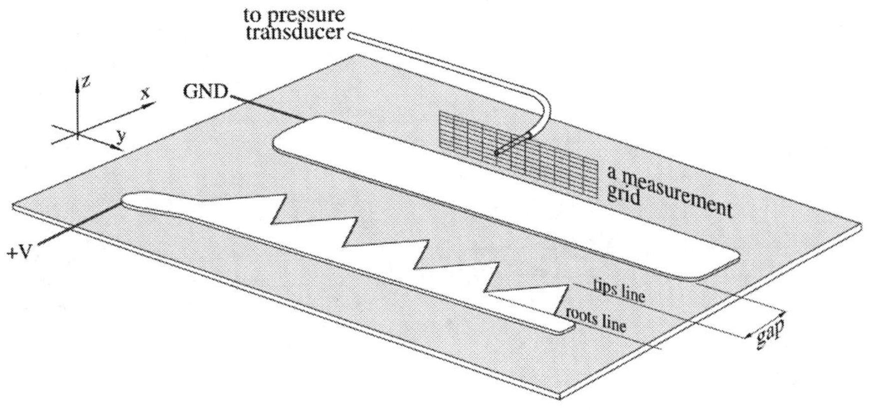

Fig. 3: Experimental setup, shown for a multi-tip corona actuator (non-scaled for clarity). The setup for DBDs is the same, but the grounded electrode is under the PMMA flat plate that serves as dielectric and is aligned with the roots.

sets. The actuators, placed on a flat plate in still air, are evaluated on the basis of the far field ionic wind velocity, measured by means of an insulating pitot probe as in figure 3, which shows also the reference system xyz. The pressure probe has an outer diameter $d_o = 0.8$ mm, an inner one $d_i = 0.4$ mm and is connected to a differential pressure transducer with range ± 200 Pa and accuracy 0.05 Pa. The induced longitudinal velocity is measured across different normal planes yz at distances $x = 17$ mm and 40 mm from the tips line of the active electrode; this ensures that the field of motion is far enough from the tips as to have a ionic wind direction close to the x-direction, and in particular within the directional error band of the probe, that is less than 1% for misalignments within 34° around the probe axis. The velocities are then obtained through the equation

$$p_m - p = \frac{1}{2} C_p \rho v^2 , \tag{1}$$

where p_m is the measured pressure, p the reference atmospheric pressure and C_p a coefficient including the experimental corrections, purposely set to maintain the 1% pressure accuracy even when the Reynolds number is very low and/or the probe touches the wall [7]. The humidity during the tests has been maintained to $22 \pm 5\%$. The gas density has been determined by the atmospheric pressure and temperature through the state law, assuming a negligible plasma heating. The measurements system described above gives uncertainties on the velocity data in the order of 0.1 m/s, excluding error sources due to small irregularities of the electrodes shapes. Furthermore, the unlikely case of large misalignments in the far field can be treated by means of the calibration curves of

Table 2: Parameters of the DBD actuators under test.

Name	h (mm)	w (mm)	$r = h/w$	$n = N/S$ (m^{-1})
D1	0	-	0	-
D2	3	10	0.3	100
D3	5	5	1	200
D4	5	10	0.5	100
D5	10	10	1	100
D6	15	10	1.5	100
D7	20	5	4	200
D8	20	10	2	100
D9	20	20	1	50
D10	20	25	0.8	40
D11	20	40	0.5	25
D12	25	10	2.5	100
D13	30	10	3	100
D14	35	10	3.5	100
D15	40	10	4	50

the probe: for angular differences between the velocity vector and the probe axis up to ±40°, the uncertainty for the longitudinal velocity component cannot exceed values of 0. 2m/s.

3 Results

3.1 Corona stability

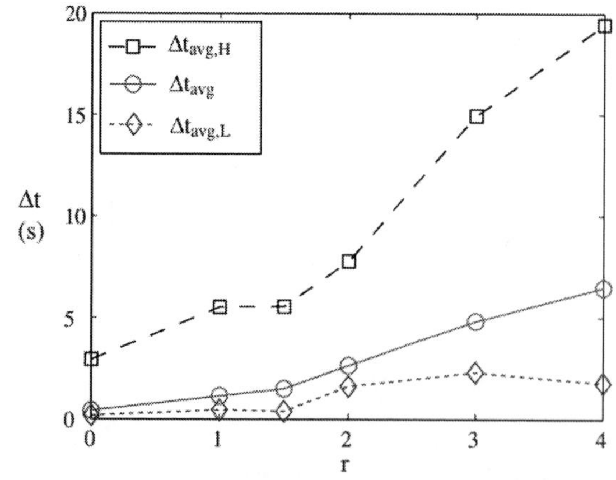

Fig. 4: Mean times between transient sparks for corona actuators (C1,4,5,9,11,12) of increasing sharpness ($r = 0$ to 4).

In this work most data about corona actuators have been acquired thanks to their improved stability, that can be related to the geometry under study, as shown in what follows. In general, the sparks are visible to the naked eye but they can also be recorded as short and strong current pulses, in the order of 1A and more. The power supply can not afford a permanent transition to the arc regime, so that this phenomenon can be described as a 'transient spark' [8], sustained only by the occasional complete discharge of the circuit capacity. The stability of a corona under steady working conditions can be quantified by reporting the time Δt between consecutive transient sparks as in figure 4: this figure is obtained from the histograms of the Δt measured along tests including hundredth of events for a set of actuators of sharpness growing from $r = 0$ to 4, namely C1,4,5,9,11 and 12. The histogram for a given actuator has in general a peak corresponding to relatively short Δt and a wide tail representing long Δt. The relevant information is shown in figure 4 by the lower and upper curves. The former is the mean time $\Delta t_{avg,L}$ of the lower part of the histograms, below the median value; the latter is the mean time $\Delta t_{avg,H}$ of the upper part of the histograms. The figure is completed by the central curve, which is the standard mean time Δt_{avg}, and shows a clear increase of the time between sparks as the tip sharpness grows. This property should be related to the anode shape, since the sparks can start only from the small tip regions, and to the relevant local field, that grows with r and makes the discharge less sensitive to perturbations. As a final remark, it is important noting that even if the trend of the data is significant, the Δt values are specifically related to the present setup through its physical parameters as the circuit capacity.

3.2 Flow velocity data

This section presents some examples of velocity data, showing the ionic wind dependence on the geometry of the tips. The flow is periodical in the span direction, with slight differences between different tips, due to very small imperfections in the realization process. However, the values presented here are obtained as an average on at least 2 different tips. The velocity curves along the z-axis for three different span locations, corresponding to a tip, a root and halfway between them, and two different distances along the x-axis (17 and 40 mm from the tips line, see figure 3) are presented in figures 5a,b and 6a,b for two representative actuators, corona C11 of table 1 and DBD D13 of table 2, that share the same geometry: $r = 3$, $n = 100$. Furthermore, the velocity maps over the cross-section yz for the same distances along the x-axis are presented in figures 5c,d and 6c,d for two tips. The measurements locations are marked with black dots. Obviously, the flow pattern is periodical over all the other tips, with minor exceptions for the extreme tips, due to border effects.

For the corona actuator in figure 5 the maximum of the velocity is measured downstream of the tips at a height $z = 3.75$ mm for $x = 17$ mm and $z = 3.9$ mm for $x = 40$ mm. The 2D velocity maps show the momentum diffusion as x grows, whereas the position of the velocity maximum remains quite far from the wall. This behaviour is similar for all the multi-tip corona tested, instead the wire-to-plate reference actuator has its maximum located nearer to the wall, at $z = 2.4$ mm for $x = 17$ mm. The velocities measured downstream of the tips are larger than downstream of the roots, because the corona discharge originates only from the anode tips and expands like a fan towards the cathode. However, due to the diffusive mechanisms, the differences between tips and roots decrease when moving along the x-axis, as it can be seen comparing figure 5a,c and b,d. The velocities induced by the multi-tip actuators are larger than for the straight wire-to-plate: the tips are regions of high electric field and may cause an increase in the ionization efficiency. Indeed, the region close to the tip is characterized by a very strong luminosity, much more than close to the wire anode in the wire-to-plate actuator, implying that a high concentration of ions is achieved. This phenomenon is observed also for the DBDs: even if in that case the discharge originates along the whole electrode perimeter, the luminosity, and consequently the ion concentration, is higher in the proximity of the tips.

Confronting the DBD curves and maps in figure 6 with the corona ones of Fig. 5, the induced flow appears more complex for the DBDs. At the nearest x distance (17 mm), once again the velocity maximum is found downstream of the tips, but at a lower height, $z = 0.65$ mm. This value is the same for all the DBD tested, with the exception of the straight edge, that gives $z = 0.9$ mm. This great difference between coronas and DBDs may be due to the different morphology of the electric field between the electrodes: the field lines are more flattened toward the wall for the DBDs, while they are more curved and extended in the z-direction for the coronas. The velocities are lower for the DBD, so that, as it will shown later, the DBDs require a larger power to produce a given ionic wind, but the location of the velocity maximum nearer to the wall may be better for the boundary layer control. Another property of the multi-tip DBDs of importance in flow control, known in literature (e.g.[5]), is the induction of a strongly three-dimensional flow since the discharge is generated along the whole active electrode perimeter and the induced velocity is locally normal to the electrode edge. This implies that in each space between adjacent tips two oppositely directed jets are generated, that collide in the middle of the region, downstream of the root; here a vertical component, directed away from the wall, is created, while in the regions behind and over the tips the flow moves toward the wall, for continuity reasons. Thus, a DBD creates a ionic wind made of a longitudinal component interacting with transverse components that produce periodic pinch-spread effects on the streamlines and a distribution of streamwise vorticity. This can be seen in the curves and maps of figure 6, where the velocity downstream of the roots has a consistent extension in the z-direction, as well as a maximum located at a larger distance from the wall. Furthermore, for increasing distances along the x-axis, the velocity maximum downstream of the roots may grow up to reach the same order of the one downstream of the tips (figure 6b and d): this phenomenon, already described in literature [9], may be due to the streamlines spreading over the tip zones and pinching over the root zones, where the higher velocities streamlines tend to coalesce while the flow develops along the x direction.

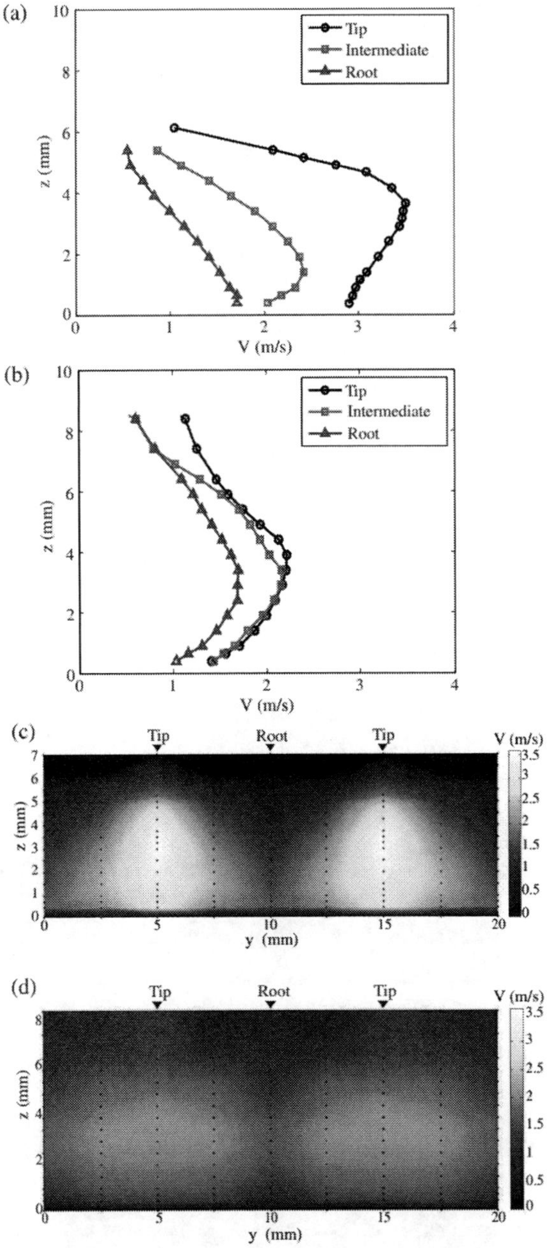

Fig. 5: Corona actuator C11: velocity curves at $x = 17$ mm (a) and 40mm (b); velocity maps at $x = 17$ mm (c) and 40 mm (d). Maps interpolated and completed by no-slip condition.

3.3 Data in the parameter space

The properties of the actuators under study can be considered as dependent on the actuator location over

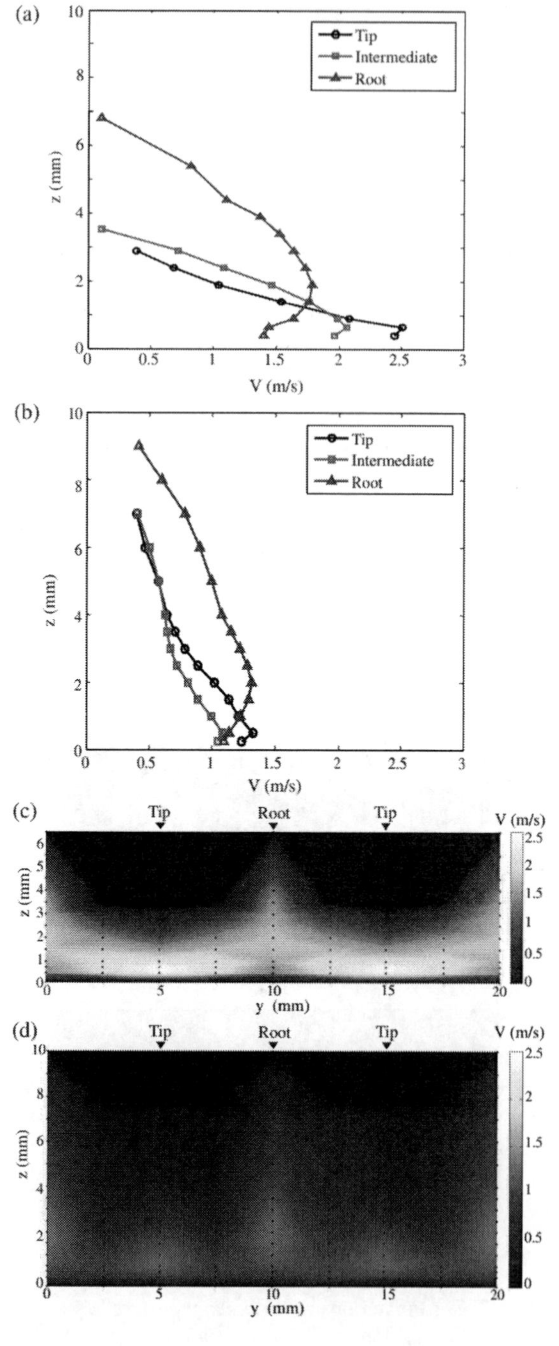

Fig. 6: DBD actuator D13: velocity curves at $x = 17$ mm (a) and 40 mm (b); velocity maps at $x = 17$ mm (c) and 40 mm (d). Maps interpolated and completed by no-slip condition.

the parameter space (n, r), i.e. as function of the parameters $r = h/w$ (tip sharpness) and $n = N/S$ (tips number per unit length) introduced in §2. Figures 7 and 8 present a comparison of 3 coronas (C6,C9, C10) and 3 DBDs (D7, D8, D11) lying in the parameter space (n, r) along the same oblique straight line $r = 0.02\,n$. This line represents actuators having the same tip length $h = 20$ mm but different widths w. It can be seen that the velocity difference downstream of tips and roots increases when w grows, i.e. when the tips are more spaced. Also the values of the maximum velocity increase. It seems that more isolated tips are more efficient in generating high velocities, whereas, if the tips are too close each other, their beneficial effects are partially lost and the behaviour may become similar to that of the straight actuator. However, it is important noting that the characterization of an actuator by means of the maximum velocity induced locally can not be considered as complete; in what follows, all the actuators will be compared on the basis of a reliable global parameter.

Figure 9 shows the global mass flows produced by all the actuators under test, obtained by integration of the velocity profiles over the measurement grids, at $x = 17$ mm. The mass flow is an important parameter, even more important than the maximum velocity, to characterize the effectiveness of a given actuator for flow control purposes. For corona actuators (figure 9a), the performance improves increasing the sharpness r: the more acute the tips are, the more intense is the electric field in their proximity and thus their performance. The optimal mass flow is achieved for the actuator C12 ($n = 100$, $r = 4$), that has very sharp tips but not so numerous. It is probable that excessively increasing the parameter n leads to very close ionization regions, and the electric field eventually becomes similar to the one of the wire-to-plate, thus losing the beneficial effects of the tips. On the other hand, the point at $n = 50$, $r = 4$ shows that tips too distant each other worsen the performance, so the best mass flow arises as a balance between the sharpness and the numerosity of the tips. For the DBDs (figure 9b), the highest mass flow is obtained for the D7 ($n = 200$, $r = 4$): also in this case an increase of the sharpness leads to better performance, but the proximity of tips is not so negative as for the corona actuators, even if it reduces the maximum induced velocity. This is confirmed by the good performance achieved by actuator D3 ($n = 200$, $r = 1$), that is also located on the right border of the figure. This may be due to the fact that the active perimeter for DBDs is very extended for a high number of tips, so that the ionic wind generated between tips compensates the lower velocity maxima with respect to the case of more isolated tips. In conclusion, coronas induce mass flows larger than DBDs, which is good for flow control purposes, but the velocity maxima are more distant from the wall, that is a negative characteristic for boundary layers control.

Electrical measurements have also been performed, in order to quantify the electrical power consumed by the actuators and the electro-mechanical efficiency, that is defined as the ratio between the available mechanical power and the consumed electrical power. In general, it is found that DBD actuators require a higher power than coronas to produce similar velocities. This reflects also on the electro-mechanical effi-

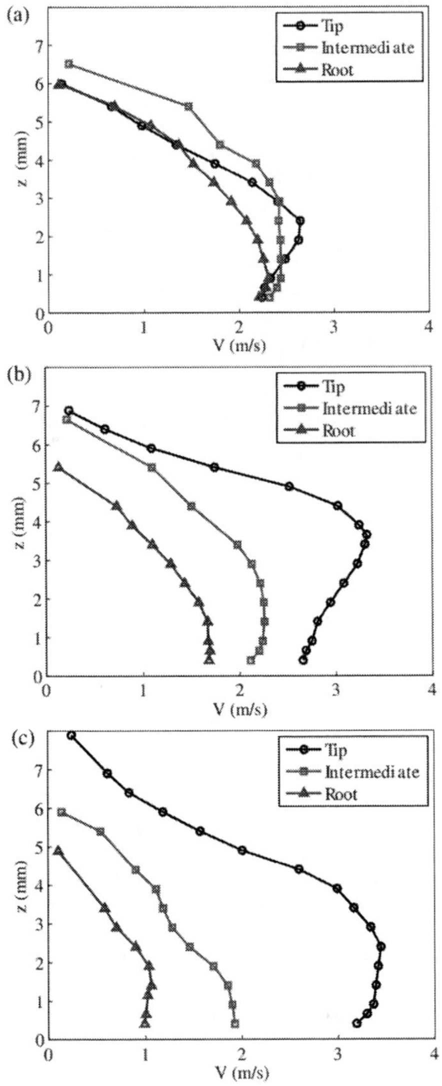

Fig. 7: Velocity curves at x = 17 mm for the corona actuators C6 (a), C9 (b), C10 (c) having the same h = 20 mm and growing $w = 4, 10, 20$ mm.

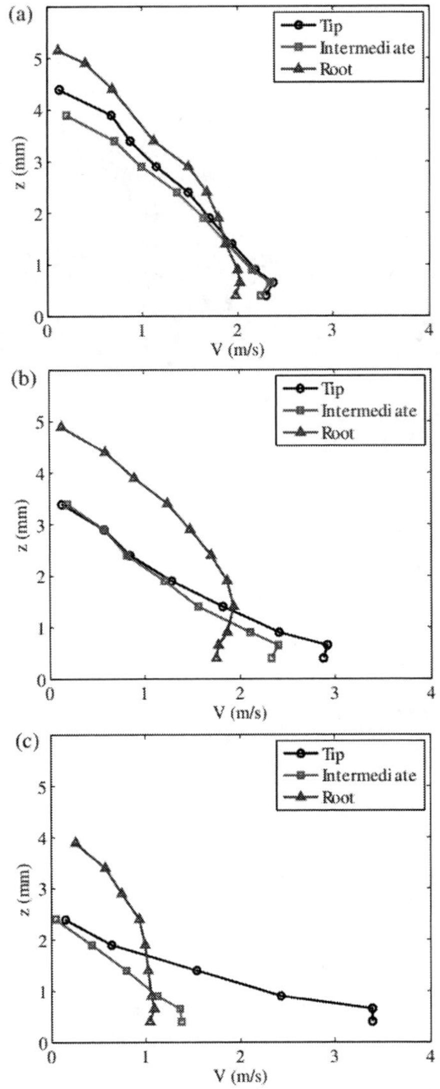

Fig. 8: Velocity curves at x = 17 mm for the DBD actuators D7 (a), D8 (b), D11 (c) having the same h = 20 mm and growing $w = 4, 10, 20$ mm.

ciency: the corona actuator C11, whose velocity curves are shown in figure 5, has an efficiency of 0.96%, while the DBD D13 of figure 6 has an efficiency of 0.05%, that is remarkably lower. Nevertheless, it must be noticed that the efficiencies of the straight reference actuators, that are about 0.05% for the corona wire-to-plate and 0.04% for the DBD plate-to-plate, are lower than the ones of almost every multi-tip actuator tested. In particular, the efficiency improvement for corona actuators is about two order of magnitude, while it is less substantial, but still present, for the DBDs.

4 Conclusions

In this work, velocity measurements downstream of two sets of multi-tip plasma actuators, both corona and DBD types, have been performed. Reference straight actuators have also been tested. The geometry of the upstream electrodes was characterized by the tip sharpness r and the tips number per unit length n. For the corona set, r has been varied between 0 and 5, while n between 50 and 250; for the DBD set, r was investigated between 0 and 4, and n between 25 and 200. The measured profiles show that the induced velocity is not uniform in the span direction: for coronas, it is periodic and higher downstream of the tips, thanks to the increased ionization intensity in the near-tip regions; for DBDs, the velocity field is also periodic

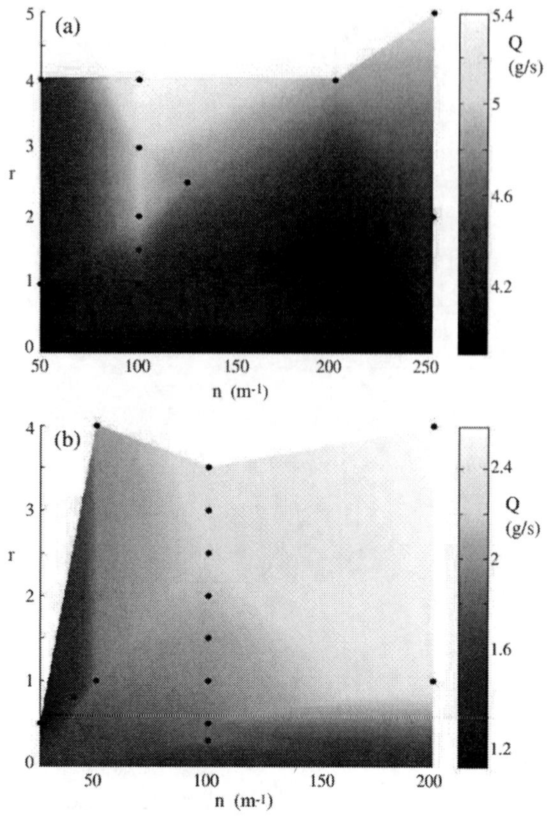

Fig. 9: Mass flow rate Q vs. tip number per unit length n and tip sharpness r for the tested corona (a) and DBD (b) actuators (interpolated data). Each actuator is represented by a dot on the map, except the straight ones, which corresponds to the horizontal axes.

but more complicated, since in the near field of these actuators the induced velocity vector is locally normal to the perimeter of the exposed electrode. A remarkable effect is the generation of a couple of streamwise counter-rotating vortices downstream of each space between adjacent tips, with possible beneficial effects for flow control purposes, as reported in [5].

Optimal geometries within those tested have also been determined in terms of induced mass flow. For the corona actuators, the maximum mass flow has been registered for very sharp, but not so numerous tips (actuator C12). The tips are regions of high electric field, and thus of high ionization rate. Generally speaking, the performance improves by increasing the tip sharpness. However, an excessive number of tips (i.e. tips very close each other) may lose these benefits, leading to a condition more similar to the straight wire-to-plate, that has the worst performance among the actuators tested. On the other hand, since the corona discharge is generated only from the anode tips, an insufficient number of tips may also be detrimental for performance. For the DBD set, the maximum mass flow has been measured for very sharp and numerous tip (actuator D7): in this case a great number of tips implies a very extended active perimeter, since the discharge is generated on the whole active edge of the exposed electrode.

In general, it is important to notice that almost all the multi-tip actuators under test have achieved better performance in terms of induced velocity and mass flow with respect to the straight reference actuators. In particular, thanks to the different characteristics of the electric fields, coronas induce the maximum velocity at a higher distance from the wall than DBDs. This may not be a positive characteristics for boundary layer control. However, multi-tip corona actuators require less power than DBDs to induce larger mass flows: the electro-mechanical efficiency is higher. This feature, in addition to the improved discharge stability with the multi-tip geometry, is very promising.

References

1. E. Moreau, J. Phys. D: Applied Physics **40**, 605–636 (2007)
2. T.C. Corke, C.L. Enloe, S.P. Wilkinson, Annual Rev. Fluid Mech. **42**, 505–529 (2010)
3. J.J. Wang, K.S. Choi, L.H. Feng, T.N. Jukes, R.D. Whalley, Prog. Aerosp. Sci. **62**, 52–78 (2013)
4. N. Benard, E. Moreau, Exp. Fluids **55**: 1846 (2014)
5. R. Joussot, A. Leroy, R. Weber, H. Rabat, S. Loyer, D. Hong, J. Phys. D: Appl. Phys. **46**: 125204 (2013)
6. F. O. Thomas, T.C. Corke, M. Iqbal, A. Kozlov, D.Schatzman, AIAA J. **47**, 2169–2178 (2009)
7. C. Tropea, A.L. Yarin, J.F.Foss (Eds.), *Springer Handbook of Experimental Fluid Mechanics*, Section 5.1.1 (Springer–Verlag, Heidelberg, 2007)
8. M. Janda, V. Martišovitš, Z. Machala, Plasma Sources Sci. Technol. **20**: 035015 (2011)
9. S. Roy, C.C. Wang, J. Phys. D: Appl. Phys. **42**: 032004 (2009)

EPJ Web of Conferences **114**, 02074 (2016)
DOI: 10.1051/epjconf/201611402074

Optimization principle of operating parameters of heat exchanger by using CFD simulation

Jozef Mičieta[1,a], Vondál Jiří[2], Jozef Jandačka[1], Richard Lenhard[1]

[1]*University of Žilina, Faculty of Mechanical Engineering, Department of Power Engineering, Univerzitná 1, 010 26 Žilina, Slovakia*
[2]*Brno University of Technology, Institute of Process and Environmental Engineering, Technická 2896/2, 616 69 Brno, Czech Republic*

Abstract. Design of effective heat transfer devices and minimizing costs are desired sections in industry and they are important for both engineers and users due to the wide-scale use of heat exchangers. Traditional approach to design is based on iterative process in which is gradually changed design parameters, until a satisfactory solution is achieved. The design process of the heat exchanger is very dependent on the experience of the engineer, thereby the use of computational software is a major advantage in view of time. Determination of operating parameters of the heat exchanger and the subsequent estimation of operating costs have a major impact on the expected profitability of the device. There are on the one hand the material and production costs, which are immediately reflected in the cost of device. But on the other hand, there are somewhat hidden costs in view of economic operation of the heat exchanger. The economic balance of operation significantly affects the technical solution and accompanies the design of the heat exchanger since its inception. Therefore, there is important not underestimate the choice of operating parameters. The article describes an optimization procedure for choice of cost-effective operational parameters for a simple double pipe heat exchanger by using CFD software and the subsequent proposal to modify its design for more economical operation.

1 Introduction

Heat exchangers belong to the most important elements of energy facilities and they are used in a wide field of industry [1]. They are devices for heat transfer among heat carrying fluids. They can be classified according to numerous criteria e.g. according to way of use (heaters, coolers, evaporators, condensers, etc.), design arrangement (heat transfer between two or more fluids, etc.), way of the heat exchange (without or with phase change) or according to contact of one and other fluid (mixer-heat exchanger – it does not have a heat exchange surface and the fluids are mixed together; regenerator – it have one heat exchange surface, which is alternatively flowed around by hot and cold fluid stream and the exchangers use heat accumulation (it is also referred to as direct transfer type); finally there is recuperator, which is referred to as indirect transfer type, because a wall separates the fluid streams) [2].

The technical level of heat exchanger construction significantly adjudicates on the effectiveness and economic return of investments. Minimization of costs are therefore an important aim both for the designers, but mainly for users. Optimal design and operation of the heat exchanger require understanding of heat transfer issues, design and operational requirements. The traditional approach to the design of the heat exchanger is based on an iterative process. It contains gradually changed design parameters until a satisfactory solution for assigned specification is reached [3]. However, these methods, besides that they are time consuming, they do not guarantee optimum economic solution. The calculation process can be supported by specific software and with some constrains also by CFD (Computational Fluid Dynamics) software. The paper shows the optimizing principle of the operating parameters, which is based on the condition of minimal operating cost.

2 Design of heat exchanger

The mathematical description of some occurring processes in the heat exchanger is so complex that there is not possible do it without similarity theory and practical experience [4]. The structural design of the heat exchanger is essentialy dependent on the experience of the engineer and the use of computational software is major advantage from timesaving point of view. Typically, the first there is selected geometry of the device. Subsequently, there are defined value of the design variables according to the specifications and several provided mechanical and thermodynamic parameters in order to obtain a satisfactory heat transfer coefficient. Individual engineer`s options are according to customer requests and depending on the iteration

[a] Corresponding author: jozef.micieta@fstroj.uniza.sk

procedure until an acceptable design of the heat exchanger complies to the specification with satisfactory compromise from efficiency point of view [3].

2.1 Sizing of heat exchanger

Calculating method of heat exchanger depends on case whether new heat exchanger is designed (primary result are design and size of the heat transfer surface) or control calculation (size of heat transfer surface is known and usually there are calculated the output or input temperature, transferred heat flux or other operating parameters). The choice of a suitable type of heat exchanger precedes the sizing of heat exchanger, which is dependent on many parameters [5]. The next paragraphs are focused on the design calculation. Thermal calculation of the heat exchanger has a number of phases:

- Gathering of initial data: kind and physical properties of hot – heating fluid (index 1) and cold – heated fluid (index 2), flow rates and temperatures of both fluids, optionally heat exchanger performance, requirements of the pressure and permissible pressure drop.
- A draft of heat exchanger type, its layout (parallel flow, contraflow, cross-flow) and dimensions.
- Thermal calculation, which is comprised of heat balance of the heat exchanger.
- Technical and economical optimization and specification of the parameters in order the heat exchanger keeps economic requirements.

Among two most common methods for calculation of heat exchanger is included LMTD method (Logarithmic Mean Temperature Difference), which is based on the known inlet and outlet temperatures of both media. However, if some temperature is unknown, there is necessary to interpolate it. The second method, which is based on heat exchanger efficiency for transferring a certain amount of heat (NTU – Number of Transfer Units) is preferable to LMTD method. NTU method is suitable for the control calculation and comparison of various types of heat exchangers [6]. In the paper, there is below, partly mentioned only LMTD method, but NTU method is not described in detail.

2.2 Thermal balance of the heat exchanger

The basic equation for thermal calculating is enthalpy balance of heat exchanger. For illustration, we consider a simple recuperator, concretely the double pipe heat exchanger (Figure 1). For the layout of the heat exchanger in Figure 2 is the conservation of energy for the heat exchanger expressed

$$\sum \dot{H}' = \sum \dot{H}'' + \sum \dot{Q}_{\text{loss}} \qquad \text{(W)}, \qquad (1)$$

where $\dot{H}$ enthalpy flow, W; $\dot{Q}_{\text{loss}}$ is heat loss to the environment, W; and superscripts $'$ and $''$ mark the inlet and outlet of fluid in the heat exchanger. When the heat loss in common systems with insulation does not exceed 5% [7], it is possible to simplify calculations by neglecting of heat loss or it can be estimated. Furthermore, equation (1) can be itemized to individual enthalpy flows as equation

$$\dot{H}_1' + \dot{H}_2' = \dot{H}_1'' + \dot{H}_2'' + \dot{Q}_{\text{loss}} \qquad \text{(W)}, \qquad (2)$$

which for two-phase flow of fluid after treatment and expression through $\dot{m}$ mass flow, kg m^{-3}; and h the specific enthalpy, J kg^{-1} has the form

$$\dot{m}_1 \left(h_1' - h_1''\right) = \dot{m}_2 \left(h_2'' - h_2'\right) + \dot{Q}_{\text{loss}} \qquad \text{(W)}. \qquad (3)$$

If the total heat flux is denoted $\dot{Q}$, W simple modification of equation (3) gives the next expression (4) for the enthalpy balance of one fluid. Its heat flux is given by the enthalpy loss of hotter fluid. The heat flux causes an increasing enthalpy of the heated fluid and increasing enthalpy of environment around the heat exchanger (heat loss to the environment).

$$\dot{m}_1\, p_{c_1} \left(T_1' - T_1''\right) = \dot{m}_2\, p_{c_2} \left(T_2'' - T_2'\right) + \dot{Q}_{\text{loss}} \qquad \text{(W)}, \qquad (4)$$

where c_p specific heat capacity, J kg^{-1} K^{-1} is considered at main temperature of fluid (hotter or colder). For the control calculation, there are three unknown variables: outlet temperatures T_1'', T_2'' and $\dot{Q}$ heat output, W. On the contrary, at design calculations, there are given these temperatures and the heat performance. There is necessary to calculate the heat exchange surface S, m^2 according to the equation for heat transfer calculation

$$\dot{Q} = k\, S\, \Delta T_{\text{mean}} \qquad \text{(W)}, \qquad (5)$$

where k heat transfer coefficient, W m^{-2} K^{-1} and the heat flux $\dot{Q}$, W can be estimated from enthalpy balance, which is expressed by the previous equation (4). The temperatures of both liquids are during fluid flow through the heat exchanger changed, thereby there is changed also the temperature difference between them. ΔT_{mean} is the mean temperature gradient, which is defined by integrals

$$\Delta T_{\text{mean}} = \frac{1}{S} \int \left(T_1 - T_2\right) \mathrm{d}S \qquad \text{(K)}, \qquad (6)$$

where the temperature difference $\left(T_1 - T_2\right)$ is the local value at the location. Mean temperature gradient, which is valid for the heat exchanger, must be obtained by integration of equation (6) over the entire surface of the heat exchanger. The result of this integration is the equation for the medium logarithmic temperature difference (it is valid not only for a parallel flow, but also for contraflow), which is marked LMTD and defined by equation

$$\Delta T_{\text{ln}} = \frac{\Delta T'' + \Delta T'}{\ln \dfrac{\Delta T''}{\Delta T'}} \qquad \text{(K)}, \qquad (7)$$

where $\Delta T'$ and $\Delta T''$ are the extreme temperature differences (gradients) at the inlet and at the outlet of heat exchanger $\left(\Delta T' > \Delta T''\right)$, which mean inlet and outlet temperature differences of the fluid flows. If there is changed the temperature gradient along the heat transfer surface a little, there is possible to calculate an average temperature gradient as the arithmetic average of the two extreme temperature differences (gradients)

$$\Delta T_{\text{mean}} = \frac{1}{2}\left(\Delta T'' + \Delta T'\right) \qquad \text{(K)}. \qquad (8)$$

The mean arithmetic gradient is always greater than the mean logarithmic temperature difference. For $\left(\Delta T'/\Delta T''\right) > 0{,}5$ is the deviation less than 4%. In contrast to the pure parallel flow or contraflow, which includes also the case of double pipe heat exchanger, the calculation of the mean temperature gradient in the heat exchangers with the cross-flow or a combination flow of both fluid is difficult. Therefore, for the case of conventional practice, the results are presented graphically or by equations. By use of these parameters

$$P = \frac{T_2'' - T_2'}{T_1' - T_2'} \quad (-); \quad P = \frac{T_2'' - T_1'}{T_2' - T_2''} \quad (-) \qquad (9)$$

is determined corrective coefficient F. It is used for multiplying ΔT_{ln} for case of the contraflow adjustment. The correction factor F is for the structural layout of the degree of deviation from the maximum possible LMTD value.

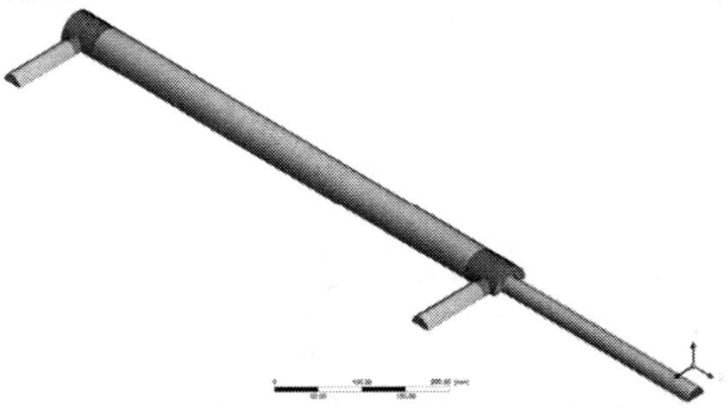

Figure 1. Model of heat exchanger with computational mesh.

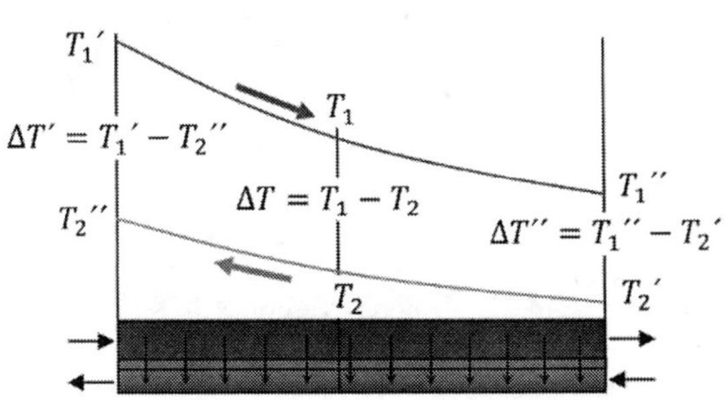

Figure 2. Temperature progress in parallel flow arrangement of heat exchanger.

The reciprocal value of the heat transfer coefficient $1/k$ is the total thermal resistance of the heat transfer from hot to cold fluid, which is expressed by the equation (10). Total thermal resistance R_{c} is calculated according to serial alignment. Thus, it is sum of the individual thermal resistances, which are caused by heat transfer on the both sides of the wall, the heat resistances of layers of heat transfer wall and the heat resistance of fouling on both sides. The total thermal resistance of wall which is composed from different layers, is solved

$$R_{\text{c}} = \frac{1}{k} = \frac{1}{\alpha_{\text{i}}} + \sum_{j=1}^{n} \frac{\delta}{\lambda_{\text{j}}} + \frac{1}{\alpha_{\text{e}}} \qquad (\text{m}^2\ \text{K W}^{-1}), \qquad (10)$$

where subscript i and e represent the internal and external heat exchange surface. Heat transfer coefficient α, W m^{-2} K^{-1} can be determined by similarity theory in the form of Nusselt number for the design of heat exchanger, the hydrodynamic conditions and the fluid characteristic. The determination of heat transfer coefficient is quite large task, but there is enough literature sources for individual assessment of different cases.

From the economy view, for design calculation, there is important the determination of the parameters that directly affect investments and operating costs. Investments depend on the size of the heat transfer surface and on the complexity of production (type) of the heat exchanger, which is reflected in the price for m^2 heat transfer surface. Operating costs include energy, maintenance and repair services. The increasing velocity of hot or cold fluid causes that the heat transfer coefficient α increases, too. Besides, it also allows to decrease investment costs due to the size reduction of the heat transfer surface, but simultaneously operating costs rise significantly, due to increased pressure loss. In design calculations, there have to be determinate the flow velocity (flow rate) so that the total costs, which can be expressed as the sum of investment and operation, were minimal.

2.3 Hydraulic calculation

In order to sizing of devices for circulation of working fluid, there is necessary the calculation of the flow resistance (pressure loss) in the heat exchanger. The total pressure drop of the fluid flow Δp_{c} is equal to the sum of friction losses Δp_{t} and the loss of local resistance Δp_{ζ}.

$$\Delta p_{\text{c}} = \Delta p_{\text{t}} + \Delta p_{\zeta} = \left(\lambda \frac{L}{d} + \Sigma\zeta\right)\frac{v^2}{2}\rho \qquad \text{(Pa)} \qquad (11)$$

There are λ coefficient of friction, –, depending on the Reynolds number and the wall roughness; L the considered length of channel through which fluid flows, m; d equivalent diameter (cross section of the channel), m; ξ shape coefficient of resistance, –, which primarily depends on the shape and less on the character of flow (Reynolds number); v mean velocity of flow, m s^{-1}; and ρ density of the fluid flow, kg m^{-3} [8]. P_{e} total demand for electric power of the circulating pump, W is determined by equation

$$P_{\text{e}} = \Delta p_{\text{c}}\,\dot{V} = \Delta p_{\text{c}_1}\dot{V}_1 + \Delta p_{\text{c}_2}\dot{V}_2 = \Delta p_{\text{c}_1}\frac{\dot{m}_1}{\rho_1} + \Delta p_{\text{c}_2}\frac{\dot{m}_2}{\rho_2}. \qquad (12)$$

3 Task of the heat exchanger calculation

Operating efficiency of heat exchanger is established on the electricity costs associated with forced circulation of heat transfer liquid and on assumption of known incomes from distribution of heat. Capital costs are neglected in this case, because there was considered a heat exchanger in operation and it could be affected only its operation setting. There was necessary to establish flows of both media so that the operation of the device was as economically as possible.

For heat exchanger was set the cold (heated) water with temperature of 30 °C flows to the straight tube and through the pipe interspace flowed the hot (heating) water with inlet temperature of 110 °C. Input parameters for the boundary conditions were flows of heat transfer fluids or amended geometry of the heat exchanger. The model of flow and heat transfer in the heat exchanger`s geometry was simulated in a commercial CFD software Ansys Fluent 15. For evaluation of the parameter`s impact was changed only one parameter while others remained constant. The main assessment detail from CFD analyses was the performance of heat exchanger, which was expressed by the ratio between incomes from the heat distribution and operating costs for circulation of heat transfer fluid. In the calculation were considered the cost of electricity 150 € MWh^{-1} [9] and income from heat distribution 100 € MWh^{-1} (approximately 27.80 € GJ^{-1}) from heat distribution [10, 11, 12].

3.1 The model of the heat exchanger and the principle of optimization

Double pipe heat exchangers are often used in special operating conditions such as low flow volume, for large temperature differences, at desired residence time, for fast temperature changes or at high pressure operation and if there is a requirement for pure parallel flow or counterflow [2]. The heat exchanger consists of simple concentric arrangement of the two pipes with different diameter. One of the fluid flows in the inner tube and the second into the created annulus. In this model case, there was not assumed a phase change of heat transfer fluid (water). Double pipe heat exchanger is one of the simplest, therefore it was selected for example. The optimization process was focused on the adjustment of cost-effective flow rate of heat transfer fluids for given dimensions of the heat exchanger and the inlet temperatures of the both streams.

The goal of paper is not to describe the settings of the model in detail because it is a very large task. Therefore, the description of the model is mentioned only briefly. Since the model heat exchanger was symmetric, there was appropriate to model only half of the heat exchanger in order to speed up the calculation. From an actual geometry could be removed insignificant volumes as heat exchanger walls, there was left only partition - heat exchanger surface and volumes of liquids. The heat transfer thorough the outer walls was included in the model. But on other hand the outer wall was not physically modelled, because temperature field in the direction of the wall thickness was not significant. Simplified computational domain was then filled by mesh, which was near the walls and in areas with expected intensive flow smoother, but on other hand, in the direction, in which there were not expected substantial changes of flow were mesh cells more stretched. Finally, this approach contributes to more accurate and less time consuming calculation. Therefore, the mesh quality was acceptable. For solution, there was implemented the Realizable k-epsilon turbulent model with Enhance wall treatment that describes the flow near the walls of heat exchanger [13]. Boundary condition on inlet was defined by mass flow rate in normal direction to boundary. Outlet boundary condition was pressure-outlet type. Calculation also includes heat loss to the environment and for roughness was assumed that height of roughness peaks does not exceed the hydrodynamic boundary layer. For chosen, above-mentioned boundary conditions at inlet were specified input values and the calculation was made subsequently. When tasks were converged, the results were plotted in temperature, speed and pressure profiles. The flow was drawn by velocity vectors and streamlines. Subsequently, the required data were shown in a table and exported for further analysis of costs in the spreadsheet program Excel. Electricity costs did not include efficiency of pumps, since there were other parameters more or less estimated. The results of this process were graphs of various analysed parameters. In the graphic data (Figure 5, Figure 6), there was found the maximum of efficiency curve which shows the optimum flow.

3.2 Data evaluation

For determination of mentioned variables were decisive data about pressure loss for calculation of consumed energy and heat fluxes in the heat exchanger for determination of the heat output. In order to determine the most economically advantageous operating parameters (flow) of fluid was sought global maximum of reference indicator (efficiency - the ratio of energy prices) by use of regression analysis. Several values, which were calculated by numerical calculation in CFD software, were for properly selected input parameters approximated by a polynomial function [14]. The emphasis in approximation of values was given on the accuracy of the maximum values. Greater variation in the limit values of the reference interval was not important then, because the approximation function was decisive for determination of the maximum. The approximation function in the limit values describes mostly the trend line.

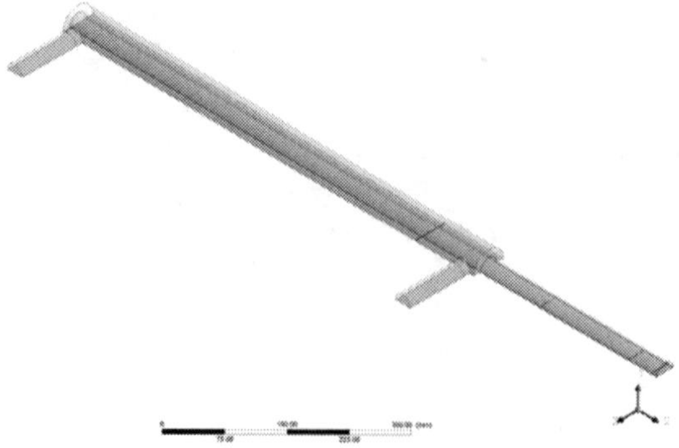

Figure 3. Lines for analysed velocity profiles.

Figure 3 shows lines in given distance from inlet, where were drawn velocity profiles. They are shown in next Figure 4.

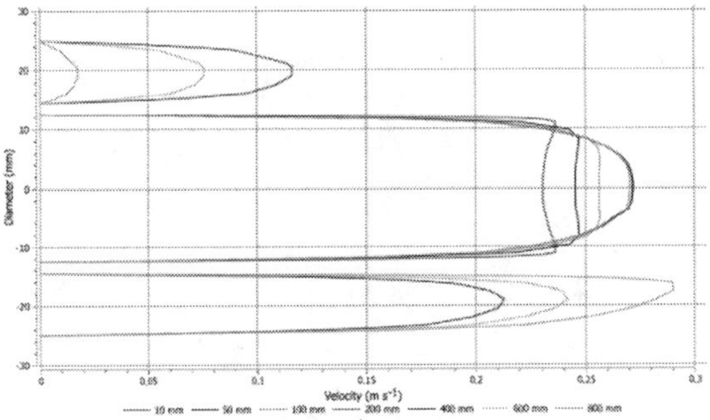
Figure 4. Velocity profiles (m s^{-1}) in symmetry plane.

From shape of velocity profiles is visible their tendency to form the parabolic profile, since in distance of about 8 times the pipe diameter leads to stable profile.

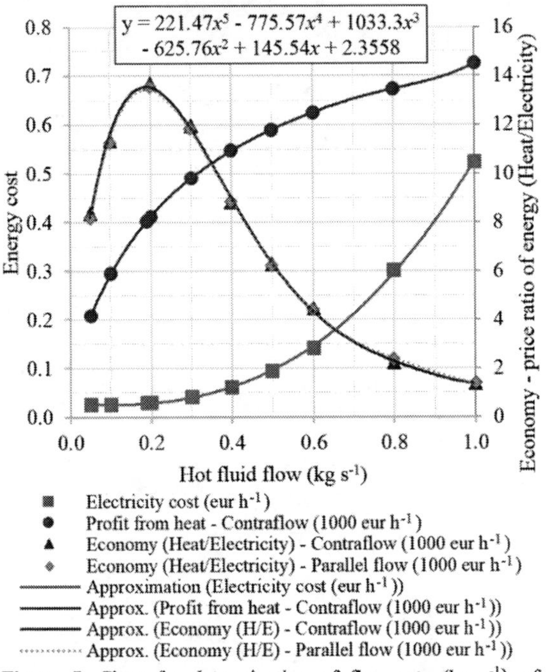

Figure 5. Chart for determination of flow rate (kg s^{-1}) of heating (hot) fluid.

Subsequently, when the flow rate of one fluid was specified (Figure 5), there was implemented the same analysis for the heated fluid in the inner pipe. For this case of the heated fluid, there was set optimum flow from previous calculation (Figure 6). For verification of the calculated optimum flows were CFD simulations realized for these values again.

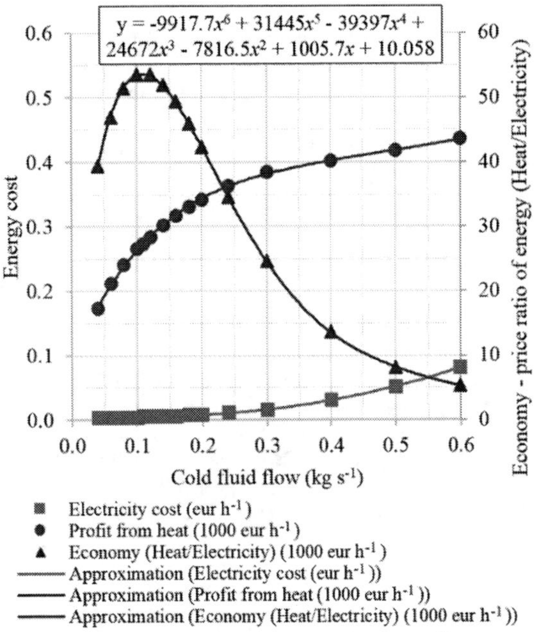

Figure 6. Chart for determination of flow rate (kg s^{-1}) of heated (cold) fluid in parallel flow adjustment.

The results in the case of parallel flow adjustment were almost identical to contraflow adjustment, but the efficiency of the contraflow adjustment should be significantly better. The reason for this discrepancy was very little difference of output temperatures in parallel flow and contraflow adjustment in Figure 7 and Figure 8.

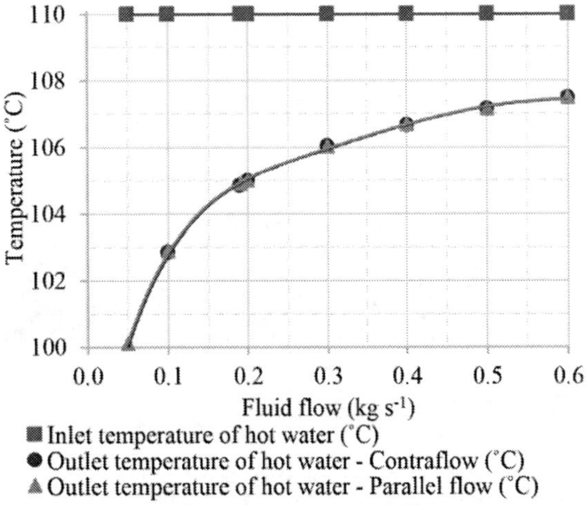

Figure 7. Inlet and outlet temperatures (°C) of heating (hot) fluid in both cases of adjustment.

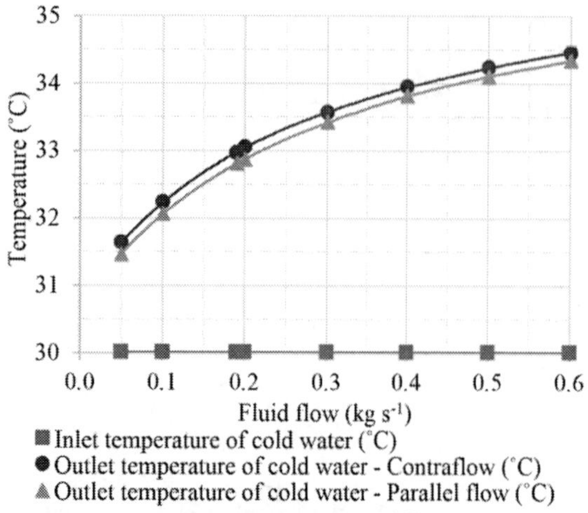

Figure 8. Inlet and outlet temperatures (°C) of heated (cold) fluid in both cases of arrangement.

Figure 9 and Figure 10 display the temperature profiles in the symmetry plane, respectively in the cross-section, approximately in the middle of the heat exchanger.

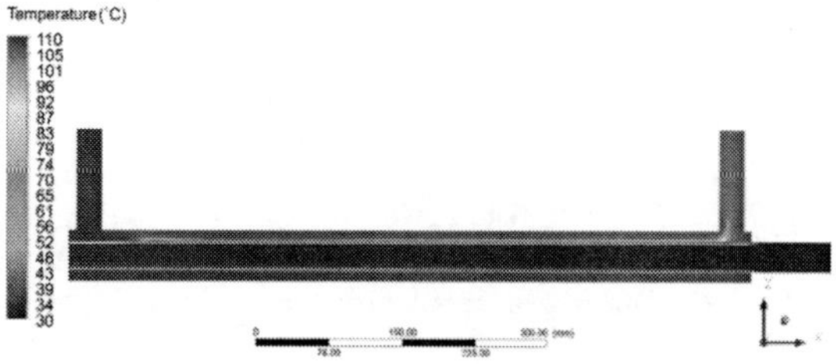

Figure 9. Temperature field (°C) in the cross-section – 24 ribs in the symmetry plane.

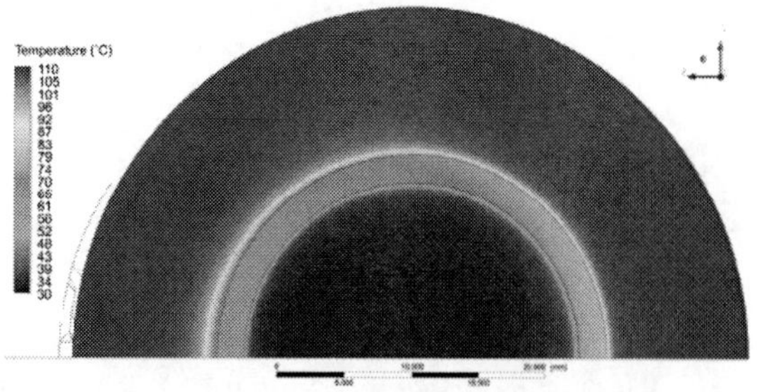

Figure 10. Temperature field (°C) in the cross-section, approximately in the middle of the heat exchanger.

The heat transfer can be increased by adding a suitable number of ribs (Figure 11, Figure 12). However, in this respect, there are structural impacts and costs of materials and manufacturing becoming more significant.

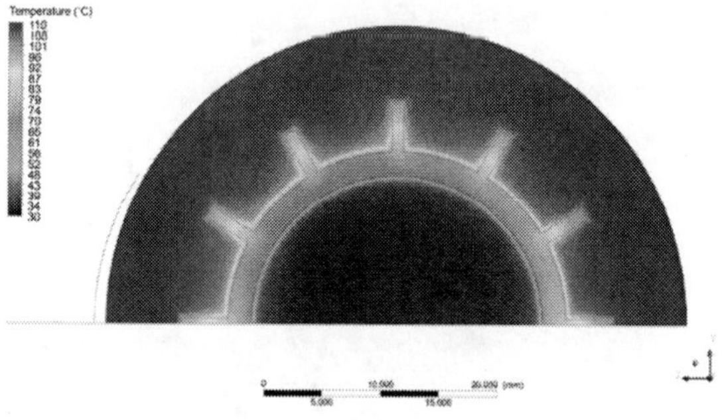

Figure 11. Temperature field (°C) in the cross-section – 12 fins.

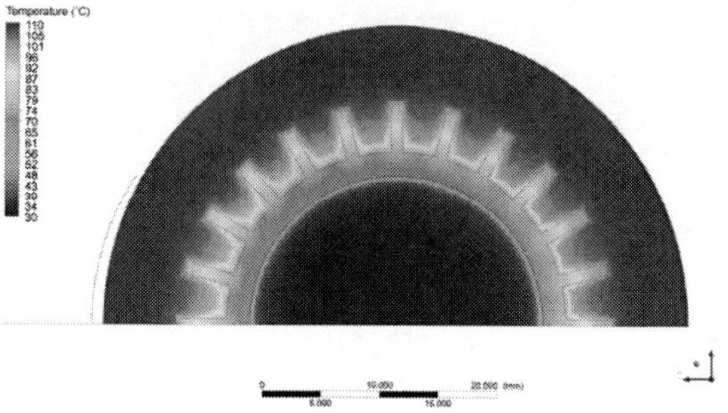

Figure 12. Temperature field (°C) in the cross-section – 24 fins.

There was achieved more guided flow by addition of fins in the longitudinal direction, as illustrate velocity profiles (Figure 13, Figure 14).

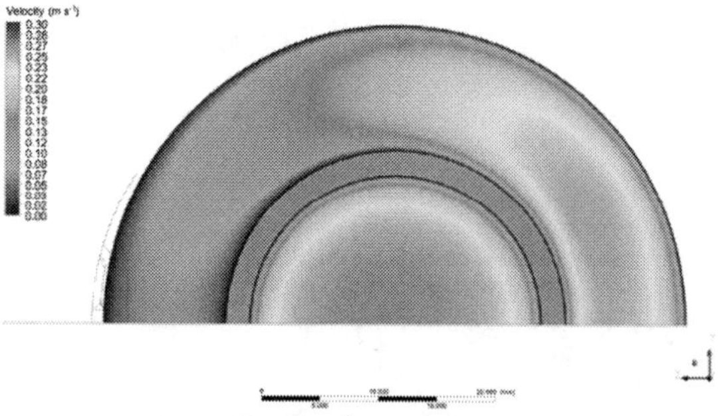

Figure 13. Velocity field (m s^{-1}) in the cross-section – without fins.

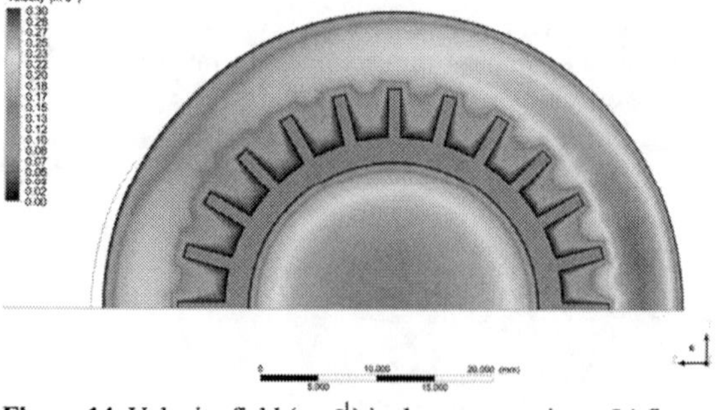

Figure 14. Velocity field (m s^{-1}) in the cross-section – 24 fins.

4 Conclusions

The calculation was unable point out more pronounced difference between parallel flow and contraflow arrangement of flows. Mentioned output temperatures showed that the contraflow arrangement is only a little advantageous as the parallel flow. The optimal flow rate of heated fluid was determined, for both flow arrangement, by a maximum of polynomial functions, which are specified by approximation of calculated values. For parallel flow arrangement was determined optimum flow rate approximately 0.191 kg s^{-1} and for the contraflow adjustment was optimal flow 0.190 kg s^{-1}, which are almost the same values. The determined flow rate in annulus space, in case contraflow arrangement of the heat exchanger, was eked by the optimized flow in a straight (inner) pipe (0.109 kg s^{-1}). In the above mentioned processes was determined the best fluid flow from the economic point of view. For a comprehensive analysis should be taken into account also other operating parameters in the calculation. Similarly, it would be appropriate to assess adding fins, which were proposed for the intensification of heat transfer. Then there was necessary to assess besides desired height, width and number of fins also costs of material and manufacture.

Acknowledgments

Moderné vzdelávanie pre vedomostnú spoločnosť/Projekt je spolufinancovaný zo zdrojov EÚ

Podpora kvality vzdelávania a rozvoj ľudských zdrojov v oblasti technického výskumu a vývoja v priestore modernej vedomostnej spoločnosti ITMS 26110230117

The research is supported by European regional development fund and Slovak state budget by the project Research centre of University of Žilina“, ITMS 26220220183.

References

1. G. V. S. Rao, C. J. Rao, N. H. Babu, International J. of Research in Aero. and Mech. **2**, 11 - 16 (2014)
2. K. Jelemenský, iDB Journal **2**, 34 - 36 (2011)
3. A. C. Caputo, M. P. Pacifico, P. Salini, Applied Thermal Engineering **28**, 1151 - 1159 (2008)
4. Z. Ibler, J. Karták, J. Mertlová, *Technický průvodce energetika*, 61 (2002)
5. K. Jelemenský, iDB Journal **3**, 38 - 40 (2011)
6. M. Baláš, *Kotle a výměníky tepla*, 119 (2013)
7. P. Stehlík, J. Kohoutek, J. Němčanský, *Tepelné pochody: výpočet výměníku tepla*, 129 (1991)
8. M. Székyová, K. Fernstl, R. Nový, *Vetranie a klimatizácia*, 422 (2004)
9. Energy online, a. s., www.cenaenergie.sk (2015)
10. I. Buranský, www.energiaweb.sk (2014)
11. P. Laube, energostat.cz (2014)
12. Energostat, www.energostat.cz (2013)
13. J. Hužvár, P. Nemec, Proc. of the Optim. Using Ex.-based Meth. and Comp. Fluid, 219 - 224 (2009)
14. A. Kapjor, J. Hužvár, B. Ftorek, M. Vantúch, Communications **16**, 54 - 61 (2014)

EPJ Web of Conferences **114**, 02075 (2016)
DOI: 10.1051/epjconf/201611402075

Wind tunnel studies of gas dispersion over complex terrain

Petr Michálek[1,a] and David Zacho[1]

[1]Aerospace Research and Test Establishment (VZLU), Department of Aerodynamics, Beranových 130, Praha 9, Czech Republic

Abstract. Wind tunnel studies of gas dispersion over complex terrain model were performed in VZLU Prague. The terrain model with a ground-level emission source was mounted in a boundary layer wind tunnel. Flow and concentration field behind the source was measured. The model presented an area of the Liberec city, 9.0 × 2.4 km in full scale. The emission source was mounted at the position of a heating plant in the model centre and concentration field was measured using flame ionisation detectors. The experimental results will be used for validation and verification of a new computational dispersion model intended for use in case of accidents with dangerous gas leakages in selected areas in Czech Republic.

1 Introduction

There are many chemical plants in the Czech Republic and some of them present a serious risk to people living in its neighbourhood. Recent gas explosion in Litvinov has proven again the need for a tool, which will be able to predict dangerous gas dispersion in urban areas in the vicinity of selected chemical factories. Such computational dispersion model is being developed in VZLU. The model uses pre-calculated results of flow and dispersion in selected urban areas together with actual meteorological situation to estimate the most endangered area. The results will be available for rescue services in the case of an accident and for local authorities for the development of emergency plans.

Experimental studies concerning gas dispersion from ground level source were made e.g. by Hall et al. [1], Lawton and Robins [2], Yassin et al. [3]. Britter and Hanna [4] reviewed flow and dispersion in different scales. In Czech Republic, Ulman et al. [5, 6] studied dispersion in topographically complicated terrain, Nosek et al. [7], Janour et al. [8], Brych et al. [9] investigated gas dispersion from a chemical factory.

The mathematical model from VZLU is based on the numerical numerical solution of the compressible gas flow in the gravitational field, uses finite volume method for the solution of the Reynolds-Averaged Navier-Stokes equations with the k-omega or EARSM turbulence models, together with the additional equation describing mass conservation of the additional gas specie. It uses special handling of the boundary conditions, based on the analysis of the Riemann problem; see Kyncl and Pelant [10, 11].

2 Experimental setup

The experiments were made in a boundary layer wind tunnel (BLWT) in VZLU, which an open-circuit wind tunnel with cross section 1.8 × 1.5 m in model section and 13.5 m long test section for boundary layer development. The tunnel is powered with 55 kW fan behind the 2 m long model section. Maximal velocity above the boundary layer is ca. 25 m s^{-1} and the ceiling is adjustable in model and test section. The simulation means can create agricultural, suburban or urban boundary layer according to Eurocode 1 [12]. The boundary layer developed over 7 m long part of the test section with boards covered with glued 4 mm gravel and a rectangular barrier 140 mm high, which served as turbulence generator. Continuous transition from boards with gravel to the model was made with increasing number of layers of boards in order to reach the level of the terrain model edge. This boundary layer corresponds with terrain type III (suburban).

The terrain model was manufactured from lightweight boards on the basis of a digital terrain model of Liberec city. The model was manufactured in scale 1:1500 and presents an area of 9.0 km × 2.4 km in full scale. The modeled area is in northwest-southeast direction along the valley of Lužická Nisa river and contains a heating plant as the possible source of gas leakage and its densely urbanized neighborhood. The terrain model was covered with the same 4 mm gravel in order to increase the aerodynamic roughness. Models of significantly high buildings and housing estates were manufactured too. Other buildings were substituted with the gravel. Windward part of the terrain model mounted in the wind tunnel is shown in Figure 1; the emission source is under the white spot. The emission source is a small chamber 25 × 25 × 10 mm covered with filtration fabric on its top

[a] Corresponding author: michalek@vzlu.cz

and built in the model. The source presents a ground level point or small area source of gas emissions.

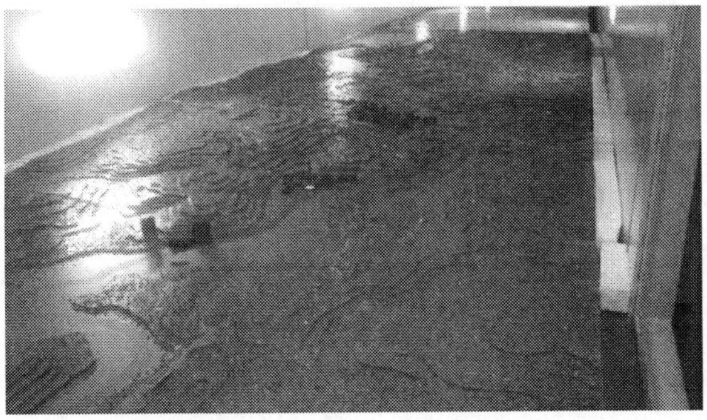

Figure 1. The terrain model of Liberec in wind tunnel

The modeled area is presented in Figure 2. The model was mounted in the wind tunnel in southeast-northwest direction, so the main flow direction was from bottom right to top left corner of the map in Figure 2.

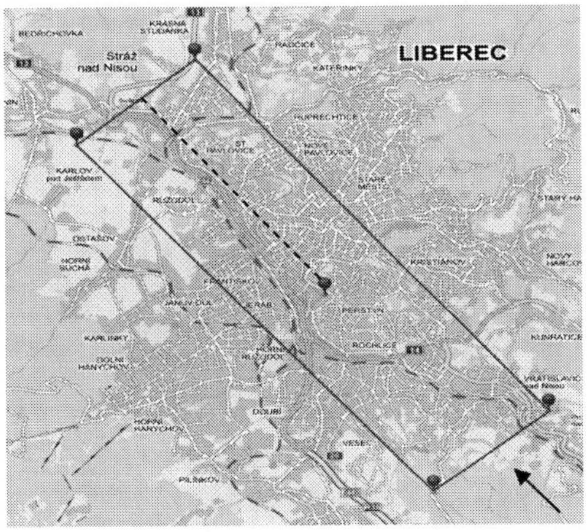

Figure 2. The modelled area of Liberec with source location

The incoming boundary layer above the emission source was measured with dual-sensor hot-wire probe and its vertical mean velocity profile is shown in Figure 3, where U (m s^{-1}) is longitudinal component of velocity vector and V (m s^{-1}) is perpendicular velocity vector component, Iu (%) is turbulence intensity of the component U, I_v (%) is turbulence intensity of the component V and z (mm) is vertical height above the emission source.

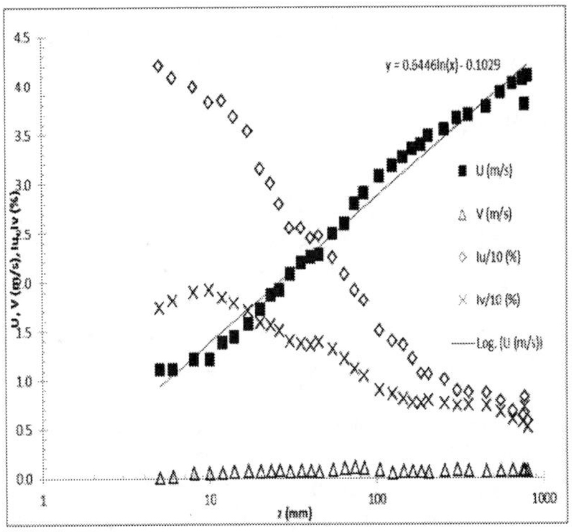

Figure 3. Vertical velocity and turbulence intensity profile

This vertical mean velocity profile was measured above the emission source and indicates that the ideal logarithmic profile differs from the actual velocity profile of component U due to high turbulence caused by the hills and dense build-up areas. The reference velocity was set to ca. 4.3 m s^{-1} above the boundary layer and this reference velocity was used in the dispersion experiments as well. For this size of the modeled area, the so-called small-scale similarity criteria applies, which means that the minimal Reynolds roughness number Re* has to be satisfied. This ensures invariance of the flow pattern above the structure. The Reynolds roughness number is defined as

$$Re^* = z_0 \times u^* / \nu = 11.9 \geq 1, \quad (1)$$

where ν is kinematic viscosity of air (m^2 s^{-1}), u* is frictional velocity (m s^{-1}) and z_0 is aerodynamic roughness (m). Values of u* = 0.26 m s^{-1} and z_0 = 1.17 mm were derived from the logarithmic fit of the mean velocity profile. The turbulence spectrum in Figure 4 was evaluated at height 300 mm above the end of boards with gravel at the transition point from boards to the model. The measured spectrum showed good agreement with Eurocode 1 wind turbulence spectrum.

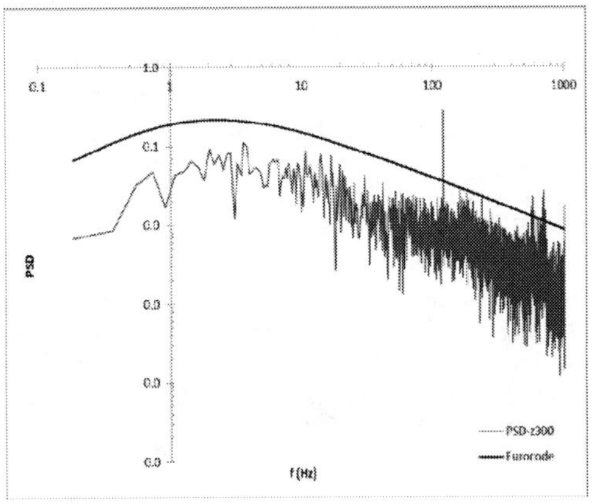

Figure 4. Power spectral density of the incoming flow

The emission source was connected to the gas tanks with tracer gas (ethane) and compressed air source via flow controllers. For neutral (non-buoyant) emission modeling, a mixture of compressed air with volume flow 5.1 l min^{-1} and ethane with flow 0.25 l min^{-1} was established. For dense gas emission simulation, sulfur hexafluoride SF_6 can be added into the emission mixture. Concentration field was measured using suction probe with four tubes mounted on a traversing device, necessary tubing, peristaltic pump and four flame ionization detectors (FID). FIDs measured on the principle of the hydrogen flame ionization, which changed according to burning of tracer gas (ethane) contained in the incoming sample. FID calibration was made with calibration gas mixture with precise concentration 100 ppm (parts per million) of ethane in air. Sampling time was set to 60 s for one point, and then the probe moved to next point and waited 30 s to clean the tubing from previous sample. The response of this system is up to 10 Hz that is why only mean concentration values were evaluated. FID signals were recorded in the computer using analog to digital input card and LabView software.

3 Results

Vertical and horizontal (longitudinal and transversal) concentration profiles were measured over the entire traversing device range to examine the extent of the concentration field. Then a network of points at certain heights above the source level, i.e. 5, 15, 25, 50 mm, etc. was measured to investigate the entire concentration field. The results were then evaluated into graphs using Excel and Origin spreadsheet software.

The measured mean concentration values c (ppm) were recalculated into dimensionless concentration

$$c^* = \frac{cU_{ref}H^2}{Q}, \qquad (2)$$

where c is measured concentration, U_{ref} is reference velocity (4.3 m s^{-1}), H is difference between lowest and highest point in the modeled area in full scale (135 m) and Q is source volume flow (5.35 l min^{-1}). The concentration was measured in a network of points indicated with crosses. The horizontal longitudinal concentration profiles measured downstream the emission source at heights from 45 to 180 m in full scale is in Figure 5. The location of the profile is depicted in Figure 2 with dashed line. The main flow direction is from right to left. The values indicate high concentration at the lowest level and rapidly decreasing concentration with increasing distance from the source. At higher level the emission concentration raises not so much, therefore our interest should be concentrated on the lowest levels below ca. 90 m in full scale above the source level.

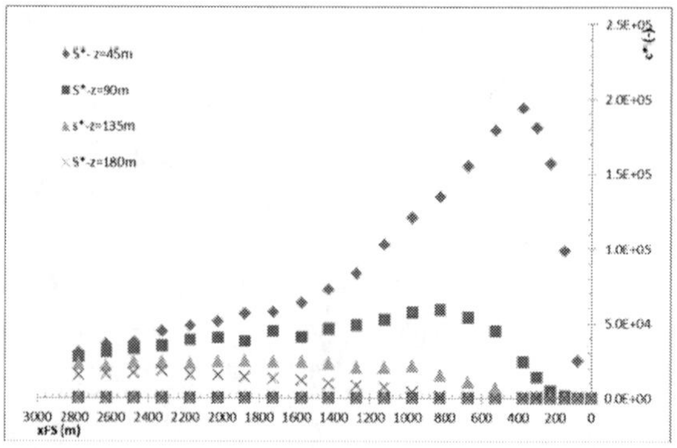

Figure 5. Horizontal longitudinal concentration profile

The resulting image from measurements within a network of points processed as contour graph at height 5 mm (7.5 m in full scale) above the source level is shown in Figure 6. The coordinates xFS (m) and yFS (m) are in full scale values. Crosses indicate measuring points. This height 5 mm was the lowest height above the model surface, where measuring was possible. Main flow direction was from right to left. The plume extension behind the source can be observed. For better evaluation of the plume in real built environment, the graph can be compared with a satellite image of the studied area within the same scale.

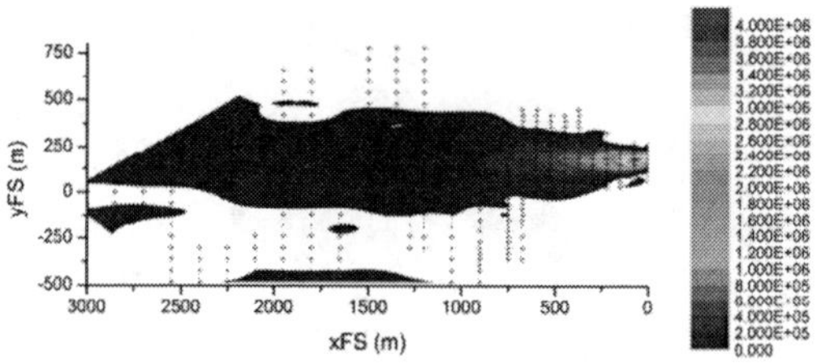

Figure 6. Concentration field c* for z = 7.5 m in full scale above the source level, neutral emission

4 Conclusion

Mean velocity profiles and concentration fields were measured over a complex terrain model of Liberec in VZLU. Mean concentration and velocity profiles and fields were measured with the use of tracer gas emission, flame ionization detectors and hot-wire probes. The extent of the plume from simulated gas leakage from a heating plant in Liberec was determined horizontally and vertically. The mean concentration fields were measured in several heights in order to use them for comparison and validation of the mathematical dispersion model. Future work includes dense gas dispersion measurements, dispersion in opposite (northwest-southeast) flow direction and field experiments with dense gas dispersion at the studied area using non-toxic tracer gas (sulfur hexafluoride). The comparison between experimental and numerical results will be made in the future as well.

Acknowledgement

Authors appreciate the support by Ministry of the Interior of Czech Republic under project no No. VG20122015098 "Scent".

References

1. D.J. Hall, V. Kukadia, S. Walker, P. Tilz, G.W. Marsland, *Proceedings of Physmod 2007* (2007)
2. T. Lawton, A. Robins, *Proceedings of Physmod 2007* (2007)
3. M.F. Yassin, M. Ohba, H. Tanaka, *Proceedings of 14th joint conference on application of air pollution meteorology with the air and waste management* Assoc, Atlanta, Georgia, USA, 1-9, (2006)
4. R.E. Britter, S.R. Hanna, *Annu. Rev. Fluid. Mech.* **35**:469-96 (2003)
5. R. Ulman, J. Drbohlav, D. Zachoval, *Proceedings of 7th UK Conference on Wind Engineering*, 201-204 (2006)
6. R. Ulman, Z. Janour, H. Sedenkova, J. Drbohlav, *Proceedings of EACWE4*, Prague (2005), paper #189
7. S. Nosek, Z. Janour, L. Kukacka, K. Jurcakova, R. Kellnerova, E. Gulikova, EPL, **67** (2014)
8. Z. Janour, K. Jurcakova, K. Brych, F. Dittrt, F. Dittrich, Proc. Saf. Environ. Prot. **88** (2010)
9. K. Brych, F. Dittrt, J. Korenar, E. Houbova, Z. Janour, *Proceedings of Konference Orlice 2001* (2001)
10. M. Kyncl, Numerical solution of the three-dimensional compressible flow. Ph.D. thesis, MFF UK, Prague (2011)
11. M. Kyncl, J. Pelant, Technical report R-5453, VZLU, Prague (2012)
12. Eurocode 1: Actions on structures – Part 1-4: General actions – Wind loads (EN 1991-1-4), European Committee for Standardisation, Brussels (2005)

EPJ Web of Conferences **114**, 02076 (2016)
DOI: 10.1051/epjconf/201611402076

Numerical and experimental study on the flow distribution in a water manifold

Gwansik Min[1,a], Pil Jong Lee[1] and Jong Hoon Kang[1]

[1]*Technical Research Laboratories, POSCO, Pohang 37859, South Korea*

Abstract. This study presents water distribution analysis of the device for spraying cooling water through specific nozzles numerically and experimentally. Numerical analysis was performed using the 3-D incompressible, multi-phase flow model, for different Reynolds numbers of 4 x 10^5, 8 x 10^5. Experimental analysis was performed at real-size, under the same conditions. The calculated results and the measured results for the distribution of flow were matched relatively well. The distribution of the nozzle flow depends on the Reynolds number.

1 Introduction

Thermal management systems that require very high heat transfer cooling with a simple equipment use the free impinging jets of liquid [1]. Run-out table temperature control is one of the processes to obtain the mechanical properties of a steel strip employed in the metal processing industry. Accelerated cooling has been implemented in hot strip mills to reduce the duration of this process in order to achieve the required coiling temperature and to obtain the desirable grain structure [2-4]. In general, impinging circular water jets are used as a rapid cooling system for hot steel plates [1].

Impinging water jets require the uniform mass flow rate to make strips with uniform properties. In two-phase flow, mal-distribution of the mass flow rate can significantly increase the various defects of the strip.

Wu and Webb [5] conducted tests in an evaporator with a horizontal header and an upward tube configuration. Lalot et al. [6] investigated fluid mal-distribution in an electrically-heated heat exchanger. They developed a numerical model to determine the effectiveness of the heat exchanger and noted a 7% reduction in the calculated heat exchanger performance due to mal-distribution in the counter-current flow heat exchangers and condensers.

Rao et al. [7,8] evaluated the thermal performance of falling film plate condensers with flow mal-distribution from the feeding tube to the channel. The heat transfer coefficient was considered as a function of mass flow rate inside the channels. They evidenced a heat transfer coefficient reduction due to mal-distribution among the parallel channels.

Marchitto et al. [9] commented that uneven two-phase flow distribution may occur inside each channel, owing to the asymmetrical parallel and diagonal flow, as well as inside the header, owing to the separation of the two-phase mixture.

This work presents the water distribution analysis of the manifold and comparison on experimental and numerical results. Figure 1. shows the manifold geometry considered in this paper.

2 Approach of problem

2.1 Experimental setup

Experiments were performed in a real-size test in order to investigate water distribution problems of the model manifold under various flow conditions in addition to the effect of baffle fitting.

An apparatus was made that consisted of a water manifold, an elevated tank, a pump, and a basin. The elevated tank was installed at 30 m height and fed the basin water through the pump. The downstream of the elevated tank flowed through a pipe with an inner diameter (D) of 0.13 m. The water manifold was connected to the flow system by a horizontal pipe with a flow meter and pressure transmitter. The instrumentation was designed to record the evolution of the flow rate and pressure inside the pipe. An array of the valves upstream of the manifold could be operated in order to control the flow rate coming from the elevated tank. The pressurized water manifold sprayed the water back into the basin. A schematic diagram of the flow system is shown in Figure 2.

The manifold consists of a horizontal channel of rectangular cross-section, two air vent halls and a system of laminar jet nozzles. The total length of the channel is

[a] Corresponding author: corundum@posco.com

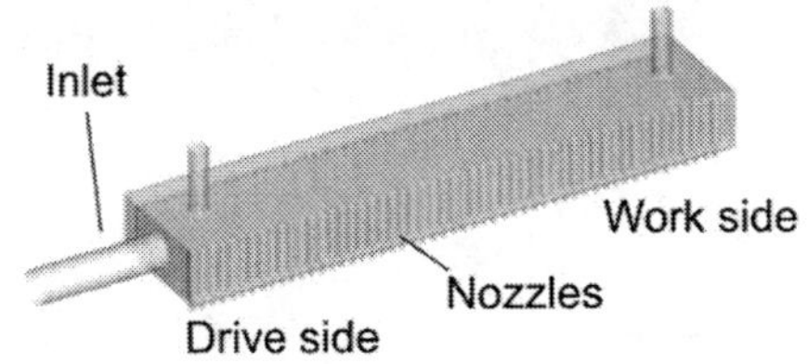

Figure 1. The model water manifold considered in this paper.

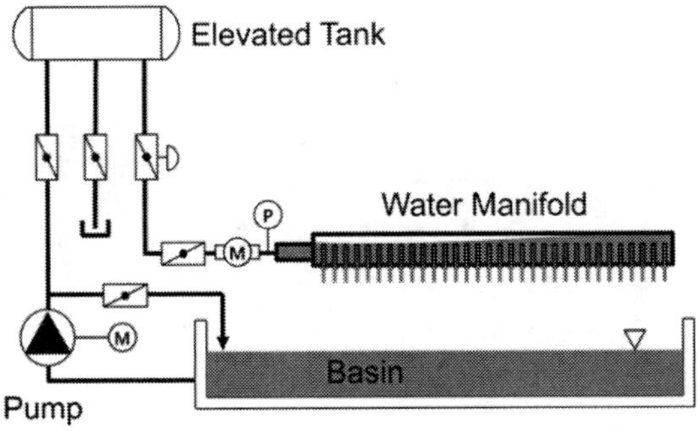

Figure 2. Schematic diagram of the flow system.

over 15D and minimum cross-sectional area (A) is 0.089 mm^2. The gap between the nozzle and manifold wall is over 0.2 D.

For this case, the Reynolds number is defined as follows:

$$\mathrm{Re} = \frac{\rho V D_h}{\mu} \tag{1}$$

where, ρ represents the density, V the mean velocity in the minimum cross-sectional area of the manifold, D_h the hydraulic diameter calculated from the cross-section of the manifold, and μ the viscosity.

The nozzle flow rate was measured by the weight of water sprayed from the bottom of the manifold, and the measured time was over 40 seconds. The measuring points were located on the drive side, center side and work side.

2.2 Numerical simulation

2.2.1 Governing equations

The mass and momentum conservations for the incompressible multi-phase flow can be written as follows:

$$\nabla \cdot \mathbf{u} = 0 \tag{2}$$

$$\rho \frac{\partial \mathbf{u}}{\partial t} + \rho \nabla(\mathbf{uu}) = -\nabla p + \mu \nabla^2 \mathbf{u} + \mathbf{F} \tag{3}$$

where ρ represents the density, **u** the velocity vector, p the local pressure, μ the viscosity and **F** any body forces acting on the fluids. The momentum equation (3) is dependent on the volume fractions of all phases through the properties ρ and μ.

Table 1. Numerical parameters.

Parameters	
Multiphase model	Volume of Fluid, Multi-fluid VOF (Eulerian)
Turbulent model	SST k-ω
Pressure-Velocity coupling scheme	SIMPLE
Working fluid	Water, Air
Inlet condition	Uniform velocity inlet
Outlet condition	Constant pressure outlet

In the two-phase model, two types of model were compared, VOF (volume of fluid) and Multi-fluid VOF.

In the VOF model, a single set of conservation equations is shared by the phases and the volume fraction of each of the phases is tracked in each computional cell thorughtout the domain [12]. Therefore VOF adopted surface-tracking techniques and the interfacial forces are replaced by smoothly varying volumetric forces.

The multi-fluid VOF model couple the VOF model and Eulerian multi-phase models. This allows the discretization schemes suited to both sharp and dispersed interface regimes while overcoming some limitations of the VOF model that arise due to the shared velocity.

For the numerical computations, the commercial CFD code ANSYS FLUENT 15.0 was used. A QUICK scheme for the spatial discretization and SIMPLE algorithm for the pressure-velocity coupling were adopted [11]. The turbulence model is considered an SST k-ω model. Numerical convergence criterion for the residuals was 1.0 x 10^{-3} for the mass components and the other quantities.

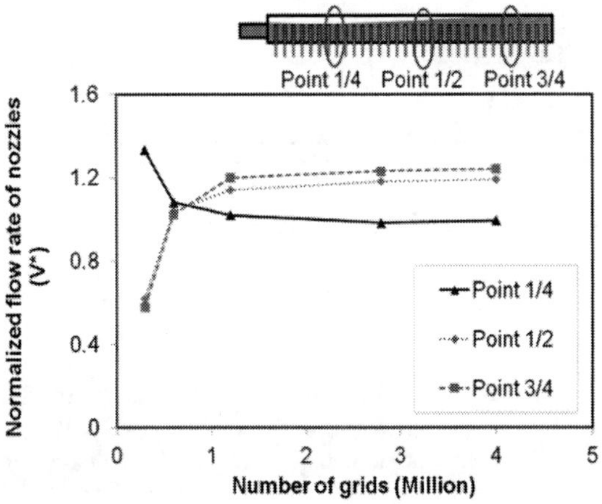

Figure 3. Normalized flow rate of simulation vs. number of grids.

2.2.2 Grid dependency

In order to determine a suitable number of grids, a preliminary grid dependency test for the geometry of the manifold was carried out to verify that the solution was grid independent. Five grids with numbers of cells that ranged from 0.3 million to 4.0 million were tested.

The normalized flow rate of nozzles depending on the number of grid shown in Figure 3. We selected 1.2 million cells for the grid system. In this case, the y+ near the wall was 51.

3 Results and Discussion

Figure 4. shows flow distribution of the water manifold from the experimental and numerical results at the Re=8.4 x 10^5. Experimental results are represented by the points; computational results are represented by the line. Single-phase flow calculation results did not describe the unbalanced flow rate distribution. In the case of the VOF model, mal-distribution was exaggerated. The Eulerian model that was calculated separately for each phase tends to be relatively consistent with the experimental results.

To compare the above two multi-phase models, the VOF model was suitable for finding the unbalanced area and the Eulerian model was appropriate for estimation of mal-distribution flow

To solve a simpler problem, the VOF model is a better choice, since it solves a smaller number of equations which requires less computational effort than the Eulerian model. If accuracy is more important than computational effort, the Eulerian model is a better choice.

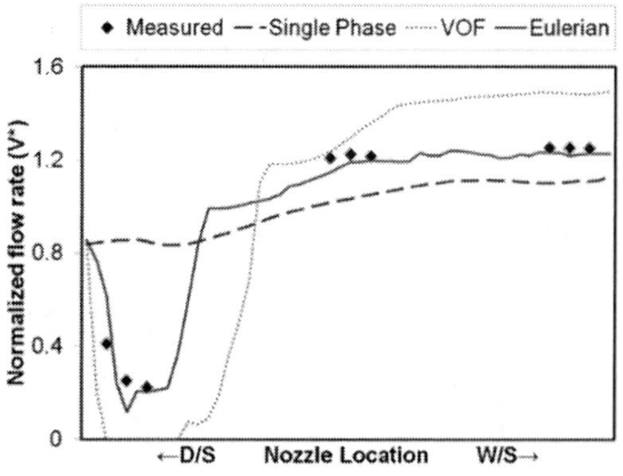

Figure 4. Comparison between the experimental and the numerical results.

The base model and modified model with slope baffles of the water manifold are shown in Figure 5. Computations and experiments were conducted for Re = 8.4 x 10^5, and only computations for Re = 4.2 x 10^5.

The effect of slope baffles on the water distribution is shown in Figure 6. For Re = 8.4 x 10^5, the model with the baffles had a better distribution compared to the base model. For Re = 4.2 x 10^5, the flow rate largely changed according to the location of baffles.

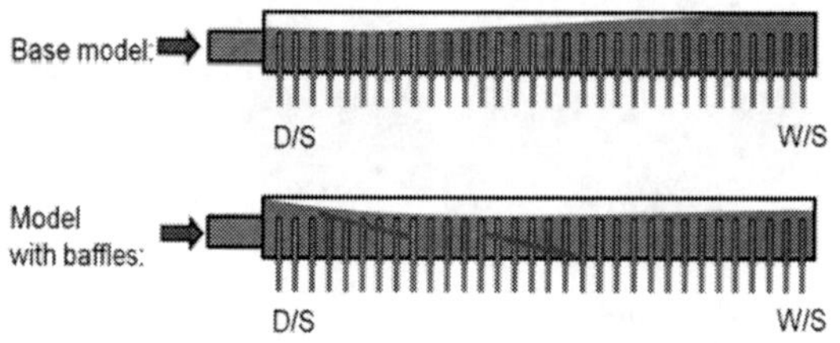

Figure 5. Schematic diagram of the base model and model with

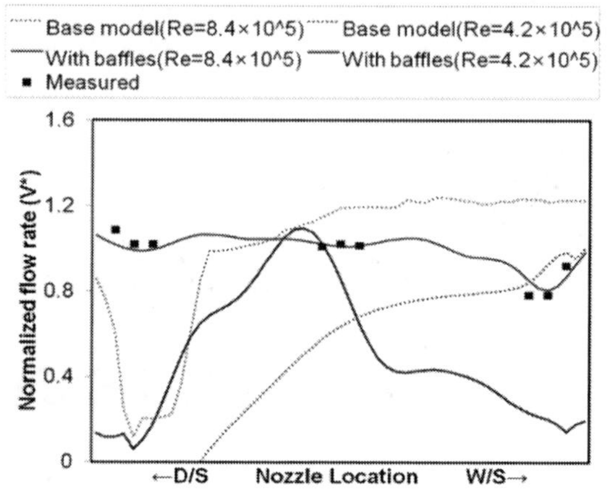

Figure 6. Water distribution comparison between base model and model with the baffles.

(a) Base model

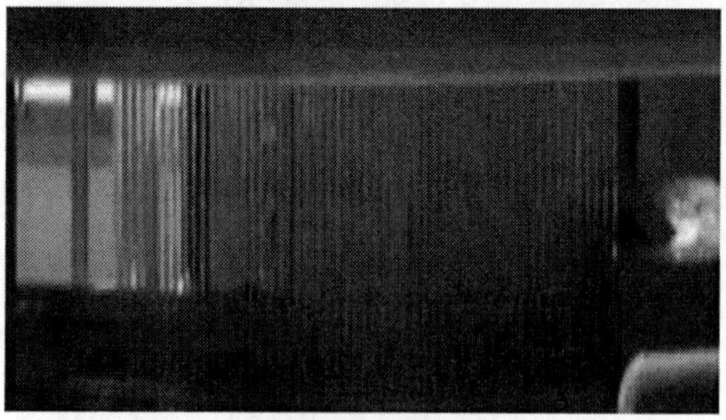

(b) Model with baffles

Figure 7. Photographs of the water distribution of the manifold: (a) Base model, (b) Model with the baffles.

The calculated results and the measured results for the distribution of flow matched relatively well.

Figure 7. represents two experimental results. The drive side of base model sprays imperfectly compared to the work side. The model with the baffles shows more even water jet columns than the base model.

4 Conclusions

An experimental and numerical study has been conducted to investigate the water distribution of the manifold for different Reynolds numbers of 4.2×10^5 and 8.4×10^5.

In the multi-phase model, the VOF model was suitable for finding the unbalanced area and the Eulerian model was appropriate for estimation of mal-distribution flow.

The distribution of the nozzle flow depends on the Reynolds number.

The installation of the baffles is effective to ease the flow distribution of a manifold under large flow condition.

References

1. P. Lee, H. Choi, S. Lee, ISIJ International **44**, 704 (2004)
2. S.D. Cox, S.J. Hardy, D.J. Parker, Ironmaking and Steelmaking, **28**, 363 (2001)
3. C. Devadas, I.V. Samarasekera, Ironmaking and Steelmaking, **13**, 311 (1986)
4. J.F. Evans, I.D. Roebuck, H.R. Watkins, Iron and Steel Engineer, **70**, 50 (1993)
5. X. Wu, R.L. Webb, Applied Thermal Engineering, **22**, 1369, (2002)
6. S. Lalot, P. Florent, S. Lang, A. Bergles, Applied Thermal Engineering, **19**, 847 (1999)
7. B.P. Rao, B. Sunden, S.K. Das, Journal of Heat Transfer, **127**, 332 (2005)
8. B.P. Rao, B. Sunden, S.K. Das, International Journal of Heat and Mass Transfer, **49**, 4966 (2006)
9. A. Marchitto, M. Fossa, G. Guglielmini, Experimental Thermal and Fluid Science, **33**, 895 (2009)
10. E.R. Dario, L. Tadrist, J.C. Passos, Applied Thermal Engineering **59**, 316 (2013)
11. Ansys Inc., *Ansys fluent 14 user's guide* (Ansys Inc., USA, 2011)
12. V.V Ranade, *Computational Flow Modeling for Chemical Reactor Engineering*, **5**, (Academic Press, USA, 2002)

EPJ Web of Conferences **114**, 02077 (2016)
DOI: 10.1051/epjconf/201611402077

Numerical simulation heat transfer by natural convection in liquid metal with a sinusoidal temperature

Abdelkrim Missoum[1,a], Mohamed Elmir[1], Mohamed Bouanini[1], Abdellah Belkacem[1] and Belkacem Draoui[1]

[1] *ENERGARID Laboratory, Tahri Mohamed University of Bechar, B.P 417, Ageria*

Abstract. This study focuses on the numerical simulation of heat transfer by natural convection in a rectangular enclosure, filled with a liquid metal (low Prandtl number) partially heated from below with a sinusoidal temperature. The value of the study lies in its involvement in the crystal growth for the manufacture of semiconductors and electronics cooling. Indeed, the occurrence of convection during crystal growth can lead to in homogeneities that lead to striations and defects that affect the quality of the crystals obtained by the Bridgman techniques or Chochrawlski. Temperature of the oscillations, due to the instabilities of the convective flow in the liquid metal, also induces non-uniform cooling in the solidification front. Convection is then studied in order to reduce it. A modelling of the problem in two dimensions was conducted using Comsol computer code that is based on the finite element method, by varying the configuration of the control parameters, namely, the Rayleigh number, the nature of fluid (Prandtl number) and amplitude of temperature on heat transfer rate (Nusselt number) on convective structures that appear.

1 Introduction

Natural convection in liquid metals is the subject of studies by metallurgists and engineers of nuclear reactors. Indeed, liquid metals have an excellent heat transfer medium and thus make it possible to cool the nuclear reactor, the electronic components and especially during the crystal growth process [1].

Many studies have been conducted in recent years and have allowed to better understand the behavior of liquid metals when stressed thermally. Stewarl and Weinberg [2] were among the first to study the natural convection in a two-dimensional cavity fluid containing low Prandtl number with insulated horizontal walls. Their results show that the behavior of the flow in the liquid metal is different from conventional fluids. This study has given researchers a starting point for research in liquid metals. [3], [4] studied this phenomenon in several conditions with comparison of the experimental results by digital. Their results confirm that the liquid metals are an excellent heat transfer medium.

Interest in the study of this problem is also its involvement in the growth of crystals for the manufacture of semiconductors. Indeed, the occurrence of convection during crystal growth can lead to in homogeneities that lead to striations and defects that affect the quality of the crystals obtained by the Bridgman techniques and Czochralski. Temperature of the oscillations due to flow instabilities in the liquid metal also induce non-uniform cooling in the forehead solidification. In this case, convection is studied in order to avoid or reduce it.

In addition to this practical aspect, the study of convective flows in liquid metals is of interest of fundamental research perspective. Indeed this type of opaque fluids at low melting temperatures (eg 302.8 K for gallium) are fluids at low Prandtl number (Pr) and behavior when subjected to temperature gradients are quite different from those observed in fluids high number of Pr (such as water and air).

Our work is to study the numerical heat transfer by natural convection in a rectangular enclosure filled with a liquid metal (low Prandlt), partially heated from below with a sinusoidal temperature. The results obtained by the Comsol software using the numerical method of finite elements.

2 Problem and mathematical model

2.1 Problem definition

A schematic of the two dimensional system, and the coordinates and boundary conditions are shown in Figure 1. The square cavity is partially heated by sinusoidal temperature and the top wall is maintained cold temperature. The vertical wall and the no-heated portions of bottom wall are adiabatic. The height of the cavity is H. The finite-sized heater is length L_1 , and here temperature strength is $T_C = 1+A*\sin(w*t)$. The

[a] Corresponding author: missoum101@yahoo.fr

dimension length the portions no-heated and of the heater are same at a value L_1=H/3. All boundaries satisfy the no-slip velocity conditions.

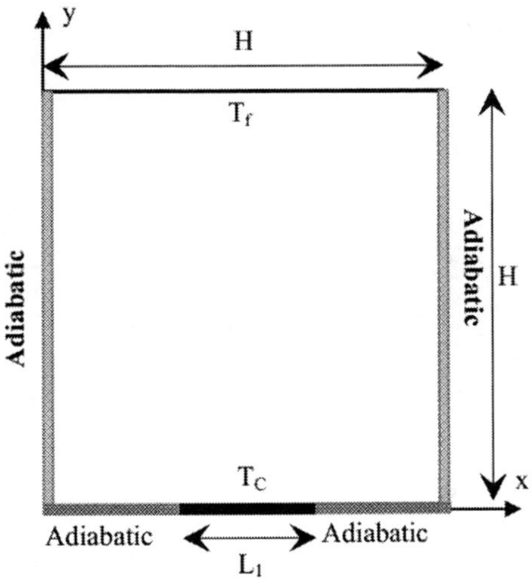

Figure 1. Schematic of the physical system.

2.2 Mathematical model

Following assumptions are made: the flow is laminar of an incompressible Newtonian fluid and two-dimensional, there is no viscous dissipation, the gravity acts in the vertical direction, fluid properties are constant and fluid density variations are neglected except in the buoyancy term (the Boussinesq approximation), and radiation heat exchange is negligible.

The unsteady state, two dimensional continuity, momentum and energy equations governing the flow and heat transfer are given by:

$$\frac{\partial u}{\partial x}+\frac{\partial v}{\partial y}=0 \quad (1)$$

$$\frac{\partial u}{\partial t}+u\frac{\partial u}{\partial x}+v\frac{\partial v}{\partial y}=-\frac{1}{\rho}\frac{\partial p}{\partial x}+\upsilon\left(\frac{\partial^2 u}{\partial x^2}+\frac{\partial^2 u}{\partial y^2}\right) \quad (2)$$

$$\frac{\partial v}{\partial t}+u\frac{\partial v}{\partial x}+v\frac{\partial v}{\partial y}=-\frac{1}{\rho}\frac{\partial p}{\partial y}+\upsilon\left(\frac{\partial^2 v}{\partial x^2}+\frac{\partial^2 v}{\partial y^2}\right)+ g.\beta.(T-T_C) \quad (3)$$

$$\frac{\partial T}{\partial t}+u\frac{\partial T}{\partial x}+v\frac{\partial T}{\partial y}=\alpha\left(\frac{\partial^2 T}{\partial x^2}+\frac{\partial^2 T}{\partial y^2}\right) \quad (4)$$

The continuity, momentum and energy equations are written. The non-dimensional governing equations are obtained as:

$$\frac{\partial U}{\partial X}+\frac{\partial V}{\partial Y}=0 \quad (5)$$

$$\frac{\partial U}{\partial \tau}+U\frac{\partial U}{\partial X}+V\frac{\partial U}{\partial Y}=-\frac{\partial P}{\partial X}+Pr\left(\frac{\partial^2 U}{\partial X^2}+\frac{\partial^2 U}{\partial Y^2}\right) \quad (6)$$

$$\frac{\partial V}{\partial \tau}+U\frac{\partial V}{\partial X}+V\frac{\partial V}{\partial Y}=-\frac{\partial P}{\partial Y}+Pr\left(\frac{\partial^2 V}{\partial X^2}+\frac{\partial^2 V}{\partial Y^2}\right)+Ra.Pr.\theta \quad (7)$$

$$\frac{\partial \theta}{\partial \tau}+U\frac{\partial \theta}{\partial X}+V\frac{\partial \theta}{\partial Y}=\frac{\partial^2 \theta}{\partial X^2}+\frac{\partial^2 \theta}{\partial Y^2} \quad (8)$$

Here, the x and y coordinates are normalized as: $X=\frac{x}{H}$; $Y=\frac{y}{H}$. U and V are the dimensionless velocity components in the X and Y directions, non-dimensionalized as: $U=\frac{u.H}{\alpha}$; $V=\frac{v.H}{\alpha}$. P is the dimensionless effective pressure, scaled as: $P=\frac{p.H^2}{\rho.\alpha^2}$. θ is the dimensionless temperature, non- dimensionalized as: $\theta=\frac{T-T_f}{T_C-T_f}$. a and W are the dimensionless respectively amplitude and period, non- dimensionalized as: $a=\frac{A}{T_C-T_f}$ and $W=\frac{w.\alpha}{H^2}$.

The steady-state solutions are obtained from unsteady state equations, Eqs (5)-(8). The local and average Nusselt numbers are calculated respectively as:

$$\left.\begin{aligned} Nu_L &= -\frac{\partial \theta}{\partial Y} \\ Nu_m &= \int_0^1 Nu_L dX \end{aligned}\right\} \quad (9)$$

The stream function is calculated from its definition as:

$$U=-\frac{\partial \psi}{\partial Y},\ V=\frac{\partial \psi}{\partial X} \quad (10)$$

2.3 Boundary conditions

The boundary conditions of the system are shown in Figure. 1. Velocities are zero on all solid surfaces, normal pressure gradient are zero on all solid surfaces at the outside boundaries. On the adiabatic boundaries, temperature gradient is zero. On the bottom centre portion horizontal boundary, a sinusoidal temperature is applied. Thus,

On all solid boundaries: $U=V=0$

$$Y=0,\quad \frac{H}{3}<X<\frac{2H}{3}:\ \theta=1+a.\sin(W.\tau)$$

$$Y=0,\quad 0<X<\frac{H}{3}\ and\ \frac{2H}{3}<X<1:\ \frac{\partial \theta}{\partial Y}=0$$

$$Y=H,\quad 0<X<H:\ \theta=0$$

$$X=0\ and\ X=H, 0<Y<H:\ \frac{\partial \theta}{\partial X}=0$$

3 Numerical techniques

Numerical results are obtained by solving the system of unsteady differential equations (1)-(4), with appropriate boundary and initial conditions, using Comsol Multiphysics based for Galerkin finite elements method. The computational domain consists of bi-quadratic elements which correspond to 590 elements and a Lagrange quadratic interpolation has been chosen.

4 Results and discussion

The square cavity with height and length H=1 here studied at Rayleigh number from 10^3 to 10^6. The partially centre portion is hot by the sinusoidal temperature with the amplitude and period are varied between respectively a=0 to 0.8 and W=0 to 1. The working fluid is chosen as liquid metal (Gallium) Pr = 0.025.

4.1 Stream functions and isotherms

The figure 2 shows the streamlines and isotherms for different Rayleigh numbers with an amplitude and a period is respectively 0.8 and 0.5 for the liquid metal. The results show that the flow of fluid within the square cavity becomes large by increasing the number of Rayleigh. Can be observed at low flow intensities (Ra = 10^3) the formation of two counter-rotating cells and subsequently turns into a single main drive unit of the fluid mass in a direction of counterclockwise rotation with the appearance of recirculation zones have low flow intensities located in the four corners of the cavity. For Ra = 10^3, the isotherms are parallel to the horizontal wall of the heating characterizes the conductive heat transfer. Nevertheless, from Ra = 10^4 the onset of transfer with convection , beyond this value of Ra the transfer is purely convective. Indeed, the intensity of the natural convection entrained the elevation large quantities of hot fluid to be subsequently transported to the opposite side (cold).

For comparison, the structure of the flow and isotherms for the air (Figure 3) are different from the liquid metal. As noted, the presence of two counter-rotating cells to Ra = 10^3 and 10^5 and a single vortex to Ra = 10^4 and 10^6, it is also noted that the intensity of the flow increases by increasing Ra. Heat transfer by convection is triggered from as Ra = 10^4. Thus, there is a symmetry of the isotherms for Ra = 10^5 and the transfer becomes fully convective for Ra = 10^6. Note that the effect found on the stream lines, affects the structure of the isotherms, This changing of the flow structure explained by the density of air is lighter than the liquid metal.

4.2 Heat transfer

Local Nusselt number profile on the cold wall for amplitude 0.8 and period 1 is shown in Figure 4. In this figure, the profile is traced for X to 0 from 1. As discussed earlier, we can see clearly that the local Nusselt number profile along X is line in Ra=10^3, axisymmetric for Ra=10^4, 10^5 and symmetric (Gaussian form) for Ra=10^6. We clearly notice that the value of the local Nusselt number believed along the horizontal axis of the cavity then it starts decreased from the center of the cavity. This is due to the increase in the central portion along the horizontal axis X of the temperature gradient between the fluids and parietal layers cooled wall. Indeed, the fluid in contact with the hot wall is heated and transported to the top of the cavity under the effect of the decrease in density of the fluid. We note the maximum values for the local Nusselt number, this can be explained by the high values of the Rayleigh number that represents the predominant effect of convective fluid movement.

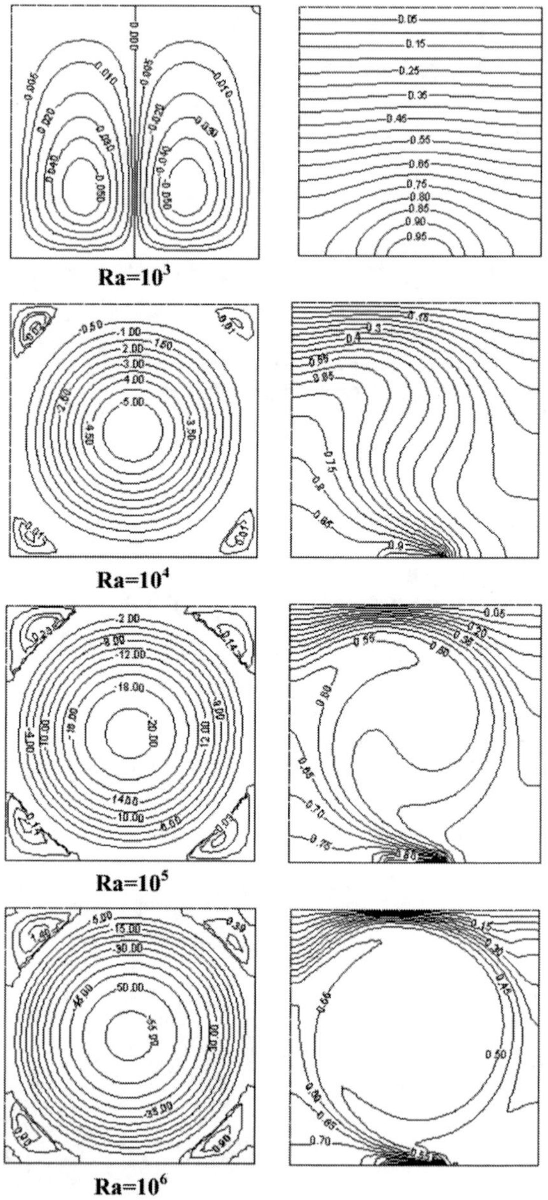

Figure 2. Stream functions (in the left) and isotherms (in the right) for a=0.8, W=0.5 with different values of Rayleigh for metal liquid.

Average Nusselt number on the cold wall is calculated by Eq. (5) and presented in the figure 5 and 6 respectively with differentes values the amplitude for period is fixed at 0.5 and differentes values the period for amplitude is fixed at 0.2. In the case of liquid metal the average Nusselt number increases with the number of Rayleigh, therefore the heat transfer becomes more intense with the increase of Rayleigh number (Figure 4) and it decreases with increase in the value of the amplitude of the heating temperature. However, for the case of air, the average transfer rate remains almost constant at low amplitude values, and then decreases with the decrease of the amplitude value of the heating temperature. The transfer rate increases with the increase of Rayleigh number and the period of the heating temperature (Figure 6) for the case of fluid.

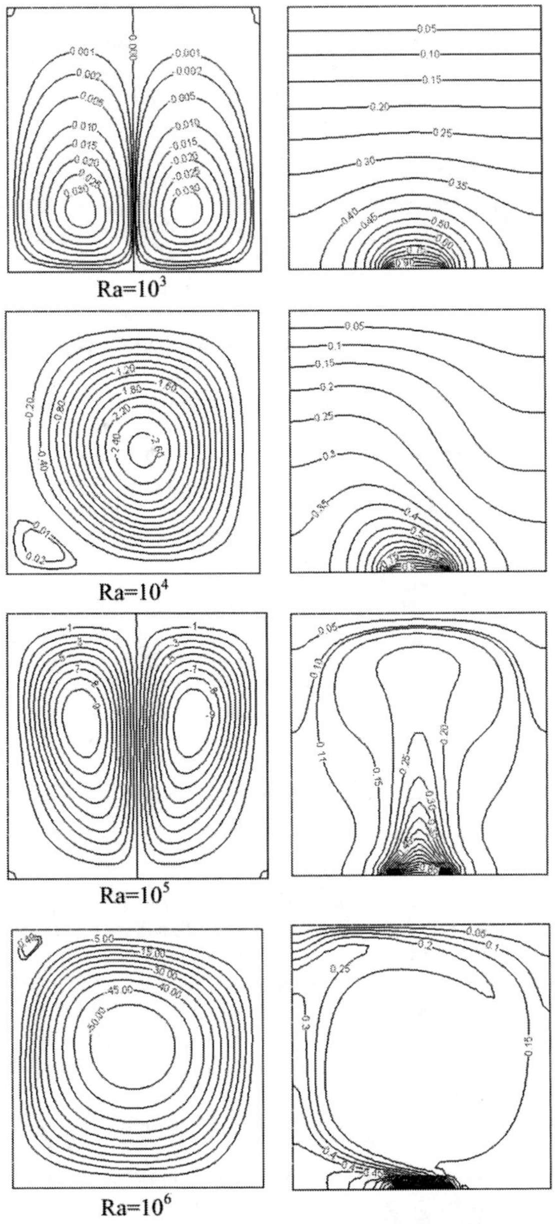

Figure 3. Stream functions (in the left) and isotherms (in the right) for a=0.8, W=0.5 with different values of Rayleigh for air.

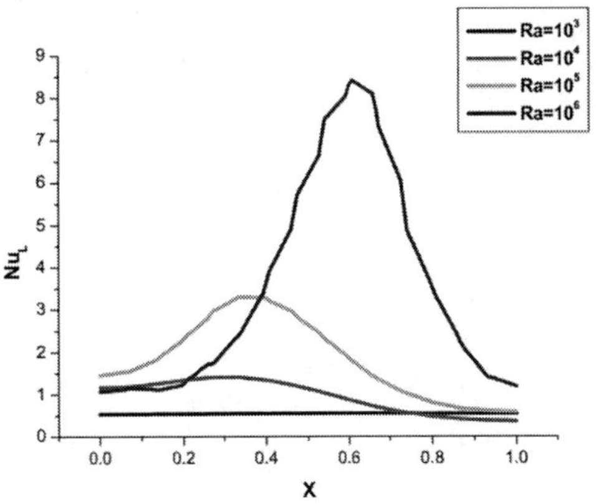

Figure 4. Local Nusselt number at (Y=1, X) along cold wall of the enclosed for a=0,8 and W=1 for various Rayleigh number

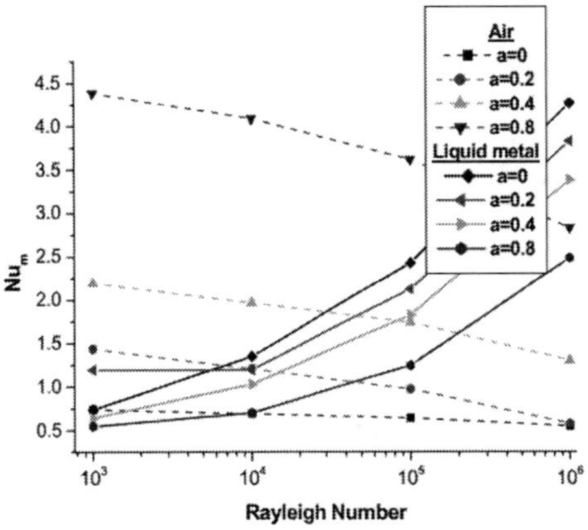

Figure 5. Average Nusselt number the liquid metal and air as a function of Rayleigh number with the amplitude value as a emperature for period fixed at 0.5

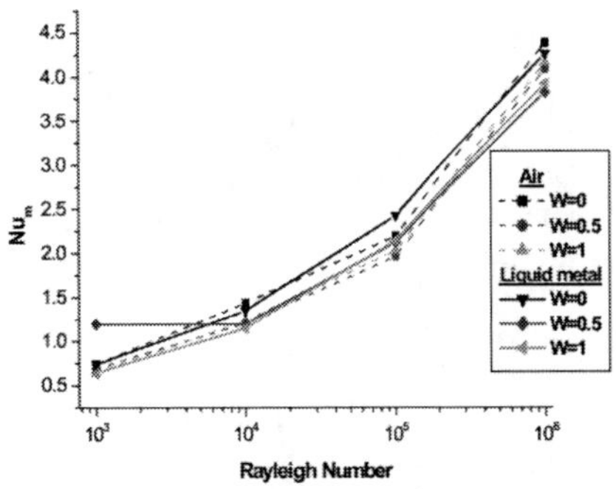

Figure 6. Average Nusselt number of the liquid metal and air as a function of Rayleigh number with the period value as a temperature for amplitude fixed at 0.2

4.3 Maximum streamt function

Figures 7 and 8 respectively show the variations of the maximum current in absolute value function as a function of time for different values of the amplitude and the period. We read on these curves the instant the heating temperature was imposed. Naturally, at this moment we can see that the steady state is well established. A transition period characterizing the fluid response is observed at the start of heating by the sinusoidal temperature.

After the transition period, the solution becomes periodically characterized by sinusoidal oscillation having the same period as the heating temperature. The oscillations of the maximum current in absolute value function are around a mean value substantially identical to that of the steady state (a = 0), which increases an amplitude of a. Nevertheless, increasing the amplitude, it is noted that the maximum value of the amplitude above the steady state (a = 0) is less than that below.

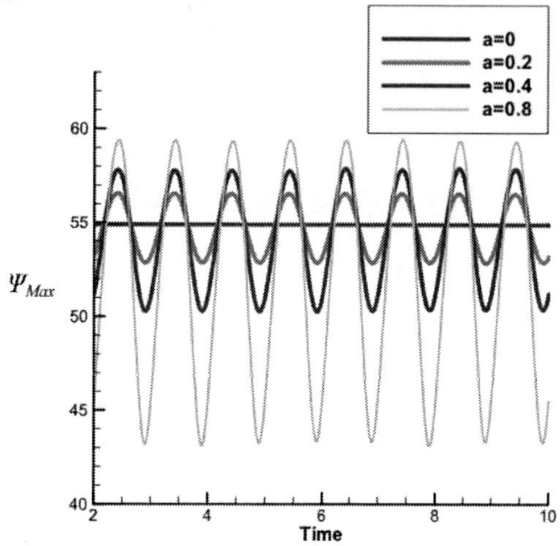

Figure 7. Variation of maximal stream function for time for different value of amplitude with period at 1

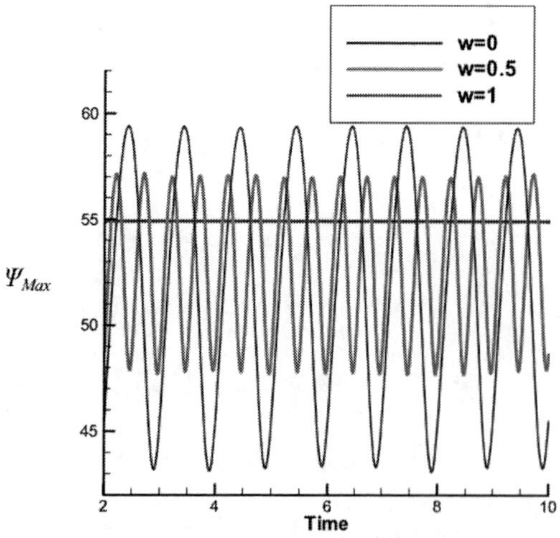

Figure 8. Variation of maximal stream function for time for different value of period with amplitude at 0.8

5 Conclusion

A numerical study has been carried out in square enclosures is partially heated by sinusoidal temperature in bottom and the top wall is maintained cold temperature. The vertical wall and the no-heated portions of bottom wall are insulated. The influence of Rayleigh number and of the amplitude, the period of the heated part on the heat transfer characteristics is examined. In view of the results and discussion presented, the following main conclusions have been drawn.

- Whatever the nature of the fluid, increasing the Rayleigh number causes a remarkable change in the structure of the flow field and the temperature field, and stabilizes the convective flow.
- The curves of the stream function and the temperature profile of a changing sinusoidally around the mean values. Comparing amplitudes of the oscillation parameters, it may be noted that these amplitudes are of different sizes with respect to that of the heating temperature, this leads to the conclusion that the oscillation amplitudes depend on the positions of the particles in the cavity, while the oscillation frequencies are the same everywhere in the fluid.
- The effect of the variation of the amplitudes of oscillation parameters has an effect on the transfer rate.
- The rate of transfer of liquid metals is important with respect to large values of the number of Prandlt such as air, it can be concluded that the liquid metal have a significant calorific value is for cooling as for heating.

References

1. A.A. Mohamad and R. Viskanta, 'Modeling of Turbulent Buoyant Flow and Heat Transfer in Liquid Metals', International Journal of Heat and Mass Transfer, Vol. 36, N°11, pp. 2815 –2826, 1993.
2. M.J. Stewart and F. Weinberg, "Fluid Flow in Liquid Metals I. Theoretical Analysis", J. Crystal Growth, 12, 217-227, (1972).
3. F. Wolff and al, "Natural convection of liquid metals cavities "Experimental Thermal Fluid Science,1, 83-91 (1988).K. Elissa, "Title of paper if known" unpublished.
4. R. Derbail and J. N. Koster, "Numerical Simulation of Natural Convection of Gallium in a Narrow Gap", Int J. Heat Mass Transfer, 40, 1169-1180, (1997).
5. S.V. Patankar, "Numerical heat transfer and fluid flow", Hemisphere, Washington DC, (1980).
6. G..Barakos and E.Mitsoulis, "Natural Convection Flow in a Square cavity Revised: Laminar and Turbulent models with wall Functions", International journal for numerical methods in fluids, Vol.18, 695-719, (1994).
7. R. J. Krane and J. Jessee, "Some Detailed Field Measurements for a Natural Convection Flow in a Vertical Square Enclosure" Proc. Isr Asme-Jsme Thermal Engineering Joint Conf, Vol. 1, pp. 323-329, (1983).

8. J. M. Aurnou and P. L. Olson, "Experiments on ayleigh-Bénard convection, magnetoconvection and rotating magnetoconvection in liquid gallium." J. Fluid Mech, 30, 283-307 (2001).
9. D. A. Kramer, "Gallium and Gallium compounds" dans "Metals Handbook" – volume 2: Properties and Selection: Nonferrous Alloys and Special-Purpose Materials, Edited by ASM International Handbook Committee, 740-749, tenth edition (1990).
10. R. Viskanta, D.M. Kim and C. Gau, "Three-dimensional natural convection heat transfer of a liquid metal in a cavity", Int. J.Heat Mass Transfer, 29, 475-485 (1986).
11. F. Wolff, C. Beckermann and R. Viskanta, "Natural convection of liquid metals in vertical cavities", Experimental Thermal and Fluid Science, 1, 83-91 (1988).
12. M. G. Braunsfurthi, A. C. Skeldon, A. Juel, T. Mullin and D. S.Riley "Free convection in liquid gallium", J. Fluid Mech, 342, 295-314. (1997).
13. A. Juel, T.Mullin, H. Ben Hadid and D. Henry, "Threedimensional free convection in molten gallium", J. Fluid Mech, 436,267-281 (2001).
14. Di Piazza and M. Ciofalo, "MHD free convection in a liquid-metal fielled cubic enclosure. I. Differential heating" Int. J.Heat Mass Transfer, 45, 1477–1492 (2002).
15. Di Piazza and M. Ciofalo, "MHD free convection in a liquid-metal fielled cubic enclosure. II. Internal heating" Int. J. Heat Mass Transfer, 45, 1493–1511 (2002).

EPJ Web of Conferences **114**, 02078 (2016)
DOI: 10.1051/epjconf/201611402078

Comparison of Y-jet and OIL effervescent atomizers based on internal and external two-phase flow characteristics

Marek Mlkvik[1,a], Matous Zaremba[2], Jan Jedelsky[2] and Miroslav Jicha[2]

[1]*FME STU in Bratislava, Nam. Slobody 17, 81231 Bratislava, Slovak Republic*
[2]*FSI BUT, Technicka 2896/2, 616 69 Brno, Czech Republic*

Abstract. Presented paper focuses on spraying of two viscous liquids (μ = 60 and 143 mPa·s) by two types of twin-fluid atomizers with internal mixing. We compared the well-known Y-jet atomizer with the less known, "outside in liquid" (OIL), configuration of the effervescent atomizer. The required liquid viscosity was achieved by using the water-maltodextrin solutions of different concentrations. Both the liquids were sprayed at two gas inlet pressures (Δp = 0.14 and 0.28 MPa) and various gas-to-liquid ratios (GLR = 2.5%, 5%, 10% and 20%). The comparison was focused on four characteristics: liquid flow-rate (for the same working regimes, defined by Δp and GLR), internal flow regimes, Weber numbers of a liquid breakup (We) and droplet sizes. A high-speed camera and Malvern Spraytec laser diffraction system were used to obtain necessary experimental data. Comparing the results of our experiments, we can state that for both the liquids the OIL atomizer reached higher liquid flow-rates at corresponding working regimes, it was typical by annular internal flow and higher We in the near-nozzle region at all the working regimes. As a result, it produced considerably smaller droplets than the second tested atomizing device, especially for GLR < 10%.

1 Introduction

Twin-fluid atomizers can be found in various industrial applications such as internal combustion and gas turbine engines [1], process industry [2], spray coating [3], fire suppression [4] or food process engineering and spray drying [5].

They are favored for their capability to process high-viscous liquids such as heavy fuels [6] or suspensions used in food processing industry. These atomizers also provide easy control of the individual spray parameters with lower atomizing gas consumption and lower input pressures than their externally mixed counterparts.

In this paper, we decided to provide a comparison of the Y-jet atomizer and OIL atomizer as the reaction to the lack of available systematic atomizers comparisons. Selected spraying devices differ in the design of their internal parts and mixing principles. Our aim was to investigate differences in the internal and external flows for spraying liquids of different viscosity under the same working parameters.

2 Experiment

Our experiments were performed on cold test bench equipped with eccentric screw pump at constant rotation speed used to drive the liquid into the atomizer. The liquid mass flow was controlled by a valve in bypass line. The atomizing gas (air) was provided by a house supply, and its pressure was controlled by pressure relief valve. This regulation was independent on the air mass flow. The test rig was also equipped with sensors and data acquisition system for measurements of flow parameters. All the experiments were performed at the room temperature.

The atomizers were operated under two input pressures of air (Δp = 0.14 and 0.28 MPa) and four ratios of gas and liquid mass flows (GLR = 2.5, 5, 10 and 20%).

The required liquid viscosity was achieved by using water–maltodextrin solutions of different concentrations (Table 1).

The high-speed camera (Olympus i-speed2) equipped with PENTAX TV lens (50 mm, fl:1.4) with extension rings of a total length of 25 mm was used to observe primary breakup of liquid in the near-nozzle area.

Droplet sizes were measured 100 mm downstream of the discharge orifice using a Malvern Spraytec laser diffraction system working on frequency 500 Hz [7].

Table 1. Physical properties of used water–maltodextrin solutions

Label	Maltodextrin powder concentration [%]	μ [mPa·s]	σ [mN/m]	ρ [kg/m^3]
liq.1	40	60	74.54	1185
liq.2	45	143	74.26	1121

* μ – viscosity, σ – surface tension, ρ – density

[a] Corresponding author: marek.mlkvik@stuba.sk

2.1 Atomizers

Investigated atomizers (Figure 1.) have different internal geometry, which leads to the different internal flows for the same working regimes and consequently to the different spray characteristics such as the primary breakup Weber number or droplet sizes.

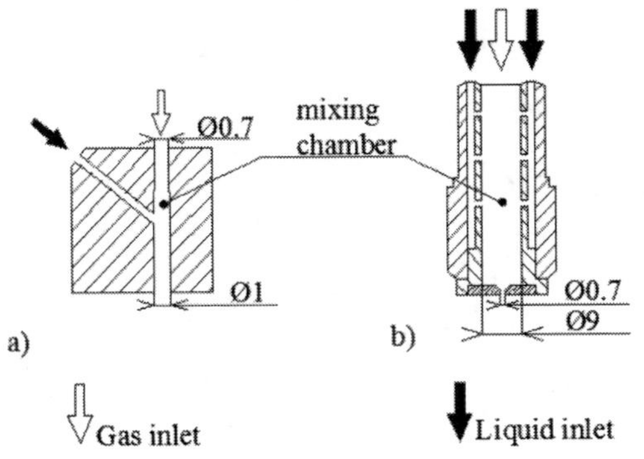

Figure 1. Schematic drawings of investigated atomizers

The mixing chamber of our Y-jet atomizer (Figure 1a) was designed according to the recommendations of Mullinger [8], and its cross-section diameter is 1 mm. The working liquid is injected under the angle 52^0 to the atomizer axis through a liquid injection port of diameter 0.4 mm. The air is injected in the axial direction from the top of the mixing chamber.

The mixing chamber of the OIL effervescent atomizer (Figure. 1b.) has a cylindrical shape with cross-section diameter of 9 mm. The internal two-phase flow is created by radial injection of the working liquid into the axial gas stream through 20 injectors with diameter 1 mm. The two-phase mixture is discharged through the orifice of diameter 0.7 mm.

The hydraulic resistance of internal design was judged by comparison of liquid flow-rates for corresponding Δp and GLR. The influence of atomizers design on the internal two-phase flow was also investigated. The internal flow pattern was determined using liquid-to-gas momentum ratio ($\Phi = m_l^2 \cdot d_l^2 \cdot \rho_g \cdot \sin\Theta / (m_g^2 \cdot d_g^2 \cdot \rho_l)$, originally used by Song for Y-jet nozzles [10]. The meaning of symbols in equation for Φ is the mass flux per surface unit, density, port diameter, and intersecting angle for "m", "ρ", "d" and "Φ". The indes "l " and " g " denote the liquid or gas phases. The Weber number ($We = \rho \cdot w^2 \cdot D / \sigma$, "w" denotes the liquid-to-gas velocity difference, where "D" stands for the ligament diameter) of observed near-nozzle flows were calculated and used for comparison of primary breakup regimes of the atomizers. Finally, the spray quality was related to the measured Sauter mean diameter of the droplets.

3 Results

In this section, we provide a comparison of atomizers based on the four characteristics: liquid flow-rate (for the same working regimes), internal flows, Weber number of liquid breakup and droplet sizes.

3.1 Internal flow: liquid flow-rates and flow pattern

The atomizers were designed to work with the same liquid flow-rates at identical working regimes, but the measured data (Figure 2.) show that the real mass flows were significantly lower for the Y-jet atomizer. This difference wasmore significant for lower Δp and GLR < 10% for the more viscous liquid. Considering the dimensions of atomizers mixing chambers, the maximum mixture velocity, calculated from maximum measured mass flows, reached 40 m/s for the Y jet atomizer, while it reached only 0.6 m/s for the OIL atomizer. The friction pressure loss was therefore dominant in the case of the Y-jet atomizer. On the other hand, the local pressure loss was significant at the discharge orifice of the OIL atomizer. It was caused by the narrow cross-section reduction (from 9 mm to 0.7 mm) while it was less significant for the Y-jet atomizer. We thus presume that the mass flow difference of the atomizers was caused by the above-described relevance of local (viscosity independent) and friction (viscosity dependent, [9]) pressure losses.

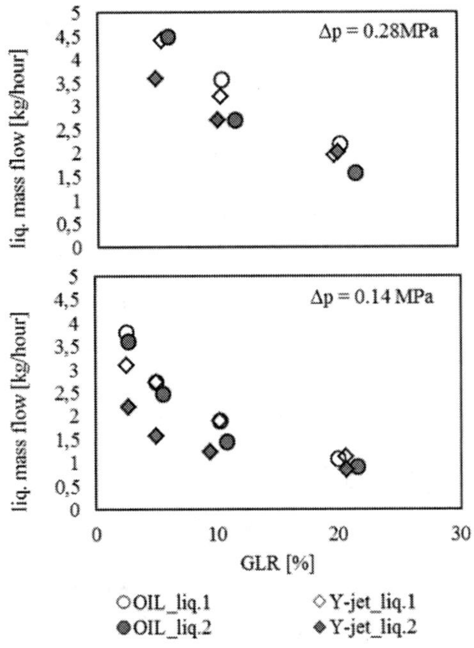

Figure 2. Measured mass flows of liquid

The design of the mixing chamber also influenced the internal flow pattern. Because the mixing principle for both the atomizers was the same — injection of liquid into the gas stream — we could estimate the internal flow patterns by the parameter Φ [10]. Figures 3 and 4 show, that the internal flow patterns were different for the corresponding working regimes. The main difference between the atomizers was that the internal flow of the OIL atomizer was annular amongst all measured regimes.

The gas velocity in the mixing chamber of the OIL atomizer was low, sothe annular liquid flow was not accelerated intensively, but the flow pattern was stable which led to the production of stable spray [7]. In the Y-jet atomizer the liquid could reach the central portion of the mixing chamber. As a result, it was intensively accelerated by the surrounding high-velocity gas. The gas-to-liquid interaction also caused the flow instabilities as referred by Song [10] and decreased the spray stability [7].

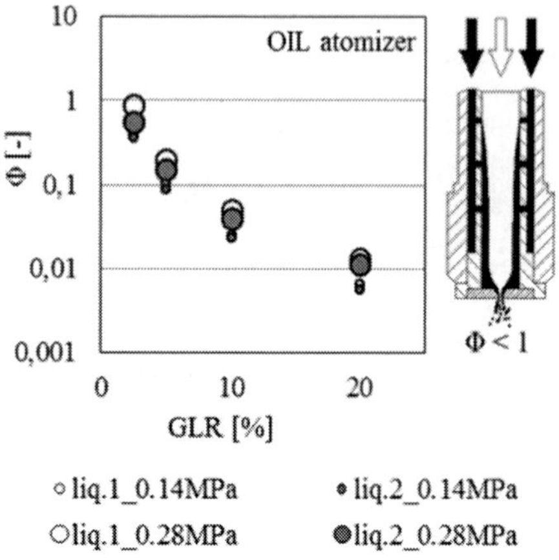

Figure 3. Estimated internal flow of the OIL atomizer

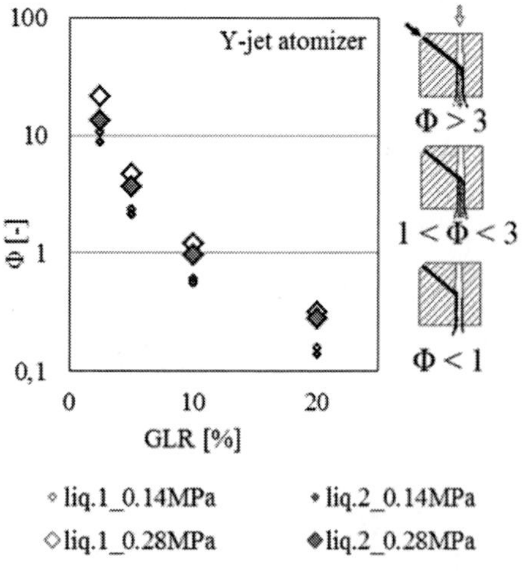

Figure 4. Estimated internal flows of the Y-jet atomizer

3.2 External flow: liquid breakup and droplet sizes

The liquid discharge of the Y-jet atomizer was characterized by low We (Figure 5). The liquid acceleration in the mixing chamber decreased the liquid-to-gas velocity slip in the discharge area. Therefore, σ became the dominant force of the droplet deformation, which caused a number of spherical droplets observed in the near-nozzle flow of the Y-jet atomizer [7]. The We observed for the OIL atomizer were considerably larger regardless to the working regime or liquid viscosity. In this case, the air drag force deformed the liquid into fibers and potentially disrupted the large liquid structures [7].

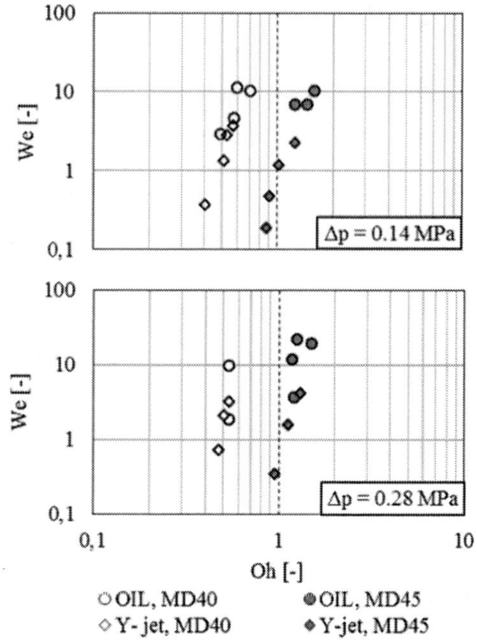

Figure 5. Weber's numbers of the near-nozzle flows

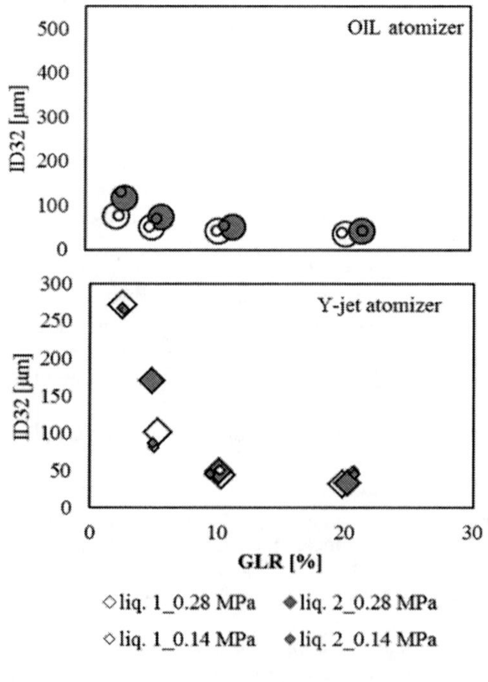

Figure 6. Measured ID32 in distance 100 mm from the discharge orifice

The liquid structures, observed in the near-nozzle region, were further fragmented and their size was reduced until they reached the stable spherical shape as the effect of surface tension. These droplets were observed in the distance 100 mm from the discharge orifice and their size were measured by the laser diffraction system. As shown in Figure 6, the general trend for both the atomizers shows a decrease of ID32 with increasing GLR, which accords to the number of previous works [11, 12, 13]. The Y-jet atomizer produced spray with ID32 < 50 μm for GLR = 10 and 20% for both the liquids, which can be considered as good quality for such viscous liquids. The ID32 rapidly increased up to 270 μm with GLR reduction to 5 and 2.5% respectively. The OIL atomizer produced spray with smaller droplets for low GLRs. The ID32 rose up at lower GLRs, but the increase was not as rapid as in previous case. This atomizer also showed higher sensitivity to liquid viscosity, especially for low GLRs.

Conclusions

We studied flow and spray characteristics of a Y-jet atomizer in comparison with OIL effervescent atomizer. Our tests demonstrate a reduction of liquid flow-rate for the Y-jet atomizer for the corresponding working regimes as the result of the atomizer mixing chamber design. The slow mixture motion in the OIL atomizer mixing chamber and its rapid acceleration in the discharge orifice produced a local (nearly viscosity) pressure loss. The pressure loss of the Y-jet atomizer was produced mainly by liquid internal friction, due to the high velocity of the internal flow.

The geometry of the mixing chamber also influenced the internal flow pattern of the atomizers. We estimated and compared internal two-phase flow types of investigated atomizers and we found that the internal flow of the OIL atomizer was annular for all measured regimes, while three types of internal flow patterns were identified for the Y-jet atomizer over the range of tested working regimes. The main influence of the internal flow regime on the atomizer performance can be found in the spray stability as published in our previous works.

The internal flow also influenced the near-nozzle flow and primary breakup of the liquid. The liquid discharge of the Y-jet atomizer was characterized by the low We, which was caused by the liquid acceleration in the mixing chamber. The We observed for the OIL atomizer were considerably larger for all of the measured cases.

Different conditions of the near-nozzle flow affected the droplet sizes of the final spray. The Y-jet atomizer produced considerably larger droplets as the result of the low effect of the air drag force in the near-nozzle region. The OIL atomizer, characterized by larger We number of the primary breakup, produced smaller droplets, especially for the low GLRs

Considering the previous analysis, we can state that the OIL atomizer provided better results for all investigated working regimes. It reached higher flow-rates at corresponding Δp and it also produced smaller droplets even for low GLRs. Thus, it fulfilled the task to atomize viscous liquids more effectively than the investigated Y-jet atomizer (for given range of liquid viscosities and working regimes).

Acknowledgements

This work has been supported by the project No. GA15-09040S funded by the Czech Science Foundation, the project LO1202 NETME CENTRE PLUS with the financial support from the Ministry of Education, Youth and Sports of the Czech Republic under the "National Sustainability Programme I" and project Reg. No. FSI-S-14-2355 funded by the Brno University of Technology.

The authors also appreciate support from Prof. Dr.-Ing. Heike P. Schuchmann, Dr. rer. nat. Dipl.-Ing. Volker Gaukel and Dipl.-Ing Philipp Stähle from Karlsruhe Institute of Technology for providing us their test rig for experiments and other support and cooperation.

References

1. A.H. Lefebvre, 1988. Indian Defence Sci. J. **38**, (1988)
2. D. Loebker, H.J. Empie, Proc.10th Annual Conference on Liquid Atomization and Spray Systems, pp. 253–257, (1997)
3. L.J. Qian, J.Z. Lin, H.B. Xiong, Int. J. Therm. Sci. **50**, 1417–1427, (2011)
4. X. Huang, X.S. Wang, G.X. Liao, Proc. Combust. Inst. **33**,pp. 2573–2579, (2011)
5. P. Stähle, V. Gaukel, H.P. Schuchmann, Food Research International, (2015)
6. J. Jedelsky, M. Jicha, J. Slama, J. Otahal, Energy Fuels **23**, pp. 6121–6130, (2009)
7. M. Mlkvik, P. Stähle, H.P. Schuchmann, V. Gaukel, J. Jedelsky, M. Jicha, Int. J. Multiphase Flow **77**, pp. 19-31, (2015)
8. P. J. Mullinger, N.A. Chigier, J. Inst. Fuel **47**, pp. 251-261 (1974)
9. S. K. Chen, A. H. Lefebvre, Atomization sprays **4**, pp. 275-290, (1994)
10. S.H. Song, S.Y. Lee, Atomization Sprays **6**, pp. 193–209, (1996)
11. J.D. Whitlow, A.H. Lefebvre, Atomization Sprays **3**, pp.137–155, (1993)
12. E. Sher, M. Koren, D. Katoshewski, V. Kholmer, Proc. ILASS-Europe, Paper II.7, (2000)
13. S. Sovani, P. Sojka, A. Lefebvre, Prog. Energ. Combust. **27**(4), (2001)

EPJ Web of Conferences **114**, 02079 (2016)
DOI: 10.1051/epjconf/201611402079

Optimization of hydraulic turbine diffuser

Prokop Moravec[1, a], Juraj Hliník[2] and Pavel Rudolf[1]

[1] *Victor Kaplan Dept. of Fluid Engineering, Faculty of Mechanical Engineering, Brno University of Technology, Czech Republic*
[2] *Dept. of Mathematical Engineering, Faculty of Mechanical Engineering, Brno University of Technology, Czech Republic*

Abstract. Hydraulic turbine diffuser recovers pressure energy from residual kinetic energy on turbine runner outlet. Efficiency of this process is especially important for high specific speed turbines, where almost 50% of available head is utilized within diffuser. Magnitude of the coefficient of pressure recovery can be significantly influenced by designing its proper shape. Present paper focuses on mathematical shape optimization method coupled with CFD. First method is based on direct search Nelder-Mead algorithm, while the second method employs adjoint solver and morphing. Results obtained with both methods are discussed and their advantages/disadvantages summarized.

1 Introduction

Swirl turbine [1] is suited for low heads and relatively large discharges. Almost a half of the available head should be exploited in the draft tube by recovery of kinetic energy into pressure energy. Velocity field in the draft tube is characterized by swirl component on its inlet. This imposes stress on the draft tube performance [2].

Figure 1. Swirl turbine [2]

Liquid, which flows through draft tube (diffuser), is affected by hydraulic losses. These losses are undesirable and there is an effort to minimize them. The term minimization means process of finding the minimum of examined function. In our case the examined function is coefficient of pressure recovery c_p, defined:

$$c_p = \frac{p_2 - p_1}{\frac{1}{2}\rho v_1^2}, \qquad (1)$$

where p_1 is static pressure at inlet, p_2 is static pressure at outlet; ρ is density of liquid, v_1 is stream velocity at inlet.

2 Nelder-Mead method

Nelder-mead method, known as a simplex method, is a very popular method that does not use derive of the examined functions.

A method is described for the minimization of a function of n variables, which depends on the comparison of function values at the $(n+1)$ vertices of a general simplex, followed by the replacement of the vertex with the highest value by another point. The simplex adapts itself to the local landscape, and contracts on to the final minimum. The method is shown to be effective and computationally compact [3].

Example of Nelder-Mead method: finding a global minimum of function:

$$f(x_1, x_2) = x_1^2 + x_2^2 - 3x_1 - x_1x_2 + 3\,.$$

Global minimum lies in $x = (2, 1)$. The path of simplex is shown in Figure 2.

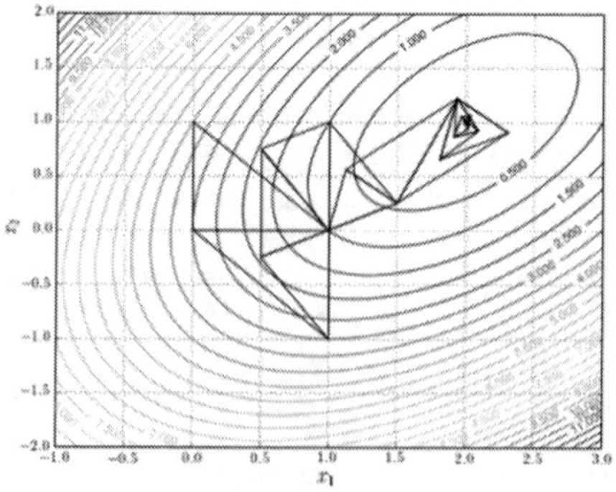

Figure 2. The path of simplex in Nelder-Mead algorithm [4]

[a] Corresponding author: Prokop.Moravec@seznam.cz

2.1 Nelder-Mead algorithm

Nelder-Mead algorithm consists of several main steps [5]:

2.1.1 Choice of initial simplex and parameters

The first step of the algorithm is a creation of regular simplex with side length *a*, using following definition:

$$x_j = x_1 + pe_j + \sum_{k \neq j} qe_k \,, \tag{2}$$

where x_1 is specified vertex, $\boldsymbol{e}_k$ are unitary basis vectors and

$$p = \frac{a}{n\sqrt{2}}\left(\sqrt{n+1} + n - 1\right);$$

$$q = \frac{a}{n\sqrt{2}}\left(\sqrt{n+1} - 1\right),$$

where n is a dimension. Important part of algorithm is also an ending tolerance criterion:

$$\left(\frac{1}{n+1}\sum_{i=1}^{n+1}[f_i - f(\bar{x})]^2\right)^2 < \epsilon \,, \tag{3}$$

where ε is required precision (tolerance).

Algorithm consists of four basic operations that depend on chosen parameters:

- Reflection $\rho > 0$,
- Expansion $\eta > 1$, $\eta > \rho$,
- Contraction $0 < \gamma < 1$,
- Reduction $0 < \sigma < 1$.

Parameter values are often chosen as follows:

$$\rho = 1;\ \eta = 2;\ \gamma = 0.5;\ \sigma = 0.5\,.$$

2.1.2 Vertices ordering

The next step involves ordering according to the values of objective function at vertices:

$$f_1 = f(x_1) \leq \cdots \leq f_{n+1} = f(x_{n+1}) \,,$$

after that comes control of tolerance criterion, if epsilon is reached, the end of algorithm will occur.

2.1.3 Reflection

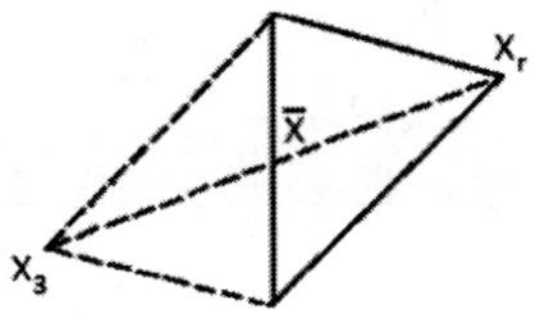

Figure 3. Reflection [5]

First, we must calculate $\bar{x}$ – the centre of gravity of the centroid (Figure 3):

$$\bar{x} = \frac{1}{n}\sum_{i=1}^{n} x_i \,, \tag{4}$$

second, coordinates of reflected vertex x_r are computed (Figure 3):

$$x_r = \bar{x} + \rho(\bar{x} - x_{n+1}) \,, \tag{5}$$

finally, if $f_1 \leq f_r \leq f_n$, then vertex x_r is accepted as replacement of x_{n+1}. Also function value in x_r is calculated $f_r = f(x_r)$.

2.1.4 Expansion

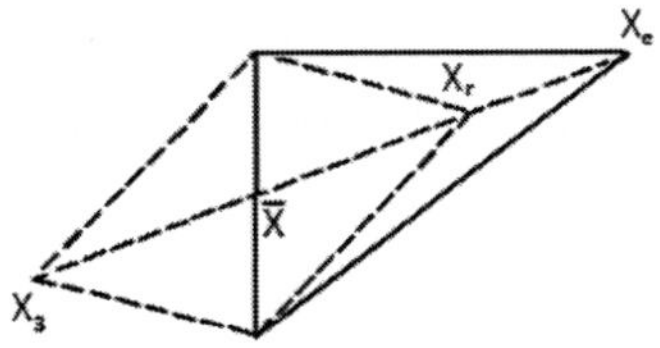

Figure 4. Expansion [5]

Expansion only takes place when $f_r \leq f_1$. Then coordinates of expanded vertex are computed:

$$x_e = \bar{x} + \eta(x_r - \bar{x}) \,, \tag{6}$$

finally, if $f_e < f_r$ then vertex x_e is accepted as replacement of x_{n+1} (else x_r is). Also function value in x_e is calculated $f_e = f(x_e)$.

2.1.5 Contraction

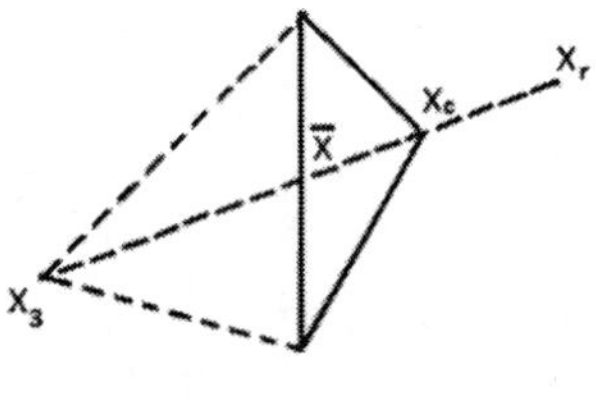

Figure 5. Contraction [5]

Contraction occurs when a given condition $f_r \geq f_n$ is satisfied. Depending on which vertex is better suited (x_r or x_{n+1}), comes external or internal contraction.

Coordinates of internally contracted vertex are computed (if $f_r \geq f_{n+1}$):

$$x_{ci} = \bar{x} + \gamma(x_{n+1} - \bar{x}) \,, \tag{7}$$

then if $f_{ci} < f_{n+1}$, computed vertex x_{ci} is accepted as replacement of x_{n+1}, else coordinates of externally contracted vertex (if $f_r < f_{n+1}$) are computed:

$$x_{ce} = \bar{x} + \gamma(x_r - \bar{x}) \,, \tag{8}$$

then if $f_{ce} < f_r$, computed vertex x_{ce} is accepted as replacement of x_{n+1}. Also function value in x_{ci} or x_{ce} is calculated $f_{ci} = f(x_{ci})$ or $f_{ce} = f(x_{ce})$.

2.1.6 Reduction

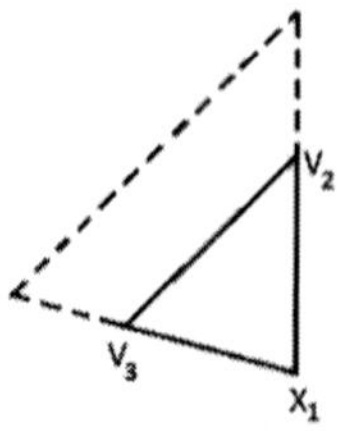

Figure 6. Reduction [5]

In this step of Nelder-Mead algorithm is an edge reduction of initial simplex:

$$v_i = x_1 + \sigma(x_i - x_1) \quad i = 1,..,n+1\,, \tag{9}$$

where v_i, ..., v_{n+1} ($v_1 = x_1$) are vertices of a new reduced simplex (Figure 6.).

Notice: After each step of Nelder-Mead algorithm the control of tolerance criterion must be checked. If is not satisfied, algorithm returns to step called vertices ordering.

3 Adjoint solver

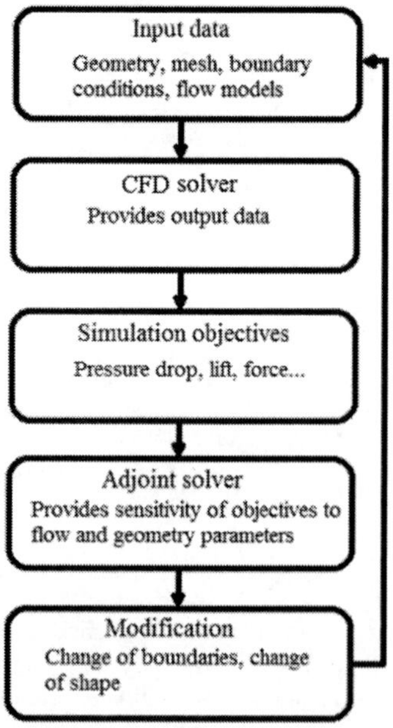

Figure 7. Adjoint solver optimization approach

Adjoint solver is a specialized tool that extends the scope of the analysis provided by a conventional flow solver by providing detailed sensitivity data for the performance of a fluid system [7].

Optimization via adjoint solver consists of several main steps – Figure 7.

There are two types of adjoint approaches – continuous and discrete. Ansys Fluent uses only the discrete one.

3.1 Discrete adjoint approach theory

A detailed description of discrete adjoint approach may be found in [6].

The adjoint solver needs a fully converged flow simulation, from which the gradients of objective function are derived.

The gradient of objective function $J = J[X_D, Q(X_D)]$ is [6]:

$$G = \frac{dJ}{dX_D} = \frac{\partial J}{\partial X_D} + \frac{\partial J}{\partial Q}\frac{dQ}{dX_D}, \tag{10}$$

where X_D is vector of design variables (in [6] is reduced to scalar) and fraction $\frac{dQ}{dX_D}$ is referred as flow sensitivities (Q is vector of conserved flow variables).

This gradient could be rewritten in form [6]:

$$G = \frac{\partial J}{\partial X_D} - \psi^T \frac{\partial R}{\partial X_D}, \tag{11}$$

where $R = R[X_D, Q(X_D)]$ is the discretized RANS residual vector, ψ represents vector of adjoint variables and is obtained by solving the linear system of equations (known as adjoint equation) [6]:

$$\frac{\partial R^T}{\partial Q}\psi = \frac{\partial J^T}{\partial Q}. \tag{12}$$

4 Overview

Initial diffuser design which was used in mathematical optimization methods (Nelder-Mead, adjoint solver) is shown in Figure 8.

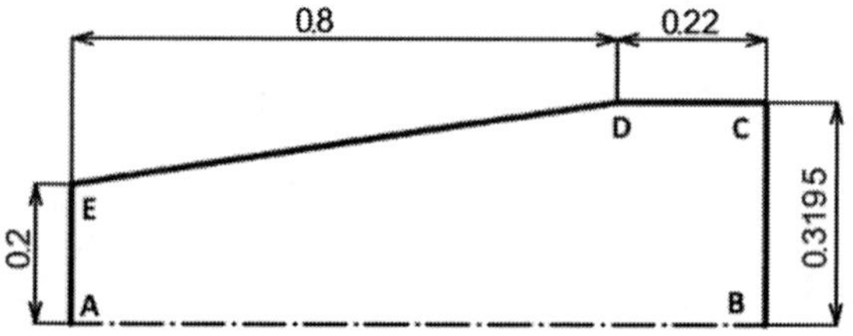

Figure 8. Domain design (dimensions are in metres) [8]

Table 1. displays types of boundary conditions, their locations, magnitudes and turbulence intensity values.

Table 1. Boundary conditions 2D

	Type	Magnitude	Turbulent intensity
A-E	Velocity inlet	2 m/s (constant)	5 %
B-C	Pressure outlet	0 Pa	7 %
A-B	Axis	-	-
E-D-C	Wall	-	-

Notice: Value of velocity at inlet is constant and does not match with real velocity profile behind the runner of swirl turbine.

For initial (Figure 8.) design was created computational mesh (Figure. 9). Value of pressure recovery coefficient for this initial shape diffuser design was **0.736824**.

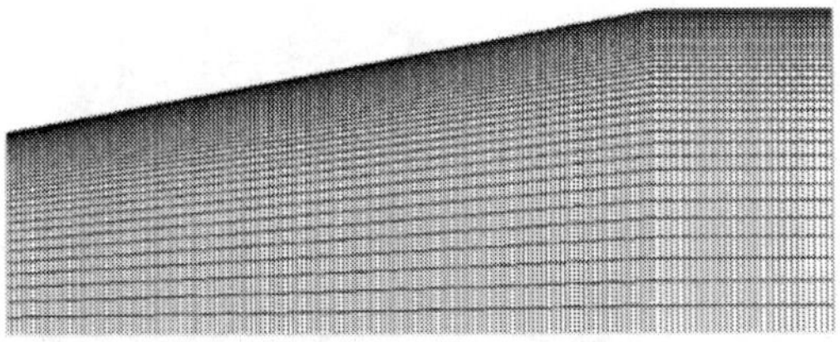

Figure 9. Initial 2D domain mesh

Complete setting of the chosen discretization schemes is listed in the following chart.

Pressure:	Standard
Momentum:	Quick
Turbulent diss. rate:	Second order upwind
Turbulent kinetic energy:	Second order upwind

4.1 Nelder-Mead method [8]

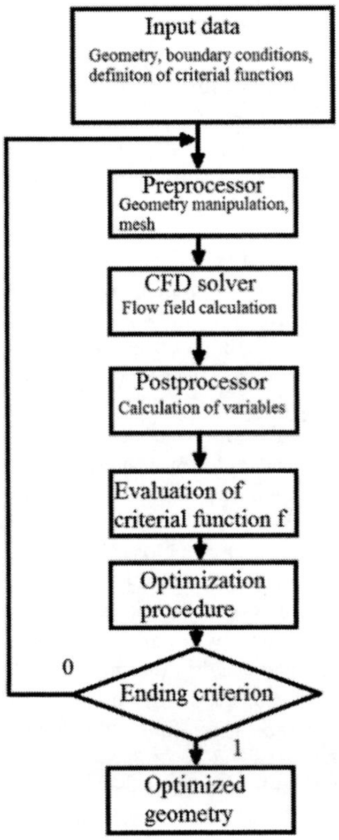

Figure 10. Algorithm of Nelder-Mead optimization [11]

Shape optimization using Nelder-Mead algorithm consists of several steps (Figure 10.).

Very important part is the first step, in which we define the criterial function (c_p in our case), geometry (Figure 8.) and boundary conditions (Table 1.).

Preprocessing involves creation of geometry and computational mesh. Everything is handled in Gambit software using programmed script.

The next two steps – CFD solver and postprocessor – take place in Ansys Fluent, which also uses programmed script. Results (Fluent outcome) are later saved in text files due to easier access to data of examined variables.

The rest of the loop is realized in software called Matlab (including script calling from Gambit and Fluent).

Notice: The geometry change occurs only between vertices *E* and *D*.

4.1.1 Piecewise linear approach

Initial simplex D_i was chosen near vertex *D*. After 75 iterations of Nelder-Mead algorithm, the value of c_p coefficient stabilized at **0.807863**. Coordinates of simplex D_i changed to

$$D_i = [0.0686020; 0.1973931] .$$

Figure 11. shows the change of diffuser geometry (red line).

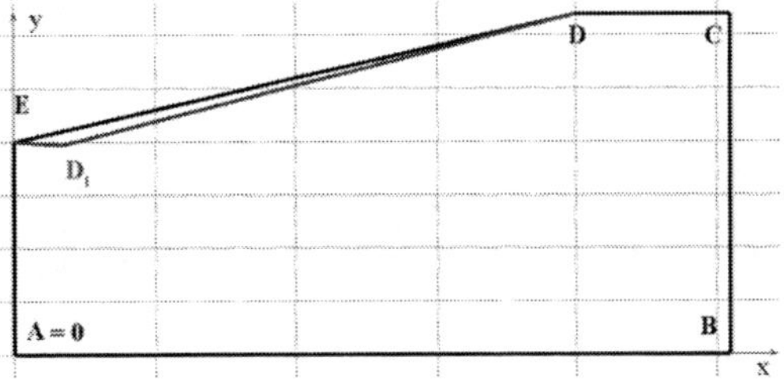

Figure 11. Piecewise linear approach - diffuser geometry

4.1.2 Optimization using Bezier curves

In the next step of optimization, Bezier curves were used as restriction of wall *E-D*.

<u>First order Bezier curve</u>

First order Bezier curve is characterized only by one point of control polygon - P_1.

Number of iterations in this approach was 52. Value of coefficient of pressure recovery c_p reached up to **0.810309**. Coordinates of control point P_1 were computed during Nelder-Mead algorithm as follows:

$$P_1 = [0.138172; 0.185848] .$$

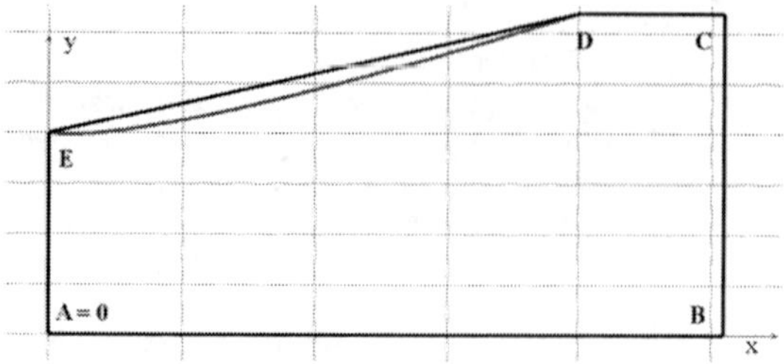

Figure 12. First order Bezier curve – diffuser geometry

Second order Bezier curve

Second order Bezier curve is characterized by two points of control polygon - P_1, P_2.

Number of iterations was 118. Value of coefficient of pressure recovery c_p reached up to **0.812146**. Coordinates of control points P_1 and P_2 were computed during Nelder-Mead algorithm as follows:

$$P_1 = [0.1725671; 0.1814290] ;$$

$$P_2 = [0.5137642; 0.3125861] .$$

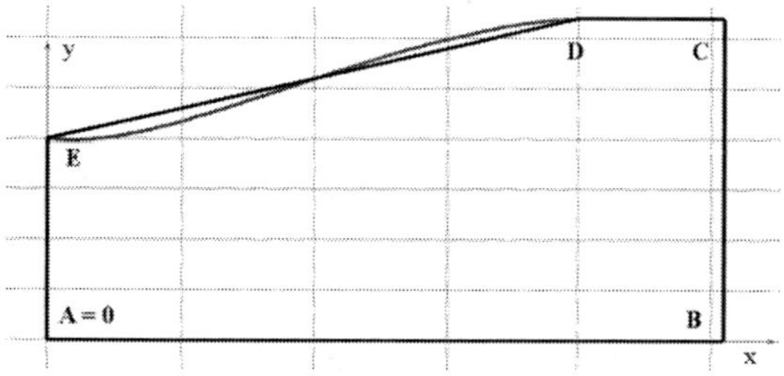

Figure 13. Second order Bezier curve– diffuser geometry

Third order Bezier curve

Third order Bezier curve is characterized by three points of control polygon - P_1, P_2 and P_3.

Number of iterations in this approach was 78. Value of coefficient of pressure recovery c_p reached up to **0.813706**. Coordinates of control points P_1, P_2 and P_3 were computed as follows:

$$P_1 = [0.128563; 0.237429] ;$$

$$P_2 = [0.186375; 0.186554] ;$$

$$P_3 = [0.427714; 0.270078] .$$

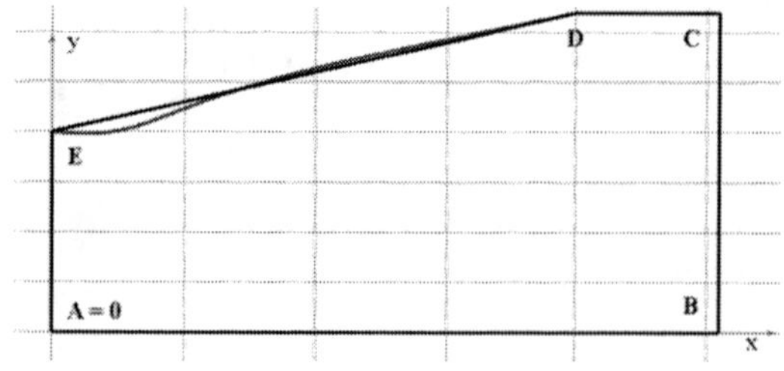

Figure 14. Third order Bezier curve – diffuser geometry

4.2 Adjoint solver

Adjoint solver is incompatible with an axis boundary condition in Fluent. Due to this fact, the working computational domain was created entirely in 3D (Figure 15.). After optimization in 3D, shape of diffuser was transformed back into 2D.

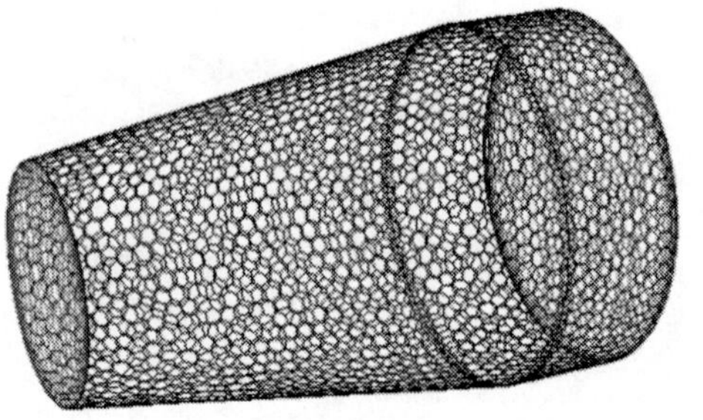

Figure 15. Initial 3D domain mesh

In [10] author recommends polyhedral type of mesh (Figure 16.) to ensure better mesh behaviour during mesh morphing.

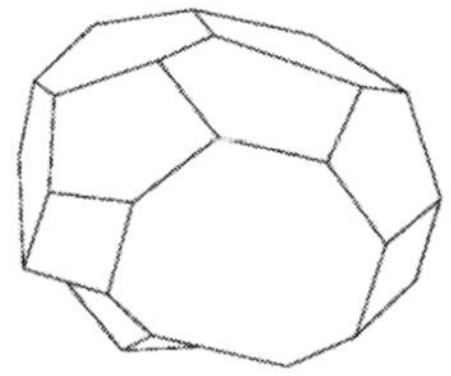

Figure 16. Polyhedral cell [9]

Types of boundary conditions, their locations, magnitudes and turbulence intensity values used in Fluent software during optimization via adjoint solver are the same as in Table 1.

Simulation objective of adjoint solver was set to minimize static pressure at inlet, which is consistent with minimization of c_p.

The next step of adjoint solver optimization approach (Figure 7.) involves shape modification – mesh morphing. Morphing takes place, when sensitivity field is computed (through adjoint equations). This morphing procedure has a double role. Firstly, it smooths the surface sensitivity field and secondly provides smooth distortion in boundary and interior mesh [10]. In adjoint solver rectangular volume encloses boundaries which will be morphed. An array of control points is then distributed in this volume. These points in combination with the local mesh coordinates will define the proper movement both on the boundary and in the interior mesh. The local coordinate system relies on the properties of Bernstein polynomials [10]. Bernstein polynomials are the basis of Bezier curves.

The most important information, which adjoint solver offers are optimal displacements (Figure 17.). Vectors of optimal displacement show location and intensity of diffuser shape change.

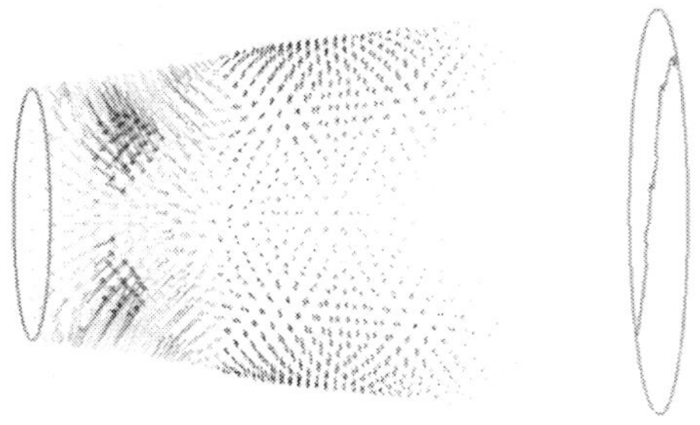

Figure 17. Optimal displacements - vectors

After 30 loops optimization was stopped, 3D geometry was transformed back to 2D and 2D geometry was recalculated using the same CFD settings as in Nelder-Mead algorithm. Coefficient of pressure recovery in this case reached up to value **0.812354** (22. loop – Figure 18.).

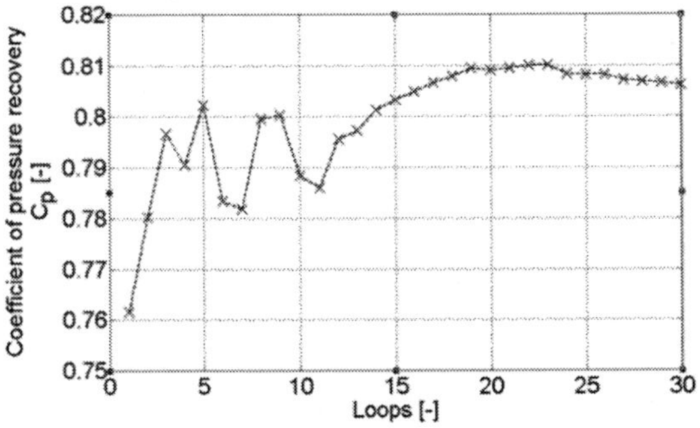

Figure 18. Adjoint solver approach

Due to the absence of better restriction of shape (for example: using only the first order Bezier curve as a wall restriction) vertices of changed wall were later approximated in Excel using the 4. order polynomial (Figure 19.). Coefficient of pressure recovery changed to **0.812614**.

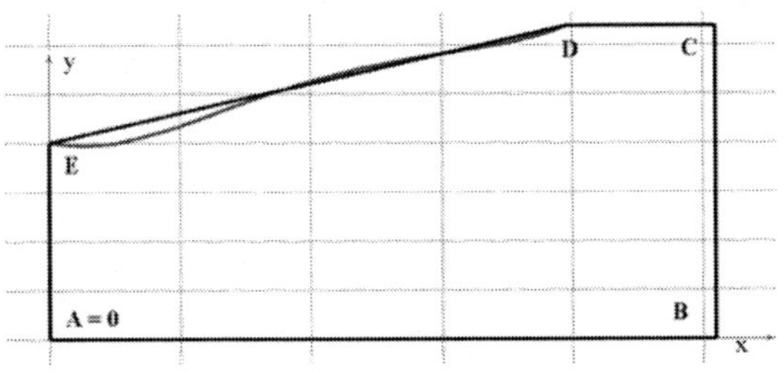

Figure 19. Adjoint solver – 4. order polynomial approx. – diffuser geometry

5 Conclusions

The values of c_p obtained in both optimization methods – Nelder-Mead and adjoint solver - are summarized in Table 2.

Optimized shape of diffuser via Nelder-Mead method has 10.43 % higher pressure recovery than the original (initial) shape. This value was achieved through the use of 3. order Bezier curve as a wall constraint. The benefits of this method include: achieved value of c_p and ability to control changed boundaries. Disadvantages include: the need to use additional software (Matlab) and the higher number of loops to achieve the final value of c_p.

Optimized shape of diffuser via adjoint solver in Fluent has 10.25 % higher pressure recovery than the original (initial) shape, but this shape had extensively corrugated boundary. Problem was solved by polynomial approximation of vertices that form the diffuser. Value of c_p increased by 0.04 %. The main advantages of this method are: the implementation in Ansys Fluent and shape modification is also present in the solver. Disadvantages include mainly the absence of better restriction of changed boundaries and a large number of settings, which adjoint solver uses.

Comparison of execution time for both methods has not been done due to use of two different computers (clusters).

It should be also mentioned that the value of velocity at inlet was constant and does not correspond with the real velocity profile behind the runner of swirl turbine.

Coefficient of pressure recovery c_p diffuser increases as a result of weaker velocity gradients and shear layers (it was also confirmed by theoretical analysis [11]).

Table 2. Overview

	c_p [-]	Change [%]
Initial geometry	0.736824	-
Nelder-Mead – Piecewise linear approach	0.807863	9.64
Nelder-Mead – Bezier curve 1. order	0.810309	9.97
Nelder-Mead – Bezier curve 2. order	0.812146	10.22
Nelder-Mead – Bezier curve 3. order	0.813706	10.43
Adjoint solver	0.812354	10.25
Adjoint solver – 4. order polynomial approximation	0.812614	10.29

References

1. F. Pochylý, M. Haluza, P. Rudolf, F. Šob: Design of a New Low-Head Turbine. In Proceedings of the XXIst Symposium on Hydraulic Machinery and Systems. September 9-12. 2002. Lausanne. pp. 29-34. ISBN 3-85545-865-0
2. P. Rudolf: Optimization Methods for Hydraulic Machines Design - Shape Optimization of Swirl Turbine Draft Tube. Conference paper – Hydroturbo. September 2006.

3. J.A. Nelder, R. Mead: A Simplex Method for Function Minimization. 1965. The Computer Journal. 7(4): 308-313. DOI: 10.1093/comjnl/7.4.308. ISSN 0010-4620. Available online: http://comjnl.oxfordjournals.org/cgi/doi/10.1093/comjnl/7.4.308
4. Nelder-mead simplex algorithm. Figure: one-minimum.png [online]. [cit. 2015-09-01]. Available online: http://www.jakubkonka.com/images/nelder-mead/one-minimum.png
5. J. Machalová, H. Netuka: Numerické metody nepodmíněné optimalizace. 1. vyd. Olomouc: Univerzita Palackého v Olomouci, 2013, 142 s. ISBN 978-80-244-3403-2
6. M. Nemec, D.W. Zingg: Newton - Krylov Algorithm for Aerodynamic Design Using the Navier-Stokes Equations, AIAA Journal, Vol. 40, No. 6, June 2002, pp. 1146-1154.
7. ANSYS FLUENT 16.1 Manual
8. J. Hliník: Shape optimization of hydraulic turbine diffuser. Bachelor thesis. 2015, Brno.
9. Converting the domain to a polyhedra. Figure: img474.gif [online]. [cit. 2015-09-01]. Available online: https://www.sharcnet.ca/Software/Fluent6/html/ug/img474.gif
10. A. Tzanakis: Duct optimization using CFD software 'ANSYS Fluent Adjoint Solver'. Master's thesis in Automotive Engineering. Department of Applied Mechanics. Division of Vehicle Engineering and Autonomous Systems. Chalmers University of Technology. Goteborg, Sweden 2014
11. P. Rudolf: Study of the shear layers for swirl draft tube optimization. PhD thesis. In Czech

Acknowledgement

Presented research was supported by centre of competence of Technology Agency of the Czech Republic TE02000232 "Rotary machines".

EPJ Web of Conferences **114**, 02080 (2016)
DOI: 10.1051/epjconf/201611402080

Application of the balancing holes on the turbine stage discs with higher root reaction

Lukáš Mrózek[1,2,a], Ladislav Tajč[2,1], Michal Hoznedl[1,2] and Martin Miczán[3,1]

[1]Doosan Škoda Power, Experimental Research of Flow, Tylova 1/57 Pilsen 301 28, Czech Republic
[2]Univerzity of West Bohemia, Department of Power System Engineering, Univerzitní 22 Pilsen 306 14, Czech Republic
[3]Czech Technical University in Prague, Department of Fluid Dynamics and Thermodynamics, Technická 4, Prague 6, Czech Republic

Abstract. The influence of the flow area of balancing holes in the discs of a steam turbine on its thermodynamic efficiency is examined for turbine stages operating in the increased root reaction regime. The interstage seal flooding with steam is modelled. The impact is given on stage reaction changes, on mass flow adjustments, and on overpressure operating in the rotor disc.

1 Introduction

The balancing holes on the discs are used at the turbine stages of the impulse pressure type. The holes enable minimization of the differences between the pressures on both sides of the disc and thus keeping the axial force acting on the rotor at the acceptable level. With the onset stages with the higher root reaction a question arises whether it is still acceptable to use the balancing holes of the original arrangement in the given number and diameter, or if it is necessary to modify suitably their flow area considering the root reaction of the stages and the amount of the steam leaked through shaft seal. It is also desirable to assess the possible leakage of the steam from the blade space to the balancing holes and to evaluate the impact on the performance deterioration. Therefore, the efficiency of the stage with the root reaction at a 13% - stage and with three different areas of the balancing holes was tested on the experimental turbine. The original arrangement with fully closed holes was considered as well as the arrangement with half the diameter of the holes and the variant with fully open holes. At the same time the influence of the seal steam on the change of the root and the velocity data, at the original level. Reaction was tested. Since the mass flow through the shaft seal is in comparison with the main flow insignificant, the decrease of the efficiency is caused mainly by the leakage of the steam through the balancing holes. The quantity of the leaked steam cannot be defined by using experiments. Thus, it is impossible to evaluate the quantity of the steam leaked through the nozzle blade cascade. In order to enable to compare the results from individual experiments there is the effort to keep all key aerodynamic parameters such as the input pressure and the steam temperature, as well as the pressure ratio at the stage.

[a] Corresponding author: lukas.mrozek@doosan.com

2 Stage implementation

Basic geometrical and operational parameters of the tested stage are given in table 1. Blades adjustment and profiles implementation is evident from figure 1.

Table 1. Geometrical and operational parameters of the tested stage

Stage	VT 5	Units
Chord	120	mm
Aspect ratio	0.528	–
Tip seal clearance	0.4	mm
Pitch to chord ratio	0.99	–
Root diameter	930	mm
Velocity ratio u/c	0.48	–
RPM	3611	1/min
Inlet temperature	152.6	°C
Inlet pressure	0.775	bar
Pressure ratio	0.697	–
Root reaction	0.15	–

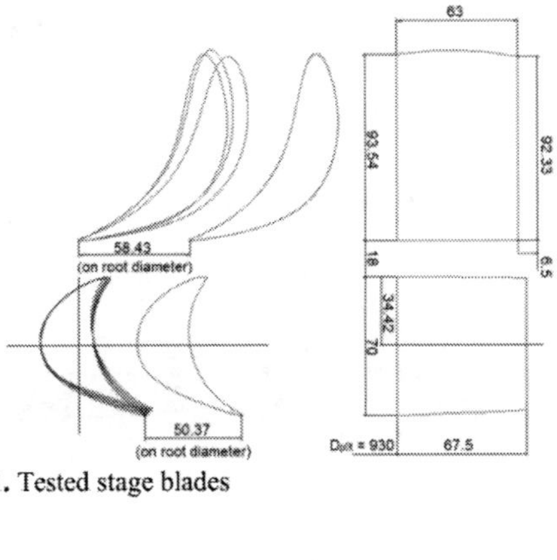

Figure 1. Tested stage blades

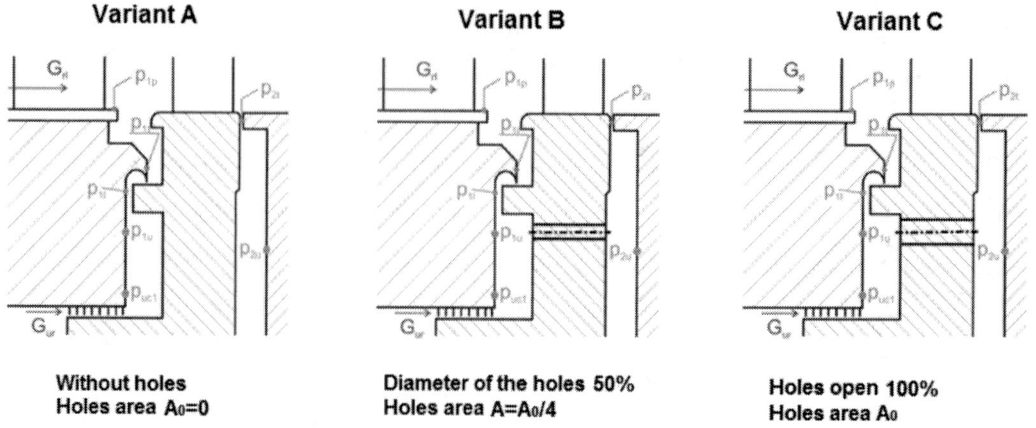

Figure 2. Variants of balancing holes

The stage is, for the experiment purposes, made in the 1.27 : 1 scale. This corresponds with rotation speed n = 3600 1 / min and root diameter $D_p = 0.93$ m. It is a stage of HP turbine part with designed root reaction $\rho_p = 0.15$.

Implementations of individual variants of balancing holes are shown in figure 2. The openings are placed under the knife-edge seal. It should limit possible steam leak from the blades to the holes. In the picture places are marked for pressure measurements and mass flows through stator blade G_{rl} and interstage seal G_{ur}. Variant A is without holes. In variant B half diameter d = 20 mm is used and in variant C the openings have the original dimension, i.e. d = 40 mm.

Marking of all measured and calculated data in the observed stage is depicted in figure 3. Using controlled steam suction G_{x1} it is possible to adjust the chamber pressure p_{uc0} and thus the influence the mass flow through the interstage seal G_{ur}. The experimental turbine is designed above all to define the thermodynamical efficiency, i.e. for operations with minimal influence of the interstage seal steam. In common stage tests the mass flow is $G_{ur} = 0$ and balancing holes are covered. Turbine arrangement enables flooding and suction of steam from the interstage steal, but only in a very limit scope.

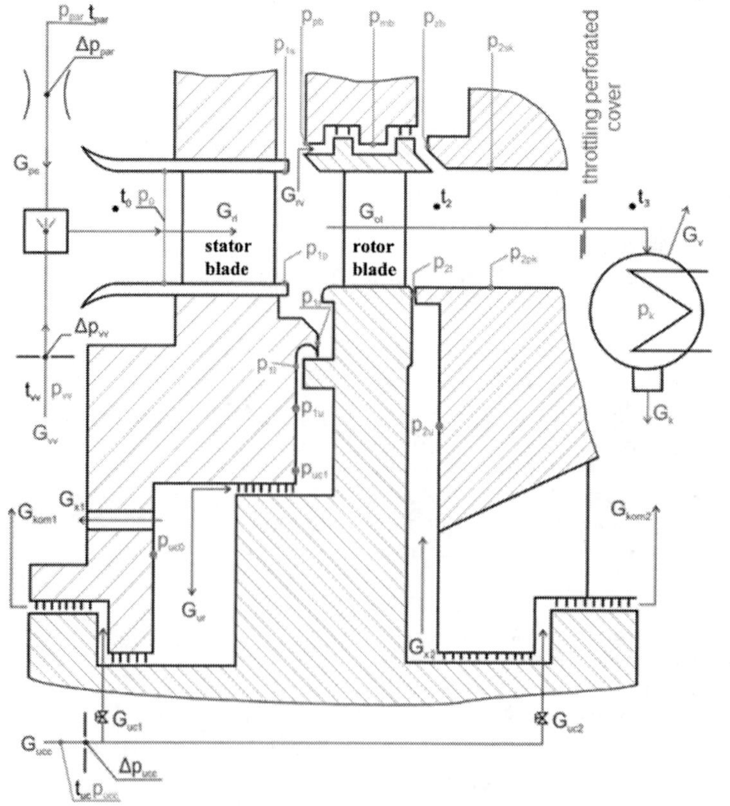

Figure 3. Scheme of measured parameters on the tested stage

3 Findings from experiment

The course of temperatures and pressures on the stage in relation to the velocity ratio u/c is given in figure 4 and 5. The pressure drop on stage is permanent for all variants of balancing holes adjustments with $\varepsilon = 0.697$. The Mach and Reynolds numbers on stage are kept at the same level.

In figure 4 steam temperatures are given before and behind the stage with the covered holes variant. Temperature t_2 is measured immediately behind the rotor blade. Temperature t_3 is recorded in the output casing near the regulator sieve above the condenser.

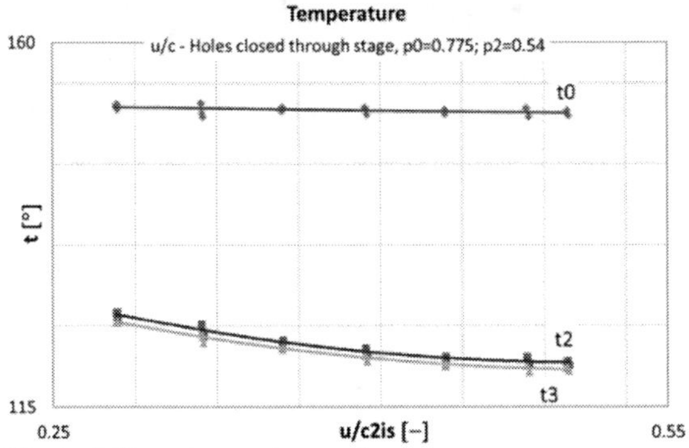

Figure 4. Steam temperature on the stage

The mean values on the stage are dependent on the velocity ratio u / c is given in figure 5. Dividing pressure p_{1mean} is already influenced by the stage reaction course and also by the size of the balancing holes area. The mass flow through the stator blades is related to the pressure change p_{1mean}.

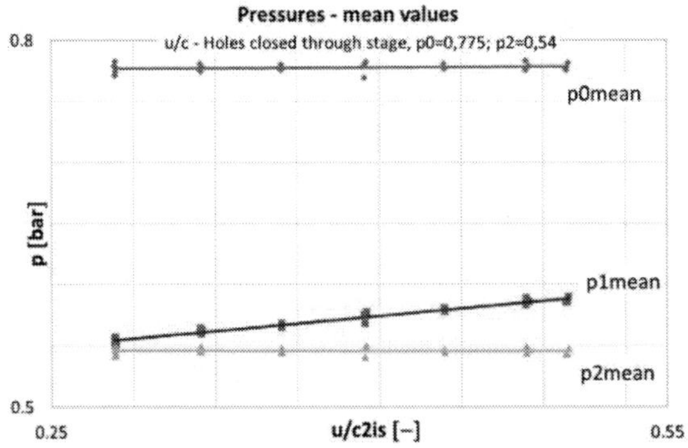

Figure 5. Pressures on stage

The mass flow through the stator blades is recorded in figure 6. With the increase of velocity ratio u / c, under the influence of increasing pressure p_{1mean}, mass flow G_{rl} is decreasing. For variant A it is possible to calculate the steam flow through the seals and thus the mass flow through the rotor blades G_{ol}. As it is impossible from the experiments to define the steam leak through the uncovered balancing holes, the steam flow through the rotor blades cannot be calculated.

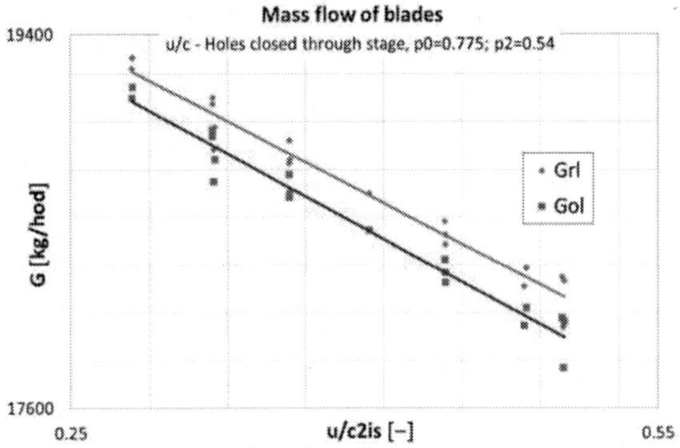

Figure 6. Steam mass flow through the stage

The scope of steam leak through the rotor seal and controlled steam flow through the interstage seal is shown in figure 7. Technical arrangement of the experimental turbine does not allow to change the steam flow in interstage seal in the scope higher than $\pm G_{ur} = 40$ kg / hour. This corresponds with proportional flow $G_{ur} / G_{ul} \approx \pm 0.002$. It is less than in the real implementation of the turbine stage, where the flow through the interstage seal could be comparable with the steam leak in the rotor seal.

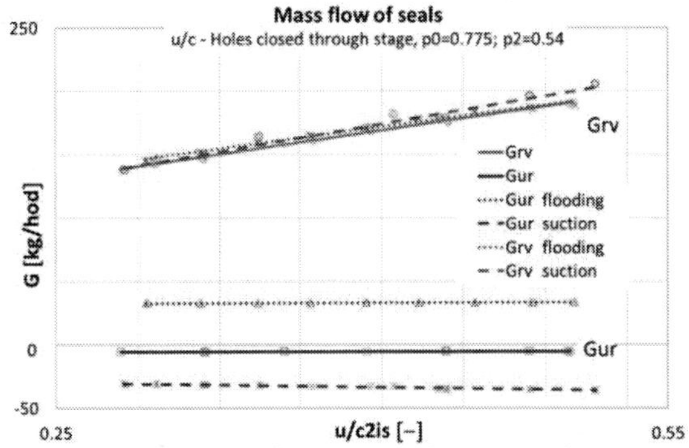

Figure 7. Mass steam flow through the rotor and interstage seals.

Comparison of proportional steam flows through both seals can be found in figure 8.

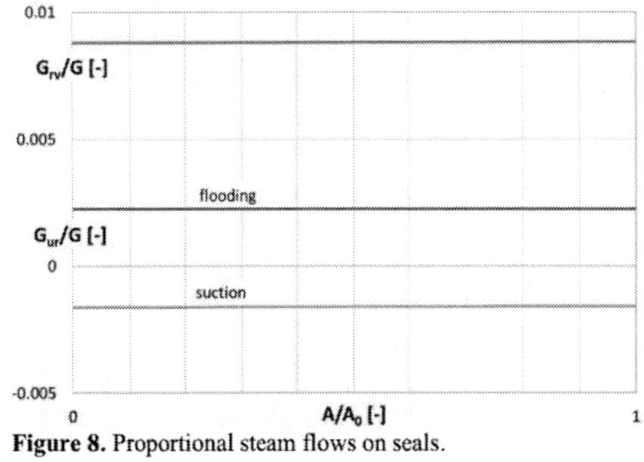

Figure 8. Proportional steam flows on seals.

It is evident that a relatively small steam leak from the interstage seal won´t have a major impact on the change of root reaction or on efficiency. Possible changes of their values rather may be related to disc openings and the steam leak from the blade area. The effect of interstage seal steam flow adjustment on the reaction course in variant A is shown in figure 9. The dispersion of values on the root and tip of the stage is in both cases insignificant.

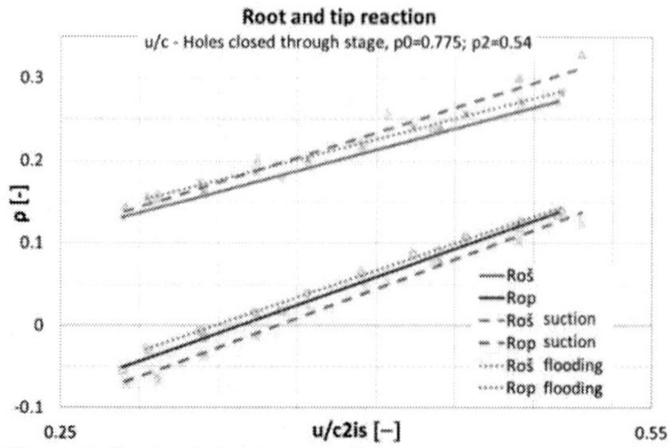

Figure 9. Root and tip stage reaction

Figure 10 shows how the root reaction changes are dependent on the amount the balancing holes are uncovered. Uncovering the holes causes drop in root reaction by as much as 4%.

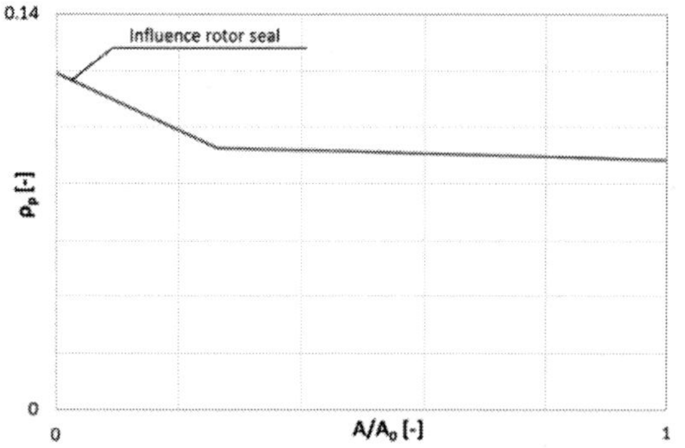

Figure 10. The influence of balancing holes area on the root reaction change for u/c_{2is}=0.49

Figure 11 shows how the steam flow through the interstage seal influences the thermodynamic efficiency of the stage.

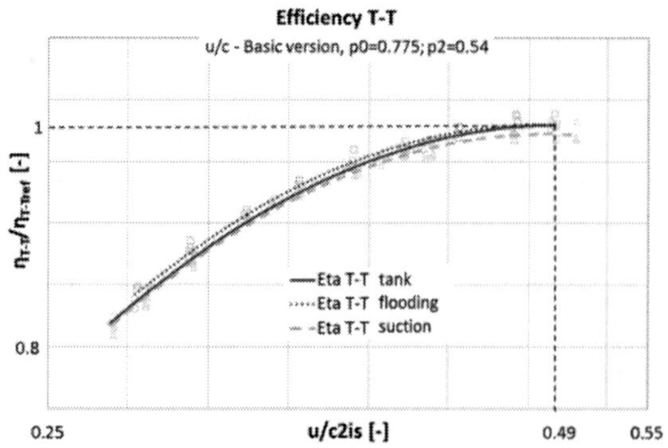

Figure 11. Turbine stage efficiency

Interstage seal flooding does not have any effect on change in efficiency. The decrease in efficiency occurs only while there is suction of steam. This phenomenon does not occur in common turbine stages. Efficiency can be calculated using temperatures measured before and behind the stage. This calculation of efficiency for variant A is shown in figure. 12.

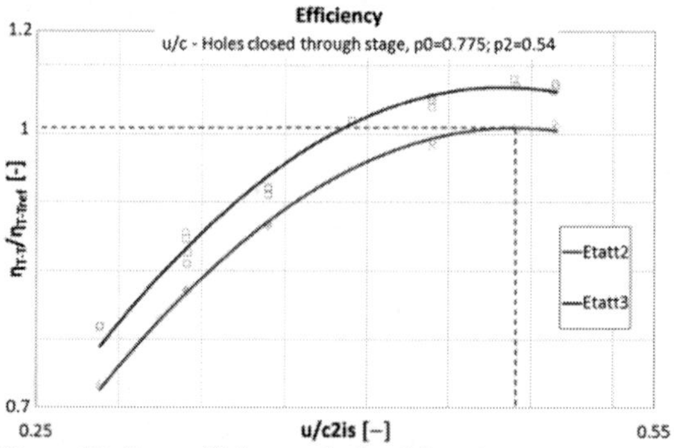

Figure 12. Stage efficiency calculated from temperatures

It is obvious that efficiency calculated from temperature t_3 is better than efficiency from temperature measured directly behind the rotor blades. Temperature t_3 can be influenced by heat transfer to the turbine body and to the condenser. Efficiency from temperature t_2 corresponds with efficiency defined using torque-meter.

Influence of covering and uncovering the balancing holes on efficiency is shown in Figure 13 and 14. In figure 13 efficiency calculated from temperature t_2 is shown.

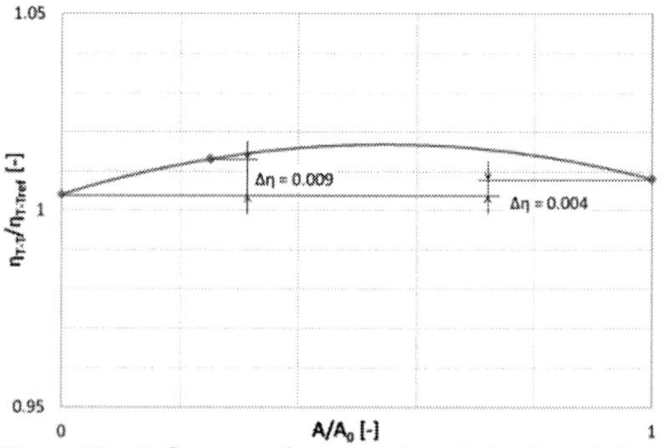

Figure 13. Influence of uncovering balancing holes on efficiency changes calculated from temperatures

Calculation of efficiency from torquemeter is shown in figure 14.

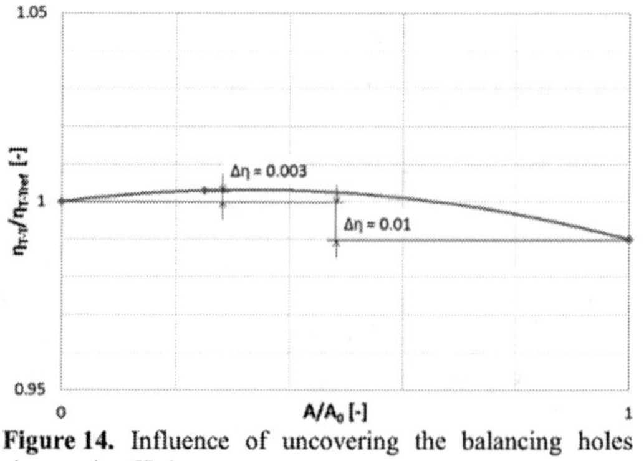

Figure 14. Influence of uncovering the balancing holes on changes in efficiency

Partial uncovering of holes slightly increased the efficiency. But the changes do not exceed $\Delta\eta = 0.5\%$. In comparison with the covered holes variant, total hole opening reduced the efficiency by only 1%. It represents only a negligible drop in efficiency. When increasing the designed root reaction to the level around 15% it is not necessary to adjust the dimensions of the balancing holes, as no distinctive steam leak from the blade area to the holes on the disc occurs. A positive effect of the sealing fin under the blade root is applied here.

Efficiency appears to be more sensitive to the Mach number level than to root reaction changes and adjustment of balancing holes. As it is evident from figure 15, with decreasing Mach number efficiency increases even by several percent. The converse effect of the loss dependent on Mach number known from experiments with blade cascades can be observed in the stage.

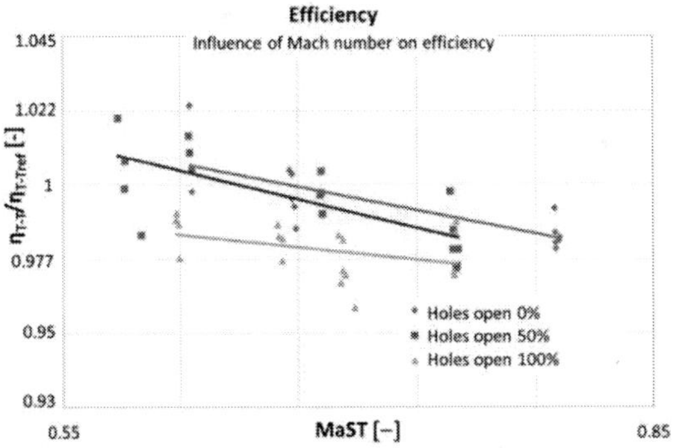

Figure 15. Influence of Mach number on efficiency

Balancing holes are supposed above all to lower the overpressure and the size of axial force influencing the rotor. It is possible to derive an estimate of the axial force from the pressure differences operating on the disc and rotor blades. The pressures before and behind the disc are recorded above and below the sealing fin and at the interstage seal level. Unfortunately while assessing the results leakage from the pressure pipes for the section around the sealing fin appeared. In figure 16 proportional pressure differences for individual variants of the experiment are calculated only from pressures on the interstage seal level and pressures at the tip and the root of the stage. Despite distorted data on pressures before and after the fin it is obvious that balancing holes perform their function and contribute to considerable drop in overpressure on the disc.

A more specific idea of pressure distribution on the disc surface can be obtained from the numerical study calculated for the stage of the same root reaction as expected in the experimentally measured stage [2]. Results of the numerical study are processed in figure 17. Variants are considered with or without the sealing fin and also the stage with or without balancing holes. The influence of holes and the sealing fin is confirmed on the pressure distribution on the disc. Both the holes and the sealing fin contribute to lowering the axial force operating on the rotor. Placing the fin nearer to the root radius of the stage can lower the final axial force even more.

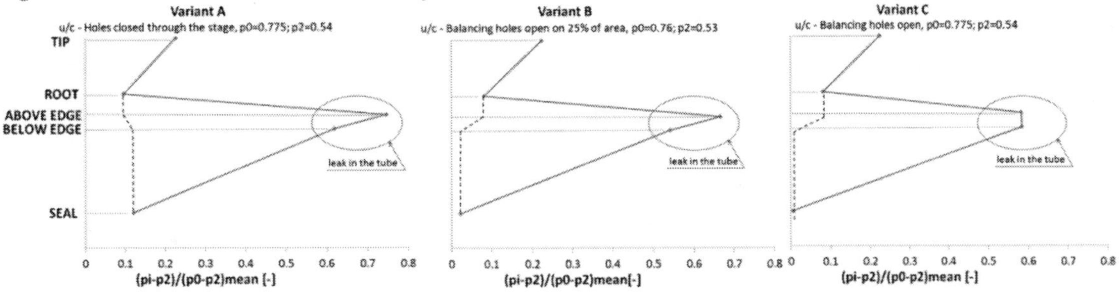

Figure 16. Pressure ratios in operational parameters

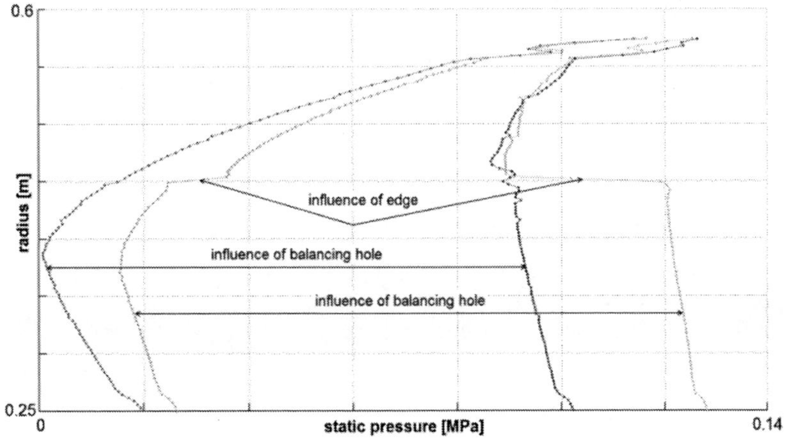

Figure 17. Numerical study of overpressure on disc with and without holes

4 Conclusions

• In stages with increased root reaction $\rho_p \approx 0.15$ for standard version with disc wheels it is not necessary to adjust the amount or size of balancing holes.

• Using the holes does not have crucial influence on the efficiency drop, but a drop in root reaction occurs.

• The size of the holes does not need adjustment for the amount of steam leak through the interstage seal.

• The influence of Mach number in the subsonic flow area on efficiency is confirmed.

• Holes in the disc contribute to lowering axial force operating on the rotor. Application of sealing fin is necessary. Also proper placing of the sealing fin is important.

References

1. L. Tajč, M. Hoznedl, L. Bednář, L. Mrózek: *Turbine Stage HP5 – Stage with Higher Root Reaction* (Turbínový stupeň VT5 – stupeň se zvýšenou reakcí na patě), Výzkumná zpráva Doosan Škoda Power, VZTP 1087, 2015
2. V. Sláma: *Numerical study Lovisa VTZ – Force Effects and the Pressure Distribution on the Rotor Disc* (Výpočtový studie Lovisa VTZ – působení sil a rozložení tlaků na disku oběžného kola), Doosan Škoda Power, 2013

EPJ Web of Conferences **114**, 02081 (2016)
DOI: 10.1051/epjconf/201611402081

Distribution of heat flux by working fluid in loop heat pipe

Patrik Nemec[1,a] and Milan Malcho[1]

[1] *University of Zilina, Faculty of Mechanical Engineering, Department of Power Engineering, Univerzitná 8215/1, 01026 Žilina, Slovakia*

Abstract. The main topics of article are construction of loop heat pipe, thermal visualization of working fluid dynamics and research results interpretation. The work deals about heat flux transport by working fluid in loop heat pipe from evaporator to condenser evolution. The result of the work give us how the hydrodynamic and thermal processes which take place inside the loop of heat pipe affect on the overall heat transport by loop heat pipe at start-up and during operation.

1 Introduction

Infrared thermography is one of the most important sensing technologies applied to the detection and monitoring of manufacturing and production equipment. Until recently this sophisticated technology was prohibitively expensive, being driven primarily by military applications; over the last few years, however, the technology has improved and it has been introduced to high volume commercial and professional applications by innovative companies. This has brought the price down to a level which is opening up a host of new applications to use infrared thermography. One of new infrared thermograpy applications is thermal imaging heat flux transferred working fluid in heat pipes to verify and test heat pipe operation [1].

Heat pipe is device which is more and more used to heat transport industry branch, such as industries using thermal processes, heat transfer processes or in power electronics cooling, because use to heat transport latent heat contained in vapour at working fluid vaporization and thus significantly increases the heat pipe thermal conductivity.

The most cases of incorrect operation heat pipes is due to, carelessness in filling and sealing the heat pipe, wrong choice working fluid amount or the formation of non-condensable gas arising on account materials and working fluid incompatibility or impurities in the working fluid and materials of the heat pipe. These aspects affect the transferred heat flux which is reflected in the increase of the temperature difference between the evaporator and condenser part of the heat pipe. Scanning the surface temperature of the heat pipe during operation by infrared camera can be monitored transferred heat flux of the working fluid and thus determine whether the heat pipe working correctly. Some works deals about heat pipe experiments [2 - 4].

2 Experiments

In this section is described experiment of thermal imaging heat flux transferred working fluid in loop heat pipe by infrared camera. The experiment deals about influence of working fluid on heat transport ability of loop heat pipe and distribution of heat flux by working fluid.

2.1 Loop heat pipe construction

Loop heat pipe consists of an evaporator with wick, a condenser, a compensation chamber, and liquid and vapor line. Only the evaporator and the compensation contain wicks; the rest of the loop is made of smooth wall tubing. The wick in the evaporator is made with fine pores for purpose of developing a capillary pressure to circulate fluid around the loop, while the wick in the compensation chamber is made with larger pores for purpose of managing fluid ingress and egress. The operating principle of the LHP is as follows. As heat is applied to the evaporator, liquid is vaporized and the menisci formed at the liquid/vapour interface in the evaporator wick develop capillary forces to push the vapour through the vapour line to the condenser. Vapour condenses in the condenser and the capillary forces continue to push liquid back to the evaporator. The waste heat from the heat source provides the driving force for the circulation of the working fluid and no external pumping power is required. The two - phase compensation chamber stores excess liquid and controls the operating temperature of the loop [5].

[a] Corresponding author: patrik.nemec@fstroj.uniza.sk

This experiment was realized with loop heat pipe model made in our workplace. The loop heat pipe model is shown in figure 1. In the figure 2 is cross section of compensation chamber and evaporator with capillary structure. The condenser of loop heat pipe model was created from finned pipe, so that heat loaded to the evaporator was dissipate to the ambient air by natural convection. The working fluid is acetone (50 % of loop heat pipe volume).

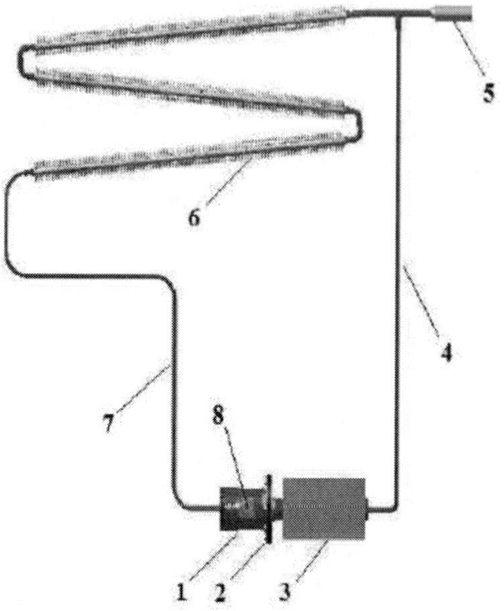

Figure 1. Model of loop heat pipe. 1 – compensation chamber, 2 – flange with isolation, 3 – evaporator, 4 – vapour line, 5 – vent valve, 6 – condenser, 7 – liquid line, 8 - filling and closing valve.

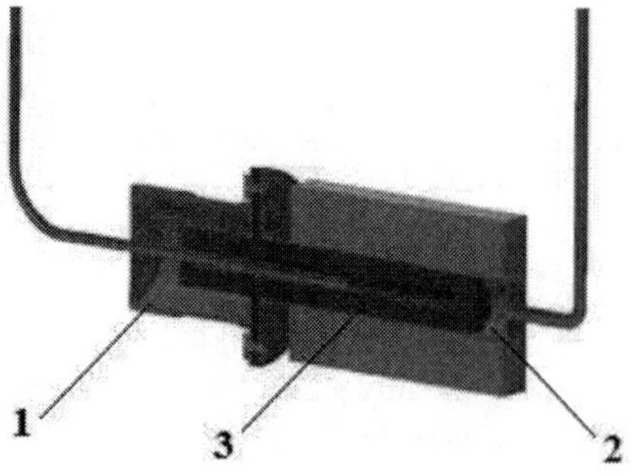

Figure 2. Cross section of compensation chamber and evaporator with capillary structure. 1 – compensation chamber, 2 – evaporator, 3 – capillary structure.

To achieve good thermal performance of loop heat pipe, capillary wicks with high permeability and porosity and fine pore radius are expected. These parameters depend mainly on the manufacturing process. The most frequently used wicks are made of sintered metal, like nickel, cooper, titanium, stainless steel or polymers (polyethylene, polypropylene, PTFE) [6]. There is known a number of research works which deal with sintered wick structure suitable in loop heat pipe [7 - 11]. According [12] the main parameters of wick are porosity, pore diameter and permeability. The optimal porosity of sintered wick is between 30- 75% regardless of the pore diameter. The sintered material porosity increases when the temperature or the forming pressure decrease. The optimal permeability is between 10^{-14} and 3.10^{-13} m^2. The capillary structure used in loop heat pipe model shown in figure 3 was made from copper powder with granularity of 50 µm by sintering in the electric oven.

Figure 3. Capillary structure.

2.2 Heat flux transport by working fluid in loop heat pipe

In the figure 4 to 10 are shown thermal images of the loop heat pipe loaded by heat 60 W and heat flux distribution by working fluid at the time. On the first four thermal images is seen start-up of loop heat pipe, where is the heat flux gradually transferred by working fluid from evaporator to the condenser part. On the last two thermal images is see fully operated loop heat pipe without no temperature changes at time. There is seen different temperature between vapour line (on the right) and condenser line (on the left), too. The vapour line is hotter than the condensation line. It means that the vapours of the working fluid transfer all heat flux from evaporator (on the bottom) to the condenser (on the top). There vapours of the working fluid condense to the liquid, heat transferred from evaporator is removed to the surrounding and working fluid return back to the evaporator.

Figure 4. Surface temperatures develop of loop heat pipe at time 2 min from heating start.

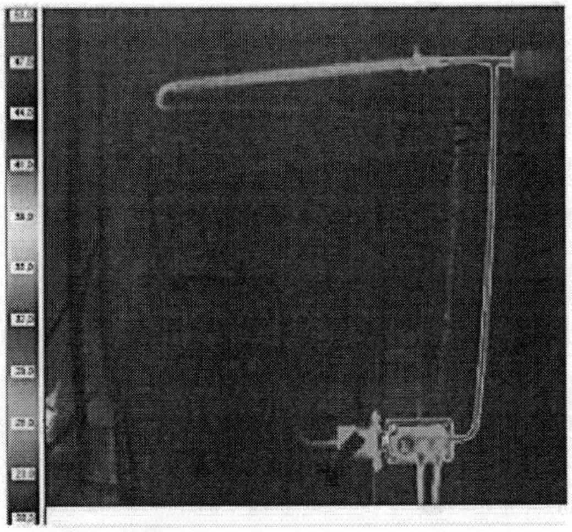

Figure 5. Surface temperatures develop of loop heat pipe at time 8 min from heating start.

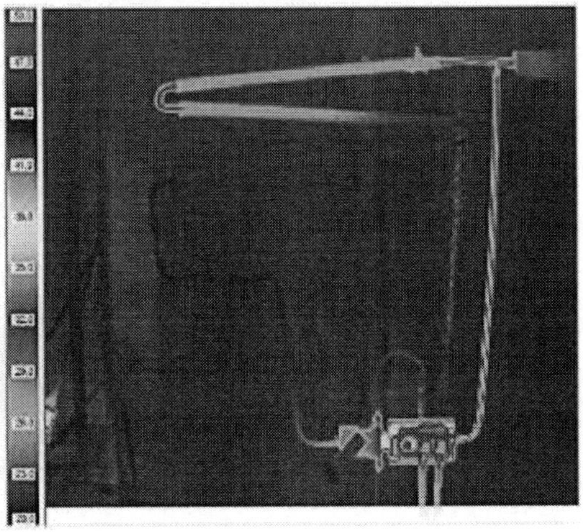

Figure 6. Surface temperatures develop of loop heat pipe at time 12 min from heating start.

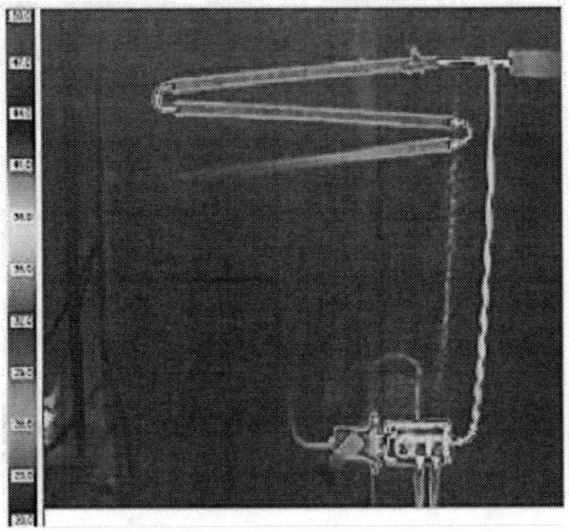

Figure 7. Surface temperatures develop of loop heat pipe at time 20 min from heating start.

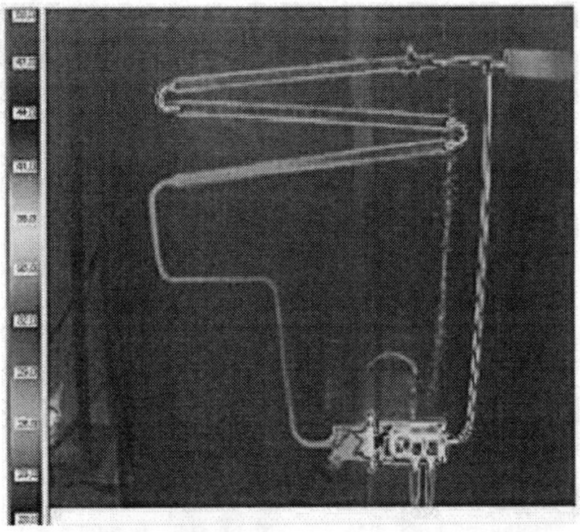

Figure 8. Surface temperatures develop of loop heat pipe at time 28 min from heating start.

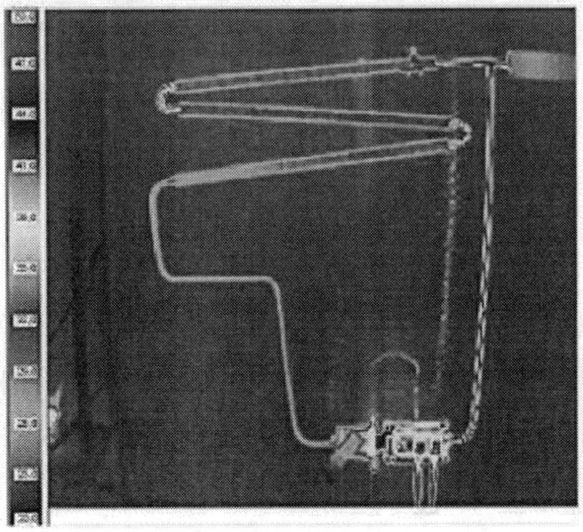

Figure 9. Surface temperatures develop of loop heat pipe at time 32 min from heating start.

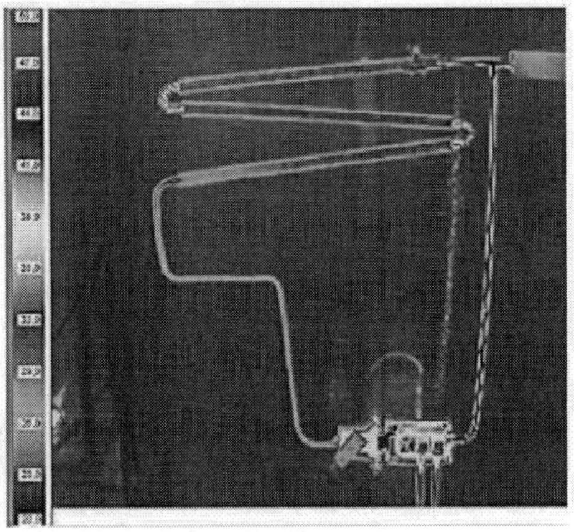

Figure 10. Surface temperatures develop of loop heat pipe at time 52 min from heating start.

3 Conclusions

The goal of this work was design and construction of the loop heat pipe with possibility removed heat fluxes to the surrounding in range approximately of 50 W. Thermal imaging of the loop heat pipe shows that the loop heat pipe loaded by heat 60 W has stabilized evaporator and condenser temperature after 30 minutes from start-up. Its operation was in equilibrium state without any temperature changes on evaporator and condenser part at time. This means that this loop heat pipe operate correctly and is able remove heat fluxes up-to 100 W.

Acknowledgment

The research is supported by the European Regional Development Fund and the Slovak state budget for the project "Research Centre of University of Zilina", ITMS 26220220183 (33 %).
This work is supported by the financial assistance of the project APVV - 0577-10 "Cooling of power electronic systems by cooling cycles without mechanical drive" (34 %).
The work is supported by Moderné vzdelávanie pre vedomostnú spoločnosť/Projekt je spolufinancovaný zo zdrojov EÚ. Podpora kvality vzdelávania a rozvoj ľudských zdrojov v oblasti technického výskumu a vývoja v priestore modernej vedomostnej spoločnosti, ITMS 26110230117 (33 %).

References

1. A. Kapjor, T. Grešák, J. Hužvár, M. Vantúch, AIP Conf. Proc. **1608**, 103-107 (2014)
2. R. Nosek, S. Gavlas, R. Lenhard, V. Sedlak, A. H. Mollerhagen, Communications. **16(3A)**, 62-66 (2014)
3. R. Lenhard, K. Kaduchová, Š. Papučík, J. Jandačka, EPJ Web of Conf. **67**, 02067 (2014)
4. A. Čaja, M. Patsch, EPJ Web of Conf. **92**, 02008 (2015)
5. J. Ku, 29th Int. Conf. on Envir. Sys. (1999)
6. J. Li, Y. Zou, L. Cheng, R. Singh, A. Akbarzadeh, Pow. Tech. **204(2-3)**, 241-248 (2010)
7. T. Kaya, J. Ku, J. of Thermophys. and Heat Transf. **17(4)**, 464-470 (2003)
8. K. C. Leong, C. Y. Liu, G.Q. Lu, J. of Por. Mat. **4**, 303-308 (1997)
9. E. G. Reimbrechta, M. C. Fredel, E. Bazzo, F. M. Pereira, Mat. Res. **2(3)**, 225-229 (1999)
10. N. J. Gernert, G. J. Baldassarre, J. M. Gottschlich, SAE Tech. Pap. 961319 (1996)
11. L. J. Orman, AIP Conf. Proc. **1608**, 169-172 (2014)
12. S. Launay, V. Sartre, J. Bonjour, Int. J. Therm. Sci. **46**, 621-636 (2007)

EPJ Web of Conferences **114**, 02082 (2016)
DOI: 10.1051/epjconf/201611402082

Experimental study of high-performance cooling system pipeline diameter and working fluid amount

Patrik Nemec[1,a], Milan Malcho[1] , Peter Hrabovsky[1] and Štefan Papučík[1]

[1]*University of Zilina, Faculty of Mechanical Engineering, Department of Power Engineering, Univerzitná 8215/1, 01026 Žilina, Slovakia*

Abstract. This work deals with heat transfer resulting from the operation of power electronic components. Heat is removed from the mounting plate, which is the evaporator of the loop thermosyphon to the condenser and by natural convection is transferred to ambient. This work includes proposal of cooling device - loop thermosyphon, with its construct and follow optimization of cooling effect. Optimization proceeds by selecting the quantity of working fluid and selection of diameters vapour line and liquid line of loop thermosyphon.

1 Introduction

Demands on the dimensions of electronic components, more precisely their reduction, are an important factor for the proper function of these components. There are many requirements on increasing electronic components performance and currently unchanged its dimension or reduce them. In most cases this result in produce of greater heat amount and thus temperature too, what may endanger the reliability operation of electronic components or even the whole system. Therefore, it is necessary the generated heat remove to other places where it does not affect the electronic system. One from the many option of the electronic components cooling may be a cooling system based on evaporation and condensation of the working medium circumfluent in a closed loop. In this work is just such a cooling system designed to remove heat from the electronic components placed on the mounting plate. The heat removal to the ambient by this cooling system takes place without additional electrical energy, which means that it is going on a heat transfer by free convection [1, 2, 3].

The closed loop thermosyphon may be visualised as a long hollow pipe, bent and the ends joined to form a continuous loop, filled with working fluid and orientated in a vertical plane. If the one side of the loop is heated and the other side cooled, the average density of the fluid in the heated side is less than in the cooled side. An essentially hydrostatic pressure difference, as a result of the thermally induced temperature gradient between the hot and the cold sides, rives the fluid flows around the loop. The 'buoyancy' force, as it is often termed, driving the fluid is in turn counteracted by an opposing frictional force that tends to retard the flow [4]. A thermosyphon loop can transfer heat from the interior of a microelectronic system to a central location where space limitations are less stringent. The advantages that a thermosyphon system enjoys over a conventional refrigeration system include: (1) absence of moving parts leading to a more reliable system operation, (2) increased choices for selecting a working fluid compatible with microelectronics chips since it does not have to go through a refrigeration cycle, (3) reducing the decomposition rate of the working fluid as the higher temperatures at the compressor discharge in a vapour compression refrigeration system are not encountered, (4) clean operation as no oil is circulated through the system. In comparison to pool boiling systems employing vapour space condensation, a thermosyphon loop offers more flexibility in terms of providing a centralized condenser with different feed lines to individual evaporator stations. Further, with the addition of a liquid circulating pump in a thermosyphon loop, higher heat transfer coefficients associated with flow boiling systems could be realized [5].

2 Cooling device construction

The cooling device has been designed as a specific type of thermosyphon - loop thermosyphon whose operation depends on the position of the evaporation and condensation section. The condensation section has to be placed above the evaporation section. This means that the return of the condensate is affected by the gravity. The proposed model of the loop thermosyphon is heat removed into the surrounding by natural convection.

[a] Corresponding author: patrik.nemec@fstroj.uniza.sk

Moreover, this type of cooling devices work on the evaporation and condensation principle of the working fluid in a closed loop, and therefore it do not need any mechanical elements or additional electric energy to drive the heat transfer medium. In the figure 1 is shown proposed model of the cooling device.

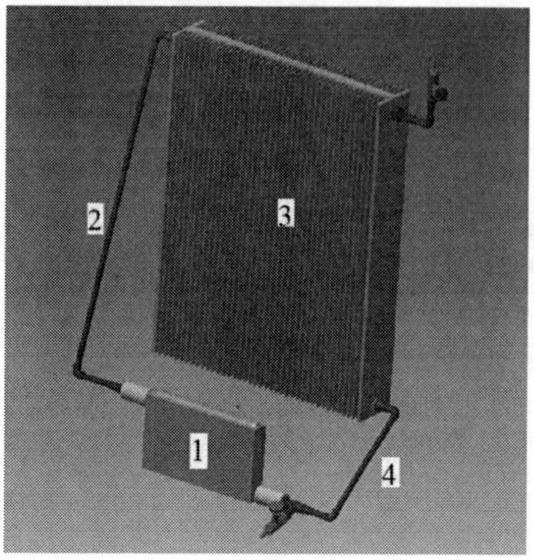

Figure 1. Model of cooling system, 1 - evaporator, 2 - vapour pipeline, 3 - condenser, 4 - liquid pipeline.

Main parts of the cooling device are:

- evaporator,
- condenser,
- pipeline,
- fill and vent valve,
- drain valve.

The evaporator enables on the base of a phase change (boiling) of the working fluid an intensive heat removal from its surface. It has to be constructed so that it will prevent the leakage of the working fluid, maintain pressure differences in all the walls and enable heat transfer from the electronic component into the working fluid as well as a suitable distribution of liquid and vapor phases of the working fluid. When choosing the material suitable for the construction of evaporator, it is necessary to pay attention to its thermokinetic characteristics. To provide a minimal temperature drop between a heat source and evaporator, the evaporator material must feature high thermal conductivity. To prevent escape of vapor, it should not be porous. The material should have high strength but, at the same time, it should be easily machineable and compatible with the working fluid [6]. The evaporator body shown in the figure 2 is a plate with dimensions 116 x 206 x 30 mm. To provide the working fluid circulation there are two 12 mm openings drilled horizontally on the plate and connected with twenty one 6 mm vertical connecting channels. They provide the transport of heated fluid vapor from the bottom to the top section of the evaporator. On the outer contact surface of the evaporator and electronic component there is a groove with a mounted temperature sensor.

The condenser shown in the figure 3 is proposed so that it was possible removed heat in to the surrounding by natural convection. It is created from two alumina fin coolers with dimensions 400 x 480 mm and fin height 45 mm. The each cooler has on the back side along the length ninety 7.5 mm deep and wide vertical grooves, and on the top and bottom two 10 mm deep and wide horizontal grooves. The cooler are soldered together on the back side and through the grooves can the working fluid flow down to the evaporator. This robust condenser construction provides to the cooling device remove high heat fluxes loaded to the evaporator.

Figure 2. Evaporator of cooling system.

Figure 3. Condenser of cooling system.

The pipelines connected evaporator with condenser was designed as simple as possible, so that when the working fluid circulating in the loop has minimum pressure losses. The pipelines are made from copper material, due to its good installation properties and compatibility with almost all mediums used in heat pipe [7, 8].

One of the construction conditions of the cooling device is that the cooling system has to be electrically isolated from the electrical components. Therefore are on the input and output pipelines to the mounting plate inserted Teflon insulators that prevent a cooling system against possible electrical voltage breakdown. The other parts on the cooling device are needle valves. One is located at the top of the condenser, which is used as a fill and vent valve. At the bottom part of the evaporator at input is the second needle valve, which is used to discharge the working fluid [9].

The main objective in the cooling device design has been focusing on the maximal the heat flux transfer at high heat load of the mounting plate with electronic components. The cooling systems proposal is based on a comparison of current work dealing with electronic components cooling device investigation. In the figure 4 is a graphical comparison of cooling effect results different types of cooling devices with heat pipe technology. After results comparison has been chosen the cooling system was able remove the highest heat flux (LT with aluminium condenser).

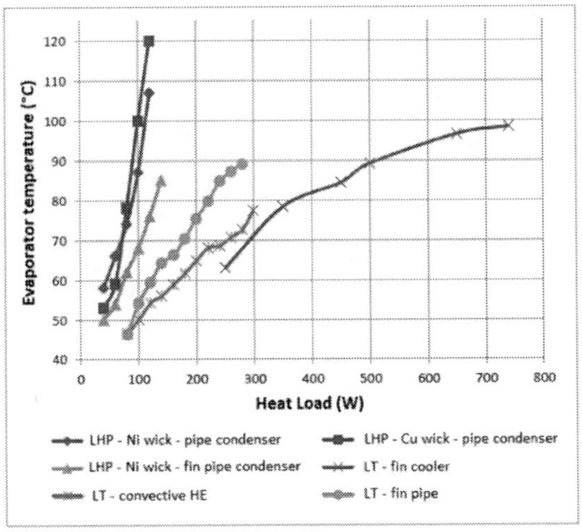

Figure 4. Comparison of cooling effect of different cooling devices with heat pipe technology.

3 Working fluid amount option

The scientific works declare that the proper percentage amount of the working fluid in loop thermosyphon is in the range of 40 - 60% of the total loop thermosyphon volume [10, 11, 12]. In this work was performed experiment to finding proper amount of the working fluid for the LT with working fluid percentage amount in the range of 20 – 70 %. As the working fluid was used ethanol. The pipeline diameter was 15 mm. In the figure 5 are shown measurement values. In the graph is seen that the evaporator temperature curve of all percentage amounts, in addition to 70 % are not very different. The differences are ± 1 ° C, sometimes even less.

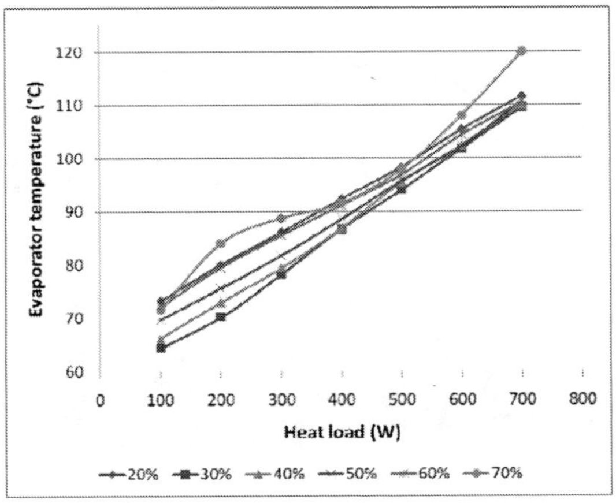

Figure 5. Influence of working fluid amount on cooling effect of cooling system.

In the figure 6 are shown results of evaporator temperature achieved at heat load 700 W. Following the results in both graphs was optimal quantity of the working fluid in the cooling device 40 and 30 % established. The choice of working fluid optimal quantity was realized on the basis of achieving the lowest evaporator temperature at the highest thermal load.

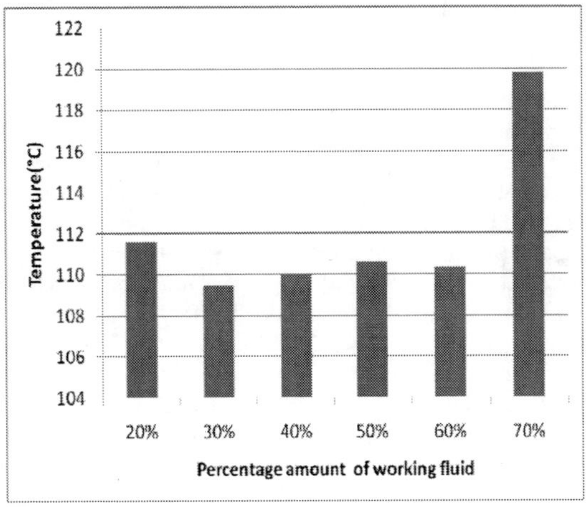

Figure 6. Evaporator temperature at heat load 700 W

4 Pipeline diameter option

One of the possibilities affected on the amount of heat transferred through the LT is the option of the pipelines diameter. In the figure 7 are compared individual tested pipelines diameters of LT filled with the working medium ethanol. From the results is seen that the ethanol is compatible mainly with the large-diameter pipes. The size pipeline diameter 15 mm appears as a best choice for the LT with working fluid ethanol, when at the maximal heat load of the LT, the evaporator temperature did not rise above 110 ° C.

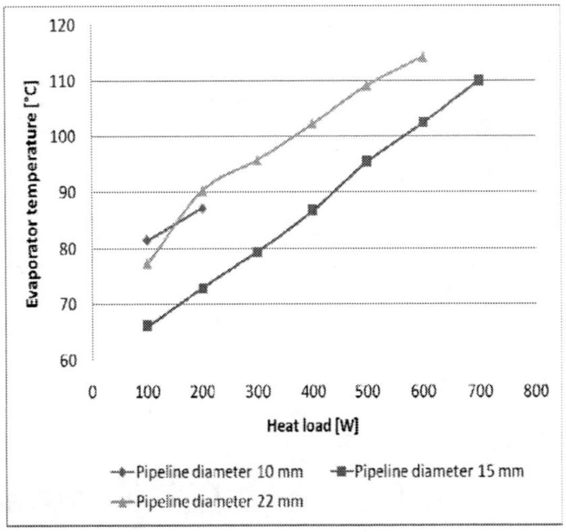

Figure 7. Influence pipeline diameter of cooling system on cooling effect.

Conclusion

The correct function of the most electronic components depends on operating temperature range. In practice this mean, that the components and the equipments too has to be cooled, therefore the heat that is produced by using them has to be removed.

Aim of the work was to construct a cooling device and optimize the size of the pipeline diameter and quantity of working fluid, due to remove as much of the heat from the mounting plate of the electronic components. It has been found that a cooling device with a pipeline diameter of at least 15 mm working correctly, and that the size of pipeline diameter 10 mm and less is not suitable for this type of device, because with such a small pipeline diameter, this type of device cannot operate. From the measurements, it was found that the quantity of working fluid ideal for this type of cooling device is 40 % or even 30 % of its total volume, also. Performed measuring the quantity of heat removed at the maximum heat load found that the device made by us can remove sufficient heat so that the temperature of the electronic components fixed to the mounting plate does not exceed the limit of 120 ° C at the heat load 700 W.

Other optimization of the designed cooling devices e.g. working fluid type option or condenser modification can overcome us detected limits.

Acknowledgment

This work is supported by the projects:
KEGA 029ŽU-4/2015 „Recovering heat from technological processes“ (33 %)
APVV 0577-10 "Cooling of power electronic systems by cooling cycles without mechanical drive" (34 %)
European Regional Development Fund and the Slovak state budget for the project “Research Centre of University of Zilina”, ITMS 26220220183 (33 %).

References

1. P. G. Peterson, *An introduction to heat pipes modelling, testing and applications*, (2010)
2. D. Reay, P. Kew and R. McGlen, *Heat pipes – Theory, design and applications*, (2014)
3. C. Silverstein, *Design and technology of heat pipes for cooling and heat exchangers*, (2011)
4. C. Ruppersberg, R. T. Dobson, JESA, **18** (3), pp. 32-40, (2007)
5. J. Hartenstine, R. Boner III, J. Montgomery, T. Semenic, *ASME InterPAC*, pp. 1-8 , (2007)
6. M. Holubčik, P. Ďurčanský, J. Jandačka and R. Nosek, The Sci. World J., **2014** (2014)
7. M. Vantúch, J. Hužvár and A. Kapjor, EPJ Web of Conf. **67**, 02122, (2014)
8. R. Lenhard, J. Jandačka, AIP Conf. Proc., **1558**, pp. 2138-2141, (2013)
9. M. Patsch, A. Čaja, EPJ Web of Conf., **95**, 02061, (2015)
10. M. Hamdan, E. Elnajjare, World Acad. of Sci., Engineer. And Tech., **3**, pp. 161-167, (2009)
11. D. Wolf, W. Bienert, *Int. Conf. on Envir. Syst.*, 941576, (1994)
12. S. W. Kang, M. Ch. Tsai, Ch. S. Hsieh and J. Y. Chen, Tamkang J. of Sci. and Engineer., **13** (3), pp. 281-288, (2010)

EPJ Web of Conferences **114**, 02083 (2016)
DOI: 10.1051/epjconf/201611402083

Working fluid flow visualization in gravity heat pipe

Patrik Nemec[1,a], Milan Malcho[1]

[1]University of Zilina, Faculty of Mechanical Engineering, Department of Power Engineering, Univerzitná 8215/1, 01026 Žilina, Slovakia

Abstract. Heat pipe is device working with phase changes of working fluid inside hermetically closed pipe at specific pressure. The phase changes of working fluid from fluid to vapour and vice versa help heat pipe to transport high heat flux. The article deal about gravity heat pipe construction and processes casing inside during heat pipe operation. Experiment working fluid flow visualization is performed with two glass heat pipes with different inner diameter (13 mm and 22 mm) and filled with water. The working fluid flow visualization explains the phenomena as a working fluid boiling, nucleation of bubbles, and vapour condensation on the wall, vapour and condensate flow interaction, flow down condensate film thickness on the wall occurred during the heat pipe operation.

1 Introduction

The thermosyphon (gravity heat pipe) is shown in figure 1. The bottom is a heating (evaporation) section, the middle an adiabatic section, and the upper part a condensation section. A small quantity of liquid is placed in a tube from which the air is then evacuated and the tube sealed. The lower end of the pipe is heated, this cause that the vapour flows from the bottom to the condensation section where it is cooled and condensed into liquid. Then the liquid flows downwards along the wall as a very thin film. However, liquid plugs may form at high heat input. This entire cycle repeats in heat pipe again and again. The Thermosyphon works on both conduction and convection mechanism. The heat from the heat source is transferred to the thermosyphons through conduction and inside thermosyphon heat convection occurs [1].

Since the latent heat of evaporation is large, considerable quantities of heat can be transported with a very small temperature difference from end to end. Thus, the structure will also have a high effective thermal conductance. One limitation of the basic thermosyphon is that in order for the condensate to be returned to the evaporator region by gravitational force, the evaporation section must be situated at the lowest point [2].

The amount of heat that can be transported by these systems is normally several orders of magnitude greater than pure conduction through a solid metal [3].

A thermosyphon needs only a temperature difference to transfer large amount of heat, and are widely used in various areas of industry, such as chemical engineering, thermal engineering or thermal management systems with limited space applications and other heat recovery systems.

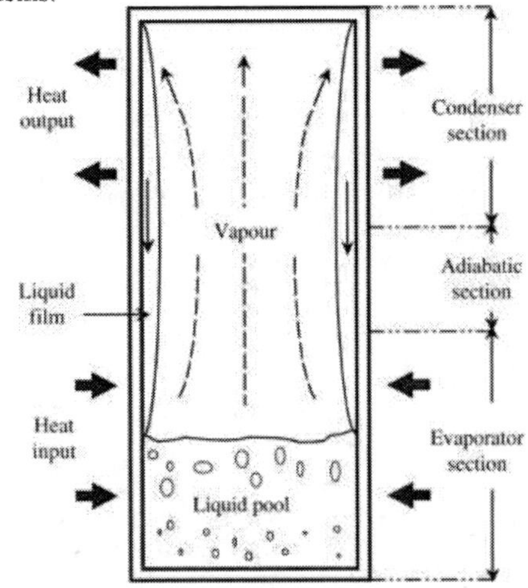

Figure 1. Working principle of the Thermosyphon [4]

2 Boiling and condensation

The theory of boiling is complex and not yet fully understood. The boiling process depends on factors such as mass flow, vapour content and the temperature difference between the working fluid and the heating surface. The available research is based on empirical experiments on smooth tubes. However, the processes of evaporation could be explained by a combination of the available research on boiling and flow regimes.

[a] Corresponding author: patrik.nemec@fstroj.uniza.sk

2.1 Boiling

The heat transfer coefficient for pool boiling has a characteristic curve shown in figure 2 that displays heat flux versus the temperature difference between the evaporating working fluid and heating surface. As the temperature difference increases, the energy flow per heat transfer area (kW/m2) will increase until an unstable maximum value for heat transfer is reached, the critical heat flux. At this point, vapour will form to such an extent that it hinders contact between the liquid of working fluid and the heat transfer area. As the temperature difference increases, a continuous gas film will form and the heat flux will reach a minimum. In the figure 2 are shown different boiling regimes.

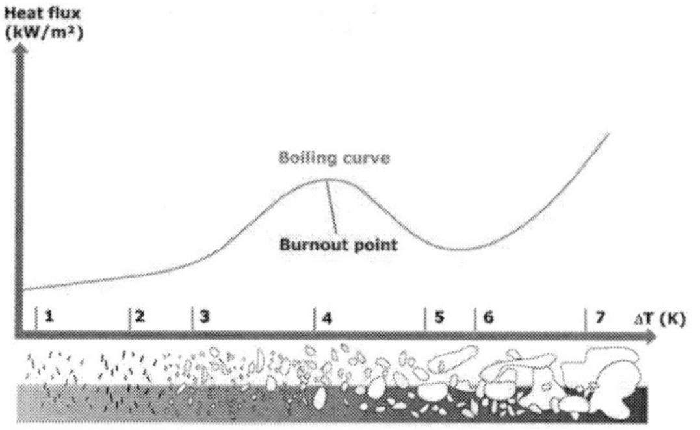

Figure 2. Boiling regimes [5]

Free convection evaporation: Heat is transferred between the wall and the working fluid without bubble formation. The liquid close to the surface wall becomes slightly superheated and evaporates in the interface between liquid and gas.

Sub-cooled nucleate boiling region: Heat transfer between the wall and the working fluid is sufficiently large to create bubbles, but they collapse in contact with the liquid bulk.

Nucleate boiling region: This is the most important boiling region for technical applications. The temperature difference required to enter this region is approximately 3K. Superheated liquid overcomes the surface tension, forming unstable vapour bubbles that collapse in contact with the sub-cooled liquid of working fluid. The additional turbulence caused by forming and collapsing bubbles increases the heat transfer. The heat transfer increases as the temperature difference between the working fluid and the heating surface wall increases to reach a maximum critical heat flux at the burnout point.

Partial film boiling: Increasing the temperature difference further, beyond the burnout point, will cause the working fluid to evaporate too quickly to allow new liquid of working fluid access to the heat transfer surface. The partial vapour blanket acts as insulation, causing the heat transfer coefficient to drop greatly and reducing the overall heat flux. This is an unstable region that should be avoided, because performance is uncertain and may fluctuate substantially.

Complete film boiling: With a very high temperature difference, a stable film of working fluid vapour will form on the heat transfer area. The film will act as an effective insulation layer hindering any direct contact between the heating surface and the liquid working fluid. Although the heat transfer coefficient reaches a minimum here, this evaporation region is preferable over the unstable partial film-boiling region, because predictions of the heat transfer coefficient are more reliable.

2.2 Condensation

Condensation is the phase transformation from vapour to liquid state. It may be carried in a space or on a solid surface and is provided with a heat and mass transfer. The heat pipe efficiency is supported by high rates of heat transfer during the condensation. The condensation on a solid surface can occur in the two different ways shown in figure 3 as a film condensation or dropwise condensation.

When the film condensation occurs, the condensate is form in a continuous liquid film that covers and wets the solid wall cold surface. The liquid film flows down the surface under gravity or other forces (e.g. capillarity). Heat transfer to the solid surface extends through the liquid film [6].

When the dropwise condensation occur, the working fluid vapour reach the wall cold surface, reduces its energy and causing the working fluid liquefaction and droplets formation on the cold surface that grow with direction of the vapour condensation. The small neighbour's drops are collecting to the bigger drops until they are tearing down from the surface by the gravity or other forces. As drops are flow down the wall, they leaving clean surface where condensation can proceed [7].

The heat transfer coefficient of the dropwise condensation is 10 times larger than the film condensation, but its duration is short and from this reason is difficult usable in practice. If the dropwise condensation has been continuous or if the thickness of the liquid film has been reduced can be achieved increasing heat transfer coefficient [8].

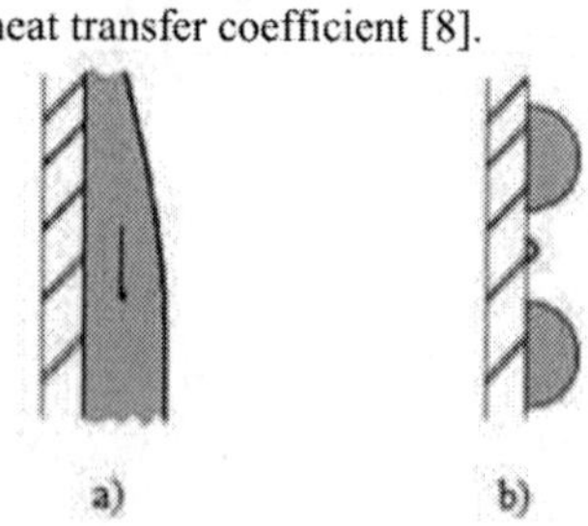

Figure 3. Condensation regimes: a) Film condensation, b) Dropwise condensation [9]

3 Experiments

Experiments deal about working fluid (water) flow visualization in glass gravity heat pipes with inner diameter 13 and 22 mm and total length 500 mm. In the figure 4 is shown positioning of the experimental apparatus. The working fluid flow inside the glass heat pipe placed on the heat source with heat output regulator was scanned by high speed video camera. The visualization of the working fluid flow in heat pipes can be helpful in solution or CFD simulation of the flow dynamics or heat and mass transfer of the heat pipes [10, 11].

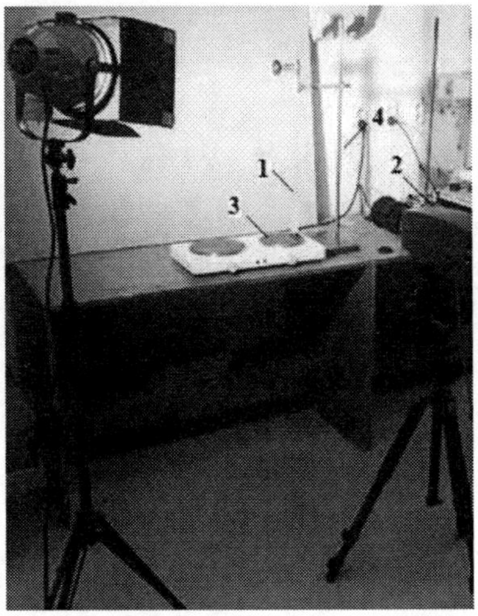

Figure 4. Measuring setup 1 – thermosyphon, 2 – video camera, 3 – heat source, 4 – stand

3.1 Heat pipe construction

In the figure 5 are containers of gravity heat pipes manufactured for experiment of working fluid visualization. The body of heat pipes is created form borosilicate glass tube. The borosilicate glass is more resistant to thermal shock than others glass and therefore is suitable for laboratory experiments. It has very low coefficients of thermal expansion and has good withstand capability to temperature differences at the same time. On the bottom is Cu cap and on the top is filling and closing valve.

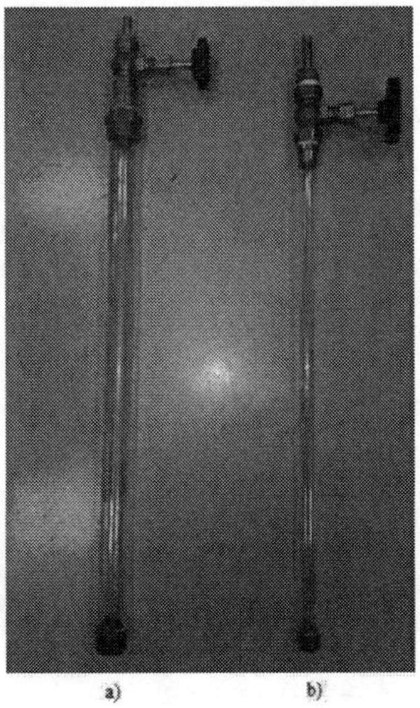

Figure 5. Glass containers of gravity heat pipes a) Inner diameter: 13 mm b) Inner diameter: 22 mm

The heat pipes manufacturing was performed by method of working fluid evaporation. This method covers filling the total volume of heat pipe with working fluid and then evaporation of working fluid to 20 % of heat pipe. This method guaranty that inside heat pipe is only pure working fluid and no air is inside.

3.2 Working fluid flow visualization

In the figure 6 and 7 is shown two-phase flow regime of working fluid in gravity heat pipe at evaporator temperature 105 °C scanned by high speed video camera.

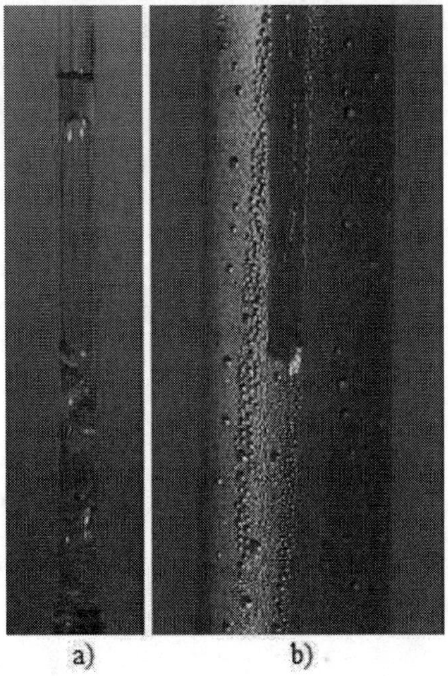

Figure 6. Glass gravity heat pipe (inner diameter: 13 mm, working fluid: water) a) Evaporation section b) Condensation section.

In the figure 6 are shown evaporation and condensation sections of glass gravity heat pipe with inner diameter 13 mm and water working fluid. In the figure 6 a) is shown film boiling and creating big bubbles of the working fluid on bottom heat pipe evaporator section. On the top is seen as a vapour pushes the liquid to the adiabatic section. In the figure 6 b) is shown dropwise condensation of the working fluid on the condensation section surface. The small droplets are collected to the big drops and they flow down to the evaporator section due to gravitation action. Thus is created free space for a new dropwise condensation on the clear surface.

In the figure 7 are shown evaporation and condensation sections of glass gravity heat pipe with inner diameter 22 mm and water working fluid. In the figure 7 a) is shown film boiling and creating big bubbles of the working fluid on bottom heat pipe evaporator section. It is similar action as in the heat pipe with inner diameter 13 mm, but in this case liquid is not pushes by vapour to the adiabatic section so high. This may caused bigger diameter of heat pipe. In the figure 7 b) is shown dropwise condensation of the working fluid on the condensation section surface. In this case this action is same as at heat pipe with inner diameter 13 mm.

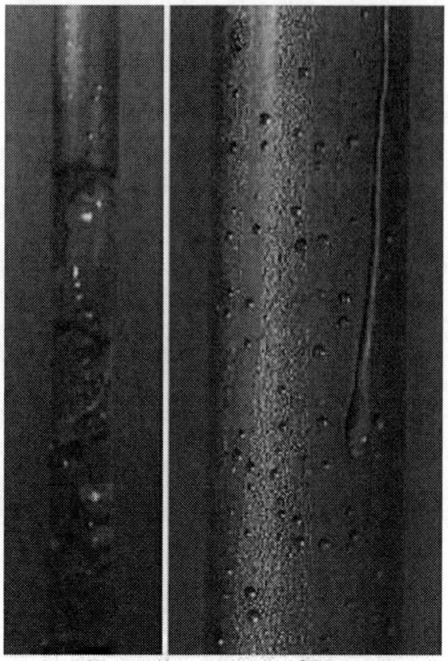

a) b)

Figure 7. Glass gravity heat pipe (inner diameter: 22 mm, working fluid: water) a) Evaporation section b) Condensation section.

3 Conclusions

Aim of this experimental work was working fluid flow visualization in the gravity heat pipes and working fluid flow comparison considering to different inner diameter of heat pipes. The experiment shows that at the both inner diameter dimensions of heat pipes film boiling with big bubbles in the evaporator section occurs. The different was at level height of liquid pushed in to the adiabatic section. In the heat pipe with inner diameter 13 mm the level height of pushed liquid was higher than in the heat pipe with inner diameter 22 mm. The working fluid flow in the condensation section was at both cases the same - dropwise condensation. The small droplets condensate of working fluid was gradually collected to the big drops, which flown down to the evaporator section due to gravitation action.

Acknowledgment

The work is supported by Moderné vzdelávanie pre vedomostnú spoločnosť/Projekt je spolufinancovaný zo zdrojov EÚ. Podpora kvality vzdelávania a rozvoj ľudských zdrojov v oblasti technického výskumu a vývoja v priestore modernej vedomostnej spoločnosti, ITMS 26110230117 (33 %).
The research is supported by the European Regional Development Fund and the Slovak state budget for the project "Research Centre of University of Zilina", ITMS 26220220183 (33 %).
This work is supported by the financial assistance of the project APVV - 0577-10 "Cooling of power electronic systems by cooling cycles without mechanical drive" (34 %).

References

1. A. S. Sundaram, A. Bhaskaran, J. of El. Cool. and Therm. Cont. **1**, 15-21 (2011)
2. D. Ray, P. Kew, *Heat pipes, fifth eddition* Elseviere, Oxford (2006)
3. P. Dunn, D. Ray, *Heat Pipes, 4th Edition* Pergamon Press, Oxford (1994)
4. S. Liu, J. Li, Q. Chen, Flow Measur. and Instrum. **18**, 216 – 222 (2007)
5. http://handbooks.swep.net/RefrigerantHandbook/chapter_06/Pages/6.2-Boiling-Regimes.aspx
6. R. Nosek, M. Holubčík, Š. Papučík, Sci. World J. **2014**, 487549 (2014)
7. M. Čarnogurská, M. Příhoda, J. Molínek, J. of Mech. Sci. and Tech. **25 (11)**, 2935-2941 (2011)
8. T. Brestovič, N. Jasminská, R. Pyszko, M. Lázár, M. Puškár, Measurement: J. of the Inter. Measur. Confed. **72**, 3380, 52-60 (2015)
9. http://ltcm.epfl.ch/files/content/sites/ltcm/files/shared/import/migration/COURSES/TwoPhaseFlowsAndHeatTransfer/lectures/Chapter_7.pdf
10. R. Lenhard, J. Jandačka, AIP Conf. Proc. **1558**, 2138-2141 (2013)
11. R. Lenhard, K. Kaduchová, Š. Papučík, J. Jandačka, EPJ Web of Conf. **67**, 02067 (2014)

EPJ Web of Conferences **114**, 02084 (2016)
DOI: 10.1051/epjconf/201611402084

Effect of shelter porosity on downwind flow characteristics

Š. Nosek[1,a], R. Kellnerová[1], K. Jurčáková[1], Z. Jaňour[1], H. Chaloupecká[1,2], and M. Jakubcová[1,3]

[1] Institute of Thermomechanics AS CR, Dolejškova 1402/5, Prague, Czech Republic

[2] Charles University in Prague, Faculty of Mathematics and Physics, Department of Meteorology and Environment Protection, V Holešovičkách 2, Prague, Czech Republic

[3] Czech University of Life Sciences Prague, Faculty of Environmental Sciences, Department of Water Resources and Environmental Modeling, Kamýcká 1176, Prague, Czech Republic

Abstract. Previous wind-tunnel studies were focused mainly on lonely standing windbreaks or wind fences with respect to their wind velocity reduction efficiency and effective shelter distance. In presented wind-tunnel study, we investigated the effects of a three different fence porosities (0.5, 0.25 and 0) embodied in a shelter-like building for coal convey by means of two-component Laser Doppler Anemometry (LDA). The turbulent flow characteristics behind the fences were compared with those performed without the fence. For characterization of the fence effectiveness we used following quantities: wind-speed and turbulence kinetic energy reduction, and time fractions of the turbulent coherent structures associated with the sediment transport (sweeps and outward interactions). Results from mentioned quantities revealed that for the case of embodied fence the shelter construction has significant impact on the flow characteristics behind. The fence of the 0.5 porosity has been indicated as the most shelter effective considering the studied quantities.

1 Introduction

The windbreaks or wind fences have been studied since the 1940th. The main objectives of these studies remain the same: shelter efficiency and the factors that help to optimize the windbreak or wind fence design with respect to control the particle sedimentation (sand, dust, snow, etc.). The fence shelter efficiency is evaluated as the reduction of the wind velocity and downwind distance for which wind velocity remains below the threshold velocity [1]. Here the threshold velocity relates to the particle movement initiation. Previous studies (e.g. [2]) observed that the shelter efficiency is mainly influenced by the fence geometry and porosity. As one might expected, lower porosity brings about higher wind velocity reduction. However, higher turbulent intensities, shear stresses and pressures are produced as well. These flow turbulent characteristics are unfavourable to fence efficiency since decreasing the fence shelter distance and enhancing the particle uplift just downstream the fence.

In this paper, we present twofold extension relating to the fence shelter efficiency analysis: construction complexity and turbulent coherent structures. Previous wind-tunnel studies were mainly focused on a lonely standing windbreaks or wind fences. However, there are many practical fence applications. Thus, in present study, we considered the fence integration into the shelter for avoiding the coal dust escape from an convey belt to the surrounding area. This will give a better insight into a practical fence utilization.

The second fence shelter efficiency analysis extension follows the experimental study on aeolian sediment transport over a sand dune [9], which revealed that the turbulent coherent structures, sweeps and outward interactions, have a dominant influence on sand transport. This finding is in contrast to research in fluvial domain where turbulent structures associated with the bursting processes, sweeps and ejections, were found to be dominant for sediment transport [3].

2 Experimental setup

The experiment was carried out in the low-speed Environmental wind-tunnel of the Institute of Thermomechanics of the Czech academy of sciences in Nový Knín. For detailed wind-tunnel description see [4]. The schema of the shelter model for convey belt of scale 1:100 positioned at the wind-tunnel test section with respect to the wind-tunnel coordinates is presented in figure 1a. Detail of the model, where the measures are normalised by the shelter model characteristic height $H = 90$ mm ($H = 9$ m at full scale), is presented in figure 1b. The fence, made of polypropylene fibres, was positioned at the leeward side of the shelter at $x/H = 0$ and $z/H = 0.17$ and runs along the entire lateral shelter leeward side (from $y/H = -8.3$ to $y/H = 8.3$). The gap that arisen between the bottom of the wind tunnel and the fence is necessary for dust explosiveness prevention. We investigated the following fence porosities $\epsilon = 0, 0.25$ and 0.5, defined according to [5] as

$$\epsilon = \left(\frac{B-D}{D}\right)^2, \quad (1)$$

where B is the fence opening width and D is the fence fibre diameter.

[a] e-mail: nosek@it.cas.cz

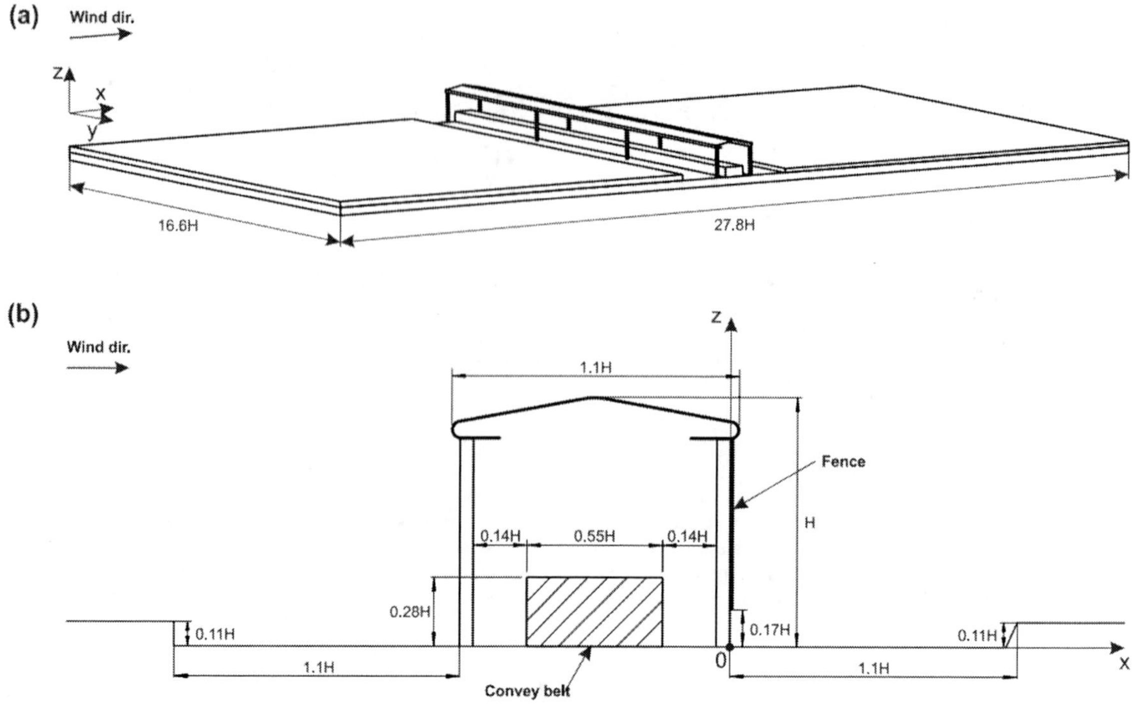

Fig. 1: Schemas of the shelter model with respect to the wind-tunnel coordinates x, y and z: (a) isometric view of the whole model incorporated in the wind-tunnel measuring section; (b) detailed side view with the origin of the coordinate system for the measurement grid (the origin of the y coordinate is in the middle of the model lateral direction, $16.6H/2$).

The wind-tunnel free stream velocity was maintained at 6 m s^{-1} during the whole experiment. For the characteristic height of the shelter model the Reynolds number was sufficiently high (Re = 34600) to fulfill the Reynolds number independence (Re_{crit} = 11000), recommended by [6]. The free stream velocity also fulfil the criterium for the fence porosity independence on pressure drop according to [7].

The turbulent boundary layer simulating the moderately rough atmospheric boundary layer for the area of the interest (e.g. moderate hills, shrubs) was developed along the 20 m long wind-tunnel development section. Four vortex generators were spaced laterally at the beginning of the development section and the remaining part was covered by the roughness elements. The boundary layer characteristics were measured at the model entrance with the two-dimensional Laser Doppler Anemometry (LDA) based on DANTEC BSA F-60 burst processor. The parameters of the longitudinal velocity profile (roughness length z_0, displacement length d_0 and friction velocity u_*) were derived from the logarithmic law and are compared with those recommended by the VDI guidelines [8] in table 1. The parameters of the longitudinal velocity profile as well as the longitudinal and vertical turbulent profiles (not shown here) were in accordance with the VDI guidelines [8].

3 Data Analysis

The data were collected from the measured grid of the vertical plane xz downstream from the fence by LDA system at a rate ranging from 1 to 2 kHz. The sampling time for each point of the measured grid was kept constant at 90 s. The measurement error was established on 0.83% from 36 measured reference velocities, U_{ref}, obtained during the experiment from the same point at dimensionless height $z/H = 5.5$.

Table 1: Approaching boundary layer parameters for moderately rough terrain

	z_0 (m)	d_0 (m)	u_* (m s^{-1})
Exp	0.15	0	0.17
VDI	0.12 – 0.18	0	–

The origin of the coordinate system for the measurement grid is depicted in figure 1b, where the scheme of the vertical cut at $y = 0$ is presented. We measured at 6 heights 13 points downstream from the fence for each fence porosity and without the fence. One extra height (z/H = 0.1), consisting of 5 downstream points, was measured to provide the data below the fence gap (z/H < 0.17, see figure 1b) for each model type as well.

For analysing the fence effectiveness, we used a dimensionless wind-speed reduction coefficient according to [2]

$$R_U = 1 - \frac{U_{dim}}{U_{dim0}}, \quad (2)$$

where $U_{dim} = U/U_{ref}$ is the dimensionless mean velocity measured at a given point and fence porosity and $U_{dim0} = U_0/U_{ref0}$ is the dimensionless mean velocity measured at the same given point without the fence. The same

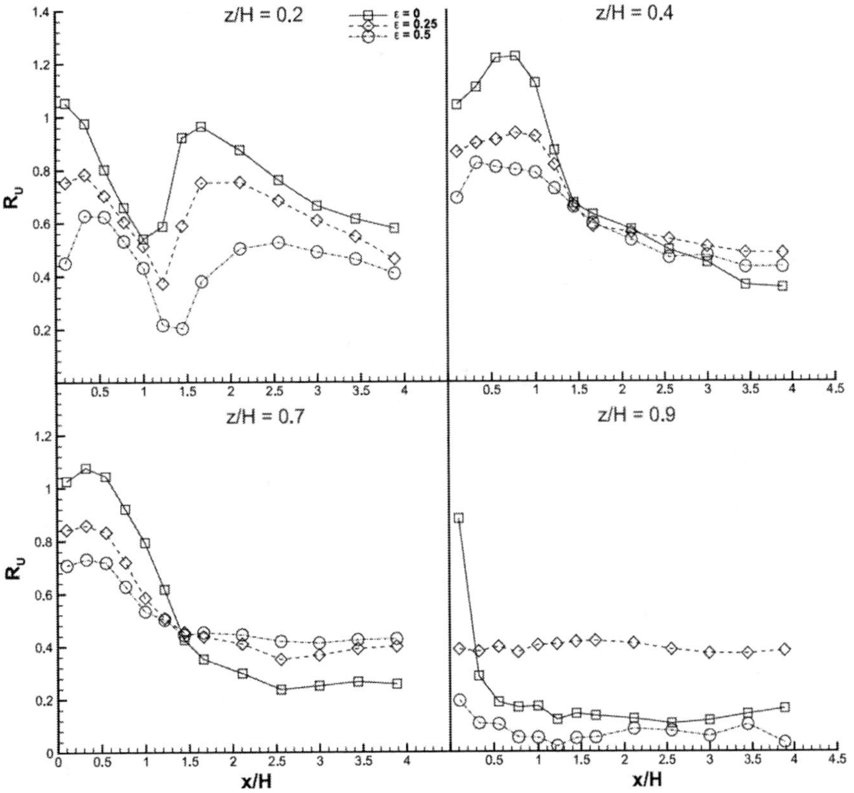

Fig. 2: Dimensionless longitudinal velocity reduction coefficient of the fence with porosity $\epsilon = 0$ (squares), $\epsilon = 0.25$ (diamonds) and $\epsilon = 0.5$ (circles) as a function of the dimensionless streamwise distance from the fence, x/H, for four different dimensionless heights $z/H = 0.2, 0.4, 0.7$ and 0.9.

was applied for the dimensionless turbulent kinetic energy reduction coefficient

$$R_{TKE} = 1 - \frac{TKE_{dim}}{TKE_{dim0}}. \quad (3)$$

Here, we assumed that the turbulent flow is a quasi two-dimensional and used the following formulas for dimensionless turbulent kinetic energies

$$TKE_{dim} = \frac{\langle u'^2 \rangle + \langle w'^2 \rangle}{2U_{ref}^2}, \quad TKE_{dim0} = \frac{\langle {u_0}'^2 \rangle + \langle {w_0}'^2 \rangle}{2U_{ref0}^2}, \quad (4)$$

where u' and w' is the fluctuation of the longitudinal and vertical velocity, respectively, $\langle \quad \rangle$ is the time averaging and the zero index denotes the case without the fence. If R_U and R_{TKE} are equal to one, the fence reduces the wind-speed and turbulent kinetic energy to zero and is 100% efficient. The negative values mean that the fence is increasing the wind-speed and turbulent kinetic energy, respectively.

Assuming the same approach for the sand transport [9], we performed the quadrant analysis according to [10] in order to identify turbulent coherent structures. This conditioning technique separates the flow into four discrete categories according to four quadrants of the scatter plots of two turbulent quantities, u' and w' : 1. outward interactions ($u' > 0$, $w' > 0$), 2. sweeps ($u' > 0$, $w' < 0$), 3. inward interactions ($u' < 0$, $w' < 0$) and 4. ejections ($u' < 0$, $w' > 0$). We used the hole quadrant analysis in order to filter insignificant events such that the attribution of the flow structure occurred if $|u'w'| > \sigma_{uw}$, where σ_{uw} is the standard deviation of the kinematic stress $u'w'$ and represents the selected "hole threshold".

We analysed the fence porosity efficiency with respect to relevant events for sediment transport (sweeps and outward interactions) by means of their time fraction to the total sampling time T defined by [11] as

$$D_i = \frac{1}{T} \int_0^T I_i(t) \mathrm{d}t, \quad (5)$$

where D_2 and D_1 is total sweep and outward interaction durations within T, respectively, and I_i is the indicator function, where $I_i = 1$, if $u'w'$ is within quadrant $i = 1, 2$ and $I_i = 0$, otherwise.

4 Results

4.1 Reduction coefficients

The dimensionless longitudinal velocity reduction coefficient, R_U, of three fence porosities as a function of the dimensionless streamwise distance from the fence, x/H, is presented in figure 2. The four graphs of R_U are plotted for dimensionless height $z/H = 0.2, 0.4, 0.7$ and 0.9, respectively.

As one expected, the solid wall, $\epsilon = 0$ (square symbols in figure 2), has the highest wind-speed reduction.

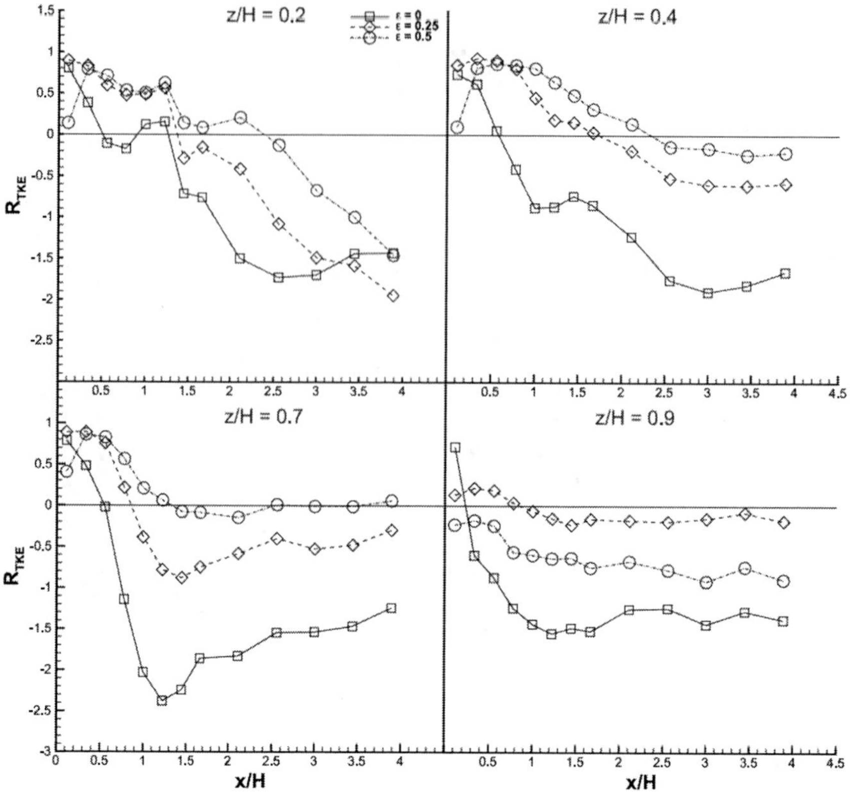

Fig. 3: The same as in figure 2 but for the dimensionless turbulent kinetic energy reduction coefficient.

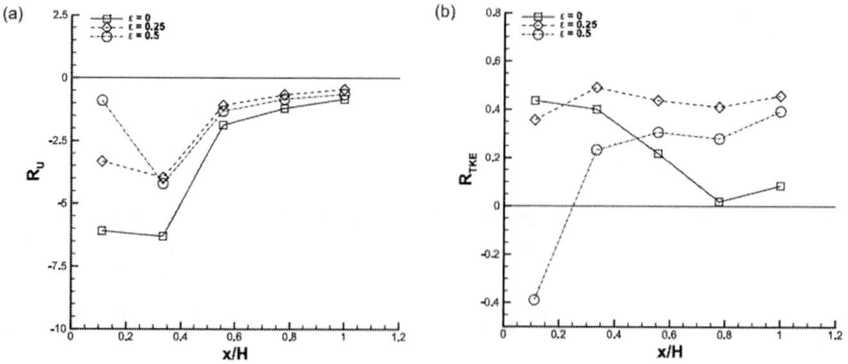

Fig. 4: Reduction coefficient of the (a) dimensionless longitudinal velocity and (b) dimensionless turbulent kinetic energy for the fence porosity $\epsilon = 0$ (squares), $\epsilon = 0.25$ (diamonds) and $\epsilon = 0.5$ (circles) as a function of the dimensionless streamwise distance from the fence, x/H, at dimensionless height $z/H = 0.1$.

However, this reduction is absolute (for all investigated heights) only at the fence vicinity ($x/H = 0.1$). For heights $z/H > 0.4$ we can observe that R_U decreases with the increasing downstream distance and height. The R_U is even lower than the R_U of the fence with the highest porosity ($\epsilon = 0.5$) for $x/H > 1.5$ and $z/H > 0.2$. The values of R_U in case of the solid wall means that the backward wind direction (vortex) is present. The wind-speed reduction is independent on the dimensionless downstream distance x/H from ceratin value of x/H. We can observe that the higher the height the shorter the downstream distance is for the wind-speed reduction independence on downstream distance.

Another parameter for determining the fence effectiveness, the dimensionless turbulent kinetic energy reduction coefficient R_{TKE}, is presented in figure 3. The meanings of the symbols and coordinates are the same as in figure 2.

From R_{TKE} point of view, the solid wall is even the worst case with respect to the sediment transport behind the shelter. It increases the turbulent kinetic energy (negative values of R_{TKE}) at downstream distances further than $x/H = 0.5$ at all investigated heights, compared to the case without the fence. Both investigated porous fences

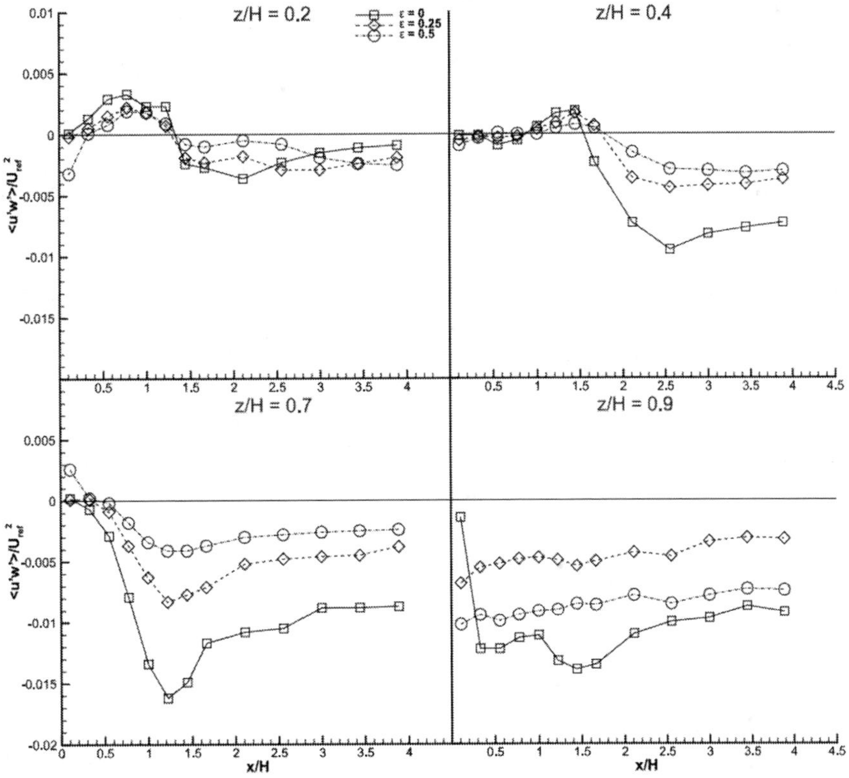

Fig. 5: The same as in figure 2 but for the dimensionless vertical momentum flux.

are more efficient compared to the solid wall. The best results of R_{TKE} are achieved with the fence porosity of $\epsilon = 0.5$ (circles in figure 3), where up to the downstream distance of $x/H = 2$ and for the lower investigated heights ($z/H = 0.2$ and 0.4) the fence still reducing the TKE, compared with the case without the fence. At the vicinity of the fence ($x/H < 1$), the TKE reductions of the fence with 0.5 porosity are very similar with those from the fence of 0.25 porosity (diamonds in figure 3).

If we observe the effect of a necessary vertical gap between the fence and the bottom of the model ($z/H = 0.1$), we can see from the figure 4a and figure 4b that strong wind-speed accelerations and not so strong turbulent kinetic energy reductions occur for all investigated porosities, respectively. Thus, the simulated vertical gap has unfavourable effect for sediment transport.

Because of better wind-speed and turbulent kinetic energy reductions are attained by fence of 0.25 and 0.5 porosity, respectively, the most favorable porosity cannot be stated. Hence, turbulent coherent structure analysis at following chapter would give a better insight to the fence porosity effectiveness.

4.2 Turbulent coherent structures analysis

Because the presented turbulent coherent structures are the partitions of the total vertical momentum flux $\langle u'w' \rangle$, we present the dimensionless momentum flux $\langle u'w' \rangle / U_{ref}^2$ in figure 5 as well. As in previous presented results, the vertical momentum flux is plotted as a function of dimensionless streamwise distance x/H at four graphs for dimensionless heights $z/H = 0.2$, 0.4, 0.7 and 0.9, respectively.

The effect of the gap is evident at the height $z/H = 0.2$ and up to downstream distance from the fence $x/H = 1.5$, where upward momentum flux occurs. At the middle height ($z/H = 0.4$), the fence momentum flux reduction effect ($\langle u'w' \rangle / U_{ref}^2 \approx 0$) up to $x/H = 1.5$ can be observed. For higher levels, the strong momentum flux downwards are present from downstream distances even close to the fence ($x/H > 0.5$). The lowest values of the momentum flux are attained by the fence of the 0.5 porosity across the downstream distance and up to $z/H = 0.7$.

The time fractions of the relevant coherent structures, sweeps (D_2) and outward interactions (D_1), with respect to the dimensionless streamwise distance x/H and height z/H are presented in figure 6. Using the hole quadrant analysis that filters the "weaker" coherent structures brings about the maximum time fraction decay approx. to 0.3 of the total momentum flux (approx. 70% of the events are filtered, not shown here). Thus, the maximum time fraction value of the sweep or outward interaction might be 0.3 from all investigated points.

The strong impact of the height on sweep time fractions can be observed from figure 6 (D_2, filled symbols). At the levels higher than the middle height of the fence $z/H > 0.4$) and at the downstream distances further than one height of the fence ($x/H > 1$), the sweep time fractions dominates over outward interaction ones. The highest values of the sweep time fractions (around 10% of sampling time) are attained by solid wall (filled squares). If we consider that the time fractions are filtered from the

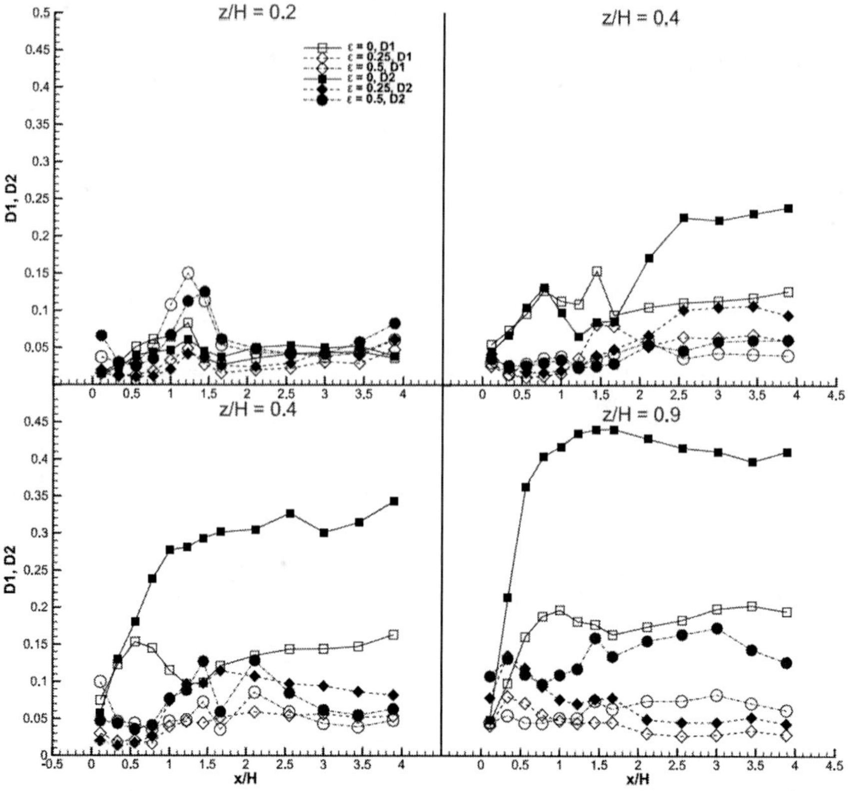

Fig. 6: The same as in figure 2 but for the sweep (D_2, filled symbols) and outward interaction (D_1, empty symbols) time fractions.

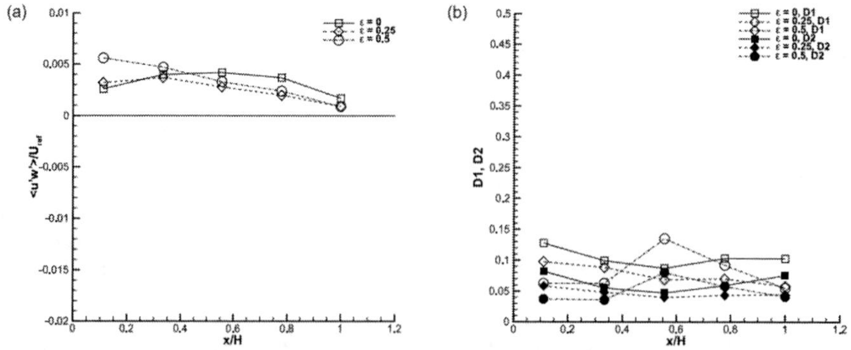

Fig. 7: (a) Dimensionless vertical momentum flux and (b) sweep (D_2, filled symbols) and outward interaction (D_1, empty symbols) time fractions for the fence porosity $\epsilon = 0$ (squares), $\epsilon = 0.25$ (diamonds) and $\epsilon = 0.5$ (circles) as a function of the dimensionless streamwise distance from the fence, x/H, at dimensionless height $z/H = 0.1$.

insignificant ones, the dominant sweep time fractions occupy approx. 30% of the relative sampling time at these higher levels. In comparison with the sweep time fractions for the fences of the 0.5 and 0.25 porosities, this represents approx. four time higher occupancy of the relative sampling time. The lowest values of the sweep time fractions between can be observed for the fences of 0.25 porosities (filled diamonds) at all investigated heights.

The outward interaction time fractions (D_1, empty symbols in figure 6) attained the highest values for the solid wall at height $z/H < 0.9$. For porous fences, the highest values can be observed the fence vicinity ($x/H < 1.5$) and lower heights ($z/H < 0.4$). This points out on the effect of the vertical gap below the fences. As for the case of the sweeps, the lowest values of the outward interaction time durations are attained by the fence of 0.25 porosity. However, these values are slightly lower than in case of the fence with 0.5 porosity. Compared with the sweeps, lesser impact of the height on outward interaction time fraction can be observed.

From figure 7a and figure 7b we can see no significant effect of the fence porosity on total momentum flux and coherent structure time fractions in the region close to the vertical gap ($z/H = 0.1$), respectively. In the case of the coherent structure time fractions the outward interactions (empty symbols in figure 7b) are more frequent than the sweep events (filled symbols in figure 7b).

5 Conclusions

The presented wind-tunnel experiment on porous shelter model revealed the impact of the geometry and porosity on downstream flow quantities. We demonstrated that not only the wind-speed and turbulent kinetic reduction coefficients but also the vertical momentum flux and its partitions should be taken into account for the shelter efficiency determination with respect to the aeolian sediment transport. According to these turbulent quantities the non-porous shelter shows the most unfavourable flow for the sediment transport, hence the worse shelter efficiency. The necessary shelter construction arrangement, comprising the vertical gap below the shelter porous/non-porous leeward wall, has unfavourable effect on shelter efficiency of all investigated shelter porosities. The best shelter performance associated with the highest turbulent kinetic reduction, the lowest total vertical momentum fluxes, and very low dominant turbulent coherent structures time fractions were observed for the shelter of 50% porosity. This porosity was worse than others only for the wind-speed reduction coefficient and have slightly lower values of the dominant coherent structures time fractions than the fence of 25% porosity. These results for the very high fence porosity suggest that the vertical gap at the shelter leeward wall would be unnecessary considering the dust explosiveness limits, hence improve sheltering effect.

Acknowledgements

This work was supported by the Czech Science Foundation GACR (project GAP15-18964S) and by the institutional support RVO: 61388998.

References

1. Z. Dong, W. Luo, G. Qian, P. Lu, H. Wang, Journal of Arid Environments **74**, (2010) 193-207
2. W. M. Cornelis, D. Gabriels, Journal of Arid Environments **61**, (2005), 315-332
3. W. Yue, C. L. Lin, V. C. Patel, J. Fluid Eng. **127**, (2005), 858-864
4. L. Kukačka, Š. Nosek, R. Kellnerová, K. Jurčáková, Z. Jaňour, TheScientificWorldJournal **2012**, (2012) 13
5. Z. Jaňour, P. Jonáš, Strojnický časopis **42**, (1991) 443-450 (in Czech)
6. W. H. Snyder, *Guidelines for fluid modelling of atmospheric diffusion* (EPA office of Air quality, USA, 1981) 199
7. P. Jonáš, Čs. Čas. Fys. **7**, (1957) 76-86 (in Czech)
8. VDI, Verein Deutcher Ingenieure, GER, 2000 36
9. G. F. S. Wiggs, C. M. Weaver, Geophysical research letters **39**, (2012)
10. S. S. Lu, W. W. Willmarth, Earth Sci. Rev. **80**, (1973)
11. G. Katul, D. Poggi, D. Cava, J. Finnigan, Boundary-Layer Meteorology, **120**, (2006) 367-375

EPJ Web of Conferences **114**, 02085 (2016)
DOI: 10.1051/epjconf/201611402085

Evaluation of thermal conditions inside a vehicle cabin

Tadeusz Orzechowski[1,a] and Zbigniew Skrobacki[1]

[1]*Kielce University of Technology 25-314 Kielce, Al. Tysiąclecia PP 7, Poland*

Abstract. There are several important factors influencing road accidents. Temperature inside the vehicle is ranked third after alcohol and seat belts. For this reason, maintaining thermal comfort in the passenger compartment is essential. Thermal comfort is provided by the air conditioning system, which consumes much energy. In the case of electrically powered vehicles, this results in a shorter range. Optimization of such systems is therefore required. This paper proposes a set of equations describing the thermal conditions inside the vehicle, which are the result of appropriate energy balances for air, interior elements, and glass. Variable transmission conditions are included for transparent materials exposed to short and long wave radiation. The study focused on unsteady air-conditioning of the vehicle interior. The measurement data was compared with the results obtained through numerical solutions of the proposed set of equations.

1 Introduction

The beginnings of air conditioning go back to 1902, when Carrier designed and constructed a spray-type air washer, which was first applied to control the temperature and humidity of a publishing company. Since then, air conditioning has been defined as a process of supplying and maintaining desirable internal atmospheric conditions for human health and comfort. In the years that followed, the process was used first to improve the comfort within buildings and later also within motor vehicles. At the beginning, it was a relatively primitive, uncontrollable system. From 1933, it was installed in American luxurious limousines with powerful engines capable of handling additional load. After World War 2, Chrysler began installing such systems in series production cars. In the late 1960s, half of the cars produced in the US were equipped with such a system. In Europe and Asia, the American invention was perceived as a luxury car feature. This century it is becoming increasingly common in popular cars. Air conditioning is required when temperature is high; in the summer, when it frequently exceeds 30 °C, air conditioning is even indispensable for the safe use of motor vehicles. These observations have been corroborated by findings on the relationship between road accidents and ambient temperature. Reference [1] considers the effects of several dozen different factors on accident frequency rates. The results show that alcohol is the most dangerous factor and temperature is reported to be the third.

Thermal comfort in enclosed spaces can be determined using, for instance, the Predicted Mean Vote (PMV) index developed by Fanger [2]. It describes human thermal impressions expressed in a seven-point scale from hot (+3) to cold (-3). It is recommended that the index value should not exceed ± 0.5, which corresponds to the percentage of occupants not satisfied with the indoor climate, known as the Predicted Percentage Dissatisfied (PPD), which, in turn, should not to be greater than 10 %. The temperature and humidity comfort is affected by the temperature and relative humidity of the air, the air flow velocity, and the temperature of the surrounding surfaces. However, these are not the only factors affecting the conditions required for thermal comfort. It is important to consider the following: the microbiological, particle and gaseous composition of air, air quality, lighting, noise, etc., as well as insulating effects of clothes, which are different in summer and winter. As shown in [3], the parameters are also dependent on the velocity of the moving vehicle.

As discussed in [4], some factors are difficult to predict; these include road conditions, environmental conditions, the mental and physical fitness of the driver, the driver's behaviour, etc. When considered, these factors relate the optimal comfort temperature with the air quality, which may lead to a decrease in energy consumption. Another study [5] shows that the local temperature and humidity comfort conditions have a considerable influence on the operation of the HVAC system in an electrically-powered vehicle. This reduces the driving range between recharging. An attempt was made to solve the problem, using the example of an electrically powered vehicle [6]. The thermal comfort model discussed in this work can be used to answer the questions concerning the effects of the environmental conditions on the range of electric vehicles, the requirements resulting from local climatic conditions, and the influence of air pre-conditioning to achieve the

[a] Corresponding author: todek@tu.kielce.pl

required comfort conditions before the journey on the vehicle range.

The effects of changes in the relative humidity of the cabin environment on the thermal comfort of the occupants were considered in [7]. From the thermodynamic and psychometric analysis it is evident that a change in the relative humidity can facilitate the process of air conditioning and reduce energy losses. The analysis results were represented graphically as PPD versus PMV. It was found that when the ambient temperature was high and the relative humidity of the air increased, there was a rise in the energy consumption of the air conditioning system. If the AC system is to be controlled effectively, it needs to take account of the relative humidity of the fresh air. However, air conditioning by evaporative cooling requires considerable changes in the system structure.

In another study [8], the tests were conducted on a group of fifty people, who were randomly selected to drive under three different climatic conditions: under cold, neutral and hot conditions, which corresponded to 5 °C, 20 °C, and 35 °C, respectively. In all the cases, the same relative humidity conditions of 50 % were maintained. The best driver's performance was observed under neutral conditions. During the tests, which lasted 30 minutes, no influence of ambient temperature on the human body temperature and the heart rate was reported. However, the driving quality was very much affected by the driver's concentration, which declined significantly when the air conditioning parameters were controlled manually. Additionally, it was found that the manual selection of the direction of the air flow affected the thermal comfort of occupants and the efficiency of the HVAC system [9].

Investigations conducted in the climatic conditions of Paris [10] showed that the important factors affecting the temperature inside a vehicle with a large area of glazing were the solar radiation, the glass type, the vehicle colour, and the radiation properties of the interior materials. The mathematical model proposed by the researchers was based on the heat balance equation. However, it did not take into consideration the influence of the long-wave radiation caused by heated interior elements, i.e. the radiation almost entirely absorbed by transparent heat barriers, on the glass temperature, which can be responsible for certain discrepancies between the calculation results and the measurement data. A similar approach was used to calculate the thermal conditions in [11]. Despite the relatively good agreement between the calculation results and the measurement data, the researchers indicated that the model was not perfect because of the insufficient accuracy of the description of the radiation heat transfer inside the vehicle and the variable heat gains from the occupants attributable to the temperature and humidity conditions in the passenger compartment.

Different materials are used for the interior elements. Their physical properties are not always known. While determining dynamic changes in thermal conditions, we can improve the calculation accuracy by taking into account the thermal capacity of the interior elements and their materials and the radiation properties of different surfaces, especially their transmission and absorption properties [12 , 13].

Air conditioning is a highly energy-consuming process, which leads to a higher consumption of fuel in conventional vehicles, and a shorter range in electric vehicles [14]. Possible solutions to this drawback include improving the efficiency of energy-storage systems and optimizing the operation of others, especially the heating and air-conditioning systems [15].

2 Energy balance

In this study it was assumed that, during unsteady-state heating or cooling of the vehicle passenger compartment, the interior was a space with uniform air flow conditions. With this assumption, the instantaneous temperature T in time t was determined from the following heat balance equation:

$$\begin{aligned} m c_p \frac{dT}{dt} &= m_{AC} c_p (T_{AC} - T) + m_V c_p (T_V - T) + \\ &+ m_{\inf} c_p (T_a - T) + \alpha_{iv} (T_g - T) A_g + \\ &+ (\alpha_{ih} A_h + \alpha_{iv} A_v)(T_b - T) + N q_{p1} \end{aligned} \quad , \quad (1)$$

where m is the cabin air mass and c_p is the air specific heat.

The next terms on the right side of the heat balance equation denote: heat transferred to the interior air when the air conditioning is on, heat supplied with the ventilation air, heat from the air removed from the interior, heat from the surrounding surfaces due to convection phenomena, and heat from the occupants, respectively. The mass and temperature of the inlet air streams, m_{AC}, m_V, m_{inf} and T_{AC}, T_V, T_{inf}, respectively, can be determined from the parameter plots and recommendations on the settings of the HVAC system. These are: the mass and temperature of the air supplied by the air conditioning system (m_{AC}, T_{AC}), the mass and temperature of the air supplied by the ventilation system (m_V, T_V), the mass and temperature of the air entering the passenger compartment due to infiltration through open air blow ducts and other leaks (m_{inf}, $T_{inf} = T_a$), the temperature of the transparent surfaces T_g with surface areas A_g, and the temperature of other interior elements marked T_b. As the heat transfer coefficients differed in value, the surface area of the walls enclosing the passenger compartment space was considered according their direction: vertical A_v and horizontal A_h. N is the number of occupants in the compartment, and q_{pl} is the total heat, which is a sum of the sensible heat and the latent heat emitted by a single person. The temperature of the solid elements inside the passenger compartment (seats, instrument panel, etc.), T_b, present in Eq. (1) can be calculated from the following equation:

$$\begin{aligned} C_b \frac{dT_b}{dt} &= Q_{cdb} + Q_{rad} = \\ &= \alpha_b (T - T_b) A_b + \varepsilon_g \sigma (T_g^4 - T_b^4) A_b \approx \alpha_b (T - T_b) A_b \end{aligned} \quad , \quad (2)$$

where C_b is the thermal capacity of the interior elements, Q_{ab} is the heat transferred by convection with the interior air, and Q_{rad} is the heat gained (or lost) due to radiation through transparent heat barriers and radiation between solid elements and transparent barriers differing in temperature, e.g. the inside surface of the vehicle roof.

The solid parts include the seats, the floor, the instrument panel, etc. Because of their different location in the passenger compartment space, they are subjected to different direct radiation of the sun, and accordingly, they may differ in temperature. This is particularly true about the instrument panel. It is thus more convenient to determine the average air temperature and the heat transfer coefficient α_b.

Radiation may affect materials. The effects are dependent on the wavelength and angle of radiation as well as the optical properties of the material. In the case of spaces enclosed with glass elements that are transparent for short-wave radiation and almost entirely absorb long-wave radiation, the thermal load caused by solar radiation is calculated for the two components separately [16, 17]: short-wavelength (SW) radiation for the spectral range $\lambda \leq 2.5$ μm and long-wave (LW) radiation for $\lambda > 2.5$ μm. Thermal effects in the long wavelength range are characteristic of diffuse and reflected radiation and low temperature bodies.

Assuming the glass thickness h_g, glass density ρ_g and glass specific heat c_g, we can calculate the change in temperature T_g over time from the following equation:

$$h_g \rho_g c_g \frac{dT_g}{dt} = \alpha_e (T_a - T_g) + \\ + \alpha_{gsw} I_{sw} + \alpha_{glw} I_{lw} + \alpha_i (T - T_g) + \\ + \alpha_{glw} \sigma T_b^4 - 2\varepsilon_g \sigma T_g^4 + \alpha_{glw} I_n \quad (3)$$

It should be noted that, for a moving vehicle, the heat transfer coefficient on the outside α_e is dependent on the actual velocity of the air. The coefficient is determined from an appropriate relationship for the Nusselt number, i.e. for forced convection, when the vehicle is moving, and for natural convection, when the vehicle is stopped [16]. I denotes radiation intensity in the SW or LW range. The radiation absorption coefficient for radiation absorbed by the glass in the appropriate spectral range was denoted by α_g. I_n is the intensity of the solar radiation, whose spectrum is in the LW range. For air with temperature T_a containing water vapour whose amount is defined by its partial pressure p_{H2O} (in hPa), this quantity can be calculated using the Stefan-Boltzmann law, in which the temperature of the sky T_n is determined from the formula:

$$T_n = T_a \sqrt[4]{0.526 + 0.065\sqrt{p_{H_2O}}} \,. \quad (4)$$

The windscreen also radiates energy. Assuming that the glass is optically grey, its value is determined by means of the specific emission coefficient ε_g. An appropriate term describing this quantity is included in the heat balance equation (3).

The proposed model takes account of the radiation effects related to the emission and absorption properties of the surrounding surfaces [13]. Additionally, it can be used to optimize the operation of the evaporator in an automotive AC system, which is achieved by microscale surface enhancement in systems where the principle of operation involves the phase change of the refrigerant [18, 19] in different boiling regimes: from nucleate to film boiling [20, 21, 22]. The model, which is based on a system of three differential equations, can be employed to optimize the control of the automotive HVAC system, and accordingly, to improve the fuel economy or increase the range [23]. This approach is particularly important for electrically-powered vehicles.

3 Test results

The tests were conducted on a Suzuki SX4 equipped with a manufacturer - installed manual single-zone air conditioning system. In this vehicle, the air flow parameters are selected with two knobs: one for turning the temperature up or down (continuous mode), and the other for changing the position of the fan speed setting (step mode). In the case of the latter, there are four positions of the fan speed corresponding to four air flow rates. The air coming from the blower can be directed only towards the windscreen, both towards the windscreen and the upper or lower zone of the internal space through a system of directional panel vents and floor vents. Their location is shown in the schematic diagram in figure 1.

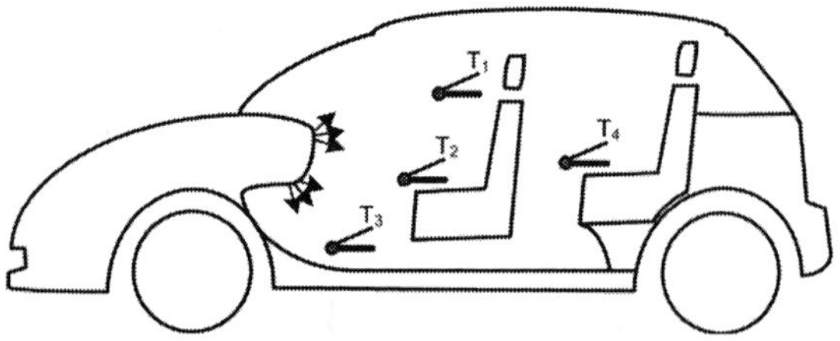

Figure 1. Location of the thermocouples in the passenger compartment.

The parameters of the cooled air supplied to the interior were measured by setting the air conditioning system at the lowest temperature with the air flow directed towards the head and the feet. The temperature was registered at four different positions of the fan speed (positions I to IV). The temperature and velocity of the air blown inside were measured with a thermocouple sensor vane anemometer. The resolution of the temperature readings obtained with the meter was 0.1 K, while its accuracy was ±0.5 K. The corresponding values for the measurement of the incoming air flow velocity were: resolution 0.01 m/s and accuracy ±0.1 m/s. The sensors were placed in the plane of two panel grilles with the air directed towards the head. Each measurement was repeated several times. Their average values are shown in Table 1.

Table 1. Air temperature and air flow velocity at four different positions of the fan speed.

Setting	T_{mean} °C	V_{mean} m/s
I	9.2	0.45
II	12.35	0.97
III	14.6	1.7
IV	17.3	2.81

The temperature of the outside air was constant at about 30.5°C before cooling. After cooling, its value was strongly dependent on the air flow velocity. At the lowest fan speed (control knob in position I), it was at about 9.2°C, while at the highest it rose to 17.3 °C.

After a longer period of non-operation at hot summer temperatures, with the vehicle parked in full sun, the interior air reaches a temperature at which the driver's performance drops. Under such circumstances, the air needs to be cooled immediately, which is achieved by switching on air conditioning. The system can operate in different modes. One of the parameters to control is the blower speed. The speed affects the temperature of the incoming fresh air. Thus, thermal conditions not only change over time but they are also dependent on the parameters set. Five thermocouples were mounted inside the passenger compartment to monitor them; a sixth one was used to measure the temperature of the outside air. The first of the thermocouples was employed to measure the temperature at the driver's head level T_1. The second determined the temperature at the driver's knee level T_2. Temperature T_3 was measured at the feet level. Two thermocouples were placed at the back to measure the temperature at the knee level for passengers occupying the rear seat (in figure 1, both marked T_4). All the sensors were linked to the data acquisition station, which was programmed to register all signals with the same frequency of 1/15 Hz (four measurements per minute). Before the tests, the vehicle was parked in full sun and then travelled about 5 km to heat the engine so that, when switched on, the air conditioning system could operate in steady state thermal conditions. Next, the vehicle was parked in the shade and the measurement of temperature commenced.

In the first series, the fan speed control knob was placed in position I, which corresponds to the minimum capacity of the fan. After about 7 minutes, the fan was turned off, the interior was aired thoroughly, and then the fan control knob was placed in position II. Measurements were taken after about 7 minutes, with all the doors closed. The test results are illustrated in figure 2.

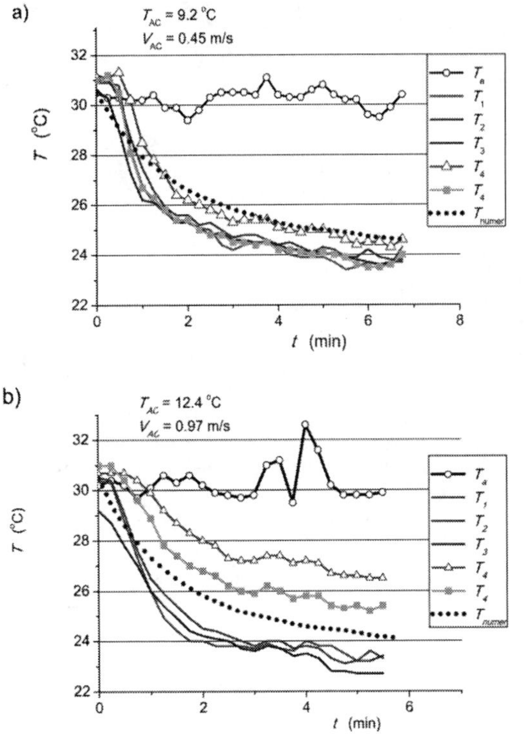

Figure 2. In-vehicle temperature plots for different points in the compartment (see figure 1) and the ambient temperature plot T_a, with the blower fan control knob placed in a) position I, b) position II.

In the first series, the fan speed control knob was placed in position I, which corresponds to the minimum capacity of the fan. After about 7 minutes, the fan was turned off, the interior was aired thoroughly, and then the fan control knob was placed in position II. Measurements were taken after about 7 minutes, with all the doors closed. The test results are illustrated in figure 2.

4 Result analysis

The flow of the air blown into the passenger compartment can be blended, low turbulent or laminar in nature. The division is dependent on the degree of turbulence of the air stream, understood as the ratio of the time-averaged pulsation velocity to the average flow velocity.

When the turbulence is higher than 20 %, extensive blending occurs at the stream threshold between the air blown into the compartment and the interior air. This leads to an increase in the amount of air in the stream with an increase in the distance from the air inlet opening. In the low turbulent flow, with the degree of turbulence between 10 and 20 %, cold air flows in and causes the warm interior air, which absorbed heat from the surrounding objects, to go up by convection. As a result, the differences in temperature in the vertical cross-section are reduced considerably. A laminar flow is characteristic of very small and steady-state flow velocities, lower than 0.4 m/s; to keep it, it is necessary to maintain steady state conditions of the inflow of fresh air. This type of the air conditioning of the vehicle interior is achieved when the blower is off, the driving velocity is constant and the conditions outside the vehicle are steady.

Figure 2 shows plots of temperature registered at different points of the passenger compartment; the measurements were taken by changing the fan speed only, which resulted in different air flow velocities. With a relatively small air flow velocity and the fan speed set at a minimum, the air stream directed towards the heads of the occupants in the front seats does not reach the heated parts of the ceiling. As the cool air is heavier than the air in the upper zones of the passenger compartment, it falls towards the floor. After being slightly heated there, it is forced up by convection to the upper parts of the compartment. The phenomena are characteristic of the low turbulent air flow, which results in a uniform distribution of temperature inside the passenger compartment.

In the second series of the measurements, when the fan speed was higher, the average velocity in the instrument panel plane doubled. This results in a strongly turbulent flow. Air is directed down towards the occupants' feet and up towards their heads. Since the tests were conducted for an empty vehicle, with no occupants in the passenger compartment, the air stream moved freely to the ceiling and then parallel to it. Its temperature was higher than the ambient temperature so the air stayed in the central part of the compartment. The temperature measured at the knee level at the back for passengers occupying the rear seat was 3 to 4 degrees higher than that reported at the front. This is likely due to the hindered flow of cool air under the front seats from the front to the rear part of the passenger compartment. In the front part of the compartment, however, no considerable changes in temperature were reported in the vertical cross-section.

Ambient temperature, plotted in figure 2, was an independently measured quantity. The measuring thermocouple was placed directly at the data acquisition station near the front door at the doorstep level. The measurements were carried out in a relatively short period of time so that the ambient temperature remained almost steady at about 30.5 °C. As it was necessary to maintain steady-state operating conditions of the A/C system, the tests were performed on a preheated engine. While running under such conditions, the engine got very hot; as a result, every now and then, the fan switched on automatically to draw the air through the radiator and blow it over the engine surface. Some of the air flew as far as the sensor. As can be seen in Fig. 2, an instantaneous increase in temperature was registered then.

Equations (1), (2) and (3) describe the time-dependent temperature of the cabin air, the cabin equipment, and the windows, respectively. For the test conditions, the system was solved numerically. For both analyzed settings of the air blower, the calculation results were plotted in figure 2. The discrepancies between the measurement data and the calculation results are due to the simplifying assumptions made in this study.

From the study it is clear that at the predetermined control settings, the most uniform and, accordingly, the most favourable distribution of the cabin air temperature occur at low turbulent flow. This corresponds to the minimum capacity of the blower in the HVAC system.

It is important to note that the measurements were performed on a parked vehicle with the engine switched off and no occupants inside. The presence of a driver and one or more passengers would have definitely changed the cabin air flow. However, it seems that the physical phenomena related to the nature of the air flow and the influence of the air flow on the temperature distribution would have been similar.

References

1. Th.-J. Zlatoper, Accid. Anal. Prev., **23**, 5 (1991)
2. P.O. Fanger, *Thermal comfort* (Arkady, Warszawa, 1974) (in Polish).
3. J. Fišer, J. Pokorný, EPJ Web of Conferences, **64** (2014)
4. H. Khayyam, Appl Therm Eng., **51** (2013)
5. K.R. Kambly, T.H. Bradley, J Power Sources, **259** (2014)
6. K.R. Kambly, T.H. Bradley, J Power Sources, **275** (2015)
7. A. Alahmer, M.A Omar, A. Mayyas, S. Dongi, Appl Therm Eng, **31** (2011)
8. H.A.M. Daanen, E. Vliert, Xu Huang, Appl Ergon, **34** (2003)
9. J. Fišer, M. Jícha, EPJ Web of Conferences, **45** (2013)
10. A. Mezrhab, M. Bouzidi, Appl Therm Eng, **26** (2006)
11. D. Marcos, F. J. Pino, C. Bordons, J. J. Guerra, Appl Therm Eng, **66** (2014)
12. J. Pokorny, J. Fišer and M. Jicha, EPJ Web of Conferences, **64** (2014)
13. T. Orzechowski, K. Ziętala, District Heating, Heating, Ventilation, **45/8** (2014)
14. K.R. Kambly, T.H. Bradley, J Power Sources, **259** (2014)
15. J. Fišer, J. Pokorný, M. Jícha, EPJ Web of Conferences, **25** (2012)
16. F. P. Incropera, D.P. DeWitt, T.L. Bergman, A.S. Lavine, *Fundamentals of Heat and Mass Transfers* (John Wiley & Sons Inc., U.S.A, 2006)
17. T. Kruczek, Energy, **91** (2015)
18. T. Orzechowski, Exp Therm Fluid Sci, **31** (2007)
19. R. Pastuszko, Int J Therm Sci, **49** (2010)
20. T. Orzechowski, A. Tyburczyk, EPJ Web of Conferences, **67** (2014)
21. T. Orzechowski, S. Wciślik, Int J Heat Mass Tran, **73** (2014)
22. R. Pastuszko, T. M. Wójcik, Exp Therm Fluid Sci, **63** (2015)
23. Z. Skrobacki, Maintenance and Reliability, **3** (2007)

EPJ Web of Conferences **114**, 02086 (2016)
DOI: 10.1051/epjconf/201611402086

Quasi-stationary phase change heat transfer on a fin

Tadeusz Orzechowski[1,a], Katarzyna Stokowiec[1]

[1] *Kielce University of Technology, 25-314 Kielce, Al. 1000-lecia P.P. 7, Poland*

Abstract. The paper presents heat transfer research basing on a long fin with a circular cross-section. Its basis is welded to the pipe where the hot liquid paraffin, having a temperature of 70oC at the inflow, is pumped. The analyzed element is a recurrent part of a refrigeration's condenser, which is immersed in a paraffin. The temperature of the inflowing liquid is higher than the temperature of the melting process for paraffin, which allows the paraffin to liquify. The temperature at the basis of the rib changes and it is assumed that the heat transfer is quasi-stationary. On this basis the estimation of the mean value of heat transfer coefficient was conducted. The unsteady thermal field of the investigated system was registered with an infrared camera V50 produced by a Polish company Vigo System. This device is equipped with a microbolometric detector with 384 × 288 elements and the single pixel size 25 × 25 μm. Their thermal resolution is lower than 70 mK at a temperature of 30 °C. The camera operates at 7,5 ÷ 14 μm long-wave infrared radiation range. For a typical lens 35 mm the special resolution is 0.7 mrad. The result of the calculations is mean heat transfer coefficient for the considered time series. It is equal to 50 W m $^{-2}$ K^{-1} and 47 W m $^{-2}$ K^{-1} on the left and right side of the fin, respectively. The distance between the experimental data and the curve approximating the temperature distribution was assessed with the standard deviation, Sd = 0.04 K.

1 Introduction

There are multiple energy saving requirements both in residential and industrial constructions. They include statutory restrictions dealing with low heat transfer coefficients in each of the building partitions on one hand and an increased necessity of heat recovery applications and absorption of mostly renewable energy on the other. Currently employed methods with sensible heat storage are not effective enough due to the possible short time of its storage. Therefore phase change materials (PCM) can be a promising alternative used in order to absorb latent energy. They seem to be a solution to the problems of sensible systems since they work during phase change process and thus stabilize the temperature [1]. Latent heat is usually produced through solidification or melting.

The multiple advantages of energy storage with phase change materials predispose this kind of installation as preferable applications. Phase change process can be applied in many areas where solar energy absorption systems with solar collectors are used [2]. The application of phase change materials allows to acquire the proper microclimate conditions inside rooms while energy is stored for heating [3] and domestic hot water [4] purposes. On the other hand cool absorption by means of PCM is also possible for air-conditioning or free-cooling purposes [5]. The processes mentioned can be delivered by filling the proper tank with PCM and storage of energy gained for long periods of time [6]. The other possibility is to build the phase change material in the building structure. It allows to decrease the influence of temperature change inside the room on microclimate [7]. Such an installation may also absorb heavy heat fluxes that may be derived from fire flames [8] and constitute the passive element of fire-fighting protection. Phase change processes can be also applied in inactive road defrost [9] decreasing the exploitation costs and eliminating the dependence on mobility of traffic maintenance workers. This kind of solution brings advantages both in economic and ecologic aspects not only to the direct user.

In order to select the PCM suitable for the required system it is necessary to learn the thermal features of the substance chosen such as heat capacity, phase change heat and melting temperature [10]. Phase change materials are diverse. There are organic substances such as fatty acids, esters or paraffin and non-organic ones such as salts or hydrated salts among the PCMs [11]. When taking into consideration the last-mentioned ones we observe the disadvantages of supercooling and phase separation during heat absorption. The issues can be limited with sodium chloride solution addition [12]. As for the organic substances the low thermal conductivity problem is the drawback and can be solved by

[a] Corresponding author: todek@tu.kielce.pl

encapsulation [13] as well as application of multiple nanoparticles additives such as boron nitrate [14] or cement composites [15]. The research presents other study areas. Shell materials are employed in order to create capsules. It contributes to the limitation of natural convection processes during the phase change in favor of enhancing the thermal conductivity [16]. The combination of paraffin with nanoparticles of iron oxide Fe_3O_4 seems to be a very promising method since it results in an increase of thermal conductivity by ca. 8 % [17]. The shape of the tank as well as the distance between PCM and the tank's wall are also significant factors since they affect the dynamics of the process [18].

The correct selection of phase change material for the process required involves the analysis of some parameters of the material chosen. First of all, it is necessary to learn the latent heat of fusion and the melting temperature. Therefore the paraffin can be applied in cooling installations as well as in order to satisfy the users' needs for domestic hot water [19].

Accumulation of heat that can be used for future purposes requires the installation of different kinds of tanks. The most effective ones employ the phase change processes. The loading of the PCM tank is conducted when the energy is available while the discharging process proceeds during the peak heat or cool demand hours. Both processes take place in a relatively short period of time. That is why the heat exchanger needs to be designed with a highly developed surface that is usually finned. In order to improve the heat transfer some surface modifications can also be applied [20, 21]. Experimental results are used in case of calculation procedures for heat exchangers working during the phase change process [22].

It is required to learn the heat transfer coefficient in order to design and optimize the construction of such installations. Unfortunately the availability of these data for phase change systems is insufficient. Therefore the aim of this paper is the study of methodology and measurement of heat transfer coefficient for a finned heat exchanger with paraffin filling.

2 Methodology

Heat transfer coefficient values vary depending on fluid physical parameters and geometry of heat transfer systems. Such a great number of factors required to be determined causes the necessity to obtain the individual measurements. That mostly involves PCM installations with heat exchangers with complex geometry. In order to measure the heat transfer coefficient during paraffin melting a test facility was set. Its scheme is presented in Figure 1.

A standard refrigerator's condenser was employed in the research (1). It is constructed with a spirally arranged steel coils and fins with round intersection transversely welded. This kind of heat exchanger can be used with PCM tank due to the high density of the fins and properly chosen geometrical parameters of the exchanger.

A 30x15 cm element was cut out of the condenser and placed in the Perspex vessel that was prepared specially for this purpose. Isolation features of this material limit the heat losses to the ambient air. On the other hand its transparency permits the observation of the phase change process directly. The vessel with the investigated element inside was immersed in the paraffin (2). The vessel was installed in a measurement system, as presented in figure 1. The liquid paraffin with an invariable flow was pumped through the exchanger coils. The oil temperature was constant by means of ultathermostat (4). The temperatures at inflow and outflow to the heat exchanger were controlled continuously during the research. Three thermocouples (3) were applied for that reason. Two of them were located at the liquid paraffin inflow and outflow to the coils while the third one registered the temperature inside the ultrathermostat. All the values were recorded by means of data acquisition station DaqLab (8). The station was connected with the PC computer (9) where the experimental data were saved continuously.

The temperature of the liquid paraffin at the inflow to the heat exchanger was kept as constant. Its stabilization, according to the ultrathermostat producer, was controlled with the precision of ≈ 0.1 K.

The liquid paraffin mass flow was assessed by

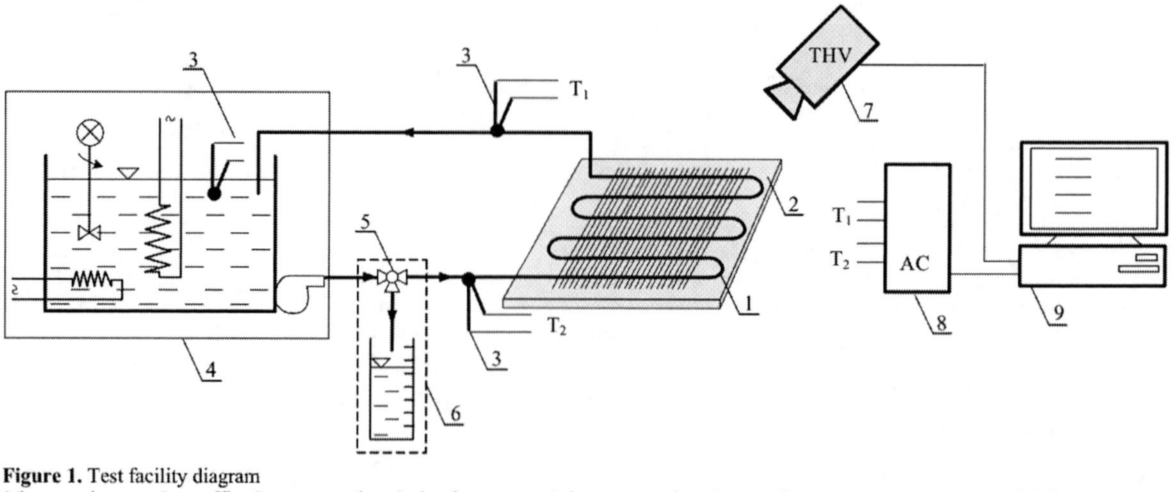

Figure 1. Test facility diagram
1-heat exchanger, 2-paraffin, 3-termocouples, 4-ultrathermostat, 5-three-way valve, 6-mass flow measurement system, 7-infrared camera, 8-data acquisition station, 9-computer

measurement of the fluid volume during the flow through the heat exchanger. In order to conduct the test the liquid paraffin flow to the heat exchanger was interrupted with a three-way valve (5) and the oil volume (6) as well as the filling time were measured. The mass flow was calculated since the dimension of the pipe was known.

The temperature distribution on the outer surface of the condenser was registered by means of infrared camera (7) V50 produced by Polish company Vigo System. The device is equipped with a microbolometric detector with 384 × 288 elements and with the single pixel size of 25 × 25 μm. Their thermal resolution is lower than 70 mK in 30 °C. The camera operates at a long-wave spectrum of infrared radiation that is 8 – 14 μm. The spatial resolution for a typical lens 35 mm is 0.7 mrad.

Each infrared camera operates according to the individual curve that is extrapolated during the device calibration process against the blackbody. The actual lenses emit not only their own radiation but also the reflected one [23], which influences the final measurement results [24]. The valid temperature reading requires the individual assessment of emission coefficient [25] whose estimation was conducted for the THV camera used in the tests.

The result of the research with the infrared camera is the temperature distribution on the outer surface of the heat exchanger. One of the examples is presented in figure 2. The thermogram shows two adjoining pipes with inflowing hot liquid paraffin. They are joined with

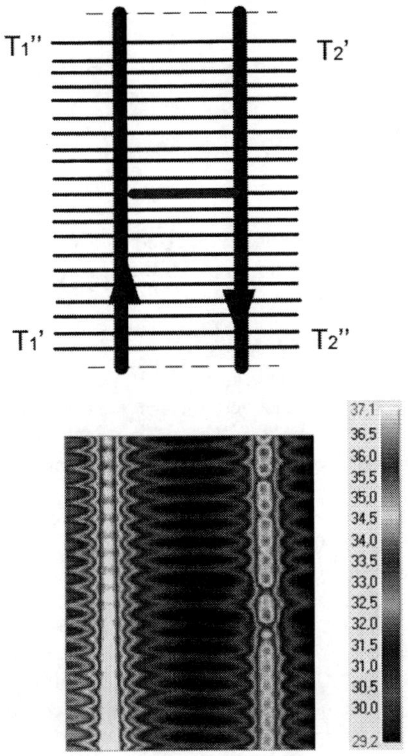

Figure 2. Thermogram of a recurrent element of the condenser.

steel elements in a fin shape, increasing the exchanger surface and intensifying the heat exchange process.

The surface temperature distribution can be read from the presented thermogram. One of the elements was selected as an example and its temperatures were presented in figure 3 where eight curves were shown. Each of them gives the temperatures on the fin every 50 seconds.

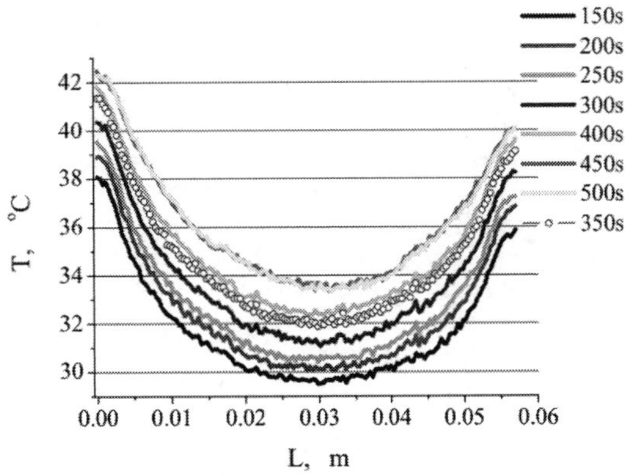

Figure 3. Temperature distribution along the fin

3 Results analysis

Heat transfer coefficient is constant in case of free convection and when the temperature change is low. Its value can be assessed by checking the measured temperature field against the analytical fit describing one-dimensional temperature distribution along the fin. In this case the analyzed element was divided in two parts so that the temperature derivative at each end equals zero. It present a fin isolated at one end. For those assumptions, in steady conditions with no internal heat sources, the dependence describing temperature T distribution along the fin with a fixed cross-section is given [26]:

$$T - T_f = (T_w - T_f)\left(\frac{e^{mx}}{1+e^{2mL}} + \frac{e^{-mx}}{1+e^{-2mL}}\right) \quad (1),$$

where: T_w temperature at the base of the fin, °C, T current temperature, °C, T_f ambient temperature, °C, L fin length, m, m coefficient for a fin with round intersection, given:

$$m^2 = \frac{\alpha \cdot P}{\lambda \cdot F} \quad (2),$$

where: α heat transfer coefficient, W m^{-2} K^{-1}, P element perimeter, m, λ heat conduction coefficient, W m^{-1} K^{-1}, F cross section area, m^2

The relation (1) can be applied in order to approximate the measured temperature distribution along the fin (see figure 3), which was calculated in each of the time series were conducted. The analyzed element was then considered as a system of two fins connected with the frontal surfaces. The temperatures at the base are different due to the heating process and gradual temperature increase of the liquid paraffin flowing through the heat exchanger. That is why we initially observe the difficulties in assessing the differences in heat transfer coefficient values for each of the fin. That is why the model (1) fit of the temperature distribution measured was conducted for each of the measurements individually applying the minimum mean square error assumption.

The approximation was performed for 0-1 system. The dimensionless proportion was assumed with the suitable temperature differences. The difference between the temperature on the fin and the minimal temperature for the fin system connected with the frontal surfaces was divided by the difference of temperature at the base of the fin and the minimal one. Next, the heat transfer coefficient was calculated for each of the measurements. An approximation for one of the time series (t = 350 s) is presented in figure 4 as an example. In this case heat transfer coefficient is 51 W m^{-2} K^{-1} on the left side of an analyzed element and 48 W m^{-2} K^{-1} on the right one. Similar calculations were performed for all the experimental data receiving approximate heat transfer coefficient values.

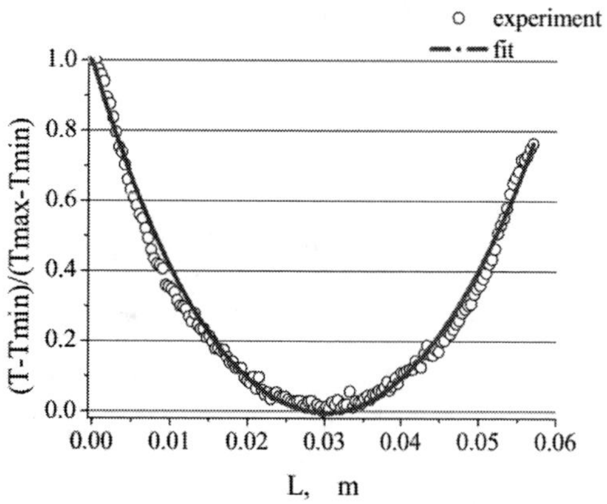

Figure 4. Temperature distribution along the fin for t = 350s.

The identical procedure was applied for all the results jointly, which is presented in figure 5. Mean heat transfer coefficient value for the investigated condenser on the left and right side of the fin are different from each other and they amount to 50 W m^{-2} K^{-1} and 47 W m^{-2} K^{-1}, respectively. As proved, the values do not diverge from analogically calculated values in each of the time series.

The method error was assessed applying the standard deviation value. It was calculated on the basis of the points distance from the approximation curve for the temperature distribution along the fin (see figure 5). The results are shown in figure 6. The calculated value of standard deviation is Sd = 0.04 K. The little value validates the physical similarity of the phenomenon on the considered element, what proves the quasi-stationary process. Therefore the given attempt to impose results in the constant system response with accurate precision. In the analyzed case the temperature increase at the fin base presents an almost parallel displacement of the temperature distribution (see figure 3).

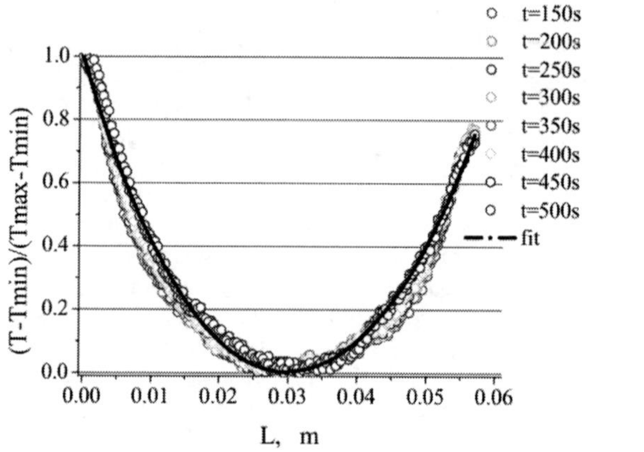

Figure 5. Temperature distribution along the fin for all the experimental data.

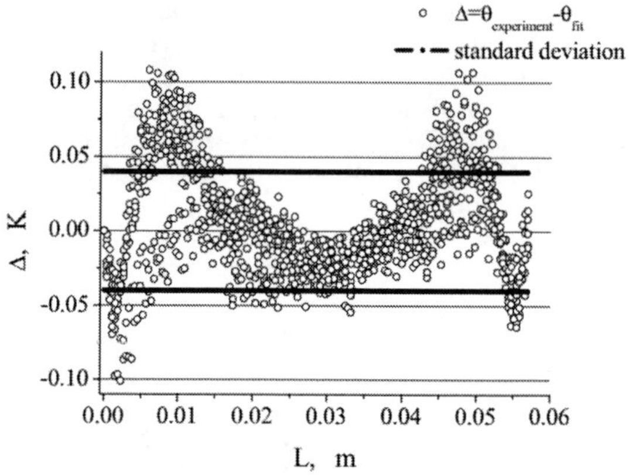

Figure 6. Comparison of measured and approximated temperatures.

4 Conclusions

This paper proposes the heat transfer coefficient measurement methodology for the refrigeration's condenser, which is immersed in a vessel with paraffin. Heat source is liquid paraffin that flows through coils in a serpentine shape. The applied heat exchanger is finned with crosswise steel rods in order to improve the heat transfer process. The measurements were conducted for a system presented in figure 1. The temperature field on the outer exchanger surface was examined with an infrared camera with a constant frequency. The temperatures on the fins' surfaces were selected basing on the temperature distribution. The authors' methodology to assess the heat transfer coefficient was developed for an analyzed element. It was discovered that the system is in quasi-stationary state after a period of time, as shown in figure 6.

The result of the calculations is mean heat transfer coefficient for the considered time series. It is equal 50 W m^{-2} K^{-1} fin 47 W m^{-2} K^{-1} on the left and right side of the fin, respectively. The distance between the experimental data and the curve describing the temperature distribution was assessed with the standard deviation. In this paper its value is low Sd = 0.04 K (see figure 6). It proves the quasi-stationary process thesis and effectiveness of the chosen methodology for heat transfer coefficient measurement.

References

1. P. Charvat, J. Stetina, O. Pech, L. Klimes and M. Ostry, EPJ Web of Conferences **67**, 02046 (2014)
2. P. Charvat, O. Pech and J. Hejcik, EPJ Web of Conferences **45**, 01127 (2013)
3. N. Soares, J.J. Costa, A.R. Gaspar, P. Santos, Energ Buildings **59** (2013)
4. M. Pomianowski, P. Heiselberg, Y. Zhang, Energ Buildings **67** (2013)
5. S. E. Kalnæs, B. Pe. Jelle, Energ Buildings **94** (2015)
6. E. Rodriguez-Ubinas, L. Ruiz-Valero, S. Vega, J. Neila, Energ Buildings **50** (2012)
7. X. Shia, Sh. Ali Memon, W. Tang, H. Cui, F. Xinga, Energ Buildings **71** (2014)
8. P. Wang, N. Li, Ch. Zhao, L. Wu, G. Han, Procedia Engineering **71** (2014)
9. A. Tyburczyk, Logistyka, **4** (2015)
10. T. Orzechowski, K. Stokowiec, COW, **44/1** (2013)
11. K. Pielichowska, K. Pielichowski, Progr Mat Sci **65** (2014)
12. Y.Yang, J. Luo, S. Li, G. Song, Y. Liu, G. Tang, . Sol Energ Mat Sol C **139** (2015)
13. L. Fan, J.M. Khodadadi, Renew Sust Energ Rev **15** (2011)
14. Z. Yu, G. Cheng, Y. Hu, Energ Convers Manage **80** (2014)
15. B. Xu, Z. Li, Appl Energ **121** (2014)
16. A. R. Archibold, M. M. Rahman, J. Gonzalez-Aguilar, D. Y. Goswami, E. K. Stefanakos, M. Romero, Energy Procedia **57** (2014)
17. N. Şahan, M. Fois, H. Paksoy, Energ Mat Sol C **137** (2015)
18. N. S. Dhaidan, J. M. Khodadadi, Renew Sust Energ Rev 43 (2015)
19. C. Barrenechea, H. Navarro, S. Serrano, L. F. Cabeza, A. I. Fernández, Energy Procedia **57** (2014),
20. R. Pastuszko, Int. J. Therm Sci, **49** (2010)
21. R. Pastuszko, T. M. Wójcik, Exp Therm Fluid Sci, **63** (2015)
22. T. Orzechowski and A. Tyburczyk, EPJ Web of Conferences, **67** (2014)
23. T. Kruczek, Energy, **91** (2015)
24. T. Kruczek, Energy, **62** (2013)
25. M. Pešek and M. Pavelek, EPJ Web of Conferences, **25** (2012)
26. F. P. Incropera, D.P. DeWitt, T.L. Bergman, A.S. Lavine, *Fundamentals of Heat and Mass Transfers* (John Wiley & Sons Inc., U.S.A, 2006)

EPJ Web of Conferences **114**, 02087 (2016)
DOI: 10.1051/epjconf/201611402087

Numerical modelling of swirling diffusive flames

Teresa Parra-Santos[1,a], Ruben Perez[2], Robert Z. Szasz[3], Artur N. Gutkowski[4] and Francisco Castro[1]

[1] *University of Valladolid, Department of Energy and Fluid Mechanics, Valladolid, Spain*
[2] *Tecnicas Reunidas, Thermo-Hydraulic Section, Madrid, Spain*
[3] *Lund University, Energy Division, Lund, Sweden*
[4] *Technical University of Lodz, Department of Heat Technology and Refrigeration, Lodz, Poland*

Abstract. Computational Fluid Dynamics has been used to study the mixing and combustion of two confined jets whose setup and operating conditions are those of the benchmark of Roback and Johnson. Numerical model solves 3D transient Navier Stokes for turbulent and reactive flows. Averaged velocity profiles using RNG swirl dominated k-epsilon model have been validated with experimental measurements from other sources for the non reactive case. The combustion model is Probability Density Function. Bearing in mind the annular jet has swirl number over 0.5, a vortex breakdown appears in the axis of the burner. Besides, the sudden expansion with a ratio of 2 in diameter between nozzle exits and the test chamber produces the boundary layer separation with the corresponding torus shape recirculation. Contrasting the mixing and combustion models, the last one produces the reduction of the vortex breakdown.

1 Introduction

An analysis was carried out to study numerically the interaction of two coaxial jets with different compositions, temperatures and axial velocities.

The geometrical set up corresponds with the case of Roback and Johnson [1], see figure 1, being the diameter of the combustion chamber the double of the outer diameter of the annular nozzle. In addition, the annual nozzle has eight fixed blades, swirler, that produce azimuthal velocity, [2].

Swirl number is defined as the ratio between the azimuthal and the axial momentum. When Swirl number is over a critical value, the radial distribution of pressure promotes reverse flow near the longitudinal axis of the chamber. This inner recirculation zone (IRZ) is known as vortex breakdown.

An increase of the swirl number usually means an increase of the IRZ's size, [3]. That is the reason of using swirling flows to promote enhanced mixture.

This work is devoted to high swirl number to study its flow pattern and its effect on the diffusion flame.

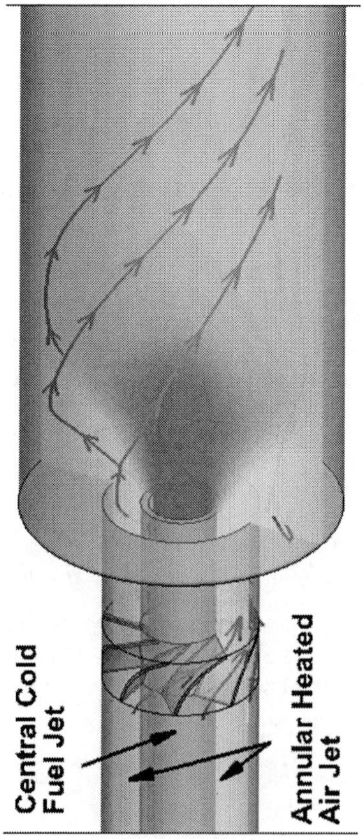

Fig. 1. Set up of the physical domain of Roback and Johnson burner

2 Numerical model

The Reynolds number based on bulk velocity at the test chamber corresponds to a transitional turbulence regime.

Navier Stokes equations are solved in a 3D computational domain with about 600 thousand cells. Figures 2 show a detail of the mesh on the swirler and in a longitudinal slice.

[a] Corresponding author: tpsantos@mail.com

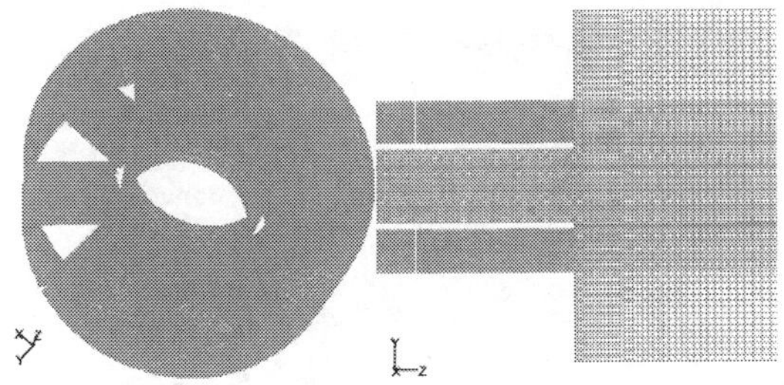

Fig. 2. Details of the mesh swirler (left) longitudinal slice (right).

The swirler has tetrahedral mesh whereas the nozzle and the chamber have a hexahedral mesh. As for the mesh quality, table 1 summarizes the statistics of equi-angle skew, i.e. the dispersion of the hexahedral cell from angle 90° and the tetrahedral cells from angle 60°. Also the aspect ratio is analysed.

Table 1. Mesh quality statistic.

		Very Good	Slightly Bad	Very Bad
Equi-Angle skew	Range	(0-0.3)	(0.5, 0.7)	(0.7, 1.0)
	% cells	88.86%	1.88%	0.84%
Aspect ratio	Range	(1, 5)	(10, 50)	(50, 200)
	% cells	73.96%	16.57%	0.42%

The numerical scheme is second order accuracy. SIMPLE is the method to coupling pressure and velocity resolution.

Two different simulations were conducted: non-reactive flow and reactive flow.

2.1 Boundary conditions

The boundary conditions and the geometrical details of the Roback and Johnson burner [1] are summarized in table 2. The nozzles discharge in the test chamber with a sudden expansion of 4 in area. The length of the chamber is 1 m but the refinement occurs in the first 300 mm where the flow patterns are complex.

Table 2. Summary of geometrical details and the inlet conditions for the nozzles.

Magnitude	Central jet	Annular inlet
Diameter (m)	0-0.025	0.0306-0.059
Velocity (ms^{-1})	0.66	1.52
Intensity of turbulence (%)	12	7.5
Temperature (K)	300	1200
Mass Fraction (-)	0.234 CH_4	0.22O_2 + 0.78N_2
Viscosity (kg m^{-1} s^{-1})	1.087 10^{-5}	1.7894 10^{-5}
Molecular Weight (kg $kmol^{-1}$)	16.043	28.966
Specific Heat Transfer (J kg^{-1} K^{-1})	Polynomial function of temperature	

2.2 Turbulence model

The turbulence model was swirl-dominated RNG k-ε model because it is specially recommended for large zones with relevant vorticity. A detailed comparison of different turbulence models is presented in [2].

Two additional conservation equations are solved. One is for the turbulent kinetic energy, eq. (1), and another for its dissipation, eq. (2). The influence of swirling is represented by the term R_ε that depends on the strain tensor, [4].

$$\frac{\partial}{\partial t}(\rho k) + \frac{\partial}{\partial x_i}(\rho k u_i) = \frac{\partial}{\partial x_j}\left(\alpha_k \mu_{eff} \frac{\partial k}{\partial x_j}\right) + G_k - \rho\varepsilon \tag{1}$$

$$\frac{\partial}{\partial t}(\rho\varepsilon) + \frac{\partial}{\partial x_i}(\rho\varepsilon u_i) = \frac{\partial}{\partial x_j}\left(\alpha_\varepsilon \mu_{eff} \frac{\partial \varepsilon}{\partial x_j}\right) + C_{1\varepsilon}\frac{\varepsilon}{k}G_k - C_{2\varepsilon}\rho\frac{\varepsilon^2}{k} - R_\varepsilon \tag{2}$$

2.3 Combustion model

The simulation of turbulent reactive flows requests large computational time when the chemical kinetic is involved, [5-6]. The model Probability Density Function (PDF) proposed by Jones & Whitelaw [7] is an excellent option. A probability function of the mixture fraction provides the temperature and composition values based on tabulated experimental or theoretical data.

$$\bar{T} = \int_0^1 p(f)T(f)df \tag{3}$$

The model solves only two conservation equations: one for the mixture fraction f and another its variance f'^2.

$$\frac{\partial}{\partial t}(\rho\bar{f}) + \frac{\partial}{\partial x_i}(\rho u_i \bar{f}) = \frac{\partial}{\partial x_i}\left(\frac{\mu_t}{\sigma_t}\frac{\partial}{\partial x_i}\bar{f}\right) \tag{4}$$

$$\frac{\partial}{\partial t}(\rho\overline{f'^2}) + \frac{\partial}{\partial x_i}(\rho u_i \overline{f'^2}) = -C_d\rho\frac{\varepsilon}{k}\overline{f'^2} + \frac{\partial}{\partial x_i}\left(\frac{\mu_t}{\sigma_t}\frac{\partial}{\partial x_i}\overline{f'^2}\right) + C_g\mu_t\left(\frac{\partial}{\partial x_i}\bar{f}\right)^2 \tag{5}$$

A beta probability function is calculated in each cell based on the mixture fraction, and its variance, eq. (6)

$$p(f) = \frac{f^{\alpha-1}(1-f)^{\beta-1}}{\int f^{\alpha-1}(1-f)^{\beta-1}df} \tag{6}$$

Being the coefficients α and β evaluated with equations (7) and (8).

$$\alpha = \bar{f}\left[\frac{\bar{f}(1-\bar{f})}{\overline{f'^2}} - 1\right] \tag{7}$$

$$\beta = (1-\bar{f})\left[\frac{\bar{f}(1-\bar{f})}{\overline{f'^2}} - 1\right] \tag{8}$$

An input of the model is the tabulated results. Figure 3 shows a sample for natural gas.

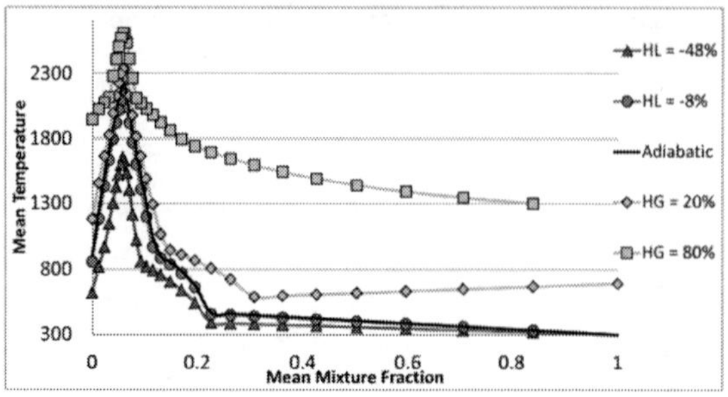

Fig. 3. Tabulated temperature based on mixture fraction and scaled heat loss (HL) and heat gain (HG).

3 Flow pattern for non-reactive case

This section intends to stablish the accuracy of the numerical model for non-reactive case. That is the validity of the turbulence model, mesh resolution and numerical algorithm.

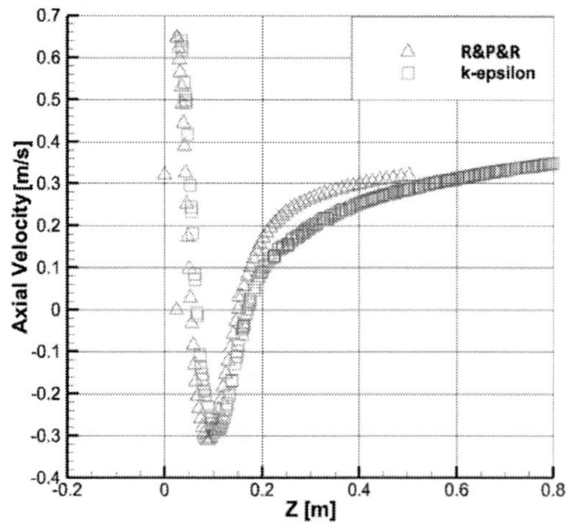

Fig. 4. Validation of the profile of axial velocity along the axis of the chamber. Swirl = 1.2. Triangles: experimental data [8], rectangles: numerical results.

Figure 4 shows the longitudinal profile of the axial velocity for swirl number equals to 1.2. The line is the axis of the chamber. Then negative values correspond with the IRZ and are limited by two stagnation points. It is clear that experimental results from [8] and numerical results provide the same IRZ's size. However the numerical results are more dissipative than the experimental ones because they provide a weaker gradient.

4 Flow pattern for reactive case

This section is devoted to the study of the reactive cases. Figures 5 depict the radial profiles of axial velocities for Swirl number 1.2 in two sections: at a distance of 25 mm and 100 mm from the nozzles' discharge.

At 25 mm it is possible to identify the momentum transfer between the two jets, the annular and the central one. At 100 mm the shear layer has suffer a radial displacement because of the IRZ.

As for the comparison of the non-reactive and the reactive cases, the combustion produces an expansion of the reaction's products and then, axial velocities are higher because mass balance must be accomplished.

Recirculation zones are evident in locations with negative axial velocity. Bearing in mind the IRZ, it is evident the combustion's tendency to minimize the IRZ's size.

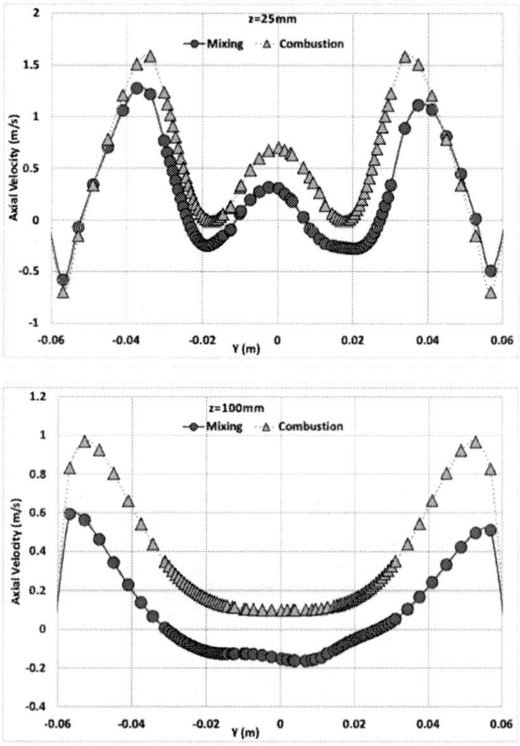

Fig. 5. Radial profile of axial velocity. Swirl = 1.2. Solid line: non-reactive case. Dash line: reactive case.

Figures 6 show a slice from the end of the swirl generator in the nozzle located in Z = -50 mm to Z = 400 mm. The combustion chamber is 1 m long to prevent the influence of the outlet in in flow field.

Both figures 6 evidence the mixture is homogenous in a length around 3 diameters of the combustion chamber.

Figure 6 left shows the contours of axial velocity in a longitudinal plane of the chamber. There are two recirculation zones: the annular ring in touch with the walls and the central one near the axis.

The first one is due to the sudden expansion of the burner whereas the second one is because to the high swirl number of the annular jet. This figure evidence a size of the IRZ for combustion lower than that of figure 4 for non-reactive case.

Figure 6 right shows the contours of temperature in a longitudinal plane. The maximum gradient of temperatures identifies the flame front. This corresponds with the shear layer between the jets, where mixture occurs. The annular ring, near the wall, and the IRZ have high temperature and help to warm the fresh mixture. It is clear the IRZ has the role to fix the location of the flame front and therefore is mainly composed by reaction products. The walls are thermal insulated that is why the lines of iso-temperature cut in perpendicular to the chamber's walls.

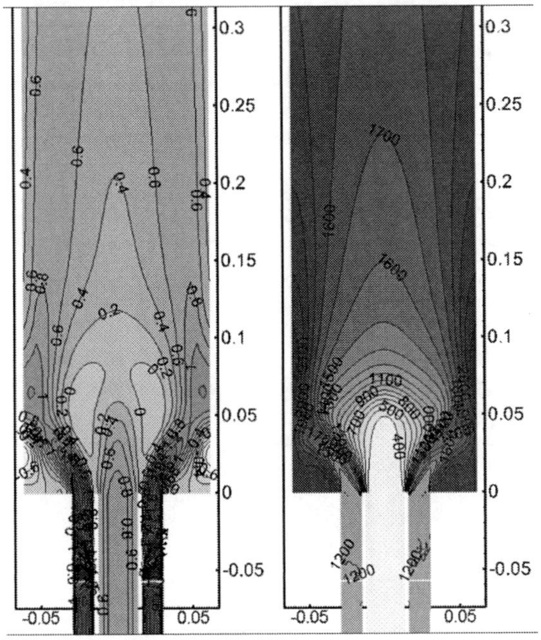

Fig. 6. Longitudinal contour of axial velocity (left) and temperature (right). Swirl = 1.2.

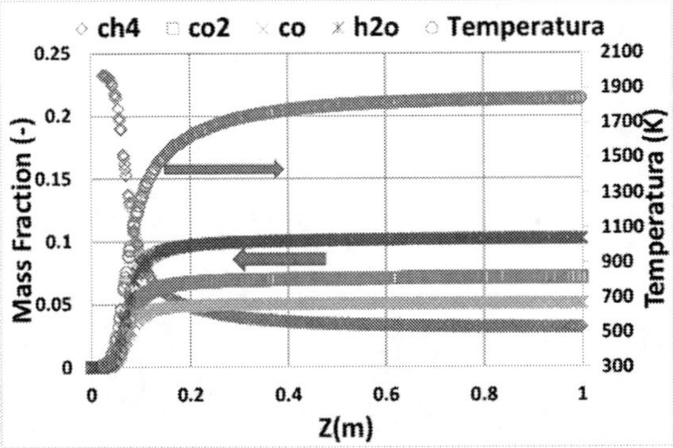

Figure 7. Axial profiles of mass fraction and temperature for Swirl = 1.2.

Figure 7 depicts the longitudinal profiles of temperature and mass fraction of different components along the axis of the chamber. The profiles are smooth because the local thermodynamic state of the flame provided by the PDF combustion model is weighted based on the beta probability function. Also the turbulence model is RANS; that means that the flow field is averaged in time.

Hence, this PDF model is not suitable for the study of instabilities associated to PVC (Precessing Vortex Core). The main advantage of this model is it requests little computational resources. The chemical kinetics is a stiff problem that needs fine temporal and spatial resolution, [9].

5 Influence of the diffuser

This section is devoted to the study of the influence of the diffuser. A conical diffuser of 120° angle was included between the nozzles' discharge and the test chamber.

Figures 8 show the contours of axial velocity and temperature. The main effect of the diffuser is that prevents the formation of the annular ring recirculation near the walls.

The case with the diffuser produces a larger IRZ. As a result, mixing is better. The mixture is homogeneous in a length around 2 diameters of the chamber.

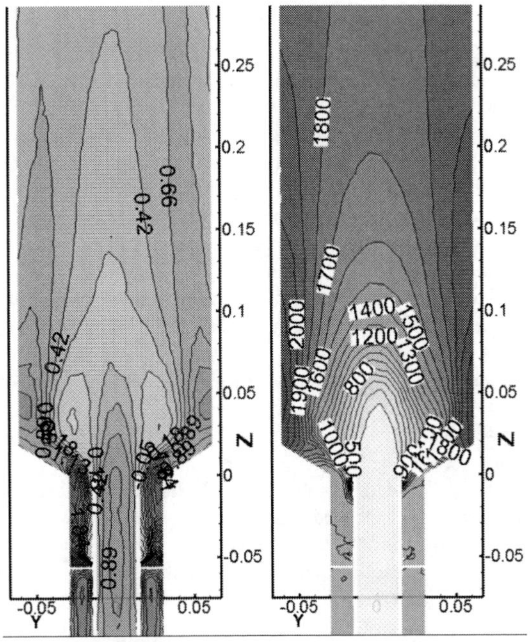

Fig. 8. Longitudinal contour of axial velocity (left) and temperature (right). Swirl = 1.2. Conical diffuser of 120°

This kind of burner has the problem of high temperatures in the walls because of the confinement of the flame, see figure 6 right and 8 right. This could produce thermal fatigue of the burner. Future works will reduce the degree of confinement of the burner.

Figures 9 show f'^{2}, the mixture fraction variance, that is a representation of the flame front location in both cases: with and without diffuser.

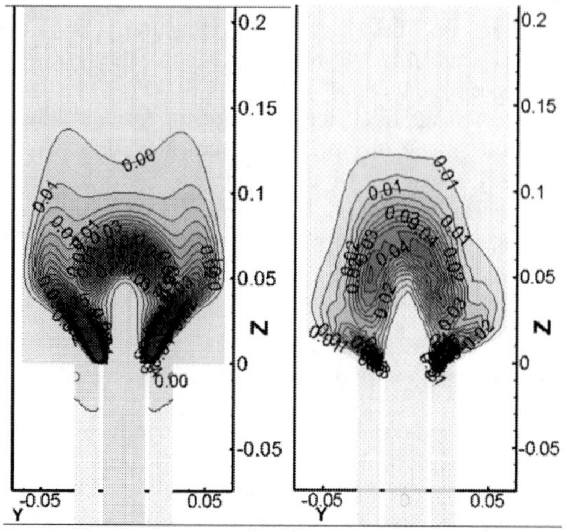

Fig. 9. Contours of mixture fraction variance Swirl = 1.2. With diffuser (right), without it (left).

Conclusions

A confined diffusive swirling flame of natural gas has been simulated using CFD.

The set up corresponds with a swirling burner where two coaxial jets are to be mixed. Because of a swirler generator, the vortex breakup appears. This Inner Recirculation Zone is responsible of an enhanced heat and mass transfer. A secondary recirculation zone is the annular ring because of the boundary layer separation.

Combustion is a challenging numerical task because the wide range of characteristic times involved with the chemical kinetics, the convective transport and the diffusive one. The aim was to find a suitable model for reactive turbulent flows with little requisites of computational resources.

The conclusion is that the PDF model based on tabulated experimental o equilibrium results provides tendencies of the flame behaviour with good convergence.

In the case of combustion, IRZ is smaller than the non-reactive case, because of the increase on axial velocities. Besides, both recirculation zones have hot products of reaction that warm the fresh mixture. Also, the IRZ is downwards the flame front and help to keep its location.

Finally, a simulation with a conical diffuser was carried out. The diffuser promotes a larger IRZ, and a shorter mixing.

Acknowledgment

The author thankfully acknowledges the Spanish Ministry of Science and Innovation for the financial resources in the framework of the project reference ENE2011-25468.

We acknowledge PRACE for awarding us access to resource Curie-GENCI@CEA based in France and MareNostrum@BSC based in Spain. Ref. 2010PA1766

References

1. R. Roback, B.V. Johnson. NASA CR-168252 (1983)
2. T. Parra-Santos; J. R. Perez-Dominguez; R. Z. Szasz; F. Castro; Engineering Computations, **32/3** (2015) pp. 668-686
3. T. Parra, J. R. Perez, R. Szasz; et al, EPJ Web of Conferences, **92** (2015) 02059
4. H. K. Versteeg W. Malalasekera. Computational Fluid Dynamics, The finite volume method (1995).
5. K. Kuo. Principles of Combustion. John Wiley and Sons (1986)
6. E. S. Oran and J. P. Boris, Numerical Simulation of Reactive Flow, Elsevier, New York, (1987).
7. W. Jones, J. Whitelaw. Calculation Methods for Reacting Turbulent Flows: A Review. Combustion and Flame **48** (1982) pp. 1-26
8. K. K. J. Ranga Dinesh, M. P. Kirkpatrick, K.W. Jenkins. Investigation of the influence of swirl on a confined coannular swirl jet. Computers & Fluids **39** (2010) pp. 756–767.
9. J. Warantz, U. Maas and R. W. Dibble. Combustion, Physical and Chemical Fundamentals, Modelling and Simulation, Experiments, Pollutant Formation. editorial Springer Verlag Berlin Heidelberg New York, ISBN 3-540-65228-0, (1999)
10. T. Parra, C. Vega, A. Gallegos; et al, EPJ Web of Conferences, **92** (2015) 02058

EPJ Web of Conferences **114**, 02088 (2016)
DOI: 10.1051/epjconf/201611402088

Numerical study on small scale vertical axis wind turbine

Teresa Parra-Santos[1,a], Armando Gallegos[2], Cristóbal N. Uzarraga[3] and Miguel A. Rodriguez[1]

[1]*University of Valladolid, Department of Energy and Fluid Mechanics, Valladolid, Spain*
[2]*University of Guanajuato, Salamanca, Mexico*
[3]*Technologic Institute of Durango, Mexico*

Abstract. The performance of a Vertical Axis Wind Turbine (VAWT) is numerically analyzed. The set-up is H-Darrieus with three straight blades airfoils NACA attached to a rotating vertical shaft. The wind turbine has solidity equals to the unity operating with wind velocity of 7 m/s. Influence of pitch angle is tested to get design tendencies. 2D, transient, Navier Stokes equations are solved using the code Ansys-Fluent. Conservation equations were solved with a Third-Order MUSCL scheme using SIMPLE to couple pressure and velocity. More than six revolutions must be simulated to get the periodic behavior. Two models of turbulence have been contrasted Realizable k-epsilon and Transition SST concluding the last one show more realistic flow features. Pitch angles of 0°, -6° and -10° have been tested with Tip Speed Ratios ranging from 0.7 and 1.6. The no null pitch angles improve the performance of the wind turbine. Instantaneous and averaged power coefficients as well as detailed flow field around the airfoils are showed.

1 Introduction

The computational model of an H- Darrieus wind turbine is built up using three air-foils NACA0015. The model is scaled down in order to match experimental requisites in a wind tunnel facility. Solidity of the wind turbine is the unity calculated as three times the ratio of chord versus diameter.

This research line is a collaboration of three teams from Universities of Valladolid (Spain), Guanajuato and Durango (Mexico). Previous works involved with this research line were carried out using k-ε turbulence models and non-arm configuration. For instance,

- The influence of the twist angle variation at the chord ends in symmetric airfoils NACA0015, [1].
- The analysis of the camber effect in non-symmetric airfoils NACA2425 and NACA7425, [2-3].
- The study of the solidity of the rotor NACA0025, [3-4].

Bearing in mind the Reynolds number is not very high, actual work uses the Transition SST turbulence model. Lanzafame states it performs well in this kind of applications, [5].

A secondary aim of this work is to analyze the influence of a fixed negative pitch angle, such as is proposed in [6].

2 Numerical model

The Reynolds number based on air velocity corresponds to a transitional turbulence regime.

Transient Navier Stokes equations are solved in a 2D computational domain with about 78 thousand cells. The numerical scheme is third order accuracy. SIMPLE is the method to coupling pressure and velocity resolution.

The domain considers two regions, ambient with poorer spatial resolution and rotor with a finer mesh. Moving reference mesh in stablished in the rotor zone, see figure 1 for a sample of the domain in a plane with arms.

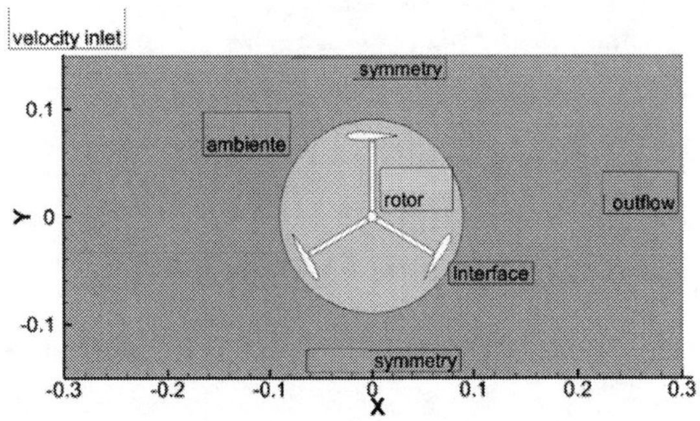

Fig. 1. Set up of the computational domain of H-Darrieus.

---

[a] Corresponding author: tpsantos@mail.com

2.1 Boundary conditions

The geometrical details of the wind turbine [1] are summarized in table 1.

The wind is considered to be 7 ms^{-1} with an intensity of turbulence of 5%.

Table 1. Summary of geometrical details and the inlet conditions for VAWT.

Airfoil	NACA0015
Shaft Radius	0.01 m
Rotor Radius (R)	0.07 m
Length (b)	0.2 m
Chord (c)	0.05 m
Area (A)	0.030 m^2
Number of Blades (Z)	3
Solidity(σ)	1.0

2.2 Sensibility to the mesh

Different meshes have been tested in order to stablish the optimal resolutions considering the computational cost constrain, see table 2 for details of cell numbers in the rotor and ambient regions.

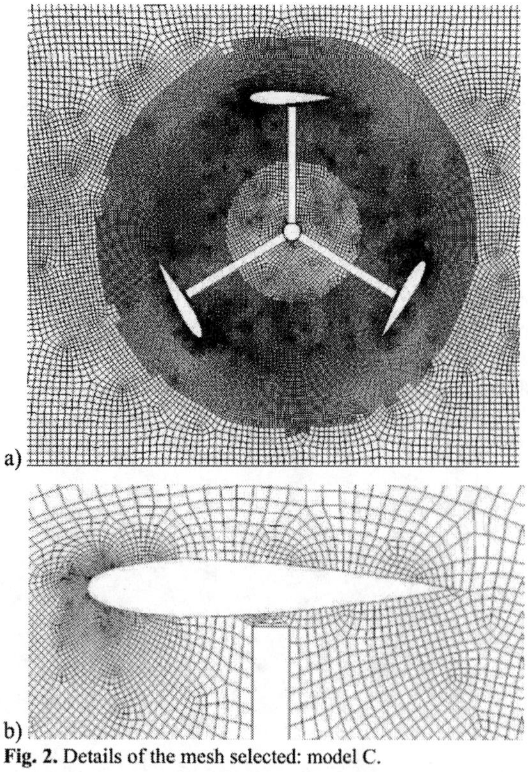

Fig. 2. Details of the mesh selected: model C.

Table 2. Details of mesh models.

Airfoil	Model A	Model B	Model C
Rotor mesh	3737	14948	54320
Ambient mesh	4473	6084	24336
Total mesh	8210	21032	78686
Y+	(5, 30)	(5, 30)	(5, 30)

Because of the imposed rotation of the mesh, the refinement criterion must warranty an axisymmetric mesh, see figure 2 for the selected mesh in a slice with arms.

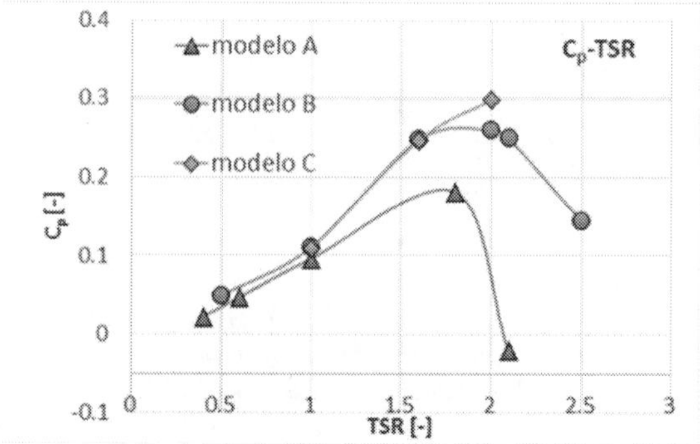

Fig. 3. Averaged power coefficient versus the Tip Speed Ratio for NACA0015, solidity 1 with arms.

Each simulation with a certain rotation speed let obtain the corresponding averaged power coefficient that is a point of the characteristic curve. Changing the rotation speed, it is possible to build the curve. See figure 3 for a sample of influence of the mesh on the power coefficient curves.

Power coefficient depends on the torque T, rotation speed ω and wind velocity V_∞, see equation 1,

$$C_p = \frac{T\cdot\omega}{\frac{1}{2}\rho\cdot v_\infty^3\cdot A} \quad (1)$$

TSR stands for Tip Speed Ratio and is defined in equation 2.

$$TSR = \frac{\omega\cdot R}{v_\infty} \quad (2)$$

2.3 Turbulence model

Transition SST is a suitable model for flow fields in the transition between laminar and turbulent regimes.

$$\frac{\partial}{\partial t}(\rho\Upsilon) + \mathrm{div}(\rho\vec{v}\Upsilon) = \mathrm{div}\left(\left(\mu + \frac{\mu_T}{\sigma_\gamma}\right)\mathrm{grad}(\gamma)\right) + S \quad (3)$$

The source term has different contributions of production, P and destruction, D, of the laminar regime. Sub-index 1 is related to the strain tensor whereas sub-index 2 is function of rotation

$$S = P_{\gamma 1} - D_{\gamma 1} + P_{\gamma 2} - D_{\gamma 2} \quad (4)$$

Control functions to stablish the laminar or turbulent regime are based on 2 definitions of Reynolds number indicated in equations (5) and (6).

$$Re_v = \frac{\rho y^2 S}{\mu} \quad (5)$$

$$R_T = \frac{\rho K}{\mu \omega} \quad (6)$$

Where S is the strain tensor, ω is the turbulent frequency scale and k is the turbulent kinetic energy.

Figure 4 shows the peak power coefficient using different turbulence models. The realizable k-ε and the transition SST provide similar results in agreement with the conclusions of [5].

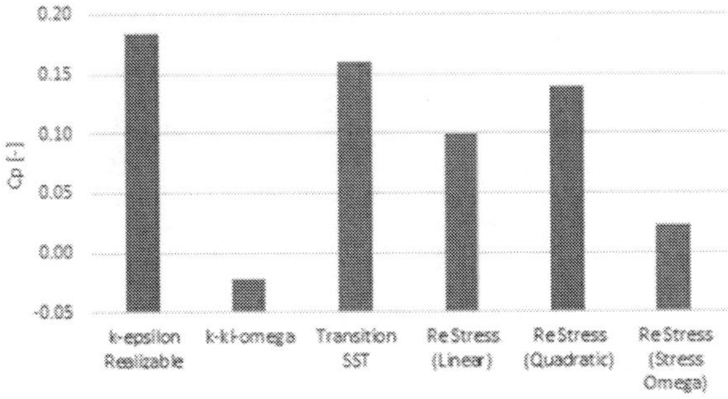

Fig. 4. Comparison of different turbulence models for design conditions.

2.4 Validation of the model

In order to validate the results using the experimental measurements by Lanzafame [5], it was used a slice of the VAWT without arms.

Figure 5 shows the non-dimensional curves of averaged power coefficient versus the Tip Speed Ratio. The numerical model fails in the prediction of the range of TSR where the wind turbine can operate.

At least, the performance of the turbine is well predicted for TSR between 0.5 and 1.25.

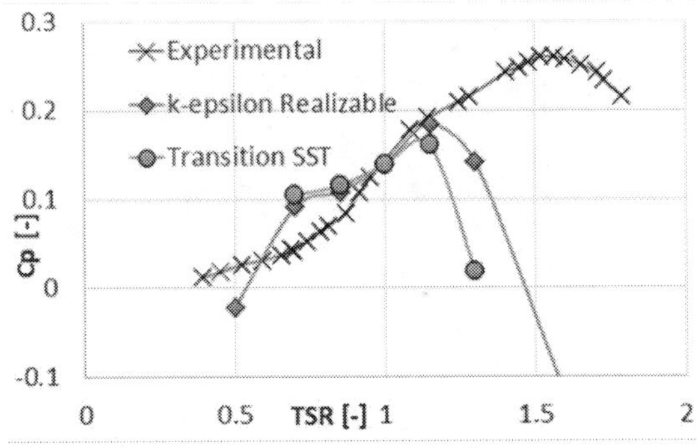

Fig. 5. Averaged power coefficient versus the Tip Speed Ratio model without arms. Comparison of k-ε Realizable, Transition SST and experimental values from Lanzafame [5].

3 Flow pattern at design conditions

This section is devoted to the analysis of the flow pattern. The simulation corresponds with the design conditions, i.e. the maximum averaged power coefficient of the predicted characteristic curve.

Because of the periodic performance of the turbine, the instantaneous power coefficient changes along the revolution of the rotor. Then, there are positions of maximum and minimum instantaneous power coefficient.

The criterion to locate the airfoils is: 0° starts in 12 hours of a watch and the sense of rotation is against clockwise.

In the case of the model without arms, the most favourable position (maximum instantaneous power coefficient) is nearly when the air-foils are in positions 60°, 180° and 300° whereas the least favourable conditions are with air-foils near 0°, 120° and 240°.

In the case of model with arms, the flow pattern is completely different. The most favourable position is 0°, 120° and 240° while the least one is 60°, 180° and 300°.

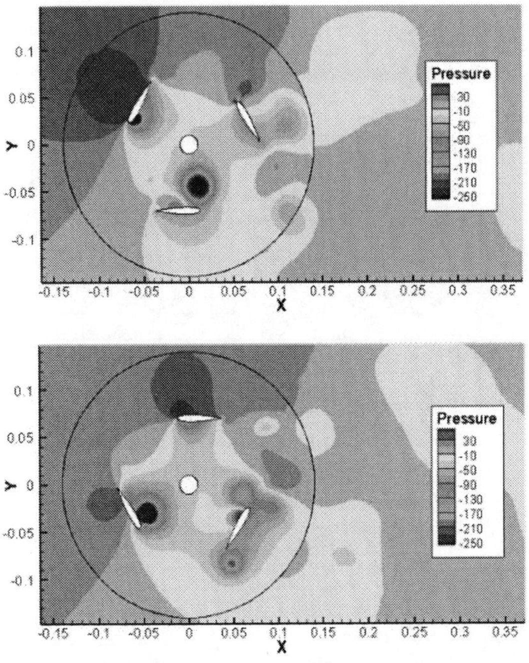

Fig. 6. Contours of pressure NACA0015 for solidity 1 (top) at maximum instantaneous power coefficient (bottom) at minimum instantaneous power coefficient.

Because the sections without arms are more representative of the performance of the rotor, this section will show the flow patterns without the effect of the arms.

Figures 6 show contours of pressure. The higher pressure zone is up-winds from the rotor whereas the lower pressure is downwards. The suction side, associated with lower static pressure, is the inner part of the air-foils at positions 0°, 60° and 120°. Air-foils at 180° and 240° have the suction side in the outer part. The performance of the air-foil at 300° does not seem to be relevant.

The local minimum of pressure corresponds with the kernel of a vortex. These structures are consequence of the separation of the boundary layer. They can be observed near the suction side of air-foils located at 60°, 120°, 180° and 240°.

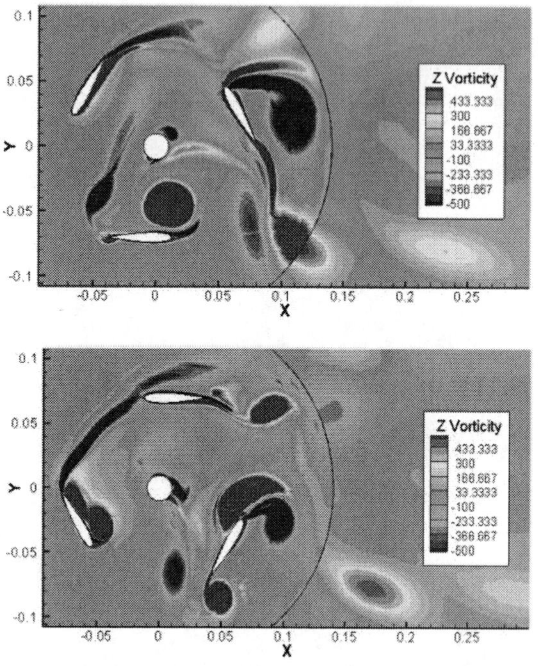

Fig. 7. Contours of vorticity NACA 0015 for solidity 1 (top) at maximum instantaneous power coefficient (bottom) at minimum instantaneous power coefficient.

Figures 7 show contours of vorticity. Positive values are assigned to vortices rotating against clockwise. It is easier to locate the vortex kernel as local minimum and maximum of vorticity. The vorticity flow field let identify the wake downstream of the turbine. At the position of minimum instantaneous power coefficient, the wake of blade at 120° interact with blade at 0°. Also the wake downstream of the turbine is larger than that for maximum instantaneous power coefficient.

4 Influence of the pitch angle

This kind of turbines is suitable for developing countries, and then the design criterion is to minimize cost. It is not cost-efficient a system of variable pitch. But the use of a fixed pitch angle can be a benefit.

This section is devoted to see the performance's tendency when the pitch angle is modified. A fixed pitch angle is used such as Lanzafame proposed in [6].

Figure 8 shows a scheme with a pitch angle negative. Mesh with angles -6° and -10° where tested.

Figures 9 show contours of pressure for different pitch angles on design conditions at the least favourable instantaneous conditions. The negative pitch angle is a precursor of lower pressure gradient on the three airfoils. Hence, smaller wakes are produced.

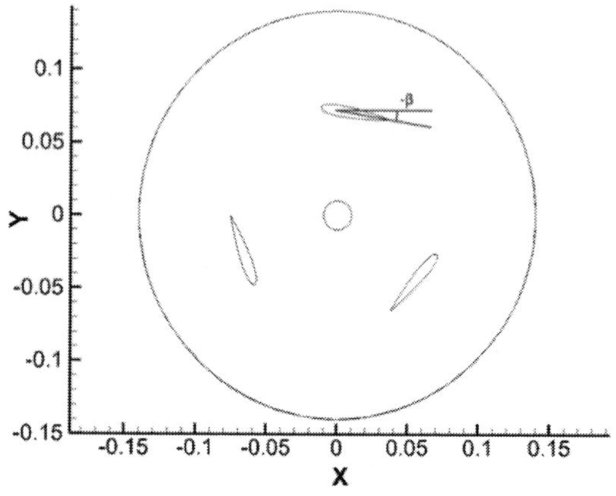

Fig. 8. Scheme of the negative pitch angle.

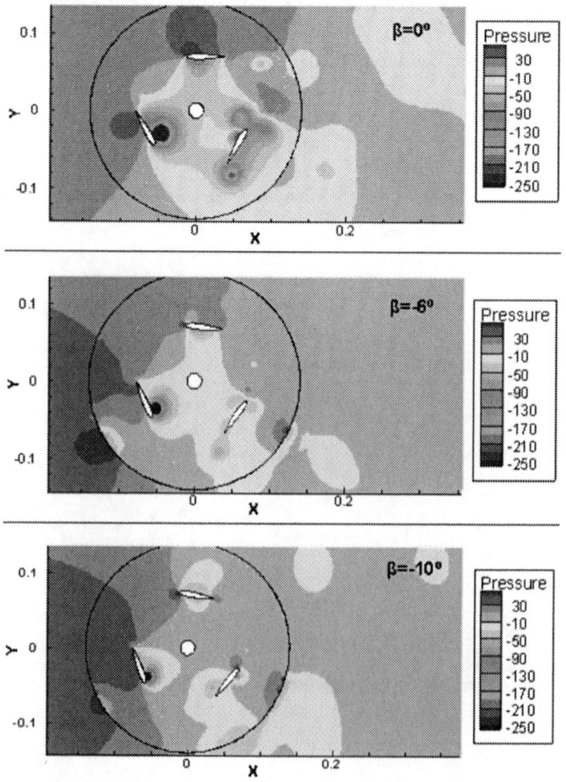

Fig. 9. Contours of pressure NACA0015 for solidity 1 at minimum instantaneous power coefficient with different pitch angles: 0, -6° and -10°.

Figure 10 shows the non-dimensional curves of averaged power coefficient versus the Tip Speed Ratio without arms and with different fixed pitch angles. It is evident there is an increase of range of TSR operating as turbine for negative pitch angles. Also the peak power coefficient is increased.

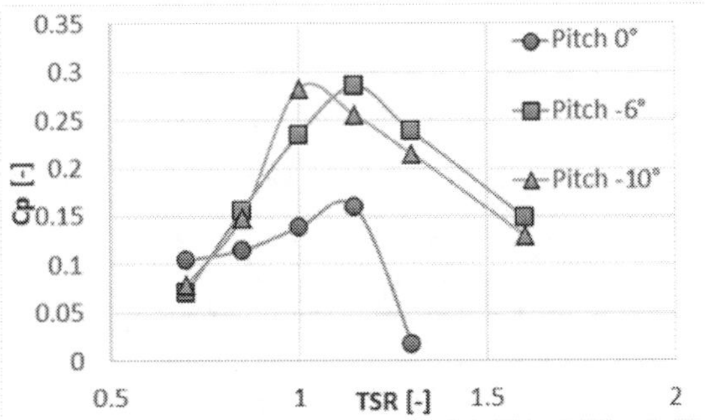

Fig. 10. Averaged power coefficient versus the Tip Speed Ratio model without arms using transition SST and different pitch angles.

Conclusions

An H-Darrieus has been simulated using CFD. The set-up corresponds with 3 NACA0015 blades. Two slices were considered: with and without arms.

The model requests the transient resolution of conservation equations for incompressible and turbulent flows. Moving reference mesh is used to impose the rotation velocity.

As for the mesh refinement, it must be axisymmetric in order to achieve the periodicity of the flow.

Because the Reynolds number does not corresponds to fully developed turbulent flow, the turbulent model was transition SST that has been successfully used in the literature. It was found the results are similar to those provided by the realizable turbulence model. Besides, the convergence is adequate. It is a suitable model for flow fields between laminar and turbulent regimes.

The flow pattern for the maximum and minimum instantaneous power coefficient were analysed based on the pressure and vorticity fields. Suction sides and separation of the boundary layer were identified for the slice without arms.

The authors are aware that the characteristic curves of the averaged power coefficient versus the TSR show tendencies because no three-dimensional effects are considered.

The use of a pitch angle of -6° or -10° shows an increase of the performance of the turbine, increasing the range of TSR and the peak of the power coefficient. This conclusion matches the results found in the literature.

References

1. N.C. Uzarraga, A. Gallegos, M. T. Parra, J. M. Beltran. ASME ESFuelCell2012-91113 (2012)
2. T. Parra, C. Uzarraga, A. Gallegos, M.A. Rodriguez, F. Castro. FEDSM2014-21356 (2014)
3. T. Parra, C. Vega, A. Gallegos; et al, EPJ Web of Conferences, **92** (2015) 02058
4. T. Parra, M. Esteban, F. Gonzalez, C. Uzarraga, A. Gallegos, F. Castro. International Conference of Advanced Technologies and Sciences (2014) pp. 770-772
5. R. Lanzafame, S. Mauro, M. Messina Energy Procedia **45**, pp 131 – 140 (2014)
6. R. Lanzafame, S. Mauro, M. Messina. Renewable Energy (2013).

EPJ Web of Conferences **114**, 02089 (2016)
DOI: 10.1051/epjconf/201611402089

On computations of temperature dependent incompressible flows by high order methods

Jan Pech[1,a]

[1]Institute of Thermomechanics AS CR, v. v. i., Dolejškova 1402/5, Praha 8, 182 00, Czech Republic

Abstract. Parallel vortex shedding from the heated cylinder as dependent on the cylinder temperature is investigated in numerical simulation. The computational scheme for the system of the incompressible Navier-Stokes equations coupled with the heat equation is presented. The scheme is based on the mixed implicit-explicit solver of the third order and the spectral element method for the spatial approximation. The results are compared with the experiments for flow of air and water. The Strouhal-Prandtl-Reynolds relationship and evolution of the separation angle as dependence on the temperature is the final result of the computations.

1 Introduction

The vortex shedding from the circular cylinder is widely studied phenomena both experimentally and in the numerical simulations. This, geometrically simple, problem demonstrates various effects in the transition of the flow from the laminar to turbulent behaviour. From the wide literature concerning this problem we have already knowledge about the dependence of the vortex shedding frequency on the velocity magnitude and vortical structures in the wake. In the present study we restrict our attention to the flow regime, where the vortex shedding parallel to the cylinder occurs, and where a 2D numerical simulation is justified. This regime is achievable in the range of Reynolds numbers $Re_c < Re < 160$, if Re_c denotes the critical Reynolds number, separating the steady and unsteady flow. In this flow regime we study the influence of the cylinder wall heating to the vortex shedding frequency (described in terms of the Strouhal number St). The Reynolds number is here defined as $Re = \frac{|\mathbf{u}_\infty|\, D\rho_\infty}{\mu_\infty}$, where D stands for the cylinder diameter, μ is the dynamic viscosity, ρ density, $|\mathbf{u}|$ the characteristic velocity magnitude and subscript ∞ denotes values in the far-field/inlet. The Strouhal number is defined as $St = \frac{fD}{|\mathbf{u}_\infty|}$, where f is the frequency of vortex shedding.

The fluids considered are water and air. For our study is most important that these two substances exhibit opposite temperature dependence of its viscosities. We expect the flow to be incompressible in case of both studied fluids, because the velocity magnitudes satisfy $M^2 \ll 1$ (M is the Mach number). Effects of the buoyancy and viscous heating are neglected, since it has a minor influence to the flow in our setting. Therefore the influence of the flow by the temperature distribution is given by the temperature variation of the dynamic viscosity μ and the thermal conductivity λ. Frequency of the vortex shedding depends mostly on the viscosity, but variability in the thermal conductivity is also important, especially for preserving the correct local value of the Prandtl number ($Pr = c_p\,\mu_\infty/\lambda_\infty$), which moderates the heat equation. For the isothermal case, when the cylinder wall has the same temperature as the inlet fluid, the results of numerical simulation using the scheme of [1] give excellent agreement with St-Re relationship (see e.g. [2]). For the study of the heating effects to the St-Re relation, the scheme for isothermal case is enhanced using technique already proposed in [3] and applied in the same manner to the heat equation as in [4]. In comparison to the previous results of [4] using the first order scheme, we present the third order scheme and optimise the computation in the spatial solver.

The flow parameters were chosen to correspond with those of [5], where an influence of heating to the Strouhal number was studied experimentally. The Strouhal number obtained from our computation in comparison to the values from the experiment then provides a basic information about the relevance of the computational results.

2 Physical model

A system of equations describing the motion and temperature distribution of an incompressible fluid with the temperature dependent viscosity and thermal conductivity may be written in the dimensionless form as

$$\frac{\partial \mathbf{u}}{\partial t} + (\mathbf{u}\cdot\nabla)\mathbf{u} = -\nabla p + \nabla\cdot\left[\frac{1}{Re}\frac{\mu(T)}{\mu_\infty}\left(\nabla\mathbf{u} + (\nabla\mathbf{u})^{\mathrm{T}}\right)\right] \quad (1)$$

$$\nabla\cdot\mathbf{u} = 0 \quad (2)$$

$$\frac{\partial T}{\partial t} + (\mathbf{u}\cdot\nabla)T = \nabla\cdot\left(\frac{1}{Pr\,Re}\frac{\lambda(T)}{\lambda_\infty}\nabla T\right) \quad (3)$$

where $\mathbf{u} \in [0 : 1]$ denotes velocity, p is dimensionless *kinematic pressure*, $\mu(T)$ is the temperature dependent dy-

[a] e-mail: jpech@it.cas.cz

Table 1. The power-law exponents of the viscosity and therm. conductivity of water and air used in the present computations.

	M	L
water	-7.0	0.71
air	0.7774	0.85

namic viscosity and $\lambda(T)$ is the thermal conductivity. Denoting by T_W the temperature of the cylinder wall we have set $T \in [1 : T_W/T_\infty]$.

To complete above mathematical system the functions $\lambda(T)$ and $\mu(T)$ are approximated as power laws

$$\lambda(T) = \lambda_\infty \left(\frac{T}{T_0}\right)^L \tag{4}$$

$$\mu(T) = \mu_\infty \left(\frac{T}{T_0}\right)^M \tag{5}$$

based on the data [6] and procedure suggested in [2], respectively. Exponents M and L, used in the computations, are listed in the table 1.

The whole system is fully coupled by the variable properties. The technique used in the numerical scheme decouples the system as the operator splitting method together with linearization and results in acceptably accurate and fast solver.

3 Numerical methods

The computational scheme for the system (1)-(3) originate from the high order splitting scheme [1] enhanced by the approach for variable viscosity ([3]) and was already presented in its first-order version in [4].

The temperature dependent properties of viscosity and thermal conductivity are divided into a constant and variable (subscript s) part

$$\mu(T) = \mu_\infty + \mu_s(T), \ \lambda(T) = \lambda_\infty + \lambda_s(T). \tag{6}$$

The terms with variable parts (μ_s, λ_s) are left to the explicit evaluation together with the convective non-linearities. The constants μ_∞ and λ_∞ stay on the place as in the constant-property system.

Introducing a notation of a function value at the n-th time step, $f(n\Delta t) = [f]^n$, we can write the explicit part of the scheme for the incompressible Navier-Stokes equations

$$\frac{\gamma_0 \hat{\mathbf{u}} - \sum_{q=0}^{2} \alpha_q [\mathbf{u}]^{n-q}}{\Delta t} = \\ = \sum_{q=0}^{2} \beta_q \left[\nabla \cdot \left(\frac{\mu_s(T)}{Re\,\mu_\infty}\left(\nabla \mathbf{u} + (\nabla \mathbf{u})^{\mathrm{T}}\right)\right) - (\mathbf{u}\cdot\nabla)\mathbf{u}\right]^{n-q}. \tag{7}$$

Evaluation of the viscosity follows $[\mu_s(T)]^n = \mu_s([T]^n)$ and coefficients α_q, β_q are listed in the table 2.

The pressure Poisson equation as known from the fractional step methods follows

$$\nabla^2 [p]^{n+1} = \nabla \cdot \left(\frac{\hat{\mathbf{u}}}{\Delta t}\right) \tag{8}$$

Table 2. The coefficients of the third-order *IMEX* scheme. For details we refer to [7], pg.264.

q	α_q	β_q	γ_0
0	3	3	11/6
1	-3/2	-3	
2	1/3	1	

and is accompanied with the high order pressure boundary condition (HOPBC), which has crucial impact on the final accuracy of the scheme

$$\frac{\partial [p]^{n+1}}{\partial \mathbf{n}} = \\ = \mathbf{n} \cdot \sum_{q=0}^{2} \beta_q \left[\nabla \cdot \left(\mu(T)\left(\nabla \mathbf{v} + (\nabla \mathbf{v})^{\mathrm{T}}\right)\right) - (\mathbf{v}\cdot\nabla)\mathbf{v}\right]^{n-q}. \tag{9}$$

The condition in the form 9 expects a constant velocity on the boundary, what is satisfied in our model. The second step (8) allows use of the spectral element method as the spatial solver. It also applies the incompressibility to the second intermediate velocity field $\hat{\hat{\mathbf{u}}}$, which satisfy

$$\frac{\hat{\hat{\mathbf{u}}} - \hat{\mathbf{u}}}{\Delta t} = -\nabla p^{n+1} \tag{10}$$

and finally forms the right hand side of the Helmholtz equation for the velocity at the new time step

$$\left(\frac{1}{Re}\nabla^2 - \frac{\gamma_0}{\Delta t}\right)[\mathbf{u}]^{n+1} = \frac{\gamma_0}{\Delta t}\hat{\hat{\mathbf{u}}} \tag{11}$$

The temperature dependent thermal conductivity results in a strongly non-linear heat equation, which is linearized and splitted in the same manner as the equation of motion. The operator splitting is applied to perform linear terms implicitly and the non-linearities, containing convective term and variable part of the diffusion term, are evaluated explicitly. The explicit step of the third-order scheme then results in

$$\frac{\gamma_0 \hat{T} - \sum_{q=0}^{2} \alpha_q [T]^{n-q}}{\Delta t} = \\ = \sum_{q=0}^{2} \beta_q \left[-(\mathbf{v}\cdot\nabla)T - \frac{1}{Pr\,Re}\nabla \cdot \left(\frac{\lambda_s(T)}{\lambda_\infty}\nabla T\right)\right]^{n-q}. \tag{12}$$

Implicit part has again the form of the Helmholtz equation

$$\left(\frac{1}{Pr\,Re}\nabla^2 - \frac{\gamma_0}{\Delta t}\right)[T]^{n+1} = -\frac{\hat{T}}{\Delta t}. \tag{13}$$

and is solved by the spectral element method.

4 Results

The computational domain with dimensions measured from the center of the cylinder had $51D$ upstream, $100D$ downstream and $51D$ on both sides from the cylinder. A refinement towards the cylinder and the area of the wake was

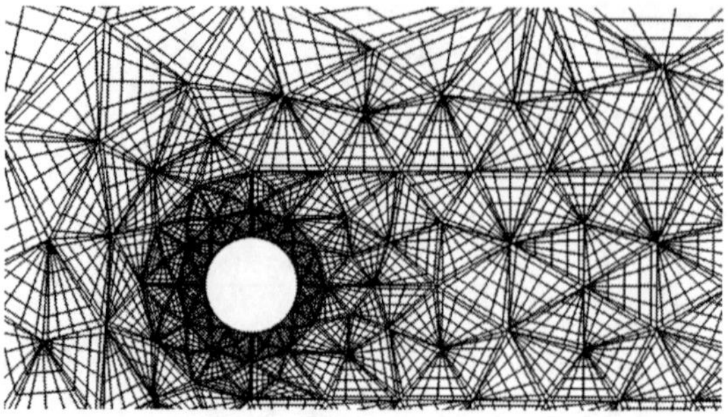

Fig. 1. Detail of the computational mesh around the cylinder. The quadrature points are connected by lines, so the particular triangular elements may be recognized. In the direction of converging lines the Gauss-Radau quadrature is used and Gauss-Lobatto rule is on the opposite side. The mesh, coarse in comparison to those of low order methods, has curved boundaries at the cylinder surface.

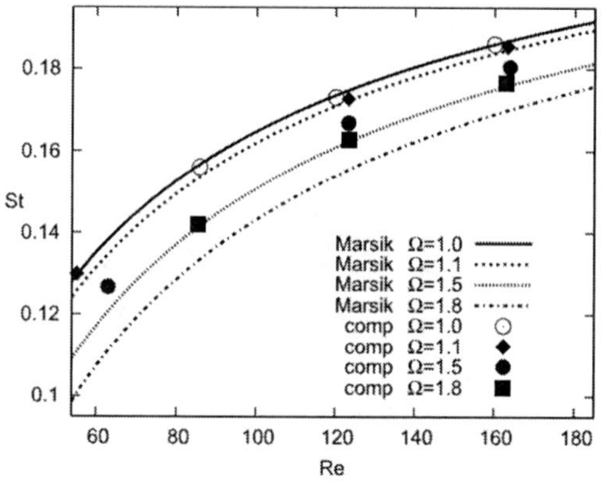

Fig. 2. The Strouhal-Reynolds number relationship for flow of air in isothermal case $\Omega = 1.0$ and three heated cases ($\Omega = T_W/T_\infty$). Results of computations are denoted by *comp*, curves *Marsik* are plots of eq.(4.4) from [2].

designed to keep a satisfactory spatial resolution. A detail of the cylinder area of the mesh consisting of 2360 triangular elements is in figure 1. Polynomial basis of order 6 was chosen as sufficient after convergence tests.

The computational scheme (7)-(13) was implemented on base of the Nektar++ library [8], a C++ library for implementation of the schemes based on the spectral element methods. Among the output quantities were positions of the zero shear stress on the cylinder boundary and values of the drag C_D and lift C_L coefficients

$$C_D = \frac{2F_1}{\rho\,|\mathbf{u}_\infty|^2 D}, \; C_L = \frac{2F_2}{\rho\,|\mathbf{u}_\infty|^2 D}. \qquad (14)$$

Here we denoted

$$\mathbf{F} = \begin{pmatrix} F_1 \\ F_2 \end{pmatrix} = \int_C (\mathbb{T}\mathbf{n})\mathrm{d}S\,,$$

C the cylinder surface and $\mathbb{T} = -p\mathbb{I} + \mu(\nabla\mathbf{u} + (\nabla\mathbf{u})^{\mathrm{T}})$ the stress tensor. The shear stress was computed as a tangential component of the velocity gradient in the direction of the normal $\mathbf{n}$ to the cylinder wall

$$\frac{\partial\mathbf{u}}{\partial\mathbf{n}} \cdot \mathbf{t}\,. \qquad (15)$$

Values of C_D, C_L and positions of all detected zeros of the shear stress were saved in every time step.

4.1 Temperature dependence of St-Re relationship

If $Re > Re_c$, a disturbances from initialisation of the computational process are not suppressed and the flow develops to the vortex shedding regime. History of C_D and C_L reflects the evolution of the flow and must be processed, since only the fully developed vortex street contains the frequency, which refers to the Strouhal number. The resulting St-Re dependencies for both air and water for various temperatures of the cylinder are in the figure 2 and figure 3. The Strouhal number decreases with temperature

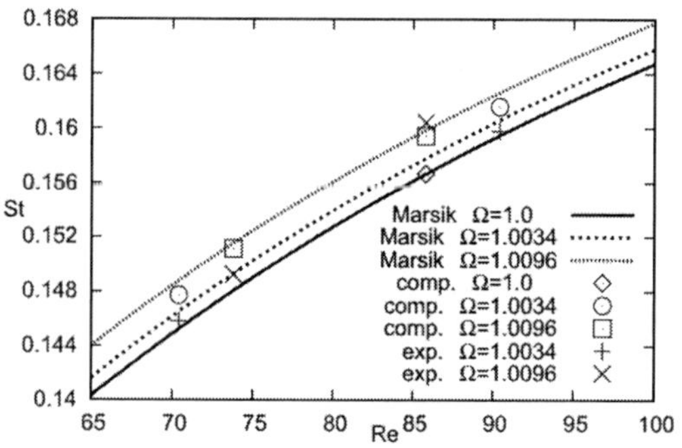

Fig. 3. Strouhal-Reynolds number relationship for flow of water. Experimental results of [5] are denoted as *exp*, other notation is adopted from figure 2.

increment in case of air and increases in case of water. Presently computed data coincide with those computed in [4] by the polynomial approximation up to order 49. The new results therefore support applicability of very high order spatial approximations and shows that distinction of the computation from the experiment is not given by the spatial approximation with mentioned settings.

4.2 Influence of heating to the separation angle

The separation angle Θ is measured on the arc defined by frontal stagnation point, cylinder center and the separation point (figure 4). There are four positions of the zero shear stress on the cylinder, which correspond to frontal and backward stagnation points and points of separation on the "top" and "bottom" side of the cylinder. Obtaining the position of any of the zero points stayed on interpolation of the zero between two quadrature points with opposite sign of the shear stress value. The position of all four points oscillates with the frequency of the vortex shedding. Re-

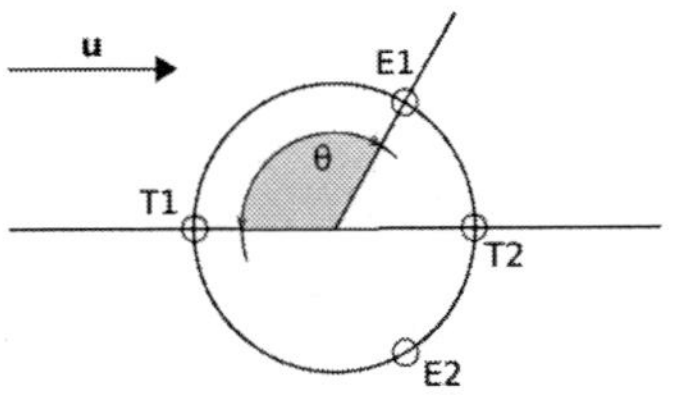

Fig. 4. The separation angle Θ is measured from the upstream zero shear stress point. Approximative positions on frontal (T1), backward (T2) stagnation points and two separation points (E1, E2) is drawn.

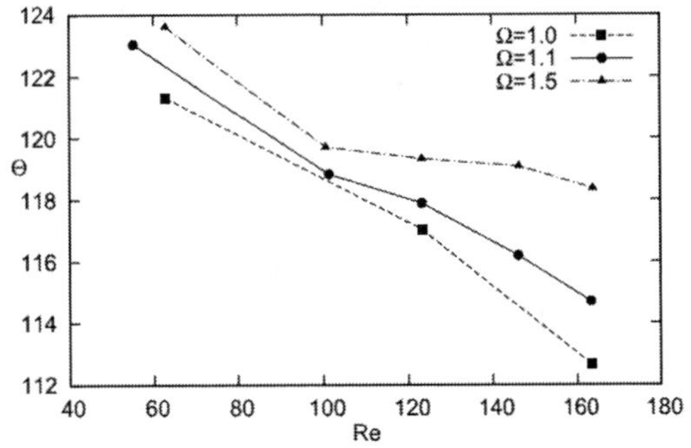

Fig. 5. Θ-Re dependence for various temperature ratios $\Omega = T_W/T_\infty$. The values of separation angles are the mean values in time. For the maximal and minimal values observed we refer to figure 6.

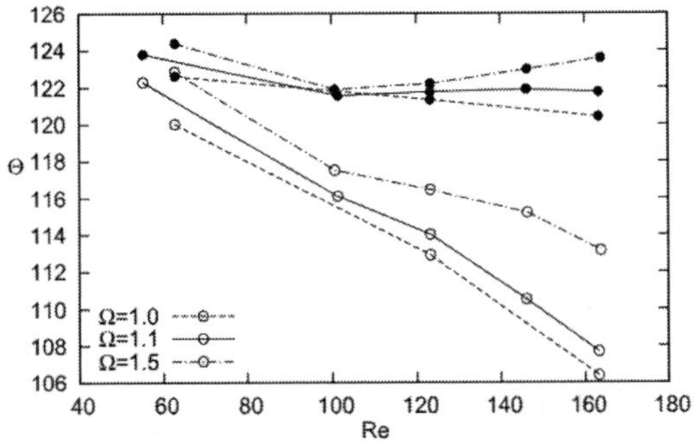

Fig. 6. Temperature dependence of amplitude of separation angle for various ratios of flow and cylinder temperatures $\Omega = T_W/T_\infty$. Filled points refers to maximal separation angles, empty points denotes the minima.

sulting angles as dependent on the Reynolds number and temperature are plotted in figure 5 and figure 6.

The frontal stagnation point very accurately preserves mean value of $\Theta = 0$ and oscillates with moderate amplitude < 2 degrees. Time-averaged positions of the separation points (E1 and E2 in figure 4) appear symmetrically ($\Theta_{S1} \simeq 360 - \Theta_{S2}$). A difference between angles evaluated on the present mesh and mesh for very high order basis of [4] was observed. This deviation in results is up to 2 angular degrees and suggests a need of improvement of the present evaluation method, whose accuracy depends on local distribution of quadrature points, which is non-uniform over the elements.

5 Conclusion

There are still various settings in the computation, which need to be revised in relation to the results of the experiment. Values of the final Strouhal number are strongly influenced by the choice of the reference temperature and dependencies of μ and λ. Even though the buoyancy effects are negligible, the model expecting a constant density is limiting for the case of the heated flows even if the velocity magnitudes in the flow are small. Successes of the present computations are in confirmation of relevance of the previous very high order computations [4] and inclusion of the variable properties into the solver capable of the fast and accurate computations of the fluid flow. Results of the computations clearly shows the impacts of heating in both settings with properties of water and air. Accuracy of the computation is comparable with tolerance of the experimental data. Observed deviation of the absolute values of computation and experiment are left for wider discussion. Computations of the separation angle shows its dependence on temperature and confirms its dependence on the flow velocity. Observation of this quantity has many complications in experimental studies and the computation clarifies the phenomenon especially in its time dependence.

Acknowledgement

Institutional support RVO: 61388998 is gratefully acknowledged.

References

1. G.E. Karniadakis, M. Israeli, S. A. Orszag, J. Comput. Phys. **97**, 414-443 (1991)
2. F. Maršík, Z. Trávníček, R.-H. Yen, W.-Y. Tu, A.-B. Wang, Proceedings of CHT-08 **2008**
3. G. Karamanos, S. Sherwin, Appl. Numer. Math. **33**, 455-462 (2000)
4. J. Pech, Programs and Algorithms of Numer. Math. **17**, 169-174
5. T. Vít, M. Ren, Z. Trávníček, F. Maršík, C. Rindt, Exp. Therm. Fluid Sci. **31**, 751-760
6. B. Gebhart, *Heat conduction and mass diffusion* (McGraw-Hill, Inc., 1993)
7. G.E. Karniadakis, S. Sherwin, *Spectral/hp Element Methods for CFD* (Oxford University Press, New York, 2005)
8. C.D. Cantwell, et al., Comput. Phys. Commun. **192**, 205-219 (2015)

EPJ Web of Conferences **114**, 02090 (2016)
DOI: 10.1051/epjconf/201611402090

Coarse particle support in turbulent flow of visco-plastic carrier

Vojtěch Pěník[1,a], Mikoláš Kesely[1], Václav Matoušek[1]

[1]Czech Technical University in Prague, Faculty of Civil Engineering, Thákurova 7,166 29 Praha 6, Czech Republic

Abstract. The paper deals with a flow behaviour of mixture composed of coarse glass-bead particles and non-Newtonian carrying liquid of Herschel-Bulkley type in a pressurized pipe. Essentially, there are two mechanisms governing support of particles in the flowing carrier: mutual inter-particle collisions and interaction of particles with carrier turbulent eddies. The turbulent support is quantified by a turbulent-diffusion model, which predicts a distribution of concentration of particles in the flow. In the model, the particle turbulent diffusivity is an important parameter dependent on a number of flow quantities, e.g. the carrier rheological parameters. In case of Newtonian carrier, the viscosity is the only rheological parameter and it is constant throughout the flow. In the non-Newtonian carrier, however, local values of the viscosity vary with the shear strain and so affect the particle diffusivity. The paper discusses experimental and analytical results for non-Newtonian mixture flows of measured rheological parameters and flow characteristics. Based on measured concentration profiles, a suitable method is used to determine the local particle diffusivity experimentally.

1 Introduction

Conveying and pumping of non-Newtonian slurries are processes which occur in many branches of industry as mining, dredging, chemistry, or food industry. Depending on slurry properties and conveying conditions, non-Newtonian slurries can flow in the laminar regime or in the turbulent regime in pressurized pipes [1,2]. While laminar flow is relatively simple and its behaviour is well understood, turbulent flow is much more complex and much less understood. A degree of complexity further increases if coarse particles are involved [3-5].

For stratified and heterogeneous flows of coarse-grain slurries with a Newtonian carrier (as water), relatively successful models are available for predicting the distribution of particles in pipe flow [6-9]. Such models are not readily available for turbulent flows of complex slurries based on a non-Newtonian carrier. As a first approximation, it can be assumed that the Newtonian models may be used for non-Newtonian conditions provided that rheology of the non-Newtonian carrier is taken into account. The aim of this paper to evaluate a capability of the Newtonian-based turbulent diffusion model appropriately modified for non-Newtonian liquid to predict a distribution of volumetric concentration of coarse particles in a vertical of a pipe cross section. The attention focusses on the key parameter of the turbulent-diffusion model - the particle diffusivity.

2 Survey of modelling of concentration distribution

A distribution of solid particles in slurry flowing through a pressurized pipe is an important property of the flow. Together with a distribution of particle velocity across the flow cross sectional area, it determines the flow internal structure resulting from mechanisms that govern particle support in flowing carrying liquid and hence affect flow friction.

Basically, there are two very different mechanisms that can support particles in flowing slurry: the interaction of a particle with turbulent eddies of the flowing carrier and the interaction of a particle with other particle through contacts. The contacts between particles are either permanent or sporadic. Sporadic contacts have a form of inter-particle collisions. Hence, particles can be dispersed and distributed throughout a slurry flow either by a diffusive action of turbulent eddies (e.g. [10]) or by collisions between particles travelling at different velocities (e.g. [11]).

A turbulent-diffusion model of the Schmidt-Rouse type have proven to be successful in predicting concentration profiles of fine particles (supported by turbulent eddies) in dilute suspensions based on Newtonian carrying liquids [12],

$$\frac{dc}{dy} = \frac{-v_t}{\varepsilon_s} \cdot c \qquad (1)$$

[a] Corresponding author: vojtech.penik@cvut.fsv.cz

in which dc/dy is the concentration gradient, v_t is the terminal settling velocity of particle, ε_s is the local turbulent diffusivity of particle at vertical position y, and c is the local volumetric concentration of particles at the position y.

A typical modification of the model in order to make it useable for concentrated slurries is an implementation of the hindered settling effect to the model equation [6,10,13],

$$\frac{dc}{dy} = \frac{-v_t}{\varepsilon_s} \cdot (1-c)^m \cdot c \qquad (2)$$

in which and m is the exponent in Richardson-Zaki equation for hindered settling velocity [14].

3 Experimental work

3.1 Experimental rig

A pipe loop (Figure 1) is located in the Water Engineering Laboratory of the Czech Technical University in Prague. It is composed of pieces of a PE pipe (I.D. 51.4mm, blank pipe in Figure 1) and piece of transparent acrylic pipe (I.D. 50.0mm, grey pipe in Figure 1). The total length of the loop is 22.96m and its volume is 45.08l. The length of the horizontal section is 6.20m. The pump EBARA 3M 40-200/7,5kW is driven by an electric motor with a variable frequency converter TECO GD100-011G-4 11kW. Pump parameters are: power 7,5kW, impeller diameter 200mm, maximum flow 11.67l/s, total head from 58m to 44m (valid for water for maximum flow).

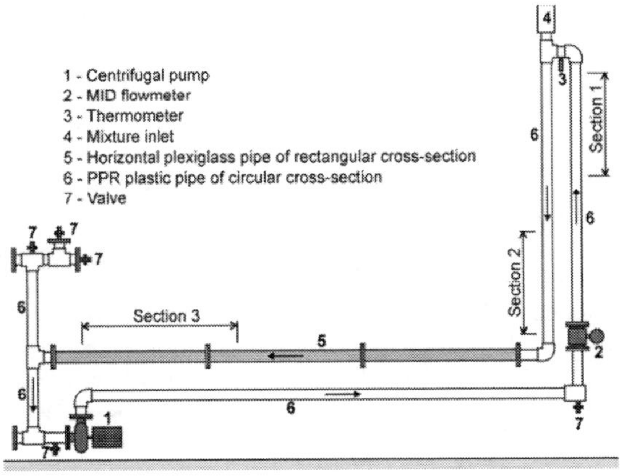

Figure 1. Test pipe loop in Water Engineering Laboratory.

Differential pressures are measured over vertical Sections 1, 2 (1.3m long) and the horizontal Section 3 (1m long) using the differential pressure transducers Fischer Rosemount DP1151 (Sections 1 and 2) and the transducer Siemens Sitrans P DSIII (Section 3). An electromagnetic flow meter Krohne Optiflux 5000 measures the flow rate in the vertical pipe mounted to the discharge outlet of the centrifugal pump. The temperature of the flowing medium is measured in the vertical invert pipe. The rheology of fluids are tested on viscometer HAAKE VT 550.

An ERT (Electrical Resistance Tomography) equipment was used to measure a distribution of particle concentration in the pipe. The equipment contained 3 rings located at 3 different locations along the transparent acrylic horizontal pipe. Each ring had 16 electrodes and was able to sense a tomographic image of conductivity distribution in one pipe cross section. The rings were connected with the UCT processing unit controlled by UCT Tomography software. The processed ERT data were further post-processed by EIDORS procedures.

A sedimentation column was exploited to test the terminal settling velocity of particles in quiescent non-Newtonian liquid of different rheologies.

3.2 Tested materials

Tested mixtures are composed of Carbopol solutions as carrying liquids and glass beads as conveyed solids. Carbopol is an acidic powder. After mixing with water and neutralizing, it forms a non-Newtonian solution of Herschel-Bulkley type described using the rheological model $\tau = \tau_y + K\gamma^n$, in which γ (s^{-1}) is the shear rate. Values of the rheological parameters (τ_y, K, n) depend on a concentration of the powder in the solution. A big advantage of Carbopol is its transparency and a quite simple preparation of a solution of various concentrations. Transparency is very useful for visual observations of solids in non-Newtonian mixtures flow.

The fraction of glass beads TK1.5 is virtually monodisperse with d_{50} = 1.5 mm, the sieving test showed that all grains were finer than 1.61 mm and all grains were coarser than 1.49 mm. Its density is ρ_s = 2488 kg/m^3. The B7 fraction is narrow graded (grain sizes from 400 to 600 micron) with the median size d_{50} = 0.57 mm, and density of particles 2450 kg/m^3.

3.3 Measured quantities and flow conditions

A set of quantities was measured and stored at each installed flow condition forming a test run. The output of the magnetic flow meter was interpreted as the average velocity of the mixture in the pipe, V_m. The differential pressure measured in the horizontal pipe (Section 3) was recalculated to the hydraulic gradient, i. Furthermore, ERT images and temperature were measured at each test run. Test runs collected for slurry of one value of the mean volumetric concentration of particles, C_v, in the loop formed one test series. C_v was determined as the ratio of the known volume of particles introduced to the loop and the total volume of the loop. Samples of flowing liquid were taken from the loop before and after the test series for viscometry tests in the rotational viscometer. For some series, rheological parameters were determined also for several runs within the series.

A typical test series collected a set of test runs for a broad range of velocities with the same slurry. Results of one series produced one flow curve as in Figure 2. Table 1 gives a list of collected test series used in the evaluation of turbulent diffusivities. Table 2 specifies flow conditions for the runs incorporated to the evaluation.

Table 1. Survey of test series

Test no.	Code of test series
1	CBP0,175b_TK1,5_20140803_a
2	CBP0,175b_TK1,5_20140803_b
3	CBP0,150e_B7_20150520_a
4	CBP0,250b_TK1,5_B7_20150812_b

Table 2. Flow conditions for test series

Test No.	C_v [-]	d_m [mm]	V_m range [m s^{-1}]	No. of profiles	τ_y [Pa]	K [Pa s^n]	n [-]
1	0.1	1.5	3.41-4.51	5	0.811	1.414	0.504
2	0.2	1.5	3.51-4.45	5	1.872	2.365	0.476
3	0.2	0.57	3.39-3.79	3	1.435	1.296	0.534
4	0.167	1.04	3.38-4.52	6	0.462	0.452	0.623

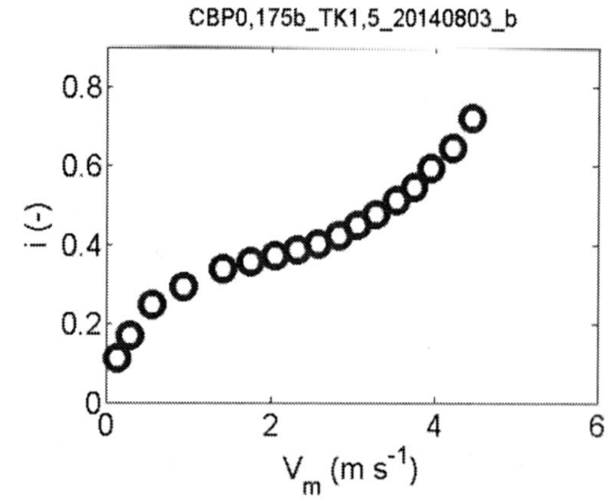

Figure 2. Flow curve (hydraulic gradient versus average velocity of mixture flow) for Test series No. 2.

4 Determination of particle diffusivity from experiments

Turbulent-regime test runs (e.g. the 5 points at the highest velocities in Figure 2) were processed to evaluate the particle turbulent diffusivity for the Rouse-Schmidt turbulent-diffusion model (Eq. 2). The evaluation process required measured concentration profiles as inputs.

4.1 Concentration profile and gradient

Measured profiles of local chord-averaged volumetric concentration (left panels of Figure 3) exhibited a non-uniform distribution (and hence non-zero concentration gradient) of conveyed particles in in the pipe cross section. In order to determine local gradients, the concentration profiles were approximated by a high order polynomic function. This smoothened a profile curve and enabled to obtain values of the local gradient *dc/dy* by differentiating the polynomic function.

4.2 Local settling velocity

An additional input to the evaluation procedure was the vertical distribution of local settling velocities of particles. In flowing non-Newtonian liquids, the local viscosity varies across the pipe cross section because it is sensitive to the local strain rate (i.e. to a velocity profile of the flow). This feature cannot be captured by the static settling velocity test and hence our velocities measured in the sedimentation column could not be directly used in the evaluation procedure.

Instead, it was assumed that the local terminal settling velocity could be calculated using a method for a Newtonian carrier [15] provided that suitable local turbulent viscosities replaced the Newtonian dynamic viscosity.

A distribution function based on results of DNS-simulations of turbulent flows of Herschel-Bulkley was employed to get the vertical distribution of chord-averaged values of the turbulent viscosity of particles. The local values of this viscosity were used to calculate the local settling velocity using the Haider-Levenspiel method [15].

4.3 Distribution of diffusivity and average diffusivity

The local settling velocity and the local concentration gradient were fed to turbulent diffusion formula (Eq. 2) to get local values of the particle diffusivity ε_s (right panels of Figure 3). The local particle diffusivity seems to be almost uniformly distributed throughout the core of the flow. Hence, the average cross-sectional value of the particle diffusivity, $\varepsilon_{s,mean}$, is a useful parameter to model particle suspension using a turbulent-diffusion model. For our purposes, values of $\varepsilon_{s,mean}$ were determined as average values from the core of flow within the range <-0.5R;0.5R>.

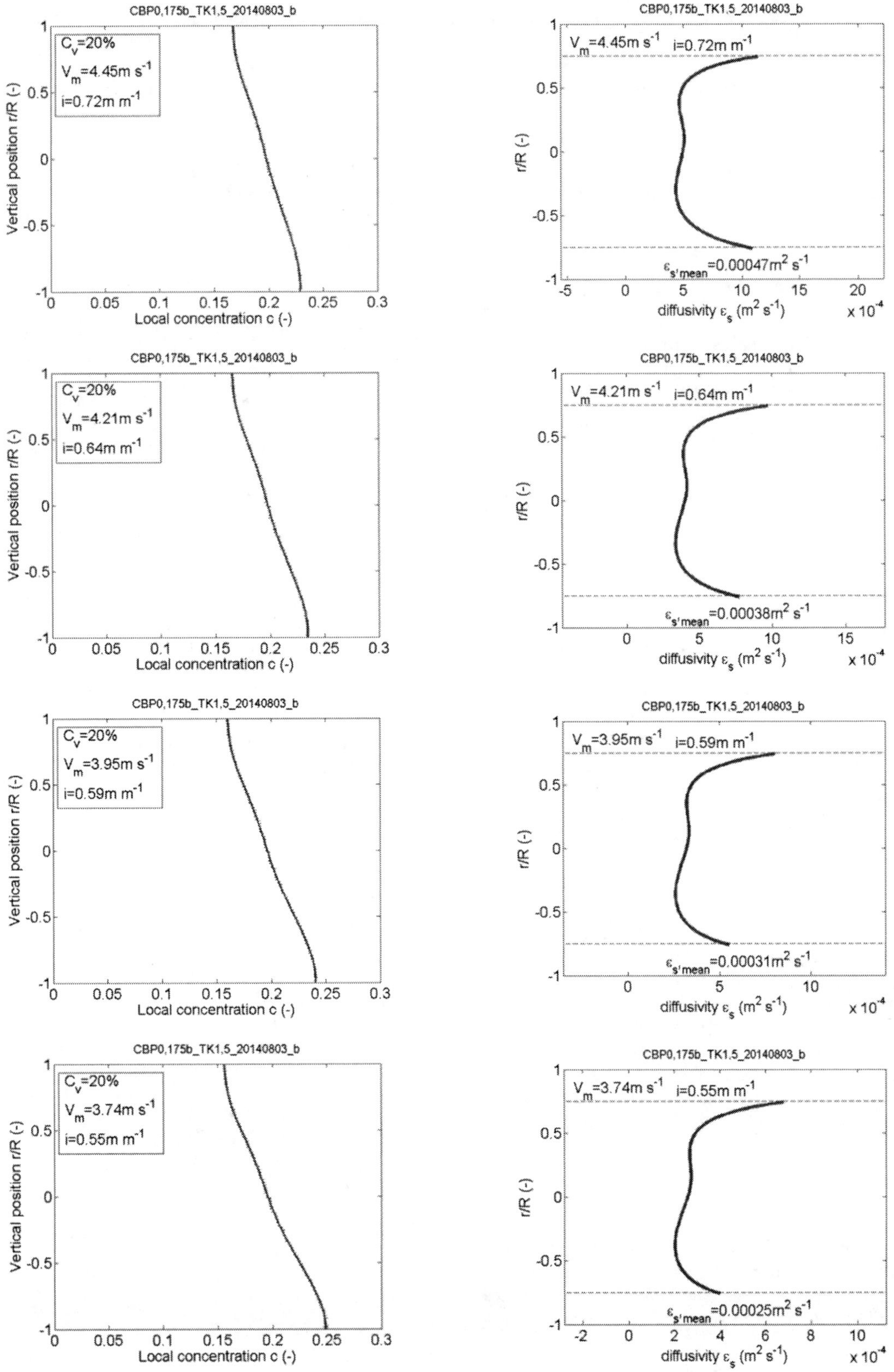

Figure 3. Measured concentration profile and evaluated particle diffusivity.

The dimensionless average particle diffusivity was obtained by normalizing $\varepsilon_{s,mean}$ with the flow shear velocity and the pipe inner radius, $\varepsilon_{s,mean}/(u_{*}{*}R)$. The shear velocity was calculated from the measured hydraulic gradient i,

$$u_{*} = \sqrt{g\frac{D}{4}i} \tag{3}$$

Figure 4 shows the dimensionless diffusivities obtained for different test series.

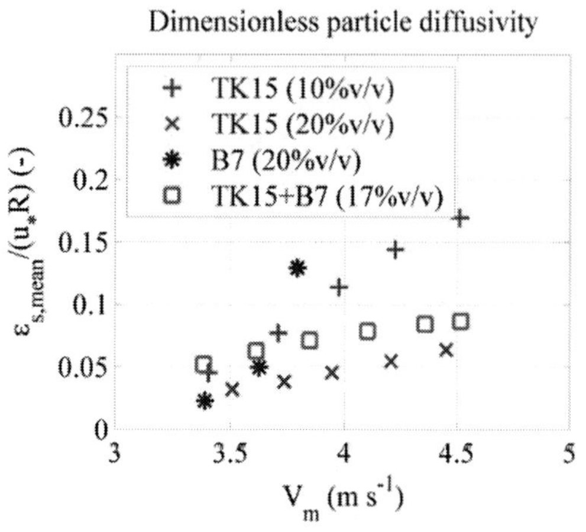

Figure 4. Dimensionless diffusivities obtained for all test series.

5 Discussion

Figure 4 indicates that the dimensionless turbulent diffusivity of particles in mildly viscous Herschel-Bulkley fluid is sensitive to flow velocity, solids concentration and fluid rheology. It also indicates that the diffusivity values are of the same order of magnitude as in a Newtonian carrier. An experimentally determined average value for the diffusivity in flows of sand-water mixtures through a 150-mm pipe was found equal to approximately 0.12 [10], other experiments suggested a value of 0.08 for the centre of the solid-water flow [6]. Compared to the Newtonian case in [10], the diffusivity in the non-Newtonian carrier seems to be considerably more sensitive to the velocity and concentration.

6 Conclusions

The turbulent diffusivity was evaluated in coarse mixture flows within Herschel-Bulkley carrier. The evaluations indicate that the dimensionless diffusivity of particles is of the same order of magnitude in flows of slurry based on a mildly viscous non-Newtonian carrier as in slurry flows with an equivalent Newtonian carrier.

The dimensionless average particle diffusivity appears to be sensitive to carrier rheology, average velocity of mixture and average concentration of coarse particles in the mixture. The data base needs to be extended further to serve as a basis for an appropriate correlation.

The evaluated particle diffusivity can be used in a turbulent diffusion model in order to predict a 1-D concentration distribution of coarse particles in turbulent flow of non-Newtonian carrying liquid.

Acknowledgements

The support by CTU in Prague through a student grant project SGS CVUT No. OHK1-089/15 "Behaviour of solid grains in non-Newtonian fluids and flows" is highly acknowledged.

The authors also would like to thank the following companies who have sponsored aspects of this work as part of the AMIRA P1087 project "Integrated tailings management", i.e. Anglo American PLC, BASF Australia Ltd., Freeport-McMoRan Inc., Gold Fields Australasia Pty. Ltd., Outotec Pty. Ltd., Nalco-Ecolab Pty Ltd., Newmont Mining Corp., Shell Canada Energy Ltd. and Total E&P Canada Ltd.

References

1. P. Vlasák, Z. Chára, *Powder Technol*, **104**(3), 200-206, (1999)
2. P. Vlasák, Z. Chára, *Particul Sci Technol*, **22**(2), 189-200, (2004)
3. P. Vlasák, Z. Chára, *Particul Sci Technol*, **27**(5), 428-443, (2009)
4. L. Pullum, L.J.W. Graham, J. Wu, *Proc. Hydrotransport 18*, 261-276, (2010)
5. L.J.W. Graham, J. Wu, L. Pullum, *Can J Chem Eng*, **89**(4), 817-824, (2011)
6. R.G. Gillies, C.A. Shook, *Particul Sci Technol*, **12**, 45-69, (1994)
7. D.R. Kaushal, Y. Tomita, *Int J Multiphas Flow*, **28**, 1697-1717, (2002)
8. D. Eskin, *Chem Eng Sci*, **82**, 84-94, (2012)
9. V. Matoušek, J. Krupička, *Powder Technol*, **260**, 42-51, (2014)
10. V. Matoušek, *J Hydrol Hydromech*, **48(3)**, 180-196, (2000)
11. V. Matoušek, P. Vlasák, Z. Chára, J. Konfršt, J., *P I Civil Eng-Mar En*, **168(2)**,93-100,(2015)
12. H. Rouse, *Trans. ASCE*, **102**, 463–505, (1937)
13. V. Matoušek, J. Krupička, V. Pěník, *Particul Sci Technol*, **32(2)**, 186-196, (2014)
14. J. F. Richardson, W.M. Zaki, *Sedimentation and fluidization, Part 1. Trans. I. Ch. E.*,**32**, 35–53, (1954)
15. Haider, O. Levenspiel, *Powder Technol*, **58(1)**, 63-70, (1989)

EPJ Web of Conferences **114**, 02091 (2016)
DOI: 10.1051/epjconf/201611402091

Heat transfer measurements and CFD simulations of an impinging jet

Karel Petera[a], Martin Dostál

Czech Technical University in Prague, Department of Process Engineering, Technická 4, Praha 6, Czech Republic

Abstract. Heat transport in impinging jets makes a part of many experimental and numerical studies because some similarities can be identified between a pure impingement jet and industrial processes like, for example, the heat transfer at the bottom of an agitated vessel. In this paper, experimental results based on measuring the response to heat flux oscillations applied to the heat transfer surface are compared with CFD simulations. The computational cost of a LES-based approach is usually too high therefore a comparison with less computationally expensive RANS-based turbulence models is made in this paper and a possible improvement of implementing an anisotropic explicit algebraic model for the turbulent heat flux model is evaluated.

1 Introduction

Impinging jets are frequently used in industrial processes because high heat transfer intensities can be achieved at relatively small surfaces. A lot of experimental, numerical and analytical data concerning the hydrodynamics and heat or mass transfer in impinging jets can be found in the literature, mainly in a configuration where the fluid (liquid) jet impinges a smooth plane surface.

In CFD simulations of an impinging jet, the computational cost of LES-based or DNS-based approach is usually too high therefore less computationally expensive RANS-based (Reynolds averaged Navier-Stokes equation) turbulence models [1] are more frequently used in practise. With a RANS-based approach, the Reynolds stress tensor in the time-averaged Navier-Stokes equation must be described by some turbulence model. The simplest definition is based on the Boussinesq eddy-viscosity approximation [2] which assumes that the Reynolds (turbulent) stress tensor is proportional to mean strain-rate tensor S_{ij} and the proportional factor is the eddy (turbulent) viscosity.

$$-\varrho\overline{u_i'u_j'} = 2\mu_t S_{ij} - \frac{2}{3}\varrho k\delta_{ij} \qquad (1)$$

Turbulence models like, for example, Spalart-Allmaras, $k-\varepsilon$ and $k-\omega$ are based on this approach and using a set of another (transport) equations, quantities like turbulence kinetic energy k and energy dissipation ε are described and the turbulent viscosity is then expressed as a scalar value.

$$\mu_t = \varrho C_\mu \frac{k^2}{\varepsilon}. \qquad (2)$$

In some situations when the turbulence anisotropy is important (highly swirling flows, stress induced secondary flows in rectangular ducts, impinging jets, etc), this approach may not give correct results and a second-order closure Reynolds-Stress Model [3] could be more useful.

Even with the RSM models, the turbulent heat flux representing similar term in the time-averaged form of the energy equation as the Reynolds stress tensor in the momentum equation (for the sake of simplicity, incompressible system and no other internal sources of heat are assumed here)

$$\varrho c_p \frac{\partial \overline{T}}{\partial t} + \varrho c_p \overline{u}_i \frac{\partial \overline{T}}{\partial x_i} = \frac{\partial}{\partial x_i}\left(\lambda \frac{\partial \overline{T}}{\partial x_i}\right) + \frac{\partial}{\partial x_i}\left(-\varrho c_p \overline{u_i'T'}\right) \qquad (3)$$

is frequently described by the scalar value of the eddy viscosity and turbulent Prandtl number (which is frequently taken as constant 0.85 [4]).

$$-\varrho c_p \overline{u_i'T'} = \frac{c_p \mu_t}{\mathrm{Pr}_t}\frac{\partial \overline{T}}{\partial x_i} \qquad (4)$$

This approach is not able to reflect the possible anisotropy in the turbulent heat flux as pointed out by [5] and also other authors. A full second-order closure transport equation for the turbulent heat flux, similar to the transport equation for the Reynolds stress tensor [6] could resolve the anisotropy of the turbulent heat flux, nevertheless with many other unknown terms which would have to be modeled somehow. Therefore, simpler explicit algebraic heat flux models can be used as proposed by Daly and Harlow [7]

$$-\overline{u_i'T'} = C_\theta \frac{k}{\varepsilon}\overline{u_i'u_j'}\frac{\partial \overline{T}}{\partial x_j}, \quad C_\theta = 0.3, \qquad (5)$$

and other authors [8,9]. Such models can improve results in complex cases where the turbulence anisotropy is important and they have relatively small impact on the total computational requirements.

2 Experimental methods

Experimental data in the literature describing the heat transfer for impinging jets on a flat plate can show relatively large scatter (20 – 30%), see Figure 1 comparing literature

[a] Corresponding author: karel.petera@fs.cvut.cz

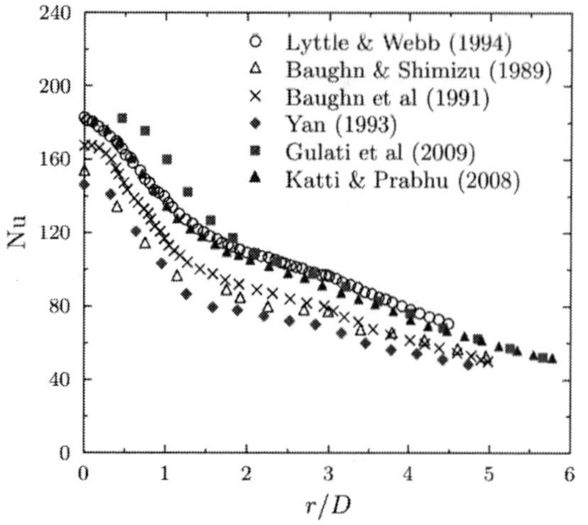

Figure 1. Nusselt number experimental data for ratio $H/D = 6$ and Reynolds number Re = 23000, see references [10–16] .

data [10–16] . The discrepancies in the available data can lie in slightly different geometrical configurations, boundary conditions, as well as in the experimental method itself and its accuracy.

Some experimental techniques are based on measuring the velocity or temperature profiles in the fluid and the quantities like the shear stress or heat transfer coefficients at the impinging surface are then indirectly evaluated using the momentum or heat transfer equations. Large group of experimental methods measures the momentum and/or heat transfer characteristics directly at the impinged surface. The shear stress (shear rate) at the impinged surface can be measured by the electrodiffusion method which is based on measuring the limiting diffusion current between two electrodes. The small working electrode is situated at the impinged surface and the size of the diffusion current is proportional to the molar flux of ion exchange and, at some conditions, it is also proportional to the heat transfer coefficient. The limiting diffusion current can be used to determine the shear stress (rate) at the location of the working electrode, see for example [17–20]. Using the analogy between mass and heat transfer, the heat transfer characteristics can be determined [21].

The heat transfer coefficients are frequently measured by the infrared-thermography methods. A stream of fluid impinging the plate heated by Joule's heat leads to a non-uniform temperature field at the plate surface which can be measured by the infrared thermography and consequently, the local heat transfer coefficients can be determined, see [16], for example. The infrared thermography was also used by Freund in his work [22] to measure the temperature response of the plate induced by an oscillating heat flux impinging the plate. This was an extension of work by [23] and it is briefly described below. See [24–26] for applications of this method (TOIRT - Temperature Oscillation Infra Red Thermography).

The principle of the TOIRT method by Wandelt and Roetzel [23] is depicted on Figure 2. It is based on measuring the time dependency of $T|_{z=\delta}$ (or other corresponding characteristic) at the surface plate of width δ induced by oscillating heat flux $q|_{z=\delta}$. The temperature response depends on thermophysical properties of the plate and heat transfer coefficients on both sides of the plate, i.e. α_0 and α_δ. Transient temperature field T in a homogeneous plate with heat diffusivity a is described by the Fourier equation

$$\frac{\partial T}{\partial t} = a\left(\cancel{\frac{\partial^2 T}{\partial x^2}} + \cancel{\frac{\partial^2 T}{\partial y^2}} + \frac{\partial^2 T}{\partial z^2}\right) . \qquad (6)$$

Neglecting the lateral heat conduction described by the first two terms on the right-hand side and adding boundary conditions on both sides of the plate, it is possible to find an analytical solution for the case of a periodically oscillating impinging heat flux. The boundary conditions of the third kind can be applied on both sides, with a periodic oscillation of the heat flux on one side. They can be described as

$$\lambda \left.\frac{\partial T}{\partial z}\right|_{z=0} = \alpha_0 \, T|_{z=0} \, , \qquad (7)$$

$$\lambda \left.\frac{\partial T}{\partial z}\right|_{z=\delta} = \hat{q} \sin \omega t - \alpha_\delta \, T|_{z=\delta} \, , \qquad (8)$$

where λ is the thermal conductivity of the plate, ω is the frequency of the oscillating heat flux with amplitude $\hat{q}$, and α_0, α_δ are the already mentioned heat transfer coefficients on both sides of the plate. Using the Laplace transform, a temperature solution on both sides of the plate for amplitude A and phase shift φ can be found [23]

$$T(z, t) = A(z) \sin (\omega t - \varphi(z)) \, . \qquad (9)$$

Then, the phase shift at the surface with induced temperature oscillations ($z = \delta$) can be described [23] as

$$\tan \varphi|_{z=\delta} = \frac{c_1 + 2\xi\psi_0 c_2 + 2\xi^2\psi_0^2 c_3}{2\xi\psi_0 (1 + r) c_0 + 2\xi^2\psi_0^2 (1 + 2r) c_1 + 4\xi^3\psi_0^3 r c_2 + c_3} \qquad (10)$$

where the following dimensionless variables and parameters are used

$$r = \alpha_\delta/\alpha_0 \, , \quad \psi_0 = \frac{\alpha_0}{\delta \varrho c_p \omega} = \frac{\alpha_0 a}{\delta \lambda \omega} \, , \quad \xi = \delta \sqrt{\frac{\omega}{2a}} \qquad (11)$$

$$\begin{aligned} c_0 &= \cosh^2 \xi \cos^2 \xi + \sinh^2 \xi \sin^2 \xi \, , \\ c_1 &= \cosh \xi \sinh \xi + \cos \xi \sin \xi \, , \\ c_3 &= \cosh^2 \xi \sin^2 \xi + \sinh^2 \xi \cos^2 \xi \, , \\ c_4 &= \cosh \xi \sinh \xi - \cos \xi \sin \xi \, . \end{aligned} \qquad (12)$$

It is maybe not clear from the complicated solution (10), but the phase shift of the temperature response depends on the size of unknown heat transfer coefficient α_0, and, on the contrary, it does not depend on the amplitude of the heat flux and, furthermore, it does not depend on the accuracy of the oscillating temperature field measurements. In this case with the neglected lateral heat conduction, it is not necessary to measure the surface temperature precisely, only the exact time dependency between the impinging heat flux and the measured temperature response is required. Figure 3 shows an example of relation between the heat transfer coefficient α_0 and phase shift $\varphi|_{z=\delta}$ for our experimental parameters and values mentioned in the following text.

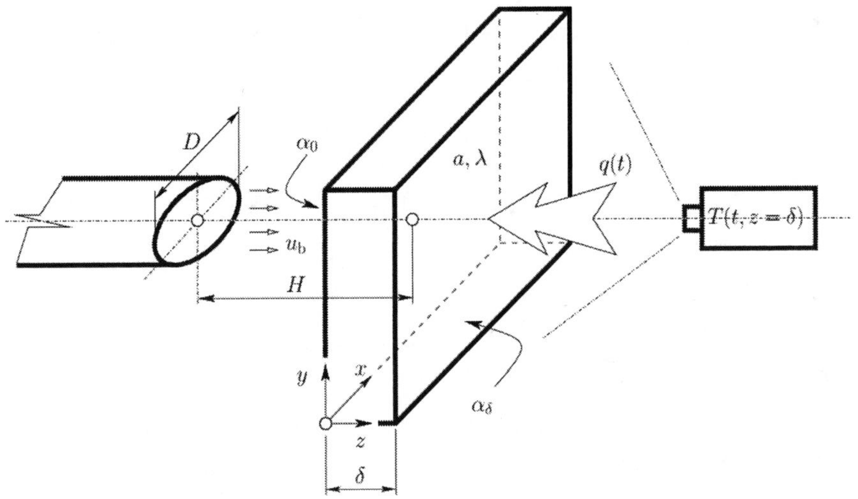

Figure 2. The principle of the TOIRT method by Wandelt and Roetzel [23]. The left side of the wall is impinged by a stream of air, the right side is heated by an oscillating heat flux and the temperature response $T(t, z = \delta)$ is then measured by an infrared camera.

3 Experimental results

The heat transfer coefficient in an impinging jet was measured on a circular plate with diameter 500 mm and thickness 1 mm. The plate was made of steel with heat conductivity 15 W m^{-1} K^{-1}, density 7628 kg m^{-3} and specific heat capacity 460 J kg^{-1} K^{-1}, that is thermal diffusivity 4.275×10^{-6} m^2 s^{-1}. The impinging jet was produced by a stream of air leaving the circular jet nozzle of inner diameter D = 45.8 mm at distance H = 91.6 mm from the impinged plate, that is $H/D = 2$, see Figure 2. The temperature of the jet air and surroundings was 25.9°C, and the corresponding properties are: kinematic viscosity 15.85×10^{-6} m^2 s^{-1}, thermal conductivity 0.026 W m^{-1} K^{-1}, and the Prandtl number 0.71. Corresponding mean volumetric (bulk) velocity u_b = 7.82 m s^{-1} used in our measurements was calculated for the Reynolds number 22596.

The oscillating heat flux impinging the other side of the plate was produced by 2 × 500 W halogen lamp. The light intensity modulation, that is the size of the impinging heat flux was maintained within range 44% – 100% of the maximum value using the voltage variation. The modulation proceeded with frequency 0.1 Hz and it was controlled by a precise sine-impulse generator. The temperature response of this plate side was measured by the infrared camera thermoIMAGER TIM 160 with sampling frequency 10 Hz (camera resolution 160 x 120 points, the real distance between the measured points corresponded to approximately 1.56 mm, the maximum sampling frequency is 120 Hz which we have not used because the measured data are stored in a proprietary file format at this frequency and cannot be processed easily).

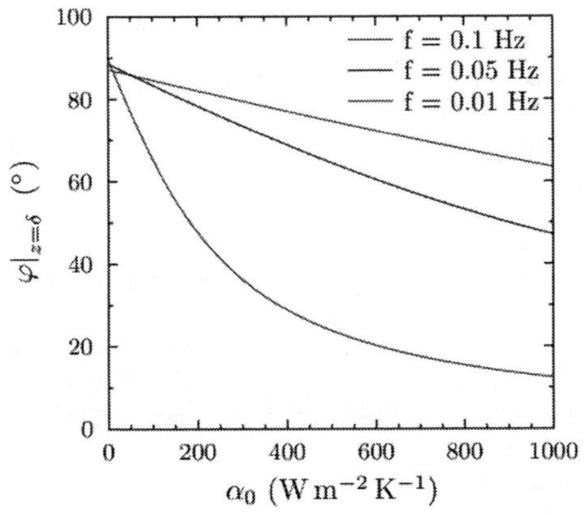

Figure 3. Dependency of phase shift $\varphi|_{z=\delta}$ on heat transfer coefficient α_0 for our experimental parameters and values. α_δ was assumed to be 0.

As mentioned above, the approximate solution (10) neglects the lateral heat conduction. Freund et al [25] wrote that if $\xi < 0.5$, the lateral conduction is negligible compared to the heat transfer rate through the wall and the spatial resolution is then expected to be limited by the resolution of the camera. In our case, the value the dimensionless thickness was $\xi = 0.27$ therefore we were in accordance with this condition.

Figure 3 presents the solution of Equation (10) for the thermophysical parameters mentioned above. The corresponding values of the Nusselt number were then expressed as

$$\mathrm{Nu} = \frac{\alpha D}{\lambda}, \qquad (13)$$

where λ is the thermal conductivity of the fluid (air) and they are presented on Figure 4 in comparison with other literature data [12, 14, 16]. The measured data do not agree very well with the literature data. The main reasons lie in relatively small values of the heat transfer coefficients (below 10 W m^{-2}K^{-1}) therefore the measured dependency between the phase shift φ and the heat transfer coefficient (see Figure 3) was at the accuracy limit of our measurements.

4 Numerical simulations

We have performed several numerical simulations to get a qualitative as well as quantitative comparison of several turbulence models available in ANSYS Fluent. Figure 5 shows our domain of interest, that is a single round jet of

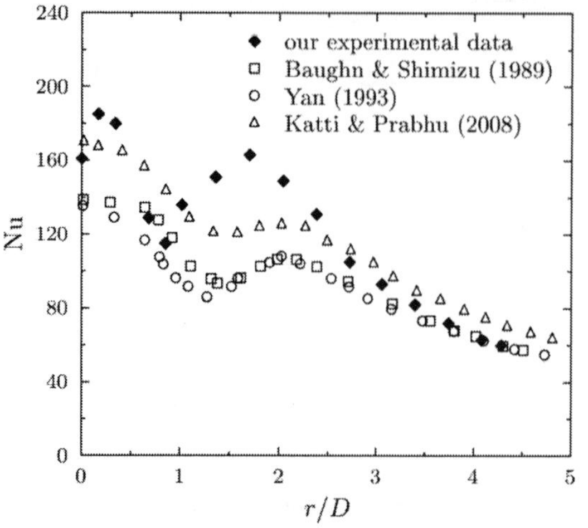

Figure 4. The Nusselt number dependency of our experimental data and comparison with the literature data [12, 14, 16], $H/D = 2$, Reynolds number Re $\approx$ 23000.

diameter D at distance H from a plate which is heated by constant heat flux q. We have solved the case as symmetric (axisymmetric), and a fully developed velocity profile obtained in a separate computation for the given Reynolds number $\mathrm{Re} = u_b D/\nu$ and corresponding turbulence model was applied at the inlet. The pressure outlet boundary condition was used at the top and the right boundary of the solution domain. The maximum radius of the outer boundary related to the jet diameter was $r/D = 10$ in order to minimize the influence of reversed (back-flow) quantities. At the bottom wall, constant heat flux $q = 1000\ \mathrm{W/m^2}$ were used in heat transfer computations. The mesh used in our simulations satisfied condition $y^+ \approx 1$ at the impinged wall boundary. A grid independence study was performed for three different mesh sizes (13000, 23000, 50000), and the grid convergency index [27] for the finest mesh was evaluated as 3.3% for the mean velocity field at $r/D = 1.0$, and 0.03% for the temperature value at the stagnation point. Figure 6 shows relatively small differences of numerical results on a full 3-D geometry (700 000 mesh elements) compared to the 2-D axisymmetric case therefore we made most of our simulations on the 2-D finest mesh with 50 000 mesh elements.

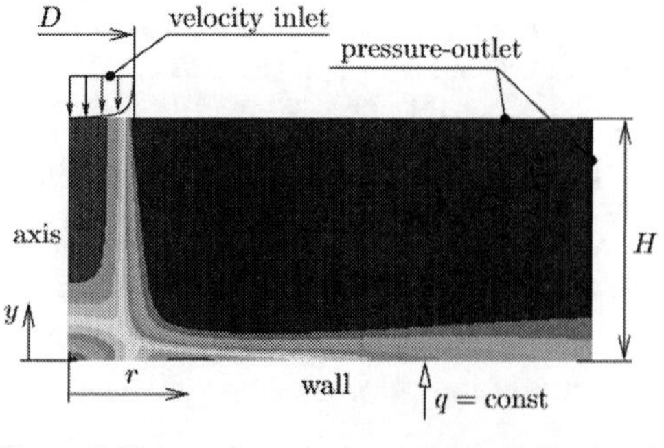

Figure 5. Picture of our single round jet solution domain with boundary conditions and dimensions marked out. The contours illustrates here the mean velocity $\sqrt{u^2 + v^2}$ for Re = 23000 and $H/D = 2$.

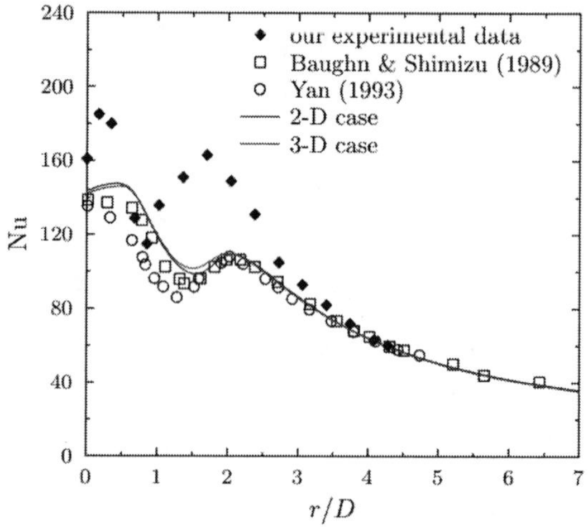

Figure 6. The Nusselt number comparison of 2-D axisymmetric case (50 000 mesh elements) and a full 3-D case (700 000 mesh elements) with experimental data [12, 14]. Height of the jet discharge above the plate is $H/D = 2$, Reynolds number Re = 23000. The Intermittency Transition turbulence model was used here.

In our previous work [28], we have described the implementation of anisotropic models for the turbulent heat flux in ANSYS Fluent. The anisotropic turbulent heat flux was implemented through a user-defined-scalar (UDS) transport equation of scalar ϕ

$$\frac{\partial \varrho\phi}{\partial t} + \frac{\partial}{\partial x_i}(\varrho u_i \phi) = \frac{\partial}{\partial x_i}\left(\Gamma_{ij}\frac{\partial \phi}{\partial x_j}\right) \tag{14}$$

It provides a possibility to use anisotropic diffusivity Γ_{ij}. The following definition based on Daly-Harlow model (5) for the turbulent heat flux

$$\Gamma_{ij} = \frac{\lambda}{c_P}\delta_{ij} + \varrho C_\theta \frac{k}{\varepsilon}\overline{u_i' u_j'}, \tag{15}$$

and $\phi = c_P T$ can give us the following equation

$$\varrho c_P u_i \frac{\partial T}{\partial x_i} = \frac{\partial}{\partial x_i}\left[\left(\lambda\delta_{ij} + \varrho c_P C_\theta \frac{k}{\varepsilon}\overline{u_i' u_j'}\right)\frac{\partial T}{\partial x_j}\right], \tag{16}$$

which, divided by specific heat capacity c_P, corresponds to Equation (3) for Reynolds averaged temperature $\overline{T}$ (steady-state case with no internal heat sources, constant density and specific enthalpy). Formally same equation as (16) and same definition (15) for diffusivity coefficient Γ_{ij} can be obtained for temperature as the user-defined scalar, that is $\phi = T$. Some differences should appear only in the definition of boundary conditions, see [28] for more details.

Implementing an anisotropic turbulent heat flux with some available RSM turbulence models in ANSYS Fluent unfortunately did not yield in a better agreement with the experimental data, the values of the Nusselt number were overpredicted especially in the stagnation region, see Figure 7. Daly-Harlow explicit algebraic heat flux model [7] was used here and qualitatively similar results were obtained with other models [8, 9]. The following code shows an example of UDF function implementing diffusivity coefficient Γ_{ij} in the user-defined scalar equation for the Daly-Harlow model in 2-D.

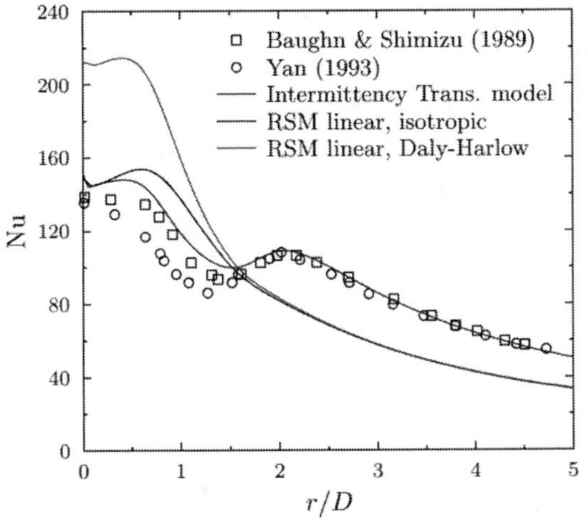

Figure 7. Comparison of simulations with isotropic and anisotropic turbulent heat flux along with experimental data [12, 14]. Linear pressure-strain RSM model was used here [4]. $H/D = 2$, Re = 23000.

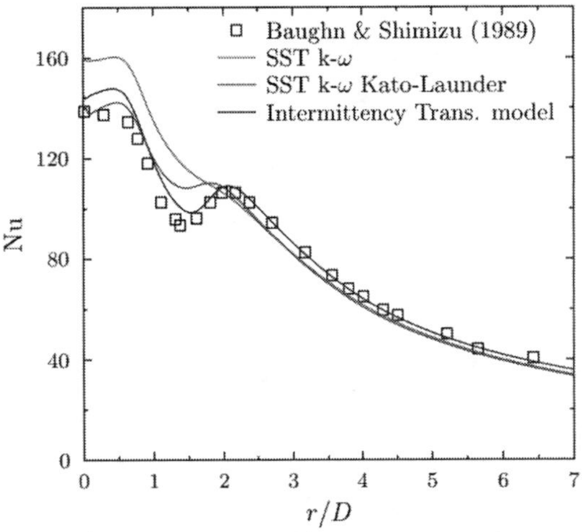

Figure 8. The influence of Kato-Launder [29] production limiter on SST $k-\omega$ turbulence model and comparison with the Intermittency Transition model (which includes the Kato-Launder limiter by default). $H/D = 2$, Re = 23000.

```
DEFINE_ANISOTROPIC_DIFFUSIVITY (gamma_DalyHarlow, c, t,
                                i, dmatrix)
{
  real Cth = 0.3;
  real k = C_K(c,t); real eps = C_D(c,t);
  dmatrix[0][0] = LAMBDA/CP+RHO*Cth*k/eps*C_RUU(c,t);
  dmatrix[1][1] = LAMBDA/CP+RHO*Cth*k/eps*C_RVV(c,t);
  dmatrix[0][1] = RHO*Cth*k/eps*C_RUV(c,t);
  dmatrix[1][0] = dmatrix[0][1];
}
```

The overprediction of the Nusselt values near the stagnation point could be resolved by a production limiter like Kato-Launder [29] which limits the excessive production of turbulence kinetic energy caused by a very high level of shear stress rate in the stagnation regions [4]. See Figure 8 comparing the results for SST $k-\omega$ turbulence model with and without the Kato-Launder limiter.

The Kato-Launder limiter [29] is also part of the ANSYS Fluent implementation of the SST $k-\omega$ Intermittency Transition model which provided the best agreement with the experimental data. It is a modified version of the Transition SST model which is also known as $\gamma - \mathrm{Re}_\theta$ transition model [30]. The Intermittency Transition model solves only one additional transport equation for intermittency γ in addition to the kinetic turbulence energy and dissipation equations. It avoids the dependency of Re_θ on the velocity [4] and makes the Intermittency Transition model Galilean invariant in contrast to $\gamma - \mathrm{Re}_\theta$ transition model which has another transport equation for momentum-thickness Reynolds number Re_θ. The coupling of the transition model with the SST turbulence model is done through the modified production and destruction terms in the equation for turbulent kinetic energy k, see [4, 30–32] for more details.

5 Conclusions

Results of numerical simulations of a round impinging jet were compared with our own experimental data as well as with the literature experimental data. The measured heat transfer coefficients were relatively small therefore the evaluated phase shift with the TOIRT method was at the accuracy limit of our measurements. A different fluid media, water instead of air, for example, could help to resolve these problems and get better agreement with the literature data.

We have implemented an anisotropic model for the turbulent heat flux through a user-defined-scalar equation in ANSYS Fluent to get better prediction of the heat transfer characteristics in an impinging jet. Nevertheless, the best agreement was provided with the SST $k-\omega$ Intermittency Transition model which belongs to the RANS turbulence models based on the scalar eddy viscosity (Boussinesq approximation). The second-order closure Reynolds Stress models with our implementation of the explicit algebraic model for the anisotropic turbulent heat flux were not able to get better agreement with experimental data. The location of the secondary peak in the Nusselt dependency was not correct and the values of the Nusselt number were overpredicted in the stagnation region substantially. A further investigation is necessary to identify the main reasons of these discrepancies.

Acknowledgment

This work has been supported by research project of the Grant Agency of Czech Republic No. 14-18955S.

Nomenclature

a	thermal diffusivity, $a = \lambda/\varrho c_\mathrm{P}$ $(\mathrm{m^2\,s^{-1}})$
$c_{0,1,\ldots,4}$	constants (parameters) used in Equation (10), see Equation (12) for their definitions (–)
c_P	specific heat capacity $(\mathrm{J\,kg^{-1}\,K^{-1}})$
D	jet nozzle diameter (m)
f	frequency $(\mathrm{Hz, s^{-1}})$
H	jet discharge height over the plate (m)
k	turbulent kinetic energy $(\mathrm{m^2 s^{-2}},\ \mathrm{J\,kg^{-1}})$
Nu	Nusselt number, $\mathrm{Nu} = \alpha D/\lambda$, where heat transfer coefficient α is evaluated as $\alpha = q/(T_\mathrm{w} - T_\mathrm{b})$ (–)

P	pressure (Pa)
$\mathrm{Pr_t}$	turbulent Prandtl number (–)
q	heat flux ($\mathrm{W\,m^{-2}}$)
$\hat{q}$	amplitude of oscillating heat flux ($\mathrm{W\,m^{-2}}$)
r	radius (m)
r	dimensionless parameter in Equation (10), see Equation (11) for its definition (–)
Re	Reynolds number, $u_b D/\nu$ (–)
Re_θ	momentum-thickness Reynolds number (–)
S_{ij}	strain-rate tensor ($\mathrm{s^{-1}}$)
t	time (s)
T	temperature (°C, K)
T_b	bulk temperature (average) at the jet nozzle (°C, K)
T_w	wall (boundary) temperature (°C, K)
T'	temperature fluctuation (°C, K)
$\overline{T}$	time-averaged temperature (°C, K)
u	velocity ($\mathrm{m\,s^{-1}}$)
u_b	bulk velocity (average velocity at the jet nozzle) ($\mathrm{m\,s^{-1}}$)
u_m	mean velocity, velocity magnitude, $\sqrt{u^2+v^2}$ in 2-D space ($\mathrm{m\,s^{-1}}$)
u'	velocity fluctuation ($\mathrm{m\,s^{-1}}$)
$\overline{u}$	time-averaged velocity ($\mathrm{m\,s^{-1}}$)
S_h	source term in the energy equation ($\mathrm{W\,m^{-3}}$)
S_ϕ	source term in the user-defined scalar equation (–)
x, y, z	spatial coordinates of the Cartesian coordinate system (m)
α	heat transfer coefficient ($\mathrm{W\,m^{-2}K^{-1}}$)
γ	intermittency (–)
Γ	diffusivity in the user-defined scalar equation ($\mathrm{kg\,m^{-1}s^{-1}}$)
δ	width of impinged plate (m)
δ_{ij}	Kronecker unity tensor (–)
ε	dissipation rate of turbulent kinetic energy ($\mathrm{m^2\,s^{-3}}$)
λ	thermal conductivity ($\mathrm{W\,m^{-1}\,K^{-1}}$)
μ	dynamic viscosity (Pa s, $\mathrm{kg\,m^{-1}s^{-1}}$)
μ_t	turbulent (eddy) dynamic viscosity (Pa s, $\mathrm{kg\,m^{-1}s^{-1}}$)
ν	kinematic viscosity, $\nu = \mu/\varrho$ ($\mathrm{m^2\,s^{-1}}$)
ξ	dimensionless width, $\xi = \delta\sqrt{\omega/2a}$ (–)
ϱ	density ($\mathrm{kg\,m^{-3}}$)
ϕ	general scalar variable in the user-defined scalar equation (–)
φ	phase shift (–)
ψ_0	dimensionless parameter used in Equation (10), see Equation (11) for its definition (–)
ω	specific dissipation rate of turbulent kinetic energy, $\omega = \varepsilon/k$ ($\mathrm{s^{-1}}$)
ω	angular frequency of oscillating heat flux, $\omega = 2\pi f$ ($\mathrm{s^{-1}}$)

References

1. D. Wilcox, *Turbulence Modeling for CFD* (DCW Industries, Inc., 2006)
2. J.O. Hinze, *Turbulence* (McGraw-Hill Publishing Co., New York, 1975)
3. B.E. Launder, G.J. Reece, W. Rodi, Journal of Fluid Mechanics **68**, 537 (1975)
4. ANSYS Fluent, *ANSYS Fluent Theory Guide*, ANSYS, Inc., Release 15.0 (2013)
5. C.F. Dietz, S.O. Neumann, B. Weigand, Numerical Heat Transfer, Part A **52**, 101 (2007)
6. B.A. Younis, C.G. Speziale, T.T. Clark, Proceedings of The Royal Society A **461**, 575 (2005)
7. B.J. Daly, F.C. Harlow, Physics of Fluids **13**, 2634 (1970)
8. K. Abe, K. Suga, International Journal of Heat and Fluid Flow **22**, 19 (2001)
9. B.A. Younis, B. Weigand, S. Spring, Journal of Heat Transfer **129**, 425 (2007)
10. M. Behnia, S. Parneix, Y. Shabany, P.A. Durbin, International Journal of Heat and Fluid Flow **20**, 1 (1999)
11. D. Lytle, B. Webb, International Journal of Heat and Mass Transfer **37**, 1687 (1994)
12. J. Baughn, S. Shimizu, ASME Journal of Heat Transfer **111**, 1096 (1989)
13. J. Baughn, A. Hechanova, X. Yan, ASME Journal of Heat Transfer **113**, 1023 (1991)
14. X. Yan, Ph.D. thesis, University of California (1993)
15. P. Gulati, V. Katti, P. S.V., International Journal of Thermal Sciences **48**, 602 (2009)
16. V. Katti, S. Prabhu, International Journal of Heat and Mass Transfer **51**, 4480 (2008)
17. A. Meslem, M. Kristiawan, I. Nastase, V. Sobolik, Journal of Physics: Conference Series **395**, 012034 (2012)
18. M.E. Hassan, H.H. Assoum, V. Sobolik, J. Vétel, K. Abed-Meraim, A. Garon, A. Sakout, Experiments in fluids **52**, 1475 (2012)
19. M. Kristiawan, A. Meslem, I. Nastase, V. Sobolik, International Journal of Heat and Mass Transfer **55**, 282 (2012)
20. M. Kristiawan, K. Sodjavi, B. Montagné, A. Meslem, V. Sobolik, Chemical Engineering Science **132**, 32 (2015)
21. M. Dostál, K. Petera, T. Jirout, I. Fořt, *Heat transfer similarities between impinging jets and axial-flow impellers*, in *15th European Conference on Mixing* (2015)
22. S. Freund, Ph.D. thesis, Hamburg (2007)
23. M. Wandelt, W. Roetzel, *Lockin thermography as a measurement technique in heat transfer*, in *QIRT 96 - Eurotherm Series 50* (1997), pp. 189–194
24. S. Freund, S. Kabelac, *Measurement of local convective heat transfer coefficients with temperature oscilation IR thermography and radiant heating*, in *Proceedings of HT2005* (2005), pp. 1–7
25. S. Freund, A.G. Pautsch, T.A. Shedd, S. Kabelac, International Journal of Heat and Mass Transfer **50**, 1953 (2007)
26. S. Freund, S. Kabelac, International Journal of Heat and Mass Transfer **53**, 3764 (2010)
27. I. Celik, U. Ghia, P. Roache, C. Freitas, H. Coleman, P. Raad, Journal of Fluids Engineering **130** (2008)
28. Petera, K., EPJ Web of Conferences **92**, 02063 (2015)
29. M. Kato, B. Launder, *The modelling of turbulent flow around stationary and vibrating square cylinders*, in *Ninth Symposium on "Turbulent Shear Flows"* (1993)
30. F.R. Menter, R.B. Langtry, S.R. Likki, Y.B. Suzen, P.G. Huang, S. Volker (2004), ASME-GT2004-53452
31. F.R. Menter, R. Langtry, S. Völker, Flow, Turbulence and Combustion **77**, 277 (2006)
32. F.R. Menter, AIAA Journal **47**, 2894 (2009)

EPJ Web of Conferences **114**, 02092 (2016)
DOI: 10.1051/epjconf/201611402092

Heat and mass transfer in a vertical channel under heat-gravitational convection conditions

Michail Petrichenko[1], Darya Nemova[1], Elisaveta Reich[1,a], Svetlana Subbotina[1] and Faina Khayrutdinova[1]

[1] *St. Petersburg State Polytechnical University, 195251 St. Petersburg, Russian Federation*

Abstract. Heat-gravitational motion of an air flow in a vertical channel with one-sided heating in an area with low Reynolds number is stated in Boussinesq approximation. Hydraulic variables field in a heat-gravitational motion is modeled with the application of ANSYS-FLUENT. It is converted to average velocity and temperature values in a cross section of the channel. The value of an average velocity is determined by rate of heat supply in a barotropic flow with a polytropic coefficient n<k. An average velocity versus flow resistance characteristic is ascertained. Largest extremum of channel flow coefficient is less than 0,707. Physical modelling of the flow is performed on a variable geometry unit. Calculation and experimental data established that an average velocity reaches a maximum in a ventilated vertical channel with free air access and in the absence of gaps. In a channel with closed air access inleakage of the cold air through gaps on an unheated side leads to decrease in an average speed at least twice in comparison to channel with free air access.

1 Introduction

Heat-gravitational convection occurs in different technical devices and natural phenomena (from the heat engine to meteorology and ocean dynamics). It also explains the movement of an air in a vertical heated channel.

JV Boussinesq has formulated the conditions for simplifying motion equation of a viscous incompressible fluid that is known as the Boussinesq approximation. He formulated key insight of the theory, allowing writing motion equation of heat-gravitational motion of viscous fluid in a gravitational field.

Initial interest in the natural convection evident in the work of physicists. Physical bases of natural convection was investigated by H. Lorentz, J. Boussinesq, E. Eckert, V. Zhukovsky and etc.

E. Eckert in his work in 1929 by expansion in power series integrates the equations of motion and energy. He obtained for the first time criterial formula for the transfer coefficients on a plate under natural convection in a semi-infinite space.

The issues of heat-gravitational convection in engineering structures are viewed in reports [1-29].

N. Mingotti, T. Chenvidyakarn, A. Woods in their paper described buoyancy flow in ventilated facades. They suggested model, that hydraulically describes flow mechanics. Thus, it became possible to estimate an influence of design feature of the façade on air flow in construction. [1].

Portuguese scientists F.M. Silva, G. Gomes, M. Rodrigues basing on in-situ testing of ventilated facades developed a mathematical model that made possible to predict heat-gravitational flow in these constructions [2].

C. Popa, D. Ospira, S. Fohannoa, C. Chereches made a numerical simulation of buoyancy flow in ventilated facades. They took in account Rayleigh number and influence of boundary layer in their paper [3].

The aim of this study is to determine discharge characteristics of heated vertical channels in a low-Reynoldsless than area. Accomplish this aim such challenges should be figured out:

1. To perform theoretical estimate of flow characteristics in conditions of heat-gravitational convection (average velocity, heat transfer rate, pressure loss, average temperature of the flow) and to evaluate rate of heat transfer, temperature and moisture in conditions of heat-gravitational convection.
2. To perform numerical simulation of plain flow in conditions of heat-gravitational convection in ANSYS Fluent.
3. To perform physical simulation of heat-gravitational convection in different types of vertical channels.

2 Deriving of calculation dependences

Fundamentally there is an Eckert model of heat-gravitational motion. It consists of equations that state motion, energy and continuity equations:

[a] Corresponding author: lisa_reich@mail.ru

$$u_z\frac{\partial u_z}{\partial z}+u_y\frac{\partial u_z}{\partial y}-\nu\frac{\partial^2 u_z}{\partial y^2}+g\theta$$
$$u_z\frac{\partial \theta}{\partial z}+u_y\frac{\partial \theta}{\partial y}=\frac{\nu}{\sigma}\frac{\partial^2\theta}{\partial y^2} \quad (1)$$
$$\frac{\partial u_z}{\partial z}+\frac{\partial u_y}{\partial y}=0$$

where, $u_{z,y}$ – longitudinal and transverse velocity components, θ - non-dimensional temperature head, $\theta:=(T-T_c)/(T_h-T_c)$, T_h, T_c – temperatures of heated surface and cold air, ν - coefficient of kinematic viscosity of an air, σ - Prandtl number. This set of equations is necessary and opportune in conditions of heat-gravitational convection due to minimum of allowances and relying on basic mechanics lows.

In Eckert model the fulfillment of conditions of hydrostatic pressure distribution in vertical direction is essential:(Boussinesq approximation). In other words, force of the flow pressure in heat-gravitational motion is counterbalanced by gravity force. The one volume force is – ascending force with mass density gθ. Set of equations is supplemented by boundary conditions, that state no-slip condition on the surface and continuity of non-dimensional temperature head on air-surface boundary. Eckert model is correct for laminar (low-Reynolds) flow. In conditions of real vertical ventilated layer thermal Grashof number:

$$Gr_{h,\theta}=gh^3/\nu^2*(T_h-T_c)/T_h \quad (2)$$

where h – width of the channel, is less than $10^5 \ldots 10^7$, that is in accord with laminar flow.

As usual hydraulic model, Eckert model determines velocity and temperature head distributions in flow space. In hydromechanics, local velocity and temperature head distributions are substituted with average velocities and temperature heads in cross-section. Therefore averaging across the width of boundary layer or channel allows to calculate hydraulic flow characteristics – average velocity and temperature head. In addition, with the knowledge of velocity and temperature head gradients on the side of the channel it is possible to calculate coefficient of hydraulic friction number and heat-transfer coefficient.

Boussinesq approximation is applied to heat-gravitational motion of viscous liquid. Basic allowances, composing Boussinesq model form the following set of conditions: liquid is a perfect gas, $p=R\rho T$; motion is barotropic (polytropic index n), static pressure p is in accord with equilibrium condition. Barotropic condition replaces differential energy equation in Eckert system.

Average velocity of the flow in vertical channel is determined.

When heat supply to air flow n≤k. Therefore:

Technical works of expansion in absence of external heat-exchange is less than technical works of polytropic expansion $l_{T,n}$, n≤k.

Excess of technical work is expended in mass motion of an air in the channel. Consequently:

$$v=\varphi\Lambda\sqrt{RT_c}\sqrt{\frac{1}{n}-\frac{1}{k}},\ \Lambda:=\frac{gL}{RT_c} \quad (3)$$

Using basic allowances it can be showed that:

$$v=\sqrt{\frac{1-\theta}{1+\zeta_r}}\sqrt{2gL}=\varphi\sqrt{\frac{T_h-T_c}{T_h}}\sqrt{2gL} \quad (4)$$

In formula (3), (4): φ - velocity coefficient, L – height of the channel.

The following functions were deduced:

1. For an average velocity:

$$v=\varphi\sqrt{1-\frac{T_c}{T_h}}\sqrt{2gL} \quad (5)$$

where:

$\varphi:=\dfrac{1}{\sqrt{1+\lambda\frac{L}{h}+\zeta}}$ - velocity coefficient

λ - coefficient of hydraulic friction,

ζ - local head loss coefficent,

2. For heat-transfer rate:

$$\mathrm{N}=\frac{h\delta}{\kappa}=\sqrt[4]{3\sigma^2} \quad (6)$$

where:

$\delta=\sqrt[4]{\dfrac{4\nu^2 z}{g}}$ - width of the «boundary layer of force of the flow pressure»

ν – coefficient of hydraulic friction,

z – vertical distance from the lower bound of ventilated layer;

h – heat transfer coefficient,

κ – air thermal conductivity,

σ – Prandtl number. For an air σ=0,707.

3. For head loss:

$$h_r=\frac{V^2}{2g\varphi^2} \quad (7)$$

where:

$\varphi:=\dfrac{1}{\sqrt{1+\lambda\frac{L}{h}+\zeta}}$ - velocity coefficient

λ - coefficient of hydraulic friction,

ζ - local head loss coefficient;

4. For an average temperature of the flow:

$$T_m=\frac{T_h-T_c}{\sqrt[4]{3}}(1-e^{-\sqrt[4]{3}\cdot\zeta})+T_c \quad (8)$$

where:

T_h - temperature of heated surface

T_c - temperature of the cold surface

$\zeta=\dfrac{y}{\delta}$ - nondimensional coordinate (transverse coordinate to width of the boundary layer ratio).

3 Numerical simulation

Numerical simulation of heat-gravitational motion in a vertical air layers is performing for the sake of maximized approximation to real design features of various vertical channels of engineering structure. Simulation was performing in ANSYS Fluent. The set of motion, energy and continuity equations should be solved. There are 3 equations and 3 unknown quantities in all(2 velocity components and temperature). Energy equation can be substitute with the condition of barotropy of the motion. It is a determined system.

In the field of varying of calculated functions there are some difficulties in calculation of distributions and their gradients. Original set of equations is written as averaging system, there are different translation models for coefficients of stress matrix.

Numerical simulation consists on two parts:
- selection and testing of model constants;
- problem solving for heat-gravitational motion.

3.1 Selection and testing of model constants

The model has been selected from a three models: k-ε standart, Girimadji, SST k-ω - the model of kinetic energy - vorticity of Saffman-Wilcoxon and its modification. Calculation was performed for a two-dimensional channel, Reynolds number is between 1500 and 8000, hydraulically smooth wall, motion, similar to uniform, Darcy number λ=0,016.

By way of the test for heat-gravitational motion two cases were considered: isothermal plate flotation and plate flotation in constant heat flow conditions. In case of a real ventilated layer none of these conditions are realized: hot surface is not isothermal, heat flow throughout the height of the wall is changeable. Therefore test calculations evaluate program results in the limit states, constraining real mode. Velocity and temperature head distributions, head gradient of the plate had been calculated. Velocity fields in the channel with a free vent and in absence of rustication joints are shown in figures 1 and 2.

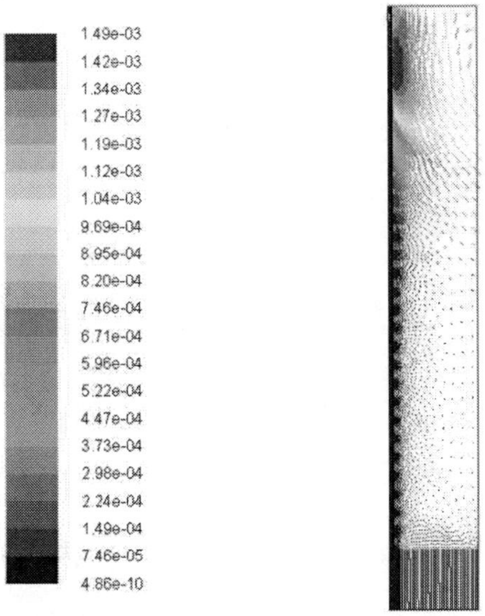

Figure 1. Velocity vectors in an ideal ventilated channel.

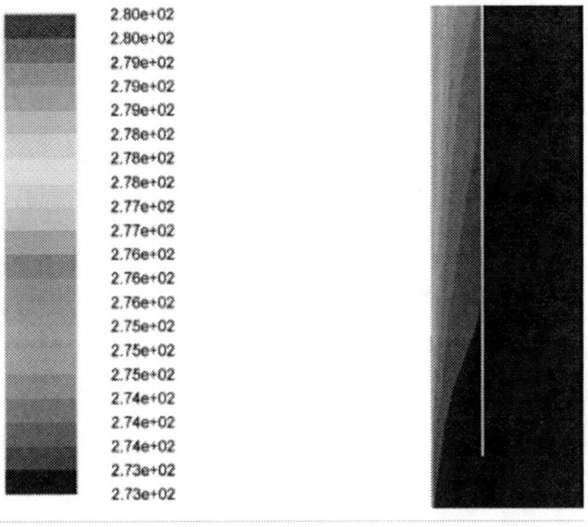

Figure 2. Temperature distribution in an ideal ventilated channel.

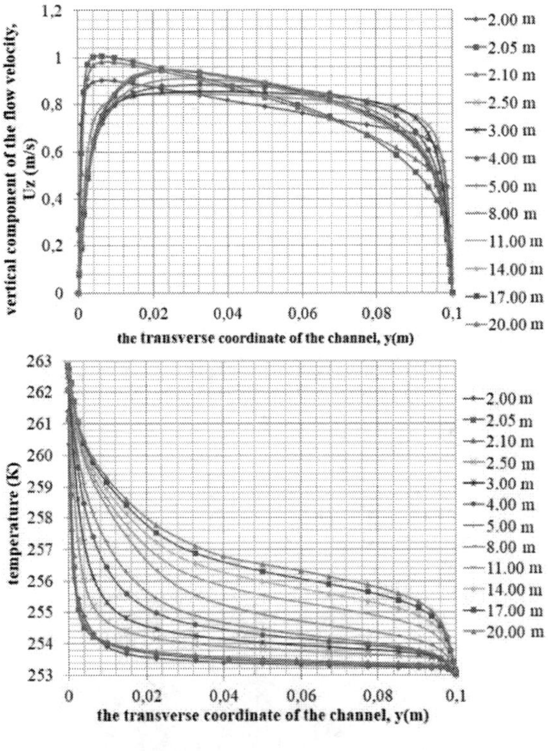

Figure 3. Distribution of the vertical component of the flow velocity in the transverse coordinate of the channel and calculated dependence temperature of the transverse coordinate channel (ideal channel).

According to results it can be concluded:

1. Maximum of the velocity and temperature head are localized along the hot surface of the vertical channel. Result does not depend on the width of the channel;
2. Stabilization of velocity fields and pressure head along the length of the flow is attended with decrease of distribution inhomogeneity;

3. Effect of width of the channel on velocity and pressure head distribution along the hot surface is significant when width of the channel is about 2δ, δ - width of the boundary layer;
4. Velocity and temperature head distribution across the width of the channel provide support for analytical estimations. Temperature distribution demonstrates quick and monotone head loss from the hot surface to flow:

$$\theta = \frac{N}{3a}\exp(-3a\zeta) \tag{9}$$

where:
a - parametr
Velocity distribution is not monotone, with the largest extremum along the hot surface:

$$f'(\zeta) = \frac{1}{9a^2}\zeta\exp(-3a\zeta) \tag{10}$$

Max is attained when $\zeta = \dfrac{1}{3a}$.

Numerical experiment had also been performed for real constructions with an absence of air access for providing their nonoptimality in hydraulic point of view. The results of an experiment are shown in figures 4-7.

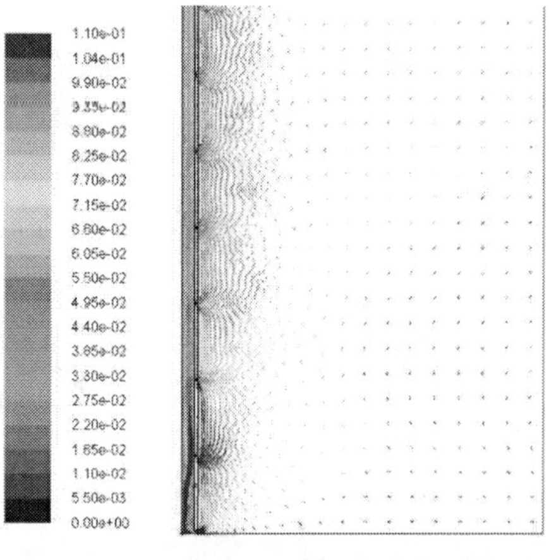

Figure 4. Velocity vectors in real ventilated channel.

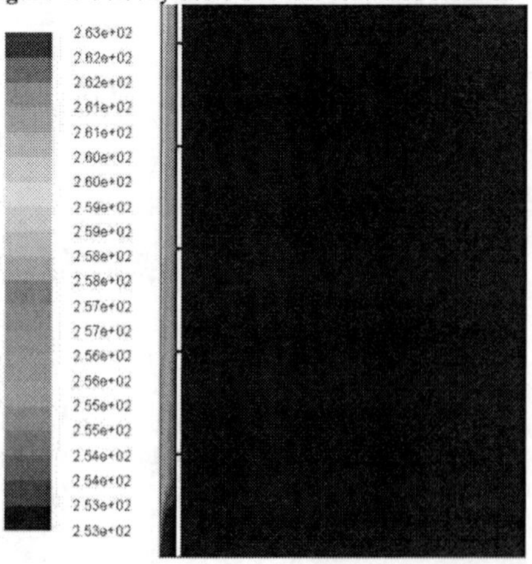

Figure 5. The temperature distribution in the ventilated real channel.

Simulation had been performed for evaluation of effect of design features on air flow in the vertical channel. Such constructions have a rustication joints and densely adjoin to pedestal, impeding airing from below to the channel. Therefore, these constructions do not provide optimal conditions for heat-gravitational convection. It means an absence of the optimal heat-to-humidity conditions in the building.

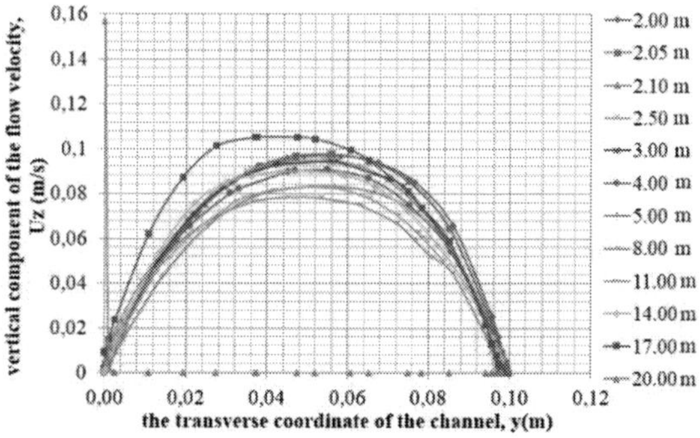

Figure 6. Distribution of the vertical component of the flow velocity in the transverse coordinate of the channel (simulation of real of the channel).

It can be seen that real velocities are far less than in perfect structure and at top point velocity is equal to zero. In perfect structures velocity distribution across the width of an air layer is uniform. In a real structure velocity attain a maximum only in the midpoint of the layer. Boundary layer distribute almost across the width of the layer, heat transfer coefficient is small, moisture removal is ineffective.

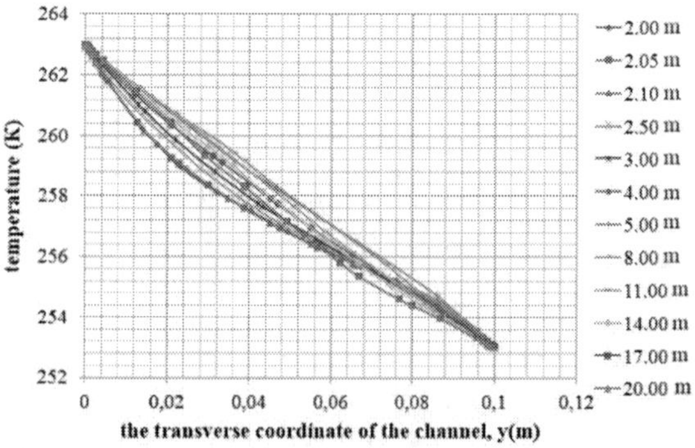

Figure 7. Estimated dependence of the temperature on the transverse coordinate of the channel (simulation of real of the channel).

Figure shows that temperature distribution in a real structures is almost linear. This is evidenced by absence of convection.

4 Physical model

There are different installations for studying of heat and moisture transfer physics and for studying of real structures (channels).

Installations of the first group are for verification of convection mathematical model. They represent lengthful vertical and inclined plates, that work in conditions of constant temperature of the surface or in case of constant heat flow on surface of the plate. Boundary layer of ascending force on this surface expanding along the plate with no limits. An absence of cold adiabatic surface distorts the results and restrict their using for real structures.

Isothermal surface and surface in case of constant heat flow are not realized in a real ventilated channels. Usually dissipation from hot surface to flow is known. These installations overlook design features, that are used for hinged ventilated façade: entry and exit conditions of the flow, existence of rustication joints and others.

Installations of the second group are applied for boiler construction, engine technology and others. For these technical fields studying of convection for capillary liquid and molten metal is of current concern Studying of heat-gravitational convection of an air in channels of HVF demands development of a special ohysical model. It should take into account real design features of the ventilated layer. It is necessary to vary with L/h, entry conditions of the flow to channel from free and impactless to its absence, presence of rustication joints, their values and frequency distribution along the hot. There must be heat source, that provides density of heat flow, according to operation of real structures. Cold surface, that imitates façade, should be adiabatic.

Physical model of the vertical ventilated layer features installation 2050 mm high and 600 mm wide, it has two edges, bounding an air gap and imitating lining layer and wall of the building. One of the edge of the installation is heated and fixed, the second one imitates cladding layer, it is unheated and moveable. It can be created different width h of an air gap in the range of 20-300 mm. Unheated edge consists on sheets of 300 mm, separated by gaps simulating rustication joints by 5 mm. It is possible to control air access from the bottom of the channel in the installation. Regulation of an air access is realized by dint of the seal, capable of restricting of an air access either in part or in whole.

Figure 8. Installation for physical experiment.

Thus, installation allows to carry out experiments by adjusting the conditions of air inlet, geometrical parameters of an air gap and the quantity of heat radiated by the heated edge.

To confirm theoretical calculations, numerical experiment and for studying the effects of rustication joints and air access series of experiments had been performed in the following sizes and configurations of the installation:

1st case: air access to the bottom of the installation is fully open, rustication joints are hermetically sealed (imitation of the perfect channel).

2nd case: air access to the bottom of the installation is fully open, rustication joints are presented (imitation of the channel, closer to real conditions).

3rd case: air access to the bottom of the installation is partially-obscured, rustication joints are presented (imitation of the channel, closer to the real conditions of design changes).

4th case: air access to the bottom of the installation is fully closed, rustication joints are presented (imitation of the real channel).

On each occasion the width of an air gap was variable and changed within the range 20 to 300 mm, the height of the air layer was constant and was equal to 2050 mm. Before starting of series of experiments, the temperature in the laboratory was equal to 293 K. During the experiment, on each occasion an average velocity and temperature of an air flow in the channel at each width of an air gap was measured by dint of hot-wire anemometer.

Air velocity varied between 0 m/s and 0.4 m/s, the temperature varied between 298.6 K and 301.2 C. Data reduction process was performed for each of the four series of experiments. As the variables there were used L/h and $v/\sqrt{gL}$. The results are shown in figure 11-12. One of experiments is considered as a probe (first case). The purpose of the test experiment - to estimate the value of the coefficient of hydraulic friction. Basic functional connection used in the processing of the experiment is as follows:

$$v/\sqrt{gL} = \varphi\sqrt{2(T_h - T_c)/T_h} \qquad (11)$$

In this case, $\varphi := 1/\sqrt{1+\lambda L/h}$. Thus, it is possible to calculate the velocity coefficient φ and coefficient of hydraulic friction $\lambda = h/L(1/\varphi^2 - 1)$.

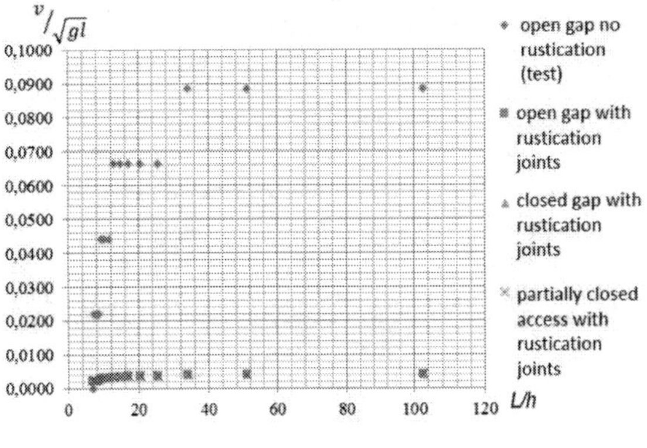

Figure 9. Dependence of froude number on geometric parameters of the channel.

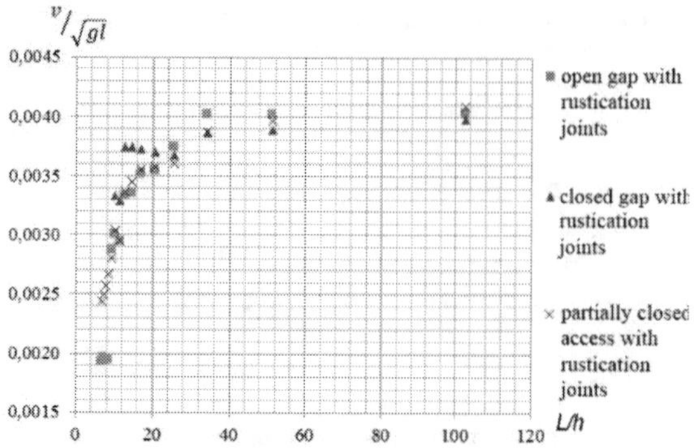

Figure 10. Dependence of froude number on geometric parameters of the channel.

From all results it follows: in the "perfect channel" (air access to the bottom of the installation is fully open, rustication joints are hermetically sealed) air velocity is maximal. This is evidenced by the results of numerical simulations. Existence of rustication joints even in the open channel reduces the maximum value of Froude number at least twice. Therefore rustication joints are an additional local hydraulic resistance, reducing velocity coefficient φ. Closed channel in presence or absence of rustication joints operates at significantly lower values Froude number. Probably, there is no directional heat-gravitational motion, determining lower value of an average velocity for all values of L/h.

5 Conclusions

1. Hydraulic model of heat-gravitational flow in a vertical ventilated layer was developed.
2. Limit values of an average velocity, heat transfer rate, head loss, average temperature of the flow ensuring removal of the dew point in the air gap were evaluated. Basic functional connection was obtained.
3. Numerical simulation of plane flow in conditions of heat-gravitational convection had been performed.
4. Calculation data and experiments showed that the maximum value of an average velocity attained in the vertical channel with fully open air access in the absence of rustication joints. In a channel with closed air access air infiltration through rustication joints provides reducing of an average velocity at least twice in comparing with an open access channel.
5. Design of the gap is offered - installation of accelerating cascade with blockage degree less than 1/2 of the input section of the channel.

References

1. N. Mingotti, T. Chenvidyakarn, A. Woods, Build. Environ. **46**, 807-823 (2011)
2. F.M. Silva, G. Gomes, M. Rodrigues, Build. Environ. **87**, 292-301 (2015)
3. C. Popaa, D. Ospir, S. Fohannoa, C. Chereches, Energ. Buildings **50**, 229–233 (2012)
4. M. Mourshed, S. Shikder, A. Price, Adv. Eng. Inform. **25**, 676–687 (2011)
5. R. Fuliotto, F. Cambuli, N. Mandas, N. Bacchin, G. Manara, Q. Chen, Energ. Build. **42**, 23–28 (2010)
6. N. Hashemi, R. Fayaz, M. Sarsha, Energ. Buildings **42**, 1823–1832 (2010)
7. S. Lee, S. Kim, Y. Na (2015) Energ. Buildings **99**, 67–74 (2015)
8. A. Chabauda, J. Eynarda, S. Grieu, Energ. Buildings **99**, 9–31 (2015)
9. F. Favoinoa, Q. Jinb, M. Overend, Energ. Proc. **62**, 289–298 (2014)
10. W. Tizani, M. Mawdesley, Adv. Eng. Inform. **25**, 569–572 (2011)
11. J. Ploennigs, A. Ahmed, B. Hensel, P. Stack, K. Menzel, Adv. Eng. Inform. **25**, 688–698 (2011)
12. L. Gracia, Navarro, A. Castell, L. Cabeza, Appl. Therm. Eng. **61**, 372-380 (2013)
13. W. Pasut, M. Carli, Appl. Therm. Eng. **37**, 267-274 (2012)
14. C-S. Park, G. Augenbroe, Automat. Constr. **30**, 50–56 (2013)
15. A. Hunter, N. Williams, J. Rayner, L. Aye, D. Hes, S. Livesleya, Ecol. Eng. **63**, 102-113 (2014)
16. A.L.S. Chan, T.T. Chow, Energ. Buildings **69** (2014)
17. N. Mingottia, T. Chenvidyakarn, A.W. Woods, Energ. Buildings **58**, 237–249 (2013)
18. H. Radhi, S. Sharples, F. Fikiry, Energ. Buildings **56**, 179–188 (2013)
19. W. Loua, M. Huanga, M. Zhang, N. Lin, Energ. Buildings **54**, 179–191 (2012)
20. T. Ayinde, S. A. M. Said, M. Habib, Int. J. Heat. Mass Trans. **44**, 1207-1216 (2008)
21. T. Yilmaz, A. Gilchrist, Int. J. Heat. Mass Trans. **43**, 707-719 (2007)
22. D. Vuksanovic, V Murgul, N. Vatin, V. Pukhkal, Appl. Mech. Mater. **725-726** (2014)
23. M. Penić, N. Vatin, V. Murgul, Appl. Mech. Mater. **680**, 534-538 (2014)
24. V. Pukhkal, N. Vatin, V. Murgul, Appl. Mech. Mater. **680**, 529-533 (2014)
25. A. de Gracia, L. Navarro, A. Castell, L. F. Cabeza, Appl. Therm. Eng. **61**, 372-380 (2013)
26. E. Abu-Nada, Int. J. Heat Fluid Fl. **29(1)**, 242-249 (2008)
27. E. Abu-Nada, Numer. Heat Tr. Part A **58**, 660-679 (2010)
28. H. Oztop, E. Abu-Nada, Int. J. Heat Fluid Fl. **29**, 1326-1336 (2008)
29. Z. Li,Y. Tang, H. Lei, B. Caswell, G. Em Karniadakis, J. Comput. Phys. **265**, 113-127 (2014)

EPJ Web of Conferences **114**, 02093 (2016)
DOI: 10.1051/epjconf/201611402093

Possibilities of radioisotope measuring in control of an unstable solid particles hydrotransport

Leszek Petryka[1,a], Marcin Zych[1], Robert Hanus[2], Jerzy Sobota[3], Pavel, Vlasak[4], Monika Śleziak[1] and Dariusz Świsulski[5]

[1]*AGH University of Science and Technology, Krakow 30-059, Poland,*
[2]*Rzeszow University of Technology, Rzeszow 35-959, Poland,*
[3]*Wrocław University of Environmental and Life Sciences, Wroclaw 50-375, Poland,*
[4]*Institute of Hydrodynamics AS CR, v. v. i. 160 00 Prague 6, Czech Republic,*
[5]*Gdansk University of Technology, 80-233 Gdansk, Poland*

Abstract. The paper presents γ-radiation proposal to control a multiphase flow, independently from circumstances. So this method may be applied even in compound industrial or environmental processes. Moreover in many cases, it is the only method for applications for dense mixture containing coarse angular grains. The constructed equipment allows continuous measurement of density as well as solid phase for both concentration and average velocity. Due to pressure loss, it gives the output digital signal convenient for cybernation of the control process. The proposed procedures were tested at a laboratory installation modeling conditions expected during planned excavation of nodules from Pacific bottom.

1 Introduction

Hydrotransport is often applied in many branches of industry. Commonly in this operation, solid particles are transported by water through a pipe or open channel. However, the management of such process is being subject to numerous physical and economicl restrictions. For example, the size of the grains may vary and decreasing of the velocity would cause sedimentation of the solid phase. Moreover, even small modification of viscosity or the wall roughness may radically change head losses and turbulence of the flow [1, 2].

For a staff, it is evident how improper concentration of solids or flow velocity might result in high energy loss, over-wearing of pipes or even their clogging [3]. For example, in the typical hydrotransport installation with dry mineral throughput of 10,000 tons per hour, where minimal velocity preventing sedimentation is 3 m / s, the operating takes place with velocity of 4 or even up to 6 m / s. In consequence all pipes are used no longer than for 2 years [4].

Necessity of such flow control demands continuous measurements of at least solid particles' concentration and velocity, but this idea is really a great challenge, especially in harsh industrial or environmental circumstances, when ultrasonic flowmeters fail. As an example, transport of tough mineral particles may quickly wear down any probe insight the flow. Bearing that in mind, the proposal of nuclear method based on gamma radiation passing through both pipe and flowing mixture is often desirable [5].

The paper presents a gamma absorption method and describes results of laboratory tests demanded to formulate the proposal of control of the Pacific nodules exploitation.

2 Nuclear method of two-phase flow control

Hydraulic transport of solid particles is an example of two-phase flow especially convenient for an analysis by gamma radiation, due to photons transmission through pipes and delivering digital signals related to density of flow and averaged velocity of solid particles [6, 7].

2.1 Density measurement

In a hydrotransport analysis, a concentration of solid phase in a flowing mixture or its density belongs to the most important parameters and is commonly applied in the largest installations [3, 4 and 8].

The idea of gamma-ray absorption measurement is presented in figure 1. This procedure is based on exponential decreasing of narrow monoenergetic photon beam intensity in function of composition and geometry of absorbent. In this figure the sealed radioactive source emits incident gamma radiation beam shaped by collimator (1). Photons pass through the pipeline with analyzed mixture (3) and achieve the probe (2) contains a scintillation detector.

[a] Corresponding author: Leszek.Petryka@fis.agh.edu.pl

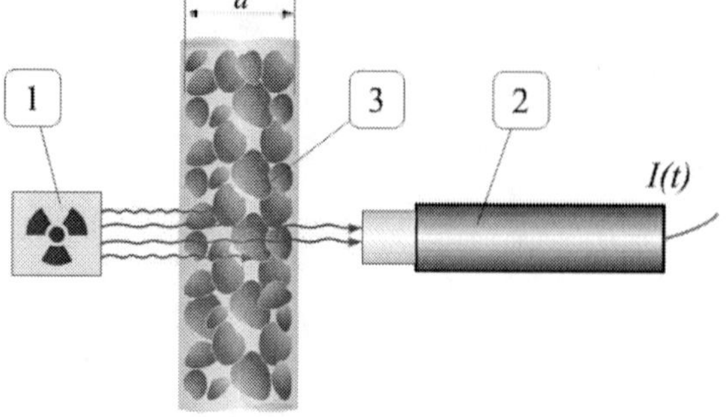

Figure 1. The principle of gamma absorption measurement: 1-gamma radiation source in collimator, 2 – scintillation probe, 3 – pipe with a hydromixture.

However in contrary to typical nuclear equipment, Authors decided to apply a twin absorption set shown in figure 2. That solution allows both, increase accuracy of density measurement and ensures solid particle's velocity determination exactly in the same part of the flow.

In result, the count rates intensities provided by both probes 1 and 2 shown in figure 2, allow determination of the intensity of photons I^* which pass through the investigated flow:

$$I^* = I - I_B \tag{1}$$

where I total pulses rate intensity emerges at a probe output, cpms; I_B background rate intensity, cpms.
For convenience, all those intensities were recorded in counts per millisecond (cpms).

The above idea is illustrated by an arbitrary selected part of the experiment Wrs70, which is an example of a nonstationary flow typical for conditions anticipated throughout nodules excavation [7].

During selected measurement:

$$I_B = const \tag{2}$$

and received data in sections x and y determined by probes shown in figure 2, are plotted in chart 3.
In this figure it is easy to find differences in the probes efficiency resulting in higher $I_x^*(t)$ intensity then $I_y^*(t)$ and small dissimilarities caused by electonic equipment noise, despite of small distance between probes, which in that experiment was equal only 90 mm. So the new signal is:

$$I_{xy}^*(t) = \sqrt{I_x^*(t) \cdot I_y^*(t)} \tag{3}$$

Consequently, that signal is more adequate for describing the content of the flow and may be used for its monitoring. Moreover, in connection with the calibration, one may obtain the density or solid phase concentration plot [2, 3].

In the described experiment, calibration was made with the same solid partiles and fragment of the original pipe brought to the Laboratory of Sedimentology, Faculty of Geology, Geophysics and Environmental Protection, AGH – University of Science, Krakow.

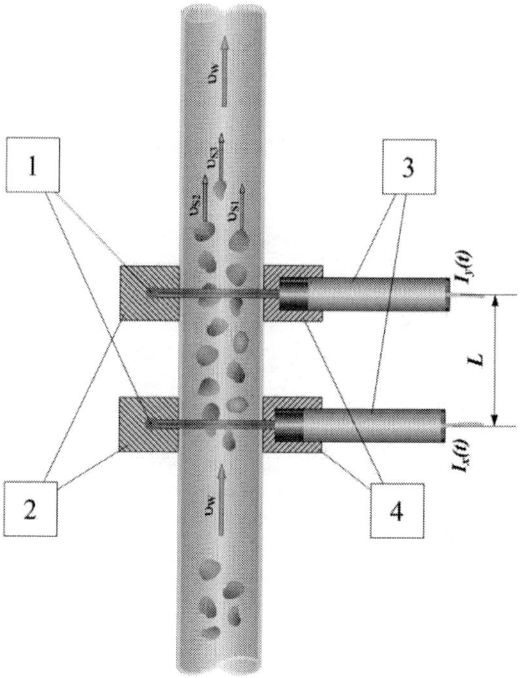

Figure 2. The twin absorption set used to both the improved density and grain's velocity measurement: 1 – sealed radioactive source, 2 – collimator of the source, 3 – scintillation probe, 4 - collimator of the detector, υ_{si} – velocity of a particular grain i, υ_w – average velocity of water.

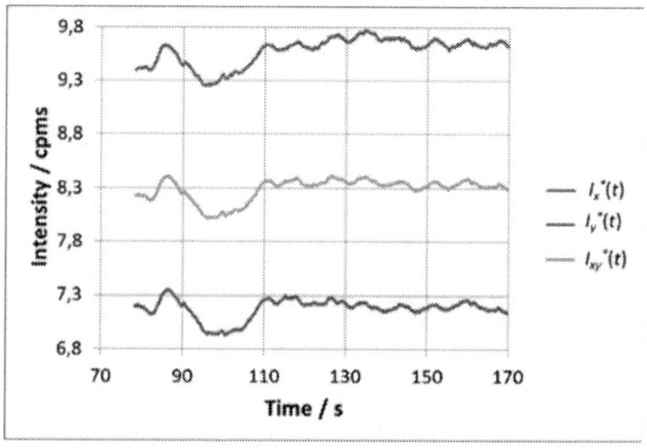

Figure 3. Signals recorded and calculated in the experiment Wrs70

In result, the density of the flow ρ_m was calculated from:

$$I^* = I_0^* \exp(-\mu_m \cdot \rho_m \cdot d) \tag{4}$$

where I^* γ radiation intensity after transfer through the pipe with investigated solid-liquid mixture, cpms; I_0^*- gmmma ray intensity after transfer through the empty pipe, cpms; μ_m - the mass attenuation coefficient, traditionally presented in $cm^2\ g^{-1}$; ρ_m - density, $g\ cm^{-3}$; d - internal diameter of the pipe, cm.

The obtained averaged mixture density distribution is presented in figure 4.

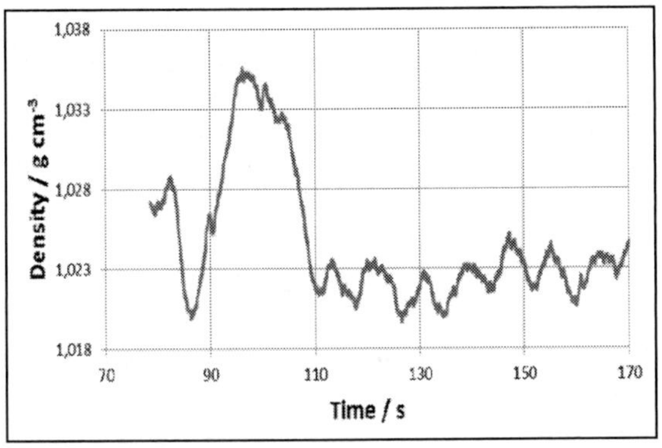

Figure 4. Mixture density estimated in the experiment Wrs70

Consequently the volumetric concentration of solid phase in the analyzed part of flow was calculated by:

$$C_V = \frac{V_S}{V_M} = \frac{V_S}{V_L + V_S} \quad (5)$$

where V_S - volume of solid phase, m^3, V_M - volume of the mixture, m^3, V_L - volume of liquid phase, m^3.

The obtained volumetric concentration distribution in the investigated part of flow is shown in figure 5.

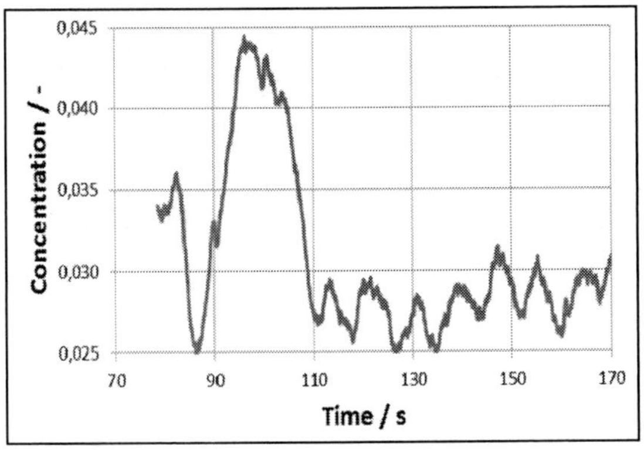

Figure 5. Volumetric concentration estimated in the experiment Wrs70

2.2 Grains velocity measurement

The two γ absorption sets shown in figure 2, illustrate the idea of the solid phase velocity measurement.

The signals delivered by detectors, after the background I_B removal, represent photons passed through two cross sections x and y of $L = 90$ mm distance. That allows for the determination of the transit time delay τ_0 used for transportation of the solid phase through the investigated part of the flow.

Averaged transit time delay τ_0 was determined by the maximum of the cross – correlation distribution $R_{xy}(\tau)$ of the I_x* and I_y* signals:

$$R_{xy}(\tau) = \lim_{T\to\infty} \frac{1}{T}\int_0^T I_x^*(t)\cdot I_y^*(t+\tau)dt \quad (6)$$

where T is approximation time sufficient for grains' transport evaluation, s; τ time delay, s [9-14].

Then, the average velocity of solid particles flow υ_S may be calculated from:

$$\upsilon_S = L / \tau_0 \quad (7)$$

During transportation by liquid, each solid particle i moves with the coaxial part of velocity υ_{Si} depending from its position in a pipe cross-section and interactions with other grains. In consequence, the velocity of grain i is function of its coordinates in cross-section and time, $\upsilon_{si}(x,y,t)$. So the average coaxial velocity:

$$\upsilon_s = \frac{\sum_{i=1}^{n} \upsilon_{si}}{n} \quad (8)$$

where n is number of grains in the volume determined by γ radiation plane passing to first probe (I_x) at the beginning of the velocity measuring time T, and similar plane passing to second probe (I_y) at end of T.

The velocity measuring time T depends from a geometry and density of the flow as well as activity of the radioactive sources and presence of such rapid fluctuations of a flow as turbulences, whirling and transitory fluctuations. In practice, its' reliable determination demands evaluation of the particular flow during a standard exploitation.

In the experiment Wrs70 the approximation time $T = 10$ s was sufficient to estimation the average velocity of solid particles distribution $\upsilon(t)$ shown in figure 6.

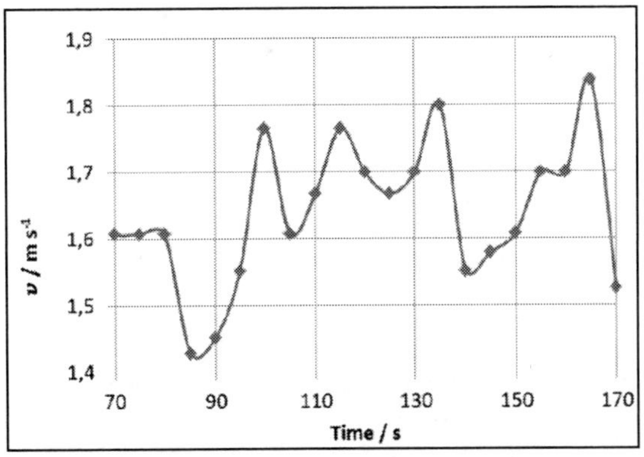

Figure 6. Velocity of solid phase estimated in the experiment Wrs70

In that view, the proposal of the nodules hydrotransport during excavation from Pacific bottom was preceded by construction of a laboratory installation suitable for modelling conditions of the full scale exploitation.

3 Laboratory tests

At Wroclaw University of Environmental and Life Sciences an installation was constructed which allows the study of phenomena expected during possible exploitation of the Pacific nodules [7].

During tests, specially designed ceramic models were applied to study manganese nodules transportation. The experimental pipe loop shown in figure 7 was equipped with pressure gauges, ultrasonic and electromagnetic flow meters as well as a special weighing scale with sampling tank to measure the flow rate of the investigated mixture.

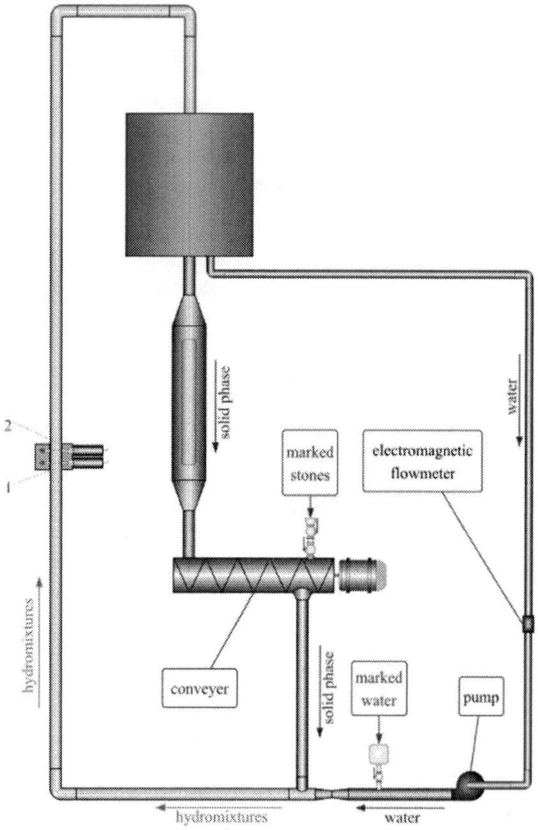

Figure 7. The experimental pipe loop [3]

In addition, the installation gives an opportunity to control flow rate of centrifugal pump with a variable speed drive and a grain feeder yield by auger revolution. Thanks to this, the system gave opportunity to observe the transport of grains in the vertical pipeline when water velocity changed from 2 up to 4 ms^{-1} and vary concentrations of solid particles to simulate the flow conditions during the expected undersea transportation.

The phases separator located in upper part of the installation allows convenient recirculation of water and solid particles in the individual loops equipped in the measuring and control facilities.

Careful analysis of nodules hydraulic transportation facilitated authors to construct the measuring equipment consisting of two radioactive sources and two detectors as shown in figure 2. In this set, such isotopes as ^{241}Am, ^{137}Cs, or ^{60}Co may be used, depending on the pipe material and diameter.

The data acquisition equipment based on scintillation detectors and the dedicated 8-channel counter connected to a PC by an USB port. Due to that, TTL voltage pulses I_x and I_y provided by probes may be counted within the sampling time $\Delta t = 1$ ms for hydrotransport and 100 µs for pneumatic flows. Duration of the data recording in this system is only limited by PC memory, and may be observe on a monitor [3].

A selected part of the experiment Wrs70 illustrates a non-stationary flow typical for anticipated Pacific nodules excavation and was convenient for the flow control evaluation.

4 Pressure loss determinations

The most convenient signal for the flow dynamic visualisation seems to be one proportional to the pressure loss recorded in the installation. This parameter may be determined from the Darcy–Weisbach equation:

$$\Delta p = f_D \cdot \frac{L_p}{d} \cdot \frac{\rho u^2}{2} = c \cdot \Delta p^* \qquad (9)$$

where f_D is Darcy friction factor, L_p / d is the ratio of the length to diameter of the pipe, ρ is the density of the fluid, u is the mean flow velocity and c is dimensionless coefficient.

Due to the above, the signal proposed for the flow control:

$$\Delta p^* = \rho_m \cdot v_s^2 \qquad (10)$$

The exemplary shape of that signal is shown in figure 8.

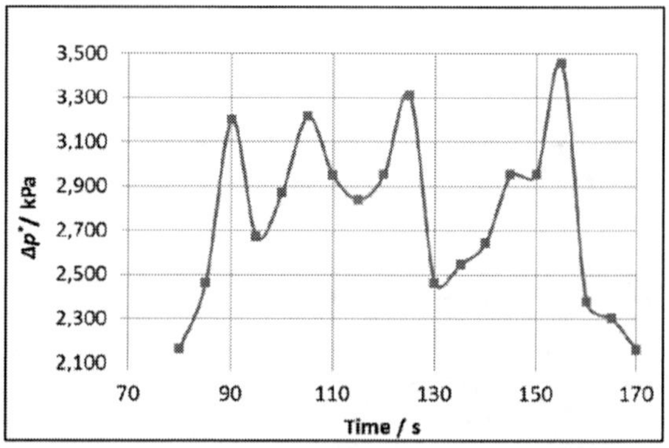

Figure 8. The signal proportional to the pressure loss estimated in the experiment Wrs70

5 Conclusions

The proposed control of a flow gives possibility to reduce velocity of solid phase transportation without hazard of the pipe clogging and promising high profit due to longer lasting safe exploitation of the industrial installation.

Moreover, the completed test confirms possibility of limited extension of nuclear methods applied only for the density examination in a particular flow. In many cases, the establishment of the second gamma absorption set, identical to the existing one, would allow realizing the offered method in limited cost.

Confirmation of the prepared solution in the laboratory installation modeling unstable conditions expected during the nodules exploitation suggests more easy application in typical industrial installations.

Acknowledgements

All presented investigations were possible due to support under the project P105/10/1574 of the Grant Agency of the Czech Republic, and RVO: 67985874 of the ASCR, as well as the Project No N N523 755340 of the National Science Centre of Poland, which are gratefully acknowledged.

The Authors would like to acknowledge the EFM Conference Organizers for arranging a promotable opportunity to discuss in detail the proposed solutions and help in their dissemination in numerous societies in industry and science.

References

1. G. Falcone, G.F. Hewitt, C. Alimonti, *Multiphase flow metering: principles and applications* (Elsevier, Amsterdam, 2009)
2. B.K. Arvoh, R. Hoffmann, M. Halstensen, Flow. Meas. Instrum. **23**, 56-65 (2012)
3. L. Petryka, et al, *EPJ WoC* **92,** 02065 (2015)
4. http://www.oilsandsmagazine.com/
5. G. Johansen, P. Jackson, *Radioisotope gauges for industrial process measurements* (John Wiley and Sons, NY, 2004)
6. M. Kutz (ed.), *Handbook of Measurement in Science and Engineering* (John Wiley and Sons, NY, 2013)
7. J. Sobota, S. Boczarski, L. Petryka, M. Zych, *Proc. of OMS-2005,* Changsha, Hunan, China, Oct. 9-13, (2005)
8. P. Vlasak, Z. Chara, J. Krupicka, J. Konfrst, J. Hydrol. Hydrom. **62**, 241-247 (2014)
9. J.S. Bendat, A.G. Piersol, *Random data - analysis and measurement procedures* (John Wiley and Sons, NY, 2010)
10. L. Petryka, M. Zych, R. Hanus, J. Sobota, P. Vlasak, *Proc. of 10th ISOPE Ocean Mining & Gas Hydrates Symp. OMS-2013*, Szczecin, Poland, Sept. 22-26, 230-233 (2013)
11. L. Petryka, M. Zych, R. Murzyn, Nukleonika **50**, 43–46 (2005)
12. L. Petryka, J. Oszajec, Nucl. Geophys. **7,** 323-333 (1993)
13. R. Hanus, L. Petryka, M. Zych, Flow. Meas. Instrum. **40**, 58–63 (2014)
14. R. Hanus, M. Zych, L. Petryka, V. Mosorov, P. Hanus, *EPJ WoC* **92,** 02020 (2015)

EPJ Web of Conferences **114**, 02094 (2016)
DOI: 10.1051/epjconf/201611402094

Comparison of two methods for contactless surface temperature measurement

Magdalena Piasecka[1,a], Dariusz Michalski[1] and Kinga Strąk[1]

[1]*Kielce University of Technology, Faculty of Mechatronics and Mechanical Engineering, 25-314 Kielce*

Abstract. The aim of the paper is to determine the temperature filed of the heating surface on the basis of temperature measurements taken by liquid crystal thermography and infrared thermography applied in boiling heat transfer research during FC-72 flow in minichannels, and to compare them. The essential part of the experimental stand is the test section with two parallel rectangular minichannels, each 1.7 mm deep, 24 mm wide and 360 mm long. It is possible to observe the channel surfaces through panes: of the first minichannel allows observing foil temperature changes on the plain side due to liquid crystal thermography (LCT), which required treating the foil surface with thermochromic liquid crystals, of the other minichannel enables detecting outer glass or foil surface temperature changes due to infrared thermography (IRT). Comparison of the results of the measurements are presented in graphical form as thermographs and as heating surface temperature vs. minichannel length. The differences between two sets of measurement data concerning the temperature of the heating surface obtained with LCT and IRT, were analyzed using: Pearson linear correlation coefficient, determination coefficient, Experimental Method Error and Method Accuracy. The comparative data analysis shows that similar values and distributions of the surface temperature were obtained from both techniques.

1 Introduction

Infrared thermography (IRT) and liquid crystal thermography (LCT) are non-invasive technique for measuring surface temperature. They are becoming increasingly common as a contactless method during heat transfer in small channels.

The application of an IR camera, which is characterized by a low response time and high sensitivity, is a better solution than the use of standard transducers. Being fully two-dimensional, the camera is able to produce not only colour images but also surface temperature fields. Its errors are easy to assess because tangential distortions can be corrected. The IR camera is frequently used to analyze complex fluid flows. It is a suitable tool to study convective heat fluxes on the heated wall or the behaviour of the surface flow field [1]. The applications of IRT have been numerous. For instance, the method was used to assess the increase in the dryout heat flux for four fluids flowing through two small-diameter silver tubes [2]. It proved to be useful to analyze the thermo-hydrodynamics of the initial phase of the hydrodynamically fully developed but thermally developing internal laminar flow of water in a square cross-sectional minichannel. Also, it was found suitable for high-resolution measurement of the local temperature on the heated wall for one-dimensional conduction in the transverse direction as well as for the determination of the heat transfer coefficient [3, 4]. Another application of IRT was to analyze the local heat transfer coefficient during the Taylor bubble train flow in a square cross-sectional minichannel [5]. The technique was also used to investigate enhanced boiling heat transfer and critical heat flux (CHF) for reduced graphene oxide, graphene, and SiC-coated surfaces. Finally, the infrared camera was employed to determine the CHF conditions and the temperature fields of heater surfaces [6].

Liquid crystal thermography (LCT) is another method for contactless measurement of surface temperature. Liquid crystals, which are available in many different temperature ranges, change colour from red at the lower end of the active temperature range through the other colours of the visible spectrum to blue at the upper end. LCT measurement of temperature requires calibration [7]. This technique has various applications. For example, it was used to determine the local heat transfer coefficient for saturated flow boiling and to study the onset of boiling and flow boiling in a silicon microchannel heat sink containing a rectangular silicon chip in which there are 45 rectangular microchannels [8]. Liquid crystal thermography was found suitable to study the distribution of surface temperature while simultaneously imaging bubble motion in flow boiling in a narrow vertical duct [9] and a horizontal minichannel [7]. Moreover, the method was employed to measure spatial steady-state distributions of wall/fluid temperatures and to assess the

[a] Corresponding author: tmpmj@tu.kielce.pl

thermo-hydrodynamics of simultaneously developing single-phase liquid flows in a semi-circular minichannel [10]. It was possible to use the technique to experimentally study boiling heat transfer in a rectangular, oriented minichannel [11-13]. Liquid crystals have proved useful to analyze not only flow boiling but also pool boiling heat transfer. Reference [14] discusses the use of liquid crystals to determine the oscillating temperature field beneath a single isolated nucleation, the nucleation dynamics, and the two-phase flow structure.

This paper presents the comparison of two contactless surface temperature measurement techniques (liquid crystal and infrared thermography) applied in flow boiling heat transfer in minichannels. The main aim is to compare both temperature results using elements of statistical analysis.

2 Experimental set up

The most important part of the system is a measurement module which consists of two parallel vertically-oriented minichannels, each 1.7 mm deep, 24 mm wide and 360 mm long, as shown in figure 1. The heating element for FC-72 flowing in each minichannel (1) is the heating foil made of super alloy Haynes 230 (2). Temperature of the outer foil surface are measured using two contactless measurement methods: liquid crystal thermography and infrared thermography. In one minichannel (LCT minichannel) a glass pane (4a) enables recording of a sequence of hues on the liquid crystal layer (3) located on the foil contacting this pane. After the calibration, it is possible to define the two-dimensional distribution of temperature on this surface by applying liquid crystal thermography [11-13]. Simultaneously, infrared thermography is used to observe changes in temperature on the outer side of the foil in the other minichannel (IRT minichannel). In this case, the surface of the foil is coated with black paint (8) to gain an emissivity of 0.83 [15]. Temperature is measured by using infrared camera in the central, axially symmetric, part of the channel (approx. 10 mm x 350 mm). The channel is reinforced on both sides with glass panels to prevent the heating foil from deforming. The other side of the minichannels can be observed through another pair of glass panes. The result are used to visualize two-phase flow in the minichannels (not discussed here). K-type thermocouples (7) and pressure converters are installed at the inlet and outlet of the minichannel. The essential part of the experimental setup is a data and image acquisition system comprising: Canon G11 digital camera, FLIR SC640 infrared camera, DaqLab2005 data acquisition station, a computer with special software and a lighting system with fluorescent lamps emitting 'cold white light', figure 2.

3 Two contactless methods of surface temperature measurements

3.1 Infrared Thermography

Infrared thermography is one of the most universal and contactless technologies that use detection of infrared radiation. Thermal radiation is emitted by objects with a temperature higher than absolute zero (-273 °C/0 K). The infrared radiation region is between visible light and microwaves within the wavelength spectrum of 0.78 – 1000 μm [16, 17].

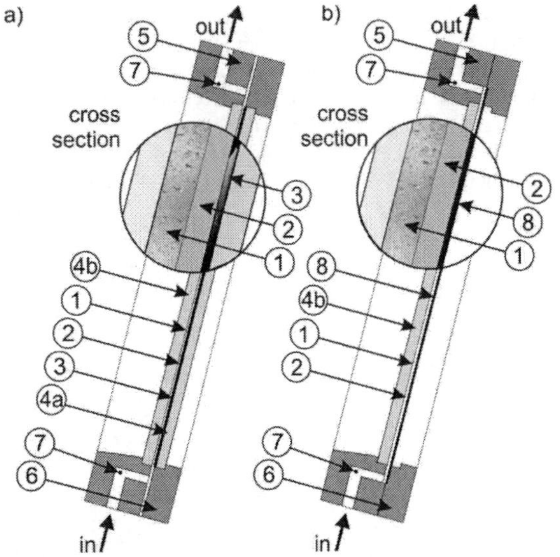

Figure 1. Schematic diagrams of the measurement module: a) with the LCT minichannel, b) with the IRT minichannel; 1 - minichannel, 2 - heating foil, 3 - liquid crystal layer, 4a,b - glass panes, 5 - channel body, 6 - front cover, 7 - thermocouple, 8 - black paint layer.

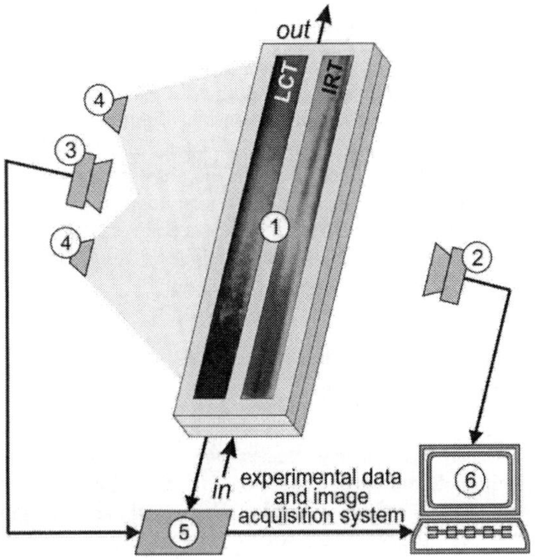

Figure 2. Schematic diagram of a data and image acquisition system: 1 - measurement module, 2 - infrared camera, 3 - digital camera, 4 - lighting system, 5 - data acquisition station, 6 – computer.

In the study of infrared thermography, the emissivity of the object is an important parameter as it determines the ratio of radiation emitted from an object (real body) to the radiation emitted from a blackbody at the same temperature. The emissivity is in the range from 0 to 1 and determines the ability of the test object to emit its own energy [18, 19].

IRT provides visual image records that are qualitative (image) and quantitative (temperature) information for deeper knowledge of thermal conditions and thermodynamic processes in the research object. The infrared camera allows measuring static conditions and monitoring dynamic processes in a laboratory. To accurately analyze and interpret thermograms recorded, a good knowledge of the properties of the object, the material, the effect of climatic conditions is required, along with the knowledge of measurement techniques. The sources of interference occurring in practice are an important issue when testing infrared cameras. These include: a hot heating element (i.e. radiator), the impact of sunlight on the test object, lamp light incident on the test surface, air flow, or the effect of moisture on the test object [20, 21].

3.2 Liquid Crystal Thermography

Liquid crystals have been widely applied in technology because of their unique properties manifested in the change of colour and/or its intensity. That can be used to detect very slight changes in temperature, stress, electromagnetic radiation or chemical agent properties. At present, the most popular liquid crystal applications make use of the above-mentioned property of light-selective reflection from planar oriented layers of the substance. LCT provides a perfect means of quantitative temperature measurement, as it enables us to detect and observe a region which is warmer or cooler than the surroundings, without measuring the absolute temperature. Liquid crystals application to the detection of surface temperature must be preceded by colour vs. temperature calibration of the liquid crystal surface. That has to be carried out under constant lighting conditions. Calibration should be repeated when a new batch of liquid crystals is used or when there are changes in any geometrical parameters of the image acquisition system. What should also be taken into account is that liquid crystals are affected by ageing [22]. Exemplary calibration curve is presented in figure 3c.

4 Results

4.1 Experimental data

The main experimental results were collected as: infrared thermographic images of the heating foil for one minichannel (tabure 3a) and liquid crystal thermographic images of the outer surface of the glass pane for the other minichannel (figure 3c).

The calibration procedure has to precede temperature measurement by liquid crystal thermography in the boiling heat transfer investigation. Its aim is to assign corresponding temperature values to the hues observed on the surface covered with liquid crystals [22]. As the result of the calibration procedure the calibration curve was obtained, figure 3c.

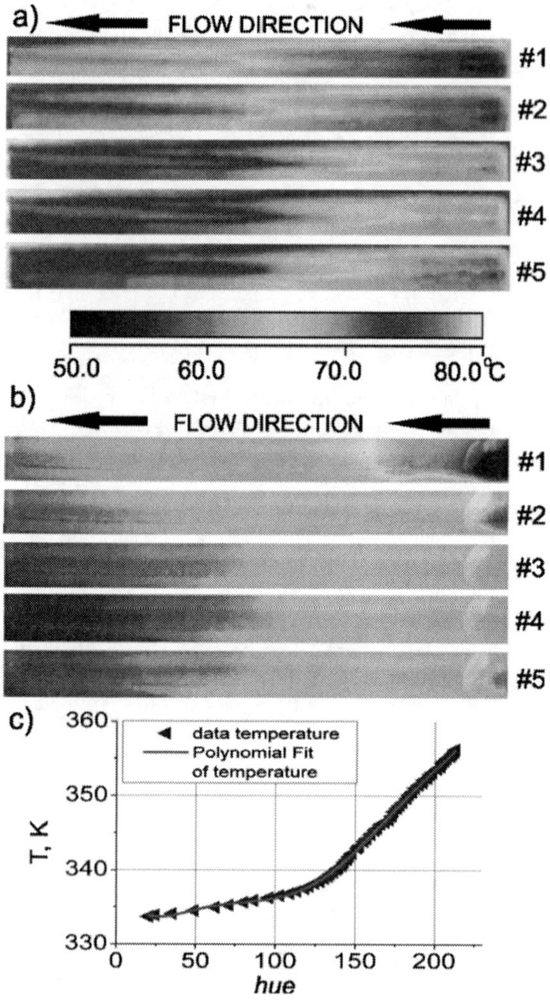

Figure 3. a) Thermograms of the heating foil surface obtained by infrared thermography, b) thermograms of the heating foil surface obtained by liquid crystal thermography; experimental parameters: average mass flux 162.5 kg $m^{-2}s^{-1}$, average inlet pressure 194 kPa, inlet liquid subcooling 51 K,
#1) $q_v = 1.22\cdot10^5$ kW m^{-3}; #2) $q_v = 1.34\cdot10^5$ kW m^{-3};
#3) $q_v = 1.45\cdot10^5$ kW m^{-3}; #4) $q_v = 1.37\cdot10^5$ kW m^{-3};
#5) $q_v = 1.15\cdot10^5$ kW m^{-3};
c) *calibration curve*- dependence of temperature as a function of *hue*, $T = 1.02524E\text{-}11 hue^6 - 7.63921E\text{-}9 hue^5 +$
$+ 2.16167E\text{-}6 hue^4 - 2.88229E\text{-}4 hue^3 + 0.01897 hue^2 -$
$+ 0.54528 hue + 339.199$.

4.2 Uncertainty and limitations of IRT and TLC

The images taken with FLIR SC640 infrared camera system operating in the spectral range from 7.5 µm to 13 µm were recorded in a digital system with image frequency of 30 Hz. The accuracy of the camera was ± 2 K (3.6 °F) or ± 2% of reading. The camera can be used for remote real time measurement of temperature with a thermal resolution ranging from 0.06 °C to 30 °C [20]. The analysis takes into account the most unfavourable value of measurement accuracy, i.e. ± 2K of the readings.

When IRT was used to measure the surface temperature, the influence of the external conditions on the measurement accuracy was significant. The major parameters affecting the measurement uncertainty in this case were: the object emissivity, the effect of the sources of interference such as the heat produced by the heating elements or by the lamps lighting the surface, the air supply, the moisture of the object, the distance from the object, and the angle of observation.

The accuracy of the measurements of the heating foil temperature by liquid crystal thermography was estimated in [22]. The mean error in the measurement of the heating foil temperature using liquid crystal thermography was approximately 0.86 K.

Liquid crystals application to the detection of two-dimensional heating surface temperature distribution in flow boiling in a minichannel must be preceded by hue vs. temperature calibration of liquid crystal surface. That has to be carried out under constant lighting conditions. LCT method seems to be less sensitive to measurement errors. Because of the limitations of the hue recording system, the method cannot be applied to the range described by *calibration curve*. When the highest or the lowest temperature from the active range of applied liquid crystal mixture are observed, that rapid change in foil temperature cannot be recognizable properly. According to the authors, the LCT data should be omitted for the distance of approx. 15 per cent of the distance from the minichannel length.

5 Comparative statistical analysis of temperature data

For comparative statistical analysis temperature data obtained from the two contactless methods: IRT and LCT, for five settings of the increasing heat flux supplied to the foil, was selected.

The procedure of comparative analyzes were as follows:

a. Graphical comparison of the experimental data concerning the measurement of the foil temperature;

b. Statistical comparative analysis of the foil temperature, including:

- histograms of the foil temperature obtained by IRT and LCT separately;
- expected normal value vs. experimental values;
- analyses of the foil temperature data obtained by IRT and LCT using coefficients of Pearson's linear correlation and the coefficient of determination;

c. Determination of the experimental error of the method for measuring temperature, according to the principles of statistical inference, taking account of the temperature:

- the procedure for the estimation the mean value of the experimental error [23, 24];
- the procedure for the estimation mean deviations of the experimental error [23, 24];
- the estimation of the confidence interval of a single method error and measurement accuracy [26].

5.1 Graphical Comparison

The data used in the graphical comparison was the heating surface temperature monitored along the central line in the axially symmetric part of IRT and LCT minichannels so that the results acquired from both methods could be adequately compared. The values obtained for: infrared thermography and liquid crystal thermography, for five settings of the increasing heat flux, are shown in figure 4, respectively. Each graph shows also two fitting curves (linear function), received for each method, separately.

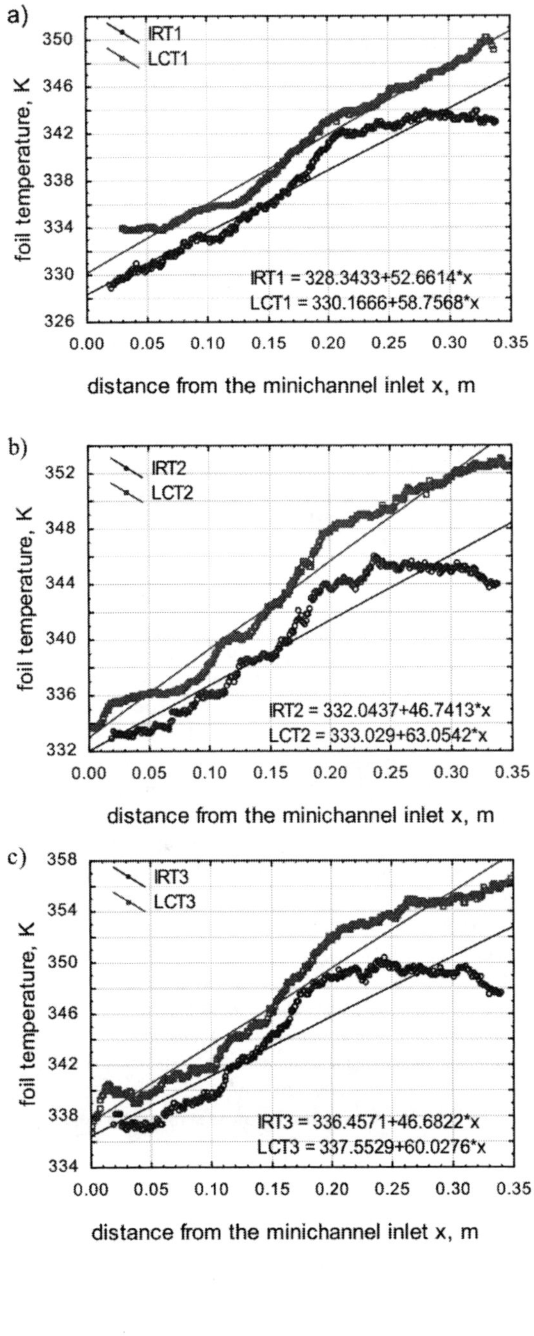

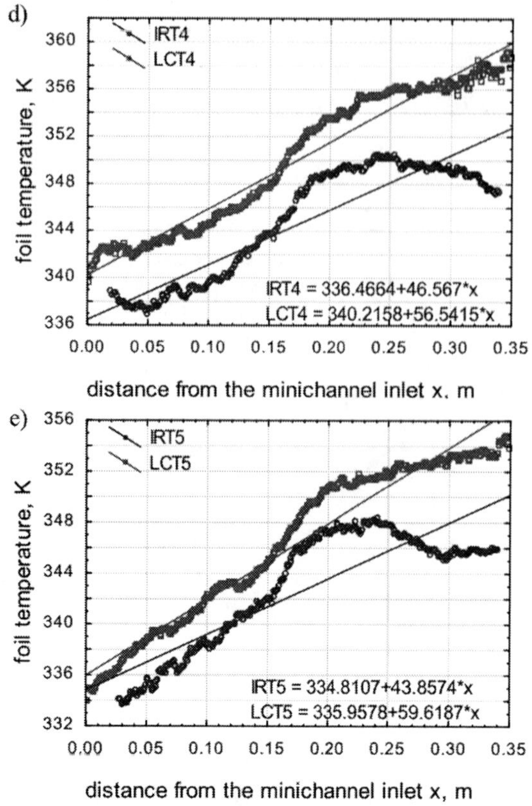

Figure 4. Temperature of the heating foil surface vs. distance from the minichannel inlet, data obtained from infrared thermography (blue points) and liquid crystal thermography (red points); experimental parameters as for figure 3, setting of the heat flux: a) #1, b) #2, c) #3, d) #4, e) #5.

It was assumed that application of IRT and LCT for the heating surface temperature measurement, gives similar results. It should be underlined that calibration curve (LCT) have to affect the limiting the measured temperature range (active band of the liquid crystal mixture used in experiment is 20 K). Data obtained by IRT technique shows that there was a large scatter of results for the foil temperature at the inlet and outlet of the channel, e.g. data for the setting #4, figure 4d. It can be result of heat transfer by conduction from heating elements to the body of the testing module or other measurement errors. Some of the IRT temperature data was reported to almost coincide with the data fitting curve. Probably local decreasing of the surface temperature are caused by the formation of nucleation centers, during development of nucleate boiling in IRT minichannel. This implies that the IR camera is able to immediately detect changes in the surface temperature, while liquid crystals are well-suited for indicating changes in the surface temperature with some delay.

5.2 Expected Normal Value

In figure 5 the dependence of expected normal values for the analyzed settings of the heat flux vs. foil temperature obtained from IRT and LCT thermographs (experimental data in Expected Normal Value method).

From the curves shown in figure 5 it is clear that the greatest deviations from the expected normal values for surface temperature registered with the two methods are found at the limits of the range. For each setting of the heat flux, the coincidence of the measurement results is very high, especially the central part of the minichannels. At the minichannel ends, extremes of temperature occur.

It was concluded that the calculated value of experimental method accuracy is very satisfactory. The IRT and LCT method can by use for heating surface temperature monitoring in the scientific laboratory.

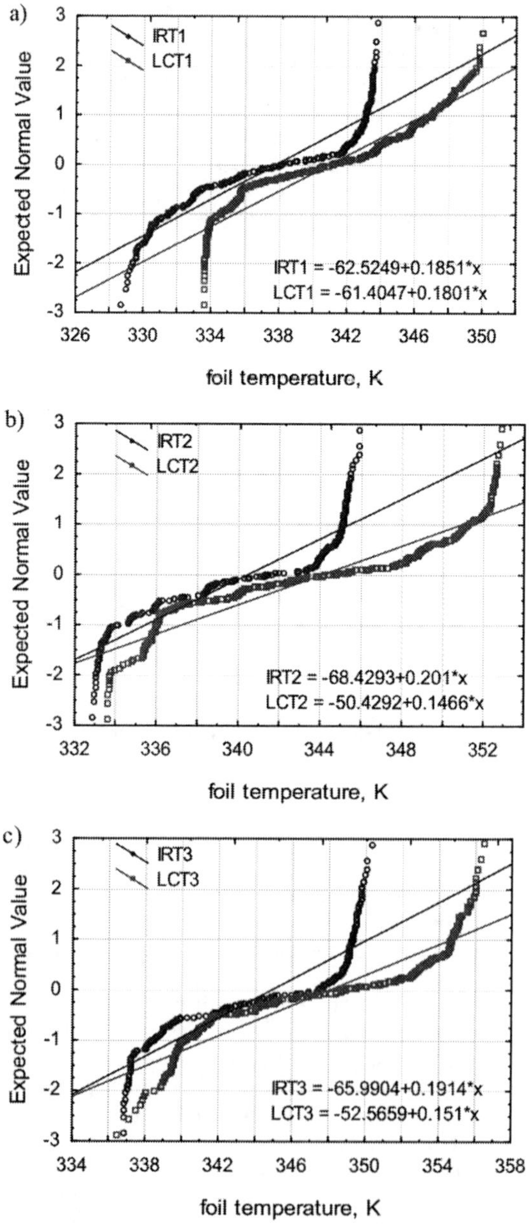

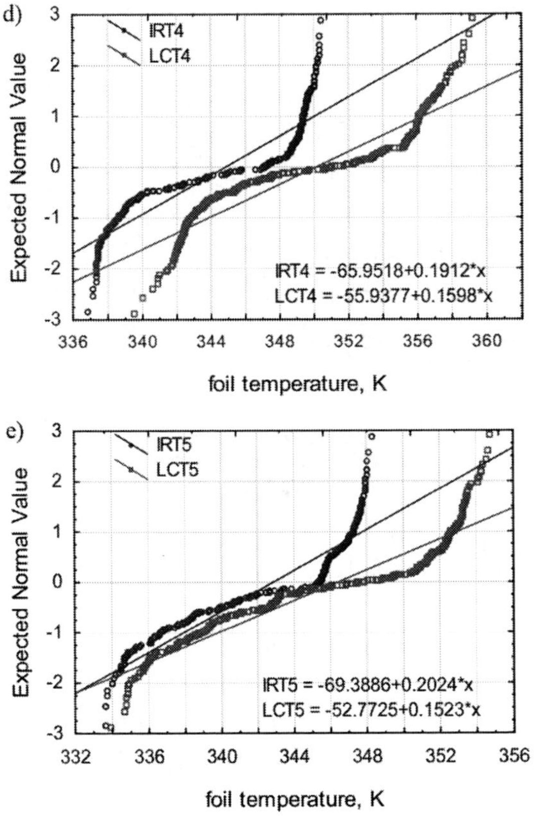

Figure 5. Expected normal values vs. foil temperature obtained from IRT and LCT thermographies, for the analyzed settings of the heat flux; data obtained from infrared thermography (blue points) and liquid crystal thermography (red points); experimental parameters as for figure 3.

5.3 Histograms

Figures 6-10 present histograms constructed for the analyzed settings of the heat flux vs. foil temperatures, for data obtained from IRT and LCT, separately. Additionally, each histogram shows the normal distribution - marked as red (LCT) or blue line (IRT).

The histograms indicate that the temperature distributions are similar in the central part of the range, especially for settings #2 to #5. The largest differences are observed at the limits of the range.

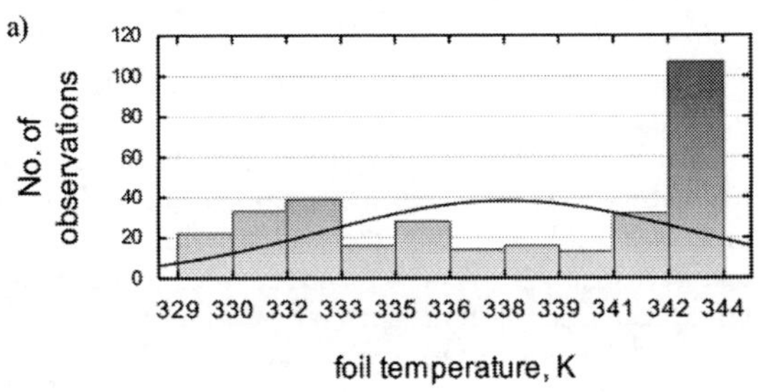

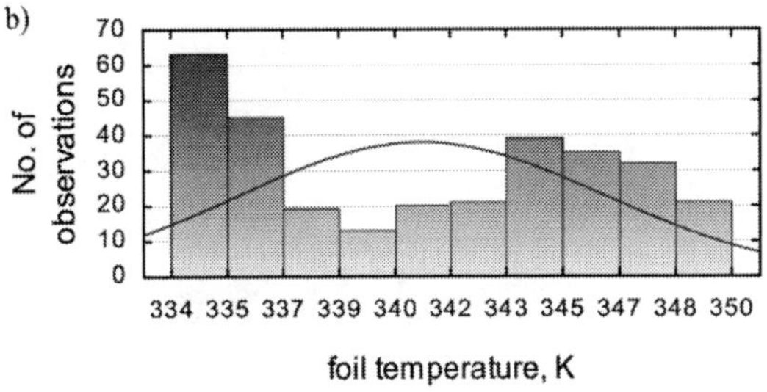

Figure 6. Histogram for foil temperature obtained from: a) IRT, b) LCT; experimental parameters as for figure 3, data for the setting #1.

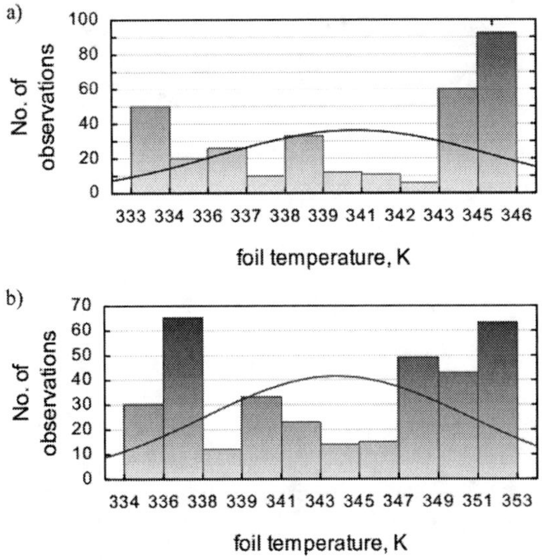

Figure 7. Histogram for foil temperature obtained from: a) IRT, b) LCT; experimental parameters as for figure 3, data for the setting #2.

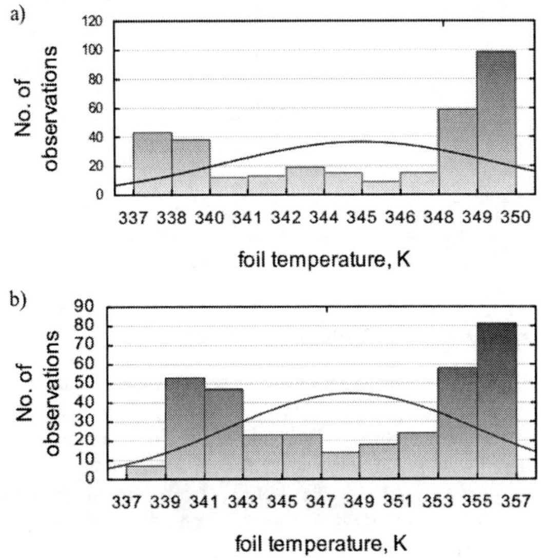

Figure 8. Histogram for foil temperature obtained from: a) IRT, b) LCT; experimental parameters as for figure 3, data for the setting #3.

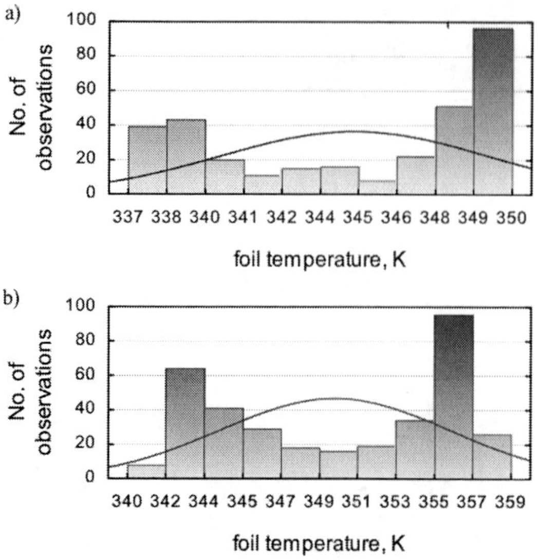

Figure 9. Histogram for foil temperature obtained from:
a) IRT, b) LCT; experimental parameters as for figure 3, data for the setting #4.

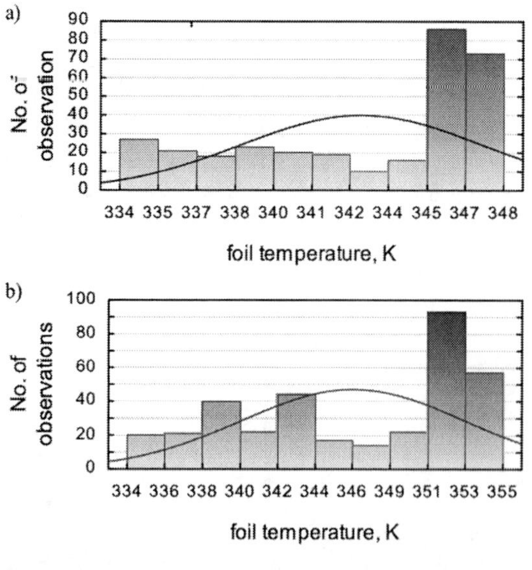

Figure 10. Histogram for foil temperature obtained from:
a) IRT, b) LCT; experimental parameters as for figure 3, data for the setting #5.

5.4. Basic statistical parameters

All basic statistical parameters for comparative statistical analysis temperature data obtained from the two contactless methods: IRT and LCT, for each analyzed settings of the increasing heat flux supplied to the foil, are summarized in table 1. They are: mean square error (MSE), mean absolute error (MAE), mean relative squared error (MRSE), mean relative absolute error (MRAE), Pearson linear correlation coefficient (R) and coefficient of determination (R2)[25].

Table 1. Basic statistical parameters

Basic parameters	IRT1 vs. LCT1	IRT2 vs. LCT2	IRT3 vs. LCT3	IRT4 vs. LCT4	IRT5 vs. LCT5
MSE	9.88705	19.0522	15.0822	32.8862	21.0597
MAE	2.91012	3.9413	3.4995	5.5355	4.2270
MRSE	0.00008	0.0002	0.0001	0.0003	0.0002
MRAE	0.00851	0.0114	0.0099	0.0158	0.0121
R	0.9757	0.9779	0.9731	0.9694	0.9521
R^2	0.9436	0.9572	0.9495	0.9455	0.9141

MSE - Mean Square Error
MAE - Mean Absolute Error
MRSE - Mean Relative Squared Error
MRAE - Mean Relative Absolute Error
R - Pearson linear correlation coefficient
R^2 - coefficient of determination

The analysis of the coefficient of Pearson's linear correlation and the coefficient of determination performed on the basis of J. P. Guilford's table indicates that there is very good correlation between the methods of temperature measurement used in the tests (IRT and LCT).

5.5. Experimental Relative Method Error

The accuracy of the measurement methods was assessed by comparing their results with those obtained by means of the reference method [24, 26].

The experimental relative method error (EME) was determined from the relation [23]:

$$EME = \frac{T_{IRT} - T_{LCT}}{T_{LCT}}, \tag{1}$$

where: T_{LCT} - foil temperature obtained from LCT (the standard method), T_{IRT} - foil temperature measured by IRT (the tested method).

Foil temperature obtained from LCT was selected as the standard method because LCT was characterized by the lowest measurement error [22]. It was assumed that IRT is the tested method in the experimental relative method error shown below.

The mean value of the experimental method error ($\overline{EME}$) was estimated using the following formula:

$$\overline{EME} = \frac{1}{n}\sum_{i=1}^{n} EME_i, \tag{2}$$

where: n - sample size, EME_i - experimental relative method error of foil temperature measurement.

The mean value of the method error in the analyzed population was assessed by means of the following procedure [26]: determine the method error for the IRT and LCT methods calculate the mean value of the method error, and determine the interval of confidence for the mean value of the method error with normal distribution.

Confidence interval (CI) for the mean value of the method error was estimated using the following relation:

$$\left(\overline{EME} - u_p \cdot s \quad \overline{EME} + u_p \cdot s\right), \tag{3}$$

where: $\overline{EME}$ – mean values of the experimental relative method error, defined in eq. (2), u_p – the expansion coefficient for probability $P = 0.95$, $u_p = 2$ in accordance with the recommendations contained in the ISO standard [27], s – mean deviation of the method error.

Data from Experimental Method Error vs. distance from the minichannel inlet, constructed for the analyzed settings of the heat flux vs. foil temperatures, is shown in figure 11.

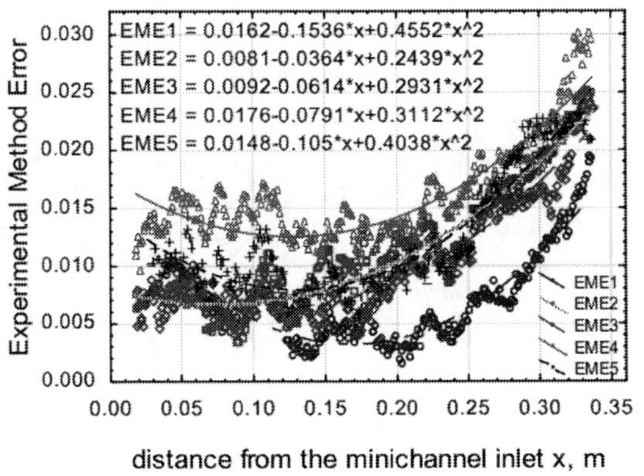

Figure 11. Data from Experimental Method Error vs. distance from the minichannel inlet.

The diagram in figure 11 suggests that the results concerning the Experimental Method Error for all the five analyzed settings are similar and can be approximated with a 2nd order polynomial. The highest values are found at the limits of the range for the minichannel inlet. In the worst case, the EME does not exceed 0.031.

5.6. Method accuracy of IRT and LCT methods

From the obtained values of the method error, an accuracy of the method can be estimated by applying the following relation:

$$MA = \left| \overline{EME} \pm u_p \cdot s \right|_{\max} \cdot 100\%, \qquad (4)$$

where: $\overline{EME}$ – mean values of the relative method error, defined in eq. (2), u_p – the expansion coefficient for probability $P = 0.95$, $u_p = 2$ according to the recommendations contained in the ISO standard [27], s – mean deviation of the method error.

The method accuracy (MA) calculated with the eq. (4) [23] is used to assess qualitatively the measurement accuracy of examined measuring device FLIR SC640 infrared camera used to measure of foil temperature.

Table 2 shows results of experimental investigation with the statistical analysis of experimental relative method error obtained while comparing the LCT and IRT methods.

Table 2. Method accuracy and statistical basic statistical parameters of the experimental method error

Basic parameters	IRT1 vs. LCT1	IRT2 vs. LCT2	IRT3 vs. LCT3	IRT4 vs. LCT4	IRT5 vs. LCT5
Sample size n	299	320	321	320	312
Min EME	0.0014	0.0033	0.0040	0.0096	0.0048
Max EME	0.0193	0.0250	0.0242	0.0300	0.0233
$\overline{EME}$	0.0068	0.0114	0.0101	0.0160	0.0123
s	0.0036	0.0052	0.0047	0.0042	0.0051
CI	0.0068± 0.0073	0.0114± 0.0105	0.0101± 0.0095	0.0160± 0.0084	0.01227± 0.0102
MA	1.40%	2.19%	1.95%	2.43%	2.25%

EME – Experimental Method Error
$\overline{EME}$ – mean values of the Experimental relative Method Error
s - mean deviation
CI - Confidence Interval
MA - Method Accuracy

The parameter Method Accuracy (MA) indicates that the two methods used for temperature measurement do not differ much; in certain ranges they can be used interchangeably. Acceptable accuracy limit of a measuring system used in scientific research conditions is 15% [24].

Table 3 shows the proposed ranges of the relative method errors and the corresponding applications based on [24].

Table 3. Proposed ranges of experimental relative method errors in surface temperature measurement and the corresponding applications based on [24].

Measurement accuracy range %	Type of application
2 ÷ 5	Measurement of standards temperature
5 ÷ 15	Scientific research
10 ÷ 25	Measurement of surface temperature under industrial conditions

The results of statistical analysis of experimental relative method error indicate that the IRT method accuracy is equal to about MA = 2.5% comparing to LCT.

6 Conclusions

The article presented two contactless measurement techniques application to investigations into boiling heat transfer in the flow through two parallel asymmetrically heated vertical minichannels. The temperature of the heated foil was measured by applying two methods: infrared thermography (IRT) and liquid crystal thermography (LCT). Based on the available literature demonstrates the use of both the contactless methods for measuring the surface temperature in many heat transfer experiments and applications.

It was assumed that application of IRT and LCT for the heating surface temperature measurement, gives similar results. Similar values and distributions of temperature fields on the heating foil were obtained from IRT and LCT thermographs. It should be underlined that LCT method calibration curve have to affect the limiting of the measured temperature range.

For each setting of the heat flux, the highest deviations from the expected normal values were observed at the limits of the range. A high goodness of fit between the measurement results was reported in the central part of the minichannels. According to the authors, the calculated value of the experimental method accuracy is very satisfactory. The IRT and LCT methods can by use for heating surface temperature monitoring in the scientific laboratory.

The analysis of the histograms shows that the distributions of temperature are very similar, especially in the central part of the range. The greatest differences occur at the limits.

From the comparison of the coefficient of Pearson's linear correlation and the coefficient of determination it is evident that their values approach unity. This suggests that there is goodness of fit between the temperature measured with IRT and that recorded by means of LCT.

The parameter Method Accuracy indicates that the differences between the compared methods are small and that in certain ranges they are interchangeable.

References

1. G. M. Carlomagno, G. Cardone, Exp. Fluids **49**, 1187–1218 (2010)
2. D. Mikielewicz, J. Wajs, M. Gliński, R.S. Zrooga, Exp. Therm.Fluid Sci. **44**, 556–564 (2014)
3. B. Mehta, A. Dhyani, S. Khandekar, *Proc. 21st National & 10th ISHMT-ASME Heat Mass Transf. Conf.*, India, paper ISHMT_IND_02_041 (2011)
4. B. Mehta, S. Khandekar, Exp. Thermal Fluid Sci. **42**, 219–229 (2012)
5. B. Mehta, S. Khandekar, Int. J. Heat Fluid Flow **45**, 41–52 (2014)
6. H. Seo, J. H. Chu, S.-Y. Kwon, I. Ch. Bang, Int. J. Heat Mass Transf. **82**, 490–502 (2015)
7. A. B. Ozer, A. F. Oncel, D. K. Hollingsworth, L. C. Witte, Exp. Therm. Fluid Sci. **35**, 1522–1529 (2011)
8. A. Megahed, Int. J. Multiphase Flow **39**, 55–65 (2012)
9. S. M. S. Murshed, K. Vereen, D. Strayer, R. Kumar, Energy **35**, 5143–5150 (2010)
10. M. Ra, S. Khandekar, *Proc. 19 th National & 8 th ISHMT-ASME Heat Mass Transf. Conf.*, JNTN Hyderabad, India (2008)
11. M. Piasecka, Ann. Nucl. Energy **73**, 282–293 (2014)
12. M. Piasecka, Heat Transf. Eng. **35**, 903–912 (2014)
13. S. Hożejowska, M. Piasecka, Heat Mass Transf. **50**, 1053–1063 (2014)
14. A. Miglani, D. Joo, S. Basu, R. Kumar, Int. J. Heat Mass Transf. **60**, 188–200 (2013)
15. T. Orzechowski, *Heat transfer on ribs with microstructured surface*, Monographs, studies, hearings **39** (in Polish) (the Publishing House of the Kielce Univ. Techn., Kielce, Poland, 2003)
16. R Gade, T.B. Moeslund, Mach. Vision Appl. **25**, 245–262 (2014)
17. A. Rogalski, K. Chrzanowski, Opto-Electron. Rev., **10** (2), 111–136 (2002)
18. T. Karwat, Izolacje **5**, 33–36 (2008) (in Polish)
19. K. Ziętala, *Proc. 11th European Conf. Young Research Sci. Workers TRANSCOM 2015*, Žilina, 347–352 (2015)
20. User's manual ThermaCam B640, P640, SC640, *Publ. No 155850 Rev.a 201-ENGLISH (EN)* (2007)
21. M. Pešek, O. Pech, EPJ Web Conf. **67**, paper 02091 (2014)
22. M. Piasecka, Metrol. Meas. Sys. **2**, 205–216 (2013)
23. S. Adamczak, D. Michalski, Metrol. Meas. Sys. **3**, 221–234 (2004)
24. S. Adamczak, *Measurement of surface texture. Form profiles, waviness and roughness* (in Polish) (WNT, Warsaw, Poland 2008)
25. G. Bohm, G. Zech, *Introduction to Statistics and Data Analysis for Physicists* (Verlag Deutsches Elektronen-Synchrotrofn, 2010)
26. M. L. Tiku, *Aust. J. Stat.* **22**, 260–275 (1980)
27. ISO 10012:2003 *Measurement management systems - Requirements for measurement processes and measuring equipment*

EPJ Web of Conferences **114**, 02095 (2016)
DOI: 10.1051/epjconf/201611402095

Effect of the heating surface enhancement on the heat transfer coefficient for a vertical minichannel

Magdalena Piasecka[1,a] and Kinga Strąk[1]

[1]Kielce University of Technology, Faculty of Mechatronics and Mechanical Engineering, 25-314 Kielce

Abstract. The aim of the paper is to estimate effect of the heating surface enhancement on FC-72 flow boiling heat transfer for a vertical minichannel 1.7 mm deep, 24 mm wide and 360 mm long. Two types of enhanced heating surfaces were used: one with minicavities distributed unevenly, and the other with capillary metal fibrous structure. It was to measure temperature field on the plain side of the heating surface by means of the infrared thermography and to observe the two-phase flow patterns on the enhanced foil side. The paper analyses mainly the impact of the microstructured heating surface on the heat transfer coefficient. The results are presented as heat transfer coefficient dependences on the distance along the minichannel length. The data obtained using two types of enhanced heating surfaces in experiments was compared with the data when smooth foil as the heating surface was used. The highest local values of heat transfer coefficient were obtained using enhanced foil with minicavities - in comparison to other cases. Local values of heat transfer coefficient received for capillary fibrous structure were the lowest, even compared with data obtained for smooth foil. Probably this porous structure caused local flow disturbances.

1 Introduction

Recently, there is a worldwide interest in compact heat exchangers with minichannels. They are characterized by many advantages such as higher efficiency and reduced size. Extensive studies are required to understand the phenomena with change of phase in such heat exchangers [1-3]. The results may be used for the designing of cooling systems with minichannels to improve their efficiency and thermal stability. Experimental investigations show that the average heat transfer coefficient in compact heat exchangers with minichannels is higher [1]. It is thus possible that compact heat exchangers with enhanced walls will have better thermal parameters. However, if the surface of the heating element is enhanced too much, i.e. it is too porous, the flow of the refrigerant and, accordingly, the whole boiling process in the minichannel, may be periodically disturbed [2]. Investigations often deals with frictional pressure drop in minichannels [3].

The use of microstructured surfaces in minichannels is known to intensify the heat transfer process. Results obtained by the application such surfaces in flow boiling in micro- or mini spaces [4-15] are not widely discussed in the specialist literature, but they are promising. The properties of a microstructured surface can be modified using chemical, thermal, mechanical or combined mechanical and thermal processes. The thermal processes include plasma spraying, electric arc spraying, sintering, laser surface texturing or electrical discharge texturing [16-24]. Enhanced heating surfaces are common in pool boiling systems. They are considered to be very efficient because of providing a large number of nucleation sites, and, in consequence, substantially increase the heat flux [15, 25-27].

Research on heat transfer in minichannels of the very small cross-section is considerably difficult to conduct because of the temperature measurement. In the contact temperature method a sensor placed in the minispace could disturb the fluid flow. Applying a contactless method for measuring temperature of the minichannel wall allows us to eliminate this problem. The methods that provide sufficient measurement accuracy are infrared thermography and liquid crystal thermography.

Infrared thermography (IRT) is becoming increasingly common as a contactless for measuring surface temperature. Using an infrared (IR) camera for determining of surface temperature is a very useful tool for researchers because such measurement are characterized by high sensitivity (down to 20 mK), low response time (down to 20 μs) and it is fully two dimensional. Error assessment is also very convenient because tangential distortions of the camera can be corrected. IR camera has become a common tool in the analysis of complex fluid flows either to evaluate convective heat fluxes on the heated wall or to investigate the behaviour of the surface flow field [28]. The IRT has been reported suitable to experimentally study the

[a] Corresponding author: tmpmj@tu.kielce.pl

thermo-hydrodynamics of laminar flow of water in a square cross-sectional minichannel (5 mm x 5 mm). It ensured high-resolution spatial measurement of the local wall temperature with the assumption of one-dimensional conduction in the transverse direction and evaluation of the heat transfer coefficient [29, 30]. The IRT has also been used to measure the wall temperature in the analysis of the increase in dryout heat flux for four fluids in two small diameter tubes (diameters of 1.15 mm and 2.3 mm) [31]. In [32] the IRT has been applied to determine the local heat transfer coefficients during the Taylor bubble train flow in a square cross-sectional minichannel (5 mm x 5 mm). IR camera was used to determine the CHF conditions and the temperature fields of the heater surfaces [33]. It has been found that is very useful technique to study enhanced boiling heat transfer and critical heat flux (CHF) for reduced graphene oxide, graphene, and SiC-coated surfaces. The IRT was also applied to visualize temperature field to study the subcooled flow boiling in a vertical square channel; it was used for the validation of CFD simulations and for the improvement and development of turbulence models under subcooled boiling conditions [34]. Researchers in [35] stated that the IRT was well suited for measurement of two-phase flow parameters in a convective boiling system. In [36] the local Nusselt numbers were obtained using IRT in two-phase non-boiling slug flows under constant flux. Researchers in [37, 38] have used IRT to study pool boiling on a fin.

2 Experimental stand, methodology and uncertainties

2.1 Experimental stand

The experimental stand (figure 1) is composed of several systems: the test loop in which Fluorinert FC-72 circulates as the working fluid, the calibration loop, the data and image acquisition system, the supply and control system and the lighting system, described in [11-16]. The test loop consists of: a rotary pump, a heat exchanger, a compensating tank, rotameters, a filter and a deaerator. The data and image acquisition system comprised of FLIR SC640 infrared camera, SLR Canon EOS 550D camera, DaqLab 2005 data acquisition station, a computer with special software and a lighting system.

2.2 The testing module with minichannels

The most important part of the system is the testing module (figure 2) which comprises two parallel vertically-oriented minichannels each 1.7 mm deep, 24 mm wide and 360 mm long, as shown in figure 2. The heating element for FC-72 flowing in minichannels is the foil (2). Infrared thermography is used to observe changes in temperature on the outer smooth side of the foil in one minichannel and this data is taken account in further calculations and analyses. Foil temperature is measured by IR camera in the central, axially symmetric, part of the channel (approx. 10 mm x 350 mm). The surface of foil is coated with black paint (9) to gain an emissivity of 0.83 [39]. The minichannel is reinforced on both sides with narrow glass panels to prevent the heating foil from deforming.

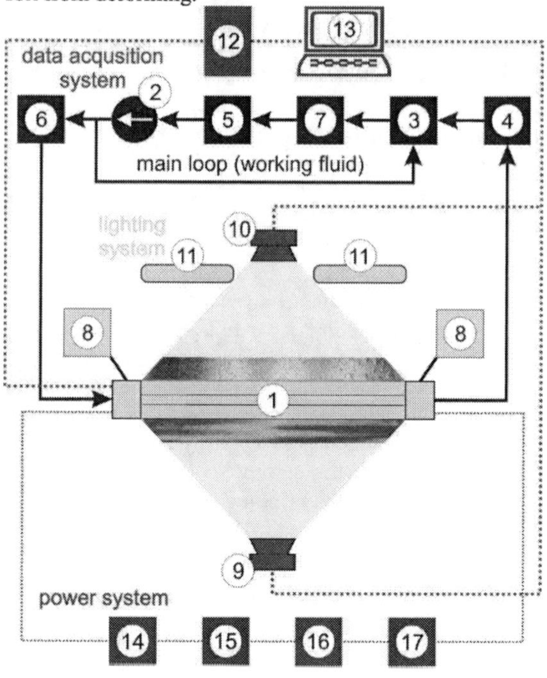

Figure 1. Schematic diagram of the main loops of the experimental set-up, 1 - testing module with minichannels; 2 - rotary pump; 3 - compensating tank/pressure regulator; 4 - tube-type heat exchanger, 5 - filter, 6 - rotameters; 7 - deaerator, 8 - pressure converter, 9 - IR camera, 10 - digital SLR camera, 11 - halogen lamp, 12 - data acquisition station, 13 - computer, 14 - inverter welder, 15 - shunt, 16 - ammeter, 17 - voltmeter.

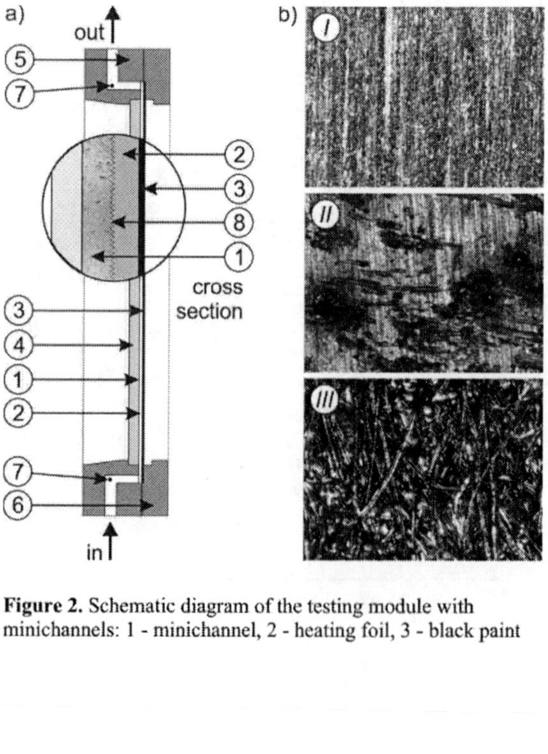

Figure 2. Schematic diagram of the testing module with minichannels: 1 - minichannel, 2 - heating foil, 3 - black paint

layer, 4 - glass pane, 5 - channel body, 6 - front cover, 7 - thermocouple, 8 - enhanced surface, b) photos of: the smooth foil (I), enhanced foils (II, III) with: minicavities (II) and capillary metal fibrous structure (III).

The other side of the minichannels, where a two-phase flow occurs on the enhanced surface of the foil in contact with the fluid, are observed through another two glass panes. K-type thermocouples and pressure converters are installed at the inlet and outlet of the minichannel.

2.3 Characteristic of the heating foil enhancement

The heating element for FC-72 flowing in the minichannel was the thin alloy foil (approx. 0.1 mm in depth) designated as Haynes-230, with the smooth surface roughness of $Ra = 0.121$ μm, see figure 2.b (I). Two types of enhanced heating surface were used: one with minicavities, and another with capillary metal fibrous structure.

The minicavities, see figure 2.b (II), produced by spark erosion over the whole surface area of the foil using an arcograph were distributed unevenly and their dimensions varied. The rims of resolidified melt from the electrode and the foil were several μm in height, with a maximum height being 5 μm. The craters were usually less than 1 μm deep [16].

The length of the capillary stainless steel fibrous structure, see figure 2.b (III), is usually up to 2 cm. They were spot welded with a microwelder and rolled to a thickness of 0.15-0.2 mm.

2.4 Experimental methodology

During experimental series, after the deaeration, the working fluid (FC-72) flows laminarly along the minichannel. When flow rate and the pressure are fixed, there is a gradual increase in the electric power supplied to the heating foil followed by an increase in the heat flux transferred to the fluid in the minichannel. This leads to the onset of nucleate boiling and the heat transfer enhancement. It is possible to measure the distribution of temperature on the smooth foil surface due to infrared thermography. The flow structure is observed simultaneously on the opposite side of the minichannel (on the smooth or enhanced foil surface in contact with the fluid).

2.5 Experimental uncertainties

The images taken with a FLIR SC640 infrared camera system operating in the spectral range from 7.5 μm to 13 μm were recorded in a digital system with an image frequency of 30 Hz. The accuracy of the camera was ± 2 K or ± 2% of reading. The camera can be used for remote real time measurement of temperature with a thermal resolution ranging from 0.06 °C to 30 °C [40].

3 Local heat transfer coefficients determination

Various methods of the determination of the local heat transfer coefficients have been proposed [41- 43]. Here, the simple one-dimensional approach was used. The resulting heat transfer coefficient for the saturated boiling was calculated according to the following equation:

$$\alpha_{IRT}(x) = \frac{q_{w\_IRT}}{T_F^{IRT}(x,\delta_G) - T_{sat}(x)} \tag{1}$$

where T_F^{IRT} – foil temperature measured by infrared thermography, q_{w_IRT} – heat flux transferred to the fluid in the IRT minichannel, $T_{sat}(x)$ – saturation temperature determined on the basis of the linear distribution of pressure along the length of the minichannel from the inlet to the outlet.

The heat transferred to the fluid was calculated as:

$$q_{w\_IRT} = \frac{I \cdot \Delta U}{A_{F\_IRT}} - q_{wl} \tag{2}$$

where I - current, ΔU - voltage drop, A_{F_IRT} - surface area of the heating foil in the IRT minichannel, q_{wl} - the heat lost to the surroundings in the IRT minichannel [43].

The central part of the heating foil where the surface temperature is measured with the infrared camera (10 mm x 350 mm) is not insulated. Therefore, the loss of heat to the surroundings is the greatest. It was estimated as 1.3 % of the heat flux density [43].

The local heat transfer coefficients obtained for the single phase convection, boiling incipience and subcooled boiling region were analysed in [41].

4 Results

Figures 3-5 show the results selected for the analysis plotted as the heat transfer coefficient versus the distance along the length of the minichannel. During the measurement series, thermal and flow parameters were nearly identical. The heat flux density was the same for experiments with minicavities on the enhanced foil surface and for the smooth foil.

Figure 3 presents the data obtained for the enhanced surface with minicavities and for the smooth foil. The local heat transfer coefficients obtained for capillary fibrous structure differ significantly from those for other foil surface types.

The heat flux supplied to the enhanced foil with capillary fibrous structure was much lower than those supplied to the smooth foil and enhanced foil with minicavities. For that reason, it is shown separately in figure 4.

Figure 5 presents local heat transfer coefficients as a function of the minichannel length for all the heating surfaces: two enhanced foils (with minicavities, with capillary stainless steel fibrous structure) and the smooth foil.

In the fully developed nucleate boiling, the heat transfer coefficient was the highest but then it decreased sharply with an increase in the distance from the channel inlet, which was accompanied by an increase in the vapour volume fraction in the two-phase flow mixture, as shown in figures 3-4. As the boiling process proceeded, partial dryout and a reduction in heat transfer were reported [13-15], as in the works of other researchers [44-48]. Finally, the heat transfer coefficient decreased steadily to much lower values and heat transfer returned to single-phase forced convection. The lowest values of the coefficient were noticed at the minichannel outlet.

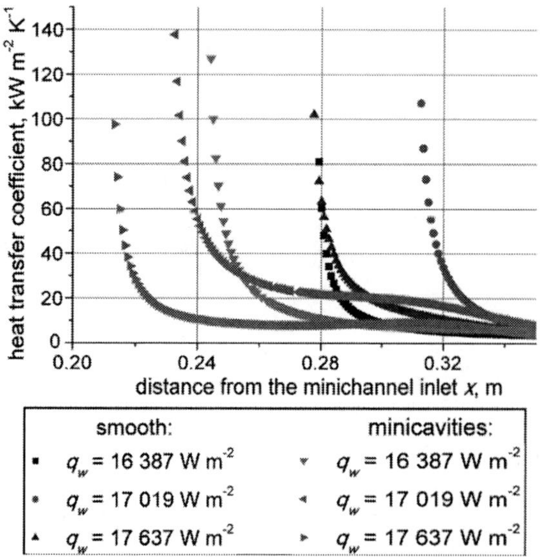

Figure 3. Heat transfer coefficient vs. the minichannel length determined for the smooth foil and the enhanced foil with minicavities.

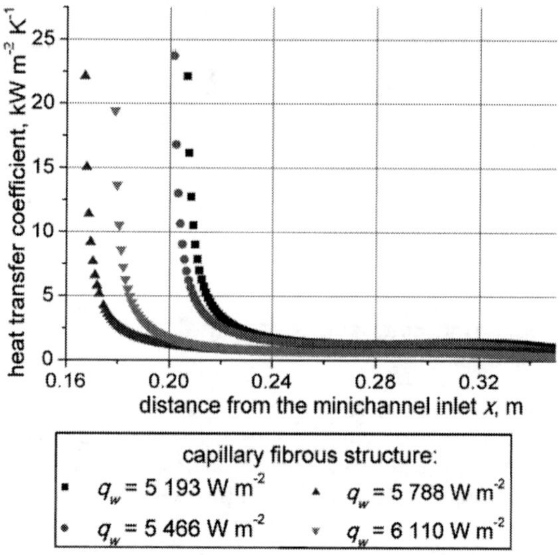

Figure 4. Heat transfer coefficient vs. the minichannel length for the enhanced foil with capillary stainless steel fibrous structure.

The highest local values of heat transfer coefficient were obtained with using enhanced foil with minicavities - in comparison to other cases. Local values of heat transfer coefficient received for capillary fibrous structure were the lowest, even compared with data obtained for smooth foil. Probably this porous structure caused local flow disturbances during fluid flow along the minichannel. The formation of vapour bubbles could also affect heat transfer process causing its deterioration.

From the results of the pool boiling experiments it occurs that for surfaces with minicavities the geometrical parameters need to be selected to match specific liquids. For example, smaller diameters of cavities are suitable for liquids with lower surface tension (refrigerants) [25].

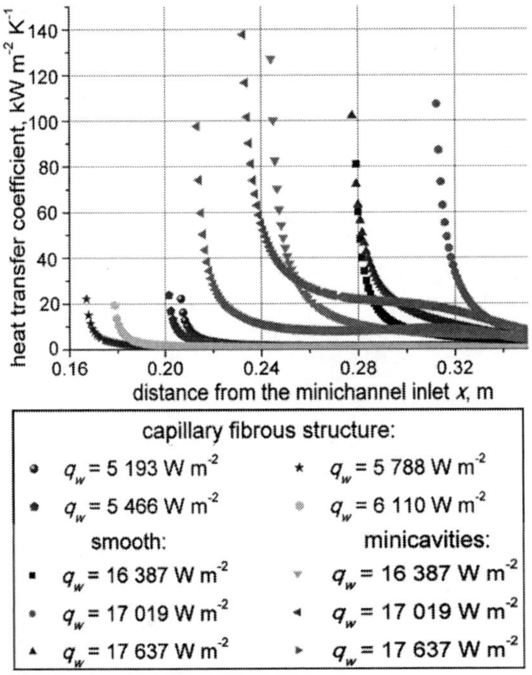

Figure 5. Heat transfer coefficient vs. the minichannel length for: the smooth foil, the enhanced foil with minicavities, the enhanced foil with capillary stainless steel fibrous structure.

5 Conclusions

The aim of the paper was to estimate effect of the heating surface enhancement on FC-72 flow boiling heat transfer for a vertical minichannel. Two types of enhanced heating surfaces were used: one with minicavities distributed unevenly, and the other with capillary metal fibrous structure. It was to measure temperature field on the plain side of the heating surface by means of the infrared thermography and to observe the two-phase flow patterns on the enhanced foil side. The impact of the microstructured heating surface on the heat transfer coefficient were analyzed. The results are presented as heat transfer coefficient dependences on the distance along the minichannel length. Data obtained using two types of enhanced heating surfaces in experiments was compared with the data when smooth foil as the heating surface was used. The highest local values of heat transfer coefficient were obtained using the enhanced foil with minicavities - in comparison to other cases. Local values of heat transfer coefficient received for capillary fibrous structure were the lowest, even compared with data obtained for smooth foil. Probably this porous structure caused local flow disturbances during fluid flow along the minichannel.

Acknowledgements

The research reported herein was supported by a grant from the National Scientific Center (No. DEC-2013/09/B/ST8/02825).

References

1. T. Bohdal, Annual Set Env. Protect. **15**, 107–126 (2013)
2. T. Bohdal, W. Kuczyński, Heat Tran. Eng. **32** (5), 359–368 (2011)
3. K. Dutkowski, Heat Tran. Eng. **31**, 321–330 (2010)
4. C. N. Ammerman, S. M. You, J. Heat Tran. **123**, 976–983 (2001)
5. A. Kosar, C.-J. Kuo, Y. Peles, Int. J. Heat Mass Tran. **48**, 4867–4886 (2005)
6. C.-J. Kuo., Y. Peles, Int. J. Heat Mass Tran. **50**, 4513–4526 (2007)
7. V. Khanikar, I. Mudawar, T. Fisher, Int. J. Heat Mass Tran. **52**, 3805–3817 (2009)
8. H. Wang, R. B. Peterson, IEEE Trans. Compon. Pack. Techn. **33**, 784–793 (2010)
9. Y. Sun, L. Zhang, H. Xu, X. Zhong, Exp. Therm. Fluid Sci. **35**, 1418–1426 (2011)
10. Y. Sun, L. Zhang, H. Xu, X. Zhong, Int. J. Therm. Sci. **50**, 881–889 (2011)
11. M. Piasecka, Exp. Heat Tran. **27**, 231–255 (2014)
12. M. Piasecka, Heat Tran. Eng. **35** (10), 903–912 (2014)
13. M. Piasecka, Int. J. Heat Mass Tran. **81**, 114-121 (2015)
14. M. Piasecka, Int. J. Refrig. **56**, 198–212 (2015)
15. M. Piasecka, Ann. Nucl. Energy **73**, 282–293 (2014)
16. M. Piasecka, Adv. Mater. Research **874**, 95–100 (2014)
17. W. Depczyński, *Proc. 23rd Int. Conf. Metallurgy Materials METAL 2014*, Poland (2014)
18. W. Depczyński, S. Spadło, P. Młynarczyk, E. Ziach, *Proc. 24th Int. Conf. Metallurgy Materials METAL 2015*, Poland (2015)
19. B. Grabas, Arch. Metall. Mater. **60** (1), 33–39 (2015)
20. B. Grabas, Exp. Therm. Fluid Sci. **68**, 499–508 (2015)
21. B. Grabas, Adv. Mater. Research **874**, 71–75 (2014)
22. N. Radek, J. Pietraszek, B. Antoszewski, Adv. Mater. Research **874**, 29–34 (2014)
23. J. Pietraszek, N. Radek, K. Bartkowiak, Solid State Phen. **197**, 198-202 (2013)
24. N. Radek, B. Antoszewski, Kovove Materialy-Metallic Materials **47** (1), 31–38 (2009)
25. R. Pastuszko, M. Piasecka, J. Physics Conf. Ser. **395**, paper No. 012137 (2012)
26. R. Pastuszko, M. Wójcik, Exp. Therm. Fluid Sci. **63**, 34–44 (2015)
27. R. Pastuszko, EPJ Web Conf. **45**, paper No. 01020 (2013)
28. G. M. Carlomagno, G. Cardone, Exp. Fluids **49**, 1187–1218 (2010)
29. B. Mehta, A. Dhyani, S. Khandekar, *Proc. 21st National & 10th ISHMT-ASME Heat Mass Tran. Conf. Madras, India, paper ISHMT_IND_02_041* (2011)
30. B. Mehta, S. Khandekar, Exp. Therm. Fluid Sci. **42**, 219–229 (2012)
31. D. Mikielewicz, J. Wajs, M. Gliński, A.-B. R.S Zrooga, Exp. Therm. Fluid Sci. **44**, 556–564 (2014)

32. B. Mehta, S. Khandekar, Int. J. Heat Fluid Flow **45**, 41–52 (2014)
33. H. Seo, J. H. Chu, S.-Y. Kwon, I. Ch. Bang, Int. J. Heat Mass Tran. **82**, 490–502 (2015)
34. C. E. Estrada-Perez, J. Yoo, Y. A. Hassan, Int. J. Multiph. Flow **73**, 17–33 (2015)
35. J. Yoo, C. E. Estrada-Perez, Y. A. Hassan, Int. J. Therm. Sci. **90**, 248–266 (2015)
36. P. A. Walsh, E. J. Walsh., Y. S. Muzychka, Int. J. Heat Mass Tran. **53**, 3193–3201 (2010)
37. T. Orzechowski, Exp. Therm. Fluid Sci. **31** (8), 947–955 (2007)
38. T. Orzechowski, A. Tyburczyk, ISI Master Journal List **87** (7), 48–51 (2011)
39. T. Orzechowski, *Heat transfer on ribs with microstructured surface*, Monographs, studies, hearings **39** (in Polish) (the Publishing House of the Kielce Univ. Techn., Kielce, Poland, 2003)
40. *User's manual ThermaCam B640, P640, SC640*, Publ. No 155850 Rev.a 201-ENGLISH (EN) (2007)
41. M. Piasecka, B. Maciejewska, Exp. Therm. Fluid Sci. **68**, 459–467 (2015)
42. S. Hożejowska, M. Piasecka, Heat Mass Tran. **50**, 1053–1063 (2014)
43. K. Ziętala, M. Piasecka, Trans. Inst. Fluid-Flow Machinery, **128** (to be published)
44. S. Hożejowska, R. Kaniowski, M. E. Poniewski, Int. J. Numer. Method Heat Fluid Flow **24**, 811–824 (2014)
45. M. E. Steinke, S. G. Kandlikar, J. Heat Tran. **126** (4), 518–526 (2004)
46. M. Cortina Dı´az, J. Schmidt, Int. J. Heat Fluid Flow **28**, 95–102 (2007)
47. E. Sobierska, R. Kulenovic, R. Mertz, M. Groll, Exp. Therm. Fluid Sci. **31**, 111–119 (2006)
48. E. Sobierska, R. Kulenovic, R. Mertz, Int. J. Therm. Sci. **46**, 1172–1181 (2007)

EPJ Web of Conferences **114**, 02096 (2016)
DOI: 10.1051/epjconf/201611402096

Numerical calculations of the thermal deformations of the rectangular minichannel walls

Magdalena Piasecka[1,a], Sławomir Błasiak[1] and Tomasz Musiał[1]

[1]Kielce University of Technology, Faculty of Mechatronics and Mechanical Engineering, Al. 1000-lecia P.P. 7, 25-314 Kielce, Poland

Abstract. This paper analyzes the thermal deformations of the walls of a rectangular minichannel. The distribution of temperature on the walls and the local coefficients of heat transfer between the heated wall and the boiling fluid were determined on the basis of the experimental data. The thermal deformations were examined for two vertical walls: a heated Hastelloy X plate and an unheated stainless steel plate. The heat conduction and displacement equations were solved numerically using Ansys Workbench software. The distributions of temperature and the thermal deformations of the walls were obtained by appropriately combining the software modules with the mathematical description of the physical phenomena occurring in the minichannel and introducing the boundary conditions and the experimental data. It was found that, for the predetermined thermal conditions and the assumed thicknesses of the minichannel walls, thermal deformations can be omitted from further analysis. However, it is important to note that, if the geometrical dimensions of the minichannel had been different, the difference in the temperature along the minichannel length would have resulted in changes in the minichannel shape, i.e. its local narrowing, a reduction in the heat transfer efficiency, and, finally, a local change in the fluid flow rate.

1 Introduction

An increase in the heat transfer efficiency is achieved by applying appropriate conditions of flow boiling heat transfer. Thermal stress is associated with non-uniform thermal expansion.

In recent years, many works on heat transfer in minichannels have been published. An example of a research paper on the practical application of heat exchangers with a minichannel is [1]; it discusses the usefulness of minichannels in the cooling of integrated circuits. In [2], the author presents a study of heat transfer for the initial phases of boiling flow in a minichannel for environmentally friendly refrigerants. The analysis and model presented in [3] refer to a boiling refrigerant in a heat exchanger; the heat transfer coefficient was determined for channels with conventional and small diameters. The authors of [4] investigated boiling heat transfer for the counter-current flow of R134a refrigerant.

The world literature on flow boiling heat transfer in minichannels does not provide any information on the thermal stress in minichannel elements. Thermal deformations of different surfaces exposed to elevated temperatures have been discussed in the specialist literature but there is no data on heat exchangers with minichannels. The authors of [5] focused on thermal stresses in a planar solid oxide fuel cell. In [6] researchers presented the influence of thermal stress on the failure of a martensitic stainless steel (CA-15M) roll determined numerically. The relationship between the level of thermal stress and the geometry of the pressure vessel was presented in [7]. Reference [8] describes a model of a boiler start-up taking account of thermal stress. In the study presented in [9], thermal stress is analyzed to optimize the heating of a steam pipeline and a T-pipe. Thermal stress observed on the minichannel surfaces may cause changes in the minichannel geometry (a reduction in the width and volume), which may influence the heat transfer and flow conditions.

2 Main aim

This paper analyzes the thermal stress and deformations accompanying flow boiling heat transfer in a rectangular minichannel of a heat exchanger. The investigations involved determining thermal deformations of two metal walls: a heated Hastelloy X plate and an unheated stainless steel plate. The elements were mounted between two fixed supports. The thermal deformations and the distributions of temperature were determined using the Ansys Workbench software. The experimental part of the study required using infrared thermography. These results were used in numerical calculations to determine the temperature distribution on the minichannel walls and the

[a] Corresponding author: tmpmj@tu.kielce.pl

local coefficients of heat transfer between the heated wall and the boiling fluid in the minichannel.

3 Experimental issues

The most important part of the experimental stand was the test module with a vertically-oriented rectangular minichannel of the following dimensions: 1.6 mm (depth), 16 mm (width), and 180 mm (length). The schematic diagram of the test module is shown in fig. 1. The heated wall, which ensured the flow of FC-72 in the minichannel (1), was a metal plate made of Hastelloy X (2) with a thickness of about 0.6 mm. The temperature was measured in the central, axially symmetric, part of the channel using infrared thermography. The channel was insulated on both sides with micanit to prevent the heated wall from deforming. The other wall of the minichannel was a stainless steel plate (3) with a thickness of 5.6 mm. The temperature of both minichannel walls was monitored using a FLIR SC640 infrared camera. The temperature distributions on the outer sides of the minichannel were also registered. Images obtained from the infrared camera were used to produce IRT thermograms (see fig. 3). Before each experimental series, the two metal plates were coated with the black paint (7) to achieve an emissivity of 0.83 [10]. In two consecutive experiments conducted at similar thermal and flow parameters, each minichannel surface was observed with an infrared camera. K-type thermocouples and pressure converters (6) were installed at the minichannel inlet and outlet.

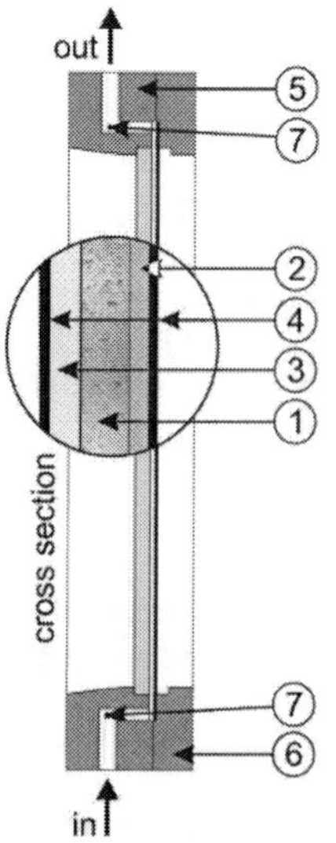

Figure 1. The schematic diagram of the test module: 1-minichannel, 2-the heated Hastelloy X plate, 3-the unheated stainless steel plate, 4-black paint layer, 5-channel body, 6-front cover, 7-thermocouple.

The experimental data was collected using: a data acquisition station, a computer and appropriate software. The supply and control system contained: an inverter welder, a shunt, an ammeter and a voltmeter. The experimental stand is discussed in detail in [11-13].

The thermal deformations of the minichannel surfaces were determined using the experimental data, which included local temperatures of the walls and the heat transfer coefficient calculated by means of a one-dimensional mathematical model.

The direction of the heat flow in the model was perpendicular to the direction of the fluid flow in the real minichannel (see fig. 2). The onset of nucleate boiling and enhancement of heat transfer in the minichannel were observed at an appropriate setting of the heat flux supplied to the heated surface. The results correspond to steady-state regimes. Details of the mathematical model are presented in [11,13].

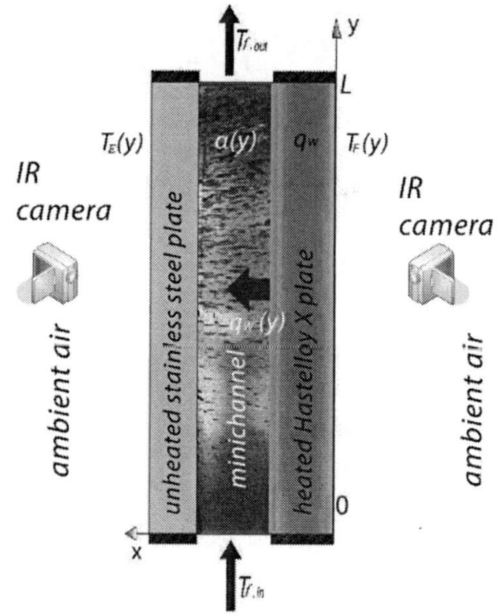

Figure 2. Schematic diagram of the main elements of the test module.

The local values of the heat transfer coefficient $\alpha(y)$ in the area between the fluid and the heated surface of the minichannel were determined using the Robin boundary condition (third-kind boundary condition) on the basis of the known local fluid temperature and the distribution of temperature on the heated plate measured by means of infrared thermography, according to the following equation [13]:

$$\alpha(y) = q_w / \left[T_F(y) - T_f(y)\right], \tag{1}$$

where q_w - heat flux density transferred to the fluid in the minichannel, $T_F(y)$ - local surface temperature measured by infrared thermography, $T_f\ (y)$ - local bulk liquid temperature determined on the basis of the linear distribution of temperature along the length of the channel from the inlet ($T_{f,in}$) to the outlet ($T_{f,out}$), y - distance from the minichannel inlet.

The results were analyzed for two settings – higher and lower values of the heat flux – supplied to the heated plate. The thermograms of the temperature distribution on the heated and unheated plates obtained at two different heat fluxes are presented in fig. 3.

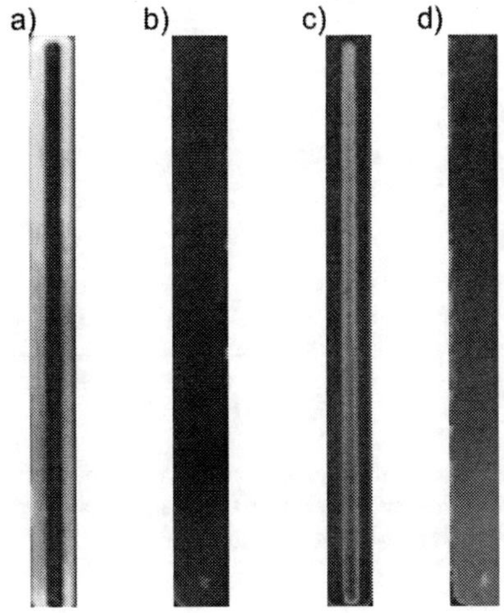

Figure 3. Thermograms of the temperature distribution on: (a,c) the heated plate, (b,d) the unheated plate, for the higher values of the heat flux (a,b) and the lower values of the heat flux (c,d).

4 Numerical calculations

The numerical calculations were performed using Ansys Workbench Academic Research software based on the Finite Element Method (FEM). The channel geometry with its main elements, i.e. the 0.6 mm thick heated plate and the 5.6 mm thick stainless steel unheated plate, was modelled using the assumptions presented in fig. 5.

Temperature difference is known to be the source of stress and strain generation. The temperature field and the stress field are difficult to determine in the steady state. Because of the model complexity, the problem is solved numerically. The calculation process is divided into two steps.

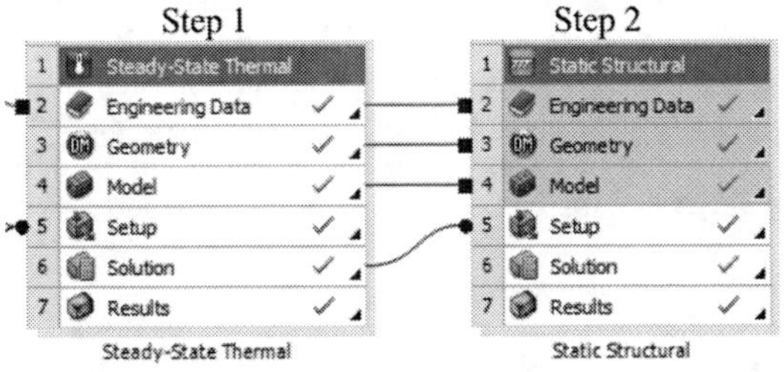

Figure 4. Block connection in Ansys Workbench.

In the first step, the distributions of temperature in the minichannel are determined using the first-kind boundary conditions (the Dirichlet conditions).

In the second step, the problem of elasticity is solved. If the solution is coupled two physical problems, it is possible to calculate the stresses and displacements associated with the changes in temperature along the minichannel walls.

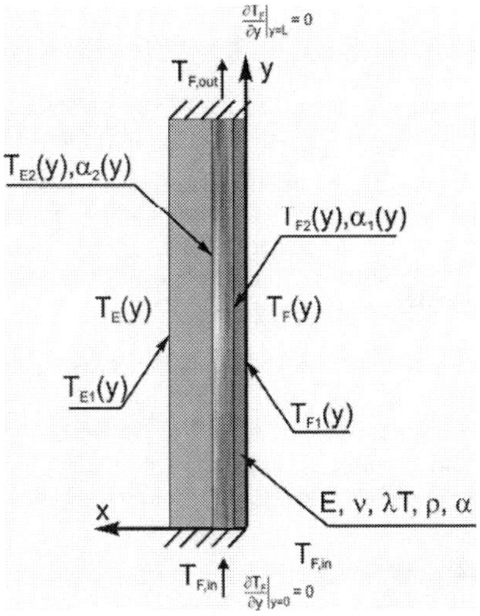

Figure 5. Numerical model assumptions.

The physical phenomena observed during heat transfer in the minichannel for the heated element [14, 15] and the unheated element can be described by means of the two-dimensional Laplace equation in the following form:

$$\frac{\partial^2 \tau_k}{\partial x^2}+\frac{\partial^2 \tau_k}{\partial y^2}=0 \quad \text{where } k=F,E \qquad (2)$$

where: x - direction coinciding with the direction of the fluid flow in the minichannel, y - direction perpendicular to the fluid flow direction, with appropriate boundary conditions based on the experimental data.

The effects of the temperature changes on the stress in the elements can be determined using Hooke's law, the general form of which is written as:

$$\sigma_{ij}=2\mu\varepsilon_{ij}+(\lambda e-\beta\tau)\delta_{ij}, \qquad (3)$$

where e - expansion $e=\varepsilon_{kk}=\varepsilon_{xx}+\varepsilon_{yy}+\varepsilon_{zz}$, λ and μ - the Lame constants, β - thermoelastic constant, σ_{ij} - stress, τ - temperature distribution, δ_{ij} - the Kronecker delta, υ - Poisson's ratio.

The relationships between the elastic constants and the thermoelastic constant can be described as:

$$\mu=\frac{E}{2(1+\upsilon)} \quad \lambda=\frac{2\upsilon\mu}{1-2\upsilon} \quad \beta=\alpha(3\lambda+2\mu), \qquad (4)$$

where α - thermal expansion coefficient, E - Young's modulus.

From the equation of equilibrium of forces $\sigma_{ji,j}+F_i=0 \quad (i,j=1,2,3)$ and the component of deformations it is possible to determine the displacement equation in index notation:

$$\mu\nabla^2 u_i+(\lambda+\mu)u_{k,ki}-\beta\tau_{,i}+F_i=0 \qquad (i=1,2,3), \qquad (5)$$

where $\nabla^2 = \frac{\partial^2}{\partial x^2} + \frac{\partial^2}{\partial y^2}$, F_i - unit forces, σ_{ij} - stress divergence, u_i - displacement, ∇^2 - the Laplacian.

The general equations, which describe the physical phenomena in the minichannel, are employed in the appropriate Ansys Workbench modules.

5 Results and discussion

The numerical analysis was conducted using a calculation grid of the wall cross-sectional area (similarly as i [4,5]) divided into 114 419 and 13 200 elements for the heated plate and the unheated stainless steel plate, respectively. Figure 6 shows the mesh employed in the numerical analysis.

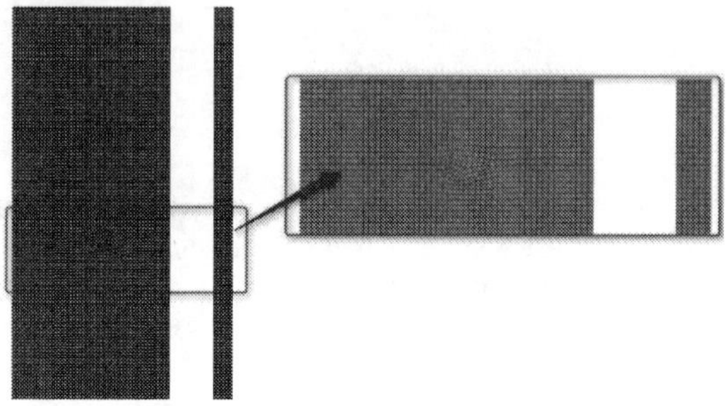

Figure 6. Mesh used in the numerical analysis.

The numerical analysis required assuming the material parameters for the heated Hastelloy X plate and the unheated stainless steel plate, the main elements of the test module. The constants for the materials studied are listed in table 1.

Table 1. Material constants used in the calculations.

	Hastelloy X	Stainless Steel
Density, ρ	8 970 [kg/m^3]	7 750 [kg/m^3]
Poisson's Ratio, ν	0.32	0.31
Thermal Conductivity, λT	8.9 [W(/m K)]	15.1 [W(/m K)]
Young's Modulus, E	2.11 10^{11} [Pa]	1.93 10^{11} [Pa]
Coefficient of Thermal Expansion, α	1.17 10^{-5} [1/K]	1 7 10^{-5} [1/K]

The simulations were based on the experimental data obtained for two heating power settings, where the heat flux supplied to the heated wall was higher or lower.

In one case, the value of the heat flux was 29.1 kW/m^2. Figure 7 shows changes in temperature along the minichannel length, while fig. 8 presents the thermal deformation along the minichannel length. The numbers 1 and 2 refer to the data obtained for the heated and unheated plates, respectively.

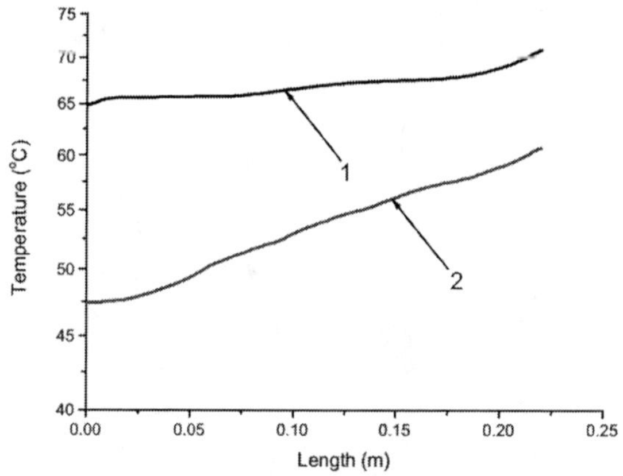

Figure 7. Distribution of temperature along the minichannel length, $q_w = 29.1$ kW/m^2; data for: 1) the heated Hastelloy X plate, 2) the stainless steel plate.

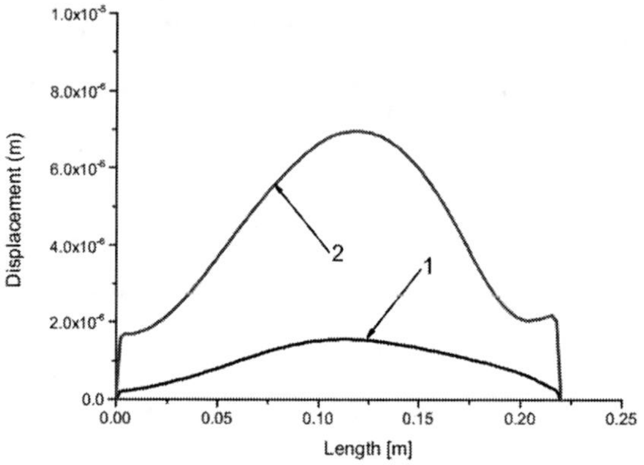

Figure 8. Thermal deformation along the minichannel length, $q_w = 29.1$ kW/m^2; data for: 1) the heated Hastelloy X plate, 2) the stainless steel plate.

It can be seen that the difference in temperature of the heated element between the inlet and the outlet is approx. 6 °C. In the case of the other surface, the difference in temperature in the minichannel between the ends is much higher, i.e. approx. 15 °C. The non-uniform temperature distributions generate thermal stress leading to changes in the minichannel geometry. As can be seen from fig. 8, the resulting deformations of the major elements are approx. 1-7 10^{-6} m and they are the highest in the central part of the minichannel length.

In the other case, the heat flux was 14.5 kW/m^2. The graphs in fig. 9 show that the differences in temperature between the analyzed surfaces did not exceed 5 °C and at a distance of two-thirds of the length from the inlet, the temperatures were almost equal.

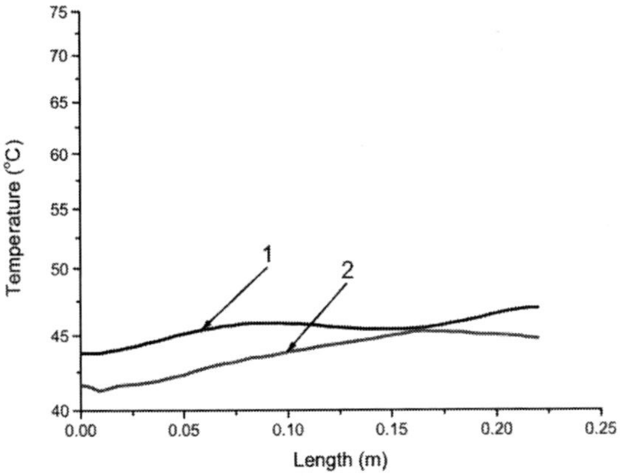

Figure 9. Distribution of temperature along the minichannel length q_w = 14.5 kW/m^2; data for: 1) the heated Hastelloy X plate, 2) the stainless steel plate.

The small differences in temperature were responsible for the deformations of the test elements (fig. 10). The deformations recorded for the stainless steel plate and the heated element made of Hastelloy X were below 1 10^{-6} m and above 2 10^{-6} m, respectively.

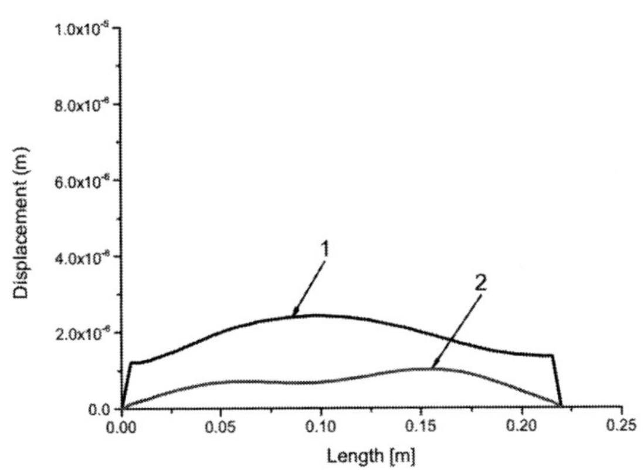

Figure 10. Thermal deformations along the minichannel length q_w = 14.5 kW/m^2; data for: 1) the heated Hastelloy X plate, 2) the stainless steel plate.

From the analysis of the numerical data concerning the thermal deformations in the minichannel it is clear that they did not exceed 0.0001 m. Their values were about three orders of magnitude smaller than the minichannel width, which, practically, had no direct impact on the refrigerant flow.

The problem of thermoelasticity was solved assuming that the heated body affected by thermal deformations and changes in temperature.In the calculations, it was assumed that the material was isotropic and that the temperature change accompanying the deformation was small, so a rise in temperature did not cause a significant change in the elastic and thermal properties of the material.

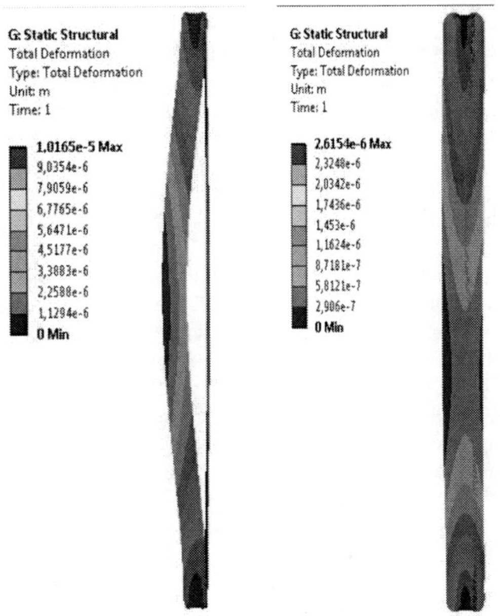

Figure 11. Thermal deformations of the minichannel walls; data for: a) q_w = 14.5 kW/m^2, b) q_w = 14.5 kW/m^2.

Figure 11 shows the thermal deformations of the minichannel walls in their cross-sections for two settings of the heat flux density. The results clearly indicate that, when the heat flux density was higher, the difference in temperature distribution was greater, which caused greater buckling. The thermal deformation was more than 1 10^{-5} m at a distance of half of the minichannel length. However, when the heat flux density was lower, expansion of the stainless steel element was observed along the OX axis. This deformation was not as large as in the other case; the maximum deformation was 2 10^{-6} m.

The major aim of the simulations was to establish whether the minichannel surfaces undergo deformations in the predetermined temperature range . It was locally change its cross-sectional area and, consequently, lead to any distortions of the fluid flow in the minichannel.

The final finding is that for the predetermined thermal conditions and the assumed thicknesses of the minichannel walls, the deformations can be omitted from further analysis.

6 Conclusions

The paper has discussed a numerical analysis of the thermal deformations of rectangular minichannel walls. The distribution of temperature on the minichannel walls and the local heat transfer coefficients were determined on the basis of the experimental results. The numerical calculations were performed to study the thermal deformations of two minichannel walls: the heated Hastelloy X plate and the stainless steel plate. The heat conduction and displacement equations were solved numerically using Ansys Workbench software. The analysis involved appropriately combining the modules with the mathematical description of the physical phenomena occurring in the minichannel, introducing the boundary conditions and the experimental data, and calculating the distributions of temperature and the thermal deformations of the walls. The results were represented graphically as a printout of Ansys analysis.

Thermoelastic deformations of the minichannel walls should be one of the most important aspects to be taken into consideration in the modelling of the physical phenomena occurring in this minichannel. The difference in temperature of the heated wall can result in changes in the minichannel shape, with its local narrowing. Moreover, the thermal deformations of the heated wall, which was relatively thick, were not greater than 0.0001 m; they were three orders of magnitude smaller than the minichannel depth and, practically, had no direct impact on the fluid flow.

Acknowledgements

The research presented here was supported by the National Scientific Center under grant No. DEC-2013/09/B/ST8/02825.

References

1. L. A. Campbell, S. G. Kandlikar, Proc. 2d Int. Conf. Microchannels and Minichannels, Rochester (2004)
2. T. Bohdal, Annual Set The Env. Protection, (in Polish) **2**, 262-279 (2000)
3. D. Mikielewicz, J. Mikielewicz, J. Tesmar, Int. J. Heat Mass Tran. **50**, 3949–3956 (2007)
4. J. Kaew-On, K. Sakamatapan, S. Wongwises, Exp. Therm. Fluid **35**, 364–374 (2011)
5. P. Fan , Y. Zeng , Y. Zhang , G. Li , Int. J. Therm. Sc. **77**, 1-10 (2014)
6. E.A. Ariza., A.P. Tschiptschin, C.R.F. Azevedo, Eng. Failure Analysis **48**, 78-93 (2015)
7. M. Ahmed, R. Khan, S. Badshah, S. Jan, J. Eng. Adv. Techn. **4**, 118-124 (2014)
8. J. Taler, P. Dzierwa, D. Taler, P. Harchut, Energy (2015)
9. J. Taler, D. Taler, S. Lubecki, J. Therm. Stresses **34**, 101-1034 (2011)
10. Orzechowski T., Heat transfer on ribs with microstructured surface (in Polish), Kielce Univ. Techn., Kielce, 2003
11. M. Piasecka, Int. J. Heat Mass Tran. **66**, 472-488 (2013)
12. M. Piasecka, Heat Tran. Eng. **35** (10), 903-912 (2014)
13. M. Piasecka, B. Maciejewska, K. Strąk, Heat Tran. Eng., to be printed
14. S. Błasiak, Int. J. Heat Mass Tran. **81** 90-102 (2015)
15. S. Błasiak, J. E. Takosoglu, P. A. Łaski, J. Therm. Sc. Techn. **9** (2), 1-8 (2014)

EPJ Web of Conferences **114**, 02097 (2016)
DOI: 10.1051/epjconf/201611402097

Measurement of airflow and pressure characteristics of a fan built in a car ventilation system

Jan Pokorný[1,a], Filip Poláček[1], Miloš Fojtlín[1], Jan Fišer[1] and Miroslav Jícha[1]

[1]*Brno University of Technology, Faculty of Mechanical Engineering, Energy Institute, Technická 2896/2, Brno, Czech Republic*

Abstract. The aim of this study was to identify a set of operating points of a fan built in ventilation system of our test car. These operating points are given by the fan pressure characteristics and are defined by a pressure drop of the HVAC system (air ducts and vents) and volumetric flow rate of ventilation air. To cover a wide range of pressure drops situations, four cases of vent flaps setup were examined: (1) all vents opened, (2) only central vents closed (3) only central vents opened and (4) all vents closed. To cover a different volumetric flows, the each case was measured at least for four different speeds of fan defined by the fan voltage. It was observed that the pressure difference of the fan is proportional to the fan voltage and strongly depends on the throttling of the air distribution system by the settings of the vents flaps. In case of our test car we identified correlations between volumetric flow rate of ventilation air, fan pressure difference and fan voltage. These correlations will facilitate and reduce time costs of the following experiments with this test car.

1 Introduction

The amount of ventilation air flowing into a car cabin is important parameter, which affects cabin indoor environment and HVAC system performance. It should be considered already in early stage of car design and has to be known before the new car series are produced to be able evaluate the efficiency of cabin's ventilation, which has close relationship to the driving comfort and safety. "The safety of occupants to reduce driver fatigue, ensure good visibility and maintain comfort is key to the successful design of such systems" [1]. To support the early stage of the HVAC system design our research team has been developed a Virtual Testing Stand of the Car Cabin [2], which is a computational model for the calculation of the car cabin heat load in various environments. The first verification was done [3], however still the model needed to improve the advection part. The amount of ventilation air and its distribution to the cabin was important task how to improve the advection scheme of the model.

In this paper the measurement of amount of sucked air into the cabin is presented for wide range of manual vents settings. The ventilation air distribution in automatic mode during various seasons is contemporary published in the paper [4]. Aim of this study was to find correlations between the setup of fan speed, volumetric flow rate and pressure difference of the fan for various ventilation setups. This study we performed on the our test and the obtained results will be used for subsequent experiments within work package Human Cabin Centred Design of the Czech Republic project Josef Božek Competence Centre for Automotive Industry.

2 Methods

There are plenty of methods how to measure air flow rates through ventilation system, which are based on various physical phenomena, see [5]. For example there are mechanical flow meters, differential pressure flowmeters (orifice plates, Venturi tubes, etc.) or velocity flow meters (Pitot tubes, anemometers: vane, hot wire, laser Doppler, etc.). In our experiment we utilized the last mentioned category velocity flow meters, which identify the flow rate by measuring the air velocity and integrating over the flow area. To measure air velocity we used a couple of hot bulb anemometer probes manufactured by Testo.

The method is more accurate when the stilling channel is used to ensure that air flow through a ducting system will be evenly distributed with developed air flow regimes. The volumetric flow rate $\boldsymbol{V}$ was afterwards calculated from the knowledge of cross-section area $\boldsymbol{S}$ of this channel and mean velocity $\overline{v}$. During the experiments we also kept knowledge of the barometric pressure and air temperature to be able to calculate air density $\boldsymbol{\rho}$ and thus also mass flow rate $\boldsymbol{m}$ through the ventilation system, see Equation 1.

$$m = V \cdot \rho = v \cdot S \cdot \rho = \frac{\sum_{i=1}^{n}(v_i \cdot S_i)}{N} \cdot \rho \quad (\text{kg·s}^{-1}) \qquad (1)$$

[a] Corresponding author: pokorny.j@fme.vutbr.cz

2.1 Experimental setup

The stilling channel added in front of the test car suction inlet was a plastic tube of diameter 150 mm and length 2000 mm. Because the suction inlet is located under the hood, we had to remove it to be able prepare the experimental setup, see Figure 1 on the left; on the right side, there is shown the ring around the cross-section area, which was located in the middle of the tube's length to locate measuring axis un . The ring we used for better

Figure 1. Experimental setup. On the left: Stilling channel during measurements. On the right: the ring around the cross-section area of tube where the probes were placed (axis location +x=45 ° +y= 135° from the horizontal plane).

During the test the car engine was idling to ensure that the fan be powered with same power as during real traffic coditions (supply voltage about 13.8 V). For the measurement of the voltage we used classical multi-meter and for the measurement of the pressure difference we used Testo 435 with a build pressure sensor. The value of pressure difference was recorded manually ten times for the each test case. In the case of air velocity we recorded measurement automatically with sampling rate 1 Hz. We used two hot bulb anemometers connected to data logger Testo 350 M/XL. The each point was measured for five seconds and other five seconds were used for the changing position of the probes in the tube. The position of the probes in the tube was setup according to the theory see [6] using the Equation 2.

$$r_i = R \cdot \sqrt{(2i-1)/2N} \quad (\mathrm{m \cdot s^{-1}}) \tag{2}$$

where R = 75 mm is diameter of the tube and N = 7 is number of traversed lines in each half-axis. Using this the approach it allows to split cross-sectional area of tube (176.7 mm^2) into 28 sectors of equal areas. For each sector was measured one air velocity at its centre and then we calculated the mean velocity $\bar{v}$ simply as the average value of these local velocities. Additionally we measured the central point [x=0, y=0], so all together it counts 29 points to measure for each test case. However central point not enter to the calculation of mean velocity, because this points represents annulus of zero area (see Figure 2).

2.2 Uncertainties

To obtain a representative view about a flow regime inside the tube the air velocity at the cross-section's traverse lines (x, y axis) was measured repeatedly. For each ventilation setup it was traversed five times to be able evaluate uncertainty components: of type A. In this case the standard uncertainty u_A can be associated with standard deviation of the mean. We applied this for the air velocity and obtained the Equation 3.

$$u_A = k_{uA} \cdot \sqrt{\frac{1}{n(n-1)} \sum_{i=1}^{n} (v_i - \bar{v})^2} \quad (\mathrm{m/s}) \tag{3}$$

where $k_{uA} = 1.4$ is safety factor for number of repetition $n = 5$. Then the expanded uncertainty for the 95 % confidence interval is $U_A = 2 \cdot u_A$, where 2 is the relevant coverage factor, see [7]. The uncertainty of type B corresponding to the accuracy of measurement devices was neglected in this study.

2.3 Measurement schedule

The measurement schedule included test cases combining four different fan speed (level 2, 3, 4, 6) and four different configurations of closing vent flaps ("all vents opened", "only central vents closed", "only central vents opened" and "all vents closed"). We manipulate only with the flaps of the dashboards vents (two at centre and two at sides) and with two tunnel vents oriented to the rear seat. These configuration were chosen with respect to the cover a wide range of ventilation system operational conditions. We measured all mentioned combination which rises matrix 4x4 of examined test cases, see Table 1.

3 Results

In the first subchapter, there is demonstrated how the volumetric and mass flow rate were evaluated through the measurement of air velocities, which was the most time consuming procedure of the experiment. The procedure is described for the test case S6-C1, see Table 1. In the second subchapter a summary of results of all test cases is presented and in the third one the resulting pressure-volume characteristics are plotted.

3.1 Test procedure for particular test case

During the test case S6-C1 (fan speed setup to 6th level and only central vents opened) the mean air temperature was 25.4 °C and barometric pressure was 102 375 Pa, which implies air density ρ = 1.194 kg/m^3. Voltage on the fan was during the test almost constant U = 11.4 V and pressure difference at fan Δp = 325.2 Pa was quite stable as well (mostly it was between 320 and 328 Pa).

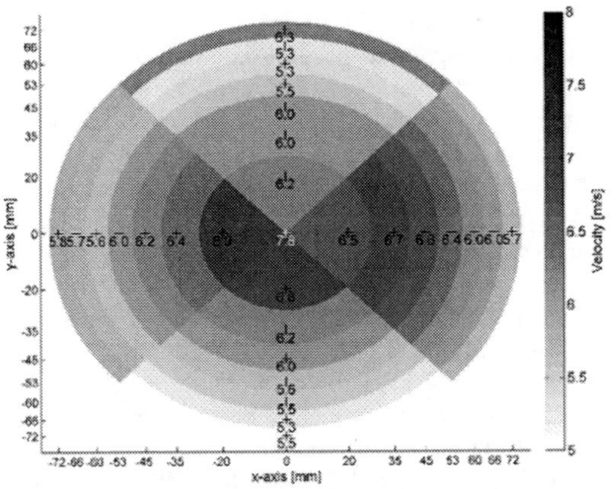

Figure 2. Test case S6-C1. Measured velocity profile in the tube with 28 equal areas of cross-section.

The measured values of air velocities in specific points of the cross-section area of stilling channel are plotted in Figure 2. There are illustrated location and measured values of all 28 sectors, which implies the mean velocity $\bar{v} = 6.00 \pm 0.60\ m/s$. Hence by using the Equation 1 the volumetric flow rate was $\dot{V}$ = 106.0 l/s and mass flow rate $\dot{m}$ =126.5 g/s.

In Figure 3, there are shown the velocity profiles for the x-axis and y-axis, with error bars showing the value of expanded standard deviation U_A for each point. It can be seen that the air flow in the stilling channel had almost fully developed turbulent flow regime with the maximal velocity in the centre. For example, in the case S6-C1, the highest value of U_A was for the point on y-axis y=+72 mm, where v = 5.71 ± 1.40 m/s. The individual time samples were [7.17, 5.91, 4.02, 5.63, 5.81]. On the other hand the lowest standard deviation was for the point on x-axis x=+66 mm, where v = 5.27 ± 0.18 m/s. The individual time samples were [5.21, 5.53, 5.22, 5.27, 5.13]. These uncertainties rather than outline some fluctuations in flow structure inside a tube, represents the precision of the measurement at given point.

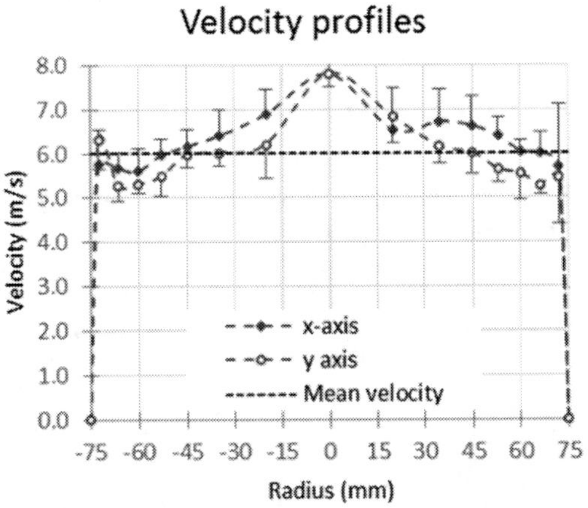

Figure 3. Test case S6-C1. Measured velocity profiles and the mean velocity (point r = 0 mm not enter to its calculation).

3.2 Fan and ducting system characteristics

All identified operating points are noted in Table 1. Each point is defined by fan setup, vent flaps configuration, value of volumetric flow rate and pressure differences. From this table were plotted graphs in Figure 4 and Figure 5, representing operating points of the fan and ducting system, respectively.

The fan characteristics expresses how the specific fan of our test car works during various intensity of ventilation, which depends on the fan speed and thus also on the voltage of the fan. We derived for the fan a regression model using the polynomials of 2nd order see Figure 4. It can be seen that in the case S2 the volumetric flow rate is very sensitive to the pressure difference. From this reason we not measure the case S1, because the range of pressure difference is in order of units of Pascals which is close to the accuracy of the Testo device. On the other hand with the higher speed of fan the range of pressure differences respect to the range of volumetric flow rate is increasing and thus the interpretation of results should be more reliable.

Table 1. Matrix of all measured test cases

Fan setup	**C1**: all opened	**C2**: only central closed	**C3**: only central opened	**C4**: all closed
S2: U = 4.8 V	42.3 l/s 99.8 Pa	39.8 l/s 110.2 Pa	35.9 l/s 112.4 Pa	23.9 l/s 124.3 Pa
S3: U = 6.1 V	53.0 l/s 148.4 Pa	48.2 l/s 160.3 Pa	45.3 l/s 165.3 Pa	31.5 l/s 192.1 Pa
S4: U = 7.6 V	75.4 l/s 208.8 Pa	69.2 l/s 214.8 Pa	65.0 l/s 227.8 Pa	47.1 l/s 269.5 Pa
S6: U = 11.4 V	106.0 l/s 322.5 Pa	97.5 l/s 368.8 Pa	91.9 l/s 395.1 Pa	76.4 l/s 492.6 Pa

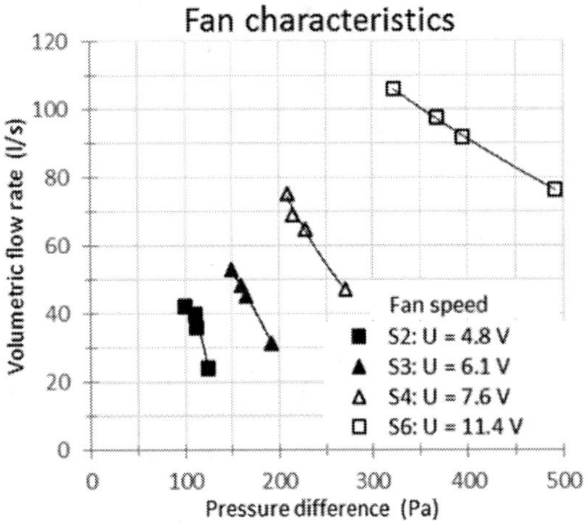

Figure 4. Fan characteristics

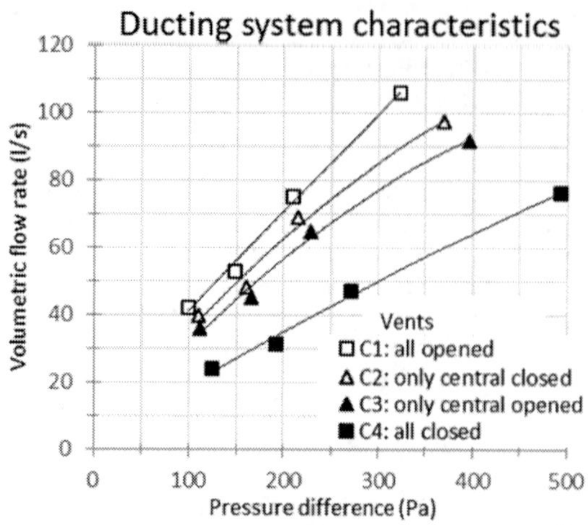

Figure 5. Ducting system characteristics

The ducting system characteristics express the relation between the difference pressure and the volumetric for the specific ducting system. The most interesting curves are C1 and C4, which defines a thresholds of possible flow conditions through the given ducting. We derived for the the ducting also a regression model based on polynomials of 2nd order see Figure 5.

The main results of this experiment are identified regression curves for the fan and ducting system inside out test car. If we plot the fan curves and ducting system curves over themselves, we obtain from their intersections the operating points, which we already measured. However the advantage is that it is possible to change the type of fan on the current ducting, or also it is possible to change ducting system for the current fan to obtain new operating points.

4 Discussion

The experiments shown that the voltage of the fan is linearly proportional to the user settings of fan controller, as well as the volumetric flow rate see Figure 6. It is evident that the speed of fan not define volumetric flow rate unambiguously, because there is strong influence of how the ventilation ducting system is throttled by vent flaps settings. This is the main benefit of our experiment, to considered pressure - volumetric flow rate characteristics of ventilator (Figure 4) and ducting system (Figure 5) to be able cover all possible settings of ventilation control (defrost, automatic mode, closing or manual setup of vent flaps, etc.)

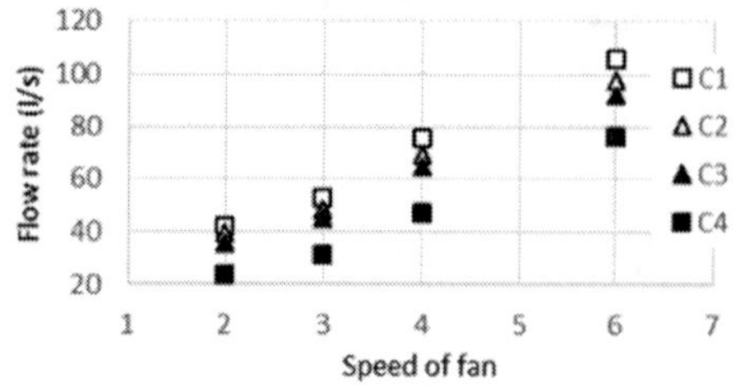

Figure 6. Speed of fan versus volumetric flow rate.

The positioning of the probes were done manually, so the error of fastening and positing could occur and slightly influence particular measured point. Still the standard deviation (confidence interval 95 %) during the measurement were acceptable generally and the anemometer based method shows a good repeatability.

The highest standard deviations of velocity were near the wall. We assume that is due to the higher velocity gradient at the wall, which caused that the measured values were more sensitive to precise probe location. The higher accuracy of the measurement could be achieved by increasing of repetition of measurement from 5 for example to 10. However we chose $n = 5$ due to the fact that the measurement would be too time consuming in case of 10 and the k_{uA} factor would not change significantly. The stilling channel improved the stability of flow a lot and reduce the fluctuations which occur directly before the suction inlet of the ventilation system. The peak in the velocity profile (Figure 3) at the centre of channel cross-section can be explained by the fact that all points were measured by two probes simultaneously, however in case of central point it was used only one probe. Nevertheless the value of central point was not considered in the calculation of the mean velocity and volumetric flow rate.

5 Conclusion

We identified fan and ventilation ducting system characteristics describing correlations between volumetric flow rate, setup of ventilation system and pressure difference at the fan. Obtained correlations are valid only for fan and ducting system of our test car, but even so they will allow evaluate volumetric flow rate of air directly from the fan pressure difference. This approach will facilitate and reduce time costs of the consequent experiments with our test car substantially. Acquired knowledge is possible to use not only for climatic chamber tests, but even for real traffic conditions measurements.

Acknowledgement

The research was supported by the project Reg. No. FSI-S-14-2355 of the Brno University of Technology; the project LO1202 Netme Centre Plus with the financial support from the Ministry of Education, Youth and Sports of the Czech Republic under the "National Sustainability Programme I" and the project of the Josef Božek Competence Centre for Automotive Industry TE01020020.

References

1. S. Daly. *Automotive air-conditioning and climate control systems*. 1st ed. Oxford: Butterworth Heinemann (2006).
2. J. Pokorny, J. Fiser, M. Jicha, Advances in engineering software **76** (2014)

3. J. Pokorny, Fiser, M. Jicha, EPJ Web of Conferences **92**, 02072 (2015)
4. M. Fojtlin, M. Planka, J. Fišer, J. Pokorný, M. Jícha. Airflow Measurement of the Car HVAC Unit Using Hot-wire Anemometry. *Proceedings of Experimental Fluid Mechanics* 2015, Prague, Czech Rep. (to be published).
5. H. D. Goodfellow, E. Tähti. *Industrial Ventilation Design Guidebook. Part 12. Experimental techniques*, (2001).
6. M. Pavelek. Experimentální metody v technice prostředí. 3. vyd. Brno : Akademické nakladatelství CERM (2007).
7. P.Němeček. *Nejistoty měření*. 1. vyd. Praha : Česká společnost pro jakost (2008).

EPJ Web of Conferences **114**, 02098 (2016)
DOI: 10.1051/epjconf/201611402098

The effects of ambient impurities on the surface tension

A. Ponce-Torres[1,a] and E. J. Vega[1,b]

[1] *Depto. de Ingeniería Mecánica, Energética y de los Materiales, Universidad de Extremadura, E-06006 Badajoz, Spain*

Abstract. A liquid bridge is a liquid column held captive between two coaxial and parallel solid disks. It is an excellent test bench where measuring the surface tension. In this paper, we used this fluid configuration to examine experimentally the effects of ambient impurities on the surface tension over time. For this purpose, the liquid bridge equilibrium shape was analyzed when the liquid bridge was surrounded by three environments: the uncontrolled ambient, and both air and argon encapsulated in a small glass cover. Ambient contamination produced a sharp decrease of the surface tension of ultra-pure water. The presence of an anionic surfactant in the free surface of an aqueous solution did not inhibit the action of impurities coming from the ambient. Impurities can influence the dynamical behavior of the free surface in flows dominated by the surface tension. Therefore, a careful control of that influence can be crucial in many applications of fluid mechanics.

1 Introduction

What role plays the ambient impurities in the surface tension of liquids? The usual answer is: it can be regarded as negligible for the first minutes of a certain experiment involving exposed liquids to an uncontrolled ambient. It is believed that liquids with low surface tensions ($20 \lesssim \sigma \lesssim 30$ mN/m) are almost insensitive to impurities. Because of the fact that the surface tension of the pure liquid is small, impurities of surfactant nature cannot lower significantly that tension. However, impurities can affect noticeably the behavior of capillary systems with high surface tensions. For example, they can cause Marangoni convection due to the gradients of surface tension over the interface. This phenomenon is particularly important in experiments with air-water interfaces.

The presence of a tainted monolayer due to the ambient impurities could produce an extra-dissipation of energy in millimetric liquid free surface [1]. In fact, surface contamination is known to play a significant role in surface wave damping [2] (and in surface wave dynamics too, see Refs. [3] and [4] when water is used). The measurement of surface tension over time can help us to know whether ambient impurities are forming a monolayer on the free surface (even if surfactants are previously added on purpose) which could affect our experimental results.

Liquid bridges are excellent test benches where analyzing varied phenomena associated with surface tension [5–8]. The large surface/volume ratios of millimeter bridges make the surface tension force dominant in most situations. Liquid bridges offer significant advantages over their close relatives: the pendant and sessile drops. When working with liquid bridges, the experimenter has at his disposal an additional adjustable parameter, the slenderness, to enlarge the range of experimental conditions. In addition, the liquid bridge shape remains sensitive to the surface tension value even in the minimum volume stability limit, while drops become spherical independently of the surface tension value for small enough volumes.

The liquid bridge configuration considered in this work (figure 1) consists of an isothermal mass of liquid of volume $\mathcal{V}$, held between two parallel coaxial circular solid supports placed a distance L apart. The radius of the two solid supports is R. Due to the sharpness of their edges, one assumes that the liquid anchors perfectly to those edges, preventing motion of the triple contact line. Hence, spreading of the liquid over the lateral surfaces of the solid supports does not occur. The liquid bridge is surrounded by another fluid medium of negligible density and viscosity, and is subjected to the action of the force of gravity of magnitude g per unit mass. The liquid bridge density and viscosity are ρ and μ, respectively, while σ is the surface tension associated with the interface. For this axisymmetric configuration, the equilibrium contour of the liquid bridge is characterized by the function $F_0(z)$, which measures the distance between a surface element and the axis of the disks (z axis). The dimensionless parameters characterizing the fluid configuration are the slenderness $\Lambda \equiv L/(2R)$, the reduced volume $V \equiv \mathcal{V}/(\pi R^2 L)$, the static Bond number $B_0 \equiv \rho g R^2/\sigma$, and the Capillary number (defined as the square root of the Ohnesorge number) $C_\mu \equiv \mu(\rho\sigma R)^{-1/2}$.

In this work, we will examine experimentally the effects of ambient impurities on the surface tension over time. For this purpose, the liquid bridge equilibrium shape will be analyzed when the liquid bridge is surrounded by

[a] e-mail: albertoponce0692@gmail.com
[b] e-mail: ejvega@unex.es

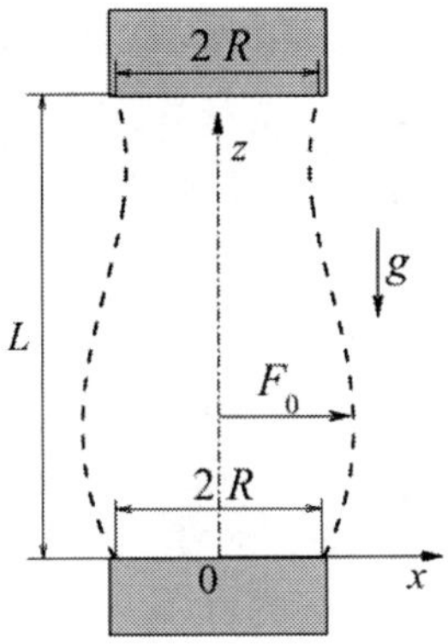

Figure 1. Liquid bridge configuration.

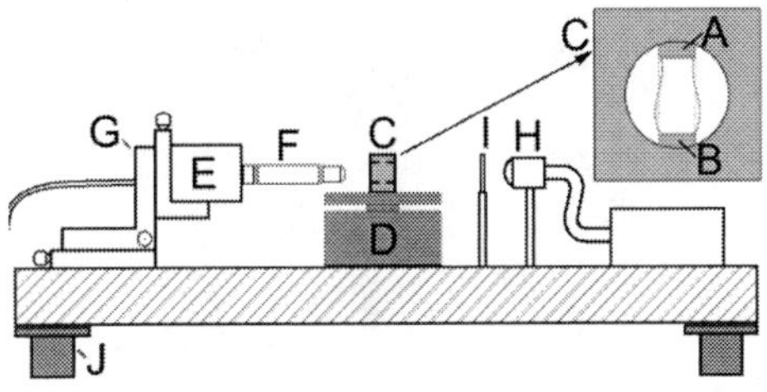

Figure 2. Experimental apparatus: upper needle (A), bottom disk (B), liquid bridge cell (C), platform (D), camera (E), optical lenses (F), micrometer screws (G), optical fiber (H), frosted diffuser (I), and optical table (J).

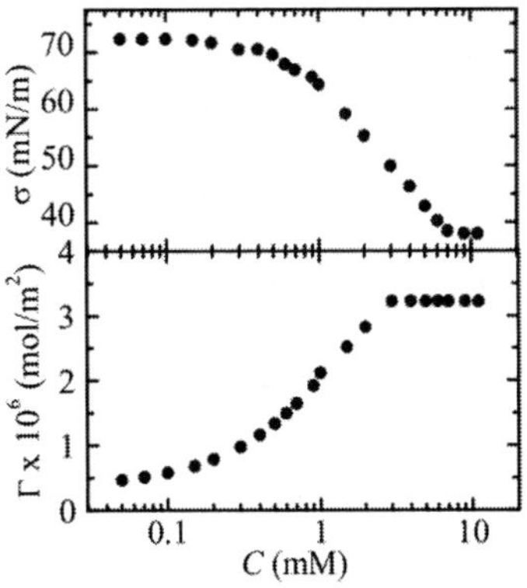

Figure 3. Surface tension and surface density for aqueous SDS solutions as a function of the bulk surfactant concentration C. The points come from the interpolation of the data in the reference [10].

three environments: the uncontrolled ambient, and both air and argon encapsulated in a small glass cover. We will analyze the effects of impurities in both pure water and an aqueous solution of an anionic surfactant. The paper is organized as follows. The experimental method used to analyze the problem is described in Sec. 2. The results are presented and discussed in Sec. 3. Finally, the paper closes with some concluding remarks in Sec. 4.

2 Experimental Method

Figure 2 shows the experimental setup used in the present study. A liquid bridge was formed between an upper needle (A) and a coaxial bottom disk (B), both about 1 mm in radius and placed in a cubic cell (C). The upper needle could be displaced along its axis, and was used to feed and remove liquid by using a syringe pump connected to a stepping motor. The cell was mounted on a platform (D). Digital images of 704×600 pixels were acquired using a CMOS camera (E), equipped with a set of optical lenses (F) providing a frame covering an area of about 4×3.4 mm. The magnification obtained was 5.73 μm/pixel. The camera could be displaced both horizontally and vertically using a triaxial translation stage (G) to focus the liquid bridge. The fluid configuration was illuminated from the back side by cool white light provided by an optical fiber (H) connected to a light source. A frosted diffuser (I) was positioned between the optical fiber and the cell to provide a uniformly lit background. All these elements were mounted on an optical table with a pneumatic anti-vibration isolation system (J) to damp the vibrations coming from the building. The experiments were performed at room temperature, approximately $T = 25 \pm 1$ °C. The images of the liquid bridge at equilibrium were processed with the Theoretical Image Fitting Analysis-Axisymmetric Interfaces (TIFA-AI) method [9] to measure the surface tension.

The test liquids used in the experiments were deionized water for synthesis (Di-water, Merck) ($\rho = 998$ kg/m^3, and $\mu = 0.001$ kg/ms), and an aqueous solution of the anionic surfactant SDS (sodium dodecyl sulfate for molecular biology, assay = 99.0% (GC, gas chromatography), Sigma Aldrich). The concentration was $C = 0.84$ mM, much smaller than the accepted value 8 mM of the cmc (critical micelle concentration), and than that leading to the maximum package surface density of aqueous solution of SDS, around 3 mM. In figure 3, one can observe the values of surface tension and surfactant surface density Γ for aqueous SDS solutions as a function of the bulk surfactant concentration C. The points come from the interpolation of the data in the reference [10], where is mentioned that both surfactant (SDS) and water were carefully purified in order to minimize the effects of impurities. Therefore, these curves can be interpreted as "ideal" curves for aqueous SDS solutions.

We conducted experimental runs with both test liquids, Di-water and 0.84 mM solution of SDS in Di-water, for three different environments: the uncontrolled ambient of air, and saturated air and argon (ultra-pure argon, purity 5.0, Linde, $\rho = 1.7$ kg/m^3) encapsulated in a small glass cover, repeating each experimental run to check the reproducibility of the results. Argon was chosen because it is the least active gas with liquids, water-insoluble, and its density is higher than air, which makes easier create the argon ambient in the cover. Figure 4 shows the small glass cover introduced in the liquid bridge cell of the experi-

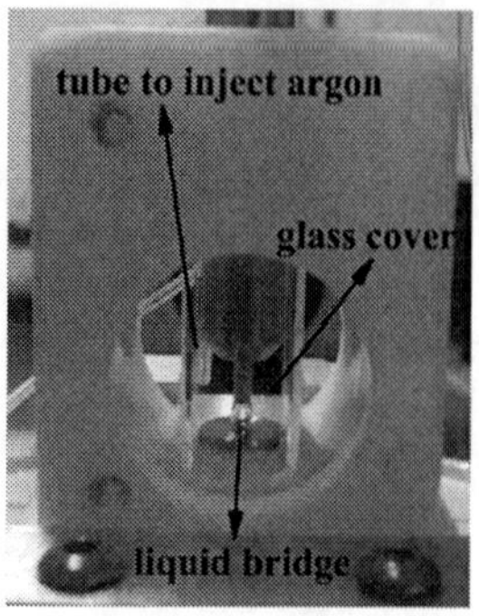

Figure 4. Image of the liquid bridge cell with a glass cover to control the ambient.

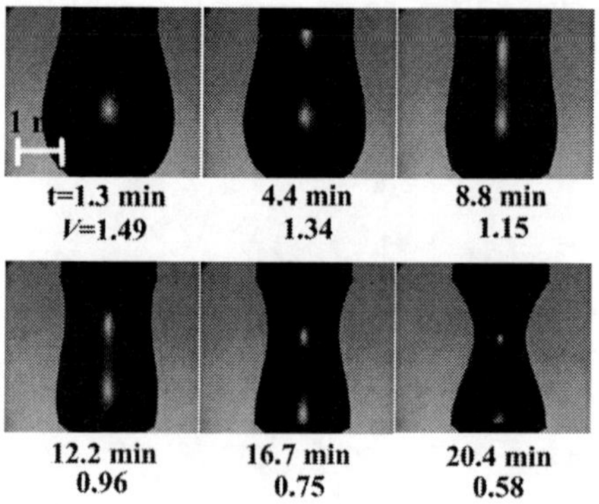

Figure 5. Images of the liquid bridge captured during the experimental run of the 0.84 mM aqueous SDS solution in the argon environment.

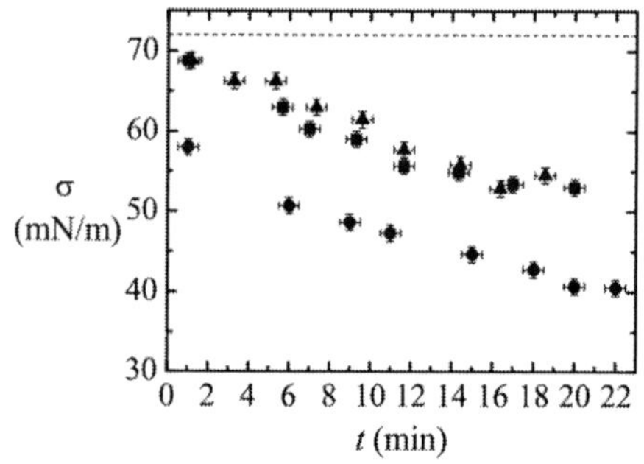

Figure 6. Surface tension as a function of the time t when a liquid bridge of Di-water is surrounded by three environments: uncontrolled ambient (solid circles), saturated air (solid squares) and argon (solid triangles). The dashed line indicates the ideal value of surface tension for Di-water, 72 mN/m. The parameters characterizing the experiments were $\Lambda = 1.45$, $B_0 = 0.135–0.25$, and $C_\mu = 3.73 - 5 \times 10^{-3}$.

mental setup to create the controlled ambients. The optical glass cover was closed hermetically, only in the bottom part, thanks to a rubber gasket. In the case of controlled ambient of air, the bottom part of the glass cover was filled with a small amount of Di-water to saturate the ambient, while argon was injected through a small plastic tube placed in the glass cover (as it can be seen in the figure) to obtain a ambient of argon. The velocity of the argon flow was low enough to neglect any effects on the liquid bridge.

In each experimental run: a liquid bridge was formed, and one image of the liquid bridge was acquired approximately every 2 minutes. We measured the surface tension over time in the first 23 minutes after forming the liquid bridge. During the experiments, the liquid bridge volume was reduced over time due to the evaporation (see figure 5), with the exception of the cases in which the liquid bridge was encapsulated in saturated air. For the experiments of 0.84 mM aqueous SDS solution, it was supposed that the evaporation did not change the concentration because the diffusion of the surfactant particles from the needle to the bulk/free surface of the liquid bridge can be regarded as immediate if it is compared with the evaporation rate. In other words, there is always a spices and thermodynamic equilibrium between the injection system and the bulk of the liquid bridge. Consequently, the bulk concentration and its associated surface density remains constant even when the volume of the liquid bridge changes. All the experiments were conducted with extreme care to avoid the contamination of the liquid during its injection and the formation of the liquid bridge.

3 Results

Figure 6 shows the results obtained for the interface contamination over time for Di-water. As can be observed, surface tension depends strongly on time in all the cases. In our experiments, only when a controlled ambient of saturated air or argon is used, it is possible to get a value next to the well-known value of 72 mN/m for pure water, over the first 5 minutes. When the liquid bridge is surrounded by an uncontrolled ambient the value of surface tension is already under 60 mN/m in the first minute of our experimental run, indicating that the impurities reach rapidly the liquid free surface in that period. Thus a controlled ambient slows down that first quick contamination. After 20 minutes exposed to the uncontrolled ambient the surface tension goes down to 40 mN/m (from 58 mN/m). In both saturated air and argon ambients encapsulated in a small glass cover, the surface tension reachs an almost constant value around 53 mN/m after 15 minutes.

Figure 7 allows one to analyze the effects of impurities in an aqueous solution of SDS surfactant. In this figure, we plot the results for a 0.84 mM aqueous SDS solution exposed over time at the same three environments used with the previous liquid. Interestingly, although the free surface of the liquid bridge is doped with a surfactant, the latter does not inhibit the action of impurities coming from the ambient. When surfactant is introduced in the problem, it can be seen only a slight difference between the values of surface tension for the different environments in the first minute, since the free surface is previously doped with the

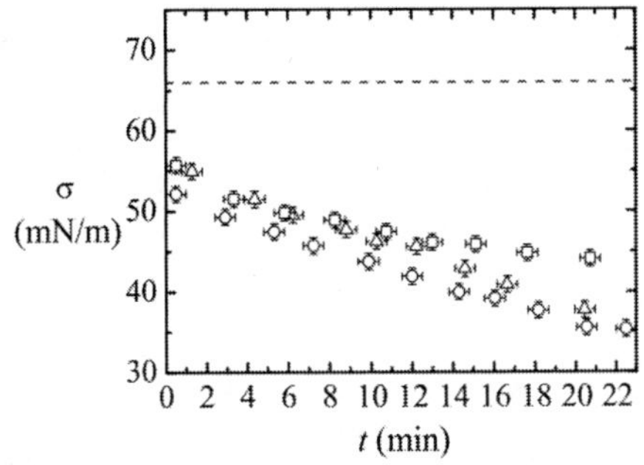

Figure 7. Surface tension as a function of the time t when a liquid bridge of 0.84 mM aqueous SDS solution is surrounded by three environments: uncontrolled ambient (open circles), saturated air (open squares) and argon (open triangles). The dashed line indicates the ideal value of surface tension for a 0.84 mM aqueous SDS solution, around 66 mN/m from the figure 3. The parameters characterizing the experiments were $\Lambda = 1.45$, $B_0 = 0.175 - 0.28$, and $C_\mu = 4.23 - 5.35 \times 10^{-3}$.

surfactant molecules. The discrepancy between the ideal value 66 mN/m (dashed line in the figure) and our results in the first minute may be due to a lack of purificaction of the components of our solution [10]. Since surfactants are by definition highly surface active, chemical impurities in the solution often tend to be surface active and may affect surface measurements. In the uncontrolled ambient, the surface tension suffers an important reduction after 20 minutes in the experiment (from 52 to 35 mN/m). On contrary, in the saturated air ambient the surface tension decreases more slowly, and after 15 minutes a transition to a constant value around 44 mN/m can be obseved. Finally, when argon is used as environment the surface tension changes sharply, as it happens in an uncontrolled ambient. However, the values of surface tension are above of the uncontrolled ambient case.

4 Conclusions

The effects of ambient impurities on the surface tension over time was analyzed experimentally. The change in the surface tesion for water and an aqueous solution of a surfactant over time was relevant in all the cases. Then, the presence of a surfactant on the free surface did not inhibit the action of impurities coming from the ambient. In general, the use of a controlled ambient allowed us to keep the surface tension around values higher than the uncontrolled case, and after around 15 minutes it seemed to reach a constant value.

Thus, in this work it was shown impurities can influence the dynamical behavior of the free surface in flows dominated by the surface tension. A careful control of that influence can be crucial in many applications of fluid mechanics. For this porpuse, an accurate knowledge of the surface tension value over time is crucial to properly analyze the experimental results.

References

[1] M.A. Herrada, J.M. Montanero, J.M. Vega, Phys. Fluids **23**, 082102 (2011)

[2] J.A. Nicolás, J.M. Vega, J. Fluid Mech. **410**, 367 (2000)

[3] D.M. Henderson, J. Fluid Mech. **365**, 89 (1998)

[4] E. Martín, J.M. Vega, J. Fluid Mech. **546**, 203 (2006)

[5] J. Eggers, Rev. Mod. Phys. **69**, 865 (1997)

[6] Q.S. Chen, W.R. Hu, Int. J. Heat Mass Transfer **41**, 825 (1998)

[7] J.C. Burton, J.E. Rutledge, P. Taborek, Phys. Rev. Lett. **92**, 244505 (2004)

[8] E.J. Vega, J.M. Montanero, C. Ferrera, M.A. Herrada, Meas. Sci. Technol. **25**, 075303 1 (2014)

[9] M.G. Cabezas, A. Bateni, J.M. Montanero, A.W. Neumann, Colloids Surf. A **255**, 193 (2005)

[10] A.J. Prosser, E.I. Franses, Colloid Surf. A-Physicochem. Eng. Asp. **178**, 1 (2001)

EPJ Web of Conferences **114**, 02099 (2016)
DOI: 10.1051/epjconf/201611402099

Stereoscopic PIV measurement of boundary layer affected by DBD actuator

Pavel Procházka[1,a], Václav Uruba[1]

[1]*Institute of Thermomechanics, Czech Academy of Sciences, Czech Republic*

Abstract. The effect of ionic wind generated by plasma actuator on developed boundary layer inside a narrow channel was investigated recently. Since the main investigated plane was parallel to the channel axis, the description of flow field was not evaluated credibly. This paper is dealing with cross-section planes downstream the actuator measured via 3D time-resolved PIV. The actuator position is in spanwise or in streamwise orientation so that ionic wind is blown in the same direction as the main flow or in opposite direction or perpendicularly. The interaction between boundary layer and ionic wind is evaluated for three different velocities of main flow and several parameters of plasma actuation (steady and unsteady regime, frequency etc.). Statistical properties of the flow are shown as well as dynamical behaviour of arising longitudinal vortices are discussed via phase-locked measurement and decomposition method.

1 Introduction

This article is devoted to investigation of the influence of plasma DBD (dielectric barrier discharge) actuator on zero-pressure gradient boundary layer (BL) developed inside a narrow channel in order to use it for active flow control of flow separation over GG (Glaubert-Goldschmiedt) body-HUMP [1]. Plasma actuators belong to the active flow control device group and have many advantages in comparison with others as synthetic jets. For example, they have no moving mechanical parts, they can be easily fastened to the surface and they have very short response time [2]. Conversely, they have quite low efficiency of energy conversion. Generally, plasma DBD actuators which are powered by alternating current, are used to control in steady regime when instantaneous ionic wind (wall-jet like flow) is produced [3,4,5]. There is also an unsteady regime [6] which occurs after that rectangular modulation is applied. Then vortex train is presented inside ionic wind, whereas the vortex parameter is given by electrical parameter as modulation frequency and duty cycle. The plasma actuator unsteady regime is more effective and is in the interest of research to investigate the behaviour which is to be used for effective control of boundary layer over GG body.

Previously, the effect of DBD actuator on BL was observed in narrow channel when the ionic wind has been oriented in the parallel direction as the channel flow- either same or opposite direction. Further, the effect of perpendicularly oriented ionic wind on BL was observed also, whereas the measurement plane was in the middle of channel width and perpendicular to the bottom side [7,8]. The biggest effect on BL was observed for lower velocities (5 or 10 m/s) and for steady actuation where velocity gain was up to 10% in normalized velocity profile. The effect of modulation frequency was obvious but only for spanwise configuration and in streamwise configuration was no effect at all which is not in agreement with previously gained results and this effect should be investigated in cross-section plane also. That is why stereoscopic TR PIV measurement was suggested in this paper for cross-section plane perpendicular to the channel flow.

This article will report about flow field behind plasma DBD actuator in base case (no actuation) , in steady and in unsteady case. Time-resolved PIV was suggested as main experimental method and no numerical simulation are not incorporated. The effect of ionic wind oriented in the same/opposite/perpendicular direction to BL is studied and huge longitudinal vortical structure is observed in two distinct planes. There are two approaches how to describe the flow. The first one is statistical approach where mean flow field is computed from long ensemble of measurement. Phase-lock measurement was used to describe several individual phases of vortex motion. Dynamical approach utilized fast acquisition frequency (500 Hz) because fast dynamical processes are under investigation via analysis algorithm (POD).

2 Description of experiment

Plasma DBD actuator is used in simple configuration (two electrodes) in wire configuration. This means that it

[a] Corresponding author: prochap@it.cas.cz

consists of two electrodes and a dielectric layer between them. Upper electrode is powered by high-voltage high-frequency waveform and is actually very thin wire (30 μm) used commonly in HWA. Lower electrode is made from gold glaze, its thickness is only a few μm and the electrode width is 18 mm. The electrode gap is 2 mm. The dielectric layer is made from silica glass with thickness of only 1 mm. Silica glass has good dielectric strength and it enables to use sufficient value of voltage. Lower electrode is grounded and is encapsulated to retain the plasma discharge only at the top side of actuator resulting in better efficiency.

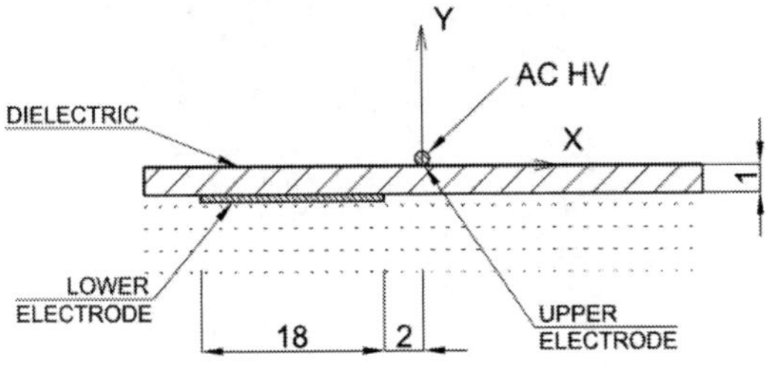

Figure 1. Plasma DBD actuator – wire type

The actuator is used in spanwise (figure 1) orientation as well as in streamwise [8] orientation. The actuator anchor points had to be replaced to the bottom part to not affect developed BL in both orientation mainly in streamwise one. Now, all anchor points and preloading mechanism of wire electrode are placed below. The platform dimension of plasma actuator is 150 x 100 mm to be easily inserted inside a channel. The position of wire electrode in spanwise orientation is in the middle of actuator length (150 mm) which means that wire electrode is oriented across the channel width. On the other hand, the position of electrode in streamwise orientation is not in the middle of the width but is shifted to one side (17 mm). The right-hand side coordinate system is oriented so that x is in the direction of channel flow and y is perpendicular to the bottom side. The power source parameter were introduced in detail in [7,8] and did not undergo any alternatives. It can generate alternating voltage waveform with maximal value around 12 kV. The frequency range is adjustable but it was used fixed value of 16 kHz. The main feature of this power source is a shut-down function which enables to use rectangular modulation of voltage signal. Then another two parameters is needed to take into account – modulation frequency and duty cycle (DC). This regime was called as unstable regime of actuation. When the signal is not modulated (steady regime), the generated ionic wind is observed as wall-jet like flow, parallel to the actuator surface oriented from powered electrode to the grounded one and maximal value gains up to one millimeter above the surface. The unstable regime is characterized by alternating existence of plasma discharge resulting in existence of series of vortices inside ionic wind, so called vortex train. This regime is mainly under investigation from dynamical point of view.

The boundary layer was developed inside a rectangular perspex channel connected to the blow-down wind tunnel [9]. The cross-section dimension was 100 x 250 mm. The total length was 3000 mm. The actuator was placed inside channel so that it created a part of bottom side and the distance from inlet edge to the actuator center was 2275 mm. The spanwise orientation generated ionic wind in the same/opposite orientation in comparison with channel flow. The streamwise orientation generates ionic wind in perpendicular direction.

The boundary layer was developed without any turbulizators under channel flow velocities 5, 10 and 20 m/s. Corresponding Re number are given in table 1. The boundary layer thickness was determined from PIV 2D data in place where actuator is located without actuation and were 54, 51 and 41 mm, respectively.

Table 1. Reynolds number for rectangular channel.

	Re_x	Re_δ
5 $m \cdot s^{-1}$	$750 \cdot 10^3$	$18 \cdot 10^3$
10 $m \cdot s^{-1}$	$1500 \cdot 10^3$	$34 \cdot 10^3$
20 $m \cdot s^{-1}$	$3000 \cdot 10^3$	$55 \cdot 10^3$

This experiment was carried out for three basic regimes. The first one, when actuator is off, is used as base case. The steady regime is characterized by continuous ionic wind. Since unsteady regime enables to tune actuator by changing two parameters, it has become as the most researched case. The modulation frequency range from 10 Hz to 50 Hz and duty cycle was set to 30%.

Stereoscopic TR-PIV was used as a measurement technique in cross-section plane, which means that all three velocity components are specified. Laser New Wave Pegasus Nd:YLF with double head and cylindrical optics was used. The wavelength is 527 nm, maximal operation frequency is 10 kHz and energy of one pulse is 10 mJ (for 1 kHz). We used two identical CCD cameras NanoSense MkIII with resolution of 1280x1024 and maximal sampling frequency 512 Hz. Total memory of camera was 4 GB. Scheinpflug criterion was fulfilled and specific target moving in x-direction was used to calibrate precisely all three dimensions. With respect to fulfill all specific features [10], the precision of this method is within 1-2 %. The software DynamicStudion ver. 3.41 was used.

The acquisition frequency differs in dependency on data treatment. The frequency was 100 Hz for at least 10 second for statistical data treatment and 500 Hz and at least 2 second for dynamical one. Phase-locked measurement was conducted for unstable case. 20 phases was taken during one period for 7.5 second resulting in averaging over 150 period for 10 Hz - case. However, modulation frequency of 20 Hz has required two times higher acquisition frequency resulting in half measurement time. For modulation frequency 50 Hz, it was necessary to capture only 10 phases per period due to limited camera memory.

3 Results

The character of longitudinal vortex structures generated behind plasma actuator oriented in streamwise position can be clearly seen from mean flow field evaluated in two positions distanced 110 mm from each other. Figure 2 shows typical image of vortex oriented in left-hand side of the channel flow. There is a position of wire electrode marked as a black spot. The position of vortex core differs for first and second position. However this difference is always equal for steady and for unsteady (for 10, 20 and 50 Hz) regimes. The vortex core is inclined about 5.5°, 3.4°and 2.6° for channel velocities 5,10 and 20 m/s, respectively. This corresponds with previously gained results, where the effect of 20 m/s was the smallest one.

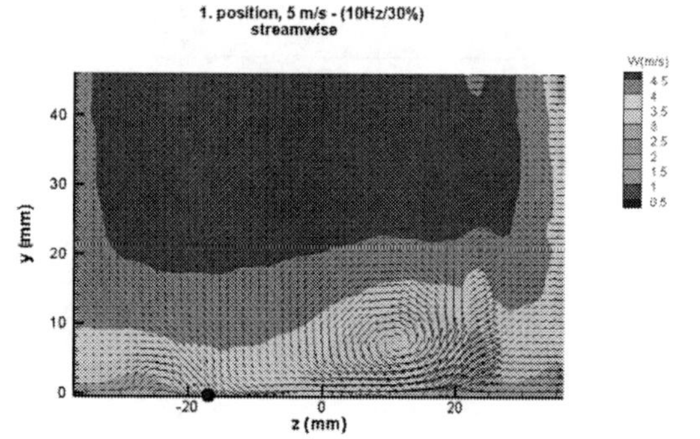

Figure 2. The cross-section of longitudinal vortex

The position of mean vortex core strongly differs in dependency on used channel velocity (see figure 2 and 3). When the steady regime is used, the position of vortex core is rather in the middle of the channel. If the unsteady regime of the actuator is applied, the vortex core is shifted toward the channel wall. The vortex core does not occur so far away from the centre with increasing value of modulation frequency (figure 4). It is obvious that lower value of modulation frequency has biggest effect.

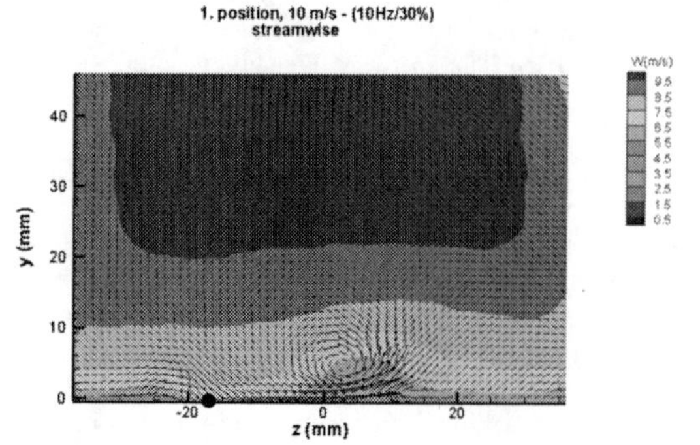

Figure 3. The cross-section of longitudinal vortex

Figures 5 and 6 show typical image of flow field generated by actuator in spanwise position. It can be seen that plasma can add momentum into boundary layer, hence the velocities are accelerated and boundary layer is more stable [7]. The flow in cross-section plane is axisymmetrical. These results perfectly correspond with results gained by 2D PIV in longitudinal plane and it shows that biggest effect on velocities has steady regime.

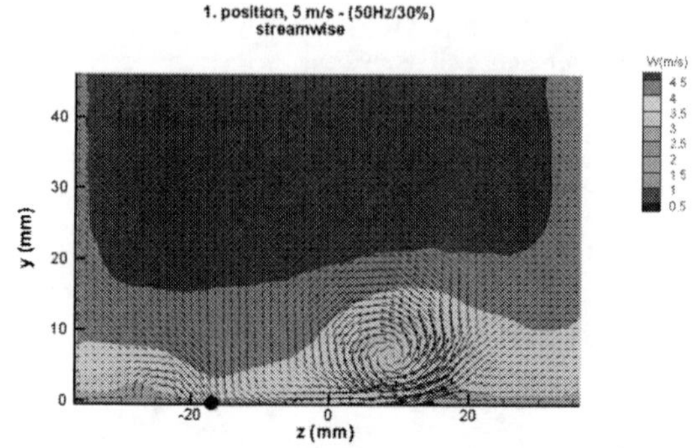

Figure 4. The cross-section of longitudinal vortex

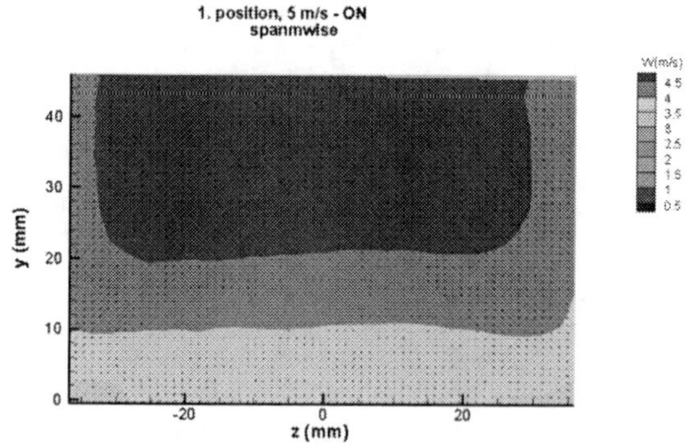

Figure 5. Boundary layer behind actuator, steady regime

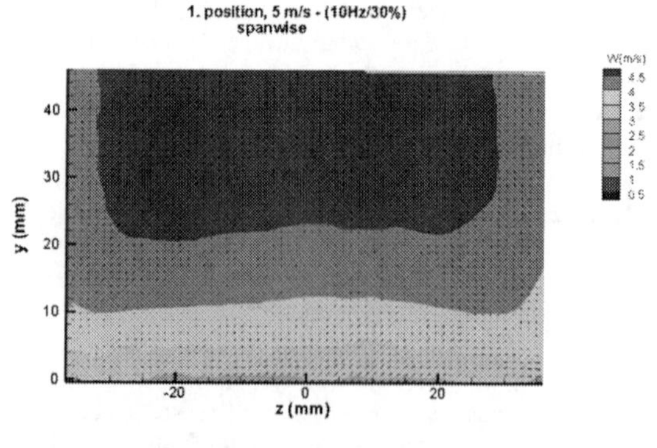

Figure 6. Boundary layer behind actuator, unsteady regime

The phase-locked measurement was done and 20 phases per period was taken for unsteady regime generated by 10 Hz. The position of vortex core (figure 7) was evaluated during one period (0.1 second) and vortex core shifting velocity was determined as approx.

0.6 m/s. The vortex core was propagating all the time toward the wall until the 8. phase, when the motion was stopped, although the plasma discharge continued until the 13. phase. Then the vortex started to fade.

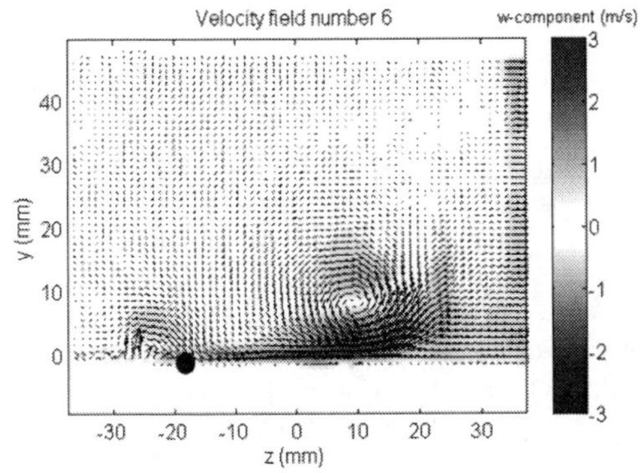

Figure 7. 6. phase of vortex core motion of one period

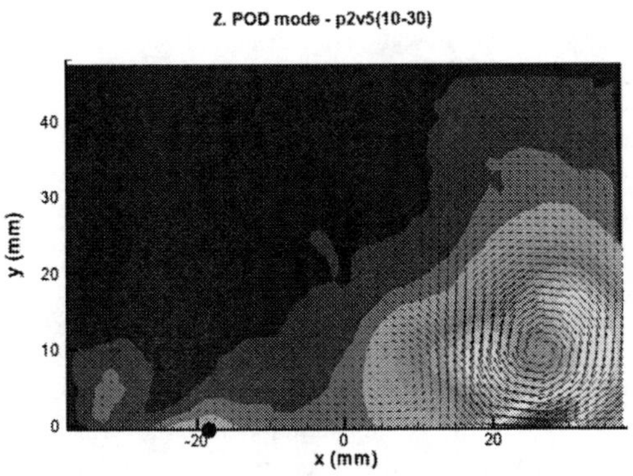

Figure 8. Second POD mode for unsteady actuation

POD analysis was applied on data taken during steady and unsteady regimes to show energy-important modes. The table 2 shows the energy distribution of first five modes. It seems that modes in second position (three right-hand side columns) are slightly more powerful. The modes with biggest energy content were found out for unsteady cases.

Table 2. Five most important modes sorted by kin. energy

p 1 steady	p 1 un-steady	p 1 base case	p 2 steady	p 2 un-steady	p 2 base case
4,48	12,86	4,72	5,44	13,97	5,1
3,73	5,38	3,94	4,85	7,74	4,17
2,64	3,01	2,39	2,83	2,71	2,54
2,03	2,59	2,29	2,32	2,21	2,29
1,86	2,27	1,97	2,18	2,16	2,13

4 Conclusion

The cross-section plane of rectangular channel was investigated experimentally by stereoscopic PIV measurement. The developed boundary layer was influenced by plasma DBD actuator which was mounted in spanwise and in streamwise orientation. The flow field was evaluated from statistically gained data, from phase-locked measured data and from dynamical analysis. The velocity distribution demonstrated better properties (e.g. stability etc.) for spanwise orientation when ionic wind was blowed in the same direction as channel flows. The streamwise orientation is characterized by perpendicularly blowed ionic wind resulting in long inclined vortical structure whose behaviour is strongly dynamical, especially in horizontal sense.

The POD analysis revealed that the most powerful POD modes were found for unsteady actuation as the interaction between boundary layer and the vortex train inside ionic wind generates biggest structures. However, to assign distinct frequency to particular mode, more advanced analysis is required e.g. OPD. To reconstruct the flow field more precisely, several additional cross-section planes should be measured as well as planes parallel to channel bottom side.

Acknowledgement

The authors gratefully acknowledge financial support of the Grant Agency of the Czech Republic, No. GP14-25354P.

References

1. J. W. Naughton, S. A. Viken, D. Greenblatt, AIAA Journal **44**, 1255-1265 (2006)
2. E. Moreau, Journal of Physics D: Applied Physics **40**, 605-636 (2007)
3. Y. Li, X. Zhang, X. Huang, Exp. Fluids **49**, 367-377 (2010)
4. N. Bénard, J. Jolibois, E. Moreau, R. Sosa, G. Artana, G. Touchard, Thin Solid Films **516**, 6660-6667 (2008)
5. G. I. Font, AIAA Journal **44**, 1572-1578 (2006)
6. N. Benard, N. Balcon, G. Touchard, E. Moreau, Exp. Fluids **45**, 333-355 (2008)
7. P. Procházka, V. Uruba, Journal of Physics: Conference Series **1** (2014)
8. P. Procházka, V. Uruba, Conference on Modelling Fluid Flow, Budapest (2015)
9. P. Jonáš, O. Mazur, V. Uruba, IV. Watermanagement conference, Praha (2004)
10. M. Raffel, C. Willert, S. Wereley, J. Kompenhans, Particle image velocimetry, Springer-Verlag (2007)

EPJ Web of Conferences 114, 02100 (2016)
DOI: 10.1051/epjconf/201611402100

Hybrid solar receiver as a source of high-temperature medium for an absorption chiller supply

Estera Przenzak[1,a] and Mariusz Filipowicz[1]

[1]*AGH University of Science and Technology, Faculty of Energy and Fuels, Department of Sustainable Energy Development, Mickiewicza Av. 30, 30-059 Krakow, Poland*

Abstract. This article discusses the problems related with the cold production, i.e. energy efficiency of the process. The idea of solar cooling systems has been presented as the solution of the problem of big electricity demand. The paper discusses the principle of the operation of absorption chillers. Disadvantages and advantages of the solar cooling systems were discussed. The installation for manufacturing high-temperature heat based on solar collectors and concentrator of solar radiation constructed in AGH in Cracow has been presented. This installation is a first stage of projected, complete solar cooling system. The special attention is paid to the dedicated solar high-temperature heat receiver as a most important element of the system. The achieved values of temperature, power and efficiency depending on the working medium flow has been presented and discussed. The intensity of solar radiation during the measurements has been taken into account. Two versions of heat receiver were investigated: non-insulated and insulated with mineral wool. The obtained efficiency of the heat receiver (less than 30%) is not satisfactory but possibility of improvements exist.

1 Introduction

In the developed countries, the clear increase in energy demand for refrigeration and air conditioning has been observed for many years. The reason for this is the general desire of the society to increase the thermal comfort throughout the year. The increasing energy consumption contributed to the formation of the currently ongoing boom to promote alternative energy sources. At the same time the steps are undertaken aimed to reduce the demand for electricity by, among others, modernising the lighting, using modern energy-efficient household and electronic devices or thermo-modernisation of buildings. An additional factor influencing such changes includes the European and global restrictions announced in Directives and legal regulations [1-3].

The production of cool requires a lot of electricity. This applies especially to the summer period, when cool is used not only in catering or medicine, but also in air conditioning. The study [4] presents the analysis of the energy demand in residential buildings with a different shape coefficient. The authors analysed, among others, the energy use for heating and cooling purposes during the year on the example of Canada (figure 1). The works has been performed based on computer simulations using the eQuest software. Calculations were performed for buildings with a different orientation of elevation relative to world sides and with varied proportions of their length. Profiles visible in chart 1 mean:

- G90 °Θ profile: a building oriented with shorter walls to north/south with wall proportions 1:4.2;
- profile C: a building oriented with longer walls to north/south with wall proportions 1.5:1.

In figure 1 can be seen that regardless of the geometry of the building the greater demand for energy needed to maintain thermal comfort occurs in the winter period.

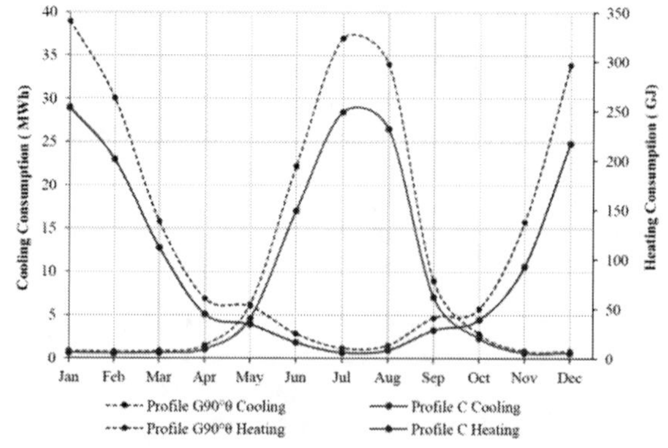

Figure 1. Comparison of energy consumption between two building profiles with different aspect ratio and orientation for Toronto simulations [4].

However, for the purpose of supplying the central heating installation the solid or gas (less often oil fuel boilers are used most commonly. Thus, the demand for

[a] Corresponding author: esterap@agh.edu.pl

electricity in the winter period is not as high. While in order to improve the quality of indoor air in the summer period most often the compressor air conditioners are used, which are powered by electricity. Hence the emerging problem of high electricity consumption during hot weather. According to [5] the demand for electricity needed to maintain the thermal comfort conditions in the summer period is much higher than in the winter period.

In the European Union, 55.1% of electricity comes from coal thermal power plants, 28% from nuclear power plants, while less than 17% from renewable energy sources of various kinds [6]. The biggest share in electricity production, not only in Europe but also around the world, comes from carbon power plants. They are characterised by high inertia, therefore energy safety during summer is at risk. The demand for electricity is not evenly distributed in time. Difficult to predict weather conditions are sometimes rapid within days, and sometimes even hours. This means that during the period of strong heat, users suddenly overload the power grid. Due to the inertia of power plants, energy suppliers cannot always ensure the continuous supply of electricity. Dangerous situations of overloads an thus shut-offs of networks because of the too high consumption of energy needed to power air conditioning devices happened, among others, in China [7] or the United States [8]. Also the effects associated with strong heat were observed in Poland in August 2015, when the amount of generated power dropped in the power system due to water shortages (and its too high temperature) for cooling power units. In addition, high electricity consumption (derived from conventional sources) for the purposes of producing cold causes the increased emission of carbon dioxide to the atmosphere [9] which in the long term has a negative impact on the environment.

Comparing the blue curve (online version) in figure 1 with the graph of the average global intensity of solar radiation within a year (figure 2) it can be seen that these distributions are similar. Data presented in graph 2 were created in order to assess different methods of modelling of weather conditions presented in [10].

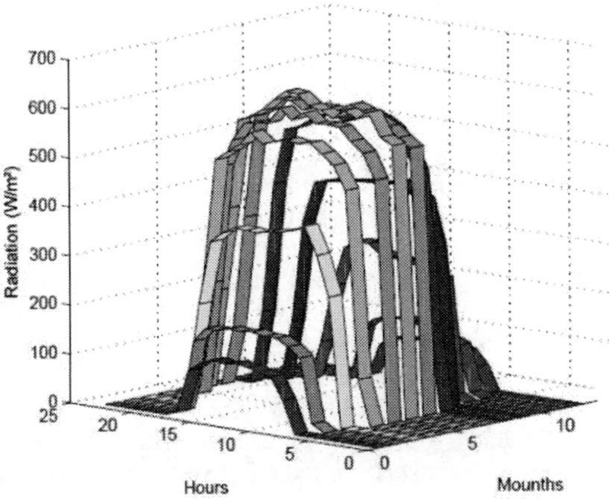

Figure 2. Measured monthly average daily global radiation [10].

Measurements of solar radiation were performed in northern France (Méteo France ground station of the Troyes-Barbery), i.e. in a place characterised by a relatively high cloudiness throughout the year. The presented coherence is the basis for research on modern cooling systems based on solar energy. It is possible by applying the sorption refrigerator, i.e., a heat-powered device.

1.1 Fundamentals of sorption refrigeration

In sorption aggregates the working medium is the two-component medium, i.e., the solution of sorbent and substances absorbed by it. In sorption chillers the absorber is in the solid form, and the adsorption process takes place at the phase boundary. In case of absorption chillers the liquid absorbent is used, so the sorption process takes place in the whole volume of the medium. Most often two types of absorption chillers are used: bromine-lithium ($LiBr$-H_2O) and ammonia (H_2O-NH_3). In the first ones, water plays the role of the refrigerant, so they are used in systems, where chilled water has the temperature not smaller than approx. 5°C. While the ammonia absorption chillers can be used to obtain lower temperatures (even up to -60°C), however they require higher investment outlay. Compared to conventional compressor devices, absorption chillers are characterised by lower noise level, lower service costs (a small number of moving parts) and a much longer period of exploitation, because even to 25-20 years (for the compressor refrigerator it is approx. 15 years) [11].

As in the compressor refrigerators, also in the absorption devices in the evaporator the absorption of heat takes place by the medium with low pressure, while in the condenser at high pressure this heat is released. The change of pressure takes place by using a throttling valve and the absorption-desorption system, replacing the compressor (figure 3).

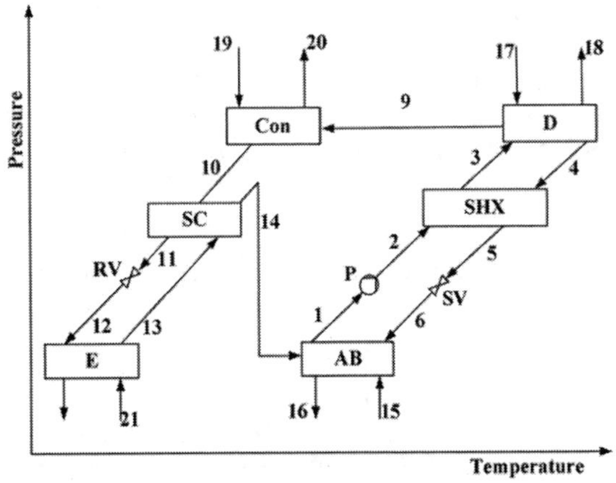

Figure 3. An exemplary scheme of the absorption refrigerator [12]: AB-absorber, P-solution pump, SHX-solution heat exchanger, SV-solution expansion valve, D-desorber, Con-condenser, SC-refrigerant subcooler, RV- refrigerant expansion valve, E-evaporator.

The refrigerant of low pressure and in the liquid form, flowing through the evaporator, takes the heat from the ice water led to it, thus reducing its temperature. Ice water can be then used to distribute cold in the rooms, e.g., using the fan coil unit. Vapour of the refrigerant created in the absorber flow to the system with the absorber, where in contact with the absorbing medium they are subject to absorption. In this process the dissolution heat is released, which must be discharged from the system using the external system with the cooling water. The rich solution created in the absorber is pumped using a circulation pump to the desorber. In this place the high-temperature heat is applied, which is usually generated using gas or electric heater. Increasingly, it is also a waste heat created as a result of other technological processes or heat produced using solar installations. In the desorber, using the high-temperature heat led to the system, the cooling medium is evaporated from the absorbent. The absorbent (weak solution) flows back to the absorber. In contrast, the refrigerant vapour goes to the condenser, where by means of cooling water from the external system flowing through the exchanger, they are condensed. The liquid working medium created this way is directed through the throttling valve in order to achieve low pressure, required to conduct the evaporation process in the evaporator.

1.2 Review of solar sorption cooling systems

Studies on such systems are carried out around the world. For example, in Poland at the Department of Engineering and Environmental Protection of the Czestochowa Technical University studies on the possibility to create conditions of thermal comfort in summer in the room with the capacity of $30 m^3$ were conducted. The research work was carried out based on adsorption refrigerator powered by heat produced by the installation of five flat solar collectors. It has been proven that it is possible to use the standard installation with solar collectors for the purposes of cold production. However, we should bear in mind that the thermal effectiveness coefficient TEC is in this case quite low and is approx. 0.6-0.7 [13, 14]. This is due to the fact that solar collectors only produce heat with low temperature (approx. 60 °C). Compared to absorption aggregates, adsorption ones are characterised by a much lower thermal effectiveness coefficient. Nevertheless, studies on solar power supplies of this type of refrigeration are conducted around the world on a large scale. The overview of modern solutions and description of future trends in solar adsorption installations was described, among others in [15, 16].

At the same time in the energy industry we observe the dynamic development of solar absorption systems. In Oberhausen in Germany in summer months the system with the absorption cooling aggregate with the power of 34 kW has been tested, powered by vacuum solar collectors with the surface of $108 m^2$. To facilitate the operation of the system, it was equipped with the cooling tower with the capacity of 134 kW. As a result, over the 5-year exploitation of the installation 25% cold was produced needed to meet the demand for rooms with a total area of $270 m^2$. At this time, the TEC coefficient values of the used cooling device were quite low and amounted to 0.37-0.81 [17].

Similar studies were carried out in Madrid, where the TEC coefficient of the bromine-lithium absorption chiller with the power of 35 kW was 0.23-0.42. This system was also powered only by the installation with solar (flat) collectors. The cooling aggregate in this case achieved the maximum of only 7.5 kW power (21% nominal) [18].

For sorption devices, and particularly absorption refrigerators, to work with high efficiency, it is necessary to provide desorption heat to the system with temperature over 85 °C. This is shown in the graph (figure 4) of the dependency of cooling efficiency of the absorption refrigerator on the nominal power of 17.6 kW on temperature of the medium flowing into the desorber. It is an example of energy characteristics included in the catalogue card of the YAZAKI aggregate [19].

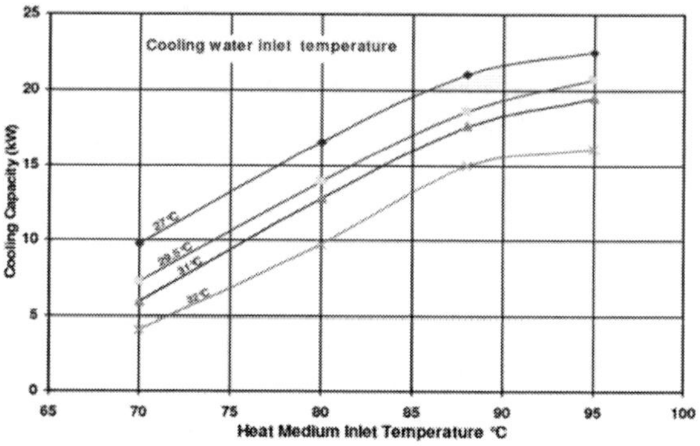

Figure 4. Cooling capacity of the absorption device depending on temperature of the heating medium [19].

From the graph shown in figure 4 it can be seen that for the absorption systems to work efficiently it is necessary to provide the high-temperature heat to the system. Therefore, in solar systems used for generating cold, the systems concentrating the solar radiation have a great potential. The concentrated light can in fact be used to increase the temperature of the working medium pre-heated in collectors, or be used as the only source of heat in this system.

Studies on focus systems used to produce cold were conducted, among others, in Seville in Spain, where the system focusing light in the using linear Fresnel lens with the total surface of $352 m^2$ has been constructed. In the discussed studies the two-stage bromine-lithium cooler was used with the power of 174 kW as well as the system for heating the working medium in the gas burners. The conducted measurements showed that in the period from May to October, i.e., when there is the largest demand for cold, the cooler works with the TEC effectiveness at the level of 1.1-1.25. Furthermore, the SHF factors (Solar Heat Fraction – the ratio of thermal energy to the one obtained from solar installation to thermal energy powering the desorber) and SCR (Solar Cooling Ratio – the efficiency of the whole system) are at a very high level, respectively 75% and 44%. In the discussion of research works also the problems, which concern the focus systems, were indicated. Because it has been studied that on cloudy days (predominance of scattered radiation) the use of gas burners in the discussed system is necessary and constitutes the production of approx. 60% energy needed for the appropriate increase of temperature of the working medium. Also the issue of the impact of the focus installation pollution on the effectiveness of the whole system has been raised (decrease of efficiency even by 50%) [20].

While in Tunisia at the Centre of Energy Research and Technologies in Bordj-Cedria the system of trough solar radiation concentrators has been constructed and tested, which powers the absorption cooler (BrLi) with the nominal power of 16 kW, which in turn was used as powering for several fan coils installed in individual rooms of the building. In this case the device used to heat the working medium was also used (electrical heater). In order to increase the SHF factor the buffer tank was used, which was applied for storing heat to be used at night. The solar degree of coverage thanks to this operation increased from the value of 0.54 up to 0.77. In the discussed system, the cooler reached the maximum power at the level of 75% of its nominal value, i.e., the maximum of 12kW has been achieved. On the basis of a number of conducted measurements and calculations it has been shown that the TEC factor of the studied absorption aggregate is in the range of 0.8-0.91 [21].

Analysing the discussed solutions it can be seen that the coefficient of thermal effectiveness in the case of applying the system focusing the solar radiation is higher than in systems powered by solar collectors. The review of systems using the solar radiation energy for powering sorption systems has been performed, among others, in [11, 22-24].

2 Solar system for high-temperature heat generation

At the Department of Sustainable Energy Development in the Faculty of Energy and Fuels at AGH in Cracow there are ongoing works on the construction of the solar cooling system analogous to the described ones. The target system consists of the hybrid connection of solar collectors with the parabolic concentrator of solar radiation. The system will also use the absorption aggregate, which has not yet been mounted.

Figure 5 presents the diagram of the proposed solution. According to the idea of the described system, the working medium (thermal oil) is pumped to solar collectors (1) by using the circulation pump (7). Then, the pre-heated medium goes to the heat exchanger (3) located near the focal point of the concentrator, where its temperature increases. The resulting heat is used to power the absorption cooler (4), which in turn supplies chilled water to the fan coil units. The remaining amount of heat is accumulated in the buffer tank (6) and is used for heating the utility water. The expansion vessel (8) (just like in classic installations) protects the installation against the excessive pressure increase. Given that this is the high-temperature installation, the expansion vessel was additionally equipped with the cooling vessel. On colder days (spring, autumn), when air conditioning devices are not used, the heated medium is directed by the three-way valves directly to the coil of the buffer tank. When the accumulation tank has the already sufficient amount of the heat, the working medium after leaving the absorption cooler, is directed directly to solar collectors (it omits the tank). The system was equipped with the fan cooler (5), which serves as the protection function in case of achieving too high temperatures at the outlet of the sorption aggregate.

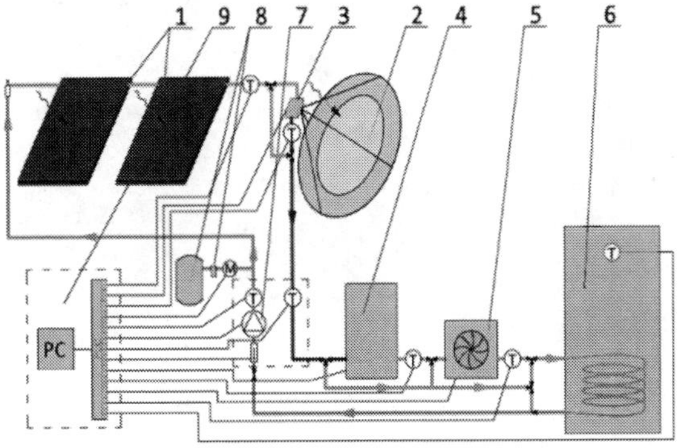

Figure 5. The diagram of solar collector system with the solar radiation concentrator for absorption cooling.

Currently, the system consists of two flat solar collectors mounted on the roof connected with the system of solar radiation concentrator (figure 6a). The concentrator is mounted on the tracker using the differential sensor built of six LED diodes. This type of tracker allows the quite accurate tracking of the sun location, what in case of focus systems has a significant meaning. The concentrator is built of the bowl with a diameter of 1.8 m lined with the reflective foil and the mirror deposited on it with a diameter of 1.2 m. On the roof there is also a fan cooler. Due to the moving elements of the installation, piping is made of corrugated steel pipes in the rubber coating.

While in the laboratory other installation elements have been mounted, which are visible in figure 6b. Only the absorption aggregate has not yet been installed.

Figure 6. View of the hybrid installation of solar collectors with a concentrator a) on the roof, b) in the laboratory.

The system was equipped with the system of control and measurement automation based on the PLC controller (figure 7). The following measurement elements were used:

- oil temperature sensors in different points of the system (also outside air temperature) (°C),
- pressure sensor (bar),
- oval-wheel flow meter (l / min),
- solar radiation intensity sensor (W / m^2).

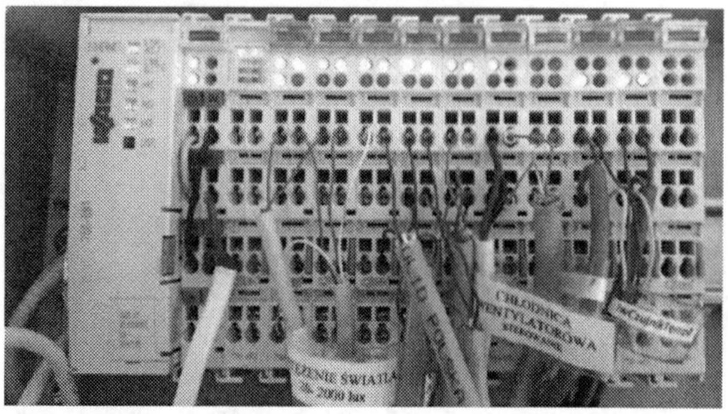

Figure 7. PLC controller of the WAGO company with connected measurement system elements.

Using the industrial programmable controller it is also possible to control the performance of the circulation pump and fan radiator. Preview of data, their recording and control of individual parameters takes place through the visualisation presented in figure 8 performed in the CoDeSys software.

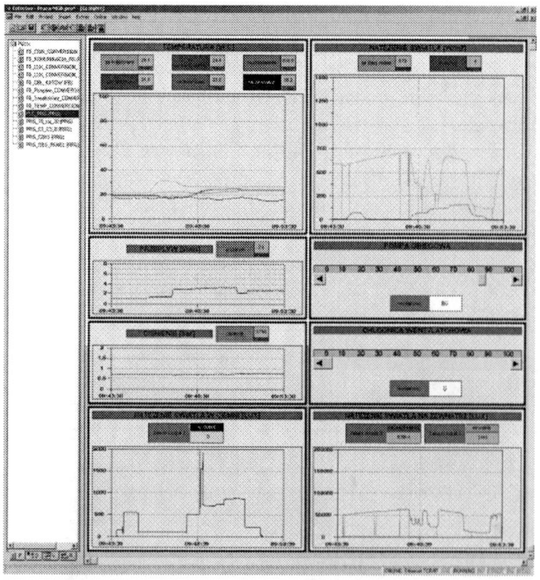

Figure 8. Control and data recording panel.

One of the most important elements of the discussed installation is the receiver of high-temperature heat absorbing the concentrated solar radiation. In the first version, the receiver is made using two interconnected aluminium blocks with the channel milled inside. The diameter of the receiver is 18 cm (figure 9). The surface, on which the concentrated solar radiation falls was dulled and blackened. Simultaneously to the experimental studies the simulation studies are performed. They serve as an aid in optimising the receiver system, namely: radiation absorption, its conversion and high-temperature heat reception. Effectiveness of the listed processes depends, among others, on the external shape and internal geometry of the receiver, type of absorption surface, receiver location in the concentrator geometrical system, as well as the proper flow values depending on the current parameters of the working medium.

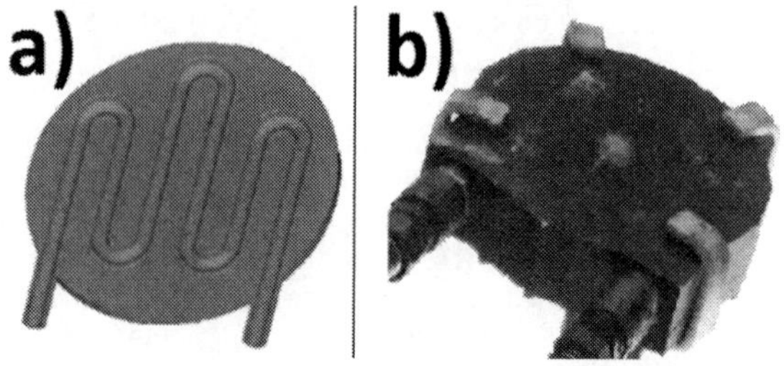

Figure 9. High-temperature heat receiver a) cross section, b) view of the receiver in the mounting holder.

3 Solar system for high-temperature heat generation

In summer 2015 studies on the possibility to produce heat using only the solar radiation concentrator were performed. During measurements, the working medium flowed also through the system with two solar collectors, but for the purposes of studies they have been covered. The effectiveness of conversion and heat absorption from the concentrator were tested using the receiver presented in figure 9. Two versions of the receiver were studied: without insulation and with insulation (coating of mineral wool). Insulation of 2cm thickness was fastened to the back and lateral surfaces of the receiver. In graphs figure 10 and 11 the change of temperature of the working medium has been presented before and behind the concentrator and with the buffer tank with utility water. The dotted line marks the change in performance of the circulation pump beginning from 100% to 40% in case of the insulated receiver. Below the graphs the solar radiation intensity during the study is also presented.

It can be seen that small decreases in solar radiation intensity are caused by the temporary cloudiness, such as, e.g., figure 10: 5 and 26 minute of measurement or figure 11 at: 1 h 17 min, 1 h 36 min, 2 h 08 min, do not affect the temperature change of the working medium. This results from the inertia of the heat absorption system, which is mainly caused by the volume of oil in the receiver and its heat capacity.

Despite shadowing of solar collectors to some extent they contribute to the heating of medium in the installation. Therefore, the oil temperature before the solar radiation concentrator is higher than its temperature behind the tank. Only when the pump capacity is at the level of 50-55% in the measurement using the insulation on the receiver the lower temperature was noted on the way between the tank and the concentrator. This is caused by the heat losses in the pipes.

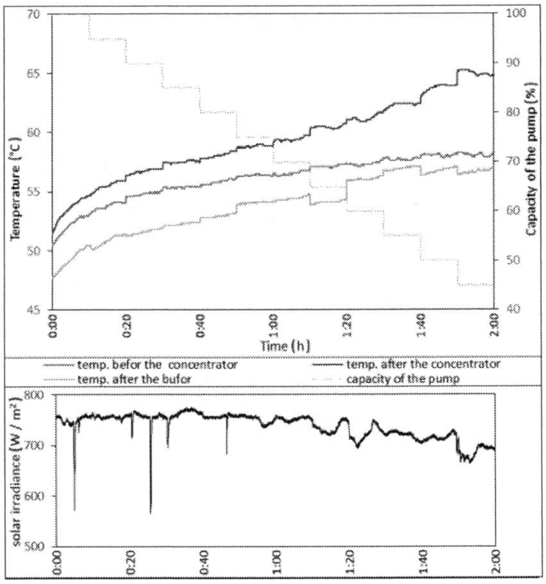

Figure 10. Medium temperature variation for the non-insulated receiver.

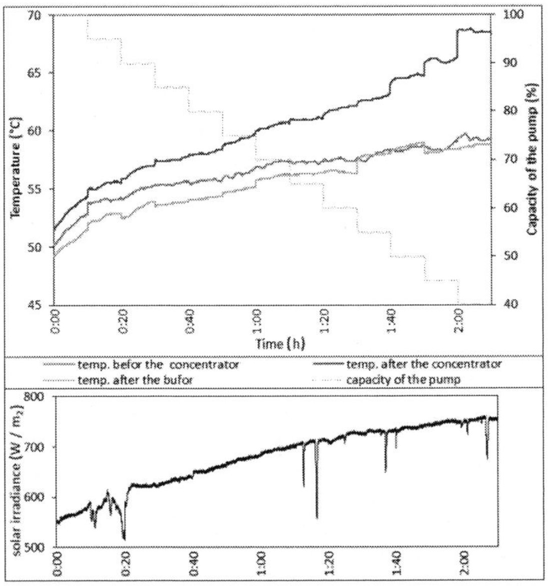

Figure 11. Medium temperature variation for the insulated receiver.

In graph in figure 10 we can see that the decrease of the flow is due to the change of pump capacity from 70% to 65% (time moment 1:10 h) caused a slightly more effective heat exchange in the buffer tank (stated based on medium temperature observation in the buffer tank). It is likely that this has caused high enough temperature increase in the tank that the temperature difference between the working medium and the utility water dropped. Therefore, in both cases (non-insulated and insulated receiver) the adjustment of pump capacity under 60% caused the temperature increase behind the tank – what means the worse heat exchange in the buffer in relation to other capacities. The presented graphs also show that significant temperature changes are mostly observed during the work of a pump with the capacity of 40%-80%. The largest increase of oil temperature behind the receiver was observed with the lowest flows. This particularly applies to the insulated tank, where the temperature increases are very clear with the pump capacity from 55% to 40%. Due to the risk of overheating, the pump cannot work with the yield lower than minimal, which is defined as a certain part of nominal performance in relation to liquid temperature – in the discussed case this is 1,5 L / min in temperature range to 80°C. Therefore, measurements were limited to the minimal efficiency of 40% which gives approx. 1,8 L / min.

Due to the fact that the measurements were started from the maximum flow rate, which was then reduced during studies, graphs in figure 12-14 present results in reference to the decreasing scale of its value. In order to compare the operation of the receiver working without insulation and with insulation, a graph of the average temperature growth depending on the average flow was made (figure 12). The arithmetic mean values were calculated from 10-minute time intervals during which the fixed value of the circular pump was set. In figure 12 there are also the average values of solar radiation intensity prevailing at the given time of measurement.

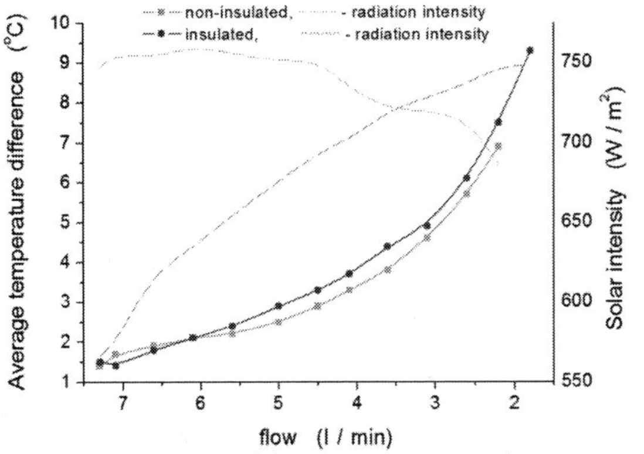

Figure 12. Comparison of the average temperature increase from the flow in case of the receiver without and with thermal insulation.

It can be seen that the greater increase of temperature was achieved in the case of the insulated receiver. However, it may be noted that the study also had the increase of the solar radiation intensity in contrast to the study marked with green curves (online version) - no insulation. Certainly different light intensity has also had a significant impact on the operation of the installation. This can be especially seen at the flow rate of 7.1 L / min, where the lower temperature increase is noted for the red curve (online version), while the radiation intensity is approx. 180 W / m^2 lower than in case of the study marked with the green curve (online version). Only when this difference reaches approx. 100 W / m^2 (with the flow rate of 5.6 L / min) the use of thermal insulation causes the achievement of higher temperatures. An important information is that despite the drop of the solar radiation intensity, we can observe a significant increase of temperature caused by the flow regulation. What is also promising is the fact that in the first version of the heat reception system presented in this article, it was possible to achieve the temperature increase already at the level of over 9 °C.

Knowing the oil temperature before and behind the receiver and the basic properties of oil the power of the receiver was calculated for different flows of the working medium. Results are presented in graph in figure 13.

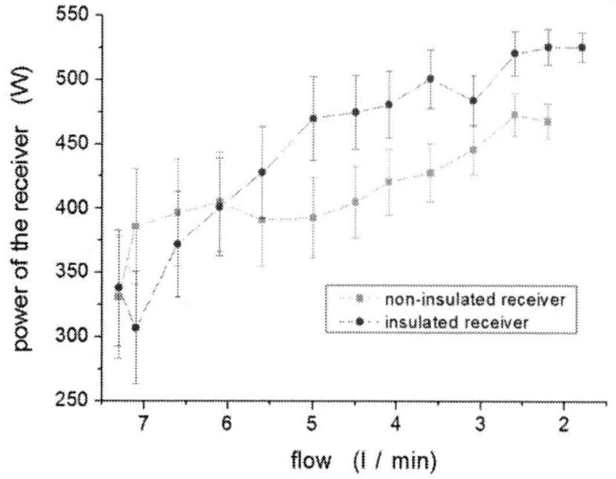

Figure 13. The power of the receiver for varied flow.

In the graph in figure 13 it can be seen that the higher power was obtained in the case of the insulated receiver. In general, the power of the receiver increases with the decrease of the flow and achieves the maximum with the lowest flow, i.e., 1.8 L / min. For comparison of studies on the non-insulated and insulated receiver the efficiency graph was made (figure 14).

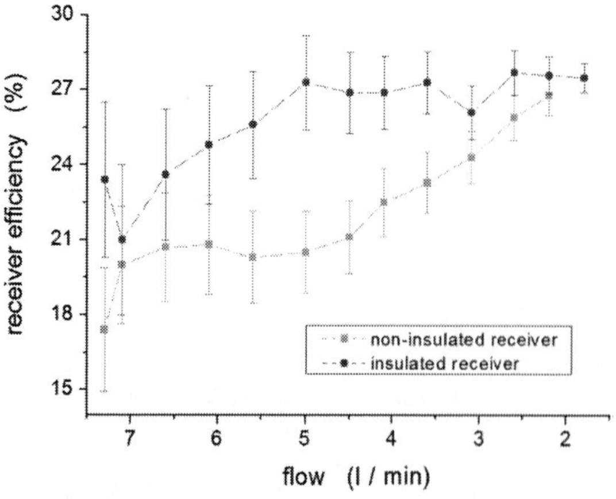

Figure 14. The efficiency of heat receiver in varied flow.

It has been assumed in the calculations that the concentrator system has the efficiency of 100%, i.e. the enhancement of light on the exchanger equals 85, what was calculated from the ratio of the concentrator surface to the receiver surface. It was assumed that the whole solar radiation falling on the concentrator bowl surface is reflected, and all rays reach the absorber surface. In reality, taking into account the efficiency of the concentrating system will contribute to the increase of the calculated values of the receiver efficiency.

In the graph in figure 14 it can be seen that the thermal insulation of mineral wool contributed to the increase of efficiency of the receiver system. Effectiveness in both cases is higher for lower flows, what is caused by achieving a greater difference of temperatures before and behind the receiver (figure 12). At the flow of 5 L / min it is visible that the use of insulation significantly increases the efficiency of heat reception. The studies did not include the impact of wind during the measurement, what also has the effect on the presented results. However, it can be noticed that the best results are obtained for the lowest flows.

Generally the efficiency of non-insulated receiver was estimated at the level of 22.9 ±0.5%, however insulated receiver reached the value of 26.6 ±0.3%. So even small thickness of insulation material led to important improvement of its efficiency.

4 Conclusions

The study results presented in this article are the basis for the continuation of research into the possibility of obtaining the high-temperature heat from the solar system for producing cold. An important element of the discussed system is the receiver of the high-temperature heat. Initially, the efficiency of the system was increased by using the insulation of mineral wool, what in average raised the effectiveness more than 15%.

To raise the efficiency of the absorption system, conversion and heat reception it is necessary to introduce next modifications. Currently, the works are ongoing about, among others, the application of the absorption surface with the lowest possible emissivity. Also the external shape of the receiver is very important. The absorption surface should be constructed in such a way so that the concentrated rays fall on it at the straight angle, what will contribute to increasing the absorption efficiency. Extremely important here is also the optimal placement of the receiver in the concentrator system. In studies on increasing the radiation absorption efficiency the simulation tools are useful, which use the "ray tracing" technique [25]. Also the efficient reception of heat is important, thus the internal geometry of the receiver. The simulation of the medium flow and different work parameters of the system are conducted at the same time to experimental works using the ANSYS software based on CFD methods.

The examples of the solar systems used in the World for producing cold quoted in the article and the presented results confirm that using only the solar power without the help of the additional heat source can be an effective way for producing cold in the summer period. The biggest advantage of this type of systems is the fact that in case of their application the coherence occurs, that the largest heat production occurs in the summer period, i.e., when the greatest demand for cold takes place.

Acknowledgments

This work was financially supported by Dean Grant of Faculty of Energy and Fuels at the AGH University of Science and Technology and in was also completed as a part of the statutory activities of the Department of Sustainable Energy Development "Studies concerning the conditions of sustainable energy development".

References

1. Directive 2009/28/EC
2. Directive 2010/31/EU
3. Directive 2008/50/EC
4. P. McKeen, A. S. Fung, Buildings, 4, 336-354 (2014)
5. A. Q. Jakhrani, S. R. Samo, A. R. H. Rigit, S. A. Kamboh, WASJ 22 (9), 1334-1343 (2013)
6. Europe's Energy Portal: https://www.energy.eu/
7. P. Shi, C. Jaeger, Q. Ye, Integrated Risk Governance, IHDP, 218 (2013)
8. B. Gradoń, S. Gil, W. Bialik, Polska Energetyka Słoneczna, 1-4, 55-59 (2012)
9. R. E. Critoph, Polska Energetyka Słoneczna, 1-4, 14-19 (2009)
10. A.B. Jemaa, S. Rafa, N. Essounbouli, A. Hamzaoui, F. Hnaien, F. Yalaoui, Energy Procedia, **42**, 406-415 (2013)
11. J. Stefaniak, Rocznik Ochrona Środowiska, **15**, 1216-1227 (2013)
12. B.H. Gebreslassie, G. Guillen-Gosalbez, L. Jimenez, D. Boer, Energy, **35**, 3849-3862 (2010)
13. M. Turski, R. Sekret, Architecture Civil Engineering Environment, **2**, 109-115 (2012)
14. M. Turski, R. Sekret, Instal, **9**, 47-50 (2013)
15. M.S. Fernandes, G.J.V.N. Brites, J.J. Costa, A.R. Gaspar, V.A.F. Costa, Renewable and Sustainable Energy Reviews, **39**, 102-123 (2014)
16. E.E. Anyanwu, Energy Conversion and Management, **44**, 2, 301-312 (2003)
17. A.H.H. Ali, P. Noeres, P. Pollerberg, Solar Energy, **82**, 1021-1030 (2008)
18. A. Syed, M. Izquierdo, P. Rodriguez, G. Maidment, J. Missenden, A. Lecuona, R. Tozer, International Journal of Refrigeration, **6/28**, 859-871 (2005)
19. Technical specification of WFC-SC5 YAZAKI absorption chiller
20. P. Bermejo, F. J. Pino, F. Rosa, Solar Energy, **84**, 1503-1512 (2010)
21. M. Balghouthi, M.H. Chahbani, A. Guizani, Applied Energy, **98**, 138-148 (2012)
22. I. Sarbu, C. Sebarchievici, Energy Conversion and Management, **105**, 403-422 (2015)
23. K.R. Ullah, R. Saidur, H.W. Ping, R.K. Akikur, N.H. Shuvo, Renewable and Sustainable Energy Reviews, **24**, 499-513 (2013)
24. I. Sarbu, C. Sebarchievici, Energy and Buildings, **67**, 286-297
25. E. Bożek, M. Filipowicz, JCEEA, **26**, 2, 19-28 (2015)

EPJ Web of Conferences **114**, 02101 (2016)
DOI: 10.1051/epjconf/201611402101

Study of blades inclination influence of gate impeller with a non-Newtonian fluid of Bingham

Lakhdar Rahmani[1], O. Seghier[1], B. Draoui[1] and E. Benachour[1]

[1]*Arids zones energetic laboratory (ENERGARID), TAHRI Mohamed University of Béchar, BP 417, 08000 Béchar, Algeria.*

Abstract. A large number of chemical operations, biochemical or petrochemical industry is very depending on the rheological fluids nature. In this work, we study the case of highly viscous of viscoplastic fluids in a classical system of agitation: a cylindrical tank with plate bottom without obstacles agitated by gate impeller agitator. We are interested to the laminar, incompressible and isothermal flows. We devote to a numerical approach carried out using an industrial code CFD Fluent 6.3.26 based on the method of finites volumes discretization of Navier - Stokes equations formulated in variables (U.V.P). The threshold of flow related to the viscoplastic behavior is modeled by a theoretical law of Bingham. The results obtained are used to compare between the five configurations suggested of power consumption. We study the influence of inertia by the variation of Reynolds number.

1 Introduction

Usually many industrial processes often face difficulties related to the implementation of non-Newtonian fluids such as polymer solutions, detergents, petroleum products, biological fluids, food, drilling fluids and emulsions to be transported, stirred and mixed. If agitation is intended to promote a physical process, such as homogenization or enhancement of heat transfer, the prediction of the power necessary for its implementation will be the primary concern. We have studied the hydrodynamics developed in an agitation in the presence of viscoplastic fluids. Hydrodynamics is a key control and process optimization. His knowledge and control are crucial to improve efficiency. To do this, we seek to determine the influence of rheological properties of the fluid on these flows.

The objective of this work is the study of mechanical agitation of highly viscous fluid by an agitator-type proximity barrier using five configurations depending on the inclination angles of blades with a non baffled cylindrical tank with flat bottom such that the geometry is close to an industrial setting. The choice of this geometry has the advantage of generating a flow that can, in first approximation considered as a two-dimensional. The flow regimes studied are laminar and implemented fluids are viscoplastic type of Bingham. The difficulty of modeling problems lies in the viscoplastic behavior of the fluid. The results enable a comparative study between five blade configurations chosen especially from the point of view of energy consumption (number of power), this allows us to make a choice of an optimal configuration (i.e.) with low energy consumption.

2 Numerical models

The mixing systems are presented in Figure 1. The tank is a flat-bottomed vessel (inside diameter: D) equipped with a two blades impellers (diameter d). H is the fluid height. Values for the impellers width L, agitator-to-wall clearance w and impairs thickness e, Alpha is the inclination of the agitator blades are reported in Table 1.

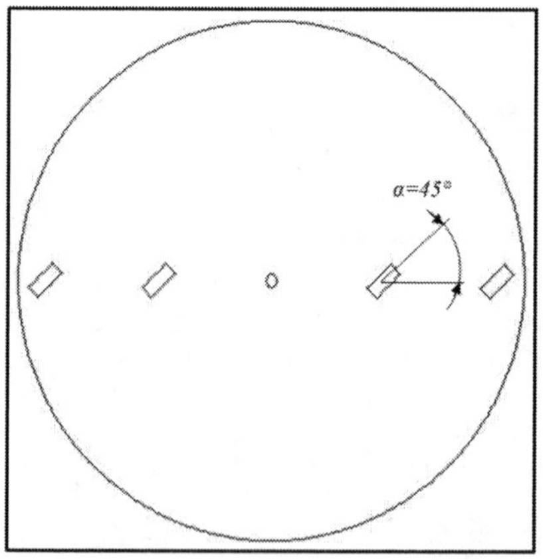

Figure 1. The mixing system model.

[a] Corresponding author: rahmani_l1@yahoo.fr

Table 1. Geometrical characteristics of agitator.

D	1
d	0.96
da	0.023
E	0.027
l	0.067
w	0.02
α	0°,15°,30°, 45°,60°

3 Comportment law of fluid

To study the yield stress fluids which constitute the purpose of this paper, mixing systems are dealt with a numerical simulation using Viscoplastic fluids modeled by Bingham law:

$$\overline{\overline{D}} = 0 \text{ for } \|\tau\| < \tau_0 \tag{1}$$

$$\overline{\overline{\tau}} = \left(\frac{\tau_0}{\dot{\gamma}} + \eta_\infty \right) \overline{\overline{D}} \text{ for} \|\overline{\overline{\tau}}\| > \tau_0 \tag{2}$$

Where $\overline{\overline{D}}$ and $\overline{\overline{\tau}}$ are, respectively, the rate of strain tensor and the stress tensor. τ_0 is the yield stress.

The shear rate is defined as $\dot{\gamma} = \sqrt{2\, tr\, \overline{\overline{D}}^2}$ where tr stands for the trace.

$\overline{\overline{D}}$ *And* $\overline{\overline{\tau}}$ *are defined as:* $\overline{\overline{D}} = 1/2\left(\nabla V + \nabla V^T\right)$

$$\left|\overline{\overline{\tau}}\right| = \sqrt{\frac{1}{2} \overline{\overline{\tau}} : \overline{\overline{\tau}}^T}$$

According to equations (1) and (2), the flow domain for Bingham fluid is characterized by two distinct regions. In the regions where $\left|\overline{\overline{\tau}}\right| \prec \tau_0$ the material behaves as a solid (rigid zone), and in the regions where $\left|\overline{\overline{\tau}}\right| \succ \tau_0$ the material flow with an apparent viscosity η_{ap}.

$$\eta_{ap} = \eta_\infty + \frac{\tau_0}{\dot{\gamma}} \tag{3}$$

The major difficulty with the constitutive equation (1) and (2) when used for numerical simulation is the discontinuity associated with infinite value of the viscosity when $|\tau|$ approaches τ_0. This is reached closed to the yield surface delimiting plug regions. Some numerical works have been published on methods adapted to the initial set of equations [(1), (2)] completed by continuity equation and momentum equation in stress formulation. But most of the published works on this topic use modified versions which consist in replacing equations. (1) + (2) By a unique and continuous equation written with a variable viscosity η_{ap} [9]:

$$\tau = 2\eta_{ap}\overline{\overline{D}} \tag{4}$$

4 Model equations and computational procedure

The equations describing fluid flow are derived from the conservation of mass and momentum which are also known as Navier–Stokes equations for fluids of constant physical properties:

$$\nabla \cdot V = 0 \tag{5}$$

$$\rho \frac{\partial V}{\partial t} + \rho(V.\nabla) - \eta_{ap}\Delta V - 2\overline{\overline{D}} \cdot \nabla \eta_{ap} + \nabla P = 0 \tag{6}$$

4.1 Boundary conditions

In the CFD code used two reference frames can be treated. One is stationary, the impeller is in rotation and the tank is fixed. The other frame, which is rotating, the impeller is kept stationary while the outer wall of the vessel is given an angular velocity equal and opposite to the velocity of the rotating frame. Constant boundary conditions have been set respecting a fixed reference frame (RRF) approach. The boundary conditions for velocity are fixed on the gate impeller and the vessel. On the gate impeller: Moving reference frame implemented in the CFD code.

Flow regime must be retained in non-Newtonian laminar flow, it can be used to define a generalized Reynolds number. From this generalized Reynolds number, an apparent viscosity ηap can be deduced. Therefore, ηap is defined as the viscosity of the Newtonian fluid providing the same power consumption as the non-Newtonian fluid for the studied system:

$$\text{Re } g = \frac{\rho.N.d^2}{\eta_{ap}} \tag{7}$$

Metzner and Otto [14] introduced the concept of apparent viscosity which is linked to an apparent shear rate $\dot{\gamma}_{ap}$ using the rheological characterization of the fluid $\eta(\dot{\gamma}_{ap})$. They suggested this apparent shear rate to be proportional to the rotation frequency N:

$$\dot{\gamma}_{ap} = \text{Ks. N} \tag{8}$$

These results for the bi-viscosity model are quite similar for $\dot{\gamma}_c = 10^{-2}$.

4.2 Validation

To verify the model of the fluid and to validate our results we compared our power number with previous works on an anchor agitator. Figure 2 gives an example of comparison between our results and a numerical work of Marouche [8]. The results show a satisfactory agreement.

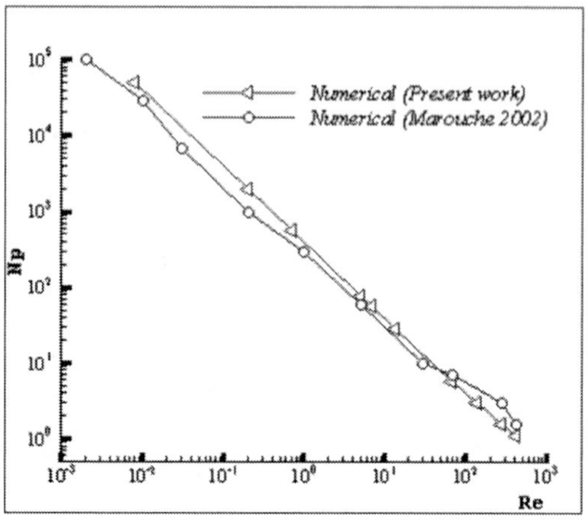

Figure 2. Variation of power number as function of Reynolds number.

5 Results and interpretations

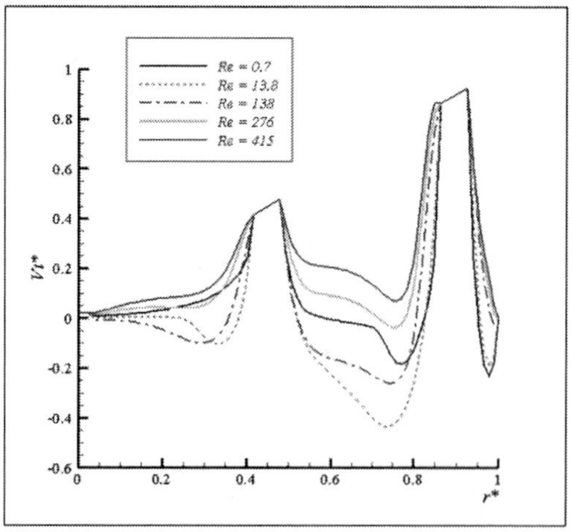

Figure 3. Influence of inertia on the tangential velocity at the blade and extension.

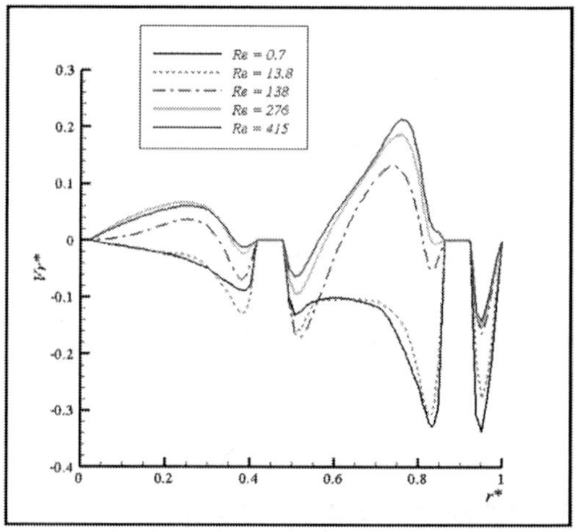

Figure 4. Influence of inertia on the radial velocity at the blade and extension.

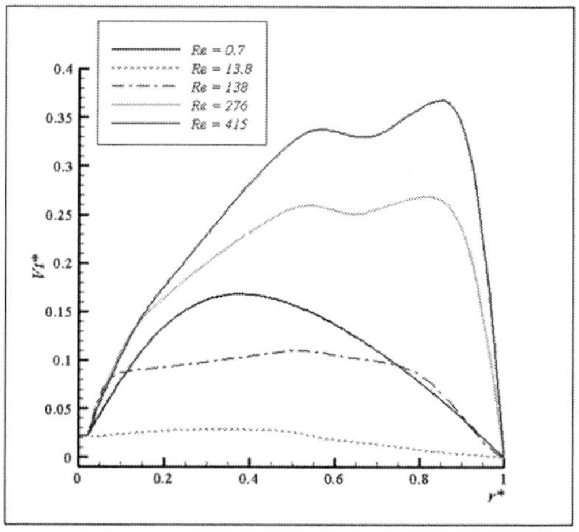

Figure 5. Influence of inertia on the tangential velocity at the mediator.

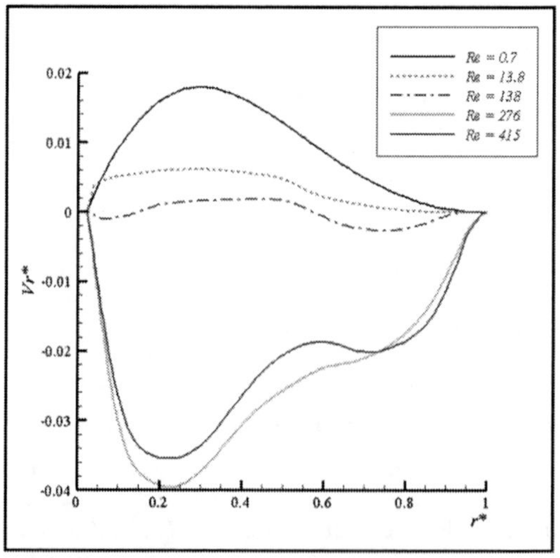

Figure 6. Influence of inertia on the radial velocity of the mediator.

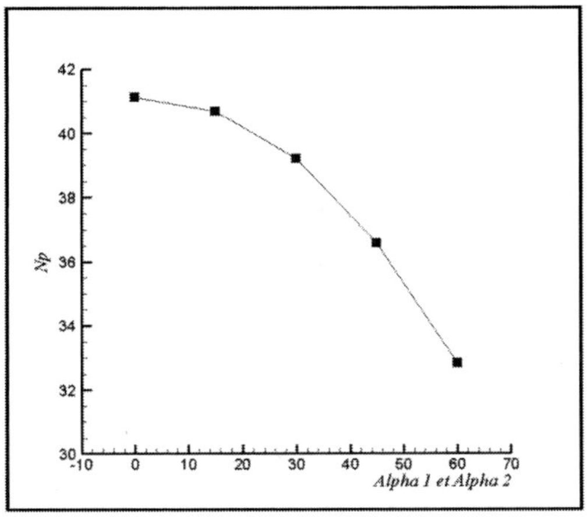

Figure 7. Inclination influence of the two blades on the power consumption.

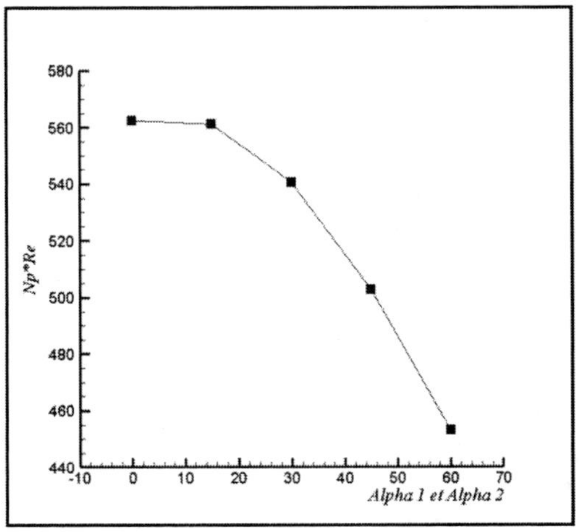

Figure 8. Inclination influence of the two blades on the Product Np * Re.

Based on the analysis of previous curves results which show the variation in energy consumption compared to the blade pitch position or one by one or both at the same time, we find that the optimal configuration that consumes less energy is the fifth configuration (i.e.) configuration where Alpha 1 and Alpha 2 equals 60 °.

Figures 3, 4, 5 and 6 show the variation of the tangential and radial velocity of blade and its extension and on the mediator according to the radial position. We studied the influence of Reynolds number on the nature of the flow by varying this parameter from 0.7 to 415 for a Bingham fluid threshold τ 0= 1 Pa.

The tangential component of the velocity varies in the following manner Figure 3 growth of the tangential velocity from the agitator shaft to the first blade and a linear increase between the two ends thereof, a decrease to a relative minimum, and growth until the inner end of the second blade and a second linear increase in the level of the second blade, once end, a sudden decrease of velocity to the vessel wall which is zero. It is noted again that the tangential velocity increases with the increase of Reynolds number in the two zones, one between the axis of the agitator and the first vane and the other between the two blades.

We note that upstream of the first blade and the second, the tangential velocity Vt* is negative for low Reynolds numbers, when Re is greater than 150, the tangential velocity is always positive, for against the radial velocity Vr* Figure 5 is always positive: this means that the inner end of each blade draws fluid from the center to the wall of the vessel, while the outer end of the second blade has a tendency of the shaving of the tank towards the center thereof.

On the mediator Figure 4, the velocity is nearly zero for low Reynolds numbers, but reaches a maximum with increasing relative inertia. Moreover this maximum is lower than the outer end of the second blade, which means that the flow in this zone is delayed beyond this maximum speed decreases as one moves towards the wall of the vessel, although to a slope greater than that of the rising part of the curve.

The influence of Reynolds number on the profile of the radial velocity Vr* on the blades Figure 5, we see that the variation of the radial velocity is noticeable Clearly between axis of the rod end and the internal blade and the first entre le two blades, this variation is Rapidly decreasing to a low Reynolds number, we noted the existence of negative velocities which results in the formation of recirculation area between the two blades and near the wall where the flow is delayed in this area. Moreover, on the mediator Figure 6, the radial velocity is zero Almost.

We note that the number of power varies linearly with the inertia (Re) on a logarithmic scale, produced by the cons of Np * Re, the constant of laminar flow is Re <100 equal to 452 for our mobile, All above this value, there is a rapid increase in the product Np * Re, which means that the power consumption increases faster than the Reynolds number.

6 Conclusions

This contribution focuses on the viscoplastic fluids, which behavior is poorly controlled and which intervene in many processes, particularly the polymerization in emulsion. During the latter, the behavior progresses through time: practically, Newtonian at the beginning of the reaction, then becomes shear thinning and finally viscoplastic as long as the concentrations increase. The hydrodynamics of viscoplastic fluids in complex geometries such as the systems of agitation is modestly approached. However it plays a significant role, in the control and the effectiveness of processing since it conditions the heat and mass transfer. The problem is analyzed under the numerical aspect. The latter is not treated by a specific code but using an existing industrial code which is the Fluent code. The numerical simulations carried out in geometries 2D made it possible to show the important significations by the viscoplastic behavior and in particular, the quasi immobilization of the fluids in a broad area when the effects of yield are dominant. It is worth nothing that this figure does not work if we consider the mixture of two substances of this type of fluid.

Characterization of laminar flow and power consumption in an agitated vessel with two blades agitators has been developed in order to provide a physical analysis of mixing in viscoplastic fluids.

Has led us to see that this type of fluid for the choice of the fifth configuration is appropriate and consumes less power compared to other configurations to be precise our simulation has shown that we can make a saving of 20%, hence the interest to use it often in the case of these fluids.

References

1. K. Yapici, B. Karasozen, M. Schafer, Y. Uludag. *Numerical investigation of the effect of the Rushton type turbine design factors on agitated tank flow characteristics*. Chem. Eng. and Proc. (2007).
2. F. Ein-Mozaffari, C.P.J. Bennington, G.A. Dumont, *Suspension yield stress and the dynamic response of agitated pulp chests*. Chem. Eng. Sci. **60** (2005).
3. T.P. Elson, *The growth of caverns formed around rotating impellers during the mixing of a yield stress fluid*, Chem. Eng. **96** 303–319 (1990).
4. D. Anne-Archard, M. Marouche, H.C. Boisson. *Hydrodynamics and Metzner–Otto correlation in stirred vessels for yield stress fluids*. Chem. Eng. J. **125**, 15–24 (2006).
5. J.N. Haque, T. Mahmud, K. Roberts, *Modeling turbulent flows with free surface in baffled agitated vessels*, Ind. Eng. Chem. Res. **45**, 2881–2891 (2006).
6. V.V. Ranade, S.M.S. Dommeti, *Computational snapshot of flow generated by axial impellers in baffled stirred vessels*, Trans. Inst. Chem. Eng. **74**, 476–484 (1996).
7. J.J. Derksen, *Numerical simulation of solid suspension in a stirred tank*, AIChE J. **49**, 2700–2714 (2003).
8. M. Marouche, *Hydrodynamique d'un système d'agitation en fluide viscoplastique et en régime laminaire inertiel*, Ph. D Theses, National Polytechnic Institut of Toulouse. (2002).
9. L. Rahmani et al. *Heat transfer to Bingham fluid during laminar flow in agitated tank*, Int. Rev. Mech. Eng. **09**, (2013).
10. E.J. O'Donovan, R.I. Tanner, *Numerical analysis of the Bingham squeeze film problem*, J. of Non-New. Flu. Mec. **15**, 75-83, (1984).
11. S.V. Patankar. *Numerical heat transfer and fluid flow*. Mc-Graw Hill. (1980).
12. G.C. Vradis, M.V. Otugen, *The axisymmetric sudden expansion flow of a non-Newtonian viscoplastic fluid*, J. Fluid Eng. **119**, 193–200 (1997).
13. M. Marouche, D. Anne-Archard, H.C. Boisson, *A numerical model of yield stress fluid dynamics in a mixing vessel*, Appl. Rheol. **12**, 182–191 (2002).
14. A.B. Metzner, R.E. Otto, *Agitation of non-Newtonian fluids*, AIChE J. **3**, 3–10 (1957).

EPJ Web of Conferences **114**, 02102 (2016)
DOI: 10.1051/epjconf/201611402102

Mathematical model of non-stationary temperature distribution in the metal body produced by induction heating process

Josef Rak[1,2,a]

[1] *University of Pardubice, Faculty of Electrical Engineering and Informatics, Studentska 95 532 10 Pardubice, Czech Republic*
[2] *Some results were obtained by author's doctoral study in Charles University in Prague, Faculty of Mathematics and Physics, Department of Numerical Mathematics*

Abstract. An induction heating problem can be described by a parabolic differential equation. For this equation, specific Joule looses must be computed. It can be done by solving the Fredholm Integral Equation of the second kind for the eddy current of density. When we use the Nyström method with the singularity subtraction, the computation time is rapidly reduced. This paper shows the method for finding non-stationary temperature distribution in the metal body with illustrative examples.

Key words. induction heating, integral equation of the second kind, non-stationary temperature distribution, Nyström method

1 Introduction

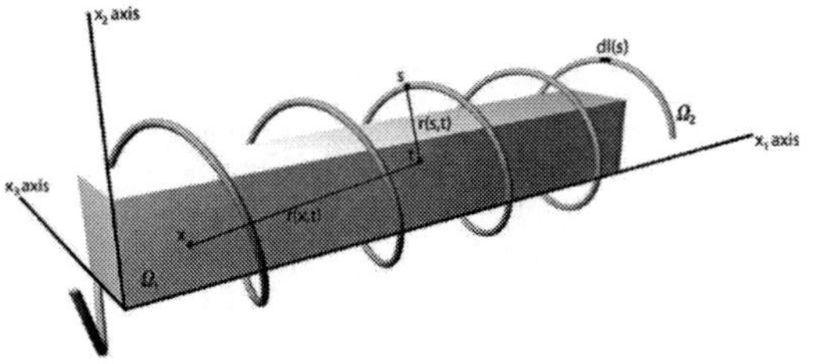

Figure 1. Heated body and coil.

A bounded metal cuboid body Ω_1 of sizes $l_1 \times l_2 \times l_3$ is heated by an external electromagnetic field produced by inductor Ω_2 (see figure 1). The inductor is formed by a conductor of general shape and position that carries the harmonic current I_{ext}.

2 Eddy current of density

Computation of the temperature distribution depends on the eddy current of density. It is a phasor

$$J_{eddy} = (J_{eddy,x_1}, J_{eddy,x_2}, J_{eddy,x_3}).$$

The x_1 component of $J_{eddy,x_1}(\boldsymbol{x})$ can be computed (see [4] and [5]) by Fredholm integral equation of the second kind

$$\iota J_{eddy,x_1}(\boldsymbol{x}) - \kappa(\boldsymbol{x}) \int_{\Omega_1} \frac{J_{eddy,x_1}(\boldsymbol{t})}{r(\boldsymbol{x},\boldsymbol{t})} dt_1 dt_2 dt_3 = I_{ext} F(\boldsymbol{x}), \quad (1)$$

where

$$F(\boldsymbol{x}) = \kappa(\boldsymbol{x}) \int_{\Omega_2} \frac{dl(s).e_{x_1}}{r(\boldsymbol{x},s)}, \quad (2)$$

$$\kappa(\boldsymbol{x}) = \frac{\omega\gamma(T(\boldsymbol{x}))\mu_0}{4\pi}, \quad (3)$$

$r(\boldsymbol{x},\boldsymbol{t})$ is the Euclidean distance, $\boldsymbol{x} = (x_1, x_2, x_3)$ and $\boldsymbol{t} = (t_1, t_2, t_3)$ are points in the metal body, $s = (s_1, s_2, s_3)$ is a point at the inductor, I_{ext} is harmonic current carried by the inductor, ω is angular frequency, $\gamma(T(\boldsymbol{x}))$ is electrical conductivity, μ_0 is permeability of vacuum, e_{x_1} is unite vector $e_{x_1} = (1, 0, 0)$ and ι is the complex unit.

For each bounded and continuous temperature distribution $T(\boldsymbol{x})$, $\kappa(\boldsymbol{x})$ is a real, positive, bounded and continuous function.

Formulas for the remaining components J_{eddy,x_2} and J_{eddy,x_3} can be obtained by mere interchanging of indices.

With the notation

$$J_R(\boldsymbol{x}) = Re J_{eddy,x_1}(\boldsymbol{x}) \quad (4)$$
$$J_I(\boldsymbol{x}) = Im J_{eddy,x_1}(\boldsymbol{x}) \quad (5)$$
$$I_R = Re(I_{ext}) \quad (6)$$
$$I_I = Im(I_{ext}) \quad (7)$$

we have system of integral equations

$$J_R(\boldsymbol{x}) - \kappa(\boldsymbol{x}) \int_{\Omega_1} \frac{J_I(\boldsymbol{t})}{r(\boldsymbol{x},\boldsymbol{t})} dt = I_I F(\boldsymbol{x}),$$
$$-J_I(\boldsymbol{x}) - \kappa(\boldsymbol{x}) \int_{\Omega_1} \frac{J_R(\boldsymbol{t})}{r(\boldsymbol{x},\boldsymbol{t})} dt = I_R F(\boldsymbol{x}). \quad (8)$$

The specific Joule losses at the point $\boldsymbol{x}$ in the body, which are needed to compute the temperature distribution are given by

$$\omega(\boldsymbol{x}) = \frac{1}{\gamma} J_e(\boldsymbol{x}) \overline{J_e(\boldsymbol{x})}, \quad (9)$$

[a] Corresponding author: josef.rak@upce.cz

where

$$J_e(\boldsymbol{x}) =$$

$$\sqrt{[ReJ_{eddy,x_1}(\boldsymbol{x})]^2 + [ReJ_{eddy,x_2}(\boldsymbol{x})]^2 + [ReJ_{eddy,x_3}(\boldsymbol{x})]^2} +$$

$$+\iota\sqrt{[ImJ_{eddy,x_1}(\boldsymbol{x})]^2 + [ImJ_{eddy,x_2}(\boldsymbol{x})]^2 + [ImJ_{eddy,x_3}(\boldsymbol{x})]^2}$$

and $\overline{J_e(\boldsymbol{x})}$ is complex conjugate to $J_e(\boldsymbol{x})$.

3 The non-stationary temperature distrubution

The non-stationary distribution of the temperature $T(\boldsymbol{x}, t)$ at point $\boldsymbol{x}$ and time t in the metal body is described by a partial differential equation of the parabolic type

$$\operatorname{div}\left[\lambda(T(\boldsymbol{x}, t)) \operatorname{grad} T(\boldsymbol{x}, t)\right] =$$

$$= \rho(T(\boldsymbol{x}, t))c(T(\boldsymbol{x}, t))\frac{\partial T(\boldsymbol{x}, t)}{\partial t} - \omega(\boldsymbol{x}, t), \qquad (10)$$

where $\lambda = \lambda(T(\boldsymbol{x}, t))$ denotes the thermal conductivity, $\rho = \rho(T(\boldsymbol{x}, t))$ the specific mass of the material, $c = c(T(\boldsymbol{x}, t))$ specific heat and $\omega(\boldsymbol{x}, t)$ the specific Joule losses given by (9) at time t.

The boundary condition along the whole surface of the body reads

$$-\lambda(T(\boldsymbol{x}, t))\frac{\partial T(\boldsymbol{x}, t)}{\partial n} = \alpha\left(T(\boldsymbol{x}, t) - T_{ext}\right), \qquad (11)$$

where α denotes the coefficient of the convective heat transfer, T_{ext} the temperature of the surrounding medium and n the direction of the outward normal.

4 Numerical solution

The metal body Ω_1 was assumed to be cuboid. Let us cover the body by n_1 subsuboides in the x_1 direction, n_2 subsuboides in the x_2 direction and n_3 subsuboides in the x_3 direction. The size of each subcuboid is $h_1 \times h_2 \times h_3$, where

$$h_1 = \frac{l_1}{n_1},\, h_1 = \frac{l_2}{n_2},\, h_3 = \frac{l_3}{n_3}.$$

We need to solve integral equation (1) to get Joule losses. Then we can put computed Joule losses to partial differential equation (10) to get the temperature distribution. The Joule losses must be recalculated at the time when changed.

For computation of Joule losses we will apply the Nyström method with the compound mid-cuboid rule. For time integration we will use the classical finite difference method for partial differential equations of the parabolic type.

4.1 Application of Nyström method

The Nyström method is based on approximation of the integral by the numerial integration rule. We will use the compound min-point rule. Let the node points be $\boldsymbol{x}_i$ defined as center of sub cuboides. For the weight let

$$w_j = w = \frac{|\Omega_1|}{N},\, j = 1, ..., N,$$

where $N = n_1 n_2 n_3$.

Here the numerical integration rule cannot be applied directly. The reason is that the function $r(\boldsymbol{x}, \boldsymbol{t})^{-1}$ is singular at $\boldsymbol{x} = \boldsymbol{t}$. One way to deal with such a problem is singularity subtraction. It was described in [3]. Let $r_N(\boldsymbol{x}, \boldsymbol{t})$ be an approximation of $r(\boldsymbol{x}, \boldsymbol{t})$ which coincide outside a certain neighborhood of $\boldsymbol{x} = \boldsymbol{t}$. The integrand in equation (1) is approximated by

$$\frac{J_{eddy,x_1}(\boldsymbol{t})}{r(\boldsymbol{x}, \boldsymbol{t})} \approx \frac{J_{eddy,x_1}(\boldsymbol{t}) - J_{eddy,x_1}(\boldsymbol{x})}{r_N(\boldsymbol{x}, \boldsymbol{t})} + \frac{J_{eddy,x_1}(\boldsymbol{x})}{r(\boldsymbol{x}, \boldsymbol{t})}. \qquad (12)$$

Since the first element of the approximation is zero when $\boldsymbol{x} = \boldsymbol{t}$, the exact constructioin of $r_N(\boldsymbol{x}, \boldsymbol{t})$ is immaterial. By using the numerical integration rule, approximation (12) and running $\boldsymbol{x}$ throung the node points we get a system of linear equations for numerical approximation of the x_1 component of the eddy current of density $\widetilde{J}_N$

$$\left[1 - \iota\kappa(\boldsymbol{x}_i)\sum_{\substack{j=1\\ j\neq i}}^{N}\frac{w}{r_N(\boldsymbol{x}_i, \boldsymbol{x}_j)} = \iota\kappa(\boldsymbol{x}_i)\int_{\Omega_1}\frac{dt_1dt_2dt_3}{r(\boldsymbol{x}_i, \boldsymbol{t})}\right]\widetilde{J}_N(\boldsymbol{x}_i) +$$

$$+\iota\kappa(\boldsymbol{x}_i)\sum_{\substack{j=1\\ j\neq i}}^{N}\frac{w\widetilde{J}_N(\boldsymbol{x}_j)}{r_N(\boldsymbol{x}_i, \boldsymbol{x}_j)} = -\iota I_{ext}F(\boldsymbol{x}_i), i = 1, ..., N. \qquad (13)$$

$\widetilde{J}_N(\boldsymbol{x}_i)$ is complex number. Let us define for each $i = 1, ..., N$

$$\widetilde{J}_i^{(R)} = \mathrm{Re}\widetilde{J}_N(\boldsymbol{x}_i) \text{ and } \widetilde{J}_i^{(I)} = \mathrm{Im}\widetilde{J}_N(\boldsymbol{x}_i). \qquad (14)$$

Then (13) is equivalent the real system of linear equations

$$\widetilde{J}_i^{(R)} + \kappa(\boldsymbol{x}_i)\left[\sum_{\substack{j=1\\ j\neq i}}^{N}\frac{w}{r_N(\boldsymbol{x}_i, \boldsymbol{x}_j)} - \int_{\Omega_1}\frac{dt_1dt_2dt_3}{r(\boldsymbol{x}_i, \boldsymbol{t})}\right]\widetilde{J}_i^{(I)} -$$

$$-\kappa(\boldsymbol{x}_i)\sum_{\substack{j=1\\ j\neq i}}^{N}\frac{w\widetilde{J}_j^{(I)}}{r_N(\boldsymbol{x}_i, \boldsymbol{x}_j)} = \mathrm{Im}I_{ext}F(\boldsymbol{x}_i), i = 1, ..., N$$

$$-\widetilde{J}_i^{(I)} + \kappa(\boldsymbol{x}_i)\left[\sum_{\substack{j=1\\ j\neq i}}^{N}\frac{w}{r_N(\boldsymbol{x}_i, \boldsymbol{x}_j)} - \int_{\Omega_1}\frac{dt_1dt_2dt_3}{r(\boldsymbol{x}_i, \boldsymbol{t})}\right]\widetilde{J}_i^{(R)} -$$

$$-\kappa(\boldsymbol{x}_i)\sum_{\substack{j=1\\ j\neq i}}^{N}\frac{w\widetilde{J}_j^{(R)}}{r_N(\boldsymbol{x}_i, \boldsymbol{x}_j)} = \mathrm{Re}I_{ext}F(\boldsymbol{x}_i), i = 1, ..., N. \qquad (15)$$

Following Nyström interpolation formula (for details see for example [1]) we get the numerical solution for eddy currents $\widetilde{J}_N$ outside the node points

$$\widetilde{J}_N(\boldsymbol{x}) = \frac{-\iota I_{ext}F(\boldsymbol{x}) - \iota\kappa(\boldsymbol{x})\sum_{j=1}^{N}\frac{w}{r_N(\boldsymbol{x}, \boldsymbol{x}_j)}\widetilde{J}_N(\boldsymbol{x}_j)}{1 - \iota\kappa(\boldsymbol{x})\sum_{j=1}^{N}\frac{w}{r_N(\boldsymbol{x}, \boldsymbol{x}_j)} + \iota\kappa(\boldsymbol{x})\int_{\Omega_1}\frac{1}{r(\boldsymbol{x}, \boldsymbol{t})}dt}, \qquad (16)$$

where

$$\widetilde{J}_N(x_i) = \widetilde{J}_i^{(R)} + \iota\widetilde{J}_i^{(I)}.$$

Since Ω_1 is a cuboid, the integral in (15) and (16) can be computed analytically. The x_2 and x_3 components of the eddy current of density can be computed in an analogous way.

5 Application of the finite difference method

Let us use a finite difference method for the numerical solution of equation (10). Since thermal conductivity is a constant that depends on the temperature we can rewrite equation (10) into the form

$$\frac{\partial T(\boldsymbol{x},t)}{\partial t} = \frac{1}{\rho(T(\boldsymbol{x},t))c(T(\boldsymbol{x},t))}\Bigg[\lambda(T(\boldsymbol{x},t))\cdot \left(\frac{\partial^2 T(\boldsymbol{x},t)}{\partial x_1^2} + \frac{\partial^2 T(\boldsymbol{x},t)}{\partial x_2^2} + \frac{\partial^2 T(\boldsymbol{x},t)}{\partial x_3^2}\right) + \omega(\boldsymbol{x},t)\Bigg]. \quad (17)$$

Let the coordinates of the node point be

$$(x_{1_i}, x_{2_j}, x_{3_k}), i = 1, ..., n_1, j = 1, ...n_2, k = 1, ..., n_3.$$

With the notation

$$x_{i,j,k} = x_{1_i}, x_{2_j}, x_{3_k}$$
$$T_{i,j,k}(t) = T(x_{i,j,k}, t)$$
$$\omega_{i,j,k}(t) = \omega(x_{i,j,k}, t)$$

the x_1, x_2, x_3 derivatives are approximated by

$$\frac{\partial^2 T(\boldsymbol{x},t)}{\partial x_1^2} \approx \frac{T_{i+1,j,k}(t) - 2T_{ijk}(t) + T_{i-1,j,k}(t)}{h_1^2}$$
$$\frac{\partial^2 T(\boldsymbol{x},t)}{\partial x_2^2} \approx \frac{T_{i,j+1,k}(t) - 2T_{ijk}(t) + T_{i,j-1,k}(t)}{h_2^2}$$
$$\frac{\partial^2 T(\boldsymbol{x},t)}{\partial x_3^2} \approx \frac{T_{i,j,k+1}(t) - 2T_{ijk}(t) + T_{i,j,k-1}(t)}{h_3^2}. \quad (18)$$

Now let us approximate of the derivation by difference in the boudary condition (11) to define T_{ijk} for $i = 0$, $i = n_1 + 1$, $j = 0$, $j = n_2 + 1$, $k = 0$ and $k = n_3 + 1$ as

$$T_{0,j,k}(t) = T_{1,j,k} - \frac{h_1\alpha}{\lambda(T_{1,j,k}(t))}(T_{1,j,k} - T_{ext})$$
$$T_{n_1+1,j,k}(t) = T_{n_1,j,k} - \frac{h_1\alpha}{\lambda(T_{n_1,j,k}(t))}(T_{n_1,j,k} - T_{ext})$$
$$T_{i,0,k}(t) = T_{i,1,k} - \frac{h_2\alpha}{\lambda(T_{i,1,k}(t))}(T_{i,1,k} - T_{ext})$$
$$T_{i,n_2+1,k}(t) = T_{i,n_2,k} - \frac{h_2\alpha}{\lambda(T_{i,n_2,k}(t))}(T_{i,n_2,k} - T_{ext})$$
$$T_{i,j,0}(t) = T_{i,j,1} - \frac{h_3\alpha}{\lambda(T_{i,j,1}(t))}(T_{i,j,1} - T_{ext})$$
$$T_{i,j,n_3+1}(t) = T_{i,j,n_3} - \frac{h_3\alpha}{\lambda(T_{i,j,n_3}(t))}(T_{i,j,1n_3} - T_{ext}). \quad (19)$$

Let us define time step Δt and approximate time derivation by

$$\frac{\partial T(\boldsymbol{x},t)}{\partial t} \approx \frac{T(\boldsymbol{x},t+\Delta t) - T(\boldsymbol{x},t)}{\Delta t}. \quad (20)$$

From (17), (18), (19) and (20) we have numerical formula for temperature evolution

$$T_{i,j,k}(t+\Delta t) = T_{i,j,k}(t) + \frac{\Delta t}{\rho(T_{i,j,k}(t))c(T_{i,j,k}(t))}\cdot \Bigg[\lambda(T_{ijk}(t))\Bigg(\frac{T_{i+1,j,k}(t) - 2T_{i,j,k}(t) + T_{i-1,j,k}(t)}{h_1^2} + \frac{T_{i,j+1,k}(t) - 2T_{i,j,k}(t) + T_{i,j-1,k}(t)}{h_2^2} + \frac{T_{i,j,k+1}(t) - 2T_{i,j,k}(t) + T_{i,j,k-1}(t)}{h_3^2}\Bigg) + \omega_{i,j,k}(t)\Bigg].$$

Joule losses need to be updated when time changes. The whole calculation is started with an initial condition. It is the starting temperature of the body and is equal to temperature of the air of the surroundig medium.

6 Example 1

A brass cuboid body with the size $0.15 \times 0.01 \times 0.01$ m is heated with a stationary inductor starting at room temperature 20 °C. The inductor has the form of a coil which turns around the heated body in the x_1-direction in 6 loops. The radius of the coil is 0.015 m, exciting current 500 A and frequency 150 kHz. The length of the coil is 0.15 m. The cuboid is partitioned by 75 elements in x_1 direction, 10 elements in x_2 and x_3 directions. Figures 2 - 4 show temperature distribution during time evolution (1, 10 and 60 s) at the surface of the heating body. Figures 5 - 7 show temperature distribution during time evolution (1, 10 and 60 s) at cuts of the body when $x_2 = 0$ and $x_3 = 0$.

Computation was made by Matlab. The integration for computing $F(x_i)$ is done by the Matlab's method *quad*. It uses an aplication of the adaptive Simpson quadrature. Improper integral is computed analytically.

7 Example 2

To see the dependence of the parameters let us show examples with different parameters. Let the parameters be the same as in example 1. Figure 8 shows temperature distribution at 30 s. Other figures have change in one parameter.

- at figure 9 the coil rotates over the heating body at 3 loops.
- at figure 10 the size of the body is $0.15 \times 0.005 \times 0.005$ m.
- at figure 11 the coil is exciting current 1000 A.
- at figure 12 the angular frequency is 300 kHz.

8 Conclusion

In example 2, the temperature changed as expected. It was shown in [5], that the time to compute the Joule losses is reduced to approximately 10 percent compared to the collocation method. Since the Joule losses need to be recalculated 3 – 10 times during time evolation to 60 s is this method a big improovement.

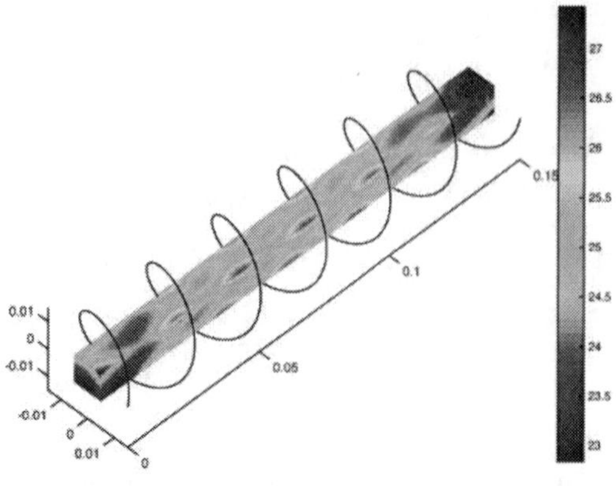

Figure 2. Example 1, t = 1 s, body

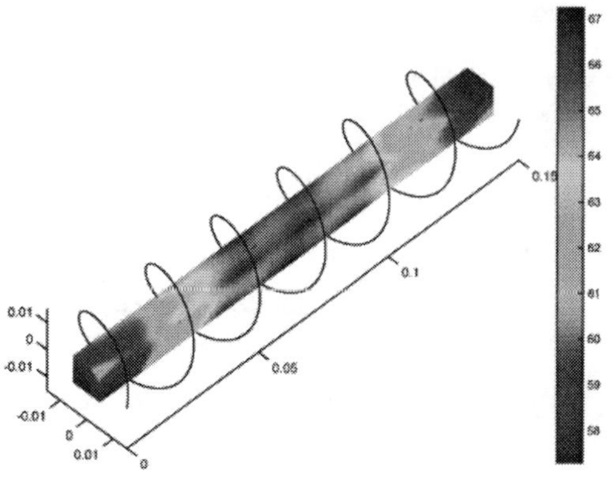

Figure 3. Example 1, t = 10 s, body

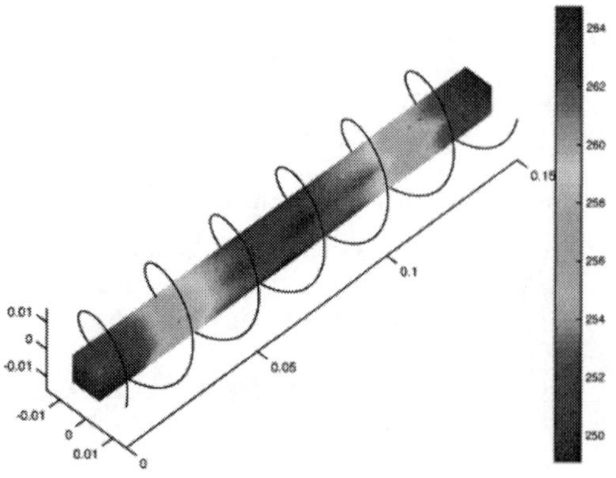

Figure 4. Example 1, t = 60 s,body

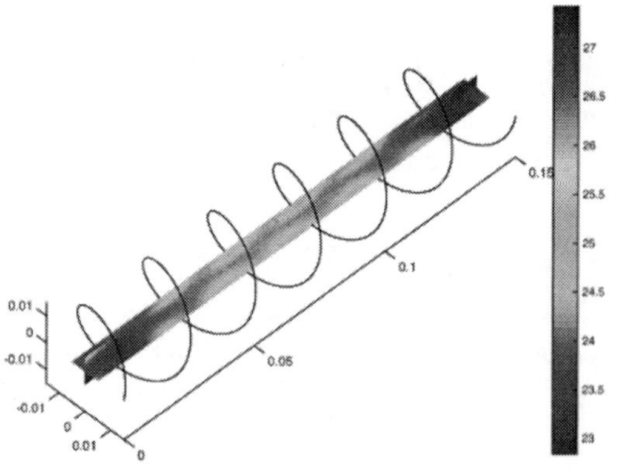

Figure 5. Example 1, t = 1 s, body cut

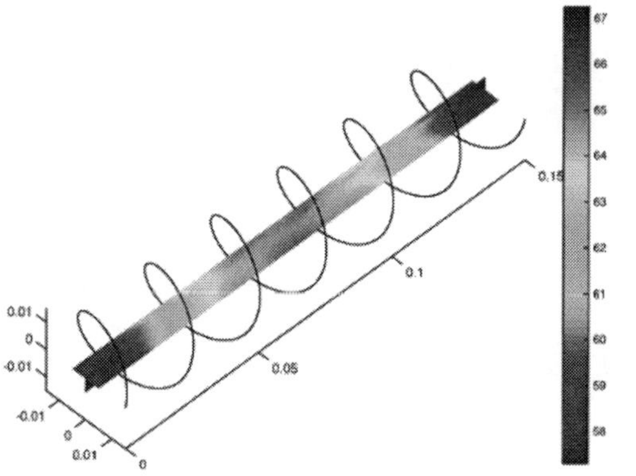

Figure 6. Example 1, t = 10 s, body cut

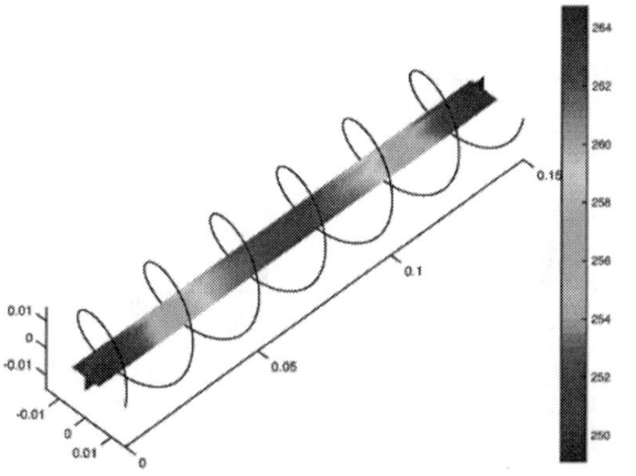

Figure 7. Example 1, t = 60 s, body cut

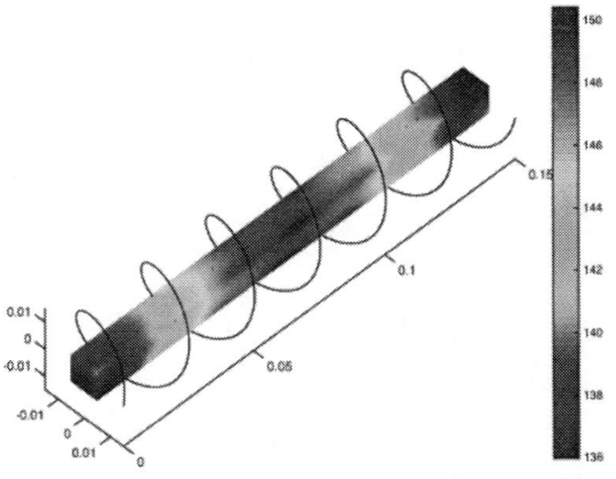

Figure 8. Example 2, t = 30 s, original parameters

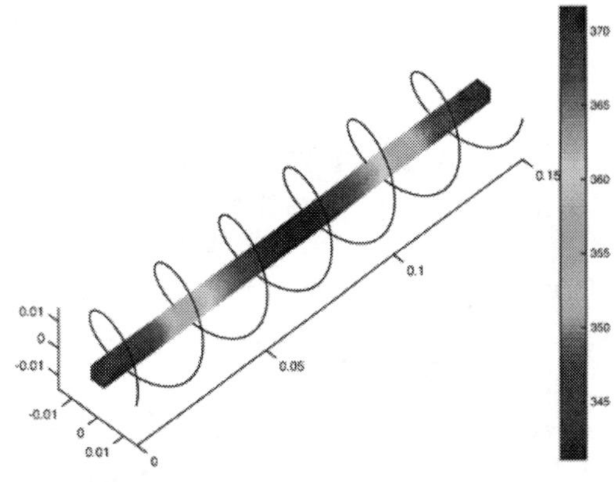

Figure 10. Example 2, t = 30 s, size 0.15 × 0.005 × 0.005 m

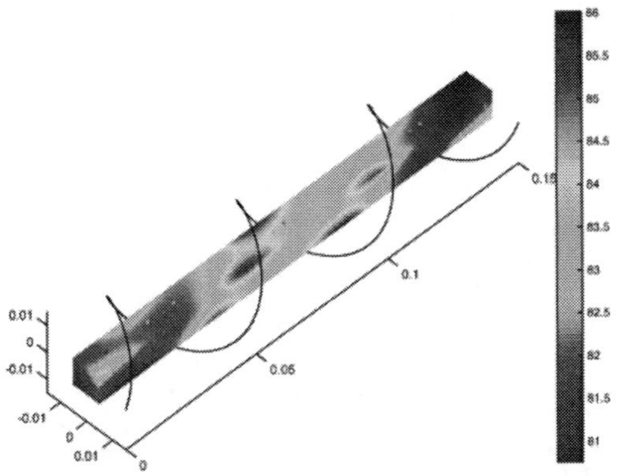

Figure 9. Example 2, t = 30 s, 3 loops of the coil

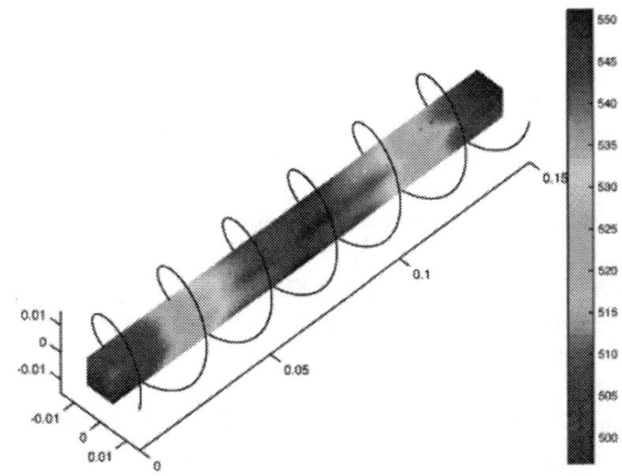

Figure 11. Example 2, t = 30 s, harmonic current 1000 A

References

1. K. E. Atkinson: *The Numerical Solution of Integral Equations of the Second Kind* (Cambridge University Press, 1997)
2. P. M. Anselone: *Collectively compact operator approximation theory and applications to integral equations* (Prentice-Hall, Inc., Englewood Cliffs, N.J. 1971)
3. L. V. Kantorovich, V. I. Krylov: *Approximate Methods of Higher Analysis*, (Interscience, New York, 1958)
4. P. Solin, I. Dolezel, M. Skopek, B. Ulrych, *SPETO 2001 International Conference*, 143-146 (2001)
5. J. Rak, EPJ web conf. **92**, 02075 (2015)

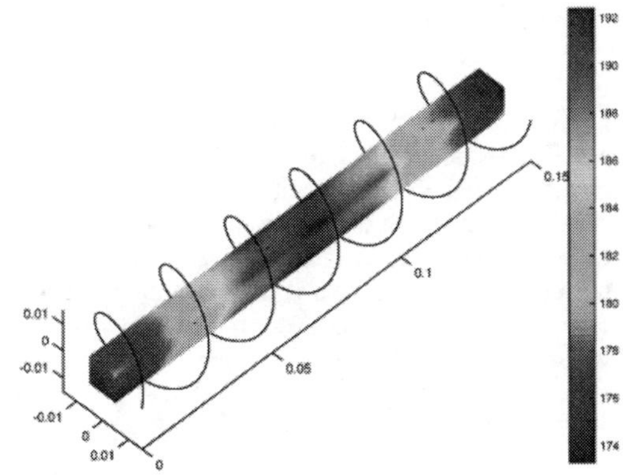

Figure 12. Example 2, t = 30 s, angular frequency 300 kHz

EPJ Web of Conferences **114**, 02103 (2016)
DOI: 10.1051/epjconf/201611402103

Experimental study of turbulence intensity influence for turbulent cross flow in staggered tube bundle using grooved cylinder

Ouissem Sahli[1], Lahouari Adjlout[1, a], Omar Ladjedel[1] and Mohamed amine Ghazi[1]

[1] *Laboratoire d'aérohydrodynamique naval Département de Génie Maritime Mechanical Engineering Faculty USTO MB Oran, Algeria*

Abstract. The present work is an experimental investigation on the effect of the turbulence intensity variation in a staggered tube bundle equipped with grooves at 90° and 270°.The experiments were carried out in a subsonic wind tunnel. Three Reynolds numbers and three turbulence levels were tested. The pressure distributions and drag forces were measured. Surface visualizations were also performed. The obtained results show that the turbulence intensity for different Reynolds number has an influence on the reduction of the drag coefficient.

1 Introduction

Cross-flow in tube bundles is an active area of research because of its wide practical application in the design for heat exchanger of tubes, offshore structure, cooling system for nuclear plants etc… There has been a lot of experimental and numerical works dedicated to the study of different aspects of flow around cylinders in tandem and to flow in arrays of cylinders. Over the past years, numerous experimental investigations have been carried out to study the complex mechanism of the flow characteristics in different configurations of tube bundles. Aiba et al. [1] performed an experimental study on square in-line tube banks for gap ratio of 1.2 and 1.6. They found that the flow field of the cylinders situated in the second row is relatively affected by the upstream ones because of the absence of vortex shedding behind the first row. Romberg and Popp [2,3] conducted experimental investigations on the stability factors of normal and rotated triangular tube arrays with different pitch-to-diameter ratios. Chen and Srikantiah [4] presented series of experiments to measure motion-dependent fluid forces for various tube arrays. Rottmann and Popp [5] further investigated the influence of approach flow on the first row of tube bundles. Žukauskas and Ulinskas [6] reported extensive experimental data for heat transfer and fluid friction during viscous cross-flow in-line and staggered banks of tubes under two thermal boundary conditions. Paul et al. [7] preformed measurements in a staggered tube bundle with fixed pitch-to-diameter ratios at sub-critical Reynolds number using the particle image velocimetry technique. Their results showed that the flow exhibits strong Reynolds number dependence in the developing region, however no significant Reynolds number effect could be observed in the spatially periodic region. Paul et al. [8] perform an experimental study on turbulent cross-flow in a staggered tube bundle with transverse and longitudinal pitch-to-diameter ratios of 3.8 and 2.1, respectively. The results reveal higher shear rates in the wake regions.

Although some interesting studies which are based on the premise of modified tube shapes have been carried out, most of the investigations have focused on the characteristics of fluidelastic instability of tube bundles with the circular cylinders. It was found that a cylinder with wavy surface can weaken, to the extent that it may even suppress the vibration comparing with a circular cylinder at the same Reynolds numbers Lam et al. [9, 10]. An optimal spanwise wavelength ratio for drag reduction and fluctuating lift suppression was found at low Reynolds number of 100 over a wide wavelength range by Lam and Lin [11, 12]. The effects of introducing a row of wavy cylinders with suitable wavelength into circular tube bundles for the control of flow-induced vibration were investigated by K. Lam et al. [13], two types of tube bundles with the first row of circular tubes replaced by different wavy cylinders are studied and the results are compared with that of purely circular cylinders at the Reynolds number of 7500. Because of the introduction of this kind of wavy tubes at the first row, the turbulent kinetic energy behind the first row tubes are reduced and the wake patterns of the second row tubes are better organized compared with the tube bundle with purely circular cylinders and the tube bundle with wavy cylinders placed in the first row. The power density spectra of the vortex shedding in the tube bundle are evidently suppressed. This implies that the turbulent forces on the second tube row are weakened. It is concluded that by introducing a suitable wavy cylinders in tube bundles, some good effects on the control of flow-

[a] Corresponding author : adjloutl@yahoo.fr

induced vibration of the tube bundles in the heat exchanger can be achieved. Ladjedel et al. [14] performed an experimental investigation using grooved cylinders in in-line tube bundles configuration. They found a reduction of 36.5% in the pressure drop for a Reynolds number of 1.33×10^4.

Traub [15], has studied the influence of turbulence intensity on the heat transfer and the pressure drop in plain tube bundles. The results of this study showed that an increase of the turbulence intensity from 0.8% to 25% leads to an increase of the Nusselt number about 42% for a single row and 14% for a tube bundle with three rows. Furthermore, Traub [15] presented the results for the enhancement of heat transfer according to the criterion of Stephan and Mitrovic [16]. The dimensionless pressure drop coefficient is related to the Reynolds number and to the turbulence intensity level of the external flow around the heat exchanger tubes. The Reynolds number of the flow through the heat exchanger is calculated using a characteristic length of the heat exchanger, which is usually either the tube diameter or the minimum distance between the tubes inside the heat exchanger and the maximum velocity of the flow passing through the heat exchanger, Umeda and Yang [17]. In previous studies it has been observed that for high Reynolds number and turbulence levels the pressure drop decreases. Beziel and Stephan [18] concluded to this observation for a range of Reynolds numbers from 4000 to 200,000 and for turbulence intensities from 0.8% up to 38.8%. Regarding the heat transfer mechanism, Sak et al. [19] studied the role of the turbulence length scale and intensity on forced convection from a heated horizontal circular cylinder. Different levels of turbulence were imposed by a grid located in the inlet section of the flow field which had the role of turbulence generator. They derived an expression correlating the Nusselt number with the Reynolds number and the turbulence intensity. They showed that an increase of Reynolds number and turbulence intensity leads to higher Nusselt numbers

Experimental investigation is performed; the forces on the cylinders are measured using a TE 81 balance and a TE44DPS differential pressure with a multi-channel scanner. The purpose of the present work is to investigate experimentally the influence of the turbulence intensity on staggered configuration equipped with grooved cylinders.

2 Experimental set-up

The experiments are carried out in closed circuit wind tunnel. The test section is of square cross-section 0.46m × 0.46m and 1.2m long. The closed circuit wind tunnel is of conventional design in which the velocity could reach 60 m/s and the turbulence intensity is less than 1%. The test section is composed of 17 rigid PVC tubes of an identical external diameter of 40 mm and 460mm length arranged in a staggered configuration as shown in Figure 1. Half tubes are also mounted along the bottom and top walls of the test model alternately to simulate an infinite tube bundle and minimize the wall boundary layer. All the tube arrays are investigated in this study. A schematic diagram of the arrangement is shown in Figure 2.

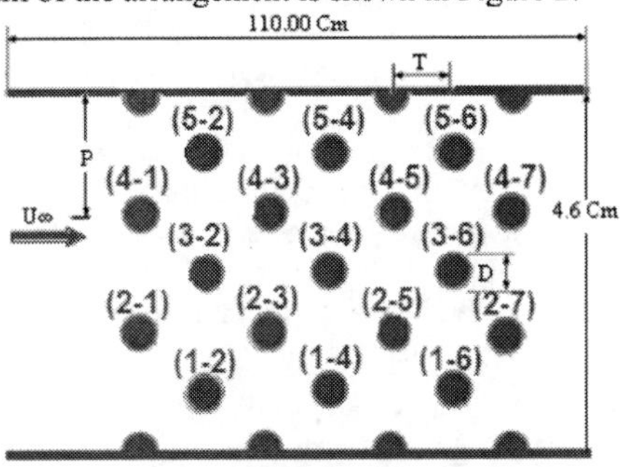

Figure 1. The geometrical features of the tube bundle.

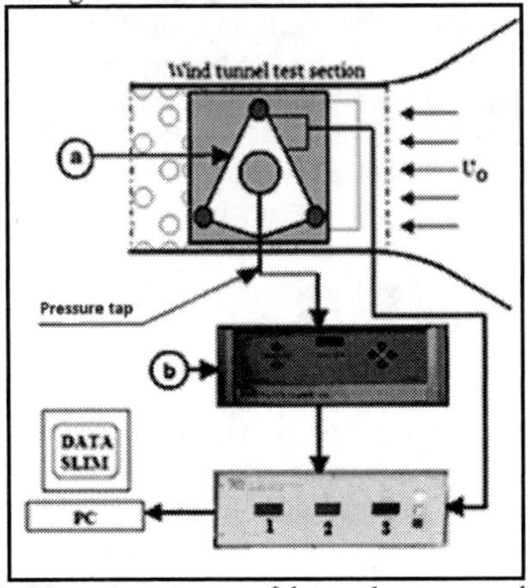

Figure 2. Measurement setup of drag and pressure. a: balance TE81, b: TE44 DPS

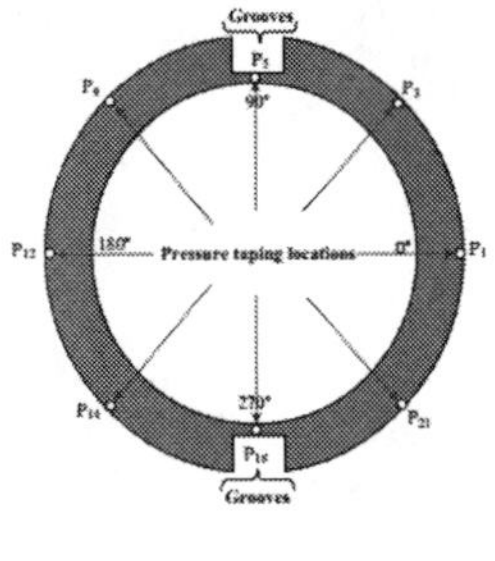

Figure 3. Pressure taping locations.

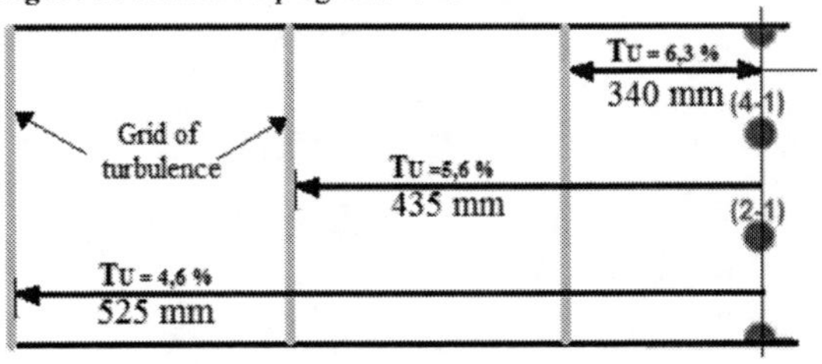

Figure 4. turbulence level generated for each position of the turbulence grid.

The experiment setup is shown in Figure 2. The tubes in the array are rigidly fixed, except for one tube which will be referred to as the instrumented cylinder equipped with 22 pressure tapings on its circumference at mid-span. The shape of the groove and the pressure tapings position are shown in Figure 3. The diameter of the pressure tapings is 0.5 mm. The pressure measurements are carried out for the 17 tubes for an azimuthal angle from 0° to 360°. The pressure distributions on the cylinder are measured using a TE44DPS differential pressure with a multi-channel scanner. The latter pressure scanning box allows sequential selection of up to 20 pressure tapings. The display unit links to a computer, loaded with DATASLIM software for data analysis and logging of the results. To generate turbulence Roach [20] method was adopted. The latter uses different turbulence grids. In this study the SMS (square mesh arrays of square bars) type is chosen (figure 4).

3 Result and discussion

In order to investigate the change of flow characteristics around staggered tube bundle by the use of the grooved cylinders the experiments are performed for the following physical parameters:

- Three Reynolds number are studied Re = 9300, Re= 13950 and Re= 20000 (based on the free stream velocity).
- Pitch to diameter ratio, P/D = 3.8 and T/D = 2.1.

However, visualizations are carried out for selected Reynolds numbers.

The measurement of pressure distribution starts from the frontal point facing the mainstream flow and ends at the same point after a complete revolution. The uncertainty in the pressure measurement was estimated to be less than ± 5% and for the drag and lift forces less than ± 2.4 %.

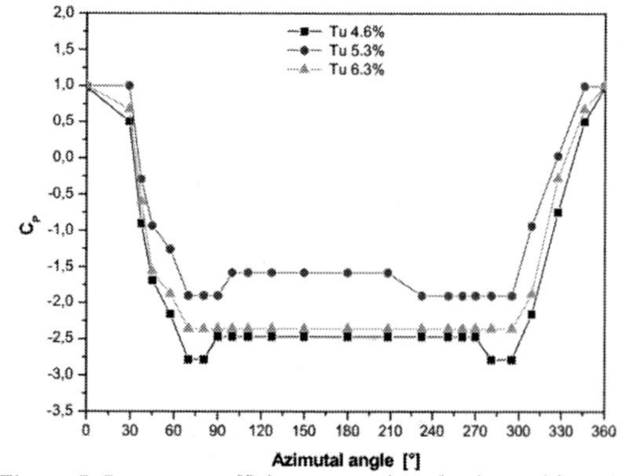

Figure 5. Pressure coefficient comparison for the position 2-1 (Re = 9300).

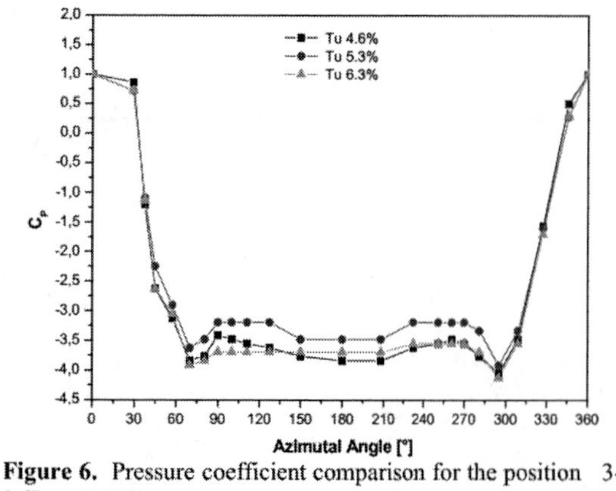

Figure 6. Pressure coefficient comparison for the position 3-2 (Re= 13950).

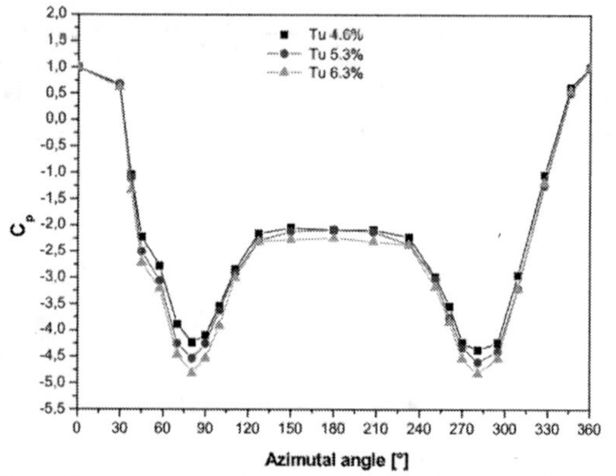

Figure 7. Pressure coefficient comparison for the position 4-3 (Re= 20000).

Figure 5, 6 and 7 show the pressure coefficient distribution around the cylinders (2-1) , (3-2) and (4-3) respectively at different Reynolds numbers and turbulence intensities (Tu). It is observed that the curves have the same trend where the flow around the cylinders starts from the stagnation point (θ =0°) in which Cp takes a value of one. The flow speeds up then starts to increase accompanied by a drop in the Cp to negative values. The flow velocity begins to reduce near θ = 60° along with an increment of pressure in the direction of the flow, which results in the adverse pressure gradient situation. The flow now has to move against the pressure force in addition to the viscous force. This leads to a reduced gradient of the velocity profile and the wall shear stress. Separation of flow occurs when the shear stress cannot overcome the adverse pressure gradient, this happens at θ = 75° for the tube 2-1, θ = 70° for the tube 3-2 and at θ = 85°for the tube 4-3. It is clearly seen in figure 5 and 6 that the pressure drag is more reduced due to the observed reduction in the pressure on the downstream of the cylinder for Tu = 5.3 % compared to the other levels .

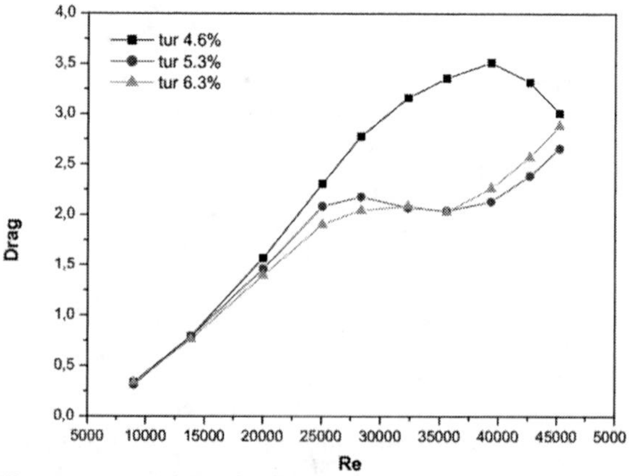

Figure 8. Drag force distribution along the tube length for different Reynolds number and turbulence intensities. (tube 5-2)

Figure 8 shows the drag force evolutions for different Reynolds number and turbulence intensities for the position 5-2. It is noted that the curves of the drag force distribution for the two levels 5,3 and 6,3% have the same trends; but at a critical Reynolds number of 25000 the trend of the level Tu = 4,6 % is not behaving in the same way. This indicates an influence of the level of turbulence. For level Tu = 4.6% arriving at Re = 40000, the strength of the drag decreases. This phenomenon is explained by the surface visualization for Re = 45000 of figure 9, where the separation point is situated after the groove. Therefore a reduced wake distance resulting in a lower drag force is obtained.

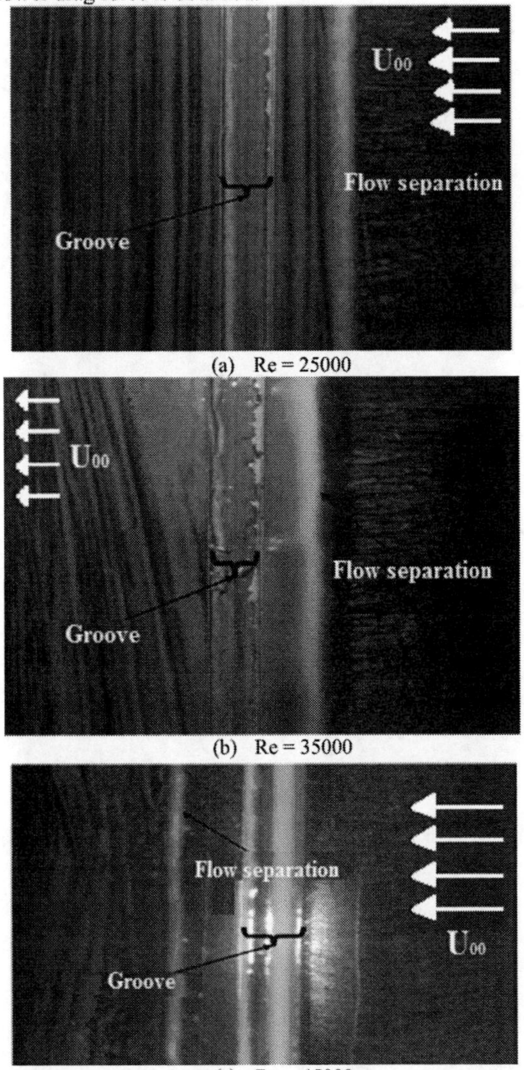

(a) Re = 25000

(b) Re = 35000

(c) Re = 45000

Figure 9. Surface flow visualization for Tu = 4.6% (tube 5-2).

From Re = 25000, the drag force decreases. Compared to Tu = 4.6%, the drag force has lower values for Tu = 6.3%. The visualization shows that when exceeding a number of Re = 25000, the flow separation is close to the groove which leads to the decrease in the drag force value.

Consequently, it is noted that by increasing the turbulence intensity the separation angle is delayed causing a decrease in the value of the drag.

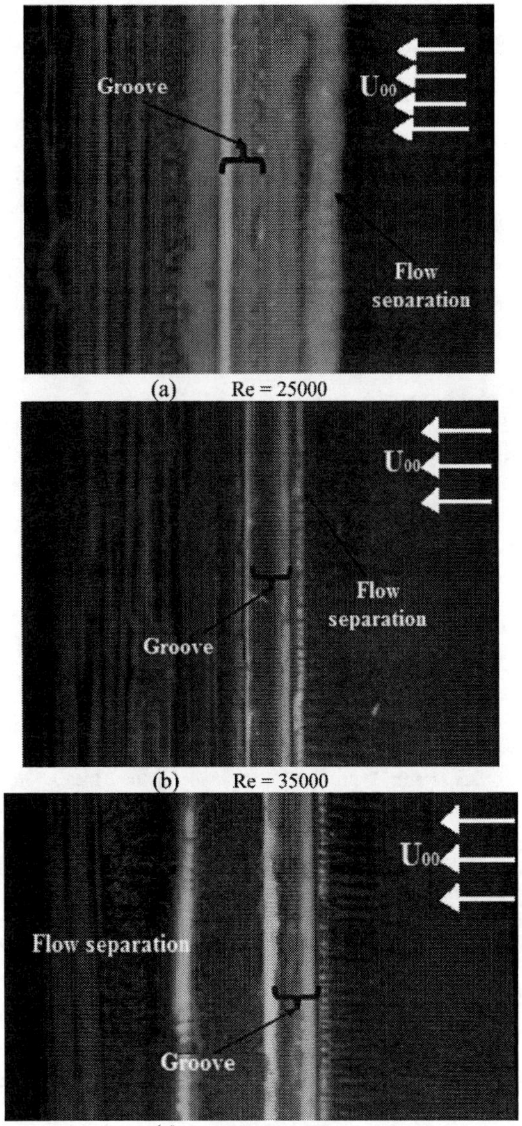

(a) Re = 25000

(b) Re = 35000

(c) Re = 45000

Figure 10. Surface flow visualization for Tu = 6.3% (tube 5-2).

Conclusion

The turbulent cross flow in staggered tube bundle arrangement with transverse and longitudinal pitch-to-diameter ratios of 3.8 and 2.1 respectively is experimentally investigated. The configuration is equipped with grooved cylinders. The obtained results seem to agree rather well with the experimental and theoretical data already gathered, in particular on the distribution of pressure around the examined cylinders. The following conclusions are drawn:

- Increasing the turbulence level delayed the separation point;
- At some critical Reynolds number, increased turbulence level leads to a decrease in the drag force in column 2;

- By increasing the Reynolds number of the separation point is delayed.
- The evaluation and analysis of the distribution of the pressure on the surface of the cylinder; it is found that the pressure drag reduction is due to the decreasing pressure in the front face of the circular cylinder and the moving aft of the separation point on the circular cylinder.

Flow visualization provided a clear picture of the separation point which can help in better understanding of the flow character.

References

1. S. Aiba, H. Tsuchida, T. Ota. Bull JSME, **25**, 919 (1982).
2. O. Romberg, K. Popp. J. Fluid Struct., **12**, 17 (1998a).
3. O. Romberg, K. Popp. J. Fluid Struct, **12**, 153 (1998a).
4. S.S. Chen, G.S. Srikantiah. ASME J. Press. Vess. Technol., **123**, 429 (2001).
5. M. Rottmann, K. Popp. J. Fluid Struct, **18**, 595 (2003).
6. A. Žukauskas,R. Ulinskas. Heat Transfer in Tube Banks in Crossflow, Hémisphère, New York. (1988).
7. S.S. Paul, M.F. Tachie, S.J. Ormiston. Int. J. Heat Fluid Flow, **28**, 441 (2007).
8. S.S. Paul, S.J. Ormiston, M.F. Tachie. Int. J. of Heat and fluid flow, **29**, 387 (2008).
9. K. Lam, F.H. Wang, J.Y. Li, R.M.C. So. J. Fluid. Struct., **19**, 321 (2004a).
10. K. Lam, F.H. Wang, R,M.C. So. J. Fluid Struct., **19**, 815 (2004b).
11. K. Lam, Y.F. Lin. J. Mech. Sci. Technol., **21**, 1331 (2007).
12. K. Lam, Y.F. Lin. Fluid Mech., **620**, 195 (2009).
13. K. Lam, F. Lin, L. Zou, Y. Liu. Int. Journal of Heat and Fluid Flow, **31**, 32 (2010).
14. O. Ladjedel, L. Adjlout, T. Yahiaoui, O. Imine. Mechanics & Industry, **14**, 287 (2013).
15. D. Traub, Chem. Eng. Process. **28**, 173 (1990).
16. K. Stephan, J. Mitrovic, Chem. Ing.Technol. **56**, 427 (1984).
17. S. Umeda, S.W. Yang. Exp. Fluids **26**, 389 (1999).
18. M. Beziel, K. Stephan. Chem. Eng. Technol. **15**, 219 (1992).
19. C. Sak, R. Liu, D.S.-K. Ting, G.W. Rankin. Exp. Thermal Fluid Sci. **31**, 279 (2007).
20. P. E. Roach ''the generation of nearly isotropic turbulence by means of grids''. **Vol 8**, N° 2, June (1987).

EPJ Web of Conferences **114**, 02104 (2016)
DOI: 10.1051/epjconf/201611402104

Effect of the duty cycle on the spark-plug plasma synthetic jet actuator

Mehmet Seyhan[1,a], Yahya Erkan Akansu[2], Fuat Karakaya[3], Cihan Yesildag[2] and Hürrem Akbıyık[4]

[1]*Karadeniz Technical University, Mechanical Engineering Department, 61080 Trabzon, Turkey*
[2]*Nigde University, Mechanical Engineering Department, 51200 Niğde, Turkey*
[3]*Nigde University, Electrical and Electronics Engineering Department, 51200 Niğde, Turkey*
[4]*Cukurova University, Mechanical Engineering Department, 01380 Adana, Turkey*

Abstract. A promising novel actuator called Spark-Plug Plasma Synthetic Jet (SPSJ) has been developed in Atmospheric Plasma Research Laboratory at Niğde University. It generates electrothermally high synthetic jet velocity by using high voltage. SPSJ actuator can be utilized to be an active flow control device having some advantages such as no moving parts, low energy consumption and easy to integrate the system. This actuator consists of two main components: semi-surface spark plug (NGK BUHW) as an anode electrode and a cap having an orifice as a cathode electrode. The cap, having a jet exit orifice diameter of 2 mm, has diameter of 4.4 mm and height of 4.65 mm. This study presents the characteristics of SPSJ actuator by using the hot wire anemometer in order to approximately determine jet velocity in quiescent air. Peak velocity as high as 180 m/s was obtained for f_c= 100 and duty cycle 50%. The flow visualization indicated that the actuator's jet velocity is enough to penetrate the developed boundary layer.

1 Introduction

In recent years, active flow control (AFC) has been greatly drawn attention due to some advantages such as no moving parts, low energy consumption and without external air supply. Researchers have developed various devices based on plasma such as dielectric barrier discharge (DBD) actuator [1], DBD plasma synthetic jet actuator [2] and plasma spark jet (PSJ) actuator [3-5]. Especially PSJ actuators have very high jet velocity in comparison with the other plasma based actuator like DBD actuator.

A PSJ actuator called as a Sparkjet was developed by Grossmann et al. [3] at Johns Hopkins University Applied Physics Laboratory. This actuator had two electrodes inside a small cavity. The periods of this actuator consists of three different steps. Respectively, these are referred to as energy deposition, gas discharge and recovery. Moreover, the PSJ actuator can be classified as a two electrode [3, 5 and 6] and a three-electrode PSJ actuator [4, 7]. Explanations of these steps will be comprehensively explained in the section 2. Narayanaswamy et al. [6] carried out the experimental study via PSJ actuator having different cavity volume and orifice diameter. They had measured temperature of actuator by using optical emission spectroscopy and also obtained jet exit velocity of about 250 m/s with consuming discharge energies of about 35 mJ per generated jet. Zong et al. [7] implemented to experimental study using three-electrode PSJ actuator having a volume of 1178 mm^3 and an orifice diameter of 3 mm. They measured jet exit velocity of about 230 m/s by using the Schlieren method. According to the above mentioned literature results, this actuator can be effectively used in both subsonic and supersonic flow applications without moving parts and external air supply. But it also has some disadvantages, such as high price components (macorTM, high voltage power supply, etc.), and quick oxidation of actuator's electrodes.

This study has two purposes. The first is to explain how to develop a promising novel plasma synthetic jet based on spark-plug so as to use in flow control applications such as military and civilian aircraft. The second is to characterize the developed novel SPSJ actuator by using hot-wire anemometer in the quiescent air. Smoke-wire flow visualization experiments were also carried out to present how the developed SPSJ actuator can be affected the boundary layer developing over the flat plate.

2 Experimental setup

2.1 Designed SPSJ Actuator

A promising novel actuator called Spark-Plug Plasma Synthetic Jet (SPSJ) Actuator has been developed in Atmospheric Plasma Research Laboratory (APRL) at Niğde University. Developed SPSJ is inspired by

[a] Corresponding author: mehmetseyhan@ktu.edu.tr

internal combustion gasoline engine. The engine cylinder with spark plug has very big volume while our developed SPSJ has very small volume as well as an orifice. As shown in figure 1, SPSJ actuator which can be generated high synthetic jet velocity consists of two main components: semi-surface spark plug (NGK BUHW) as an anode electrode and a cap having an orifice that is made of aluminum as a cathode electrode. This cap, having orifice diameter of 2 mm, has diameter of 4.4 mm and height of 4.65 mm. The cavity volume is almost 70 mm^3.

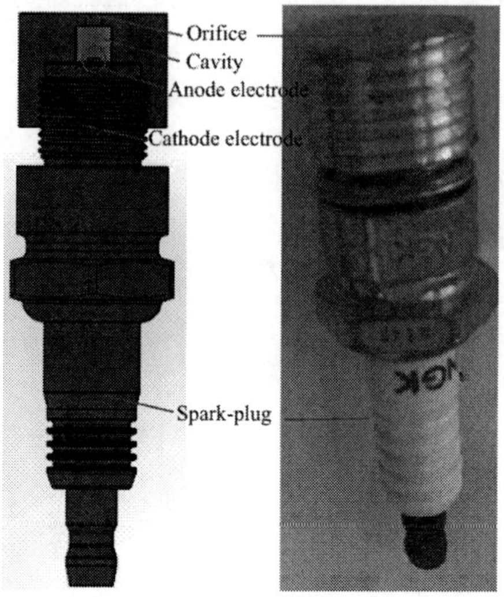

(a) (b)

Figure 1. (a) Schematic view of developed SPSJ actuator, (b) Photo of developed SPSJ actuator.

The process of this actuator consists of three different steps which is same as the PSJ actuator [3]. As illustrated in figure 2, these are referred to as energy deposition, gas discharge and recovery respectively. The cap creates the cavity volume so as to heat rapidly the air in the cavity and then bring about suddenly augmentation pressure of the air, when applied pulse DC high voltage between these electrodes. End of this period, pressure of the air is very high compared to the ambient air and this pressurized air is expelled at high velocity through the orifice. Afterwards, the air pressure, which is dropped under the ambient pressure in the cavity, generates vacuum inside the cavity. Therefore, the ambient air is drawn into the cavity and the SPSJ actuator is ready for another cycle.

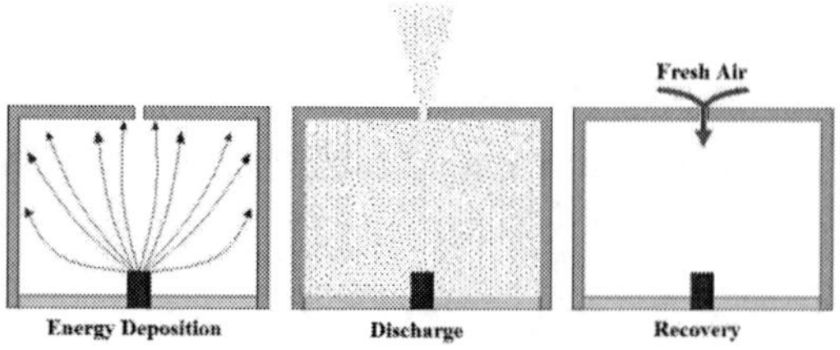

Figure 2. Process of Spark-Plug plasma synthetic jet.

2.2 Power Supply

Belinger et al. [8] mentioned about two type power supply that can work the actuator. One is a high voltage power supply to commute high voltage directly during stage of discharge. The other is a low voltage power supply associated with a high voltage transformer and a transistor. According to this, we design a low voltage power supply associated with the high voltage transformer in this study. Designed power supply consists of a DC power supply, a high-voltage transformer and a MOSFET switch. DC power supply (PACIFIC^{TM}), which has output 30 V and 5 A, is connected to developed low cost DC high voltage transformer, which is worked between 2 - 6 V input, having maximum 400kV output. The MOSFET switch is also inserted in the ground line between DC power supply and the high voltage transformer. The MOSFET switch is controlled by an arbitrary wave generator (AATech-AWG-1010) which can be adjusted with the actuator excitation to get desired frequency. As shown in figure 3, input current was measured by Fluke 80i-110a AC/DC current probe which is monitored with the help of Tektronix model TDS2012B oscilloscope. Another oscilloscope was also used for signal visualization generated by function generator. Input mean voltage(V_m) and the current (I_{pp}) was measured as V_m=2 V and I_{pp}=2.48 A respectively.

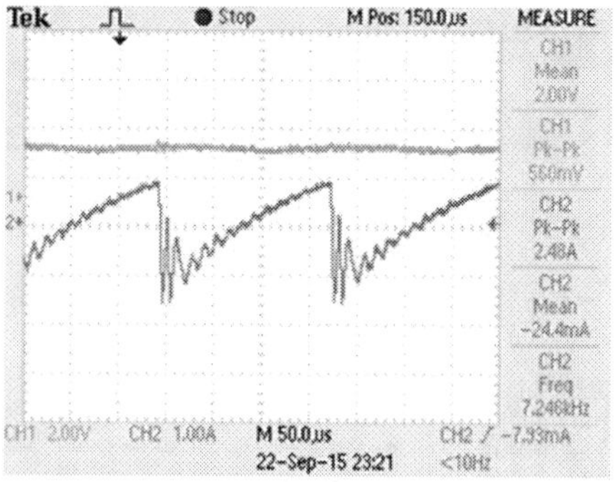

Figure 3. Captured image from oscilloscope in order to present input mean voltage and peak to peak current

2.3 Measurement Devices

The experiments were conducted to determine the characteristic of the developed novel SPSJ actuator by using hot-wire anemometer in the quiescent air. Hot-wire measurement, which is measured the instantaneous air velocity, was carried out by using the Dantec 55P11 single wire probe. This probe operates together with Dantec Multichannel Constant Temperature Anemometer (CTA). The hot-wire probe was calibrated in a nozzle calibrator against a micromanometer. For each measurement, 32000 data was collected from the hot-wire at a sample frequency of 4 kHz with the help of NI PCIe-6323 DAQ card. The probe was positioned at jet axis. It was placed on computer controlled two axis traverse mechanism so as to drive sensitively to the desired location on x and y axis. Secondly, Smoke-wire flow visualization experiment was implemented to present the effects of the SPSJ actuator on flow over the flat plate with the free stream velocity of 1.5 m/s. The SPSJ actuator was vertically placed on the flat plate in the open-type wind tunnel having a square cross-section 57 cm x 57 cm in the Aerodynamics Research Laboratory of the Mechanical Engineering department at Niğde University. Both the leading edge and the trailing edge of the flat plate are beveled.

3 Results

3.1 Velocity measurements by using hot-wire anemometer

In order to present characteristic of the developed SPSJ actuators, instantaneous velocity measurement is being carried out with the help of the hot wire anemometer in quiescent air at the probe location above 6 mm from the top surface of the cap. As illustrated in figure 4, the instantaneous velocity distribution as a function of time was presented for different duty cycle changing between 10% and 90% in driving frequency of 0.5 Hz. Peak instantaneous velocity is as high as 175 m/s which was obtained at duty cycle of 90%. On the other hand, in the velocity time history, some of the jet velocities weren't exactly acquired rightly because of insufficient sampling frequency of the hot-wire anemometer. Therefore, the peak jet velocities obtained randomly due to very small jet formation time according to efficient sampling period. As clearly shown in figure 4, the used low-cost high voltage power supply has instable characteristic during the elapsed time. It sometimes provides enough energy to SPSJ actuator but it doesn't. In this case, instantaneous velocities acutely change. As duty cycle decrease from 90% to 10%, peak velocities significantly diminish. At the low duty cycle, the discharge time is too short in order to obtain enough energy for producing plasma in comparison with other duty cycle rate; therefore the produced plasma by giving energy to the cavity volume generates the low jet velocity.

Figure 5 shows effect of the variation of duty cycle on the instantaneous velocity distribution at f_e= 1 Hz. Peak jet velocity is as high as 145 m/s which was obtained at the duty cycle of 90% and 75%. At f_e= 1 Hz, there is no significant effect of different duty cycles over peak jet velocity for the range of duty cycle from 25% to 90%. For duty cycle 10%, the first pulse jet was obtained as high as 90 m/s but the actuator doesn't produce strong enough jet at the other periods. Aforementioned problems are valid for this case such as instability of power supply and sampling sensitivity of the measurement.

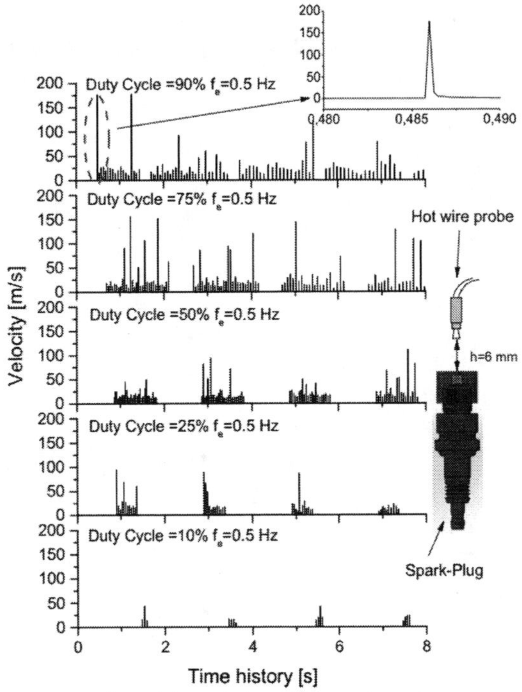

Figure 4. Effect of the variation of duty cycle on the instantaneous velocity distribution at f_e= 0.5 Hz.

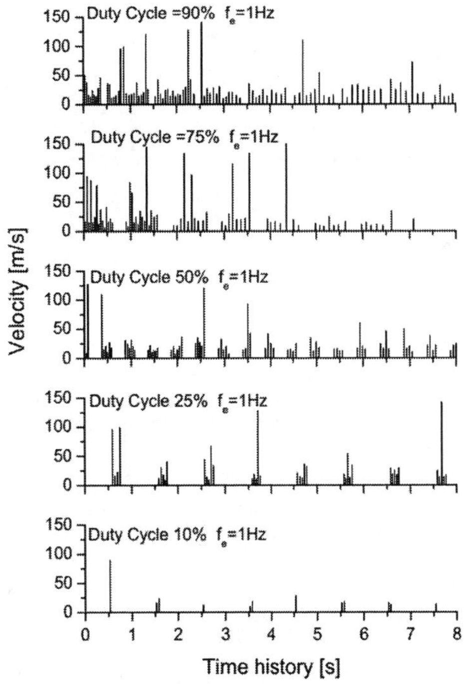

Figure 5. Effect of the variation of duty cycle on the instantaneous velocity distribution at f_e= 1 Hz.

As illustrated in figure 6, the instantaneous velocity distribution as a function of time was presented for different excitation frequency, changing from f_e=0.2 Hz to f_e=100 Hz, at duty cycle of 50%. As clearly seen from figure 6, there is no significant change in peak value of the instantaneous velocity but the repetition of the jet decreases by the increasing of the excitation frequency due to characteristic of power supply at the high excitation frequency. Peak instantaneous velocity is obtained 180 m/s which are highest velocity in comparison with the other excitation frequencies for f_e at 100 Hz.

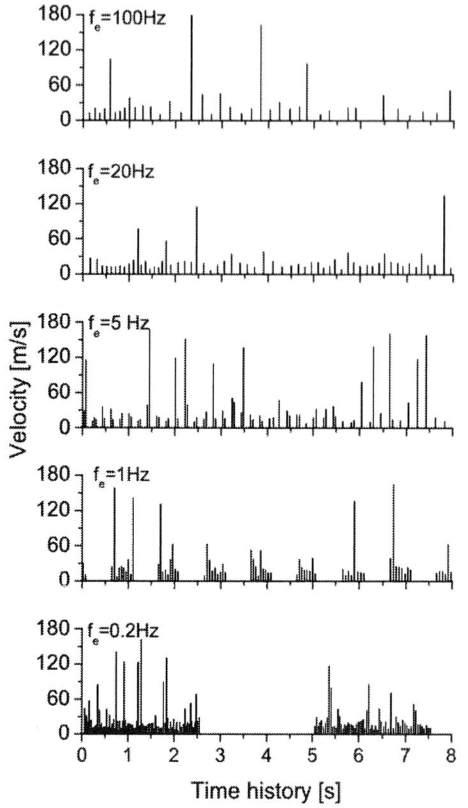

Figure 6. Effect of the variation of excitation frequency on the instantaneous velocity distribution at duty cycle 50%.

3.2 Flow visualization

As shown in figure 7, the smoke-wire flow visualization experiments were carried out to present how the developed SPSJ actuator, which generates high jet velocity, can affect the boundary layer developing over the flat plate from leading edge to trailing edge. The first captured photo shows velocity field in the actuator off situation. The other captured photos show the perturbation effect of the SPSJ actuator's jet on the flow. These photos present the evolution of jet at different time inside the developed boundary layer. It should be noted that this visualization also shows the ability of the actuator in order to use the active flow control applications. Periodically generated jets are shown as a short line during the exposure time of the camera in the frame. Produced jets having different velocity submerged into the boundary layer and some of them, which are high velocity in nature, penetrate the shear layers.

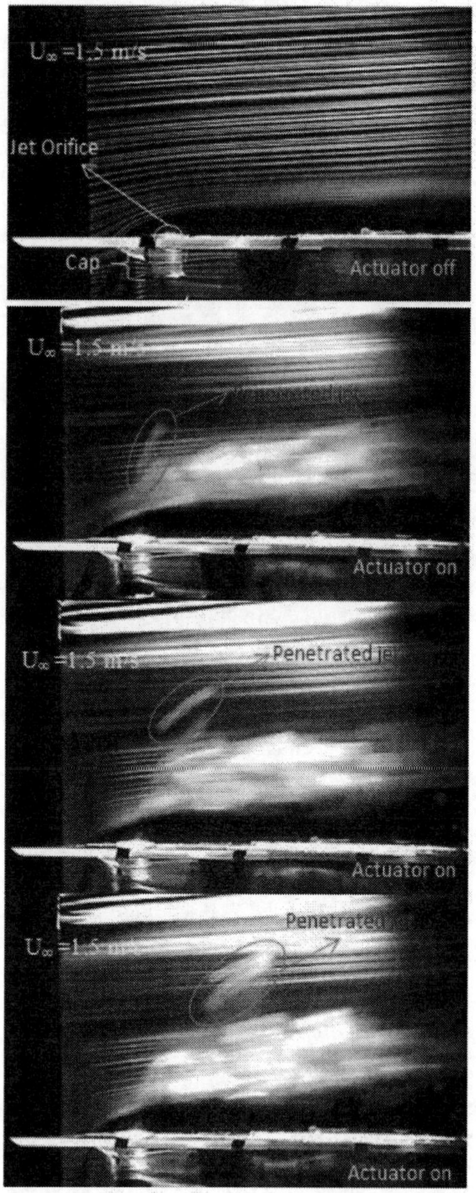

Figure 7. Effects of evolution of jet inside boundary layer developing over the flat plate.

4 Conclusions

A promising novel SPSJ actuator was developed so as to use in AFC applications. Characteristics of the developed SPSJ actuator was also investigated by using hot-wire anemometer in the quiescent air. Peak velocity was obtained as high as 180 m/s for f_e= 100 and duty cycle 50%. Smoke-wire flow visualization experiment was implemented to present the effects of the SPSJ actuator on flow around the flat plate. The flow visualization indicates that the actuator's jet velocity is enough to penetrate the developed boundary layer. Evolution of the jet inside boundary layer is clearly presented. These results revealed that the SPSJ actuator is a promising and satisfying for the AFC application but we haven't enough qualified equipment such as high speed Schlieren system in order to make a precise measurement of this developed actuator.

Acknowledgements

The authors would like to acknowledge the financial support of this work by the Scientific and Technological Research Council of Turkey (TUBITAK) under the Contract Number of 213M179.

Nomenclature

V	applied plasma voltage
f_e	excitation frequency of the actuator
U_∞	freestream velocity
h	distance between the probe and the cap
AC	alternating current
DC	direct current

References

1. E. Moreau, J.Phys. D: Appl. Phys., **40**, 605-636 (2007)
2. A. Santhanakrishnan, J.D. Jacob, *44th AIAA Aerospace Sciences Meeting and Exhibit*, pp.317, (2006)
3. K.R. Grossman, B.Z. Cybyk, M.VanWie, AIAA, **57**, (2003)
4. S.J. Haack, T. Taylor, J. Emhoff, B. Cybyk, *5th Flow Control Conferance*, **28**, (2010)
5. B. Caruana, P. Barricau, P. Hardy, J.P. Cambronne, A. Belinger, AIAA, **1307**, (2009)
6. V. Narayanaswamy, L.L. Raja, N.T. Clemens, AIAA J. **48**, 297-305 (2010)
7. H.H. Zong, W. Cui, Y. Wu, Z.B. Zhang, H. Liang, M. Jia, Y.H. Li, Sensor Actuat. A-Phys. **222**, 114-121 (2015)
8. A. Belinger, P. Hardy, P. Barricau, J.P. Cambronne, D. Caruana, J.Phys. D: Appl. Phys., **44**, 365201, (2011)

EPJ Web of Conferences **114**, 02105 (2016)
DOI: 10.1051/epjconf/201611402105

CFD simulation of flow-induced vibration of an elastically supported airfoil

Petr Šidlof[1,a]

[1] *Technical University of Liberec, NTI FM, Studentská 2, 461 17 Liberec 1, Czech Republic*

Abstract. Flow-induced vibration of lifting or control surfaces in aircraft may lead to catastrophic consequences. Under certain circumstances, the interaction between the airflow and the elastic structure may lead to instability with energy transferred from the airflow to the structure and with exponentially increasing amplitudes of the structure. In the current work, a CFD simulation of an elastically supported NACA0015 airfoil with two degrees of freedom (pitch and plunge) coupled with 2D incompressible airflow is presented. The geometry of the airfoil, mass, moment of inertia, location of the centroid, linear and torsional stiffness was matched to properties of a physical airfoil model used for wind-tunnel measurements. The simulations were run within the *OpenFOAM* computational package. The results of the CFD simulations were compared with the experimental data.

1 Introduction

In aerospace engineering, fluid-structure interaction can play a very important and potentially dangerous role: under certain circumstances, the coupling between flow and structure may lead to unstable exponentially increasing oscillations. The classical example is the flutter instability of airfoils, which occurs for systems with two degrees of freedom when the critical flow velocity is surpassed [1]. Flutter is a dynamic instability of an elastic structure coupled to airflow, caused by the interaction between elastic, inertial and aerodynamic forces. Flutter may occur at certain flow velocities and structural natural vibration frequencies, when the energy is transferred from the airflow to the structure and the internal damping is not able to absorb it. This can lead to vibration of an aircraft component with exponentially increasing amplitudes and catastrophic consequences. Thus, the aircraft components (wings, flaps and ailerions, stabilators, elevons and rudders) must be designed and tested not only to sustain dynamic loads induced by mechanical accelerations and air turbulence during take-off, cruise and landing, but also to ensure that within design flight conditions, the aircraft components may never encounter dynamic flow-induced instability.

In general, flutter instability is a complex nonlinear phenomenon which is not yet fully understood in all the aspects. The first fundamental type of flutter is the coupled-mode flutter, which can occur for elastic structures with two degrees of freedom. If the natural frequencies of the respective modes (without airflow) are not far apart, when the object is subjected to airflow and the flow velocity increases, the frequencies of the modes approach. For a critical flow velocity the frequencies merge and coupled-mode flutter instability occurs. The second common type of aeroelastic instability is the stall flutter. Here, the physical mechanism is different. The stall flutter of airfoils is triggered for high angles of attack, when the airflow separates from the airfoil surface. In this case, the ratio of the vortex shedding frequency and natural frequency of the airfoil vibration become important.

The current study reports on a 2D CFD model of flow-induced vibration of a rigid airfoil with two degrees of freedom – pitch (rotation about the elastic axis) and plunge (vertical deflection), supported by a linear and torsional spring. Such system can model a real aircraft wing, with pitching corresponding to torsional vibration of the elastic aircraft wing structure, and plunge corresponding to vibrations in the bending mode. The CFD model was designed according to parameters of the physical model used for wind tunnel measurements [2-5], so that the results of the CFD simulations can be compared with the experimental results.

2 Methods

The computational model tries to fit as well as possible the geometry, mechanical and aerodynamic parameters of the physical airfoil model depicted in Figure 1. The physical setup in Figure 1 is a redesigned and improved version of a model reported in [2]. The airfoil, supported by a linear and torsional springs, is fixed in a suction-type wind tunnel with optical access through the side walls for

[a] Corresponding author: petr.sidlof@tul.cz

Schlieren visualizations and interferometric measurements of the flow field.

Figure 1. Physical model of the pitch-plunge airfoil in the wind tunnel (with front wall dismantled).

2.1 Geometry and mesh

The airfoil has a symmetric NACA0015 profile with a chord length of 59.5 mm. The airfoil is located in a 2D channel with a height of 210 mm and length of 500 mm (measured from the trailing edge).

The computational grid for the CFD simulations is a C-type structured mesh generated in *OpenFOAM's* structured mesh generator *blockMesh* using an octave script provided in [6]. The mesh is refined towards the airfoil surface. For the purpose of first testing of the fluid-structure interaction simulations, a coarse mesh composed of only 10k quadrilateral elements was used in order to keep the simulation time as low as possible. The mesh is depicted in Figure 2.

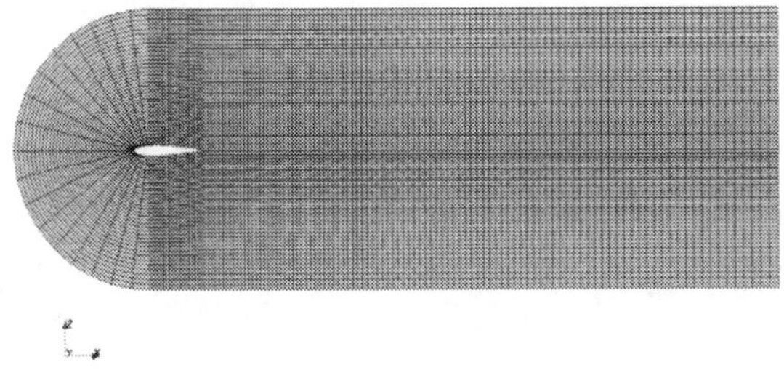

Figure 2. Geometry of the computational domain and mesh used for CFD simulations.

2.2 Mechanical parameters

All the mechanical parameters of the computational model were set according to the values directly measured or identified from the physical airfoil model. The elastic axis of the model is located in 1/3 of the chord length – slightly upstream of the centroid, located at $x_T = 26.4$ mm. The mass of the airfoil and the moving parts of the frame (leading rods) is $m_t = 1.498$ kg, the mass of the airfoil itself $m = 148$ g. The natural frequencies of the plunging and pitching mode are $f_1 = 16.8$ Hz and $f_2 = 14.5$ Hz, respectively. The linear stiffness of the support was measured as $k_t = 16\,500$ N/m. The computational model does not consider the total mass m_t, but the mass of the airfoil only. Thus, a reduced stiffness $k = 1629$ N/m preserving the same natural frequency f_1 was used in the computational model. The moment of inertia of the airfoil was identified from the torsional natural frequency and torsional stiffness as

$$J = \frac{k_T}{(2\pi f_T)^2} = 4.78710^{-5}\ \mathrm{kg\,m^2}\ . \qquad (1)$$

A problematic issue is the damping of the respective modes. During parameter identification, the damping ratios were identified as $b_{1id} = 5.606$ N s/m and $b_{2id} = 0.000785$ N m s/rad. However, the mechanical design of physical model does not ensure that the damping of both modes is precisely linear, which resulted in discrepancies in vibration attenuation when using these values of the damping ratios in the computational model. Eventually, the ratios were set to $b_1 = 1.5$ N s/m and $b_2 = 0.0001$ N m s/rad to match the attenuation of the physical model for zero flow velocity.

2.3 CFD model for the fluid-structure interaction, initial and boundary conditions

The fluid flow is modeled by incompressible Navier-Stokes equations on the 2D computational domain

$$\begin{aligned} &\nabla \cdot \mathbf{u} = 0 \\ &\frac{\partial}{\partial t}\mathbf{u} + \nabla \cdot (\mathbf{u} \otimes \mathbf{u}) - \nu \Delta \mathbf{u} + \frac{1}{\rho}\nabla p = 0 \end{aligned} \qquad (2)$$

for the flow velocity $\boldsymbol{u}$ and pressure p, with the kinematic viscosity of air $\upsilon = 1.58\ 10^{-5}$ m^2/s. No turbulence modeling was used in the CFD simulations published in this paper.

The boundary conditions imposed on the pressure and velocity fields were chosen to approximate the wind tunnel conditions. The air flow is driven by a constant velocity at inlet, which was measured by a Pitot tube in the physical setup. Due to the presence of large vortical structures convected downstream up to the outlet of the computational domain and consequent possible backflow to the domain, which destabilizes the numerical solution, a stabilized outflow condition was specified at the outlet of the domain: $\delta\mathbf{u}/\delta\mathbf{n} = 0$ if the velocity direction points outward from the domain, and $\mathbf{u} = 0$ otherwise. On the top and bottom walls a no-slip boundary condition was specified, on the airfoil surface, the flow velocity was set equal to the velocity of the structure. The pressure $p = 0$ at outlet, on the other parts of the boundary $\delta p/\delta\mathbf{n} = 0$.

The initial conditions were zero pressure and constant velocity in the whole computational domain. The initial position of the airfoil was set to $z = 3$ mm, simulating the initial vertical deflection imposed to the physical model during wind-tunnel measurements. Additionally, the airfoil was given an initial angular velocity $\omega = -15.76$ rad/s (corresponding to harmonic rotation with an amplitude of 10 degrees) to trigger the rotational mode.

The computational domain is deformed in time due to airfoil motion. The position of the airfoil and the coordinates of the element vertices in a new timestep are resolved using the following algorithm:

- With the knowledge of the pressure and velocity fields, the aerodynamic drag and lift force and the aerodynamic moment on the airfoil surface (both viscous and pressure parts) are calculated.
- The new position of the airfoil is found by solving the 2-DOF equation of motion of the system, including the external aerodynamic forces.
- The new position of the mesh vertices is calculated by a specialized *OpenFOAM* mesh motion solver based on spherical linear interpolation, which is designed to prevent cell shearing.

2.4 Numerical solution

The numerical solution was implemented in *OpenFOAM*, an object-oriented open-source set of C++ libraries for CFD simulations. *OpenFOAM* is based on the finite-volume method in a collocated cell-centered approach on unstructured meshes. The incompressible Navier-Stokes equations on a moving computational mesh were solved numerically using a modified PISO algorithm [7]. In contrast to the standard PISO algorithm, it has a substep iteration loop: multiple cycles over the same timestep with the last iteration results (optionally under-relaxed) used as an initial guess for the next substep iteration.

The discretization schemes were as follows: implicit first-order Euler for the time derivative, linear upwind scheme for the convective term, and central differencing scheme (CDS) with nonorthogonal correction for the diffusion term. The timestep Δt is adjusted automatically during the transient solution so that the maximum local Courant number is kept below a predefined limit. The Courant number on unstructured 3D meshes is calculated as

$$Co = \frac{1}{2}\frac{\sum_f |\Phi_f|}{\Delta V}\Delta t \;, \qquad (3)$$

where $\Phi_f = A_f(\mathbf{u}_f \cdot \mathbf{n}_f)$ is the velocity flux normal to face f with the surface A_f of a cell with a volume ΔV.

Due to significant element distortion and consequent loss of element orthogonality in the proximity of the moving airfoil, two outer loops of nonorthogonal correctors were utilized within the modified PISO algorithm to guarantee stability of the computation. Unlike the discretization of the temporal, convective and diffusive terms, the nonorthogonal correctors are treated explicitly. The numerical scheme thus acquires explicit character and the timestep has to be limited so that the CFL condition is satisfied. In the current simulations, the Courant number was limited to Co < 0.9.

The resulting linear system for momentum was solved using a Gauss-Seidel smoother. For the pressure predictor and corrector steps and for the mesh motion, a geometric multigrid solver was used with an *OpenFOAM*-specific cell agglomeration algorithm and a conjugate-gradient-type method for solution of the coarsest level matrix.

The simulations were run in parallel using the domain decomposition method. The current simulations on a coarse mesh consisting of 10k elements proved efficient parallel speedup up to four CPU cores on a SGI Altix UV-100 shared-memory supercomputer.

3 Results

To reveal the stability boundaries of the system and determine the critical flow velocity for the onset of instable exponentially increasing airfoil vibration, the numerical simulations were run for inflow velocities $u = 30 - 90$ m/s (see Table 1).

Table 1. Inlet flow velocity u [m/s] in numerical simulations s520a .. s520g. The airfoil oscillations are stable for s520a-e, unstable for s520f-g.

s520	a	b	c	d	e	f	g
***u* [m/s]**	30	40	50	60	70	80	90

For the velocities $u = 30 - 70$ m/s, the oscillations remain stable and exponentially decay due to the mechanical damping in the system (see Figure 3). For inflow velocities exceeding 80 m/s, however, the system is instable and the amplitudes exponentially increase in time. The CFD simulations are limited by the extent of airfoil motion, which induces mesh deformation. Under very large amplitudes, especially in the angle of attack, the computational mesh deforms too much, the elements distort and the numerical simulation becomes unstable. This happened in s520f and s520g for pitching angles exceeding about 50 degrees.

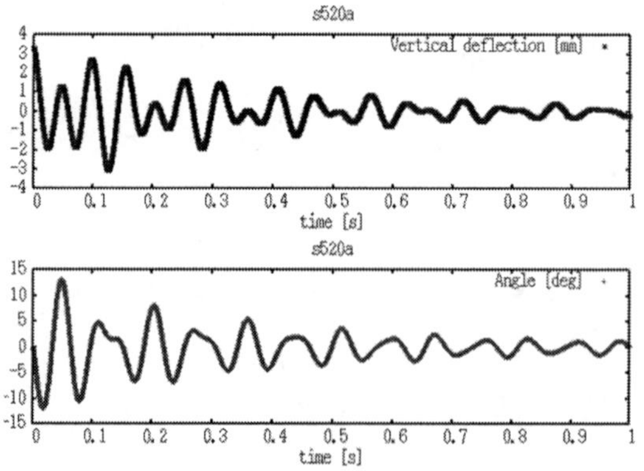

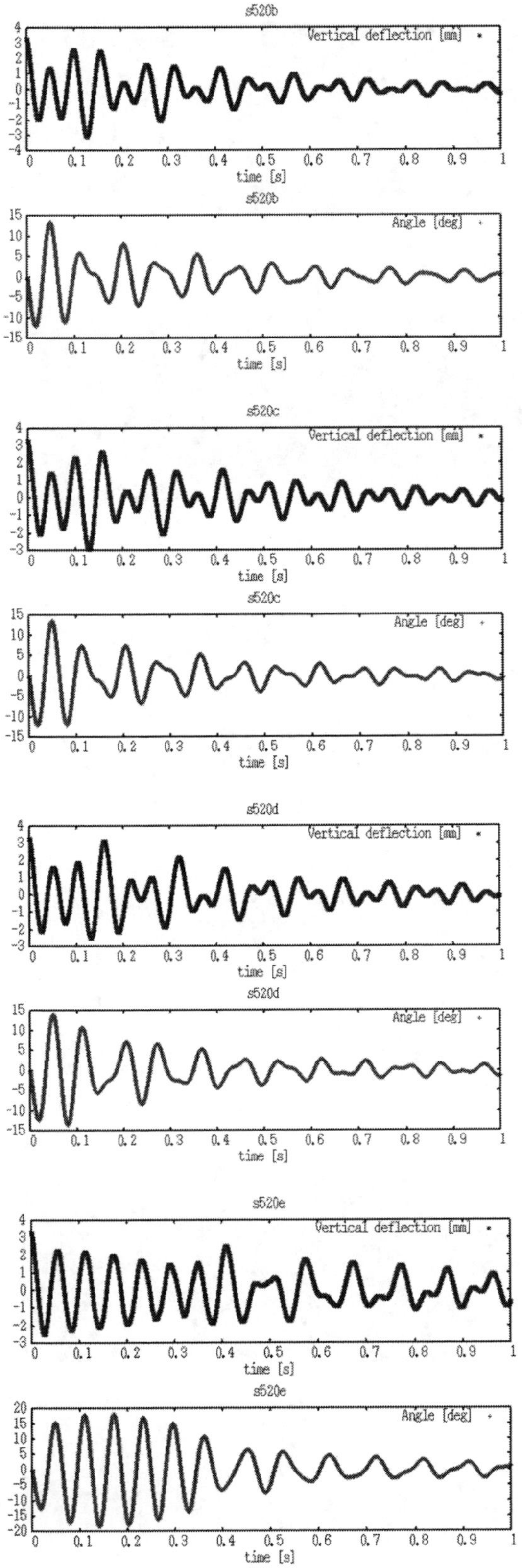

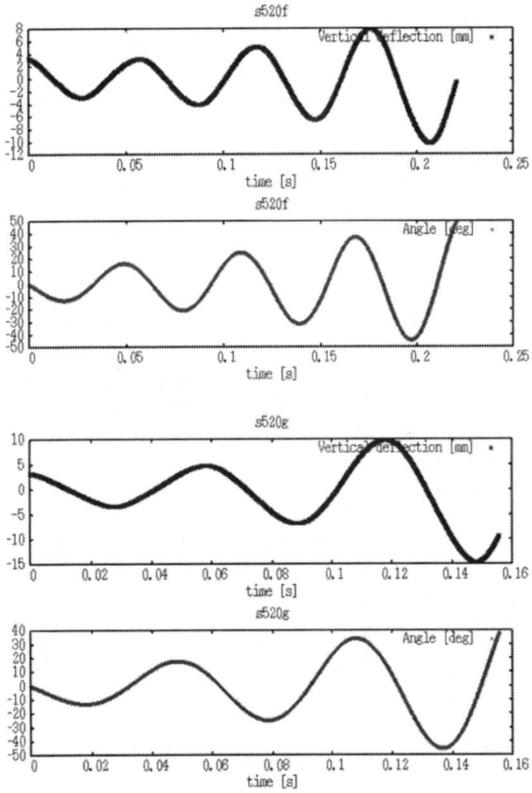

Figure 3. Time history of plunge and pitch in numerical simulations s520. Stable damped oscillations (s520a..s520e), instable exponentially increasing amplitudes (stall flutter, s520f and s520g).

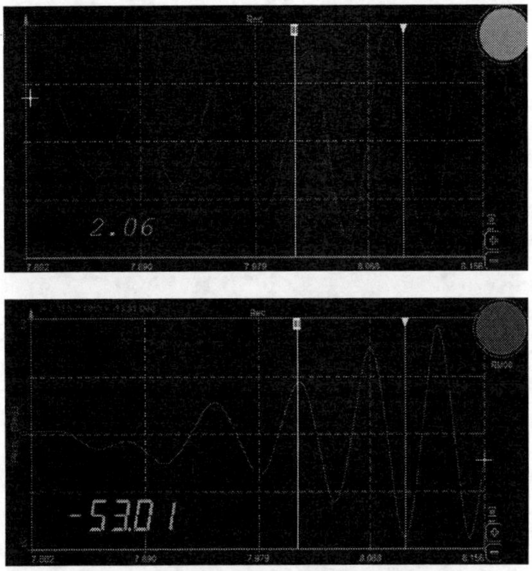

Figure 4. Time history of plunge and pitch in wind-tunnel measurement No. 2915-22 (inlet Mach number Ma = 0.228, u = 78.2 m/s. After about two seconds of the transient development with amplitudes reaching 10 mm and 56 deg, the amplitudes stabilize at 5.9 mm and 42 deg (not shown in figure).

The simulated time history of pitch and plunge in Figure 3 can be compared with values measured on the physical model in the wind tunnel. Figure 4 shows the time development of the vertical deflection and angle of attack in measurement 2915-22, where the inlet Mach number was Ma = 0.228 (inflow velocity u = 78.2 m/s), corresponding approximately to numerical simulation s520f. The numerical simulation is in quite good agreement with the measurement, only the amplitudes in the measurement are slightly higher.

The velocity fields from numerical simulation s520f are shown in Figure 5 in time instants corresponding to zero, minimum and maximum pitching angles. The simulation shows the vortex street generated in the wake of the airflow. For the higher angles of attack, the flow is massively separated with vortex structures shed from the shear layer.

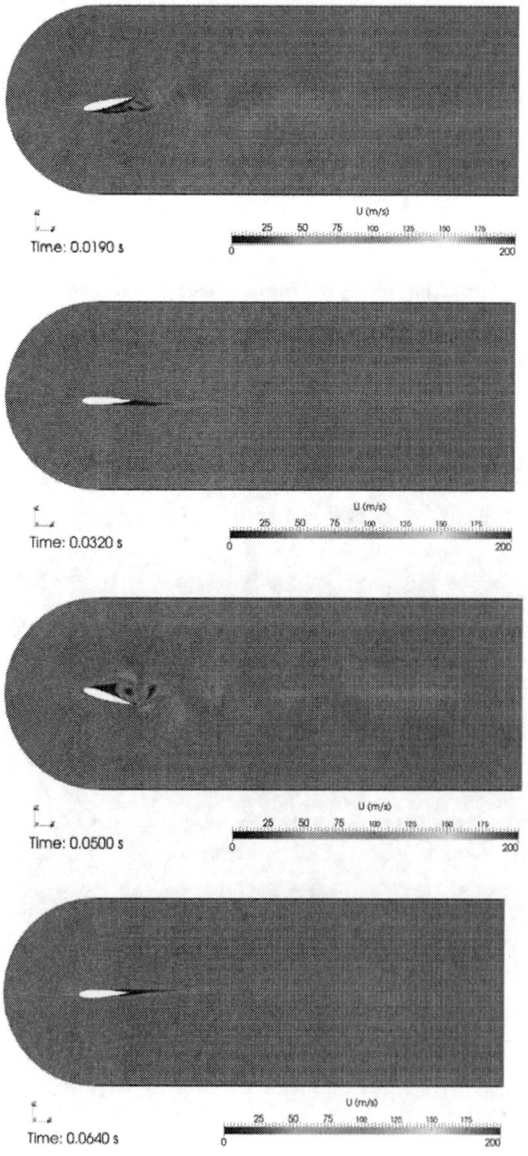

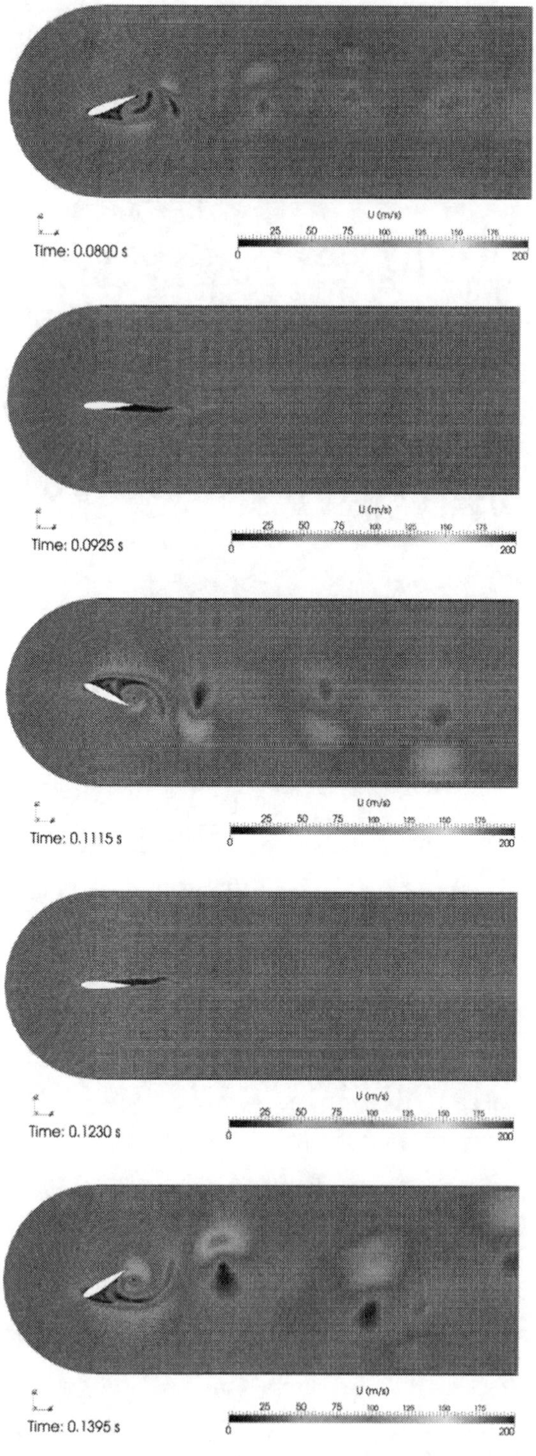

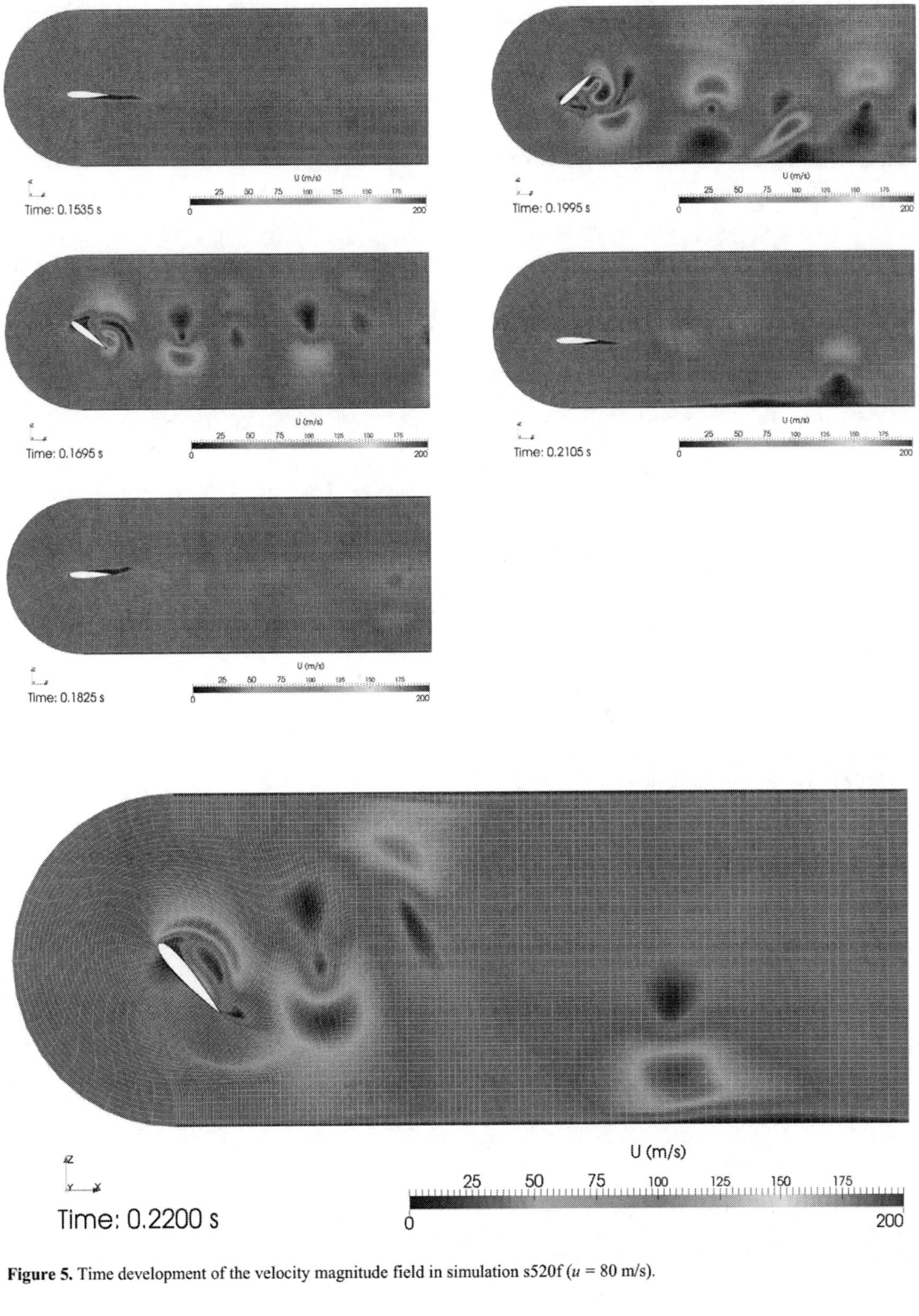

Figure 5. Time development of the velocity magnitude field in simulation s520f (u = 80 m/s).

4 Discussion

Taking into account that the numerical simulation is performed on a very coarse 2D mesh and that it contains numerous simplifications, the results compare to the experimental data unexpectedly well. Without any artificial parameter tuning, the critical flow velocity predicted by the numerical simulation lies between 70 – 80 m/s, while the measurements revealed the flutter boundary at about 65 m/s. The time histories of the pitching and plunging motion also match very well.

4.1 Limitations of the current approach

The incompressible flow assumption holds reasonably for flow velocities up to Ma = 0.3, and thus the incompressible flow model used in the simulations is on the verge on its validity. In the current approach, no turbulence modeling was considered. Further, the computational grid is too coarse to capture accurately the boundary layer on the airfoil. Due to drastic differences between the computational requirements of 2D and 3D simulations, the first tests were run in 2D only. Especially in the downstream section of the geometry, the predicted 2D flow fields necessarily depart from the 3D reality due to fundamentally different vortex dynamics in two and three dimensions.

4.2 Future steps

The computational model is ready for parallel numerical simulations in 3D. In previous parallel speedup testing [8], the code proved good scaling up to 40 CPU cores for meshes consisting of about three million elements. The k-ω SST turbulence model and the compressible flow model have already been successfully used in previous work [9], and can be included in further simulations.

5 Conclusions

Both experiments and simulations show that the instability onset is sensitive to the initial conditions. The instability is clearly of the stall-flutter type, which is driven by the flow separation on the suction surface of the airfoil. To trigger the pitching motion and thus flow separation, the numerical simulations needed an initial impulse in angular velocity, otherwise the pitching angles remained too small and the instability did not occur even for higher flow velocities.

Acknowledgements

The research has been supported by the Czech Science Foundation, project 13-10527S "Subsonic flutter analysis of elastically supported airfoils using interferometry and CFD".

The parallel scaling tests were run on the facilities of the Supercomputing center at the Czech Technical University in Prague. The experimental data were acquired during measurements in the wind tunnel of the Institute of Thermomechanics, Academy of Sciences of the Czech Republic in Nový Knín, in cooperation with Martin Štěpán, Václav Vlček, Martin Luxa and David Šimurda. The natural frequencies and damping ratios of the mechanical model were measured by Štěpán Chládek.

References

1. E.H. Dowell, *A Modern Course in Aeroelasticity* (Kluwer Academic Publishers, Dodrecht, 1995)
2. P. Šidlof, V. Vlček, M. Štěpán, J. Horáček, M. Luxa, D. Šimurda, J. Kozánek, Proceedings of the ASME PVP 2014 conference, ASME, 2014
3. P. Šidlof, M. Štěpán, V. Vlček, V. Řidký, D. Šimurda, J. Horáček, EPJ Web of Conferences **67**, 02108 (2014)
4. M. Štěpán, Proceedings of Colloquium DYNAMESI 2014, Institute of Thermomechanics, (2014)
5. V. Vlček, M. Štěpán, I. Zolotarev, J. Kozánek, *Proceedings of Engineering Mechanics 2015*, pp. 350-351 (2015)
6. https://www.hpc.ntnu.no/display/hpc/OpenFOAM+-+Airfoil+Calculations, accessed 2015-30-07
7. J.H. Ferziger, M. Peric, *Computational Methods for Fluid Dynamics* (Springer, Berlin, 2002)
8. P. Šidlof, V. Řidký, EPJ Web of Conferences **92**, 02080 (2015)
9. Řidký V., Šidlof P., *Engineering Mechanics 2014*, pp. 548-551, (2014)

EPJ Web of Conferences **114**, 02106 (2016)
DOI: 10.1051/epjconf/201611402106

Numerical and Experimental Study of a Cooling for Vanes in a Small Turbine Engine

Jan Šimák[1,a] and Jan Michálek[1]

[1] *Aerospace Research and Test Establishment, Prague*

Abstract. This paper is concerned with a cooling system for inlet guide vanes of a small turbine engine which are exposed to a high temperature gas leaving a combustion chamber. Because of small dimensions of the vanes, only a simple internal cavity and cooling holes can be realized. The idea was to utilize a film cooling technique. The proposed solution was simulated by means of a numerical method based on a coupling of CFD and heat transfer solvers. The numerical results of various scenarios (different coolant temperature, heat transfer to surroundings) showed a desired decrease of the temperature, especially on the most critical part - the trailing edge. The numerical data are compared to results obtained by experimental measurements performed in a test facility in our institute. A quarter segment model of the inlet guide vanes wheel was equipped with thermocouples in order to verify an effect of cooling. Despite some uncertainty in the results, a verifiable decrease of the vane temperature was observed.

1 Introduction

Modern turbine engines are very sophisticated machines designed to transform energy into mechanical work as efficiently as possible. A desired high level of the efficiency is connected with high temperatures and pressures that thermally and mechanically strain turbine components. The present-day aircraft engines often operate close or above melting point of the used material and therefore require an intensive heat protection.

The most common and simplest way to reduce temperature to safe values is the use of cold air [1]. A combination of an internal and film cooling is a typical example of this approach. An amount of the air used for the active cooling may reach up to twenty percent of the entire engine core mass flow [2], which in turn somewhat reduces achieved thermal efficiency. However, a gain from the increased working temperature is still higher than losses. The mentioned approach is sometimes combined with a thermal barrier coating (TBC) applied on the most stressed surfaces to extend the component lifespan.

The internal cooling system usually consists of a complicated labyrinth of channels, turbulizators, pins and cavities inside the vane, which distributes the coolant through the body in order to transfer the excessive heat away. The film cooling is achieved through a formation of a thin layer of 'cold' air uniformly distributed over the surface which provides an insulating thermal barrier against the hot gas of the bulk flow. This is usually accomplished by a large quantity of shaped and optimized holes connecting the internal channels with the vane surface.

[a]Corresponding author: simak@vzlu.cz

The problem of cooling is even more challenging in small engines due to their dimensions and compact design. Moreover, the most stressed parts, such as vanes and blades of a gas generator, are often surrounded by an annular combustion chamber which prevents the heat to conduct away. The inlet guide vanes (IGV) of the studied engine are not presently cooled and suffer from an extensive heat degradation mainly in the vicinity of the trailing edge. An initial attempt considered a simple vane film cooling introduced from a slit tube in the leading edge region [3, 4]. However, the obtained results were found questionable and this concept was abandoned. For the above reasons, a new IGV profile was proposed taking in account an internal cavity and several cooling holes deployed at the rear part of the pressure side. This novel concept was examined within the presented work.

The problem was at first studied by means of three-dimensional numerical simulations, which provided a good preview of the situation and of the impact of the suggested solution. The simulations were performed by a CFD solver based on the Reynolds averaged Navier-Stokes equations (RANS) and a solver for a heat conduction in a solid. The obtained results showed a feasibility of the adopted concept which was further verified by conducted experiments.

2 Description of the problem

The problem considered in this work is an initial study of an IGV cooling system for a gas generator of turbine engines intended for light aircraft. The small dimensions of the engine (maximum outer diameter is about

270 mm) and of the IGV wheel (the hub and shroud radii are 50 mm and 69.75 mm) and also its counter flow combustion chamber arrangement (see figure 1) strongly limit any practicable design variants of a cooling system. Moreover, due to the small size of the vanes, it would be very difficult to create a sophisticated system of internal channels that distribute cooling air. Limiting were also the aspects regarding the cost and technological complexity related to an application of a TBC. The engine parts are manufactured from nickel-chrome super-alloys (e.g. Inconel 713 or 718).

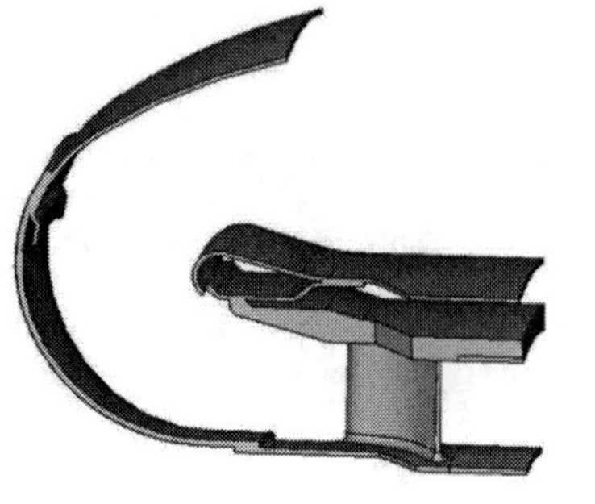

Figure 1. A complete CAD model of the IGV section with an exit from a combustion chamber.

From these reasons, a simple concept of an active fluid cooling was introduced. A practical experience showed that the most critical part of the vane regarding the thermal strain was the trailing edge region mainly at the vane mid-height. Therefore, an emphasis was put on this area during a design phase. According to preliminary numerical simulations performed with a uniform temperature distribution of the incoming air, temperatures at the trailing edge region exceeded 930 °C. In a case of a non-uniform distribution, temperature could possibly grow even higher.

Because of the above mentioned reasons, it was decided to create a simple internal cavity situated in the thickest part of the vane with seven cooling holes leading to the pressure side. The main idea was to decrease the vane temperature by the heat exchange with the cold air inside the cavity and by forming a thin film layer protecting the most critical part on the trailing edge. This solution was however allowed only by a redesign of the vane profile which was supplied by the producing company. The new profile was enlarged to allow the cavity and the trailing edge was thickened to reduce the heat corrosion. As a result, the redesigned IGV wheel consists of only 17 vanes, while the original one has 26 vanes.

An overview of the basics concerning the film cooling can be found e.g. in [5]. It is well known that the position, direction and shape of holes influence the overall efficiency but we tried to suggest a simple solution. Thus the holes are of a cylindrical shape and in a small number. The tested cooling system consists of an oval shaped cavity which opens into an annular channel next to the shroud. A cooling medium, which is a cold compressed air diverted before reaching the combustion chamber, flows through this channel. Further, this cold air is brought from inside of the cavity to the surface of the pressure side of the vane by seven cooling holes. These holes are parallel to each other, each of them has a circular cross-section with a diameter d = 0.8 mm and are spaced two diameters apart. The central cooling hole is positioned slightly above the mean radius of the wheel and the openings are located on the surface in about one third of the vane from the trailing edge. The cooling channels are designed such that the middle one is inclined 30° to the surface. The axes of the cooling channels are also tilted with respect to the stator wheel axis with a compound angle of 30°. The configuration is shown in figure 2.

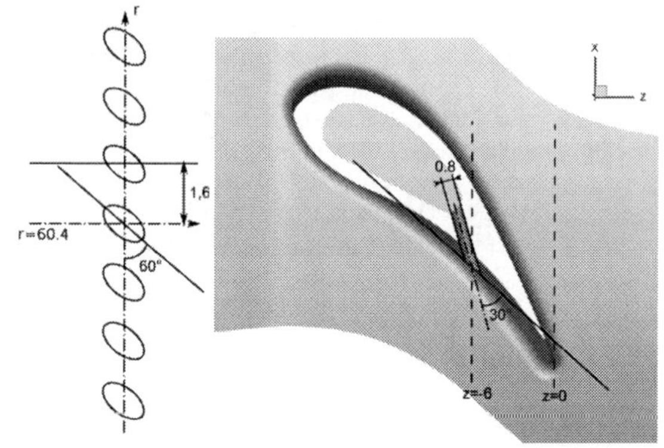

Figure 2. Geometry of the vane cooling channels: meridional plane (left), tangential plane (right).

3 Numerical methods

From a numerical point of view, the problem was solved as two separated problems coupled by a common boundary. An iterative process (called outer in this text) then provided an exchange of information between a CFD solver and a heat solver.

3.1 Solution of the flow problem

We considered a steady, compressible flow of a perfect gas (for simplicity air), described by the three-dimensional RANS equations. The problem was solved using the CFD software Edge v5.2, which is based on a finite volume method on dual meshes. The presented results were obtained by the use of an upwind scheme with the Roe averaging of state variables on cell faces. To increase spatial order of accuracy, a reconstruction of variables on the cell faces (symmetric TVD scheme) was applied. Since the flow was transonic and contained shock waves, the *minmod* limiter was used to suppress undesirable numerical oscillations near flow discontinuities.

The governing equations were integrated in a time variable to a stationary solution using an explicit three-stage Runge-Kutta method with a local time step. To accelerate the convergence, a multigrid and an implicit residual smoothing were involved.

The flow was modelled as fully turbulent with an EARSM k-ω turbulence model proposed by Hellsten [6], which is also the default choice of Edge. More information about the solver can be found in [7].

A weak pressure condition, which prescribes the overall state variables - total pressure, total temperature and flow direction, was chosen as a boundary condition at the main inlet and also at the entrance to the cavity. At the outlet boundary there was prescribed an integral average of a static pressure. Walls were treated either as adiabatic or with a prescribed wall temperature obtained from the heat solver.

The source code of the software used in this study was slightly modified compared to the original version, namely the boundary conditions were adapted for problems of internal aerodynamics and procedures, which were necessary for coupling with a heat solver, were added.

A flow field with adiabatic walls was taken as an initial condition for the heat transfer problem. Further, at each step of the outer iteration process inner iterations (up to 20) were performed to get a new thermal boundary profile that corresponded to new wall temperatures.

3.2 Mesh generation

The shape of a computational domain was derived from a simplified geometry without considering a rotor. The domain was a periodic part containing one vane and relevant parts of channels between vanes. The outlet boundary was extended further behind the trailing edge from reasons to prevent possible reflections from this boundary. The shape of hub and shroud walls was also slightly simplified.

The computational domain was discretized by a multiblock structured grid, created by the GridPro software. The computational domain thus consisted of blocks with six-sided cells. The grids for the flow and heat conduction problem were generated separately. Since cells of the inner and outer grids are not necessarily linked one to one, an interpolation matrix was constructed, which maps surface vertices from one grid to the other grid and vice versa. The shape of periodic boundaries was determined by the grid generator automatically.

In order to properly capture a profile of a boundary layer in the simulation, the height of the first row of cells on walls was chosen such that $y^+ \leq 1$ based on the expected velocity. The number of cells in the grid was approximately $7{\cdot}10^6$ for a simple vane and $8.5{\cdot}10^6$ for the vane with the cooling system.

3.3 Solution of the heat conduction

The heat conduction in the vane and in connecting hub and shroud walls was performed using an in-house software for a solution of a heat equation. This equation has the form of

$$c\rho\frac{\partial T}{\partial t} - \operatorname{div}(\lambda \nabla T) = 0, \tag{1}$$

where T is the temperature, c heat capacity, ρ density and λ thermal conductivity. The equation was discretized by a finite volume method on a multi-block structured grid and integrated towards a steady state by the backward Euler scheme.

At parts of the boundary, which are common to the solid body and the flow field, a heat transfer boundary condition was considered. This condition is based on the Fourier's law $\mathbf{q} = -\lambda\nabla T$ for a local heat flux $\mathbf{q}$ and equality of the heat fluxes from both sides of the boundary. So the boundary condition can be formulated as

$$\lambda\frac{\partial T}{\partial n} = \lambda_a\frac{T_a - T}{d_a}, \tag{2}$$

where $\mathbf{n}$ is the outer unit normal vector, λ_a the thermal conductivity of the air, T_a temperature of the air in the first row of nodes and finally d_a denotes the distance of these nodes to the wall.

The periodic boundary condition ensures then the portability of obtained results to the whole IGV wheel geometry.

4 Numerical results

4.1 Setting up the problem

The engine IGV wheel was made from a nickel-chrome alloy Inconel 713LC which has the thermal conductivity $\lambda = 20\ \mathrm{W\,m^{-1}\,K^{-1}}$ at 760 °C. The heat capacity at the same temperature is $c = 625\ \mathrm{J\,kg^{-1}\,K^{-1}}$, the values of λ and c are changing with temperature. The melting point of the material is around 1260 °C.

The total pressure and total temperature at the inlet to the IGV wheel were

$$P_{01} = 476.481\ \text{kPa},\ T_{01} = 950\ °\text{C}.$$

The coolant total pressure and total temperature were set to

$$P_{0c} = 495.015\ \text{kPa},\ T_{0c} = 210\ °\text{C} - 470\ °\text{C}.$$

The integral average of the static pressure at the exit boundary was set to

$$P_2 = 265.553\ \text{kPa}.$$

The turbulence intensity of the incoming air was set to $Tu = 3\%$. The flow field entering the computational domain was assumed homogeneous and perpendicular to the boundary. The isentropic Mach number behind the IGV was $M_{2,is} = 0.95$.

4.2 Vane cooling without heat transfer

The first step was to evaluate the effect of the proposed cooling system and to find the best configuration. Hence the problem with adiabatic walls (e.g. no heat transfer to the vane) was solved. The temperature of the cooling air was set T_{0c} = 210 °C. The effectiveness was evaluated in terms of the film effectiveness, η, defined as follows:

$$\eta = \frac{T_{01} - T_{aw}}{T_{01} - T_{0c}}, \quad (3)$$

where T_{aw} refers to the adiabatic wall temperature. The distribution of η on the vane surface is shown in figure 3. It is clearly seen that the holes do not create a continuous film of cold air that would completely insulate the surface from the hot air. To do this, one would need to use more holes in more rows or change the shape of the holes, which is in a contradiction with our requirements of simplicity and robustness. However, the area of the surface influenced by the cold air seemed to provide sufficient decrease of the trailing edge temperature. The mass flow rate of the cooling air entering to one vane was evaluated as $\dot{m}_c$ = 0.00173 kg s^{-1}, the mass flow rate at the main inlet was $\dot{m}$ = 0.1074 kg s^{-1} per one period. That means the cooling air represented 1.59% of the total flow. The mass flow $\dot{m}_c$ is given by the number of holes and by the pressure drop between the entrance to the cavity and the vane pressure side near the hole exit location.

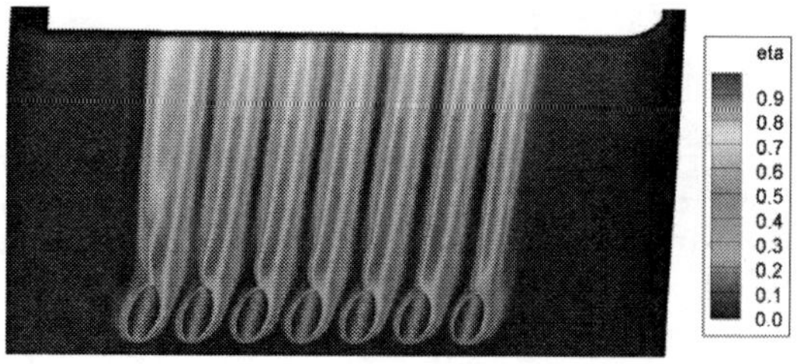

Figure 3. The film effectiveness on the adiabatic vane surface

4.3 Vane cooling with heat transfer

Further, the heat transfer model was accompanied into the simulations. It is very difficult to model the same situation as in a real engine because we have no information about heat fluxes through outer boundaries of our model. We do not even know the temperatures of these parts. Thus, we assumed a zero heat transfer on these outer boundaries. This means that we supposed the solid body was isolated from the surroundings. This approach is thought to be satisfactory for our purpose of evaluation of the cooling efficiency. However, the comparison with experimental data is more difficult. In the later simulations, this approach was partly abandoned to find out the influence of the surroundings.

Using the described method, a temperature distribution was obtained at first on a vane with no cooling (NC). At the trailing edge, an average temperature was 931.8 °C, while temperature of the incoming flow was 950 °C. This result was then compared to a case when the vane is actively cooled with the inner cavity and cooling holes. When using the cooling air of a temperature 210 °C, an average temperature at the trailing edge decreased to a value of 838.8 °C. That means reduction of 93 °C.

Comparing the results to a variant with only the inner cooling (with the same performance) present, we can express the effectiveness of the cold film. Figure 4 shows a decrease of the temperature up to 34 °C on the trailing edge. Moreover, this result is for the case with higher cooling air temperature, T_{01} = 910 °C and T_{0c} = 470 °C.

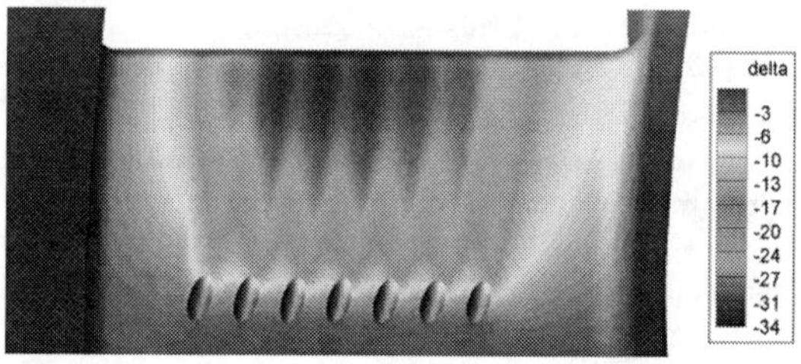

Figure 4. A temperature decrease due to the film cooling.

The maps of surface temperature for NC and cooling (T_{0c} = 210 °C) variants are presented in figure 5. The injection is usually described by the blowing ratio defined as

$$BR = \frac{\rho_c V_c}{\rho V} \quad (4)$$

where ρ and V are the mainstream density and velocity above the holes, the momentum flux ratio

$$I = \frac{\rho_c V_c^2}{\rho V^2} \quad (5)$$

and the density ratio

$$DR = \frac{\rho_c}{\rho}. \quad (6)$$

The values for this case are BR = 1.098, I = 0.58 and DR = 2.07. The cooling jets are located on a concave surface with a non-dimensional parameter $2R/d = -69$ where R denotes the curvature radius. These values indicate a regime that could lead to a decrease of the efficiency. The following figures 6 and 7 show shapes of the cooling jets. Because of the small number of holes and their spacing, the resulting cooling jets do not form a continuous thin film, but the individual jets remain isolated. The pictures show that the cross section of the jet is not homogeneous, but it has a shape of a horseshoe. That means the jet is not fully attached to the vane surface and the efficiency is lower than in a case of a fully attached flow. A more appropriate choice of the coolant holes configuration would certainly eliminate this imperfection.

The mass flow rate of the cooling air per one vane is $\dot{m}_c$ = 0.00144 kg s^{-1} (corrected mass flow rate Q_c = 1.09 kg s^{-1} K$^{1/2}$ MPa^{-1}). In comparison with the mass flow rate through the inlet boundary, which for one blade to blade channel attained a value of $\dot{m}$ = 0.1074 kg s^{-1} (Q_c = 134.01 kg s^{-1} K$^{1/2}$ MPa^{-1}), it represented approximately 1.32% of the total mass flow.

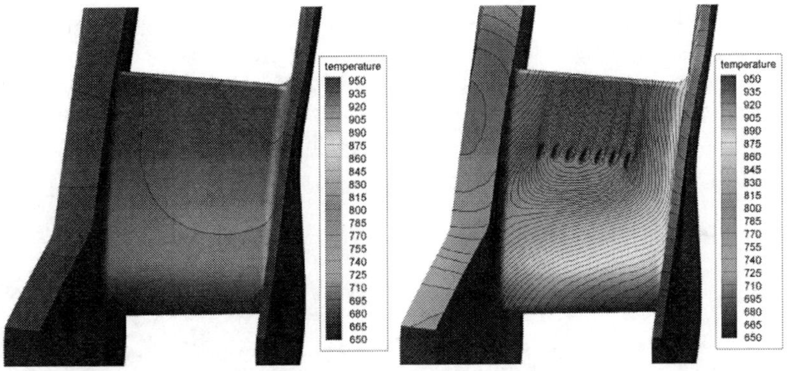

Figure 5. A distribution of temperature on the vane surface; uncooled (left), cooled (right); the coolant temperature 210 °C.

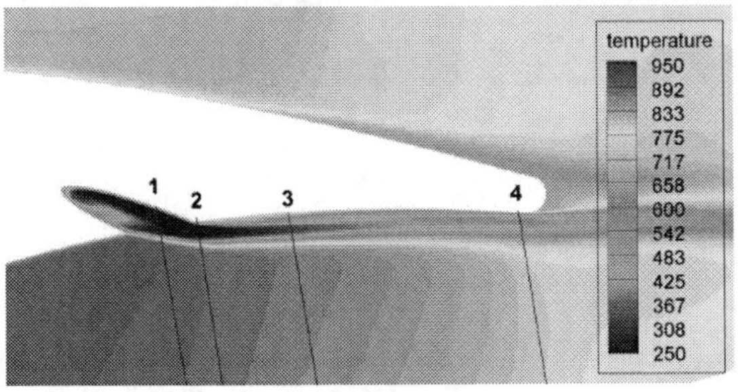

Figure 6. A section along the cool jet from the middle cooling hole with indicated positions of cross-sections.

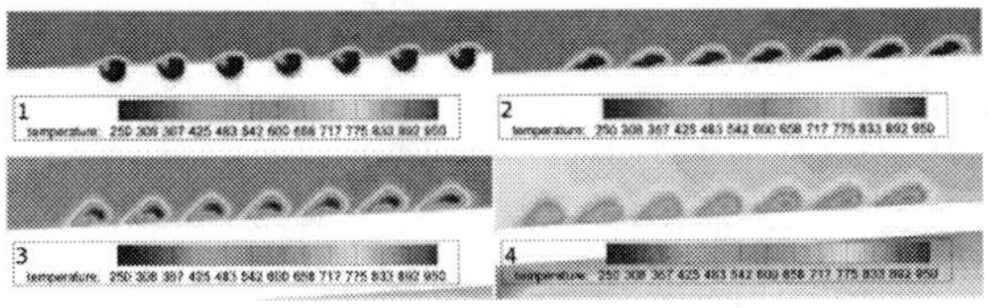

Figure 7. Cross-sections of the cool jets at given positions.

The calculated mass-weighted average values of the kinetic energy loss ζ and total pressure loss ξ, given as follows:

$$P_{0,mix} = \frac{\dot{m}_c}{\dot{m}_c + \dot{m}} P_{0c} + \frac{\dot{m}}{\dot{m}_c + \dot{m}} P_{01}, \quad (7)$$

$$M_{is} = \sqrt{\left[\left(\frac{P_{0,mix}}{P}\right)^{\frac{\kappa-1}{\kappa}} - 1\right]\frac{2}{\kappa - 1}}, \quad (8)$$

$$\zeta = 1 - M^2/M_{is}^2, \quad (9)$$

$$\xi = 1 - P_0/P_{0,mix}, \quad (10)$$

where M denotes the Mach number. The values at a distance of 5 mm behind the trailing edge are presented in table 1. The resulting temperature on the trailing edge is presented in figure 8.

Table 1. The loss coefficients of the uncooled and cooled IGV.

	ζ	ξ
NC	0.0518	0.0286
Cooling T_{0c} = 210° C	0.0521	0.0289
Cooling T_{0c} = 410° C	0.0532	0.0294

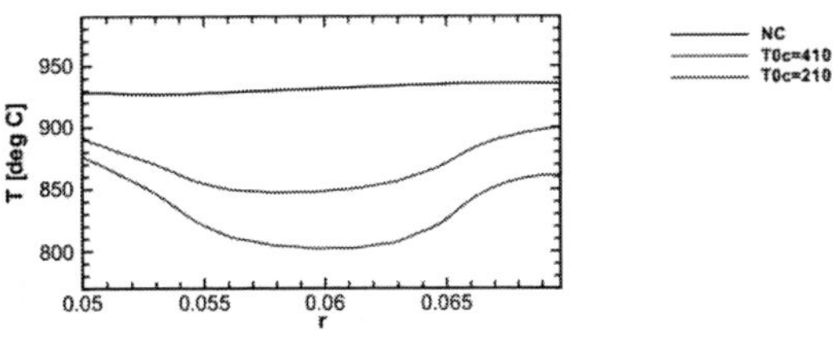

Figure 8. A distribution of the trailing edge temperature plotted along the radius.

4.4 Vane cooling with heat transfer through tip shroud

Looking at the figure 1 it is possible to take notice of the annular channel bringing the cooling air to the vane. Since the coolant flows around this wall, it is correct to expect some heat transfer through this boundary. The variants with a prescribed coolant temperature of 210, 310, 410 and 470 °C were evaluated. The latter variant for coolant temperature 470 °C was evaluated with a different inlet boundary condition. Profiles of the total pressure and total temperature were prescribed on the inlet boundary according to data obtained by the experiments. The inlet temperature at the inner radius was 876.6 °C and at the outer radius 932.75 °C. The average total temperature T_{01} = 910 °C, the total pressure remained the same. This variant was thus closest to the experimental data. However, the numerical model still differs from the experimental model because we have no information about the heat transfer through outer boundaries.

When we therefore considered a heat transfer through the tip shroud, we observed a decrease of the temperature in this area. A comparison of a temperature on the trailing edge for the considered variants is in figure 9.

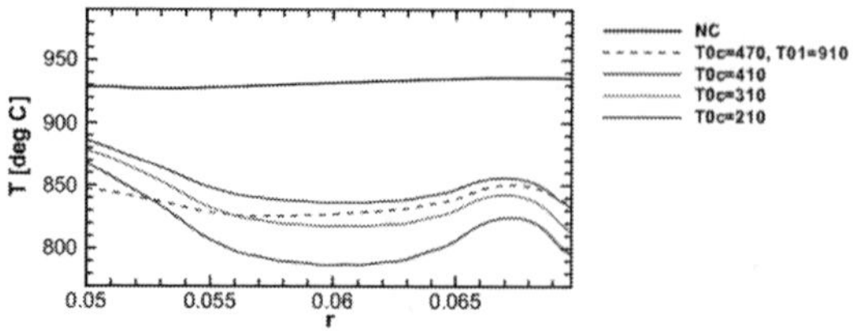

Figure 9. A distribution of the trailing edge temperature plotted along the radius. Considered is a heat transfer through the tip shroud; $T_{01} = 950$ °C.

5 Experimental evaluation

5.1 Test facility

An experimental verification of the proposed cooling system was carried out on a test rig consisted of three principal parts: radial compressor (RC), combustion chamber (CC) and tested IGV segment. An amount of the air supplied by the radial compressor at desired pressure was restricted by its working regime and therefore only a quarter segment of the combustion chamber and of the IGV model was considered in the test rig. Considering the proposed number of the vanes (17), the test model resulted in an annular segment with only four blade to blade channels, i.e. three full vanes. Due to the technological purposes a nickel-chrome alloy Inconel 713 LC, used in the serial production, was substituted with Inconel 718 which has similar physical properties as the original material. Three alternative IGV segments were used within the experimental campaign: a variant *without* cooling (IGV-NC), one with an *internally* delivered cooling (IGV-IC) and one with an *externally* delivered cooling (IGV-EC). The IGV-NC segment served as a reference for a comparison with the other two cooled segments. A scheme of the test rig with alternative cooling arrangements supplied with measured parameters is presented in figure 10.

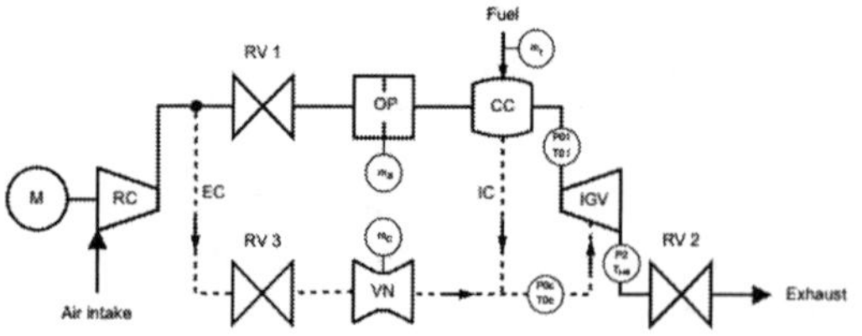

Figure 10. A scheme of the test rig with two alternative cooling arrangements; a) internally delivered coolant (IC), b) externally delivered coolant (EC).

The IGV-IC (figure 11) variant represented a real engine configuration with three hollow vanes and seven cooling holes located on the pressure side of each vane. The variant utilized the coolant (mostly compressed air) from a space between the combustion chamber and the outer diameter of the IGV wheel. The mass flow rate of the coolant was given by prescribed flow parameters and by an effective flow area of the cooling holes.

Figure 11. The IGV-IC segment. An inlet view with the cooling channels on the vanes pressure sides.

Finally, the IGV-EC configuration allowed us to control an amount of the coolant air delivered to the vane internal cavity from an external source. The air was taken at the compressor exhaust and delivered to each of the internal vane cavity through a single air distributor welded to the hub block. The opposite ends of the vane internal cavities were blinded. An amount of the cooling air was controlled through the regulating valve and the actual value was measured by means of a calibrated Venturi nozzle (figure 10).

5.2 IGV probes instrumentation

All the three IGV segments were equipped with thermocouple probes and pressure probes. The vane metal temperature was monitored by means of three unshielded K-type thermocouples installed into the trailing edge region of the three individual vanes. They were flush-mounted into the pre-machined groves at vanes mid-height using a high-temperature resistant cement. The thermal conductivity of the cement was slightly higher than the base material Inconel 718. The total pressure (P_{0c}) and total temperature (T_{0c}) of the coolant were in the case of IGV-EC arrangement monitored by means of a Pitot probe and a T-type thermocouple probe placed in the supply pipe just before the IGV entry. In the case of IGV-IC configuration, the total temperature and pressure readings were provided by means of a shielded K-type thermocouple and a stainless-steel tube installed inside one of the vane internal cavities.

6 Experimental results

6.1 IGV segment without cooling

Tests on the uncooled IGV segment were performed for a range of flow regimes characterized by the exit isentropic

Mach number. The inlet total temperature was held approximately constant on a value of 850 °C.

The air mass flow rate, as measured on the orifice plate, increased together with the higher Mach number until eventually conditions close to the choking were reached for desired pressure ratio $P_2/P_{01} = 0.557$, $P_{01} = 476.481$ kPa, $P_2 = 265.553$ kPa.

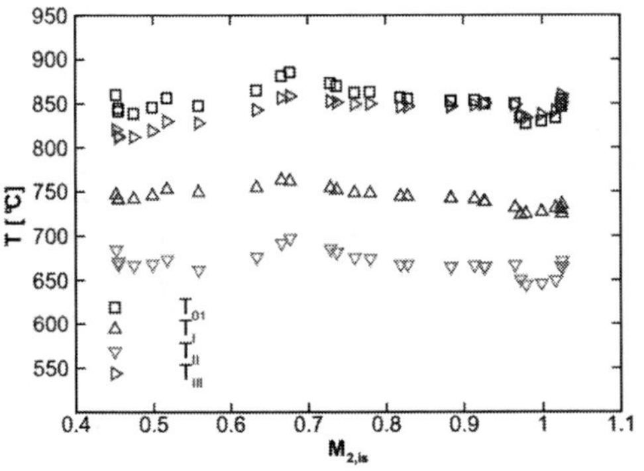

Figure 12. IGV-NC – the inlet total temperature and the trailing edge metal blade temperatures as a function of the exit isentropic Mach number.

Relatively large temperature differences were measured between the thermocouple probes installed in the individual vanes as presented in figure 12. While the two probes placed at the same vane (III) showed virtually same temperatures (T_{01}, T_{III}), the other two neighbouring vanes exhibited significantly lower values. A temperature difference between the vane III and II reached values as large as 200 °C despite the fact that preliminary tests in the duct behind the combustion chamber showed a satisfactorily uniform distribution of temperatures in the circumferential direction with an only slight variation along the vane height.

Given the above information and a fact that similar pattern was found also for the tests performed on the other two types of the IGV it was concluded that the temperature variation was systematic due to a temperature gradient and consequent heat transfer within the solid of the test section. The real engine configuration with an annular IGV shielded by a combustion chamber is, however, qualitatively different and it should be taken in account while interpreting the presented data.

6.2 IGV segment with external coolant delivery

The measurements on a variant of the IGV with the external coolant delivery (IGV-EC) were carried out in a narrow range of the exit isentropic Mach number close to unity. The pressures prescribed during the tests attained approximately the required boundary conditions (section 4.1). The total temperature at the inlet was set to a value near to 910 °C in the beginning of the test.

An amount of the coolant through the IGV was during the test controlled by the RV3 valve from a fully closed state until a point when the mass flow did not change anymore. A ratio of the coolant and the main flow total pressures was ranging roughly between 0.88 and 1.02 and thus the maximal reached value was slightly below the required conditions, i.e. 1.038.

A similar nature, as observed in the case of IGV-NC, regarding a temperature variation between individual vanes was found also for this type of the IGV segment (figure 13). For zero coolant mass flow the trailing edge temperature of the vane III was again akin to the total temperature which was measured at the leading edge of the same vane. The other two adjacent vanes I and II exhibited much lower values with a difference about 150 °C with respect to the vane III.

When the coolant is introduced, an effect of the cooling is clearly evident throughout the entire range of the imposed mass flow rate. A rate of change of the vanes temperatures was much higher for a small mass flow of the coolant and then asymptotically tended to a saturated state which was not fully reached during the present experiment. A difference in the vane temperatures between conditions without cooling and with a maximum mass flow of the coolant was approximately 150 °C. It should be noted, however, that the coolant temperature was lower than required.

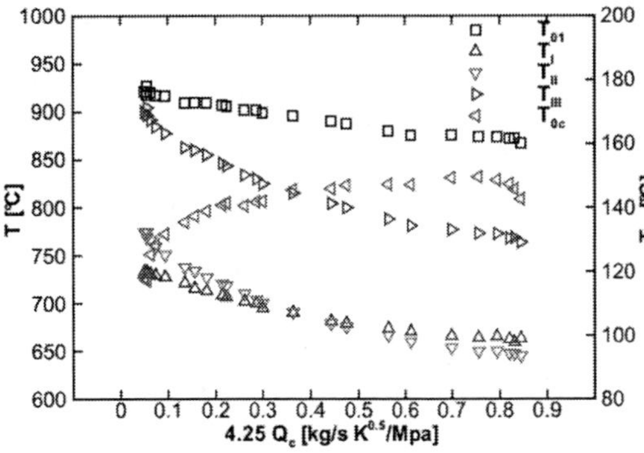

Figure 13. IGV-EC – the measured temperatures with respect to a variation of the coolant mass flow rate capacity.

6.3 IGV segment with internal coolant delivery

The tests on the IGV-IC segment were carried out for measurements with a constant exit Mach number ($M_{2,is} = 1$) while varying inlet total temperatures. An evolution of the blade temperatures with respect to a change of the total conditions at the inlet is presented in figure 14. The same pattern in a distribution of the vanes temperatures was observed as for the previously tested configurations with the highest temperature indicated for the vane III and the lowest on the vane II. The observed temperature differences between the individual vanes grew almost linearly with an increase of the total temperature which further supports the assumption of spatially non-uniform heat conduction through the test section. For the highest attained temperature at the inlet (910 °C) a difference between the vanes II and III metal temperatures was again about 150 °C.

The temperature of the coolant air measured in the vane cavity is also depicted in figure 14. It attained a maximum value above 450 °C while the temperature at the inlet to the combustion chamber was held nearly unchanged at 310 °C.

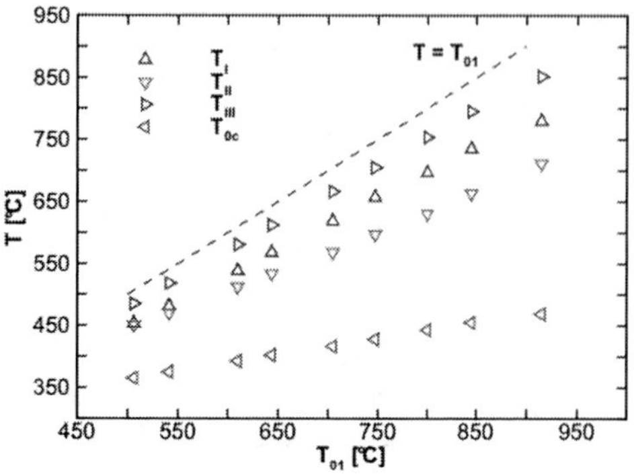

Figure 14. IGV-IC – an evolution of the blade metal temperatures with a variation of the inlet total temperature; $M_{2,is} \approx 1$.

A temperature drop obtained for the vane III was about 60 °C when the highest inlet temperature about 910 °C was imposed. It is a significantly lower value then one reached with the IGV-EC arrangement. Nevertheless, the coolant temperature in the present case was also much higher (≈ 450 °C) which in turn resulted in a reduced cooling efficiency.

The other two vanes, labelled as I and II, showed somewhat lower metal temperatures compared to the vane III (figure 14), but it was likely only due to their overall lower temperature level as similarly indicated for the IGV-NC and IGV-EC experiments (figure 12 and 13).

7 Conclusion

The paper presents results of a work focused on a design and verification of a novel cooling system for the inlet guide vanes (IGV) of small turbine engines. The proposed cooling concept was firstly verified using numerical simulations and later tested during an experimental campaign.

The cooling air was brought to the hollow vane from a space between the combustion chamber and the outer diameter of the IGV and then discharged back to the main flow through seven holes uniformly positioned along the vane height in the rear part of the pressure side.

The performed flow simulations, coupled with a heat transfer to the solid parts, indicated a significant reduction of the solid temperature of the entire vane and particularly in the trailing edge region. An achieved drop of the temperature was about 80 °C for nominal flow conditions with approximately 1.5% of a total amount of the air considered for cooling.

The experimental results obtained for an IGV quarter segment with the internally delivered coolant (a real engine configuration) showed a somewhat lower cooling efficiency than the one predicted by the numerical simulations. The measured temperature drop was approximately 60 °C when the flow parameters close to nominal conditions were prescribed. It should be noted, however, that the measured coolant temperature was a little bit higher than one assigned in the prediction (460 °C vs. 410 °C). On the contrary, when the externally delivered coolant with a desired mass flow and a comparable temperature were considered, a measured temperature drop of the solid reached a similar value as predicted by the numerical simulation.

In general, the presented results proved a feasibility of the proposed cooling concept adopted for a small turbine engine. An extended life span of the cooled IGV should therefore be expected while maintaining the high temperature at the inlet. Additional tests on a real engine should be however performed in order to verify the concept with more realistic (heat transfer) boundary conditions and with a longer operational period.

Acknowledgement
Presented results were obtained during the work in project ESPOSA – Efficient Systems and Propulsion for the Small Aircraft, GA284859 co-funded by European Commission within 7th Framework Program.

References

1. F. Bayley, A.R.C. Technical Report 3110, Aeronautical Research Council (1959)
2. R. Bunker, *Turbine Engine Film Cooling Design and Applications*, in *VKI Lecture Series, Film Cooling Science and Technology: State-Of-The-Art Experimental And Computational Knowledge* (Von Karman Institute for Fluid Dynamics, 2007)
3. P. Straka, R - 5052, VZLU - Aerospace Research and Test Establishment (2011)
4. P. Straka, R - 5074, VZLU - Aerospace Research and Test Establishment (2011)
5. R.A. Dennis, ed., *The Gas Turbine Handbook* (U.S. Department of Energy-National Energy Technology Laboratory (NETL), 2006), `http://www.netl.doe.gov/research/coal/energy-systems/turbines/publications/handbook`
6. A. Hellsten, Ph.D. thesis, Helsinki University of Technology (2004)
7. *Edge - Theoretical formulation, ver. 5.2*, FOI (2010)

EPJ Web of Conferences **114**, 02107 (2016)
DOI: 10.1051/epjconf/201611402107

Investigation of a corrugated channel flow with an open source PIV software

Deniz Sivas[1, 2], A. Bahadır Olcay[2, a], Hojin Ahn[2]

[1]*Dalgakıran Makine Inc.,34885 İstanbul, Turkey*
[2]*Department of Mechanical Engineering, Yeditepe University,34775 İstanbul, Turkey*

Abstract. In this study, the corrugated channel flow was investigated by using an open-source particle image velocimetry (PIV) software. The open-source software called OpenPIV was first verified by using images of an earlier experimental work of a vortex ring formation. The corrugated channel flow images were taken with 200 W power LED light source and a high speed camera and those images were analysed with these spatial and temporal tools of OpenPIV. Laminar, transient and turbulent flow regimes were identified when Reynolds number was below 1100, in between 1100 and 2000 and higher than 2000, respectively. The velocity vectors were found to be about 20% lower than the previous study results. The flow inside the grooves was also investigated with OpenPIV and flow characteristics at the grooves were captured when interrogation window size was lowered. The visualization of the flow was presented for different Reynolds numbers with the relative scale values. As a result of this study, OpenPIV software was determined as promising open source PIV analysis software.

1 Introduction

Fluid motion analysis is one of the most difficult fields of mechanical engineering. Most of the equations which are being used to understand and to predict fluid motion are complex and difficult to solve. Additionally, sometimes it becomes so hard to find an equation which helps to understand fluid motion just by analytical approach. As a result of these difficulties, experimental studies became crucial and inevitable for this field. One method for understanding fluid motion experimentally is flow visualization technique. A fundamental method to visualize the flow is use of dyes. When a certain colour of dye is used to mark fluid particles, successful qualitative results can be obtained; however, these results would not possess any qualitative findings. Besides, when dye is used for flow visualization, it is known that dye would diffuse after certain time and dye would no longer mark the fluid particles in fluid flow. On the other hand, along with the technological development of optics, advanced visualization techniques have been developed. Planar Laser Induced Fluorescence (PLIF) is an optical diagnostic technique being used widely for flow visualization. In this technique, naturally buoyant particles are mixed with the liquid which is desired to be observed. If the fluid is in gas form, then smoke particles are used. As a different approach, hydrogen bubbles generated from electrolysis can be used for water flow applications [1]. The commercial particle image velocimetry (PIV) products use high-tech lasers for illumination and high speed cameras with a data hub to control image capture process. Because of this high-tech setup, these analysis tools require huge investments. Since the PIV technique mainly depends on the images, low quality images taken with a high-speed camera can also be analysed to see if they provide satisfying results. There are promising studies about development of open-source PIV software [2].

The most complicated flow regime to analyse is turbulent flow. Turbulent flow has advantages when mixing of fluid streams or exchange of heat is required. On the other hand, the heat transfer rate mainly depends on temperature difference and heat transfer area so that different geometrical approaches have been studied to increase the heat transfer area. Corrugated wall channels are one of the examples of these studies. These are used widely as heat exchangers in heat transfer systems. The flow around corrugation which is called bulk flow and the flow inside of the corrugations which is called groove flow should be analysed to see the effect of different geometries on the flow.

The objective of this study is the validation of an open-source PIV software and then using it to analyse different flow types in corrugated pipes.

[a] Corresponding author: bahadir.olcay@yeditepe.edu.tr

2 Previous studies

In the present study, the corrugated channel flow is intensely investigated. These investigations were focused on geometry of the corrugation, pressure drop around the corrugation and vortex formation. Since other properties can be derived from velocity, the flowrate estimation due to geometry was important. There are studies about estimation flowrate due to the corrugated pipe geometry [3]. The pressure drop needs to be investigated for proper heat transfer rates and proper implementation of corrugated pipes. To understand this, particle image velocimetry is also used [4]. The vortices in the grooves enhance the turbulent flow. If the vortex movement is analysed, then the geometry or dimensions of corrugations can be changed to optimize vortex formation and locations of them. There are experimental studies about the location change of vortices inside the groove and the magnitude of it [5]. Setting up an experiment is costly and tedious way to analyse flow around corrugation so that different approaches have been used. Computational analysis is a way to analyse the fluid motion around a corrugation and with a successful meshing; it can provide accurate results [6].

3 Experimental setup

The data used in the present study was taken from an experimental setup as shown in Figure 1. The experimental setup includes a corrugated surface section with Plexiglas walls, a pump for circulation of water, a tank for steady flow and three valves for control of the flow. The section of interest for the images is the corrugated surface section. Its schematic representation which was taken from a previous study can be seen below.

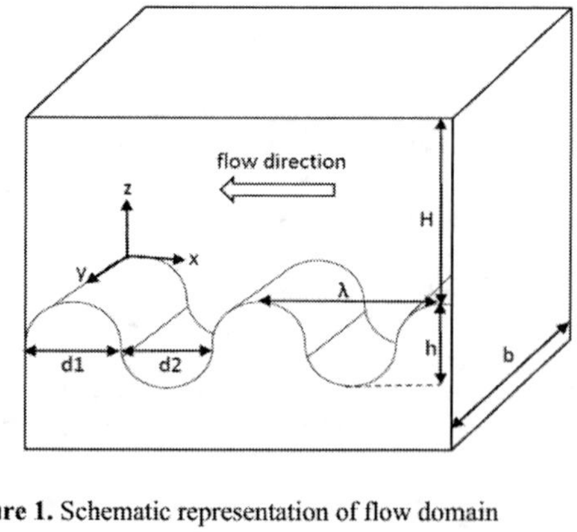

Figure 1. Schematic representation of flow domain

In here, H stands for channel height and it is given as 15.86 mm. The corrugation height is defined as h and it is given as 4.15 mm. The wave length of the corrugation is defined with λ and it is given as 6.6 mm. d1 stands for the cavity size and it is given as 3.0 mm. The corrugation size which is defined as d2 is 3.6 mm. The channel width which is defined with b is 20 mm. A high speed camera was placed perpendicular and a light source was used to illuminate the seeding particles. The Reynolds number is defined as

$$\mathrm{Re} = \frac{\rho \cdot u \cdot D_h}{\mu} \quad (1)$$

Where ρ stands for density of water, u stands for mean velocity of water, D_h stands for hydraulic diameter of channel and μ stands for dynamic viscosity of water. A stopwatch and a container were used for measuring mass flow rate and thereby velocity. The hydraulic diameter was calculated as 17.69 mm.

4 Digital Particle Image Velocimetry Technique

4.1 Typical setup

The typical PIV setup includes a double-pulsed laser for illumination, laser sheet optics to transform laser beam into a light sheet, a transparent walled flow setup, buoyant seeding particles and a high-speed camera to capture images. The laser should be synchronised with the camera so precisely because the time interval between two consecutive images is crucial for good PIV data. The seeding also plays a very important role since the seeding marking the fluid particles must follow the flow. Their diameter should be in between 10-100 μm. They should be naturally buoyant so that they decrease the time lag between accelerations of fluid and the particles. The material of the particles could be reflective or fluorescent.

4.2 Image processing

The method to analyse PIV images is mainly based on cross-correlation technique. The cross-correlation technique is used to determine the similarity between two series as a function of time difference between them. PIV software subdivides each image pair into interrogation windows then looks for the intensity correlation between each interrogation window by using cross-correlation. The location of the peak in this correlation function indicates the average displacement of particles in the two related interrogation windows in image A and image B. Error reduction is done by windows shifting and post-processing of the results. In windows shifting, the software uses the displacement field found from cross-correlation of images A and B. The image A is subdivided into interrogation windows like the first cross-correlation process then the known displacement vector is used to determine the corresponding interrogation window in image B. A new cross-correlation is done between these two interrogation windows to get a correction to the previously found displacement vector. In post processing, outlier removal is crucial and easy to perform. To find the outlier, a threshold deviation is defined. The software checks magnitudes of vectors, and if a vector exceeds the threshold value times standard deviation of the whole vector field then it is identified as outlier. The outlier vectors are replaced with interpolated vectors from neighbours. The rest is the conversion process of the pixel displacement field to velocity field.

5 Validation of OpenPIV software

5.1 Introduction

In order to use OpenPIV as an analysis tool, its output needs to be verified by using a known previous work or by comparison with a professional commercial product output. The validation of the results for this study was done by using a previously captured and published vortex ring formation experiment images [7]. These high quality professional PIV images were analysed with OpenPIV and the resultant vectors were post-processed via Spatial and Temporal Analysis Toolbox which is complementary software for OpenPIV. The validation was conducted for three different time instants. These points were selected according to the previous work. An example set of analysed consecutive images can be seen below. (See Figure 2)

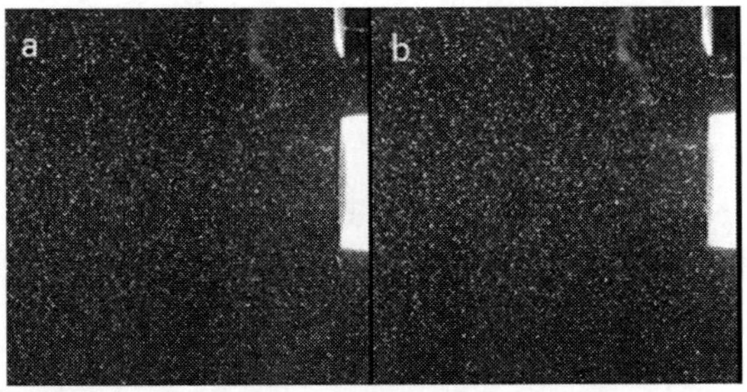

Figure 2. Consecutive validation images

This analysis was conducted with 32x32 interrogation window size and 32x32 overlap. The result of this analysis can be seen in Figure 3.

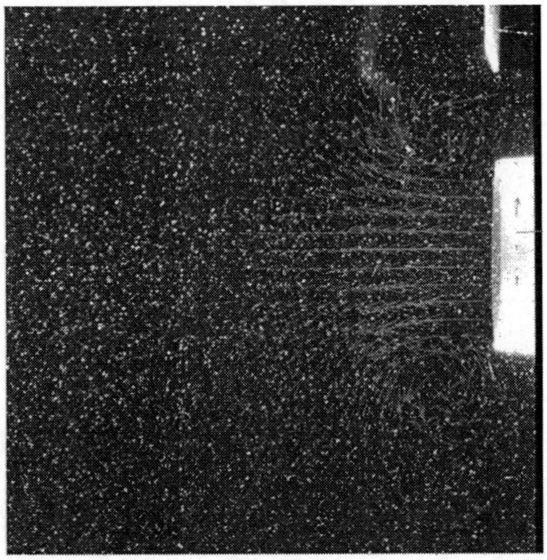

Figure 3. Validation Run 1 Result

It can be clearly seen that there are some extraordinary long and misdirected vectors in front of the vortex ring generator. These vectors are called outliers. These outliers are reduced with outlier filtering. Experimental apparatus also has remarkable effect on outliers. After getting the vector files, filtered and interpolated results were imported to Spatial and Temporal Analysis Toolbox. An example view of the vectors before the analysis process can be seen below (See Figure 4).

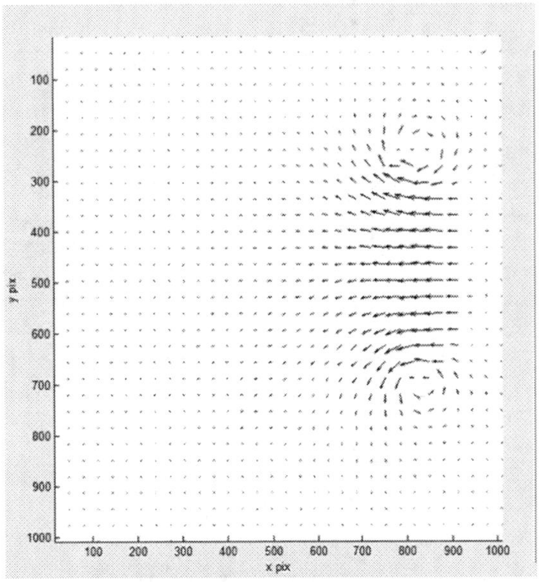

Figure 4. Vector field in spatial analysis toolbox

5.2 Results and discussion

For three different time instants three image pairs were analysed. OpenPIV analysis mode was selected as instantaneous and the contour quantity selected as vorticity. The results of OpenPIV can be seen in below figures. (See Figures 5-7)

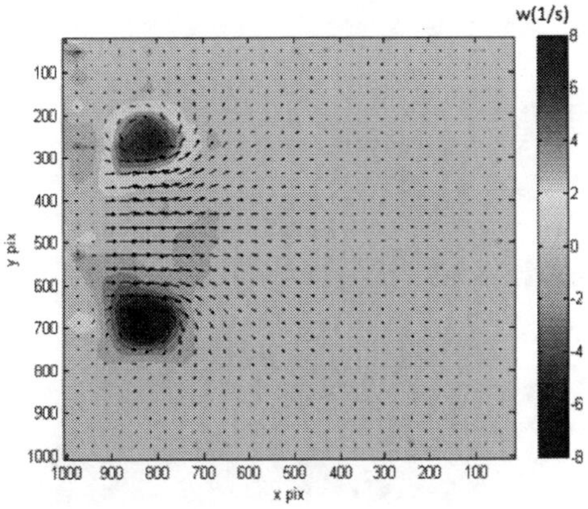

Figure 5. Vorticity analysis result at t = 2s

To be sure about the position of the vortices, a pixel conversion was also performed. Since the earlier work has used x/D as a coordinate axis, the pixel numbers corresponding to these values had to be calculated. The D value which represents the inner diameter of the cylinder was given as 3.73 cm. The images have 1024 x 1024 pixel spatial resolution and the field of view is 4.65D x 4.65D. So there are 59.0389 pixels for 1 cm. The bottom vortex location is around 0.65. The pixel number corresponds to this position was calculated as 143.14 px. The starting position is 1024 px, the cylinder head black section should be taken into account, then the starting pixel coordinate becomes 1000 px, the end position can be found easily by subtraction of the x distance. So that the vortex position was calculated as 856.86 px.

The above plot which was generated according to the OpenPIV result shows that the vortices are around 860 pixels. So it can be certified that OpenPIV results are quite accurate when compared to results of a commercial product. Since the validation only at one point is not statistically satisfying, the other points were also analysed by following the same procedure. The results were presented below. (See Figures 6 and 7)

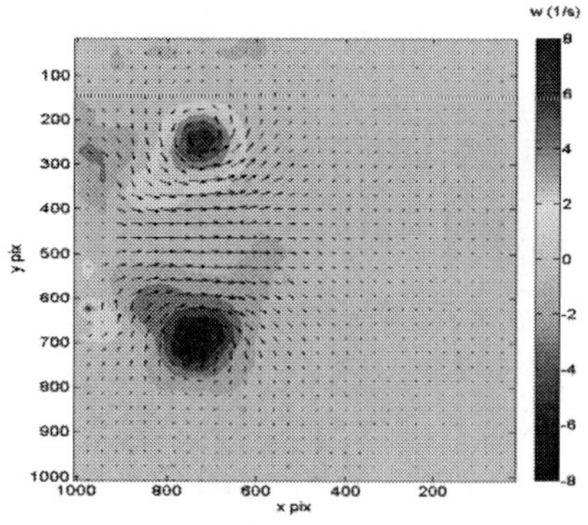

Figure 6. Vorticity analysis result at t=3s

The most important visual property of this analysis was the stopping vortices. These vortices can be seen next to the main vortices. Small and reverse directional vortices shown in Figure 7 are called stopping vortices.

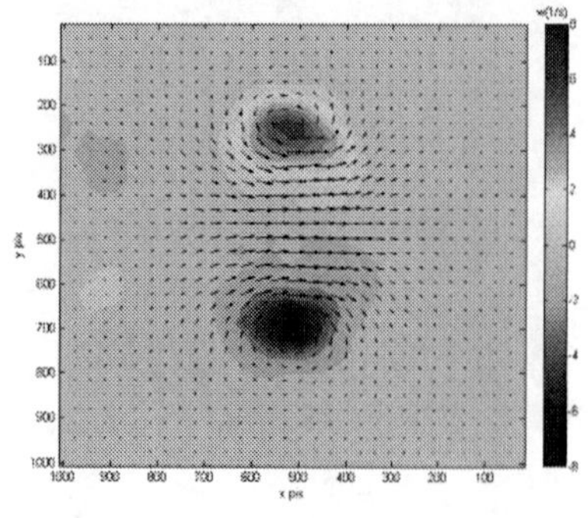

Figure 7. Vorticity analysis result at t = 5

The magnitudes of the vorticity were quite accurate but the problem about these results is that when the position verification applied to these images, vortex position for t = 2s was roughly 60 pixels ahead of the paper result and position for t = 3s was roughly 40 pixels ahead of the paper result. The vortex at t = 5s also had a positioning problem; it was roughly 100 pixels ahead of the paper result.

The accuracy of the software was checked by using a known experimental data and the results were satisfying. The position problem of the vortices was tried to be solved by changing the scale or the analysis type but the result was failure. Analysis of the corrugated pipe flow was decided to be done to see the error of the software when analysing velocity values.

6 Velocity analysis of bulk flow

6.1. Introduction

The visualization of the flow around corrugated pipes with a high speed camera gives rough information about the flow characteristic but it needs to be supported qualitatively. This section is about velocity analysis of corrugated channel flow by using OpenPIV software. In order to do that, images of a previous graduation thesis which were taken with a high speed camera and a 200 W LED source were used. These image pairs were selected for five different Reynolds numbers and three different flow regimes.

6.2 Results and discussion

In order to get a statistically satisfying result three different time instants were analysed for each Reynolds number. For Re < 1200, the flow was named as laminar, for 1200 < Re < 2000 the flow was named as transient and for Re >2000 the flow was named as turbulent. The image pairs were selected as one from the start of the experiment, one from the middle and the last one from nearly the end of recording. Those images were imported to OpenPIV and the scaling was set to 1. The analysis was conducted with 32x32 interrogation window size and with 32x32 overlap. One representative result for each flow was presented below (See Figures 8-12).

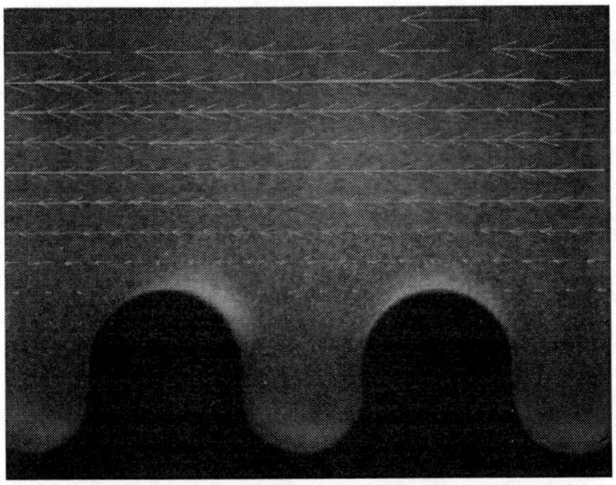

Figure 8. Result for laminar flow with Re = 664.92

Figure 9. Result for laminar flow with Re = 1017.59

Figure 10. Result for transient flow with Re = 1483.92

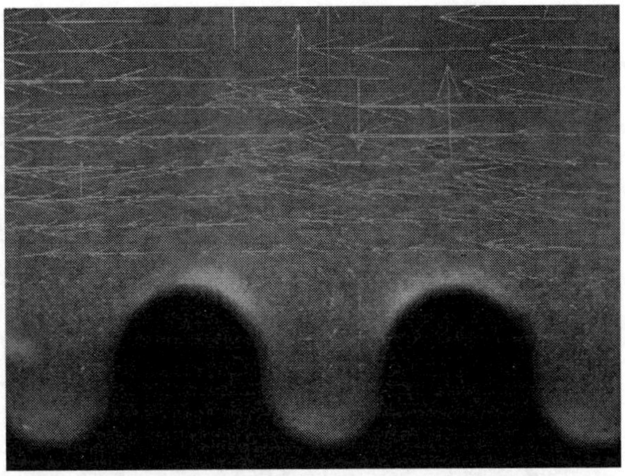

Figure 11. Result for turbulent flow with Re = 3243.87

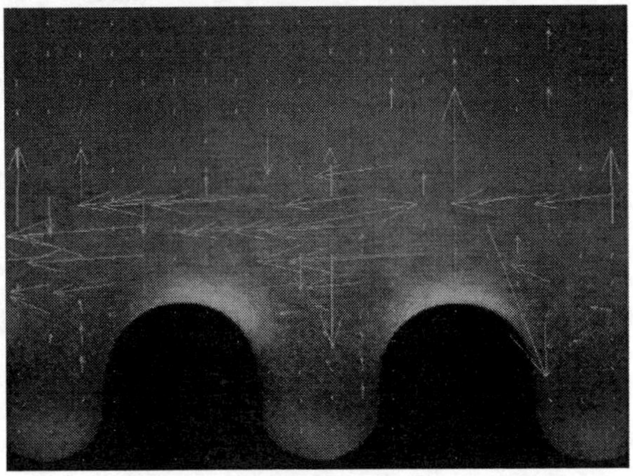

Figure 12. Result for turbulent flow with Re = 7340.02

For laminar flow regime and for turbulent flow with Re = 3243.87, the software gave reasonable result. For transient and turbulent flow with Re = 7340.02 analysis, the accuracy was lower. As it can be clearly seen from results, the upper part of the corresponding images could not be analysed accurately. These areas are lack of velocity vectors because of the quality of the images. To visualize it better, original frames were presented. (See Figure 13)

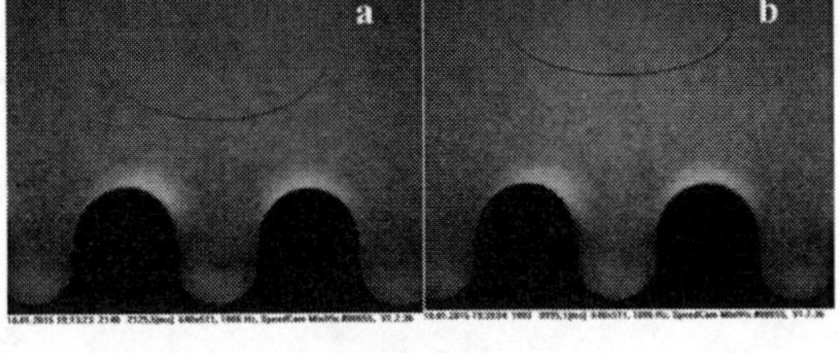

Figure 13. a) Non-processed image for turbulent flow with Re = 7340.02 b) Non-processed image for transient flow with Re = 1483.92

The red circles show the hardest locations to analyse since there are no significant intensity points. As stated before, PIV technique works with the light intensity of points. The lesser the significant intensity points, the lower the accuracy of the analysis.

The analysis was conducted with scale value 1. Hence, velocity vectors have the unit of pixels/dt. Where dt represents the time interval between images. To get a physical result, a basic conversion was done. In order to do this, two main parameters were needed to be known, these parameters were dt in seconds and the meter-to-pixel ratio of the images. The information of the pixels in the image was taken from the camera itself. The images have 640 x 511 pixels and 1006 Hz. Some of the images were taken for different fps rates so that these changes were taken into account while converting the px/dt values to physical units. As it can be seen from the schematic representation of flow domain (See Figure 1) the diameter of the crest (d2) is 3.6 mm. This physical length was taken as reference and the meter-to-pixel ratio was approximated by using an image manipulation program called GIMP. The original image was analysed with GIMP and the pixel ratio of original image and analysis PC's display was found as 1.2. It means 640 pixels are represented with 535 pixels in the GIMP GUI.

The diameter of the crest was measured with "Measure" tool and it was found as 129 pixels. The proportion between 129 pixels and 535 pixels was calculated as 4.1473. So the physical length of the image was calculated as 14.93 mm by multiplying crest's physical length with the proportion. This means 640 pixels in the image corresponds to 14.93 mm. So, for 1 m there are 42867 pixels.

The formula used for conversion is the following

$$velocity\left[\frac{m}{s}\right] = \frac{u_x\,[px]}{dt[s]} \cdot \frac{1[m]}{42867[px]} \quad (2)$$

The vector components in the x-direction were selected with the MIN function which returns the minimum value of a column of data in MS Excel. Since the flow direction is –x, the vectors are negative. The values were then entered to a MATLAB script to calculate the velocity values. The returned value was double-checked if it is an outlier or not. This was done by inspecting its position and the neighbour vector lengths.

The theoretical velocity versus calculated velocity results for different Reynolds numbers were tabulated and presented (See Table 1). Re is the Reynolds number, FPS is the frame-per-second rate, u_x stands for the velocity vector length in the x-direction. V_c is the calculated velocity of the flow and V_t stands for the theoretical velocity of the flow due to mass flow rate.

Table 1. Velocity analysis results for all Reynolds numbers

	Re	FPS	u_x(px)	V_c(m/s)	V_t (m/s)	Error
Laminar	664,92	300	6,9707	0,050	0,042	0,183
	664,92	300	6,2045	0,045	0,042	0,053
	664,92	300	6,2545	0,045	0,042	0,061
	1017,59	300	11,0514	0,080	0,065	0,222
	1017,59	300	10,4853	0,075	0,065	0,159
	1017,59	300	10,5458	0,076	0,065	0,166
Transient	1483,92	1006	4,7727	0,115	0,095	0,217
	1483,92	1006	4,0866	0,099	0,095	0,042
	1483,92	1006	4,3566	0,105	0,095	0,111
Turbulent	3243,87	505	15,9799	0,194	0,206	-0,062
	3243,87	505	15,7341	0,191	0,206	-0,076
	3243,87	505	16,5771	0,201	0,206	-0,027
	7340,02	1006	14,0398	0,339	0,468	-0,276
	7340,02	1006	12,0364	0,291	0,468	-0,380
	7340,02	1006	12,0877	0,292	0,468	-0,377

6.3 Conclusion

The results show that using OpenPIV to analyze experimental findings highly depends on the image quality, especially when analysing turbulent and transitional flow. When the all velocity values are compared with the theoretical ones, one can say that the error increases as the flow becomes turbulent. The main reason for the error increment is the lack of vector fields around the top of the image due to the blurriness. To be able to analyse corrugated pipe flow with OpenPIV software, the illumination and seeding methods need to be changed. The accuracy of this software is mainly based on the illumination and seeding processes rather than the camera quality.

7 Visualization of flow inside the grooves

The flow was analysed by using region of interest (ROI) option and the visualization of the flow inside the groove for different Reynolds numbers were presented below.

Figure 14. Flow inside the grooves with Re = 664.52 (Scale 20)

Figure 15. Flow inside the grooves with Re = 1017.59 (Scale 20)

Figure 16. Flow inside the grooves with Re = 1483.92 (Scale 20)

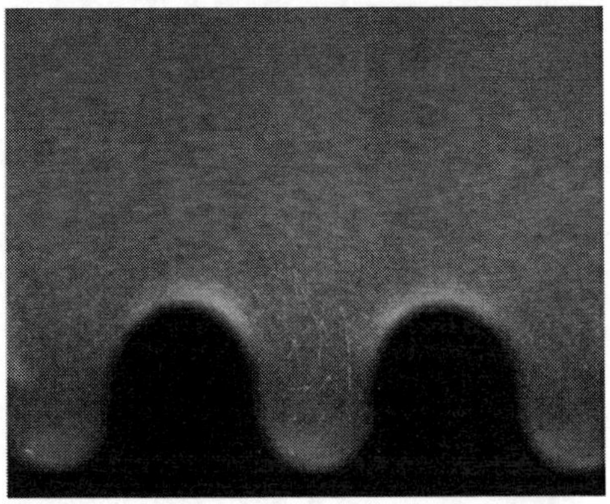

Figure 17. Flow inside the grooves with Re = 3243.87 (Scale 5)

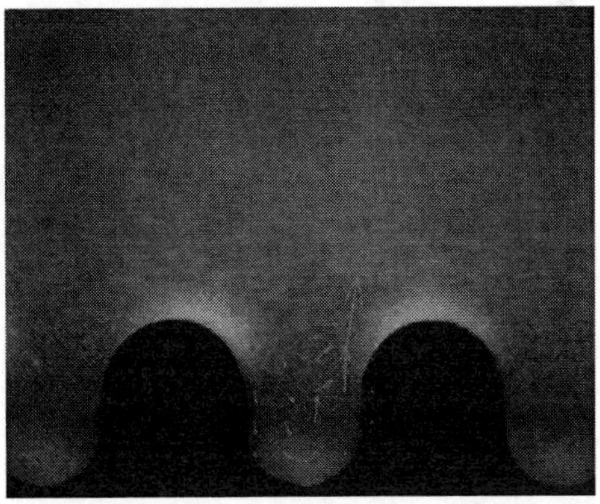

Figure 18. Flow inside the grooves with Re = 7340.02 (Scale 5)

8 Conclusions

In conclusion, OpenPIV was verified as promising open source PIV analysis software. The vorticity results came out to be slightly different for location but the magnitude of the vortices were agreed with the earlier study results when validation test was performed. The velocity analysis result does not follow a pattern. For laminar flow, the error is in between 5% and 22%. For transient case the error is in between, 4% and 22%. For turbulent flow, it is in between 3% and 38%. The reasons for that are probably related to the image quality and selection of seeding. The visualization experiment needs to be reconducted by using different seeding methods and illumination methods to obtain more accurate velocity vector fields. Also, the velocity was calculated by using a stopwatch so that the human reaction time delay should be taken into account while considering the error of the results.

The visualization of the images with OpenPIV provided a rough idea about the flow behaviour. When bulk flow results are considered, flow regimes could be identified based on the fluctuating velocity components with OpenPIV. When flow in the grooves are analysed, OpenPIV results provided sense of flow rotation in the grooves.

References

1. Shraub, F.A., Kline, S.J., Henry, J., Runstadler, P.W., Littell , JR. A., J. of Basic E. , **87**, 429-444 (1965)
2. Taylor, Z.J., Gurka, R., Kopp, G.A., Liberzon, A., IEEE Tr., **59**, 3262-3269 (2010)
3. Irwin, W., J. of Irr. and Dr. Eng, **110**, 237-241(1984)
4. Ahn H, Uslu I, IMECE 66061 (2013)
5. Unal E., Ahn H., Sorguven E., IMECE66063 (2013)
6. Unal E., Sorguven E.,Ahn H., IMECE66062 (2013)
7. Olcay, A. B.,Krueger, P. S., Exp. In Fld., **44**, 235-247, (2008)

EPJ Web of Conferences **114**, 02108 (2016)
DOI: 10.1051/epjconf/201611402108

Influence of balancing holes and disc sealing in turbine stages on thermodynamic efficiency and rotor forces

Václav Sláma[1,a]

[1]R&D Specialist Sciences, Thermodynamics Department, Doosan Škoda Power s.r.o., Tylova 1/57, 301 28 Plzeň, Czech Republic

Abstract. This paper deals with CFD analysis of a particular turbine impulse stage including the surrounding parts such as diaphragm seals, shroud seals and a disc axial gap, where balancing holes and a disc sealing is designed. The aim is to investigate effects on efficiency and axial forces by evaluation of a pressure distribution on the rotor disc and respective mass flow balances. As a result of this, the possible methodology of design optimization could be found. The other part of the project is to find the way for a more exact preliminary computation which is essential for increasing efficiency and the reduction of the axial forces.

1 Introduction

1.1 Motivation

An increasing of efficiency is the main challenge for each research and development department in the companies of power engineering. Although a lot of significant improvements have been made such as the development of blade profiles or the optimization of exhaust hoods, there are still many possibilities how the efficiency can be increased. One of them is a reduction of the negative influence of steam which cannot produce an output because it does not flow through blade cascades (nozzles and buckets), where the output is produced, but through the surrounding parts such as diaphragm seals or shroud seals. Owing to construction requirements, a small but not insignificant amount of the steam always has to go through these parts. We can reduce this amount, but never completely. An approach how to estimate this mass flow and possibilities of the reduction is described, for example, in reference 1. The rest of this steam should be used in an appropriate way so as not to disturb the main flow.

In addition, it is very desirable to find ways how to decrease the turbine price. For example the price could be decreased by using optimal bearings whose size determines construction costs. If the influence of the particular turbine parts on the flow phenomena is known, it could allow us to estimate an axial force in the rotor. Accordingly, it is possible to use smaller bearings which could be considerably cheaper.

The mentioned possibility of increasing the efficiency and the accurate estimation of forces is closely dependent on the turbine sections around a rotor disc. By using a CFD analysis, which is described in this paper, the influence of important construction parts in the rotor disc have been identified and described. The results have been implemented in a homemade loss model and have also been considered during a standard design procedure.

1.2 Wheel stage design

One of the widely used designs for turbine stages is a wheel stage design which is shown in figure 1. It was created particularly for impulse stages but the similar principle could be used in various modern stages.

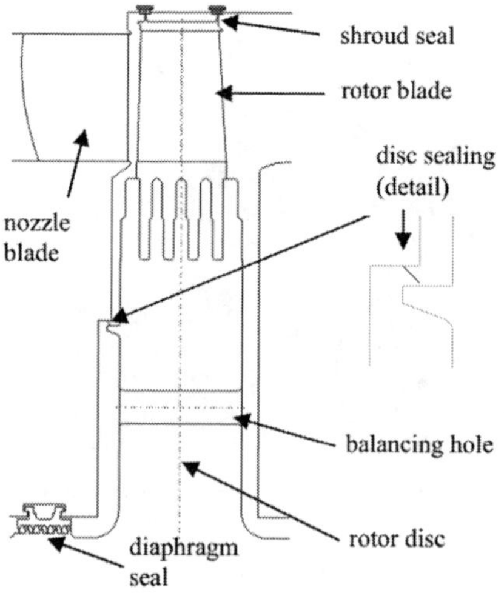

Figure 1. Typical wheel stage design.

In figure 1, are shown typical parts of all turbine stages – the nozzle and the rotor blade and the diaphragm and the shroud seal. There are also the parts which are specifically in the wheel stage design – the balancing

[a] Corresponding author: vaclav.slama@doosan.com

hole and the disc sealing. These two parts are main subjects of this study.

2.1.1 Balancing hole

One purpose of the balancing hole is to allow the steam going from the diaphragm seal to flow through the rotor disc to the next turbine stage so that this steam would not be mixed with the main flow upstream of the rotor blades. This mixing produces losses because the direction of the steam going through the diaphragm seal and adjacent rotor disc parts is different from the direction of the steam going from the stator blades. These losses are often included into the secondary losses. A lot of detailed researches and studies have been carried out. (For examples see references 2, 3 and 4). On the whole, the balancing holes could significantly increase the efficiency. Another purpose of using the balancing holes is to decrease an axial force in the rotor. This effect is sometimes considered to be more important than the efficiency increasing.

2.1.2 Disc sealing

The other typical construction part in the wheel stage designs is the disc sealing. It usually consists of a thin strip which is situated against the rotor part (it can be seen in detail in figure 1). The main purpose of the disc sealing is to decrease the amount of steam going from the cascade. This effect is described in detail in the results section. As a consequence, it helps to increase the efficiency. On the other hand, by using the disc sealing the axial force could increase.

2 Computational model

2.1 Domains and meshes

A package of ANSYS tools was used for carrying out the CFD analysis. Turbo Grid 15.0 was used for creating meshes of the rotor and the stator domain while ICEM 15.0 was used for other domains, where a blocking strategy was largely used. An appropriate boundary layer was used so that an average value of y+ would be lower than 1.

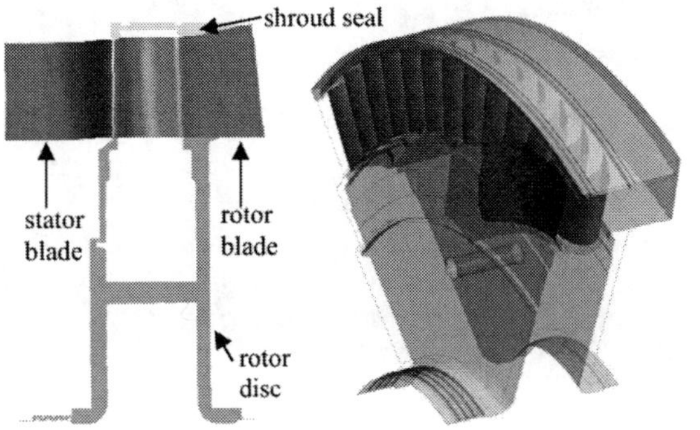

Figure 2. Computation model.

The used model, which is depicted in figure 2, consists of four domains. The number of rotor blades had to be adjusted according to the balancing hole (there are thirteen blades for one hole) so that flow phenomena could be simulated correctly in the rotor disc part.
What is more, because we need to compare the influence of the mentioned parts, the model was adapted so that the balancing hole and the disc sealing could or could not be used. Thanks to this, four variants could be computed and the influence of each observed part could be described. The names of the variants are in table 1.

Table 1. Used variants.

name	balancing hole	disc sealing
A	no	no
B	no	yes
C	yes	no
D	yes	yes

2.2 Numerical solution and boundary conditions

A numerical solver CFX 15.0 was chosen as the main CFD software tool. A numerical computation using the steady state RANS analysis was performed by using a compressible turbulent flow with the k-ω SST turbulent model. The medium was considered as a real gas.

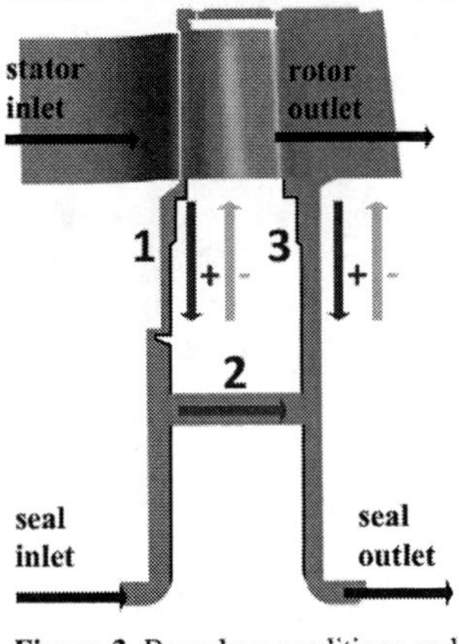

Figure 3. Boundary conditions and monitoring places.

Boundary conditions with the flow directions are shown in figure 3. A total pressure and temperature distribution as well as a flow direction was defined at the stator inlet and an average static pressure with a radial equilibrium condition was defined at the rotor outlet. A total pressure according to the computed mass flow was defined at the seal inlet and a mass flow outlet was defined at the seal outlet (there is in fact an inlet to the diaphragm seal of the next turbine stage). Rotational periodicity conditions were applied to corresponding free side boundaries of the computation domains. Connections between the domains were defined as interfaces which were specified as a stage or a frozen rotor.

2.3 Monitored parameters

Due to the fact that the analysis is very complex, many parameters had to be monitored during the computation

in order to check that the solution was converged enough. The monitored parameters were mass flow balances, axial forces acting to parts connected with the rotor (the rotor disc, the rotor blades and the shroud seal) and static pressure distributions on rotor disc surfaces.
The mass flows were monitored especially in the rotor disc part (defined as 1, 2 and 3 sections in figure 3). The monitoring of mass flow balances through the other parts, which are the shroud seal, the stator inlet and the rotor outlet were carried out too, but these results are not important for our further evaluations. The pressure distribution was monitored on the disc surfaces (this surface is represented by the bold black line in figure 3). The inlet part of the disc as well as the outlet part was considered. The other essential monitored parameter was the stage efficiency which was evaluated as total to total.

3 Results

As was mentioned before, the computations of four different models, which differ in either having the balancing hole and the disc sealing or not, were carried out. According to the preliminary assumptions, the results are strongly dependent on the computed variant.

3.1 Efficiency and axial forces

The following graphs in figure 4 present the comparison of the efficiency and of the axial rotor force.

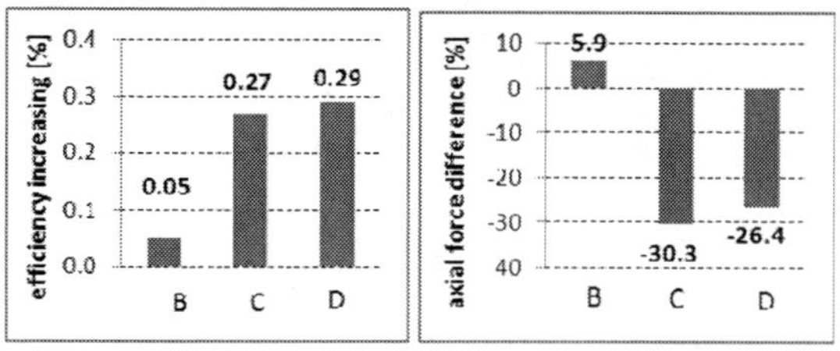

Figure 4. Efficiency and forces comparison.

If we compare the variants where there is the balancing hole (C and D) with the variants without the hole (A and B), the positive effect can be seen. The efficiency was increased by about 0.3%. In addition, the force was significantly reduced by about 30%. The effect of the disc sealing (the difference between 0.29 and 0.27) seems to be negligible but in the wider terms each increment of the efficiency is very desirable. The other fact is that this increment should be significantly greater in other turbine stages.

3.2 Mass flow balances

A reason why the balancing hole considerably influences the efficiency can be explained by comparing the mass flows which are in table 2 (the evaluating positions of the flows 1, 2 and 3 were described in figure 3). In this picture, we can also see plus and minus directions. If the steam goes from the cascade to the adjacent rotor disc part, the direction is positive because this effect increases the efficiency. On the contrary, the minus direction means that the flow goes to the cascade which has a negative effect. The optimal is when the mass flow through the part 1 is very closed to zero but this case can hardly be reached in practice. The greater the leakage going from the cascade, the greater losses are produced. For this reason, the disc sealing is commonly used. However, the positive effect is not always too substantial. Although the difference of the mass flow between the variants C and D is at about 0.3 kg s^{-1}, the efficiency was increased only by 0.02%. This result shows that if we want to increase the efficiency, we need to reduce the leakage more significantly. In a practical way, it can for example mean to design a smaller clearance in the disc sealing.

Table 2. Mass flows through observed parts (in kg s^{-1}).

name	flow 1	flow 2	flow 3
A	-1.217	-	0.867
B	-1.217	-	0.867
C	0.902	2.119	-1.252
D	0.507	1.781	-0.914

Regarding the facts mentioned before, it seems to be not difficult to generalize that by using balancing holes and disc sealing is always of benefit in each turbine impulse stage. However, the influence of the disc sealing is small in this study. In addition, the question as to how significant is the increase in efficiency has to be answered for future using because the accurate estimation of the efficiency is very important. Therefore, many other detailed studies, including CFD, and experiments have been done. See references 4 and 5. As a result, a homemade efficiency prediction procedure has been considerably increased.

3.3 Pressure distribution

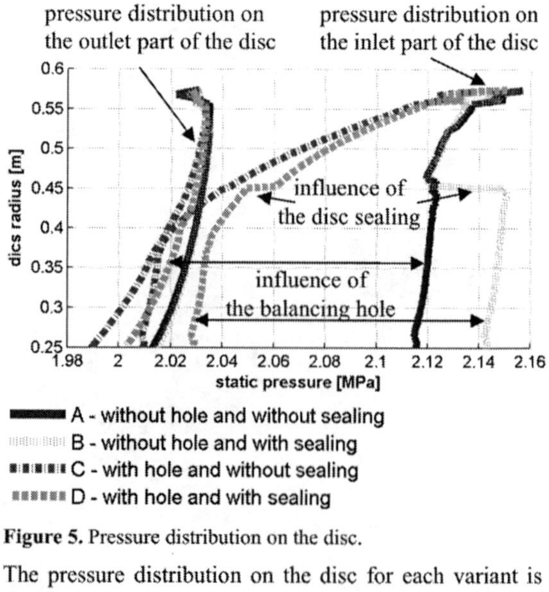

Figure 5. Pressure distribution on the disc.

The pressure distribution on the disc for each variant is depicted in figure 5. An understanding of these

distributions is essential for an axial forces estimation. There can be clearly seen the influence of the balancing holes as well as the influence of the disc sealing.
The significant pressure difference between the variants with the balancing holes (C and D) and the variants without the balancing holes (A and B) shows the important impact of the holes. We would like to avoid having the big pressures which are in the variants A and B. Therefore, the balancing holes should be used in order to have the lower axial force in the rotor. The pressure is the lowest in the variant C where the pressure on the inlet disc part is almost the same as in the outlet disc part, particularly in the middle section where the balancing hole is situated. It is the best solution regarding the axial forces but in order to increase the efficiency the disc sealing is used. Its effect can be visible as horizontal parts in the variants of B and D, where the pressure is sharply changed. It means that if the disc sealing is added, the pressure difference between the inlet and outlet part is greater which shows the disc sealing impact that is not desirable. Fortunately, this pressure increase, which is in fact the axial force increase, is not as significant as the impact of the balancing hole.

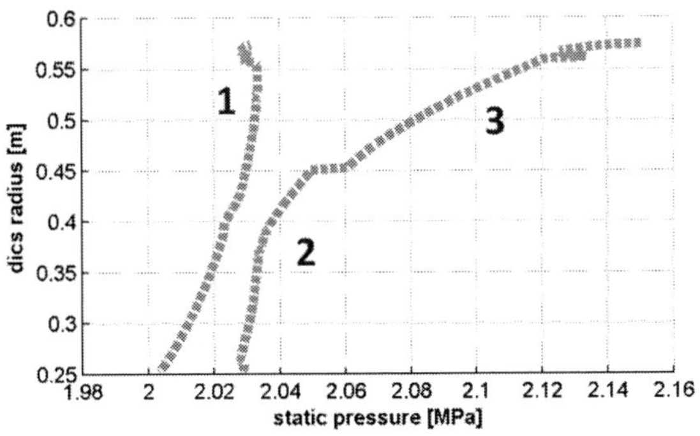

Figure 6. Pressure distribution on the disc in the variant D.

The result of the pressure distribution of D variant is used in this paper to briefly describe how to deal with the generalisation of results. This generalization is essential owing to the fact that it is absolutely not economic to use CFD for computing each turbine stage which one is going to design. As it was said, the estimation of the pressure distribution on the disc is vital for computing the rotor axial force. In order to create this estimation, we can use general equations of flow in the disc (see reference 6) and add the particular coefficients according to the knowledge from CFD computations.
As is shown in figure 6, the pressure distribution is not linear. It is caused by the influence of whirls, rotor revolutions and the amount of steam going through the rotor disc parts. As we can see, the curve could be divided into different parts. While the pressure distribution on the outlet part of the disc could be described as one curve, the pressure distribution on the inlet part has to be described by two curves at least. As a result, the final estimation of the pressure distribution could be defined as a function of the revolutions, a radial position, a clearance between rotor and stator parts, mass flow through the relevant parts and pressures and temperatures at the rotor hub as well as in the seals. In addition, this function has to be multiplied by the coefficients which are different according to the position on the disc (1, 2 or 3 in figure 6).
As we can imagine, the computed coefficients are dependent on a turbine design, therefore they would be different for variant turbine stages. It means that the different values, which have been either computed or estimated, have to be used during the stage design procedure. For this reason, not only the turbine stage presented here was computed but also some experiments have been carried out.

4 Conclusion and further work

The numerical analysis of the flow phenomena in the rotor disc section representing the wheel stage design has provided general and valuable information about the dependence of the balancing holes and the disc sealing on the efficiency and the axial force in a rotor. It was clearly shown that the balancing holes have considerable effect on increasing the efficiency and in reducing the axial force. The influence of the disc sealing was also described by using the comparison of mass flow balances.
The importance was placed on the analysis of the pressure distribution on the rotor disc. It was said that the static pressure distribution could be described as dependent on the basic stage parameters and by using the appropriate partition of the showed pressure distribution curves.
The results, which were presented in this paper, have been used as the references for similar computations and for the improvement of the current prediction of the efficiency and for the estimation of the axial force in the rotor. This endeavour leads to the fact that the efficiency of a steam turbine can be improved and the production costs can be reduced by using a modernised estimation computational approach.

References

1. V. Sláma, P. Epikaridis, V. Vaník, M. Hajšman: *Study of Mass Flow in Steam Turbine Labyrinth Seals.* (29th Conference of Computational Mechanics, Špičák 2013)
2. L. Bednář, M. Hoznedl, L. Tajč: *Účinnost turbínových stupňů III. upravené vydání.* (Doosan Škoda Power, Plzeň 2010)
3. M. Hoznedl, J. Linhart, L. Tajč: *Vliv vefukování do mezní vrstvy na ztrátový součinitel výstupního hrdla.* (pp. 53-56, Praha 2006)
4. V. Sláma, L. Tajč, K. Yun: *Vliv hřídelové ucpávkové páry na účinnost turbínového stupně.* (Turbostroje, Praha 2015)
5. K. Yun, L. Tajč: *Vliv geometrických úprav oběžných lopatek a přívodního kanálu pro páru z hřídelové ucpávky na aerodynamické parametry proudu páry.* (Doosan Škoda Power, Plzeň 2014)
6. A. V. Ščegljajev: *Parní turbíny 1. Svazek.* (SNTL, Praha 1983)

EPJ Web of Conferences **114**, 02109 (2016)
DOI: 10.1051/epjconf/201611402109

Numerical solution of inviscid and viscous laminar and turbulent flow around the airfoil

Martin Slouka[1,a], Karel Kozel[1,2]

[1]Dept. of Technical Mathematics, Faculty of Mechanical Eng., CTU in Prague, 121 35 Prague 1, Czech Rep.
[2]Institute of Thermomechanics, Academy of Science CR, 182 00 Prague 8, Czech Rep.

Abstract. This work deals with the 2D numerical solution of inviscid compressible flow and viscous compressible laminar and turbulent flow around the profile. In a case of turbulent flow algebraic Baldwin-Lomax model is used and compared with Wilcox k-omega model. Calculations are done for NACA 0012 and RAE 2822 airfoil profile for the different angles of upstream flow. Numerical results are compared and discussed with experimental data.

1 Introduction

Mathematical models used in this work are based on the solution of the inviscid compressible flow and viscous compressible laminar and turbulent flow around the airfoil.

Computational domain represents NACA 0012 and RAE 2822 airfoil profiles. The results show differences between the uses of each numerical model and experimental results.

2 Governing equations

The two-dimensional viscous compressible flow is prescribed by the system of Navier-Stokes equations in following conservation form

$$\mathbf{W}_t + \mathbf{F}_x + \mathbf{G}_y = \mathbf{R}_x + \mathbf{S}_x, \quad (1)$$

where

$$\begin{aligned} \mathbf{W} &= (\rho, \rho u, \rho v, e)^T, \\ \mathbf{F} &= (\rho u, \rho u^2 + p, \rho uv, (e+p)u)^T, \\ \mathbf{G} &= (\rho v, \rho uv, \rho v^2 + p, (e+p)v)^T, \\ \mathbf{R} &= (0, \tau_{xx}, \tau_{xy}, u\tau_{xx} + v\tau_{xy} - q_x)^T, \\ \mathbf{S} &= (0, \tau_{xy}, \tau_{yy}, u\tau_{xy} + v\tau_{yy} - q_y)^T \end{aligned} \quad (2)$$

and

$$\begin{aligned} \tau_{ij} &= \mu\left(\frac{\partial u_i}{\partial x_j} + \frac{\partial u_j}{\partial x_i}\right) - \frac{1}{3}\delta_{ij}\frac{\partial u_k}{\partial x_k}, \\ q_i &= -\frac{\kappa}{\kappa-1}\frac{\mu}{Pr}\frac{\partial}{\partial x_i}\left(\frac{p}{\rho}\right). \end{aligned} \quad (3)$$

This system is closed by following equation

$$p = (\kappa - 1)\left[e - \frac{1}{2}\rho\left(u^2 + v^2\right)\right]. \quad (4)$$

In the equations above ρ denotes density, (u, v) are components of local velocity in x and y direction respectively, e denotes total energy per unit volume, p is pressure given by previous equation of state, τ_{ij} is shear stress, q_i represents heat flux, μ represents dynamical viscosity.

Following constants were used in previous equations: κ is isentropic exponent equalled to 1.4, Pr is Prandtl number equalled to 0.7.

Viscous compressible turbulent flow represents system of Reynolds averaged Navier-Stokes equations (RANS) which is formally the same as (1) with enclosed used model of turbulence which is algebraic Baldwin-Lomax or two-Wilcox k-ω model [1] in this work.

Inviscid flow is represented by the system of Euler equations which is simply reached after neglecting of viscosity in viscous fluxes on the right hand side in the system (1).

[a] Corresponding author: m.slouka@seznam.cz

3 Models of turbulence

3.1 Baldwin-Lomax model

Baldwin-Lomax (B-L) model is the modification of basic Cebeci and Smith algebraic turbulent model. The boundary layer is divided into two regions, inner region (close to the wall) and outer region. Turbulent viscosity in the inner part is given by

$$\mu_{ti} = \rho F_d^2 \kappa^2 y^2 |\Omega| \tag{5}$$

where vorticity is in 2D case defined as

$$\Omega = \frac{\partial u}{\partial y} - \frac{\partial v}{\partial x}, \tag{6}$$

van Driest function given as follow

$$F_d = 1 - \exp\left(-\frac{1}{A}\frac{u_\tau y}{\nu_w}\right) \tag{7}$$

and friction velocity by formula

$$u_\tau = \sqrt{\frac{\tau_w}{\rho}} = \left(\nu \frac{\partial u}{\partial y}\right)_w^{1/2}. \tag{8}$$

Turbulent viscosity in outer region is given by

$$\mu_{to} = \alpha C_{cp} F_w F_k. \tag{9}$$

Function F_w is determined by the relation

$$F_w = y_{max} F_{max}. \tag{10}$$

where F_w is the maximum of the function bellow

$$F = y|\Omega| F_d \tag{11}$$

and y_{max} is the distance from the wall in which $F(y_{max}) = F_{max}$ holds and

$$F_k = \left[1 + 5.5\left(C_{KL}\frac{y}{y_{max}}\right)^6\right]^{-1}. \tag{12}$$

In Baldwin-Lomax turbulent model following constants are used: $\kappa = 0.4$, $A = 26$, $\alpha = 0.0168$, $C_{cp} = 1.6$ and $C_{KL} = 0.3$.

3.2 Wilcox k-ω model

The two-equation model is given by transport equations for two characteristic scales of turbulent motion, in this case turbulent energy k and specific dissipation rate ω. Then turbulent viscosity is defined as

$$\mu_t = \rho \frac{k}{\omega}. \tag{13}$$

The Wilcox k-ω model [1] was used as follow

$$\begin{aligned} &\frac{\partial(\rho k)}{\partial t} + \frac{\partial(\rho u_j k)}{\partial x_j} = P_k + \frac{\partial}{\partial x_j}\left[\left(\mu + \sigma^* \mu_t\right)\frac{\partial k}{\partial x_j}\right] \\ &- \beta^* \rho k \omega, \\ &\frac{\partial(\rho\omega)}{\partial t} + \frac{\partial(\rho u_j \omega)}{\partial x_j} = \gamma \frac{\omega}{k} P_k + \frac{\partial}{\partial x_j}\left[(\mu + \sigma\mu_t)\frac{\partial \omega}{\partial x_j}\right] \\ &- \beta\rho k \omega^2 + C_D, \end{aligned} \tag{14}$$

where

$$P_k = \tau_{ij}\frac{\partial u_i}{\partial x_j} \tag{15}$$

represents production of turbulent energy and

$$C_D = \frac{\rho}{2\omega}\max\left(\frac{\partial u_i}{\partial x_j}\frac{\partial u_i}{\partial x_j}, 0\right). \tag{16}$$

Following model coefficients are used: $\alpha = 5/9$, $\beta = 3/40$, $\beta^* = 9/100$ and $\sigma = \sigma^* = 1/2$.

4 Numerical method

4.1 Numerical scheme

Finite volume method was applied on cell centred non-orthogonal structured grid with quadrilateral cells.

For numerical solutions Lax-Wendroff scheme in McCormack modification was used with predictor step

$$\mathbf{W}_{i,j}^{n+1/2} = \mathbf{W}_{i,j}^{n} - \Delta t \mathrm{Res} \mathbf{W}_{i,j}^{n} \tag{17}$$

and corrector step as follow

$$\mathbf{W}_{i,j}^{n+1} = \frac{1}{2}\left(\mathbf{W}_{i,j}^{n} + \mathbf{W}_{i,j}^{n+1/2} - \Delta t \mathrm{Res}\, \mathbf{W}_{i,j}^{n+1/2}\right). \tag{18}$$

The Jameson's artificial dissipation model was applied to damp the oscillations.

$$\widetilde{\mathbf{W}}_{i,j}^{n+1} = \mathbf{W}_{i,j}^{n+1} + AD\left(\mathbf{W}_{i,j}^{n}\right), \tag{19}$$

$$\begin{aligned} AD\left(\mathbf{W}_{i,j}^{n}\right) &= C_1\psi_1\left(\mathbf{W}_{i+1,j}^{n} - 2\mathbf{W}_{i,j}^{n} + \mathbf{W}_{i-1,j}^{n}\right) \\ &+ C_2\psi_2\left(\mathbf{W}_{i,j-1}^{n} - 2\mathbf{W}_{i,j}^{n} + \mathbf{W}_{i,j-1}^{n}\right), \end{aligned} \tag{20}$$

$$\psi_1 = \frac{\left|p^n_{i+1,j} - 2p^n_{i,j} + p^n_{i-1,j}\right|}{\left|p^n_{i+1,j}\right| + \left|p^n_{i,j}\right| + \left|p^n_{i-1,j}\right|}, \tag{21}$$

$$\psi_2 = \frac{\left|p^n_{i,j+1} - 2p^n_{i,j} + p^n_{i,j-1}\right|}{\left|p^n_{i,j+1}\right| + \left|p^n_{i,j}\right| + \left|p^n_{i,j-1}\right|}. \tag{22}$$

4.2 Boundary conditions

Inlet boundary conditions were realized for inviscid flow as follow: inlet velocity Ma_∞ together with angle of attack α, density ρ_∞ and total energy per volume e_∞ were set; pressure p_∞ was extrapolated from the flow field. In a case of viscous flow the inlet pressure p_∞ was set.

Outlet boundary conditions were the same for the both type of flow. Outlet pressure was set and other variables extrapolated from the flow field.

Solid wall condition was realized by the adding of virtual cells. In a case of inviscid flow velocity components were prescribed so that sum of velocity vectors equals zero in its tangential component. In a case of viscous flow velocity components were prescribed so that sum of velocity vectors equals zero on the wall.

5 Numerical results and discussion

Mach number isolines are shown in all pictures together with inlet flow parameters below where Ma_∞ corresponds to the inlet Mach number and α is attack angle of inlet flow. Numerical results were validated for different setting of artificial dissipation to damp the oscillations and in a case of viscous model the influence on the physical viscosity was checked.

The set of the following six figures corresponds with results of the calculation of transonic inviscid flow for NACA 0012 and RAE 2822 airfoil respectively. This configuration was set as a reference for the testing of behaviour of the inviscid model with regarding to the proper setting of the artificial dissipation. Mach number isolines and pressure coefficient C_p are compared with [2].

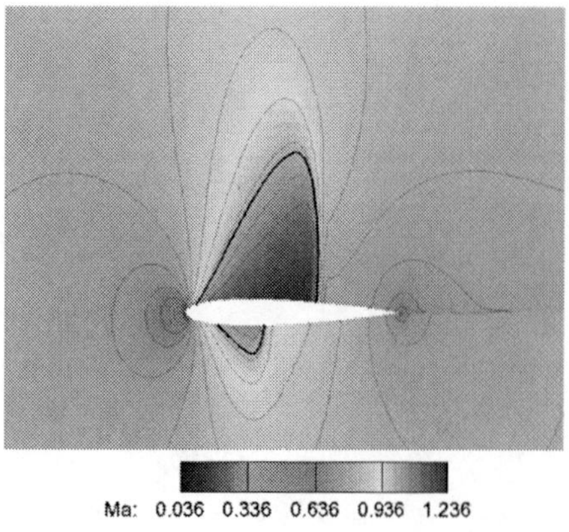

Figure 1. NACA 0012 airfoil, inviscid flow, $Ma_\infty = 0.8$, $\alpha = 1.25°$, Mach number isolines

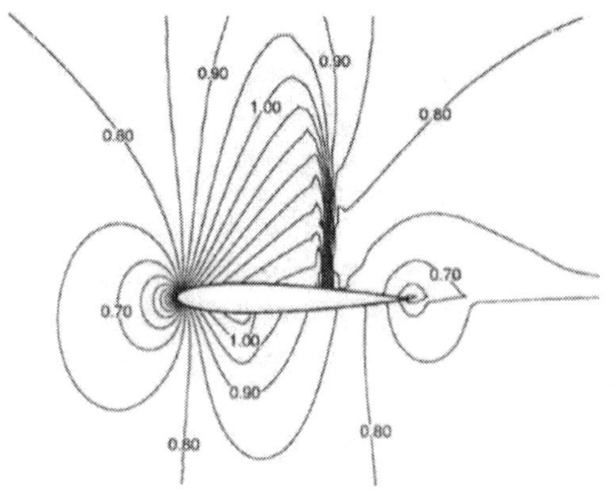

Figure 2. NACA 0012 airfoil, inviscid flow, $Ma_\infty = 0.8$, $\alpha = 1.25°$, Mach number isolines by [2]

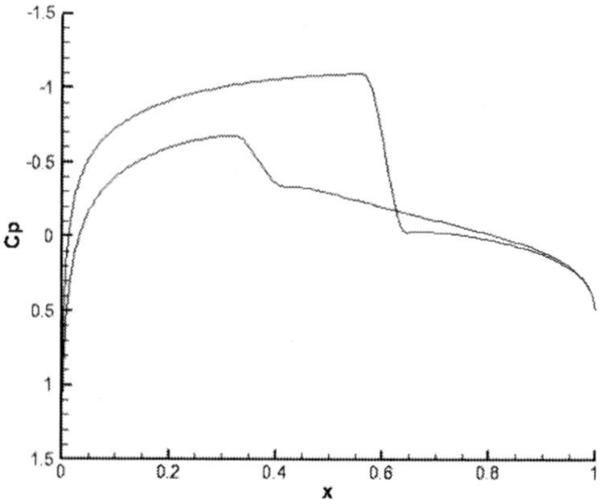

Figure 3. NACA 0012 airfoil, inviscid flow, $Ma_\infty = 0.8$, $\alpha = 1.25°$, pressure coefficient C_p along the profile

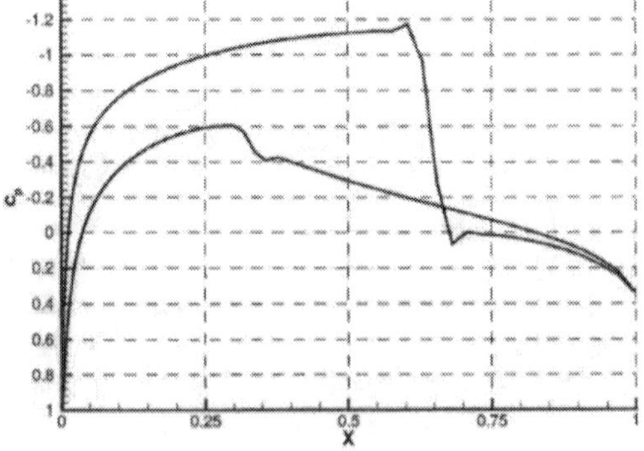

Figure 4. NACA 0012 airfoil, inviscid flow, $Ma_\infty = 0.8$, $\alpha = 1.25°$, pressure coefficient C_p along the profile by [2]

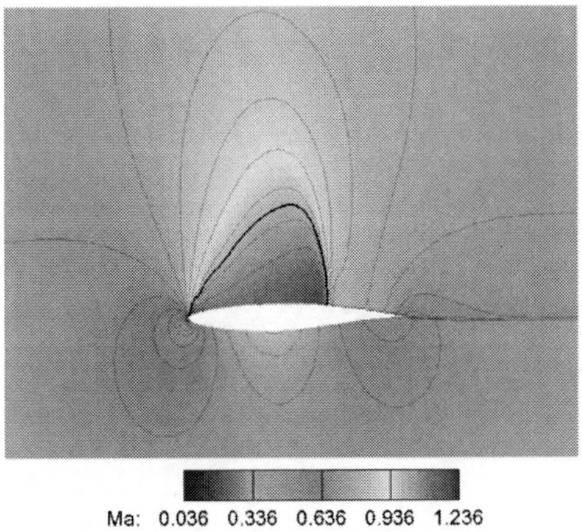

Figure 5. RAE 2822 airfoil, inviscid flow, $Ma_\infty = 0.73$, $\alpha = 2.79°$, Mach number isolines

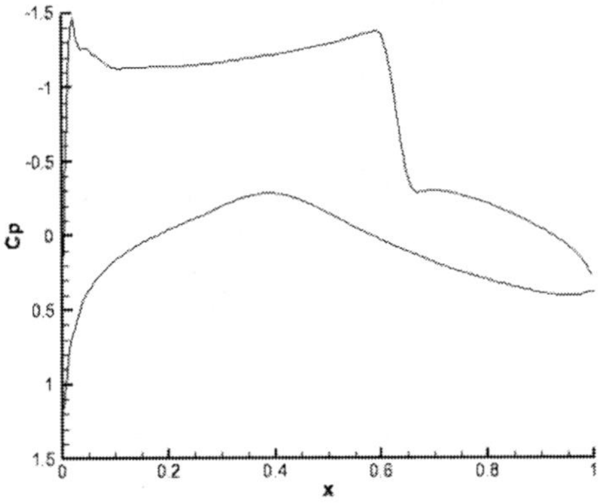

Figure 6. RAE 2822 airfoil, inviscid flow, $Ma_\infty = 0.73$, $\alpha = 2.79°$, pressure coefficient C_p along the profile

The next set of figures shows some test cases of viscous laminar and turbulent flow. First set of figures shows laminar flow around the NACA 0012 airfoil for the inlet Mach number $Ma_\infty = 0.85$ with angle of attack $\alpha = 0.0°$ and Reynolds number $Re = 500$. The case is quite far from the real physical conditions but it is generally used as the test case for the laminar model validation. Quite good result is clear and can be compared e.g. with results in [2]. The other figures show comparison between the used two different models of turbulence and experimental data. Inlet Mach number is $Ma_\infty = 0.15$ with angle of attack $\alpha = 0.0°$ and Reynolds number $Re = 2.88\times10^6$. It is seen good correlation to the each other.

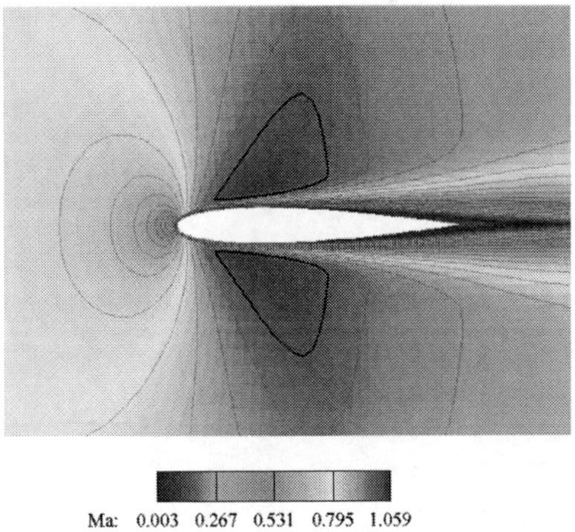

Figure 7. NACA 0012 airfoil, viscous laminar flow, $Ma_\infty = 0.85$, $\alpha = 0.0°$, $Re = 5\times10^2$, Mach number isolines

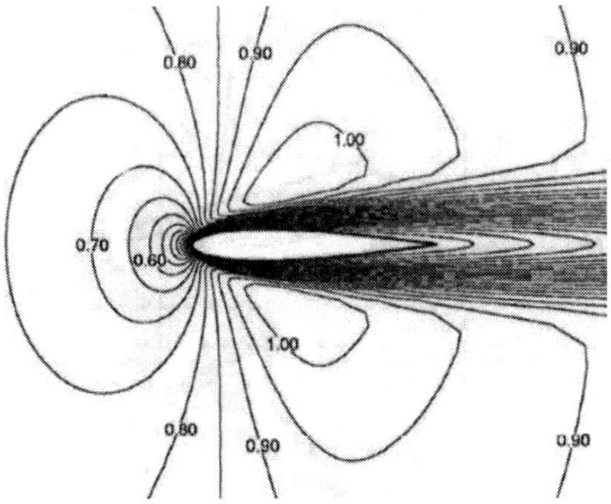

Figure 8. NACA 0012 airfoil, viscous laminar flow, $Ma_\infty = 0.85$, $\alpha = 0.0°$, $Re = 5\times10^2$, Mach number isolines by [2]

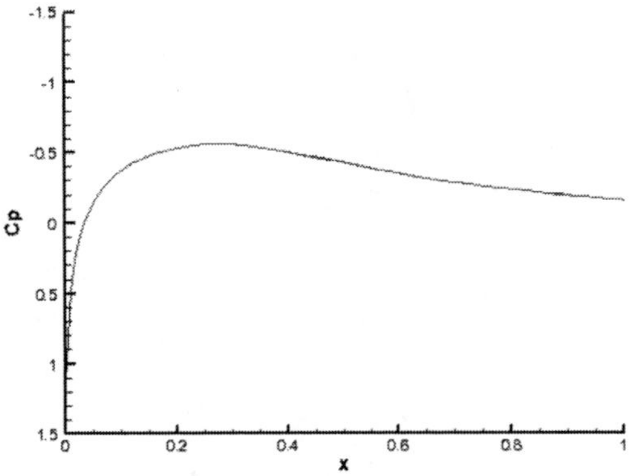

Figure 9. NACA 0012 airfoil, viscous laminar flow, $Ma_\infty = 0.85$, $\alpha = 0.0°$, $Re = 5\times10^2$, pressure coefficient C_p along the profile

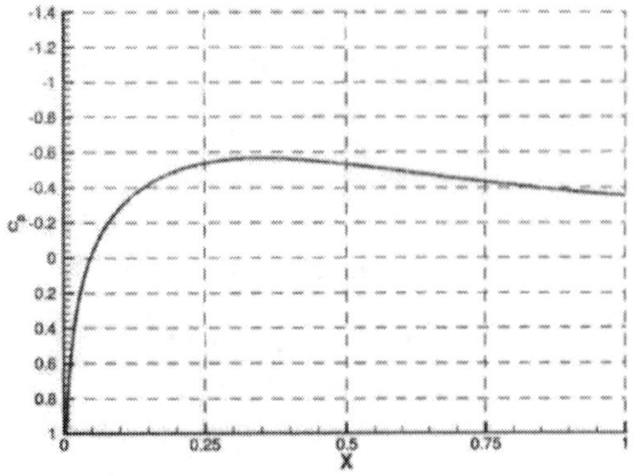

Figure 10. NACA 0012 airfoil, viscous laminar flow, $Ma_\infty = 0.85$, $\alpha = 0.0°$, $Re = 5\times10^2$, pressure coefficient C_p along the profile by [2]

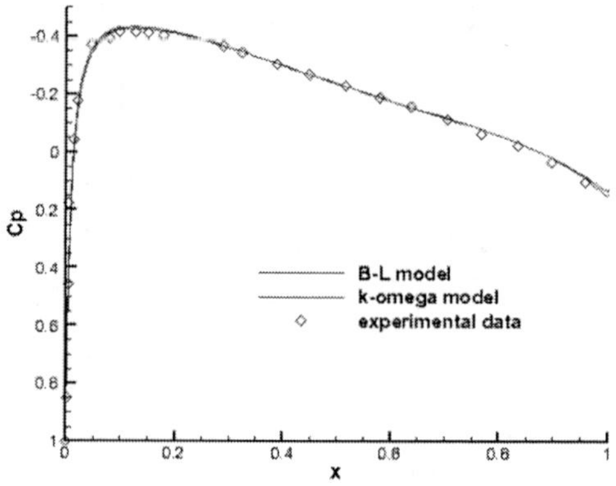

Figure 13. NACA 0012 airfoil, viscous turbulent flow, $Ma_\infty = 0.15$, $\alpha = 0.0°$, $Re = 2.88\times10^6$, pressure coefficient C_p along the profile

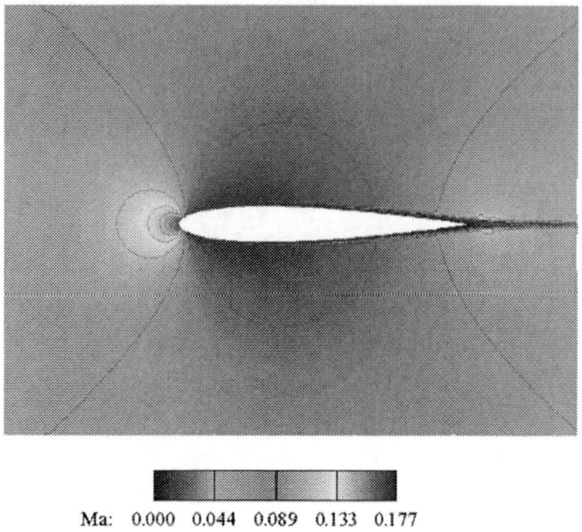

Figure 11. NACA 0012 airfoil, viscous turbulent flow, B-L model, $Ma_\infty = 0.15$, $\alpha = 0.0°$, $Re = 2.88\times10^6$, Mach number isolines

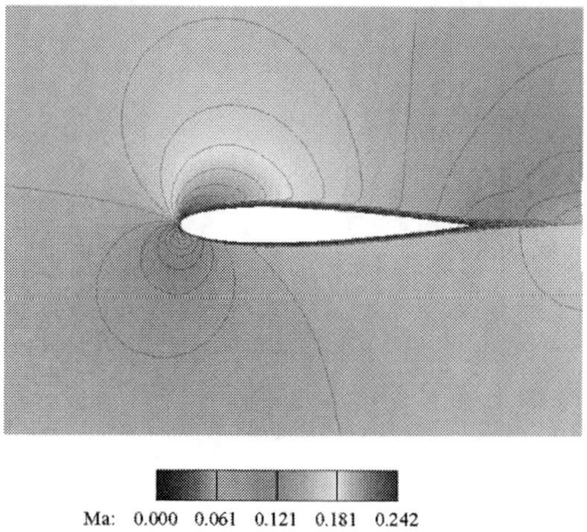

Figure 14. NACA 0012 airfoil, viscous turbulent flow, Wilcox k-ω model, $Ma_\infty = 0.15$, $\alpha = 10.0°$, $Re = 2.88\times10^6$, Mach number isolines

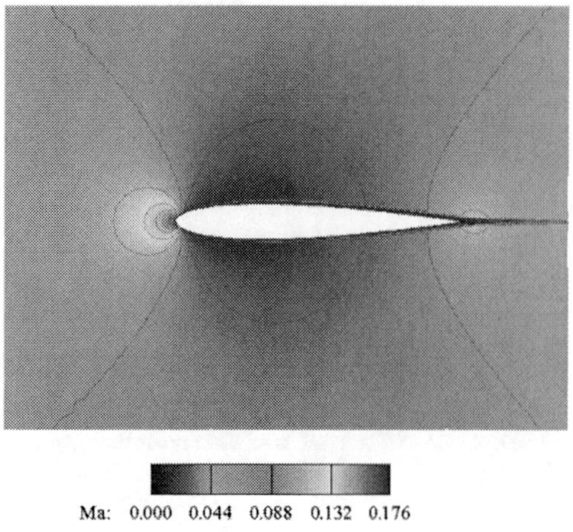

Figure 12. NACA 0012 airfoil, viscous turbulent flow, Wilcox k-ω model, $Ma_\infty = 0.15$, $\alpha = 0.0°$, $Re = 2.88\times10^6$, Mach number isolines

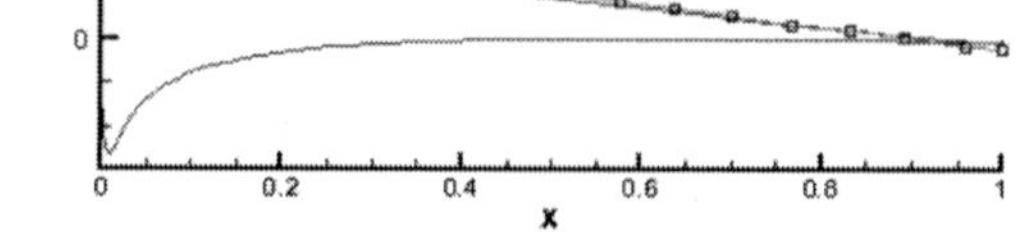

Figure 15. NACA 0012 airfoil, viscous turbulent flow, Wilcox k-ω model, $Ma_\infty = 0.15$, $\alpha = 10.0°$, $Re = 2.88\times10^6$, pressure coefficient C_p along the profile

The results of turbulent flow around the NACA 0012 profile are shown also at the figures above. In this case inlet Mach number was set to $Ma_\infty = 0.15$ and angle of attack α = 10.0°, Reynolds number Re = 2.88×10^6. Pressure coefficient is compared with experimental data by [3] for the upper surface with free transition. The deviation close the leading edge is visible and needs to be more investigated. It can be caused e.g. by coarse grid spacing in x axis around the leading edge. Also the problem of artificial dissipation model needs to be solved in the next study to reduce the influence of additional numerical viscosity that could be also the source of incorrect behaviour.

6 Conclusions

The developed software was successfully tested for some basic types of the calculations of inviscid and viscous turbulent flow around the NACA 0012 and RAE 2822 airfoil. Deeper testing still needs to be done in a case of viscous turbulent flow to verify the possibilities of different numerical approaches as well as the testing of another modification of two-equation k-ω model. Also the deeper investigation will be done at a field of the transition between the laminar and turbulent flow.

Acknowledgements

The research has been financed in part by grants P 101/10/1329 and P 101/12/1271 and by the Education for Competitiveness Operation Programme.

References

1. D. C. Wilcox, AIAA J., **46**, 2823, (2008)
2. J. Fort, K. Kozel, J. Fürst, J. Halama, J Dobes *Num. solution of flow problems I* (CTU in Prague, Prague 2004)
3. N. Gregory, C. L. O'Reilly, NASA R&M, **3726**, (1970)
4. R. Dvorak, K. Kozel, *Mathematical modelling in aerodynamics* (CTU in Prague, Prague, 1996)
5. R. Dvorak, *Transonic Flows* (Academia, Prague, 1986)
6. M. Slouka, K. Kozel, *Topical Problems of Fluid Mech. 2014,* 127-130 (2014)
7. M. Slouka, K. Kozel, *Topical Problems of Fluid Mech. 2013,* 57-60, (2013)
8. M. Slouka, K. Kozel, *Fluid Dynamics 2013,* 29-30, (2013)

EPJ Web of Conferences **114**, 02110 (2016)
DOI: 10.1051/epjconf/201611402110

Investigation of possible causes of the additional torque on the yacht's rudder

Soukup Lubomir[1,a], Stigler Jaroslav[1], and Kharicha Abdellah[2]

[1]Victor Kaplan Dept. of Fluid Engineering, Faculty of Mechanical Engineering, Brno University of Technology, Czech Republic
[2]Dept. of Simulation and Modelling of Metallurgical Process, Montanuniversität Leoben, Austria

Abstract. The present article deals with investigation of possible causes of the additional torque on the yacht's rudder. One of the most important aspect for design of yachts are the symmetric conditions of all parts, which are located under water level and concentricity of the ship's screw, rudder and keel relative to the hull. These symmetric and concentricity conditions have a major impact on the resulting dynamic properties of the ships. They have either substantial impact on the overall efficiency of installed engine and ship's screw. As the result of poorly designed above mentioned parts, there can be an unsolicited additional torque on the yacht's rudder and higher consumption of the fuel. Last but not least of these problems leads to poor controllability and discomfort within steering. This article is focused on the investigation of possible causes of the additional torque on the yacht's rudder.

1 Introduction

Several studies have been carried out on the problematic of the symmetric shapes of rudders or the shapes of wings in connection with other parts as are a hull and keel of the yacht or a ship's screw shaft. Some of them were conducted by using an analytical approaches, others by using experiments and in some of them were used numerical simulations as for example CFD or CFX. German physicist Ludwig Prandtl (1875 – 1935) was one of the founders of the modern aerodynamics and the theory of boundary layer. It is possible to find many theories in Prandtl's publications including a flow around symmetric and asymmetric profiles. The foundations aerodynamic theories are described for example in [1] or in [2]. There are other researches of aerodynamic problems around different profiles made by Munson and collective [3] and research of CFD simulation of propeller and rudder performance when using additional thrust fins [4]. Other research was performed about the flow around the rudder and hull of sailing yacht, where the results from CFD simulation are compared with experiments [5]. An aerodynamic analysis was conducted for special yacht, where is described a comparison between experiment in towing tank and CFD results which considered free surface waves [6].

Yachts are generally divided into two groups. First group are yachts without combustion engine (often sport's yachts) and second group are yachts with combination of the sails and a combustion engine, which can be used in case of needs to sail a yacht regardless of the wind. In this work is described the yacht having the combination of combustion engine and sails. There are very important steps for design of yachts, especially the symmetric conditions of all parts which are located under water level and concentricity of the ship's screw, rudder and keel relative to the hull. If one of these parts is not symmetric or not in concentricity, it can among other occur an additional unsolicited torque acted on the rudder when the crew uses the combustion engine to sail the yacht. This work describes a yacht which has one above described problem or combination of these problems, namely when the crew uses the engine to sail the yacht at the specific speed, the maximal additional torque acted on the rudder is $T_r=218.16\ N\cdot m$. As a consequence of this additional torque, there is torque acting on steering wheel $T_{sw}=31.88\ N\cdot m$. This value means additional weight 5 kg on the diameter of the steering wheel.

For all above mentioned parts of the yacht, it is very difficult to find out the symmetric and concentricity conditions directly on the yacht via some measurement. There is only way to find the exact shape, symmetric conditions and concentricity of all above described parts, when the yacht would be picked up from the water and it would be made 3-D scan or very precise measurement of all problematic parts, which are located under water level. However this process is very expensive and time-consuming. Experimental measurements is also almost unfeasible since to measure all the necessary parameters (water flow, velocity and pressure) directly on the sailing yacht is almost impossible, since the measurement can be

[a] Corresponding author: lubomirsoukup@seznam.cz

affected by many factors in case of natural conditions on the open sea.

An alternative way is the numerical simulation of problematic cases. The possible impacts of poorly designed parts will therefore be determined by numerical simulation in program Ansys Fluent. An approximate shape of rudder, keel and hull was obtained from the technical documentation of the yacht. There were provided also the data of the measured torque on the steering wheel and transmission ratio of the rudder system.

2 Possible problems of the yacht and technical parameters

If the yacht sails without using the combustion engine then the additional torque on rudder has almost zero value. It is possible to make decision that the additional torque has been caused only by parts related to the ship's screw and rudder. The most likely problems of this examined yacht are therefore:

- Asymmetric shape of the rudder
- Misalignment of the rudder relative to the hull (concentricity of the ship's shaft screw relative to the hull)
- Combination of misalignment of the rudder relative to the hull and deviation of the ship's shaft screw

Technical parameters were obtained or estimated from the technical documentation. Since the technical documentation was incomplete particularly with regard to the dimensions, the unknown proportions were estimated from the ratio of pixels and known dimensions in technical drawings. Some of the dimensions were completely unknown so in these cases were used the data measured and provided by the owner of the yacht. The ship's screw in real case is about 1000 mm behind the hull. With regard to this distance it was suggested to simulate the water stream from ship's screw as a stream flowing out from a "tunnel". Although this suggestion is not correct, we consider this simplification as negligible with respect to 2-D approach of the entire problem. Important data which were used in the following simulations are:

- Diameter of the ship's screw $D_{screw} = 508\ mm$
- Speed of yacht $v_{yacht} = 5.8\ knots = 3\ m/s$

Other physical values used for simulation

- Density of the sea water $\rho_w = 1024\ kg/m^3$
- Density of the air $\rho_{air} = 1.23\ kg/m^3$
- Kinematic viscosity of the sea water $\nu_w = 1.15 \cdot 10^{-6}\ m^2/s$

3 Measured values of the torque on the rudder

Two sets of torque measurements on the rudder were provided by the yacht owner. Each set was performed under different conditions. First measurement was performed for a headwind v_{1_hwind}=3-5 knots. (Tab 1).

The second set was performed for the headwind v2_hwind=5-7 knots (Tab 2). The dependences of torque on the yacht velocity for both cases are also drawn in Figure 1.

Tab 1. Values of torque for headwind 3-5 kn

Revolutions [min^{-1}]	Velocity of the yacht [kn]	[m.s^{-1}]	Traction on the steering wheel [kg]	T_{kk} [Nm]	T_k [Nm]
750	2,1	1,167	0	0	0
1000	3,5	1,944	1	6,38	43,72
1400	4,9	2,722	1,3	8,29	56,84
1800	5,9	3,278	2	12,75	87,45
2200	6,8	3,778	3	19,13	131,17
2600	7,7	4,278	4	25,51	174,9

Tab 2. Values of torque for headwind 5-7 kn

Revolutions [min^{-1}]	Velocity of the yacht [kn]	[m.s^{-1}]	Traction on the steering wheel [kg]	T_{kk} [Nm]	T_k [Nm]
1000	2,6	1,444	0,2	1,28	8,74
1400	4,1	2,278	1	6,38	43,7
1800	5	2,778	1,8	11,48	78,70
2200	6	3,333	2,3	14,67	131,1
2600	6,9	3,833	3	19,13	131,1
3100	7,7	4,278	5	31,88	218,6

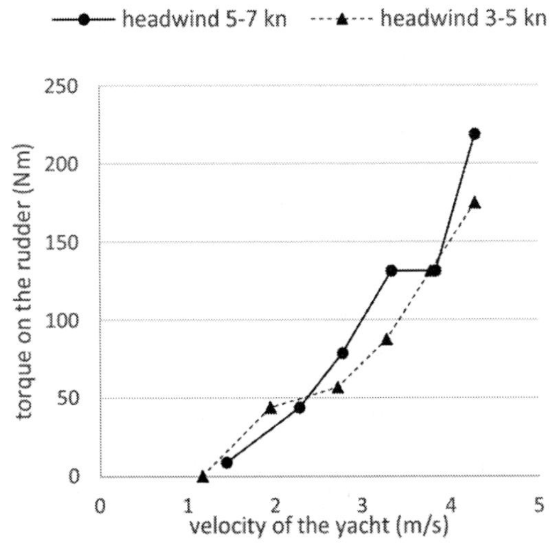

Figure 1. Graph of the additional torque on the rudder shaft

4 Solving of problem by software and solving process

The increasing of torque during the acceleration of the yacht can be determined numerically. Since there were no exact dimensions of some parts of the yacht, the main aim of the numerical simulation cannot be to determine the exact torque acting on the rudder of yacht, but to determine whether is possible to obtain additional torque with similar values or with similar trends which were measured on the rudder in real case. A geometry of the yacht was created in software Gambit. Numerical simulation was done by software Ansys Fluent. This software is based on the finite volume method.

The solving process was completely solved in 2-D dimension with respect to impossibility to create the exact 3-D model from the provided technical documentation and drawings. This process was divided into three separate parts namely: simulation of the asymmetric shape of the rudder, simulation of misalignment of the rudder relative to the hull (concentricity of the ship's screw shaft relative to the hull) and last simulation was performed for misalignment of the rudder relative to the hull considering the deviation of the ship's shaft screw relative to the hull.

The asymmetric geometry is based on a rough measurement of the real rudder on the yacht. Since the measurement cannot be considered as very accurate with regard to the fact that the measurement was carried out under water surface. It was suggested to perform the simulation with a small deviation 5 mm in the location of the largest width of the rudder (Figure 2), since it was found out with certainty that the least measured deviation was 5 mm.

All cases of misalignment and deviation in rudder shaft have initial position as in ideal case, it means misalignment 0 mm and deviation 0°. The step in misalignment is 10 mm and a maximum misalignment of rudder is 40 mm. The deviation of the ship's shaft screw is 3° for all cases of misalignment. For all 2-D simulations was used the top cross-section geometry from the location of the ship's screw shaft as is depicted in Figure 3 and Figure 5.

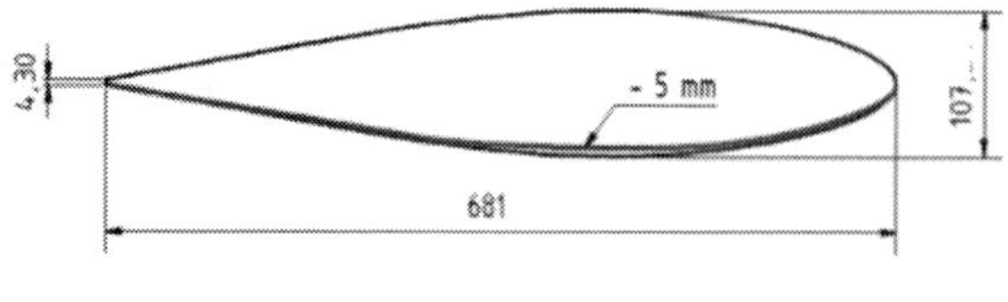

Figure 2. Asymmetric geometry of the rudder in *mm* (top cross-section)

5 Mathematical simulation

Computational meshes for mathematical simulations were made in software Gambit. It was necessary to prepare the mesh of asymmetric rudder (the shape of the asymmetric rudder is in Figure 2), five meshes for the misalignment of the rudder relative to the hull (concentricity of the ship's shaft screw relative to the hull – Figure 3) and five meshes for the misalignment of the rudder relative to the hull considering deviation of the ship's shaft screw relative to the hull (Figure 4).

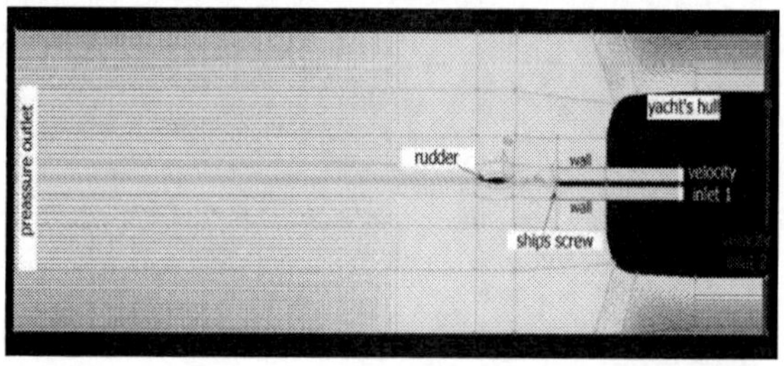

Figure 3. 40 mm misalignment of the rudder relative to the hull, the hull and ship's screw shaft are concentric (top cross-section)

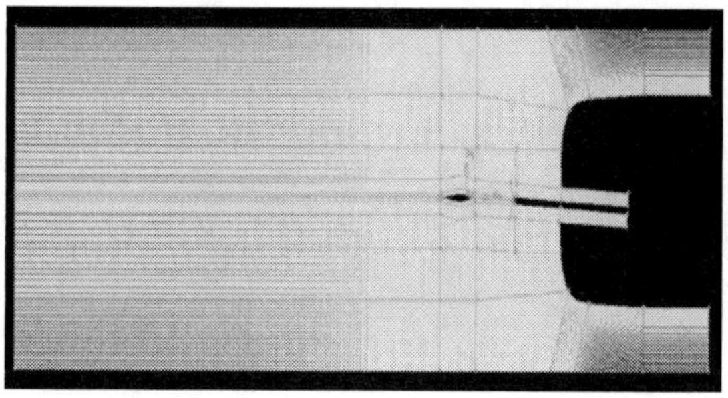

Figure 4. 40 mm misalignment of the rudder relative to the hull, deviation of the ship's screw shaft relative to the hull (top cross-section)

Numbers of cells for each mesh is 490 thousand and all meshes were made by quad cells and a map type. In meshing processes the qualitiy as EquiSize skew were checked, the worst element had a value of 0.62. Difficulties of whole meshing were related to a fine meshing process in area behind the hull and around the rudder. The boundary conditions were set as velocity inlet on all inlets and as pressure outlet on all outlets – the boundary conditions are displayed in the Figure 3. There are two velocity inlets with different velocity values. *Velocity inlet 2* (vi_2) represents the water stream from the water flowing around the hull and *velocity inlet 1* (vi_1) represents the water flow from the ship's screw. All simulations were conducted with the same velocity at *velocity inlet 2* $vi_2 = 3\,m/s$ (this is an average velocity of the sailing yacht in relation to see) and with different velocity at *velocity inlet 1* $vi_1 = 0, 5, 10 \; and \; 15\,m/s$. The progress of velocity on *velocity inlet 1* was specific for case considering asymmetric rudder shape as is described in a Chapter 6. The maximal velocity of the water stream from the ship's screw can be approximately 15 – 20 m/s (this information was obtained from the producer of ship's screw website for engine with output 75 kW by which is the yacht fitted).

In program Ansys Fluent were simulated almost 60 different cases. It was elected turbulent K-ε model with non-equilibrium wall function. The values of Wall Y^+ were checked to ensure the correct results. The torque was evaluated directly from the Ansys Fluent for cases of steady flow and for cases of unsteady flow was the torque evaluated from the time-averaged maximum values of torque coefficient. The torque was related to the shaft of the rudder. The height of the rudder was chosen to be a partial height $h=508\ mm$ to have possibility to evaluate the particularly impact of the fluid flow from the combustion engine, respectively from the ship's screw. This height was implemented into simulation only for expressions of the value of the additional torque from 2-D dimension. This assumption can be considered as assumption close to real case since the rudder has almost the same top cross-section's shape in the area behind the ship's screw [7],[8].

6 Results

With regard to the many numerical simulations, it was decided to make an evaluation for each problem separately. At first, partial results were evaluated for asymmetric rudder shape. It is clear from the graph in the Figure 5 that for the velocities from 0 to approximately 18 m/s the additional torque on the rudder's shaft is negligible however the additional torque increases significantly for the velocities near to 18 – 20 m/s. There were made more simulations in this case in order to acquire necessary velocity of the water stream form ship's screw to find a region of increment of the additional torque. Figures 6 and 7 show the velocity and pressure contours in an area around the rudder.

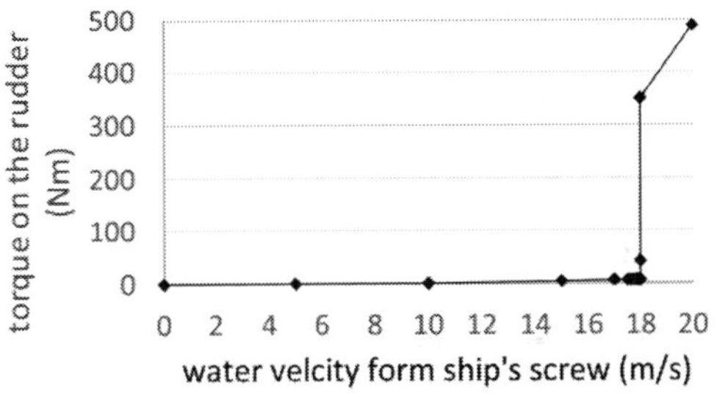

Figure 5. Graph of additional torque on the asymmetric rudder

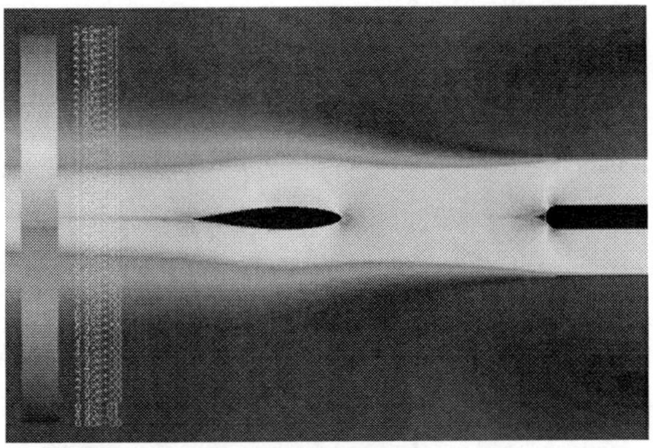

Figure 6. Contours of asymmetric rudder - velocity field

Further partial results were evaluated for the case of the misalignment of the rudder relative to the hull (considering concentricity of the ship's shaft screw relative to the hull).

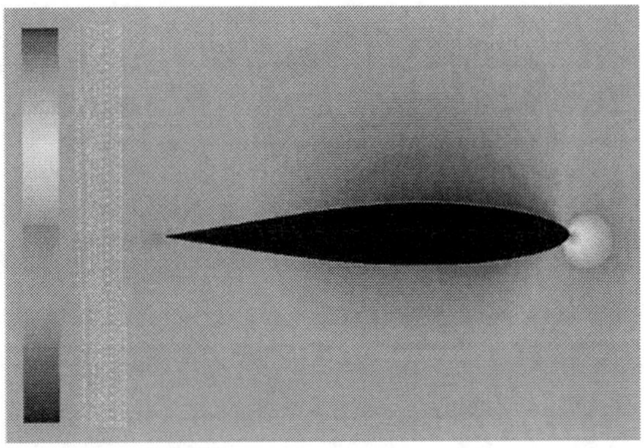

Figure 7. Contours of asymmetric rudder - static pressure field

It is evident in the graph in Figure 8 that the additional torque on the rudder shaft is increasing slowly and from the velocity 10 to 15 m/s the torque increases significantly. It was not necessary to continue with simulation to the velocity 20 m/s, since the value of the maximum torque is in all cases higher or very close to the value of the maximum torque from the measured data in the graph in Figure 1. The fields of the velocity and pressure contours are depicted in Figures 9 and 10.

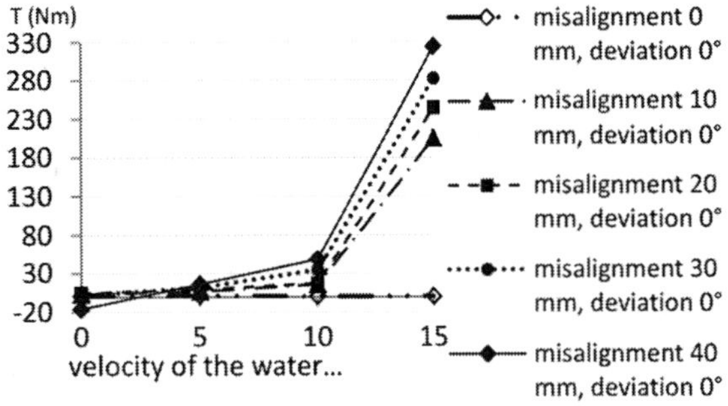

Figure 8. Additional torque for misalignment of the rudder relative to the hull considering concentricity of the ship's screw shaft relative to the hull

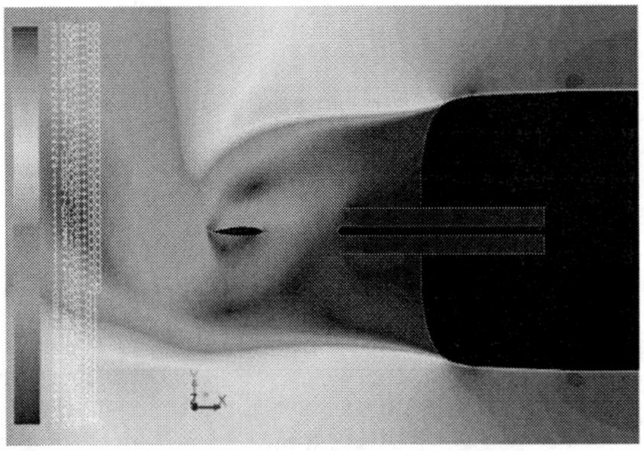

Figure 9. Contours of zero misalignment - velocity field

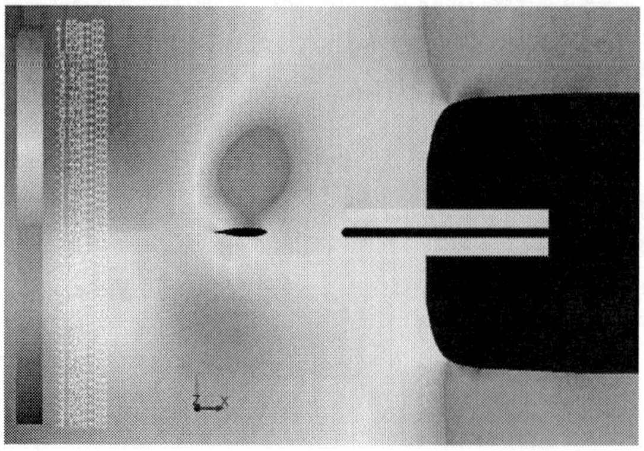

Figure 10. Contours of zero misalignment - static pressure field

Last set of the results were evaluated for the misalignment of the rudder relative to the hull considering deviation angle 3° of the ship's screw shaft relative to the hull. It is obvious from the graph in Figure 11 that the additional torque on the rudder shaft has negligible or negative value for velocities from 0 to 10 m/s and for velocities from 10 to 15 m/s the torque increases (instead of case of misalignment 0 mm and deviation 3°). It is apparent that this trend does not respond to the trend from the graph in Figure 1 and therefore it was not necessary to continue in this simulation to the velocity 20 m/s.

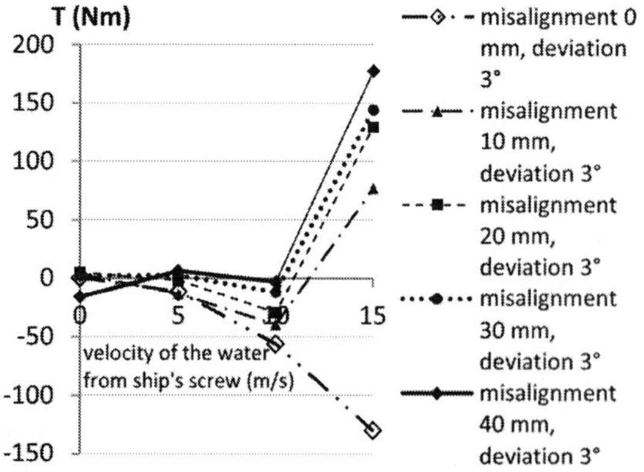

Figure 11. Additional torque for misalignment of the rudder relative to the hull considering deviation of the ship's shaft screw relative to the hull

7 Assessment of the results and comparison with measured data

The direct comparison of the measured and simulated data is not possible since there is no dependence in measured data on the velocity of the water flow from ship's screw. The assessment must be performed from a trend of the additional torque. As it is shown in the graph in Figure 11 the trend for solution of the additional torque for misalignment of the rudder relative to the hull considering deviation of the ship's shaft screw relative to the hull does not correspond to measured data. From this fact it can be argued that this problematic is the least likely problem from all three cases.

The most similar results to measured data are represented by solution of additional torque for the misalignment of the rudder relative to the hull (concentricity of the ship's screw shaft relative to the hull). The trend in this case is the most similar trend to the measured data. However, it should be taken into account either the set of results of asymmetric shape of the rudder. In both cases (misalignment of rudder and asymmetric shape of rudder) is the fundamental reason of additional torque a non-uniform velocity field around the rudder. This non-uniform velocity filed causes action of buoyancy force on the rudder. If the asymmetric shape of the rudder has higher deviation then it is sure that the additional torque will grow rapidly for lower velocities and the trend of the torque would have more similar progress to the measured data.

8 Proposed changes and their effects

First proposed change is to remove the rudder from the yacht and perform a precise measurement of the potential deviation from the ideal symmetric rudder shape. If the first adjustment shows that the ruder is asymmetric, then it would be necessary to use special sealants which have a long durability in seawater and modify the shape of the rudder to the accurate symmetric shape. If the first modification shows that the rudder has symmetric shape then it has to be performed the exact measurement of the concentricity of the rudder and ship's screw shaft relatively to the hull.

Effect of first proposed change was fortunately successful since there was found out a larger deviation than 5 mm in the symmetry of the yacht's rudder. After precise modification of the shape of the rudder to precise symmetric shape the additional torque decreased to the negligible value.

9 Conclusion

At the beginning of this article the problem of the yacht is described. There are three possible cases, which may constitute a reason of the additional torque on the rudder shaft namely:
- Asymmetric shape of the rudder
- Misalignment of the rudder relative to the hull
- Deviation of the ship's shaft screw relative to the hull.

It would be possible to make 3-D scan of all parts located under water level to find out the exact geometry of the yacht, or there could be performed the exact measurement directly on the yacht. However both these methods are very time-consuming and expensive. The mathematic simulation can be made as a further method to find out the reason of the additional torque.

Subsequently the simulation mesh was created in the software Gambit and the quality of all meshes were checked (Equi Size Skew worst element had quality 0.62). The number of cells was 490 thousand for all meshes. A turbulent model was elected as K-ε with non-equilibrium wall function. Boundary conditions were set as velocity inlet on all inlets and pressure outlet on all outlets for all simulations.

Then the results obtained from the mathematical simulations were presented. It was impossible to perform the direct comparison of simulation results with the measured data since the measured data has no dependence on the velocity of the water flow from the ship's screw. The trend of additional torque was used as a comparative benchmark to decide which case of simulated problem could be the reason of the additional torque. The comparison of results shows that the additional torque could occur by the misalignment of the rudder relative to the hull, or by asymmetric shape of the rudder.

Some modifications were suggested as a consequence of the results of numerical simulations. First modification was to remove the rudder from the yacht and perform a precise measurement of the potential deviation from the ideal symmetric shape. This proposal proved to be as successful. It was found out that the deviation of the rudder shape was more than 5 mm. The additional torque gained negligible values after modification of the rudder shape by special sealant.

Investigation of the influence of geometrical changes of the rudder to the additional torque can be done as a future work. There can be included the dependence of asymmetric shape of the rudder together with the overall size of the hull to the resultant additional torque. It could show how precise must be the production process of the rudders in relation to the overall size of the ship's hull and to the output of combustion engine respective to the ship's screw.

Acknowledgment

This research was financially supported under the project "Innovative fluid machines ", number: FSI-S-14-2480

References

1. Prandtl, L, Tietjens, O. Hydro- und Aeromechanics *I*. 238 p. S.l.: s.n., 1929
2. Prandtl, L, Tietjens, O. Fundamentals of *Hydro- and aeromechanics*. New York: Dover Publications, 2003, 280 p. ISBN 0486603741
3. Munson, Bruce Roy, Donald F Young a Theodore H Okiishi. *Fundamentals of fluid mechanics*. 5th. ed. Hoboken: Wiley, 2006, 770 s. ISBN 0-471-67582-2.
4. Huang, S, Xiang-Yuan, Z, Chun-Yu, G, Xin, Ch. *CFD simulation of propeller and rudder performance when using additional thrust fins.* Journal of Marine Science and Application. Online ISSN 1993-9433
5. Wu-Yoan K, Hye-Ryoun, Ch, *CFD application for turbulent flow analysis around the sails and hull of a sailing yacht.* Journal of Marine Science and Application. ISBN 978-1-880653-68-5
6. Wu-Yoan K, Jaehoon, Y, Zhengshou, Ch, Shin Hyung, R, Hye-Ryoun, Ch, Haeseong, A. *Hydro- and aerodynamic analysis for the design of sailing yacht.* Journal of Marine Science and Application. Online ISSN 1437-8213
7. Scott, Thomas E. *Power transmission: mechanical, hydraulic, pneumatic, and electrical.* Upper Saddle River: Prentice Hall, c2000, xiii, 319 s. ISBN 0130953865.
8. Pletcher, Richard H, John C Tannehill, Dale A, Anderson. *Computational fluid mechanics and heat transfer.* 3rd ed. Boca Raton: CRC Press, c2013, xx, 753 s. ISBN 978-1-59169-037-5.

EPJ Web of Conferences **114**, 02111 (2016)
DOI: 10.1051/epjconf/201611402111

Estimation of the drift eliminator efficiency using numerical and experimental methods

Jiří Stodůlka[1,a], Rut Vitkovičová[1]

[1]CTU in Prague, FME, Department of Fluid Dynamics and Thermodynamics, Prague, Czech Republic

Abstract. The purpose of the drift eliminators is to prevent water from escaping in significant amounts the cooling tower. They are designed to catch the droplets dragged by the tower draft and the efficiency given by the shape of the eliminator is the main evaluation criteria. The ability to eliminate the escaping water droplets is studied using CFD and using the experimental IPI method.

1 Introduction

This paper deals with the simulation of the flow in the optimized multifunctional cooling tower drift eliminator in order to obtain it´s functional parameters. The flow field which is directly connected with the efficiency estimation are both simulated using CFD code ANSYS Fluent which was previously validated by the PIV visualization method and by experimental pressure loss evaluation on the prototype model. A further analysis in terms of the efficiency itself has been done based on the validated flow field using again the CFD methods and then compared with experimental IPI method.

2 Drift eliminators

Beside the water distribution system and cooling panels are the eliminators another important functional part of the cooling tower [1], whose arrangement is schematically described in the figure 1.

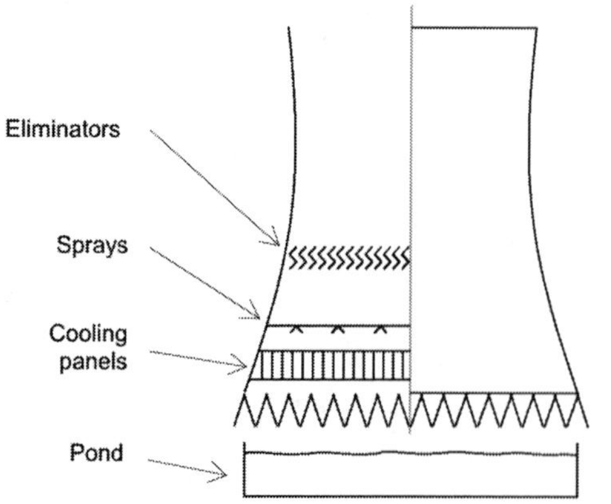

Figure 1. Natural draft cooling tower scheme.

The main purpose of these curved shape panels is to rapidly reduce the total mass flow carried away by air exiting the tower (drift) using their shape to change the flow direction causing enough inertia forces, so that the fluid droplets can't follow the air stream and hit the wall tickeling down as a film and drop back to the tower. The amount of water contained in both the form of vapor generated during the cooling process [2] and liquid droplets generated by condensation, but also very small diameter droplets drifted away from the spraying zone [3] escaping the tower, was estimated at approx. 0.001 m^3/hr/MW back in the year 2000. This number may seem small, but taking into account high power outputs of modern energetic open recirculation systems, the total mass flow escaping is significant and that plays an important role in economical and also ecological function of the system. To restrict all the unpleasant side effects, the eliminator manufacturers have to guarantee nowadays to reduce the drift according to standards to maximum of 0.0005%

The most commonly used types are the grid eliminators, made of thin profiled slats connected together. Example of such eliminator is shown in the figure 2.

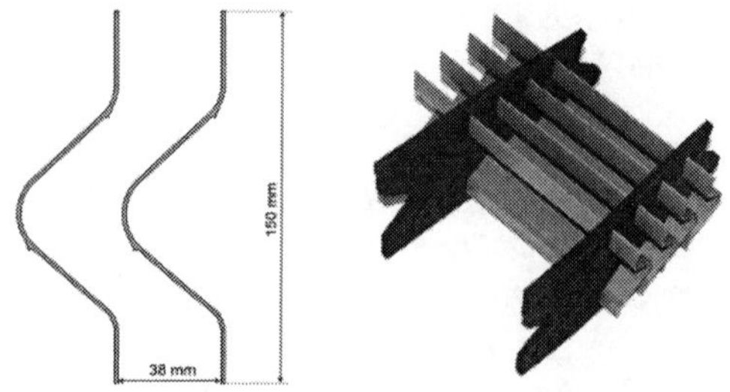

Figure 2. Grid type eliminator.

[a] Corresponding author: jiri.stodulka@fs.cvut.cz

Another relatively new type is the eliminator with moisture recovery system. This type combines two principles; the first is the elimination of liquid phase similarly like the grid type eliminator and simultaneously a function of heat exchanger. These eliminators are bigger and hollow, to allow outer air with lower temperature pass through to cool down the escaping moist air ending in condensation of the moist air. A lot of attention is paid nowadays to improve the behaviour of these multifunctional eliminators in order to save more and more amount of water escaping from the evaporative cooling systems.

3 Design

Shape of every eliminator has to fulfil three main requirements. These are maximal efficiency in terms of caught water droplets, minimal pressure loss and minimal amount of used material. Unfortunately all three requirements are against each other, so it is necessary to find appropriate balance between them. The shape of investigated eliminator row is in the figure 3.

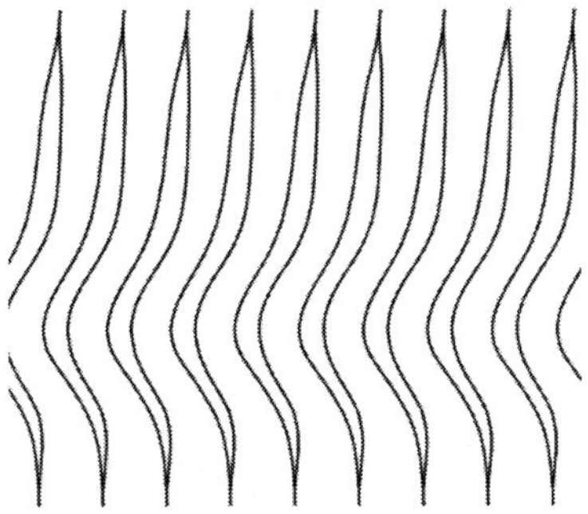

Figure 3. Eliminator shape.

The overall height of the eliminator is 450 mm, while distance between neighbouring ones is 60 mm.

4 Flow simulation and validation

For the numerical simulations the commercial CFD code ANSYS Fluent was used. Schematic view on the model and boundary conditions setup is in the figure 4.

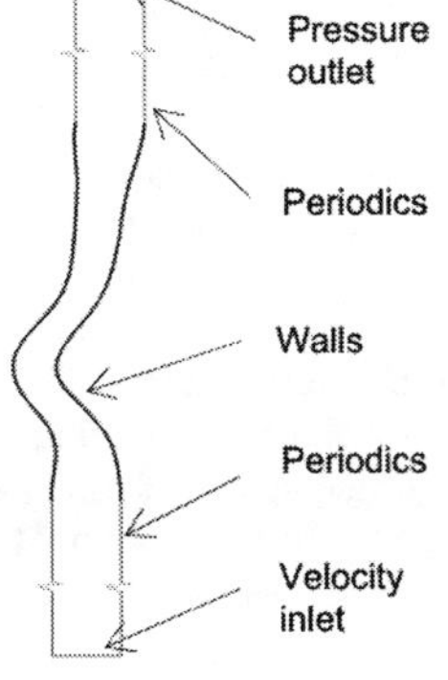

Figure 4. Computational model description.

Velocity inlet values vary depending on concrete cooling tower conditions from 1.5 to 4 m/s with the turbulent intensity of 2%. Pressure at the outlet is set to be atmospheric. Due to relatively simple geometry the mapped 2d quad mesh was used for the whole domain with total count of approx. 100000 elements. Implemented sst k-ω turbulence model was used, what leads to demand for fine mesh close to the walls. The values of Y+ oscillate around 0.5 for all velocity modes. The detail of the mesh is shown in figure 5.

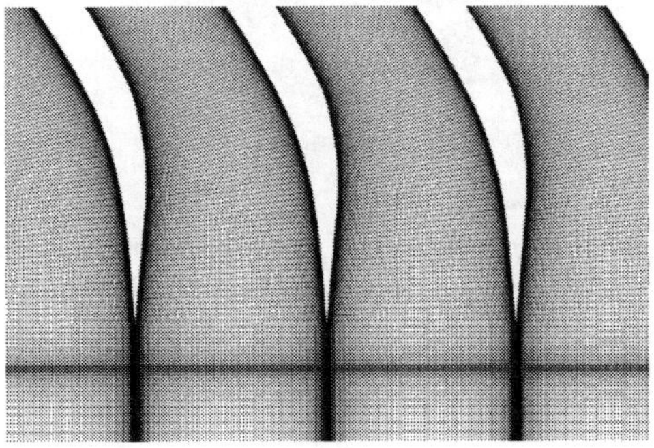

Figure 5. Quad computational mesh.

Results in form of velocity contours for inlet mode 2.6 m s^{-1} are shown in the figure 6. Except the trailing edge of the eliminator, no other separation occurs resulting in reasonable pressure loss. The flow is also significantly accelerated, so that droplets elimination can be expected in curved areas.

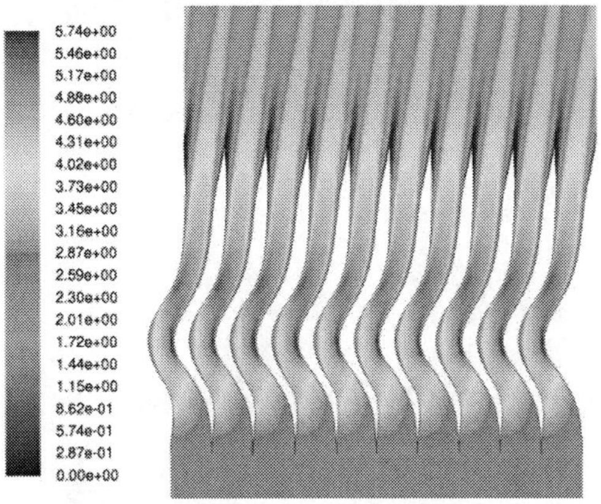

Figure 6. Contours of velocity magnitude (m s^{-1})

For experimental evaluation of the flow field the PIV method was used. Models of eliminator were placed in wind tunnel prepared for optical visualisation methods; see figure 7. The inlet velocity was approximately 3.2 m s^{-1} and the CFD boundary condition for comparison simulation was adjusted according to that. Pressure loss was simply evaluated by pressure difference in front of and behind the eliminator in bigger classical open aerodynamic tunnel to minimize the effect of tunnel itself.

Figure 7. Experimental model.

For a direct comparison between experimental and numerical results, the data from both were analyzed using Matlab. Data in form of velocity component contours were plotted on the same grid and then subtracted from each other to see the differences. The evaluation of the difference between CFD and PIV for the x(right) and y(left)-velocity components is shown in the figure 8.

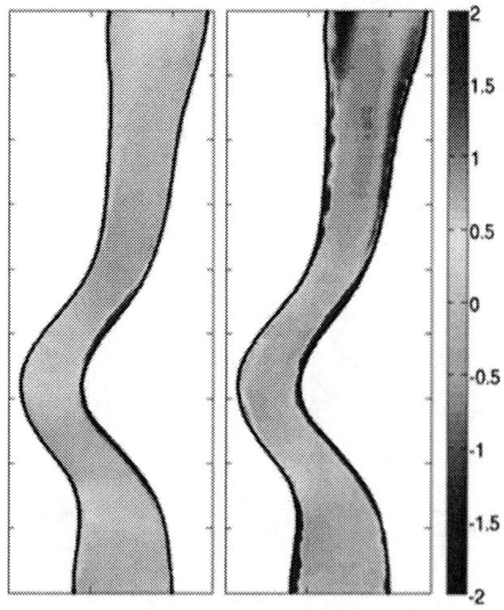

Figure 8. Velocity components comparison

From this figure is obvious a good accordance between these two approaches, at least in the main centre flow, where the difference ranges only in few tenths of m/s, what was also confirmed from velocity profiles. Near walls is the difference more dramatic, but that can be due to limits of the PIV near strongly reflective boundary and direct real and model geometry comparison. Major inconsistency appears only at the upper part of the eliminator, where mainly y-components exhibit significant disparity. That was probably caused by inconsistence between numerical and physical eliminator model and the finite dimension of the wind tunnel can play its role as well. However, this area is no more important in terms of the main function of the eliminator, while all possible droplet separation happens in the curved section.Considering the geometry differeces, the CFD is slightly underpredicting in direct comparison, but the numerical results are still within the errorbars. In terms of pressure loss, the maximal value for highest velocity mode (4 m s^{-1}) is 5.9 Pa, for the lowest mode (1.5 m s^{-1}) is 1.2 Pa. The experimental pressure loss estimation confirmed these values as well as all the numbers from the numerical simulation were inside the error tolerance of the experimental setup. The further details can be also found in [4].

5 Efficiency estimation

According to previous results it can be said, that in terms of the flow filed the numerical model predicts the behavior sufficiently. The efficiency then can be calculated using the same code with built-in Discrete phase modeler (DPM). This model computes the trajectory of solid or liquid particles in main fluid flow according to the forces equilibrium:

$$m_p \frac{du_p}{dt} = F_{drag} + F_{pressure} + F_{virtual\,mass} + F_{gravity} + F_{other}. \tag{1}$$

Where the particular forces are the drag forces

$$F_{drag} = \frac{\rho A_p}{2} C_d \|u - u_p\| (u - u_p), \tag{2}$$

the pressure forces

$$F_{pressure} = \int -p dA, \tag{3}$$

the forces of virtual mass

$$F_{virtual\,mass} = \frac{m_p}{2} \frac{\rho}{\rho_p} \frac{d(u - u_p)}{dt} \tag{4}$$

and the gravity forces.

$$F_{gravity} = m_p \frac{\rho - \rho_p}{\rho_p} g. \tag{5}$$

The individual particles cannot interact between each other but it is possible to simulate the interaction between particles and the main flow, but in this case, this option was not used. The particles with properties of water were released from the inlet boundary condition and their behavior on the walls was set to "trap" option, that means that when the particle touches the wall, it simply disappears. On the outlet, the particles are allowed to escape from the domain. The overall efficiency is then the ratio of number of trapped to number of released. Calculated trajectories for different particle diameter for inlet velocity 2.6 m/s are shown in the figure 9.

The particular efficiencies are marked in the efficiency-diameter graph for all investigated velocities. In order to obtain continual distribution, the points in the graph were fitted with the funcion

$$\eta = 1 - \exp(a\, D^b). \tag{6}$$

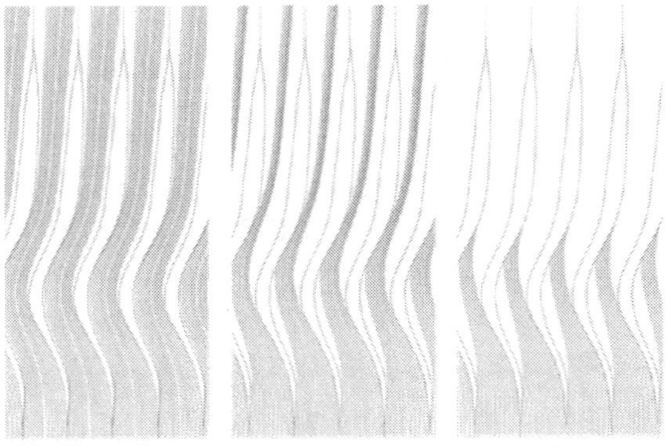

Figure 9. Particle trajectories.

The final efficiency diagram for different velocities is shown in the figure 10.

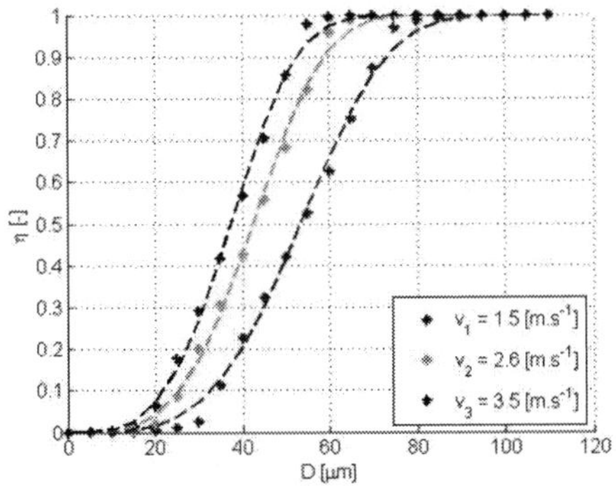

Figure 10. Eliminator efficiency.

This graph shows that for any inlet velocity, almost all particles smaller than 10 microns in diameter are dragged away. After this point, the eliminators starts to eliminate accordingly to the velocity magnitude - for higher velocities is the efficiency higher. For the slowest regime the biggest particles should not exceed 100 microns. Considering the fact that the absolute majority of the particles in the cooling tower are in range of tenths of millimeters or millimeters, is the efficiency for this special low drag eliminator still very good.

These numbers were also verified using experimental methods, concretely using the Interferometric particle imaging or IPI. This optical method is based on capturing the interfering beams reflected from transparent particle. More about this method and the particular setup can be found in [5]. The beam of coherent light incidents on the particle and is partly reflected and partly passes through and refracts. These two beams then interfere in the illuminated area and if the particle lies out of the object of the camera we can see an interferogram image. The example of such obtained image of captured particles by this method is shown in the figure 11. The diameter of the interferogram corresponds to the particle towards the object plane, the number of lanes in the interferogram, or their spatial frequency corresponds to the particle size.

Figure 11. Interferogram image of photographed particles.

Due to the size of the experimental tunnel which was not originally designed for this particular case properties are not the data from IPI method for small particles very accurate, but for bigger diameters it has been observed that the diameter of escaping particles was not higher than approximately 80 microns for the velocity regime of 2.6 m/s. This result corresponds sufficiently with the numbers obtained from the CFD simulations.

6 Conclusions

A profound comparison between numerical and experimental results of the mixed flow in the cooling tower drift eliminator has been done in order to validate the CFD methods for this application. Modern CFD is a very fast and effective tool that can significantly help with design or optimization in this field and the reliability of this approach has been proved in terms of flow field, pressure loss and finally with the efficiency as well. Considering problems with exact geometry fabrication and experimental setup limitations, the results of this comparison are pretty reasonable.

References

1. J. Čížek, L. Nováková, J. Nožička, *Reducing Liquid Phase Drift from Cooling Towers*, (Gradient, Prague 2009)
2. T. Hyhlík, EPJ Web of Conferences, **92**, (2015), 02027
3. P. Vitkovič, V. Syrovátka, Journal of Flow Visualization and Image Processing, **16(4)**, (2009), pp. 367-375
4. J. Stodůlka, R. Vitkovičová, EPJ Web of Conferences, **92**, (2015), 02086
5. R. Vitkovičová, P. Vitkovič, J. Čížek, *Modifikace metody IPI pro určování velikosti úletu z chladicích věží*, SKMTaT, Ústí nad Labem, 2015

Acknowledgement

This work has been supported by *Technology Agency of the Czech Republic* under the project *Advanced Technologies for Heat and Electricity Production* – TE01020036

EPJ Web of Conferences **114**, 02112 (2016)
DOI: 10.1051/epjconf/201611402112

Measurement of noise and its correlation to performance and geometry of small aircraft propellers

Vít Štorch[1,a], Jiří Nožička[1], Martin Brada[1], Jiří Gemperle[1], Jakub Suchý[1]

[1]*Department of Fluid Dynamics and Thermodynamics, Faculty of Mechanical Engineering, CTU in Prague, Czech Republic*

Abstract. A set of small model and UAV propellers is measured both in terms of aerodynamic performance and acoustic noise under static conditions. Apart from obvious correlation of noise to tip speed and propeller diameter the influence of blade pitch, blade pitch distribution, efficiency and shape of the blade is sought. Using the measured performance data a computational model for calculation of aerodynamic noise of propellers will be validated. The range of selected propellers include both propellers designed for nearly static conditions and propellers that are running at highly off-design conditions, which allows to investigate i.e. the effect of blade stall on both noise level and performance results.

1 Introduction

The effort in designing aircraft propellers is usually focused to obtaining highest efficiency possible at the design point or throughout a specific design point range. For both fixed and variable pitch propellers, determining the correct operation conditions for a specific application plays also an important role. The propeller can be tailored for maximum efficiency at cruise speed or for a relatively flat efficiency curve that suits aerobatic manoeuvres. According to the required operating range the distribution of geometric properties along the blade are chosen and optimized using an algorithm based on computed airfoil polars and 3D panel method based wake system. Using this technique small fixed pitch model aircraft and UAV propellers have been designed at the Department of Fluid Mechanics and Thermodynamics, CTU in Prague.

Since there is demand on the market for quiet fans and propellers, their noise is one of the priorities of current research. As the first step an experimental measurement is performed on a set of propellers with diameters between 20 and 24 inches in static conditions ($v_\infty = 0$). Although the results cannot be directly used for determining propeller sound power level at operating conditions, their purpose is to validate computation schemes and develop a better understanding of how and if properties like surface roughness, material, chord length distribution, chosen airfoil, pitch angle distribution and overall efficiency affect the sound pressure level, perceived noise and noise spectrum.

The material and geometry of the propeller will affect its aeroelastic behavior. In previous studies it was assumed that the propellers do not deform, in the present work it was decided to perform a simple optical measurement of bending and twisting of the blades due to centrifugal and aerodynamic loads.

[a] Corresponding author: vit.storch@fs.cvut.cz

2 Propeller noise

There are many different mechanisms of noise generation in turbomachinery and also different ways of their classification [1,2] based on the source type, periodicity and generation mechanism.

The most important frequency in turbomachine acoustics is the blade passing frequency (*BPF*) of a rotor with N blades rotating with frequency f.

$$BPF = Nf \tag{1}$$

Blade thickness noise is a periodic monopole noise source due to the fluctuating volume of fluid displaced by the blade. It becomes relevant as the tip speed and blade thickness increases. For tip speeds not approaching sonic speed the blade thickness noise is small compared to other sources [2].

Periodic noise due to aerodynamic forces is caused partly by steady pressure distribution relative to the blade which appears as fluctuations at *BPF* frequency to the observer and its higher harmonics and partly by a non-uniform rotor inflow.

Interaction effects due to the blade passing through stator wakes or its own wake are not important in case of propeller.

Vortex noise belongs to the broadband type of noise as it is caused by vortex shedding at the trailing edge with different periodicity along the blade and blade tip depending on the velocity at each blade span position. A dipole source is produced with strength proportional to the sixth power of section velocity [2].

Turbulence noise is produced in the shear layers, such as the boundary layer, due to small scale turbulence. It has a broadband quadrupole characteristic.

Propeller is a relatively simple type of open turbomachine which makes it easier to analyse. In studies focusing on propeller noise [2,3,4], the aerodynamic force induced noise is considered as the most important, together with vortex noise that plays an important role in noise reduction efforts [2].

All acoustic measurement results in this paper are presented using no weighting filter, since the main interest lies in the noise generation regardless of frequency sensitivity of human hearing (represented by A filter).

3 Propeller selection

Between the samples selected for the measurement were the propellers previously designed at the Department and labelled with D12112 as the manufacturer. For each of these propellers an exact geometry file is available. Commercially available propellers were also selected from several brands. For these propellers, some basic properties, like blade planform and pitch distribution was obtained by processing images. The geometry including airfoil shape can be also scanned using a manual 3D scanner developed in-house, which is however a very lengthy process and has not been used in the current work. The list of selected propellers can be found in table 1.

Table 1. Selected propellers for measurement

Make	Designation	Measured pitch [inch/rev]	Surface finish
D12112	20x13E3	16.8	Raw CF
D12112	20x13E4	15.9	Raw CF
D12112	22x10E	14.1	Raw CF
PT	20x12WE	13.5	Coated CF
PT	20x13E	16.2	Coated CF
PT	21x13E	15.1	Coated CF
PT	22x8HP	11.8	Coated CF
PT	22x10E I.	10.8	Coated CF
PT	22x10E II.	11.3	Coated CF
APC	22x10E I.	12.3	Fiber plastic composite
Mejzlik	20,5x12WEL	15.1	Coated CF
Mejzlik	24x8 EVO2	12.0	Coated CF
RS	21x14EL	15.4	Coated CF

The main parameters of a propeller is the diameter and pitch. Since the distribution of pitch is usually not constant along the blade, an effective pitch is considered as being the pitch of the blade at 70% of radius. However, our experience has shown that manufacturers of small propellers tend to indicate lower values, perhaps to gain an advantage over competition in thrust at a specific pitch. Some past geometry measurements of the authors have shown up to 3 inch/revolution difference in measured vs. indicated pitch so it was decided to process the images to obtain the pitch distribution and calculate a representative value of pitch taken as the average value between 40% and 90% of blade span.

The propellers measured are between 20 ÷ 24 inches (508 ÷ 609.6 mm) in diameter with pitch in range of 8 ÷ 14 inches per revolution. They have three different surface finishes, with coated carbon fiber being the finest and raw carbon fiber the coarsest. Letter E in the designation stands for electrical as the recommended type of motor. W is used by some manufacturers to indicate wider blades, if they make two versions of the same diameter and pitch. Also the number after E in case of D12112 propellers is the prototype version, meaning that E3 and E4 propellers have slightly different geometry.

To get an idea about the precision and reliability of the measurement, one propeller is represented by a pair of identical samples denoted by I. and II. (identical except for manufacturing imperfections).

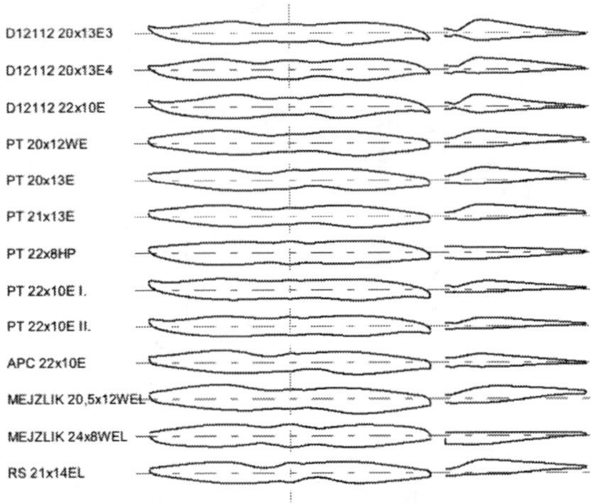

Figure 1. Planforms and side views of propellers (not rotating) obtained by processing images.

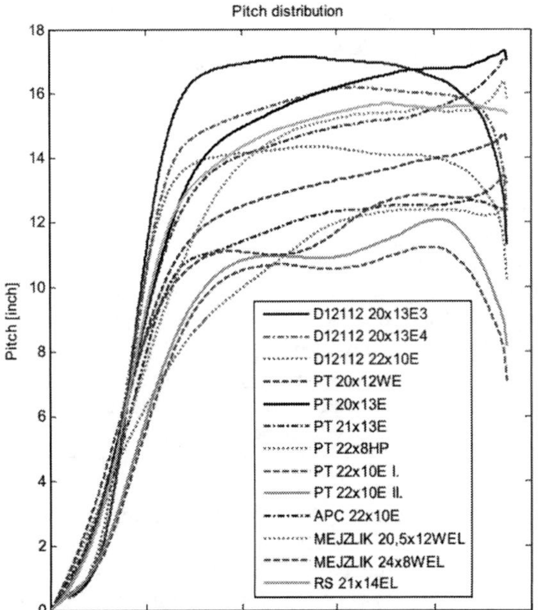

Figure 2. Pitch distribution based on image processing

Last but not least it must be noted that none of the selected propellers was designed to work under static conditions and the results provided cannot be used for judging the quality of each propeller. On the other hand, noise of propellers working in more or less off-design conditions promise more variable results.

4 Measurement

4.1 Experimental stand and data acquisition

The available wind tunnel for propeller measurement in dynamic conditions could not be used for acoustic measurement, due to the noise of the tunnel fan. It was decided to build a simple stand for static measurement. The stand consists of a welded hexagonal cage with motor mount.

Figure 3. The measurement stand consists of a welded hexagonal frame with protective net on the sides.

The motor mount contains in total 4 load cells for measurement of thrust force and torque. An optical gate together with perforated ring mounted on the motor's casing provides the measurement of rotational speed and also provides synchronization signal for the flash. All these devices are connected to an in-house designed electronic circuit with ADCs and the data read into MATLAB. For the purpose of studying aeroelastic behaviour of the propeller a DSLR camera is mounted on a tripod and positioned to capture the blade from the side. The exposure time was set at 1/10 s, which allowed for several blade passes during one exposure. After opening the shutter, the camera sends a signal to the flash to fire. This signal is delayed until the optic gate confirms correct position of the blades and the flash can fire. The logical circuit also limits the flashes to only one during one exposure. Canon EX430II was used as the flash and gives approximately 0.05 ms long light impulse at the lowest power settings. For the highest measured rotational speed of 105Hz the blade rotates aproximately 2° during the flash duration.

The acoustic quantities such as the overall sound pressure level and FFT spectrum are measured using NTI XL2 analyzer connected to a NTI M2211 omni-directional, free-field microphone. The microphone is placed 550mm from the axis of rotation and 100mm below the propeller disc plane.

4.2 Experiment procedure

Due to the large set of propellers, it was decided to measure the performance of each propeller at 4 different rpms equivalent to 40, 60, 80 and 100% of electronic throttle. Because the measurement is done under static conditions the results should be similar (when expressed in dimensionless coefficients) with the exception of small deviations due to different Reynolds numbers.

At each measurement point the propeller was rotating at a fixed speed and the values of rpm, thrust force and torque were recorded simultaneously for 5 seconds at a 1000Hz sampling rate and averaged to a single value. Two images were captured by the camera and the FFT spectra were recorded by the sound analyzer in the ranges: 7 Hz ÷ 215 Hz, 58 Hz ÷ 1.7 kHz and 484 ÷ 20 kHz.

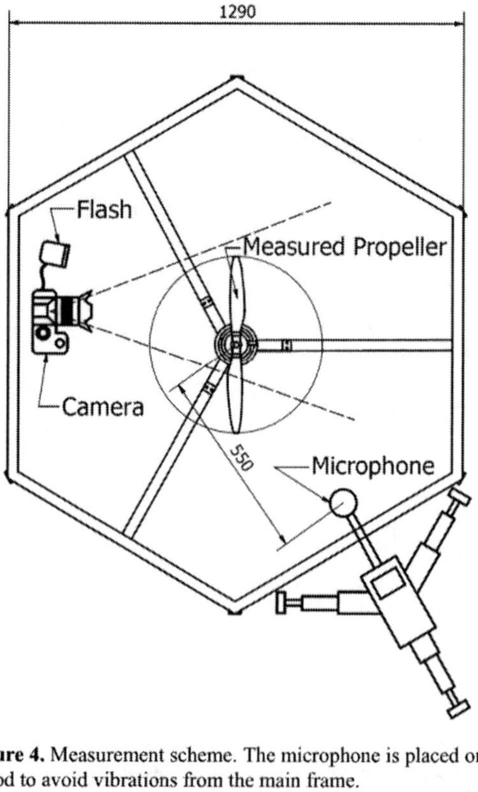

Figure 4. Measurement scheme. The microphone is placed on a tripod to avoid vibrations from the main frame.

5 Aeroelasticity measurement results

After processing the images several curves normalized to the tip radius R were obtained for each propeller and each measurement point. Based on the side view, the actual thickness $C_T = C_T$ (r/R) and midline $C_M = C_M$ (r/R) curves were calculated. Midline was defined as the curve averaging the leading edge and trailing edge curves. By subtracting the midline of the propeller at 0 rpm from each midline of rotating propeller, bending curves were obtained. By combining the actual blade thickness and the blade width distribution $C_W = C_W$ (r/R) obtained from the planform photos, a formula for twist C_Φ distribution at i-th measured rotational velocity could have been derived:

$$C_\Phi(i) = atan\left(\frac{C_{T(i)}}{C_w}\right) - atan\left(\frac{C_{T(0rpm)}}{C_w}\right) \qquad (2)$$

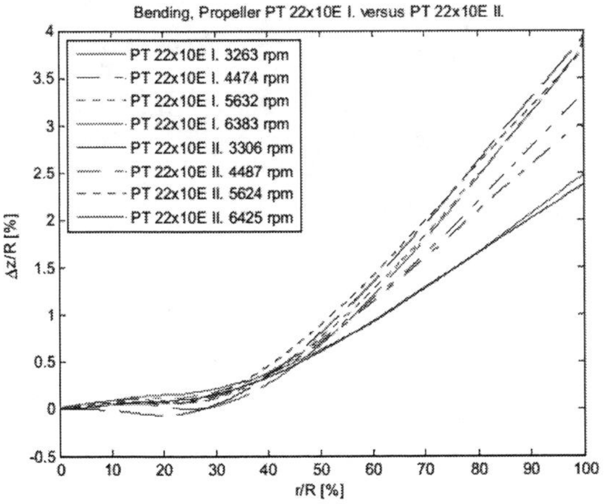

Figure 5. Typical shape of bending curves demonstrated on the two propellers PT 22x10E sample I. and II.

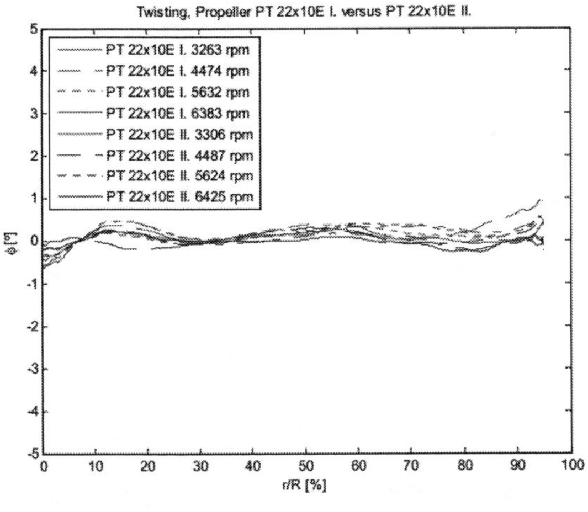

Figure 6. Typical results for twisting curves demonstrated on the two propellers PT 22x10E sample I. and II.

After running the algorithm on all samples, it was shown, that all propellers bend quite noticeably forward. By comparing the results of two samples of the same propeller model (figure 5) the measurement method produces very similar bending curves. It is interesting to note, that after a certain rotational speed the bending displacement seem to stop increasing, which is probably due to centrifugal force preventing further bending.

The angle of twist due to aerodynamic loading on the other hand stays within 0.5 degrees from zero for most blades, which seem to be also the accuracy of the method, as can be seen in figure 6.

The main purpose of this part of the measurement – to get information about changes in geometry due to aerodynamic loading was accomplished. The twisting angle falls below measurable values, so it can be assumed that the blades do not twist. Bending, on the other hand, is well measurable and depending on the material and construction of the blade the blade tip will bend forward by 0.5 ÷ 4 % of the tip radius. The results were stored for current and future numerical calculations, however some test calculations have confirmed the assumption that bending of the propeller has much smaller influence on the performance of propeller then its twisting.

6 Noise measurement results

The measurement could not be performed in an isolated anechoic chamber so an attention had to be payed to the microphone position and background noise levels. An example of measured sound spectrum is in figure 6. The difference in overall *SPL* (sound pressure level) between propeller noise and background is in this case 53 dB which is more than sufficient to neglect the background noise.

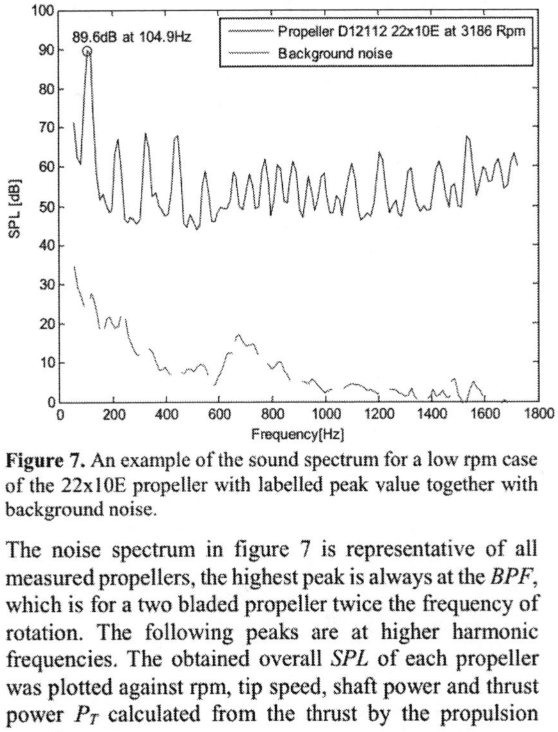

Figure 7. An example of the sound spectrum for a low rpm case of the 22x10E propeller with labelled peak value together with background noise.

The noise spectrum in figure 7 is representative of all measured propellers, the highest peak is always at the *BPF*, which is for a two bladed propeller twice the frequency of rotation. The following peaks are at higher harmonic frequencies. The obtained overall *SPL* of each propeller was plotted against rpm, tip speed, shaft power and thrust power P_T calculated from the thrust by the propulsion equation:

$$P_T = \sqrt{\frac{(F_T)^3}{2\rho A}} \qquad (3)$$

Where F_T is the thrust and A the propeller disc area.

The resulting curves are presented in figures 8-11. The *SPL* grows with rpm and tip speed in a linear manner, whereas the SPL is roughly proportional to the cubic root of shaft and thrust power, which can be expected from well-known relation between fan power and rpm. Figure 8 shows the expected fact that at constant revolutions, propellers of higher diameter and pitch tend to produce higher *SPL* and when constant tip speed is used for comparison, the higher pitch (and therefore greater section loading) produces higher noise levels.

More interesting are the plots of *SPL* vs. shaft power and thrust power. The data in figure 10 form a relatively narrow band, which means that the shaft power is the parameter that influences noise levels the most. On the other hand the thrust power curves are quite scattered, which confirms that the aerodynamic efficiency plays a crucial role also in noise reduction efforts.

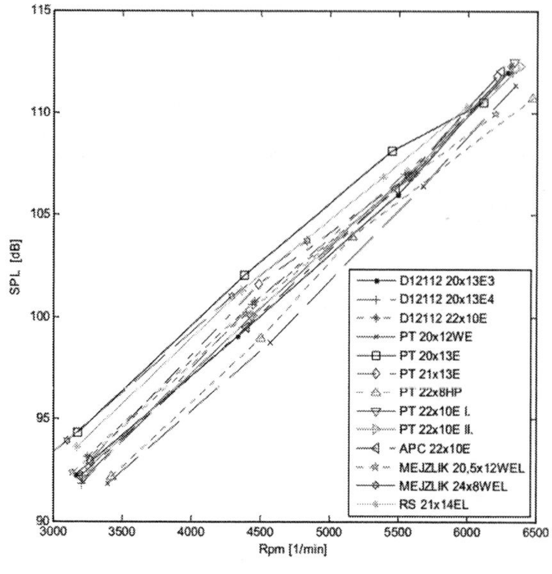

Figure 8. Sound pressure levels plotted against rpm

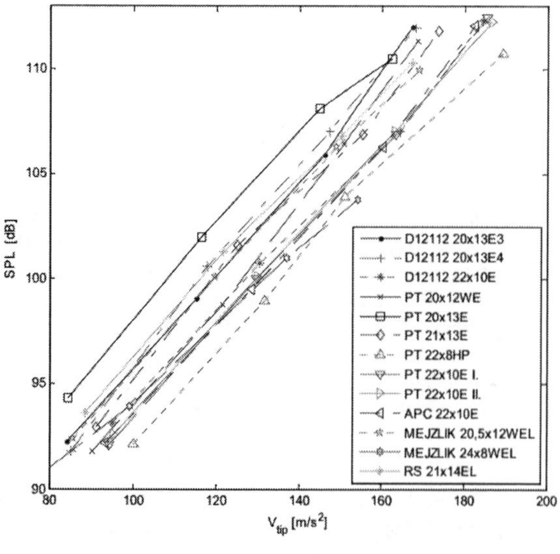

Figure 9. Sound pressure levels plotted against the tip speed.

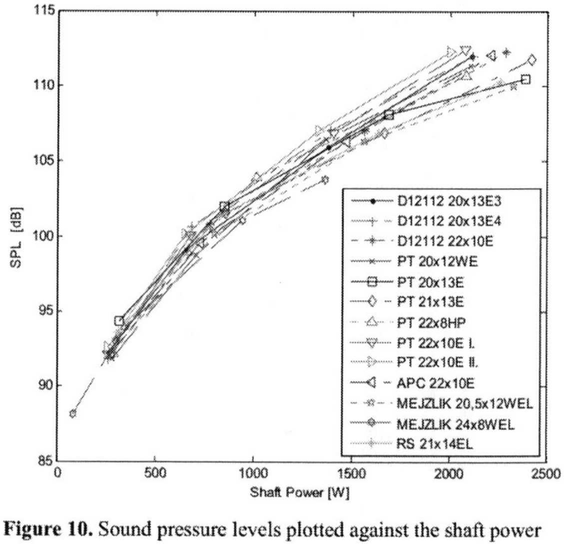

Figure 10. Sound pressure levels plotted against the shaft power

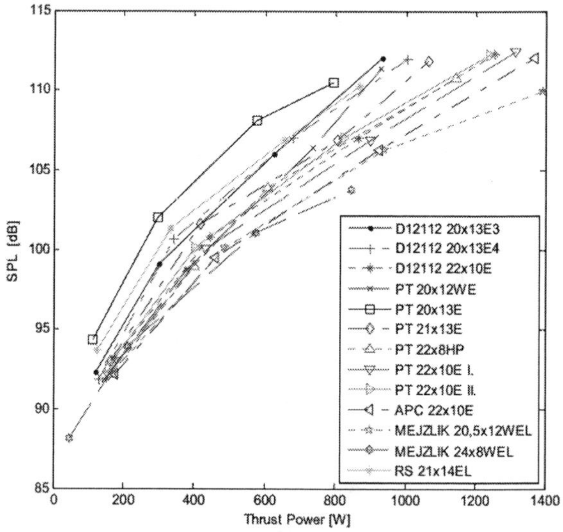

Figure 11. Sound pressure levels plotted against the thrust power

In order to be able to compare the measured propellers more efficiently values of *SPL* were interpolated from the graphs at constant chosen parameters and visualised using bar graphs (figures 12 ÷ 15). Interesting comparison of propellers in figure 14 shows that when constant shaft power is provided, more efficient propellers (note: efficiency at static conditions) tend to produce more noise. This is not true in all cases, but the tendency is noticeable. This can be explained by a hypothesis that the noise due to dissipation in a less efficient propeller (stalled and large vorticity flow) does not reach the same level of noise as a propeller with efficient aerodynamic loading (large pressure gradients of well streamlined sections) when both propellers are powered by the same shaft power.

The *SPL* at constant thrust power (figure 15) shows, that at constant thrust power, the noise is almost inversely proportional to the efficiency. This can be expected from the close tie between *SPL* and shaft power.

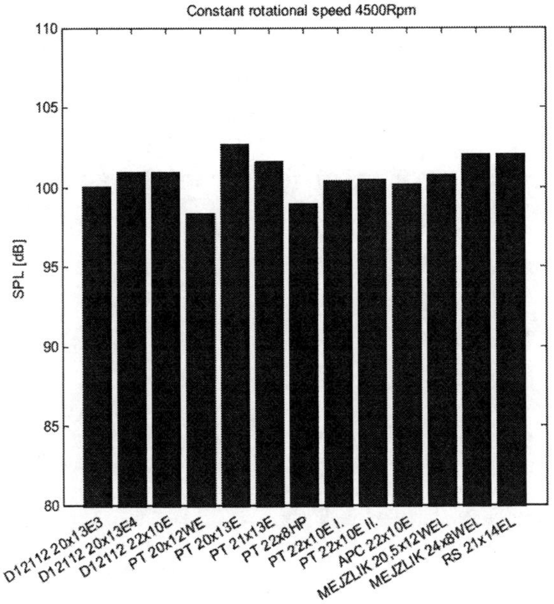

Figure 12. Sound pressure levels at 4500 rpm

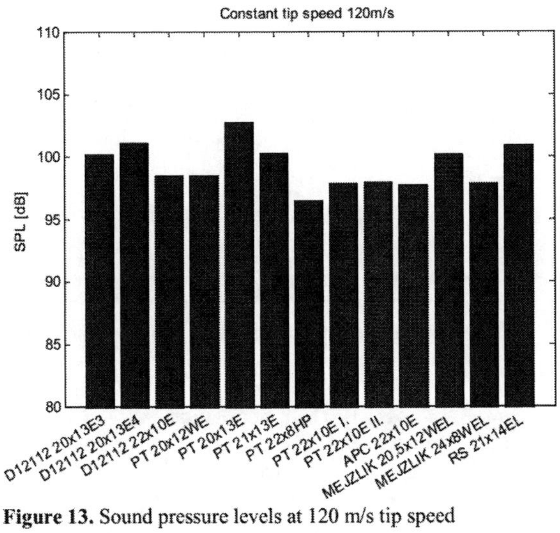

Figure 13. Sound pressure levels at 120 m/s tip speed

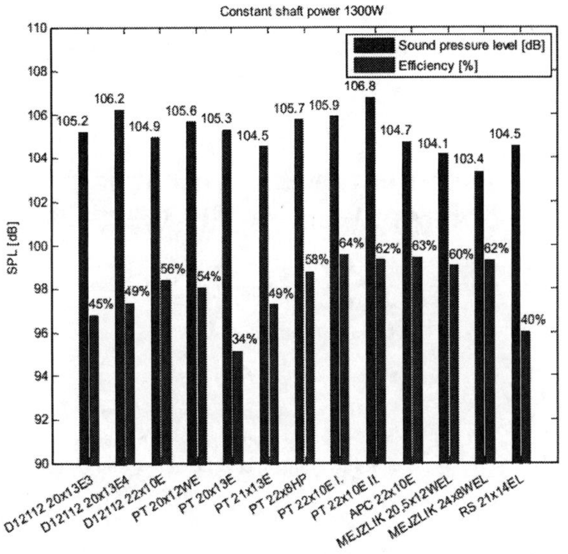

Figure 14. Sound pressure levels and efficiency at 1300W shaft power

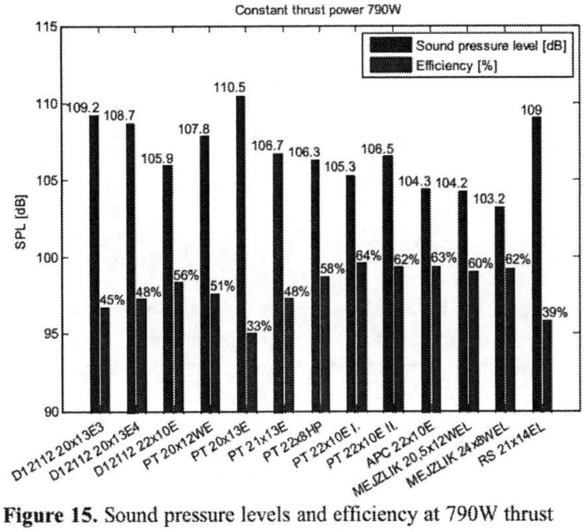

Figure 15. Sound pressure levels and efficiency at 790W thrust power

7 Conclusion

The presented work focuses on three main areas regarding small aircraft propellers – geometry, aerodynamics and acoustics. By inspecting the blade geometry it was confirmed, that the average pitch angle does not correspond to the designation of the propellers, and the pitch and chord distributions for future calculations were obtained. Image acquisition during blade rotation revealed little or no blade twisting but significant bending of the blades.

The aerodynamic performance measurement was done using proven setup for obtaining thrust, torque and rpm data. For each measurement point, the noise spectrum was recorded. The presented results will be used for the current development of computational model of propeller performance and acoustics.

Acknowledgement

This work was supported by the Grant Agency of the Czech Technical University in Prague, grant No. SGS 15/065/OHK2/1T/12

References

1. W. Cory, *Fans and Ventilation*, (Elsevier Science, Amsterdam Boston, 2005)
2. J. Marte, D. Kurtz, *A Review of Aerodynamic Noise From Propellers, Rotors and Lift Fans*, (Jet Propulsion Laboratory, Pasadena, 1970)
3. M. Carley, J. Sound Vib. **233** (2000) 255-277
4. G. Sinibaldi, L. Marino, Appl. Acoust. **74** (2013) 79-88

EPJ Web of Conferences **114**, 02113 (2016)
DOI: 10.1051/epjconf/201611402113

The effect of Bond number on pool boiling for mini-fin surfaces and different working fluids

Ewelina Strąk[1,a], Robert Pastuszko[1,b]

[1] *Kielce University of Technology, Al. 1000-lecia P.P. 7, 25-314 Kielce, Poland*

Abstract. Experimental nucleate pool boiling data were collected for structures in the form of extended surfaces sintered with perforated foil. The article describes experimental investigations for two kinds of surfaces: smooth and mini-fins with sintered perforated foil (MFP). The MFP surfaces were manufactured out of perforated copper foil (pore diameters: 0.05 – 0.3 mm) sintered with the mini-fins, uniformly spaced on the base surface.
The experiments were carried out at atmospheric pressure for four kinds of the fluids: water, ethanol, FC-72 and Novec 649. The results for these working fluids were compared in terms of the Bond number and other dimensionless parameters. For all working fluids, the heat transfer coefficients obtained for the MFP surfaces were from 2 to 5 times higher than those for the smooth surface.
Using a regression analysis with reference to selected physical properties and several characteristic dimensions, it was possible to develop a correlation for the Nusselt number.

1 Introduction

Pool boiling heat transfer has generally been associated with a combination of microconvection, vapor-liquid exchange, evaporation, latent heat and transient conduction. Vapor bubble dynamics, that is, bubble nucleation, growth, and departure as well as rewetting of the heater surface is the main pool boiling heat transfer mechanism.

This article describes the influence of the similarity numbers on pool boiling from mini-fin surfaces with water, ethanol, FC-72 and Novec 649 as working fluids.

The correlation formulas can also be found in the literature below.

In [1], the correlation is similar to that proposed by Rohsenow. The experimental data have been obtained for seven kinds of fluids (water, ethanol, CCl_4, acetone, n-hexane, R113, R141b) and for the plain surfaces without capillary structures under variable pressure. These surfaces include ordinary metal, metal with a coating and nonmetals of high thermal conductivity.

The correlation in [2] has been obtained for five fluids (R-113, n-pentane, ethanol, water and R-11) boiling on the plain, low finned surfaces and on the Gewa-T surfaces at atmospheric pressure.

Moita et al. in [3] reports the results of a detailed heat transfer and bubble dynamics analysis for pool boiling with three fluids (water, ethanol and HFE 7000) boiling on smooth and micro-structured surfaces. The bubble departure diameter is described as a function of the Jakob number.

E. van Rooyen and J.R. Thome [4] studied Turbo-B5 and Gewa-B5 enhanced surfaces. The experiment were carried out with used refrigerants R-134a, R-236fa and R-1234ze(E). Their correlation was based on a boiling heat transfer mechanism in the near-wall region.

The authors of this paper presented the effect of Bond number on boiling heat transfer coefficient and proposed their own correlation equation for the Nusselt number related to five dimensionless numbers.

2 Experimental research

2.1 Investigated surfaces

2.1.1 Mini-fins with porous structure (MFP) surfaces

The test surfaces, presented in figure 1, were fabricated out of copper to dimensions 26.5 x 26.5 mm^2. The mini-fins tips were added to the smooth surface and the perforated copper foil was sintered to the mini-fin tips.

[a] Corresponding author: enadstawna@tu.kielce.pl

In MFP surfaces the tunnel width in each direction was constant.

The tests were carried out for different diameters of pores in the copper foil (d_p = 0.05, 0.1, 0.2 and 0.3 mm). The parameters of the investigated surfaces and their designations are shown in table 1. In the figure 1, p_{p1} means the pitch thickness in one direction in the copper foil and p_{p2} means the pitch thickness in the second direction. In table 1, δ means the mini-fin thickness, p_{tun} – is the tunnel pitch, w_{tun} – is the tunnel width, s – is the distance between the mini-fins and h – is the mini-fin height.

Table 1. Parameters of surfaces tested.

Surface	δ	p_{tun}	s	w_{tun}	h	d_p	p_{p1}/p_{p2}
	[mm]	[mm]	[mm]	[mm]	[mm]	[mm]	[mm]
MFP-1.0-0.6-0.05	2.0	2.0	1.4	0.6	1.0	0.05	0.1/0.4
MFP-1.0-0.6-0.1	2.0	2.0	1.4	0.6	1.0	0.1	0.2/0.4
MFP-1.0-0.6-0.2	2.0	2.0	1.4	0.6	1.0	0.2	0.4/0.4
MFP-1.0-0.6-0.3	2.0	2.0	1.4	0.6	1.0	0.3	0.6/0.4
Smooth	-	-	-	-	-	-	-

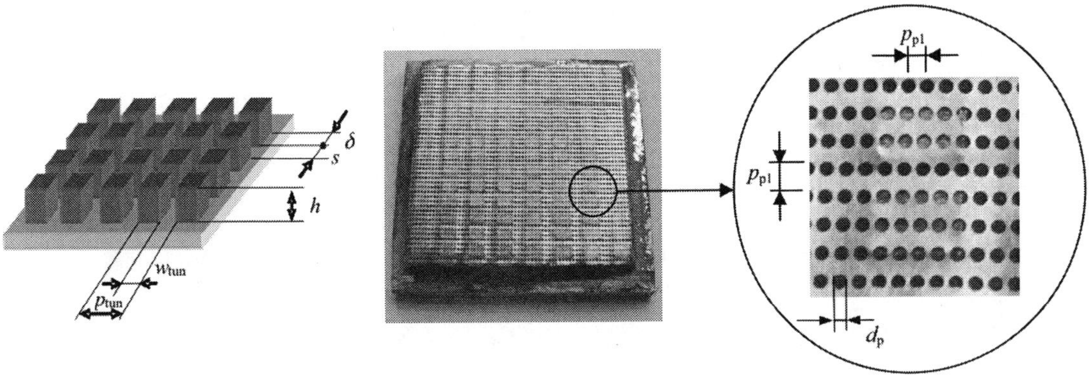

Figure 1. Dimension symbols and a photograph of mini-fin surface sintered with perforated foil (MFP)[5].

2.2 Experimental set-up

The measurement system, presented in figure 2, was designed for determining boiling curves and heat transfer coefficients.

The main device in the measurement system is the basic module supplied with alternating current by an autotransformer. The input power is measured by a wattmeter. Signals from thermocouples are collected by data logger and transferred to the PC.

The main module is instrumented with 7 thermocouples (figure 2): 2 for measuring the saturation temperature (items I, II), 2 for measuring the temperature of the base of the sample (items III, IV), and 3 for determining the heat flux (items V, VI, VII).

Selected thermodynamic properties of four working fluids used in the experiments, at the saturation temperature, are summarized in table 2.

From the experimental data, the heat flux, temperature superheating and heat transfer coefficient were calculated. The obtained results were compared with the Bond number and geometric parameters of surfaces after appropriate transformations.

Table 2. Thermodynamic parameters of saturated fluids [5, 6].

Parameters at p=101.3 kPa	Water	Ethanol	FC-72	Novec 649
T_{sat}, °C	100	78.3	56.4	49.0
ρ_l, kg/m^3	959	757	1 602	1 610
ρ_v, kg/m^3	0.597	1.43	13.24	4.33
λ_l, W/mK	0.68	0.17	0.054	0.059
σ_l, N/m	0.0589	0.0177	0.0081	0.0108
c_l, J/kgK	4 220	723	1 103	1 103
h_{lv}, kJ/kg	2 257	963	88	88
μ_l, Pa·s	0.00028	0.00044	0.00043	0.00043
L, mm	2.5	1.55	0.72	0.83

where:

L – capillary length (eq. 2), mm

T_{sat} – saturation temperature, °C

ρ_l – saturated liquid density, kg/m^3
ρ_v – saturated vapor density, kg/m^3
σ_l – liquid surface tension, N/m
λ_l – liquid thermal conductivity, W/m·K
c_l – specific heat, J/kg·K
h_{lv} – latent heat, J/kg
μ_l – dynamic viscosity, Pa·s.

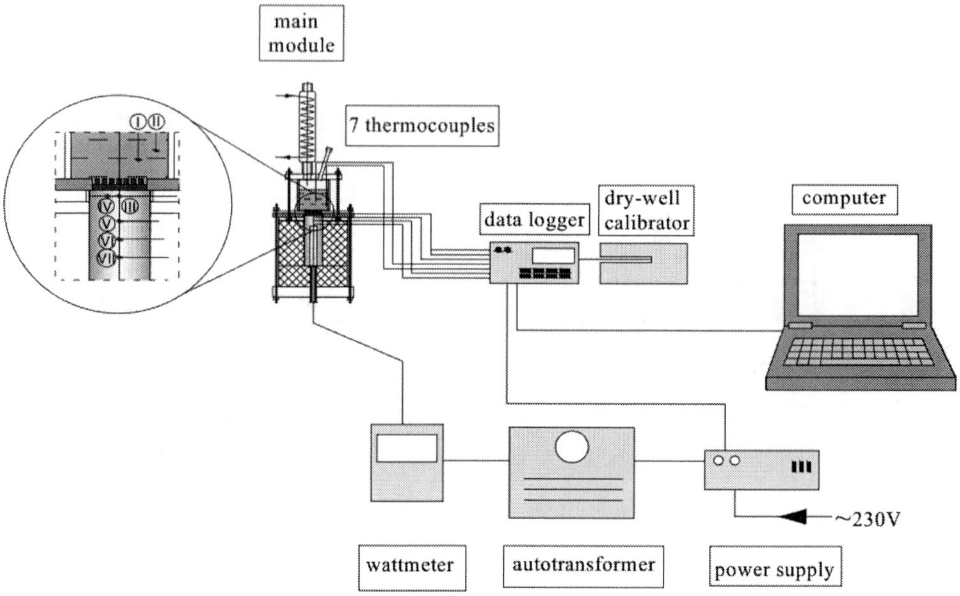

Figure 2. Measurement system.

3 Results

3.1 Bond number

The Bond number is the ratio of characteristic dimension (pore diameter, d_p) to capillary length. For MFP surfaces this similarity number was calculated from the following formula

$$Bo_d = \left(\frac{d_p}{L}\right)^2 \qquad (1)$$

The capillary length which characterizes the potential for bubble departure and coalescence is defined by the equation

$$L = \sqrt{\frac{\sigma_l}{g(\rho_l - \rho_v)}} \qquad (2)$$

where
g – gravitational acceleration, m/s^2.

3.2 Generalized results

The experimental data helped to generalize the results in the form of the relationship between the MFP surface heat transfer coefficient/smooth flat surface coefficient ratio on the Bond number and varying geometric parameters.

For mini-fin surfaces, the proposed relationship is as follows [6,7]

$$\frac{\alpha}{\alpha_s} = f\left(Bo_d^{\,0,5}\,\frac{w_{tun}}{h}\right) = f\left(\frac{d_p}{L}\,\frac{w_{tun}}{h}\right) \qquad (3)$$

where
α – heat transfer coefficient for the MFP surfaces, W/m^{2}K
α_s – heat transfer coefficient for smooth surfaces, W/m^{2}K
d_p – pore diameter, m.

The calculation results for four working fluids are presented in figure 3. Boiling heat transfer coefficients for the smooth surfaces were determined on the basis of the power-law data approximation:

- for ethanol: $\alpha_s = 0.18 \cdot q^{0.68}$
- for water: $\alpha_s = 0.39 \cdot q^{0.70}$
- for FC-72: $\alpha_s = 0.11 \cdot q^{0.72}$
- for Novec 649: $\alpha_s = 0.20 \cdot q^{0.59}$

where:
q – heat flux, W/m^2.

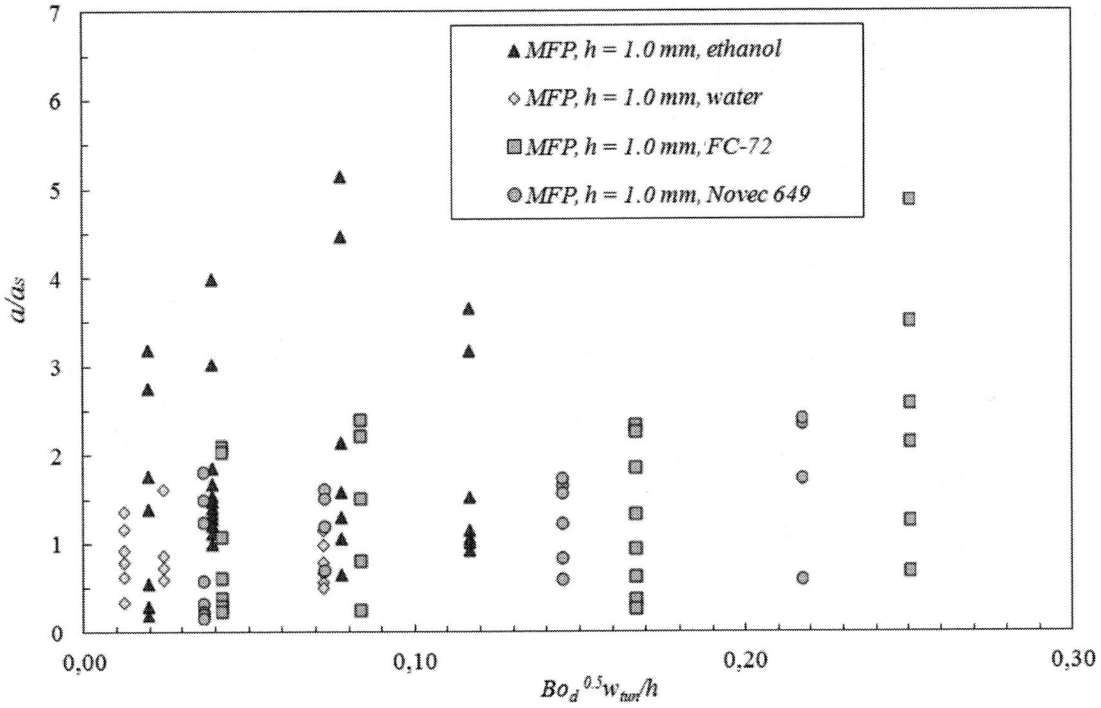

Figure 3. Generalized results for the MFP surface.

From the graph it can be concluded that the highest heat transfer coefficients for the mini-fin surfaces are obtained for the following ranges of similarity numbers:

• for ethanol

$$0.04 < Bo_d^{0.5}\frac{w_{tun}}{h} < 0.08 \quad (4)$$

• for water

$$0.013 < Bo_d^{0.5}\frac{w_{tun}}{h} < 0.025 \quad (5)$$

• for FC-72

$$Bo_d^{0.5}\frac{w_{tun}}{h} \approx 0.25 \quad (6)$$

• for Novec 649

$$Bo_d^{0.5}\frac{w_{tun}}{h} \approx 0.22 \quad (7)$$

The highest ratio between the MFP surface heat transfer coefficients and the smooth flat surface coefficients was observed for the boiling of ethanol and FC-72.

3.3 Correlation

A correlation was proposed in the form of a relationship between the Nusselt number and five similarity numbers. The Reynolds and Bond numbers were associated with the cooper foil pore diameter. The dimensionless number K_w expressed the relationships between the tunnel width and the copper foil pore diameter. Number K_λ shows the ratio of the mini-fin material (copper) thermal conductivity λ_{Cu} to the thermal conductivity of the boiling liquid λ_l.

The following correlation was proposed

$$Nu_{theor} = 0.007 Re^{0.025} Pr^{0.148} K_\lambda^{0.858} Bo_d^{-1.808} K_w^{-3.280} \quad (8)$$

The dimensionless numbers were calculated from the following equations:

- Nusselt number $Nu_{exp} = \frac{\alpha\, d_p}{\lambda_l}$,
- Reynolds number $Re = \frac{q\, d_p \rho_l}{h_{lv} \rho_v \mu_l}$,
- Prandtl number $Pr = \frac{\mu_l\, c_l}{\lambda_l}$,
- Bond number $Bo_d = \left(\frac{d_p}{L}\right)^2$,
- K_w number $K_w = \frac{w_{tun}}{d_p}$,
- K_λ number $K_\lambda = \frac{\lambda_{Cu}}{\lambda_l}$.

The estimated standard deviation of the calculated Nusselt number amounted to 1.5, and the coefficient of determination, being the measure of accuracy of simple regression matching, was R^2 = 0.8343. Relative errors were 0.2. The correlation was found for 98 observations conducted for water, ethanol, FC-72 and Novec 649. The Nusselt number calculated on the basis of experimental data, Nu_{exp}, was compared with theoretical values, Nu_{theor} (figure 4). The correlation was presented with ± 40 % tolerance.

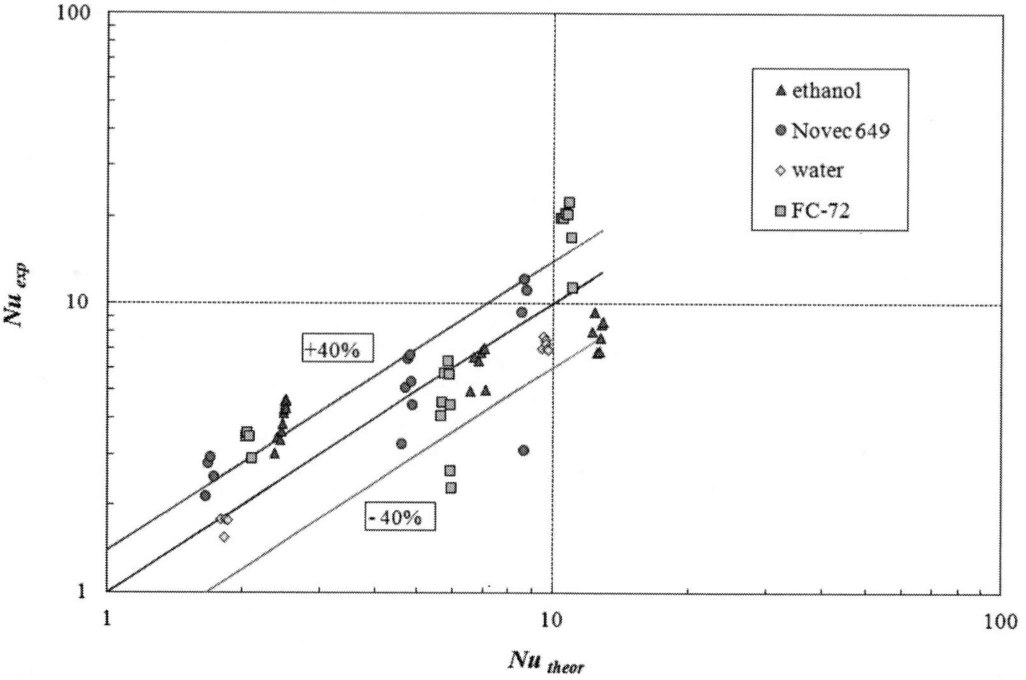

Figure 4. Correlation for pool boiling of water, ethanol FC-72 and Novec 649 for mini-fins with perforated foil.

4 Conclusions

It is necessary to match the geometrical dimensions of the tunnels and the diameters of the pores with the characteristic capillary dimension to obtain the highest possible increase in the pool boiling heat transfer coefficient value on the MFP surfaces.

From the present study the following conclusions can be drawn:

- For boiling FC-72 and ethanol, the heat transfer coefficients obtained from the MFP surfaces were more than 5 times higher than those for the smooth surface.
- For boiling water, the heat transfer coefficients obtained from the MFP surfaces were almost 2 times higher than those for the smooth surface and for boiling Novec 649 – 2.5 times higher than those for the smooth surface.
- Low values of $Bo_d^{0.5}$ (w_{tun} / h_f) should be used for the best heat transfer coefficient enhancement for the boiling liquids with high value of the capillary length (water, ethanol).
- Using a regression analysis with reference to selected physical properties and several characteristic dimensions, it was possible to develop correlations to calculate the heat transfer coefficient with ±40% tolerance.

References

1. Y.-Y. Li, Y.-J. Chen, Z.-H. Liu, *Int. J. Heat Mass Transfer* **77**, 809-817 (2014)
2. A. H. Tarrad, *JJMIE* vol. **5**, nr **2**, 139-144, (2011)
3. A. S. Moita, E. Teodori, A. L. N. Moreira, *Int. J. Heat Fluid Flow* **52,** 50-63, (2015)
4. E. Rooyen, J.R. Thome, *Int. J. Refrigeration* **36**, 447-455 (2013)
5. E. Nadstawna, R. Pastuszko, MATEC Web of Conferences **18**, 01008 (2014)
6. R. Pastuszko, EPJ Web of Conferences**, 67**, 02090 (2014)
7. R. Pastuszko, T. Wójcik, *Exp. Therm. Fluid Sci.* **63**, 34-44 (2015)

EPJ Web of Conferences **114**, 02114 (2016)
DOI: 10.1051/epjconf/201611402114

Extension of the algebraic transition model for the wall roughness effect

Petr Straka[1,a], Jaromír Příhoda[2]

[1]Aerospace Research and Test Establishment, Plc, Prague, Czech Republic
[2]Institute of Thermomechanics AS CR, v.v.i., Prague, Czech Republic

Abstract. The contribution deals with the simulation of the laminar/turbulent transition taking into account the effect of wall roughness. The correlation for the transition onset proposed by Straka and Příhoda [1] was modified for the effect of the wall roughness using the correlation according to Boyle and Stripf [2]. This correlation derived for the wall roughness formed by regularly distributed truncated cones was modified for flows over the distributed wall roughness. The algebraic transition model proposed by Straka and Příhoda [1] with the modified relation for the transition onset was verified by means of the incompressible flat-plate boundary-layer and the compressible flow through the turbine blade cascade with rough blades.

1 Introduction

Besides the pressure gradient, the laminar-turbulent transition in turbomachinery flows is mostly influenced by the free-stream turbulence and wall roughness. Most models of the bypass transition take into account the effect of the free-stream turbulence and the pressure gradient only. The simulation of compressible flows through turbine blade cascades has to take into account many factors, especially the effect of pressure gradient, free-stream turbulence, and wall roughness. The mathematical model should not be completed by the turbulence model only, but also by adequate models of the laminar/turbulent transition and turbulent heat transfer.

The majority of inner surfaces are rough due to erosion and deposits during operation. A typical example is the surface of turbine blades. An extensive review on the effect of wall roughness on flows in turbines was given e.g. by Bons [3]. Typical roughness of operational turbine blades was measured and studied by Bogard et al. [4].

While the wall roughness effect on the modelling of turbulent flow is sufficiently described, the reliable relations for the laminar/turbulent transition on rough walls are very rare. Many publications on the effect of roughness on the transition onset deal with the acceleration of the transition by means of locally placed roughness elements as wires and/or strips of sandpaper near the leading edge of airfoils. Much less experimental results were accomplished for flows over walls with distributed roughness.

The contribution deals with a short review of existing correlations for the transition onset including a modified criterion of Boyle and Stripf [2] and with its testing by means of selected test cases – incompressible flat plate flow and compressible flows through a linear turbine blade cascade.

2 Modelling of wall roughness effect

For turbulent flow, wall roughness is mostly model-led by the k-ω turbulence models with wall boundary condition for the specific dissipation rate ω proposed by Wilcox [5] and lately modified by Hellsten [6] and/or by a two/layer k-L/k-ε turbulence model according to Cebeci and Chang [7], see e.g. Stripf et al. [8].

The roughness of various types is usually described by the centreline averaged roughness Ra or the average peak to valley roughness Rz. The equivalent sand roughness k_s introduced by Nikuradse [9] is mainly used for numerical simulations. According to Bogard et al. [4], roughness parameters of operational turbine blades were in the range $Ra = 8 \div 46$ µm in contrast to new blades with typical value of $Ra = 1.5$ µm. The correlations $k_s \approx (5 \div 6)\, Ra$ can be used for blade surface roughness.

Most of correlations for the wall roughness effect on the transition onset are extensions or modifications of existing relations for transition on smooth walls. Besides, another approach is based on the discrete element model. This roughness model combined with a two-layer turbulence model was used for transition modelling by Stripf et al. [8, 10].

One of first correlations for the transition onset was proposed by Mayle [11] in the form

$$Re_{\theta tr} = 100 + 0.43\, exp\left[7 - 0.77\left(k_s / \theta_t\right)\right] \qquad (1)$$

A very extensive investigation of the wall roughness effect on the transition was carried out by Stripf et al. [12] and Stripf [13] on rough turbine blades with heat

[a] Corresponding author: straka@vzlu.cz

transfer. Wall roughness on turbine blades was modelled by regularly distributed truncated cones with various distances. Several correlations were proposed based on these experiments. Stripf et al. [8] proposed a correlation for the transition onset on rough walls with the height of cones k and the density parameter Λ_R used for the description of roughness. Stripf et al. [10] slightly modified the initial correlation to the form

$$Re_{\theta tr} = \left[\frac{1}{Re_{\theta t}} + 0.0061 f_{\Lambda} \left(\frac{k}{\delta_t^*} - 0.01 \right)^{f_{Tu}} \right]^{-1} \tag{2}$$

valid for $0.01 < k/\delta_t^* \leq 3$. The relation for the transition onset on a smooth wall is given by the relation

$$Re_{\theta t} = 500 Tu_{ef}^{-0.75} \tag{3}$$

The effect of free-stream turbulence level is expressed by the function f_{Tu} with the maximum $f_{Tumax} = 0.9$ and the effect of the element spacing by the function f_Λ. The maximal value $f_{\Lambda max} = 1$ is recommended for the distributed roughness.

Similarly Langtry and Menter [14] presented a modification of the γ-Re transition model for flow over rough walls based on experiments of Stripf et al. [12]. The proposed correlation

$$Re_{\theta tr} = Re_{\theta t}\, min \left[1; \frac{0.0554 + 1.005(\theta_t / k)^{0.87}}{1.4629 + (\theta_t / k)^{0.87}} \right] \tag{4}$$

was tested by means of experimental data of Stripf et al. [12] and measurement of Pimenta et al. [15] on a flat plate with tightly packed spheres.

Boyle and Stripf [2] came out from the simple criterion for the transition onset due to free-stream turbulence

$$Re_{\theta t} = 400 Tu^{-0.625} \tag{5}$$

proposed by Mayle [11]. The effect of wall roughness on the transition onset was expressed by the correlation

$$Re_{\theta tr} = \frac{Re_{\theta t}}{1 + Tu^{-0.625} \left[0.05 \left(k_s^+ - 5 \right) \right]^{1.25}} \tag{6}$$

According to the investigation of turbulent boundary layer on a rough wall, the surface is hydraulically smooth for the equivalent sand-grain roughness $k_s^+ < 5$. The proposed correlation was used for the modelling of flow through the turbine blade cascade with roughness formed by truncated cones according to experiments of Stripf et al. [12] by means of the two-layer k-ε/k-L turbulence model.

The latest correlation of the effect of wall roughness and free stream turbulence on the transition onset was proposed by Lorenz et al. [16]. However, this correlation is as well based on the investigation of the heat transfer in turbine blade cascades with roughness formed by truncated cones.

The empirical correlations for the transition onset due to wall roughness should be generally applicable for various types of wall roughness. Turbulence models modified for flow on rough walls are based on the non-dimensional equivalent sand grain roughness k_s^+. The application of the friction velocity as the parameter is not suitable for the numerical simulation especially in the transitional flow. The more convenient parameter is the ratio k_s/θ_t introduced by Mayle [11]. Provided that the Blasius boundary layer exists in the laminar part of the boundary layer, the non-dimensional equivalent sand-grain roughness can be correlated to the ratio k_s/θ by the relation $k_s^+ = 0.4695\, Re_\theta^{1/2} k_s/\theta$.

The transition criterion of Boyle and Stripf [2] modified to the relation $Re_{\theta tr} = f(Tu, k_s/\theta_t)$ is compared in figure 1 with the accessible experimental data. The lower limit of rough wall $k_s^+ = 5$ is denoted by the dashed line.

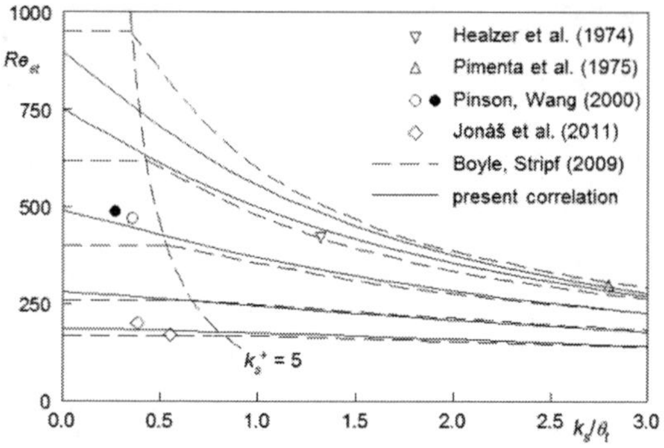

Figure 1. Modified transition correlation for rough walls

Nevertheless, results of experimental investigation of the transition on a rough flat plate with sand paper carried out by Pinson and Wang [17] and Jonáš et al. [18] show that the wall roughness influences the transition onset for values $k_s^+ < 5$. Therefore, the transition correlation of Boyle and Stripf [2] was modified to the form

$$Re_{\theta tr} = \frac{Re_{\theta t}}{1 + F_r \left(Tu, k_s / \theta_t \right)}, \tag{7}$$

where the function F_r is approximated by the relation

$$F_r = a \left(k_s / \theta_t \right)^b \qquad \text{for} \quad k_s / \theta_t \leq 3 \tag{8}$$

$$F_r = c_0 + c_1 \frac{k_s}{\theta_t} + c_2 \left(\frac{k_s}{\theta_t} \right)^2 \qquad \text{for} \quad k_s / \theta_t > 3 \tag{9}$$

Coefficients a, b, c_0, c_1 and c_2 are dependent on the free-stream turbulence level. The correlation for the transition onset on a smooth wall $Re_{\theta t} = (Tu, \lambda)$ was used according to Straka and Příhoda [1].

3 Mathematical model

The Favre-averaged Navier-Stokes equations are closed by the EARSM model according to Hellsten [19] connected with the algebraic transition model proposed by Straka and Příhoda [1]. The Reynolds stress in the EARSM model is expressed by the anisotropy tensor

$$a_{ij} = \frac{\tau_{ij}}{k} - \frac{2}{3} \delta_{ij} \tag{10}$$

Transport equations for the turbulent energy k and the specific dissipation rate ω are used in the two-layer form with two sets of model coefficients and with the blending function according to Menter [20] in the form

$$\frac{D(\rho k)}{Dt} = P_k - \beta^* \rho \omega k + \frac{\partial}{\partial x_j} \left[(\mu + \sigma_k \mu_t) \frac{\partial k}{\partial x_j} \right] \tag{11}$$

$$\frac{D(\rho\omega)}{Dt} = \frac{\gamma\omega}{k}P_k - \beta\rho\omega^2 + \frac{\partial}{\partial x_j}\left[(\mu+\sigma_\omega\mu_t)\frac{\partial\omega}{\partial x_j}\right] + C_D \quad (12)$$

Turbulent viscosity is given by the relation

$$\mu_t = C_\mu \rho k \tau_t \quad (13)$$

with the turbulent time scale

$$\tau_t = \max\left(\frac{1}{\beta^*\omega}; C_\tau\sqrt{\frac{\nu}{\beta^*\omega k}}\right), \quad (14)$$

where the Kolmogorov viscous time scale is used near the wall. The variable coefficient C_μ is obtained by

$$C_\mu = -(\beta_1 + II_\Omega \beta_6)/2 \quad (15)$$

For the reduction of the undesirable overproduction of the turbulent energy in the stagnation region, the production term in the turbulent energy equation was modified into form

$$P_k = \psi\left(\tau_{ij}\frac{\partial \bar{U}_i}{\partial x_j}\right) + (1-\psi)\mu_t\Omega^2, \quad (16)$$

where τ_{ij} is the Reynolds stress tensor and Ω is the absolute value of the vorticity tensor. The parameter ψ is given by the relation

$$\psi = 1 - \max\left[\min\left(C_\psi \tau_t \left|S^2 - \Omega^2\right|;1\right);0\right] \quad (17)$$

with C_ψ = 0.05 ÷ 0.2. The parameter ψ is equals to 1 nearly in the all computational domain except the vicinity of the stagnation point where decreases to zero.

The turbulent heat transfer is modelled according to Launder [21] by the generalized gradient hypothesis in the form

$$-\overline{u_i t} = C_t \frac{\tau_{ij}}{\omega}\frac{\partial \bar{T}}{\partial x_j} \quad (18)$$

with constant $C_t = 0.3$.

For the prediction of the transitional flows, the production and destruction terms in the k-equation are multiplied by the intermittency coefficient γ. Similarly, the effective viscosity is given by $\mu_{ef} = \mu + \gamma\mu_t$ in the transition region.

The algebraic transition model is based on the concept of different values of the intermittency coefficient in the boundary layer γ_i and in the free stream γ_e. The intermittency coefficient in the boundary layer is given by the relation

$$\gamma_i = 1 - \exp\left[-\hat{n}\sigma(Re_x - Re_{xt})^2\right] \quad (19)$$

proposed by Narasimha [22]. The transition onset is given by the momentum Reynolds number Re_{xt} usually depending on the free-stream turbulence level Tu (%) and the pressure-gradient parameter $\lambda_t = (\theta_t^2/\nu)(dU_e/dx)$ at the transition onset. The used empirical correlation is given by the relation

$$Re_{\theta t} = Re_{\theta t o}\left[1 + F(Tu)\frac{1-\exp(-40\lambda)}{1+0.4\exp(-40\lambda)}\right], \quad (20)$$

where the transition onset on the flat plate without pressure gradient is given by the relation

$$Re_{\theta t o} = 975.8 - 497.2Tu + \frac{11.4}{Tu} \quad \text{for } Tu \le 1\%$$

$$Re_{\theta t o} = 96.7 + \frac{340}{Tu} + \frac{53.3}{Tu^2} \quad \text{for } Tu > 1\% \quad (21)$$

The function $F(Tu)$ is given by the relation

$$F(Tu) = 0.29\left[1 - 0.54\exp(-3.5Tu)\right]\exp(-Tu) \quad (22)$$

The length of the transition region given by the spot generation rate $\hat{n}$ and the spot propagation rate σ is expressed using the parameter $N = \hat{n}\sigma\, Re_{\theta t}^3$ introduced by Narasimha [23]. The effect of the free-stream turbulence and the pressure gradient on the parameter N is correlated by an empirical relation N = f(Tu, λ_t) proposed for the attached flow by Solomon et al. [24].

The onset of transition in separated flow is given by the correlation proposed by Mayle [11] in the form

$$Re_{xt} = 300 Re_{\theta s}^{0.7} + Re_{xs}, \quad (23)$$

where $Re_{\theta s}$ is the momentum Reynolds number at the separation and Re_{xs} is the Reynolds number related to the distance of the separation position from the leading edge. The transition length is expressed in the form

$$\hat{n}\sigma = 2.28\times10^{-5}\, Re_{\theta s}^{-1.4} \quad (24)$$

and so the same approach can be applied as in the attached flow.

According to Langtry and Menter [25], the maximum of the vorticity Reynolds number $Re_{\nu max}$ is used to avoid the calculation of the momentum Reynolds number Re_θ in complex flow geometries. The vorticity Reynolds number is given by the relation

$$Re_V = y^2|\Omega|/\nu, \quad (25)$$

where y is the distance from the wall and Ω is the absolute value of the vorticity tensor. The link between both Reynolds numbers is expressed by the relation $Re_\theta = Re_{\nu max}/C$ where the parameter C depends on the pressure gradient. Free-stream turbulence was taken at the boundary layer thickness δ estimated by means of the position of the maximum vorticity Reynolds number using the similar solutions of Falkner and Skan [26].

The algebraic model was implemented into the in-house numerical code. The code is based on the finite volume method of the cell-centered type with the Osher's-Solomon's approximation of the Riemann solver and the two-dimensional linear reconstruction with the Van Albada's limiter. The governing equations are discretized using a multi-block quadrilateral structured grid with a block overlapping implementation. The algebraic transition model is described in detail by Fürst et al. [27].

4 Results

The modified algebraic transition model was tested by means of incompressible flow over a flat plate covered by sand paper. Further, the transition model was applied for modelling of compressible flow through the linear turbine blade cascade with wall roughness formed by the carborundum powder.

Experiments of Pinson and Wang [28] on a heated flat plate were focused on the effect of wall roughness on the boundary layer development on the heat transfer and skin friction at various configurations of the leading edge and wall roughness. The experiments were conducted on smooth wall and on rough wall covered by the 100-grit sand paper. Measurements were carried out at the free-stream velocity $U_e = 8.6$ and 16.1 m/s on the smooth wall and $U_e = 8.6$ m/s on the rough wall. The free-stream turbulence is about $Tu = 0.7$ % in the test section. The wall roughness is characterized by the centreline averaged roughness $R_a = 37$ μm for the grit 100. The equivalent sand roughness was established by means of the relation $k_s = 6\ R_a$ according to Koch and Smith [29]. The difference between wall and free-stream temperature $T_w - T_e = 8$ deg was used for heat transfer modelling.

The computational domain corresponds to the experimental arrangements. Standard inlet and outlet boundary conditions were applied. The predicted skin friction coefficient C_f is compared in figure 2 with experimental data for smooth and rough walls. A quite good agreement was achieved for both cases even though values of C_f for the turbulent boundary layer are higher than standard values. Due to the wall roughness, the transition onset moves upstream from $Re_{xt} \approx 9.2\times10^5$ on smooth wall to $Re_{xt} \approx 4.8\times10^5$ on rough wall at same flow conditions.

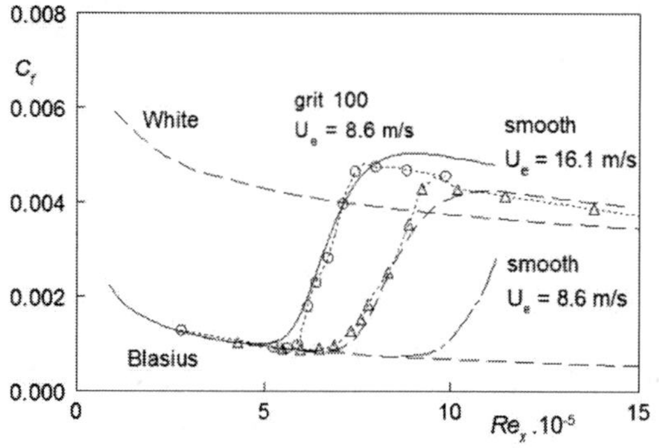

Figure 2. Skin friction distribution on the flat plate

Mean velocity profiles typical for the laminar, transitional and turbulent part of the boundary layer are demonstrated in figure 3 for smooth wall and in figure 4 for rough wall. The agreement of predicted laminar and turbulent velocity profiles with the experimental data is quite good for both cases.

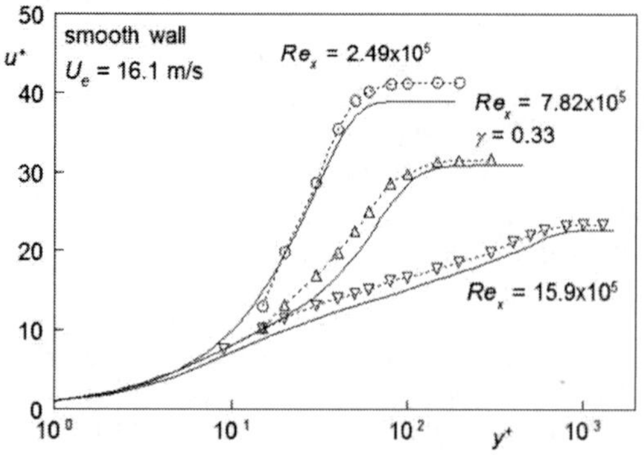

Figure 3. Mean velocity profiles on a smooth wall

For transitional flows the velocity profiles were chosen for Reynolds numbers Re_x where the measured inter-mittency coefficient γ is about 0.3 ÷ 0.4. The predicted velocity profiles correspond to the higher predicted inter-mittency coefficient γ as can be seen in figure 5.

The distribution of the intermittency coefficient γ in the inner part of the boundary layer is shown for both cases in figure 5. It follows for the distribution of the intermittency coefficient that measured transition length is somewhat shorter than the predicted one.

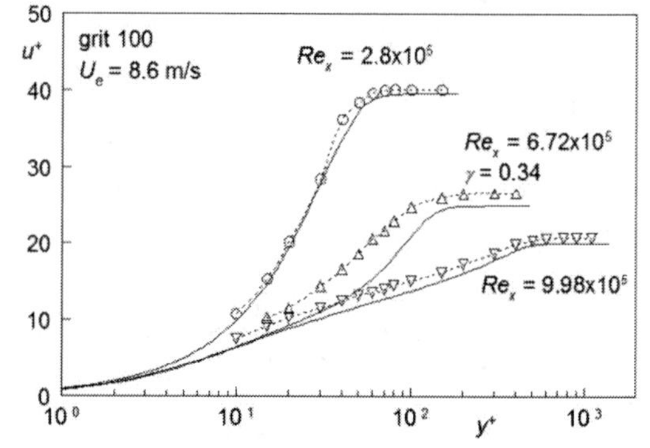

Figure 4. Mean velocity profiles on a rough wall

According to Boyle and Stripf [2], the correlation for the transition length was not modified for roughness effects. This assumption should be probably revised.

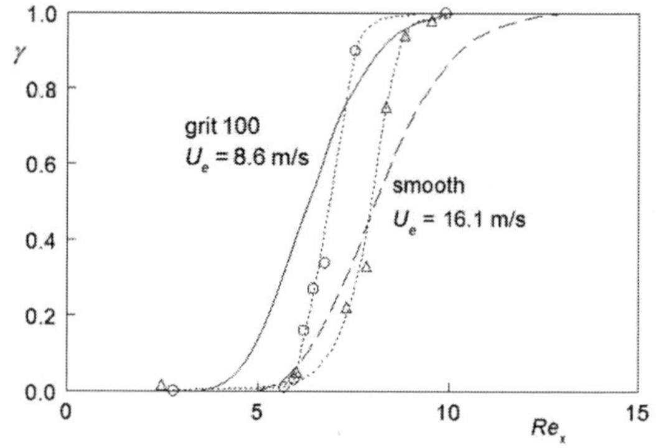

Figure 5. Distribution of the intermittency coefficient

The numerical simulation of 2D compressible flow through the linear turbine blade cascade VS-33 with the relative spacing $t/c = 0.7$ and the inlet angle $\alpha_1 = 0°$ (smooth blades) and 6° (rough blades) was carried out for the isentropic Reynolds number $Re_{2is} = 8.5\times10^5$ and for isentropic Mach numbers M_{2is} from 0.5 up to 0.9. Results were compared with experimental data according to Ulrych et al. [30] and Kladrubský [31]. The wall roughness was created by carborundum powder with Ra = 10 and/or 30 deposited on the blade surface. The relation $k_s = 4.8\ Ra$ was used for modelling of wall roughness.

For the subsonic flow, the total pressure, the total temperature, the inlet turbulence intensity $Tu \approx 4.8\%$ and the viscosity ratio $\mu_t/\mu \approx 100$ were given. At the outlet boundary, the constant static pressure determined according to the outlet isentropic Mach number was prescribed.

The flow field is shown by means of Mach number isolines in figure 6 for values $Re_{2is} = 8.5\times10^5$ and M_{2is} = 0.9 for the smooth wall and for the rough wall $Ra = 30$

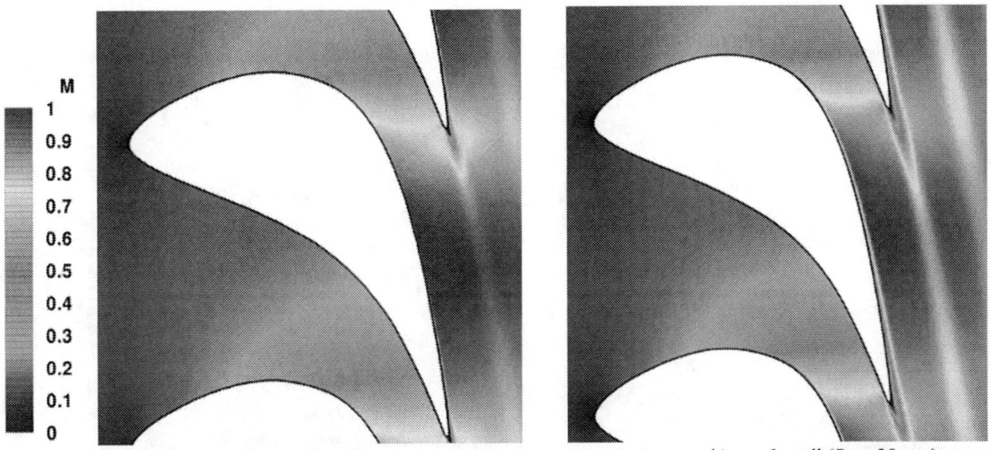

a) smooth wall b) rough wall (Ra = 30 μm)

Figure 6. Mach number isolines for M_{2is} = 0.9

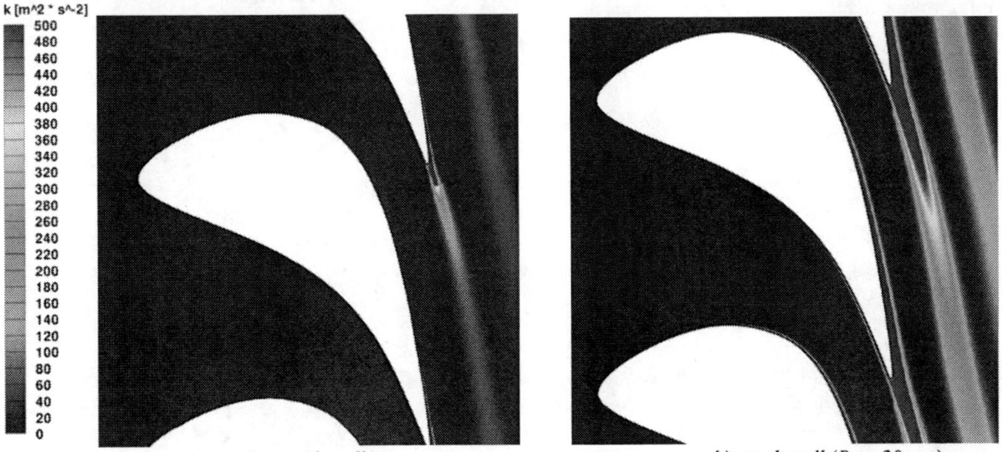

a) smooth wall b) rough wall (Ra = 30 μm)

Figure 7. Field of turbulent energy for M_{2is} = 0.9

μm. The effect of wall roughness is evident especially on the blade suction side and in the wake behind blades.

The field of turbulent energy in the turbine blade cascade is shown in figure 7 for the smooth wall and for the rough wall Ra = 30 μm. The thickness of shear layers on the suction side and the wake is much higher for the rough wall than for the smooth wall.

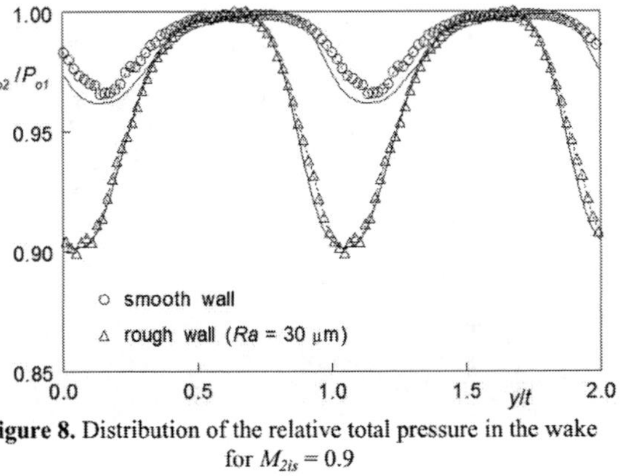

Figure 8. Distribution of the relative total pressure in the wake for M_{2is} = 0.9

The distribution of the relative total pressure in the wake behind the cascade is shown in figure 8 for the Mach number M_{2is} = 0.9. This distribution corresponds to mean velocity profiles behind blades. Numerical results are compared with experimental data obtained in the traversing plane 15 mm behind the cascade. The agreement with experimental data is quite good.

Finally, predicted loss coefficient ζ was compared with experimental data for the cascade with smooth and rough blades. The mean values of the loss coefficient across the blade spacing were obtained by the data reduction method, see Amecke and Šafařík [32]. The loss coefficient ζ is defined by the relation

$$\zeta = 1 - \left(\frac{U_2}{U_{2is}} \right)^2 , \quad (26)$$

where U_2 is the mean outlet velocity and U_{2is} is the mean outlet isentropic velocity. The variation of the loss coefficient with the isentropic Mach number M_{2is} for the isentropic Reynolds number Re_{2is} = 8.5×10^5 is shown in figure 9. The loss coefficient in the subsonic region slowly decreases with the increasing Mach number. The agreement of the prediction of energy losses carried out by the modified transition model connected with the EARSM model is very good up to M_{2is} = 0.9 for smooth and rough blades as well. In the transonic region, the agreement of predicted and measured data is worse.

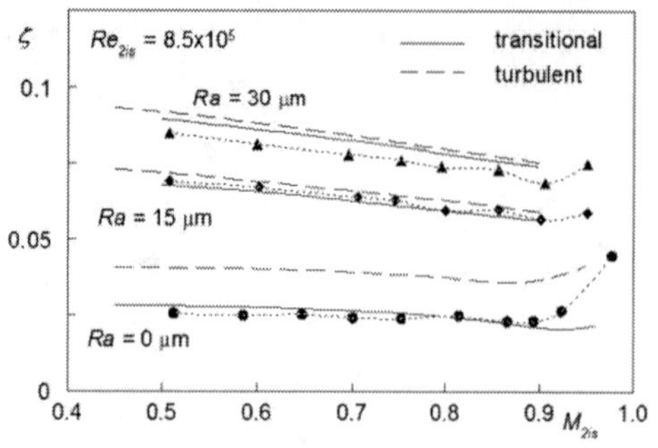

Figure 9. Variation of the loss coefficient with M_{2is}

In case of smooth blades, the application of the turbulence model only gives energy losses about 50% higher than the prediction with the transition model. The effect of the transition model on predicted losses is much lesser for rough blades. It is caused to some extent by combined effects of wall roughness and relatively high free-stream turbulence about $Tu \approx 4.8\%$.

Conclusions

The algebraic transition model was extended for flows over rough walls. The modified model was verified by means of the incompressible flat-plate boundary-layer and the compressible flow through the turbine blade cascade with rough blades. The agreement of numerical results with experimental data is in the both cases quite good. Nevertheless, the measured transition length is somewhat shorter than the predicted one for the flat-plate flow. According to other transition models, the correlation for the transition length was not modified for the effect of roughness. This assumption should be probably revised. As well, the modified model should be further tested for the transition in various boundary conditions.

Acknowledgement

The work was supported by the institutional support RVO 61388998 and by grants TA CR TA03020277 and GA CR P101/12/1271.

References

1. P. Straka, J. Příhoda, *Proc. Conf. Experimental Fluid Mechanics*, Liberec, 636-641 (2010)
2. R. J. Boyle, M. Stripf, *Jour. Turbomachinery*, **131**, 041020 (2009)
3. J. P. Bons, *Jour. Turbomachinery*, **132**, 021004, (2010)
4. D. G. Bogard, D. L. Schmidt, M. Tabbita, *Jour. Turbomachinery*, **120**, 337-342 (1998)
5. D. C. Wilcox, *AIAA Jour.*, **32**, 247-255 (1994)
6. A. Hellsten, *Some improvements in Menter's k-ω SST turbulence model*, AIAA Paper 28-2554 (1998)
7. T. Cebeci, K. C. Chang, *AIAA Jour.*, **16**, 730-735 (1978)
8. M. Stripf, A. Schulz, H.-J. Bauer, *Jour. Turbomachinery*, **130**, 021003 (2008)
9. J. Nikuradse, *Forsch.-Arb. Ing.-Wes.*, **361**, (1933)
10. M. Stripf, A. Schulz, H.-J. Bauer, S. Wittig, *Jour. Turbomachinery*, **131**, 031016, 031017 (2009)
11. R. E. Mayle, *Jour. Turbomachinery*, **113**, 509-537 (1991)
12. M. Stripf, A. Schulz, S. Wittig, *Jour. Turbomachinery*, **127**, 200–208 (2005)
13. M. Stripf, *Einfluss der Oberflächenrauigkeit auf die transitionale Grenzschicht an Gasturbinen-schaufeln*, Dissertation, Universität Karlsruhe (2007)
14. R. Langtry, F. R. Menter, *Overview of industrial transition modelling in CFX*, Technical Report ANSYS, 172 p. (2006)
15. J. J. Pimenta, R. J. Moffat, W. M. Kays, *The turbulent boundary layer: An experimental study of the transport of momentum and heat with the effect of roughness*, Report HMT-21, Dept. of Mech. Eng., Stanford University (1975)
16. M. Lorenz, A. Schulz, H.-J. Bauer, *ASME Paper GT2012-68910* (2012)
17. M. V. Pinson, T. Wang, *Jour. Turbomachinery*, **122**, 301-307, 308-316 (2000)
18. P. Jonáš, O. Hladík, O. Mazur, V. Uruba, *IOP Publishing, Journal of Physics, Conference Series*, **318**, 032030 (2011)
19. A. Hellsten, *New two-equation turbulence model for aerodynamics applications*, PhD. Thesis, Helsinki University of Technology (2004)
20. F. R. Menter, *AIAA Jour.*, **32**, 1598-1605 (1994)
21. B. E. Launder, *Jour. Heat Transfer*, **110**, 1112–1128 (1988)
22. R. Narasimha, *Jour. Aerospace Science*, **24**, 711-712 (1957)
23. R. Narasimha, *Prog. Aerospace Science*, **22**, 29-80, (1985)
24. W. J. Solomon, G. J. Walker, J. P. Gostelow, *Jour. Turbomachinery*, **118**, 744-751 (1996)
25. R. Langtry, F. R. Menter, *AIAA Jour.*, **47**, 2894-2906 (2009)
26. V. M. Falkner, S. W. Skan, *Phil. Mag.*, **12**, 865-896 (1930)
27. J. Fürst, J. Příhoda, P. Straka, *Computing*, **95**, S163-S182 (2013)
28. M. V. Pinson, T. Wang, *Jour. Turbomachinery*, **122**, 301-307, 308-316 (2000)
29. C. C. Koch, L. H. Smith Jr., *Jour. Engineering for Power*, **98**, 411-424 (1976)
30. J. Ulrych, J. Benetka, T. Jelínek, R. Valenta, L. Tajč, *Proc. XVIII Symposium on Measuring Techniques in Turbomachinery*, Thessaloniki, CD-ROM, 5 p. (2006)
31. M. Kladrubský, *Measurement of linear blade cascade VS-33 along the blade height at the spacing 0.7*, VZLÚ Praha, Research Report R–5166 (2011)
32. J. Amecke, P. Šafařík, *Data reduction of wake flow measurements with injection of an other gas*, DLR-Forschungbericht 95-32, Göttingen (1995)

EPJ Web of Conferences **114**, 02115 (2016)
DOI: 10.1051/epjconf/201611402115

Numerical simulation of high-swirl flow in axial turbine stage

Petr Straka[1,a]

[1]*Aerospace Research and Test Establishment, a.s., Beranových 130, Prague–Letňany, Czech Republic*

Abstract. This article deals with numerical modelling of flow in an axial turbine stage with prismatic blades which are not equipped with shroud. In this case flow through a radial gap above and below ends of blades leads to generation of strong secondary vortices which significantly affect an efficiency of the stage. The main attention is paid to turbulence modelling. Besides a linear model based on Boussinesq approximation a nonlinear model based on higher order closure formula is used. Also an effect of a curvature correction is taken into account.

1 Introduction

The axial steam turbine of low power (around hundreds kilowatts to units megawatts) are typically designed in configuration with the drum-type rotor. It means that the rotor blades are carried directly by the rotor of large diameter. In case of low power turbines, in order to reduce a production costs a blades with free ends that are not equipped with the shroud are still in use. For more reduction of the production cost the prismatic blades are sometimes used as it is in the present work. In this configuration it is necessary to maintain the radial gap under the hub-end of the stator blades and above the tip-end of the rotor blades. However, flow through the radial gap generates an intensive secondary vortices which interacts with following blade wheel. This interaction leads to strong swirl of flow field in the peripheral parts of the blade span.

This work focuses on numerical modelling of intensively swirl flow in case of the axial turbine stage using the URANS (Unstaedy Reynolds Averaged Navier–Stokes equations) framework. The main attention is paid on the turbulence model closure formula.

Presented work was carried out under the project: *TA02021336 - Research and Development of Small Sized Turbine Stage* (see [1]). Project deals with research of an aerodynamics and energy characteristics of the axial stage of small sized and low power steam turbine.

Also some previous works carried out under above mentioned project deal with numerical modelling of flow in the axial stage with blades that are not equipped with the shroud. In [2] there is discussed an effect of the radial gap dimension on steady flow in stator wheel without interaction with the rotor blades. Results of numerical and experimental investigation are compared in [3]. There is also discussed an influence of the interface between stator and rotor part of the computational domain. Two different kinds of the interface are compared in [3]. Paper [4] deals with clarification the mechanism of the rotor blade interaction with secondary vortices generated behind the radial gap under the hub-end of the stator blade. Also in that paper computational results are compared with experimental data. There is also discussed the possibility of reduction the computational domain by the method of scaling, which technique is applied also in present work.

Note that in [2–4] there the two-equation $k - \omega$ turbulence model with first order closure formula based on common Boussinesq approximation was used. However in general the turbulence models based on first order closure formula are not able to sufficiently well predict the secondary flow structures. Therefore present work is focussed on turbulence model based on nonlinear second order closure formula.

2 Geometric setup

Figure 1 shows the computational domain with marked inlet and outlet boundaries and sliding mesh interface between stator and rotor parts of the computational domain. It is seen that the computational domain contains only one stator and one rotor blade. Such a reduction of the actual number of blades is achieved by method of scaling. The scale factor of the stator blade is given by $2N_s/(N_s + N_r)$, where N_s is real number of the stator blades and N_r is real number of the rotor blades. The scale factor of the rotor blade is defined analogously. In our case the scale factor is $0.\overline{909}$ for the stator blade and $1.\overline{09}$ for the rotor blade. Note that scaling has no effect on static aerodynamics loading of the blades, because the pitch to chord ratio of blades is preserved. Scaling of the blade profile causes little change of the chord based Reynolds number, but it is minor problem in case of values of scale factors near 1. The method of scaling has, however, major impact on prediction of dynamic aerodynamic loading of the blades. Therefore this method is unusable in cases where such that prediction is a main scope. In our case we want to

[a]e-mail: straka@vzlu.cz

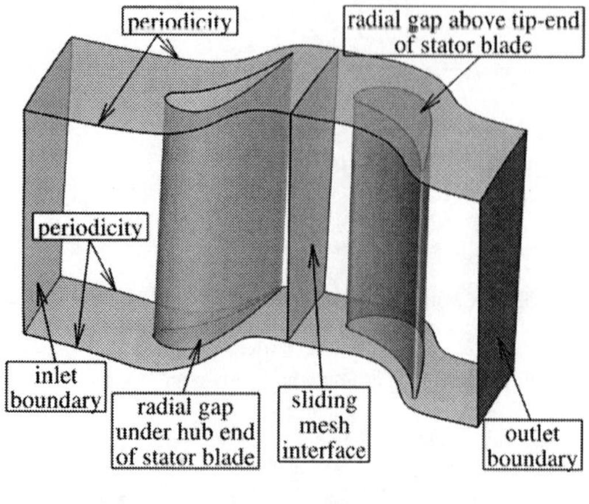

Figure 1. Scheme of the computational domain.

compare prediction of the rotor blade interaction with secondary vortices generated behind the radial gap under the hub-end of the stator blade using turbulence models with different closure formulas. After scaling, the chord of the stator blade is 29.0 mm and the chord of the rotor blade is 28.1 mm. The hub diameter is 330 mm and diameter of the outer casing is 430 mm. Between the hub-end of the stator blade and the hub-wal there is the radial gap of 1 mm. Also between the tip-end of the rotor blade and the outer casing there is the radial gap of 1.4 mm. The axial distance between the stator trailing edge and the rotor leading edge is 15 mm. The inlet boundary is placed 24 mm before the stator leading edge and the outlet boundary is placed 15 mm behind the rotor trailing edge. The sliding mesh interface between stator and rotor parts of the computational domain is placed in the middle of the axial gap between the stator and the rotor blades.

3 Flow conditions

Presented results were calculated for the total/static expansion ratio 1.4, the velocity ratio $u/c_{is} = 0.6$ (u is the circumferential velocity at middle diameter and c_{is} is the isentropic velocity at the hub diameter behind the rotor trailing edge). It corresponds with the inlet total pressure 120 kPa, the outlet pressure 85.7 kPa, speed of revolutions 7430 RPM, the total inlet temperature of 303 K. The isentropic outlet Mach number is $M_{is} = 0.7$, while the outlet isentropic Reynolds number based on blade chord is $Re_{is} \approx 3 \times 10^5$.

4 Numerical aparatus

Flow through the axial turbine stage is modeled as unsteady, 3D, compressible, viscous, fully turbulent flow of the perfect gas. Simulation is performed using the in-house numerical code based on solution of system of the RANS equation and two-equation $k-\omega$ turbulence model. Used numerical code [5, 6] is focused on turbomachinery application. The interface between stator and rotor part of the computational domain is performed via the "sliding mesh" method, where communication between stationary stator domain and rotating rotor domain is based on interpolation.

System of the governing equations reads:

$$\frac{\partial \rho}{\partial t} + \frac{\partial}{\partial x_j}\left(\rho u_j\right) = 0\,, \tag{1}$$

$$\frac{\partial}{\partial t}\left(\rho u_i\right) + \frac{\partial}{\partial x_j}\left(\rho u_j u_i + p\delta_{ij}\right) - \frac{\partial \tau_{ij}}{\partial x_j} = 0\,, \tag{2}$$

$$\frac{\partial e}{\partial t} + \frac{\partial}{\partial x_j}\left(\rho u_j(e+p) - u_k\tau_{kj} + q_j - d^t_j\right) = 0\,, \tag{3}$$

$$\frac{\partial}{\partial t}\left(\rho k\right) + \frac{\partial}{\partial x_j}\left(\rho u_j k - d^t_j\right) = P_k - f_{\beta^*}\rho k\omega\,, \tag{4}$$

$$\frac{\partial}{\partial t}\left(\rho\omega\right) + \frac{\partial}{\partial x_j}\left(\rho u_j\omega - o^t_j\right) = P_\omega - \beta\rho\omega^2 + C_D\,, \tag{5}$$

where

$$p = (\varkappa - 1)\left[e - \frac{\rho}{2}\left(u^2 + v^2 + w^2\right) - \rho k\right]\,, \tag{6}$$

$$q_j = -\left(\frac{\mu}{\mathrm{Pr}} + \frac{\mu_t}{\mathrm{Pr_t}}\right)\frac{\varkappa r}{\varkappa - 1}\frac{\partial T}{\partial x_j}\,, \tag{7}$$

$$\tau_{ij} = \tau^{lam}_{ij} + \tau^t_{ij}\,, \tag{8}$$

$$\tau^{lam}_{ij} = 2\mu\left(S_{ij} - \frac{1}{3}\frac{\partial u_k}{\partial x_k}\delta_{ij}\right)\,, \tag{9}$$

$$d^t_j = \left(\mu + \sigma_k\mu^{evm}_t\right)\frac{\partial k}{\partial x_j}\,, \tag{10}$$

$$o^t_j = \left(\mu + \sigma_\omega\mu^{evm}_t\right)\frac{\partial \omega}{\partial x_j}\,, \tag{11}$$

$$P_k = \min\left(2\mu_t S^2,\, 2\mu_t\Omega^2,\, 20 f_{\beta^*}\rho k\omega\right)\,, \tag{12}$$

$$P_\omega = \gamma\frac{\omega}{k}P_k\,, \tag{13}$$

$$C_D = \sigma_d\frac{\rho}{\omega}\max\left(\frac{\partial k}{\partial x_j}\frac{\partial \omega}{\partial x_j},\, 0\right)\,, \tag{14}$$

$$f_{\beta^*} = \beta^*\frac{1 + 680\chi_k^2}{1 + 400\chi_k^2}\,, \tag{15}$$

$$\chi_k = \frac{1}{\omega^3}\max\left(\frac{\partial k}{\partial x_j}\frac{\partial \omega}{\partial x_j},\, 0\right)\,, \tag{16}$$

$$\mu = 1.72 \times 10^{-5}\left(\frac{T}{273.15}\right)^{0.754}\,, \tag{17}$$

$$\mu^{evm}_t = \rho\frac{k}{\omega}\,, \tag{18}$$

$$S = \sqrt{2S_{ij}S_{ij}}\,, \tag{19}$$

$$\Omega = \sqrt{2\Omega_{ij}\Omega_{ij}}\,, \tag{20}$$

$$S_{ij} = \frac{1}{2}\left(\frac{\partial u_i}{\partial x_j} + \frac{\partial u_j}{\partial x_i}\right)\,, \tag{21}$$

$$\Omega_{ij} = \frac{1}{2}\left(\frac{\partial u_i}{\partial x_j} - \frac{\partial u_j}{\partial x_i}\right)\,. \tag{22}$$

The turbulent viscosity μ_t and the Reynolds stress term τ^t_{ij} will be described later in Sect. 5.

System of equations (1) to (5) we can formally rewrite in vector form as follows:

$$\frac{\partial \mathbf{W}}{\partial t} + \frac{\partial \mathbf{F}_j}{\partial x_j} = \mathbf{Q}\,, \tag{23}$$

where $\mathbf{W}$ is vector of the conservative variables, $\mathbf{F}$ is vector of the inviscid and viscous fluxes and $\mathbf{Q}$ is vector of the source terms. System (23) is discretized using the cell-centered finite-volume method on multi-block structured

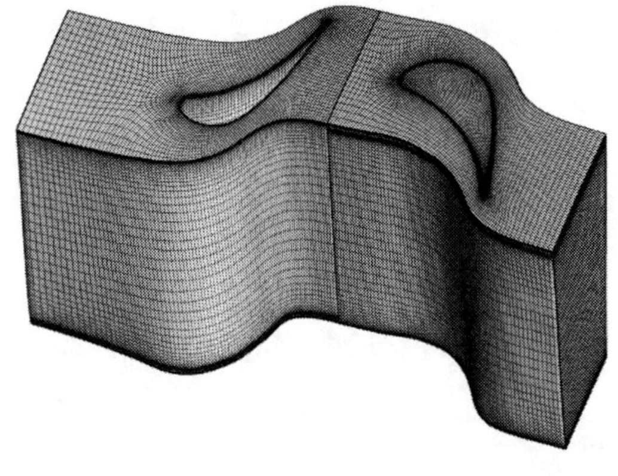

Figure 2. Computational mesh.

mesh of hexahedral cells. Temporal discretization is performed using the second-orded backward Euler formula in implicit form, which is realized through a dual iterative process:

$$\left(\frac{3}{2}\mathbf{I} + \Delta t\,\frac{\partial \mathbf{R}_i^{n,\nu}}{\partial \mathbf{W}_i^{n,\nu}}\right)\Delta \mathbf{W}_i^{n,\nu+1/2} + \Delta t \sum_{j=1}^{6} \frac{\partial \mathbf{R}_i^{n,\nu}}{\partial \mathbf{W}_j^{n,\nu}}\Delta \mathbf{W}_j^{n,\nu+1/2} =$$

$$= -\frac{3\mathbf{W}_i^{n,\nu} - 4\mathbf{W}_i^{n} + \mathbf{W}_i^{n-1}}{2} - \Delta t\,\mathbf{R}_i^{n,\nu}, \qquad (24)$$

where $\mathbf{W}^{n,0} := \mathbf{W}^n$, $\Delta\mathbf{W}^{n,\nu+1/2} = \Delta\mathbf{W}^{n,\nu+1} - \Delta\mathbf{W}^{n,\nu}$ and $\mathbf{W}^{n+1} := \mathbf{W}^{n,\nu^*}$, where $\mathbf{W}^{n,\nu^*}$ is a steady solution in the dual iterative process. In scheme (24) there n means the time level, Δt is the time step, ν is index of the dual iteration, i is index of current cell in the computational mesh, j is index of the neighbouring cells and $\mathbf{R} \approx \partial \mathbf{F}_j/\partial x_j$ is the numerical approximation of the inviscid and the viscous fluxes. The inviscid numerical fluxes are calculated using the exact solution of the 1D Rieman problem in normal direction to the cell edges. The viscous numerical fluxes are calculated using the central scheme using the Green-Gauss theorem on a dual cells. Higher order of accuracy in space is obtained using linear reconstruction with the Van Leer's slope limiter.

The computational domain was discretized by structured multiblock mesh of 1.5 millions hexahedral cells which is displayed in figure 2.

5 Closure formula

As it was mentioned above, this work is focused on using of the turbulence model based on nonlinear second order closure formula. Among the many turbulence models suitable for prediction of secondary flows a quadratic explicit algebraic Reynolds stress model proposed by Rumsey and Gatski [7] was chosen. This model takes into account the production-to-dissipation rate ratio as well as the stress tensor anisotropyo. On the other hand, because the stress tensor is modeled via algebraic formula, this model has comparable computational cost as the linear eddy-viscosity turbulence models based on the Boussinesq approximation. Because the algebraic stress models are in general derived based on weak-equilibrium assumption, it partially lose sensitivity to streamline curvature. This may be problem in case of strong swirl flow field. This deficiency can be removed by modelling the convection of the anisotropy by some means. The curvature correction method based on the strain rate tensor, that is described in [8], is used in this work. This turbulence model will be referred here as EASM (Explicit Algebraic Stress Model).

Results calculated using models EASM are in this work compared with those from [4]. Because the results published in [4] were computed using the turbulence model based on the first order closure formula, this model is also described here. Turbulence model based on the first order closure formula will be here referred as EVM (Eddy Viscosity Model).

Note that used turbulence models are based on two transport equations for turbulent kinetic energy k and specific dissipation rate of turbulent energy ω given by eq. (4) and (5).

5.1 EVM turbulence model

This model is based on the Boussinesq approximation. The Reynolds stress tensor τ^t_{ij} is modeled as a linear function of the strain rate tensor S_{ij}:

$$\tau^t_{ij} = 2\mu_t\left(S_{ij} - \frac{1}{3}\frac{\partial u_k}{\partial x_k}\delta_{ij}\right) - \frac{2}{3}\rho k \delta_{ij}, \qquad (25)$$

where the turbulent viscosity $\mu_t = \mu_t^{evm}$ is given by eq. (18). The model constants: $\beta^* = 0.09$, $\beta = 0.075$, $\gamma = 0.5532$, $\sigma_k = 0.5$, $\sigma_\omega = 0.5$ and $\sigma_d = 0.5$ corresponds with the TNT $k-\omega$ turbulence model proposed by Kok [9].

5.2 EASM turbulence model

In this model the Reynolds stress tensor τ^t_{ij} is modeled as a quadratic function of the strain rate tensor S_{ij}:

$$\begin{aligned}\tau^t_{ij} &= 2\mu_t\left(S_{ij} - \tfrac{1}{3}\frac{\partial u_k}{\partial x_k}\delta_{ij}\right) - \tfrac{2}{3}\rho k \delta_{ij} \\ &+ 2\mu_t\left[a_2 a_4\left(S_{ik}\Omega_{kj} - \Omega_{ik}S_{kj}\right)\right. \qquad (26) \\ &- \left.2a_3 a_4\left(S_{ik}S_{kj} - \tfrac{1}{3}S_{kl}S_{lk}\delta_{ij}\right)\right].\end{aligned}$$

It is clear that eq. (26) extends relation (25) about anisotropy part (term in square brackets). The turbulent viscosity μ_t is now given as

$$\mu_t = C_\mu \rho k \tau, \qquad (27)$$

where the turbulent time scale τ is defined as

$$\tau = \max\left(\frac{1}{\beta^*\omega}, C_\tau\sqrt{\frac{\mu}{\beta^*\rho k\omega}}\right), \qquad (28)$$

where $C_\tau = 6.0$ is a model constant. The nominal level of the variable coefficient C_μ in a zero-pressure-gradient log-layer is approximately 0.09. In general case value of C_μ is obtained by solving following cubic equation:

$$\left(-C_\mu\right)^3 + q_1\left(-C_\mu\right)^2 + q_2\left(-C_\mu\right) + q_3 = 0, \qquad (29)$$

where

$$q_1 = -\frac{\gamma_1^*}{II_S \tau^2 \gamma_0^*}, \qquad (30)$$

$$q_2 = \frac{1}{\left(2II_S\tau^2\gamma_0^*\right)^2}\left(\gamma_1^{*2} - 2II_S\tau^2\gamma_0^* a_1 - \tfrac{2}{3}II_S\tau^2 a3_2 - 2II_\Omega\tau^2 a_2^2\right), \qquad (31)$$

$$q_3 = \frac{\gamma_1^* a_1}{\left(2II_S\tau^2\gamma_0^*\right)^2}, \qquad (32)$$

$$II_S = S_{kl}S_{lk}, \quad II_\Omega = \Omega_{kl}\Omega_{lk}. \qquad (33)$$

The correct root to choose from this equation is the root with the lowest real part. See [7] for more details. Other parameters are gived by:

$$a_1 = \tfrac{1}{2}\left(\tfrac{4}{3} - C_2\right), \quad a_2 = \tfrac{1}{2}(2 - C_4), \qquad (34)$$

$$a_3 = \tfrac{1}{2}(2 - C_3), \quad a_4 = \tau / \left(\gamma_1^* + 2\gamma_0^* C_\mu S^2 \tau^2\right), \qquad (35)$$

$$\gamma_0^* = C_1^1/2, \quad \gamma_1^* = \tfrac{1}{2}C_1^0 + \left(\tfrac{C_{\varepsilon2} - C_{\varepsilon1}}{C_{\varepsilon1} - 1}\right), \qquad (36)$$

where $C_{\varepsilon1} = 1.44$, $C_{\varepsilon2} = 1.83$, $C_1^0 = 3.4$, $C_1^1 = 1.8$, $C_2 = 0.36$, $C_3 = 1.25$ and $C_4 = 0.4$.

The model constants are now: $\beta^* = 0.09$, $\beta = 0.0747$, $\gamma = 0.53$, $\sigma_k = 1$, $\sigma_\omega = 0.5339$ and $\sigma_d = 0.5$.

5.3 Curvature correction

The rotation rate tensor Ω_{ij} is defined now as:

$$\Omega_{ij} = \frac{1}{2}\left(\frac{\partial u_i}{\partial x_j} - \frac{\partial u_j}{\partial x_i}\right) - \frac{1}{A_0}\Omega_{ij}^{(r)}, \qquad (37)$$

where

$$\Omega_{ij}^{(r)} = -\varepsilon_{ijk} B_{km} S_{pr} \frac{DS_{rq}}{Dt} \varepsilon_{pqm}, \qquad (38)$$

$$B_{km} = \frac{II_S^2 \delta_{km} + 12III_S S_{km} + 6II_S S_{kl} S_{lm}}{2II_S^3 - 12III_S^2} \qquad (39)$$

and $A_0 = -0.72$, $III_S = S_{kl}S_{lm}S_{mk}$ and ε_{ijk} is the Levi-Civita symbol.

In eq. (38) $\frac{DS_{rq}}{Dt} = \frac{\partial S_{rq}}{\partial t} + \frac{\partial(u_k S_{qr})}{\partial x_k}$ is the material derivative of the strain rate tensor. The temporal derivative of the strain rate tensor is approximated using first order backward differencing scheme, while the convective part is discretized by standard manner of finite volume method. Following [10] a spatial filter for computed tensor $\Omega_{ij}^{(r)}$ is applied to prevent some problems with numerical oscillations of DS/Dt.

6 Computational results

The effect of flow through the radial gap under the stator blade is such that stream flowing from the radial gap generates (in interaction with main stream in blade channel) massive vortex above the stream and beside the vertical surface of the suction side as illustrates a scheme in Fig. 3. This upper vortex is in given regimes and geometrical configurations intensive enough to deform the stream flowing

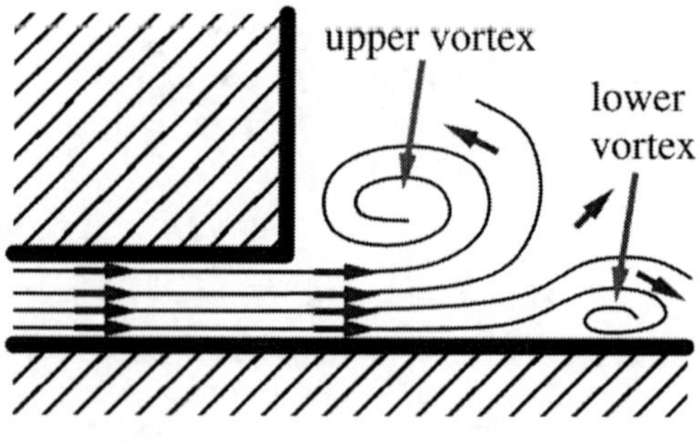

Figure 3. Scheme of flow behind the radial gap under the stator blade.

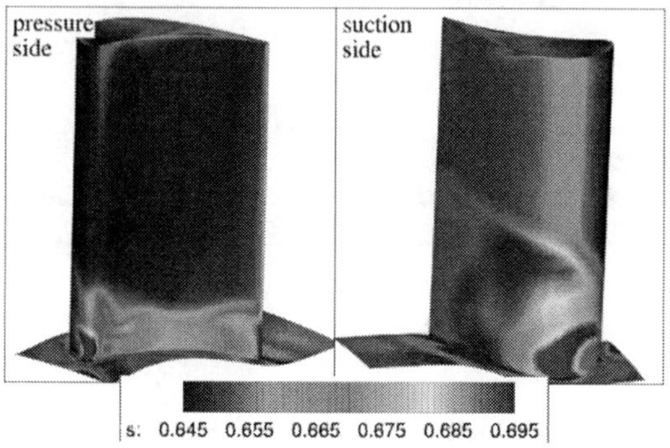

Figure 4. Distribution of the dimensionless entropy index $s = p/\rho^\kappa$ on the rotor blade surface.

from the radial gap which leads to separation of the stream from the wall and to generation of the other vortex under the stream. Both vortices then interact with the rotor blade so that they deforms in upward direction and the are carried on the suction side almost up to half of the span. This causes increase of the dissipation rate on the lower half of span on the suction side, as illustrates distribution of the dimensionless entropy index $s = p/\rho^\kappa$ (computed from the normalized pressure and density) on the rotor blade surface that is shown in figure 4.

The difference between the linear EVM model and the nonlinear EASM model, which takes the anisotropy into account, lies in prediction of forming of the upper and lower vortices. The EVM model predicts this phenomena much simpler than the EASM model. The EVM model predicts the lower vortex in parallel along the upper vortex up to the point of interaction with the leading edge of the rotor blade, while the EASM model predicts distortion of the lower vortex that is being wound around the upper vortex as is depicted schematically in figure 5 and in form of isosurface of $(|\Omega|^2 - |S|^2)$ in figure 6. This difference between EVM and EASM models has its origin in anisotropic part of the Reynolds stress tensor, which is ignored in the EVM model.

Another difference between both turbulent models lies in production of the turbulent kinetic energy. Figure 7 shows field of the turbulent kinetic energy in circumferential cross section at 10, 50 and 90 percent of span. The

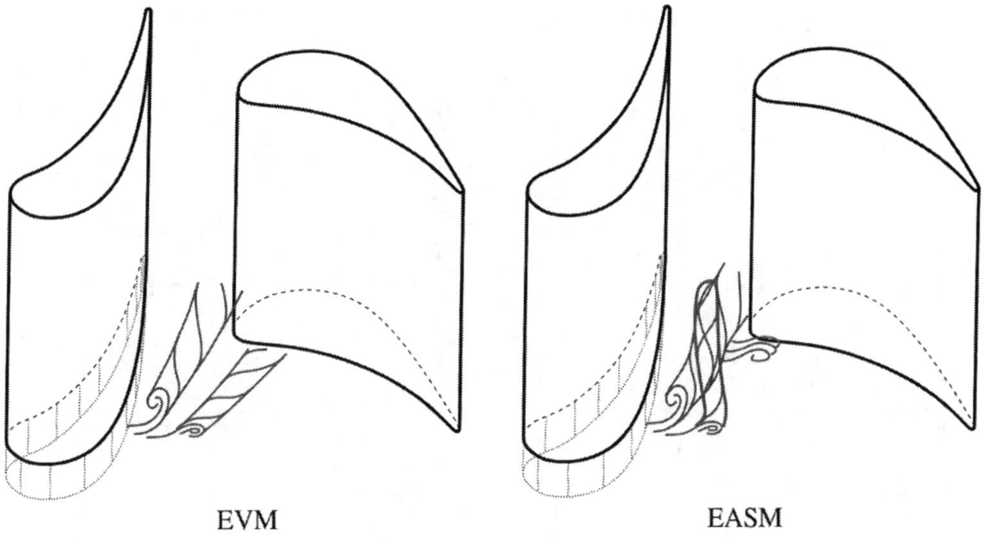

Figure 5. Scheme of formation of the secondary vortices behind the radial gap under the stator blade.

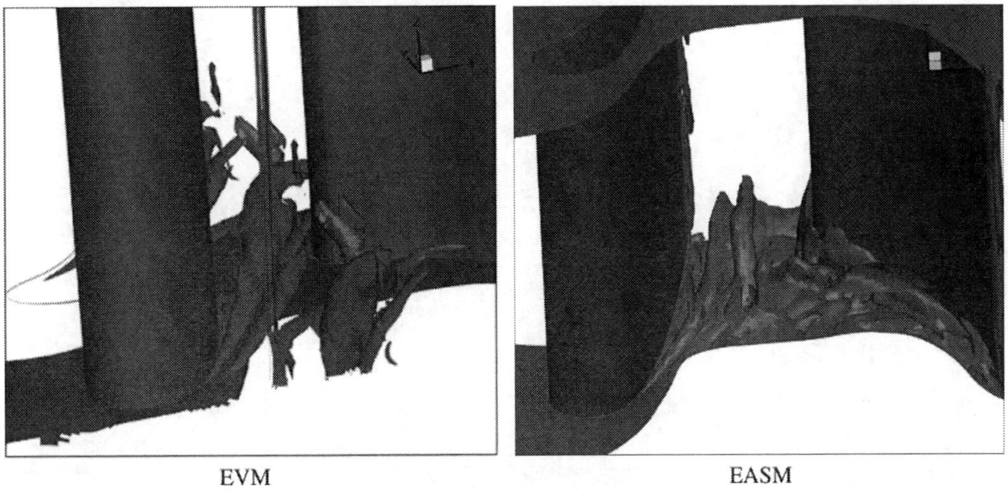

Figure 6. Isosurface of $(|\Omega|^2 - |S|^2)$.

cross section at 10 percent of span lies in region of secondary vortices generated behind the radial gap under the stator blade, while the cross section at 90 percent of span lies in region of secondary vortices generated behind the radial gap above the rotor blade. From figure 7 it is clear that EVM model predicts thinner wakes with lower production of the turbulent energy than EASM model. On the other hand the EVM model predicts higher production rate of the turbulent energy in regions of the secondary vortices and also in region around the stagnation poin at the leading edge of the rotor blade. This difference between both turbulent models is due to formulation of the turbulent viscosity, where in case of the EASM model it depends on the second invariants of the vorticity and strain tensors via variable parameter C_μ. Another reason is in different value of the diffusion coefficient: $\sigma_k = 0.5$ for the EVM model and $\sigma_k = 1$ for the EASM model. It means that the turbulend diffusion of the turbulent enery is enhanced in case of the EASM model.

Figures 8 and 9 show fields of the absolute Mach number and the angle of the velocity vector from the axial direction in tangential plane.

7 Conclusion

This paper continues the works [2–4]. The nonlinear explicit algebraic Reynolds stress model including anisotropy and curvature correction was implemented into the in-house numerical code. Results of the nonlinear EASM model were compared with previous results of the linear EVM model. It was shown that the anisotropy part of the Reynolds stress tensor influences formation and in-

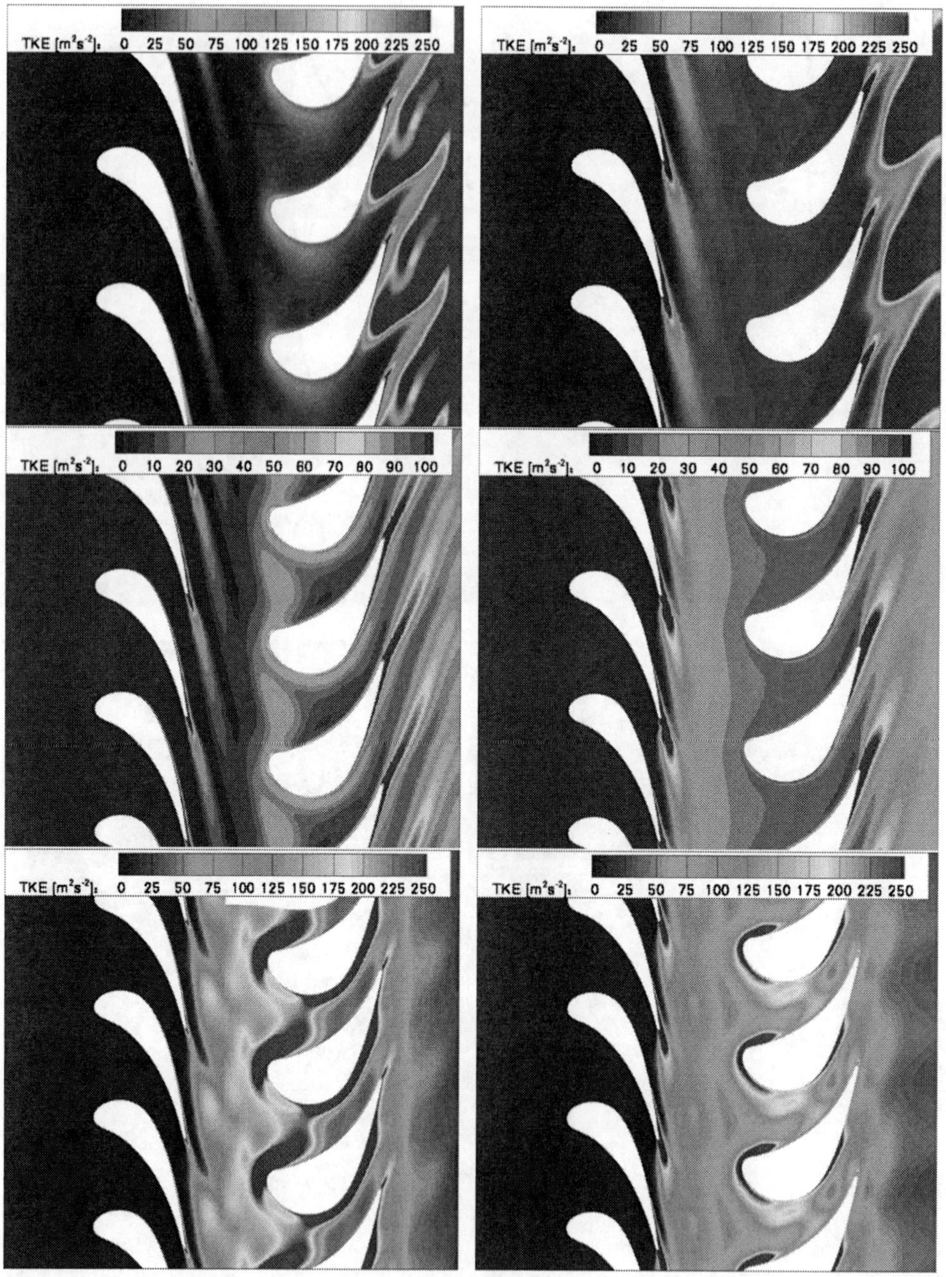

Figure 7. Field of the turbulent kinetic energy. Left column – EVM model, right column – EASM model, top – 90% of span, middle – 50_ of span, bottom – 10% of span.

teraction of the secondary vortices which are generated behind the radial gap under the hub-end of the stator blade.

Acknowledgement

This work was supported by the grant project TA02021336 of the Technology Agency of the Czech Republic and by the Long-term Framework Advancement Plan provided by the Ministry of Industry and Trade of the Czech Republic.

References

1. http://www.isvav.cz/projectDetail.do?rowId=TA02021336
2. P. Straka, EPJ Web of Conf., **67**, 02112 (2014)
3. P. Straka, M. Němec, T. Jelínek, EPJ Web of Conf., **92**, 02088 (2015)
4. P. Straka, M. Němec, J. Appl. Mechanics and Materials (in publishing process)

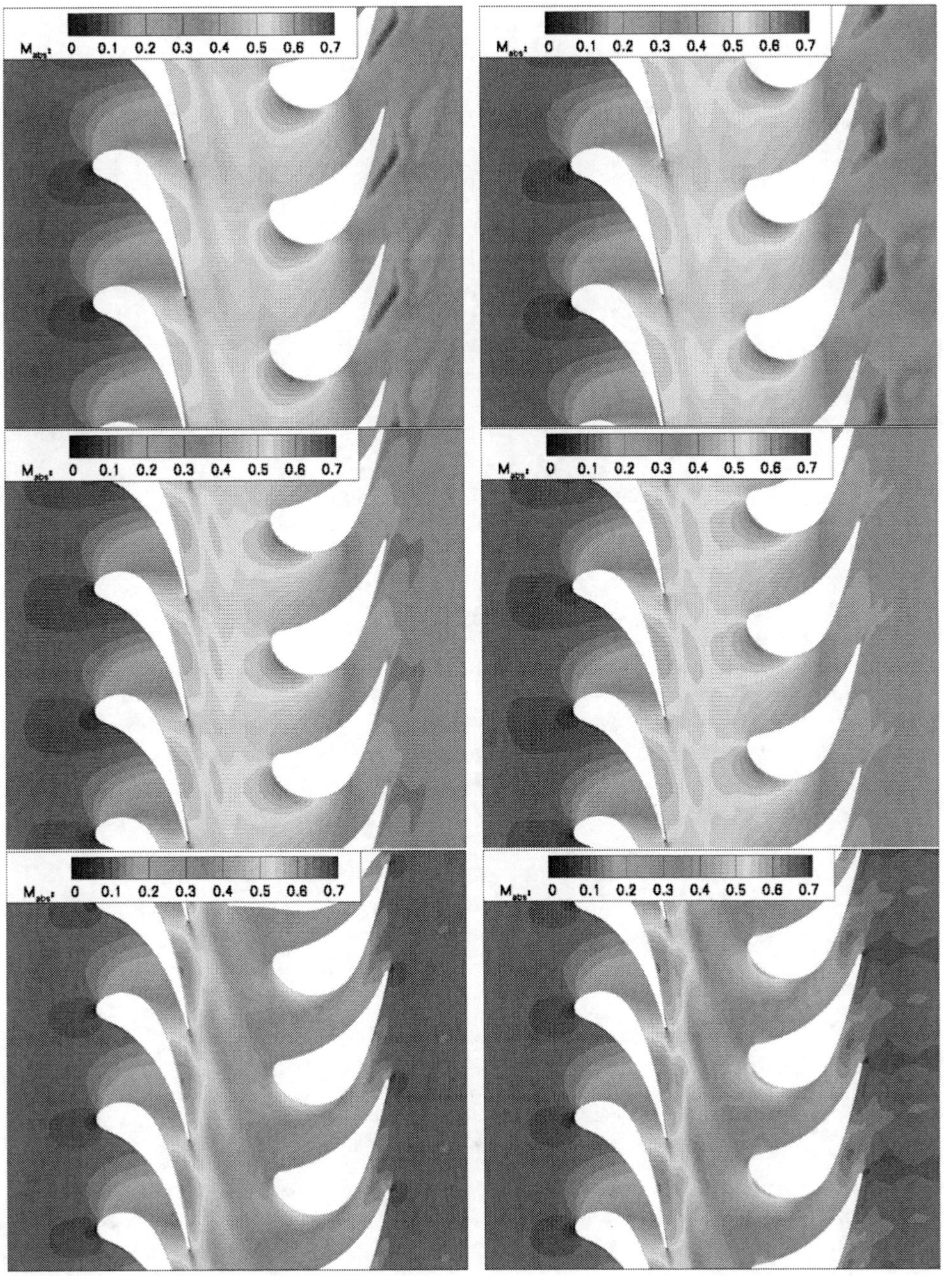

Figure 8. Field of the absolute Mach number. Left column – EVM model, right column – EASM model, top – 90% of span, middle – 50_ of span, bottom – 10% of span.

5. P. Straka, Proc. Conf. Exp. Fluid Mech. 979-987, (2011)
6. P. Straka, Technical Report VZLÚ R-4910, Prague, (2010) (in Czech)
7. C. L. Rumsey, T. B. Gatski, J. of Aircraft, **38**, No. 5, 904-910, (2001)
8. A. Hellsten, AIAA J., **40**, 1909-1911, (2002).
9. J. C. Kok, AIAA Journal **38**, No. 7, (2000)
10. S. Wallin, A. Hellsten, M. Schatz, T. Rung, D. Peshkin, A. V. Johansson, In 3rd Int. Symp. on Turbulence and Shear Flow Phenomena, Sendai, Japan, (2003)

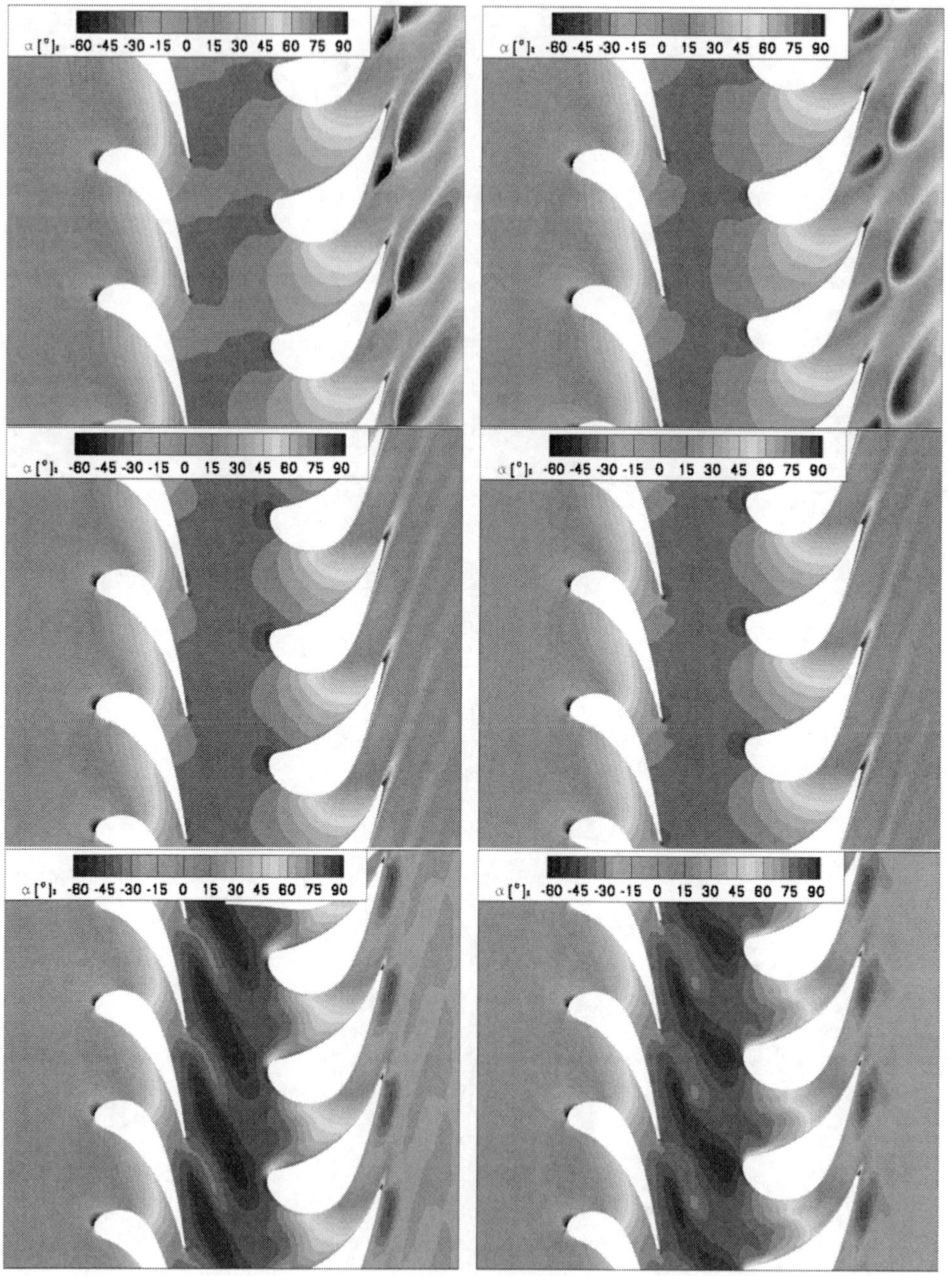

Figure 9. Angle of the velocity vector from the axial direction in tangential plane. Left column – EVM model, right column – EASM model, top – 90% of span, middle – 50_ of span, bottom – 10% of span.

EPJ Web of Conferences **114**, 02116 (2016)
DOI: 10.1051/epjconf/201611402116

An experimental investigation of flow around a vehicle passing through a tornado

Masahiro Suzuki[1,a], Kouhei Obara[1] and Nobuyuki Okura[1]

[1]*Meijo University, Department of Vehicle and Mechanical Engineering, 468-8502 Nagoya, Japan*

Abstract. Flow around a vehicle running through a tornado was investigated experimentally. A tornado simulator was developed to generate a tornado-like swirl flow. PIV study confirmed that the simulator generates two-celled vortices which are observed in the natural tornadoes. A moving test rig was developed to run a 1/40 scaled train-shaped model vehicle under the tornado simulator. The car contained pressure sensors, a data logger with an AD converter to measure unsteady surface pressures during its run through the swirling flow. Aerodynamic forces acting on the vehicle were estimated from the pressure data. The results show that the aerodynamic forces change its magnitude and direction depending on the position of the car in the swirling flow. The asymmetry of the forces about the vortex centre suggests the vehicle itself may deform the flow field.

1 Introduction

The tornado is a violent meteorological phenomenon. It sometimes damages infrastructures severely. It rarely hits moving ground vehicles. However, once it occurs it would bring a huge damage. Actually, several train-overturned accidents with casualties by tornadoes have been reported in Japan [1, 2].

The tornado has been studied by means of field observation, analytical approach and computational simulation in terms of meteorology and civil engineering. Meteorologists have tried to clarify the structure of the tornado and its mechanism, while civil engineers have managed to reduce the damage of the infrastructure. Recently, Hann et al. have developed a tornado simulator. It generates tornado-like vortices to quantify aerodynamic forces on civil engineering structures [3].

The crosswind stability of the ground vehicles has been vigorously studied in the past. Those works mainly concern steady/unsteady winds coming from one direction. The tornado like swirling flow has been barely investigated in spite of its risk to our society.

Therefore we have been investigated the effect of the tornado on the moving ground vehicle. In this paper, we describe our experimental approach by use of a tornado simulator based on Hann's study.

2 Method

Our experimental setup was composed of the tornado simulator and a moving model rig. Figure 1 shows the setup. The tornado simulator had a fan, an inner and an outer duct, 18 guide vanes and a stage (Fig. 2). The fan produced an upward flow in the inner duct. Then the flow passed between the guide vanes. The guide vanes that were attached at an angle created a twisted downward flow in the outer duct and a tornado-like swirling flow on the stage. The swirl ratio, S, is an

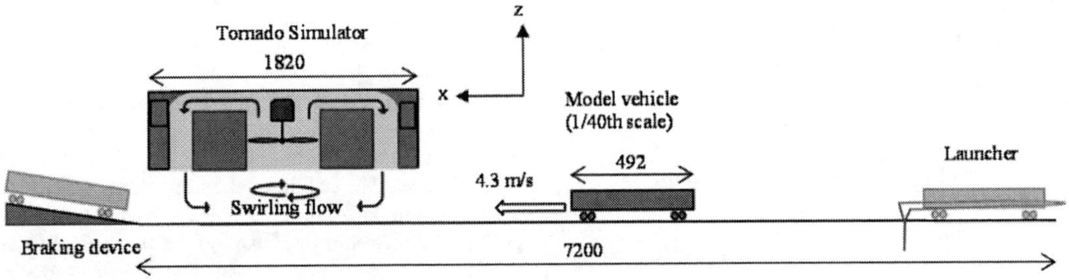

Figure 1. Experimental setup (unit: mm).

[a] Corresponding author: msuzuki@meijo-u.ac.jp

important parameter and has a close relation with the flow structure of the tornado. It is the ratio of the angular momentum to the updraft momentum and is defined as follows:

$$S = \frac{r}{2H}\tan\theta \quad (1)$$

Here, r is the radius of the inner duct, H is the height of the convergence layer (see Fig. 2(b)) and θ is the guide vane angle. In our experiment, the swirl ratio can be adjusted by changing the guide vane angle. The guide vane angle was set at 30°, 40°, 50° and 60°.

The moving model rig had a straight track, a model vehicle, a launcher and a breaking device. The track was placed in such a way the vehicle passed through the center of the vortex. The model vehicle had a train-like shape with a scale of 1/40 (Fig. 3). The vehicle had 72 pressure ports on its surface. Pressure sensors and data logger with AD converter were loaded inside the vehicle. The launcher accelerated the vehicle to 4.3 m/s by using an elastic force provided by rubber bands. A motor inside the vehicle rotated wheels to keep the speed constant. The braking device stopped the vehicle softly and safely.

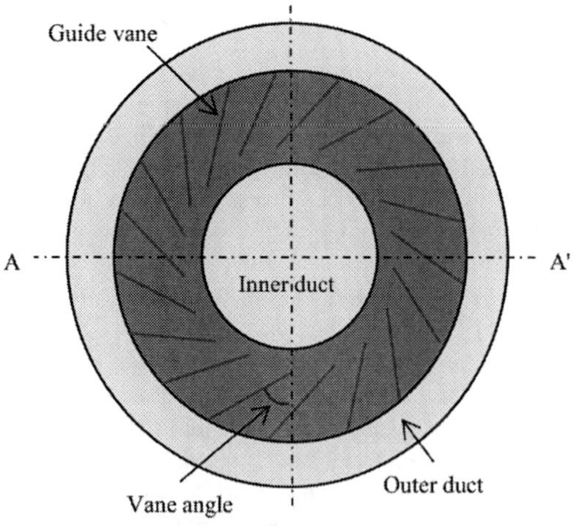

(a) Top view.

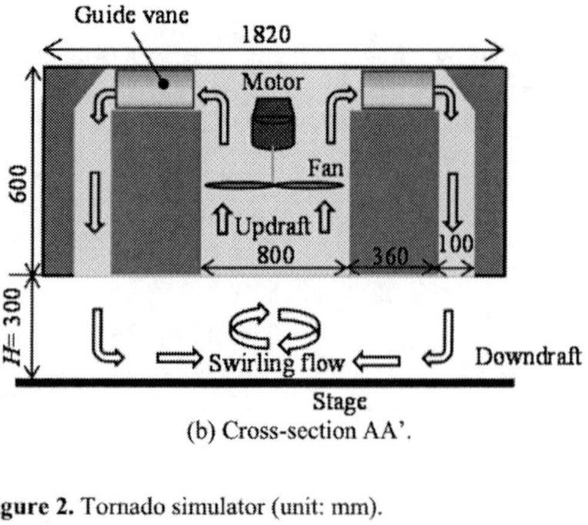

(b) Cross-section AA'.

Figure 2. Tornado simulator (unit: mm).

The flow field generated by the tornado simulator was measured by PIV. Note there were no vehicles in this measurement. The cross-sectional velocity distributions on the horizontal plane at z = 0.54R (the centre height of the vehicle body) and the vertical planes at the centre of the swirling flow were examined. The pressure distribution on the stage was also measured.

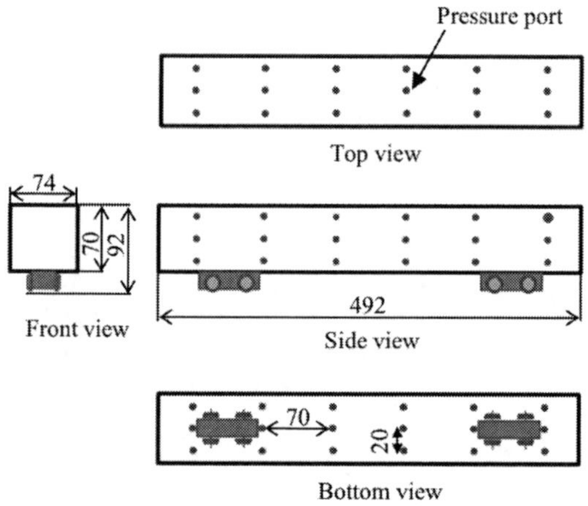

Figure 3. Model vehicle (unit: mm).

3 Results

3.1 Flow field generated by the tornado simulator

Figure 4 shows an example of the velocity vectors obtained by the PIV measurement. The swirl ratio, S, is 0.79 in this case. Colours of the vectors indicate the magnitude of the velocity on each plane. The velocity vectors on the horizontal plane clearly show the swirling flow field. The velocity vectors on the vertical plane indicate that the flow coming from the far field near the stage moves up while a weak downdraft appears in the centre area. This is flow structure is called a two-celled vortex [3]. By using the velocity field obtained by PIV, the maximum tangential velocity, *Vmax*, and the radius of vortex core, R, at which the tangential velocity becomes maximum were obtained in each case of the vane angle (Table 1). As the Swirl ratio increases, the radius of vortex core increases. The tangential velocity also becomes larger at $\theta < 60°$.

Figure 5 shows normalized tangential velocity profiles at the horizontal plane. The horizontal axis indicates normalized distance from the vortex centre. Doppler radar data from full-scale tornados at Spencer, South Dakota and Mulhall, Oklahoma [3] are also plotted in the figure. The height of the horizontal plane of these full-scale tornado data is z = 0.52R, while ours is z = 0.54R. A black thin line displays the theoretical value of the Rankin vortex model, which a widely-used simple model of the tornado [4]. The model defines that a tangential velocity in the inner region is proportional to

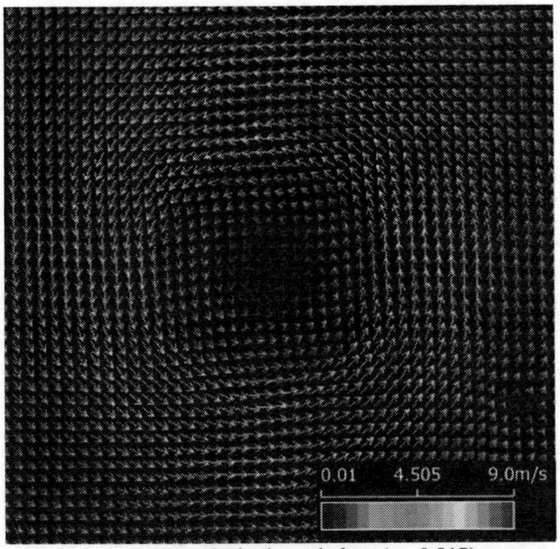

(a) Velocity vectors on the horizontal plane ($z = 0.54R$).

(b) Velocity vectors on the vertical plane at the vortex centre.

Figure 4. Velocity vectors of the vortex.

Table 1. Maximum tangential velocity and radius of vortex core.

θ (degree)	S	*Vmax* (m/s)	*R* (mm)
30	0.38	6.1	74
40	0.56	6.8	89
50	0.79	7.4	96
60	1.15	7.0	115

the distance from the centre, x, while it is inversely proportional to the distance in the outer region as follows:

$$V(x) = \begin{cases} V_{max}\frac{|x|}{R} & , |x| \le R \\ V_{max}\frac{R}{|x|} & , |x| > R \end{cases} \quad (2)$$

While our results disagree with the Rankin vortex mode at $r/R > 1$, they agree well with the radar data irrespective of the vane angle.

Figure 6 illustrates the pressure distribution on the stage in the case of $\theta = 50°$. A line indicates the Rankin vortex model. Pressure distribution of the Rankin vortex model is as follows:

$$p(x) = \begin{cases} \frac{1}{2}p_{min}\left[2 - \frac{x^2}{R^2}\right] & , |x| \le R \\ \frac{1}{2}p_{min}\frac{R^2}{x^2} & , |x| > R \end{cases} \quad (3)$$

Here, p_{min} is the minimum value of the pressure. Our experimental result agrees well with the Rankin vortex model except near the vortex centre. Hann et al. also observed this flattened profile near the vortex centre [3]. They suggested that this is due to the downdraft near the vortex centre that was also observed in our PIV measurement.

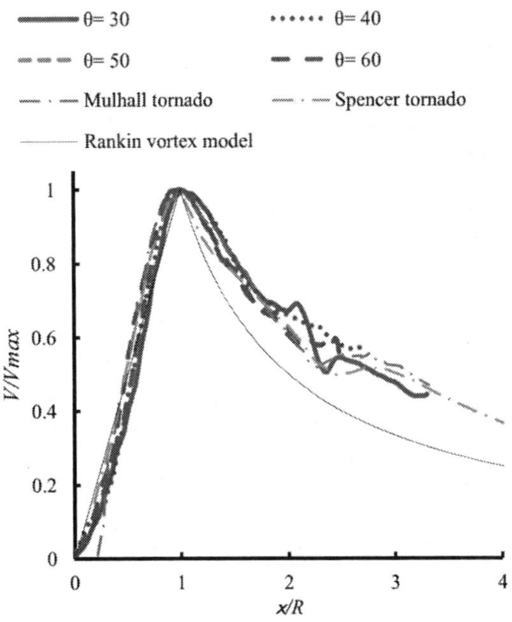

Figure 5. Tangential velocity profiles.

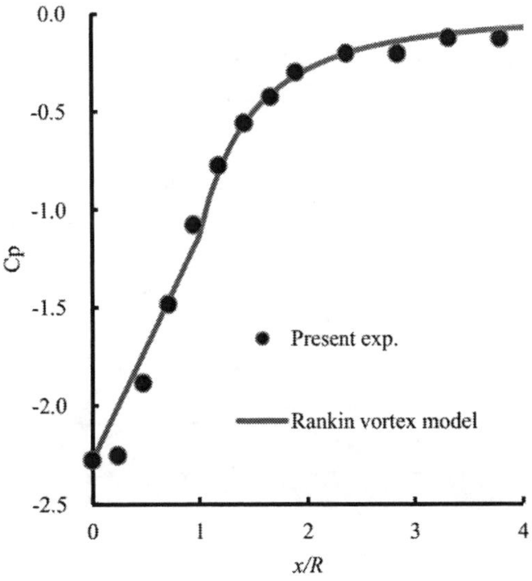

Figure 6. Pressure distribution on the stage ($\theta = 50$).

3.2 Aerodynamic forces acting on the moving vehicle

When the measurement of surface pressure on the vehicle running in the tornado was conducted, the vane angle of the tornado simulator was set at 50 degrees. The corresponding Reynolds number based on the maximum tangential velocity and the vehicle length was 2.2×10^5.

The aerodynamic forces were obtained by using the ensemble averaged surface pressure data of 30 runs. Figure 7 explains the definition of the aerodynamic forces and moment acting on the vehicle. Figure 8 presents time histories of the side coefficient, Cs, the lift force coefficient, Cl, and the yawing moment coefficient, Cy, of the moving vehicle during its run through the swirling flow generated by the tornado simulator. The horizontal axis indicates the position of the vehicle centre. The each coefficient was calculated by using the maximum tangential velocity, *Vmax*. Due to the nature of the swirling flow, the shape of the side force should be symmetric with respect to the origin, and those of the lift force and yawing moment should be symmetric with respect to the vertical axis. As seen the figure, our results show this tendency to a certain degree. However, the side force becomes positive and the yawing moment become maximum before the vehicle reaches the vortex centre. The yawing moment does not decrease to negative values at $x/R > 0$ while it does at $x/R < 0$. The lift force has a second peak at $x/R = -2$, but it does not at $x/R = 2$. These suggest that the vehicle deforms the flow structure when it runs through the swirling flow.

4 Concluding remarks

The experimental method was developed to clarify the flow around the moving vehicle in the tornado. The tornado simulator and the moving model rig were developed to perform the experiment. The flow field generated by the tornado simulator was verified by comparing with the radar data of the full-scale tornado and the Rankin vortex model. The aerodynamic forces acting on the vehicle in the swirling flow were successfully obtained. The result suggests the vehicle affects the flow structure.

PIV measurements are planned to clarify the flow structure around the static/moving vehicle in the swirling flow.

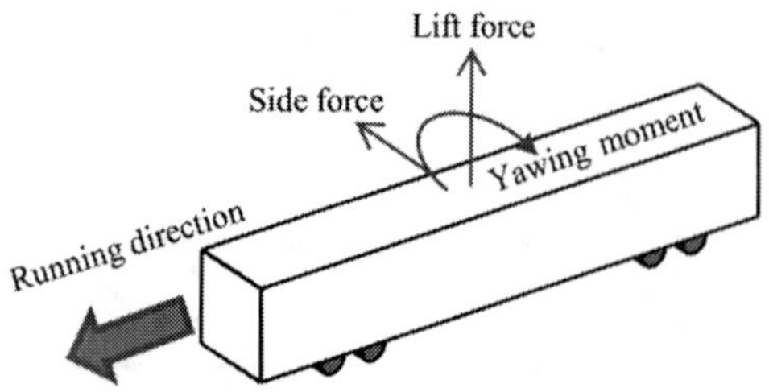

Figure 7. Definition of aerodynamic forces and yawing moment acting on the vehicle.

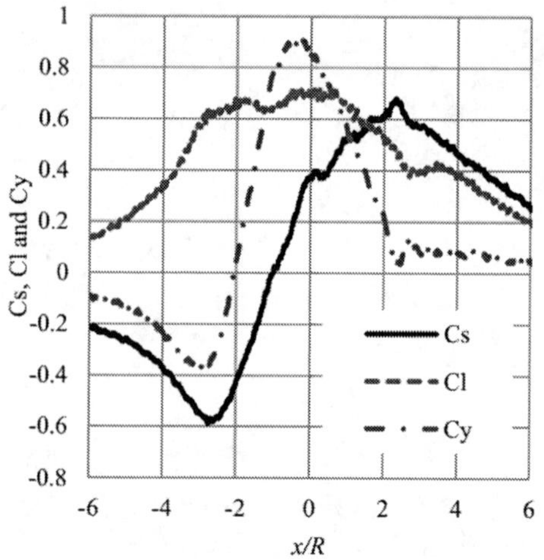

Figure 8. Ensemble averaged time histories of side and lift forces and yawing moment acting on the vehicle.

Acknowledgements

This work was supported by JSPS KAKEN Grant Number 26350491.

References

1. Aircraft and Railway Accidents Investigation Commission, Railway accident analysis report RA2008-4, (2008) (in Japanese)
2. Aircraft and Railway Accidents Investigation Commission, Railway accident analysis report RA2008-6, (2008) (in Japanese)
3. F. L. Hann, P. P. Sarkar and W. A. Gallus, Eng. Structures **30**, 1146-1159 (2008)
4. American Nuclear Society, An American national standard, ANSI/ANS-2.3-2011 (2011)

EPJ Web of Conferences **114**, 02117 (2016)
DOI: 10.1051/epjconf/201611402117

Finite element method application for turbulent and transitional flow

Petr Sváček

Czech Technical University in Prague, Faculty of Mechanical Engineering, Dep. of Technical Mathematics, Karlovo nám. 13, Praha 2, Czech Republic

Abstract. This paper is interested in numerical simulations of the interaction of the fluid flow with an airfoil. Particularly, the problem of the turbulent flow around the airfoil with elastic support is considered. The main attention is paid to the numerical approximation of the flow problem using the finite element approximations. The laminar - turbulence transition of the flow on the surface airfoil is considered. The chois of the transition model is discussed. The transition model based on the two equation $k-\omega$ turbulence model is used. The structure motion is described with the aid of two degrees of freedom. The motion of the computational domain is treated with the aid of the arbitrary Lagrangian-Eulerian method. Numerical results are shown.

1 Introduction

Recently the mathematical modelling and numerical approximation play important role in the engineering practice in the civil, aerospace and mechanical engineering (see e.g. [1], [2]). The mutual interactions of air flow and vibrating structure is the main subject of aero-elasticity, where the flow-induced vibrations may affect negatively the operation and stability of aircrafts, bladed machines, bridges, and many other structures. Typically, the main goal of aeroelasticity is the prediction of the bounds of the structure stability, see, e.g. the monographs [1], [3]. The mathematical simulation of fluid and structure interaction requires to consider the viscous, usually turbulent flow, changes of the flow domain in time, nonlinear behaviour of the elastic structure and to solve simultaneously the evolution systems for the fluid flow and for the oscillating structure. The changes fluid domain cannot be neglected and a moving mesh method ([4], [5]) must be employed.

The subject of our attention is the numerical analysis of the interaction of viscous turbulent flow with a vibrating airfoil, where the influence of the transition on the aeroelastic response of the airfoil is studied. In order to include the transition model into aeroelastic simulations and avoid using the non-local operations, the transition model based on two transport equations for intermittency and the onset momentum-thickness Reynolds number were used, see [6], [7]. In the present paper we are concerned with the numerical simulation of the aeroelastic problem of 2D viscous incompressible flow past a moving airfoil. The main attention is paid to the description of the application of the $k-\omega$ turbulence model (see [8], [9]) together with the transition model included, see [6].

2 Mathematical description

The mathematical formulation of the problem consists of the flow model, the structure model and the interface conditions.

2.1 Fluid flow

The fluid flow is described in the two-dimensional time dependent computational domain $\Omega_t \subset \mathbb{R}^2$ with the Lipschitz continuous boundary $\partial\Omega_t = \Gamma_D \cup \Gamma_O \cup \Gamma_{Wt}$. Here, Γ_D denotes the inlet part of the computational domain, Γ_O denotes the outlet part and Γ_{Wt} denotes the surface of the airfoil at time t. The Arbitrary Lagrangian-Eulerian (ALE) method is used to treat the moving domain, see [10], [11]. The differentiable ALE mapping $\mathcal{A} = \mathcal{A}(\xi, t) = \mathcal{A}_t(\xi)$ is assumed to be defined for all $t \in (0, T)$ and $\xi \in \Omega_0^{ref} = \Omega_0$. Then the domain velocity w_D reads $w_D(x,t) = \partial_t\mathcal{A}(\xi, t)$ for any $x = \mathcal{A}(\xi, t)$ and the derivative with respect to a fixed point of the reference configuration Ω_0^{ref} is called the ALE derivative denoted by $D^{\mathcal{A}}/Dt$, see [12].

The fluid motion in the domain Ω_t is modelled using the Navier-Stokes system of equations in Ω_t written in the ALE form ($i = 1, 2$)

$$\frac{D^{\mathcal{A}}u}{Dt} + ((u - w_D)\cdot\nabla)u = \nabla\cdot\sigma, \qquad \text{div}\, u = 0, \tag{1}$$

where $u = (u_1, u_2)$ is the fluid velocity vector, and the stress tensor σ with components σ_{ij} is given by

$$\sigma = -p\mathbb{I} + 2\nu_{eff}S(u),$$

where $S = S(u) = (\nabla u + \nabla^T u)/2$ is the symmetric part of the gradient with the components $S_{i,j} = (\frac{\partial u_i}{\partial x_j} + \frac{\partial u_j}{\partial x_i})/2$, p is the kinematic pressure (i.e., the pressure divided by the constant fluid density ρ), $\nu_{eff} = \nu + \gamma\nu_T$ is the effective kinematic viscosity, ν is the constant fluid kinematic viscosity, ν_T is a turbulent viscosity obtained by an additional model, see [8], and γ is the intermittency coefficient.

Further, the system (1) is equipped with an initial condition $u(x, 0) = u_0(x)$, $x \in \Omega_0$. and with the boundary conditions

a) $u = u_D$ on Γ_D,

b) $\boldsymbol{u} = \boldsymbol{w}_D$ on Γ_{Wt},

c) $-2\nu_{eff}\boldsymbol{S}(\boldsymbol{u})\cdot\boldsymbol{n} + p\boldsymbol{n} + \frac{1}{2}(\boldsymbol{u}\cdot\boldsymbol{n})^-\boldsymbol{u} = 0$ on Γ_O,

where $\alpha^- = \min(0,\alpha)$ denotes the negative part of the number $\alpha \in \mathbb{R}$ and $\boldsymbol{n} = (n_1, n_2)$ denotes the unit outward normal to $\partial\Omega_t$.

2.2 Structure motion

The flow is interacting with a flexibly supported airfoil, which is allowed to be vertically displaced by h (downwards positive) and rotated by the angle α (clockwise positive). Its motion is then described by the nonlinear equations of motion

$$m\ddot{h} + S_\alpha\ddot{\alpha}\cos\alpha - S_\alpha\dot{\alpha}^2\sin\alpha + k_h h = -L(t), \tag{2}$$
$$S_\alpha\ddot{h}\cos\alpha + I_\alpha\ddot{\alpha} + +k_\alpha\alpha = M(t).$$

where m is the mass of the airfoil, S_α is the static moment around the elastic axis (EA), and I_α is the inertia moment around EA, see [12]. The parameters k_h and k_α denote the bending and torsional spring stiffness coefficients, respectively.

2.3 Interface conditions

The aerodynamical lift force $L(t)$ is given by

$$L = -l\int_{\Gamma_{Wt}}\sum_{j=1}^{2}\sigma_{2j}n_j\,dS, \tag{3}$$

and the aerodynamical torsional moment $M(t)$ (clockwise positive) is given by

$$M = l\int_{\Gamma_{Wt}}\sum_{j=1}^{2}\sigma_{ij}n_j r_i^{\mathrm{ort}}\,dS, \tag{4}$$

where l denotes the depth of the airfoil section, δ_{ij} is the Kronecker's delta, $r_1^{\mathrm{ort}} = -(x_2 - x_2^{\mathrm{EA}})$, $r_2^{\mathrm{ort}} = x_1 - x_1^{\mathrm{EA}}$, and $x^{\mathrm{EA}} = (x_1^{\mathrm{EA}}, x_2^{\mathrm{EA}})$ is the position of EA of the airfoil at the time instant t.

3 Numerical approximation

In order to discretize the problem in time the equidistant partition $t_k = k\Delta t$ of the time interval I is considered with a time step $\Delta t > 0$. We denote the approximations $\boldsymbol{u}^k \approx \boldsymbol{u}(\cdot, t_k)$ and $p^k \approx p(\cdot, t_k)$. Moreover we approximate the domain velocity $\boldsymbol{w}_D$ at time level t_k by $\boldsymbol{w}_D^k$ and focus on the description of the discretization at an arbitrary fixed time instant $t = t_{n+1}$. For the sake of simplicity we shall omit the subscripts t or t_{n+1} in what follows. We shall consider all the function spaces $\boldsymbol{X}, \boldsymbol{W}, Q$ defined for the time instant $t = t_{n+1}$ on the domain $\Omega := \Omega_{t_{n+1}}$. Then the ALE time derivative in the weak formulation of (1) is approximated at the time $t = t_{n+1}$ by the second order backward difference formula, i.e.

$$\frac{D^{\mathcal{A}}\boldsymbol{u}}{Dt}\Big|_{t=t_{n+1}} \approx \frac{3\boldsymbol{u}^{n+1} - 4\tilde{\boldsymbol{u}}^n + \tilde{\boldsymbol{u}}^{n-1}}{2\Delta t},$$

where by $\tilde{\boldsymbol{u}}^i = \boldsymbol{u}^i \circ \mathcal{A}_{t_i} \circ \mathcal{A}_{t_{n+1}}^{-1}$ the transformation of $\boldsymbol{u}^i$ from Ω_{t_i} on Ω for $i = n-1$ and $i = n$ is denoted.

In order to spatially discretize the problem (1) by the finite element method, the spaces $\boldsymbol{X}, \boldsymbol{W}$ and Q are approximated by finite element subspaces $\boldsymbol{X}_\Delta, \boldsymbol{W}_\Delta$ and Q_Δ, respectively. The Taylor-Hood family of finite element spaces defined over a triangulation $\mathcal{T}_\Delta$ of the computational domain $\Omega = \Omega_{t_{n+1}}$ are used, i.e. the continuous piecewise quadratic velocities and the continuous piecewise linear pressures are used. Moreover for the involved high Reynolds numbers the fully stabilized scheme is used, which consists of streamline-upwind/Petrov-Galerkin(SUPG) and pressure-stabilizing/Petrov-Galerkin(PSPG) stabilization combined with the div-div stabilization, see [13]. The non-linear problem is then solved using the Oseen linearization process, where the solution of each of the linear problems is performed using the efficient direct solver, see [14].

4 Turbulent flow

For modelling of the turbulent flow with the transition the Menter's SST $k-\omega$ turbulence model, see [15] is used, with the $\gamma - \overline{Re}_{\theta t}$ transition model introduced by [6]. The viscosity coefficient ν in the equation (1) is replaced by the effective viscosity $\nu_{\mathrm{eff}} = \nu + \nu_T$, and the turbulent viscosity ν_T is modelled using the turbulent kinetic energy $k = k(x,t)$ and the turbulent specific dissipation rate $\omega = \omega(x,t)$ satisfying for any $t \in (0,T)$ in Ω_t equations

$$\begin{aligned}\frac{D^{\mathcal{A}}k}{Dt} + ((\boldsymbol{u}-\boldsymbol{w}_D)\cdot\nabla)k &= \gamma_{\mathrm{eff}}P_k - \beta^*\omega k\overline{\gamma_{\mathrm{eff}}} + \nabla\cdot(\varepsilon_k\nabla k),\\ \frac{D^{\mathcal{A}}\omega}{Dt} + ((\boldsymbol{u}-\boldsymbol{w}_D)\cdot\nabla)\omega &= P_\omega - \beta\omega^2 + \nabla\cdot(\varepsilon_\omega\nabla\omega) + C_D,\end{aligned} \tag{5}$$

where $\varepsilon_k = \nu + \sigma_k\nu_T$, $\varepsilon_\omega = \nu + \sigma_\omega\nu_T$ and the source terms P_k, P_ω are defined by

$$P_k = \nu_T S,\ S = \boldsymbol{S}(\boldsymbol{u}) : \boldsymbol{S}(\boldsymbol{u}),\ P_\omega = \frac{\alpha_\omega\omega}{k}P_k,$$

and C_D reads

$$C_D = 2(1-F_1)\frac{\sigma_D}{\omega}(\nabla k\cdot\nabla\omega)^+.$$

The turbulent viscosity is then given by

$$\nu_T = \min\left(\frac{k}{\max(\omega, SF_2/a_1)}, \frac{0.6k}{\sqrt{3}S}\right). \tag{6}$$

The closure coefficients $\beta, \beta^*, \sigma_k, \sigma_\omega, \alpha_\omega$ are chosen according to [15], i.e. $a_1 = 0.3, \kappa = 0.41, \beta^* = 0.09, \sigma_D = 0.5$ and the coefficients $\sigma_k, \sigma_\omega, \beta$ and α_ω are calculated using the blending function F_1 as $\phi = F_1\phi_1 + (1-F_1)\phi_2$, where $\sigma_{k1} = 0.85, \sigma_{\omega 1} = 0.65, \beta_1 = 0.075, \sigma_{k2} = 1, \sigma_{\omega 2} = 0.856$, $\beta_2 = 0.0828$ and $\alpha_\omega = \beta/\beta^* - \sigma_\omega\kappa^2/\sqrt{\beta^*}$. For the sake of brevity, the blending functions F_1, F_2 are not specified here, see e.g., [15]. In order to capture the transition, the modification of the SST model in the production and the destruction terms of (5) is used, using the effective intermittency γ_{eff}, $\overline{\gamma}_{\mathrm{eff}} = \max(\min(\gamma_{\mathrm{eff}}, 1), 0.1)$.

The effective intermittency is then modelled using the equation for the intermittency coefficient γ written in the ALE form

$$\frac{D^{\mathcal{A}}\gamma}{Dt} + ((\boldsymbol{u}-\boldsymbol{w}_D)\cdot\nabla)\gamma = P_\gamma - E_\gamma + \nabla\cdot((\nu + \nu_T/\sigma_f)\nabla\gamma), \tag{7}$$

where P_γ and E_γ are the transition source and destruction terms given by $P_\gamma = F_{\text{length}} c_{a1} S \sqrt{\gamma F_{\text{onset}}}(1 - c_{e1}\gamma)$, and $E_\gamma = c_{a2}\Omega\gamma F_{\text{turb}}(c_{e2}\gamma - 1)$, where S and Ω are the strain rate and vorticity magnitudes, respectively. Further, $F_{\text{onset}} = (F_{\text{onset2}} - F_{\text{onset3}})^+$, $F_{\text{onset1}} = Re_V/(2.193 Re_{\theta c})$, $Re_V = y^2 S/\nu$ $R_T = k/(\nu\omega)$, $F_{\text{turb}} = e^{-(R_T/4)^4}$, y denotes the wall distance, and $Re_{\theta t}$ is the transition Reynolds number and

$$F_{\text{onset2}} = \min\left(\max\left(F_{\text{onset1}}, F^4_{\text{onset1}}\right), 2\right),$$

$$F_{\text{onset3}} = \max\left(1 - \left(\frac{R_T}{2.5}\right)^3, 0\right).$$

The following constants for the intermittency equation were used $c_{e1} = 1$, $c_{a1} = 2$, $c_{e2} = 50$, $c_{a2} = 0.06$, $\sigma_f = 1$. Further, $Re_{\theta c}$ is the critical Reynolds number given by an empirical correlation, and another empirical correlation is used for the function F_{length}, which controls the length of the transition region. The correlations are based on newly defined transported unknown $\overline{Re}_{\theta t}$ governed by the equation written in the ALE form

$$\frac{D^{\mathcal{A}}\overline{Re}_{\theta t}}{Dt} + ((\boldsymbol{u} - \boldsymbol{w}_D)\cdot\nabla)\overline{Re}_{\theta t} = P_{\theta t} + \nabla\cdot(\sigma_{\theta t}\nu_{\text{eff}}\nabla\overline{Re}_{\theta t}), \quad (8)$$

where the source term $P_{\theta t} = c_{\theta t}\frac{\rho}{t_\infty}\left(Re_{\theta t} - \overline{Re}_{\theta t}\right)(1 - F_{\theta t})$, where $c_{\theta t} = 0.03$, $\sigma_{\theta t} = 2$, $t_\infty = 500\nu/U^2$ is the time scale, U is the magnitude of velocity $U = \|\boldsymbol{u}\|_2$ and the blending function $F_{\theta t}$ is defined as

$$F_{\theta t} = \min\left(1, \max\left(F_{\text{wake}} e^{-(y/\delta)^4}, 1 - \left(\frac{\gamma - 1/c_{e2}}{1 - 1/c_{e2}}\right)\right)\right),$$

with $\delta = (375\Omega y\theta)/U$, $\theta = \overline{Re}_{\theta t}\nu/U$, $F_{\text{wake}} = e^{-(Re_\omega/10^5)^2}$, and $Re_\omega = \omega y^2/\nu$. The right hand side includes the Reynolds number $Re_{\theta t} = \theta_t\nu/U$ given by the empirical correlation, see [16]. The equations (7) and (8) are equipped with the Dirichlet boundary conditions ($\gamma = 1$, $\overline{Re}_{\theta t}$ obtained by the previously specified correlation) at the inlet and Neumann boundary conditions at the outlet and on the airfoil surface ($\Gamma_O \cup \Gamma_{Wt}$). For the turbulence and transition models the SUPG stabilization is used and the resulting system of linear equations is solved numerically using the direct solver.

5 Numerical results

In the paper we present the numerical simulation of the coupled problem. The inlet velocity is taken as $U_\infty = 40m/s$ (post-critical regime), the structural parameters are taken same as in [12]. In that case the divergence type of instability occurs (with further increase of the far field velocity the flutter instability should be observed). Here, the we investigated the response of the system depending on the different values of the initial values of α and h. The comparison of the aeroelastic response is shown in Figures 1-2, where the divergence type of instability can be observed. The instability type changes with the further increase of the inlet velocity in Figure 3, where flutter type of instability was observed.

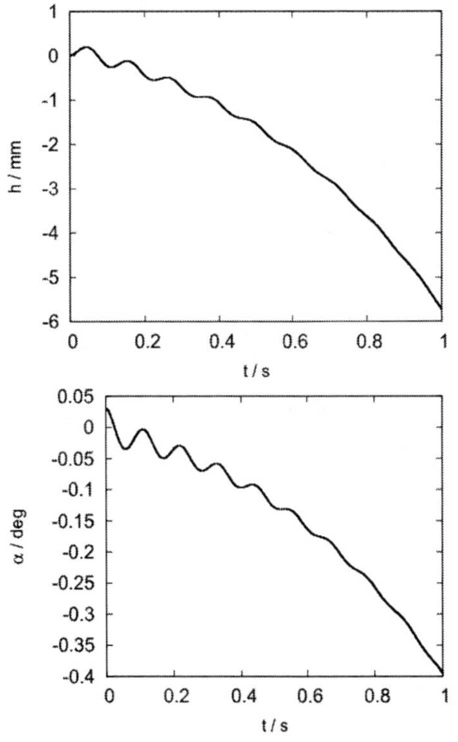

Fig. 1. The aeroelastic response of the airfoil NACA 0012. The inlet velocity is taken as $U_\infty = 40m/s$ and the instability of type divergence can be observed. The initial condition is $\alpha(0) = 0.03^o$ and $h(0) = 0$ mm.

6 Conclusion

The present paper describes the solution of FSI problem of flow induced vibrations of an elastically supported airfoil in the post-critical regime. The problem was discretized in time and in space using the finite element method. The main attention was paid to the the description of the transition model, originally proposed by [6]. The numerical results were shown.

This work was supported by the grant No. P101/12/1271 of the Czech Science Foundation.

References

1. E.H. Dowell, R.N. Clark, *A modern course in aeroelasticity*, Solid mechanics and its applications (Kluwer Academic Publishers, Dordrecht, Boston, 2004), ISBN 1-402-02039-2
2. M. Paidoussis, *Fluid-Structure Interactions. Slender structures and axial flow* (Academic Press, Elsevier, 2013), 2nd Edition
3. E. Naudasher, D. Rockwell, *Flow-Induced Vibrations* (A.A. Balkema, Rotterdam, 1994)
4. C. Farhat, M. Lesoinne, N. Maman, International Journal for Numerical Methods in Fluids **21**, 807 (1995)
5. P. Le Tallec, J. Mouro, Computer Methods in Applied Mechanics and Engineering **190**, 3039 (2001)

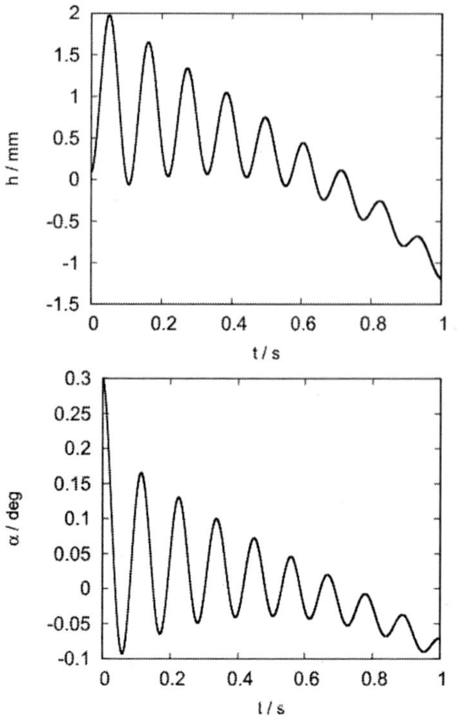

Fig. 2. The aeroelastic response of the airfoil NACA 0012. The inlet velocity is taken as $U_\infty = 40m/s$ and the instability of type divergence can be observed. The initial condition is $\alpha(0) = 0.03^o$ and $h(0) = 0$ mm.

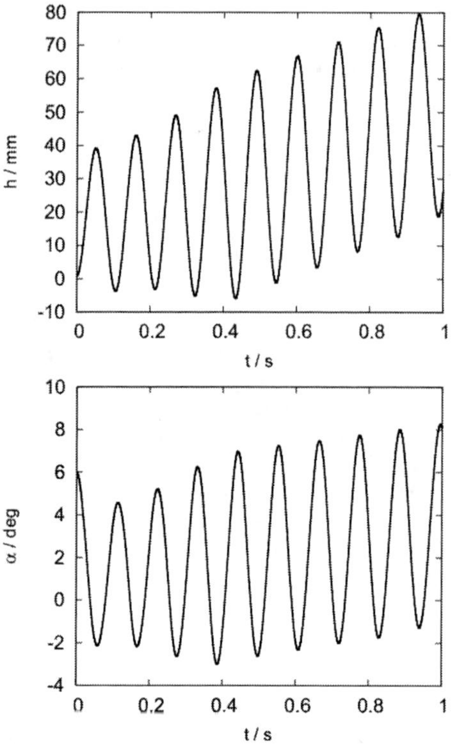

Fig. 3. The aeroelastic response of the airfoil NACA 0012. The inlet velocity is taken as $U_\infty = 40m/s$ and the instability of type divergence can be observed. The initial condition is $\alpha(0) = 6^o$ and $h(0) = 0$ mm.

6. F. Menter, R. Langtry, S. Völker, Flow, Turbulence and Combustion **77**, 277 (2006)
7. Y. Suzen, P. Huang, L.S. Hultgren, D.E. Ashpis, Journal of turbomachinery **125**, 455 (2003)
8. D.C. Wilcox, *Turbulence Modeling for CFD* (DCW Industries, 1993)
9. J.C. Kok, Tech. rep., National Aerospace Laboratory NLR (1999)
10. F. Nobile, Ph.D. thesis, Ecole Polytechnique Federale de Lausanne (2001)
11. T. Nomura, T.J.R. Hughes, Computer Methods in Applied Mechanics and Engineering **95**, 115 (1992)
12. P. Sváček, M. Feistauer, J. Horáček, Journal of Fluids and Structures **23**, 391 (2007)
13. T. Gelhard, G. Lube, M.A. Olshanskii, J.H. Starcke, Journal of Computational and Applied Mathematics **177**, 243 (2005)
14. T.A. Davis, I.S. Duff, ACM Transactions on Mathematical Software **25**, 1 (1999)
15. F.R. Menter, AIAA Journal **32**, 1598 (1994)
16. R.B. Langtry, F.R. Menter, AIAA JOURNAL **47**, 2894 (2009)

EPJ Web of Conferences **114**, 02118 (2016)
DOI: 10.1051/epjconf/201611402118

Numerical simulation of fluid-structure interactions with stabilized finite element method

Petr Sváček

Czech Technical University in Prague, Faculty of Mechanical Engineering, Dep. of Technical Mathematics, Karlovo nám. 13, Praha 2, Czech Republic

Abstract. This paper is interested to the interactions of the incompressible flow with a flexibly supported airfoil. The bending and the torsion modes are considered. The problem is mathematically described. The numerical method is based on the finite element method. A combination of the streamline-upwind/Petrov-Galerkin and pressure stabilizing/Petrov-Galerkin method is used for the stabilization of the finite element method. The numerical results for a three-dimensional problem of flow over an airfoil are shown.

1 Introduction

There are various technical or scientific applications, where an interaction of flowing fluids and vibrating structures can play an important role, see, e.g. the monographs [1], [2]. The mathematical simulation of fluid and structure interaction is a challenging problem, it requires to consider the viscous, usually turbulent flow, changes of the flow domain in time, nonlinear behaviour of the elastic structure. Moreover, the coupled system for the fluid flow and for the oscillating structure needs to solved simultaneously. The changes of the fluid domain cannot be neglected and the methods with moving meshes must be employed, see e.g. ([3], [4]) .

In this paper the numerical analysis of an interaction of the incompressible viscous flow with a vibrating airfoil is considered. The numerical analysis of 2D interactions of the incompressible flow with an airfoil was published in [5], [6]. The method allowing the solution of large amplitude flow-induced vibrations of an airfoil with 3 degrees of freedom (3-DOF) was developed and tested in [7]. The approximation of the Navier-Stokes equations in the case of 3D flow problem is much more complicated, see also [8], [9]. The incompressibility can be treated by using the pressure projection methods originated by Patankar, see [10], or Chorin's artificial compressibility method [11]. The other possibility is to use coupled solution for both pressure-velocity unknownstogether with a multilevel method. For the case of high Reynolds numbers anisotropically refined meshes need to be used in order to capture correctly the boundary layers, wakes, etc. In this paper the attention to the problem of mutual interaction of the airflow with a solid. The motion of the solid is described with the aid of two degrees of freedom (2DOF) (the bending and the torsion modes), which is equivalent to previous two-dimensional (2D) numerical results published, e.g., in [5]. The numerical method based on the fully stabilized finite element method is described.

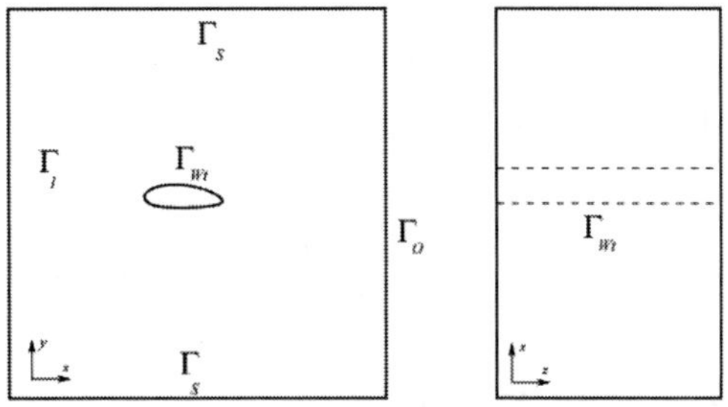

Fig. 1. The sketch of the computational domain Ω_t shown in xy and xz planes. The mutually disjoint parts of its boundary $\partial\Omega_t$ are shown.

The combination of the streamline–upwind and the pressure-stabilizing Petrov–Galerkin (SUPG/PSPG) and the grad–div stabilization is used. Numerical results are shown.

2 Mathematical model

Flow model. In order to practically treat the motion of the fluid computational domain Ω_t, the Arbitrary Lagrangian-Eulerian (ALE) method is used, see [12]. The ALE mapping $\mathcal{A} = \mathcal{A}(\xi, t) = \mathcal{A}_t(\xi)$ defined for all $t \in (0, T)$ and $\xi \in \Omega_0^{\mathrm{ref}} = \Omega_0$ is assumed to be diffeomorphism of $\overline{\Omega_0}$ onto $\overline{\Omega_t}$ at any $t \in (0, T)$. The domain velocity $\boldsymbol{w}_g(x, t)$ is then defined by $\boldsymbol{w}_g(x, t) = \frac{\partial \mathcal{A}}{\partial t}(\xi, t)$, where $x = \mathcal{A}(\xi, t)$, $x \in \Omega_t$, $\xi \in \Omega_0$. The time derivative with respect to the reference configuration Ω_0^{ref} is called the ALE derivative, denoted by $D^{\mathcal{A}}/Dt$ and satisfies (see [5], [12])

$$\frac{D^{\mathcal{A}} f}{Dt}(x, t) = \frac{\partial f}{\partial t}(x, t) + \boldsymbol{w}_g(x, t) \cdot \nabla f(x, t). \qquad (1)$$

The incompressible viscous flow in the computational domain $\Omega_t \subset \mathbb{R}^3$ is governed by the Navier-

Stokes equations written in the ALE form

$$\frac{D^{\mathcal{A}}\boldsymbol{u}}{Dt} + ((\boldsymbol{u}-\boldsymbol{w}_g)\cdot\nabla)\boldsymbol{u} = \nabla\cdot\boldsymbol{\sigma} \qquad (2)$$
$$\nabla\cdot\boldsymbol{u} = 0,$$

where $\boldsymbol{u} = \boldsymbol{u}(x,t)$ denotes the velocity vector, $\boldsymbol{u} = (u_1, u_2, u_3)$, $\boldsymbol{\sigma}$ denotes the Cauchy stress tensor

$$\boldsymbol{\sigma} = -p\mathbb{I} + \nu(\nabla\boldsymbol{u} + \nabla^T\boldsymbol{u}),$$

with components σ_{ij} given by

$$\sigma_{ij} = -p\delta_{ij} + \nu\left(\frac{\partial u_i}{\partial x_j} + \frac{\partial u_j}{\partial x_i}\right),$$

where $p = p(x,t)$ denotes the kinematic pressure and ν is the constant kinematic viscosity (i.e. the viscosity divided by the constant fluid density ρ), t denotes time and δ_{ij} is the Kronecker's delta. System of equations (2) is equipped with the initial

$$\mathbf{u}(x,0) = \mathbf{u}^0(x) \quad x\in\Omega, \qquad (3)$$

and with boundary conditions

$$\begin{aligned} &\text{a) } \boldsymbol{u} = \boldsymbol{u}_D && \text{on } \Gamma_I,\\ &\text{b) } \boldsymbol{u} = \boldsymbol{0} && \text{on } \Gamma_{Wt},\\ &\text{c) } \boldsymbol{\sigma}\cdot\boldsymbol{n} = 0 && \text{on } \Gamma_O,\\ &\text{d) } \boldsymbol{u}\cdot\boldsymbol{n} = 0, && \text{on } \Gamma_S, \end{aligned} \qquad (4)$$

where $\boldsymbol{n}$ denotes the unit outward normal vector to the Lipschitz continuous boundary $\partial\Omega$. The boundary condition (4c) is a modification of the so-called 'do-nothing' boundary condition, cf. [13].

Structure motion. The flow motion results in the aerodynamic forces acting on the flexibly supported airfoil, which can be vertically displaced by h (downwards positive) and rotated by the angle α (clockwise positive). In this case the airfoil motion is described with the aid of (nonlinear) motion equations

$$\begin{aligned} m\ddot{h} + S_\alpha\cos\alpha\ddot{\alpha} + k_h h &= -L(t), \\ S_\alpha\cos\alpha\ddot{h} + I_\alpha\ddot{\alpha} + k_\alpha\alpha &= M(t). \end{aligned} \qquad (5)$$

where m is the mass of the airfoil, S_α is the static moment around the elastic axis (EA), and I_α is the inertia moment around EA. The parameters k_h and k_α denote the stiffness coefficients. On the right-hand side the aerodynamical lift force $L(t)$ and aerodynamical torsional moment $M(t)$ are involved, which satisfy

$$L = -\int_{\Gamma_{Wt}} \rho\sigma_{2j}n_j\, dS, \qquad (6)$$
$$M = \int_{\Gamma_{Wt}} \varepsilon_{3ij}r_i\rho\sigma_{jk}n_k\, dS,$$

where ε_{3ij} is the Levi-Civita symbol and $r_i = x_i - x_i^{\mathrm{EA}}$ and x^{EA} is the position of the EA at the time instant t.

3 Numerical approximation

In this section the time discretization of the flow problem is introduced, linearized and the resulting problem is spatially approximated by the finite element method based on continuous piecewise linear (or trilinear) functions is used for approximation of both the velocity and the pressure. The application of the finite element method for the incompressible Navier-Stokes equations needs to overcome two difficulties. First, the velocity-pressure finite element pair needs to be properly chosen in order to guarantee the stability of the scheme, see, e.g., [14], or the PSPG can be used to overcome the instability, see [15]. The second source of the instability is the dominating convection flows. In this case the SUPG method is applied, cf. [16], [17], [18].

3.1 Time discretization

The time interval $[0,T]$ is divided with a constant time step $\Delta t > 0$, i.e. we denote $t_k = k\Delta t$. Further, we approximate $\boldsymbol{u}(t_n)$, $p(t_n)$, $\alpha(t_n)$, $h(t_n)$, $\dot{\alpha}(t_n)$, $\dot{h}(t_n)$ and $\boldsymbol{w}_g(t_n)$ by $\boldsymbol{u}^n$, p^n, α^n, h^n, $\dot{\alpha}^n$, $\dot{h}^n$ and $\boldsymbol{w}_g{}^n$, respectively. Here, the attention is paid only to the discretization on a fixed time level t_{n+1}.

The ALE derivative in (2) is then approximated by second order backward difference formula (BDF2), i.e.

$$\left.\frac{D^{\mathcal{A}}\boldsymbol{u}}{Dt}\right|_{t=t_{n+1}} \approx \frac{3\boldsymbol{u}^{n+1} - 4\tilde{\boldsymbol{u}}^n + \tilde{\boldsymbol{u}}^{n-1}}{2\Delta t}, \qquad (7)$$

where $\tilde{\boldsymbol{u}}^i(x) = \boldsymbol{u}^i(\mathcal{A}(t_i,\xi))$ with $x = \mathcal{A}(t_{n+1},\xi)$ is the transformation of the velocity from the domain Ω_{t_i} on the domain $\Omega_{t_{n+1}}$. The convective term is linearized by

$$\left.((\boldsymbol{u}-\boldsymbol{w}_g)\cdot\nabla)\boldsymbol{u}\right|_{t_{n+1}} \approx (2\tilde{\boldsymbol{u}}^n - \tilde{\boldsymbol{u}}^{n-1} - \boldsymbol{w}_g{}^{n+1})\cdot\nabla\boldsymbol{u}^{n+1}, \qquad (8)$$

where we shall write $\overline{\boldsymbol{w}}^n = 2\tilde{\boldsymbol{u}}^n - \tilde{\boldsymbol{u}}^{n-1} - \boldsymbol{w}_g{}^{n+1}$.

The system of equations (5) is transformed to the first order system and discretized in time using the BDF2, i.e. the following approximations for $\dot{h}$ and $\dot{\alpha}$ are used

$$\dot{\alpha}(t_{n+1}) \approx \frac{3\alpha^{n+1} - 4\alpha^n + \alpha^n}{2\Delta t},$$
$$\dot{h}(t_{n+1}) \approx \frac{3h^{n+1} - 4h^n + h^n}{2\Delta t},$$

and similarly also for the first derivatives

$$\ddot{\alpha}(t_{n+1}) \approx \frac{3\dot{\alpha}^{n+1} - 4\dot{\alpha}^n + \dot{\alpha}^n}{2\Delta t},$$
$$\ddot{h}(t_{n+1}) \approx \frac{3\dot{h}^{n+1} - 4\dot{h}^n + \dot{h}^n}{2\Delta t}.$$

Further, the aerodynamical lift force and the aerodynamical moment in equations (5) are extrapolated from the previous time levels, i.e. the approximations $L^n, M^n, L^{n-1}, M^{n-1}$ are assumed to be computed using (6) and the values of $\boldsymbol{u}^n, p^n, \boldsymbol{u}^{n-1}, p^{n-1}$. Then we extrapolate in (5) the aerodynamical lift force and the aerodynamical moment by

$$L(t_{n+1}) \approx 2L^n - L^{n-1}, \quad M(t_{n+1}) \approx 2M^n - M^{n-1}.$$

3.2 Spatial approximation

Now, the system of equation (2) is time discretized and linearized with the aid of (7) and (8), respectively. Then, the equations are formulated weakly and the solution is sought on the couple of finite element spaces $\boldsymbol{W}_h \subset \boldsymbol{H}^1(\Omega^{n+1})$ and $Q_h \subset L^2(\Omega^{n+1})$ for the approximation of the velocity components and pressure. The domain Ω is assumed to be a polyhedral domain and the finite element spaces are defined using an admissible triangulation $\mathcal{T}_h$ of the domain Ω, cf. [19], such that $\overline{\Omega} = \bigcup_{K\in\mathcal{T}_h} K$, $\mathcal{T}_h$ is formed by a finite number of closed hexahedral, tetrahedral, pyramidal or prism elements. Further, on each of reference elements K the local set of shape functions is denoted by P_K. For the reference hexahedral element the tri-linear functions are used whereas for the reference tetrahedral element the space P_K contains linear functions on K. The consistent definition of P_K on the pyramidal or the prism elements can be found in [20]. The finite element spaces are then defined by

$$Q_h = \left\{\varphi \in C(\overline{\Omega}) : \varphi\Big|_K \in P_K, \forall K \in \mathcal{T}_h\right\}, \; \boldsymbol{W}_h = [Q_h]^3$$

$$\boldsymbol{X}_h = \left\{\boldsymbol{\varphi} \in \boldsymbol{W}_h : \boldsymbol{\varphi} = 0 \text{ on } \Gamma_I \cup \Gamma_{Wt}, \; \boldsymbol{\varphi}\cdot\boldsymbol{n} = 0 \text{ on } \Gamma_S\right\}.$$

In order to introduce the stabilized weak formulation we start with the definition of the Galerkin terms for any $U = (\boldsymbol{u}, p) \in \boldsymbol{W}_h \times Q_h$, $V = (\boldsymbol{\varphi}, q) \in \boldsymbol{X}_h \times Q_h$ by

$$\begin{aligned} a(U,V) &= \frac{3}{2\Delta t}(\boldsymbol{u},\boldsymbol{\varphi})_\Omega + (\nu \boldsymbol{S}(\boldsymbol{u}), \boldsymbol{S}(\boldsymbol{\varphi}))_\Omega \\ &\quad + (\overline{\boldsymbol{w}}^n \cdot \nabla \boldsymbol{u}, \boldsymbol{\varphi})_\Omega - (p, \nabla\cdot\boldsymbol{\varphi})_\Omega + (\nabla\cdot\boldsymbol{u}, q)_\Omega, \\ f(\boldsymbol{u},\boldsymbol{\varphi}) &= \frac{1}{2\Delta t}(4\tilde{\boldsymbol{u}}^n - \tilde{\boldsymbol{u}}^{n-1}, \boldsymbol{\varphi})_\Omega, \end{aligned}$$

where by $(\cdot,\cdot)_\Omega$ the scalar product in $L^2(\Omega)$ or $\boldsymbol{L}^2(\Omega)$ is denoted. Further, the SUPG/PSPG and div-div stabilization terms are used defined by

$$\begin{aligned} \mathcal{L}(U,V) &= \sum_{K\in\mathcal{T}_h} \delta_K \\ &\quad \left(\frac{3\boldsymbol{u}}{2\Delta t} - \boldsymbol{\sigma}(\boldsymbol{u}) + (\overline{\boldsymbol{w}}^n\cdot\nabla)\boldsymbol{u}, (\overline{\boldsymbol{w}}^n\cdot\nabla)\boldsymbol{\varphi} + \nabla q\right)_K, \\ \mathcal{F}(V) &= \sum_{K\in\mathcal{T}_h} \delta_K \left(\frac{4\tilde{\boldsymbol{u}}^n - \tilde{\boldsymbol{u}}^{n-1}}{2\Delta t}, (\overline{\boldsymbol{w}}^n\cdot\nabla)\boldsymbol{\varphi} + \nabla q\right)_K, \\ \mathcal{P}(U,V) &= \sum_{K\in\mathcal{T}_h} \tau_K (\nabla\cdot\boldsymbol{u}, \nabla\cdot\boldsymbol{\varphi})_K. \end{aligned}$$

where $(\cdot,\cdot)_K$ denotes the scalar product in $L^2(K)$ or $\boldsymbol{L}^2(K)$. The choice of the parameters δ_K and τ_K is carried out according to [18] or [21] on the basis of the local element length h_K, i.e.

$$\tau_K = \nu\left(1 + Re^{\mathrm{loc}} + \frac{h_K^2}{\nu\cdot\Delta t}\right), \qquad \delta_K = \frac{h_K^2}{\tau_K}, \quad (9)$$

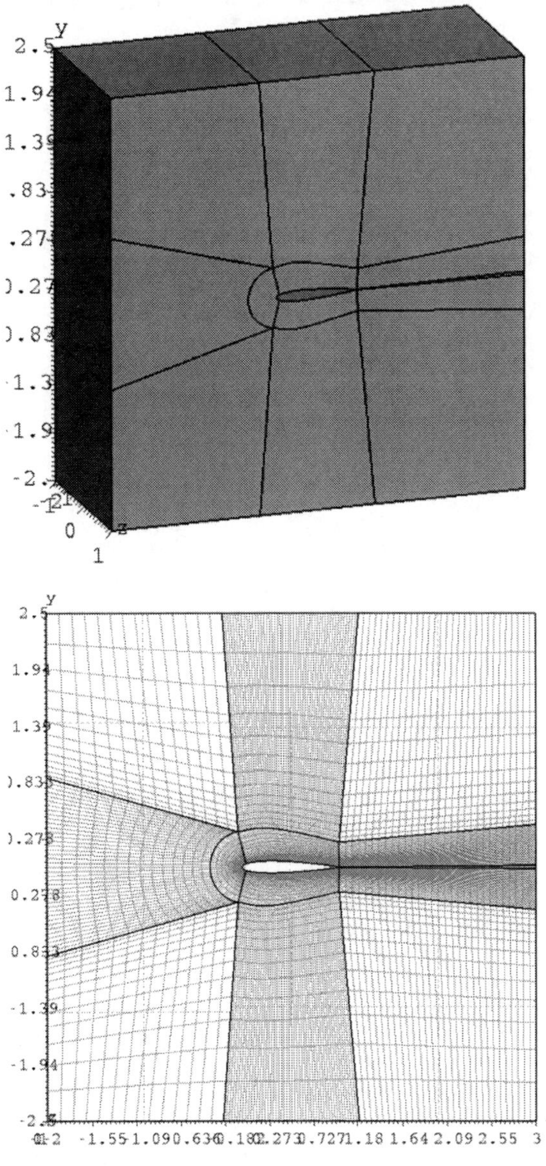

Fig. 2. The computational domain and the mesh detail around the airfoil shown in the xy-plane.

where the local element Reynolds number is given by

$$Re^{\mathrm{loc}} = \frac{h_K \|\overline{\boldsymbol{w}}^n\|_K}{2\nu}.$$

Then the *stabilized discrete problem* at a time instant $t = t^{n+1}$ reads: Find $U = (\boldsymbol{u}, p) \in \boldsymbol{W}_h \times Q_h$, $p := p^{n+1}$, $\boldsymbol{u} := \boldsymbol{u}^{n+1}$, such that $\boldsymbol{u}$ satisfies approximately the conditions (4 a-b) and

$$a(U,V) + \mathcal{L}(U,V) + \mathcal{P}(U,V) = f(V) + \mathcal{F}(V),$$

holds for all $V = (\boldsymbol{\varphi}, q) \in \boldsymbol{X}_h \times Q_h$. The solution of the arising system of linear equations is realized by a preconditioned iterative method based on inexact LU-factorization.

4 Numerical results

The stabilized finite element method was applied for the approximation of flow over NACA 0015 and NACA 0012 airfoils in the experimental channel, see [22] or [23]. The mutual interaction of the flowing air caused lead to the aeroelastic instability of flutter type. We consider the depth of the domain to be 180 mm (z-direction in Figure 1), the height of the domain (y-direction in Figure 1) to be equal to 210 mm which is the height of the experimental channel of the measurement. The artificial length of the computational domain was chosen to be 7 times the chord of the airfoil. The airfoil chord was equal to $c = 18\,\mathrm{mm}$. The computations were performed on hexahedral meshes. The air viscosity was considered to be equal to $\nu = 1.5 \times 10^{-5}\,\mathrm{m}^2\,/\,\mathrm{s}$. The inlet velocity was set equal to $U_\infty = 8.33\,\mathrm{m}\,/\,\mathrm{s}$. The Reynolds number based on the airfoil chord is then equatl to $Re = 10^4$.

For the numerical simulation two grids were used. First, the coarse grid consisting of approximately 2×10^5 hexahedral elements was used, where the discrete system with approximately 9×10^5 unknowns was solved using the iterative solver based on the incomplete blockwise LU factorization. Second, the grid with approximately 700000 hexahedral elements was used leading to the system with approximately 2.8×10^6 unknowns. The iterative solver was successful in both cases, although the numerical solution on the fine grid required much more computer time (as expected). The further improvement, optimization and parallelization of the iterative solver needs to be done.

The solution of the 3D problem was compared to the numerical approximation of the flow in 2D domain, where the grid consisting of the approximately 10000 quadrilateral elements was used. For the 2D problem the solution was realized using the direct solver. The fluid 3D domain is shown shown in Figure 2. The same grid (a slice of the 3D grid) was used for the computation of the 2D flow problem. The results in terms of flow velocity field for 2D and the central plane of 3D computations nearby the airfoil surface are shown in Figure 3.

5 Conclusion

This paper focused on the problem of numerical approximation of the incompressible flow with a flexibly supported airfoil. A model problem of channel flow around vibrating NACA 0015 airfoil was considered and mathematically described. The numerical method based on the finite element method was succesfully used for the approximation of the flow problem. This numerical method is the extension of the stabilized finite element method applied succesfully for 2D simulations. Here, however the method was tested and verified by application on a channel flow over the NACA 0012 airfoil. The results were compared to 2D resuls. Furthermore, the numerical approximation based on fully stabilized finite element method was presented and the numerical results for a three-dimensional problem of flow over an airfoil were shown.

Furthermore, the iterative method for solution of the discrete problem in 3D was succesfully used. The iterative method based on the blockwise incomplete LU factorization was used. The method was found to be efficient, but for large meshes the convergence was slow (as expected). The improvement of the iterative method will be a subject of future work.

Acknowledgment

The financial support for the present project was partly provided by the *Czech Science Foundation* under the *Grants No. P101/11/0207 and 13-00522S*.

References

1. E.H. Dowell, R.N. Clark, *A modern course in aeroelasticity*, Solid mechanics and its applications (Kluwer Academic Publishers, Dordrecht, Boston, 2004), ISBN 1-402-02039-2
2. E. Naudasher, D. Rockwell, *Flow-Induced Vibrations* (A.A. Balkema, Rotterdam, 1994)
3. C. Farhat, M. Lesoinne, N. Maman, International Journal for Numerical Methods in Fluids **21**, 807 (1995)
4. P. Le Tallec, J. Mouro, Computer Methods in Applied Mechanics and Engineering **190**, 3039 (2001)
5. P. Sváček, M. Feistauer, J. Horáček, Journal of Fluids and Structures **23**, 391 (2007)
6. P. Sváček, Mathematics and Computers in Simulation **80**, 1713 (2010)
7. M. Feistauer, J. Horáček, M. Růžička, P. Sváček, Computers & Fluids **49**, 110 (2011)
8. P. Sváček, P. Louda, K. Kozel, Journal of Computational and Applied Mathematics **270**, 451 (2014)
9. V. John, Int. J. Num. Meth. Fluids **40**, 775 (2002)
10. S. Patankar, *Numerical Heat Transfer and Fluid Flow* (McGraw-Hill, New York, 1980)
11. A.J. Chorin, Journal of Computational Physics **2**, 12 (1967)
12. F. Nobile, Ph.D. thesis, Ecole Polytechnique Federale de Lausanne (2001)
13. C.H. Bruneau, P. Fabrie, International Journal for Numerical Methods in Fluids **19**, 693 (1994)
14. R. Temam, *Navier-Stokes equations. Theory and numerical analysis.s* (North-Holland, Amsterdam, 1978)
15. T.J.R. Hughes, M. Mallet, A. Mizukami, Comput. Methods Appl. Mech. Eng. **54**, 341 (1986)
16. M. Olshanskii, G. Lube, T. Heister, J. Lwe, Computer Methods in Applied Mechanics and Engineering **198**, 3975 (2009)
17. G. Lube, G. Rapin, Computer Methods in Applied Mechanics and Engineering **195**, 4124 (2006)
18. T. Gelhard, G. Lube, M.A. Olshanskii, J.H. Starcke, Journal of Computational and Applied Mathematics **177**, 243 (2005)
19. P.G. Ciarlet, *The Finite Element Methods for Elliptic Problems* (North-Holland Publishing, 1979)

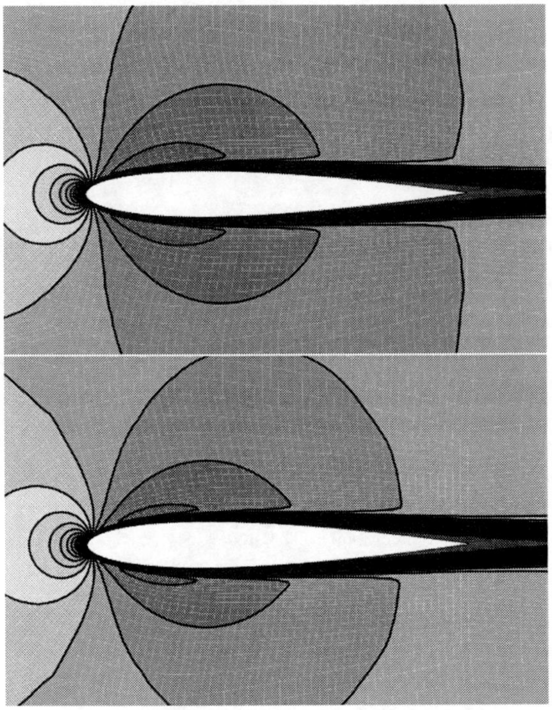

Fig. 3. The comparison of the flow velocity pattern for 3D (on the left) and 2D (on the right) computations.

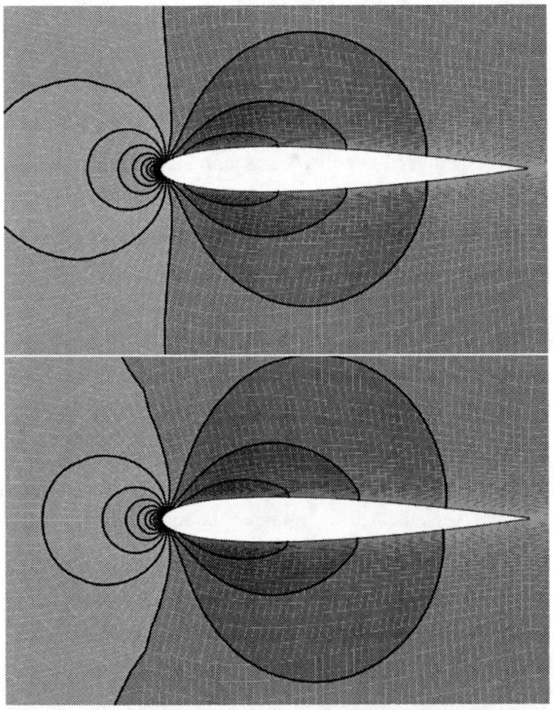

Fig. 4. The comparison of the pressure distribution for 3D (on the left) and 2D (on the right) computations.

20. C. Wieners, Preprint, University of Stuttgart (1997)
21. P. Sváček, M. Feistauer, *Application of a Stabilized FEM to Problems of Aeroelasticity*, in *Numerical Mathematics and Advanced Application* (Springer, Berlin, 2004), pp. 796–805, ISBN 3-540-21460-7
22. J. Kozánek, V. Vlček, I. Zolotarev, Journal of Applied Nonlinear Dynamics **3** (2014)
23. J. Kozánek, V. Vlček, I. Zolotarev, International Journal of Structural Stability and Dynamics **13**, 1 (2013)

EPJ Web of Conferences **114**, 02119 (2016)
DOI: 10.1051/epjconf/201611402119

Analysis of selected problems of biomass combustion process in batch boilers - experimental and numerical approach

Mateusz Szubel[1,a]

[1]*AGH University of Science and Technology, Faculty of Energy and Fuels, al. Mickiewicza 30, 30-059 Krakow, Poland*

Abstract. It is possible to list numerous groups of heating units that are used in households, such as boilers, stoves and units used as supporting heat sources, namely fireplaces. In each case, however, the same operational problems may be evoked [1]. To understand the causes of energy losses in a boiler system, a proper definition of significant elements of the unit's heat balance is necessary. In the group of energy losses, the flue gas loss and the incomplete combustion loss are the most significant factors. The problem with the loss resulting from incomplete combustion, which is related to the presence of combustible substances in the exhaust, is especially significant in case of biomass boilers [2, 3]. The paper presents results of the research and the optimisation of the biomass combustion process in the 180 kW batch boiler. The studies described have been focused on the reduction of the pollutants emission, which was primarily realised by the modifications of the air feeding system. Results of the experiments and the CFD simulations have been compared and discussed. Both in case of the model as well as the experiment, positive influence of the modifications on the emission have been observed.

1 Introduction

From the physicochemical point of view, the biomass combustion process can be divided into four stages: heating and drying of the fuel, its thermal decomposition, combustion of released volatiles and combustion of solid residues [4].

Fuel heating proceeds as a result of the radiation of the flame and the walls of the combustion chamber. Then, the fuel is dried in the temperature of 100 °C (evaporated water is carried away with the exhaust gas to the chimney).

The process of thermal decomposition depends on the composition of the compounds, such as hemicellulose, cellulose and lignin. Decomposition of hemicellulose occurs in temperatures 150-260 °C. Because cellulose and lignin are more stable, thermal decomposition of these compounds occurs in higher temperatures (240-350 °C and 280-500 °C, respectively) [4].

The composition of the gaseous substance released from the biomass fuel depends on the ratio of oxygen available and the amount of oxygen that would allow to complete the burning process, which is called "the equivalence ratio". For equivalence ratios below 0.1, pyrolysis occurs. In such a way, a significant fraction of the biomass energy is found in the char and oily residues. In case of the equivalence ratios between 0.2 and 0.4, the process is called the proper gasification (maximum energy is transferred to the gas) [5].

Carbon monoxide constitutes a major component of the biomass gasification products. It is oxidized to carbon dioxide (CO_2) in the secondary combustion region. It should be noted, that combustion of high purity carbon monoxide (CO), without moisture, hydrogen and hydrocarbons occurs in a slow manner – the oxidation of CO is catalysed even by trace amounts of hydrogen and its compounds [6].

Currently, attention of the producers of biomass batch boilers is mostly focused on the problem of CO emissions as well as particle matters emissions. Due to the European Union regulations, the emissions referred to above have to be kept at certain levels due to health considerations, but also because of the necessity to increase the efficiency of the devices.

Heat loss related to the presence of combustible substances in the gas mixture leaving the combustion chamber is referred to as the incomplete combustion loss. Incomplete combustion may be a result of a number of factors, such as insufficient penetration of the combustion chamber by air, insufficient mixing of the volatiles and the oxidator, low temperature in the combustion area, etc. [2]. In each of the cases listed above, concentration of carbon monoxide, hydrogen and volatile hydrocarbons in the exhaust increases.

Some countries such as Denmark have extensive experiences in the field of practical and theoretical studies regarding low-power biomass batch boilers [7]. In Poland, many heating units such as straw-fired units

[a] Corresponding author: mszubel@agh.edu.pl

are based on Danish designs. Despite the robust structure, reliability and simple handling, improvement in the field of CO emission is still necessary.

Both experimental and numerical methods are broadly applied in the studies of the solid biomass combustion process and, as a consequence, are used to achieve higher levels of energy conversion in the exploitation of heating devices.

Authors of [8] present experimental studies of straw and two herbaceous crops combustion in a fixed bed reactor. A wide range of air flow-rates has been used to determine the optimal process conditions. Based on the measured temperatures, gas composition and mass loss, combustion characteristics in terms of ignition front speed, burning rate, percentage of mass loss and the equivalence ratio have been determined. The relation between the air flow-rate and the stages of the combustion process has been established.

The paper [9] presents the results of the experimental studies of a batch boiler designed to burn pellets in order to evaluate the influence of the air staging used in the combustion process on the volume of NO_x emissions. Temperatures have been measured in selected points of the combustion chamber and the analysis of the CO content in the exhaust has been performed for different conditions. The results enabled the development of a set of recommendations for optimizing the operation of the unit.

Another approach is to study the process of biomass combustion independently from the operation of the device, in a laboratory. An adequate example is provided in [10], where a laboratory evaluation of four different pulverized biomass types was described. Single particles of sugarcane bagasse, pine sawdust, torrefied pine sawdust and olive residue were burned in a drop-tube furnace, set at 1400 K, in both air and O_2/CO_2 atmospheres containing different oxygen mole fractions. High-speed and high resolution images of single particles were recorded and temperature-time histories were obtained pyrometrically. Striking differences were observed between the combustion behavior of the biomass particles burned in that study and those of coal particles of all ranks investigated in previous studies in this laboratory. The biomass particles released large amounts of volatiles that burned in the form of spherical envelope flames.

One of the essential tools used to analyze the dynamics of combustion of this type of biomass is the thermogravimetric analysis (TGA). Among other things, such method allows to determine the energy consumed in the thermal decomposition of the fuel components, as well as to establish the rate of the process in a given temperature. Moreover, the TGA provides information related to the influence of the fuel composition on the combustion and thermal stability of the biomass. Based on the results obtained by means of this method, it is also possible to develop a combination of the reactions used in the combustion mechanisms included in the numerical models [11 - 13].

In works described in [14], an Euler–Euler multiphase CFD model is presented as a method of simulation of continuous fast pyrolysis of the biomass in a fluidized bed reactor. Multi-component, multi-stage kinetic model is applied to describe the pyrolysis of the biomass particle. Direct relation with the experimental (TGA) works is considering the biomass as a mixture of three pseudocomponents (cellulose, hemicellulose, and lignin) in case of the numerical model.

Authors of [14] present a three-dimensional numerical model of a straw boiler which has been developed to allow for the optimization of the structure and the operation conditions. Combustion in oxygen-enriched or conventional air atmospheres has both been considered. In the study, a combination of FLIC (Fluid Dynamic Incinerator Code) code and FLUENT commercial software has been applied. The goal of using the oxygen-enriched atmospheres was to promote the destruction of most pollutants due to the high oxygen partial pressures and temperatures which is reflected by very low residual amounts of organic combustion by-products in the bottom ash and flue gas of the straw-fired boiler unit.

Based on a review of the literature, it must be stated that both experimental and numerical methods may be used as essential tools in the optimization of design and operation parameters of biomass-fired heating devices. The research described in this article is devoted to the impact of the air feeding system design on the CO emission in case of the 180 kW straw boiler. Experimental and numerical (CFD) methods have been applied to investigate the CO emissions in a state-of-the-art biomass boiler and to develop a more efficient design of the air supply system without the necessity to introduce significant modifications in the boiler design or to use a second fan.

2 The tested heating device and the test stand

The main element of the system used for the experimental works described in the paper is the solid biomass batch boiler (Ekopal RM40 model).

Figure 1. Cross section of the studied biomass batch boiler: 1 – primary combustion chamber, 2 – secondary combustion chamber, 3 – ash separator, 4 – exhaust heat exchanger, 5 – secondary air nozzle, 6 – primary air nozzles.

The EKOPAL RM40 boiler (figure 1) is a 180 kW unit classified in the group of gasification boilers. The dimensions of the primary combustion chamber are: 2405/3075/1880 mm (height/width/depth). The device is designed to burn straw in the form of cubes or bales (in the latter case the bale is loaded into the chamber using a forklift truck). Co-firing of straw with other biomass fuels and combustion of alternative fuels such as wood wastes is also possible [15, 16].

The combustion of fuel proceeds in two stages: in the primary combustion chamber and the secondary combustion chamber. In the first stage, gasification of the fuel in the primary chamber takes place under controlled air deficiency. Combustible gaseous products of the first stage flow into the secondary chamber, where the post-combustion occurs. The next element behind the secondary combustion chamber and the ash separator is a heat exchanger, consisting of tubes arranged in three rows.

The heat exchanger is responsible for the removal of the heat from the exhaust gas (from the process of the combustion of the gasification products). The separation of the combustion stages as well as the application of the exhaust heat exchanger causes an increase in the total efficiency of energy conversion – in case of the heating unit used in the experiments it is approximately 84 % (for the nominal conditions of operation) [17].

It is important to provide a sufficient distribution of air in the areas of the two combustion chambers. The boiler system is equipped with an air supply system, which uses a centrifugal fan and an air manifold in the form of pipes arranged in series (figure 2.), connected with the main pipe responsible for the transport of the air provided by the fan to each duct terminated by the nozzle. Efficient biomass combustion is realized by the proper location of the nozzles of the air manifold (figure 3). The air distributor of the examined boiler consist of seven pipes (R1-R7), welded on the front of the main duct, connected with the area of the primary combustion chamber. The last two ducts (D1, D2) are responsible for the feeding of air to the secondary combustion chamber [18]. This is the so called "counter-combustion system", which assures the correct process of straw gasification and combustion of the generated syngas.

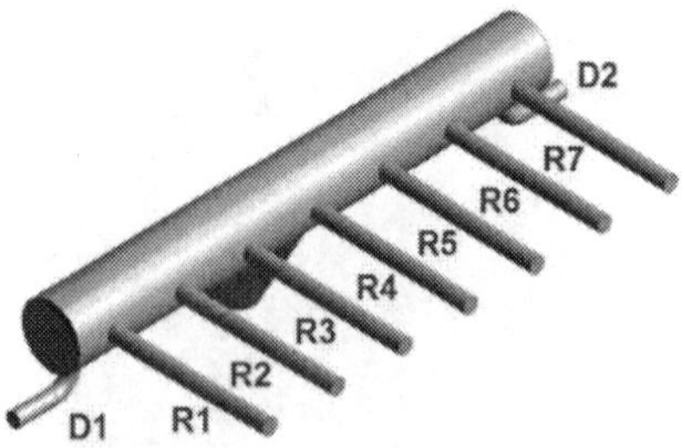

Figure 2. Air manifold applied in the EKOPAL RM40 boiler: R1-R7 – primary air ducts, D1-D2 – secondary air ducts.

The combustion chambers are surrounded by a water jacket filled with water. The jacket is responsible for the transfer of the heat resulting from the combustion and its transportation to the buffer tank. Boilers of this type operate only in open hydraulic systems.

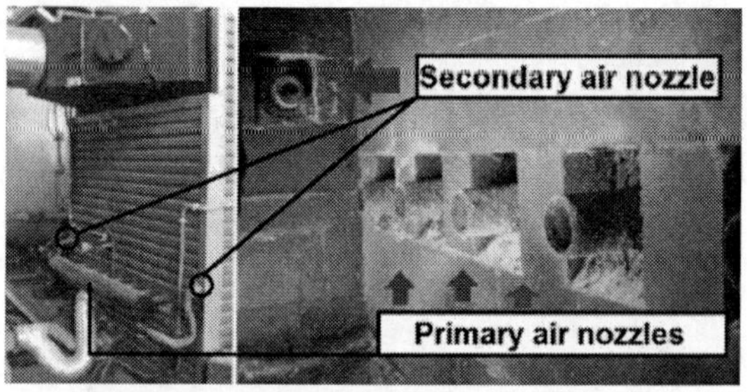

Figure 3. Location of the air manifold at the rear wall of the boiler and the view of the air pipes from the inside of the combustion chamber.

The boiler is fitted to the heating installation via a storage tank. The system of the heating unit connected to the buffer tank is presented in figure 4. The operating medium, which flows through the water jacket and between the boiler and the buffer tank, is pumped by a pump equipped with an inverter controlled by PLC. The maximum volume flow rate is 8.5 m^3/h. The water volume of the primary loop (including the water jacket, buffer tank and pipes) is about 5.5 m^3 [19].

Figure 4. General view of the measurement installation with the batch boiler and the buffer tank.

A storage tank of a proper size is very important in terms of the efficient operation of the heating system due to the requirement of the appropriate heat capacity of the system and the safety of operation (risk of overheating) [20].

To perform measurements of the temperature inside the combustion chambers, PTTK thermocouples have been applied. Five sensors have been selected to control the temperature during the operation of the boiler. The locations of the temperature measurement points (figure 5) has been provided below: DR – the centre of the door of the primary combustion chamber, PR1 – the centre of the right wall of the primary combustion chamber, LE – the centre of the left wall of the primary combustion chamber, TY – the rear wall of the boiler, at about 1/3rd of the height of the vertical area of the secondary combustion chamber (inlet to the secondary combustion chamber), PR2 – top part of the right wall of the secondary combustion chamber.

A MRU ECO 3000 exhaust analyzer has been applied during the experimental tests. The device is dedicated to the measurements performed in combustion installations, especially in power plants, chemical industries, refineries and laboratories. The analyzer is an essential tool in studying the emissions in case of boilers, hearths, engines and turbines. The device is adapted to analyze all popular solid fuels, including biomass.

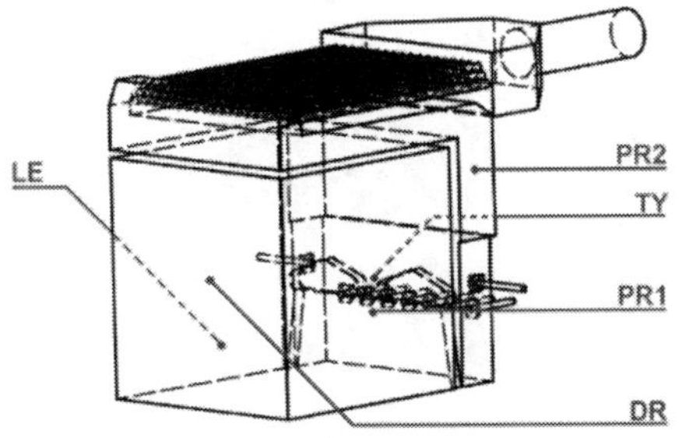

Figure 5. Locations of the temperature sensors in the boiler.

The exhaust analyzer allows to conduct a simultaneous measurement of 10 compounds. Additional parameters are also calculated based on the information from the measuring system. Moreover, the measuring system is equipped with a set of sensors not directly used in the current studies, such as resistance thermometers dedicated to measure the temperature in selected points of the water jacket, the pipes and in the flue. A flow meter and a heat meter are installed in the system. The current heating power of the device can be calculated based on the information provided by the elements of the system mentioned above.

A PLC set is the heart of the control and measurement system. It incorporates a controller connected to a PC and a set of inverters (responsible for the modulation of the operation of the fan and the pumps). The controller consist of modules which receive (from the sensors) and send (to the PC and other devices in the system) an analog and digital signal.

3 Experimental and numerical analysis of the current state of the heating device

Experimental tests of the combustion in case of the current state of the heating device have been carried out as the first stage of the study in order to analyze the dynamics and repeatability of the process. Three series of experiments have been carried out. Before each measurement, warming of the boiler has been carried out to provide appropriate conditions during the proper test.

Straw in form of cubes (700x500x500 mm) has been used. Estimated fuel weight for the individual test was 40 kg (about $1/4^{th}$ ~ $1/3^{th}$ of the nominal load). Moisture content in each straw cube has been measured using a device equipped with a bayonet probe. The average moisture content for the fuel used in the experiment carried out for the current state of the boiler (first stage of the study) as well as for the modified air supply nozzles are presented later in the paper.

In each series of measurements, the straw cubes were arranged in the same way. Ignition has been initiated through a special orifice located at the rear bottom corner of the boiler side wall. To ensure the appropriate operation of the heating device, the fan efficiency has been modulated in the same way in each of the experiments. After a given time (about thirty seconds after the ignition) necessary for the flame propagation on the rear wall of the fuel load, the fan efficiency was set to a minimum value – 20 % (~50 dm^3/s).

Figure 6 presents selected results of two experiments performed for the current design of the device. Temperature curves allow to determine the stages of the combustion process. Ignition at the rear wall of the fuel load occurs approximately in the 5th minute of the process. In the first 10 minutes of the experiments, the increase of the temperature at the inlet to the secondary combustion chamber ("TY" sensor) is observed. At the same time, an increase of the temperature at the top of the secondary chamber ("PR2" sensor) occurs. The CO emission curve behaves in a manner analogous to the "TY" temperature curve until the moment when the temperature reaches the highest level – between 700 and 800 °C.

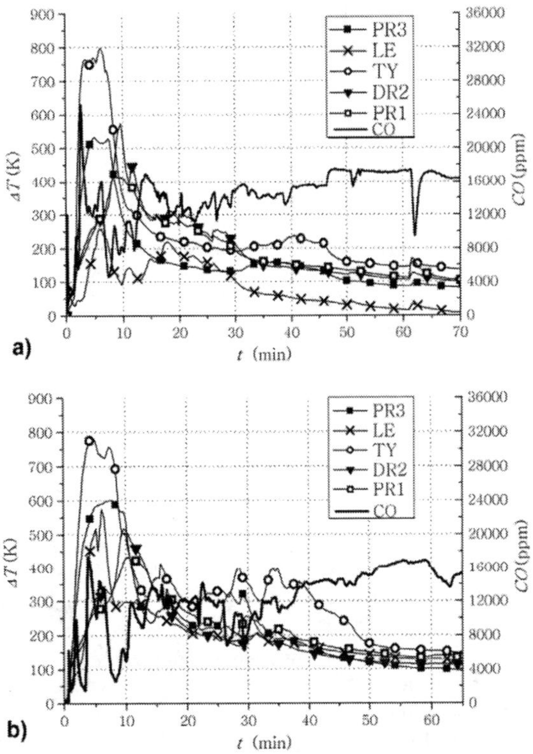

Figure 6. Results obtained in the first (a) and the second measurement (b) of the current design of the boiler.

The high content of CO in the first stage of the combustion is due to the fact that the fuel is a natural barrier closing access to the primary combustion chamber. Consequently, the area of the oxidation of the gasification products is reduced almost exclusively (depending on the fuel load) to the region of the secondary chamber. In the time of the highest temperature in the secondary combustion chamber the emission level decreases rapidly, which is the result of the conditions appropriate for the CO post-combustion. The temperature in the secondary chamber is relatively high, the combustion area moves in the direction to the door and the temperatures in the center of the reactor (especially "PR1" sensor) increase. Moreover, the combustion proceeds in a larger part of the primary combustion chamber.

The emission of CO increases slightly in the third stage of the combustion, when the temperatures in the measurement points decrease to the level below approximately 250~300 °C. Based on the temperature curves for the fuel load used in the analyses in concern, the following stages of the process can be distinguished:
- ignition stage till the 5~10th minute of the measurement,
- nominal (proper) combustion stage till the 30~35th minute of the measurement,
- afterburning of the solid residues after 35 minutes.

A numerical CFD model has been developed to analyze the phenomenon occurring at the beginning of the combustion in the area of the secondary combustion chamber. Fluid dynamics as well as the simple combustion model allowed to refer the experimental results to the air flow characteristics obtained by means of computation. Commercial tools included in the ANSYS Workbench 15 software have been applied for the geometry edition, computational grid generation as well as in problem solving.

Due to the fact that the first stage of the combustion proceeds generally in the area of the secondary combustion chamber, 3D geometry designed for the purposes of the analysis was reduced just to that region of the device (thus the primary combustion chamber has not been considered). In case of the spatial geometry corresponding to the current (existing) design of the boiler, it was also possible to reduce the geometry to that half of the device, due to the presence of the symmetry plane.

Spatial geometry has been divided into sections, allowing for the efficient discretization with the predominance of hexagonal grid elements. Subsequently, the geometry has been exported to the ANSYS Meshing. Discretization has been carried out using the "Multizone", the "Hex dominant" and the automatic method. The total number of grid elements was $1.3 \cdot 10^6$. The computational domain has been defined as the continuous fluid and the reacting mixture. Due to the fact that the goal of the study was to investigate the possibility of reducing CO emission by modifying the air feeding system, the volatiles have been defined as a mixture of CO and H_2O. Air was represented by a mixture of O_2 and N_2. The kinetics of the chemical reaction were based on the "Finite Rate Chemistry and Eddy Dissipation" model, which gives consideration to turbulences as well as the temperature in the reacting mixture. In such approach, the level of mixing of the fuel and the oxidizer is also significant. The mechanism of the CO post-combustion has been based on the Arrhenius equation. K-epsilon model of turbulence has been applied.

Heat transfer was taken into account in the model. An appropriate heat transfer coefficient has been defined for the walls of the combustion chamber – 80 [W/m2K] for the bottom part covered by the chamotte bricks and 150 [W/m2K] for the top part in direct contact with the water jacket (8 mm steel wall). The Discrete Transfer option for the thermal radiation has been set with the so-called "surface to surface" mode and the "Multigray" spectral model.

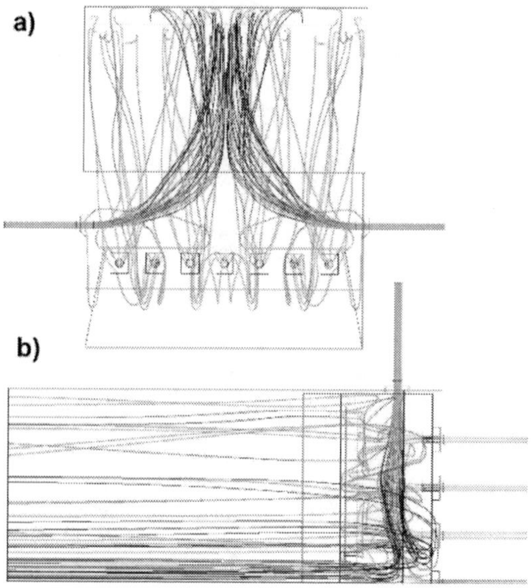

Figure 7. Streamlines of the primary (rainbow scale) and the secondary (grayscale) air in the secondary combustion chamber in case of the existing boiler.

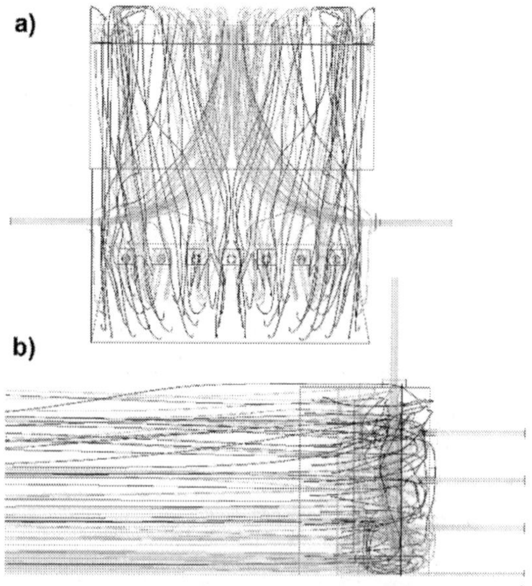

Figure 8. Streamlines of the primary and the secondary air (rainbow scale) and the CO (grayscale) in the secondary combustion chamber in case of the existing boiler.

Because of the non-homogeneous air distribution along the pipes of the air manifold, the primary and the secondary air inlets have been set as individual boundary conditions [18]. The highest value of the mass flow was set for the central pipe, located on the symmetry plane of the geometry and for the secondary air nozzles, 0.0237 kg/s and 0.0239 kg/s, respectively. The mass flow for other ducts oscillates around 0.0235 kg/s. Based on the manufacturer's recommendations regarding the exploitation of the studied device, the chimney draft for the model has been set to -40 Pa ("opening" type outlet boundary condition).

Steady state analysis has been carried out. The residues for the governing equations has been set on the level $1 \cdot 10^{-5}$.

The analysis has been carried out especially to determine the characteristics of the air fluxes in the area of the secondary combustion chamber.

Figure 7 presents the so-called streamlines generated using the options available in the post-processor of the ANSYS CFX. The lines correspond to the primary and secondary air streams in the aforementioned space. The configuration of the nozzles results in a laminar flow of the air – especially in case of the secondary air which is responsible for the post-combustion of the CO.

The location of the secondary air nozzles opposite to each other suggests that the level of the turbulences in the place of the collision of the streams has to be high, but due to the width of the chamber and the impact of the chimney draft the air flows up to the outlet from the post combustion area.

Due to the characteristics of the air flow, the mixing of the CO and the oxygen in the considered case is much more difficult. The relation between the fluxes of the air and the volatiles is shown in figure 8. The characteristics of the air flow result in high concentration of the CO, especially in the regions close to the side walls. CO emission calculated within the numerical simulation is equal to 2.6% (averaged mass fraction at the outlet of the computational domain). The effect can be reduced by the change of the air flow directions, which in some cases does not require any major modification of the heating device design and can also be applied in already existing boilers.

4 Experimental and numerical analysis of the heating device equipped with the modified secondary air feeding system

Many methods are available to achieve mixing improvement and, consequently, a better post combustion of the CO. According to the presented results of experimental measurements, the temperature in the combustion area is a significant factor which has direct impact on the chemical reactions in this region. However, the application of additional insulation surfaces on the wall of the secondary combustion chamber requires a modification of the water jacket design due to a significant decrease of the efficiency of the heat transfer through the internal walls to the operating medium.

On the other hand, in [19] significant problems related to the overheating of some areas of the studied type of biomass boilers have been described. Therefore, the insulation of the parts of the device which are prone to overheating is reasonable from the point of view of the CO emission and the overheating risk. It has to be noted that this approach is possible only in case of new devices. For the existing heating units, it is necessary to develop technically reasonable and cost-effective methods of optimization of the CO post-combustion of the boilers.

Due to the fact that the source of the problems related to the CO emission is strongly connected with the inefficient mixing of the secondary air and the volatiles, a modification of the secondary air feeding system seems to be interesting method to achieve the increase of the efficiency of the energy conversion.

Of course, the implementation of the construction elements responsible for the changing of the air stream direction can cause an increase of the resistance to flow. However, minor changes of the angles of the secondary air nozzles relative to the walls can cause a significant increase of the turbulences in the secondary combustion chamber.

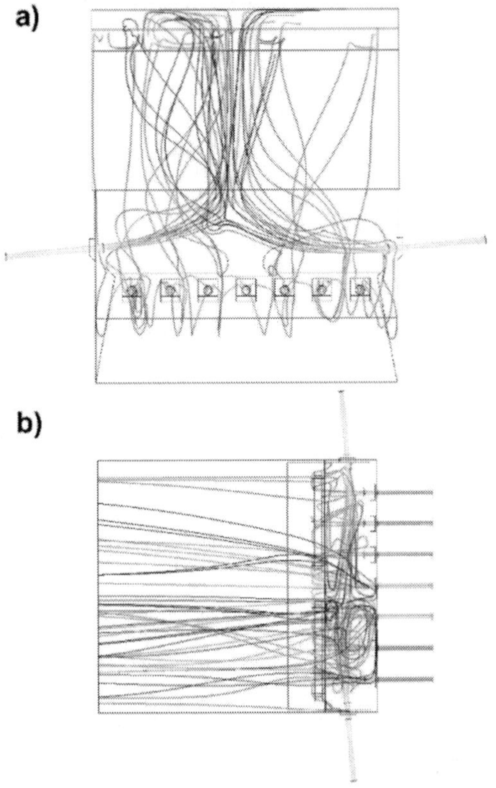

Figure 9. Streamlines of the primary (rainbow scale) and the secondary (grayscale) air in the secondary combustion chamber in case of the first modification of the secondary air nozzles.

As a first stage of the optimization of the combustion process, the proposed design changes have been applied in the numerical model (based on the model developed to study the current state of the considered heating device) to investigate the reasonability of the modifications of the existing heating device and the performance of the experimental studies. It was decided that two configurations will be tested. In the first case: a deflection of the secondary air nozzles of 5° vertically and horizontally, in the second case: an analogous deflection of the secondary air nozzles of 10°.

Figs. 9 to 12 show the results of the simulations that have been performed. It was noted that in case of the first modification (deflection of 5°), the relation between both air fluxes as well as between the air fluxes and the CO flux is much more preferred that in the case of the existing air feeding system.

The angled air fluxes generate a vortex in the centre of the secondary combustion chamber. The turbulent flow region is moved slightly towards the left sidewall of the chamber, which is a result of the manner of vertical deflection of the nozzles. Compared to the existing device, in case of the nozzles modification a much larger part of the secondary combustion chamber is characterized by the preferred process conditions. In case of the considered change of the design of the secondary air feeding system, a reduction in the CO emission by 15% has been achieved as the result of the numerical simulation.

a)

b)

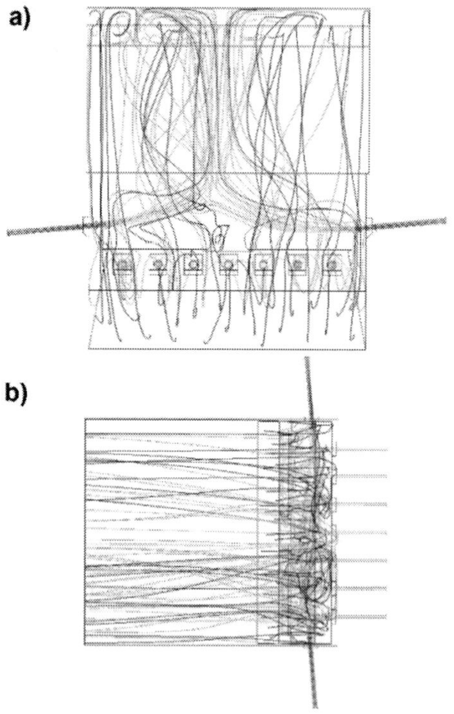

Figure 10. Streamlines of the primary and the secondary air (rainbow scale) and the CO (grayscale) in the secondary combustion chamber in case of the first modification of the secondary air nozzles.

As a second stage of the numerical analysis, secondary air nozzles with an inclination of 10° have been implemented in the secondary combustion chamber. The manner of the nozzles deflection was similar to the first case described above.

The increase of the deflection angle resulted in the passing of the secondary air streams without generating any significant turbulences in the area of the most intensive post-combustion of the CO. Moreover, as a result of the air flow, volatiles were transported to the top part of the combustion chamber. The positive effect of the second change of the design, however, is visible in the bottom corners.

The relatively low level of the turbulences in the centre of the chamber allows for a better air penetration in the areas near to the sidewalls, which was more difficult to achieve in the cases of the existing device and the first modification.

Vortices appear also at the outlet of the considered area, but the relatively low temperature in this part of the chamber significantly impedes the chemical reaction of the fuel and the oxygen.

a)

b)

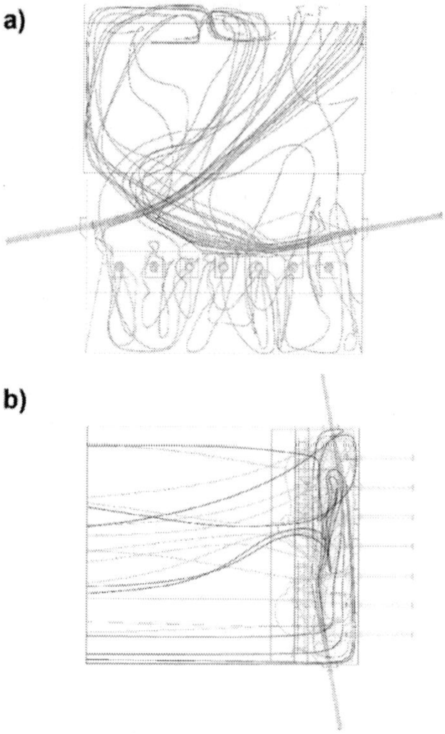

Figure 11. Streamlines of the primary (rainbow scale) and the secondary (grayscale) air in the secondary combustion chamber in case of the second modification of the secondary air nozzles.

The calculated emission level for such case was worse than in the case of the basic design (without any modifications) by about 8%.

Due to the promising results of the simulation devoted to the investigation of the system equipped with the secondary air nozzles deflected by 5°, it was decided to produce a set of steel nozzles based on the design presented in figure 13. One may note that the nozzles have been flattened at their outlets to increase the air velocity and to provide a wider air flux.

The elements deflected by 5° as well as 10° have been prepared to compare the results of the experiments with both the numerical simulations performed. The implementation of the newly designed elements of the air feeding system has been carried out by placing them at the ends of the existing secondary air pipes.

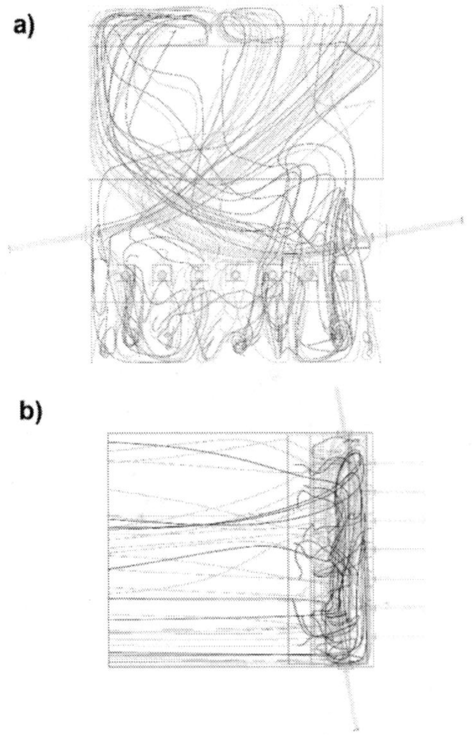

Figure 12. Streamlines of the primary and the secondary air (rainbow scale) and the CO (grayscale) in the secondary combustion chamber in case of the second modification of the secondary air nozzles.

Due to the fact that the first prototypes of the directional nozzles have been produced using simple methods and tools, the resulting components are slightly different from these assumed in the design (smaller extension at the ends of the nozzles).

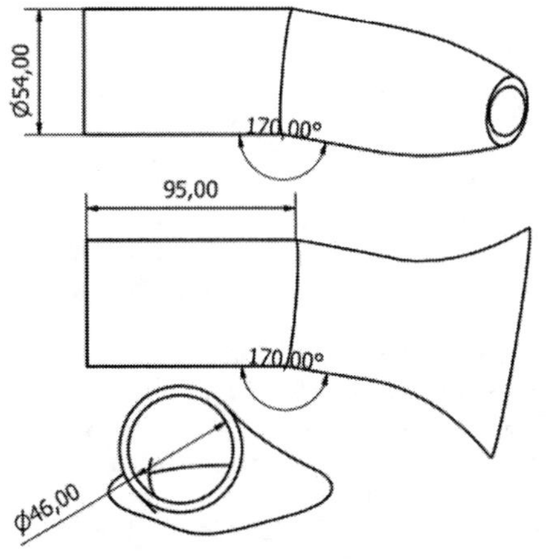

Figure 13. Design of the nozzles produced for the experimental tests (deflection by10°).

Figure 14 shows the directional nozzles which were produced for the experimental tests. The picture presents the elements removed from the secondary air pipes of the boiler after a series of measurements. No thermal deformation or other damages as well as no general wear of the nozzles has been noted.

Figure 14. The nozzles produced for the experimental tests removed from the combustion chamber after the measurements.

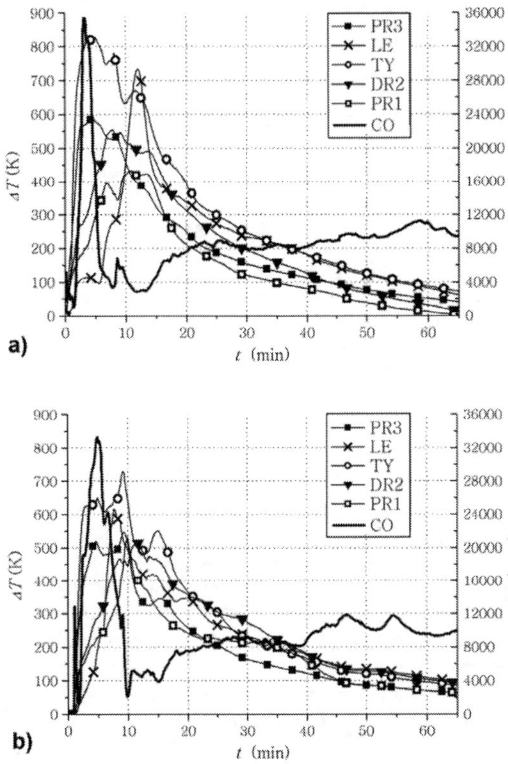

Figure 15. Results from the first (a) and the second measurement (b) of the boiler equipped with the directional nozzles (deflection by 5°).

Similarly to the studies of the heating unit before the modifications, three series of experiments have been carried out both for the operation of the boiler with the first and the second design of the directional nozzles (5 and 10 degrees). Each series included the warming up of the device and the proper experiment. Figure 15 shows the results of the measurement for two sample experiments performed for the nozzles deflected by 5 degrees.

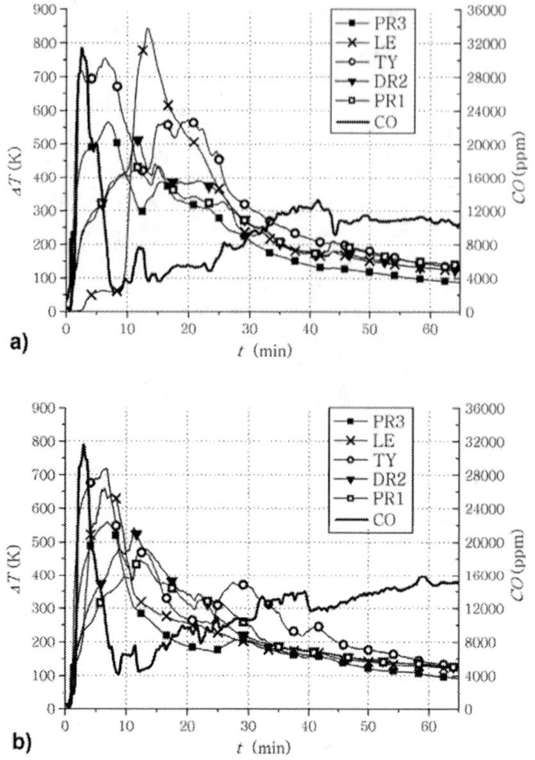

Figure 16. The results of the first (a) and the second (b) measurement of the boiler equipped with the directional nozzles (deflection of 10°).

As in the case of the basic design, the process was very repetitive due to the similar arrangement of the straw bales inside the primary combustion chamber. According to the conclusions drawn from the numerical modeling, the application of the directional nozzles results in a significant reduction of the CO concentration in the exhaust gas. However, in comparison with the basic design (lack of the nozzles deflection), a much higher emission of CO at the ignition stage has been noted in each series of experiments. Such a situation may be explained by the dynamics of the temperature changes. The temperatures at the measurement points located in the secondary combustion chamber were comparable to the values achieved in the case of the basic design.

On the other hand, it should be noted that the temperatures measured by the sensors located on the left and/or right wall of the primary combustion chamber increase rapidly from the beginning of the process and achieve much higher values already at the beginning of the stage of the proper combustion. This provides information that the intensity of the process at the beginning is greater and that it proceeds at a wider area than in case of the basic design. Consequently, the release of the products of the biomass thermal decomposition (including the carbon monoxide) is higher.

In case of the experimental data obtained in the analysis of the second set of nozzles (10°), the emission of CO and its relation with the dynamics of the temperature changes is analogous to the case of the first modification. The curve of the CO emission for the first series of measurements is similar to the one in the previous case. However, the averaged results obtained for the stage of the proper combustion and for the last part of the process are worse than in case of the smaller angle of the deflection of the nozzles. Moreover, a slight randomness in the experiments related to fuel parameters such as local moisture content has to be taken into account. Averaged moisture contents for the straw used in all series of the experimental studies presented in the paper are summarized in table 1.

Table 1. The results of the measurement of the moisture content in the fuel used in the experiments.

Measurement series	Current state (no nozzles)	Nozzles 5°	Nozzles 10°
1	12.6 %	11.4 %	10.1 %
2	12.8 %	9.5 %	8.6 %

5 Discussion

In case of biomass batch boilers, it is possible to improve the degree of the energy conversion by means of the application of additional elements of the air supply system. The directional nozzles allow to generate turbulences in the secondary combustion chamber, which is significant from point of view of the post-combustion of the gasification products. Due to the possibility of installation of the nozzles on the existing air ducts, the solution can be easily applied in the heating devices already in operation. In newly designed boilers, such nozzles could become direct structural elements.

However, to provide the most efficient oxidation of CO, it is required to determine the optimal angle of the vertical and the horizontal deflection of the nozzle. Figure 17 shows the comparison of the CO emission curves for all cases of the air feeding system considered within the studies described in the paper. Each curve has been generated as an average from the results obtained in three series of experiments.

The effect of the significant reduction of CO content in the exhaust gas has been achieved both in the case of the first and the second deflection angle at the stage of the proper combustion. Greater reduction of the CO emission has been noted for the nozzles bent at the angle of 5 degrees. In case of the deflection by 10 degrees, results of the experimental studies are inconsistent with the numerical model. The fact that in the case of the CFD model the considerations have been conducted for a steady state, where the CO source was defined as present in the same, invariable area, could be the cause of the discrepancies between the computation and the experiment.

Temperature curves generated based on the experimental data indicate that the dynamics of the flame propagation were different in the subsequent tests.

Consequently, the intensity of the CO emission has varied, which explains the discrepancies in the results.

Moreover, in case of the numerical model, the primary combustion chamber was completely excluded from the consideration.

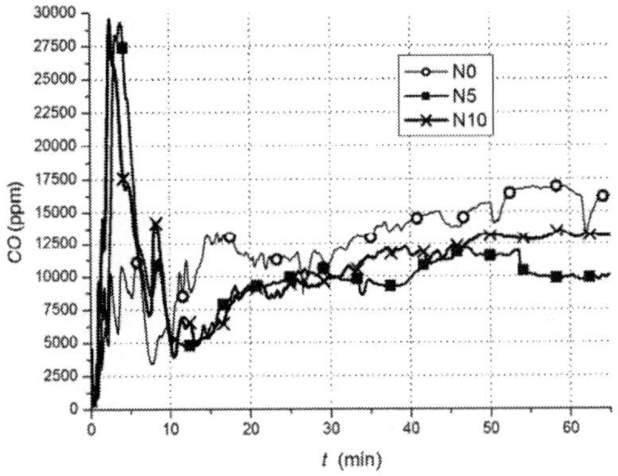

Figure 17. Mean CO emission curves for all of the considered cases: N0 – basic case, N5 – directional nozzles 5°, N10 – directional nozzles 10°.

Experiments were performed for the fuel load smaller than the nominal load and the chemical reactions could take place at the rear region of the primary combustion chamber (the chamber has not been completely cut off).

The location of biomass batch boilers in confined spaces such as the laboratory (the boilers are often installed outside because of their relatively large sizes) may cause exploitation problems related to the possibility of periodic extraction of smoke. For this reason, the ignition of the fuel has to be carried out gradually, which is achieved by appropriately increasing the fan power. The modulation of fan efficiency occurs based on the measurement of temperature curves and the general observation of the boiler operation. Because of this, it is much more difficult to conduct the combustion in the same way in each of the experiments.

Another difficulty related to the numerical modeling of the gasification is the inhomogeneous release of the gasification products from the defined source. While comparing the results of the simulation and the experimental data, randomness related to certain factors such as local moisture content or even fuel density need to be taken into account.

Despite all the problems, a simple CFD model developed within the described works has provided information regarding the characteristics of the gaseous fluxes in the considered area, which allowed to construct a prototype of efficiently operating directional nozzles. Studies carried out using the experimental system of the biomass batch boiler show that the application of the developed elements of the secondary air supply system is economically and technically reasonable.

The increased emission of CO at the ignition stage suggests that a modification of the process control at that stage has to be considered. It is recommended to apply additional air nozzles at approximately 2/3rd of the height of the secondary combustion chamber, which could provide additional secondary air fluxes during the first few minutes of the combustion without the risk of cooling of the bottom part of the combustion chamber, where the process of the biomass gasification takes place. In case of applying the solution referred to above, it is also recommended to reduce the heat transfer to the water jacket in the area of the bottom part of the secondary combustion chamber, to homogenize the process conditions in the entire vertical part of the chamber space.

Acknowledgment

This study was carried out under the "BioORC: Construction of cogeneration system with small to medium size biomass boilers" project.

References

1. S. B. Kang, J. J. Kim, K. S. Choi, B. S. Sim, Renew. En., **54,** 2-7 (2013)
2. R. Saidur, E. A. Abdelaziz, A. Demirbas, M. S. Hossain, S. Mekhilef, Renew. and Sust. En. Rev., **15**, 2262-2289 (2011)
3. A. Demirbas, Progr. in En. and Comb. Sci., **30**, 219-230 (2004)
4. B. Janković, Cellulose, **21**, 2285-2314 (2014)
5. B. Sørensen, *Renewable Energy. Its physics, engineering, environmental impacts, economics & planning*, 473-479 (2000)
6. W. Kordylewski, *Spalanie i paliwa (in Polish)*, 30-31 (2008)
7. E. F. Kristensen, J. K. Kristensen, Biom. and Bioen., **26**, 561-569 (2004)
8. A. Khor, C. Ryu, Y. Yang, V. N. Sharifi, J. Swithenbank, Fuel, **86**, 152-160 (2007)
9. H. Liu, J. Chaney, J. Li, C. Sun, Fuel, **103**, 792-798 (2013)
10. J. Riaza, R. Khatami, Y. A. Levendis, L. Alvarez, M. V. Gil, C. Pevida, F. Rubiera, J. J. Pis, Biom. and Bioen., **64**, 162-174 (2014)
11. C. Tenoiro, R. Moya, Therm. Acta, **563**, 12-21 (2013)
12. A. Garcia-Maravel, D. Salvachúa, M. J. Martinez, L. F. Diaz, M. Zamorano, Waste Man., **33**, 2245-2249 (2013)
13. C. Branca, C. Di Balsi, En.&F., **17**, 1609-1615 (2003)
14. Q. Xue, T. J. Heindel, R.O. Fox, Chem. Eng. Sci., 2440-2452 (2011)
15. J. I. Miguez, J. C. Morán, E. Granada, J. Porteiro, Renew. and Sust. En. Rev., **16**, 3867-3875 (2012)
16. S. Werle, R. K. Wilk, Renew. En., **35**, 1914-1919 (2010)
17. K. Sornek, M. Tomski, M. Szubel, M. Filipowicz, En. i Bud. (in Polish), **3**, 18-25 (2012)
18. M. Szubel, T. Siwek, Tech. Trans., **19**, 255-260 (2014)
19. M. Szubel, M. Filipowicz, EFM, 595-602 (2014)
20. U. Stritih, V. Butala, Renew. En., **29**, 2011-2012 (2004)

EPJ Web of Conferences **114**, 02120 (2016)
DOI: 10.1051/epjconf/201611402120

Comparison of two inlet boundary approaches in numerical simulation of car ventilation outlet flow

Tomáš Talanda[1a], Pavel Niedoba[1], František Lízal[1] and Miroslav Jícha[1]

[1]*Brno University of Technology, Faculty of Mechanical Engineering, Technická 2, 616 69 Brno*

Abstract. The paper is concerned with the comparison of two inlet boundary approaches of numerical simulations. The first approach is based on the simulation of a sufficiently long duct upstream the area of interest with known volumetric flow rate. The second approach rests on the experimental measurement of the section closer to the area of interest. The experimental measurement provides velocity profile and in addition turbulent intensity compared to the first approach where only a velocity profile can be computed according to known volumetric flow rate. A simplified model of car ventilation outlet was chosen as a test case. The model consists of circular and rectangular duct, area of interest (closing flap, vertical slats and horizontal slats) and outlet box. We have compared the two mentioned inlet boundary approaches for two distinct values of volumetric flow rate for which the experimental data of the section upstream of the area of interest are available. The velocity and the turbulent kinetic energy profiles downstream of the area of interest were chosen as comparative characteristics.

1 Outline

The development in computer technology still increases the advantage of numerical simulations compared to experimental measurements. The advantage is meant in the sense of time savings. The experimental measurements are still necessary for tuning of numerical models.

Computational fluid dynamics (CFD) is a tool for simulation of fluid flow [1]. The simulations can be divided into two categories. The first involves simulations which solve Navier-Stokes equations directly, e.g. Direct Numerical Simulation (DNS) and Large Eddy Simulation (LES). Due to enormous computational demands, these methods are still unsuitable for most engineering applications. More feasible is the second category of Reynolds Averaged Navier-Stokes (RANS) simulations which solve time-averaged Navier-Stokes equations and has to be closed with a turbulence model [2]. The most popular turbulence model for a wide area of applications is k-ω SST model of Menter [3].

2 Methodology

As a computation method for the CFD simulations, RANS simulation closed with the k-ω SST turbulence model was chosen. This turbulence model combines the k-ω turbulence model and k-ε turbulence model (where k is turbulent kinetic energy, ε is turbulent dissipation, and ω is specific dissipation rate) such that the k-ω model is used in the inner region of the boundary layer and switches into the k-ε model in the free shear flow region. A special blending function based on rates of turbulent kinetic energy, cell size, specific dissipation and viscosity is used for the switching. In addition, the k-ω turbulence model is suitable mostly for internal flows, while the k-ε turbulence model fits well to external flows [1]. In our case, internal flow occurs in the pipe, whereas free shear flow is present downstream of the ventilation outlet.

3 Results

In this section, the test cases as well as the numerical setup are described. Also the identification sign of each case is introduced for the comparison purpose in the last part of this section.

3.1 Test case

The benchmark ventilation outlet with one closing flap and two rows of slats was chosen as a test case (see figure 2). Airflow through this ventilation outlet was experimentally investigated by Šeda [4].

Altogether four subcases of different pipe geometries (short and long) and flow rates (33.5 m^3/h and 80.6 m^3/h) were investigated (see table 1).

[a] Corresponding author: t.talanda@seznam.cz

Table 1. Case identification sign.

		Flow rate	
		33.5 $m^3 h^{-1}$	80.6 $m^3 h^{-1}$
Duct	Long	**L33**	**L80**
	Short	**S33**	**S80**

3.1.1 Long duct geometry

The geometry of the long duct is displayed in the figure 1. It consists of circular duct, on which the rectangular part is connected. At the end of the rectangular duct, the area of interest with flap and slats is situated. The overall length of the duct is 1280 mm [4]. The outlet part of the geometry is modelled by a sufficiently large box representing free space.

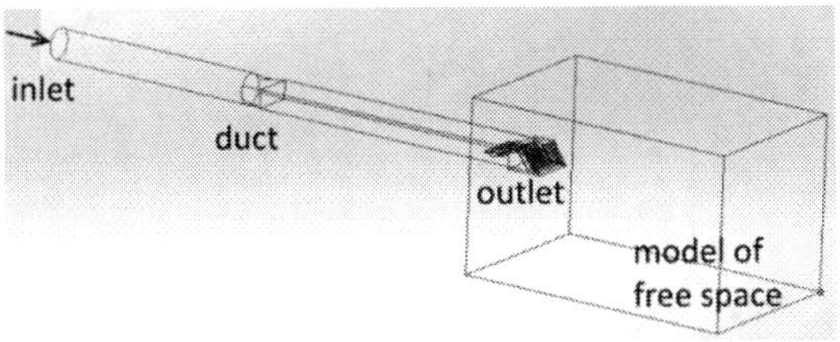

Figure 1. Long duct geometry.

3.1.2 Short duct geometry

The geometry of the short duct is depicted in figure 2, as well as the duct dimensions and flap and slats positions. In this case, the duct was shortened to 276 mm [4].

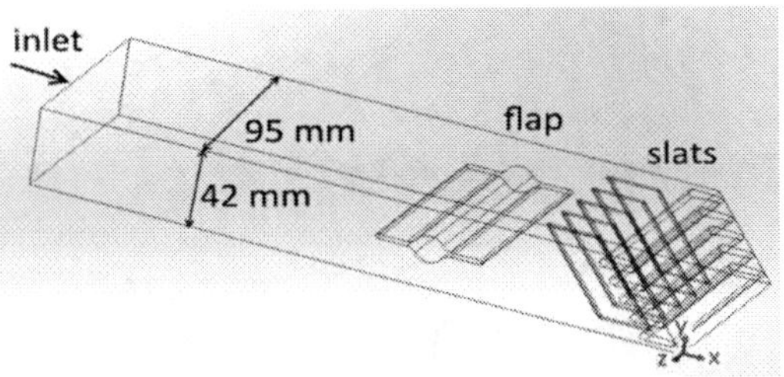

Figure 2. Short duct geometry, flap and slats position.

3.2 Numerical setup

The CFD simulation was performed with open source CFD software package OpenFOAM, where the RANS simpleFoam solver was selected. As a turbulence model, the kOmegaSST model was applied using the following wall functions, i.e., kqRWallFunction, nutkWallFunction and omegaWallFunction.

3.2.1 Cases S33 and S80

In cases S33 and S80 (see table 1), inlet values of velocity, turbulent kinetic energy k and specific dissipation rate ω at discrete points were set up according to Šeda experimental data [4]. The values of k and ω had to be calculated from the experimentally measured root mean square turbulent velocity according to the following formulas:

turbulent kinetic energy $$k = \frac{3}{2}(u'_{RMS})^2 \quad (1)$$

specific dissipation rate $$\omega = \frac{\sqrt{k}}{0.038 D_h} \quad (2)$$

hydraulic diameter $$D_h = \frac{2ab}{a+b} \quad (3)$$

where a and b are channel dimensions.

3.2.2 Cases L33 and L80

In these cases, uniform inlet profiles of velocity, turbulent kinetic energy and specific dissipation rate were prescribed. The calculation of the inlet velocity profile was based on experimentally measured volumetric flow rate. The turbulent kinetic energy and specific dissipation rate were calculated according to the formulas 1 and 2 with an assumption of turbulence intensity $I = 14$ % ($u'_{RMS} = \bar{u}I$).

3.1.2 Computational mesh

A slice of the computational mesh in the vertical symmetry plane is shown in figure 3.

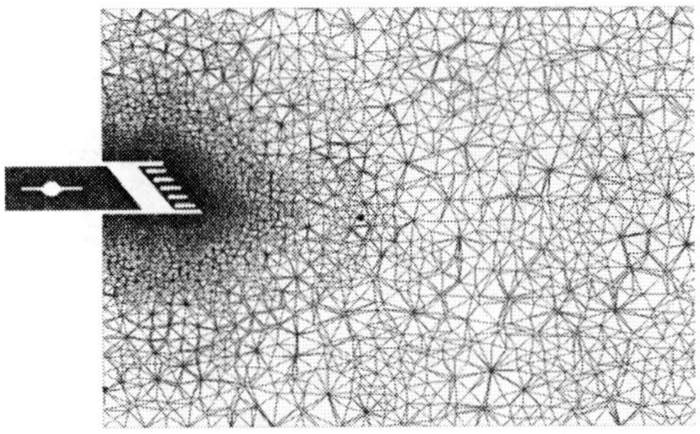

Figure 3. A slice of the computational mesh.

The outlet part was simulated by a large box (figure 1), using zeroGradient boundary condition. The computation mesh was created with an open source Salome mesher containing 4,238,467 tetrahedral cells for cases S33 and S80, and 6,682,931 cells for cases L33 and L80. Parallel computation was performed using a cluster of 32 processors. As a convergence criterion, the 10^{-4} residual was considered.

3.3 Results comparison

The computational results were compared along lines 1, 2, and 3 (see figure 4), where line 1 is parallel with the axis of the duct and line 2 represents the axis of the outflow. Line 3 is placed perpendicularly to the duct axis and 0.1 m downstream from the lower trailing edge of the outlet.

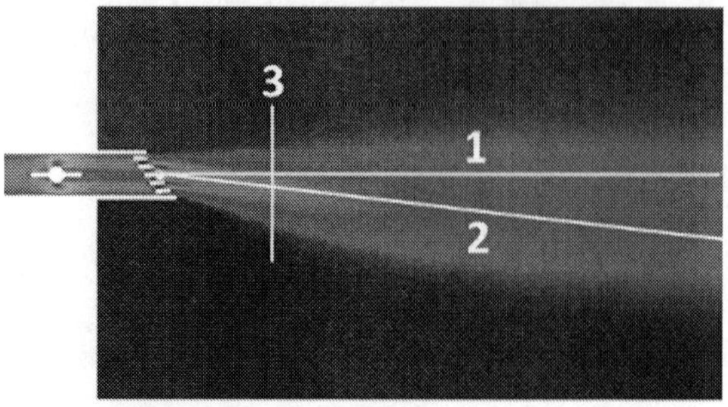

Figure 4. Lines for the results comparison.

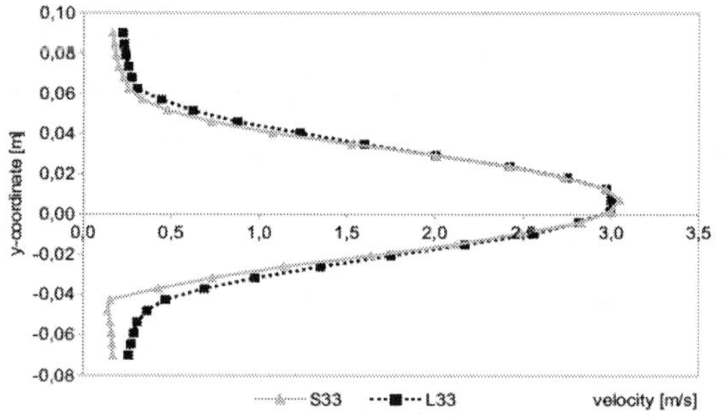

Figure 7. Plot of axial velocity in cases S33 and L33 over line 3

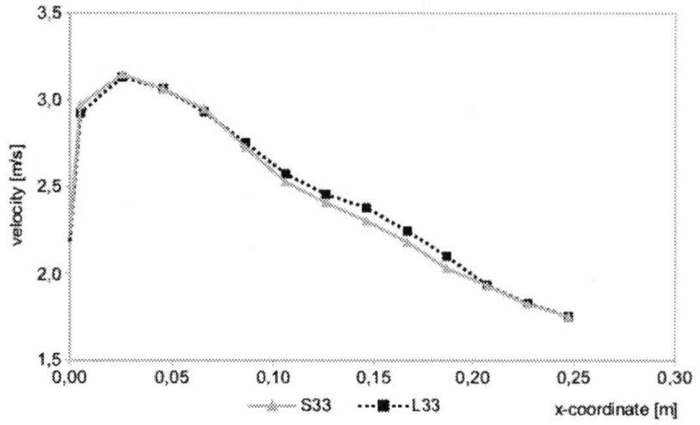

Figure 5. Plot of axial velocity in cases S33 and L33 over line 1

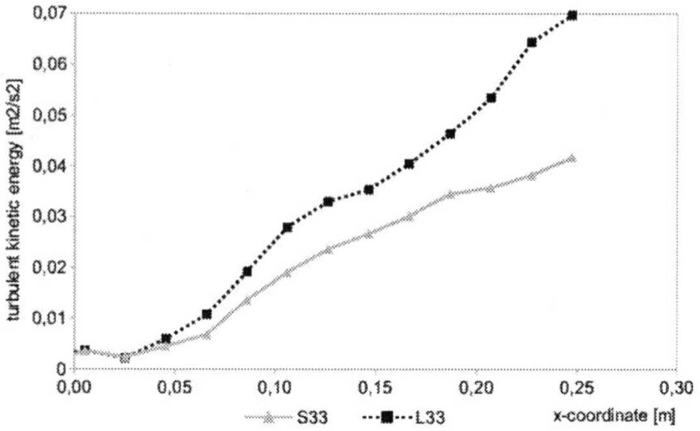

Figure 8. Plot of turbulent kinetic energy *k* in cases S33 and L33 over line 1.

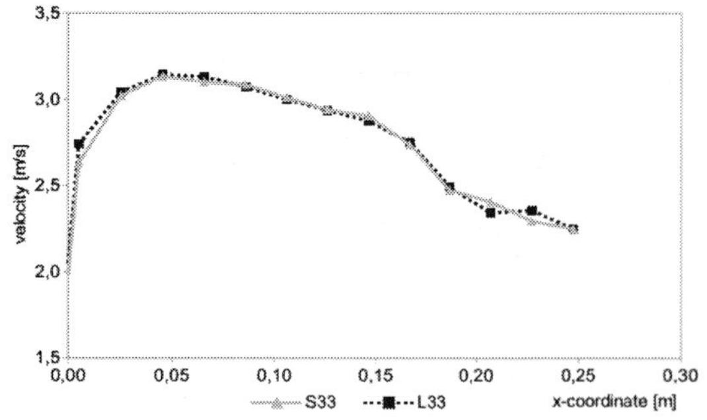

Figure 6. Plot of axial velocity in cases S33 and L33 over line 2

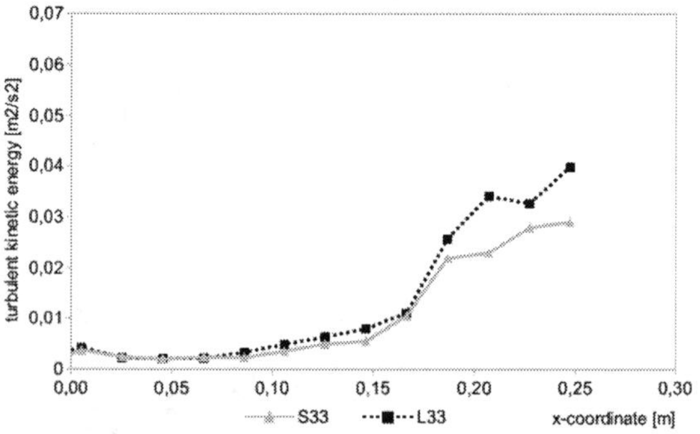

Figure 9. Plot of turbulent kinetic energy *k* in cases S33 and L33 over line 2.

The comparison of axial velocity profiles are shown in figures 5 – 7 for the cases S33 and L33 over lines 1 – 3. The same is shown for the cases S80 and L80 in figures 8 – 10.

The comparison of turbulent kinetic energy *k* profiles for the cases S33 and L33 over lines 1 and 2 can be seen in figures 11 and 12, and for the cases S80 and L80 in figures 13 and 14.

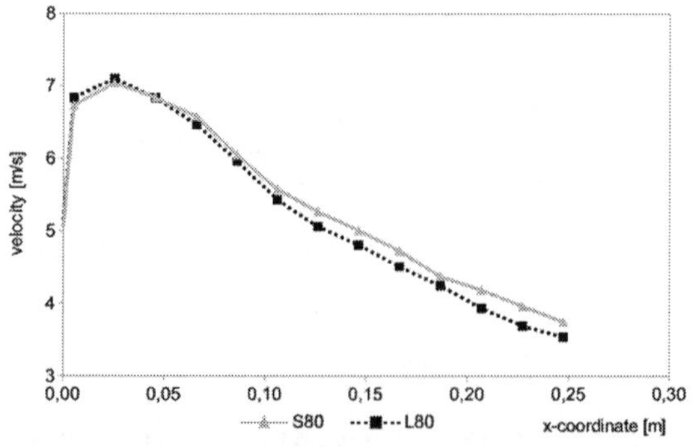

Figure 10. Plot of axial velocity in cases S80 and L80 over line 1.

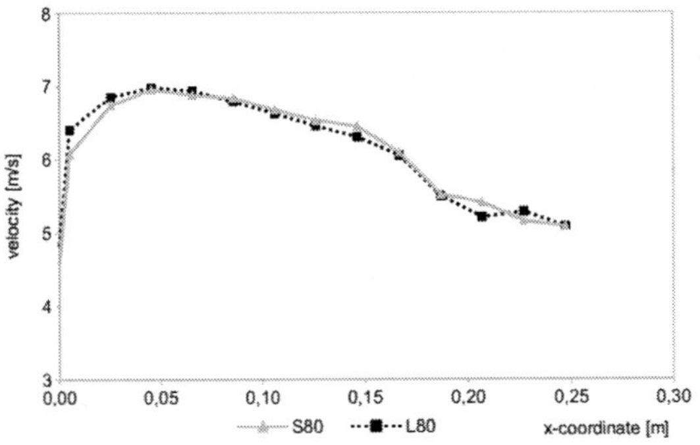

Figure 11. Plot of axial velocity in cases S80 and L80 over line 2.

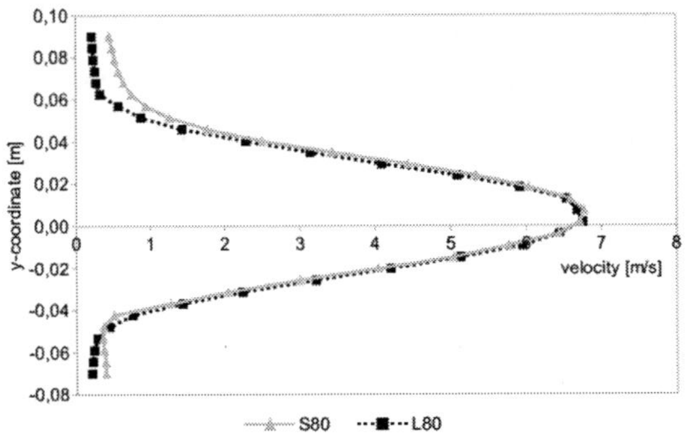

Figure 12. Plot of axial velocity in cases S80 and L80 over line 3.

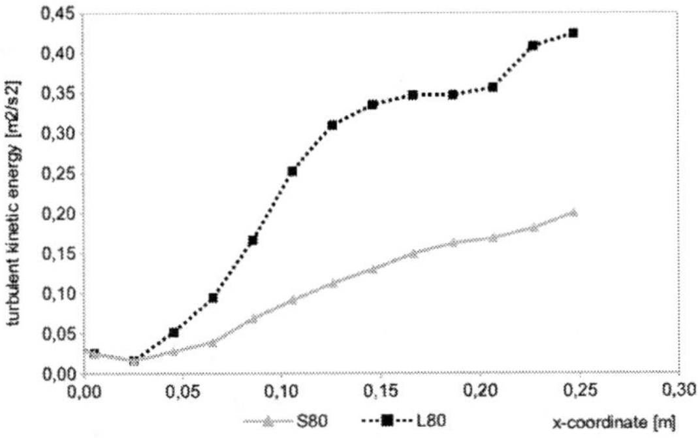

Figure 13. Plot of turbulent kinetic energy *k* in cases S80 and L80 over line 1.

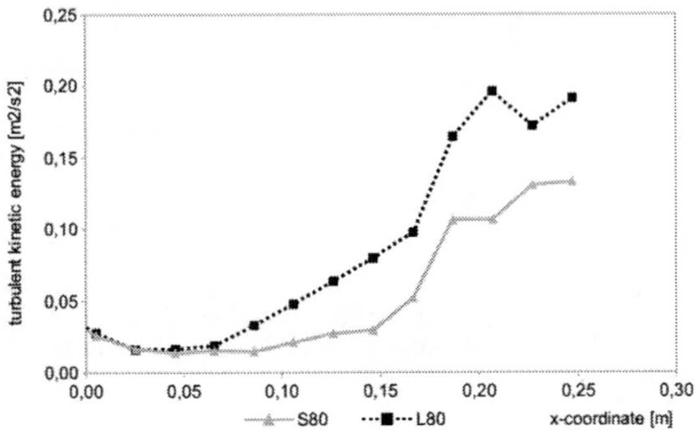

Figure 14. Plot of turbulent kinetic energy *k* in cases S80 and L80 over line 2.

4 Conclusion

In this paper, the axial velocity and the turbulent kinetic energy profiles downstream of the area of interest were compared for two different inlet boundary approaches and two distinct values of volumetric flow rate. The first approach takes into account a sufficiently long duct upstream the area of interest unlike the second approach where the inlet boundary profiles close to the area of interest in upstream direction were provided by experimental measurements. For the first approach, the inlet turbulent kinetic energy had to be computed because only the volumetric flow rate was known.

Resulted axial velocity profiles downstream of the area of interest are in very good agreement. As expected, the slight differences can be seen in the turbulent kinetic energy profiles. The differences may be caused according to the computation of inlet turbulent kinetic energy in the case of the first approach. Despite of the simple geometry, the computation of inlet turbulent kinetic energy has a significant impact on the downstream turbulent kinetic energy.

Thus, for the correct examination of the flow behaviour in the area of interest and downstream of this area, the experimentally measured inlet data sufficiently close to this area are desired. In addition, the first approach requires more computational performance relating to the extended geometry upstream of the area of interest.

Acknowledgement

This research was supported by the Czech Grant Agency under the grant GA 13-27505S and by the project FV 15-19 of the Faculty of Mechanical Engineering at Brno University of Technology.

References

1. S. B. Pope, *Turbulent Flows*, Cambridge University Press, 11th printing (2011). ISBN 0521598869
2. H. K. Versteeg and W. Malalasekera, *An introduction to computational fluid dynamics: The finite volume method*, Pearson Education Ltd., Harlow, UK, 520p (2007).
3. F. R. Menter, AIAA journal 32 (8), 1598 - 1605 (1994).
4. L. Šeda, *Analysis of flow from the car vent using thermoanemometric probe*, diploma thesis (2015)

EPJ Web of Conferences **114**, 02121 (2016)
DOI: 10.1051/epjconf/201611402121

Fluidic low-frequency oscillator consisting of load-switched diverter and a pair of vortex chambers

Václav Tesař[1], *Kazimierz Peszynski*[2], *Emil Smyk*[2]

[1]*Institute of Thermomechanics v.v.i., Academy of Sciences of the CR, Praha, Czech Republic*
[2]*UTP University of Science and Technology in Bydgoszcz, Faculty of Mechanical Engineering, Control Group, Poland*

Abstract. Paper discusses a new configuration of fluidic oscillators, a subject of recent Patent application. There is some similarity with the standard Warren oscillator with its bistable jet-deflection diverter and two feedbacks – which is not suitable for situations demanding very low oscillation frequency. For these conditions the new design replaces jet-deflection switching in the diverter by load-switching effects, with the gradually increased loading by spin-up of fluid in the vortex chambers. The spin-up time also provides the needed time delays. Behaviour is characterised by the oscillation frequency increasing with increasing fluid flow rate – for which was derived a surprisingly simple theoretical solution.

1 Introduction

Fluidic oscillator discussed in this paper generates self-exited pulsation in fluid flows in its two output terminals. It was developed in a response to demands for very low oscillation frequency that were encountered in generation of gas microbubbles in liquids. Already considerable time ago it has been recognised that the very small bubbles can bring significant advantages in a number of industrial processes (such as, e.g. [1, 2]) – but known methods of the microbubbles generation (such as, e.g., fragmentation by ultrasonics) were all uneconomical. The promising solution found in oscillating the gas flow into a percolation type aerator by inexpensive and reliable fluidic oscillators [2] was unfortunately found not dependable – sometimes it did produce the small bubbles and sometimes not. The hindrance to further progress in this direction was absence of knowledge about the microbubble formation mechanism. Initial approaches were based upon the idea of fragmentation of larger bubbles. If this were really the mechanism, it would be necessary for the fluidic oscillator to operate at very high frequencies. This is obvious since the decomposition into fragments would be the consequence of gradually growing excrescences on the surface of the original larger bubble – and effective formation of the excrescences would naturally require use of higher-mode resonant conditions. Because the "larger" bubbles are in fact quite small and their surface tension forces are consequently large, the resonant frequencies needed for the fragmentation are of the order of kHz [3]. This is higher than the usual oscillation frequencies of fluidic oscillators [4, 5]. Attempts at achieving the seemingly necessary high frequencies resulted in several extraordinary oscillator designs, like e.g. those described in [6] and [7]. A change in the approach to development was the caused by information obtained in recording the microbubble formations using a high-speed camera and very special optics [8]. The size of the bubbles generated with gas flow oscillations was found to be dictated by phenomenon of repeated conjunctions. The studies described in [8] have also indicated the ways to be followed for suppressing this phenomenon. To the general surprise, the required pulsation frequency was discovered to be the very low, i.e. the very opposite of what used to be believed before. Instead of the high frequencies the search become directed towards obtaining very low frequency.

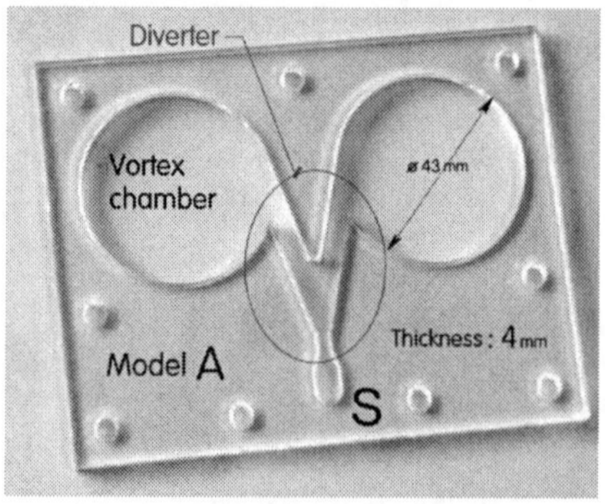

Figure 1. Photograph of the main plate of one of early oscillator models. The load-switched diverter delivers the fluid flow in an alternating manner into one of the two vortex chambers.

[a] Corresponding author: tesar@it.cas.cz

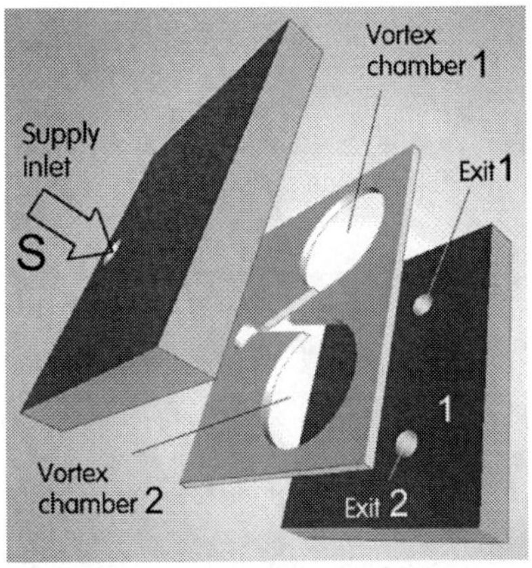

Figure 2. View of the three basic components of authors' oscillator models. The cavities for fluid flow in which is generated the oscillation are in the thin main plate (Fig. 1), clamped between the two thick cover plates, the latter containing the holes for supply (inlet) and outlet flows.

This new direction in the development of the fluidic oscillator encountered several difficulties. It became obvious that standard oscillator designs, which follow the idea of Warren [9], is unsuitable for generation of very low frequencies. The oscillators according to [9] are based on the classical concept of an amplifier with negative feedback. The amplifier is almost always of the bistable fluidic diverter layout, with jet of fluid deflected alternatively to one of two attachment walls by the Coanda effect phenomenon. Each wall leads the jet into one of two output terminals. The deflection is by an action of control flow from one of two control nozzles. This amplifier is converted into the oscillator by providing it with two feedback loop channels connecting the output terminal with the control on the same side nozzle. A part of the fluid flow leaving the amplifier output terminal is fed back into the control nozzle. There it causes separation from the attachment wall and switching of the jet to the opposite side. There the mirror image of the process takes place, resulting in periodic switching. Frequency of oscillation is primarily determined by the length of the feedback channels (which define the necessary time delay between the flow deflection and its return). Secondary influence on the frequency has the supplied fluid flow rate. This dependence on flow is a characteristic property of all no-moving-part oscillatory flows in fluids

This type of oscillator layouts cause no particular problems with standard-sized amplifiers – let us say with supply nozzle widths ~2 mm — generating oscillation at frequencies from 50 Hz to 300 Hz. Earlier, where the solution of the microbubble generation problem was sought at high frequencies, development was directed towards minimising the length of the feedback path [6]. Nowadays, with the understanding of the microbubble generating mechanism calling for the very opposite task of obtaining frequencies roughly decimal order of magnitude ***lower,*** the obvious approach is to have the feedback channels very long - as long as is physically possible. Typically the lengths are metres or tens of metres.

On one hand, such extreme lengths are difficult to stow in a microbubble generator of reasonable size. More important, on the other hand, is the fact that with such lengths the oscillators tend to stop oscillation. The reason is the hydraulic friction loss in the very long channels and weakening of the feedback signal. The signal may be so weak – and immersed in chaotic turbulence – that is may not suffice for really regular switching.

2 New layout

The reason why there is the necessity of large feedback channel lengths is the time delay needed for the deflected jet to move from one attachment wall to the opposite one – and to stay attached there for a part of the oscillation period. Apart from the channel lengths, the parameter that dictates the oscillation frequency is the signal propagation velocity in the channel. Because of the difficulties encountered with the classical layout, it became useful to consider other means of no-moving-part fluidics for obtaining the time delay. Some of the alternative approaches seemed to lead – in principle at least - to more convenient solutions of the stowage problem as well as to elimination of switching irregularity caused by the fluid energy dissipation in the long channels.

An obvious example of an alternative time delay is the accumulation of fluid for a part of the cycle in an accumulation chamber, made possible by fluid compressibility [10, 11]. An example of a fluidic oscillator using this capacitance mechanism is the Zalmanzon's design, presented in Fig. 4 of ref. [6]. Experience has shown, unfortunately, that this principle of generating time delay is suitable only for high frequency oscillation. Obtaining the desirable long

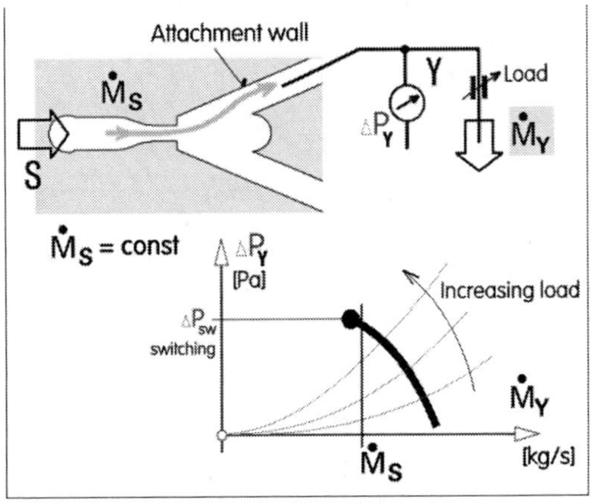

Figure 3. Principle of the load-switching. As dissipance of the load gradually increases the conditions in the output terminal Y move along the loading characteristic curve up to the switching point. — where the Coanda effect can no more keep the flow deflected into the terminal Y.

delay time in this way at the low frequencies of late interest would call for extremely large accumulation chambers. Another alternative solution, one that is discussed in this paper, is the use of the spin-up time of a vortex in a chamber with tangential inlet and centrally located outlet. It takes some time for the rotation to reach its final intensity. The idea of using this mechanism evolved from the use of vortex diodes, first discussed by Zobel [12], in valve-less pumps [13]. Despite the advantage of high achievable steady-state diodity of the Zobel diodes, they were found not suitable for use in microfluidic pumps because of the spin-up and spin-down delays. This made possible operating the vortex diodes only at very low frequencies (and hence with large size of the displacement volumes of the pump). This property, while a drawback in pumping, may be very useful for the here discussed problem of low-frequency oscillators.

The new layout may be explained using the photograph presented in Fig. 1. This oscillator layout is covered by the recent Patent application [14]. In similarity to the Warren's oscillator, the supplied fluid flow enters a diverter valve – with two Coanda-effect attachment walls and corresponding two output channels. Downstream from these diverter outputs are two vortex chambers, each in a configuration closely corresponding to the Zobel [12] diode (with small detail differences demanded my the different manufacturing

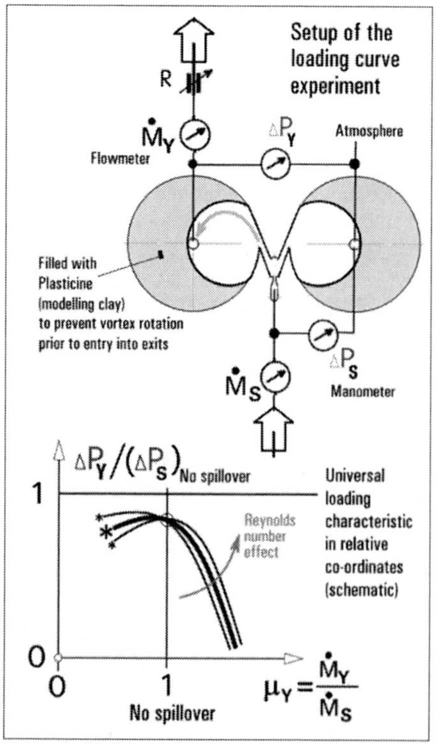

Figure 4. Schematic representation of the setup used in experimental investigation of the loading characteristic. Filling of substantial part of the vortex chambers by Plasticine converted them into settling cavities without centrifugal effects.

methods). In principle, it might be possible to switch the diverter by alternating flows issuing from jet-deflecting control nozzles. Nevertheless, in the configuration according to [14] preference is given to a simpler configuration based on the load-switching mechanism.

3 Load-switching diverter

The vortex type rotation of fluid takes place in an alternating manner in one of the two vortex chambers. The rotation is caused by the tangential orientation of the chamber entrance from the diverter. Initially, when the flow is switched to enter the vortex chamber, the fluid may is (or may be considered to be) non-rotating – and, hence, the pressure drop across the chamber is relatively low. The diverter thus does not experience any significant resistance to its outflow. The Coanda-effect causing the fluid flow to attach to the active attachment wall in the diverter can easily keep it moving into the load that is here represented by the vortex chamber.

After the initial inertia of the fluid inside the chamber is overcome, the rotation starts and is gradually increasing in intensity. The diverter output flow then meets an increasing resistance. This is represented in the loading characteristic diagram in the bottom part of Fig. 3. On the horizontal co-ordinate in this diagram is plotted the output mass flow rate from the diverter into the location Y -- while on the vertical axis is plotted the pressure difference between the location Y and the atmosphere (downstream from the exit from the vortex chamber). Represented in the diagram ion Fig. 3 is the fact that the jet issuing from the supply nozzle entrains fluid flow coming through the inactive output terminal (the bottom one in the picture) – by the jet-pumping effect. As a result, the output flow rate through Y is higher than the supply flow rate through S. The thin parabolic lines in the diagram show the characteristics of the load – the vortex chamber. The instantaneous states in the diverter output are determined by the intersections of these load characteristic (thin) and the loading characteristic of the diverter (thick line). As the rotation speed inside the vortex chamber increases, these characteristics are gradually more and more steep. The intersection moves upwards. The diverter has to overcome the higher hydraulic resistance associated with the centrifugal forces. One of the consequences of this process is the decreasing amount of jet-pumped fluid flow. The conditions inside the location Y may, as is the case in Fig. 3, reach the no-spillover regime indicated by the vertical straight line. Further increase of the rotation intensity in the vortex chamber reverses the conditions: instead of the inflow due to the jet pumping, some fluid from the supply terminal S is actually spilled over as $\dot{M}_S - \dot{M}_Y > 0$ into the inactive output from the diverter and into atmosphere.

Finally, the conditions reach the state in which the Coanda effect can no more hold the fluid flow attached to the upper (active) attachment wall. This limiting condition is indicated in the diagram in Fig. 3 by the

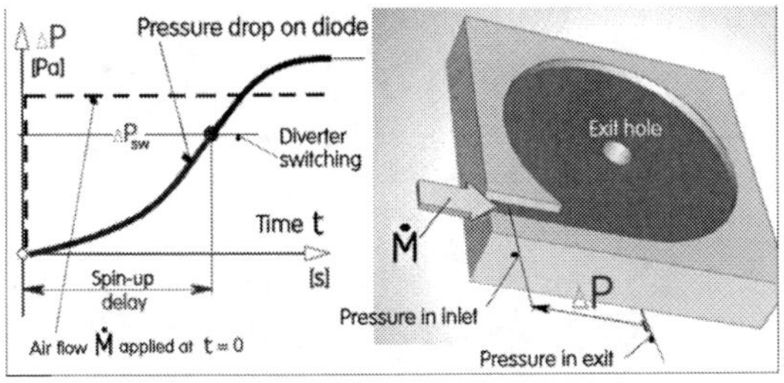

Figure 5. Response of a vortex diode to sudden input of flow $\dot{M}$ into the tangential inlet – after which is this flow kept constant. The vortex spin-up process takes a considerably long time before the pressure difference ΔP (between the tangential inlet and central outlet) reaches its final level.

horizontal line – the line of the switching pressure difference. The crossing of this line is marked in the diagram Fig. 3 by a large round black symbol: the fluid flow there separates from the upper attachment wall and is switched to the other, lower wall.

Figure 4 shows schematically the details of the experiments performed for investigating the load-switching effect of the diverter. Note there the fact that the vortical motion in the vortex chambers in these investigations would distort the diverter behaviour results. To eliminate them, the vortex chambers had to be converted into low flow velocity settling cambers, without the centrifugal effects caused by the vortical rotation. Another difference worth being noted is the fact that instead of the single loading characteristic shown in Fig. 3 were now in Fig. 4 measured three such characteristics, each for a different magnitude of the supply mass flow rate. The graphical presentation in the bottom part of Fig. 4 uses not the absolute co-ordinates (as it does in Fig. 3) but instead relative co-ordinates: on the vertical axis it is the ratio of the output mass flow rate to the supply mass flow rate — while on the vertical co-ordinate are plotted the measured pressure differences between the location Y and the atmosphere, divided by the pressure difference between the supply terminal S measured in the no-spillover states. Ideally, in these co-ordinates all three loading characteristics might be expected to squeeze into a single universal line (as a consequence of the Eulerian similarity). The

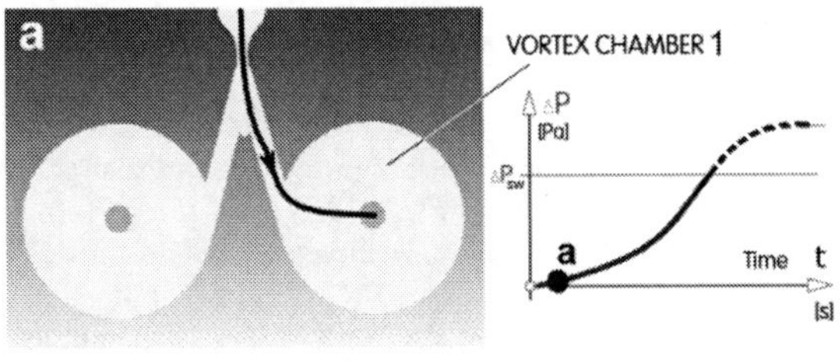

Figure 6. Conditions in the oscillator immediately after the admission of the air flow $\dot{M}$ into the vortex chamber 1. The pressure drop across this chamber is very small since there is no rotation and no centrifugal action to overcome.

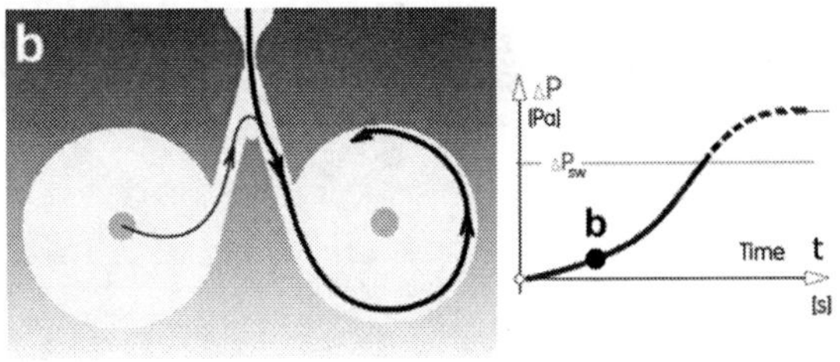

Figure 7. At a later stage "b" than in the previous Fig. 6 the pressure drop across the diode is still small since the spin-up time is quite long. The increased flow rate causes some air to be entrained by jet-pumping effect through the vortex chamber 2.

characteristics, however, represent among other influences the friction loss component, which is Reynolds-number dependent. Thus the three curves, each measured at different Reynolds number, became slightly dissimilar. Also the position of the load-switching point is Re-dependent.

4 Delayed feedback: spin-up

The switching of the diverter by the increasing load is in the new oscillator made by the increasing pressure drop across a vortex chamber the fluid inside which starts to rotate. At the same time, the inertial response to the starting rotation ensures the desirable time delay.

The vortex chambers are, as is seen in Fig. 5, very simple and therefore easy to manufacture. Their main component is a large circular cavity made in the thin main plate. The rotation is generated by the tangentially directed inlet at the chamber circumference – connected to the exits from the diverter.. In the centre of the vortex chamber is, in the bottom support (cover) plate an exit hole through which the fluid leaves the oscillator.

To get a large time delay it is advisable to make the vortex chamber relatively large (cf. Fig. 1) – because, as is discussed below, the magnitude of the delay effect is directly proportional to the mass of the fluid inside the chamber. Also the very shallow geometry (small vertical difference between the top and bottom cover plates) is desirable. The large exposed cover plate surfaces slow down rapidly by fluid friction the vortical motion once the tangential inflow is switched off. This secures turn-down effect is a necessity because it makes possible starting the spin-up process anew in each

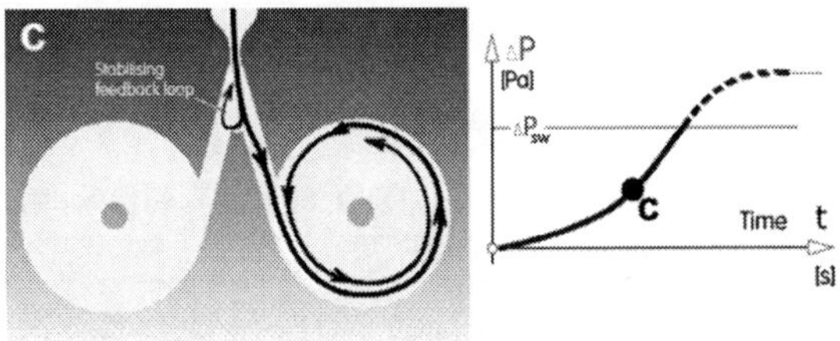

Figure 8. A short time later the rotation in the vortex chamber 1 intensifies. Increased pressure difference to be overcome by the jet causes the jet-pumping to cease. The jet nevertheless remains attached to the ON-side attachment wall.

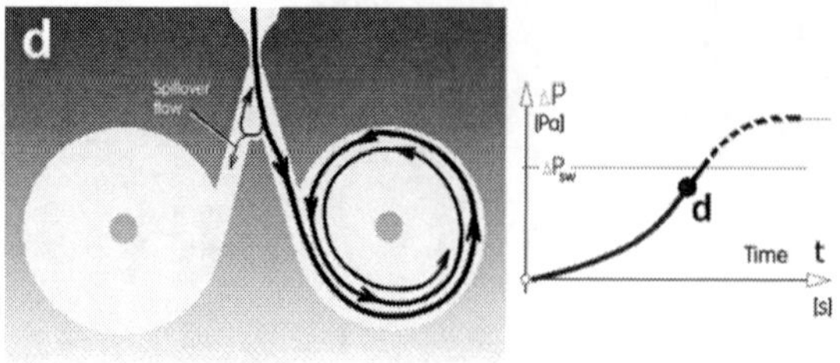

Figure 9. The intensive rotation in the vortex chamber 1 has increased the pressure drop across it substantially. There is some small spillover flow $\dot{M}_S - \dot{M}_Y > 0$ (cf. Fig. 3)

oscillation period (or at least accelerating the rotation if it is not completely stopped by the spin-down effect).

The diagram in the left part of Fig. 5 represents schematically the pressure-drop response of fluid in the vortex chamber to a sudden admission of the constant mass flow rate $\dot{M}$ into the tangential inlet. Important fact presented there is the time delay between the instant of admission and the time at which is reached the diverter switching limit. The pressure response curve in Fig. 5 is drawn as continuing above the diverter switching level. This would be found in experiments with the vortex chamber alone – when the vortex chamber is a part of the oscillator, the switching of the diverter delivers the fluid into the other chamber, as will be schematically presented below in Fig. 15.

Individual phases of the spin-up processes are presented in Figures 6 to 10. As is indicated by the arrows showing the local direction of motion, the processes are complicated by secondary effects – such as the jet-pumping effect that may be present in the very initial stages (Fig. 7) and the stabilising vortical motion created due to the bi-cuspid shape of the splitter nose (-note the 3 mm radius in Fig. 12).

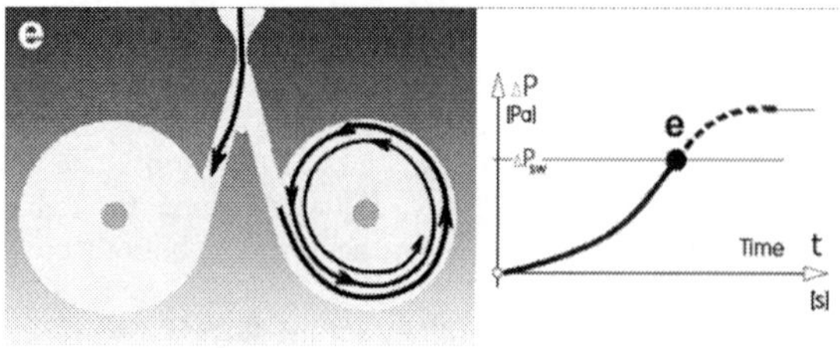

Figure 10. When the load-switching limit is reached, the jet is switched to the other side of the diverter interaction cavity. Due to inertia, some rotation continues in the vortex chamber 1.

5 Laboratory model

The model used in the laboratory investigations is shown in the photograph in Fig. 11. It is a larger version among a few initial designs (one ob them the version A, is seen in the photograph in Fig. 1). The larges size B – and hence low oscillation frequency — was chosen for the tests because of the expected less difficult

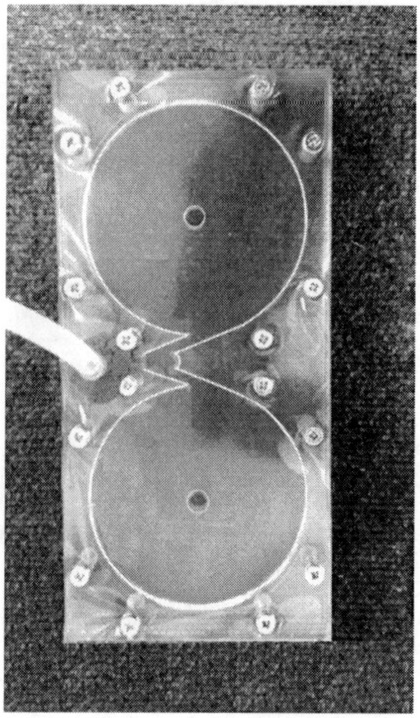

Figure 11. Photograph of the oscillator model taken during test run in the laboratory. Note the sixteen clamping screws and the plastic tube (at left) supplying the air flow. The hot-wire anemometer probe (here it was removed and is not seen) was used to measure and record instantaneous velocity samples at one of the vortex chamber exits.

measurements at very low oscillating frequencies. The vortex chamber diameters of the tested model were 100 mm. Geometry of the critical component of the model – the main plate - is presented in Fig. 2 The dominant (i.e. smallest) cross section of the oscillator, the exit of the supply nozzle, was of dimensions 3 mm x 7.8 mm – the latter dimension being the thickness of the PMMA (polymethylmethacrylate) plate in which the cavities were made by numerically controlled laser cutting. The apex angle of the two attachment walls in the diverter was 45 deg The exist from the vortex chambers were simple holes of 8 mm diameter.

6 Measurements and results

Experiments were performed with air as the working fluid. The supplied air flow rate was measured by digital mass flow meter FMA 1727A — range 0 to 40 standard litres per minute, accuracy ±1.5% of full scale, with an integral display. The oscillator output was measured by hot-wire MiniCTA anemometer manufactured and supplied by Dantec Dynamics. The probe used was 55 P 161, its hot wire positioned downstream from one of the 8 mm dia. exits from the vortex chamber. The velocities were measured at sampling frequency 2.048 kHz, The amplified electric

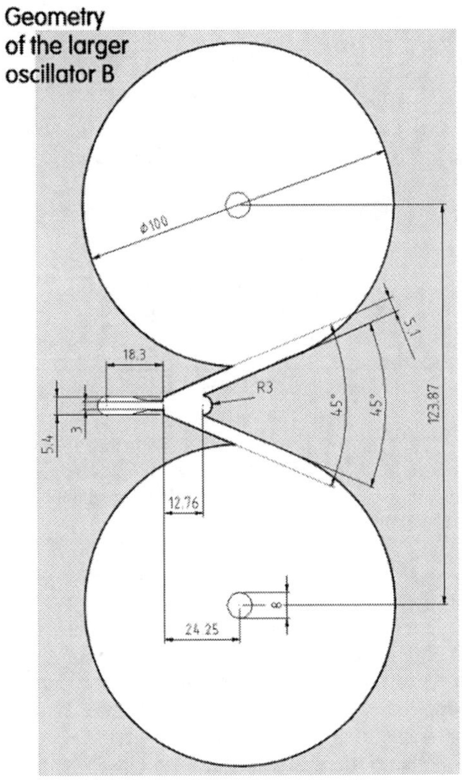

Figure 12. Geometry of the main plate (7.8 mm thick) of the oscillator model B with which were made detailed investigations of oscillatory behaviour.

output from the CTA bridge was linearised and stored in packets, each packed containing 4096 instantaneous velocity data values to be used in subsequent processing. In particular, the data were used for evaluation of time-mean velocity and by computing the velocity spectra by fast Fourier transform.

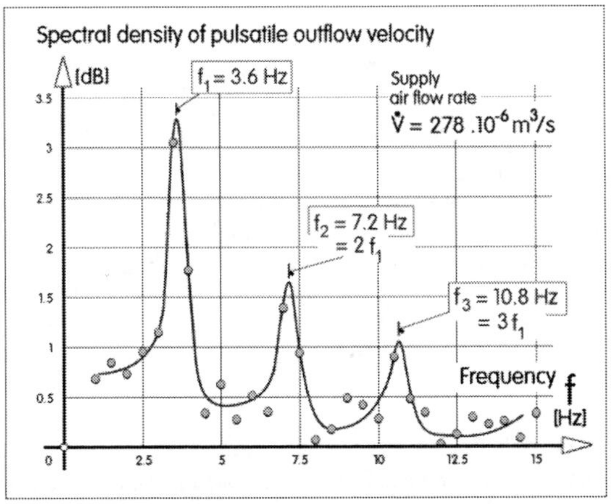

Figure 13. Typical frequency spectrum of the velocity measured at the exit from one vortex chamber at a relatively small air flow rate. Apart from the main frequency f_1 there are significant higher harmonics.

The next Fig. 13 shows a typical example of the results obtained by the Fourier transformation: a spectral density diagram. This one was obtained at a rather small air flow rate. The most important fact are the really very low frequency levels, which would be difficult to obtain with the usual fluidic oscillators. The main frequency in the case presented in Fig. 13 is mere f_1 = 3.6 Hz . Its amplitude is very much higher than all other velocity components, like the quite significant harmonics.

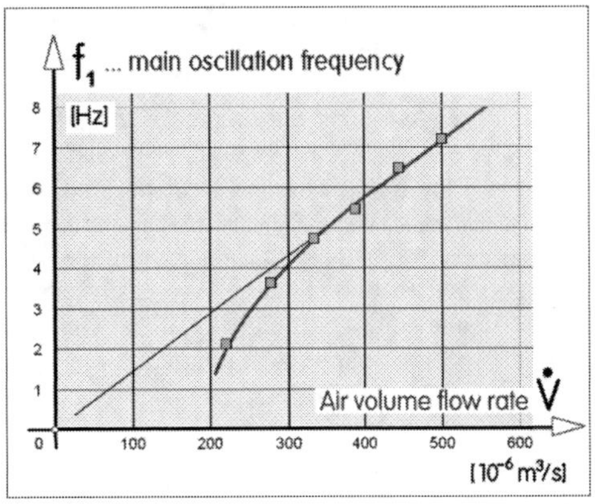

Figure 14. Dependence of the main oscillation frequency f_1 on the supplied air flow rate. Apart from the deviation at very small flow rates this behaviour indicates constancy of Strouhal number.

Similar diagrams were obtained for 8 other magnitudes of the supply flow rate. In each of them was identified the main frequency f_1 . In the next Fig. 14 is their dependence on the flow rate. It is immediately obvious that the oscillation in the developed oscillator are not a result of acoustic resonance (which would be unlikely anyway because of the size of the resonator needed for so low frequencies) which would be characterised by constant frequency independent on the flow rate. Instead, there is a monotonous growth of the frequency with the increased flow rate. A straight line fitted to the data points above the lower limit volume flow rate 0.3 10^{-3} m^3/s passes through the origin of the diagram, which means constancy of the dimensionless parameter known now as Strouhal number Sh.

$$Sh = f_1\ b\ /\ w \qquad (1)$$

- a typical property of no-moving-part aerodynamic oscillation associated with shedding of vortices. As usual in fluidics, the Strouhal number in eq. (1) is defined by evaluation from the conditions in the supply nozzle (- using the supply nozzle width b = 3 mm and bulk velocity w). The values of w were in the experiments varied between 4.7 m/s and 23,7 m/s. thus the Strouhal number values were all very near to Sh = 0.001. The corresponding Reynolds number values were in the range from Re = 942.6 to 4 713. Since the operation of the diverter depends on entrainment of outer air into the jet flow, the non-existence of the oscillation at velocities lower than the indicated value w = 4.7 m/s is obvious the dominating laminar character of the jet makes the Coanda effect attachment too weak.

7 Characteristic time – and a simple theory of oscillator functioning

There are two parameters of key importance in fluidic oscillators. The first one is the supplied mass flow rate $\dot{M}$ (the volume flow rate, an in Figs. 13 and 14, is an analogous but not conserved parameter, acceptable as alternative in situations with negligible compressibility) while the other one is the frequency of generated oscillation. The latter is in the discussed oscillator principle dependent upon the time delay caused by the spin-up process – and this, in turn, is dependent upon the inertia of the fluid which has to be put into rotation inside the vortex chamber. A very rough approximation providing an information about the magnitude of this inertia is provided by the mass M_{ch}, easily evaluated from the diameter D and height h of the chamber as well as the specific volume v of the fluid:

$$M_{ch} = \frac{\pi D^2 h}{4 v} \quad (2)$$

Obviously, the ratio of the two parameters

$$t_C = M_{ch}/\dot{M} \quad (3)$$

is the characteristic time of the spin-up process. It is interesting to compare it with the duration of the oscillation period – or, better, with the half-period

$$t_p = 1/(2 f_1) \quad (4)$$

- which is the time (compare with the schematic representation of the period in Fig. 15) taken by the pressure difference between the tangential inlet and the central outlet to increase from the initial condition (in Fig. 15 is assumed a condition of completely stopped rotation) to the switching limit. Of course, the spin-down time may be different from the spin-up, but this is of no consequence because of the symmetry of the oscillator, whereby the same start-up periods in the vortex chamber 1 and vortex chamber 2 are summed to represent the period.

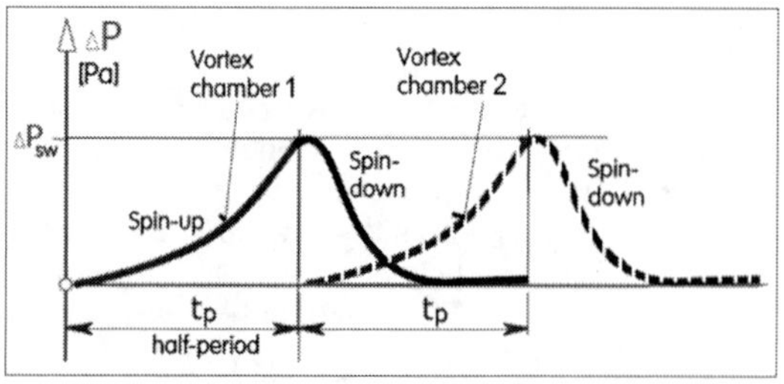

Figure 15. Schematic representation of the mutual relation between the processes in two vortex chambers. The half-period lasts $t_p = 1 / (2 f_1)$.

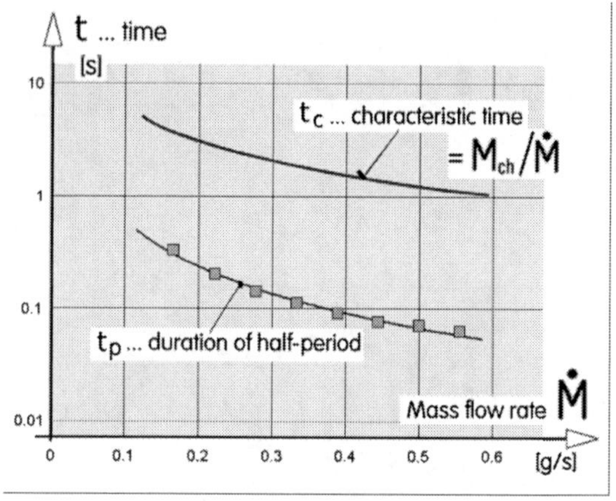

Figure 16. Comparison of the characteristic time (evaluated from the air mass M_{ch} inside the vortex chamber and the supplied air mass flow rate) with the half-period computed from the measured data.

Because the frequency f_1 in eq. (4) is known from the measurements (Fig. 14), the two elapsed times eq.(3) and eq. (4) may be compared, as is done in Fig. 16. Significantly, while the magnitudes of the two times are different, the character of their dependence on the supplied air mass flow rate is the same. When the characteristic time eq.(3) and the half-period eq. (4) are divided, their ratio was found to be an integer constant

$$t_c = 17\, t_p \quad (5)$$

- with the exception of the deviation at small flow rates, already discussed as insignificant because of the too low values of Reynolds number. By substitution of eq. (5) and eq. (2) into eq. (3) one obtains a surprisingly simple expression for the oscillator behaviour

$$f_1 = \frac{34\, v\, \dot{M}}{\pi D^2 h} \quad (6)$$

The importance of such a simple expression is obvious for predicting the performance and designing a system with the oscillator as a component. Of course, the numerical value in the expression eq.(5) is valid only for the same oscillator geometry as presented above in Fig. 12. Perhaps with considerable cautiousness, the results may be extended to slightly different geometries, with different vortex chamber diameters D and different supply nozzle widths b because of the ensuing expression for the Strouhal number — eq. (1)

$$Sh = \frac{34\, b^2}{\pi D^2} \quad (7)$$

- which indeed leads for the geometry from Fig. 12 to the values that were in experiments foud to be all very near to Sh = 0.001.

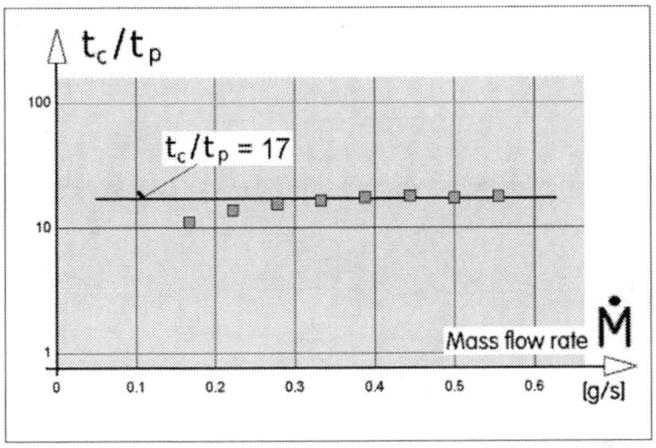

Figure 17. An important fact discovered by analysis of the experimental data: the oscillation period is a simple constant multiple of the characteristic time.

The question about the meaning of the constant 17 in eq. (5) has simple answers. The fact that it is an integer is very probably just an accident. Why it is so large is due to the simplifications introduced in its derivation. It involves the effects of rather complex velocity profiles inside the vortex chamber. Also, it includes the switching time of the diverter (neglected in the derivation completely). Finally it contains the temporal dependences of the secondary flows into (jet pumping) and out from (spillover) the diverter.

8 Conclusions

Paper presents interesting results obtained in the development of very-low-frequency fluidic oscillator, originally considered for generation of microbubbles in fluids but obviously suitable for a wide spectrum of other uses. The oscillator is of planar configuration, suitable for modern manufacturing methods. Its main advantage is compactness, which with currently standard oscillators (especially with their necessity of extremely long feedback channels) is a problem.

Apart from demonstration of realisability of the new layout, of considerable value in this paper is the development of a simple and yet very useful theory.

Acknowledgements

First author obtained support from GAČR - Grant Agency of the Czech Republic, by research grant Nr. 13-23046S. There was also institutional support RVO:61388998.

Third author obtained support from Ministry of Science and Higher Educating of Poland in form of "Diamond Grant" 0201/DIA/2015/44 for young researchers.

References

1. *Microbubbles Revolutionize Industrial Processes*, www.mitsubishielectric.com/company/environment/.../ microbubbles/

2. W. J. B. Zimmerman et al., *Applied Energy* **88** 3357 (2011)

3. V. Tesař, *Chemical Engineering Journal*. **235**. 368 (2014)

4. V. Tesař, S. Zhong, F. Rasheed, *AIAA Journal* **51** 397 (2013)

5. Raghu S., *Experiments in Fluids* **54**, 1455 (2013)

6. V. Tesař, *Chemical Engineering Research and Design* **92** 1603 (2014)

7. V. Tesař, *Sensors and Actuators A: Physical*, **234** 158 (2015)

8. V. Tesař, *Chemical Engineering Science* **116** 849 (2014)

9. R.W. Warren, *Negative feedback oscillator*, US Patent 3158166 (1962)

10. V. Tesař , *Fluidics Quarterly* **11** 77 (1979)

11. V. Tesař, Peszynski K., *Pneumatyka* **58** 48 (2006)

12. R. Zobel, *Mitt. Hydraulischen Instituts der Technischen Hochschule Munchen*, No. 8, R. Oldenbourg, Munich–Berlin (1936).

13. V. Tesař: "*Valve-Less Rectification Pumps*" Entry on p. 2132 in "Encyclopedia of microfluidics and Nanofluidics", Ed.: Dongqing Li, publ. by Springer Science+Business Media, (2008)

14. V. Tesař, *Fluidic oscillator*, Czech Republic Patent Application PV 2014-903, December 2014

15. V. Tesař, *Proc. of FLUCOME '94*, p. 741, Toulouse, France (1994)

16. Section 4.4 in Tesař V., *Pressure-driven microfluidics*, published by Artech House Inc., Norwood, MA, U.S.A. (2007)

17. V. Tesař, H.C.H. Bandalusena, *Experiments in Fluids* **50** 1225 (20110

18. V. Tesař, E. Smyk, *Chemical Engineering and Processing: Process Intensification* **90**, 6 (2015)

19. A. Ogawa, H. Murakami H., *Internal Flows*, P. Doerffler ed., Instytut Maszyn Przeplywowych PAS, Gdansk, Poland, p. 825 (2001)

20. S.-G. Lim, S.-H. Lee, Kim H.-G., Nuclear Engineering and Design **240** 1139 (2010)

21. J. Yin, L. Jiao, L. Wang, *Nuclear Engineering and Design* **240** 970 (2010)

22. A. A. Kulkarni, et al., *Chemical Engineering Science* **64** 1285 (2009)

23. Jiao., et al., *Science China – Technologiocal Sciences* **54** 332 (2011)

24. F. Haakh, *Journal of Hydraulic Research* **41** 53 (2003)

EPJ Web of Conferences **114**, 02122 (2016)
DOI: 10.1051/epjconf/201611402122

Fluid mechanics of additive manufacturing of metal objects by accretion of droplets – a survey

Václav Tesař[1,a]

[1]Institute of Thermomechanics v.v.i., Academy of Sciences of the Czech Republic
Dolejškova 5, 182 00 Praha-Kobylisy, Czech Republic

Abstract. Paper presents a survey of principles of additive manufacturing of metal objects by accretion of molten metal droplets, focusing on fluid-mechanical problems that deserve being investigated. The main problem is slowness of manufacturing due to necessarily small size of added droplets. Increase of droplet repetition rate calls for basic research of the phenomena that take place inside and around the droplets: ballistics of their flight, internal flowfield with heat and mass transfer, oscillation of surfaces, and the ways to elimination of satellite droplets.

1 Introduction

A new approach to manufacturing of objects has been recently gaining momentum. In contrast to traditional subtractive methods of removing not needed material from the processed product, the new method is additive – the basic idea is gradual addition of small particles to the machined product. Historically, the underlying idea of production by addition has never been completely absent – gluing together pieces of wood has always been used to make furniture and bricklayers have always made walls by adding together bricks. In the new approach, however, there are three essential novelties:

- accretion of very small, generally sub-millimetre material volumes,
- moving the accretion location by a servomechanism
- computer control of the process

The main advantage of the additive manufacturing is its universality. There is no need for jigs and special tools or similar auxiliary devices. Each individual product may be specially adapted – by a change in the computer program - to become a unique original.

The additive technology, also called "*3D printing*", has been already developed during the past two decades but its use so far was mostly limited to mere demonstrations of the principle. Activities concentrated on developing the computer software for the purpose and demonstrating example products made from soft polymer materials. These materials, of course, made the process easy requiring no high temperatures and no high power levels. Indeed, one of the early demonstrations used a cool process: the traversed laser light induced local polymerisation inside a bath of liquid monomer.

[a] Corresponding author: tesar@it.cas.cz

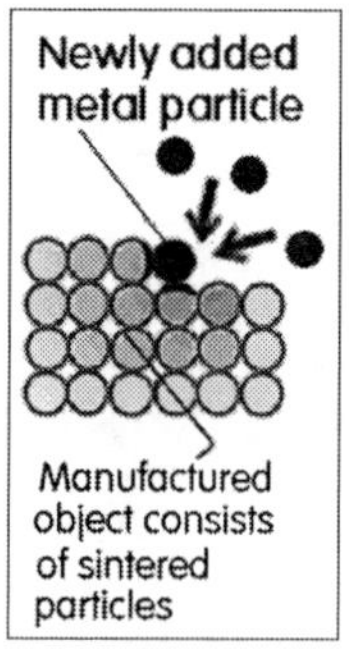

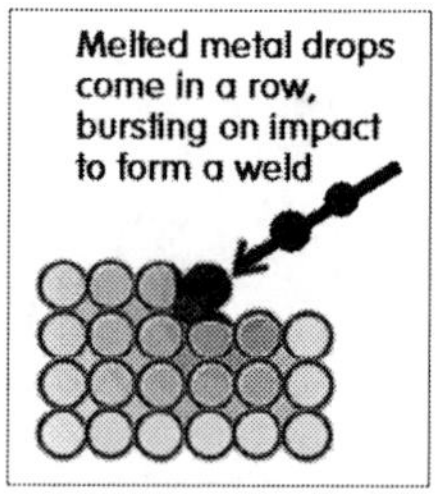

Figure 1 (left) Schematic representation of the additive manufacturing by sintering solid particles. Gaps between the particles are not necessarily a disadvantage – they decrease weight, a factor important e.g. in aeronautical engineering.
Figure 2 (right) Schematic representation of the additive manufacturing by melted metal liquid drops that burst upon impact and fill gaps. The surface expecting an impact of the droplet may by also partly melted by local application of heat.

Contemporary mechanical engineering is, of course, heavily dependent on use of metals as the construction materials. If used in the additive manufacturing, metal materials bring quite a number of technological challenges – mainly, of course, associated with the high melting temperatures. There are solutions of these problems, but they have been mostly covered by patents and also need a development of some details. It has to be said that one of the reasons leading to the recent surge of interest in the additive manufacturing is the approaching expiration of patents that were mostly filed at about the turn of the millennium by pioneer inventors.

There are additive production methods that have already gained the status of practical production processes, but almost all of them follow the easier approach: the sintering of added solid metal particles (Fig. 1). The particles are supplied in excess to the location where a power beam – mostly laser light - is applied for their partial melting and forming mutual bonds. Advantage of the sintering approach is lower demand on the input power. On the other hand, the sintering bonds between particles are generally less reliable and may result in an incomplete continuity of the inner structure of the product, with the consequence of lower acceptable mechanical loading. Since metals are chosen as the building material mainly to make components that have to carry considerable stresses, the sintering approach has its limitations.

Interest is thus gradually turning towards the more difficult additive manufacturing processes by accretion of sequentially deposited impinging melted metal drops, Fig. 2. The drops are made in a central source and fly to their accretion position by a controlled ballistic curve. They are fixed in position by fusion – the processr encountering problems mainly caused by the high melting temperatures. Some – if not most - among the problems waiting for being solved are of melted metal droplets fluid mechanics.

Particularly challenging tasks are:

- Aero/thermodynamics of droplet flight ballistics,
- Dynamics of surfaces, governed by surface tension,
- Droplet deformation upon hitting its target location,
- Formation of the droplet from a pendant liquid exposed to heating.

2 Additive method: advantages and problems

The new additive manufacturing offers a number of attractive advantages. Already mentioned was its universality. Another advantage is much less limits on the complexity of what may be manufactured in a single operation. A particular complexity example in Fig. 3 shows a cross section of a beam with thin curved walls surrounding a half-closed cavity (which may be, without problems, even fully closed. In fact, the method allows producing simultaneously a set of several,

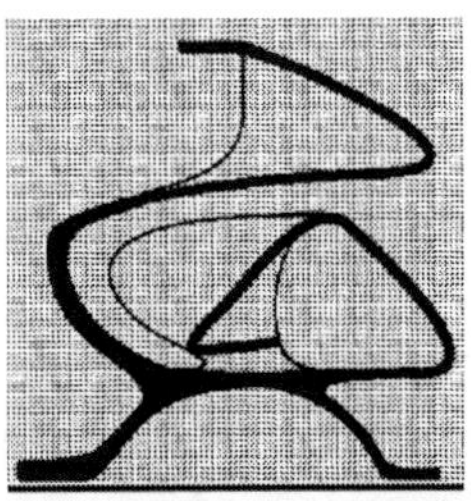

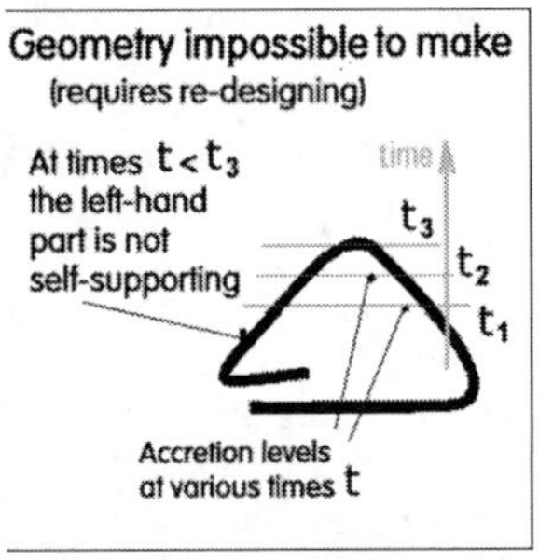

Figure 3 (left) An example of a manufactured component with a very complex cross-section shape. It would be extremely difficult to make it by classical subtractive methods but brings no problem to the additive approach.

Figure 4 (right) The limit to realisability is absence of support during the early stages of additive production.

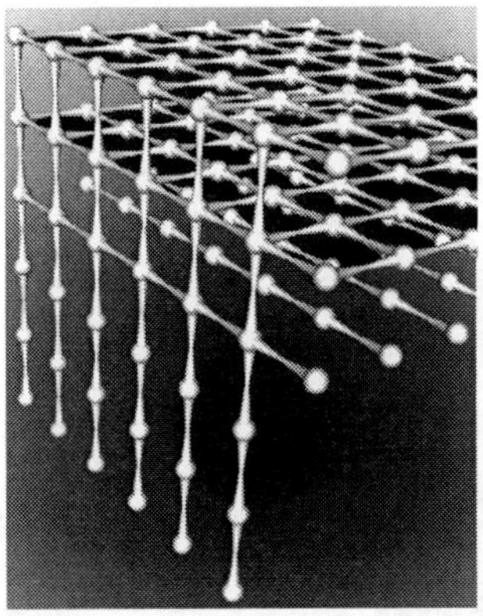

Figure 5 (left) Grids – like this spatial filigree periodic structure - are typical objects that may be quite easily and fast manufactured by the additive method.

Figure 6 (right) An example of a mechanical component with cross section not suitable for the additive method. The simple and wide cross section requires much material the transfer of which makes the additive method unable of competing with the traditional production approaches

mutually disconnected components – such as, e.g., a lock and its key, perhaps with the key positioned partly inside the lock. The number of simultaneously made mutually disconnected components placed inside one another is unlimited: already demonstrated was producing in a single operation a complete gearbox including all its internal gears inside. Obviously, this calls for a completely new approach to mechanical component design, not needing access windows to the internal space or fixing by screws.

On the other hand, it must not be forgotten that additive manufacturing is generally very costly – and is very likely to remain so in foreseeable future. Lasers and other generators of controlled power transfer are often extremely expensive. The other problem is the manufacturing cannot be fast. It would be therefore uneconomical to apply the method to making simply shaped objects, like the girder beam the cross section of which is shown in Fig. 6. Products suitable for the new approach have typically relatively thin curved walls - perhaps warped or twisted (Fig. 3) with little limitation to complexity – and objects with internal cavities. Making thin walls, of course, is preferable as it requires a smaller total number of deposited drops and hence a faster production. Adding tiny droplets to cover a large cross-section area in Fig. 6 would be wastage of time and effort.

The general request for increasing production speed leads to development activities concentrating on high-frequency of molten metal formation. Very much depends on requested surface smoothness of produced objects. The higher is the required surface quality, the smaller must be the accreted droplets and the longer takes the manufacturing process. At the current top speed, additive manufacturing has already demonstrated capability to add 200 cubic inches of metal per hour – in other words, $16\ 10^3\ mm^3/h$ (or $4.5\ mm^3/s$). As a general rule, by decreasing the size of the droplets to one half the

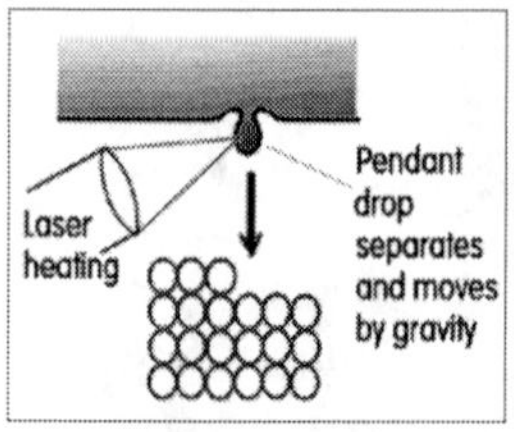

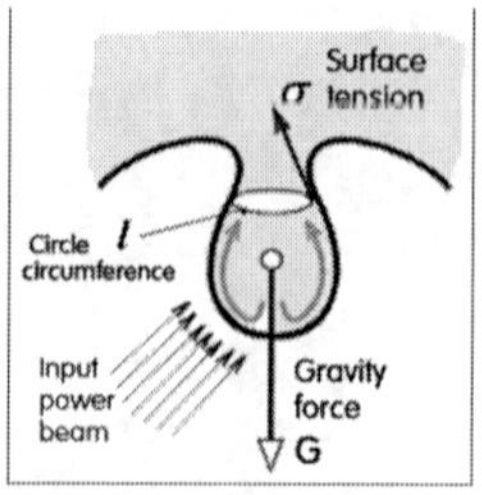

Figure 7 (left) The drop of molten metal is often produced by focused high power beam – in this case a beam of laser light.

Figure 8 (right) Detail of the effects that influence the formed pendant droplet generated by irradiation of a small part at the bottom of supplied metal body.

manufacturing time increases 8-times. Some improvement in the productivity may be obtained by operating two (or more) droplet generating units simultaneously - but apart from the obvious considerable increase in the system price, not all manufactured geometries are suited for the simultaneous operation of two sub-systems without mutual interference of the deposition trajectories or servomechanism components.

3 Producing the molten metal droplet

There are two approaches to droplet formation:

- The metal may be melted in a larger quantity in advance and stored in a vessel from which a tiny amount is taken by some ON-OFF valve when needed (Fig. 11). Because of the energy demand for melting and maintaining the molten state – as well as difficulties associated with the valve design - this approach is chosen for metals with lower melting temperatures.
- Otherwise the drop is produced "on demand" – as represented schematically in Fig. 7 – by high power heating pulse applied to the metal material at an end of a wire or at an edge of a strip. Some mechanism has to translate the material to the irradiation location.

Because of the high density of most metals, it is common to generate the drop somewhere above the produced object, so that transporting the molten droplet to its final location may be in principle by gravity. However, small droplet size, high requested repetition rate, and high speed of droplet flight make the gravity action insufficient. It may perhaps dominate the separation of the droplet, Fig. 9, but the subsequent ballistics usually uses other forces. Velocities of droplet flight are nowadays in the range 2 - 4 m/s – with a strong trend towards higher values. In principle the heating associated with impact on target surface might be used for melting the arriving particle, but this is rarely if ever used because the velocities necessary for significant melting effect would call for velocities at least a decimal order of magnitude higher than the above mentioned contemporary values.

Conditions governing the droplet formation "on demand" by input heating power are presented in Fig. 8. Most important is the surface tension which has to be overcome to separate the droplet and let it move.

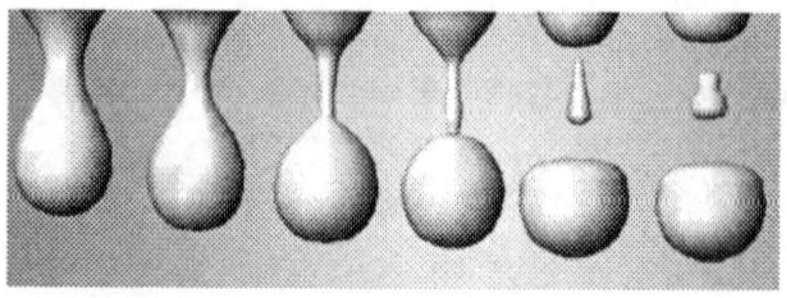

Figure 9. Typical drop formation process (time increasing from left to right). Top of the drop contracts into ligature which, after the drop has pinched off, becomes the unwelcome satellite droplet.

The surface tension, characterised by quantity σ [N/m], is identical to mechanical energy per unit of surface that must be input to cause the separation. The simplest expression for the force balance of the gravity force *G* [N] and surface tension force, neglecting unsteady effects, viscous, and other forces would be

$$G = \sigma\, l \tag{1}$$

- where *l* [13] is the circumference of the horizontal cross-section through the "neck" of the formed droplet. The balance is unstable and the separation is facilitated by the surface tension tending to squeeze the liquid into the spherical shape characterized by the least surface energy.

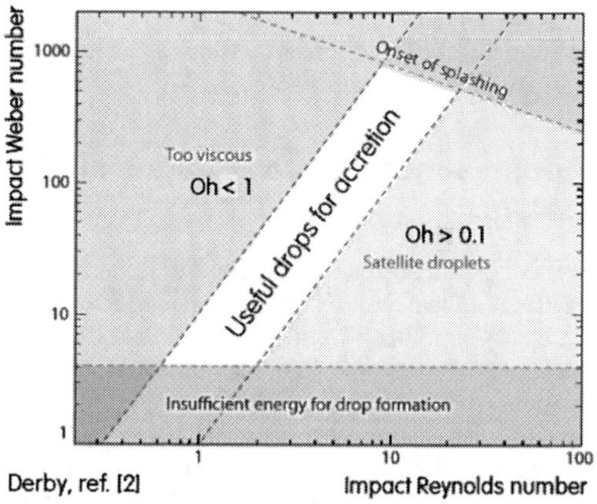

Figure 10. Parameters of metal drops suitable for use in additive manufacturing have to be adjusted to the relatively narrow range of formation conditions.

As shown in Fig. 9, the "neck" contracts into a tread-like ligature, which finally (fourth drop from left in Fig. 9) breaks so that the droplet may begin its ballistic flight. This picture also shows one of the aspects of drop formation that has been a subject of detailed research: the broken ligature under the effect of the minimised surface energy gets transformed into a smaller droplet, called a "satellite". Because of its smaller size, the character of satellite trajectory diffes from that of the main droplet. This flight to a different spot on the impact surface deteriorates the surface quality. Formation of satellite droplets depends on the magnitude of Ohnesorge number

$$Oh = \frac{\nu}{\sqrt{\sigma v d}} \tag{2}$$

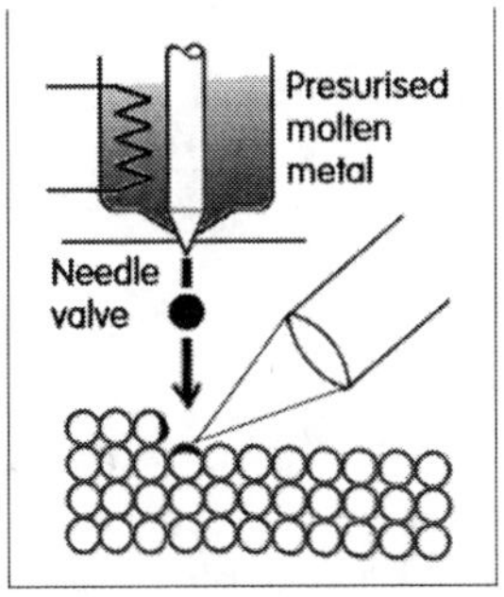

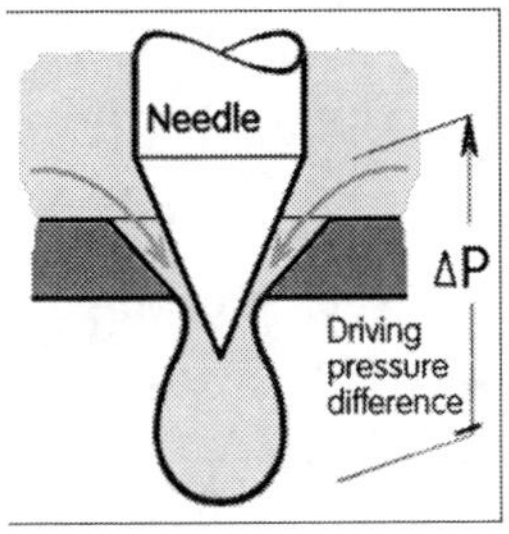

Figure 11 (Left) Generation of a molten metal drop by opening an exit from a bath of metal kept there in the liquid state.
Figure 12 (Right) Detail of producing the drop from the molten metal bath. Hydrodynamics of the flow past the needle tip is discussed in ref. [1].

- where ν [m^2/s] is molten metal viscosity (kinematic), v [m^3/kg] is its specific volume, and d [13] is drop diameter. This dimensionless parameter eq. (2) was introduced by Ohnesorge in 1936 in his PhD thesis [2]. As a general rule, its is recommended to adjust the drop formation conditions so that the Oh values are within the limits Oh > 0.1 and Oh < 1 as presented in Fig. 10 (it should be noted that these limit values are based on ref. [3] where they were actually derived for inkjet printing., where the droplet formation conditions may be different). In the diagram Fig. 10 there are also other limits, derived from experimental data on droplet impact behaviour.

A specific study deserves the drop formation in the case of metal melted in advance and stored in the melted phase (Fig. 11 – [4, 5]). Drop is produced on demand by brief opening of a valve. Figure 12 presents a typical example of the valve performing the flow control by axial motion of the needle. The force balance on the drop is here complicated by the additional pressure forces which ref. [1] shows to be far from simple.

4 High-energy beams

4.1 Flame: a beam of molecules produced by combustion

The easiest way how to generate the required beam of energetic particles – in this case consisting of molecules produced by a chemical reaction – is to use a burning flame. It may have its merits, such as low cost. It may be particularly suitable – as shown schematically in Fig. 13 – for pre-heating the target are before the droplet impact. Here it is, however, discussed mainly for completeness of the survey, because for the additive manufacturing it has considerable disadvantages.

The highest available temperature, up to 3.5 kK (kilokelvins) is found in the primary flame, using the supplied O_2. This generates roughly 2/3 of total heat production. The rest ~ 1/3 is provided by the outer, secondary flame. One of the disadvantages is flames in general do not reach very far. The length of the very hot region in front of the nozzle is comparable with the nozzle exit diameter. As shown in Fig. 13, the fuel gas is usually brought in by the central nozzle and the oxygen through the outer annular nozzle. The secondary flame uses additional oxygen from the atmosphere. This mixing with atmospheric air decreases the temperature, often to only ~ 1 kK at the tip of the secondary (outer) flame. Also, the flame widths are generally large compared with the size of typical droplets to be melted. This makes this approach to heat generation rather ineffective for the discussed purposes. A maximum achievable value of thermal power density on the heated surfaces is approximately ~ 0.1 GW/m^2.

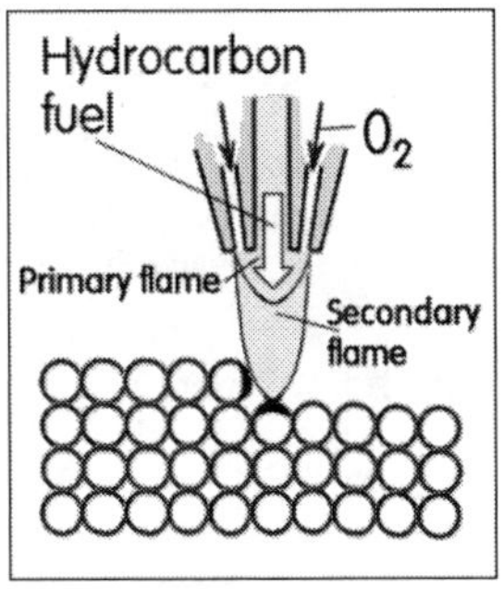

Figure 13. A relatively low-intensity power transfer beam may be generated by exothermic chemical reactions – burning gas in a torch, generally with poor focusing on a small droplet.

4.2 Plasma torch: a beam of ionised gases

More powerful than the chemical reaction – and generally better suited for producing the thermal input into drops (and droplets) for their melting are plasma torches. They use energy supplied from an external source in the form of electricity. There are too many alternative variants of plasma torch designs to discuss them all in this survey in detail (most of them were developed, anyway, for a different purpose - metal cutting and welding).

In essence, the impinging heating flow is here plasma (or hot gas with a certain plasma contents) generated by the voltage difference between two electrodes (Fig. 14). One of the electrodes, the cathode, is positioned upstream. Downstream, at or near the exit is a hollow anode. Plasma generated by arc was introduced in 1957 by Dr. R. Gage whose Patent [6] gave for 17 years a virtual monopoly to Union Carbide, Gage's employer. Meanwhile, Prof. Browning [7] and his graduate student Thorpe were developing their own plasma torches concentrating on the task of obtaining high temperature. They managed to achieve 5.6 kK, more than the temperature of sun photosphere. The gas that is ionised to become the power carrier for metal heating is mostly an inert one, such as Ar, because even tungsten electrodes would erode rapidly in the presence of oxygen. Much welcome was discovery by Soviet researchers [8, 9] of hafnium and zirconium electrodes, capable of working with air. The jet, issuing from the orifice in the anode, is narrow due to electric gradient. Since this means the temperature may be concentrated in a small cross-section area on jet axis, plasma torches may reach high power density, generally as high as 5 GW/m^2.

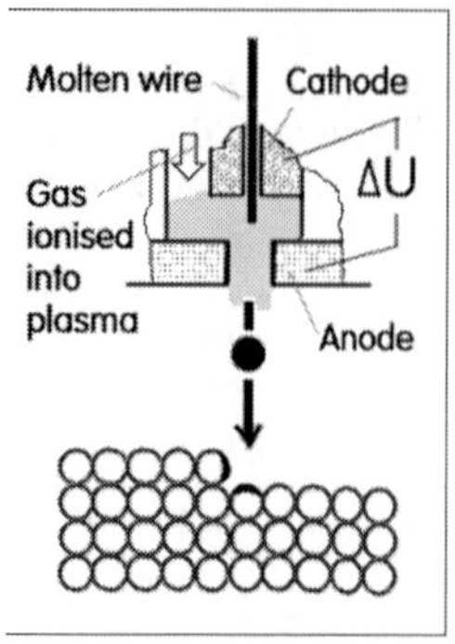

Figure 14. Schematic representation of typical plasma torch. A beam of ions is produced by voltage difference ΔU exceeding the dielectric breakdown strength of the supplied gas. The hot flow generated by the arc may be easily accelerated either by the gas pressure and/or by the voltage difference.

4.3 Laser light: photon beams

Lasers generate beams of photons emitted by a process of optical amplification. Because of the high coherence, the generated laser light beam may remain narrow over long distances – and may be effectively focused by an optical system to a very small illuminated spot. Also here are many existing variants. The best solution for industrial uses is offered by CO_2 lasers, invented by Patel [10]. They can continuously emit many hundreds of watts in a single spatial mode in the thermal infrared at 10.6 μm. It is possible to reach power density values of the order 10 GW/m^2. Important is the fact that efficiency of CO_2 lasers is quite high, over 30%.

Very important is the capability of lasers to operate in the mode of extremely short pulses. The table-top sized multi-pass laser described in [11] generates at infra-red in picosecond pulses a power 1.4 kW. Ref. [12] discusses picosecond and femtosecond regime (1 fs = 10^{-15}s) pulsed lasers processing components by the additive manufacturing of Ni- and Ti- base alloys. Due to the short pulse duration, power density may easily reach values above 10 TW/m^2. Lasers thus can easily melt metal droplets so as to form a dense chain (Figs. 17, 18) – the drop density limited only by the physical dimensions of the drops. Unfortunately, not only the laser itself but also its necessary accompanying facilities tend to be very expensive.

4.4 Electron beams

The inventor of this approach to generation of power carrying beams - in this case beams of electrons – was Pirani [13] working for Siemens in Germany. Electrons possess extremely small mass m = 9.1 · 10^{-31} kg and a negative electric charge e = 1.6 · 10^{-19} C. Free electrons in vacuum may be thus very much accelerated by high voltage, usually applied in the range from ΔU = 10 kV to 1 MV, in narrow beams carrying high kinetic energy transformed into heat upon the impact on solids.

The extraordinary values of handled energy, difficult to compare with common phenomena, has led in this field to introduce an auxiliary unit (not SI) eV = electron volt. It is approximately equal to 1.6 10^{-19} Ws. For comparison, photons of visible light carry 1.5 – 3.5 eV, the electron gun in a cathode ray display tube operate with approximately 20 keV, and energy of medical X-rays is typically 0.2 MeV. Electrons are emitted from cathode (Fig. 16). They are the conduction electrons in cathode metal (i.e. not bound to atoms). Leaving the cathode requires getting kinetic energy higher than the potential barrier — which e.g. in tungsten is 4.5 eV (often used cathode material is lanthanum hexaboride LaB_6 , with the low barrier level around 2.5 eV). Thermionic cathodes use the fact that following the Richardson's rule, the number of electrons possessing the required kinetic energy increases with metal temperature. Another approach is field emission electron gun applying extremely high electric field gradient, typically 10^9 V/m. The beam of electrons may be shaped into a narrow cone by magnetic "lenses" – suitably configured coils.

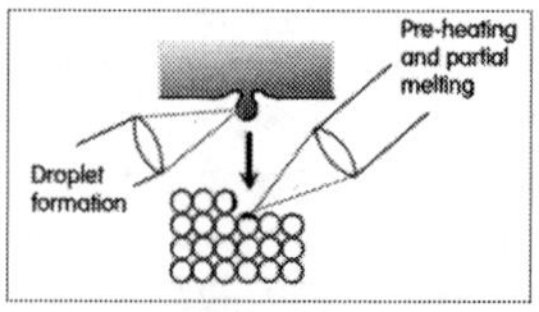

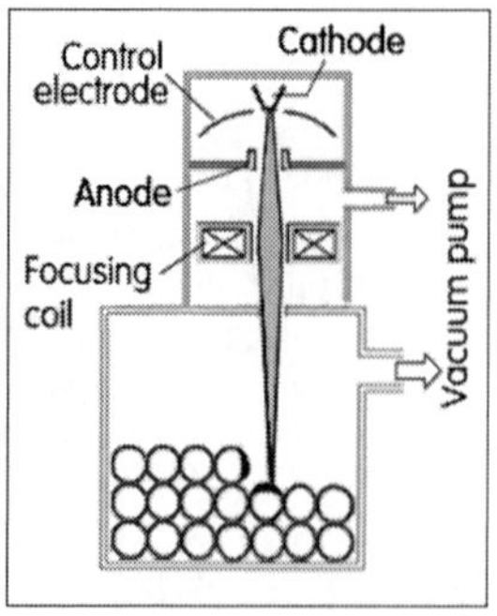

Figure 15 (left) Formation of strong bond with the newly added drop may require re-heating of the target area – perhaps with partial melting, cf. Fig. 2, so that the drop impact may be a case of liquid-liquid interaction.
Figure 16 (right) A particularly high heating power can be carried by a beam of electrons, with the advantageous capability of exact focusing to very small droplet surface area. A disadvantage is the need to operate in vacuum.

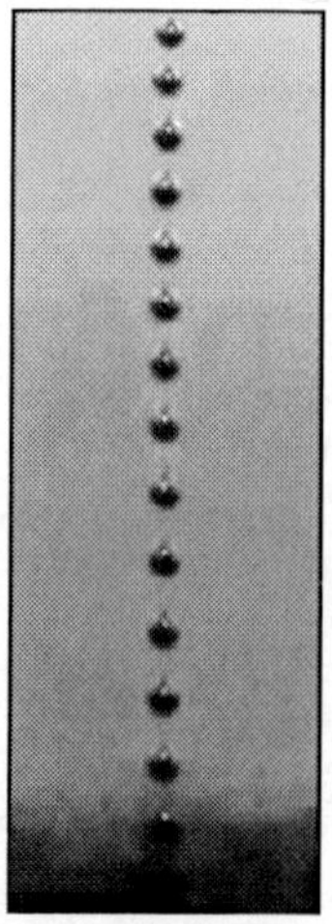

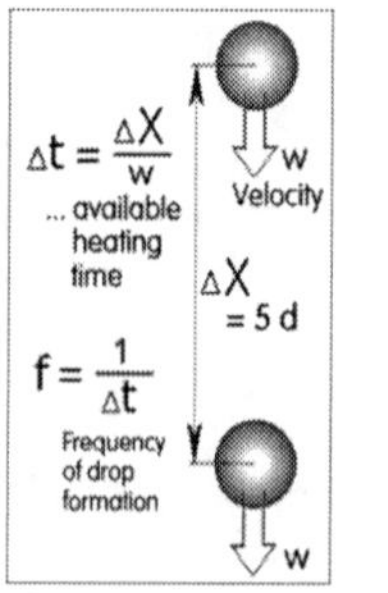

Figure 17 (left) Droplets falling in a chain. Requirements of increasing the manufacturing speed may necessitate working with chains of similar density.
Figure 18 (Right) Basic relations of a droplet chain. The distance of five drop diameters shown in this picture between the chain members may be typical.

Of course, considerable practical disadvantage of the electron gun is the fact that the electrons can move over the distances of interest only in vacuum – which complicates and slows down the placing of the processed object into its position.

5. Kinetics of metal drop motion

Behaviour of molten metal drops in their use for additive manufacturing presents many unanswered questions. Obtaining a deeper understanding is obviously necessary for making the "*3D printing*" more efficient and therefore more widely applicable. There is also the question of higher surface quality of the manufactured object without decreasing excessively the drop size.

Because of the complexity and at the same time difficulties associated with experimental investigations of high temperature molten metals, it may be convenient -- at least in the initial stages – to simplify the analysis by separating two aspects:

- The fluid mechanics of metal drop motion, and
- Thermomechanics of heat and mass transfer between the drops, their environment (either gas atmosphere or vacuum), and finally the surface on which they are deposited.

5.1 Ballistics of droplet flight

Fluid mechanics of drop motion must take into account several important specific facts. One of them is the oscillation that was observed to take place immediately after the drop separation from its source. This may endanger the required high precision of drop motion trajectory – in a similar manner as the oscillation influences the motion of the complementary objects – bubbles in liquid [14, 15]. The importance of the initial part of the droplet trajectory increases with the demands of productivity resulting in the trajectories being rather short. In spite of the rapid viscous damping, because of the generally low Reynolds numbers, the oscillation can significantly influence the initial stages of the separated drop – and these stages are very important because of the general demand to shorten the length of the droplet flight trajectory due to the demands of increased productivity.

Also because of the demands of higher productivity, there is a strong tendency to produce and eject the drops at very high repetition rates. The presence of neighbor drops may influence significantly their trajectories. The drops are likely in near future to move in chains, with rather short distances between the chain members, as is shown in Fig. 15. This means the studies of an individual drop flight, while certainly a useful starting point, are no more than just a preliminary to studies of the motion in chains [4] – especially in those cases where the drops fly in gas atmosphere and thus move in the wake of the predecessor drop. In present-day additive manufacturing systems the frequency of drop generation (Fig. 19) is not particularly high – typically of the order from 2 Hz to 10 Hz - but already foreseen are requirements of working with frequencies higher than these by a decimal order of magnitude. This results in very short times Δt, less than 0.1 s , that is available for delivering into the forming drop the heat for its melting.

Admittedly, at present the velocities of drop ballistic motion in additional manufacturing facilities are not very high; characteristic values being let us say between 2 m/s and 4 m/s. This motion speed determines the pitch distance ΔX between the drops flying in a chain – the value ΔX = 5 d in Figs. 18 and 19 is here chosen as very probably characteristic for the conditions in the very near future. Later condition have to be expected. even higher values The hydrodynamic drag of droplets in a dense chain differ significantly from the usually considered Stokes' law.

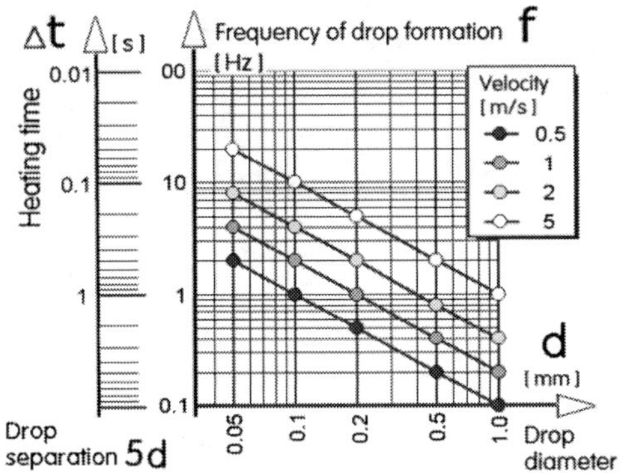

Figure 19. Diagram useful for providing an idea about the drop velocity, frequency of their generation, and the time available for their heating and melting.

5.2 Dynamics of impact

Important mechanism influencing proper deposition of the metal from the drop to the target area is the dynamics of droplet impact on its target location. This is a complex problem calling for deeper fluid mechanical studies. There are some not very numerous papers (such as, e.g. [3] and [16]) analyzing the drop impact processes – but their interest was directed towards a different branch of technology: the impact of ink drops on paper, investigated because of the needs of developed inkjet

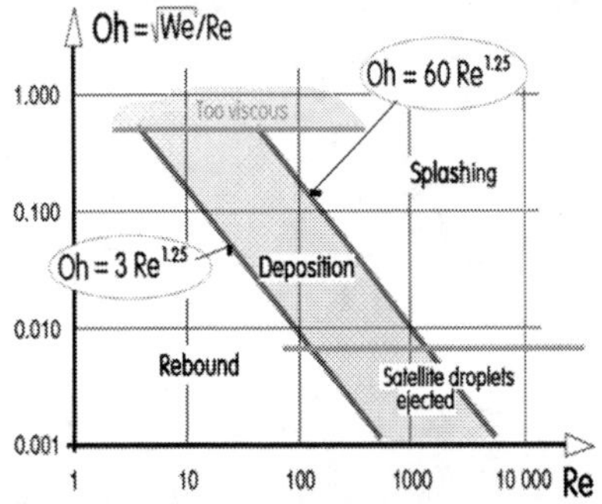

Figure 20. Conditions at the impact should be adjusted so that its dimensionless parameters fit into the deposition region shaded in this diagram. The splashing regime is particularly unwelcome since the impinging droplet there is deformed into a thin layer on the impingement surface rather than a blob suitable for the role of building a structure.

printers. Some, predominantly qualitative information may be transferable from there, but the need of detailed analysing the process for the purposes of manufacturing of metal objects is obvious.

One of the aspects that may significantly change the character of the impact is the quite often applied heating of the target area, Fig. 16, especially if the heating is at the (quite desirable) level at which the target surface becomes melted. The impact can then get the character of liquid/liquid interaction [17], with the splattered droplet generating the well known "milk crown" configurations, many times reproduced in early high-speed photography.

On the other boundary in Fig. 20, the one opposite to the splashing limit, the proper deposition of the drops may be influenced again by the shape oscillation, Fig 20 – some information about which is known from ref. [14].

It is now well established that the character of the drop impact is determined by the pair of dimensionless characterisation number, the impact Reynolds number:

$$Re = \frac{w_{imp} d}{\nu} \tag{3}$$

where w_{imp} [m/s] is the velocity of the drop prior to the impact, and the Weber number of impact:

$$We = \frac{w_{imp}^2 d}{\sigma v} \tag{4}$$

Substantial role in the dynamics of the motions – such as, e.g., of the rebound oscillation (Fig. 2) – has the characteristic impact time:

$$t_{imp} = \frac{w_{imp}}{d} \tag{5}$$

The splattering or splashing is an unwelcome phenomenon for the additive manufacturing because it reduces the height of the added material and thus reduces the manufacturing speed. It is also likely to aggravate the quality of generated surface, especially at high Reynolds numbers of the impinging motion. As shown in Fig. 20 that combines some aspects of available information about the impact, there is a regime suitable for the drop deposition, the boundaries of which are mainly dependent on the value of Ohnesorge number, eq. (1).

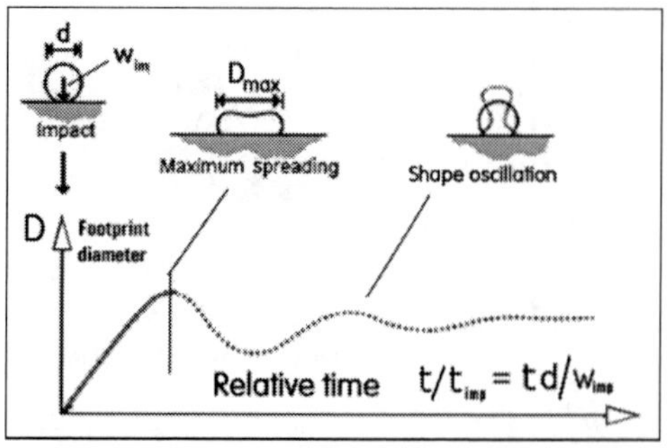

Figure 21 Shape oscillation in the rebound regime is governed by surface energy (and also by viscosity that determines damping of the oscillation). Conditions correspond to those discussed in [14].

6 Droplet thermomechanics

The unusual term "thermomechanics" is here used due to the fact that the unsteady thermodynamics of generated drop, cooled during flight and finally solidified involves several aspects of the process, involving also thermokinetics of heat transfer across the droplet surface. Moreover, unless the processes take place in vacuum, there are present also chemical changes because the drops tend to be very soon covered by a layer of oxides.

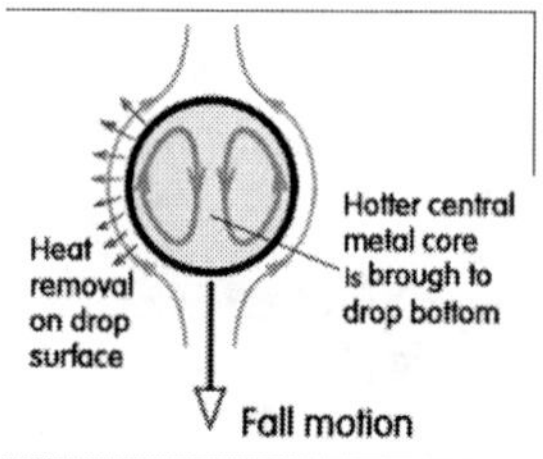

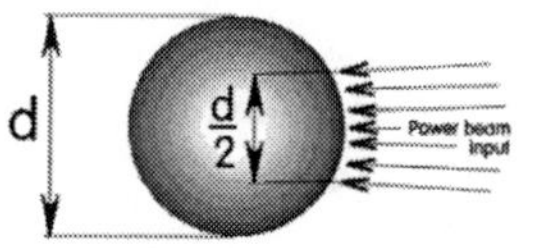

Figure 22 (left) Thermal processes inside a falling drop are influenced by the internal flowfield. Also important is the fact that many (perhaps most) most molten metal drops in air atmosphere are rapidly covered by a layer of oxides.

Figure 23 (right) Simple model of irradiation of a drop used for plotting the diagram Fig. 25

The heat transfer, as suggested in Fig. 22, is also strongly dependent on the difficult to access internal flowfield inside the drop – partly perhaps generated by the skin friction and partly by the inertia of the internal flows in the drop during its formation and separation from the solid mother metal.

Of particular importance are also the extraordinary magnitudes of the radiation heat transfer, involving the extreme devices discussed in Sect. 4. Their high intensities means the governing laws are quite likely to involve nonlinear terms, usually neglected in the relations of standard thermokinetics.

In this paper of survey character, details of these problems cannot be discussed. It is, nevertheless, useful to list them and to provide some basic data for order-of-magnitude considerations. The essential question that may be asked is the magnitude of thermal energy that has to be delivered into the metal particle so that it is melted and becomes the droplet. The essential information is presented in the diagram Fig. 24. For a range of drop diameters and several representative metals are there plotted thermal input data evaluated from the quantities

	Melting temperature [°C]	Specific thermal capacity [J/kg K]	Specific volume [m³/kg]
Aluminium	660	910	0.000369
Beryllium	1285	1830	0.000543
Copper	1084	390	0.000112
Steel, stainless	1510	450	0.000134
Titanium	1670	540	0.000222

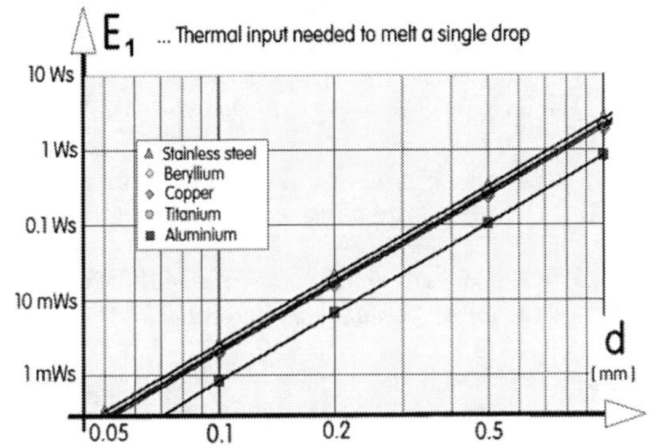

Figure 24. Order-of-magnitude diagram for evaluation of energy to be delivered into solid spheres of various size so as to convert them into the liquid drops. Interestingly, the requirements are nearly the same for many metals.

presented in the accompanying table. A somewhat surprising fact visible in Fig. 24 is the critical dependence of heat transfer on the metal droplet diameter d — while the dependence on the sort of the metal is rather small, almost negligible for the metals of interest (despite the quite wide range of values of the properties in the table on the previous page). Also the order of magnitude spread of the necessary thermal input into the droplet may be interesting, from less than 1 mWs on one side to full one joule (J = Ws) and more for larger drops.

It is interesting to note in the next Fig. 25 how large power densities are actually needed for melting a modelled stainless steel spherical drop partly irradiated, as described in Fig. 23. To overcome the problem of some of the power transfer radiation impinging at a very small incidence angle (and hence reflected rather than absorbed), the model used for the evaluation according to Figs. 23 and 25 considers only absorption on surface area diameter equal to only one half of the overall droplet diameter d. The irradiated surface is considered ideally black, absorbing all incident radiation. The heating is assumed applied in a pulse lasting for the whole droplet formation time (determined by the formation frequency f). As seen in the diagram Fig. 25, the required thermal power density values for melting the metal may be very high – but fortunately lower than the limits discussed above in Sect. 4 and summarised in Fig. 26.

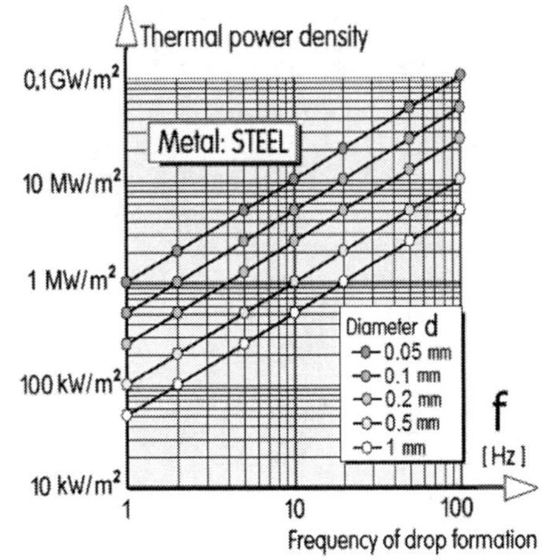

Figure 25. Surface density of irradiation power needed to melt a steel drop of various diameters generated at a particular formation frequency – using the model of Fig. 23.

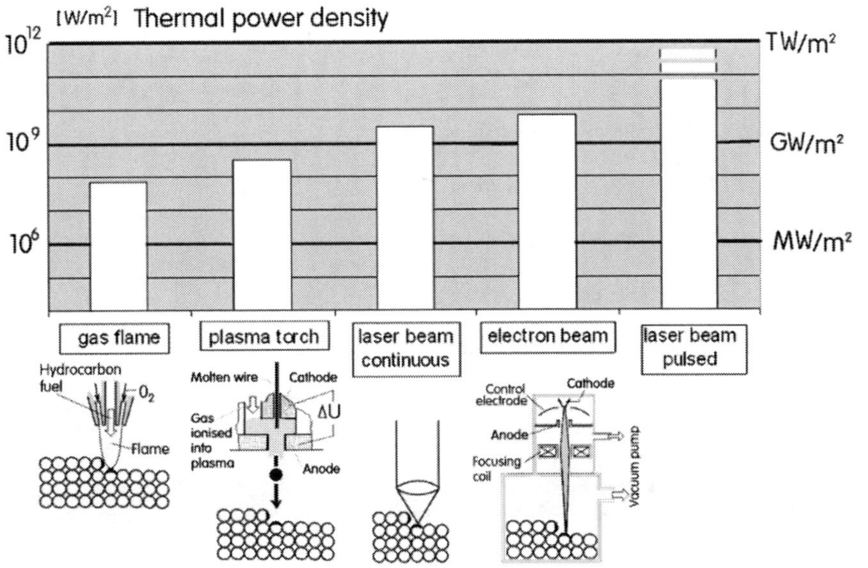

Figure 26. Thermal power densities obtainable with various heat delivering beams. Laser light and electron beams can react sufficiently fast to increase the power by concentrating the output into short pulses.

Conclusions

This paper presents, at an introductory level, a general survey of the principles on which is based the recently popular additive manufacturing. The discussion focuses on problem areas that call for deeper investigation. The key problem is low production speed due to the small size of the added droplets and limited repetition rate of their formation. The small droplet size, of course, is necessary for obtaining good quality of produced surfaces – and is also desirable as it decreases the necessary thermal power delivered into the melted droplet. Improvement in productivity needs paying attention to deeper understanding of the processes in the droplets. Some useful information is obtained from earlier studies directed at inkjet printing; nevertheless the molten metal brings its own problems. Some research areas mentioned in this paper are ballistics of controlled flight, undesirable formation of satellite droplets, melted surface oscillation, internal flowfield inside the droplet, and the heat and mass evaporative loss. Most of these problems are of fluid-mechanical and thermokinetic nature.

Acknowledgements

Author obtained support from GAČR - Grant Agency of the Czech Republic, by research grant Nr. 13-23046S. There was also institutional support RVO:61388998.

References

1. V. Tesař, *Sensors and Actuators A: Physical* **220** 1 (2014)
2. W. Ohnesorge, *Journal of Applied Mathematics and Mechanics* **16,** 355 (1936)
3. B. Derby, *Annual Review of Material Research* **40** 395 (2010)
4. B. M. Michaelis et al., *International Journal of Heat and Mass Transfer* **50** 4554 (2007)
5. A. Amirzadeh, M. Raessi, S. Chandra, *Experimental Thermal and Fluid Science* **47** 26 (2013)
6. Gage R., *US Patent* No. US2806124, filed July 1955
7. J. A. Browning, *US Patent* No. 4370538, filed May 1980
8. M. N. Butova, *US Patent* No. 3930139, filed May 1974
9. V. I. Savin et al., *Izvestiya Akademii Nauk SSSR, Metally*, Nov.-Dec. 1979
10. C.K.N. Patel, *Physical Review* **136** A1187 (1964)
11. J.-P. Negel J.-P., *SPIE Newsroom*, (Aug. 2014)
12. I. Mingareev et al., *Proc. SPIE8970 Conf. on Laser 3D Manufacturing*, (Mar. 2014)
13. M. von Pirani, *Journal of Physics D: Applied Physics* **45** 175202 (2012)
14. V. Tesař V., *Chemical Engineering Journal*, **235** 368 (2013)
15. J. M. Hoey et al., *Journal of Nanotechnology*, **2012** Article ID 324380 (2012)
16. Š. Šikalo , E.N. Ganič, *Experimental Thermal and Fluid Science* **31** 97 (2006)
17. K.-L. Pan, C.-Y. Hung, *Journal of Colloid and Interface Science* **352** 186 (2010)
18. B. Bachmann, E. Siewert, J. Schein, *Journal of Physics D: Applied Physics*, **45** 175202 (2012)
19. Y. S. Oh et al., *International Journal of Heat and Fluid Flow* **50** 27 (2014)
20. E. J. Vega et al., *International Journal of Advanced Manufacturing Technology* **70** 1395 (2014)
21. M.P. Mughal, H. Fawad, R. Mufti, *Heat and Mass Transfer*, **42** 226 (2006)
22. G. Trapaga et al., *Metallurgical Transactions B* **23B** 7101 (1992)
23. N. Lass et al., *Micromachines* **4** 49 (2013)
24. V. Tesař, *Chemical Engineering Science* **116** 843 (2014)

EPJ Web of Conferences **114**, 02123 (2016)
DOI: 10.1051/epjconf/201611402123

Stochastic regimes in very-low-frequency fluidic oscillator

Václav Tesař[1,a]

[1]Institute of Thermomechanics v.v.i., Academy of Sciences of the Czech Republic, Dolejškova 5, 182 00 Praha-Kobylisy, Czech Republic

Abstract. Paper discusses interesting unexpected stochastic regimes discovered in a fluidic oscillator designed for operation at very low oscillation frequencies - without the inconvenience of the long feedback loops needed in standard low-frequency oscillator designs. The new oscillator contains a pair of bistable turn-down active valves operating in anti-parallel — essentially analogous to Abraham & Bloch electric "*multibrateur*" invented in 1919. Three different self-excited oscillation regimes were found. In the order of increasing supplied flow rate, these regimes are characterised by: (A) generation of stochastic-duration multi-pulse packs, (B) generation of individual pulses with a degree of periodicity, and (C) regime with randomly appearing flow pulses separated by intervals of the order of seconds.

1 Introduction

This paper describes an unusual oscillator supplied with steady air flow and generating flow pulses in its output terminal. Unusual property of this oscillator is an extremely low frequency - and significant pulse randomness. Laboratory model of the oscillator was investigated for potential use in producing very small, sub-millimetre size gas microbubbles in water. There is a wide range of engineering applications in which the microbubbles bring significant advantages over standard larger size bubbles. It may suffice to name just the use in supplying CO_2 to algae cultivated in a bioreactor for producing biofuel [1]. It is an important process, pursued with the ultimate target of reducing the present-day automobile fuel dependence on fossil oil imported from politically unstable regions. The CO_2 from the microbubbles is needed for algal photosynthesis and an additional task of O_2 removal. The small size increases the area of total collective gas/liquid interface and thus increases the rate of diffusion transport. The fact that no-moving-part fluidic oscillators can generate smaller than usual bubbles by pulsating the supplied gas flow was demonstrated already ~10 years ago [2], but for a considerable time the details of the microbubble generation mechanism remained unexplained.

Fluidic oscillators, possessing no movable or deformable components, are particularly attractive for the role of the gas flow pulsation in microbubble generation because of their reliability, no need of maintenance, and low cost (since they are nothing more than just a specially shaped gas inlet cavities). Until recently, development of oscillators for this purpose was characterised by demand of very high oscillation frequency. This seemed quite logical because the microbubbles, due to their small size and dominance of relatively strong surface tension forces, posses very high natural frequency of surface oscillation, of the order of kilohertz [3]. It was believed that the small size of the microbubbles is the result of pulsation-promoted fragmentation of initially larger

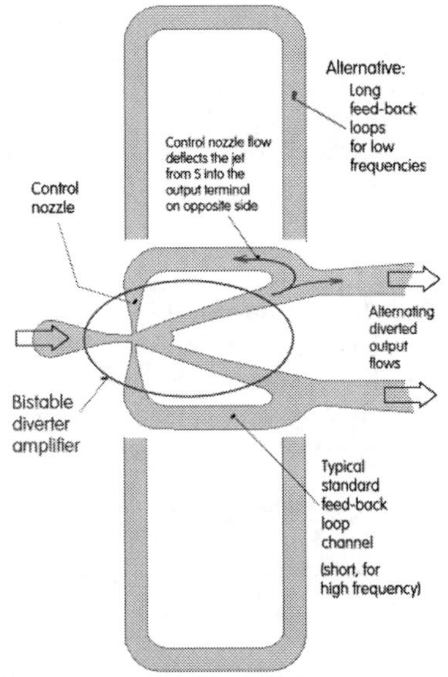

Figure 1. Standard configurations of present-day fluidic oscillators [6, 7]. The main component is a bistable diverter amplifier. It is provided with feedback loop channels, connecting amplifier outputs with control nozzles. Low oscillation frequency would need very long feedback loops.

bubbles — with the oscillator obviously operating in resonance with this natural frequency.

[a] Corresponding author: tesar@it.cas.cz

The problem was difficulty in obtaining the high frequency with typical fluidic oscillators operating very much below the required kilohertz range. The present-day oscillators, as shown in Fig. 1, consist of two parts: **a)** a fluidic amplifier, usually of the bistable jet-diverting type, and **b)** feedback loops. Typical achievable high frequencies with the version resembling Fig. 1 are near to f ~ 300 Hz. Somewhat higher frequencies, perhaps up to f ~ 500 Hz, may be obtained with the recently introduced configuration described in [4], with the loop feedback replaced by signal returned from a resonator. Of course, both frequency values mentioned above are very much below the requested kilohertz regime. A very small oscillator size was necessary, calling for special fabrication techniques and limiting the production rate of the microbubbles. Solutions were sought in special high-frequency versions, e.g., designs facilitating generation of higher harmonics [5].

Quite recently, the target of the development was completely reversed. Performed detailed studies of microbubble formation, using very high speed camera and special objective lenses with very long working distances, have revealed that the key problem is the growth of slowly moving microbubbles at the aerator exit by repeated multiple conjunctions – discussed in Part I of [8]. In the next Part II of [8] it was demonstrated that the conjunctions may be suppressed by reversing, for a part of the oscillation period, the direction of motion in the aerator passages. The microbuble protected inside the passage is thus prevented from contact with other, previously generated microbubbles. It was also found that this working mode, with reversing flow direction of water columns in aerator channels, actually necessitates low driving frequency.

The classical feedback oscillator configuration of Figure 1 is unsuitable for pursuit of the newly emerged target of very low frequencies. It would be necessary to have the feedback loop channels of lengths of the order of metres or even tens of metres [6]. This would make the oscillator loops difficult to stow in the available spaces near the aerator. Also the length – as well as the multiple channel bends, with flow direction changes made in an effort towards compactness — would lead to large hydraulic losses inside the loop. The losses may cause the oscillator to cease operating.

Absence of any success in the attempts to adapt the configuration from Figure 1 to very low frequency regimes has led to recognition of the need to develop new oscillators, based on completely different operation principles.

An inspiration was found in early history of electronics. Design of fluidic circuits has, quite naturally, followed the pattern of electronic circuits as its historical predecessor. Of course, there is a large difference in the final steps of actual realisations. The conceptual similarity of the de-stabilising feedback on one hand – and yet different realisation of the idea on the other hand may be seen in the example of the feedback presented in Fig. 1.

2 Electronic circuits with anti-parallel, mutually blocked valves

If a designer of an electronic circuit is requested to generate a signal output in the form of a train of rectangular pulses, he has as an option the use of an astable flip-flop circuit. This consists of a pair of electric current-controlling devices operated in anti-parallel, with mutual blockage. This is set up by interconnecting the output terminals of one device to the input terminal of the other one – and vice versa. The result is a circuit switching between two regimes. In one regime the OPEN state in one device is simultaneous with CLOSED state in the other one. In the next regime their role are exchanged.

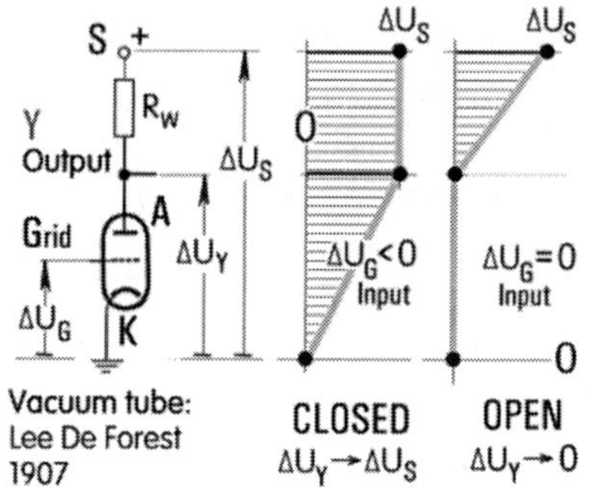

Figure 2. Principle of the current control valve used by Abraham and Bloch in their electronic oscillator. Negative input control voltage ΔU_G applied to grid closes the current of electrons from the cathode K to anode A. The consequent absence of voltage drop on the working resistor R_w increases the output voltage ΔU_Y.

The idea of the anti-parallel circuits appeared in early history of vacuum-tube electronics towards the end of the first world war. The current control in vacuum tubes operates as shown in Fig. 2, by switching between. the two alternative regimes described as OPEN and CLOSED state. The current passing through the vacuum valve is controlled by admission of negative voltage ΔU_G to the grid positioned inside the valve between its cathode K and anode A. If there is no negative voltage applied on the grid, the vacuum is open for motion to the anode A of electrons emitted from the heated cathode K. The motion is driven by voltage difference ΔU_S between K and S. The vacuum tube is connected in series with working

Figure 3. Eugene Bloch (1878–1944), one of the two inventors of the electronic *"multivibrateur"* – the ancestor of oscillators with two anti-parallel active devices.

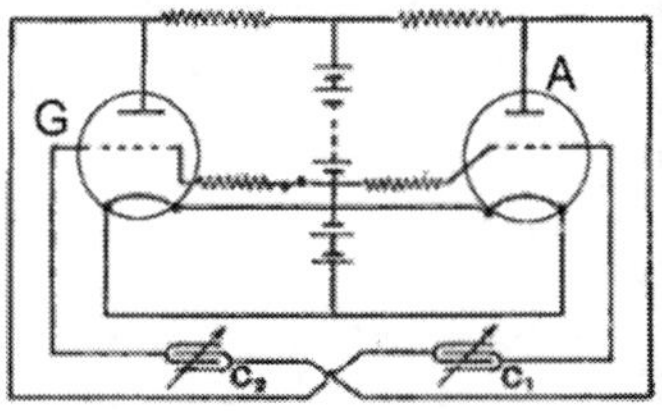

Figure 4 (Left) The original circuit diagram of the "multivibrateur" as it appeared in the 1919 publication ref. [21]. Anode A of one valve is coupled to the grid G of the other valve by means of the capacitors C_1 and C_2 which cause time delays.
Figure 5 (Right) Henri Abraham (1868-1943) – the other one of the two inventors of the idea of anti-parallel valves.

resistor R_w. The output signal voltage ΔU_Y is taken from the location between this resistor and the valve. As shown in the idealised voltage distributions at the left-hand part of Fig. 2, in the OPEN regime the voltage difference between the output terminal Y and the cathode K is practically zero. The whole supplied voltage difference ΔU_S is present as voltage drop on the working restrictor R_w. The output signal voltage ΔU_Y is zero.

Once the control voltage ΔU_G is applied, the motion of electrons in the resultant CLOSED regime is blocked. Since no current then passes through the working restrictor R_w, there is practically no voltage trop on it. The whole voltage difference ΔU_S is now available as the output signal between Y and K.

This description of vacuum tube behaviour helps in understanding what experienced the inventors E. Bloch (Fig. 3) and H. Abraham (Fig. 5), in the study of their interesting circuit with two vacuum tubes in anti-parallel, i.e. one in the OPEN and the other CLOSED regimes and mutually connected, as is apparent from Fig. 4. Note that the anode of one valve is connected to the grid on the other side. This connection is through capacitors C_1 and C_2. These cause a time delay (the time needed for charging the capacitor, so that a particular configuration of currents and voltages has a chance to exits briefly before it is switched off). Condensers, unable to transfer DC current, prevent the circuit from remaining in one of the regimes permanently.

Abraham and Bloch called their circuit a "*multi-vibrateur*". Priority of their invention was sometimes considered doubtful because they made the tests of this circuit during the first world war while working for the French *Ministère de la Guerre*. This, naturally enough during the war, demanded keeping the invention secret. It could be first described in open literature ref. [7] later, after the war in 1919. Prior to this publication, however, two British electrical engineers, Prof. Eccless (Fig. 7) and a technician in his laboratory, W. Jordan (Fig. 8) have applied in 1918 for patent on essentially the same circuit as presented in Fig. 6 (illustration taken from ref. [6]). Their circuit, also with the pair of mutually blocking vacuum tubes, became known as "*flip-flop*" because, in contrast to the Abraham-Bloch "*multivibrateur*" the signal transfer was by DC connection. This made the circuit bistable, remaining permanently in one of the two regimes until the arrival of a switching input signal. Nevertheless, the monostable version (with one regime

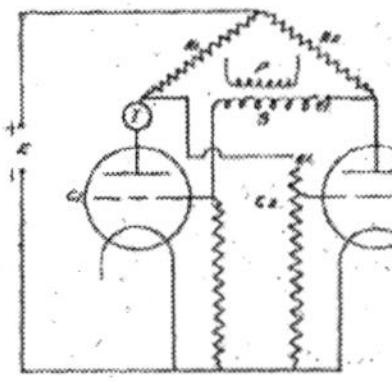

Figure 6 (Left) The original circuit diagram of the "*flip-flop*" from the 1918 patent. It remains stably in one of two alternative regimes.
Figure 7 (Right) Professor William H. Eccless (1875-1966) who, with his technician Frank Jordan, invented essentially a year before the same circuit as Abraham and Bloch – except that in its standard version the circuit was not set up to oscillate.

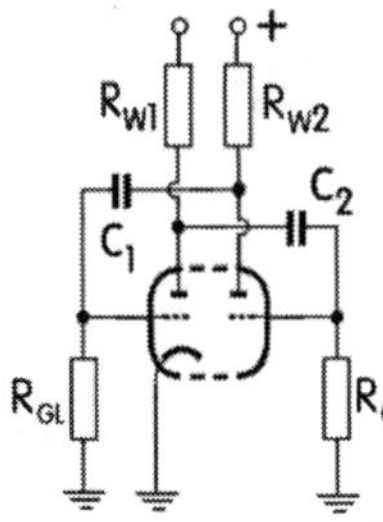

Figure 8 (Left) Frank Wilfred Jordan (1882-?) – the researcher who by toying with electronic circuits in Eccless' laboratory discovered the interesting behaviour of anti-parallel valve connection.
Figure 9 (Right) Eccles-Jordan flip-flop circuit in the oscillator version (note the capacitors C_1 and C_2)

preferred) and "astable", i.e. oscillating, were also soon tested. The bistable version was later used in large numbers in the earliest digital computers - in the natural realisation of binary algebra operations. The large amount of these units deserved producing for this purpose special vacuum tubes containing two separate electrode systems in the same glass bulb, as shown in the schematic representation in Fig. 9.

3 Fluidic bistable turn-down valves

The electronic circuits, presented schematically in Figs. 6 and 9, are complicated by added auxiliary components. These, e.g. such as the "grid-leak" resistors R_{GL} in Fig. 9 (they serve for removal of electrons accumulated on the grid), are irrelevant for the adaptation of the basic idea to the fluidic "astable" oscillator. Rather than a closer study of the electronic circuits with these details, it is advisable to follow the basics of the function logic as presented in Figs. 10 and 11, showing the basic logic element and the mutual blockage connection of two such elements A and B, respectively.

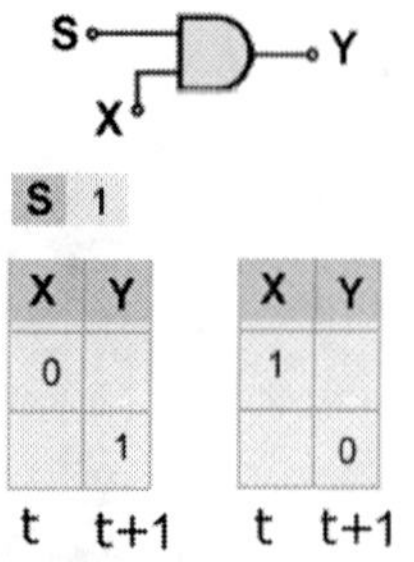

Figure 10. General representation of the basic element performing logical negation after a time step from t to t+1. Note the constant supply S =1.

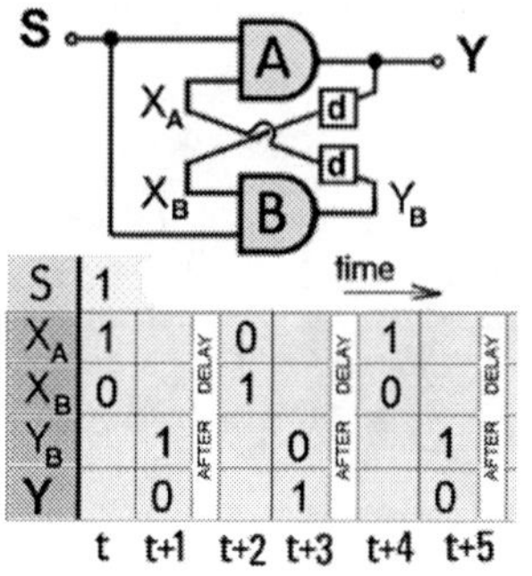

Figure 11. Schematic principle of the Abraham-Bloch oscillating circuit. The output signal in terminal Y switches periodically between the states Y=0 and Y=1 in each pair of time steps.

The basic problem encountered in an attempt to translate the circuits into the fluidic version is the need of the fluidic valve devices possessing a direct behaviourial analogy to the vacuum tubes. Usual active fluidic devices that might be considered for the role, bistable amplifiers, behave differently. They are not capable of being switched between the OPEN and fully blocked CLOSED regimes. Instead of closing the flow in one their stable regime they divert the fluid flow into another output terminal. If the turning down (not a complete blockage) of a fluid flow is required in no-moving-part fluidics, the solution is to use a vortex amplifier, e.g. [9, 10, 11]. These, however, in their standard version operate not by switching between two regimes. but instead turn the flow down continuously. Although they suffer from two significant limitations:

- Since the basic principle of no-moving-part fluidics is use of inertial forces in fluid (for example the inertia is used to keep a diverted jet flowing into the deflected direction even if it is opposed by the load connected to the outlet), the complete turning down (i.e. closure) of the flow passing through the device is obviously out of question. Stopping the flow would mean an absence of the inertial force. Pure fluidic devices without moved or deformed mechanical components thus can only decrease the flow but cannot stop it. A way out from this dilemma, little known, does exist. It is an addition of a component producing a jet-pump effect. An example is the radial-diffuser jet pump mentioned in [12], with which this limitation is not an unsurmounable one.
- The other until recently serious problem associated with known vortex amplifiers is the fact that their control input pressure must be higher than the pressure in the supply terminal [13, 14]. This would obviously lead to substantial complication in the design of the anti-parallel circuit, at least if solutions are sought among circuits exhibiting the configuration similarity $\Delta U_G < \Delta U_S$ of the electronic circuits discussed in Sect. 2.

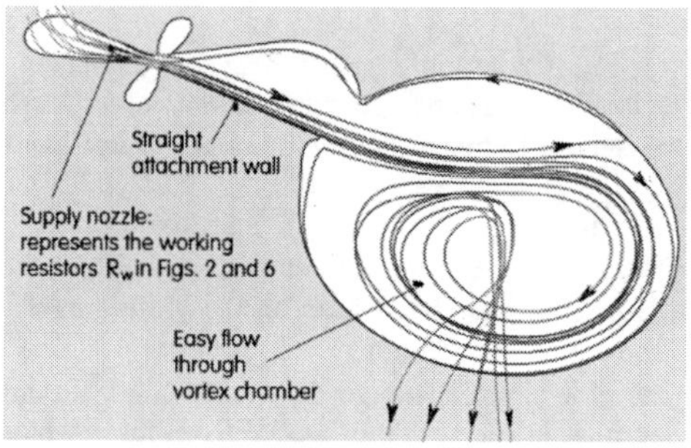

Figure 12. Perspective view of computed 3D pathlines in the author's bistable fluidic device in the OPEN regime of attachment to the straight attachment wall. The pressure drop across the vortex chamber is in this regime very small.

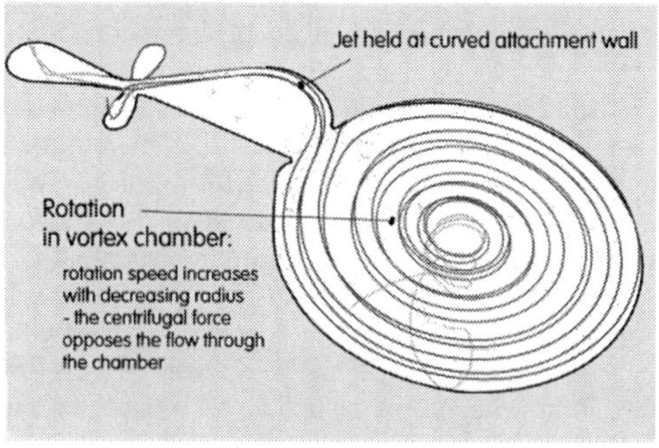

Figure 13. Computed in the same active device as in Fig. 12, is here the fluid flow in the CLOSED regime. The jet from the supply nozzle is attached to the curved attachment wall in the pre-chamber. Large pressure drop across the vortex chamber is due to the necessity of overcoming the centrifugal forces.

Fortunately, a solution to this second, more serious problem has been found in the device discussed in [12] – the bistable turn-down valve. From one point of view this valve may be seen as an integrated module consisting of two devices: a bistable diverter upstream and a vortex chamber [15, 16] downstream. Another point of view is to see it as a development of the idea of induced rotation [17] in two cavities: here the roughly triangular pre-chamber and the large round vortex chamber. It is this later point of view that explains the elimination of collectors and diffusers (otherwise unthinkable in a diverter amplifier). Also, this point of view explains the curvature of one of the attachment walls (AW1 in Fig.15) and the presence of the single communicating window between the pre-chamber and the vortex chamber.

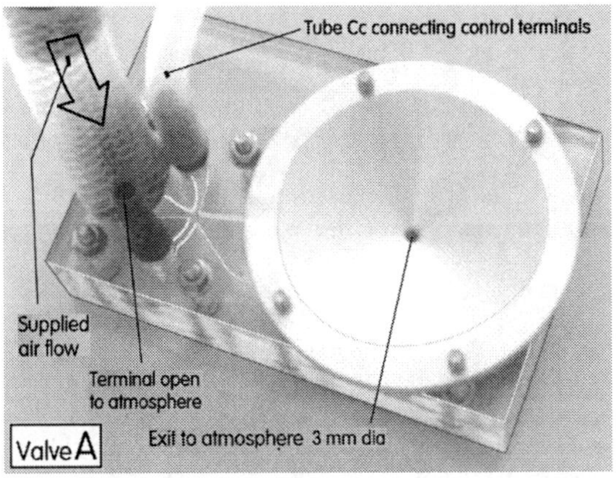

Figure 14. Photograph of the bistable turn-down valve during the performed tests. The terminal C_1 is open to atmosphere so that the CLOSED state, with attachment to the curved attachment wall AW1 (figure 15), is possible only under the action of the control inflow into C_2.

As shown in Fig. 12, in the regime that may be described as OPEN the fluid flow from the supply inlet through this valve manages to get easily into the central exit hole from the vortex chamber lid (the lid is the round aluminium component seen on top of the chamber in the photograph Fig. 14) — with minimal pressure loss. In the other, CLOSED regime – Fig. 13 - the jet issuing form the supply inlet into the pre-camber attaches to the curved attachment wall. This guides it to enter the vortex chamber tangentially. There it begins to rotate.

As the rotation radius decreases, its speed increases (because of conservation – though only approximate – of the fluid moment of momentum). The faster rotation is

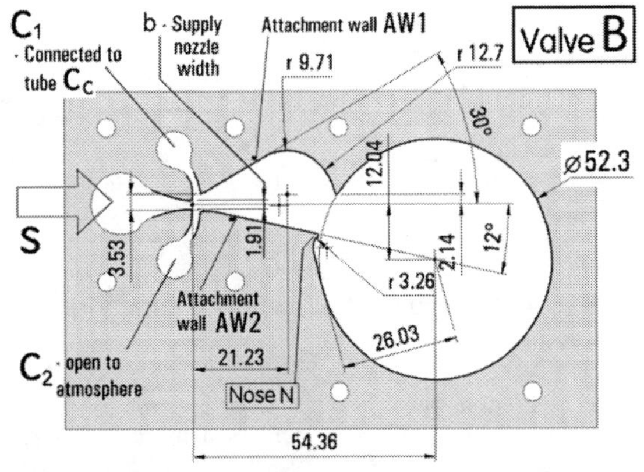

Figure 15. Internal geometry of the two active turn-down bistable valves. Note that A and B valves (Figs. 14, 16, 17) are mutual mirror images.

accompanied with larger centrifugal force acting on the rotating fluid. This force grows to a degree at which it opposes the flow towards the central exit almost completely. Some flow into the output nevertheless remains – it gets to the exit through the less intensively rotating boundary layers on the vortex chamber bottom and lid.

4 Fluidic circuit with anti-parallel turn-down valves

To form the anti-parallel fluidic circuit, the two active valves corresponding to Fig. 14 were connected in the manner discussed in the previous Sect. 3. The valves, described as A and B in Fig. 16, were made by laser cutting in a 2 mm thin PMMA plate, covered from top and bottom by simple flat cover plates. The laser-cut cavities in the two main plates were actually made identical. The required mirror image configuration of Fig. 16 was simply obtained by reversing (top to bottom) one of the main plates.

Both valves were connected to the common supply inlet S, with the exit terminals open into the atmosphere. Also one of their control terminals — in Fig. 16 it is the left one in both valves — was open to atmosphere. The other control terminals are mutually connected by

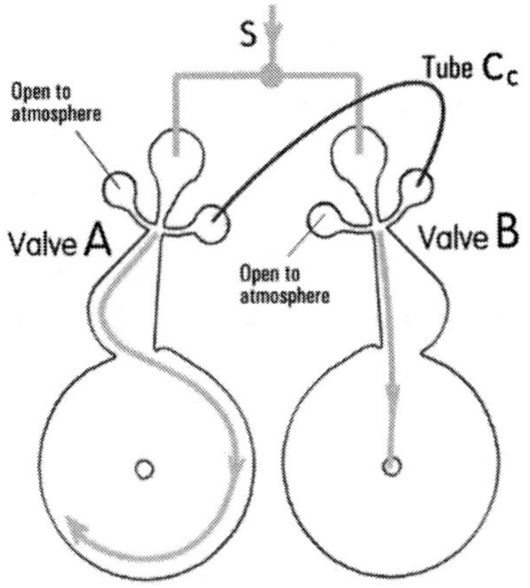

Figure 16. Schematic representation of mutual positions of the two bistable vortex turn-down valves in the tested anti-parallel circuit. In contrast to the flip-flops of Sect. 2 there is an immediately apparent asymmetry of the control signal transfer tube C_c.

Both valves are connected to the common supply terminal S – in the experiments this was an adjustable source of compressed air. The pressure drop across the supply nozzles corresponds to the voltage decrease on the working resistors R_w in the electric analogy in Sect. 2. No such fluidic resistors as separate components are therefore necessary. The pressure in the pre-chamber is led through the connecting tube C_c to the control nozzle of the other valve. The silicone rubber tube C_c is also clearly visible in figure 17. It was of 620 mm length and 6 mm internal diameter. In contrast to the classical "*multivibrateur*", no such second connection by a tube was in the discussed experiments applied to the other pair of control nozzles, which were both open to the atmosphere (which may be perhaps considered to be also a certain form of the interconnection). Of course, the unrestricted inflow of atmospheric air made the corresponding Coanda-effect attachment untenable, so that both devices actually in the present test series were practically monostable. Other variants of control terminal connections were used later in other experimental investigations — the present configuration is remarkable just because of the discovered stochastic regimes.

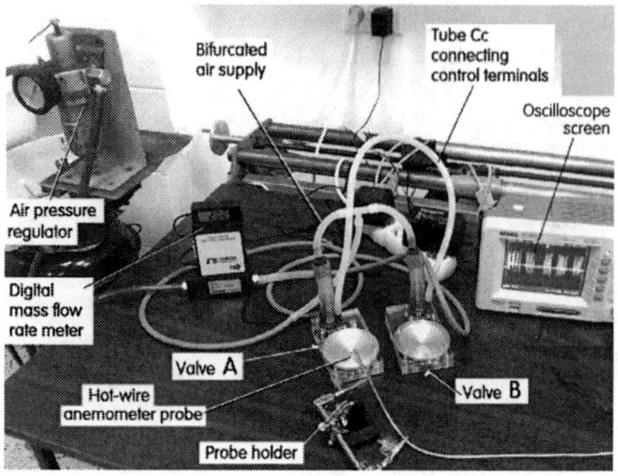

Figure 17. Schematic representation of mutual positions of the two bistable vortex turn-down valves in the anti-parallel circuit in the here discussed experiments.

While the intensity of air flow rotation in the vortex chamber served in the discussed circuit as the variable resistor, increasing or decreasing the fluid flow, the vortex also simultaneously served another purpose. It represented the delay devices (note the necessity of the delays in Fig. 11). The delay was simply caused by the spin-up and spin-down times of the rotating vortex [12, 18].

5 Experiments

The setup for the tests performed in the laboratory is presented in figure 17. At extreme left in the picture is the pressure regulator used to adjust the compressed air flow. Downstream from it is seen digital mass flow meter FMA 1727A with integral display, capable of measurements in the range 0 to 40 standard litres per minute, calibrated for nitrogen, accuracy ±1.5% of full scale guaranteed by supplier Omega Engineering Inc. Conversion to values for air, together with kg/s units replacing SLM, were made in data processing. The metered air flow then entered the bifurcation into the two supply terminals S.

Availability of only one flowmeter, connected upstream from the bifurcation, caused impossibility of evaluating actual distribution of flow rates between the two devices. After all, the fast changes of the distribution during each period of the investigated oscillation would eould make such measurement impossible anyway, due to the limitations of the flowmeter response (the supplier sytates the settling time of the meter is 800 ms). The flow rate was actually used only for indicating the order of magnitude values of Reynolds number and for this purpose it was considered sufficient to work with nominal mean velocity $\overline{w}$ computed under the equipartition condition:

$$\overline{w} = \frac{\dot{M}\, v}{2\, b\, h}, \tag{1}$$

where $\dot{M}$ [kg/s] is the mass flow rate, v [m^3/kg] is specific volume of the gas (air), b = 1.91 mm is the width of each supply nozzles (cf. figure 15), and h = 2 mm is the thickness of the plate with the air flow cavities.

The two tested valves A and B are recognisable near the middle of the photograph Fig. 17. The oscillator output was measured by hot-wire MiniCTA anemometer supplied by Dantec. Its probe 55 P 161 was positioned downstream from the exit of the valve A. The anemometer signal was displayed on the screen of Rigol oscilloscope, visible at the extreme right of the photograph figure 17.

6 Results and discussion

The onset of oscillation was recognised by audible response at a certain magnitude of the gradually increased supply airflow. There were, in response to further increase, other changes in the character of the air flow. These may be divided into the three regimes presented in the accompanying table, characterised by the values of the nominal Reynolds number:

$$\overline{Re} = \frac{\overline{w}\, b}{\nu}, \tag{2}$$

with ν [m^2/s] kinematic viscosity of air.

Ⓐ	At small flow rates, - just above oscillation threshold	Pulse packets, irregulatly appearing Irregular number of pulses in packet
Ⓑ	Medium flow rates $\overline{Re} \sim 6.5\ 10^3$	Individual pulses a degree of regularity
Ⓒ	High flow rates	individual pulses appearing at random

Typical appearances of the oscilloscope screen records in these three regimes are presented in Figs. 18, 19, and 20.

6.1 Low $\overline{Re}$ regime: stochastically appearing pulse packets

The oscillation signal became audible at roughly $\overline{Re}$ ~ ~2.5 10^3 and its character remained more or less the same for increased Reynolds numbers up to $\overline{Re}$ ~ 6 10^3. The example of appearance of the anemometer signal on oscilloscope screen presented in Fig. 18 is inside this range. Dominant feature there are packets of pulses. Typically there were 5 pulses in a packet. Sometimes (especially at lower $\overline{Re}$) the number is increased to 8 and (exceptionally) even more. Sometimes there was the other exception, only a single pulse. Frequency of the pulsation in a packet was of the order f ~20 Hz (note the value f = 17.3 Hz in Fig. 18) and remained quite constant for given air flow rate. Repetition rate of the packets was quite stochastic. Evaluation from many recorded cases indicated mean period 0.589 s, corresponding to frequency 1.7 Hz.

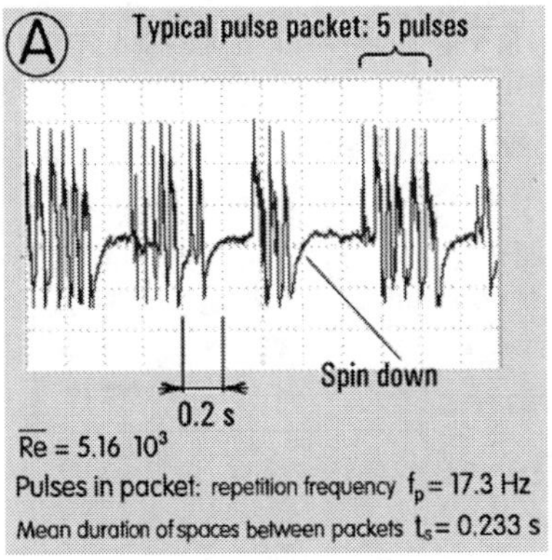

Figure 18. Typical oscilloscope screen record of oscillator output at small flow rates – the regime **(A)**. The oscillation, more or less regular, stops after several pulses to re-appear later. The repetition frequency as well as the no-oscillation space between the pulse packs are quite irregular.

An interesting feature found at the end of each packet is the exponential character of the variation of the output flow velocity signal (labelled "spin-down" in Fig. 18)

The irregularity of pulse generation may be obviously inconvenient in some applications that might be possibly considered for this oscillator, but there is no particular request for regularity in the generation of microbubbles in a bioreactor, which was the *raison d'etre* of the performed investigation. On the other hand, the capability to attain low oscillation frequencies with repeated sudden clearing bursts may be a distinct advantage – especially when obtained in a very compact oscillator device.

6.2 Middle $\overline{Re}$ regime: single flow pulses

The record presented in the next Fig. 19 shows a phenomenon observed with continuing increase of the supplied air flow rate. It is a distinct decrease of the number of pulses in a packet – finally leading to there being only a single pulse. Another feature there is again the remarkable spin-down, apparent at the end of each pulse. Although there is no exact regularity in the repetition of the pulses, the deviations from the mean period in this regime were quite small.

In the case of the record presented in Fig. 19 the mean period between the generated pulses was 0.325 s, which means the frequency of repetition f ~3 Hz. This is higher than was the frequency of repetition rate of the packets in the low $\overline{Re}$ regime of Fig. 18, nevertheless still very low indeed.

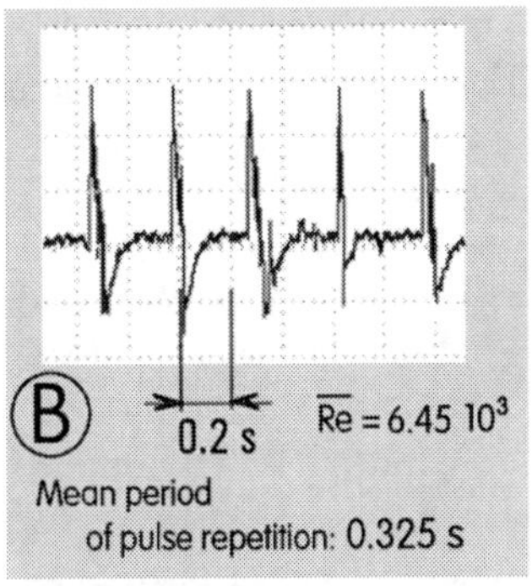

Figure 19. Typical oscillogram recorded in the regime **(B)**. The pulse packets of Fig. 18 are here reduced to single pulses. There is some visible regularity in their repetition.

6.3 High $\overline{Re}$: pulses at very long intervals and exponential dependence

A result of the experiments that was the least expected one was discovered at the highest supplied air flow rates. The more or less regular repetition rate of Fig. 20 disappeared and was replaced by isolated pulses mutually separated by very long intervals. The example presented in Fig. 20 shows one of these intervals lasting more than one second – but followed by a much shorter one, persisting for only one half of a second.

To evaluate the governing law of the pulse appearances, this obviously stochastic behaviour was investigated applying the classical statistical processing. Time axis was divided into "windows" lasting 0.2 s and in a very large number of recorded samples was counted the number of pulses in each such "window".

What was expected as a result of the plotting the number of pulse duration occurrences as a function the

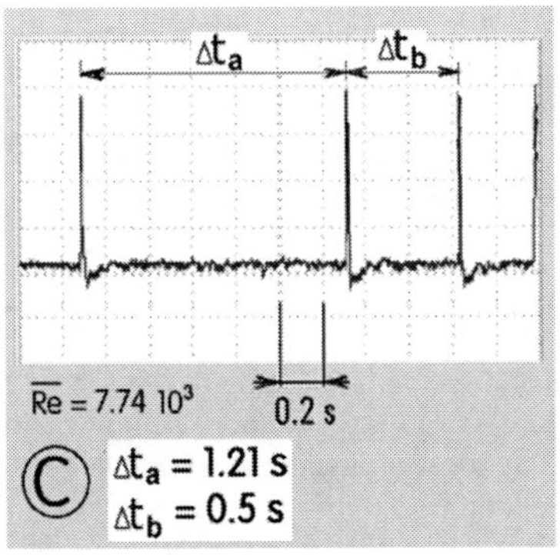

Figure 20. Typical oscillogram of output flow pulsations appearing in the highest $\overline{Re}$ regime **(C)**. Characteristic are the wide differences in time delays between the pulses.

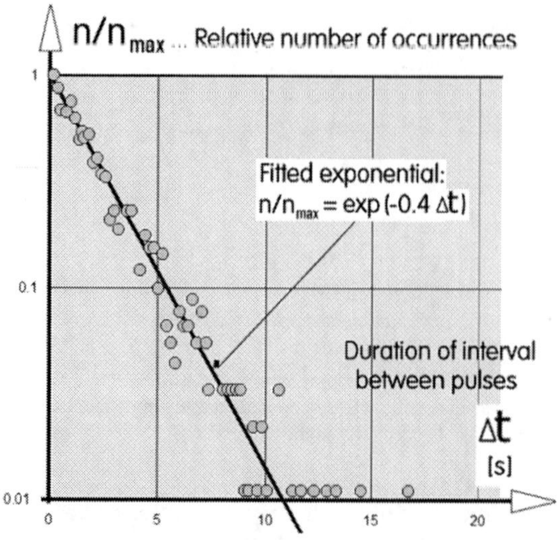

Figure 21. Interesting result of the statistical processing of the lengths (durations) of the interval between the pulses in the regime (**C**). The number n of occurrences of the interval lengths inside the 200 ms "windows" was counted in sample ensembles consisting of more than 50 oscilloscope records.

time interval between the pulses was a curve similar to the Gauss' distribution, with a maximum at the most probable interval lengths.

What was actually found is shown plotted in semi-logarithmic scale diagram in Fig. 21. The most probable interval between pulses is the shortest one. Number of occurrences then decreases in a monotonic manner — so that were at the end of the diagram also pulses that appeared after more than 10 seconds of waiting for them (the resolution, determined by the 0.2 s width of the "windows", is seen to be too crude for details of this part of the diagram). The semi-logarithmic scale in Fig. 21 made possible as the nest fitting through the data points the exponential decrease line. The slope of the lines depended on the air mass flow rate. In the particular case of Fig. 21, the fitted line indicates the characteristic time 1/0.4 = 2. 5 seconds.

7 Conclusions

In an investigation inspired by the search for microbubble generating, very low frequency fluidic oscillator was tested the idea of a circuit with two controlled flow valves operating in anti-parallel, with mutually blocking interconnection of their control inlets. In the basic principle this corresponds to the idea of Abraham & Bloch "*multibrateur*", invented the very beginnings of electronics. The fluidic valves used are of the little-known bistable amplifiers version with attachment walls pre-chamber and flow turning down vortex chamber.

Three different self-excited oscillation regimes were found, dependent upon the supplied fluid flow rate. The low flow rate regime (A) is characterised by multi-pulse packs with stochastic appearance and duration. More regular behaviour was found in the regime (B), with individual pulses and a degree of periodicity. The most surprising is the regime (C), with pulses seemingly appearing at random – but the interval separation between them, of the order of seconds, revealing extraordinary exponential dependence.

While the idea of a stochastic bubble generator may sound strange, algae do not care for the regularity in the CO_2 supply so that the behaviour may be acceptable. There seem to be even some other used of this idea.

Acknowledgements

Author obtained support from GAČR - Grant Agency of the Czech Republic, by research grant Nr. 13-23046S. There was also institutional support RVO:61388998.

References

1. *Biofuel from algae*, eds: A. Padney et al., Elsevier, 2014
2. W. Zimmerman, V. Tesař, British Patent GB20060021561 filed Oct. 2006
3. V. Tesař, *Chemical Engineering Journal,* **235**, 368, (2014)
4. V. Tesař, S. Zhong, R. Fayaz, *AIAA Journal*, **51** 397 (2013)
5. V. Tesař, *Chemical Engineering Research and Design* **92** 1603 (2014)
6. V. Tesař, C.-H. Hung, W. Zimmerman, *Sensors and* Actuators A: Physical **125** 159 (2006)
7. Raghu S., Experiments in Fluids **54**, 1455 (2013)
8. V. Tesař, *Chemical Engineering Science* **116** Part I: 843, Part II: 849 (2014)
9. D. N. Wormley, H. H. Richardson, *Journal of Basic Eng.* **92** 369 (1970)
10. R. J. Woolhouse, J. R. Tippetts, S. B. M. Beck, *Proc. Inst. Mech. Eng., Part C: J. Mech. Eng. Sci.*, **215** 893
11. N. Syred, J. K. Royle,*Second IFAC Symposium on Fluidics*, Prague, paper B5 (1971)
12. V. Tesař , E. Smyk E., *Chemical Engineering and Processing: Process Intensification* **90**, 6 (2015)
13. D. Parker, M. J. Birch, J. Francis, *J. Fluids Eng.* **133,** 041103 (2011)
14. J. Francis, M. J. Birch, D. Parker, *J. Fluids Eng* **134,** 021103 (2012)
15. V. Tesař V. *Messen-steuern-regeln* 26, Nr. 4 (1983)
16. V. Tesař V., *Proc.gs of the 9-th International Fluidics 'Jablonna' Conf.*, Jablonna, Poland, Paper A-5, Sept. 1982
17. V. Tesař., *EPJ Web of Conferences* **92,** article 02096 (2015)
18. V. Tesař, paper #206, *Proc. 21st Internat. Conf. Engineering Mechanics 2015*, p. 326, Svratka, Czech Republic, May 11 (2015)
19. V. Tesař, *Pressure-Driven Microfluidics,* Artech House, Boston, USA (2007)
20. W.H. Eccles, F.W. Jordan, *British Patent Nr. 148582,* filed June 1918
21. H. Abraham, E. Bloch, *Annales de Physique* **9** 237 (1919)

EPJ Web of Conferences **114**, 02124 (2016)
DOI: 10.1051/epjconf/201611402124

DESIGN OF THEORETICALLY OPTIMAL THERMOACOUSTIC COOLING DEVICE

Tomáš Tisovský[a,1], Tomáš Vít[b,1]

[1]*Department of Power Engineering Equipment, Faculty of Mechanical Engineering, Technical University of Liberec, Czech Republic*

Abstract. The aim of this article is to design theoretically optimal thermoacoustic cooling device. The opening chapter gives the reader brief introduction to thermoacoustic, specializing in the thermoacoustic principle in refrigerator regime. Subsequent part of the article aims to explain the principle on which thermoacoustic is simulated in DeltaEC. Numbers of executed numerical simulations are listed and the resulting thermoacoustic cooling device design is presented along with its main operation characteristics. In conclusion, recommendations for future experimental work are given and the results are discussed.

1 Thermoacoustic as an alternative cooling technology

Thermoacoustics was defined by Rott [1] as a subject dealing generally with effects in acoustics in which heat conduction and entropy variations of a medium are of importance.

Acoustic wave spreading throughout a fluid is a phenomenon composed out of variations of pressure and velocity. These pressure variations are usually small enough for the wave to be considered an isoentropic process. If the fluid under the influence of acoustic wave is left in thermal contact with constant temperature solid boundary, heat transfer between solid and fluid will take place resulting in the thermoacoustic effect.

It was found out that device can be made in which the termoacoustic effect can be used to transfer heat from low potential source into high potential heat sink. Energy used to pump the heat is supplied by the acoustic wave induced in the fluid by external source such as loudspeaker.

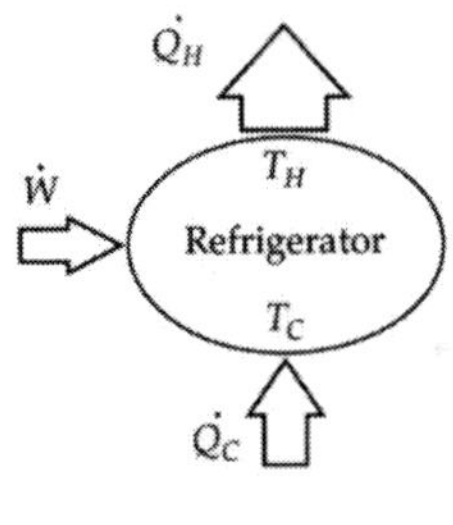

Figure 1. Flows of work and heat inside a thermoacoustic refrigerator [2]

[a] Corresponding author: tomas.tisovsky@tul.cz
[b] Corresponding author: tomas.vit@tul.cz

2 Introduction to thermoacoustics

To explain the thermoacoustic effect in more detail it is convenient to imagine a thin stationary solid plate surrounded by volume of fluid subjected to oscillatory boundary condition in a way that the resulting acoustic wave propagates in the direction of the longest plate dimension.

Such simple device will perform as a refrigerator (i.e. transferring heat in an initially uniform temperature plate from one end to the other) only as long as the temperature gradient alongside the plate is small enough. The process which single parcel of working fluid undergoes is shown in Figure 2a.

The process which every parcel of fluid undergoes within a piston movement period consists of two reversible adiabatic steps 1-2 and 3-4 and two isobaric heat transfer steps 2-3 and 4-1.

At the beginning of the cycle, the piston moves to the right and compresses the parcel which consequently warms up. In thermoacoustic refrigeration cycle, the parcel is now warmer that the adjacent part of the solid plate and heat flows irreversibly from parcel to plate. Next the piston moves to the left to its original position and the parcel expands and cools in process. Again, if the temperature gradient along the plate is small enough, heat flows irreversibly, this time from the solid plate into the parcel. After that the parcel is in its initial position and the cycle starts again.

The amount of heat which is taken from/given to the plate at during one cycle is quite small and many cycles are needed for a substantial temperature gradient to build up along the plate.

It is also notable that the acoustic work consumed to compress the gas in step 1-2 is of the same amount as work which is returned by expanding gas in step 3-4. Therefore no net work is consumed in those steps. Net work is consumed only in steps 2-3 and 4-1 in which piston has to move a little bit to keep the pressure throughout the gas constant.

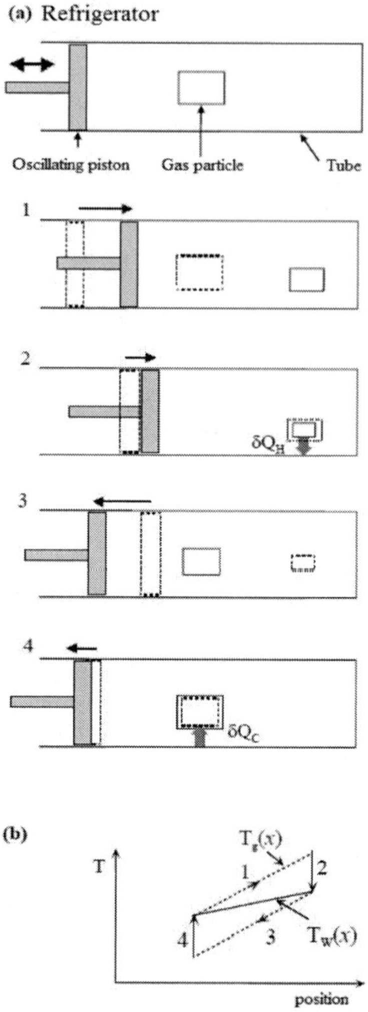

Figure 2. a) a typical gas parcel executing the four step cycle in a thermoacoustic refrigerator
b) gas and wall temperature plotted versus the position in resonator, [2]

The time lag between displacement and heat transfer is necessary for the operation of thermoacoustic device and is caused by the imperfect thermal contact between solid and gas. Sound wave at greater distances from solid boundary is, as is mentioned above, adiabatic process and if the gas and plate were at a zero distance from each other, the parcel would simply be compressed and expanded isothermally. Parcel that are the most effective to the thermoacoustic effect are located approximately in a distance about the thermal penetration depth δ_k away from the solid boundary.

The thermal penetration depth is defined as

$$\delta_k = \sqrt{\frac{k_g}{\pi f \rho_m c_p}}, \tag{1}$$

where k_g is the thermal conductivity of a working gas, ρ_m is its mean density and c_p is its specific heat capacity. It is clear from previous discussion that it is desirable to have as much of a working gas volume as possible in appropriate, relatively small, distance away from solid boundary on which temperature gradient is imposed. Way to achieve this is to place a component called stack downstream of the source from which the acoustic wave is propagating. Stack consists of many parallel channels aligned in a direction of the acoustic wave motion giving the passing wave large surface to interact with.

The main parts combining the thermoacoustic cooling device are therefore: resonator, loudspeaker, stack, hot heat exchanger and cold heat exchanger. The loudspeaker at one end of the resonator gives rise to the acoustic wave which then propagates through resonator to its other end where it reflects. Characteristics of the reflected wave are given by the characteristics of the incident wave as well as by the type of boundary at the point of reflection (closed end / open end).The incident wave and the reflected wave interfere in the resonator resulting in the standing wave pattern. Generally, the incident and the reflected wave do not form the standing wave pattern. This has to be achieved by correctly choosing oscillation frequency for a given apparatus.

In standing wave based thermoacoustic devices, the heat flow due to thermoacoustic effect is always in the direction towards the nearest pressure antinode.

The remaining parts of the thermoacoustic cooling device, which have not been discussed, are the heat exchangers. The hot heat exchanger is located closer to the loudspeaker and its purpose is to maintain proper temperature on the hot end of a stack, usually the ambient temperature of about 300K should suffice. The cold heat exchanger, on the other hand, is design to heat up the working gas enough to maintain the desired cooling temperature.

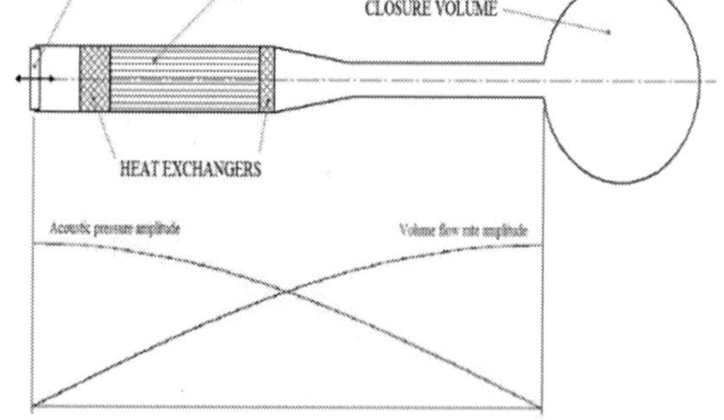

Figure 3. Resonator scheme

3 Design of thermoacoustic cooling device in DeltaEC

DeltaEC - Design Environment for Low–Amplitude Thermoacoustic Energy Conversion, is a computer program that can calculate details of how thermoacoustic equipment performs, or can help the user to design equipment to achieve desired performance. [4]
DeltaEC integrates in one spatial dimension using low-amplitude acoustic approximation and sinusoidal time dependence. It integrates up to four, two of which are complex, equations in gas in a geometry given by the user as a sequence of segments. The four equations can are integrated to obtain complex acoustic pressure amplitude, complex volume flow rate amplitude, mean temperature and total power flow.
Apart from few smaller exceptions DeltaEC does not include any nonlinear effects which arise at high pressure amplitudes. As a consequence care must be taken to choose the maximum pressure amplitude, which occurs at the loudspeaker, such that two conditions hold.

a) Condition (2) for the acoustic Mach number, given by means of maximum acoustic pressure amplitude p_{01} and the square of velocity of sound in gas a, is

$$M_T = \frac{p_{01}}{\rho_m a^2} = 0{,}1. \tag{2}$$

b) The Reynolds number based on Stokes boundary layer thickness δ_v is less than a value given by equation (3), which is the condition for transition to turbulence in boundary layer

$$Re^{\delta_v} = \frac{u_1 \delta_v}{\nu} = 550. \tag{3}$$

Figure 4 shows a simplified model of thermoacoustic cooling device. In reality, model also includes many RPN (Reversed Polish Notation) math segments which are used for better control of the model behaviour. Main parts of the model are listed in top left corner of the figure 4. The begin segment is used to define 6 real boundary conditions needed for an unique solution of the equations (2) to (5) to exist. Model below was designed so that few boundary conditions were given elsewhere from the begin segment hence a shooting method needs to take place. Shooting method algorithm is adjusting variables called guesses which are selected by the user to gain a correct value of variable called targets (i.e. the boundary conditions).

The guesses are inlet volume flow rate amplitude, phase shift and magnitude of the loudspeaker source voltage, heat flow from/to both heat exchangers and length of two resonator ducts.

The targets are resonance condition, temperatures of the working gas at exchangers' levels, value of the maximum acoustic pressure amplitude, position of the center of the stack and the conditions that ensure closure and isolation of the resonator at the compliance end.

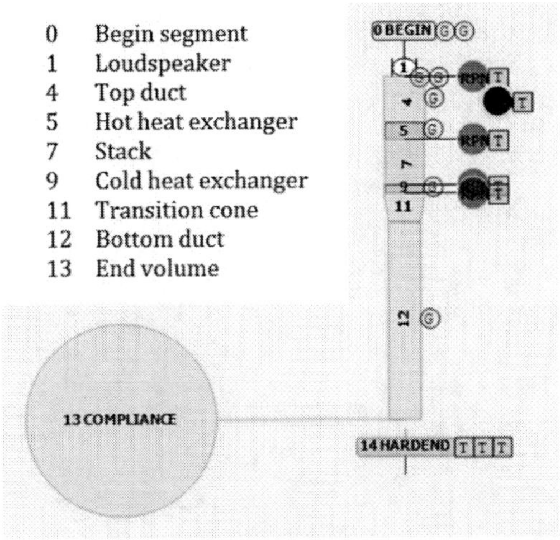

Figure 4. Composition of numerical model in DeltaEC

Series of simulations was carried out, namely the analysis of the appropriate end volume diameter, analysis of the stack plate half-spacing magnitude, analysis of the optimal stack center position in the resonator, comparison of different stack materials, analysis of the influence of various resonator duct diameter, optimal stack length analysis and analysis of the appropriate narrowing of the resonator duct. Most of these analyses were evaluated with respect to the highest possible coefficient of performance.

Table 1. Parameters of the resulting device

p_m [MPa]	f [Hz]	p_{01} [kPa]	$L_{resonator}$ [mm]	$\frac{x_{stack\ center}}{L_{resonator}}$ [1]
1,2	500	120	416	0,25

$D_{resonator}$ [mm]	$\frac{d_{duct}}{D_{duct}}$ [1]	$D_{compliance}$ [mm]	$\frac{y_0}{\delta_k}$ [1]	y_0 [mm]
34	0,78	105	2,64	0,23

T_C [K]	COP [1]	$\frac{COP}{COP_{ideal}}$ [1]	η_{ea} [1]	$\dot{W}_{el}$ [W]	$\dot{E}$ [W]
260	0,79	0,12	0,54	428	229

$\dot{Q}_C$ [W]	$\dot{Q}_H$ [W]	$Re^{\delta_v}{}_{max\_stack}$ [1]	$Re^{\delta_v}{}_{max\_res}$ [1]
182	610	235	594

Table 1 shows the resulting resonator dimensions and its working parameters. Variable η_{ea} stands for the efficiency of transformation of electric energy of loudspeaker source to acoustic energy and it is highest possible for given device. Variable y_0 is the stack plate half-spacing.

One of the greatest advantages of thermoacoustic refrigerator is the ease in which it can be adjusted to work optimally in various cooling temperature modes. Cooling power corresponding with given temperature can be than easily adjusted by imposing different acoustic pressure amplitude. Tables 2 and 3 show the difference between simple adjustment of a stack position when cooling to different temperature and leaving the original resonator unchanged.

Table 2. Resonator parameters corresponding to various cooling temperature – original resonator

	original resonator		
T_C [K]	f [Hz]	ε_{ch} [1]	$\dot{Q}_C$ [W]
290	526	1,26	358
270	509	0,97	239
250	491	0,59	126
230	472	0,10	17

Table 3. Resonator parameters corresponding to various cooling temperature – adjusted resonator

	resonator with optimal stack position					
T_C [K]	$\frac{x_s}{L}$ [%]	L_{rez} [mm]	x_s [mm]	f [Hz]	ε_{ch} [1]	$\dot{Q}_C$ [W]
290	38	434	166	500	1,60	495
270	28	422	120	500	1,01	251
250	22	411	90	500	0,62	133
230	18	379	71	500	0,36	70

4 Conclusions

Designed thermoacoustic cooling device has several advantages compared to conventional cooling devices, mainly it is the ease of proportional control, very simple structure and the fact that they are environmentally friendly. Moreover such device would occupy a relatively small space and surprisingly enough, it would be silent. This is due to the fact that the working gas inside the thermoacoustic cooling device is pressurized and therefore the ratio of the acoustic pressure to the mean pressure would be small even for amplitudes of acoustic pressure that would be considered high in atmospheric conditions. The only significant source of noise can therefore only be the mechanical vibrations of the whole device and such problem can be easily solved by more suitable connection of the device to its base.

For their advantages, thermoacoustic cooling devices can be suitable to use in relatively extreme conditions such are in space or in Third world countries.

On the other hand, there are also some disadvantages of thermoacoustic cooling technology. The main disadvantage is the poor efficiency exhibited by thermoacoustic cooling devices. This might be resolved once a better understanding of involved physics is obtained, mainly the understanding of non-linear effects accompanying coupled heat transfer and acoustics. There are also technological and construction aspects of improving efficiency of thermoacoustic devices since present heat exchangers and stack are still hardly optimal. The device proposed in this article could be used, for example, in food merchandising, where toxicity of a conventionally used refrigerant can be a problem.

At the end, it is worth mentioning that thermoacoustic cooling devices can be constructed so that they can work without a loudspeaker or similar device (acoustic wave can be generated only by stack along which is maintained a proper temperature gradient), or they can be constructed so that they can directly cool the gas which requires cooling (no additional heat exchanger is required).

References

[1] Rott, N. (1980). "Thermoacoustics". *Adv. Appl. Mech.* **20** (135): 135. doi:10.1016/S0065-2156(08)70233-3

[2] In't Panhuis P. H. M. W. *Mathematical Aspects of Thermoacoustics* [online]. Universiteit Eindhoven: Print Service Technische, 2003, 190 s. ISBN 978-90-386-1862-3.

[3] Tijani, Moulay El Hassan. *Loudspeaker-driven thermo-acoustic refrigeration.* Eindhoven: Technische Universiteit Eindhoven, 2001, 170 s. ISBN 90-386-1829-8.

EPJ Web of Conferences **114**, 02125 (2016)
DOI: 10.1051/epjconf/201611402125

Verification of the boundary condition at the porous medium–fluid interface

Kirill Tsiberkin[1,a], Ekaterina Kolchanova[2], and Tatyana Lyubimova[3]

[1] Perm State University, Theoretical Physics Department, 614990 Perm, Bukirev 15 street, Russia
[2] Perm State University, General Physics Department, 614990 Perm, Bukirev 15 street, Russia
[3] Institute of Continuous Media Mechanics UB RAS, CFD Laboratory, 614013 Perm, Acad. Korolev 1 street, Russia

Abstract We study the hydrodynamic stability of liquid flowing down over the inclined layer of uniform porous medium and compare the results obtained in the different frameworks. The flow in porous medium is described by the Brinkman model and by the Darcy model with corresponding boundary conditions at the interface between the homogeneous fluid and porous medium. It is shown the critical Reynolds number is calculated for the Darcy model much lower than in case of the Brinkman model, while the flow velocity is the same in the both models. It is a pure mathematical effect, which can be used to verify the models and to determine the empirical coefficients in the boundary conditions from an experimental study of flow instability.

1 Introduction

The fluid flow over the saturated porous medium enthralls the fluid inside it. The drag force in a porous medium leads to the rapid decrease of the tangential velocity, and it permits the flow instability development. The coupled flows have been studied since 1950s [1–7]. The problem has many applications in the environmental science and technology: contaminant transport in the rivers and atmosphere [8, 9], solidification of the liquid metals [10–12], heat and mass transfer between the contacting media [13], optimization of the flows in the fuel cells [5], etc.

There is also a fundamental interest. The problem of the conditions at the interface between the coupled flows is not solved until the end. Many authors offer the different kinds of the boundary condition being applied to the different filtration models [1, 3, 10, 14–18]. The porous media models by Darcy and Brinkman are the most frequently used in known studies [1].

The Darcy model usually uses the interface boundary conditions by Beavers and Joseph [3]. These conditions were obtained from the experimental data assuming the Darcy law determines the filtration. The flow velocity is discontinuous in that case. An alternative for the Darcy model is the approximate conditions by le Bars and Worster [10]. It is more precise than the first, because the velocity stays continuous in that model. This approach postulates the transition layer exists between the flows. In spite of the transition zone is located in the porous layer, the flow velocity inside it is assumed to have a parabolic profile which continues the overlying fluid velocity.

Another description of the filtration is given by the Brinkman model [19]. It was introduced as empirical, but the further studies show that the Brinkman equations can be obtained from the Navier–Stokes equations by averaging at the microscopic scale [4, 14, 15, 20]. Moreover, it gives the boundary conditions at the interface between the porous medium and homogeneous fluid. Ochoa-Tapia and Whitaker obtained one of the most used condition for the Brinkman model [14, 15]. Despite it is obtained by the strict formal way, these conditions include other empirical parameter sets a tangential stress jump at the interface. Other conditions are proposed in the elder and the latest studies [16–18], but these variants are not well known yet.

It is shown the flow structure and the velocity magnitude obtained in the different models may be matched by tuning of the empirical parameters in the interface boundary conditions [13]. The velocities are the same in the overlying layer and in the porous medium. The difference exists only in the thin transition zone between the flows. Thus, the models cannot be distinguished by the stationary flow measurements. However, we have found the difference of the calculated flow structure makes a great influence on the results of the theoretical analysis of flow stability. The critical Reynolds number and wave number of the flow in the framework of the Darcy model are significantly different from results for the Brinkman model. We suppose the comparison of the theoretical and experimental data about the flow stability can verify the filtration model and interface boundary conditions, can determine the empirical parameters of these models, and may clarify their applicability area.

In this study, we present a comparative, theoretical analysis of the coupled flow stability using the Darcy model with the interface boundary conditions by Beavers and Joseph, and the Brinkman model with the conditions by Ochoa-Tapia and Whitaker. In addition, we analyse an influence of the heavy admixture dissolved in the porous layer on flow stability.

The paper is organized as follows. After the Introduction, Section 2 describes the problem geometry and flow models. The stationary solutions are presented in Section 3, while the results of numerical linear stability analysis and comparison between the models are given in Section 4. Section 5 contains a brief description of the linear stabil-

[a] e-mail: kbtsiberkin@psu.ru

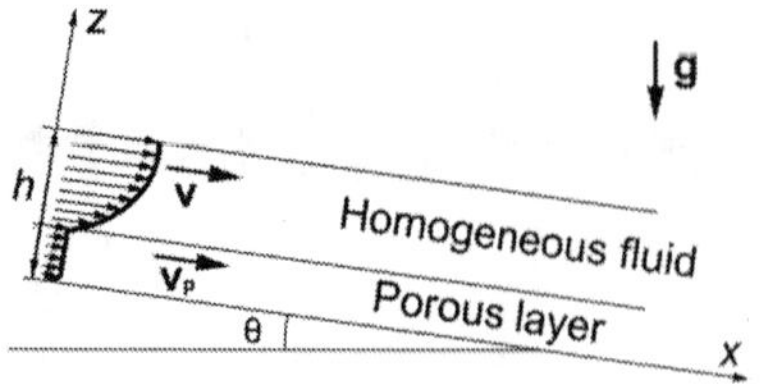

Figure 1. Geometry of the two-layer system and schematic structure of the flow.

ity analysis of the fluid with a heavy admixture extracting from the porous layer. Conclusion includes the discussion and final remarks.

2 Two-layer system

2.1 Problem geometry

We consider the layer of incompressible homogeneous fluid is flowing down over the inclined uniform porous medium, which is saturated with the same fluid (Figure 1). The overlying fluid is bounded by a free rigid surface from above, and the porous layer has an impermeable rigid wall from below.

The porous layer thickness is h_p, and the full thickness of the system is h. The layers are inclined at angle θ to the horizon. The permeability and porosity of the porous medium are K and φ, respectively. The fluid has the density ρ and kinematic viscosity ν.

The fluid moves only because of the gravity force, and there is no additional longitudinal pumping. Besides, the system is isothermal, and the fluid does not contain any dissolved or weighted admixture excluding the problem considered in Section 5.

2.2 Flow models

2.2.1 Governing equations

The overlying fluid layer is described by the Navier–Stokes equations:

$$\frac{\partial \mathbf{v}}{\partial t} + \mathrm{Re}(\mathbf{v}\cdot\nabla)\mathbf{v} = -\nabla p + \nabla^2\mathbf{v} - \frac{\gamma}{\sin\theta}, \qquad (1)$$

$$\nabla\cdot\mathbf{v} = 0, \qquad (2)$$

where $\mathbf{v}$ is flow velocity, p is pressure, Re is the Reynolds number, θ is the slope angle of the layers, and γ is a unit vector contrary the gravity acceleration $\mathbf{g}$. We present all the equations in dimensionless form. The variable scales and parameters are listed in Section 2.2.3.

We use the Darcy model and the Brinkman model to describe the filtration [1, 19]. The Darcy equation sets the fluid velocity is proportional the pressure gradient:

$$\frac{\partial \mathbf{v}_p}{\partial t} + \mathrm{Re}(\mathbf{v}_p\cdot\nabla)\mathbf{v}_p = -\nabla p_p - q^2\varphi\mathbf{v}_p - \frac{\gamma}{\sin\theta}, \qquad (3)$$

$$\nabla\cdot\mathbf{v}_p = 0, \qquad (4)$$

where $\mathbf{v}_p$ is fluid velocity in pores, φ is medium porosity, and q replaces the Darcy number and determines the medium permeability (see Section 2.2.3).

We save the inertial terms in the filtration equations. They are usually negligible, but we consider the medium with high permeability and porosity, thus the inertial terms are significant because of the pore-scale Reynolds number is large [1]. It equals $10^2 \div 10^3$ in the water plants, and $10^1 \div 10^2$ in the foametal layer. The time-derivative is in the equation because of flow instability is a rapid process. The two-layer flow instability is oscillatory, and the fast travelling waves occur in the system. Thus, we save the inertial terms to describe these factors correctly.

The Brinkman equations are as follows:

$$\frac{\partial \mathbf{v}_p}{\partial t} + \mathrm{Re}(\mathbf{v}_p\cdot\nabla)\mathbf{v}_p = -\nabla p_p - q^2\varphi\mathbf{v}_p + \nabla^2\mathbf{v}_p - \frac{\gamma}{\sin\theta}, \qquad (5)$$

$$\nabla\cdot\mathbf{v}_p = 0. \qquad (6)$$

It includes an additional effective viscous term. The Brinkman model works well if the medium has the large porosity and permeability. This model was proposed as an empirical [19], but it may be also obtained theoretically by the averaging of the Navier–Stokes equation at the pore-scale level [4, 14, 15, 20].

2.2.2 Boundary conditions

The filtration models include the equations of a different order and require the different boundary conditions at the walls and at the interface between homogeneous flow and porous medium.

The top surface $z = 1$ is rigid and free, and the conditions are as follows:

$$v_z = 0, \quad \frac{\partial v_x}{\partial z} + \frac{\partial v_z}{\partial x} = 0. \qquad (7)$$

It includes the undeformability condition and absence of the tangential viscous stress. The bottom wall $z = 0$ is rigid and impermeable. The condition for the Darcy model sets only the wall is impermeable:

$$v_{p,z} = 0, \qquad (8)$$

while the Brinkman model requires no-slip condition:

$$\mathbf{v}_p = 0. \qquad (9)$$

At the same time, many different kinds of the interface boundary conditions were developed. These conditions usually include the empirical coefficients connected with the properties of the porous medium, and one may vary the parameters to equate the stationary flow velocity [13]. A unified approach to describe the coupled flows is still not developed [1, 13, 16].

We apply the two most frequently used kinds of the interface boundary conditions at $z = d$. The conditions by Beavers and Joseph developed empirically for the Darcy model are as follows [3]:

$$v_z = \varphi v_{p,z}, \quad p = p_p, \quad \frac{\partial v_x}{\partial z} = \alpha q(v_x - \varphi v_{p,x}), \qquad (10)$$

where α is the parameter should be found from the experiment. The conditions by Beavers and Joseph determine the jump of the tangential flow velocity. The normal velocity component and pressure are continuous.

The Brinkman model uses the conditions by Ochoa-Tapia and Whitaker obtained by the direct averaging of the Navier–Stokes equations at the pore-scale level [14, 15]:

$$\mathbf{v} = \varphi \mathbf{v}_p, \quad p - 2\frac{\partial v_z}{\partial z} = p_p - 2\varphi\frac{\partial v_{p,z}}{\partial z}, \tag{11}$$

$$\frac{\partial v_x}{\partial z} - \frac{\partial v_{p,x}}{\partial z} = -\beta q \varphi v_{p,x}. \tag{12}$$

Here, β is another empirical parameter. These conditions set the continuity of the flow velocity and normal stresses, while the tangential viscous stress has a jump at the interface.

2.2.3 Dimensionless parameters

We use the following units of the variables:

$$[\mathbf{r}] = h, \ [t] = \frac{h^2}{\nu}, \ [U] = \frac{gh^2 \sin\theta}{\nu}, \ [p] = \rho g h \sin\theta. \tag{13}$$

Namely, the length is measured in the units of the whole system thickness. The time-scale is a characteristic viscous time. The velocity unit is doubled maximum velocity of the flow over the inclined plane. The pressure scale is determined by the longitudinal hydrostatic gradient.

The dimensionless equations and boundary conditions include the following parameters:

$$\mathrm{Re} = \frac{gh^3}{\nu^2}\sin\theta, \ d = \frac{h_p}{h}, \ q = \frac{h}{K^{1/2}} = \frac{1}{\mathrm{Da}^{1/2}}, \tag{14}$$

where Re is the Reynolds number, d is a relative thickness of the porous layer and q is an equivalent of the Darcy number Da. Here, the Reynolds number is larger than it is usually determined for the channel flow. To compare the results of the current study with the channel flow stability [21], one should multiply Re by $(1 - d)^3$.

3 Stationary flow

The equations (1)–(6) permit the stationary solution for the plane-parallel flow. The velocity profiles calculated in the Darcy as well as Brinkman model frameworks are presented in Figure 2 and Figure 3, and the structure of flow near the "fluid–porous medium" interface is magnified. The velocity magnitudes are the same in the both layers. Nevertheless, the boundary conditions (10) and (11) produce the difference between the solutions.

The Brinkman flow has a continuous profile at the interface (Figure 2). It reaches the filtration velocity determined by the Darcy law within the boundary layer with thickness is of order q^{-1}, which becomes wider in the high permeable medium. If q is low, the velocity profile tends to the parabolic Poiseuille-like flow in the whole system. Besides, the stationary velocity profile has a break at the interface is connected to the tangential stress jump.

On the other hand, the Darcy model gives a constant value of the velocity in the porous layer, thus the tangential velocity is discontinuous at the interface (Figure 3). That makes a significant influence on the results of the theoretical flow stability analysis.

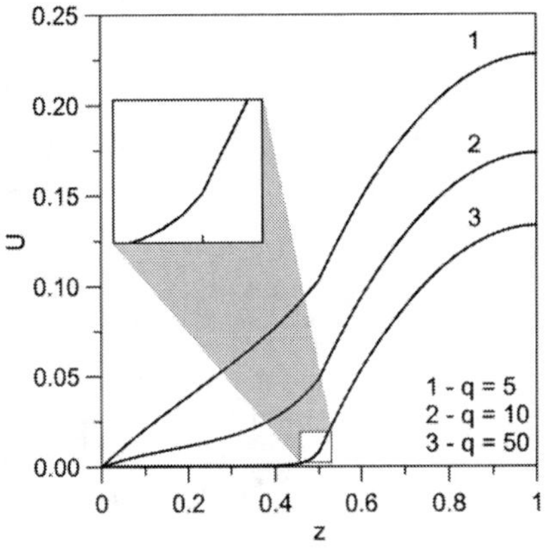

Figure 2. Profiles of the stationary plane-parallel flow in the Brinkman model.

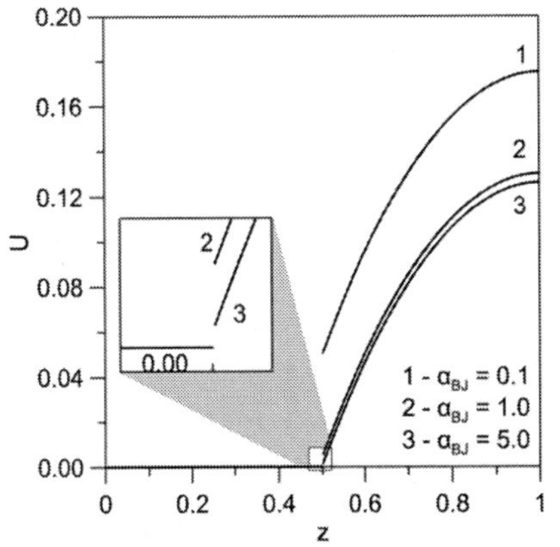

Figure 3. Profiles of the stationary plane-parallel flow in the Darcy model.

4 Linear stability analysis

4.1 Shooting method

In this paragraph, we present a brief review of the method applied to the linear stability analysis. First, we introduce the small perturbations of the stationary velocity and pressure. Substitution

$$\mathbf{v} \to U\mathbf{e}_x + \mathbf{v}, \quad p \to P + p, \tag{15}$$

where $|\mathbf{v}| \ll U$ and $p \ll P$, and linearization of the full equations and boundary conditions (1)–(11) produce the system of linear ordinary differential equations for the perturbations. Assuming they are longitudinal waves growing exponentially, i.e.

$$\mathbf{v}, p \sim \exp(\lambda t + ikx), \tag{16}$$

and expressing the pressure through the velocity, we obtain the linear system of ordinary differential equations for the amplitude functions depending on z. It includes the Orr–Sommerfeld equation for the homogeneous fluid and its' equivalent for the porous medium [21]. The equation system is of the sixth order if the Darcy model applied, and

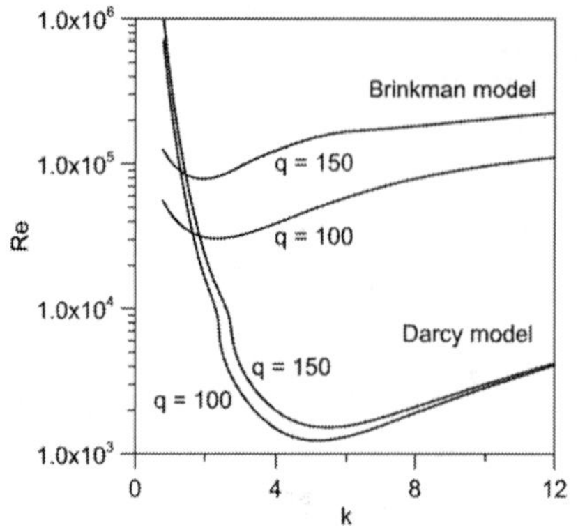

Figure 4. Neutral curves calculated in the both models at d = 0.50, α = 1 and β = 0.

Table 1. Critical Re and k for the both flow models at different permeability of the porous layer, d = 0.50, α = 1 and β = 0.

q	k		Re, $\times 10^4$	
	Darcy	Brinkman	Darcy	Brinkman
30	4.62	2.40	0.0821	0.546
50	4.85	2.38	0.0946	1.03
80	5.07	2.37	0.112	2.07
100	5.21	2.37	0.124	3.07
120	5.39	2.24	0.135	4.44
150	5.51	1.88	0.152	7.87
170	5.69	1.42	0.164	13.2
200	5.75	2.45	0.181	19.6

eighth order in the case of the Brinkman model. It has no an analytic solution, and stability analysis is provided by the numerical shooting method [22–24].

We solve the equations in the both fluid layers numerically by the adaptive Runge–Kutta–Merson 4th order method, and try to satisfy the interface boundary conditions. It is possible only with some specific Re, k and other parameters. Thus, the semi-automatic numerical algorithm (namely, two-dimensional secant method) varies the Re, k and $\lambda = i\omega$ until the boundary conditions are not satisfied with some precision. The method has some additional features, such as forced orthogonalization of the solution vectors must be linearly independent [24]. The described method is well proven by the studied of many different problems of hydrodynamic and convective instability.

4.2 Results

The numerical linear stability analysis shows the huge difference between the flow models we use. We set $d = 0.50$, $\alpha = 1$ and $\beta = 0$, and $q = 100$ and 150. The relative difference of stationary flow velocity calculated in the both frameworks is about 0.5%. Nevertheless, the critical Reynolds number for the Darcy model is almost one hundred times lower than for the Brinkman model (Figure 4). Table 1 contains the values of the wave number and critical Reynolds number calculated for the both models at the different q and fixed d, α and β.

The structure of the neutral curves is also different. The long-wave minimum of the curve dominates in the Brinkman model, while the short-wave perturbations are critical in the Darcy model. The great difference is a result of the flow structure changes at the interface between the overlying fluid and the porous medium. The velocity jump in the Darcy model determines the instability mechanism closes the Kelvin–Helmholtz. It also makes an influence on the Brinkman model, but it is mitigated here. To compare the stability threshold with the well-known data for the channel flow (Re $\approx$ 5700) [21], the critical Reynolds number may be scaled to the overlying fluid thickness (see Section 2.2.3). It gives the critical Re for the Brinkman model is comparable to the known value for the channel flow, while it is much lower for the Darcy model.

One should remember the obtained difference is the pure theoretical result, and it shows the way to verify the filtration models and boundary conditions. Comparing the critical Re and k with experimental data, one may check the reliability of the mathematical models and determine the empirical parameter in the boundary conditions.

The bimodality of the neutral curves is a specific feature of the multilayer systems [25, 26]. It means there are two physical mechanisms of the instability. The first generates the large-scale vortices cover the upper layer and have a relatively large longitudinal wavelength, while the second produces the short wave vortices localized near the interface. The detailed study of the two-layer system instability, its' main peculiarities and mechanisms of instability developlment are presented in papers [27, 28].

5 Admixture influence on flow stability

5.1 Admixture transport model

In addition, we consider an influence of the heavy dissolved admixture in a porous layer on the flow stability. The contaminant is initially located in the porous medium. It forms the diffusional boundary layer near the interface, and the flow enthralls the admixture even at the stable state.

We use the Brinkman model (1), (2), (5) and (6) with the boundary conditions (11) to describe the fluid motion with an extended gravitational force term:

$$\gamma \to \gamma(1 + \beta_C C), \tag{17}$$

where β_C characterises the fluid density dependence on concentration:

$$\rho(C) = \rho_0(1 + \beta_C C). \tag{18}$$

The dimensionless admixture transport equations are as follows:

$$\frac{\partial C}{\partial t} + \mathrm{Re}\mathbf{v} \cdot \nabla C = \frac{1}{\mathrm{Sc}} \nabla^2 C, \tag{19}$$

$$\frac{\partial C_p}{\partial t} + \mathrm{Re}\mathbf{v}_p \cdot \nabla C_p = \frac{\varphi}{\mathrm{Sc}} \nabla^2 C_p. \tag{20}$$

Here, the Schmidt number $\mathrm{Sc} = \nu / D$ is introduced, where D is a molecular diffusion coefficient. The top and bottom boundaries of the system are impermeable for the contaminant:

$$\left(\frac{\partial C}{\partial z}\right)_{z=1} = 0, \quad \left(\frac{\partial C_p}{\partial z}\right)_{z=0} = 0, \tag{21}$$

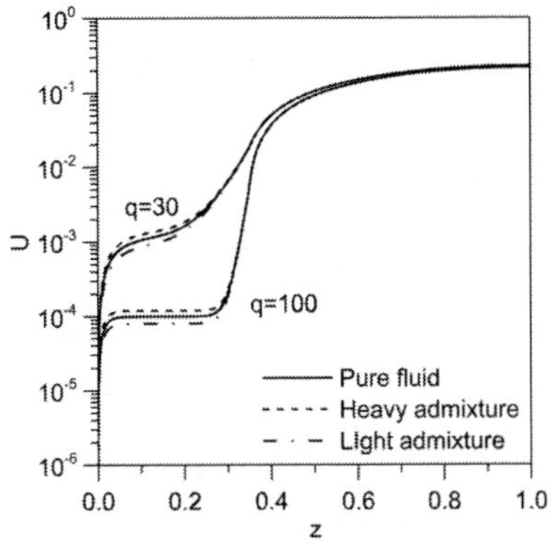

Figure 5. Quasistatic velocity profile of the plane-parallel flow at $q = 30$ and $q = 100$.

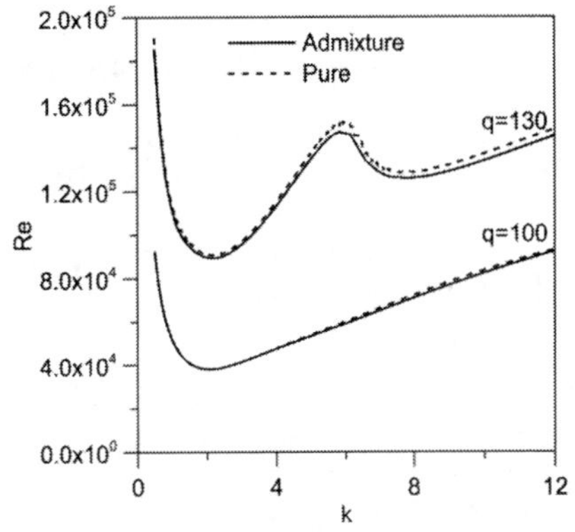

Figure 6. Comparison of the neutral curves of the flow without and with the heavy admixture dissolved in the porous layer, at $d = 0.35$, and $q = 100$ and $q = 130$.

while the concentration and mass flux are continuous at the interface $z = d$:

$$C = C_p, \quad \frac{\partial C}{\partial z} = \varphi \frac{\partial C_p}{\partial z}. \qquad (22)$$

The admixture diffuses from the porous layer into the external flow and the diffusional boundary layers occur at the interface. Their dimensionless widths δ and δ_p are as follows:

$$\delta = 2\left(\frac{t}{\mathrm{Sc}}\right)^{1/2}, \quad \delta_p = 2\left(\frac{\varphi t}{\mathrm{Sc}}\right)^{1/2}. \qquad (23)$$

We do not consider the light admixture. In that case, a convective instability may occur at the time moment when the concentration gradient in the diffusional boundary layers becomes large enough, even if there is no longitudinal flow.

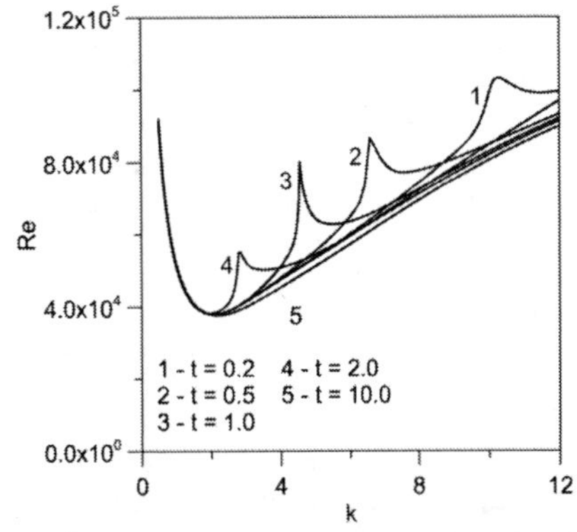

Figure 7. Neutral curves of the flow with heavy admixture dissolved in the porous layer at the different time moments, $q = 100$ and $d = 0.35$.

5.2 Velocity profile and flow stability

A typical diffusion time is several thousand times larger than the characteristic viscous time. The instability development is the most rapid process in the two-layer system. Thus, we may provide the linear stability analysis using quasi-static concentration profile and corresponding velocity profile as the basic state. The analytical structure of the exact solution for the plane-parallel flow is huge, therefore we apply a piecewise approximation of the concentration profile and the basic profile of the velocity corresponding to it (Figure 5). Unfortunately, the approximate solution is also huge, and we do not present it here.

The velocities of the pure fluid and flow with admixture have no large difference, therefore the critical Reynolds numbers close each other too (Figure 6). The contaminant presence makes a little flow acceleration in the porous medium and slight destabilization of the basic state.

The main effect of the heavy admixture on the flow stability is change of the neutral curve structure. The additional minimum occurs on the curves (Figure 7). It corresponds to a damping of the perturbation with some specific wavelength. The critical wavelength of that minimum obtained from the numerical data rises by the $t^{1/2}$-law like the diffusional boundary layer width (23) until the both minima of the curve do not merge.

Conclusion

We have studied the linear stability of the two-layer system of finite thickness homogeneous fluid overlying the finite porous medium. The problem has been solved in the two frameworks. The first is the Darcy filtration model with interface boundary conditions by Beavers and Joseph, and the second is the Brinkman model with conditions by Ochoa-Tapia and Whitaker. The magnitudes of the stationary flow coincide each other in the both approaches excluding the thin transition zone near the interface. The Darcy model gives a discontinuity of the tangential velocity at the interface, while the velocity profile calculated in the Brinkman model is continuous, but has a break. It means the tangential stresses are discontinuous at the interface, and the break point is an analog of the inflection point.

The linear stability analysis shows the great difference between the models. The critical Reynolds number calculated in the Darcy model is one hundred times lower than in the Brinkman model. It is a result of the tangential velocity jump in the flow profile is determined by the boundary conditions. The Brinkman model gives the continuous profile, and the instability occurs because of the velocity break at the interface in accordance with Rayleigh's inflection point theorem. The neutral curves are bimodal at the both frameworks.

Furthermore, we study an influence of the heavy admixture dissolved in the porous layer on flow stability. The critical Reynolds number and wave number do not change significantly, while the new minimum occurs at the neutral curve. Its' wavelength changes in time like the diffusional boundary layer width, and it means the additional perturbation damping occurs within the diffusional boundary layer near the interface. The heavy admixture does not produce a new instability mechanism, and only effect on the pure hydrodynamics mechanisms described previously.

Contrary to that, the light admixture causes a concentrational convection even if there is no flow. Here, the instability occurs when the concentration gradient becomes large enough. These results prove it is possible to use the heavy visualizing admixtures to study the stationary plane-parallel flow stability and structure of the critical motions without large error.

We suppose the empirical parameters of the models can be determined from the experimental study of the flow stability. The flow structure in the transition zone cannot be defined with good precision. Therefore, the stability study can be more effective than the velocity measurements. The study of instability development does not require any information about the flow within the porous medium. The wave number and critical Reynolds number can be found from the flow structure only in a homogeneous fluid.

Another possible method to determine the parameters of the models is a measurement of the viscous friction force acting on porous medium. It is determined by the velocity magnitude in the main flow and by its' gradients in the transition zone. Thus, the calculating of the viscous force in the both models we consider and the result comparison with the experiments can be effective to obtain the empirical parameter values. Nevertheless, it seems to be not so effective to specify the filtration model because it is based on the stationary flow magnitude.

Acknowledgements

The study in financially supported by the Russian Science Foundation (Grant 14-21-00090).

References

1. D.A. Nield, A. Bejan, *Convection in porous media* (Springer, New-York, 2013)
2. A.S. Berman, J. Appl. Phys. **24**, 1231-1235, (1953)
3. G.S. Beavers, D.D. Joseph, J. Fluid Mech. **30**, 197-207 (1967)
4. N. Tilton, L. Cortelezzi, J. Fluid Mech. **604**, 411-445 (2008)
5. A.A. Hill, B. Straughan, Adv. Water Resour. **32**, 1609-1614 (2009)
6. Q. Liu, A. Prosperetti, J. Fluid Mech. **679**, 77-100 (2011)
7. P.D. Antonladis, M.V. Papalexandris, Phys. Fluids **27**, 014104 (2015)
8. M. Ghisalberti, H. Nepf, Environ. Fluid Mech. **6**, 277-301 (2006)
9. K. Gavrilov, G. Accary, D. Morvan, D. Lyubimov, S. Méradji, O. Bessonov, Flow Turbul. Combust. **86**, 89-111 (2011)
10. M. le Bars, M.G. Worster, J. Fluid Mech. **550**, 149-173 (2006)
11. S. Govender, Transport Porous Med. **90**, 393-402 (2011)
12. E.A. Kolchanova, D.V. Lyubimov, T.P. Lyubimova, Transport Porous Med. **97**, 25-42 (2013)
13. B. Alazmi, K. Vafai, Int. J. Heat Mass Transfer **44**, 1735-1749 (2001)
14. J.A. Ochoa-Tapia, S. Whitaker, Int. J. Heat Mass Transfer **38**, 2635-2646 (1995)
15. J.A. Ochoa-Tapia, S. Whitaker, Int. J. Heat Mass Transfer **38**, 2647-2655 (1995)
16. P. Angot, Appl. Math. Lett. **24**, 803-810 (2011)
17. M. Minale, Phys. Fluids **26**, 123101 (2014)
18. M. Minale, Phys. Fluids **26**, 123102 (2014)
19. H.C. Brinkman, Appl. Sci. Res. A **1**, 27-34 (1947)
20. S. Whitaker, Transport Porous Med. **25**, 27-61 (1996)
21. C.C. Lin, *The theory of hydrodynamic stability* (Cambridge University Press, 1966)
22. S.D. Conte, SIAM Review **8**, 309-321 (1966)
23. S.K. Godunov, Uspekhi Mat. Nauk **16**, 171-174 (1966)
24. G.Z. Gershuni, E.M. Zhukhovitskij, A.A. Nepomnyashchij, *Stability of convective flows* (Nauka, Moscow, 1989)
25. D.V. Lyubimov, I.D. Muratov, Hydrodynamics **10**, 38-46 (1977)
26. F. Chen, C.F. Chen, J. Heat Trans. **110**, 403-409 (1988)
27. T.P. Lyubimova, D.T. Baydina, D.V. Lyubimov, Nonlinear Proc. Geophysics **20**, 543-547 (2013)
28. T.P. Lyubimova, D.V. Lyubimov, D.T. Baydina, E.A. Kolchanova, K.B. Tsiberkin, Phys. Fluids (to be published)

EPJ Web of Conferences **114**, 02126 (2016)
DOI: 10.1051/epjconf/201611402126

Scour depth estimation using an equation based on wind tunnel experiments

Takayuki Tsutsui[1,a]

[1]Department of Mechanical Engineering, The National Defense Academy, 1-10-20 Hashirimizu, Yokosuka, Kanagawa 239-8686, Japan

Abstract. Scour is the result of degradation and aggradation by wind or moving fluid in the front and back of a pole standing in sand, respectively, and is often observed at the bottom of bridge piers in rivers. In this study, we propose a method of estimating the scour depth around a cylindrical structure standing in sand. The relationships among the depth of the scour, the aspect ratio of the structure (= height/diameter), the fluid velocity, and the sand properties (particle size and density) were determined experimentally using a wind tunnel. The experiments were carried out under clear-water scour conditions. In the experiments, the aspect ratio of the cylindrical structure, the fluid velocity, and the sand particle size were varied systematically. The diameters of the structure were 20, 40, and 60 mm, and the aspect ratio was varied from 0.25 to 3.0. Sand particles of four sizes (200, 275, 475, and 600 μm) were used in the experiment, and the velocity was varied from 4 to 11 m/s. The depth and radius of the scour were measured. As a result, we have developed an equation for estimating the scour depth that uses the aspect ratio, fluid velocity, and sand particle size as parameters.

1 Introduction

Scour is the result of the erosive action of flowing water, which excavates and carries away material from riverbeds and around the piers of bridges. Scour can be found not only in rivers but also in sand [1] and snow [2]; the basic mechanism of scour is the same in these three cases.

The scour mechanism is caused by three vortex systems: horseshoe, wake, and trailing [3, 4]. In particular, studies have investigated wake vortices, which are also called roll-up vortices. A roll-up vortex is a spiral vortex that can be observed behind a pole [5]. In this vortex mechanism, sand is flung up by the roll-up vortex and falls down behind the pole. Graf and Istiarto [6] and Pasiok and Stilger-Szydlo [7] call this type of vortex the wake vertical vortex.

Numerous studies have also estimated the scour depth around bridge piers in rivers (e.g., [8]). Such estimates are valuable in the field of civil engineering. Therefore, we propose a method for estimating the scour depth in an air–sand system. Water–sand and air–sand systems are defined as environments in which scour is caused by water and air, respectively.

In this study, to determine the important factors that influence scour in the air–sand system, we experimentally investigated the scour depth around a cylindrical structure in the sand of a riverbed by using a wind tunnel. We systematically varied the diameter and aspect ratio of the cylindrical structure, the particle size of the bed material, and the fluid velocity and optically measured the scour depth and radius. The main factors that influenced the scour were determined by analyzing the data, and we define a simplified equation for estimating the scour depth.

2 Experimental apparatus and methods

The experimental coordinate system and a schematic of the low-speed wind tunnel used in this study are shown in figures 1 and 2, respectively. The experiments were conducted in a low-speed wind tunnel with dimensions of 1000 mm × 400 mm × 300 mm. The diameter and height of the cylindrical structure are denoted by D and H, respectively. The scour depth is denoted by Z and is measured from the ground level to the deepest point in the scour. The radius R is the distance from the front surface of the cylindrical structure to the front edge of the scour. Cylindrical structures made of acrylic resin with diameters of 20, 40, and 60 mm were used in this study. The aspect ratio H / D of the cylindrical structures was varied from 0.25 to 3.0, and the fluid velocity U from 4 to 11 m/s. The bed materials were sands of varying particle size. The properties of the different bed materials (particle size d_s, density ρ_s, and angle of repose θ_s) are given in table 1.

Experiments were designed and conducted for different velocities, diameters, aspect ratios, and sand particle sizes. The scour radius and depth were measured by a video camera and a laser rangefinder, as shown in figure 3.

[a] Corresponding author : tsutsui@nda.ac.jp

3 Results and discussion

3.1 Differences between air–sand and water–sand systems

Figure 4 shows an example of the scour depth and radius as functions of time t. Both have the same profile; that is, they first change rapidly and subsequently increase monotonically, ultimately reaching their maximum (equilibrium) values. To compare the air–sand and water–sand systems, the experimental data under the conditions of the water–sand system in a flume are plotted against time t in figure 5. The profile is the same as that in figure 4. These data are transformed into dimensionless values and plotted on a single graph (figure 6). Figure 6 shows the logarithmically nondimensional scour depth Z / D and radius R / D as functions of the nondimensional time U t / D. The progress in the depth and radius changes at approximately $U t / D = 3 \times 10^4$, which is the deflection point. This point defines the earlier ($U t / D \leq 3 \times 10^4$) and later ($U t / D \geq 3 \times 10^4$) periods.

The scour depth and radius are related to the angle of repose of the bed material. Figure 7 shows the slope of the scour [$= \tan^{-1}(Z / R)$]. The slope is 35°, which is the angle of repose of 200 μm sand. We can estimate the radius from the depth; therefore, we consider only the depth for the remainder of this paper.

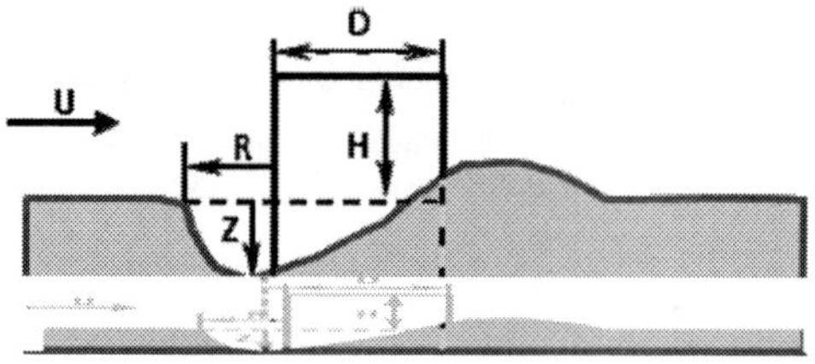

Figure 1. Flow geometry and coordinate system

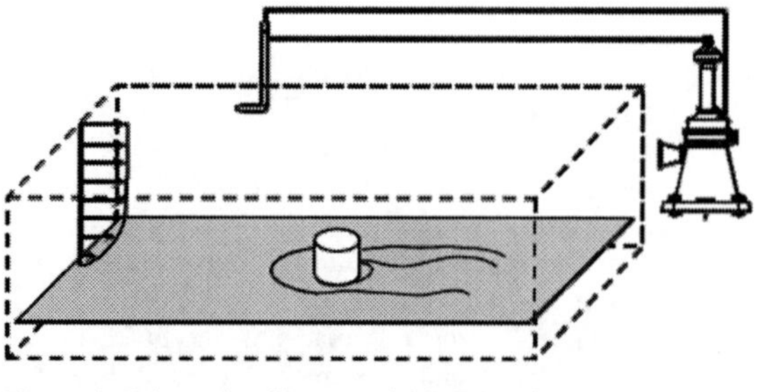

Figure 2. Schematic of law speed wind tunnel

Table 1. Property of bed material

mean size d_s [μm]	size range d_s [μm]	density ρ_s [kg/m^3]	angle of repose θ_s [deg.]
200	180 - 212	2600	35
275	250 - 300	2600	32
475	450 - 500	2300	35
600	575 - 625	2300	34

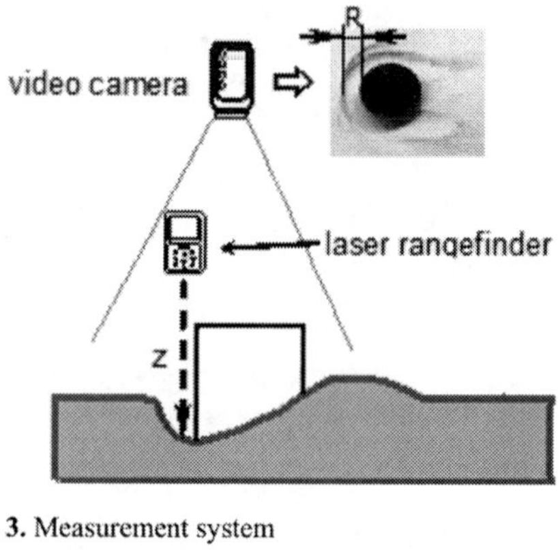

Figure 3. Measurement system

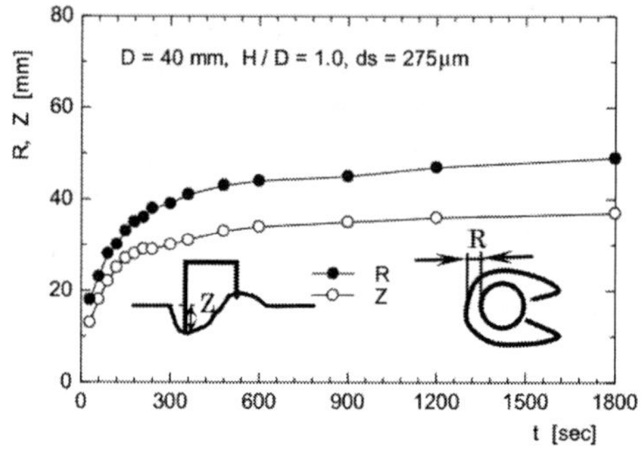

Figure 4. Example of scour depth and radius in air – sand field by wind tunnel (D = 40 mm, U = 7 m/s, ds = 275 mm)

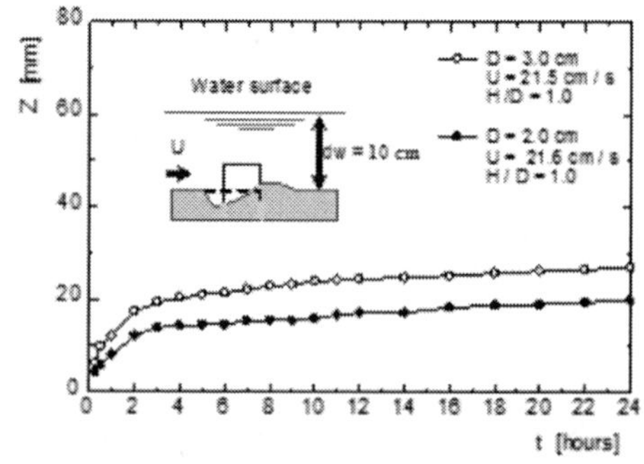

Figure 5. Example of scour depth in water – sand field by flume (modified after T. Euler and J. Herget, 2012, p. 41)

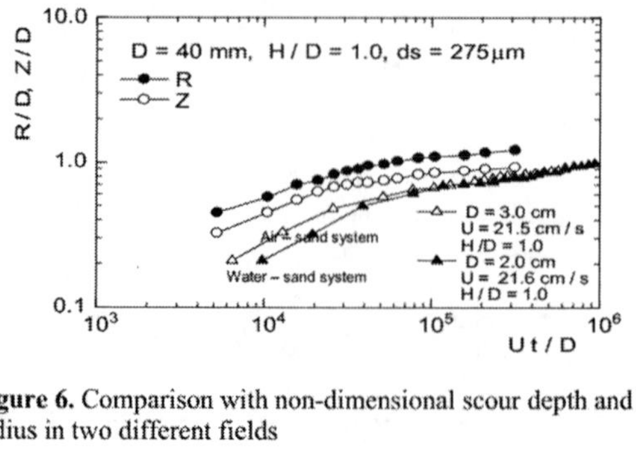

Figure 6. Comparison with non-dimensional scour depth and radius in two different fields

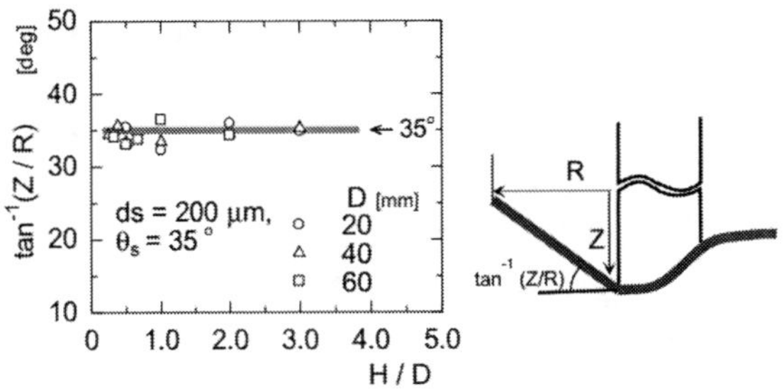

Figure 7. Slope of the scour (ds = 200 mm)

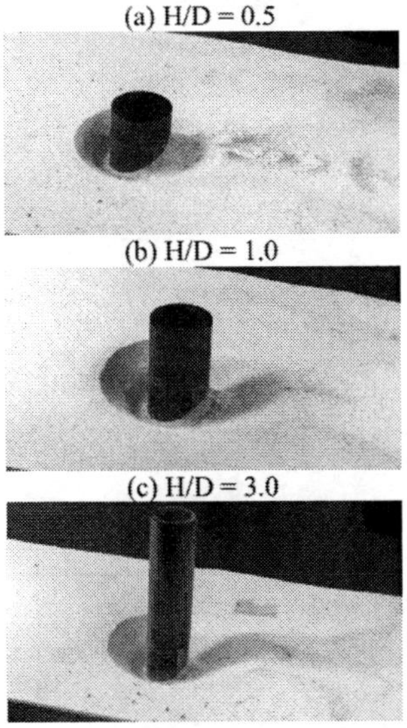

Figure 8. Photograph of scour hole changed by aspect ratio (D = 40 mm, U = 6 m/s, 300 second elapsed)

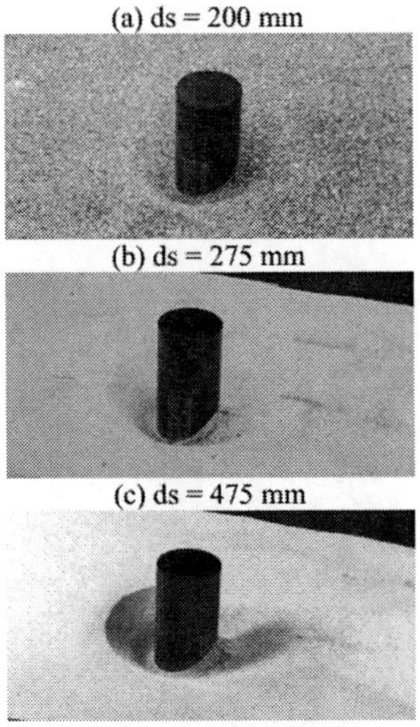

Figure 9. Photograph of scour hole changed by particle size (D = 40 mm, U = 6 m/s, H/D = 1.0, 300 second elapsed)

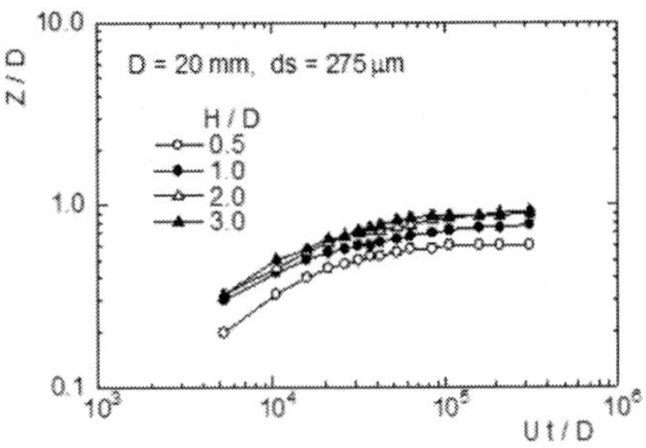

Figure 10. Scour depth changed by aspect ratio (D = 20 mm, ds = 275 mm)

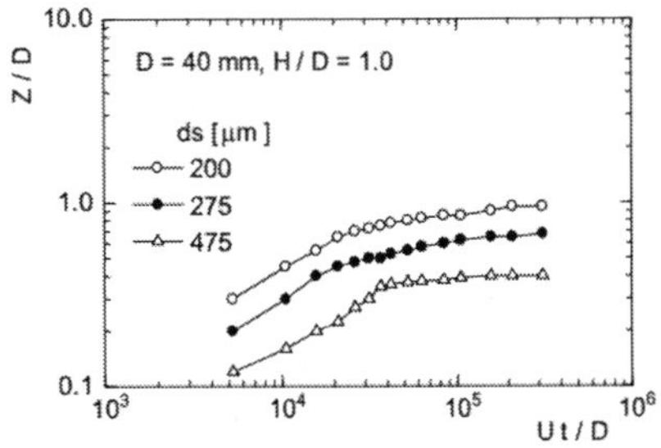

Figure 11. Scour depth changed by particle size (D = 40 mm, H / D = 1.0)

3.2 Progress of scour

Figure 8 shows photographs of scour holes for different aspect ratios (H / D = 0.5, 1.0, and 3.0) obtained 300 seconds after turning on the wind tunnel fan. The degradation around the cylinder increases with increasing aspect ratio. Figure 9 shows photographs of scour holes for the different particle sizes investigated in this study. The scour depth increases with decreasing particle diameter. The photographs show the influence of the particle size on the scour hole characteristics.

According to figures 10 and 11, the aspect ratio and particle size have a large influence on the scour depth

3.3 Scour depth estimation by dimensionless analysis

Dimensionless quantities are introduced for the purpose of discussing the proposed scour depth estimation method. The physical quantities that determine the equilibrium scour depth Z_s are the diameter D and height H of the cylindrical structure, the fluid velocity U, the sediment particle size d_s, the acceleration due to gravity g, the viscosity of air μ, the density of air ρ, and the density of the sediment ρ_s. The dimensions of these quantities are defined in terms of mass M, length L, and time T, and the dimensional equations given in terms of M, L, and T are called MLT equations. The dimensions of the given quantities are $Z_s = [L]$, $D = [L]$, $H = [L]$, $U = [L\,T^{-1}]$, $d_s = [L]$, $g = [L\,T^{-2}]$, $\mu = [M\,T^{-1}\,L^{-1}]$, $\rho = \rho_s = [M\,L^{-3}]$.
The dimensionless equilibrium scour depth Z_s / D is

$$Z_s / D = K\, D^a H^b U^c d_s^d g^e \mu^f \rho^g \rho_s^h, \qquad (1)$$

where K is a constant and a, b, c, d, e, f, g, and h are exponents. Writing this as an MLT equation, we have

$$0 = [L]^a [L]^b [L T^{-1}]^c [L]^d [L T^{-2}]^e [M T^{-1} L^{-1}]^f [M L^{-3}]^g [M L^{-3}]^h . \qquad (2)$$

Next, we equate the powers of each dimension on the right and left-hand sides of equation (2), where M and L^{-1} are always shown together as ML^{-1}. We equate powers of ML^{-1}, length L, and time T.

$$M L^{-1}: f + g + h = 0, \qquad (3)$$

$$L: a + b + c + d + e - 2g - 2h = 0, \qquad (4)$$

$$T: - c - 2e - f = 0. \qquad (5)$$

The numbers of the reference and basic dimensions are 8 and 3. We thus obtain the following five dimensionless quantities:

$$a = -b - d + e - f, \qquad (6)$$

$$c = -2e - f, \qquad (7)$$

$$g = -f - h, \qquad (8)$$

$$\frac{Z_s}{D} = K\left(\frac{H}{D}\right)^b \left(\frac{d_s}{D}\right)^d \left(\frac{Dg}{U^2}\right)^e \left(\frac{\mu}{UD\rho}\right)^f \left(\frac{\rho_s}{\rho}\right)^h . \qquad (9)$$

We can combine these dimensionless quantities to simplify them:

$$\left(\frac{d_s}{D}\right) \Big/ \left(\frac{\mu}{U\rho\rho}\right) = \left(\frac{Ud_s}{\mu/\rho}\right) = Re_s \qquad (10)$$

$$1/\sqrt{\left(\frac{d_s}{D}\right)\left(\frac{Dg}{U^2}\right)\left(\frac{\rho_s}{\rho}\right)} = \frac{U}{\sqrt{\frac{\rho_s}{\rho} d_s g}} \Rightarrow \frac{U}{\sqrt{\frac{(\rho_s - \rho)}{\rho} d_s g}} = \frac{U}{\sqrt{(\rho_s/\rho - 1) d_s g}} = N_s, \qquad (11)$$

where H / D, Re_s, and N_s are the aspect ratio, sediment Reynolds number, and sediment number, respectively. Furthermore, ρ_s / ρ is the sediment specific gravity, which we modified as $(\rho_s - \rho) / \rho$ to allow the introduction of N_s. As a result, Z_s / D can be represented as a function of three dimensionless quantities:

$$Z_s / D = f(H / D, Re_s, N_s). \qquad (12)$$

Figure 12 shows the correlation between the equilibrium scour depth Z_s / D and the aspect ratio H / D. The log of Z_s / D is proportional to the log of H / D. We obtain the relation

$$Z_s / D \propto (H / D)^{0.35}. \qquad (13)$$

The correlation between Z_s / D and N_s and that between Z_s / D and Re_s are shown in figures 13 and 14, respectively. From these figures, the following relations are obtained:

$$Z_s / D / (H / D)^{0.35} \propto N_s^{3.4}, \qquad (14)$$

$$Z_s / D / [(H / D)^{0.35} N_s^{3.4}] \propto Re_s^{-0.43}, \qquad (15)$$

$$Z_s / D = K\, (H/D)^{0.35} N_s^{3.4} Re_s^{-0.43}, \qquad (16)$$

where K is a constant.

According to figure 15, K is 0.1. Thus, for the variable ranges of H / D = 0.5 – 3.0, N_s = 2.2 – 3.6, and Re_s = 78 – 440, the following equation is obtained:

$$Z_s / D = 0.1 (H / D)^{0.35} N_s^{3.4} Re_s^{-0.43} . \qquad (17)$$

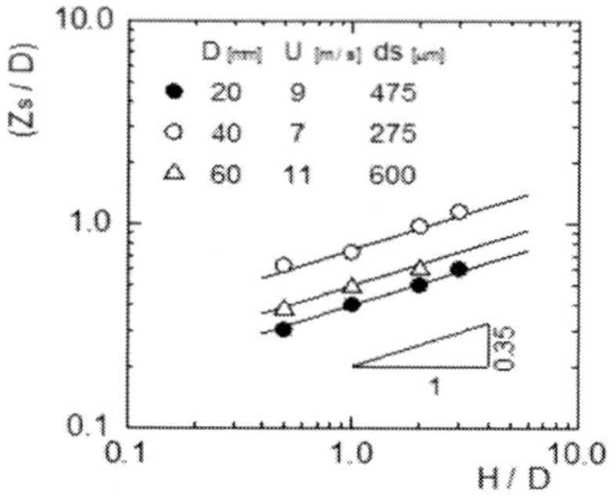

Figure 12. Correlation between scour depth and aspect ratio

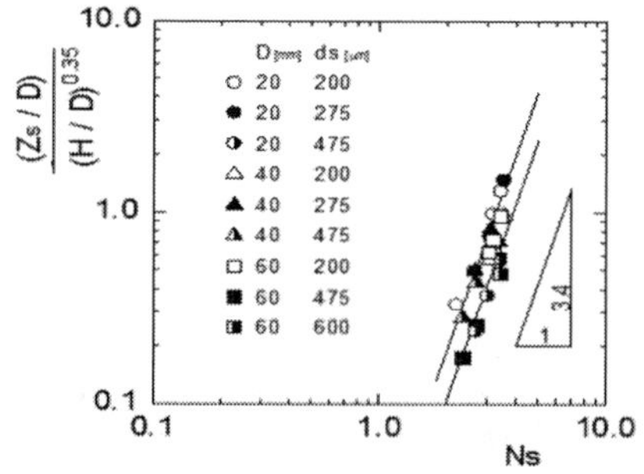

Figure 13. Correlation among scour depth, aspect ratio and sediment number

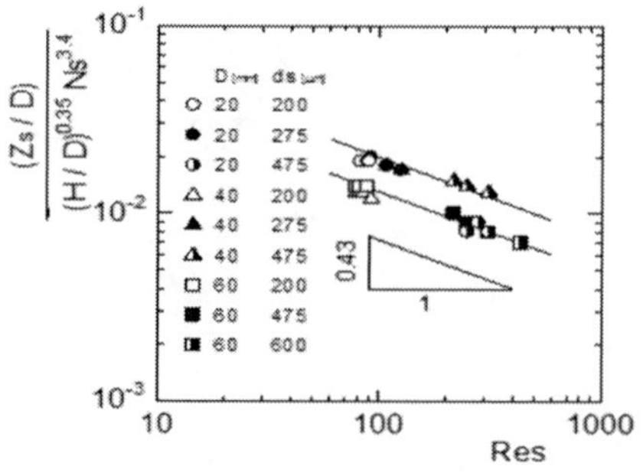

Figure 14. Correlation among scour depth, aspect ratio, sediment number and sediment Reynolds number

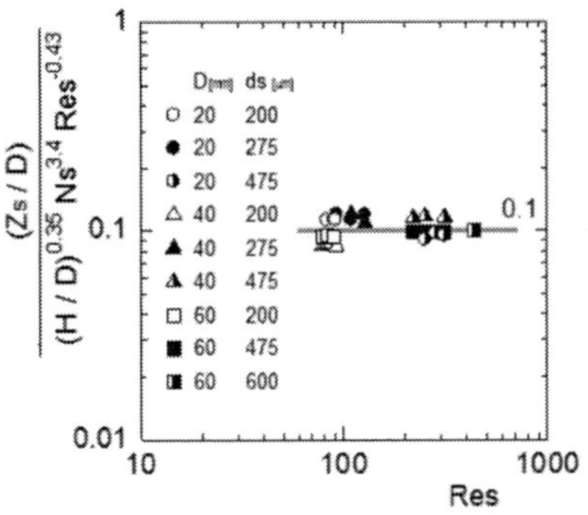

Figure 15. Constant value K

4 Conclusions

We measured the scour depth around a cylindrical structure in the sand of a riverbed by using a wind tunnel. The diameter and aspect ratio of the cylindrical structure, the particle size of the bed material, and the fluid velocity were systematically varied. The results are summarized below.

(1) The tendency of the change in the scour depth and radius is the same in air–sand and water–sand systems.

(2) The aspect ratio and particle size significantly influence the scour characteristics.

(3) The relationship between Z_s / D, N_s and Re_s is obtained as $Z_s / D = 0.1(H / D)^{0.35} N_s^{3.4} Re_s^{-0.43}$.

References

1. H. Mochizuki, Y. Okada, S. Kumar, S. Sinha, J. of Arid Land Studies, **22**-1 267-270 (2012)
2. J.H.M. Beyers, P.A. Sundsbφ, T.M. Harms, J.Wind Eng. Ind.Aerodyn. **92** 725-747 (2004).
3. H. W. Shen, V. R. Schneider, S. Karaki, Proc. ASCE, J. Hyd. Div., **95**, HY 6, 1919-1940 (1969)
4. H. N. C. Breusers, G. Nicollet, H. W. Shen, J. Hyd. Research, **15**-3, 211-252 (1977)
5. T. Tsutsui, Proceeding of 6th Int. Colloquium on Bluff Bodies Aerodynamics and Applications (2008), pp.408-411.
6. W. H. Graf, I. Istiarto, J. Hyd. Research **40** (1), 13-20 (2002)
7. R. Pasiok, E. Stilger-szydlo, Archives of Civil and Mechanical Engineering **10** (2), 67-79 (2010)
8. T. Euler, J. Herget, Catena **91**,35-46 (2012)

EPJ Web of Conferences **114**, 02127 (2016)
DOI: 10.1051/epjconf/201611402127

Passive heating of the ground surface

Anna Tyburczyk[1,a]

[1]*Kielce University of Technology, 25-314 Kielce, Al. 1000-lecia P.P. 7, Poland*

Abstract. The phenomenon of phase change is one of the most important contemporary issues of thermal engineering. In particular, this applies to all kinds of heat exchanger systems, which should achieve the highest possible efficiency while reducing investment and operating costs. Some of these systems are heat pipes or thermosyphons, which, among others, are used for the heat transfer, temperature stabilization and the regulation of heat flux density. Additionally, they are passive systems, and therefore do not require an external power supply. Heat pipes can be used to stabilize the surface temperature of roads and driveways. Large heat tubes can be applied for heating the surface of bridges and overpasses, which become icy in unfavorable climatic conditions. The paper presents research on the test facility, whose main component is a long vertical copper fin. The temperature at the base of the fin was kept constant for a given series of measurements. Heat receiving fluid was ethanol at atmospheric pressure. The measurement methodology and the results of investigations were discussed. The surface temperature distribution was measured with the infrared camera, and on this basis the local values of heat flow and the heat transfer coefficient were determined. The results were presented as boiling curves for both the fin with the smooth surface and the one covered with a metal capillary-porous structure. The results obtained are useful in the design of heat exchangers, including passive heating of the ground.

1 Introduction

The phenomenon of phase change in economy sectors is one of the most important contemporary issues of thermal engineering. It refers in particular to different kinds of exchanger systems, which should achieve the highest possible efficiency while reducing investment and operating costs.

Regarding this, it is advisable to pay attention to systems based on heat tubes, which are used i.a. in heat transfer, temperature stabilization and the regulation of heat flux density. In the construction industry they are applied in solar collectors, heat exchangers, wall heating systems and radiant wall panels.

Heat tubes or heat pipes [1,2] are simple devices without movable parts, filled with a working fluid and the vapours of this fluid. They can successfully transport large amounts of heat over long distances with a nearly constant temperature along the length. In addition, their effective operation requires no external energy supply, as opposed to other exchangers where the flow of the agent is forced by the operation of the pump or fan.

The heat pipe, schematically shown in figure 1, is a slender tube lined with a structure designed with carefully selected geometrical/physical parameters, containing small amounts of a pressurized fluid, whose boiling temperature is selected in view of the heat exchange conditions [3].

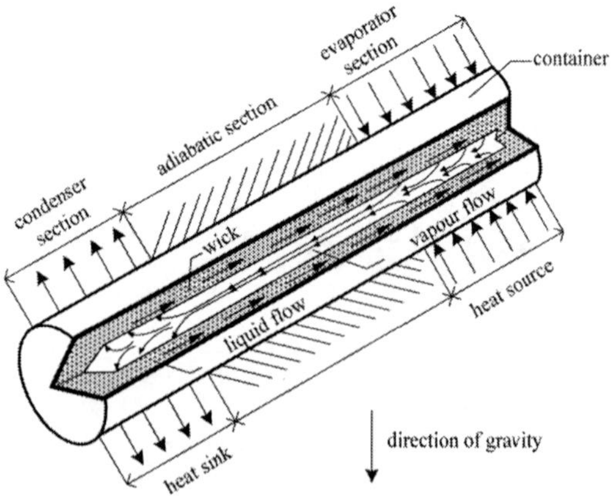

Figure 1. Diagram of a typical heat pipe with a schematic principle of operation

Heat pipes can be divided primarily with regard to the way in which the condensate is transported to the evaporator. One can distinguish here thermosyphons, rotating heat pipes and tubes with capillary structures in

[a] Corresponding author: annazw@tu.kielce.pl

which the transport of the condensate occurs as a result of capillary pressure.

Owing to the capillary-porous filling, heat pipes can work both in horizontal and vertical position. The reason for this is the high value of capillary forces as compared to the forces of gravity. A greater efficiency of such systems can be achieved by maximizing the capillary forces, with the research in this area focused on the use of various kinds of porous fillings [4].

One of the simplest and most common methods of developing the surface is the application of mesh structures such as the metal-fiber or powder ones, including those that are thermally sprayed [5]. The metal-fiber capillary-porous and mesh structures are of special interest in this group [6]. Metal-fiber coverings are made of thin fibers of various metals and their alloys. Careful application of such structures results in a high value of the heat transfer coefficient, with appropriately selected parameters such as thickness, porosity and skeletal conductivity. Mesh structures are commercially available and capable of forming capillary-porous structures with desired geometrical parameters, defined by volume porosity and height.

Inside the heat pipe a capillary-porous lining is applied, performing various functions depending on the location. One end thereof is an evaporator in which the fluid evaporation occurs as a result of absorbing heat from the surrounding environment. The condensation combined with heat emission to the surrounding atmosphere occurs at the opposite end. The operation of exchangers equipped in heat tubes is limited by crisis phenomena [7]. Excessive overheating of the evaporator results in the stagnation phenomenon, and restarting the operation is associated with lowering the surface temperature, which proceeds in film boiling conditions.

Transport and construction industry take advantage primarily of heat tubes with capillary structures [8]. They are used for:

- ground temperature stabilization around foundations and supports in arctic areas,
- defrosting of icy roads,
- the preparation of domestic hot water in heat exchangers,
- accumulation of solar energy in solar collectors,
- wall and underfloor heating,
- heating ventilation air in combination with ecological energy sources.

Heat pipes have also been used to stabilize the temperature of the road surface [9]. For example, during the construction of the railway between China and Tibet they were placed in the permafrost sections. During thaw the ground under the railway sleepers thawed, which resulted in the collapsed tracks and broken rails. For these reason the pipes were designed to transfer the heat in one direction only, i.e. from the ground to the environment. The condensing section of the tubes was finned and projected above the surface of the ground, where it emits the heat extracted from the ground to the air. This caused the lowering of the ground temperature, which remains negative even during the thaw.

A similar solution was applied in the construction of the pipeline which transports oil from the north to the south of Alaska. A number of heat pipes that protect the ground from thawing and prevent pipe damage were installed along the route of the pipeline.

In the Nordic countries heat tubes are used for heating the road [10]. They are installed directly beneath the road surface to maintain a constant, positive temperature of asphalt. Such solutions take advantage of geothermal energy available throughout the year [11], which can be accumulated to cover the peak demand [12].

Another application of large-size heat pipes is using them for heating the surface of bridges and overpasses, which in unfavorable climatic conditions become icy [13]. Such a system operates as a thermosyphon, where the supplied heat is associated with the convection movement. In this case there is no need for capillary-porous layer between the section of the evaporator and the condenser, as the liquid phase condensed in the upper part flows down to the evaporator by gravity [14].

There are two types of solutions available here: passive and active systems. The passive system is the one in which the evaporator section is inserted in the ground at the depth where the temperature is constantly above 0°C. By contrast, the condenser section is placed directly under the surface, which should be protected from freezing. It is worth noting that this system further lowers the carrier temperature of the roadway in summer, increasing its durability. A simplified diagram of such a system is shown in figure 2. This system works spontaneously, without further supply of energy, which advantageously reduces the cost of operation [15, 16].

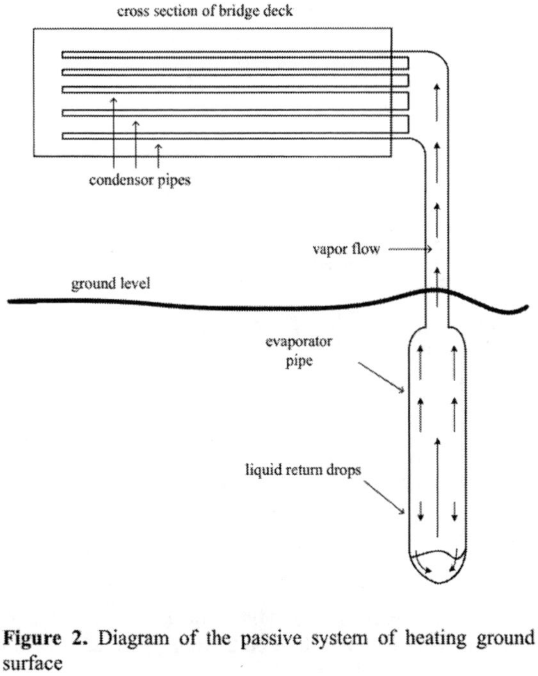

Figure 2. Diagram of the passive system of heating ground surface

In active systems operating with an external heat source e.g. a boiler, the systems equipped with heat pipes provide the stabilization of temperature, its fine adjustment and the reduction of the maximum value (see figure 3).

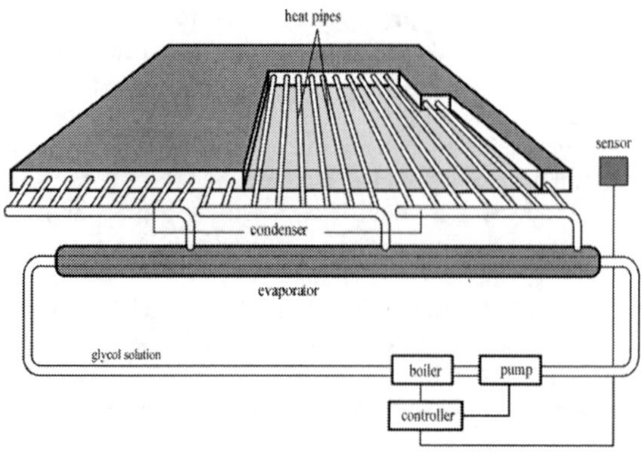

Figure 3. Active system diagram. Ground heating system that uses heat tubes and glycolic indirect system

The objective of this paper is to analyze the heat exchange at the boiling of ethanol as illustrated by non-isothermal surface of a long copper fin coated with one layer of mesh and metal-fiber capillary-porous structure with the predetermined microstructure parameters.

2 Test facility

The tests were conducted on the stand, whose simplified diagram is presented in figure 4.

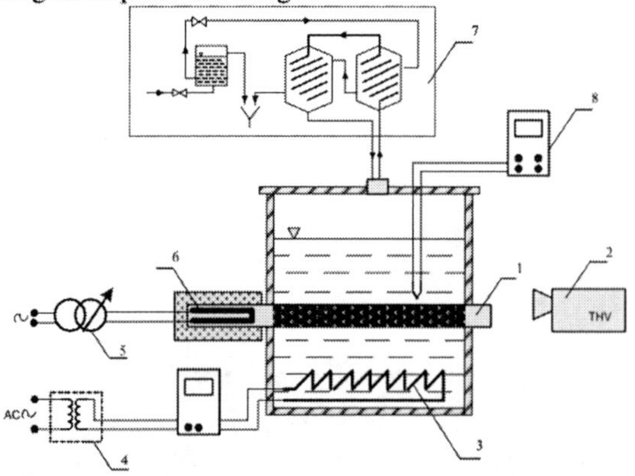

Figure 4. Diagram of the test facility for non-isothermal samples 1 – the examined element, 2 – the thermovision camera, 3 – the auxiliary heater, 4 – the separation system with the control of supplied electric power, 5 – the main heater system with the electric power measurement, 6 – the main heater, 7 – the liquid supply and condensate recovery system, 8 – temperature measurement and control system

Its main part is a flat fin with a rectangular cross section. It is mounted between two side bakelite plates, which provide a housing for the container filled with a boiling liquid. The fin is one-sidedly covered with the structure intensifying the heat exchange (mesh layer with the metal-fiber capillary-porous structure) on the side of the liquid. The tests were performed in stationary conditions at the set temperature, different, however, for each series of measurements at the base of the fin. The necessary heat was supplied by the main electric heater powered from the autotransformer. In order to maintain boiling throughout the vessel, the auxiliary heater system was applied, with the heating part in the form of a resistance wire placed spirally at the bottom. Constant electric power supplied in the auxiliary heater system was maintained for all the measurement series. The vapour cooling and the condensate recovery were performed in a separate system of two radiators whose temperature was stabilized with district heating water in an open system.

The object of measurement is the surface temperature distribution across the fin from the outside, which was measured with an infrared camera VarioCAM®hr. The camera is equipped with an uncooled bolometric matrix with the 640 x 480 pixels image format. The device is designed to work in a long-wave range of infrared radiation of 7.5 ÷ 14 μm. The study was conducted with a standard camera lens (30 x 23)°. The studied surface was evenly covered with black paint in order to ensure the optical uniformity of the outer surface of the tested samples.

The assembled measuring system permits assuming a single-dimensional phenomenon, i.e. the assumption that the change in temperature in relation to the fin thickness is negligibly small. This assumption is valid for the numbers of $Bi < 0.1$ [17]. The result of the study is a one-dimensional temperature distribution alongside the length of the fin along its axis.

Real objects emit not only their own, but also reflected radiation [18], which have a large impact on the final outcome measure [19]. Correct temperature reading requires an individual determination the emissivity of the investigated facility [20]. Therefore, before starting the measurement, calibration was performed examined surface radiation properties.

3 Research results and methodology

The tests were performed for a copper sample. A mesh structure with a metallic-fiber capillary-porous layer of the following parameters: height $h_w = 0.8$ mm and a volume porosity of ~ 65% was sintered on the sample. The structure was made from cut wire with the diameter of 0.05 mm and the fiber length of ~ 3 mm, sintered in reducing hydrogen atmosphere [21]. Ethanol at atmospheric pressure was the heat-receiving agent. For comparative and calibration purposes a copper fin with a smooth surface was tested in identical conditions [22].

The thermal field was obtained with the help of an infrared camera as a result of examining the outer surface of the analyzed part.

Figure 5 shows an example of a post-axial temperature distribution along the studied fin at the constant power of the main heater for two samples: the one with the smooth surface and the one coated with a mesh and capillary-porous structure. The Figure below shows that the application of the fibrous coating enhances the heat exchange as compared to a smooth surface. Due to a correspondingly large size, the cooling to the liquid boiling temperature occurs below the half-height point of the sample. This is observed for all the measurement series performed at a different electric power levels applied at the base of the main heater.

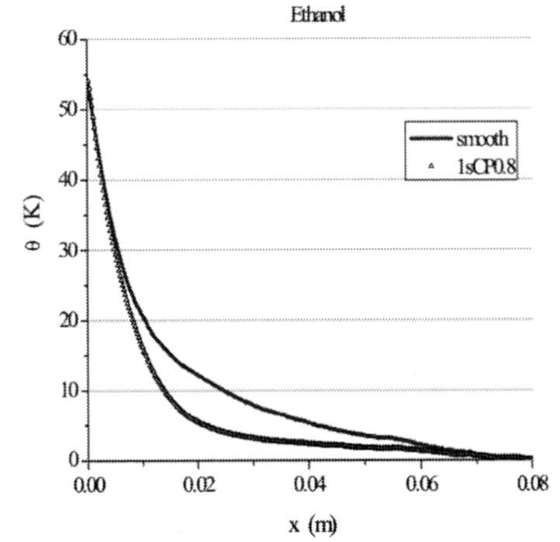

Figure 5. Temperature distribution on the copper fin along the center line at the constant power of the main heater for a smooth and capillary-porous sample with a layer height of 0.8 mm

The use of the infrared camera for measurement results in a large number of measurement points, permitting the application of the numerical differentiation combined with smoothing. Using this method, a derivative for the two examined samples, presented in figure 6, was determined with respect to the fin length. It is proportional to the amount of the transposed heat along the length of the element.

It should be noted that the value of the derivative is almost zero at a point below the half-height of the fin.

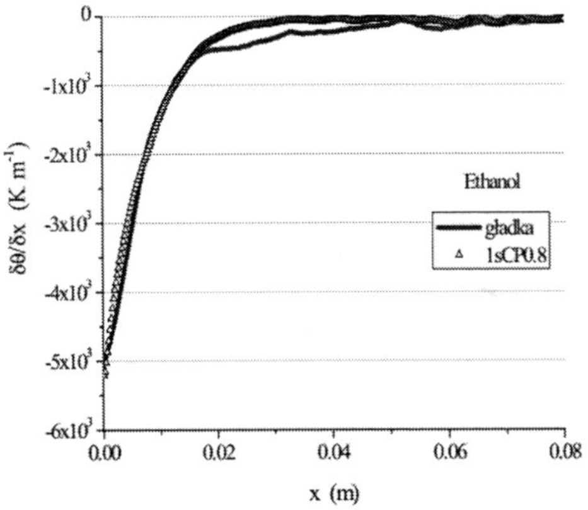

Figure 6. Temperature gradient versus length for the smooth sample and the sample coated with a single mesh layer with the capillary-porous structure

The derivative shown in figure 6 is used to calculate the locally transferred heat fluxes, and consequently to determine the boiling curve for the analyzed element with a non-isothermal surface. The result of the calculations for the smooth sample and the sample coated with mesh with the capillary-porous structure is presented in figure 7. The detailed calculation procedure is described in the paper [23].

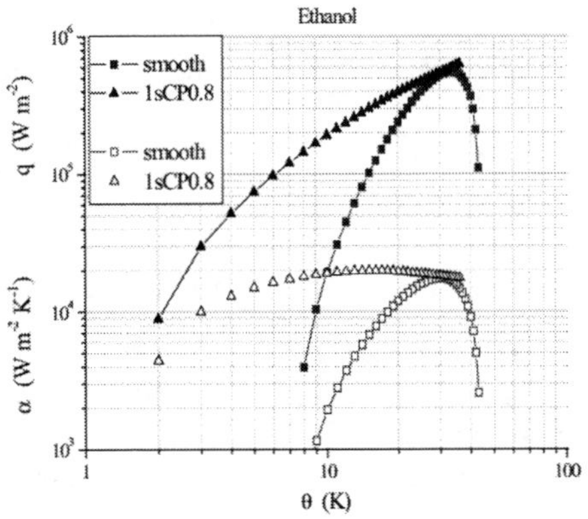

Figure 7. Boiling curve for the smooth sample and the sample coated with a mesh structure with a metallic-fibrous capillary-porous layer

4 Conclusions

The final result of the conducted research is the relation of the locally transferred fluxes and the heat transfer coefficients for fins: with a smooth surface and with mesh and capillary-porous coating. As shown in diagram 7 nucleate pool boiling starts at the superheating of ~ 8 K for a smooth surfaced sample and at only ~ 2 K for a sample with a structural coating. A similar shift towards lower values of overheating is observed for maximum values of heat flux. Table 1 contains local values of heat flux density at constant superheating. The difference between those values was calculated for both samples. While the difference initially grows, it decreases for higher values of superheating. The maximum difference is noted in the case of 14 K superheat, for which $\Delta q = q_{\text{capillary-porous}} - q_{\text{smooth}} = 197.3 \text{ kW m}^{-2}$.

Tab. 1. Heat flux density at constant superheating

q (kW m^{-2})	θ = 10 (K)	θ = 14 (K)	θ = 19 (K)	θ = 26 (K)	θ = 29 (K)
Q_{smooth}	19.3	80.3	206.9	425.0	502.8
$Q_{capillary\text{-}porous}$	190.0	277.6	374.3	487.1	530.7
Δq	**170.7**	**197.3**	**167.4**	**62.1**	**27.9**

References

1. D. Reay, A. Harvey, Appl. Therm. Eng., **57** (2013)
2. L.L. Vasiliev, Appl. Therm. Eng., **25** (2005)
3. H.N. Chaudhry, B.R. Hughes, S.A. Ghani, Renew. Sust. Energ. Rev., **16** (2012)
4. T.M. Wójcik, EPJ Web of Conferences, **25** (2012)
5. M.G. Semena, A.N. Geršuni, V.K. Zaripov, Višč. Škola, Kiev, 1984
6. R. Pastuszko, EPJ Web of Conferences, **45** (2013)

7. T. Orzechowski, S. Wciślik, Int. J. Heat Mass Tran., **73** (2014)
8. T.S. Wiśniewski, *Wymiana ciepła*, WNT, Warszawa, 1995
9. http://en.sunpower.com.cn/, [online 15.06.2015]
10. http://www.alyeska-pipeline.com/, [online 15.06.2015]
11. A. Manzella, Geothermal energy, EPJ Web of Conferences, **98** (2015)
12. T. Orzechowski, K. Stokowiec, COW, **44/1** (2013)
13. http://www.smartbridge.okstate.edu/, [online 15.06.2015]
14. P. Jakóbowski, TChiK, **4 – 5** (2009)
15. A. Nurpeiis, EPJ Web of Conferences, **76** (2014)
16. P. Nemec, A. Čaja, M. Malcho, EPJ Web of Conferences, **45** (2013)
17. I.L. Pioro, W. Rohsenow, S.S. Doerffer, Int. J. Heat Mass Tran., **47** (2004)
18. T. Kruczek, Energy, **91** (2015)
19. T. Kruczek, Energy, **62** (2013)
20. T. Orzechowski, Exp. Therm. Fluid Sci., **31** (2007)
21. T. Orzechowski, A. Zwierzchowska, S. ZwierzchowskA, Inż. Ap. Chem., **6** (2009)
22. T. Orzechowski, A. Zwierzchowska, S. Zwierzchowska, 8-th European Conference of Young Research and Scientific Workers, 2009
23. T. Orzechowski, Exp. Therm. Fluid Sci., **8** (2007)

EPJ Web of Conferences **114**, 02128 (2016)
DOI: 10.1051/epjconf/201611402128

A novel technique to neutralize the Yawing moment due to asymmetric thrust in a hybrid buoyant aircraft

Anwar U Haque[1], Waqar Asrar[1,a], Ashraf A Omar[2], Erwin Sulaeman[1] and Mohamed J.S Ali[1]

[1]*International Islamic University Malaysia (IIUM), 50728 Kuala Lumpur, Malaysia*
[2]*University of Tripoli (UOT), P.O. Box 13154 Tripoli-Libya*

Abstract. Dorsal fin is used in swimming animals like shark for the generation of thrust as well as to meet the requirement of the lateral stability. In the case of aircraft, rudders are normally used for the said requirement. In the present work, this nature inspired idea is explored for its application to neutralize the unavoidable asymmetric thrust produced by the twin engines of a hybrid buoyant aircraft. First, the estimation of asymmetric thrust is obtained with the help of analytical techniques for maximum thrust condition at 4 degree angle of attack. The moment generated by it is utilized for the sizing of a dorsal fin which looks similar to a tapered wing and is placed aft of the center of gravity. Wind tunnel testing at subsonic speed is carried out to explore the design features of this rotatable dorsal fin. It is found that a small rotation of 5 degree can generate the required moment. However, such rotation requires a complete pneumatic/electro-mechanical system and an alternative of it is to use a cambered airfoil for the dorsal fin installed at fixed location. Such a flow controlling device can also be used as an antenna mast, which is commonly installed out the fuselage of the aircraft for communication purposes. Moreover, by incorporating this technique, a pilot doesn't have to put an extra effort to make the aircraft stable in the presence of side wind.

1 Introduction

Studies related to static lateral stability of conventional and unconventional fuel powered twin engine aircraft, based on well-known design methods, are routinely done in industry, academia, and research labs and their results are mostly based on the analytical and experimental techniques. Such techniques can be employed to analyze the issues related to the asymmetric thrust of twin engines and its effect on the static lateral stability. Thrust produced by engine affects the overall static stability in three different ways [1]. First, the thrust produced by the propellers generated a moment about the center of gravity *CG*. Second, the influence of the propwash upon the aft portion of fuselage and on the tail surfaces. This effect is not so dominant if the engines are placed far away from the fuselage and the tail. The third one is a destabilizing moment due to the side force generated by the propellers [1], which is more critical for twin engine aircraft. The origin of this unbalanced torque in twin engine aircraft is not that both the engines turning in the same direction but it is due to the *"P-Factor"* [2]. The lift distribution over the propeller disk is always asymmetric, which tends to turn the aircraft to the right, if assuming an anti-clockwise propeller [3]. This always requires rudder input for cancellation of the propellers' torque and for the said condition; the pilot has to turn the rudder to the left [4]. The turning angle of the rudder depends of the flying speed of aircraft and hence a function of the engine's power and *RPM* of the propeller. Thrust loading of propeller produces a normal force when an aircraft is given an angle of attack.

Asymmetric thrust loading is always about the center of the thrust forces. It generates a small moment about the *CG*, if the engines are placed close of the fuselage. But the magnitude of the moment will be considerable for a short take-off and landing aircraft in which moment arm of thrust of engine is quite large due to voluminous space accommodated by the fuselage. Hybrid buoyant (HB) aircraft is one of the examples of such an aircraft in which partial takeoff weight is balanced by the aerostatic lift [5]. A complete detail about the anatomy of such an aircraft has been discussed in detail in section 3. Although, such an aircraft is statically stable in longitudinal and lateral direction with power off [6]. But power on effects is yet to be explored, specially the influence of propulsion system in yaw stability. Asymmetric thrust is always a fatique for the pilot [4] as additional efforts has to be put by pressing the rudder pedal to counter balance the reading shown by the turn and bank indicator in the cockpit. Such a forced turning causes an increase in drag as well due to more frontal area of the aircraft exposed to the incoming air in flight.

[a] Corresponding author: waqar@iium.edu.my

This requires special attention to explore a suitable option to minimize such affects.

Issue of asymmetric thrust is also quite critical in the climb and descent segment of takeoff and landing segment respectively, especially when the cross winds on the runway can trigger the asymmetric thrust loading. Also from the operational point of view, it requires additional assembly to deflect the rudder. For instance, in the case of the large aircraft, a pneumatically powered *"rudder boost system"* [7] is usually used, which require more operational and maintenance cost along with the weight penalty.

Flying altitude of a *HB* aircraft is limited due to pressure height and in such an aircraft; the rate of climb is limited due to the constraint of ballonet's inflation [8]. Therefore, such aircraft does not leave a room for the pilot to neglect the said moment even during flight. It is infact of great importance to come up with some novel concept to counter measure the side force due to twin engines. In this regard, to the authors' best knowledge, no effort has been done in past for the control of such asymmetric side force by other alternative ways of active flow control. In the present research work, a novel concept is proposed to implant a small dorsal fin of the fuselage and behind the *CG*, which can generate an equivalent yaw force. This concept is perhaps derived from the nature by applying the potential advantages of dorsal fin in Shark fish, which lack swim bladders and any automatic neutral buoyancy [9]. This work is motivated due to huge amount of good work has gone on in past years on swimming animal, especially of sharks in which the dorsal fin/fins is movable [10-21]. Such motion becomes more interesting, when the tip of the fin is also deflectable to have additional component of lift in longitudinal direction. Such an aerodynamic component is hardly seen in bird in which both wings perform the same down stroke and up stroke simultaneously to generate the required thrust. Similar to a shark fish, a dorsal fin made up of a symmetric airfoil can be rotated in the lateral direction. But similar to a multi-element airfoil wing, it will require additional adds-on for the electro mechanical control. Also it will be a problem for breakage in communication, if such a fin is housing an antenna for communication, usually known as antenna mast [22].

Inplane forces due to the thrust produced by propeller engines can be derived by using Blade Element Theory, in particular for unconventional propeller shape. For conventional shapes, a *NACA* report contains theory for the estimation of propeller side force that can be used for either for angle of attack as well as yaw angle [23]. This work was placed into helpful curves by Seckel [24], which was further elaborated with the help of an example of fuel powered engine's propeller by [25]. These analytical relationships are used to get a first order approximation of yaw force, which is later utilized in designing a dorsal fin. Vertical tail itself has destabilizing effect on the longitudinal stability [1] and such an affect can be minimized by defining an anhederal angle to get a partial component of lift to balance such an affect. Anhederal angle is defined here as the tilt angle, once looking from the front view of the aircraft. Effect of anhederal and chamber of airfoil of such a dorsal fin is analyzed in the light of wind tunnel experiments for a dorsal fin made up of symmetric airfoil with a small anhederal angle. Although at certain angle of attack, the desired force requirement can be met but it requires a mechanical system to replace the muscles anatomy, which exist in swimming animals to turn the dorsal fin. However, a dorsal fin with a cambered one airfoil can do this job without additional weight penalty and it can meet the requirement of momentum to neutralize the counter moment of asymmetric thrust. This novel idea is discussed in this paper for its future application to keep the *HB* aircraft to its original state, without any input by the pilot.

2 Estimation of Asymmetric Thrust

A three step methodology is adopted for the estimation of asymmetric thrust, Tab. 1. All the important input parameters required for three variables are highlighted in it.

Table 1. Methodology for the estimation of Asymmetric Thrust [25]

Side Force Factor and Advance Ratio	
Area of Propeller's disk Number of propellers	Testing velocity RPM of propeller
Propeller Normal Force Derivative and its Factor	
Coefficient of thrust	Horse power of engine
Propeller Normal Force Derivative and upwash/downwash Gradient	
Dynamic pressure	Number of engines

Equation 1 represents an empirical method [25] for the estimation of the propeller normal force which is based upon propeller charts [23-24]. In this equation, *Ne* is the number of the blades per propeller, A_p is the area of one propeller, q is the dynamic pressure and α is the angle of attack. The function $f(T)$ adjusts for non-zero thrust and is given as function of coefficient of thrust, C_T:

$$T_N = q\left[N_e A_p \frac{dC_{NB}}{d\alpha}\frac{d\beta}{d\alpha} f(T)\right].\alpha \qquad (1)$$

In order to evaluate the value of T_N against q equal to 750.3 N/m^2, value of $\frac{dC_{NB}}{d\alpha}$ and $\frac{d\beta}{d\alpha}$ were calculated first along with the value of *SFF*, which is equal to 14. 82 against value of A_p. After doing an iterative calculation with a number of inputs, the equired value of against angle of attack equal to 4 degree comes out to be equal to 42.2 N and the moment generated by it is 304 N.m. Hence, based on the reference area of wing, the required coefficient of lift is equal to 0.0029.

3 Details of the Configuration

The configuration of HB aircraft consists of a voluminous fuselage to fulfill the requirement of volume for aerostatic lift and ballonets to control the expansion of the lifting gas. The fineness ratio of the aircraft is 5.5 which resembles that of a *California Sea Lion*; a hybrid buoyant aircraft by nature [26]. Two fuel powered engines are installed on the ultralight weight wings of span 20 m, reference area of the wing equal to 27.5 m^2 and 1.5 m is its mean geometric chord. The *CG* of the aircraft is about 15.1 m, which is far ahead of the wing. An *H- tail* is selected for such a configuration to meet the requirement of longitudinal and lateral stability [27]. A pictorial view of the configuration is shown in figure 1(a). This figure also shows the direction of the moments generated by the propellers of the engines as well that produced by the lift force of the dorsal fin. The distance between the aerodynamic center of the dorsal fin and CG i.e. X_T is about *CG*. Geometry of the fin is shown in figure 1(b).

(a) HB aircraft

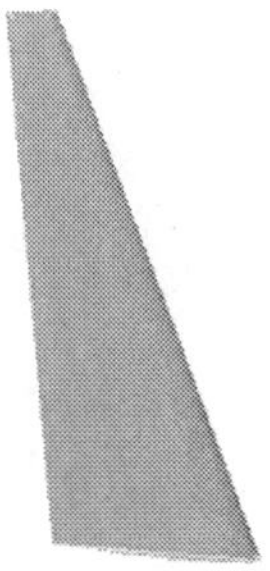

(b) Dorsal fin

Figure 1. Pictorial view of HB aircraft and dorsal fin

It is believed that the shape of dorsal fin in swimming animals is symmetrical and is perpendicular to the body. Such a shape is of interest for HB aircraft as well. This is due to interesting augmentation of the lift produced by its deflection in longitudinal as we all in in lateral direction. Aircraft always have some technological constrains due to a movable part. However, as mentioned earlier in the introduction part that a dorsal fin can be placed at constant incidence angle and the additional lift contribution in the longitudinal direction may be utilized for generating a negative pitching moment about the CG. However, the sizing of the dorsal fin depend on the required lift, which is calculated based on the asymmetric force generated by the propellers of the twin engines, discussed earlier in the section 2. The dorsal fin under study has span of 0.9 m and its mean aerodynamic chord is 0.28 m. The root sand tip chord is equal to 0.40 m and 0.10 m.

4 Design of Experiment

In the case of aircraft, the main contributor towards directional stability is vertical tail. When the airplane experiences side slip, the angle of attack of the vertical tail changes the lift so obtained is utilized to produce a yawing moment about center of gravity. The same concept is adopted here and a full model of dorsal fin, discussed in previous section is tested in *IIUM* wind tunnel. A rounded tip fairing was added to the wing tip and mounted vertically to the wind tunnel turn table, it has sweepback and sweep angles of leading and trailing edge are 30° and 5° respectively, with a tapered *NACA 23012* profile and span of 0.9 m. The wing taper ratio equal to λ=0.25 and aspect ratio of 7.2. The model is designed such that it has an inherent anhederal angel of 5 degree.

Model forces are measured with an external six-component balance and the variation of the lift coefficient with angle of attack is shown in figure 2. The maximum lift coefficient is found to be equal to 0.887. Due to reasonably small blockage ratio of 0.012 (1.2%) at 4 degree angle of attack, no blockage correction was applied.

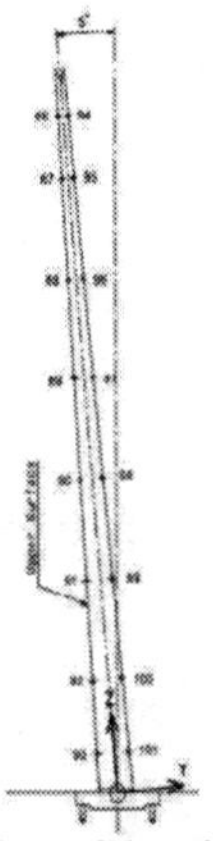

Figure 2. A Frontal view of the wing

5 Results and Discussion

The required value of C_L is basically function of the reference area of the wing and the local value C_L of the dorsal fin is based on its own reference area. Its estimated value is derived by plotting as function of angle of attack, Fig. 3. Against α equal to 8.5 degree, experimental value of C_L is equal to about 0.45, which is further non-dimensionalized by the reference area of the wing to get the required C_L equal to 0.0029. This value is found to be consistent with that obtained by using panel method software. The distance between the aerodynamic center of the dorsal fin is taken at its quarter chord from the leading edge. In this way, the moment arm of the force generated by the dorsal fin is calculated to be equal to 6.2 m. Plot of C_L as function of α is shown in figure 3. The plots of coefficient of pressure of this dorsal fin at α equal to zero degree and 8.5 degree are shown in figure 4(a) and 4(b), respectively.

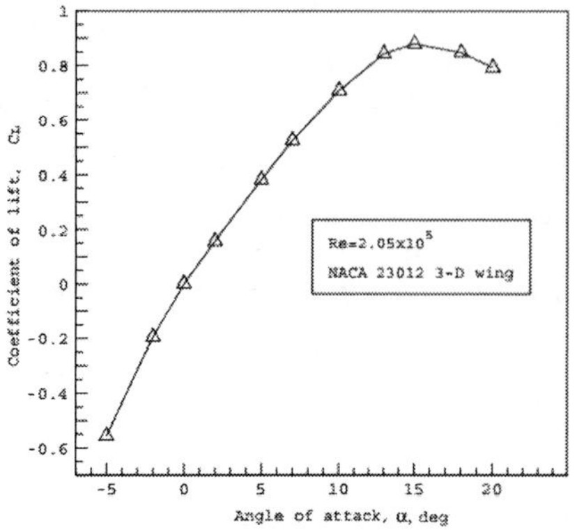

Figure 3. Plot of C_L as function of α

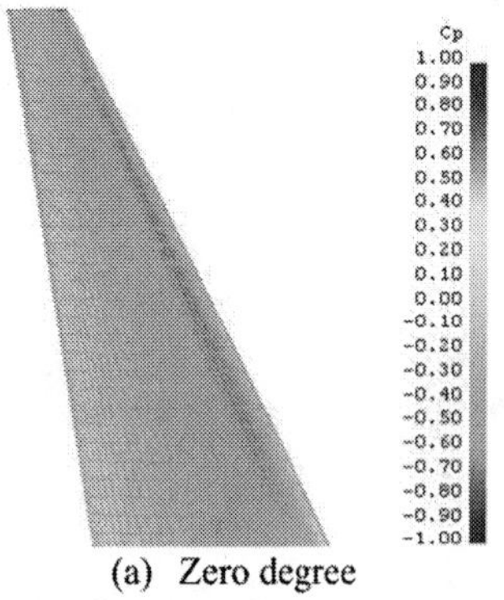

(a) Zero degree

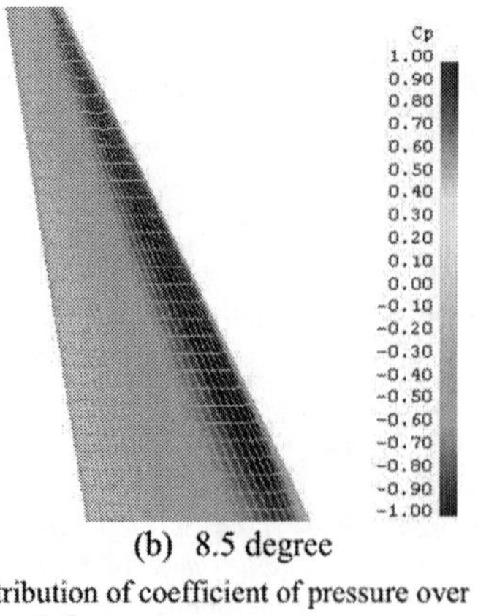

(b) 8.5 degree

Figure 4. Distribution of coefficient of pressure over cambered dorsal fin

For the option two case, *NACA 4412* airfoil is selected after an iterative calculation to meet the requirement of the C_L for zero degree angle of attack, for the same dimensions of the dorsal fin. It is important to note that the dorsal fins can be of a shape similar to that of marine animals as well. In this regard, pictorial view of two such fins is shown in figure 5 for quick reference. Such fins will have less span and bigger root chord. Only design requirement for it is to meet the said requirement of asymmetric thrust.

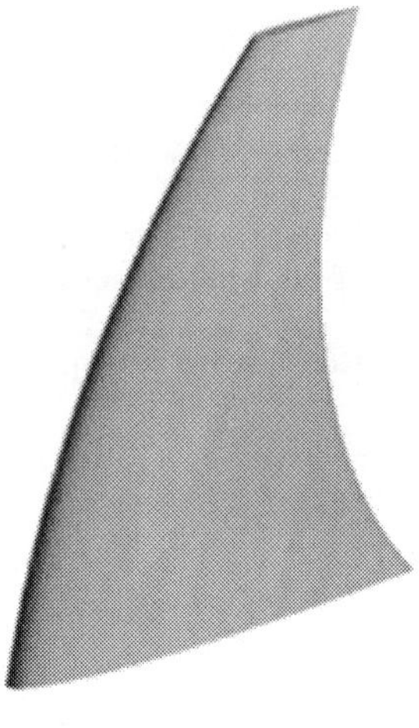

(a)

(b)

Figure 5. Alternative profiles of dorsal fins

The results so obtained will be applied in the conceptual design of such aircraft [26-27] and later, wind tunnel testing on its scaled down model for power-on condition. Furthermore, as the scope of this research is only limited to the yaw force generated by the propellers. Therefore, no calculations have been done to estimate the component of lift force generated by the dorsal fin, which contributes towards longitudinal stability.

The proposed concept does identify clearly any implications for research work for other applications. For example it can be applied to buoyant aerial vehicle, which utilizes DC electric brushless motors which has additional weight penalty due to massive battery to give power for the operation of motors. It can also be applied to under water hull as well for which a pin obstacle was recently proposed to act as a source of asymmetric normal force to improve the longitudinal stability. This phenomenon is similar to the asymmetric vertical flow which exists in aircraft which fly at high angle of attack.

Although no results were provided in this work to justify that the asymmetric normal force of the cambered shaped dorsal fin can be used in the longitudinal control of the HB aircraft, through induced pitching moment. But it is believed that an engineering method can be developed to validate the said concept. Furthermore, it is perceived that when the certification of such an aircraft will be done in future than such nature inspired concept will ease of the pilot in flight operation.

6. Conclusion

The proposed concept of dorsal fin is quite different than that traditionally used in aircraft with vertical tail to enhance the performance of vertical tail by generating the vortex flow ahead of it. As per the results of the generic model, it can be concluded that the proposed dorsal fin can intrinsically balance the asymmetric thrust for the condition that there is no offset of the thrust vector. Moreover, it will give a road map for future wind tunnel testing with bilateral propulsion system on such aircraft in which long moment arm for the dorsal fin is available.

Acknowledgment

The support of the Ministry of Science, Technology and Innovation (MOSTI), Malaysia, under the grant 06-01-08-SF0189 is gratefully acknowledged. Authors are thankful to Senior Engineer Azliza Binti Embong, Nor Hafiz Adnan and Firdauz Hakem for providing assistance during the experimental work.

References

1. Raymer, D., Aircraft Design: A Conceptual Approach, Washington, DC: American Institute of Aeronautics and Astronautics, Inc., (2012)
2. Cover, A.B, Engine Response to Distorted Inflow Conditions, AGARD Report, (1986)
3. Shin, H. J., Lee, J. S., Lee, K. H., Han, M. R., Hur, E. B., & Shin, S. C. Numerical and experimental investigation of conventional and un-conventional pre swirl duct for VLCC. *International Journal of Naval Architecture and Ocean Engineering*, **5** 3, 414-430 (2013)
4. McRuer, D. T., Graham, D., and Ashkenas, I. *Aircraft dynamics and automatic control*. Princeton University Press. (2014)
5. Haque, A. U., Asrar, W., Omar, A. A., Sulaeman, E., and Ali, J. S. M., Effect of Wing Loading on Gross Takeoff Mass of a Hybrid Buoyant Aircraft, International Aerospace Engineering Conference, 27-28 August, 2015 – Vancouver, Canada (2015).
6. Haque, A. U., Asrar, W., Omar, A. A., Sulaeman, E., and Ali, J. S. M., Power-off static stability analysis of a clean configuration of a hybrid buoyant aircraft, 7th Ankara International Aerospace Conference, METU, Ankara Turkey, 11-13 September (2015)
7. Wallace, F. R., & Hein, J. N. U.S. Patent No. 5,238,207. Washington, DC: U.S. Patent and Trademark Office, (1993)
8. Haque, A. U., Asrar, W., Omar, A. A., Sulaeman, E., and Ali, J. S. M., Assessment of Engine's Power Budget for Hydrogen Powered Hybrid Buoyant Aircraft, Journal of Propulsion and Power Research, Elesvier (in press), (2015)
9. Nakamura, I., Meyer, C. G., & Sato, K. (2015). Unexpected Positive Buoyancy in Deep Sea Sharks, Hexanchus griseus, and a Echinorhinus cookei. PloS one, 10(6).
10. J. E. Harris, The role of the fins in the equilibrium of the swimming fish. I. Wind-tunnel tests on a model of *Mustelus canis* (Mitchill). *Journal of Experimental Biology* **13**, 476-493, (1936)
11. J. E. Harris, The mechanical significance of the position and movements of the paired fins in the Teleostei.*Papers from Tortugas Laboratory* **31**, 173-189, (1973)
12. J. E. Harris, 1938. The role of the fins in the equilibrium of the swimming fish. II. The role of the pelvic fins. Journal of Experimental Biology **15**, 32-47, (1938)
13. J. E. Harris, (1953). Fin patterns and mode of life in fishes. *In* S. M. Marshall and P. Orr (eds), *Essays in Marine Biology*, Oliver & Boyd, Edinburgh, (1953)
14. P.W. Webb, (1975). Hydrodynamics and energetics of fish propulsion. *Bulletin of the Fisheries Research Board of Canada* **190** 1-158, (1975)
15. P. W. Webb, Simple physical principles and vertebrate aquatic locomotion. *American Zoologist* **28** 709-725, (1988).
16. P. W. Webb, Designs for stability and maneuverability in aquatic vertebrates: What can we learn. In: Tenth International. Symp. Unmanned Untethered Submersible Tech.: Proc. Sp. Ses. Bio-Eng Res. Related to Autonomous Underwater Vehicles, Durham, NH, 85-108, (1997).
17. P. W. Webb, Control of posture, depth, and swimming trajectories of fishes. Integrative and Comparative Biology **42** 94-101, (2002).
18. P.W. Webb, and Keyes, R, Swimming kinematics of sharks. Fish. Bull. **80** 803-812, (1982)
19. D. Weihs, (1989). Design features and mechanics of axial locomotion in fish. American Zoologist 29:151-160.
20. D. Weihs, Stability of aquatic animal locomotion. *Cont. Math.* **141** 443-461, (1993)
21. D. Weihs, (2002). *Stability versus maneuverability in aquatic locomotion. Integrative and Comparative Biology* Experimental Biology, **42** 127-134, (2002)
22. Burnside, W. D., & Marhefka, R. J. *Antennas on aircraft, ships, or any large, complex environment. An Antenna handbook*, Springer US, 1435-1534, (1988)

23. Rebneb, B. H. S., Formulas for propellers in yaw and charts of the sede-force derivative, (1943)
24. Seckel, E., & Morris, J. J., The Stability Derivatives of the Navion Aircraft Estimated by Various Methods and Derived from Flight Test Data (No. AMS-923). Princeton University, Dept. of Aerospace and mechanical sciences, (1971)
25. Gudmundsson, S., *General aviation aircraft design: Applied Methods and Procedures*. Butterworth-Heinemann, (2013)
26. Haque, A. U., Asrar, W., Omar, A. A., Sulaeman, E., and Ali, J. S. M., A Novel Design of a Hybrid Buoyant Aircraft- A Potential Greener Solution for Inter Connectivity of Malaysian Islands", AER0 15, 19-21 April, 2015, Monterial Canada, (2015)
27. Haque, A. U., Asrar, W., Omar, A. A., Sulaeman, E., and Ali, J. S. M., (2015) "Pugh Analysis for Configuration Selection of a Hybrid Buoyant Aircraft", SAE 2015 AeroTech Congress & Exhibition, September 22-24, Seattle, Washington, USA, (2015)

EPJ Web of Conferences **114**, 02129 (2016)
DOI: 10.1051/epjconf/201611402129

Comparison of data correction methods for blockage effects in semi-span wing model testing

Anwar U Haque[1], Waqar Asrar[1,a], Ashraf A Omar[2], Erwin Sulaeman[1] and Mohamed J.S Ali[1]

[1]*International Islamic University Malaysia (IIUM), 50728 Kuala Lumpur, Malaysia*
[2]*University of Tripoli (UOT), P.O. Box 13154 Tripoli-Libya*

Abstract. Wing alone models are usually tested in wind tunnels for aerospace applications like aircraft and hybrid buoyant aircraft. Raw data obtained from such testing is subject to different corrections such as wall interference, blockage, offset in angle of attack, dynamic pressure and free stream velocity etc. Since the flow is constrained by wind tunnel walls, therefore special emphasis is required to deliberate the limitation of correction methods for blockage correction. In the present research work, different aspects of existing correction methods are explored with the help of an example of a straight semi-span wing. Based on the results of analytical relationships of standard methods, it was found that although multiple variables are involved in the standard methods for the estimation of blockage, they are based on linearized flow theory such as source sink method and potential flow assumption etc, which have intrinsic limitations. Based on the computed and estimated experimental results, it is recommended to obtain the corrections by adding the difference in results of solid walls and far-field condition in the wind tunnel data. Computational Fluid Dynamics technique is found to be useful to determine the correction factors for a wing installed at zero spacer height/gap, with and without the tunnel wall.

1 Introduction

One of the major problems associated with the wind tunnel test for a wing is the blockage effects produced by the wind tunnel walls [1]. The presence of tunnel-induced lift increments, as well as the increment in drag count is due to wall bounded conditions [2]. Reliable aerodynamic and static stability coefficients of an aircraft's wing are usually obtained from wind tunnel testing, which needs to be corrected for such affects. The effect of the sidewalls of the tunnel is only applicable to 3-D models where trailing vortices exist. In such cases, the tunnel walls induce a span-wise variation in up-wash, which decreases the normal downwash [3]. Thus, wind tunnel testing on a wing or lifting body will have too little downwash, as the wing appears to have a larger effective aspect ratio than it would have if tested in free air. These effects are small if the span is much less than the tunnel width, but large span models produce serious distortions in lift distribution which also affect stalling characteristics [4]. To alleviate such interference, a correction scheme to cater the blockage effects is always required. But deficiencies in such correction procedure are likely to be most severe when the model is mounted on the tunnel's floor and the interference flow created by such an arrangement has significant non-uniformity in the region of the model [5].

Till now, significant work has been done, to have a better understanding about the blockage corrections due to wall effects applied to the raw data of wind tunnel [5]. Such research efforts are mostly related to the development of analytical relationships for data correction method, especially for blockage correction equations [6], but the validation of the same through experiments is quite limited for half model testing of wing alone models. Furthermore, to the authors' best knowledge; very little effort has been made to correlate the raw data of wing alone half model testing with the numerical simulation techniques in the subsonic regime.

It is well known that wind-tunnel-induced flow distortion around a test model causes drag, lift and pitching moment increments that are absent in free air [6]. Necessary correction must be made before using this data for engineering design. In the present work, results of two data correction schemes were compared with those obtained through CFD. For the proof of concept, qualitative as well as quantitative comparison is done for straight level flight condition.

All the empirical relationships and relative curves used for the estimation of horizontal buoyancy, wake blockage, total blockage and corrections to aerodynamic and static longitudinal stability parameters are taken from Pope and Harper [7] and Herriot [8]. The experimental testing is done on a generic model of a straight semi-span

[a] Corresponding author: waqar@iium.edu.my

wing made up of NACA-65(3)-218 wing and the experimental results are utilized to compare the standard methods of blockage correction. Flow physics related to the creation of horse shoe vortex due to the unavoidable interaction of the tunnel's floor with the base of wing's model is quite matured. Therefore, it has not been discussed here as it a well-known phenomenon in past [9] as well as in recent studies [10].

2 Data Correction Methods

It has been mentioned earlier that the conditions under which a wing's model is tested in a wind tunnel are not the same as those in free air. The walls of a closed test section produces changes in flow patterns, which in turn changes the flow physics. To check this, blockage corrections is one of the prospective options for correcting the raw data obtained in a wind tunnel.

2.1 Blockage

The purpose of the blockage corrections is to yield estimates for the aerodynamic forces and moments experienced by the actual test model in free stream, rather than for the original model [5]. Such a correction falls into the following two categories:

2.1.1 Solid Blockage

The presence of a model in the test section reduces the flow area and thereby increases the velocity of the air through the test section. The increase in velocity due to solid blockage at the scaled down model is usually quite small than that calculated from a direct area reduction since the streamlines near the tunnel walls are displaced more than those near the model [7]. Solid blocking causes an increase in dynamic pressure, which in turn increases all forces and moments [7,11].

2.1.2 Wake Blockage

Any scaled down model tested in a wind tunnel will always have a wake behind it which is basically due to the detachment of the boundary layer from the model. This wake has a mean velocity lower than that of the far field condition, which perhaps keeps the mass flow rate constant from the test section entrance to exit. This results in a lower wake pressure. This effect increases with wake size, hence is highest when large flow separation occurs. The wake blocking also creates an increase of dynamic pressure on the model [7].

2.2 Calculations of Blockage Effects

Available analytical relationships are briefly reviewed are to get first-hand knowledge about the blockage correction techniques for a closed circuit wind tunnel. These relationships are applicable only for the condition when the size of the scaled down model is smaller than the test section size. All such relationships are grouped into two categories, a brief detail of which is given below. Formulas for the first group are taken for incompressible flow [7] and the second group with consideration of the effect of compressibility [8]. Existing *ESDU* standard for half model testing is perhaps based on the potential flow method and some deficiencies in this method have recently been diagnosed by Alvaro [12] . This includes an offset between the angle of attack for wall corrections and the slope of the angle-of-attack correction versus lift coefficient curve [12]. Hence this method has not been discussed and used for the comparison study.

2.2.1 Conventional Method of Correction

Blockage is the sum of wake blockage and horizontal buoyancy. In this way; total correction factor ϵ can be written [7] as follows, Eq. (1):

$$\varepsilon = \varepsilon_{sb} + \varepsilon_{wb} \qquad (1)$$

Where, ε_{sb} is the wake blockage and ε_{wb} is the wake blockage. The analytical relationship [7] for solid blockage ε_{sb} is shown below:

$$\varepsilon_{sb} = \frac{k_1 \tau_1 \omega_v}{C^{3/2}} \qquad (2)$$

Almost all wind tunnels with closed test sections have a variation in static pressure, even with no model attached with it. This is caused by the thickening of the wall boundary layer, which in turn causes a decrease in the effective test section flow area [7]. This effect can be reduced by having the cross-section area of the tunnel increase towards the test section exit. However, with fixed walls, it is only possible to achieve a uniform axial test section pressure at one flow rate. For all other flows, the test section pressure normally decreases as the end of the test section is approached. This pressure variation over the model produces an additional drag force. Such an effect on a wing's scaled down model can analytically be estimated by using Eq. (3) [7]:

$$\varepsilon_{wb} = \frac{S_{ref}}{4C} C_{DU} \qquad (3)$$

Where, C_{D_U} is uncorrected drag coefficient, S_{ref} is reference area of wing and term C is the tunnel's cross sectional area.

The corrected i.e. true angle of attack due to blockage is expressed in Eq. (4) [7]. Similar to the case of a pitot-static tube of an aircraft, we have difference in indicated and true air speed. The same is true for the angle of attack for the wind tunnel testing.

$$\alpha = \alpha_U + \left[\delta \frac{S}{C} C_L (57.3)\right](1 + \tau_2) \qquad (4)$$

In the above equation, terms δ and τ_2 are shape factor of a tunnel and wing respectively. The change in angle of attack causes the lift to increase, the corrected lift is estimated by using Eq. (5)-(8) [7]:

$$\Delta\alpha = \Delta\alpha_1 + \tau_2 \Delta\alpha_1 \quad (5)$$

$$\Delta\alpha_1 = \frac{6\sigma}{\pi^2} C_1 \quad (6)$$

$$\sigma = \frac{\pi^2}{48}\left(\frac{\bar{c}}{\sqrt{C}}\right)^2 \quad (7)$$

$$C_L = C_{LU}(1 - 2\varepsilon) - \tau_2 \Delta\alpha . a \quad (8)$$

Similarly, the change in coefficient of drag and the corrected drag coefficient is estimated by using Eq. (9)-(10) [7]:

$$\Delta C_{DW} = \frac{k_1 \tau_1 \omega_v}{C^{3/2}} c_{DU} \quad (9)$$

$$C_D = C_{DU}(1 - 2\varepsilon) - \Delta C_{DW} \quad (10)$$

Pitching moment is usually referred at the quarter chord of the wing. This is actually the location of the aerodynamic centre of the wing which may change or may not change in the wing-fuselage case. This is because of the fact that, all analytical methods except the Diehl method state that their results are the location of the wing fuselage aerodynamic centre ahead of the original wing aerodynamic centre, which does not necessary have to be 0.25c, [13]. Since the case under discussion is only of the wing, therefore the coefficient of moment is taken at the quarter chord of the wing, Eq. (11):

$$C_{M1/4} = C_{M1/4U}(1 - 2\varepsilon) + 0.25\tau_2 \Delta\alpha . a \quad (11)$$

It is important to note that all uncorrected coefficients are defined in terms of an uncorrected dynamic pressure q_u; which is the q that would exist if no model was present in the test section [7].

$$C_{L_U} = \frac{Lift}{q_u S} \quad (12)$$

$$C_{D_U} = \frac{Drag - \Delta D_B}{q_u S} \quad (13)$$

$$C_{M_U} = \frac{Moment}{q_u Sc} \quad (14)$$

Where S, c and ΔD_B are reference area of the wing model, its characteristic length and change in drag due to horizontal buoyancy drag respectively.

2.2.2. Correction with constraint of Compressibility and Test Section of Rectangular shape

Herriot's work [8] is based on the estimation of blockage correction for closed-test-section wind tunnels of different cross-sectional shapes. He has also provided the experimental data for blockage estimation of different shapes of wings with tabulation of the appropriate constants. In order to account for the effect of the compressibility of flow, a correction for Mach number 'M' and for the free stream velocity 'U' is suggested in his work:

$$M = M'\left\{1 + \left[1 + 0.2(M')^2\right]K\right\} \quad (15)$$

$$U = U'(1 + K) \quad (16)$$

And the corrected dynamic pressure 'q' can be obtained by using Eq. (17), [8]:

$$q = q'\left\{1 + \left[1 + 0.2(M')^2\right] K\right\} \quad (17)$$

$$K = K_\omega + K_{\omega k} \quad (18)$$

In Eq. (18), K_ω and $K_{\omega k}$ have the same meaning as that of ε_{sb} and ε_{wb} as described in Eq. (2) and (3), respectively.

$$K_\omega = \frac{1}{\left[1 - (M')^2\right]^{3/2}} \frac{K_1 . \tau . \nabla_\omega}{C^{3/2}} \quad (19)$$

Where, value of K_1 is estimated by Eq. (20):

$$K_1 = \frac{\pi^{3/2}}{16} \frac{\Delta}{t/c} \frac{1}{\kappa_1} \quad (20)$$

Here t/c is the thickness to chord ratio of the wing and $K_{\omega k}$ is the wake blockage correction factor, which is given by:

$$K_{\omega k} = \frac{\left[1 + 0.4(M')^2\right]}{\left[1 - (M')^2\right]} \times \frac{C_D' S}{4C} \quad (21)$$

The corrected coefficient of drag is:

$$C_D = C_D'\left\{1 - \left[2 - (M')^2\right] K\right\} - \Delta C_{D_W}' \quad (22)$$

Herriot [14] also provided the analytical relationship for an additive wake-induced incremental drag adjustment, shown below in Eq. (23):

$$\Delta C_{D_W}' = \frac{\left[1 + 0.4(M')^2\right]}{\left[1 - (M')^2\right]^{3/2}} \times \frac{K_1 . \tau . C_D' . V_\omega}{C^{3/2}} \quad (23)$$

It is important to note that C_{D_U} given in Eq. (13) and C_D' of Eq. (23) have the same meanings. The analytical relationships described in sub section 2.2.1 and 2.2.2 are used to correct the data for the test on the semi-span wing conducted at a free stream velocity equal to 50 m/s. The corrected results are discussed in section 4.2.

3 CFD vis-`a-vis Wind Tunnel Testing

To allow reprocessing of the available wind-tunnel data, with available data correction schemes and comparison of results with CFD, a methodology used earlier for transonic flow [14] is adopted and applied to the available testing data of wing model [15] at a subsonic speed.

3.1 Methodology

The incompatible requirements in scaling down the model, complex interaction of the flow from the walls and its interference with the model raise a question on the reliability of the existing data correction schemes [7]- [8] for half model testing of straight wings. Under such an unpredictable scenario; confidence in the results can be achieved by numerically simulating the flow with and without the walls. In this way, flow field information can easily be obtained. It is well-known that the wall mounted model faces additional interference due to the boundary layer of the tunnel's floor but the assessment of such interference due to the complicated boundary condition can only be achieved by testing the model at different spanner heights/gaps. Moreover, there is no universal school of thought for defining any fixed spanner height/gap. The effect of the walls on the model can be obtained by performing a numerical simulation of the test with no physical gap between the wing and tunnel's floor. Results so obtained can further be compared with those obtained by applying the farfield boundary condition at the outer computational domain Complete details of both the simulations have to be described in section 4.1. The difference so obtained is perhaps a representation of flow interaction due to the wall and can be added to the wind tunnel results to get the data for the free stream condition. This methodology taken from reference [14] is shown in figure 1 for quick reference. Numerical wind tunnel testing for the specified size of the test section has already been in practice in transonic regime for ventilated [16] as well as for solid wall [17] and has been applied earlier for different aerospace applications [14].

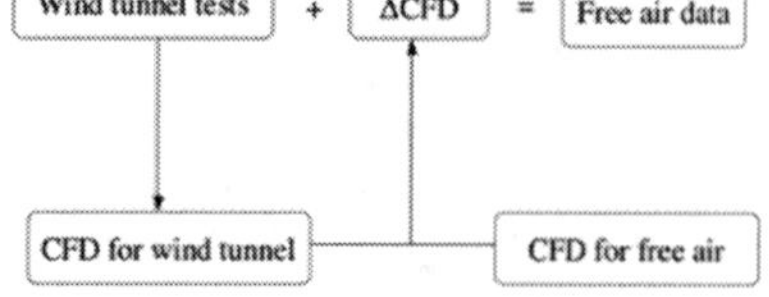

Figure 1. Role of CFD for data correction [14]

In order to validate this numerical approach for subsonic flow, CFD cases are run and the difference so obtained is used later to correct the wind tunnel measurements. These results alongwith those obtained by using the existing analytical formulas are discussed in the following section:

4 Results and Discussion

This section details the analytical and computational calculations used to correct the raw data for a wing tested in a closed test section of *IIUM-LSWT* [15]. Complete details of the experimental setup and flow conditions have been provided in reference [15]. Due to the weight and attachment constraints, no additional gap is considered between the model and the floor of the tunnel. Moreover, no data interpolation technique was applied on the raw data of the wind tunnel which is only corrected for tare and balance affects.

Figure 2. Model of wing tested in IIUM-LSWT [15]

4.1 Numerical Wind Tunnel - with and without wall

Fluent®, a commercial *CFD* software is used to solve the time averaged Navier-Stokes equations with SST turbulence model. As the scope of the work is limited to find the correction factor by using CFD. Hence, less effort has been put for correlating the numerical results with the raw data of the wind tunnel. Two different cases alongwith the grid generations for CFD analysis are described below to find the correction factor, described earlier in section 3.1.

4.1.1 Numerical Wind Tunnel

The geometry of the wing alongwith the geometrical details of the test section is modeled using CATIA ® CAD software, figure 3(a). Few necessary simplifications in the model are made for the ease of grid generation. As the flow over the wing is analyzed in the clean configuration therefore it is modeled without modeling the nuts and bolts. These few small external components which are ignored, mainly add to the drag in small amount but their contribution towards the lift and the moments is negligible. Two cases with different grids are used so that flow should be aligned with the grid. In the first case the walls of the tunnel are defined as per the size of the test section and the grid is made keeping in view the flow physics so extensive grid cells are distributed on and near the tunnel's floor where the wing is physically attached with it as shown in figure 3(b). Moreover, velocity inlet condition is defined at the inlet and outflow condition is used at the outlet. The presence of a low pressure zone due to flow separation at the junction of tunnel's floor and the root chord of the wing is obvious from figure 3(c). The quantitative results so obtained alongwith the experimental results for level flight condition are discussed at the end of sub section 4.2.2.

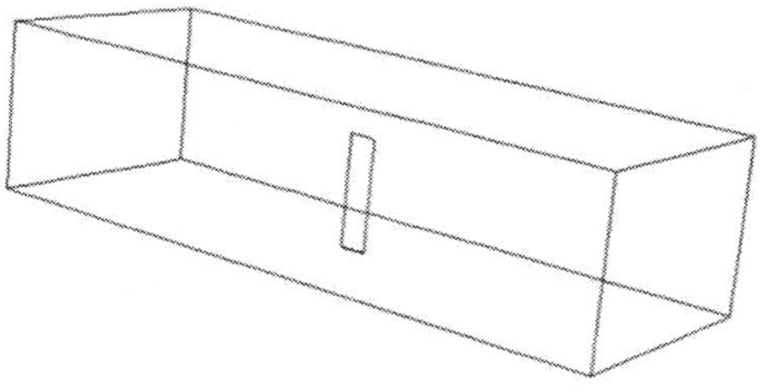

(a) Computational domain for solid wall test

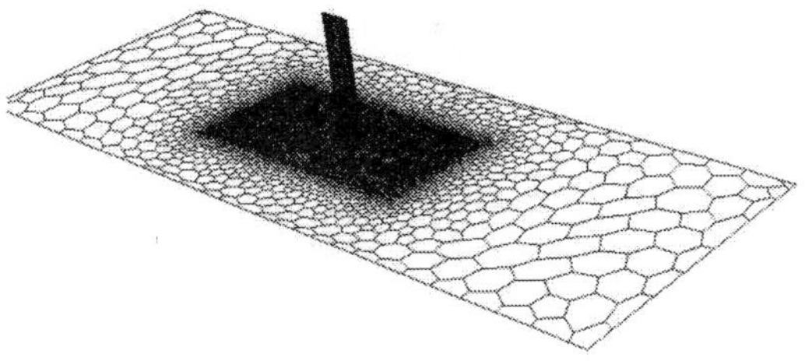

(b) Grid clustering near the junction of wing

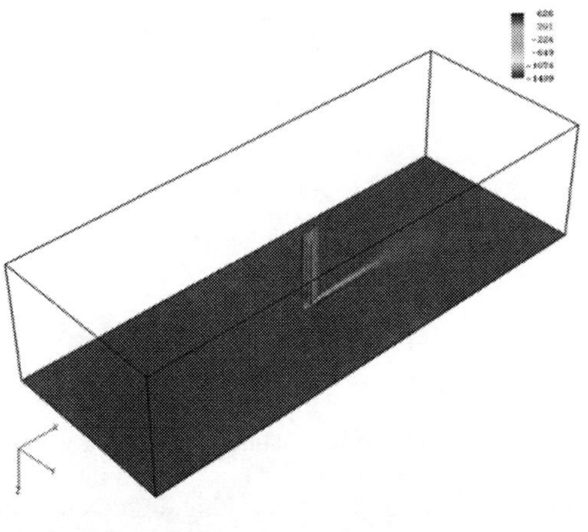

(c) Pressure contour at the tunnel's floor

Figure 3. Results of numerical wind tunnel

4.1.2 Numerical Case of Farfield Condition

For the second case of numerical simulation, pressure farfield boundary condition is used and no-slip wall condition is applied at the wall of the model. The outer domain of farfield boundary is about 30 times the span of the model. This is because of the fact that although the effect of compressibility are quite small for $M \leq 0.3$. But in nature, the flow is compressible and it is important to run the numerical simulation for the said condition. Free stream Mach number corresponding to the free stream velocity of 50 m/s is taken equal to 0.145. This value of Mach number along with the appropriate value of atmospheric pressure is prescribed in the boundary condition panel. Density is calculated by using the ideal gas law and Sutherland's law is employed to calculate the viscosity in the numerical simulations. For this case, the domain extents are kept at twenty times the length of the body in order to avoid the flow variations near the wing from affecting the boundary conditions. The numbers of grid cells are 0.4 million, out of which 40 % of the cells are on the tunnel's floor and in the vicinity of the test article. Grid is kept coarse in the farfield domain as well as for the sidewalls of the tunnel. The grid clustering can be seen in the sectional plots of grid at the center plane as well as at the inlet, shown in figure 4(a).

Since the temperature problem is not of interest, therefore adiabatic wall conditions with no slip boundary condition is employed on the wing surface. Contours of the pressure distribution on the surface of the wing at α equal to zero are shown in figure 4(b). Grid independent studies are also carried out to obtain grid independent numerical solutions. Values of the estimated aerodynamic coefficients for V=50 m/s are tabulated in Tab. 1.

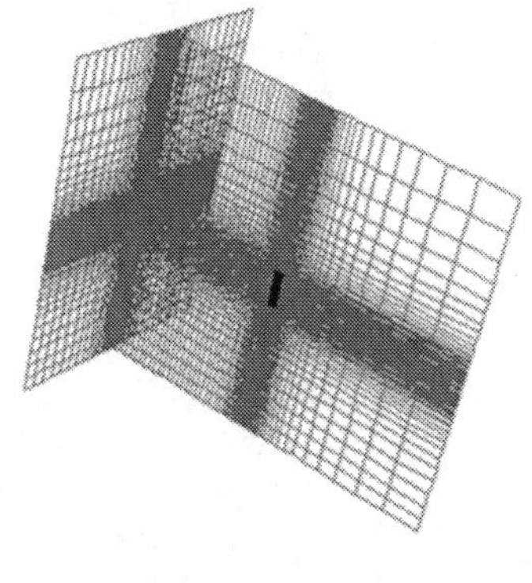

(a) Sectional planes of computational grid

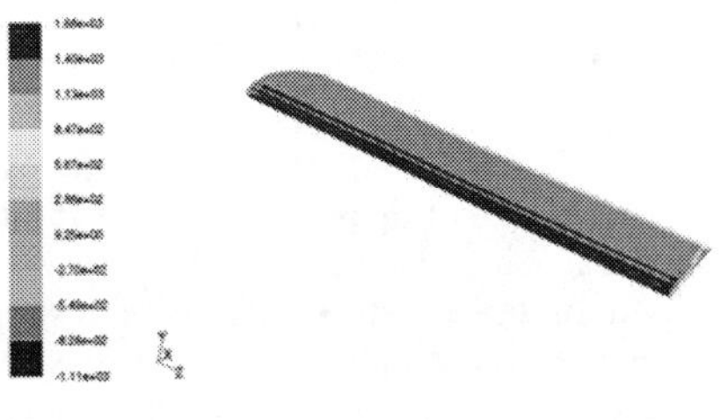

(b) Contours of total pressure (Pascal) on the model

Figure 4. Grid planes and pressure distribution on the wing for the case of unbounded wall condition

Table 1. Comparison of CFD results for two different cases at $\alpha = 0^0$ and at V=50 m/s

	Farfield	*Solid wall*	*Δ CFD*
C_L	0.175	0.183	0.008
C_D	0.0209	0.032	0.01
C_m	0.0019	0.0026	0.007

From above mentioned table, the walls' influence on the aerodynamic results in numerical wind tunnel is considerable. This is because of the fact that when the flow is constrained by the wall than it gives blockage effects to the model placed inside the test section. Due to which aerodynamic coefficients increases, when compared to a a farfield condition. For the case under discussion, this increase in C_L, C_D and C_m is 0.008, 0.01 and 0.007 respectively. Although the presented results are only for $\alpha = 0^0$, but it motivates to carry out a parametric study in future, to determine the sensitivity of the numerical results at higher angle of attack as well. It is perceived that the interference flow field will be quite non- uniform due to the massive flow separation at the floor-model junction and *CFD* will provide the interference factor for data correction.

4.2 Comparison of Results

Analytical relationships for blockage correction are applied ow data of the wind tunnel, figure 5. In this figure the corrected coefficients are nomenclatured as per references [7] and [8], which are overlapping each other due to no significant difference in the corresponding values for the defined α sweep. It can be observed from this figure that the difference between the corrected values and the raw data of the wind tunnel increases with increases in the angle of attack (positive as well as negative). For zero degree angle of attack, the impact of data correction scheme is quite less i.e. the values of ΔC_L, ΔC_D and ΔC_M are 0.0016, 0.001 and 0.0001 respectively. Furthermore, both the blockage correction schemes predicted the same results. This is perhaps due to small difference in the estimated values of k_1 and τ_1, which is 0.01 and 0.033 respectively. Moreover, wind tunnel over predicted the stall angle and hence the coefficient of maximum lift, figure 5(a). But there is negligible difference in the value of coefficient of drag at zero lift, figure 5(a). It can be observed from the qualitative comparison of plots of figure 5 that after $\alpha > 6^0$, the effect of data correction method is considerable. Moreover, a comparison of these results with those highlighted in Tab.1 for at $\alpha = 0^0$ condition revealed that the estimated values of ΔC_L, ΔC_D and ΔC_M is more than that predicted by the analytical relationships. *CFD* results for the said condition are not overlapped on figure 5 as the difference in the values will not be obvious due to other larger values at high α.

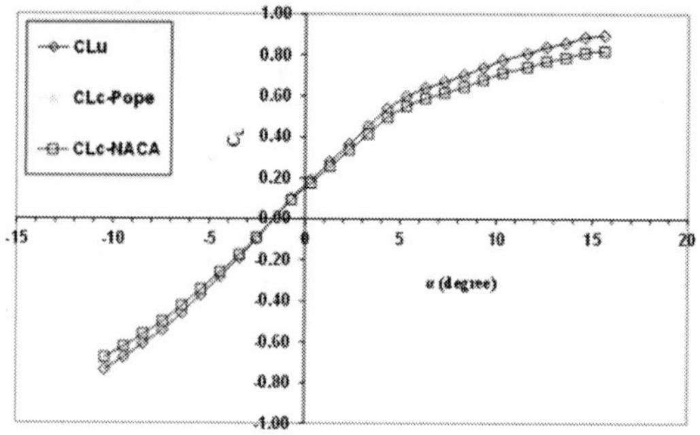

(a) C_L vs α

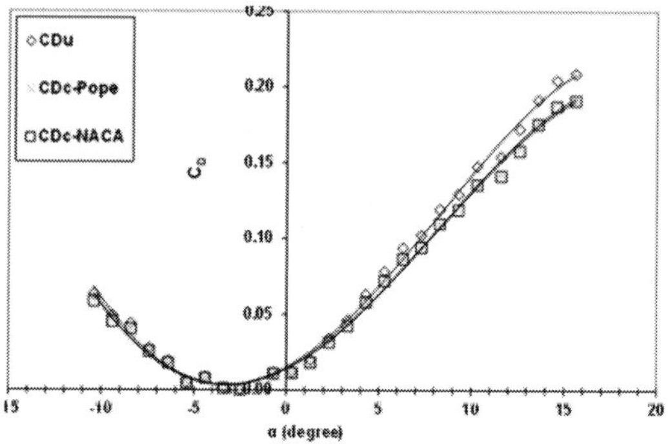

(b) C_D vs α

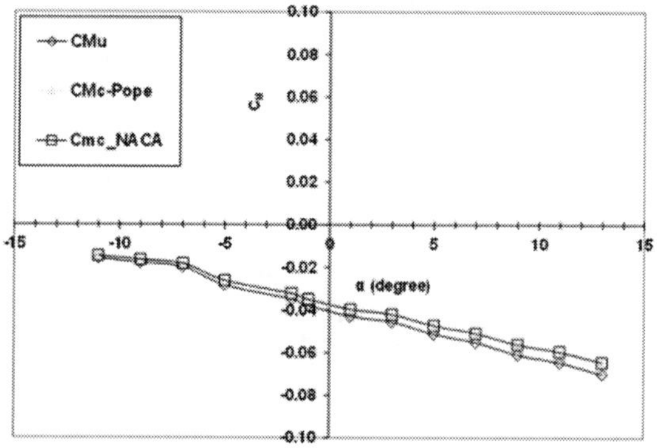

(c) C_m vs α

Figure 5. Comparison of raw and corrected data

at *V=50 m/s*

5 Conclusion

Preliminary *CFD* results obtained from solid-walls and free stream tests showed a difference of values at zero degree angle of attack. Though this difference is quite small but due to strong interaction of the flow with the wall, it will increase with increase in angle of attack. Such an inconsistent offset can be observed between the raw data of wind tunnel and those corrected by standard methods. Compressibility effects on the data correction method are found to be negligible for the defined flow conditions. Moreover, both the analytical schemes for blockage correction decrease the values of aerodynamic coefficient for positive angle of attack and vice versa.

Acknowledgment

The support of the Ministry of Science, Technology and Innovation (MOSTI), Malaysia, under the grant 06-01-08-SF0189 is gratefully acknowledged. Authors are thankful to Senior Engineer Azliza Binti Embong, Nor Hafiz Adnan and Firdauz Hakem for providing assistance during the experimental work.

References

1. R. Crites and F.W. Steinle, Wall Interference Reduction Methods for Subsonic Wind Tunnels 33rd Aerospace Sciences Meeting and Exhibition, AIAA, January 9-12 (1995).
2. R.U.E. Ancelle, A Selection of Experimental Test Cases for the Validation of CFD Codes, AGARD Report No. AR-303 Vol. II, August (1994).
3. J.M. Macha, R.J. Buffington, J.F. Henflingj, Slotted-Wall Blockage Corrections for Disks and Parachutes, Journal of Aircraft, **28** 9, (1990).
4. N. Saha, Gap Size Effect on Low Reynolds Number Wind Tunnel Experiments, Doctoral dissertation, Virginia Polytechnic Institute and State University, (1990).
5. M. M., Freestone, Upwash interference for wings in solid-liner wind tunnels using subsonic linearised-theory, ESDU Data Item 95014 (1995).
6. J. E. Hackett, K.R. Cooper, M.L. Perry, Drag , lift and pitching moment increments due to wind tunnel wall constraint: extension to three dimensions, ICAS Congress (2000).
7. A. Rae, W.H, Pope, Low Speed Wind Tunnel Testing, 3rdd ed., John Wiley & Sons, Ltd, New York, NY, (1999).
8. J.G. Herriot, Blockage corrections for the three-dimensional flow closed throat wind tunnels, with consideration of the effect of compressibility, NACA Report No. 995 (1950).
9. G. H. Tidbury, Wind-Tunnel Wall Interference, Corrections to the Airflow in closed Rectangular Interference, Aircraft Engineering, pp. 192–194 (1945).
10. A. Malik, Suppression of junction flow effects in half model wind tunnel testing, Doctoral dissertation, Loughborough University, (2012).
11. Y. Abudaram, B.Stanford, P.Ifju, Wind Tunnel Testing of Load-Alleviating Membrane Wings at Low Reynolds Numbers, AIAA 47th Aerosp. Science Meeting, (2009).
12. A. Toledano, C. Broughton, J. Weiss, T. Nrc, T.W. Tunnel, I. Wind-tunnel, Improvement of Subsonic Wall Corrections in an Industrial Wind, CASI 62nd Aeronautics Conference and AGM, Montréal, (2015).
13. F. M. Catalano, E. De Engenharia, D.S. Carlos, A.J. Dagnone, J.S. Angelina, S. Carlos, Theoretical and experimental analysis of the fuselage influence on the wing aerodynamic center position at low speed conditions, in 22nd International Congress Mechanical Engineering (COBEM 2013), Ribeirão Preto, SP, Brazil, pp. 2089–2100, November. 3-7 (2010).
14. N. Qin, CFD for Better Understanding of Wind Tunnel Tests, Integrating Experiments and CFD, Glasgow (2003)
15. A. U. Haque; W. Asrar; A.Omar ; E. Sulaeman; J.S.M.Ali, "Half Model Testing for Canard of a Hybrid Buoyant Aircraft", World Acad. Sci. Eng. Technol. Int. Sci. Index, Aerosp. Mech. Eng. **2** 10 (2015).
16. M. Bouriga, F. Morency, and J. Weiss, Numerical Investigation of Wall Mounting Effects in Semi-Span Wind-Tunnel Tests, 53rd AIAA Aerospace Science Meeting, American Institute of Aeronautics and Astronautics, AIAA, Kissimmee, Florida, (2015).
17. S. E. Rogers, K. Roth, S. M. Nash,. Validation of computed high-lift flows with significant wind-tunnel effects. AIAA journal, **39** 10, 1884-1892, (2001).

EPJ Web of Conferences **114**, 02130 (2016)
DOI: 10.1051/epjconf/201611402130

Numerical solution of fluid-structure interaction represented by human vocal folds in airflow

J. Valášek[1,a], P. Sváček[1], and J. Horáček[2]

[1] Department of Technical Mathematics, Faculty of Mechanical Engineering, Czech Technical University in Prague, Karlovo nám. 13, Praha 2, 121 35, Czech Republic
[2] Institute of Thermomechanics, Czech Academy of Sciences, Dolejškova 5, 182 00 Praha 8

Abstract. The paper deals with the human vocal folds vibration excited by the fluid flow. The vocal fold is modelled as an elastic body assuming small displacements and therefore linear elasticity theory is used. The viscous incompressible fluid flow is considered. For purpose of numerical solution the arbitrary Lagrangian-Euler method (ALE) is used. The whole problem is solved by the finite element method (FEM) based solver. Results of numerical experiments with different boundary conditions are presented.

1 Introduction

The fluid-structure interaction is a complex and very interesting problem with many, not only technical applications. Similar physical description of interaction between fluid flows and elastic bodies can be used in various studies dealing for example with airfoil or bridge stability as well as in investigation of blood flow in compliant arteries or in biomechanics of human voice production. This paper is concentrated on the biomechanics of human vocal folds, see e.g. [1]. Human voice is one of basic human being's characteristics and it plays an important role in the quality of a human life. The air flow from lungs excites the vibrations of the human vocal folds which in reverse influence air flow and vice versa. This problem was studied in many papers, see e.g. [2], [3]. Therefore we speak about coupled problem. The human voice originates in basic sound produced by the vocal fold vibration. The attention of this paper is focused on fluid-structure interaction (FSI) problem and comparison of different boundary conditions. First, the interaction problem is described in the terms of linear elasticity theory and the incompressible viscous fluid flow modelled by the Navier-Stokes equations. Then the mathematical model is derived with the help of the arbitrary Lagrangian-Eulerian (ALE) method which enables to handle with the time-dependent domain. The numerical algorithm is based on the finite element method (FEM). For the fluid solver the P1-bubble/P1 elements and for the elastic part piecewise linear P1 elements are implemented. For the coupled problem the fully implicit scheme is used. This implemented algorithm is verified on few test cases when different boundary conditions are studied.

2 Mathematical model

For the sake of simplicity our study is restricted to a 2D model problem. Scheme of the model domain is shown in figure 1. Quantities connected with fluid and structure are labelled by indeces ^{f} and ^{s}, respectively. The boundary of fluid domain $\partial\Omega_t^f$ at arbitrary time t consists of the inlet part Γ_{In}^f, the outlet part Γ_{Out}^f and $\Gamma_{\mathrm{Dir}}^f \cup \Gamma_{\mathrm{W}_t}$. The part Γ_{Dir}^f represents fixed walls of the glottal channel and the time dependent boundary Γ_{W_t} representing the interface on the vibrating vocal folds. The undeformed (reference) shape of the domain is marked by index $_{ref}$. The fluid flow is described within the Eulerian coordinates and later solved using the arbitrary Lagrangian Eulerian coordinates, see [4]. The Lagrange coordinates are used for solution of the structure motion.

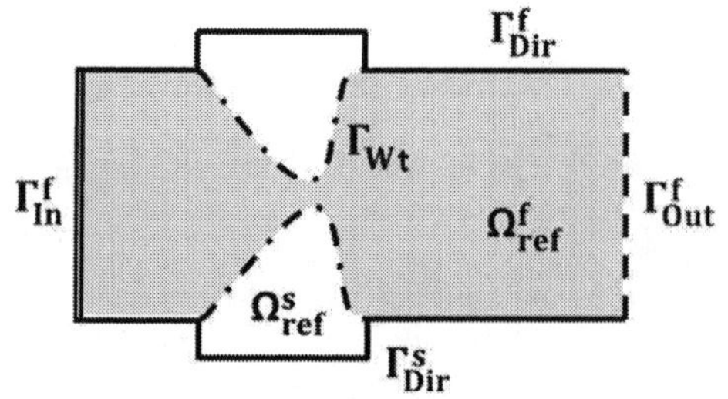

Fig. 1. Schema of vocal folds model with boundaries of the glottal channel.

Flow model. Just for this moment we consider firmly given fluid domain Ω^f independent on time. Then the motion of the viscous incompressible Newtonian fluid in Eulerian coordinates is described by the Navier-Stokes equations

$$\frac{\partial \mathbf{v}}{\partial t} + (\mathbf{v} \cdot \nabla)\mathbf{v} - \nu^f \Delta \mathbf{v} + \nabla p = \mathbf{g}^f \quad \text{in } \Omega^f, \qquad \operatorname{div} \mathbf{v} = 0 \quad \text{in } \Omega^f, \tag{1}$$

where $\mathbf{v}$ denotes the fluid velocity and p is the kinematic pressure, ν^f is the kinematic viscosity of the fluid and $\mathbf{g}^f$ is the force vector.

[a] e-mail: jan.valasek1@fs.cvut.cz

ALE method. During the interaction the vocal fold changes it's shape, thus the fluid domain is also deformed and the ALE method is suitable approach to resolve it. It is the generalization of the Eulerian and the Lagrangian description. The basic assumption of this method is the existence of diffeomorphism $A_t : \overline{\Omega}^f_{ref} \longmapsto \overline{\Omega}^f_t$, e.g. $X \in \overline{\Omega}^f_{ref} \longmapsto x = A_t(X) \in \overline{\Omega}^f_t$. These conditions ensure that mapping A_t maps deformed fluid domain in time t back to reference shape. It follows thereof the similarity with Lagrangian method, but on contrary ALE method does not describe the real motion of the particles.

In addition mapping A_t must meet the conditions

$$\frac{\partial A_t}{\partial t} \in C(\overline{\Omega}^f_{ref}), \quad A_t(\partial \Omega^f_{ref}) = \partial \Omega^f_t, \quad t \in (0, \mathrm{T}). \quad (2)$$

This method allows us to define velocity of deformation of the fluid domain $\mathbf{w}_D$ as

$$\mathbf{w}_D(x,t) = \hat{\mathbf{w}}_D(A_t^{-1}(x), t), \quad t \in (0,\mathrm{T}),\ x \in \Omega^f_t, \quad (3)$$

where $\hat{\mathbf{w}}_D$ is quantity defined on Ω^f_{ref} as $\hat{\mathbf{w}}_D(X,t) = \frac{\partial}{\partial t} A_t(X)$, $t \in (0,\mathrm{T})$, $X \in \Omega^f_{ref}$. Afterwards we can establish so called ALE derivative

$$\frac{D^A}{Dt} f(x,t) = \frac{\partial f}{\partial t}(A_t(x), t) = \frac{\partial f}{\partial t}(x,t) + \mathbf{w}_D(x,t) \cdot \nabla f(x,t), \quad (4)$$

where $\mathbf{w}_D(x,t)$ is given by (3). For more information see e.g. [4].

Then the substitution of the ALE derivative into Eg. (1) leads to Navier-Stokes equations in ALE form, see e.g. [5]

$$\frac{D^A \mathbf{v}}{Dt} + ((\mathbf{v} - \mathbf{w}_D) \cdot \nabla)\mathbf{v} - \nu^f \Delta \mathbf{v} + \nabla p = \mathbf{g}^f \quad \text{in } \Omega^f_t,$$
$$\operatorname{div} \mathbf{v} = 0 \quad \text{in } \Omega^f_t. \quad (5)$$

The formulation of the fluid problem is completed by the initial conditions

$$\mathbf{v}(x,0) = \mathbf{v}_0(x) \qquad \text{for } x \in \Omega^f_{ref}, \quad (6)$$

and boundary conditions, specially for solid walls and the interface

$$\mathbf{v}(x,t) = \mathbf{0} \qquad \text{for } x \in \Gamma^f_{\mathrm{Dir}},\ t \in (0,\mathrm{T}), \quad (7)$$
$$\mathbf{v}(x,t) = \mathbf{w}_D(x,t) \qquad \text{for } x \in \Gamma_{\mathrm{W}_t},\ t \in (0,\mathrm{T}), \quad (8)$$

for the inlet we consider two possibilities:

$$\text{a)} \quad \mathbf{v}(x,t) = \mathbf{v}_{\mathrm{Dir}}(x,t),$$
$$\text{or} \quad (9)$$
$$\text{b)} \quad (p(x,t) - p_{ref})\mathbf{n}^f - \nu^f \frac{\partial \mathbf{v}}{\partial \mathbf{n}^f}(x,t) = \mathbf{0},$$

where $x \in \Gamma^f_{\mathrm{In}}$, $t \in (0,\mathrm{T})$ and $\mathbf{n}^f$ denotes outer unit normal to $\partial\Omega^f_t$.

And finally boundary condition at the outlet is

$$\text{a)} \quad (p(x,t) - p_{ref})\mathbf{n}^f - \nu^f \frac{\partial \mathbf{v}}{\partial \mathbf{n}^f}(x,t) = \mathbf{0},$$
$$\text{or} \quad (10)$$
$$\text{b)} \quad (p(x,t) - p_{ref})\mathbf{n}^f - \nu^f \frac{\partial \mathbf{v}}{\partial \mathbf{n}^f}(x,t) + \frac{1}{2}(\mathbf{v} \cdot \mathbf{n}^f)^- \, \mathbf{v} = \mathbf{0},$$

where $x \in \Gamma^f_{\mathrm{Out}}$, $t \in (0,\mathrm{T})$ and p_{ref} is a reference pressure. Only one version of each condition by Eqs. (9) and (10) will be chosen for simulations.

Elastic structure. Vocal folds are modelled as an 2D compliant structure with the assumption of small displacements. It's deformation is given by, see e.g. [6]

$$\rho^s \frac{\partial^2 \mathbf{u}}{\partial t^2} - \frac{\partial \tau^s_{ij}(\mathbf{u})}{\partial X_j} = \mathbf{f}^s \quad \text{in } \Omega^s \times (0,\mathrm{T}), \quad (11)$$

where $\mathbf{u} = (u_1, u_2)$ is the displacement vector, ρ^s is the structure density, the tensor τ^s_{ij} is the Cauchy stress tensor, the vector $\mathbf{f}^s$ is the volume density of an acting force and $X = (X_1, X_2)$ are the reference coordinates. Using the generalized Hook law for an isotropic body, the stress tensor τ^s_{ij} can be expressed as

$$\tau^s_{ij} = \lambda^s(\operatorname{div} \mathbf{u})\mathbb{I} + 2\mu^s e^s_{ij}, \quad (12)$$

where $\mathbb{I}$ is the unit matrix and e^s_{jk} is the strain tensor for small displacements

$$e^s_{jk} = \frac{1}{2}\left(\frac{\partial u_j}{\partial X_k} + \frac{\partial u_k}{\partial X_j}\right) \quad (13)$$

and λ^s, μ^s are Lame coefficients given by the Young modulus of elasticity E^s and the Poisson's ratio σ^s, see [6].

Eq. (11) is supplied with the following initial and boundary conditions

$$\text{a)} \quad \mathbf{u}(X,0) = \mathbf{u}_0(X) \qquad \text{for } X \in \Omega^s,$$
$$\text{b)} \quad \frac{\partial \mathbf{u}}{\partial t}(X,0) = \mathbf{u}_1(X) \qquad \text{for } X \in \Omega^s, \quad (14)$$
$$\text{c)} \quad \mathbf{u}(X,t) = \mathbf{u}_{\mathrm{Dir}}(X,t) \qquad \text{for } X \in \Gamma^s_{\mathrm{Dir}},\ t \in (0,\mathrm{T}),$$
$$\text{d)} \quad \tau^s_{ij}(X,t)\, n^s_j(X) = q^s_i(X,t), \qquad \text{for } X \in \Gamma_{\mathrm{W_{ref}}},\ t \in (0,\mathrm{T}),$$

where the boundaries $\Gamma_{\mathrm{W}_t}, \Gamma^s_{\mathrm{Dir}}$ are shown in figure 1, $\mathbf{n}^s(X) = (n^s_1, n^s_2)$ is the unit outer normal to $\partial\Omega^s_{ref}$ and $\mathbf{q}^s(X,t) = (q^s_1, q^s_2)$ is the aerodynamic force vector described later.

Coupling conditions. The fluid-structure interaction is a coupled problem. It means that solutions in the fluid and the structure domain are connected via boundary conditions. Moreover, the location of the interface Γ_{W_t} at time t is not known and it depends on establishing force equilibrium between aerodynamic and elastic forces, which are dependent on the position and the velocity of the interface.

The shape of the interface Γ_{W_t} in time t is the result of force equilibrium between fluid and structure. It's position is given by the deformation $\mathbf{u}$, also

$$\Gamma_{\mathrm{W}_t} = \left\{x \in \mathbb{R}^2 \,|\, x = X + \mathbf{u}(X,t),\ X \in \Gamma_{\mathrm{W_{ref}}}\right\}. \quad (15)$$

So called dynamic condition expresses the aerodynamic force $\mathbf{q}^s$ acting on the structure interface

$$q^s_i(X,t) = -\sum_{j=1}^{2} \sigma^f_{ij}(x)\, n^f_j(x), \quad x = X + \mathbf{u}(X,t),\ X \in \Gamma_{\mathrm{W_{ref}}}, \quad (16)$$

where $\sigma^f_{ij} = -\rho^f p \mathbb{I} + \rho^f \nu^f (\frac{\partial v_i}{\partial x_j} + \frac{\partial v_j}{\partial x_i})$ for $i, j \in \{1, 2\}$ are components of the fluid stress tensor and $\mathbf{n}^f(x) = (n^f_1, n^f_2)$ denotes the components of the unit normal (here to the interface Γ_{W_t}) oriented out of Ω^f_t.

On the other hand the kinematic condition given by Eq. (8) enforces the correspondence between fluid motion and elastic body deformation on boundary Γ_{W_t}.

3 Numerical approximation

The presented mathematical model represented by the partial differential equations (11), (5) is discretized in space by the finite element method and in time by the Newmark and the backward differentiation formula of second order (BDF2), respectively. The time interval [0, T] is divided into the equidistant partition with a constant time steps Δt.

Elastic structure. The solution $\mathbf{u}$ is sought in the space $\mathbf{V} = V \times V$, where $V = \{f \in W^{1,2}(\Omega^s) | f = 0 \text{ on } \Gamma^s_{\text{Dir}}\}$, and $W^{k,p}(\Omega)$ denotes the Sobolev's space. The equation (11) can be reformulated in the weak sense

$$(\rho^s \frac{\partial^2 \mathbf{u}}{\partial t^2}, \boldsymbol{\Phi})_{\Omega^s} + (\lambda^s (\text{div}\, \mathbf{u}) \mathbb{I} + 2\mu^s e^s_{ij}(\mathbf{u}), e^s_{ij}(\boldsymbol{\Phi}))_{\Omega^s} = = (\mathbf{f}^s, \boldsymbol{\Phi})_{\Omega^s} + (\mathbf{q}^s, \boldsymbol{\Phi})_{L^2(\Gamma^s_{\text{Neu}})}, \tag{17}$$

where $(\cdot,\cdot)_\Omega$ denotes scalar product of space $L^2(\Omega)$ or $[L^2(\Omega)]^2$. The solution $\mathbf{u}$ of Eq. (17) must satisfy that equation for all $\boldsymbol{\Phi} \in \mathbf{V}$. For further details see [7].

Furthermore after approximation of the space $\mathbf{V}$ by finite dimensional space $\mathbf{V}_h \subset \mathbf{V}$ the discrete solution $\mathbf{u}_h(x, t)$ can be written as a linear combination of basis functions $\boldsymbol{\Phi}_j(x) \in \mathbf{V}_h$ which yields

$$\mathbb{M}^T \ddot{\alpha} + \mathbb{K}^T \alpha = \mathbf{b}(t), \tag{18}$$

where $\alpha(t) = (\alpha_j)(t)$ denotes unknown time dependent coefficient of linear combination of basis function $\boldsymbol{\Phi}_j$ and the elements of the matrices $\mathbb{K} = (k_{ij}), \mathbb{M} = (m_{ij})$ and vector $\mathbf{b}(t) = (b_i)(t)$ are given by

$$k_{ij} = (\lambda^s (\text{div}\, \boldsymbol{\Phi}_j) \mathbb{I} + 2\mu^s e^s_{ij}(\boldsymbol{\Phi}_j), e^s_{ij}(\boldsymbol{\Phi}_i))_{\Omega^s}, \tag{19}$$
$$m_{ij} = (\rho^s \boldsymbol{\Phi}_j, \boldsymbol{\Phi}_i)_{\Omega^s}, \quad b_i(t) = (\mathbf{f}^s, \boldsymbol{\Phi}_i)_{\Omega^s} + (\mathbf{q}^s, \boldsymbol{\Phi}_i)_{L^2(\Gamma^s_{\text{Neu}})}.$$

The Lagrange finite elements of the first order are used based on a triangulation, which give the first order of accuracy in space. The Newmark's method is used for numerical solution of ordinary differential equations (19), which results for a proper parameter choice to the second order accurancy in time, for application details see [7].

ALE method. In every time step the ALE mapping must be constructed. For that purpose the pseudoelastic solver is used. It is elastic body deformation solver with unphysical Lame parameters, which ensures better distribution of mesh deformation in the vicinity of very deformed part of the mesh, e.g. around the interface. See [8].

The discrete version of mesh deformation is dealt with BDF2.

Flow model. Firstly we discretize the equations (5) in time and then in space. For time discretization we used BDF2 scheme suitable for ordinary differential equations of first order. So the ALE derivative is approximated by

$$\frac{D^A \mathbf{v}}{Dt} \approx \frac{3\mathbf{v}^{n+1} - 4\mathbf{v}^n + \mathbf{v}^{n-1}}{2\Delta t}, \tag{20}$$

where for the fixed time instant t_{n+1} we denote $\overline{\mathbf{v}}^i(x_{n+1}) = \mathbf{v}^i(x_i)$ for $x_i = A_{t_i}(A^{-1}_{t_{n+1}}(x_{n+1}))$, $i = \{n-1, n\}$, $x_{n+1} \in \Omega^f_{n+1}$.
In next we omit time index and we shall write $\Omega^f := \Omega^f_{t_{n+1}}$. The application of the approximation in Eq. (5) leads to the scheme

$$\frac{3\mathbf{v}^{n+1} - 4\overline{\mathbf{v}}^n + \overline{\mathbf{v}}^{n-1}}{2\Delta t} + ((\mathbf{v}^{n+1} - \mathbf{w}^{n+1}_D) \cdot \nabla)\mathbf{v}^{n+1} - -\nu^f \Delta \mathbf{v}^{n+1} + \nabla p^{n+1} = \mathbf{g}^{f,n+1}, \tag{21}$$
$$\nabla \cdot \mathbf{v}^{n+1} = 0.$$

Then we continue with the weak formulation, where the velocity solution $\mathbf{v}$ in time t_{n+1} is sought in the functional space $\mathbf{W}^{1,2}(\Omega^f)$ and $q \in M = L^2(\Omega^f)$. Further, the space of the test function $\mathbf{X} = X \times X$ depends on choice of boundary condition. For Eq. (9 a)) the space is defined by

$$X = \{f \in W^{1,2}(\Omega^f) | f = 0 \text{ on } \Gamma^f_{\text{Dir}} \cup \Gamma^f_{\text{In}} \cup \Gamma^f_{\text{W}_{t_{n+1}}}\} \subset W^{1,2}(\Omega^f).$$

If alternative condition Eq. (9 b)) is chosen then test functions can be nonzero on Γ^f_{In}. In the next it is written for better arrangement $\mathbf{v}^{n+1}$ as $\mathbf{v}$. The weak formulation in space is acquired by multiplication of the first equation (21) by $\varphi \in \mathbf{X}$ and integration over the whole domain Ω^f, that results in the following equations

$$\left(\frac{3\mathbf{v} - 4\overline{\mathbf{v}}^n + \overline{\mathbf{v}}^{n-1}}{2\Delta t}, \varphi\right)_{\Omega^f} + (((\mathbf{v} - \mathbf{w}_D) \cdot \nabla)\mathbf{v}, \varphi)_{\Omega^f} - -\nu^f (\Delta \mathbf{v}, \varphi)_{\Omega^f} + (\nabla p, \varphi)_{\Omega^f} = (\mathbf{g}^f, \varphi)_{\Omega^f}, \tag{22}$$

With help of Green's theorem third and fourth term can be modified

$$\nu^f (\Delta \mathbf{v}, \varphi)_{\Omega^f} = -\nu^f (\nabla \mathbf{v}, \nabla \varphi)_{\Omega^f} + \nu^f \left(\frac{\partial \mathbf{v}}{\partial \mathbf{n}^f}, \varphi\right)_{L^2(\partial\Omega^f)}, \tag{23}$$
$$(\nabla p, \varphi)_{\Omega^f} = -(p, \text{div}\, \varphi)_{\Omega^f} + (p, \varphi \cdot \mathbf{n}^f)_{L^2(\partial\Omega^f)}.$$

The first terms of right side of Eq. (23) will be used in weak formulation and the sum of second ones form the boundary condition

$$(p, \varphi \cdot \mathbf{n}^f)_{L^2(\partial\Omega^f)} - \nu^f \left(\frac{\partial \mathbf{v}}{\partial \mathbf{n}^f}, \varphi\right)_{L^2(\partial\Omega^f)} = (p_{ref}, \varphi \cdot \mathbf{n}^f)_{L^2(\partial\Omega^f)}, \tag{24}$$

where the definition of test space and the knowledge that pressure is exact up to arbitrary constant was used. This condition can be prescribed on Γ^f_{Out} or also on Γ^f_{In} if the simulation driven by pressure gradient is requested.

The weak formulation then reads

$$\left(\frac{3\mathbf{v} - 4\overline{\mathbf{v}}^n + \overline{\mathbf{v}}^{n-1}}{2\Delta t}, \varphi\right)_{\Omega^f} + (((\mathbf{v} - \mathbf{w}_D) \cdot \nabla)\mathbf{v}, \varphi)_{\Omega^f} + +\nu^f (\nabla \mathbf{v}, \nabla \varphi)_{\Omega^f} - (p, \text{div}\, \varphi)_{\Omega^f} = (\mathbf{g}^f, \varphi)_{\Omega^f} - (p_{ref}, \varphi \cdot \mathbf{n})_{L^2(\Gamma^f_{\text{Out}})},$$
$$(q, \text{div}\, \mathbf{v})_{\Omega^f} = 0. \tag{25}$$

For derivation of condition (10 b)) we divide term $((\mathbf{v} - \mathbf{w}_D) \cdot \nabla)\mathbf{v}, \varphi)_{\Omega^f}$ in two halves. Then the Green's theorem gives

$$\frac{1}{2}(((\mathbf{v} - \mathbf{w}_D) \cdot \nabla)\mathbf{v}, \varphi)_{\Omega^f} = \frac{1}{2}(((\mathbf{v} - \mathbf{w}_D) \cdot \mathbf{n}^f)\mathbf{v}, \varphi)_{\partial\Omega^f} - -\frac{1}{2}(\mathbf{v}, \nabla \cdot ((\mathbf{v} - \mathbf{w}_D)\varphi))_{\Omega^f} = \frac{1}{2}\left(((\mathbf{v} \cdot \mathbf{n})\mathbf{v}, \varphi)_{\Gamma^f_{\text{Out}}}\right)^+ - \tag{26}$$
$$-\frac{1}{2}\left(((\mathbf{v} \cdot \mathbf{n})\mathbf{v}, \varphi)_{\Gamma^f_{\text{Out}}}\right)^- - \frac{1}{2}(((\mathbf{v} - \mathbf{w}_D) \cdot \nabla)\varphi, \mathbf{v})_{\Omega^f},$$

where the boundary term was split to positive and negative part: $(\cdot)^+ = \max(\cdot, 0)$, $(\cdot)^- = -\min(\cdot, 0)$, together with $\mathbf{w}_D = \mathbf{0}$ on $\partial\Omega^f \setminus \Gamma_{W_t}$ and $\boldsymbol{\varphi} = \mathbf{0}$ on Γ_{W_t}. This explains an additional term in Eq. (10 b)) and changes Eq. (25)

$$\begin{aligned}
&\left(\frac{3\mathbf{v} - 4\overline{\mathbf{v}}^n + \overline{\mathbf{v}}^{n-1}}{2\Delta t}, \boldsymbol{\varphi}\right)_{\Omega^f} + \frac{1}{2}(((\mathbf{v} - \mathbf{w}_D) \cdot \nabla)\mathbf{v}, \boldsymbol{\varphi})_{\Omega^f} - \\
&-\frac{1}{2}(((\mathbf{v} - \mathbf{w}_D) \cdot \nabla)\boldsymbol{\varphi}, \mathbf{v})_{\Omega^f} + \frac{1}{2}\left(((\mathbf{v} \cdot \mathbf{n})\mathbf{v}, \boldsymbol{\varphi})_{\Gamma^f_{\mathrm{Out}}}\right)^+ + \\
&+\nu^f(\nabla\mathbf{v}, \nabla\boldsymbol{\varphi})_{\Omega^f} - (p, \operatorname{div}\boldsymbol{\varphi})_{\Omega^f} = (\mathbf{g}^f, \boldsymbol{\varphi})_{\Omega^f} - (p_{ref}, \boldsymbol{\varphi} \cdot \mathbf{n})_{L^2(\Gamma^f_{\mathrm{Out}})}, \\
&\qquad (q, \operatorname{div}\mathbf{v})_{\Omega^f} = 0. \qquad (27)
\end{aligned}$$

This formulation suppresses backward intake of flow on the outlet part of the boundary and the term $\frac{1}{2}\left(((\mathbf{v} \cdot \mathbf{n})\mathbf{v}, \boldsymbol{\varphi})_{\Gamma^f_{\mathrm{Out}}}\right)^+$ increases stability of final scheme. For details see [9].

The FEM then approximates spaces $\mathbf{X}$ and M by the finite dimension spaces $\mathbf{X}_h$ and M_h, so the solution $\mathbf{v} \approx \mathbf{v}_h$ can be expressed again as linear combination of basis functions leading from Eq. (25) to the system

$$\begin{pmatrix} \mathbb{A}(\mathbf{v}_h^*) & \mathbb{B} \\ \mathbb{B}^T & 0 \end{pmatrix} \begin{pmatrix} \boldsymbol{\beta} \\ \boldsymbol{\gamma} \end{pmatrix} = \begin{pmatrix} \mathbf{g} \\ \mathbf{0} \end{pmatrix}, \qquad (28)$$

where $\boldsymbol{\beta}, \boldsymbol{\gamma}$ are vectors of linear combination coefficients and $\mathbb{A}(\mathbf{v}_h) = \frac{1}{\Delta t}\mathbb{M} + \mathbb{C}(\mathbf{v}_h^*) + \mathbb{D}$. The elements of the matrices $\mathbb{M} = (m_{ij}), \mathbb{C} = (c_{ij}), \mathbb{D} = (d_{ij})$ are given by

$$\begin{aligned}
&m_{ij} = \frac{3}{2}(\boldsymbol{\varphi}_j, \boldsymbol{\varphi}_i)_{\Omega^f}, \qquad c_{ij} = ((\mathbf{v}_h^* - \mathbf{w}_D) \cdot \nabla)\boldsymbol{\varphi}_j, \boldsymbol{\varphi}_i)_{\Omega^f}, \\
&d_{ij} = \nu^f(\nabla\boldsymbol{\varphi}_j, \nabla\boldsymbol{\varphi}_i)_{\Omega^f}, \quad b_{ij} = (-q_j, \operatorname{div}\boldsymbol{\varphi}_i)_{\Omega^f}, \qquad (29) \\
&g_i = (\mathbf{g}^f, \boldsymbol{\varphi}_i)_{\Omega^f} - (p_{ref}, \boldsymbol{\varphi}_i \cdot \mathbf{n})_{L^2(\Gamma^f_{\mathrm{Out}})} + (\frac{4\overline{\mathbf{u}}^n - \overline{\mathbf{u}}^{n-1}}{2\Delta t}, \boldsymbol{\varphi}_i)_{\Omega^f}.
\end{aligned}$$

The system of equations (28) is non-linear. For its solution the linearization $\mathbf{v}_h^* = \mathbf{v}^n$ is used. and the mathematical library UMFPACK is employed, see [10]. In this article P1-bubble/P1 elements are used, which according to [11] satisfy the well-known Babuška-Brezzi condition.

Coupled problem. The aerodynamic forces are evaluated from the velocity and pressure values at adjacent triangles with the help of numerical quadrature.The algorithm solving the coupled problem is implemented in fully implicit form, sometimes called strongly coupled. It means that new time step is solved only after previous time step converges, see [7].

4 Numerical results

The numerical results for fluid flow interacting with the vocal fold model M5 suggested by paper [12] are presented. The model M5 together with the triangulation is shown in figure 2. Here, only one half of the channel was used as the computational domain with symmetric boundary conditions prescribed at the top of the fluid domain ($y = 0.003$ m).

Flow in the fixed channel without interaction. The flow solver was tested for flow prescribed by pressure gra-

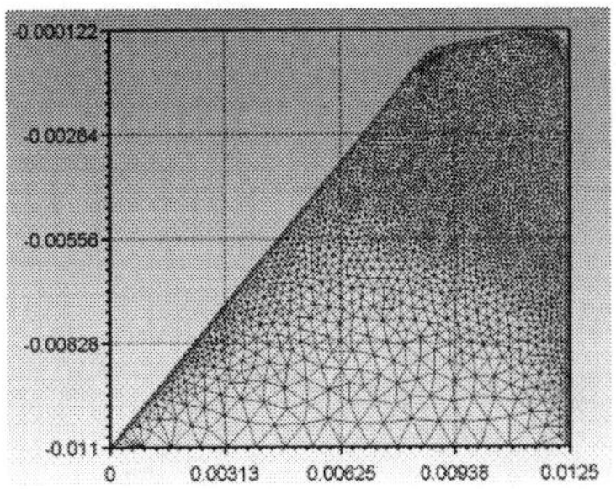

Fig. 2. The triangulation of the computational domain Ω_0^f of the vocal fold model M5 (dimensions in [m]). The triangulation of the computational domain Ω_0^s of the vocal fold model M5 (dimensions in [m]). Domain Ω_0^f is bounded by rectangle $(-0.02, 0.05) \times (-0.011, 0.003)$ m.

dient in the domain Ω^f_{ref} without any interaction. So condition (9 b)) together with (10 a)) was selected. The kinematic viscosity was set $\nu^f = 1{,}5 \cdot 10^{-5}\,\mathrm{m^2/s}$, the difference between inlet and outlet kinematic pressure was set to $2\,\mathrm{Pa\,m^3/kg}$ and $\rho^f = 1.7\,\mathrm{kg/m^3}$. Figure 3 shows the development of the flow velocity magnitude. After short time corresponding to the first figure the periodic process established where acceleration behind glottis can be observed repeatedly. The flow pattern is not too much turbulent so condition (10 a)) is sufficient. The kinetic energy of fluid

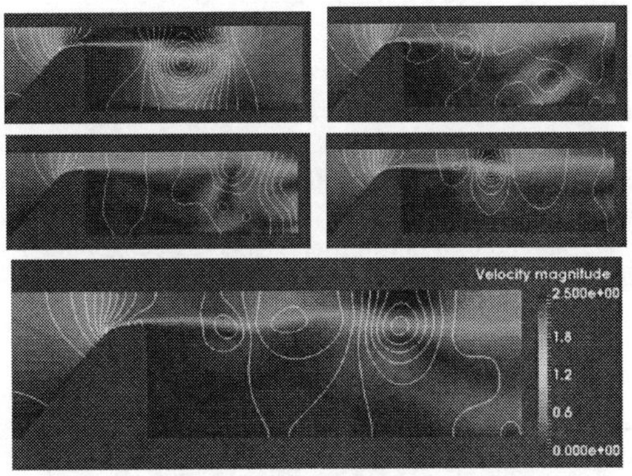

Fig. 3. The magnitude of velocity pattern around the fixed model at four different time instants $t = 0.05 + j \cdot 0.075$ s, $(j = 0, 1, 2, 3)$ and detail at 0.35 s are shown. Figures are ordered from the left to the right. At the same time 25 pressure contour lines in range from -2.6 to $2.0\,\mathrm{Pa\,m^3/kg}$ are depicted.

is given as

$$E^f_{kin} = \frac{1}{2}\int_{\Omega^f_t} \rho^f \mathbf{v}^2 \,\mathrm{d}x \qquad (30)$$

and it can be calculated with use of numerical quadrature. It's time development and Fourier analysis is shown in ff-figure 4 and 5.

It is interesting that even without interaction the motion of fluid demonstrates periodic behaviour, here with dominant frequency 42 Hz. It is the frequency of the boundary layer separation and the formation of vortices. This phenomena is more obvious if the simulation is driven by pressure gradient because then the inlet velocity is not constant but also oscillatory.

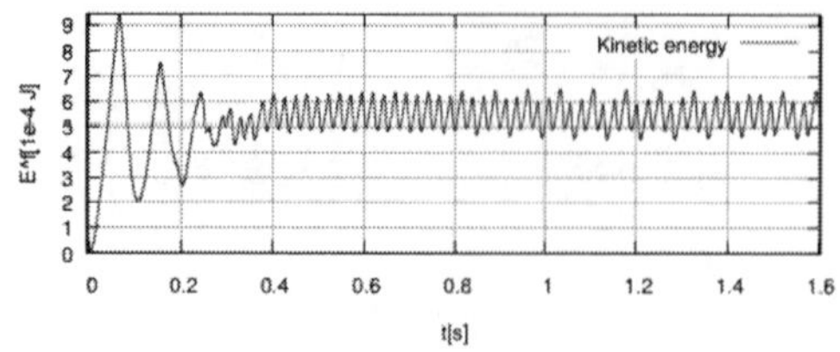

Fig. 4. The kinetic energy of fluid in 10^{-4} [J] in dependence on time in [s].

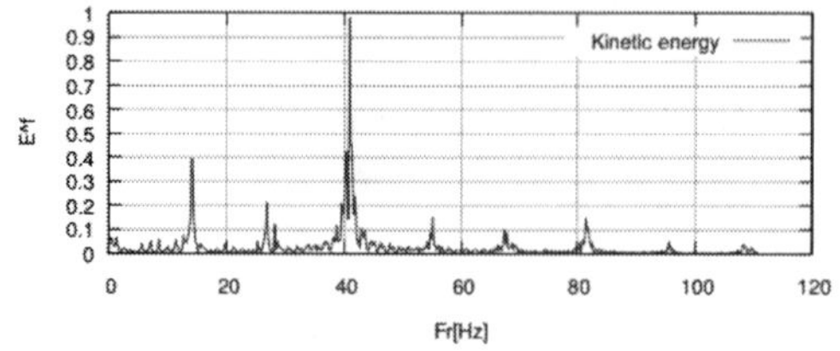

Fig. 5. Fourier transformation of periodic section of E^f_{kin}, bottom axis is in [Hz].

FSI test. We simulated interaction between fluid and elastic body in two tests. In both cases all initial conditions were prescribed equal zero. The elastic body constants were set: Young's modulus of elasticity $E^s = 100$ kPa in a thin layer along the interface $\Gamma_{\mathrm{W_{ref}}}$ (epithelium) and $E^s = 12$ kPa otherwise inside Ω^s_{ref} (muscle), $\sigma^s = 0.4$ everywhere in Ω^s_{ref}. In the first test the vibration of vocal fold was excited by prescribed parabolic inlet fluid velocity with maximum value $\mathbf{v}_{Dir} = 1.5\,\mathrm{m/s}$, i.e. by condition (9 a)) together with classic do-nothing condition (10 a)).

The interaction was enabled after time $t = 0.1$ s, when the velocity profile behind the vocal fold was developed. The flow excited a periodic vibration of vocal fold with small amplitudes, but in time 0.45 s the computation fell down due to backward inlet on Γ_{Out}, see figure 6. The usage of condition (10 b)) can prevent it.

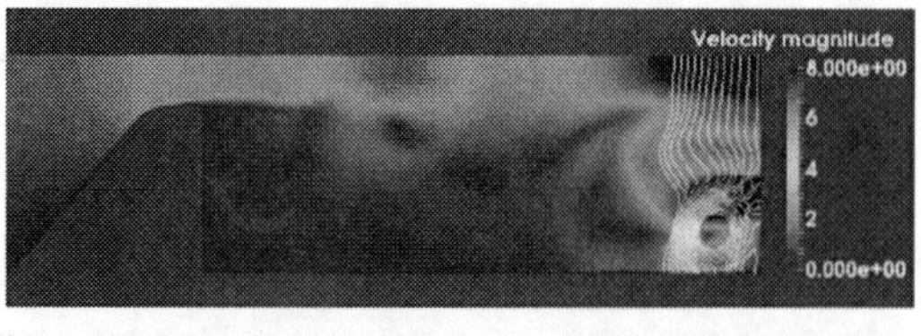

Fig. 6. Flow velocity magnitude in time 0.447 s a few steps before solver failure. Maximum of speed 15 m/s is out of color scale and 20 pressure isolines are spread between −20 and 30 Pa m^3/kg.

We also controlled the total energy E^s of vibrating structure which consists of kinetic and potential energy – $E^s = E^s_{kin} + E^s_{pot}$. These energies are approximated in discrete form as

$$E^s_{kin} = \frac{1}{2}\int_{\Omega^s} \rho^s \dot{\mathbf{u}}^2 \,\mathrm{d}x \approx \frac{1}{2}\dot{\mathbf{u}}^T \mathbb{M}\dot{\mathbf{u}}, \tag{31}$$

$$E^s_{pot} = \frac{1}{2}\int_{\Omega^s} (\lambda^s div\, \boldsymbol{\Phi}_j \mathbb{I} + 2\mu^s e^s_{kl}(\boldsymbol{\Phi}_j)) : e^s_{kl}(\boldsymbol{\Phi}_i)\,\mathrm{d}x \approx \frac{1}{2}\mathbf{u}^T \mathbb{K}\mathbf{u}.$$

Figure 7 shows the time development of E^s and figure 8 it's Fourier analysis. There are two peaks at the frequencies 54.54 Hz and 278.80 Hz. The first one corresponds to the first eigenfrequency of the vocal folds, see [7]. The second one is excited just in case of higher inlet flow velocities and it agrees with seventh eigenfrequency (285.6 Hz).

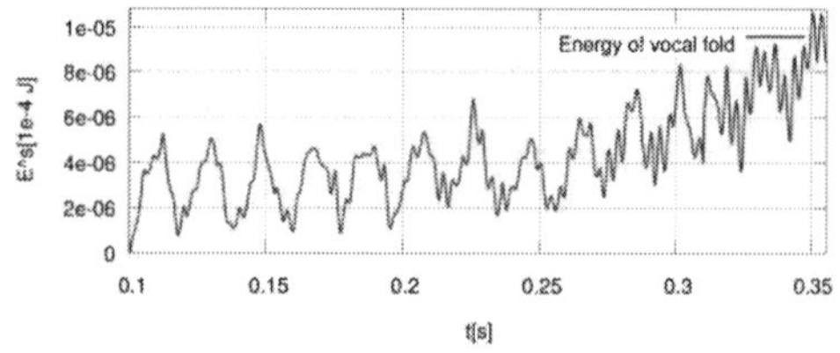

Fig. 7. Time development of E^s, time is in [s] and energy in [10^{-4}J].

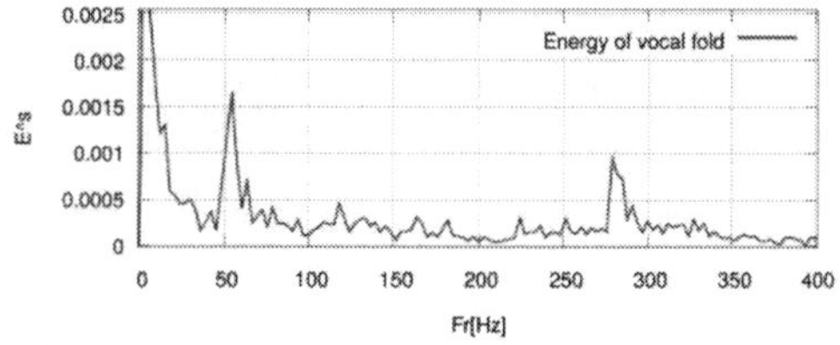

Fig. 8. Fourier analysis of E^s with two dominant frequencies 54.55 Hz and 278.80 Hz.

In the second test the pressure gradient was set to 4 Pa m^3/kg and the condition (9 b)) together with non-linear outlet condition (10 b)) were used. This setting resulted in stable periodic vibrations of the vocal fold despite higher outlet velocities. The detail of flow field in time 0.8 s is shown in figure 9.

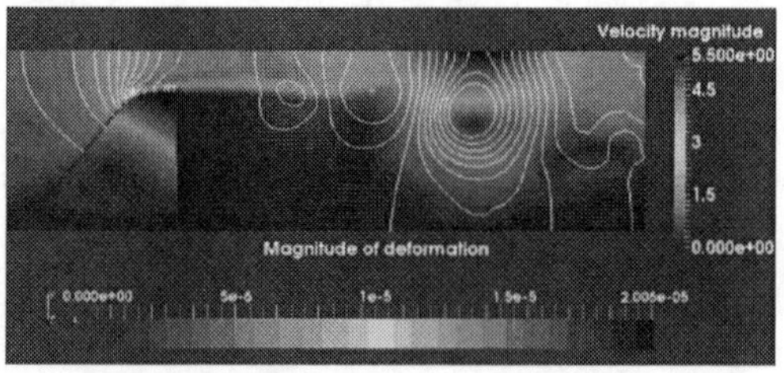

Fig. 9. Flow velocity magnitude and pressure isolines in time 0.808 s.

The total energy of elastic body E^s was monitored in figure 10 and it's dominant frequency was found equal to 52.0 Hz, see figure 11. It is almost the same as in the first test. The absence of the higher frequency can be caused by slower maximum velocity of fluid flow.

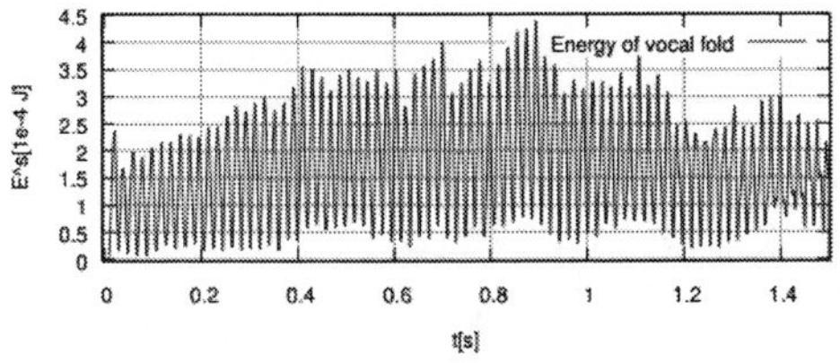

Fig. 10. Time development of total energy E^s.

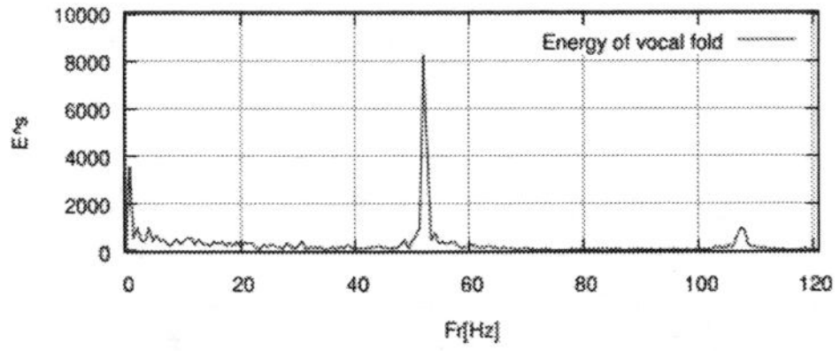

Fig. 11. Fourier analysis of total energy E^s with dominant frequency 52.0 Hz.

5 Conclusion

The paper presented the formulation of the FSI problem focused on the example of human fold vibration in airflow. The formulation is based on FEM and the ALE method and the developed numerical scheme was implemented in an own program. Special attention was devoted to the type of the boundary conditions used. When the fluid flow was driven by the pressure gradient the simulated interaction showed the vocal fold vibration with nearly the same frequencies as for the inlet Dirichlet boundary condition. In addition the considered non-linear version of the do-nothing condition gives more robust solution especially suitable for problems with higher flow velocities at the outlet of the channel.

The financial support for the present project was partly provided by the *Czech Science Foundation* under the *Grant No. 101/11/0207* and project *SGS* 13/174/OHK2/3T/12.

References

1. I. R. Titze and F. Alipour. *The myoelastic aerodynamic theory of phonation*. National Center for Voice and Speech, 2006.
2. A. Kosík, M. Feistauer, J. Horáček, and P. Sváček. Numerical simulation of interaction of human vocal folds and fluid flow. *Vibration Problems ICOVP 2011*, 139:765–771, 2011.
3. P. Svacek and J. Horacek. Numerical simulation of glottal flow in interaction with self oscillating vocal folds: comparison of finite element approximation with a simplified model. *Communications in Computational Physics*, 12:789–806, 2012.
4. N. Takashi and T. J. R. Hughes. An arbitrary Lagrangian-Eulerian finite element method for interaction of fluid and a rigid body. *Computer Methods in Applied Mechanics and Engineering*, 95:115–138, 1992.
5. M. Feistauer, P. Sváček, and J. Horáček. Numerical simulation of fluid-structure interaction problems with applications to flow in vocal folds. In T. Bodnár, G. P. Galdi, and S. Nečasová, editors, *Fluid-structure Interaction and Biomedical Applications*, pages 312–393. Birkhauser, 2014.
6. M. Brdička, L. Samek, and B. Sopko. *Continuum mechanics*. Academia, 2000.
7. J. Valášek, P. Sváček, and J. Horáček. On numerical approximation of fluid-structure interactions of air flow with a model of vocal folds. *Topical Problems of Fluid Mechanics 2015*, pages 245–254, 2015.
8. M. Hadrava, M. Feistauer, J. Horáček, and A. Kosík. Discontinuous Galerkin method for the problem of linear elasticity with applications to the fluid-structure interaction. *AIP Conference Proceedings*, 1558:2348–2351, 2013.
9. M. Braack and P. B. Mucha. Directional do-nothing condition for the Navier-Stokes equations. *Journal of Computational Mathematics*, 32:507–521, 2014.
10. T. A. Davis. *Direct methods for sparse linear systems*. SIAM, 2006.
11. V. Girault and P. A. Raviart. *Finite element methods for Navier-Stokes equations*. Springer-Verlag, 1986.
12. R. C. Scherer, D. Shinwari, K. J. De Witt, Ch. Zhang, B. R. Kucinschi, and A. A. Afjeh. Intraglottal pressure profiles for a symmetric and oblique glottis with a divergence angle of 10 degrees. *The Journal of the Acoustical Society of America*, 109:1616–1630, 2001.

EPJ Web of Conferences **114**, 02131 (2016)
DOI: 10.1051/epjconf/201611402131

Eigenfrequency of Hydraulic Systems of Loading Device

Martin Vašina[1,a], Lumír Hružík[1] and Adam Bureček[1]

[1]VŠB-Technical University of Ostrava, Faculty of Mechanical Engineering, Department of Hydromechanics and Hydraulic Equipment, 708 33 Ostrava, Czech Republic

Abstract. Eigenfrequency of hydraulic systems belongs to important dynamic quantities. If the excitation frequency of a given hydraulic system is equal to the system eigenfrequency, high-amplitude pressure and flow pulsations can arise. It has a negative influence on load of hydraulic elements, system tightness etc. For this reason it is necessary to eliminate the operation of the hydraulic system at its eigenfrequency. The paper deals with experimental determination of the system eigenfrequency in various operating modes of the investigated loading device.

1 Introduction

Hydraulic systems are applied in many areas at the present time. They are used as drives of heavy duty equipment (e.g. cranes, loaders and excavators) [1]. Furthermore hydraulic systems are characterized by relatively small dimensions in comparison with different drives. It is their big advantage in comparison to other types of drives.

There are different requirements for hydraulic systems, e.g. energy costingness, purchase costs, reliability, dynamic properties, safety, weight, dimensions and service life. It is necessary to consider the requirements during the design process of a given hydraulic system [2].

This paper is focused on dynamics of hydraulic systems. Dynamical properties of hydraulic systems are also characterized by the system eigenfrequency, which depends on many parameters. The eigenfrequency can be determined by means of experimental measurements, mathematical simulations or simplified mathematical formulas.

The objective of this work is to experimentally investigate the eigenfrequency of a hydraulic system of the WM 185 hydraulic manipulator. The eigenfrequency was obtained under different working conditions of the manipulator.

2 Eigenfrequency of hydraulic systems

The eigenfrequency of hydraulic systems belongs to significant dynamic quantities. If the excitation frequency of a given hydraulic system is consistent with its eigenfrequency, high-amplitude pressure and flow pulsations are generated in the system. These pulsations have a negative influence on the system operation (e.g. on system tightness and cyclic loading of hydraulic elements). It is possible to determine the eigenfrequency of hydraulic systems by means of mathematical simulations, mathematical formulas or experimental measurements.

Mathematical models of hydraulic systems are described by a system of algebraic and differential equations. These equations are subsequently solved by Laplace transformation [3] or by computer programs (e.g. using Matlab, Mathematica or Mathcad software).

Simplified mathematical formulas [4] are used to an approximate determination of the system eigenfrequency, e.g. in the design phase. For example, the eigenfrequency f_0 of a rotary hydraulic motor is given by the formula:

$$f_0 = \frac{1}{2 \cdot \pi} \cdot \sqrt{\frac{k_\varphi}{J_M}}, \tag{1}$$

where k_φ static stiffness of the rotary hydraulic motor, N m; J_M imoment of inertia of rotating masses, kg m^2. Similarly, the eigenfrequency f_0 of a linear hydraulic motor is given by the equation:

$$f_0 = \frac{1}{2 \cdot \pi} \cdot \sqrt{\frac{k_h}{m_{red}}}, \tag{2}$$

where k_h static stiffness of the linear hydraulic motor, N m^{-1}; m_{red} reduced mass of moving parts, kg. In the case of the linear hydraulic motor with a single piston rod and with minimum pressure (i.e. $p_2 \rightarrow 0$) at its output (see figure 1), it is possible to express the eigenfrequency f_0 as follows:

$$f_0 = \frac{S_1}{2 \cdot \pi} \cdot \sqrt{\frac{K}{m_{red} \cdot (S_1 \cdot x + V_1)}}, \tag{3}$$

[a] Corresponding author: martin.vasina@vsb.cz

where S_1 piston area on its input side with the input working pressure p_1, m^2; x piston extension ($x \in$ <0, h>), m; h piston stroke, m; V_1 liquid volume in an inlet transmission channel (i.e. between the linear hydraulic motor and the control valve CV), m^3. As shown in figure 1, Q_1 is input flow rate of pressure liquid into the motor, Q_2 is output flow rate of the pressure liquid from the motor and v is speed of movement of the motor.

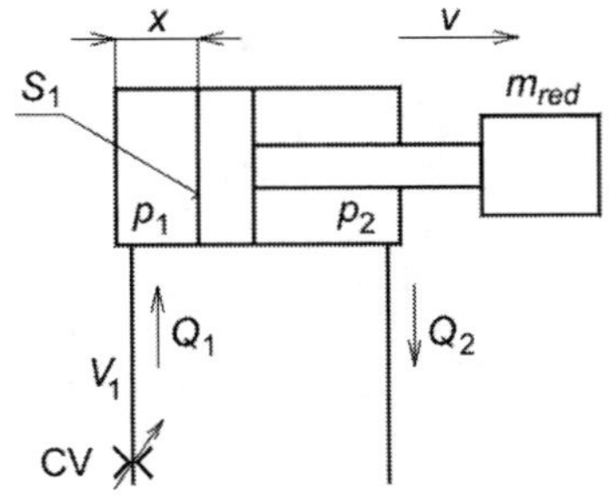

Figure 1. Schematic representation of linear hydraulic motor.

Experimental determination of the system eigenfrequency is realized by means of time-response characteristics, e.g. on basis of sudden changes of certain quantities. The time-response characteristics of a measured function $y(t)$ can be in general periodically damped or stochastic.

If the time dependence of the function $y(t)$ is periodically damped around the mean value y_m (see figure 2), the system eigenfrequency is subsequently determined by the ratio [5, 6]:

$$f_0 = \frac{1}{T}, \tag{4}$$

where T period of oscillation (see figure 2) of the function $y(t)$, s.

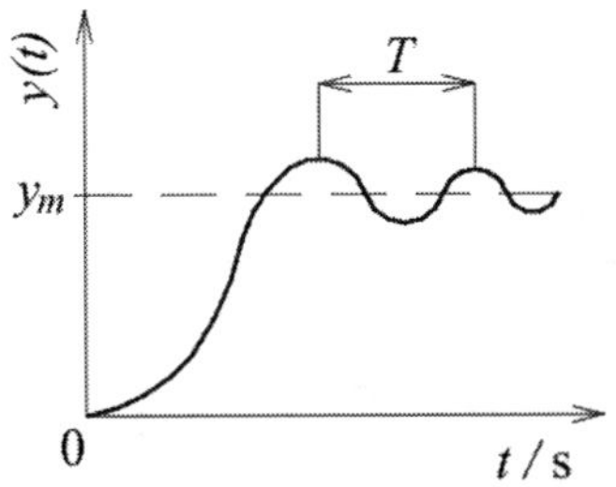

Figure 2. Example of time dependence of the periodically damped function $y(t)$.

In the case of the stochastic course (see figure 3) of the function $y(t)$, the system eigenfrequency is determined by means of the power spectral density [7]. Then the system eigenfrequency is obtained at the maximum value of the power spectral density [8].

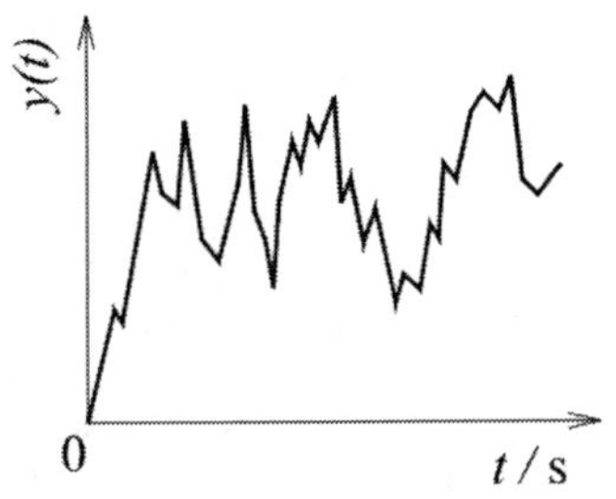

Figure 3. Example of stochastic course of the function $y(t)$.

3 Investigated hydraulic system

The system eigenfrequency was experimentally measured on the WM 185 hydraulic manipulator (see figure 4). The manipulator is used to loading and unloading of timber and other materials. The manually controlled WM 185 hydraulic manipulator is lifting equipment with the maximum lifting moment of 185 kN·m and the maximum radius of 7.99 m. The hydraulic manipulator consists of the following hydraulic motors [8]:

- Hydraulic motors of left and right stabilizing supports;
- Hydraulic motor for extension of supports;
- Turning hydraulic motor (for rotation of swivelling post);
- Hydraulic lifting motor (HLM);
- Two tilting hydraulic cylinders in order to achieve a pendulous motion of a telescopic boom;
- Hydraulic motor of the telescopic boom (HTM);
- Hydraulic motor of pliers (for opening or closing of the pliers) and
- Rotator (for turning the pliers).

Figure 4. View of the investigated WM 185 hydraulic manipulator.

The hydraulic motors of both stabilizing supports and the hydraulic motor for extension of supports are used to ensure lateral stability of the manipulator. The other hydraulic motors are used for material handling (i.e. for lifting, lowering, loading and transloading of different materials).

The investigated hydraulic system belongs to closed center load sensing systems and is characterized by higher energy savings. The hydraulic system also consists of a hydraulic pump that is a pressure energy source of the system. The working principle of the WM 185 hydraulic manipulator and the schematic diagram of the hydraulic system are described in detail in [8].

4 Experimental determination of system eigenfrequency

The eigenfrequency of the investigated WM 185 hydraulic manipulator was experimentally measured under different working conditions, which were obtained on the basis of transient changes in behaviour of the hydraulic lifting motor (i.e. at a sudden stop of the lifting motor HLM in a given piston position of the motor). The other hydraulic motors did not move. They were in given positions at the same time. Furthermore the hydraulic motor of the telescopic boom HTM was located at several piston positions of this boom and the manipulator was loaded with different reduced masses during the measurements. The experimental measurement of the system eigenfrequency was purposely chosen for the lifting motor HLM due to the highest reduced masses m_{red} of the lifting motor compared to the other hydraulic motors (see figure 4). As indicated in the Eq. (3), the reduced mass significantly reduces the system eigenfrequency f_0. For this reason the system eigenfrequency was determined under these working conditions of the hydraulic manipulator.

The time-response characteristics of the input pressure p_1 and the output pressure p_2 (see figure 1) of the lifting motor HLM were measured by means of the M5000 Hydrotechnik measuring equipment. The connection of this equipment on the WM 185 hydraulic manipulator is shown in figure 5.

Figure 5. Connection of the M5000 Hydrotechnik measuring equipment

The time-response characteristic (see figure 6) of the investigated hydraulic manipulator was firstly measured during a sudden stop of the hydraulic lifting motor HLM (see figure 4) in the middle of its stroke (i.e. in the motor position $x = h/2$). At the same time, the telescopic boom HTM with the piston position $x = 0$ was unloaded. It is evident that the input pressure p_1 is subsequently changed during the time-response characteristic. The time dependence of the input pressure is periodically damped around the mean value $p_1 \cong 64 \cdot 10^5$ Pa with the period of oscillation $T \cong 0.55$ s. Then it is possible to determine the system eigenfrequency from the Eq. (4):

$$f_0 = \frac{1}{T} = \frac{1}{0.55} = 1.82 \text{ Hz}. \tag{5}$$

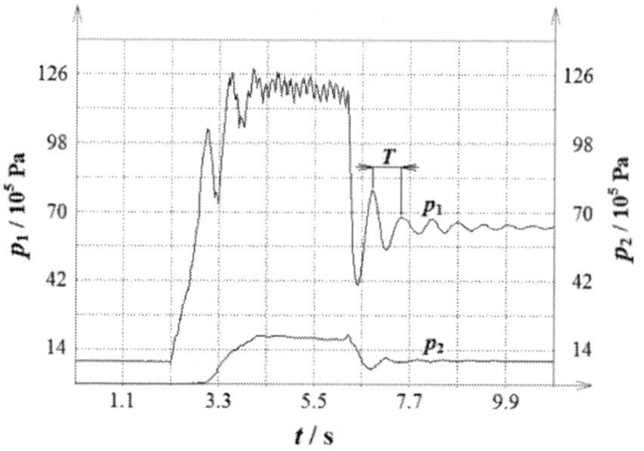

Figure 6. Time dependencies of the pressures p_1 and p_2 [8].

The system eigenfrequency was subsequently determined during a sudden stop of the hydraulic lifting motor HLM in the position $x = 2\,h/3$. In this case the telescopic boom HTM with the piston position $x = 0$ was loaded. It is visible (see figure 7) that the input pressure p_1 is periodically damped with the period of oscillation $T \cong 1.17$ s. For this reason the system eigenfrequency is determined as the reciprocal of the period:

$$f_0 = \frac{1}{T} = \frac{1}{1.17} = 0.85 \text{ Hz}. \tag{6}$$

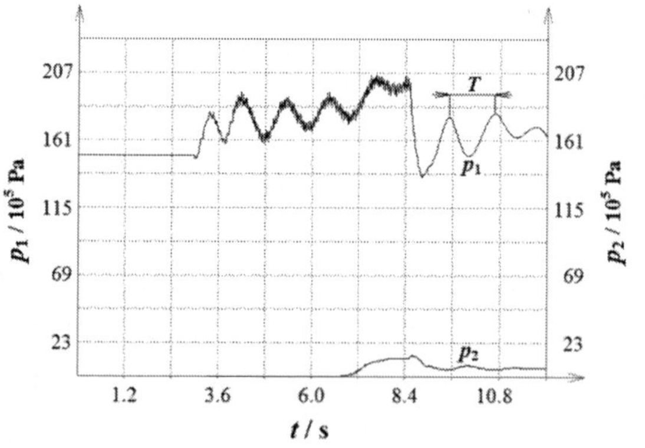

Figure 7. Time dependencies of the pressures p_1 and p_2 [8].

Figure 8 demonstrates the time dependencies of the pressures p_1 and p_2 during a sudden stop of the hydraulic lifting motor HLM in the position $x = 2\,h/3$. At the same time the telescopic boom HTM with the piston position $x = h$ was again loaded. It is obvious that the input pressure p_1 oscillates around the mean value

$p_1 \cong 210 \cdot 10^5$ Pa with the period of oscillation $T \cong 1.34$ s. Then, the eigenfrequency of the investigated hydraulic manipulator is determined as follows:

$$f_0 = \frac{1}{T} = \frac{1}{1.34} = 0.75\ \text{Hz}. \quad (7)$$

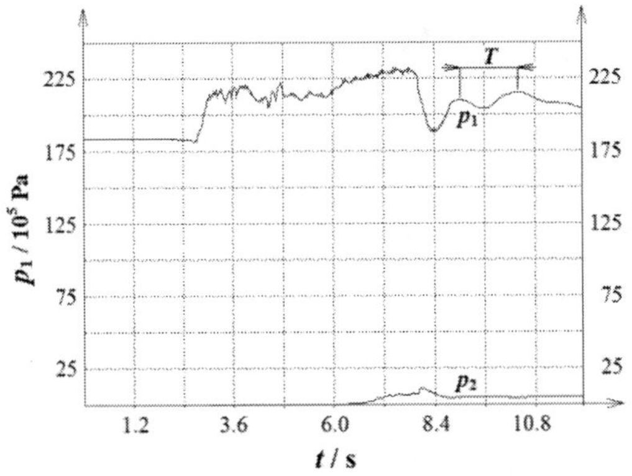

Figure 8. Time dependencies of the pressures p_1 and p_2 [8].

It is evident from the above-mentioned time-response characteristics (see figures 6–8) that the eigenfrequency of the investigated WM 185 hydraulic manipulator is very low under the given working conditions. Nevertheless, it is recommended to ensure a minimum value of the system eigenfrequency f_{0min} during operation of the hydraulic system. In the case of the investigated hydraulic manipulator, which is represented as a load compensation system [4], the minimum eigenfrequency $f_{0min} = 4$ Hz. It is evident that this frequency criterion is not fulfilled. This fact can lead to many complications during operation of the hydraulic system. It can be reflected in increased load of hydraulic elements, decreased system tightness and excessive noise and vibration. At worst, low values of the eigenfrequency can also lead to an accident of the system. It is also evident from the experimental measurements that the system eigenfrequency decreases with increasing piston extension x of the motors HLM and HTM and with increasing reduced mass of moving parts m_{red}. It is in accordance with the Eq. (3).

It can be concluded that low values of the system eigenfrequency were probably caused by system aeration. Air bubbles in liquids have a negative influence on the system stiffness and dynamic behaviour of the investigated hydraulic manipulator. For this reason it is necessary to properly deaerate the hydraulic system. It can be also caused by a relatively large reduced mass m_{red} of moving parts. Furthermore, it is possible to perform suitable precautions when designing the system. According to the Eq. (3), it is suitable to ensure the minimum liquid volume V_1 in the applied hydraulic system and to increase the piston area S_1 in the design phase.

5 Conclusions

The purpose of this paper was to experimentally determine the eigenfrequency of the WM 185 hydraulic manipulator that is used to loading and unloading of timber and other materials. The eigenfrequency was determined under different working conditions from time-response characteristics.

It was found in this study that the experimentally determined values of the system eigenfrequency were very low. It can lead to different failures during operation of the hydraulic manipulator. For this reason it is necessary to increase the system eigenfrequency through appropriate measures. It is suitable to properly deaerate the hydraulic system in order to increase its stiffness. Furthermore it is possible to reduce the reduced mass of moving parts and to perform different precautions when designing the system.

It is also desirable that the excitation frequency of a hydraulic pump in a given hydraulic system did not correspond to the system eigenfrequency under given working conditions. If the excitation frequency and the system eigenfrequency are consistent, it can have a negative influence on dynamic behaviour of the hydraulic system.

References

1. M. Ochiai, S. Ryu, 7th JFPS 2008, *Hybrid in construction machinery*, (Toyama, 2008)
2. P. Drexler, H. Faatz, F. Feith, H. Geis, J. Morlok, E. Wiesmann, *The Hydraulic Trainer Volume* 3 – Planning and Design of Hydraulic Power Systems (Mannesmann Rexroth GmbH, Lohr am Main, 1988)
3. A. Furman, S. P. Neumam, Advances in Water Resources **26**, 12 (2003)
4. H. Dörr, R. Ewald, J. Hutter, D. Kretz, F. Liedhegener, A. Schmitt, *The Hydraulic Trainer Volume* 2 – Proportional and Servo Valve Technology (Mannesmann Rexroth GmbH, Lohr am Main, 1986)
5. M. S. Chandrasekar, M. Pushpavanam, Electrochimica Acta **53**, 8 (2008)
6. M. D. Bootman, M. J. Berridge, P. Lipp Cell **91**, 3 (1997)
7. P. Laguna, G. B. Moody, R. G. Mark, IEEE Transactions on Biomedical Engineering **45**, 6 (1998)
8. M. Vašina, *Energy saving hydraulic systems of lifting and loading equipment mounted on lorries* (VŠB – Technical University of Ostrava, Czech Republic, 2000)

Acknowledgements

The work presented in this paper was supported by a grant SGS "Investigation of dynamics of fluid systems" SP2015/95.

EPJ Web of Conferences **114**, 02132 (2016)
DOI: 10.1051/epjconf/201611402132

Experimental study of effect of stenosis geometry on pressure loss for periodic flow

Ondřej Veselý[1,a], Ludmila Nováková[1] and Josef Adamec[1]

[1]*CTU in Prague, Faculty of Mechanical Engineering, Department of Fluid Dynamics and Thermodynamics, Technická 4, 16607, Prague 6, Czech Republic*

Abstract. A stenosis is a narrowing in a tubular organ. In medicine, vessel stenosis poses health risk for people. In the last work, experimental investigation of pressure loss coefficient for varying stenosis eccentricity and shape for steady flow were performed. In this work, experimental investigation of pressure loss for varying stenosis eccentricity and shape under periodic flow were performed. Four models of different geometry were studied, two models are axisymmetric stenoses and two models are eccentric stenoses. All models were stenosis of 75% area reduction. The periodic flow, generated by a controllable pump, has sinus shape in an inlet. The measuring range of medium Reynolds number was from 500 to 1500, range of ratio between an amplitude and medium flow rate was from 0.2 to 0.6 and range of frequency was from 0.2 to 1 Hz. The pressure loss for each conditions was quantified by mean value, amplitude and phase shift against flow rate.

1 Introduction

Stenosis is a narrowing of vessel. Stenosis significantly affect the blood flow, which has implication for the body system. [1] The characteristic feature of the stenoses is atherosclerosis [2].

The relationship between flow parameters and occurrence of the atherosclerosis has been studied by many studies [3], [4], [5]. The complexity of this issue, high demands on measurement accuracy and ethical limitations transfer research from the medical field to the field of technology. In laboratory for measuring flow parameters is used technically advanced method (for example PIV).

In fluid dynamics, the blood flow is an unsteady flow of the non-Newtonian fluid through an elastic tube. For certain experiments conditions the non-Newtonian characteristics can be neglected and the use of Newtonian fluid is sufficient [6], [7]. Many research is focused on models with a solid [8], [9] as well as elastic wall [10], [11]. This work is focused on model with solid wall.

The effect of unsteadiness is described in several experimental and numeric work. For example Siouffi et al. [12] compared different types of pulsatile flows with physiological one. This experimental work is focused on effect of stenosis geometry on pressure loss for unsteady flow, which have sinus shape in an inlet.

For the description of unsteady flow is used a dimensionless number to be called the Womersley number. It is an expression of the pulsatile flow frequency in relation to viscous effect. The Womersley number is defined by the relation:

$$\alpha = r\sqrt{\frac{\omega}{\nu}}, \tag{1}$$

where r is the diameter of unobstructed vessel and ν is the kinematic viscosity.

In this work four models were studied; two symmetric stenoses and two eccentric stenoses. All models have circular cross-section and they were stenosis with 75% area reduction. The measuring range of Womersley number was from 8 to 18.

From obtained data, the local pressure loss is evaluated from equation:

$$\Delta p_l = \Delta p_t - \Delta p_f, \tag{2}$$

where Δp_t is total pressure loss and Δp_f is major pressure loss. The total pressure loss is sum of major and local pressure loss. From obtained data amplitude, mean value and phase shift of local pressure loss was evaluated. Amplitude of local pressure loss is given by equation:

$$\Delta p_{a,l} = \frac{\Delta p_{max,l} - \Delta p_{min,l}}{2}, \tag{3}$$

where $\Delta p_{max,l}$ is maximum of the local pressure loss and $\Delta p_{min,l}$ is minimum of the local pressure loss. Mean value is given by equation.

$$\Delta p_{m,l} = \frac{\Delta p_{max,l} + \Delta p_{min,l}}{2}. \tag{4}$$

Phase shift of local pressure loss is given by equation:

[a] Corresponding author: O.Vesely@fs.cvut.cz

$$\Delta\varphi_l = \varphi_{max,fr} - \varphi_{max,l}, \qquad (5)$$

where $\varphi_{max,fr}$ is phase of maximum value of flow rate and $\varphi_{max,l}$ is phase of maximum value of local pressure loss.

The geometry of stenosis models is shown in Figure 1, 2, 3 and 4. All models have the diameter of unobstructed cross-section 14 mm, the diameter of obstructed cross-section 7 mm, the length of model 50 mm, the length of stenosis 5 mm.

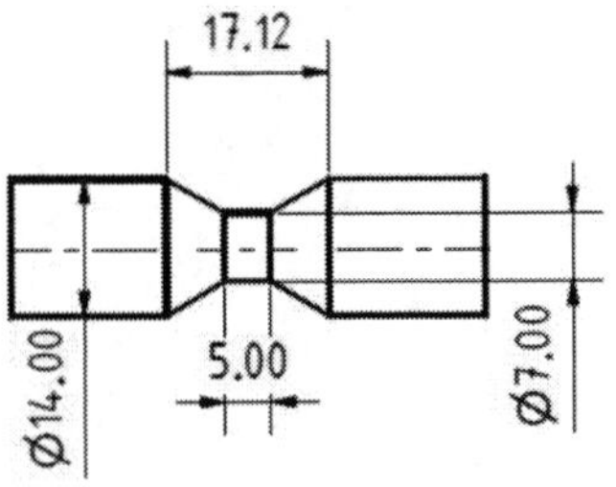

Figure 1. Symmetric stenosis number 1, narrowing angle 45°.

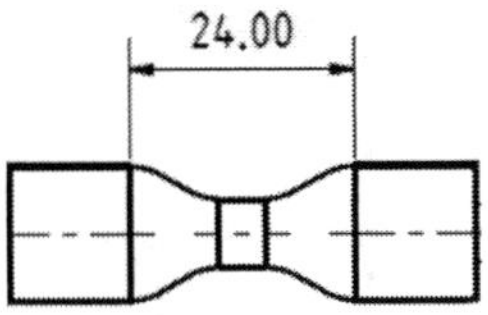

Figure 2. Symmetric stenosis number 2, smooth transition portion.

Figure 3. Eccentric stenosis number 1, narrowing angle 45°.

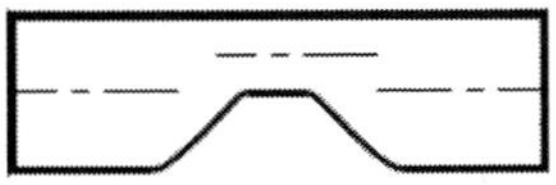

Figure 4. Eccentric stenosis number 2, smooth transition portion

The aim of this experimental study is determine, if pressure loss was affected more by shape of stenosis or by eccentricity.

2 Method

Schematic diagram of the experimental setup for measuring the pressure loss is shown in Figure 5.

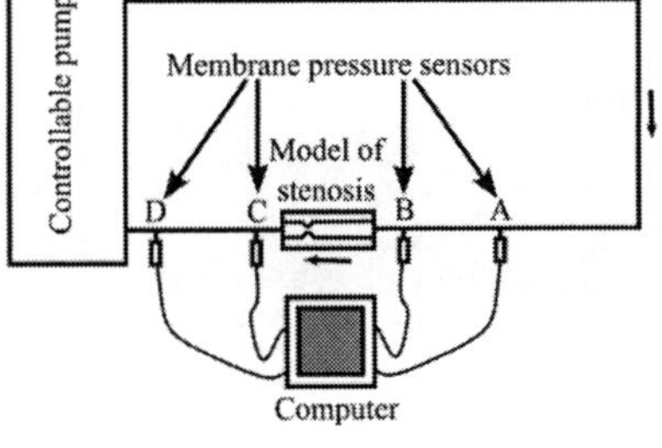

Figure 5. Experimental setup.

The unsteady flow was established in the model. The pressure loss was measured by four membrane pressure sensors. The distance between sensors is 70 cm. The sensors were calibrated before measurement using water column.

For each flow regime the pressure gradient between sensors was recorded. Pressure gradient between sensor A and sensor B (sensors are labelled in Figure 5) is major pressure loss in pipe with length 70 cm.

Between sensor B and C was measured total pressure loss, which is combination of major and local pressure loss. Sensor D is used to check the correct functioning of the sensors. The control is performed by comparison of data obtained between sensor A and B and between sensor C and D.

3 Results

Table 1 summarises the set of experimental flow regime. For each flow regime, measurement for five value of frequency was performed.

Table 1. Flow regimes

Number of regime	Mean Reynolds number (1)	Amplitude of Reynolds number (1)	Range of measured frequency (Hz)
1	500	100	0.2 – 1.0
2	500	200	
3	500	300	
4	1000	200	
5	1000	400	
6	1000	600	
7	1500	300	
8	1500	600	
9	1500	900	

Figures 6 – 8 plot results for symmetric stenosis number 1. The results show that the difference of maximum and minimum value of mean and amplitude local pressure loss increases with increasing Reynolds number. The value of phase shift increase with Reynolds number and maximum value has for Re = 1500.

Figures 9 – 11 plot results for symmetric stenosis number 2. This stenosis has smoother transition portion then symmetric stenosis number 1. The results show that the value of mean and amplitude of local pressure loss is approximately same with stenosis number 1.

Figures 12 – 14 plot results for eccentric stenosis number 1. Comparing with symmetric stenosis number 1, the eccentric stenosis has higher difference of maximum and minimum value of mean local and amplitude pressure loss. The value of phase shift increase with Reynolds number and maximum value has for Re = 1000 and 1500.

Figures 15 – 17 plot results for eccentric stenosis number 2. This stenosis has smoother transition portion then eccentric stenosis number 1. The results show that this stenosis has lower the difference of maximum and minimum value of mean and amplitude local pressure loss is lower than in the eccentric stenosis number 1. The phase shift for Re = 500 is lower than eccentric stenosis number 1, for higher Re the value is same.

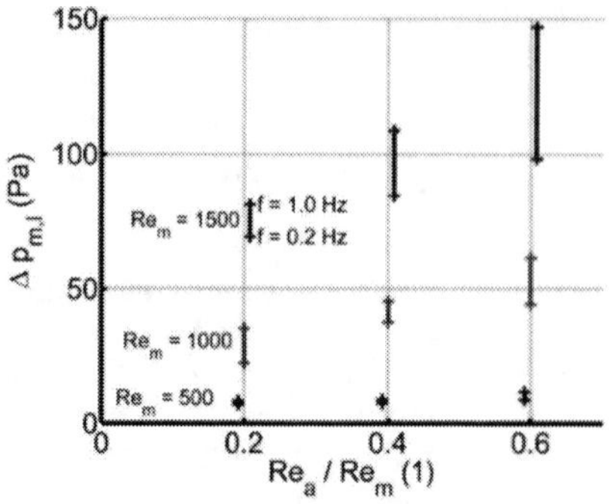

Figure 6. The mean value of local pressure loss for symmetric stenosis number 1.

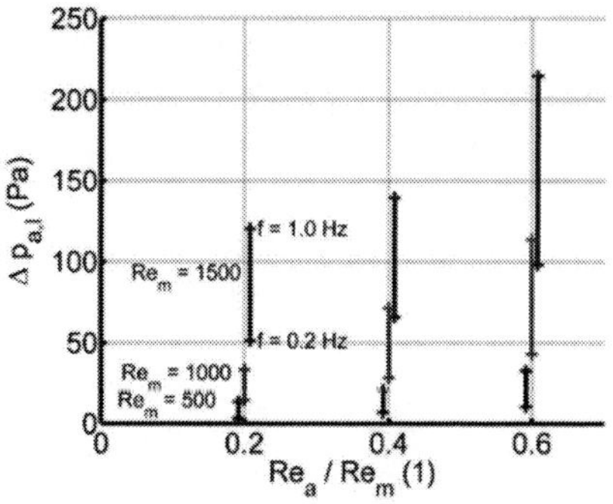

Figure 7. The amplitude of local pressure loss for symmetric stenosis number 1.

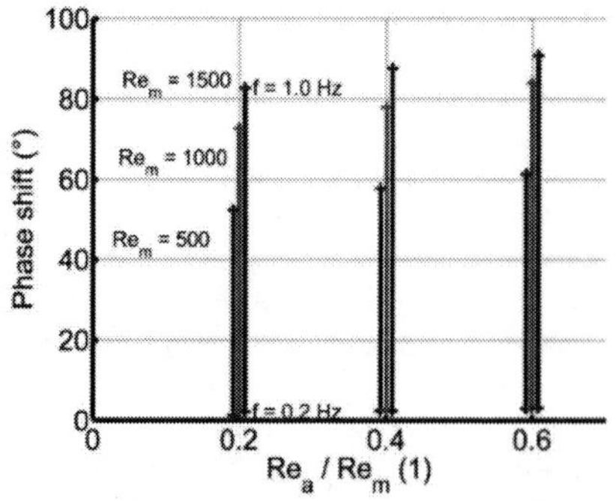

Figure 8. The phase shift of local pressure loss for symmetric stenosis number 1.

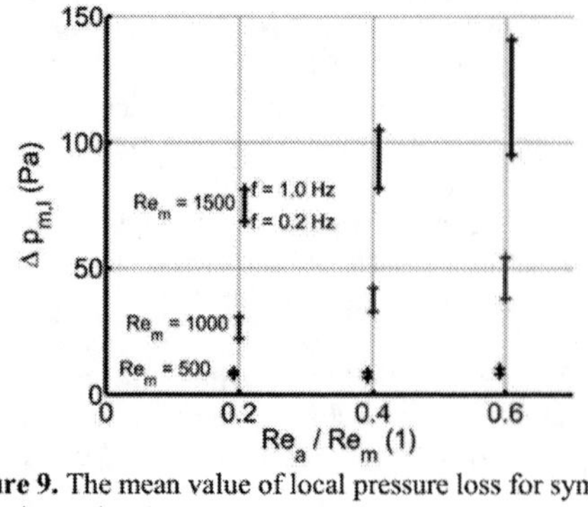

Figure 9. The mean value of local pressure loss for symmetric stenosis number 2.

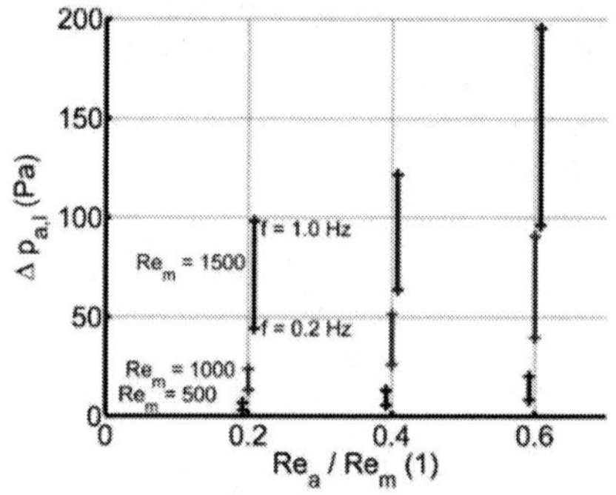

Figure 10. The amplitude of local pressure loss for symmetric stenosis number 2.

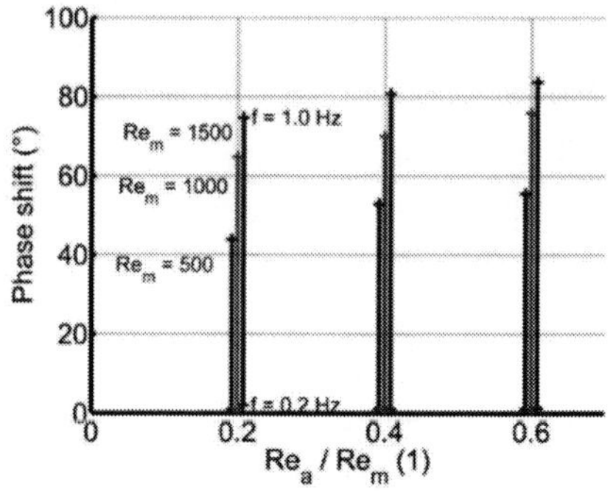

Figure 11. The phase shift of local pressure loss for symmetric stenosis number 2.

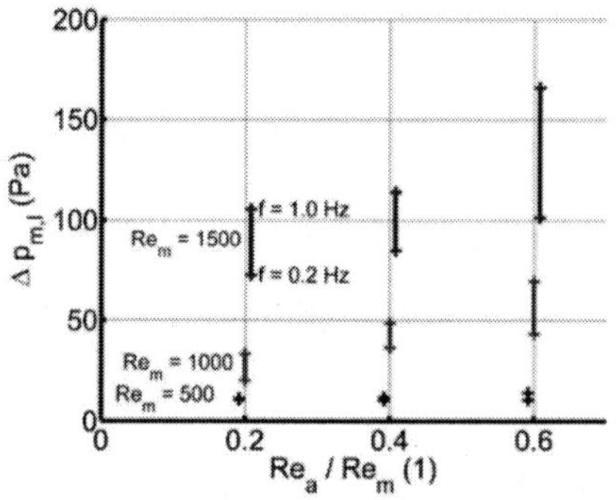

Figure 12. The mean value of local pressure loss for eccentric stenosis number 1.

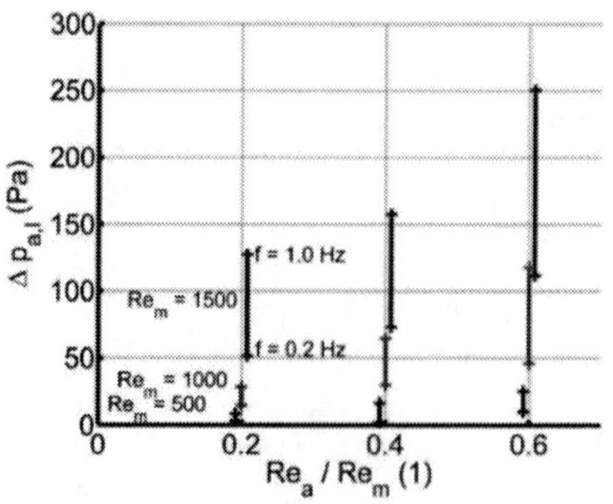

Figure 13. The amplitude of local pressure loss for eccentric stenosis number 1.

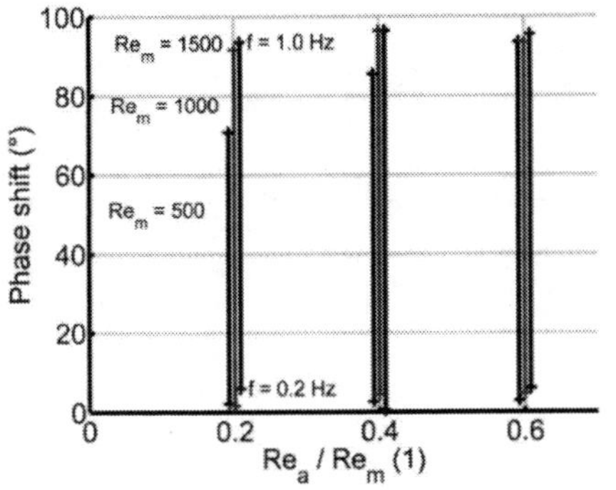

Figure 14. The phase shift of local pressure loss for eccentric stenosis number 1.

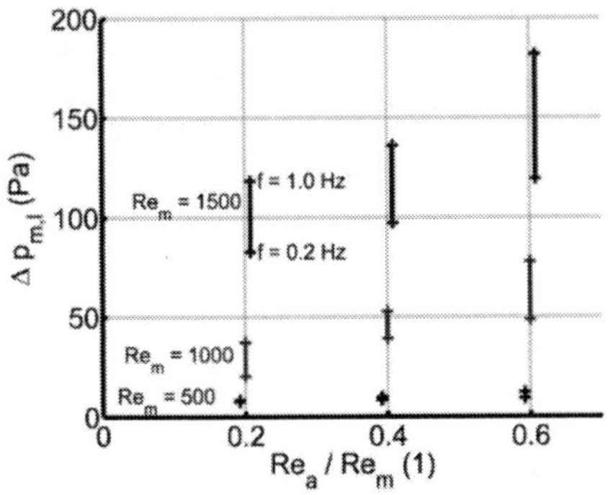

Figure 15. The mean value of local pressure loss for eccentric stenosis number 2.

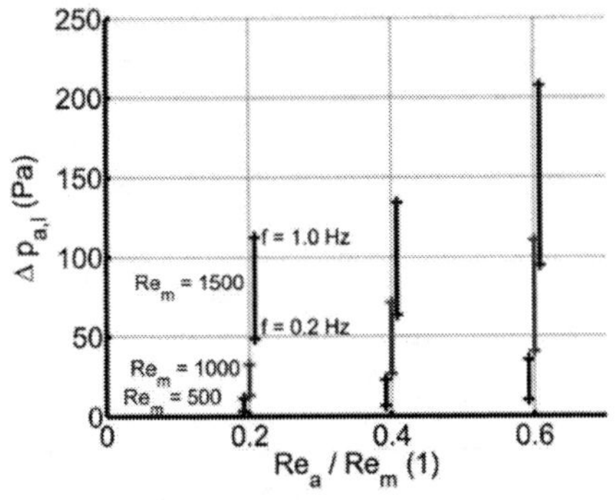

Figure 16. The amplitude of local pressure loss for eccentric stenosis number 2.

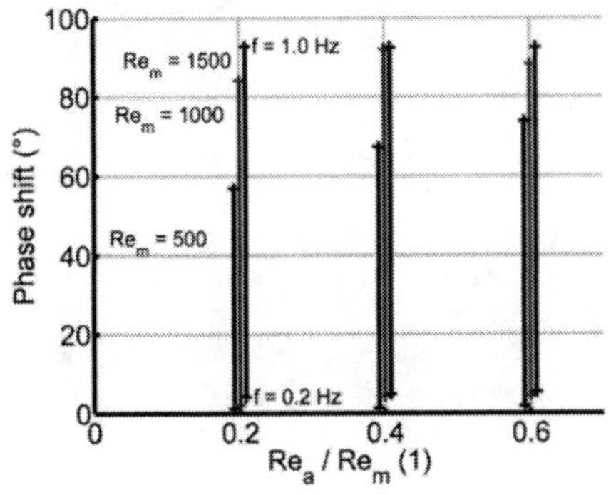

Figure 17. The phase shift of local pressure loss for eccentric stenosis number 2.

4 Conclusion

In this work, experimental investigation of pressure loss of stenosis for the four models were performed: two symmetric stenoses and two eccentric stenoses. All models were stenosis of 75% area reduction. The results prove the relationship between pressure loss and varying eccentricity and shape of stenosis.

For all models, difference of maximum and minimum value of mean and amplitude local pressure loss increases with increasing Reynolds number. The eccentric stenoses have higher value of difference than symmetric stenoses. The phase shift for Re = 500 is higher for eccentric stenoses. The symmetric stenoses have maximum value of phase shift in Re = 1500, the eccentric stenoses in Re = 1000. The maximum value is approximately same for all models.

Investigation of the local pressure loss indicates that eccentric stenoses create higher pressure loss and higher difference between maximum and minimum value. The loss is little affected by shape of stenosis, but significantly by the eccentricity.

References

1. L. Caplain, J Am Med Association, **300**, 81 – 90 (2008)
2. A. Malek, S. Alper, S. Izumo, J Am Med Association, **282**, 2035 – 2042 (1999)
3. S. Glagov, C. K. Zarins, D. P. Giddens, D. N. Ku, Arch. Pathol. Lab. Med., **316**, 1371 – 1375 (1988)
4. Y. Jiang, K. Kohara, K. Hiwada, Am Jounar of Hypertension, **13**, 892 – 898 (2000)
5. C. G. Caro, Arteriscler Thromb Vasc Biol., **29**, 158 – 161 (2009)
6. T. M. Liou, S. N. Liou, In Proc. Natl. Sci. Counc. ROC(B), **23**, 133 – 148 (1999)
7. D. N. Ku, Ann Rev Fluid Mech., **29**, 399 – 434 (1997)
8. A. S. Shuib, P. R. Hoskins, W. J. Easson, Engineering and Technology, **60**, 56 – 69 (2011)
9. B. Liu, Medical Engineering and Physics, **29**, 868 – 876 (2007)
10. K. Pielhop, M. Klaas, W. Scroder, European Journal of Mechanics of Fluid, **35**, 102 – 110 (2012)
11. J. Kolínský, J. Matěcha, J. Novotný, In Proc. Of. ASME 2009 Fluids Engineering Division, New York, (2009)
12. M. Siouffi, V. Deplano, R. Pélisser, Journal of Biomechanics, **31**, 11 – 19 (1998)

Acknowledgement

The research has been supported by Faculty of Mechanical Engineering of the Czech Technical University in Prague through the student grant project No. SGS15/191/OHK2/3T/12 (The Unsteady flow in Pipe with Variable Area Cross Section).

EPJ Web of Conferences **114**, 02133 (2016)
DOI: 10.1051/epjconf/201611402133

Dependence of the isobaric specific heat capacity of water vapor on the pressure and temperature

Magda Vestfálová[1,a], Pavel Šafařík[2]

[1]KEZ, TU Liberec, Studentská 1402/2, 461 17 Liberec 1, Czech Republic
[2]U12112, CTU Prague, Technická 4, 166 07 Praha 6, Czech Republic

Abstract. The fundamental base for the calculation of the thermodynamic properties of substances is the thermal equation of state and the dependence of some of the basic specific heat capacities on temperature. Dependence of isobaric specific heat capacity on the pressure can already be deduced from these relations. International standards of the properties of water and steam are based on the new scientific formulation IAPWS-95. The equation is in the form of Helmholtz dimensionless function with very much parameters. The aim of this paper is to design the simple dependence of the isobaric specific heat capacity of water vapor on the pressure and temperature in the range in which the steam occurs in the atmospheric moist air.

1 Introduction

Water and water vapor plays an important role in many natural phenomena. They are also the usual work media or process media in various industrial sectors, especially in technical or chemical fields. Therefore, their properties are explored in detail, measured and tabulated.

Since 1929, research and standardization of the properties of water and steam has been organized internationally. The aim was to provide a reliable and uniform thermophysical data. At first, it was a loose association of professionals engaged in measuring, describing, or using thermophysical properties of water and steam. Since 1972, a loose association has been replaced by of a permanent organization, the International Association for the Properties of Water and Steam - IAPWS. One of the member countries of IAPWS is also Czech Republic.

During a period of international cooperation to standardization the properties of water and steam, IAPWS has published a series of documents based on a thorough verification of the available experimental data on the properties of water [1], on the analysis of variance of data files, on data classification according to plausibility, on interpolation of the data [2] and on appropriate expression of the properties of water vapor, especially thermodynamic properties and transport properties.

In 1995, the Pruß-Wagner equation [3] was adopted as an international formulation of thermodynamic properties of water and steam for scientific purposes IAPWS-95. It was evaluated as the best equation of state for water and steam for a wide range of parameters, which represents the available experimental data very well. The equation is in the form of Helmholtz dimensionless function according on temperature and reduced density in the form of a uniform state equation valid for the liquid phase and gas phase too. International standards of the properties of water and steam are based on the new scientific formulation IAPWS-95 and they are available for example in [4]. Document IAPWS-95 was recently revised in 2014 [5].

In 1997, it was received a document stating the properties of water and steam in the formulation for industrial purposes - IAPWS IF-97. Its importance lies in the fact that it is binding international basis for energy (particularly for tenders and guarantee tests). Formulation of the properties of water and steam for industrial purposes is compatible with scientific standards with sufficient accuracy and reduces significantly the computational time to achieve the values of thermodynamic parameters of water and steam in five defined areas of parameters. In these areas, fundamental equations are defined - in three fields these are specific equations for the Gibbs free energy, in one area this is the equation for the Helmholtz free energy and in one area it is implicit quadratic equation as a relationship between the pressure and temperature of saturation. International Standards of properties of water and steam based on the industrial formulation IAPWS IF-97 are available for example in the form of tables in [6]. Document IAPWS IF-97 is gradually completed [7], for example by the reverse functions to further reducing the computational time required values of thermodynamic parameters of water and steam, and by other cues from the current industrial applications.

The aim of the presented paper is to derive the simple dependence of the isobaric specific heat capacity of water

[a] Corresponding author: pavel.safarik@fs.cvut.cz

vapor on the pressure and temperature in the range in which the steam occurs in the atmospheric moist air. The dependence is based on the tabulated data according to formulation for scientific purposes IAPWS-95 and it was derived for using in calculations of the properties of the atmospheric moist air.

2 Specific heat capacity at constant pressure of real gases

Let dependency of states parameters (pressure p, temperature T and the specific volume v) of real gas is expressed by some thermal equation of state

$$f(p,T,v)=0. \tag{1}$$

Let the isobaric specific heat capacity of a real gas depends on the temperature and pressure

$$c_p=\left(\frac{\partial q}{\partial T}\right)_p=\left(\frac{\partial h}{\partial T}\right)_p=c_p(T,p), \tag{2}$$

where q is the specific heat and h is the specific enthalpy.

The full differential of isobaric specific heat capacity is expressed

$$dc_p=\left(\frac{\partial c_p}{\partial T}\right)_p dT+\left(\frac{\partial c_p}{\partial p}\right)_T dp. \tag{3}$$

The dependence of the specific heat capacity on the pressure can be derived from thermal state equation (1) (see further). Its temperature dependence can not be deduced from other properties of the substance and thus it belongs (together with the thermal equation of state) to the necessary initial foundations for building computational relationships of thermodynamic properties of substances.

In the theory of ideal gas, specific heat capacity has a constant value. For model of partly ideal gas, it is usually worked with dependency of specific heat capacity on temperature in the polynomial form

$$c_p=A+BT+CT^2+DT^3+\dots. \tag{4}$$

where A $(J\cdot kg^{-1}\cdot K^{-1})$, B $(J\cdot kg^{-1}\cdot K^{-2})$, C $(J\cdot kg^{-1}\cdot K^{-3})$, D $(J\cdot kg^{-1}\cdot K^{-4})$, ... are constants of polynomial. In the calculations taking into account the real properties of substances, the relationship (4) expresses the limit approximation of the real gas for $p\to 0$, it is dependence of the isobaric specific heat capacity on the temperature in an idealized state of gas.

The dependence of the isobaric specific heat capacity on the pressure can be obtained from the thermal state equation by the following procedure:

Equation (2) is differentiated by pressure at a constant temperature

$$\left(\frac{\partial c_p}{\partial p}\right)_T=\left[\frac{\partial}{\partial p}\left(\frac{\partial h}{\partial T}\right)_p\right]_T=\left[\frac{\partial}{\partial T}\left(\frac{\partial h}{\partial p}\right)_T\right]_p. \tag{5}$$

The derivative of enthalpy can be obtained by differentiating of the Gibbs equation ($dh=Tds+vdp$, where s is the specific entropy)

$$\left(\frac{\partial h}{\partial p}\right)_T=T\left(\frac{\partial s}{\partial p}\right)_T+v. \tag{6}$$

Substituting the Maxwell equation

$$\left(\frac{\partial h}{\partial p}\right)_T=-T\left(\frac{\partial v}{\partial T}\right)_p+v. \tag{7}$$

Substituting equation (7) into equation (5)

$$\left(\frac{\partial c_p}{\partial p}\right)_T=\left[\frac{\partial}{\partial T}\left\{-T\left(\frac{\partial v}{\partial T}\right)_p+v\right\}\right]_v=-T\left(\frac{\partial^2 v}{\partial T^2}\right)_p. \tag{8}$$

If the thermal equation of state of technical type (it is in the form $v=v(p,T)$) is available (and it describes the properties of steam with sufficient accuracy) then it would be possible to obtain a derivative $\left(\frac{\partial v}{\partial T}\right)_p$ and to solve the equation (8) and to obtain the dependence of the isobaric specific heat capacity at pressure.

3 Specific heat capacity of water vapor - regression of IAPWS-95 data

Another way to obtaining the functional dependence of the specific heat capacity on pressure and temperature, is the simple regression of data, for example from IAPWS-95.

The paper is focused on the properties of water vapor occurring in the unsaturated moist air at "normal" temperatures, it is at temperatures of *0°C* to *200°C*. The values of isobaric specific heat capacity of water vapor by IAPWS-95 under these conditions are shown in Table 1 [4]. Graphical dependence of the isobaric specific heat capacity on the temperature with parameter - pressure is shown in Figure 1. It is evident that the course of this dependence is not monotonic. But the dependence of the isobaric specific heat capacity on the pressure with parameter - temperature, as it is shown in Figure 2, appears to be linear. Therefore, the relationship for dependence of isobaric specific heat capacity on the pressure in the shape

$$c_p=E+Fx, \tag{9}$$

Table 1. The isobaric specific heat capacity of water vapor in $kJ \cdot kg^{-1} \cdot K^{-1}$ - dependence on pressure and temperature by IAPWS-95 - selected values [4].

p [Pa] t [°C]	611.657	1000	2000	5000	10000	20000	50000	101325
5	1.875							
10	1.872	1.884						
15	1.87	1.878						
20	1.87	1.875	1.896					
25	1.87	1.874	1.888					
30	1.871	1.874	1.884					
40	1.872	1.875	1.881	1.905				
50	1.874	1.876	1.881	1.896	1.928			
60	1.877	1.878	1.882	1.894	1.914			
70	1.880	1.881	1.884	1.893	1.909	1.942		
80	1.883	1.884	1.887	1.894	1.906	1.931		
90	1.887	1.888	1.890	1.895	1.905	1.925	1.991	
100	1.891	1.891	1.893	1.898	1.906	1.922	1.974	2.080
120	1.899	1.900	1.901	1.904	1.910	1.921	1.956	2.021
140	1.909	1.909	1.910	1.912	1.916	1.924	1.949	1.994
160	1.919	1.919	1.919	1.921	1.924	1.930	1.948	1.980
180	1.929	1.929	1.930	1.931	1.933	1.938	1.951	1.976
200	1.940	1.940	1.941	1.942	1.943	1.947	1.957	1.976

where E and F are the parameters of linear dependence will be investigated.

Parameter E (it is the value of the isobaric specific heat capacity c_p at the lowest "known" pressure) depends on

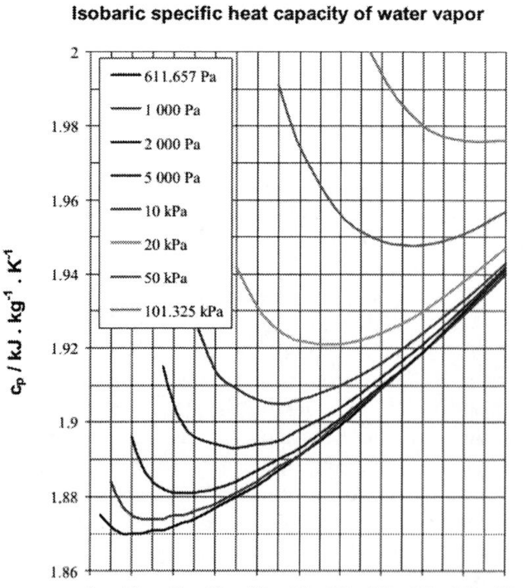

Figure 1. The isobaric specific heat capacity of water vapor - dependence on pressure and temperature by IAPWS-95 [4] (parameter - pressure).

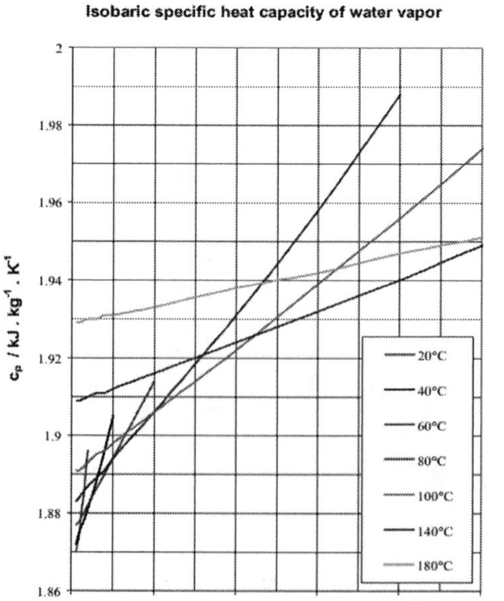

Figure 2. The isobaric specific heat capacity of water vapor - dependence on pressure and temperature by IAPWS-95 [4] (parameter - temperature).

the temperature (see Figure 2). As low "known" pressure is adopted pressure *611.657 Pa* (the pressure of the triple point of water). Therefore parameter E corresponds with the isobaric specific heat capacity at a pressure equal to the pressure of the triple point of water. In the linear dependence (9) will not be direct the value of the pressure, but the value of difference between the pressure and the pressure of the triple point of water. The pressure of the triple point can be considered as the pressure "very low", so, it can be assumed that this is an idealized state of water vapor, it is water vapor acts as an ideal gas at this pressure. Dependence of the parameter E on temperature is therefore for us the dependence of isobaric specific heat capacity in an idealized state of the gas on temperature according to equation (4)

$$E=E(t)=(c_p)_{p=611.657\,Pa}=A_E+B_E\,t+C_E\,t^2+D_E\,t^3+\ldots\,, \quad (10)$$

where t (°C) is a temperature, A_E, B_E, C_E, D_E, ... are the coefficients of polynomial regression.

By regression of data IAPWS-95 by 2nd order polynomial, we obtained the course of the isobaric specific heat capacity c_p at a pressure of *611.657 Pa* with a deviation less than *0.1 %* (Figure 3).

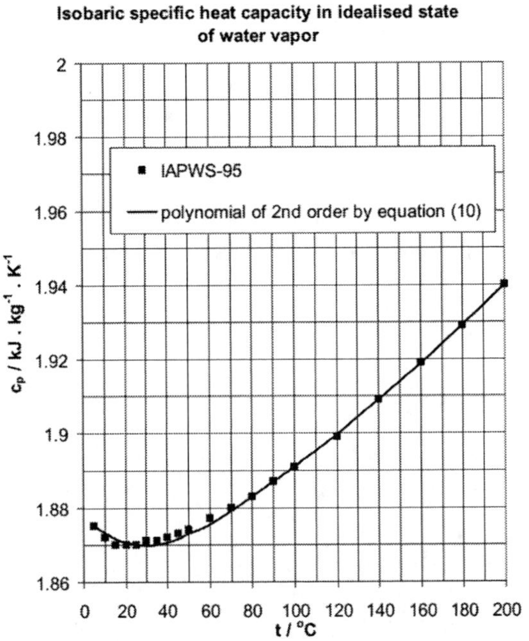

Figure 3. The isobaric specific heat capacity of water vapor at low pressure (p_{tr}) according IAPWS-95 [4], and by regression polynomial, equation (10).

The parameter F is a slope of the growth of the isobaric specific heat capacity c_p when the pressure increases. It is temperature-dependent function. It decreases with temperature (the higher is the temperature, the slower is the growth of c_p when the pressure increases, as Figure 2 shows).

Polynomial dependence of inverted value of slope F in linear relationship in equation (9) has the form

$$\frac{1}{F}=\frac{1}{F(t)}=A_F+B_F\,t+C_F\,t^2+D_F\,t^3+\ldots\,. \quad (11)$$

It is achieved the compliance with data IAPWS - 95 (see Figure 4) already using 2nd order polynomial in equation (11).

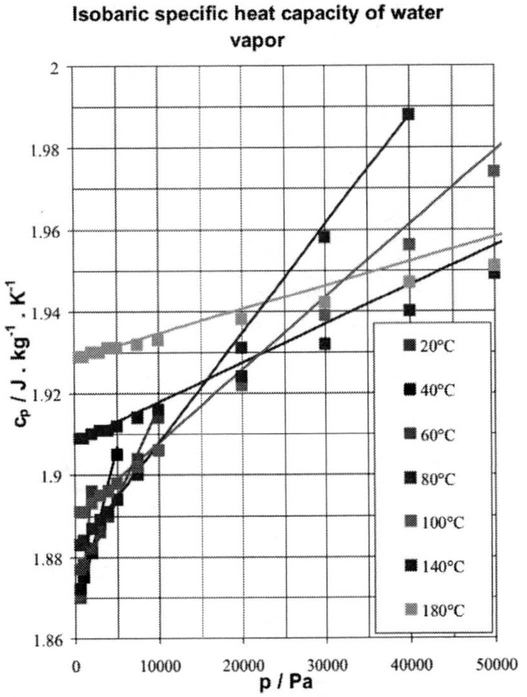

Figure 4. The isobaric specific heat capacity of water vapor in dependence on pressure and temperature (parameter - temperature) according IAPWS-95 [4], and by regression polynomial (11).

Figure 5 shows the dependence of isobaric specific heat capacity c_p on the temperature and on the pressure determined by a simplified relationship using equation (10) and (11) in the form

$$c_p(p,t)=E(t)+F(t)(p-p_{tr})= \quad (12)$$

$$A_E+B_E\,t+C_E\,t^2+\frac{1}{A_F+B_F\,t+C_F\,t^2}(p-p_{tr}).$$

Table 2. Coefficients of polynomials regression of isobaric specific heat capacity c_p of water vapor for the temperature range *0°C - 200°C* and pressures to *200 kPa*.

	t < 50°C	**t > 50°C**
A_E [m^2 . s^{-2} . K^{-1}]	1877.2	1856.1
B_E [m^2. s^{-2}. K^{-2}]	-0.49545	0.28056
C_E [m^2. s^{-2}. K^{-3}]	$8.1818 \cdot 10^{-3}$	$6.9444 \cdot 10^{-4}$
A_F [kg . K . m^{-3}]	22.537	22.537
B_F [kg . m^{-3}]	0.49321	0.49321
C_F [kg . K^{-1}. m^{-3}]	0.048927	0.048927

Table 3. Variations of the isobaric specific heat capacity c_p of water vapor calculated by the regression polynomial (12) from the values IAPWS-95 in % in the temperature range *0°C - 200°C* and pressures to *200 kPa*.

p [Pa] t [°C]	611.657	1000	2000	5000	10000	20000	50000	100000	101325	200000
5	$2\cdot10^{-13}$									
10	0.061	0.060								
15	0.090	0.169								
20	0.034	0.166	0.071							
25	$1\cdot10^{-13}$	0.103	0.170							
30	0.066	0.029	0.150							
40	0.078	0.066	0.056	0.102						
50	0.056	0.040	0.008	0.150	0.018					
60	0.083	0.046	0.025	0.035	0.134					
70	0.045	0.029	0.009	0.050	0.093	0.128				
80	$5\cdot10^{-14}$	0.002	0.016	0.037	0.106	0.191				
90	0.001	0.010	0.001	0.077	0.118	0.199	0.131			
100	0.006	0.042	0.031	0.049	0.097	0.191	0.260	0.422	0.452	
120	0.041	0.014	0.029	0.071	0.090	0.179	0.337	0.356	0.340	
140	$9\cdot10^{-14}$	0.019	0.017	0.062	0.101	0.180	0.360	0.583	0.545	0.421
160	0.012	0.003	0.042	0.053	0.089	0.160	0.372	0.615	0.665	0.881
180	0.006	0.018	0.004	0.036	0.085	0.131	0.370	0.643	0.631	1.068
200	$1\cdot10^{-13}$	0.010	0.017	0.006	0.078	0.120	0.346	0.649	0.631	1.136

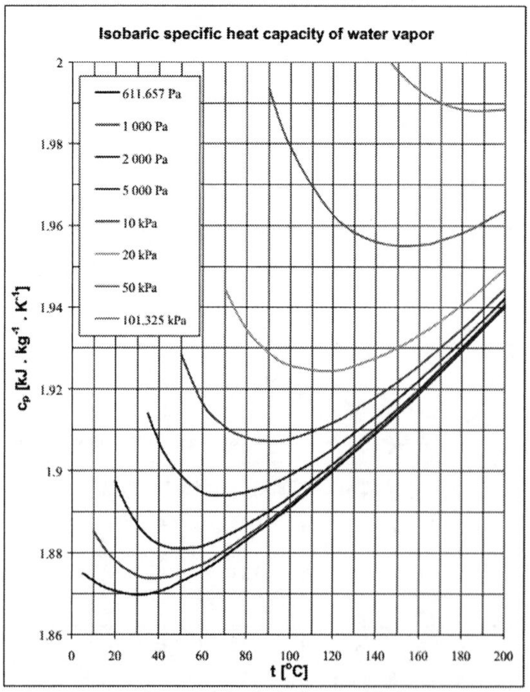

Figure 5. Dependence of the isobaric specific heat capacity of water vapor on the pressure and on the temperature (parameter - pressure) calculated according to the regression equation (12).

To calculation, the coefficients of polynomial regression according to Table 2 were used.

The values shown in Figure 5 are compared with the data IAPWS-95 (see Figure 1) for temperatures up to *200°C* and pressures up to *200 kPa* and differences are less than *1.2%* (see Table 3).

4 Conclusion

The relationship for calculation the isobaric specific heat capacity of water vapor in the states in which the water vapor founds in humid atmospheric air under normal conditions, is derived, equation (12). Values obtained with derived relationship deviates from the values IAPWS-95 by less than *1.2 %* (while this deviation for isobaric specific heat capacity of water vapor as an ideal gas reaches for example at a temperature of *200°C* and a pressure of *200 kPa* values around *35%*). Such variation is acceptable for most engineering calculations in the humid air.

Acknowledgements

The first author, M.Vestfálová, express his gratitude for the support of project SGS No.21000 „Experimental and Numerical Investigation in Fluid Mechanics, Thermodynamics and Heat Transfer“ .

The second author, P. Šafařík, expresses his gratitude for the support of research from TAČR Competence

Centre, "Advanced Technologies for the Production of Heat and Electricity," Reg. No. TE01020036.

References

1. H. Sato, K. Watanabe, J.M.H. Levelt Sengers, J.S. Gallagher, P.G. Hill, J. Straub, W. Wagner, Sixteen Thousands Evaluated Experimental Thermodynamic Property Data for Water and Steam, Journal of Physical and Chemical Reference Data, **20**, 1023-1044 (1991)
2. J. Straub, K. Scheffler, *Water and Steam, Their Properties and Current Industrial Applications* (1980)
3. W. Wagner, A. Pruß, The IAPWS Formulation 1995 for the Thermodynamic Properties of Ordinary Water Substance for General and Scientific Use, Journal of Physical and Chemical Reference Data, **31**, 387-535 (2002)
4. O. Šifner, J. Klomfar, *International Standards of Thermophysical Properties of Water and Steam* (1996)
5. Revised Release on the IAPWS Formulation 1995 for the Thermodynamic Properties of Ordinary Water Substance for General and Scientific Use (2014)
6. R. Mareš, O. Šifner, J. Kadrnožka, *Tables of Properties of Water and Steam* (1999)
7. Revised Release on the IAPWS Industrial Formulation 1997 for the Thermodynamic Properties of Water and Steam (2007)

EPJ Web of Conferences **114**, 02134 (2016)
DOI: 10.1051/epjconf/201611402134

Thermal equation of state BWR for water and water steam

Magda Vestfálová[1,a], Markéta Petříková[1]

[1]KEZ, TU Liberec, Studentská 1402/2, 461 17 Liberec 1, Czech Republic

Abstract. The fundamental bases for the calculation of the thermodynamic properties of substances are the thermal equation of state and the dependence of some of the basic specific heat capacities on temperature. Any other thermodynamic properties of a substance can already be deduced from these relations. The aim of this paper is to design the simple thermal equation of state for water steam based on original BWR thermal equation of state, which describes the behavior of H_2O in the gaseous phase in region of usual pressures and temperatures with sufficient accuracy.

1 Introduction

Water and water vapor plays an important role in many phenomena occurring in nature. They are also the customary working substances in various industrial fields (for example in engineering industry, chemistry field, food industry, ...). Therefore, their properties are explored, measured and tabulated in detail. [1] Research of properties of water and water steam began to organize on an international level since 1929 to provide reliable and consistent data on their properties. In 1972 there was founded the International Association for the Properties of Water and Steam - IAPWS which deals with verification and interpolation of experimental data and with expressing of the thermodynamic and transport properties of water and steam. In 1995, there was accepted Pruß - Wagner equation as the international formulation of thermodynamic properties of water and steam for scientific purposes IAPWS-95 [2]. The equation is in the form of Helmholtz dimensionless function depending on temperature and reduced density. The equation is valid for the liquid and gas in a wide range of temperatures and pressures. It contains a large number of constants. International Standards properties of water and steam for example [3] are based on Scientific formulation IAPWS-95.

In many cases, there is necessary to use some "simple" equation for calculation of thermodynamic properties of water steam. "Simple" form of thermal equation of state will be applicable only on a limited range of temperatures and pressures, and it will have less accuracy. The aim of this paper is to present, if possible, a "simple" thermal equation of state and to define its application area, where agreed accuracy is maintained.

In 1940 Manson Benedict, G. B. Webb and L. C. Rubin constructed "quite simple" so called original Benedict - Webb - Rubin thermal equation of state (hereinafter BWR) [4].

$$p = rT\rho + rT\rho^2 B_0 + rT\rho^3 b - A_0\rho^2 - a\rho^3 + \\ + a\alpha\rho^6 - \frac{\rho^2 C_0}{T^2} + \frac{c\rho^3}{T^2}\left(1+\gamma\rho^2\right)e^{-\gamma\rho^2}. \tag{1}$$

Equation has some physical basics in the state behavior of gases. It contains an addition to the specific constant r still eight parameters: A_0, B_0, C_0, a, b, c, α and γ, that characterize the individual properties of the gas. They must be determined from experimental dates by regression. So, as well as other multi-parameter equations, BWR equation can be only used to description of properties of substances for which there is sufficient amount of experimental data in wide range of pressures and temperatures. This requirement is fulfilled for water and water steam undoubtedly.

In 1973 modification of the Benedict–Webb–Rubin equation of state by Starling was made [5]

$$p = rT\rho + rT\rho^2 B_0 + rT\rho^3 b - A_0\rho^2 - \\ - a\rho^3 + a\alpha\rho^6 + \frac{\alpha d\rho^6}{T} - \frac{\rho^2 C_0}{T^2} + \\ + \frac{\rho^2 D_0}{T^3} + \frac{\rho^2 E_0}{T^4} + \frac{c\rho^3}{T^2}\left(1+\gamma\rho^2\right)e^{-\gamma\rho^2}. \tag{2}$$

Values of the various parameters for 15 substances can be found in work [5].

Another modification of the Benedict–Webb–Rubin equation of state was made in 1987 by Jacobsen and Stewart [6], it is noted as mBWR

$$p = \sum_{n=1}^{9} a_n \rho^n + \sum_{n=10}^{15} a_n \rho^{2n-17} e^{-\gamma\rho^2}. \tag{3}$$

[a] Corresponding author: magda.vestfalova@tul.cz

The mBWR equation subsequently evolved into a 32 term version with numerical parameters determined by fitting the equation to empirical data for a reference fluid [7].

At first BWR equation was suggested for calculation of behavior of hydrocarbons. For these substances values obtained from BWR equation are very close to the experimental date [8]. For these substances BWR equation describes the behavior of materials in single phase areas (it is in superheated steam and in the super cooled liquid) very well. In heterogeneous areas (below the limit curve), as expected, there is a local maximum and a local minimum and the results are totally unrealistic. Parameters of BWR equation for hydrocarbons are presented for example in work [9] too.

We want to obtain a "simple" equation of state for water and water steam, so we try to use the original BWR equation.

In [10] there was shown the using of BWR thermal equation of state for H_2O. There was described derivation of the isochoric specific heat capacity and enthalpy from the thermal equation BWR. It has been shown that in the area where the error of *p-v-T* values obtained from BWR thermal state equations is small, the values of enthalpy, calculated from the relation derived using equation BWR are sufficiently precise too. The disadvantage was that the application area (the area with small errors) is very small.

The aim of this paper is to provide the values of constants of BWR equation A_0, B_0, C_0, a, b, c, α and γ so that this BWR equation describes with a "sufficient" accuracy the behavior of water seam, it means H_2O in gas phase in commonly used ranges of pressures and temperatures, and if possible tills to limit curve. A "reasonable" accuracy we will consider the deviation of the values obtained from the solution BWR and tabulated data according to formulation for scientific purposes IAPWS-95 was less than *1 %*.

2 Model of ideal gas

The first term on the right side of BWR equation (1) - $(r \cdot T \cdot \rho)$, which is proportional to the first power of density, corresponds to the state equation of ideal gas. Other members, proportional to the higher powers of density, represent the correction on the real behavior of gas. Obviously, the higher is the density, the greater is the influence of correction terms on the calculated pressure value.

In the "bigger" distances from the limit curve, it means in a state of much diluted gases (at "high" temperatures and "low" pressures, it means the "big" specific volumes, respectively "small" densities) the behavior of real materials approaches to the behavior of ideal gases. Correction terms in the BWR equation with higher power of density is unlikely to be strongly applied. In the case of water vapor we can illustrate this fact on several isotherms: error to *1 %* is for temperature *1000°C* to the pressure *15 MPa*, for temperature *500°C* to the pressure *1 MPa*, for temperature *200°C* to the pressure *0.1 MPa*, for temperature *100°C* to the pressure *0.05 MPa* (in Figure 1. - red line). For temperature *50°C* is the pressure of saturated steam very low - 0.012351 *MPa*. In this point (on the limit curve) we obtain values with accuracy *0.4 %* by using the model of ideal gas. The accuracy to *1 % is* maintained on the limit curve until the temperature *80°C*. In triple point (temperature *0.01°C*, pressure 0.000611657MPa), the state of saturated steam is calculated with accuracy *0.07 %* using the model of ideal gas.

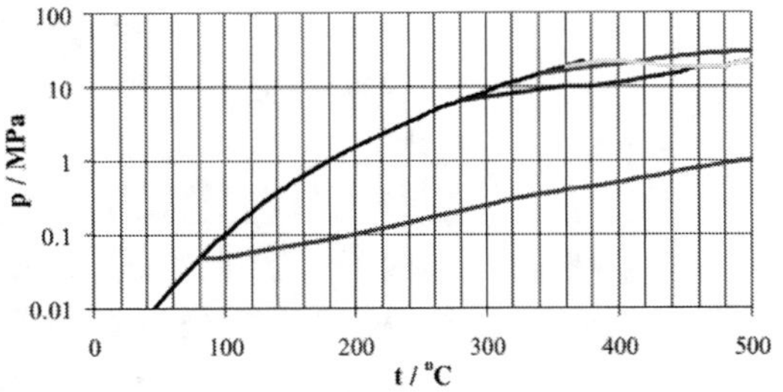

Figure 1. Field of application of the different models with the accuracy to 1 %: thick line - the limit curve; red line - limit of applicability of ideal gas model; blue line - limit of applicability of model with correction A_0, B_0 and C_0; green line - limit of applicability of model with correction A_0, B_0, C_0, a and b; yellow line - limit of application of BWR equation.

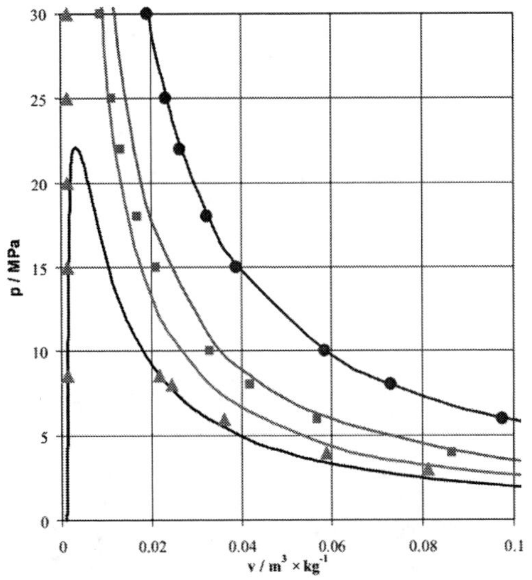

Figure 2. Comparison of isotherms of an ideal gas with real isotherms: thick line - the limit curve; green - isotherm *300°C*; pink - isotherm *500°C*; blue - isotherm *1000°C*; lines - ideal gas isotherms; points - real values.

Figure 2. shows the real isotherms *300°C*, *500°C* and *1000°C* and corresponding isotherms calculated using the ideal gas model. With increasing specific volume is he ideal isotherms and the real isotherms close to each other, conversely a decreasing specific volume increases the difference between the ideal and the real isotherm. We obtain higher pressure using the model of ideal gas than its actual value at low specific volumes.

Table 1. Specific volume $10^3 \cdot v / m^3\ kg^{-1}$ of water steam depending on the pressure and temperature; yellow highlighted numbers - accuracy of model of ideal gas less then 1 %; pink highlighted numbers - accuracy of model with correction A_0, B_0 and C_0 less then 1 %; green highlighted numbers - accuracy of model with correction A_0, B_0, C_0, a and b less then 1 %.

p / MPa t / °C	18	15	10	8	6	4	2	1	0.5	0.1	0.01
1000	32.282	38.808	58.390	73.079	97.560	146.52	293.42	587.21	1174.80	5875	58757
500	16.810	20.827	32.811	41.767	56.671	86.44	175.68	354.11	710.94	3565	35679
480	15.990	19.923	31.627	40.362	54.891	83.90	170.85	344.66	692.26	3473	34736
460	15.116	18.973	30.407	38.926	53.083	81.34	165.98	335.19	673.56	3380	33833
440	14.171	17.694	29.144	37.451	51.240	78.74	161.09	325.69	654.84	3288	32909
420	13.122	16.874	27.826	35.928	49.355	76.11	156.17	316.17	636.09	3195	31986
400	11.915	15.671	26.436	34.344	47.419	73.43	151.21	306.61	617.30	3103	31063
380	10.419	14.289	24.950	32.681	45.418	70.70	146.21	297.00	598.47	3010	30139
360	8.111	12.582	23.325	30.912	43.333	67.90	141.15	287.35	579.60	2917	29216
340			21.487	28.992	41.135	65.02	136.03	277.64	560.67	2824	28292
320			19.270	26.840	38.780	62.02	130.82	267.86	541.68	2732	27369
300				24.279	36.189	58.87	125.51	257.98	522.61	2639	26445
280					33.199	55.50	120.05	248.00	503.42	2546	25521
260						51.78	114.41	237.88	484.12	2453	24598
240							108.50	227.30	464.66	2360	23674
220							102.18	216.98	444.96	2266	22749
200								206.02	425.03	2172	21825
180								194.44	404.63	2078	20901
160									383.66	1984	19976
140										1889	19050
120										1793	18124
100										1696	17196
50											14866

In Table 1., the area in which the ideal gas model gives accuracy to *1 % is* "highlighted" in yellow.

3 Correction using members proportional to the square of density

If the state of gas comes "near" to the limit curve, it is (due to the independent variables in the BWR equation) if specific volume decreases (resp. density increases), the deviation of the ideal gas model from the fair value increases. In the BWR equation, the influence of the correction members with higher power density grows. The second, the fourth and the seventh term on the right side are proportional to the square of density (other members depend on even higher powers of density) their effects are thus manifested first. The fourth member - $\left(-A_0 \cdot \rho^2\right)$ is directly proportional only to the square of the density and it corresponds to the absolute deviation between the ideal and the real isotherm apparently. Whereas it is necessary to make such a correction, which "reduce" the ideal pressure in the BWR equation this member is subtracted. Deviation between ideal and real isotherm is not the same for all temperatures. At high temperatures (the substance is closer to the model of an ideal gas) the deviation is less, therefore the first correction (reduction of the ideal pressure due to fourth member) is "mitigated" due to the second member - $\left(r \cdot T \cdot \rho^2 \cdot B_0\right)$, which depends directly proportional to temperature, it is its influence increases with increasing of temperature. In the temperature range where the substance is describable due to ideal gas model, the fourth and the second member are mutually compensated. Conversely, at low temperatures, where the substance approaches the limit of liquefaction, is the absolute difference between the ideal and actual pressure greater - correction from ideal behavior is reinforced due to seventh member - $\left(-\frac{\rho^2 \cdot C_0}{T^2}\right)$, which is inversely proportional to the square of the temperature and therefore it significantly increases if the temperature falls.

[a] Corresponding author: magda.vestfalova@tul.cz

To estimate the coefficients A_0, B_0, C_0 was used IAPWS-95 data of water steam at different temperatures and such densities at which the maximum error of the model of an ideal gas from real data was 5 % approximately.

If the thermodynamic properties of water vapor was calculated from the ideal gas state equation adjusted to the real behavior of water steam through the 2nd, 4th and 6th member in BWR equation, the area, where the error of calculated values was less than *1%,* significantly increased: for temperature *1000°C* to the pressure *50 MPa*, for temperature *500°C* to the pressure *22 MPa.* On the limit curve, the accuracy *1 % is* maintained at temperatures up to *290°C.* In Table 1. the region is indicated by a pink color, in the Figure 1. by blue color.

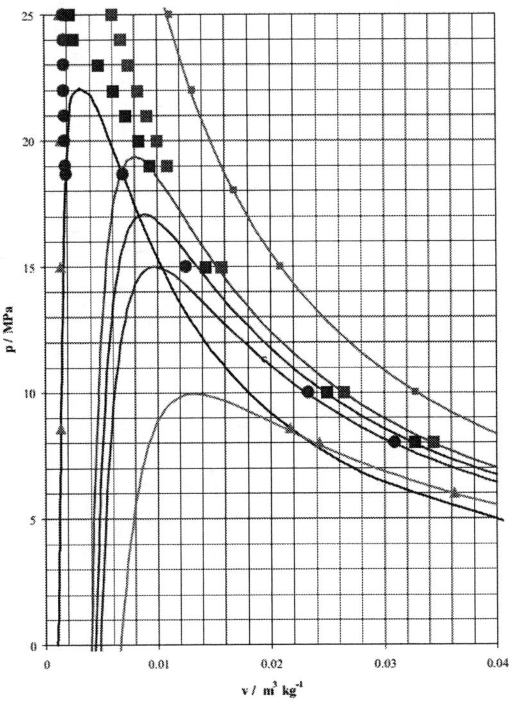

Figure 3. Comparison of isotherms of model with correction A_0, B_0 and C_0 with real isotherms: thick line - the limit curve; green - isotherm *300°C*; brown - isotherm *360°C*; bleu - isotherm *380°C;* red - isotherm *400°C*; pink - isotherm *500°C*; lines - model with correction A_0, B_0 and C_0 isotherms; points - real values.

4 Correction using members proportional to the third power of density

At low values of specific volume, however, the error is still too big in gaseous state too, see Figure 3. At low specific volumes, the model gives the lower value of pressure than its real value. Obviously this is the area in which begin to apply the correcting members with higher than the square power of the density. The third and the fifth member are proportional to the third power of density. The fifth member $\left(-a\cdot\rho^3\right)$ is directly proportional only to the third power of density and apparently corresponds to the absolute deviation between the calculated and real isotherm. Whereas it is necessary to make such a correction that "increases" calculated pressure, in BWR equation that member will be added (negative value of *a*). The deviation of calculated pressures from the real values of pressure is not the same for all temperatures - the higher the temperature, the smaller the difference between the actual and the calculated pressure. The original correction (fifth member) is therefore necessary to "alleviate" (decrease) by correction using the third member $\left(T\cdot\rho^3\cdot b\right)$, whose influence is greater the higher the temperature of the gas. To estimate the coefficients *a* and *b* was used IAPWS-95 data of steam water at different temperatures and such densities at which the maximum error of the model of an ideal gas with the previous correction (2nd, 4th and 7th member in a BWR equation) from the real data was approximately *5 %*.

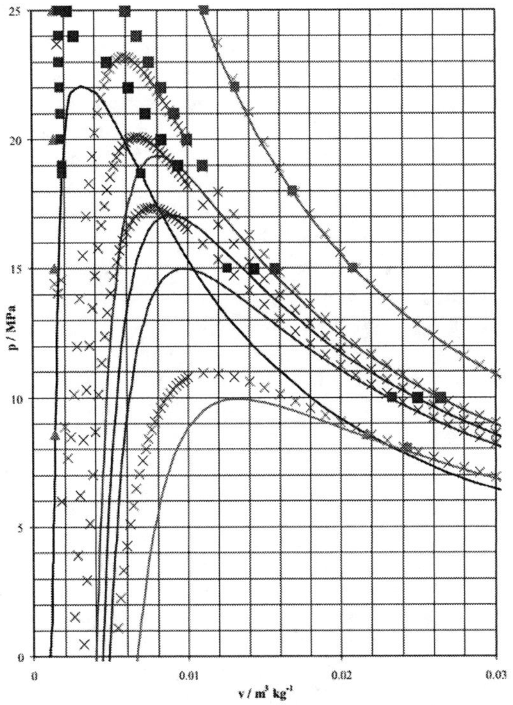

Figure 4. Comparison of isotherms of model with correction A_0, B_0, C_0, *a* and *b* (lines) with isotherms of model with correction only A_0, B_0 and C_0 (crosses) and with real isotherms (points): thick line - the limit curve; green - isotherm *300°C*; brown - isotherm *360°C*; bleu - isotherm *380°C;* red - isotherm *400°C;* pink - isotherm *500°C*.

If the thermodynamic properties of water vapor was calculated from the ideal gas state equation adjusted to the real behavior of water steam through the 2nd, 3rd, 4th, 5th and 6th member in BWR equation, the area, where the error of calculated values was less than *1 %,* increased again: on the limit curve, the accuracy *1 % is* maintained at temperatures up to *340°C.* In Table 1., the region is indicated by a green color, in the Figure 1. by green color too.

5 Correction using members with special dependence on density

Isotherms calculated using such equations have already the known shape of Van der Waals isotherm (with the local maximum and minimum in the area of wet steam), see Figure 4. The big mistake, however, still occurs near the critical point, it is at much lower levels of specific volume. Obviously, this is the area where it would be needed used further correction in the state equation.

The third power density is found in the eighth member $\left(\frac{c\cdot\rho^3}{T^2}\left(1+\gamma\cdot\rho^2\right)e^{-\gamma\cdot\rho^2}\right)$, but the dependence of this correction term on the density is more complicated. To illustrate the effect of the eighth member look at Figure 5., where dependence $\frac{\rho^3}{T^2}\left(1+\rho^2\right)e^{-\rho^2}$ on $\frac{1}{\rho}$ (it is on specific volume) is plotted for two different temperatures. It is evident, that this member neither monotonically decreases nor monotonically increases with increasing density (it is with decreasing specific volume), but its influence is reflected only in a certain range of densities. Its dependence has a marked maximum for concretely value of density. This member is inversely proportional to the square of temperature. His influence, therefore, probably will decrease significantly with increasing temperature. This member is significantly exhibit to correcting the shape of isotherms in the occurrence of local maxima and minima; it is below the limit curve (area of phase change). The constant γ should be determined so that the region of influence of the eighth member coincides to the densities corresponds with the start and end of the phase transformation of water vapor.

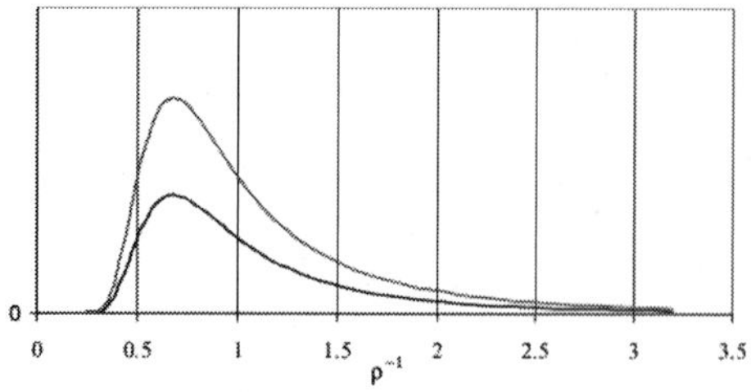

Figure 5. Dependence of $\frac{\rho^3}{T^2}\left(1+\rho^2\right)e^{-\rho^2}$ on $\frac{1}{\rho}$ for temperature 500°C (blue line) and for temperature 300°C (pink line).

The last unused member of the BWR equation is the 7th member - $\left(a\cdot\alpha\cdot\rho^6\right)$. It is proportional to the sixth power of density and thus it has influence only at significantly high densities. This member does not depend on the temperature, which means that adjusts value of pressure independently on the temperature - for the some density it add to the calculated values of pressure (see the previous correction) still the same value.

When parameter c and parameter α are estimated, we need for small values of specific volumes (where correction using the 7th member and the 8th member may have effect on the calculated value of pressure) to ensure: the applicability of the equation on the saturation line to the higher temperatures, refinement of equation near the critical point, the existence of a critical point - the inflection point on the critical isotherm, maintaining sufficient accuracy equation at high temperatures.

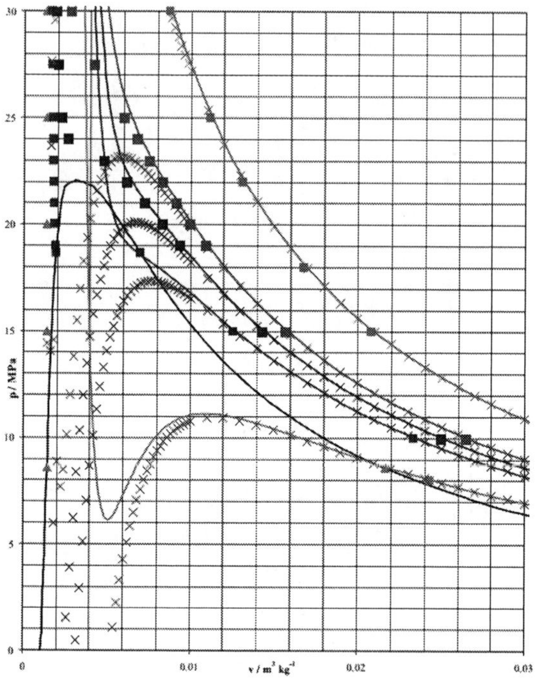

Figure 6. Comparison of isotherms of model with correction A_0, B_0, C_0, a, b, c *and* α (lines) with isotherms of model with correction only A_0, B_0, C_0, a and b (crosses) and with real isotherms (points): thick line - the limit curve; green - isotherm *300°C*; brown - isotherm *360°C*; bleu - isotherm *380°C;* red - isotherm *400°C*; pink - isotherm *500°C*.

It turns out that it is not possible to satisfy simultaneously these requirements for H_2O using the original BWR equation. It is possible to extend the application area of the equation on the limit curve to higher temperatures, but at the cost of a reduced pressure into which an equation is usable in gas phase at high temperatures (see the yellow line in Figure 1.). Figure 6. shows some isotherms calculated by equation with correcting members with A_0, B_0, C_0, a and b (crosses) and same isotherms calculated using the BWR equation with such choice of parameters c and α, that we obtained the good accuracy on the limit curve to the high temperatures (see isotherm *360°C*). We see that for this temperature and pressure of saturation, BWR equation gives a more accurate value, but the shape of this isotherm is bad - it looks like supercritical isotherms (no inflection point). If we use such combination of correction parameters that on the critical isotherm (it is on the isotherm *373.946°C) is* an inflection point, we get the critical point shifted significantly to the left (to higher values of specific volumes than corresponds with data IAPWS-95) and the error of calculated values on the boundary curve is much more even for lower temperatures.

It turns out that if we want to calculate the thermodynamic properties of steam using "simple" equation of state based on the original BWR equation, it is useless to use its full form - 7th and 8th correction terms have not brought any significant refinement of calculated values of pressure. For routine engineering calculations of water vapor it is thus possible to use a simplified BWR equation of the form

$$p = rT\rho + rT\rho^2 B_0 + rT\rho^3 b - A_0\rho^2 - a\rho^3 - \frac{\rho^2 C_0}{T^2} \quad (4)$$

This equation gives sufficiently accurate results. When we use simplified BWR equation with coefficients, which we obtained by shown procedure, we can calculate the properties of water vapor to a state of beginning of condensation to the pressure *15 MPa* (temperature 340°C). For high temperature, equation can be used for even higher pressures (see Figure 1.), but pressures up to *30 MPa*.

6 Conclusions

Simplified BWR equation using coefficients derived in this paper, is less accurate than the equations used in IAPWS, near the critical point and in liquid areas even is not usable at all. The great advantage is the simplicity of equation, and thus easier possibility of calculating of the properties of the water vapor. Another advantage is the possibility of derivation of the relationship for other thermodynamic properties (isobaric specific heat capacity c_p, enthalpy h, ...) [10] and thus the simple possibility of calculating of the thermodynamic processes in the superheated steam without the using of thermodynamic tables of water and steam.

Acknowledgment

Work was supported by the project SGS 21000.

References

1. M.Vestfálová, P.Šafařík, Experimental Fluid Mechanics 2015 (to be published)
2. W.Wagner, A.Pruß, *The IAPWS Formulation 1995 for the Thermodynamic Properties of Ordinary Water Substance for General and Scientific Use*, Journal of Physical and Chemical Reference Data, **31**: 387-535 (2002)
3. O.Šifner, J.Klomfar, *Mezinárodní standardy termofyzikálních vlastností vody a vodní páry*, (1996)
4. M. Benedict, G. Webb, L. C. Rubin, Journal of Chemical Physics **10** (12): 747-758 (1942)
5. K. E. Starling, *Fluid Properties for Light Petroleum Systems*, (1973)
6. R. C. Reid, J. M. Prausnitz, B. E. Poling, *The Properties of Gases & Liquids* (4th ed.) (1987)
7. B. A. Younglove, J. F. Ely, Journal of Physical and Chemical Reference Data **16** (4): 577 (1987)
8. M. Benedict, G. Webb, L. C. Rubin, Chemical Engineering Progress (CEP) **47** (8): 419–422 (1951)
9. J. Novák, *Termodynamické vlastnosti plynů*, (2007)
10. M. Vestfálová, Experimental Fluid Mehcanics (2014)

EPJ Web of Conferences **114**, 02135 (2016)
DOI: 10.1051/epjconf/201611402135

Improvements of the experimental apparatus for measurement of the surface tension of supercooled liquids using horizontal capillary tube

Václav Vinš[1,a], Jan Hošek[1], Jiří Hykl[1] and Jan Hrubý[1]

[1]Institute of Thermomechanics of the CAS, v. v. i., Dolejškova 1402/5, 182 00 Prague 8, Czech Republic

Abstract. An experimental apparatus with a horizontal capillary tube for measurement of the surface tension of supercooled liquids, i.e. liquids in a metastable state below the equilibrium freezing point, was designed and tested in the previous study [V. Vinš *et al.*, EPJ Web Conf. **92**, 02108 (2015)]. In this work, recent modifications of both the experimental setup and the measurement analysis are described. The main aim is to improve the accuracy and the reproducibility of measured surface tension and to achieve higher degrees of supercooling. Temperature probes measuring the temperature of cooling medium near the horizontal capillary tube were calibrated in the relevant temperature range from − 31 °C to + 45 °C. An additional pressure transducer was installed in the helium distribution setup at the position close to the capillary tube. The optical setup observing the liquid meniscus at the open end of the horizontal capillary tube together with the video analysis were thoroughly revised. The red laser illuminating the liquid meniscus, used at the original apparatus, was replaced by a fiber optic light source, which significantly improved the quality of the meniscus image. The modified apparatus was used for the measurement of surface tension of supercooled water at temperatures down to − 11 °C. The new data have a lower scatter compared to the previous horizontal measurements and show a good agreement with the other data obtained with a different measurement technique based on the modified capillary rise method.

1 Introduction

An experimental apparatus developed for the measurement of surface tension of supercooled liquids, i.e. metastable liquids at temperatures below the equilibrium freezing point, using a horizontal capillary tube was introduced in the previous study [1]. The apparatus is based on a measuring technique developed by Ferguson and Kennedy [2] in 1932. In this method, a short thread of liquid sample is placed inside a horizontal capillary tube. One end of the capillary tube is connected to the pressure setup which allows tuning of a small counterpressure of a gas. The other end is left open to the ambient. It is illuminated by the light source and observed by optics. The gas counterpressure compensates the surface forces at the two menisci of the liquid thread. At low counterpressure, the meniscus at the open end of the capillary tube is concave, i.e. inside the capillary tube. With increasing counterpressure, the outer meniscus changes its shape from concave to planar and subsequently to convex. In case of the planar shape of outer meniscus, the surface forces become equal to zero. Consequently, the counterpressure of gas is in equilibrium only with the surface forces at the second meniscus inside the capillary tube. In this case, the surface tension σ at the inner meniscus can be determined from the following equation

$$\sigma = \frac{\Delta p \cdot d}{4\cos\theta}, \qquad (1)$$

where Δp denotes the counterpressure of gas at the planar outer meniscus, d is the inner diameter of the capillary tube, and θ is the contact angle between the liquid and the capillary tube wall.

The horizontal technique by Ferguson and Kennedy [2] was later improved by Hacker and used for the measurement of surface tension of supercooled water in 1951 [3]. Hacker's data measured down to − 22 °C show a significant change in the temperature trend of the surface tension of supercooled water. This anomaly, called the second inflection point (SIP) in the surface tension of water [4], was trusted to be a real behavior of the supercooled water. The reason is also the fact that the data by Hacker have relatively low scattering and good internal consistency. However, recent experiments performed with another technique based on the modified capillary rise method show that the surface tension of water does not show any anomaly at temperatures down to − 26 °C [4-6]. The surface tension of supercooled water can be predicted by the IAPWS (International Association for Properties of Water and Steam) standard [6] extrapolated below 0.01 °C. The IAPWS standard has following form

$$\sigma = 235.8\,\tau^{1.256}\left(1-0.625\,\tau\right), \qquad (2)$$

[a] Corresponding author: vins.vaclav@seznam.cz

where σ is the surface tension in mN m^{-1}, τ denotes a dimensionless temperature $1-T/T_c$ with the critical temperature of water $T_c = 647.096$ K. Equation (2) was developed by Vargaftik *et al.* [8] in 1970's and approved by IAPWS as an international standard [6]. The IAPWS standard is currently being revised with the use of recent experimental data in order to improve the estimates of its uncertainties and to extend the range of validity also to the supercooled region [8].

In the atmospheric research focused on nucleation processes, i.e. formation of droplets and ice crystals in the upper atmosphere, the correlation for the surface tension of supercooled water by Viisanen *et al.* [10] is often employed

$$\sigma = -2.75\cdot 10^{-4} T^2 + 9.133\cdot 10^{-3} T + 93.6635 \ , \quad (3)$$

where σ is the surface tension in mN m^{-1} and T is the thermodynamic temperature. Equation (3) given as a simple quadratic fit of temperature provides similar results as the extrapolated IAPWS correlation in the temperature range down to − 25 °C. However at higher temperatures within the stable region, equation (3) shows significant deviation from the IAPWS standard.

In this study, the experimental apparatus with the horizontal capillary tube introduced in the previous work [1] was further modified in order to improve the reproducibility and the accuracy of measured surface tension of supercooled water. The improved setup will allow achieving higher degrees of supercooling of up to 20 °C. Main motivation for this work is to obtain new valuable data for supercooled water with an experimental method different from the recent measurements with the modified capillary rise technique [4-6]. Moreover, the new measurements with the horizontal capillary tube may clarify the SIP anomaly in the Hacker's data [3], as Hacker used also the horizontal technique. New data from the preliminary measurements on the improved apparatus at temperatures from − 11 °C to + 15 °C are also provided.

2 Experimental apparatus

The measurement principle is similar to that in the experiments by Ferguson and Kennedy [2] and Hacker [3]. A simplified scheme of the experimental setup is shown in Figure 1. The horizontal capillary tube is partly placed inside the temperature–controlled chamber with flowing ethanol. The liquid thread inside the capillary tube is long such that the inner meniscus is located inside the temperature–controlled chamber. The outer meniscus at the open end is kept at an ambient temperature. Knowing the inner diameter of capillary tube, the temperature of inner meniscus, and the counterpressure of gas corresponding to the planar meniscus at the open end, the surface tension can be determined from equation (1). A constant contact angle $\theta = 3°$ [11] can be considered for the water and the fused silica capillary tube at temperatures below 200 °C [12]. Helium with purity of 99.996% was used as a working gas due to its low solubility in water and negligible adsorption on the liquid surface of water.

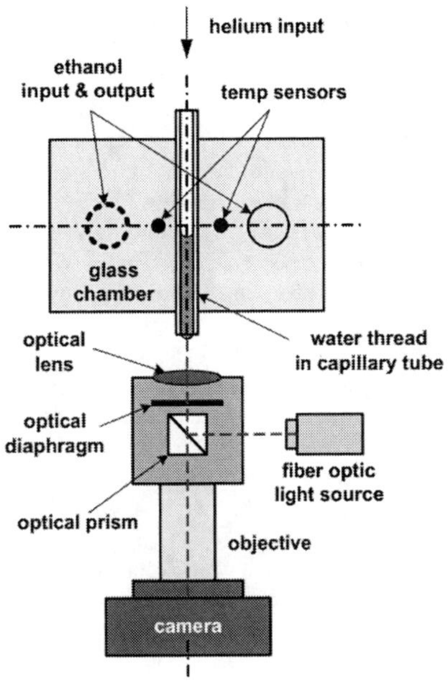

Figure 1. Scheme of experimental apparatus with the horizontal capillary tube; view from top

Similarly to our previous measurements with the modified capillary rise method [4-6], the horizontal measurement presented in this work is a relative measurement. The inner diameter of capillary tube d is evaluated as an average from the measurements at the reference temperature $T_{ref} = 15$ °C and the surface tension calculated from the IAPWS correlation (2)

$$d = \frac{1}{N}\sum_{i=1}^{N} \frac{\Delta p\left(T_{\mathrm{ref},i}\right)\cdot \sigma_{\mathrm{IAPWS}}\left(T_{\mathrm{ref},i}\right)}{4\cos\theta} \ . \quad (4)$$

The average inner diameter of capillary tube used in this study had value $d = 0.3353 \pm 0.0022$ mm.

All the experiments were performed with ultrapure water with a constant resistivity of 18.2 MΩ cm, total organic carbon content below 5 ppb, and free of particles larger than 0.2 μm.

3 Improvements of the setup

Preliminary measurements performed in the previous study [1] proved the functionality of main components of the new experimental apparatus with the horizontal capillary tube. However, only relatively low degree of supercooling of − 7 °C was achieved and the data for the surface tension of water were quite scattered. Consequently, several improvements of the entire apparatus were applied in this study. The temperature sensors were carefully calibrated in the relevant temperature range. An additional pressure transducer was installed in the helium distribution setup at a position closer to the capillary tube. Most importantly, the optical setup together with the subsequent analysis of recorded video were thoroughly revised, which markedly improved the accuracy and the consistency of measured data for the surface tension of water.

3.1 Measurement of temperature

Resistive temperature sensors Pt100 of Omega PR-11-3 series measuring the temperature of liquid ethanol inside the temperature–controlled chamber were carefully calibrated in the relevant temperature range from – 31 °C to + 45 °C. A secondary standard platinum resistance thermometer Isotech 909Q (25.5 ohm at 0 °C) was taken as the reference thermometer. All sensors were connected to the ASL F500 precision thermometer with accuracy better than ± 0.005 °C and ± 5 mK in the whole range from – 200 °C to + 962 °C. The calibrated sensors together with the reference thermometer were submerged in the Lauda Proline RP855 thermostatic bath with liquid ethanol. The accuracy of the thermostatic bath stated by the manufacturer is better than ± 0.01 °C. The standard deviation of the reference temperature did not exceed 0.0061 °C during the calibration.

Calibration curves were correlated with the Callendar–van Dusen (CvD) equation

$$R_t = R_0\left[1 + At + Bt^2 + C(t-100)t^3\right], \qquad (5)$$

where t denotes temperature in degrees of Celsius, R_t is resistance in ohm at temperature t, R_0 is the reference resistance at 0 °C, and A, B, and C are correlated coefficients. Coefficient C is equal to zero for $t > 0$ °C. Calibration parameters for the two most important temperature sensors positioned near the capillary tube are given in Table 1. The value for C was taken from the norm DIN 90.

Table 1. CvD coefficients of the calibrated temperature probes.

	Probe 22	Probe 23
R_0 / ohm	99.9862	99.9656
A / ohm °C^{-1}	$3.91442\cdot10^{-3}$	$3.91446\cdot10^{-3}$
B / ohm °C^{-2}	$-5.67342\cdot10^{-7}$	$-5.67737\cdot10^{-7}$
C / ohm °C^{-4}	$-4.1830\cdot10^{-12}$	$-4.1830\cdot10^{-12}$

3.2 Measurement of pressure

The analysis of uncertainties of measurements performed in the previous study [1] showed that the measurement of counterpressure Δp had the major influence on the uncertainty budget of measured surface tension. Therefore, new pressure transducer Omega PX 653 with a pressure range of 1245 Pa and accuracy 0.25% FS was additionally installed inside the helium distribution setup described in section 2.2 in the previous work [1]. The additional pressure transducer was placed after the mechanical filter shown in Figure 3 in the previous study [1]. Its position is markedly closer to the horizontal capillary tube than the original pressure transducer Furness Controls FCO 332 installed behind the high-precision needle valve on the pressure manifold. The additional transducer should help to detect possible differences between the pressure measured by the Furness Controls transducer on the manifold and the pressure near the capillary tube. Nevertheless, no significant differences or time delays, when varying the flow rate of helium through the distribution setup, were observed between the pressures measured at the manifold and near the capillary tube. The output signals of both pressure transducers were logged with a constant time step of 125 ms using the data acquisition (DAQ) unit Agilent 34970A with fast multiplexer 34902A.

3.3 Optical setup

The modifications of the optical setup and the subsequent analysis of recorded video represent the most important improvements compared to the previous study [1]. The original design of optical setup was described by Hošek *et al.* [13] in detail. In the current study, the red laser light source used for the illumination of liquid meniscus was replaced with a fiber optic light source Zeiss KL 200. An additional optical diaphragm was placed behind the optical prism shown in Figure 1 in order to exclude nonparallel beams which negatively influenced the final image of liquid meniscus. We note that the diaphragm was not used in the original setup with the red laser even though it was shown in Figure 1 in the previous study [1].

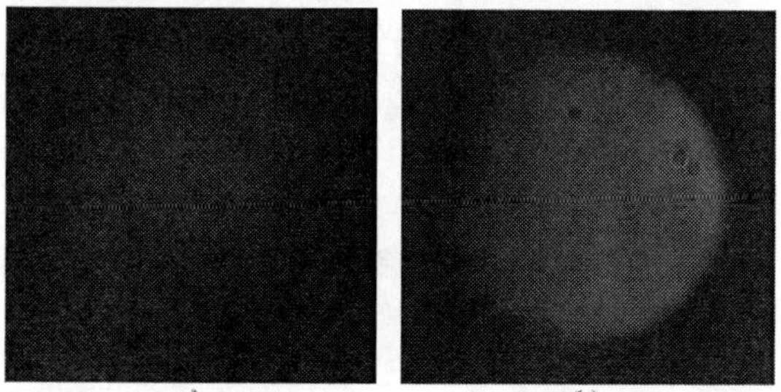
a) b)

Figure 2. Planar liquid meniscus illuminated by the red laser (a) and the fiber optic light source with diaphragm (b)

Figure 2 compares the pictures of planar liquid meniscus illuminated by the red laser and the fiber optic light source taken by the digital camera Thorlabs DCC1645C-HQ. As can be seen the picture taken with the fiber optic light source is more uniform compared to the picture obtained with the laser light source. The difference becomes more distinct when the meniscus has more convex or concave shape. In case of the laser light source, disturbing reflections appear in the image, which negatively affect the subsequent analysis of the recorded video.

A constant frame rate of 8 frames per second, corresponding to the time step of 125 ms used also on DAQ for the pressure transducers, was set for the video record. Even though the digital camera and DAQ worked with the same time step, a difference in the lengths of the recorded video and the data logged by DAQ was observed. The video record was usually slightly shorter than the DAQ record. With the kind help of Thorlabs GmbH helpdesk, it was found that the reason for the shorter video record was caused by the failures in the video transmission between the camera and the measuring computer. Luckily, the Thorlabs software enables detection of such errors. An example of the log file containing the transmission errors is shown in Figure 3. Knowing the exact time of the missing frames allowed correcting the video in the subsequent analysis of the measurement such that the video record and the pressure record became fully consistent with each other.

```
24  -  15:05:32.291  -  Transfer failed: Capture Error (178).
25  -  15:05:51.858  -  Transfer failed: Capture Error (178).
26  -  15:06:57.758  -  Transfer failed: Capture Error (178).
27  -  15:07:33.334  -  Transfer failed: Capture Error (178).
```

Figure 3. Example of the log file containing the flags for errors in the video transfer

4 Description of measurement

The principle of current measurement is similar to the previous study [1]. As it was discussed in section 3, the most important improvement lies in better analysis of the recorded video. An important fact that needs to be mentioned is that the open end of the horizontal capillary tube had to be precisely polished and grinded as described by Hošek and Studenovský [14]. The outer surface of the capillary tube has to be planar and perpendicular to the main capillary tube axis in order the well visible and reproducible planar meniscus can be achieved.

4.1 Methodology

The video record and the pressure record were synchronized using a white light-emitting diode (LED) illuminating the open end of the capillary tube. The power voltage of LED of 5 V was logged as an additional channel on DAQ for this purpose. The video record and the pressure record were synchronized by turning off and turning on the diode. No difference between the number of video frames and the number of data points logged by DAQ was detected after the video transfer analysis discussed in section 3.3. The measurements with only one missing frame were still taken as good points. All measurements with the difference higher than one frame were treated as bad points and excluded from the further analysis.

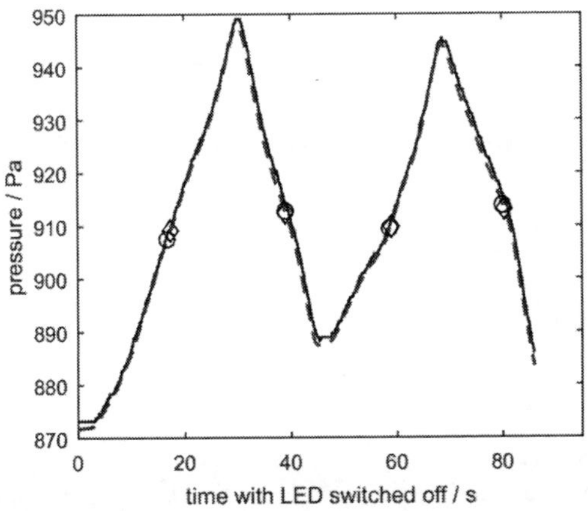

Figure 4. Two pressure loops measured over time with turned off LED for the measurement at – 6.9 °C. Red - -, pressure transducer Furness Controls FCO 332; blue –, pressure transducer Omega PX653. Pressures corresponding to the planar meniscus measured by, black o, Furness Controls FCO 332 and, black ◊, Omega PX653.

Similarly as in the previous study [1], each measurement was taken as two subsequently increasing and decreasing pressure loops. The final value of counterpressure corresponding to the planar meniscus was taken as an overall average of two points at the increasing pressure and two points at the decreasing pressure. No systematic difference between the four values was observed in any of the current experiments. Figure 4 shows an example of the typical pressure course with two loops for the measurement at – 6.9 °C.

Two to three independent measurements were carried out for each temperature. The uncertainty of measured temperature was better than ± 0.06 °C in all cases. The uncertainty of measured counterpressure of helium was better than ± 2.3 Pa, i.e. better than in the previous study [1] where $u(\Delta p)$ was better than ± 3.0 Pa.

4.2 Analysis of the video

Figure 5 shows a typical picture of capillary tube end taken by the digital camera. The capillary tube is illuminated by LED in this case. A small spot inside the capillary tube in the center of the figure is the concave meniscus illuminated by the fiber optic light source. The recorded video is analyzed inside the measuring rectangle containing the hole of capillary tube and its close surroundings. Yellow point on the right side of the figure is an arbitrary chosen point for the detection of light intensity of LED.

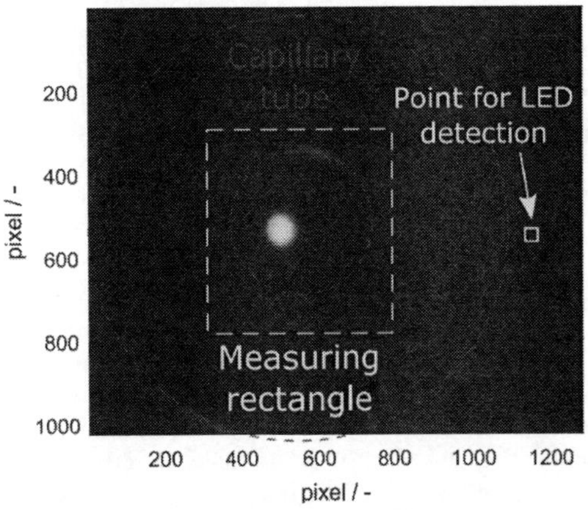

Figure 5. Photograph of the concave meniscus inside the capillary tube with the low counterpressure of helium at the beginning of measurement. LED is turned on.

The varying shape of liquid meniscus can be detected from the sum of light intensity inside the measuring rectangle. At the concave or convex shape, the meniscus behaves as a curved mirror. Consequently, only a small central part of the meniscus is visible in the camera and the sum of light intensity is relatively low. On the other hand, the planar meniscus behaves as a plane mirror. Most of the pixels inside the hole of capillary tube are illuminated in this case and the sum of light intensity within the measuring rectangle reaches its maximum.

Figure 6 shows an example of relative light intensity, i.e. the light intensity divided by its value at the start of measurement, over time for the measurement at – 6.9 °C. Clear four maxima in the light intensity were found, which correspond to the two pressure loops shown in Figure 4. The light intensity at the point for the detection of LED is also shown. Both the light intensity and the power voltage of LED change within one time step when turning LED on or off. Points for the synchronization of the recorded video and the recorded pressure are marked with purple circles.

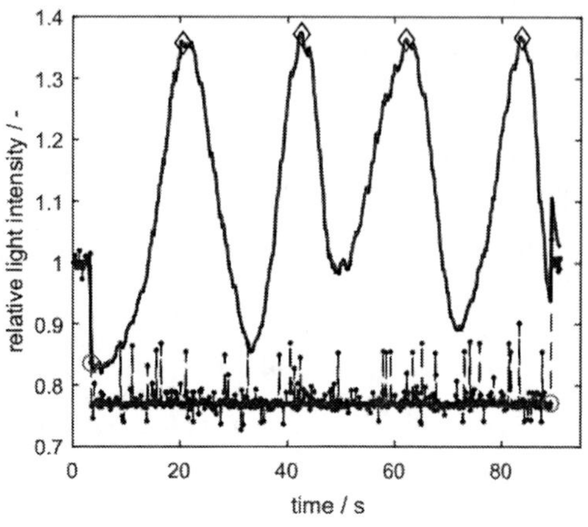

Figure 6. Relative light intensity I_i/I_1 over time for the measurement at – 6.9 °C. Blue -•-, light intensity at the point for detection of LED; purple ○, time of turning LED off/on; black –, sum of light intensity within the measuring rectangle; red –, moving average of light intensity within the rectangle; black ◊, maxima of sum of light intensity corresponding to the planar meniscus.

5 Results

The improved experimental apparatus with the horizontal capillary tube was successfully used for the measurement of surface tension of supercooled water. Preliminary measurements were performed at temperatures down to – 11 °C. Main reason for the presented measurements was to test the improved apparatus and to validate the video analysis. The relatively low degrees of supercooling were measured due to a potential risk of crystallization of the liquid sample. Unlike in the modified capillary rise method [4-6], the horizontal capillary tube is open to ambient, which increases the risk of abrupt solidification of the water sample.

Figure 7 shows the experimental data for the surface tension of supercooled water. The new measurements at temperatures between – 11 °C and + 15 °C (red diamonds) are in good agreement with both the capillary rise data [4-6] and the IAPWS correlation (2) extrapolated below 0.01 °C. The reference temperature T_{ref} for the evaluation of the inner diameter from equation (4) was + 15 °C.

The difference between the experimental data and the IAPWS standard extrapolated below the triple point of water is shown in Figure 8. As can be seen, the new data obtained with the improved apparatus are more consistent than the previous measurements [1]. Moreover, the new data support the previous measurements of our group with the modified capillary rise method [4-6]. A linear extrapolation of the IAPWS correlation [6] from 0.01 °C and the correlation by Viisanen *et al.* [10] are also shown. The uncertainty of surface tension measured with the horizontal method is better than 0.62 mN m^{-1}. This value is 50% higher than the uncertainty of the capillary rise measurements [6], which have the uncertainty of 0.41 mN m^{-1} at temperatures around – 12 °C. Nevertheless, similarly to the capillary rise measurements, the main output quantity of the presented experiment shall be the relative surface tension $Y = \sigma/\sigma_{corr}(T_{ref})$. The relative surface tension together with its uncertainty will be determined in the next step.

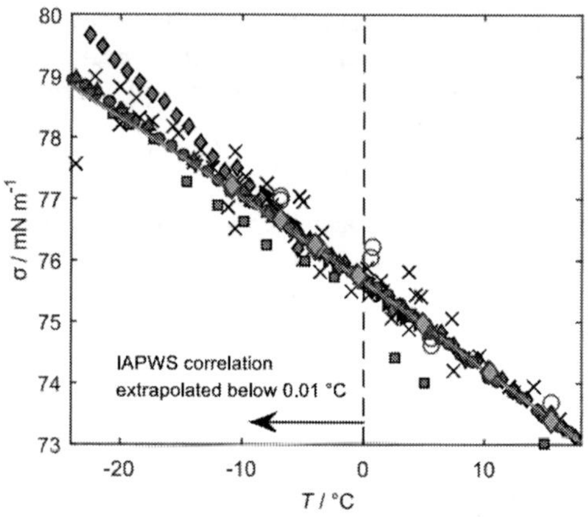

Figure 7. Experimental data for the surface tension of supercooled water. Green –, IAPWS correlation extrapolated below 0.01 °C [6]; purple - -, correlation by Viisanen *et al.* [10]; red ◊, this study; red ○, previous study [1]; black ▶, Humphreys and Mohler [15]; black ♦, Hacker [3]; black x, Floriano and Angell [16]; black ■, Trinh and Ohsaka [17]; blue •, Hrubý *et al.* [4]; blue ▲ and ▼, Vinš *et al.* [6] *p*-data and *h*-data, respectively.

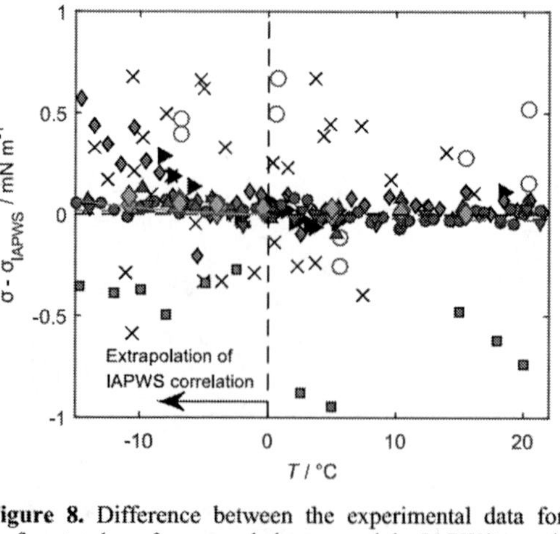

Figure 8. Difference between the experimental data for the surface tension of supercooled water and the IAPWS correlation extrapolated below 0.01 °C [6]. Green –, linear extrapolation of IAPWS correlation from the triple point; purple - -, correlation by Viisanen *et al.* [10]; red ◊, this study; red ○, previous study [1]; black ▶, Humphreys and Mohler [15]; black ♦, Hacker [3]; black x, Floriano and Angell [16]; black ■, Trinh and Ohsaka [17]; blue •, Hrubý *et al.* [4]; blue ▲ and ▼, Vinš *et al.* [6] *p*-data and *h*-data, respectively.

6 Conclusions

In this study, an experimental apparatus for measurement of the surface tension of supercooled liquids was improved compared to its original design introduced in the previous work [1]. Temperature sensors were calibrated in the range of temperatures relevant for the supercooled aqueous liquids. An additional pressure transducer was installed closer to the capillary tube and used for the control of the helium distribution setup. The optical setup and the analysis of recorded video was thoroughly revised. The new data for the surface tension of supercooled water measured down to – 11 °C are significantly less scattered than the previous measurements. Moreover, the data obtained with the horizontal technique agree with the capillary rise measurements [4-6], which represent a very important result of this study. The data for the surface tension of supercooled water obtained both with the capillary rise method and the horizontal technique will be used in the revision of the current IAPWS standard for the surface tension of water [6,8].

The SIP anomaly detected by Hacker [3], using also the horizontal capillary tube, was not observed in the new measurements in the temperature range from – 11 °C to + 15 °C. Nevertheless, further measurements at higher degrees of supercooling are planned in order to definitely clarify the SIP anomaly. The improved apparatus will allow achieving temperatures down to – 20 °C. A detailed analysis of the uncertainty budget and the evaluation of the relative surface tension are also under preparation.

Acknowledgement

The study was supported by the Czech Science Foundation grant GJ15-07129Y and by the institutional support RVO:61388998 from IT CAS. Many thanks go to Prof. Karel Studenovský from the Czech Technical University in Prague for his help with the capillary tube grinding and polishing and to Wolfgang Skopalik from Thorlabs GmbH for his help with the detection of errors in the video transfer from the digital camera.

References

1. V. Vinš, J. Hošek, J. Hykl, J. Hrubý, EPJ Web Conf. **92**, 02108 (2015)
2. A. Ferguson, S.J. Kennedy, Proc. Phys. Soc. **44**, 511 (1932)
3. P.T. Hacker, NACA TN **2510** (1951)
4. J. Hrubý, V. Vinš, R. Mareš, J. Hykl, J. Kalová, J. Phys. Chem. Lett. **5**, 425 (2014)
5. V. Vinš, J. Hrubý, J. Hykl, J. Blaha, B. Šmíd, EPJ Web Conf. **45**, 01094 (2013)
6. V. Vinš, M. Fransen, J. Hykl, J. Hrubý, J. Phys. Chem. B **119**, 5567 (2015)
7. IAPWS revised release on surface tension of ordinary water substance. 2014; http://www.iapws.org
8. N.B. Vargaftik, B.N. Volkov, L.D. Voljak. J. Phys. Chem. Ref. Data **12**, 817 (1983)
9. Minutes from 2014 IAPWS Meeting, TPWS/SCSW Working Group. 2014; http://www.iapws.org
10. Y. Viisanen, R. Strey, H. Reiss, J. Chem. Phys. **99**, 4680 (1993)
11. V.G. Stepanov, L.D. Volyak, Y.V. Tarlakov, J. Eng. Phys. Thermophys. **32**, 646 (1977). Translated from Inzhenerno-Fizicheskii Zhurnal **32**, 1000 (1977)
12. L.D. Volyak, V.G. Stepanov, Y.V. Tarlakov, Zh. Fiz. Khim. **49**, 2931 (1975)
13. J. Hošek, V. Vinš, J. Hykl, Proc. of SPIE **9442**, 94420E (2015)
14. J. Hošek, K. Studenovský, EPJ Web Conf. **48**, 00005 (2013)
15. W.J. Humphreys, J.F. Mohler, Phys. Rev. **2**, 387 (1895)
16. M.A. Floriano, C.A. Angell, J. Phys. Chem. **94**, 4199 (1990)
17. E. Trinh, K. Ohsaka, Int. J. Thermophys. **16**, 545 (1995)

EPJ Web of Conferences **114**, 02136 (2016)
DOI: 10.1051/epjconf/201611402136

Molecular Simulations of the Vapor–Liquid Phase Interfaces of Pure Water Modeled with the SPC/E and the TIP4P/2005 Molecular Models

Václav Vinš[1], David Celný[1,2], Barbora Planková[1,2,a], Tomáš Němec[1], Michal Duška[1], and Jan Hrubý[1]

[1] Institute of Thermomechanics of the CAS, v. v. i., Dolejškova 1402/5, Prague, 182 00, Czech Republic
[2] Faculty of Nuclear Sciences and Physical Engineering, Czech Technical University in Prague, Břehová 78/7, Prague, Czech Republic

Abstract. In our previous study [Planková et al., EPJ Web. Conf. **92**, 02071 (2015)], several molecular simulations of vapor-liquid phase interfaces for pure water were performed using the DL_POLY Classic software. The TIP4P/2005 molecular model was successfully used for the modeling of the density profile and the thickness of phase interfaces together with the temperature dependence of the surface tension. In the current study, the extended simple point charge (SPC/E) model for water was employed for the investigation of vapor-liquid phase interfaces over a wide temperature range from 250 K to 600 K. The TIP4P/2005 model was also used with the temperature step of 25 K to obtain more consistent data compared to our previous study. Results of the new simulations are in a good agreement with most of the literature data. TIP4P/2005 provides better results for the saturated liquid density with its maximum close to 275 K, while SPC/E predicts slightly better saturated vapor density. Both models give qualitatively correct representation for the surface tension of water. The maximum absolute deviation from the IAPWS standard for the surface tension of ordinary water is 10.4 $mN \cdot m^{-1}$ and 4.1 $mN \cdot m^{-1}$ over the temperature range from 275 K to 600 K in case of SPC/E and TIP4P/2005, respectively.

1 Introduction

Molecular simulations (MS) represent a computational technique that allows to study a large variety of physical and chemical processes on the molecular scale. MS works particularly well for fluid systems as the liquid phase and the gas phase can be treated simultaneously. The only difference between the phases is given by the local density.

There are two different methods of MS, the molecular dynamics (MD) and Monte Carlo. MD solves a simple Newton equation of motion for every atom in discretized time-steps while Monte Carlo uses the probability nature of the statistical mechanics. Although all the methods in MS stand on relatively simple basis, their more wide-spread use has emerged only in recent years. The main limitation of all MS methods are their enormous computational demands. However, as computers are getting faster and more powerful, MS can gradually solve larger systems over longer time periods and more complex physical models.

One of the most important substances investigated using the approaches of MS is water. Water is essential in daily life, industry, and biological and natural processes. However, it has a typical non-standard behavior and many anomalies, which are still hard to model theoretically. The non-trivial phenomena are caused especially by its polar character and by the formation of hydrogen bonds.

The pioneers in successful simulations of water were Barker and Watts [1] and Rahman and Stillinger [2] around 1970. For many problems which include modeling complex aqueous systems, relatively simple water models such as SPC or TIPXP-type (TIP3P, TIP4P, or TIP5P) can be used. Currently, the modifications of the original SPC and TIP4P models are considered as the most convenient water models; namely the SPC/E model [3,4] and the TIP4P/2005 model [5,6].

In this study, we continue MD simulations of water [7] focused on the investigation of properties at the vapor–liquid phase interface. A series of MS was performed with both the SPC/E and TIP4P/2005 models in order to evaluate the density profiles, the thicknesses of the phase interface, and the surface tensions. A "slab" geometry with water molecules was simulated at various temperatures from the triple point of water to the temperatures close to the critical point. The conditions and settings of MS were inspired by the works of Vega and Miguel [8] and Sakamaki *et al.* [9]. All simulations were performed using DL_POLY Classic 1.9 [10,11] running on a cluster with 4 × 2 Intel(R) Xeon(R) CPU E5645 (12 cores effectively for each CPU) @ 2.53GHz CPUs. New results were compared to both MS from the literature and the experimental data.

2 Water models

Despite the fact that water molecule is flexible and polarizable, simple rigid models are used regularly in many MS. Therefore, systems with large numbers of molecules can be modeled for a longer times. The most widespread simple models belong to the SPC and TIPXP families. A common feature of these models is that they locate the Lennard-Jones interaction site on the position of the oxygen

[a] e-mail: : barbora.plankova@gmail.com

and the positive charge on the hydrogen atoms. Nevertheless, they differ in the target properties, i.e. the charge distribution and the angle of H-O-H bonds. The basic three-site models TIP3P [12] and SPC [13] were developed in 1980's. An extended modification of the simple bond charge, i.e. the SPC/E model by Berendsen *et al.* [14], including empirical corrections to the polarization energy is very successful and reproduces vapor–liquid data very well. A modification of TIP3P is the TIP4P model, which tries to improve the electrostatic distribution within the water molecule [12] by adding an auxiliary atom. This atom is located close to the oxygen atom (only 0.15 Å distant), it has no mass and carries a negative charge. A further modification, called TIP4P/2005, was proposed by Vega and Abascal [15] in 2005. They tried to combine the good phase diagram of TIP4P with target properties of SPC/E improving the melting point. There are many other modifications of the TIP4P model, e.g., the extension TIP5P by Mahoney *et al.* [16] carrying the negative charge on two auxiliary atoms. However, the general performance of most of these models is not better than TIP4P/2005.

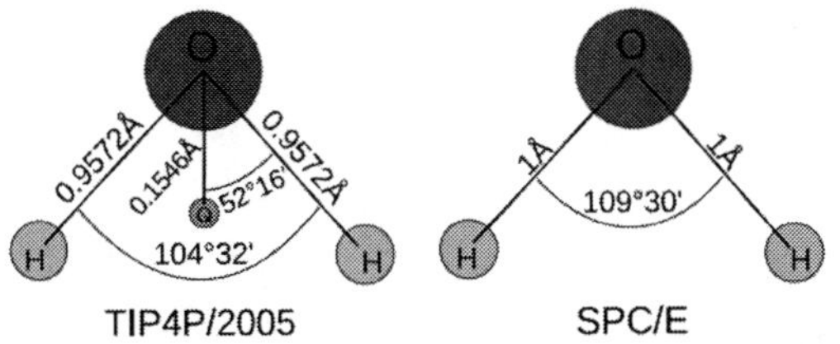

Fig. 1. Schematic representation of the SPC/E and TIP4P/2005 molecular models for water.

In the present study, the two models TIP4P/2005 and SPC/E are employed to investigate vapor–liquid interfacial properties of water. Both the TIP4P/2005 and SPC/E models consider the Lennard-Jones interactions only between the oxygen atoms, because hydrogen atoms have a negligible mass compared to oxygen. The electrostatic interaction occurs between the hydrogen atoms and oxygen atom in the SPC/E model and between hydrogen atoms and the massless atom Q with the negative charge in the TIP4P/2005 model, respectively. A schematic representation of the SPC/E and TIP4P/2005 molecular models is shown in figure 1. The parameters of the models are summarized in Table 1. Both models provide a very good description of the vapor–liquid equilibria compared to the other molecular models popular for water, e.g., SPC, TIP3P, TIP4P/Ice [8]. Moreover, TIP4P/2005 gives a reasonable description also for the compressibility, the maximum in liquid density (near room temperature), and thermal expansion coefficient.

Table 1. Parameters of the molecular models SPC/E and TIP4P/2005 for water: oxygen O, hydrogen H, massless atom with charge Q. M denotes molar mass, q is charge in units of elementary charge, ϵ and σ are Lennard–Jones parameters.

Model / atom	M (g/mol)	q (e)	ϵ (kJ/mol)	σ (Å)
SPC/E				
O	15.99940	−0.8476	0.6500	3.1660
H	1.00794	0.4238	–	–
TIP4P/2005				
O	15.99940	0.0	0.7749	3.1589
H	1.00794	0.5564	–	–
Q	0.0	−1.1128	–	–

3 Simulation methods

A series of MD simulations was performed with the NVT ensemble, i.e. for a constant number of molecules, a constant total volume of the system, and a constant temperature. The ensemble consisted of 1372 molecules, which is considerably more than in other studies, e.g., both Vega and Miguel [8] and Sakamaki *et al.* [9] considered 1024 molecules. The simulations were carried out in two steps. At first, a cubic box consisting only of liquid phase was simulated for 50 ps with 1 ps equilibration time. Unlike in other studies, the size of the cubic box was not constant as it varied with temperature. The size of the box was determined from the temperature dependence of the saturated liquid density calculated from the TREND package [17] using the reference equation of state for water by Wagner and Pruss [18]. The liquid densities of the SPC/E and TIP4P/2005 models significantly differ from the experimental values at high temperatures close to the critical point. Consequently, at temperatures above 450 K, the box size had to be evaluated from an estimate of the liquid density from previous MS and not from the reference equation of state. After a homogeneous liquid phase was formed in the cubic box, the z-size of the simulation box was symmetrically expanded by a factor of 3. As a result, the liquid phase forms a slab in the centre of the simulation box and the expanded empty volumes on each side of the slab allow subsequent formation of the vapor phase. The slab configuration was simulated for 10 ns in all cases. An additional equilibration was not required as the liquid phase formed from the cubic box had convenient initial physical conditions. A constant time step of 2 fs was considered in all simulations. This value was also used in similar simulations by Chen and Smith [4] and Sakamaki *et al.* [9].

An initial configuration of molecules within the cubic box had to be prepared as an input for the DL_POLY program before the simulation. The configuration was prepared using our code developed in MATLAB [19]. The initial cubic box was generated as a grid of oxygen atoms with face-centered cubic unit cell. The remaining atoms of the water molecules, i.e. the hydrogen atoms and in case of TIP4P/2005 also the massless charge atom Q, were afterwards placed with consideration of the molecular structure (figure 1) and then randomly rotated. A thin void of 1 Å was added around the cubic box on each side to prevent overlapping of atoms from the simulation box randomly. The last configuration of the simulation within the cubic box was subsequently used as the input configuration for the slab simulation. In both simulations (in the cubic box and in the slab geometry), the periodic boundary conditions were imposed on all walls of the simulation box. However, the periodic boundary conditions can cause problems when extending the simulation box in the z-direction, because the molecules close to the walls whose atoms are periodically reflected on the other side of the cubic box can split up. Consequently, each molecule influenced by the periodic conditions on the boundaries had to be joint together after the expansion of the simulation box such that the molecule would not be pulled apart. We developed another MATLAB code for this purpose.

The MD simulations do not fix the temperature implicitly, therefore a thermostat is needed to correct kinetic energies of the molecules and thus regulate the temperature of the system. For this purpose, we used the Nosé–Hoover thermostat with the relaxation constant of 100 fs. The Verlet velocity integrator was used. The cutoffs of the Lennard–Jones interactions and the van der Waals forces were set to 14.5 Å. The direct Ewald method with an automatic parameter optimization constant set to 10^{-5} was used for electrostatic interactions. This value gave a convergence parameter of 0.192 $Å^{-1}$ and k-vector indexes of $6 \times 6 \times 6$ for the cubic box and $6 \times 6 \times 18$ for the slab geometry. For higher temperatures closer to the critical point, k-vector indexes increased up to $7 \times 7 \times 7$ and $7 \times 7 \times 29$, respectively.

4 Results

A series of 15 simulations at various temperatures spaced by 25 K was performed for each of the two water models. One simulation of the slab geometry running on 22 CPU cores took on average 48 hours in the case of the SPC/E model and 62 hours in the case of the TIP4P/2005 model. The difference in the computational time is caused by the additional massless atom Q that needed to be included in the TIP4P/2005 model.

The density profile $\rho(z)$ at a given temperature was computed as an average from the histograms of positions of water molecules in the z-direction calculated at each time step. Properties of the phase interface can, according to Chapela *et al.* [20], be determined from the correlation of the averaged density profile by a hyperbolic tangent function given in the following manner

$$\rho(z) = \frac{\rho_L + \rho_V}{2} - \frac{\rho_L - \rho_V}{2} \tanh\left(\frac{z - z_0}{d}\right). \quad (1)$$

In equation (1), z marks the coordinate perpendicular to the phase interface, $\rho(z)$ denotes the density in $Å^{-3}$ and z_0, d, ρ_L, and ρ_V are adjustable parameters. z_0 is the position of the Gibbs dividing surface of the interface, d is a parameter of the phase interface thickness, ρ_L denotes the saturated liquid density, and ρ_V is the saturated vapor density. We note that Chapela *et al.* [20] used the term $2(z - z_0)/d$ in the hyperbolic tangent, while the subsequent studies, e.g., [3, 8, 9], considered only $(z - z_0)/d$. To assure compatibility with other MS studies of water, we decided to use the later definition in equation (1) as well.

Figure 2 shows the saturated liquid and vapor densities obtained with the SPC/E model compared to other MS and to experimental data. The experimental data are represented by predictions of the reference equation of state IAPWS-95 [18] obtained from the TREND package [17]. The new simulations are in good agreement with other studies [3, 8, 9]. The saturated vapor density ρ_V obtained with the SPC/E model is in a quite good agreement with the IAPWS-95 equation of state, while the saturated liquid density ρ_L shows a clear deviation. The SPC/E liquid density is lower by 0.6% and 5.3% than the experimental data at temperatures 250 K and 500 K, respectively. We note that the difference becomes less pronounced when comparing the densities related to the reduced temperature $T_r = T/T_c$. As will be discussed later, near Table 2, both molecular models have lower critical temperature T_c than the physically correct value employed in IAPWS-95.

The melting point temperature of both water models considered in this work is below the real triple point of water 273.16 K. The triple point values are 213 K for the SPC/E model and 249 K for the TIP4P/2005 model [9]. Consequently, the vapor–liquid equilibrium could be calculated down to 250 K for both models.

The densities for the TIP4P/2005 model are shown in figure 3. The saturated liquid density obtained with TIP4P/

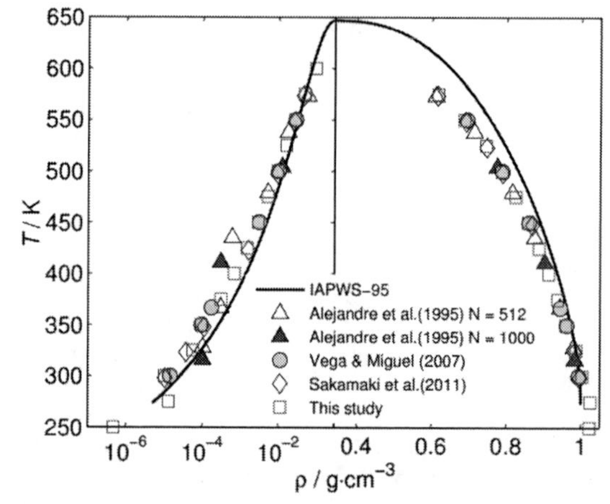

Fig. 2. Saturated liquid and vapor densities for water obtained from MS with the SPC/E water model. Comparison of results from this study with the literature data [3, 8, 9] and the reference equation of state IAPWS-95 [18].

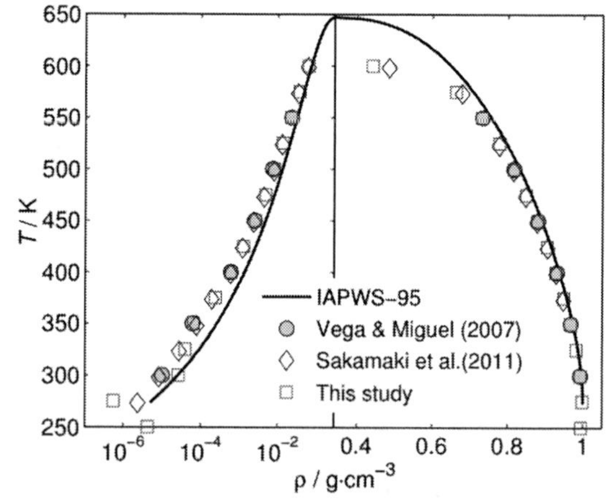

Fig. 3. Saturated liquid and vapor densities for water obtained from MS with the TIP4P/2005 water model. Comparison of results from this study with the literature data [8, 9] and the reference equation of state IAPWS-95 [18].

2005 lies closer to IAPWS-95 than the SPC/E model data (see figures 2 and 3). The TIP4P/2005 liquid density is lower only by 0.2% and 1.9% than the experimental data at temperatures 250 K and 500 K, respectively. Moreover, the TIP4P/2005 model provides relatively good representation of the maximum in the density of liquid water at 277 K. The maximum of approximately 0.999 g/cm^3 lies close to 275 K in our simulations. The saturated vapor density obtained with TIP4P/2005 is slightly worse than in the case of SPC/E.

The temperature dependence of the 10–90 thickness of the phase interface t related to thickness parameter d from equation (1) as $t = 2.1972d$ is shown in figure 4. The 10–90 thicknesses calculated with SPC/E and TIP4P/2005 are in good mutual agreement over the whole temperature range from 250 K to 600 K. The results of the new simulations are compared to other MS and the experimental data by Matsumoto and Kataoka [21]. As can be seen, the new results agree with the previous simulations at low temperatures below 400 K. However, with the increasing temperature, the thickness from the new simulations becomes considerably higher than in other MS, studies which are in relatively good mutual agreement over the entire temperature range. The reason for this difference might be the higher amount of molecules. Our ensemble consists of 1372 molecules compared to Alejandre *et al.* [3] with 1000 molecules or Vega and Miguel [8] with 1020 molecules in ensemble. As will be shown later in Table 3, number of molecules in the simulated ensemble has a systematic influence on the calculated results for the saturated liquid density and the interface thickness.

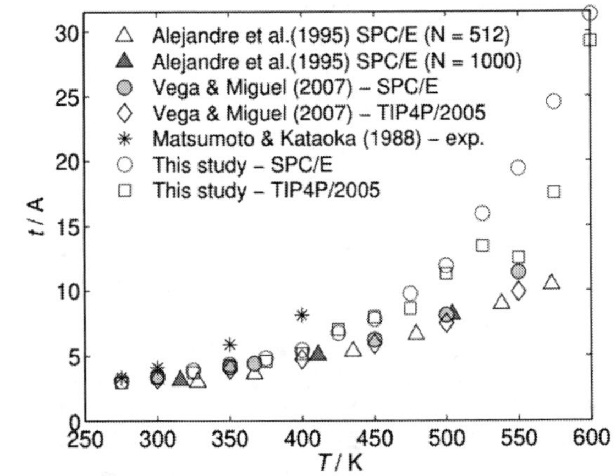

Fig. 4. The 10–90 thickness of the interface given as $t = 2.1972d$, where the thickness parameter d was determined from relation (1). Results of the new simulation compared to the literature data [3,8] and experimental data [21].

The experimental data for the interface thickness t that is shown in figure 4 were obtained by Matsumoto and Kataoka [21] using elipsometry, a measuring technique that allows to determine information about the interface thickness and the dielectric constant from the polarization of the light reflected at the phase interface. As can be seen in figure 4, the MS agree well with the experimental data at temperatures below 300 K. At higher temperatures of 350 K and 450 K, both the new simulations and the previous MS studies underpredict the experimental data for t by several Ångströms.

The surface tension was computed in a conventional way as a sum of the pressure tensor (or initial) term γ_{pt} and the tail correction to the long-range Lennard–Jones interactions γ_{tail} that corrects the cutoff truncation error,

$$\gamma = \gamma_{pt} + \gamma_{tail}. \tag{2}$$

The pressure tensor term can be determined in the following manner

$$\gamma_{pt} = \frac{L_z}{2}\left(P_{zz} - \frac{P_{xx} + P_{yy}}{2}\right), \tag{3}$$

where L_z denotes the box size in z direction, P_{ii} is the ii-th diagonal component of the pressure tensor. The tail correction γ_{tail} is given by

$$\gamma_{tail} = 12\pi\epsilon\sigma^6(\rho_L - \rho_V)^2 \times \int_0^1 \int_{r_c}^{\infty} \coth\left(\frac{rs}{d}\right)\frac{3s^3 - s}{r^3}\mathrm{d}r\mathrm{d}s, \tag{4}$$

where ϵ and σ are the Lennard–Jones parameters for oxygen atom, r_c is the cutoff for the Lennard–Jones potential and ρ_L and ρ_V are the bulk densities determined from equation (1).

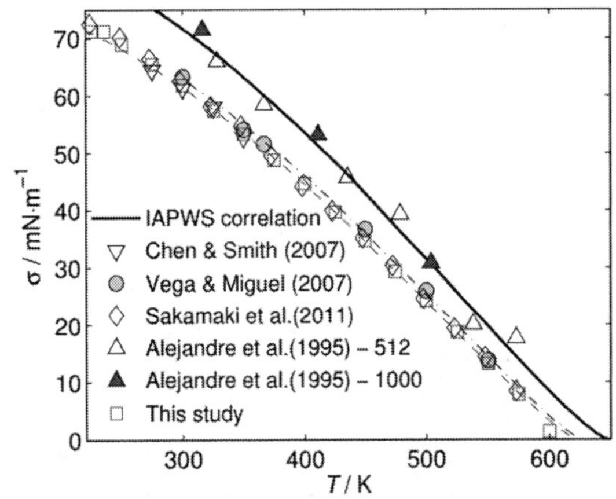

Fig. 5. Surface tension of water obtained from MS with the SPC/E water model. Comparison of results from this study with the literature data [3,4,8,9] and the IAPWS standard [22]. Dash-and-dotted black line and dashed red line are results of relation (5) with parameters taken from Table 2 for Vega and Miguel [8] and this study, respectively.

The results for the surface tension obtained with the SPC/E model are shown in figure 5. The new simulations are in a very good agreement with the previous studies by Chen and Smith [4], Vega and Miguel [8], and Sakamaki *et al.* [9]. The surface tension predicted in 1995 by Alejandre *et al.* [3] is much higher compared to other MS studies and lies remarkably close to the experimental data represented by the IAPWS standard for the surface tension of ordinary water [22]. Moreover, the data by Alejandre *et al.* [3] show relatively large scatter compared to other studies. Alejandre *et al.* [3] performed rather short simulations with 250 ps equilibration time and subsequent production runs of 250 ps and 125 ps for samples with 512 and 1000 molecules, respectively. The time step was set to 2.5 fs, i.e. a similar value as in other simulations. However, the overall simulation time between 375 and 500 ps seems to be insufficient in order to obtain correct and reproducible results. For example, Vega and Miguel [8] stated that a minimum time for the surface tension simulations is above 0.5 ns. They performed their simulations for 1.5 to 2.0 ns, which is still at least five times less than in the case of Sakamaki *et al.* [9] and this study. Chen and Smith [4] concluded that a minimum time for an accurate surface tension simulations is around 2 to 5 ns. The longer simulation time of 10 ns in the new simulations may also explain the differences in the increasing 10–90 interface thickness t shown in figure 4. Unfortunately, Sakamaki *et al.* [9] did not provide the results for t from their 10 ns simulations, which would be an interesting comparison with our results.

The TIP4P/2005 model gives a better prediction for the surface tension of water compared to the SPC/E model. The difference between the models can be seen from the deviations of the SPC/E and TIP4P/2005 results from the black line representing the IAPWS standard [22] in figures 5 and 6. The new results for TIP4P/2005 are in good agreement with our previous simulations [7] and the results by Vega and Miguel [8] and Sakamaki *et al.* [9].

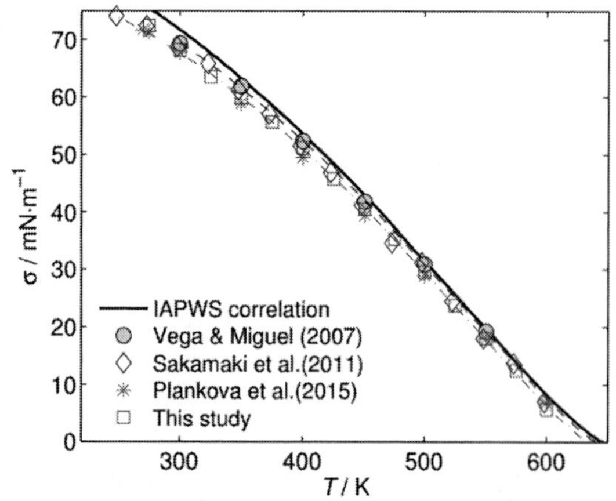

Fig. 6. Surface tension of water obtained from MS with the TIP4P/2005 water model. Comparison of results from this study with the literature data [7–9] and the IAPWS standard [22]. Dash-and-dotted black line and dashed red line are results of correlation (5) with parameters taken from Table 2 for Vega and Miguel [8] and this study, respectively.

The surface tension data shown in figures 5 and 6 were correlated by the following equation

$$\gamma(T) = B(1 - T/T_c)^{11/9}[1 - b(1 - T/T_c)], \qquad (5)$$

where B and b are adjustable parameters and T_c is the critical temperature of the molecular model. We continue using this equation for comparison purposes. We note that Eq. (5) is of the same form as the IAPWS standard [22], except for the exponent, whose IAPWS value 1.256 is

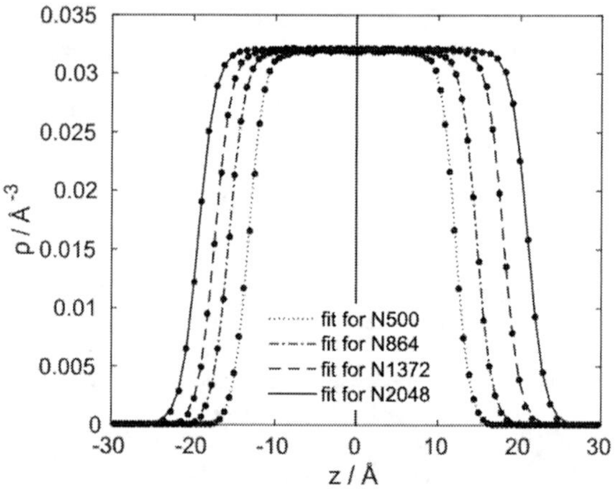

Fig. 7. Density profiles of the SPC/E model at a constant temperature 350 K for simulations with various amount of water molecules N. Black points are results of MS, lines represent correlation (1).

Table 2. Critical temperature T_c and coefficients B and b from equation (5) correlated to the surface tension data obtained from MS.

Model / reference	T_c (K)	B (mN/m)	b (–)
SPC/E			
Vega & Miguel [8]	625.7	205.3	0.613
Sakamaki et al. [9]	625.3	193.2*	0.548*
This study	621.6	199.8	0.603
TIP4P/2005			
Vega & Miguel [8]	641.4	227.9	0.641
Sakamaki et al. [9]	636.3	227.8*	0.654*
This study	635.5	226.7	0.660
IAPWS** [22]	647.096	235.8	0.625

* correlated in this study to the data by Sakamaki et al. [9]

** exponent has different value of 1.256 in the IAPWS standard

closer to the current estimate 1.260 of the universal critical exponent [23]. The values for T_c, B, and b are given in Table 2. The critical temperature for SPC/E is approximately 4 K lower than the values provided by Vega and Miguel [8] and Sakamaki *et al.* [9]. Parameters B and b from the new simulations lie between the values of the other two MS studies. The values for T_c, B, and b for the TIP4P/2005 model are very close for all three studies. The only difference can be seen in the critical temperature, where the value by Vega and Miguel [8] is 5 K above by Sakamaki *et al.* [9] and this study. We note that values in Table 2 are rather sensitive to the amount and quality of the correlated data points. The values for T_c, B, and b given in the IAPWS standard [22] are also given in Table 2 for comparison.

The influence of the ensemble size was thoroughly investigated with the SPC/E model at a constant temperature of 350 K. Four simulations were performed with the numbers of molecules varying from 500 to 2048. Figure 7 shows the density profiles close to the centre of the simulation box for all four simulations. The thickness of the

Table 3. Influence of the number of molecules N in the simulation with the SPC/E model at a constant temperature 350 K on the following results: ρ_L liquid density, ρ_V vapor density, t 10–90 interface thickness, γ surface tension.

Property / N	500	864	1372	2048
ρ_L (g/cm^3)	0.954	0.957	0.959	0.960
$\rho_V \cdot 10^5$ (g/cm^3)	9.85	10.41	9.11	11.54
t (Å)	3.879	4.256	4.332	4.550
γ (mN/m)	53.38	53.71	53.35	53.65

liquid slab is continuously increasing as the amount of water molecules gets higher.

Table 3 contains results for the saturated liquid and vapor densities, the thickness of the phase interface, and the surface tension. The liquid density slightly increases in a larger ensemble. The vapor density and the surface tension show small fluctuations with no obvious dependency on the number of simulated molecules. The only significant influence of the ensemble size can be seen for the 10–90 thickness of the phase interface t, which continuously increases approximately by 17% with increasing the number of molecules from 500 to 2048. The observed dependence of the interface thickness on the ensemble size can also explain differences in t shown in figure 4. The new simulations were made with 1372 molecules, while Alejandre *et al.* [3] and Vega and Miguel [8] used a maximum number of molecules of 1000 and 1024, respectively.

5 Conclusions

Properties of the vapor–liquid phase interfaces of pure water were studied using the approaches of MD. Two commonly used molecular models of water, SPC/E and TIP4P/2005, were used to model the density profiles and the surface tensions of water at temperatures ranging from 250 K to 600 K. Results of our new simulations are in good agreement with most of the previous MS studies. Both models were found to provide relatively good results for the vapor-liquid phase equilibria and the densities of water at temperatures up to 500 K. At higher temperatures, both models start to deviate from the experimental data as they underpredict the critical point of water. SPC/E gives slightly better reproduction of the vapor density while TIP4P/2005 is more accurate on the liquid side. The TIP4P/2005 model gives higher values for the surface tension lying closer to the experimental data correlated by IAPWS [22] than the SPC/E model. An influence of the number of molecules in the ensemble on the 10–90 thickness of the phase interface was observed. The interface thickness increases with the larger amount of molecules in the simulation.

On the basis of the previous studies by Vega and Miguel [8] and Sakamaki *et al.* [9] and our new results we conclude that the TIP4P/2005 molecular model is convenient for modelling the interfacial properties and surface tension of pure water in the temperature range from 250 K to 550 K, although a fully quantitative prediction cannot be expected of these relatively simple models.

Acknowledgement

The authors would like to thank Dr. Martin Lísal from the Institute of Chemical Process Fundamentals of the CAS for his valuable advice about DL_POLY program. The project has been supported by Czech Science Foundation grants GAP101/11/1593 and GJ15-07129Y and the institutional support RVO:61388998.

References

1. J. Barker, R. Watts, Chem. Phys. Lett. **3**, 144 (1969)
2. A. Rahman, F.H. Stillinger, J. Chem. Phys. **55**, 3336 (1971)
3. J. Alejandre, D.J. Tildesley, G.A. Chapela, J. Chem. Phys. **102**, 4574 (1995)
4. F. Chen, P.E. Smith, J. Chem. Phys. **126**, 221101 (2007)
5. C. Vega, J.L. Abascal, M. Conde, J. Aragones, Farad. discuss. **141**, 251 (2009)
6. C. Vega, J.L. Abascal, Phys. Chem. Chem. Phys. **13**, 19663 (2011)
7. B. Planková, V. Vinš, J. Hrubý, M. Duška, T. Němec, D. Celný, EPJ Web. Conf. **92**, 02071 (2015)
8. C. Vega, E. De Miguel, J. Chem. Phys. **126**, 154707 (2007)
9. R. Sakamaki, A.K. Sum, T. Narumi, K. Yasuoka, J. Chem. Phys. **134** (2011)
10. I. Todorov, W. Smith, K. Trachenko, M. Dove, J. Mater. Chem. **16**, 1911 (2006)
11. W. Smith, T. Forester, I.T.Todorov, The DL_POLY Classic User Manual, Daresbury Laboratory, United Kingdom (2012)
12. W.L. Jorgensen, J. Chandrasekhar, J.D. Madura, R.W. Impey, M.L. Klein, The Journal of chemical physics **79**, 926 (1983)
13. H.J. Berendsen, J.P.M. Postma, W.F. van Gunsteren, A. DiNola, J. Haak, J. Chem. Phys. **81**, 3684 (1984)
14. H. Berendsen, J. Grigera, T. Straatsma, J. Phys. Chem. **91**, 6269 (1987)
15. J.L. Abascal, C. Vega, J. Chem. Phys. **123**, 234505 (2005)
16. M.W. Mahoney, W.L. Jorgensen, J. Chem. Phys. **112**, 8910 (2000)
17. R. Span, T. Eckermann, S. Herrig, S. Hielscher, A. Jäger, M. Thol, Tech. rep., TREND. Thermodynamic Reference and Engineering Data 2.0, Lehrstuhl fuer Thermodynamik, Ruhr-Universitaet Bochum (2015)
18. W. Wagner, A. Pruss, J. Phys. Chem. Ref. Data **31**, 387 (2002)
19. MATLAB R2013a, The MathWorks Inc. (2013), `www.mathworks.com`
20. G. Chapela, G. Saville, S. Thomson, J. Rowlinson, J. Chem. Soc. Faraday Trans. II pp. 1133 – 1144 (1977)
21. M. Matsumoto, Y. Kataoka, J. Chem. Phys. **88**, 3233 (1988)
22. IAPWS revised release on the surface tension of ordinary water substance (2014), `http://www.iapws.org/`
23. M.A. Anisimov, C.E. Bertrand, in Applied Thermodynamics of Fluids (The Royal Society of Chemistry, 2010)

EPJ Web of Conferences **114**, 02137 (2016)
DOI: 10.1051/epjconf/201611402137

Droplet Measurement below Single-Layer Grid Fill

Pavol Vitkovic[1, a]

[1] *CTU in Prague, Department of Fluid Dynamics and Thermodynamics, 166 07 Prague, Czech Republic*

Abstract. The main part of the heat transfer in a cooling tower is in a fill zone. This one is consist of a cooling fill. For the cooling tower is used a film fill or grid fill or splash fill in the generally. The grid fill has lower heat transfer performance like film fill usually. But their advantage is high resistance to blockage of the fill. The grid fill is consisted with independent layers made from plastic usually. The layers consist of several bars connected to the different shapes. For experiment was used the rhombus shape. The drops diameter was measured above and below the Grid fill.

1 Introduction

The wet cooling towers [1] are used wherever need transfer the low potential heat to the atmosphere through the cooling water by the air. It is basically a mixed heat exchanger [2]. The wet cooling tower are consist of a shell [1], a water distribution system with a nozzles [3], a fills [4], [5] and a drift eliminators [6]. The nozzle creates a spray zone [3], [7] and the fill create rain zone [8]. These are two important zones in this experimental study of the Grid fill. The spray zone is above the fill and the rain zone is below the fill. The purpose of this paper is to measure the droplets size at the entry to the Grid fill and the droplet size generated by the Grid fill with the rhombus shape bar.

2 Measurement method

The Laser-Induced Fluorescence (LIF) method is used for obtaining an experimental data. LIF has excellent detection sensitivity because a signal is observed against a dark background [9]. Fluorescent dyes (Rhodamine B) in water can absorb light (green light) at one frequency and subsequently reemit light at a different frequency. Because the fluoresced light is of a different frequency than the excitation light can be filtered out. The laser beam is expanded into a sheet using two cylindrical lenses. The cylindrical lens produces a light sheet of few millimeter thick. Rhodamine B as a tracer dye after absorbing light with the wavelength of 532 nm emitted light of wavelength between 500 and 700 nm. Excitation light was filtered by a red filter with the cut-on wavelength about 600 nm. The red filter was installed at the lens of a camera.

3 Experimental setup

The test facility has a modular construction. It is consist of a two sections made from a plexiglass. A cross-section is 300 x 288 mm. The total height of the test facility is 900 mm. The first section is consist of a collecting basin with a pump. The pump has maximum flow rate $0.01083\ m^3\ s^{-1}$. The water flowrate of the pump is controlled by a voltage. At a bottom of the basin is placed a piece of drift eliminator. Because it blocks secondary water droplets from a drop impact on the water surface. The second section has a water distribution system.

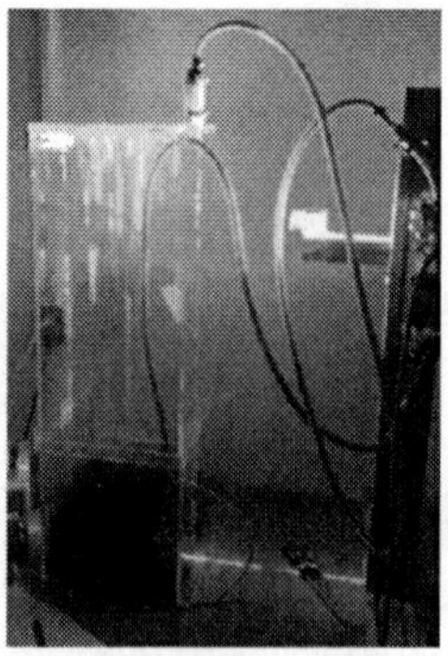

Figure 1. Test facility – experiment setup

A droplets generator was made from a plexiglass sheets. It's consist of a three layers of a plexiglass cut by a laser. The last layer has holes with 1 mm diameter arranged to the net. The middle layer has a rectangular distribution channels. And the third layer has a connection tube for pipelines. All the layers are glued together into the one piece.

[a] Corresponding author: pavol.vitkovic@fs.cvut.cz

Figure 2. Droplets generator

We can expect about 3.5 mm diameter water drops. For predicted calculation was used equation (1) of an approximate static drops formation [10]:

$$D = \left(\frac{6 \cdot d_o \cdot \sigma}{\rho_L \cdot g}\right)^{\frac{1}{3}} \tag{1}$$

The measurement of the spray zone [7] shows that 60% of the total drops has the size larger than 3.5 mm and the largest drops are 7.8 mm. And the corresponding Sauter mean diameter is 3.25 mm.

The Sauter mean drop diameter is defined as

$$d_{32} = \frac{\sum_{i=1}^{m} n_i d_i^3}{\sum_{i=1}^{m} n_i d_i^2} \tag{2}$$

For measurement was used the full range pump operate (from 5 V to 12 V with 1 V step). Below 5 V the pump can't pull the water to the droplets generator.

Table 1. Measurement points setup

Voltage (V)	Water flowrate ($m^3 s^{-1}$)
5	0.00176
6	0.00327
7	0.00422
8	0.00493
9	0.00569
10	0.00627
11	0.00696
12	0.00747

The drop size distribution based on the volume of the droplets generator is at Figure. 3. All measurement point has similar drop size distribution (triangle mark) except point with lowest water flow rate (0.00176 $m^3 s^{-1}$ – star mark) and the second lowest (0.00327 $m^3 s^{-1}$ – circle mark). The mean drop size is lower as expected from formula (1).

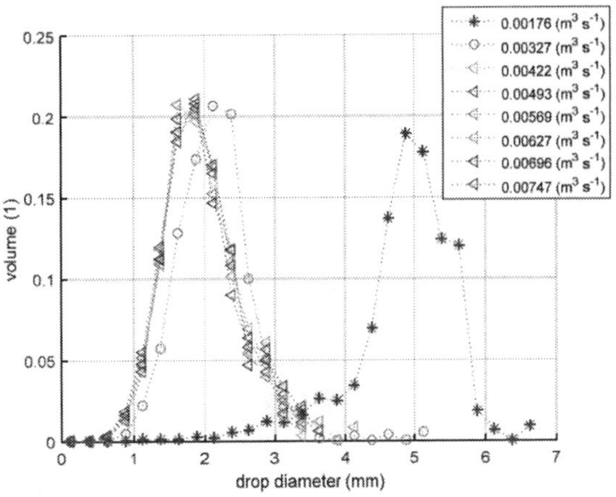

Figure 3. Droplet generator drop size distribution based on volume.

The experimental equipment consists of the GigE Sony camera XCG-H280E and the continual laser 1000 mW with the wavelength of 532 nm (green light) for the sheet generation. The exposition time of the camera is possible to be set up between 0.0002 s and 2 s. The experiment was conducted at exposition time 0.0005 s.

For experiment, it was used the Grid fill with the rhombus shape bar (Figure 4). The side length of the rhombus is 53 mm and the thickness of the bar is 5 mm.

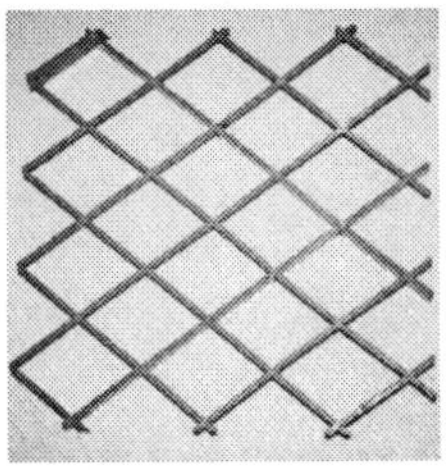

Figure 4. Grid fill with rhombus shape bar.

4 Experimental Results

Before measuring it was necessary to a calibrated measurement. For this purpose, graph paper was chosen as a calibration target. After calibration, it was possible to calculate droplets diameter in millimetre size. The signal from the fluorescent light is converted to white blobs on a black background. From the drop, area is calculated an equivalent spherical drop diameter. The drop size distribution based on volume are at following figures.

For all measurements points the Sauter mean drop diameter was computed. The Sauter mean drop diameter is calculated from (2).

Figure 6. provides the Sauter mean drop diameter for all measured points. The Grid fill generates the Sauter mean drop diameter about $d_{32} = 7.5$ mm except the point with lowest water flow rate ($d_{32} = 11.5$ mm).

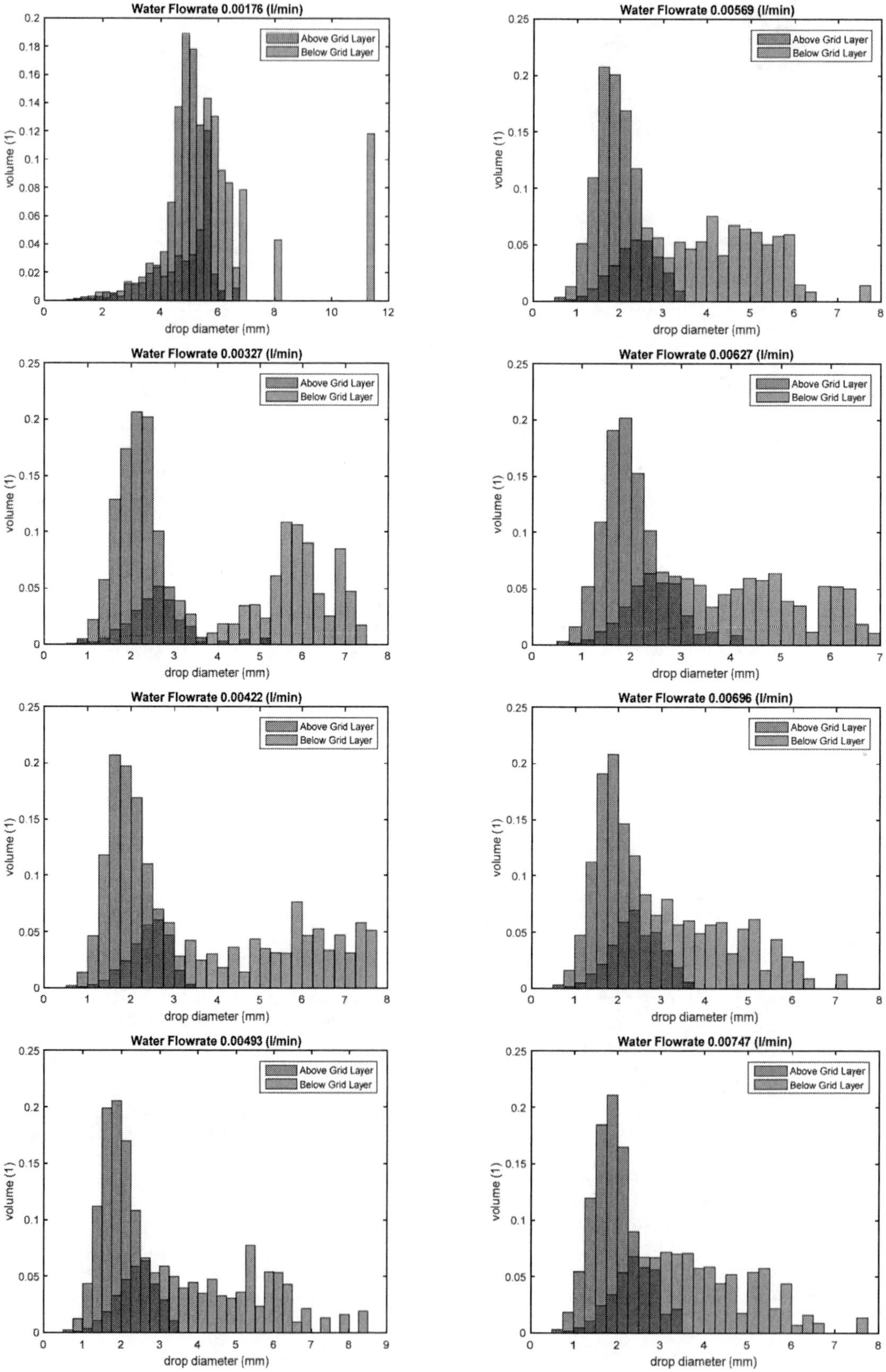

Figure 5. Drop size histogram based on volume.

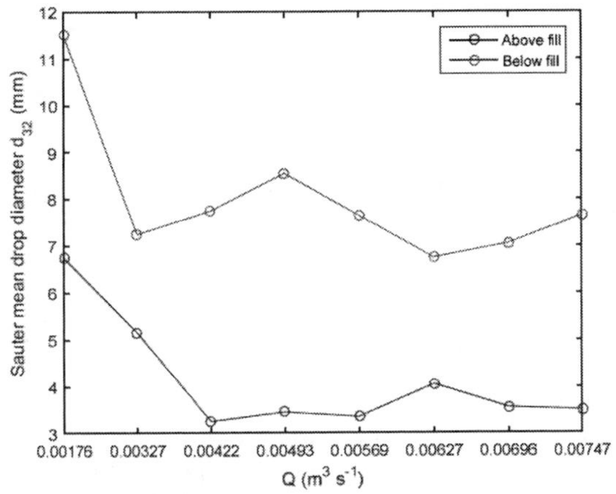

Figure 6. Sauter mean drop diameter for a different water flow rate.

5 Conclusion

The drop size distribution below and above the Grid fill with rhombus shaped bar was obtained. What is interesting in this data is generating bigger droplets by the Grid fill as droplets before coming to the Grid fill. The areas from where large drops drip is a bar intersection [4].

The section between the two layers of Grid fill is the rain zone basically. It can be described by the mathematical model of the rain zone [11], [12]. The results of this study show that we can obtain the size of droplets between the Grid fill as the boundary conditions for the mathematical model of the rain zone [11]. This is an important issue for future research. In the future investigation, it might be possible to use a different drops size at the entry of the Grid fill.

Acknowledgments

The author acknowledges the support received from Advanced technologies for heat and electricity production TE01020036 and Centre for research of multiphase flow and thermodynamics processes in renewable sources and energetics – NEW ENERGY supported by the European Union (CZ.2.16/3.1.00/22130).

References

1. D.G. Kröger, *Air-Cooled Heat Exchangers and Cooling Tower*, (2004)
2. VDI Heat Atlas, (2010)
3. P. Vitkovic, EPJ Web of Conferences, 92, 02109, (2015)
4. R. Terblanche, H.C.R. Reuter, D.G. Kröger, Applied Thermal Engineering, 29, 1552-1560, (2009)
5. L. Dvorák, J. Nožička, EPJ Web of Conferences, **67**, 02024, (2014)
6. J. Stodůlka, R. Vitkovicova, EPJ Web of Conferences, **92**, 02086, (2015)
7. Reuter, H.C.R., Viljoen, D.J., Kröger, D.G., IHTC 14, 4, pp. 619-628 (2010).
8. L. Dvořák, J. Čížek, J. Nožička, EPJ Web of Conferences, 92, 02014, (2015)
9. R. N. Zare, Annual Review of Analytical Chemistry, **5**, 1-14, (2012)
10. A. H. Lefebvre, *Atomization and sprays*, (1989)
11. T. Hyhlík, Engineering Mechanics, **21**, 240-243, (2014)
12. T. Hyhlík, ICCT 2014 International Conference, 219-226, (2014)

EPJ Web of Conferences **114**, 02138 (2016)
DOI: 10.1051/epjconf/201611402138

Interaction of two circular cylinders in the flow field at low Reynolds numbers

Rut Vitkovičová[1a], Vladislav Skála[1]

[1] *CTU in Prague, FME, Department of Fluid Dynamics and Thermodynamics, Technická 4, 166 07 Prague, Czech Republic*

Abstract. A set of points of a velocity fluctuation was measured in the wake behind the two circular cylinders by CTA method. Measurements were carried out for several positions of the secondary cylinder in the wake of primary cylinder and for a set of modes of low Reynolds number (Re max 2600). The goal was to obtain a set of data and information about the influence of the secondary cylinder on the primary cylinder.

1 Introduction

Investigating the mutual influence of two or more cylinders in the flow field is the theme of many works in the field of fluid mechanics and related fields, but nevertheless it is still an open problem which is not fully resolved. Understanding the mutual influence of cylinders in the flow field allows us among other things, to understand the phenomena taking place in geometrically similar applications in technology, e.g. construction (e.g. the chimney towers), engineering etc., and also allows to find areas affecting the wake of the dominant cylinder.

M. M. Zdravkovich in his review [1] deals with an observation of an influent of two cylinders in a configuration tandem and a staggered arrangement. Hwang and Choi introduce in [2] with works that deal with a control of wake on the one hand passively and other actively. They comparing the area of influence of the secondary cylinder on primary cylinder in research Strykowski and Sreenivasan in [3] with the results of Hwang and Choi [4]. D. Sumner in his article [5] clearly summarizes several studies by authors engaged in research of wake behind a cylinder in a number of settings and gives an overview of the performed measurements, measuring apparatuses and measuring stands. Overview of the measured dependence Strouhal number ratio gap between the cylinders in a parallel arranged and comparing them with the results of other authors is given in a paper from Yokoi [6].

This article describes an experimental measurement of the influence of cylinders in the flow field in staggered configuration using the CTA method and presents the results of these measurements. This paper follows the measurements published in [7].

2 Measurements

2.1 Experimental tunnel and measurement techniques

The measurement by method CTA carried out in an Eifffel type wind tunnel which has a speed range of approximately 0-20 m / s. The intensity of turbulence in the tunnel is around 1.1%. A test section of the tunnel is closed. The length of the test section is 750 mm and its cross section is 300 mm x 200 mm. The test section is equipped with a movable system which enables automatic 3D positioning of the probe. Tunnel control, monitoring and data logging and control of movable system are performed by software written in LabView programming environment. The tunnel was fitted with a measuring system for the hot wire anemometers MiniCTA 54T30 (Dantec Dynamics) with 55P16 HW-probes for the measurements. The hot wire and cylinders are shown in Figure 1.

Figure 1. The hot wire probe and cylinders

[a] Corresponding author: rut.vitkovicova@fs.cvut.cz

2.2 Experimental set-up

Measuring method CTA was carried out for two pairs of cylinders, D/d = 1 and D/d = 2.5. The interaction of these cylinders was investigated for 7 (resp. 6) relative positions in the flow field, which were selected according to [2]. In Table 1 are given their coordinates and pitch ratios (R). The secondary cylinder was always located behind the primary cylinder. For each pair of cylinders were measured 16 modes of Re in the range of approximately from 400 to 2600. For each mode Re was measured from 120 to 149 points placed in the wake of cylinders. Presented frequencies of vortex shedding was measured at a distance of 55 - 85 mm behind the primary cylinder. Sampling frequency of hot-wire probe was set to 30 kHz. In a Figure 2 is shown the set-up of the cylinders and coordinate system.

Table 1. Measured positions of secondary cylinder

Position No.	x/D	y/D	R (mm)
1	4.3	-1.25	44.8
2	3.2	-0.8	33.0
3	2.4	-0.8	25.3
4	1.95	-0.8	21.1
5	2	-0.6	20.9
6	1.5	-0.35	15.4
7	1.5	0	15.0

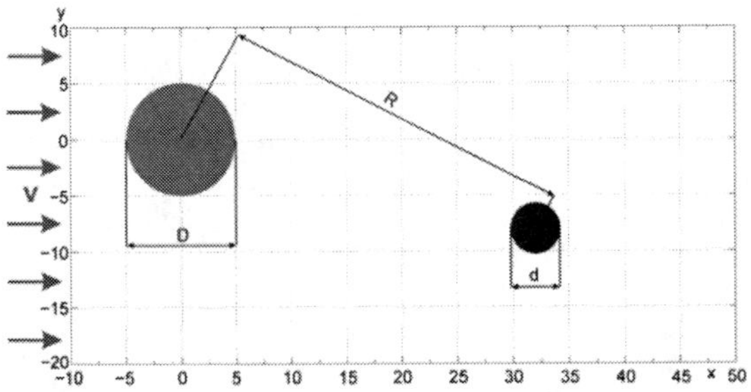

Figure 2. The set-up of the cylinders and coordinate system

3 Results

In the first step was obtained dominant vortex shedding frequencies from the measurement data for each mode of Re and the mutual position of the cylinders. In Figures 3 - 10 are compared a dependence of the frequency of vortex shedding on the R / D for the cylinders ratios 1 and 2.5. From these graphs it is clear that the vortex shedding frequency for ratio D / d = 1 and thus the corresponding Strouhal number are lower than for the ratio of 2.5, except for positions 1, 2 and 7, which are in most cases approximately the same frequency. Positions 1 and 6 located outside of the areas of attenuating instabilities in [3] and [4], the positions 2, 3, 4 and 7 is located within these areas, and position 5 is located in the area defined in [4].

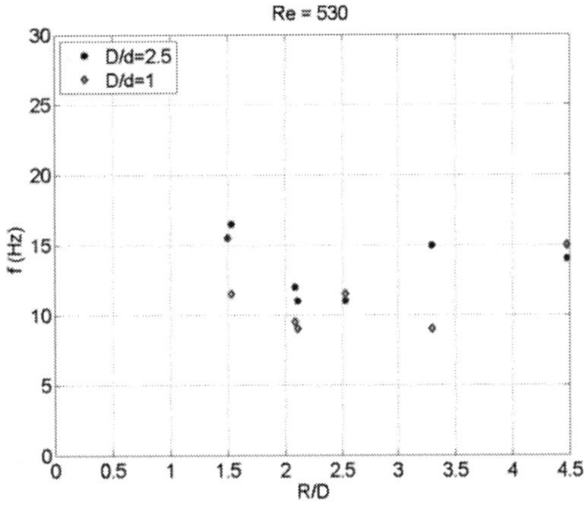

Figure 3. Relationship between the frequency and the pitch distance for Re 530

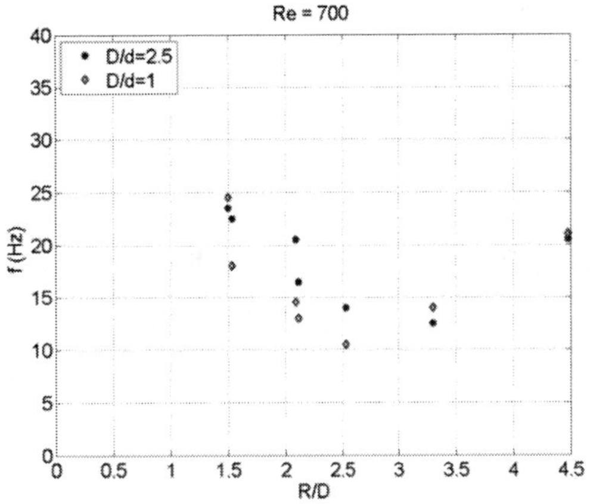

Figure 4. Relationship between the frequency and the pitch distance for Re 700

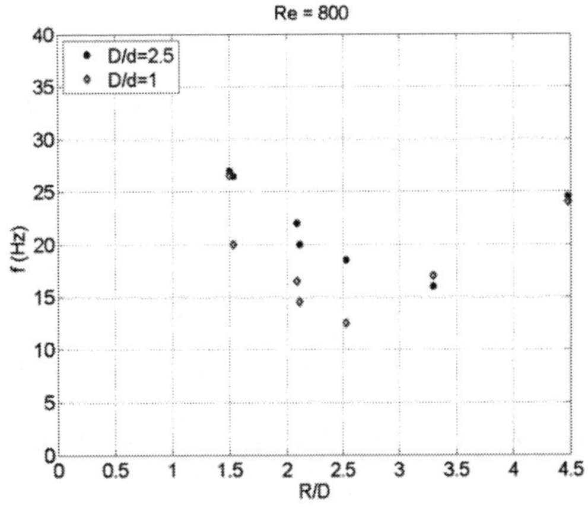

Figure 5. Relationship between the frequency and the pitch distance for Re 800

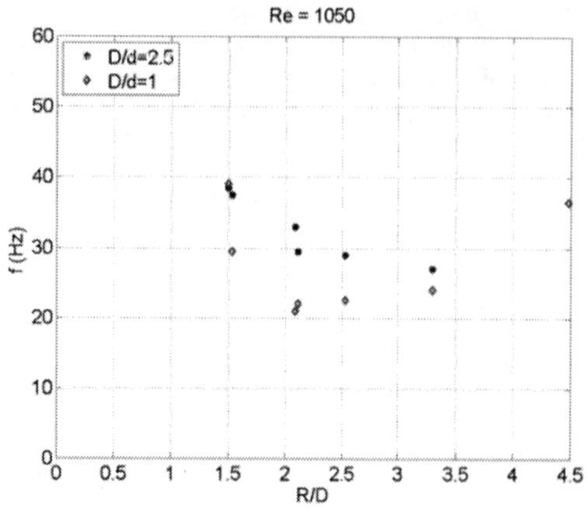

Figure 6. Relationship between the frequency and the pitch distance for Re 1050

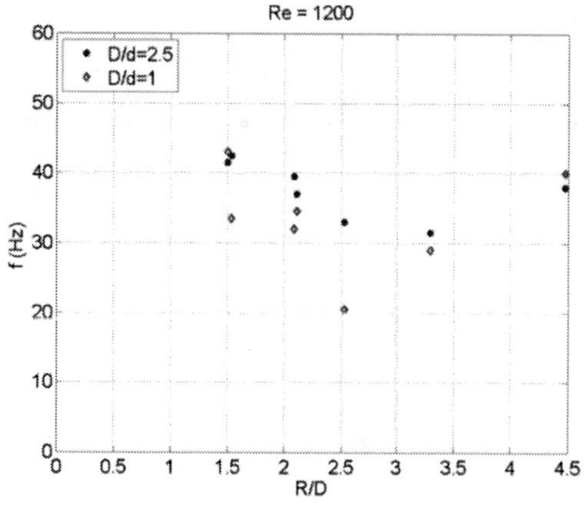

Figure 7. Relationship between the frequency and the pitch distance for Re 1200

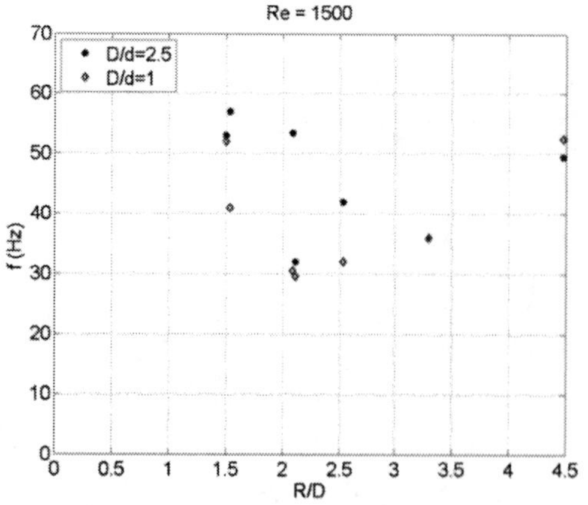

Figure 8. Relationship between the frequency and the pitch distance for Re 1500

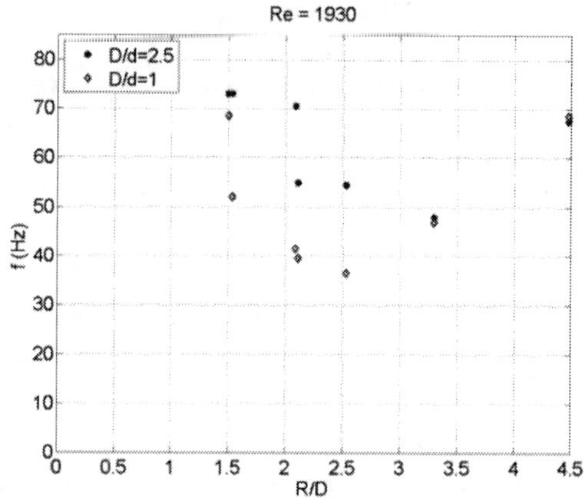

Figure 9. Relationship between the frequency and the pitch distance for Re 1930

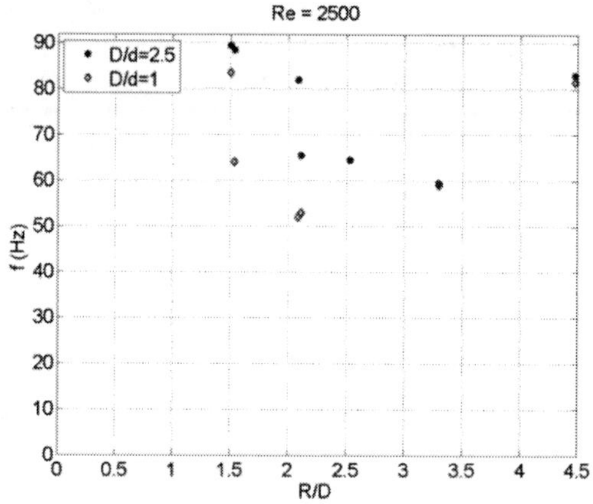

Figure 10. Relationship between the frequency and the pitch distance for Re 2500

Figures 11 and 12 compares the position of the secondary cylinder for a D / d ratio for all measurement modes Re. The obtained data show that for the positions 3 and 4 the smallest instabilities occur in the wake for both cylinders ratios. The next position with a low shedding frequency is position 2 for a ratio D / d = 2.5, and for D / d = 1 it is the position 5.

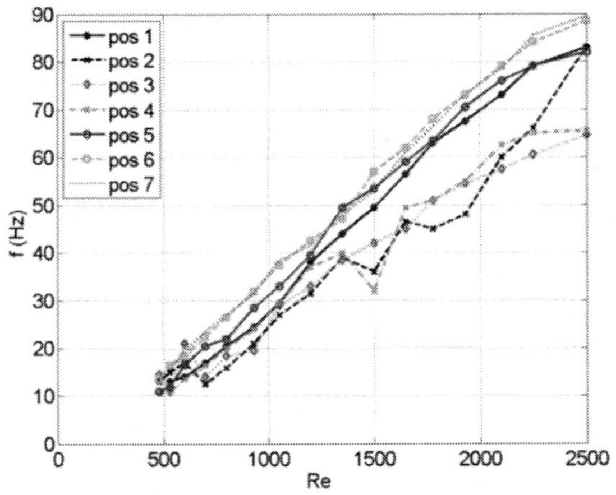

Figure 11. Relationship between the frequency and Reynolds numbers for D/d = 2.5

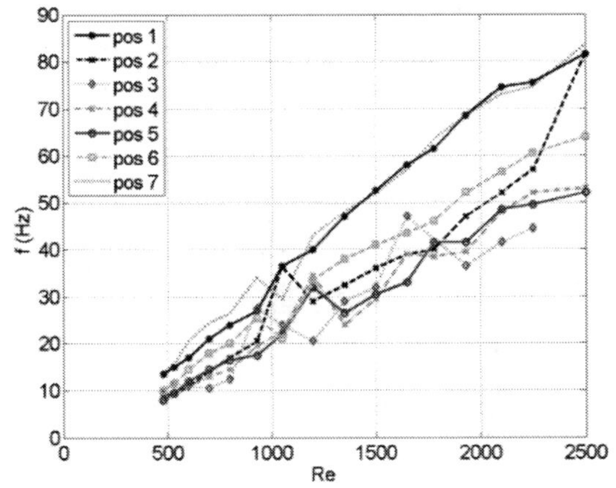

Figure 12. Relationship between the frequency and Reynolds numbers for D/d = 1

4 Conclusion

Set of measurements of frequencies of vortex shedding were performed for combination of cylinders in staggered configuration and for 16 modes of Reynolds number. The main aim was to compare the individual positions of the secondary cylinder in the wake of the primary cylinder, and compare these positions with research [3] and [4], which were made, however, for lower Re and other D / d ratio.

The measured results show that for the D / d = 1 the position with most suppressing instabilities are 3, 4 and 5, which corresponds to position 3: x / D = 2.4 and y / D = -8, position 4: x / D = 1.95 and y / D = -0.8 and position 5: X / D = 2 and y / D = -0.6. These points lie in the area defined in [4]. Positions 1 and 6, which cause greater disruption in the wake, respectively significantly do not affect the flow, also agrees with the research in [3] and [4]. Only the results for the 7-position will vary, which may be due to the fact that the primary and secondary cylinder behave as a single bluff body, as indicated by various authors in [5].

For D / d = 2.5, it was found that the most suppressive instability positions are 2, 3 and 4 where position 2 corresponding to x / D = 3.2 and y / D = -8. These points also located in the area defined in [4]. For the other positions the same applies as for D / d = 1. If the results are compared for D / d = 1 and D / d = 2.5 among them we find that the D / d = 1 suppresses more formation of instability than D / d = 2.5.

Acknowledgements

The work was supported by the Student Grant Competition of Czech Technical University in Prague and the Centre SGS14/058/OHK2/1T/12 for Research on multiphase flow and thermodynamic phenomena in the field of renewable sources and energy - NEW ENERGY (CZ.2.16/3.1.00/22130).

References

1. M. M. Zdravkovich, J. Fluids Eng. 99(4): 618-633 (1977)
2. H. Choi, I. Lee, W.P. Jeon., J. Kim, Annu. Rev. Fluid Mech. 40:113-39 (2008)
3. P. J. Strykowski, K. R. Sreenivasan, J. Fluid Mech. 218: 71-107 (1990)
4. Y. Hwang, H. Choi, J. Fluid Mech., 560: 465-475 (2006)
5. D. Sumner, J. Fluids and Structures, 26: 849-899 (2010)
6. Y. Yokoi, K. Hirao, EPJ Web of Conf. 45, 01098 (2013)
7. R. Vitkovicova, J. Nozicka, J. Cizek, V. Skala, EPJ Web of Conf. 92, 02110 (2015)

EPJ Web of Conferences **114**, 02139 (2016)
DOI: 10.1051/epjconf/201611402139

Numerical experiment about the arrangement determination of multiple bluff cylinders which obtain the most suitable new type stir

Yoshifumi Yokoi[1,2,a]

[1]*National Defense Academy, Department of Mechanical Engineering, 1-10-20 Hashirimizu, Yokosuka 239-8686, Japan*
[2]*Techniche Universität Berlin, Müller-Breslau-Str. 8, 10623 Berlin, Germany*

Abstract. When developing the stirring rod for mixing two liquid with which the densities into which gas melted differ in little stirring operation, verification of the stirring effect was performed in the numerical simulation using a vortex method. Since it is calculation for specification examination of the preceding paragraph which creates a producing model, it becomes important to obtain a calculation result early and to feed back to the design. Two dimension calculation of the double file circular cylinder group which carried out staggered arrangement of the nine circular cylinders was performed. The aspects of the flow which strayed off from the design point or the design point were shown. The mixed degree was judged from the degree of the complexity of the flow pattern. In a design point, the complicated flow by a vortex is formed with oscillating 1 period, and interference stirring occurs. Adaptation ranges other than a design point are that the ranges of the value of P/L are 0.8-1.2.

1 Introduction

In the restaurant in Japan, various alcoholic beverages are offered like the countries of Europe. Beer, sake, white distilled liquor, wine, whiskey, etc. are prepared there. Consumption of plum wine with soda is increasing in such an alcoholic beverage. This is because plum wine with soda passes along a throat smoothly. In women or the younger age group, it is reported that the amount of consumption of plum wine with soda is increasing [1]. In response to this matter, alcoholic beverage brewing industry companies vied in production increase-in-quantity sale of a "soda cocktail" drink mutually, and planned it. In order to attain differentiation on sale, the improvement idea of packaging, addition of a sales promotion article, etc. were planned. Development of a new type "muddler" was desired as one of the plans of differentiation on sale. Here, although the "muddler" is "cocktail stir" in use, it is the popular name currently called in usually in the food-and-drink industry of Japan. The popular name was used in this study. Generally, the "muddler" of the stirring use of the cocktail drink is a long and slender bar configuration. The method for use agitates by making it circle horizontally in the liquid in a glass. Since the liquid in a glass also rotates synchronizing with a muddler, in the shape of an ordinary muddler, stirring efficiency is low. Therefore, in order to obtain enough stirring effect, it is necessary to continue rotating a muddler over a long time. In a "soda cocktail", if it continues rotating a muddler over a long time, the demerit in which carbon dioxide falls off from the liquid by the centrifugal force produced in a motion of the muddler will occur. The good method of raising the efficiency of stirring should just make a muddler reciprocating movement within the glass horizontally. However, the method is not efficiently mixed over the overall fluid in the glass.

The author is observing that mix as the symmetrical vortex street discharged from two circular cylinders which are oscillating in the direction of a flow flows and goes down through study of a lock-in phenomenon until now, and rearrangement occurs as a vortex of another kind [2, 3]. By using the lock-in phenomenon and phenomenon of vortical rearrangement, the author conjectured what can promote liquid stirring. Then, the author performed creation development of the muddler (type shaken longitudinal) of the new shape based on interference stirring by vortices. And the muddler was used as the product and sent out to the market [1].

This paper reports the details which are the numerical simulations performed in the early stage of the creation development.

2 Numerical calculations

2.1 Test design of the muddler for numerical simulations

Mixture of ume liquor and carbonated water was assumed with the cylinder glass with a capacity of 460 ml (the

[a] Corresponding author: yokoi@nda.ac.jp

diameter of 70 mm, the depth of 120 mm) containing ice. In order to make it mix moderately in a little operation, a motion of the muddler makes it reciprocating movement in the perpendicular direction of the glass. In order to enlarge the contact surface product when moving in one operation in the inside of a liquid, the tip section of a stick made shape of a muddler the shape rolled spirally. It was considered as the shape of a coil. The shape is shown in figure 1. The circular cylinder which became this coil shape plans making a vortex, discharging and agitating surrounding fluid by the mutual interference of the discharged vortices. The muddler is inconvenient with the metal which rusts on a health concept target in order to touch a drink. It becomes a cost overrun although what is necessary is just to use stainless steel for the material. The cast made of resin which can be mass-produced at a low price also from the field of manufacture expense was assumed. In consideration of intensity and productivity, the diameter d of a material round bar was determined as 4 mm.

As for a vortex shedding occurring independently mutually, it is known for parallel two circular cylinder that the ratio (G/d = (L - d)/d) of the circular cylinder interval G and the circular cylinder diameter d is 1.0 or more. In tandem two circular cylinders, it is known that the separating shear layer discharged from the upstream side circular cylinder at the time of about P/d (ratio of circular cylinder center interval P and circular cylinder diameter d) = 2.5 will carry out a reattachment to the downstream side circular cylinder [4]. In consideration of this thing, the design of the coil centre distance L and the design of a coil pitch P were performed. The value of P/d was set to 3.5 in order to secure the space which has been rolled before the circular cylinder by the side of downstream by the separating shear layer discharged from the circular cylinder by the side of the upstream. The value of space ratio P/L which is the ratio of a coil pitch P and the coil centre distance L was decided to be 1.0 (design point). As a result, it was decided that it would be $P = L$ = 14 mm. The length of the coil section was made into the 4 times roll ($4P$ = 56 mm) in consideration of the appearance as the muddler. The conditions of oscillation were decided as follows. Half oscillating amplitude a was set to 30 mm so that oscillating amplitude might become equivalent to the length of the coil section. Frequency f of oscillation was set to 1 Hz in consideration of the cardiac beats rate of the human body at the time of drinking.

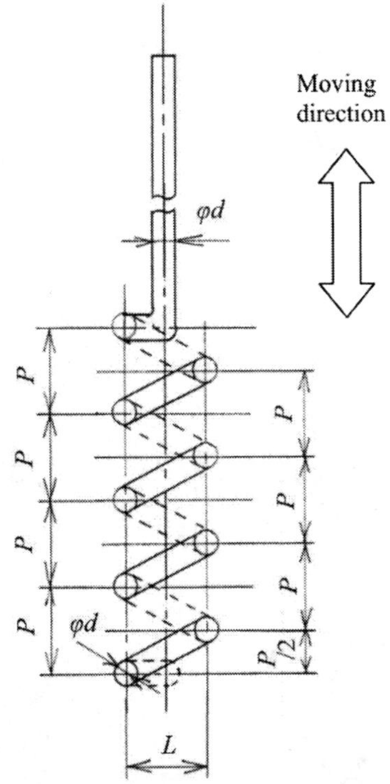

Figure 1. Muddler tip section shape

2.2 The calculation method and apparatus

The numerical simulation was performed by commercial fluid analysis tools and the commercial note type computer. The fluid analysis tools of the marketing used "Uzu Cruise 2D Ver.1.1.3 Rev.H" by College Master Hands, Inc. The numerical computation technique of the used fluid analysis tools is a vortex method which introduced the surface of wall vortex element method which is a type the vortex method application level 3 which used together the boundary element method and the vortex method developed. The details as the calculation technique of the vortex method are written to reference [5] and reference [6]. The computer used the commercial note type personal computer (NEC LaVie LL850/L).

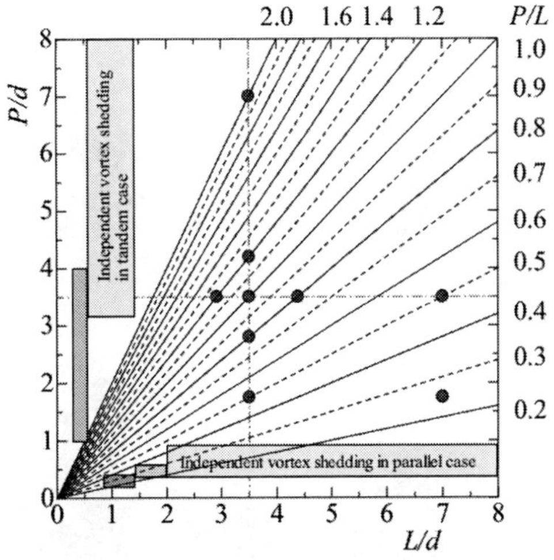

Figure 2. The diagram of the calculation points

2.3 Calculation conditions

This calculation is calculation for specification examination of the preceding paragraph which creates a producing model. So, it is important to obtain the calculation result early and to feed back to a design. In this study, two dimensional calculation of the circular cylinder group which treated the section including the main axis of the coil, then carried out staggered arrangement of the nine circular cylinders was performed. The mixed degree was judged from the degree of the complexity of the flow pattern. Two kinds of liquids with which density differs from specific gravity in fact are mixed. However, in order to verify the process of mixture by the vortex, it was considered as single fluid and the physical-properties value of water was used for calculation here. Here, the spatial restriction supposing a glass, ice, etc. is not provided. The state of early fluid is stationary. Half oscillating amplitude was set to a = 30 mm, and the oscillation frequency was set to f = 1 Hz. The Reynolds number ($Re = u_{max} \cdot d/v$) based on the maximum movement speed of body ($u_{max} = 2\pi af$) was about 660. If the Strouhal number was calculated using the body oscillation frequency and amplitude, the value of the Strouhal number was about 0.02.

In such the state, in order to investigate the applicability of space ratio (P/L), the value of space ratio (P/L) was varied and the simulation was performed. The value of space ratio (P/L) was changed with 0.25, 0.5, 0.8, 1.0, 1.2, and 2.0. The total number of times of repetition calculation is 1000 times, and this corresponds by 2.5 times the oscillating period. Figure 2 shows a design point and each calculating point in each space ratio with the aspect of the flow of parallel two circular cylinders, and the aspect of the flow of tandem two circular cylinders. An abscissa is the distance ratio in the case of parallel two circular cylinders, and an ordinate is the distance ratio in the case of tandem two circular cylinders. The straight lines in the figure mean the space ratio (P/L).

3 Result and discussion

3.1 Aspect of the flow in a design point

The aspect of the flow for every time progress in a design point (P/L = 1.0, L/d = 3.5, P/d = 3.5) is shown in figure 3. Here, "Origin" is in the state of the calculation start. "T.D.C." and "B.D.C." mean the top dead center of oscillation, and the bottom dead center of oscillation, respectively. The "cycle" expresses one period and the numeric character in a parenthesis shows the number of calculation steps. Since preservation of calculated data was performed every ten steps, the flow pattern of each period, the top dead center, and the bottom dead center used the thing of the step near it. It is imagined that mixture of flow is promoted at the time of the turn to the bottom dead center from the top dead center and the turn to the top dead center from the bottom dead center. When it sees through oscillating one period, it turns out that the complicated flow by vortices was formed and interference stirring has occurred. Furthermore, the flow increased complexity by continuing oscillation. After

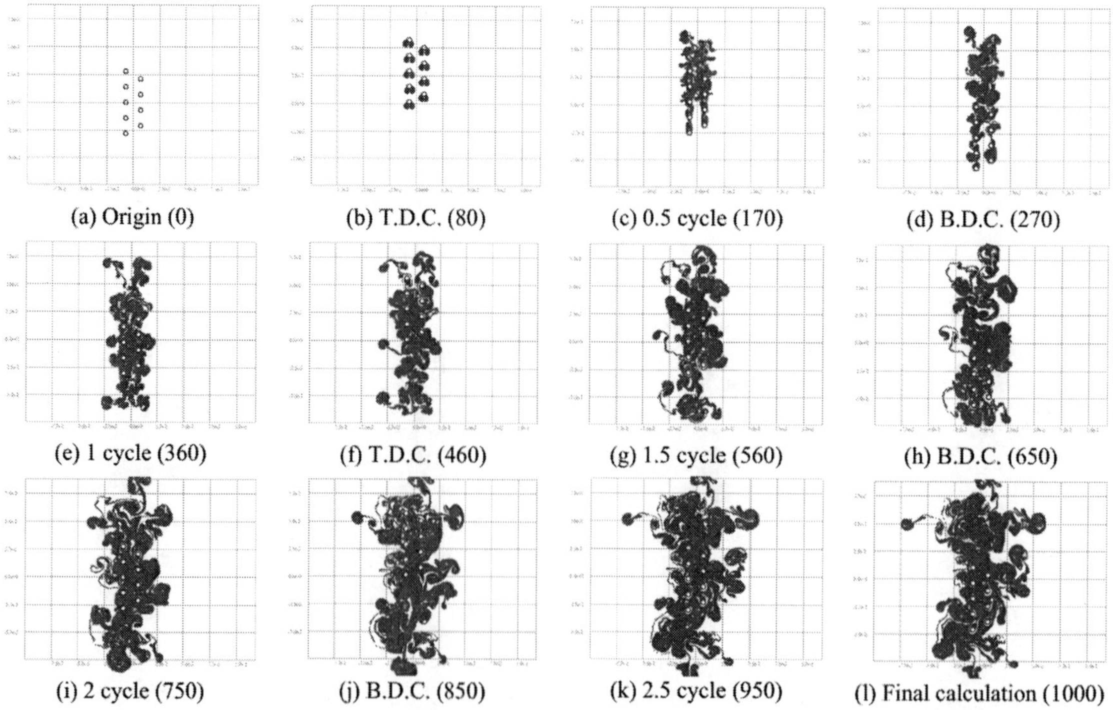

(a) Origin (0) (b) T.D.C. (80) (c) 0.5 cycle (170) (d) B.D.C. (270)
(e) 1 cycle (360) (f) T.D.C. (460) (g) 1.5 cycle (560) (h) B.D.C. (650)
(i) 2 cycle (750) (j) B.D.C. (850) (k) 2.5 cycle (950) (l) Final calculation (1000)

Figure 3. The time history of the flow pattern in a design point (P/L = 1.0, L/d = 3.5, P/d = 3.5)

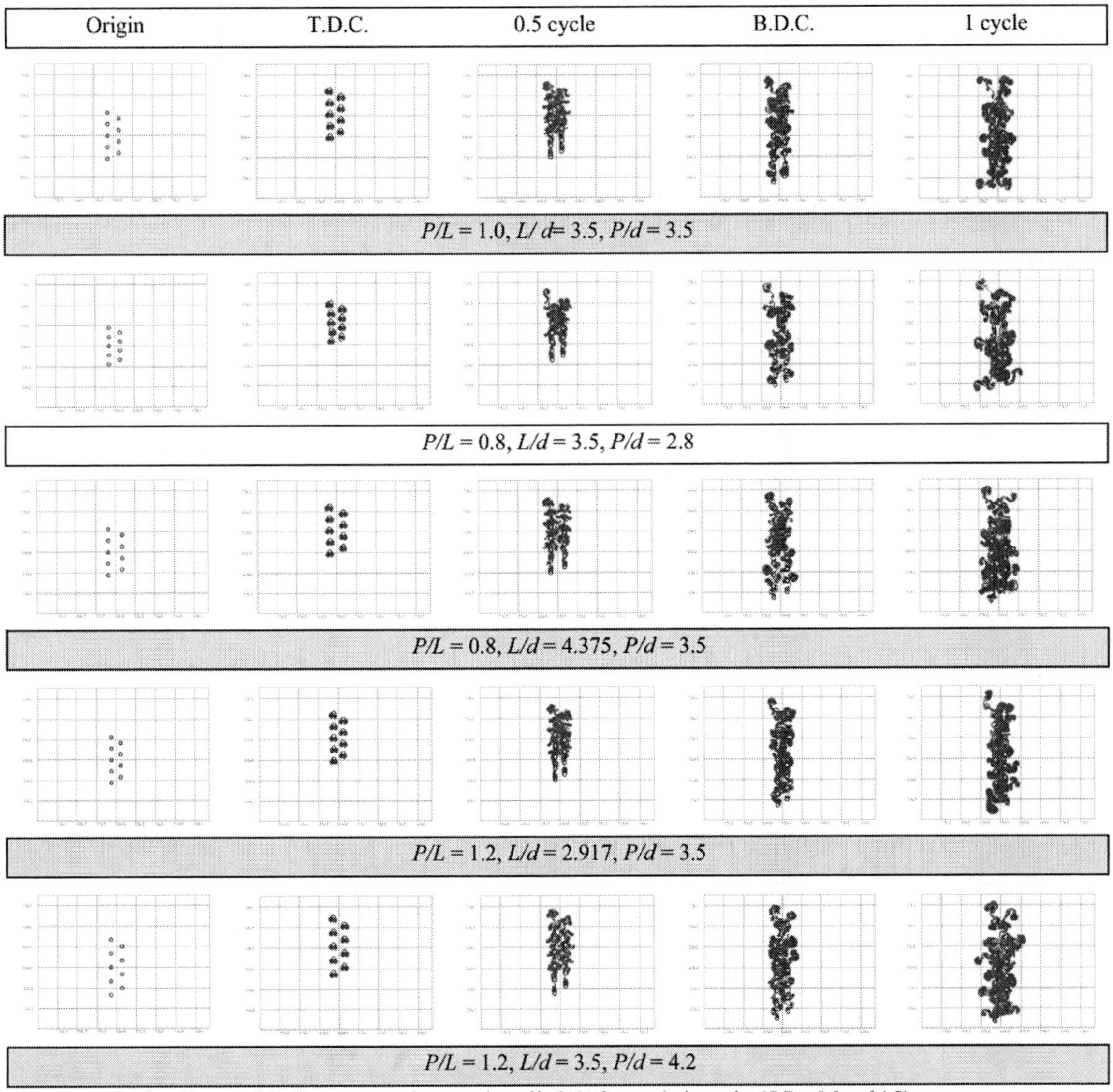

Figure 4. The time history of the flow pattern when straying off ±20% from a design point (*P/L* = 0.8 and 1.2)

passing over oscillating two periods, mixture by large-scale vortices can be seen. This thing suggests that "good stirring" is obtained by little movement.

3.2 Examination of the space ratio suitable for stirring

The aspects of the flow of space ratios other than a design point are shown in figure 4 and figure 5. The aspect of the flow of the design point is also displayed on both figures for comparison. In the design point, the complicated developed flow is formed oscillating one period. Therefore, the flow patterns of the "initial state", the "top dead center", the "half cycle", the "bottom dead center", and the "one period" are shown here, respectively. The conditions currently colored by yellow all over the figure show the case where the vortices which became independent of each circular cylinder are discharged. Figure 4 shows the flow patterns of the conditions used as *P/L* = 0.8 and *P/L* = 1.2 which strayed off from the space ratio in a design point ±20 %. On every condition, the complicated flow by a vortex was formed with oscillating one period, and interference stirring has occurred, and it can grasp so that it may scarcely change to the aspect of the flow in a design point. On the other hand, figure 5 shows the flow pattern in the case of the space ratio which strayed off from the design point further. When a result is seen, it turns out that enough on any conditions stirring is not obtained. These matters show that the ranges of *P/L* of space ratio suitable for stirring are 0.8-1.2. Here, when it is on the line of the same space ratio, the one where the value of *L/d* or *P/d* is smaller is considered to be a desirable thing on the size of the product.

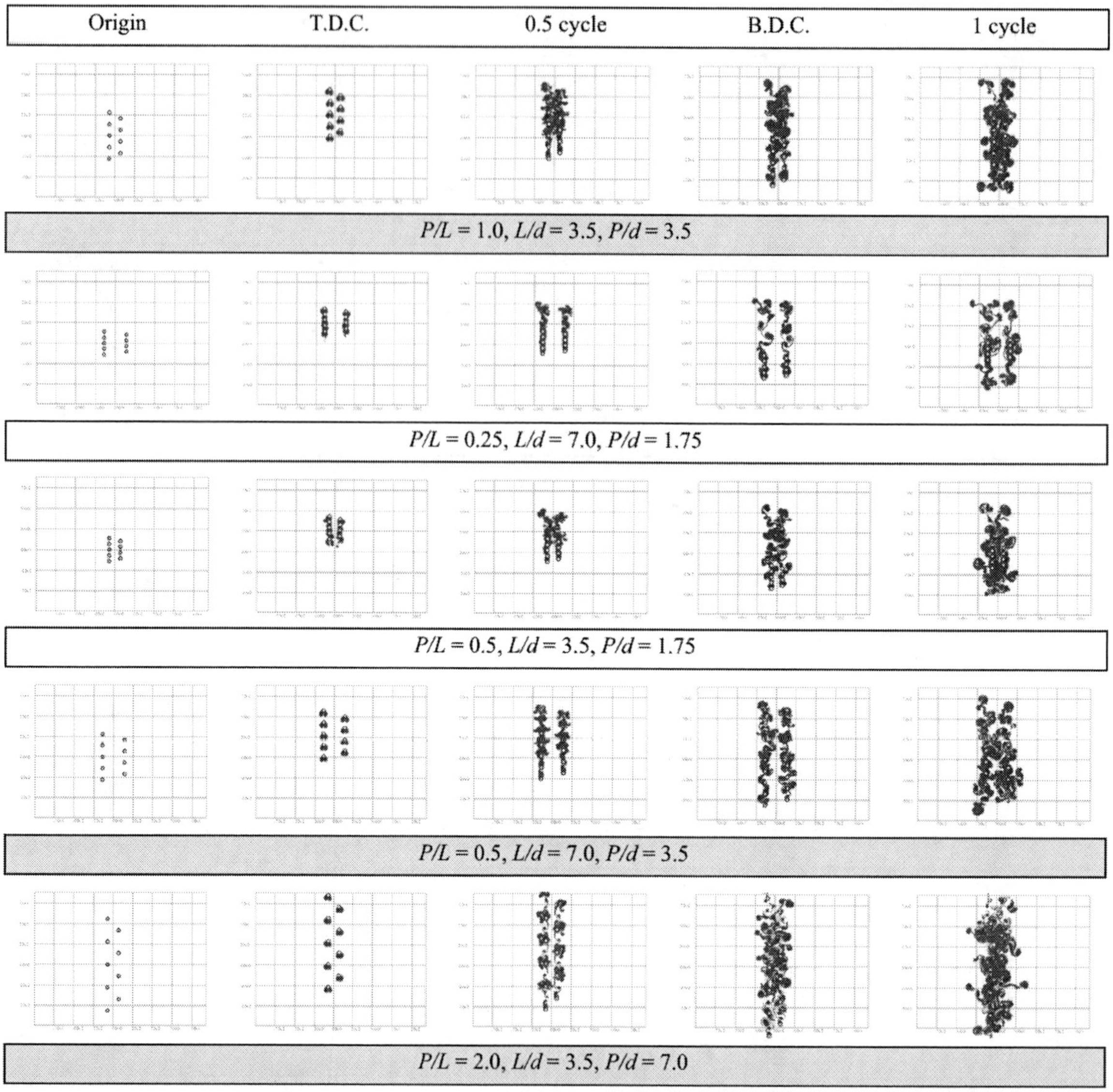

Figure 5. The time history of the flow pattern when straying off further from a design point (P/L = 0.25, 0.5 and 2.0)

4 Conclusions

In order to develop the stirring rod for mixing two liquid with which densities called sodium carbonate and ume liquor differ in little stirring operation, verification of the stirring effect was performed in the numerical simulation using the vortex method. The following conclusions were obtained.

(1) In a design point, a complicated flow is made from oscillation of the one period by vortices. And interference stirring occurs.

(2) Adaptation ranges other than the design point are that the ranges of the value of P/L are 0.8-1.2.

(3) The calculation using the vortex method is the simple technique whose product development is possible while making it feed back to the design.

Acknowledgement

An author expresses gratitude to Sapporo Breweries, Ltd., Daiko Advertising, Inc., and Avalanche, Inc. The opportunity to which an author carries out research from the staff of those companies was obtained. And they sent out the product borne by this research to the market. An author expresses deep gratitude to the staff members anew.

References

1. T. Monna, Y. Yokoi, *Proc. 50th all Japan packing technical research convention*, 265-268 (2012)
2. Y. Yokoi, K. Hirao, Trans. Jpn Soc. Mech. Eng. **74**-748, 2466-2475 (2008)
3. Y. Yokoi, K. Hirao, Trans. Jpn Soc. Mech. Eng. **77**-775, 590-601 (2011)
4. JSME ed., *JSME Standard S012*, B78(1998)
5. K. Kamemoto, Computational Fluid Dynamics, **2**-1, 20-29 (1993)
6. K. Kamemoto, Computational Fluid Dynamics, **2**-2, 28-39 (1994)

Appendix

The data of "the new type muddler" which the author developed are attached. Figure 6 shows the blueprint and its finished goods. The material of parts is plastic resin. This article is the goods developed for advertisement sales promotion of ume liquor. Figure 7 shows the muddler attached to the package of ume liquor. The sale was performed by such packaging. Figure 8 shows the covering letter attached to the muddler. Since the campaign was performed in Japan, it is written in Japanese. The muddler is shown in the beer jug which a model woman (right-hand side) has in a hand. The author (left-hand side) is also introduced as a designer. Since "the cheers" was called "Kanpai" in Japan, this muddler was named the "Kanpai-muddler." Figure 9 is the back of the covering letter, and the method for use of the muddler is written. Here, the busy condition of the muddler was also shown.

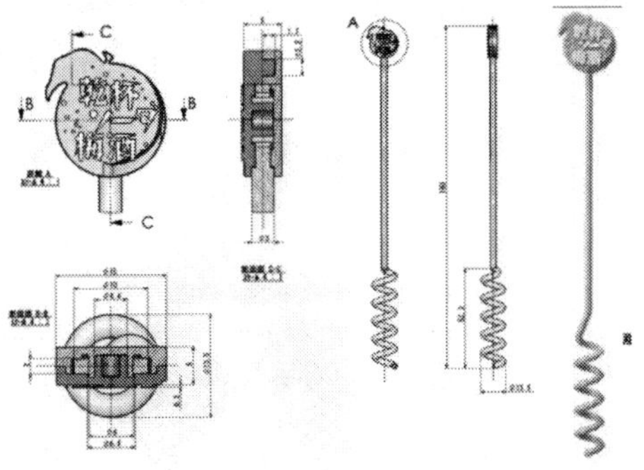

Figure 6. The blueprint and finished goods of the muddler

Figure 7. Packaging of goods

The "Kanpai-muddler" was the schedule of 20,000 piece creation in the first plan. However, since it was reputable, additional creation of the " Kanpai-muddler " was performed, and 80,000 totals were created.

Figure 8. The surface of the covering letter attached to the muddler

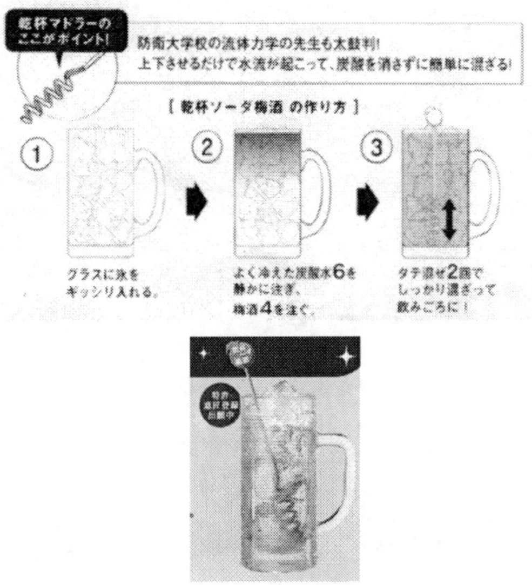

Figure 9. The back of the covering letter attached to the muddler and the busy condition of the muddler

EPJ Web of Conferences **114**, 02140 (2016)
DOI: 10.1051/epjconf/201611402140

A shape modification of the balancing slot of the impulse turbine stage with the drum rotor arrangement

Kukchol Yun[1,a], L. Tajč[2] and M. Kolovratník[1]

[1]Czech Technical University in Prague, Department of Mechanical Engineering, Technická 4 Praha 6, Czech Republic
[2]University of West Bohemia, Department of Mechanical Engineering, Univerzitní 22 Plzeň, Czech Republic

Abstract. At the previous conference in Český Krumlov it was presented how the balancing slots may influence the efficiency of the impulse turbine stage with the drum rotor arrangement. A few options for shape modification of the balancing slot were also suggested. Implementation of the balancing slots and their possible modifications are subject to various design and technological limitations. During the consultation with the turbine designers and technologists, we were able to determine the modification of the balancing slot which, in terms of their design and production technology, will be the most simple and feasible. It is predicted that the suggested shape modification of the balancing slot will improve the efficiency of the turbine stage. This prediction is then supported by CFD simulations.

1 Introduction

The drum rotor arrangement in the impulse turbine is one of the original concepts of Doosan Škoda Power Company for the construction of industrial steam turbines. This new concept enables the low-output turbines to extend the rotor blades and in this way lowers their boundary losses.

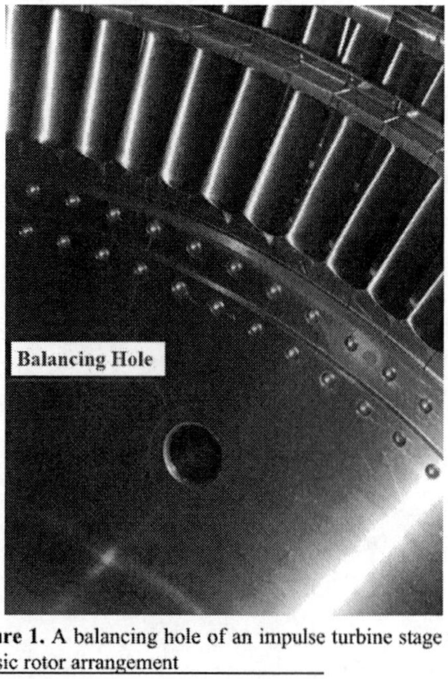

Figure 1. A balancing hole of an impulse turbine stage with classic rotor arrangement

[a] Kukchol Yun: kukchol.yun@gmail.com

As is well-known, rotor blades of a classical impulse turbine stage are fitted into a large rotor disc, where sizeable balancing holes are placed (see figure. 1). These balancing holes primarily serve to balance pressures in front of and behind the rotor. Because of this, the axial force affecting the disc decreases. The pressure balancing in front of and behind the rotor disc is accomplished by extracting the seal steam by properly dimensioned balancing slots. Unlike in a traditional rotor arrangement of an impulse turbine, the blades in the drum rotor arrangement are placed directly on the drum rotor without using the rotor disc. This constructional design does not allow the application of the balancing holes in the way used in the traditional impulse turbine stage. In other words, the seal steam in a turbine stage with the drum rotor arrangement cannot be extracted by the balancing holes.

In the new concept, there are two possibilities how to transfer the seal steam behind the turbine stage.

The first possibility is to have all the seal steam blown into the gap between the stator (nozzle) blades and rotor (bucket) blades. In this case a fusion of the main steam (steam coming from the nozzle blade cascades) and the seal steam occurs. The seal steam has a velocity and flow direction that differs from that of the main steam. Its temperature is also higher than that of the main steam. Their mixing causes a change in aerodynamic and thermodynamic parameters in front of the rotor blades, specifically near their hub area. The influence of this change in these parameters on the efficiency of the corresponding turbine stage is described in detail in our paper from the previous conference [1]. It can be stated with certainty that the blowing of the seal steam always causes a decrease in the efficiency of the impulse turbine stage.

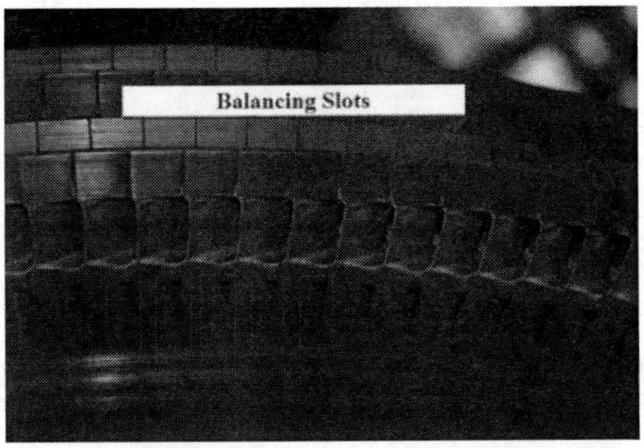

Figure 2. Balancing slots of an impulse turbine stage with the drum rotor arrangement

Another possibility how to solve the problem with seal steam in the drum rotor arrangement is to find an alternative for the classical balancing holes. Specifically, in our case the balancing holes are replaced by slots with small cross-section areas, which are a part of every rotor blade. These slots are called balancing slots (see figure 2). The results of numerical calculations clearly show that the extracting of the seal steam through the balancing slots has a positive influence on aerodynamic and thermodynamic parameters of the steam flow upstream from the rotor blades, and in this way leads to an increase in the efficiency of the turbine stage (see figure 3).

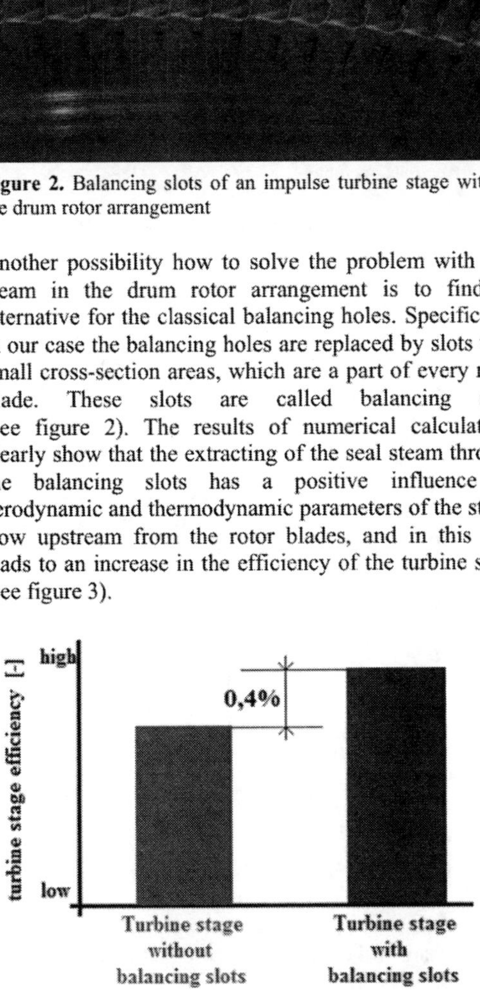

Figure 3. Comparison of the efficiencies of the turbine stage with and without balancing slots

2 Balancing Slots and Their Shape Modification

As mentioned above, in an impulse turbine with the drum rotor arrangement, it is necessary to implement the balancing slots for an improvement in efficiency.

These balancing slots are essentially replacements for the balancing holes of a classical impulse turbine stage. Thus these slots must meet all the requirements that are typical for the balancing holes. In addition, they have to satisfy another specific requirement which is derived from the constructional particularity of the new drum concept.

The balancing slots in the drum rotor arrangement are placed nearer to the shaft seal than the balancing holes in the traditional rotor arrangement (see figure 4).

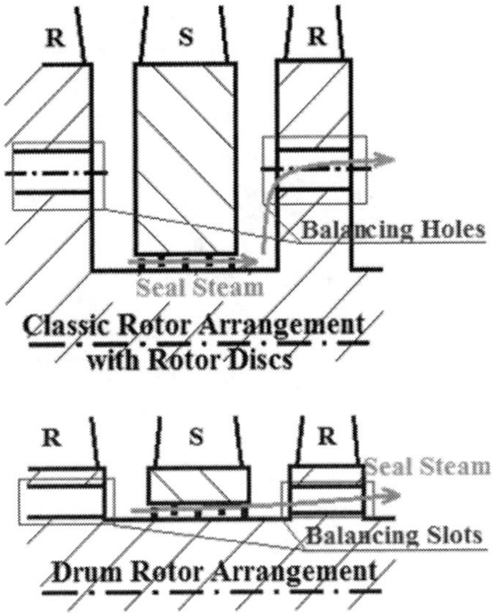

Figure 4. Balancing slots of an impulse turbine stage with the drum rotor arrangement

Thanks to this arrangement, the seal steam moves immediately to the balancing slots and thus its kinetic energy is not "wasted". Because of this, it is possible to get greater output from the seal steam flow through the balancing slots (The balancing slots could have mechanical output).

The amount of output gained in this way will depend on the ability of the balancing slots to effectively use the kinetic energy of the seal steam. For this reason, to maximize the utilisation of the kinetic energy of the seal steam is another requirement, which the balancing slots have to satisfy.

When the steam flows along the turbine blade profile, different pressures occur on its pressure and suction side. The pressure on the pressure side is generally higher than the pressure on the suction side. The output of the steam flowing through the blade cascade corresponds to this pressure gradient.

Similarly, the seal steam flow through the balancing slots will produce output if the pressure on the pressure side of every balancing slot is higher than the pressure on its suction side.

In the opposite case, the seal steam flow will produce "negative" output, i.e. it will slow down the rotor.

To depict clearly the pressures on both sides of the balancing slot, a few control lines have been created (see figure 5).

The pressures on these lines are shown in figure 6.

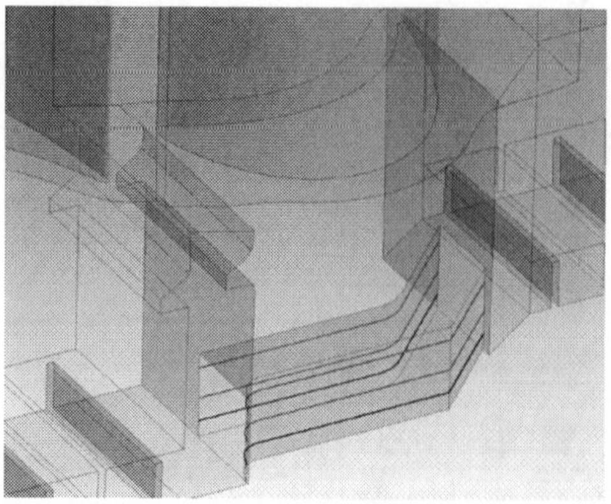

Figure 5. Control lines created on the side walls of the balancing slot

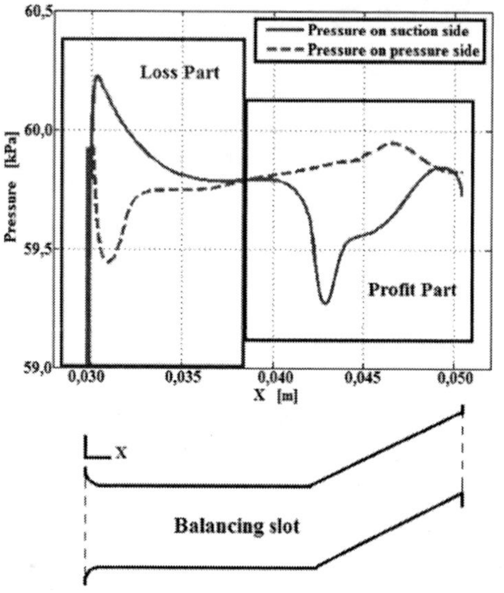

Figure 6. Pressure distribution on the control lines

The production of output in the balancing slot can be divided into two parts - the profit and the loss. It is obvious that the amount of output produced by the steam flow through the balancing slot depends on its profit and loss part ratio. Specifically, the larger the profit part of the balancing slot, the higher its mechanical output and vice versa. As described in our paper from the previous conference [1], an expansion of the profit part of the balancing slot can be reached by a change in the outlet angle of the seal steam flow and by acceleration of the steam flow at its outlet. In both cases a reduction in the tangential velocity of the steam flow on the balancing slot outlet occurs and it leads to an increase in the mechanical output of the balancing slot (see figure 7).

As mentioned before, the balancing slot is a part of each rotor blade. For this reason, its shape modification certainly leads to a geometric change of the rotor blade. To change a blade´s geometry is very complicated both in terms of construction and technology. It is also very expensive.

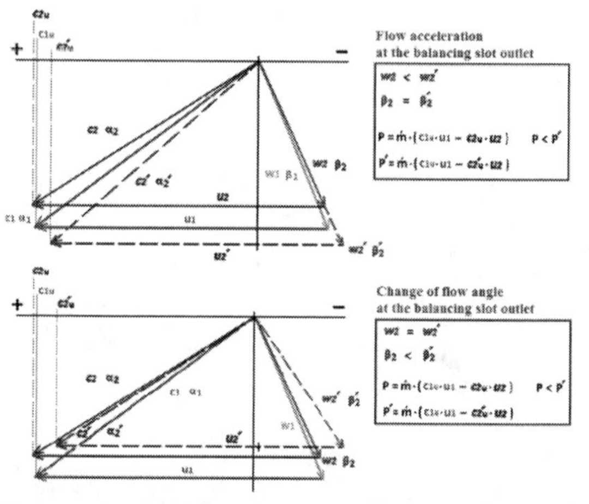

Figure 7. Steam flow acceleration and change of flow angle at the balancing slots outlet

That´s why we tried to discover a shape modification of the balancing slot, which causes minimal interference with the rotor blade while at the same time enabling the change in the outlet angle and acceleration of the steam flow at its outlet.

One of the modifications which meets the above mentioned criteria is gradual narrowing of the balancing slot´s outlet channel (see figure 8).

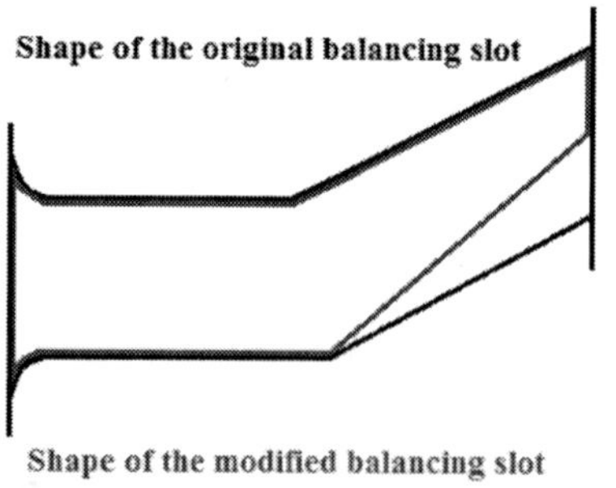

Figure 8. Shapes of the original and modified balancing slots

This shape modification of the balancing slot clearly minimizes the geometric impact of the rotor blade. Gradual narrowing of the balancing slot´s inner channel should lead to the change in the outlet angle and to the acceleration of the steam flow at its outlet. The question, whether this modification of the balancing slot´s shape will really bring about the increase in its mechanical output and the improvement of corresponding turbine stage efficiency respectively remains.

To answer this question, CFD calculations of two-stage steam turbines with the original as well as modified balancing slots have been carried out.

3 Calculation Model

The calculation model includes 2 complete turbine stages including their shaft seals and tip seals (see figure 9).
This complexity of the calculation model will guarantee the exact condition for the steam flow through the balancing slot.

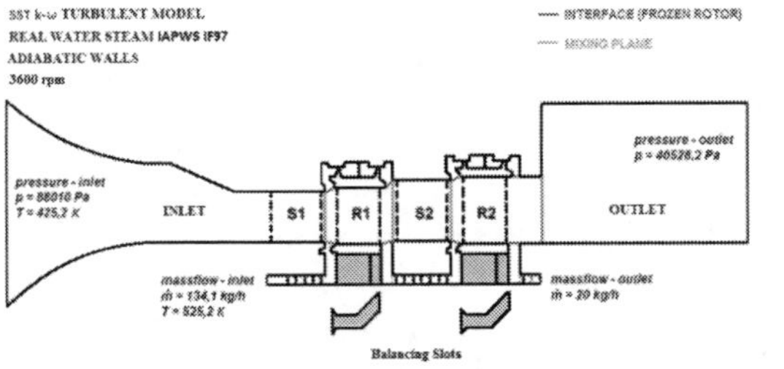

Figure 9. Calculation model

The mesh of the calculation model is prepared either in ANSYS TurboGrid 15.0 (stator and rotor domains) or in ANSYS Meshing 15.0 (shaft and tip seal domains). The mesh of the calculation model including several details in selected places is shown in figure 10.

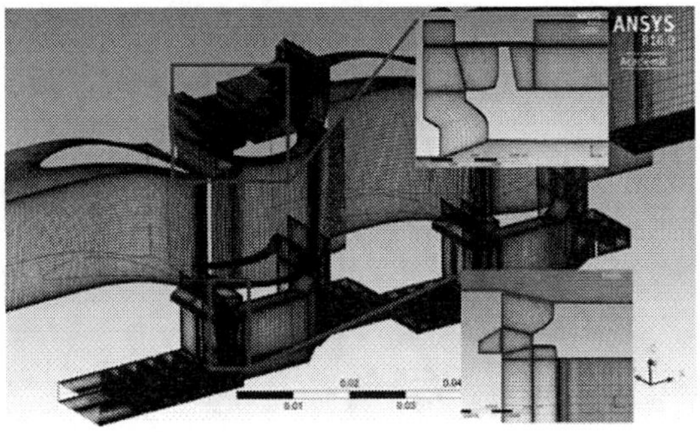

Figure 10. Mesh of the calculation model

The inlet and outlet boundary conditions of the calculation model are set by pressures, temperature and mass flux which were measured in the experiments. All calculations are accomplished in ANSYS CFX 15.0 using database IAPWS IF97 describing the real steam properties.
Two calculation variants (variant0 with the original balancing slots and variant1 with modified balancing slots) are carried out and their results are subsequently compared.

4 Results

In figure 11 the pressure distribution on the control lines is compared. In this picture it is possible to observe that the profit part of the variant1 (variant with modified balancing slots) is larger than that of the variant0 (variant with original balancing slots). This expansion of the profit part in the variant1 brought about considerable increase in its mechanical output.

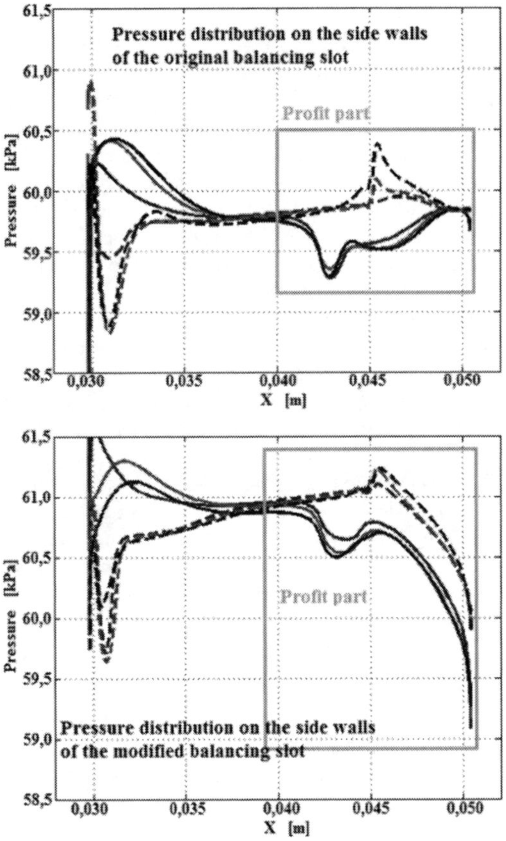

Figure 11. Comparison of the profit part of the balancing slot of variant0 and variant1

As is seen in figure 12, mechanical output of the modified balancing slot is about 2 times higher than the output of the original slot.

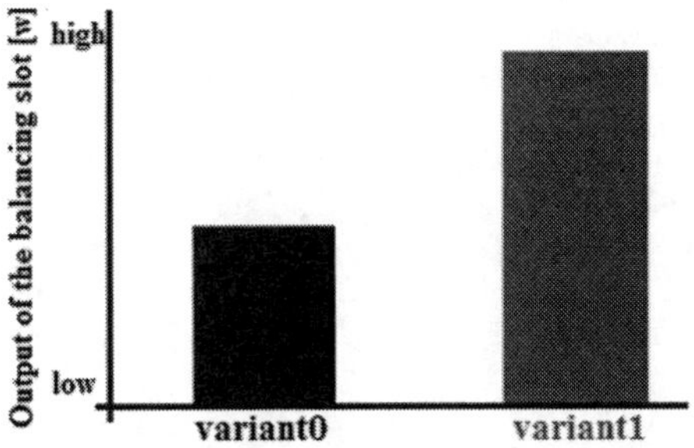

Figure 12. Comparison of the output of the balancing slot of variant0 and variant1

This increase in mechanical output of the balancing slot in variant1 contributes to the improvement of its efficiency. Specifically, the efficiency of the variant1 with the modified balancing slots is 0.52% higher than the efficiency of the variant 0 with the original balancing slots.

Conclusion

After detailed analysis of the steam flow through the balancing slot, it was discovered that the change in the outlet angle and the acceleration of the steam flow at the balancing slot's outlet can cause an increase in its mechanical output. Each shape modification of the balancing slot is related to a geometric change of the rotor blade, which is very demanding concerning both construction and technology. Therefore we had to find the shape modification of the balancing slot, which could be most easily implemented and at the same time enable the change of the outlet angle and acceleration of the steam flow at its outlet.

After consultation with the designers, the new shape of the balancing slot with the narrowing outlet channel was suggested.

Using CFD calculations the benefit of the newly designed balancing slots was verified and its positive influence on the turbine efficiency was confirmed. Replacing the original balancing slots by the modified ones caused improvement in the turbine efficiency by 0.52%.

Acknowledgement

The authors of the paper would like to thank Czech Technical University in Prague for the financial support within the project with the grant number SGS 14/183.

References

1. K. Yun, M. Kolovratník, L. Tajč, Proceeding, Experimental Fluid Dynamics 2014, **748-752** (2014)
2. J. Synáč, *The steam blowing to the shear layers and their extruding from the side walls of the turbine stage* (1992)
3. M. Hoznedl, L. Bednář, L. Tajč, *Analysis of the steam flow in the two-stage steam turbine with the drum rotor arrangement*, Technical report, Doosan Škoda Power, (2011)
4. K. Yun, Z. Jůza, M. Hoznedl, L. Bednář, L. Tajč, Proceeding, 17th Conference Engineering Mechanics, **671-674**, (2011)
5. M. Hoznedl, L. Tajč, Proceeding, 10th European Conference on Turbomachinery, **48-53**, (2013)

EPJ Web of Conferences **114**, 02141 (2016)
DOI: 10.1051/epjconf/201611402141

A CFD Analysis of Steam Flow in the Two-Stage Experimental Impulse Turbine with the Drum Rotor Arrangement

Kukchol Yun[1,a], L. Tajč[2] and M. Kolovratník[1]

[1]CTU, Department of Mechanical Engineering, 16607 Technická 4 Praha 6, Czech Republic
[2]WBU, Department of Mechanical Engineering, 30614 Univerzitní 22 Plzeň, Czech Republic

Abstract. The aim of the paper is to present the CFD analysis of the steam flow in the two-stage turbine with a drum rotor and balancing slots. The balancing slot is a part of every rotor blade and it can be used in the same way as balancing holes on the classical rotor disc. The main attention is focused on the explanation of the experimental knowledge about the impact of the slot covering and uncovering on the efficiency of the individual stages and the entire turbine. The pressure and temperature fields and the mass steam flows through the shaft seals, slots and blade cascades are calculated. The impact of the balancing slots covering or uncovering on the reaction and velocity conditions in the stages is evaluated according to the pressure and temperature fields. We have also concentrated on the analysis of the seal steam flow through the balancing slots. The optimized design of the balancing slots has been suggested.

1 Introduction

At present a considerable increase in demand of industrial steam turbines exists. The Doosan Škoda Power tries to satisfy its customers also in this area and develops its own conception for industrial turbine construction. One of these conceptions is the drum rotor arrangement. Industrial turbines are generally characterized by a small output and it leads frequently to a problem of unacceptable increase in secondary loss of rotor blades. In the new conception this problem is restrained by the replacement of the classical rotor with big discs by the drum rotor (see Fig. 1).

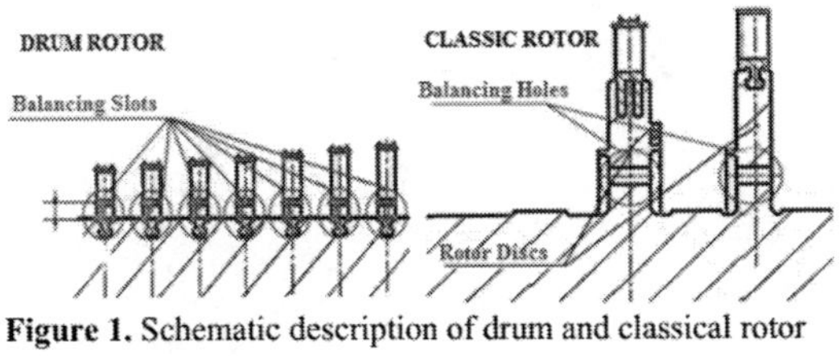

Figure 1. Schematic description of drum and classical rotor

This new concept of the rotor arrangement enables even for low output to use longer rotor blades, and in this way to reach lowering of their secondary losses. For construction reasons in the drum rotor arrangement large balancing holes, which are a common part of the classical impulse stage, cannot be realized. The balancing holes are made in the rotor disc and primarily serve to balance pressures in front of and behind the disc. Thus, the axial force affecting the rotor disc decreases. Another task of the balancing holes in the classical impulse stage is to extract the seal steam behind the rotor disc. The seal steam has generally higher temperature than the main steam flowing through the blade cascade. If this warmer seal steam is blown to the main flow part, the steam flow in the blade cascade is strongly influenced, mainly in the hub part of rotor blades. The change of the turbine stage reaction and deformation of a velocity triangle occurs.
The influence of blowing seal steam on the turbine stage efficiency was explored in detail by J. Synáč in his PhD thesis. From the findings it is evident that blowing seal steam always causes decrease in the turbine stage efficiency.

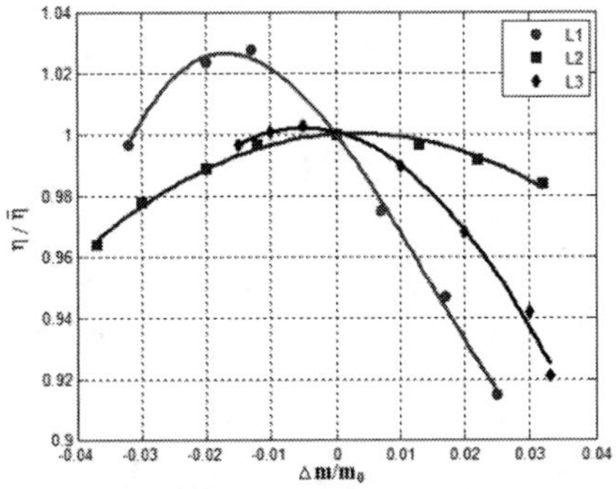

Figure 2. Dependence of efficiency in blown or extracted amount of steam for blades L1, L2 and L3

[a] Kukchol Yun: kukchol.yun@gmail.com

In Fig.2 the dependences of the turbine stage efficiency on the blown or extracted amount of steam Δm for various rotor blades L1, L2 and L3 are shown. These blades differ mainly in length. Specifically, blade L1 is the shortest and L3 the longest one.

As mentioned before, due to the construction limitations in the rotor with the drum arrangement it is not possible to make the large balancing holes. That's why, in the new conception these holes are replaced by slots of a small diameter, which are called balancing slots. These slots are part of each rotor blade and serve to extract seal steam behind the stage. Moreover, the shape of these slots is designed so that the flowing seal steam could produce a part of shaft power.

2 Incomprehensibility of Experimental Results

In Doosan Škoda Power laboratories many experiments have been carried out on the two stage impulse turbine with the drum rotor arrangement. While processing their results some difficulties were appeared. They were caused mainly by certain limitations while getting experimental data, uncertainty around the mass flow of steam through the balancing slots and by evaluations of representative pressure and temperature values in the individual turbine stages to calculate their efficiency. In addition, analysing experimental results brought controversial finding about the functionality of the balancing slots.

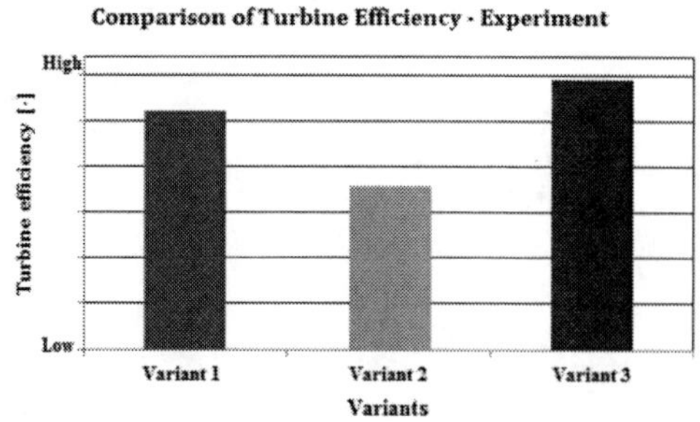

Figure 3. Comparison of turbine efficiency-Experiment

In Fig. 3 the turbine efficiencies are compared for variant 1, 2 and 3, which differ in various operational setting of the 1st and 2nd stage of experimental turbine balancing slots (see Tab. 1).

Table 1. Operational setting of balancing slots for various calculation variants

variant	Balancing slots	
	1st stage	2nd stage
variant 1	uncovered	uncovered
variant 2	covered	covered
variant 3	covered	covered

Turbine efficiency for each variant is evaluated using measured values of pressures, temperatures, mass flow of steam and torque moments at experiments. It is obvious that the highest efficiency is measured for variant 3, where all balancing slots are covered and, thus, all seal steam is blown into the main flow parts of both stages.

This fact is in direct contrast with Dr. Synáč's experimental results, which show negative influence of blowing of seal steam on the turbine stage efficiency. It is also noticeable that the lowest efficiency is achieved in variant 2, where the balancing slots are covered only in the 1st stage. Incomprehensible influence of the slot covering on the turbine efficiency is one of the reasons for implementing numerical analyses.

3 CFD Calculations

For better understanding of the steam flow not only through the balancing slots, but through the entire turbine, 3 CFD calculations modelling the same operational conditions that were applied at experimental turbine with the drum rotor arrangement were implemented. The calculation model with detailed description of the boundary conditions is shown schematically in Fig.4.

All calculations were realized in the ANSYS CFX using IAPWS IF97, describing the real steam properties.

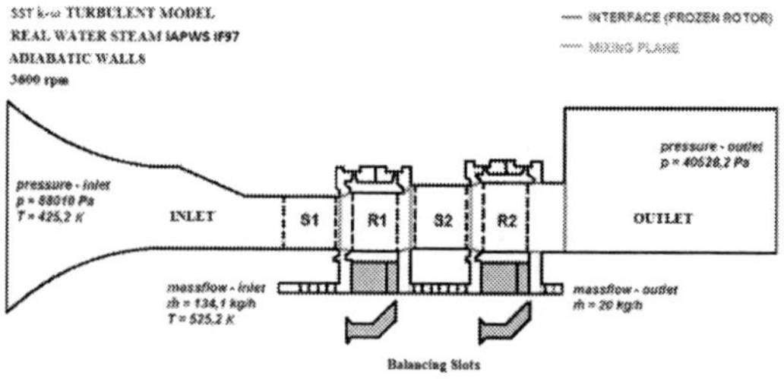

Figure 4. Calculation model

In Fig.5 the turbine efficiencies evaluated using results obtained from the CFD calculations are compared with ones measured from experiments. The comparison shows the same dependence of the turbine efficiencies on the operational settings of the 1st and 2nd stage balancing slots.

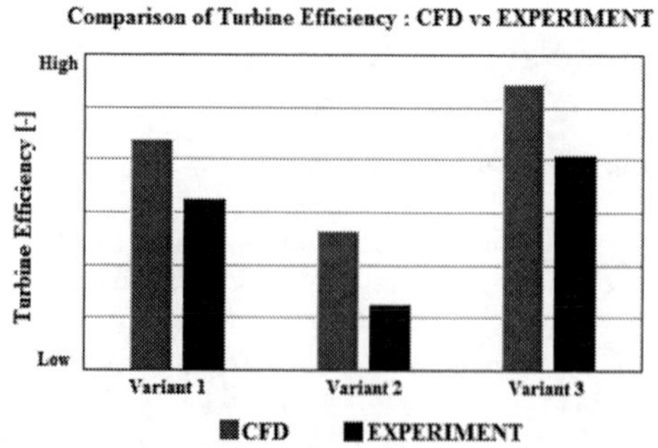

Figure 5. Comparison of turbine efficiency – CFD vs EXPERIMENT

Detailed analysis of CFD calculation results also shows negative influences of the seal steam blowing on aerodynamic parameters of the steam flow in the main flow part. As shown in Fig.6 and 7, seal steam blowing leads to increasing static pressure in front of the rotor and to slowing down the steam flow in the rotor cascade. This slowdown is more considerable mainly at the rotor blade hub and it is connected with the change of the angle α and β respectively (see Fig. 8 and 9).

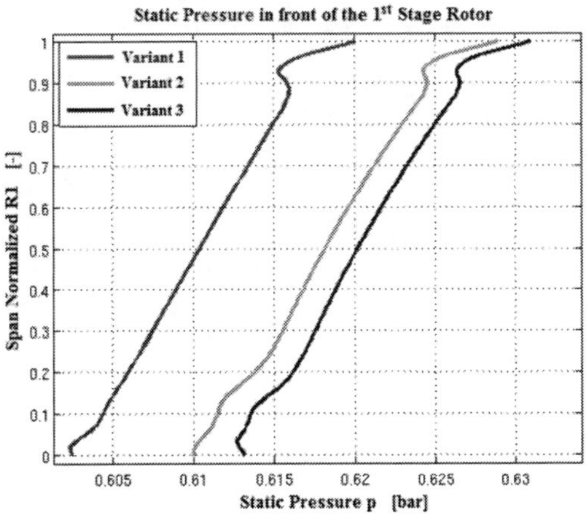

Figure 6. Static pressure in front of the 1st stage rotor

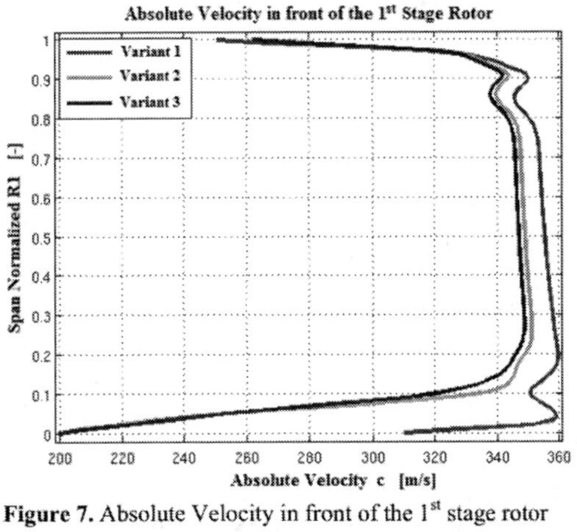

Figure 7. Absolute Velocity in front of the 1st stage rotor

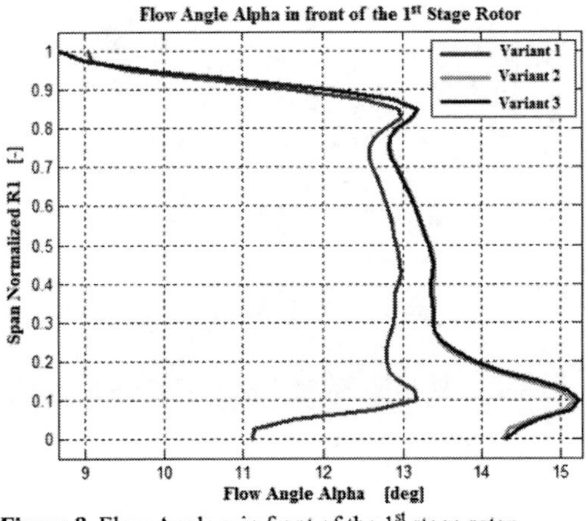

Figure 8. Flow Angle α in front of the 1st stage rotor

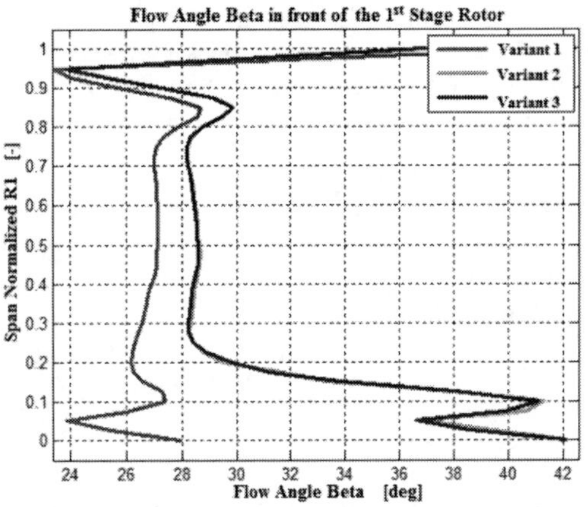

Figure 9. Flow Angle β in front of the 1st stage rotor

Seal steam has higher temperature than steam in the main flow part. Because of that, while blowing the seal steam into the main flow part, a change (increase) in reaction at the rotor blade hub occurs (see Fig. 10).

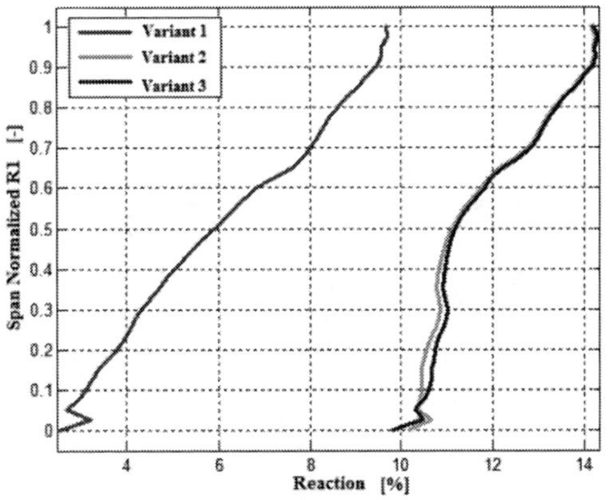

Figure 10. Reaction of the 1st turbine stage

As the experimental turbine is an impulse turbine and has prismatic cylindrical blades, the above mentioned changes negatively influence on the turbine stage output. Thus, there are two contradictory conclusions.

The blowing seal steam leads to worsening of the aerodynamic parameters of steam in the main flow part and it will cause reduction of the turbine stage mechanical output. However, the highest efficiency is recorded in variant 3, where seal steam is blown into the main flow part.

After a more detailed analysis of all implemented CFD calculation results it is evident, that the reason for this contradiction could be the fact that, at experiments the seal steam was considerably warmer than the main steam entering the turbine stage. Due to the technical reasons it was not possible to regulate the source of seal steam, its temperature was by 100 K higher than the main steam, which enters the turbine stage. Because of this considerable temperature difference, blowing seal steam represents a non-negligible „intake" of heat into the main flow part, even if its flux is relatively small comparing to the main steam flux. The in-this-way "heated" steam is processed in the following stage in variant 3 and in this way the loss of output in the previous stage, caused by worsening of aerodynamic parameters of the main steam, is compensated. In variant 2, where only the balancing slots of the 1st stage are covered, a part of the heated steam is extracted by uncovered balancing slots of the 2nd stage without using its potential to produce shaft power. This extracting of better quality steam can provide answers to the question, why the efficiency of this variant is calculated as the lowest.

Such extremely high temperature of seal steam is not practically common. As a matter of the fact, the seal steam temperature is almost as high as the steam temperature in the main flow part.

To confirm the correctness of the given assumption, calculations are done of other three variants (variant 1a, 2a a 3a). These new calculation variants differ from the previous ones by boundary condition for the seal steam inlet. In this case the seal steam has the same pressure and temperature as the main steam entering the turbine stage.

The subsequent processing of these CFD calculations confirms the correctness of the assumption. A direct proof is provided by comparison of the turbine efficiency for variants 1a, 2a and 3a (see Fig. 11).

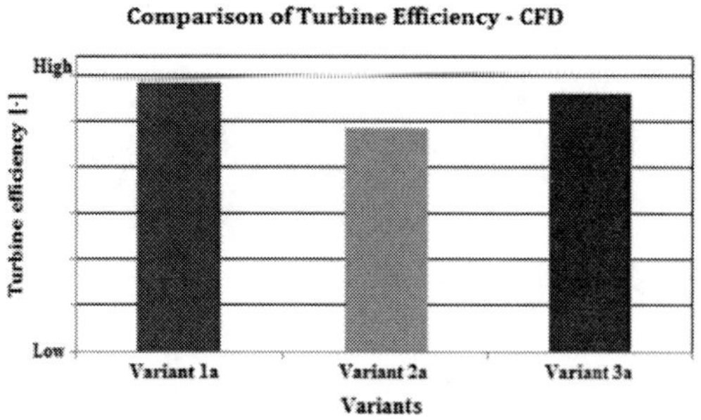

Figure 11. Comparison of turbine efficiency-CFD

This time the highest efficiency is reached in variation 1a, where all balancing slots of both stages are uncovered. It explains why the results gained from experiments do not correspond with the older theoretical findings.

4 Steam Flow through Balancing Slots

The results of CFD calculations show that extracting seal steam through balancing slots is, from the viewpoint of turbine stage efficiency, more favourable than steam blowing into the main flow steam. That´s why considerable attention is devoted to examination of steam flow through the balancing slots.

In Fig. 12 steam flow through the balancing slots is presented.

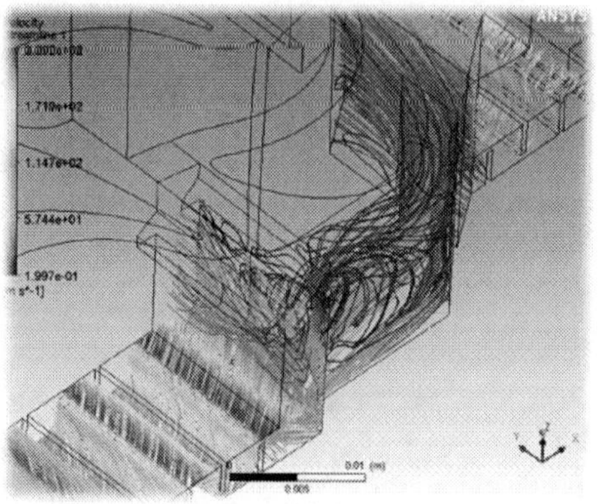

Figure 12. Steam flow through balancing slots

A detailed analysis shows that flow cross-sectional areas of existing balancing slots are oversized. This means that not only the seal steam, but also a part of the main flow steam goes through the balancing slots. Since balancing slots are not able to effectively gain energy from the steam flow like the blade grid, extracting main steam is not desirable. Moreover, we found out that the velocity ratio in front of and behind the balancing slots is very different from the ratio in front of and behind the rotor cascade of the turbine stage.

In Fig. 13 velocity triangles for the balancing slots are compared with velocity triangles for rotor cascade.

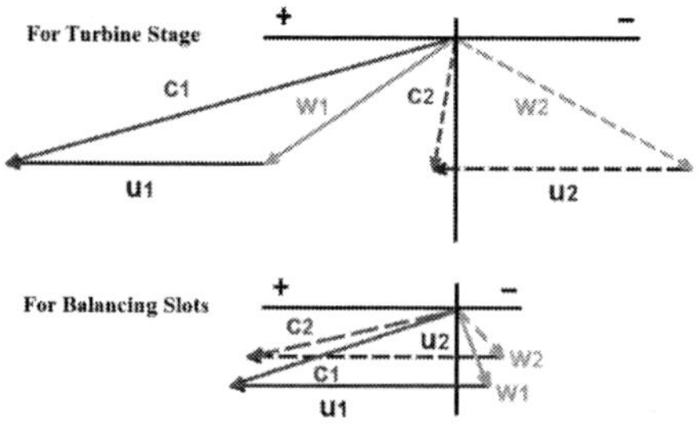

Figure 13. Velocity triangles for the turbine stage and balancing slots

Due to this difference local increase of pressure occurs on the suction side of balancing slots (see Fig. 14).

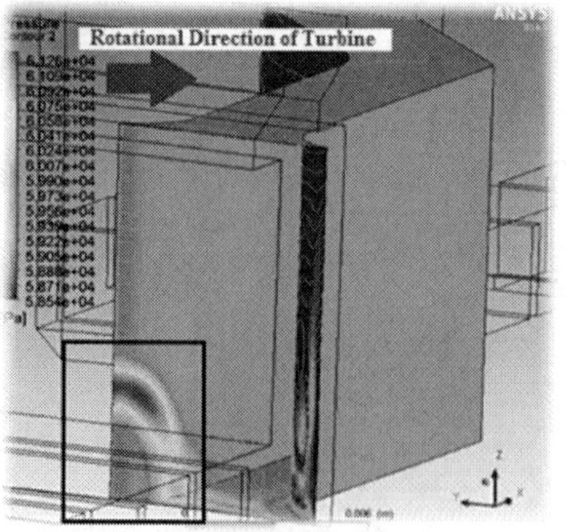

Figure 14. Occurrence of local increase of pressure on the suction side of balancing slot

Several possibilities to improve functioning of the balancing slots are founded.

The first possibility is optimization of flow cross-sectional area of the balancing slots. The steam flow going through the balancing slots mainly depends on their flow cross-sectional area. Even more optimization of the flow cross-sectional area can prevent extraction of steam from the main flow part, which leads to reduction of turbine stage output.

Another possibility can be the design of more suitable inlet shape of the balancing slots. Inlet shape of the balancing slots that fit better the flow ratios can reduce the occurrence of local increase of pressure on their suction side. This reduction could cause the steam, which flows through the balancing slots to produce shaft power more effectively (see Fig. 15).

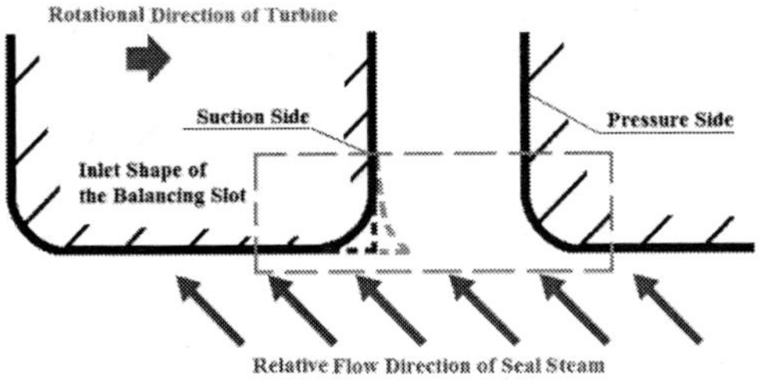

Figure 15. Possible inlet shapes of balancing slot

Acceleration and change of the flow angle at the balancing slots outlet can also lead to improvement of gaining shaft power from the seal steam and increase in the turbine stage efficiency. As shown in Fig.13, acceleration and change of flow angle at cause lowering of the peripheral component of the absolute velocity C_{2u}. It is obvious from the Euler Turbine Equation that decreasing this component leads to increase of mechanical output.

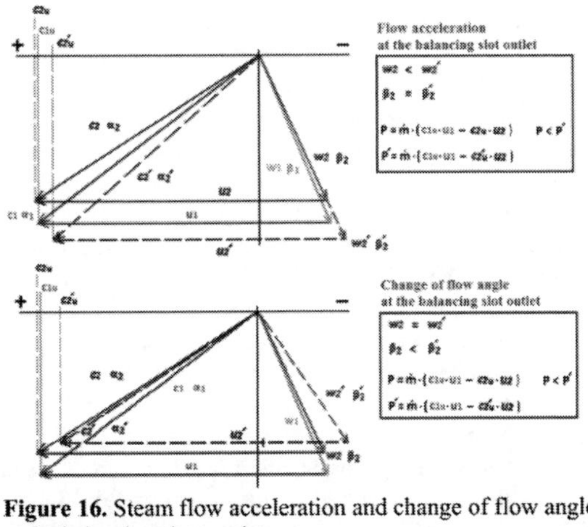

Figure 16. Steam flow acceleration and change of flow angle at the balancing slots outlet

Conclusion

To clarify some uncertainties of experimental results and to understand better the steam flow through the entire experimental turbine, a large range of CFD calculations was implemented.

Detailed processing of obtained results brought a hypothesis about the cause of these uncertainties and consequently confirmed it.

The results of the CFD calculations further showed that extracting seal steam through balancing slots in impulse turbine with drum rotor arrangement is, from the viewpoint of its efficiency, more favourable than it´s blowing into the main flow part.

After a detailed analysis of steam flow through the existing balancing slots some options for their shape optimization were suggested.

Final dimension and shape of the optimized balancing slots will be a result of the combination of these proposed options of modification.

References

1. A. Mecke, I. Lee, J.R. Baker jr., M.M. Banaszak Holl, B.G. Orr, Eur. Phys. J. E **14**, 7 (2004)
2. M. Ben Rabha, M.F. Boujmil, M. Saadoun, B. Bessaïs, Eur. Phys. J. Appl. Phys. (to be published)

EPJ Web of Conferences **114**, 02142 (2016)
DOI: 10.1051/epjconf/201611402142

Comparison of different configurations of Phase Doppler Analyser

Matouš Zaremba[1,a], Milan Malý[1], Jan Jedelský[1] and Miroslav Jícha[1]

[1]Brno University of Technology, Faculty of Mechanical Engineering, Department of the thermodynamics and environmental engineering, Technická 2896/2, 616 69 Brno, Czech Republic

Abstract. A phase Doppler anemometry (PDA) technique is widely used in experimental fluid mechanics to measure size and velocity of particles in the fluid flow. Even though this method is common in experimental fluid mechanics, there are only few techniques that might serve for the purpose of the evaluation of the PDA system. To examine results of the PDA visualizations techniques are usually used. However, this approach suffers from several aspects. Mainly, it is difficult to determine the exact position of the measurement volume of PDA system. Then it is complicated to determine which particles are passing through the measuring volume. Another way how to examine performance of the PDA system is to use two PDA systems simultaneously. By using one laser for both systems we can avoid previously mentioned aspects. In our experiments, we use fiber based PDA system and classical PDA system both made by Dantec Dynamics. The aim of this paper is to compare results from various configurations and highlight crucial parameters that influence measurements.

1 Introduction

Non-intrusive optical measuring techniques play an important role in the field of the experimental fluid mechanics. Especially, in the application of measuring velocity and size of droplets or bubbles in two-phase flows. These parameters are of importance in several aspects in industrial applications. Mainly in the field of spray combustion systems such as automotive and aircraft engines, combustion of liquid fuels in power plants or spray drying processes in the food industry [1].

Several techniques were developed to measure characteristics of the spray, for example: particle image velocimetry (PIV), laser Doppler anemometry (LDA), phase Doppler anemometry (PDA), Malvern laser particle size analyser and others. These measuring techniques give us valuable and relatively precise results about droplet velocity or droplet size or both. However, determining the uncertainty of the measuring system is a difficult task [2, 3]. Even though several workers tried to calculate precision of PDA system a precise method has not yet been made. Some publications investigated different measuring techniques, for example: [4] however, the comparison between two system which works on the same principle but have different post processing tools or different configuration was examined in just few publications, for example [5]. Moreover, various commercial PDA systems were developed in the past, and their performance is usually assumed to be precise. It is common to compare results from measurements with previous observations. However, the difference between measuring systems should be taken into account. Thus, in this paper we focus on the examination of two different phase Doppler analysers in three configurations. The aim is to find crucial parameters or components of the PDA systems that influence measured results.

[a] Corresponding author: zaremba@fme.vutbr.cz

2 Material and Methods

Experiments were conducted on the cold test bench at still ambient condition and room temperature at the Brno University of Technology in Sprays laboratory.

2.1 Atomizer

The PDA systems were tested in the fully developed spray. The spray was generated by the twin-fluid atomizer, so called outside-in liquid injection atomizer (OIL). This type of atomizer works on a principle of injecting liquid and gas into the mixing chamber where two-phase flow is created. Then this mixture is pushed downstream the atomizer and through the exit orifice into the ambient atmosphere. Detailed description of the atomizer can be found in [6]. Description of the test bench can be found in [7]. Light heating oil (LHO) and pressurized air were used as operating fluids. Physical properties of these fluids are presented in Table 1.

Operating regime of the atomizer is determined by the operating overpressure inside the atomizer and liquid and gas mass flows. These parameters were captured during the experiment. Measured values are written in the Table 2. Values of mass flows result in gas to liquid ratio (GLR) of 20% which indicates the annular two-phase flow inside the atomizer [6].

Table 1. Physical properties of used fluids.

Name	Dynamics viscosity (kg/(m·s))	Density (kg/m³)	Surface tension (kg/s²)
LHO	0.0185	874	0.0297
Air	$1.82 \cdot 10^{-5}$	1.23	-

Table 2. Operating regime of the atomizer.

LHO mass flow (kg/h)	LHO pressure (kPa)	Air mass flow (kg/h)	Air pressure (kPa)
2.55	66	0.13	69

2.2 Description of examined systems

In this work, we examined two phase Doppler analysers in three configurations. The first configuration (system A) is fiber based 2D phase Doppler analyser made by Dantec Dynamics. This system consists of:

- Spectra Physics Stabilite 2017 Argon laser: max. power output 6 Watts,
- 60X41 Transmitter, which splits the beam into its individual color components (488.0 nm, 514.5 nm) and divides each color into two beams.
- Brag cell is implemented in the transmitter which gives frequency shift 40 MHz to each beam from given pair.
- 60X81 2D 85 mm transmitting optics with 50X82 beam translator and focal length of the final lens of 500 mm,
- 57X50 112 mm diameter fiber PDA receiver optics with spatial filter and focal length of the lens of 800 mm,
- Fiber PDA Detector unit and
- BSA P80 flow and particle processor that was set to measure velocity within the range from 0 to 25 m/s and droplet size in the range from 0 to 274 microns.

The transmitting line (laser, transmitter and transmitting optics) of the system A was used for all measurements to keep the same parameters of measuring volume (m.v.). The second configuration (system B) is classic 1D Dantec phase/Doppler particle analyser. Transmitting optics and processor were used from the system A. So the receiving line of the system B consists of:

- Dantec 57X10 receiving optics with focal length of a final lens of 500 mm and
- BSA P80 flow and particle processor.

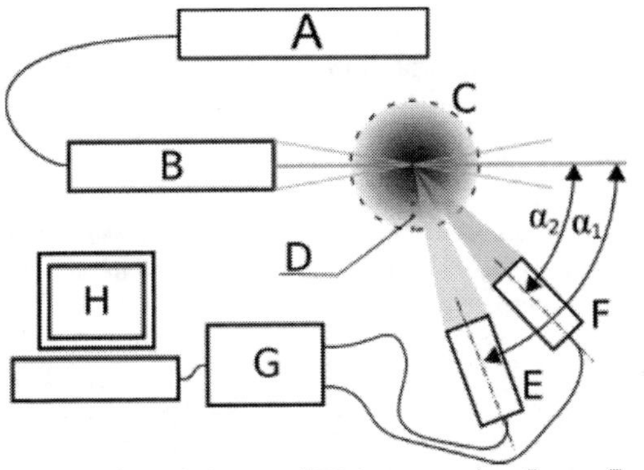

Figure 1. Schematic layout of PDA system. A – Laser, B – transmitting optics, C – Spray, D – measured points, E – Fiber PDA receiving optics, F – classic PDA receiving optics, G – processor, H – computer.

Third configuration (system C) uses again transmitting line from the system A but with the processor from classic PDA system. Then, the receiving line of system C consists of:

- Dantec 57X10 receiving optics with focal length of a final lens of 500 mm and
- Dantec Model 58N50 signal processor that was set to measure velocity within the range from 0 to 25 m/s and droplet size in the range from 0 to 236 microns.

2.2.1. Parameters of the measuring volume

Two intersecting laser beams coming out from the transmitting optics create light objects in the intersecting point, so-called measuring volume (m.v.). This ellipsoid consists of layers of high and low intensities of light i.e. fringes. The optical configuration of examined system results in fringe separation of 2.14 μm and dimensions of m.v. 0.08 × 0.08 × 0.7 mm.

The effective volume of m.v. has a shape of an ellipsoid truncated by a slit shaped by a spatial filter placed in the receiver optics [8] as shown in the figure 2. We assume that m.v. is symmetric among the main axis. Effective length of the m.v. is determined by the slit used in the receiving optics. In system A, the slit was set to 200 μm which results into the length L_S = 0.52 mm. The slit of systems B and C result in effective m.v. length L_S = 0.26 mm.

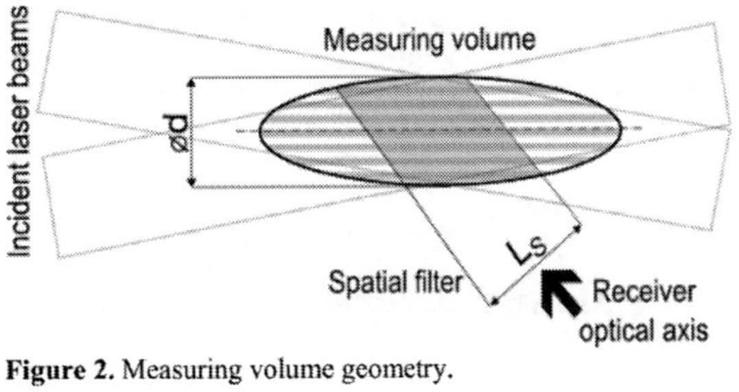

Figure 2. Measuring volume geometry.

2.2.2. Experimental setup

Measurements were performed in the axial distance of 100 mm from the atomizer's exit orifice. This distance was chosen because we predict that only spherical droplet should be presented in this area according to our previous findings [9, 10]. We also calculated Weber number (We) of droplets, and we found out that 97.5% of droplets have lower We than the critical value (11 in our case [1]). The mean value of We is then equal to 2, so we assume that all atomization processes are finished. Thus, PDA system, which can measure only spherical droplets, is able to measure with high validation (70-90% depends on the position in the spray). Due to the symmetry of the spray we measured only one half of the radial profile: 5 points by the step of 5 mm. Systems were set to measure 50 000 samples or one minute in low dense regions. Only axial velocity component was measured due to the fact that systems B and C are 1D only.

The PDA setup was optimized to a given spray conditions to get the highest validation and data rate. The velocity and drop size measuring spans were set according to preliminary measurements to be able to detect all droplet sizes and velocities. Systems B and C were set to get the similar velocity and droplet size measuring spans as it was in the system A. However due to different construction of PDA receiving optics the exact same values cannot be reached.

The PDA receiving optics is ideally set into the Brewster's angle where only one scattering mode is dominant. In our case, it is 69 degrees to the main axes of transmitting optics (shown as angle α in figure 1). Due to limited optical access to the spray, consequently to the m.v., systems were set into two different scattering angles. System A was put in the angle of $\alpha_1 = 75$ degrees and systems B and C were set in the angle of $\alpha_2 = 65$ degrees, see figure 1. The difference in measured values from various scattering angles was checked before the measurements. Variations in the mean axial velocity were in the range of 1.5% from measured range. Mean diameter differences were in the range of 1.4%. Thus, we assume that deviation from the ideal scattering angle has only negligible influence on our results.

2.2.3. Repeatability of the measurements

For this purpose only system A was used in a standard configuration as mentioned above. Measurements were performed in the same positions in the spray and results are plotted in graphs as error bars showing measured span of values. These results serve for the purpose of prediction of an error caused by the repetition of measurements.

3 Results and discussion

As the first examination of the measurements, we investigate velocity and drop size profiles. The upper graph in figure 4 shows axial velocity profiles with error bars obtained by the repetitive measurements and typical data rate profile.

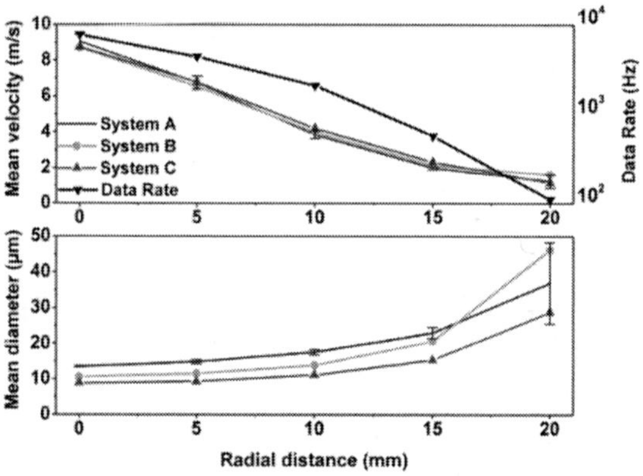

Figure 3. Mean axial velocity and mean diameter profiles. The zero position corresponds to the centre of the spray.

The differences between velocities are in the range from minimum 3.7% to maximum of 22% from a measured value which are 1% and 1.8% from measured range. Minimal difference value was found in the radial distance of 5 mm and maximal in the distance of 20 mm. The maximal difference is probably caused by a low data rate (see figure 3) and by an instability of the spray at the edge [11]. Nevertheless, considering error bars we see that the majority of measured points fall into the error range. Thus, we assume that difference in mean velocity might be caused by the uncertainty of repetitive measurements. We assume that systems were set correctly and further in this paper we focus on an examination of one point only (in the centre of the spray).

Drop size distribution profiles are shown in figure 3 at the bottom. It indicates that there is a demonstrable influence of the receiving optics and used processor to the resulting drop size distribution. The system A measured largest droplets except of value at the edge of the spray. The system B measured smaller droplets and the system A the smallest droplets. Only one point in a profile falls into predicted range of the error. This point is at the edge of the spray where the largest error was observed. We assume that this behaviour was caused again by low data rate and instabilities of the spray.

One of the ways how to examine PDA data it is to plot transit time (TT) of droplets through the measuring volume versus droplet size or velocity. This approach was recommended in PDA system manuals [12]. Ideally, when droplet flows through the centre of the m.v. magnitude of TT should be linearly dependent on droplet size and velocity. When particle is not flowing through the centre of the m.v. then TT would not be directly proportional to the droplet size and velocity. However, as mentioned in [12] linear relationship should be observed for the majority of measured droplets. Thus for evaluation of droplet size measurements we plot diameter of droplets versus TT, see figure 4.

Results in figure 4 show that systems A and B perform similarly. However, system C somehow rounds the values of TT. This is an effect of maximal sampling frequency. Systems were set to have the highest measuring resolution. Minimal TT that can be measured for systems A and B was 0.1 μs and for the system C was 2 μs. Thus, the fringe pattern, presented in figure 4 for the system C, is a consequence of its resolution.

Relationship of TT and droplet size show that droplet size is relatively well correlated with TT for most particles. However, there are large droplets presented in the low values of TT (mainly in the cases of system A and B) and small ones in the range of high TT (system C). It means that those large particles flew probably on the edge of measuring volume. Assuming that laser beams have Gaussian intensity profile then Gaussian beam effect might occur [13]. This effect causes bias in PDA measurements. There is an approach how to distinguish false particles generated by this effect [14]. This approach uses signal amplitude analysis which is not possible in our systems.

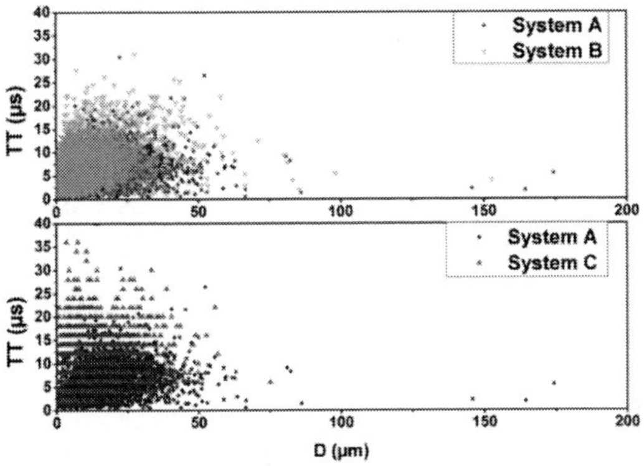

Figure 4. Droplet size-transit time correlation. Only 5000 samples are shown for better visibility.

When examining small droplets we calculated ideal TT from a given velocity, the diameter of m.v. and droplet diameter. Ideal TT in this case means that we predict a droplet which flew through the centre of the m.v. We observed that 99% of small particles with TT higher than 25 μs had higher measured TT than computed idealised value. It means that they travelled longer trajectory in a given time than is a diameter of the m.v.. Thus, we assume that small particles with the TT should not be real. However, our approach is simplified and we do not take into account variations in signal to noise ratio, the precision of the laser beam alignment, changes in droplet velocity and other factors.

Conclusions

We examined two PDA systems in three configurations in this paper. Results show that under the same conditions these systems measure velocity precisely, but values of mean droplet diameter are different. Variations in droplet size and TT were found for the system C where 58N50 processor was used. Moreover, we observed that systems A and B measure more large particles (larger than 100 microns) whereas the system C measures more small particles that have very small transit time. Thus, caution has to be made when comparing results from other measuring systems.

An approach, recommended by the manufacturer, points to the fact that probably several invalid particles were measured. However, the method presented in this paper is not precise enough, and it should be extended by evaluation of signal amplitude, or it should be supported by the high-speed imaging of the m.v. during the measurements. This, however, requires self-made PDA system or camera with long distance microscope synchronized with PDA system.

Acknowledgements

The presented research was supported by project No. GA15-09040S, which was funded by the Czech Grant Agency, the project LO1202 NETME CENTRE PLUS with the financial support from the Ministry of Education, Youth and Sports of the Czech Republic under the "National Sustainability Programme I" and project Reg. No. FSI-S-14-2355 funded by the Brno University of Technology.

References

1. Lefebvre, A., *Edtion ed.: Taylor & Francis*, ISBN 9780891166030, (1988)
2. Calvo, E., J. Garcia, I. Garcia and L. Aisa *Experiments in Fluids*, **47**(3), 489-508, (2009)
3. Czarske, J. W. *Optics Letters*, **21(7)**, 522-524, (1996)
4. Dullenkopf, K., M. Willmann, S. Wittig, F. Schone, et al., *Particle & Particle Systems Characterization*, **15(2)**, 81-89, (1998)
5. Wigley, G., J. Heath, G. Pitcher and A. Whybrew, *Particle and Particle Systems Characterization*, **18(4)**, 169-178, (2001)
6. Mlkvik, M., P. Stahle, H. P. Schuchmann, V. Gaukel, et al. *International Journal of Multiphase Flow*, **77**, 19-31, (2015)
7. Jedelský, J., M. Zaremba, M. Malý and M. Jícha., *EPJ Web of Conferences*, **67**, (2014)
8. Albrecht, H. E., N. Damaschke, M. Borys and C. Tropea, *Springer Berlin Heidelberg*, ISBN 9783662051658, (2013)
9. Jedelský, J., M. Jícha, J. Sláma and J. Otáhal *Energy & Fuels*, **23**, 6121-6130, (2009)
10. Jedelský, J. and M. Jícha. In *WIT Transactions on Engineering Sciences*. **82**, 227-238, (2014)
11. Jedelský, J. and M. Jícha, *Atomization and Sprays*, **18(1)**, 49-83, (2008)
12. Dantec Dynamics, *Edtion ed. Tonsbakken*, **18**, DK-2740 Skovlunde, Denmark, 119-122, (2001)
13. Albreacht, H., M. Borys, M. Wenzel and T. Wriedt, *Particle & Particle Systems Characterization*, **11(4)**, 339-344, (1994)
14. Bachalo, W. D., patent, US4986659 A, (1991).

EPJ Web of Conferences **114**, 02143 (2016)
DOI: 10.1051/epjconf/201611402143

Experimental turbine VT-400

Zitek Pavel[1,a], Petr Milčák[2], Noga Tomáš[1]

[1]*UWB in Pilsen, Department of Power System Engineering, Univerzitni 8, 306 14 Pilsen, Czech Republic*
[2]*Doosan Škoda Power, Tylova 1/57, 301 00 Pilsen, Czech Republic*

Abstract. The experimental air turbine VT400 is located in hall laboratories of the Department of Power System Engineering. It is a single-stage air turbine located in the suction of the compressor. It is able to solve various problems concerning the construction solution of turbine stages. The content of the article will deal mainly with the description of measurements on this turbine. The up-to-now research on this test rig will be briefly mentioned, too, as well as the description of the ongoing reconstruction.

1 Introduction

Research of flow through the flowpath of a steam turbine is highly important for increasing efficiency and reliability of a blading machine of this type. In general, it can be implemented by various ways.

The most suitable would be to carry out the research on real steam turbines in power plants. However, possibilities of such experiments are quite limited; usually there are extended guarantee tests providing the basic capacity data.

The second option is to measure on experimental research steam turbines in laboratories of their producers. Their construction currently enables more detailed measurements of performance characteristics. Yet, operation of these test rigs is fairly expensive, and operational parameters often do not provide the option of more detailed measurement inside a flow part.

Air turbine models located at specialized sites are the last option of experimental research. A construction of these turbines allows to inspect flow fields in various operational modes, and to compare results with a design state. Nevertheless, when we want to obtain such detailed data, it brings about a necessary issue of how to apply the results directly on real machines (geometrical and physical similarity).

2 Description of the experimental device

It is a single-stage air turbine located in the compressor suction. The turbine is a model of a high-pressure part of a steam turbine in the scale of 1:2. This layout provides, apart from almost constant air input parameters, also an easy access to the turbine and its measuring points. The scheme of the experimental loop is on figure 1, whereas figure 2 shows the whole device.

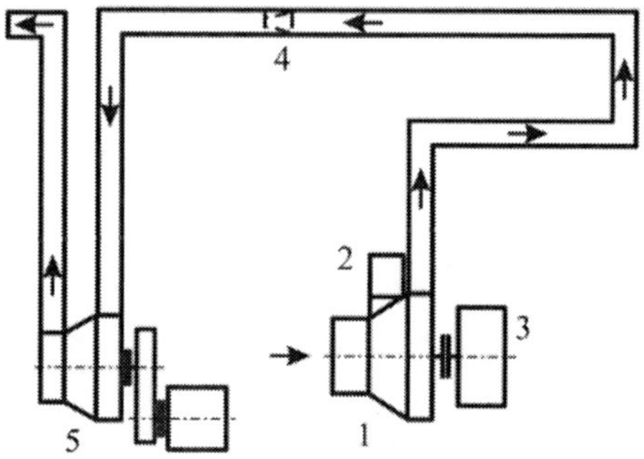

Figure 1 Scheme of experimental line

Figure 2 Experimental device

As apparent from the scheme, the rig consists of 5 main parts in total. The base is the turbine VT400 of a disc construction without a shaft gland (marked as No. 1 on the scheme). Various problems concerning turbine stages constructional solution are possible to solve on the air turbine. A hub diameter is 400 mm, the largest blade is 47 mm long. Maximum speed is 4000 1/min. A turbine section can be viewed on figure 3.

[a] Corresponding author: zitek@kke.zcu.cz

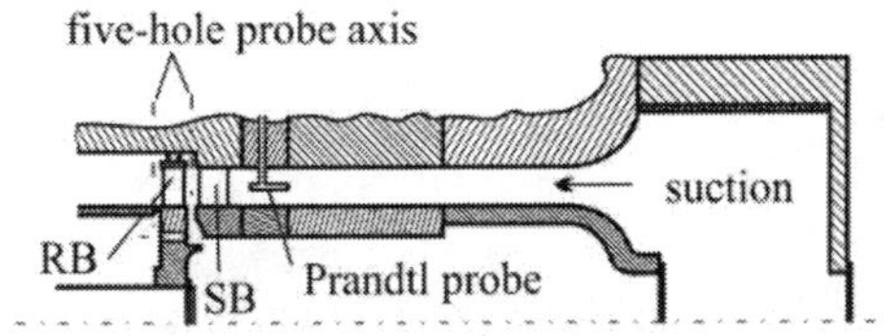

Figure 3 Section of experimental turbine

The second device of the experimental rig is a traversing device, marked as No. 2 on figure 1. The traverser enables the pneumatic probe to shift radially (under the hub as well as above the tip), peripherally (over two spacings of stationary blades), and also its automatic flow-wise turn. It allows a thorough measuring of the flow field behind both stationary and moving blades. A five-hole pneumatic probe , designed specially for this purpose, is used for such measurements on the air turbine. The probe is 700 mm long, and the holder diameter is 10 mm. The end piece of the probe is adjusted in such a manner, that it has minimum influence on the flow, and the probe sensor is placed in its rotating axis (see the model on figure 4)

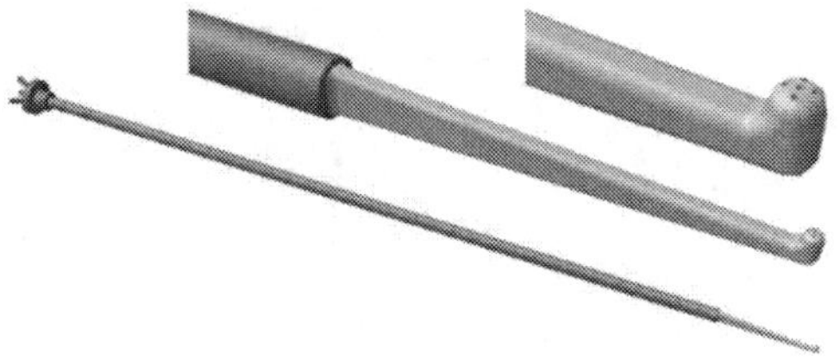

Figure 4 Model of 5-hole probe

The number three on the scheme represents a dynamometer, which provides information on speed and the moment of torsion, and at the same time, it keeps one of those values constant. It is a product of MEZ Vsetín; its maximum speed is 5000 1/min. and maximum capacity 150 kW.

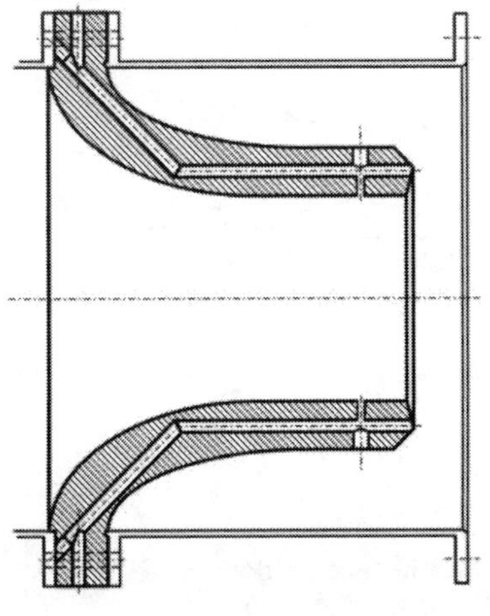

Figure 5 Nozzle for flow rate measurement

Another part of the rig is a nozzle for measuring the air flow (figure 1 number 4). The nozzle was made according to the ASME norm; the nozzle section is on figure 5.

A radial single-stage compressor represents the last part of the experimental loop (see No. 5 on figure 1). Minimum speed of the compressor is 1850 1/min, and maximum pressure drop, which can be reached on the turbine due to the compressor, is 6500 Pa.

3 Measurements

In the past, the turbine served, for instance, to determine the influence of partial admission on the stage efficiency. Since 2000, the research has been focused mainly on the influence of blade shaping on the stage efficiency [1], [2], [3], [4].

Currently, the testing of the experimental turbine VT400 stages consists of two parts – measuring integral characteristics and stationary measurements of flow fields behind stationary (SB) and moving blades (MB).

The purpose of integral characteristics measurements is to obtain the efficiency dependence on the ratio u/c. A change of pressure drop on the stage appears during this measurement at constant turbine speed. Due to the experimental device, the inlet pressure in front of the turbine is determined by the atmospheric pressure, and thus only outlet pressure and pressure drop on the stage can be measured. Pressure drop can be changed by the compressor performance, or more precisely, by the performance of the engine which powers the compressor. Changes are made in steps and manually, which complicates to set up a certain required mode and maintain a constant state. It is also possible to change the velocity ratio u/c by a change in the turbine speed at the constant pressure drop.

Apart from speed, moment of torsion and pressure drop, the flow rate and several static pressures at the hub and the tip in front of and behind the stage are measured as well. There are also pressure extractions at the outlet from SB (figure 6), which provide information regarding the size of the hub and tip reaction. Inlet and outlet velocities are calculated from the measured mass flow and the continuity equation. The total states in front of and behind are calculated from velocities and measured static pressures on walls.

Figure 6 Static pressured extractions at SB

Flow fields stationary measurements behind both stationary (SB) and moving blades (MB) grant very important information on flow in the stage. They provide pressure, or rather, velocity image of a flow field, and distribution of several variables along the blade height, after suitable peripheral data averaging, which is very valuable for designers. It is not only about loss values, but also velocities, in other words, angles to calculate deviations and flow incidences; furthermore e.g. reaction distribution and flow rate etc. Such data provide substantial information on behavior and influence of various shape modifications of blades or influence of off-design operation, when compared with theoretical design.

Flow fields measurement (wake traversing) takes place at a selected operational mode (pressure drop and ratio u/c), which is preserved at the constant value within the tunnel possibilities. It is measured by a five-hole pneumatic probe (figure 7) of the size of approximately 4mm, which represents roughly 9% of the blade height. Data acquired by this can be partially distorted. Due to the probe size (and its holder), the axial gap between SB and MB was extended. The probe is fixed in the traversing device. It can be manually shifted in two positions – for measuring behind SB or MB.

Figure 7 A 5-hole probe detail

Apart from static pressures and pressures from a five-hole probe, we can measure the velocity profile at the inlet into the stage using the Prandtl probe. We receive values of speed and moments of torsion from the dynamometer. Moreover, we measure air temperatures to determine the flow rate. We also extract static pressures in front of and behind the nozzle located in the pipeline to determine the flow rate. We use the following procedure to determine the flow rate.

Air density in front of and behind the nozzle is

$$\rho_{d1} = \frac{p_{d1}}{r_v T_{d1}}(1 - k_1\varphi) = \frac{p_{d1}}{287{,}139 \cdot T_2} \cdot 0{,}995 \qquad (1)$$

$$\rho_{d2} = \frac{p_{d2}}{r_v T_{d2}}(1 - k_1\varphi) = \frac{p_{d2}}{287{,}139 \cdot T_2} \cdot 0{,}995 \qquad (2)$$

where

$k_1\varphi$ is a correction on relative air humidity

T_{d1} is temperature in front of the nozzle [K]

r_v is individual gas constant; for air it is $r_v = 287{,}139 \frac{\mathrm{J}}{\mathrm{kgK}}$

note.: we consider $T_2 = T_{d1} = T_{d2}$

A ratio of the minimum nozzle section D_{dMin} and pipeline section D_p is

$$m = \left(\frac{D_{dMin}}{D_p}\right)^2 = \left(\frac{0{,}18}{0{,}4}\right)^2 = 0{,}2025 \qquad (3)$$

Air velocity in the narrowest nozzle section

$$w_2 = \frac{1}{\sqrt{1 - m^2 \cdot \left(\frac{p_{d2}}{p_{d1}}\right)^{\frac{2}{\kappa}}}} \cdot \sqrt{2 \cdot \frac{\kappa}{\kappa - 1} \cdot \frac{p_{1d}}{\rho_{d1}} \cdot \left[1 - \left(\frac{p_{d2}}{p_{d1}}\right)^{\frac{\kappa-1}{\kappa}}\right]} \qquad (4)$$

Reynolds number of the nozzle is

$$Re_{d2} = \frac{w_2 \cdot D_{d2} \cdot \rho_{d2}}{\mu_2} \qquad (5)$$

A substitution of $D_{d2} = 0{,}18$ m and ρ_{d2} from the equation (2)

$$Re_{d2} = 6{,}269 \cdot 10^{-4} \cdot \frac{p_{d2}(1 - k_1\varphi)}{T_2 \cdot \mu_2} \cdot w_2 \qquad (6)$$

Nozzle flow coefficient is defined

$$\alpha = \frac{c_q}{\sqrt{1 - m^2}} \qquad (7)$$

The coefficient c_q is independent on Re_{d2} and equals to 0,996 for values $Re_{d2} > 0{,}5 \cdot 10^6$, which can be practically considered for measuring on the air line. The flow coefficient is $\alpha = 1{,}017$ after a substitution for m in the equation (7).

Expansion coefficient

$$\varepsilon = \sqrt{\frac{1 - m^2}{1 - m^2 \cdot \left(\frac{p_{d2}}{p_{d1}}\right)^{\frac{2}{\kappa}}}} \cdot \sqrt{\frac{1}{1 - \frac{p_{d2}}{p_{d1}}} \cdot \frac{\kappa}{\kappa - 1} \cdot \left[\left(\frac{p_{d2}}{p_{d1}}\right)^{\frac{2}{\kappa}} - \left(\frac{p_{d2}}{p_{d1}}\right)^{\frac{\kappa+1}{\kappa}}\right]} \qquad (8)$$

Air volume

$$m_v = \alpha \cdot \varepsilon \cdot S_{d2} \cdot \sqrt{2 \cdot \rho_{d2}(p_{d1} - p_{d2})} \ , \qquad (9)$$

where

α is the nozzle flow coefficient according to (7); $\alpha = 1{,}017$

ε is expansion coefficient according to (8)

S_{d2} is the outlet nozzle section; for $D_{d2} = 0{,}18$ m is $S_{d2} = 0{,}02543\ \mathrm{m}^2$

4 Device modernization

Currently, the whole device is undergoing a reconstruction, which should be finished soon. The reason for the reconstruction is to increase accuracy and measurement speed, and to expand measurement posibilities. The renovation concerns three parts of the loop.

The first change refers to measurements of flow fields behind stationary and moving blades. The current traverser allows to measure by one probe only, i.e. we can measure either behind stationary or moving rings. The current traverser controlling mechanism is not sufficient either. The control is not fully automatized, the probe shift in a radial direction is possible to be made in a step mode only. The new traverser (model on figure 8) will not only be fully automatized, but it will also enable measuring by two probes simultaneously, i.e. we will measure the flow field both behind SB and MB at the same time. Furthermore, the turbine casing will be adapted, and the range of circumferential movement of the probe will be enlarged.

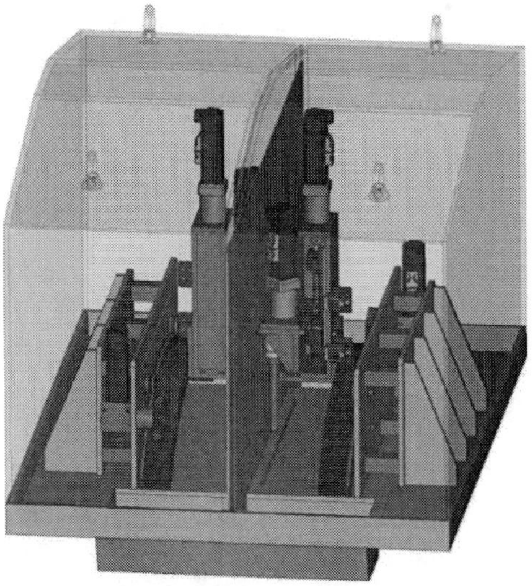

Figure 8 A new traverser model

A purchase of new pneumatic probes is related to the traverser exchange. We will receive two five-hole probes with a head diameter of 1,6 mm and a frequency response of 0,5 kHz. We shall also receive one five-hole probe with a head diameter of 1,6 mm and a frequency response of 2,0 kHz meant for non-stationary measurement, along with the above mentioned new probes. We will buy new pressure scanners for data collection, not only from those probes, but for static pressure extractions as well. It will result in a growth of velocity and measurement accuracy, and, moreover, we shall be able to measure the pressure at more points at the same time.

The next innovated part will be a source of pressurized air. The technical condition of the current compressor corresponds to the year of its production and is completely insufficient. It does not allow a fluent automatic regulation, but only a step regulation. The compressor also does not enable to measure the pressure drop; it is determined from measuring on the turbine. The objective of the pressurized air source modernization is to improve the operational stability, and mainly, to enhance the possibility of regulation, including the performance program adaptation. The new compressor performance is required to be up to 50 kW, pressure drop up to 12000 Pa and flow rate of 2-2,5 kg/s (the compressor characteristic on figure 9). The new compressor shall allow both the increase in pressure drop and the connection to the control system of the whole loop; this will allow automatic regulation of the compressor as well as data registration from the compressor.

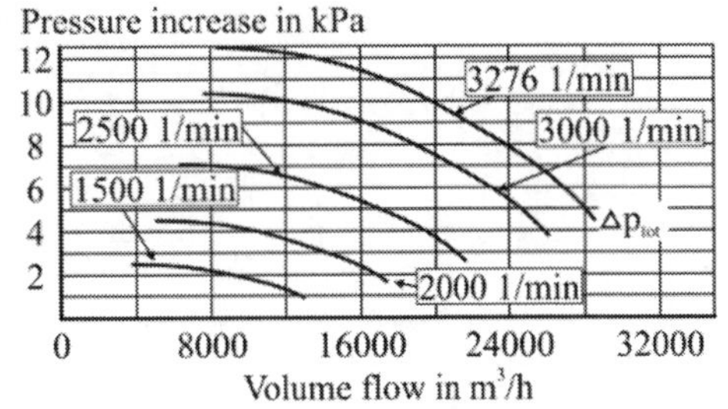

Figure 9 Compressor characteristics

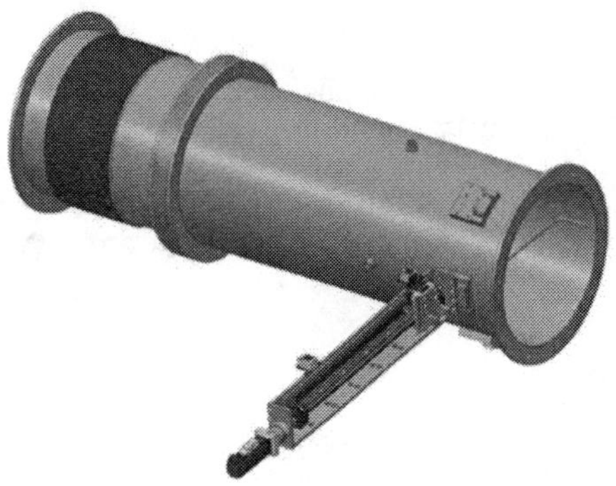

Figure 10 Location of 1D traverser

Another change will appear in defining the air flow rate. Currently, a nozzle located in the pipeline between the turbine and the compressor (see Chapter 3) is used for measuring the flow rate. To know the air temperature is, among others, necessary to analyze the flow. As for now, the air temperature is measured at the turbine outlet, i.e. in a distance of several dozens of meters from the nozzle itself. Also the pressure transducer, which evaluates the pressure extractions from the probe, is currently located several tens of metres from the nozzle, therefore, some inaccuracies in measurements may occur. The temperature sensor and pressure transducer will be newly placed at the point of the flow measurement. In addition, the flow rate measuring will be extended by traversing the velocity profile in the pipeline, near the nozzle by a newly installed 1D traverser, including a new pressure probe, which will allow the nozzle calibration. The 1D traverser model can be viewed on figure 10.

5 Conclusions

The knowledge of detailed flow flied in the turbine flow component is a necessary condition for a correct blading design. An acquisition of this data is a subject of testing on model air turbines. VT-400 is one of two such turbines in the Czech Republic. The reconstruction of the whole apparatus will allow a continuing research of flowing in the air turbine stage, in cooperation with our industrial partner Doosan Skoda Power.

We would like to continue the research in the 3D blade shaping influence on the stage efficiency in the future; we intent to study the turbine stage behaviour during off-desing, and we would also like to pursue the inspection of the influence of balancing holes onto the efficiency, or the influence of axial gap flows on hub between moving and stationary rings.

We would also like to concentrate on the air turbine, which has been only slightly affected by the reconstruction. At present, we are exploring options of VT-400 construction adjustments with the objective to change the hub diameter or rings which would extend / shorten the blade length, and thus influence the L/D ratio. Furthermore, we would like to propose such adjustments, which would allow to research the influence of tip seals.

Acknowledgements

This work has been supported by the SUSEN Project CZ.1.05/2.1.00/03.0108 realized in the framework of the European Regional Development Fund (ERDF) and student project SGS-2014-070 (Zvyšování účinnosti, spolehlivosti a životnosti energetických strojů a zařízení 3).

References

1. J. Vomela, P. Milčák, *The development of high-efficiency turbine blades.* Cieplne maszyny przeplywowe. Turbomachinery. Lodz : Technical University of Lodz, 2005, 557-564. ISSN 0137-2661
2. P. Milčák, M. Hoznedl, D. Krivánka, *Testování turbínového stupně s 3D lopatkováním – úvodní experimenty.* Strojárstvo, Mechanical Engineering Journal, 2009, ISSN 1335-2938.
3. P. Milčák, M. Hoznedl, P. Žitek, *Comparison of results from experimental testing of three variants of turbine stage with modern 3D blades.* In EPJ Web of Conferences. DOI: 10.1051/epjconf/20134501063
4. M. Klimko, D. Okresa, *Measuring on the air turbine VT-400,* in Acta Polytechnica - Journal of Advanced Engineering (to be published).

EPJ Web of Conferences **114**, 02144 (2016)
DOI: 10.1051/epjconf/201611402144

Distribution of velocity and turbulent characteristics in coarse-sediment laden flows above erodible plane bed in open channel

Štěpán Zrostlík[1,a], Vojtěch Bareš[1], Jan Krupička[1], Tomáš Picek[1], and Václav Matoušek[1]
[1] *Department of Hydraulics and Hydrology, Czech Technical University in Prague, 166 29 Prague 6*

Abstract. Extreme conditions as flash floods in steep mobile-bed streams induce flows which carry a huge amount of sediment. Typically, the flow with the intense transport of sediment is stratified with a layered structure above an erodible plane bed. The paper deals with local flow characteristics in the layered structure of the two-phase flow composed of water and sediment. We present experimental data and results for two different fractions of plastic particles obtained using two acoustic Doppler methods and a Prandtl probe. The results contain vertical profiles of longitudinal velocity component and local turbulent characteristics of the flow as the Reynolds stress and turbulent intensity. The effect of the layered character of the flow on the flow characteristics is discussed. Special attention is paid to the distribution of Reynolds stress in the sediment-rich transport layer above the bed and to the evaluation of ability of carrier turbulent eddies to support sediment grains in and above the transport layer.

1 Introduction

The two-phase flow of coarse-sediment mixture occurs in mobile-bed open channels at extreme conditions as flash floods in mountain streams. A huge amount of transported material changes properties of the flowing medium and this needs to be considered while computing the flow rate and resistance of the flow [1] as well as [2]. The coarse-sediment flow occurs in the regime with bed forms or with no bed forms. In the bed-form free regime called the upper plane bed (UPB) regime, the flow resistance is significantly affected by the transported material. Flow in the UPB regime can be considered as a layered flow with three zones. The uppermost layer contains clear water and occasionally saltating particles. The transport layer is situated between the uppermost layer and the layer with deposit material. In the transport layer, a vast majority of the total load is transported. A UPB-based flow is described in details in [3] as well as [4] and its structure is shown in Figure 1. The logarithmic shape of a velocity profile is expected in the water layer and the linear shape is expected in the transport layer. Figure 1 shows also the expected linear concentration profile across the transport layer. This assumption are based on previous experimental observations in the UPB regime [5] as well as [6].

Distributions of local velocity and turbulent characteristics of flow are inevitable for better understanding of mechanisms governing flow structure and resistance. Measuring the distributions using classical optical methods as LDA or PIV is impossible because of high concentration of particles which does not allow to evaluate the point velocity in the center of a channel. Capart and Fraccarollo [6] used the modified optical method PTV (Particle Tracking Velocity) for measuring velocity profiles of transported particles near the side wall of the rectangular channel. Other authors used Prandtl probe for measuring the local velocity [7] and considered the measured velocities as velocities of water. Ultrasonic probes are utilized frequently. Rajesh et al. [8] and Song et al. [9] used Acoustic Doppler Velocimeter for measuring turbulence statistics above the erodible bed at relatively low transport concentration of sediment. Revil-Baudard et al. [10] measured velocity and concentration profiles with ACVP (Acoustic Concentration and Velocity Profiler) in mixture of bakelite grains and water. Bakhtyar et al. [11] used UVP (Ultrasonic Velocity Profiling) for measuring velocity characteristics above the sand bed under the condition of oscillatory sheet flow.

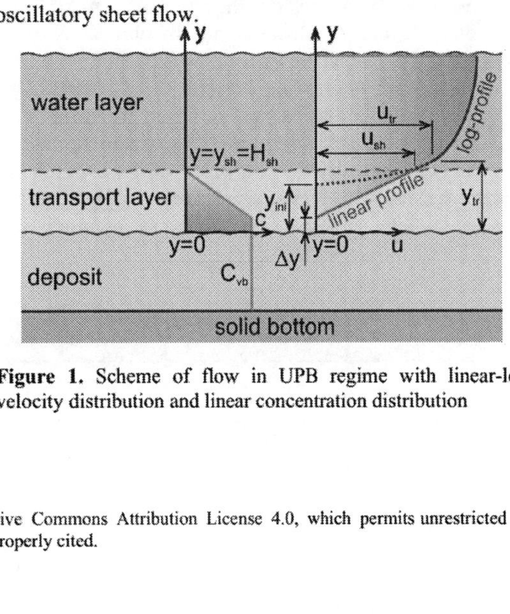

Figure 1. Scheme of flow in UPB regime with linear-log velocity distribution and linear concentration distribution

[a] Corresponding author: stepan.zrostlik@fsv.cvut.cz

The work presented here is a continuation of our previous investigations on velocity distribution in stratified flows with both intense and weak transport of sediment. In this paper, we exploit our new experimental data to evaluate velocity profiles and focus on an evaluation of turbulent characteristics in the transport layer.

2 Methods

The experiments were carried out in the Water Engineering Laboratory of the Czech Technical University (CTU) in Prague. The experimental tilting flume and measured variables have been described elsewhere [3, 4].

2.1 Measuring instruments

We used three different measurement techniques to measure a vertical velocity profile of the longitudinal component of the velocity in the axis of the flume. The first technique is Prandtl probe (PT). It is a robust standard method and we used it primarily to validate velocity profiles obtained from the other methods. The second method used a UVP probe directed opposite to the flow. A disadvantage of the method in this particular application is that particles are in sporadic contact with the measuring membrane of the probe. Compared to the other two methods, the down-looking ADVP (Acoustic Doppler Velocity Profiler) is non-invasive while measuring within the transport layer. The probe senses the instantaneous 3-D velocity information along 32 mm long measuring section below the probe. The measuring instruments are described in more details and an estimation of their measuring errors is given in [3]. Processing of measured data is described in [3] as well as [4].

2.2 Materials

Experiments were carried out with two different fractions of plastic particles. Particles of both fractions are of a shape of a cylinder with one concave base and the other convex base. The diameter of the cylinders is 4.7 mm and the height for the coarser fraction is 4.9 mm (TLT50), the height for the smaller fraction is 2.2 mm (TLT25). Measured terminal settling velocity v_t of particles and diameters d_{50} of equivalent spherical particles of the same volume are presented in Table 1. Both particles have a similar density (ρ_s), they are made of PVC (TIULIT). Particles showed no signs of mechanical damage during the experiments. Overall, 44 runs were measured with TLT25 and 55 runs with TLT50.

Table 1. Properties of particles

Properties	TLT25	TLT50
d_{50} (mm)	4.23	5.41
v_t (m.s^{-1})	0.106	0.149
ρ_s (kg . m^{-3})	1381	1307

2.3 Velocity profile

A velocity profile of flow of water above the hydraulically rough boundary can be described using the logarithmic law of the wall [3],

$$\frac{u}{u_b^*} = 2.5\ln\left(B\frac{y - y_{ini}}{k_s}\right) \quad (1)$$

where u is longitudinal velocity at vertical position y, u_b* is the bottom shear velocity, y_{ini} is position of origin of logarithmic profile, k_s is the boundary roughness and B is an integration constant for open channel flow ($B = 30$). The velocity distribution is linear in the transport layer and it can be described by following equation [3],

$$\frac{u}{u_{tr}} = \left(\frac{y - \Delta y}{y_{tr} - \Delta y}\right)^n \quad (2)$$

where y_{tr} is vertical position, at which a profile smoothly transforms its shape from linear to logarithmic, u_{tr} is velocity at position y_{tr}, Δy is position of zero velocity and n is an exponent (n=1).

Taking the above assumptions into account, we carried out an optimization procedure on measured velocity profiles resulting in values of five parameters (u_b*, y_{ini}, k_s, y_{tr}, Δy).

2.4 Turbulent characteristics

Knowledge of turbulent characteristics is vital for better understanding of particle movements in sediment-laden flows. By definition, the instantaneous velocity can be described as the time average velocity $\bar{u}$ plus the fluctuation component u_i'. For the longitudinal component of local velocity one can write,

$$u_i = \bar{u} + u_i' \quad (3)$$

The same holds for the velocity in vertical direction, w. We consider the intensity of turbulence in the longitudinal direction as $\sqrt{u_i'^2}/u_*$. The local Reynolds stress in turbulent flow of water is defined as

$$\tau_{x,z} = -\rho_w \overline{u'w'} \quad (4)$$

3 Evaluation of velocity characteristics

3.1 Velocity distribution

The velocity profiles, obtained using the different measuring methods, correspond with each other quite well (Figures 2 and 3). The lin-log shape of a velocity profile describes the measured vertical distribution of velocity very well, which is in an agreement with the earlier measurements with other sediment materials under the same condition [4]. Based on experimental data processed using the optimization procedure containing Equations (1) and (2), it is confirmed that the position of the transition in the trend of the velocity profile approximately corresponds with the position of the interface between the transport layer and the water layer. The slope of the linear profile remains approximately constant at very high shear stresses. The position of the virtual origin of the logarithmic part of a velocity profile is at of about 65 per cent of the height of the transport layer.

The upper plots of Figures 2 and 3 show examples of measured and processed velocity profiles at relatively low high stress, the lower plots show the profile at high shear stress. Experimental data from all three methods are plotted. The UVP data are from the distance 20, 30, 40 mm upstream the probe (UVP20, UVP30 and UVP40, respectively) and data from ADVP are from different installed vertical positions of the probe (ADVP1, ADVP2 etc.).

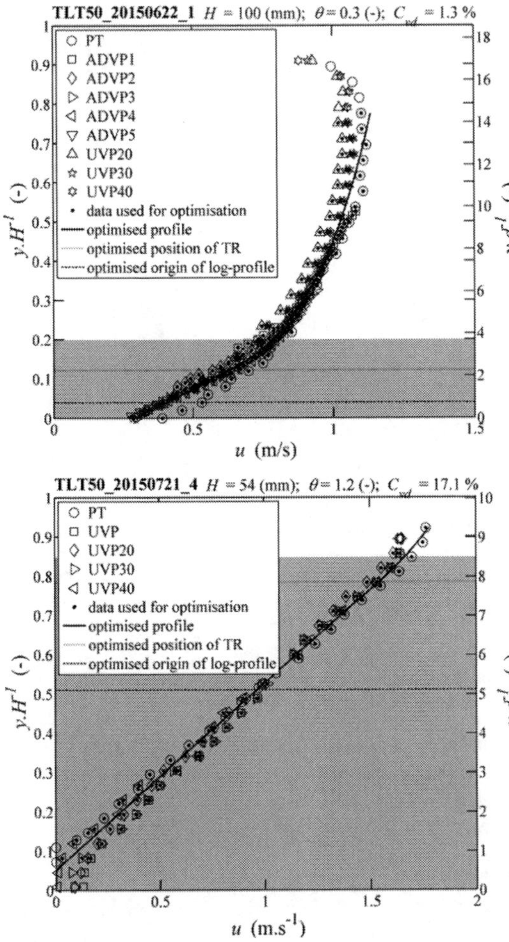

Figure 2. Measured and processed velocity profiles for TLT50. Legend: C_{vd} – delivered concentration of sediment, θ – Shields parameter, H – depth of flow

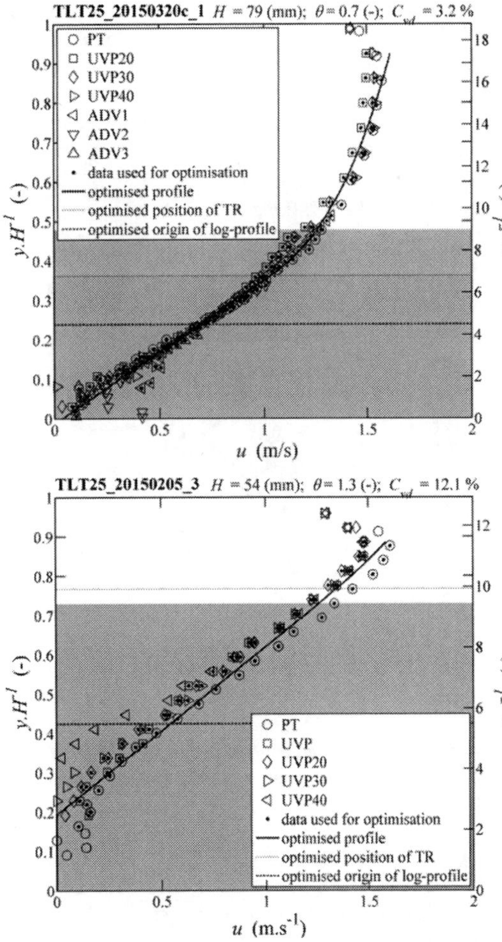

Figure 3. Measured and processed velocity profiles for TLT25. Legend: C_{vd} – delivered concentration of sediment, θ – Shields parameter, H – depth of flow

We have employed two methods to evaluate the shear velocity (u_*). The first approach estimates the shear velocity from the integral experimental data,

$$u_{*\text{int}} = \sqrt{g \cdot R_b \cdot \sin \beta} \tag{5}$$

where g is the gravitational acceleration, β is the angle of the longitudinal slope of uniform flow and R_b is the hydraulic radius associated with bed. The second method gives the shear velocity, $u_{*\text{opt}}$, as a result of the optimization procedure to fit the theoretical velocity profile (Eq. 1) with the measured data in the water layer (i.e. in the theoretical inner region of the turbulent boundary layer). Figure 4 compares results of the two methods. The shear velocity from the velocity profile tends to be bigger than the shear velocity from the integral data and the difference increases with the increasing value of the shear velocity. It must be mentioned that there are some important differences which cause that the two velocities cannot be considered the same parameters. The first difference is caused by the spatial distribution of shear velocity along the width of the channel. The u_{*int} data represent a width-averaged value of the shear velocity, whereas u_{*opt} is evaluated for the plane of the vertical velocity profile and must be considered a local value. The second difference is caused by the effect of the vertical position. The u_{*opt} is optimized for the position of the origin of a logarithmic profile while u_{*in} is determined for the top of the bed, i.e. the bottom of the transport layer.

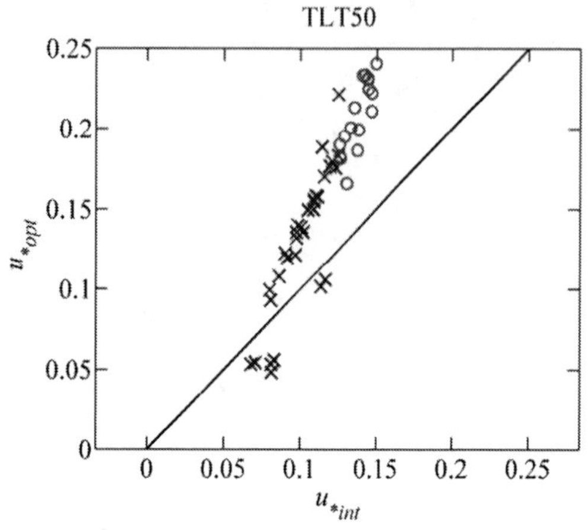

Figure 4. Comparison of shear velocity from integral data (u_{*int}) and shear velocity from velocity profile (u_{*opt}). Legend: x - data for Shields number lower than 1, o - data for Shields number higher than 1, solid line - line of agreement

3.2 Turbulent characteristics

The turbulent characteristics were evaluated for all experimental data sets (corresponding with measured profiles) obtained using ADVP: 14 sets for TLT25 and 21 sets for TLT50. For longitudinal turbulent intensities, UVP measurements were used as well. Measurements from Prandtl probe were not considered for an analysis of turbulent characteristics.

3.2.1 Longitudinal turbulent intensity

Figure 5 shows evaluated longitudinal intensities of turbulence (the bed shear velocity from integral data is used to normalize the velocity fluctuation). The intensities obtained from the different measuring methods may differ in values, but exhibit the same trend. The intensity of turbulence increases approximately linearly from the water surface (the zero-value at the water surface) to the top of the transport layer. In the transport layer, the turbulence intensity starts to decrease with the increasing depth below the top of the layer. Apparently, damping of intensities is due primarily to the presence of particles and due to the increase of particle concentration with the depth within the transport layer. The presence of particles and their mutual collisions attenuate the longitudinal fluctuation associated with flow turbulence.

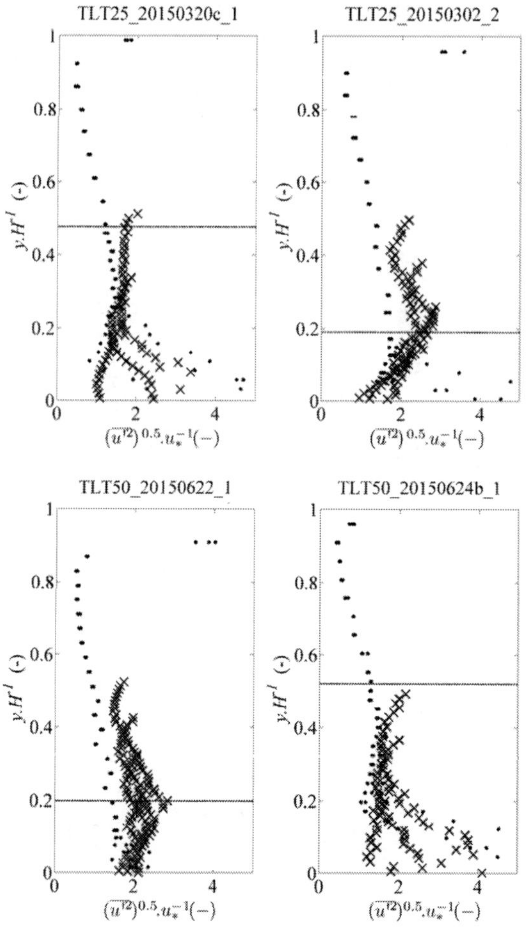

Figure 5. Distribution of longitudinal turbulent intensity (fluctuation normalized by bed shear velocity from integral data). Upper panels - TLT25, lower panels - TLT50. Legend: dots – UVP, x – ADVP, line - visually observed top of transport layer.

3.2.2 Vertical turbulent intensity

Instantaneous vertical velocity components were obtained by the ADVP probe only. Vertical pulsations were evaluated in the same way as the longitudinal ones. In Figure 6, the hindered settling velocity (v_{th}) is used to normalise the square root of the mean vertical fluctuation velocity. In the transport layer, the hindered settling velocity was determined from the Richardson-Zaki equation,

$$v_{th} = v_t \cdot (1 - c_v)^m \quad (6)$$

where c_v is the local volumetric concentration of particles at given vertical position. The exponent m was computed from the Wallis equation,

$$m = \frac{4.7 \cdot \left(1 + 0.15 \cdot \mathrm{Re}_p^{\,0.687}\right)}{1 + 0.253 \cdot \mathrm{Re}_p^{\,0.687}} \quad (7)$$

where Re_p is the particle Reynolds number. Figure 6 shows that the time-averaged vertical velocity fluctuation is considerably smaller than the terminal settling velocity of particles at the top of the transport layer. This suggests that support of particles comes from interparticle collisions rather than turbulent vortices at the top of the transport layer. The observed increase in values of the velocity ratio within the transport layer is due to the decreasing local hindered settling velocity of particles.

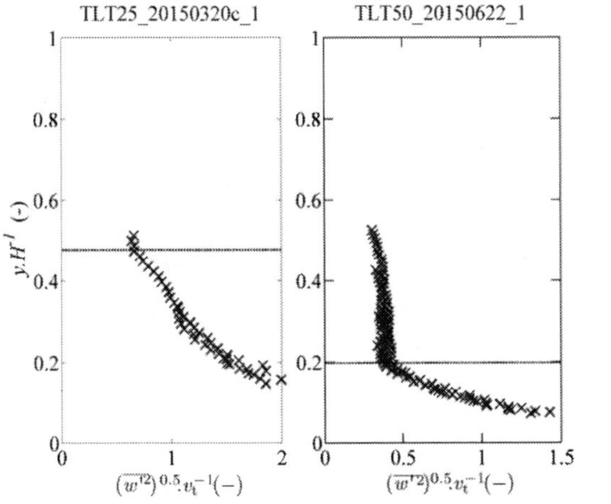

Figure 6. Distribution of vertical turbulent intensity (fluctuation normalized by local hindered settling velocity). Legend: x – ADVP, line - visually observed top of transport layer.

3.2.3 Reynolds stress

The local Reynolds stress could be evaluated from the ADVP measurements only just as the vertical turbulence intensity. Plotted in Figure 7, the Reynolds stress obtained from measured velocity components using Eq. (4) is normalized by the square of u_{*int} and the density of water. In Figure 7, the inclined line delineates the theoretical distribution of total applied shear stress in uniform flow above the bed [9]. The linear course and the maximum value of 1 at the flow boundary follows from the integral force balance on the control volume of flow. In the right-side panels of Figure 7, experimentally determined Reynold stresses exhibit a linear distribution at vertical positions above the transport layer and a change in the trend at the top of the transport layer. In the transport layer, the Reynolds stress decreases towards a zero value at the top of the bed. The linear trend of the data above the transport layer indicates that the Reynolds stress is the exclusive local resisting stress. This is not the case within the transport layer, where transported particles through their mutual collisions take care of a part of the local stress. Frequency of collisions increases with increasing local concentration of particles and so increases the local particulate stress. Hence the momentum transfer due to collisions takes a more important role in the total shear stress transfer. This is indicated by the increasing difference between the measured Reynolds stress and the total shear stress with the increasing depth in the transport layer. This decrease in the Reynolds stress can be interpreted as a result of damping of turbulence in the layer due to particles. The left-side panels of Figure 7 confirm the observed trends for flow of the smaller depth and higher bed shear stress. A value of u_{*opt} is indicated by the red dot in Figure 7. The horizontal ordinate of the dot gives $(u_{*opt} / u_{*int})^2$. The vertical ordinate of the dot indicates the virtual origin of the log-profile, y_{ini}/H.

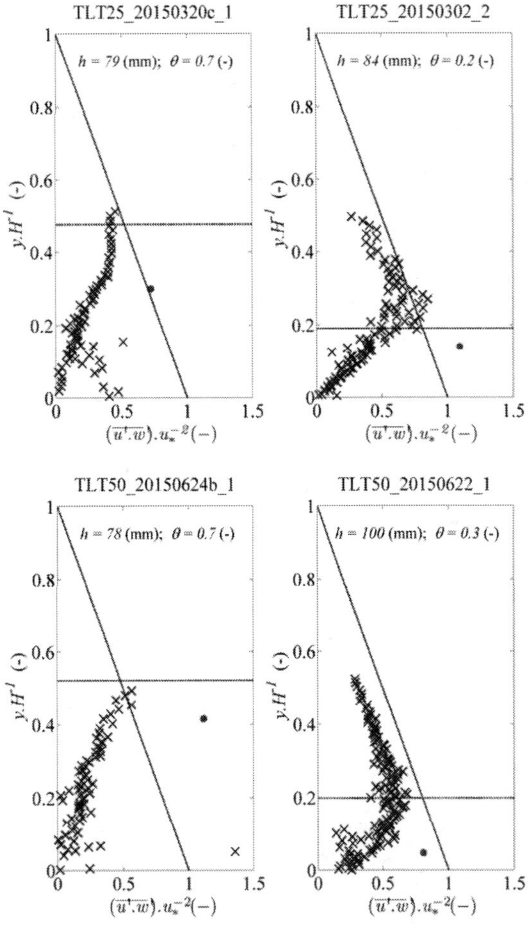

Figure 7. Distribution of shear stresses (stress normalised by square of shear velocity evaluated from integral data). Legend: crosses – Reynolds stress measured by ADVP, dot – value obtained by optimization of parameters of theoretical velocity profile, inclined line - theoretical distribution of total shear stress, horizontal line - visually observed position of top of transport layer.

3.2.4 Particulate stress in transport layer

As discussed above, the difference between the local Reynolds stress and the local total shear stress is the local particulate shear stress transferred by interparticle collisions. This collisional stress is plotted in Figure 8 (green dots) and it seems to exhibit an approximately linear distribution across the transport layer (see the fit by a red line). The position of the zero collisional stress (i.e. the position of the intersection of the red line with the y-axis) marks a position above which no particles are supported by mutual contacts. Since there are no particles supported by turbulent eddies at this position either (Fig. 6), no particles are expected to be present above the position of the zero collisional stress and hence this position should be the vertical position of the top of the transport layer. This is a method of a determination of the top of the transport layer alternative to the method of the visual determination of the top of the layer.

The Reynolds stress decreases towards zero at the bottom of the transport layer indicating that the applied shear stress is resisted exclusively by the particulate stress at the top of the bed. Also at the top of the bed, the collisional particulate shear stress τ_0 is related to the interparticle normal stress σ_0 through the Coulomb criterion (e.g. Capart and Fraccarollo [6])

$$\tau_0 = \tan(\alpha)\cdot\sigma_0 \qquad (8a)$$

where α is the quasi-static internal angle of friction (Bagnold [12] reported values of $\tan(\alpha)$ to vary from 0.32 to 0.75 in his experiments) and the normal stress is expressed using parameters associated with the transport layer,

$$\sigma_0 = (\rho_s - \rho_w)\cdot H_{sh}\cdot g\cdot C_{IL}\cdot\cos(\beta) \qquad (8b)$$

in which C_{TL} is mean volumetric concentration of particles in the transport layer (typically one half of local concentration at the bottom of the transport layer) and H_{sh} is thickness of transport layer. A combination of Eqs. (8a) and (8b) gives a formula for $\tan(\alpha)$,

$$\tan(\alpha) = \frac{\tau_0}{H_{sh}\cdot(\rho_s - \rho_w)\cdot g\cdot C_{TL}\cdot\cos(\beta)} \qquad (9)$$

Based on results of our experiments and estimated C_{TL} = 0.27, we obtained values of $\tan(\alpha)$ in the range from 0.2 to 0.44 from Eq. (9).

4 Conclusions

A large experimental data set of vertical velocity profiles was analysed for layered flows with both relatively weak and intense transport of sediment. The data confirm that a velocity profile can be considered composed of a linear part and a logarithmic part in flow with a developed transport layer.

Turbulent characteristics evaluated from acoustic Doppler-based measurements indicated that turbulent eddies might not be able to support sediment particles at the top of the transport layer, i.e. the sediment was transported predominantly as bed load. Furthermore, an analysis of turbulent velocity fluctuations in the transport layer suggested that an interaction of turbulent eddies with colliding particles leads to attenuation of both the longitudinal turbulent intensity and the Reynolds stress. The measure of attenuation increases with the increasing local concentration of particles. At the bottom of the transport layer, the Reynolds stress is negligible and the applied shear stress is resisted exclusively by the particulate shear stress.

The angle of internal friction of colliding particles was evaluated to vary from 0.22 to 0.44 at the bottom of the transport layer.

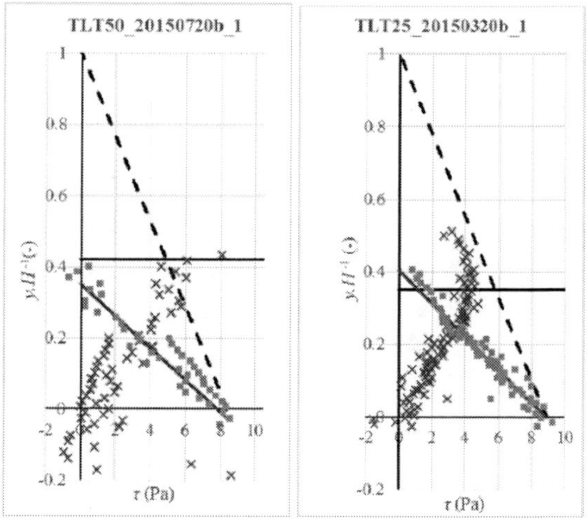

Figure 8. Distribution of shear stresses. Legend: crosses – Reynolds stress, squares – calculated collisional shear stress, dashed line – theoretical distribution of total shear stress, red line – linear fit of collisional shear stress, horizontal line - the visually observed interface between water and transport layer.

Acknowledgments

The research has been supported by the Czech Science Foundation through the grant project No. P105/12/1082 and by Faculty of Civil Engineering of the Czech Technical University in Prague through the student grant project No. SGS14/179/OHK1/3T/11.

References

1. D. Rickenman, A. Koschni, *Hydrol. Process.*, **24**, 93-1007, (2010)
2. V. Matoušek, J. Krupička, *Proc. River Flow 2014*, Lausanne, Switzerland, p. 917-922, (2014).
3. V. Matoušek, V. Bareš. J. Krupička, T. Picek, Š. Zrostlík, *J. Hydrol. Hydromech.*, **63**(4), in press, (2015)
4. Š. Zrostlík, V. Bareš, J. Krupička, T. Picek, V. Matoušek, *EPJ Web of Conferences 92*, 02120 (2015)
5. V. Matoušek, *J. Hydraul. Eng.*-ASCE, **135**(12), 1101-1106, (2009)
6. H. Capart, L. Fraccarollo, *Geophysi. Res. Lett.*, **38,** (2011)
7. A. Yeganeh, H. Gotoh, T. Sakai, *J. Hydraul. Res.*, **38**(5), 389-398, (2000)
8. J. K. Rajesh, A. Kumar, U. Kothyari, *Journal of Hydro-environment Research*, in press, (2015)

9. T. Song, H. W. Graf, U. Lemin, *J. Hydraul. Res.*, **32**(6), 861-876, (1994)
10. T. Revil-Baudard, J. Chauchat, D. Hurther, P-A. Barraud, *J. Fluid Mech.*, **767**, 1-30, (2015)
11. P. Bakhtyar, A. Yeganeh-Bakhtiary, D.A. Barry, A. Ghaheri, *Advances in Water Resources*, **32**, 1267-1283, (2009)
12. R. A. Bagnold, *U.S. Geological Survey Professional Paper*, **422-I**, p. 37, (1966)

EPJ Web of Conferences **114**, 02145 (2016)
DOI: 10.1051/epjconf/201611402145

Application of nuclear techniques in two-phase liquid-solid particles hydrotransport investigations

Marcin Zych[1,a], Robert Hanus[2], Pavel Vlasak[3], Leszek Petryka[1] and Marek Jaszczur[1]

[1]*AGH University of Science and Technology, Krakow 30-059, Poland*
[2]*Rzeszow University of Technology, Rzeszow 35-959, Poland*
[3]*Institute of Hydrodynamics AS CR, v. v. i., 160 00 Prague, Czech Republic*

Abstract. The paper presents gamma radiation application to two-phase flow investigation in a vertical pipeline, where the flow of solid particles transported by water was examined by use of both: radiotracers and gamma-absorption method. The simultaneous use of two methods allows analyzing of important parameters of solid particles hydrotransport. In the described experiments as solid phase the ceramic models representing natural polymetallic ocean nodules were used. Radiotracers allow to track the movements of selected models, representing specified grain size and the designation of its velocity. However gamma-absorption method enables measurement of average solid-phase velocity. For analysis of electrical signals obtained from scintillation detectors the cross-correlation method has been applied.

1 Introduction

Two-phase liquid-solid particles flow occurs often in the mining industry. An example would be vertical pipeline hydrotransport of minerals, e.g. polymetallic nodules. Nodules are porous organic and mineral compositions which contain a various metals (mostly Mn, Si, Fe, Al, Na, Mg, Ni, K, Cu) [1-4]. They occur on the bottom of seas and oceans, but the richest deposits can be found at the Pacific Ocean bed in depth about 5000 m in the so called Clarion-Clipperton zone. The ocean nodules usually have the form of irregular grains with the diameter of several mm up to 0.5 m and density of about 2 g / cm^3 in wet state. The mining of nodules using the hydraulic method requires nodules vertical transport by water to sea level in extremely hard and varying environment. Determination of the velocity of nodules of various sizes in the vibrating pipeline at a significant depth is a very difficult task and requires the use non-invasive measurement techniques [5, 6].

One of these techniques, which are employed for many years in measurements of two-phase flows in pipelines and open channels, is a method using radioactive isotopes. In nuclear techniques radiotracer is injected under certain conditions into the flow and/or sealed radioactive sources are used [7-19]. In both cases as detectors of radiation the scintillation probes are mounted outside of the analyzed stream.

Typically in such measurements the mutually delayed stochastic signals are provided by detectors located on the outer walls of the pipeline. Measured time delay of signals is applied to determine the velocity of the marked phase (radiotracers method) or averaged velocity of the solid phase (absorption method). In the carried out experiments solid phase has been modeled using ceramic particles representing three grain sizes of natural polymetallic ocean nodules.

The processing of stochastic signals from scintillation probes requires the use of statistical methods in the time and frequency domain [20-26]. The cross-correlation function (*CCF*) is the most known methods of time delay estimation applied for stationary random signals.

2 Application of radioisotopes to investigations of nodules hydro-transport in a vertical pipeline

The basic principle of gamma-ray absorption measurement is based on fact that the intensity of a collimated gamma beam decreases as it passes through matter [5-7, 10-17].

Radiotracer idea is based on marked particle tracking using a scintillation probes placed along the pipeline [7, 8, 18].

The principle of both methods and typical gamma-radiation measuring equipment for liquid–solid particles flow evaluation in a vertical pipeline is presented in figure 1.

[a] Corresponding author: zych@geol.agh.edu.pl

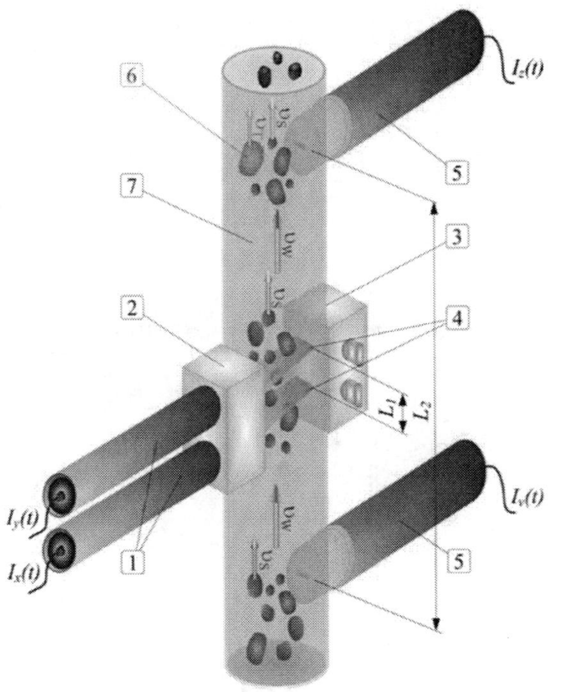

Figure 1. The gamma radiation measuring equipment: 1 – scintillation probes of gamma-absorption, 2 – detector's collimator, 3 – two gamma-radiation sources with collimator, 4 – gamma-rays, 5 – scintillation probe for radiotracer measurements, 6 – marked particle, 7 – pipeline with two-phase flow, υ_S, υ_W – velocities of solid particles and water respectively, υ_T – velocity of marked particles.

Two sealed radioactive sources in collimator (3) emit gamma radiation beams shaped by the collimator. Photons pass through the pipeline (7) with the multiphase mixture. The changes in the intensity of radiation are recorded by the scintillation probes of the absorption set (1) with collimators (2) and next converted into output electrical impulses [7]. Count rates I_x and I_y at the outputs of probes depend on the flowing medium content in the test cross section. In the present investigations a linear ^{241}Am γ-ray sources with an activity of 100 mCi and probes with 2" NaI(Tl) scintillation crystal were used. Two such sets were mounted on the pipe with the distance of $L_1 = 90$ mm between them. Based on the time delay of signals I_x and I_y and the distance L_1 one can calculate the average velocity of the solid phase.

Probes for radiotracer measurements (5) are based on the longer distance of $L_2 = 1882$ mm. These probes record the radiation intensity I_v and I_z from the marked particle (6). Velocity of the marked particle can be determined as before by determining time delay between signals.

The current study has been performed for a mixture of water and artificial nodules at volumetric concentration 0.05. The density of the particles representing the nodules was 2000 kg / m^3.

Table 1 presents the characteristics of solid particles and figure 2 shows image of particles before experiments.

Table 1. Characteristics of the artificial nodules.

Particle size grade	Dimensions (mm)	Weight, (g)
I - small	29.5x26.9x20.5	18.5
II - medium	31.2x47.7x27.0	47.9
III - big	59.6x43.7x54.1	128.3

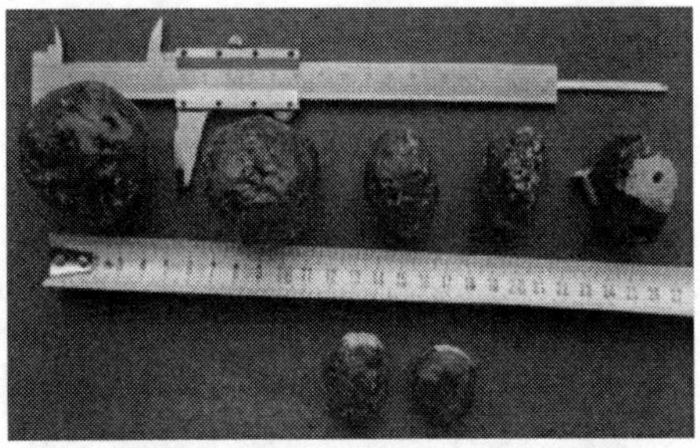

Figure 2. The image of the particles (right-hand side: the particle with a bore for the isotope Tc-99m).

3 Experimental set-up

The measuring system presented in figure 1 has been installed in the experimental set-up located in the Water Laboratory of the Wroclaw University of Environmental and Life Sciences, Poland. The entire experimental installation presented in figure 3 has been used for investigation of the polymetallic nodules hydrotransport.

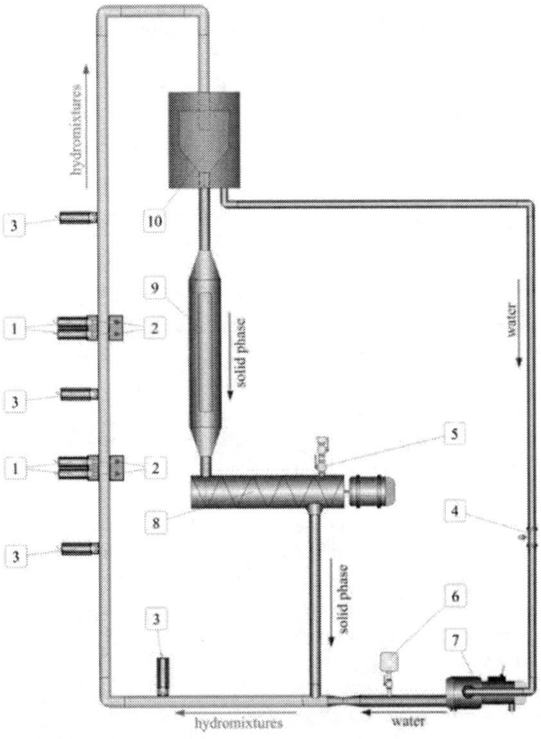

Figure 3. Diagram of experimental set-up: 1 – absorption probes, 2 – sealed sources, 3 – probes recording radiotracers, 4 – electromagnetic flow meter, 5 – marked particles feeder, 6 – feeder of marked water, 7 – pump, 8 – conveyer, 9 – container for solid particles, 10 – separating container.

The key part of the installation is a vertical pipeline of length 7.75 m and with an internal diameter of 150 mm made of acrylic glass. The laboratory set-up has been described in details in papers [5, 13, 16]. The view of some parts of the experimental installation together with measuring section and gamma-absorption set as well as probe for radiotracer experiments is shown in figure 4.

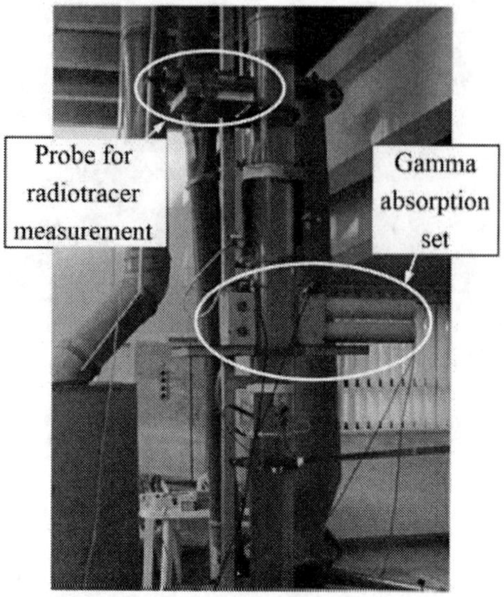

Figure 4. Test section view of the experimental set-up.

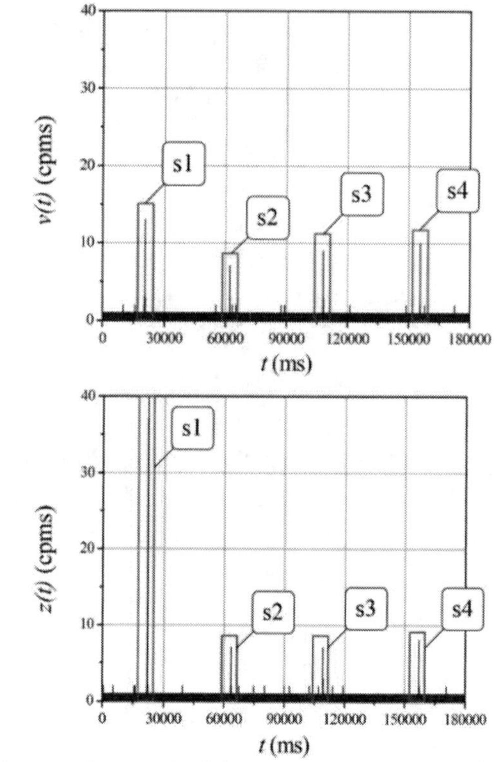

Figure 5. Time records of signal $v(t)$ and $z(t)$ obtained in the WRS0057 experiment.

The data acquisition equipment was comprised of the dedicated 8-channel EC Electronics counter of HSC 8000 connected to PC using a USB port.

4 The analysis of signals

Signal pulses I_x, I_y, I_v and I_z from the four probes are counted within the selected sampling time $\Delta t = 1$ ms and create mutually delayed discrete stochastic signals $x(t)$, $y(t)$, $v(t)$ and $z(t)$.

An examples of the time records $v(t)$ and $z(t)$ obtained in the WRS0057 experiment (particle size grade I) are presented in figure 5, where s1, s2, s3, s4 denote consecutive marked particle passes through a measuring section of the pipe.

For comparison, the signals $x(t)$ and $y(t)$ from the probes of absorption set recorded at the same experiment is shown in figure 6.

The signals recorded at two spatially separated location within the distance L_1 allow determination of the τ_{0S} time delay (transit time) necessary for transportation of the solid particles through the measuring section of the pipe. The particles average velocity υ_S can be calculated from the following formula:

$$\upsilon_S = L_1 / \tau_{0S} \tag{1}$$

The velocity of the marked particles υ_T is calculated as follows:

$$\upsilon_T = L_2 / \tau_{0T} \tag{2}$$

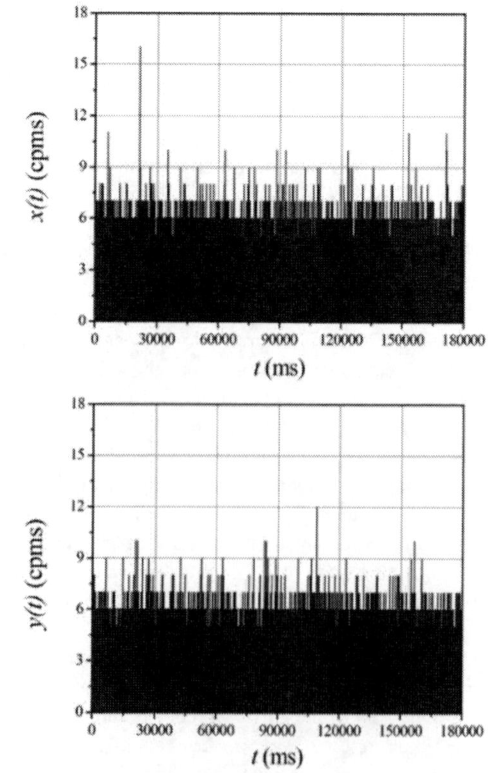

Figure 6. Time records of signal $x(t)$ and $y(t)$ in obtained from absorption set in the WRS0057 experiment

The frequently used method of time delay estimation of the ergodic random signals $x(t)$ and $y(t)$ is based on the cross-correlation function (CCF) $R_{xy}(\tau)$, defined by equation [20-24]:

$$R_{xy}(\tau) = \lim_{T\to\infty} \frac{1}{T}\int_0^T x(t)\cdot y(t+\tau)dt \qquad (3)$$

where T is the averaging time, τ – time delay.
The transportation time delay τ_0 is determined base on position of the CCF global maximum.

5 Results

The graphs of the cross-correlation functions obtained in the WRS0057 experiment are shown in figure 7. Figure 7a presents CCF for $v(t)$ and $z(t)$ signals obtained for marked particle and figure 7b presents CCF for signals from absorption set.

a)

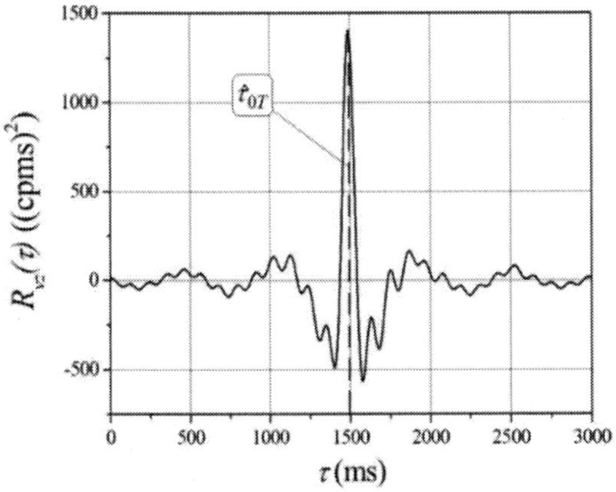

b)

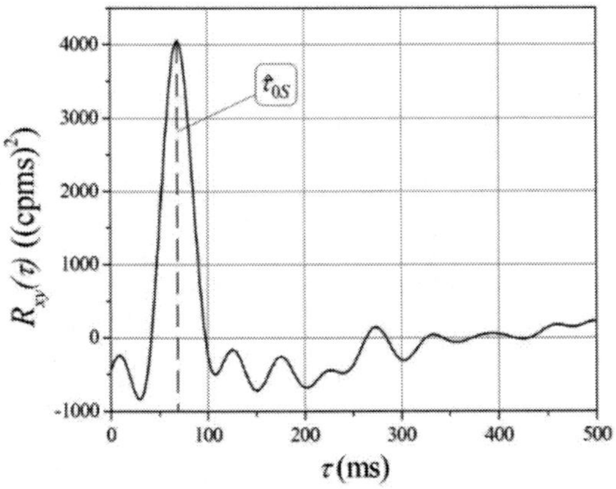

Figure 7. *CCF* calculated in the WRS0057 experiment; (a) for signals from radiotracers (b) for signals from absorption set.

For correct interpretation of absorption and radiotracers experiments simultaneously conducted for different distances L_1 and L_2, the *CCFs* were converted to the function of inverse velocity. The converted *CCFs*, which include amplitude adjustment, are shown in figure 8.

The combined standard uncertainty $u_c(\upsilon)$ of the solids velocity, with negligible small uncertainties of the acquisition set, depends on an inaccuracy of uncorrelated L and τ_0 determination [27]:

$$u_c(\upsilon) = \left[\left(\frac{\partial \upsilon}{\partial L}\right)^2 \cdot u^2(L) + \left(\frac{\partial \upsilon}{\partial \tau_0}\right)^2 \cdot u^2(\tau_0)\right]^{1/2} \qquad (4)$$

where $u(L)$ is the standard uncertainty of the distance between detectors.

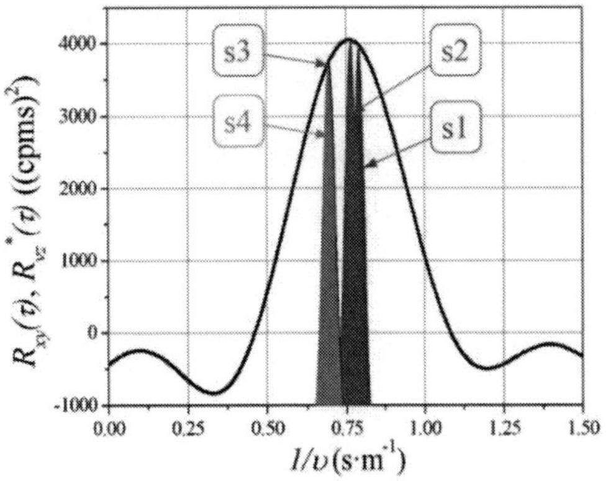

Figure 8. *CCF* for signals obtained for marked particles (colored areas) and for signals from absorption set (solid line) in the WRS0057 experiments.

Table 2 shows the results of the marked solid particle velocity υ_T obtained from radiotracer measurements and its uncertainty. The average particle velocity υ_S calculated using absorption method is $\upsilon_S = 1.323$ m / s with uncertainty $u_c(\upsilon_S) = 0.031$ m / s.

Table 2. Results of the marked solid particle velocity measurement in the WRS0057 experiment.

Type of experiment	υ_T (m/s)	$u_c(\upsilon_T)$ (m/s)
tracer, s1	1.2632	0.0023
tracer, s2	1.3078	0.0026
tracer, s3	1.4373	0.0030
tracer, s4	1.4435	0.0013
average	1.3630	0.0048

From the analysis of figure 8 and Table 2 it can be seen that velocity of the marked particles are in the range of greater υ_S values of the CCF distribution determined for absorption measurement. Velocities for s1, s2, s3, s4 are arranged within $\pm 1\sigma$ (where σ is standard deviation) aforementioned *CCF*.

The tracer experiment gives the ability to track individual selected grains whose velocity depends on the position in the flow. In the case of large number of radiotracer experiments when the number of passes indicated in the installation of stones is over 50, the calculated velocity values should be close to the *CCF* distribution obtained from absorption measurements.

6 Conclusions

Radiotracer experiments allow tracking the behavior of the selected grain size classes of the solid phase in the liquid-solids particle flow in vertical pipeline. This is important for the development of technology of hydrotransport of polymetallic nodules from the seabed.

The absorption measurements give an average velocity of the entire population of solid particles. The resulting distribution of the *CCF* can be considered as distribution of all possible time delays (velocities) of the moving grain population.

Comparison of the results of both methods indicates that the velocities of marked particles obtained for s1, s2, s3, s4 are arranged within $\pm 1\sigma$ aforementioned *CCF*.

Measurements of radiotracers, due to the use of open radioisotopes must be performed under conditions of radiological protection. However, in many cases of flow measurements there is no alternative method that gives similar opportunities.

Acknowledgment

The authors would like to thank team of prof. Jerzy Sobota for their cooperation during the measurements undertaken at Wroclaw University of Environmental and Life Sciences.

All presented investigations were possible due to support under the project P105/10/1574 of the Grant Agency of the Czech Republic, and RVO: 67985874 of the ASCR, as well as the Project No NN523755340 of the National Science Centre of Poland, which are gratefully acknowledged.

This publication is funded by AGH University of Science and Technology (No 11.11.140.645).

References

1. D.S. Cronan (ed.), *Handbook of marine mineral deposits* (CRC Press, Boca Raton, 2000)
2. R. Kotliński, http://www.iom.gov.pl/uploads/iom_attachment-95635.pdf
3. S. Depowski, et al, *Mineral resources of the seas and oceans* (SCHOLAR, Warsaw, 1998)
4. http://www.iom.gov.pl/
5. L. Petryka, et al., *EPJ WoC* **92,** 02065 (2015)
6. R. Hanus, et al., *EPJ WoC* **92,** 02020 (2015)
7. G. Falcone, G.F. Hewitt, C. Alimonti, *Multiphase flow metering: principles and applications* (Elsevier, Amsterdam, 2009)
8. G.A. Johansen, P. Jackson, *Radioisotope gauges for industrial process measurements* (Wiley, New York, 2004)
9. M. Zych, et al., Flow Meas. Instrum. **35,** 11–15 (2014)
10. Z. Yu, B. Qincheng, H. Richa, Appl. Therm. Eng. **60,** 398-410 (2013)
11. M. Zych, et al., *EPJ WoC* **92,** 02122 (2015)
12. W.A.S. Kumara, B.M. Halvorsen, M.C. Melaaen, Int. J. Multiphas. Flow **36,** 467-480 (2010)
13. R. Hanus, L. Petryka, M. Zych, Flow Meas. Instrum. **40,** 58–63 (2014)
14. M. Zych, et al., *MATEC WoC* **18,** 02004 (2014)
15. B.K. Arvoh, R. Hoffmann, M. Halstensen, Flow Meas. Instrum. **23,** 56-65 (2012)
16. M. Zych, et al., Prz. Elektrotech. **88** (10a), 98-100 (2012)
17. S.H. Jung, et al., Appl. Rad. Isotop. **67,** 1254-1258 (2009)
18. H.J. Pant, et al., Appl. Radiat. Isotop. **67,** 1609–1615 (2009)
19. P. Vlasak et al., J. Hydrol. Hydrom. **62,** 241-247 (2014)
20. M.S. Beck, A. Plaskowski, *Cross correlation flowmeters - their design and application.* (Adam Hilger, Bristol, 1987)
21. R. Hanus, Int. J. Multiphas. Flow **72,** 210–217 (2015)
22. J.S. Bendat, A.G. Piersol, *Random data - analysis and measurement procedures* (Wiley, New York, 2010)
23. R. Hanus, et al., Math. Probl. Eng. **2014,** 475735 (2014)
24. S.L. Soo (ed.), *Instrumentation for fluid-particle flow* (Noyes Publications, New Jersey, 1999)
25. R. Hanus, M. Zych, L. Petryka, *JPCS* **530,** 012042 (2014)
26. R. Hanus, Prz. Elektrotech. **86** (6), 232-235 (2010)
27. *Guide to the expression of uncertainly in measurement* (International Organisation for Standardisation, 1995)

EPJ Web of Conferences **114**, 02146 (2016)
DOI: 10.1051/epjconf/201611402146

Determination of volumetric concentration of solids in vertical pipeline hydrotransport

Marcin Zych[1,a], Robert Hanus[2], Leszek Petryka[1], Anna Strzępowicz[1] and Piotr Zych[3]

[1]*AGH University of Science and Technology, Krakow 30-059, Poland*
[2]*Rzeszow University of Technology, Rzeszow 35-959, Poland*
[3]*Warsaw University of Technology, Warszawa 00-661, Poland*

Abstract. This paper presents an application of radioisotopes with reference to the determination of the solid phase volumetric concentration in a hydromixture by calibration of the measuring set. It shows how the gamma absorption equipment consisting of radioactive isotopes ^{241}Am and scintillation probe, may be applied to the measurement of solid particles volumetric concentration in a flow. It is based on fact that the intensity of a gamma beam decreases as it passes through matter. In the described experiments as solid phase the ceramic models representing natural polymetallic ocean nodules were used. The especially constructed calibration stand and obtained relation between the related intensity of radiation and mean volumetric concentration of the solid phase are presented.

1 Introduction

Polymetallic nodules occur on the bottom of seas and oceans. They are solid structures of organic and mineral compositions which contain a various metals (mostly Mn, Si, Fe, Al, Na, Mg, Ni, K, Cu) [1-4]. The mining of nodules using the hydraulic method requires grain vertical transport by water. To determination of the solid phase volumetric flow rate in pipeline it is necessary to know the volumetric concentration of nodules and their velocity. For this purpose the method based on gamma ray absorption was proposed [5-10].

The idea of gamma-ray absorption measurement is presented in figure 1.

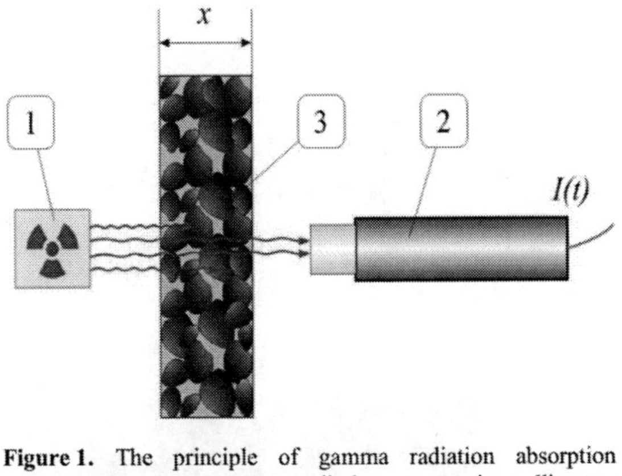

Figure 1. The principle of gamma radiation absorption measurement: 1 – gamma radiation source in collimator, 2 – scintillation probe, 3 – absorbent.

The gamma absorption is based on exponential decreasing of a monoenergetic gamma beam in function of composition and geometry of absorbent:

$$I = I_0 \cdot exp(-\eta \cdot \mu \cdot x) \qquad (1)$$

where I_0 is the inlet to absorbent radiation intensity, I is the outlet intensity detected after the beam has traveled a distance x through the absorbing materials, η and μ represents consequently the density and mass absorption coefficient of these materials [11-14].

The sealed radioactive sources emit gamma radiation beams shaped by collimators (1). Photons pass through the pipeline with analyzed mixture (3) and are achieved by the detector (2). In the presented experiment linear ^{241}Am source with energy of 59.5 keV and probe with NaI(Tl) scintillation detectors are used.

Using two sets of source - probe presented in figure 1 we can measure the averaged velocity of the solid phase [6-10]. At the same time one of these sets can be used to determination of the volumetric concentration of solid particles.

In this work we use data recorded on the experimental laboratory stand built up in the Water Laboratory of the Wroclaw University of Environmental and Life Sciences (Poland) for investigation of the polymetallic nodules hydrotransport in a vertical pipe with a inner diameter 150 mm. In the described experiments as solid phase the ceramic models representing three grain sizes (up to 70 mm) of natural polymetallic ocean nodules was used. The density of the particles representing the nodules was 2000 kg / m^3.

The ceramic models of nodules are shown in figure 2.

[a] Corresponding author: zych@geol.agh.edu.pl

Figure 2. The ceramic models of ocean nodules used in experiments.

2 Determination of volumetric concentration of solid particles

The volumetric concentration in liquid-solid particles flow is defined as the ratio of the solid phase volume to the total volume of the mixture:

$$C_V = \frac{V_S}{V} = \frac{V_S}{V_W + V_S} \tag{2}$$

where V_W is water volume, $V_S = V_{S1} + V_{S2} + \ldots$ is the sum of the volume of solid particles in the analyzed section dz of the pipe. Figure 3 presents the essential parameters for determination of volume concentration.

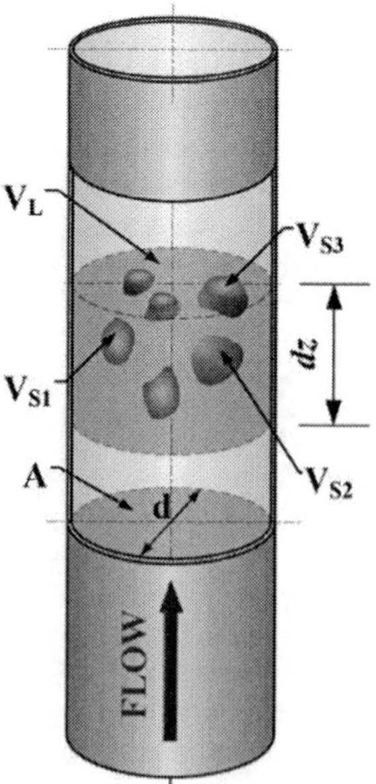

Figure 3. The essential parameters for determination of volume concentration.

Passing to the limit $dz \rightarrow 0$, the volumetric concentration of cross-section is defined as [15]:

$$C_{VP} = \frac{A_S}{A} = \frac{A_S}{A_W + A_S} \tag{3}$$

where: $A = \pi \cdot d^2 / 4$ – cross sectional surface of the pipe, d – inner diameter of the pipe, A_S and A_W – cumulative cross sectional surfaces of the solid phase and liquid phase respectively.

The weakening of radiation in a mixture of concentration C_{VP} transported through pipeline can be described by the differential equation:

$$\frac{dI}{dC_{VP}} + \mu_C \cdot I = 0 \tag{4}$$

where μ_C is the relative volumetric absorption coefficient. For the boundary conditions that occur during the measurement the solution of equation (4) is:

$$I = I_0 \cdot exp(-\mu_C \cdot C_{VP}) \tag{5}$$

where I_0 is the intensity of radiation recorded by the probe, where the pipe is filled with water only ($C_{VP} = 0$).

Because it is difficult to consider all the phenomena of absorption of radiation and its detection, experimental calibration on the laboratory stand is necessary.

3 Experimental setup and calibration procedure

Determination of the relationship between concentration and intensity of gamma radiation weakened in the liquid-solid particles mixture was carried out on an experimental calibration stand built in the Sedimentological Laboratory in Faculty of Geology, Geophysics and Environmental Protection AGH University of Science and Technology in Krakow. Diagram of the calibration stand is shown in figure 4.

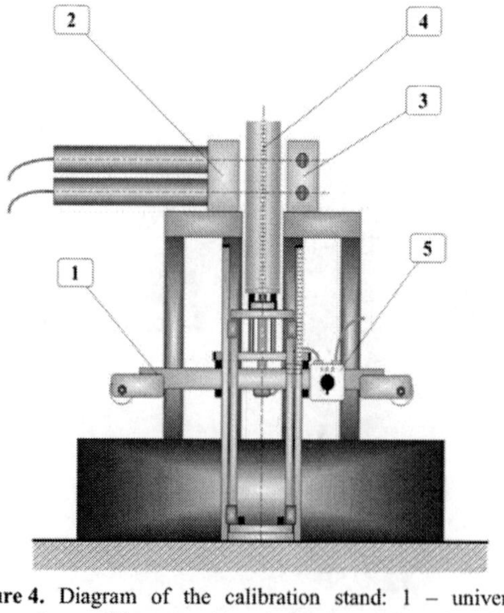

Figure 4. Diagram of the calibration stand: 1 – universal trolley, 2 – collimator with scintillation probes, 3 – lead collimator with gamma ray sources, 4 – calibration section of the pipeline, 5 – switch to control of cross-section high.

The stand consists of a universal measuring trolley (1) with the possibility of vertical travel, the controlled switch (5) and the calibration section of the pipeline (4). The fragment containing the test pipe with liquid-solids mixture was placed in the same way as during the nodule hydrotransport measurements. The gamma beams of radiation formed through collimator (3) and (2) pass through the pipeline at a predetermined height.

The mixture was prepared in such way that in the test section of pipe a real nodules concentration was reconstructed:

- for C_{VP} range of 0.400 ÷ 0.700 models filled the entire volume of the pipe,
- for C_{VP} range of 0.180 ÷ 0.400 the additional thin-walled plastic container with dimensions of 110 mm × 120 mm × 200 mm was used,
- for C_{VP} range of 0.010 ÷ 0.180 the fragment of the cone with diameters of Ø 80 to 85 mm and a height of 250 mm was used.

In order to determine the concentration, the pipe was progressively filled with water. The thickness of each layer of water was measured by using a laser range finder and caliper gauge. During refilling with water, the test section of the pipeline was placed on the scale, which allowed measurement of the mass increase of a further layer of the liquid.

During calibration the volumetric concentration of the solid phase C_{VP} was determined from the relationship:

$$C_{VP} = 1 - C_{VW} \qquad (6)$$

where

$$C_{VW} = \frac{4 \cdot m_W}{\rho_W \cdot \pi \cdot d^2 \cdot dz} \qquad (7)$$

where m_W is the weight of the respective water layer having a thickness dz, and ρ_W is density of water.

For each layer a series of 18000 measurements of radiation intensity was made. Each time the background radiation was measured before and after experiment. To properly map the distribution of solid particles in the pipeline, container was additionally rotated of angles from 45° to 180° during the measurement.

As a result of the study the following calibration curve was obtained using a linear least squares fitting technique:

$$\ln\left(\frac{I}{I_0}\right) = a \cdot C_{VP} + b \qquad (8)$$

where regression coefficients are: $a = -2.58 \pm 0.06$ and $b = -0.23 \pm 0.03$ at coefficient of determination $R^2 = 0.99$.

The graph of the obtained straight line is shown in figure 5.

Dependence (8) allows to determining the average volume concentration of solid phase in nodules hydrotransport measurements conducted at the research installation.

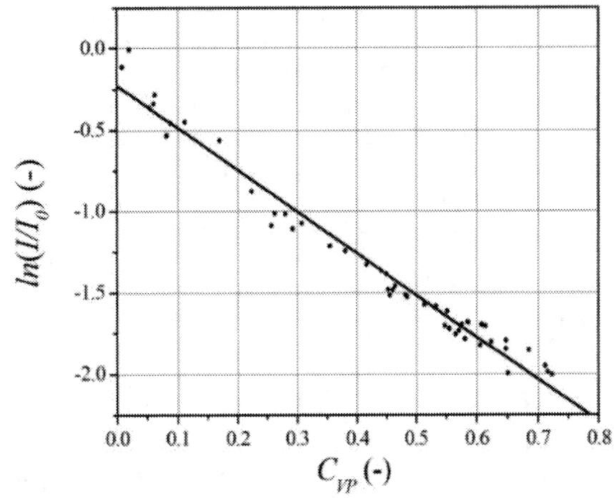

Figure 5. The relation between I/I_0 and the volumetric concentration C_{VP}.

4 Exemplary results

To use the calibration curve in practice, they are essential reference measurements, which are carried out to the water itself. This allows to find the point for which the value $C_{VP} = 0$. Thus this enables temperature compensation and taking into account the stability of the apparatus. It should be noted that the slope of the function is unchanged, because it is associated solely with the composition of the transported mixture.

The curve shown in figure 5 were used in the studies of hydrotransport of ceramic models of polymetallic nodules conducted in the Water Laboratory of the Wroclaw University of Environmental and Life Sciences. For example, figure 6 presents the results obtained in the experiment WRS039. The line with points shows the evolution of the concentration of the solid phase for the averaging in the time interval of 10 seconds. Whereas the horizontal straight line shows the average value for the whole measurement (300 seconds).

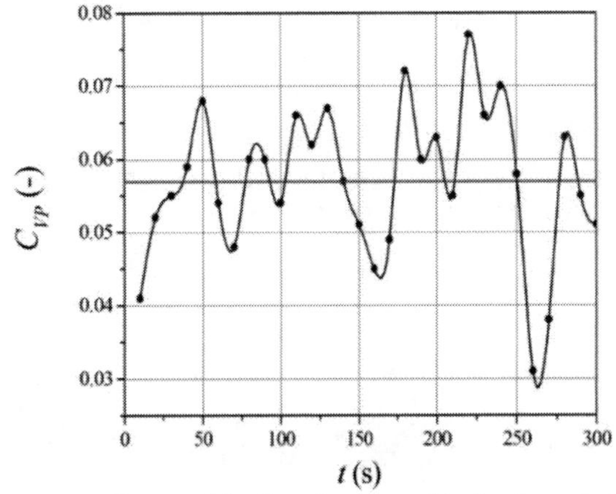

Figure 6. Changes of the C_{VP} value averaged in time interval of 10 seconds (line with points) and an average C_{VP} value for the whole measurement (horizontal straight line) in experiment WRS039.

The average value C_{VP} for the whole measurement presented in figure 6 is equal to 0.057 ± 0.003. The C_{VP} values averaged in time interval of 10 seconds are characterized by relatively high volatility in the range from 0.030 to 0.080. This means that the grains of the solid phase not flowed continuously but formed clusters, between which there was only water.

The measurement accuracy of C_{VP} was determined with law of propagation of uncertainty [16].

In the case of use of the calibration line the fluctuations in the intensity of radiation I should be considered. They are the result of the statistical nature of radioactive decay and are described by the standard uncertainty [17]:

$$u(I) = \sqrt{I} \qquad (9)$$

On the basis of the analysis we can be stated that the uncertainty of the average volumetric concentration of solid phase in the flowing mixture does not exceed 10%.

5 Conclusions

For determination of volumetric concentration C_{VP} using gamma absorption it is necessary to calibrate a measuring set of source of radiation - scintillation probe.

Implementation of the calibration procedure required to build a laboratory stand for which reconstruction of fragment of pipeline and modeling of the flow structures were conducted. The resulting calibration dependence allows the determination of the average volumetric concentration of solid phase for the measured radiation intensity. On the basis of the C_{VP} value and the velocity of the solid phase we can determine the volumetric flow rate of the nodules [15].

The advantage of the presented solution is to use the same absorption set to measure of both: the volumetric concentration of the solid phase and their average velocity [6-10]. This allows the determination of the volumetric flow rate of the solid phase additional detectors or independent measurements.

Described in the paper calibration procedure can be in the future replaced by simulations, using e.g. MCNP software.

Acknowledgment

The authors would like to thank teams of prof. Jerzy Sobota and prof. Pavel Vlasak for their cooperation during the measurements undertaken at Wroclaw University of Environmental and Life Sciences.

All presented investigations were possible due to support under the project P105/10/1574 of the Grant Agency of the Czech Republic, and RVO: 67985874 of the ASCR, as well as the Project No NN523755340 of the National Science Centre of Poland, which are gratefully acknowledged.

This publication is funded by AGH University of Science and Technology (No 11.11.140.645).

References

1. D.S. Cronan (ed.), *Handbook of marine mineral deposits* (CRC Press, Boca Raton, 2000)
2. R. Kotliński, http://www.iom.gov.pl/uploads/iom_attachment-95635.pdf
3. S. Depowski, et al, *Mineral resources of the seas and oceans* (SCHOLAR, Warsaw, 1998)
4. http://www.iom.gov.pl/
5. L. Petryka, M. Zych, R. Murzyn, Nukleonika **50**, 43–46 (2005)
6. R. Hanus, L. Petryka, M. Zych, Flow Meas. Instrum. **40**, 58–63 (2014)
7. L. Petryka, et al, *EPJ WoC* **92,** 02065 (2015)
8. R. Hanus, et al, *EPJ WoC* **92,** 02020 (2015)
9. R. Hanus, et al., Math. Probl. Eng. **2014**, 475735 (2014)
10. L. Petryka, et al, *Proc. 10th ISOPE Ocean Mining & Gas Hydrates Symp. OMS-2013*, 230-233 (2013)
11. G. Falcone, G.F. Hewitt, C. Alimonti, *Multiphase flow metering: principles and applications* (Elsevier, Amsterdam, 2009)
12. G.A. Johansen, P. Jackson, *Radioisotope gauges for industrial process measurements* (Wiley, New York, 2004)
13. M. Zych, et al., Flow Meas. Instrum. **35,** 11–15 (2014)
14. Z. Yu, B. Qincheng, H. Richa, Appl. Therm. Eng. **60,** 398-410 (2013)
15. M. Zych, et al., MAaM **56** (4), 318-321 (2010)
16. *Guide to the expression of uncertainly in measurement* (International Organisation for Standardisation, 1995)
17. G.F Knoll, *Radiation detection and measurement* (Wiley, New York, 1986)

EPJ Web of Conferences 114, 02147 (2016)
DOI: 10.1051/epjconf/201611402147

An experimental study of the glottal jet

Darina Jasikova[1,a], Petr Sidlof[1], Michal Kotek[1] and Vaclav Kopecky[1]

[1] *Institute of new technologies and applied informatics, Faculty of mechatronics, informatics and interdisciplinary studies, Technical university of Liberec, Studentska 1402/2, 461 17 Liberec, Czech Republic*

Abstract. The paper presents results of the flow field analysis in a static scaled model of human larynx. Here we are focused on the effect of the nozzle gap parameter and the incoming flow velocity. The study is performed in the aerodynamic channel using the PIV technique.

1 Introduction

Human voice is created by expiring air from the lungs through a narrow constriction called glottis. This constriction is formed by the vocal folds, located in the larynx. The vocal folds are two symmetric soft tissue structures fixed between the thyroid cartilage and arytenoid cartilages. Under certain conditions (subglottal pressure, glottal width, and longitudinal tissue tension) the vocal folds can start to oscillate, thus modulating the flow field. The acoustic signal is created by complex aeroacoustic mechanisms in glottis and supraglottal areas, filtered by the vocal tract, radiated from the mouth, and perceived as voice.

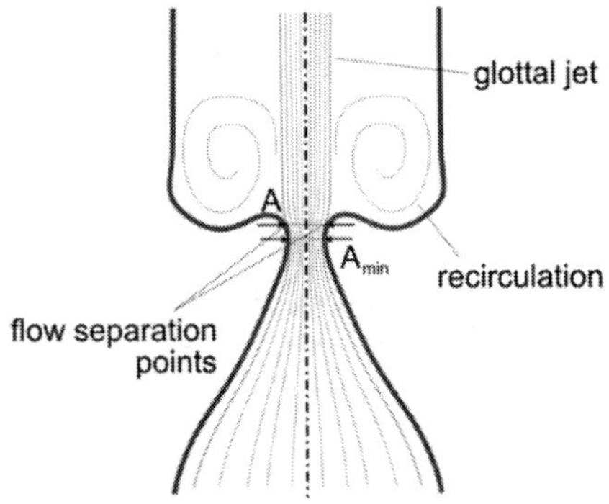

Figure 1. Idealized symmetric airflow in the glottis. Physiological orientation – flow in the inferior-to-superior direction.

From the fluid-mechanical point of view, the human larynx can be seen as a nearly planar nozzle with time-varying clearance. In the convergent part the airflow accelerates. Near the narrowest cross-section, airflow separates due to adverse pressure gradient and forms a jet (see Fig.1). The shear layer of the glottal jet generates significant vorticity in the flow field, with periodic vortex shedding creating large-scale coherent turbulent structures in the supraglottal field. Due to the interaction of the glottal jet with the vortices, the glottal jet rarely remains axisymmetric and rather teds to deflect to either of the supraglottal walls.

The concept of fluid-structure-acoustic interaction between the airflow, elastic vocal folds, and sub- and supraglottal acoustic spaces relies on knowledge of aerodynamics in the larynx. However, due to periodic closure of the glottal channel during vocal fold vibration and inherent unsteadiness of the airflow, the aerodynamic effects in the larynx are very complex. In spite of the progress in fundamental research of human voice production during recent years, some features of the glottal flow are not yet fully understood.

First glottal flow visualizations and PIV measurements were published by Triep et al. [1], who used a three times scaled hydrodynamic setup, and Erath and Plesniak [2, 3] on a static 7.5 times life-size vocal fold model. Kucinschi et al. [4] confronted his Fluent computations with pressure and flow rate measurements on a mechanically driven physical model, but did not assess velocity fields. Li et al. [5] used a similar technique (with a static physical model) and tried to evaluate the flow separation points, although only qualitatively. Krane et al. [6] made measurements on an externally driven model of the human glottis in a water channel, which operated at lower frequencies.

An extensive PIV dataset on glottal flow was published by Neubauer et al. [7], who used a life-sized model of Thomson et al. [8] and quasi phase-locked PIV to measure near-field flow structures. The paper provides detailed data on jet core velocity, jet inclination angle, and also on the flow separation point, however only qualitatively. The paper of Becker et al. [9] is focused on elucidation of the mechanisms of sound production in the larynx. The authors used a synthetic life-sized self-oscillating vocal fold model. Erath and Plesniak [10] published a study on asymmetric flow features in the

[a] Corresponding author: author@email.org

glottis. Using an externally driven 7.5 times life-sized model precisely mimicking vocal fold oscillation, they quantify jet skewing in the divergent part of the glottis and deduce implications of flow asymmetries on sound production. Triep et al. [11] used an improved experimental setup equipped with time-resolved PIV to show that the supraglottal flow field is highly 3D. Their results also show that the presence of the ventricular folds decreases the pressure loss and stabilizes the jet during the divergent phase. Sidlof et al. [12] measured the position of the flow separation point during self-induced vocal fold oscillation on a scaled physical model.

Recently, Lodermeyer et al. [13] investigated the glottal jet behaviour downstream of self-oscillating synthetic vocal folds by means of phase-locked PIV technique.

In a computational study, Sidlof et al. [14] have shown that the main aeroacoustic sources are located inside glottis, and in places where the glottal jet interacts with the channel wall. The current study aims at high-resolution PIV measurements of the velocity fields in domains, which are crucial for source sound generation. A scaled transparent static model of symmetric vocal folds with the "M5" shape [15] was fabricated and used for supraglottal velocity field measurements.

2 Experimental setup

For the purpose of this study we build the simplified model directly in the air tunnel. The air circuit was set with the narrow part of cross section (45x45) mm. In the middle of this part has been placed and fixed two segments that create the nozzle. The experimental setup of the nozzle is seen in figure 2.

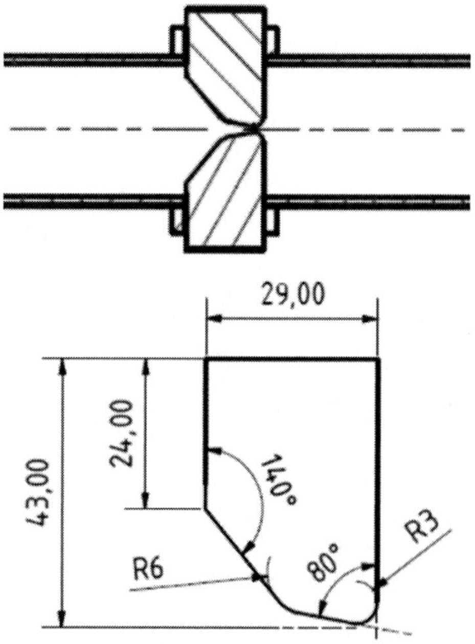

Figure 2. Experimental setup of the nozzle and the segments parameters.

Both segments were manufactured from the polycarbonate plastic and polished. The model was mounted into the chamber in the air tunnel. The nozzle gap can be adjusted. Here we used three setups: 1 mm, 2 mm, and 3 mm. On the inlet to the chamber was fixed hot wire anemometry probe. The constant flow on inlet to the chamber was set by regulation on the ventilation system on the air tunnel.

The glottal jet was visualized using PIV method. The laser sheet was brought to the chamber using optical arm and the camera was fixed to the traversing unit. We recorded the flow in two different positions of camera to reach the complex view of the flow. The first view enabled capture area of interest of (140x50) mm and the second one enabled field of view (20x20) mm to obtain macroscopic view of the flow between the nozzles.

The PIV system consisted of NewWave Gemini pulsed laser Nd: YAG energy of 120 mJ per pulse length of 10 ns, which operates at a wavelength of 532 nm. The images were captured with Neo CMOS chip 5,5 MPixel size of 6.5 microns camera. Laser and camera system is controlled from a computer and synchronized to external TimerBox.

The flow on the inlet to the chamber was set in two steps: firstly on 0,375 m/s and the second flow was 0,75 m/s. This velocity setup was scaled according to the scaling of the glottis model.

3 Results and discussion

In terms of visualization and measurement using PIV method in this task we are facing with three problems: the glottal jet represents changes in dynamic range, the laser light on the wall reflects and the nozzle represents the task of measurement in the narrow fissure.

The first problem - accelerated glottal jets against to slower moving vortex structures close to steps, has been resolved by changing of the dynamic range by adjustment method. The flow was firstly captured with the time between laser pulses optimized to slower flow that supposed to reach maximum speed of 2 m/sec. This ensured a record of slow vortex structures. In the second part of the measurement the time between laser pulses was optimized on accelerated glottal jet, wherein the expected velocity of about 8 m/s. All records were separately calculated, validated and represented as vector maps and statistics. Numerical data were sent to Matlab.

The designed algorithm combines suitable values from both vector maps. The threshold for correct values could be set by the experienced operator. More vector maps with several "time between pulses" may be combined into the resulted vector map. This procedure enlarges the low dynamical range, one of the most important limits of the PIV method.

The second problem was dealing with sufficient lightening inside the nozzle. The light was reflected to the nozzle using two inclined optical mirrors. This setup had to be precisely in geometry and the reflected light should lead directly to the source sheet, otherwise the reflection created second focusing plane that would bring the errors in the cross correlation, because of comparing different particles in two different measuring laser sheets. This effect was magnified in the nozzle area. The figure 3 shows the illumination of the nozzle area.

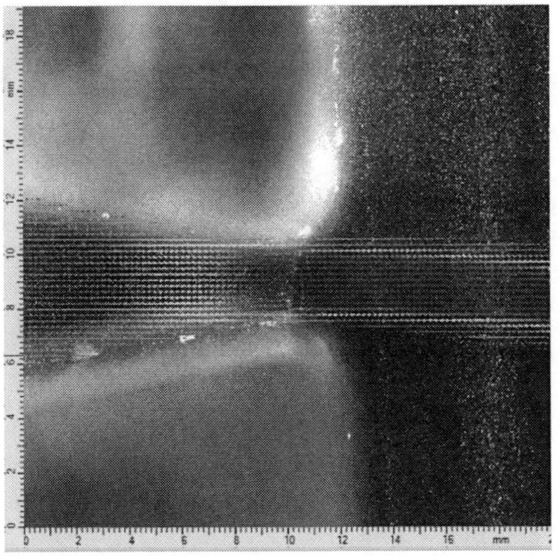

Figure 3. The magnified nozzle area under laser sheet illumination with seeding particles and calculated vector map of accelerated glottal jet.

The third problem of visualization is dealing with the illumination in the narrow structures. The segments are manufactured from the polished material that reflects and refracts the laser light, although the laser sheet is precisely focused in the sheet plane, the backward refraction bringing noise to the cross correlated vector maps. This can be solved by using image masking algorithm and image processing function that deduct all constantly refracted noise.

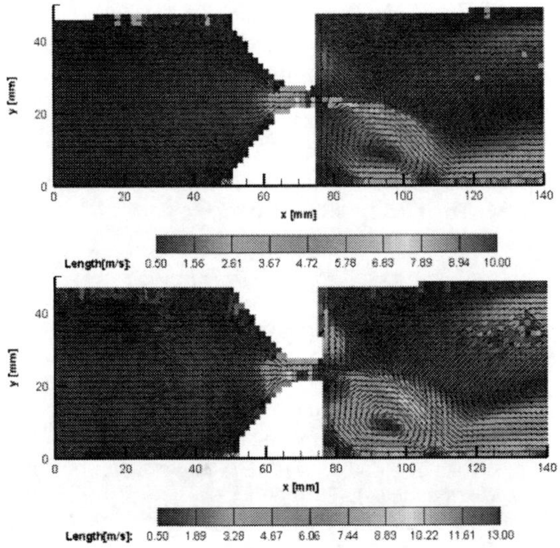

Figure 4. Vector map optimized with algorithm in Matlab – position 1 with nozzle fissure 1mm and incoming flow velocity 0,375 m/s and 0,75 m/s.

Vector maps on Fig. 4 describe the flow field in the model of the glottal jet, close to the fissure with inlet velocities 0,375 m/s and 0,75 m/s. The character of the flow stays similar in the full range of the fissure dimension. Behind the glottal nozzle the flow does not expand symmetrically, the main stream adherences to the lower wall. According the theoretical assumption the adherence should be random (upper or lower wall). Design of the model forces the stream to adhere always down. That phenomenon facilitates the statistical analysis.

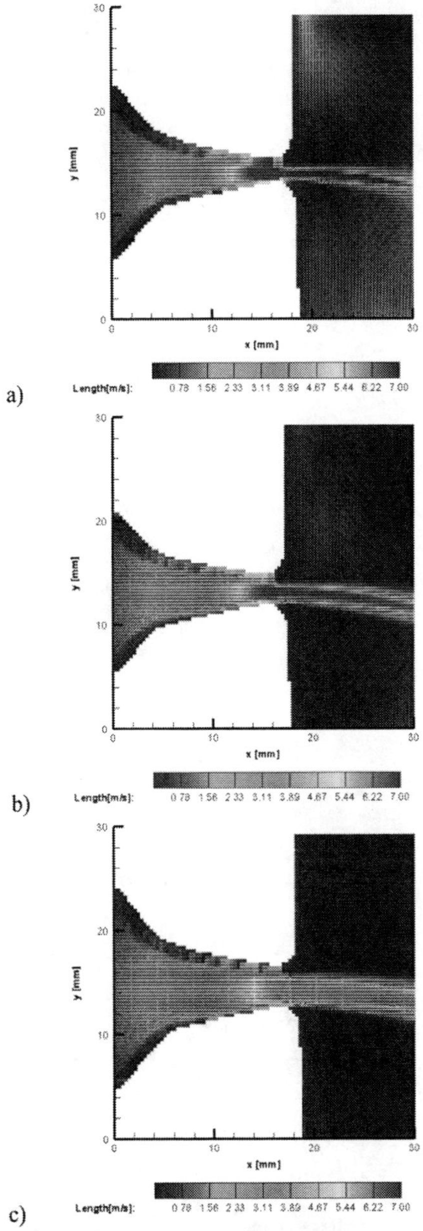

Figure 5. Vector map processed in the second position on maximal magnification. A) Calculation of the nozzle 1 mm, b) for the nozzle fissure 2 mm and c) the nozzle 3 mm for the inlet flow) 0,37 5m/s.

The stagnation point (point of the flow adherence to the wall) is found 30 mm (for inlet velocity 0,375 m/s) and 32 mm (for inlet velocity 0,75 m/s).

The series of vector maps on Fig. 5 compares the flow behind the vocal nozzle with different dimension in detail. This study proves the ability of the PIV to measure the flow in the thin vocal fissure. Using Macro Nikkon lens and HiSense Neo camera the flow profile consists of 12 vectors in 1 mm. Analysed data are suitable to verify the numerical simulation and to set the initial conditions of the model. To get real glottal jet data, the model will be in the next step expanded with a mechanism to make the fissure oscillate.

There is seen the comparison of the velocity profiles taken in the stagnation point of the glottal jet in the figure 6. There is seen the acceleration of the jet and maximum velocities that are increasing with the narrowing of the nozzle.

A)

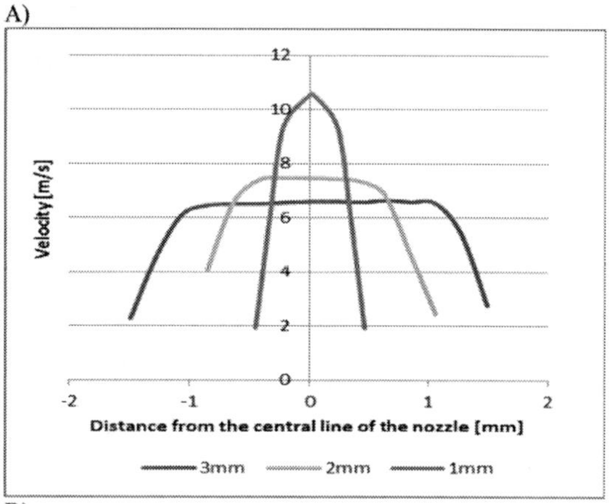

B)

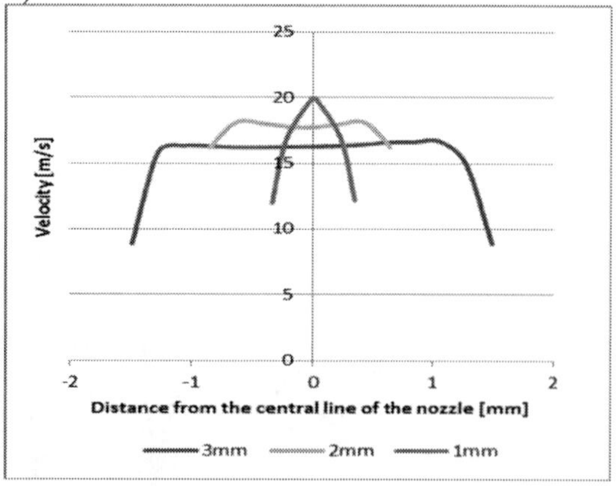

Figure 6. The comparison of the velocity profiles in the stagnation point in the nozzle for inlet flow a) 0,375 m/s and b) 0,75 m/s.

4 Conclusions

The concept of fluid-structure-acoustic interaction between the airflow, elastic vocal folds, and sub- and supraglottal acoustic spaces relies on knowledge of aerodynamics in the larynx. Analytical and numerical studies are run by many research teams, but the experimental data for validation and numerical model optimization are still not sufficiently provided. There are many input parameters and boundary conditions that have to be obtained. Here we started the research on the simplified scaled static model of human glottal and we applied the PIV method for accelerated glottal jet in narrow space. Following the problem of dynamic range changes we draw up the Matlab algorithm that takes in account the requirements on slow vortex structures motion estimation as well as the accelerated glottal jet. We confirmed the possibility of evaluating the flow in narrow gap. In the next step, it will be extended of the segments movement and synchronized measurement with segments oscillation in time.

Acknowledgements

The authors acknowledge the institutional support of the Faculty of Mechatronics, Informatics and Interdisciplinary Studies of the Technical University of Liberec.

References

1. M. Triep, C. Brücker, W. Schröder, *Exp. in Fluids* **39** (2005)
2. B. Erath, M. Plesniak, *Exp. in Fluids* **41** (2006)
3. B. Erath, M. Plesniak, *J. of the Acoustical Society of America* **120**(2) (2006)
4. B. Kucinschi, R. Scherer, K. Dewitt, T. Ng, *J. of the Acoustical Society of America* **119** (2006)
5. S. Li, R. Scherer, M. Wan, S. Wang, H. Wu, *J. of the Acoustical Society of America* **119** (2006)
6. M. Krane, M. Barry, T. Wei, *J. of the Acoustical Society of America* **122** (2007)
7. J. Neubauer, Z. Zhang, R. Miraghaie, D. Berry, *J. of the Acoustical Society of America* **121** (2007)
8. S. Thomson, L. Mongeau, S. Frankel, *J. of the Acoustical Society of America* **118** (2005)
9. S. Becker, S. Kniesburges, S. Müller, A. Delgado, G. Link, et al., *J. of the Acoustical Society of America* **125**, (2009)
10. B. Erath, M. Plesniak, *Exp. in Fluids* **49** (1), (2010)
11. M. Triep, C. Brücker, *J. of the Acoustical Society of America* **127**(3), (2010)
12. P. Šidlof, O. Doaré, O. Cadot, A. Chaigne, *Exp. in Fluids* **51**(1), (2011)
13. A. Lodermeyer, S. Becker, M. Döllinger, S. Kniesburges, *Exp. in Fluids* **56** (4), (2015)
14. P. Šidlof, S. Zörner, A. Hüppe, *Biomechanics and Modeling in Mechanobiology* **14** (3), (2015)
15. R. Scherer, D. Shinwari, K.D. Witt, C. Zhang, B. Kucinschi et al., *J. of the Acoustical Society of America* **109**(4), (2001)
16. F. De Lillo, F. Cecconi, G. Lacorata, A. Vulpiani, EPL, 84 (2008)
17. L. T. De Luca, Propulsion physics (EDP Sciences, Les Ulis, 2009)

EPJ Web of Conferences **114**, 02148 (2016)
DOI: 10.1051/epjconf/201611402148

Air flow phenomena in the model of the blind drift

Marek Jaszczur[1,a], Michał Karch[1], Marcin Zych[1], Robert Hanus[2], Leszek Petryka[1] and Dariusz Świsulski[3]

[1]*AGH University of Science and Technology, 30-059 Krakow, Poland*
[2]*Rzeszow University of Technology, 35-959 Rzeszow, Poland*
[3]*Gdansk University of Technology, 80-233 Gdansk, Poland*

Abstract. In the presented paper, Particle Image Velocimetry (PIV) has been used to investigate flow pattern and turbulent structure in the model of blind drift. The presented model exist in mining, and has been analyzed to resolve ventilation issues. Blind region is particularly susceptible to unsafe methane accumulation. The measurement system allows us to evaluate all components of the velocity vector in channel cross-section simultaneously. First order and second order statistic of the velocity fields from different channel cross-section are computed and analyzed.

1 Introduction

The researchers from AGH University of Science and Technology (traditional university name is AGH Mining and Metallurgy) in the last few decades carried out large number of studies related to the safety and comfort of people working in extremely difficult climatic conditions that occur in underground mines [1,2]. One of the most important element of this research are issues related to ensuring proper ventilation in the people workplace. In order to ensure the ventilation designers should have the knowledge, experience and tools that will help solve emerging questions and problems [3-6]. Proper ventilation requires good knowledge about velocity field in blind drift [7-8]. Maintained for technological reasons, the blind chamber is exposed particularly to dangerous accumulations of methane. A methane-air mixture flows into the cave from a goaf space. This flow is not limited, because it should push away the zone of high methane concentrations from the conveyor drive located at the terminal section of the wall. The existing system of duct connections causes this to be the least intensively ventilated part of the typical longwall.

In literature there are only few experimental studies related to similar topics. Most of works concentrate on numerical analysis [9-12] in which depending on the considered variant of the exploitation system, various numerical models has been tried and tested for the best way to reproduce the real flow. In order to use any new or existing turbulence model, this model has to be validated base on theory, experimental measurement based on real or laboratory scale model [13-14].

In the typical copper mines chamber-pillared system has been used. The chamber is ventilated by fans placed at the inlets. The length of the chambers is typically 25-30 m, in some cases ventilated chamber length up to 60 m. The geometry investigated in this study differs significantly from those analyzed in literature. It arranges the connections between the main ventilation ducts and separately ventilated bind chamber.

[a] Corresponding author: marek.jaszczur@agh.edu.pl

2 Experimental setup

The experimental model of the blind drift set-up is shown in figure 1. It is assumed, that dimensions of the real object are: the blind drift 4 m x 2 m, and the length of the blind drift is 20 m. The main fan diameter is 0.8 m. The geometrical scale of the physical model is 1:10.

The outlet of the fan has been installed 15 cm from the beginning of the channel. In real flow the Reynolds number (base on the hydraulic diameter and mean velocity) excess of 100 000.

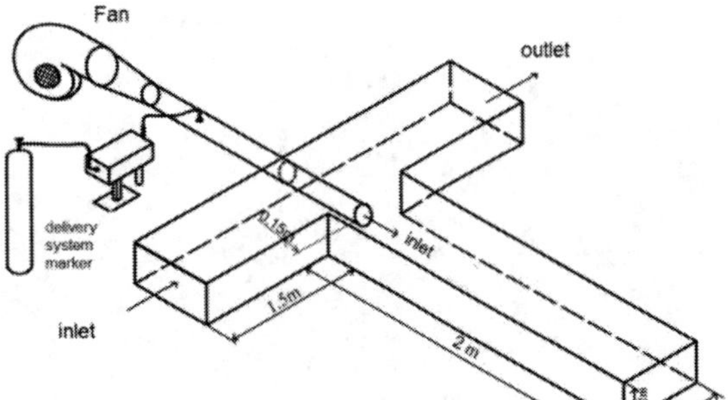

Figure 1. Experimental set-up of the blind drift.

Measurements were performed for the average airflow velocity in the inlet duct equal 35.4 m / s, which gives a Reynolds number of about 180 000.

In order to stabilize the air jet and to obtain the axisymmetric flow velocity profile downstream from the fan, a 1.35 m jet-stabilized section was provided. This section comprised a diffuser, a confuser and a pipe 0.2 m in diameter with a force-mode fan at the inlet. The channel was made entirely of plexiglas (but in some cases side and bottom walls were cover by real stones to make the wall more realistic - see figure 4) to meet the requirement of transparency which is imposed by the selected method of measurement.

3 Measurement

Particle Image Velocimetry (PIV) and Stereoscopic –PIV method was used to evaluate the velocity vector components [15].

Velocity measurement in this method is based on a statistical estimate of the particle motion marker mixed with the air in the measurement area [16]. The seeding for the PIV method was DEHS oil sprayed through an atomiser, to obtain a particle diameter of about 1 μm. The seeding particles were introduced to the inlet air stream, immediately behind the flow-stabilizing section. During the measurement, the test area was illuminated with a laser beam (ie. the knife light). Particles injected to the flow works as a tracer particles and reflect laser light. Reflected light is recorded by a CCD camera placed perpendicular (or in the same of stereoscopic measurement inclined) to the plane of the light. In order to determine the distribution of the velocity field of the flow, tracer particles are recorded by the camera on two frames. The measuring equipment was subjected to testing and calibration procedures under flow condition to yield the accurate statistics. The Davis software ver. 8.0 [17] was used to analyze the images to evaluate the velocity vector components as well as other statistics.

Measuring section is presented in figure 2. The particles were illuminated with a double-pulse Nd:YAG laser of energy of about 60 mJ per pulse. The digital images were acquired by 4 Mpx monochromatic CCD camera.

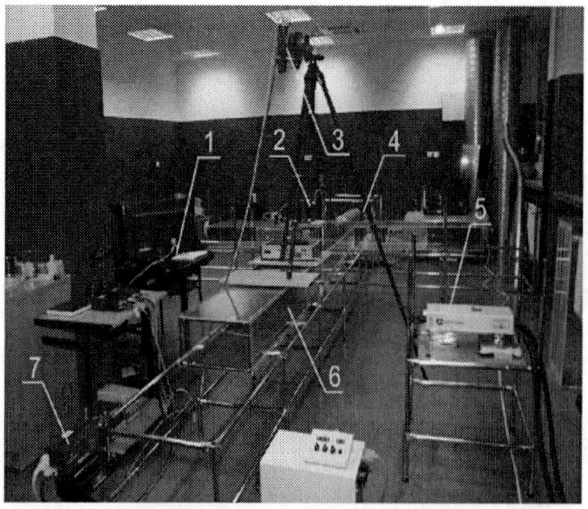

Figure 2. The photo of a measuring section: 1-measuring computer, 2-seeds system, 3-camera no.1, 4-centrifugal fan, 5- double-pulse Nd:YAG laser, 6-blind drift, 7-camera No. 2.

In each experiment, up to 1000 double frame images were recorded with the camera recording at a frequency of 5-7 Hz, which resulted in an overall time of one measure of around 3 minutes. Time Δt between two subsequent frames varied from about 20 μs to 200 μs. However, the measurements at the end of the blind drift were taken in the range of 3000-4000 μs, because (in some cases) the velocities there were much lower than in the other sections. During the calculations, the size of interrogation windows that exhibit satisfying results was not constant but varied from 16x16 to 64x64 pixels.

All fan arrangements in the model blind drift study in this paper are presented in the figure 3. In the first case (1) the fan tube is parallel to the canal walls and the distance between the axis of the fan and the top and left wall was 5 cm. In the case (2) the fan tube is also parallel to the channel walls but the distance between was 8 cm.

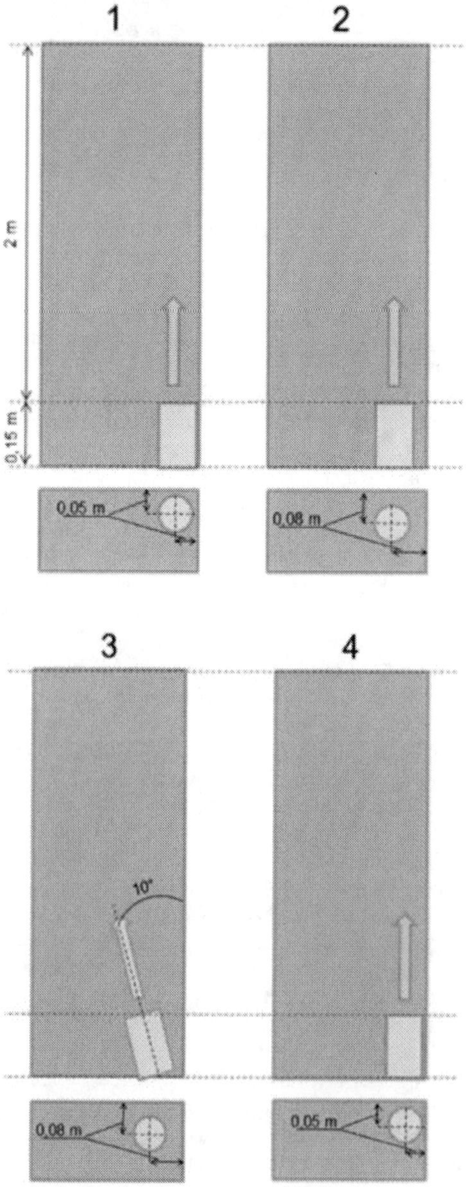

Figure 3. Schematic diagram of four different fan outlet location and inclination in the model of the blind drift.

For case (3) the fan tube is inclined with an angle of 10 degrees to the side wall of the channel and located as for case (2). Case (4) - fan is set parallel to the channel walls. The distance between the axis of the fan and the rough wall was 5 cm. For cases (1)-(3) all channel walls were smooth while in case (4) two of four walls have a high wall roughness (see figure 4).

The high wall roughness was obtained using real small size stones. Size of applied stones contain from 1 to 1.5 cm. The use of walls with high surface roughness was to bring actual conditions in the blind drift.

Figure 4. The channel with two walls with a high roughness.

As was mentioned, only in the first case, the air flow stream goes to the end of the channel parallel to the wall. In the other three settings the air flow stream breaks away from the wall and turns toward the opposite wall. In the current research optimal position and inclination has been evaluated in order to get correct and efficient ventilation of the blind drift.

4 Results

Contours of steam-wise velocity component in the blind drift for all cases considered here is presented in figure 5 in the x-y cross section located in the axis at the fan (high equal to the fan tube axis position). It is interesting to compare the results obtained for case (1) and (4) for which fan tube is located in the same place but the channel walls are smooth or have high wall roughness. Qualitative difference is observed between velocity fields obtained from these two scenarios. In case (4), when the walls are rough main flow stream turned into the opposite wall. Flow pattern for case (4) is similar to case (3) where fan tube is inclined and main flow stream in a natural way turned left.

Figures 6-8 shown profiles of the stream-wise velocity component along the channel horizontal line at the axis fan and for y = 30 cm from the fan outlet.

Figure 6 shows the measured velocity for cases (1), (3) and (4) at the distance y = 30 cm from the fan outlet. The main flow stream of air is directed at an angle for case (3). Velocity distributions obtained from cases (1) and (4) are similar. Velocity distributions obtained from case (3) is shifted to the left wall.

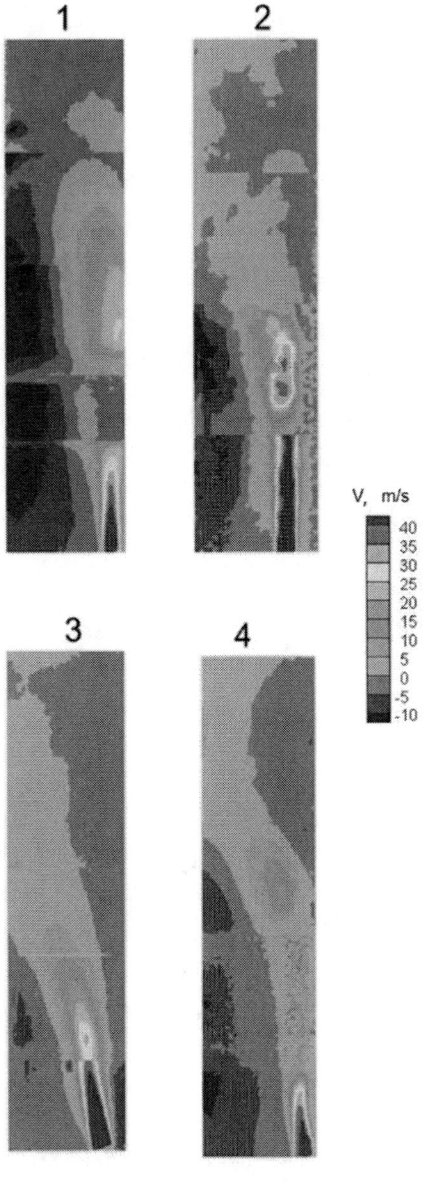

Figure 5. Contours of stream-wise velocity in the blind drift in the x-y cross section.

Figure 7 shows the measured velocity for case (1) and (4) at a distance y = 80 cm from the fan outlet. At this point, you can see that in case (4) (the walls are rough) air flow main stream begins to turn toward the opposite wall. There is slight difference (about 0.5 m / s) between maximum velocities for presented cases.

Figure 8 shows the measured velocity in case 3 in at a distance of 120 cm from the fan outlet. Figures 5 (bottom) show a reversal of the air flow in the duct. Such a situation occurred in all cases studied, except case 1. For different settings of the fan turning air stream in the direction of the opposite wall is followed at different points in the channel. It can be seen in figure 5.

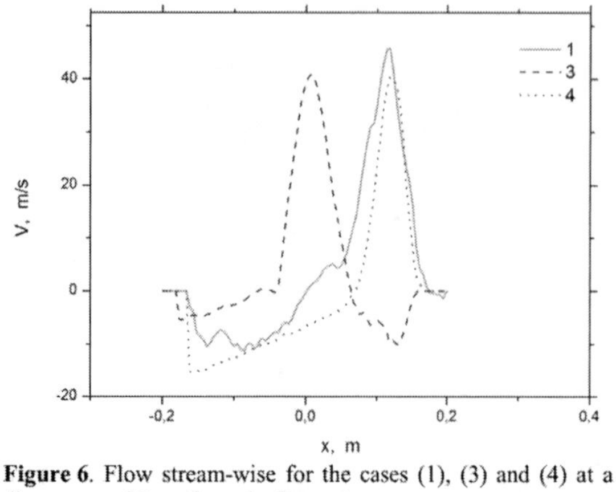

Figure 6. Flow stream-wise for the cases (1), (3) and (4) at a distance y = 30 cm from the fan outlet.

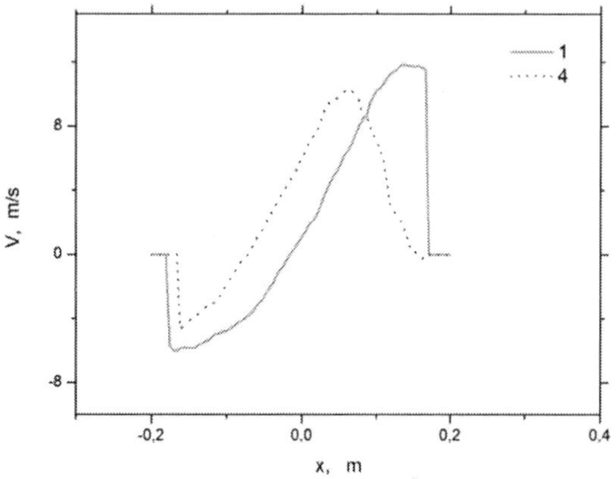

Figure 7. Flow stream-wise for the cases (1), (4) at a distance y = 80 cm from the fan outlet.

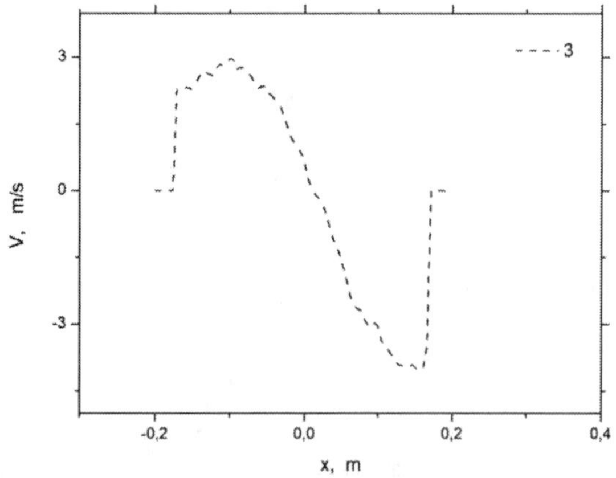

Figure 8. Flow stream-wise for the cases (3) at a distance y = 120 cm from the fan outlet.

Figure 9 shows flow fields obtained for case (1) and (3). The flow field is shown in the x-z cross-section plane of the channel. This data has been obtained using stereoscopic measurement which allow to determine all velocity vectors components.

In figure 9, the presented measurement was taken at a distance of 30 cm counted from the end of the channel. As can be seen for case (1) main air flow stream hits the end wall while for case (4) situation is different and largest positive velocity is observed close to the left wall and negative velocity close to the right wall. This is related to the large recirculation areal existing at the end of the blind drift which directly influence ventilation processes.

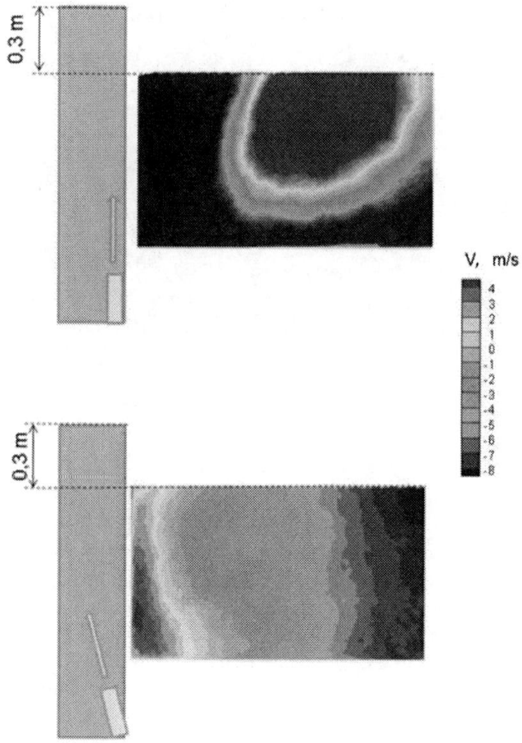

Figure 9. The velocity flow field obtained in the cave using 3D stereoscopic measurement (Stereo-PIV) in the x-z cross section at the distance 30 cm from the channel end.

Figure 9 also shows, that in case (1) when the axis of the fan tube is at a distance of 5 cm from the channel walls the main flow stream of air is not reversing before it reach the end of the channel. While the fan tube is set at an angle of 10 degrees as for the case (3) the air flow is fully reversed and the zone of negative stream-wise velocity component covers right part of the channel (at presented distance). It is also worth to notice, that in case (3) the area with velocity close to or equal 0 m / s (no air movement) is bigger than for case 1. This may also suggest, that this area may be less ventilated.

5 Conclusions

In this paper preliminary measurement of velocities in blind drift result are presented. Roughness and angle of fan have influence on velocity fields in the blind chamber. The laboratory setup represents a simplified model of the crossing of longwall with ventilation gallery. Despite of the simple geometry of the flow domain, the structure of the velocity field is very complex. The examined flow is characterized by such flow features as separation, air stream impingement on the wall, stress-driven flow and strong streamline curvature.

Depending on the configuration, one large or two small zones with recirculating flow can be distinguished in the flow due to the shape of the velocity field created by the inlet air jet. This directly influences ventilation processes at the end of the blind drift. In the case when one large vortex exists, the whole cavity is well ventilated and fresh air is delivery to the end wall. When the flow forms two separated recirculation zones or two partially mixed zones as in case (3) and case (4) the ventilation conditions become much worse and only a small or very small amount of primary fresh air is delivery to the end of cavity.

Stereoscopic PIV offers a good insight into ventilation processes in the blind drift aired with jet fans. The system is able to capture all key features of the real flow with accuracy sufficient for the needs of mine ventilation practices.

Acknowledgment

Present work was supported by the Grant AGH No. 11.11.210.312.

References

1. M. Jaszczur, R. Nowak, J. Szmyd, M. Branny, M. Karch, W. Wodziak, *JPCS* **318,** 052043 (2011)
2. S.M. Aminossadati, K. Hooman, *12th U.S/North AMVS* (2009)
3. B. Kuan, W. Yang, M.P. Schwarz, *Chemical Engineering Science* **62**, 2068–2088 (2007)
4. R. Mossad, W. Yang, M.P. Schwarz, *7th International Conference on CFD in the Minerals and Process Industries* (CSIRO, Melbourne, Australia 2009)
5. H. Nakayama, M. Hirota, K. Shinoda, S. Koide, *Thermal Science and Engineering* **13**, 17-23 (2005)
6. S.A. Silvester, PhD thesis, The University of Nottingham (2002)
7. A.M. Wala, S. Vytla, C.D. Taylor, G. Huang. *SME ME* **59** (10), 49 (2007)
8. A.M. Wala, J.R. Stoltz, J.D. Jacob, *Proceedings of the 7th IMVC*, 411-418 (2001)
9. M. Jaszczur, L. Portela, *ERCOFTAC SER.* **12**, 343-354 (2008)
10. J. Szmyd, M. Branny, M. Karch, W. Wodziak, M. Jaszczur, R. Nowak, *Arch. Min. Sci.* **58** (2), 333-348 (2013)
11. M. Branny, W. Filipek, *Arch. Min. Sci.* **53** (2), 221-234 (2008)
12. J. Krawczyk, *Arch. Min. Sci.* Monograph **2** (2007)
13. R.D. Keane, R.J. Adrian, *Appl. Sci. Res.* **49** (3), 191-215 (1992)
14. R.J. Adrian, *Exp. Fluids* **39** (2), 159-169 (2005)
15. M. Raffel et al., *Particle Image velocimetry: A Practical Guide* (Springer-Verlag, 2007)
16. J. Westerweel, *Digital Particle Image Velocimetry, Theory and applications* (Delft University Press, 1993)
17. *Flow Master Manual for Davis 8.0* (LaVision Gmbh, 2007)

EPJ Web of Conferences **114**, 02149 (2016)
DOI: 10.1051/epjconf/201611402149

Research of vibration asymmetry of fluidic oscillator with vortex chambers

Kazimierz Peszynski[1,a], Emil Smyk[1], Sylwester Wawrzyniak[1], and Daniel Perczynski[1]

[1]*UTP University of Science and Technology in Bydgoszcz, Faculty of Mechanical Engineering, Control Group, Poland*

Abstract. The paper presents an analysis of the test results of a new type of fluidic oscillator. The oscillator design is based on the internal feedback in the form of two vortex chambers. The results were transformed into a dimensionless form. Then, the effect of deliberately introduced asymmetry into one of two input channels of the feedback chambers on the device oscillation asymmetry was examined.

1 Introduction

In the fluidic oscillators, periodic oscillation of fluid passing through a device can be induced without moving elements. The effect is obtained by suitable shaping of channels and chambers where fluid (mostly air) goes through. The fluidic oscillator is characterised by the input feedback.

The typical oscillator (Fig. 1) consists of the following elements: the supply channel, nozzle, separator chamber with specified inclination of walls, splitter etc. The feedback circuit is usually a channel which connects one of the output channels (in case of oscillators with two outputs) or working chamber (in case of oscillators with one output) with the interaction region.

Oscillators with the feedback have already been described in the 1980s (Belsterling, 1971, Davies, 1970, Tippetts et al., 1973) – Fig. 1. The jet from the nozzle attracts to one of two walls of the operating cavity as a result of the Coanda effect and flows in the direction of the output channel. Before leaving the cavity, the stream is divided into two parts on the splitter of suitable shaped separator. Most of the fluid flows out through the output channel, the rest of it flows into the feedback channel. After passing through the channel, the stream gets the required speed and affects the adjacent input stream and moves it towards the second wall. The process of the stream separation repeats in the second part of chamber, which results in relocation of the stream towards the other wall. The phenomenon leads to the pulsation of pressure in both output channels (Fig.1a) or one output channel (Fig. 1b). It is common for this type of oscillator to have two output channels and the stream flows out alternately from one or the other channel.

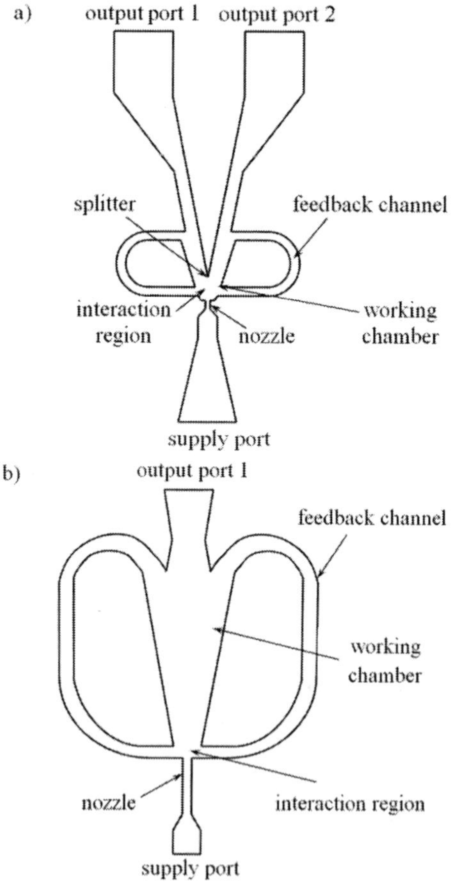

Figure 1. Characteristic parts of the classic oscillator with feedback: a) with two outputs, b) with one output

[a] Corresponding author: kazimierz.peszynski@utp.edu.pl

The following characteristic numbers play the key role in the design, analysis and mathematical modelling of the fluidic elements: Strouhal number (*Sr*), Reynolds number (*Re*), Euler number (*Eu*), and the dimensionless parameter of the feedback circuit (*L'*), which can be calculated from the equations:

$$Sr = \frac{fh}{u} \quad (1)$$

$$Re = \frac{uh}{\upsilon} \quad (2)$$

$$Eu = \frac{2\Delta p}{\rho u^2} \quad (3)$$

$$L' = \frac{4l'n}{\pi d'^2} \quad (4)$$

where h is the characteristic dimension of the oscillator, f – the frequency of pressure oscillation, u – the speed of fluid at the entry, v – the kinematic viscosity, Δp– the pressure drop, and ρ – the fluid density. Furthermore, $l'=l/d_n$ and $d'=d/d_n$, where l and d are the length and diameter of the feedback channel. d_n and n are the width and shape factor of the output channel of the circuit.

2 Research subject and test stand

The subject of the study was a new type of fluidic oscillator without a classic feedback circuit (Figure 2).

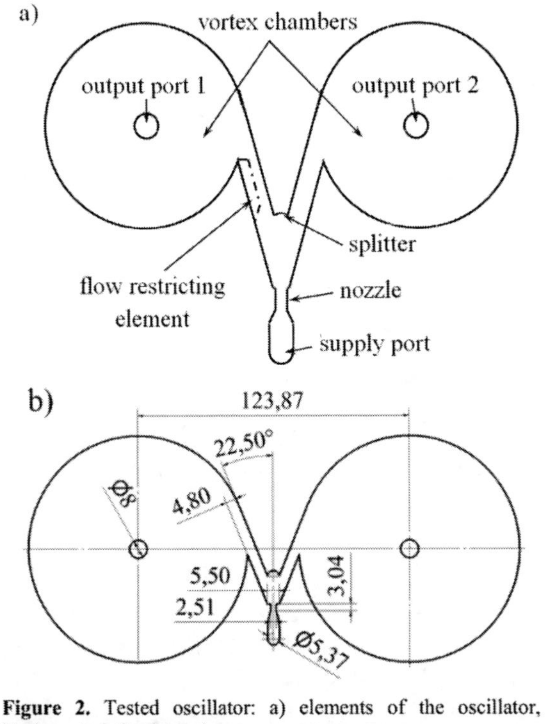

Figure 2. Tested oscillator: a) elements of the oscillator, b) characteristic dimensions

The preliminary test results of the oscillator, designed in the Institute of Thermomechanics of Academy of Science of Czech Republic by Professor Tesař, has been presented in the paper by Tesař and Smyk (2015b). The feedback channels were replaced with the vortex chambers, which are located at the end of each output channel of interaction region. Increasing pressure in the vortex chambers plays the role of the feedback. As a result of Coanda effect, the stream from the oscillator attracts to one of the walls and goes to the vortex chamber. It starts to move in rotational motion, which causes the accumulation of fluid mass and increase of pressure. As a result of increased resistance, the stream is moved to the second vortex chamber with a lower pressure. During the filling of the second chamber, some of accumulated fluid in the first chamber leaves it through the output port located in the middle. The phenomenon repeats in the second chamber, resulting in an oscillation in both output ports of the oscillator.

The shape of tested oscillator is presented in Figure 2b. The oscillator chamber was made of PMMA plate, 7.8 mm thick. The fluid stream flows into the device through the supply port with a connection tube, 5.37 mm diameter. At the end of the input channel, there is a nozzle of rectangular cross-section, 2.51 mm width. After flowing through the nozzle, the gas attracts to one of the working chamber walls and flows through the channel, 4.80 mm width, and gets into one the vortex chambers.

The air was delivered from a tank with a capacity of $28 \cdot 10^{-3}$ m^3 DURR-DENTAL reciprocating compressor. The tank suppressed pressure pulsation of the supply air. The measurement and setting of the gas flow rate was made by VEB MLW PRUFGERATER-WERK float (glass) rotameter with an operating distance of 0.2 ÷ 2 and measurement accuracy of 0.010 m^3·h^{-1}. The speed of gas at the oscillator outlet was measured by HWA (Hot-Wire Anemometer) thermo-anemometer with the use of Constant Temperature Anemometer (CTA) single-sensor wire probe, No. 55P16, DANTEC company. The tests were taken for the following parameters of the gas: absolute number $T = 295$ K, ambient pressure $p = 102500$ Pa, and relative humidity $\varphi = 28\%$.

3 Research results

Before the main tests, there were carried out preliminary tests of the oscillator without elements which limit the flow of one of the channels behind the separation chamber. The tests were used to verify actions of the oscillator and to test the methods of results development.

The next step was to place in the left channel behind the separation funnel, the elements which limit A_k cross-section.

3.1 Preliminary studies

Figure 3 shows the process of outflow gas speed at the oscillator output port for selected rate of the supply stream velocity. The measured flow is unsettled. However, the tendency to form the output sinusoidal stream can be noted. The Fast Fourier Transform (FFT) was used in order to determine the approximating frequency of the oscillator. 4096 samples were treated for

8 flow rates in the range of $(166.7 \div 555.6) \cdot 10^{-6}\,\mathrm{m^3/s}$, graduated at $66.7 \cdot 10^{-6}\,\mathrm{m^3/s}$. The selected time and sampling frequency allowed determining the frequency of approximating signal with accuracy of $f_{\min} = 0.5\,\mathrm{Hz}$. For the case presented in Figure 3, the period was obtained as $T = 0.182\,\mathrm{s}$ and the amplitude as $A = 2.63\,\mathrm{m/s}$.

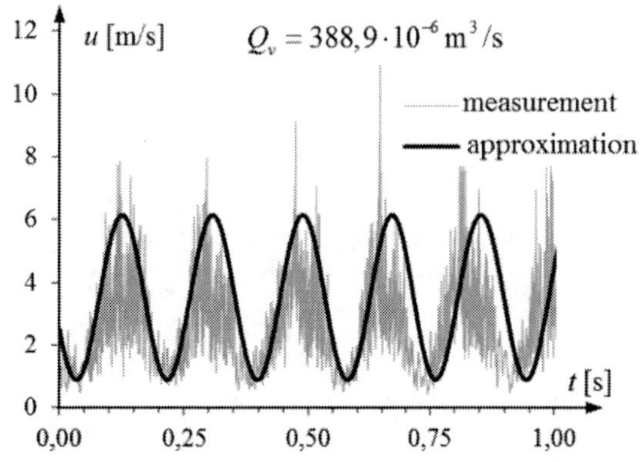

Figure 3. The course of speed at the oscillator outlet with sinusoidal process

For a given range of the flow rates of the supply stream, characteristic was drawn and presented in Figure 4.

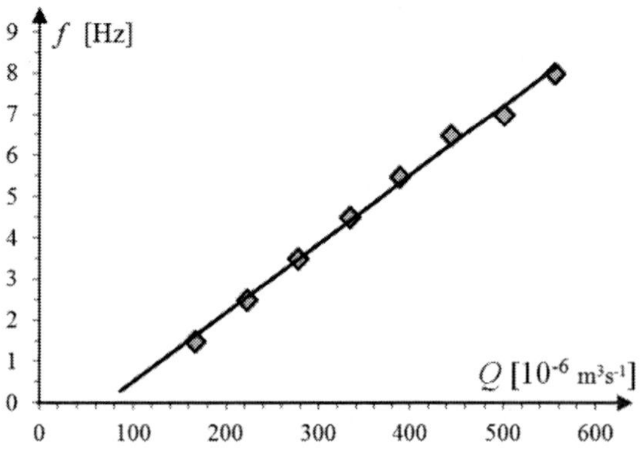

Figure 4. Approximated and real frequencies versus flow rate function

The obtained dependence is assumed to be linear, but it does not go through the beginning of the coordinate system, so:

$$f = aQ_V + b = a\,A_k u_{sr} + b \tag{5}$$

After linear approximation, the slope of the straight line was determined as $a = 16714$ and the intercept of a line $b = -1.1607$.

In order to generalise the results of research, the results were presented in the form of dimensionless numbers $Sr = f(Re)$. To calculate Strouhal number (*Sr*) and Reynolds number (*Re*), the characteristic dimension of the channel width was used with the initial value of $h = 4.8\,\mathrm{mm}$, as well as average speed of the stream at the channel input of the oscillator interaction region, which was determined from the equation $Q_V = A_k u_{sr}$.

The kinematic viscosity was calculated from the equation

$$\nu = \frac{\mu}{\rho} \tag{6}$$

where the dynamic viscosity μ was given from the equation

$$\mu = \frac{1.458 \cdot 10^{-6} \cdot T_L^{0.5}}{1 + \dfrac{110.4}{T_P}} \tag{7}$$

It was assumed that the temperature of the working medium T_P is approximately equal to the temperature in the laboratory.

The specific mass ρ was determined from the equation

$$\rho = \frac{p_{bar}}{r \cdot T_P} \tag{8}$$

where the individual gas constant for air was

$$r = \frac{0.426 \cdot \psi + 287}{1 + 0.001 \cdot \psi} \tag{9}$$

the specific humidity

$$\psi = \frac{622 \cdot E}{p_{bar} - 0.378 \cdot E} \tag{10}$$

the saturated vapour pressure was calculated from Mangus equation

$$E = 610 \cdot 10^{\frac{7.5 \cdot (T_L - 273.15)}{T_L - 38.15}} \tag{11}$$

When multiplied the equation (5) by the factor h / u_{sr}, we obtain:

$$\frac{f\,h}{u_{sr}} = \frac{(a\,A_k u_{sr} + b)\,h}{u_{sr}} = a\,A_k h + bh\frac{1}{u_{sr}} \tag{12}$$

Based on the equation (2), it is possible to determine the average air speed u_{sr} as $u_{sr} = \frac{\nu Re}{h}$. After substituting the equation (12) on the left side with the equation (1) and $u_{sr} = \frac{\nu Re}{h}$ on the right side, we obtain

$$Sr = a\,A_k h + \frac{b\,h^2}{\nu\,Re} \tag{13}$$

Substituting the given numerical values of $A_k = 2{,}827 \cdot 10^{-5}\,\mathrm{m^2}$ and $\nu = 1{,}518 \cdot 10^{-5}\,\mathrm{m^2 s^{-1}}$ into equation (13), we obtain

$$Sr = 2{,}2684 \cdot 10^{-3} - 1{,}76148\frac{1}{Re} \tag{14}$$

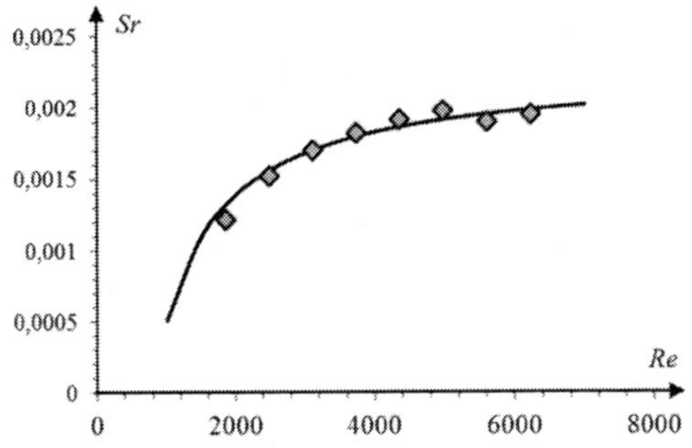

Figure 5. Characteristics of the oscillator with the vortex chamber – dependence between Strouhal number and Reynolds number

The range of possible to achieve frequencies is limited in practical solution. The increase of the flow rate is impossible, because hydraulic resistance increases rapidly, which causes significant pressure drop. The phenomenon is unfavourable in series connections of the stream elements.

3.2 Asymmetric oscillations

The research of the oscillator asymmetry was carried out because of three reasons:

- determining the influence of the asymmetry on the output frequency of the oscillator,
- determining the possible impact of the errors on the actions of the oscillator,
- determining the dependence between the cross-section asymmetry of the feedback input channels and the output frequency.

The oscillator asymmetry was obtained by reducing the cross-section of one of the input channels to the vortex chambers (Figure 2a by permanently placing an element which reduces the cross-section in the channel. The minimum geometric dimensions of the channel in the plane of reduced cross-section, for different cases, are shown in Table 1.

Table 1. Dimension of the channel for different settings of experiment

Percentage of cross-section	Channel dimensions
100% of cross-section	4.80 x 7.80 mm
80% of cross-section	3.84 x 7.80 mm
60% of cross-section	2.88 x 7.80 mm
40% of cross-section	1.92 x 7.80 mm
20% of cross-section	0.96 x 7.80 mm
0% of cross-section	Channel fully blocked

Figure 6 presents the characteristic of the oscillator in the output of the vortex chamber, supplied by the channel with unchanged cross-section. The cross-section of the other channel was changed in accordance to the Table 1.

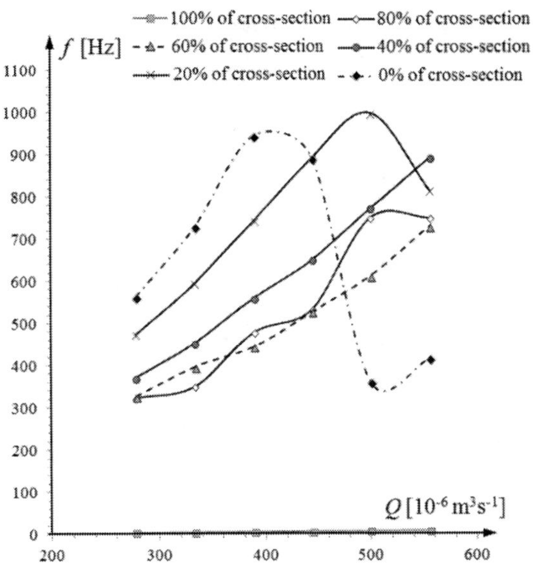

Figure 6. Oscillator asymmetry for the output channel with unchanged input cross-section

On the output of the other chamber (Figure 2a), the oscillation frequency significantly increases with the implementation of the asymmetry. The oscillation frequency dependence from the power flow rate $f_2 = f(Q_V, x)$ is approximately linear, at least for several ranges of the volumetric flow rate. This allows the possibility of precise control of the output signal. The oscillation frequencies f of the output signals for two channels (one with the reduced cross-section and the other one with unchanged cross-section) have different values, but they are different in both cases, which is related directly to the difference of the flow rate in each channel. It should be noted that reducing the cross-section area of the channel is equivalent to its throttling, which means reduction of the flow rate.

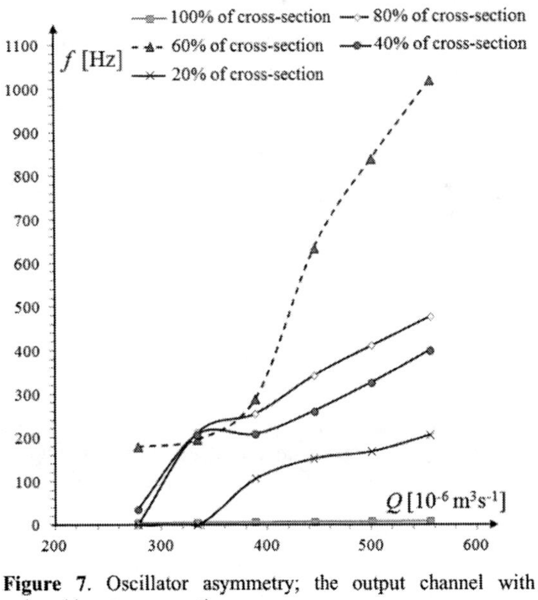

Figure 7. Oscillator asymmetry; the output channel with reduced input cross-sections

3.3 Frequency spectrum analysis

The frequency spectrum analysis, obtained by the fast Fourier transform (FFT), allows analysing the quality of received signals. Figure 8 presents the spectrum obtained for the output signal in case of the oscillator with unchanged input cross-section (Figure 8a), as well as the spectrum obtained on the output of the channel with an element which limits the cross-section of the other channel of 40% of initial cross-section (Figure 8b).

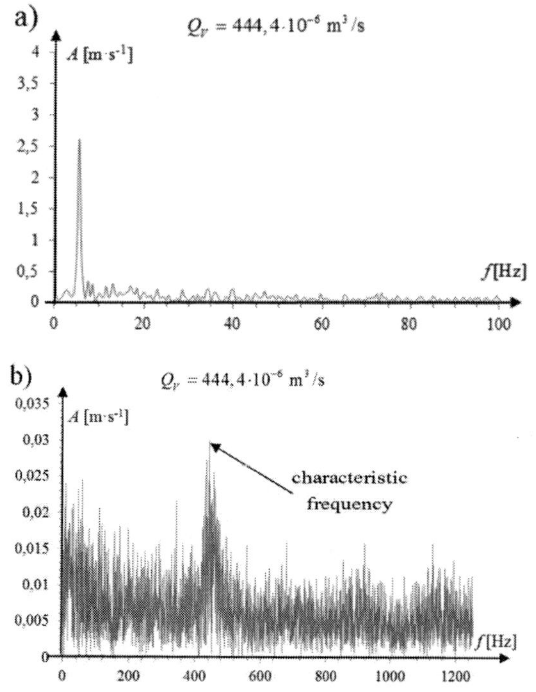

Figure 8. Analysis of the frequency spectrum, input value, oscillator, a) for the channel without damping mass, b) for the channel without intalled mass in the cross-section of the second channel of 40% of initial cross-section

The frequency spectrum in the first case shows the oscillation frequency equal $f = 5,5\,\text{Hz}$, without significant noise for other frequencies. In the case of installing mass, which limits the cross-section, in the oscillator, the frequency analysis shows a considerable share of the noise in the output signal. This phenomenon is unwanted. Moreover, the characteristic frequency amplitude is slight (approx. $A = 0,04\,\text{m/s}$). In these cases, the signal should be filtered and strengthened, which includes other operations.

4 Conclusions

The results confirm the possibility of changing the characteristics of the proposed oscillator by changing the cross-sectional area of one of the channels. The operation helped to increase the oscillation frequency of the oscillator output signal, without increasing the volumetric flow rate. Despite the increase in the noise level, the characteristic oscillation frequency can be found for selected parameters which limit the cross-section of one of two input channels. An important result of the study is the oscillation frequency of the output significantly increased.

Acknowledgements

Second author obtained support from Ministry of Science and Higher Educating of Poland in form of "Diamond Grant" 0201/DIA/2015/44 for young researchers.

There was also institutional support BS 16/2013 for all authors.

References

1. C.A. Belsterling, *Fluidic Systems Design*. John Wiley & Sons, (1971)
2. R.C. Davies, *Functional characteristics of fluid elements*, Fluidics Quarterly 2.2, 1-43. (1970)
3. E.W. Simôes, R. Furlan, R.E.B. Leminski, M.R. Gongora-Rubio, M.T. Pereira, N.I. Morimoto, J.J.S. Avilés, *Microfluidic oscillator for gas flow control and measurement.* Flow Measurement and Instrumentation, 16(1), 7-12, (2005)
4. V. Tesař, C.H. Hung, W. B. Zimmerman, *No-moving-part hybrid-synthetic jet actuator.* Sensors and Actuators A: Physical, *125*(2), 159-169 (2006)
5. V. Tesař, *Configurations of fluidic actuators for generating hybrid-synthetic jets.* Sensors and Actuators A: Physical, 138(2), 394-403 (2007)
6. V. Tesař, E. Smyk, *Fluidic low-frequency oscillator with vortex spin-up time delay.* Chemical Engineering and Processing: Process Intensification, *90*, 6-15. (2015)
7. V. Tesař, E. Smyk, *Infrasonic fluidic oscillator for use in anti-terrorist warfare.* Safety and Security Engineering VI, *151*, 179, (2015)
8. J.R. Tippetts, H.K. Ng, and J.K. Royle, *An oscillating bistable fluid amplifier for use as a flowmeter,* Fluidics Quarterly 5.1, 28-42, (1973)
9. J.T. Yang, C.K. Chen, K.J. W.Z. Tsai Lin,. H.J. Sheen, *A novel fluidic oscillator incorporating step-shaped attachment walls.* Sensors and Actuators A: Physical, *135*(2), 476-483, (2007)

EPJ Web of Conferences **114**, 02150 (2016)
DOI: 10.1051/epjconf/201611402150

A new technique for investigating the induced and profile drag coefficients of a smooth wing and a tubercled wing

Michael Bolzon[1], Richard Kelso[1] and Maziar Arjomandi[1]

[1] *The School of Mechanical Engineering, The University of Adelaide, South Australia, Australia, 5005.*

Abstract. The induced and profile drag coefficients of a wing are typically determined through a complex experimental technique, such as wake surveying. Such a technique requires measurement of all three orthogonal components of the downstream velocity to find the components of drag, which results in the necessary usage of a sophisticated and costly measurement device, such as multi-hole pressure probe. However, in this paper data is presented which demonstrate that the relative changes in the induced and profile drag coefficients can largely be determined through the sole measurement of the downstream, streamwise velocity. To demonstrate this, the induced and profile drags of two NACA 0021 wings, one with a smooth leading edge and the other wing a tubercled leading edge for comparison, are determined through the measurement of the three orthogonal velocities. The downstream, streamwise velocity distribution of each wing is then constructed and relationships can be determined. The wings were surveyed at 3°, 9°, and 12°. It has been found that the relative magnitude of the profile drag coefficient can be found for all considered angles of attack, while the relative magnitude of the induced drag coefficient can be found at 9° and 12°. These findings produce an innovative, simpler, and more cost effective experimental technique in determining the components of drag of a wing, and reduces the burdensome requirement of a sophisticated measurement device for such an experiment. Further investigation is required to determine the induced drag at 3°.

1 Introduction

The drag of a wing is an important parameter as it largely determines the wing's efficiency. While it is a relatively simple task to determine a wing's drag, such as through the use of load measurements, designing a wing to produce less drag is far more complex. The drag can be divided into two components; the induced and profile drags. In order to reduce the drag of a wing one of these two components can be targeted, however, a positive effect on one component can often lead to a negative one for the other, for example tripping a boundary layer may result in a reduced profile drag, but this also results in an increased induced drag. For some flow control devices the total drag coefficient may increase. Therefore, in order to reduce the overall drag of a wing both the induced and profile drags must be considered during the design phase, and consequently these two components of drag must be measured. The profile drag is greatly affected by flow separation, as a greater flow separation results in a greater pressure drag, which is a constituent of profile drag. Therefore, determining to what extent the flow over a wing has separated will provide valuable insight into the relative magnitude of the profile drag. Furthermore, for swept wings the flow tends to separate first at the wingtip, which reduces the circulation in this region, and in turn reduces the induced drag. Therefore, determining the location and extent of flow separation will also indicate the relative induced drag.

A typical method used to investigate the induced and profile drags of a wing is to perform a wake survey, whereby the wing is placed in a wind tunnel and a plane downstream of the wing is designated as the "wake plane", as detailed in figure 1. This wake plane is then divided into a grid of nodes and the three orthogonal velocities are measured at each node. From these three orthogonal velocities the induced and profile drags can be computed from Eqs. 1 and 2 [1]. To measure the necessary three orthogonal velocities requires a sophisticated tool, such as a multi-hole pressure probe, which is typically expensive to purchase or labour intensive to manufacture and calibrate. The ability to measure induced and profile drags then becomes limited to a select few, and hence impacts the progress of wing design. Naturally, a question arises as to whether the induced and profile drags can be investigated without knowing three orthogonal velocities. By reducing the number of parameters that must be measured a simpler methodology that can be utilised by a greater number of researchers, may result, thereby accelerate the wing design process.

[a] Corresponding author: michael.bolzon@adelaide.edu.au

The drag created by a wing will manifest itself as a reduction in the downstream, streamwise velocity component, as such by studying the streamwise velocity distribution in the wake plane the drag distribution over a wing can be ascertained. Over a conventional wing the vast majority of the induced drag originates at the wingtip, whereas the profile drag generally occurs over the entire span. Therefore, by considering the downstream, streamwise velocity in certain regions it may be possible to qualitatively deduce the induced and profile drags of a given wing. While the streamwise velocity profile will only serve as a qualitative indication of the induced and profile drags, this method is well suited to preliminary investigations into wings with new flow control devices or planform shapes, as the effects on the induced and profile drags will be evident when comparing two wings. If the induced and profile drags can be investigated from the downstream, streamwise velocity profile then a tool as common and inexpensive as a Pitot-static probe will suffice. The aim of this investigation is to determine whether any patterns exist between the downstream, streamwise velocity distribution and the induced or profile drag coefficients, or the flow separation over a wing. To do so, a comparison of two wings, one with a smooth leading edge and the other with a tubercled leading edge, is performed as it is known that the flow physics over a tubercled wing is different to that over the smooth wing [2].

2 Method

In order to determine whether the induced and profile drag coefficients can be investigated through the downstream, streamwise velocity distribution, a wake survey has been conducted on two wings. The wings are shown in figure 2 and are identical except that one has a smooth leading edge while the other has a new flow control device known as tubercles. The wings have a NACA 0021 profile, a quarter chord sweep angle of 35°, a taper ratio of 0.4, a root chord of 175mm, and the tubercles have a constant amplitude of 10.5mm and a constant wavelength of 75mm (A10.5λ75) [2]. The wings were CNC machined from aluminium and the tubercles were implemented to preserve a constant thickness-to-chord ratio along the entire wingspan. This resulted in chordwise "bumps" and "valleys" forming over the peaks and troughs, respectively, along the chord. These features washed out towards the trailing edge. The wake survey was conducted with a Turbulent Flow Instruments brand multi-hole pressure probe known as a Cobra probe, from which three orthogonal velocities can be found and the induced and profile drags can be computed from Eqs. 1 and 2. The downstream, streamwise velocity distribution can also be found from the same measurements. The wake plane was located 390mm downstream of the wing's root trailing edge, as shown in figure 1. The wake plane size differed for each angle of attack, as expected, because the wake size grew with increasing angle of attack. Therefore, to reduce experimental time the wake plane size was smaller for lower angles of attack, however, it was kept consistent between the smooth and tubercled wings for any given angle of attack. The wake plane was divided into nodes spaced 6mm x 6mm apart. The wings were placed on a JR3 load cell, which was placed on a Vertex brand rotary table, in turn. The load cell was used to determine the 0° angle of attack of the wing, while the rotary table was used to change the wing's angle of attack. The experimental setup can be seen in figure 1. The wake survey was conducted in The University of Adelaide's "KC" open-return wind tunnel with a freestream velocity of the wind tunnel was 27.5m/s, which resulted in a Mean Aerodynamic Chord, MAC, Reynolds number of 225,000. The turbulence intensity of the wind tunnel is 0.6% ~ 0.8%. The wings were placed at 3°, 9°, and 12°. At 3° both wings produce induced drag and were performing at a regular low angle

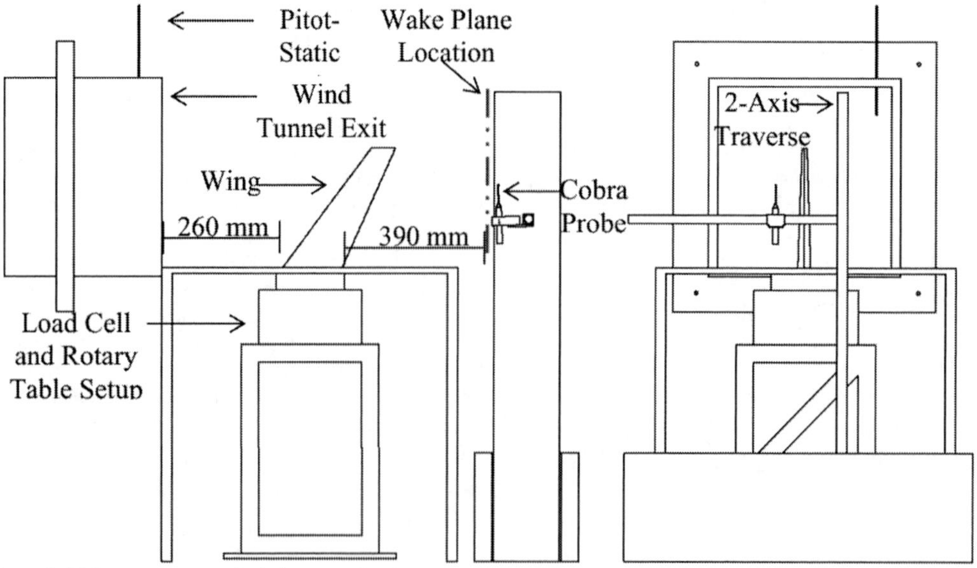

Figure 1. Wake survey setup depicting smooth wing.

of attack. At 9° the flow over both wings was still mostly attached, whereas at 12° it was found by Bolzon et al. [3] that the flow over the tubercled wing began to separate near the wingtip greatly, which resulted in an increase in the profile drag coefficient and reduction in the induced drag coefficient [4]. However, this flow separation was localised to the final trough [5]. Therefore, comparing 9° to 12° is of high interest as it marked an appreciable change in the flow patterns over the tubercled wing, therefore, comparing the induced and profile drag coefficient distributions of the smooth and tubercled wings at these angles of attack may reveal potential relationships between the induced and profile drag coefficient distributions, and the extent of flow separation, and the downstream, streamwise velocity distribution.

$$D_i = \iint_S \tfrac{1}{2}\, \rho_\infty (v^2 + w^2 - u'^2).dy.dz, \quad (1)$$

$$D_p = \iint_S P_{t\infty} - P_t + \tfrac{1}{2}\, \rho_\infty (u^* - u)(u^* - u - u_\infty).dy.dz, \quad (2)$$

where

$$u^* = (u^2 + (2/\rho)\,(P_{t\infty} - P_t))^{1/2}, \quad (3)$$

and

$$u' = u^* - u_\infty. \quad (4)$$

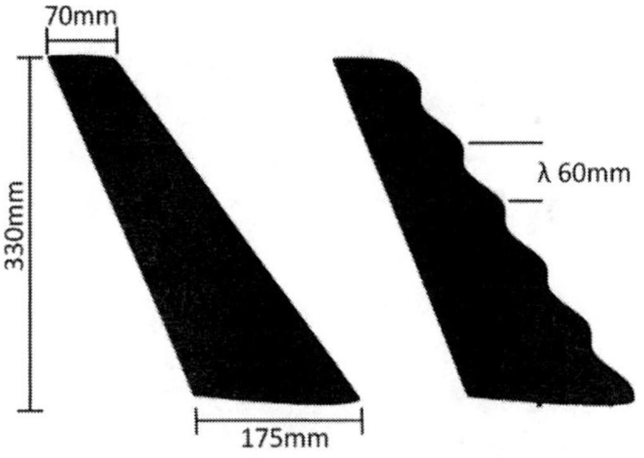

Figure 2. The smooth and tubercled wings manufactured, dimensioned.

3 Results

The induced and profile drag coefficients, and the downstream, streamwise velocity distributions of the smooth and tubercled wings at 3°, 9°, and 12° can be seen in figures 3 to 20. The velocity gradients of the downstream, streamwise velocity distributions have been included on the figures.

3.1 The Effects of the Induced Drag Coefficient on the Downstream, Streamwise Velocity Distribution

At an angle of attack of 3°, as shown in figures 3 and 4, the vast majority of the induced drag coefficient of either wing occurred near the wingtip, as expected. However, while this finding is promising in that the induced drag coefficient was concentrated in a relatively small area, and as such any changes in the induced drag coefficient should translate directly to the downstream, streamwise velocity distribution, the profile drag coefficient in the wingtip region was substantially higher, as shown in figure 5 and 6. The reason why the profile coefficient drag was high in the wingtip region was because there was significant flow separation over the wingtip of each wing, which caused an increase in pressure drag [5]. Therefore, using the superficial idea that any change in the downstream, streamwise velocity in the wingtip region is indicative only of the induced drag coefficient is false, and a better criterion must be made in order to decouple the induced drag coefficient effects from the profile drag coefficient effects.

Up to 9° the flow over the wings was relatively consistent, with the flow staying largely attached. However, increasing the tubercled wing's angle of attack from 9° to 12° resulted in the wingtip to stall, whereas the flow over the smooth wing's wingtip continued to stay largely attached (Bolzon et al., 2015, b, c). This resulted in the tubercled wing producing less lift near the wingtip, which resulted in a decreased induced drag coefficient and an increased profile drag coefficient compared to the smooth wing (Bolzon et al., 2015, b). This range of angles of attack is ideal for investigating the effects of induced and profile drag coefficients on the downstream, streamwise velocity distribution because of the change in the tubercles' effects on the induced and profile drag coefficients. Figures 13 and 19 depict the downstream, streamwise velocity distributions for the smooth wing at 9° and 12° angles of attack, respectively. The higher magnitude contours have increased in size and the velocity gradient has increased, and in addition Bolzon et al. [4] found that increasing the smooth wing's angle of attack from 9° to 12° resulted in an increase in both the induced and profile drag coefficients. Conversely, when the tubercled wing was pitched from 9° to 12° the induced drag coefficient reduced in the wingtip region, but the profile drag coefficient increased dramatically, because of flow separation [4]. From the downstream, streamwise velocity distributions of the tubercled wing at 9° and 12° presented in figures 14 and 20, respectively, it can be seen that increasing the angle of attack of the tubercled wing resulted in an increase in the lower magnitude contour sizes, but the velocity gradients of these lower velocity contours decrease dramatically, as labelled on figure 20. Therefore, it can be deduced that this expansion of the lower velocity contours coupled with a reduced velocity gradient was caused by flow separation. As this occurred near the wingtip, a reduction in the lift and hence induced drag coefficient resulted.

3.2 The Effects of the Profile Drag Coefficient on the Downstream, Streamwise Velocity Distribution

Decoupling the effects of the profile drag coefficient on the downstream, streamwise velocity is relatively simple compared to decoupling the induced drag coefficient

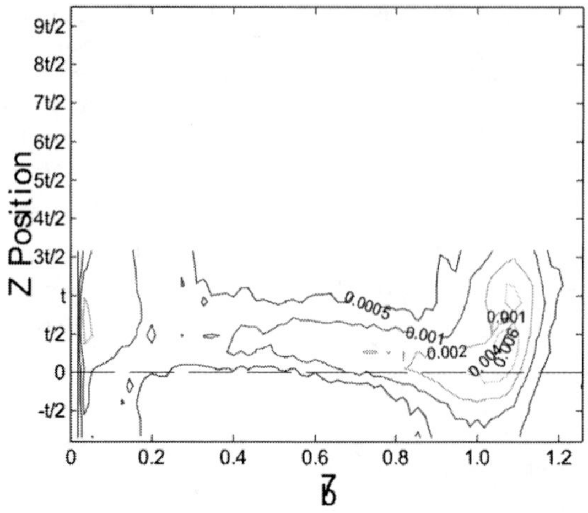

Figure 3. Induced drag coefficient distribution over the smooth wing at 3°.

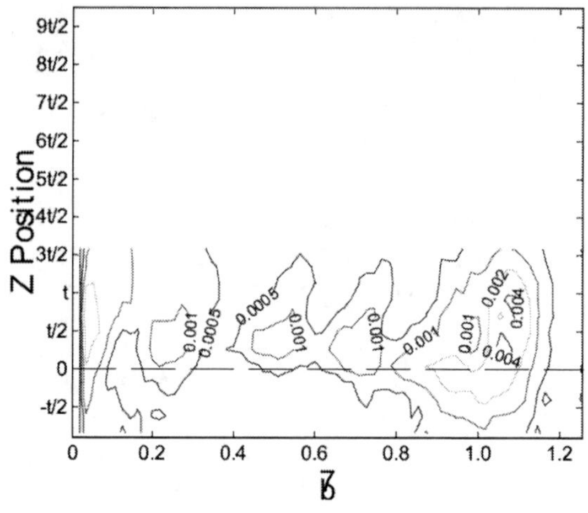

Figure 4. Induced drag coefficient distribution over the tubercled wing at 3°.

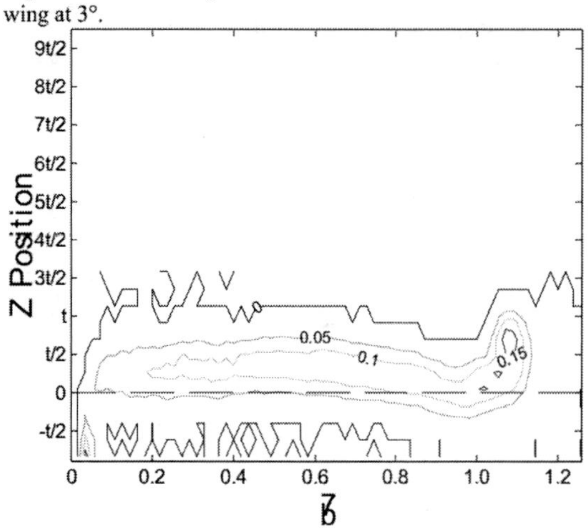

Figure 5. Profile drag coefficient distribution over the smooth wing at 3°.

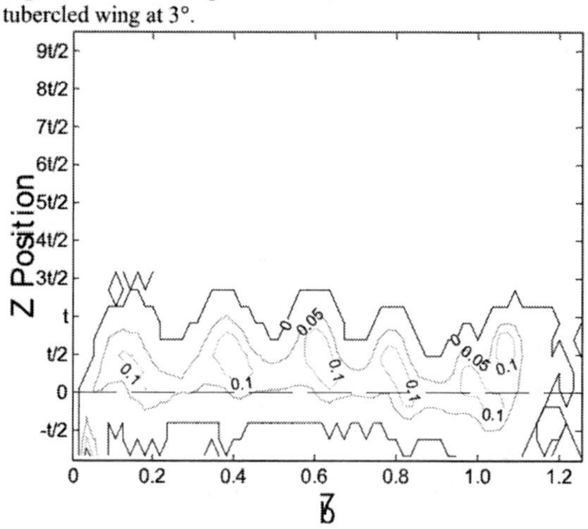

Figure 6. Profile drag coefficient distribution over the tubercled wing at 3°.

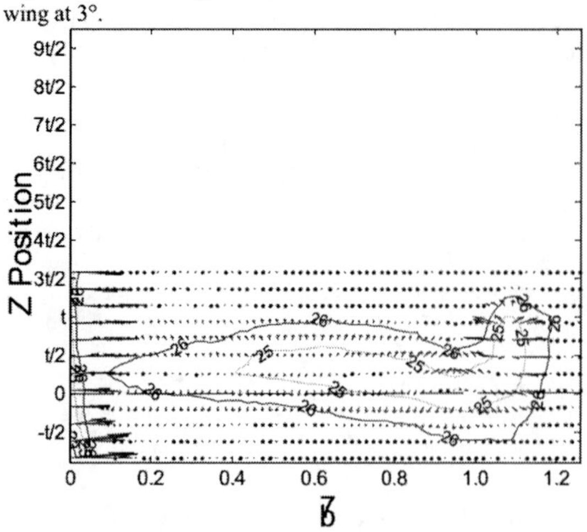

Figure 7. Downstream, streamwise velocity distribution over the smooth wing at 3°.

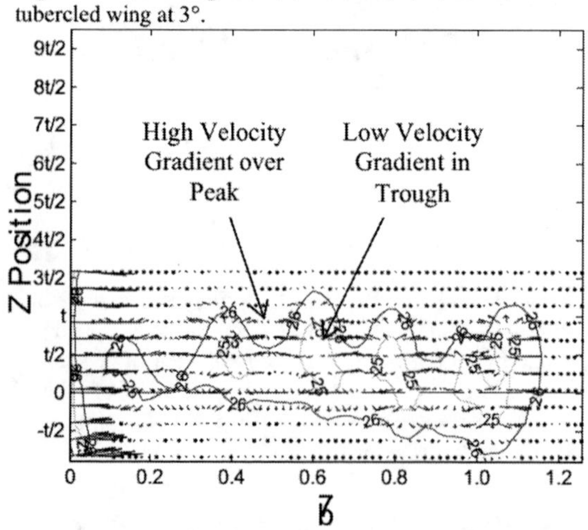

Figure 8. Downstream, streamwise velocity distribution over the tubercled wing at 3°.

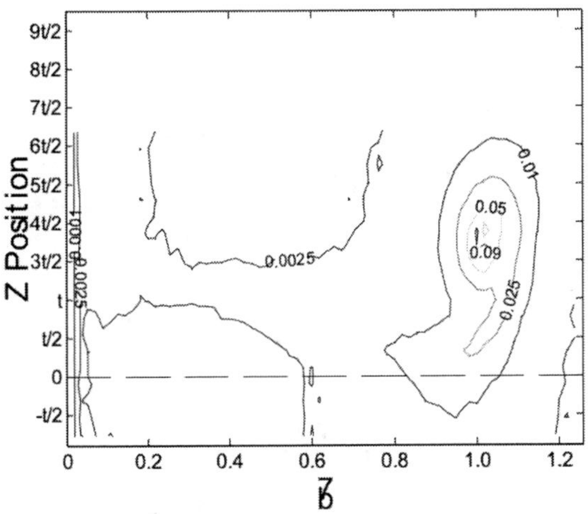

Figure 9. Induced drag coefficient distribution over the smooth wing at 9°.

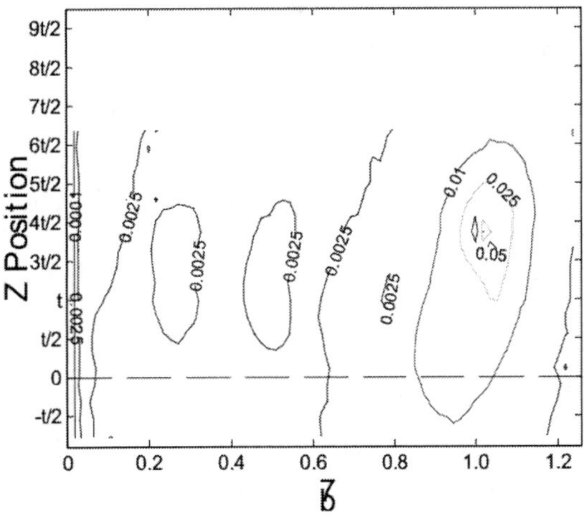

Figure 10. Induced drag coefficient distribution over the tubercled wing at 9°.

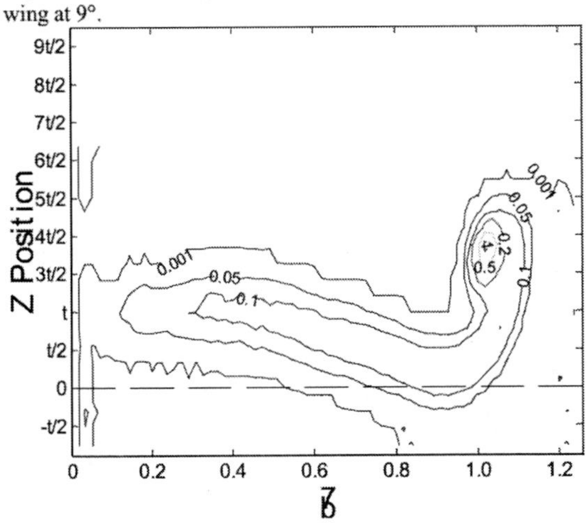

Figure 11. Profile drag coefficient distribution over the smooth wing at 9°.

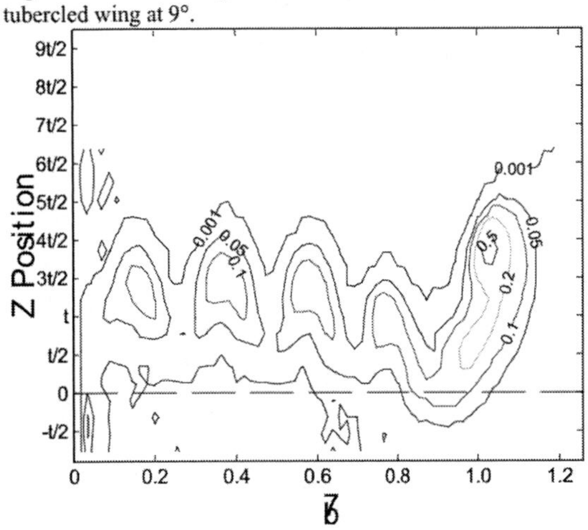

Figure 12. Profile drag coefficient distribution over the tubercled wing at 9°.

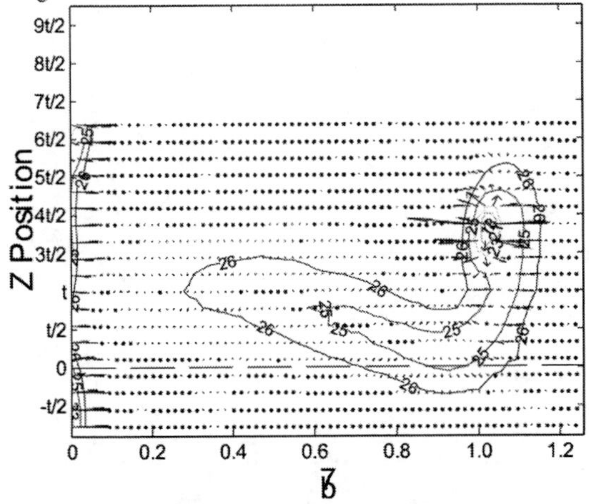

Figure 13. Downstream, streamwise velocity distribution over the smooth wing at 9°.

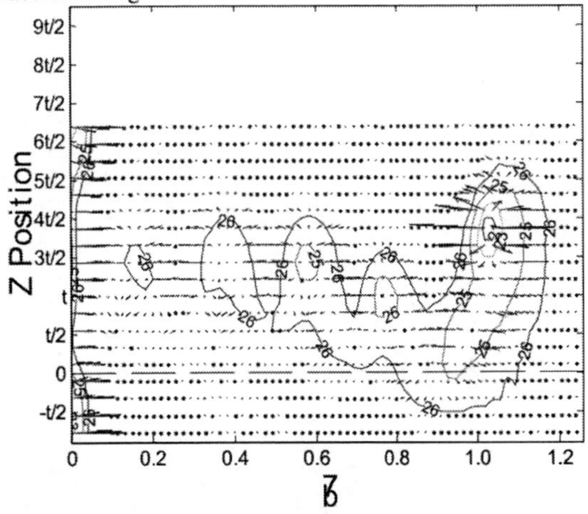

Figure 14. Downstream, streamwise velocity distribution over the tubercled wing at 9°.

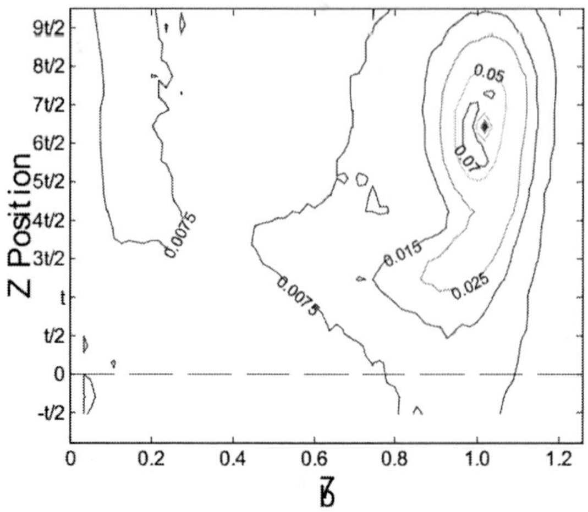

Figure 15. Induced drag coefficient distribution over the smooth wing at 12°.

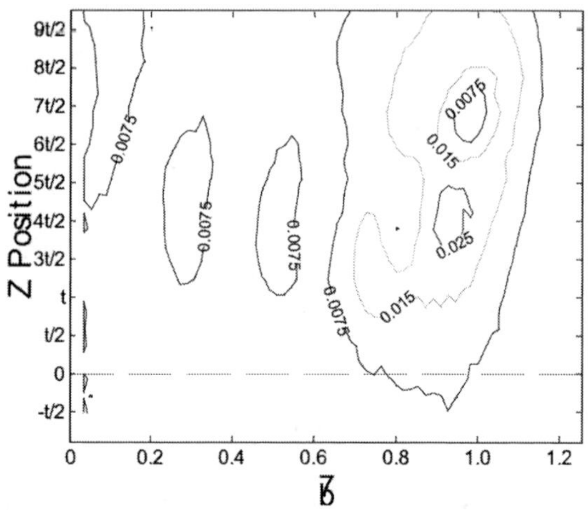

Figure 16. Induced drag coefficient distribution over the tubercled wing at 12°.

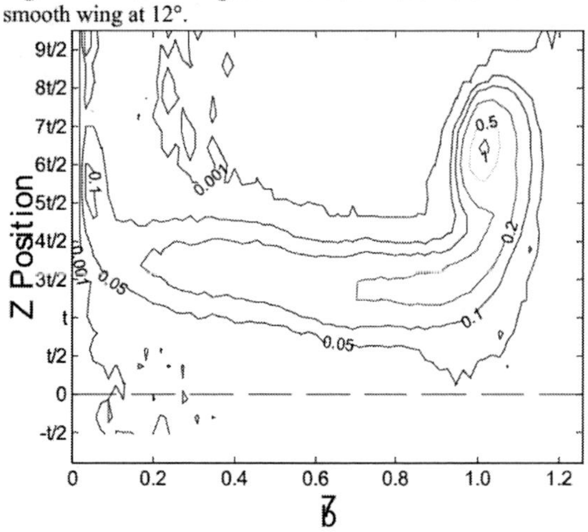

Figure 17. Profile drag coefficient distribution over the smooth wing at 12°.

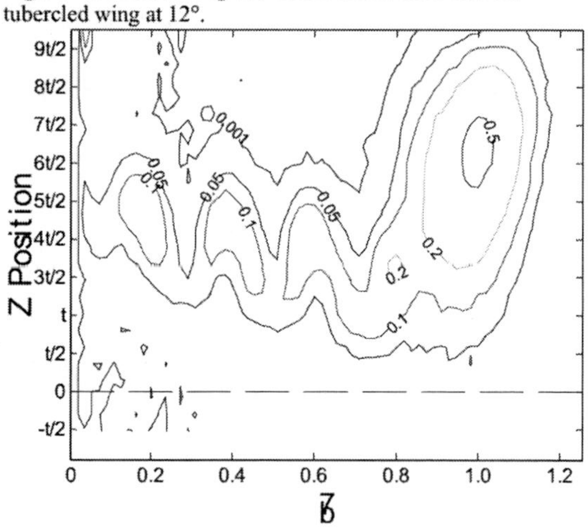

Figure 18. Profile drag coefficient distribution over the tubercled wing at 12°.

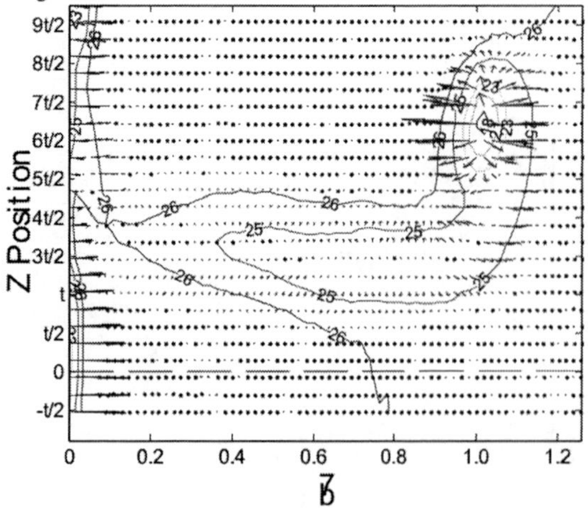

Figure 19. Downstream, streamwise velocity distribution over the smooth wing at 12°.

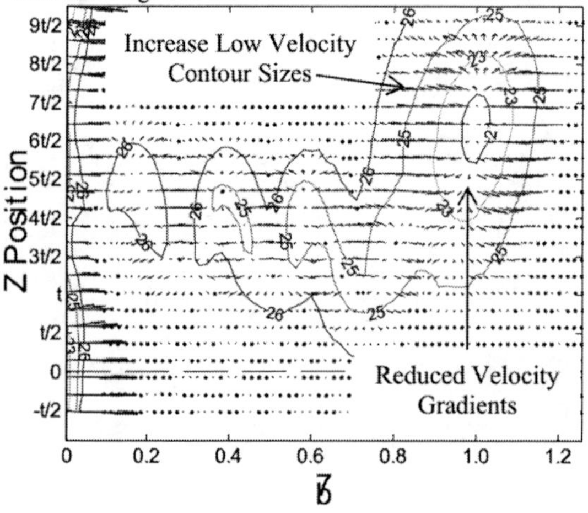

Figure 20. Downstream, streamwise velocity distribution over the tubercled wing at 12°.

effects. It was already pointed out that the induced drag coefficient was largely localised to the wingtip region, therefore, the downstream, streamwise velocity over the span of a wing will largely reflect the profile drag coefficient. Evidence for this claim can be seen from the tubercled wing at any of the angles of attack presented. For example, at 3° the tubercled wing modulated the profile drag coefficient into local maxima and minima, as shown in figure 6, and Bolzon et al. [4] found that these local maxima and minima correspond to the tubercle troughs and peaks, respectively. It was hypothesized that the local maxima were produced in the troughs as the flow separates more readily [4]. The downstream, streamwise velocity shown in figure 8 exhibits the same modulation. It should be noted that a reduction in the downstream, streamwise velocity will indicate an increase in the profile drag coefficient.

As has been explained previously, as the induced and profile drag coefficients are relatively large in the wingtip region, it is much more difficult to investigate the profile drag coefficient in this region purely from the downstream, streamwise velocity distribution. However, it was deduced that an increase in the downstream, streamwise velocity contour size in the wingtip region coupled with a reduction in the velocity gradient indicates flow separation, which subsequently resulted in an increase in the profile drag coefficient. While this relationship is useful, it is only applicable to angles of attack when the wing is stalling, hence a question arises as to whether the profile drag in the wingtip region can be deduced from the downstream, streamwise velocity at lower angles of attack.

For all of the angles of attack considered the profile drag coefficient in the wingtip region of either wing is significantly greater than the induced drag coefficient, therefore, the effect of the profile drag coefficient on the downstream, streamwise velocity will be substantially higher. As such, changes in the downstream, streamwise velocity in the wingtip region at lower angles of attack will typically indicate a change in the profile drag coefficient. This means that the effect of any change in the induced drag coefficient in the wingtip region on the downstream, streamwise velocity distribution at lower angles of attack is overshadowed by the profile drag coefficient effect.

3.3 Investigating Stall from the Downstream, Streamwise Velocity Distribution

Bolzon et al. [5] found that this tubercled wing begins to severely stall near the wingtip at an angle of attack of 12°, which results in an increase in profile drag coefficient and decrease in induced drag coefficient, as explained previously. The flow separation at the wingtip is easily identified from the downstream, streamwise velocity distribution in figure 20 where, as previously discussed, the expansion of the velocity deficit and reduction in the velocity gradient indicates an increased drag and lower lift production near the wingtip, which is arguably caused by flow separation.

For the smooth wing the flow separation follows a common pattern, where flow begins to separate at the trailing edge and progresses towards the leading edge with increasing angle of attack [5]. However, the flow separation pattern over the tubercled wing is far more interesting, as not only does the flow separate over the wingtip region, but in the troughs [5]. The known flow separation in the troughs is the cause of the localised increases in the profile drag coefficient that can be seen in figure 6, 12, and 18. While it is easy to determine that these increases in the profile drag coefficient are caused by flow separation given the flow visualisation data obtained by Bolzon et al. [5], an aim of this study is to determine whether flow separation can be deduced solely from the downstream, streamwise velocity profile. In the wingtip region, a relationship between the size and gradient of the velocity deficit and the flow separation exists, and perhaps this same relationship can be extended to the flow separation in the troughs of the tubercled wing.

As highlighted in figure 8, the downstream, streamwise velocity gradient is much higher over the peaks than in the troughs. Furthermore, the lower velocity contours are much larger in the troughs than over the peaks, which can be likened to the wingtip expansion discussed previously. Therefore, similar conditions exist in the troughs as over the wingtip region when flow separation occurs, hence the criteria used to determine if the flow has separated in the wingtip region can also be used for the flow over the rest of the wing, namely increased lower velocity contour sizes and reduced velocity gradients in these contours. This suggests that the extent of flow separation over a wing can be investigated from the downstream, streamwise velocity distribution alone.

4 Conclusion

A wake survey has been conducted on two swept wings, one with a smooth leading edge and the other with a tubercled leading edge, at 3°, 9°, and 12°. The induced and profile drag coefficient distributions of each wing were subsequently determined. The downstream, streamwise velocity profiles of each wing have also been found and patterns emerging from these distributions have been compared to the induced and profile drag coefficient distributions to ascertain potential relationships.

It was found that at angles of attack nearing stall, an expansion of the velocity contours in the wingtip region coupled with a reduction in the downstream, streamwise velocity gradient indicates a reduction in the induced drag coefficient and an increase in the profile drag coefficient. It was also found that if these two criteria are fulfilled then it can be deduced that the flow has separated regardless of the location over the wing. This is a particularly important result as flow separation typically results in an increase in the local profile drag coefficient.

At angles of attack where the flow is still mostly attached the induced and profile drag coefficients were not decoupled from the downstream velocity distribution alone. However, for all angles of attack considered the profile drag coefficient is considerably greater than the induced drag coefficient along the span of the wing, therefore, changes in the downstream, streamwise velocity distribution largely represent the profile drag coefficient, and as such changes in the profile drag coefficient can be determined. As the induced drag coefficient is overshadowed by the profile drag coefficient, determining the relative induced drag value from the downstream, streamwise velocity when the flow was still attached was not possible.

This study has found several interesting trends between the induced and profile drag coefficients and the downstream, streamwise velocity distribution, which suggests that valuable information about these components of drag, and whether the flow is attached, can be determined from a more simplistic approach. Further investigation is required to attempt to decouple the effects of the induced drag coefficient from the profile in the wingtip region on the downstream, streamwise velocity distribution when the flow is still attached.

Acknowledgements

Research undertaken for this report has been assisted with a grant from the Sir Ross and Sir Keith Smith Fund (Smith Fund) (www.smithfund.org.au). The support is acknowledged and greatly appreciated.

The Smith Fund by providing funding for this project does not verify the accuracy of any findings or any representations contained in it. Any reliance on the findings in any written report or information provided to you should be based solely on your own assessment and conclusions.

The Smith Fund does not accept any responsibility or liability from any person, company or entity that may have relied on any written report or representations contained in this report if that person, company or entity suffers any loss (financial or otherwise) as a result.

References

1. G.W. Brune, J. Aircraft **Vol. 31**, 7 (1994)
2. K.L. Hansen, Ph.D. Dissertation, The University of Adelaide (2012)
3. M.D. Bolzon, R.M. Kelso, M. Arjomandi, AIAA Journal (under review)
4. M.D. Bolzon, R.M. Kelso, M. Arjomandi, AIAA Journal (under review)
5. M.D. Bolzon, R.M. Kelso, M. Arjomandi, Phys. Fluids (under review)

EPJ Web of Conferences **114**, 02151 (2016)
DOI: 10.1051/epjconf/201611402151

Mathematical modeling of control system for the experimental steam generator

Szymon Podlasek[1], Krzysztof Lalik[2], Mariusz Filipowicz[1], Krzysztof Sornek[1,a], Robert Kupski[3] and Anita Raś[3]

[1]*AGH University, Faculty of Energy and Fuels, Mickiewicza Av. 30, 30-059 Krakow, Poland*
[2]*AGH University, Faculty of Mechanical Engineering and Robotics, Mickiewicza Av. 30, 30-059 Krakow, Poland*
[3]*Dom Project Sp. z o.o., ul. Willowa 19, 32-085 Modlniczka, Poland*

Abstract. A steam generator is an essential unit of each cogeneration system using steam machines. Currently one of the cheapest ways of the steam generation can be application of old steam generators came from army surplus store. They have relatively simple construction and in case of not so exploited units – quite good general conditions, and functionality of mechanical components. By contrast, electrical components and control systems (mostly based on relay automatics) are definitely obsolete. It is not possible to use such units with cooperation of steam bus or with steam engines. In particular, there is no possibility for automatically adjustment of the pressure and the temperature of the generated steam supplying steam engines. Such adjustment is necessary in case of variation of a generator load. The paper is devoted to description of improvement of an exemplary unit together with construction of the measurement-control system based on a PLC. The aim was to enable for communication between the steam generator and controllers of the steam bus and steam engines in order to construction of a complete, fully autonomic and maintenance-free micro-cogeneration system.

1 Introduction

According to the new trends, traditional heat sources are going to be replaced by combined heat and power systems (CHP). Such units integrate the production of heat and electricity in one highly efficient technology process. Up to now, cogeneration systems were used almost only in the professional power industry, but nowadays, there are more and more popular also in distributed energy solutions. Due to installed power, distributed energy systems can be divided as follows:

- Micro scale systems – up to 50 kWe,
- Small scale systems – from 50 kWe to 1 MWe,
- Medium scale systems – from 1 MWe to 50 MWe,
- Large scale systems – from 50 MWe to 150 MWe.

Currently available microcogeneration devices use several different technologies, including internal and external combustion engines (Stirling engines, Otto engines etc.), steam expansion engines and fuel cells. Many of developed solutions are still in the research and application stage and only a small part of them will be commercially available in the future. Such units are mainly fired by natural gas or oil. On the other hand, if we take into account steam turbines and steam engines (devices suitable to cooperate with external heat sources), we can use biomass fired boilers as heat sources. The huge potential is connected with the use of configured in such way CHP units [1].

The operation of steam units is based on the Rankine Cycle (the same cycle, which is used in the professional power plants). The main part of such systems is a steam generator (e.g. a steam boiler). Heat generated during the combustion process allows to evaporate water, which is used as a working medium. Steam goes from boiler to the turbine and expands passing the blades disposed on the rotor (transfers to the rotor the part of its energy). Mechanical energy is converted to the power using a dedicated power generator. In the typical power plant, partially expanded medium goes next to the condenser what allows to utilize the heat of the condensation.

An extension of classical Rankine cycle is the Organic Rankine cycle (ORC). In ORC cycle the working medium (water) is replaced by low-boiling fluid, such as silicone oil, benzene, toluene etc. The use of organic working fluids results in the ORC cycle having several advantages over the steam cycle. Higher molecular weight than water increases the mass flow rate of the fluid for the same size of turbine. The result is a better power generation efficiency (about 10-20%). Most importantly, the boiling point of ORC fluids is much lower than that of water, so they can be applied in lower temperatures [2].

[a] Corresponding author: ksornek@agh.edu.pl

The idea of a small cogeneration system based on the use of steam turbines (or steam engines) was shown in figure. 1. The comparison of classical steam cycle and ORC cycle was shown in the T-S diagram in figure. 2.

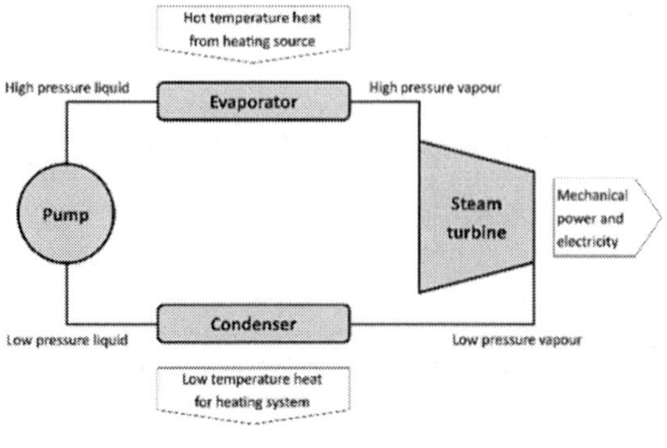

Figure 1. The idea of a small cogeneration system (source: [2]).

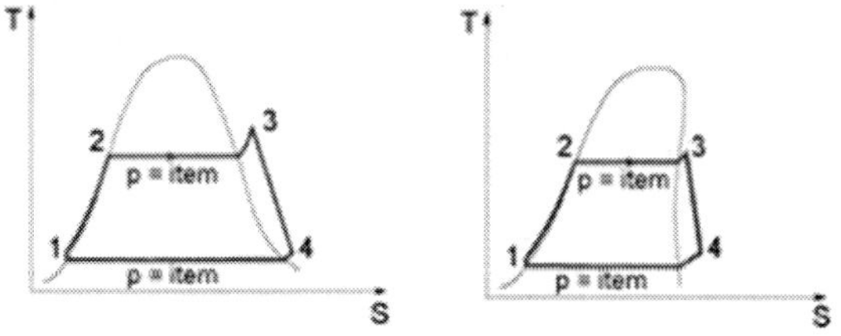

Figure 2. Comparison of classic steam and ORC cycle [3].

Biomass boilers as a heat sources for CHP systems

According to assumptions presented in the *EU Energy, Transport and GHG Emission Trends to 2050* (document published by the European Commission), the overall share of renewable energy sources in the electricity production will achieve a level of 50% in 2050 [4]. The increase in electricity production from RES will be connected with the dynamic development of distributed energy systems. Power will be generated e.g. in classical Rankine Cycle and Organic Rankine Cycle based systems powered by biomass-fired boilers. The expected share of biomass in local CHP systems and network heating up to 2050 was presented in figure 3.

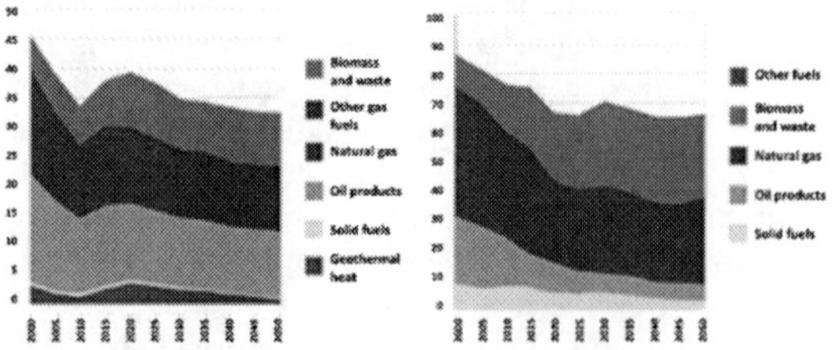

Figure 3. Predicted biomass share in local CHP systems and heating systems (own work based on [4]).

The use of cogeneration systems equipped with the biomass-fired boilers is possible both in the case of residential buildings, public buildings and farms [5]. It is really important especially in the cases, where the problems with access to the grid occur. In such facilities the oil-fired external combustion engines, which operation is really expensive are mostly used. Replacing such devices by biomass-fired CHP systems will allow reducing energy costs and obtaining full or partial energy independence.

2 Testing version of the cogeneration system

All tests of the steam generator were done using simple set-up shown in figure 4. The installation includes steam engines and steam turbine with power of range 1-10 kW. Some auxiliary elements are also indicated.

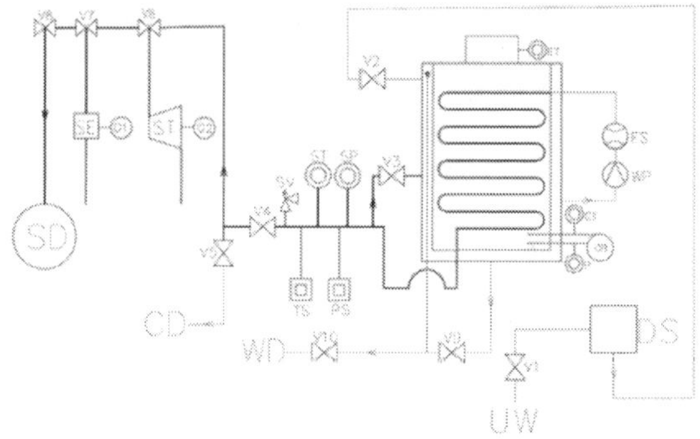

Figure 4. Diagram of simple version of the cogeneration system including steam machines, steam bus and steam generator. Description of symbols in the text.

The meaning of symbols in figure 4 is the following:

V – Ball Vavle
SV – Safety Vavle
UW – Untreated Water
DS – Demineralizing Station
ET – Exhaust Temperature
FS – Flow Sensor
WP – Water Pump
CT – Control Temperature
P – Photoresistor
WD – Water Discharge
ST – Steam Temperature
SP – Steam Pressure
TS – Temperature Switch
PS – Pressure Switch
CD – Condensate Discharge
SD – Steam Discharge
SE – Steam Engine
ST – Stem Turbine
G – Generator.

It is visible that the steam machines are mounted on the common steam bus and each engine is equipped with a cut-off valve. Also, the cut-off valves are mounted between the steam generators and the steam bus.

In the considered system certain parameters of working medium in the close loop have to be secured. Application of some number of the auxiliary elements such as dehumidifiers, separators, reduction valves etc. influence on proper operation of the system and allows to obtain steam with given parameters, allowing then for safe and efficient operation of the engines. Those factors also may reduce possibility of mechanical damages in the whole system. Very efficient control system cooperating with the other elements allows for generation steam with constant parameters of pressure and temperature. It increases security of the system operations and also increase period of the continuous unit operation.

In the installation, role of actuators perform the steam machines and the microturbine. This complete steam system will enable to obtain characteristics of the installed machines in the operational conditions.

The unit made in 1988 and modernized in 2015 was used as a steam generator. The nominal power is ca. 130 kW_t and steam output – ca. 180 kg/h. It is example of flow water-tube boiler. Heating and evaporation of the medium takes place in the coil directly heated by the fuel flame. A small volume of the water/steam inside the coil makes some difficultness regarding securing constant parameters of the generated stem. Hence, advanced control system of burner is required to eliminate this disadvantage. The fuel and air supply to the burner should be strictly controlled. In the considered unit the fuel pump and the fan are placed on the common shaft, what makes that these elements are dependent. It means that only rotational speed of the shaft can be used to control the burner power. Unfortunately, the measurements of the temperature directly in the coil are impossible. In this way some delays in the system have to be taken into account in a control algorithm.

Visualization of the steam generator is shown in figure 5.

Figure 5. The steam generator – outlook view (left) nd the cross section (right). Place of the maximal temperature T_{max} and measuring point T_{sen} are indicated.

It is visible that the steam temperature sensor is placed in some distance from point of maximal temperature. The photo of the coil is presented in figure 6.

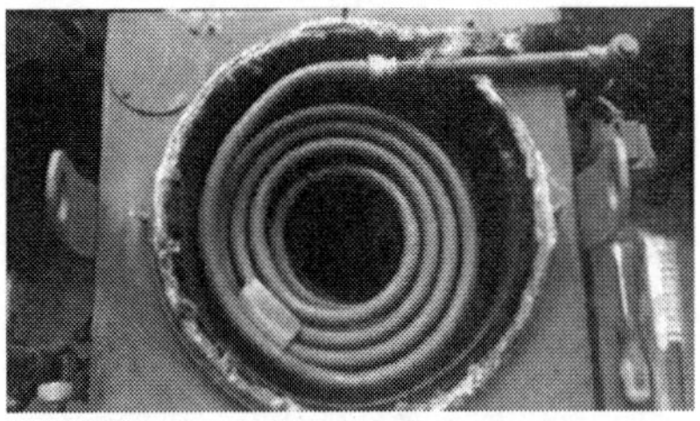

Figure 6. Upper view of the coil, the upper cover with smoke conduit is removed. The water volume of the coil is ca. 30 liters.

The biggest part of the generator is the combustion chamber made of two layers of the boiler steel separated by the insulation layer. Inside the combustion chamber the coil is placed. The water jacket prevents the generator overheating and initially heats the water. Before filling the water jacked, the municipal water has to be treated to prevent appearance of limescale, and then is directed to the water jacket. The formation of limescale has to be reduced because, due to the small diameter of the coil tube, the water flow may be strongly limited or even stopped. Also the heat transfer from the flame to the coil is decreased. The water is sucked from the water jacket by positive displacement pump and flows through the filter to the coil. During its flow, the water is heated and evaporation took place. The set of manual valves allows for initial circulation of the water inside the steam generator, and after reaching required parameters, the steam is directed to the steam bus.

In this way, the coil in such arrangement plays triple role: heater, evaporator and superheater. Due to volume extension of the working medium, the diameter of the coil tube is larger in the "steam part".

The operation of a typical old steam generator with the original automatic do not allow for stable properties of the generated steam. Strong pressure and temperature fluctuations unfortunately exist.

To solve this problem, a dedicated PLC (programmable logic controller) and software is required. Description of this controller, its configuration and tuning and some results of initial test are discussed in the further part of the paper.

3 Construction of dedicated control system

An electrical control circuit was updated during modernization of the steam generator. Most of the actuators and sensors were replaced. The outlook diagram of the measurement stand is presented in figure 7. The variation of the ventilator's rotational velocity results in variation of air amount supplied to the combustion chamber. The shaft of the ventilator is coupled by a flexible clutch with the shaft of the fuel pomp. It results also with variation of the fuel dose delivered to the combustion chamber. Therefore, the thermal power of the generator is strictly correlated with the speed of the ventilator.

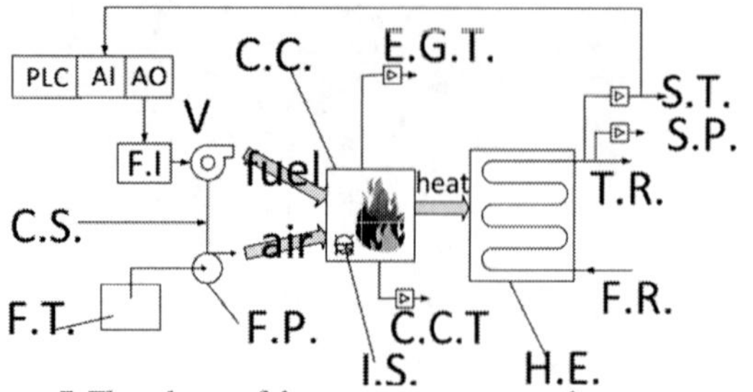

Figure 7. The scheme of the steam generator stand.

The meaning of symbols in figure 7 is the following:

C.C. - combustion chamber
C.C.T - combustion chamber temperature
C.S. - coupled shafts
E.G.T. - exhaust gas temperature
F.I. - frequency inverter
F.P. - fuel pomp
F.R. - from receiver
F.T. - fuel tank
H.E. - heat exchanger
I.S. - ignition system
S.P. - steam pressure
S.T. - steam temperature
T.R. - to receiver
V. – ventilator.

The diagram shown in figure 8 presents the structure of the control system. The heating coil temperature sensor is used as a feedback signal and is connected to the analog input module in the PLC. The value of this temperature is subtracted from the setpoint, which is entered from graphical user interface. This calculated error is subsequently transformed by the microprocessor and the appropriate control signal is obtained from its value. This signal is afterwards physically generated from the analog output module and directed to the frequency inverter. The inverter regulates the speed of the ventilator according to this calculated control value. With the variation of the ventilator speed, the amount of the energy generated in the combustion chamber and temperature of working medium vary.

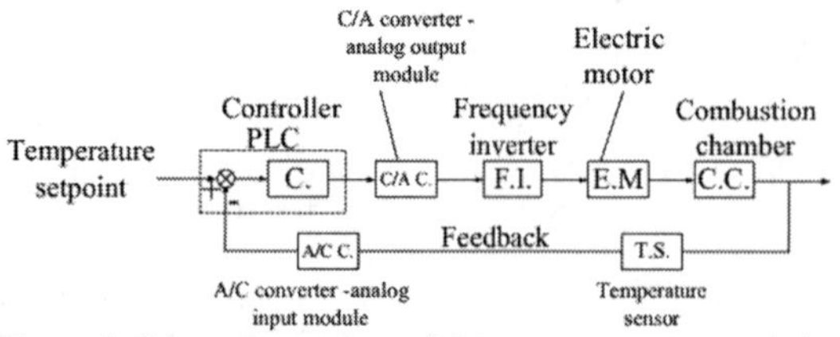

Figure 8. Schematic structure of the steam generator regulation system.

3.1 Structure of regulatory system

In systems with large time constant value, like objects such as heating systems, even the simplest type of the control system, like relay control can be applied [6,7]. However, in this case these solutions were not applied. This is mainly due to the possibility of control the speed of the asynchronous motor continuously in the entire velocity range by using frequency inverter. The second reason is the highly non-stationariness of the system. Object parameters variability occurs from previously discussed modes of operation of the steam generator and various energy retrieve from it [8]. On the one hand, setting the wide range set of the hysteresis control will occur in oscillations of the process variable around the set point, but the amplitude of those oscillations may be unacceptable by the expected parameters of the steam generator operation. On the other hand, if the range of settings of the hysteresis control loop would be to narrow, then it would be causing an excessive frequent switching of the actuator, which will reduce its lifespan. Additionally, for certain set of control operating parameters, this relay control method may result in rise of the self-excited oscillations and lead to loss of the stability of the process [9].

For these reasons, it was decided to realize the control of the system via a classical PID controller in the form given by the equation 1.

$$m(t) = K_P e(t) + K_I \int_0^t e(t)dt + K_D \frac{de(t)}{dt}, \quad (1)$$

where *m(t)* is an control signal and also the controller output, e(t) is an error signal, K_P is a proportional gain, K_I is an integral gain, a K_D is a derivative gain.

3.2 Numerical model

In order to obtain the most optimal tuning parameters, the offline methods were selected. This choice is made because the most popular method used to find the tuned PID parameters, which is the classic Ziegler–Nichols method, requires from object under examination to be on its stability limit. It might lead even to the destruction of the system.

The offline method requires from designer of the control system a knowledge about the structure of the system. Therefore, an identification experiment was carried out. It involves an examination of the system step response. For this purpose, the open system, where the feedback from the temperature of the collector was disconnected, was tested. The system was stimulated by setting the ventilator motor at 65% of its rated speed. Then both the motor speed and the temperature of the collector were measured. The experimental results of the identification are shown in figure 9.

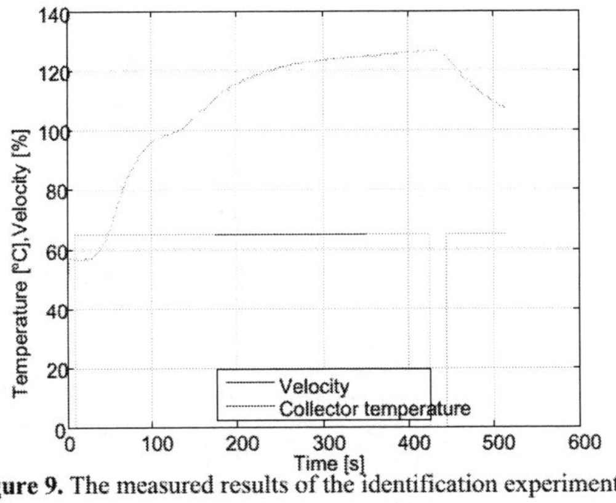

Figure 9. The measured results of the identification experiment

Some trends could be noticed by analyzing the charts showed in figure 9. Firstly, for both kind of medium: the water and the steam, a delay between the reference signal and the system response exists. It occurs because the temperature sensor is located not directly in the heat exchanger, but a little bit behind it, inside the steam collector. Therefore, a classical transport delay is present in the system. This statement is important for building a numerical model of the generator, because such a delay must not be ignored.

The second conclusion, which comes from analysis of the charts showed in figure 9, is the existence of a distinct point of inflection for the temperature of 100°C. This point is a phase change of the medium from its liquid to gas state. This phenomenon causes a large nonstationarity of the system. However, given the fact that the operating point of the system is above this value, the designer of the control system can build the mathematical model of the system regardless the lower temperatures.

The step response above the inflection point suggests, that the system is an inertial object, like most thermal systems. However, the row of this inertia is still unknown. Therefore, the Nyquist characteristics will be used to investigate its value. According to the control theory, the row of inertia is equal to the number of quarters, through which its spectral characteristics is crossing. The first step to obtain this characteristic is a transition from time domain to frequency domain using power spectral density function S given by equation 2.

$$S(\omega) = \int_{-\infty}^{\infty} R(\tau)e^{-i\omega\tau}d\tau, \quad (2)$$

where ω is an angular frequency of the signal, τ is time, and $R(\tau)$ is an signal autocorrelation function.

For input signal x and system step response y, the spectral transmittance G(jω) is given by equation 3.

$$G(j\omega) = \frac{S_{xy}}{S_{xx}}, \quad (3)$$

where S_{xy} is a power cross-spectral density function for input and output signal, and S_{xx} a power cross-spectral density function for input signal.

The spectral transmittance of the object obtained using this method is shown in figure 10.

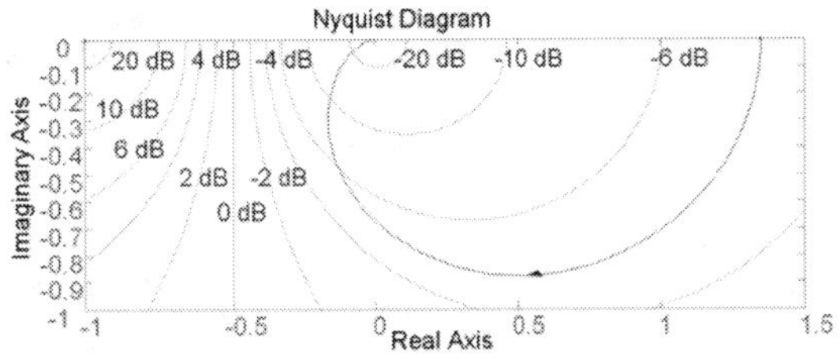

Figure 10. The spectral transmittance of the steam generator obtained as a result of the identification experiment.

As it is shown in figure 10, the Nyquist characteristic is present both in the left and right half of the complex plane in quadrant III and IV. Thus, the row of the system inertia may be defined as the second. Considering also the previously described delay in the structure, the entire system of the steam generator may therefore be described as inertial object of the second row with transport delay.

Knowing the structure of the system, the designer can therefore begin to determine the parameters of the system. For this purpose, an iterative method of minimizing the mean square error of the predicted output can be used. This method uses an optimization to minimize the cost function VN, which is given by equation 4.

$$V_N = \sum_{t=1}^{N} e^2(t), \quad (4)$$

where e(t) is the difference between the measured output and the predicted model output.

The parameters of the object model were calculated using Matlab® package. Its transfer function G(s) is given by equation 5. According to the trends presented in figure 11, the model response reaches 94.74 % fit to the real measurement stand. It is therefore sufficient to the task of fine-tuning of the PID controller in the offline conditions.

$$G(s) = \frac{1.349}{31.153s+1}\cdot\frac{1}{31.108s+1}e^{-14.307t}, \quad (5)$$

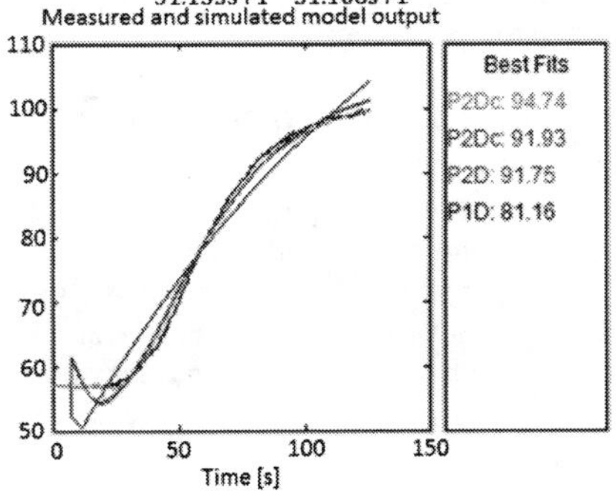

Figure 11. The fit between the model and the measured response.

3.3 Controller fine-tuning

The numerical model in Matlab Simulink® was built using mathematical model of the steam generator given by equation 5. This model is shown in figure 12. The signal Constraint toolbox was used to obtain the tuned parameters of the controller. It is using again the algorithm, which minimalize the cost function to obtain the optimum controller settings at the system operating point.

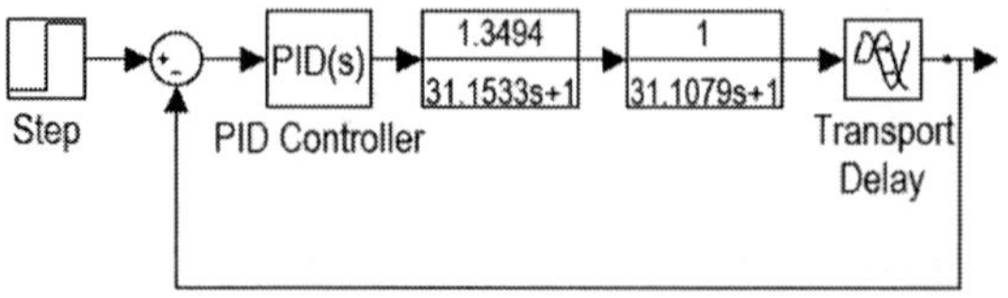

Figure 12. Numerical model of the steam generator with the PID controller.

The setpoint value was determined as 130°C. For this operating point and presented model the obtained PID controller parameters are shown in Table 1.

Table 1. Optimal PID parameters for operating point T=130 C.

Controller Part	Value
Proportional Gain K_P	1.27
Integral gain K_I	0.02
Derivative gain K_D	61.99

Taking into account the fact, that the object is highly non-stationary, the parameter K_I was reduced to value 0.01. This would reduce the overshoot of the process in other operation point with different temperature.

The biggest impact on the stationarity of the physical objects has the change of its load. It is represented in case of the steam generator as a change or switch between devices, which are consuming the steam. Thus, the simulation studies over the control quality ratios were conducted. For this purpose, an integral quality factor, defined by the formula 6 was chosen. Factor called Integral Square Error (ISE) has a very pragmatic meaning. It represents the energy, which is lost at a particular type of control.

$$I = \int_0^{\infty} e^2(t)dt. \qquad (6)$$

After all of this assumption, the simulation studies were carried out. The results of those simulations for different operating points were presented in Table 2. According to those results, the best control quality is achieved around the assumed operating point. The integral quality factor reached the lowest value for this operating point.

Table 2. Value of ISE for different temperatures.

Temperature [C]	ISE
110	4.766
120	4.563
130	4.415
140	5.123
150	5.877
160	6.687
170	7.549
180	8.463
190	9.439
200	10.460
210	11.533

For higher temperatures selected settings are therefore worse, but in contrast to the hysteresis regulation type, they do not cause oscillations and instability. The output error is eliminated and the overshoot, due to reduced integral gain, is small. For other temperature operating point, only the control time may be increased. Adaptive control can be applied in order to overcome this disadvantage. However, it is much more computationally expensive than the classic PID algorithm.

Calculated tuned parameters were implemented into the PLC controller, and the results of the experiment on the real system are shown in figure 13.

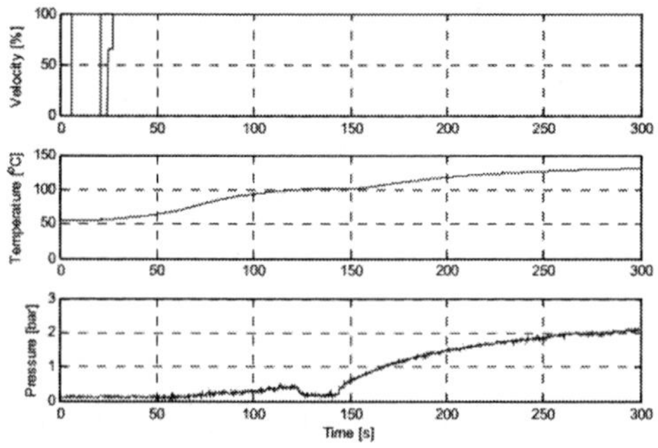

Figure 13. Control results for a real steam generator with tuned PID controller.

4 Characterization of the generator operation

The modified steam generator according described procedure was tested. The control and data acquisition was performed using dedicated software. The main panel is presented in figure. 14.

The blocks responsible for control of the circulating pump, the burner ignitron system, the fuel pump/air fan frequency inverter are visible. Users are allowed to set working temperature and pressure of the medium and limits of those parameters. The control lights indicate status of the pump and burner. The variation of the parameters may be observed in real time on the graphs. Exemplary results of the generator operation are presented in figure. 15.

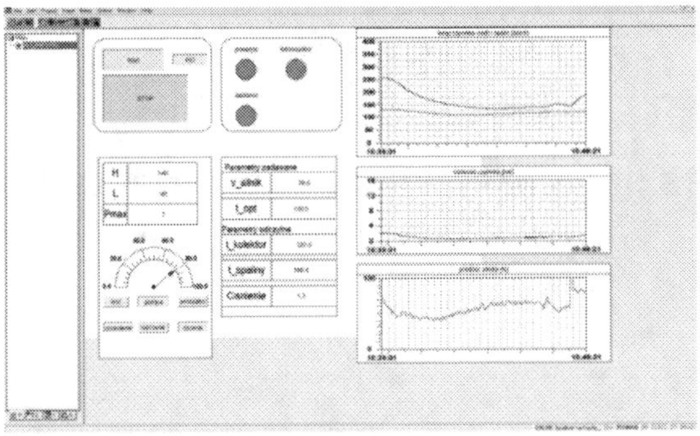

Figure 14. Main panel of the dedicated control software.

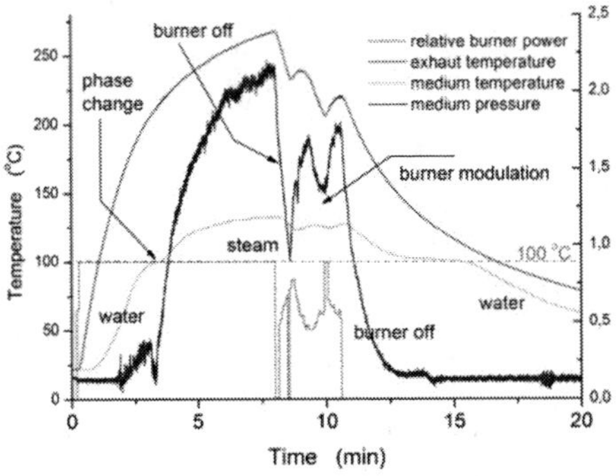

Figure 15. Main characteristics of the steam generator operation.

The test was performed for the partially open circuit – steam, via the steam bus was directed to container with 100 liters volume. Part of the steam condensed and the rest was released to the atmosphere. This arrangement resulted relatively low steam pressure during the test – maximally ca. 2 bars.

The initial phase of the test is clearly visible – the water temperature increase from the ambient temperature to 100oC – to beginning of the evaporation. This phase took about 3 minutes. The inverter works in this period with nominal frequency (100%). Then the evaporation starts and the steam temperature increase till reaching setting value. The small temperature fluctuations around setting value are visible.

5 Conclusions

It was show that adaptations of simple and currently obsolete military steam generators for purpose of small, home cogeneration systems are possible. The possibility of creation of a full control model of such units allows for further, deeper modifications – e.g. burner replacement or modification of the combustion chamber. In such a case a repetition of the whole tuning procedure should be performed, what is related with different dynamics of burners and heat generation area.

Additionally, this paper presents the procedure of the fine-tuning of the PID controller parameters. A concise procedure of the offline method was presented. The conditions of the identification procedure were given, and subsequently the reverse model of the real object was delivered. Designed numerical model was used to determine the optimal value of the controller parameters. This paper presents also discussion about the optimization of selected settings for other operating points. For this purpose, the integral quality factors were used.

Disadvantages of the most frequently used type of control system, which is the relay controller, are also discussed in the paper. Some disadvantages of classical continuous control using PID controller were also indicated. Those can be eliminated using its modification, called adaptive control. It allows to optimum process control in the full range of operational conditions, but unfortunately it is also much more computationally complex. This algorithm may therefore require increasing a memory of computing unit. However, in further development of this project it will be implemented.

Acknowledgements

The works have been conducted in collaboration between Dom Project Sp. z o.o. and AGH UST in Krakow (under statutory activities of the Faculty of Energy and Fuels “Studies concerning the conditions of sustainable energy development”). Purchase of the steam generator and its modifications were financed by Dom Project Sp. Z o.o. Components, used in the tests, were purchased with the funds of the “BioORC: Construction of cogeneration system with small a medium size biomass boiler” project.

References

1. K. Sornek, M. Filipowicz, K. Izdebski. Journal of Civil Engineering, Environment and Architecture, **62**, 2 (2015)
2. K. Sornek, M. Filipowicz, M. Szubel, E. Bożek, K. Izdebski, MATEC Web of Conferences, **18** (2014)
3. R. Wróblewski, Energia Gigawat, **10-11** (2014)
4. EU Energy, *Transport and GHG Emissions Trends to 2050*, Reference Scenario 2013
5. K. Sornek, M. Szubel, W. Goryl, E. Bożek, M. Filipowicz, Przemysł Chemiczny, **93**, 12 2014)
6. Y. Li, K.H. Ang, G.C.Y. Chong, Control Systems, IEEE, **26**, 1 (2006)
7. Y. Li, K.H. Ang, G.C.Y. Chong, IEEE Control Systems Magazine, **26**, 1 (2006)
8. J. Wołoszyn, A. Gołaś, Energy Conversion and Management, **87** (2014)
9. J. Kwasniewki, I. Dominik, K. Lalik, Journal Of Vibroengineering, **14** (2012)
10. I. Dominik, Solid State Phenomena, **164** (2010)

EPJ Web of Conferences **114**, 02152 (2016)
DOI: 10.1051/epjconf/201611402152

Maximization of integral outlet quantities of an axisymmetric synthetic jet actuator based on a loudspeaker

J. Kordík[1,a] and Z. Trávníček[1]

[1]Institute of Thermomechanics, AS CR

Abstract. The goal of this paper is to find an optimal nozzle size of an axisymmetric synthetic jet actuator based on a loudspeaker. The desirable maximized output quantities are: volumetric flow, momentum flux, and kinetic energy flux. To evaluate these quantities velocity profiles were measured using a hot-wire probe at the actuator nozzle exit. Six different nozzle diameters and three supplied real power levels were tested to find the maxima of the quantities. The actuator operated always at resonance during experiments. It was found out that the momentum flux and the kinetic energy flux reach distinguishable local maxima at particular diameters of the nozzle. Besides, the maxima of the particular quantities do not coincide and the best nozzle size slightly increases with the supplied real power to the actuator.

1 Introduction

Synthetic jets (SJs), or zero-net-mass-flux jets, are fluid flows that are generated by fluid pulsations and are formed in a free space behind a nozzle. The fluid is pushed and pulled through the nozzle, whose inner part exits into a cavity. The source of the pulsations is an oscillating diaphragm that bounds one side of the cavity. The nozzle is the only inlet/outlet in/from the cavity, therefore time mean mass flux through the nozzle is zero. The momentum flux along the nozzle axis is, however, non-zero, therefore the flow continues ahead downstream from the nozzle exit and forms similar flow structures as a continuous jet.

Since the end of the last century, the topic of synthetic jets has been intensively investigated [1,2] and many useful applications of SJs have been developed. The applications of SJs have been found especially in boundary-layer separation control [3–10], jet vectoring [11–16], heat transfer enhancement [17–22], and mixing [23–25]. A variant of the SJ, namely a non-zero-net-mass-flux jet (or hybrid SJ), has been investigated more recently [26–28,21,29–31].

It is a known fact that the most convenient operating condition from the energetic point of view is a resonance (if we consider the synthetic jet actuators based on elastic diaphragms e.g. based on a loudspeaker or a piezoceramic membrane etc.). Therefore, many authors concentrated their efforts on location and prediction of the SJ actuator resonances (see e.g. [32–44]). Having a SJ actuator operating at a resonance frequency, there are still a lot of ways how to enhance its exploitable outlet quantities such as volumetric flow, momentum or kinetic energy fluxes. This will be the topic of this work. Particularly, this paper investigates the axisymmetric SJ actuator based on a loudspeaker and working at the first resonance. Under these conditions the goal is to find such a nozzle diameter of the SJ actuator to achieve a) the highest volumetric flow, b) the highest momentum flux, and c) the highest kinetic energy flux. The optimization is made here experimentally with the following constrictions: the actuator works with an incompressible fluid, the diaphragm (loudspeaker) properties and the nozzle diameter to length ratio are constant, and the diaphragm excitation is sinusoidal (harmonic).

[a] e-mail: kordik@it.cas.cz

2 Observed Parameters

First of all we define a characteristic velocity of the SJ, which is used in the present auxiliary experiments (including measurements of velocity resonance curves):

$$U_0 = \frac{1}{T S_N} \int_0^{T_E} \int_{S_N} u \mathrm{d}S \, \mathrm{d}t, \qquad (1)$$

where S_N is the nozzle exit cross-sectional area ($\mathrm{d}S$ is the differential element of S_N), u denotes the velocity component aligned with the nozzle axis, T_E is the extrusion stroke duration time, T is the actuation period, and the area-averaged velocity $\frac{1}{S_N}\int_{S_N} u \mathrm{d}S$ has the origin at $t = 0$ (t is time).

Now, we can define the observed quantities, whose maxima are desirable. The definitions of the quantities follow in this order: the time-average volumetric flow (Q_0), momentum flux (M_0), and kinetic energy flux (E_0):

$$Q_0 = \frac{1}{T} \int_0^{T_E} \int_{S_N} u \mathrm{d}S \, \mathrm{d}t = U_0 S_N, \qquad (2)$$

$$M_0 = \frac{1}{T} \int_0^{T_E} \int_{S_N} \rho u^2 \mathrm{d}S \, \mathrm{d}t, \qquad (3)$$

$$E_0 = \frac{1}{T} \int_0^{T_E} \int_{S_N} \frac{\rho u^3}{2} \mathrm{d}S \, \mathrm{d}t, \qquad (4)$$

where ρ is the working fluid density.

Note, that the interval $(0, T_E)$ can be a bit different for each quantity Q_0, M_0, and E_0. Generally, the interval represents a time spans where the surface integrals from Eqs. (2,3,4) are greater or equal to zero.

3 Experimental Setup and Methods

3.1 The Synthetic Jet Actuators

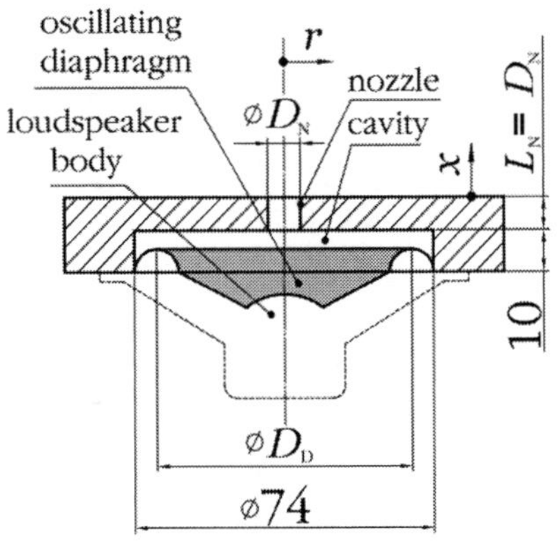

Fig. 1. Schema of the current actuator.

Figure 1 shows the schematic view of the tested SJA that operates at standard room conditions (temperature 20 – 25 °C and barometric pressure 970 – 990 hPa) with air as the working fluid. The present actuator is based on a Aurasound NS3-193-8A loudspeaker, whose diaphragm generates the fluid oscillations in the nozzle.

During all experiments, the loudspeaker of the actuator was supplied by a harmonic electric current generated by Agilent 33220A function generator and gained in Pioneer A-209R amplifier. The experiments were performed with the following set of replaceable nozzles: D_N = 6, 8, 12, 15, 20, and 30 mm. The length of all nozzles was always $L_N = D_N$.

Each nozzle was attached to the actuator body and the velocities were measured by a hot-wire probe at the nozzle outer exit. The measurements were performed at actuator velocity resonance and for three input power levels: P_e = 0.05, 0.24, and 1.18 W. The desirable quantities Q_0, M_0, and E_0 were evaluated numerically from the velocity measurements using corresponding Eqs. (2,3,4).

3.2 Data Acquisition and Hardware Setup

As was mentioned above the quantities Q_0, M_0, and E_0 will be investigated at the actuator velocity resonance. To locate the velocity resonances auxiliary experiments were performed primarily. These experiments included the measurements of velocity resonance curves. The curves are represented by the characteristic velocity, U_0, as function of the actuating frequency, f, and were acquired at constant power input into the actuator.

The velocities U_0 for the resonance curves were evaluated from a velocity waveform obtained at the centre of the nozzle ($r = 0$ and $x = 0$, for the coordinate system see figure 1). The MiniCTA 54T30 DANTEC hot-wire anemometer with a single wire probe 55P16 and the NI-PCI 6023E data acquisition device were used for this measurement. The hot-wire signal was sampled with a sampling frequency of $f \cdot 64$ and the number of samples was 16384. The temperature of the working fluid (air) was measured using a fast-response Pt100 sensor (PT100-SMD0805) at the same time as the velocity samples. The measured temperature was used for the temperature correction of the hot-wire data. To evaluate and check the constant power input into the actuator, the supplied electric current and voltage were acquired concurrently with velocity and temperature samples.

After the identification of the velocity resonance frequencies (for each combination of the nozzle diameter and input power level), the velocity profiles were measured at the resonances along nozzle diameter at the distance of x = 0.3 mm. The velocity profiles were acquired at 14 discrete positions on the nozzle diameter. The same experimental equipment and hardware setup as during auxiliary experiments was used.

For all measurements the hot-wire probe was calibrated in the velocity range of 0.2 – 53 m/s. The maximum relative uncertainty of a single velocity sample was below 5% for velocities in range of 0.4 – 53 m/s. For very small velocities in the range of 0.2 – 0.4 m/s the uncertainty was 14.2%. A more detailed description of the anemometer settings, probe calibration method, and uncertainty evaluation is available in [42].

4 Experimental Results

4.1 Auxiliary Experiments

The measured resonance curves for identification of particular resonance frequencies are shown in figure 2. The set of curves in figure 2 was acquired at constant power input into the actuator: P_e = 0.24W and for the full set of replaceable nozzle diameters. Other power levels were tested, as well, but the locations of the resonance frequencies for the same nozzles did not change substantially.

Figure 2 reveals the fact that the resonance frequency gets higher with the increasing nozzle diameter. Moreover, we can notice that the resonance curve for the smallest diameter (D_N = 6 mm) is quite flat and the resonance peak is hardly identifiable. This is caused by a strong damping having an origin in local losses at cavity/nozzle or nozzle/ambiance transitions.

The experimentally located velocity resonance frequencies, f_{res}, are listed in table 1; they can be, however, evaluated easily theoretically, as well. The appropriate formula can be found e.g. in [44,31]:

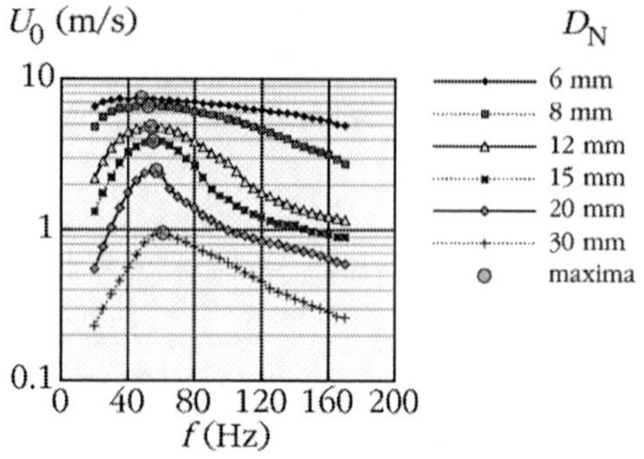

Fig. 2. Resonance curves and their maxima acquired at P_e = 0.24*W*.

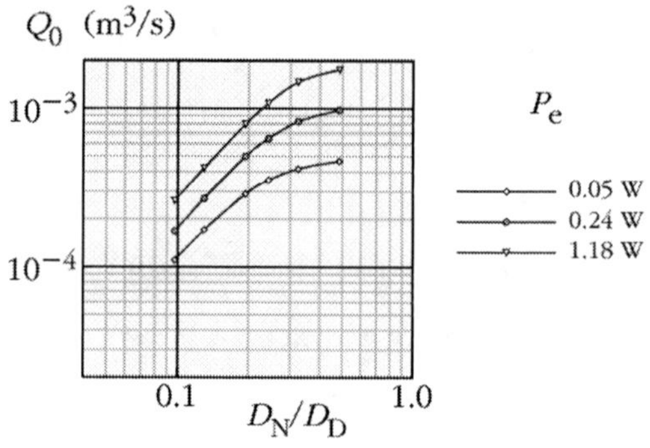

Fig. 3. Volumetric flows.

Table 1. The parameters of experiments, found resonances.

D_N (mm)	f_{res} (Hz)		P_e (W)
	Experiment	Theory	
6	48	47.0	
8	52	50.3	
12	54	54.4	0.05, 0.24, 1.18
15	55	56.3	
20	57	58.4	
30	61	60.9	

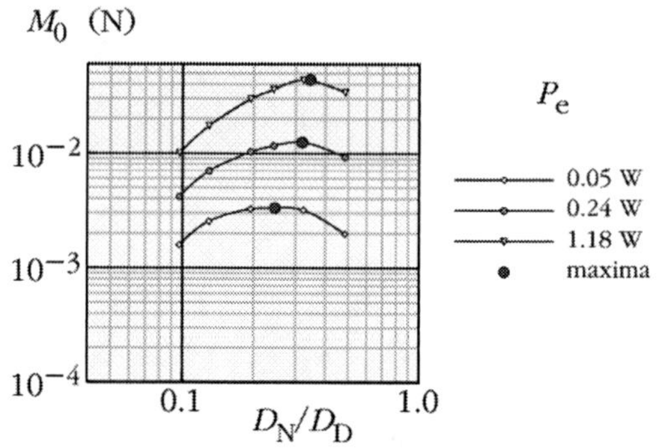

Fig. 4. Momentum flux.

$$f_{res} = \frac{1}{2\pi}\sqrt{\frac{K_D}{m_D + D_D^4/D_N^4 m_N}}, \text{ where :}$$

$$m_N = \frac{\rho\pi D_N^2}{4}(L_N + 0.85D_N), \tag{5}$$

and where K_D = 704 N/m is the diaphragm spring constant, m_D = 3.9 g is its mass, and D_D = 61.8 mm is the diaphragm effective diameter (the values of these parameters were identified in [43]). The fluid mass moving in the nozzle and its vicinity is assigned m_N. The evaluation of m_N suggested in Eq. (5) is taken from [45]. Comparing the experimental and theoretical results in table 1 we can state that there is an acceptable agreement.

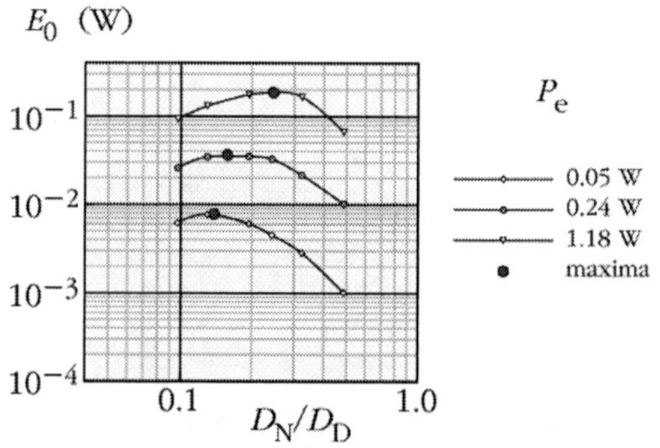

Fig. 5. Kinetic energy flux.

4.2 Main Experiments and Discussion

The next results were obtained only at resonance frequencies summarized in table 1. Figure 3 shows the results evaluated for the volumetric flows, Q_0. The flows, Q_0, are drawn here as functions of the ratio D_N/D_D. A monotonous character of all curves in figure 3 implies the following: to achieve the highest value of Q_0 the optimal nozzle diameter should be large as possible. By other words, the diameter for maximal volumetric flows from the SJA probably does not exist for the current configurations. This phenomenon can be explained from the theory of linear damped harmonic oscillator. An increasing damping of the forced oscillator always decreases its displacements and velocities in the whole frequency range (and especially at resonance frequency). The narrower the nozzle is, the higher damping is imposed onto the oscillating diaphragm. Therefore, due to the incompressibility conditions the highest volumetric flows can be achieved with the largest nozzles (or without any nozzle and the actuator body, but in this case no SJ would be produced, see [46]).

Results obtained for the momentum flux can be seen in figure 4. Clear maxima were found for each input power level at the ratio of D_N/D_D = 0.25 – 0.35. Moreover, we can notice that for a maximum momentum flux, the optimal nozzle diameter slightly increases with increasing input power, P_e.

Very similar results as for the momentum flux were found for the kinetic energy flux – see figure 5. The local maxima were found for each input power level at the ratio of $D_N/D_D = 0.14 - 0.25$. Therefore, we can conclude that smaller nozzle diameters are necessary to operate the actuator with the maximal kinetic energy flux than with the maximal momentum flux.

It is worth to note here that actuators working at resonance and having the optimal nozzle diameters with respect to the kinetic energy flux work with maximal energetic efficiency defined by [47, 31].

5 Conclusion

A synthetic jet actuator based on a loudspeaker was tested with various nozzles and input powers at the first resonance frequency. The nozzle diameter to length ratio was constant ($D_N/L_N \approx 1$) and operating medium was air at incompressible conditions. Outlet velocity profiles were measured by hot-wire probe at the actuator nozzle exit. The values of the volumetric flow, momentum flux, and kinetic energy flux were evaluated from the measured velocity profiles.

The maximal volumetric flow was found for the largest tested nozzle diameter. The volumetric flow as function of the nozzle diameter appeared to be monotonic and had no local maximum. On the other hand, momentum and kinetic energy fluxes had their local maxima at particular nozzle diameters. An optimal geometry of the actuator, represented by a ratio of the nozzle diameter to the diaphragm diameter, slightly depended on the actuator input power. In both cases (momentum and kinetic energy fluxes) the best nozzle diameter slightly increased with the increasing power input and was larger for momentum flux than for the kinetic energy flux.

Acknowledgement

We gratefully acknowledge the support of the Grant Agency CR (project number 14-08888S) and the institutional support RVO:61388998.

References

1. B.L. Smith, A. Glezer, Physics of Fluids **10**, 2281 (1998)
2. J.E. Cater, J. Soria, Journal of Fluid Mechanics **472**, 167 (2002)
3. G. Hong, Sensor Actuat. A-Phys. **132**, 607 (2006)
4. J. Dandois, E. Garnier, P. Sagaut, J. Fluid Mech. **574**, 25 (2007)
5. D. You, P. Moin, J. Fluid Struct. **24**, 1349 (2008)
6. S.G. Mallinson, J.A. Reizes, R. Hillier, Flow Turbulence and Combustion **66**, 1 (2001)
7. M. Amitay, A. Glezer, International Journal of Heat and Fluid Flow **23**, 690 (2002)
8. R. Rathnasingham, K.S. Breuer, Journal of Fluid Mechanics **495**, 209 (2003)
9. S. Lardeau, M.A. Leschziner, Journal of Fluid Mechanics **683**, 172 (2011)
10. C. Bernardini, M. Carnevale, M. Manna, F. Martelli, D. Simoni, P. Zunino, Journal of Thermal Science **21**, 404 (2012)
11. B.L. Smith, A. Glezer, Journal of Fluid Mechanics **458**, 1 (2002)
12. Z.B. Luo, Z.X. Xia, Modern Physics Letters B **19**, 1619 (2005)
13. Z.B. Luo, Z.X. Xia, Y.G. Xie, Chinese Journal of Aeronautics **20**, 193 (2007)
14. Z.X. Xia, Z.B. Luo, Applied Mathematics and Mechanics **28**, 907 (2007)
15. D.A. Tamburello, M. Amitay, International Journal of Heat and Fluid Flow **29**, 967 (2008)
16. M. Ben Chiekh, M. Ferchichi, J.C. Béra, Experiments in Fluids **53**, 1797 (2012)
17. M.B. Gillespie, M.S. thesis, Georgia Institute of Technology (1998)
18. Z. Trávníček, V. Tesař, International Journal of Heat and Mass Transfer **46**, 3291 (2003)
19. M.B. Gillespie, W.Z. Black, C. Rinehart, A. Glezer, Journal of Heat Transfer **128**, 990 (2006)
20. M. Arik, Applied Thermal Engineering **27**, 1483 (2007)
21. Z. Trávníček, V. Tesař, J. Kordík, Journal of Visualization **11**, 221 (2008)
22. M. Chaudhari, B. Puranik, A. Agrawal, International Journal of Heat and Mass Transfer **53**, 1057 (2010)
23. H. Wang, S. Menon, AIAA Journal **39**, 2308 (2001)
24. M. Al-Atabi, Journal of Fluids Engineering **133**, 094503 (2011)
25. Q. Xia, S. Zhong, International Journal of Heat and Fluid Flow **37**, 64 (2012)
26. Z. Trávníček, A.I. Fedorchenko, A.B. Wang, Sensors and Actuators A: Physical **120**, 232 (2005)
27. Z. Trávníček, T. Vít, V. Tesař, Physics of Fluids **18**, 081701 (2006)
28. V. Tesař, Sensors and Actuators A **138**, 394 (2007)
29. J. Kordík, Z. Trávníček, Journal of Fluids Engineering **135**, 101101 (2013)
30. Z. Trávníček, T. Vít, International Journal of Heat and Mass Transfer **85**, 473 (2015)
31. J. Kordík, Z. Trávníček, M. Pavelka, Experimental Thermal and Fluid Science **69**, 119 (2015)
32. Q. Gallas, M.S. thesis, University of Florida, Gainesville (2002)
33. Q. Gallas, J. Mathew, A. Kaysap, R. Holman, T. Nishida, B. Carroll, M. Sheplak, L. Cattafesta, *Lumped element modeling of piezoelectric-driven synthetic jet actuators*, in *40th AIAA Aerospace Sciences Meeting & Exhibit* (AIAA Paper 2002-0125, Reno, 2002)
34. Q. Gallas, R. Holman, T. Nishida, B. Carroll, M. Sheplak, L. Cattafesta, AIAA Journal **41**, 240 (2003)
35. Q. Gallas, M. Sheplak, L. Cattafesta, Final report for NAG-1-03031, University of Florida, Department of Mechanical and Aerospace Engineering, Gainesville (2005)

36. Q. Gallas, R. Holman, R. Raju, R. Mittal, M. Sheplak, L. Cattafesta, *Low dimensional modeling of zero-net mass flux actuators*, in *2nd AIAA Flow Control Conference* (AIAA Paper 2004-2413, Portland, 2004)
37. Q. Gallas, PhD. thesis, University of Florida, Gainesville (2005)
38. B.H. Kim, D.R. Williams, S. Emo, M. Acharya, AIAA Journal **43**, 314 (2005)
39. H. Tang, S. Zhong, *Lumped element modelling of synthetic jet actuators*, in *The 3rd AIAA Flow Control Conference* (AIAA Paper 2006-3696, San Francisco, 2006)
40. H. Tang, S. Zhong, M. Jabbal, L. Garcillan, F. Guo, N. Wood, C. Warsop, Flow, Turbulence and Combustion **78**, 309 (2007)
41. H. Tang, S. Zhong, Aerospace Science and Technology **13**, 331 (2009)
42. J. Kordík, *Synthetic and Hybrid Synthetic Jet Actutators – Theoretical and Experimental Analysis* (LAP LAMBERT Academic Publishing, 2013), ISBN 978-3-659-32910-4
43. J. Kordík, Z. Trávníček, AIAA Journal **51**, 2862 (2013)
44. J. Kordík, Z. Broučková, T. Vít, M. Pavelka, Z. Trávníček, Experiments in Fluids **55**, 1757 (2014)
45. L.E. Kinsler, A.R. Frey, A.B. Coppens, J.V. Sanders, *Fundamentals of acoustics*, 4th edn. (John Wiley & Sons, Inc. New York, 2000)
46. R. Holman, Y. Utturkar, R. Mittal, B.L. Smith, L. Cattafesta, AIAA Journal **43**, 2110 (2005)
47. Z. Broučková, Z. Trávníček, Journal of Visualization (2015)

AUTHOR INDEX

AUTHOR INDEX

AUTHOR INDEX

AUTHOR INDEX